Contents—1996 HVAC SYSTEMS AND EQUIPMENT

AIR-CONDITIONING AND HEATING SYSTEMS

Chapter
1. Air-Conditioning System Selection and Design
2. All-Air Systems
3. Air-and-Water Systems
4. All-Water Systems
5. Unitary Refrigerant-Based Systems for Air Conditioning
6. Panel Heating and Cooling
7. Cogeneration Systems and Engine and Turbine Drives
8. Applied Heat Pump and Heat Recovery Systems
9. Air Distribution Design for Small Heating and Cooling Systems
10. Steam Systems
11. District Heating and Cooling
12. Hydronic Heating and Cooling System Design
13. Condenser Water Systems
14. Medium- and High-Temperature Water Heating Systems
15. Infrared Radiant Heating

AIR-HANDLING EQUIPMENT

Chapter
16. Duct Construction
17. Air-Diffusing Equipment
18. Fans
19. Evaporative Air Cooling
20. Humidifiers
21. Air-Cooling and Dehumidifying Coils
22. Desiccant Dehumidification and Pressure Drying Equipment
23. Air-Heating Coils
24. Air Cleaners for Particulate Contaminants
25. Industrial Gas Cleaning and Air Pollution Control

HEATING EQUIPMENT

Chapter
26. Automatic Fuel-Burning Equipment
27. Boilers
28. Furnaces
29. Residential In-Space Heating Equipment
30. Chimney, Gas Vent, and Fireplace Systems
31. Unit Ventilators, Unit Heaters, and Makeup Air Units
32. Hydronic Radiators
33. Solar Energy Equipment

GENERAL COMPONENTS

Chapter
34. Compressors
35. Condensers
36. Cooling Towers
37. Liquid Coolers
38. Centrifugal Pumps
39. Motors and Motor C
40. Pipes, Tubes, and Fittings
41. Valves
42. Air-to-Air Energy Recovery

UNITARY EQUIPMENT

Chapter
43. Room Air Conditioners and Dehumidifiers
44. Unitary Air Conditioners and Unitary Heat Pumps
45. Applied Packaged Equipment
46. Codes and Standards

Contents—1995 HVAC APPLICATIONS

COMFORT AIR CONDITIONING AND HEATING

Chapter
1. Residences
2. Retail Facilities
3. Commercial and Public Buildings
4. Places of Assembly
5. Domiciliary Facilities
6. Educational Facilities
7. Health Care Facilities
8. Surface Transportation
9. Aircraft
10. Ships

INDUSTRIAL AND SPECIAL AIR CONDITIONING AND VENTILATION

Chapter
11. Industrial Air Conditioning
12. Enclosed Vehicular Facilities
13. Laboratory Systems
14. Engine Test Facilities
15. Clean Spaces
16. Data Processing System Areas
17. Printing Plants
18. Textile Processing
19. Photographic Materials
20. Environmental Control for Animals and Plants
21. Drying and Storing Farm Crops
22. Air Conditioning of Wood and Paper Products Facilities
23. Nuclear Facilities
24. Ventilation of the Industrial Environment
25. Mine Air Conditioning and Ventilation
26. Industrial Exhaust Systems
27. Industrial Drying Systems
28. Kitchen Ventilation

ENERGY SOURCES

Chapter
29. Geothermal Energy
30. Solar Energy Utilization
31. Energy Resources

BUILDING OPERATION AND MAINTENANCE

Chapter
32. Energy Management
33. Owning and Operating Costs
34. Testing, Adjusting, and Balancing
35. Operation and Maintenance Management
36. Computer Applications
37. Building Energy Monitoring
38. Building Operating Dynamics and Strategies
39. Building Commissioning

GENERAL APPLICATIONS

Chapter
40. Thermal Storage
41. Control of Gaseous Indoor Air Contaminants
42. Automatic Control
43. Sound and Vibration Control
44. Corrosion Control and Water Treatment
45. Service Water Heating
46. Snow Melting
47. Evaporative Air Cooling
48. Smoke Management
49. Radiant Heating and Cooling
50. Seismic Restraint Design
51. Codes and Standards

1998 ASHRAE® HANDBOOK

REFRIGERATION

SI Edition

American Society of Heating, Refrigerating and Air-Conditioning Engineers, Inc.
1791 Tullie Circle, N.E., Atlanta, GA 30329
(404) 636-8400   http://www.ashrae.org

Copyright ©1998 by the American Society of Heating, Refrigerating and Air-Conditioning Engineers, Inc. All rights reserved.

DEDICATED

TO THE ADVANCEMENT OF

THE PROFESSION

AND ITS ALLIED INDUSTRIES

No part of this book may be reproduced without permission in writing from ASHRAE, except by a reviewer who may quote brief passages or reproduce illustrations in a review with appropriate credit; nor may any part of this book be reproduced, stored in a retrieval system, or transmitted in any form or by any means—electronic, photocopying, recording, or other—without permission in writing from ASHRAE.

ASHRAE has compiled this publication with care, but ASHRAE has not investigated, and ASHRAE expressly disclaims any duty to investigate, any product, service, process, procedure, design, or the like which may be described herein. The appearance of any technical data or editorial material in this publication does not constitute endorsement, warranty, or guaranty by ASHRAE of any product, service, process, procedure, design, or the like. ASHRAE does not warrant that the information in this publication is free of errors. The entire risk of the use of any information in this publication is assumed by the user.

Comments, criticisms, and suggestions regarding the subject matter are invited. Any errors or omissions in the data should be brought to the attention of the Editor. If required, an errata sheet will be issued at approximately the same time as the next Handbook. Notice of any significant errors found after that time will be published in the *ASHRAE Journal*.

ISBN 1-883413-54-0

CONTENTS

Contributors

ASHRAE Technical Committees and Task Groups

Preface

REFRIGERATION SYSTEM PRACTICES

Chapter
1. **Liquid Overfeed Systems** (TC 10.1, Custom Engineered Refrigeration Systems)
2. **System Practices for Halocarbon Refrigerants** (TC 10.3, Refrigerant Piping, Controls, and Accessories)
3. **System Practices for Ammonia Refrigerant** (TC 10.3)
4. **Secondary Coolants in Refrigeration Systems** (TC 10.1)
5. **Refrigerant System Chemistry** (TC 3.2, Refrigerant System Chemistry)
6. **Control of Moisture and Other Contaminants in Refrigerant Systems** (TC 3.3, Refrigerant Contaminant Control)
7. **Lubricants in Refrigerant Systems** (TC 3.4, Lubrication)

FOOD STORAGE AND EQUIPMENT

Chapter
8. **Thermal Properties of Foods** (TC 10.9, Refrigeration Application for Foods and Beverages)
9. **Cooling and Freezing Times of Foods** (TC 10.9)
10. **Commodity Storage Requirements** (TC 10.5, Refrigerated Distribution and Storage Facilities)
11. **Food Microbiology and Refrigeration** (TC 10.9)
12. **Refrigeration Load** (TC 10.8, Refrigeration Load Calculations)
13. **Refrigerated Facility Design** (TC 10.5)
14. **Methods of Precooling Fruits, Vegetables, and Cut Flowers** (TC 10.9)
15. **Industrial Food Freezing Systems** (TC 10.9)

FOOD REFRIGERATION

Chapter
16. **Meat Products** (TC 10.9)
17. **Poultry Products** (TC 10.9)
18. **Fishery Products** (TC 10.9)
19. **Dairy Products** (TC 10.9)
20. **Eggs and Egg Products** (TC 10.9)
21. **Deciduous Tree and Vine Fruit** (TC 10.9)
22. **Citrus Fruit, Bananas, and Subtropical Fruit** (TC 10.9)
23. **Vegetables** (TC 10.9)
24. **Fruit Juice Concentrates and Chilled Juice Products** (TC 10.9)
25. **Beverages** (TC 10.9)
26. **Processed, Precooked, and Prepared Foods** (TC 10.9)
27. **Bakery Products** (TC 10.9)
28. **Candies, Nuts, Dried Fruits, and Dried Vegetables** (TC 10.9)

DISTRIBUTION OF CHILLED AND FROZEN FOOD

Chapter
29. **Trucks, Trailers, and Containers** (TC 10.6, Transport Refrigeration)
30. **Marine Refrigeration** (TC 10.6)
31. **Air Transport** (TC 10.6)

INDUSTRIAL APPLICATIONS

Chapter
32. **Insulation Systems for Refrigerant Piping** (TC 10.3)
33. **Ice Manufacture** (TC 10.2, Automatic Icemaking Plants and Skating Rinks)
34. **Ice Rinks** (TC 10.2)
35. **Concrete Dams and Subsurface Soils** (TC 10.1)
36. **Refrigeration in the Chemical Industry** (TC 10.1)

LOW-TEMPERATURE APPLICATIONS

Chapter
37. **Environmental Test Facilities** (TC 9.2, Industrial Air Conditioning)
38. **Cryogenics** (TC 10.4, Ultra-Low Temperature Systems and Cryogenics)
39. **Low-Temperature Refrigeration** (TC 10.4)
40. **Biomedical Applications of Cryogenic Refrigeration** (TC 10.4)

REFRIGERATION EQUIPMENT

Chapter
41. **Absorption Cooling, Heating, and Refrigeration Equipment** (TC 8.3, Absorption and Heat Operated Machines)
42. **Forced-Circulation Air Coolers** (TC 8.4, Air-to-Refrigerant Heat Transfer Equipment)
43. **Liquid Chilling Systems** (TC 8.1, Positive Displacement Compressors, and TC 8.2, Centrifugal Machines)
44. **Component Balancing in Refrigeration Systems** (TC 10.1)
45. **Refrigerant-Control Devices** (TC 8.8, Refrigerant System Controls and Accessories)
46. **Factory Dehydrating, Charging, and Testing** (TC 8.1)

UNITARY REFRIGERATION EQUIPMENT

Chapter
47. **Retail Food Store Refrigeration and Equipment** (TC 10.7, Commercial Food and Beverage Cooling Display and Storage)
48. **Food Service and General Commercial Refrigeration Equipment** (TC 10.7)
49. **Household Refrigerators and Freezers** (TC 7.1, Residential Refrigerators and Food Freezers)
50. **Automatic Ice Makers** (TC 10.2)
51. **Codes and Standards**

INDEX

Composite index to the 1995 Applications, 1996 Systems and Equipment, 1997 Fundamentals, and 1998 Refrigeration volumes.

CONTRIBUTORS

In addition to the Technical Committees, the following individuals contributed significantly to this volume. The appropriate chapter numbers follow each contributor's name.

Thomas K. O'Donnell (1, 25, 35, 44)
Gallo Winery

David F. Ward (1, 44)
HEC Energy Services

Robert A. Jones (2)
Sporlan Valve Company

M. David Gardner (3)
Process Engineering, Inc.

George C. Briley (4)
Technicold Services Inc.

Martin L. Timm (4)
APV Heat Transfer Tech

Robert G. Doerr (5)
Trane Co.

Scott T. Jolley (5)
Lubrizol Corp

Raymond H.P. Thomas (5)
Allied Signal

Richard E. Cawley (6)
The Trane Company

Alan P. Cohen (6)
UOP Research Center

Danny M. Halel (6)
ASHRAE

Kenneth W. Manz (6)
InSource Technologies

Shelvin Rosen (6)

Ward D. Wells (6)
E.I. DuPont de Nemours and Co.

James A. Cancila (7)
Elf Lubricants North America

Barry Greig (7)
Castrol International

David R. Henderson (7)
Spauschus Associates

Thomas E. Rajewski (7)
CPI Engineering Services, Inc.

Carl F. Speich (7)
The Trane Company

Bryan R. Becker (8, 9, 27)
University of Missouri

Brian A. Fricke (8, 9, 27)
University of Missouri

Gordon E. Follette (10, 13, 14, 15, 19, 23, 26, 28)
Follette Engineering, Inc.

Edward W. Fuhrmann (10, 13, 18, 39)
Hendon-Lurie & Associates, Inc.

William W. Humm (10, 13)
FES, Inc.

Patricia A. Curtis (11, 16, 17)
North Carolina State University

Godan P. Nambudiripad (11)
The Pillsbury Company

William H. Sperber (11)
Cargill, Inc.

Katherine M.J. Swanson (11)
The Pillsbury Company

Gideon Zeidler (11, 16, 17, 20)
University of California

Ronald A. Cole (12)
R.A. Cole & Associates, Inc.

Jon M. Edmonds (12)
Edmonds Engineering Co.

George R. Smith (12)
HCR, Inc.

Clifford Studman (14, 21)
Massey University

Christopher Watkins (14, 21)
Cornell University

Terry D. Barber (15)
Northfield

Robert L. Hendrickson (16)
Oklahoma State University

Ronald P. Vallort (16)
The Haskell Company

Benjamin C. Smith (18)
National Sea Products Ltd.

Kenneth E. Anderson (20)
North Carolina State University

Mo Samimi (20)

William J. Stadelman (20)
Purdue University

Tony Yang (20)
Papetti Egg Products

Giustino N. Mastro (21)
University of Vermont

William M. Miller (22)
University of Florida

Steve Sargent (23)
University of Florida

Joseph Bene (24)
Bene Engineering Company

Evans J. Lizardos (25)
Lizardos Engineering Associates

Joseph G. Ponte, Jr. (27)
Kansas State University

Will F. Stoecker (27, 39)
University of Illinois

Ronald H. Zelch (27)
American Institute of Baking

William J. Hannett (29, 30)
UTC Carrier (retired)

Sung Lim Kwon (29)
Thermo King Corporation

James J. Bushnell (31)
HVAC Consulting Services

Alan H. Benton (32)
Dow Chemical

Robert S. Burdick (32)
Bassett Mechanical

Robert P. Connolly (32)
Connolly Associates

Paul A. Hough (32)
Armstrong World Industries, Inc.

Kathleen M. Posteraro (32)
Pittsburgh Corning Corporation

Kelly Huang (33)
Trico Refrigeration

Ronald H. Strong (33)
R H Strong & Associates Inc.

John Topliss (33)
Refrigeration Components Canada Ltd.

James L. Carver (34, 47)
Webb Technologies

James J. Shepherd (35)
Toromont Processing Systems

Brent J. Allardyce (36)
Startec Refrigeration

Earl M. Clark (36, 39)
DuPont Fluorochemical

Richard Evans (37)
Westinghouse Hanford Company

Laszlo Emho (38)
IPARTERV Rt.

Klaus D. Timmerhaus (38)
University of Colorado at Boulder

CONTRIBUTORS (Concluded)

Arthur P. Garbarino (39)
Air Service Inc.

K. Ted Hartwig (39)
Texas A&M University

Predrag Hrnjak (39)
University of Illinois

Stephen A. Mowrer (39)
Liebert Corporation

Rudolph Stegmann (39, 46)
The Enthalpy Exchange

Chaun Weng (39)
Revco Scientific, Inc.

Kenneth R. Diller (40)
University of Texas at Austin

Eckhard A. Groll (41)
Purdue University

Joe G. Murray (41)
JEM Associates

Rodney L. Osborne (41)
Battelle Memorial Institute

William J. Plzak (41)
The Trane Company

Roland Ares (42)
Ares Corporation

Michel Lecompte (42)
RefPlus, Inc.

Alfi Helmy Malek (42)
CETIM

Robert L. Bates (43)
E.I. du Pont de Nemours and Co.

Richard C. Niess (43)
Gilbert & Associates

John H. Roberts (43)
The Trane Company

Donald K. Miller (44)
MDK Engineering Corporation

Allan N. Podhorodeski (44)
P-B Engineering Inc.

Robert R. Bittle (45)
Texas Christian University

Richard J. Buck (45)
Sporlan Valve Company

Larry D. Cummings (45)
Delphi Harrison Thermal Division (GMC)

Piotr A. Domanski (45)
National Institute of Standards and Technology

Richard Krause (45)
Henry Valve Company

Duane A. Wolf (45)
J.L. Hall Engineering Services

Nicholas G. Zupp (45)
Delphi Harrison Thermal Division (GMC)

Atma Advani (47)

Albert A. Domingorena (47)
Hussmann Corporation

Carl E. Laverrenz (47, 48)
Tyler Refrigeration

David R. Menninger (47)
The Kroger Company

Robert W. Parkes (47, 48)
Parkes Associates

Fayez F. Ibrahim (48)
Kysor/Bangor

Michael J. Palladino (48)
Victory Refrigeration

Thomas H. Davis (49)
General Electric Appliances

Richard Devos (49)
Frigidaire

John T. Dieckmann (49)
Arthur D. Little, Inc.

ASHRAE HANDBOOK COMMITTEE

Harold G. Lorsch, Chair

1998 Refrigeration Volume Subcommittee: **Evans J. Lizardos,** Chair

Donald L. Fenton **Arthur P. Garbarino** **Stephen A. Mowrer** **Thomas K. O'Donnell** **Gideon Zeidler**

ASHRAE HANDBOOK STAFF

Robert A. Parsons, Editor   **Christina D. Tate,** Associate Editor

Joy M. Moses, Editorial Assistant

Scott A. Zeh, Nancy F. Thysell, and **Jayne E. Jackson,** Publishing Services

W. Stephen Comstock
Director, Communications and Publications
Publisher

ASHRAE TECHNICAL COMMITTEES AND TASK GROUPS

SECTION 1.0—FUNDAMENTALS AND GENERAL
- 1.1 Thermodynamics and Psychrometrics
- 1.2 Instruments and Measurements
- 1.3 Heat Transfer and Fluid Flow
- 1.4 Control Theory and Application
- 1.5 Computer Applications
- 1.6 Terminology
- 1.7 Operation and Maintenance Management
- 1.8 Owning and Operating Costs
- 1.9 Electrical Systems
- 1.10 Energy Resources

SECTION 2.0—ENVIRONMENTAL QUALITY
- 2.1 Physiology and Human Environment
- 2.2 Plant and Animal Environment
- 2.3 Gaseous Air Contaminants and Gas Contaminant Removal Equipment
- 2.4 Particulate Air Contaminants and Particulate Contaminant Removal Equipment
- 2.6 Sound and Vibration Control
- 2.7 Seismic Restraint Design
- TG Buildings' Impacts on the Environment
- TG Global Climate Change

SECTION 3.0—MATERIALS AND PROCESSES
- 3.1 Refrigerants and Brines
- 3.2 Refrigerant System Chemistry
- 3.3 Refrigerant Contaminant Control
- 3.4 Lubrication
- 3.5 Desiccant and Sorption Technology
- 3.6 Corrosion and Water Treatment
- 3.8 Refrigerant Containment

SECTION 4.0—LOAD CALCULATIONS AND ENERGY REQUIREMENTS
- 4.1 Load Calculation Data and Procedures
- 4.2 Weather Information
- 4.3 Ventilation Requirements and Infiltration
- 4.4 Thermal Insulation and Moisture Retarders
- 4.5 Fenestration
- 4.6 Building Operation Dynamics
- 4.7 Energy Calculations
- 4.9 Building Envelope Systems
- 4.10 Indoor Environmental Modeling
- 4.11 Smart Building Systems
- TG Integrated Building Design

SECTION 5.0—VENTILATION AND AIR DISTRIBUTION
- 5.1 Fans
- 5.2 Duct Design
- 5.3 Room Air Distribution
- 5.4 Industrial Process Air Cleaning (Air Pollution Control)
- 5.5 Air-to-Air Energy Recovery
- 5.6 Control of Fire and Smoke
- 5.7 Evaporative Cooling
- 5.8 Industrial Ventilation
- 5.9 Enclosed Vehicular Facilities
- 5.10 Kitchen Ventilation

SECTION 6.0—HEATING EQUIPMENT, HEATING AND COOLING SYSTEMS AND APPLICATIONS
- 6.1 Hydronic and Steam Equipment and Systems
- 6.2 District Heating and Cooling
- 6.3 Central Forced Air Heating and Cooling Systems
- 6.4 In-Space Convection Heating
- 6.5 Radiant Space Heating and Cooling
- 6.6 Service Water Heating
- 6.7 Solar Energy Utilization
- 6.8 Geothermal Energy Utilization
- 6.9 Thermal Storage
- 6.10 Fuels and Combustion

SECTION 7.0—PACKAGED AIR-CONDITIONING AND REFRIGERATION EQUIPMENT
- 7.1 Residential Refrigerators and Food Freezers
- 7.4 Unitary Combustion-Engine-Driven Heat Pumps
- 7.5 Room Air Conditioners and Dehumidifiers
- 7.6 Unitary Air Conditioners and Heat Pumps

SECTION 8.0—AIR-CONDITIONING AND REFRIGERATION SYSTEM COMPONENTS
- 8.1 Positive Displacement Compressors
- 8.2 Centrifugal Machines
- 8.3 Absorption and Heat Operated Machines
- 8.4 Air-to-Refrigerant Heat Transfer Equipment
- 8.5 Liquid-to-Refrigerant Heat Exchangers
- 8.6 Cooling Towers and Evaporative Condensers
- 8.7 Humidifying Equipment
- 8.8 Refrigerant System Controls and Accessories
- 8.10 Pumps and Hydronic Piping
- 8.11 Electric Motors and Motor Control

SECTION 9.0—AIR-CONDITIONING SYSTEMS AND APPLICATIONS
- 9.1 Large Building Air-Conditioning Systems
- 9.2 Industrial Air Conditioning
- 9.3 Transportation Air Conditioning
- 9.4 Applied Heat Pump/Heat Recovery Systems
- 9.5 Cogeneration Systems
- 9.6 Systems Energy Utilization
- 9.7 Testing and Balancing
- 9.8 Large Building Air-Conditioning Applications
- 9.9 Building Commissioning
- 9.10 Laboratory Systems
- 9.11 Clean Spaces
- TG Combustion Gas Turbine Inlet Air Cooling Systems
- TG Tall Buildings

SECTION 10.0—REFRIGERATION SYSTEMS
- 10.1 Custom Engineered Refrigeration Systems
- 10.2 Automatic Icemaking Plants and Skating Rinks
- 10.3 Refrigerant Piping, Controls, and Accessories
- 10.4 Ultra-Low Temperature Systems and Cryogenics
- 10.5 Refrigerated Distribution and Storage Facilities
- 10.6 Transport Refrigeration
- 10.7 Commercial Food and Beverage Cooling Display and Storage
- 10.8 Refrigeration Load Calculations
- 10.9 Refrigeration Application for Foods and Beverages

PREFACE

This Handbook covers the refrigeration equipment and systems used for applications other than human comfort. This book includes information on cooling, freezing, and storing food; industrial applications of refrigeration; and low-temperature refrigeration. While this Handbook is primarily a reference for the practicing engineer, it is also a useful reference for anyone involved in the cooling and storage of food products.

Most of the chapters from the 1994 *Refrigeration Handbook* were revised for this volume. Two new chapters and two chapters from the 1993 *Fundamentals Handbook*, which were rewritten, have been added to this book. Some of the revisions that have been made are as follows:

- Chapter 5, Refrigerant System Chemistry, includes more refrigerants and blends in tables listing ozone depletion potential and global warming potential. The information on hydrocarbons, refrigerant analysis, and compatibility of refrigerants, lubricants, and additives has been expanded.
- Chapter 6, Control of Moisture and Other Contaminants in Refrigerant Systems, now describes the advantage of molecular sieves in excluding refrigerant molecules, which increases the capacity of desiccants. Additional information has been included on residuals in polyol ester lubricants and refrigerant recovery, recycling, and reclamation. A new section on contaminant control during retrofit has been added.
- Chapter 7, Lubricants in Refrigerant Systems, continues to include information on CFC refrigerants for the benefit of developing countries. The information on synthetic lubricants has been expanded, and the chapter briefly describes preparation for refrigerant conversion.
- Chapter 8, Thermal Properties of Foods, has been moved from the 1993 *Fundamentals Handbook* and rewritten. It includes tables listing physical and thermal properties of various food products and describes methods for calculating those properties.
- Chapter 9, Cooling and Freezing Times of Foods, has also been moved from the 1993 *Fundamentals Handbook*. It suggests theoretical models that can be used to calculate cooling and freezing times of individual products shaped as spheres, slabs, or cylinders.
- Chapter 10, Commodity Storage Requirements. The thermal properties of produce formerly listed in this chapter have been updated and moved to Chapter 8. Recommended storage periods remain the same as listed in the 1994 volume.
- Chapter 11, Food Microbiology and Refrigeration, now includes information on HACCP, a preventive program that establishes procedures to control microorganisms.
- Chapter 12, Refrigeration Load, has been expanded, and information on heat gain from motors, internal loads, equipment-related loads, and loads from open doors has been clarified.
- Chapter 13, Refrigerated Facility Design, now includes more information on controls and control architecture. Also included is a description of a construction method that incorporates an excellent vapor-retarder system.
- Chapter 15, Industrial Food Freezing Systems, shows a specialized contact freezer and a new section on refrigeration systems.
- Chapter 16, Meat Products, now shows typical steps in meat processing. Also, information on sanitation has been expanded.
- Chapter 17, Poultry Products, includes a figure that summarizes the process flow. Information on sanitation has been expanded.
- Chapter 20, Eggs and Egg Products, has additional information on sanitation and HACCP.
- Chapter 26, Processed, Precooked, and Prepared Foods, includes additional information on freezing fruit.
- Chapter 27, Bakery Products, has additional information on uses of refrigeration in the bakery industry. Sections on ingredient storage, mixing, bread cooling and freezing, and freezing of other bakery products are expanded and updated.
- Chapter 28, Candies, Nuts, Dried Fruits, and Dried Vegetables, has an updated description of refrigeration equipment used in candy making plants.
- Chapter 32, Insulation Systems for Refrigerant Piping, is a new chapter that discusses the materials and methods used to insulate piping for low-temperature applications.
- Chapter 38, Cryogenics, was rewritten. It discusses properties of materials and gases at cryogenic temperatures, basic refrigeration cycles for achieving cryogenic temperatures, cryocoolers, and the separation and purification of gases. The equipment, insulation, storage, transfer systems, instrumentation, and hazards involved with low-temperature installations are also covered.
- Chapter 39, Low-Temperature Refrigeration, is a new chapter that discusses refrigeration between -50 to $-100°C$. It describes cascade refrigeration systems. It also updates information on metallurgy and adds information on plastics, adhesives, insulation systems, and properties of refrigerants and secondary coolants. Heat transfer characteristics in this temperature range are also discussed.
- Chapter 41, Absorption Cooling, Heating, and Refrigeration Equipment, now has a section (and diagram) that covers single-effect heat transformers.
- Chapter 42, Forced-Circulation Air Coolers, has been updated and includes additional information on types of coolers. It also discusses recent studies on the effect of frost accumulation on evaporator performance and defrosting methods.
- Chapter 45, Refrigerant-Control Devices, has been updated and includes a new method and equations for designing capillary tubes.
- Chapter 46, Factory Dehydrating, Charging, and Testing, includes an expanded table listing typical factory dehydrating and moisture-measuring methods for refrigeration systems.
- Chapter 47, Retail Food Store Refrigeration and Equipment, combines the retail food store refrigeration and equipment chapters in the 1994 volume.

Each Handbook is published in two editions. One edition contains inch-pound (I-P) units of measurement, and the other contains the International System of Units (SI).

Look for corrections to the 1995, 1996, and 1997 volumes of the Handbook on the Internet at http://www.ashrae.org. Any changes to this volume will be reported in the 1999 *ASHRAE Handbook* and on the Internet.

If you have suggestions for improving a chapter or you would like more information on how you can help revise a chapter, e-mail bparsons@ashrae.org; write to Handbook Editor, ASHRAE, 1791 Tullie Circle, Atlanta, GA 30329; or fax (404) 321-5478.

Robert A. Parsons
ASHRAE Handbook Editor

CHAPTER 1

LIQUID OVERFEED SYSTEMS

Overfeed System Operation 1.1	Evaporator Design 1.6
Refrigerant Distribution 1.2	Refrigerant Charge 1.6
Oil in System 1.3	Start-Up and Operation 1.6
Circulating Rate 1.4	Line Sizing 1.7
Pump Selection and Installation 1.4	Low-Pressure Receiver Sizing 1.7
Controls 1.5	

OVERFEED systems are those in which excess liquid is forced, either mechanically or by gas pressure, through organized-flow evaporators, separated from the vapor, and returned to the evaporators.

Terminology

Low-pressure receiver. Sometimes referred to as an **accumulator**, this vessel acts as the separator for the mixture of vapor and liquid returning from the evaporators. A constant refrigerant level is usually maintained by conventional control devices.

Pumping unit. One or more mechanical pumps or gas-operated liquid circulators arranged to pump the overfeed liquid to the evaporators. The pumping unit is located below the low-pressure receiver.

Wet returns. Connections between the evaporator outlets and the low-pressure receiver through which the mixture of vapor and overfeed liquid is drawn.

Liquid feeds. Connections between the pumping unit outlet and the evaporator inlets.

Flow control regulators. Devices used to regulate the overfeed flow into the evaporators. They may be needle-type valves, fixed orifices, calibrated manual regulating valves, or automatic valves designed to provide a fixed liquid rate.

Advantages and Disadvantages

The main advantages of liquid overfeed systems are high system efficiency and reduced operating expenses. These systems have lower energy cost and fewer operating hours because

1. The evaporator surface is used efficiently through good refrigerant distribution and completely wetted internal tube surfaces.
2. The compressors are protected. Liquid slugs resulting from fluctuating loads or malfunctioning controls are separated from suction gas in the low-pressure receiver.
3. Low-suction superheats are achieved where the suction lines between the low-pressure receiver and the compressors are short. This causes a minimum discharge temperature, preventing lubrication breakdown and minimizing condenser fouling.
4. With simple controls, evaporators can be hot-gas defrosted with little disturbance to the system.
5. Refrigerant feed to evaporators is unaffected by fluctuating ambient and condensing conditions. The flow control regulators do not need to be adjusted after the initial setting because the overfeed rates are not generally critical.
6. Flash gas resulting from refrigerant throttling losses is removed at the low-pressure receiver before entering the evaporators. This gas is drawn directly to the compressors and eliminated as a factor in the design of the system low side. It does not contribute to increased pressure drops in the evaporators or overfeed lines.

7. Refrigerant level controls, level indicators, refrigerant pumps, and oil drains are generally located in the equipment rooms, which are under operator surveillance or computer monitoring.
8. Because of ideal entering suction gas conditions, compressors last longer. There is less maintenance and fewer breakdowns. The oil circulation rate to the evaporators is reduced as a result of the low compressor discharge superheat and separation at the low-pressure receiver (Scotland 1963).
9. Overfeed systems have convenient automatic operation.

The following are possible disadvantages:

1. In some cases, refrigerant charges are greater than those used in other systems.
2. Higher refrigerant flow rates to and from evaporators cause the liquid feed and wet return lines to be larger in diameter than the high-pressure liquid and suction lines for other systems.
3. Piping insulation, which is costly, is generally required on all feed and return lines to prevent condensation, frost formation, or heat gain.
4. The installed cost may be greater, particularly for small systems or those having fewer than three evaporators.
5. The operation of the pumping unit requires added expenses that are offset by the increased efficiency of the overall system.
6. The pumping units may require maintenance.
7. Pumps sometimes have cavitation problems to due low available net positive suction pressure.

Generally, the more evaporators used, the more favorable are the initial costs for liquid overfeed compared to a gravity recirculated or flooded system (Scotland 1970). Liquid overfeed systems compare favorably with thermostatic valve feed systems for the same reason. For small systems, the initial cost for liquid overfeed may be higher than for direct expansion.

Ammonia Systems. Easy operation and lower maintenance are attractive features for even small ammonia systems. However, for ammonia systems operating below −20°C evaporating temperature, some manufacturers do not supply direct-expansion evaporators due to unsatisfactory refrigerant distribution and control problems.

OVERFEED SYSTEM OPERATION

Mechanical Pump

Figure 1 shows a simplified pumped overfeed system in which a constant liquid level is maintained in a low-pressure receiver. A mechanical pump circulates liquid through the evaporator(s). The two-phase return mixture is separated in the low-pressure receiver. The vapor is directed to the compressor(s). The makeup refrigerant enters the low-pressure receiver by means of a refrigerant metering device.

Figure 2 shows a horizontal low-pressure receiver with a minimum pump pressure, two service valves in place, and a strainer on the suction side of the pump. Valves from the low-pressure receiver to the pump should be selected to have a minimal pressure drop. The

---

The preparation of this chapter is assigned to TC 10.1, Custom Engineered Refrigeration Systems.

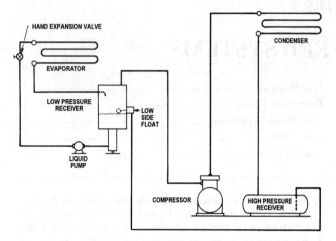

Fig. 1 Liquid Overfeed with Mechanical Pump

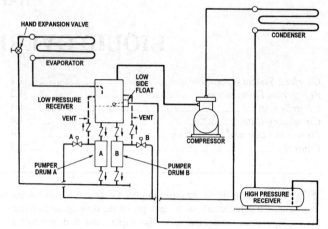

Fig. 3 Double Pumper Drum System

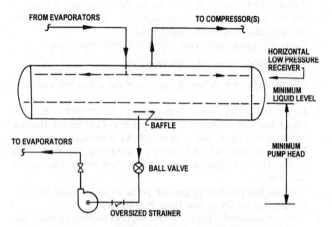

Fig. 2 Pump Circulation, Horizontal Separator

strainer protects hermetic pumps when oil is miscible with the refrigerant. It should have a free area twice the transverse cross-sectional area of the line in which it is installed. With ammonia, use of a suction strainer should be evaluated. Open drive pumps do not require strainers. If no strainer is used, a dirt leg should be used to reduce the risk of solids getting into the pump.

Generally, the minimum pump pressure should be at least double the net positive suction pressure to avoid cavitation. The liquid velocity to the pump should not exceed 0.9 m/s. Net positive suction pressure and flow requirements vary with pump type and design. The pump manufacturer should be consulted for specific requirements. The pump should be evaluated over the full range of operation at low and high flow conditions. Centrifugal pumps have a "flat curve" and have difficulty with systems in which discharge pressure fluctuates.

Gas Pump

Figure 3 shows a basic gas-pumped liquid overfeed system, in which the pumping power is supplied by gas at condenser pressure. In this system, a level control maintains the liquid level in the low-pressure receiver. There are two pumper drums; one is filled by the low-pressure receiver, while the other is drained as hot gas pushes liquid from the pumper drum to the evaporator. Pumper drum B drains when hot gas enters the drum through valve B. To function properly, the pumper drums must be correctly vented so they can fill during the fill cycle.

Another common arrangement is shown in Figure 4. In this system, the high-pressure liquid is flashed into a controlled-pressure receiver that maintains constant liquid pressure at the evaporator

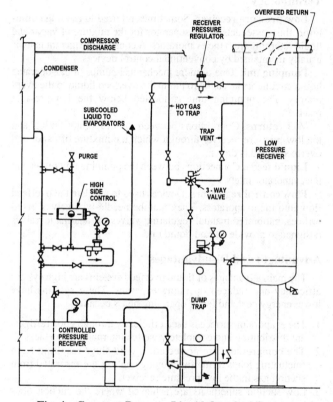

Fig. 4 Constant-Pressure Liquid Overfeed System

inlets, resulting in continuous liquid feed at constant pressure. The flash gas is drawn into the low-pressure receiver through a receiver pressure regulator. Excess liquid drains into a liquid dump trap from the low-pressure receiver. Check valves and a three-way equalizing valve transfer the liquid into the controlled-pressure receiver during the dump cycle. Refinements of this system are used for multistage systems.

REFRIGERANT DISTRIBUTION

To prevent underfeeding and excessive overfeeding of refrigerants, metering devices regulate the liquid feed to each evaporator and/or evaporator circuit. An automatic regulating device continuously controls refrigerant feed to the design value. Other devices commonly used are hand expansion valves, calibrated regulating valves, orifices, and distributors.

Liquid Overfeed Systems

It is time-consuming to adjust hand expansion valves to achieve ideal flow conditions. However, they have been used with some success in many installations prior to the availability of more sophisticated controls. One factor to consider is that standard hand expansion valves are designed to regulate flows caused by the relatively high pressure differences between condensing and evaporating pressure. In overfeed systems, large differences do not exist, so valves with larger orifices may be needed to cope with the combination of the increased quantity of refrigerant and the relatively small pressure differences. Caution must be exercised when using larger orifices because controllability decreases as orifice size increases.

Calibrated, manually operated regulating valves reduce some of the uncertainties involved in using conventional hand expansion valves. To be effective, the valves should be adjusted to the manufacturer's recommendations. Because the refrigerant in the liquid feed lines is above saturation pressure, the lines should not contain flash gas. However, liquid flashing can occur if excessive heat gains by the refrigerant and/or high pressure drops build up in the feed lines.

Orifices should be carefully designed and selected; once installed they cannot be adjusted. They are generally used only for top- and horizontal-feed multicircuit evaporators. Foreign matter and congealed oil globules can cause flow restriction; a minimum orifice of 2.5 mm is recommended. With ammonia, the rate of circulation may have to be increased beyond that needed for the minimum orifice size because of the small liquid volume normally circulated. Pumps and feed and return lines larger than minimum may be needed. This does not apply to halocarbons because of the greater liquid volume circulated as a result of fluid characteristics.

Conventional multiple outlet distributors with capillary tubes of the type usually paired with thermostatic expansion valves have been used successfully in liquid overfeed systems. Capillary tubes may be installed downstream of a distributor with oversized orifices to achieve the required pressure reduction and efficient distribution.

Existing gravity-flooded evaporators with accumulators can be connected to liquid overfeed systems. Changes may be needed only for the feed to the accumulator, with suction lines from the accumulator connected to the system wet return lines. An acceptable arrangement is shown in Figure 5. Generally, gravity-flooded evaporators have different circuiting arrangements from overfeed evaporators. In many cases, the circulating rates developed by thermosiphon action are greater than the circulating rates used in conventional overfeed systems.

Example 1. Find the orifice diameter of an ammonia overfeed system with a refrigeration load per circuit of 4.47 kW and a circulating rate of 7. The evaporating temperature is −35°C, the pressure drop across the orifice is 55 kPa, and the coefficient of discharge for the orifice is 0.61. The circulation per circuit is 33.3 mL/s.

Solution: Orifice diameter may be calculated as follows:

$$d = \left(\frac{Q}{C_d}\right)^{0.5}\left(\frac{\rho}{p}\right)^{0.25} \quad (1)$$

where

d = orifice diameter, mm
Q = discharge through orifice, mL/s
p = pressure drop through orifice, Pa
ρ = density of fluid at −35°C
  = 683.7 kg/m³
C_d = coefficient of discharge for orifice

$$d = \left(\frac{33.3}{0.61}\right)^{0.5}\left(\frac{683.7}{55 \times 1000}\right)^{0.25} = 2.47 \text{ mm}$$

Note: As noted in the text, use a 2.5 mm diameter orifice to avoid clogging.

OIL IN SYSTEM

In spite of reasonably efficient compressor discharge oil separators, oil finds its way into the system low-pressure sides. In the case of ammonia overfeed systems, the bulk of this oil can be drained from the low-pressure receivers with suitable oil drainage facilities. In low-temperature systems, a separate valved and pressure-protected, noninsulated oil drain pot can be placed in a warm space at the accumulator. The oil/ammonia mixture flows into the pot, and the refrigerant evaporates. This arrangement is shown in Figure 6. At pressures lower than atmospheric, high-pressure vapor must be piped into the oil pot to force oil out. Because of the low solubility of oil in liquid ammonia, thick oil globules circulate with the liquid and can restrict flow through strainers, orifices, and regulators. To maintain high efficiency, oil should be removed from the system by regular draining.

Except at low temperatures, halocarbons are miscible with oil. Therefore, positive oil return to the compressor must be ensured. There are many methods, including oil stills using both electric heat and heat exchange from high-pressure liquid or vapor. Some arrangements are discussed in Chapter 2. At low temperatures, oil skimmers must be used because oil migrates to the top of the low-pressure receiver.

Buildup of excessive oil in evaporators must not be allowed because it causes efficiency to decrease rapidly. This is particularly critical in evaporators with high heat transfer rates associated with low volumes, such as flake-type ice makers, ice cream freezers, and scraped-surface heat exchangers. Because the refrigerant flow rate through such evaporators is high, excessive oil can accumulate and rapidly reduce efficiency.

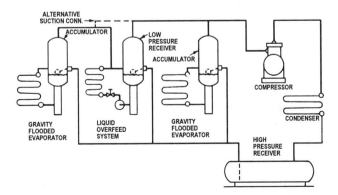

Fig. 5 Liquid Overfeed System Connected on Common System with Gravity-Flooded Evaporators

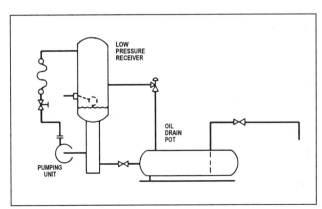

Fig. 6 Oil Drain Pot Connected to Low-Pressure Receiver

CIRCULATING RATE

In a liquid overfeed system, the **circulating number or rate** is the mass ratio of liquid pumped to amount of vaporized liquid. The amount of liquid vaporized is based on the latent heat for the refrigerant at the evaporator temperature. The **overfeed rate** is the ratio of liquid to vapor returning to the low-pressure receiver. When vapor leaves an evaporator at saturated vapor conditions with no excess liquid, the circulating rate is 1 and the overfeed rate is 0. With a circulating rate of 4, the overfeed rate at full load is 3; at no load, it is 4. Most systems are designed for steady flow conditions. With few exceptions, the load conditions may vary, causing fluctuating temperatures outside and within the evaporator. Evaporator capacities vary considerably; with constant refrigerant flow to the evaporator, the overfeed rate fluctuates.

For each evaporator, there is an ideal circulating rate for every loading condition that will result in the minimum temperature difference and the best evaporator efficiency (Lorentzen 1968, Lorentzen and Gronnerud 1967). With few exceptions, it is impossible to predict ideal circulating rates or to design a plant for automatic adjustment of the rates to suit fluctuating loads. The optimum rate can vary with heat load, pipe diameter, circuit length, and number of parallel circuits to achieve the best performance. High circulating rates can cause excessively high pressure drops through evaporators and wet return lines. The sizing of these return lines, discussed in the section on Line Sizing, can have a bearing on the ideal rates. Many evaporator manufacturers specify recommended circulating rates for their equipment. The rates shown in Table 1 agree with these recommendations.

Because of distribution considerations, higher circulating rates are common with top feed evaporators. In multicircuit systems, the refrigerant distribution must be adjusted to provide the best possible results. Incorrect distribution can cause excessive overfeed in some circuits, while others may be starved. Manual or automatic regulating valves can be used to control flow for the optimum or design value.

Halocarbon densities are about twice that of ammonia. If halocarbons R-22, R-134a, and R-502 are circulated at the same rate as ammonia, the halocarbons require 6 to 8.3 times more energy for pumping to the same height than the less dense ammonia. Because this pumping energy must be added to the system load, halocarbon circulating rates are usually lower than those for ammonia. Ammonia has a relatively high latent heat of vaporization, so for equal heat removal, much less ammonia mass must be circulated compared to halocarbons.

Although halocarbons circulate at lower rates than ammonia, the wetting process in the evaporators is still efficient because of the liquid and vapor volume ratios. For example, at –40°C evaporating temperature, with constant flow conditions in the wet return connections, similar ratios of liquid and vapor are experienced with a circulating rate of 4 for ammonia and 2.5 for R-22, R-502, and R-134a. With halocarbons, some additional wetting is also experienced because of the solubility of the oil in these refrigerants.

When bottom feed is used for multicircuit coils, a minimum feed rate per circuit is not necessary because orifices or other distribution devices are not required. The circulating rate for top feed and horizontal feed coils may be determined by the minimum rates from the orifices or other distributors in use.

Figure 7 provides a method for determining the liquid refrigerant flow (Niederer 1964). The charts indicate the amount of refrigerant vaporized in a 1 kW system with circulated operation having no flash gas in the liquid feed line. The value obtained from the chart may be multiplied by the desired circulating rate and by the total refrigeration to determine total flow.

The pressure drop through the flow control regulators is usually 10 to 50% of the available feed pressure. The pressure at the outlet of the flow regulators must be higher than the vapor pressure at the low-pressure receiver by an amount equal to the total pressure drop of the two-phase mixture through the evaporator, any evaporator pressure regulator, and wet return lines. This pressure loss could be 35 kPa in a typical system. When using recommended liquid feed sizing practices, assuming a single-story building, the frictional pressure drop from the pump discharge to the evaporators is about 70 kPa. Therefore, a pump for 140 to 170 kPa should be satisfactory in this case, depending on the lengths and sizes of feed lines, the quantity and types of fittings, and the vertical lift involved.

PUMP SELECTION AND INSTALLATION

Types of Pumps

Mechanical pumps, gas pressure pumping systems, and injector systems are available for liquid overfeed systems.

Types of mechanical pump drives include open, semihermetic, magnetic clutch, and hermetic. Rotor arrangements include positive rotary, centrifugal, and turbine vane. Positive rotary and gear-type pumps are generally operated at slow speeds up to 900 rpm. Whatever type of pump is used, care should be taken to prevent flashing at the pump suction and/or within the pump itself.

Centrifugal pumps are typically used for larger volumes, while semihermetic pumps are best suited for halocarbons at or below atmospheric refrigerant saturated pressure. Regenerative turbines are used with relatively high pressure and large swings in discharge pressure.

Open-type pumps are fitted with a wide variety of packing or seals. For continuous duty, a mechanical seal with an oil reservoir or a liquid refrigerant supply to cool, wash, and lubricate the seals is commonly used. Experience with the particular application or the recommendations of an experienced pump supplier are the best guide for selecting the packing or seal. A magnetic coupling between the motor and the pump can be used instead of shaft coupling to eliminate shaft seals. A small immersion-type electric heater within the oil reservoir can be used with low-temperature systems to ensure that the oil remains fluid. Motors should have a service factor that compensates for drag on the pump if the oil is cold or stiff.

Considerations should include ambient temperatures, heat leakage, fluctuating system pressures from compressor cycling, internal bypass of liquid to pump suction, friction heat, motor heat conduction, dynamic conditions, cycling of automatic evaporator liquid and suction stop valves, action of regulators, gas entrance with liquid, and loss of subcooling by pressure drop. Another factor to consider is the time lag caused by the heat capacity of pump suction, cavitation, and net positive suction pressure factors (Lorentzen 1963).

The motor and stator of hermetic pumps are separated from the refrigerant by a thin nonmagnetic membrane. The metal membrane should be strong enough to withstand system design pressures. Normally, the motors are cooled and the bearings lubricated by liquid refrigerant bypassed from the pump discharge. It is good practice to use two pumps, one operating and one standby.

Installing and Connecting Mechanical Pumps

Because of the sensitive suction conditions of mechanical pumps operating on overfeed systems, the manufacturer's application and installation specifications must be followed closely. Suction connections should be as short as possible, without restrictions, valves, or elbows. Angle or full-flow ball valves should be used. Using

Table 1 Recommended Minimum Circulating Rate

Refrigerant	Circulating Rate[a]
Ammonia (R-717)	
Downfeed (large-diameter tubes)	6 to 7
Upfeed (small-diameter tubes)	2 to 4
R-22—upfeed	3
R-134a	2

[a]Circulating rate of 1 equals evaporating rate.

Liquid Overfeed Systems

and drained from each system. This comparison determines whether oil is accumulating in systems. Oil should not be drained in halocarbon systems. Due to the miscibility of oil with halocarbons at high temperatures, it may be necessary to add oil to the system until an operating balance is achieved (Stoecker 1960, Soling 1971).

Operating Costs and Efficiency

Operating costs for overfeed systems are generally lower than for other systems. Operating costs may not be lower in all cases due to the variety of inefficiencies that exist from system to system and from plant to plant. However, in cases where existing dry expansion plants were converted to liquid overfeed, the operating hours, power, and maintenance costs were reduced. The efficiency of the early gas pump systems has been improved by using high-side pressure to circulate the overfeed liquid. This type of system is indicated in the controlled pressure system shown in Figure 4. Refinements of the double pumper drum arrangement (shown in Figure 3) have also been developed.

Gas-pumped systems, which use refrigerant gas to pump liquid to the evaporators or to the controlled-pressure receiver, require additional compressor volume, from which no useful refrigeration is obtained. These systems consume 4 to 10% or more of the compressor power to maintain the refrigerant flow.

If the condensing pressure is reduced as much as 70 kPa, the compressor power per unit of refrigeration drops by about 7%. Where outdoor dry- and wet-bulb conditions allow, a mechanical pump can be used to pump the gas with no effect on evaporator performance. Gas-operated systems must, however, maintain the condensing pressure within a much smaller range to pump the liquid and maintain the required overfeed rate.

LINE SIZING

The liquid feed line to the evaporator and the wet return line to the low-pressure receiver cannot be sized by the method described in Chapter 33 of the 1997 *ASHRAE Handbook—Fundamentals*. Figure 7 can be used to size liquid feed lines. The circulating rate from Table 1 is multiplied by the evaporating rate. For example, an evaporator with a circulating rate of 4 that forms vapor at a rate of 50 g/s needs a feed line sized for $4 \times 50 = 200$ g/s.

Alternative methods that may be used to design wet returns include the following:

1. Use one pipe size larger than calculated for vapor flow alone.
2. Use a velocity selected for dry expansion reduced by the factor $\sqrt{1/\text{Circulating Rate}}$. This method suggests that the wet return velocity for a circulating rate of 4 is $\sqrt{1/4} = 0.5$ or half that of the acceptable dry vapor velocity.
3. Use the design method described by Chaddock et al. (1972). The report includes tables of flow capacities at 0.036 K drop per metre of horizontal lines for R-717 (ammonia), R-12, R-22, and R-502.

When sizing refrigerant lines, the following design precautions should be taken:

1. Carefully size overfeed return lines with vertical risers because more liquid is held in risers than in horizontal pipe. This holdup increases with reduced vapor flow and increases pressure loss because of gravity and two-phase pressure drop.
2. Use double risers with halocarbons to maintain velocity at partial loads and to reduce liquid static pressure loss (Miller 1979).
3. Add the equivalent of a 100% liquid static height penalty to the pressure drop allowance to compensate for liquid holdup in ammonia systems that have unavoidable vertical risers.
4. As alternatives in severe cases, provide traps and a means of pumping liquids, or use dual-pipe risers.
5. Install low pressure drop valves so the stems are horizontal or nearly so (Chisolm 1971).

LOW-PRESSURE RECEIVER SIZING

Low-pressure receivers are also called liquid separators, suction traps, accumulators, liquid-vapor separators, flash-type coolers, gas and liquid coolers, surge drums, knock-out drums, slop tanks, or low-side pressure vessels, depending on their function and the preference of the user.

The sizing of low-pressure receivers is determined by the required liquid holdup volume and the allowable gas velocity. The volume must accommodate the fluctuations of liquid in the evaporators and overfeed return lines as a result of load changes and defrost periods. It must also handle the swelling and foaming of the liquid charge in the receiver, which is caused by boiling during temperature increase or pressure reduction. At the same time, a liquid seal must be maintained on the supply line for continuous circulation devices. A separating space must be provided for gas velocity low enough to cause a minimum entrainment of liquid drops into the suction outlet. Space limitations and design requirements result in a wide variety of configurations (Miller 1971; Stoecker 1960; Lorentzen 1966; Niemeyer 1961; Scheiman 1963, 1964; Sonders and Brown 1934; Younger 1955).

In selecting a gas-and-liquid separator, adequate volume for the liquid supply and a vapor space above the minimum liquid height for liquid surge must be provided. This requires an analysis of operating load variations. This, in turn, determines the **maximum operating liquid level**. Figures 8 and 9 identify these levels and the important parameters of vertical and horizontal gravity separators.

Vertical separators maintain the same separating area with level variations, while separating areas in horizontal separators change with level variations. **Horizontal separators** should have inlets and outlets separated horizontally by at least the vertical separating distance. A useful arrangement in horizontal separators distributes the inlet flow into two or more connections to reduce turbulence and horizontal velocity without reducing the residence time of the gas flow within the shell (Miller 1971).

In horizontal separators, as the horizontal separating distance is increased beyond the vertical separating distance, the residence time of the vapor passing through is increased so that higher velocities than allowed in vertical separators can be tolerated. As the separating distance is reduced, the amount of liquid entrainment from gravity separators increases. Table 2 shows the gravity separation velocities. For surging loads or pulsating flow associated with large step changes in capacity, the maximum steady flow velocity should be reduced to a value achieved by a suitable multiplier such as 0.75.

The gas-and-liquid separator may be designed with baffles or eliminators to separate liquid from the suction gas returning from the top of the shell to the compressor. More often, sufficient separation space is allowed above the liquid level for this purpose. Such a design is usually of the vertical type, with a separation height above the liquid level of from 600 to 900 mm. The shell diameter is sized to keep the suction gas velocity at a value low enough to allow the liquid droplets to separate and not be entrained with the returning suction gas off the top of the shell.

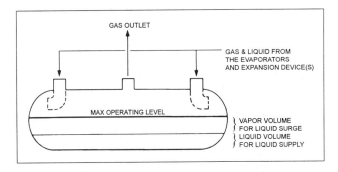

Fig. 8 Basic Horizontal Gas-and-Liquid Separator

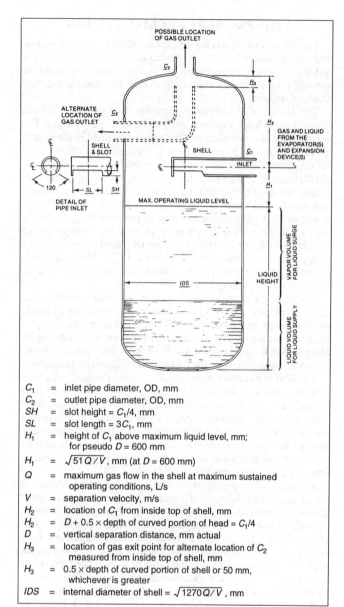

Fig. 9 Basic Vertical Gravity Gas-and-Liquid Separator

C_1 = inlet pipe diameter, OD, mm
C_2 = outlet pipe diameter, OD, mm
SH = slot height = $C_1/4$, mm
SL = slot length = $3C_1$, mm
H_1 = height of C_1 above maximum liquid level, mm; for pseudo D = 600 mm
H_1 = $\sqrt{51Q/V}$, mm (at D = 600 mm)
Q = maximum gas flow in the shell at maximum sustained operating conditions, L/s
V = separation velocity, m/s
H_2 = location of C_1 from inside top of shell, mm
H_2 = D + 0.5 × depth of curved portion of head = $C_1/4$
D = vertical separation distance, mm actual
H_3 = location of gas exit point for alternate location of C_2 measured from inside top of shell, mm
H_3 = 0.5 × depth of curved portion of shell or 50 mm, whichever is greater
IDS = internal diameter of shell = $\sqrt{1270Q/V}$, mm

Although separators are made with length-to-diameter (L/D) ratios of 1/1 increasing to 10/1, the least expensive separators usually have L/D ratios between 3/1 and 5/1. Vertical separators are normally used for systems with reciprocating compressors. Horizontal separators may be preferable where vertical height is critical and/or where large volume space for liquid is required. The procedures for designing vertical and horizontal separators are different.

A vertical gas-and-liquid separator is shown in Figure 9. The end of the inlet pipe C_1 is capped so that flow dispersion is directed downward toward the liquid level. The suggested opening is four times the transverse internal area of the pipe. The height H_1 with a 120° dispersion of the flow reaches to approximately 70% of the internal diameter of the shell.

An alternate inlet pipe with a downturned elbow or mitered bend can be used. However, the jet effect of entering fluid must be considered to avoid undue splashing. The outlet of the pipe must be a minimum distance of IDS/5 above the maximum liquid level in the shell. H_2 is measured from the outlet to the inside top of the shell. It equals D + 0.5 times the depth of the curved portion of the head.

Table 2 Maximum Effective Separation Velocities for R-717, R-22, R-12, and R-502, with Steady Flow Conditions

Temp., °C	Vertical Separation Distance, mm	Maximum Steady Flow Velocity, m/s			
		R-717	R-22	R-12	R-502
+10	250	0.15	0.07	0.08	0.06
	610	0.64	0.31	0.36	0.25
	910	0.71	0.39	0.43	0.32
−7	250	0.21	0.10	0.11	0.08
	610	0.87	0.44	0.49	0.35
	910	0.99	0.52	0.58	0.42
−23	250	0.31	0.14	0.16	0.11
	610	1.29	0.61	0.69	0.49
	910	1.43	0.72	0.81	0.59
−40	250	0.48	0.21	0.24	0.17
	610	1.99	0.88	1.01	0.71
	910	2.17	1.04	1.17	0.84
−57	250	0.80	0.33	0.37	0.25
	610	3.30	1.36	1.54	1.08
	910	3.54	1.57	1.78	1.25

Source: Adapted from Miller (1971).

For the alternate location of C_2, determine IDS from the following equation:

$$IDS = \sqrt{\frac{1270Q}{V} + C_2^2} \qquad (2)$$

The maximum liquid height in the separator is a function of the type of system in which the separator is being used. In some systems this can be estimated, but in others, previous experience is the only guide for selecting the proper liquid height. The accumulated liquid must be returned to the system by a suitable means at a rate comparable to the rate at which it is being collected.

With a horizontal separator, the vertical separation distance used is an average value. The top part of the horizontal shell restricts the gas flow so that the maximum vertical separation distance cannot be used. If H_t represents the maximum vertical distance from the liquid level to the inside top of the shell, the average separation distance as a fraction of IDS is as follows:

H_t/IDS	D/IDS	H_t/IDS	D/IDS
0.1	0.068	0.6	0.492
0.2	0.140	0.7	0.592
0.3	0.215	0.8	0.693
0.4	0.298	0.9	0.793
0.5	0.392	1.0	0.893

The suction connection(s) for refrigerant gas leaving the horizontal shell must be located at or above the location established by the average distance for separation. The maximum cross-flow velocity of gas establishes the residence time for the gas and any entrained liquid droplets in the shell. The most effective removal of entrainment occurs when the residence time is at a maximum practical value. Regardless of the number of gas outlet connections for uniform distribution of gas flow, the cross-sectional area of the gas space is

$$A_x = \frac{2000DQ}{VL} \qquad (3)$$

where

A_x = minimum transverse net cross-sectional area or gas space, mm²
D = average vertical separation distance, mm
Q = total quantity of gas leaving vessel, L/s
L = inside length of shell, mm
V = separation velocity for separation distance used, m/s

For nonuniform distribution of gas flow in the horizontal shell, determine the minimum horizontal distance for gas flow from point of entry to point of exit as follows:

$$RTL = \frac{1000QD}{VA_x} \quad (4)$$

where

RTL = residence time length, mm
Q = maximum flow for that portion of the shell, L/s

All connections must be sized for the flow rates and pressure drops permissible and must be positioned to minimize liquid splashing. Internal baffles or mist eliminators can reduce the diameter of vessels; however, test correlations are necessary for a given configuration and placement of these devices.

An alternate formula for determining separation velocities that can be applied to separators is

$$v = k\sqrt{\frac{\rho_l - \rho_v}{\rho_v}} \quad (5)$$

where

v = velocity of vapor, m/s
ρ_l = density of liquid, kg/m^3
ρ_v = density of vapor, kg/m^3
k = factor based on experience without regard to vertical separation distance and surface tension for gravity separators

In gravity liquid/vapor separators that must separate heavy entrainment from vapors, use a k of 0.03. This gives velocities equivalent to those used for 300 to 350 mm vertical separation distance for R-717 and 350 to 400 mm vertical separation distance for halocarbons. In knockout drums that separate light entrainment, use a k of 0.06. This gives velocities equivalent to those used for 900 mm vertical separation distance for R-717 and for halocarbons.

REFERENCES

Chaddock, J.B., D.P. Werner, and C.G. Papachristou. 1972. Pressure drop in the suction lines of refrigerant circulation systems. *ASHRAE Transactions* 78(2):114-23.

Chisholm, D. 1971. Prediction of pressure drop at pipe fittings during two-phase flow. *Proceedings* I.I.R., Washington, D.C.

Lorentzen, G. 1963. Conditions of cavitation in liquid pumps for refrigerant circulation. *Progress Refrigeration Science Technology* I:497.

Lorentzen, G. 1965. How to design piping for liquid recirculation. *Heating, Piping & Air Conditioning* (June):139.

Lorentzen, G. 1966. On the dimensioning of liquid separators for refrigeration systems. *Kältetechnik* 18:89.

Lorentzen, G. 1968. Evaporator design and liquid feed regulation. *Journal of Refrigeration* (November-December):160.

Lorentzen, G. and R. Gronnerud. 1967. On the design of recirculation type evaporators. *Kulde* 21(4):55.

Miller, D.K. 1971. Recent methods for sizing liquid overfeed piping and suction accumulator-receivers. *Proceedings* I.I.R., Washington, D.C.

Miller D.K. 1974. Refrigeration problems of a VCM carrying tanker. *ASHRAE Journal* 11.

Miller, D.K. 1979. Sizing dual suction risers in liquid overfeed refrigeration systems. *Chemical Engineering* 9.

Niederer, D.H. 1964. Liquid recirculation systems—What rate of feed is recommended. *The Air Conditioning & Refrigeration Business* (December).

Niemeyer, E.R. 1961. Check these points when designing knockout drums. *Hydrocarbon Processing and Petroleum Refiner* (June).

Scheiman, A.D. 1964. Horizontal vapor-liquid separators. *Hydrocarbon Processing and Petroleum Refiner* (May).

Scheiman, A.D. 1963. Size vapor-liquid separators quicker by nomograph. *Hydrocarbon Processing and Petroleum Refiner* (October).

Scotland, W.B. 1963. Discharge temperature considerations with multicylinder ammonia compressors. *Modern Refrigeration* (February).

Scotland, W.B. 1970. Advantages, disadvantages and economics of liquid overfeed systems. *ASHRAE Symposium Bulletin* KC-70-3, Liquid overfeed systems.

Soling, S.P. 1971. Oil recovery from low temperature pump recirculating hydrocarbon systems. *ASHRAE Symposium Bulletin* PH-71-2, Effect of oil on the refrigeration system.

Sonders, M. and G.G. Brown. 1934. Design of fractionating columns, entrainment and capacity. *Industrial & Engineering Chemistry* (January).

Stoecker, W.F. 1960. How to design and operate flooded evaporators for cooling air and liquids. *Heating, Piping & Air Conditioning* (December).

Younger, A.H. 1955. How to size future process vessels. *Chemical Engineering* (May).

BIBLIOGRAPHY

Chaddock, J.B. 1976. Two-phase pressure drop in refrigerant liquid overfeed systems—Design tables. *ASHRAE Transactions* 82(2):107-33.

Chaddock, J.B., H. Lau, and E. Skuchas. 1976. Two-phase pressure drop in refrigerant liquid overfeed systems—Experimental measurements. *ASHRAE Transactions* 82(2):134-50.

Geltz, R.W. 1967. Pump overfeed evaporator refrigeration systems. *Air Conditioning, Heating & Refrigeration News* (January 30, February 6, March 6, March 13, March 20, March 27).

Lorentzen, G. and A.O. Baglo. 1969. An investigation of a gas pump recirculation system. *Proceedings* of the Xth International Congress of Refrigeration, p. 215. International Institute of Refrigeration, Paris.

Richards, W.V. 1959. Liquid ammonia recirculation systems. *Industrial Refrigeration* (June):139.

Richards, W.V. 1970. Pumps and piping in liquid overfeed systems. *ASHRAE Symposium Bulletin* KC-70-3, Liquid overfeed systems.

Slipcevic, B. 1964. The calculation of the refrigerant charge in refrigerating systems with circulation pumps. *Kältetechnik* 4:111.

Thompson, R.B. 1970. Control of evaporators in liquid overfeed systems. *ASHRAE Symposium Bulletin* KC-70-3, Liquid overfeed systems.

Watkins, J.E. 1956. Improving refrigeration systems by applying established principles. *Industrial Refrigeration* (June).

CHAPTER 2

SYSTEM PRACTICES FOR HALOCARBON REFRIGERANTS

Refrigerant Flow 2.1	*Piping at Multiple Compressors* 2.19
Refrigerant Line Sizing 2.3	*Piping at Various System Components* 2.20
Discharge (Hot-Gas) Lines 2.14	*Refrigeration Accessories* 2.23
Defrost Gas Supply Lines 2.16	*Pressure Control for Refrigerant Condensers* 2.27
Receivers 2.16	*Keeping Liquid from Crankcase During Off Cycles* 2.28
Air-Cooled Condensers 2.18	*Hot-Gas Bypass Arrangements* 2.29

REFRIGERATION is the process of moving heat from one location to another by use of refrigerant in a closed cycle. Oil management; gas and liquid separation; subcooling, superheating, and piping of refrigerant liquid and gas; and two-phase flow are all part of refrigeration. Applications include air conditioning, commercial refrigeration, and industrial refrigeration.

Desired characteristics of a refrigeration system may include

- Year-round operation, regardless of outdoor ambient conditions
- Possible wide load variations (0 to 100% capacity) during short periods without serious disruption of the required temperature levels
- Frost control for continuous-performance applications
- Oil management for different refrigerants under varying load and temperature conditions
- A wide choice of heat exchange methods (e.g., dry expansion, liquid overfeed, or flooded feed of the refrigerants) and the use of secondary coolants such as salt brine, alcohol, and glycol
- System efficiency, maintainability, and operating simplicity
- Operating pressures and pressure ratios that might require multistaging, cascading, and so forth

A successful refrigeration system depends on good piping design and an understanding of the required accessories. This chapter covers the fundamentals of piping and accessories in halocarbon refrigerant systems. Hydrocarbon refrigerant pipe friction data can be found in petroleum industry handbooks. Use the refrigerant properties and information in Chapters 2, 18, and 19 of the 1997 *ASHRAE Handbook—Fundamentals* to calculate friction losses.

For information on refrigeration load, see Chapter 12.

Piping Basic Principles

The design and operation of refrigerant piping systems should

- Ensure proper refrigerant feed to evaporators
- Provide practical refrigerant line sizes without excessive pressure drop
- Prevent excessive amounts of lubricating oil from being trapped in any part of the system
- Protect the compressor at all times from loss of lubricating oil
- Prevent liquid refrigerant or oil slugs from entering the compressor during operating and idle time
- Maintain a clean and dry system

REFRIGERANT FLOW

Refrigerant Line Velocities

Economics, pressure drop, noise, and oil entrainment establish feasible design velocities in refrigerant lines (see Table 1).

Higher gas velocities are sometimes found in relatively short suction lines on comfort air-conditioning or other applications where the operating time is only 2000 to 4000 h per year and where low initial cost of the system may be more significant than low operating cost. Industrial or commercial refrigeration applications, where equipment runs almost continuously, should be designed with low refrigerant velocities for most efficient compressor performance and low equipment operating costs. An owning and operating cost analysis will reveal the best choice of line sizes. (See Chapter 33 of the 1995 *ASHRAE Handbook—Applications* for information on owning and operating costs). Liquid lines from condensers to receivers should be sized for 0.5 m/s or less to ensure positive gravity flow without incurring backup of liquid flow. Liquid lines from receiver to evaporator should be sized to maintain velocities below 1.5 m/s, thus minimizing or preventing liquid hammer when solenoids or other electrically operated valves are used.

Table 1 Gas Line Velocities for R-22, R-134a, and R-502

Suction line	4.5 to 20 m/s
Discharge line	10 to 18 m/s

Refrigerant Flow Rates

Refrigerant flow rates for R-22, R-134a, and R-502 are indicated in Figures 1 through 3. To obtain the total system flow rate, select the proper rate value and multiply by the system capacity. Enter curves using saturated refrigerant temperature at the evaporator outlet and actual liquid temperature entering the liquid feed device (including subcooling in condensers and liquid-suction interchanger, if used).

Because Figures 1 through 3 are based on a saturated evaporator temperature, they may indicate slightly higher refrigerant flow rates than are actually in effect when the suction vapor is superheated in excess of the conditions mentioned in the last paragraph. Refrigerant flow rates may be reduced approximately 3% for each 5.5 K increase in superheat in the evaporator.

Suction line superheating downstream of the evaporator due to line heat gain from external sources should not be used to reduce evaluated mass flow. This suction line superheating due to line heat gain increases volumetric flow rate and line velocity per unit of evaporator capacity, but not mass flow rate. It should be considered when evaluating a suction line size for satisfactory oil return up risers.

Suction gas superheating from the use of a liquid-suction heat exchanger has an effect on oil return similar to that of suction line superheating. The liquid cooling that results from the heat exchange reduces mass flow rate per kilowatt of refrigeration. This can be seen in Figures 1 through 3 because the reduced temperature of the liquid supplied to the evaporator feed valve has been taken into account.

The preparation of this chapter is assigned to TC 10.3, Refrigerant Piping, Controls, and Accessories.

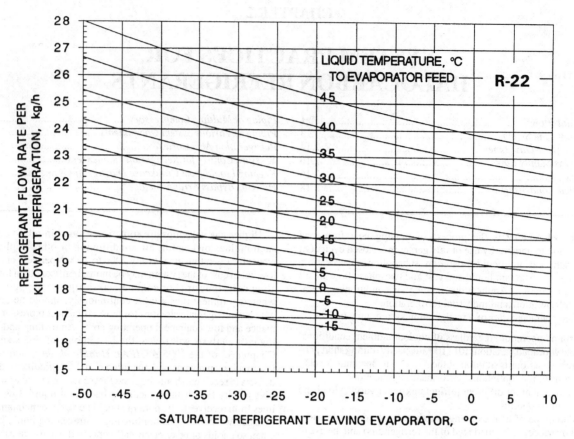

Fig. 1 Flow Rate per Kilowatt of Refrigeration for Refrigerant 22

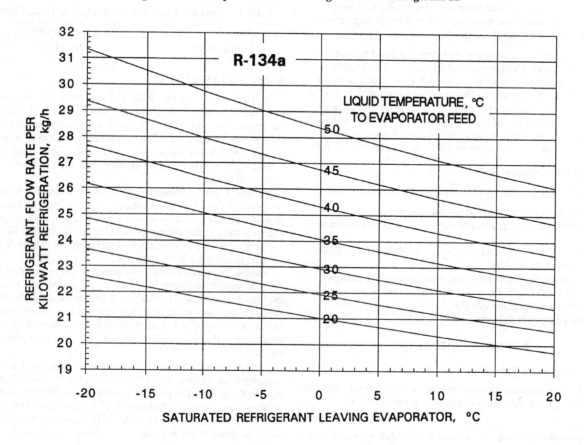

Fig. 2 Flow Rate per Kilowatt of Refrigeration for Refrigerant 134a

System Practices for Halocarbon Refrigerants

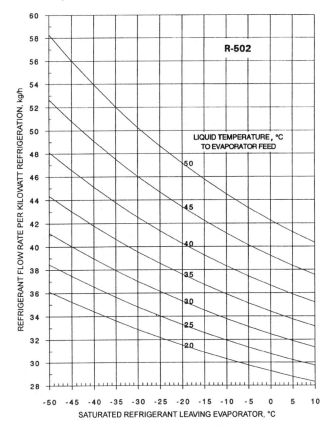

Fig. 3 Flow Rate per Kilowatt of Refrigeration for Refrigerant 502

Superheat due to heat in a space not intended to be cooled is always detrimental because the volumetric flow rate increases with no compensating gain in refrigerating effect.

REFRIGERANT LINE SIZING

In sizing refrigerant lines, cost considerations favor keeping line sizes as small as possible. However, suction and discharge line pressure drops cause loss of compressor capacity and increased power usage. Excessive liquid line pressure drops can cause the liquid refrigerant to flash, resulting in faulty expansion valve operation. Refrigeration systems are designed so that friction pressure losses do not exceed a pressure differential equivalent to a corresponding change in the saturation boiling temperature. The primary measure for determining pressure drops is a given change in saturation temperature.

Pressure Drop Considerations

Pressure drop in refrigerant lines causes a reduction in system efficiency. Correct sizing must be based on minimizing cost and maximizing efficiency. Table 2 indicates the approximate effect of refrigerant pressure drop on an R-22 system operating at a 5 °C saturated evaporator temperature with a 40 °C saturated condensing temperature.

Pressure drop calculations are determined as normal pressure loss associated with a change in saturation temperature of the refrigerant. Typically, the refrigeration system will be sized for pressure losses of 1 K or less for each segment of the discharge, suction, and liquid lines.

Liquid Lines. Pressure drop should not be so large as to cause gas formation in the liquid line, insufficient liquid pressure at the liquid feed device, or both. Systems are normally designed so that the pressure drop in the liquid line, due to friction, is not greater than

Table 2 Approximate Effect of Gas Line Pressure Drops on R-22 Compressor Capacity and Power[a]

Line Loss, K	Capacity, %	Energy, %[b]
Suction Line		
0	100	100
1	96.8	104.3
2	93.6	107.3
Discharge Line		
0	100	100
1	99.2	102.7
2	98.4	105.7

[a]For system operating at 5 °C saturated evaporator temperature and 40 °C saturated condensing temperature.
[b]Energy percentage rated at kW (power)/kW (cooling).

that corresponding to about a 0.5 to 1 K change in saturation temperature. See Tables 3 through 6 for liquid line sizing information. Liquid pressure losses for a change of 0.5 K saturation at 40 °C condensing pressure are approximately as follows:

Refrigerant	Change, kPa
R-22	18.7
R-134a	13.6
R-502	19.4

Liquid subcooling is the only method of overcoming the liquid line pressure loss to guarantee liquid at the expansion device in the evaporator. If the subcooling is insufficient, flashing will occur within the liquid line and degrade the efficiency of the system.

Friction pressure drops in the liquid line are caused by accessories such as solenoid valves, filter driers, and hand valves, as well as by the actual pipe and fittings between the receiver outlet and the refrigerant feed device at the evaporator.

Liquid line risers are a source of pressure loss and add to the total loss of the liquid line. The loss due to risers is approximately 11.3 kPa per metre of liquid lift. The total loss is the sum of all friction losses plus the pressure loss from liquid risers.

The following example illustrates the process of determining the liquid line size and checking for total subcooling required.

Example 1. An R-22 refrigeration system using copper pipe operates at 5 °C evaporator and 40 °C condensing. Capacity is 14 kW, and the liquid line is 50 m equivalent length with a riser of 6 m. Determine the liquid line size and total required subcooling.

Solution: From Table 3, the size of the liquid line at 1 K drop is 15 mm OD. Use the equation in Note 3 of Table 3 to compute actual temperature drop. At 14 kW,

Actual temperature drop	$= \dfrac{(50 \times 0.02)}{(14.0/21.54)^{1.8}}$	= 0.46 K
Estimated friction loss	$= 0.46 \times 18.7$	= 8.6 kPa
Loss for the riser	$= 6 \times 11.3$	= 67.8 kPa
Total pressure losses	$= 67.8 + 8.6$	= 76.4 kPa
Saturation pressure at 40 °C condensing		= 1534.1 kPa
Initial pressure at beginning of liquid line		1534.1 kPa
Total liquid line losses		− 76.4 kPa
Net pressure at expansion device		= 1457.7 kPa

The saturation temperature at 1457.7 kPa is 37.9 °C.
Required subcooling to overcome the liquid losses = (40.0 − 37.9) or 2.1 K

Refrigeration systems that have no liquid risers and have the evaporator below the condenser/receiver benefit from a gain in pressure due to liquid weight and can tolerate larger friction losses without flashing. Regardless of the routing of the liquid lines when

Table 3 Suction, Discharge, and Liquid Line Capacities in Kilowatts for Refrigerant 22 (Single- or High-Stage Applications)

Nominal Line OD, mm	Suction Lines (Δt = 0.04 K/m)					Discharge Lines (Δt = 0.02 K/m, Δp = 74.90)			Liquid Lines See notes a and b	
	Saturated Suction Temperature, °C					Saturated Suction Temperature, °C			Velocity = 0.5 m/s	Δt = 0.02 K/m Δp = 749
	−40	−30	−20	−5	5	−40	−20	5		
	Corresponding Δp, Pa/m									
	196	277	378	572	731					
TYPE L COPPER LINE										
12	0.32	0.50	0.75	1.28	1.76	2.30	2.44	2.60	7.08	11.24
15	0.61	0.95	1.43	2.45	3.37	4.37	4.65	4.95	11.49	21.54
18	1.06	1.66	2.49	4.26	5.85	7.59	8.06	8.59	17.41	37.49
22	1.88	2.93	4.39	7.51	10.31	13.32	14.15	15.07	26.66	66.18
28	3.73	5.82	8.71	14.83	20.34	26.24	27.89	29.70	44.57	131.0
35	6.87	10.70	15.99	27.22	37.31	48.03	51.05	54.37	70.52	240.7
42	11.44	17.80	26.56	45.17	61.84	79.50	84.52	90.00	103.4	399.3
54	22.81	35.49	52.81	89.69	122.7	157.3	167.2	178.1	174.1	794.2
67	40.81	63.34	94.08	159.5	218.3	279.4	297.0	316.3	269.9	1415.0
79	63.34	98.13	145.9	247.2	337.9	431.3	458.5	488.2	376.5	2190.9
105	136.0	210.3	312.2	527.8	721.9	919.7	977.6	1041.0	672.0	4697.0
STEEL LINE										
10	0.47	0.72	1.06	1.78	2.42	3.04	3.23	3.44	10.66	15.96
15	0.88	1.35	1.98	3.30	4.48	5.62	5.97	6.36	16.98	29.62
20	1.86	2.84	4.17	6.95	9.44	11.80	12.55	13.36	29.79	62.55
25	3.52	5.37	7.87	13.11	17.82	22.29	23.70	25.24	48.19	118.2
32	7.31	11.12	16.27	27.11	36.79	46.04	48.94	52.11	83.56	244.4
40	10.98	16.71	24.45	40.67	55.21	68.96	73.31	78.07	113.7	366.6
50	21.21	32.23	47.19	78.51	106.4	132.9	141.3	150.5	187.5	707.5
65	33.84	51.44	75.19	124.8	169.5	211.4	224.7	239.3	267.3	1127.3
80	59.88	90.95	132.8	220.8	299.5	373.6	397.1	422.9	412.7	1991.3
100	122.3	185.6	270.7	450.1	610.6	761.7	809.7	862.2	711.2	4063.2

Notes:
1. Table capacities are in kilowatts of refrigeration.
   Δp = pressure drop per unit equivalent length of line, Pa/m
   Δt = corresponding change in saturation temperature, K/m
2. Line capacity for other saturation temperatures Δt and equivalent lengths L_e
   $$\text{Line capacity} = \text{Table capacity} \left(\frac{\text{Table } L_e}{\text{Actual } L_e} \times \frac{\text{Actual } \Delta t}{\text{Table } \Delta t} \right)^{0.55}$$
3. Saturation temperature Δt for other capacities and equivalent lengths L_e
   $$\Delta t = \text{Table } \Delta t \left(\frac{\text{Actual } L_e}{\text{Table } L_e} \right) \left(\frac{\text{Actual capacity}}{\text{Table capacity}} \right)^{1.8}$$
4. Values in the table are based on 40°C condensing temperature. Multiply table capacities by the following factors for other condensing temperatures.

Condensing Temperature, °C	Suction Line	Discharge Line
20	1.18	0.80
30	1.10	0.88
40	1.00	1.00
50	0.91	1.11

[a] The sizing shown is recommended where any gas generated in the receiver must return up the condensate line to the condenser without restricting condensate flow. Water-cooled condensers, where the receiver ambient temperature may be higher than the refrigerant condensing temperature, fall into this category.

[b] The line pressure drop Δp is conservative; if subcooling is substantial or the line is short, a smaller size line may be used. Applications with very little subcooling or very long lines may require a larger line.

flashing takes place, the overall efficiency is reduced, and the system may malfunction.

The velocity of liquid leaving a partially filled vessel (such as a receiver or shell-and-tube condenser) is limited by the height of the liquid above the point at which the liquid line leaves the vessel, whether or not the liquid at the surface is subcooled. Because the liquid in the vessel has a very low (or zero) velocity, the velocity V in the liquid line (usually at the vena contracta) is $V^2 = 2gh$, where h is the height of the liquid in the vessel. Gas pressure does not add to the velocity unless gas is flowing in the same direction. As a result, both gas and liquid flow through the line, limiting the rate of liquid flow. If this factor is not considered, excess operating charges in receivers and flooding of shell-and-tube condensers may result.

No specific data are available to precisely size a line leaving a vessel. If the height of the liquid above the vena contracta produces the desired velocity, the liquid will leave the vessel at the expected rate. Thus, if the level in the vessel falls to one pipe diameter above the bottom of the vessel from which the liquid line leaves, the capacity of copper lines for R-22 at 6.4 g/s per kilowatt of refrigeration is approximately as follows:

OD, mm	kW
28	49
35	88
42	140
54	280
67	460
79	690
105	1440

The whole liquid line need not be as large as the leaving connection. After the vena contracta, the velocity is about 40% less. If the line continues down from the receiver, the value of h increases. For a 700 kW capacity with R-22, the line from the bottom of the receiver should be about 79 mm. After a drop of 1300 mm, a reduction to 54 mm is satisfactory.

Suction Lines. Suction lines are more critical than liquid and discharge lines from a design and construction standpoint. Refrigerant lines should be sized to (1) provide a minimum pressure drop at full load, (2) return oil from the evaporator to the compressor under minimum load conditions, and (3) prevent oil from draining

System Practices for Halocarbon Refrigerants

Table 4 Suction, Discharge, and Liquid Line Capacities in Kilowatts for Refrigerant 134a (Single- or High-Stage Applications)

Nominal Line OD, mm	Suction Lines (Δt = 0.04 K/m)					Discharge Lines (Δt = 0.02 K/m, Δp = 538 Pa/m)			Liquid Lines See notes a and b	
	Saturated Suction Temperature, °C					Saturated Suction Temperature, °C				
	−10	−5	0	5	10				Velocity = 0.5 m/s	Δt = 0.02 K/m Δp = 538 Pa/m
	Corresponding Δp, Pa/m									
	318	368	425	487	555	−10	0	10		
TYPE L COPPER LINE										
12	0.62	0.76	0.92	1.11	1.33	1.69	1.77	1.84	6.51	8.50
15	1.18	1.45	1.76	2.12	2.54	3.23	3.37	3.51	10.60	16.30
18	2.06	2.52	3.60	3.69	4.42	5.61	5.85	6.09	16.00	28.40
22	3.64	4.45	5.40	6.50	7.77	9.87	10.30	10.70	24.50	50.10
28	7.19	8.80	10.70	12.80	15.30	19.50	20.30	21.10	41.00	99.50
35	13.20	16.10	19.50	23.50	28.10	35.60	37.20	38.70	64.90	183.00
42	21.90	26.80	32.40	39.00	46.50	59.00	61.60	64.10	95.20	304.00
54	43.60	53.20	64.40	77.30	92.20	117.00	122.00	127.00	160.00	605.00
67	77.70	94.60	115.00	138.00	164.00	208.00	217.00	226.00	248.00	1080.00
79	120.00	147.00	177.00	213.00	253.00	321.00	335.00	349.00	346.00	1670.00
105	257.00	313.00	379.00	454.00	541.00	686.00	715.00	744.00	618.00	3580.00
STEEL LINE										
10	0.87	1.06	1.27	1.52	1.80	2.28	2.38	2.47	9.81	12.30
15	1.62	1.96	2.36	2.81	3.34	4.22	4.40	4.58	15.60	22.80
20	3.41	4.13	4.97	5.93	7.02	8.88	9.26	9.64	27.40	48.20
25	6.45	7.81	9.37	11.20	13.30	16.70	17.50	18.20	44.40	91.00
32	13.30	16.10	19.40	23.10	27.40	34.60	36.10	37.50	76.90	188.00
40	20.00	24.20	29.10	34.60	41.00	51.90	54.10	56.30	105.00	283.00
50	38.60	46.70	56.00	66.80	79.10	100.00	104.00	108.00	173.00	546.00
65	61.50	74.30	89.30	106.00	126.00	159.00	166.00	173.00	246.00	871.00
80	109.00	131.00	158.00	288.00	223.00	281.00	294.00	306.00	380.00	1540.00
100	222.00	268.00	322.00	383.00	454.00	573.00	598.00	622.00	655.00	3140.00

Notes:
1. Table capacities are in kilowatts of refrigeration.
   Δp = pressure drop per equivalent line length, Pa/m
   Δt = corresponding change in saturation temperature, K/m
2. Line capacity for other saturation temperatures Δt and equivalent lengths L_e

$$\text{Line capacity} = \text{Table capacity}\left(\frac{\text{Table } L_e}{\text{Actual } L_e} \times \frac{\text{Actual } \Delta t}{\text{Table } \Delta t}\right)^{0.55}$$

3. Saturation temperature Δt for other capacities and equivalent lengths L_e

$$\Delta t = \text{Table } \Delta t \left(\frac{\text{Actual } L_e}{\text{Table } L_e}\right)\left(\frac{\text{Actual capacity}}{\text{Table capacity}}\right)^{1.8}$$

4. Values in the table are based on 40°C condensing temperature. Multiply table capacities by the following factors for other condensing temperatures.

Condensing Temperature, °C	Suction Line	Discharge Line
20	1.239	0.682
30	1.120	0.856
40	1.0	1.0
50	0.888	1.110

[a] The sizing shown is recommended where any gas generated in the receiver must return up the condensate line to the condenser without restricting condensate flow. Water-cooled condensers, where the receiver ambient temperature may be higher than the refrigerant condensing temperature, fall into this category.

[b] The line pressure drop Δp is conservative; if subcooling is substantial or the line is short, a smaller size line may be used. Applications with very little subcooling or very long lines may require a larger line.

from an active evaporator into an idle one. A pressure drop in the suction line reduces a system's capacity because it forces the compressor to operate at a lower suction pressure to maintain a desired evaporating temperature in the coil. The suction line is normally sized to have a pressure drop from friction no greater than the equivalent of about a 1 K change in saturation temperature. See Tables 3 through 9 for suction line sizing information. The equivalent pressure loss at 5°C saturated suction temperature is approximately

Refrigerant	Suction Loss, K	Pressure Loss, kPa
R-22	1	18.1
R-134a	1	12.2
R-502	1	19.7

At suction temperatures lower than 5°C, the pressure drop equivalent to a given temperature change decreases. For example, at −40°C suction with R-22, the pressure drop equivalent to a 1 K change in saturation temperature is about 4.9 kPa. Therefore, low-temperature lines must be sized for a very low pressure drop, or higher equivalent temperature losses, with resultant loss in equipment capacity, must be accepted. For very low pressure drops, any suction or hot-gas risers must be sized properly to assure oil entrainment up the riser so that the oil is always returned to the compressor.

Where pipe size must be reduced to provide sufficient gas velocity to entrain oil up vertical risers at partial loads, greater pressure drops are imposed at full load. These can usually be compensated for by oversizing the horizontal and down run lines and components.

Discharge Lines. Pressure loss in hot-gas lines increases the required compressor power per unit of refrigeration and decreases the compressor capacity. Table 2 illustrates the power losses for an R-22 system at 5°C evaporator and 40°C condensing temperature. Pressure drop is kept to a minimum by generously sizing the lines

Table 5 Suction, Discharge, and Liquid Line Capacities in Kilowatts for Refrigerant 502 (Single- or High-Stage Applications)

Nominal Line Size, mm	Suction Lines ($\Delta t = 0.04$ K/m)					Discharge Lines ($\Delta t = 0.02$ K/m, $\Delta p = 465$)			Liquid Lines See notes a and b	
	Saturated Suction Temperature, °C					Saturated Suction Temperature, °C			Velocity = 0.5 m/s	$\Delta t = 0.02$ K/m $\Delta p = 779$
	−40	−30	−20	−5	5	−40	−20	50		
	Corresponding Δp, Pa/m									
	230	320	431	635	799					
TYPE L COPPER LINE										
12	0.26	0.42	0.63	1.10	1.53	1.70	1.91	2.14	4.48	7.39
15	0.51	0.80	1.21	2.10	2.92	3.25	3.64	4.08	7.27	14.13
18	0.88	1.39	2.10	3.64	5.05	5.62	6.30	7.07	11.02	24.60
22	1.56	2.45	3.70	6.40	8.89	9.88	11.07	12.42	16.87	43.34
28	3.09	4.85	7.31	12.62	17.47	19.43	21.77	24.43	28.20	85.71
35	5.67	8.89	13.41	23.08	32.00	35.54	39.81	44.67	44.62	157.4
42	9.43	14.77	22.22	38.25	52.90	58.77	65.84	73.88	65.45	261.1
54	18.77	29.34	44.09	75.78	104.7	116.2	130.2	146.0	110.2	518.3
67	33.46	52.31	78.44	134.5	185.7	206.0	230.8	259.0	170.8	922.9
79	51.86	81.03	121.4	208.0	287.2	318.2	356.5	400.0	238.2	1429.9
105	111.1	173.2	259.3	443.4	611.4	677.5	759.0	851.8	425.2	3057.7
STEEL LINE										
10	0.38	0.59	0.87	1.47	2.01	2.22	2.49	2.79	6.75	10.33
15	0.71	1.09	1.61	2.72	3.72	4.12	4.61	5.17	10.74	19.15
20	1.49	2.29	3.39	5.73	7.83	8.65	9.70	10.88	18.85	40.33
25	2.82	4.34	6.41	10.79	14.76	16.32	18.28	20.52	30.49	76.24
32	5.84	8.96	13.24	22.29	30.48	33.65	37.69	42.30	52.87	157.6
40	8.76	13.45	19.87	33.45	45.66	50.50	56.57	63.48	71.96	236.5
50	16.94	25.96	38.29	64.46	88.01	97.24	108.9	122.2	118.6	455.7
65	26.98	41.36	61.01	102.5	140.3	154.8	173.4	194.6	169.1	726.0
80	47.66	73.07	107.8	181.2	247.3	273.6	306.5	344.0	261.2	1282.7
100	97.35	148.9	219.8	369.4	504.3	556.6	623.6	699.8	450.0	2614.6

Notes:
1. Table capacities are in kilowatts of refrigeration.
   Δp = pressure drop per equivalent line length, Pa/m
   Δt = corresponding change in saturation temperature, K/m
2. Line capacity for other saturation temperatures Δt and equivalent lengths L_e

   $$\text{Line capacity} = \text{Table capacity} \left(\frac{\text{Table } L_e}{\text{Actual } L_e} \times \frac{\text{Actual } \Delta t}{\text{Table } \Delta t}\right)^{0.55}$$

3. Saturation temperature Δt for other capacities and equivalent lengths L_e

   $$\Delta t = \text{Table } \Delta t \left(\frac{\text{Actual } L_e}{\text{Table } L_e}\right) \left(\frac{\text{Actual capacity}}{\text{Table capacity}}\right)^{1.8}$$

4. Values in the table are based on 40°C condensing temperature. Multiply table capacities by the following factors for other condensing temperatures.

Condensing Temperature, °C	Suction Line	Discharge Line
20	1.26	0.87
30	1.17	0.94
40	1.00	1.00
50	0.86	1.05

[a] The sizing shown is recommended where any gas generated in the receiver must return up the condensate line to the condenser without restricting condensate flow. Water-cooled condensers, where the receiver ambient temperature may be higher than the refrigerant condensing temperature, fall into this category.

[b] The line pressure drop Δp is conservative; if subcooling is substantial or the line is short, a smaller size line may be used. Applications with very little subcooling or very long lines may require a larger line.

for low friction losses, but still maintaining refrigerant line velocities to entrain and carry oil along at all loading conditions. Pressure drop is normally designed not to exceed the equivalent of a 1 K change in saturation temperature. Recommended sizing tables are based on a 0.02 K/m change in saturation temperature.

Location and Arrangement of Piping

Refrigerant lines should be as short and direct as possible to minimize tubing and refrigerant requirements and pressure drops. Plan piping for a minimum number of joints using as few elbows and other fittings as possible, but provide sufficient flexibility to absorb compressor vibration and stresses due to thermal expansion and contraction.

Arrange refrigerant piping so that normal inspection and servicing of the compressor and other equipment is not hindered. Do not obstruct the view of the oil level sight glass or run piping so that it interferes with the removal of compressor cylinder heads, end bells, access plates, or any internal parts. Suction line piping to the compressor should be arranged so that it will not interfere with removal of the compressor for servicing.

Provide adequate clearance between pipe and adjacent walls and hangers or between pipes for insulation installation. Use sleeves that are sized to permit installation of both pipe and insulation through floors, walls, or ceilings,. Set these sleeves prior to pouring of concrete or erection of brickwork.

Run piping so that it does not interfere with passages or obstruct headroom, windows, and doors. Refer to ASHRAE *Standard* 15, Safety Code for Mechanical Refrigeration, and other governing local codes for restrictions that may apply.

Protection Against Damage to Piping

Protection against damage is necessary, particularly for small lines, which have a false appearance of strength. Where traffic is heavy, provide protection against impact from carelessly handled hand trucks, overhanging loads, ladders, and fork trucks.

Piping Insulation

All piping joints and fittings should be thoroughly leak tested before insulation is sealed. Suction lines should be insulated to prevent sweating and heat gain. Insulation covering lines on which moisture can condense or lines subjected to outside conditions must be vapor sealed to prevent any moisture travel through the insulation or condensation in the insulation. Many commercially available types are provided with an integral waterproof jacket for this purpose. Although the liquid line ordinarily does not require insulation, the suction and liquid lines can be insulated as a unit on installations

System Practices for Halocarbon Refrigerants

Table 6 Suction, Discharge, and Liquid Line Capacities in Kilowatts for Refrigerant 22 (Intermediate- or Low-Stage Duty)

Nominal Type L Copper Line OD, mm	Suction Lines (Δt = 0.04 K/m)					Discharge Lines[a]	Liquid Lines
	Saturated Suction Temperature, °C						
	−70	−60	−50	−40	−30		
	Corresponding Δp, Pa/m						
	31.0	51.3	81.5	121	228		
12	0.09	0.16	0.27	0.47	0.73	0.74	
15	0.17	0.31	0.52	0.90	1.39	1.43	
18	0.29	0.55	0.91	1.57	2.43	2.49	
22	0.52	0.97	1.62	2.78	4.30	4.41	
28	1.05	1.94	3.22	5.52	8.52	8.74	
35	1.94	3.60	5.95	10.17	15.68	16.08	
42	3.26	6.00	9.92	16.93	26.07	26.73	See Table 3
54	6.54	12.03	19.83	33.75	51.98	53.28	
67	11.77	21.57	35.47	60.38	92.76	95.06	
79	18.32	33.54	55.20	93.72	143.69	174.22	
105	39.60	72.33	118.66	201.20	308.02	316.13	
130	70.87	129.17	211.70	358.52	548.66	561.89	
156	115.74	210.83	344.99	583.16	891.71	915.02	

Notes:
1. Table capacities are in kilowatts of refrigeration.
   Δp = pressure drop per equivalent line length, Pa/m
   Δt = corresponding change in saturation temperature, K/m

2. Line capacity for other saturation temperatures Δt and equivalent lengths L_e

$$\text{Line capacity} = \text{Table capacity} \left(\frac{\text{Table } L_e}{\text{Actual } L_e} \times \frac{\text{Actual } \Delta t}{\text{Table } \Delta t} \right)^{0.55}$$

3. Saturation temperature Δt for other capacities and equivalent lengths L_e

$$\Delta t = \text{Table } \Delta t \left(\frac{\text{Actual } L_e}{\text{Table } L_e} \right) \left(\frac{\text{Actual capacity}}{\text{Table capacity}} \right)^{1.8}$$

4. Values in the table are based on −15°C condensing temperature. Multiply table capacities by the following factors for other condensing temperatures.

Condensing Temperature, °C	Suction Line	Discharge Line
−30	1.08	0.74
−20	1.03	0.91
−10	0.98	1.09
0	0.91	1.29

5. Refer to the refrigerant property tables (Chapter 19 of the 1997 *ASHRAE Handbook—Fundamentals*) for the pressure drop corresponding to Δt.

[a] See the section on Pressure Drop Considerations.

where the two lines are clamped together. When it passes through an area of higher temperature, the liquid line should be insulated to minimize heat gain. Hot-gas discharge lines usually are not insulated; however, they should be insulated if the heat dissipated is objectionable or to prevent injury from high-temperature surfaces. In the latter case, it is not essential to provide insulation with a tight vapor seal because moisture condensation is not a problem unless the line is located outside. Hot-gas defrost lines are customarily insulated to minimize heat loss and condensation of gas inside the piping.

While all joints and fittings should be covered, it is not advisable to do so until the system has been thoroughly leak tested.

Vibration and Noise in Piping

Vibration transmitted through or generated in refrigerant piping and the resulting objectionable noise can be eliminated or minimized by proper piping design and support.

Two undesirable effects of vibration of refrigerant piping are (1) physical damage to the piping, which results in the breaking of brazed joints and, consequently, loss of charge; and (2) transmission of noise through the piping itself and through building construction with which the piping may come into direct physical contact.

In refrigeration applications, piping vibration can be caused by the rigid connection of the refrigerant piping to a reciprocating compressor. Vibration effects are evident in all lines directly connected to the compressor or condensing unit. It is thus impossible to eliminate vibration in piping; it is only possible to mitigate its effects.

Flexible metal hose is sometimes used to absorb vibration transmission along smaller pipe sizes. For maximum effectiveness, it should be installed parallel to the crankshaft. In some cases, two isolators may be required, one in the horizontal line and the other in the vertical line at the compressor. A rigid brace on the end of the flexible hose away from the compressor is required to prevent vibration of the hot-gas line beyond the hose.

Flexible metal hose is not as efficient in absorbing vibration on larger sizes of pipe because it is not actually flexible unless the ratio of length to diameter is relatively great. In practice, the length is often limited, so flexibility is reduced in larger sizes. This problem is best solved by using flexible piping and isolation hangers where the piping is secured to the structure.

When piping passes through walls, through floors, or inside furring, it must not touch any part of the building and must be supported only by the hangers (provided to avoid transmitting vibration to the building); this eliminates the possibility of walls or ceilings acting as sounding boards or diaphragms. When piping is erected where access is difficult after installation, it should be supported by isolation hangers.

Vibration and noise from a piping system can also be caused by gas pulsations from the compressor operation or from turbulence in the gas, which increases at high velocities. It is usually more apparent in the discharge line than in other parts of the system.

When gas pulsations caused by the compressor create vibration and noise, they have a characteristic frequency that is a function of the number of gas discharges by the compressor on each revolution. This frequency is not necessarily equal to the number of cylinders, since on some compressors two pistons operate together. It is also varied by the angular displacement of the cylinders, such as in V-type compressors. Noise resulting from gas pulsations is usually objectionable only when the piping system amplifies the pulsation by resonance. On single-compressor systems, resonance can be reduced by changing the size or length of the resonating line or by installing a properly sized hot-gas muffler in the discharge line immediately after the compressor discharge valve. On a paralleled compressor system, a harmonic frequency from the different speeds of multiple compressors may be apparent. This noise can sometimes be reduced by installing mufflers.

When noise is caused by turbulence, and isolating the line is not effective enough, the installation of a larger diameter pipe to reduce the gas velocity is sometimes helpful. Also, changing to a line of heavier wall or from copper to steel to change the pipe natural frequency may help.

Table 7 Suction Line Capacities in Kilowatts for Refrigeration 22 (Single- or High-Stage Applications) for Pressure Drops of 0.02 and 0.01 K/m Equivalent

Nominal Line OD, mm	Saturated Suction Temperature, °C									
	−40		−30		−20		−5		5	
	Δt = 0.02 Δp = 97.9	Δt = 0.01 Δp = 49.0	Δt = 0.02 Δp = 138	Δt = 0.01 Δp = 69.2	Δt = 0.02 Δp = 189	Δt = 0.01 Δp = 94.6	Δt = 0.02 Δp = 286	Δt = 0.01 Δp = 143	Δt = 0.02 Δp = 366	Δt = 0.01 Δp = 183
TYPE L COPPER LINE										
12	0.21	0.14	0.34	0.23	0.51	0.34	0.87	0.59	1.20	0.82
15	0.41	0.28	0.65	0.44	0.97	0.66	1.67	1.14	2.30	1.56
18	0.72	0.49	1.13	0.76	1.70	1.15	2.91	1.98	4.00	2.73
22	1.28	0.86	2.00	1.36	3.00	2.04	5.14	3.50	7.07	4.82
28	2.54	1.72	3.97	2.70	5.95	4.06	10.16	6.95	13.98	9.56
35	4.69	3.19	7.32	4.99	10.96	7.48	18.69	12.80	25.66	17.59
42	7.82	5.32	12.19	8.32	18.20	12.46	31.03	21.27	42.59	29.21
54	15.63	10.66	24.34	16.65	36.26	24.88	61.79	42.43	84.60	58.23
67	27.94	19.11	43.48	29.76	64.79	44.48	110.05	75.68	150.80	103.80
79	43.43	29.74	67.47	46.26	100.51	69.04	170.64	117.39	233.56	161.10
105	93.43	63.99	144.76	99.47	215.39	148.34	365.08	251.92	499.16	344.89
STEEL LINE										
10	0.33	0.23	0.50	0.35	0.74	0.52	1.25	0.87	1.69	1.18
15	0.61	0.42	0.94	0.65	1.38	0.96	2.31	1.62	3.15	2.20
20	1.30	0.90	1.98	1.38	2.92	2.04	4.87	3.42	6.63	4.65
25	2.46	1.71	3.76	2.62	5.52	3.86	9.22	6.47	12.52	8.79
32	5.11	3.56	7.79	5.45	11.42	8.01	19.06	13.38	25.88	18.20
40	7.68	5.36	11.70	8.19	17.16	12.02	28.60	20.10	38.89	27.35
50	14.85	10.39	22.65	14.86	33.17	23.27	55.18	38.83	74.92	52.77
65	23.74	16.58	36.15	25.30	52.84	37.13	87.91	61.89	119.37	84.05
80	42.02	29.43	63.95	44.84	93.51	65.68	155.62	109.54	211.33	148.77
100	85.84	60.16	130.57	91.69	190.95	134.08	317.17	223.47	430.77	303.17
125	155.21	108.97	235.58	165.78	344.66	242.47	572.50	403.23	776.67	547.16
150	251.47	176.49	381.78	268.72	557.25	391.95	925.72	652.73	1255.93	885.79
200	515.37	362.01	781.63	550.49	1141.07	803.41	1895.86	1336.79	2572.39	1813.97
250	933.07	656.12	1413.53	996.65	2063.66	1454.75	3429.24	2417.91	4646.48	3280.83
300	1494.35	1050.57	2264.54	1593.85	3305.39	2330.50	5477.74	3867.63	7433.20	5248.20

Δp = pressure drop per unit equivalent line length, Pa/m
Δt = corresponding change in saturation temperature, K/m

Table 8 Suction Line Capacities in Kilowatts for Refrigeration 134a (Single- or High-Stage Applications) for Pressure Drops of 0.02 and 0.01 K/m Equivalent

Nominal Line OD, mm	Saturated Suction Temperature, °C									
	−10		−5		0		5		10	
	Δt = 0.02 Δp = 159	Δt = 0.01 Δp = 79.3	Δt = 0.02 Δp = 185	Δt = 0.01 Δp = 92.4	Δt = 0.02 Δp = 212	Δt = 0.01 Δp = 106	Δt = 0.02 Δp = 243	Δt = 0.01 Δp = 121	Δt = 0.02 Δp = 278	Δt = 0.01 Δp = 139
TYPE L COPPER LINE										
12	0.42	0.28	0.52	0.35	0.63	0.43	0.76	0.51	0.91	0.62
15	0.81	0.55	0.99	0.67	1.20	0.82	1.45	0.99	1.74	1.19
18	1.40	0.96	1.73	1.18	2.09	1.43	2.53	1.72	3.03	2.07
22	2.48	1.69	3.05	2.08	3.69	2.52	4.46	3.04	5.34	3.66
28	4.91	3.36	6.03	4.13	7.31	5.01	8.81	6.02	10.60	7.24
35	9.05	6.18	11.10	7.60	13.40	9.21	16.20	11.10	19.40	13.30
42	15.00	10.30	18.40	12.60	22.30	15.30	26.90	18.40	32.10	22.10
54	30.00	20.50	36.70	25.20	44.40	30.50	53.40	36.70	63.80	44.00
67	53.40	36.70	65.40	44.90	79.00	54.40	95.00	65.40	113.00	78.30
79	82.80	56.90	101.00	69.70	122.00	84.30	147.00	101.00	176.00	122.00
105	178.00	122.00	217.00	149.00	262.00	181.00	315.00	217.00	375.00	260.00
STEEL LINE										
10	0.61	0.42	0.74	0.52	0.89	0.62	1.06	0.74	1.27	0.89
15	1.13	0.79	1.38	0.96	1.65	1.16	1.97	1.38	2.35	1.65
20	2.39	1.67	2.91	2.03	3.49	2.44	4.17	2.92	4.94	3.47
25	4.53	3.17	5.49	3.85	6.59	4.62	7.86	5.52	9.33	6.56
32	9.37	6.57	11.40	7.97	13.60	9.57	16.30	11.40	19.30	13.60
40	14.10	9.86	17.10	12.00	20.50	14.40	24.40	17.10	28.90	20.40
50	27.20	19.10	32.90	23.10	39.50	27.70	47.00	33.10	55.80	39.40
65	43.30	30.40	52.50	36.90	62.90	44.30	75.00	52.70	88.80	62.70
80	76.60	53.80	92.80	65.30	111.00	78.30	133.00	93.10	157.00	111.00
100	156.00	110.00	189.00	133.00	227.00	160.00	270.00	190.00	320.00	226.00

Δp = pressure drop per unit equivalent line length, Pa/m
Δt = corresponding change in saturation temperature, K/m

Table 9 Suction Line Capacities in Kilowatts for Refrigeration 502 (Single- or High-Stage Applications) for Pressure Drops of 0.02 and 0.01 K/m Equivalent

Nominal Line OD, mm	Saturated Suction Temperature, °C									
	−40		−30		−20		−5		5	
	$\Delta t = 0.02$ $\Delta p = 115$	$\Delta t = 0.01$ $\Delta p = 57.5$	$\Delta t = 0.02$ $\Delta p = 160$	$\Delta t = 0.01$ $\Delta p = 80$	$\Delta t = 0.02$ $\Delta p = 215$	$\Delta t = 0.01$ $\Delta p = 108$	$\Delta t = 0.02$ $\Delta p = 317$	$\Delta t = 0.01$ $\Delta p = 159$	$\Delta t = 0.02$ $\Delta p = 399$	$\Delta t = 0.01$ $\Delta p = 200$
TYPE L COPPER LINE										
12	0.18	0.12	0.28	1.19	0.43	0.29	0.75	0.51	1.05	0.72
15	0.34	0.23	0.54	0.37	0.83	0.56	1.44	0.98	2.00	1.37
18	0.60	0.41	0.95	0.65	1.44	0.98	2.50	1.71	3.48	2.38
22	1.06	9.72	1.67	1.14	2.53	1.73	4.40	3.01	6.11	4.20
28	2.11	1.44	3.32	2.27	5.02	3.44	8.68	5.97	12.06	8.29
35	3.88	2.65	6.11	4.17	9.21	6.32	15.92	10.95	22.08	15.22
42	6.46	4.42	10.16	6.95	15.31	10.50	26.39	18.17	36.58	25.25
54	12.90	8.83	20.20	13.86	30.41	20.90	52.36	36.11	72.55	50.15
67	23.02	15.78	36.00	24.78	54.19	37.30	93.16	64.36	128.96	89.18
79	35.69	24.53	55.79	38.40	83.93	57.89	144.18	99.69	199.33	137.96
105	76.67	52.64	119.66	82.48	179.67	123.91	208.06	213.05	425.46	295.15
STEEL LINE										
10	0.27	0.18	0.41	0.29	0.61	0.43	1.03	0.72	1.41	0.99
15	0.49	0.34	0.76	0.53	1.13	0.79	1.91	1.34	2.62	1.84
20	1.04	0.73	1.61	1.13	2.38	1.67	4.03	2.83	5.52	3.88
25	1.98	1.38	3.04	2.13	4.51	3.16	7.61	5.34	10.40	7.32
32	4.09	2.87	6.29	4.42	9.31	6.55	15.70	11.04	21.48	15.13
40	6.15	4.31	9.45	6.64	14.00	9.82	23.56	16.59	32.23	22.70
50	11.88	8.34	18.24	12.82	26.96	18.99	45.40	31.97	62.10	43.83
65	18.96	13.30	29.06	20.45	42.96	30.24	72.34	50.94	98.98	69.69
80	33.55	23.57	52.44	36.21	76.04	53.53	127.82	90.19	174.90	123.40
100	68.50	48.10	104.93	73.85	155.04	109.08	260.59	183.83	355.77	251.27
125	123.60	86.98	189.39	122.41	279.53	196.87	469.33	331.39	642.34	453.07
150	200.11	140.76	306.60	215.68	452.01	318.75	758.96	535.98	1038.85	732.68
200	409.66	288.50	627.19	442.16	925.80	652.64	1552.5	1097.8	2125.3	1498.7
250	741.87	522.36	1134.4	800.86	1672.3	1180.4	2812.4	1982.9	3839.5	2711.0
300	1184.9	835.42	1814.6	1279.0	2675.2	1888.2	4492.8	3172.2	6133.7	4337.2

Δp = pressure drop per unit equivalent length of line, Pa/m
Δt = corresponding change in saturation temperature, K/m

Refrigerant Line Capacity Tables

Tables 3 through 6 show capacities for R-22, R-134a, and R-502 at specific pressure drops. The capacities shown in the tables are based on the refrigerant flow that develops a friction loss, per metre of equivalent pipe length, corresponding to a 0.02 K change in the saturation temperature (Δt) for discharge and liquid lines. Suction lines are based on a 0.04 K change. Tables 7, 8, and 9 show suction line capacities for a 0.02 and 0.01 K/m change in the saturation suction temperature. Pressure drops are given in kelvins because this pipe sizing method is convenient and accepted throughout the industry. Corresponding pressure drops are also shown.

The refrigerant line sizing capacity tables are based on the Darcy-Weisbach relation and friction factors as computed by the Colebrook function (Colebrook 1938, 1939). Tubing roughness height is 1.5 μm for copper and 46 μm for steel pipe. Viscosity extrapolations and adjustments for pressures other than 101.325 kPa were based on correlation techniques as presented by Keating and Matula (1969). Discharge gas superheat was 45 K for R-134a and R-502 and 60 K for R-22.

The refrigerant cycle for determining capacity is based on saturated gas leaving the evaporator. The calculations neglect the presence of oil and assume nonpulsating flow.

For additional charts and discussion of line sizing refer to Timm (1991), Wile (1977), and Atwood (1990).

Equivalent Lengths of Valves and Fittings

Refrigerant line capacity tables are based on unit pressure drop per metre length of straight pipe or per combination of straight pipe, fittings, and valves with friction drop equivalent to a metre of straight pipe.

Generally, pressure drop through valves and fittings is determined by establishing the equivalent straight length of pipe of the same size with the same friction drop. Line sizing tables can then be used directly. Tables 10, 11, and 12 give equivalent lengths of straight pipe for various fittings and valves, based on nominal pipe sizes.

The following example illustrates the use of various tables and charts to size refrigerant lines.

Example 2. Determine the line size and pressure drop equivalent (in degrees) for the suction line of a 105 kW R-22 system, operating at 5°C suction and 40°C condensing temperatures. The suction line is copper tubing, with 15 m of straight pipe and six long-radius elbows.

Solution: Add 50% to the straight length of pipe to establish a trial equivalent length. Trial equivalent length is 15 × 1.5 = 22.5 m. From Table 3 (for 5°C suction, 40°C condensing), 122.7 kW capacity in 54 mm OD results in a 0.04 K loss per metre equivalent length.

Straight pipe length	= 15.0 m
Six 50 mm long-radius elbows at 1.0 m each (Table 10)	= 6.0 m
Total equivalent length	= 21.0 m

$$\Delta t = 0.04 \times 21.0(105/122.7)^{1.8} = 0.63 \text{ K}$$

Since 0.63 K is below the recommended 1 K, recompute for the next smaller (42 mm) tube; i.e., Δt = 2.05 K. But this temperature drop is too large; therefore the 54 mm tube is recommended.

Oil Management in Refrigerant Lines

Oil Circulation. All compressors lose some lubricating oil during normal operation. Because oil inevitably leaves the compressor with the discharge gas, systems using halocarbon refrigerants must return this oil at the same rate at which it leaves (Cooper 1971).

Table 10 Fitting Losses in Equivalent Metres of Pipe
(Screwed, Welded, Flanged, Flared, and Brazed Connections)

Nominal Pipe or Tube Size, mm	Smooth Bend Elbows						Smooth Bend Tees			
	90° Std[a]	90° Long-Radius[b]	90° Street[a]	45° Std[a]	45° Street[a]	180° Std[a]	Flow Through Branch	Straight-Through Flow		
								No Reduction	Reduced 1/4	Reduced 1/2
10	0.4	0.3	0.7	0.2	0.3	0.7	0.8	0.3	0.4	0.4
15	0.5	0.3	0.8	0.2	0.4	0.8	0.9	0.3	0.4	0.5
20	0.6	0.4	1.0	0.3	0.5	1.0	1.2	0.4	0.6	0.6
25	0.8	0.5	1.2	0.4	0.6	1.2	1.5	0.5	0.7	0.8
32	1.0	0.7	1.7	0.5	0.9	1.7	2.1	0.7	0.9	1.0
40	1.2	0.8	1.9	0.6	1.0	1.9	2.4	0.8	1.1	1.2
50	1.5	1.0	2.5	0.8	1.4	2.5	3.0	1.0	1.4	1.5
65	1.8	1.2	3.0	1.0	1.6	3.0	3.7	1.2	1.7	1.8
80	2.3	1.5	3.7	1.2	2.0	3.7	4.6	1.5	2.1	2.3
90	2.7	1.8	4.6	1.4	2.2	4.6	5.5	1.8	2.4	2.7
100	3.0	2.0	5.2	1.6	2.6	5.2	6.4	2.0	2.7	3.0
125	4.0	2.5	6.4	2.0	3.4	6.4	7.6	2.5	3.7	4.0
150	4.9	3.0	7.6	2.4	4.0	7.6	9	3.0	4.3	4.9
200	6.1	4.0	—	3.0	—	10	12	4.0	5.5	6.1
250	7.6	4.9	—	4.0	—	13	15	4.9	7.0	7.6
300	9.1	5.8	—	4.9	—	15	18	5.8	7.9	9.1
350	10	7.0	—	5.5	—	17	21	7.0	9.1	10
400	12	7.9	—	6.1	—	19	24	7.9	11	12
450	13	8.8	—	7.0	—	21	26	8.8	12	13
500	15	10	—	7.9	—	25	30	10	13	15
600	18	12	—	9.1	—	29	35	12	15	18

[a] R/D approximately equal to 1. [b] R/D approximately equal to 1.5.

Table 11 Special Fitting Losses in Equivalent Metres of Pipe

Nominal Pipe or Tube Size, mm	Sudden Enlargement, d/D			Sudden Contraction, d/D			Sharp Edge		Pipe Projection	
	1/4	1/2	3/4	1/4	1/2	3/4	Entrance	Exit	Entrance	Exit
10	0.4	0.2	0.1	0.2	0.2	0.1	0.5	0.2	0.5	0.3
15	0.5	0.3	0.1	0.3	0.3	0.1	0.5	0.3	0.5	0.5
20	0.8	0.5	0.2	0.4	0.3	0.2	0.9	0.4	0.9	0.7
25	1.0	0.6	0.2	0.5	0.4	0.2	1.1	0.5	1.1	0.8
32	1.4	0.9	0.3	0.7	0.5	0.3	1.6	0.8	1.6	1.3
40	1.8	1.1	0.4	0.9	0.7	0.4	2.0	1.0	2.0	1.5
50	2.4	1.5	0.5	1.2	0.9	0.5	2.7	1.3	2.7	2.1
65	3.0	1.9	0.6	1.5	1.2	0.6	3.7	1.7	3.7	2.7
80	4.0	2.4	0.8	2.0	1.5	0.8	4.3	2.2	4.3	3.8
90	4.6	2.8	0.9	2.3	1.8	0.9	5.2	2.6	5.2	4.0
100	5.2	3.4	1.2	2.7	2.1	1.2	6.1	3.0	6.1	4.9
125	7.3	4.6	1.5	3.7	2.7	1.5	8.2	4.3	8.2	6.1
150	8.8	6.7	1.8	4.6	3.4	1.8	10	5.8	10	7.6
200	—	7.6	2.6	—	4.6	2.6	14	7.3	14	10
250	—	9.8	3.4	—	6.1	3.4	18	8.8	18	14
300	—	12.4	4.0	—	7.6	4.0	22	11	22	17
350	—	—	4.9	—	—	4.9	26	14	26	20
400	—	—	5.5	—	—	5.5	29	15	29	23
450	—	—	6.1	—	—	6.1	35	18	35	27
500	—	—	—	—	—	—	43	21	43	33
600	—	—	—	—	—	—	50	25	50	40

Note: Enter table for losses at smallest diameter *d*.

System Practices for Halocarbon Refrigerants

Table 12 Valve Losses in Equivalent Metres of Pipe

Nominal Pipe or Tube Size, mm	Globe[a]	60° Wye	45° Wye	Angle[a]	Gate[b]	Swing Check[c]	Lift Check
10	5.2	2.4	1.8	1.8	0.2	1.5	Globe and vertical lift same as globe valve[d]
15	5.5	2.7	2.1	2.1	0.2	1.8	
20	6.7	3.4	2.1	2.1	0.3	2.2	
25	8.8	4.6	3.7	3.7	0.3	3.0	
32	12	6.1	4.6	4.6	0.5	4.3	
40	13	7.3	5.5	5.5	0.5	4.9	
50	17	9.1	7.3	7.3	0.73	6.1	
65	21	11	8.8	8.8	0.9	7.6	
80	26	13	11	11	1.0	9.1	
90	30	15	13	13	1.2	10	
100	37	18	14	14	1.4	12	
125	43	22	18	18	1.8	15	
150	52	27	21	21	2.1	18	
200	62	35	26	26	2.7	24	Angle lift same as angle valve
250	85	44	32	32	3.7	30	
300	98	50	40	40	4.0	37	
350	110	56	47	47	4.6	41	
400	125	64	55	55	5.2	46	
450	140	73	61	61	5.8	50	
500	160	84	72	72	6.7	61	
600	186	98	81	81	7.6	73	

Note: Losses are for valves in fully open position and with screwed, welded, flanged, or flared connections.

[a] These losses do not apply to valves with needlepoint seats.

[b] Regular and short pattern plug cock valves, when fully open, have same loss as gate valve. For valve losses of short pattern plug cocks above 150 mm, check with manufacturer.

[c] Losses also apply to the in-line, ball-type check valve.

[d] For Y pattern globe lift check valve with seat approximately equal to the nominal pipe diameter, use values of 60° wye valve for loss.

Oil that leaves the compressor or oil separator reaches the condenser and dissolves in the liquid refrigerant, enabling it to pass readily through the liquid line to the evaporator. In the evaporator, the refrigerant evaporates, and the liquid phase becomes enriched in oil. The concentration of refrigerant in the oil depends on the evaporator temperature and types of refrigerant and oil used. The viscosity of the oil/refrigerant solution is determined by the system parameters. Oil separated in the evaporator is returned to the compressor by gravity or by the drag forces of the returning gas. The effect of oil on pressure drop is large, increasing the pressure drop by as much as a factor of 10 in some cases (Alofs et al. 1990).

One of the most difficult problems in low-temperature refrigeration systems using halocarbon refrigerants is returning lubrication oil from the evaporator to the compressors. With the exception of most centrifugal compressors and rarely used nonlubricated compressors, refrigerant continuously carries oil into the discharge line from the compressor. Most of this oil can be removed from the stream by an oil separator and returned to the compressor. Coalescing oil separators are far better than separators using only mist pads or baffles; however, they are not 100% effective. The oil that finds its way into the system must be managed.

Oil mixes well with halocarbon refrigerants at higher temperatures. As the temperature decreases, miscibility is reduced, and some of the oil separates to form an oil-rich layer near the top of the liquid level in a flooded evaporator. If the temperature is very low, the oil becomes a gummy mass that prevents refrigerant controls from functioning, blocks flow passages, and fouls the heat transfer surfaces. Proper oil management is often the key to a properly functioning system.

In general, direct-expansion and liquid overfeed system evaporators have fewer oil return problems than do flooded system evaporators because refrigerant flows continuously at velocities high enough to sweep oil from the evaporator. Low-temperature systems using hot-gas defrost can also be designed to sweep oil out of the circuit each time the system defrosts. This reduces the possibility of oil coating the evaporator surface and hindering heat transfer.

Flooded evaporators can promote oil contamination of the evaporator charge because they may only return dry refrigerant vapor back to the system. Skimming systems must sample the oil-rich layer floating in the drum, a heat source must distill the refrigerant, and the oil must be returned to the compressor. Because flooded halocarbon systems can be elaborate, some designers avoid them.

System Capacity Reduction. The use of automatic capacity control on compressors requires careful analysis and design. The compressor is capable of loading and unloading as it modulates with the system load requirements through a considerable range of capacity. A single compressor can unload down to 25% of full-load capacity, while multiple compressors connected in parallel can unload to a system capacity of 12.5% or lower. System piping must be designed to return oil at the lowest loading, yet not impose excessive pressure drops in the piping and equipment at full load.

Oil Return up Suction Risers. Many refrigeration piping systems contain a suction riser because the evaporator is at a lower level than the compressor. Oil circulating in the system can return up gas risers only by being transported by the returning gas or by auxiliary means such as a trap and a pump. The minimum conditions for oil transport correlate with buoyancy forces (i.e., the density difference between the liquid and the vapor, and the momentum flux of the vapor) (Jacobs et al. 1976).

The principal criteria determining the transport of oil are gas velocity, gas density, and pipe inside diameter. The density of the oil-refrigerant mixture plays a somewhat lesser role because it is almost constant over a wide range. In addition, at temperatures somewhat lower than –40°C, oil viscosity may be significant. Greater gas velocities are required as the temperature drops and the gas becomes less dense. Higher velocities are also necessary if the pipe diameter increases. Table 13 translates these criteria to minimum refrigeration capacity requirements for oil transport. Suction risers must be sized for minimum system capacity. Oil must be returned to the compressor at the operating condition corresponding to the minimum displacement and minimum suction temperature at which the compressor will operate. When suction or evaporator pressure regulators are used, suction risers must be sized for actual gas conditions in the riser.

For a single compressor with capacity control, the minimum capacity is the lowest capacity at which the unit can operate. For multiple compressors with capacity control, the minimum capacity is the lowest at which the last operating compressor can run.

Riser Sizing. The following example demonstrates the use of Table 13 in establishing maximum riser sizes for satisfactory oil transport down to minimum partial loading.

Example 3. Determine the maximum size suction riser that will transport oil at the minimum loading, using R-22 with a 120 kW compressor with a capacity in steps of 25, 50, 75, and 100%. Assume the minimum system loading is 30 kW at 5°C suction and 40°C condensing temperatures with 10 K superheat.

Solution: From Table 13, a 54 mm OD pipe at 5°C suction and 30°C liquid temperature has a minimum capacity of 23.1 kW. From the chart at the bottom of Table 13, the correction multiplier for 40°C suction temperature is about 1. Therefore, the 54 mm OD pipe is suitable.

Based on Table 13, the next smaller line size should be used for marginal suction risers. When vertical riser sizes are reduced to provide satisfactory minimum gas velocities, the pressure drop at full load increases considerably; horizontal lines should be sized to keep the total pressure drop within practical limits. As long as the

Table 13 Minimum Refrigeration Capacity in Kilowatts for Oil Entrainment up Suction Risers
(Copper Tubing, ASTM B 88M Type B, Metric Size)

Refrigerant	Saturated Temp., °C	Suction Gas Temp., °C	Tubing Nominal OD, mm											
			12	15	18	22	28	35	42	54	67	79	105	130
22	−40	−35	0.182	0.334	0.561	0.956	1.817	3.223	5.203	9.977	14.258	26.155	53.963	93.419
		−25	0.173	0.317	0.532	0.907	1.723	3.057	4.936	9.464	16.371	24.811	51.189	88.617
		−15	0.168	0.307	0.516	0.880	1.672	2.967	4.791	9.185	15.888	24.080	49.681	86.006
	−20	−15	0.287	0.527	0.885	1.508	2.867	5.087	8.213	15.748	27.239	41.283	85.173	147.449
		−5	0.273	0.501	0.841	1.433	2.724	4.834	7.804	14.963	25.882	39.226	80.929	140.102
		5	0.264	0.485	0.815	1.388	2.638	4.680	7.555	14.487	25.058	37.977	78.353	135.642
	−5	0	0.389	0.713	1.198	2.041	3.879	6.883	11.112	21.306	36.854	55.856	115.240	199.499
		10	0.369	0.676	1.136	1.935	3.678	6.526	10.535	20.200	34.940	52.954	109.254	189.136
		20	0.354	0.650	1.092	1.861	3.537	6.275	10.131	19.425	33.600	50.924	105.065	181.884
	5	10	0.470	0.862	1.449	2.468	4.692	8.325	13.441	25.771	44.577	67.560	139.387	241.302
		20	0.440	0.807	1.356	2.311	4.393	7.794	12.582	24.126	41.731	63.246	130.488	225.896
		30	0.422	0.774	1.301	2.217	4.213	7.476	12.069	23.141	40.027	60.665	125.161	216.675
134a	−10	−5	0.274	0.502	0.844	1.437	2.732	4.848	7.826	15.006	25.957	39.340	81.164	140.509
		5	0.245	0.450	0.756	1.287	2.447	4.342	7.010	13.440	23.248	35.235	72.695	125.847
		15	0.238	0.436	0.732	1.247	2.370	4.206	6.790	13.019	22.519	34.129	70.414	121.898
	−5	0	0.296	0.543	0.913	1.555	2.956	5.244	8.467	16.234	28.081	42.559	87.806	152.006
		10	0.273	0.500	0.840	1.431	2.720	4.827	7.792	14.941	25.843	39.168	80.809	139.894
		20	0.264	0.484	0.813	1.386	2.634	4.674	7.546	14.468	25.026	37.929	78.254	135.471
	5	10	0.357	0.655	1.100	1.874	3.562	6.321	10.204	19.565	33.843	51.292	105.823	183.197
		20	0.335	0.615	1.033	1.761	3.347	5.938	9.586	18.380	31.792	48.184	99.412	172.098
		30	0.317	0.582	0.978	1.667	3.168	5.621	9.075	17.401	30.099	45.617	94.115	162.929
	10	15	0.393	0.721	1.211	2.063	3.921	6.957	11.232	21.535	37.250	56.456	116.479	201.643
		25	0.370	0.679	1.141	1.944	3.695	6.555	10.583	20.291	35.098	53.195	109.749	189.993
		35	0.358	0.657	1.104	1.881	3.576	6.345	10.243	19.640	33.971	51.486	106.224	183.891
502	−40	−35	0.129	0.236	0.397	0.676	1.284	2.279	3.679	7.054	12.201	18.492	38.152	66.048
		−25	0.125	0.229	0.385	0.657	1.248	2.215	3.575	6.855	11.858	17.972	37.079	64.190
		−15	0.121	0.223	0.374	0.638	1.212	2.151	3.472	6.658	11.516	17.453	36.009	62.337
	−20	−15	0.210	0.385	0.647	1.102	2.096	3.718	6.003	11.510	19.909	30.173	62.253	107.769
		−5	0.204	0.374	0.628	1.070	2.033	3.607	5.823	11.166	19.314	29.272	60.392	104.549
		5	0.198	0.363	0.611	1.041	1.978	3.510	5.666	10.865	18.793	28.482	58.763	101.728
	−5	0	0.288	0.528	0.887	1.510	2.871	5.094	8.224	15.770	27.277	41.341	84.292	147.655
		10	0.279	0.511	0.859	1.464	2.783	4.937	7.970	15.282	26.434	40.063	82.656	143.091
		20	0.271	0.496	0.834	1.421	2.701	4.793	7.737	14.835	25.661	38.891	80.239	138.907
	5	10	0.347	0.637	1.071	1.824	3.467	6.151	9.931	19.041	32.936	49.917	102.986	178.286
		20	0.336	0.617	1.036	1.765	3.356	5.954	9.613	18.431	31.881	48.318	99.688	172.577
		30	0.326	0.598	1.005	1.713	3.256	5.777	9.326	17.882	30.932	46.880	96.721	167.439

Notes:
1. Refrigeration capacity in kilowatts is based on saturated evaporator as shown in table and condensing temperature of 40°C. For other liquid line temperatures, use correction factors in the table to the right.
2. These tables have been computed using an ISO 32 mineral oil for R-22 and R-502. R-134a has been computed using an ISO 32 ester-based oil.

Refrigerant	Liquid Temperature, °C		
	20	30	50
22	1.17	1.08	0.91
134a	1.20	1.10	0.89
502	1.26	1.12	0.86

horizontal lines are level or pitched in the direction of the compressor, oil can be transported with normal design velocities.

Because most compressors have multiple capacity reduction features, gas velocities required to return oil up through vertical suction risers under all load conditions are difficult to maintain. When the suction riser is sized to permit oil return at the minimum operating capacity of the system, the pressure drop in this portion of the line may be too great when operating at full load. If a correctly sized suction riser imposes too great a pressure drop at full load, a double suction riser should be used.

Oil Return up Suction Risers—Multistage Systems. The movement of oil in the suction lines of multistage systems requires the same design approach as that for single-stage systems. For oil to flow up along a pipe wall, a certain minimum drag of the gas flow is required. Drag can be represented by the friction gradient; Table 14 shows values for minimum friction gradients. Use the following sizing data for refrigerants other than those listed in Tables 13 and 14.

Saturation Temperature, °C	Line Size	
	50 mm or less	Above 50 mm
−18	80 Pa/m	45 Pa/m
−46	100 Pa/m	57 Pa/m

Double Suction Risers. Figure 4 shows two methods of double suction riser construction. Oil return in this arrangement is accomplished at minimum loads, but it does not cause excessive pressure

System Practices for Halocarbon Refrigerants

Table 14 Minimum Refrigeration Capacity in Kilowatts for Oil Entrainment up Hot-Gas Risers
(Copper Tubing, ASTM B 88M Type B, Metric Size)

Refrigerant	Saturated Discharge Temp., °C	Discharge Gas Temp., °C	Tubing Diameter, Nominal OD, mm											
			12	15	18	22	28	35	42	54	67	79	105	130
22	20	60	0.563	0.032	0.735	2.956	5.619	9.969	16.094	30.859	43.377	80.897	116.904	288.938
		70	0.549	1.006	1.691	2.881	5.477	9.717	15.687	30.078	52.027	48.851	162.682	281.630
		80	0.535	0.982	1.650	2.811	5.343	9.480	15.305	29.346	50.761	76.933	158.726	173.780
	30	70	0.596	1.092	1.836	3.127	5.945	10.547	17.028	32.649	56.474	85.591	176.588	305.702
		80	0.579	1.062	1.785	3.040	5.779	10.254	16.554	31.740	54.901	83.208	171.671	297.190
		90	0.565	0.035	1.740	2.964	5.635	9.998	16.140	30.948	53.531	81.131	167.386	289.773
	40	80	0.618	1.132	1.903	3.242	6.163	10.934	17.653	33.847	58.546	88.732	183.069	316.922
		90	0.601	1.103	1.853	3.157	6.001	10.647	17.189	32.959	47.009	86.403	178.263	308.603
		100	0.584	1.071	1.800	3.067	5.830	10.343	16.698	32.018	55.382	83.936	173.173	299.791
	50	90	0.630	1.156	1.943	3.310	6.291	11.162	18.020	34.552	59.766	90.580	186.882	323.523
		100	0.611	1.121	1.884	3.209	6.100	10.823	17.473	33.503	57.951	87.831	181.209	313.702
		110	0.595	1.092	1.834	3.125	5.941	10.540	17.016	32.627	56.435	85.532	176.467	305.493
134a	20	60	0.469	0.860	1.445	2.462	4.681	8.305	13.408	25.709	44.469	67.396	139.050	240.718
		70	0.441	0.808	1.358	2.314	4.399	7.805	12.600	24.159	41.788	63.334	130.668	226.207
		80	0.431	0.790	1.327	2.261	4.298	7.626	12.311	23.605	40.830	61.881	127.671	221.020
	30	70	0.493	0.904	1.519	2.587	4.918	8.726	14.087	27.011	46.722	70.812	145.096	252.916
		80	0.463	0.849	1.426	2.430	4.260	8.196	13.232	25.371	43.885	66.512	137.225	237.560
		90	0.452	0.829	1.393	2.374	4.513	8.007	12.926	24.785	42.870	64.974	134.052	232.066
	40	80	0.507	0.930	1.563	2.662	5.061	8.979	14.496	27.794	48.075	72.863	150.328	260.242
		90	0.477	0.874	1.469	2.502	4.756	8.439	13.624	26.122	45.184	68.480	141.285	244.588
		100	0.465	0.852	1.432	2.439	4.637	8.227	13.281	25.466	44.048	66.759	137.735	238.443
	50	90	0.510	0.936	1.573	2.679	5.093	9.037	14.589	27.973	48.385	73.332	151.296	261.918
		100	0.479	0.878	1.476	2.514	4.779	8.480	13.690	26.248	45.402	68.811	141.969	245.772
		110	0.467	0.857	1.441	2.454	4.665	8.278	13.364	25.624	44.322	67.173	138.590	239.921
502	20	60	0.453	0.831	1.397	2.380	4.524	8.027	12.959	24.848	42.980	65.141	134.396	232.661
		70	0.440	0.807	1.357	2.311	4.393	7.795	12.585	24.130	41.737	63.257	130.509	225.933
		80	0.429	0.788	1.324	2.255	4.286	7.605	12.278	23.542	40.720	61.715	127.329	220.427
	30	70	0.459	0.841	1.414	2.409	4.580	8.125	13.118	25.152	43.506	65.937	136.038	235.504
		80	0.446	0.818	1.375	2.343	4.454	7.902	12.757	24.461	42.311	54.126	132.302	229.036
		90	0.435	0.798	1.341	2.285	4.343	7.706	12.441	23.854	41.260	62.534	129.017	233.350
	40	80	0.451	0.827	1.389	2.367	4.499	7.983	12.888	24.711	42.743	64.780	133.652	231.374
		90	0.439	0.804	1.352	2.303	4.378	7.767	12.540	24.044	41.589	63.031	130.044	225.127
		100	0.427	0.783	1.316	2.241	4.260	7.559	12.203	23.398	40.472	61.340	126.554	219.085
	50	90	0.432	0.791	1.330	2.266	4.307	7.641	12.336	23.652	40.912	62.006	127.927	221.463
		100	0.418	0.767	1.289	2.196	4.174	7.406	11.956	22.925	39.654	60.100	123.996	214.657
		110	0.406	0.745	1.253	2.134	2.056	7.197	11.619	22.279	38.536	58.404	120.498	208.602

Notes:
1. Refrigeration capacity in kilowatts is based on saturated evaporator at −5 °C, and condensing temperature as shown in table. For other liquid line temperatures, use correction factors in the table to the right.
2. These tables have been computed using an ISO 32 mineral oil for R-22 and R-502. R-134a has been computed using an ISO 32 ester-based oil.

	Saturated Suction Temperature, °C						
Refrigerant	−50	−40	−30	−20	0	5	10
22	0.87	0.90	0.93	0.96	—	1.02	—
134a	—	—	—	—	1.02	1.04	1.06
502	0.77	0.83	0.88	0.93	—	1.04	—

drops at full load. The sizing and operation of a double suction riser are as follows:

1. Riser A is sized to return oil at the minimum load possible.
2. Riser B is sized for satisfactory pressure drop through both risers at full load. The usual method is to size riser B so that the combined cross-sectional area of A and B is equal to or slightly greater than the cross-sectional area of a single pipe sized for an acceptable pressure drop at full load without regard for oil return at minimum load. The combined cross-sectional area, however, should not be greater than the cross-sectional area of a single pipe that would return oil in an upflow riser under maximum load conditions.
3. A trap is introduced between the two risers, as shown in both methods. During part-load operation, the gas velocity is not sufficient to return oil through both risers, and the trap gradually fills up with oil until riser B is sealed off. The gas then travels up riser A only with enough velocity to carry oil along with it back into the horizontal suction main.

The oil holding capacity of the trap is limited to a minimum by close-coupling the fittings at the bottom of the risers. If this is not done, the trap can accumulate enough oil during part-load operation to lower the compressor crankcase oil level. Note in Figure 4 that riser lines A and B form an inverted loop and enter the horizontal suction line from the top. This prevents oil drainage into the risers, which may be idle during part-load operation. The same purpose

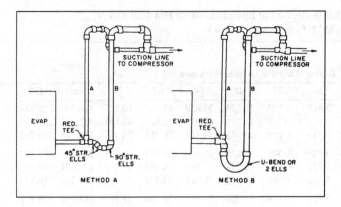

Fig. 4 Double-Suction Riser Construction

can be served by running the risers horizontally into the main, provided that the main is larger in diameter than either riser.

Often, double suction risers are essential on low-temperature systems that can tolerate very little pressure drop. Any system using these risers should include a suction trap (accumulator) and a means of returning oil gradually.

For systems operating at higher suction temperatures, such as for comfort air conditioning, single suction risers can be sized for oil return at minimum load. Where single compressors are used with capacity control, minimum capacity will usually be 25 or 33% of maximum displacement. With this low ratio, pressure drop in single suction risers designed for oil return at minimum load is rarely serious at full load.

When multiple compressors are used, one or more may shut down while another continues to operate, and the maximum-to-minimum ratio becomes much larger. This may make a double suction riser necessary.

The remaining portions of the suction line are sized to permit a practical pressure drop between the evaporators and compressors because oil is carried along in horizontal lines at relatively low gas velocities. It is good practice to give some pitch to these lines toward the compressor. Traps should be avoided, but when that is impossible, the risers from them are treated the same as those leading from the evaporators.

Preventing Oil Trapping in Idle Evaporators. Suction lines should be designed so that oil from an active evaporator does not drain into an idle one. Figure 5A shows multiple evaporators on different floor levels with the compressor above. Each suction line is brought upward and looped into the top of the common suction line to prevent oil from draining into inactive coils.

Figure 5B shows multiple evaporators stacked on the same level, with the compressor above. Oil cannot drain into the lowest evaporator because the common suction line drops below the outlet of the lowest evaporator before entering the suction riser.

Figure 5C shows multiple evaporators on the same level, with the compressor located below. The suction line from each evaporator drops down into the common suction line so that oil cannot drain into an idle evaporator. An alternate arrangement is shown in Figure 5D for cases where the compressor is above the evaporators.

Figure 6 illustrates typical piping for evaporators above and below a common suction line. All horizontal runs should be level or pitched toward the compressor to ensure oil return.

The traps shown in the suction lines after the evaporator suction outlet are recommended by various thermal expansion valve manufacturers to prevent erratic operation of the thermal expansion valve. The expansion valve bulbs are located on the suction lines between the evaporator and these traps. The traps serve as drains and help prevent liquid from accumulating under the expansion valve bulbs during compressor off cycles. They are useful only where straight runs or risers are encountered in the suction line leaving the evaporator outlet.

DISCHARGE (HOT-GAS) LINES

Hot-gas lines should be designed to

- Avoid trapping oil at part-load operation
- Prevent condensed refrigerant and oil in the line from draining back to the head of the compressor
- Have carefully selected connections from a common line to multiple compressors
- Avoid developing excessive noise or vibration from hot-gas pulsations, compressor vibration, or both

Oil Transport up Risers at Normal Loads. Although a low pressure drop is desired, oversized hot-gas lines can reduce gas velocities to a point where the refrigerant will not transport oil. Therefore, when using multiple compressors with capacity control, hot-gas risers must transport oil at all possible loadings.

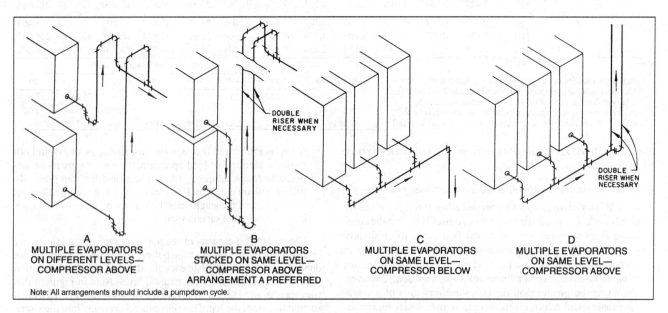

Fig. 5 Suction Line Piping at Evaporator Coils

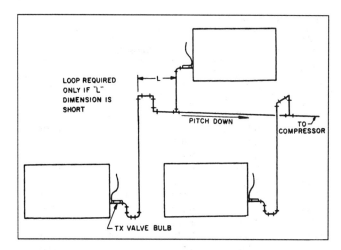

Fig. 6 Typical Piping from Evaporators Located above and below Common Suction Line

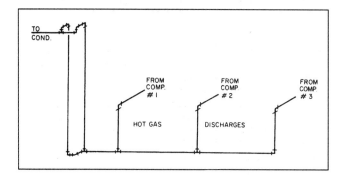

Fig. 7 Double Hot-Gas Riser

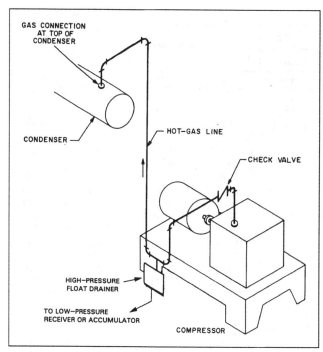

Fig. 8 Hot-Gas Loop

Minimum Gas Velocities for Oil Transport in Risers. Minimum capacities for oil entrainment in hot-gas line risers are shown in Table 14. On multiple-compressor installations, the lowest possible system loading should be calculated and a riser size selected to give at least the minimum capacity indicated in the table for successful oil transport.

In some installations with multiple compressors and with capacity control, a vertical hot-gas line, sized to transport oil at minimum load, has excessive pressure drop at maximum load. When this problem exists, either a double riser or a single riser with an oil separator can be used.

Double Hot-Gas Risers. A double hot-gas riser can be used the same way it is used in a suction line. Figure 7 shows the double riser principle applied to a hot-gas line. Its operating principle and sizing technique are described in the section on Double Suction Risers.

Single Riser and Oil Separator. As an alternative, an oil separator located in the discharge line just before the riser permits sizing the riser for a low pressure drop. Any oil draining back down the riser accumulates in the oil separator. With large multiple compressors, the capacity of the separator may dictate the use of individual units for each compressor located between the discharge line and the main discharge header. Horizontal lines should be level or pitched downward in the direction of gas flow to facilitate travel of oil through the system and back to the compressor.

Piping to Prevent Liquid and Oil from Draining to Compressor Head. Whenever the condenser is located above the compressor, the hot-gas line should be trapped near the compressor before rising to the condenser, especially if the hot-gas riser is long. This minimizes the possibility that refrigerant, condensed in the line during off cycles, will drain back to the head of the compressor. Also, any oil traveling up the pipe wall will not drain back to the compressor head.

The loop in the hot-gas line (Figure 8) serves as a reservoir and traps liquid resulting from condensation in the line during shutdown, thus preventing gravity drainage of liquid and oil back to the compressor head. A small high-pressure float drainer should be installed at the bottom of the trap to drain any significant amount of refrigerant condensate to a low-side component such as a suction accumulator or low-pressure receiver. This float prevents an excessive buildup of liquid in the trap and possible liquid hammer when the compressor is restarted.

For multiple-compressor arrangements, each discharge line should have a check valve to prevent gas from active compressors from condensing on the heads of the idle compressors.

For single-compressor applications, a tightly closing check valve should be installed in the hot-gas line of the compressor whenever the condenser and the receiver ambient temperature are higher than that of the compressor. The check valve prevents refrigerant from boiling off in the condenser or receiver and condensing on the compressor heads during off cycles.

This check valve should be a piston type, which will close by gravity when the compressor stops running. The use of a spring-loaded check may incur chatter (vibration), particularly on slow-speed reciprocating compressors.

For compressors equipped with water-cooled oil coolers, a water solenoid and water-regulating valve should be installed in the water line so that the regulating valve maintains adequate cooling during operation, and the solenoid stops flow during the off cycle to prevent localized condensing of the refrigerant.

Hot-Gas (Discharge) Mufflers. Mufflers can be installed in hot-gas lines to dampen the discharge gas pulsations, reducing vibration and noise. Mufflers should be installed in a horizontal or downflow portion of the hot-gas line immediately after it leaves the compressor.

Because gas velocity through the muffler is substantially lower than that through the hot-gas line, the muffler may form an oil trap. The muffler should be installed to allow oil to flow through it and not be trapped.

Table 15 Refrigerant Flow Capacity for Defrost Lines

Nominal Pipe Size, mm		R-22 Mass Flow, kg/s			R-134a Mass Flow, kg/s			R-502 Mass Flow, kg/s		
		Velocity			Velocity			Velocity		
Copper		5 m/s	10 m/s	15 m/s	5 m/s	10 m/s	15 m/s	5 m/s	10 m/s	15 m/s
12		0.012	0.024	0.035	0.016	0.032	0.049	0.024	0.049	0.073
15		0.019	0.038	0.057	0.026	0.053	0.079	0.040	0.079	0.119
18		0.029	0.058	0.087	0.040	0.080	0.119	0.060	0.120	0.180
22		0.044	0.088	0.133	0.061	0.122	0.183	0.092	0.184	0.276
28		0.074	0.148	0.222	0.102	0.204	0.305	0.154	0.307	0.461
35		0.120	0.230	0.350	0.160	0.320	0.480	0.240	0.490	0.730
42		0.170	0.340	0.510	0.240	0.470	0.710	0.360	0.710	1.070
54		0.290	0.580	0.870	0.400	0.800	1.190	0.600	1.200	1.800
67		0.450	0.890	1.340	0.620	1.230	1.850	0.930	1.860	2.790
79		0.620	1.250	1.870	0.860	1.720	2.580	1.300	2.590	3.890
105		1.110	2.230	3.340	1.530	3.070	4.600	2.310	4.630	6.940
130		1.730	3.460	5.180	2.380	4.760	7.140	3.590	7.180	10.800
156		2.500	5.010	7.510	3.450	6.900	10.300	5.200	10.400	15.600
206		4.330	8.660	13.000	5.970	11.900	17.900	9.000	18.000	27.000
257		6.730	13.500	20.200	9.280	18.600	27.800	14.000	28.000	42.000
Steel										
mm	SCH									
10	80	0.018	0.035	0.053	0.024	0.049	0.073	0.037	0.074	0.110
15	80	0.028	0.056	0.084	0.039	0.078	0.116	0.059	0.120	0.180
20	80	0.049	0.099	0.148	0.068	0.136	0.204	0.103	0.210	0.310
25	80	0.080	0.160	0.240	0.110	0.220	0.330	0.166	0.330	0.500
32	40	0.139	0.280	0.420	0.191	0.382	0.570	0.290	0.580	0.860
40	40	0.190	0.380	0.570	0.260	0.520	0.780	0.390	0.780	1.180
50	40	0.310	0.620	0.930	0.430	0.860	1.280	0.650	1.290	1.940
65	40	0.440	0.890	1.330	0.610	1.220	1.830	0.920	1.840	2.760
80	40	0.680	1.370	2.050	0.940	1.890	2.830	1.420	2.840	4.270
100	40	1.180	2.360	3.540	1.620	3.250	4.870	2.450	4.900	7.350
125	40	1.850	3.700	5.550	2.550	5.100	7.650	3.850	7.690	11.500
150	40	2.680	5.350	8.030	3.690	7.370	11.100	5.560	11.100	16.700
200	40	4.630	9.260	13.900	6.380	12.800	19.100	9.630	19.300	28.900
250	40	7.300	14.600	21.900	10.100	20.100	30.200	15.200	30.300	45.500
300	ID	10.500	20.900	31.400	14.400	28.900	43.300	21.800	43.500	65.300

Note: Refrigerant flow data based on saturated condensing temperature of 21°C.

DEFROST GAS SUPPLY LINES

Sizing refrigeration lines to supply defrost gas to one or more evaporators has not been an exact science. The parameters associated with sizing the defrost gas line are related to allowable pressure drop and refrigerant flow rate during defrost.

Engineers have used an estimated two times the evaporator load for effective refrigerant flow rate to determine line sizing requirements. The pressure drop is not as critical during the defrost cycle, and many engineers have used velocity as the criterion for determining line size. The effective condensing temperature and average temperature of the gas must be determined. The velocity determined at saturated conditions will give a conservative line size.

Some controlled testing (Stoecker 1984) has shown that in small coils with R-22, the defrost flow rate tends to be higher as the condensing temperature is increased. The flow rate is on the order of two to three times the normal evaporator flow rate, which supports the estimated two times used by practicing engineers.

Table 15 (R-22, R-134a, and R-502) provides guidance on selecting defrost gas supply lines based on velocity at a saturated condensing temperature of 21°C. It is recommended that initial sizing be based on twice the evaporator flow rate and that velocities from 5 to 10 m/s be used for determining the defrost gas supply line size.

Gas defrost lines must be designed to continuously drain any condensed liquid.

RECEIVERS

Refrigerant receivers are vessels used to store excess refrigerant circulated throughout the system. Receivers perform the following functions:

1. Provide pumpdown storage capacity when another part of the system must be serviced or the system must be shut down for an extended time. In some water-cooled condenser systems, the condenser also serves as a receiver if the total refrigerant charge does not exceed its storage capacity.

2. Handle the excess refrigerant charge that occurs with air-cooled condensers using the flooding-type condensing pressure control (see the section on Pressure Control for Refrigerant Condensers).

3. Accommodate a fluctuating charge in the low side and drain the condenser of liquid to maintain an adequate effective condensing surface on systems where the operating charge in the evaporator and/or condenser varies for different loading conditions. When an evaporator is fed with a thermal expansion valve, hand expansion valve, or low-pressure float, the operating charge in the evaporator varies considerably depending on the loading. During low load, the evaporator requires a larger charge since the boiling is not as intense. When the load increases, the operating charge in the evaporator decreases, and the receiver must store excess refrigerant.

System Practices for Halocarbon Refrigerants

4. Hold the full charge of the idle circuit on systems with multicircuit evaporators that shut off the liquid supply to one or more circuits during reduced load and pump out the idle circuit.

Connections for Through-Type Receiver. When a through-type receiver is used, the liquid must always flow from the condenser to the receiver. The pressure in the receiver must be lower than that in the condenser outlet. The receiver and its associated piping provide free flow of liquid from the condenser to the receiver by equalizing the pressures between the two so that the receiver cannot build up a higher pressure than the condenser.

If a vent is not used, the piping between condenser and receiver (condensate line) is sized so that liquid flows in one direction and gas flows in the opposite direction. Sizing the condensate line for 0.5 m/s liquid velocity is usually adequate to attain this flow. Piping should slope at least 20 mm/m and eliminate any natural liquid traps. See Figure 9 for this configuration.

The piping between the condenser and the receiver can be equipped with a separate vent (equalizer) line to allow receiver and condenser pressures to equalize. This external vent line can be piped either with or without a check valve in the vent line (see Figures 11 and 12). If no check valve is installed in the vent line, prevent the discharge gas from discharging directly into the vent line; this should prevent a gas velocity pressure component from being introduced on top of the liquid in the receiver. When the piping configuration is unknown, install a check valve in the vent with the direction of flow toward the condenser. The check valve should be selected for minimum opening pressure (i.e., approximately 3.5 kPa). When determining the condensate drop leg height, allowance must be made to overcome both the pressure drop across this check valve and the refrigerant pressure drop through the condenser. This ensures that there will be no liquid backup into an operating condenser on a multiple-condenser application when one or more of the condensers is idle. The condensate line should be sized so that the velocity does not exceed 0.75 m/s.

The vent line flow is from receiver to condenser when the receiver temperature is higher than the condensing temperature. Flow is from condenser to receiver when the air temperature around the receiver is below the condensing temperature. The rate of flow depends on this temperature difference as well as on the receiver surface area. Vent size can be calculated from this flow rate.

Connections for Surge-Type Receiver. The purpose of a surge-type receiver is to allow liquid to flow to the expansion valve without exposure to refrigerant in the receiver, so that it can remain subcooled. The receiver volume is available for liquid that is to be removed from the system. Figure 10 shows an example of connections for a surge-type receiver. The height h must be adequate for a liquid pressure at least as large as the pressure loss through the condenser, liquid line, and vent line at the maximum temperature difference between the receiver ambient and the condensing temperature. The condenser pressure drop at the greatest expected heat rejection should be obtained from the manufacturer. The minimum value of h can then be calculated and a decision made as to whether or not the available height will permit the surge-type receiver.

Multiple Condensers. Two or more condensers connected in series or in parallel can be used in a single refrigeration system. If the condensers are connected in series, the pressure losses through each must be added. Condensers are more often arranged in parallel. The pressure loss through any one of the parallel circuits is always equal to that through any of the others, even if it results in filling much of one circuit with liquid while gas passes through another.

Figure 11 shows a basic arrangement for parallel condensers with a through-type receiver. The condensate drop legs must be long enough to allow liquid levels in them to adjust to equalize pressure losses between condensers at all operating conditions. The drop legs should be 150 to 300 mm higher than calculated to ensure that liquid outlets remain free-draining. This height provides a liquid pressure to offset the largest condenser pressure loss. The liquid seal prevents gas blow-by between condensers.

Large single condensers with multiple coil circuits should be piped as though the independent circuits were parallel condensers. For example, assume the left condenser in Figure 11 has 14 kPa more pressure drop than the right condenser. The liquid level on the left side will be about 1.2 m higher than that on the right. If the condensate lines do not have enough vertical height for this level difference, the liquid will back up into the condenser until the pressure drop is the same through both circuits. Enough surface may be covered to reduce the condenser capacity significantly.

The condensate drop legs should be sized based on 0.75 m/s velocity. The main condensate lines should be based on 0.5 m/s. Depending on prevailing local and/or national safety codes, a relief device may have to be installed in the discharge piping.

Figure 12 shows a piping arrangement for parallel condensers with a surge-type receiver. When the system is operating at reduced load, the flow paths through the circuits may not be symmetrical. Small pressure differences would not be unusual; therefore, the liquid line junction should be about 600 to 900 mm below the bottom of the condensers. The exact amount can be calculated from pressure loss through each path at all possible operating conditions.

When condensers are water-cooled, a single automatic water valve for the condensers in one refrigeration system should be used. Individual valves for each condenser in a single system would not be able to maintain the same pressure and corresponding pressure drops.

With evaporative condensers (Figure 13), the pressure loss may be high. If parallel condensers are alike and all are operated, the

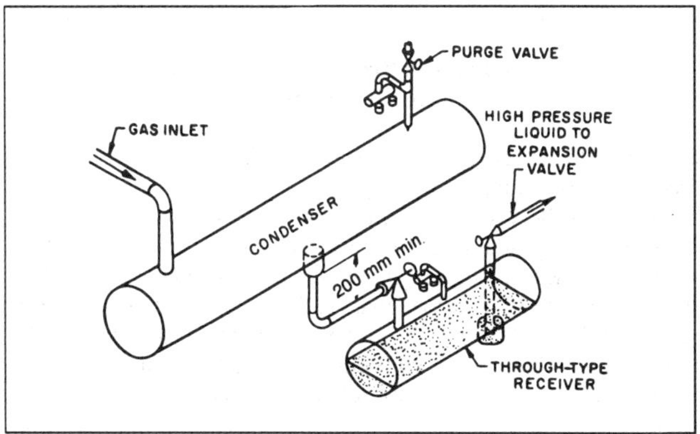

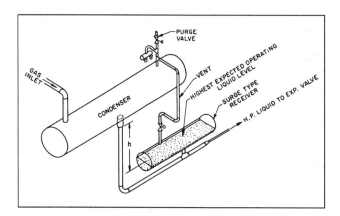

Fig. 9 Shell-and-Tube Condenser to Receiver Piping (Through-Type Receiver)

Fig. 10 Shell-and-Tube Condenser to Receiver Piping (Surge-Type Receiver)

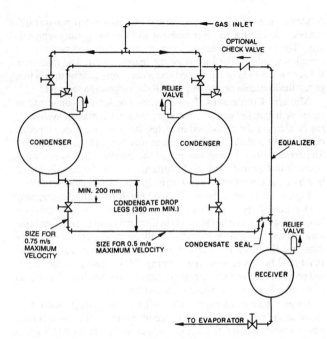

Fig. 11 Parallel Condensers with Through-Type Receiver

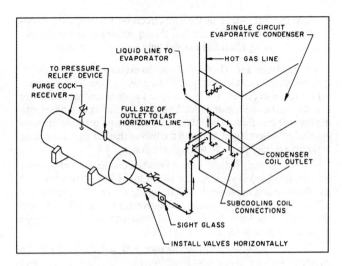

Fig. 13 Single-Circuit Evaporative Condenser with Receiver and Liquid Subcooling Coil

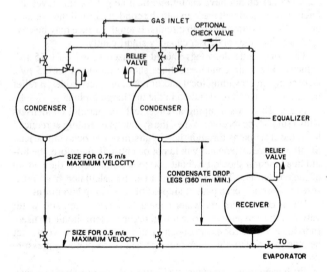

Fig. 12 Parallel Condensers with Surge-Type Receiver

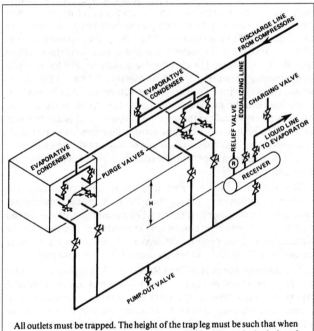

All outlets must be trapped. The height of the trap leg must be such that when one or more units are idle (fan or pump stopped), liquid may rise in the leg of the operating unit so that the static head equals the pressure drop in the operating unit at all conditions. The trap height should be 150 to 300 mm greater than the calculated height to assure that the liquid outlets remain free-draining.

Fig. 14 Multiple Evaporative Condensers with Equalization to Condenser Inlets

differences may be small, and the height of the condenser outlets above the liquid line junction need not be more than 600 to 900 mm. If the fans on one condenser are not operated while the fans on another condenser are, then the liquid level in the one condenser must be high enough to compensate for the pressure drop through the operating condenser.

When the available level difference between condenser outlets and the liquid line junction is sufficient, the receiver may be vented to the condenser inlets (see Figure 14). In this case, the surge-type receiver can be used. The level difference must then be at least equal to the greatest loss through any condenser circuit plus the greatest vent line loss when the receiver ambient is greater than the condensing temperature.

AIR-COOLED CONDENSERS

The refrigerant pressure drop through air-cooled condensers must be obtained from the supplier for the particular unit at the specified load. If the refrigerant pressure drop is low enough and it is practical to so arrange the equipment, parallel condensers can be connected to allow for capacity reduction to zero on one condenser without causing liquid backup in active condensers (Figure 15). Multiple condensers with high pressure drops can be connected as shown in Figure 15, provided that (1) the ambient at the receiver is equal to or lower than the inlet air temperature to the condenser; (2) capacity control affects all units equally; (3) all units operate when one operates, unless valved off at both inlet and outlet; and (4) all units are of equal size.

A single condenser with any pressure drop can be connected to a receiver without an equalizer and without trapping height if the condenser outlet and the line from it to the receiver can be sized for

System Practices for Halocarbon Refrigerants

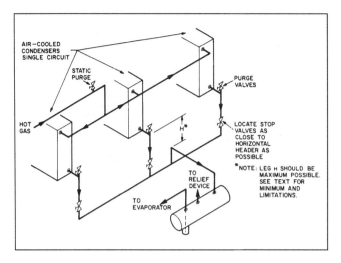

Fig. 15 Multiple Air-Cooled Condensers

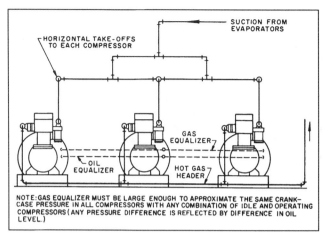

Fig. 16 Suction and Hot-Gas Headers for Multiple Compressors

sewer flow without a trap or restriction, using a maximum velocity of 0.5 m/s. A single condenser can also be connected with an equalizer line to the hot-gas inlet if the vertical drop leg is sufficient to balance the refrigerant pressure drop through the condenser and the liquid line to the receiver.

If unit sizes are unequal, additional liquid height H, equivalent to the difference in full-load pressure drop, is required. Usually, condensers of equal size are used in parallel applications.

If the receiver cannot be located in an ambient temperature below the inlet air temperature for all operating conditions, sufficient extra height of drop leg H is required to overcome the equivalent differences in saturation pressure of the receiver and the condenser. The subcooling formed by the liquid leg tends to condense vapor in the receiver to reach a balance between rate of condensation, at an intermediate saturation pressure, and heat gain from ambient to the receiver. A relatively large liquid leg is required to balance a small temperature difference; therefore, this method is probably limited to marginal cases. The liquid leaving the receiver will nonetheless be saturated, and any subcooling to prevent flashing in the liquid line must be obtained downstream of the receiver. If the temperature of the receiver ambient is above the condensing pressure only at part-load conditions, it may be acceptable to back liquid into the condensing surface, sacrificing the operating economy of lower part-load pressure for a lower liquid leg requirement. The receiver must be adequately sized to contain a minimum of the backed-up liquid so that the condenser can be fully drained when full load is required. If a low-ambient control system of backing liquid into the condenser is used, consult the system supplier for proper piping.

PIPING AT MULTIPLE COMPRESSORS

Multiple compressors operating in parallel must be carefully piped to ensure proper operation.

Suction Piping

Suction piping should be designed so that all compressors run at the same suction pressure and so that oil is returned in equal proportions. All suction lines should be brought into a common suction header in order to return the oil to each crankcase as uniformly as possible. Depending on the type and size of compressors, oil may be returned by designing the piping in one or more of the following schemes.

1. Oil returned with the suction gas to each compressor
2. Oil contained with a suction trap (accumulator) and returned to the compressors through a controlled means
3. Oil trapped in a discharge line separator and returned to the compressors through a controlled means (see the section on Discharge Piping)

The suction header is a means of distributing the suction gas equally to each compressor. The design of the header can be to freely pass the suction gas and oil mixture or to provide a suction trap for the oil. The header should be run above the level of the compressor suction inlets so that oil can drain into the compressors by gravity.

Figure 16 shows a pyramidal or yoke-type suction header to maximize pressure and flow equalization at each of three compressor suction inlets piped in parallel. This type of construction is recommended for applications of three or more compressors in parallel. For two compressors in parallel, a single feed between the two compressor takeoffs is acceptable. Although not as good with regard to equalizing flow and pressure drops to all compressors, one alternative is to have the suction line from the evaporators enter at one end of the header instead of using the yoke arrangement. Then the suction header may have to be enlarged to minimize pressure drop and flow turbulence.

Suction headers designed to freely pass the gas and oil mixture should have the branch suction lines to the compressors connected to the side of the header. The return mains from the evaporators should not be connected into the suction header to form crosses with the branch suction lines to the compressors. The header should be full size based on the largest mass flow of the suction line returning to the compressors. The takeoffs to the compressors should either be the same size as the suction header or be constructed in such a manner that the oil will not trap within the suction header. The branch suction lines to the compressors should not be reduced until the vertical drop is reached.

Suction traps are recommended wherever any of the following are used: (1) parallel compressors, (2) flooded evaporators, (3) double suction risers, (4) long suction lines, (5) multiple expansion valves, (6) hot-gas defrost, (7) reverse-cycle operation, and (8) suction pressure regulators.

Depending on the size of the system, the suction header may be designed to function as a suction trap. The suction header should be large enough to provide a region of low velocity within the header to allow for the suction gas and oil to separate. Refer to the section on Low-Pressure Receiver Sizing in Chapter 1 to arrive at recommended velocities for separation. The suction gas flow for individual compressors should be taken off the top of the suction header. The oil can be returned to the compressor directly or through a vessel equipped with a heater to boil off the refrigerant and then allow the oil to drain to the compressors or other devices used to feed oil to the compressors.

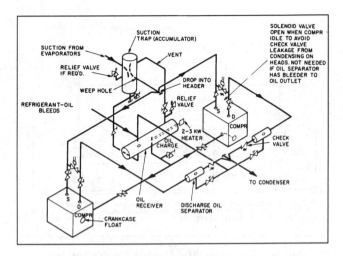

Fig. 17 Parallel Compressors with Gravity Oil Flow

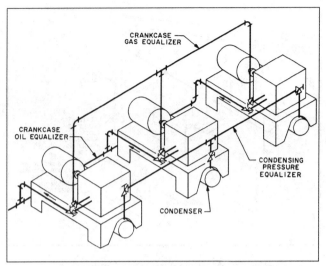

Fig. 18 Interconnecting Piping for Multiple Condensing Units

The suction trap must be sized for effective gas and liquid separation. Adequate liquid volume and a means of disposing of it must be provided. A liquid transfer pump or heater may be used. Chapter 1 has further information on separation and liquid transfer pumps.

An oil receiver equipped with a heater effectively evaporates liquid refrigerant accumulated in the suction trap. It also assumes that each compressor receives its share of oil. Either crankcase float valves or external float switches and solenoid valves can be used to control the oil flow to each compressor.

A gravity feed oil receiver should be elevated to overcome the pressure drop between it and the crankcase. The oil receiver should be sized so that a malfunction of the oil control mechanism cannot overfill an idle compressor.

Figure 17 shows a recommended hookup of multiple compressors, suction trap (accumulator), oil receiver, and discharge line oil separators. The oil receiver also provides a reserve supply of oil for the compressors where the oil in the system external to the compressor varies with system loading. The heater mechanism should always be submerged.

Discharge Piping

The piping arrangement shown in Figure 16 is suggested for discharge piping. The piping must be arranged to prevent refrigerant liquid and oil from draining back into the heads of idle compressors. A check valve in the discharge line may be necessary to prevent refrigerant and oil from entering the compressor heads by migration. It is recommended that, after leaving the compressor head, the piping be routed to a lower elevation so that a trap is formed to allow for drainback of refrigerant and oil from the discharge line when flow rates are reduced or the compressors are off. If an oil separator is used in the discharge line, it may suffice as the trap for drainback for the discharge line.

A bullheaded tee at the junction of two compressor branches and the main discharge header should be avoided because it causes increased turbulence, increased pressure drop, and possible hammering in the line.

When an oil separator is used on multiple compressor arrangements, the oil must be piped to return to the compressors. This can be done in a variety of methods depending on the oil management system design. The oil may be returned to an oil receiver that is the supply for control devices feeding oil back to the compressors.

Interconnection of Crankcases

When two or more compressors are to be interconnected, a method must be provided to equalize the crankcases. Some compressor designs do not operate correctly with simple equalization of the crankcases. For these systems, it may be necessary to design a positive oil float control system for each compressor crankcase. A typical system allows the oil to collect in an oil receiver that, in turn, supplies oil to a device that meters oil back into the compressor crankcase to maintain a proper oil level (Figure 17).

Compressor systems that can be equalized should be placed on foundations so that all oil equalizer tapping locations are exactly level. If crankcase floats (as shown in Figure 17) are not used, an oil equalization line should connect all of the crankcases to maintain uniform oil levels. The oil equalizer may be run level with the tapping, or, for convenient access to the compressors, it may be run at the floor (Figure 18). It should never be run at a level higher than that of the tapping.

For the oil equalizer line to work properly, equalize the crankcase pressures by installing a gas equalizer line above the oil level. This line may be run to provide head room (Figure 18) or run level with the tapping on the compressors. It should be piped so that oil or liquid refrigerant will not be trapped.

Both lines should be the same size as the tapping on the largest compressor and should be valved so that any one machine can be taken out for repair. The piping should be arranged to absorb vibration.

PIPING AT VARIOUS SYSTEM COMPONENTS

Flooded Fluid Coolers

For a description of flooded fluid coolers, see Chapter 37 of the 1996 *ASHRAE Handbook—Systems and Equipment*.

Shell-and-tube flooded coolers designed to minimize liquid entrainment in the suction gas require a continuous liquid bleed line (Figure 19) installed at some point in the cooler shell below the liquid level to remove trapped oil. This continuous bleed of refrigerant liquid and oil prevents the oil concentration in the cooler from getting too high. The location of the liquid bleed connection on the shell depends on the refrigerant and oil used. For refrigerants that are highly miscible with the refrigeration oil, the connection can be anywhere below the liquid level.

Refrigerants 22 and 502 can have a separate oil-rich phase floating on a refrigerant-rich layer. This becomes more pronounced as the evaporating temperature drops. When R-22 and R-502 are used with mineral oil, the bleed line is usually taken off the shell just slightly below the liquid level, or there may be more than one valved bleed connection at slightly different levels so that the optimum point can

System Practices for Halocarbon Refrigerants

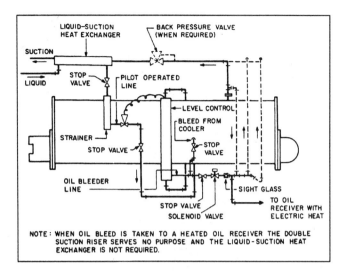

Fig. 19 Typical Piping at Flooded Fluid Cooler

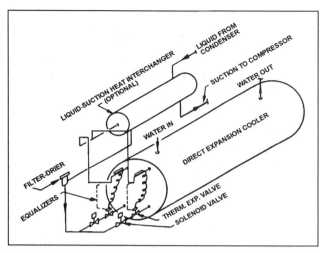

Fig. 20 Two-Circuit Direct-Expansion Cooler Connections
(for Single-Compressor System)

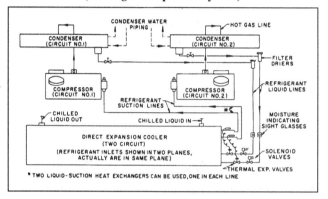

Fig. 21 Typical Refrigerant Piping in Liquid Chilling
Package with Two Completely Separate Circuits

be selected during operation. With alkyl benzene lubricants, oil-refrigerant miscibility may be high enough that the oil bleed connection can be anywhere below the liquid level. The solubility charts in Chapter 7 give specific information.

Where the flooded cooler design requires an external surge drum to separate the liquid carryover from the suction gas off the tube bundle, the richest oil concentration may or may not be in the cooler. In some cases, the surge drum will have the highest concentration of oil. Here, the refrigerant and oil bleed connection is taken from the surge drum. The refrigerant and oil bleed from the cooler by gravity. The bleed sometimes drains into the suction line so that the oil can be returned to the compressor with the suction gas after the accompanying liquid refrigerant is vaporized in a liquid-suction heat interchanger. A better method is to drain the refrigerant-oil bleed into a heated receiver that boils the refrigerant off to the suction line and drains the oil back to the compressor.

Refrigerant Feed Devices

For further information on refrigerant feed devices, see Chapter 45. The pilot-operated low-side float control (Figure 19) is sometimes selected for flooded systems using halocarbon refrigerants. Except for small capacities, direct-acting low-side float valves are impractical for these refrigerants. The displacer float controlling a pneumatic valve works well for low-side liquid level control; it allows the cooler level to be adjusted within the instrument without disturbing the piping.

High-side float valves are practical only in systems having one evaporator because distribution problems result when multiple evaporators are used.

Float chambers should be located as near to the liquid connection on the cooler as possible because a long length of liquid line, even if insulated, can pick up room heat and give an artificial liquid level in the float chamber. Equalizer lines to the float chamber must be amply sized to minimize the effect of heat transmission. The float chamber and its equalizing lines must be insulated.

Each flooded cooler system must have a way of keeping oil concentration in the evaporator low, both to minimize the bleedoff needed to keep oil concentration in the cooler low and to reduce system losses from large stills. A highly efficient discharge gas/oil separator can be used for this purpose.

At low temperatures, periodic warm-up of the evaporator permits recovery of oil accumulation in the chiller. If continuous operation is required, dual chillers may be needed to permit deoiling an oil-laden evaporator, or an oil-free compressor may be used.

Direct-Expansion Fluid Chillers

For further information on these chillers, see Chapter 43. Figure 20 shows typical piping connections for a multicircuit direct-expansion chiller. Each circuit contains its own thermostatic expansion and solenoid valves. One of the solenoid valves can be wired to close at reduced system capacity. The thermostatic expansion valve bulbs should be located between the cooler and the liquid-suction interchanger, if used. Locating the bulb downstream from the interchanger can cause excessive cycling of the thermostatic expansion valve because the flow of high-pressure liquid through the interchanger ceases when the thermostatic expansion valve closes; consequently, no heat is available from the high-pressure liquid, and the cooler must starve itself to obtain the superheat necessary to open the valve. When the valve does open, excessive superheat causes it to overfeed until the bulb senses liquid downstream from the interchanger. Therefore, the remote bulb should be positioned between the cooler and the interchanger.

Figure 21 shows a typical piping arrangement that has been successful in packaged water chillers having direct-expansion coolers. With this arrangement, automatic recycling pumpdown is needed on the lag compressor to prevent leakage through compressor valves, permitting migration to the cold evaporator circuit. It also prevents liquid from slugging the compressor at start-up.

On larger systems, the limited size of thermostatic expansion valves may require use of a pilot-operated liquid valve controlled by a small thermostatic expansion valve (Figure 22). The small thermostatic expansion valve pilots the main liquid control valve. The

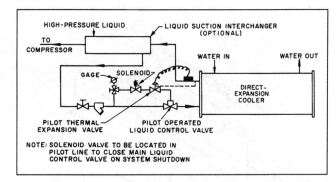

Fig. 22 Direct-Expansion Cooler with Pilot-Operated Control Valve

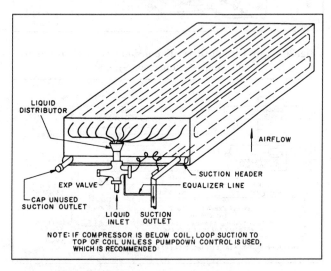

Fig. 23 Direct-Expansion Evaporator (Top-Feed, Free-Draining)

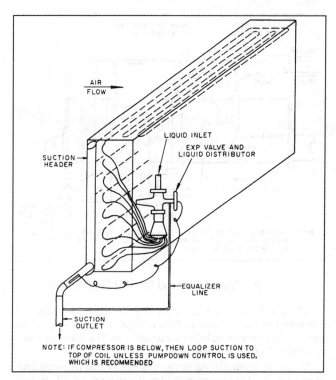

Fig. 24 Direct-Expansion Evaporator (Horizontal Airflow)

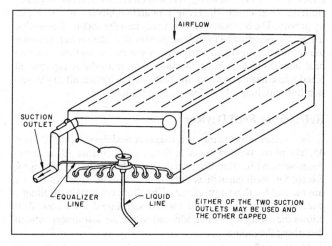

Fig. 25 Direct-Expansion Evaporator (Bottom-Feed)

equalizing connection and the bulb of the pilot thermostatic expansion valve should be treated as a direct-acting thermal expansion valve. A small solenoid valve in the pilot line will shut off the high side from the low during shutdown. However, the main liquid valve will not open and close instantaneously.

Direct-Expansion Air Coils

For further information on these coils, see Chapter 21 of the 1996 *ASHRAE Handbook—Systems and Equipment*. The most common ways of arranging direct-expansion coils are shown in Figures 23 and 24. The method shown in Figure 24 provides the necessary superheat to operate the thermostatic expansion valve and is effective for heat transfer because the leaving air contacts the coldest evaporator surface. This arrangement is advantageous on low-temperature applications, where the coil pressure drop represents an appreciable change in evaporating temperature.

Direct-expansion air coils can be located in any position as long as proper refrigerant distribution and continuous oil removal facilities are provided.

Figure 23 shows top-feed, free-draining piping with a vertical up airflow coil. In Figure 24, which illustrates a horizontal airflow coil, the suction is taken off the bottom header connection, providing free oil draining. Many coils are supplied with connections at each end of the suction header so that a free-draining connection can be used regardless of which side of the coil is up; the other end is then capped.

In Figure 25, a refrigerant upfeed coil is used with a vertical downflow air arrangement. Here, the coil design must provide sufficient gas velocity to entrain oil at lowest loadings and to carry it into the suction line.

Pumpdown compressor control is desirable on all systems using downfeed or upfeed evaporators in order to protect the compressor against a liquid slugback in cases where liquid can accumulate in the suction header and/or the coil on system off cycles. Pumpdown compressor control is described later.

Thermostatic expansion valve operation and application are described in Chapter 45. Thermostatic expansion valves should be sized carefully to avoid undersizing at full load and oversizing at partial load. The refrigerant pressure drops through the system (distributor, coil, condenser, and refrigerant lines, including liquid lifts) must be properly evaluated to determine the correct pressure drop available across the valve on which to base the selection. Variations in condensing pressure greatly affect the pressure available across the valve, and hence its capacity.

System Practices for Halocarbon Refrigerants

Oversized thermostatic expansion valves result in a cycling condition that alternates flooding and starving the coil. This occurs because the valve attempts to throttle at a capacity below its capability, which causes periodic flooding of the liquid back to the compressor and wide temperature variations in the air leaving the coil. Reduced compressor capacity further aggravates this problem. Systems having multiple coils can use solenoid valves located in the liquid line feeding each evaporator or group of evaporators to close them off individually as compressor capacity is reduced.

For information on defrosting, see Chapter 42.

Flooded Evaporators

Flooded evaporators may be desirable when a small temperature differential is required between the refrigerant and the medium being cooled. A small temperature differential is an advantageous feature in low-temperature applications.

In a flooded evaporator, the coil is kept full of refrigerant when cooling is required. The refrigerant level is generally controlled through a high- or low-side float control. Figure 26 represents a typical arrangement showing a low-side float control, oil return line, and heat interchanger.

Circulation of refrigerant through the evaporator depends on gravity and a thermosiphon effect. A mixture of liquid refrigerant and vapor returns to the surge tank, and the vapor flows into the suction line. A baffle installed in the surge tank helps prevent foam and liquid from entering the suction line. A liquid refrigerant circulating pump (Figure 27) provides a more positive way of obtaining a high rate of circulation.

Where the suction line is taken off the top of the surge tank, difficulties arise if no special provisions are made for oil return. For this reason, the oil return lines shown in Figure 26 should be installed. These lines are connected near the bottom of the float chamber and also just below the liquid level in the surge tank (where an oil-rich liquid refrigerant exists). They extend to a lower point on the suction line to allow gravity flow. Included in this oil return line is (1) a solenoid valve that is open only while the compressor is running and (2) a metering valve that is adjusted to allow a constant but small volume return to the suction line. A liquid line sight glass may be installed downstream from the metering valve to serve as a convenient check on the liquid being returned.

Oil can be returned satisfactorily by taking a bleed of refrigerant and oil from the pump discharge (Figure 27) and feeding it to the heated oil receiver. If a low-side float is used, a jet ejector can be used to remove oil from the quiescent float chamber.

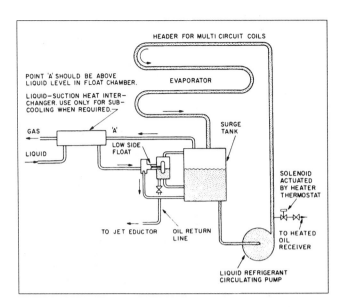

Fig. 27 Flooded Evaporator (Forced Circulation)

REFRIGERATION ACCESSORIES

Liquid-Suction Heat Exchangers

Generally, liquid-suction heat exchangers subcool the liquid refrigerant and superheat the suction gas. They are used for one or more of the following functions:

- *Increasing the efficiency of the refrigeration cycle.* The efficiency of the thermodynamic cycle of certain halocarbon refrigerants can be increased when the suction gas is superheated by removing heat from the liquid. This increased efficiency must be evaluated against the effect of pressure drop through the suction side of the exchanger, which forces the compressor to operate at a lower suction pressure. Liquid-suction heat exchangers are most beneficial at low suction temperatures. The increase in cycle efficiency for systems operating in the air-conditioning range (down to about $-1°C$ evaporating temperature) usually does not justify their use. The heat exchanger can be located wherever convenient.

- *Subcooling the liquid refrigerant to prevent flash gas at the expansion valve.* The heat exchanger should be located near the condenser or receiver to achieve subcooling before pressure drop occurs.

- *Evaporating small amounts of expected liquid refrigerant returning from evaporators in certain applications.* Many heat pumps incorporating reversals of the refrigerant cycle include a suction line accumulator and liquid-suction heat exchanger arrangement to trap liquid floodbacks and vaporize them slowly between cycle reversals.

If the design of an evaporator makes a deliberate slight overfeed of refrigerant necessary, either to improve evaporator performance or to return oil out of the evaporator, a liquid-suction heat exchanger is needed to evaporate the refrigerant.

A flooded water cooler usually incorporates an oil-rich liquid bleed from the shell into the suction line for returning oil. The liquid-suction heat exchanger boils the liquid refrigerant out of the mixture in the suction line. Exchangers used for this purpose should be placed in a horizontal run near the evaporator. Several types of liquid-suction heat exchangers are used.

Liquid and Suction Line Soldered Together. The simplest form of heat exchanger is obtained by strapping or soldering the suction and liquid lines together to obtain counterflow and then insulating the lines as a unit. To obtain the greatest capacity, the

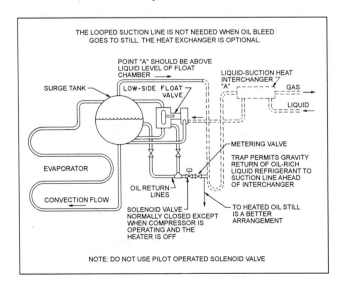

Fig. 26 Flooded Evaporator (Gravity Circulation)

liquid line should always be on the bottom of the suction line, since liquid in a suction line runs along the bottom (Figure 28). This arrangement is limited by the amount of suction line available.

Shell-and-Coil or Shell-and-Tube Heat Exchangers (Figure 29). These units are usually installed so that the suction outlet drains the shell. When the units are used to evaporate liquid refrigerant returning in the suction line, the free-draining arrangement is not recommended. Liquid refrigerant can run along the bottom of the heat exchanger shell, having little contact with the warm liquid coil, and drain into the compressor. By installing the heat exchanger at a slight angle to the horizontal (Figure 30) with the gas entering at the bottom and leaving at the top, any liquid returning in the line is trapped in the shell and held in contact with the warm liquid coil, where most of it is vaporized. An oil return line, with a metering valve and solenoid valve (open only when the compressor is running), is required to return oil that collects in the trapped shell.

Concentric Tube-in-Tube Heat Exchangers. The tube-in-tube heat exchanger is not as efficient as the shell-and-finned-coil type. It is, however, quite suitable for cleaning up small amounts of excessive liquid refrigerant returning in the suction line. Figure 31 shows typical construction with available pipe and fittings.

Plate-Type Heat Exchangers. Plate-type heat exchangers provide high-efficiency heat transfer. They are very compact, have low pressure drop, and are lightweight devices. They are good for use as liquid subcoolers.

For air-conditioning applications, heat exchangers are recommended for liquid subcooling or for clearing up excess liquid in the suction line. For refrigeration applications, heat exchangers are recommended to increase cycle efficiency, as well as for liquid subcooling and removing small amounts of excess liquid in the suction line. Excessive superheating of the suction gas should be avoided.

Two-Stage Subcoolers

To take full advantage of the two-stage system, the refrigerant liquid should be cooled to a temperature near the interstage temperature to reduce the amount of flash gas handled by the low-stage compressor. The net result is a reduction in total system power requirements. The amount of gain from cooling to near interstage conditions varies among refrigerants.

Figure 32 illustrates an open or flash-type cooler. This is the simplest and least costly type, which has the advantage of cooling liquid to the saturation temperature of the interstage pressure. One disadvantage is that the pressure of the cooled liquid is reduced to interstage pressure, leaving less pressure available for liquid transport. Although the liquid temperature is reduced, the pressure is correspondingly reduced, and the expansion device controlling flow to the cooler must be large enough to pass all the liquid refrigerant flow. Failure of this valve could allow a large flow of liquid to the upper-stage compressor suction, which could seriously damage the compressor.

Liquid from a flash cooler is saturated, and liquid from a cascade condenser usually has little subcooling. In both cases, the liquid temperature is usually lower than the temperature of the surroundings. Thus, it is important to avoid heat input and pressure losses that would cause flash gas to form in the liquid line to the expansion device or to recirculating pumps. The cold liquid lines should be insulated, as expansion devices are usually designed to feed liquid, not vapor.

Figure 33 shows the closed or heat exchanger type of subcooler. It should have sufficient heat transfer surface to transfer heat from the liquid to the evaporating refrigerant with a small final temperature difference. The pressure drop should be small, so that full

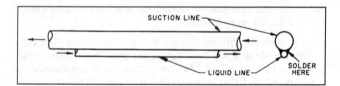

Fig. 28 Soldered Tube Heat Exchanger

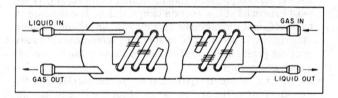

Fig. 29 Shell-and-Finned-Coil Heat Exchanger

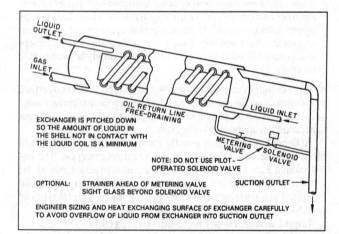

Fig. 30 Shell-and-Finned-Coil Exchanger Installed to Prevent Liquid Floodback

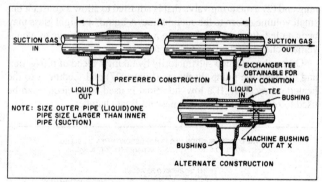

Fig. 31 Tube-in-Tube Heat Exchanger

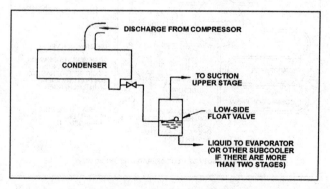

Fig. 32 Flash-Type Cooler

System Practices for Halocarbon Refrigerants

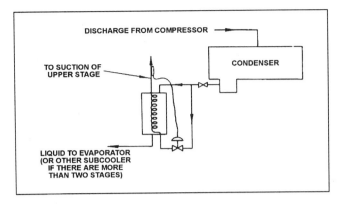

Fig. 33 Closed-Type Subcooler

pressure is available for feeding liquid to the expansion device at the low-temperature evaporator. The subcooler liquid control valve should be sized to supply only the quantity of refrigerant required for the subcooling. This prevents a tremendous quantity of liquid from flowing to the upper-stage suction in the event of a valve failure.

Discharge Line Oil Separators

Oil is always in circulation in systems using halocarbon refrigerants. Refrigerant piping is designed to ensure that this oil passes through the entire system and returns to the compressor as fast as it leaves. Although well-designed piping systems can handle the oil in most cases, an oil separator can have certain advantages in some applications (see Chapter 45). The following are some of the applications where discharge line oil separators can be useful:

- In systems where it is impossible to prevent substantial absorption of refrigerant in the crankcase oil during shutdown periods. When the compressor starts up with a violent foaming action, oil will be thrown out at an accelerated rate, and the separator will immediately return a large portion of this oil to the crankcase. Normally, the system should be designed with pumpdown control or crankcase heaters to minimize liquid absorption in the crankcase.
- In systems using flooded evaporators, where refrigerant bleedoff is necessary to remove oil from the evaporator. Oil separators reduce the amount of bleedoff from the flooded cooler needed for operation.
- In direct-expansion systems using coils or tube bundles that require bottom feed for good liquid distribution and where refrigerant carryover from the top of the evaporator is essential for proper oil removal.
- In low-temperature systems, where it is advantageous to have as little oil as possible going through the low side.
- In screw-type compressor systems, where an oil separator is necessary for proper operation. The oil separator is usually supplied with the compressor unit assembly directly from the compressor manufacturer.
- In multiple compressors operating in parallel. The oil separator can be an integral part of the total system oil management system.

In applying oil separators in refrigeration systems, the following potential hazards must be considered:

- Oil separators are not 100% efficient, and they do not eliminate the need to design the complete system for oil return to the compressor.
- Oil separators tend to condense out liquid refrigerant during compressor off cycles and on compressor start-up. This is true if the condenser is in a warm location, such as on a roof. During the off cycle, the oil separator cools down and acts as a condenser for refrigerant that evaporates in warmer parts of the system. A cool oil separator may condense discharge gas and, on compressor start-up, automatically drain it into the compressor crankcase. To minimize this possibility, the drain connection from the oil separator can be connected into the suction line. This line should be equipped with a shutoff valve, a fine filter, hand throttling and solenoid valves, and a sight glass. The throttling valve should be adjusted so that the flow through this line is only a little greater than would normally be expected to return oil through the suction line.
- The float valve is a mechanical device that may stick open or closed. If it sticks open, hot gas will be continuously bypassed to the compressor crankcase. If the valve sticks closed, no oil is returned to the compressor. To minimize this problem, the separator can be supplied without an internal float valve. A separate external float trap can then be located in the oil drain line from the separator preceded by a filter. Shutoff valves should isolate the filter and trap. The filter and traps are also easy to service without stopping the system.

The discharge line pipe size into and out of the oil separator should be the full size determined for the discharge line. For separators that have internal oil float mechanisms, allow enough room for removal of the oil float assembly for servicing requirements.

Depending on system design, the oil return line from the separator may feed to one of the following locations:

- Directly to the compressor crankcase.
- Directly into the suction line ahead of the compressor.
- Into an oil reservoir or device used to collect oil. This reservoir is used as a source of oil for a specifically designed oil management system.

When a solenoid valve is used in the oil return line, the valve should be wired so that it is open when the compressor is running. To minimize the entrance of condensed refrigerant from the low side, a thermostat may be installed and wired to control the solenoid in the oil return line from the separator. The thermostat sensing element should be located on the oil separator shell below the oil level and set high enough so that the solenoid valve will not open until the separator temperature is higher than the condensing temperature. A superheat-controlled expansion valve can perform the same function. If a discharge line check valve is used, it should be downstream of the oil separator.

Surge Drums or Accumulators

A surge drum is required on the suction side of almost all flooded evaporators to prevent liquid slopover to the compressor. Exceptions include shell-and-tube coolers and similar shell-type evaporators, which provide ample surge space above the liquid level or contain eliminators to separate gas and liquid. A horizontal surge drum is sometimes used where headroom is limited.

The drum can be designed with baffles or eliminators to separate liquid from the suction gas. More often, sufficient separation space is allowed above the liquid level for this purpose. Usually, the design is vertical, with a separation height above the liquid level of 600 to 750 mm and with the shell diameter sized to keep the suction gas velocity low enough to allow the liquid droplets to separate. Because these vessels are also oil traps, it is necessary to provide oil bleed.

Although separators may be fabricated with length-to-diameter (L/D) ratios of 1/1 increasing to 10/1, the lowest cost separators will usually be for L/D ratios between 3/1 and 5/1.

Compressor Floodback Protection

Certain systems periodically flood the compressor with excessive amounts of liquid refrigerant. When periodic floodback through the suction line cannot be controlled, the compressor must be protected against it.

The most satisfactory method appears to be a trap arrangement that catches the liquid floodback and (1) meters it slowly into the suction line, where the floodback is cleared up with a liquid-suction heat interchanger; (2) evaporates the liquid 100% in the trap itself by using a liquid coil or electric heater, and then automatically returns oil to the suction line; or (3) returns it to the receiver or to one of the evaporators. Figure 30 illustrates an arrangement that handles moderate liquid floodback, disposing of the liquid by a combination of boiling off in the exchanger and a limited bleedoff into the suction line. This device, however, would not have sufficient trapping volume for most heat pump applications or hot-gas defrost systems employing reversal of the refrigerant cycle.

For heavier floodback, a larger volume is required in the trap. The arrangement shown in Figure 34 has been applied successfully in reverse-cycle heat pump applications using halocarbon refrigerants. It consists of a suction line accumulator with sufficient volume to hold the maximum expected floodback and a large enough diameter to separate liquid from suction gas. The trapped liquid is slowly bled off through a properly sized and controlled drain line into the suction line, where it is boiled off in a liquid-suction heat exchanger between cycle reversals.

With the alternate arrangement shown, the liquid-oil mixture is heated to evaporate the refrigerant, and the remaining oil is drained into the crankcase or the suction line.

Refrigerant Driers and Moisture Indicators

The effect of moisture in refrigeration systems is discussed in Chapters 5 and 6. Using a permanent refrigerant drier is recommended on all systems and with all refrigerants. It is especially important on low-temperature systems to prevent ice from forming at expansion devices. A **full-flow drier** is always recommended in hermetic compressor systems to keep the system dry and to prevent the products of decomposition from getting into the evaporator in the event of a motor burnout.

Replaceable element filter-driers are preferred for large systems because the drying element can be replaced without breaking any refrigerant connections. The drier is usually located in the liquid line near the liquid receiver. It may be mounted horizontally or vertically with the flange at the bottom, but it should never be mounted vertically with the flange on top because any loose material would then fall into the line when the drying element was removed.

A three-valve bypass is usually used, as shown in Figure 35, to provide a means for isolating the drier for servicing. The refrigerant charging connection should be located between the receiver outlet valve and the liquid line drier so that all refrigerant added to the system passes through the drier.

Reliable moisture indicators can be installed in refrigerant liquid lines to provide a positive indication of when the drier cartridge should be replaced.

Strainers

Strainers should be used in both liquid and suction lines to protect automatic valves and the compressor from foreign material, such as pipe welding scale, rust, and metal chips. The strainer should be mounted in a horizontal line with the orientation so that the screen can be replaced without loose particles falling into the system.

A liquid line strainer should be installed before each automatic valve to prevent particles from lodging on the valve seats. Where multiple expansion valves with internal strainers are used at one location, a single main liquid line strainer will protect all of these. The liquid line strainer can be located anywhere in the line between the condenser (or receiver) and the automatic valves, preferably near the valves for maximum protection. Strainers should trap the particle size that could affect valve operation. With pilot-operated valves, a very fine strainer should be installed in the pilot line ahead of the valve.

Filter driers perform the dual function of drying the refrigerant and filtering out particles far smaller than those trapped by mesh strainers. No other strainer is needed in the liquid line if a good filter drier is used.

Refrigeration compressors are usually equipped with a built-in suction strainer, which is adequate for the usual system with copper piping. The suction line should be piped at the compressor so that the built-in strainer is accessible for servicing.

Both liquid and suction line strainers should be adequately sized to ensure sufficient foreign material storage capacity without excessive pressure drop. In steel piping systems, an external suction line strainer is recommended in addition to the compressor strainer.

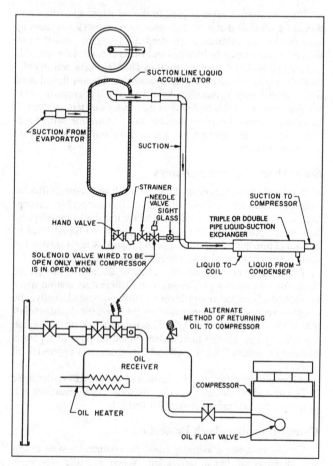

Fig. 34 Compressor Floodback Protection Using Accumulator with Controlled Bleed

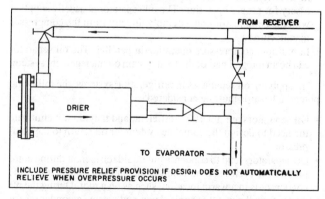

Fig. 35 Drier with Piping Connections

System Practices for Halocarbon Refrigerants

Liquid Indicators

Every refrigeration system should have a way to check for sufficient refrigerant charge. The common devices used are liquid line sight glass, mechanical or electronic indicators, and an external gage glass with equalizing connections and shutoff valves. A properly installed sight glass shows bubbling when the charge is insufficient.

Liquid indicators should be located in the liquid line as close as possible to the receiver outlet or to the condenser outlet if no receiver is used (Figure 36). The sight glass is best installed in a vertical section of the line, a sufficient distance downstream from any valve that the resulting disturbance does not appear in the glass. If the sight glass is installed too far away from the receiver, the line pressure drop may be sufficient to cause flashing and bubbles in the glass, even if the charge is sufficient for a liquid seal at the receiver outlet.

When sight glasses are installed near the evaporator, often no amount of system overcharging will give a solid liquid condition at the sight glass because of pressure drop in the liquid line or lift. Subcooling is required here. An additional sight glass near the evaporator may be needed to check on the condition of the refrigerant at that point.

Sight glasses should be installed full size in the main liquid line. In very large liquid lines, this may not be possible; the glass can then be installed in a bypass or saddle mount that is arranged so that any gas in the liquid line will tend to move to it. A sight glass with double ports (for back lighting) and seal caps, which provide added protection against leakage, is preferred. Moisture-liquid indicators large enough to be installed directly in the liquid line serve the dual purpose of liquid line sight glass and moisture indicator.

Oil Receivers

Oil receivers serve as reservoirs for replenishing crankcase oil pumped by the compressors and provide the means to remove refrigerant dissolved in the oil. They are selected for systems having any of the following components:

- Flooded or semiflooded evaporators with large refrigerant charges
- Two or more compressors operated in parallel
- Long suction and discharge lines
- Double suction line risers

A typical hookup is shown in Figure 34. Outlets are arranged to prevent oil from draining below the heater level to avoid heater burnout and to prevent scale and dirt from being returned to the compressor.

Purge Units

Noncondensable gas separation using a purge unit is useful on most large refrigeration systems where suction pressure may fall below atmospheric pressure (see Figure 11 of Chapter 3).

PRESSURE CONTROL FOR REFRIGERANT CONDENSERS

For further information regarding pressure control, see Chapter 35 of the 1996 *ASHRAE Handbook—Systems and Equipment*.

Water-Cooled Condensers

With water-cooled condensers, pressure controls are used both for maintaining the condensing pressure and for conserving water. On cooling tower applications, they are used only where it is necessary to maintain condensing temperatures.

Condenser Water Regulating Valves

The shutoff pressure of the valve must be set slightly higher than the saturation pressure of the refrigerant at the highest ambient temperature expected when the system is not in operation. This ensures that the valve will not pass water during off cycles. These valves are usually sized to pass the design quantity of water at about a 170 to 200 kPa difference between design condensing pressure and valve shutoff pressure. Chapter 45 has further information.

Water Bypass

In cooling tower applications, a simple bypass with a manual or automatic valve responsive to pressure change can also be used to maintain condensing pressure. Figure 37 shows an automatic three-way valve arrangement. The valve divides the water flow between the condenser and the bypass line to maintain the desired condensing pressure. This maintains balanced flow of water on the tower and pump.

Evaporative Condensers

Among the methods used for condensing pressure control with evaporative condensers are (1) cycling the spray pump motor; (2) cycling both fan and spray pump motors; (3) throttling the spray water; (4) bypassing air around duct and dampers; (5) throttling air via dampers, on either inlet or discharge; and (6) combinations of these methods. For further information, see Chapter 35 of the 1996 *ASHRAE Handbook—Systems and Equipment*.

In water pump cycling, a pressure control at the gas inlet starts and stops the pump in response to pressure changes. The pump sprays water over the condenser coils. As the pressure drops, the pump stops and the unit becomes an air-cooled condenser.

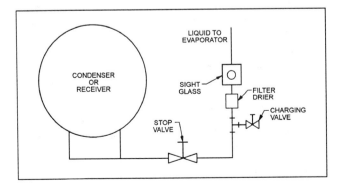

Fig. 36 Sight Glass and Charging Valve Locations

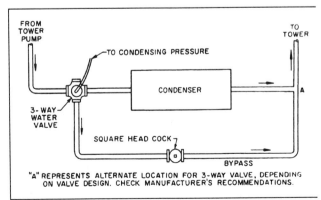

Fig. 37 Head Pressure Control for Condensers Used with Cooling Towers (Water Bypass Modulation)

Constant pressure is difficult to maintain with coils of prime surface tubing because as soon as the pump stops, the pressure goes up and the pump starts again. This occurs because these coils have insufficient capacity when operating as an air-cooled condenser. The problem is not as acute with extended surface coils. Short-cycling results in excessive deposits of mineral and scale on the tubes, decreasing the life of the water pump.

One method of controlling pressure is using cycle fans and pumps. This minimizes water-side scaling. In northern climates, an indoor water sump with a remote spray pump(s) is required. The fan cycling sequence follows:

Upon dropping pressure

- Stop fans.
- If pressure continues to fall, stop pumps.

Upon rising pressure

- Start fans.
- If pressure continues to rise, start pumps.

Damper control (see Figure 38) may be incorporated in systems requiring more constant pressures (e.g., some systems using thermostatic expansion valves). One drawback of dampers is the formation of ice on the dampers and linkages.

Figure 39 incorporates an air bypass arrangement for controlling pressure. A modulating motor, acting in response to a modulating pressure control, positions dampers so that the mixture of recirculated and cold inlet air maintains the desired pressure. In extremely cold weather, most of the air is recirculated.

Air-Cooled Condensers

Methods used for condensing pressure control with air-cooled condensers include (1) cycling fan motor, (2) air throttling or bypassing, (3) coil flooding, and (4) fan motor speed control. The first two methods are described in the section on Evaporative Condensers.

The third method holds the condensing pressure up by backing liquid refrigerant up in the coil to cut down on effective condensing surface. When the pressure drops below the setting of the modulating control valve, it opens, allowing discharge gas to enter the liquid drain line. This restricts liquid refrigerant drainage and causes the condenser to flood enough to maintain the condenser and receiver pressure at the control valve setting. A pressure difference must be available across the valve to open it. Although the condenser would impose sufficient pressure drop at full load, pressure drop may practically disappear at partial loading. Therefore, a positive restriction must be placed parallel with the condenser and the control valve. Systems using this type of control require extra refrigerant charge.

In multiple-fan air-cooled condensers, it is common to cycle fans off down to one fan and then to apply air throttling to that section or modulate the fan motor speed. Consult the manufacturer before using this method, as not all condensers are properly circuited for it.

Using ambient temperature change (rather than condensing pressure) to modulate air-cooled condenser capacity prevents rapid cycling of condenser capacity. A disadvantage of this method is that the condensing pressure is not closely controlled.

KEEPING LIQUID FROM CRANKCASE DURING OFF CYCLES

The control of reciprocating compressors should prevent excessive accumulation of liquid refrigerant in the crankcase during off cycles. Any one of the following control methods accomplishes this.

Automatic Pumpdown Control (Direct-Expansion Air-Cooling Systems)

The most effective way to keep liquid out of the crankcase during system shutdown is to operate the compressor on automatic pumpdown control. The recommended arrangement involves the following devices and provisions:

- A liquid-line solenoid valve in the main liquid line or in the branch to each evaporator.
- Compressor operation through a low-pressure cutout providing for pumpdown whenever this device closes, regardless of whether or not the balance of the system is operating.
- Electrical interlock of the liquid solenoid valve with the evaporator fan, so that the refrigerant flow will be stopped when the fan is out of operation.
- Electrical interlock of the refrigerant solenoid valve with the safety devices (such as the high-pressure cutout, oil safety switch, and motor overloads), so that the refrigerant solenoid valve closes when the compressor stops.

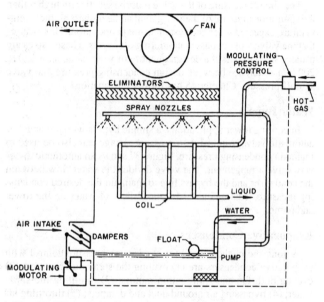

Fig. 38 Head Pressure Control for Evaporative Condenser (Air Intake Modulation)

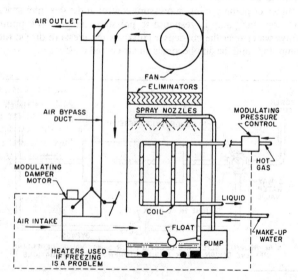

Fig. 39 Head Pressure for Evaporative Condenser (Air Bypass Modulation)

System Practices for Halocarbon Refrigerants

- Low-pressure control settings such that the cut-in point will correspond to a saturated refrigerant temperature lower than any expected compressor ambient air temperature. If the cut-in setting of the low-pressure switch is any higher, liquid refrigerant can accumulate and condense in the crankcase at a pressure corresponding to the ambient temperature. Then, the crankcase pressure would not rise high enough to reach the cut-in point, and effective automatic pumpdown would not be obtained.

Crankcase Oil Heater (Direct-Expansion Systems)

A crankcase oil heater with or without single (nonrecycling) pumpout at the end of each operating cycle does not keep liquid refrigerant out of the crankcase as effectively as automatic pumpdown control, but many compressors equalize too quickly after stopping automatic pumpdown control. Crankcase oil heaters maintain the crankcase oil at a temperature higher than that of other parts of the system, minimizing the absorption of the refrigerant by the oil.

Operation with the single pumpout arrangement is as follows. Whenever the temperature control device opens the circuit, or the manual control switch is opened for shutdown purposes, the crankcase heater is energized, and the compressor keeps running until it cuts off on the low-pressure switch. Because the crankcase heater remains energized during the complete off cycle, it is important that a continuous live circuit be available to the heater during the off time. The compressor cannot start again until the temperature control device or manual control switch closes, regardless of the position of the low-pressure switch.

This control method requires

- A liquid-line solenoid valve in the main liquid line or in the branch to each evaporator
- Use of a relay or the maintained contact of the compressor motor auxiliary switch to obtain a single pumpout operation before stopping the compressor
- A relay or auxiliary starter contact to energize the crankcase heater during the compressor off cycle and deenergize it during the compressor on cycle
- Electrical interlock of the refrigerant solenoid valve with the evaporator fan, so that the refrigerant flow will be stopped when the fan is out of operation
- Electrical interlock of the refrigerant solenoid valve with the safety devices (such as the high-pressure cutout, oil safety switch, and motor overloads), so that the refrigerant flow valve will close when the compressor stops

Control for Direct-Expansion Water Chillers

Automatic pumpdown control is not desirable for direct-expansion water chillers because freezing is possible if excessive cycling occurs. A crankcase heater is the best solution, with a solenoid valve in the liquid line that closes when the compressor stops.

Effect of Short Operating Cycle

With reciprocating compressors, oil will leave the crankcase at an accelerated rate immediately after starting. Therefore, each start should be followed by a sufficiently long operating period to permit the regain of the oil level. Controllers used for compressors should not produce short-cycling of the compressor. Refer to the compressor manufacturer's literature for guidelines on maximum or minimum cycles for a specified period.

HOT-GAS BYPASS ARRANGEMENTS

Most large reciprocating compressors are equipped with unloaders that allow the compressor to start with most of its cylinders unloaded. However, it may be necessary to further unload the compressor to (1) reduce starting torque requirements so that the compressor can be started both with low starting torque prime movers and on low-current taps of reduced voltage starters and (2) permit capacity control down to 0% load conditions without stopping the compressor.

Full (100%) Unloading for Starting

Starting the compressor without load can be done with a manual or automatic valve in a bypass line between the hot-gas and suction lines at the compressor.

To prevent overheating, this valve is open only during the starting period and closed after the compressor is up to full speed and full voltage is applied to the motor terminals.

In the control sequence, the unloading bypass valve is energized on demand of the control calling for compressor operation, equalizing pressures across the compressor. After an adequate delay, a timing relay closes a pair of normally open contacts to start the compressor. After a further time delay, a pair of normally closed timing relay contacts opens, deenergizing the bypass valve.

Full (100%) Unloading for Capacity Control

Where full unloading is required for capacity control, hot-gas bypass arrangements can be used in ways that will not overheat the compressor. In using these arrangements, hot gas should not be bypassed until after the last unloading step.

A hot-gas bypass should (1) give acceptable regulation throughout the range of loads, (2) not cause excessive superheating of the suction gas, (3) not cause any refrigerant overfeed to the compressor, and (4) maintain an oil return to the compressor.

A hot-gas bypass for capacity control is an artificial loading device that maintains a minimum evaporating pressure during continuous compressor operation, regardless of the evaporator load. This is usually done by an automatic or manual pressure-reducing valve that establishes a constant pressure on the downstream side.

Four of the more common methods of using hot-gas bypass are shown in Figure 40. Figure 40A illustrates the simplest type of hot-gas bypass. It will dangerously overheat the compressor if used for protracted periods of time. Figure 40B shows the use of hot-gas bypass to the exit of the evaporator. The expansion valve bulb should be placed at least 1.5 m downstream from the bypass point of entrance, and preferably further, to ensure good mixing.

In Figure 40D, the hot-gas bypass enters after the evaporator thermostatic expansion valve bulb. Another thermostatic expansion valve supplies liquid directly to the bypass line for desuperheating purposes. It is always important to install the hot-gas bypass far enough back in the system to maintain sufficient gas velocities in suction risers and other components to ensure oil return at any evaporator loading.

Figure 40C shows the most satisfactory hot-gas bypass arrangement. Here, the bypass is connected into the low side between the expansion valve and the entrance to the evaporator. If a distributor is used, the gas enters between the expansion valve and distributor. Refrigerant distributors are commercially available with side inlet connections that can be used for hot-gas bypass duty to a certain extent. Pressure drop through the distributor tubes must be evaluated to determine how much gas bypassing can be effected. This arrangement provides good oil return.

Solenoid valves should be placed before the constant pressure bypass valve and before the thermal expansion valve used for liquid injection desuperheating, so that these devices cannot function until they are required.

The control valves for the hot gas should be close to the main discharge line because the line preceding the valve usually fills with liquid when closed.

The hot-gas bypass line should be sized so that its pressure loss is only a small percentage of the pressure drop across the valve.

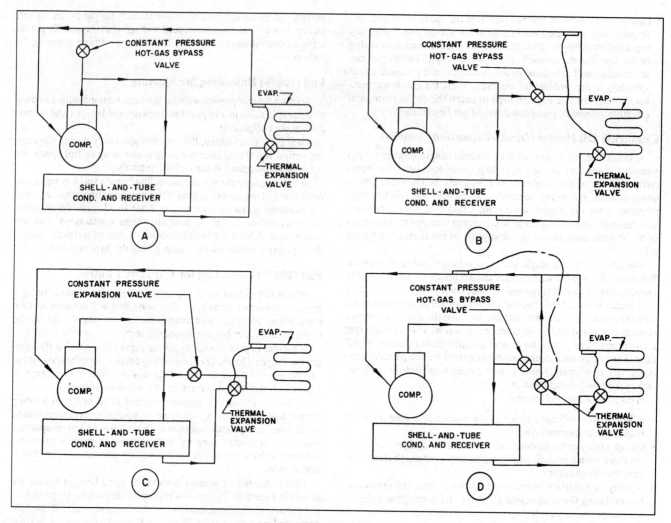

Fig. 40 Hot-Gas Bypass Arrangements

Usually, it is the same size as the valve connections. When sizing the valve, consult a control valve manufacturer to determine the minimum compressor capacity that must be offset, the refrigerant used, the condensing pressure, and the suction pressure.

When unloading (Figure 40C), the pressure control requirements increase considerably because the only heat delivered to the condenser is that caused by the motor power delivered to the compressor. The discharge pressure should be kept high enough that the hot gas bypass valve can deliver gas at the required rate. The condenser pressure control must be capable of meeting this condition.

Safety Requirements

ASHRAE *Standard* 15, Safety Code for Mechanical Refrigeration, and ASME *Standard* B31.5, Refrigeration Piping, should be used as guides for safe practice because they are the basis of most municipal and state codes. However, some ordinances require heavier piping and other features. The designer should know the specific requirements of the installation site. Only A106 Grade A or B or A53 Grade A or B should be considered for steel refrigerant piping.

REFERENCES

Alofs, D.J., M.M. Hasan, and H.J. Sauer, Jr. 1990. Influence of oil on pressure drop in refrigerant compressor suction lines. *ASHRAE Transactions* 96:1.

ASHRAE. 1994. Safety code for mechanical refrigeration. ANSI/ASHRAE *Standard* 15.

ASME. 1992 and 1994 addendum. Refrigeration piping. ANSI/ASME *Standard* B31.5. American Society of Mechanical Engineers, New York.

Atwood, T. 1990. Pipe sizing and pressure drop calculations for HFC-134a. *ASHRAE Journal* 32(4):62-66.

Colebrook, D.F. 1938, 1939. Turbulent flow in pipes. *Journal of the Institute of Engineers* 11.

Cooper, W.D. 1971. Influence of oil-refrigerant relationships on oil return. *ASHRAE Symposium Bulletin* PH71(2):6-10.

Jacobs, M.L., F.C. Scheideman, F.C. Kazem, and N.A. Macken. 1976. Oil transport by refrigerant vapor. *ASHRAE Transactions* 81(2):318-29.

Keating, E.L. and R.A. Matula. 1969. Correlation and prediction of viscosity and thermal conductivity of vapor refrigerants. *ASHRAE Transactions* 75(1).

Stoecker, W.F. 1984. Selecting the size of pipes carrying hot gas to defrosted evaporators. *International Journal of Refrigeration* 7(4):225-228.

Timm, M.L. 1991. An improved method for calculating refrigerant line pressure drops. *ASHRAE Transactions* 97(1):194-203.

Wile, D.D. 1977. *Refrigerant line sizing.* ASHRAE.

CHAPTER 3

SYSTEM PRACTICES FOR AMMONIA REFRIGERANT

System Selection 3.1	Condenser and Receiver Piping 3.15
Equipment 3.2	Evaporative Condensers 3.17
Controls 3.6	Evaporator Piping 3.19
Piping 3.7	Multistage Systems 3.21
Reciprocating Compressors 3.10	Liquid Recirculation Systems 3.22
Rotary Vane, Low-Stage Compressors 3.12	Safety Considerations 3.27
Screw Compressors 3.13	

CUSTOM-ENGINEERED ammonia (R-717) refrigeration systems often have design conditions that span a wide range of evaporating and condensing temperatures. Examples are (1) a food freezing plant operating at temperatures from 10 to −45°C; (2) a candy storage requiring 15°C dry bulb with precise humidity control; (3) a beef chill room at −2 to −1°C with high humidity; (4) a distribution warehouse requiring multiple temperatures for storage of ice cream, frozen food, meat, and produce and for docks; or (5) a chemical process requiring multiple temperatures ranging from 15 to −50°C. Ammonia is the refrigerant of choice for many industrial refrigeration systems.

For safety and minimum design criteria for ammonia systems, refer to ASHRAE *Standard* 15, Safety Code for Mechanical Refrigeration; IIAR *Bulletin* 109, Minimum Safety Criteria for a Safe Ammonia Refrigeration System; IIAR *Standard* 2, Equipment, Design and Installation of Ammonia Mechanical Refrigeration Systems; and applicable state and local codes.

See Chapter 12 for information on refrigeration load calculations.

Ammonia Refrigerant for HVAC Systems

The use of ammonia for HVAC systems has received renewed interest, due in part to the scheduled phaseout and increasing costs of chlorofluorocarbon (CFC) and hydrochlorofluorocarbon (HCFC) refrigerants. Ammonia secondary systems that circulate chilled water or other secondary refrigerant are a viable alternative to halocarbon systems, although ammonia is inappropriate for direct refrigeration systems (ammonia in the air unit coils) for HVAC applications. Ammonia packaged chilling units are available for HVAC applications. As with the installation of any air-conditioning unit, all applicable codes, standards, and insurance requirements must be followed.

SYSTEM SELECTION

In selecting an engineered ammonia refrigeration system, several design decisions must be considered, including whether to use (1) single-stage compression, (2) economized compression, (3) multistage compression, (4) direct-expansion feed, (5) flooded feed, (6) liquid recirculation feed, and (7) secondary coolants.

Single-Stage Systems

The basic single-stage system consists of evaporator(s), a compressor, a condenser, a refrigerant receiver (if used), and a refrigerant control device (expansion valve, float, etc.). Chapter 1 of the 1997 ASHRAE *Handbook—Fundamentals* discusses the compression refrigeration cycle.

The preparation of this chapter is assigned to TC 10.3, Refrigerant Piping, Controls, and Accessories

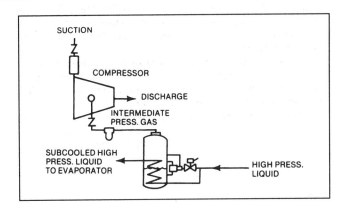

Fig. 1 Shell-and-Coil Economizer Arrangement

Economized Systems

Economized systems are frequently used with rotary screw compressors. Figure 1 shows an arrangement of the basic components. Subcooling the liquid refrigerant before it reaches the evaporator reduces its enthalpy, resulting in a higher net refrigerating effect. Economizing is beneficial since the vapor generated during the subcooling is injected into the compressor part way through its compression cycle and must be compressed only from the economizer port pressure (which is higher than suction pressure) to the discharge pressure. This produces additional refrigerating capacity with less increase in unit energy input. Economizing is most beneficial at high pressure ratios. Under most conditions, economizing can provide operating efficiencies that approach that of two-stage systems, but with much less complexity and simpler maintenance.

Economized systems for variable loads should be selected carefully. At approximately 75% capacity, most screw compressors revert to single-stage performance as the slide valve moves such that the economizer port is open to the compressor suction area.

A flash-type economizer, which is somewhat more efficient, may often be used instead of the shell-and-coil economizer (Figure 1). However, ammonia liquid delivery pressure is reduced to economizer pressure.

Multistage Systems

Multistage systems compress the gas from the evaporator to the condenser in several stages. They are used to produce temperatures of −25°C and below. This is not economical with single-stage compression.

Single-stage reciprocating compression systems are generally limited to between 35 and 70 kPa (gage) suction pressure. With lubricant-injected economized rotary screw compressors, where the discharge temperatures are lower because of the lubricant cooling,

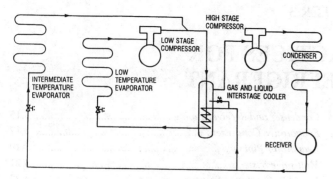

Fig. 2 Two-Stage System with High- and Low-Temperature Loads

the low-suction temperature limit is about –40°C, but efficiency is very low. Two-stage systems are used down to about −60°C evaporator temperatures. Below this temperature, three-stage systems should be considered.

Two-stage systems consist of one or more compressors that operate at low suction pressure and discharge at intermediate pressure and have one or more compressors that operate at intermediate pressure and discharge to the condenser (Figure 2).

Where either single- or two-stage compression systems can be used, two-stage systems require less power and have lower operating costs, but they can have a higher initial equipment cost.

EQUIPMENT

Compressors

Compressors available for single- and multistage applications include the following:

Reciprocating
    Single-stage (low-stage or high-stage)
    Internally compounded
Rotary vane
Rotary screw (low-stage or high-stage, with or without
    economizing)

The reciprocating compressor is the most common compressor used in small, 75 kW or less, single-stage or multistage systems. The screw compressor is the predominant compressor above 75 kW, both in single- and multistage systems. Various combinations of compressors may be used in multistage systems. Rotary vane and screw compressors are frequently used for the low-pressure stage, where large volumes of gas must be moved. The high-pressure stage may be a reciprocating or screw compressor.

Factors to be considered in selecting a compressor include the following:

- System size and capacity requirements.
- Location, such as indoor or outdoor installation at ground level or on the roof.
- Equipment noise.
- Part- or full-load operation.
- Winter and summer operation.
- Pulldown time required to reduce the temperature to desired conditions for either initial or normal operation. The temperature must be pulled down frequently for some applications for a process load, while a large cold storage warehouse may require pulldown only once in its lifetime.

Lubricant Cooling. When a reciprocating compressor requires lubricant cooling, an external heat exchanger using a refrigerant or secondary cooling is usually added. Screw compressor lubricant cooling is covered in detail in the section on Screw Compressors.

Compressor Drives. The correct electric motor size(s) for a multistage system is determined by the pulldown load. When the final low-stage operating level is −75°C, the pulldown load can be three times the operating load. Positive displacement reciprocating compressor motors are usually selected for about 150% of operating power requirements for 100% load. The compressor's unloading mechanism can be used to prevent motor overload. Electric motors should not be overloaded, even when a service factor is indicated. For screw compressor applications, motors should be sized by adding 10% to the operating power. Screw compressors have built-in unloading mechanisms to prevent motor overload. The motor should not be oversized because an oversized motor has a lower power factor and lower efficiency at design and reduced loads.

Steam turbines or gasoline, natural gas, propane, or diesel internal combustion engines are used when electricity is unavailable, or if the selected energy source is cheaper. Sometimes they are used in combination with electricity to reduce peak demands. The power output of a given engine size can vary as much as 15% depending on the fuel selected.

Steam turbine drives for refrigerant compressors are usually limited to very large installations where steam is already available at moderate to high pressure. In all cases, torsional analysis is required to determine what coupling must be used to dampen out any pulsations transmitted from the compressor. For optimum efficiency, a turbine should operate at a high speed that must be geared down for reciprocating and possibly screw compressors. Neither the gear reducer nor the turbine can tolerate a pulsating backlash from the driven end, so torsional analysis and special couplings are essential.

Advantages of turbines include variable speed for capacity control and low operating and maintenance costs. Disadvantages include higher initial costs and possible high noise levels. The turbine must be started manually to bring the turbine housing up to temperature slowly and to prevent excess condensate from entering the turbine.

The standard power rating of an engine is the absolute maximum, not the recommended power available for continuous use. Also, torque characteristics of internal combustion engines and electric motors differ greatly. The proper engine selection is at 75% of its maximum power rating. For longer life, the full-load speed should be at least 10% below maximum engine speed.

Internal combustion engines, in some cases, can reduce operating cost below that for electric motors. Disadvantages include (1) higher initial cost of the engine, (2) additional safety and starting controls, (3) higher noise levels, (4) larger space requirements, (5) air pollution, (6) requirement for heat dissipation, (7) higher maintenance costs, and (8) higher levels of vibration than with electric motors. A torsional analysis must be made to determine the proper coupling if engine drives are chosen.

Condensers

Condensers should be selected on the basis of total heat rejection at maximum load. Often the heat rejected at the start of pulldown is several times the amount rejected at normal, low-temperature operating conditions. Some means, such as compressor unloading, can be used to limit the maximum amount of heat rejected during pulldown. If the condenser is not sized for pulldown conditions, and compressor capacity cannot be limited during this period, condensing pressure might increase enough to shut down the system.

Evaporators

Several types of evaporators are used in ammonia refrigeration systems. Fan-coil, direct-expansion evaporators can be used, but they are not generally recommended unless the suction temperature is −18°C or higher. This is due in part to the relative inefficiency of the direct-expansion coil, but more importantly, the low mass flow rate of ammonia is difficult to feed uniformly as a liquid to the coil.

Instead, ammonia fan-coil units designed for recirculation (overfeed) systems are preferred. Typically in this type of system, high-pressure ammonia from the system high stage flashes into a large vessel at the evaporator pressure from which it is pumped to the evaporators at an overfeed rate of 2.5 to 1 to 4 to 1. This type of system is standard and very efficient. See Chapter 1 for more details.

Flooded shell-and-tube evaporators are often used in ammonia systems in which indirect or secondary cooling fluids such as water, brine, or glycol must be cooled.

Some problems that can become more acute at low temperatures include changes in lubricant transport properties, loss of capacity caused by static pressure from the depth of the pool of liquid refrigerant in the evaporator, deterioration of refrigerant boiling heat transfer coefficients due to lubricant logging, and higher specific volumes for the vapor.

The effect of pressure losses in the evaporator and suction piping is more acute in low-temperature systems because of the large change in saturation temperatures and specific volume in relation to pressure changes at these conditions. Systems that operate near zero absolute pressure are particularly affected by pressure loss.

The depth of the pool of boiling refrigerant in a flooded evaporator causes a liquid pressure that is exerted on the lower part of the heat transfer surface. Therefore, the saturation temperature at this surface is higher than the saturated temperature in the suction line, which is not affected by the liquid pressure. This temperature gradient must be considered when designing the evaporator.

Spray-type shell-and-tube evaporators, while not commonly used, offer certain advantages. In this design, the liquid depth penalty for the evaporator can be eliminated since the pool of liquid is below the heat transfer surface. A refrigerant pump sprays the liquid over the surface. The pump energy is an additional heat load to the system, and more refrigerant must be used to provide the net positive suction pressure required by the pump. The pump is also an additional item that must be maintained. This evaporator design also reduces the refrigerant charge requirement compared to a flooded design (see Chapter 1).

Vessels

High-Pressure Receivers. Industrial systems generally incorporate a central high-pressure refrigerant receiver, which serves as the primary refrigerant storage location in the system. It handles the refrigerant volume variations between the condenser and the system's low side during operation and pumpdowns for repairs or defrost. Ideally, the receiver should be large enough to hold the entire system charge, but this is not generally economical. An analysis of the system should be made to determine the optimum receiver size. Receivers are commonly equalized to the condenser inlet and operate at the same pressure as the condenser. In some systems, the receiver is operated at a pressure between the condensing pressure and the highest suction pressure to allow for variations in condensing pressure without affecting the system's feed pressure.

If additional receiver capacity is needed for normal operation, extreme caution should be exercised in the design. Designers usually remove the inadequate receiver and replace it with a larger one rather than install an additional receiver in parallel. This procedure is best because even slight differences in piping pressure or temperature can cause the refrigerant to migrate to one receiver and not to the other.

Smaller auxiliary receivers can be incorporated to serve as sources of high-pressure liquid for compressor injection or thermosiphon, lubricant cooling, high-temperature evaporators, and so forth.

Intercoolers (Gas and Liquid). An intercooler (subcooler/desuperheater) is the intermediate vessel between the high stage and low stage in a multistage system. One purpose of the intercooler is to cool the discharge gas of the low-stage compressor to prevent overheating the high-stage compressor. This can be done by

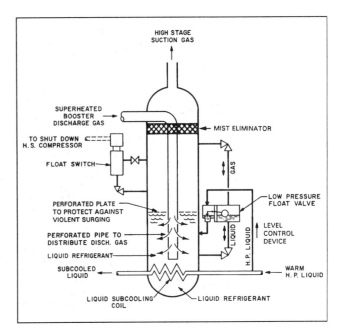

Fig. 3 Intercooler

bubbling the discharge gas from the low-stage compressor through a bath of liquid refrigerant or by mixing liquid normally entering the intermediate vessel with the discharge gas as it enters above the liquid level. The heat removed from the discharge gas is absorbed by the evaporation of part of the liquid and eventually passes through the high-stage compressor to the condenser. Disbursing the discharge gas below a level of liquid refrigerant separates out any lubricant carryover from the low-stage compressor. If the volume of liquid in the intercooler is to be used for other purposes, such as liquid makeup or feed to the low stage, periodic lubricant removal is important.

Another purpose of the intercooler is to lower the temperature of the liquid being used in the system low stage. Lowering the refrigerant temperature increases the refrigeration effect and reduces the low-stage compressor's required displacement, thus reducing its operating cost.

Two types of intercoolers for two-stage compression systems are shell-and-coil and flash-type intercoolers. Figure 3 depicts a shell-and-coil intercooler incorporating an internal pipe coil for subcooling the high-pressure liquid before it is fed to the low stage of the system. Typically, the coil subcools the liquid to within 6 K of the intermediate temperature.

Vertical **shell-and-coil intercoolers** with float valve feed perform well in many applications using ammonia refrigerant systems. The vessel must be sized properly to separate the liquid from the vapor that is returning to the high-stage compressor. The superheated gas inlet pipe should extend below the liquid level and have perforations or slots to distribute the gas evenly in small bubbles. The addition of a perforated baffle across the area of the vessel slightly below the liquid level protects against violent surging. A float switch that shuts down the high-stage compressor when the liquid level gets too high should always be used with this type of intercooler in case the feed valve fails to control properly.

The **flash-type intercooler** is similar in design to the shell-and-coil intercooler with the exception of the coil. The high-pressure liquid is flash cooled to the intermediate temperature. Caution should be exercised in selecting a flash-type intercooler because all the high-pressure liquid is flashed to intermediate pressure. While colder than that of the shell-and-coil intercooler, the liquid in the flash-type intercooler is not subcooled and is susceptible to flashing due to system pressure drop. Two-phase liquid feed to control

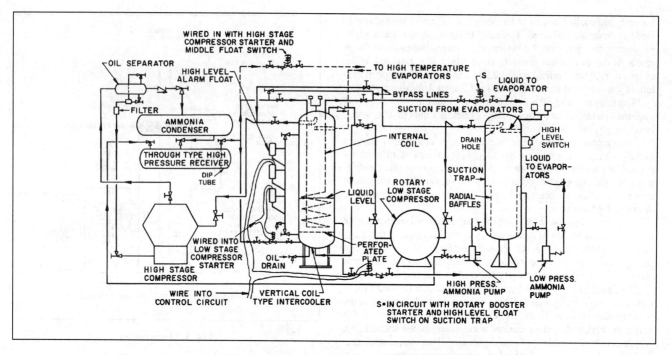

Fig. 4 Arrangement for Compound System with Vertical Intercooler and Suction Trap

valves may cause premature failure due to the wire drawing effect of the liquid/vapor mixture.

Figure 4 shows a vertical shell-and-coil intercooler as piped into the system. The liquid level is maintained in the intercooler by a float that controls the solenoid valve feeding liquid into the shell side of the intercooler. Gas from the first-stage compressor enters the lower section of the intercooler, is distributed by a perforated plate, and is then cooled to the saturation temperature corresponding to intermediate pressure.

When sizing any intercooler, the designer must consider (1) the low-stage compressor capacity; (2) the vapor desuperheating, liquid makeup requirements for the subcooling coil load, or vapor cooling load associated with the flash-type intercooler; and (3) any high-stage side loading. The volume required for normal liquid levels, liquid surging from high-stage evaporators, feed valve malfunctions, and liquid/vapor must also be analyzed.

Accessories necessary are the liquid level control device and the high-level float switch. While not absolutely necessary, an auxiliary oil pot should also be considered.

Suction Accumulator. A suction accumulator (also known as a knockout drum, suction trap, pump receiver, recirculator, etc.) prevents liquid from entering the suction of the compressor, whether on the high stage or low stage of the system. Both vertical and horizontal vessels can be incorporated. Baffling and mist eliminator pads can enhance liquid separation.

Suction accumulators, especially those not intentionally maintaining a level of liquid, should have a means of removing any buildup of ammonia liquid. Gas boil-out coils or electric heating elements are costly and inefficient.

Although it is one of the more common and simplest means of liquid removal, a liquid boil-out coil (Figure 5) has some drawbacks. Generally, the warm liquid flowing through the coil is the source of liquid being boiled off. Liquid transfer pumps, gas-powered transfer systems, or basic pressure differentials are a more positive means of removing the liquid (Figures 6 and 7).

Accessories should include a high-level float switch for compressor protection along with additional pump or transfer system controls.

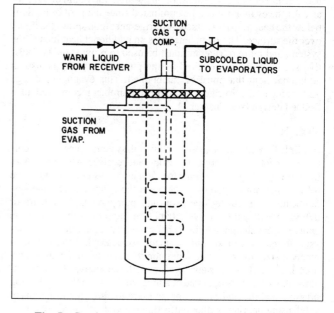

Fig. 5 Suction Accumulator with Warm Liquid Coil

Vertical Suction Trap and Pump. Figure 8 shows the piping of a vertical suction trap that uses a high-pressure ammonia pump to transfer liquid from the system's low-pressure side to the high-pressure receiver. Float switches piped on a float column on the side of the trap can start and stop the liquid ammonia pump, sound an alarm in case of excess liquid, and sometimes stop the compressors.

When the liquid level in the suction trap reaches the setting of the middle float switch, the liquid ammonia pump starts and reduces the liquid level to the setting of the lower float switch, which stops the liquid ammonia pump. A check valve in the discharge line of the ammonia pump prevents gas and liquid from flowing backward through the pump when it is not in operation. Depending on the type of check valve used, some installations have two valves in a series as an extra precaution against pump "backspin."

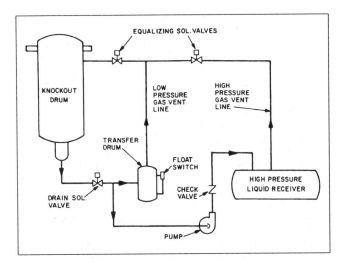

Fig. 6 Equalized Pressure Pump Transfer System

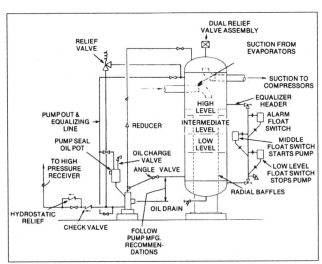

Fig. 8 Piping for Vertical Suction Trap and High-Pressure Pump

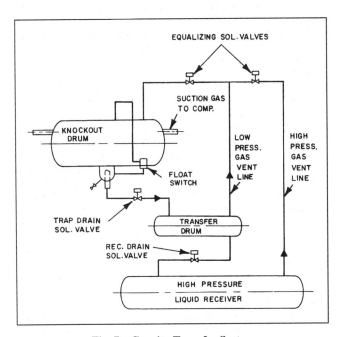

Fig. 7 Gravity Transfer System

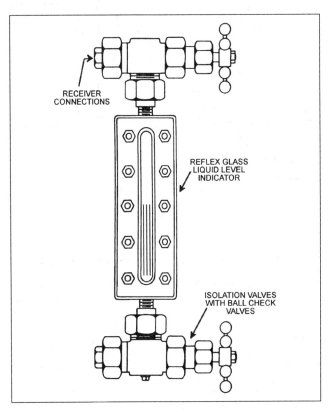

Fig. 9 Gage Glass Assembly for Ammonia

Compressor controls adequately designed for starting, stopping, and capacity reduction result in minimal agitation, which aids in separating the vapor and liquid in the suction trap. Increasing the compressor capacity slowly and in small increments reduces the boiling of liquid in the trap, which is caused by the refrigeration load of cooling the refrigerant and metal mass of the trap. If another compressor is started when plant suction pressure increases, it should be brought on line slowly to prevent a sudden pressure change in the suction trap.

A high level of liquid in a suction trap should activate an alarm or stop the compressors. Although eliminating the cause is the most effective way to reduce a high level of excess surging liquid, a more immediate solution is to stop part of the compression system and raise the plant suction pressure slightly. Continuing high levels indicate insufficient pump capacity or suction trap volume.

Liquid Level Indicators. Liquid level can be indicated by visual indicators, electronic sensors, or a combination of the two. Visual indicators include individual circular reflex level indicators (bull's-eyes) mounted on a pipe column or stand-alone linear reflex glass assemblies (Figure 9). For operation at temperatures below the frost point, transparent plastic frost shields covering the reflex surfaces are necessary. Also, the pipe column must be insulated, especially when control devices are attached.

Electronic level sensors can continuously monitor the liquid level. Digital or graphic displays of the liquid level can be locally or remotely monitored (Figure 10).

Adequate isolation valves for the level indicators should be employed. High-temperature glass tube indicators should incorporate stop check or excess flow valves for isolation and safety.

Purge Units. A noncondensable gas separator (purge unit) is useful in most plants, especially when suction pressure is below atmospheric pressure. Purge units on ammonia systems are piped to

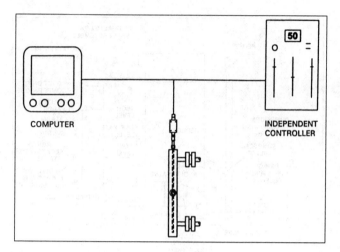

Fig. 10   Electronic Liquid Level Control

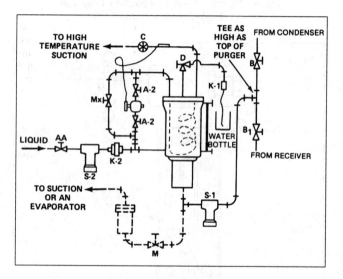

Fig. 11   Purge Unit and Piping for Noncondensable Gas

carry noncondensables (air) from the receiver and condenser to the purger, as shown in Figure 11. High-pressure liquid expands through a coil in the purge unit, providing a cold spot in the purge drum. The suction from the coil should be taken to one of the low-temperature suction mains. Ammonia vapor and noncondensable gas are drawn into the purge drum, and the ammonia condenses on the cold surface. When the drum fills with air and other noncondensables, a float valve within the purger opens and permits them to leave the drum and pass into the open water bottle. Purge units are available for automatic operation.

Lubricant Management

Most lubricants are immiscible in ammonia and separate out of the liquid easily when flow velocity is low or when temperatures are lowered. Normally, lubricants can be easily drained from the system. However, if the temperature is very low and the lubricant is not properly selected, it becomes a gummy mass that prevents refrigerant controls from functioning, blocks flow passages, and fouls the heat transfer surfaces. Proper lubricant selection and management is often the key to a properly functioning system.

In two-stage systems, proper design usually calls for lubricant separators on both the high- and low-stage compressors. A properly designed coalescing separator can remove almost all the lubricant that is in droplet or aerosol form. Lubricant that reaches its saturation vapor pressure and becomes a vapor cannot be removed by a separator. Separators equipped with some means of cooling the discharge gas will condense much of the vapor for consequent separation. Selection of lubricants that have very low vapor pressures below 80°C can minimize carryover to 2 or 3 mg/kg. Care must be exercised, however, to ensure that refrigerant is not condensed and fed back into the compressor or separator, where it can lower lubricity and cause compressor damage.

In general, direct-expansion and liquid overfeed system evaporators have fewer lubricant return problems than do flooded system evaporators because refrigerant flows continuously at good velocities to sweep lubricant from the evaporator. Low-temperature systems using hot-gas defrost can also be designed to sweep lubricant out of the circuit each time the system defrosts. This reduces the possibility of coating the evaporator surface and hindering heat transfer.

Flooded evaporators can promote lubricant buildup in the evaporator charge because they may only return refrigerant vapor back to the system. In ammonia systems, the lubricant is simply drained from the surge drum. At low temperatures, this procedure is difficult if the lubricant selected has a pour point above the evaporator temperature.

Lubricant Removal from Ammonia Systems. Most lubricants are miscible with liquid ammonia only in very small proportions. The proportion decreases with the temperature, causing lubricant to separate. The evaporation of ammonia increases the lubricant ratio, causing more lubricant to separate. Increased density causes the lubricant (saturated with ammonia at the existing pressure) to form a separate layer below the ammonia liquid.

Unless lubricant is removed periodically or continuously from the point where it collects, it can cover the heat transfer surface in the evaporator, reducing performance. If gage lines or branches to level controls are taken from low points (or lubricant is allowed to accumulate), these lines will contain lubricant. The higher lubricant density will be at a lower level than the ammonia liquid. Draining lubricant from a properly located collection point is not difficult unless the temperature is so low that the lubricant does not flow readily. In this case, maintaining the lubricant receiver at a higher temperature may be beneficial. Alternatively, a lubricant with a lower pour point can be selected.

Lubricant in the system is saturated with ammonia at the existing pressure. When the pressure is reduced, the ammonia vapor separates, causing foaming.

Draining lubricant from ammonia systems requires special care. Ammonia in the lubricant foam normally starts to evaporate and produces a smell. Operators should be made aware of this. On systems where lubricant is drained from a still, a spring-loaded drain valve should be installed. This type of valve will close if the valve handle is released.

CONTROLS

Refrigerant flow controls are discussed in Chapter 45. The following precautions are necessary in the application of certain controls in low-temperature systems.

Liquid Feed Control

Many controls available for single-stage, high-temperature systems may be used with some discretion on low-temperature systems. If the liquid level is controlled by a low-side float valve (with the float in the chamber where the level is controlled), low pressure and temperature have no appreciable effect on operation. External float chambers, however, must be thoroughly insulated to prevent heat influx that might cause boiling and an unstable level, affecting the float response. Equalizing lines to external float chambers, particularly the upper line, must be sized generously so that liquid can reach

System Practices for Ammonia Refrigerant

the float chamber, and gas resulting from any evaporation may be returned to the vessel without appreciable pressure loss.

The superheat-controlled (thermostatic) expansion valve is generally used in direct-expansion evaporators. This valve operates on the difference between the bulb pressure, which is responsive to the suction temperature, and the pressure below the diaphragm, which is the actual suction pressure.

The thermostatic expansion valve is designed to maintain a preset superheat in the suction gas. Although the pressure-sensing part of the system responds almost immediately to a change in conditions, the temperature-sensing bulb must overcome thermal inertia before its effect is felt on the power element of the valve. Thus when compressor capacity decreases suddenly, the expansion valve may overfeed before the bulb senses the presence of liquid in the suction line and reduces the feed. Therefore, a suction accumulator should be installed on direct-expansion low-temperature systems with multiple expansion valves.

Controlling Load During Pulldown

System transients during pulldown can be managed by controlling compressor capacity. Proper load control reduces the compressor capacity so that the energy requirements stay within the capacities of the motor and the condenser. On larger systems using screw compressors, a current-sensing device reads motor amperage and adjusts the capacity control device appropriately. Cylinders on reciprocating compressors can be unloaded for similar control.

Alternatively, a downstream, outlet, or crankcase pressure regulator can be installed in the suction line to throttle the suction flow should the pressure exceed a preset limit. This regulator limits the compressor's suction pressure during pulldown. The disadvantage of this device is the extra pressure drop it causes when the system is at the desired operating conditions. To overcome some of this pressure drop, the designer can use external forces to drive the valve, causing it to be held fully open when the pressure is below the maximum allowable. Systems incorporating downstream pressure regulators and compressor unloading must be carefully designed so that the two controls complement each other.

Operation at Varying Loads and Temperatures

Compressor and evaporator capacity controls are similar for multi- and single-stage systems. Control methods include compressor capacity control, hot-gas bypass, or evaporator pressure regulators. Low pressure can affect control systems by significantly increasing the specific volume of the refrigerant gas and the pressure drop. A small pressure reduction can cause a large percentage capacity reduction.

System load usually cannot be reduced to near zero, since this would result in little or no flow of gas through the compressor and consequent overheating. Additionally, high pressure ratios would be detrimental to the compressor if it were required to run at very low loads. If the compressor cannot be allowed to cycle off during low load, an acceptable alternative is a **hot-gas bypass**. The high-pressure gas is fed into the low-pressure side of the system through a downstream pressure regulator. The gas should be desuperheated by injecting it at a point in the system where it will be in contact with expanding liquid, such as immediately downstream of the liquid feed to the evaporator. Otherwise, extremely high compressor discharge temperatures can result. The artificial load supplied by the high-pressure gas can fill the gap between the actual load and the lowest stable compressor operating capacity. Figure 12 shows such an arrangement.

Electronic Control

Microprocessor and computer-based control systems are becoming the norm for control systems on individual compressors as well as for entire system control. Almost all screw compressors use microprocessor control systems to monitor all safety functions and operating conditions. These machines are frequently linked together with a programmable controller or computer for sequencing multiple compressors so that they load and unload in response to system fluctuations in the most economical manner. Programmable controllers are also used to replace multiple defrost time clocks on larger systems for more accurate and economical defrosting. Communications and data logging permit systems to operate at optimum conditions under transient load conditions even when operators are not in attendance.

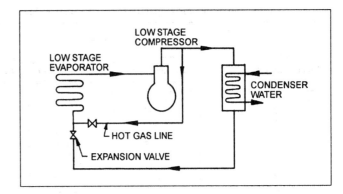

Fig. 12 Hot-Gas Injection Evaporator for Operations at Low Load

PIPING

The following recommendations are given for ammonia piping. Local codes or ordinances governing ammonia mains should also be complied with.

Recommended Material

Because copper and copper-bearing materials are attacked by ammonia, they are not used in ammonia piping systems. Steel piping, fittings, and valves of the proper pressure rating are suitable for ammonia gas and liquid.

Ammonia piping should conform to ASME *Standard* B31.5, Refrigeration Piping, and IIAR *Standard* 2, which states the following:

1. Liquid lines 40 mm and smaller shall be not less than Schedule 80 carbon steel pipe.
2. Liquid lines 50 through 150 mm shall be not less than Schedule 40 carbon steel pipe.
3. Liquid lines 200 through 300 mm shall be not less than Schedule 20 carbon steel pipe.
4. Vapor lines 150 mm and smaller shall be not less than Schedule 40 carbon steel pipe.
5. Vapor lines 200 through 300 mm shall be not less than Schedule 20 carbon steel pipe.
6. Vapor lines 350 mm and larger shall be not less than Schedule 10 carbon steel pipe.
7. All threaded pipe shall be Schedule 80.
8. Carbon steel pipe shall be ASTM *Standard* A 53 Grade A or B, Type E (electric resistance welded) or Type S (seamless); or ASTM *Standard* A 106 (seamless), except where temperature-pressure criteria mandate a higher specification material. *Standard* A 53 Type F is not permitted for ammonia piping.

Fittings

Couplings, elbows, and tees for threaded pipe are for a minimum of 21 MPa design pressure and constructed of forged steel. Fittings for welded pipe should match the type of pipe used (i.e., standard

Table 1 Suction Line Capacities in Kilowatts for Ammonia with Pressure Drops of 0.005 and 0.01 K/m Equivalent

Steel Nominal Line Size, mm	Saturated Suction Temperature, °C					
	−50		−40		−30	
	Δt = 0.005 K/m Δp = 12.1 Pa/m	Δt = 0.01 K/m Δp = 24.2 Pa/m	Δt = 0.005 K/m Δp = 19.2 Pa/m	Δt = 0.01 K/m Δp = 38.4 Pa/m	Δt = 0.005 K/m Δp = 29.1 Pa/m	Δt = 0.01 K/m Δp = 58.2 Pa/m
10	0.19	0.29	0.35	0.51	0.58	0.85
15	0.37	0.55	0.65	0.97	1.09	1.60
20	0.80	1.18	1.41	2.08	2.34	3.41
25	1.55	2.28	2.72	3.97	4.48	6.51
32	3.27	4.80	5.71	8.32	9.36	13.58
40	4.97	7.27	8.64	12.57	14.15	20.49
50	9.74	14.22	16.89	24.50	27.57	39.82
65	15.67	22.83	27.13	39.27	44.17	63.77
80	28.08	40.81	48.36	69.99	78.68	113.30
100	57.95	84.10	99.50	143.84	161.77	232.26
125	105.71	153.05	181.16	261.22	293.12	420.83
150	172.28	248.91	294.74	424.51	476.47	683.18
200	356.67	514.55	609.20	874.62	981.85	1 402.03
250	649.99	937.58	1 107.64	1 589.51	1 782.31	2 545.46
300	1 045.27	1 504.96	1 777.96	2 550.49	2 859.98	4 081.54

Steel Nominal Line Size, mm	Saturated Suction Temperature, °C					
	−20		−5		+5	
	Δt = 0.005 K/m Δp = 42.2 Pa/m	Δt = 0.01 K/m Δp = 84.4 Pa/m	Δt = 0.005 K/m Δp = 69.2 Pa/m	Δt = 0.01 K/m Δp = 138.3 Pa/m	Δt = 0.005 K/m Δp = 92.6 Pa/m	Δt = 0.01 K/m Δp = 185.3 Pa/m
10	0.91	1.33	1.66	2.41	2.37	3.42
15	1.72	2.50	3.11	4.50	4.42	6.37
20	3.66	5.31	6.61	9.53	9.38	13.46
25	6.98	10.10	12.58	18.09	17.79	25.48
32	14.58	21.04	26.17	37.56	36.94	52.86
40	21.99	31.73	39.40	56.39	55.53	79.38
50	42.72	61.51	76.29	109.28	107.61	153.66
65	68.42	98.23	122.06	174.30	171.62	245.00
80	121.52	174.28	216.15	308.91	304.12	433.79
100	249.45	356.87	442.76	631.24	621.94	885.81
125	452.08	646.25	800.19	1 139.74	1 124.47	1 598.31
150	733.59	1 046.77	1 296.07	1 846.63	1 819.59	2 590.21
200	1 506.11	2 149.60	2 662.02	3 784.58	3 735.65	5 303.12
250	2 731.90	3 895.57	4 818.22	6 851.91	6 759.98	9 589.56
300	4 378.87	6 237.23	7 714.93	10 973.55	10 810.65	15 360.20

Note: Capacities are in kilowatts of refrigeration resulting in a line friction loss per unit equivalent pipe length (Δp in Pa/m), with corresponding change in saturation temperature per unit length (Δt in K/m).

fittings for standard pipe and extra heavy fittings for extra heavy pipe).

Tongue and groove or ANSI flanges should be used in ammonia piping. Welded flanges for low-side piping can have a minimum 1 MPa design pressure rating. On systems located in high ambients, low-side piping and vessels should be designed for 1.4 to 1.6 MPa. The high side should be 1.7 MPa if the system uses water-cooled or evaporative cooled condensing. Use 2.1 MPa minimum for air-cooled designs.

Pipe Joints

Joints between lengths of pipe or between pipe and fittings can be threaded if the pipe size is 32 mm or smaller. Pipe 40 mm or larger should be welded. An all-welded piping system is superior.

Threaded Joints. Many sealants and compounds are available for sealing threaded joints. The manufacturer's instructions cover compatibility and application method. Do not use excessive amounts or apply on female threads because any excess can contaminate the system.

Welded Joints. Pipe should be cut and beveled before welding. Use pipe alignment guides to align the pipe and provide a proper gap between pipe ends so that a full penetration weld is obtained. The weld should be made by a qualified welder, using proper procedures such as the Welding Procedure Specifications, prepared by the National Certified Pipe Welding Bureau (NCPWB).

Gasketed Joints. A compatible fiber gasket should be used with flanges. Before tightening flange bolts to valves, controls, or flange unions, properly align the pipe and bolt holes. When flanges are used to straighten pipe, they put stress on adjacent valves, compressors, and controls, causing the operating mechanism to bind. To prevent leaks, flange bolts are drawn up evenly when connecting the flanges. Flanges at compressors and other system components must not move or indicate stress when all bolts are loosened.

Union Joints. Steel (21 MPa) ground joint unions are used for gage and pressure control lines with screwed valves and for joints up to 20 mm. When tightening this type of joint, the two pipes must be axially aligned. To be effective, the two parts of the union must match perfectly. Ground joint unions should be avoided if at all possible.

Pipe Location

Piping should be at least 2.3 m above the floor. Locate pipes carefully in relation to other piping and structural members, especially when the lines are to be insulated. The distance between insulated lines should be at least three times the thickness of the insulation for screwed fittings, and four times for flange fittings. The space between the pipe and adjacent surfaces should be three-fourths of these amounts.

Hangers located close to the vertical risers to and from compressors keep the piping weight off the compressor. Pipe hangers should be placed no more than 2.5 to 3 m apart and within 0.6 m of a change

System Practices for Ammonia Refrigerant

Table 2 Suction, Discharge Line, and Liquid Capacities in Kilowatts for Ammonia (Single- or High-Stage Applications)

Steel Nominal Line Size, mm	Suction Lines (Δt = 0.02 K/m)					Discharge Lines Δt = 0.02 K/m, Δp = 684.0 Pa/m			Steel Nominal Line Size, mm	Liquid Lines	
	Saturated Suction Temperature, °C					Saturated Suction Temp., °C				Velocity = 0.5 m/s	Δp = 450.0
	−40 Δp = 76.9	−30 Δp = 116.3	−20 Δp = 168.8	−5 Δp = 276.6	+5 Δp = 370.5	−40	−20	+5			
10	0.8	1.2	1.9	3.5	4.9	8.0	8.3	8.5	10	3.9	63.8
15	1.4	2.3	3.6	6.5	9.1	14.9	15.3	15.7	15	63.2	118.4
20	3.0	4.9	7.7	13.7	19.3	31.4	32.3	33.2	20	110.9	250.2
25	5.8	9.4	14.6	25.9	36.4	59.4	61.0	62.6	25	179.4	473.4
32	12.1	19.6	30.2	53.7	75.4	122.7	126.0	129.4	32	311.0	978.0
40	18.2	29.5	45.5	80.6	113.3	184.4	189.4	194.5	40	423.4	1 469.4
50	35.4	57.2	88.1	155.7	218.6	355.2	364.9	374.7	50	697.8	2 840.5
65	56.7	91.6	140.6	248.6	348.9	565.9	581.4	597.0	65	994.8	4 524.8
80	101.0	162.4	249.0	439.8	616.9	1 001.9	1 029.3	1 056.9	80	1 536.3	8 008.8
100	206.9	332.6	509.2	897.8	1 258.6	2 042.2	2 098.2	2 154.3	—	—	—
125	375.2	601.8	902.6	1 622.0	2 271.4	3 682.1	3 783.0	3 884.2	—	—	—
150	608.7	975.6	1 491.4	2 625.4	3 672.5	5 954.2	6 117.4	6 281.0	—	—	—
200	1 252.3	2 003.3	3 056.0	5 382.5	7 530.4	12 195.3	12 529.7	12 864.8	—	—	—
250	2 271.0	3 625.9	5 539.9	9 733.7	13 619.6	22 028.2	22 632.2	23 237.5	—	—	—
300	3 640.5	5 813.5	8 873.4	15 568.9	21 787.1	35 239.7	36 206.0	37 174.3	—	—	—

Notes:
1. Table capacities are in kilowatts of refrigeration.

   Δp = pressure drop due to line friction, Pa/m

   Δt = corresponding change in saturation temperature, K/m

2. Line capacity for other saturation temperatures Δt and equivalent lengths L_e

   $$\text{Line capacity} = \text{Table capacity} \left(\frac{\text{Table } L_e}{\text{Actual } L_e} \times \frac{\text{Actual } \Delta t}{\text{Table } \Delta t} \right)^{0.55}$$

3. Saturation temperature Δt for other capacities and equivalent lengths L_e

   $$\Delta t = \text{Table } \Delta t \left(\frac{\text{Actual } L_e}{\text{Table } L_e} \right) \left(\frac{\text{Actual capacity}}{\text{Table capacity}} \right)^{1.8}$$

4. Values in the table are based on 30°C condensing temperature. Multiply table capacities by the following factors for other condensing temperatures:

Condensing Temperature, °C	Suction Lines	Discharge Lines
20	1.04	0.86
30	1.00	1.00
40	0.96	1.24
50	0.91	1.43

5. Liquid line capacities are based on −5°C suction.

Table 3 Liquid Ammonia Line Capacities in Kilowatts

Nominal Size, mm	Pumped Liquid Overfeed Ratio			High-Pressure Liquid at 21 kPa[a]	Hot-Gas Defrost[a]	Equalizer High Side[b]	Thermosiphon Lubricant Cooling Lines Gravity Flow[c]		
	3:1	4:1	5:1				Supply	Return	Vent
40	513	387	308	1 544	106	791	59	35	60
50	1 175	879	703	3 573	176	1 055	138	88	106
65	1 875	1 407	1 125	5 683	324	1 759	249	155	187
80	2 700	2 026	1 620	10 150	570	3 517	385	255	323
100	4 800	3 600	2 880	—	1 154	7 034	663	413	586
125	—	—	—	—	2 089	—	1 041	649	1 062
150	—	—	—	—	3 411	—	1 504	938	1 869
200	—	—	—	—	—	—	2 600	1 622	3 400

Source: Wile (1977).
[a] Hot-gas line sizes are based on 0.34 kPa pressure drop per equivalent metre of pipe at 690 kPa (gage) discharge pressure and 3 times the evaporator refrigeration capacity.
[b] Line sizes are based on experience using total system evaporator kilowatts.
[c] From Frick Co. (1995). Values for line sizes above 100 mm are extrapolated.

in direction of the piping. Hangers should be designed to bear on the outside of insulated lines. Sheet metal sleeves on the lower half of the insulation are usually sufficient. Where piping penetrates a wall, a sleeve should be installed; and where the pipe penetrating the wall is insulated, it must be adequately sealed.

Piping to and from compressors and to other components must provide for expansion and contraction. Sufficient flange or union joints should be located in the piping that components can be assembled easily during initial installation and also disassembled for servicing.

Pipe Sizing

Table 1 presents practical suction line sizing data based on 0.005 K and 0.01 K differential pressure drop equivalent per metre total equivalent length of pipe. For data on equivalent lengths of valves and fittings, refer to Tables 10, 11, and 12 in Chapter 2. Table 2 lists data for sizing suction and discharge lines at 0.02 K differential pressure drop equivalent per metre equivalent length of pipe, and for sizing liquid lines at 0.5 m/s. Charts prepared by Wile (1977) present pressure drops in saturation temperature equivalents.

For a complete discussion of the basis of these line sizing charts, see Timm (1991). Table 3 presents line sizing information for pumped liquid lines, high-pressure liquid lines, hot-gas defrost lines, equalizing lines, and thermosiphon lubricant cooling ammonia lines.

Valves

Stop Valves. These valves should be placed in the inlet and outlet lines to all condensers, vessels, evaporators, and long lengths of pipe so that they can be isolated in case of leaks and to facilitate "pumping out" by evacuation. Sections of liquid piping that can be valved off and isolated must be protected with a relief device.

Installing globe-type stop valves with the valve stems horizontal lessens the chance (1) for dirt or scale to lodge on the valve seat or disk and cause it to leak or (2) for liquid or lubricant to pocket in the area below the seat. Wet suction return lines (recirculation system) should employ angle valves to reduce the possibility of liquid pockets and to reduce pressure drop.

Welded flanged or weld-in-line valves are desirable for all line sizes; however, screwed valves may be used for 32 mm and smaller lines. Ammonia globe and angle valves should have the following features:

- Soft seating surfaces for positive shutoff (no copper or copper alloy)
- Back seating to permit repacking the valve stem while in service
- Arrangement that allows packing to be tightened easily
- All-steel construction (preferable)
- Bolted bonnets above 25 mm, threaded bonnets for 25 mm and smaller

Consider seal cap valves in refrigerated areas and for all ammonia piping. To keep pressure drop to a minimum, consider angle valves (as opposed to globe valves).

Control Valves. Pressure regulators, solenoid valves, and thermostatic expansion valves should be flanged for easy assembly and removal. Valves 40 mm and larger should have welded companion flanges. Smaller valves can have threaded companion flanges.

A strainer should be used in front of self-contained control valves to protect them from pipe construction material and dirt. A ceramic filter installed in the pilot line to the power piston protects the close tolerances from foreign material when pilot-operated control valves are used.

Solenoid Valves. Solenoid valve stems should be upright with their coils protected from moisture. They should have flexible conduit connections, where allowed by codes, and an electric pilot light wired in parallel to indicate when the coil is energized. A manual opening stem is useful for emergencies.

Solenoid valves for high-pressure liquid feed to evaporators should have soft seats for positive shutoff. Solenoid valves for other applications, such as in suction lines, hot-gas lines, or gravity feed lines, should be selected for the pressure and temperature of the fluid flowing and for the pressure drop available.

Relief Valves. Safety valves must be provided in conformance with ASHRAE *Standard* 15 and Section VIII, Division 1, of the ASME *Boiler and Pressure Vessel Code*. For ammonia systems, IIAR *Bulletin* 109 also addresses the subject of safety valves.

Dual relief valve arrangements enable testing of the relief valves (Figure 13). The three-way stop valve is constructed so that it is always open to one of the relief valves if the other is removed to be checked or repaired.

Isolated Line Sections

Sections of piping that can be isolated between hand valves or check valves can be subjected to extreme hydraulic pressures if cold liquid refrigerant is trapped in them and subsequently warmed. Additional safety valves for such piping must be provided.

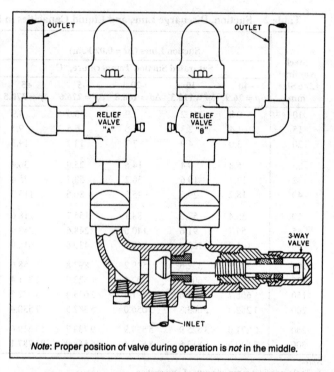

Fig. 13 Dual Relief Valve Fitting for Ammonia

Insulation and Vapor Retarders

Chapter 32 covers insulation and vapor retarders. Insulation and effective vapor retarders on low-temperature systems are very important. At low temperatures, the smallest leak in the vapor retarder can allow ice to form inside the insulation, which can totally destroy the integrity of the entire insulation system. The result can cause a significant increase in load and power usage.

RECIPROCATING COMPRESSORS

Piping

Figure 14 shows a typical piping arrangement for two compressors operating in parallel off the same suction main. Suction mains should be laid out with the objective of returning only clean, dry gas to the compressor. This usually requires a suction trap sized adequately for gravity gas and liquid separation based on permissible gas velocities for specific temperatures. A dead-end trap can usually trap only scale and lubricant. As an alternative to the dead-end trap, a shell-and-coil accumulator with a warm liquid coil may be considered. Suction mains running to and from the suction trap or accumulator should be pitched toward the trap at 10 mm per metre for liquid drainage.

In sizing the suction mains and the takeoffs from the mains to the compressors, consider how the pressure drop in the selected piping affects the compressor size required. First costs and operating costs for compressor and piping selections should be optimized.

Good suction line systems have a total friction drop of 0.5 to 1.5 K pressure drop equivalent. Practical suction line friction losses should not exceed 0.01 K equivalent per metre equivalent length.

A well-designed discharge main has a total friction loss of 7 to 15 kPa. Generally, a slightly oversized discharge line is desirable to hold down discharge pressure and, consequently, discharge temperature and energy costs. Where possible, discharge mains should be pitched (10 mm/m) toward the condenser, without creating a liquid trap; otherwise, pitch should be toward the discharge line separator.

System Practices for Ammonia Refrigerant

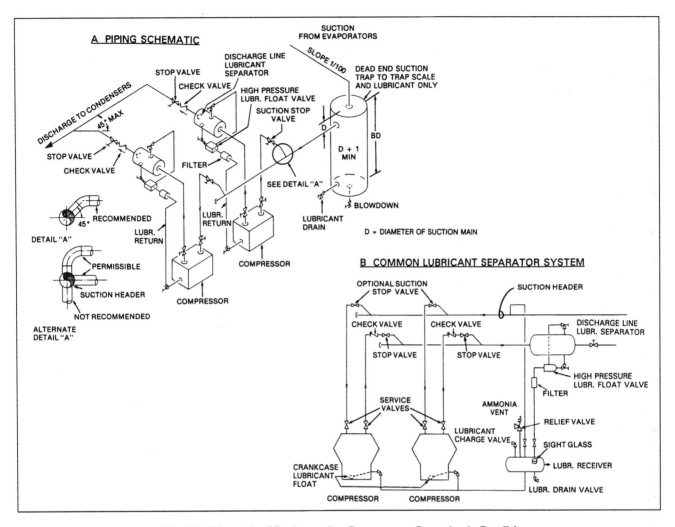

Fig. 14 Schematic of Reciprocating Compressors Operating in Parallel

High- and low-pressure cutouts and gages and lubricant pressure failure cutout are installed on the compressor side of the stop valves to protect the compressor.

Lubricant Separators. Lubricant separators are located in the discharge line of each compressor (Figure 14A). A high-pressure float valve drains the lubricant back into the compressor crankcase or lubricant receiver. The separator should be placed as far from the compressor as possible, so that the extra pipe length can be used to cool the discharge gas before it enters the separator. This reduces the temperature of the ammonia vapor and makes the separator more effective.

Liquid ammonia must not reach the crankcase. Often, a valve (preferably automatic) is installed in the drain from the lubricant separator, open only when the temperature at the bottom of the separator is higher than the condensing temperature. Some manufacturers install a small electric heater at the bottom of a vertical lubricant trap instead. The heater is actuated when the compressor is not operating. Separators installed in cold conditions must be insulated to prevent ammonia condensation.

A filter is recommended in the drain line on the downstream side of the high-pressure float valve.

Lubricant Receivers. Figure 14B illustrates two compressors on the same suction line with one discharge line lubricant separator. The separator float drains into a lubricant receiver, which maintains a reserve supply of lubricant for the compressors. Compressors should be equipped with crankcase floats to regulate the lubricant flow to the crankcase.

Discharge Check Valves and Discharge Lines. Discharge check valves on the downstream side of each lubricant separator prevent high-pressure gas from flowing into an inactive compressor and causing condensation (Figure 14A).

The discharge line from each compressor should enter the discharge main at a 45° maximum angle in the horizontal plane so that the gas flows smoothly.

Unloaded Starting. Unloaded starting is frequently needed to stay within the torque or current limitations of the motor. Most compressors are unloaded either by holding the suction valve open or by external bypassing. Control can be manual or automatic.

Suction Gas Conditioning. Suction main piping should be insulated, complete with vapor retarder to minimize thermal losses, to prevent sweating and/or ice buildup on the piping, and to limit superheat at the compressor. Additional superheat results in increased discharge temperatures and reduces compressor capacity. Low discharge temperatures in ammonia plants are important to reduce lubricant carryover and because the compressor lubricant can carbonize at higher temperatures, which can cause cylinder wall scoring and lubricant sludge throughout the system. Discharge temperatures above 120°C should be avoided at all times. Lubricants should have flash point temperatures above the maximum expected compressor discharge temperature.

Cooling

Generally, ammonia compressors are constructed with internally cast cooling passages along the cylinders and/or in the top heads. These passages provide space for circulating a heat transfer medium, which minimizes heat conduction from the hot discharge gas to the incoming suction gas and lubricant in the compressor's crankcase. An external lubricant cooler is supplied on most reciprocating ammonia compressors. Water is usually the medium circulated through these passages (**water jackets**), and the lubricant cooler at a rate of about 2 mL/s per kilowatt of refrigeration. The lubricant in the crankcase (depending on type of construction) is about 50°C. Temperatures above this level reduce the lubricant's lubricating properties.

For compressors operating in ambients above 0°C, water flow is sometimes controlled entirely by hand valves, although a solenoid valve in the inlet line is desirable to make the system automatic. When the compressor stops, water flow must be stopped to keep the residual gas from condensing and to conserve water. A water-regulating valve, installed in the water supply line with the sensing bulb in the water return line, is also recommended. This type of cooling is shown in Figure 15.

The thermostat in the water line leaving the jacket serves as a safety cutout to stop the compressor if the temperature becomes too high.

For compressors installed where ambient temperatures below 0°C may exist, a means for draining the jacket on shutdown to prevent freeze-up must be provided. One method is shown in Figure 16. Water flow is through the normally closed solenoid valve, which is energized when the compressor starts. The water then circulates through the lubricant cooler and the jacket, and out through the water return line. When the compressor stops, the solenoid valve in the water inlet line is deenergized and stops the water flow to the compressor. At the same time, the solenoid valve opens to drain the water out of the low point to wastewater treatment. The check valves in the air vent lines open when pressure is relieved and allow the jacket and cooler to be drained. Each flapper check valve is installed so that water pressure closes it, but absence of water pressure allows it to swing open.

When compressors are installed in spaces below 0°C or where water quality is very poor, cooling is best handled by using an inhibited glycol solution or other suitable fluid in the jackets and lubricant cooler and cooling with a secondary heat exchanger. This method for cooling reciprocating ammonia compressors eliminates the fouling of the lubricant cooler and jacket normally associated with city water or cooling tower water.

ROTARY VANE, LOW-STAGE COMPRESSORS

Piping

Rotary vane compressors have been used extensively as low-stage compressors in ammonia refrigeration systems. Now, however, the screw compressor has largely replaced the rotary vane compressor for ammonia low-stage compressor applications. Piping requirements for rotary vane compressors are the same as for reciprocating compressors. Most rotary vane compressors are lubricated by injectors since they have no crankcase. In some designs, a lubricant separator, lubricant receiver, and cooler are required on the discharge of these compressors; a pump recirculates the lubricant to the compressor for both cooling and lubrication. In other rotary vane compressor designs, a discharge lubricant separator is not used, and the lubricant collects in the high-stage suction accumulator or intercooler, from which it may be drained. Lubricant for the injectors must periodically be added to a reservoir.

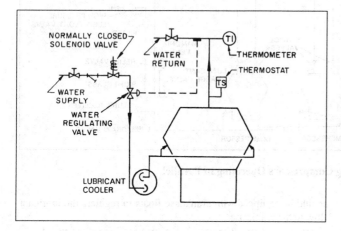

Fig. 15 Jacket Water Cooling for Ambient Temperatures above Freezing

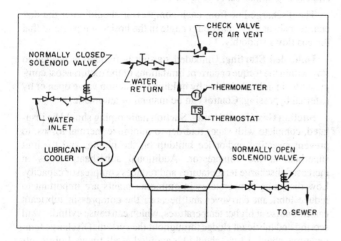

Fig. 16 Jacket Water Cooling for Ambient Temperatures below Freezing

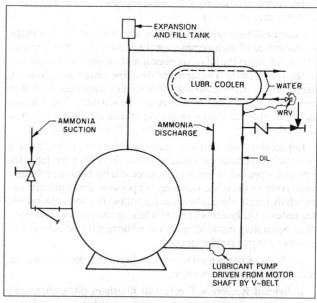

Fig. 17 Rotary Vane Booster Compressor Cooling with Lubricant

System Practices for Ammonia Refrigerant

Cooling

The compressor jacket is cooled by circulating a cooling fluid, such as water or lubricant. Lubricant is recommended because it will not freeze and can be the same lubricant used for lubrication (Figure 17).

SCREW COMPRESSORS

Piping

Helical screw compressors are the choice for most industrial refrigeration systems. All helical screw compressors have a constant volume (displacement) design. The volume index (V_i) refers to the internal volume ratio of the compressor. There are three types of screw compressors:

1. Fixed V_i with slide valve
2. Variable V_i with slide valve and slide stop
3. Fixed V_i with bypass ports in lieu of slide valve

When V_i is fixed, the compressor functions most efficiently at a certain absolute compression ratio (CR). In selecting a fixed V_i compressor, the average CR rather than the maximum CR should be considered. A guide to proper compressor selection is based on the equation $V_i^k = CR$, where $k = 1.4$ for ammonia.

For example, for a screw compressor at $-12°C$ (268 kPa) and 36°C (1390 kPa) with CR = 5.19, $V_i^{1.4} = 5.19$ and $V_i = 3.24$. Thus, a compressor with $V_i = 3.6$ might be the best choice. If the ambient conditions are such that the average condensing temperature is 24°C (973 kPa), then the CR is 3.63 and the ideal V_i is 2.51. Thus, a compressor with $V_i = 2.4$ is the proper selection to optimize efficiency.

Fixed V_i compressors with bypass ports in lieu of a slide valve are often applied as booster compressors, which normally have a V_i requirement of less than 2.9.

A variable V_i compressor makes compressor selection simpler because it can vary its volume index from 2.0 to 5.0; thus, it can automatically match the internal pressure ratio within the compressor with the external pressure ratio.

Typical flow diagrams for screw compressor packages are shown in Figures 18 and 19. Figure 18 is for indirect cooling, and Figure 19 is for direct cooling with refrigerant liquid injection. Figure 20 illustrates a variable V_i compressor that does not require a full-time lube pump but rather a pump to prelube the bearings. Full-time lube pumps are required when fixed or variable V_i compressors are used as low-stage compressors. Lubrication systems require at least a 500 kPa pressure differential for proper operation.

Lubricant Cooling

The lubricant in screw compressors may be cooled three ways:

1. Liquid refrigerant injection
2. Indirect cooling with glycol or water in a heat exchanger
3. Indirect cooling with boiling high-pressure refrigerant used as the coolant in a thermosiphon process

Refrigerant injection cooling is shown schematically in Figures 19 and 21. Depending on the application, this cooling method usually decreases compressor efficiency and capacity but lowers equipment cost. Most screw compressor manufacturers publish a derating curve for this type of cooling. Injection cooling for low-stage compression has little or no penalty on compressor efficiency or capacity. However, the system efficiency can be increased by using an indirectly cooled lubricant cooler. With this configuration, the heat from the lubricant cooler is removed by the evaporative condenser or cooling tower and is not transmitted to the high-stage compressors.

The refrigerant liquid for liquid injection oil cooling must come from a dedicated supply source. The source may be the system receiver or a separate receiver; a 5 min uninterrupted supply of refrigerant liquid is usually adequate.

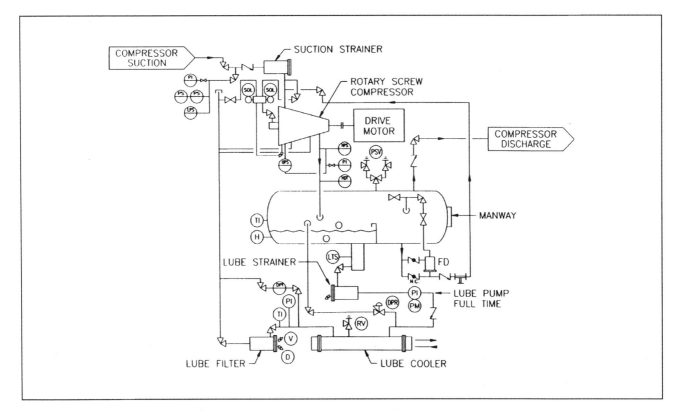

Fig. 18 Fixed V_i Screw Compressor Flow Diagram with Indirect Lubricant Cooling

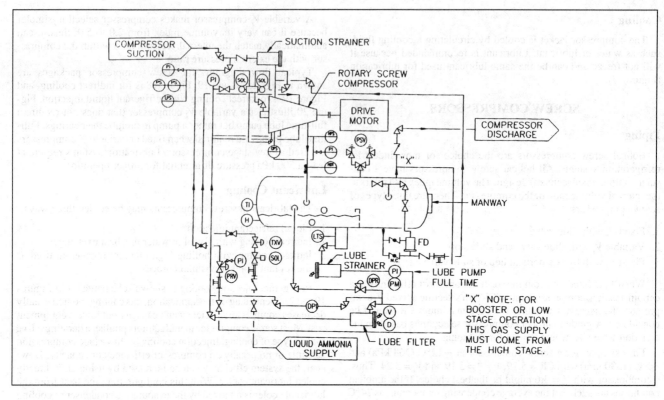

Fig. 19 Fixed V_i Screw Compressor Flow Diagram with Liquid Injection Cooling

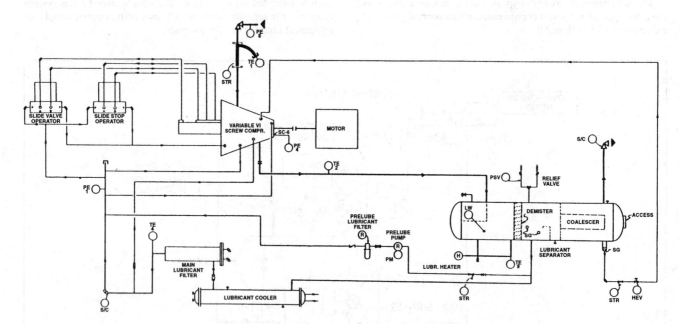

Fig. 20 Flow Diagram for Variable V_i Screw Compressor High-Stage Only

Indirect or thermosiphon lubricant cooling for low-stage screw compressors rejects the lubricant cooling load to the condenser or auxiliary cooling system—this load is not transferred to the high-stage compressor, which improves the system efficiency. Indirect lubricant cooling systems using glycol or water reject the lubricant cooling load to a section of an evaporative condenser, a separate evaporative cooler, or a cooling tower. A three-way lubricant control valve should be used to control lubricant temperature.

Thermosiphon lubricant cooling is the industry standard. In this system, high-pressure refrigerant liquid from the condenser, which boils at condensing temperature/pressure (usually 32 to 35°C design), cools the lubricant in a tubular heat exchanger. Typical thermosiphon lubricant cooling arrangements are shown in Figures 18, 20, 22, 23, and 24. Note on all figures that the refrigerant liquid supply to the lubricant cooler receives priority over the feed to the system low side. It is important that the gas equalizing line (vent) off

System Practices for Ammonia Refrigerant

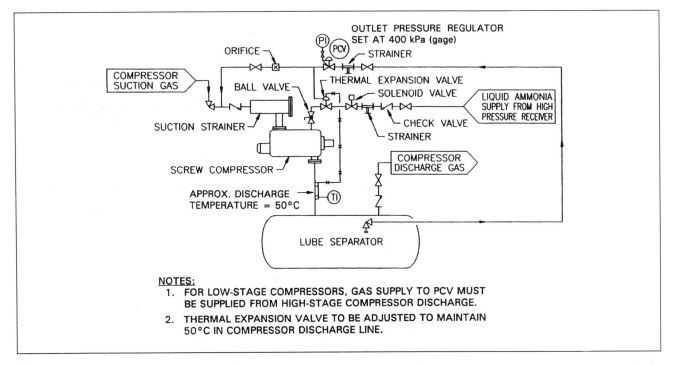

Fig. 21 Flow Diagram for Screw Compressors with Refrigerant Injection Cooling

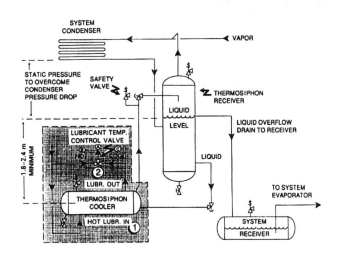

Fig. 22 Typical Thermosiphon Lubricant Cooling System with Thermosiphon Accumulator

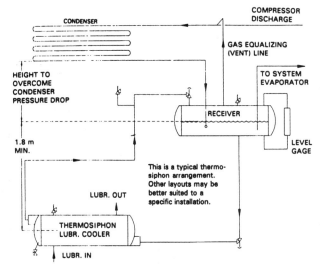

Fig. 23 Thermosiphon Lubricant Cooling System with Receiver Mounted above Thermosiphon Lubricant Cooler

the top of the thermosiphon receiver be adequately sized to match the lubricant cooler load to prevent the thermosiphon receiver from becoming gas bound.

Figure 25 shows a typical capacity control system for a fixed V_i screw compressor. The four-way valve controls the slide valve position and thus the compressor capacity from typically 100 to 10% with a signal from an electric, electronic, or microprocessor controller. The slide valve unloads the compressor by bypassing vapor back to the suction of the compressor.

Figure 26 shows a typical capacity and volume index control system in which two four-way control valves take their signals from a computer controller. One four-way valve controls the capacity by positioning the slide valve in accordance with the load, and the other positions the slide stop to adjust the compressor internal pressure ratio to match the system suction and discharge pressure. The slide valve works the same as that on fixed V_i compressors. Volume index is varied by adjusting the slide stop on the discharge end of the compressor.

Screw compressor piping should generally be installed in the same manner as for reciprocating compressors. Although screw compressors can ingest some liquid refrigerant, they should be protected against liquid carryover. Screw compressors are furnished with both suction and discharge check valves.

CONDENSER AND RECEIVER PIPING

Properly designed piping around the condensers and receivers keeps the condensing surface at its highest efficiency by draining liquid ammonia out of the condenser as soon as it condenses and keeping air and other noncondensables purged.

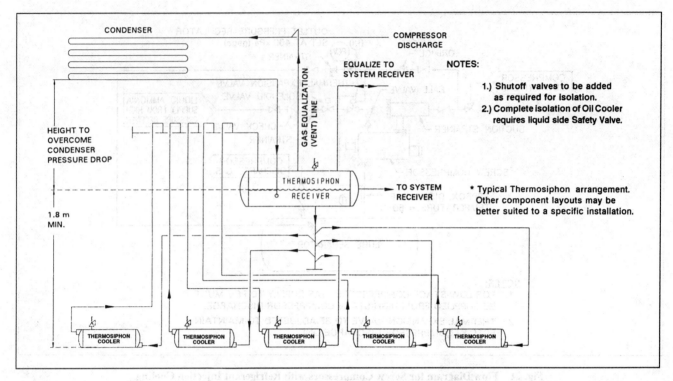

Fig. 24 Typical Thermosiphon System with Multiple Oil Coolers

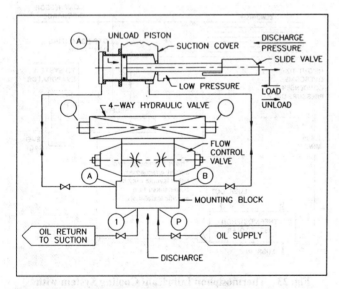

Fig. 25 Typical Hydraulic System for Slide Valve Capacity Control for Screw Compressor with Fixed V_i

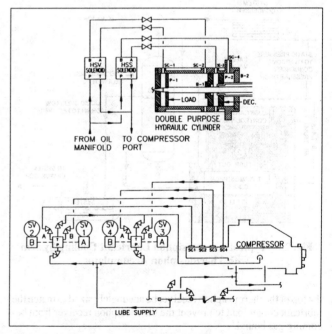

Fig. 26 Typical Positioning System for Slide Valve and Slide Stop for Variable V_i Screw Compressor

Horizontal Shell-and-Tube Condenser and Through-Type Receiver

Figure 27 shows a horizontal water-cooled condenser draining into a through-type (top inlet) receiver. Ammonia plants do not always require controlled water flow to maintain pressure. Usually, pressure is adequate to force the ammonia to the various evaporators without water regulation. Each situation should be evaluated by comparing water costs with input power cost savings at lower condenser pressures.

Water piping should be arranged so that condenser tubes are always filled with water. Air vents should be provided on condenser heads and should have hand valves for manual purging.

Receivers must be below the condenser so that the condensing surface is not flooded with ammonia. The piping should provide (1) free drainage from the condenser and (2) static height of ammonia above the first valve out of the condenser greater than the pressure drop through the valve.

The drain line from condenser to receiver is designed on the basis of maximum velocity to allow gas equalization between condenser and receiver. Refer to Table 2 for sizing criteria.

System Practices for Ammonia Refrigerant

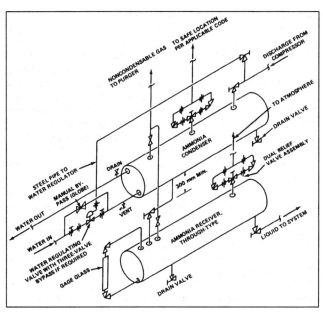

Fig. 27 Horizontal Condenser and Top Inlet Receiver Piping

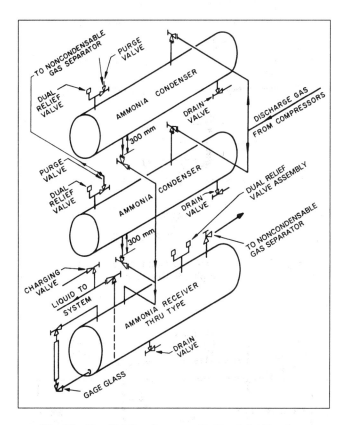

Fig. 28 Parallel Condensers with Top Inlet Receiver

Parallel Horizontal Shell-and-Tube Condensers

Figure 28 shows two condensers operating in parallel with one through-type (top inlet) receiver. The length of horizontal liquid drain lines to the receiver should be minimized, with no traps permitted. Equalization between the shells is achieved by keeping the liquid velocity in the drain line less than 0.5 m/s. The drain line can be sized from Table 2.

EVAPORATIVE CONDENSERS

Evaporative condensers are selected based on the wet-bulb temperature in which they operate. The 1% design wet bulb is that wet-bulb temperature that will be equalled or exceeded 1% of the months of June through September, or 29.3 h. Thus, for the majority of industrial plants that operate at least at part load all year, the wet-bulb temperature will be below design 99.6% of the operating time. The resultant condensing pressure will only equal or exceed the design condition during 0.4% of the time if the design wet-bulb temperature and peak design refrigeration load occur coincidentally. This peak condition is more a function of how the load is calculated, what load diversity factor exists or is used in the calculation, and what safety factor is used in the calculations, than of the size of the condenser.

Location

If an evaporative condenser is located with insufficient space for air movement, the effect is the same as that imposed by an inlet damper, and the fan may not deliver enough air. In addition, evaporative condenser discharge air may recirculate, which adds to the problem. The high inlet velocity causes a low-pressure region to develop around the fan inlet, inducing flow of discharge air into that region. If the obstruction is from a second condenser, the problem can be even more severe because discharge air from the second condenser flows into the air intake of the first.

Prevailing winds can also contribute to recirculation. In many areas, the winds shift with the seasons; wind direction during the peak high-humidity season is the most important consideration.

The tops of the condensers should always be higher than any adjacent structure to eliminate downdrafts that might induce recirculation. Where this is impractical, discharge hoods can be used to discharge air far enough away from the fan intakes to avoid recirculation. However, the additional static pressure imposed by a discharge hood must be added to the fan system. The fan speed can be increased slightly to obtain proper air volume.

Installation

A single evaporative condenser used with a through-type (top inlet) receiver can be connected as shown in Figure 29. The receiver must always be at a lower pressure than the condensing pressure. Design ensures that the receiver is cooler than the condensing temperature.

Installation in Freezing Areas. In areas having ambient temperatures below 0°C, the water in the evaporative condenser drain pan and water circuit must be kept from freezing at light plant loads. When the temperature is at freezing, the evaporative condenser can operate as a dry-coil unit, and the water pump(s) and piping can be drained and secured for the season.

Another method of preventing the water from freezing is to place the water tank inside and install it as illustrated in Figure 30. When the outdoor temperature drops, the condensing pressure drops, and a pressure switch with its sensing element in the discharge pressure line stops the water pump; the water is then drained into the tank. An alternative is to use a thermostat that senses the water temperature or outdoor ambient temperature and stops the pump at low temperatures. The exposed piping and any trapped water headers in the evaporative condenser should be drained into the indoor water tank.

Air volume capacity control methods include inlet, outlet, or bypass dampers; two-speed fan motors; or fan cycling in response to pressure controls.

Liquid Traps. Because all evaporative condensers have a substantial pressure drop in the ammonia circuit, liquid traps are needed at the outlets when two or more condensers or condenser coils are installed (Figure 31). Also, an equalizer line is necessary to maintain a stable pressure in the receiver to ensure free drainage from the condensers. For example, assume a 10 kPa pressure drop in the operating condenser in Figure 31, which is producing a lower pressure (1290 kPa) at its outlet compared to the idle condenser (1300 kPa) and the receiver (1300 kPa). The trap creates a liquid seal so that a liquid height h of 1700 mm (equivalent to 10 kPa) builds up in the vertical drop leg and not in the condenser coil.

The trap must have enough height above the vertical liquid leg to accommodate a liquid height equal to the maximum pressure drop that will be encountered in the condenser. The example illustrates the extreme case of one unit on and one off; however, the same phenomenon occurs to a lesser degree with two condensers of differing pressure drops when both are in full operation. Substantial differences in pressure drop can also occur between two different brands of the same size condenser or even different models produced by the same manufacturer.

The minimum recommended height of the vertical leg is 1500 mm for ammonia. This vertical dimension h is shown in all evaporative condenser piping diagrams. This height is satisfactory for operation within reasonable ranges around normal design conditions and is based on the maximum condensing pressure drop of the coil. If service valves are installed at the coil inlets and/or outlets, the pressure drops imposed by these valves must be accounted for by increasing the minimum 1500 mm drop-leg height by an amount equal to the valve pressure drop in height of liquid refrigerant (Figure 32).

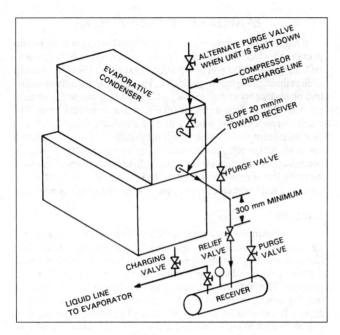

Fig. 29 Single Evaporative Condenser with Top Inlet Receiver

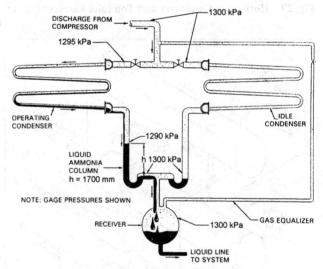

Fig. 31 Two Evaporative Condensers with Trapped Piping to Receiver

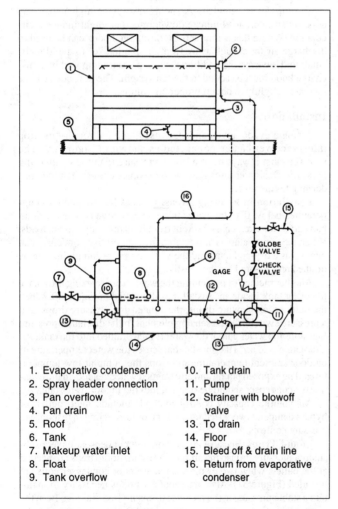

1. Evaporative condenser
2. Spray header connection
3. Pan overflow
4. Pan drain
5. Roof
6. Tank
7. Makeup water inlet
8. Float
9. Tank overflow
10. Tank drain
11. Pump
12. Strainer with blowoff valve
13. To drain
14. Floor
15. Bleed off & drain line
16. Return from evaporative condenser

Fig. 30 Evaporative Condenser with Inside Water Tank

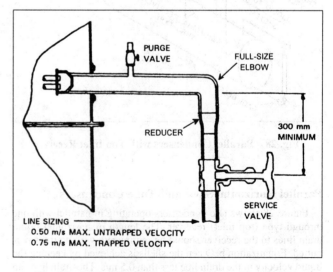

Fig. 32 Method of Reducing Condenser Outlet Sizes

System Practices for Ammonia Refrigerant

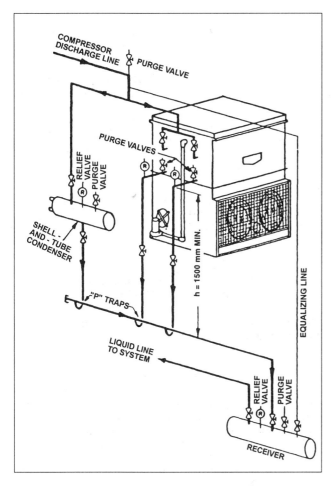

Fig. 33 Piping for Shell-and-Tube and Evaporative Condensers with Top Inlet Receiver

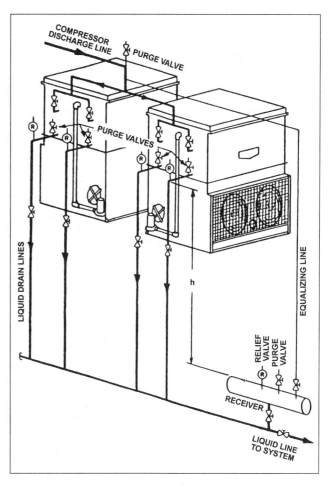

Fig. 34 Piping for Parallel Condensers with Surge-Type Receiver

Figures 33, 34, and 35 illustrate various piping arrangements for evaporative condensers.

EVAPORATOR PIPING

Proper evaporator piping and control are necessary to keep the cooled space at the desired temperature and also to adequately protect the compressor from surges of liquid ammonia out of the evaporator. The evaporators illustrated in this section show some methods used to accomplish these objectives. In some cases, combinations of details shown on several illustrations have been used.

When using hot gas or electric heat for defrosting, the drain pan and drain line must be heated to prevent the condensate from refreezing. With hot gas, a heating coil is embedded in the drain pan. The hot gas flows first through this coil and then into the evaporator coil. With electric heat, an electric heating coil is used under the drain pan. Wraparound or internal electric heating cables are used on the condensate drain line when the room temperature is below 0°C.

Figure 36 illustrates a thermostatic expansion valve on a unit cooler using hot gas for automatic defrosting. Because this is an automatic defrosting arrangement, hot gas must always be available at the hot-gas solenoid valve near the unit. The system must contain multiple evaporators so that the compressor will be running when the evaporator to be defrosted is shut down. The hot-gas header must be kept in a space where ammonia will not condense in the pipe. Otherwise, the coil receives liquid ammonia at the start of defrosting and is unable to take full advantage of the latent heat of hot-gas condensation entering the coil. This can also lead to severe hydraulic shock loads. If the header must be in a cold space, the insulated hot-gas main must be drained to the suction line by a high-pressure float.

The liquid line and suction line solenoid valves are open during normal operation only and are closed during the defrost cycle. When the defrost cycle starts, the hot-gas solenoid valve is opened. Refer to IIAR *Bulletin* 116 for information on possible hydraulic shock when the hot-gas defrost valve is opened after a defrost.

A defrost pressure regulator maintains a gage pressure of about 480 to 550 kPa in the coil.

Unit Cooler—Flooded Operation

Figure 37 illustrates a flooded evaporator with a close coupled low-pressure vessel for feeding ammonia into the coil and automatic water defrost.

The lower float switch on the float column at the vessel controls the opening and closing of the liquid line solenoid valve, regulating ammonia feed into the unit to maintain a liquid level. The hand expansion valve downstream of the solenoid valve should be adjusted so that it will not feed ammonia into the vessel at a rate higher than the vessel can accommodate while raising the suction pressure of gas from the vessel no more than 6 to 14 kPa.

The static height of liquid in the vessel should be sufficient to flood the coil with liquid under normal loads. The higher float switch should be wired into an alarm circuit and possibly a compressor shutdown circuit for when the liquid level in the vessel is too high. With flooded coils having horizontal headers, distribution between the multiple circuits is accomplished without distributing orifices.

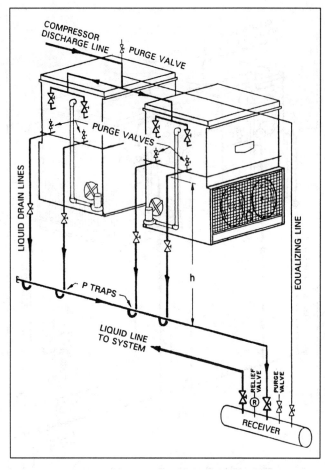

Fig. 35 Piping for Parallel Condensers with Top Inlet Receiver

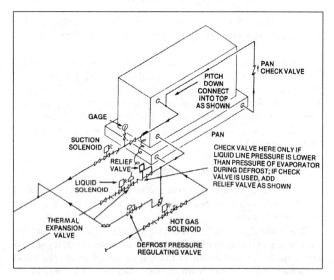

Fig. 36 Piping for Thermostatic Expansion Valve Application for Automatic Defrost on Unit Cooler

A combination evaporator pressure regulator and stop valve is used in the suction line from the vessel. During operation, the regulator maintains a nearly constant back pressure in the vessel. A solenoid coil in the regulator mechanism closes it during the defrost cycle. The liquid solenoid valve should also be closed at this time.

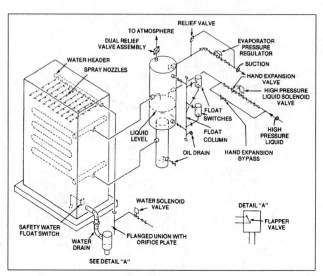

Fig. 37 Arrangement for Automatic Defrost of Air Blower with Flooded Coil

One of the best means of controlling room temperature is a room thermostat that controls the effective setting of the evaporator pressure regulator.

A spring-loaded relief valve is used around the suction pressure regulator and is set so that the vessel is kept below 860 kPa (gage).

A solenoid valve unaffected by downstream pressure is used in the water line to the defrost header. The defrost header is constructed so that it drains at the end of the defrost cycle and the downstream side of the solenoid valve drains through a fixed orifice.

Unless the room is maintained above 0°C, the drain line from the unit should be wrapped with a heater cable or provided with another heat source and then insulated to prevent the defrost water from refreezing in the line.

The length of the water line within the space leading up to the header and the length of the drain line in the cooled space should be kept to a minimum. A flapper or pipe trap on the end of the drain line prevents warm air from flowing up the drain pipe and into the unit.

An air outlet damper may be closed during defrosting to prevent thermal circulation of air through the unit, which would affect the temperature of the cooled space. The fan is stopped during defrost.

This type of defrosting requires a drain pan float switch for safety control. If the drain pan fills with water, the switch overrides the time clock to stop the flow into the unit by closing the water solenoid valve.

There should be a 5 min delay at the end of the water spray part of the defrosting cycle so that the water can drain from the coil and pan. This limits the ice buildup in the drain pan and on the coils after the cycle is completed.

On completion of the cycle, the pressure in the low-pressure vessel may be about 500 kPa (gage). When the unit is opened to the much lower pressure suction main, some liquid surges out into the main; therefore, it may be necessary to gradually bleed off this pressure before fully opening the suction valve in order to prevent thermal shock. Generally, a suction trap in the engine room removes this liquid before the gas stream enters the compressors.

The type of refrigerant control shown in Figure 37 can be used on brine spray-type units where brine is sprayed over the coil at all times to pick up the condensed water vapor from the airstream. The brine is reconcentrated continually to remove the water absorbed from the airstream.

High-Side Float Control

When a system has only one evaporator, a high-pressure float control can be used to keep the condenser drained and to provide a

System Practices for Ammonia Refrigerant

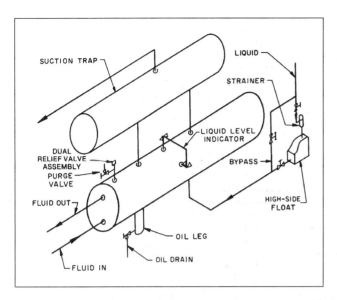

Fig. 38 Arrangement for Horizontal Liquid Cooler and High-Side Float

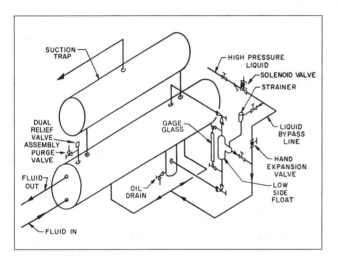

Fig. 39 Piping for Evaporator and Low-Side Float with Horizontal Liquid Cooler

liquid seal between the high side and the low side. Figure 38 illustrates a brine or water cooler with this type of control. The high-side float should be located near the evaporator to avoid insulating the liquid line.

The amount of ammonia in this type of system is critical because the charge must be limited so that liquid will not surge into the suction line under the highest loading in the evaporator. Some type of suction trap should be used. One method is to place a horizontal shell above the cooler, with the suction gas piped into the bottom and out of the top. The reduction of gas velocity in this shell causes the liquid to separate from the gas and draw back into the chiller.

Coolers should include a liquid indicator. A reflex glass lens with a large liquid chamber and vapor connections for boiling liquids and with a plastic frost shield to determine the actual level should be used. A refrigeration thermostat measuring the temperature of the chilled fluid as it exits the cooler should be wired into the compressor starting circuit to prevent freezing.

A flow switch or differential pressure switch should prove flow before the compressor starts. The fluid to be cooled should be piped into the lower portion of the tube bundle and out of the top portion.

Low-Side Float Control

For multiple evaporator systems, low-side float valves are used to control the refrigerant level in flooded evaporators. The low-pressure float shown in Figure 39 has an equalizer line from the top of the float chamber to the space above the tube bundle and an equalizer line out of the lower side of the float chamber to the lower side of the tube bundle.

For positive shutoff of liquid feed when the system stops, a solenoid valve in the liquid line is wired so that it is only energized when the brine or water pump motor is operating and the compressor is running.

A reflex glass lens with large liquid chamber and vapor connections for boiling liquids should be used with a plastic frost shield to determine the actual level and with front extensions as required.

Usually a high-level float switch is installed above the operating level of the float to shut the liquid solenoid valve if the float should overfeed.

MULTISTAGE SYSTEMS

As pressure ratios increase, single-stage ammonia systems encounter problems including (1) high discharge temperatures on reciprocating compressors causing the lubricant to deteriorate, (2) loss of volumetric efficiency as high pressure leaks back to the low-pressure side through compressor clearances, and (3) excessive stresses on compressor moving parts. Thus, manufacturers usually limit the maximum pressure ratios for multicylinder reciprocating machines to approximately 7 to 9. For screw compressors, which incorporate cooling, compression ratio is not a limitation, but efficiency deteriorates at high ratios.

When the overall system pressure ratio (absolute discharge pressure divided by absolute suction pressure) begins to exceed these limits, the pressure ratio across the compressor must be reduced. This is usually accomplished by employing a multistage system. A properly designed two-stage system exposes each of the two compressors to a pressure ratio approximately equal to the square root of the overall pressure ratio. In a three-stage system, each compressor is exposed to a pressure ratio approximately equal to the cube root of the overall ratio. When screw compressors are used, this calculation does not always guarantee the most efficient system.

Another advantage to multistaging is that successively subcooling the liquid at each stage of compression increases the overall system operating efficiency. Additionally, multistaging can be used to accommodate multiple loads at different suction pressures and temperatures in the same refrigeration system. In some cases, two stages of compression can be contained in a single compressor, such as an internally compounded reciprocating compressor. In these units, one or more cylinders are isolated from the others so that they can act as independent stages of compression. Internally compounded compressors are economical for small systems that require low temperature.

Two-Stage Screw Compressor System

A typical two-stage, two-temperature system using screw compressors provides refrigeration for high- and low-temperature loads (Figure 40). For example, the high-temperature stage will supply refrigerant to all process areas operating between −2 and 10°C. A −8°C intermediate suction temperature is selected. The low-temperature stage requires a −37°C suction temperature for blast freezers and continuous or spiral freezers.

The system employs a flash-type intercooler that doubles as a recirculator for the −8°C load. It is the most efficient system available if the screw compressor uses indirect lubricant cooling. If refrigerant injection cooling is used, system efficiency is decreased. This system is efficient for several reasons:

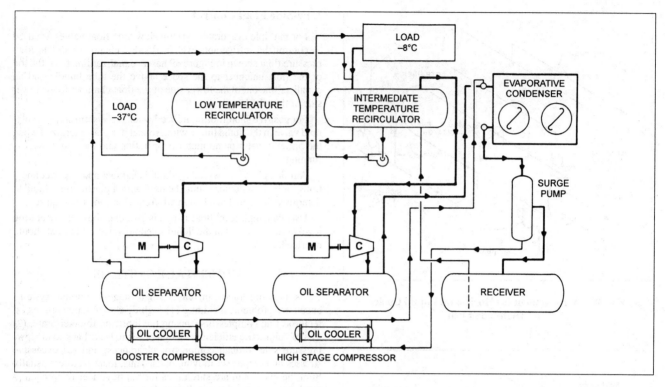

Fig. 40 Compound Ammonia System with Screw Compressor Thermosiphon Cooled

1. Approximately 50% of the booster (low-stage) motor heat is removed from the high-stage compressor load by the thermosiphon lubricant cooler.

   Note: In any system, thermosiphon lubricant cooling for booster and high-stage compressors is about 10% more efficient than injection cooling. Also, plants with a piggyback, two-stage screw compressor system without intercooling or injection cooling can be converted to a multistage system with indirect cooling to increase system efficiency approximately 15%.

2. Flash-type intercoolers are more efficient than shell-and-coil intercoolers by several percent.

3. Thermosiphon lubricant cooling of the high-stage screw compressor provides the highest efficiency available. Installing indirect cooling in plants with liquid injection cooling of screw compressors can increase the compressor efficiency by 3 to 4%.

4. Thermosiphon cooling saves 20 to 30% in electric energy during the low-temperature months. When outside air temperature is low, the condensing pressure can be decreased to 600 to 700 kPa (gage) in most ammonia systems. With liquid injection cooling, the condensing pressure can only be reduced to approximately 850 to 900 kPa (gage).

5. Variable V_i compressors with microprocessor control require less total energy when employed as high-stage compressors. The controller tracks the compressor operating conditions to take advantage of ambient conditions as well as variations in load.

Converting Single-Stage into Two-Stage Systems

When plant refrigeration capacity must be increased and the system is operating below about 70 kPa (gage) suction pressure, it is usually more economical to increase capacity by adding a compressor to operate as the low-stage compressor of a two-stage system than to implement a general capacity increase. The existing single-stage compressor then becomes the high-stage compressor of the two-stage system. The following are some items to consider when converting:

- The motor on the existing single-stage compressor may have to be increased in size when used at a higher suction pressure.
- The suction trap should be checked for sizing at the increased gas flow rate.
- An intercooler should be added to cool the low-stage compressor discharge gas and to cool high-pressure liquid.
- A condenser may have to be added to handle the increased condensing load.
- A means of purging air should be added if plant suction gage pressure is below zero.
- A means of automatically reducing compressor capacity should be added so that the system will operate satisfactorily at reduced system capacity points.

LIQUID RECIRCULATION SYSTEMS

The following discussion gives an overview of liquid recirculation (liquid overfeed) systems. See Chapter 1 for more complete information. For additional engineering details on liquid overfeed systems, refer to Stoecker (1988).

In a liquid ammonia recirculation system, a pump circulates the ammonia from a low-pressure receiver to the evaporators. The low-pressure receiver is a shell for storing refrigerant at low pressure and is used to supply evaporators with refrigerant, either by gravity or by a low-pressure pump. It also takes the suction from the evaporators and separates the gas from the liquid. Because the amount of liquid fed into the evaporator is usually several times the amount that actually evaporates there, liquid is always present in the suction return to the low-pressure receiver. Frequently, three times the evaporated amount is circulated through the evaporator (see Chapter 1).

Generally, the liquid ammonia pump is sized by the flow rate required and a pressure differential of about 170 kPa. This is satisfactory for most single-story installations. If there is a static lift on the pump discharge, the differential is increased accordingly.

The low-pressure receiver should be sized by the cross-sectional area required to separate liquid and gas and by the volume between

System Practices for Ammonia Refrigerant

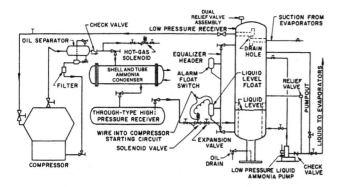

Fig. 41 Piping for Single-Stage System with Low-Pressure Receiver and Liquid Ammonia Recirculation

the normal and alarm liquid levels in the low-pressure receiver. This volume should be sufficient to contain the maximum fluctuation in liquid from the various load conditions (see Chapter 1).

The liquid at the pump discharge is in the subcooled region. A total pressure drop of about 35 kPa in the piping can be tolerated.

The remaining pressure is expended through the control valve and coil. The pressure drop and heat pickup in the liquid supply line should be low enough to prevent flashing in the liquid supply line.

Provisions for liquid relief from the liquid main back to the low-pressure receiver are required, so that when the liquid line solenoid valves at the various evaporators are closed, either for defrosting or for temperature control, the excess liquid can be relieved back to the receiver. Generally, relief valves used for this purpose are set at about 275 kPa differential when positive displacement pumps are used. When centrifugal pumps are used, a hand expansion valve or a minimum flow orifice is acceptable to ensure that the pump is not dead-headed.

The suction header between the evaporators and the low-pressure receiver should be pitched 1% to permit excess liquid flow back to the low-pressure receiver. The header should be designed to avoid traps.

Liquid Recirculation in Single-Stage System. Figure 41 shows the piping of a typical single-stage system with a low-pressure receiver and liquid ammonia recirculation feed.

Hot-Gas Defrost

This section was taken from a technical paper by Briley and Lyons (1992). Several methods are used for defrosting coils in areas below 2°C room temperature. These include

- Hot refrigerant gas (the predominant method)
- Water
- Air
- Combinations of hot gas, water, and air

The evaporator (air unit) in a liquid recirculation system is circuited so that the refrigerant flow provides maximum cooling efficiency. The evaporator can also work as a condenser if the necessary piping and flow modifications are made. When the evaporator is operated as a condenser and the fans are shut down, the hot refrigerant vapor raises the surface temperature of the coil enough to melt any ice and/or frost on the surface so that it drains off. Although this method is effective, it can be troublesome and inefficient if the piping system is not properly designed.

Even when the fans are not operating, up to 50% or more of the heat given up by the refrigerant vapor is lost to the space. Because the rate of heat transfer varies with the temperature difference between the coil surface and the room air, the temperature/pressure of the refrigerant during defrost should be minimized.

Another reason to maintain the lowest possible defrost temperature/pressure, particularly in freezers, is to keep the coil from steaming. Steam increases the refrigeration load, and the resulting icicle or frost formation must be dealt with. Icicles increase maintenance during cleanup; ice formed during defrost tends to collect at the fan rings, which sometimes restricts fan operation.

Defrosting takes slightly longer at lower defrost pressures. The shorter the time heat is added to the space, the more efficient is the defrost. However, with slightly extended defrost times at lower temperature, the overall defrosting efficiency is much greater than at higher temperature/pressure because refrigeration requirements are reduced.

Another loss during defrost can occur when hot, or uncondensed, gas blows through the coil and the relief regulator and vents back to the compressor. Some of this gas load cannot be contained and must be vented to the compressor through the wet return line. It is most energy-efficient to vent this hot gas to the highest suction possible; an evaporator defrost relief should be vented to the intermediate or high-stage compressor if the system is two-stage. See Figure 42 for a conventional hot-gas defrost system for evaporator coils of 50 kW of refrigeration and below. Note that the wet return is above the evaporator and that a single riser is employed.

Defrost Control. Because the efficiency of defrosting is low, frequency and duration of defrosting should be kept to the minimum necessary to keep the coils clean. Less defrosting is required during the winter than during hotter, more humid periods. An effective energy-saving measure is to reset defrost schedules in the winter.

Several methods are used to initiate the defrost cycle. **Demand defrost**, actuated by a pressure device that measures the air pressure drop across the coil, is a good way of minimizing total daily defrost time. The coil is defrosted automatically only when necessary. Demand initiation, together with a float drainer to dump the liquid formed during defrost to an intermediate vessel, is the most efficient defrost system available. See Figure 43 for details.

The most common defrost control method, however, is **time-initiated, time-terminated**; it includes adjustable defrost duration and an adjustable number of defrost cycles per 24 h period. This control is commonly provided by a defrost timer.

Estimates indicate that the load placed on a refrigeration system by a coil during defrost is up to three times the operating design load. Thus, it is important to properly engineer hot-gas defrost systems.

Designing Hot-Gas Defrost Systems. Several approaches are followed in designing hot-gas defrost systems. Figure 43 shows a typical demand defrost system for both upfeed and downfeed coils. This design returns the defrost liquid to the system's intermediate pressure. An alternative is to direct the defrost liquid into the wet suction. A float drainer or thermostatic trap with a hot-gas regulator installed at the hot-gas inlet to the coil is much better than the relief regulator (see Figure 43).

Most defrost systems installed today (Figure 42) use a time clock to initiate defrost; the demand defrost system shown in Figure 43 uses a low differential pressure switch to sense the air pressure drop across the coil and actuate the defrost. A thermostat terminates the defrost cycle. A timer is used as a backup to make sure the defrost terminates.

Sizing and Designing Hot-Gas Piping. Hot gas is supplied to the evaporators in two ways:

1. The preferred method is to install a pressure regulator set at approximately 700 kPa (gage) in the equipment room at the hot-gas takeoff and size the piping accordingly.
2. The alternative is to install a pressure regulator at each evaporator or group of evaporators and size the piping for minimum design condensing pressure, which should be 500 to 600 kPa (gage).

A maximum of one-third of the coils in a system should be defrosted at one time. If a system has 900 kW of refrigeration capacity, the main hot-gas supply pipe could be sized for 300 kW of

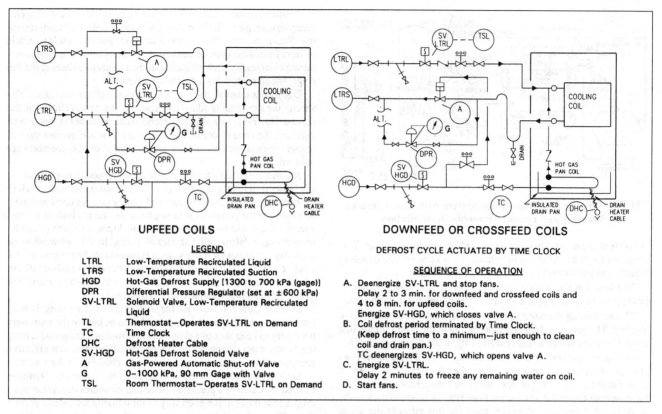

Fig. 42 Conventional Hot Gas Defrost Cycle
(For coils with 50 kW refrigeration capacity or below)

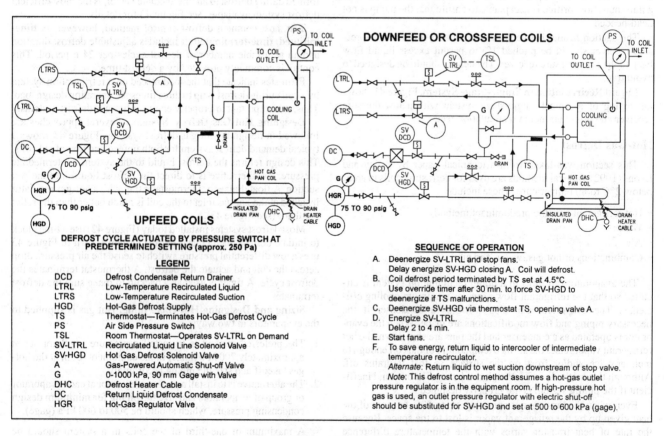

Fig. 43 Demand Defrost Cycle
(For coils with 50 kW refrigeration capacity or below)

System Practices for Ammonia Refrigerant

refrigeration. The outlet pressure-regulating valve should be sized in accordance with the manufacturer's data.

Reducing the defrost hot-gas pressure in the equipment room has advantages, notably that less liquid will condense in the hot-gas line as the condensing temperature is reduced to 11 to 18°C. A typical equipment room hot-gas pressure control system is shown in Figure 44. If hot-gas lines in the system are trapped, a condensate drainer must be installed at each trap and at the low point in the hot-gas line (Figure 45). Defrost condensate liquid return piping from coils where a float or thermostatic valve is used should be one size larger than the liquid feed piping to the coil.

Demand Defrost. The following are advantages and features of demand defrost:

1. It uses the least energy for defrost.
2. It increases total system efficiency because coils are off-line for a minimum amount of time.
3. It imposes less stress on the piping system because there are fewer defrost cycles.
4. Regulating the hot gas to approximately 700 kPa (gage) in the equipment room gives the gas less chance of condensing in the supply piping. Liquid in hot-gas systems may cause problems due to the hydraulic shock created when the liquid is forced into an evaporator (coil). Coils in hot-gas pans may rupture as a result.
5. Draining the liquid formed during defrost with a float or thermostatic drainer eliminates hot-gas blowby normally associated with pressure-regulating valves installed around the wet suction return line pilot check valve.
6. Returning the ammonia liquid to the intercooler or high-stage recirculator saves considerable energy in a system. A 70 kW refrigeration coil defrosting for 12 min can condense up to 11 kg/min of ammonia, or 132 kg total. The enthalpy difference between returning to the low-stage recirculator (−40°C) and the intermediate recirculator (−7°C) is 148 kJ/kg or 19.5 MJ total or 27 kW of refrigeration removed from the −40°C booster for 12 min. This assumes that only liquid is drained and is the saving when liquid is drained to the intermediate point, not the total cost to defrost. If a pressure-reducing valve is used around the pilot check valve, this rate could double or triple because hot gas flows through these valves in unmeasurable quantities.

Soft Hot-Gas Defrost System. The soft hot-gas defrost system is particularly well suited to large evaporators and should be used on all coils of 50 kW of refrigeration or over. The system eliminates the valve clatter, pipe movements, and noise associated with large coils during hot-gas defrost. Soft hot-gas defrost can be used for upfeed or downfeed coils; however, the piping systems differ (Figure 46). Coils operated in the horizontal plane must be orificed. Vertical coils that usually are cross-fed are also orificed.

The soft hot-gas defrost system is designed to increase coil pressure gradually as defrost is initiated. This is accomplished by a small hot-gas feed having a capacity of about 25 to 30% of the estimated duty with a solenoid and a hand expansion valve adjusted to bring the pressure up to about 275 kPa (gage) in 3 to 5 min. (Refer to Sequence of Operation in Figure 46.) After defrost, a small suction line solenoid is opened so that the coil can be brought down to operation pressure gradually before liquid is introduced and the fans started. The system can be initiated by a pressure switch; however, for large coils in spiral or individual quick freezing systems, manual initiation is preferred.

The soft hot-gas defrost system eliminates check valve chatter and most, if not all, liquid hammer (i.e., hydraulic problems in the piping). In addition, features 4, 5, and 6 listed in the section on Demand Defrost apply to the soft hot-gas defrost system.

Double Riser Designs for Large Evaporator Coils

Static pressure penalty is the pressure/temperature loss associated with a refrigerant vapor stream bubbling through a liquid bath.

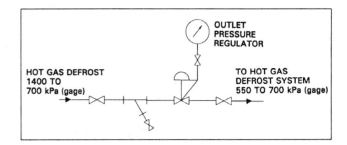

Fig. 44 Equipment Room Hot-Gas Pressure Control System

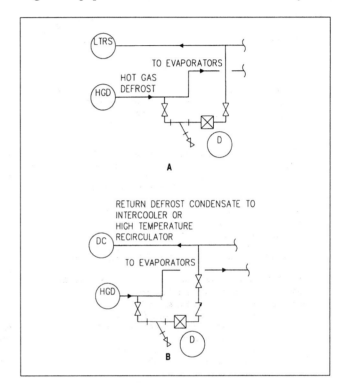

Fig. 45 Hot-Gas Condensate Return Drainer

If the velocity in the riser is high enough, it will carry over a certain amount of liquid, thus reducing the penalty. For example, at −40°C ammonia has a density of 689.9 kg/m^3, which is equivalent to a pressure of 689.9(9.807 m/s^2)/1000 = 6.77 kPa per metre of depth. Thus, a 5 m riser has a column of liquid that exerts 5 × 6.77 = 33.8 kPa. At −40°C, ammonia has a saturation pressure of 71.7 kPa. At the bottom of the riser then, the pressure is 33.8 + 71.7 = 105.5 kPa, which is the saturation pressure of ammonia at −33°C. This 7 K difference amounts to a 1.4 K penalty per metre of riser. If a riser were oversized to the point that the vapor did not carry the liquid to the wet return, the evaporator would be at −33°C instead of −40°C. This problem can be solved in several ways:

1. Install the low-temperature recirculated suction (LTRS) line below the evaporator. This method is very effective for downfeed evaporators. The suction from the coil should not be trapped. This piping arrangement also ensures lubricant return to the recirculator.
2. Where the LTRS is above the evaporator, install a liquid return system below the evaporator (Figure 47). This arrangement eliminates static penalty, which is particularly advantageous for plate, individual quick freeze, and spiral freezers.
3. Use double risers from the evaporator to the LTRS (Figure 48).

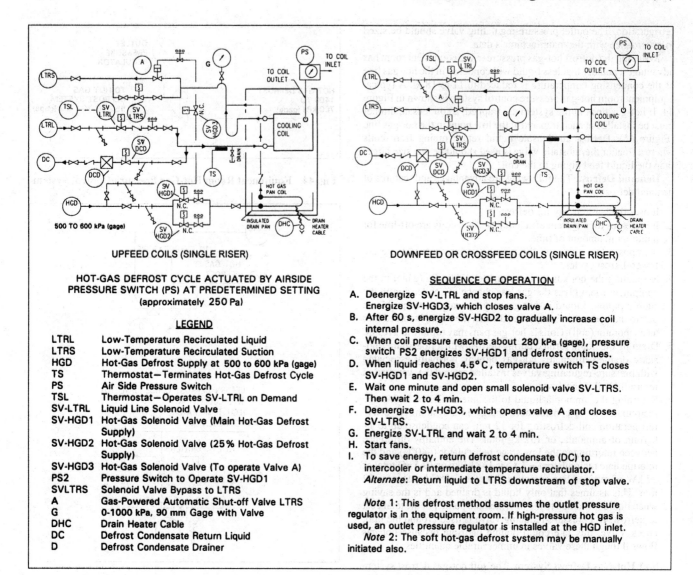

Fig. 46 Soft Hot-Gas Defrost Cycle
(For coils with 50 kW refrigeration capacity or above)

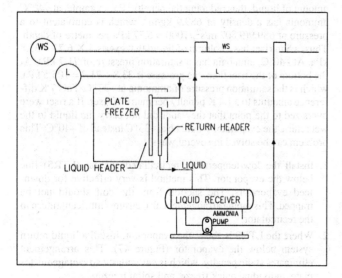

Fig. 47 Recirculated Liquid Return System

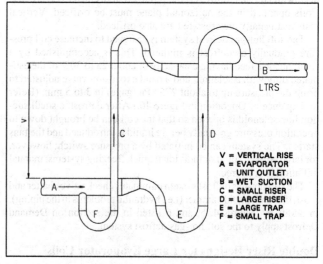

Fig. 48 Double Low-Temperature Suction Risers

System Practices for Ammonia Refrigerant

If a single riser is sized for minimum pressure drop at full load, the static pressure penalty is excessive at part load, and lubricant return could be a problem. If the single riser is sized for minimum load, then the pressure drop in the riser is excessive and counterproductive.

Double risers solve these problems (Miller 1979). Figure 48 shows that when maximum load occurs, both risers return vapor and liquid to the wet suction. At minimum load, the large riser is sealed by the liquid ammonia in the large trap, and the refrigerant vapor flows through the small riser. A small trap on the small riser ensures that some lubricant and liquid return to the wet suction.

The risers should be sized so that the pressure drop, calculated on a dry gas basis, is at least 70 Pa/m. The larger riser is designed for approximately 65 to 75% of the flow and the small one for the remainder. This design results in a velocity of approximately 25 m/s or higher. Some coils may require three risers (large, medium, and small).

Over the years, freezers have continued to grow in capacity. As they became larger, so did the evaporators (coils). Where these freezers are in line and the product to be frozen is wet, the defrost cycle can be every 4 or 8 h. Many production lines limit defrost duration to 30 min. If the coils are large (some coils have a refrigeration capacity of 700 to 1000 kW), it is difficult to design a hot-gas defrost system that can complete a safe defrost in 30 min. Sequential defrost systems, where the coils are defrosted alternately during production, are feasible but require special treatment.

SAFETY CONSIDERATIONS

Ammonia is an economical choice for industrial systems. While ammonia has superior thermodynamic properties, it is considered toxic at low concentration levels of 35 to 50 mg/kg. Large quantities of ammonia should not be vented to enclosed areas near open flames or heavy sparks. Ammonia at 16 to 25% by volume burns and can explode in air in the presence of an open flame.

The importance of ammonia piping is sometimes minimized when the main emphasis is on selecting major equipment pieces. Mains should be sized carefully to provide low pressure drop and avoid capacity or power penalties caused by inadequate piping.

Rusting pipes and vessels in older systems containing ammonia can create a safety hazard. Oblique X-ray photographs of welded pipe joints and ultrasonic inspection of vessels may be used to disclose defects. Only vendor-certified parts for pipe, valving, and pressure-containing components according to designated assembly drawings should be used to reduce hazards. Cold liquid refrigerant should not be confined between closed valves in a pipe where the liquid can warm and expand to burst piping components. Rapid multiple pulsations of ammonia liquid in piping components (e.g., those developed by cavitation forces or hydraulic hammering from compressor pulsations with massive slugs of liquid carryover to the compressor) must be avoided to prevent equipment and piping damage and injury to personnel.

Most service problems are caused by inadequate precautions during design, construction, and installation (IIAR *Standard* 2, ASHRAE *Standard* 15). Ammonia is a powerful solvent that removes dirt, scale, sand, or moisture remaining in the pipes, valves, and fittings during installation. These substances are swept along with the suction gas to the compressor, where they are a menace to the bearings, pistons, cylinder walls, valves, and lubricant. Most compressors are equipped with suction strainers and/or additional disposable strainer liners for the large quantity of debris that can be present at initial start-up.

Moving parts are often scored when a compressor is run for the first time. Damage starts with minor scratches, which increase progressively until they seriously affect the operation of the compressor or render it inoperative.

A system that has been carefully and properly installed with no foreign matter or liquid entering the compressor will operate satisfactorily for a long time. As piping is installed, it should be power rotary wire brushed and blown out with compressed air. The piping system should be blown out again with compressed air or nitrogen before evacuation and charging. See ASHRAE *Standard* 15 for system piping test pressure.

REFERENCES

ASHRAE. 1994. Safety code for mechanical refrigeration. ANSI/ASHRAE *Standard* 15-1994.

ASME. 1992. Refrigeration piping. ANSI/ASME *Standard* B31.5-92. American Society of Mechanical Engineers, New York.

ASME. 1995. Rules for construction of power boilers. *Boiler and pressure vessel code*, Section VIII, Division 1-95.

ASTM. 1995. Specification for seamless carbon steel pipe for high-temperature service. ANSI/ASTM *Standard* A 106-95. American Society for Testing and Materials, West Conshohocken, PA.

ASTM. 1996. Specification for pipe, steel, black and hot-dipped, zinc-coated, welded and seamless. ANSI/ASTM *Standard* A 53-96.

Briley, G.C. and T.A. Lyons. 1992. Hot gas defrost systems for large evaporators in ammonia liquid overfeed systems. IIAR *Technical Paper* 163.

Frick Co. 1995. Thermosyphon oil cooling. *Bulletin* E70-900Z (August). Frick Company, Waynesboro, PA.

IIAR. 1988. Minimum safety criteria for a safe ammonia refrigeration system. *Bulletin* 109. International Institute of Ammonia Refrigeration, Washington, DC.

IIAR. 1992. Equipment, design, and installation of ammonia mechanical refrigeration systems. ANSI/IIAR *Standard* 2-1992.

IIAR. 1992. Guidelines for avoiding component failure in industrial refrigeration systems caused by abnormal pressure or shock. *Bulletin* 116.

Miller, D.K. 1979. Sizing dual-suction risers in liquid overfeed refrigeration systems. *Chemical Engineering* (September 24).

NCPWB. Welding procedure specifications. National Certified Pipe Welding Bureau, Rockville, MD.

Stoecker, W.F. 1988. Chapters 8 and 9 in *Industrial refrigeration*. Business News Publishing Company, Troy, MI.

Timm, M.L. 1991. An improved method for calculating refrigerant line pressure drops. *ASHRAE Transactions* 97(1):194-203.

Wile, D.D. 1977. *Refrigerant line sizing*. ASHRAE RP 185.

BIBLIOGRAPHY

BAC. 1983. *Evaporative condenser engineering manual*. Baltimore Aircoil Company, Baltimore, MD.

Bradley, W.E. 1984. Piping evaporative condensers. In Proceedings of IIAR meeting. International Institute of Ammonia Refrigeration, Chicago.

Cole, R.A. 1986. Avoiding refrigeration condenser problems. *Heating/Piping/and Air-Conditioning*, Parts I and II, 58(7, 8).

Loyko, L. 1989. Hydraulic shock in ammonia systems. IIAR *Technical Paper* T-125. International Institute of Ammonia Refrigeration, Washington, DC.

Nuckolls, A.H. The comparative life, fire, and explosion hazards of common refrigerants. Miscellaneous Hazard No. 2375. Underwriters Laboratory, Northbrook, IL.

Strong, A.P. 1984. Hot gas defrost—A one-a-more-a-time. IIAR *Technical Paper* T-53. International Institute of Ammonia Refrigeration, Washington, DC.

CHAPTER 4

SECONDARY COOLANTS IN REFRIGERATION SYSTEMS

Coolant Selection 4.1
Design Considerations 4.2
Applications 4.6

A SECONDARY coolant is a liquid that is used as a heat transfer fluid and that changes temperature as it gains or loses heat energy without changing into another phase. For the lower temperatures of refrigeration, this requires a coolant with a freezing point below that of water. In this chapter, the design considerations for components, system performance requirements, and applications for secondary coolants are discussed. Related information can be found in Chapters 2, 3, 19, 20, and 33 of the 1997 *ASHRAE Handbook—Fundamentals*.

COOLANT SELECTION

A secondary coolant must be compatible with the other materials in the system at the pressures and temperatures encountered for maximum component reliability and operating life. The coolant should also be compatible with the environment and the applicable safety regulations, and it should be economical to use and replace.

The coolant should have a minimum freezing point of 3 K below and preferably 8 K below the lowest temperature to which it will be exposed. When subjected to the lowest temperature in the system, the viscosity of the coolant should be low enough to allow satisfactory heat transfer and reasonable pressure drop.

The vapor pressure of the coolant should not exceed that allowed at the maximum temperature encountered. To avoid a vacuum in a low vapor pressure secondary coolant system, the coolant can be pressurized with pressure-regulated dry nitrogen in the expansion tank. However, some special secondary coolants such as those used for computer circuit cooling have a high solubility for nitrogen and must therefore be isolated from the nitrogen with a suitable diaphragm.

Load Versus Flow Rate

The secondary coolant pump is usually in the return line upstream of the chiller. Therefore, to be accurate, the pumping rate in gallons per minute is based on the density at the return temperature. The mass flow rate for a given heat load is based on the desired temperature range and required coefficient of heat transfer at the average bulk temperature.

To determine heat transfer and pressure drop, the density, specific heat, viscosity, and thermal conductivity are based on the average bulk temperature of the coolant in the heat exchanger, noting that film temperature corrections are based on the average film temperature. Trial solutions of the secondary coolant side coefficient compared to the overall coefficient and the total log mean temperature difference (LMTD) determine the average film temperature. Where the secondary coolant is cooled, the more viscous film reduces the heat transfer rate and raises the pressure drop compared to what can be expected at the bulk temperature. Where the secondary coolant is heated, the less viscous film approaches the heat transfer rate and pressure drop expected at the bulk temperature.

The greater the amount of turbulence and mixing of the bulk and film, the better the heat transfer and the higher the pressure drop.

Where secondary coolant velocity in the tubes of a heat transfer device results in laminar flow, the heat transfer can be improved by inserting spiral tapes or spring turbulators that promote mixing the bulk and film. This usually increases pressure drop. The inside surface can also be spirally grooved or augmented by other devices. Since the state of the art of heat transfer is constantly improving, use the most cost-effective heat exchanger to provide optimum heat transfer and pressure drop. Energy costs for pumping the secondary coolant must be considered when selecting the fluid to be used and the heat exchangers to be installed.

Pumping Cost

Pumping costs are a function of the secondary coolant selected, the load and temperature range where energy is transferred, the pump head required by the system pressure drop (including that of the chiller), the mechanical efficiencies of the pump and driver, and the electrical efficiency and power factor where the driver is an electric motor. Small centrifugal pumps, operating in the range of approximately 3 L/s at 240 kPa to 9 L/s at 210 kPa, for 60 Hz applications, typically have 45 to 65% efficiency, respectively. Larger pumps, operating in the range of 30 L/s at 240 kPa to 95 L/s at 210 kPa, for 60 Hz applications, typically have 75 to 85% efficiency, respectively.

A pump should operate near its peak operating efficiency for the flow rate and head that usually exist. The secondary coolant temperature increases slightly from the energy expended at the pump shaft. If a semihermetic electric motor is used as the driver, the motor inefficiency is added as heat to the secondary coolant, and the total kilowatt input to the motor must be considered in establishing load and temperatures.

Performance Comparisons

Assuming that the total refrigeration load at the evaporator includes the pump motor input and brine line insulation heat gains, as well as the delivered beneficial cooling, tabulating typical secondary coolant performance values assists in the coolant selection. A 27-mm ID smooth steel tube evaluated for pressure drop and internal heat transfer coefficient at the average bulk temperature of $-6.7°C$ and a temperature range of 5.6 K for 2.1 m/s tube-side velocity provides comparative data (see Table 1) for some typical coolants. Table 2 ranks the same coolants comparatively, using data from Table 1.

For a given evaporator configuration, load, and temperature range, select a secondary coolant that gives satisfactory velocities, heat transfer, and pressure drop. At the $-6.7°C$ level, hydrocarbon and halocarbon secondary coolants must be pumped at a rate of 2.3 to 3.0 times the rate of water-based secondary coolants for the same temperature range.

Higher pumping rates require larger coolant lines to keep the pressure and brake power requirement for the pump within reasonable limits. Table 3 lists approximate ratios of pump power for secondary coolants. The heat transferred by a given secondary coolant affects the cost and perhaps the configuration and pressure drop of a chiller and other heat exchangers in the system; therefore, Tables 2 and 3 are only guides of the relative merits of each coolant.

The preparation of this chapter is assigned to TC 10.1, Custom Engineered Refrigeration Systems.

Table 1 Secondary Coolant Performance Comparisons

Secondary Coolant	Concentration (by Mass), %	Freeze Point, °C	L/(s·kW)[c]	Pressure Drop,[a] kPa	Heat Transfer Coefficient[b] h_i, W/(m²·K)
Propylene glycol	39	−20.6	0.0459	20.064	1164
Ethylene glycol	38	−21.6	0.0495	16.410	2305
Methanol	26	−20.7	0.0468	14.134	2686
Sodium chloride	23	−20.6	0.0459	15.858	3169
Calcium chloride	22	−22.1	0.0500	16.685	3214
Aqua-ammonia	14	−21.7	0.0445	16.823	3072
Trichloroethylene	100	−86.1	0.1334	14.548	2453
d-Limonene	100	−96.7	0.1160	10.204	1823
Methylene chloride	100	−96.7	0.1146	12.824	3322
R-11	100	−111.1	0.1364	14.341	2430

[a]Based on one length of 4.9 m tube with 26.8 mm ID and use of Moody Chart (1944) for an average velocity of 2.13 m/s. Input/output losses equal $V^2\rho/2$ for 2.13 m/s velocity. Evaluations are at a bulk temperature of −6.7°C and a temperature range of 5.6 K.

[b]Based on a curve fit equation for Kern's adaptation (1950) of Sieder and Tate heat transfer equation (1936) using a 4.9-m tube for $L/D = 181$ and a film temperature of 2.8 K lower than average bulk temperature with 2.13 m/s velocity.

[c]Based on inlet secondary coolant temperature at the pump of 3.9°C.

Table 2 Comparative Ranking of Heat Transfer Factors at 2 m/s[a]

Secondary Coolant	Heat Transfer Factor
Propylene glycol	1.000
d-Limonene	1.566
Ethylene glycol	1.981
R-11	2.088
Trichloroethylene	2.107
Methanol	2.307
Aqua-ammonia	2.639
Sodium chloride	2.722
Calcium chloride	2.761
Methylene chloride	2.854

[a]Based on Table 1 values using 27-mm ID tube 4.9 m long. The actual ID and length vary according to the specific loading and refrigerant applied with each secondary coolant, tube material, and surface augmentation.

Table 3 Relative Pumping Energy Required[a]

Secondary Coolant	Energy Factor
Aqua-ammonia	1.000
Methanol	1.078
Propylene glycol	1.142
Ethylene glycol	1.250
Sodium chloride	1.295
Calcium chloride	1.447
d-Limonene	2.406
Methylene chloride	3.735
Trichloroethylene	4.787
R-11	5.022

[a]Based on the same pump pressure, refrigeration load, −6.7°C average temperature, 6 K range, and the freezing point (for water-based secondary coolants) 11 to 13 K below the lowest secondary coolant temperature.

Other Considerations

Corrosion must be considered when selecting the coolant, an inhibitor, and the system components. The effect of secondary coolant and inhibitor toxicity on the health and safety of plant personnel or consumers of food and beverages must be considered. The flash point and explosive limits of secondary coolant vapors must also be evaluated.

Examine the secondary coolant stability for anticipated moisture, air, and contaminants at the temperature limits of materials used in the system. The skin temperatures of the hottest elements determine the secondary coolant stability.

If defoaming additives are necessary, their effect on the thermal stability and toxic properties of the coolant must be considered for the application.

DESIGN CONSIDERATIONS

The secondary coolant vapor pressure at the lowest operating temperature determines whether a vacuum could exist in the secondary coolant system. To keep air and moisture out of the system, pressure-controlled dry nitrogen can be applied to the top level of secondary coolant (e.g., in the expansion tank or a storage tank). The gas pressure over the coolant plus the pressure created at the lowest point in the system by the maximum vertical height of coolant determine the minimum internal pressure for design purposes. The coincident highest pressure and lowest secondary coolant temperature dictate the design working pressure (DWP) and material specifications for the components.

To select proper relief valve(s) with settings based on the system DWP, the highest temperatures to which the secondary coolant could be subjected should be considered. This temperature would occur in case of heat radiation from a fire in the area or the normal warming of the valved-off sections. Normally, a valved-off section is relieved to an unconstrained portion of the system and the secondary coolant can expand freely without loss to the environment.

Safety considerations for the system are found in ASHRAE *Standard* 15, *Safety Code for Mechanical Refrigeration*. The design standards for pressure piping can be found in ASME *Standard* B31.5, and the design standards for pressure vessels can be found in Section VIII of the ASME *Boiler and Pressure Vessel Code*.

Piping and Control Valves

Piping should be sized for reasonable pressure drop using the calculation methods in Chapters 2 and 33 of the 1997 *ASHRAE Handbook—Fundamentals*. Balancing valves or orifices in each of the multiple feed lines help distribute the secondary coolant. A reverse-return piping arrangement balances the flow. Control valves that vary the flow are sized for 20 to 80% of the total friction pressure drop through the system for proper response and stable operation. Valves sized for pressure drops smaller than 20% may respond too slowly to a control signal for a flow change. Valves sized for pressure drops in excess of 80% can be too sensitive, causing control cycling and instability.

Storage Tanks

Storage tanks can shave peak loads for brief periods, limit the size of the refrigeration equipment, and reduce energy costs. In off-peak hours, a relatively small refrigeration plant cools a secondary coolant stored for later use. A separate circulating pump sized for the maximum flow needed by the peak load is started to satisfy the peak load. Energy cost savings are enhanced if the refrigeration equipment is used to cool secondary coolant at night, when the cooling medium for heat rejection is generally at the lowest temperature.

Secondary Coolants in Refrigeration Systems

The load profile over 24 h and the temperature range of the secondary coolant determine the minimum net capacity required for the refrigeration plant, the sizes of the pumps, and the minimum amount of secondary coolant to be stored. For maximum use of the storage tank volume at the expected temperatures, choose inlet velocities and locate the connections and the tank for maximum stratification. Note, however, that maximum use will probably never exceed 90% and, in some cases, may equal only 75% of the tank volume.

Example 1. Figure 1 depicts the load profile and Figure 2 shows the arrangement of a refrigeration plant with storage of a 23% (by mass) sodium chloride secondary coolant at a nominal −6.7°C. During the peak load of 176 kW, a range of 4.4 K is required. At an average temperature of −4.4°C, with a range of 4.4 K, the specific heat of the coolant c_p is 3.314 kJ/(kg·K). At −2.2°C, the density of coolant at the pump (ρ_L) is 1183 kg/m^3; at −6.7°C, the ρ_L is 1185 kg/m^3.

Determine the minimum size storage tank for 90% use, the minimum capacity required for the chiller, and the sizes of the two pumps. The chiller and the chiller pump run continuously. The secondary coolant storage pump runs only during the peak load. A control valve to the load source diverts all coolant to the storage tank during a zero load condition, so that the initial temperature of −6.7°C is restored in the tank. During the low load condition, only the required flow rate for a range of 4.4 K at the load source is used; the balance returns to the tank and restores the temperature to −6.7°C.

Solution: If x is the minimum capacity of the chiller, determine the energy balance in each segment by subtracting the load in each segment from x. Then multiply the result by the time length of the respective segments, and add as follows:

$$6(x - 0) + 4(x - 176) + 14(x - 31.7) = 0$$
$$x = 47.8 \text{ kW}$$

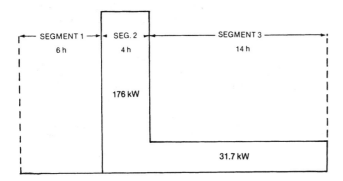

Fig. 1 Load Profile of Refrigeration Plant Where Secondary Coolant Storage Can Save Energy

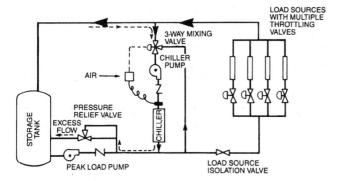

Fig. 2 Arrangement of System with Secondary Coolant Storage

Calculate the secondary coolant flow rate W at peak load:

$$W = 176/(3.314 \times 4.4) = 12.07 \text{ kg/s}$$

For the chiller at 52.8 kW, the secondary coolant flow rate is

$$W = 52.8/(3.314 \times 4.4) = 3.62 \text{ kg/s}$$

Therefore, the coolant flow rate to the storage tank pump is 12.07 − 3.62 = 8.48 kg/s. The chiller pump size is determined by

$$1000 \times 3.62/1183 = 3.06 \text{ L/s}$$

Calculate the storage tank pump size as follows:

$$1000 \times 8.48/1185 = 7.16 \text{ L/s}$$

Using the concept of stratification in the storage tank, the interface between warm return and cold stored secondary coolant falls at the rate pumped from the tank. Since the time segments fix the total amount pumped and the storage tank pump operates only in segment 2 (see Figure 1), the minimum tank volume V at 90% use is determined as follows:

Total mass = 8.48 kg/s × 4 h × 3600 s/h/0.9 = 135 700 kg

and

$$V = 135\,700/1185 = 114.5 \text{ m}^3$$

A larger tank (e.g., 190 m^3) provides flexibility for longer segments at peak load and accommodates potential mixing. It may be desirable to insulate and limit heat gains to 2.3 kW for the tank and lines. Energy use for pumping can be limited by designing for 160 kPa. With the smaller pump operating at 51% efficiency and the larger pump at 52.5% efficiency, the pump heat added to the secondary coolant would be 970 and 2190 W, respectively.

For cases with various time segments and their respective loads, the maximum load for segment 1 or 3 with the smaller pump operating cannot exceed the net capacity of the chiller minus insulation and pump heat gain to the secondary coolant. For various combinations of segment time lengths and cooling loads, the recovery or restoration rate of the storage tank to the lowest temperature required for satisfactory operation should be considered.

Figure 2 depicts a system arrangement with secondary coolant storage as described in Example 1.

As load source circuits shut off, the excess flow is bypassed back to the storage tank. The temperature setting of the 3-way valve is the normal return temperature for full flow through the load sources.

When only the storage tank requires cooling, the flow is as shown by the dotted lines with the load source isolation valve closed. When the storage tank temperature is at the desired level, the load isolation valve can be opened to allow cooling of the piping loops to and from the load sources for full restoration of storage cooling capacity.

Expansion Tanks

Figure 3 shows a typical closed secondary coolant system without a storage tank; it also illustrates different control strategies. The reverse-return piping assists flow balance. Figure 4 shows a secondary coolant strengthening unit for salt brines. The secondary coolant expansion tank volume is determined by considering the total coolant inventory and the differences in coolant density at the lowest temperature of coolant pumped to the load location (t_1) and the maximum temperature. The expansion tank is sized to accommodate a residual volume with the system coolant at t_1, plus an expansion volume and vapor space above the coolant. A vapor space equal to 20% of the expansion tank volume should be adequate. A level indicator, used to prevent overcharging, is calibrated at the residual volume level versus lowest system secondary coolant temperature.

Example 2. Assume a 190 m^3 charge of 23% sodium chloride secondary coolant at t_1 of −6.7°C in the system. If 37.8°C is the maximum temperature, determine the size of the expansion tank required. Assume

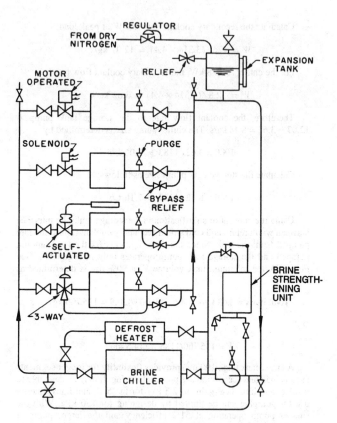

Fig. 3 Typical Closed Salt Brine System

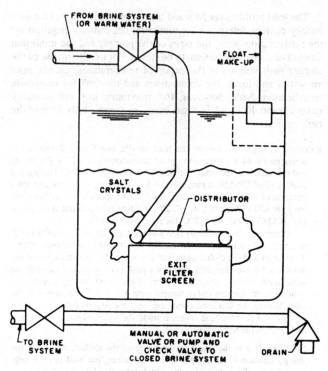

Fig. 4 Brine Strengthening Unit for Salt Brines Used as Secondary Coolants

that the residual volume is 10% of the total tank volume and that the vapor space at the highest temperature is 20% of the total tank volume.

$$\text{ETV} = \frac{V_S[(\rho_1/\rho_2) - 1]}{1 - (R_F + V_F)}$$

where

- ETV = expansion tank volume
- V_S = system secondary coolant volume at t_1 temperature
- ρ_2 = density at t_1
- ρ_1 = density at maximum temperature
- R_F = residual volume of tank liquid (low level) at t_1, expressed as a fraction
- V_F = volume of vapor space at highest temperature, expressed as a fraction

If the density of the secondary coolant is 1185 kg/m³ at −6.7°C and 1155 kg/m³ at 38°C, the tank volume is

$$\text{ETV} = \frac{190[(1185/1155) - 1]}{1 - (0.10 + 0.20)} = 7.05 \text{ m}^3$$

Pulldown Time

Example 1 is based on a static situation of secondary coolant temperature at two different loads—normal and peak. The length of time for pulldown from 37.8°C to the final −6.7°C may need to be calculated. For a graphical solution, required heat extraction versus secondary coolant temperature is plotted. Then, by iteration, the pulldown time is solved by finding the net refrigeration capacity for each increment of coolant temperature change. A mathematical method may also be used.

The 52.8 kW refrigeration system cited in the examples has a 105.7 kW capacity at a maximum of 10°C saturated suction temperature (STP). For pulldown, a compressor suction pressure regulator (holdback valve) is sometimes used. The maximum secondary coolant temperature must be determined when the holdback valve is wide open and the STP is at 10°C. For Example 1, this is at 21°C coolant temperature. As the coolant temperature is further reduced with a constant 3.0 L/s, the capacity of the refrigeration system is gradually reduced until a 52.8 kW capacity is reached with −3.3°C coolant in the tank. Further cooling to −6.7°C will be at reduced capacity.

Temperatures of the secondary coolant mass, storage tanks, piping, cooler, pump, and insulation must all be reduced. In Example 1, as the coolant is reduced in temperature from 37.8 to −6.7°C, the total heat removed from these items is as follows:

Brine Temperature, °C	Total Heat Removed, GJ
37.8	33.27
26.7	24.92
21.1	20.76
15.6	16.59
4.4	8.29
−6.7	0

From a secondary coolant temperature of 37.8 to 21.1°C, the refrigeration system capacity is fixed at 105.7 kW, and the time for pulldown is essentially linear (system net kW for pulldown is less than the compressor capacity because of heat gain through insulation and added pump heat). In Example 1, the pump heat was not considered. When recognizing the variable heat gain for a 35°C ambient, and the pump heat as the secondary coolant temperature is reduced, the following net capacity is available for pulldown at the various secondary coolant temperatures:

Brine Temperature, °C	Net Capacity, kW
37.8	105.1
26.7	104.1
21.1	103.6
15.6	89.0
4.4	62.7
−3.3	49.6
−6.7	44.7

Secondary Coolants in Refrigeration Systems

A curve fit shows capacity is a straight line between the values for 37.8 and 21.1°C. Therefore, the pulldown time for this interval is

$$\theta = \frac{(33.27 - 20.76)10^6}{0.5(105.1 + 103.6)3600} = 33.3 \text{ h}$$

From 21.1 to −6.7°C, the capacity curve fits a second-degree polynomial equation as follows:

$$q = 54.868 + 1.713t + 0.02880t^2$$

where

t = secondary coolant temperature, °C
q = capacity for pulldown, kW

Using the arithmetic average pulldown net capacity from 21.1 to −6.7°C, the time interval would be

$$\theta = \frac{20.76 \times 10^6}{0.5(103.6 + 44.7)(3600)} = 77.8 \text{ h}$$

If the logarithmic (base e) mean average net capacity for this temperature interval is used, the time is

$$\theta = \frac{20.76 \times 10^6}{70.08 \times 3600} = 82.3 \text{ h}$$

This is a difference of over 4.5 h and neither solution is correct. A more exact calculation uses a graphical analysis or calculus. One mathematical approach determines the heat removed per degree of secondary coolant temperature change per kilowatt of capacity. Because the coolant's heat capacity and the heat leakage change as the temperature drops, the amount of heat removed is best determined by first fitting a curve to the data for total heat removed versus secondary coolant temperature. Then a series of iterations for secondary coolant temperature ±1 K is made as the temperature is reduced. The polynomial equations may be solved by computer or hand-held calculator with a suitable program or spreadsheet. The time for pulldown will be less if supplemental refrigeration is available for pulldown or if less secondary coolant is stored.

The correct answer is 88.2 h, which is 7% greater than the logarithmic mean average capacity and 13% greater than the arithmetic average capacity over the temperature range.

Therefore, total time for temperature pulldown from 37.8 to −6.7°C is

$$\theta = 33.3 + 88.2 = 121.5 \text{ h}$$

System Costs

Various alternatives may be evaluated to justify a new project or system modification. Means (1988) lists the installed cost of various projects. Park and Jackson (1984) and NBS (1978) discuss engineering and life-cycle cost analysis. Using the various time value of money formulas, the payback for storage tank handling of peak loads compared to large refrigeration equipment and higher energy costs can be evaluated. The trade-offs in these costs—initial, maintenance, insurance, increased secondary coolant, loss of space, and energy escalation—all must be considered.

Corrosion Prevention

Corrosion prevention requires choosing proper materials and inhibitors, routine testing for pH, and eliminating contaminants. Because potentially corrosive calcium chloride and sodium chloride salt brine secondary coolant systems are widely used, test and adjust the brine solution monthly. To replenish salt brines in a system, a concentrated solution may be better than a crystalline form, because it is easier to handle and mix.

A brine should not be allowed to change from an alkaline to an acid condition. Acids rapidly corrode the metals ordinarily used in refrigeration and ice-making systems. Calcium chloride usually contains sufficient alkali to render the freshly prepared brine slightly alkaline. When any brine is exposed to air, it gradually absorbs carbon dioxide and oxygen, which eventually make the brine slightly acid. Dilute brines dissolve oxygen more readily and generally are more corrosive than concentrated brines. One of the best preventive measures is to make a closed rather than open system, using a regulated inert gas over the surface of a closed expansion tank (see Figure 2). However, many systems, such as ice-making tanks, brine spray unit coolers, and brine spray-type carcass chill rooms, cannot be closed.

A brine pH of 7.5 for a sodium or calcium chloride system is ideal, since it is safer to have a slightly alkaline rather than a slightly acid brine. Brine system operators should check pH regularly.

If a brine is acid, the pH can be raised by adding caustic soda dissolved in warm water. If a brine is alkaline (indicating ammonia leakage into the brine), carbonic gas or chromic, acetic, or hydrochloric acid should be added. Ammonia leakage must be stopped immediately so that the brine can be neutralized.

In addition to controlling the pH, an inhibitor should be used. Generally, sodium dichromate is the most effective and economical for salt brine systems. The dichromate has a bright orange color, a granular form, and readily dissolves in warm water. Since it dissolves very slowly in cold brine, it should be dissolved in warm water and added to the brine far enough ahead of the pump so that only a dilute solution reaches the pump. The quantities recommended are: 2 kg/m^3 of calcium chloride brine, and 3.2 kg/m^3 of sodium chloride brine.

Adding sodium dichromate to the salt brine does not make it noncorrosive immediately. The process is affected by many factors, including water quality, density of the brine, amount of surface and kind of material exposed in the system, age, and temperature. Corrosion stops only when protective chromate film has built up on the surface of the zinc and other electrically positive metals exposed to the brine. No simple test is available to determine the chromate concentration. Since the protection afforded by the sodium dichromate treatment depends greatly on maintaining the proper chromate concentration in the brine, brine samples should be analyzed annually. The proper concentration for calcium chloride brine is 0.13 g/L (as $Na_2Cr_2O_7 \cdot 2H_2O$); for sodium chloride brine, it is 0.21 g/L (as $Na_2Cr_2O_7 \cdot 2H_2O$).

Since crystals and concentrated solutions of sodium dichromate can cause severe skin rash, avoid contact. If contact does occur, wash the skin immediately. *Warning: Sodium dichromate should not be used for brine spray decks, spray units, or immersion tanks where food or personnel may come in contact with the spray mist or the brine itself.*

Polyphosphate-silicate and orthophosphate-boron mixtures in water-treating compounds are useful for sodium chloride brines in open systems. However, where the rate of spray loss and dilution is very high, any treatment other than density and pH control is not economical. For the best protection of spray unit coolers, housings and fans should be of a high quality, hot-dipped galvanized construction. Stainless steel fan shafts and wheels, scrolls, and eliminators are desirable.

While the nonsalt secondary coolants described in this chapter are generally noncorrosive when used in systems for long periods, recommended inhibitors should be used, and a pH check should be performed occasionally.

Steel, iron, or copper piping should not be used to carry the salt brines. Use copper nickel or suitable plastic. Use all-steel and iron tanks if the pH is not ideal. Similarly, calcium chloride systems usually have all-iron and steel pumps and valves to prevent electrolysis in the presence of acidity. Sodium chloride systems usually have all-iron or all-bronze pumps. When the pH can be controlled in a system, brass valves and bronze fitted pumps may be satisfactory. A stainless steel pump shaft is desirable. Consider salt brine composition and temperature to select the proper rotary seal or, for dirtier systems, the proper stuffing box.

APPLICATIONS

Applications for secondary coolant systems are extensive (see Chapters 10 through 36). A glycol coolant prevents freezing in solar collectors and outdoor piping. Secondary coolants heated by solar collectors or by other means can be used to heat absorption cooling equipment, to melt a product such as ice or snow, or to heat a building. Process heat exchangers can use a number of secondary coolants to transfer heat between locations at various temperature levels. Using secondary coolant storage tanks increases the availability of cooling and heating and reduces peak demands for energy.

Each supplier of refrigeration equipment that uses secondary coolant flow has specific ratings. Flooded and direct-expansion coolers, dairy plate heat exchangers, food processing, and other air, liquid, and solid chilling devices come in various shapes and sizes. Refrigerated secondary coolant spray wetted-surface cooling and humidity control equipment has an open system that absorbs moisture while cooling and then continuously regenerates the secondary coolant with a concentrator. Although this assists the cooling, dehumidifying, and defrosting process, it is not strictly a secondary coolant flow application for refrigeration, unless the secondary coolant also is used in the coil. Heat transfer coefficients can be determined from vendor rating data or by methods described in Chapter 3 of the 1997 *ASHRAE Handbook—Fundamentals* and appropriate texts.

A primary refrigerant may be used as a secondary coolant in a system by being pumped at a flow rate and pressure high enough that the primary heat exchange occurs without evaporation. But the refrigerant is then subsequently flashed at a low pressure, with the resulting flash gas being drawn off to a compressor in the conventional manner.

REFERENCES

ASHRAE. 1994. Safety code for mechanical refrigeration. ANSI/ASHRAE *Standard* 15-1994.

ASME. 1992. Refrigeration piping. ANSI/ASME *Standard* B31.5-92. American Society of Mechanical Engineers, New York.

ASME. 1995. Rules for construction of pressure vessels. *Boiler and pressure vessel code*, Section VIII-95.

Kern, D.Q. 1950. *Process heat transfer*, p. 134. McGraw-Hill, New York.

Means. Updated annually. *Means mechanical cost data*, 11th ed. Section 2, U.C.I. Division 15.5. Robert Snow Means Co., Kingston, MA.

Moody, L.F. 1944. Frictional factors for pipe flow. *ASME Transactions* (November):672-73.

Park, W.R. and D.E. Jackson. 1984. *Cost engineering analysis*, 2nd ed. John Wiley and Sons, New York.

NBS. 1978. National Bureau of Standards Building Science Series 113, Life cycle costing. SD Catalog Stock No. 003-003-01980-1, U.S. Government Printing Office, Washington, D.C.

Sieder, E.N. and G.E. Tate. 1936. Heat transfer and pressure drop of liquids in tubes. *Industrial and Engineering Chemistry* 28(12):1429.

CHAPTER 5

REFRIGERANT SYSTEM CHEMISTRY

Chemical Evaluation Techniques 5.1
Refrigerants 5.1
Compatibility of Materials 5.4
Chemical Reactions 5.6
Refrigerant Database and ARTI/MCLR Research Projects 5.10

A GOOD understanding of the chemical interactions of the materials in a refrigeration system is necessary for designing systems that are reliable and have a long service life. This chapter covers the chemical aspects of both old refrigerants and new, environmentally acceptable refrigerants. Physical aspects such as measurement and contaminant control (including moisture) are discussed in Chapter 6. Physical properties of lubricants are discussed in Chapter 7.

CHEMICAL EVALUATION TECHNIQUES

Chemical problems can often be attributed to inadequate testing of a new material, improper application of a previously tested material, or inadvertent introduction of contaminants into the system. Three techniques are used to chemically evaluate materials: (1) sealed tube material tests, (2) component tests, and (3) system tests.

Sealed Tube Material Tests

The glass sealed tube test, as described by ASHRAE *Standard 97*, is widely used to assess the stability of refrigerant system materials. It is also used to identify chemical reactions that are likely to occur in operating units.

Generally, glass tubes are charged with refrigerant, oil, metal strips, and other materials to be tested. The tubes are then sealed and aged at elevated temperatures for a specified time. The tubes are inspected for color and appearance and compared to control tubes that are processed identically to the specimen tubes, but might contain a reference material rather than the test material. The contents of the tubes can be analyzed for changes in the test materials by methods such as gas chromatography, ion chromatography, liquid chromatography, infrared spectroscopy, and specific ion electrode. Wet methods, such as total acid number analysis, are also used.

The sealed tube test was originally designed to compare lubricants, but it is effective in testing other materials as well. For example, Huttenlocher (1972) evaluated zinc die castings, Guy et al. (1992) reported on the compatibilities of motor insulation materials and elastomers, and Mays (1962) studied the decomposition of R-22 in the presence of 4A-type molecular sieve desiccants.

Although the sealed tube is very useful, it has some disadvantages. Because chemical reactions likely to occur in a refrigeration system are greatly magnified, results can be misinterpreted. Also, reactions in which mechanical energy plays a role (e.g., in a failing bearing) are not easily studied in a static sealed tube.

The sealed tube test, in spite of its proven utility, is only a screening tool and not a full simulation of a refrigeration system. Sealed tube tests alone should not be used to predict field behavior. Material selection for refrigerant systems requires follow-up with component or system tests or both.

The preparation of this chapter is assigned to TC 3.2, Refrigerant System Chemistry.

Component Tests

Component tests carry material evaluations a step beyond sealed tube tests; materials are tested not only in the proper environment, but also under dynamic conditions. Motorette (enameled wire, ground insulation, and other motor materials assembled into a simulated motor) tests used to evaluate hermetic motor insulation are described in Underwriters Laboratories (UL) *Standard 984* and are a good example of this type of test. Component tests are conducted in large pressure vessels or autoclaves in the presence of a lubricant and a refrigerant. Unlike sealed tube tests, in which temperature and pressure are the only means of accelerating the aging process, autoclave tests can include external stresses that may accelerate the phenomena likely to occur in an operating system. These stresses can include mechanical vibration, on-off electrical voltages, and refrigerant liquid floodback.

System Tests

System tests can be divided into two major categories:

1. Testing a sufficient number of systems under a broad spectrum of operating conditions to obtain a good, statistical reference base. Failure rates of units containing the new materials can be compared to those of units containing proven materials.
2. Testing under well-controlled conditions. Temperatures, pressures, and other operating conditions are continuously monitored. Chemical analysis of the refrigerant and lubricant is done before, during, and after the test.

In most cases, the tests are conducted under severe operating conditions to obtain results quickly. Analyzing the lubricant and refrigerant samples during the test and inspecting the components after teardown can yield information on (1) the nature and rate of chemical reactions taking place in the system, (2) the products formed by these reactions, and (3) possible effects on system life and performance. Accurate interpretation of these data determines system operating limits that keep chemical reactions at an acceptable level.

REFRIGERANTS

Environmental Acceptability

The common chlorine-containing refrigerants contribute to the depletion of the ozone layer. The ozone depletion potential (ODP) of a material is a measure of its ability to destroy stratospheric ozone.

Halocarbon refrigerants also contribute to global warming and are considered greenhouse gases. The global warming potential (GWP) of a greenhouse gas is an index describing its relative ability to trap radiant energy. This index is based on carbon dioxide, which has a very long atmospheric lifetime. The GWP, therefore, is connected to a particular time scale, for example 100 or 500 years. Appliances using a given refrigerant also produce indirect (energy-related) emissions that contribute to global warming; this indirect effect may be much larger than the direct effect of the refrigerants.

The total equivalent warming impact (TEWI) of an appliance is based on the direct warming potential of the refrigerant and the indirect effect of the energy used by the appliance.

Environmentally preferred refrigerants (1) have low or zero ODP, (2) provide good system efficiency, and (3) have low GWP. Hydrogen-containing compounds such as the hydrochlorofluorocarbon HCFC-22 or the hydrofluorocarbon HFC-134a have shorter atmospheric lifetimes than chlorofluorocarbons (CFCs) because they are largely destroyed in the lower atmosphere by reactions with OH radicals, resulting in lower ODP and GWP values.

Tables 1 and 2 show the boiling points, atmospheric lifetimes, ODPs, GWPs, and flammability of new refrigerants and the refrigerants being replaced. The ODP values have been established through the Montreal Protocol and are unlikely to change. ODP values calculated using the latest scientific information are sometimes lower but are not used for regulatory purposes. Because HFCs do not contain chlorine atoms, their ODP values would be essentially zero (Ravishankara et al. 1994). The lifetimes and GWPs were last reviewed by the International Panel on Climate Change (IPCC) in 1995 and are subject to review in the future.

Compositional Groups

Chlorofluorocarbons. CFC refrigerants such as R-12, R-11, R-114, and R-115 have been used extensively in the air-conditioning and refrigeration industries. Because of their chlorine content, these materials have nonzero ODP values. The Montreal Protocol, which governs the elimination of ozone-depleting substances, was strengthened at the London meeting in 1990 and confirmed at the Copenhagen meeting in 1992. In accordance with this international agreement, production of CFCs in industrialized countries was totally phased out as of January 1, 1996. Production in developing countries will be phased out in 2010.

Hydrochlorofluorocarbons. These refrigerants have shorter atmospheric lifetimes (and lower ODP values) than CFCs. Nevertheless, the Montreal Protocol calls for limiting the production of HCFCs by January 1, 1996. This limit is based on a consumption cap of 2.8% of the ODP of the CFCs plus the HCFCs consumed in a country in 1989. It will be followed by a 35% reduction by January 1, 2004, a 65% reduction by January 1, 2010, a 90% reduction by January 1, 2015, a 99.5% reduction by January 1, 2020, and finally a total phaseout by January 1, 2030.

HCFC-22 is the most widely used hydrochlorofluorocarbon. R-410A is now the leading alternative for HCFC-22 for new equipment. R-407C is another HCFC-22 replacement and can be used in retrofits as well as in new equipment. HCFC-123 is now used commercially as a replacement for CFC-11.

Hydrofluorocarbons. These refrigerants contain no chlorine atoms, so their ODP is zero. HFC methanes, ethanes, and propanes have been extensively considered for use in air conditioning and refrigeration.

Fluoromethanes. Mixtures that include R-32 (difluoromethane, CH_2F_2) are being considered as a replacement for R-22 and R-502. For very low temperature applications, R-23 (trifluoromethane, CHF_3) is being considered as a replacement for R-13 and R-503 (Atwood and Zheng 1991).

Fluoroethanes. Refrigerant 134a (CF_3CH_2F) of the fluoroethane series is used extensively as a direct replacement for R-12 and as a replacement for R-22 in higher temperature applications. R-125 and

Table 1 Refrigerant Properties

Refrigerant Number	Structure	Boiling Point, °C	Atmospheric Lifetime,[a] Years	ODP[b]	GWP,[c] 100 Year ITH	Flammable[d]
E125	CHF_2OCF_3	−34.6	~160[e]	0	—	No
E143	CHF_2OCH_2F	29.9	—	0	—	Yes
E143a	CF_3OCH_3	−24.1	~5[e]	0	—	Yes
11	CCl_3F	23.9	50	1.0	3 800	No
12	CCl_2F_2	−29.8	102	1.0	8 100	No
22	$CHClF_2$	−40.8	12.1	0.055	1 500	No
23	CHF_3	−82.1	264	0	11 700	No
32	CH_2F_2	−51.7	5.6	0	650	Yes
113	CCl_2FCClF_2	[47.6]	85	0.8	4 800	No
114[f]	$CClF_2CClF_2$	3.8	300	1.0	9 000	No
115[f]	$CClF_2CF_3$	−39.1	1 700	0.6	9 000	No
116	CF_3CF_3	−78.1	10 000	0	9 200	No
123	$CHCl_2CF_3$	27.9	1.4	0.02	90	No
124	$CHClFCF_3$	−12.1	6.1	0.022	470	No
125	CHF_2CF_3	−48.6	32.6	0	2 800	No
134a	CH_2FCF_3	−26.2	14.6	0	1 300	No
142b	$CClF_2CH_3$	−9.8	18.4	0.065	1 800	Yes
143	CH_2FCHF_2	3.8	41	0	300	Yes
143a	CF_3CH_3	−47.8	48.3	0	3 800	Yes
152a	CHF_2CH_3	−25.0	1.5	0	140	Yes
218	$CF_3CF_2CF_3$	−31.5	2 600	0	7 000	No
227ea	CF_3CHFCF_3	−18.3	36.5	0	2 900	No
236ea	$CF_3CHFCHF_2$	6.5	8.0[g]	0	710[g]	No
236fa	$CF_3CH_2CF_3$	−0.7	209	0	6 300	No
245ca	$CHF_2CF_2CH_2F$	25.0	6.6	0	560	Yes
245fa	$CF_3CH_2CHF_2$	15.3	7.6[h]	0	820[h]	No

ODP = ozone depletion potential
GWP = global warming potential; ITH = integrated time horizon
[a] 1995 IPCC Report: HFCs Table 2.9; CFCs and HCFCs Table 2.2 (Houghton et al. 1996).
[b] Ozone Secretariat UNEP (1996).
[c] 1995 IPCC Report: HFCs Table 2.9; CFCs and HCFCs Table 2.8 last column (Houghton et al. 1996).
[d] ASHRAE *Standard* 34.
[e] Estimated using the OH rate constant recommended by a NASA panel (DeMore et al. 1997).
[f] Values for the lifetime listed in the 1995 IPCC Report (Houghton et al. 1996). The 100 year GWP values were calculated by scaling the 1994 IPCC values (Houghton et al. 1995) to reflect changes in lifetime and CO_2 decay function used in the 1995 IPCC Report.
[g] Gierczak et al. (1996).
[h] Ko et al. (1997).

Refrigerant System Chemistry

Table 2 Properties of Refrigerant Blends

Refrigerant Number	Composition (Mass Percent)	Bubble Point, °C	ODP	GWP,[a] 100 year
401A	22/152a/124 (53/13/34)	−33.1	0.037	970
401B	22/152a/124 (61/11/28)	−34.7	0.04	1 060
401C	22/152a/124 (33/15/52)	−28.4	0.03	760
402A	125/290/22 (60/2/38)	−49.2	0.021	2 250
402B	125/290/22 (38/2/60)	−47.4	0.033	1 960
403A	C_2H_6/22/218 (5/75/20)	−50.0	0.041	2 530
403B	C_2H_6/22/218 (5/56/39)	−49.5	0.031	3 570
404A	125/143a/134a (44/52/4)	−46.5	0	3 260
405A	22/152a/142b/C318 (45/7/5.5/42.5)	−27.3	0.028	4 480
406A	22/600a/142b (55/4/41)	−32.3	0.057	1 560
407A	32/125/134a (20/40/40)	−45.5	0	1 770
407B	32/125/134a (10/70/20)	−47.3	0	2 290
407C	32/125/134a (23/25/52)	−43.7	0	1 530
407D	32/125/134a (15/15/70)	−39.5	0	1 430
408A	125/143a/22 (7/46/47)	−43.5	0.026	2 650
409A	22/124/142b (60/25/15)	−34.2	0.048	1 290
410A	32/125 (50/50)	−52.7	0	1 730
411A	1270/22/152a (1.5/87.5/11.0)	−38.6	0.048	1 330
411B	1270/22/152a (3/94/3)	−41.6	0.052	1 410
412A	22/218/142b (70/5/25)	−38.5	0.055	1 850
500	12/152a (73.8/26.2)	−34	0.74	6 010
501	22/12 (75/25)	−41.0	0.29	3 150
502	22/115 (48.8/51.2)	−45.4	0.33	5 260
503	23/13 (40.1/59.9)	−88.7	0.6	11 350
504	32/115 (48.2/51.8)	−57.2	0.31	4 890
507A	125/143a (50/50)	−46.7	0	3 300
508A	23/116 (39/61)	−122.3	0	10 200
508B	23/116 (46/54)	−124.4	0	10 400
509A	22/218 (44/56)	−47.1	0.032	4 580

[a]GWPs are the mass fraction average for the GWP values for individual components obtained from the 1995 IPCC Report (Houghton et al. 1996).

R-143a are used in azeotropes or zeotropic blends with R-32 and/or R-134a as replacements for R-22 or R-502. R-152a is flammable and less efficient than R-134a in applications using suction line heat exchangers (Sandvordenker 1992), but it is still being considered as a candidate for R-12 replacement. R-152a is also being considered as a component, with R-22 and R-124, in zeotropic blends (Bateman et al. 1990, Bivens et al. 1989) that can be R-12 and R-500 alternatives.

Fluoropropanes. Desmarteau et al. (1991) identified a number of fluoropropanes as potential refrigerants. R-245ca is being considered as a chlorine-free replacement for R-11. Evaluation by Doerr et al. (1992) showed that R-245ca is stable and compatible with key components of the hermetic system. However, Smith et al. (1993) demonstrated that R-245ca is slightly flammable in humid air at room temperature. Keuper et al. (1996) investigated the performance of R-245ca in a centrifugal chiller; they found that the refrigerant might be useful in new equipment but posed some problems when used as a retrofit for R-11 and R-123 machines. R-245fa is being developed as a foam-blowing agent and may also be considered as a refrigerant. R-236fa is currently under investigation as a replacement for R-114 in naval centrifugal chillers.

Fluoroethers. Booth (1937), Eiseman (1968), Kopko (1989), O'Neill and Holdsworth (1990), O'Neill (1992), and Wang et al. (1991) proposed these compounds as refrigerants. The fluoroethers are usually more reactive than the fluorinated hydrocarbons, both physiologically and chemically. Fluorinated ethers have found use as anesthetics and convulsants (Krantz and Rudo 1966, Terrell et al. 1971a, 1971b). Reactivity with glass is characteristic of some fluoroethers (Doerr et al. 1993, Simons et al. 1977, Gross 1990). Misaki and Sekiya (1995, 1996) investigated 1-methoxyperfluoropropane (boiling point 34.2°C) and 2-methoxyperfluoropropane (boiling point 29.4°C) as potential low-pressure refrigerants.

Hydrocarbons. Hydrocarbons such as propane, *n*-butane (R-600), isobutane (R-600a), and blends of these are being used as refrigerants. Hydrocarbons have zero ODP and low GWP. However, they are very flammable, which is a serious obstacle to their widespread use as refrigerants. Hydrocarbons are commonly used in small proportions in mixtures with nonflammable halogenated refrigerants and in small equipment requiring low refrigerant charges. Hydrocarbons are currently being used in air-conditioning and refrigeration equipment in Europe and China (Powell 1996; Lohbeck 1996; Mianmiam 1996).

Ammonia. Used extensively in large, open-type compressors for industrial and commercial applications, ammonia (R-717) has high refrigerating capacity per unit displacement, low pressure losses in connecting piping, and low reactivity with refrigeration lubricants (mineral oils). See Chapter 3 for detailed information.

The toxicity and flammability of ammonia offset its advantages. Ammonia is such a strong irritant to the human nose (detectable below 5 mg/kg) that people automatically avoid exposure to it. Ammonia is considered toxic at 35 to 50 mg/kg. Ammonia-air mixtures are flammable, but only within a narrow range of 15.2 to 27.4% by volume. These mixtures can explode but are difficult to ignite because they require an ignition source of at least 650°C.

Refrigerant Analysis

With the introduction of many new pure refrigerants and refrigerant mixtures, interest in the analysis of refrigerants has increased. Refrigerant analysis is addressed in ARI *Standard* 700 and its Analytical Procedures Appendix. Gas chromatographic methods have been published for the purity determination of R-134a and R-141b (Gehring et al. 1992a, 1992b). Gehring (1995) deals with the measurement of water in refrigerants Bruno and Caciari (1994) and Bruno et al. (1995) have done extensive work developing chromatographic methods for the analysis of refrigerants using a graphitized carbon black column with a coating of hexafluoropropene. Bruno et al. (1994) have also published the refractive indices of some of the alternative refrigerants. There is interest in developing methods for the field analysis of refrigerant systems. Systems for field analysis of both oils and refrigerants are commercially available.

Flammability and Combustibility

Refrigerant flammability testing is defined in UL *Standard* 2182, Section 7. For many refrigerants, flammability is enhanced by increase in temperature and humidity. These factors must be controlled accurately to obtain reproducible, reliable data.

Fedorko et al. (1987) studied the flammability envelope of R-22/air as a function of pressure (up to 1.4 MPa) and fuel (R-22)-to-oxygen ratio. They found that R-22 was nonflammable under 500 kPa. In addition, the flammable compositions between 30% and 45% generated maximum heats of reaction. Their results were in general agreement with those of Sand and Andrjeski (1982), who found that pressurized mixtures of R-22 and air containing at least 50% air are combustible. R-11 and R-12 did not ignite under similar conditions.

Reed and Rizzo (1991) and Lindley (1992), using different experimental arrangements, studied the combustibility of R-134a at high temperature and pressure. Lindley notes that the results depend on the nature of the equipment used in the tests. Reed and Rizzo showed that R-134a is combustible at pressures above 100 kPa (gage) at room temperature and air concentrations greater than 80% by volume. At 177°C, combustibility was observed at pressures above 36 kPa (gage) and air concentrations above 60% by volume. Lindley found flammability limits of 8 to 22% by volume in air at 170°C and 700 kPa. Both researchers found R-134a to be nonflammable at ambient conditions and under the likely operating conditions of air-conditioning and refrigeration equipment. Blends of R-22/152a/114 showed combustion at temperatures above 82°C at atmospheric pressure and above, with air concentrations above 80% by volume (Reed and Rizzo 1991).

Richard and Shankland (1991) followed the ASTM *Standard* E 681 method to study the flammability of R-32, R-141b, R-142b, R-152a, R-152, R-143, R-161, methylene chloride, 1,1,1-trichloroethane, propane, pentane, dimethyl ether, and ammonia. They used several ignition methods, including the electrically activated match ignition source as specified in ASHRAE *Standard* 34. They also reported on the critical flammability ratio of mixtures such as R-32/125, R-143a/134a, R-152a/125, propane/R-125, R-152a/22, R-152a/124, and R-152a/134a. The **critical flammability ratio** is the maximum amount of flammable component that a mixture can contain and still be nonflammable, regardless of the amount of air. These data are important because mixtures containing flammable components are being considered as refrigerants.

Zhigang et al. (1992) published data on the flammability of R-152a/22 mixtures. Their measured lower flammability limit in air of R-152a is 11.4% by volume. They state that values reported in the literature range from 4.7 to 16.8% by volume. Richard and Shankland (1991) reported an average flammable range of 4.1 to 20.2% by mass for R-152a. Zhigang et al. (1992) also provide data on the length of the flame as a function of R-22 concentration. They found that the flame no longer existed somewhere between 17 and 40% R-22 by mass in the mixture. This is in apparent disagreement with the data of Richard and Shankland (1991) who found a critical flammability ratio of 57.1% R-22 by mass. Comparison of results is difficult because results depend on the apparatus and methods used. Grob (1991) reported on the flammabilities of R-152a, R-141b, and R-142b. He describes R-152a as having "the lowest flammable mixture percentage, highest explosive pressure and highest potential for ignition of the refrigerants studied." Womeldorf and Grosshandler (1995) used an opposed-flow burner to evaluate the flammability limits of refrigerants.

COMPATIBILITY OF MATERIALS

Electrical Insulation

The insulation on electric motors is affected by the refrigerant and/or the lubricant through two primary modes: extraction of insulation polymer into the refrigerant or absorption of refrigerant by the polymer.

Extraction of insulation material causes embrittlement, delamination, and general degradation of the material. In addition, extracted material can separate from solution, deposit out, and cause components to stick or passages such as capillary tubes to clog.

Absorption of refrigerant can change the dielectric strength or physical integrity of the material through softening or swelling. Rapid desorption (off-gassing) of the refrigerant by internal heating can have a more serious effect than refrigerant absorption. Rapid desorption results in high internal pressures that cause blistering or voids within the insulation, causing a decrease in dielectric or physical strength.

In compatibility studies of 10 refrigerants and seven lubricants with 24 motor materials in various combinations, Doerr and Kujak (1993) showed that R-123 was absorbed to the greatest extent, but R-22 caused more damage because of more rapid desorption and higher internal pressures. They also observed damage to insulation after desorption of R-32, R-134, and R-152a in a 150°C oven, but not to the extent exhibited by R-22.

Compatibility studies of motor materials were also conducted under retrofit conditions in which materials were exposed to the original refrigerant/mineral oil followed by exposure to the alternative refrigerant/polyester lubricant (Doerr and Waite 1995b, 1996). Alternative refrigerants included R-134a, R-407C, R-404A,
R-123, and R-245ca. Most of the motor materials were unaffected except for increased brittleness in polyethylene terephthalate (PET) due to moisture and blistering between layers of sheet insulation from the adhesive. Exposure of many of the same materials to ammonia resulted in complete destruction. The magnet wire enamel was degraded, and the PET sheet insulation completely disappeared, having been being converted to a terephthalic acid diamide precipitate (Doerr and Waite 1995a).

Ratanaphruks et al. (1996) determined the compatibility of metals, desiccants, motor materials, plastics, and elastomers with the HFCs R-245ca, R-245fa, R-236ea, and R-236fa, and HFE-125. Most metals and desiccants were compatible. Plastics and elastomers were compatible except for excessive absorption of refrigerant or lubricant (resulting in unacceptable swelling) observed with fluoropolymers, hydrogenated nitrile butyl rubber, and natural rubber. Corr et al. (1994) tested compatibility with R-22 and R-502 replacements. Kujak and Waite (1994) studied the effect on motor materials of HFC refrigerants with polyol ester lubricants containing elevated levels of moisture and organic acids. They concluded that a 500 ppm moisture level in the polyol ester lubricant had a greater effect on the motor materials than an organic acid level of 2 mg KOH/g. Exposure to R-134a/polyol ester with a high moisture level had less effect than exposure to R-22/mineral oil with a low moisture level.

Ellis et al. (1996) developed an accelerated test to determine the life of motor materials in alternative refrigerants using a simulated stator unit. Hawley-Fedder (1996) studied the breakdown products of a simulated motor burnout in HFC refrigerant atmospheres.

Magnet Wire Insulation. The magnet wire is coated with enamels, which are cured by heating. The most common insulation is a polyester base coat followed by a polyamide-imide top coat. A polyester-imide base coat is also used. Acrylic and polyvinyl formal enamels are found on older motors. An enameled wire with an outer layer of polyester-glass is used in larger hermetic motors for greater wire separation and thermal stability.

The magnet wire insulation is the primary source of electrical insulation and the most critical in regard to compatibility with refrigerants. Most electrical tests (NEMA *Standard* MW 1000) are conducted in air and may not be valid for hermetic motors. For example, wire enamels absorb R-22 up to 15 to 30% by mass (Hurtgen 1971) and at different rates depending on their chemical structure and degree of cure and the conditions of exposure to the refrigerant. Refrigerant permeation is shown by changes in electrical, mechanical, and physical properties of the wire enamels. Fellows et al. (1991) measured the dielectric strength, Paschen curve minimum, dielectric constant, conductivity, and resistivity for 19 HFCs in order to predict electrical properties in the presence of these refrigerants.

Wire enamels in refrigerant vapor typically exhibit **dielectric loss** with increasing temperature, as shown in Figure 1. Depending on the atmosphere and degree of cure, each wire enamel or enamel/varnish combination exhibits a characteristic temperature t_{max} above which dielectric losses increase sharply. Table 3 shows values of t_{max} for several hermetic enamels. Continued heating above t_{max} causes aging, evidenced by the irreversible alteration of dielectric properties and an increase in the conductance of the insulating material.

Table 3 Maximum Temperature t_{max} for Hermetic Wire Enamels in R-22 at 450 kPa

Enamel Type	t_{max}, °C
Acrylic	108
Polyvinyl formal	136
Isocyanate-modified polyvinyl formal	151
Polyamide-imide	183
Polyester-imide	215
Polyimide	232

Refrigerant System Chemistry

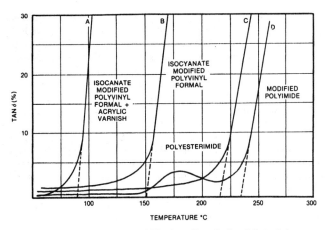

Fig. 1 Loss Curves of Various Insulating Materials
(Spauschus and Sellers 1969)

Spauschus and Sellers (1969) showed that the change rate in conductance is a quantitative measure of aging in a refrigerant environment. They proposed an aging rate for varnished and unvarnished enamels at two levels of R-22 pressure typical of high- and low-side hermetic motor operation.

Apart from the effects on long-term aging, R-22 can also affect the **short-term insulating properties** of certain wire enamels. Beacham and Divers (1955) demonstrated that the resistance of polyvinyl formal drops drastically when the material is submerged in liquid R-22. A parallel experiment using R-12 showed a much smaller drop, followed by quick recovery to the original resistance. The relatively rapid permeation of R-22 into polyvinyl formal, coupled with the low volume resistivity of liquid R-22 and other electrical properties of the two refrigerants, explains the phenomenon.

Softening of the wire coating, which can cause the insulation to fail, occurs with certain combinations of coatings and refrigerants. Table 4 shows data on softening measured in terms of abrasion resistance for a number of wire enamels exposed to R-22. At the end of the shortest soaking period, the urethane-modified polyvinyl formal had lost all its abrasion resistance. All the other insulations, except polyimide, lost their abrasion resistance at a slower rate, which over a 3-month period approached the rate of the urethane-polyvinyl formal. The polyimide showed only a minimal effect, although its abrasion resistance was originally among the lowest.

Because of the time dependency of softening, which is related to the rate of R-22 permeation into the enamel, Sanvordenker and Larime (1971) proposed that comparative tests on magnet wire be made only after the enamel is completely saturated with refrigerant, so that the effect on enamel properties of long-term exposure to R-22 can be evaluated.

Table 4 Effect of Liquid R-22 on Abrasion Resistance

Magnet Wire Insulation	Average Cycles to Failure			
	As Received	After Time in Liquid R-22		
		7 to 10 Days	One Month	Three Months
Urethane-polyvinyl formal batch 1	40	3	2	2
Urethane-polyvinyl formal batch 2	42	2	2	7
Polyester-imide batch 1	44	15	18	6
Polyester-imide batch 2	24	10	5	6
Dual coat amide-imide top coat, polyester base	79	35	23	11
Dual coat, polyester	35	5	5	9
Polyimide	26	25	23	21

Source: Sanvordenker and Larime (1971).

The second consequence of R-22 permeation is **blistering**, caused by the rapid change in pressure and temperature after a wire enamel is exposed to R-22. Heating greatly increases the internal pressure as the dissolved R-22 expands; because the polymer film has already been softened, portions of the enamel lift up in the form of blisters. Although blistered wire has a poor appearance, field experience indicates that mild blistering is not cause for concern, as long as the blisters do not break and the enamel film remains flexible. Currently used wire enamels have the characteristics mentioned previously and maintain dielectric strength even after blistering. However, hermetic wire enamel with strong resistance is preferred.

Varnishes. After the stator of an electric motor is wound, it is usually treated with a varnish by a vacuum-and-pressure impregnation process for form-wound, high-voltage motors or a dip-and-bake process for low-voltage, random wound motors. The varnished motor is cured in a 135 to 177 °C oven. The varnish holds the windings together in the magnetic field and acts as a secondary source of electrical insulation. The windings have a tendency to move, and independent movement of the wires causes abrasion and wear of the insulation. High-voltage motors contain form-wound coils wrapped with a porous fiberglass, which is saturated with varnish and cured as an additional layer.

Many different chemicals are used as motor varnishes. The most common are epoxies, polyesters, phenolics, and modified polyimides. Characteristics important to a varnish are good adhesion and bond strength to the wire enamel; flexibility and strength under both hot and cold conditions; thermal stability; good dielectric properties; and chemical compatibility with the wire enamel, sheet insulation, and refrigerant/lubricant mixture.

The compatibility of a varnish is determined by exposing the cured varnish (in the form of a section of a thin disk and varnished magnet wire in single strands, helical coils, and twisted pairs) to a refrigerant at elevated temperatures. The properties of the varnish are then compared to samples not exposed to refrigerant and to other exposed samples placed in a hot oven to rapidly remove absorbed refrigerant. The disk sections are evaluated for absorption, extraction, degradation, and changes in flexibility. The single strands are wound around a mandrel, and the varnish is examined for flexibility and effect on the wire enamel. In many cases, the varnish does not flex as well as the enamel; if bound tightly to the enamel, the varnish removes the enamel from the copper wire. The helical coils are evaluated for bond strength (ASTM *Standard* D 2519) before and after exposure to a refrigerant/lubricant mixture. The twisted pair is tested for dielectric breakdown voltage, or burnout time, while subjected to resistance heating (ASTM *Standard* D 1676).

During the compatibility testing of motor materials with alternative refrigerants, researchers observed that the varnish could absorb considerable amounts of refrigerant, especially R-123. Doerr (1992) studied the effects of time and temperature on the rate of absorption and desorption of R-123 and R-11 by epoxy motor varnishes. Absorption of refrigerant was faster at higher temperatures. Desorption of refrigerant at temperatures as high as 120°C was slow. The equilibrium absorption value for R-123 was linearly dependent on temperature, with higher absorption at lower temperatures. Absorption of R-11 remained the same at all test temperatures.

Ground Insulation. Sheet insulation material is used in slot liners, phase insulation, and wedges in hermetic motors. The sheet material is usually a PET film or an aramid (aromatic polyamide) mat. These can be used singly or laminated together. PET or aramid films possess excellent dielectric properties and good chemical resistance to refrigerants and oils.

The PET film selected must contain little of the low relative molecular mass polymers that exhibit a temperature-dependent solubility in mineral lubricants and tend to precipitate as noncohesive granules at temperatures lower than those of the motor. Another limitation of this film is that, like most polyesters, it is susceptible to degradation by hydrolysis; however, the quantity of water required is more than that generally found in refrigerant systems. Sundaresan and Finkenstadt (1991) discuss the effect of synthetic lubricants on PET films.

Elastomers

Refrigerants, oils, or mixtures of both can, at times, extract enough filler or plasticizer from an elastomer to change its physical or chemical properties. This extracted material can harm the refrigeration system by increasing its chemical reactivity or by clogging screens and expansion devices. Many elastomers are unsuitable for use with refrigerants because of excessive swell or shrinkage. Some neoprenes tend to shrink in HFC refrigerants, and nitriles swell in R-123. Hamed and Seiple (1993a, 1993b) determined swell data on 95 elastomers in 10 refrigerants and seven lubricants. Compatibility data on general classifications of elastomers such as neoprenes or nitriles should be used with caution because results depend on the particular formulation. Users should be aware that elastomeric behavior is strongly affected by the specific formulation of an elastomer as well as by the general type of elastomer.

Plastics

The effect of refrigerants on plastics usually decreases as the amount of fluorine in the molecule increases. For example, R-12 has less effect than R-11, while R-13 is almost entirely inert. Cavestri (1993) studied the compatibility of 23 engineering plastics with alternative refrigerants and lubricants.

Each type of plastic material should be tested for compatibility with the refrigerant before it is used. Two samples of the same type of plastic might be affected differently by the refrigerant because of differences in polymer structure, relative molecular mass, and plasticizer.

CHEMICAL REACTIONS

Halocarbons

Thermal Stability in the Presence of Metals. All common halocarbon refrigerants have excellent thermal stability, as shown in Table 5. Bier et al. (1990) studied R-12, R-134a, and R-152a. For R-134a in contact with metals, traces of hydrogen fluoride (HF) were detected after 10 days at 200°C. This decomposition did not increase much with time. R-152a showed traces of HF at 180°C after five days in a steel container. Bier et al. suggested that vinyl fluoride is formed during the thermal decomposition of R-152a. Vinyl fluoride can then react with water to form acetaldehyde. Hansen and Finsen (1992) conducted lifetime tests on small hermetic compressors with a ternary mixture of R-22/152a/124 and an alkyl benzene lubricant. In agreement with Bier et al., they found that vinyl fluoride and acetaldehyde were formed in the compressor. Aluminum, copper, and brass and solder joints lower the temperature at which decomposition begins. Decomposition also increases with time.

Under extreme conditions, such as above red heat or with molten metal temperatures, refrigerants react exothermically to produce metal halides and carbon. Extreme temperatures may occur in devices such as centrifugal compressors if the impeller rubs against the housing when the system malfunctions. Using R-12 as the test refrigerant, Eiseman (1963) found that aluminum was most reactive, followed by iron and stainless steel. Copper is relatively unreactive. Using aluminum as the reactive metal, Eiseman reported that R-14 causes the most vigorous reaction, followed by R-22, R-12, R-114, R-11, and R-113. Dekleva et al. (1993) studied the reaction of various CFCs, HCFCs, and HFCs in vapor tubes at very high temperatures in the presence of various catalysts and measured the onset temperature of decomposition. These data also showed the HFCs to be more thermally stable than the CFCs and HCFCs. The researchers found that when molten aluminum is in contact with R-134a, a layer of unreactive aluminum fluoride forms and inhibits further reaction.

Hydrolysis. The halogenated refrigerants are susceptible to reaction with water (hydrolysis), but the rates of reaction are so slow that they are negligible (see Table 6). Desiccants, which are discussed in Chapter 6, are used to keep refrigeration systems dry. Cohen (1993) investigated the compatibility of desiccants with R-134a and refrigerant blends.

Ammonia

Reactions involving ammonia, oxygen, oil degradation acids, and moisture are common factors in the formation of ammonia compressor deposits. Sedgwick (1966) suggested that ammonia or ammonium hydroxide reacts with organic acids produced by oxidation of the compressor oil to form ammonium salts (soaps), which can decompose further to form amides (sludge) and water. The reaction may be described as follows:

$$NH_3 + RCOOH \Leftrightarrow RCOONH_4 \Leftrightarrow RCONH_2 + H_2O$$

Water may be consumed or released during the reaction, depending on system temperature, metallic catalysts, and chemical state (acidic or basic condition). Compressor deposits can be minimized by keeping the system clean and dry, preventing entry of air, and maintaining proper compressor temperatures. Care should be taken to ensure that ester lubricants and ammonia are not used together, as large quantities of soaps and sludges would be produced.

At atmospheric pressure, ammonia starts to dissociate into nitrogen and hydrogen at about 300°C in the presence of active catalysts such as nickel and iron. However, because these high temperatures are unlikely to occur in open-type compression systems, thermal stability is not a problem. Ammonia attacks copper in the presence of even small amounts of moisture; therefore, except for some specialty bronzes, copper-bearing materials are excluded in ammonia

Table 5 Inherent Thermal Stability of Halocarbon Refrigerants

Refrigerant	Formula	Decomposition Rated at 200°C in Steel, % per yr[a]	Temperature at Which Decomposition is Readily Observed in Laboratory,[b] °C	Temperature at Which 1%/Year Decomposes in Absence of Active Materials, °C	Major Gaseous Decomposition Products[c]
22	$CHClF_2$	—	430	250	CF_2CF_2,[d] HCl
11	CCl_3F	2	590	300[e]	R-12, Cl_2
114	$CClF_2CClF_2$	1	590	380	R-12
115	$CClF_2CF_3$	—	630	390	R-13
12	CCl_2F_2	Less than 1	760	500	R-13, Cl_2
13	$CClF_3$	—	840	540[f]	R-14, Cl_2, R-116

Sources: Norton (1957), Du Pont (1959, 1969), and Borchardt (1975).
[a]Data from UL Standard 207.
[b]Decomposition rate is about 1% per min.
[c]Data from Borchardt (1975).
[d]A variety of side products are also produced, here and with the other refrigerants, some of which may be quite toxic.
[e]Conditions were not found where this reaction proceeds homogeneously.
[f]Rate behavior too complex to permit extrapolation to 1% per year.

Refrigerant System Chemistry

Table 6 Rate of Hydrolysis in Water
(Grams per Litre of Water per Year)

Refrigerant	Formula	101.3 kPa at 30°C Water Alone	101.3 kPa at 30°C With Steel	Saturation Pressure at 50°C with Steel
113	CCl_2FCClF_2	<0.005	50	40
11	CCl_3F	<0.005	10	28
12	CCl_2F_2	<0.005	1	10
21	$CHCl_2F$	<0.01	5	9
114	$CClF_2\text{-}CClF_2$	<0.005	1	3
22	$CHClF_2$	<0.01	0.1	—

Source: Du Pont (1959, 1969).

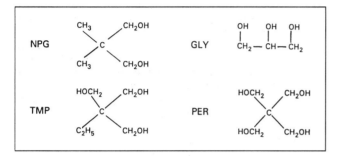

Fig. 2 Types of Alcohols Used for Ester Synthesis

Table 7 Influence of Type of Alcohol on Ester Viscosity

Type of Alcohol	Ester Viscosity at 40°C, mm²/s
NPG	13.3
GLY	31.9
TMP	51.7
PER	115

Note: Ester derived using the same carboxylic acid.

Table 8 R-134a Miscibility and Viscosity of Several Pentaerythritol-Based Esters

Acid Used	R-134a Miscibility at 20% Ester, °C	Ester Viscosity at 40°C, mm²/s
5 carbon, linear	<−70	15.6
6 carbon, linear	−47	18.5
7 carbon, linear	1	21.2
8 carbon, linear	>65	26.7
9 carbon, linear	>65	31.0
5 carbon, branched	<−70	25.2
8 carbon, branched	−15	44.4
9 carbon, branched	−27	112.9

Source: Jolley (1997).

systems. A corollary to this design practice is that copper plating (see the section on Copper Plating) is nonexistent in ammonia systems.

Lubricants

Lubricants now in use and under consideration for the new refrigerants are mineral oils, alkyl benzenes, polyol esters, polyalkylene glycols, modified polyalkylene glycols, and polyvinyl ethers. Gunderson and Hart (1962) give an excellent introduction to synthetic lubricants, including polyglycols and esters.

Polyol Esters. Commercial esters (Jolley 1991) are manufactured from four types of alcohols. They are neopentyl glycol (NPG) with two OH reaction sites; glycerin (GLY) with three OH sites; trimethylolpropane (TMP) with three OH sites; and pentaerythritol (PER) with four OH sites. The formulas for the four alcohol types are shown in Figure 2. The viscosities of the esters formed by the reaction of a given acid with each of the four alcohol types are given in Table 7.

Polyol esters are widely used as lubricants in HCFC refrigerant systems, due mainly to their physical properties. Because they are made from a wide variety of materials, polyol esters can be designed to optimize desirable physical characteristics. The system chemistry of the lubricant can be significantly influenced by the type and chain length of the carboxylic acid used to prepare the ester.

Table 8 gives R-134a miscibility and viscosity data for several esters based on pentaerythritol. From these data it can be seen that polyol ester lubricants rapidly lose refrigerant miscibility when linear carbon chain lengths exceed six carbons. The data also show that the use of branched chain acids in the preparation of these lubricants can greatly enhance refrigerant miscibility. Chain branching also enables the preparation of higher viscosity esters, which are needed in some industrial refrigeration applications.

The thermal stability of polyol esters is well known. Those esters made from polyols that posses a central "neo" structure, which consists of a carbon atom attached to four other carbon atoms (i.e., structures corresponding to NPG, TMP, and PER in Figure 2), have demonstrated outstanding thermal stability. Gunderson and Hart (1962) have reviewed research measuring the thermal stability of various polyol esters and dibasic acid esters at 260°C by heating them in evacuated tubes for up to 250 hours. Such tests demonstrated the increased thermal stability expected from "neo" ester structure, with the dibasic acid esters decomposing at a rate three times faster than the polyol esters.

Hydrolysis of Esters. An alcohol and an organic acid react to produce an organic ester and water; this reaction is called esterification, and it is reversible. The reverse reaction of an ester and water to produce an alcohol and an organic acid is called **hydrolysis**:

$$\underset{\text{Ester}}{RCOOR'} + \underset{\text{Water}}{HOH} \Leftrightarrow \underset{\text{Acid}}{RCOOH} + \underset{\text{Alcohol}}{R'OH}$$

Hydrolysis may be the most important chemical stability issue associated with esters. The degree to which esters are subject to hydrolysis is related to their processing parameters and their structure. Important processing parameters are total acid number (TAN), degree of esterification, the nature of the catalyst used during production, and the catalyst level remaining in the polyol ester after processing. Dick et al. (1996) have demonstrated that (1) the use of polyol esters prepared with acids known as α-branched acids significantly reduces ester hydrolysis and (2) the use of α-branched esters with certain additives can eliminate hydrolysis.

Hydrolysis is undesirable in refrigeration systems because free carboxylic acid can react with metal surfaces to cause corrosion. The metal carboxylate soaps that may be produced by hydrolysis can also cause capillary tube blockage. Davis et al. (1996) have reported that polyol ester hydrolysis proceeds via autocatalytic reaction. They were able to determine reaction rate constants for hydrolysis through sealed tube tests. Jolley et al. (1996) and others have used compressor testing, along with variations of the ASHRAE *Standard* 97 sealed tube test, to examine the potential for lubricant hydrolysis in operating systems. Compressor tests run with lubricant saturated with water (2000 ppm) have gone 2000 hours with no significant capillary tube blockage, indicating that under normal, much drier operating conditions, little or no detrimental ester hydrolysis will occur with the use of polyolester lubricants. Hansen and Snitkjær (1991) demonstrated the hydrolysis of esters in compressor life tests run without desiccants and in sealed tubes. They detected hydrolysis by measuring the total acid number and showed that desiccants can reduce the extent of hydrolysis in a compressor. They concluded that with filter driers, refrigeration systems using esters and R-134a can be very reliable.

Greig (1992) ran the thermal and oxidation stability test (TOST) by heating an oil/water emulsion to 95°C and bubbling oxygen through the emulsion in the presence of steel and copper. Greig shows that the hydrolysis of esters can be suppressed by appropriate additives. While agreeing that esters can be successfully used in refrigeration, Jolley et al. (1996) point out that some additives are themselves subject to hydrolysis. Cottington and Ravner (1969) and Jones et al. (1969) studied the effect of tricresyl phosphate, a common antiwear agent, on the decomposition of esters.

Field and Henderson (1998) studied the effect of elevated levels of organic acids and moisture on the corrosion of metals in the presence of R-134a and POE lubricant. Copper, brass, and aluminum showed little corrosion, but cast iron and steel were severely corroded. At 200°C, iron caused the POE lubricant to break down, even in the absence of additional acid and moisture.

Polyalkylene Glycols. Polyalkylene glycols (PAGs) are of the general formula RO-[CH_2-CHR'-O]-R'. They are used as lubricants in automotive applications that use R-134a. Linear PAGs can have one or two terminal hydroxyl groups. Modified PAG molecules have both ends capped by various groups. Sundaresan and Finkenstadt (1990) discuss the use of PAGs and modified PAGs in refrigeration compressors. Short and Cavestri (1992) present data on PAGs.

These lubricants and their additive packages may (1) oxidize, (2) degrade thermally, (3) react with system contaminants such as water, and/or (4) react with the refrigerant or system materials such as polyester films.

Oxidation is usually not a problem in hermetic systems using hydrocarbon oils because no oxygen is available to react with the lubricant. However, if a system is not adequately evacuated or if air is allowed to leak into the system, organic acids and sludges can be formed. Lockwood and Klaus (1981) and Clark et al. (1985) found that iron and copper catalyze the oxidative degradation of esters. These reaction products are detrimental to the refrigeration system and can cause failure. Some have suggested that the oxidative breakdown products of PAG lubricants (Komatsuzaki et al. 1991) and perhaps of esters are volatile, while those of mineral oils are more likely to include sludges.

Sanvordenker (1991) studied the thermal stability of PAG and ester lubricants and found that above 204°C, water is one of the decomposition products of esters (in the presence of steel) and of PAG lubricants. He recommends that polyol esters be used with metal passivators to enhance their stability when in contact with the metallic bearing surfaces, which can experience 204°C temperatures. Sanvordenker presents data on the kinetics of the thermal decomposition of polyol esters and PAGs. These reactions are catalyzed by metal surfaces in the following order: low carbon steel > aluminum > copper (Naidu et al. 1988).

Lubricant Additives

Additives are often used to improve the performance of lubricants in refrigeration systems. Additives have become more important with the increased use of HFC refrigerants. The chlorine found in CFC refrigerants acted as an antiwear agent, so the mineral oil lubricants needed minimal or no additives to provide wear protection. HFC refrigerants such as R-134a do not contain chlorine and thus do not provide this antiwear benefit. Additives such as antioxidants, detergents, dispersants, rust inhibitors, and so forth, are not normally used because the conditions they treat are absent in most refrigeration systems. Many HFC/polyol ester refrigeration systems function well without lubricant additives. However, some systems that have aluminium wear surfaces require an additive to supplement wear protection. Antiwear protection is likely to be necessary in future systems with lower viscosity lubricants to improve energy efficiency, especially if branched acid polyol esters are used. Randles et al. (1996) discuss the advantages and disadvantages of using additives in polyol ester lubricants for refrigeration systems.

The active ingredient in antiwear additives is typically phosphorous, sulfur, or both. Organic phosphates, phosphites, and phosphonates are typical phosphorous-containing antiwear agents. Tricresylphosphate (TCP) is the best known of these. Sulfurized olefins and disulfides are typical of sulfur-containing additives for wear protection. Zinc dithiophosphates are the best examples of mixed additives. Vinci and Dick (1995) have shown that additives containing phosphorous can perform well as antiwear agents and that sulfur-containing additives are not thermally stable as determined by the ASHRAE *Standard* 97 sealed tube stability test.

Other additives finding some use in HFC/polyol ester combinations are foam-producing agents (compressor start-up noise reduction) and hydrolysis inhibitors. Vinci and Dick (1995) show that a combination of antiwear additive and hydrolysis inhibitor can produce exceptional performance with respect to both wear and capillary tube blockage in bench testing and long-term compressor endurance tests. Sanvordenker (1991) has shown that iron surfaces can catalyze the decomposition of esters at 200°C. He proposed using a metal passivator as an additive to minimize this effect in systems where high temperatures are possible. Schmitz (1996) describes the use of a siloxane ester foaming agent for noise reduction. Swallow et al. (1995, 1996) suggested the use of additives to control the release of refrigerant vapor from polyol ester lubricants.

System Reactions

Average strengths of carbon-chlorine, carbon-hydrogen, and carbon-fluorine bonds are 328, 412, and 441 kJ/mole, respectively (Pauling 1960). The relative stabilities of refrigerants that contain chlorine, hydrogen, and fluorine bonded to carbon can be understood by considering these bond strengths. The CFCs have characteristic reactions that depend largely on the presence of the C-Cl bond. Spauschus and Doderer (1961) concluded that R-12 can react with a hydrocarbon oil by exchanging a chlorine for a hydrogen. This reaction is characteristic of chlorine-containing refrigerants. In the reaction, R-12 forms the reduction product R-22, R-22 forms R-32 (Spauschus and Doderer 1964), and R-115 forms R-125 (Parmelee 1965). For R-123, Carrier (1989) demonstrated that the reduction product R-133a is formed at high temperatures.

Factor and Miranda (1991) studied the reaction between R-12, steel, and oil sludge. They concluded that it can proceed by a predominantly Friedel-Crafts mechanism in which Fe^{3+} compounds are key catalysts. They also concluded that oil sludge can be formed by a pathway that does not generate R-22. They suggest that, except for the initial formation of Fe^{3+} salts, the free-radical mechanism plays only a minor role. Further work is needed to clarify this mechanism.

Huttenlocher (1992) tested 23 refrigerant-lubricant combinations for stability in sealed glass tubes. HFC refrigerants were shown to be very stable even at temperatures much higher than normal operating temperatures. HCFC-124 and HCFC-142b were slightly more reactive than the HFCs, but less reactive than CFC-12. HCFC-123 was less reactive than CFC-11 by a factor of approximately 10.

Fluoroethers were studied as alternative refrigerants. Sealed glass tube and Parr bomb stability tests with E-245 (CF_3-CH_2-O-CHF_2) showed evidence of an autocatylic reaction with glass that proceeds until either the glass or the fluoroether is consumed (Doerr et al. 1993). High pressures (about 14 MPa) usually cause the sealed glass tubes to explode.

The breakdown of CFCs and HCFCs can usually be tracked by observing the concentration of the reaction products formed. Alternatively, the amount of fluoride and chloride formed in the system can be observed. For HFCs, no chloride will be formed, and reaction products are highly unlikely because the C-F bond is strong. Decomposition of HFCs is usually tracked by measuring the fluoride ion concentration in the system (Thomas and Pham 1989, Spauschus 1991, Thomas et al. 1993). This test showed that R-125, R-32, R-143a, R-152a, and R-134a are quite stable.

Refrigerant System Chemistry

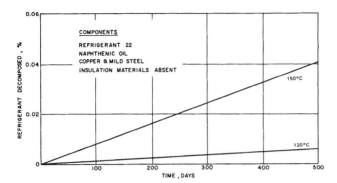

Fig. 3 Stability of Refrigerant 22 Control System
(Kvalnes and Parmelee 1957)

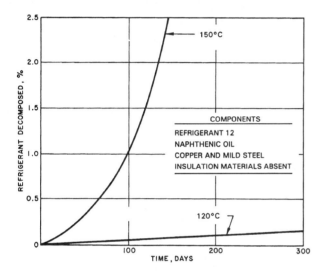

Fig. 4 Stability of Refrigerant 12 Control Systems
(Kvalnes and Parmelee 1957)

The possibility that hydrogen fluoride released by the breakdown of the refrigerants being studied will react with glass of the sealed tube has been a concern. Sanvordenker (1985) confirmed this possibility with R-12. Spauschus et al. (1992) found no evidence of fluoride on the glass surface of sealed tubes with R-134a.

Figures 3 and 4 show sealed tube test data for reaction rates of R-22 and R-12 with oil in the presence of copper and mild steel. The formation of chloride ion was taken as a measure of decomposition. These figures show the extent to which temperature accelerates reactions; it also shows that R-22 is much less reactive than R-12. The data only illustrate the chemical reactivities involved and do not represent actual rates in refrigeration systems.

The chemistry that occurs in CFC systems retrofitted to use HFC refrigerants and their lubricants is an area of growing interest. Corr et al. (1992) point out that a major problem is the effect of chlorinated residues in the new system. Komatsuzaki et al. (1991) showed that R-12 and R-113 degrade PAG lubricants. Powers and Rosen (1992) performed sealed tube tests and concluded that the threshold of reactivity for R-12 in R-134a and PAG lubricant is between 1 and 3%.

Copper Plating

Copper plating is the formation of a copper film on steel surfaces in refrigeration and air-conditioning compressors. A blush of copper is often discernible on compressor bearing and valve surfaces when machines are cut apart. After several hours of exposure to air, this thin film becomes invisible, probably because metallic copper is converted to copper oxide. In severe cases, the copper deposit can build up to substantial thickness and interfere with compressor operation. Extreme copper plating can cause compressor failure.

Thomas and Pham (1989) compared copper plating in R-12/mineral oil and R-134a/PAG systems. They showed that R-134a/PAG systems produced much less total copper (in solution and as precipitate) than R-12/mineral oil systems. They also showed that water did not significantly affect the amount of copper produced. They found that in the R-134a/PAG system, the copper was largely precipitated. In the R-12/mineral oil system, the copper was found in solution when dry and precipitated when wet. Walker et al. (1960) found that water below the saturation level had no significant effect on copper plating for R-12/mineral oil systems. Reyes-Gavilan (1993) observed copper plating in reciprocating compressors charged with R-134a and PAG lubricants. Spauschus (1963) observed that copper plating in sealed glass tubes was more prevalent with medium-refined naphthenic pale oil than with a highly refined white oil. He concluded that the refrigerant-lubricant reaction was an essential precursor for gross copper plating. The excess acid produced by the decomposition of the refrigerant had little effect on copper solubility but facilitated the plating process. Herbe and Lundqvist (1996, 1997) examined a large number of systems retrofitted from R-12 to R-134a for contaminants and copper plating. They reported that copper plating did not occur in retrofitted systems where the level of contaminants was low.

Contaminant Generation by High Temperature

Hermetic motors can overheat well beyond design levels under adverse conditions such as line voltage fluctuations, brownouts, or inadequate airflow over the condenser coils. Under these conditions, motor winding temperatures can exceed 150°C. Prolonged exposure to these thermal excursions can damage the motor insulation, depending on the thermal stability of the insulation materials, their reactivity with the refrigerant and lubricant, and the temperature levels encountered.

Another potential for high temperatures is in the bearings. Oil film temperatures in hydrodynamically lubricated journal bearings are usually not much higher than the bulk oil temperature; however, in elastohydrodynamic films in bearings with a high slide/roll ratio, the temperature can be several hundred degrees above the bulk oil temperature (Keping and Shizhu 1991). Local hot spots in boundary lubrication can reach very high temperatures. Fortunately, the amount of material exposed to these temperatures is usually very small. The appearance of methane or other small hydrocarbon molecules in the refrigerant indicates cracking of the lubricant due to high bearing temperatures.

Thermal decomposition of organic insulation materials and some types of lubricants produces noncondensable gases such as carbon dioxide and carbon monoxide. These gases circulate with the refrigerant, increasing the discharge pressure and lowering unit efficiency. At the same time, the compressor temperature and the rate of deterioration of the insulation or lubricant increase. Liquid decomposition products circulate with the lubricating oil either in solution or as colloidal suspensions. Dissolved and suspended decomposition products circulate throughout the refrigeration system, where they clog oil passages; interfere with the operation of expansion, suction, and discharge valves; or plug capillary tubes.

Appropriate control mechanisms in the refrigeration system minimize exposure to such high temperatures. Identifying potential reactions, performing adequate laboratory tests to qualify the materials before using them in the field, and finding means to remove the contaminants generated by high-temperature excursions are equally important (see Chapter 6).

REFRIGERANT DATABASE AND ARTI/MCLR RESEARCH PROJECTS

The U.S. Department of Energy (DOE) funded research to accelerate the phaseout of CFCs and the conversion to alternative refrigerants and lubricants. The funds supported about 40 materials compatibility and lubricants research (MCLR) projects and a refrigerants database. This database is maintained by the Air-Conditioning and Refrigeration Technology Institute (Calm 1995). The computer-searchable database covers over 5000 references with abstracts or summaries of presentations and technical papers. Performance, physical property, toxicity, compatibility, and flammability data for refrigerants are also included.

REFERENCES

ARI. 1995. Specifications for fluorocarbons and other refrigerants. *Standard* 700-95. Air-Conditioning and Refrigeration Institute, Arlington, VA.

ARI. 1995. Analytical procedures for ARI *Standard* 700-95.

ASHRAE. 1983. Sealed glass tube method to test the chemical stability of material for use within refrigerant systems. ANSI/ASHRAE *Standard* 97-1983 (RA 89).

ASHRAE. 1992. Number designation and safety classification of refrigerants. ANSI/ASHRAE *Standard* 34-1992.

ASTM. 1994. Test method for concentration limits of flammability of chemicals. *Standard* E 681-94. American Society for Testing and Materials, West Conshohocken, PA.

ASTM. 1995. Test method for acid number of petroleum products by potentiometric titration. ANSI/ASTM *Standard* D 664-95.

ASTM. 1995. Test methods for film-insulated magnet wire. ANSI/ASTM *Standard* D 1676-95.

ASTM. 1996. Test method for bond strength of electrical insulating varnishes by the helical coil test. *Standard* D 2519-96.

Atwood, T. and J. Zheng. 1991. Cascade refrigeration systems: The HFC-23 solution. International CFC and Halon Alternatives Conference, sponsored by The Alliance for Responsible CFC Policy, Arlington, VA.

Bateman, D.J., D.B. Bivens, R.A. Gorski, W.D. Wells, R.A. Lindstrom, R.A. Morse, and R.L. Shimon. 1990. Refrigerant blends for the automotive air conditioning aftermarket. SAE *Technical Paper* Series 900216. Society of Automotive Engineers, Warrendale, PA.

Beacham, E.A. and R.T. Divers. 1955. Some practical aspects of the dielectric properties of refrigerants. *Refrigerating Engineering* (July):33.

Bier, K., M. Crone, M. Tuerk, W. Leuckel, M. Christill, and B. Leisenheimer. 1990. Studies of the thermal stability and ignition behavior and combustion properties of the refrigerants R-152a and R-134a. *DKV-Tagungsbericht* 17:169-91.

Bivens, D.B., R.A. Gorski, W.D. Wells, A. Yokozeki, R.A. Lindstrom, and R.L. Shimon. 1989. Evaluation of fluorocarbon blends as automotive air conditioning refrigerants. SAE *Technical Paper* Series 890306. Society of Automotive Engineers, Warrendale, PA.

Booth, H.S. 1937. Halogenated methyl ethers. U.S. Patent 2,066,905.

Borchardt, H.J. 1975. *Du Pont innovation* G(2).

Bruno, T.J. and M. Caciari. 1994. Retention of halocarbons on a hexafluoropropylene epoxide modified graphitized carbon black, Part 1: Methane based fluids. *Journal of Chromatography A* 672:149-158.

Bruno, T.J., M. Wood, and B.N. Hansen. 1994. Refractive indices of alternative refrigerants. *Journal of Research of the National Institute of Standards and Technology* 99(3):263-66.

Bruno, T.J., M. Caciari, and K.H. Wertz. 1995. Retention of halocarbons on a hexafluoropropylene epoxide modified graphitized carbon black, Part 4: Propane based fluids. *Journal of Chromatography A* 708:293-302.

Calm, J.M. 1995. Refrigerants and lubricants—Data for screening and application. Proceedings of the International CFC and Halon Alternatives Conference, Washington, D.C., pp. 169-78.

Carrier. 1989. Decomposition rates of R-11 and R-123. Carrier Corporation, Syracuse, NY. Available from ARTI *Refrigerant Database*. Air-Conditioning and Refrigeration Technology Institute, Arlington, VA.

Cavestri, R.C. 1993. Compatibility of refrigerants and lubricants with plastics. *Final Report* No. DOE/CE 23810-15. ARTI *Refrigerant Database* (December). Air-Conditioning and Refrigeration Technology Institute, Arlington, VA.

Clark, D.B., E.E. Klaus, and S.M. Hsu. 1985. The role of iron and copper in the degradation of lubricating oils. *Journal of ASLE* (May):80-87. Society of Tribologists and Lubrication Engineers, Park Ridge, IL.

Cohen, A.P. 1993. Test methods for the compatibility of desiccants with alternative refrigerants. *ASHRAE Transactions* 99(1):408-12.

Corr, S., R.D. Gregson, G. Tompsett, A.L. Savage, and J.A. Schukraft. 1992. Retrofitting large refrigeration systems with R-134a. Proceedings of the International Refrigeration Conference—Energy Efficiency and New Refrigerants 1:221-30. Purdue University, West Lafayette, IN.

Corr, S., P. Dowdle, G. Tompsett, R. Yost, T. Dekleva, J. Allison, and R. Brutsch. 1994. Compatibility of non-metallic motor components with R-22 and R-502 replacements. ASHRAE Winter Meeting, New Orleans, LA. ARTI *Refrigerant Database* 4216. Air-Conditioning and Refrigeration Technology Institute, Arlington, VA.

Cottington, R.L. and H. Ravner. 1969. Interactions in neopentyl polyol ester-tricresyl phosphate-iron systems at 500°F. *ASLE Transactions* 12:280-86. Society of Tribologists and Lubrication Engineers, Park Ridge, IL.

Davis, K.E., S.T. Jolley, and J.R. Shanklin. 1996. Hydrolytic stability of polyolester refrigeration lubricants. ARTI *Refrigerant Database*. Air-Conditioning and Refrigeration Technology Institute, Arlington, VA.

Desmarteau, D., A. Beyerlein, S. Hwang, Y. Shen, S. Li, R. Mendonca, K. Naik, N.D. Smith, and P. Joyner. 1991. Selection and synthesis of fluorinated propanes and butanes as CFC and HCFC alternatives. International CFC and Halon Alternatives Conference. See Atwood and Zheng (1991).

Dekleva, T.D., A.A. Lindley, and P. Powell. 1993. Flammability and reactivity of select HFCs and mixtures. *ASHRAE Journal* (12):40-47.

DeMore, W.B., S.P. Sander, D.M. Golden, R.F. Hampson, M.J. Kurylo, C.J. Howard, A.R. Ravishankara, C.E. Kolb, and M.J. Molina. 1997. Chemical kinetics and photochemical data for use in stratospheric modeling. *Evaluation Number 12*, January 15, 1997. NASA and JPL, California Institute of Technology.

Dick, D.L., J.N. Vinci, K.E. Davis, G.R. Malone, and S.T. Jolley. 1996. ARTI *Refrigerant Database*. Air-Conditioning and Refrigeration Technology Institute, Arlington, VA.

Doerr, R.G. 1992. Absorption of HCFC-123 and CFC-11 by epoxy motor varnish. *ASHRAE Transactions* 98(2):227-34.

Doerr, R.G. and S.A. Kujak. 1993. Compatibility of refrigerants and lubricants with motor materials. *ASHRAE Journal* 35(8):42-47.

Doerr, R.G., D. Lambert, R. Schafer, and D. Steinke. 1992. Stability and compatibility studies of R-245ca, $CHF_2\text{-}CF_2\text{-}CH_2F$, a potential low-pressure refrigerant. International CFC and Halon Alternatives Conference. See Atwood and Zheng (1991).

Doerr, R.G., D. Lambert, R. Schafer, and D. Steinke. 1993. Stability studies of E-245 fluoroether, $CF_3\text{-}CH_2\text{-}O\text{-}CHF_2$. *ASHRAE Transactions* 99(2):1137-40.

Doerr, R.G. and T.D. Waite. 1995a. Compatibility of ammonia with motor materials. ASHRAE Annual Meeting, San Diego CA, June. ARTI *Refrigerant Database*. Air-Conditioning and Refrigeration Technology Institute, Arlington, VA.

Doerr, R.G. and T.D. Waite. 1995b. Compatibility of refrigerants and lubricants with motor materials under retrofit conditions. Proceedings of the International CFC and Halon Alternatives Conference, Washington, D.C., pp. 159-68.

Doerr, R.G. and T.D. Waite. 1996. Compatibility of refrigerants and lubricants with motor materials under retrofit conditions. *Final Report* No. DOE/CE 23810-63. ARTI *Refrigerant Database*. Air-Conditioning and Refrigeration Technology Institute, Arlington, VA.

Du Pont. 1959. Properties and application of the "Freon" fluorinated hydrocarbons. *Technical Bulletin* B-2. Freon Products Division, E.I. du Pont de Nemours and Co.

Du Pont. 1969. Stability of several "Freon" compounds at high temperatures. *Technical Bulletin* XIA, Freon Products Division, E.I. du Pont de Nemours and Co.

Eiseman, B.J. 1968. Chemical processes. U.S. Patent 3,362,180.

Eiseman, B.J, Jr. 1963. Reactions of chlorofluoro-hydrocarbons with metals. *ASHRAE Journal* 5(5):63.

Ellis, P.F., A.F. Ferguson, and K.T. Fuentes. 1996. Accelerated test methods for predicting the life of motor materials exposed to refrigerant-lubricant mixtures. *Report* No. DOE/ CE 23810-69. ARTI *Refrigerant Database*. Air-Conditioning and Refrigeration Technology Institute, Arlington, VA.

Factor, A. and P.M. Miranda. 1991. An investigation of the mechanism of the R-12-oil-steel reaction. *Wear* 150:41-58.

Fedorko, G., G. Fredrick, and J.G. Hansel. 1987. Flammability characteristics of chlorodifluoromethane (R-22)-oxygen-nitrogen mixtures. *ASHRAE Transactions* 93(2):716-24.

Fellows, B.R., R.C. Richard, and I.R. Shankland. 1991. Electrical characterization of alternative refrigerants. XVIIIth International Congress of Refrigeration, Montreal, 45:398.

Field, J.E. and D.R. Henderson. 1998. Corrosion of metals in contact with new refrigerants/lubricants at various moisture and organic acid levels. *ASHRAE Transactions* 104(1).

Gehring, D.G. 1995. How to determine concentration of water in system refrigerants. *ASHRAE Journal* 37(9):52-55.

Gehring, D.G., D.J. Barsotti, and H.E. Gibbon. 1992a. Chlorofluorocarbons alternatives analysis, Part I: The determination of HFC-134a purity by gas chromatography. *Journal of Chromatographic Science* 30:280.

Gehring, D.G., D.J. Barsotti, and H.E. Gibbon. 1992b. Chlorofluorocarbons alternatives analysis, Part II: The determination of HCFC-141b purity by gas chromatography. *Journal of Chromatographic Science* 30:301.

Gierczak, T., R.K. Talukdar, J.B. Burkholder, R.W. Portman, J.S. Daniel, S. Solomon, and A.R. Ravishankara. 1996. Atmospheric fate and greenhouse warming potentials of HFC 236fa and HFC 236ea. *Journal of Geophysical Research* 101(D8):12,905-11.

Greig, B.D. 1992. Formulated polyol ester lubricants for use with HFC-134a. The role of additives and conversion of existing CFC-12 plant to HFC-134a. International CFC and Halon Alternatives Conference. See Atwood and Zheng (1991).

Grob, D.P. 1991. Summary of flammability characteristics of R-152a, R-141b, R-142b and analysis of effects in potential applications. House hold Refrigerators. 42nd Annual International Appliance Technical Conference, Batavia, IL.

Gross, T.P. 1990. Sealed tube tests—Grace ether (E-134). ARTI *Refrigerant Database* RDB0904. Air-Conditioning and Refrigeration Technology Institute, Arlington, VA.

Gunderson, R.C. and A.W. Hart. 1962. *Synthetic lubricants*. Reinhold Publishing Corporation, New York.

Guy, P.D., G. Tompsett, and T.W. Dekleva. 1992. Compatibilities of non-metallic materials with R-134a and alternative lubricants in refrigeration systems. *ASHRAE Transactions* 98(1):804-16.

Hamed, G.R. and R.H. Seiple. 1993a. Compatibility of elastomers with refrigerant/lubricant mixtures. *ASHRAE Journal* 35(8):173-76.

Hamed, G.R. and R.H. Seiple. 1993b. Compatibility of refrigerants and lubricants with elastomers. *Final Report* No. DOE/CE 23810-14. ARTI *Refrigerant Database*. Air-Conditioning and Refrigeration Technology Institute, Arlington, VA.

Hansen, P.E. and L. Finsen. 1992. Lifetime and reliability of small hermetic compressors using a ternary blend HCFC-22/HFC-152a/HCFC-124. International Refrigeration Conference—Energy Efficiency and New Refrigerants. D. Tree, ed. Purdue University, West Lafayette, IN.

Hansen, P.E. and L. Snitkjær. 1991. Development of small hermetic compressors for R-134a. XVIIIth International Congress of Refrigeration, Montreal, 223:1146.

Hawley-Fedder, R. 1996. Products of motor burnout. *Report* No. DOE/CE 23810-74. ARTI *Refrigerant Database*. Air-Conditioning and Refrigeration Technology Institute, Arlington, VA.

Herbe, L. and P. Lundqvist. 1996. Refrigerant retrofit in Sweden—Field and laboratory studies. Proceedings of the International Refrigeration Conference at Purdue University, West Lafayette, IN, pp. 71-76.

Herbe, L. and P. Lundqvist. 1997. CFC and HCFC refrigerants retrofit—Experiences and results. *International Journal of Refrigeration* 20(1): 49-54.

Houghton, J.T., L.G. Filho, B.A. Callander, N. Harris, A. Kattenberg, and K. Maskell. 1995. *Climate change 1994: The science of climate change*. Contribution of WGI to the First Assessment Report of the International Panel on Climate Change, Cambridge University Press, Cambridge, UK.

Houghton, J.T., L.G. Filho, B.A. Callander, N. Harris, A. Kattenberg, and K. Maskell. 1996. *Climate change 1995: The science of climate change*. Contribution of WGI to the Second Assessment Report of the International Panel on Climate Change, Cambridge University Press, Cambridge, UK.

Hurtgen, J.R. 1971. R-22 blister testing of magnet wire. Proceedings of the 10th Electrical Insulation Conference, Chicago. pp. 183-85.

Huttenlocher, D.F. 1972. Accelerated sealed-tube test procedure for Refrigerant 22 reactions. Proceedings of the 1972 Purdue Compressor Technology Conference. Purdue University, West Lafayette, IN.

Huttenlocher, D.F. 1992. Chemical and thermal stability of refrigerant-lubricant mixtures with metals. *Final Report* No. DOE/CE 23810-5. ARTI *Refrigerant Database*. Air-Conditioning and Refrigeration Technology Institute, Arlington, VA.

Jolley, S.T. 1991. The performance of synthetic ester lubricants in mobile air conditioning systems. U.S. DOE Grant No. DE-F-G02-91 CE 23810. ARTI *Refrigerant Database*. Air-Conditioning and Refrigeration Technology Institute, Arlington, VA.

Jolley, S.T. 1997. Polyester lubricants for use in environmentally friendly refrigeration applications. American Chemical Society National Meeting, San Francisco, CA. *Preprints of Papers, Division of Petroleum Chemistry* 42(1):238-41.

Jolley, S.T., K.E. Davis, and G.R. Malone. 1996. The effect of desiccants in HFC refrigeration systems using ester lubricants. ARTI *Refrigerant Database*. Air-Conditioning and Refrigeration Technology Institute, Arlington, VA.

Jones, R.L., H.L. Ravner, and R.L. Cottington. 1969. Inhibition of iron-catalyzed neopentyl polyol ester thermal degradation through passivation of the active metal surface by tricresyl phosphate. Paper presented at ASLE/ASME Lubrication Conference, Houston, TX.

Keping, H. and W. Shizhu. 1991. Analysis of maximum temperature for thermo elastohydrodynamic lubrication in point contacts. *Wear* 150:1-10.

Keuper, E.F., F.B. Hamm, and P.R. Glamm. 1996. Evaluation of R-245ca for commercial use in low pressure chiller. *Final Report* DOE/CE23810-67. ARTI *Refrigerant Database* (March). Air-Conditioning and Refrigeration Technology Institute, Arlington, VA.

Keuper, E.F. 1996. Performance characteristics of R-11, R-123 and R-245ca in direct drive low pressure chillers. Proceedings of the International Compressor Conference at Purdue, pp.749-54.

Ko, M., R.L. Shia, and N.D. Sze. 1997. Report on calculations of global warming potentials. Prepared for AFEAS, Contract P97-134, July.

Komatsuzaki, S., Y. Homma, K. Kawashima, and Y. Itoh. 1991. Polyalkylene glycol as lubricant for HCFC-134a compressors. *Lubrication Engineering* (Dec.):1018-25.

Kopko, W.L. 1989. Extending the search for new refrigerants. Proceedings of CFC Technology Conference, Gaithersburg, MD.

Krantz, J.C. and F.G. Rudo. 1966. The fluorinated anesthetics. In *Handbook of Experimental Pharmacology* XX(1):501-64. Eichler, O., H. Herken, and A.D. Welch, eds. Springer Verlag, Berlin.

Kvalnes, D.E. and H.M. Parmelee. 1957. Behavior of Freon-12 and Freon-22 in sealed-tube tests. *Refrigerating Engineering* (November):40.

Kujak, S.A., and T.D. Waite. 1994. Compatibility of motor materials with polyolester lubricants: Effect of moisture and weak acids. Proceedings of the International Refrigeration Conference at Purdue University, pp. 425-29. West Lafayette, IN.

Lindley, A.A. 1992. KLEA 134a flammability characteristics at high-temperatures and pressures. ARTI *Refrigerant Database*. Air-Conditioning and Refrigeration Technology Institute, Arlington, VA.

Lockwood, F. and E.E. Klaus. 1981. Ester oxidation—The effect of an iron surface. *ASLE Transactions* 25(2):236-44. Society of Tribologists and Lubrication Engineers, Park Ridge, IL.

Lohbeck, W. 1996. Proceedings of International Conference on Ozone Protection Technologies, pp. 247-51. Washington, D.C.

Mays, R.L. 1962. Molecular sieve and gel-type desiccants for refrigerants. *ASHRAE Transactions* 68:330.

Mianmiam, Y. 1996. Proceedings of International Conference on Ozone Protection Technologies, pp. 260-66. Washington, D.C.

Misaki, S. and A. Sekiya. 1995. Development of a new refrigerant. Proceedings of International CFC and Halon Alternatives Conference, pp. 278-85. Washington, D.C.

Misaki, S. and A. Sekiya. 1996. Update on fluorinated ethers as alternatives to CFC refrigerants. Proceedings of International Conference on Ozone Protection Technologies, pp. 65-70. Washington, D.C.

Naidu, S.K., B.E. Klaus, and J.L. Duda. 1988. Thermal stability of esters under simulated boundary lubrication conditions. *Wear* 121:211-22.

NEMA. 1987. Magnet wire. *Standard* MW 1000-87. National Electrical Manufacturer's Association, Washington, D.C.

Norton, F.J. 1957. Rates of thermal decomposition of $CHClF_2$ and Cl_2F_2. *Refrigerating Engineering* (September):33.

O'Neill, G.J. 1992. Synthesis of fluorinated dimethyl ethers. U.K. Patent Application 2,248,617.

O'Neill, G.J. and R.S. Holdsworth. 1990. Bis(difluoromethyl) ether refrigerant. U.S. Patent 4,961,321.

Ozone Secretariat UNEP. 1996. *Handbook for the international treaties for the protection of the ozone layer*, 4th ed., pp. 275 and 303.

Parmelee, H.M. 1965. Sealed-tube stability tests on refrigerant materials. *ASHRAE Transactions* 71(1):154.

Pauling, L. 1960. The nature of the chemical bond and the structure of molecules and crystals. In *An introduction to modern structural chemistry*. Cornell University Press, Ithaca, NY.

Powell, L. 1996. Field experience of HC's in commercial applications. Proceedings of International Conference on Ozone Protection Technologies, pp. 237-46. Washington, D.C.

Powers, S. and S. Rosen. 1992. Compatibility testing of various percentages of R-12 in R-134a and PAG lubricant. International CFC and Halon Alternatives Conference, See Atwood and Zheng (1991).

Randles, S.J., P.J. Tayler, S.H. Colmery, R.W. Yost, A.J. Whittaker, and S. Corr. 1996. The advantages and disadvantages of additives in polyol esters: An overview. ARTI *Refrigerant Database*. Air-Conditioning and Refrigeration Technology Institute, Arlington, VA.

Ratanaphruks, K., M.W. Tufts, A.S. Ng, and N.D. Smith. 1996. Material compatibility evaluations of HFC-245ca, HFC-245fa, HFE-125, HFC-236ea, and HFC-236fa. Proceedings of the International Conference on Ozone Protection Technologies, Washington, D.C., pp. 113-22.

Ravishankara, A.R., A.A. Turnipseed, N.R. Jensen, and R.F. Warren. 1994. Do hydrofluorocarbons destroy stratospheric ozone? *Science* 248: 1217-19.

Reed, P.R. and J.J. Rizzo. 1991. Combustibility and stability studies of CFC substitutes with simulated motor failure in hermetic refrigeration equipment. Proceedings of the XVIIIth International Congress of Refrigeration, Montreal, 2:888-91

Reed, P.R. and H.O. Spauschus. 1991. HCFC-124: Applications, properties and comparison with CFC-114. *ASHRAE Journal* 40(2).

Reyes-Gavilan, J.L. 1993. Performance evaluation of naphthenic and synthetic oils in reciprocating compressors employing R-134a as the refrigerant. *ASHRAE Transactions* 99(1):349.

Richard, R.G. and I.R. Shankland. 1991. Flammability of alternative refrigerants. Proceedings of the XVIIIth Congress of Refrigeration, Montreal, 384.

Sand, J.R. and D.L. Andrjeski. 1982. Combustibility of chlorodifluoromethane. *ASHRAE Journal* 24(5):38-40.

Sanvordenker, K.S. 1985. Mechanism of oil-R12 reactions—The role of iron catalyst in glass sealed tubes. *ASHRAE Transactions* 91(1A):356-63.

Sanvordenker, K.S. 1991. Durability of HFC-134a compressors—The role of the lubricant. Proceedings of the 42nd Annual International Appliance Technical Conference, University of Wisconsin, Madison.

Sanvordenker, K.S. 1992. R-152a versus R-134a in a domestic refrigerator-freezer, energy advantage or energy penalty. Proceedings of the International Refrigeration Conference—Energy Efficiency and New Refrigerants, Purdue University, West Lafayette, IN.

Sanvordenker, K.S. and M.W. Larime. 1971. Screening tests for hermetic magnet wire insulation. Proceedings of the 10th Electrical Insulation Conference, pp. 122-36. Institute of Electrical and Electronics Engineers, Piscataway, NJ.

Schmitz, R. 1996. Method of making foam in an energy efficient compressor. U.S. Patent 549908.

Sedgwick, N.V. 1966. *The organic chemistry of nitrogen*, 3rd ed. Clarendon press, Oxford, UK.

Short, G.D. and R.C. Cavestri. 1992. High viscosity ester lubricants for alternative refrigerants. *ASHRAE Transactions* 98(1):789-95.

Simons, G.W., G.J. O'Neill, and J.A. Gribens. 1977. New aerosol propellants for personal products. U.S. Patent 4,041,148.

Smith, N.D., K. Ratanaphruks, M.W. Tufts, and A.S. Ng. 1993. R-245ca: A potential far-term alternative for R-11. *ASHRAE Journal* 35(2):19-23.

Spauschus, H.O. 1963. Copper transfer in refrigeration oil solutions. *ASHRAE Journal* 5:89.

Spauschus, H.O. 1991. Stability requirements of lubricants for alternative refrigerants. *Paper* No. 148. XVIIIth International Congress of Refrigeration, Montreal.

Spauschus, H.O. and G.C. Doderer. 1961. Reaction of Refrigerant 12 with petroleum oils. *ASHRAE Journal* 3(2):65.

Spauschus, H.O. and G.C. Doderer. 1964. Chemical reactions of Refrigerant 22. *ASHRAE Journal* 6(10).

Spauschus, H.O. and R.A. Sellers. 1969. Aging of hermetic motor insulation. IEEE *Transactions* E1-4(4):90.

Spauschus, H.O., G. Freeman, and T.L. Starr. 1992. Surface analysis of glass from sealed tubes after aging with HFC-134a. ARTI *Refrigerant Database* RDB2729. Air-Conditioning and Refrigeration Technology Institute, Arlington, VA.

Sundaresan, S.G. and W.R. Finkenstadt. 1990. Status report on polyalkylene glycol lubricants for use with HFC-134a in refrigeration compressors. Proceedings of the 1990 USNC/IIR-Purdue Conference, ASHRAE-Purdue Conference, 138-44.

Sundaresan, S.G. and W.R. Finkenstadt. 1991. Degradation of polyethylene terephthalate films in the presence of lubricants for HFC-134a: A critical issue for hermetic motor insulation systems. *International Journal of Refrigeration* 14:317.

Swallow, A., A. Smith, and B. Greig. 1995. Control of refrigerant vapor release from polyol ester/halocarbon working fluids. *ASHRAE Transactions* 101(2):929-34.

Swallow, A.P., A.M. Smith, and D.G.V. Jones. 1996. Control of refrigerant vapor release from polyolester/halocarbon working fluids. Proceedings of the International Conference on Ozone Protection Technologies, Washington, D.C., pp. 123-32.

Terrell, R.C., L. Spears, A.J. Szur, J. Treadwell, and T.R. Ucciardi. 1971a. General anesthetics. 1. Halogenated methyl ethers as anesthetics agents. *Journal of Medical Chemistry* 14:517.

Terrell, R.C., L. Spears, T. Szur, T. Ucciardi, and J.F. Vitcha. 1971b. General anesthetics. 3. Fluorinated methyl ethyl ethers as anesthetic agents. *Journal of Medical Chemistry* 14:604.

Thomas, R.H. and H.T. Pham. 1989. Evaluation of environmentally acceptable refrigerant/lubricant mixtures for refrigeration and air conditioning. SAE Passenger Car Meeting and Exposition. Society of Automotive Engineers, Warrendale, PA.

Thomas, R.H., W.T. Wu, and R.H. Chen. 1993. The stability of R-32/125 and R-125/143a. *ASHRAE Transactions* 99(2).

UL. 1993. Refrigerant-containing components and accessories, nonelectrical, 6th ed. ANSI/UL *Standard* 207-93. Underwriters Laboratories, Northbrook, IL.

UL. 1994. Refrigerants. *Standard* 2182-94.

UL. 1996. Hermetic refrigerant motor-compressors, 7th ed. *Standard* 984-96.

Vinci, J.N. and D.L. Dick. 1995. Polyol ester lubricants for HFC refrigerants: A systematic protocol for additive selection. International CFC and Halon Alternatives Conference. See Atwood and Zheng (1991).

Walker, W.O., S. Rosen, and S.L. Levy. 1960. A study of the factors influencing the stability of mixtures of Refrigerant 22 and refrigerating oils. *ASHRAE Transactions* 66:445.

Wang, B., J.L. Adcock, S.B. Mathur, and W.A. Van Hook. 1991. Vapor pressures, liquid molar volumes, vapor non-idealities and critical properties of some fluorinated ethers: $CF_3-O-CF_2-O-CF_3$, $CF_3-O-CF_2-CF_2H$, $c-CF_2-CF_2-CF_2-O-$, CF_3-O-CF_2H, and CF_3-O-CH_3 and of CCl_3F and CF_2ClH. *Journal of Chemical Thermodynamics* 23:699-710.

Womeldorf, C. and W. Grosshandler. 1995. Lean flammability limits as a fundamental refrigerant property. *Final Report* No. DOE/CE 23810-68. ARTI *Refrigerant Database*. Air-Conditioning and Refrigeration Technology Institute, Arlington, VA.

Zhigang, L., L. Xianding, Y. Jianmin, T. Xhoufang, J. Pingkun, C. Zhehua, L. Dairu, R. Mingzhi, Z. Fan, and W. Hong. 1992. Application of HFC-152a/HCFC-22 blends in domestic refrigerators. ARTI *Refrigerant Database* RDB2514. Air-Conditioning and Refrigeration Technology Institute, Arlington, VA.

CHAPTER 6

CONTROL OF MOISTURE AND OTHER CONTAMINANTS IN REFRIGERANT SYSTEMS

Moisture 6.1
Other Contaminants 6.6
Refrigerant Recovery, Recycling, and Reclamation 6.8
System Cleanup Procedure After Hermetic Motor Burnout 6.11
Contaminant Control During Retrofit 6.12

THIS chapter covers the following topics: (1) control of moisture, which is an important and universal contaminant in refrigeration systems; (2) control of other contaminants in refrigeration systems; (3) recovery, recycling, and reclamation of refrigerants; (4) service procedures typically used in cleaning a refrigeration system following a hermetic motor burnout; and (5) control of contaminants during retrofit to alternative refrigerants.

MOISTURE

The amount of moisture in a refrigerant system must be kept below an allowable maximum to provide satisfactory operation. Moisture must be removed from components during manufacture and assembly to minimize the amount of moisture in the completed assembly. Any moisture that enters during installation or servicing should be removed as quickly as is feasible.

Sources of Moisture

Moisture in a refrigerant system results from:

1. Inadequate equipment drying in factories and service operations
2. Introduction during installation or service operations in the field
3. Low-side leaks, resulting in entrance of moisture-laden air
4. Leakage of water-cooled heat exchangers
5. Oxidation of certain hydrocarbon lubricants to produce moisture
6. Wet lubricant, refrigerant, or desiccant
7. Decomposing cellulose insulation in hermetically sealed units
8. Moisture entering a nonhermetic refrigerant system via permeation of nonmetallic hoses and seals

Drying equipment in the factory is discussed in Chapter 43. Proper installation and service procedures as given in ASHRAE *Guideline* 3 minimize the sources listed in items 2, 3, and 4. Lubricants are discussed in Chapter 7. If purchased refrigerants and lubricants meet specifications and are properly handled, the moisture content generally remains satisfactory. See the section on hermetic motor insulation in Chapter 5 and the section on motor burnouts in this chapter.

Effects of Moisture

Excess moisture in a refrigerating system can cause one or all of the following undesirable effects:

1. Ice formation in expansion valves, capillary tubes, or evaporators
2. Corrosion of metals
3. Copper plating
4. Chemical damage to motor insulation in hermetic compressors or other system materials
5. Hydrolysis of lubricants and other materials

Ice or solid hydrate separates from refrigerants if the water concentration is high enough and the temperature low enough. Solid hydrate, a complex molecule of refrigerant and water, can form at temperatures higher than those required to separate ice. Liquid water forms at temperatures above those required to separate ice or solid hydrate. Ice forms during refrigerant evaporation when the relative saturation of vapor reaches 100% at temperatures of 0°C or below.

The separation of water as ice or liquid also is related to the solubility of water in a refrigerant. This solubility varies for different refrigerants and with temperature (see Table 1). Various investigators have obtained different results on water solubility in R-134a and R-123. The data presented here are the best available. The greater the solubility of water in a refrigerant, the less the possibility

Table 1 Solubility of Water in the Liquid Phase of Certain Refrigerants

Temp., °C	Solubility, mg/kg								
	R-11	R-12	R-13	R-22	R-113	R-114	R-123	R-134a	R-502
71.1	460	700	—	4100	460	450	2600	4200	1780
65.5	400	560	—	3600	400	380	2300	3600	1580
60.0	340	440	—	3150	344	320	2000	3200	1400
54.4	290	350	—	2750	290	270	1800	2800	1220
48.9	240	270	—	2400	240	220	1600	2400	1080
43.3	200	210	—	2100	200	180	1400	2000	930
37.8	168	165	—	1800	168	148	1200	1800	810
32.2	140	128	—	1580	140	120	1000	1500	690
26.7	113	98	—	1350	113	95	900	1300	580
21.1	90	76	—	1140	90	74	770	1100	490
15.6	70	58	44	970	70	57	660	880	400
10.0	55	44	—	830	55	44	560	730	335
4.4	44	32	26	690	44	33	470	600	278
−1.1	34	23.3	—	573	34	25	400	490	225
−6.7	26	16.6	14	472	26	18	330	390	180
−12.2	20	11.8	—	384	20	13	270	320	146
−17.8	15	8.3	7	308	15	10	220	250	115
−23.3	11	5.7	—	244	11	7	180	200	90
−28.9	8	3.8	3	195	8	5	140	150	69
−34.4	6	2.5	—	152	6	3	110	120	53
−40	4	1.7	1	120	—	2	90	89	40
−45.6	3	1.1	—	91	—	1.5	70	66	30
−51.1	2	0.7	—	68	—	1	53	49	22
−56.7	1	0.4	—	50	—	0.6	40	35	16
−62.2	0.8	0.3	—	37	—	0.4	30	25	11
−67.8	0.5	0.1	—	27	—	0.2	22	18	8
−73.3	0.3	0.1	—	19	—	0.1	16	12	5

Data on R-134a adapted from Thrasher et al. (1993) and Allied-Signal Corporation.
Data on R-123 adapted from Thrasher et al. (1993) and E.I. duPont de Nemours & Company. Remaining data adapted from E.I. duPont de Nemours & Company and Allied-Signal Corporation. Data used by permission.

The preparation of this chapter is assigned to TC 3.3, Refrigerant Contaminant Control.

Table 2 Distribution of Water Between the Vapor and Liquid Phases of Certain Refrigerants

Temperature, °C	Water in Vapor/Water in Liquid, mass %/mass %						
	R-11	R-12	R-22	R-114	R-123	R-134a	R-502
37.8	30.1	5.5	0.400	13.3	4.25	0.542	0.63
32.2	—	6.1	0.397	14.4	4.57	0.571	0.64
26.7	34.8	6.3	0.405	16.0	4.53	0.573	0.65
21.1	—	7.5	0.404	17.6	4.69	0.585	0.66
15.6	43.1	8.2	0.401	19.7	4.82	0.625	0.66
10.0	—	9.0	0.391	21.7	4.97	0.637	0.66
4.4	50.5	9.9	0.390	24.7	5.16	0.650	0.66
−1.1	—	11.2	0.378	26.9	5.17	0.654	0.65
−6.7	57.9	11.9	0.351	29.5	5.09	0.639	0.61
−17.8	62.4	13.1	0.301	32.1	4.91	0.584	0.53
−28.9	71.0	15.3	0.251	37.4	4.78	0.543	0.47
−40	—	17.1	0.203	52.2	4.43	0.484	0.40

Data adapted from E.I. duPont de Nemours & Company, Inc. Used by permission.

that ice or liquid water will separate in a refrigerating system. The solubility of water in ammonia, carbon dioxide, and sulfur dioxide is so high that ice or liquid water separation is not a problem.

Elsey and Flowers (1949) recognized that the concentration of water by mass at equilibrium is greater in the gas phase than in the liquid phase of Refrigerant 12. The opposite is true for Refrigerants 22 and 502. The ratio of mass concentrations differs for each refrigerant; it also varies with temperature. Table 2 shows the distribution ratios of water in the vapor phase to water in the liquid phase for common refrigerants. It can be used to calculate the equilibrium water concentration of the liquid phase refrigerant if the gas phase concentration is known, and vice versa. The water content in the vapor phase is determined by:

$$W = \left[\frac{P_w}{(P_w)^0}\right]\left[\frac{(d_w)^0}{(d_R)^0}\right] \quad (1)$$

where

W = mass water/mass refrigerant
P_w = partial pressure of water vapor
$(P_w)^0$ = partial pressure of water vapor at saturation
$(d_w)^0$ = density of water vapor at saturation
$(d_R)^0$ = density of refrigerant vapor at saturation

Freezing at expansion valves or capillary tubes can occur when excessive moisture is present in a refrigerating system. Formation of ice or hydrate in evaporators can partially insulate the evaporator. Walker et al. (1962) showed that excess moisture can cause corrosion and enhance copper plating. Other factors affecting copper plating are discussed in Chapter 5.

The moisture required for freeze-up is a function of the amount of flash gas formed during expansion and the distribution of water between the liquid and gas phases downstream of the expansion device. For example, in an R-12 system with a 43.3°C liquid temperature and a −28.9°C evaporator temperature, the refrigerant after expansion is 41.3% vapor and 58.7% liquid (by mass). The percentage of vapor formed is determined by:

$$\% \text{ Vapor} = 100\frac{h_{L(\text{liquid})} - h_{L(\text{evap})}}{h_{fg(\text{evap})}} \quad (2)$$

where

$h_{L(\text{liquid})}$ = saturated liquid enthalpy for refrigerant at liquid temperature
$h_{L(\text{evap})}$ = saturated liquid enthalpy for refrigerant at evaporating temperature
$h_{fg(\text{evap})}$ = latent heat of vaporization of refrigerant at evaporating temperature

Table 1 lists the saturated water content of the R-12 liquid phase at −28.9°C as 3.8 mg/kg. Table 2 is used to determine the saturated vapor phase water content as:

3.8 mg/kg × 15.3 = 58 mg/kg

When the vapor contains more than the saturation quantity (100% rh), free water will be present as a third phase. If the temperature is below 0°C, ice will form. Using the saturated moisture values and the liquid-vapor ratios, the critical water content of the circulating refrigerant can be calculated as:

3.8 × 0.587 = 2.2
58.0 × 0.413 = 24.0
                 ———
                 26.2 mg/kg

Maintaining moisture levels below critical value keeps free water from the low side of the system.

The previous analysis can be applied to all refrigerants and applications. An R-22 system with 43.3°C liquid and −28.9°C evaporating temperatures reaches saturation when the moisture circulating is 139 mg/kg. Note that this value is less than the liquid solubility, 195 mg/kg at −28.9°C.

Excess moisture causes paper or polyester motor insulation to become brittle, which can cause premature motor failure. However, not all motor insulations are affected adversely by moisture. The amount of water present in a refrigerant system must be small enough to avoid ice separation, corrosion, and insulation breakdown.

Polyol ester lubricants (POEs), which are used largely with hydrofluorocarbons (HFCs), absorb substantially more moisture than mineral oils and do so very rapidly on exposure to the atmosphere. Once present, the moisture is difficult to remove. Hydrolysis of these POEs can lead to the formation of acids and alcohols that, in turn, can have a negative impact on system durability and performance (Griffith 1993). For these reasons, POEs should not be exposed to ambient air except for very brief periods required for compressor installation. Also, adequate driers are particularly important elements for equipment containing POEs.

Exact experimental data on the maximum permissible moisture level in refrigerant systems are not known because so many factors are involved. Table 3 shows the results of moisture analysis of refrigerants taken from equipment in normal service from 1 to 16 years (ARI 1991a).

Table 3 Data on Moisture in Refrigeration Systems

Application	Range of Moisture Content Observed in Refrigerant, mg/kg			
	R-11	R-12	R-22	R-502
Centrifugal chillers	0 – 30	0 – 25	—	—
Reciprocating and screw chillers	—	—	0 – 56	—
Refrigeration systems	—	0 – 2	1 – 122	3 – 33
Large unitary systems ≥ 35 kW	—	—	0 – 75	—
Small unitary systems < 35 kW	—	—	3 – 127	—

Drying Methods

Equipment in the field is dried by evacuation and driers. Prior to opening the equipment for service, refrigerant must be isolated or recovered into an external storage container. After installation or service, noncondensables (air) should be removed with a vacuum pump connected preferably to both suction and discharge service ports. The absolute pressure should be reduced to 130 Pa or less, which is below the vapor pressure of water at ambient temperature. External heat may be required to vaporize water in the system. Care should be taken not to overheat the equipment. Even with these procedures, small amounts of moisture trapped under an oil film,

Control of Moisture and Other Contaminants in Refrigerant Systems

adsorbed by the motor windings, or located far from the vacuum pump are difficult to remove.

It is good practice to install a drier. On larger systems a drier with a replaceable core is frequently used. The core may need to be changed several times before the proper degree of dryness is obtained. A moisture indicator in the liquid line can indicate when the system has been dried satisfactorily.

Special techniques are required to remove free water in a refrigeration or air-conditioning system due to a burst tube or water chiller leak. The refrigerant should be transferred to a pumpdown receiver or recovered in a separate storage tank. Parts of the system may have to be disassembled and the water drained from system low points. In some large systems, the semihermetic or open-drive compressor may need to be cleaned by disassembling and hand-wiping the various parts. After reassembly, it should be dried further by passing dry nitrogen through the system and by heating and evacuation. Drying may take an extended period and require frequent changes of the vacuum pump oil. The liquid line driers should be replaced and temporary suction line driers installed. During the initial operating period, driers will need to be changed often.

If the refrigerant in the pumpdown receiver is to be reused, it must be thoroughly dried before being reintroduced into the system. One way to dry it is to draw a liquid refrigerant sample and record the ambient temperature. If a chemical analysis of the sample by a qualified laboratory reveals a moisture content at or near the water solubility of Table 1 at the recorded temperature, then free water is probably present. In that case, a recovery unit with a suction filter-drier and/or a moisture/oil trap must be used to transfer the bulk of the refrigerant from the receiver liquid port to a separate tank. When the free water reaches the tank liquid port, most of the remaining refrigerant can be recovered through the receiver vapor port. The water can then be drained from the pumpdown receiver.

Moisture Indicators

Moisture-sensitive elements that change color according to moisture content can gage the moisture level in the system; the color changes at a low enough level to be safe. Manufacturer's instructions must be followed since the color change point is also affected by the liquid line temperature and the refrigerant used.

Moisture Measurement

Techniques for measuring the amount of moisture in a compressor, or in an entire system, are discussed in Chapter 45. The following methods are used to measure the moisture content of various halocarbon refrigerants. The moisture content to be measured is generally in the milligram per kilogram, and the procedures require special laboratory equipment and techniques.

The **Karl Fischer Method** is suitable for measuring the moisture content of a refrigerant, even if it contains mineral oil. Although different firms have slightly different ways of performing this test and get somewhat varying results, the method remains the common industry practice for determining moisture content in refrigerants. The refrigerant sample is bubbled through predried methyl alcohol in a special sealed glass flask; any water present remains with the alcohol. In the **volumetric titration method**, Karl Fischer reagent is added, and the solution is immediately titrated to a "dead stop" electrometric end point. The reagent reacts with any moisture present so that the amount of water in the sample can be calculated from a previous calibration of the Karl Fischer reagent.

In the **coulometric titration method** (ARI *Standard* 700, Appendix Part 2) water is titrated with iodine that is generated electrochemically. The instrument measures the quantity of electric charge used to produce the iodine and titrate the water and calculates the amount of water present.

These titration methods, considered among the most accurate, are also suitable for measuring the moisture content of pure oil or other liquids. Special instruments designed for this particular analysis are available from laboratory supply companies. Haagen-Smit et al. (1970) describe improvements in the equipment and technique that significantly reduce analysis time.

The **gravimetric method** for measuring moisture content of refrigerants is described in ASHRAE *Standards* 63.1 and 35. It is not widely used in the industry. In this method, a measured amount of refrigerant vapor is passed through two tubes in series, each containing phosphorous pentoxide (P_2O_5). Moisture present in the refrigerant reacts chemically with the P_2O_5 and appears as an increase in mass in the first tube. The second tube is used as a tare. This method is satisfactory when the refrigerant is pure, but the presence of oil produces inaccurate results, because it is weighed as moisture. Approximately 200 g of refrigerant are required for accurate results. Because the refrigerant must pass slowly through the tube, analysis requires many hours to complete.

DeGeiso and Stalzer (1969) discuss the **electrolytic moisture analyzer**, which is suitable for high purity refrigerants. Other electronic hygrometers are available that sense moisture by the adsorption of water on an anodized aluminum strip with a gold foil overlay (Dunne and Clancy 1984). Calibration of such instruments is critical to obtain maximum accuracy. These hygrometers give a continuous moisture reading and respond rapidly enough to monitor changes. Data showing drydown rates can be gathered with these instruments (Cohen 1994). Brisken (1955) used this method in a study of moisture migration in hermetic equipment.

Thrasher (1993) used nuclear magnetic resonance spectroscopy to determine the moisture solubilities in R-134a and R-123. Another method, the infrared spectrophotometer, is used for moisture analysis, but requires a large sample for precise results and is subject to interference if oil is present in the refrigerant.

Desiccants

Desiccants used in refrigeration systems adsorb or react chemically with the moisture contained in a liquid or gaseous refrigerant-oil mixture. Solid desiccants, used widely as dehydrating agents in refrigerant systems, remove moisture from both new or field-installed equipment. The desiccant is contained in a device called a drier (also spelled dryer) or filter-drier and can be installed in either the liquid or the suction line of a refrigeration system.

Desiccants must remove water and not react unfavorably with any other materials in the system. Activated alumina, silica gel, and molecular sieves are the most widely used desiccants acceptable for refrigerant drying. Water is physically adsorbed on the internal surfaces of these highly porous desiccant materials.

Activated alumina and silica gel have a wide range of pore sizes. The pores are large enough to adsorb refrigerant, lubricant, additives, and water molecules. The pore sizes of molecular sieves, however, are uniform with an aperture of approximately 0.3 nm for a type 3A molecular sieve or 0.4 nm for a type 4A molecular sieve. The uniform openings exclude lubricant molecules from the adsorption surfaces. Molecular sieves can be selected to exclude refrigerant molecules as well. This property gives the molecular sieve the advantage of increasing the water capacity and improving the chemical compatibility between refrigerant and desiccant (Cohen 1993, 1994; Cohen and Blackwell 1995). The drier or desiccant manufacturer can provide information about which desiccant adsorbs or excludes a particular refrigerant.

Drier manufacturers offer combinations of desiccants that can be used in a single drier and may have certain advantages over a single desiccant because they can adsorb a greater variety of refrigeration contaminants. Two combinations are activated alumina with molecular sieves and silica gel with molecular sieves. Activated carbon is also used in some mixtures.

Desiccants are available in granular, bead, and block forms. Solid core desiccants, or block forms, consist of desiccant beads, granules, or both held together by a binder (Walker 1963). The

Table 4 Reactivation of Desiccants

Desiccant	Temperature, °C
Activated alumina	200 to 310
Silica gel	180 to 310
Molecular sieves	260 to 350

binder is usually a nondesiccant material. Suitable filtering action, adequate contact of the desiccant with the refrigerant, and low pressure drop are obtained by properly sizing the desiccant particles used to make up the core, and by the proper geometry of the core with respect to the flowing refrigerant. Beaded molecular sieve desiccants offer higher water capacity per unit mass than solid core desiccants. The composition and form of the desiccant is varied by drier manufacturers to achieve the desired properties.

Desiccants that take up water by chemical reaction are not recommended. Calcium chloride reacts with water to form a corrosive liquid. Barium oxide is known to cause explosions. Magnesium perchlorate and barium perchlorate are powerful oxidizing agents, which are potential explosion hazards in the presence of oil. Phosphorous pentoxide is an excellent desiccant, but its fine powdery form makes it difficult to handle and produces a high resistance to gas and liquid flow. A mixture of calcium oxide and sodium hydroxide has limited use, and should not be used as a desiccant.

Desiccants readily adsorb moisture and must be protected against it until ready for use. If a desiccant has picked up moisture, it can be reactivated under laboratory conditions by heating for about 4 h at a suitable temperature, preferably with a dry air purge or in a vacuum oven (see Table 4). Only adsorbed water is driven off at the temperatures listed, and the desiccant is returned to its initial activated state. Care must be taken against repeated reactivation and excessive temperatures during reactivation as this may damage the desiccant. The desiccant in a refrigerating equipment drier should not be reactivated for reuse, because of oil and other contaminants in the drier as well as possible damage caused by overheating the drier shell.

Equilibrium Conditions of Desiccants. Desiccants in refrigeration and air-conditioning systems function on the equilibrium principle. If an activated desiccant contacts a moisture-laden refrigerant, the water is adsorbed from the refrigerant-water mixture onto the desiccant surface until the vapor pressures of the adsorbed water (i.e., at the desiccant surface) and the water remaining in the refrigerant are equal. Conversely, if the vapor pressure of the water on the desiccant surface is higher than that in the refrigerant, water is released into the refrigerant-water mixture, and a new equilibrium point is established.

Adsorbent desiccants function by holding (adsorbing) moisture on their internal surfaces. The amount of water adsorbed from a refrigerant by an adsorbent at equilibrium is influenced (1) by the pore volume, pore size, and surface characteristics of the adsorbent; (2) by the temperature and moisture content of the refrigerant; and (3) by the solubility of water in the refrigerant.

Figures 1, 2, and 3 are equilibrium curves for various adsorbent desiccants with R-12 and R-22. These curves are representative of commercially available materials. They (adsorption isotherms) are based on the technique developed by Gully et al. (1954), as modified by ASHRAE *Standard* 35. ASHRAE *Standards* 35 and 63.1 define the moisture content of the refrigerant as Equilibrium Point Dryness (EPD), and the moisture held by the desiccant as water capacity. The curves show that for any specified amount of water in a particular refrigerant, the desiccant holds a corresponding specific quantity of water.

Figure 1 and Figure 2 show moisture equilibrium curves for three common adsorbent desiccants in drying R-12 and R-22 at 24°C. As shown, the capacity of a desiccant can vary widely for different refrigerants when the same EPD is required. Generally, a refrigerant in which moisture is more soluble requires more desiccant for adequate drying than one that has less solubility.

Figure 3 shows the effect of temperature on moisture equilibrium capacities of activated alumina and R-12. Much higher water capacities are obtained at lower temperatures, demonstrating the advantage of locating the alumina driers at relatively cool spots in the system. When using molecular sieves, the effect of temperature on water capacity is much less. ARI *Standard* 710 requires determining the water capacity for R-12 at an EPD of 15 mg/kg, and for R-22 at 60 mg/kg. Each determination must be made at 24°C (see Figures 1 and 2) and 52°C.

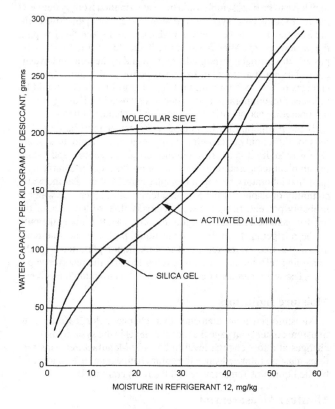

Fig. 1 Moisture Equilibrium Curves for R-12 and Three Common Desiccants at 24°C

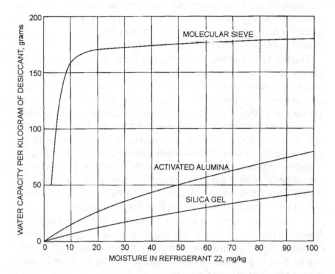

Fig. 2 Moisture Equilibrium Curves for R-22 and Three Common Desiccants at 24°C

Control of Moisture and Other Contaminants in Refrigerant Systems

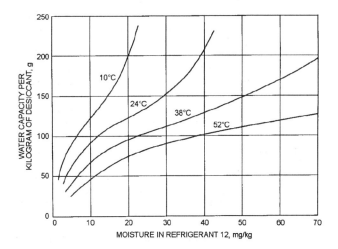

Fig. 3 Moisture Equilibrium Curves for Activated Alumina at Various Temperatures in R-12

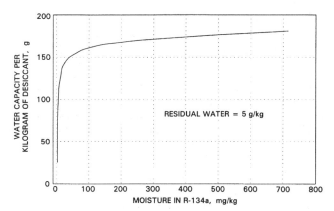

Fig. 4 Moisture Equilibrium Curves for Molecular Sieve in R-134a at 52°C
(Courtesy UOP, Reprinted with permission.)

Figure 4 shows water capacity of a molecular sieve in liquid R-134a at 52°C. This data was obtained using the Karl Fischer method similar to that described in Dunne and Clancy (1984).

Although the figures show that molecular sieves have higher capacities than activated alumina or silica gel at the indicated EPD, all three desiccants are suitable if sufficient quantities are used. Cost, operating temperature, other contaminants present, and equilibrium capacity at the desired EPD must be considered when choosing a desiccant for refrigerant drying. The desiccant manufacturer has information and equilibrium curves for specific desiccant-refrigerant systems.

Desiccant Applications

In addition to removing water, desiccants may be capable of adsorbing or reacting with acids, dyes, chemical additives, and refrigerant oil reaction products.

Acids. Generally, acids can harm refrigerant systems. The quantity of acid that a refrigerant system tolerates depends in part on the size, the mechanical design and construction of the system, the type of motor insulation, the type of acid, and the amount of water in the system.

The acid removal capacity of desiccants is difficult to determine because the environment is complex. Hoffman and Lange (1962) and Mays (1962) showed that desiccants remove acids from refrigerants and oils by adsorption and/or chemical reaction. Hoffman also showed that the concentration of water in the desiccant, the type of desiccant, and the type of acid play a major role in a desiccant's ability to remove acids from refrigerant systems. In addition, acids formed in these systems can be inorganic, such as HCl and HF, or a mixture of organic acids. All factors must be considered to establish acid capacities of desiccants.

Colors. Colored materials frequently are adsorbed by activated alumina and silica gel and occasionally by calcium sulfate and molecular sieves. Leak detector dyes may lose their effectiveness when used in systems containing desiccants. The interaction of the dye and drier should be evaluated before putting a dye in the system.

Oil Deterioration Products. Oils can react chemically to produce substances that are adsorbed by desiccants. Some of these are hydrophobic and, when adsorbed by the desiccant, reduce the rate at which it can adsorb liquid water. The rate and capacity of the desiccant to remove water dissolved in the refrigerant are not significantly impaired, however (Walker et al. 1955). Frequently, the reaction products are sludges or powders that can be filtered out mechanically by the drier.

Chemicals. Refrigerants that can be adsorbed by desiccants cause the drier temperature to rise considerably when the refrigerant is first admitted. This temperature rise is not the result of moisture in the refrigerant, but the adsorption heat of the refrigerant. Oil additives may be adsorbed by silica gel and activated alumina. Methanol is coadsorbed with moisture and competes with it for adsorption capacity in desiccants. Because of small pore size, molecular sieves generally do not adsorb additives or the oil.

Driers

A drier is a device containing a desiccant. It collects and holds moisture, but it also acts as a filter and adsorber of acids and other contaminants.

To prevent moisture from freezing in the expansion valve or capillary tube, a drier is installed in the liquid line close to these devices. Hot locations should be avoided. Driers can function on the low-pressure side of expansion devices, but this is not the preferred location (Jones 1969).

Moisture is reduced as liquid refrigerant passes through a drier. However, Krause et al. (1960) showed that considerable time is required before moisture equilibrium is reached in a refrigeration unit. The moisture is usually distributed throughout the entire system, and time is required for the circulating oil-refrigerant mixture to carry the moisture to the drier. Cohen and Dunne (1987) and Cohen (1994) treat the kinetics of drying refrigerants in circulating systems.

Driers are also used effectively to clean up systems severely contaminated due to hermetic motor burnouts and mechanical failures (see the System Cleanup Procedure after Hermetic Motor Burnout section).

Drier Selection

The drier manufacturer selection chart lists the amount of desiccants, flow capacity, filter area, water capacity, and a specific recommendation covering the type and refrigeration capacity of the drier for various applications.

The equipment manufacturer must consider the following factors when selecting a drier:

1. The **desiccant** is the heart of the drier and its selection is most important. The section on desiccants has further information.
2. The drier's **water capacity** is measured by methods described in ARI *Standard* 710. The reference points are set arbitrarily to prevent confusion arising from determinations made at other points. The specific refrigerant, the amount of desiccant, and the effect of temperature are all considered in the statement of water capacity.
3. The **liquid line** flow capacity is listed at 14kPa pressure drop across the drier by the official procedures of ASHRAE *Standard* 63.1 and ARI *Standard* 710. Jones (1964) developed a gravity flow method for determining flow capacities that uses R-113 and

converts the results to R-12 and R-22. Rosen et al. (1965) described a closed-loop method for evaluating filtration and flow characteristics of liquid line refrigerant driers. The flow capacity of suction line filters and filter-driers is determined according to ASHRAE *Standard* 78-1985 and ARI *Standard* 730-86. The latter standard gives recommended pressure drops for selecting suction line filter-driers for permanent and temporary installations. Flow capacity may be reduced quickly when critical quantities of solids and semisolids are filtered out by the drier. Whenever flow capacity drops below the machine's requirements, the drier should be replaced.

4. Although limits for particle size vary with refrigerant system size and design, and with the geometry and hardness of the particles, manufacturers publish **filtration capabilities** for comparison.

Testing and Rating

Desiccants and driers are tested according to the procedures of ASHRAE *Standards* 35 and 63.1, respectively. Driers are rated under ARI *Standard* 710. Minimum standards for listing of refrigerant driers can be found in UL *Standard* 207. Filtration ratings and test standards have not been developed.

OTHER CONTAMINANTS

Refrigerant filter-driers are the principal devices used to remove contaminants from refrigeration systems. The filter-drier is not a substitute for poor workmanship or design, but a maintenance tool necessary for continued and proper system performance. Contaminants removed by filter-driers include moisture, acids, hydrocarbons with a high molecular mass, oil decomposition products, and insoluble material, such as metallic particles and copper oxide.

Metallic Contaminants and Dirt

Small contaminant particles frequently left in refrigerating systems during manufacture or servicing include chips of copper, steel, or aluminum; copper or iron oxide; copper or iron chloride; welding scale; brazing or soldering flux; sand; and other dirt. Some of these contaminants, such as copper chloride, develop from normal wear or chemical breakdown during system operation. Solid contaminants vary widely in size, shape, and density. Solid contaminants create problems by:

- Scoring cylinder walls and bearings
- Lodging in the motor insulation of a hermetic system, where they act as conductors between individual motor windings or abrade the wire coating when flexing of the windings occurs
- Depositing on terminal blocks and serving as a conductor
- Plugging expansion valve screen or capillary tubing
- Depositing on suction or discharge valve seats, significantly reducing compressor efficiency
- Plugging oil holes in compressor parts, leading to improper lubrication
- Increasing the rate of chemical breakdown. At elevated temperatures, R-22, for example, decomposes more readily when in contact with iron powder, iron oxide, or copper oxide (Norton 1957)
- Plugging driers

Liquid line filter-driers, suction filters, and strainers isolate contaminants from the compressor and expansion valve. Filters minimize the return of particulate matter to the compressor and expansion valve; but the capacity of permanently installed liquid and/or suction filters must accommodate this particulate matter without causing excessive, energy-consuming pressure losses. Equipment manufacturers should consider the following procedures to ensure proper operation during the design life:

1. Develop cleanliness specifications that include a reasonable value for maximum residual matter. Some manufacturers specify allowable quantities in terms of internal surface area. ASTM *Standard* B 280 allows a maximum of 37 mg of contaminants per square metre of internal surface.
2. Multiply the factory contaminant level by a factor of five to allow for solid contaminants that will be added during installation. This factor depends on the type of system and the previous experience of the installers, among other considerations.
3. Determine a value for maximum pressure drop to be incurred by the suction or liquid filter when loaded with the quantity of solid matter calculated in Step 2.
4. Conduct pressure drop tests according to ASHRAE *Standard* 63.2.
5. Select driers for each system according to its capacity requirements and test data. In addition to contaminant removal capacity, tests can evaluate filter efficiency, maximum escaped particle size, and average escaped particle size.

Very small particles passing through filters tend to accumulate in the crankcase. Most compressors tolerate a small quantity of these particles without allowing them into the oil pump inlet, where they can damage running surfaces.

Organic Contaminants—Sludge, Wax, and Tars

Organic contaminants in a refrigerating system with a mineral oil lubricant can appear when organic materials, such as oil, insulation, varnish, gaskets, and adhesives decompose. As opposed to inorganic contaminants, these materials are mostly carbon, hydrogen, and oxygen. Organic materials may be partially soluble in the refrigerant-lubricant mixture or may become so through the action of heat. They then circulate in the refrigerating system and can plug small orifices. Organic contaminants in a refrigerating system using a synthetic polyol ester lubricant may also generate sludge. The following contaminants should be avoided:

Paraffin—typically found in mineral oil lubricants
Silicone—found in some machine lubricants
Phthalate—found in some machine lubricants

Whether mineral oil or synthetic lubricants are used, some organic contaminants remain in a new refrigerating system during manufacture or assembly. For example, excessive brazing paste introduces a wax-like contaminant into the refrigerant stream. Certain cutting lubricants, corrosion inhibitors, or drawing compounds frequently contain paraffin based compounds. These lubricants can leave a layer of paraffin on a component that may be removed by the refrigerant-lubricant combination and generate insoluble material in the refrigerant stream. Organic contamination also results during the normal method of fabricating return bends. The die used during forming is lubricated with these organic materials, and afterwards the return bend is brazed to the tubes to form the evaporator and/or condenser. During brazing, residual lubricant inside the tubing and bends can be baked to a resinous deposit.

If organic materials are handled improperly, certain contaminants remain. Resins used in varnishes, wire coating, or casting sealers may not be cured properly and can dissolve in the refrigerant-lubricant mixture. Solvents used in washing stators may be adsorbed by the wire film and later, during compressor operation, carry chemically reactive organic extractables. Chips of varnish, insulation, or fibers can detach and circulate in the system. Portions of improperly selected or cured rubber parts or gaskets can dissolve in the refrigerant.

Refrigeration grade oil decomposes under adverse conditions to form a resinous liquid or a solid frequently found on refrigeration filter-driers. These oils decompose noticeably when exposed for as little as 2 h to temperatures as low as 120°C in an atmosphere of air or oxygen. The compressor manufacturer should perform all high-temperature dehydrating operations on the machines prior to adding the oil charge. In addition, equipment manufacturers should

Control of Moisture and Other Contaminants in Refrigerant Systems

not expose compressors to processes requiring high temperatures unless the compressors contain refrigerant or inert gas.

The result of organic contamination is frequently noticed at the expansion device. Materials dissolved in the refrigerant-lubricant mixture, under liquid line conditions, may precipitate at the lower temperature in the expansion device, resulting in restricted or plugged capillary tubes or sticky expansion valves. A few milligrams of these contaminants can render a system inoperative. These materials have physical properties that range from a fluffy powder to a solid resin entraining inorganic debris. If the contaminant is dissolved in the refrigerant-lubricant mixture in the liquid line, it may not be removed by a filter-drier.

Chemical identification of these organic contaminants is very difficult. Infrared spectroscopy can characterize the type of organic groups present in contaminants. Materials found in actual systems vary from wax-like aliphatic hydrocarbons to resin-like materials containing double bonds, carbonyl groups and carboxyl groups. In some cases, organic compounds of copper and/or iron have been identified.

These contaminants can be eliminated by carefully selecting materials and strictly controlling cleanliness during manufacture and assembly of the components as well as the final system. Because heat degrades most organic materials and enhances chemical reactions, operating conditions with excessively high discharge temperatures must be avoided to prevent formation of degradation products.

Residual Cleaning Agents for Mineral Oil Systems

Solvents used for cleaning compressor parts are likely contaminants if left in refrigerating equipment. Solvents in this category are considered pure liquids without additives. If additives are present, they are reactive materials and should not be in a refrigerating system. Some solvents are relatively harmless to the chemical stability of the refrigerating system, while others initiate or accelerate degradation reactions. For example, the common mineral spirits solvents are considered harmless. Other common compounds react rapidly with hydrocarbon lubricating oils (Elsey et al. 1952).

Residual Cleaning Agents for Polyol Ester Lubricated Systems

Typical solvents used in cleaning mineral oil systems are not compatible with polyol ester lubricants. Several chemicals must be avoided to reduce or eliminate possible contamination and sludge generation. For example, a small amount of the following contaminants can cause a refrigerating system to fail:

Chlorides—typically found in chlorinated solvents
Acid or alkali—found in some water-based cleaning fluids
Water—a component of water-based cleaning fluids

Noncondensable Gases

Gases, other than the refrigerant, are another contaminant frequently found in refrigerating systems. These gases result: (1) from incomplete evacuation, (2) when functional materials release sorbed gases or decompose to form gases at an elevated temperature during system operation, (3) through low-side leaks, and (4) from chemical reactions during system operation. Chemically reactive gases, such as hydrogen chloride, attack other components, and, in extreme cases, the refrigerating unit fails.

Chemically inert gases, which do not liquefy in the condenser, reduce cooling efficiency. The quantity of inert, noncondensable gas that is harmful depends on the design and size of the refrigerating unit and on the nature of the refrigerant. Its presence contributes to higher than normal head pressures and resultant higher discharge temperatures, which speed up undesirable chemical reactions.

Gases found in hermetic refrigeration units include nitrogen, oxygen, carbon dioxide, carbon monoxide, methane, and hydrogen. The first three gases originate from incomplete air evacuation or a low-side leak. Carbon dioxide and carbon monoxide usually form when organic insulation is overheated. Hydrogen has been detected when a compressor is experiencing serious bearing wear. These gases are also found where a significant refrigerant-lubricant reaction has occurred. Only trace amounts of these gases are present in well-designed, properly functioning equipment.

Spauschus and Olsen (1959), Doderer and Spauschus (1966), and Gustafsson (1977) developed sampling and analytical techniques for establishing the quantities of contaminant gases present in refrigerating systems. Parmelee (1965), Spauschus and Doderer (1961, 1964), and Kvalnes (1965) applied gas analysis techniques to sealed tube tests to yield information on stability limitations of refrigerants, in conjunction with other materials used in hermetic systems.

Motor Burnouts

Motor burnout is the final result of hermetic motor insulation failure. During burnout, high temperatures and arc discharges can severely deteriorate the insulation, producing large amounts of carbonaceous sludge, acid, water, and other contaminants. In addition, a burnout can chemically alter the lubricating oil, and/or thermally decompose refrigerant in the vicinity of the burn. The products of burnout escape into the system, causing severe cleanup problems. If decomposition products are not removed, replacement motors fail with increasing frequency.

While RSES (1988) has chosen to differentiate between mild and severe burnouts, many compressor manufacturer's service bulletins treat all burnouts alike. A rapid burn from a spot failure in the motor winding results in a mild burnout with little oil discoloration and no carbon deposits. A severe burnout occurs when the compressor remains on line and burns over a longer period, resulting in highly discolored oil, carbon deposits, and acid formation.

Because the condition of the lubricant can be used to indicate the amount of contamination, the lubricant should be examined during the cleanup process. Wojtkowski (1964) stated that acid in R-22/mineral oil systems should not exceed 0.05 acid number (milligrams KOH per kilogram refrigerant). Commercial acid test kits can be used for this analysis. An acceptable acid number for other lubricants has not been established.

Various methods are recommended for cleaning a system after hermetic motor burnout (RSES 1988). However, the suction line filter-drier method is commonly used (see the section on System Cleanup Procedure after Hermetic Motor Burnout).

Field Assembly

Proper field assembly and maintenance are essential for contaminant control in refrigerating systems and to prevent undesirable refrigerant emissions to the atmosphere. Driers may be too small or carelessly handled so that drying capacity is lost. Improper tube-joint soldering is a major source of water, flux, and oxide scale contamination. Copper oxide scale from improper brazing is one of the most frequently observed contaminants. Careless tube cutting and handling can introduce excessive quantities of dirt and metal chips. Care should be taken to minimize these sources of internal contamination. In addition, because an assembled system can not be dehydrated easily, oversized driers should be installed. Even if components are delivered sealed and dry, weather and the time the unit is open during assembly can introduce large amounts of moisture.

In addition to internal sources, external factors can cause a unit to fail. Too small or too large transport tubing, mismatched or misapplied components, fouled air condensers, scaled heat exchangers, inaccurate control settings, failed controls, and improper evacuation are some of these factors.

REFRIGERANT RECOVERY, RECYCLING, AND RECLAMATION

Studies have shown that chlorofluorocarbon (CFC) and hydrochlorofluorocarbon (HCFC) refrigerants deplete ozone in the stratosphere when released to the atmosphere. Various international and federal regulations require actions to contain and properly handle refrigerants. The procedures involved in removing contaminants when recycling refrigerants are similar to those discussed earlier in this chapter. Service techniques, proper handling and storage, and possible mixing of refrigerants are of concern. Building owners, equipment manufacturers, and contractors are concerned about reintroducing refrigerants with unacceptable levels of contaminants into refrigeration equipment.

Installation and Service Practices

Proper installation and service procedures, including proper evacuation and leak checking, are essential to minimize major equipment repairs. Service lines should be made of hose material with low permeability and should include shutoff valves. Larger systems should include isolation valves and pumpdown receivers. ASHRAE *Guideline* 3 gives further detail on equipment, installation, and service requirements.

Recovering refrigerant to an external storage container and then returning the refrigerant for cleanup inside the refrigeration system is similar to the procedure described in item C in the section on System Cleanup Procedure after Hermetic Motor Burnout. Some additional air and moisture contamination may be introduced in the service procedure. In general, the refrigeration system must be cleaned whether the refrigerant is isolated in the receiver, recovered into a storage container, recycled, reclaimed, or replaced with new refrigerant. This cleanup is necessary because contaminants are distributed throughout the system. The advantage of new, reclaimed, or recycled refrigerant is that a properly cleaned system is not recontaminated by the addition of impure refrigerant.

Contaminants

The contaminants encountered in recovered refrigerants are covered in previous sections of this chapter. The main contaminants are moisture, acid, non-condensables, particulates, high boiling residue (oil and sludge), and other condensable gases (Manz 1995). The characteristics of these contaminants are as follows:

1. Moisture is normally dissolved in the refrigerant or lubricant, but sometimes free water is present. Moisture is removed by passing the refrigerant through a filter-drier. Some moisture is also removed by oil separation.
2. Acid consists of organic and inorganic types. Organic acids are normally contained in the oil and are removed in the oil separator and in the filter-drier. Inorganic acids, like hydrochloric acid, are removed by the non-condensable purge process, reaction with metal surfaces, and by the filter-drier.
3. Non-condensable gases consist primarily of air. These gases can come from the refrigeration equipment or can be introduced during servicing procedures. The control method consists of minimizing infiltration through proper equipment construction and installation (ASHRAE *Guideline* 3). Proper service equipment construction, connection techniques, and maintenance procedures (such as during filter-drier change) also reduce air contamination. Typically, a vapor purge is used to remove air.
4. Particulates are removed by suction filters, oil separators, and filter-driers.
5. High boiling residues consist primarily of refrigerant oil and sludge. Because different refrigeration systems use different oils and because oil is a collection point for other contaminants, the oil is considered a contaminant. High boiling residues are removed by separators designed to extract the oil from the vapor phase refrigerant, or by a distillation process.
6. Other condensable gases consist mainly of other refrigerants. They can be generated in small quantities by high-temperature operation or during a burnout. In rare cases, refrigerants may be mixed intentionally for performance or to top off with substitutes. Refrigerants should not be mixed in order to maintain the purity of the used refrigerant supply as well as the performance and durability of the particular system. In general, separation of other condensable gases, if possible, can only be done at a fully equipped reclamation center.
7. Mixed refrigerants are a special case of other condensable gases in that the refrigerant would not meet product specifications even if all moisture, acids, particulates, oil, and non-condensables were removed. Inadvertent mixing may occur because of a failure to

- Dedicate and clearly mark containers for specific refrigerants
- Clear hoses or recovery equipment before switching to a different refrigerant
- Test the suspect refrigerant before consolidating it into large batches, or
- Use proper retrofit procedures

Refrigerant Recovery

To **recover** means to remove refrigerant in any condition from a system and to store it in an external container. Recovery reduces refrigerant emissions to the atmosphere and is a necessary first or concurrent step to either recycling or reclamation. The largest potential for service-related emissions of refrigerant occurs during recovery. These emissions consist of refrigerant left in the system (recovery efficiency) and losses due to service connections (Manz 1995).

The key to reducing emissions is proper recovery equipment and techniques. The recovery equipment manufacturer and the technician must share this responsibility to minimize refrigerant loss to the atmosphere. Training in handling halocarbon refrigerants is required to learn the proper techniques (RSES 1991).

Important: *Recover refrigerants into a suitable container and keep containers for different refrigerants separate. Do not overfill containers as liquid expansion with rising temperature could cause loss of refrigerant through the pressure relief valve or even rupture of the container.*

Medium- and high-pressure refrigerants are commonly recovered using a compressor-based recovery unit to pump the refrigerant directly into a storage container (Manz 1995). Such a system is shown in Figure 5. Minimum functions include the processes of evaporating, compressing, condensing, storing, and controlling. Where possible, the recovery unit should be connected to both the high and low side ports to speed up the process (Manz 1995). Removal of the refrigerant as a liquid, especially where the refrigerant is to be reclaimed, greatly speeds the process (Clodic 1994). As a variation, a refrigeration unit may be used to cool the storage container to directly transfer the refrigerant. For low pressure refrigerants (R-11), a compressor or vacuum pump may be used to lower the pressure in the storage container and raise the pressure in the vapor space of the refrigeration system so that the liquid refrigerant will flow without evaporation. An alternative is to use a liquid pump to transfer the refrigerant (Manz 1995). A pumpdown unit such as a condensing unit may be required to remove the remaining vapor refrigerant after liquid removal is complete. Recovery systems for use at a factory for charging or leak testing operations will probably be larger in size and of specialized construction to meet the specific needs of the manufacturer (Parker 1988).

Components in which liquid could be trapped, such as an accumulator, may need to be gently heated (with a thermostatically-controlled heating blanket or a warm air gun) to remove all the

Control of Moisture and Other Contaminants in Refrigerant Systems

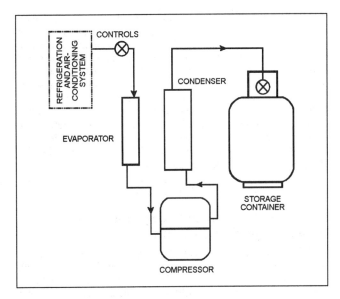

Fig. 5 Recovery Functions

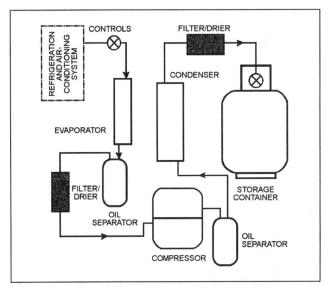

Fig. 6 Single Pass Recycling

refrigerant. Good practice requires watching for a pressure rise after recovery is completed to determine if the recovery unit needs to be restarted to remove all refrigerant. Where visual inspection is possible, such components would be identified by frosting on external surfaces to the level of the liquid refrigerant inside.

For fast refrigerant transfer, the entire liquid phase must be recovered without evaporating it or evaporating only a very small fraction. Depending on the particular refrigeration circuit, special methods must be developed, access may have to be created, and components may need to be modified; and these modifications must be simple and fast (Clodic 1994). Oil separation is essential in systems where used refrigerant is to be introduced without reclaiming. It may take longer to pump out vapor and separate oil, but clean recovery units, clean storage containers, and clean refrigeration systems are usually worth the extra time (Manz 1995).

Refrigerant Recycling

To **recycle** means to reduce contaminants in used refrigerants by separating oil, removing noncondensables, and using devices such as filter-driers to reduce moisture, acidity, and particulate matter. The term usually applies to procedures implemented at the field job site or at a local service shop. Industry guidelines (ARI [IRG-2] 1994) and federal regulations (EPA 1996) specify maximum contaminant levels in recycled refrigerant for certified recycling equipment under ARI *Standard* 740.

Recycling conserves limited supplies of regulated refrigerants (e.g., R-12). A single pass recycling schematic is shown in Figure 6 (Manz 1995). In the single pass recycling unit, refrigerant is processed by oil separation and filter-drying in the recovery path. Typically, air and noncondensables are not removed during the recovery process and are handled at a later time.

In a multiple pass recycling unit, as shown in Figure 7, the refrigerant is typically processed through an oil separation process during recovery. The filter-drier may be placed in the compressor suction line or in a bypass recycling loop or both. During a continuous recycling loop, refrigerant is withdrawn from the storage tank and processed through filter-driers and returned to the storage container. Noncondensable purge is accomplished during this recycling loop.

The primary function of the filter-drier is to remove moisture while the secondary function is to remove acid, particulate, and sludge (Manz 1995). The capability of the filter-drier to remove moisture and acid from used refrigerants is improved if the oil is separated prior to passing through the filter-drier (Kauffman 1992).

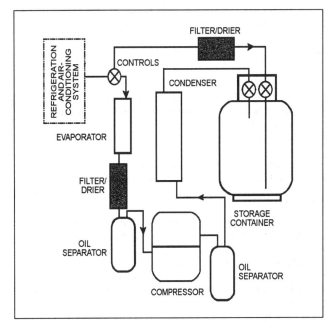

Fig. 7 Multiple Pass Recycling

Moisture indicators are typically used to indicate when a filter-drier change is required. For some refrigerants, these devices cannot indicate a moisture level as low as the purity level required by ARI *Standard* 700. Devices such as an in-situ mass flowmeter can be used to accurately determine when to change the filter-drier and still meet the purity requirements (Manz 1995).

Important: *The service technician must change recovery/recycle unit filter-driers at frequent intervals as directed by the indicators and manufacturer's instructions.*

The primary advantage of recycling is performing this operation at the job site or at a local service shop and avoiding transportation costs. The likelihood of mixing refrigerants is reduced if recycling is done at the service shop compared to consolidating refrigerant batches for shipment to a reclamation facility. Recycling equipment cannot separate mixed refrigerants to bring them back to product specifications.

A preliminary investigation of recycling refrigerants R-404A, R-410A, and R-507 showed that the refrigerant blend compositions changed less than 1% after 23 repetitions (Manz 1996). The study showed similar moisture removal capabilities as for R-22 under the test conditions. The most difficult contaminants to remove were non-condensable gases (air).

Industry Guidelines

An industry recycling guide (IRG-2), "Handling and Reuse of Refrigerants in the United States," (ARI 1994) includes a flow chart that outlines the following options:

Option 1: Put refrigerant back into the system without recycling it.
Option 2: Recycle refrigerant and put it back into the system from which it was removed or back into a system with the same owner.
Option 3: Recycle the refrigerant, test to verify conformance to ARI *Standard* 700 prior to reuse in a different owner's equipment provided that the refrigerant remains in the contractor's custody and control at all times from recovery through recycling to reuse.
Option 4: Send refrigerant to a certified reclaimer

IRG-2 states, "Used refrigerants shall not be sold, or used in a different owner's equipment, unless the refrigerant has been analyzed and found to meet requirements of ARI *Standard* 700 (latest edition), 'Specifications for Fluorocarbon and Other Refrigerants'."

IRG-2 provides maximum contaminant levels of recycled refrigerants in same owner's equipment and lists the following reasons for concern for mixed refrigerant:

- Effect on performance and operating characteristics that may affect the capacity and efficiency of the equipment
- Effect on materials compatibility, lubrication, equipment life, and warranty costs
- Increased service and repair requirements and higher operating costs
- High cost or inability to separate refrigerants
- High cost of disposal and loss of refrigerant for future service

Equipment Standards

Recovery and recycling equipment comes in a variety of sizes, shapes, and functions (Manz 1995). ARI *Standard* 740 establishes methods of testing for rating performance of equipment for type of equipment, designated refrigerants, liquid recovery rate, vapor recovery rate, final recovery vacuum, recycle rate, and trapped refrigerant. The standard requires that refrigerant emissions due to oil draining, non-condensable purging, and clearing between refrigerant types not exceed 3% by mass.

In the test method, contaminated refrigerant (sometimes called "dirty cocktail") is processed to determine the contaminant levels of the recycled refrigerant. Measured contaminant levels of recycled refrigerant include moisture content, chloride ions, acidity, high boiling residue, particulates/solids, and non-condensables. Each refrigerant is sampled at the time of the first filter-drier change when levels are expected to be highest. United States regulations (EPA 1996) require that recycling equipment meet the maximum contaminant levels in IRG-2 for recycled refrigerant (Option 2).

The basic distinction between recycling and reclamation is best illustrated by associating recycling with certification (ARI 740) and reclamation with analysis (ARI 700). ARI *Standard* 740 covers certification testing of the recycling equipment using a "standard contaminated refrigerant sample" in lieu of chemical analysis of each batch. It provides a means of comparing equipment performance under controlled conditions. By contrast, as described later, ARI *Standard* 700 is based on chemical analysis of a refrigerant sample from each batch after contaminant removal. ARI 700 provides a means of analyzing refrigerant and comparing the contaminant levels to product specifications.

ARI *Standard* 740 applies to single-refrigerant systems and their normal contaminants. It does not apply to refrigerant systems or storage containers with mixed refrigerants. No attempt is made to rate the equipment's ability to remove different refrigerants and other condensable gases from recovered refrigerant. The standard places responsibility on the equipment operator to identify those situations and to treat them accordingly. One of the uncertainties associated with recycled refrigerants is describing the purity levels when offering the refrigerant for resale. The purity of recycled refrigerant is uncertain because appropriate field measurement techniques do not exist for all contaminants listed in ARI 700. The Industry Recycling Guide (IRG-2) discusses possible options.

The Society of Automotive Engineers (SAE) has developed standards for mobile air-conditioning equipment.

Special Considerations and Equipment for Handling Multiple Refrigerants

Different refrigerants must be kept separate. Storage containers should meet applicable standards for transportation and use with that refrigerant as specified in ARI *Guideline* K. Disposable cylinders must not be used (RSES 1991). Containers should be filled per ARI *Guideline* K and marked with the refrigerant type. Container colors for recovered refrigerants and for new and reclaimed refrigerants are specified in ARI *Guidelines* K and N, respectively.

Recovery/recycling (R/R) equipment capable of handling more than one refrigerant is readily available and often preferred. The R/R equipment should only be used for labelled refrigerants. At the time the technician desires to switch refrigerants, a significant amount of the previous refrigerant will be contained in the R/R equipment, particularly in the condenser section (Manz 1991). This trapped refrigerant must be removed preferably by using isolation/bypass valves to connect the condenser section to the compressor suction and by connecting the compressor discharge directly to the storage container (Manz 1995). After the bulk of the refrigerant has been removed, the system should be evacuated before changing to the appropriate storage container for the new refrigerant. This procedure should also include all lines and connecting hoses and may include replacement of the filter-driers.

The need for purging noncondensables is determined by comparing the refrigerant pressure to the saturation pressure of pure refrigerant at the same temperature. Circulation to achieve thermal equilibrium may be required to eliminate the effect of the temperature difference on pressures. A sealed bulb is often used to determine the saturation pressure for a single refrigerant system. When purging air from R/R equipment capable of handling multiple refrigerants, the difference in saturation pressures between refrigerants far exceeds any allowable partial pressure due to noncondensables. Special equipment and/or techniques are required (Manz 1991, 1995).

Refrigerant to be recovered may be in vapor or liquid states. To optimize the recovery, R/R equipment must have the ability to handle each of these states. For some equipment, this may involve one hookup or piece of R/R equipment for liquid and a separate hookup or second piece of R/R equipment to recover vapor. In general, a single hookup is desired. When handling multiple refrigerants, traditional liquid flow control devices such as capillary tubes or expansion valves either compromise performance or simply do not work. Possible solutions include (1) the operator watching a sight glass for liquid flow and switching a valve; (2) multiple flow control devices with a refrigerant selection switch; and (3) a two-bulb expansion valve, which controls temperature differential across the evaporator (Manz 1991).

Refrigerant Reclamation

To **reclaim** means to process used refrigerant to new product specifications. Chemical analysis of the refrigerant is required to

Control of Moisture and Other Contaminants in Refrigerant Systems

determine that appropriate product specifications have been met. This term usually implies the use of processes or procedures available only at a reprocessing or manufacturing facility. United States regulations (EPA 1996) require that refrigerants must meet ARI *Standard* 700 contaminant levels in order to be sold using one of the following options discussed in IRG-2 (ARI 1994):

Option 3: Recycle the refrigerant, test to verify conformance to ARI *Standard* 700 prior to reuse in a different owner's equipment provided that the refrigerant remains in the contractor's custody and control at all times from recovery through recycling to reuse

Option 4: Send refrigerant to a certified reclaimer

United States regulations (EPA 1996) call for use of third party certified laboratories for option 3 and third party certified reclaimers for option 4 both based on ARI *Standard* 700.

Some equipment warranties, especially those for smaller consumer appliances, may not permit the use of refrigerants reclaimed to purity levels specified in ARI 700. For small appliances (refrigerators and freezers), the manufacturer's literature should be consulted before charging with reclaimed refrigerants.

Reclamation has traditionally been used for systems containing more than 50 kg refrigerant (O'Meara 1988). Assistance is often provided by the reclaimer in furnishing shipping containers and labeling instructions. Many reclaimers use air-conditioning and refrigeration wholesalers as collection points for refrigerant. Mixing of refrigerants at the consolidation points is possible. If the refrigerant is contaminated beyond limits, the price paid for the refrigerant may be reduced or the shipment may be refused. One of the advantages associated with reclaimed refrigerants is in describing the purity levels when offering the refrigerant for resale.

Purity Standards

ARI *Standard* 700 covers halocarbon refrigerants, regardless of source, and defines acceptable levels of contaminants, which are the same as Federal Specifications for Fluorocarbon Refrigerants BB-F-1421B. It specifies laboratory analysis methods for each contaminant. Only fully equipped laboratories with trained personnel are currently capable of performing the analysis.

Because ARI 700 is based on chemical analysis of a sample from each batch after contaminant removal, it is not concerned with the level of contaminants before contaminant removal (Manz 1995). This disassociation from the "standard contaminated refrigerant sample" required in ARI 740 is the basic distinction between analysis/reclamation and certification/recycling.

SAE *Standards* J1991 and J2099 contain recycled refrigerant purity levels for mobile air-conditioning systems using R-12 and R-134a, respectively.

SYSTEM CLEANUP PROCEDURE AFTER HERMETIC MOTOR BURNOUT

Introduction

This procedure is limited to positive-displacement hermetic compressors. Centrifugal compressor systems are highly specialized and are frequently designed for a particular application. A centrifugal system should be cleaned according to the manufacturer's recommendations. All or part of the procedure can be used depending on such factors as severity of the burnout and size of the refrigeration system.

After a hermetic motor burnout, the system must be cleaned thoroughly to remove all contaminants. Otherwise a repeat burnout will *likely* occur. Failure to follow these minimum cleanup recommendations as quickly as possible increases the potential for repeat burnout.

Procedure

A. **Make sure a burnout has occurred.** Although a motor that will not start appears to be a motor failure, the problem may be improper voltage, starter malfunction, or a compressor mechanical fault (RSES 1988).

1. Check for proper voltage.
2. Check that the compressor is cool to the touch. An open internal overload could prevent the compressor from starting.
3. Check the compressor motor for improper grounding using a megohmmeter or a precision ohmmeter.
4. Check the external leads and starter components.
5. Obtain a small sample of oil from the compressor, examine it for discoloration, and analyze it for acidity.

B. **Safety.** In addition to electrical hazards, service personnel should be aware of the hazard of acid burns. If the oil or sludge in a burned-out compressor must be touched, wear rubber gloves to avoid a possible acid burn.

C. **Cleanup after a burnout.** Just as proper installation and service procedures are essential to prevent compressor and system failures, proper system cleanup and installation procedures when installing the replacement compressor are also essential to prevent repeat failures. Key elements of the recommended procedures are:

1. In the United States, federal regulations require that the refrigerant be isolated in the system or recovered into an external storage container to avoid discharge into the atmosphere. Prior to opening any portion of the system for inspection or repairs, refrigerant should be recovered from that portion until the vapor pressure has been reduced to less than 87.6 kPa (absolute) for R-22 or 67.3 kPa (absolute) for CFC or other HCFC systems.
2. Remove the burned-out compressor and install the replacement. Save a sample of the new compressor oil that has not been exposed to refrigerant and store in a sealed glass bottle. This will be used later for comparison.
3. Inspect all system controls such as expansion valves, solenoid valves, check valves, etc. Clean or replace if necessary.
4. Install an oversized drier in the suction line to protect the replacement compressor from any contaminants remaining in the system. Install a pressure tap upstream of the filter-drier. This tap permits measuring the pressure drop from the tap to the service valve during the first hours of operation to determine if the suction line drier needs to be replaced.
5. Remove the old liquid line drier, if one exists, and install a replacement drier of the next larger capacity than is normal for this system. Install a moisture indicator in the liquid line if the system does not have one.
6. Evacuate and leak check the system or portion opened to the atmosphere according to the manufacturer's recommendations.
7. Recharge the system and begin operations according to the manufacturer's startup instructions. Typically:

    a. Observe pressure drop across the suction line drier for the first 4 hours. Follow the manufacturer's guide; otherwise compare to pressure drop curve in Figure 8 and replace driers as required.

    b. After 24 to 48 hours, check pressure drop and replace driers as required. Take an oil sample and check with an acid test kit. Compare the oil sample to the initial sample saved at the time the replacement compressor was installed. Cautiously smell the oil sample. Replace oil if acidity persists or if color or odor indicates.

    c. After 7 to 10 days or as required, repeat step b.

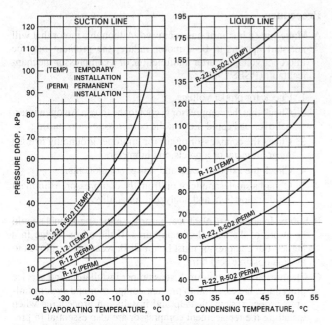

Fig. 8 Maximum Recommended Filter-Drier Pressure Drop

D. Additional suggestions

1. If sludge or carbon has backed up into the suction line, swab it out or replace that section of the line.
2. If a change in the suction line drier is required, change the oil in the compressor each time the cores are changed, if the compressor design permits.
3. Remove the suction line drier after several weeks of system operation to avoid excessive pressure drop in the suction line. This problem is particularly significant on commercial refrigeration systems.
4. In some cases, noncondensable gases are produced during the burnout. With the system off, compare the head pressure to the saturation pressure after stabilization at ambient temperature. Adequate time must be allowed to ensure stabilization. If required, purge the charge by recycling it or submit the purged material for reclaim.

Special System Characteristics and Procedures

Because of unique system characteristics, the procedures described here may require adaptations.

A. If an oil sample cannot be obtained from the new compressor, determine another method to get a sample from the system.

1. Install a tee and a trap in the suction line. An access valve at the bottom of the trap permits easy oil drainage. Only 15 mL of oil is required for an acid analysis. Be certain the oil sample represents oil circulating in the system. It may be necessary to drain the trap and discard the first amount of oil collected, before collecting the sample to be analyzed.
2. Make a trap from 35 mm copper tubing and valves. Attach this trap to the suction and discharge gage port connections with a charging hose. By blowing discharge gas through the trap and into the suction valve, enough oil will be collected in the trap for analysis. This trap becomes a tool that can be used repeatedly on any system that has suction and discharge service valves. Be sure to clean the trap after every use to avoid cross contamination.

B. On semihermetic compressors, remove the cylinder head to determine the severity of burnout. Dismantle the compressor for solvent cleaning and hand wiping to remove contaminants. Consult the manufacturer's recommendations on compressor rebuilding and motor replacement.

C. In rare instances on a close-coupled system, where it is not feasible to install a suction line drier, the system can be cleaned by repeated changes of the cores in the liquid line drier and repeated oil changes.

D. On heat pump systems, the four-way valve and the compressor should be carefully inspected after a burnout. In cleaning a heat pump after a motor burnout, it is essential to remove any drier originally installed in the liquid line. These driers may be replaced for cleanup, or a biflow drier may be installed in the common reversing liquid line.

E. Systems with a critical charge require a particular effort for proper operation after cleanup. If an oversized liquid line drier is installed, an additional charge must be added. Check with the drier manufacturer for specifications. However, no additional charge is required for the suction line drier that may be added.

F. The new compressor should not be used to pull a vacuum. Refer to the manufacturer's recommendations for evacuation. Normally, the following method is used, after determining that there are no refrigerant leaks in the system.

Pull a high vacuum to an absolute pressure of less than 65 mPa for several hours. Allow the system to stand several hours to be sure the vacuum is maintained. This requires a good vacuum pump and an accurate high-vacuum gage.

CONTAMINANT CONTROL DURING RETROFIT

Because of the phase-out of CFCs, existing refrigeration and air-conditioning systems are commonly retrofitted to alternative refrigerants. The term "refrigerant" in this section refers to a fluorocarbon working fluid offered as a possible replacement for a CFC, whether that replacement consists of one chemical, an azeotrope of two chemicals, or a blend of two or more chemicals. The terms "retrofitting" and "conversion" are used interchangeably to mean the modification of an existing refrigeration or air-conditioning system designed to operate on a CFC so that it can safely and effectively operate on an HCFC or HFC refrigerant. This section only covers the contaminant control aspects of such conversions. Equipment manufacturers should be consulted for guidance regarding the specifics of actual conversion. Industry standards and manufacturers' literature are also available which contain supporting information (UL 1993A, UL 1993B, UL 1993C).

From a contaminant control point of view, retrofitting a CFC system to an alternative refrigerant fall into the following categories:

- **Cross contamination of old and new refrigerants.** This should be avoided even though there are usually no chemical compatibility problems between the CFCs and their replacement refrigerants. One problem with mixing refrigerants is that it is difficult to determine system performance after retrofit. Pressure-temperature relationships are different for a blend of two refrigerants than for each refrigerant individually. A second concern with mixing refrigerants is that if the new refrigerant charge must be removed in the future, the mixture may not be reclaimable (DuPont 1992).
- **Cross-contamination of old and new lubricant.** Equipment manufacturers generally specify that the existing lubricant be replaced with the lubricant they consider suitable for use with a given HFC refrigerant. In some cases, the new lubricant is incompatible with the old one or with chlorinated residues present. In other cases, the old lubricant is insoluble with the new refrigerant and tends to collect in the evaporator, interfering with heat transfer. When mineral oil is replaced by, for example, a polyol ester oil during retrofit to an HFC refrigerant, a typical recommendation is to reduce the old oil content to 5% or less of the nominal oil charge (Castrol 1992). Some retrofit recommendations specify lower levels of acceptable contamination in the case of polyol ester lubricant/HFC retrofits,

Control of Moisture and Other Contaminants in Refrigerant Systems

so original equipment manufacturers recommendations should be obtained before attempting a conversion.

- **Chemical compatibility of old system components with new fluids.** One of the preparatory steps in a retrofit is to confirm that either the existing materials in the system are acceptable or that replacement materials are on hand to be installed in the system during the retrofit. Fluorocarbon refrigerants generally have solvent properties, and some are very aggressive. This characteristic can lead to swelling and extrusion of polymer "O" rings, undermining their sealing capabilities. Material can also be extracted from polymers, varnishes, and resins used in hermetic motor windings. These extracts can then collect in expansion devices, interfering with system operation. Residual manufacturing fluids such as those used to draw wire for compressor motors can be extracted from components and deposited in areas where they can interfere with operation. Suitable materials of construction have been identified by equipment manufacturers for use with HFC refrigerant systems.

Drier media must also be chemically compatible with the new refrigerant and effective in removing moisture, acid, and particulate in the presence of the new refrigerant. Drier media commonly used with CFC refrigerants tend to accept small HFC refrigerant molecules and lose moisture retention capability (Cohen and Blackwell 1995). Drier media have been developed that minimize this tendency.

REFERENCES

ARI. 1986. Flow-capacity rating and application of suction-line filters and filter-driers. *Standard* 730-86. Air-Conditioning and Refrigeration Institute, Arlington, VA.

ARI. 1986. Liquid-line driers. *Standard* 710-86. Air-Conditioning and Refrigeration Institute, Arlington, VA.

ARI. 1990. Containers for recovered fluorocarbon refrigerants. *Guideline* K-90. Air-Conditioning and Refrigeration Institute, Arlington, VA.

ARI. 1991. ARI refrigerant contaminant study—Final report. Air-Conditioning and Refrigeration Institute, Arlington, VA.

ARI. 1994. Handling and reuse of refrigerants in the United States. *Industry Recycling Guide* (IRG-2). Air-Conditioning and Refrigeration Institute, Arlington, VA.

ARI. 1995. Specifications for fluorocarbon and other refrigerants. *Standard* 700-95. Air-Conditioning and Refrigeration Institute, Arlington, VA.

ARI. 1995. Refrigerant recovery, recycling equipment. *Standard* 740-95. Air-Conditioning and Refrigeration Institute, Arlington, VA.

ARI. 1995. Guideline for assignment of refrigerant container colors. *Guideline* N-95. Air-Conditioning and Refrigeration Institute, Arlington, VA.

ASHRAE. 1985. Method of testing flow capacity of suction line filter driers. ASHRAE *Standard* 78-1985 (RA 90).

ASHRAE. 1992. Method of testing desiccants for refrigerant drying. ASHRAE *Standard* 35-1992.

ASHRAE. 1995. Method of testing liquid line refrigerant driers. ASHRAE *Standard* 63.1-1995.

ASHRAE. 1996. Method of testing liquid line filter-drier filtration capability. ASHRAE *Standard* 63.2-1996.

ASHRAE. 1996. Reducing emission of fully halogenated refrigerants in refrigeration and air-conditioning equipment and systems. ASHRAE *Guideline* 3-1996.

ASTM. 1995. Standard specification for seamless copper tube for air conditioning and refrigeration field service. *Standard* B 280 A-95. American Society for Testing and Materials, W. Conshohocken, PA.

Brisken, W.R. 1955. Moisture migration in hermetic refrigeration systems as measured under various operating conditions. *Refrigerating Engineering* (July):42.

Castrol. 1992. *Technical Bulletin No. 2*. Castrol Industrial North America, Specialty Products Division, Irvine, CA.

Cohen, A.P. 1993. Test methods for the compatibility of desiccants with alternative refrigerants. *ASHRAE Transactions* 99(1).

Cohen, A.P. 1994. Compatibility and performance of molecular sieve desiccants with alternative refrigerants. *Proceedings* International Conference, CFCs, The Day After. International Institute of Refrigeration, Paris.

Cohen, A.P. and C.S. Blackwell. 1995. Inorganic fluoride uptake as a measure of relative compatibility of molecular sieve desiccants with fluorocarbon refrigerants. *ASHRAE Transactions* 101(2).

Cohen, A.P. and S.R. Dunne. 1987. Review of automotive air-conditioning drydown rate studies—The kinetics of drying Refrigerant 12. *ASHRAE Transactions* 93(2).

Clodic, D. and F. Sauer. 1994. *The refrigerant recovery book*. ASHRAE, Atlanta.

DeGeiso, R.C. and R.F. Stalzer. 1969. Comparison of methods of moisture determination in refrigerants. *ASHRAE Journal* (April).

Doderer, G.C. and H.O. Spauschus. 1966. A sealed tube-gas chromatograph method for measuring reaction of Refrigerant 12 with oil. *ASHRAE Transactions* 72(2):IV, 4.1.

Dunne, S.R. and T.J. Clancy. 1984. Methods of testing desiccant for refrigeration drying. *ASHRAE Transactions* 90(1A):164.

DuPont. 1992. Acceptance specification for used refrigerants. *Bulletin* H-31790-1. E.I. duPont de Nemours and Company, Wilmington, DE.

Elsey, H.M. and L.C. Flowers. 1949. Equilibria in Freon-12—Water systems. *Refrigerating Engineering* (February):153.

Elsey, H.M., L.C. Flowers, and J.B. Kelley. 1952. A method of evaluating refrigerator oils. *Refrigerating Engineering* (July):737.

EPA. 1996. Regulations governing sale of refrigerant. United States Federal Register.

Griffith, R. 1993. Polyolesters are expensive, but probably a universal fit. *Air Conditioning, Heating & Refrigeration News* (May 3):38.

Gully, A.J., H.A. Tooke, and L.H. Bartlett. 1954. Desiccant-refrigerant moisture equilibria. *Refrigerating Engineering* (April):62.

Gustafsson, V. 1977. Determining the air content in small refrigeration systems. Purdue Compressor Technology Conference.

Haagen-Smit, I.W., P. King, T. Johns, and E.A. Berry. 1970. Chemical design and performance of an improved Karl Fischer titrator. *American Laboratory* (December).

Hoffman, J.E. and B.L. Lange. 1962. Acid removal by various desiccants. *ASHRAE Journal* (February):61.

Jones, E. 1964. Determining pressure drop and refrigerant flow capacities of liquid line driers. *ASHRAE Journal* (February):70.

Jones, E. 1969. Liquid or suction line drying? *Air Conditioning and Refrigeration Business* (September).

Kauffman, R.E. 1992. Chemical analysis and recycling of used refrigerant from field systems. *ASHRAE Transactions* 98(1).

Krause, W.O., A.B. Guise, and E.A. Beacham. 1960. Time factors in the removal of moisture from refrigerating systems with desiccant type driers. *ASHRAE Transactions* 66:465.

Kvalnes, D.E. 1965. The sealed tube test for refrigeration oils. *ASHRAE Transactions* 71(1):138.

Manz, K.W. 1988. Recovery of CFC refrigerants during service and recycling by the filtration method. *ASHRAE Transactions* 94(2).

Manz, K.W. 1991. How to handle multiple refrigerants in recovery and recycling equipment. *ASHRAE Journal* 33(4).

Manz, K.W. 1995. The challenge of recycling refrigerants. Business News Publishing, Troy, MI.

Manz, K.W. 1996. Recycling alternate refrigerants R-404A, R-410A, and R-507. *Proceedings* International Conference on Ozone Protection Technologies. Frederick, MD 411-419.

Mays, R.L. 1962. Molecular sieve and gel-type desiccants for Refrigerants 12 and 22. *ASHRAE Journal* (August):73.

McCain, C.A. 1991. Refrigerant reclamation protects HVAC equipment investment. *ASHRAE Journal* 33(4).

Norton, F.J. 1957. Rates of thermal decomposition of CHClF2 and CF2Cl2. *Refrigerating Engineering* (September):33.

O'Meara, D.R. 1988. Operating experiences of a refrigerant recovery services company. *ASHRAE Transactions* 94(2).

Parker, R.W. 1988. Reclaiming refrigerant in OEM plants. *ASHRAE Transactions* 94(2).

Parmelee, H.M. 1965. Sealed tube stability tests on refrigeration materials. *ASHRAE Transactions* 71(1):154.

Rosen, S., A.A. Sakhnovsky, R.B. Tilney, and W.O. Walker. 1965. A method of evaluating filtration and flow characteristics of liquid line driers. *ASHRAE Transactions* 71(1):200.

RSES. 1988. Standard procedure for replacement of components in a sealed refrigerant system (compressor motor burnout). Refrigeration Service Engineers Society, Des Plaines, IL.

RSES. 1991. Refrigerant service for the 90's, first edition. Refrigeration Service Engineers Society, Des Plaines, IL.

SAE. 1989. Standard of purity for use in mobile air-conditioning systems. *Standard* 1991-89. Society of Automotive Engineers, Warrendale, PA.

SAE. 1991. Standard of purity for recycle HFC-134a for use in mobile air-conditioning systems. *Standard* J2099-91. Society of Automotive Engineers, Warrendale, PA.

Spauschus, H.O. and G.C. Doderer. 1961. Reaction of Refrigerant 12 with petroleum oils. *ASHRAE Journal* (February):65.

Spauschus, H.O. and G.C. Doderer. 1964. Chemical reactions of Refrigerant 22. *ASHRAE Journal* (October):54.

Spauschus, H.O. and R.S. Olsen. 1959. Gas analysis—A new tool for determining the chemical stability of hermetic systems. *Refrigerating Engineering* (February):25.

Thrasher, J.S., R. Timkovich, H.P.S. Kumar, and S.L. Hathcock. 1993. Moisture solubility in Refrigerant 123 and Refrigerant 134a. *ASHRAE Transactions* 100(1).

UL. 1993a. Field conversion/retrofit of products to change to an alternative refrigerant—Construction and operation. *Standard* 2170-93. Underwriters Laboratories. Northbrook, IL.

UL. 1993b. Field conversion/retrofit of products to change to an alternative refrigerant—Insulating material and refrigerant compatibility. *Standard* 2171-93. Underwriters Laboratories. Northbrook, IL.

UL. 1993c. Field conversion/retrofit of products to change to an alternative refrigerant—Procedures and methods. *Standard* 2172. Underwriters Laboratories. Northbrook, IL.

UL. 1993d. UL standard for safety refrigerant-containing components and accessories, nonelectrical, sixth ed. *Standard* 207-93. Underwriters Laboratories, Northbrook, IL.

Walker, W.O. 1963. Latest ideas in use of desiccants and driers. *Refrigerating Service & Contracting* (August):24.

Walker, W.O., J.M. Malcolm, and H.C. Lynn. 1955. Hydrophobic behavior of certain desiccants. *Refrigerating Engineering* (April):50.

Walker, W.O., S. Rosen, and S.L. Levy. 1962. Stability of mixtures of refrigerants and refrigerating oils. *ASHRAE Journal* (August):59.

Wojtkowski, E.F. 1964. System contamination and cleanup. *ASHRAE Journal* (June):49.

BIBLIOGRAPHY

Boing, J. 1973. Desiccants and driers. RSES Service Manual, Section 5, 620-16B. Refrigeration Service Engineers Society, Des Plaines, IL.

Byrne, J.J., M. Shows, M.W. Abel. 1996. Investigation of flushing and clean-out methods for refrigeration equipment to ensure system compatibility. Air-Conditioning and Refrigeration Technology Institute, Arlington, VA. DOE/CE/23810-73.

Burgel, J., N. Knaup, and H. Lotz. 1988. Reduction of CFC-12 emission from refrigerators in the FRG. *International Journal of Refrigeration* 11(4).

Du Pont. 1976. Mutual solubilities of water with fluorocarbons and fluorocarbon-hydrate formation. E.I duPont de Nemours and Company, Wilmington, DE.

Guy, P.D., G. Tompsett, T.W. Dekleva. 1992. Compatibilities of nonmetallic materials with R-134a and alternative lubricants in refrigeration systems. *ASHRAE Transactions* 98(1).

Kauffman, R.E. 1992. Sealed tube tests of refrigerants from field systems before and after recycling. *ASHRAE Transactions* 99(2).

Kitamura, K., T. Ohara, S. Honda, and H. SakaKibara. 1993. A new refrigerant-drying method in the automotive air conditioning system using HFC-134a. *ASHRAE Transactions* 99(1).

Sundaresan, S.G. 1989. Standards for acceptable levels of contaminants in refrigerants. CFCs—Time of Transition, pp. 220-23. ASHRAE.

Walker, W.O. 1960. Contaminating gases in refrigerating systems. RSES Service Manual, Section 5, 620-15. Refrigeration Service Engineers Society, Des Plaines, IL.

Walker, W.O. 1985. Methyl alcohol in refrigeration. RSES Service Manual, Section 5, 620-17A. Refrigeration Service Engineers Society, Des Plaines, IL.

Zahorsky, L.A. 1967. Field and laboratory studies of wax-like contaminants in commercial refrigeration equipment. *ASHRAE Transactions* 73(1):II, 1.1.

Zhukoborshy, S.L. 1984. Application of natural zeolites in refrigeration industry. Proceedings of the International Symposium on Zeolites, Portoroz, Yugoslavia (September).

CHAPTER 7

LUBRICANTS IN REFRIGERANT SYSTEMS

Tests for Boundary Lubrication 7.1	Wax Separation (Floc Tests) 7.15
Refrigeration Lubricant Requirements 7.2	Solubility of Hydrocarbon Gases 7.19
Mineral Oil Composition 7.2	Solubility of Water in Lubricants 7.19
Component Characteristics 7.3	Solubility of Air in Lubricants 7.20
Synthetic Lubricants 7.3	Foaming and Antifoam Agents 7.20
Lubricant Additives 7.4	Oxidation Resistance 7.20
Lubricant Properties 7.4	Chemical Stability 7.20
Lubricant-Refrigerant Solutions 7.8	Conversion from CFC Refrigerants
Lubricant Return from Evaporators 7.14	to Other Refrigerants 7.23

THE primary function of a lubricant is to reduce friction and minimize wear. A lubricant achieves this by interposing a film between moving surfaces that reduces direct solid-to-solid contact or lowers the coefficient of friction.

Understanding the role of a lubricant requires an analysis of the surfaces to be lubricated. While bearing surfaces and other machined parts may appear and feel smooth, close examination reveals microscopic peaks (asperities) and valleys. With a sufficient quantity of lubricant, a layer is provided that has a thickness greater than the maximum height of the mating asperities, so that moving parts ride on a lubricant cushion.

These dual conditions are not always easily attained. For example, when the shaft of a horizontal journal bearing is at rest, the static loads squeeze out the lubricant, producing a discontinuous film with metal-to-metal contact at the bottom of the shaft. When the shaft begins to turn, there is no layer of liquid lubricant separating the surfaces. As the shaft picks up speed, the lubricating fluid is drawn into the converging clearance between the bearing and the shaft, generating a hydrodynamic pressure that eventually can support the load on an uninterrupted fluid film (Fuller 1984).

Various regimes or conditions of lubrication can exist when surfaces are in motion with respect to one another. Regimes of lubrication are defined as follows:

- *Full fluid film or hydrodynamic.* Mating surfaces are completely separated by the lubricant film.
- *Boundary.* Gross surface-to-surface contact occurs because the bulk lubricant film is too thin to separate the mating surfaces.
- *Mixed fluid film or quasi-hydrodynamic.* Occasional or random surface contact occurs.

Many materials can be used to separate and lubricate contacting surfaces. Separation can be maintained by a boundary layer on a metal surface, a fluid film, or a combination of both.

The function of a lubricant extends beyond preventing surface contact. It also removes heat, provides a seal to keep out contaminants or to retain pressures, inhibits corrosion, and carries away debris created by wear. Lubricating oils are best suited to meet these various requirements.

Viscosity is the most important property to consider in choosing a lubricant under full fluid film conditions. Under boundary conditions, the asperities are the contact points and support much, if not all, of the load; and the contact pressures are usually sufficient to cause welding and surface deformation. However, even under these conditions, wear can be controlled effectively with nonfluid, multimolecular films formed on the surface. These films must be strong enough to resist rupturing, yet have acceptable frictional and shear characteristics to reduce surface fatigue, adhesion, abrasion, and corrosion, which are the four major sources (either singularly or together) of rapid wear under boundary conditions. The slightly active constituents left in commercially refined mineral oils give them their natural film-forming properties.

Additives have also been developed to improve lubrication under boundary conditions. These materials are characterized by terms such as oiliness agents, lubricity improvers, and antiwear additives. They form a film on the metal surface through polar attraction or chemical action. These films or coatings have lower coefficients of friction under the loads imposed during boundary conditions. In chemical action, the temperature increase brought about by friction-generated heat brings about a reaction between the additive and the metal surface. Films such as iron sulfide and iron phosphate can be formed depending on the additives and the energy available for the reaction. In some instances, organic phosphates and phosphites are used in refrigeration oils to improve boundary lubrication. The nature of the metal and the condition of the metal surfaces are important. Refrigeration compressor designers often treat ferrous pistons, shafts, and wrist pins with phosphating processes that impart a crystalline, discontinuous film of metal phosphate to the surface. This film aids boundary lubrication during the break-in period.

TESTS FOR BOUNDARY LUBRICATION

Film strength or load carrying ability are terms often used to describe lubricant lubricity characteristics under boundary conditions. Laboratory tests that measure the degree of scoring, welding, or wear have been developed to evaluate lubricants. However, bench-type tests cannot be expected to accurately simulate actual field performance in a given compressor and are, therefore, merely screening devices. Some of these tests have been standardized by ASTM and other organizations.

In the **four-ball extreme-pressure method** (ASTM D 2783), the antiwear property is determined from the average scar diameter on the stationary balls and is stated in terms of a load-wear index. The smaller the scar, the better the load-wear index. The maximum load carrying capability is defined in terms of a weld point, i.e., the load at which welding by frictional heat occurs.

The **Falex** method (ASTM D 2670) allows measurement of wear during the test itself, and the scar width on the V-blocks and/or the mass loss of the pin can be used as a measure of the antiwear properties. The load carrying capability is determined from a failure, which can be caused by excess wear or extreme frictional resistance. The **Timken** method (ASTM D 2782) determines the load at which rupture of the lubricant film occurs, and the **Alpha** LFW-1 machine (ASTM D 2714) measures frictional force and wear.

The FZG gear test facility can provide useful information on how a lubricant performs in a gear box. Specific applications include

---

The preparation of this chapter is assigned to TC 3.4, Lubrication.

gear-driven centrifugal compressors in which the dilution of the lubricant by a refrigerant is expected to be quite low.

However, because all these machines operate in air, available data may not apply to a refrigerant environment. Divers (1958) questioned the validity of tests in air because several of the oils that performed poorly in Falex testing have been used successfully in refrigerant systems. Murray et al. (1956) suggest that halocarbon refrigerants can aid in boundary lubrication. Refrigerant 12, for example, when run hot in the absence of oil, reacted with steel surfaces to form a lubricating film. These studies emphasize the need for laboratory testing in a simulated refrigerant environment.

In Huttenlocher's (1969) method of simulation, refrigerant vapor is bubbled through the lubricant reservoir before the test to displace the dissolved air. The refrigerant is bubbled continually during the test to maintain a blanket of refrigerant on the lubricant surface. Using the Falex tester, Huttenlocher showed the beneficial effect of R-22 on the load carrying capability of the same lubricant compared with air or nitrogen. Sanvordenker and Gram (1974) describe a further modification of the Falex test using a sealed sample system.

Both R-12 (a CFC) and R-22 (a HCFC) atmospheres had beneficial effects on a lubricant's boundary lubrication characteristics when compared with tests in air. HFC refrigerants, which are chlorine-free, provide beneficial effects to boundary lubrication when compared to tests conducted in air. However, they contribute to increased wear as compared to a chlorinated refrigerant with the same lubricant.

Komatsuzaki and Homma (1991) used a modified four-ball tester to determine the antiseizure and antiwear properties of R-12 and R-22 in mineral oil and R-134a in a propylene glycol.

Test parameters must simulate as closely as possible the system conditions; namely the base material from which the test specimens are made, their surface condition, the processing methods, and the operating temperature. There are several bearings or rubbing surfaces in a refrigerant compressor, each of which may use different materials and may operate under different conditions. A different test may be required for each bearing. Moreover, bearings in hermetic compressors have very small clearances. Permissible bearing wear is minimal because wear debris remains in the system and can cause other problems even if the clearances stay within working limits. Compressor system mechanics must be understood to perform and interpret simulated tests.

Some aspects of compressor lubrication are not suitable for laboratory simulation. One aspect, the return of liquid refrigerant to the compressor, can cause the lubricant to dilute or wash away from the bearings, creating conditions of boundary lubrication. Tests using operating refrigerant compressors have also been considered, and one such wear test has been proposed as a German *Standard* (DIN 8978). The test is functional for a given compressor system and may permit comparison of lubricants within that class of compressors. However, it is not designed to be a generalized test for the boundary lubricating capability of a lubricant. Other tests using radioactive tracers in refrigerant systems have given useful results (Rembold and Lo 1966).

REFRIGERATION LUBRICANT REQUIREMENTS

Refrigeration compressors are classified as continuous or dynamic and positive displacement. Dynamic types such as the centrifugal compressor depend on energy transfer from a rotating set of blades to the gas. Momentum imparted to the gas is converted to useful pressure by decelerating the gas. Positive-displacement compressors can be either reciprocating or rotary. Both designs confine discrete volumes of gas within a closed space and then elevate pressure by reducing the volume.

Refrigeration systems require lubricant to do more than lubricate. Oil seals compressed gas between the suction and discharge sides, and acts as a coolant to remove heat from the bearings and to transfer heat from the crankcase to the compressor exterior. Oil also reduces noise generated by moving parts inside the compressor. Generally, the higher the viscosity of the oil, the better the sealing and noise insulation.

Although the compression components of centrifugal compressors require no internal lubrication, rotating shaft bearings, seals, and couplings must be adequately lubricated. Turbine or other types of lubricants can be used when the lubricant is not in contact or circulated with refrigerant gas. Chapter 34 of the 1996 *ASHRAE Handbook—Systems and Equipment* describes how reciprocating and rotary compressors are lubricated.

Hermetic systems, in which the motor is exposed to the lubricant, require a lubricant with electrical insulating properties. The refrigerant gas carries some lubricant with it into the condenser and evaporator. This lubricant must return to the compressor in a reasonable time and must have adequate fluidity at low temperatures. The lubricant should remain miscible with the refrigerant for good heat transfer in the evaporator and for good lubricant return. The lubricant must be free of suspended matter or components such as wax that might clog the expansion tube or deposit in the evaporator and interfere with heat transfer. In hermetic refrigeration systems, the lubricant is charged only once, so it must function for the lifetime of the compressor. The chemical stability required of the lubricant in the presence of refrigerants, metals, motor insulation, and extraneous contaminants is perhaps the most important characteristic distinguishing refrigeration lubricants from those used for all other applications (see Chapter 5).

As expected, an ideal lubricant does not exist; a compromise must be made to balance the requirements. A high-viscosity lubricant seals the gas pressure best, but may offer more frictional resistance. Slight foaming can reduce noise, but excessive foaming can carry too much lubricant into the cylinder and cause structural damage. Lubricants that are most stable chemically are not necessarily good lubricants. The lubricant should not be considered alone, because it functions as a lubricant-refrigerant mixture.

The precise relationship between composition and performance is not well defined. Standard ASTM bench tests can provide such information as (1) viscosity, (2) viscosity index, (3) color, (4) specific gravity, (5) refractive index, (6) pour point, (7) aniline point, (8) oxidation resistance, (9) dielectric breakdown voltage, (10) foaming tendency in air, (11) moisture content, (12) wax separation, and (13) volatility. Other properties, particularly those involving interactions with a refrigerant, must be determined by special tests described in the refrigeration literature. Among these nonstandard properties are (1) mutual solubility with various refrigerants, (2) chemical stability in the presence of refrigerants and metals, (3) chemical effects of contaminants or additives that may be in the oils, (4) boundary film-forming ability, (5) solubility of air, and (6) viscosity, vapor pressure, and density of oil-refrigerant mixtures.

MINERAL OIL COMPOSITION

For typical applications, the numerous compounds in refrigeration oils of mineral origin can be grouped into the following structures: (1) paraffins, (2) naphthenes (cycloparaffins), (3) aromatics, and (4) nonhydrocarbons. Paraffins consist of all straight chain and branched carbon chain saturated hydrocarbons. N-Pentane and isopentane are examples of paraffinic hydrocarbons. **Naphthenes** are also completely saturated but consist of cyclic or ring structures; cyclopentane is a typical example. **Aromatics** are unsaturated cyclic hydrocarbons containing one or more rings characterized by alternate double bonds; benzene is a typical example. The **nonhydrocarbons** are molecules containing atoms such as sulfur, nitrogen, or oxygen in addition to carbon and hydrogen.

The preceding structural components do not necessarily exist in pure states. In fact, a paraffinic chain frequently is attached to a

Lubricants in Refrigerant Systems

naphthenic or aromatic structure. Similarly, a naphthenic ring to which a paraffinic chain is attached may in turn be attached to an aromatic molecule. Because of such complications, mineral oil composition is usually described by carbon-type and molecular analysis.

In carbon-type analysis, the number of carbon atoms on the paraffinic chains, naphthenic structures, and aromatic rings is determined and represented as a percentage of the total. Thus, % C_P, the percentage of carbon atoms having a paraffinic configuration, includes not only the free paraffins but also those paraffinic chains attached to naphthenic or to aromatic rings.

Similarly, % C_N includes the carbon atoms on the free naphthenes as well as those on the naphthenic rings attached to the aromatic rings, and % C_A represents the carbon atoms on the aromatic rings. Carbon analysis describes a lubricant in its fundamental structure and correlates and predicts many physical properties of the lubricant. However, direct methods of determining carbon composition are laborious. Therefore, common practice uses a correlative method, such as the one based on the refractive index-density-relative molecular mass (n-d-m) (Van Nes and Weston 1951) or one standardized by ASTM D 2140 or ASTM D 3288. Other methods are ASTM D 2008, which uses ultraviolet absorbency and a rapid method using infrared spectrophotometry and calibration from known oils.

Molecular analysis is based on methods of separating the structural molecules. For refrigeration oils, important structural molecules are (1) saturates or nonaromatics, (2) aromatics, and (3) nonhydrocarbons. All the free paraffins and naphthenes (cycloparaffins), as well as mixed molecules of paraffins and naphthenes are included in the saturates. However, any paraffinic and naphthenic molecules attached to an aromatic ring are classified as aromatics. This representation of lubricant composition is less fundamental than carbon analysis. However, many properties of the lubricant relevant to refrigeration can be explained with this analysis, and the chromatographic methods of analysis are fairly simple (ASTM D 2549, ASTM D 2007, Mosle and Wolf 1963, Sanvordenker 1968).

The traditional classification of oils as paraffinic or naphthenic refers to the amount of paraffinic or naphthenic molecules in the refined lubricant. Paraffinic crudes contain a higher proportion of paraffin wax, and, thus, have a higher viscosity index and pour point relative to naphthenic crudes.

COMPONENT CHARACTERISTICS

Saturates have excellent chemical stability, but poor solubility with polar refrigerants, such as R-22; they are also poor boundary lubricants. Aromatics are somewhat more reactive but have very good solubility with refrigerants and good boundary lubricating properties. Nonhydrocarbons are the most reactive but are beneficial for boundary lubrication, although the amounts needed for that purpose are small. The reactivity, solubility, and boundary lubricating properties of a refrigeration lubricant are affected by the relative amounts of these components in the lubricant.

The saturate and aromatic fractions separated from a lubricant do not have the same viscosity as the parent lubricant. The saturate fraction is much less viscous, while the aromatic fraction is much more viscous than the parent lubricant. Both fractions have the same boiling range. Thus, for this range, the aromatics are more viscous than the saturates. For the same viscosity, the aromatics have a higher volatility than the saturates. Also, the saturate fraction has a lower density and a lower refractive index, but a higher viscosity index and molecular mass than the aromatic fraction of the same lubricant.

Among the saturates, the straight chain paraffins are undesirable for refrigeration applications because they precipitate as wax crystals when the lubricant is cooled to its pour point and tend to form flocs in certain refrigerant solutions (see the section on Wax Separation). The branched chain paraffins and naphthenes are less viscous at low temperatures and have extremely low pour points.

Nonhydrocarbons are mostly removed during the refining of refrigeration oils. Those that remain are expected to have little effect on the physical properties of the lubricant, except, perhaps, on its color, stability and lubricity. Since all the nonhydrocarbons (e.g., sulfur compounds) are not dark, even a colorless lubricant does not necessarily guarantee the absence of nonhydrocarbons. Kartzmark et al. (1967) and Mills and Melchoire (1967) found indications that nitrogen-bearing compounds cause or act as catalysts toward the deterioration of oils. The sulfur and oxygen compounds are thought to be less reactive, with some types considered to be natural inhibitors and lubricity enhancers.

Solvent refining, hydrofinishing, or acid treatment followed by a separation of the acid tar formed are often used to remove more thermally unstable aromatic and unsaturated compounds from the base stock. These methods also produce refrigeration oils that are free from carcinogenic materials sometimes found in crude oil stocks.

The properties of the components naturally are reflected in the parent oil. An oil with a very high saturate content, as is frequently the case with paraffinic oils, also has a high viscosity index, low specific gravity, high relative molecular mass, low refractive index, and low volatility. In addition, it would have a high aniline point and would be less miscible with polar refrigerants. The reverse is true of naphthenic oils. Table 1 lists typical properties of several mineral-based refrigeration oils.

SYNTHETIC LUBRICANTS

The limited solubility of mineral oils with R-22 and R-502 originally led to the investigation of synthetic lubricants for refrigeration use. In more recent times, the lack of solubility of mineral oils in nonchlorinated fluorocarbon refrigerants, such as R-134a and R-32, has led to the commercial use of some synthetic lubricants. Gunderson and Hart (1962) describe a number of commercially available synthetic lubricants such as synthetic paraffins, polyglycols, dibasic acid esters, neopentyl esters, silicones, silicate esters, and fluorinated compounds. Many have properties suited to refrigeration purposes. Sanvordenker and Larime (1972) describe the properties of these synthetic lubricants, alkylbenzenes, and phosphate esters in regard to refrigeration applications using chlorinated fluorocarbon refrigerants. The phosphate esters are unsuitable for refrigeration use because of their poor thermal stability. Although very stable and compatible with refrigerants, the fluorocarbon lubricants are expensive. Among others, only the synthetic paraffins have poor miscibility relations with R-22. Dibasic acid esters, neopentyl esters, silicate esters, and polyglycols all have excellent viscosity temperature relations and remain miscible with R-22 and R-502 to very low temperatures. At this time, the three synthetic lubricants seeing the greatest use are alkylbenzene for R-22 and R-502 service and polyglycols and polyol esters for use with R-134a and refrigerant blends using R-32.

There are two basic types of alkylbenzenes, namely, branched and linear. The products are synthesized by reacting an olefin or chlorinated paraffin with benzene in the presence of a catalyst. Catalysts commonly used for this reaction are aluminum chloride and hydrofluoric acid. After the catalyst is removed, the product is distilled into fractions. The relative size of these fractions can be changed by adjusting the relative molecular mass of the side chain (olefin or chlorinated paraffin) and by changing other variables. The quality of alkylbenzene refrigeration lubricant varies, depending on the type (branched or linear) and manufacturing scheme. In addition to good solubility with refrigerants, such as R-22 and R-502, these lubricants have better high-temperature and oxidation stability than mineral oil-based refrigeration oils. Typical properties for a branched alkylbenzene are shown in Table 1.

Polyalkylene glycols (PAGs) derive from ethylene oxide or propylene oxide. The polymerization is usually initiated with either an alcohol, such as butyl alcohol, or by water. Initiation by an alcohol

Table 1 Typical Properties of Refrigerant Lubricants

Property	ASTM	Mineral Lubricants				Synthenic Lubricants				
		Naphthenic			Paraffinic	Alkyl-benzene	Ester		Glycol	
Viscosity, mm^2/s at 38°C	D 445	33.1	61.9	68.6	34.2	31.7	30	100	29.9	90
Viscosity index	D 2270	0	0	46	95	27	111	98	210	235
Density, kg/m^3	D 1298	913	917	900	862	872	995	972	990	1007
Color	D 1500	0.5	1	1	0.5					
Refractive index	D 1747	1.5015	1.5057	1.4918	1.4752					
Relative molecular mass	D 2503	300	321	345	378	320	570	840	750	1200
Pour point, °C	D 97	−43	−40	−37	−18	−46	−48	−30	−46	−40
Floc point, °C	ASHRAE 86	−56	−51	−51	−35	−73				
Flash point, °C	D 92	171	182	204	202	177	234	258	204	168
Fire point, °C	D 92	199	204	232	232	185				
Composition Carbon-type							Branched acid penta-erythritol	Branched acid penta-erythritol	PP monol mono-functional poly-propylene glycol	PP diol di-functional poly-propylene glycol
%C$_A$	Van Nes and Weston (1951)	14	16	7	3	24				
%C$_N$		43	42	46	32	None				
%C$_P$		43	42	47	65	76				
Molecular composition	D 2549									
% Saturates		62	59	78	87	None				
% Aromatics		38	41	22	13	100				
Aniline point, °C	D 611	71	74	92	104	52				
Critical solution temp. with R-22, °C	—	−3.9	1.7	23	27	−73				

results in a monol while initiation by water results in a diol. PAGs are commonly used as lubricants in automotive air-conditioning systems using R-134a. PAGs have excellent lubricity, low pour points, good low-temperature fluidity, and good compatibility with most elastomers. Major concerns are that these oils are somewhat hygroscopic, are immiscible with mineral oils, and require additives for good chemical and thermal stability (Short 1990).

Polyalphaolefins (PAOs) are normally manufactured from linear α-olefins. The first step in manufacture is the synthesis of a mixture of oligomers in the presence of a BF$_3$·ROH catalyst. Several parameters can be varied to control the distribution of the oligomers so formed. The second step involves a hydrogenation processing of the unsaturated oligomers in the presence of a metal catalyst (Shubkin 1993). PAOs have good miscibility with R-12 and R-114. Some R-22 applications have been tried but are limited by the low miscibility of the fluid in R-22. PAOs are immiscible in R-134a (Short 1990). PAOs are mainly used as an immiscible oil in ammonia systems.

Neopentyl esters (polyol esters) are derived from a reaction between an alcohol and a normal or branched carboxylic acid. The most common alcohols used are pentaerythritol, trimethylolpropane, and neopentyl glycol. For higher viscosities, a dipentaerythritol is often used. The acids are usually selected to give the correct viscosity and fluidity at low temperatures matched to the miscibility requirements of the refrigerant. Complex neopentyl esters are derived by a sequential reaction of the polyol with a dibasic acid followed by a reaction with mixed monoacids (Short 1990). This results in a lubricant with a higher relative molecular mass, high viscosity indices, and higher ISO viscosity grades. Polyol ester lubricants are used commercially with HFC refrigerants in all types of compressors.

LUBRICANT ADDITIVES

Additives are used to enhance certain lubricant properties or impart new characteristics. They generally fall into three groups, namely polar compounds, polymers, and compounds containing active elements such as sulfur or phosphorous. Additive types include (1) pour point depressants for mineral oil, (2) floc point depressants for mineral oil, (3) viscosity index improvers for mineral oil, (4) thermal stability improvers, (5) extreme pressure and antiwear additives, (6) rust inhibitors, (7) antifoam agents, (8) metal deactivators, (9) dispersants, and (10) oxidation inhibitors.

Some additives offer performance advantages in one area but are detrimental in another. For example, anti-wear additives can reduce the wear on compressor components; but because of the chemical reactivity of these materials, the additives can reduce the overall stability of the lubricant. Some additives work best when combined with other additives. They must be compatible with materials in the system (including the refrigerant) and be present in the optimum concentration; too little may be ineffective, while too much can be detrimental or offer no incremental improvement.

In general, additive-type lubricants are not required to lubricate a refrigerant compressor. However, lubricants that contain additives give highly satisfactory service and some additives, such as anti-wear additives, offer some performance advantages over straight mineral oils. Their use is justified as long as the user knows of their presence and provided the additives are not significantly degraded with use. Additives can often be used with synthetic lubricants to reduce wear because, unlike mineral oil, they do not contain sulfur.

An additive-type lubricant is only used after thorough testing to determine whether the additive material (1) is removed by system dryers, (2) is inert to system components, (3) is soluble in refrigerants at low temperatures so as not to cause deposits in capillary tubes or expansion valves, and (4) is stable at high temperatures to avoid adverse chemical reactions such as harmful deposits. This can best be done by sealed tube and compressor testing using the actual additive/base lubricant combination intended for field use.

LUBRICANT PROPERTIES

Viscosity and Viscosity Grades

Viscosity defines a fluid's resistance to flow. It can be expressed as absolute or dynamic viscosity (mPa·s), or kinematic viscosity (mm^2/s). In the United States, dynamic viscosity is expressed in Saybolt Seconds Universal viscosity (abbreviated SSU or SUS),

Lubricants in Refrigerant Systems

Table 2 Recommended Viscosity Ranges

Small and Commercial Systems

Refrigerant	Type of Compressor	Lubricant Viscosities at 38°C, mm²/s
Ammonia	Screw	60-65
Ammonia	Reciprocating	32-65
Carbon dioxide	Reciprocating	60-65[a]
Refrigerant 11	Centrifugal	60-65
Refrigerant 12	Centrifugal	60-65
Refrigerant 12	Reciprocating	32-65
Refrigerant 12	Rotary	60-65
Refrigerant 123	Centrifugal	60-65
Refrigerant 22	Centrifugal	60-86
Refrigerant 22	Reciprocating	32-65
Refrigerant 22	Scroll	60-65
Refrigerant 22	Screw	60-173
Refrigerant 134a	Scroll	22-68
Refrigerant 134a	Screw	32-100
Refrigerant 134a	Centrifugal	60-65
Refrigerant 407C	Scroll	22-68
Refrigerant 407C	Reciprocating	32-687
Refrigerant 410A	Scroll	22-68
Halogenated refrigerants	Screw	32-800

Industrial Refrigeration[b]

Type of Compressor	Lubricant Viscosities at 38°C, mm²/s
Where lubricant may enter refrigeration system or compressor cylinders	32-65
Where lubricant is prevented from entering system or cylinders	
In force-feed or gravity systems	108-129
In splash systems	32-34
Steam-driven compressor cylinders when condensate is reclaimed for ice-making	High viscosity lubricant (30-35 mm²/s at 100°C)

[a]Some applications may require lighter lubricants of 14-17 mm²/s; others, heavier lubricants of 108-129 mm²/s.
[c]Ammonia and carbon dioxide compressors with splash, force-feed, or gravity circulating systems.

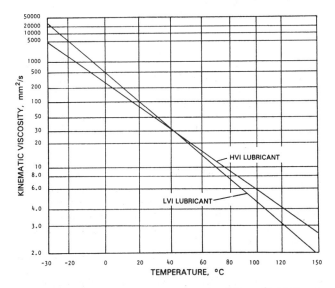

Fig. 1 Viscosity-Temperature Chart for 108 mm²/s HVI and LVI Lubricants

which is the time required for 60 cm³ to gravity flow through a Saybolt universal viscometer. ASTM *Standard* D 2161 contains tables to convert SSU to kinematic viscosity. The density must be known to convert kinematic viscosity to absolute viscosity; that is, absolute or dynamic viscosity (mPa·s) equals density (g/cm³) times kinematic viscosity (mm²/s). Refrigeration oils are sold in viscosity grades, and ASTM has proposed a system of standardized viscosity grades (D 2422).

In selecting the proper viscosity grade, the environment to which the lubricant will be exposed should be considered. The viscosity of the lubricating fluid decreases if temperatures rise or if the refrigerant dissolves appreciably in the lubricant. Synthetic oils, with the exception of alkyl benzenes, are less affected by temperature change than mineral oils.

A large reduction in the viscosity of the lubricating fluid may affect the lubricity and, more likely, the sealing function of the lubricant, depending on the nature of the machinery. The design of some types of hermetically sealed units, such as the single-vane, rotary units, requires the lubricating fluid to act as an efficient sealing agent. In reciprocating compressors, the lubricant film is spread over the entire area of contact between the piston and the cylinder wall, providing a very large area to resist leakage from the high- to the low-pressure side. In a single-vane rotary type, however, the critical sealing area is a line contact between the vane and a roller. In this case, viscosity reduction is serious.

The lubricant with the lowest viscosity that gives the necessary sealing properties with the refrigerant used for the entire range of temperatures and pressures encountered should be chosen to minimize power consumption. A practical method for determining the minimum safe viscosity is to calculate the total volumetric efficiency of a given compressor using several lubricants of widely varying viscosities. The lubricant of lowest viscosity that gives satisfactory volumetric efficiency should be selected. Tests should be run at a number of ambient temperatures, for example, 20, 30, and 40°C. As a guideline, Table 2 lists the viscosity ranges recommended for various refrigeration systems.

The International Organization for Standardization (ISO) has established a series of viscosity levels as a standard for specifying or selecting lubricant for industrial applications. This system, covered in the United States by ASTM D 2422, is designed to eliminate intermediate or unnecessary viscosity grades while providing enough viscosity grades for operating equipment. The system reference point is kinematic viscosity at 40°C, and each viscosity grade with suitable tolerances is identified by the kinematic viscosity at this temperature. Therefore an ISO VG 32 grade lubricant would identify a lubricant with a viscosity of 32 mm²/s at 40°C. Table 3 lists various standardized viscosity grades of lubricants.

Viscosity Index

Lubricant viscosity decreases as temperature increases and increases as temperature decreases. The relationship between temperature and kinematic viscosity is represented by the following equation (ASTM D 341):

$$\log \log(\nu + 0.7) = A + B \log T$$

where

ν = kinematic viscosity, mm²/s, $2 \leq \nu \leq 2 \times 10^7$
T = thermodynamic temperature, K
A, B = constants for each lubricant

This relationship is the basis for the viscosity-temperature charts published by ASTM and permits a straight line plot of viscosity over a wide temperature range. Figure 1 shows a plot for two different lubricants, such as a naphthenic mineral oil (LVI) and a synthetic lubricant (HVI). This plot is applicable over the temperature range in which the oils are homogenous liquids.

Table 3 Viscosity System for Industrial Fluid Lubricants (ASTM D 2422)

Viscosity System Grade Identification	Midpoint Viscosity, mm²/s at 40°C	Kinematic Viscosity Limits, mm²/s at 40°C	
		Minimum	Maximum
ISO VG 2	2.2	1.98	2.42
ISO VG 3	3.2	2.88	3.52
ISO VG 5	4.6	4.14	5.06
ISO VG 7	6.8	6.12	7.48
ISO VG 10	10	9.00	11.00
ISO VG 15	15	13.50	16.50
ISO VG 22	22	19.80	24.20
ISO VG 32	32	28.80	35.20
ISO VG 46	46	41.40	50.60
ISO VG 68	68	61.20	74.80
ISO VG 100	100	90	110
ISO VG 150	150	135	165
ISO VG 220	220	198	242
ISO VG 320	320	288	352
ISO VG 460	460	414	506
ISO VG 680	680	612	748
ISO VG 1000	1000	900	1100
ISO VG 1500	1500	1350	1650

The slope of the viscosity-temperature lines is different for different lubricants. The viscosity-temperature relationship of a lubricant is described by an empirical number called the viscosity index (VI) (ASTM D 2270). A lubricant with a high viscosity index (HVI) shows less change in viscosity over a given temperature range than a lubricant with a low viscosity index (LVI). In the example shown in Figure 1, both oils possess equal viscosities (32 mm²/s) at 40°C. However, the viscosity of the LVI lubricant (a naphthenic mineral oil), as shown by the steeper slope of the line, increases to 520 mm²/s at 0°C, whereas the HVI lubricant (a synthetic lubricant) viscosity increases only to 280 mm²/s.

The viscosity index is related to the composition of the mineral oil. Generally, an increase in cyclic structure, aromatic and naphthenic, decreases the viscosity index. Paraffinic oils usually have a high viscosity index and low aromatic content. Naphthenic oils, on the other hand, have a lower viscosity index and are usually higher in aromatics. For the same base lubricant, the viscosity index decreases with increasing aromatic content. Generally, among common synthetic lubricants, polyalphaolefins, polyalkylene glycols, and polyol esters will have high viscosity indices. As shown in Table 1, alkylbenzenes have lower viscosity indices.

Density

Figure 2 shows published values for pure lubricant densities over a range of temperatures. These density-temperature curves all have approximately the same slope and appear merely to be displaced from one another. If the density of a particular lubricant is known at one temperature but not over a range of temperatures, a reasonable estimate at other temperatures can be obtained by drawing a line paralleling those in Figure 2.

Density indicates the composition of a lubricant for a given viscosity. As shown in Figure 2, naphthenic oils are usually more dense than paraffinic oils and synthetic lubricants are generally more dense than mineral oils. Also, the higher the aromatic content, the higher the density. For equivalent compositions, higher viscosity oils have higher densities, but the change in density with aromatic content is greater than it is with viscosity.

Relative Molecular Mass

In refrigeration applications, the relative molecular mass of a lubricant is often needed. Albright and Lawyer (1959) showed that,

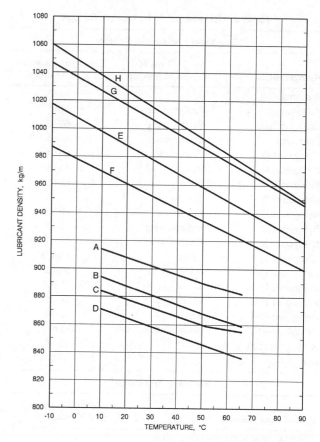

	Lubricant	Viscosity at 38°C, mm²/s	Ref.
A	Naphthene	64.7	1
B	Naphthene	15.7	1
C	Paraffin	64.7	1
D	Paraffin	32.0	1
E	Branched acid POE	32	2
F	Branched acid POE	100	2
G	Polypropylene glycol mono butyl ether	32	2
H	Polyoxypropylene diol	80	2

References: 1. Albright and Lawyer (1959)   2. Cavestri (1993)

Fig. 2 Variation of Refrigeration Lubricant Density with Temperature

on a molar basis, Refrigerants 22, 115, 13, and 13B1 have about the same viscosity-reducing effects on a paraffinic lubricant.

For most mineral oils, a reasonable estimate of the average molecular mass can be obtained by a standard test (ASTM D 2502), based on kinematic viscosities at 40°C and 100°C; or from viscosity-gravity correlations of Mills et al. (1946). Direct methods (ASTM D 2503) can also be used when greater precision is needed or when the correlative methods are not applicable.

Pour Point

Any lubricant intended for low-temperature service should be able to flow at the lowest temperature that it will encounter. This requirement is usually met by specifying a suitably low pour point. The pour point of a lubricant is defined as the lowest temperature at which it will pour or flow, when tested according to the standard method prescribed in ASTM D 97.

The loss of fluidity at the pour point may manifest itself in two ways. Naphthenic oils and synthetic lubricants usually approach the pour point by a steady increase in viscosity. Paraffinic oils, unless heavily dewaxed, tend to separate out a rigid network of wax crystals, which may prevent the flow while still retaining unfrozen

Lubricants in Refrigerant Systems

liquid in the interstices. Pour points can be lowered by adding chemicals known as pour point depressants. These chemicals are believed to modify the wax structure, possibly by depositing a film on the surface of each wax crystal, so that the crystals no longer adhere to form a matrix, and do not interfere with the lubricant's ability to flow. Pour point depressants are not suitable for use with halogenated refrigerants.

Standard pour test values are significant in the selection of oils for ammonia and carbon dioxide systems, and any other system in which refrigerant and lubricant are almost totally immiscible. In such a system, any lubricant that gets into the low side is essentially refrigerant free; therefore, the pour point of the lubricant itself determines whether loss of fluidity, congealment, or wax deposition will occur at low-side temperatures.

Because oil in the low-pressure side of halogenated refrigerant systems contains significant amounts of dissolved refrigerant, the pour point test, which is conducted on pure oils and in air, is of little significance. The viscosity of the lubricant-refrigerant solutions at the low-side conditions and the wax separation (or floc test) are important considerations.

The pour point for a lubricant should not be confused with the freezing point. Pour point is determined by exposing the lubricant to a low temperature for a short time. Refrigeration lubricants will solidify after long-term exposure to a low temperature, even if the temperature is higher than the pour point.

Some refrigerant lubricants stored at a low temperature become unstable. Typically, components of the lubricant separate and form crystals or deposits on the container.

Volatility—Flash and Fire Points

Because the boiling ranges and vapor pressure data on lubricants are not readily available, an indication of the volatility of a lubricant is obtained from the flash and fire points (ASTM D 92). These properties are normally not significant in refrigeration equipment. However, some refrigerants, such as sulfur dioxide, ammonia, and methyl chloride, have a high ratio of specific heats (c_p/c_v) and consequently have a high adiabatic compression temperature. These refrigerants frequently carbonize oils with low flash and fire points when operating in high ambient temperatures. Carbonization of the lubricant can also occur in some applications that use halogenated refrigerants and require high compression ratios (such as domestic refrigerator-freezers operating in high ambient temperatures). Because such carbonization or coking of the valves is not necessarily accompanied by the general lubricant deterioration, the tendency of a lubricant to carbonize is referred to as thermal instability as opposed to chemical instability. Some manufacturers circumvent such problems by using paraffinic oils, which in comparison to naphthenic oils, have higher flash and fire points. Others prevent them through appropriate design.

Vapor Pressure

Vapor pressure is the pressure at which the vapor phase of a substance is in equilibrium with the liquid phase at a specified temperature. The composition of the vapor and liquid phases (when not pure) influences the equilibrium pressure. With refrigeration lubricants, the type, boiling range, and viscosity are also factors influencing vapor pressure; naphthenic oils of a specific viscosity grade generally show higher vapor pressures than paraffinic oils.

The vapor pressure of a lubricant increases with increasing temperature, as shown in Table 4. In practice, the vapor pressure of a refrigeration lubricant at an elevated temperature is negligible compared with the refrigerant at that temperature. The vapor pressure of narrow boiling petroleum fractions can be plotted as straight line functions. If the lubricant's boiling range and its type are known, standard tables may be used to determine the lubricant's vapor pressure up to 101.3 kPa at any given temperature (API 1970).

Aniline Point

Aniline, an aromatic compound, is more soluble in oils containing a greater quantity of similar compounds. The temperature at which a lubricant and aniline are mutually soluble is the lubricant's aniline point (ASTM D 611). Therefore, the relative aromaticity of a mineral oil can be determined by its solubility in aniline. In comparing mineral oils, lower aniline points correspond to the presence of more naphthenic and/or aromatic molecules.

Aniline point can also predict a mineral oil's effect on elastomer seal materials. Generally, a highly naphthenic lubricant swells a specific elastomer material more than a paraffinic lubricant. This is caused by the greater solvency of aromatic and naphthenic compounds present in a naphthenic-type lubricant. However, aniline point gives only a general indication of lubricant-elastomer compatibility. Within a given class of elastomer material, lubricant resistance varies widely because of differences in compounding practiced by the elastomer manufacturer. Finally, in some retrofit applications, a high aniline point mineral oil may cause elastomer shrinkage and possible seal leakage.

Elastomers in synthetic lubricants, such as alkylbenzenes, polyalkylene glycols, and polyol esters behave differently than in mineral oils. For example, an alkylbenzene has an aniline point lower than that of a mineral oil of the same viscosity grade. However, the amount of swell in a chloroneoprene type O ring is generally less than that found with the mineral oil. For these reasons, lubricant-elastomer compatibility needs to be tested under conditions anticipated in actual service.

Solubility of Refrigerants in Oils

All gases are soluble to some extent in lubricants, and many refrigerant gases are highly soluble. The amount dissolved depends on the pressure of the gas and the temperature of the lubricant, on the nature of the gas, and on the nature of the lubricant. Because refrigerants are much less viscous than lubricants, any appreciable amount in solution causes a marked reduction in viscosity.

Two refrigerants usually regarded as poorly soluble in mineral oil are ammonia and carbon dioxide. Data showing the slight absorption of these gases by mineral oil are given in Table 5. The

Table 4 Increase in Vapor Pressure and Temperature

	Vapor Pressure 32 mm²/s Oil	
Temperature, °C	Alkylbenzene, kPa	Naphthene Base, kPa
149	0.10	0.12
163	0.21	0.26
177	0.45	0.50
191	0.89	0.95
204	1.73	1.74
218	3.23	3.06
232	5.83	5.24

Table 5 Absorption of Low Solubility Refrigerant Gases in Oil

	Ammonia[a] (Percent by Mass)				
Absolute Pressure, kPa	Temperature, °C				
	0	20	65	100	140
98	0.246	0.180	0.105	0.072	0.054
196	0.500	0.360	0.198	0.144	0.108
294	0.800	0.540	0.304	0.228	0.166
393	—	0.720	0.398	0.300	0.222
979	—	—	1.050	0.720	0.545

	Carbon Dioxide[b] (Percent by Mass)			
Absolute Pressure, kPa	Temperature, °C			
	0	20	65	100
101	0.26	0.19	0.13	0.10

[a]Type of oil: Not given (Steinle 1950)
[b]Type of oil: HVI oil, 34.8 mm²/s at 38°C (Baldwin and Daniel 1953)

amount absorbed increases with increasing pressure and decreases with increasing temperature. In ammonia systems, where pressures are moderate, the 1% or less refrigerant that dissolves in the lubricant should have little, if any, effect on lubricant viscosity. However, operating pressures in CO_2 systems tend to be much higher (not shown in Table 5), and in that case, the quantity of gas dissolved in the lubricant may be enough to substantially reduce viscosity. At 2.7 MPa, for example, Beerbower and Greene (1961) observed a 69% reduction when a 32 mm^2/s lubricant (HVI) was tested under CO_2 pressure at 27°C.

LUBRICANT-REFRIGERANT SOLUTIONS

Many chlorinated refrigerants are highly soluble in oils at any temperature likely to be encountered. The only limit to the amount of these refrigerants that the lubricant can dissolve is established by the refrigerant pressure at a given temperature. Chlorinated refrigerants such as R-22 and R-114 may show limited solubilities with some lubricants at evaporator temperatures (exhibited in the form of phase separation) and unlimited solubilities in the higher temperature regions of a refrigerant system. In some systems using HFC refrigerants, a second, distinct two-phase region may occur at high temperatures. For such refrigerants, solubility studies must, therefore, be carried out over an extended temperature range.

Because halogenated refrigerants have such high solubilities, the lubricating fluid can no longer be treated as a pure lubricant, but rather as a lubricant-refrigerant solution whose properties are markedly different from those of pure lubricant. The amount of refrigerant dissolved in a lubricant depends on the pressure and temperature. Therefore, the composition of the lubricating fluid is different in different sections of a refrigeration system operating at steady state and changes from the time of start-up until the system attains the steady state. The most pronounced effect is on viscosity.

The crankcase of a compressor can be used as an example. For this case, the refrigerant and lubricant are assumed to be in equilibrium and the viscosity is as shown in Figure 37. If the lubricant in the crankcase at start-up is 24°C, the viscosity of pure 32 ISO VG branched acid polyol ester is about mm^2/s. Under operating conditions, the lubricant in the crankcase is typically about 52°C. At this temperature the viscosity of the pure lubricant is about 20 mm^2/s. If R-134a is the refrigerant and the pressure in the crankcase is 352 kPa, the viscosity of the lubricant-refrigerant mixture at start-up is about 10 mm^2/s and decreases to 9 mm^2/s at 52°C.

Hence, if only lubricant properties are considered, an erroneous picture of the system is obtained. As another example, when the lubricant returns from the evaporator to the compressor, the highest viscosity does not occur at the lowest temperature, because the lubricant contains a large amount of dissolved refrigerant. As the temperature increases, the lubricant loses some of the refrigerant and the viscosity reaches a maximum at a point away from the coldest spot in the system.

Similar to the lubricating fluid, the properties of the working fluid (a high refrigerant concentration solution) are also affected. The vapor pressure of a lubricant-refrigerant solution is markedly lower than that of the pure refrigerant. One result of this property is that the evaporator temperature is higher than if the refrigerant is pure. Another result is what is sometimes called the flooded start-up. When the crankcase and the evaporator are at about the same temperature, the fluid in the evaporator (which is mostly refrigerant) has a higher vapor pressure than the fluid in the crankcase (which is mostly lubricant). This difference in vapor pressures drives the refrigerant to the crankcase and it is absorbed in the lubricant until the pressures are equalized. At times, the moving parts in the crankcase may be completely immersed in this lubricant-refrigerant solution. At start-up, the change in pressure and turbulence can cause excessive amounts of liquid to enter the cylinders, causing damage to the valves and starving the crankcase of lubricant.

The use of crankcase heaters to prevent such problems caused by highly soluble refrigerants is discussed in Chapter 2 and by Neubauer (1958). The problems associated with rapid outgassing from the lubricant is more pronounced with synthetic oils than with mineral oils. Synthetic oils release absorbed refrigerant more quickly and have a lower surface tension, which results in a lack of stable foam found with mineral oils (Swallow et al. 1995).

Density

When estimating the density of a lubricant-refrigerant solution, the solution is assumed ideal so that the specific volumes of the components are additive. The formula for calculating the ideal density (ρ_{id}) is:

$$\rho_{id} = \frac{\rho_o}{1 + W(\rho_o/\rho_R - 1)}$$

where

ρ_o = density of pure lubricant at the solution temperature
ρ_R = density of refrigerant liquid at the solution temperature
W = mass fraction of refrigerant in solution

Depending on the refrigerant, the actual density of a lubricant-refrigerant solution may deviate from the ideal by as much as 8%. The solutions are usually more dense than calculated, but sometimes they are less. For example, R-11 forms ideal solutions with oils, whereas R-12 and R-22 show significant deviations. Density correction factors for R-12 and R-22 solutions are depicted in Figure 3. The corrected densities can be obtained from the relation:

$$\text{Mixture Density} = \rho_m = \rho_{id}/A$$

where A is the density correction factor read from Figure 3 at the desired temperature and refrigerant concentration.

Van Gaalen et al. (1990, 1991ab) provide values of density for four refrigerant/lubricant pairs, namely, R-22/mineral oil, R-22/alkylbenzene, R-502/mineral oil, and R-502/alkylbenzene. Figures 4, 5, 6, and 7 provide data on the variation of density with temperature for R-134a in combination with a 32 ISO VG polyol ester, a 100 ISO VG polyol ester, a 32 ISO VG polyalkylene glycol, and an 80 ISO VG polyalkylene glycol, respectively (Cavestri 1993).

Thermodynamics and Transport Phenomena

Dissolving lubricant in liquid refrigerant affects the thermodynamic properties of the working fluid. The vapor pressures of refrigerant-lubricant solutions at a given temperature are always less than the vapor pressure of pure refrigerant at that temperature. Therefore, dissolved lubricant in an evaporator leads to lower suction pressures and higher evaporator temperatures than expected from pure refrigerant tables. Bambach (1955) gives an enthalpy diagram for R-12/lubricant solutions over the range of compositions from 0 to 100% lubricant and temperatures from −40°C to 115°C. Spauschus (1963) developed general equations for calculating thermodynamic functions of refrigerant-lubricant solutions and applied them to the special case of R-12/mineral oil solutions.

Pressure-Temperature-Solubility Relations

When a refrigerant is in equilibrium with a lubricant, a fixed amount of refrigerant is present in the lubricant at a given temperature and pressure. This is evident if the Gibbs phase rule is applied to basically a two-phase, two-component mixture. The lubricant, although a mixture of several compounds, may be considered one component, and the refrigerant the other. The two phases are the liquid phase and the vapor phase. The phase rule defines this mixture as having two degrees of freedom. Normally, the variables involved are pressure, temperature, and the compositions of the liquid and of

Lubricants in Refrigerant Systems

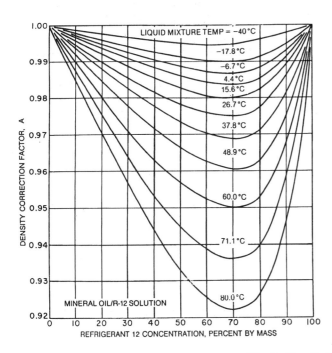
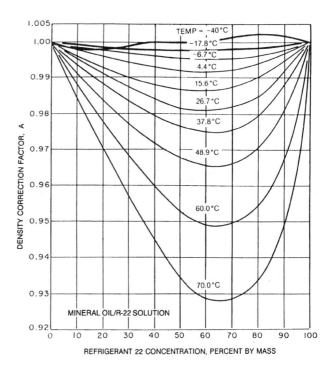

Fig. 3 Density Correction Factors
(Loffler 1959)

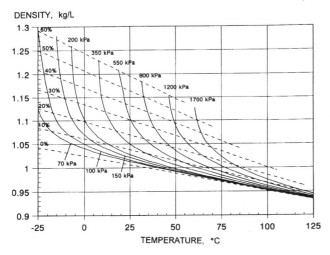

Fig. 4 Density as Function of Temperature and Pressure for Mixture of R-134a and 32 ISO VG Branched Acid Polyol Ester Lubricant

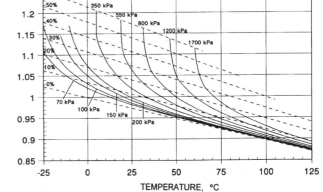

Fig. 5 Density as Function of Temperature and Pressure for Mixture of R-134a and 100 ISO VG Branched Acid Polyol Ester Lubricant

the vapor. Since the vapor pressure of the lubricant is negligible compared with that of the refrigerant, the vapor phase is essentially pure refrigerant, and only the composition of the liquid phase needs to be considered. If the pressure and temperature are defined, the system is invariant, i.e., the liquid phase can have only one composition. This is a different but more precise way of stating that a lubricant-refrigerant mixture having a known composition exerts a certain vapor pressure at a certain temperature. If the temperature is changed, the vapor pressure also changes.

Pressure-temperature-solubility relations are usually presented in the form shown in Figure 8. On this graph, P_1^0 and P_2^0 represent the saturation pressures of the pure refrigerant at temperatures t_1 and t_2, respectively. Point E_1 represents an equilibrium condition, where one and only one composition of the liquid, represented by W_1, is possible at the pressure P_1. If the temperature of this system is increased to t_2, some of the liquid refrigerant evaporates, and the equilibrium point shifts to E_2, corresponding to a new pressure P_2. In either case, the lubricant-refrigerant solution exerts a vapor pressure less than that of the pure refrigerant at the same temperature.

Mutual Solubility

In a compressor, the lubricating fluid is a solution of refrigerant dissolved in lubricant. In other parts of the refrigerant system, the solution is a lubricant in liquid refrigerant. In both instances, either the lubricant or the refrigerant could exist alone as a liquid if the other were not present; therefore, any distinction between the dissolving and dissolved component merely reflects a point of view. Either of the liquids can be considered as dissolving the other. This relationship is termed mutual solubility.

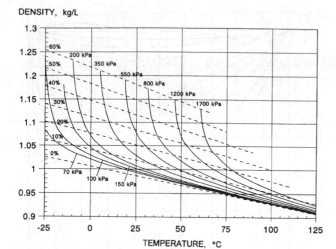

Fig. 6 Density as Function of Temperature and Pressure for Mixture of R-134a and 32 ISO VG Polypropylene Glycol Butyl Ether Lubricant

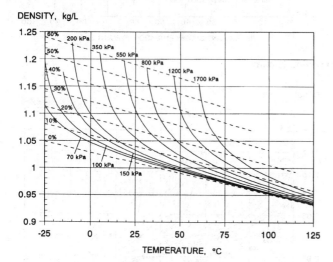

Fig. 7 Density as Function of Temperature and Pressure for Mixture of R-134a and 80 ISO VG Polyoxypropylene Glycol Diol Lubricant

Table 6 Mutual Solubility of Refrigerants and Mineral Oil

Completely Miscible	Partially Miscible			Immiscible
	High Miscibility	Intermediate Miscibility	Low Miscibility	
R-11	R-123	R-22	R-13	Ammonia
R-12		R-114	R-14	CO_2
R-113			R-115	R-134a
			R-152a	R-407C
			R-C318	R-410A
			R-502	

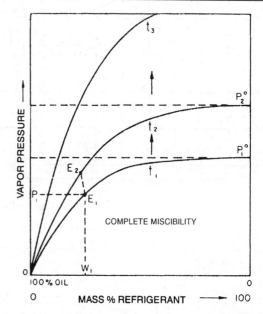

Fig. 8 P-T-S Diagram for Completely Miscible Refrigerant-Lubricant Solutions

refrigerant-rich, depending on the predominant component. Each phase may contain substantial amounts of the leaner component, and these two solutions are themselves immiscible with each other.

The importance of this concept is best illustrated by R-502, which is considered a low miscibility refrigerant exhibiting a high CST as well as a broad immiscibility range. However, even at −20°C, the lubricant-rich phase contains about 20 mass % of dissolved refrigerant (see Figure 11). Examples of partially miscible systems, in addition to R-502, are R-22, R-114, and R-13 with mineral oils.

The basic properties of the immiscible region can be recognized by applying the phase rule. With three phases (two liquid and one vapor) and two components, there can be only one degree of freedom. Therefore, either the temperature or the pressure automatically determines the composition of both liquid phases. If the system pressure is changed, the temperature of the system changes and the two liquid phases assume somewhat different compositions determined by the new equilibrium conditions.

Figure 9 illustrates the behavior of partially miscible mixtures. Point C on the graph represents the critical solution temperature t_3. There are three separate regions below this temperature on the diagram. Reading from left to right, a family of the smooth solid curves represents a region of completely miscible lubricant-rich solutions. These curves are followed by a wide break representing a region of partial miscibility in which there are two immiscible liquid phases. On the right side, the partially miscible region disappears into a second completely miscible region of refrigerant-rich solutions. A dome-shaped envelope (broken line curve OCR) encloses the partially miscible region; everywhere outside this dome the refrigerant and lubricant are completely miscible. In a

Refrigerants are classified as completely miscible, partially miscible, or immiscible, according to their mutual solubility relations with mineral oils. Because several commercially important refrigerants are partially miscible, further designation as having high, intermediate, or low miscibility is shown in Table 6.

Completely miscible refrigerants and lubricants are mutually soluble in all proportions at any temperature encountered in a refrigeration or air-conditioning system. This type of mixture always forms a single liquid phase under equilibrium conditions, no matter how much refrigerant or lubricant is present.

Partially miscible refrigerant-lubricant solutions are mutually soluble to a limited extent. Above the critical solution temperature (CST) or consolute temperature, many refrigerant/lubricant mixtures in this class are completely miscible, and their behavior is identical to that just described. As mentioned previously, R-134a and some synthetic lubricants exhibit a region of immiscibility at higher temperatures.

Below the critical solution temperature, however, the liquid may separate into two phases. Such phase separation does not mean that the lubricant and the refrigerant are insoluble in each other. Each liquid phase is a solution; one is lubricant-rich and the other

Lubricants in Refrigerant Systems

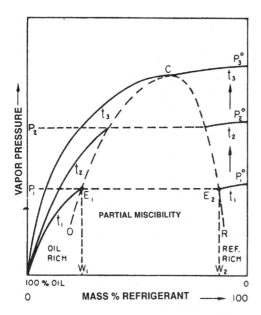

Fig. 9 P-T-S Diagram for Completely Miscible Refrigerant-Oil Solutions

sense, Figure 9 is a variant of Figure 8 in which the partial miscibility dome (OCR) blots out a substantial portion of the continuous solubility curves. Under the dome, i.e., in the immiscible region, the two points E_1 and E_2 on the temperature line t_1 represent the two phases coexisting in equilibrium.

These two phases differ considerably in composition (W_1 and $W2$) but have the same refrigerant pressure P_1. The solution pressure P_1 lies not far below the saturation pressure of pure refrigerant P_1^0. Commonly, refrigerant-lubricant solutions near the partial miscibility limit show less reduction in refrigerant pressure than observed at the same lubricant concentration with completely miscible refrigerants.

Totally immiscible lubricant-refrigerant solutions are defined in this chapter as only very slightly miscible. In such mixtures, the immiscible range is so broad that mutual solubility effects can be ignored. Critical solution temperatures are seldom found in mixtures of the totally immiscible type. Examples are ammonia and lubricant, and carbon dioxide and mineral oil.

Effects of Partial Miscibility in Refrigerant Systems

Evaporator. The evaporator is the coldest part of the system and the most likely part in which immiscibility or phase separation will occur. If the evaporator temperature is below the critical solution temperature, phase separation is likely to occur in some part of the evaporator. The fluid entering the evaporator is mostly liquid refrigerant containing a small fraction of lubricant, whereas the liquid leaving the evaporator is mostly lubricant, since the refrigerant is in vapor form. No matter how little lubricant the entering refrigerant carries, the liquid phase, as it progresses through the evaporator, passes through the critical composition which usually lies in 15% to 20% lubricant in the total liquid phase.

Phase separation in the evaporator can sometimes cause problems. In a dry-type evaporator, there is usually enough turbulence to cause the phases to emulsify. In this case, the heat transfer characteristics of the evaporator may not be significantly affected. In flooded-type evaporators, however, the working fluid may separate into layers, and the lubricant-rich phase may float on top of the boiling liquid. In addition to the heat transfer, partial miscibility may also affect the lubricant's return from the evaporator to the crankcase. Usually the lubricant is moved by high velocity suction gas

transferring momentum to the droplets of lubricant on the return line walls.

If a lubricant-rich layer separates at the evaporator temperatures, this viscous, nonvolatile liquid can migrate and collect in pockets or blind passages not easily reached by the high velocity suction gas. The lubricant return problem may be magnified and, in some cases, an oil-logging can occur. The design of the system should take into account all these possibilities, and evaporators should be designed to promote entrainment (see Chapter 2). Oil separators are frequently required in the discharge line to minimize lubricant circulation when refrigerants of poor solvent power are used or in systems involving very low evaporator temperatures (Soling 1971).

Crankcase. With certain refrigerant and lubricant pairs, such as R-502 and mineral oil, or even with R-22 in applications such as heat pumps, phase separation sometimes occurs in the crankcase when the system is shut down. When this happens, the refrigerant-rich layer settles to the bottom, often completely immersing the pistons, bearings, and other moving parts. At start-up, the fluid that lubricates these moving parts is mostly refrigerant with little lubricity, and severe bearing damage may result. The turbulence at start-up may cause the liquid refrigerant to enter the cylinders, carrying large amounts of lubricant with it. Precautions in design will prevent such problems in partially miscible systems.

Condenser. Partial miscibility is not a problem in the condenser, since the liquid flow lies in the turbulent region and the temperatures are relatively high. Even if phase separation occurs, there is little danger of separation of layers, the main obstacle to efficient heat transfer.

Solubility Curves and Miscibility Diagrams

Figure 10 shows mutual solubility relations of partially miscible refrigerant-lubricant mixtures. More than one curve of this type can be plotted on a miscibility diagram. Each single dome then represents the immiscible ranges for one lubricant and one refrigerant. Miscibility curves for R-13, R-13B1, R-502 (Parmelee 1964), R-22, and mixtures of R-12 and R-22 (Walker et al. 1957) are shown in Figure 11. Miscibility curves for R-13, R-22, R-502, and R-503 in an alkylbenzene refrigeration lubricant are shown in Figure 12. A

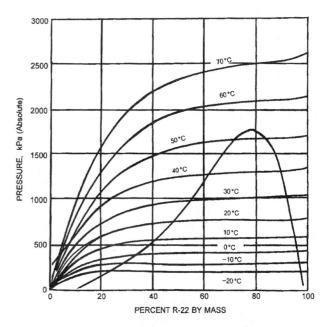

Fig. 10 P-T-S Relations of R-22 with 43 mm²/s White Oil (0% C_A, 55% C_N, 45% C_P)
(Spauschus 1964)

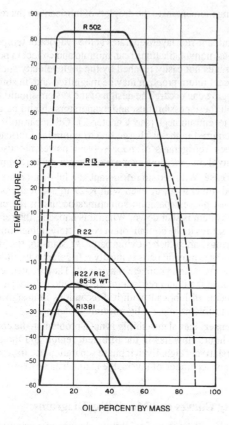

Fig. 11 Critical Solubilities of Refrigerants with 32 mm²/s Naphthenic Lubricant (C_A 12%, C_N 44%, C_P 44%)

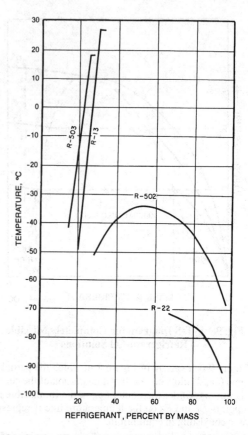

Fig. 12 Critical Solubilities of Refrigerants with 32 mm²/s Alkylbenzene Lubricant

comparison with Figure 11 illustrates the greater solubility of refrigerants in this type of lubricant.

Effect of Lubricant Type on Solubility and Miscibility

When compared on a mass basis, low viscosity oils absorb more refrigerant than high viscosity oils. Also, naphthenic oils absorb more than paraffinic oils. However, when compared on a mol basis, some confusion arises. Paraffinic oils absorb more refrigerant than naphthenic oils, i.e., reversal of the mass basis, and there is little difference between a 15.7 mm²/s and a 64.7 mm²/s naphthenic lubricant (Albright and Lawyer 1959, Albright and Mandelbaum 1956). The differences on either basis are small, i.e., within 20% of each other. Comparisons of oils by their carbon-type analyses are not available, but in view of the data on naphthenic and paraffinic types, differences between oils with different carbon-type analyses, except perhaps for extreme compositions, are unlikely.

The effect of lubricant type and composition on miscibility is better defined than solubility. When the critical solution temperature (CST) is used as the criterion of miscibility, oils with higher aromatic contents show a lower CST. Higher viscosity grade oils show a higher CST than lower viscosity grade oils, and paraffinic oils show a higher CST than naphthenic oils (see Figures 13 and 25). When the entire dome of immiscibility is considered, a similar result is noticeable. Oils that exhibit a lower CST, usually show a narrowing of the immiscibility range, i.e., the mutual solubility is greater at any given temperature.

Miscibility of R-22 with Lubricants

Parmelee (1964) showed that polybutyl silicate improves miscibility with R-22 (and also R-13) at low temperatures. Alkylbenzenes, by themselves or mixed with mineral oils, also have better miscibility with R-22 than do mineral oils alone (Seeman and Shellard 1963). Polyol esters, which are HFC miscible, are completely miscible with R-22 irrespective of viscosity grade.

Among the mineral oils, Walker et al. (1962) provide detailed miscibility diagrams of 12 brand name oils commonly used for refrigeration systems. Walker's data show that in every case, higher viscosity lubricant of the same base and type has a higher critical solution temperature.

Loffler (1957) provides complete miscibility diagrams of R-22 and 18 oils. A portion of the properties of the oils used and the critical solution temperatures are summarized in Table 7. Although precise correlations are not evident in the table, certain trends are clear. For the same viscosity grade and base, the effect of aromatic carbon content is seen in oils 2, 3, 7, and 8 and between oils 4 and 6. Similarly, for the same viscosity grades, the effect of paraffinic structure (with essentially the same % C_A) is noticeable between oils 6 and 17 and between oils 8 and 18.

According to Loffler, the most pronounced effect on the critical solution temperature is exerted by the aromatic content of the lubricant; the table indicates that the paraffinic structure reduces the miscibility compared with naphthenic structures. Sanvordenker (1968) reported the miscibility relations of the saturated fractions and the aromatic fractions of mineral oils as a function of their physical properties. The critical solution temperatures with R-22 increase with increasing viscosities for the saturates, as well as for the aromatics. For equivalent viscosities, aromatic fractions with naphthenic linkages show lower critical solution temperatures than aromatics with only paraffinic linkages.

Pate et al. (1993) developed miscibility data for 10 refrigerants and 14 lubricants. Table 8 lists lower and upper critical solution temperatures for several of the refrigerant-lubricant pairs studied.

Lubricants in Refrigerant Systems

Table 7 Critical Miscibility Values of R-22 with Different Oils

Oil No.	Oil Base Type[a]	Approximate Viscosity Grade, mm²/s	Viscosity at 50°C (Converted), mm²/s	Carbon-Type Composition			Critical Solution Temperature, °C
				%C_A	%C_N	%C_P	
2	N	15	11.2	23	34	43	−37
3	N	15	10.2	2.5	48.5	49	−6
1	N	32	18.5	13	43	44	3
8	N	46	24.6	0.6	45	55	24
7	N	46	26.7	2.8	44	54	20
5	N	46+	27.9	22	30	47	−16
4	N	46	28.6	26	28	46	−20
6	N	46	29.7	4	45	51	17
13	N	100	54.5	1.9	41	56	None[b]
12	N	100	60.7	4	41	55	None[b]
11	N	150	69.1	7	40	53	None[b]
10	N	220	93.2	21	27	52	16
9	N	220	109.0	27	24	50	9
18	P	46+	29.3	0.5	33	67	None[b]
17	P	46	31.6	3.5	34	63	None[b]
16	P	68	35.2	6.4	30	63	None[b]
15	P	68	45.2	14.3	25	61	None[b]
14	P	100	50.0	18.1	22	60	44[c]

[a]P = Paraffinic, N = Naphthenic
[b]Never completely miscible at any temperature
[c]A second (inverted) miscibility dome was observed above 58°C. Above this temperature, the oil/R-22 mixture again separated into two immiscible solutions.

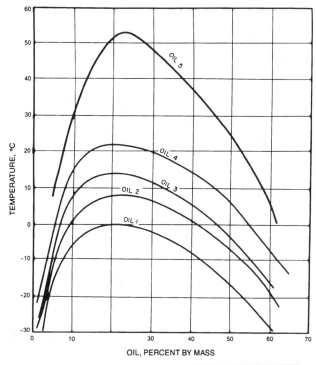

Fig. 13 Effect of Oil Properties on Miscibility with R-22

Oil No.	Viscosity at 38°C, mm²/s	Compositions, %			Ref.
		C_A	C_N	C_P	
1	34.0	12	44	44[a]	1
2	33.5	7	46	47[a]	1
3	63.0	12	44	44[a]	1
4	67.7	7	46	47[a]	1
5	41.3	0	55	45	2

References: 1. Walker et al. (1957) 2. Spauschus (1964)
[a]Estimated composition, not in original reference.

Table 8 Critical Solution Temperatures for Selected Refrigerant/Lubricant Pairs

Refrigerant	Lubricant	Critical Solution Temperature, °C	
		Lower	Upper
R-22	ISO 32 Naphthenic mineral oil	−5	>60
R-22	ISO 32 Modified polyglycol	−12	>60
R-22	ISO 68 Naphthenic mineral oil	15	>60
R-123	ISO 68 Naphthenic mineral oil	−39	>60
R-123	ISO 58 Polypropylene glycol butyl monoether	−50	14
R-134a	ISO 58 Polypropylene glycol butyl monoether	−50	56
R-134a	ISO 32 Modified polyglycol	10	>90
R-134a	ISO 22 Penta erythritol, mixed-acid ester	−42	>90
R-134a	ISO 58 Polypropylene glycol butyl monoether	−46	6
R-134a	ISO 100 Polypropylene glycol diol	−46	11
R-134a	ISO 100 Penta erythritol, mixed-acid ester	−35	>32
R-134a	ISO 100 Penta erythritol, branched-acid ester	−46	12

Solubilities and Viscosities of Lubricant-Refrigerant Solutions

Although the differences are small on a mass basis, naphthenic oils are better solvents than paraffinic oils. When considering the viscosity of lubricant-refrigerant mixtures, naphthenic oils show greater viscosity reduction than paraffinic oils for the same mass percent of dissolved refrigerant. When the two effects are compounded, under the same conditions of temperature and pressure, a naphthenic lubricant in equilibrium with a given refrigerant shows a significantly lower viscosity than a paraffinic lubricant.

Refrigerants also differ in their viscosity-reducing effects when the solution concentration is measured in mass percent. However, when the solubility is plotted in terms of mol percent, the reduction in viscosity is approximately the same, at least for Refrigerants 13, 13B1, 22, and 115 (Figure 14).

Spauschus (1964) reports numerical vapor pressure data on a R-22/white oil system; solubility-viscosity graphs on naphthenic and paraffinic oils have been published by Loffler (1960), Little (1952), and Albright and Mandelbaum (1956). Some discrepancies, particularly at high R-22 contents, have been shown in data on

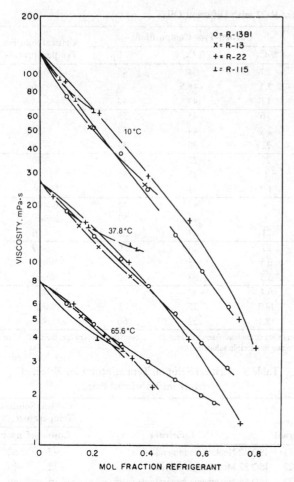

Fig. 14 Viscosity of Mixtures of Various Refrigerants and 32 mm²/s Paraffinic Oil
(Albright and Lawyer 1959)

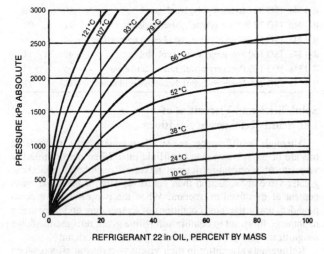

Fig. 15 Solubility of R-22 in 32 mm²/s Naphthenic Oil

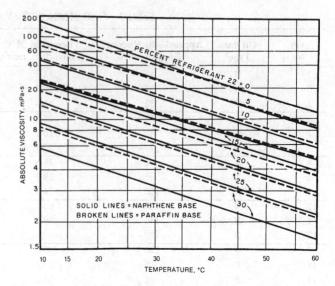

Fig. 16 Viscosity-Temperature Chart for Solutions of R-22 in 32 mm²/s Naphthene and Paraffin Base Oils

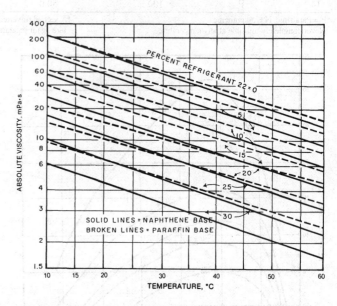

Fig. 17 Viscosity-Temperature Chart for Solutions of R-22 in 65 mm²/s Naphthene and Paraffin Base Oils

viscosities, which apparently could not be attributed to the properties of the lubricant and remain unexplained. However, general plots reported by the aforementioned authors are satisfactory for engineering and design purposes.

Spauschus and Speaker (1987) have compiled references of solubility and viscosity data. Selected solubility-viscosity data are summarized in Figures 10 and 15 through 27.

Where possible, solubilities have been converted to mass percent to provide consistency among the various charts. Figures 10 and 15 through 19 contain data on R-22 and oils, Figure 20 on R-502, Figures 21 and 22 on R-11, Figures 23 and 24 on R-12, and Figures 25 and 26 on R-114. Figure 27 contains data on the solubility of various refrigerants in alkylbenzene lubricant. Viscosity-solubility characteristics of mixtures of R-13B1 and lubricating oils have been investigated by Albright and Lawyer (1959). Similar studies on R-13 and R-115 are covered by Albright and Mandelbaum (1956).

LUBRICANT RETURN FROM EVAPORATORS

Regardless of the miscibility relations of a lubricant with refrigerants, for a refrigeration system to function properly, the lubricant must return adequately from the evaporator to the crankcase. Parmelee (1964) showed that the viscosity of the lubricant, saturated with refrigerant under low-pressure and low-temperature

Lubricants in Refrigerant Systems

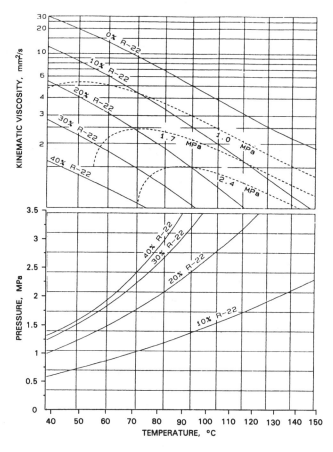

Fig. 18 Viscosity-Temperature Chart for Solutions of R-22 in 32 mm²/s Naphthenic Oil
(Van Gaalen, et al. 1990 and 1991a)

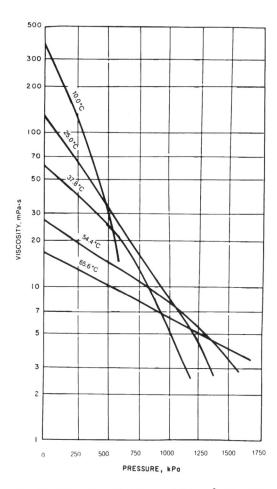

Fig. 19 Viscosity of Mixtures of 65 mm²/s Paraffin Base Oil and R-22
(Albright and Mandelbaum 1956)

conditions, is important in providing good lubricant return. The viscosity of the lubricant-rich liquid that accompanies the suction gas changes as it sees rising temperatures on its way back to the compressor. Two opposing factors then come into play. First, the increasing temperature tends to decrease the viscosity of the fluid. Second, since the pressure remains unchanged, the increasing (Loffler 1960) temperature also tends to drive off some of the dissolved refrigerant from the solution, thereby increasing its viscosity.

Figures 28 through 30 show the variation in viscosity with temperature and pressure for three lubricant-refrigerant solutions ranging from −40°C to 21°C. In all cases, the viscosities of the solutions passed through maximum values as the temperature was changed at constant pressure, a finding that was also consistent with previous data obtained by Bambach (1955) and Loffler (1960). According to Parmelee, the existence of a viscosity maximum is significant, because the lubricant-rich solution becomes most viscous, not in the coldest regions in the evaporator, but at some intermediate point where much of the refrigerant has escaped from the lubricant. This condition is possibly in the suction line. The velocity of the return vapor, which may be high enough to move the lubricant-refrigerant solution in the colder part of the evaporator, may be too low to achieve the same result at the point of maximum viscosity. The designer must consider this factor to minimize any lubricant return problems. Chapters 2 and 3 have further information on velocities in return lines.

Another aspect of viscosity data at the evaporator conditions is shown in Figure 31, which compares a synthetic alkylbenzene lubricant with a naphthenic mineral oil. The two oils are the same viscosity grade, but the highly aromatic alkylbenzene lubricant has a much lower viscosity index in the pure state and shows a higher viscosity at low temperatures. However, at 135 kPa or approximately −40°C evaporator temperature, the viscosity of the lubricant/R-502 mixture is considerably lower for alkylbenzene than for naphthenic lubricant. In spite of the lower viscosity index, alkylbenzene returns more easily than naphthenic lubricant.

Estimated viscosity-temperature-pressure relationships for a naphthenic lubricant with R-502 is shown in Figure 32. Figures 33 and 34 show viscosity-temperature-pressure plots of alkylbenzene and R-22 and R-502, respectively. These data are based on experimental data from Van Gaalen et al. (1991ab). Figures 35 and 36 show viscosity-temperature-pressure data for mixtures of R-134a and a 32 ISO VG polyalkylene glycol and an 80 ISO VG polyalkylene glycol, respectively. Figures 37 and 38 show similar data for R-134a and a 32 ISO VG polyol ester and a 100 ISO VG polyol ester, respectively (Cavestri 1993).

Sundaresan and Radermacher (1996) observed oil return in a small, air-to-air heat pump. Three refrigerant lubricant pairs (R-22/mineral oil, R-407C/mineral oil, and R-407C/polyol ester) were studied under four conditions—steady-state cooling, steady-state heating, cyclic operation, and a simulated lubricant pumpout situation. The lubricant returned rapidly to the compressor in the R-22/mineral oil and R-407C/polyol ester tests, but oil return was unreliable in the R-407C/mineral oil test.

WAX SEPARATION (FLOC TESTS)

Wax separation properties are of little importance with synthetic lubricants because they do not contain wax or wax-like molecules.

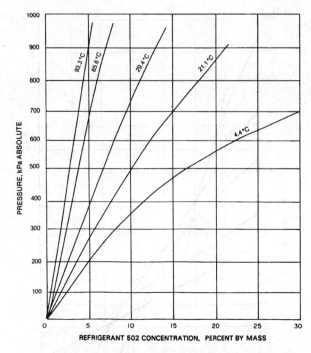

Fig. 20 Solubility of R-502 in 32 mm²/s Naphthenic Oil
(C_A 12%, C_N 44%, C_P 44%)

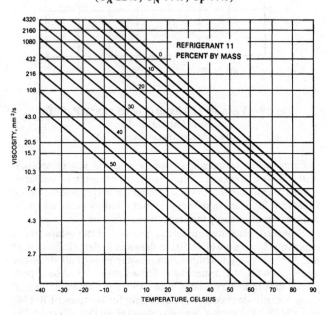

Fig. 21 Viscosity-Temperature Curves for Solutions of R-11 in 65 mm²/s Naphthene Base Oil

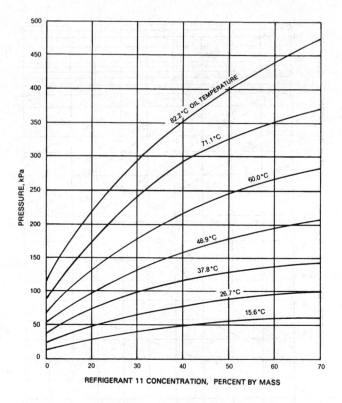

Fig. 22 Solubility of R-11 in 65 mm²/s Oil

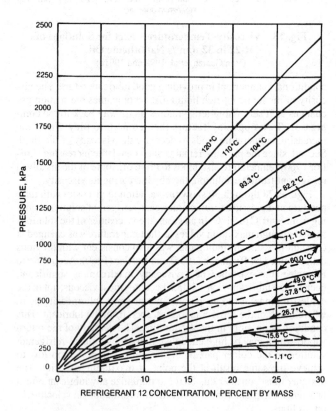

Solid lines: 32 mm²/s naphthene base oil.
 Also 52 mm²/s German base oil (Bambach 1955).
Broken lines: 32 mm²/s and 70 mm²/s mixed base oils.

Fig. 23 Solubility of R-12 in Refrigerant Oils

However, petroleum-derived lubricating oils are mixtures of large numbers of chemically distinct hydrocarbon molecules. At low temperatures in the low-pressure side of refrigeration units, some of the larger molecules separate from the bulk of the lubricant, forming wax-like deposits. This wax can clog capillary tubes and cause expansion valves to stick, which is undesirable in refrigeration systems. Bosworth (1952) describes other wax separation problems.

In selecting a lubricant to use with completely miscible refrigerants, the wax-forming tendency of the lubricant can be determined by the floc test. The floc point is the highest temperature at which wax-like materials or other solid substances precipitate when a mixture of 10% lubricant and 90% R-12 is cooled under specific

Lubricants in Refrigerant Systems

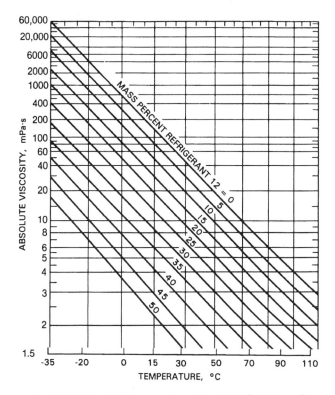

Fig. 24 Viscosity-Temperature Chart for Solutions of R-12 in 32 mm^2/s) Naphthene Base Oil

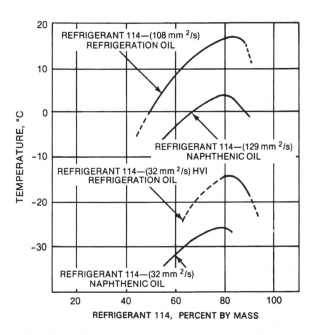

Fig. 25 Critical Solution Temperatures of R-114/Oil Mixtures

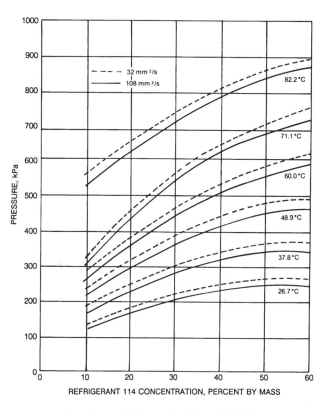

Fig. 26 Solubility of R-114 in HVI Oils

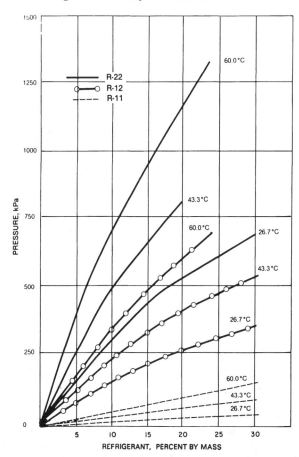

Fig. 27 Solubility of Refrigerants in 32 mm^2/s Alkylbenzene Oil

conditions. Because different refrigerant and lubricant concentrations are encountered in actual equipment, test results cannot be used directly to predict performance. The lubricant concentration in the expansion devices of most refrigeration and air-conditioning systems is considerably less than 10%, resulting in significantly lower temperatures at which wax separates from lubricant-refrigerant mixture. ASHRAE *Standard* 86 describes a standard method of determining the floc characteristics of refrigeration oils in the presence of R-12.

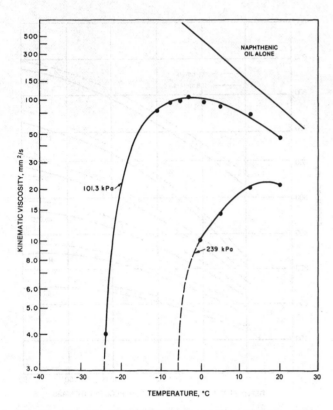

Fig. 28 Viscosity of R-12/Oil Solutions at Low-Side Conditions
(Parmelee 1964)

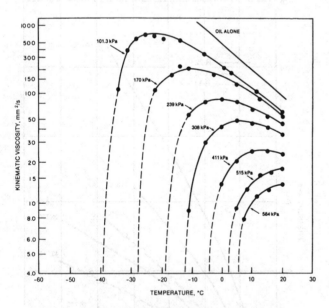

Fig. 29 Viscosity of R-22/Naphthenic Oil Solutions at Low-Side Conditions
(Parmelee 1964)

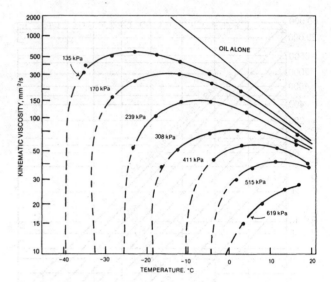

Fig. 30 Viscosity of R-502/Naphthenic Oil Solutions at Low-Side Conditions

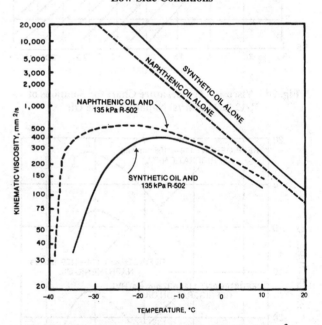

Fig. 31 Viscosities of Solutions of R-502 with 32 mm²/s Naphthenic Oil (C_A 12%, C_N 44%, C_P 44%) and 32 mm²/s Synthetic Alkylbenzene Oil

Attempts to develop a test for the floc point of partially miscible refrigerants for use with R-22 have not been successful. The solutions being cooled often separate into two liquid phases. Once phase separation occurs, the components of the lubricant distribute themselves into the lubricant-rich phase and the refrigerant-rich phase in such a way that the highly soluble aromatics concentrate into the refrigerant phase, while the less soluble saturates concentrate into the lubricant phase. The waxy materials stay dissolved in the refrigerant-rich phase only to the extent of their solubility limit. On further cooling, any wax that separates from the refrigerant-rich phase migrates into the lubricant-rich phase. Therefore, a significant floc point cannot be obtained with partially miscible refrigerants once phase separation has occurred. However, lack of flocculation does not mean lack of wax separation. Wax may separate in the lubricant-rich phase causing it to congeal. Parmelee (1964) reported such phenomena with a paraffinic lubricant and R-22.

Floc point might not be reliable when applied to used oils. A part of the original wax may already have been deposited, and the used lubricant may contain extraneous material from the operating equipment.

Good design practice suggests selecting oils that do not deposit wax on the low-pressure side of a refrigeration system, regardless of single-phase or two-phase refrigerant-lubricant solutions. Mechanical design affects how susceptible equipment is to wax deposition. Wax deposits at sharp bends and the suspended wax particles build up on the tubing walls by impingement. Careful design avoids bends and materially reduces the tendency to deposit wax.

Lubricants in Refrigerant Systems

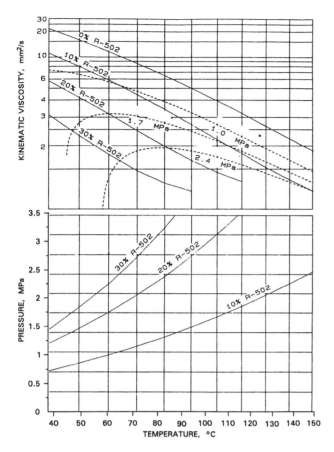

Fig. 32 Viscosity-Temperature-Pressure Chart for Solutions of R-502 in 32 mm²/s Naphthenic Oil

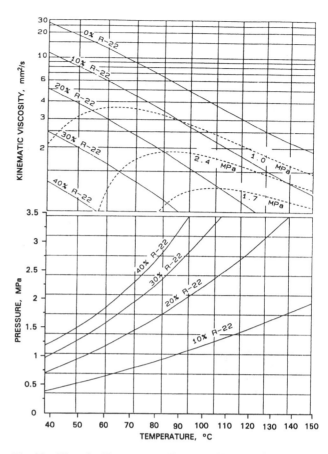

Fig. 33 Viscosity-Temperature-Pressure Chart for Solutions of R-22 in 32 mm²/s Alkylbenzene Oil

SOLUBILITY OF HYDROCARBON GASES

Hydrocarbon gases such as propane (R-290) and ethylene (R-1150) are miscible with the compressor lubricating oil and are absorbed by the lubricant. The lower the boiling point or critical temperature, the less soluble the gas, all other values being equal. Gas solubility increases with decreasing temperature and increasing pressure (Figures 39 and 40). As with other lubricant miscible refrigerants, absorption of the hydrocarbon gas reduces lubricant viscosity.

SOLUBILITY OF WATER IN LUBRICANTS

Refrigerant systems must be dry internally because high moisture content can cause ice to form in the expansion valve or capillary tube, corrosion of bearings, reactions that affect the lubricant/refrigerant stability, or other operational problems. Synthetic polyol ester and polyalkylene oils will bind with water so that ice does not form, even at very low temperatures.

As with other components, the refrigeration lubricant must contain the least amount of moisture. Normal manufacturing and refinery handling practices result in a moisture content of about 30 mg/kg for almost all types of lubricants. Polyalkylene glycols are an exception and generally contain several hundred ppm of water. However, this amount may increase between the time of shipment from the refinery and the time of actual use, unless proper preventive measures are taken. Small containers are usually sealed. Tank cars are not normally pressure sealed or nitrogen blanketed except for shipment of synthetic polyol ester and polyalkylene lubricants, which are quite hygroscopic.

During transit, changes in ambient temperatures cause the lubricant to expand and contract and draw in humid air from outside. Depending on the extent of such cycling, the moisture content of the lubricant may be significantly higher than at the time of shipment. Users of large quantities of refrigeration oils frequently dry the lubricant before use. Chapter 43 discusses the methods of drying lubricants. Normally, removing any moisture present will also deaerate the lubricant.

Because POE and PAG lubricants are quite hygroscopic, when they are in a refrigeration system they should circulate through a filter/dryer designed for liquids. A filter/dryer can be installed in the line carrying liquid refrigerant or in a line returning lubricant to the compressor. The material in the filter/dryer must be compatible with the lubricant. Also, desiccants can remove some additives in the lubricant.

Spot checks show that water solubility data for transformer oils obtained by Clark (1940) also apply to refrigeration oils (Figure 41).

A simple method, widely used in industry to detect free water in refrigeration oils, is the dielectric breakdown voltage (ASTM D 877), which is designed to control moisture and other contaminants in electrical insulating oils. The method does not work with polyester and polyalkylene glycol oils, however. According to Clark (1940), the dielectric breakdown voltage decreases with increasing moisture content at the same test temperature and increases with temperature for the same moisture content. At 27°C, when the solubility of water in a 32 mm²/s naphthenic lubricant is between 50 to 70 mg\kg, a dielectric breakdown voltage of about 25 kV indicates that no free water is present in the lubricant. However, the lubricant may contain dissolved water up to the solubility limit. Therefore, a dielectric breakdown voltage of 35 kV is commonly specified to indicate that the moisture content is well below saturation.

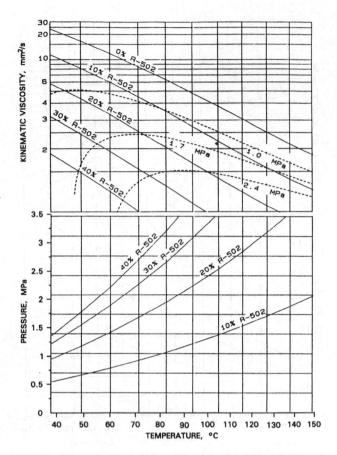

Fig. 34 Viscosity-Temperature-Pressure Chart for Solutions of R-502 in 32 mm²/s Alkylbenzene Oil

The ASTM D 877 test is not sensitive below about 60% saturation. General practice has been to measure directly the total moisture content by procedures such as the Karl Fischer (ASTM D 1533) method for low moisture levels.

SOLUBILITY OF AIR IN LUBRICANTS

Refrigerant systems should not contain excessive amounts of air or other noncondensable gases. Oxygen in air can react with the lubricant to form oxidation products. More importantly, nitrogen in the air (which does not react with lubricant) is a noncondensable gas that can interfere with performance. In some systems, the tolerable volume of noncondensables is very low. Therefore, if the lubricant is added after the system is evacuated, it must not contain an excessive amount of dissolved air or other noncondensable gas. Dissolved air is removed when a vacuum is used to dry the lubricant. However, if the deaerated lubricant is stored under pressure in dry air, it will reabsorb air in proportion to the pressure (Baldwin and Daniel 1953).

FOAMING AND ANTIFOAM AGENTS

Excessive foaming of the lubricant is undesirable in refrigeration systems. Brewer (1951) suggests that abnormal refrigerant foaming reduces the lubricant's effectiveness in cooling the motor windings and removing heat from the compressor. Too much foaming also can cause too much lubricant to pass through the pump and enter the low-pressure side. Foaming in a pressure oiling system can result in starved lubrication under some conditions.

However, moderate foaming is beneficial in refrigeration systems, particularly for noise suppression. A foamy layer on top of the lubricant level dampens the noise created by the moving parts of the compressor. Moderate foam also lubricates effectively, yet it is pumpable, which minimizes the risk of vapor lock of the oil pump at start-up. There is no general agreement on what constitutes excessive foaming or how it should be prevented. Some manufacturers add small amounts of an antifoam agent, such as silicone fluid, to refrigerator oils. Others believe that foaming difficulties are more easily corrected by equipment design.

Goswami et al. (1997) observed the foaming characteristics of R-32, R-125, R-134a, R-143a, R-404A, R-407C, and R-410A with two ISO 68 polyol ester lubricants. They compared them to R-12 and R-22 paired with both an ISO 32 and ISO 68 mineral oil and found that the foamability and foam stability of the HFC/POE pairs were much lower than the R-12 and R-22/mineral oil pairs.

OXIDATION RESISTANCE

Refrigeration oils are seldom exposed to oxidizing conditions in hermetic systems. Once a system is sealed against air and moisture, the oxidation resistance of a lubricant is not significant unless it reflects the chemical stability. Handling and manufacturing practices include elaborate care to protect lubricants against air, moisture, or any other contaminant. Oxidation resistance by itself is rarely included in refrigeration lubricant specifications.

Nevertheless, oxidation tests are justified, because oxidation reactions are chemically similar to the reactions between oils and refrigerants. An oxygen test, using power factor as the measure, correlates with established sealed-tube tests. However, such oxidation resistance tests are not used as primary criteria of chemical reactivity, but rather to support the claims of chemical stability determined by sealed-tube and other tests.

Oxidation resistance may become a prime requirement during manufacture. The small amount of lubricant used during compressor assembly and testing is not always completely removed before the system is dehydrated. If the subsequent dehydration process is carried out in a stream of hot, dry air, as is frequently the case, the hot oxidizing conditions can cause the residual lubricant to become gummy, leading to stuck bearings, overheated motors, and other operating difficulties. For these purposes, the lubricant should have high oxidation resistance. However, the lubricant used under such extreme conditions should be classed as a specialty process lubricant rather than a refrigeration lubricant.

CHEMICAL STABILITY

Refrigeration lubricants must have excellent chemical stability. Within the enclosed refrigeration environment, the lubricant must resist chemical attack by the refrigerant in the presence of all the materials encountered, including the various metals, motor insulation, and any unavoidable contaminants trapped in the system. A lubricant reacts with the refrigerant at elevated temperatures, and the reaction is catalyzed by metals. Methods for evaluating the chemical stability of lubricant-refrigerant mixtures are covered in Chapter 5.

Various phenomena in an operating system, such as sludge formation, carbon deposits on valves, gumming, and copper plating of bearing surfaces have been attributed to lubricant decomposition in the presence of the refrigerant. In addition to the direct reactions of the lubricant and refrigerant, the lubricant may also act as a medium for reactions between the refrigerant and the motor insulation, particularly when the refrigerant extracts the lighter components of the insulation. Factors affecting the stability of various components such as wire insulation materials in hermetic systems are also covered in Chapter 5.

Effect of Lubricant Type

Mineral oils differ in their ability to withstand chemical attack by a given refrigerant. In an extensive laboratory sealed-tube test program, Walker et al. (1960, 1962) showed that color darkening,

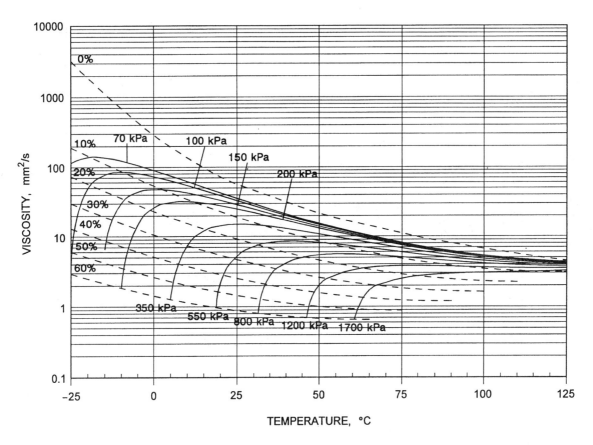

Fig. 35 Viscosity-Temperature-Pressure Plot for 32 ISO VG Polypropylene Glycol Butyl Mono Ether with R-134a

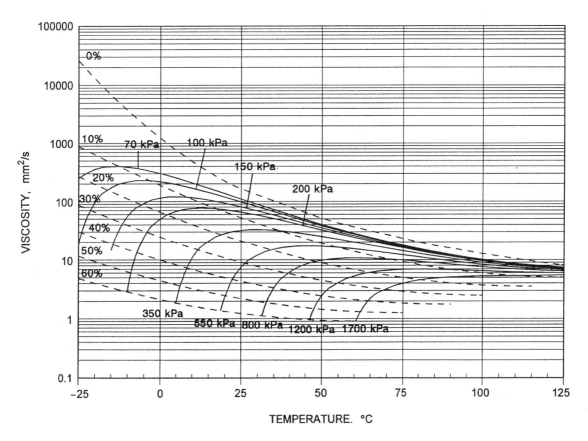

Fig. 36 Viscosity-Temperature-Pressure Plot for 80 ISO VG Polyoxypropylene Diol with R-134a

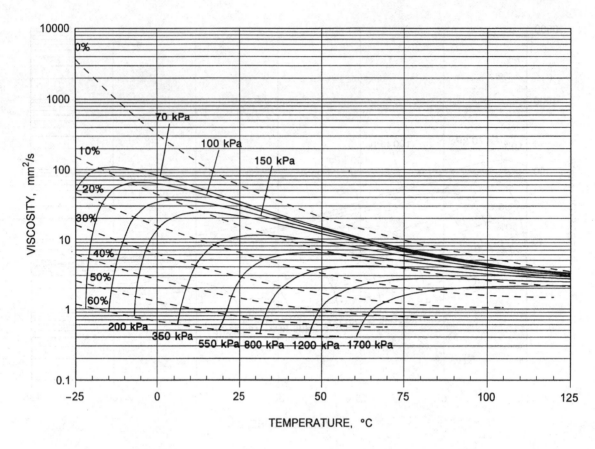

Fig. 37 Viscosity-Temperature-Pressure Plot for 32 ISO VG Branched Acid Polyol Ester with R-134a

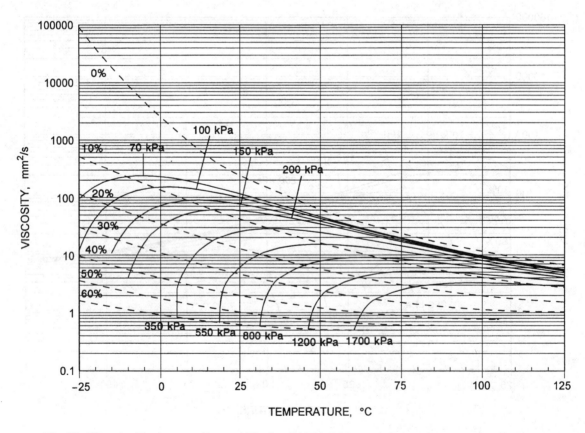

Fig. 38 Viscosity-Temperature-Pressure Plot for 100 ISO VG Branched Acid Polyol Ester with R-134a

Lubricants in Refrigerant Systems

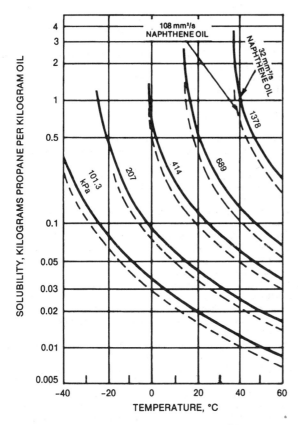

Fig. 39 Solubility of Propane in Oil

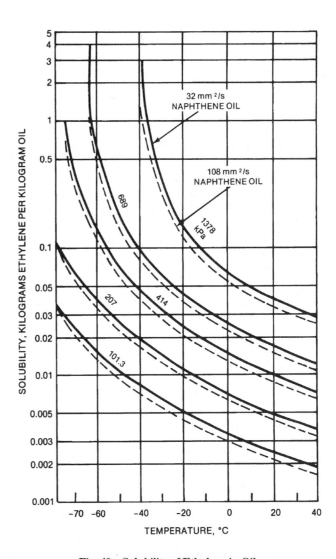

Fig. 40 Solubility of Ethylene in Oil
(Witco)

corrosion of metals, deposits, and copper plating occur less in paraffinic oils than in naphthenic oils. Using gas analysis, Spauschus and Doderer (1961) and Doderer and Spauschus (1965) show that a white oil containing only saturates and no aromatics is considerably more stable in the presence of R-12 and R-22 than a medium-refined lubricant. Steinle (1950) reported the effect of oleoresin (nonhydrocarbons) and sulfur content on the reactivity of the lubricant, using the Philipp test. A decrease in the oleoresin content, accompanied by a decrease in sulfur and aromatic content, showed an improvement in the chemical stability with R-12, while the oil's lubricating properties became poorer. Schwing's (1968) study on a synthetic polyisobutyl benzene lubricant reports that it is not only chemically stable but also has good lubricating properties.

HFC refrigerants are chemically very stable and show little tendency to degrade with any lubricant. Hygroscopic synthetic POE and PAG lubricants are less chemically stable with chlorinated refrigerants than mineral oil due to the interaction of moisture with the refrigerant at high temperatures.

CONVERSION FROM CFC REFRIGERANTS TO OTHER REFRIGERANTS

The most common conversion from a CFC refrigerant to another refrigerant is retrofitting to use HCFC or HFC refrigerants. A conversion may require draining each component to completely remove an incompatible lubricant. Often, flushing in some manner is the only way to remove old lubricant. The flushing medium may be liquid refrigerant, an intermediate fluid, or the lubricant that will be charged with the alternate refrigerant. Liquid CFC refrigerants circulated through the entire system are preferable for flushing, although other refrigerants are used. The refrigerant is recovered with equipment modified for this use.

The refrigeration equipment must be operated if intermediate fluids and lubricants are used for flushing. The system is charged

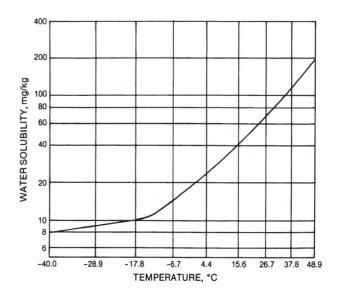

Fig. 41 Solubility of Water in Mineral Oil

with the flushing material and CFC refrigerant and operated for at least eight hours. After operation, the lubricant charge is drained from the compressor. This process is repeated until the lubricant in the drained material is reduced to a specified level. Chemical test kits or portable refractometers are available to determine the amount of old lubricant that is mixed with the flushed material.

REFERENCES

Albright, L.F. and J.D. Lawyer. 1959. Viscosity-solubility characteristics of mixtures of Refrigerant 13B1 and lubricating oils. *ASHRAE Journal* (April):67.

Albright, L.F. and A.S. Mandelbaum. 1956. Solubility and viscosity characteristics of mixtures of lubricating oils and "Freon-13 or -115." *Refrigerating Engineering* (October):37.

API 1970. *Technical data book-Petroleum refining*, 2nd ed. American Petroleum Institute, Washington, DC.

ASHRAE. 1994. Methods of testing the floc point of refrigeration grade oils. *Standard 86*.

Baldwin, R.R. and S.G. Daniel. 1953. *Journal of the Institute of Petroleum* 39:105.

Bambach, G. 1955. The behavior of mineral oil-F12 mixtures in refrigerating machines. *Abhandlungen des Deutschen Kältetechnischen Vereins*, No. 9. (Translated by Carl Demrick.) Also see abridgement in *Kältetechnik* 7(7):187.

Beerbower, A. and D.F. Greene. 1961. The behavior of lubricating oils in inert gas atmospheres. *ASLE Transactions* 4(1):87.

Bosworth, C.M. 1952. Predicting the behavior of oils in refrigeration systems. *Refrigerating Engineering* (June):617.

Brewer, A.F. 1951. Good compressor performance demands the right lubricating oil. *Refrigerating Engineering* (October):965.

Cavestri, R.C. 1993. Measurement of the solubility, viscosity and density of synthetic lubricants with HFC-134a. Final Report for ASHRAE 716-RP.

Clark, F.M. 1940. Water solution in high-voltage dielectric liquids. *Electrical Engineering Transactions* 59(8):433.

Divers, R.T. 1958. Better standards are needed for refrigeration lubricants. *Refrigeration Engineering* (October):40.

Doderer, G.C. and H.O. Spauschus. 1965. Chemical reactions of R-22. *ASHRAE Transactions* 71(I):162.

Fuller, D.D. 1984. *Theory and practice of lubrication for engineers*. John Wiley and Sons, Inc., New York.

Goswami, D.Y., D.O. Shah, C.K. Jotshi, S. Bhagwat, M. Leung, and A.S. Gregory. 1997. Foaming characteristics of HFC refrigerants. *ASHRAE Journal* 39(6):39-44.

Gunderson, R.C. and A.W. Hart. 1962. *Synthetic lubricants*. Reinhold Publishing Corp., New York.

Huttenlocher, D.F. 1969. A bench scale test procedure for hermetic compressor lubricants. *ASHRAE Journal* (June):85.

Kartzmark, R., J.B. Gilbert, and L.W. Sproule. 1967. Hydrogen processing of lube stocks. *Journal of the Institute of Petroleum* 53:317.

Komatsuzaki, S. and Y. Homma. 1991. Antiseizure and antiwear properties of lubricating oils under refrigerant gas environments. *Lubrication Engineering* 47(3):193.

Little, J.L. 1952. Viscosity of lubricating oil-Freon-22 mixtures. *Refrigerating Engineering* (Nov.):1191.

Loffler, H.J. 1957. The effect of physical properties of mineral oils on their miscibility with R-22. *Kältetechnik* 9(9):282.

Loffler, H.J. 1959. Density of oil-refrigerant mixtures. *Kältetechnik*. 11(3):70.

Loffler, H.J. 1960. Viscosity of oil-refrigerant mixtures. *Kältetechnik*. 12(3):71.

Mills, I.W., et al. 1946. Molecular weight-physical property correlations for petroleum fractions. *Industrial and Engineering Chemistry* 38:442.

Mills, I.W. and J.J. Melchoire. 1967. Effect of aromatics and selected additives on oxidation stability of transformer oils. *Industrial and Engineering Chemistry* (Product Research and Development) 6:40.

Mosle, H. and W. Wolf. 1963. *Kältetechnik* 15:11.

Murray, S.F., R.L. Johnson, and M.A. Swikert. 1956. Difluoro-dichloromethane as a boundary lubricant for steel and other metals. *Mechanical Engineering* 78(3):233.

Neubauer, E.T. 1958. Compressor crankcase heaters reduce oil foaming. *Refrigerating Engineering* (June):52.

Parmelee, H.M. 1964. Viscosity of refrigerant-oil mixtures at evaporator conditions. *ASHRAE Transactions* 70:173.

Pate, M.B., S.C. Zoz, and L.J. Berkenbosch. 1993. Miscibility of lubricants with refrigerants. Report No. DOE/CE/23810-18. U.S. Department of Energy.

Rembold, U. and R.K. Lo. 1966. Determination of wear of rotary compressors using the isotope tracer technique. *ASHRAE Transactions* 72:VI.1.1.

Sanvordenker, K.S. 1968. Separation of refrigeration oil into structural components and their miscibility with R-22. *ASHRAE Transactions* 74(I):III.2.1.

Sanvordenker, K.S. and W.J. Gram. 1974. Laboratory testing under controlled environment using a falex machine. Compressor Technology Conference, Purdue University.

Sanvordenker, K.S. and M.W. Larime. 1972. A review of synthetic oils for refrigeration use. *ASHRAE Symposium*, Lubricants, Refrigerants and Systems—Some Interactions.

Schwing, R.C. 1968. Polyisobutyl benzenes and refrigeration lubricants. *ASHRAE Transactions* 74(I):III.1.1.

Seeman, W.P. and Shellard, A.D. 1963. Lubrication of Refrigerant 22 machines. IX International Congress of Refrigeration, Paper III-7, Munich, Germany.

Short, G.D. 1990. Synthetic lubricants and their refrigeration applications. *Lubrication Engineering* 46(4):239.

Shubkin, R.L. 1993. Polyalphaolefins, in *Synthetic lubricants and high performance functional fluids*. Marcel Dekker, Inc., New York.

Soling, S.P. 1971. Oil recovery from low temperature pump recirculating halocarbon systems. *ASHRAE Symposium* PH-71-2.

Spauschus, H.O. and G.C. Doderer. 1961. Reaction of Refrigerant 12 with petroleum oils. *ASHRAE Journal* (February):65.

Spauschus, H.O. 1963. Thermodynamic properties of refrigerant-oil solutions. *ASHRAE Journal* (April):47; (October):63.

Spauschus, H.O. 1964. Vapor pressures, volumes and miscibility limits of R-22-oil solutions. *ASHRAE Transactions* 70:306.

Spauschus, H.O. and L.M. Speaker. 1987. A review of viscosity data for oil-refrigerant solutions. *ASHRAE Transactions* 93(2):667.

Steinle, H. 1950. *Kaltemaschinenole*. Springer-Verlag, Berlin, Germany, 81.

Sundaresan, S.G. and R. Radermacher. 1996. Oil return characteristics of refrigerant oils in split heat pump system. *ASHRAE Journal* 38(8):57.

Swallow, A., A. Smith, and B. Greig. 1995. Control of refrigerant vapor release from polyol ester/halocarbon working fluids. *ASHRAE Transactions* 101(2):929.

Van Gaalen, N.A., M.B. Pate, and S.C. Zoz. 1990. The measurement of solubility and viscosity of oil/refrigerant mixtures at high pressures and temperatures: Test facility and initial results for R-22/naphthenic oil mixtures. *ASHRAE Transactions* 96(2):183.

Van Gaalen, N.A., S.C. Zoz, and M.B. Pate. 1991a. The solubility and viscosity of solutions of HCFC-22 in naphthenic oil and in alkylbenzene at high pressures and temperatures. *ASHRAE Transactions* 97(1):100.

Van Gaalen, N.A., S.C. Zoz, and M.B. Pate. 1991b. The solubility and viscosity of solutions of R-502 in naphthenic oil and in an alkylbenzene at high pressures and temperatures. *ASHRAE Transactions* 97(2):285.

Van Nes, K. and H.A. Weston. 1951. *Aspects of the constitution of mineral oils*. Elsevier Publishing Co., Inc., New York.

Walker, W.O., A.A. Sakhanovsky, and S. Rosen. 1957. Behavior of refrigerant oils and Genetron-141. *Refrigerating Engineering* (March):38

Walker, W.O., S. Rosen, and S.L. Levy. 1960. A study of the factors influencing the stability of the mixtures of Refrigerant 22 and refrigerating oils. *ASHRAE Transactions* 66:445.

Walker, W.O., S. Rosen, and S.L. Levy. 1962. Stability of mixtures of refrigerants and refrigerating oils. *ASHRAE Transactions* 68:360.

Witco. Sonneborn Division, *Bulletin* 8846.

CHAPTER 8

THERMAL PROPERTIES OF FOODS

Thermal Properties of Food Constituents 8.1	*Enthalpy* 8.8
Thermal Properties of Food 8.1	*Thermal Conductivity* 8.9
Water Content 8.1	*Thermal Diffusivity* 8.17
Initial Freezing Point 8.1	*Heat of Respiration* 8.17
Ice Fraction 8.2	*Transpiration of Fresh Fruits and Vegetables* 8.19
Density 8.7	*Surface Heat Transfer Coefficient* 8.24
Specific Heat 8.7	*Nomenclature* 8.27

THERMAL properties of foods and beverages must be known to perform the various heat transfer calculations involved in the design of storage and refrigeration equipment and the estimation of process times for refrigerating, freezing, heating, or drying of foods and beverages. Because the thermal properties of foods and beverages strongly depend on chemical composition and temperature; and, because many food items are available, it is nearly impossible to experimentally determine and tabulate the thermal properties of foods and beverages for all possible conditions and compositions. However, composition data for foods and beverages are readily available from sources such as Holland et al. (1991) and USDA (1975). These data consist of the mass fractions of the major components found in food items. The thermal properties of food items can be predicted by using this composition data in conjunction with temperature dependent mathematical models of the thermal properties of the individual food constituents.

Thermophysical properties of foods and beverages that are often required for heat transfer calculations include density, specific heat, enthalpy, thermal conductivity, and thermal diffusivity. In addition, if the food item is a living organism, such as a fresh fruit or vegetable, it generates heat via respiration and loses moisture via transpiration. Both of these processes should be included in heat transfer calculations. This chapter summarizes prediction methods for estimating these thermophysical properties and includes examples on the use of these prediction methods. In addition, tables of measured thermophysical property data for various foods and beverages are provided.

THERMAL PROPERTIES OF FOOD CONSTITUENTS

Constituents commonly found in food items include water, protein, fat, carbohydrate, fiber, and ash. Choi and Okos (1986) developed mathematical models for predicting the thermal properties of these food components as functions of temperature in the range of −40°C to 150°C (see Table 1). Choi and Okos also developed models for predicting the thermal properties of water and ice (Table 2). Table 3 lists the composition of various food items, including the mass percentage of moisture, protein, fat, carbohydrate, fiber, and ash (USDA 1996)

THERMAL PROPERTIES OF FOOD

In general, the thermophysical properties of a food or beverage are well behaved when the temperature of the food is above its initial freezing point. However, below the initial freezing point, the thermophysical properties vary greatly due to the complex processes involved during freezing.

The preparation of this chapter is assigned to TC 10.9, Refrigeration Application for Foods and Beverages.

The initial freezing point of a food is somewhat lower than the freezing point of pure water due to dissolved substances in the moisture in the food. At the initial freezing point, a portion of the water within the food crystallizes and the remaining solution becomes more concentrated. Thus, the freezing point of the unfrozen portion of the food is further reduced. The temperature continues to decrease as the separation of ice crystals increases the concentration of the solutes in solution and depresses the freezing point further. Thus, the ice and water fractions in the frozen food depend on temperature. Because the thermophysical properties of ice and water are quite different, the thermophysical properties of frozen foods vary dramatically with temperature. In addition, the thermophysical properties of the food above and below the freezing point are drastically different.

WATER CONTENT

Because water is the predominant constituent in most foods, water content significantly influences the thermophysical properties of foods. Average values of moisture content (percent by mass) are given in Table 3. For fruits and vegetables, water content varies with the cultivar as well as with the stage of development or maturity when harvested, the growing conditions, and the amount of moisture lost after harvest. In general, the values given in Table 3 apply to mature products shortly after harvest. For fresh meat, the values of water content given in Table 3 are at the time of slaughter or after the usual aging period. For cured or processed products, the water content depends on the particular process or product.

INITIAL FREEZING POINT

Foods and beverages do not freeze completely at a single temperature but rather they freeze over a range of temperatures. In fact, foods which are high in sugar content or foods packed in high syrup concentrations may never be completely frozen, even at typical frozen food storage temperatures. Thus, there is not a distinct freezing point for foods and beverages, but an initial freezing point at which the crystallization process begins.

The initial freezing point of a food or beverage is important not only for determining the proper storage conditions for the food item, but also for the calculation of thermophysical properties. During the storage of fresh fruits and vegetables, for example, the commodity temperature must be kept above its initial freezing point to avoid freezing damage. In addition, because there are drastic changes in the thermophysical properties of foods as they freeze, knowledge of the initial freezing point of a food item is necessary to accurately model its thermophysical properties. Experimentally determined values. of the initial freezing point of foods and beverages are given in Table 3.

Table 1 Thermal Property Models for Food Components ($-40°C \leq t \leq 150°C$)

Thermal Property	Food Component	Thermal Property Model
Thermal conductivity, W/(m·K)	Protein	$k = 1.7881 \times 10^{-1} + 1.1958 \times 10^{-3}t - 2.7178 \times 10^{-6}t^2$
	Fat	$k = 1.8071 \times 10^{-1} - 2.7604 \times 10^{-3}t - 1.7749 \times 10^{-7}t^2$
	Carbohydrate	$k = 2.0141 \times 10^{-1} + 1.3874 \times 10^{-3}t - 4.3312 \times 10^{-6}t^2$
	Fiber	$k = 1.8331 \times 10^{-1} + 1.2497 \times 10^{-3}t - 3.1683 \times 10^{-6}t^2$
	Ash	$k = 3.2962 \times 10^{-1} + 1.4011 \times 10^{-3}t - 2.9069 \times 10^{-6}t^2$
Thermal diffusivity, m²/s	Protein	$\alpha = 6.8714 \times 10^{-2} + 4.7578 \times 10^{-4}t - 1.4646 \times 10^{-6}t^2$
	Fat	$\alpha = 9.8777 \times 10^{-2} - 1.2569 \times 10^{-4}t - 3.8286 \times 10^{-8}t^2$
	Carbohydrate	$\alpha = 8.0842 \times 10^{-2} + 5.3052 \times 10^{-4}t - 2.3218 \times 10^{-6}t^2$
	Fiber	$\alpha = 7.3976 \times 10^{-2} + 5.1902 \times 10^{-4}t - 2.2202 \times 10^{-6}t^2$
	Ash	$\alpha = 1.2461 \times 10^{-1} + 3.7321 \times 10^{-4}t - 1.2244 \times 10^{-6}t^2$
Density, kg/m³	Protein	$\rho = 1.3299 \times 10^{3} - 5.1840 \times 10^{-1}t$
	Fat	$\rho = 9.2559 \times 10^{2} - 4.1757 \times 10^{-1}t$
	Carbohydrate	$\rho = 1.5991 \times 10^{3} - 3.1046 \times 10^{-1}t$
	Fiber	$\rho = 1.3115 \times 10^{3} - 3.6589 \times 10^{-1}t$
	Ash	$\rho = 2.4238 \times 10^{3} - 2.8063 \times 10^{-1}t$
Specific heat, kJ/(kg·K)	Protein	$c_p = 2.0082 + 1.2089 \times 10^{-3}t - 1.3129 \times 10^{-6}t^2$
	Fat	$c_p = 1.9842 + 1.4733 \times 10^{-3}t - 4.8008 \times 10^{-6}t^2$
	Carbohydrate	$c_p = 1.5488 + 1.9625 \times 10^{-3}t - 5.9399 \times 10^{-6}t^2$
	Fiber	$c_p = 1.8459 + 1.8306 \times 10^{-3}t - 4.6509 \times 10^{-6}t^2$
	Ash	$c_p = 1.0926 + 1.8896 \times 10^{-3}t - 3.6817 \times 10^{-6}t^2$

Source: Choi and Okos (1986)

Table 2 Thermal Property Models for Water and Ice ($-40°C \leq t \leq 150°C$)

	Thermal Property	Thermal Property Model
Water	Thermal conductivity, W/(m·K)	$k_w = 5.7109 \times 10^{-1} + 1.7625 \times 10^{-3}t - 6.7036 \times 10^{-6}t^2$
	Thermal diffusivity, m²/s	$\alpha_w = 1.3168 \times 10^{-1} + 6.2477 \times 10^{-4}t - 2.4022 \times 10^{-6}t^2$
	Density, kg/m³	$\rho_w = 9.9718 \times 10^{2} + 3.1439 \times 10^{-3}t - 3.7574 \times 10^{-3}t^2$
	Specific heat, kJ/(kg·K) (For temp. range of −40°C to 0°C)	$c_w = 4.0817 - 5.3062 \times 10^{-3}t + 9.9516 \times 10^{-4}t^2$
	Specific heat, kJ/(kg·K) (For temp. range of 0°C to 150°C)	$c_w = 4.1762 - 9.0864 \times 10^{-5}t + 5.4731 \times 10^{-6}t^2$
Ice	Thermal conductivity, W/(m·K)	$k_{ice} = 2.2196 - 6.2489 \times 10^{-3}t + 1.0154 \times 10^{-4}t^2$
	Thermal diffusivity, m²/s	$\alpha_{ice} = 1.1756 - 6.0833 \times 10^{-3}t + 9.5037 \times 10^{-5}t^2$
	Density, kg/m³	$\rho_{ice} = 9.1689 \times 10^{2} - 1.3071 \times 10^{-1}t$
	Specific heat, kJ/(kg·K)	$c_{ice} = 2.0623 + 6.0769 \times 10^{-3}t$

Source: Choi and Okos (1986)

ICE FRACTION

To predict the thermophysical properties of frozen foods, which depend strongly on the fraction of ice within the food, the mass fraction of water that has crystallized must be determined. For temperatures below the initial freezing point, the mass fraction of water that has crystallized in a food item is a function of temperature.

In general, food items consist of water, dissolved solids and undissolved solids. During the freezing process, as some of the liquid water crystallizes, the solids dissolved in the remaining liquid water become increasingly more concentrated, thus lowering the freezing temperature. This unfrozen solution can be assumed to obey the freezing point depression equation given by Raoult's law (Pham 1987). Thus, based on Raoult's law, Chen (1985) proposed the following model for predicting the mass fraction of ice x_{ice} in a food item:

$$x_{ice} = \frac{x_s R T_o^2}{M_s L_o} \frac{(t_f - t)}{t_f t} \tag{1}$$

where

x_s = mass fraction of solids in food item
M_s = relative molecular mass of soluble solids
R = universal gas constant = 8.314 kJ/(kg mol·K)
T_o = freezing point of water = 273.2 K
L_o = latent heat of fusion of water at 273.2 K = 333.6 kJ/kg
t_f = initial freezing point of food, °C
t = food temperature, °C

The relative molecular mass of the soluble solids within the food item may be estimated as follows:

$$M_s = \frac{x_s R T_o^2}{-(x_{wo} - x_b) L_o t_f} \tag{2}$$

where x_{wo} is the mass fraction of water in the unfrozen food item and x_b is the mass fraction of "bound water" in the food (Schwartzberg 1976). Bound water is that portion of the water in a food item that is bound to solids in the food, and thus is unavailable for freezing. The mass fraction of bound water may be estimated as follows:

$$x_b = 0.4 x_p \tag{3}$$

where x_p is the mass fraction of protein in the food item.

Thermal Properties of Foods

Table 3 Unfrozen Composition Data, Initial Freezing Point and Specific Heats of Foods[a]

Food Item	Moisture Content, % x_{wo}	Protein, % x_p	Fat, % x_f	Carbohydrate, % x_c	Fiber, % x_{fb}	Ash, % x_a	Initial Freezing Point, °C	Specific Heat Above Freezing, kJ/(kg·K)	Specific Heat Below Freezing, kJ/(kg·K)
Vegetables									
Artichokes, Globe	84.9	3.27	0.15	10.51	5.40	1.13	−1.2	3.64	1.88
Artichokes, Jerusalem	78.0	2.00	0.01	17.44	1.60	2.54	−2.5	3.48	1.84
Asparagus	92.4	2.28	0.20	4.54	2.10	0.57	−0.6	3.94	2.01
Beans, Snap	90.3	1.82	0.12	7.14	3.40	0.66	−0.7	3.94	2.39
Beans, Lima	70.2	6.84	0.86	20.16	4.90	1.89	−0.6	3.06	1.68
Beets	87.6	1.61	0.17	9.56	2.80	1.08	−1.1	3.77	1.68
Broccoli	90.7	2.98	0.35	5.24	3.00	0.92	−0.6	3.85	1.97
Brussels sprouts	86.0	3.38	0.30	8.96	3.80	1.37	−0.8	3.68	1.67
Cabbage	92.2	1.44	0.27	5.43	2.30	0.71	−0.9	3.94	1.97
Carrots	87.8	1.03	0.19	10.14	3.00	0.87	−1.4	3.77	1.93
Cauliflower	91.9	1.98	0.21	5.20	2.50	0.71	−0.8	3.89	1.97
Celeriac	88.0	1.50	0.30	9.20	1.80	1.00	−0.9	3.81	1.93
Celery	94.6	0.75	0.14	3.65	1.70	0.82	−0.5	3.98	2.01
Collards	90.6	1.57	0.22	7.11	3.60	0.55	−0.8	3.85	1.94
Corn, Sweet, yellow	76.0	3.22	1.18	19.02	2.70	0.62	−0.6	3.32	1.77
Cucumbers	96.0	0.69	0.13	2.76	0.80	0.41	−0.5	4.10	2.05
Eggplant	92.0	1.02	0.18	6.07	2.50	0.71	−0.8	3.94	2.01
Endive	93.8	1.25	0.20	3.35	3.10	1.41	−0.1	3.94	2.01
Garlic	58.6	6.36	0.50	33.07	2.10	1.50	−0.8	3.31	1.76
Ginger, root	81.7	1.74	0.73	15.09	2.00	0.77	—	3.85	1.94
Horseradish	78.7	9.40	1.40	8.28	2.00	2.26	−1.8	3.27	1.76
Kale	84.5	3.30	0.70	10.01	2.00	1.53	−0.5	3.73	1.93
Kohlrabi	91.0	1.70	0.10	6.20	3.60	1.00	−1.0	3.85	1.98
Leeks	83.0	1.50	0.30	14.15	1.80	1.05	−0.7	3.98	1.91
Lettuce, iceberg	95.9	1.01	0.19	2.09	1.40	0.48	−0.2	4.02	2.01
Mushrooms	91.8	2.09	0.42	4.65	1.20	0.89	−0.9	3.89	1.97
Okra	89.6	2.00	0.10	7.63	3.20	0.70	−1.8	3.85	1.93
Onions	89.7	1.16	0.16	8.63	1.80	0.37	−0.9	3.77	1.93
Onions, dehydrated flakes	3.9	8.95	0.46	83.28	9.20	3.38	—	—	—
Parsley	87.7	2.97	0.79	6.33	3.30	2.20	−1.1	3.62	1.91
Parsnips	79.5	1.20	0.30	17.99	4.90	0.98	−0.9	3.52	1.93
Peas, green	78.9	5.42	0.40	14.46	5.10	0.87	−0.6	3.31	1.76
Peppers, freeze dried	2.0	17.90	3.00	68.70	21.30	8.40	—	—	—
Peppers, sweet, green	92.2	0.89	0.19	6.43	1.80	0.30	−0.7	3.94	1.97
Potatoes, main crop	79.0	2.07	0.10	17.98	1.60	0.89	−0.6	3.63	1.82
Potatoes, Sweet	72.8	1.65	0.30	24.28	3.00	0.95	−1.3	3.14	1.68
Pumpkins	91.6	1.00	0.10	6.50	0.50	0.80	−0.8	3.85	1.97
Radishes	94.8	0.60	0.54	3.59	1.60	0.54	−0.7	3.98	2.01
Rhubarb	93.6	0.90	0.20	4.54	1.80	0.76	−0.9	4.02	2.01
Rutabaga	89.7	1.20	0.20	8.13	2.50	0.81	−1.1	3.81	1.97
Salsify (vegetable oyster)	77.0	3.30	0.20	18.60	3.30	0.90	−1.1	3.48	1.84
Spinach	91.6	2.86	0.35	3.50	2.70	1.72	−0.3	3.94	2.01
Squash, Summer	94.2	0.94	0.24	4.04	1.90	0.58	−0.5	4.02	2.02
Squash, Winter	87.8	0.80	0.10	10.42	1.50	0.90	−0.8	3.81	1.91
Tomatoes, mature green	93.0	1.20	0.20	5.10	1.10	0.50	−0.6	3.98	2.01
Tomatoes, ripe	93.8	0.85	0.33	4.64	1.10	0.42	−0.5	3.98	2.01
Turnip greens	91.1	1.50	0.30	5.73	3.20	1.40	−0.2	3.93	1.97
Turnip	91.9	0.90	0.10	6.23	1.80	0.70	−1.1	3.89	1.97
Watercress	95.1	2.30	0.10	1.29	1.50	1.20	−0.3	4.00	2.01
Yams	69.6	1.53	0.17	27.89	4.10	0.82	—	3.53	1.77
Fruits									
Apples, fresh	83.9	0.19	0.36	15.25	2.70	0.26	−1.1	3.60	1.84
Apples, dried	31.8	0.93	0.32	65.89	8.70	1.10	—	2.27	1.14
Apricots	86.3	1.40	0.39	11.12	2.40	0.75	−1.1	3.68	1.93
Avocados	74.3	1.98	15.32	7.39	5.00	1.04	−0.3	3.81	2.05
Bananas	74.3	1.03	0.48	23.43	2.40	0.80	−0.8	3.35	1.76

Table 3 Unfrozen Composition Data, Initial Freezing Point and Specific Heats of Foods[a] (Continued)

Food Item	Moisture Content, % x_{wo}	Protein, % x_p	Fat, % x_f	Carbohydrate, % x_c	Fiber, % x_{fb}	Ash, % x_a	Initial Freezing Point, °C	Specific Heat Above Freezing, kJ/(kg·K)	Specific Heat Below Freezing, kJ/(kg·K)
Blackberries	85.6	0.72	0.39	12.76	5.30	0.48	−0.8	3.68	1.68
Blueberries	84.6	0.67	0.38	14.13	2.70	0.21	−1.6	3.60	1.88
Cantaloupes	89.8	0.88	0.28	8.36	0.80	0.71	−1.2	3.94	2.01
Cherries, Sour	86.1	1.00	0.30	12.18	1.60	0.40	−1.7	3.65	1.89
Cherries, Sweet	80.8	1.20	0.96	16.55	2.30	0.53	−1.8	3.65	1.89
Cranberries	86.5	0.39	0.20	12.68	4.20	0.19	−0.9	3.77	1.93
Currants, European Black	82.0	1.40	0.41	15.38	0.00	0.86	−1.0	3.80	1.91
Currants, Red and White	84.0	1.40	0.20	13.80	4.30	0.66	−1.0	3.80	1.91
Dates, Cured	22.5	1.97	0.45	73.51	7.50	1.58	−15.7	1.51	1.09
Figs, Fresh	79.1	0.75	0.30	19.18	3.30	0.66	−2.4	3.43	1.80
Figs, Dried	28.4	3.05	1.17	65.35	9.30	2.01	—	1.63	1.13
Gooseberries	87.9	0.88	0.58	10.18	4.30	0.49	−1.1	3.77	1.93
Grapefruit	90.9	0.63	0.10	8.08	1.10	0.31	−1.1	3.81	1.93
Grapes, American	81.3	0.63	0.35	17.15	1.00	0.57	−1.6	3.60	1.84
Grapes, European Type	80.6	0.66	0.58	17.77	1.00	0.44	−2.1	3.60	1.84
Lemons	87.4	1.20	0.30	10.70	4.70	0.40	−1.4	3.85	1.93
Limes	88.3	0.70	0.20	10.54	2.80	0.30	−1.6	3.73	1.93
Mangos	81.7	0.51	0.27	17.00	1.80	0.50	−0.9	3.77	1.93
Melons, Casaba	92.0	0.90	0.10	6.20	0.80	0.80	−1.1	4.00	2.01
Melons, Honeydew	89.7	0.46	0.10	9.18	0.60	0.60	−0.9	3.94	2.01
Melons, Watermelon	91.5	0.62	0.43	7.18	0.50	0.26	−0.4	4.06	2.01
Nectarines	86.3	0.94	0.46	11.78	1.60	0.54	−0.9	3.77	2.05
Olives	80.0	0.84	10.68	6.26	3.20	2.23	−1.4	3.35	1.76
Oranges	82.3	1.30	0.30	15.50	4.50	0.60	−0.8	3.77	1.93
Peaches, Fresh	87.7	0.70	0.90	11.10	2.00	0.46	−0.9	3.77	1.93
Peaches, Dried	31.8	3.61	0.76	61.33	8.20	2.50	—	2.30	1.16
Pears	83.8	0.39	0.40	15.11	2.40	0.28	−1.6	3.75	1.89
Persimmons	64.4	0.80	0.40	33.50	0.00	0.90	−2.2	3.52	1.80
Pineapples	86.5	0.39	0.43	12.39	1.20	0.29	−1.0	3.68	1.88
Plums	85.2	0.79	0.62	13.01	1.50	0.39	−0.8	3.52	1.92
Pomegranates	81.0	0.95	0.30	17.17	0.60	0.61	−3.0	3.68	2.01
Prunes, Dried	32.4	2.61	0.52	62.73	7.10	1.76	—	2.37	1.19
Quinces	83.8	0.40	0.10	15.30	1.90	0.40	−2.0	3.68	1.88
Raisins, Seedless	15.4	3.22	0.46	79.13	4.00	1.77	—	1.97	1.07
Raspberries	86.6	0.91	0.55	11.57	6.80	0.40	−0.6	3.56	1.88
Strawberries	91.6	0.61	0.37	7.02	2.30	0.43	−0.8	3.89	1.14
Tangerines	87.6	0.63	0.19	11.19	2.30	0.39	−1.1	3.89	2.09
Whole Fish									
Cod	81.2	17.81	0.67	0.0	0.0	1.16	−2.2	3.77	2.05
Haddock	79.9	18.91	0.72	0.0	0.0	1.21	−2.2	3.43	1.80
Halibut	77.9	20.81	2.29	0.0	0.0	1.36	−2.2	3.35	1.80
Herring, Kippered	59.7	24.58	12.37	0.0	0.0	1.94	−2.2	3.18	1.72
Mackerel, Atlantic	63.6	18.60	13.89	0.0	0.0	1.35	−2.2	2.76	1.55
Perch	78.7	18.62	1.63	0.0	0.0	1.20	−2.2	3.52	1.84
Pollock, Atlantic	78.2	19.44	0.98	0.0	0.0	1.41	−2.2	3.48	1.84
Salmon, Pink	76.4	19.94	3.45	0.0	0.0	1.22	−2.2	2.97	1.63
Tuna, Bluefin	68.1	23.33	4.90	0.0	0.0	1.18	−2.2	3.18	1.72
Whiting	80.3	18.31	1.31	0.0	0.0	1.30	−2.2	3.60	1.84
Shellfish									
Clams	81.8	12.77	0.97	2.57	0.0	1.87	−2.2	—	—
Lobster, American	76.8	18.80	0.90	0.50	0.0	2.20	−2.2	3.48	1.84
Oysters	85.2	7.05	2.46	3.91	0.0	1.42	−2.2	3.48	1.84
Scallop, Meat	78.6	16.78	0.76	2.36	0.0	1.53	−2.2	3.52	1.84
Shrimp	75.9	20.31	1.73	0.91	0.0	1.20	−2.2	3.48	1.88
Beef									
Brisket	55.2	16.94	26.54	0.0	0.0	0.80	—	—	—
Carcass, Choice	57.3	17.32	24.05	0.0	0.0	0.81	−2.2	—	—
Carcass, Select	58.2	17.48	22.55	0.0	0.0	0.82	−1.7	—	—

Thermal Properties of Foods

Table 3 Unfrozen Composition Data, Initial Freezing Point and Specific Heats of Foods[a] *(Continued)*

Food Item	Moisture Content, % x_{wo}	Protein, % x_p	Fat, % x_f	Carbohydrate, % x_c	Fiber, % x_{fb}	Ash, % x_a	Initial Freezing Point, °C	Specific Heat Above Freezing, kJ/(kg·K)	Specific Heat Below Freezing, kJ/(kg·K)
Liver	69.0	20.00	3.85	5.82	0.0	1.34	−1.7	3.43	1.72
Ribs, Whole (Ribs 6-12)	54.5	16.37	26.98	0.0	0.0	0.77	—	—	—
Round, Full Cut, Lean and Fat	64.8	20.37	12.81	0.0	0.0	0.97	—	3.35	1.68
Round, Full Cut, Lean	70.8	22.03	4.89	0.0	0.0	1.07	—	3.35	1.68
Sirloin, Lean	71.7	21.24	4.40	0.0	0.0	1.08	−1.7	3.08	1.55
Short Loin, Porterhouse Steak, Lean	69.6	20.27	8.17	0.0	0.0	1.01	—	—	—
Short Loin, T−Bone Steak, Lean	69.7	20.78	7.27	0.0	0.0	1.27	—	—	—
Tenderloin, Lean	68.4	20.78	7.90	0.0	0.0	1.04	—	—	—
Veal, Lean	75.9	20.20	2.87	0.0	0.0	1.08	—	3.35	1.93
Pork									
Backfat	7.7	2.92	88.69	0.0	0.0	0.70	—	2.60	0.94
Bacon	31.6	8.66	57.54	0.09	0.0	2.13	—	2.09	1.26
Belly	36.7	9.34	53.01	0.0	0.0	0.49	—	2.42	1.22
Carcass	49.8	13.91	35.07	0.0	0.0	0.72	—	2.60	1.31
Ham, Cured, Whole Lean	68.3	22.32	5.71	0.05	0.0	3.66	—	3.10	1.56
Ham, Country Cured Lean	55.9	27.80	8.32	0.30	0.0	7.65	—	2.72	1.37
Shoulder, Whole, Lean	72.6	19.55	7.14	0.0	0.0	1.02	−2.2	2.90	1.46
Sausage									
Braunschweiger	48.0	13.50	32.09	3.13	0.0	3.27	—	—	—
Frankfurter	53.9	11.28	29.15	2.55	0.0	3.15	−1.7	3.60	2.35
Italian	51.1	14.25	31.33	0.65	0.0	2.70	—	—	—
Polish	53.2	14.10	28.72	1.63	0.0	2.40	—	3.03	1.52
Pork	44.5	11.69	40.29	1.02	0.0	2.49	—	3.73	2.35
Smoked Links	39.3	22.20	31.70	2.10	0.0	4.70	—	3.60	2.35
Poultry Products									
Chicken	66.0	18.60	15.06	0.0	0.0	0.79	−2.8	3.31	1.55
Duck	48.5	11.49	39.34	0.0	0.0	0.68	—	3.40	1.71
Turkey	70.4	20.42	8.02	0.0	0.0	0.88	—	3.31	1.55
Egg									
White	87.8	10.52	0.0	1.03	0.0	0.64	−0.6	3.88	1.95
White, Dried	14.6	76.92	0.04	4.17	0.0	4.25	—	1.90	0.95
Whole	75.3	12.49	10.02	1.22	0.0	0.94	−0.6	3.18	1.68
Whole, Dried	3.1	47.35	40.95	4.95	0.0	3.65	—	1.05	0.89
Yolk	48.8	16.76	30.87	1.78	0.0	1.77	−0.6	2.81	1.48
Yolk, Salted	50.8	14.00	23.00	1.60	0.0	10.60	−17.2	2.93	1.47
Yolk, Sugared	51.2	13.80	22.75	10.80	0.0	1.40	−3.9	2.95	1.48
Lamb									
Composite of Cuts, Lean	73.4	20.29	5.25	0.0	0.0	1.06	−1.9	3.20	1.61
Leg, Whole, Lean	74.1	20.56	4.51	0.0	0.0	1.07	—	3.30	1.66
Dairy Products									
Butter	17.9	0.85	81.11	0.06	0.0	0.04	—	1.38	1.05
Cheese									
Camembert	51.8	19.80	24.26	0.46	0.0	3.68	—	2.98	1.50
Cheddar	36.8	24.90	33.14	1.28	0.0	3.93	−12.9	2.60	1.31
Cottage, Uncreamed	79.8	17.27	0.42	1.85	0.0	0.69	−1.2	3.65	1.84
Cream	53.8	7.55	34.87	2.66	0.0	1.17	—	2.93	1.88
Gouda	41.5	24.94	27.44	2.22	0.0	3.94	—	—	—
Limburger	48.4	20.05	27.25	0.49	0.0	3.79	−7.4	2.93	1.68
Mozzarella	54.1	19.42	21.60	2.22	0.0	2.62	—	—	—
Parmesan, Hard	29.2	35.75	25.83	3.22	0.0	6.04	—	—	—
Roquefort	39.4	21.54	30.64	2.00	0.0	6.44	−16.3	2.72	1.34
Swiss	37.2	28.43	27.45	3.38	0.0	3.53	−10.0	2.68	1.51
Processed American	39.2	22.15	31.25	1.30	0.0	5.84	−6.9	2.68	1.34
Cream									
Half and Half	80.6	2.96	11.50	4.30	0.0	0.67	—	3.68	1.85
Table	73.8	2.70	19.31	3.66	0.0	0.58	−2.2	3.48	1.75
Heavy Whipping	57.7	2.05	37.00	2.79	0.0	0.45	—	3.56	1.68

Table 3 Unfrozen Composition Data, Initial Freezing Point and Specific Heats of Foods[a] (*Continued*)

Food Item	Moisture Content, % x_{wo}	Protein, % x_p	Fat, % x_f	Carbohydrate, % x_c	Fiber, % x_{fb}	Ash, % x_a	Initial Freezing Point, °C	Specific Heat Above Freezing, kJ/(kg·K)	Specific Heat Below Freezing, kJ/(kg·K)
Ice Cream									
Chocolate	55.7	3.80	11.0	28.20	1.20	1.00	−5.6	3.27	1.88
Strawberry	60.0	3.20	8.40	27.60	0.30	0.70	−5.6	3.27	1.88
Vanilla	61.0	3.50	11.00	23.60	0.0	0.90	−5.6	3.27	1.88
Milk									
Canned, Condensed Sweetened	27.2	7.91	8.70	54.40	0.0	1.83	−15.0	2.35	1.18
Evaporated	74.0	6.81	7.56	10.04	0.0	1.55	−1.4	3.53	1.77
Skim	90.8	3.41	0.18	4.85	0.0	0.76	—	4.00	2.51
Skim, Dried	3.2	36.16	0.77	51.98	0.0	7.93	—	1.75	0.88
Whole	87.7	3.28	3.66	4.65	0.0	0.72	−0.6	3.85	1.94
Whole, Dried	2.5	26.32	26.71	38.42	0.0	6.08	—	1.72	0.87
Whey, Acid, Dried	3.5	11.73	0.54	73.45	0.0	10.77	—	1.80	0.90
Whey, Sweet, Dried	3.2	12.93	1.07	74.46	0.0	8.35	—	1.80	0.90
Nuts, Shelled									
Almonds	4.4	19.95	52.21	20.40	10.90	3.03	—	1.80	0.90
Filberts	5.4	13.04	62.64	15.30	6.10	3.61	—	1.82	0.92
Peanuts, Raw	6.5	25.80	49.24	16.14	8.50	2.33	—	1.82	0.92
Peanuts, Dry Roasted with Salt	1.6	23.68	49.66	21.51	8.00	3.60	—	1.72	0.87
Pecans	4.8	7.75	67.64	18.24	7.60	1.56	—	1.75	0.88
Walnuts, English	3.6	14.29	61.87	18.34	4.80	1.86	—	1.78	0.89
Candy									
Fudge, Vanilla	10.9	1.10	5.40	82.30	0.0	0.40	—	1.92	0.97
Marshmallows	16.4	1.80	0.20	81.30	0.10	0.30	—	2.10	1.05
Milk Chocolate	1.3	6.90	30.70	59.20	3.40	1.50	—	1.70	0.85
Peanut Brittle	1.8	7.50	19.10	69.30	2.00	1.50	—	1.72	0.87
Juice and Beverages									
Apple Juice, Unsweetened	87.9	0.06	0.11	11.68	0.10	0.22	—	3.79	1.95
Grapefruit Juice, Sweetened	87.4	0.58	0.09	11.13	0.10	0.82	—	3.77	1.94
Grape Juice, Unsweetened	84.1	0.56	0.08	14.96	0.10	0.29	—	3.66	1.90
Lemon Juice	92.5	0.40	0.29	6.48	0.40	0.36	—	3.94	2.01
Lime Juice, Unsweetened	92.5	0.25	0.23	6.69	0.40	0.31	—	3.94	2.01
Orange Juice	89.0	0.59	0.14	9.85	0.20	0.41	−0.4	3.82	1.96
Pineapple Juice, Unsweetened	85.5	0.32	0.08	13.78	0.20	0.30	—	3.71	1.92
Prune Juice	81.2	0.61	0.03	17.45	1.00	0.68	—	3.56	1.86
Tomato Juice	93.9	0.76	0.06	4.23	0.40	1.05	—	3.99	2.02
Cranberry-Apple Juice Drink	82.8	0.10	0.0	17.10	0.10	0.0	—	3.61	1.88
Cranberry-Grape Juice Drink	85.6	0.20	0.10	14.00	0.10	0.10	—	3.71	1.92
Fruit Punch Drink	88.0	0.0	0.0	11.90	0.10	0.10	—	3.79	1.95
Club Soda	99.9	0.0	0.0	0.0	0.0	0.10	—	4.19	2.10
Cola	89.4	0.0	0.0	10.40	0.0	0.10	—	3.83	1.97
Cream Soda	86.7	0.0	0.0	13.30	0.0	0.10	—	3.74	1.93
Ginger Ale	91.2	0.0	0.0	8.70	0.0	0.0	—	3.90	1.99
Grape Soda	88.8	0.0	0.0	11.20	0.0	0.10	—	3.81	1.96
Lemon-Lime Soda	89.5	0.0	0.0	10.40	0.0	0.10	—	3.84	1.97
Orange Soda	87.6	0.0	0.0	12.30	0.0	0.10	—	3.77	1.94
Root Beer	89.3	0.0	0.0	10.60	0.0	0.10	—	3.83	1.97
Chocolate Milk, 2% fat	83.6	3.21	2.00	10.40	0.50	0.81	—	3.64	1.89
Miscellaneous									
Honey	17.1	0.30	0.0	82.40	0.20	0.20	—	—	—
Maple Syrup	32.0	0.00	0.20	67.20	0.0	0.60	—	2.05	1.30
Popcorn, Air-Popped	4.1	12.00	4.20	77.90	15.10	1.80	—	—	—
Popcorn, Oil-Popped	2.8	9.00	28.10	57.20	10.00	2.90	—	—	—
Yeast, Baker's, Compressed	69.0	8.40	1.90	18.10	8.10	1.80	—	3.22	1.72

[a] Composition data from USDA (1996). Initial freezing point data from ASHRAE (1993). Specific heats from Polley et al. (1980) and ASHRAE (1993).

Thermal Properties of Foods

Substitution of Equation (2) for relative molecular mass into Equation (1) yields a simple method for predicting the ice fraction as follows (Miles 1974):

$$x_{ice} = (x_{wo} - x_b)\left[1 - \frac{t_f}{t}\right] \quad (4)$$

Because Equation (4) underestimates the ice fraction at temperatures near the initial freezing point and overestimates the ice fraction at lower temperatures, Tchigeov (1979) proposed an empirical relationship to estimate the mass fraction of ice:

$$x_{ice} = \frac{1.105 x_{wo}}{1 + \frac{0.8765}{\ln(t_f - t + 1)}} \quad (5)$$

Fikiin (1996) notes that Equation (5) applies to a wide variety of food items and provides satisfactory accuracy.

Example 1. A 150 kg beef carcass is to be frozen to a temperature of −20°C. What is the mass of the frozen water and the mass of the unfrozen water at −20°C?

Solution:
From Table 3, the mass fraction of water in the beef carcass is 0.58 and the initial freezing point for the beef carcass is −1.7°C. Using Equation (5), the mass fraction of ice is:

$$x_{ice} = \frac{1.105 \times 0.58}{1 + \frac{0.8765}{\ln(-1.7 + 20 + 1)}} = 0.52$$

The mass fraction of unfrozen water is:

$$x_u = x_{wo} - x_{ice} = 0.58 - 0.52 = 0.06$$

The mass of frozen water at −20°C is:

$$x_{ice} \times 150 \text{ kg} = 0.52 \times 150 = 78 \text{ kg}$$

The mass of unfrozen water at −20°C is:

$$x_u \times 150 \text{ kg} = 0.06 \times 150 = 9.0 \text{ kg}$$

DENSITY

Modeling the density of foods and beverages requires knowledge of the food porosity, as well as the mass fraction and density of the food components. The density ρ of foods and beverages can be calculated accordingly:

$$\rho = \frac{(1 - \varepsilon)}{\sum x_i / \rho_i} \quad (6)$$

where ε is the porosity, x_i is the mass fraction of the food constituents, and ρ_i is the density of the food constituents. The porosity ε is required to model the density of granular food items stored in bulk, such as grains and rice. For other food items, the porosity is zero.

SPECIFIC HEAT

Specific heat is a measure of the energy required to change the temperature of a food item by one degree. Therefore, the specific heat of foods or beverages can be used to calculate the heat load imposed on the refrigeration equipment by the cooling or freezing of foods and beverages. In unfrozen foods, specific heat becomes slightly lower as the temperature rises from 0°C to 20°C. For frozen foods, there is a large decrease in specific heat as the temperature decreases. Table 3 lists experimentally determined values of the specific heats for various foods above and below freezing.

Unfrozen Food

The specific heat of a food item, at temperatures above its initial freezing point, can be obtained from the mass average of the specific heats of the food components. Thus, the specific heat of an unfrozen food item c_u may be determined as follows:

$$c_u = \sum c_i x_i \quad (7)$$

where c_i is the specific heat of the individual food components and x_i is the mass fraction of the food components.

A simpler model for the specific heat of an unfrozen food item is presented by Chen (1985). If detailed composition data is not available, the following expression for the specific heat of an unfrozen food item can be used:

$$c_u = 4.19 - 2.30 x_s - 0.628 x_s^3 \quad (8)$$

where c_u is the specific heat of the unfrozen food item in kJ/(kg·K) and x_s is the mass fraction of the solids in the food item.

Frozen

Below the freezing point of the food item, the sensible heat due to temperature change and the latent heat due to the fusion of water must be considered. Because latent heat is not released at a constant temperature, but rather over a range of temperatures, an apparent specific heat must be used to account for both the sensible and latent heat effects. A common method to predict the apparent specific heat of food items is that of Schwartzberg (1976):

$$c_a = c_u + (x_b - x_{wo})\Delta c + E x_s \left[\frac{R T_o^2}{M_w t^2} - 0.8 \Delta c\right] \quad (9)$$

where

- c_a = apparent specific heat
- c_u = specific heat of food item above initial freezing point
- x_b = mass fraction of bound water
- x_{wo} = mass fraction of water above initial freezing point
- Δc = difference between specific heats of water and ice = $c_w - c_{ice}$
- E = ratio of relative molecular masses of water M_w and food solids M_s ($E = M_w / M_s$)
- R = universal gas constant = 8.314 kJ/(kg mol·K)
- T_o = freezing point of water = 273.2 K
- t = food temperature

The specific heat of the food item above the freezing point may be estimated with Equation (7) or Equation (8).

Schwartzberg (1981) expanded on his earlier work and developed an alternative method for determining the apparent specific heat of a food item below the initial freezing point as follows:

$$c_a = c_f + (x_{wo} - x_b)\left[\frac{L_o (t_o - t_f)}{t_o - t}\right] \quad (10)$$

where

- c_f = specific heat of fully frozen food item (typically at −40°C)
- t_o = freezing point of water = 0°C
- t_f = initial freezing point of food, °C
- t = food temperature, °C
- L_o = latent heat of fusion of water = 333.6 kJ/kg

Experimentally determined values of the specific heat of fully frozen food items are given in Table 3.

A slightly simpler apparent specific heat model, which is similar in form to that of Schwartzberg (1976), was developed by Chen

(1985). Chen's model is an expansion of Siebel's equation (Siebel 1892) for specific heat and has the following form:

$$c_a = 1.55 + 1.26 x_s + \frac{x_s R T_o^2}{M_s t^2} \qquad (11)$$

where

c_a = apparent specific heat, kJ/(kg·K)
x_s = mass fraction of solids
R = universal gas constant
T_o = freezing point of water = 273.2 K
M_s = relative molecular mass of soluble solids in food item
t = food temperature, °C

If the relative molecular mass of the soluble solids is unknown, Equation (2) may be used to estimate the molecular mass. Substitution of Equation (2) into Equation (11) yields:

$$c_a = 1.55 + 1.26 x_s - \frac{(x_{wo} - x_b) L_o t_f}{t^2} \qquad (12)$$

Example 2. A 150 kg lamb is to be cooled from 10°C to 0°C. Using the specific heat, determine the amount of heat which must be removed from the lamb.

Solution:
From Table 3, the composition of lamb is given as follows:

x_{wo} = 0.7342 x_f = 0.0525
x_p = 0.2029 x_a = 0.0106

Evaluate the specific heat of lamb at an average temperature of (0 + 10)/2 = 5°C. From Tables 1 and 2, the specific heat of the food constituents may be determined as follows:

$c_w = 4.1762 - 9.0864 \times 10^{-5}(5) + 5.4731 \times 10^{-6}(5)^2$
     $= 4.1759$ kJ/(kg·K)

$c_p = 2.0082 + 1.2089 \times 10^{-3}(5) - 1.3129 \times 10^{-6}(5)^2$
     $= 2.0142$ kJ/(kg·K)

$c_f = 1.9842 + 1.4733 \times 10^{-3}(5) - 4.8008 \times 10^{-6}(5)^2$
     $= 1.9914$ kJ/(kg·K)

$c_a = 1.0926 + 1.8896 \times 10^{-3}(5) - 3.6817 \times 10^{-6}(5)^2$
     $= 1.1020$ kJ/(kg·K)

The specific heat of lamb can be calculated with Equation (7):

$c = \sum c_i x_i = (4.1759)(0.7342) + (2.0142)(0.2029)$
          $+ (1.9914)(0.0525) + (1.1020)(0.0106)$

$c = 3.59$ kJ/(kg·K)

The heat to be removed from the lamb is as follows:

$Q = mc\Delta T = 150 \times 3.59 (10 - 0) = 5390$ kJ

ENTHALPY

The change in enthalpy of a food item can be used to estimate the energy that must be added or removed to effect a temperature change. Above the freezing point, enthalpy consists of sensible energy, while below the freezing point, enthalpy consists of both sensible and latent energy. Enthalpy may be obtained from the definition of constant pressure specific heat:

$$c_p = \left(\frac{\partial H}{\partial T}\right)_p \qquad (13)$$

where c_p is constant pressure specific heat, H is enthalpy, and T is temperature. Mathematical models for enthalpy may be obtained by integrating expressions of specific heat with respect to temperature.

Unfrozen Food

For food items that are at temperatures above their initial freezing point, enthalpy may be obtained by integrating the corresponding expression for specific heat above the freezing point. Thus, the enthalpy of an unfrozen food item H may be determined by integrating Equation (7) as follows:

$$H = \sum H_i x_i = \sum \int c_i x_i dT \qquad (14)$$

where H_i is the enthalpy of the individual food components and x_i is the mass fraction of the food components.

In the case of the method of Chen (1985), the enthalpy of an unfrozen food may be obtained by integrating Equation (8):

$$H = H_f + (t - t_f)(4.19 - 2.30 x_s - 0.628 x_s^3) \qquad (15)$$

where

H = enthalpy of food item, kJ/kg
H_f = enthalpy of food at initial freezing temperature, kJ/kg
t = temperature of food item, °C
t_f = initial freezing temperature of food item, °C
x_s = mass fraction of food solids

The enthalpy at the initial freezing point H_f may be estimated by evaluating either Equation (17) or (18) at the initial freezing temperature of the food as discussed in the following section.

Frozen Foods

For food items below the initial freezing point, mathematical expressions for enthalpy may be obtained by integrating the previously mentioned apparent specific heat models. Integration of Equation (9) between a reference temperature T_r and the food temperature T leads to the following expression for the enthalpy of a food item (Schwartzberg 1976):

$$H = (T - T_r) \times \left\{ c_u + (x_b - x_{wo})\Delta c + Ex_s\left[\frac{RT_o^2}{18(T_o - T_r)(T_o - T)} - 0.8\Delta c\right]\right\} \qquad (16)$$

Generally, the reference temperature T_r is taken to be 233.2 K (−40°C) at which point the enthalpy is defined to be zero.

By integrating Equation (11) between a reference temperature T_r and the food temperature T, Chen (1985) obtained the following expression for enthalpy below the initial freezing point:

$$H = (t - t_r)\left(1.55 + 1.26 x_s + \frac{x_s R T_o^2}{M_s t t_r}\right) \qquad (17)$$

where

H = enthalpy of food item
R = universal gas constant
T_o = freezing point of water = 273.2 K

Substitution of Equation (2) for the relative molecular mass of the soluble solids M_s simplifies Chen's method as follows:

$$H = (t - t_r)\left[1.55 + 1.26 x_s + \frac{(x_{wo} - x_b) L_o t_f}{t_r t}\right] \qquad (18)$$

As an alternative to the enthalpy models developed by integration of specific heat equations, Chang and Tao (1981) developed empirical correlations for the enthalpy of food items. Their enthalpy correlations are given as functions of water content, initial and final

Thermal Properties of Foods

temperatures, and food type (meat, juice or fruit/vegetable). The correlations at a reference temperature of (−45.6°C) have the following form:

$$H = H_f(y\bar{T} + (1-y)\bar{T}^z) \qquad (19)$$

where

H = enthalpy of food item, kJ/kg
H_f = enthalpy of food item at initial freezing temperature, kJ/kg
$\bar{T}$ = reduced temperature, $\bar{T} = (T - T_r)/(T_f - T_r)$
T_r = reference temperature (zero enthalpy) = 227.6 K (−45.6°C)
y, z = correlation parameters

By performing regression analysis on experimental data available in the literature, Chang and Tao (1981) developed the following correlation parameters y and z used in Equation (19):

Meat Group:

$$\begin{aligned} y &= 0.316 - 0.247(x_{wo} - 0.73) - 0.688(x_{wo} - 0.73)^2 \\ z &= 22.95 - 54.68(y - 0.28) - 5589.03(y - 0.28)^2 \end{aligned} \qquad (20)$$

Fruit, Vegetable, and Juice Group:

$$\begin{aligned} y &= 0.362 + 0.0498(x_{wo} - 0.73) - 3.465(x_{wo} - 0.73)^2 \\ z &= 27.2 - 129.04(y - 0.23) - 481.46(y - 0.23)^2 \end{aligned} \qquad (21)$$

They also developed correlations to estimate the initial freezing temperature T_f for use in Equation (19). These correlations give T_f as a function of water content:

Meat Group:

$$T_f = 271.18 + 1.47x_{wo} \qquad (22)$$

Fruit/Vegetable Group:

$$T_f = 287.56 - 49.19x_{wo} + 37.07x_{wo}^2 \qquad (23)$$

Juice Group:

$$T_f = 120.47 + 327.35x_{wo} - 176.49x_{wo}^2 \qquad (24)$$

In addition, the enthalpy of the food item at its initial freezing point is required in Equation (19). Chang and Tao (1981) suggest the following correlation for determining the enthalpy of the food item at its initial freezing point H_f

$$H_f = 9.79246 + 0.405096x_{wo} \qquad (25)$$

Table 4 presents experimentally determined values for the enthalpy of some frozen foods at a reference temperature of −40°C as well as the percentage of unfrozen water in these foods.

Example 3. A 150 kg beef carcass is to be frozen to a temperature of −20°C. The initial temperature of the beef carcass is 10°C. How much heat must be removed from the beef carcass during this process?

Solution:
From Table 3, the mass fraction of water in the beef carcass is 0.5821, the mass fraction of protein in the beef carcass is 0.1748 and the initial freezing point of the beef carcass is −1.7°C. The mass fraction of solids in the beef carcass is:

$$x_s = 1 - x_{wo} = 1 - 0.5821 = 0.4179$$

The mass fraction of bound water is given by Equation (3):

$$x_b = 0.4x_p = 0.4 \times 0.1748 = 0.0699$$

The enthalpy of the beef carcass at −20°C is given by Equation (18) for frozen foods:

$$H_{-20} = [-20 - (-40)]\left\{1.55 + (1.26)(0.4179) \right.$$
$$\left. - \frac{(0.5821 - 0.0699)(333.6)(-1.7)}{(-40)(-20)}\right\} = 48.79 \text{ kJ/kg}$$

The enthalpy of the beef carcass at the initial freezing point is determined by evaluating Equation (18) at the initial freezing point:

$$H_f = [-1.7 - (-40)]\left\{1.55 + (1.26)(0.4179) \right.$$
$$\left. - \frac{(0.5821 - 0.0699)(333.6)(-1.7)}{(-40)(-1.7)}\right\} = 243.14 \text{ kJ/kg}$$

The enthalpy of the beef carcass at 10°C is given by Equation (15) for unfrozen foods:

$$H_{10} = 243.14 + [10 - (-1.7)]$$
$$\times [4.19 - (2.30)(0.4179) - (0.628)(0.4179)^3]$$
$$= 280.38 \text{ kJ/kg}$$

Thus, the amount of heat removed during the freezing process is:

$$Q = m\Delta H = m(H_{10} - H_{-20})$$
$$= 150(280.38 - 48.79) = 34{,}700 \text{ kJ}$$

THERMAL CONDUCTIVITY

Thermal conductivity relates the conduction heat transfer rate to the temperature gradient. The thermal conductivity of a food depends on such factors as composition, structure, and temperature. Early work in the modeling of thermal conductivity of foods and beverages includes Eucken's adaption of Maxwell's equation (Eucken 1940). This model is based on the thermal conductivity of dilute dispersions of small spheres in a continuous phase:

$$k = k_c \frac{1 - [1 - a(k_d/k_c)]b}{1 + (a-1)b} \qquad (26)$$

where

k = conductivity of mixture
k_c = conductivity of continuous phase
k_d = conductivity of dispersed phase
$a = 3k_c/(2k_c + k_d)$
$b = V_d/(V_c + V_d)$
V_d = volume of dispersed phase
V_c = volume of continuous phase

In an effort to account for the different structural features of foods, Kopelman (1966) developed thermal conductivity models for homogeneous and fibrous food items. The differences in thermal conductivity parallel and perpendicular to the food fibers are accounted for in Kopelman's fibrous food thermal conductivity models.

For an isotropic, two-component system composed of continuous and discontinuous phases, in which the thermal conductivity is independent of the direction of heat flow, Kopelman (1966) developed the following expression for thermal conductivity k:

$$k = k_c\left[\frac{1 - L^2}{1 - L^2(1-L)}\right] \qquad (27)$$

Table 4 Enthalpy of Frozen Foods

Food Item	Water Content (% by mass)		−40	−30	−20	−18	−16	−14	−12	−10	−9	−8	−7	−6	−5	−4	−3	−2	−1	0
Fruits/Vegetables																				
Applesauce	82.8	Enthalpy (kJ/kg)	0	23	51	58	65	73	84	95	102	110	120	132	152	175	210	286	339	343
		% water unfrozen	—	6	9	10	12	14	17	19	21	23	27	30	37	44	57	82	100	—
Asparagus, peeled	92.6	Enthalpy (kJ/kg)	0	19	40	45	50	55	61	69	73	77	83	90	99	108	123	155	243	381
		% water unfrozen	—	—	—	—	5	6	—	7	8	10	12	15	17	20	29	58	100	
Bilberries	85.1	Enthalpy (kJ/kg)	0	21	45	50	57	64	73	82	87	94	101	110	125	140	167	218	348	352
		% water unfrozen	—	—	—	7	8	9	11	14	15	17	18	21	25	30	38	57	100	—
Carrots	87.5	Enthalpy (kJ/kg)	0	21	46	51	57	64	72	81	87	94	102	111	124	139	166	218	357	361
		% water unfrozen	—	—	—	7	8	9	11	14	15	17	18	20	24	29	37	53	100	—
Cucumbers	95.4	Enthalpy (kJ/kg)	0	18	39	43	47	51	57	64	67	70	74	79	85	93	104	125	184	390
		% water unfrozen	—	—	—	—	—	—	—	5	—	—	—	11	14	20	37	100		
Onions	85.5	Enthalpy (kJ/kg)	0	23	50	55	62	71	81	91	97	105	115	125	141	163	196	263	349	353
		% water unfrozen	—	5	8	10	12	14	16	18	19	20	23	26	31	38	49	71	100	—
Peaches, without stones	85.1	Enthalpy (kJ/kg)	0	23	50	57	64	72	82	93	100	108	118	129	146	170	202	274	348	352
		% water unfrozen	—	5	8	9	11	13	16	18	20	22	25	28	33	40	51	75	100	—
Pears, Bartlett	83.8	Enthalpy (kJ/kg)	0	23	51	57	64	73	83	95	101	109	120	132	150	173	207	282	343	347
		% water unfrozen	—	6	9	10	12	14	17	19	21	23	26	29	35	43	54	80	100	—
Plums, without stones	80.3	Enthalpy (kJ/kg)	0	25	57	65	74	84	97	111	119	129	142	159	182	214	262	326	329	333
		% water unfrozen	—	8	14	16	18	20	23	27	29	33	37	42	50	61	78	100	—	
Raspberries	82.7	Enthalpy (kJ/kg)	0	20	47	53	59	65	75	85	90	97	105	115	129	148	174	231	340	344
		% water unfrozen	—	—	7	8	9	10	13	16	17	18	20	23	27	33	42	61	100	—
Spinach	90.2	Enthalpy (kJ/kg)	0	19	40	44	49	54	60	66	70	74	79	86	94	103	117	145	224	371
		% water unfrozen	—	—	—	—	—	—	6	7	—	9	11	13	16	19	28	53	100	
Strawberries	89.3	Enthalpy (kJ/kg)	0	20	44	49	54	60	67	76	81	88	95	102	114	127	150	191	318	367
		% water unfrozen	—	—	5	—	6	7	9	11	12	14	16	18	20	24	30	43	86	100
Sweet cherries, without stones	77.0	Enthalpy (kJ/kg)	0	26	58	66	76	87	100	114	123	133	149	166	190	225	276	317	320	324
		% water unfrozen	—	9	15	17	19	21	26	29	32	36	40	47	55	67	86	100	—	
Tall peas	75.8	Enthalpy (kJ/kg)	0	23	51	56	64	73	84	95	102	111	121	133	152	176	212	289	319	323
		% water unfrozen	—	6	10	12	14	16	18	21	23	26	28	33	39	48	61	90	100	—
Tomato pulp	92.9	Enthalpy (kJ/kg)	0	20	42	47	52	57	63	71	75	81	87	93	103	114	131	166	266	382
		% water unfrozen	—	—	—	—	5	—	6	7	8	10	12	14	16	18	24	33	65	100
Fish/Meat																				
Cod	80.3	Enthalpy (kJ/kg)	0	19	42	47	53	59	66	74	79	84	89	96	105	118	137	177	298	323
		% water unfrozen	10	10	11	12	12	13	14	16	17	18	19	21	23	27	34	48	92	100
Haddock	83.6	Enthalpy (kJ/kg)	0	19	42	47	53	59	66	73	77	82	88	95	104	116	136	177	307	337
		% water unfrozen	8	8	9	10	11	11	12	13	14	15	16	18	20	24	31	44	90	100
Perch	79.1	Enthalpy (kJ/kg)	0	19	41	46	52	58	65	72	76	81	86	93	101	112	129	165	284	318
		% water unfrozen	10	10	11	12	12	13	14	15	16	17	18	20	22	26	32	44	87	100
Beef, lean, fresh[b]	74.5	Enthalpy (kJ/kg)	0	19	42	47	52	58	65	72	76	81	88	95	105	113	138	180	285	304
		% water unfrozen	10	10	11	12	13	14	15	16	17	18	20	22	24	31	40	55	95	100
Beef, lean, dried	26.1	Enthalpy (kJ/kg)	0	19	42	47	53	62	66	70	72	74	—	79	—	84	—	89	—	93
		% water unfrozen	96	96	97	98	99	100	—	—	—	—	—	—	—	—	—	—	—	—
Eggs																				
Egg white	86.5	Enthalpy (kJ/kg)	0	18	39	43	48	53	58	65	68	72	75	81	87	96	109	134	210	352
		% water unfrozen	—	—	10	—	—	—	13	—	—	—	18	20	23	28	40	82	100	
Egg yolk	50.0	Enthalpy (kJ/kg)	0	18	39	43	48	53	59	65	68	71	75	80	85	91	99	113	155	228
		% water unfrozen	—	—	—	—	—	—	—	16	—	—	—	21	22	27	34	60	100	
Egg yolk	40.0	Enthalpy (kJ/kg)	0	19	40	45	50	56	62	68	72	76	80	85	92	99	109	128	182	191
		% water unfrozen	20	—	—	22	—	24	—	27	28	29	31	33	35	38	45	58	94	100
Whole egg, w/shell[c]	66.4	Enthalpy (kJ/kg)	0	17	36	40	45	50	55	61	64	67	71	75	81	88	98	117	175	281
Bread																				
White bread	37.3	Enthalpy (kJ/kg)	0	17	35	39	44	49	56	67	75	83	93	104	117	124	128	131	134	137
Whole wheat	42.4	Enthalpy (kJ/kg)	0	17	36	41	48	56	66	78	86	95	106	119	135	150	154	157	160	163

Source: Adapted from Dickerson (1968) and Riedel (1951, 1956, 1957, 1959).
[b]Data for chicken, veal, and venison nearly matched the data for beef of the same water content (Riedel 1957)
[c]Calculated for a mass composition of 58% white (86.5% water) and 32% yolk (50% water).

Thermal Properties of Foods

where k_c is the thermal conductivity of the continuous phase and L^3 is the volume fraction of the discontinuous phase. In Equation (27), the thermal conductivity of the continuous phase is assumed to be much larger than the thermal conductivity of the discontinuous phase. However, if the thermal conductivity of the discontinuous phase is much larger than the thermal conductivity of the continuous phase, the following expression is used to calculate the thermal conductivity of the isotropic mixture:

$$k = k_c \left[\frac{1-M}{1-M(1-L)} \right] \qquad (28)$$

where $M = L^2(1 - k_d/k_c)$ and k_d is the thermal conductivity of the discontinuous phase.

For an anisotropic, two-component system in which the thermal conductivity depends on the direction of heat flow, such as in fibrous food materials, Kopelman (1966) developed two expressions for thermal conductivity. For heat flow parallel to the food fibers, Kopelman proposed the following expression for thermal conductivity $k_=$:

$$k_= = k_c \left[1 - N^2 \left(1 - \frac{k_d}{k_c} \right) \right] \qquad (29)$$

where N^2 is the volume fraction of the discontinuous phase in the fibrous food product. If the heat flow is perpendicular to the food fibers, then the following expression for thermal conductivity $k_\perp$ applies:

$$k_\perp = k_c \left[\frac{1-P}{1-P(1-N)} \right] \qquad (30)$$

where $P = N(1 - k_d/k_c)$.

Levy (1981) introduced a modified version of the Maxwell-Eucken equation. Levy's expression for the thermal conductivity of a two-component system is as follows:

$$k = \frac{k_2[(2 + \Lambda) + 2(\Lambda - 1)F_1]}{(2 + \Lambda) - (\Lambda - 1)F_1} \qquad (31)$$

where Λ is the thermal conductivity ratio ($\Lambda = k_1/k_2$), k_1 is the thermal conductivity of component 1, and k_2 is the thermal conductivity of component 2. The parameter F_1, introduced by Levy is given as follows:

$$F_1 = 0.5 \left\{ \left(\frac{2}{\sigma} - 1 + 2R_1 \right) - \left[\left(\frac{2}{\sigma} - 1 + 2R_1 \right)^2 - \frac{8R_1}{\sigma} \right]^{0.5} \right\} \qquad (32)$$

where

$$\sigma = \frac{(\Lambda - 1)^2}{(\Lambda + 1)^2 + (\Lambda/2)} \qquad (33)$$

and R_1 is the volume fraction of component 1, or:

$$R_1 = \left[1 + \left(\frac{1}{x_1} - 1 \right) \left(\frac{\rho_1}{\rho_2} \right) \right]^{-1} \qquad (34)$$

Here, x_1 is the mass fraction of component 1, ρ_1 is the density of component 1, and ρ_2 is the density of component 2.

To use Levy's method, follow these steps:

1. Calculate the thermal conductivity ratio Λ
2. Determine the volume fraction of constituent 1 using Equation (34)
3. Evaluate σ using Equation (33)
4. Determine F_1 using Equation (32)
5. Evaluate the thermal conductivity of the two-component system via Equation (31)

When foods consist of more than two distinct phases, the previously mentioned methods for the prediction of thermal conductivity must be applied successively to obtain the thermal conductivity of the food product. For example, in the case of frozen food, the thermal conductivity of the ice and liquid water mix is calculated first by using one of the earlier methods mentioned. The resulting thermal conductivity of the ice/water mix is then combined successively with the thermal conductivity of each remaining food constituent to determine the thermal conductivity of the food product.

Numerous researchers have proposed the use of parallel and perpendicular (or series) thermal conductivity models based on analogies with electrical resistance (Murakami and Okos 1989). The parallel model is the sum of the thermal conductivities of the food constituents multiplied by their volume fractions:

$$k = \sum x_i^v k_i \qquad (35)$$

where x_i^v is the volume fraction of constituent i. The volume fraction of constituent i can be found from the following equation:

$$x_i^v = \frac{x_i/\rho_i}{\sum (x_i/\rho_i)} \qquad (36)$$

The perpendicular model is the reciprocal of the sum of the volume fractions divided by their thermal conductivities:

$$k = \frac{1}{\sum (x_i^v/k_i)} \qquad (37)$$

These two models have been found to predict the upper and lower bounds of the thermal conductivity of most food items.

Tables 5 and 6 list the thermal conductivities for many food items (Qashou et al. 1972). Data in these tables have been averaged, interpolated, extrapolated, selected, or rounded off from the original research data. Tables 5 and 6 also include ASHRAE research data on foods of low and intermediate moisture content (Sweat 1985).

Example 4. Determine the thermal conductivity and density of lean pork shoulder meat which is at a temperature of –40°C. Use both the parallel and perpendicular thermal conductivity models.

Solution:
From Table 3, the composition of lean pork shoulder meat is:

$x_{wo} = 0.7263$ $x_f = 0.0714$
$x_p = 0.1955$ $x_a = 0.0102$

In addition, the initial freezing point of lean pork shoulder meat is –2.2°C. Because the temperature of the pork is below the initial freezing point, the fraction of ice within the pork must be determined. Using Equation (4), the ice fraction becomes:

$$x_{ice} = (x_{wo} - x_b)\left[1 - \frac{t_f}{t}\right] = (x_{wo} - 0.4x_p)\left[1 - \frac{t_f}{t}\right]$$

$$= (0.7263 - (0.4)(0.1955))\left(1 - \frac{-2.2}{-40}\right) = 0.6125$$

The mass fraction of unfrozen water is then:

$$x_w = x_{wo} - x_{ice} = 0.7263 - 0.6125 = 0.1138$$

Table 5 Thermal Conductivity of Foods

Food Item[a]	Thermal Conductivity W/(m·K)	Temperature, °C	Water Content, % by mass	Reference[b]	Remarks
Fruits, Vegetables					
Apples	0.418	8	—	Gane (1936)	Tasmanian French crabapple, whole fruit; 140 g
Apples, dried	0.219	23	41.6	Sweat (1985)	Density = 0.86 g/cm^3
Apple juice	0.559	20	87	Riedel (1949)	Refractive index at 20°C = 1.35
	0.631	80	87		
	0.504	20	70		Refractive index at 20°C = 1.38
	0.564	80	70		
	0.389	20	36		Refractive index at 20°C = 1.45
	0.435	80	36		
Apple sauce	0.549	29	—	Sweat (1974)	
Apricots, dried	0.375	23	43.6	Sweat (1985)	Density = 1.32 g/cm^3
Beans, runner	0.398	9	—	Smith et al. (1952)	Density = 0.75 g/cm^3; machine sliced, scalded, packed in slab
Beets	0.601	28	87.6	Sweat (1974)	
Broccoli	0.385	−6	—	Smith et al. (1952)	Density = 0.56 g/cm^3; heads cut and scalded
Carrots	0.669	−16	—	Smith et al. (1952)	Density = 0.6 g/cm^3; scraped, sliced and scalded
Carrots, puree	1.26	−8	—	Smith et al. (1952)	Density = 0.89 g/cm^3; slab
Currants, black	0.310	−17	—	Smith et al. (1952)	Density = 0.64 g/cm^3
Dates	0.337	23	34.5	Sweat (1985)	Density = 1.32 g/cm^3
Figs	0.310	23	40.4	Sweat (1985)	Density = 1.24 g/cm^3
Gooseberries	0.276	−15	—	Smith et al. (1952)	Density = 0.58 g/cm^3; mixed sizes
Grapefruit juice vesicle	0.462	30	—	Bennett et al. (1964)	Marsh, seedless
Grapefruit rind	0.237	28	—	Bennett et al. (1964)	Marsh, seedless
Grape, green, juice	0.567	20	89	Riedel (1949)	Refractive index at 20°C = 1.35
	0.639	80	89		
	0.496	20	68		Refractive index at 20°C = 1.38
	0.554	80	68		
	0.396	20	37		Refractive index at 20°C = 1.45
	0.439	80	37		
	0.439	25	—	Turrell and Perry (1957)	Eureka
Grape jelly	0.391	20	42.0	Sweat (1985)	Density = 1.32 g/cm^3
Nectarines	0.585	8.6	82.9	Sweat (1974)	
Onions	0.575	8.6	—	Saravacos (1965)	
Orange juice vesicle	0.435	30	—	Bennett et al. (1964)	Valencia
Orange rind	0.179	30	—	Bennett et al. (1964)	Valencia
Peas	0.480	−13	—	Smith et al. (1952)	Density = 0.70 g/cm^3; shelled and scalded
	0.395	−3	—		
	0.315	7	—		
Peaches, dried	0.361	23	43.4	Sweat (1985)	Density = 1.26 g/cm^3
Pears	0.595	8.7	—	Sweat (1974)	
	0.550	20	85		Refractive index at 20°C = 1.36
	0.629	80	85		
	0.475	20	60		Refractive index at 20°C = 1.40
	0.532	80	60		
	0.402	20	39		Refractive index at 20°C = 1.44
	0.446	80	39		
Plums	0.247	−16	—	Smith et al. (1952)	Density = 0.61 g/cm^3; 40 mm dia.; 50 mm long
Potatoes, mashed	1.09	−13	—	Smith et al. (1952)	Density = 0.97 g/cm^3; tightly packed slab
Potato salad	0.479	2	—	Dickerson and Read (1968)	Density = 1.01 g/cm^3
Prunes	0.375	23	42.9	Sweat (1985)	Density = 1.22 g/cm^3
Raisins	0.336	23	32.2	Sweat (1985)	Density = 1.38 g/cm^3
Strawberries	1.10	−14	—	Smith et al. (1952)	Mixed sizes, density = 0.80 g/cm^3, slab
	0.96	−15	—		Mixed sizes in 57% sucrose syrup, slab
Strawberry jam	0.338	20	41.0	Sweat (1985)	Density = 1.31 g/cm^3
Squash	0.502	8	—	Gane (1936)	
Meat and Animal Byproducts					
Beef brain	0.496	35	77.7	Poppendick et al. (1966)	12% fat; 10.3% protein; density = 1.04 g/cm^3
Beef fat	0.190	35	0.0	Poppendick et al. (1966)	Melted 100% fat; density = 0.81 g/cm^3
	0.230	35	20		Density = 0.86 g/cm^3
Beef fat ⊥[a]	0.217	2	9	Lentz (1961)	89% fat
	0.287	−9	9		
Beef kidney	0.524	35	76.4	Poppendick et al. (1966)	8.3% fat; 15.3% protein; density = 1.02 g/cm^3
Beef liver	0.488	35	72	Poppendick et al. (1966)	7.2% fat, 20.6% protein
Beef, lean =[a]	0.506	3	75	Lentz (1961)	Sirloin; 0.9% fat
	1.42	−15	75		
Beef, lean =[a]	0.430	20	79	Hill et al. (1967)	1.4% fat
	1.43	−15	79		
Beef, lean =[a]	0.400	6	76.5	Hill et al. (1967), Hill (1966)	2.4% fat
	1.36	−15	76.5		
Beef, lean ⊥[a]	0.480	20	79	Hill et al. (1967)	Inside round; 0.8% fat
	1.35	−15	79		
Beef, lean ⊥[a]	0.410	6	76	Hill et al. (1967), Hill (1966)	3% fat
	1.14	−15	76		
Beef, lean ⊥[a]	0.471	3	74	Lentz (1961)	Flank; 3 to 4% fat
	1.12	−15	74		
Beef, ground	0.406	6	67	Qashou et al. (1970)	12.3% fat; density = 0.95 g/cm^3
	0.410	4	62		16.8% fat; density = 0.98 g/cm^3
	0.351	6	55		18% fat; density = 0.93 g/cm^3

Thermal Properties of Foods

Table 5 Thermal Conductivity of Foods (*Continued*)

Food Item[a]	Thermal Conductivity W/(m·K)	Temperature, °C	Water Content, % by mass	Reference[b]	Remarks
	0.364	3	53		22% fat; density = 0.95 g/cm^3
Beefstick	0.297	20	36.6	Sweat (1985)	Density = 1.05 g/cm^3
Bologna	0.421	20	64.7	Sweat (1985)	Density = 1.00 g/cm^3
Dog food	0.319	23	30.6	Sweat (1985)	Density = 1.24 g/cm^3
Cat food	0.326	23	39.7	Sweat (1985)	Density = 1.14 g/cm^3
Ham, country	0.480	20	71.8	Sweat (1985)	Density = 1.03 g/cm^3
Horse meat⊥[a]	0.460	30	70	Griffiths and Cole (1948)	Lean
Lamb⊥[a]	0.456	20	72	Hill et al. (1967)	8.7% fat
	1.12	−15	72		
Lamb =[a]	0.399	20	71	Hill et al. (1967)	9.6% fat
	1.27	−15	71		
Pepperoni	0.256	20	32.0	Sweat (1985)	Density = 1.06 g/cm^3
Pork fat	0.215	3	6	Lentz (1961)	93% fat
	0.218	−15	6		
Pork, lean flank	0.460	2.2	—	Lentz (1961)	3.4% fat
	1.22	−15	—		
Pork, lean leg =[a]	0.478	4	72	Lentz (1961)	6.1% fat
	1.49	−15	72		
Pork, lean =[a]	0.453	20	76	Hill et al. (1967)	6.7% fat
	1.42	−13	76		
Pork, lean leg ⊥[a]	0.456	4	72	Lentz (1961)	6.1% fat
	1.29	−15	72		
Pork, lean ⊥[a]	0.505	20	76	Hill et al. (1967)	6.7% fat
	1.30	−14	76		
Salami	0.311	20	35.6	Sweat (1985)	Density = 0.96 g/cm^3
Sausage	0.427	25	68	Woodams (1965), Nowrey and	Mixture of beef and pork; 16.1% fat, 12.2% protein
	0.385	25	62	Woodams (1968)	Mixture of beef and pork; 24.1% fat, 10.3% protein
Veal ⊥[a]	0.470	20	75	Hill et al. (1967)	2.1% fat
	1.38	−15	75		
Veal =[a]	0.445	28	75	Hill et al. (1967)	2.1% fat
	1.46	−15	75		
Poultry and Eggs					
Chicken breast⊥[a]	0.412	20	69−75	Walters and May (1963)	0.6% fat
Chicken breast with skin	0.366	20	58−74	Walters and May (1963)	0−30% fat
Turkey breast ⊥[a]	0.496	3	74	Lentz (1961)	2.1% fat
	1.38	−15	74		
Turkey leg ⊥[a]	0.497	4	74	Lentz (1961)	3.4% fat
	1.23	−15	74		
Turkey breast = ⊥[a]	0.502	3	74	Lentz (1961)	2.1% fat
	1.53	−15	74		
Egg white	0.558	36	88	Spells (1960−61), Spells (1958)	
Egg, whole	0.960	−8	—	Smith et al. (1952)	Density = 0.98 g/cm^3
Egg yolk	0.420	31	50.6	Poppendick et al. (1966)	32.7% fat; 16.7% protein, density = 1.02 g/cm^3
Fish and Sea Products					
Fish, Cod ⊥[a]	0.534	3	83	Lentz (1961)	0.1% fat
	1.46	−15	83		
Fish, Cod	0.560	1	—	Long (1955), Jason and Long (1955)	
	1.69	−15	—	Long (1955)	
Fish, Herring	0.80	−19	—	Smith et al. (1952)	Density = 0.91 g/cm^3; whole and gutted
Fish, Salmon⊥[a]	0.531	3	67	Lentz (1961)	12% fat; *Salmo salar* from Gaspe peninsula
	1.24	−15	67		
Fish, Salmon ⊥[a]	0.498	5	73	Lentz (1961)	5.4% fat; *Oncorhynchus tchawytscha* from
	1.13	−15	73		British Columbia
Seal blubber ⊥[a]	0.197	5	4.3	Lentz (1961)	95% fat
Whale blubber ⊥[a]	0.209	18	—	Griffiths and Cole (1948)	Density = 1.04 g/cm^3
Whale meat	0.649	32	—	Griffiths and Hickman (1951)	Density = 1.07 g/cm^3
	1.44	−9	—		
	1.28	−12	—	Smith et al. (1952)	0.51% fat; density = 1.00 g/cm^3
Dairy Products					
Butterfat	0.173	6	0.6	Lentz (1961)	
	0.179	−15	0.6		
Butter	0.197	4	—	Hooper and Chang (1952)	
Buttermilk	0.569	20	89	Riedel (1949)	0.35% fat
Milk, whole	0.580	28	90	Leidenfrost (1959)	3% fat
	0.522	2	83	Riedel (1949)	3.6% fat
	0.550	20	83		
	0.586	50	83		
	0.614	80	83		
Milk, skimmed	0.538	2	90	Riedel (1949)	0.1% fat
	0.566	20	90		
	0.606	50	90		
	0.635	80	90		
Milk, evaporated	0.486	2	72	Riedel (1949)	4.8% fat
	0.504	20	72		
	0.542	50	72		
	0.565	80	72		
Milk, evaporated	0.456	2	62	Riedel (1949)	6.4% fat

Table 5 Thermal Conductivity of Foods (Continued)

Food Item[a]	Thermal Conductivity W/(m·K)	Temperature, °C	Water Content, % by mass	Reference[b]	Remarks
	0.472	20	62		
	0.510	50	62		
	0.531	80	62		
Milk, evaporated	0.472	23	67	Leidenfrost (1959)	10% fat
	0.504	41	67		
	0.516	60	67		
	0.527	79	67		
	0.324	26	50	Leidenfrost (1959)	15% fat
	0.340	40	50		
	0.357	59	50		
	0.364	79	50		
Whey	0.540	2	90	Riedel (1949)	No fat
	0.567	20	90		
	0.630	50	90		
	0.640	80	90		
Sugar, Starch, Bakery Products, and Derivatives					
Sugar beet juice	0.550	25	79	Khelemskii and Zhadan (1964)	
	0.569	25	82		
Sucrose solution	0.535	0	90	Riedel (1949)	Cane or beet sugar solution
	0.566	20	90		
	0.607	50	90		
	0.636	80	90		
	0.504	0	80		
	0.535	20	80		
	0.572	50	80		
	0.600	80	80		
	0.473	0	70		
	0.501	20	70		
	0.536	50	70		
	0.563	80	70		
	0.443	0	60		
	0.470	20	60		
	0.502	50	60		
	0.525	80	60		
	0.413	0	50		
	0.437	20	50		
	0.467	50	93–80		
	0.490	80	93–80		
	0.382	0	40		
	0.404	20	40		
	0.434	50	40		
	0.454	80	40		
Glucose solution	0.539	2	89	Riedel (1949)	
	0.566	20	89		
	0.601	50	89		
	0.639	80	89		
	0.508	2	80		
	0.535	20	80		
	0.571	50	80		
	0.599	80	80		
	0.478	2	70		
	0.504	20	70		
	0.538	50	70		
	0.565	80	70		
	0.446	2	60		
	0.470	20	60		
	0.501	50	60		
	0.529	80	60		
Corn syrup	0.562	25	—	Metzner and Friend (1959)	Density = 1.16 g/cm^3
	0.484	25	—		Density = 1.31 g/cm^3
	0.467	25	—		Density = 1.34 g/cm^3
Honey	0.502	2	80	Reidy (1968)	
	0.415	69	80		
Molasses syrup	0.346	30	23	Popov and Terentiev (1966)	
Angel food cake	0.099	23	36.1	Sweat (1985)	Density = 0.15 g/cm^3, porosity: 88%
Applesauce cake	0.079	23	23.7	Sweat (1985)	Density = 0.30 g/cm^3, porosity: 78%
Carrot cake	0.084	23	21.6	Sweat (1985)	Density = 0.32 g/cm^3, porosity: 75%
Chocolate cake	0.106	23	31.9	Sweat (1985)	Density = 0.34 g/cm^3, porosity: 74%
Pound cake	0.131	23	22.7	Sweat (1985)	Density = 0.48 g/cm^3, porosity: 58%
Yellow cake	0.110	23	25.1	Sweat (1985)	Density = 0.30 g/cm^3, porosity: 78%
White cake	0.082	23	32.3	Sweat (1985)	Density = 0.45 g/cm^3, porosity: 62%
Grains, Cereals, and Seeds					
Corn, yellow	0.140	32	0.9	Kazarian (1962)	Density = 0.75 g/cm^3
	0.159	32	14.7		Density = 0.75 g/cm^3
	0.172	32	30.2		Density = 0.68 g/cm^3
Flax seed	0.115	32	—	Griffiths and Hickman (1951)	Density = 0.66 g/cm^3
Oats, white English	0.130	27	12.7	Oxley (1944)	
Sorghum	0.131	5	13	Miller (1963)	Hybrid Rs610 grain

Thermal Properties of Foods

Table 5 Thermal Conductivity of Foods (*Continued*)

Food Item[a]	Thermal Conductivity W/(m·K)	Temperature, °C	Water Content, % by mass	Reference[b]	Remarks
Wheat, No. 1 Northern hard spring	0.150	—	22	Moote (1953)	Values taken from plot of series of values given by authors
	0.135	34	2	Babbitt (1945)	
	0.149	—	7		
	0.155	—	10		
	0.168	—	14		
Wheat, soft white winter	0.121	31	5	Kazarian (1962)	Values taken from plot of series of values given by author; Density = 0.78 g/cm^3
	0.129	31	10		
	0.137	31	15		
Fats, Oils, Gums, and Extracts					
Gelatin Gel	0.522	5	94–80	Lentz (1961)	Conductivity did not vary with concentration in range tested (6, 12, 20%)
	2.14	−15	94		6% gelatin concentration
	1.94	−15	88		12% gelatin concentration
	1.41	−15	80		20% gelatin concentration
Margarine	0.233	5	—	Hooper and Chang (1952)	Density = 1.00 g/cm^3
Almond oil	0.176	4	—	Wachsmuth (1892)	Density = 0.92 g/cm^3
Cod liver oil	0.170	35	—	Spells (1960-61), Spells (1958)	
Lemon oil	0.156	6	—	Weber (1880)	Density = 0.82 g/cm^3
Mustard oil	0.170	25	—	Weber (1886)	Density = 1.02 g/cm^3
Nutmeg oil	0.156	4	—	Wachsmuth (1892)	Density = 0.94 g/cm^3
Olive oil	0.175	7	—	Weber (1880)	Density = 0.91 g/cm^3
Olive oil	0.168	32	—	Kaye and Higgins (1928)	Density = 0.91 g/cm^3
	0.166	65	—		
	0.160	151	—		
	0.156	185	—		
Peanut oil	0.168	4	—	Wachsmuth (1892)	Density = 0.92 g/cm^3
Peanut oil	0.169	25	—	Woodams (1965)	
Rapeseed oil	0.160	20	—	Kondrat'ev (1950)	Density = 0.91 g/cm^3
Sesame oil	0.176	4	—	Wachsmuth (1892)	Density = 0.92 g/cm^3

[a]The symbol ⊥ indicates heat flow perpendicular to the grain structure and the symbol = indicates heat flow parallel to the grain or structure.
[b]References quoted are those on which given data are based, although actual values in this table may have been averaged, interpolated, extrapolated, selected, or rounded off.

Table 6 Thermal Conductivity of Freeze-Dried Foods

Food Item	Thermal Conductivity, W/(m·K)	Temperature, °C	Pressure, Pa	Reference[b]	Remarks
Apple	0.0156	35	2.66	Harper (1960, 1962)	Delicious; 88% porosity; 5.1 tortuosity factor; measured in air
	0.0185	35	21.0		
	0.0282	35	187		
	0.0405	35	2880		
Peach	0.0164	35	6.0	Harper (1960, 1962)	Clingstone; 91% porosity; 4.1 tortuosity factor; measured in air
	0.0185	35	21.5		
	0.0279	35	187		
	0.0410	35	2670		
	0.0431	35	51000		
Pears	0.0186	35	2.13	Harper (1960, 1962)	97% porosity; measured in nitrogen
	0.0207	35	19.5		
	0.0306	35	187		
	0.0419	35	2150		
	0.0451	35	68900		
Beef =[a]	0.0382	35	1.46	Harper (1960, 1962)	Lean; 64% porosity; 4.4 tortuosity factor; measured in air
	0.0412	35	22.7		
	0.0532	35	238		
	0.0620	35	2700		
	0.0652	35	101 000		
Egg albumin gel	0.0393	41	101 000	Saravacos and Pilsworth (1965)	2% water content; measured in air
	0.0129	41	4.40	Saravacos and Pilsworth (1965)	Measured in air
Turkey =[a]	0.0287	—	5.33	Triebes and King (1966)	Cooked white meat; 68 to 72% porosity; measured in air
	0.0443	—	15.0		
	0.0706	—	467		
	0.0861	—	2130		
	0.0927	—	98 500		
Turkey ⊥[a]	0.0170	—	5.60	Triebes and King (1966)	Cooked white meat; 68 to 72% porosity; measured in air
	0.0174	—	18.9		
	0.0221	—	133		
	0.0417	—	1250		
	0.0586	—	87 600		
Potato starch gel	0.0091	—	4.3	Saravacos and Pilsworth (1965)	Measured in air
	0.0144	—	181		
	0.0291	—	2210		
	0.0393	—	102 700		

[a]The symbol ⊥ indicates heat flow perpendicular to the grain structure and the symbol = indicates heat flow parallel to the grain or structure.
[b]References quoted are those on which given data are based, although actual values in this table may have been averaged, interpolated, extrapolated, selected, or rounded off.

Using the equations presented in Tables 1 and 2, the density and thermal conductivity of the food constituents are calculated at the given temperature −40°C:

$\rho_w = 9.9718 \times 10^2 + 3.1439 \times 10^{-3}(-40) - 3.7574 \times 10^{-3}(-40)^2$
$= 991.04 \text{ kg/m}^3$

$\rho_{ice} = 9.1689 \times 10^2 - 1.3071 \times 10^{-1}(-40)$
$= 922.12 \text{ kg/m}^3$

$\rho_p = 1.3299 \times 10^3 - 5.1840 \times 10^{-1}(-40)$
$= 1350.6 \text{ kg/m}^3$

$\rho_f = 9.2559 \times 10^2 - 4.1757 \times 10^{-1}(-40)$
$= 942.29 \text{ kg/m}^3$

$\rho_a = 2.4238 \times 10^3 - 2.8063 \times 10^{-1}(-40)$
$= 2435.0 \text{ kg/m}^3$

$k_w = 5.7109 \times 10^{-1} + 1.7625 \times 10^{-3}(-40) - 6.7036 \times 10^{-6}(-40)^2$
$= 0.4899 \text{ W/(m·K)}$

$k_{ice} = 2.2196 - 6.2489 \times 10^{-3}(-40) + 1.0154 \times 10^{-4}(-40)^2$
$= 2.632 \text{ W/(m·K)}$

$k_p = 1.7881 \times 10^{-1} + 1.1958 \times 10^{-3}(-40) - 2.7178 \times 10^{-6}(-40)^2$
$= 0.1266 \text{ W/(m·K)}$

$k_f = 1.8071 \times 10^{-1} - 2.7604 \times 10^{-3}(-40) - 1.7749 \times 10^{-7}(-40)^2$
$= 0.2908 \text{ W/(m·K)}$

$k_a = 3.2962 \times 10^{-1} + 1.4011 \times 10^{-3}(-40) - 2.9069 \times 10^{-6}(-40)^2$
$= 0.2689 \text{ W/(m·K)}$

Using Equation (6), the density of the lean pork shoulder meat at −40°C can be determined:

$$\sum \frac{x_i}{\rho_i} = \frac{0.6125}{922.12} + \frac{0.1138}{991.04} + \frac{0.1955}{1350.6} + \frac{0.0714}{942.29} + \frac{0.0102}{2435.0}$$

$$= 1.0038 \times 10^{-3}$$

$$\rho = \frac{1-\varepsilon}{\sum x_i/\rho_i} = \frac{1-0}{1.0038 \times 10^{-3}} = 996 \text{ kg/m}^3$$

Using Equation (36), the volume fractions of the constituents can be found:

$$x^v_{ice} = \frac{x_{ice}/\rho_{ice}}{\sum x_i/\rho_i} = \frac{0.6125/922.12}{1.0038 \times 10^{-3}} = 0.6617$$

$$x^v_w = \frac{x_w/\rho_w}{\sum x_i/\rho_i} = \frac{0.1138/991.04}{1.0038 \times 10^{-3}} = 0.1144$$

$$x^v_p = \frac{x_p/\rho_p}{\sum x_i/\rho_i} = \frac{0.1955/1350.6}{1.0038 \times 10^{-3}} = 0.1442$$

$$x^v_f = \frac{x_f/\rho_f}{\sum x_i/\rho_i} = \frac{0.0714/942.29}{1.0038 \times 10^{-3}} = 0.0755$$

$$x^v_a = \frac{x_a/\rho_a}{\sum x_i/\rho_i} = \frac{0.0102/2435.0}{1.0038 \times 10^{-3}} = 0.0042$$

Using the parallel model, Equation (35), the thermal conductivity becomes:

$k = \sum x^v_i k_i = (0.6617)(2.632) + (0.1144)(0.4899)$
$\quad + (0.1442)(0.1266) + (0.0755)(0.2908) + (0.0042)(0.2689)$
$k = 1.84 \text{ W/(m·K)}$

Using the perpendicular model, Equation (37), the thermal conductivity becomes:

$$k = \frac{1}{\sum x^v_i/k_i} = \left[\frac{0.6617}{2.632} + \frac{0.1144}{0.4899} + \frac{0.1442}{0.1266} + \frac{0.0755}{0.2908} + \frac{0.0042}{0.2689}\right]^{-1}$$

$k = 0.527 \text{ W/(m·K)}$

Example 5. Determine the thermal conductivity and density of lean pork shoulder meat which is at a temperature of −40°C. Use the isotropic model developed by Kopelman (1966).

Solution:
From Table 3, the composition of lean pork shoulder meat is:

$x_{wo} = 0.7263 \qquad x_f = 0.0714$
$x_p = 0.1955 \qquad x_a = 0.0102$

In addition, the initial freezing point of lean pork shoulder is −2.2°C. Because the temperature of the pork is below the initial freezing point, the fraction of ice within the pork must be determined. From Example 4, the ice fraction was found to be:

$x_{ice} = 0.6125$

The mass fraction of unfrozen water is then:

$x_w = x_{wo} - x_{ice} = 0.7263 - 0.6125 = 0.1138$

Using the equations presented in Tables 1 and 2, the density and thermal conductivity of the food constituents are calculated at the given temperature, −40°C (refer to Example 4):

$\rho_w = 991.04 \text{ kg/m}^3 \qquad k_w = 0.4899 \text{ W/(m·K)}$
$\rho_{ice} = 922.12 \text{ kg/m}^3 \qquad k_{ice} = 2.632 \text{ W/(m·K)}$
$\rho_p = 1350.6 \text{ kg/m}^3 \qquad k_p = 0.1266 \text{ W/(m·K)}$
$\rho_f = 942.29 \text{ kg/m}^3 \qquad k_f = 0.2908 \text{ W/(m·K)}$
$\rho_a = 2435.0 \text{ kg/m}^3 \qquad k_a = 0.2689 \text{ W/(m·K)}$

Now, determine the thermal conductivity of the ice/water mixture. This requires the volume fractions of the ice and the water in the two component ice/water mixture:

$$x^v_{ice} = \frac{x_{ice}/\rho_{ice}}{\sum \frac{x_i}{\rho_i}} = \frac{0.6125/922.12}{\frac{0.1138}{991.04} + \frac{0.6125}{922.12}} = 0.8526$$

Note that the volume fractions calculated for the two component ice/water mixture are different from those calculated in Example 4 for the lean pork shoulder meat. Because the ice has the largest volume fraction in the two component ice/water mixture, consider the ice to be the "continuous" phase. Then, L from Equation (27) becomes:

$L^3 = x^v_w = 0.1474$
$L^2 = 0.2790$
$L = 0.5282$

Because $k_{ice} > k_w$ and the ice is the continuous phase, the thermal conductivity of the ice/water mixture is calculated using Equation (27):

$$k_{ice/water} = k_{ice}\left[\frac{1-L^2}{1-L^2(1-L)}\right]$$

$$= 2.632\left[\frac{1-0.2790}{1-0.2790(1-0.5282)}\right] = 2.1853 \text{ W/(m·K)}$$

The density of the ice/water mixture then becomes:

$\rho_{ice/water} = x^v_w \rho_w + x^v_{ice} \rho_{ice}$
$\quad = (0.1474)(991.04) + (0.8526)(922.12)$
$\quad = 932.28 \text{ kg/m}^3$

Thermal Properties of Foods

Next, find the thermal conductivity of the ice/water/protein mixture. This requires the volume fractions of the ice/water and the protein:

$$x_p^v = \frac{x_p/\rho_p}{\sum \frac{x_i}{\rho_i}} = \frac{0.1955/1350.6}{\frac{0.1955}{1350.6} + \frac{0.7263}{932.28}} = 0.1567$$

$$x_{ice/water}^v = \frac{x_{ice/water}/\rho_{ice/water}}{\sum \frac{x_i}{\rho_i}} = \frac{0.7263/932.28}{\frac{0.1955}{1350.6} + \frac{0.7263}{932.28}} = 0.8433$$

Note that these volume fractions are calculated based on a two component system composed of ice/water as one constituent and protein as the other. Because protein has the smaller volume fraction, consider it to be the discontinuous phase.

$$L^3 = x_p^v = 0.1567$$
$$L^2 = 0.2907$$
$$L = 0.5391$$

Thus, the thermal conductivity of the ice/water/protein mixture becomes:

$$k_{ice/water/protein} = k_{ice/water}\left[\frac{1-L^2}{1-L^2(1-L)}\right]$$
$$= 2.1853\left[\frac{1-0.2907}{1-0.2907(1-0.5391)}\right]$$
$$= 1.7898 \text{ W/(m·K)}$$

The density of the ice/water/protein mixture then becomes:

$$\rho_{ice/water/protein} = x_{ice/water}^v \rho_{ice/water} + x_p^v \rho_p$$
$$= (0.8433)(932.28) + (0.1567)(1350.6)$$
$$= 997.83 \text{ kg/m}^3$$

Next, find the thermal conductivity of the ice/water/protein/fat mixture. This requires the volume fractions of the ice/water/protein and the fat:

$$x_f^v = \frac{x_f/\rho_f}{\sum \frac{x_i}{\rho_i}} = \frac{0.0714/942.29}{\frac{0.0714}{942.29} + \frac{0.9218}{997.83}} = 0.0758$$

$$x_{i/w/p}^v = \frac{x_{i/w/p}/\rho_{i/w/p}}{\sum \frac{x_i}{\rho_i}} = \frac{0.9218/997.83}{\frac{0.0714}{942.29} + \frac{0.9218}{997.83}} = 0.9242$$

$$L^3 = x_f^v = 0.0758$$
$$L^2 = 0.1791$$
$$L = 0.4232$$

Thus, the thermal conductivity of the ice/water/protein/fat mixture becomes:

$$k_{i/w/p/f} = k_{i/w/p}\left[\frac{1-L^2}{1-L^2(1-L)}\right]$$
$$= 1.7898\left[\frac{1-0.1791}{1-0.1791(1-0.4232)}\right]$$
$$= 1.639 \text{ W/(m·K)}$$

The density of the ice/water/protein/fat mixture then becomes:

$$\rho_{i/w/p/f} = x_{i/w/p}^v \rho_{i/w/p} + x_f^v \rho_f$$
$$= (0.9242)(997.83) + (0.0758)(942.29)$$
$$= 993.62 \text{ kg/m}^3$$

Finally, the thermal conductivity of the lean pork shoulder meat can be found. This requires the volume fractions of the ice/water/protein/fat and the ash:

$$x_a^v = \frac{x_a/\rho_a}{\sum \frac{x_i}{\rho_i}} = \frac{0.0102/2435.0}{\frac{0.0102}{2435.0} + \frac{0.9932}{993.62}} = 0.0042$$

$$x_{i/w/p/f}^v = \frac{\frac{x_{i/w/p/f}}{\rho_{i/w/p/f}}}{\sum \frac{x_i}{\rho_i}} = \frac{\frac{0.9932}{993.62}}{\frac{0.0102}{2435.0} + \frac{0.9932}{993.62}} = 0.9958$$

$$L^3 = x_a^v = 0.0042$$
$$L^2 = 0.0260$$
$$L = 0.1613$$

Thus, the thermal conductivity of the lean pork shoulder meat becomes:

$$k_{pork} = k_{i/w/p/f}\left[\frac{1-L^2}{1-L^2(1-L)}\right]$$
$$= 1.639\left[\frac{1-0.0260}{1-0.0260(1-0.1613)}\right]$$
$$= 1.632 \text{ W/(m·K)}$$

The density of the lean pork shoulder meat then becomes:

$$\rho_{pork} = x_{i/w/p/f}^v \rho_{i/w/p/f} + x_a^v \rho_a$$
$$= (0.9958)(993.62) + (0.0042)(2435.0)$$
$$= 999 \text{ kg/m}^3$$

THERMAL DIFFUSIVITY

For transient heat transfer, the important thermophysical property is thermal diffusivity α, which appears in the Fourier equation:

$$\frac{\partial T}{\partial \theta} = \alpha\left[\frac{\partial^2 T}{\partial x^2} + \frac{\partial^2 T}{\partial y^2} + \frac{\partial^2 T}{\partial z^2}\right] \quad (38)$$

where x, y, z are rectangular coordinates, T is temperature, and θ is time. Thermal diffusivity can be defined as follows:

$$\alpha = \frac{k}{\rho c} \quad (39)$$

where α is thermal diffusivity, k is thermal conductivity, ρ is density, and c is specific heat.

Experimentally determined values of the thermal diffusivity of foods are scarce. However, thermal diffusivity can be calculated using Equation (39), with appropriate values of thermal conductivity, specific heat, and density. A few experimental values are given in Table 7.

HEAT OF RESPIRATION

All living food products respire. During the respiration process, sugar and oxygen are combined to form CO_2, H_2O, and heat as follows:

Table 7 Thermal Diffusivity of Foods

Food Item	Thermal Diffusivity, mm²/s	Water Content, % by mass	Fat Content, % by mass	Apparent Density, kg/m³	Temperature, °C	Reference
Fruits and Vegetables						
Apple, Red Delicious, whole[a]	0.14	85	—	840	0 to 30	Bennett et al. (1969)
Apple, dried	0.096	42	—	856	23	Sweat (1985)
Applesauce	0.11	37	—	—	5	Riedel (1969)
	0.11	37	—	—	65	Riedel (1969)
	0.12	80	—	—	5	Riedel (1969)
	0.14	80	—	—	65	Riedel (1969)
Apricots, dried	0.11	44	—	1323	23	Sweat (1985)
Bananas, flesh	0.12	76	—	—	5	Riedel (1969)
	0.14	76	—	—	65	Riedel (1969)
Cherries, flesh[b]	0.13	—	—	1050	0 to 30	Parker and Stout (1967)
Dates	0.10	35	—	1319	23	Sweat (1985)
Figs	0.096	40	—	1241	23	Sweat (1985)
Jam, strawberry	0.12	41	—	1310	20	Sweat (1985)
Jelly, grape	0.12	42	—	1320	20	Sweat (1985)
Peaches[b]	0.14	—	—	960	2 to 32	Bennett (1963)
Peaches, dried	0.12	43	—	1259	23	Sweat (1985)
Potatoes, whole	0.13	—	—	1040 to 1070	0 to 70	Minh et al. (1969) Mathews and Hall (1968)
Potatoes, mashed, cooked	0.12	78	—	—	5	Riedel (1969)
	0.15	78	—	—	65	Riedel (1969)
Prunes	0.12	43	—	1219	23	Sweat (1985)
Raisins	0.11	32	—	1380	23	Sweat (1985)
Strawberries, flesh	0.13	92	—	—	5	Riedel (1969)
Sugar beets	0.13	—	—	—	0 to 60	Slavicek (1962)
Meats						
Codfish	0.12	81	—	—	5	Riedel (1969)
	0.14	81	—	—	65	Riedel (1969)
Halibut[c]	0.15	76	1	1070	40 to 65	Dickerson and Read (1975)
Beef, chuck[d]	0.12	66	16	1060	40 to 65	Dickerson and Read (1975)
Beef, round[d]	0.13	71	4	1090	40 to 65	Dickerson and Read (1975)
Beef, tongue[d]	0.13	68	13	1060	40 to 65	Dickerson and Read (1975)
Beefstick	0.11	37	—	1050	20	Sweat (1985)
Bologna	0.13	65	—	1000	20	Sweat (1985)
Corned beef	0.11	65	—	—	5	Riedel (1969)
	0.13	65	—	—	65	Riedel (1969)
Ham, country	0.14	72	—	1030	20	Sweat (1985)
Ham, smoked	0.12	64	—	—	5	Riedel (1969)
Ham, smoked[d]	0.13	64	14	1090	40 to 65	Dickerson and Read (1975)
Pepperoni	0.093	32	—	1060	20	Sweat (1985)
Salami	0.13	36	—	960	20	Sweat (1985)
Cakes						
Angel food	0.26	36	—	147	23	Sweat (1985)
Applesauce	0.12	24	—	300	23	Sweat (1985)
Carrot	0.12	22	—	320	23	Sweat (1985)
Chocolate	0.12	32	—	340	23	Sweat (1985)
Pound	0.12	23	—	480	23	Sweat (1985)
Yellow	0.12	25	—	300	23	Sweat (1985)
White	0.10	32	—	446	23	Sweat (1985)

[a]Data are applicable only to raw whole apple.
[b]Freshly harvested.
[c]Stored frozen and thawed prior to test.
[d]Data are applicable only where the juices exuded during heating remain in the food samples.

Table 8 Commodity Respiration Coefficients (Becker et al. 1996b)

Commodity	f	g	Commodity	f	g
Apples	5.6871×10^{-4}	2.5977	Onions	3.668×10^{-4}	2.538
Blueberries	7.2520×10^{-5}	3.2584	Oranges	2.8050×10^{-4}	2.6840
Brussels Sprouts	0.0027238	2.5728	Peaches	1.2996×10^{-5}	3.6417
Cabbage	6.0803×10^{-4}	2.6183	Pears	6.3614×10^{-5}	3.2037
Carrots	0.050018	1.7926	Plums	8.608×10^{-5}	2.972
Grapefruit	0.0035828	1.9982	Potatoes	0.01709	1.769
Grapes	7.056×10^{-5}	3.033	Rutabagas (swedes)	1.6524×10^{-4}	2.9039
Green Peppers	3.5104×10^{-4}	2.7414	Snap Beans	0.0032828	2.5077
Lemons	0.011192	1.7740	Sugar Beets	8.5913×10^{-3}	1.8880
Lima Beans	9.1051×10^{-4}	2.8480	Strawberries	3.6683×10^{-4}	3.0330
Limes	2.9834×10^{-8}	4.7329	Tomatoes	2.0074×10^{-4}	2.8350

Thermal Properties of Foods

$$C_6H_{12}O_6 + 6O_2 \rightarrow 6CO_2 + 6H_2O + 2667 \text{ kJ} \quad (40)$$

In most stored plant products, little cell development takes place, and the greater part of respiration energy is released in the form of heat, which must be taken into account when cooling and storing these living commodities (Becker et al. 1996a). The rate at which this chemical reaction takes place varies with the type and temperature of the commodity.

Becker et al. (1996b) developed correlations that relate a commodity's rate of carbon dioxide production to its temperature. The carbon dioxide production rate can then be related to the commodity's heat generation rate due to respiration. The resulting correlation gives the commodity's respiratory heat generation rate W in W/kg as a function of temperature t in °C:

$$W = \frac{10.7f}{3600}\left(\frac{9t}{5} + 32\right)^g \quad (41)$$

The respiration coefficients f and g for various commodities are given in Table 8.

Fruits, vegetables, flowers, bulbs, florists' greens, and nursery stock are storage commodities with significant heats of respiration. Dry plant products, such as seeds and nuts, have very low respiration rates. Young, actively growing tissues, such as asparagus, broccoli and spinach, have high rates of respiration as do immature seeds such as green peas and sweet corn. Fast developing fruits such as strawberries, raspberries, and blackberries, have much higher respiration rates than do fruits that are slow to develop, such as apples, grapes, and citrus fruits.

In general, most vegetables, other than root crops, have a high initial respiration rate for the first one or two days after harvest. Within a few days, the respiration rate quickly lowers to the equilibrium rate (Ryall and Lipton 1972).

Fruits, however, are different from most vegetables. Those fruits that do not ripen during storage, such as citrus fruits and grapes, have fairly constant rates of respiration. Those that ripen in storage, such as apples, peaches, and avocados, exhibit an increase in the respiration rate. At low storage temperatures, around 0°C, the rate of respiration rarely increases because no ripening takes place. However, if fruits are stored at higher temperatures (10°C to 15°C), the respiration rate increases due to ripening and then decreases. Soft fruits, such as blueberries, figs, and strawberries, show a decrease in respiration with time at 0°C. If they become infected with decay organisms, however, respiration increases.

Table 9 lists the heats of respiration as a function of temperature for a variety of commodities while Table 10 shows the change in respiration rate with time. Most of the commodities in Table 9 have a low and a high value for heat of respiration at each temperature. When no range is given, the value is an average for the specified temperature and may be an average of the respiration rates for many days.

When using Table 9, select the lower value for estimating the heat of respiration at the equilibrium storage state and use the higher value for calculating the heat load for the first day or two after harvest, including precooling and short-distance transport. During the storage of fruits between 0°C and 5°C, the increase in the respiration rate due to ripening is slight. However, for fruits, such as mangoes, avocados, or bananas, stored at temperatures above 10°C, significant ripening occurs and the higher rates listed in Table 9 should be used. Vegetables, such as onions, garlic, and cabbage, can exhibit an increase in heat production after a long storage period.

TRANSPIRATION OF FRESH FRUITS AND VEGETABLES

The most abundant constituent in fresh fruits and vegetables is water, which exists as a continuous liquid phase within the fruit or vegetable. Transpiration is the process by which fresh fruits and vegetables lose some of this water. This process consists of the transport of moisture through the skin of the commodity, the evaporation of this moisture from the commodity surface, and the convective mass transport of the moisture to the surroundings (Becker et al. 1996b).

The rate of transpiration in fresh fruits and vegetables affects product quality. Moisture transpires continuously from commodities during handling and storage. Some moisture loss is inevitable and can be tolerated. However, under many conditions, the loss of moisture may be sufficient to cause the commodity to shrivel. The resulting loss in mass not only affects appearance, texture, and flavor of the commodity, but also reduces the salable mass (Becker et al. 1996a).

Many factors affect the rate of transpiration from fresh fruits and vegetables. Moisture loss from a fruit or vegetable is driven by a difference in water vapor pressure between the product surface and the environment. Becker and Fricke (1996a) state that the product surface may be assumed to be saturated, and thus, the water vapor pressure at the commodity surface is equal to the water vapor saturation pressure evaluated at the product's surface temperature. However, they also report that dissolved substances in the moisture of the commodity tend to lower the vapor pressure at the evaporating surface slightly.

Evaporation that occurs at the product surface is an endothermic process that cools the surface, thus lowering the vapor pressure at the surface and reducing transpiration. Respiration within the fruit or vegetable, on the other hand, tends to increase the product's temperature, thus raising the vapor pressure at the surface and increasing transpiration. Furthermore, the respiration rate is itself a function of the commodity's temperature (Gaffney et al. 1985). In addition, factors such as surface structure, skin permeability, and air flow also effect the transpiration rate (Sastry et al. 1978).

Becker et al. (1996c) performed a numerical, parametric study to investigate the influence of bulk mass, air flow rate, skin mass transfer coefficient, and relative humidity on the cooling time and moisture loss of a bulk load of apples. They found that relative humidity and skin mass transfer coefficient had little effect on cooling time while bulk mass and airflow rate were of primary importance to cooling time. Moisture loss was found to vary appreciably with relative humidity, airflow rate, and skin mass transfer coefficient while bulk mass had little effect. They reported that an increase in airflow results in a decrease in moisture loss. The increased airflow reduces the cooling time, which quickly reduces the vapor pressure deficit, thus lowering the transpiration rate.

The driving force for transpiration is a difference in water vapor pressure between the surface of a commodity and the surrounding air. Thus, the basic form of the transpiration model is as follows:

$$\dot{m} = k_t(p_s - p_a) \quad (42)$$

where $\dot{m}$ is the transpiration rate expressed as the mass of moisture transpired per unit area of commodity surface per unit time. This rate may also be expressed per unit mass of commodity rather than per unit area of commodity surface. The transpiration coefficient k_t is the mass of moisture transpired per unit area of commodity, per unit water vapor pressure deficit, per unit time. The transpiration coefficient may also be expressed per unit mass of commodity rather than per unit area of commodity surface. The quantity $(p_s - p_a)$ is the water vapor pressure deficit. The water vapor pressure at the commodity surface p_s is the water vapor saturation pressure evaluated at the commodity surface temperature while the water vapor pressure in the surrounding air p_a is a function of the relative humidity of the air.

In its simplest form, the transpiration coefficient k_t is considered to be a constant for a particular commodity. Table 11 lists

Table 9 Heat of Respiration for Fresh Fruits and Vegetables at Various Temperatures[a]

Commodity	Heat of Respiration (mW/kg)						Reference
	0°C	5°C	10°C	15°C	20°C	25°C	
Apples							
Yellow, transparent	20.4	35.9	—	106.2	166.8	—	Wright et al. (1954)
Delicious	10.2	15.0	—	—	—	—	Lutz and Hardenburg (1968)
Golden Delicious	10.7	16.0	—	—	—	—	Lutz and Hardenburg (1968)
Jonathan	11.6	17.5	—	—	—	—	Lutz and Hardenburg (1968)
McIntosh	10.7	16.0	—	—	—	—	Lutz and Hardenburg (1968)
Early cultivars	9.7-18.4	15.5-31.5	41.2-60.6	53.6-92.1	58.2-121.2	—	IIR (1967)
Late cultivars	5.3-10.7	13.6-20.9	20.4-31.0	27.6-58.2	43.6-72.7	—	IIR (1967)
Average of many cultivars	6.8-12.1	15.0-21.3	—	40.3-91.7	50.0-103.8	—	Lutz and Hardenburg (1968)
Apricots	15.5-17.0	18.9-26.7	33.0-55.8	63.0-101.8	87.3-155.2	—	Lutz and Hardenburg (1968)
Artichokes, globe	67.4-133.4	94.6-178.0	16.2-291.5	22.9-430.2	40.4-692.0	—	Sastry et al. (1978), Rappaport and Watada (1958)
Asparagus	81.0-237.6	162.0-404.5	318.1-904.0	472.3-971.4	809.4-1484.0	—	Sastry et al. (1978), Lipton (1957)
Avocados	*[b]	*[b]	—	183.3-465.6	218.7-1029.1	—	Lutz and Hardenburg (1968), Biale (1960)
Bananas, green	*[b]	*[b]	†[b]	59.7-130.9	87.3-155.2	—	IIR (1967)
Bananas, ripening	*[b]	*[b]	†[b]	37.3-164.9	97.0-242.5	—	IIR (1967)
Beans							
Lima, unshelled	31.0-89.2	58.2-106.7	—	296.8-369.5	393.8-531.5	—	Lutz and Hardenburg (1968), Tewfik and Scott (1954)
Lima, shelled	52.4-103.8	86.3-180.9	—	—	627.0-801.1	—	Lutz and Hardenburg (1968), Tewfik and Scott (1954)
Snap	*[b]	101.4-103.8	162.0-172.6	252.2-276.4	350.6-386.0	—	Ryall and Lipton (1972), Watada and Morris (1966)
Beets, red, roots	16.0-21.3	27.2-28.1	34.9-40.3	50.0-68.9	—	—	Ryall and Lipton (1972), Smith (1957)
Berries							
Blackberries	46.6-67.9	84.9-135.8	155.2-281.3	208.5-431.6	388.0-581.9	—	IIR (1967)
Blueberries	6.8-31.0	27.2-36.4	—	101.4-183.3	153.7-259.0	—	Lutz and Hardenburg (1968)
Cranberries	*[b]	12.1-13.6	—	—	32.5-53.8	—	Lutz and Hardenburg (1968), Anderson et al. (1963)
Gooseberries	20.4-25.7	36.4-40.3	—	64.5-95.5	—	—	Lutz and Hardenburg (1968), Smith (1966)
Raspberries	52.4-74.2	91.7-114.4	82.4-164.9	243.9-300.7	339.5-727.4	—	Lutz and Hardenburg (1968), IIR (1967), Haller et al. (1941)
Strawberries	36.4-52.4	48.5-98.4	145.5-281.3	210.5-273.5	303.1-581.0	501.4-625.6	Lutz and Hardenburg (1968), IIR (1967), Maxie et al. (1959)
Broccoli, sprouting	55.3-63.5	102.3-474.8	—	515.0-1008.2	824.9-1011.1	1155.2-1661.0	Lutz and Hardenburg (1968), Morris (1947), Scholz et al. (1963)
Brussels Sprouts	45.6-71.3	95.5-144.0	187.2-250.7	283.2-316.7	267.2-564.0	—	Sastry et al. (1978), Smith (1957)
Cabbage							
Penn State[c]	11.6	28.1-30.1	—	66.4-94.1	—	—	Van den Berg and Lentz (1972)
White, Winter	14.5-24.2	21.8-41.2	36.4-53.3	58.2-80.0	106.7-121.2	—	IIR (1967)
White, Spring	28.1-40.3	52.4-63.5	86.3-98.4	159.1-167.7	—	—	Sastry et al. (1978), Smith (1957)
Red, Early	22.8-29.1	46.1-50.9	70.3-824.2	109.1-126.1	164.9-169.7	—	IIR (1967)
Savoy	46.1-63.0	75.2-87.3	155.2-181.9	259.5-293.4	388.0-436.5	—	IIR (1967)
Carrots, Roots							
Imperator, Texas	45.6	58.2	93.1	117.4	209.0	—	Scholz et al. (1963)
Main Crop, U.K.	10.2-20.4	17.5-35.9	29.1-46.1	86.8-196.4 at 18°C	—	—	Smith (1957)
Nantes, Can.[d]	9.2	19.9	—	64.0-83.9	—	—	Van den Berg and Lentz (1972)
Cauliflower, Texas	52.9	60.6	100.4	136.8	238.1	—	Scholz et al. (1963)
Cauliflower, U.K.	22.8-71.3	58.2-81.0	121.2-144.5	199.8-243.0	—	—	Smith (1957)
Celery, N.Y., White	21.3	32.5	—	110.6	191.6	—	Lutz and Hardenburg (1968)
Celery, U.K.	15.0-21.3	27.2-37.8	58.2-81.0	115.9-124.1 at 18°C	—	—	Smith (1957)
Celery, Utah, Can.[e]	15.0	26.7	—	88.3	—	—	Van den Berg and Lentz (1972)
Cherries, sour	17.5-39.3	37.8-39.3	—	81.0-148.4	115.9-148.4	157.6-210.5	Lutz and Hardenburg (1968), Hawkins (1929)

Thermal Properties of Foods

Table 9 Heat of Respiration for Fresh Fruits and Vegetables at Various Temperatures[a] (*Continued*)

Commodity	Heat of Respiration (mW/kg)						Reference
	0°C	5°C	10°C	15°C	20°C	25°C	
Cherries, sweet	12.1-16.0	28.1-41.7	—	74.2-133.4	83.4-94.6	—	Lutz and Hardenburg (1968), Micke et al. (1965), Gerhardt et al. (1942)
Corn, sweet with husk, Texas	126.1	230.4	332.2	483.0	855.5	1207.5	Scholz et al. (1963)
Cucumbers, Calif.	*[b]	*[b]	68.4-85.8 at 13°C	71.3-98.4	92.1-142.6	—	Eaks and Morris (1956)
Figs, Mission	—	23.5-39.3	65.5-68.4	145.5-187.7	168.8-281.8	252.2-281.8	Lutz and Hardenburg (1968), Claypool and Ozbek (1952)
Garlic	8.7-32.5	17.5-28.6	27.2-28.6	32.5-81.0	29.6-53.8	—	Sastry et al. (1978), Mann and Lewis (1956)
Grapes							
Labrusca, Concord	8.2	16.0	—	47.0	97.0	114.4	Lutz and Hardenburg (1968), Lutz (1938)
Vinifera,							
Emperor	3.9-6.8	9.2-17.5	2.42	29.6-34.9	—	74.2-89.2	Lutz and Hardenburg (1968), Pentzer et al. (1933)
Thompson seedless	5.8	14.1	22.8	—	—	—	Wright et al. (1954)
Ohanez	3.9	9.7	21.3	—	—	—	Wright et al. (1954)
Grapefruit, Calif. Marsh	*[b]	*[b]	*[b]	34.9	52.4	64.5	Haller et al. (1945)
Grapefruit, Florida	*[b]	*[b]	*[b]	37.8	47.0	56.7	Haller et al. (1945)
Horseradish	24.2	32.0	78.1	97.0	132.4	—	Sastry et al. (1978)
Kiwi fruit	8.3	19.6	38.9	—	51.9-57.3	—	Saravacos and Pilsworth (1965)
Kohlrabi	29.6	48.5	93.1	145.5	—	—	Sastry et al. (1978)
Leeks	28.1-48.5	58.2-86.3	159.1-202.2	245.4-346.7	—	—	Sastry et al. (1978), Smith (1957)
Lemons, Calif., Eureka	*[b]	*[b]	*[b]	47.0	67.4	77.1	Haller (1945)
Lettuce							
Head, Calif.	27.2-50.0	39.8-59.2	81.0-118.8	114.4-121.2	178.0	—	Sastry et al. (1978)
Head, Texas	31.0	39.3	64.5	106.7	168.8	2.4 at 27°C	Watt and Merrill (1963), Lutz and Hardenburg (1968)
Leaf, Texas	68.4	86.8	116.9	186.7	297.8	434.5	Scholz et al. (1963)
Romaine, Texas	—	61.6	105.2	131.4	203.2	321.5	Scholz et al. (1963)
Limes, Persian	*[b]	*[b]	7.8-17.0	17.5-31.0	20.4-55.3	44.6-134.8	Lutz and Hardenburg (1968)
Mangos	*[b]	*[b]	—	133.4	222.6-449.1	356.0	Lutz and Hardenburg (1968), Gore (1911), Karmarkar and Joshe (1941)
Melons							
Cantaloupes	*[b]	25.7-29.6	46.1	99.9-114.4	132.4-191.6	184.8-211.9	Lutz and Hardenburg (1968), Sastry et al. (1978), Scholz et al. (1963)
Honeydew	—	*[b]	23.8	34.9-47.0	59.2-70.8	78.1-102.3	Lutz and Hardenburg (1968), Scholz et al. (1963), Pratt and Morris (1958)
Watermelon	*[b]	*[b]	22.3	—	51.4-74.2	—	Lutz and Hardenburg (1968), Scholz et al. (1963)
Mint[m]	23.8-44.5	89.0	225.6-270.1	311.6-403.6	492.7-673.7	762.7-940.8	Hruschka and Want (1979)
Mushrooms	83.4-129.5	210.5	—	—	782.2-938.9	—	Lutz and Hardenburg (1968), Smith (1964)
Nuts, (kind not specified)	2.4	4.8	9.7	9.7	14.5	—	IIR (1967)
Okra, Clemson	*[b]	—	259.0	432.6	774.5	1024 at 29°C	Scholz et al. (1963)
Onions							
Dry, Autumn Spice[f]	6.8-9.2	10.7-19.9	—	14.7-28.1	—	—	Van den Berg and Lentz (1972)
Dry, White Bermuda	8.7	10.2	21.3	33.0	50.0	83.4 at 27°C	Scholz et al. (1963)
Green, N.J.	31.0-65.9	51.4-202.2	107.2-174.6	195.9-288.6	231.6-460.8	290.0-622.2	Lutz and Hardenburg (1968)
Olives, Manzanillo	*[b]	*[b]	—	64.5-115.9	114.4-145.5	121.2-180.9	Maxie et al. (1959)
Oranges, Florida	9.2	18.9	36.4	62.1	89.2	105.2 at 27°C	Haller (1945)
Oranges, Calif., W. Navel	*[b]	18.9	40.3	67.4	81.0	107.7	Haller (1945)
Oranges, Calif., Valencia	*[b]	13.6	34.9	37.8	52.4	62.1	Haller (1945)
Papayas	*[b]	*[b]	33.5	44.6-64.5	—	115.9-291.0	Pantastico (1974), Jones (1942)
Parsley[m]	98.0-136.5	195.9-252.3	388.8-486.7	427.4-661.9	581.7-756.8	914.1-1012.0	Hruschka and Want (1979)
Parsnips, U.K.	34.4-46.1	26.2-51.9	60.6-78.1	95.5-127.1	—	—	Smith (1957)
Parsnips, Canada Hollow Crown[g]	10.7-24.2	18.4-45.6	—	64.0-137.2	—	—	Van den Berg and Lentz (1972)

Table 9 Heat of Respiration for Fresh Fruits and Vegetables at Various Temperatures[a] (*Continued*)

Commodity	0°C	5°C	10°C	15°C	20°C	25°C	Reference
Peaches, Elberta	11.2	19.4	46.6	101.8	181.9	266.7 at 27°C	Haller et al. (1932)
Peaches, several cultivars	12.1-18.9	18.9-27.2	—	98.4-125.6	175.6-303.6	241.5-361.3	Lutz and Hardenburg (1968)
Peanuts							
Cured[h]	0.05 at 1.7°C	—	—	—	—	0.5 at 30°C	Thompson et al. (1951)
Not cured, Virginia Bunch[i]	—	—	—	—	—	42.0 at 30°C	Schenk (1959, 1961)
Dixie Spanish	—	—	—	—	—	24.5 at 30°C	Schenk (1959, 1961)
Pears							
Bartlett	9.2-20.4	15.0-29.6	—	44.6-178.0	89.2-207.6	—	Lutz and Hardenburg (1968)
Late ripening	7.8-10.7	17.5-41.2	23.3-55.8	82.4-126.1	97.0-218.2	—	IIR (1967)
Early ripening	7.8-14.5	21.8-46.1	21.9-63.0	101.8-160.0	116.4-266.7	—	IIR (1967)
Peas, green-in-pod	90.2-138.7	163.4-226.5	—	530.1-600.4	728.4-1072.2	1018.4-1118.3	Lutz and Hardenburg (1968), Tewfik and Scott (1954)
Peas, shelled	140.2-224.1	234.7-288.7	—	—	1035-1630	—	Lutz and Hardenburg (1968), Tewfik and Scott (1954)
Peppers, sweet	*[b]	*[b]	42.7	67.9	130.0	—	Morris (1947)
Persimmons	—	17.5	—	34.9-41.7	59.2-71.3	86.3-118.8	Lutz and Hardenburg (1968), Gore (1911)
Pineapple, mature green	*[b]	*[b]	165	38.3	71.8	105.2 at 27°C	Scholz et al. (1963)
Pineapple, ripening	*[b]	*[b]	22.3	53.8	118.3	185.7	Scholz et al. (1963)
Plums, Wickson	5.8-8.7	11.6-26.7	26.7-33.9	35.4-36.9	53.3-77.1	82.9-210.5	Claypool and Allen (1951)
Potatoes							
Calif. White, Rose,							
Immature	*[b]	34.9	41.7-62.1	41.7-91.7	53.8-133.7	—	Sastry et al. (1978)
Mature	*[b]	17.5-20.4	19.7-29.6	19.7-34.9	19.7-47.0	—	Sastry et al. (1978)
Very mature	*[b]	15.0-20.4	20.4	20.4-29.6	27.2-35.4	—	Sastry et al. (1978)
Katahdin, Can.[k]	*[b]	11.6-12.6	—	23.3-30.1	—	—	Van den Berg and Lentz (1972)
Kennebec	*[b]	10.7-12.6	—	12.6-26.7	—	—	Van den Berg and Lentz (1972)
Radishes, with tops	43.2-51.4	56.7-62.1	91.7-109.1	207.6-230.8	368.1-404.5	469.4-571.8	Lutz and Hardenburg (1968)
Radishes, topped	16.0-17.5	22.8-24.2	44.6-97.0	82.4-97.0	141.6-145.5	199.8-225.5	Lutz and Hardenburg (1968)
Rhubarb, topped	24.2-39.3	32.5-53.8	—	91.7-134.8	118.8-168.8	—	Hruschka (1966)
Rutabaga, Laurentian, Can.[l]	5.8-8.2	14.1-15.1	—	31.5-46.6	—	—	Van den Berg and Lentz (1972)
Spinach							
Texas	—	136.3	328.3	530.5	682.3	—	Scholz et al. (1963)
U.K., Summer	34.4-63.5	81.0-95.5	173.6-222.6	—	549.0-641.6 at 18°C	—	Smith (1957)
U.K., Winter	51.9-75.2	86.8-186.7	202.2-306.5	—	578.1-722.6 at 18°C	—	Smith (1957)
Squash							
Summer, yellow, straight-neck	†[b]	†[b]	103.8-109.1	222.6-269.6	252.2-288.6	—	Lutz and Hardenburg (1968)
Winter Butternut	*[b]	*[b]	—	—	—	219.7-362.3	Lutz and Hardenburg (1968)
Sweet Potatoes							
Cured, Puerto Rico	*[b]	*[b]	†[b]	47.5-65.5	—	—	Lewis and Morris (1956)
Cured, Yellow Jersey	*[b]	*[b]	†[b]	65.5-68.4	—	—	Lewis and Morris (1956)
Noncured	*[b]	*[b]	*[b]	84.9	—	160.5-217.3	Lutz and Hardenburg (1968)
Tomatoes							
Texas, mature green	*[b]	*[b]	*[b]	60.6	102.8	126.6 at 27°C	Scholz et al. (1963)
Texas, ripening	*[b]	*[b]	*[b]	79.1	120.3	143.1 at 27°C	Scholz et al. (1963)
Calif., mature green	*[b]	*[b]	*[b]	—	71.3-103.8	88.7-142.6	Workman and Pratt (1957)
Turnip roots	25.7	28.1-29.6	—	63.5-71.3	71.3-74.2	—	Lutz and Hardenburg (1968)
Watercress[l]	44.5	133.6	270.1-359.1	403.6-581.7	896.3-1032.8	1032.9-1300.0	Hruschka and Want (1979)zz

[a]Column headings indicate temperatures at which respiration rates were determined, within 1°C, except where the actual temperatures are given.

[b]The symbol * denotes a chilling temperature. The symbol † denotes the temperature is borderline, not damaging to some cultivars, if exposure is short.

[c]Rates are for 30 to 60 days and 60 to 120 days storage, the longer storage having the higher rate, except at 0°C, where they were the same.

[d]Rates are for 30 to 60 days and 120 to 180 days storage, respiration increasing with time only at 15°C.

[e]Rates are for 30 to 60 days storage.

[f]Rates are for 30 to 60 days and 120 to 180 days storage; rates increased with time at all temperatures as dormancy was lost.

[g]Rates are for 30 to 60 days and 120 to 180 days; rates increased with time at all temperatures.

[h]Shelled peanuts with about 7% moisture. Respiration after 60 hours curing was almost negligible, even at 30°C.

[i]Respiration for freshly dug peanuts, not cured, with about 35 to 40% moisture. During curing, peanuts in the shell were dried to about 5 to 6% moisture, and in roasting are dried further to about 2% moisture.

[j]Rates are for 30 to 60 days and 120 to 180 days with rate declining with time at 5°C but increasing at 15°C as sprouting started.

[k]Rates are for 30 to 60 days and 120 to 180 days; rates increased with time, especially at 15°C where sprouting occurred.

[l]Rates are for 1 day after harvest.

Thermal Properties of Foods

Table 10 Change in Respiration Rates with Time

Commodity	Days in Storage	Heat of Respiration, mW/kg of Produce 0°C	Heat of Respiration, mW/kg of Produce 5°C	Reference	Commodity	Days in Storage	Heat of Respiration, mW/kg of Produce 0°C	Heat of Respiration, mW/kg of Produce 5°C	Reference
Apples, Grimes	7	8.7	38.8 at 10°C	Harding (1929)	Garlic	10	11.6	26.7	Mann and Lewis (1956)
	30	8.7	51.9			30	17.9	44.6	
	80	8.7	32.5			180	41.7	97.9	
Artichokes, globe	1	133.3	177.9	Rappaport and Watada (1958)	Lettuce, Great Lakes	1	50.4	59.2	Pratt et al. (1954)
	4	74.2	103.8			5	26.7	0.4	
	16	44.6	77.1			10	23.8	44.6	
Asparagus, Martha Washington	1	237.6	31.2	Lipton (1957)	Olives, Manzanillo	1	—	115.9 at 15°C	Maxie et al. (1960)
	3	116.9	193.0			5	—	85.8	
	16	82.9	89.2			10	—	65.5	
Beans, lima, in pod	2	88.7	106.7	Tewfik and Scott (1954)	Onions, red	1	4.8	—	Karmarkar and Joshe (1941)
	4	59.6	85.8			30	7.3	—	
	6	52.4	78.6			120	9.7	—	
Blueberries, blue crop	1	21.3	—		Plums, Wickson	2	5.8	11.6	Claypool and Allen (1951)
	2	7.9	—			6	5.8	20.8	
		17.0	—			18	8.7	26.7	
Broccoli, Waltham 29	1	—	216.7		Potatoes	2	—	17.9	
	4	—	130.4			6	—	23.8	
	8	—	97.9			10	—	20.8	
Corn, sweet, in husk	1	152.3	—	Scholz et al. (1963)	Strawberries, Shasta	1	52.1	84.9	Maxie et al. (1959)
	2	109.1	—			2	39.3	91.2	
	4	91.2	—			5	39.3	97.9	
Figs, Mission	1	38.8	—	Claypool and Ozbek (1952)	Tomatoes, Pearson, mature green	5	—	95.0 at 20°C	Workman and Pratt (1957)
	2	35.4	—			15	—	82.9	
	12	35.4	—			20	—	71.3	

values for the transpiration coefficients k_t of various fruits and vegetables (Sastry et al. 1978). Because of the many factors that influence transpiration rate, not all the values in Table 11 are reliable. They are to be used primarily as a guide or as a comparative indication of various commodity transpiration rates obtained from the literature.

Fockens and Meffert (1972) modified the simple transpiration coefficient to model variable skin permeability and to account for air flow rate. Their modified transpiration coefficient takes the following form:

$$k_t = \frac{1}{\frac{1}{k_a} + \frac{1}{k_s}} \quad (43)$$

where k_a' is the air film mass transfer coefficient and k_s is the skin mass transfer coefficient. The air film mass transfer coefficient k_a describes the convective mass transfer which occurs at the surface of the commodity and is a function of air flow rate. The skin mass transfer coefficient k_s describes the skin's diffusional resistance to moisture migration.

The air film mass transfer coefficient k_a can be estimated by using the Sherwood-Reynolds-Schmidt correlations (Becker et al. 1996b). The Sherwood number is defined as follows:

$$\text{Sh} = \frac{k_a' d}{\delta} \quad (44)$$

where k_a' is the air film mass transfer coefficient, d is the diameter of the commodity, and δ is the coefficient of diffusion of water vapor in air. For convective mass transfer from a spherical fruit or vegetable, Becker and Fricke (1996b) recommend the following Sherwood-Reynolds-Schmidt correlation, which was taken from Geankoplis (1978):

$$\text{Sh} = 2.0 + 0.552 \text{Re}^{0.53} \text{Sc}^{0.33} \quad (45)$$

In the equation Re is the Reynolds number (Re = $u_\infty d/\nu$) and Sc is the Schmidt number (Sc = ν/δ) where u_∞ is the free stream air velocity and ν is the kinematic viscosity of air. The driving force for k_a' is concentration. However, the driving force in the transpiration model is vapor pressure. Thus, the following conversion from concentration to vapor pressure is required:

Table 11 Transpiration Coefficients for Fruits and Vegetables

Commodity and Variety	Transpiration Coefficient, ng/(kg·s·Pa)	Commodity and Variety	Transpiration Coefficient, ng/(kg·s·Pa)	Commodity and Variety	Transpiration Coefficient, ng/(kg·s·Pa)
Apples		**Leeks**		**Pears**	
Jonathan	35	Musselburgh	1040	Passe Crassane	80
Golden Delicious	58	*Average for all varieties*	**790**	Beurre Clairgeau	81
Bramley's Seedling	42	**Lemons**		*Average for all varieties*	**69**
Average for all varieties	**42**	Eureka			
Brussels Sprouts		Dark green	227	**Plums**	
Unspecified	3300	Yellow	140	Victoria	
Average for all varieties	**6150**	*Average for all varieties*	**186**	Unripe	198
Cabbage		**Lettuce**		Ripe	115
Penn State Ballhead		Unrivalled	8750	Wickson	124
Trimmed	271	*Average for all varieties*	**7400**	*Average for all varieties*	**136**
Untrimmed	404	**Onions**			
Mammoth		Autumn Spice		**Potatoes**	
Trimmed	240	Uncured	96	Manona	
Average for all varieties	**223**	Cured	44	Mature	25
Carrots		Sweet White Spanish		Kennebec	
Nantes	1648	Cured	123	Uncured	171
Chantenay	1771	*Average for all varieties*	**60**	Cured	60
Average for all varieties	**1207**	**Oranges**		Sebago	
Celery		Valencia	58	Uncured	158
Unspecified varieties	2084	Navel	104	Cured	38
Average for all varieties	**1760**	*Average for all varieties*	**117**	*Average for all varieties*	**44**
Grapefruit		**Parsnips**			
Unspecified varieties	31	Hollow Crown	1930		
Marsh	55			**Rutabagas**	
Average for all varieties	**81**	**Peaches**		Laurentian	469
Grapes		Redhaven			
Emperor	79	Hard mature	917	**Tomatoes**	
Cardinal	100	Soft mature	1020	Marglobe	71
Thompson	204	Elberta	274	Eurocross BB	116
Average for all varieties	**123**	*Average for all varieties*	**572**	*Average for all varieties*	**140**

Note: Sastry et al. (1978) gathered this data as part of a literature review. The averages reported are the average of *all* published data found by Sastry et al. for each commodity. Sastry et al. selected specific varietal data because they considered it to be highly reliable data.

$$k_a = \frac{1}{R_{H_2O}T} k_a' \tag{46}$$

where R_{H_2O} is the gas constant for water vapor and T is the absolute mean temperature of the boundary layer.

The skin mass transfer coefficient k_s, which describes the resistance to moisture migration through the skin of a commodity, is based upon the fraction of the product surface covered by pores. Although it is difficult to theoretically determine the skin mass transfer coefficient, experimental determination has been performed by Chau et al. (1987) and Gan and Woods (1989). These experimental values of k_s are given in Table 12, along with estimated values of the skin mass transfer coefficient for grapes, onions, plums, and potatoes. Note that three values of skin mass transfer coefficient are tabulated for most commodities. These values correspond to the spread of the experimental data.

SURFACE HEAT TRANSFER COEFFICIENT

Although the surface heat transfer coefficient is not a thermal property of a food or beverage, it is needed to design heat transfer equipment for the processing of foods and beverages where convection is involved. Newton's law of cooling defines the surface heat transfer coefficient h as follows:

$$q = hA(t_s - t_\infty) \tag{47}$$

Table 12 Commodity Skin Mass Transfer Coefficient

	Skin Mass Transfer Coefficient k_s, μg/(m²·s·Pa)			
Commodity	Low	Mean	High	Standard Deviation
Apples	0.111	0.167	0.227	0.03
Blueberries	0.955	2.19	3.39	0.64
Brussels Sprouts	9.64	13.3	18.6	2.44
Cabbage	2.50	6.72	13.0	2.84
Carrots	31.8	156.	361.	75.9
Grapefruit	1.09	1.68	2.22	0.33
Grapes	—	0.4024	—	—
Green Peppers	0.545	2.159	4.36	0.71
Lemons	1.09	2.08	3.50	0.64
Lima Beans	3.27	4.33	5.72	0.59
Limes	1.04	2.22	3.48	0.56
Onions	—	0.8877	—	—
Oranges	1.38	1.72	2.14	0.21
Peaches	1.36	14.2	45.9	5.2
Pears	0.523	0.686	1.20	0.149
Plums	—	1.378	—	—
Potatoes	—	0.6349	—	—
Rutabagas (swedes)	—	116.6	—	—
Snap Beans	3.46	5.64	10.0	1.77
Sugar Beets	9.09	33.6	87.3	20.1
Strawberries	3.95	13.6	26.5	4.8
Tomatoes	0.217	1.10	2.43	0.67

Source: Becker and Fricke (1996a)

Thermal Properties of Foods

Table 13 Surface Heat Transfer Coefficients for Food Products

1	2	3	4	5	6	7	8	9	10
Product	Shape Length, mm[a]	Transfer Medium	Δt and/or Temp. t of Medium, °C	Velocity of Medium, m/s	Reynolds Number Range[b]	h, W/(m²·K)	Nu-Re-Pr Correlation[c]	Reference	Comments
Apple Jonathan	Spherical 52	Air	$t = 27$	0.0	N/A	11.1	N/A	Kopelman et al. (1966)	N/A indicates that data were not reported in original article
				0.39		17.0			
				0.91		27.3			
				2.0		45.3			
				5.1		53.4			
	58			0.0		11.2			
				0.39		17.0			
				0.91		27.8			
				2.0		44.8			
				5.1		54.5			
	62			0.0		11.4			
				0.39		15.9			
				0.91		26.1			
				2.0		39.2			
				5.1		50.5			
Apple Red Delicious	63	Air	$\Delta t = 22.8$ $t = -0.6$	1.5	N/A	27.3	N/A	Nicholas et al. (1964)	Thermocouples at center of fruit.
				4.6		56.8			
	72			1.5		14.2			
				4.6		36.9			
	76			0.0		10.2			
				1.5		22.7			
				3.0		32.9			
				4.6		34.6			
	57	Water	$\Delta t = 25.6$ $t = 0$	0.27		90.9			
	70					79.5			
	75					55.7			
Beef carcass	64.5 kg* 85 kg*	Air	$t = -19.5$	1.8 0.3	N/A	21.8 10.0	N/A	Fedorov et al. (1972)	*For size indication.
Cucumbers	Cylinder 38	Air	$t = 4$	1.00	N/A	18.2	$Nu = 0.291 Re^{0.592} Pr^{0.333}$	Dincer (1994)	Diameter = 38 mm. Length = 160 mm.
				1.25		19.9			
				1.50		21.3			
				1.75		23.1			
				2.00		26.6			
Eggs, Jifujitori	34	Air	$\Delta t = 45$	2-8	6000-15000	N/A	$Nu = 0.46 Re^{0.56} \pm 1.0\%$	Chuma et al. (1970)	5 points in correlation.
Eggs, Leghorn	44	Air	$\Delta t = 45$	2-8	8000-25000	N/A	$Nu = 0.71 Re^{0.55} \pm 1.0\%$	Chuma et al. (1970)	5 points in correlation.
Figs	Spherical 47	Air	$t = 4$	1.10	N/A	23.8	$Nu = 1.560 Re^{0.426} Pr^{0.333}$	Dincer (1994)	
				1.50		26.2			
				1.75		27.4			
				2.50		32.7			
Fish Pike, perch, sheatfish	N/A	Air	N/A	0.97-6.6	5000-35000	N/A	$Nu = 4.5 Re^{0.28} \pm 10\%$	Khatchaturov (1958)	32 points in correlation.
Grapes	Cylinder 11	Air	$t = 4$	1.00	N/A	30.7	$Nu = 0.291 Re^{0.592} Pr^{0.333}$	Dincer (1994)	Diameter = 11 mm. Length = 22 mm.
				1.25		33.8			
				1.50		37.8			
				1.75		40.7			
				2.00		42.3			
Hams, boneless processed	$G^* = 0.4-0.45$ * G = Geometrical factor for shrink-fitted plastic bag	Air	$\Delta t = 132$ $t = 150$	N/A	1000-86000	N/A	$Nu = 0.329 Re^{0.564}$	*Ref*: Clary et al. (1968)	$G = 1/4 + 3/(8A^2) + 3/(8B^2)$ $A = a/Z, B = b/Z$ A = characteristic length = 0.5 min. dist. $\perp$ to airflow a = minor axis b = major axis Correlation on 18 points Recalc with min. distance $\perp$ to airflow Calculated Nu with 1/2 char. length
Hams processed	N/A	Air	$t = -23.3$	0.61	N/A	20.39	N/A	Van den Berg and Lentz (1957)	38 points total. Values are averages.
			$t = -48.3$			20.44			
			$t = -51.1$			19.70			
			$t = -56.7$			19.99			
			$t = -62.2$			18.17			

Table 13 Surface Heat Transfer Coefficients for Food Products (Continued)

1	2	3	4	5	6	7	8	9	10
Product	Shape Length, mm[a]	Transfer Medium	Δt and/or Temp. t of Medium, °C	Velocity of Medium, m/s	Reynolds Number Range[b]	h, W/(m²·K)	Nu-Re-Pr Correlation[c]	Reference	Comments
Meat	Slabs 23	Air	$t = 0$	0.56 1.4 3.7	N/A	10.6 20.0 35.0	N/A	Radford et al. (1976)	
Oranges Grapefruit Tangelos bulk packed	Spheroids 58 80 53	Air	$\Delta t = 39$ to 31 $t = -9$	0.11–0.33	35000–135000	*66.4	$Nu = 5.05 Re^{0.333}$	*Ref*: Bennett et al. (1966)	Bins 1070 × 1070 × 400 mm. 36 points in correlation. Random packaging. Interstitial velocity. *Average for oranges
Oranges Grapefruit bulk packed	Spheroids 77 107	Air	$\Delta t = 32.7$ $t = 0$	0.05–2.03	180–18000	N/A	$Nu = 1.17 Re^{0.529}$	Baird and Gaffney (1976)	20 points in correlation. Bed depth: 670 mm
Peas fluidized bed	Spherical N/A	Air	$t = -26$ to -37	1.5–7.2 ±0.3	1000–4000	N/A	$Nu = 3.5 \times 10^{-4} Re^{1.5}$	Kelly (1965)	Bed: 50 mm deep
Peas bulk packed	Spherical N/A	Air	$t = -26$ to -37	1.5–7.2 ±0.3	1000–6000	N/A	$Nu = 0.016 Re^{0.95}$	Kelly (1965)	
Pears	Spherical 60	Air	$t = 4$	1.00 1.25 1.50 1.75 2.00	N/A	12.6 14.2 15.8 16.1 19.5	$Nu = 1.560 Re^{0.426} Pr^{0.333}$	Dincer (1994)	
Potatoes Pungo, bulk packed	Ellipsoid N/A N/A	Air	$t = 4.4$	0.66 1.23 1.36	3000–9000	*14.0 (at top of bin) 19.1 20.2	$Nu = 0.364 Re^{0.558} Pr^{1/3}$	*Ref*: Minh et al. (1969)	Use interstitial velocity to calculate Re. Bin is 760 × 510 × 230 mm. *Each h value is average of 3 reps with airflow from top to bottom.
Poultry Chickens and turkeys	1.18 to 9.43 kg*	**	$\Delta t = 17.8$	*	N/A	420 to 473	N/A	Lentz (1969)	Vacuum packaged. * Moderately agitated. Chickens 1.1 to 2.9 kg Turkeys 5.4 to 9.5 kg **CaCl₂ Brine, 26% by mass.
Soybeans	Spherical 65	Air	N/A	6.8	1200–4600	N/A	$Nu = 1.07 Re^{0.64}$	Otten (1974)	8 points in correlation. Bed depth: 32 mm.
Squash	Cylinder 46	Water	0.5 1.0 1.5	0.05	N/A	272 205 166	N/A	Dincer (1993)	Diameter = 46 mm. Length = 155 mm.
Tomatoes	Spherical 70	Air	$t = 4$	1.00 1.25 1.50 1.75 2.00	N/A	10.9 13.1 13.6 14.9 17.3	$Nu = 1.560 Re^{0.426} Pr^{0.333}$	Dincer (1994)	
Karlsruhe substance	Slab 75	Air	$\Delta t = 53$ $t = 38$	N/A	N/A	16.4	N/A	Cleland and Earle (1976)	Packed in aluminum foil and brown paper
Milk container	Cylinder 70 × 100 70 × 150 70 × 250	Air	$\Delta t = 5.3$	N/A	$Gr = 10^6$–5×10^7	N/A	$Nu = 0.754 Gr^{0.264}$	Leichter et al. (1976)	Emissivity = 0.7 300 points in correlation. L = characteristic length. All cylinders 70 mm dia.
Acrylic	Ellipsoid 76 (minor axis) $G = 0.297$–1.0	Air	$\Delta t = 44.4$	2.1–8.0	12000–50000	N/A	$Nu = aRe^b$ $a = 0.32 - 0.22G$ $b = 0.44 + 0.23G$	*Ref*: Smith et al. (1971)	$G = 1/4 + 3/(8A^2) + 3/(8B^2)$ A = minor length / char. length B = major length / char. length Char. length = 0.5 × minor axis Use twice char. length to calculate Re.
Acrylic	Spherical 76	Air	$t = -4.4$	0.66 1.23 1.36 1.73	3700–10000	15.0* 14.5 22.2 21.4	$Nu = 2.58 Re^{0.303} Pr^{1/3}$	Minh et al. (1969)	Random packed. Intersticial velocity used to calculate Re. Bin dimensions: 760 × 455 × 610 mm. *Values for top of bin.

[a]Characteristic length is used in Reynolds number and illustrated in the Comments column 10 where appropriate.
[b]Characteristic length is given in column 2, free stream velocity is used, unless specified otherwise in the Comments column 10.
[c]Nu = Nusselt number, Re = Reynolds number, Gr = Grashoff number, Pr = Prandtl number.

Thermal Properties of Foods

where q is the heat transfer rate, t_s is the surface temperature of the food, t_∞ is the surrounding fluid temperature, and A is the surface area of the food through which the heat transfer occurs.

The surface heat transfer coefficient h depends on the velocity of the surrounding fluid, product geometry, orientation, surface roughness and packaging, as well as other factors. Therefore, for most applications h must be determined experimentally. Experimentalists have generally reported their findings as correlations, which give the Nusselt number as a function of the Reynolds number and the Prandtl number.

Experimentally determined values of the surface heat transfer coefficient are given in Table 13. The first two columns of the table describe the product used in the experiment and its shape. Columns 3 through 6 describe the experimental conditions used to determine the surface heat transfer coefficient. Column 7 gives the experimentally determined values of the surface heat transfer coefficient while Column 8 contains the reported Nusselt-Reynolds-Prandtl correlation, if any, and its associated error. Columns 9 and 10 state the source from which the surface heat transfer coefficient data and/or correlation was obtained as well as additional comments.

The following guidelines are important for the use of Table 13:

1. Use a Nusselt-Reynolds-Prandtl correlation or a value of the surface heat transfer coefficient that applies to the Reynolds number called for in the design
2. Avoid extrapolations
3. Use data for the same heat transfer medium, including temperature and temperature difference, which are similar to the design conditions. The proper characteristic length and fluid velocity, either free stream velocity or interstitial velocity, should be used in calculating the Reynolds number and the Nusselt number.

NOMENCLATURE

- a = parameter in Equation (26): $a = 3k_c/(2k_c + k_d)$
- A = surface area
- b = parameter in Equation (26): $b = V_d/(V_c + V_d)$
- c = specific heat
- c_a = apparent specific heat
- c_f = specific heat of fully frozen food
- c_i = specific heat of ith food component
- c_p = constant pressure specific heat
- c_u = specific heat of unfrozen food
- d = commodity diameter
- D = characteristic dimension
- E = ratio of relative molecular masses of water and solids: $E = M_w/M_s$
- f = respiration coefficient
- F_1 = parameter given by Equation (32)
- g = respiration coefficient
- h = surface heat transfer coefficient
- H = enthalpy
- H_f = enthalpy at initial freezing temperature
- H_i = enthalpy of the ith food component
- k = thermal conductivity
- k_1 = thermal conductivity of component 1
- k_2 = thermal conductivity of component 2
- k_a' = air film mass transfer coefficient (driving force: vapor pressure)
- k_a = air film mass transfer coefficient (driving force: concentration)
- k_c = thermal conductivity of continuous phase
- k_d = thermal conductivity of discontinuous phase
- k_i = thermal conductivity of the ith component
- k_s = skin mass transfer coefficient
- k_t = transpiration coefficient
- $k_=$ = thermal conductivity parallel to food fibers
- $k_\perp$ = thermal conductivity perpendicular to food fibers
- L^3 = volume fraction of discontinuous phase
- L_o = latent heat of fusion of water at 0°C = 333.6 kJ/kg
- m = mass
- $\dot{m}$ = transpiration rate
- M = parameter in Equation (28) = $L^2(1 - k_d/k_c)$
- M_s = relative molecular mass of soluble solids
- M_w = relative molecular mass of water
- n = normal surface vector
- Nu = Nusselt number
- N^2 = volume fraction of discontinuous phase
- P = parameter in Equation (30) = $N(1 - k_d/k_c)$
- Pr = Prandtl number
- p_a = water vapor pressure in air
- p_s = water vapor pressure at commodity surface
- q = heat transfer rate
- Q = heat transfer
- R = universal gas constant = 8.314 kJ/(kg mol·K)
- R_1 = volume fraction of component 1
- Re = Reynolds number
- R_{H_2O} = universal gas constant for water vapor
- Sc = Schmidt number
- Sh = Sherwood number
- t = food temperature, °C
- t_f = initial freezing temperature of food, °C
- t_r = reference temperature = –40°C
- t_s = surface temperature, °C
- t_∞ = ambient temperature, °C
- T = food temperature,
- T_f = initial freezing point of food item,
- T_o = freezing point of water; T_o = 233.2 K
- T_r = reference temperature = 233.2 K
- $\bar{T}$ = reduced temperature
- u_∞ = free stream air velocity
- V_c = volume of continuous phase
- V_d = volume of discontinuous phase
- W = rate of heat generation due to respiration, W/kg
- x_1 = mass fraction of component 1
- x_a = mass fraction of ash
- x_b = mass fraction of bound water
- x_f = mass fraction of fat
- x_{fb} = mass fraction of fiber
- x_i = mass fraction of ith food component
- x_{ice} = mass fraction of ice
- x_p = mass fraction of protein
- x_s = mass fraction of solids
- x_{wo} = mass fraction of water in unfrozen food
- x_i^v = volume fraction i^{th} food component
- y = correlation parameter in Equation (19)
- z = correlation parameter in Equation (19)
- α = thermal diffusivity
- δ = diffusion coefficient of water vapor in air
- Δc = difference in specific heats of water and ice = $c_{water} - c_{ice}$
- ΔH = enthalpy difference
- ΔT = temperature difference
- ε = porosity
- θ = time
- Λ = thermal conductivity ratio = k_1/k_2
- ν = kinematic viscosity
- ρ = density of food item
- ρ_1 = density of component 1
- ρ_2 = density of component 2
- ρ_i = density of ith food component
- σ = parameter given by Equation (33)

REFERENCES

Acre, J.A., and V.E. Sweat. 1980. Survey of published heat transfer coefficients encountered in food processes. *ASHRAE Transactions* 86(2): 235-260.

Anderson, R.E., R.E. Hardenburg, and H.C. Baught. 1963. Controlled atmosphere storage studies with cranberries. *ASHS* 83: 416.

ASHRAE. 1993. *Fundamentals handbook*. Table 1, Chapter 30.

Babbitt, J.D. 1945. The thermal properties of wheat in bulk. *Canadian Journal of Research* 23F: 338.

Baird, C.D., and J.J. Gaffney. 1976. A numerical procedure for calculating heat transfer in bulk loads of fruits or vegetables. *ASHRAE Transactions* 82: 525-535.

Becker, B.R., A. Misra, and B.A. Fricke. 1996a. A numerical model of moisture loss and heat loads in refrigerated storage of fruits and vegetables. Frigair '96 Congress and Exhibition, Johannesburg.

Becker, B.R., A. Misra, and B.A. Fricke. 1996b. Bulk refrigeration of fruits and vegetables, Part I: Theoretical considerations of heat and mass transfer. *Int. J. of HVAC&R Research* 2(2): 122-134.

Becker, B.R., A. Misra, and B.A. Fricke. 1996c. Bulk refrigeration of fruits and vegetables, Part II: Computer algorithm for heat loads and moisture loss. *Int. J. of HVAC&R Research* 2(3): 215-230.

Becker, B.R., and B.A. Fricke. 1996a. Transpiration and respiration of fruits and vegetables. *New Developments in Refrigeration for Food Safety and Quality*. International Institute of Refrigeration, Paris, France, and American Society of Agricultural Engineers, St. Joseph, MI, pp. 110-121.

Becker, B.R., and B.A. Fricke. 1996b. Simulation of moisture loss and heat loads in refrigerated storage of fruits and vegetables. *New Developments in Refrigeration for Food Safety and Quality*. International Institute of Refrigeration, Paris, France, and American Society of Agricultural Engineers, St. Joseph, MI, pp. 210-221.

Bennett, A.H. 1963. Thermal characteristics of peaches as related to hydrocooling. *Technical Bulletin Number 1292* Washington, D.C.: USDA.

Bennett, A.H., W.G. Chace, and R.H. Cubbedge. 1964. Thermal conductivity of Valencia orange and Marsh grapefruit rind and juice vesicles. *ASHRAE Transactions* 70: 256-259.

Bennett, A.H., J. Soule, and G.E. Yost. 1966. Temperature response of Florida citrus to forced-air precooling. *ASHRAE Journal* 8(4): 48-54.

Bennett, A.H., W.G. Chace, and R.H. Cubbedge. 1969. Heat transfer properties and characteristics of Appalachian area, red delicious apples. *ASHRAE Transactions* 75(2): 133.

Bennett, A.H., W.G. Chace, and R.H. Cubbedge. 1970. Thermal properties and heat transfer characteristics of Marsh grapefruit. *Technical Bulletin Number 1413* Washington, D.C.: USDA.

Biale, J.B. 1960. Respiration of fruits. *Encyclopedia of Plant Physiology* 12: 536.

Chang, H.D., and L.C. Tao. 1981. Correlations of enthalpies of food systems. *Journal of Food Science* 46: 1493.

Chau, K.V., R.A. Romero, C.D. Baird, and J.J. Gaffney. 1987. Transpiration coefficients of fruits and vegetables in refrigerated storage. *ASHRAE Report 370-RP*. Atlanta: ASHRAE.

Chen, C.S. 1985. Thermodynamic analysis of the freezing and thawing of foods: Enthalpy and apparent specific heat. *Journal of Food Science* 50: 1158.

Choi, Y., and M.R. Okos. 1986. Effects of temperature and composition on the thermal properties of foods. In *Food Engineering and Process Applications*. ed. M. LeMaguer and P. Jelen. 1: 93-101. London: Elsevier Applied Science Publishers.

Chuma, Y., S. Murata, and S. Uchita. 1970. Determination of heat transfer coefficients of farm products by transient method using lead model. *Journal of the Society of Agricultural Machinery* 31(4): 298-302.

Clary, B.L., G.L. Nelson, and R.E. Smith. 1968. Heat transfer from hams during freezing by low temperature air. *Transactions of the ASAE* 11: 496-499.

Claypool, L.L., and F.W. Allen. 1951. The influence of temperature and oxygen level on the respiration and ripening of Wickson plums. *Hilgardea* 21: 129.

Claypool, L.L., and S. Ozbek. 1952. Some influences of temperature and carbon dioxide on the respiration and storage life of the mission fig. *Proc. ASHS* 60: 266.

Cleland, A.C., and R.L. Earle. 1976. A new method for prediction of surface heat transfer coefficients in freezing. *Bulletin de L'Institut International du Froid* Annexe 1976-1: 361-368.

Dickerson, R.W., Jr. 1968. Thermal properties of food. *The Freezing Preservation of Foods*, 4th ed., Vol. 2, AVI Publishing Co., Westport, CT.

Dickerson R.W., Jr., and R.B. Read, Jr. 1968. Calculation and measurement of heat transfer in foods. *Food Technology* 22: 37.

Dickerson, R.W., Jr. 1969. Thermal properties of food. In *The Freezing Preservation of Foods*, 4th Edition, ed. D.K. Tressler, W.B. Van Arsdel, and M.J. Copley. 2: 27. Westport, CT: AVI Publishing Co.

Dickerson, R.W., and R.B. Read. 1975. Thermal diffusivity of meats. *ASHRAE Transactions* 81(1): 356.

Dincer, I. 1993. Heat-transfer coefficients in hydrocooling of spherical and cylindrical food products. *Energy* 18(4): 335-340.

Dincer, I. 1994. Development of new effective Nusselt-Reynolds correlations for air-cooling of spherical and cylindrical products. *Int. J. of Heat and Mass Transfer* 37(17): 2781-2787.

Eaks, J.L., and L.L. Morris. 1956. Respiration of cucumber fruits associated with physiological injury at chilling temperatures. *Plant Physiology* 31: 308.

Eucken, A. 1940. Allgemeine Gesetzmassigkeiten fur das Warmeleitvermogen verschiedener Stoffarten und Aggregatzustande. *Forschung auf dem Gebiete des Ingenieurwesens, Ausgabe A* 11(1): 6.

Fedorov, V.G., D.N. Il'Inskiy, O.A. Gerashchenko, and L.D. Andreyeva. 1972. Heat transfer accompanying the cooling and freezing of meat carcasses. *Heat Transfer—Soviet Research* 4: 55-59.

Fikiin, K.A. 1996. Ice content prediction methods during food freezing: A Survey of the Eastern European Literature. In *New Developments in Refrigeration for Food Safety and Quality*, International Institute of Refrigeration, Paris, France, and American Society of Agricultural Engineers, St. Joseph, Michigan, pp. 90-97.

Fockens, F.H., and H.F.T. Meffert. 1972. Biophysical properties of horticultural products as related to loss of moisture during cooling down. *J. Science of Food and Agriculture* 23: 285-298.

Gaffney, J.J., C.D. Baird, and K.V. Chau. 1985. Influence of airflow rate, respiration, evaporative cooling, and other factors affecting weight loss calculations for fruits and vegetables. *ASHRAE Transactions* 91(1B): 690-707.

Gan, G., and J.L. Woods. 1989. A deep bed simulation of vegetable cooling. In *Agricultural Engineering*, ed. V.A. Dodd and P.M. Grace, pp. 2301-2308. Rotterdam: A.A. Balkema.

Gane, R. 1936. The thermal conductivity of the tissue of fruits. *Annual Report*, Food Investigation Board, Great Britain, 211.

Geankoplis, C.J. 1978. *Transport processes and unit operations*. Boston: Allyn and Bacon.

Gerhardt, F., H. English, and E. Smith. 1942. Respiration, internal atmosphere, and moisture studies of sweet cherries during storage. *Proc. ASHS* 41: 119.

Gore, H.C. 1911. Studies on fruit respiration. *USDA Bur. Chem Bulletin*, Vol. 142.

Griffiths, E. and D.H. Cole. 1948. Thermal properties of meat. *Society of Chemical Industry Journal* 67: 33.

Griffiths, E. and M.J. Hickman. 1951. *The thermal conductivity of some nonmetallic materials*. Institute of Mechanical Engineers, London, England, 289.

Haller, M.H., P.L. Harding, J.M. Lutz, and D.H. Rose. 1932. The respiration of some fruits in relation to temperature. *Proc. ASHS* 28: 583.

Haller, M.H., D.H. Rose, and P.L. Harding. 1941. Studies on the respiration of strawberry and raspberry fruits. USDA *Circular*, Vol. 613.

Haller, M.H., et al. 1945. Respiration of citrus fruits after harvest. *J. of Agricultural Research* 71(8): 327.

Harding, P.L. 1929. Respiration studies of grimes apples under various controlled temperatures. *Proceedings of the American Society for Horticultural Science*. 26: 319.

Harper, J.C. 1960. Microwave spectra and physical characteristics of fruit and animal products relative to freeze-dehydration. Report No. 6, Army Quartermaster Food and Container Institute for the Armed Forces, ASTIA AD 255 818, 16.

Harper, J.C. 1962. Transport properties of gases in porous media at reduced pressures with reference to freeze-drying. *American Institute of Chemical Engineering Journal* 8(3): 298.

Hawkins, L.A. 1929. Governing factors in transportation of perishable commodities. *Refrigerating Engineering* 18: 130.

Hill, J.E. 1966. *The thermal conductivity of beef*. Georgia Institute of Technology, Atlanta, GA, 49.

Hill, J.E., J.D. Leitman, and J.E. Sunderland. 1967. Thermal Conductivity of Various Meats. *Food Technology* 21(8): 91.

Holland, B., A.A. Welch, I.D. Unwin, D.H. Buss, A.A. Paul, and D.A.T. Southgate. 1991. *McCance and Widdowson's—The Composition of Foods*. Cambridge, U.K.: Royal Society of Chemistry and Ministry of Agriculture, Fisheries and Food.

Hooper, F.C., and S.C. Chang. 1952. Development of the Thermal Conductivity Probe. *Heating, Piping and Air Conditioning*, ASHVE 24(10): 125.

Hruschka, H.W. 1966. Storage and Shelf Life of Packaged Rhubarb. USDA *Marketing Research Report*, 771.

Hruschka, H.W., and C.Y. Want. 1979. Storage and Shelf Life of Packaged Watercress, Parsley, and Mint. USDA *Marketing Research Report*, 1102.

International Institute of Refrigeration. 1967. *Recommended Conditions for the Cold Storage of Perishable Produce*, 2nd ed., Paris, France.

Jason, A.C., and R.A.K. Long. 1955. The Specific Heat and Thermal Conductivity of Fish Muscle. Proceedings of the 9th International Congress of Refrigeration, Paris, France, 1: 2160.

Jones, W.W. 1942. Respiration and Chemical Changes of Papaya Fruit in Relation to Temperature. *Plant Physiology* 17: 481.

Karmarkar, D.V., and B.M. Joshe. 1941a. Respiration of Onions. *Indian Journal of Agricultural Science* 11: 82.

Karmarkar, D.V., and B.M. Joshe. 1941b. Respiration Studies on the Alphonse Mango. *Indian Journal of Agricultural Science* 11: 993.

Kaye, G.W.C., and W.F. Higgins. 1928. The Thermal Conductivities of Certain Liquids. *Proceedings of the Royal Society of London*, A117: 459.

Kazarian, E.A. 1962. *Thermal Properties of Grain*. Michigan State University, East Lansing, MI, 74.

Kelly, M.J. 1965. Heat Transfer in Fluidized Beds. *Dechema Monographien* 56: 119.

Khatchaturov, A.B. 1958. Thermal Processes During Air-Blast Freezing of Fish. *Bulletin of the IIR* Annexe 1958-2: 365-378.

Khelemskii, M.Z., and V.Z. Zhadan. 1964. Thermal Conductivity of Normal Beet Juice. *Sakharnaya Promyshlennost* 10: 11.

Kondrat'ev, G.M. 1950. Application of the Theory of Regular Cooling of a Two-Component Sphere to the Determination of Heat Conductivity of Poor Heat Conductors (method, sphere in a sphere). *Otdelenie Tekhnicheskikh Nauk, Isvestiya Akademii Nauk*, USSR 4(April): 536.

Kopelman, I.J. 1966. Transient Heat Transfer and Thermal Properties in Food Systems. Ph.D. Thesis. Michigan State University. East Lansing, MI.

Kopelman, I., J.L. Blaisdell, and I.J. Pflug. 1966. Influence of Fruit Size and Coolant Velocity on the Cooling of Jonathan Apples in Water and Air. *ASHRAE Transactions* 72(1): 209-216.

Leichter, S., S. Mizrahi, and I.J. Kopelman. 1976. Effect of Vapor Condensation on Rate of Warming Up of Refrigerated Products Exposed to Humid Atmosphere: Application to the Prediction of Fluid Milk Shelf Life. *Journal of Food Science* 41: 1214-1218.

Leidenfrost, W. 1959. Measurements on the Thermal Conductivity of Milk. ASME Symposium on Thermophysical Properties, Purdue University, 291.

Lentz, C.P. 1961. Thermal Conductivity of Meats, Fats, Gelatin Gels, and Ice. *Food Technology* 15(5): 243.

Lentz, C.P. 1969. Calorimetric Study of Immersion Freezing of Poultry. *Journal of the Canadian Institute of Food Technology* 2(3): 132-136.

Levy, F.L. 1981. A Modified Maxwell-Eucken Equation for Calculating the Thermal Conductivity of Two-Component Solutions or Mixtures. *Int. J. Refrig.* 4: 223-225.

Lewis, D.A., and L.L. Morris. 1956. Effects of Chilling Storage on Respiration and Deterioration of Several Sweet Potato Varieties. *Proceedings of the American Society for Horticultural Science* 68: 421.

Lipton, W.J. 1957. *Physiological changes in Harvested Asparagus (Asparagus Officinales) as Related to Temperature*. University of California at Davis.

Long, R.A.K. 1955. Some Thermodynamic Properties of Fish and Their Effect on the Rate of Freezing. *Journal of the Science of Food and Agriculture* 6: 621.

Lutz, J.M. 1938. Factors Influencing the Quality of American Grapes in Storage. USDA *Technical Bulletin*, Vol. 606.

Lutz, J.M. and R.E. Hardenburg. 1968. The Commercial Storage of Fruits, Vegetables, and Florists and Nursery Stocks. USDA *Handbook* 66.

Mann, L.K. and D.A. Lewis. 1956. Rest and Dormancy in Garlic. *Hilgardia* 26: 161.

Mathews, F.W., Jr., and C.W. Hall. 1968. Method of Finite Differences Used to Relate Changes in Thermal and Physical Properties of Potatoes. *ASAE Transactions* 11(4): 558.

Maxie, E.C., F.G. Mitchell, and A. Greathead. 1959. Studies on Strawberry Quality. *California Agriculture* 13(2): 11, 16.

Maxie, E.C., P.B. Catlin, and H.T. Hartmann. 1960. Respiration and Ripening of Olive Fruits. *Proceedings of the American Society for Horticultural Science* 75: 275.

Metzner, A.B., and P.S. Friend. 1959. Heat Transfer to Turbulent non-Newtonian Fluids. *Industrial and Engineering Chemistry* 51: 879.

Micke, W.C., F.G. Mitchell, and E.C. Maxie. 1965. Handling Sweet Cherries for Fresh Shipment. *California Agriculture* 19(4): 12.

Miles, C.A. 1974. In *Meat Freezing—Why and How?* Proceedings of the Meat Research Institute, Symposium No. 3, Bristol, 15.1-15.7.

Miller, C.F. 1963. *Thermal Conductivity and Specific Heat of Sorghum Grain*. Texas Agricultural and Mechanical College, College Station, TX, 79.

Minh, T.V., J.S. Perry, and A.H. Bennett. 1969. Forced-Air Precooling of White Potatoes in Bulk. *ASHRAE Transactions* 75(2): 148-150.

Moote, I. 1953. The Effect of Moisture on the Thermal Properties of Wheat. *Canadian Journal of Technology* 31(2/3): 57.

Morris, L.L. 1947. A Study of Broccoli Deterioration. *Ice and Refrigeration* 113(5): 41.

Nicholas, R.C., K.E.H. Motawi, and J.L. Blaisdell. 1964. Cooling Rate of Individual Fruit in Air and in Water. *Quarterly Bulletin*, Michigan State University Agricultural Experiment Station 47(1): 51-64.

Nowrey, J.E., and E.E. Woodams. 1968. Thermal Conductivity of a Vegetable Oil-in-Water Emulsion. *Journal of Chemical and Engineering Data* 13(3): 297.

Otten, L. 1974. Thermal Parameters of Agricultural Materials and Food Products. *Bulletin of the IIR* Annexe 1974-3: 191-199.

Oxley, T.A. 1944. The Properties of Grain in Bulk; III—The Thermal Conductivity of Wheat, Maize and Oats. *Society of Chemical Industry Journal* 63: 53.

Pantastico, E.B. 1974. Handling and Utilization of Tropical and Subtropical Fruits and Vegetables. *Postharvest Physiology*, Westport, CT: AVI Publishing Co., Inc.

Parker, R.E., and B.A. Stout. 1967. Thermal Properties of Tart Cherries. *Transactions of the ASAE* 10(4): 489-491, 496.

Pentzer, W.T., C.E. Asbury, and K.C. Hamner. 1933. The Effect of Sulfur Dioxide Fumigation on the Respiration of Emperor Grapes. *Proceedings of the American Society for Horticultural Science* 30: 258.

Pham, Q.T. 1987. Calculation of Bound Water in Frozen Food. *Journal of Food Science* 52(1): 210-212.

Polley, S.L., O.P. Snyder, and P. Kotnour. 1980. A Compilation of Thermal Properties of Foods. *Food Technology* 34(11): 76-94.

Popov, V.D., and Y.A. Terentiev. 1966. Thermal Properties of Highly Viscous Fluids and Coarsely Dispersed Media. *Teplofizicheskie Svoistva Veshchestv, Akademiya Nauk, Ukrainskoi SSSR, Respublikanskii Sbornik* 18: 76.

Poppendick, H.F. et al. 1965-66. Annual Report on Thermal and Electrical Conductivities of Biological Fluids and Tissues. *ONR Contract* 4094 (00), A-2, GLR-43 Geoscience Ltd., 39.

Pratt, H.K., L.L. Morris, and C.L. Tucker. 1954. Temperature and Lettuce Deterioration. *Proceedings of the Conference on Transportation of Perishables*, University of California at Davis, 77.

Pratt, H.K., and L.L. Morris. 1958. Some Physiological Aspects of Vegetable and Fruit Handling. *Food Technology in Australia* 10: 407.

Qashou, M.S., G. Nix, R.I. Vachon, and G.W. Lowery. 1970. Thermal Conductivity Values for Ground Beef and Chuck. *Food Technology* 23(4): 189.

Qashou, M.S., R.I. Vachon, and Y.S. Touloukian. 1972. Thermal Conductivity of Foods. *ASHRAE Transactions* 78(1): 165-183.

Radford, R.D., L.S. Herbert, and D.A. Lorett. 1976. Chilling of Meat—A Mathematical Model for Heat and Mass Transfer. *Bulletin de L'Institut International du Froid* Annexe 1976-1: 323-330.

Rappaport, L., and A.E. Watada. 1958. Effects of Temperature on Artichoke Quality. *Proceedings of the Conference on Transportation of Perishables*, University of California at Davis, 142.

Riedel, L. 1949. Thermal Conductivity Measurements on Sugar Solutions, Fruit Juices and Milk. *Chemie-Ingenieur-Technik* 21(17): 340-341.

Riedel, L. 1951. The Refrigeration Effect Required to Freeze Fruits and Vegetables. *Refrigeration Engineering* 59: 670.

Riedel, L. 1956. Calorimetric Investigation of the Freezing of Fish Meat. *Kaltetechnik* 8: 374-377.

Riedel, L. 1957. Calorimetric Investigation of the Meat Freezing Process. *Kaltetechnik* 9(2): 38-40.

Riedel, L. 1957. Calorimetric Investigation of the Freezing of Egg White and Yolk. *Kaltetechnik* 9: 342.

Riedel, L. 1959. Calorimetric Investigations of the Freezing of White Bread and Other Flour Products. *Kaltetechnik* 11(2): 41.

Riedel, L. 1969. Measurements of Thermal Diffusivity on Foodstuffs Rich in Water. *Kaltetechnik* 21(11): 315-316.

Reidy, G.A. 1968. Values for Thermal Properties of Foods Gathered from the Literature. Ph.D. Thesis, Michigan State University, East Lansing.

Ryall, A.L., and W.J. Lipton. 1972. Vegetables as Living Products. Respiration and Heat Production. In *Transportation and Storage of Fruits and Vegetables*, Vol. 1, Westport, CT: AVI Publishing Co.

Saravacos, G.D. 1965. Freeze-drying Rates and Water Sorption of Model Food Gels. *Food Technology* 19(4): 193.

Saravacos, G.D., and M.N. Pilsworth. 1965. Thermal Conductivity of Freeze-Dried Model Food Gels. *Journal of Food Science* 30: 773.

Sastry, S.K., C.D. Baird, and D.E. Buffington. 1978. Transpiration Rates of Certain Fruits and Vegetables. *ASHRAE Transactions* 84(1).

Sastry, S.K., and D.E. Buffington. 1982. Transpiration Rates of Stored Perishable Commodities: A Mathematical Model and Experiments on Tomatoes. *ASHRAE Transactions* 88(1): 159-184.

Schenk, R.U. 1959. Respiration of Peanut Fruit During Curing. *Proceedings of the Association of Southern Agricultural Workers* 56: 228.

Schenk, R.U. 1961. Development of the Peanut Fruit. *Georgia Agricultural Experiment Station Bulletin N.S.*, Vol. 22.

Scholz, E.W., H.B. Johnson, and W.R. Buford. 1963. Heat Evolution Rates of Some Texas-Grown Fruits and Vegetables. *Rio Grande Valley Horticultural Society Journal* 17: 170.

Schwartzberg, H.G. 1976. Effective heat capacities for the freezing and thawing of food. *Journal of Food Science* 41(1): 152-156.

Schwartzberg, H.G. 1981. Mathematical analysis of the freezing and thawing of foods, Tutorial presented at the AIChE Summer Meeting, Detroit, Michigan.

Siebel, J.E. 1892. Specific heat of various products. *Ice and Refrigeration* 256.

Slavicek, E., K. Handa, and M. Kminek. 1962. Measurements of the thermal diffusivity of sugar beets. *Cukrovarnicke Listy*, Vol. 78, Czechoslovakia, 116.

Smith, F.G., A.J. Ede, and R. Gane. 1952. The Thermal Conductivity of Frozen Foodstuffs. *Modern Refrigeration* 55: 254.

Smith, R.E., A.H. Bennett, and A.A. Vacinek. 1971. Convection Film Coefficients Related to Geometry for Anomalous Shapes. *Transactions of the ASAE* 14(1): 44-47.

Smith, R.E., G.L. Nelson, and R.L. Henrickson. 1976. Analyses on Transient Heat Transfer from Anomalous Shapes. *ASAE Transactions* 10(2): 236.

Smith, W.H. 1957. The Production of Carbon Dioxide and Metabolic Heat by Horticultural Produce. *Modern Refrigeration* 60: 493.

Smith, W.H. 1964. The Storage of Mushrooms. *Ditton and Covent Garden Laboratories Annual Report*, Great Britain Agricultural Research Council, 18.

Smith, W.H. 1966. The Storage of Gooseberries. *Ditton and Covent Garden Laboratories Annual Report*, Great Britain Agricultural Research Council, 13.

Spells, K.E. 1958. The Thermal Conductivities of Some Biological Fluids. Flying Personnel Research Committee, Institute of Aviation Medicine, Royal Air Force, Farnborough, England, FPRC-1071 AD 229 167, 8.

Spells, K.E. 1960-61. The Thermal Conductivities of Some Biological Fluids. *Physics in Medicine and Biology* 5: 139.

Sweat, V.E. 1974. Experimental Values of Thermal Conductivity of Selected Fruits and Vegetables. *Journal of Food Science* 39: 1080.

Sweat, V.E. 1985. Thermal Properties of Low- and Intermediate-Moisture Food. *ASHRAE Transactions* 91(2): 369-389.

Tchigeov, G. 1979. *Thermophysical Processes in Food Refrigeration Technology*. Food Industry, Moscow, 272 pp.

Tewfik, S., and L.E. Scott. 1954. Respiration of Vegetables as Affected by Postharvest Treatment. *Journal of Agricultural and Food Chemistry* 2:415.

Thompson, H., S.R. Cecil, and J.G. Woodroof. 1951. Storage of Edible Peanuts. *Georgia Agricultural Experiment Station Bulletin*, Vol. 268.

Triebes, T.A., and C.J. King. 1966. Factors Influencing the Rate of Heat Conduction in Freeze-Drying. *I and EC Process Design and Development* 5(4): 430.

Turrell, F.M., and R.L. Perry. 1957. Specific Heat and Heat Conductivity of Citrus Fruit. Proceedings of the American Society for Horticultural Science 70: 261.

USDA. 1975. *Composition of Foods. Agricultural Handbook* No. 8. U.S. Department of Agriculture, Washington, DC.

USDA. 1996. *Nutrient Database for Standard Reference, Release* 11. U.S. Department of Agriculture, Washington, DC.

Van den Berg, L., and C.P. Lentz. 1957. Factors Affecting Freezing Rates of Poultry Immersed in Liquid. *Food Technology* 11(7): 377-380.

Van den Berg, L. and C.P. Lentz. 1972. Respiratory Heat Production of Vegetables During Refrigerated Storage. *Journal of the American Society for Horticultural Science* 97: 431.

Wachsmuth. R. 1892. Untersuchungen auf dem Gebiet der inneren Warmeleitung. *Annalen der Physik* 3(48): 158.

Walters, R.E., and K.N. May. 1963. Thermal Conductivity and Density of Chicken Breast Muscle and Skin. *Food Technology* 17(June): 130.

Watada, A.E., and L.L. Morris. 1966. Effect of Chilling and Nonchilling Temperatures on Snap Bean Fruits. *Proceedings of the American Society for Horticultural Science* 89: 368.

Watt, B.K., and A.L. Merrill. 1963. Composition of Foods. *USDA Handbook 8*.

Weber, H.F., VII. 1880. Untersuchungen Uber die Warmeleitung in Flussigkeiten. *Annael der Physik* 10(3): 304.

Weber, H.F. 1886. The Thermal Conductivity of Drop Forming Liquids. *Exner's Reportorium* 22: 116.

Woodams, E.E. 1965. *Thermal Conductivity of Fluid Foods*. Cornell University, Ithaca, NY, 95.

Workman, M., and H.K. Pratt. 1957. Studies on the Physiology of Tomato Fruits; II, Ethylene Production at 20°C as related to Respiration, Ripening and Date of Harvest. *Plant Physiology* 32: 330.

Wright, R.C., D.H. Rose, and T.H. Whiteman. 1954. The Commercial Storage of Fruits, Vegetables, and Florists and Nursery Stocks. USDA *Handbook* 66.

CHAPTER 9

COOLING AND FREEZING TIMES OF FOODS

Thermodynamics of Cooling and Freezing 9.1
Cooling Times of Foods and Beverages 9.1
Sample Problems for Estimating Cooling Time 9.6
Freezing Times of Foods and Beverages 9.7
Sample Problems for Estimating Freezing Time 9.13
Nomenclature 9.15

PRESERVATION of food is one of the most significant applications of refrigeration. Cooling and freezing of food effectively reduces the activity of micro-organisms and enzymes, thus retarding deterioration. In addition, crystallization of water reduces the amount of liquid water in food items and inhibits microbial growth (Heldman 1975).

In order for cooling and freezing operations to be cost-effective, refrigeration equipment should fit the specific requirements of the particular cooling or freezing application. The design of such refrigeration equipment requires estimation of the cooling and freezing times of foods and beverages, as well as the corresponding refrigeration loads.

Numerous methods for predicting the cooling and freezing times of foods and beverages have been proposed, including those based on numerical, analytical, and empirical analysis. The designer is faced with the challenge of selecting an appropriate estimation method from the many available methods. This chapter reviews selected procedures available for estimating the cooling and freezing times of foods and beverages, and presents examples of these procedures.

THERMODYNAMICS OF COOLING AND FREEZING

The cooling and freezing of food is a complex process. Prior to freezing, sensible heat must be removed from the food to decrease its temperature from the initial temperature to the initial freezing point of the food. This initial freezing point is somewhat lower than the freezing point of pure water due to dissolved substances in the moisture within the food. At the initial freezing point, a portion of the water within the food crystallizes and the remaining solution becomes more concentrated. Thus, the freezing point of the unfrozen portion of the food is further reduced. As the temperature continues to decrease, the formation of ice crystals increases the concentration of the solutes in solution and depresses the freezing point further. Thus, the ice and water fractions in the frozen food depend on temperature.

The cooling and freezing of foods and beverages can be described via the Fourier heat conduction equation:

$$\frac{\partial T}{\partial \theta} = \frac{1}{\rho c}\left[\frac{\partial}{\partial x}\left(k\frac{\partial T}{\partial x}\right) + \frac{\partial}{\partial y}\left(k\frac{\partial T}{\partial y}\right) + \frac{\partial}{\partial z}\left(k\frac{\partial T}{\partial z}\right)\right] \quad (1)$$

where T is temperature, θ is time, ρ is the density of the food, c is the specific heat of the food, k is the thermal conductivity of the food and x, y, and z are the coordinate directions. For regularly shaped food items with constant thermophysical properties, uniform initial conditions, constant external conditions and a prescribed surface temperature or a convection boundary condition,

exact analytical solutions for Equation (1) exist. However, for practical cooling and freezing processes, foods are generally irregularly shaped with temperature dependent thermophysical properties. Therefore, exact analytical solutions cannot be derived for the cooling and freezing times of foods and beverages.

Accurate numerical estimations of the cooling and freezing times of foods can be obtained using appropriate finite element or finite difference computer programs. However, the effort required to perform this task makes it impractical for the design engineer. In addition, two-dimensional and three-dimensional simulations require time consuming data preparation and significant computing time. Hence, most of the research effort has been in the development of semi-analytical/empirical cooling and freezing time prediction methods that make use of simplifying assumptions.

COOLING TIMES OF FOODS AND BEVERAGES

Before a food item can be frozen, the temperature of the food must be reduced to its initial freezing point. This cooling process, also known as precooling or chilling, removes only sensible heat and thus, no phase change occurs.

The cooling time of foods and beverages is influenced by the ratio of the external heat transfer resistance to the internal heat transfer resistance. This ratio, called the Biot number, is defined as

$$\text{Bi} = hL/k \quad (2)$$

where h is the convective heat transfer coefficient, L is the characteristic dimension of the food item, and k is the thermal conductivity of the food item.

When the Biot number approaches zero (Bi < 0.1), the internal resistance to heat transfer is much less than the external resistance, and the lumped-parameter approach can be used to determine the cooling time of a food item (Heldman 1975). When the Biot number is very large (Bi > 40), the internal resistance to heat transfer is much greater than the external resistance, and the surface temperature of the food item can be assumed to be equal to the temperature of the cooling medium. For this latter situation, series solutions of the Fourier heat conduction equation are available for simple geometric shapes. When the Biot number falls within the range 0.1 < Bi < 40, both the internal resistance to heat transfer and the convective heat transfer coefficient must be considered. In this case, series solutions, which incorporate transcendental functions to account for the influence of the Biot number, are available for simple geometric shapes.

Simplified methods for predicting the cooling times of foods and beverages are applicable to regularly and irregularly shaped food items over a wide range of Biot numbers. In this chapter, these simplified cooling time estimation methods are grouped into two main categories: (1) methods based on f and j factors, and (2) methods based on equivalent heat transfer dimensionality. Furthermore, the methods based on f and j factors are divided into two subgroups: (1) methods for regular shapes, and (2) methods for irregular shapes.

The preparation of this chapter is assigned to TC 10.9, Refrigeration Application for Foods and Beverages.

Cooling Time Estimation Methods Based on f and j Factors

All cooling processes exhibit similar behavior. After an initial lag, the temperature at the thermal center of the food item decreases exponentially (Cleland 1990). As shown in Figure 1, a cooling curve depicting this behavior can be obtained by plotting, on semilogarithmic axes, the fractional unaccomplished temperature difference versus time. The fractional unaccomplished temperature difference Y is defined as follows:

$$Y = \frac{T_m - T}{T_m - T_i} = \frac{T - T_m}{T_i - T_m} \qquad (3)$$

where T_m is the cooling medium temperature, T is the product temperature, and T_i is the initial temperature of the product.

This semilogarithmic temperature history curve consists of one initial curvilinear portion, followed by one or more linear portions. Empirical formulas, which model this cooling behavior, incorporate two factors, f and j, which represent the slope and intercept, respectively, of the temperature history curve. The j factor is a measure of the lag between the onset of cooling and the exponential decrease in the temperature of the food. The f factor represents the time required to obtain a 90% reduction in the non-dimensional temperature difference. Graphically, the f factor corresponds to the time required for the linear portion of the temperature history curve to pass through one log cycle. The f factor is a function of the Biot number while the j factor is a function of the Biot number and the location within the food item.

The general form of the cooling time model is

$$Y = \frac{T_m - T}{T_m - T_i} = j e^{-2.303 \theta / f} \qquad (4)$$

where t is the cooling time. This equation can be rearranged to give the cooling time explicitly as

$$\theta = \frac{-f}{2.303} \ln\left(\frac{Y}{j}\right) \qquad (5)$$

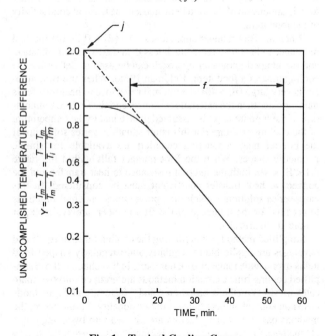

Fig. 1 Typical Cooling Curve

Determination of f and j Factors for Slabs, Cylinders, and Spheres

From analytical solutions, Pflug et al. (1965) developed charts for determining f and j factors for food items shaped either as infinite slabs, infinite cylinders, or spheres. For this development, they assumed a uniform initial temperature distribution in the food item, a constant surrounding medium temperature, convective heat exchange at the surface, and constant thermophysical properties. Figure 2 can be used to determine f values while Figures 3, 4, and 5 can be used to determine j values. Because the j factor is a function of location within the food item, Pflug et al. presented three charts for determining j factors: one corresponding to center temperature, one corresponding to volumetric average temperature and one corresponding to surface temperature.

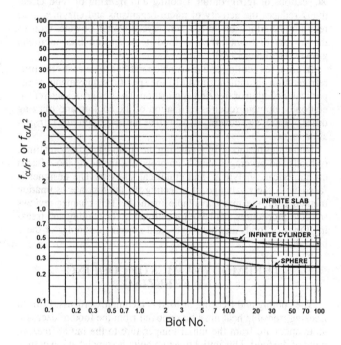

Fig. 2 Relationship Between $f\alpha/r^2$ and Biot number for Infinite Slab, Infinite Cylinder, and Sphere

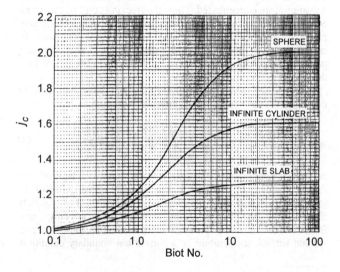

Fig. 3 Relationship Between j_c Value for Thermal Center Temperature and Biot Number for Various Shapes

Cooling and Freezing Times of Foods

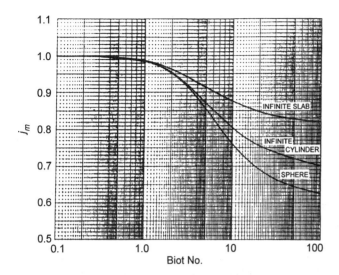

Fig. 4 Relationship Between j_m Value for Mass Average Temperature and Biot Number for Various Shapes

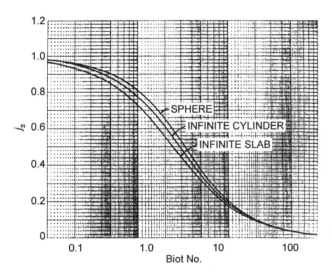

Fig. 5 Relationship Between j_s Value for Surface Temperature and Biot Number for Various Shapes

Lacroix and Castaigne (1987a) presented expressions for estimating the f and j_c factors for the thermal center temperature of infinite slabs, infinite cylinders, and spheres. These expressions, which depend upon geometry and Biot number, are summarized in Tables 1, 2 and 3. In these expressions, α is the thermal diffusivity of the food item and L is the characteristic dimension, defined to be the shortest distance from the thermal center of the food item to its surface. For an infinite slab, L is the half thickness. For an infinite cylinder or a sphere, L is the radius.

By using various combinations of infinite slabs and infinite cylinders, the f and j factors for infinite rectangular columns, finite cylinders, and rectangular bricks may be estimated. Each of these shapes can be generated by intersecting infinite slabs and infinite cylinders: two infinite slabs of proper thickness for the infinite rectangular column, one infinite slab and one infinite cylinder for the finite cylinder, or three infinite orthogonal slabs of proper thickness for the rectangular brick. The f and j factors of these composite bodies can be estimated by:

$$\frac{1}{f_{comp}} = \sum_i \left(\frac{1}{f_i}\right) \quad (6)$$

Table 1 Expressions for Estimating f and j_c Factors for the Thermal Center Temperature of Infinite Slabs

Biot Number Range	Equations for f and j factors
Bi ≤ 0.1	$\dfrac{f\alpha}{L^2} = \dfrac{\ln 10}{\text{Bi}}$ $j_c = 1.0$
0.1 < Bi ≤ 100	$\dfrac{f\alpha}{L^2} = \dfrac{\ln 10}{u^2}$ $j_c = \dfrac{2\sin u}{u + \sin u \cos u}$ where $u = 0.860972 + 0.312133\ln(\text{Bi})$ $\quad + 0.007986[\ln(\text{Bi})]^2 - 0.016192[\ln(\text{Bi})]^3$ $\quad - 0.001190[\ln(\text{Bi})]^4 + 0.000581[\ln(\text{Bi})]^5$
Bi > 100	$\dfrac{f\alpha}{L^2} = 0.9332$ $j_c = 1.273$

Source: Lacroix and Castaigne, 1987a

Table 2 Expressions for Estimating f and j_c factors for the Thermal Center Temperature of Infinite Cylinders

Biot Number Range	Equations for f and j factors
Bi ≤ 0.1	$\dfrac{f\alpha}{L^2} = \dfrac{\ln 10}{2\,\text{Bi}}$ $j_c = 1.0$
0.1 < Bi ≤ 100	$\dfrac{f\alpha}{L^2} = \dfrac{\ln 10}{v^2}$ $j_c = \dfrac{2J_1(v)}{v[J_0^2(v) - J_1^2(v)]}$ where $w = 1.257493 + 0.487941\ln(\text{Bi})$ $\quad + 0.025322[\ln(\text{Bi})]^2 - 0.026568[\ln(\text{Bi})]^3$ $\quad - 0.002888[\ln(\text{Bi})]^4 + 0.001078[\ln(\text{Bi})]^5$ and $J_0(v)$ and $J_1(v)$ are zero and first order Bessel functions, respectively
Bi > 100	$\dfrac{f\alpha}{L^2} = 0.3982$ $j_c = 1.6015$

Source: Lacroix and Castaigne, 1987a

$$j_{comp} = \prod_i j_i \quad (7)$$

where the subscript i represents the appropriate infinite slab(s) or infinite cylinder. To evaluate the f_i and j_i of Equations (6) and (7), the Biot number must be defined, corresponding to the appropriate infinite slab(s) or infinite cylinder.

Determination of f and j Factors for Irregular Shapes

In an effort to determine f and j factors for irregularly shaped food items, Smith et al. (1968) developed, for the case of Biot number approaching infinity, a shape factor called the geometry index G, which is obtained as follows:

$$G = 0.25 + \frac{3}{8B_1^2} + \frac{3}{8B_2^2} \quad (8)$$

Table 3 Expressions for Estimating f and j_c factors for the Thermal Center Temperature of Spheres

Biot Number Range	Equations for f and j factors
$Bi \leq 0.1$	$\dfrac{f\alpha}{L^2} = \dfrac{\ln 10}{3\,Bi}$ $j_c = 1.0$
$0.1 < Bi \leq 100$	$\dfrac{f\alpha}{L^2} = \dfrac{\ln 10}{w^2}$ $j_c = \dfrac{2(\sin w - w\cos w)}{w - \sin w \cos w}$ where $w = 1.573729 + 0.642906\ln(Bi)$ $\quad + 0.047859[\ln(Bi)]^2 - 0.03553[\ln(Bi)]^3$ $\quad - 0.004907[\ln(Bi)]^4 + 0.001563[\ln(Bi)]^5$
$Bi > 100$	$\dfrac{f\alpha}{L^2} = 0.2333$ $j_c = 2.0$

Source: Lacroix and Castaigne, 1987a

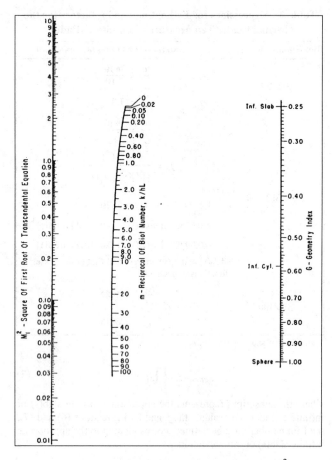

Fig. 6 Nomograph for Estimating Value of M_1^2 from Reciprocal of Biot Number and Smith's (1966) Geometry Index

where B_1 and B_2 are related to the cross sectional areas of the food:

$$B_1 = \frac{A_1}{\pi L^2} \quad B_2 = \frac{A_2}{\pi L^2} \tag{9}$$

where L is the shortest distance between the thermal center of the food and its surface, A_1 is the minimum cross sectional area containing L and A_2 is the cross sectional area that is orthogonal to A_1.

This geometry index is used in conjunction with the inverse of the Biot number m and a nomograph (shown in Figure 6) to obtain the characteristic value M_1^2. Smith et al. showed that the characteristic value M_1^2 can be related to the f factor by

$$f = \frac{2.303 L^2}{M_1^2 \alpha} \tag{10}$$

where α is the thermal diffusivity of the food. In addition, an expression for estimating a j_m factor which is used to determine the mass average temperature is given as

$$j_m = 0.892 e^{-0.0388 M_1^2} \tag{11}$$

As an alternative to estimating the value of M_1^2 from the nomograph developed by Smith et al. (1968), Hayakawa and Villalobos (1989) obtained regression formulae for estimating M_1^2. For Biot numbers approaching infinity, their regression formula is given as follows:

$$\begin{aligned}\ln(M_1^2) = &\ 2.2893825 + 0.35330539 X_g - 3.8044156 X_g^2 \\ &- 9.6821811 X_g^3 - 12.0321827 X_g^4 \\ &- 7.1542411 X_g^5 - 1.6301018 X_g^6\end{aligned} \tag{12}$$

where $X_g = \ln(G)$. For finite Biot numbers, Hayakawa and Villalobos (1989) gave the following:

$$\begin{aligned}\ln(M_1^2) = &\ 0.92083090 + 0.83409615 X_g - 0.78765739 X_b \\ &- 0.04821784 X_g X_b - 0.0408987 X_g^2 \\ &- 0.10045526 X_b^2 + 0.01521388 X_g^3 \\ &+ 0.00119941 X_g X_b^3 + 0.00129982 X_b^4\end{aligned} \tag{13}$$

where $X_g = \ln(G)$ and $X_b = \ln(1/Bi)$.

Cooling Time Estimation Methods Based on Equivalent Heat Transfer Dimensionality

The product geometry can also be considered using a shape factor called the **equivalent heat transfer dimensionality** (Cleland and Earle (1982a), which compares the total heat transfer to the heat transfer through the shortest dimension. Cleland and Earle developed an expression for estimating the equivalent heat transfer dimensionality of irregularly shaped food items as a function of Biot number. This overcomes the limitation of the geometry index developed by Smith et al. (1968), which was derived for the case of Biot number approaching infinity. However, the cooling time estimation method developed by Cleland and Earle, requires the use of a nomograph. Lin et al. (1993, 1996a, 1996b) expanded on the work of Cleland and Earle to eliminate the need for a nomograph.

In the method of Lin et al., the cooling time of a food or beverage is estimated by a first term approximation to the analytical solution for convective cooling of a sphere:

$$\theta = \frac{3\rho c L^2}{\omega^2 k E} \ln\left(\frac{j}{Y}\right) \tag{14}$$

where θ is the cooling time, ρ is the density of the food, c is the specific heat of the food, L is the radius or half-thickness of the food

Cooling and Freezing Times of Foods

item, k is the thermal conductivity of the food, j is the previously discussed lag factor, E is the equivalent heat transfer dimensionality and ω is the first root of the following transcendental function:

$$\omega \cot\omega + Bi - 1 = 0 \quad (15)$$

In Equation (14), the equivalent heat transfer dimensionality E is given as a function of Biot number:

$$E = \frac{Bi^{4/3} + 1.85}{\dfrac{Bi^{4/3}}{E_\infty} + \dfrac{1.85}{E_o}} \quad (16)$$

E_o and E_∞ are the equivalent heat transfer dimensionalities for the limiting cases of $Bi = 0$ and $Bi \to \infty$, respectively. The definitions of E_o and E_∞ make use of the dimensional ratios β_1 and β_2:

$$\beta_1 = \frac{\text{second shortest dimension of food item}}{\text{shortest dimension of food item}} \quad (17)$$

$$\beta_2 = \frac{\text{longest dimension of food item}}{\text{shortest dimension of food item}} \quad (18)$$

For two-dimensional, irregularly shaped food items, E_o, which is the equivalent heat transfer dimensionality for $Bi = 0$, is given by:

$$E_o = \left(1 + \frac{1}{\beta_1}\right)\left[1 + \left(\frac{\beta_1 - 1}{2\beta_1 + 2}\right)^2\right] \quad (19)$$

For three-dimensional, irregularly shaped food items, E_o is:

$$E_o = \frac{2[\beta_1 + \beta_2 + \beta_1^2(1+\beta_2) + \beta_2^2(1+\beta_1)]}{2\beta_1\beta_2(1+\beta_1+\beta_2)} - \frac{[(\beta_1-\beta_2)^2]^{0.4}}{15} \quad (20)$$

For finite cylinders, bricks and infinite rectangular rods, E_o may be determined as follows:

$$E_o = 1 + \frac{1}{\beta_1} + \frac{1}{\beta_2} \quad (21)$$

For spheres, infinite cylinders, and infinite slabs, $E_o = 3$, $E_o = 2$, and $E_o = 1$, respectively.

For both two-dimensional and three-dimensional food items, the general form for E_o at $Bi \to \infty$ is given as:

$$E_\infty = 0.75 + p_1 f(\beta_1) + p_2 f(\beta_2) \quad (22)$$

where

$$f(\beta) = \frac{1}{\beta^2} + 0.01 p_3 \exp\left[\beta - \frac{\beta^2}{6}\right] \quad (23)$$

with β_1 and β_2 as previously defined. The geometric parameters, p_1, p_2, and p_3, are given in Table 4 for various geometries.

Lin et al. (1993, 1996a, 1996b) also developed an expression for the lag factor applicable to the thermal center of a food item j_c as:

$$j_c = \frac{Bi^{1.35} + \dfrac{1}{\lambda}}{\dfrac{Bi^{1.35}}{L_\infty} + \dfrac{1}{\lambda}} \quad (24)$$

Table 4   Geometric Parameters

Shape	N	p_1	p_2	p_3	γ_1	γ_2	λ
Infinite slab ($\beta_1 = \beta_2 = \infty$)	1	0	0	0	∞	∞	1
Infinite rectangular rod ($\beta_1 \geq 1, \beta_2 = \infty$)	2	0.75	0	−1	$4\beta_1/\pi$	∞	γ_1
Brick ($\beta_1 \geq 1, \beta_2 \geq \beta_1$)	3	0.75	0.75	−1	$4\beta_1/\pi$	$1.5\beta_2$	γ_1
Infinite Cylinder ($\beta_1 = 1, \beta_2 = \infty$)	2	1.01	0	0	1	∞	1
Infinite ellipse ($\beta_1 > 1, \beta_2 = \infty$)	2	1.01	0	1	β_1	∞	γ_1
Squat cylinder ($\beta_1 = \beta_2, \beta_1 \geq 1$)	3	1.01	0.75	−1	$1.225\beta_1$	$1.225\beta_2$	γ_1
Short cylinder ($\beta_1 = 1, \beta_2 \geq 1$)	3	1.01	0.75	−1	β_1	$1.5\beta_2$	γ_1
Sphere ($\beta_1 = \beta_2 = 1$)	3	1.01	1.24	0	1	1	1
Ellipsoid ($\beta_1 \geq 1, \beta_2 \geq \beta_1$)	3	1.01	1.24	1	β_1	β_2	γ_1

Source: Lin et al. (1996b)

where L_∞ is as follows:

$$L_\infty = 1.271 + 0.305\exp(0.172\gamma_1 - 0.115\gamma_1^2) \\ + 0.425\exp(0.09\gamma_2 - 0.128\gamma_2^2) \quad (25)$$

and the geometric parameters λ, γ_1 and γ_2 are given in Table 4.

For the mass average temperature, Lin et al. gave the lag factor j_m as follows:

$$j_m = \mu j_c \quad (26)$$

where

$$\mu = \left[\frac{1.5 + 0.69 Bi}{1.5 + Bi}\right]^N \quad (27)$$

and N is the number of dimensions of a food item in which heat transfer is significant. The value of N for various geometries is also given in Table 4.

Algorithms for Estimating Cooling Time

The following suggested algorithm for estimating cooling time of foods and beverages is based on the equivalent heat transfer dimensionality method presented by Lin et al. (1993, 1996a, 1996b).

1. Determine thermal properties of the food item (see Chapter 8).
2. Determine surface heat transfer coefficient for the cooling process (see Chapter 8).
3. Determine characteristic dimension L and dimensional ratios β_1 and β_2 using Equations (17) and (18).
4. Calculate Biot number using Equation (2).
5. Calculate equivalent heat transfer dimensionality E for food geometry using Equation (16). This calculation requires evaluation of E_o and E_∞ using Equations (19) through (23).
6. Calculate lag factor corresponding to thermal center and/or mass average of food item using Equations (24) through (27).
7. Calculate root of transcendental equation given in Equation (15).
8. Calculate cooling time using Equation (14).

The following alternative algorithm for estimating the cooling time of foods and beverages is based on the use of f and j factors.

1. Determine thermal properties of food item (see Chapter 8).
2. Determine surface heat transfer coefficient for cooling process (see Chapter 8).
3. Determine characteristic dimension L of food item.
4. Calculate Biot number using Equation (2).
5. Calculate f and j factors by one of the following methods:
    (a) Method of Pflug et al. (1965): Figures 2 through 5.
    (b) Method of Lacroix and Castaigne (1987a): Tables 1, 2, and 3.
    (c) Method of Smith et al. (1968): Equations (8) through (11) and Figure 6.
    (d) Method of Hayakawa and Villalobos (1989): Equations (12) and (13) in conjunction with Equations (8) through (11).
6. Calculate cooling time using Equation (4).

SAMPLE PROBLEMS FOR ESTIMATING COOLING TIME

Example 1. A piece of ham, initially at 70°C, is to be cooled in an airblast freezer. The air temperature within the freezer is −1°C and the surface heat transfer coefficient is estimated to be 48.0 W/(m²·K). The overall dimension of the ham is 0.102 m by 0.165 m by 0.279 m. Estimate the time required for the mass average temperature of the ham to reach 10°C. The thermophysical properties for ham are given as follows:

c = 3.74 kJ/(kg·K)
k = 0.379 W/(m·K)
ρ = 1080 kg/m³

Solution: Use the algorithm based on the method of Lin et al. (1993, 1996a, 1996b).

Step 1: Determine the thermal properties (c, k, ρ) of the ham.
In this example the thermal properties of the ham are given above.

Step 2: Determine the heat transfer coefficient h.
The heat transfer coefficient is given as h = 48.0 W/(m²·K)

Step 3: Determine the characteristic dimension L and the dimensional ratios β_1 and β_2.
For cooling time problems, the characteristic dimension is the shortest distance from the thermal center of a food item to its surface. Assuming that the thermal center of the piece of ham coincides with its geometric center, the characteristic dimension becomes:

$$L = (0.102 \text{ m})/2 = 0.051 \text{ m}$$

The dimensional ratios then become [Equations (17) and (18)]:

$$\beta_1 = \frac{0.165 \text{ m}}{0.102 \text{ m}} = 1.62$$

$$\beta_2 = \frac{0.279 \text{ m}}{0.102 \text{ m}} = 2.74$$

Step 4: Calculate the Biot number.

$$Bi = hL/k = (48.0)(0.051)/0.379 = 6.46$$

Step 5: Calculate the heat transfer dimensionality.
Using Equation (20), E_o becomes:

$$E_o = \frac{2[1.62 + 2.74 + 1.62^2(1+2.74) + 2.74^2(1+1.62)]}{(2)(1.62)(2.74)(1+1.62+2.74)}$$
$$- \frac{[(1.62-2.74)^2]^{0.4}}{15} = 1.35$$

Assuming the ham to be ellipsoidal, the geometric factors can be obtained from Table 4:

p_1 = 1.01 p_2 = 1.24 p_3 = 1

From Equation (23):

$$f(\beta_1) = \frac{1}{1.62^2} + (0.01)(1)\exp\left[1.62 - \frac{1.62^2}{6}\right] = 0.414$$

$$f(\beta_2) = \frac{1}{2.74^2} + (0.01)(1)\exp\left[2.74 - \frac{2.74^2}{6}\right] = 0.178$$

From Equation (22):

$$E_\infty = 0.75 + (1.01)(0.414) + (1.24)(0.178) = 1.39$$

Thus, using Equation (16), the equivalent heat transfer dimensionality becomes:

$$E = \frac{6.46^{4/3} + 1.85}{\frac{6.46^{4/3}}{1.39} + \frac{1.85}{1.35}} = 1.38$$

Step 6: Calculate the lag factor applicable to the mass average temperature.
From Table 4, $\lambda = \beta_1$, $\gamma_1 = \beta_1$ and $\gamma_2 = \beta_2$. Using Equation (25), L_∞ becomes:

$$L_\infty = 1.271 + 0.305\exp[(0.172)(1.62) - (0.115)(1.62)^2]$$
$$+ 0.425\exp[(0.09)2.74 - (0.128)(2.74)^2] = 1.78$$

Using Equation (24), the lag factor applicable to the center temperature becomes:

$$j_c = \frac{6.46^{1.35} + \frac{1}{1.62}}{\frac{6.46^{1.35}}{1.78} + \frac{1}{1.62}} = 1.72$$

Using Equations (26) and (27), the lag factor for the mass average temperature becomes:

$$j_m = \left[\frac{1.5 + (0.69)(6.46)}{1.5 + 6.46}\right]^3 (1.72) = 0.721$$

Step 7: Find the root of transcendental Equation (15):

$$\omega\cot\omega + Bi - 1 = 0$$
$$\omega\cot\omega + 6.46 - 1 = 0$$
$$\omega = 2.68$$

Step 8: Calculate the cooling time.
The unaccomplished temperature difference is:

$$Y = \frac{T_m - T}{T_m - T_i} = \frac{-1-10}{-1-70} = 0.1549$$

Using Equation (14), the cooling time becomes

$$\theta = \frac{(3)(1.08 \times 10^6)(3.74)(0.051)^2}{(2.68)^2(0.379)(1.38)} \ln\left[\frac{0.721}{0.1549}\right]$$

$$\theta = 12900 \text{ s} = 3.58 \text{ h}$$

Example 2. Repeat the cooling time calculation of Example 1, but use the cooling time estimation algorithm based on the use of f and j factors.

Solution: Use the algorithm based on the method of Hayakawa and Villalobos (1989) for f and j factors.

Step 1: Determine the thermal properties of the ham.
The thermal properties of the ham are given in Example 1.

Step 2: Determine the heat transfer coefficient.
The heat transfer coefficient is given from Example 1 as

$$h = 48.0 \text{ W/(m}^2\cdot\text{K)}$$

Step 3: Determine the characteristic dimension L and the dimensional ratios β_1 and β_2.
From Example 1, L = 0.051 m, β_1 = 1.62, β_2 = 2.74

Cooling and Freezing Times of Foods

Step 4: Calculate the Biot number.

From Example 1, Bi = 6.46.

Step 5: Calculate the f and j factors using the method of Hayakawa and Villalobos (1989).

For simplicity, assume the cross-sections of the ham to be ellipsoidal in shape. The area of an ellipse is the product of π times the minor axis times the major axis, or:

$$A_1 = \pi L^2 \beta_1 \qquad A_2 = \pi(\beta_1 L)^2 \frac{\beta_2}{\beta_1}$$

Using Equations (8) and (9), calculate the geometry index G:

$$B_1 = \frac{A_1}{\pi L^2} = \frac{\pi L^2 \beta_1}{\pi L^2} = \beta_1 = 1.62$$

$$B_2 = \frac{A_2}{\pi L^2} = \frac{\pi(\beta_1 L)^2 \beta_2}{\pi L^2 \beta_1} = \beta_1 \beta_2 = (1.62)(2.74) = 4.44$$

$$G = 0.25 + \frac{3}{(8)(1.62)^2} + \frac{3}{(8)(4.44)^2} = 0.412$$

Using Equation (13), determine the characteristic value M_1^2:

$$X_g = \ln(G) = \ln(0.412) = -0.887$$

$$X_b = \ln(1/\text{Bi}) = \ln(1/6.46) = -1.87$$

$$\ln(M_1^2) = 0.92083090 + 0.83409615(-0.887) - 0.78765739(-1.87)$$
$$- 0.04821784(-0.887)(-1.87) - 0.0408987(-0.887)^2$$
$$- 0.10045526(-1.87)^2 + 0.01521388(-0.887)^3$$
$$+ 0.00119941(-0.887)(-1.87)^3$$
$$+ 0.00129982(-1.87)^4 = 1.20$$

$$M_1^2 = 3.32$$

From Equation (10), the f factor becomes:

$$f = \frac{2.303 L^2}{M_1^2 \alpha} = \frac{2.303 L^2 \rho c}{M_1^2 k}$$

$$f = \frac{(2.303)(0.051)^2(1.08 \times 10^6)(3.74)}{(3.32)(0.379)} = 19230 \text{ s} = 5.24 \text{ h}$$

From Equation (11), the j factor becomes:

$$j_m = 0.892 e^{(-0.0388)(3.32)} = 0.784$$

Step 6: Calculate the cooling time.

From Example 1, the unaccomplished temperature difference was found to be $Y = 0.1549$. Using Equation (5), the cooling time becomes:

$$\theta = -\frac{19230}{2.303} \ln\left(\frac{0.1549}{0.784}\right) = 13500 \text{ s} = 3.75 \text{ h}$$

FREEZING TIMES OF FOODS AND BEVERAGES

As discussed at the beginning of this chapter, the freezing of foods and beverages is not an isothermal process but rather occurs over a range of temperatures. In the following section, the basic freezing time estimation method developed by Plank is discussed first, followed by a discussion of those methods that modify Plank's equation. The discussion then focuses on those freezing time estimation methods in which the freezing time is calculated as the sum of the precooling, phase change, and subcooling times. The last section deals with freezing time estimation methods for irregularly shaped food items. These methods are divided into three subgroups: (1) equivalent heat transfer dimensionality, (2) mean conducting path, and (3) equivalent sphere diameter.

Plank's Equation

One of the most widely known simple methods for estimating freezing times of foods and beverages is that developed by Plank (1913, 1941). In this method, convective heat transfer is assumed to occur between the food item and the surrounding cooling medium. In addition, the temperature of the food item is assumed to be at its initial freezing temperature and that this temperature is constant throughout the freezing process. Furthermore, a constant thermal conductivity for the frozen region is assumed. Plank's freezing time estimation is as follows:

$$\theta = \frac{L_f}{T_f - T_m}\left[\frac{PD}{h} + \frac{RD^2}{k_s}\right] \qquad (28)$$

where L_f is the volumetric latent heat of fusion, T_f is the initial freezing temperature of the food, T_m is the freezing medium temperature, D is the thickness of the slab or the diameter of the sphere or infinite cylinder, h is the convective heat transfer coefficient, k_s is the thermal conductivity of the fully frozen food, and P and R are geometric factors. For the infinite slab, $P = 1/2$ and $R = 1/8$. For a sphere, $P = 1/6$ and $R = 1/24$; and for an infinite cylinder, $P = 1/4$ and $R = 1/16$.

Plank's geometric factors indicate that an infinite slab of thickness D, an infinite cylinder of diameter D and a sphere of diameter D, if exposed to the same conditions, would have freezing times in the ratio of 6:3:2. Hence, a cylinder freezes in half the time of a slab and a sphere freezes in one-third the time of a slab.

Modifications to Plank's Equation

Various researchers have noted that Plank's method does not accurately predict freezing times of foods and beverages. This is due, in part, to the fact that Plank's method assumes that freezing of foods takes place at a constant temperature, and not over a range of temperatures as is the case in actual food freezing processes. In addition, the thermal conductivity of the frozen food is assumed to be constant; but in reality, the thermal conductivity varies greatly during freezing. Another limitation of Plank's equation is that it neglects the removal of sensible heat above the freezing point, and thus, precooling times must be calculated by using one of the methods discussed in the Cooling Times of Foods and Beverages section of this chapter. Furthermore, Plank's method only applies to infinite slabs, infinite cylinders, and spheres. Subsequently, researchers have developed improved semi-analytical/empirical cooling and freezing time estimation methods that account for precooling and subcooling times, non-constant thermal properties, irregular geometries, and phase change over a range of temperatures.

Cleland and Earle (1977, 1979a,b) incorporated corrections to account for the removal of sensible heat both above and below the initial freezing point of the food as well as temperature variation during freezing. Regression equations were developed to estimate the geometric parameters P and R for infinite slabs, infinite cylinders, spheres, and rectangular bricks. In these regression equations, the effects of surface heat transfer, precooling, and final subcooling are accounted for by means of the Biot number, the Plank number, and the Stefan number, respectively.

In this section the Biot number is defined as

$$\text{Bi} = \frac{hD}{k} \qquad (29)$$

where h is the convective heat transfer coefficient, D is the characteristic dimension, and k is the thermal conductivity. The characteristic dimension D is defined to be twice the shortest distance from the thermal center of a food item to its surface. For an infinite slab, D is the thickness. For an infinite cylinder or a sphere, D is the diameter.

In general, the Plank number is defined as follows:

$$\text{Pk} = \frac{C_l(T_i - T_f)}{\Delta H} \quad (30)$$

where C_l is the volumetric specific heat of the unfrozen phase and ΔH is the volumetric enthalpy change of the food between T_f and the final food temperature. The Stefan number is similarly defined as

$$\text{Ste} = \frac{C_s(T_f - T_m)}{\Delta H} \quad (31)$$

where C_s is the volumetric specific heat of the frozen phase.

In Cleland and Earle's method, Plank's original geometric factors P and R are replaced with the modified values given in Table 5, and the latent heat L_f is replaced with the volumetric enthalpy change of the food ΔH_{10} between the freezing temperature T_f and the final center temperature, assumed to be $-10°C$. As shown in Table 5, the geometric factors P and R are functions of the Plank number and the Stefan number. Both of these parameters should be evaluated using the enthalpy change ΔH_{10}. Thus, the modified Plank equation takes the form:

$$\theta = \frac{\Delta H_{10}}{T_f - T_m}\left[\frac{PD}{h} + \frac{RD^2}{k_s}\right] \quad (32)$$

where k_s is the thermal conductivity of the fully frozen food.

Equation (32) is based on curve-fitting of experimental data in which the product final center temperature was $-10°C$. Cleland and Earle (1984) noted that this prediction formula does not perform as well when applied to situations with final center temperatures other than $-10°C$. Cleland and Earle proposed the following modified form of Equation (32) to account for different final center temperatures:

$$\theta = \frac{\Delta H_{10}}{T_f - T_m}\left(\frac{PD}{h} + \frac{RD^2}{k_s}\right)\left[1 - \frac{1.65\,\text{Ste}}{k_s}\ln\left(\frac{T_c - T_m}{T_{ref} - T_m}\right)\right] \quad (33)$$

where T_{ref} is $-10°C$, T_c is the final product center temperature, and ΔH_{10} is the volumetric enthalpy difference between the initial freezing temperature T_f and $-10°C$. The values of P, R, Pk, and Ste should be evaluated using ΔH_{10}, as previously discussed.

Hung and Thompson (1983) also improved upon Plank's equation to develop an alternative freezing time estimation method for infinite slabs. Their equation incorporates the volumetric change in enthalpy ΔH_{18} for the freezing process as well as a weighted average temperature difference between the initial temperature of the food and the freezing medium temperature. This weighted average temperature difference ΔT is given as follows:

$$\Delta T = (T_f - T_m) + \frac{(T_i - T_f)^2 \frac{C_l}{2} - (T_f - T_c)^2 \frac{C_s}{2}}{\Delta H_{18}} \quad (34)$$

where T_c is the final center temperature of the food and ΔH_{18} is the enthalpy change of the food between the initial temperature and final center temperature, assumed to be $-18°C$. Empirical equations were developed to estimate P and R for infinite slabs as follows:

$$P = 0.7306 - 1.083\,\text{Pk} + \text{Ste}\left(15.40U - 15.43 + 0.01329\frac{\text{Ste}}{\text{Bi}}\right) \quad (35)$$

$$R = 0.2079 - 0.2656U(\text{Ste}) \quad (36)$$

where $U = \Delta T/(T_f - T_m)$. In these expressions Pk and Ste should be evaluated using the enthalpy change ΔH_{18}. The freezing time prediction model is:

$$\theta = \frac{\Delta H_{18}}{\Delta T}\left[\frac{PD}{h} + \frac{RD^2}{k_s}\right] \quad (37)$$

Cleland and Earle (1984) applied a correction factor to the Hung and Thompson model [Equation (37)] and improved the prediction accuracy of the model for final temperatures other than $-18°C$. The correction to Equation (37) is as follows:

$$\theta = \frac{\Delta H_{18}}{\Delta T}\left(\frac{PD}{h} + \frac{RD^2}{k_s}\right)\left[1 - \frac{1.65\,\text{Ste}}{k_s}\ln\left(\frac{T_c - T_m}{T_{ref} - T_m}\right)\right] \quad (38)$$

where T_{ref} is $-18°C$, T_c is the product final center temperature and ΔH_{18} is the volumetric enthalpy change between the initial temperature T_i and $-18°C$. The weighted average temperature difference ΔT, Pk, and Ste should be evaluated using ΔH_{18}.

Precooling, Phase Change, and Subcooling Time Calculations

Total freezing time θ is as follows:

$$\theta = \theta_1 + \theta_2 + \theta_3 \quad (39)$$

where θ_1, θ_2, and θ_3 are the precooling, phase change, and subcooling times, respectively.

DeMichelis and Calvelo (1983) suggested that the equivalent heat transfer dimensionality method of Cleland and Earle (1982), discussed in the Cooling Times of Foods and Beverages section of this chapter, be used to estimate the precooling and subcooling times of foods and beverages. They also suggested that the phase change time calculation be made with Plank's equation. However, they suggested that in Plank's equation, the thermal conductivity of the frozen food be evaluated at the temperature, $(T_f + T_m)/2$, where T_f is the initial freezing temperature of the food and T_m is the temperature of the cooling medium. The use of Plank's equation limits the applicability of this method to infinite slabs, infinite cylinders and spheres.

Lacroix and Castaigne (1987a, 1987b, 1988) suggested the use of f and j factors to determine precooling and subcooling times of foods and beverages. They presented equations, given in Tables 1, 2, and 3, for estimating the values of f and j for infinite slabs, infinite cylinders, and spheres. Note that Lacroix and Castaigne based the Biot number on the shortest distance between the thermal center of the food item and its surface—not twice that distance.

Lacroix and Castaigne (1987a, 1987b, 1988) gave the following expression for estimating the precooling time θ_1:

$$\theta_1 = f_1 \log\left[j_1\frac{T_m - T_i}{T_m - T_f}\right] \quad (40)$$

where T_m is the coolant temperature, T_i is the initial temperature of the food, and T_f is the initial freezing point of the food. The f_1 and j_1 factors are determined from a Biot number that is calculated using an average thermal conductivity. This average thermal conductivity is based on the thermal conductivity of the unfrozen food, and the thermal conductivity of the frozen food evaluated at $(T_f + T_m)/2$.

The expression for estimating the subcooling time θ_3 is:

$$\theta_3 = f_3 \log\left[j_3 \frac{T_m - T_f}{T_m - T_c}\right] \quad (41)$$

where T_c is the final temperature at the center of the food item. The f_3 and j_3 factors are determined from a Biot number that is calculated using the thermal conductivity of the frozen food evaluated at the temperature $(T_f + T_m)/2$.

Lacroix and Castaigne model the phase change time θ_2 with Plank's equation:

$$\theta_2 = \frac{L_f D^2}{(T_f - T_m)k_c}\left[\frac{P}{2\text{Bi}_c} + R\right] \quad (42)$$

where L_f is the volumetric latent heat of fusion of the food, P and R are the original Plank geometric shape factors, k_c is the thermal conductivity of the frozen food at $(T_f + T_m)/2$, and Bi_c is the Biot number for the subcooling period ($\text{Bi}_c = hL/k_c$).

Lacroix and Castaigne (1987a, 1987b) adjusted P and R to obtain better agreement between predicted freezing times and experimental data. Using regression analysis, Lacroix and Castaigne suggested the following geometric factors:

For **infinite slabs**:

$$P = 0.51233 \quad (43)$$

$$R = 0.15396 \quad (44)$$

For **infinite cylinders**:

$$P = 0.27553 \quad (45)$$

$$R = 0.07212 \quad (46)$$

For **spheres**:

$$P = 0.19665 \quad (47)$$

$$R = 0.03939 \quad (48)$$

For **rectangular bricks**:

$$P = P'\left[-0.02175\frac{1}{\text{Bi}_c} - 0.01956\frac{1}{\text{Ste}} - 1.69657\right] \quad (49)$$

$$R = R'\left[5.57519\frac{1}{\text{Bi}_c} + 0.02932\frac{1}{\text{Ste}} + 1.58247\right] \quad (50)$$

For rectangular bricks, the values for P' and R' are calculated using the expressions given in Table 5 for the P and R of bricks.

Pham (1984) also devised a freezing time estimation method, similar to Plank's equation, in which sensible heat effects were considered by calculating precooling, phase change, and subcooling times separately. In addition, Pham suggested the use of a mean freezing point, which is assumed to be 1.5 K below the initial freezing point of the food, to account for freezing that takes place over a range of temperatures. Pham's freezing time estimation method is stated in terms of the volume and surface area of the food item and is, therefore, applicable to food items of any shape. This method is given as:

$$\theta_i = \frac{Q_i}{hA_s \Delta T_{mi}}\left(1 + \frac{\text{Bi}_i}{k_i}\right) \quad i = 1, 2, 3 \quad (51)$$

where θ_1 is the precooling time, θ_2 is the phase change time, θ_3 is the subcooling time, and the remaining variables are defined as shown in Table 6.

Pham (1986a) significantly simplified the previous freezing time estimation method to yield:

$$\theta = \frac{V}{hA_s}\left(\frac{\Delta H_1}{\Delta T_1} + \frac{\Delta H_2}{\Delta T_2}\right)\left(1 + \frac{\text{Bi}_s}{4}\right) \quad (52)$$

in which

$$\Delta H_1 = C_l(T_i - T_{fm})$$
$$\Delta H_2 = L_f + C_s(T_{fm} - T_c) \quad (53)$$

$$\Delta T_1 = \frac{T_i + T_{fm}}{2} - T_m$$
$$\Delta T_2 = T_{fm} - T_m \quad (54)$$

where C_l and C_s are volumetric specific heats above and below freezing, respectively, T_i is the initial food temperature, L_f is the volumetric latent heat of freezing, and V is the volume of the food item.

Pham suggested that the mean freezing temperature T_{fm} used in Equations (53) and (54) mainly depended on the cooling medium temperature T_m and the product center temperature T_c. By curve fitting to existing experimental data, Pham (1986a) proposed the following equation to determine the mean freezing temperature for use in Equations (53) and (54):

$$T_{fm} = 1.8 + 0.26T_c + 0.105T_m \quad (55)$$

where all temperatures are in °C.

Pham (1986b) subsequently extended the applicability of the simplified freezing time estimation model by providing provisions for variations in environmental conditions. Equations were presented to calculate the location of the freezing front for cases when step changes in boundary conditions exist. In addition, Pham (1987) developed equations to predict the location of the freezing front for the case of asymmetric heat transfer coefficients and asymmetric cooling medium temperatures.

Geometric Considerations

Equivalent Heat Transfer Dimensionality. Similar to their work involving cooling times of foods, Cleland and Earle (1982b) also introduced a geometric correction factor, called the **equivalent heat transfer dimensionality** E to calculate the freezing times of irregularly shaped food items. The freezing time of an irregularly shaped object θ_{shape}, was related to the freezing time of an infinite slab θ_{slab} via the equivalent heat transfer dimensionality as:

$$\theta_{shape} = \theta_{slab}/E \quad (56)$$

The freezing time of the infinite slab is then calculated from one of the many suitable freezing time estimation methods available for infinite slabs.

Using data collected from a large number of freezing experiments, Cleland and Earle (1982) developed empirical correlations for the equivalent heat transfer dimensionality applicable to rectangular bricks and finite cylinders. For rectangular brick shapes with dimensions D by $\beta_1 D$ by $\beta_2 D$, the equivalent heat transfer dimensionality was given as follows:

Table 5 Expressions for P and R

Shape	P and R Expressions
Infinite slab	$P = 0.5072 + 0.2018\,\text{Pk} + \text{Ste}\left[0.3224\,\text{Pk} + \dfrac{0.0105}{\text{Bi}} + 0.0681\right]$ $R = 0.1684 + \text{Ste}(0.2740\,\text{Pk} - 0.0135)$
Infinite cylinder	$P = 0.3751 + 0.0999\,\text{Pk} + \text{Ste}\left[0.4008\,\text{Pk} + \dfrac{0.0710}{\text{Bi}} - 0.5865\right]$ $R = 0.0133 + \text{Ste}(0.0415\,\text{Pk} + 0.3957)$
Sphere	$P = 0.1084 + 0.0924\,\text{Pk} + \text{Ste}\left[0.231\,\text{Pk} - \dfrac{0.3114}{\text{Bi}} + 0.6739\right]$ $R = 0.0784 + \text{Ste}(0.0386\,\text{Pk} - 0.1694)$
Brick	$P = P_2 + P_1[0.1136 + \text{Ste}(5.766 P_1 - 1.242)]$ $R = R_2 + R_1[0.7344 + \text{Ste}(49.89 R_1 - 2.900)]$ where $P_2 = P_1\left[1.026 + 0.5808\,\text{Pk} + \text{Ste}\left(0.2296\,\text{Pk} + \dfrac{0.0182}{\text{Bi}} + 0.1050\right)\right]$ $R_2 = R_1[1.202 + \text{Ste}(3.410\,\text{Pk} + 0.7336)]$ and $P_1 = \dfrac{\beta_1 \beta_2}{2(\beta_1\beta_2 + \beta_1 + \beta_2)}$ $R_1 = \dfrac{Q}{2}\left[(r-1)(\beta_1 - r)(\beta_2 - r)\ln\left(\dfrac{r}{r-1}\right) - (s-1)(\beta_1 - s)(\beta_2 - s)\ln\left(\dfrac{s}{s-1}\right)\right] + \dfrac{1}{72}(2\beta_1 + 2\beta_2 - 1)$ in which $\dfrac{1}{Q} = 4[(\beta_1 - \beta_2)(\beta_1 - 1) + (\beta_2 - 1)^2]^{1/2}$ $r = \dfrac{1}{3}\{\beta_1 + \beta_2 + 1 + [(\beta_1 - \beta_2)(\beta_1 - 1) + (\beta_2 - 1)^2]^{1/2}\}$ $s = \dfrac{1}{3}\{\beta_1 + \beta_2 + 1 - [(\beta_1 - \beta_2)(\beta_1 - 1) + (\beta_2 - 1)^2]^{1/2}\}$

Cleland and Earle (1977, 1979a, 1979b)

$$E = 1 + W_1 + W_2 \qquad (57)$$

where

$$W_1 = \left(\frac{\text{Bi}}{\text{Bi}+2}\right)\frac{5}{8\beta_1^3} + \left(\frac{2}{\text{Bi}+2}\right)\frac{2}{\beta_1(\beta_1 + 1)} \qquad (58)$$

and

$$W_2 = \left(\frac{\text{Bi}}{\text{Bi}+2}\right)\frac{5}{8\beta_2^3} + \left(\frac{2}{\text{Bi}+2}\right)\frac{2}{\beta_2(\beta_2 + 1)} \qquad (59)$$

For the case of finite cylinders where the diameter is smaller than the height, the equivalent heat transfer dimensionality was given as:

$$E = 2.0 + W_2 \qquad (60)$$

In addition, Cleland et al. (1987a, 1987b) developed expressions for determining the equivalent heat transfer dimensionality of infinite slabs, infinite and finite cylinders, rectangular bricks, spheres, and two- and three-dimensional irregular shapes. Numerical methods were used to calculate the freezing or thawing times for these various shapes. A non-linear regression analysis of the resulting numerical data yielded the following form for the equivalent heat transfer dimensionality:

$$E = G_1 + G_2 E_1 + G_3 E_2 \qquad (61)$$

where

$$E_1 = X(2.32/\beta_1^{1.77})\frac{1}{\beta_1} + [1 - X(2.32/\beta_1^{1.77})]\frac{0.73}{\beta_1^{2.50}} \qquad (62)$$

$$E_2 = X(2.32/\beta_2^{1.77})\frac{1}{\beta_2} + [1 - X(2.32/\beta_2^{1.77})]\frac{0.73}{\beta_2^{2.50}} \qquad (63)$$

$$X(x) = x/(\text{Bi}^{1.34} + x) \qquad (64)$$

where the geometric constants G_1, G_2, and G_3 are given in Table 7.

Using the freezing time prediction methods for infinite slabs and various multi-dimensional shapes developed by McNabb et al. (1990), Hossain et al. (1992a) derived infinite series expressions for E of infinite rectangular rods, finite cylinders, and rectangular bricks. For most practical freezing situations, only the first term of

Cooling and Freezing Times of Foods

Table 6 Definition of Variables for the Freezing Time Estimation Method

Process	Variables
Precooling	$i = 1$ $k_1 = 6$ $Q_1 = C_l(T_i - T_{fm})V$ $Bi_1 = (Bi_l + Bi_s)/2$ $\Delta T_{m1} = \dfrac{(T_i - T_m) - (T_{fm} - T_m)}{\ln\left[\dfrac{T_i - T_m}{T_{fm} - T_m}\right]}$
Phase change	$i = 2$ $k_2 = 4$ $Q_3 = C_s(T_{fm} - T_c)V$ $Q_2 = L_f V$ $Bi_2 = Bi_s$ $\Delta T_{m2} = T_{fm} - T_m$
Subcooling	$i = 3$ $k_3 = 6$ $Bi_3 = Bi_s$ $\Delta T_{m3} = \dfrac{(T_{fm} - T_m) - (T_o - T_m)}{\ln\left[\dfrac{T_{fm} - T_m}{T_o - T_m}\right]}$

Source: Pham (1984)

Notes:
A_s = area through which heat is transferred
Bi_l = Biot number for unfrozen phase
Bi_s = Biot number for frozen phase
Q_1, Q_2, Q_3 = heats of precooling, phase change, and subcooling, respectively
$\Delta T_{m1}, \Delta T_{m2}, \Delta T_{m3}$ = corresponding log-mean temperature driving forces
T_c = final thermal center temperature
T_{fm} = mean freezing point, assumed 1.5 K below initial freezing point
T_o = mean final temperature
V = volume of food item

Table 7 Geometric Constants

Shape	G_1	G_2	G_3
Infinite slab	1	0	0
Infinite cylinder	2	0	0
Sphere	3	0	0
Squat cylinder	1	2	0
Short cylinder	2	0	1
Infinite rod	1	1	0
Rectangular brick	1	1	1
Two-dimensional irregular shape	1	1	0
Three-dimensional irregular shape	1	1	1

Source: Cleland et al. (1987a)

these series expressions is significant. The resulting expressions for E are given in Table 8.

Hossain et al. (1992b) also presented a semi-analytically derived expression for the equivalent heat transfer dimensionality of two-dimensional, irregularly shaped food items. An equivalent "pseudo-elliptical" infinite cylinder was used to replace the actual two-dimensional, irregular shape in the calculations. A pseudo-ellipse is a shape that depends on the Biot number. As the Biot number approaches infinity, the shape closely resembles an ellipse. As the Biot number approaches zero, the pseudo-elliptical infinite cylinder approaches an infinite rectangular rod. Hossain et al. (1992b) stated that for practical Biot numbers, the pseudo-ellipse is very similar to a true ellipse. This model pseudo-elliptical infinite cylinder has the same volume per unit length and characteristic dimension as the actual food item. The resulting expression for E is given as follows:

$$E = 1 + \dfrac{1 + \dfrac{2}{Bi}}{\beta^2 + \dfrac{2\beta}{Bi}} \qquad (65)$$

In Equation (65), the Biot number is based on the shortest distance from the thermal center to the surface of the food item; not twice that distance. Using this expression for E, the freezing time of two-dimensional, irregularly shaped food items θ_{shape} can be calculated via Equation (56).

Hossain et al. (1992c) extended this analysis to the prediction of freezing times of three-dimensional, irregularly shaped food items. In this work, the irregularly shaped food item was replaced with a model ellipsoid shape having the same volume, characteristic dimension and smallest cross sectional area orthogonal to the characteristic dimension, as the actual food item. An expression was presented for E of a pseudo-ellipsoid as follows:

$$E = 1 + \dfrac{1 + \dfrac{2}{Bi}}{\beta_1^2 + \dfrac{2\beta_1}{Bi}} + \dfrac{1 + \dfrac{2}{Bi}}{\beta_2^2 + \dfrac{2\beta_2}{Bi}} \qquad (66)$$

In Equation (66), the Biot number is based upon the shortest distance from the thermal center to the surface of the food item; not twice that distance. With this expression for E, the freezing time of three-dimensional, irregularly shaped food items θ_{shape} may be calculated using Equation (56).

The method of Lin et al. (1996a, 1996b), which was discussed in the Cooling Times of Foods and Beverages section, Equations (16) through (23) and Table 4, may also be used to determine equivalent heat transfer dimensionality for freezing time calculations. It applies to all the geometric shapes given in Table 4.

Table 9 summarizes the numerous methods that have been discussed for determining the equivalent heat transfer dimensionality of various geometries. These methods can be used in conjunction with Equation (14) to calculate cooling times or in conjunction with Equation (56) to calculate freezing times.

Mean Conducting Path. Pham's freezing time formulas, given in Equations (51) and (52), require knowledge of the Biot number. To calculate the Biot number of a food, its characteristic dimension must be known. Because it is difficult to determine the characteristic dimension of an irregularly shaped food, Pham (1985) introduced the concept of the **mean conducting path**. The mean conducting path is the mean heat transfer length from the surface of the food item to its thermal center, or $D_m/2$. Thus, the Biot number becomes:

$$Bi = \dfrac{hD_m}{k} \qquad (67)$$

where D_m is twice the mean conducting path.

For rectangular blocks of food, Pham (1985) found that the mean conducting path was proportional to the geometric mean of the block's two shorter dimensions. Based on this result, Pham (1985) presented an equation to calculate the Biot number for rectangular blocks of food:

$$\dfrac{Bi}{Bi_o} = 1 + \left\{[1.5\sqrt{\beta_1} - 1]^{-4} + \left[\left(\dfrac{1}{\beta_1} + \dfrac{1}{\beta_2}\right)\left(1 + \dfrac{4}{Bi_o}\right)\right]^{-4}\right\}^{-0.25} \qquad (68)$$

where Bi_o is the Biot number based on the shortest dimension of the block D_1, or $Bi_o = hD_1/k$. The Biot number calculated with Equation (68) can then be substituted into a freezing time estimation method to calculate the freezing time for rectangular blocks.

Table 8 Expressions for Equivalent Heat Transfer Dimensionality

Shape	Expressions for Equivalent Heat Transfer Dimensionality, E
Infinite rectangular rod ($2L$ by $2\beta_1 L$)	$E = \left(1 + \dfrac{2}{\text{Bi}}\right) \left\{ \left(1 + \dfrac{2}{\text{Bi}}\right) - 4 \sum_{n=1}^{\infty} \left[\dfrac{(\sin z_n)}{z_n^3 \left(1 + \dfrac{\sin^2 z_a}{\text{Bi}}\right) \left(\dfrac{z_n}{\text{Bi}} \sinh(z_n \beta_1) + \cosh(z_n \beta_1)\right)} \right]^{-1} \right\}$ where z_n are roots of $\text{Bi} = z_n \tan(z_n)$ and $\text{Bi} = hL/k$ where L is the shortest distance from the center of the rectangular rod to the surface
Finite cylinder, height exceeds diameter (radius L and height $2\beta_1 L$)	$E = \left(2 + \dfrac{4}{\text{Bi}}\right) \left\{ \left(1 + \dfrac{2}{\text{Bi}}\right) - 8 \sum_{n=1}^{\infty} \left[y_n^3 J_1(y_n) \left(1 + \dfrac{y_n^2}{\text{Bi}^2}\right) \left(\cosh(\beta_1 y_n) + \dfrac{y_n}{\text{Bi}} \sinh(\beta_1 y_n)\right) \right]^{-1} \right\}^{-1}$ where y_n are roots of $y_n J_1(y_n) - \text{Bi} J_0(y_n) = 0$; J_0 and J_1 are Bessel functions of the first kind, order zero and one, respectively; and $\text{Bi} = hL/k$ where L is the radius of the cylinder.
Finite cylinder, diameter exceeds height (radius $\beta_1 L$ and height $2L$)	$E = \left(1 + \dfrac{2}{\text{Bi}}\right) \left\{ \left(1 + \dfrac{2}{\text{Bi}}\right) - 4 \sum_{n=1}^{\infty} \dfrac{\sin z_n}{z_n^2 (z_n + \cos z_n \sin z_n) \left(I_0(z_n \beta_1) + \dfrac{z_n}{\text{Bi}} I_1(z_n \beta_1)\right)} \right\}^{-1}$ where z_n are roots of $\text{Bi} = z_n \tan(z_n)$; I_0 and I_1 are Bessel function of the second kind, order zero and one, respectively; and $\text{Bi} = hL/k$ where L is the radius of the cylinder.
Rectangular Brick ($2L$ by $2\beta_1 L$ by $2\beta_2 L$)	$E = \left(1 + \dfrac{2}{\text{Bi}}\right) \Bigg\{ \left(1 + \dfrac{2}{\text{Bi}}\right) - 4 \sum_{n=1}^{\infty} \left[\dfrac{\sin z_n}{z_n^3 \left(1 + \dfrac{\sin^2 z_n}{\text{Bi}}\right) \left[\dfrac{z_n}{\text{Bi}} \sinh(z_n \beta_1) + \cosh(z_n \beta_1)\right]} \right]$ $- 8 \beta_2^2 \sum_{n=1}^{\infty} \sum_{m=1}^{\infty} \left[\sin z_n \sin z_m \left(\cosh(z_{nm}) + \dfrac{z_{nm}}{\text{Bi} \beta_2} \sinh(z_{nm}) \right) \right.$ $\left. z_n z_m z_{nm}^2 \left(1 + \dfrac{1}{\text{Bi}} \sin^2 z_n\right)\left(1 + \dfrac{1}{\text{Bi}\beta_1} \sin^2 z_m\right) \right]^{-1} \Bigg\}^{-1}$ where z_n are roots of $\text{Bi} = z_n \tan(z_n)$, z_m are the roots of $\text{Bi}\beta_1 = z_m \tan(z_m)$, $\text{Bi} = hL/k$ where L is the shortest distance from the thermal center of the rectangular brick to the surface, and z_{nm} is given as: $z_{nm}^2 = z_n^2 \beta_2^2 + z_m^2 \left(\dfrac{\beta_2}{\beta_1}\right)^2$

Source: Hossain et al. (1992a)

Table 9 Summary of Methods for Determining Equivalent Heat Transfer Dimensionality

Shape				
Slab		Cleland et al. (1987a, 1987b) Equations (61) – (64)		Lin et al. (1996a, 1996b) Equations (16) – (23)
Infinite Cylinder		Cleland et al. (1987a, 1987b) Equations (61) – (64)		Lin et al. (1996a, 1996b) Equations (16) – (23)
Sphere		Cleland et al. (1987a, 1987b) Equations (61) – (64)		Lin et al. (1996a, 1996b) Equations (16) – (23)
Squat cylinder		Cleland et al. (1987a, 1987b) Equations (61) – (64)	Hossain et al. (1992a) Table 8	Lin et al. (1996a, 1996b) Equations (16) – (23)
Short cylinder	Cleland and Earle (1982) Equations (59) and (60)	Cleland et al. (1987a, 1987b) Equations (61) – (64)	Hossain et al. (1992a) Table 8	Lin et al. (1996a, 1996b) Equations (16) – (23)
Infinite rod		Cleland et al. (1987a, 1987b) Equations (61) – (64)	Hossain et al. (1992a) Table 8	Lin et al. (1996a, 1996b) Equations (16) – (23)
Rectangular brick	Cleland and Earle (1982) Equations (57), (58), and (59)	Cleland et al. (1987a, 1987b) Equations (61) – (64)	Hossain et al. (1992a) Table 8	Lin et al. (1996a, 1996b) Equations (16) – (23)
2-D Irregular shape (infinite ellipse)		Cleland et al. (1987a, 1987b) Equations (61) – (64)	Hossain et al. (1992b) Equation (65)	Lin et al. (1996a, 1996b) Equations (16) – (23)
3-D Irregular shape (ellipsoid)		Cleland et al. (1987a, 1987b) Equations (61) – (64)	Hossain et al. (1992b) Equation (66)	Lin et al. (1996a, 1996b) Equations (16) – (23)

Cooling and Freezing Times of Foods

Pham (1985) noted that for squat shaped food items the mean conducting path $D_m/2$ could be reasonably estimated as the arithmetic mean of the longest and shortest distances from the surface of the food item to its thermal center.

Equivalent Sphere Diameter. Ilicali and Hocalar (1990) and Ilicali and Engez (1990) introduced the **equivalent sphere diameter** concept to calculate the freezing time of irregularly shaped food items. In this method, a sphere diameter is calculated that is based on the volume and the volume to surface area ratio of the irregularly shaped food. This equivalent sphere is then used to calculate the freezing time of the food item.

Considering an irregularly shaped food item where the shortest and longest distances from the surface to the thermal center were designated as D_1 and D_2, respectively, Ilicali and Hocalar (1990) and Ilicali and Engez (1990) defined the volume-surface diameter D_{vs} as the diameter of a sphere having the same volume to surface area ratio as the irregular shape:

$$D_{vs} = 6V/A_s \qquad (69)$$

where V is the volume of the irregular shape and A_s is the surface area of the irregular shape. In addition, the volume diameter D_v is defined as the diameter of a sphere having the same volume as the irregular shape:

$$D_v = (6V/\pi)^{1/3} \qquad (70)$$

Because a sphere is the solid geometry which has minimum surface area per unit volume, the equivalent sphere diameter $D_{eq,s}$ must be greater than D_{vs} and smaller than D_v. In addition, the contribution of the volume diameter D_v has to decrease as the ratio of the longest to the shortest dimensions D_2/D_1 increases, because the object will be essentially two dimensional if $D_2/D_1 \gg 1$. Therefore, the equivalent sphere diameter $D_{eq,s}$ is defined as follows:

$$D_{eq,s} = \frac{1}{\beta_2 + 1} D_v + \frac{\beta_2}{\beta_2 + 1} D_{vs} \qquad (71)$$

Thus, the prediction of the freezing time of the irregularly shaped food item is reduced to predicting the freezing time of a spherical food item with diameter $D_{eq,s}$. Any of the previously discussed freezing time methods for spheres may then be used to calculate this freezing time.

Algorithms for Freezing Time Estimation

The following suggested algorithm for estimating the freezing time of foods and beverages is based on the modified Plank equation presented by Cleland and Earle (1977, 1979a, 1979b). This algorithm is applicable to simple food geometries, including infinite slabs, infinite cylinders, spheres, and three-dimensional rectangular bricks.

1. Determine thermal properties of food item (see Chapter 8).
2. Determine surface heat transfer coefficient for the freezing process (see Chapter 8).
3. Determine characteristic dimension D and dimensional ratios β_1 and β_2 using Equations (17) and (18).
4. Calculate Biot number, Plank number, and Stefan number using Equations (29), (30), and (31), respectively.
5. Determine geometric parameters P and R given in Table 5.
6. Calculate freezing time using Equation (32) or Equation (33) depending on the final temperature of the frozen food.

The following algorithm for estimating the freezing time of foods and beverages is based on the method of equivalent heat transfer dimensionality. This algorithm is applicable to many food geometries, including infinite rectangular rods, finite cylinders, three-dimensional rectangular bricks, and two- and three-dimensional irregular shapes.

1. Determine thermal properties of the food item (see Chapter 8).
2. Determine surface heat transfer coefficient for the freezing process (see Chapter 8).
3. Determine characteristic dimension D and dimensional ratios β_1 and β_2 using Equations (17) and (18).
4. Calculate Biot number, Plank number, and Stefan number using Equations (29), (30), and (31), respectively.
5. Calculate freezing time of an infinite slab using a suitable method. Suitable methods include:
   (a) Equation (32) or (33) in conjunction with the geometric parameters P and R given in Table 5.
   (b) Equation (37) or (38) in conjunction with Equations (34), (35), and (36).
6. Calculate equivalent heat transfer dimensionality for the food item. Refer to Table 9 to determine which equivalent heat transfer dimensionality method is applicable to the particular food geometry.
7. Calculate the freezing time of the food item using Equation (56).

SAMPLE PROBLEMS FOR ESTIMATING FREEZING TIME

Example 3. A rectangular brick shaped package of beef (lean sirloin) with dimension 0.04 m by 0.12 m by 0.16 m is to be frozen in an air blast freezer. The initial temperature of the beef is 10°C and the freezer air temperature is −30°C. It is estimated that the surface heat transfer coefficient is 40 W/(m²·K). Calculate the time required for the thermal center of the beef to reach a temperature of −10°C.

Solution: Use the algorithm based on the modified Plank equation by Cleland and Earle (1977, 1979a, 1979b).

Step 1: Determine the thermal properties of lean sirloin.

Using the methods described in Chapter 8, the thermal properties can be calculated as follows:

Property	At −40°C (Fully Frozen)	At −10°C (Final Temp.)	At −1.7°C (Initial Freezing Point)	At 10°C (Initial Temp.)
Density, kg/m³	$\rho_s = 1018$	$\rho_s = 1018$	$\rho_l = 1075$	$\rho_l = 1075$
Enthalpy, kJ/kg	—	$H_s = 83.4$	$H_l = 274.2$	—
Specific heat, kJ/(kg·K)	$c_s = 2.11$	—	—	$c_l = 3.52$
Thermal conductivity, W/(m·K)	$k_s = 1.66$	—	—	—

Volumetric enthalpy difference between the initial freezing point and −10°C:

$$\Delta H_{10} = \rho_l H_l - \rho_s H_s$$

$$\Delta H_{10} = (1075)(274.2) - (1018)(83.4) = 210 \times 10^3 \text{ kJ/m}^3$$

Volumetric specific heats:

$$C_s = \rho_s c_s = (1018)(2.11) = 2148 \text{ kJ/(m}^3\cdot\text{K)}$$

$$C_l = \rho_l c_l = (1075)(3.52) = 3784 \text{ kJ/(m}^3\cdot\text{K)}$$

Step 2: Determine the surface heat transfer coefficient.
The surface heat transfer coefficient is estimated to be 40 W/(m²·K).

Step 3: Determine the characteristic dimension D and the dimensional ratios β_1 and β_2.

For freezing time problems, the characteristic dimension D is twice the shortest distance from the thermal center of the food item to its surface. For this example, $D = 0.04$ m.

Using Equations (17) and (18), the dimensional ratios then become:

$$\beta_1 = 0.12/0.04 = 3$$
$$\beta_2 = 0.16/0.04 = 4$$

Step 4: Using Equations (29) through (31), calculate the Biot number, the Plank number, and the Stefan number.

$$\text{Bi} = \frac{hD}{k_s} = \frac{(40.0)(0.04)}{1.66} = 0.964$$

$$\text{Pk} = \frac{C_l(T_i - T_f)}{\Delta H_{10}} = \frac{(3784)[10 - (-1.7)]}{210 \times 10^3} = 0.211$$

$$\text{Ste} = \frac{C_s(T_f - T_m)}{\Delta H_{10}} = \frac{(2148)[-1.7 - (-30)]}{210 \times 10^3} = 0.289$$

Step 5: Determine the geometric parameters P and R for the rectangular brick.
Determine P from Table 5.

$$P_1 = \frac{(3)(4)}{2[(3)(4) + 3 + 4]} = 0.316$$

$$P_2 = 0.316 \left\{ 1.026 + (0.5808)(0.211) \right.$$
$$\left. + 0.289\left[(0.2296)(0.211) + \frac{0.0182}{0.964} + 0.1050\right]\right\}$$

$$P_2 = 0.379$$

$$P = 0.379 + 0.316\{0.1136 + 0.289[(5.766)(0.316) - 1.242]\}$$

$$P = 0.468$$

Determine R from Table 5.

$$\frac{1}{Q} = 4[(3-4)(3-1) + (4-1)^2]^{1/2} = 10.6$$

$$r = \frac{1}{3}\{3 + 4 + 1 + [(3-4)(3-1) + (4-1)^2]^{1/2}\} = 3.55$$

$$s = \frac{1}{3}\{3 + 4 + 1 - [(3-4)(3-1) + (4-1)^2]^{1/2}\} = 1.78$$

$$R_1 = \frac{1}{(10.6)(2)} \left\{(3.55 - 1)(3 - 3.55)(4 - 3.55)\ln\left[\frac{3.55}{3.55 - 1}\right]\right.$$
$$\left. - (1.78 - 1)(3 - 1.78)(4 - 1.78)\ln\left[\frac{1.78}{1.78 - 1}\right]\right\}$$
$$+ \frac{1}{72}[(2)(3) + (2)(4) - 1] = 0.0885$$

$$R_2 = 0.0885\{1.202 + 0.289[(3.410)(0.211) + 0.7336]\} = 0.144$$

$$R = 0.144 + 0.0885\{0.7344 + 0.289[(49.89)(0.0885) - 2.900]\}$$
$$= 0.248$$

Step 6: Calculate the freezing time of the beef.
Because the final temperature at the thermal center of the beef is given to be $-10°C$, use Equation (32) to calculate the freezing time:

$$\theta = \frac{2.10 \times 10^8}{-1.7 - (-30)}\left[\frac{(0.468)(0.04)}{40.0} + \frac{(0.248)(0.04)^2}{1.66}\right] = 5250 \text{ s} = 1.46 \text{ h}$$

Example 4. Orange juice in a cylindrical container, 0.30 m diameter by 0.45 m tall, is to be frozen in an air blast freezer. The initial temperature of the juice is 5°C and the freezer air temperature is −35°C. It is estimated that the surface heat transfer coefficient is 30 W/(m²·K).

Calculate the time required for the thermal center of the juice to reach a temperature of −18°C.

Solution: Use the algorithm based on the method of equivalent heat transfer dimensionality.

Step 1: Determine the thermal properties of orange juice.
Using the methods described in Chapter 8, the thermal properties of orange juice are calculated as follows:

Property	At −40°C (Fully Frozen)	At −18°C (Final Temp.)	At 5°C (Initial Temp.)
Density, kg/m³	$\rho_s = 970$	$\rho_s = 970$	$\rho_l = 1038$
Enthalpy, kJ/kg	—	$H_s = 40.8$	$H_l = 381.5$
Specific Heat, kJ/(kg·K)	$c_s = 1.76$	—	$c_l = 3.89$
Thermal Cond., W/(m·K)	$k_s = 2.19$	—	—
Initial Freezing Temperature: $T_f = -0.4°C$			

Volumetric enthalpy difference between $T_i = 5°C$, and $-18°C$:

$$\Delta H_{18} = \rho_l H_l - \rho_s H_s$$

$$\Delta H_{18} = (1038)(381.5) - (970)(40.8) = 356 \times 10^3 \text{ kJ/m}^3$$

Volumetric specific heats:

$$C_s = \rho_s c_s = (970)(1.76) = 1707 \text{ kJ/(m}^3\cdot\text{K)}$$

$$C_l = \rho_l c_l = (1038)(3.89) = 4038 \text{ kJ/(m}^3\cdot\text{K)}$$

Step 2: Determine the surface heat transfer coefficient.
The surface heat transfer coefficient is estimated to be 30 W/(m²·K).

Step 3: Determine the characteristic dimension D and the dimensional ratios β_1 and β_2.
For freezing time problems, the characteristic dimension is twice the shortest distance from the thermal center of the food item to its surface. For the cylindrical sample of orange juice, the characteristic dimension is equal to the diameter of the cylinder:

$$D = 0.30 \text{ m}$$

Using Equations (17) and (18), the dimensional ratios then become:

$$\beta_1 = \beta_2 = \frac{0.45 \text{ m}}{0.30 \text{ m}} = 1.5$$

Step 4: Using Equations (29) through (31), calculate the Biot number, the Plank number, and the Stefan number.

$$\text{Bi} = \frac{hD}{k_s} = \frac{(30.0)(0.30)}{2.19} = 4.11$$

$$\text{Pk} = \frac{C_l(T_i - T_f)}{\Delta H_{18}} = \frac{(4038)[5 - (-0.4)]}{356 \times 10^3} = 0.0613$$

$$\text{Ste} = \frac{C_s(T_f - T_m)}{\Delta H_{18}} = \frac{(1707)[-0.4 - (-35)]}{356 \times 10^3} = 0.166$$

Step 5: Calculate the freezing time of an infinite slab.
Use the method of Hung and Thompson (1983). First, find the weighted average temperature difference given by Equation (34).

$$\Delta T = [-0.4 - (-35)]$$
$$+ \frac{[5 - (-0.4)]^2(4038/2) - [-0.4 - (-18)]^2(1707/2)}{356 \times 10^3} = 34.0 \text{ K}$$

Determine the parameter U:

$$U = \frac{34.0}{-0.4 - (-35)} = 0.983$$

Determine the geometric parameters, P and R, for an infinite slab using Equations (35) and (36):

Cooling and Freezing Times of Foods

$$P = 0.7306 - (1.083)(0.0613)$$
$$+ (0.166)\left[(15.40)(0.983) - 15.43 + \frac{(0.01329)(0.166)}{4.11}\right] = 0.616$$

$$R = 0.2079 - (0.2656)(0.983)(0.166) = 0.165$$

Determine the freezing time of the slab using Equation (37):

$$\theta = \frac{3.56 \times 10^8}{34.0}\left[\frac{(0.616)(0.30)}{30.0} + \frac{(0.165)(0.30)^2}{2.19}\right] = 135000 \text{ s} = 37.5 \text{ h}$$

Step 6: Calculate the equivalent heat transfer dimensionality for a finite cylinder.

Use the method presented by Cleland et al. (1987a, 1987b), Equations (61) through (64), to calculate the equivalent heat transfer dimensionality. From Table 7, the geometric constants for a cylinder are:

$$G_1 = 2 \quad G_2 = 0 \quad G_3 = 1$$

Calculate E_2:

$$E_2 = \frac{2.32}{\beta_2^{1.77}} = \frac{2.32}{1.5^{1.77}} = 1.132$$

$$X(1.132) = \frac{1.132}{4.11^{1.34} + 1.132} = 0.146$$

$$E_2 = \frac{0.146}{1.5} + (1 - 0.146)\frac{0.73}{1.5^{2.50}} = 0.324$$

Thus, the equivalent heat transfer dimensionality E becomes:

$$E = G_1 + G_2 E_1 + G_3 E_2$$
$$E = 2 + (0)(E_1) + (1)(0.324) = 2.324$$

Step 7: Calculate freezing time of the orange juice using Equation (56):

$$\theta_{shape} = \theta_{slab}/E = 135000/2.324 = 58100 \text{ s} = 16.1 \text{ h}$$

NOMENCLATURE

A_1 = cross sectional area in Equation (9), m²
A_2 = cross sectional area in Equation (9), m²
A_s = surface area of food item, m²
B_1 = parameter in Equation (8)
B_2 = parameter in Equation (8)
Bi = Biot number
Bi_1 = Biot number for precooling = $(Bi_l + Bi_s)/2$
Bi_2 = Biot number for phase change = Bi_s
Bi_3 = Biot number for subcooling = Bi_s
Bi_c = Biot number evaluated at $k_c = hD/k_c$
Bi_l = Biot number for unfrozen food = hD/k_l
Bi_o = Biot number based on shortest dimension = hD_1/k
Bi_s = Biot number for fully frozen food = hD/k_s
c = specific heat of food item, J/(kg·K)
C_l = volumetric specific heat of unfrozen food, J/(m³·K)
C_s = volumetric specific heat of fully frozen food, J/(m³·K)
D = slab thickness or cylinder/sphere diameter, m
D_1 = shortest dimension, m
D_2 = longest dimension, m
$D_{eq,s}$ = equivalent sphere diameter, m
D_m = twice the mean conducting path, m
D_v = volume diameter, m
D_{vs} = volume-surface diameter, m
E = equivalent heat transfer dimensionality
E_o = equivalent heat transfer dimensionality at Bi = 0
E_1 = parameter given by Equation (62)
E_2 = parameter given by Equation (63)
E_∞ = equivalent heat transfer dimensionality at Bi → ∞
f = cooling time parameter
f_1 = cooling time parameter for precooling
f_3 = cooling time parameter for subcooling
f_{comp} = cooling parameter for a composite shape
G = geometry index
G_1 = geometric constant in Equation (61)
G_2 = geometric constant in Equation (61)
G_3 = geometric constant in Equation (61)
h = heat transfer coefficient, W/(m²·K)
$I_o(x)$ = Bessel function of the second kind, order zero
$I_1(x)$ = Bessel function of the second kind, order one
j = cooling time parameter
j_1 = cooling time parameter for precooling
j_3 = cooling time parameter for subcooling
j_c = cooling time parameter applicable to thermal center
j_{comp} = cooling time parameter for a composite shape
j_m = cooling time parameter applicable to mass average
$J_o(x)$ = Bessel function of the first kind, order zero
$J_1(x)$ = Bessel function of the first kind, order one
k = thermal conductivity of food item, W/(m·K)
k_c = thermal conductivity of food evaluated at $(T_f + T_m)/2$, W/(m·K)
k_l = thermal conductivity of unfrozen food, W/(m·K)
k_s = thermal conductivity of fully frozen food, W/(m·K)
L = half thickness of slab or radius of cylinder/sphere, m
L_f = volumetric latent heat of fusion, J/m³
L_∞ = lag factor parameter given by Equation (25)
m = inverse of Biot number
M_1^2 = characteristic value of Smith et al. (1968)
N = number of dimensions
p_1 = geometric parameter from Lin et al. (1996b)
p_2 = geometric parameter from Lin et al. (1996b)
p_3 = geometric parameter from Lin et al. (1996b)
P = Plank's geometry factor
P' = geometric factor for rectangular bricks calculated using method in Table 5
P_1 = intermediate value of Plank's geometric factor
P_2 = intermediate value of Plank's geometric factor
Pk = Plank number = $C_l(T_i - T_f)/\Delta H$
Q = parameter given in Table 5
Q_1 = volumetric heat of precooling, J/m³
Q_2 = volumetric heat of phase change, J/m³
Q_3 = volumetric heat of subcooling, J/m³
r = parameter given in Table 5
R = Plank's geometry factor
R' = geometric factor for rectangular bricks calculated using method in Table 5
R_1 = intermediate value of Plank's geometric factor
R_2 = intermediate value of Plank's geometric factor
s = parameter given in Table 5
Ste = Stefan number = $C_s(T_f - T_m)/\Delta H$
T = product temperature, °C
T_c = final center temperature of food item, °C
T_f = initial freezing temperature of food item, °C
T_{fm} = mean freezing temperature, °C
T_i = initial temperature of food item, °C
T_m = cooling or freezing medium temperature, °C
T_o = mean final temperature, °C
T_{ref} = reference temperature for freezing time correction factor, °C
u = parameter given in Table 1
U = parameter in Equations (35) and (36) = $\Delta T/(T_f - T_m)$
v = parameter given in Table 2
V = volume of food item, m³
w = parameter given in Table 3
W_1 = parameter given by Equation (58)
W_2 = parameter given by Equation (59)
x = coordinate direction
$X(x)$ = function given by Equation (64)
X_b = parameter in Equation (13)
X_g = parameter in Equations (12) and (13)
y = coordinate direction
Y = fractional unaccomplished temperature difference
y_n = roots of transcendental equation; $y_n J_1(y_n) - Bi J_0(y_n) = 0$
z = coordinate direction
z_m = roots of transcendental equation; $Bi\beta_1 = z_m\tan(z_m)$
z_n = roots of transcendental equation; $Bi = z_n\tan(z_n)$
z_{nm} = parameter given in Table 8
α = thermal diffusivity of food, m²/s

β_1 = ratio of second shortest dimension to shortest dimension, Equation (17)
β_2 = ratio of longest dimension to shortest dimension, Equation (18)
γ_1 = geometric parameter from Lin et al. (1996b)
γ_2 = geometric parameter from Lin et al. (1996b)
θ = cooling or freezing time, s
θ_1 = precooling time, s
θ_2 = phase change time, s
θ_3 = tempering time, s
θ_{shape} = freezing time of an irregular shaped food item, s
θ_{slab} = freezing time of an infinite slab shaped food item, s
ΔH = volumetric enthalpy difference, J/m³
ΔH_1 = volumetric enthalpy difference = $C_l(T_i - T_{fm})$, J/m³
ΔH_2 = volumetric enthalpy difference = $L_f + C_s(T_{fm} - T_c)$, J/m³
ΔH_{10} = volumetric enthalpy difference between the initial freezing temperature T_f and $-10°C$, J/m³
ΔH_{18} = volumetric enthalpy difference between initial temperature T_i and $-18°C$, J/m³
ΔT = weighted average temperature difference given by Equation (34), °C
ΔT_1 = temperature difference = $(T_i + T_{fm})/2 - T_m$, °C
ΔT_2 = temperature difference = $T_{fm} - T_m$, °C
ΔT_{m1} = temperature difference for precooling, °C
ΔT_{m2} = temperature difference for phase change, °C
ΔT_{m3} = temperature difference for subcooling, °C
λ = geometric parameter from Lin et al. (1996b)
μ = parameter given by Equation (27)
ρ = density of food item, kg/m³
ω = first root of Equation (15)

REFERENCES

Cleland, A.C., and R.L. Earle. 1977. A comparison of analytical and numerical methods of predicting the freezing times of foods. *Journal of Food Science* 42(5):1390-1395.

Cleland, A.C., and R.L. Earle. 1979a. A comparison of methods for predicting the freezing times of cylindrical and spherical foodstuffs. *Journal of Food Science* 44(4):958-963, 970.

Cleland, A.C., and R.L. Earle. 1979b. Prediction of freezing times for foods in rectangular packages. *Journal of Food Science* 44(4):964-970.

Cleland, A.C., and R.L. Earle. 1982a. A simple method for prediction of heating and cooling rates in solids of various shapes. *International Journal of Refrigeration* 5(2):98-106.

Cleland, A.C., and R.L. Earle. 1982b. Freezing time prediction for foods—A simplified procedure. *International Journal of Refrigeration* 5(3):134-140.

Cleland, A.C., and R.L. Earle. 1984. Freezing time predictions for different final product temperatures. *Journal of Food Science* 49(4):1230-1232.

Cleland, A.C. 1990. *Food refrigeration processes: Analysis, design and simulation*. London: Elsevier Science Publishers.

Cleland, D.J., A.C. Cleland, and R.L. Earle. 1987a. Prediction of freezing and thawing times for multi-dimensional shapes by simple formulae—Part 1: Regular shapes. *International Journal of Refrigeration* 10(3):156-164.

Cleland, D.J., A.C. Cleland, and R.L. Earle. 1987b. Prediction of freezing and thawing times for multi-dimensional shapes by simple formulae—Part 2: Irregular shapes. *International Journal of Refrigeration* 10(4):234-240.

DeMichelis, A., and A. Calvelo. 1983. Freezing time predictions for brick and cylindrical-shaped foods. *Journal of Food Science* 48: 909-913, 934.

Hayakawa, K., and G. Villalobos, 1989. Formulas for estimating Smith et al. parameters to determine the mass average temperature of irregularly shaped bodies. *J. of Food Process Engineering* 11(4):237-256.

Heldman, D.R. 1975. *Food process engineering*. Westport, CT: AVI Publishing Co.

Hossain, Md.M., D.J. Cleland, and A.C. Cleland. 1992a. Prediction of freezing and thawing times for foods of regular multi-dimensional shape by using an analytically derived geometric factor. *International Journal of Refrigeration* 15(4):227-234.

Hossain, Md.M., D.J. Cleland, and A.C. Cleland. 1992b. Prediction of freezing and thawing times for foods of two-dimensional irregular shape by using a semi-analytical geometric factor. *International Journal of Refrigeration* 15(4):235-240.

Hossain, Md.M., D.J. Cleland, and A.C. Cleland. 1992c. Prediction of freezing and thawing times for foods of three-dimensional irregular shape by using a semi-analytical geometric factor. *International Journal of Refrigeration* 15(4):241-246.

Hung, Y.C., and D.R. Thompson. 1983. Freezing time prediction for slab shape foodstuffs by an improved analytical method. *Journal of Food Science* 48(2):555-560.

Ilicali, C., and S.T. Engez. 1990. A simplified approach for predicting the freezing or thawing times of foods having brick or finite cylinder shape. In *Engineering and Food*. ed. W.E.L. Speiss and H. Schubert. 2:442-451. London: Elsevier Applied Science Publishers.

Ilicali, C., and M. Hocalar. 1990. A simplified approach for predicting the freezing times of foodstuffs of anomalous shape. In *Engineering And Food*. ed. W.E.L. Speiss and H. Schubert. 2:418-425. London: Elsevier Applied Science Publishers.

Lacroix, C., and F. Castaigne. 1987a. Simple method for freezing time calculations for infinite flat slabs, infinite cylinders and spheres. *Canadian Inst. of Food Science and Technology Journal* 20(4):252-259.

Lacroix, C., and F. Castaigne. 1987b. Simple method for freezing time calculations for brick and cylindrical shaped food products. *Canadian Institute of Food Science and Technology Journal* 20(5):342-349.

Lacroix, C., and F. Castaigne. 1988. Freezing time calculation for products with simple geometrical shapes. *Journal of Food Process Engineering* 10(2):81-104.

Lin, Z., A.C. Cleland, G.F. Serrallach, and D.J. Cleland. 1993. Prediction of chilling times for objects of regular multi-dimensional shapes using a general geometric factor. *Refrigeration Science and Technology* 1993-3:259-267.

Lin, Z., A.C. Cleland, D.J. Cleland, and G.F. Serrallach. 1996a. A simple method for prediction of chilling times for objects of two-dimensional irregular shape. *Int. Journal of Refrigeration* 19(2):95-106.

Lin, Z., A.C. Cleland, D.J. Cleland, and G.F. Serrallach. 1996b. A simple method for prediction of chilling times: Extension to three-dimensional irregular shaped. *Int. Journal of Refrigeration* 19(2):107-114.

McNabb, A., G.C. Wake, and M.M. Hossain. 1990. Transition times between steady states for heat conduction: Part I—General theory and some exact results. *Occas Pubs Math Stat*, No. 20, New Zealand: Massey University.

Pflug, I.J., J.L. Blaisdell, and J. Kopelman. 1965. Developing temperature-time curves for objects that can be approximated by a sphere, infinite plate, or infinite cylinder. *ASHRAE Transactions* 71(1):238-248.

Pham, Q.T. 1984. An extension to Plank's equation for predicting freezing times for foodstuffs of simple shapes. *International Journal of Refrigeration* 7:377-383.

Pham, Q.T. 1985. Analytical method for predicting freezing times of rectangular blocks of foodstuffs. *International Journal of Refrigeration* 8(1):43-47.

Pham, Q.T. 1986a. Simplified equation for predicting the freezing time of foodstuffs. *Journal of Food Technology* 21(2):209-219.

Pham, Q.T. 1986b. Freezing of foodstuffs with variations in environmental conditions. *International Journal of Refrigeration* 9(5):290-295.

Pham, Q.T. 1987. A converging-front model for the asymmetric freezing of slab-shaped food. *Journal of Food Science* 52(3):795-800.

Pham, Q.T. 1991. Shape factors for the freezing time of ellipses and ellipsoids. *Journal of Food Engineering* 13:159-170.

Plank, R. 1913. Die Gefrierdauer von Eisblocken. *Zeitschrift für die gesamte Kälte Industrie* (20) 6:109-114.

Plank, R. 1941. Beitrage zur Berechnung und Bewertung der Gefriergeschwindigkeit von Lebensmitteln. *Zeitschrift für die gesamte Kälte Industrie* 3(10):1-24.

Smith, R.E. 1966. Analysis of transient heat transfer from anomalous shape with heterogeneous properties. Ph.D. thesis, Oklahoma State University, Stillwater, OK.

Smith, R.E., G.L. Nelson, and R.L. Henrickson. 1968. Applications of geometry analysis of anomalous shapes to problems in transient heat transfer. *Transactions of the ASAE* 11(2): 296-302.

CHAPTER 10

COMMODITY STORAGE REQUIREMENTS

Refrigerated Storage 10.1
Refrigerated Storage Plant Operation 10.4
Storage of Frozen Foods 10.5
Other Products 10.5

THIS chapter presents information on the storage requirements of many perishable foods that enter the market on a commercial scale. Also included is a short discussion on the storage of furs and fabrics. The data are based on the storage of fresh, high quality commodities that have been properly harvested and handled and that have been properly cooled.

Table 1 presents recommended storage requirements for various products. Some products require a curing period before storage. Other products, such as Irish potatoes, require different storage conditions depending on their intended use.

The recommended temperatures are optimum for long storage and are commodity temperatures, not air temperatures. For short storage, higher temperatures are often acceptable. Conversely, products subject to chilling injury can sometimes be held at a lower temperature for a short time without injury. Exceptions include bananas, cranberries, cucumbers, eggplant, melons, okra, pumpkins, squash, white potatoes, sweet potatoes, and tomatoes. The minimum recommended temperature for these products should be followed.

The listed storage lives are based on typical commercial practice. Special treatments can, in certain instances, extend storage life significantly.

Thermal properties of many of these products including water content, freezing point, specific heat, and latent heat of fusion are listed in Table 1 in Chapter 8. Also, because fresh fruits and vegetables are living products, they generate heat that should be included as part of the storage refrigeration load. The approximate heat of respiration for various fruits and vegetables is listed in Table 2 in Chapter 8.

REFRIGERATED STORAGE

Cooling

Because products deteriorate much faster at warm than at low temperatures, rapid removal of field heat by cooling to the storage temperature substantially increases the market life of the product. Chapter 14 describes various cooling methods.

Deterioration

The environment in which harvested produce is placed may greatly influence not only the respiration rate but other changes and products formed in related chemical reactions. In fruits, these changes are described as ripening. In many fruits, such as bananas and pears, the process of ripening is required to develop the maximum edible quality. However, as ripening continues, deterioration begins and the fruit softens, loses flavor, and eventually undergoes tissue breakdown.

In addition to deterioration after harvest by biochemical changes within the product, desiccation and diseases caused by microorganisms are also important.

Deterioration rate is greatly influenced by temperature and is reduced as temperature is lowered. The specific relationships between temperature and deterioration rate vary considerably among commodities and diseases. A generalization, assuming a deterioration rate of 1 for a fruit at $-1°C$, is as follows.

Approximate Deterioration Rate of Fresh Produce

Temperature, °C	Deterioration Rate
20	8 to 10
10	4 to 5
5	3
3	2
0	1.25
−1	1

For example, fruit that remains marketable for 12 days when stored at $-1°C$ may last only $12/3 = 4$ days when stored at 5°C. The best temperature to slow down deterioration is the lowest temperature that can safely be maintained without freezing the commodity, which is 0.5 to 1 K above the freezing point of the fruit or vegetable.

Some produce will not tolerate low storage temperatures. Severe physiological disorders that develop because of exposure to low but not freezing temperatures are classed as **chilling injury**. The banana is a classic example of a fruit displaying chilling injury symptoms, and storage temperatures must be elevated accordingly. Certain apple varieties exhibit this characteristic, and prolonged storage must be held at a temperature well above that usually recommended. The degree of susceptibility of an apple variety to chilling may vary with climatic and cultural factors. Products susceptible to chilling injury, its symptoms, and the lowest safe temperature are discussed in Chapters 8, 21, 22, and 23.

Desiccation

Water loss, which causes a product to shrivel, is a physical factor related to the evaporative potential of air, and can be expressed as follows:

$$p_D = \frac{p(100 - \phi)}{100}$$

where

p_D = vapor pressure deficit, indicating combined influence of temperature and relative humidity on evaporative potential of air
p = vapor pressure of water at given temperature
ϕ = relative humidity, percent

For example, comparing the evaporative potential of air in storage rooms at 0°C and 10°C dry bulb, with 90% rh in each room, the vapor pressure deficit at 0°C is 60 Pa, while at 10°C, it is 120 Pa. Thus, if all other factors are equal, commodities tend to lose water twice as fast at 10°C dry bulb as at 0°C at the same rh values. For equal water loss at the two temperatures, the rh has to be maintained at 95% at 10°C in comparison to 90% at 0°C. These comparisons are not precise because the water in fruits and vegetables contains a sufficient quantity of dissolved sugars and other chemical materials to cause the water to be in equilibrium with water vapor in the air at 98

The preparation of this chapter is assigned to TC 10.5, Refrigerated Distribution and Storage Facilities.

Table 1 Storage Requirements of Perishable Products

	Storage Temperature, °F	Relative Humidity, %	Approximate Storage Life[a]
Vegetables			
Artichokes			
Globe	32	95 to 100	2 weeks
Jerusalem	31 to 32	90 to 95	4 to 5 months
Asparagus	32 to 35	95 to 100	2 to 3 weeks
Beans			
Snap or Green	40 to 45	95	7 to 10 days
Lima	34 to 40	95	3 to 5 days
Dried	50	70	6 to 8 months
Beets			
Roots	32	95 to 100	4 to 6 months
Bunch	32	98 to 100	10 to 14 days
Broccoli	32	95 to 100	10 to 14 days
Brussels Sprouts	32	95 to 100	3 to 5 weeks
Cabbage, late	32	98 to 100	5 to 6 months
Carrots			
Topped-immature	32	98 to 100	4 to 6 weeks
Topped-mature	32	98 to 100	7 to 9 months
Cauliflower	32	95 to 98	3 to 4 weeks
Celeriac	32	95 to 100	6 to 8 months
Celery	32	98 to 100	2 to 3 months
Collards	32	95 to 100	10 to 14 days
Corn, Sweet	32	95 to 98	4 to 8 days
Cucumbers	45 to 50	95	10 to 14 days
Eggplant	45 to 54	90 to 95	7 to 10 days
Endive (Escarole)	32	95 to 100	2 to 3 weeks
Frozen Vegetables[b]	−10 to 0		6 to 12 months
Garlic, dry	32	65 to 70	6 to 7 months
Greens, leafy	32	95 to 100	10 to 14 days
Horseradish	30 to 32	95 to 100	10 to 12 months
Jicama	55 to 65	65 to 70	1 to 2 months
Kale	32	95 to 100	2 to 3 weeks
Kohlrabi	32	98 to 100	2 to 3 months
Leeks, green	32	95 to 100	2 to 3 months
Lettuce			
head	32 to 34	95 to 100	2 to 3 weeks
Mushrooms	32	95	3 to 4 days
Okra	45 to 55	90 to 95	7 to 10 days
Onions			
Green	32	95 to 100	3 to 4 weeks
Dry, and onion sets	32	65 to 75	1 to 8 months
Parsley	32	95 to 100	1 to 2 months
Parsnips	32	98 to 100	4 to 6 months
Peas			
Green	32	95 to 98	1 to 2 weeks
Dried	50	70	6 to 8 months
Peppers			
Dried	32 to 50	60 to 70	6 months
Sweet	45 to 50	90 to 95	2 to 3 weeks
Potatoes			
Early	38 to 40	90 to 95	4 to 5 months
Main crop	38 to 40	90 to 95	5 to 8 months
Sweet	50 to 55	85 to 90	4 to 7 months
Pumpkins	50 to 55	50 to 75	2 to 3 months
Radishes			
Spring	32	95 to 100	3 to 4 weeks
Winter	32	95 to 100	2 to 4 months
Rhubarb	32	95 to 100	2 to 4 weeks
Rutabagas	32	98 to 100	4 to 6 months
Salsify	32	98 to 100	2 to 4 months
Seed, vegetable	32 to 50	50 to 65	10 to 12 months
Spinach	32	95 to 98	10 to 14 days
Squash			
Acorn	45 to 50	70 to 75	5 to 8 weeks
Summer	41 to 50	95	5 to 14 days
Winter	50 to 55	50 to 75	4 to 6 months
Tamarillos	37 to 40	85 to 95	10 weeks
Tomatoes			
Mature green	55 to 60	90 to 95	1 to 3 weeks
Vegetables (*Continued*)			
Tomatoes, Firm, ripe	45 to 50	90 to 95	4 to 7 days
Turnips			
Roots	32	95	4 to 5 months
Greens	32	95 to 100	10 to 14 days
Watercress	32	95 to 100	2 to 3 weeks
Yams	61	85 to 90	3 to 6 months
Fruit and Melons			
Apples	32 to 38	90 to 95	3 to 8 months
Apples, dried	41 to 48	55 to 60	5 to 8 months
Apricots	30 to 32	90 to 95	1 to 3 weeks
Avocados	40 to 55	85 to 90	2 to 8 weeks
Bananas	56 to 58	85 to 95	
Blackberries	31 to 32	90 to 95	3 days
Blueberries	31 to 32	90 to 95	2 weeks
Cantaloupes	36 to 40	95	5 to 15 days
Casaba melons	45 to 50	85 to 95	4 to 6 weeks
Cherries			
Sour	32	90 to 95	3 to 7 days
Sweet	30 to 31	90 to 95	2 to 3 weeks
Coconuts	32 to 35	80 to 85	1 to 2 months
Cranberries	36 to 40	90 to 95	2 to 4 months
Crenshaw melons	45 to 50		
Currants	31 to 32	90 to 95	1 to 4 weeks
Dates, cured	0 to 32	75 or less	6 to 12 months
Dewberries	31 to 32	90 to 95	2 to 3 days
Elderberries	31 to 32	90 to 95	1 to 2 weeks
Figs			
Dried	32 to 40	50 to 60	9 to 12 months
Fresh	31 to 32	85 to 90	7 to 10 days
Frozen fruits	−12 to 0	90 to 95	18 to 24 months
Gooseberries	31 to 32	90 to 95	2 to 4 weeks
Grapefruit	58 to 60	85 to 90	6 to 8 weeks
Grapes[c]			
American	31 to 32	85 to 90	2 to 8 weeks
Vinifera	30 to 31	90 to 95	3 to 6 months
Guavas	40 to 50	90	2 to 3 weeks
Honeydew melons	41 to 50	90 to 95	2 to 3 weeks
Kiwifruit	31 to 32	90 to 95	3 to 5 months
Lemons	52 to 55[d]	85 to 90	1 to 4 months
Limes	45 to 48	85 to 90	6 to 8 weeks
Loganberries	31 to 32	90 to 95	2 to 3 days
Loquats	32	90	3 weeks
Lychees	35	90 to 95	3 to 5 weeks
Mangoes	50 to 55	85 to 90	2 to 3 weeks
Nectarines	31 to 32	90 to 95	2 to 4 weeks
Olives, fresh	41 to 50	85 to 90	4 to 6 weeks
Oranges			
CA and AZ	38 to 48	85 to 90	3 to 6 weeks
FL and TX	32 to 34	85 to 90	8 to 12 weeks
Papayas	45	85 to 90	1 to 3 weeks
Peaches	31 to 32	90 to 95	2 to 4 weeks
Peaches, dried	32 to 41	55 to 60	5 to 8 months
Pears[c]	29 to 31	90 to 95	2 to 7 months
Persian melons	45 to 50	90 to 95	2 weeks
Persimmons	30	90	3 to 4 months
Pineapples, ripe	45	85 to 90	2 to 4 weeks
Plums	31 to 32	90 to 95	2 to 4 weeks
Pomegranates	40	90 to 95	2 to 3 months
Prunes			
Fresh	31 to 32	90 to 95	2 to 4 weeks
Dried	32 to 41	55 to 60	5 to 8 months
Quinces	31 to 32	90	2 to 3 months
Raisins			
Raspberries, Black	31 to 32	90 to 95	2 to 3 days
Raspberries, Red	31 to 32	90 to 95	2 to 3 days
Strawberries	31 to 32	90 to 95	5 to 7 days
Tangerines	40	90 to 95	2 to 4 weeks
Watermelons	40 to 50	90	2 to 3 weeks

Commodity Storage Requirements

Table 1 Storage Requirements of Perishable Products

	Storage Temperature, °F	Relative Humidity, %	Approximate Storage Life[a]
Fish			
Haddock, Cod, Perch	31 to 34	95 to 100	12 days
Hake, Whiting	32 to 34	95 to 100	10 days
Halibut	31 to 34	95 to 100	18 days
Herring			
Kippered	32 to 36	80 to 90	10 days
Smoked	32 to 36	80 to 90	10 days
Mackerel	32 to 34	95 to 100	6 to 8 days
Menhaden	34 to 41	95 to 100	4 to 5 days
Salmon	31 to 34	95 to 100	18 days
Tuna	32 to 36	95 to 100	14 days
Frozen fish	−20 to −4	90 to 95	6 to 12 months
Shellfish[a]			
Scallop meat	32 to 34	95 to 100	12 days
Shrimp	31 to 34	95 to 100	12 to 14 days
Lobster			
American	41 to 50	In sea water	Indefinitely
Oysters, Clams			
(meat and liquid)	32 to 36	100	5 to 8 days
Oyster in shell	41 to 50	95 to 100	5 days
Frozen shellfish	−30 to −4	90 to 95	3 to 8 months
Beef			
Beef, fresh, average	28 to 34	88 to 95	1 week
Beef carcass			
Choice, 60% lean	32 to 39	85 to 90	1 to 3 weeks
Prime, 54% lean	32 to 34	85	1 to 3 weeks
Sirloin cut (choice)	32 to 34	85	1 to 3 weeks
Round cut (choice)	32 to 34	85	1 to 3 weeks
Dried, chipped	50 to 59	15	6 to 8 weeks
Liver	32	90	5 days
Veal, lean	28 to 34	85 to 95	3 weeks
Beef, frozen	−10 to 0	90 to 95	6 to 12 months
Pork			
Pork,			
Fresh, average	32 to 34	85 to 90	3 to 7 days
Carcass, 47% lean	32 to 34	85 to 90	3 to 5 days
Bellies, 35% lean	32 to 34	85	3 to 5 days
Backfat, 100% fat	32 to 34	85	3 to 7 days
Shoulder, 67% lean	32 to 34	85	3 to 5 days
Pork, frozen	−10 to 0	90 to 95	4 to 8 months
Ham			
74% lean	32 to 34	80 to 85	3 to 5 days
Light cure	37 to 41	80 to 85	1 to 2 weeks
Country cure	50 to 59	65 to 70	3 to 5 months
Frozen	−10 to 0	90 to 95	6 to 8 months
Bacon			
Medium fat class	37 to 41	80 to 85	2 to 3 weeks
Cured, farm style	61 to 64	85	4 to 6 months
Cured, packer style	34 to 39	85	2 to 6 weeks
Frozen	−10 to 0	90 to 95	2 to 4 months
Sausage			
Links or bulk	32 to 34	85	1 to 7 days
Country, smoked	32	85	1 to 3 weeks
Frankfurters,			
average	32	85	1 to 3 weeks
Polish style	32	85	1 to 3 weeks
Lamb			
Fresh, average	28 to 34	85 to 90	3 to 4 weeks
Choice, lean	32	85	5 to 12 days
Leg, choice,			
83% lean	32	85	5 to 12 days
Frozen	−10 to 0	90 to 95	8 to 12 months
Poultry			
Poultry			
fresh, average	28 to 32	95 to 100	1 to 3 weeks
Chicken, all classes	28 to 32	95 to 100	1 to 4 weeks
Turkey, all classes	28 to 32	95 to 100	1 to 4 weeks
Turkey breast roll	−4 to −1		6 to 12 months
Turkey frankfurters	0 to 15		6 to 16 months
Duck	28 to 32	95 to 100	1 to 4 weeks
Poultry, frozen	−10 to 0	90 to 95	12 months
Meat (Miscellaneous)			
Rabbits, fresh	32 to 34	90 to 95	1 to 5 days
Dairy Products			
Butter	32	75 to 85	2 to 4 weeks
Butter,			
frozen	−10	70 to 85	12 to 20 months
Cheese, Cheddar,			
long storage	32 to 34	65	12 months
short storage	40	65	6 months
processed	40	65	12 months
grated	40	65	12 months
Ice cream,			
10% fat	−20 to −15	90 to 95	3 to 23 months
premium	−30 to −40	90 to 95	3 to 23 months
Milk			
Fluid, pasteurized	39 to 43		7 days
Grade A (3.7% fat)	32 to 34		2 to 4 months
Raw	32 to 39		2 days
Dried, whole	70	Low	6 to 9 months
Dried, nonfat	45 to 70	Low	16 months
Evaporated	40		24 months
Evaporated,			
unsweetened	70		12 months
Condensed,			
sweetened	40		15 months
Whey, dried	70	Low	12 months
Eggs			
Eggs			
Shell	29 to 32[c]	80 to 90	5 to 6 months
Shell, farm cooler	50 to 55	70 to 75	2 to 3 weeks
Frozen,			
Whole	0		1 year plus
Yolk	0		1 year plus
White	0		1 year plus
Whole egg solids	35 to 40	Low	6 to 12 months
Yolk solids	35 to 40	Low	6 to 12 months
Flake albumen solids		Low	1 year plus
Dry spray albumen			
solids		Low	1 year plus
Candy			
Milk chocolate	0 to 34	40	6 to 12 months
Peanut brittle	0 to 34	40	1.5 to 6 months
Fudge	0 to 34	65	5 to 12 months
Marshmallows	0 to 34	65	3 to 9 months
Miscellaneous			
Alfalfa meal	0	70 to 75	1 year plus
Beer			
Keg	35 to 40		3 to 8 weeks
Bottles and cans	35 to 40	65 or below	3 to 6 months
Bread	0		3 to 13 weeks
Canned goods	32 to 60	70 or lower	1 year
Cocoa	32 to 40	50 to 70	1 year plus
Coffee, green	35 to 37	80 to 85	2 to 4 months

Table 1 Storage Requirements of Perishable Products

	Storage Temperature, °F	Relative Humidity, %	Approximate Storage Life[a]
Miscellaneous (*Continued*)			
Fur and fabrics	34 to 40	45 to 55	Several years
Honey	50		1 year plus
Hops	28 to 32	50 to 60	Several months
Lard (without antioxidant)	45	90 to 95	4 to 8 months
Maple syrup	0	90 to 95	12 to 14 months
Nuts	32 to 50	65 to 75	8 to 12 months
Oil, vegetable, salad	70		1 year plus
Oleomargarine	35	60 to 70	1 year plus
Orange juice	30 to 35		3 to 6 weeks
Popcorn, unpopped	32 to 40	85	4 to 6 weeks
Yeast, baker's compressed	31 to 32		
Tobacco			
Hogshead	50 to 65	50 to 65	1 year
Bales	35 to 40	70 to 85	1 to 2 years
Cigarettes	35 to 46	50 to 55	6 months
Cigars	35 to 50	60 to 65	2 months

Note: The text in this chapter or the appropriate commodity chapter gives additional information on many of the commodities listed.

[a]Storage life is not based on maintaining nutritional value.
[b]For a complete listing of frozen food practical storage life, see *Recommendations for the Processing and Handling of Frozen Foods*, 3rd ed. International Institute of Refrigeration, 1986.
[c]For a more complete listings of grapes and pears, see *Recommendations for Chilled Storage of Perishable Foods*, International Institute of Refrigeration, 1979.
[d]Lemons stored in production areas for conditioning are held at 13 to 14°C, but sometimes they are held at 0°C.
[e]Eggs with weak albumen freeze just below −1°C.

to 99% rh instead of 100% rh. Lowering the vapor pressure deficit by lowering the air temperature is an excellent means of reducing water loss during storage.

Other important factors in desiccation include product size, surface-to-volume ratio, the kind of protective surface on the product, and air movement. Of these, the storage operator can control only the last, and this control is greatly influenced by the container, kind of pack, and stacking arrangement (i.e., the ability of the air to move past individual fruits and vegetables).

As a rule, shrivelling does not become a serious market problem until fruits lose about 5% of their mass, but any loss reduces the salable amount. Moisture losses of 3 to 6% are enough to cause a marked loss of quality for many kinds of produce. A few kinds may lose 10% moisture and still be marketable, although some trimming may be necessary, as for stored cabbage.

The vapor pressure deficit cannot be kept at a zero level, but it should be maintained as low as possible. A maximum of about 60 Pa, which corresponds to 90% rh at 0°C, is recommended. Some compromise is possible for short storage periods. In many instances, the refrigerated storage operator may find it desirable to add moisture, or, in special cases, the owner of the produce may find it desirable to use moisture barriers such as film liners.

REFRIGERATED STORAGE PLANT OPERATION

Checking Temperatures and Humidity

To maintain top product quality, temperature in the cold storage room must be accurately maintained. Variations of 1 to 1.5 K in the product temperature above or below the desired temperature are too large in most cases. Storage rooms should be equipped either with accurate thermostats or with manual controls that receive frequent attention.

In refrigerated storage rooms, thermometers are usually placed at a height of about 1.5 m for convenience in reading. Temperatures should be monitored where they might be undesirably high or low—one or two aisle temperatures is not enough. A record of both product and air temperatures is necessary to determine performance of the storage plant. A thermometer or recording device of good quality that is periodically calibrated is essential.

Temperature in less accessible storage locations, such as the middle of stacks, can be obtained conveniently with distant reading thermometer equipment such as thermocouples or electrical-resistance thermometers.

Storage instructions or recommendations usually specify a relative humidity within 3 to 5% of the desired levels. An ordinary sling psychrometer at temperatures of 0°C or lower cannot be read that closely. An error of 0.3 K in reading either the wet- or the dry-bulb thermometer will cause an error of 5% rh. Carefully calibrated thermometers graduated to 0.05 K with a range of −5°C to +5°C are best adapted for this purpose in fruit storage. A convenient device for measuring humidity consists of a pair of these thermometers, mounted in a short length of metal casing attached to a spring or motor-operated fan that draws air past the thermometers at a speed of 0.9 m/s or faster. The thermometers should be placed so that they will not be heated by the fan motor, and they should be read quickly to prevent warming. The advantage of this instrument over the sling psychrometer is that it can be left in the room long enough to get a true wet-bulb reading. This may require 15 minutes or more if ice is formed on the wet bulb. Under these conditions, a thin coating of ice is preferable to a thick one in getting accurate readings. Hair hygrometers are satisfactory if not subjected to sudden large changes in humidity and temperature and if checked regularly with a psychrometer.

Perhaps a more accurate method of determining the relative humidity is to electrically record the dew-point temperature of the air and to use a resistance thermometer of suitable sensitivity to record the ambient or dry-bulb temperature. From these temperature records, the relative humidity can be calculated.

Air Circulation

Air must be circulated to keep refrigerated storage rooms at an even temperature throughout. Commodity temperatures in a storage room may vary because the air temperature rises as the air passes through the room and absorbs heat from the commodity; also, heat leakage may vary in different parts of the storage. In a duct system, the air near the return ducts will be warmer than the air near delivery ducts. In many storages, refrigeration units are installed over the center aisle. Air circulates from the center of the rooms outward to the walls, down through the rows of produce, and back up through the center of the room.

Rapid air circulation is needed most during removal of field heat. Sometimes this is best done in separate cooling rooms that have more refrigerating and air-moving capacity than regular refrigerated storage rooms (see Chapter 14). After field heat is removed, a high air velocity is usually undesirable. Air movement is needed only to remove respiratory heat and heat entering the room through exterior surfaces and doorways. Excessive air movement can increase moisture loss from a product, with a resulting loss of both mass and quality. However, some air circulated by fans or blowers must be uniformly distributed through all parts of the room. Also, if the air circulation or refrigeration is turned off, some precaution is needed to ensure that the stored product does not become too warm.

The type of container and the manner of stacking are important factors that influence cooling performance. An elaborate system for air distribution is useless if poor stacking prevents airflow. If spacing is irregular, the wider spaces will get a greater volume of air than narrower ones. If some spaces are partially blocked, dead air zones will occur with resultant higher temperatures.

Commodity Storage Requirements

Sanitation and Air Purification

The refrigerated storage and the product containers must be clean. Fruits and vegetables coming into the storage are generally contaminated with mold spores, which can enter through punctures or breaks in the skin of the product. Commodities so contaminated can foster a rapid development of the spores, which are carried by air currents throughout the storage. Removing decaying raw material from the storage and sanitizing the product container will reduce the problem. If mold contamination is excessive, the storage develops a musty or moldy odor that is quickly picked up by the fruits and vegetables, a fault of many apple varieties and other products held for several months in refrigerated storage. This problem can best be controlled by means of special cleaners, sanitizers, and deodorizers (see Chapters 9, 21, 22, and 23 for details on product diseases).

During several months of storage, even at −0.5°C, molds may grow on the surface of packages and on the walls and ceilings of rooms under high relative humidity conditions. These surface molds generally will not rot fruits and vegetables. However, because surface molds are unsightly, storage warehouses should have a thorough cleaning at least once a year. Good air circulation alone is of considerable value in minimizing growth of surface molds. If floors and walls become moldy, they can be scrubbed with a cleaner containing sodium hypochlorite or trisodium phosphate, then rinsed, and aired. Field boxes and equipment can be cleaned with 0.25% calcium hypochlorite solutions or by exposing to steam for 2 min.

All inspected plants and warehouses operate under regulations with sanitation requirements clearly set forth in inspection service orders. Plants should be constructed to prevent the entrance of insects and rodents. This involves rat-proof building construction and adequate screening. For doors frequently opened, special measures must be taken to prevent the entrance of insects.

The frequency of cleanup operations and the detergents and sanitizing agents used should be specified by the quality control leader. A representative from quality control should inspect all areas after cleanup and determine whether or not a suitable job has been done. Waste and miscellaneous trash accumulation areas must receive special attention around warehouses, as they become breeding places for rodents and insects. In any food warehouse, the successful quality control and sanitation program depends on the cooperation and vigilance of management.

Air may be purified in storage rooms where odors or volatiles may contribute to off-flavors and hasten deterioration. Air may be cleaned with trays or canisters containing 6 to 14 mesh (2 to 4 mm spacing of mesh), activated coconut shell carbon. Pinewood volatiles are removed by activated carbon air-purifying units. Some produce volatiles are also removed, but ethylene, a ripening gas, is not removed by activated carbon alone.

Air washing with water to remove volatiles does not retard fruit ripening. Air washing may increase the relative humidity and thus aid in maintaining good fruit appearance by reducing mass loss.

Removal of Produce from Storage

When produce is removed from storage, undue warming and condensation of moisture, which promote decay and deterioration, must be prevented. Because most storages are built on railroad sidings, canvas tunnels should be installed between the car and the storage through which the produce will be conveyed to minimize warming and condensation of moisture.

When produce is removed from storage for distribution to wholesale and retail markets, the storage operator can do little to prevent undesirable condensation. Warming the packages until they are above the dew point of the air would prevent it, but this takes time and space and is seldom practicable. Deterioration in flavor and condition proceeds rapidly after long storage periods. Therefore, the produce should be moved to consumers as rapidly as possible.

STORAGE OF FROZEN FOODS

Frozen foods deteriorate during the period between production and consumption. The extent of deterioration depends mainly on storage temperature and storage time, although other factors such as protection provided by the package are important. Bacteria in frozen foods may be killed during freezing and frozen storage, but all the bacteria present are never completely destroyed. When defrosting, frozen foods are still subject to bacterial decomposition.

OTHER PRODUCTS

Beer

Since beer in bottles or cans is either pasteurized or filtered to destroy or remove the living yeast cells, it does not require as low a storage temperature as keg beer. Bottled beer may be stored at ordinary room temperature 21 to 24°C, but for convenience, it is often stored with keg beer at a lower temperature of 2 to 4°C. The bottled product should be protected from strong light, especially direct sunlight. The storage life will vary from 3 to 6 months, depending largely on the method of processing and packaging. Keg beer, usually stored at 2 to 4°C, has a storage life of 3 to 6 weeks.

Canned Foods

Canned foods that are heat processed in hermetically sealed containers do not benefit from refrigerated storage if they are to be stored for no longer than 2 or 3 months at temperatures that rarely exceed 24°C. However, seasonal commodities produced to provide an inventory for an entire year or longer do benefit from reduced temperature and humidity storage, which delays the onset of considerable color, texture, and flavor changes, loss of nutrients, and container corrosion. Notable examples are canned asparagus, cherries, and catsup. Reduced temperature and humidity storage for canned goods is essential in environments where ambient temperatures and humidities regularly exceed 30°C and 70% rh.

Dried Foods

Dried foods and feeds, particularly those expected to supply a high level of protein, such as dehydrated milk or alfalfa meal, should be protected against high temperatures and humidities. Good packaging, such as canning in vacuum, can maintain dried food nutritive value and quality for a year or longer if ambient temperatures do not exceed 30°C regularly. When stored in bulk or in bags that are not good water vapor barriers, the storage life at 40% rh and 21°C should be limited to less than one year, and at 30°C to less than 6 months. At 60% rh and 21°C, the storage life should be limited to 6 months, and at 30°C to not more than 3 to 4 months. The only process by which dried foods can maintain quality and nutritive values for a year and longer at elevated temperatures is by packaging in zero oxygen. Similar or better results can be obtained with −20°C storage regardless of adequacy of the package or relative humidity.

Furs and Fabrics

Refrigerated storage effectively protects furs, floor coverings, garments, and other materials containing wool against insect damage. The commonly used refrigerated storage temperatures do not kill the insects, but inactivate them, preventing insect damage while the susceptible items are in storage (Table 2). However, if insects are present, the article is susceptible to damage as soon as it is removed from refrigerated storage.

Articles should be free of any possible infestation before placement in refrigerated storage. Those items that can be cleaned should be so treated. Others can be either fumigated or mothproofed.

Furs and garments should be stored at 1 to 4°C. A temperature of 4°C is most widely used commercially. The low temperature not only inactivates fabric insects but also preserves the vitality and luster of furs and the tensile strength of fabrics.

Table 2 Temperature and Time Requirements for Killing Moths in Stored Clothing

Storage Temperature, °C	All Eggs Dead After, Days	All Larvae Dead After, Days	All Adults Dead After, Days
−18 to −15	1	2	1
−15 to −12	2	21[a]	1
−12 to −9	4	—	1
−9 to −7	—	—	1
−7 to −4	21	67	4
−4 to −1	21	125[b]	7
−1 to +2	—	283[c]	—

Table adapted from USDA *Publication* AMS-57 (1955).
[a] 50 to 95% of larvae may be killed in 2 days.
[b] A few larvae survived this period.
[c] Larvae survived this period.

Continuous storage below 1 to 4°C is a wasteful expense as far as protection from insect damage is concerned. Food should not be stored with fur garments. Some storage firms maintain constant temperatures in their fur vaults between −10 and 0°C and claim excellent results. However, no research indicates that temperatures in this range are required for storing dressed furs or fabrics. Cured raw furs (but not processed) should be stored at −23 to −12°C with 45 to 60% rh and will keep up to 2 years.

Honey

Both extracted (liquid) and comb honey can be held satisfactorily in common dry storage for about a year. The slow darkening and flavor deterioration at ordinary room temperatures becomes objectionable after this time. Although cold storage is not necessary, temperatures below 10°C will maintain original quality for several years and retard or prevent fermentation. The range between 10 and 18°C should be avoided, as it promotes granulation; this increases the probability of fermentation of raw (unheated) honey. As storage temperature increases in the 27 to 38°C range, deterioration is accelerated; temperatures constantly above 30°C are unsuitable and above 32°C, quite damaging.

Honey for export is best kept in cold storage, since the half-life for honey diastase at 25°C is about 17 months.

Raw honey of greater than 20% moisture is always in danger of fermentation; the likelihood is much less at or below 18.6% moisture. Below 17% moisture, raw honey will not ordinarily ferment. Granulation increases the possibility of fermentation of raw honey by increasing the moisture content of the liquid portion. Properly pasteurized honey will not ferment at any moisture content. Granulated honey can be reliquefied by warming to 50 to 60°C.

Comb honey should not be stored above 60% rh to avoid moisture absorption through the wax, which leads to fermentation.

Finely granulated honey (honey spread, Dyce process honey, and honey cream) must not be stored above about 25°C. Higher temperatures will, in time, cause partial liquefaction and destroy the texture. Any subsequent regranulation by lower temperatures will produce an undesirable coarse texture. For holding more than 4 months, cold storage is required.

Maple Syrup

Maple syrup keeps indefinitely at room temperatures without darkening or losing flavor, if packed hot (at or within a few degrees of its boiling point) and in clean containers, which are promptly closed airtight and laid on their sides or inverted to self-sterilize the closure, and then cooled. Refrigerated storage is not necessary. However, once opened, the syrup in a bottle, can, or drum may become contaminated by organisms in the air. Mold or yeast spores, which may be present in improperly pasteurized syrup, though unable to germinate in full-density syrup, may grow in the thin syrup on the surface caused by condensed water. Small packages not completely sterile, containing spores, can be kept free of vegetative growth by periodically inverting the containers to redisperse any thin syrup on the surface caused by condensed water. Maple syrup should never be packaged at temperatures below 82°C. After pasteurizing, cool the syrup as quickly as possible to prevent stack burn, which darkens the syrup and causes a lowering of its grade.

Nursery Stock and Cut Flowers

Low temperature (0°C ± 0.5°C) and dry packaging prevent, or at least greatly retard, flower disintegration and extend the storage life. The temperature and approximate storage life given in Table 3 for cut flowers allow for a reasonable shelf life after removal from storage; however, the storage period may be extended beyond that listed. These conditions, while not widely used commercially, are recommended. Proper dry packing requires a moisture-vaporproof container in which flowers can be sealed. No free water is added because the package prevents almost all water loss.

Flowers held in water should not be crowded in the containers and should be arranged on shelves or racks to allow good air circulation. Forced air circulation should be provided, but flowers must be kept out of a direct draft. Use clean water and clean containers.

Many kinds of nursery stock can also be stored at temperatures of −0.5 to 2°C. Open packages and harden flowers before marketing if the blooms have been stored for long periods. Flowers conditioned at about 10°C following storage regain full turgidity most rapidly. Cut or crush stem ends and then place in water or a food solution at 27 to 38°C for 6 to 8 h.

Many kinds of cut flowers and greens are injured if stored in the same room with certain fruits, principally apples and pears, which give off gases (such as ethylene) during ripening. These gases cause premature aging of blooms and may defoliate greens. Greens should not be stored in the same room with cut flowers because the greens, acting in the same way as fruit, can hasten bloom deterioration.

Greens, bulbs, and certain nursery stock are usually packaged or crated when stored. Some bulbs and nursery stock are packed in damp moss or similar material, and low temperatures are required to keep them dormant. Polyethylene wraps or box liners are very effective for maintaining quality of strawberry plants, bare-root rose bushes, and certain cuttings and other nursery stock in storage. Strawberry plants can be stored up to 10 months in polyethylene-lined crates at −1 to 0°C.

Popcorn

Store popcorn at 0 to 4°C and about 85% rh. This relative humidity yields the optimum popping condition and the desired moisture content of about 13.5%.

Vegetable Seeds

Seeds generally benefit from low temperatures and low humidity storage. High temperatures and high humidity favor loss of viability. Most vegetable seeds undergo no significant decrease in germination during one season when stored at 10°C and 50% rh. Full viability is retained far longer than one year as temperature and humidity are reduced. A temperature of −7°C and 15 to 25% rh are considered ideal but rarely necessary unless seed viability must be maintained for many years. Aster, pepper, tomato, and lettuce seeds stored under these conditions had equal or better viability after 13 years than did the fresh seed.

Low moisture content of the seed is important for germination. Hemp seed containing 9.5% moisture was mostly dead after 12 years storage at 10°C, but lost only 12% viability when moisture content was 5.7%. If stored at about −20°C, moisture percentage should not exceed 10% for many species. Seed with higher moisture content, when stored at −20°C, eventually equilibrates at a lower moisture level but may initially suffer frost damage. At higher temperatures, if it is impossible to keep humidity low enough, seeds must be stored in moistureproof containers.

Commodity Storage Requirements

Table 3 Storage Conditions for Cut Flowers and Nursery Stock

Commodity	Storage Temperature, °C	Relative Humidity, %	Approximate Storage Life	Method of Holding	Highest Freezing Point, °C
Cut Flowers					
Calla Lily	4.4	90 to 95	1 week	Dry pack	
Camellia	7	90 to 95	3 to 6 days	Dry pack	−0.8
Carnation	−0.6 to 0	90 to 95	3 to 4 weeks	Dry pack	−0.7
Chrysanthemum	−0.6 to 0	90 to 95	3 to 4 weeks	Dry pack	−0.8
Daffodil (Narcissus)	0 to 0.6	90 to 95	1 to 3 weeks	Dry pack	−0.1
Dahlia	4.4	90 to 95	3 to 5 days	Dry pack	
Gardenia	0 to 1	90 to 95	2 weeks	Dry pack	−0.6
Gladiolus	2 to 5.5	90 to 95	1 week	Dry pack	−0.3
Iris, tight buds	−0.6 to 0	90 to 95	2 weeks	Dry pack	−0.8
Lily, Easter	0 to 1.7	90 to 95	2 to 3 weeks	Dry pack	−0.5
Lily-of-the-valley	−0.6 to 0	90 to 95	2 to 3 weeks	Dry pack	
Orchid	7 to 13	90 to 95	2 weeks	Water	−0.3
Peony, tight buds	0 to 1.7	90 to 95	4 to 6 weeks	Dry pack	−1.1
Rose, tight buds	0	90 to 95	2 weeks	Dry pack	−0.4
Snapdragon	4.4 to 5.5	90 to 95	1 to 2 weeks	Dry pack	−0.9
Sweet Peas	−0.6 to 0	90 to 95	2 weeks	Dry pack	−0.9
Tulips	−0.6 to 0	90 to 95	2 to 3 weeks	Dry pack	
Greens					
Asparagus (plumosus)	1.7 to 4	90 to 95	2 to 3 weeks	Polylined cases	−3.3
Fern, dagger and wood	−1 to 0	90 to 95	2 to 3 months	Dry pack	−1.7
Fern, leatherleaf	1 to 4	90 to 95	1 to 2 months	Dry pack	
Holly	0	90 to 95	4 to 5 weeks	Dry pack	−2.8
Huckleberry	0	90 to 95	1 to 4 weeks	Dry pack	−2.9
Laurel	0	90 to 95	2 to 4 weeks	Dry pack	−2.4
Magnolia	1.7 to 4.4	90 to 95	2 to 4 weeks	Dry pack	−2.8
Rhododendron	0	90 to 95	2 to 4 weeks	Dry pack	−2.4
Salal	0	90 to 95	2 to 3 weeks	Dry pack	−2.9
Bulbs					
Amaryllis	3 to 7	70 to 75	5 months	Dry	−0.7
Caladium	21	70 to 75	2 to 4 months		−1.3
Crocus	9 to 17		2 to 3 months		−1.3
Dahlia	4 to 9	70 to 75	5 months	Dry	−1.8
Gladiolus	7 to 10	70 to 75	5 to 8 months	Dry	−2.1
Hyacinth	17 to 20		2 to 5 months		−1.5
Iris, Dutch, Spanish	20 to 25	70 to 75	4 months	Dry	−1.5
Gloriosa	10 to 17	70 to 75	3 to 4 months	Polyliner	
Candidum	−0.6 to 0.6	70 to 75	1 to 6 months	Polyliner and peat	
Croft	−0.6 to 0.6	70 to 75	1 to 6 months	Polyliner and peat	
Longiflorum	−0.6 to 0.6	70 to 75	1 to 10 months	Polyliner and peat	−1.7
Speciosum	−0.6 to 0.6	70 to 75	1 to 6 months	Polyliner and peat	
Peony	0.6 to 1.7	70 to 75	5 months	Dry	
Tuberose	4.4 to 7	70 to 75	4 to 12 months	Dry	
Tulip	17	70 to 75	2 to 6 months	Dry	−2.4
Nursery Stock					
Trees and shrubs	0 to 2	95	4 to 5 months	a	
Rose bushes	−0.5 to 2	85 to 95	4 to 5 months	Bare rooted with polyliner	
Strawberry plants	−1 to 0	80 to 85	8 to 10 months	Bare rooted with polyliner	−1.2
Rooted cuttings	−0.5 to 2	85 to 95		Polywrap	
Herbaceous perennials	−2.8 to −2.2	80 to 85	4 to 8 months	a	
	−0.6 to 1.7	80 to 85	3 to 7 months		
Christmas trees	−5.5 to 0	80 to 85	6 to 7 weeks		

Data from USDA *Agricultural Handbook* No. 66.
a For details for various trees, shrubs, and perennials, see Storage of Nursery Stock (AAN 1960).

BIBLIOGRAPHY

Hardenburg, R.E., A.E. Watada, and C.Y. Wang. 1986. The commercial storage of fruits, vegetables, and florist and nursery stocks. USDA *Handbook* No. 66. USDA, Washington, D.C.

Hunt Ashby, B. 1987. Protecting perishable foods during transport by truck. USDA *Handbook* No. 669. USDA, Washington, D.C.

IIR. 1993. *Cold Stores Guide*. International Institute of Refrigeration, Paris.

IIR. 1990. Principles of refrigerated preservation of perishable foodstuffs. Manual of refrigerated storage in the warmer developing countries. International Institute of Refrigeration, Paris.

IIR. 1986. General principles for the freezing, storage and thawing of foodstuffs. Recommendations for the processing and handling of frozen foods. International Institute of Refrigeration, Paris.

ISO. 1974. Avocados—Guide for storage and transport. International Organization for Standardization, Geneva.

Kader, A.A. 1986. Biochemical basis for effect of controlled and modified atmospheres on fruits and vegetables. *Food Technology*.

Scott, V.N. 1989. Interaction of factors to control microbial spoilage of refrigerated foods. *Journal of Food Protection*.

Schlimme, D.V., M.A. Smith, and L.M. Ali. 1991. Influence of freezing rate, storage temperature, and storage duration on the quality of cooked turkey breast roll. ASHRAE *Transactions* 97(1).

Schlimme, D.V., M.A. Smith, and L.M. Ali. 1991. Influence of freezing rate, storage temperature, and storage duration on the quality of turkey frankfurters. ASHRAE *Transactions* 97(1).

Tressler, D.K. and C.F. Evers. 1957. *The freezing preservation of foods*, 3rd ed. AVI Publishing Co., Inc., Westport, CT.

Ulrich, R. 1979. Fruits and vegetables. Recommendations for chilled storage of perishable produce. International Institute of Refrigeration, Paris.

USDA. 1955. Protecting stored furs from insects. USDA *Publication* AMS-57.

Valenzuela Segura, G., D.D. Delgado, and D.R. Ramirez. 1972. Handling, storage and transport systems for exported refrigerated perishable foods. Revista del Instituto de Investigaciones Technologicas, Bogota, Columbia.

Webb, B.H., A.H. Johnson, and J.A. Alford. 1973. *Fundamentals of dairy chemistry*. AVI Publishing Co., Inc., Westport, CT.

CHAPTER 11

FOOD MICROBIOLOGY AND REFRIGERATION

Basic Microbiology 11.1
Critical Microbial Growth Requirements 11.1
Design for Control of Microorganisms 11.3
The Role of HACCP 11.4
Sanitation 11.4
Regulations and Standards 11.5

REFRIGERATION'S largest overall application is the prevention or retardation of microbial, physiological, and chemical changes in foods. Even at temperatures near the freezing point, foods may deteriorate through the growth of microorganisms, through changes caused by enzymes, or through chemical reactions. Holding foods at low temperatures merely reduces the rate at which these changes take place.

Refrigeration also plays a major role in maintaining a safe food supply. Overall, the leading factor causing food-borne illness is improper holding temperatures for food. Another important factor is improperly sanitized equipment. Engineering directly impacts the safety and stability of the food supply in design of cleanable equipment and facilities, as well as maintenance of environmental conditions that inhibit microbial growth. This chapter briefly discusses the microbiology of foods and the impact of design decisions on the production of safe and wholesome foods. Methods of applying refrigeration to specific foods may be found in other chapters.

BASIC MICROBIOLOGY

Microorganisms play several roles in a food production facility. They can contribute to food spoilage, producing off-odors and flavors, or altering product texture or appearance through slime production and pigment formation. Certain organisms cause disease; others are beneficial and are required to produce foods such as cheese, wine, and sauerkraut through fermentation.

Microorganisms fall into four categories—bacteria, yeasts, molds, and viruses. Bacteria are the most common food-borne pathogens. Bacterial growth rates, under optimum conditions, are generally faster than those of yeasts and molds, making bacteria a prime cause of spoilage, especially in refrigerated, moist foods. Bacteria have many shapes, including spheres (cocci), rods, or spirals, that are usually between 0.3 and 5 to 10 mm in size. Bacteria are capable of growth in a wide range of environments. Some bacteria, notably *Clostridium* and *Bacillus* spp., form endospores, i.e., resting states with extensive temperature, desiccation, and chemical resistance.

Yeasts and molds become important in situations that restrict the growth of bacteria, such as in acidic or dry products. Yeasts can cause gas formation in juices and slime formation on fermented products. Mildew (black mold) on humid surfaces and mold formation on spoiled foods are also common.

Viruses are obligate, intracellular parasites, which are specific to an individual host. Human viruses, such as Hepatitis A, cannot multiply outside the human body. Design features must include facilities for good employee handwashing and sanitation practices to minimize potential for product contamination. Bacterial viruses (phage), however, may contribute to starter culture failure in bacterial fermentations if proper isolation, ventilation, and sanitation procedures are not followed. The use of commercial concentrated cultures, selected for phage resistance, has greatly reduced this problem.

The preparation of this chapter is assigned to TC 10.9, Refrigeration Application for Foods and Beverages.

Sources of Microorganisms

Bacteria, yeasts, and molds are widely distributed in water, soil, air, plant materials, and the skin and intestinal tracts of humans and animals. Practically all unprocessed foods will be contaminated with a variety of spoilage and, sometimes, pathogenic microorganisms. Food processing environments that contain residual food material will naturally select the microorganisms that are most likely to spoil the particular product.

Microbial Growth

All microbial populations follow a generalized growth curve (Figure 1). An initial lag phase occurs as organisms start to grow and adapt to new environmental conditions. The lag phase is very important because the maximum extension of shelf life and length of production runs are directly related to the length of the lag phase. Once adaptation has occurred, the culture enters into the maximum (logarithmic) growth rate, and control of microbial growth is not possible without major sanitation or other drastic measures. Numbers can double as fast as every 20 to 30 min under optimum conditions.

Toxin production and spore maturation, if possible, usually occur at the end of the exponential phase as the culture enters into a stationary phase. At this time, essential nutrients are depleted and/or inhibitory by-products are accumulated. Eventually the culture dies; the rate depends on the organism, the medium, and other environmental characteristics. While refrigeration prolongs generation time and reduces enzyme activity and toxin production, in most cases, refrigeration will not reverse a predeveloped situation.

CRITICAL MICROBIAL GROWTH REQUIREMENTS

Factors that influence microbial growth can be divided into two categories: (1) intrinsic factors that are a function of the food itself and (2) extrinsic factors that are a function of the environment in which a food is held.

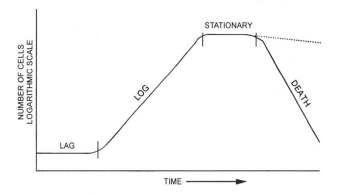

Fig. 1 Typical Microbial Growth Curve

Intrinsic Factors

Intrinsic factors affecting microbial growth include nutrients, inhibitors, biological structures, water activity, pH, and presence of competing microorganisms in a food. While engineering practices have little impact on these parameters, an understanding of how intrinsic factors influence growth is useful in predicting the types of microorganisms that may be present.

Nutrients. Like other living organisms, microorganisms require food to grow. Carbon and energy sources are usually supplied in the form of sugars and starches. Nitrogen requirements are met by the presence of protein. Vitamins and minerals are also necessary. Lactic acid bacteria have rather exacting nutritional requirements, while many aerobic spore formers have tremendous enzymatic capabilities that allow growth on a wide variety of substrates. Cleanable systems facilitate the removal of residual food material and deprive microorganisms of the nutrients required for growth, thus preventing a buildup of organisms in the environment.

Inhibitors. Either naturally occurring or added as preservatives, inhibitors may be present in food. Preservatives are not substitutes for hygienic practices and, with time, microorganisms may develop resistance. A cleanable processing system is still essential in preventing the development of a resistant population.

Competing Microorganisms. The presence of other microorganisms also affects the organisms in foods. Some organisms produce inhibiting compounds or grow faster; others are better able to use the available nutrients in a food matrix.

Water Activity. All life-forms require water for growth. Water activity (a_w) refers to the availability of water within a food system and is defined at a given temperature as:

$$a_w = \frac{\text{Vapor pressure of solution (food)}}{\text{Vapor pressure of solute (water)}}$$

The minimum water activities for growth of a variety of microorganisms, along with representative foods, are listed in Table 1. These a_w minima are also factors in environmental humidity control discussed in the section on Extrinsic Factors.

When food is enclosed in airtight packaging or in a chamber with limited air circulation, an equilibrium a_w is achieved that is equal to the a_w of the food. In these situations, the a_w of the food determines which organism will be capable of growth. If the same foods are exposed to reduced environmental relative humidity, such as meat carcasses hanging in a controlled aging room or vegetables displayed in an open case, surface dehydration acts as an inhibitor to microbial growth. Likewise, if a dry product, such as bread, is exposed to a moist environment, mold growth may occur on the surface as moisture is absorbed. Environmental relative humidity thus has a significant impact on product shelf life.

Table 1 Approximate Minimum Water Activity for Growth of Microorganisms

Organism	a_w	Foods
Pseudomonads	0.98	Fresh fruits, vegetables, meats
Salmonella spp., *E. coli*	0.95	Many processed foods
Listeria monocytogenes	0.93	
Bacillus cereus	0.92	Salted butter, fermented sausage
Staphylococcus aureus	0.86	
Molds	0.84	Soft, moist pet food
	0.80	Pancake syrup, jam
	0.70	Corn syrup
Xerotrophic molds	0.65	Caramels
Osmophilic yeasts	0.62	
Limit of microbial growth	0.60	Wheat flour
	0.40	Nonfat dry milk

pH. For most microorganisms, optimal growth occurs at neutral pH, or 7.0. Few organisms grow under alkaline conditions, while some organisms, such as yeasts, molds, and lactic acid bacteria, are acid tolerant. Figure 2 depicts pH values of a variety of foods and limiting pH values for microorganisms.

Extrinsic Factors

Extrinsic factors that influence the growth of microorganisms include temperature, environmental relative humidity, and oxygen levels. Refrigeration and ventilation systems play a major role in controlling these factors.

Temperature. Microorganisms are capable of growth over a wide range of temperatures. Minimum growth temperatures for a variety of spoilage and pathogenic bacteria of significance in foods are summarized in Table 2. Previously, 7°C was thought to be sufficient to control the growth of pathogenic organisms. However, the emergence of psychrotrophic pathogens, such as *Listeria monocytogenes*, has demonstrated the need for lower temperatures. In the United States, 5°C is now recognized as the upper limit for safe refrigeration temperature, while 1°C or lower may be more appropriate. Foods that will support the growth of pathogenic microorganisms may not be held between 5 and 60°C for more than 4 h.

Temperature is used to categorize microorganisms. Those capable of growth above 45°C, with optimum growth at 55 to 65°C, are **thermophiles**. Thermophilic growth can be extremely rapid, with generation times of 10 to 20 min. Thermophiles can become a problem in blanchers and other equipment that maintains food at elevated temperatures for extended periods. These organisms die or do not grow at refrigeration temperatures.

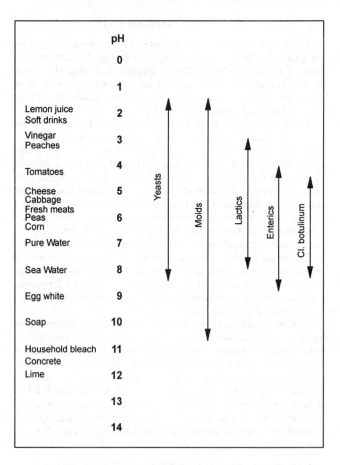

Fig. 2 pH Ranges for Microbial Growth and Representative Examples

Food Microbiology and Refrigeration

Table 2 Minimum Growth Temperatures for Some Bacteria in Foods

Organism	Possible Significance	Approximate Minimum Growth Temperature, °C
Staphylococcus aureus	Food-borne disease	10
Salmonella spp.	Food-borne disease	5.5
Clostridium botulinum, proteolytic nonproteolytic	Food-borne disease	10
Lactobacillus and *Leuconostoc*	Spoilage of cooked sausage	
Listeria monocytogenes	Food-borne disease	
Acinetobacter spp.	Spoilage of precooked foods	−1
Pseudomonads	Spoilage of raw fish, meats, poultry, and dairy products	−1

Mesophiles are organisms that grow best between 20 and 45°C. Most pathogens are in this group, with optimum growth temperatures around 37°C, i.e., body temperature. Mesophiles also include a number of spoilage organisms. Growth of mesophiles is quite rapid, with typical generation times of 20 to 30 min. Because mesophiles grow so rapidly, perishable foods must be cooled as fast as possible to prevent spoilage or potential unsafe conditions. Also, slower cooling rates cause mesophiles to adapt and grow at lower temperatures. With mild temperature abuse, prolific growth leading to spoilage or a potential health hazard can occur.

Psychrotrophs are organisms capable of growth at 5°C; some are able to grow at temperatures as low as −5°C and are a primary cause of spoilage of perishable foods. Psychrotrophic growth is slow in comparison to mesophilic and thermophilic growth, with maximum growth rates of 1 to 2 h or longer. However, control of psychrotrophic growth is a major requirement in products with extended shelf life. Because many psychrotrophs have optimum temperatures in the mesophilic range, what may seem to be an insignificant increase in temperature can have a major impact on the growth rate of spoilage organisms. Growth is roughly twice as fast with each 3 K increase in temperature.

For all the critical growth factors, the range over which growth can occur is characteristic for a given organism. The range for growth is narrower than that for survival. For example, the maximum temperature for growth is slightly above the optimum, and death usually occurs just slightly above the maximum. This is not the case at the lower end of the temperature range. Survival of psychrotrophic and most mesophilic microorganisms is enhanced by storage at low temperatures. Freezing of microorganisms is not an effective lethal process. Some organisms, notably gram-negative bacteria, are damaged by freezing and may die slowly, but others are extremely resistant. In fact, freezing is used as an effective means of preserving microbial cultures at extremely low temperatures, e.g., −80°C.

Environmental Relative Humidity. Water, previously discussed as an essential intrinsic growth factor, is also a major extrinsic factor. Environmental water acts as a vector for transmission of microorganisms from one location to another through foot traffic or aerosols. Refrigeration drain pans and drip coils have been identified as significant contributors of *L. monocytogenes* contamination in food processing environments. Aerosols have also transmitted the agent that causes Legionnaires' disease. High relative humidity in cold rooms is a particular problem and leads to black mold buildup on walls and ceilings as well as growth of organisms in drains and other reservoirs of water. Condensation that forms on ceilings supports microbial growth and can subsequently drip onto product contact surfaces. Inadequately drained equipment collects stagnant water and supports microbial growth that is easily transported throughout a production facility when people walk through puddles. It is extremely important to control environmental relative humidity in food production environments. Control measures are discussed further in the section on Regulations and Standards.

Oxygen. Microorganisms are frequently classified by their oxygen requirement. Strictly aerobic microorganisms, such as molds and pseudomonads, require oxygen for growth. Conversely, strict anaerobes, such as *Clostridium* spp., cannot grow in the presence of oxygen. Facultatively anaerobic microorganisms (e.g., coliforms) grow with or without oxygen present, and microaerophiles, such as lactobacilli, grow best in conditions with reduced oxygen levels. Controlled atmospheric chambers for fruit storage use lower oxygen levels to prolong storage life by retarding growth of spoilage organisms in addition to influencing ripening processes. Vacuum packing of foods also uses this extrinsic growth factor by inhibiting the growth of strict aerobes, such as molds and pseudomonads.

DESIGN FOR CONTROL OF MICROORGANISMS

Microorganisms can be controlled by one of three mechanisms—prevention of contamination, prevention of growth, or destruction of the organisms themselves. Design of refrigeration and ventilation systems can impact all these areas, sometimes in combination.

Prevention of Contamination

To prevent the entry of microorganisms into food production areas, ventilation systems must provide adequately clean air. Because bacteria are generally transported through air on dust particles, 95% filters (as defined in ASHRAE *Standard* 52.1, Gravimetric and Dust-Spot Procedures for Testing Air-Cleaning Devices Used in General Ventilation for Removing Particulate Matter) are sufficient to remove most microorganisms. High-efficiency particulate air (HEPA) filters provide sterile air and are used for cleanrooms.

Air-filtering materials must remain dry. Wet filters in ventilation systems support microbial growth, and organisms are transported throughout the production facility in the air. All ventilation systems must also be protected from water and the formation of condensate to prevent mold growth. This may require increased airflow or dehumidification systems. Positive pressure in the production environment prevents the entry of airborne contamination from sources other than ventilation ducts. Air intakes for production areas must not be positioned toward areas that are prone to contamination, such as puddles on roofs, nesting sites for birds, and so forth. Microbial contamination from airborne sources has been recorded, but quantitative assessments are not available to specify design parameters.

Refrigeration drip pans are a significant source of *L. monocytogenes* contamination. Condensation drip pans should be plumbed directly to drain to prevent contamination of floors and subsequent transport of organisms throughout a production facility. Drip pans must be easily accessible to allow scheduled cleaning, thus preventing organism growth. Air defrost should be avoided in critical areas. Continuous glycol-sprayed evaporative coils offer advantages, since glycol has been found to trap and kill microorganisms. Being hygroscopic, glycol depresses the dew point of the air providing a drier environment.

Traffic flow through a production facility should be planned to minimize contact between raw and cooked products. This is mandated through USDA regulations in plants that cook meat products. Straight line flow of a raw product from one end of a facility to the other prevents cross contamination. Walls separating raw product from cooked (or dirty from clean), with positive pressure in the cooked area, should be considered, as this provides the best means of protection. Provide adequate storage facilities to allow separate storage of raw ingredients from processed products, especially in facilities that handle meat products, which are a significant source of *Salmonella*. Raw meat must not be stored with cooked meats and/or vegetables or dairy products.

Prevention of Growth

Water control is one of the most effective and most frequently overlooked means of inhibiting microbial growth. All ventilation systems, piping, equipment, and floors must be designed to drain completely. Residual standing water supports rapid microbial growth, and foot and forklift traffic transports organisms from puddles throughout the production facility.

Condensation on ceilings and chilled pipes also supports microbial growth and may drip onto product contact surfaces that are not adequately protected. Efforts to prevent condensation are essential to prevent contamination. Insulation of pipe and/or dehumidifying systems may be necessary, particularly in chilled rooms. Increased airflow may also be useful in removing residual moisture. Maintenance of 70% rh prevents the growth of all but the most microorganisms; less than 60% rh prevents all microbial growth on facility surfaces (Table 1).

Sanitation procedures use much water and leave the facility and surfaces wet. Adequate dehumidification should be provided for removal of moisture during and after sanitation.

Controlling relative humidity is not always possible. For example, the aging of meat carcasses requires relative humidities of 90 to 95% to prevent excessive drying of the tissue. In these cases, temperatures just above the product freezing point should be used to inhibit microbial deterioration. Temperatures below 5°C inhibit the most common organisms that cause food-borne illness; however, 1°C is required to inhibit *L. monocytogenes*. Airflow, relative humidity, and temperature must be finely balanced to achieve maximum shelf life with limited deterioration of quality.

Freezing is also an effective means of microbial control. Limited death may occur during freezing, especially during slow freezing of gram-negative bacteria. However, freezing is not a reliable means of inactivating microorganisms. Since no microbial growth occurs in frozen foods, as long as a product remains below its freezing point, microbial safety issues are nonexistent. Frozen foods must be stored below −20°C for legal and quality reasons.

Destruction of Organisms

High temperature is an effective means of inactivating microorganisms and is used extensively in blanching, pasteurization, and canning. Moist heat is far more effective than dry heat. High temperatures (77°C) may also be used for sanitation when chemicals are not used. While hot water sanitation is effective against vegetative forms of bacteria, spores are not affected.

In addition to heat, gases such as ethylene oxide and methyl bromide, irradiation, ultraviolet light, and sanitation chemicals are effective in destroying microorganisms.

THE ROLE OF HACCP

Many of the procedures for the control of microorganisms are managed by the Hazard Analysis and Critical Control Point (HACCP) system of food safety. Developed in the food industry since the 1960s, HACCP is now accepted by food manufacturers and regulators. It is a preventive system that builds safety control features into the food product's design and the process by which it is produced. The HACCP system is used to manage physical and chemical hazards as well as biological hazards. The approach to HACCP is described in the seven principles developed by the National Advisory Committee on Microbiological Criteria for Foods (NACMCF 1992):

1. Conduct a hazard analysis and identify control measures.
2. Identify critical control points.
3. Establish critical limits.
4. Establish monitoring procedures.
5. Establish corrective actions.
6. Establish verification procedures.
7. Establish record keeping and documentation procedures.

Each food manufacturing site should have a HACCP team to develop and implement its HACCP plan. The team is multi-disciplinary, with members experienced in plant operations, product development, food microbiology, etc. Because of their knowledge of the manufacturing facility and the process equipment, engineers are important members of the HACCP team. They help identify the potential hazards and respective control measures, implement the HACCP plan, and verify its effectiveness.

SANITATION

Cleaning and sanitation are key elements that incorporate all three strategies for control of microorganisms. The cleaning phase controls microbial growth by removing the residual food material required for proliferation. Sanitizing kills most of the bacteria that remain on surfaces. This prevents subsequent contamination of food being produced. Most microbial issues that occur in food processing environments are caused by unclean equipment, sometimes due to design. Therefore, equipment and facilities designed with cleaning and sanitizing in mind maximize the effectiveness of control.

Products that are frozen prior to packaging are particularly vulnerable to contamination. Many freezing tunnels in food processing facilities are difficult or impossible to clean because of limited access and poor drainage. Although freezing temperatures control microbial growth, proliferation of organisms does occur during downtime, such as on weekends. The following points should be considered during design phases to minimize potential problems:

- Provide good access for the cleaning crew to facilitate cleaning and adequate lighting (540 lx) to allow inspection of all surfaces.
- Eliminate inaccessible parts and features that permit product accumulation.
- Design equipment that is easy to dismantle using few tools, especially for areas that are difficult to clean. Design air-handling ducts for ease of cleaning. Provide removable spools or access doors. Washable fabric ducts designed and approved for such use are another option.
- Use smooth and nonporous construction materials to prevent product accumulation. Materials must tolerate common cleaning and sanitizing chemicals listed in Table 3. Consult sanitation personnel to determine chemicals likely to be used. Give special attention to insulation materials, many of which are porous. Insulation must be protected from water to avoid saturation and resultant microbial growth. An effective method is a well-sealed PVC or stainless steel cover. Avoid using fiberglass batts in food processing plants.
- All equipment must drain completely.
- Consult references and regulations on sanitary design principles.

Innovation is needed in the area of drying after cleaning is complete. Provision of adequately sloped surfaces and sufficient drains to handle water is important. Dehumidification systems and/or increased airflow in new and existing systems could greatly reduce problems associated with water, especially in the cleaning of chilled production environments.

Table 3 Common Cleaning and Sanitizing Chemicals

Cleaning Compounds	Sanitizers
Caustic	Chlorine
Chlorinated alkaline detergents	Iodophors
Acid cleaners	Quaternary ammonium compounds
	Acid sanitizers

Food Microbiology and Refrigeration

Standard water washing procedures are not appropriate for certain food production facilities such as dry mix, chocolate, peanut butter, or flour milling operations. Refrigeration or ventilation systems for plants of this type must be made to facilitate dry cleaning, reduce condensation, and restrict water to a very confined area if it is absolutely necessary.

REGULATIONS AND STANDARDS

Facilities and equipment should be designed and installed for minimizing microbial growth and maximizing the ease of sanitation. Care should be taken to employ material that can withstand moisture and chemicals.

The food industry has developed several equipment installation standards. In the United States, examples are International Association of Milk, Food and Environmental Sanitarians (IAMFES, Ames, IA) 3-A Dairy standards, Baking Industry Sanitation Standards Committee (BISSC, Chicago, IL) bakery standards, and a select group of U.S. Department of Agriculture (USDA) standards for the meat industry. The USDA enforces the Federal Meat Inspection Act and Federal Poultry Inspection Act and requires approval of building and equipment plans.

Chapter VII, Section 701(A) of the Federal Food, Drug and Cosmetic Act as amended establishes current Good Manufacturing Practices (GMPs) in manufacturing, processing, packaging, or holding human food. These GMPs are listed in Section 21 of the Code of Federal Regulations (21 CFR), Part 110.

BIBLIOGRAPHY

Bibek, R. 1996. *Fundamental food microbiology*. CRC Press, Boca Raton, FL.

NACMCF. 1992. Hazard analysis and critical control point system. *Int. J. Food Microbiol.* 16: 1-23.

British Association of Chemical Specialties. 1989. *The control of Legionellae by safe and effective operation of cooling systems. A code of practice.*

Chartered Institute of Building Services Engineers. 1987. Technical Memorandum TM13.

FDA. 1995. *Food code 1995*. U.S. Public Health Service, Washington, DC.

Graham, D.J. 1991-1992. *Sanitary design—A mind set. A nine-part serial in Dairy, Food and Environmental Sanitation.*

Imholte, T.J. 1984. *Engineering for food safety and sanitation*. Technical Institute for Food Safety, Crystal, MN.

IIR. 1979. *Recommendations for chilled storage of perishable produce*. International Institute of Refrigeration, Paris.

IIR. 1986. *Recommendations for the processing and handling of frozen foods*, 3rd ed. International Institute of Refrigeration, Paris.

Jay, J.M. 1992. *Modern food microbiology*, 4th ed. Van Nostrand Reinhold, New York.

Marriott, N.G. 1989. *Principles of food sanitation*, 2nd ed. Van Nostrand Reinhold, New York.

NFPA. 1989. Guidelines for the development, production, distribution and handling of refrigerated foods. National Food Processors Association, Washington, D.C.

Todd, E. 1990. Epidemiology of food-borne illness: North America. *The Lancet* 336:788-93.

CHAPTER 12

REFRIGERATION LOAD

Transmission Load 12.1
Product Load 12.2
Internal Load 12.2
Infiltration Air Load 12.3
Equipment Related Load 12.5
Safety Factor 12.6
Total Refrigeration Load 12.6

THE segments of total refrigeration load are (1) transmission load, which is heat transferred into the refrigerated space through its surface; (2) product load, which is heat removed from and produced by products brought into and kept in the refrigerated space; (3) internal load, which is heat produced by internal sources, e.g., lights, electric motors, and people working in the space; (4) infiltration air load, which is heat gain associated with air entering the refrigerated space; and (5) equipment-related load.

The first four segments of load constitute the net heat load for which a refrigeration system is to be provided; the fifth segment consists of all heat gains created by the refrigerating equipment. Thus, net heat load plus equipment heat load is the total refrigeration load for which a compressor must be selected.

This chapter contains load calculating procedures and data for the first four segments and load determination recommendations for the fifth segment. Information needed for the refrigeration of specific foods can be found in Chapters 14, 16 through 26.

TRANSMISSION LOAD

Sensible heat gain through walls, floor, and ceiling is calculated at steady state as

$$q = UA\Delta t \tag{1}$$

where

q = heat gain, W
A = outside area of section, m²
Δt = difference between outside air temperature and air temperature of the refrigerated space, °C

The overall coefficient of heat transfer U of the wall, floor, or ceiling can be calculated by the following equation:

$$U = \frac{1}{1/h_i + x/k + 1/h_o} \tag{2}$$

where

U = overall heat transfer coefficient, W/(m²·K)
x = wall thickness, m
k = thermal conductivity of wall material, W(m·K)
h_i = inside surface conductance, W/(m²·K)
h_o = outside surface conductance, W/(m²·K)

A value of 9.3 W/(m²·K) for h_i and h_o is frequently used for still air. If the outer surface is exposed to 24 km/h wind, h_o is increased to 34 W/(m²·K).

With thick walls and low conductivity, the resistance x/k makes U so small that $1/h_i$ and $1/h_o$ have little effect and can be omitted from the calculation. Walls are usually made of more than one material; therefore, the value x/k represents the composite resistance of the materials. The U-factor for a wall with flat parallel surfaces of materials 1, 2, and 3 is given by the following equation:

$$U = \frac{1}{x_1/k_1 + x_2/k_2 + x_3/k_3} \tag{3}$$

The thermal conductivity of several cold storage insulations are listed in Table 1. These values decrease with age due to factors discussed in Chapter 22 of the 1997 *ASHRAE Handbook—Fundamentals*). Chapter 24 of the 1997 *ASHRAE Handbook—Fundamentals* includes more complete tables listing the thermal properties of various building and insulation materials.

Table 2 lists minimum insulation thicknesses of expanded polyisocyanurate board recommended by the refrigeration industry. These thicknesses may need to be increased to offset heat gain caused by building components such as wood and metal studs, webs in concrete masonry, and metal ties that bridge across the insulation and reduce the thermal resistance of the wall or roof. Chapter 24 of the 1997 *ASHRAE Handbook—Fundamentals* describes how to calculate heat gain through walls and roofs with thermal bridges. The metal surfaces of prefabricated or insulated panels have a negligible effect on thermal performance and should not be considered in calculating the U-factor.

Table 1 Thermal Conductivity of Cold Storage Insulation

Insulation	Thermal Conductivity k, W/(m·K)
Polyurethane board (R-11 expanded)	0.023 to 0.026
Polyisocyanurate, cellular (R-141b expanded)	0.027
Polystyrene, extruded (R-142b)	0.035
Polystyrene, expanded (R-142b)	0.037
Corkboard[b]	0.043
Foam glass[c]	0.044

a Values are for a mean temperature of 75°F and insulation is aged 180 days.
b Seldom used insulation. Data is only for reference.
c Virtually no effects due to aging.

Table 2 Minimum Insulation Thickness

Storage Temperature	Expanded Polyisocyanurate Thickness	
	Northern U.S.	Southern U.S.
°C	mm	mm
10 to 16	50	50
4 to 10	50	50
−4 to 4	50	75
−9 to −4	75	75
−18 to −9	75	100
−26 to −18	100	100
−40 to −26	125	125

The preparation of this chapter is assigned to TC 10.8, Refrigeration Load Calculations.

Table 3 Allowance for Sun Effect

Typical Surface Types	East Wall °C	South Wall °C	West Wall °C	Flat Roof °C
Dark colored surfaces				
Slate roofing	5	3	5	11
Tar roofing				
Black paint				
Medium colored surfaces				
Unpainted wood	4	3	4	9
Brick				
Red tile				
Dark cement				
Red, gray, or green paint				
Light colored surfaces				
White stone	3	2	3	5
Light colored cement				
White paint				

Note: Add °C to the normal temperature difference for heat leakage calculations to compensate for sun effect—do not use for air-conditioning design.

In most cases the temperature difference (Δt) can be adjusted to compensate for solar effect on the heat load. The values given in Table 3 apply over a 24-h period and are added to the ambient temperature when calculating wall heat gain.

Latent heat gain due to moisture transmission through walls, floors, and ceilings of modern refrigerated facilities is negligible. Data in Chapter 24 of the 1997 *ASHRAE Handbook—Fundamentals* may be used to calculate this load if moisture permeable materials are used.

Chapter 26 of the 1997 *ASHRAE Handbook—Fundamentals* gives outdoor design temperatures for major cities; values for 0.4% should be used.

Additional information on thermal insulation may be found in Chapter 22 and 23 of the 1997 *ASHRAE Handbook—Fundamentals*. Chapter 28 in the 1997 *ASHRAE Handbook* discusses load calculation procedures in greater detail.

PRODUCT LOAD

The primary refrigeration load from products brought into and kept in the refrigerated space are (1) the heat that must be removed to reduce the product temperature to storage temperature and (2) the heat generated by products in storage, mainly fruits and vegetables. The quantity of heat to be removed can be calculated as follows:

1. Heat removed to cool from the initial temperature to some lower temperature above freezing:

$$Q_1 = mc_1 (t_1 - t_2) \quad (4)$$

2. Heat removed to cool from the initial temperature to the freezing point of the product:

$$Q_2 = mc_1 (t_1 - t_f) \quad (5)$$

3. Heat removed to freeze the product:

$$Q_3 = mh_{if} \quad (6)$$

4. Heat removed to cool from the freezing point to the final temperature below the freezing point:

$$Q_4 = mc_2 (t_f - t_3) \quad (7)$$

where

Q_1, Q_2, Q_3, Q_4 = heat removed, kJ
m = mass of product, kg
c_1 = specific heat of product above freezing, kJ/(kg·K)
t_1 = initial temperature of product above freezing, °C
t_2 = lower temperature of product above freezing, °C
t_f = freezing temperature of product, °C
h_{if} = latent heat of fusion of product, kJ/kg
c_2 = specific heat of product below freezing, kJ/(kg·K)
t_3 = final temperature of product below freezing, °C

The refrigeration capacity required for products brought into storage is determined from the time allotted for heat removal and assumes that the product is properly exposed to remove the heat in that time. The calculation is:

$$q = \frac{Q_2 + Q_3 + Q_4}{3600\,n} \quad (8)$$

where

q = average cooling load, kW
n = allotted time, h

Equation (8) only applies to uniform entry of the product into storage. The refrigeration load created by nonuniform loading of a warm product may be much greater over a short period. See Chapter 14 for information on calculating the cooling load of warm product.

Specific heats above and below freezing for many products are listed in Table 3 of Chapter 8. A product's latent heat of fusion may be estimated by multiplying the water content of the product (expressed as a decimal) by the latent heat of fusion of water, which is 334 kJ/kg. Most food products freeze in the range of −3 to −0.5°C. When the exact freezing temperature is not known, assume that it is −2°C.

Example 1. 100 kg of lean beef is to be cooled from 18 to 4°C, then frozen and cooled to −18°C. The moisture content is 70%, so the latent heat is estimated as 334 × 0.70 = 234 kJ/kg. Estimate the cooling load.

Solution:
Specific heat of beef before freezing is listed in Table 3, Chapter 8 as 3.35 kJ/(kg·K); after freezing, 1.68 kJ/(kg·K).

To cool from 18 to 4°C in a chilled room:

$$100 \times 3.35\,(18 - 4) = 4690 \text{ kJ}$$

To cool from 4°C to freezing point in freezer:

$$100 \times 3.35\,[4 - (-2)] = 2010 \text{ kJ}$$

To freeze: $\quad 100 \times 234 = 23\,400 \text{ kJ}$

To cool from freezing to storage temperature:

$$100 \times 1.68\,[(-2) - (-18)] = 2690 \text{ kJ}$$

Total: $\quad 4690 + 2010 + 23\,400 + 2690 = 32\,790 \text{ kJ}$

(Example 3 in chapter 8 shows an alternative calculation method.)

Fresh fruits and vegetables respire and release heat during storage. This heat produced by respiration varies with the product and its temperature; the colder the product, the less the heat of respiration. Table 9 in Chapter 8 gives heat of respiration rates for various products.

Calculations in Example 1 do not cover heat gained from product containers brought into the refrigerated space. When pallets, boxes, or other packing materials are a significant portion of the total mass introduced, this heat load should be calculated.

Equations (4) through (8) are used to calculate the total heat gain. Any moisture removed appears as latent heat gain. The amount of moisture involved is usually provided by the end-user as a percentage of product mass; so, with such information, the latent heat component of the total heat gain may be determined. Subtracting the latent heat component from the total heat gain determines the sensible heat component.

INTERNAL LOAD

Electrical Equipment. All electrical energy dissipated in the refrigerated space (from lights, motors, heaters, and other equipment) must be included in the internal heat load. Heat equivalents of electric motors are listed in Table 4.

Refrigeration Load

Table 4 Heat Gain from Typical Electric Motors

Motor Nameplate or Rated Horsepower	(kW)	Motor Type	Nominal rpm	Full Load Motor Efficiency, %	Location of Motor and Driven Equipment with Respect to Conditioned Space or Airstream		
					A Motor in, Driven Equipment in, Watt	B Motor out, Driven Equipment in, Watt	C Motor in, Driven Equipment out, Watt
0.05	(0.04)	Shaded pole	1500	35	105	35	70
0.08	(0.06)	Shaded pole	1500	35	170	59	110
0.125	(0.09)	Shaded pole	1500	35	264	94	173
0.16	(0.12)	Shaded pole	1500	35	340	117	223
0.25	(0.19)	Split phase	1750	54	346	188	158
0.33	(0.25)	Split phase	1750	56	439	246	194
0.50	(0.37)	Split phase	1750	60	621	372	249
0.75	(0.56)	3-Phase	1750	72	776	557	217
1	(0.75)	3-Phase	1750	75	993	747	249
1.5	(1.1)	3-Phase	1750	77	1453	1119	334
2	(1.5)	3-Phase	1750	79	1887	1491	396
3	(2.2)	3-Phase	1750	81	2763	2238	525
5	(3.7)	3-Phase	1750	82	4541	3721	817
7.5	(5.6)	3-Phase	1750	84	6651	5596	1066
10	(7.5)	3-Phase	1750	85	8760	7178	1315
15	(11.2)	3-Phase	1750	86	13 009	11 192	1820
20	(14.9)	3-Phase	1750	87	17 140	14 913	2230
25	(18.6)	3-Phase	1750	88	21 184	18 635	2545
30	(22.4)	3-Phase	1750	89	25 110	22 370	2765
40	(30)	3-Phase	1750	89	33 401	29 885	3690
50	(37)	3-Phase	1750	89	41 900	37 210	4600
60	(45)	3-Phase	1750	89	50 395	44 829	5538
75	(56)	3-Phase	1750	90	62 115	55 962	6210
100	(75)	3-Phase	1750	90	82 918	74 719	8290
125	(93)	3-Phase	1750	90	103 430	93 172	10 342
150	(110)	3-Phase	1750	91	123 060	111 925	11 075
200	(150)	3-Phase	1750	91	163 785	149 135	14 738
250	(190)	3-Phase	1750	91	204 805	186 346	18 430

Table 5 Heat Equivalent of Occupancy

Refrigerated Space Temperature, °C	Heat Equivalent/Person, W
10	210
5	240
0	270
−5	300
−10	330
−15	360
−20	390

Note: Heat equivalent may be estimated by $q_p = 272 - 6t$ (°C)

adjusted. A conservative adjustment would be to multiply the values in calculated in Equation (9) by 1.25.

Latent Load. The latent heat component of the internal load is usually very small compared to the total refrigeration load and is customarily regarded as all sensible heat in the total load summary. However, the latent heat component should be calculated where water is involved in processing or cleaning.

INFILTRATION AIR LOAD

Heat gain from infiltration air and associated equipment loads can amount to more than half the total refrigeration load of distribution warehouses and similar applications.

Infiltration by Air Exchange

Infiltration most commonly occurs because of air density differences between rooms (see Figures 1 and 2). For a typical case where the air mass flowing in equals the air mass flowing out minus any condensed moisture, the room must be sealed except at the opening in question. If the cold room is not sealed, air may flow directly through the door (discussed in the following section).

Heat gain through doorways from air exchange is as follows:

$$q_t = qD_t D_f (1 - E) \quad (10)$$

where

q_t = average heat gain for the 24-h or other period, kW
q = sensible and latent refrigeration load for fully established flow, kW
D_t = doorway open-time factor
D_f = doorway flow factor
E = effectiveness of doorway protective device

Fork Lifts. Fork lifts in some facilities can be a large and variable contributor to the load. While many fork lifts may be in a space at one time, they do not all operate at the same energy level. For example, the energy used by a fork lift while it is elevating or lowering forks is different than when it is moving.

Processing Equipment. Grinding, mixing, or even cooking equipment may be in the refrigerated areas of food processing plants. Other heat sources include equipment for packaging, glue melting, or shrink wrapping. Another possible load is the makeup air for equipment that exhausts air from a refrigerated space.

People. People add to the heat load, and this load varies depending on such factors as room temperature, type of work being done, type of clothing worn, and size of the person. Heat load from a person q_p may be estimated as

$$q_p = 272 - 6t \quad (9)$$

where t is the temperature of the refrigerated space in °C. Table 5 shows the average load from people in a refrigerated space as calculated from Equation (9).

When people first enter a storage they bring in additional surface heat. As a result, when many people enter and leave every few minutes the load is greater than that listed in Table 5 and must be

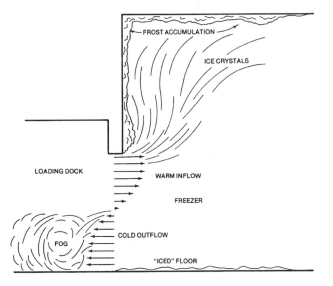

Fig. 1 Flowing Cold and Warm Air Masses that Occur for Typical Open Freezer Doors

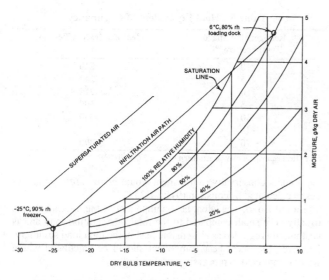

Fig. 2 Psychrometric Depiction of Air Exchange for Typical Freezer Doorway

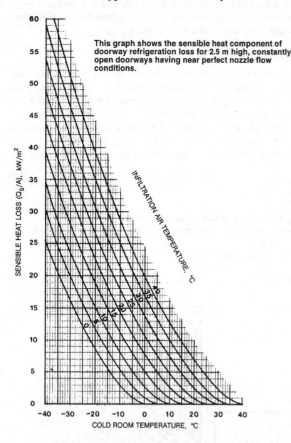

Fig. 3 Sensible Heat Gain by Air Exchange for Continuously Open Door with Fully Established Flow

Gosney and Olama (1975) developed the following air exchange equation for fully established flow:

$$q = 0.221 A (h_i - h_r) \rho_r (1 - \rho_i/\rho_r)^{0.5} (gH)^{0.5} F_m \qquad (11)$$

where

q = sensible and latent refrigeration load, W
A = doorway area, m²
h_i = enthalpy of infiltration air, kJ/kg

Table 6 Sensible Heat Ratio R_s for Infiltration from Outdoors to Refrigerated Spaces

Outdoor Cond.			Cold Space at 90% rh Dry-Bulb Temperature, °C								
DB °C	WB °C	rh, %	−30	−25	−20	−15	−10	−5	0	5	10	15
30	19.7	30	0.76	0.75	0.74	0.73	0.72	0.72	0.73	0.77	0.87	—
	21.8	40	0.71	0.69	0.68	0.66	0.65	0.63	0.63	0.64	0.68	0.83
	23.9	50	0.66	0.64	0.62	0.60	0.59	0.57	0.56	0.55	0.56	0.62
	25.8	60	0.62	0.60	0.58	0.56	0.54	0.52	0.50	0.48	0.48	0.49
35	19.0	20	0.80	0.79	0.78	0.77	0.77	0.77	0.79	0.84	0.96	—
	21.6	30	0.72	0.71	0.69	0.68	0.67	0.66	0.67	0.68	0.72	0.86
	24.0	40	0.66	0.64	0.63	0.61	0.59	0.58	0.57	0.57	0.58	0.63
	26.3	50	0.61	0.59	0.57	0.55	0.53	0.52	0.50	0.49	0.48	0.50
	28.3	60	0.56	0.54	0.53	0.51	0.49	0.47	0.45	0.43	0.42	0.41
40	20.7	20	0.76	0.75	0.74	0.73	0.72	0.72	0.73	0.75	0.82	0.98
	23.6	30	0.68	0.66	0.65	0.63	0.62	0.61	0.60	0.61	0.62	0.68
	26.2	40	0.61	0.59	0.58	0.56	0.54	0.53	0.52	0.51	0.50	0.52
	28.6	50	0.55	0.54	0.52	0.50	0.48	0.47	0.45	0.43	0.42	0.42

h_r = enthalpy of refrigerated air, kJ/kg
ρ_i = density of infiltration air, kg/m³
ρ_r = density of refrigerated air, kg/m³
g = gravitational constant = 9.81 m/s²
H = doorway height, m
F_m = density factor

$$F_m = \left(\frac{2}{1 + (\rho_r/\rho_i)^{1/3}} \right)^{1.5} \qquad (12)$$

(Chapter 6 of the 1997 *ASHRAE Handbook—Fundamentals* and the ASHRAE Psychrometrics Chart list air enthalpy and density values.)

Equation (13), when used with Figure 3, is a simplification of Equation (11):

$$q = 0.577 W H^{1.5} \left(\frac{Q_s}{A} \right) \left(\frac{1}{R_s} \right) \qquad (13)$$

where

q = sensible and latent refrigeration load, kW
Q_s/A = sensible heat load of infiltration air per square foot of doorway opening as read from Figure 3, kW/m²
W = doorway width, m
R_s = sensible heat ratio of the infiltration air heat gain, from Tables 6 or 7 (or from a psychrometric chart)

The values of R_s in Tables 6 and 7 are based on 90% rh in the cold room. A small error occurs where these values are used for 80 or 100% rh. This error together with loss of accuracy due to simplification results in a maximum error for Equation (13) of approximately 4%.

For cyclical, irregular, and constant door usage, alone or in combination, the doorway open-time factor can be calculated as follows:

$$D_t = \frac{(P\theta_p + 60\theta_o)}{3600 \, \theta_d} \qquad (14)$$

where

D_t = decimal portion of time doorway is open
P = number of doorway passages
θ_p = door open-close time, seconds per passage
θ_o = time door simply stands open, min
θ_d = daily (or other) time period, h

Refrigeration Load

Table 7 Sensible Heat Ratio R_s for Infiltration from Warmer to Colder Refrigerated Spaces

Warm Space		Cold Space at 90% rh Dry-Bulb Temperature, °C								
Temp. °C	rh, %	−40	−35	−30	−25	−20	−15	−10	−5	0
20	100	0.62	0.60	0.57	0.55	0.53	0.50	0.44	0.44	0.41
	80	0.67	0.65	0.63	0.61	0.58	0.56	0.53	0.51	0.48
	60	0.73	0.71	0.69	0.68	0.65	0.63	0.61	0.60	0.59
	40	0.80	0.79	0.78	0.76	0.75	0.73	0.73	0.73	0.76
	20	0.89	0.88	0.88	0.87	0.87	0.87	0.80	0.93	—
15	100	0.67	0.65	0.63	0.60	0.58	0.55	0.51	0.48	0.45
	80	0.72	0.70	0.68	0.66	0.63	0.61	0.58	0.55	0.53
	60	0.77	0.76	0.74	0.72	0.70	0.68	0.66	0.65	0.66
	40	0.84	0.83	0.81	0.80	0.79	0.78	0.78	0.79	0.87
	20	0.91	0.91	0.90	0.90	0.90	0.91	0.93	—	—
10	100	0.72	0.70	0.68	0.65	0.62	0.59	0.56	0.52	0.48
	80	0.76	0.75	0.73	0.70	0.68	0.65	0.63	0.60	0.59
	60	0.81	0.80	0.78	0.76	0.75	0.73	0.71	0.71	0.77
	40	0.87	0.86	0.85	0.84	0.83	0.82	0.83	0.88	—
	20	0.93	0.93	0.92	0.92	0.92	0.94	0.99	—	—
5	100	0.77	0.75	0.73	0.70	0.67	0.64	0.60	0.55	0.49
	80	0.81	0.79	0.77	0.75	0.72	0.70	0.67	0.65	0.68
	60	0.85	0.83	0.82	0.80	0.79	0.77	0.77	0.79	—
	40	0.89	0.88	0.88	0.87	0.86	0.86	0.89	—	—
	20	0.95	0.94	0.94	0.94	0.95	0.98	—	—	—
0	100	0.81	0.79	0.77	0.74	0.71	0.67	0.63	0.56	—
	80	0.84	0.83	0.81	0.79	0.76	0.74	0.71	0.71	—
	60	0.88	0.86	0.85	0.84	0.82	0.81	0.83	0.98	—
	40	0.92	0.91	0.90	0.89	0.89	0.91	0.98	—	—
		0.96	0.96	0.96	0.96	0.98	—	—	—	—
−5	100	0.85	0.83	0.81	0.79	0.75	0.71	0.65	—	—
	80	0.88	0.86	0.85	0.83	0.81	0.78	0.78	—	—
	60	0.91	0.90	0.88	0.87	0.87	0.87	0.98	—	—
	40	0.94	0.93	0.93	0.92	0.93	0.98	—	—	—
	20	0.97	0.97	0.97	0.98	—	—	—	—	—
−10	100	0.88	0.87	0.85	0.82	0.79	0.74	—	—	—
	80	0.91	0.89	0.88	0.86	0.85	0.84	—	—	—
	60	0.93	0.92	0.91	0.91	0.91	0.97	—	—	—
	40	0.95	0.95	0.95	0.95	0.98	—	—	—	—
	20	0.98	0.98	0.99	—	—	—	—	—	—
−15	100	0.91	0.90	0.88	0.85	0.81	—	—	—	—
	80	0.93	0.92	0.91	0.89	0.89	—	—	—	—
	60	0.95	0.94	0.94	0.94	0.98	—	—	—	—
	40	0.97	0.97	0.97	0.99	—	—	—	—	—
	20	0.99	0.99	1.00	—	—	—	—	—	—

The typical time θ_p for conventional pull-cord operated doors ranges from 15 to 25 s per passage. The time for high speed doors ranges from 5 to 10 s, although it can be as low as 3 s. The time for θ_o and θ_d should be provided by the user. Hendrix et al. (1989) found that steady-state flow becomes established 3 s after the cold room door is opened. This fact may be used as a basis to reduce θ_p in Equation (14), particularly for high speed doors, which may significantly reduce infiltration.

The doorway flow factor D_f is the ratio of actual air exchange to fully established flow. Fully established flow occurs only in the unusual case of an unused doorway standing open to a large room or to the outdoors, and where the cold outflow is not impeded by obstructions (such as stacked pallets within or adjacent to the flow path either within or outside the cold room). Under these conditions, D_f is 1.0.

Hendrix et al. (1989) found that a flow factor D_f of 0.8 is conservative for a 16 K temperature difference when traffic flow equals one entry and exit per minute through fast-operating doors. Tests by Downing and Meffert (1993) at temperature differences of 7 K and 10 K found a flow factor of 1.1. Based on these results, the recommended flow factor for cyclically operated doors with temperature differentials less than 11 °C is 1.1, and the recommended flow factor for higher differentials is 0.8.

The effectiveness E of open-doorway protective devices is 0.95 or higher for newly installed strip doors, fast fold doors, and other non-tight-closing doors. However, depending on the traffic level and door maintenance, E may quickly drop to 0.8 on freezer doorways and to about 0.85 for other doorways. Airlock vestibules with strip doors or push-through doors have an effectiveness ranging between 0.95 and 0.85 for freezers and between 0.95 and 0.90 for other doorways. The effectiveness of air curtains range from very poor to more than 0.7.

Infiltration by Direct Flow Through Doorways

A negative pressure created elsewhere in the building because of mechanical air exhaust without mechanical air replenishment is a common cause of heat gain from infiltration of warm air. In refrigerated spaces equipped with constantly or frequently open doorways or other through-the-room passageways, this air flows directly through the doorway. The effect is identical to that of open doorways exposed to the wind and the heat gain may be very large. Equation (15) for heat gain from infiltration by direct inflow provides the basis for either correcting the negative pressure or adding to refrigeration capacity.

$$q_t = VA(h_i - h_r)\rho_r D_t \quad (15)$$

where

q_t = average refrigeration load, kW
V = average air velocity, m/s
A = opening area, m^2
h_i = enthalpy of infiltration air, kJ/kg
h_r = enthalpy of refrigerated air, kJ/kg
ρ_r = density of refrigerated air, kg/m^3
D_t = decimal portion of time doorway is open

The area A is the smaller of the inflow and outflow openings. If the smaller area has leaks around truck loading doors in well maintained loading docks, the leakage area can vary from 0.03 m^2 to over 0.1 m^2 per door. For loading docks with high merchandise movement, the facility manager should estimate the time these doors are fully or partially open.

To evaluate velocity V, the magnitude of negative pressure or other flow-through force must be known. If differential pressure across a doorway can be determined, velocity can be predicted by converting static head to velocity head. However, attempting to estimate differential pressure is usually not possible; generally, the alternative is to assume a commonly encountered velocity. The typical air velocity through a door is 0.3 to 1.5 m/s.

The effectiveness of nontight closing devices on doorways subject to infiltration by direct airflow cannot be readily determined. Depending on the pressure differential, its tendency to vary, and the ratio of inflow area to outflow area, the effectiveness of these devices can be very low.

Sensible and Latent Heat Components

When calculating q_t for infiltration air, the sensible and latent heat components may be obtained by plotting the infiltration air path on the appropriate ASHRAE psychrometric chart, determining the air sensible heat ratio R_s from the chart, and calculating as follows:

Sensible heat: $q_s = q_t R_s$ (16)

Latent heat: $q_l = q_t(1 - R_s)$ (17)

where $R_s = \Delta h_s / \Delta h_t$.

EQUIPMENT RELATED LOAD

Heat gain associated with the operation of the refrigeration equipment consist essentially of the following:

- Fan motor heat where forced air circulation is used
- Reheat where humidity control is part of the cooling

- Heat from defrosting where the refrigeration coil operates at a temperature below freezing and must be defrosted periodically, regardless of the room temperature

Fan motor heat must be computed based on the actual electrical energy consumed during operation. The fan motor is mounted in the air stream on many cooling units with propeller fans because the cold air extends the power range of the motor. For example, a standard motor in a −25°C freezer operates satisfactorily at a 25% overload to the rated (nameplate) power. The heat gain from the fan motors should be based on the actual run time. Generally, fans on cooling units are operated continuously except during the defrost period. But, increasingly, fans are being cycled on and off to control temperature and save energy.

Cole (1989) characterized and quantified the heat load associated with defrosting using hot gas. Other common defrost methods use electricity or water. Generally, the heat gain from a cooling unit with electric defrost is greater than the same unit with hot gas defrost, and heat gain from a unit with water defrost is even less. The moisture that evaporates into the space during the defrost cycle must also be added to the refrigeration load.

Some of the heat from defrosting is added only to the refrigerant and the rest is added to the space. To accurately select refrigeration equipment, a distinction should be made between those equipment heat loads that are in the refrigerated space and those that are introduced directly to the refrigerating fluid.

Equipment heat gain is usually small at space temperatures above approximately −1°C. Where reheat or other artificial loads are not imposed, total equipment heat gain is about 5% or less of the total load. However, equipment heat gain becomes a major portion of the total load at freezer temperatures. For example, at −30°C the theoretical contribution to total refrigeration load due to fan power and coil defrosting alone can exceed, for many cases, 15% of the total load. This percentage assumes proper control of defrosting so that the space is not heat excessively.

SAFETY FACTOR

Generally, the calculated load is increased by a factor of 10% to allow for possible discrepancies between the design criteria and actual operation. This factor should be selected in consultation with the facility user and should be applied individually to the first four heat load segments.

A separate factor should be added to the coil-defrosting portion of the equipment load for freezer applications that use dry-surface refrigerating coils. However, little data are available to predict heat gain from coil defrosting. For this reason, the experience of existing similar facilities should be sought to obtain an appropriate defrosting safety factor. Similar facilities should have similar room sensible heat ratios.

The nature of frost accumulation on the cooling coils also affects the performance of the cooling units and, therefore, the refrigeration load. A very low density frost forms under certain conditions, particularly where the room sensible heat ratio is more than a few points below 1.0. This type of frost is difficult to remove and tends to block the airflow through the cooling coils more readily. Removing this type of frost requires more frequent and longer periods of defrosting of the cooling units, which increases the refrigeration load.

TOTAL REFRIGERATION LOAD

A load calculation is performed to determine the proper size of the equipment required to provide the cooling, to effectively operate the system, and to estimate operating costs. It is primarily used to select the refrigeration equipment. Several approaches can be taken to make the final selection of the equipment depending on the nature of the loads.

Peak Load Calculation. In this calculation method, all load elements are added together to establish the total load. In this method of calculation load diversity is not considered. That is, the possibility, in fact the high probability, that all maximum loads do not occur at the same time is not considered. The equipment is selected on the basis that all of the maximum load will occur at the same time to ensure that the design temperature will never be exceeded.

Hour-By-Hour Calculation. Where design load data are reasonably well defined in terms of both magnitude and occurrence, an hour-by-hour load calculation may be performed (Ballard 1992). This method accounts for the diversity of operation of refrigerated buildings, particularly large ones. Typically, the sum of the elements of the total load thus determined is smaller than that determined by the peak load method and results in smaller equipment being selected. However, with the hour-by-hour method of calculation, equipment may have insufficient capacity to handle changes in diversity from normal operation and product temperatures may rise. Estimates of the duration of peak load should be considered in the design.

BIBLIOGRAPHY

Ballard, R.N. 1992. Calculating refrigeration loads on an hour-by-hour basis: Part I—Building envelope and Part II—Infiltration and internal heat sources. *ASHRAE Transactions* 98(2).

Cole, R.A. 1989. Refrigeration loads in a freezer due to hot gas defrost, and their associated costs. *ASHRAE Transactions* 95(2):1149-54.

Cole, R.A. 1987. Infiltration load calculations for refrigerated warehouses. *Heating/Piping/Air Conditioning* (April).

Cole, R.A. 1984. Infiltration: A load calculation guide. Proceedings of the International Institute of Ammonia Refrigeration, 6th annual meeting, San Francisco (February).

Dickerson, R.W. 1972. Computing heating and cooling rates of foods. *ASHRAE Symposium Bulletin* No-72-03.

Downing, C.C. and W.A. Meffert. 1993. Effectiveness of cold-storage door infiltration protective devices. *ASHRAE Transactions* 99(2).

Fisher, D.V. 1960. Cooling rates of apples packed in different bushel containers, stacked at different spacings in cold storage. *ASHRAE Transactions* 66.

Gosney, W.B. and H.A.L. Olama. 1975. Heat and enthalpy gains through cold room doorways. Paper presented before The Institute of Refrigeration at the Faculty of Environmental Science and Technology, The Polytechnic of the South Bank, London (December).

Hamilton, J.J., D.C. Pearce, and N.B. Hutcheon. 1959. What frost action did to a cold storage plant. *ASHRAE Journal* 1(4).

Haugh, C.G., W.J. Standelman, and V.E. Sweat. 1972. Prediction of cooling/freezing times for food products. *ASHRAE Symposium Bulletin* NO-72-03.

Hendrix, W.A., D.R. Henderson, and H.Z. Jackson. 1989. Infiltration heat gains through cold storage room doorways. *ASHRAE Transactions* 95(2).

Hovanesian, J.D., H.F. Pfost, and C.W. Hall. 1960. An analysis of the necessity to insulate floors of cold storage rooms at 35°F. *ASHRAE Transactions* 66.

Jones, B.W., B.T. Beck, and J.P. Steele. 1983. Latent loads in low humidity rooms due to moisture. *ASHRAE Transactions* 89(1).

Kayan, C.F. and J.A. McCague. 1959. Transient refrigeration loads as related to energy-flow concepts. *ASHRAE Journal* 1(3).

Meyer, C.S. 1964. "Inside-out" design developed for low-temperature buildings. *ASHRAE Journal* 5(4).

Pham, O.T. and D.W. Oliver. 1983. Infiltration of air into cold stores. Meat Industry Research Institute of New Zealand. Presented at IIF-IIR 16th International Congress of Refrigeration, Paris.

Pichel, W. 1966. Soil freezing below refrigerated warehouses. *ASHRAE Journal* 8(10).

Powell, R.M. 1970. Public refrigerated warehouses. *ASHRAE Journal* 12(8).

CHAPTER 13

REFRIGERATED FACILITY DESIGN

Initial Building Considerations 13.1	Refrigeration Systems 13.7
Building Design 13.2	Insulation Techniques 13.11
Specialized Storage Facilities 13.3	Applying Insulation 13.12
Construction Methods 13.4	Other Considerations 13.14

A REFRIGERATED facility is any building or section of a building that achieves controlled storage conditions using refrigeration. Two basic storage facilities are (1) coolers that protect commodities at temperatures usually above 0°C and (2) low-temperature rooms (freezers) operating under 0°C to prevent spoilage or to maintain or extend product life.

The conditions within a closed refrigerated chamber must be maintained to preserve the stored product. This refers particularly to seasonal, shelf life, and long-term storage. Specific items for consideration include

- Uniform temperatures
- Length of air blow and impingement on stored product
- Effect of relative humidity
- Effect of air movement on employees
- Controlled ventilation, if necessary
- Product entering temperature
- Expected duration of storage
- Required product outlet temperature
- Traffic in and out of storage area

In the United States, the U.S. Public Health Service Food and Drug Administration (FDA) developed the *Food Code*, which consists of model requirements for safeguarding public health and ensuring that food is unadulterated. The code is a guide for establishing standards for all phases of handling refrigerated foods. It treats receiving, handling, storing, and transporting refrigerated foods and calls for sanitary as well as temperature requirements. These standards must be recognized in the design and operation of refrigerated storage facilities.

Regulations of the Occupational Safety and Health Administration (OSHA), Environmental Protection Agency (EPA), U.S. Department of Agriculture (USDA), and other standards must also be incorporated in warehouse facility and procedures.

Refrigerated facilities may be operated for or by a private company for storage or warehousing of their own products, as a public facility where storage services are offered to many concerns, or both. Important locations for refrigerated facilities, public or private, are (1) point of processing, (2) intermediate points for general or long-term storage, and (3) final distributor or distribution point.

The five categories for the classification of refrigerated storage for preservation of food quality are

- Controlled atmosphere for long-term storage of fruits and vegetables
- Coolers at temperatures of 0°C and above
- High-temperature freezers at −2 to −3°C
- Low-temperature storage rooms for general frozen products, usually maintained at −23 to −29°C
- Low-temperature storages at −23 to −29°C, with a surplus of refrigeration for freezing products received at above −18°C

It should be noted that due to ongoing research, the trend is toward lower temperatures for frozen foods. Refer to Chapters 22 and 23 of the 1997 *ASHRAE Handbook—Fundamentals* and Chapter 10 of this volume for further information.

INITIAL BUILDING CONSIDERATIONS

Location

Private refrigerated space is usually adjacent to or in the same building with the owner's other operations.

Public space should be located to serve a producing area, a transit storage point, a large consuming area, or various combinations of these to develop a good average occupancy. It should also have the following:

- Convenient location for producers, shippers, and distributors, considering the present tendency toward decentralization and avoidance of congested areas
- Good railroad switching facilities and service with minimum switching charges from all trunk lines to plant tracks if a railhead is necessary for the profitable operation of the business
- Easy access from main highway truck routes as well as local trucking, but avoiding location on congested streets
- Ample land for trucks, truck movement, and plant utility space plus future expansion
- Location with a reasonable land cost
- Adequate power and water supply
- Provisions for surface, waste, and sanitary water disposal
- Consideration of zoning limitations and fire protection
- Location away from residential areas, where noise of outside operating equipment (i.e., fans and engine-driven equipment on refrigerated vehicles) would be objectionable
- External appearance that is not objectionable to the community
- Minimal tax and insurance burden
- Plant security
- Favorable undersoil bearing conditions and good surface drainage

Plants are often located away from congested areas or even outside city limits where the cost of increased trucking distance is offset by better plant layout possibilities, a better road network, better or lower priced labor supply, or other economies of operation.

Configuration and Size Determination

Building configuration and size of a cold storage facility are determined by the following factors:

- Is the receipt and shipment of goods to be primarily by rail or by truck? Shipping practices affect the platform areas and internal traffic pattern.
- What relative percentages of merchandise are for cooler and for freezer storage? Products requiring specially controlled conditions, such as fresh fruits and vegetables, may justify or demand several individual rooms. Seafood, butter, and nuts also require special treatment. Where overall occupancy may be reduced because of seasonal conditions, consideration should be given to providing multiple-use spaces.

The preparation of this chapter is assigned to TC 10.5, Refrigerated Distribution and Storage Facilities.

- What percentage is anticipated for long-term storage? Products that are stored long term can usually be stacked more densely.
- Will the product be primarily in small or large lots? The drive-through rack system or a combination of pallet racks and a mezzanine have proved effective in achieving efficient operation and effective use of space. Mobile or moving rack systems are also plausible.
- How will the product be palletized? Dense products such as meat, tinned fruit, drums of concentrate, and cases of canned goods can be stacked very efficiently. Palletized containers and special pallet baskets or boxes effectively hold meat, fish, and other loose products.

The slip sheet system, which requires no pallets, eliminates the waste space of the pallet and can be used effectively for some products.

Pallet stacking racks make it feasible to use the full height of the storage and palletize any closed or boxed merchandise.

- Will rental space be provided for tenants? Rental space usually requires special personnel and office facilities. An isolated area for tenant operations is also desirable. These areas are usually leased on a unit area basis, and plans are worked into the main building layout.

The owner of a prospective refrigerated facility may want to obtain advice from specialists in product storage, handling, and movement systems.

Stacking Arrangement

Typically, the height of refrigerated spaces varies between 8.5 and 10.5 m or more clear space between the floor and structural steel to allow forklift operation. Pallet rack systems use the greater height. The practical height for stacking pallets without racks is 4.5 to 5.5 m. The clear space above the pallet stacks is used for air distribution, lighting, and sprinkler lines. Overhead space is inexpensive, and because the refrigeration requirement for the extra height is not significant in the overall plant cost, a minimum of 6 m clear height is desirable. Greater heights are valuable if automated or mechanized material handling equipment is contemplated. The effect of high stacking arrangements on insurance rates should be investigated.

The floor area in a facility where a diversity of merchandise is to be stored can be calculated on the basis of 130 to 160 kg/m³ to allow about 40% for aisles and space above the pallet stacks. In special-purpose or production facilities, products can be stacked with less aisle and open space, with an allowance factor of about 20%.

BUILDING DESIGN

Most refrigerated facilities are single-story structures. Small columns on wide centers permit palletized storage with minimal lost space. This type of building usually provides additional highway truck unloading space. The following characteristics of single-story design must be considered: (1) horizontal traffic distances, which to some extent offset the vertical travel required in a multistory building; (2) difficulty of using the stacking height with many commodities or with small-lot storage and movement of goods; (3) necessity for treatment of the floor below freezers to give economical protection against possible ground heaving; and (4) high land cost for the capacity of the building. A one-story facility with moderate or low stacking heights has a high cost per cubic metre because of the high ratio of construction costs and added land cost to product storage capacity. However, the first cost and operating cost are usually lower than that of a multistory facility.

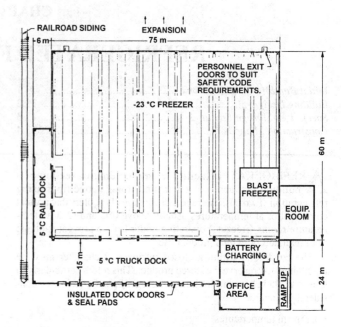

Fig. 1 Typical Plan for One-Story Refrigerated Facility

One-Story Configuration

Figure 1 shows the layout of a one-story −23°C freezer that complies with current practices. The following essential items and functions are considered:

- Refrigeration machinery room
- Refrigerated shipping docks with seal-cushion closures on the doors
- Automatic doors
- Batten doors or strip curtains
- Low-temperature storage held at −23°C or lower
- Pallet-rack systems to facilitate the handling of small lots and to comply with first-in, first-out inventory, which is required for some products
- Blast freezer or separate sharp freezer room for isolation of products being frozen
- Cooler or convertible space
- Space for brokerage offices and for distributors
- Space for empty pallet storage and repair
- Space for shop and battery charging
- Automatic sprinklers in accordance with National Fire Protection Association (NFPA) regulations

Other areas that must be in a complete operable facility are

- Electrical area
- Shipping office
- Administration office
- Personnel welfare facilities

A modified one-story design is sometimes used to reduce horizontal traffic distances and land costs. An alternative is to locate nonproductive services (including offices and the machinery room) on a second-floor level, usually over the truck platform work area to permit full use of the ground floor for production work and storage. However, potential vibration of the second floor from equipment below must be considered.

One-story design or modification thereof gives the maximum capacity per unit of investment with a minimum overall operating expense, including amortization, refrigeration, and labor. Mechanization must be considered as well. In areas where land availability

Refrigerated Facility Design

or cost is a concern, a high-rise refrigerated storage building may be a viable option.

Designs that provide minimum overall costs restrict office facilities and utility areas to a minimum. They also include ample dock area to ensure efficiency in loading and unloading merchandise.

Shipping and Receiving Docks

Regulations on temperature control during all steps of product handling have led to designing the trucking dock as a refrigerated anteroom to the cold storage area. Dock refrigeration is an absolute necessity in humid and warm climates. Typically, loading and unloading of transport vehicles is handled by separate work crews. One crew moves the product in and out of the vehicles, and a warehouse crew moves the product in and out of the refrigerated storage. This procedure may allow the merchandise to accumulate on the shipping dock. Maintaining the dock at 2 to 7°C offers the following advantages:

- Refrigeration load in the low-temperature storage area, where energy demand per unit capacity of refrigeration is higher, is reduced.
- Less ice or frost forms in the low-temperature storage because less warm air infiltrates into the area.
- Refrigerated products held on the dock maintain a more favorable temperature.
- Packaging remains in good condition because it stays dry. Facility personnel are more comfortable because temperature differences are smaller.
- Less maintenance on forklifts and other equipment is required because condensation is reduced.
- Need for anterooms or vestibules to the freezer space is reduced or eliminated.

Utility Space

Space for a general office, locker room, and machinery room is needed. A superintendent's office and a warehouse records office should be located near the center of operations, and a checker's office should be in view of the dock and traffic arrangement. Rented space should be isolated from warehouse operations.

The machinery room should include ample space for refrigeration equipment and maintenance, adequate ventilation, standby capacity for emergency ventilation, and adequate segregation from other areas. Separate exits are required by most building codes. A maintenance shop and space for parking, charging, and servicing warehouse equipment should be located adjacent to the machinery room. Electrically operated material handling equipment is used to eliminate inherent safety hazards of combustion-type equipment. Battery-charging areas should be designed with high roofs and must be ventilated due to the potential for combustible fumes resulting from the charging activity.

SPECIALIZED STORAGE FACILITIES

Material handling methods and storage requirements often dictate design of specialized storage facilities. Automated material handling within the storage, particularly for high stack piling, may be an integral part of the structure or require special structural treatments. Controlled atmosphere rooms and minimal air circulation rooms require special building designs and mechanical equipment to achieve design requirements. Drive-in and/or drive-through rack systems can improve product inventory control and can be used in combination with stacker cranes, narrow-aisle high stacker cranes, and automatic conveyors. Mobile racking systems may be considered where space is at a premium.

In general, specialized storage facilities may be classified as follows:

- Public refrigerated facility with several chambers designed to handle all commodities. Storage temperatures may range from 2 to 15°C (with humidity control) and to −29°C (without humidity control).
- Refrigerated facility area for case and break-up distribution, automated to varying degrees. The area may incorporate racks with pallet spaces to facilitate distribution.
- Facility designed for a processing operation with bulk storage for frozen ingredients and rack storage for palletized outshipment of processed merchandise. An efficient adaptation frequently seen in practice is to adjoin the refrigerated facility to the processing plant.
- Public refrigerated facility serving several production manufacturers for storing and inventorying products in lots and assembling outshipments.
- Mechanized refrigerated facility with stacker cranes, racks, infeed and outfeed conveyors, and conveyor vestibules. Such a facility may have an interior ceiling 18 to 30 m high. It is desirable to mount the evaporators in the highest internal area to help remove moisture from outside air infiltration. A penthouse to house the evaporators provides easy access through the roof for maintenance, a means of controlling condensate drip, and added rack storage space.

Controlled Atmosphere Storage Rooms

Controlled atmosphere storage rooms may be required for storage of some commodities, particularly fresh fruits and vegetables that respire, consuming oxygen (O_2) and producing carbon dioxide (CO_2) in the process. The storage life of such products may be greatly lengthened by a properly controlled environment, which includes control of temperature, humidity, and concentration of noncondensable gases (O_2, CO_2, and nitrogen). Hermetically sealing the room to provide such an atmosphere is a challenging undertaking, often requiring special gastight seals. Although information is available for some commodities, the desired atmosphere usually must be determined experimentally for the commodity as produced in the specific geographic location that the storage room is to serve.

Commercial application of controlled atmosphere storage has historically been limited to fresh fruits and vegetables that respire. The storage spaces may be classified as having either (1) product-generated atmospheres, in which the room is sufficiently well sealed that the natural oxygen consumption of the fruit balances the infiltration of O_2 into the space; or (2) externally generated atmospheres, in which nitrogen generators or O_2 consumers assist the normal respiration of the fruit. The second type of system can cope with a poorly sealed room, but the cost of operation may be high; even with the external gas generator system, a hermetically sealed room is desired.

In most cases, a CO_2 scrubber is required; the exception is the case where the total desired O_2 and CO_2 content is 21%, which is the normal balance between O_2 and CO_2 during respiration. Carbon dioxide may be removed by (1) passing the room air over dry lime that is replaced periodically; (2) passing the air through wet caustic solutions in which the caustic (typically sodium hydroxide) is periodically replaced; (3) water scrubbers in which CO_2 is absorbed from the room air by a water spray and then desorbed from the water by outdoor air passed through the water in a separate compartment; (4) monoethanolamine scrubbers in which the solution is regenerated periodically by a manual process or continuously by automatic equipment; or (5) dry adsorbents automatically regenerated on a cyclic basis.

Among the systems of room sealing to prevent outside air infiltration are (1) galvanized steel lining the walls and ceiling of the room and interfaced into a floor sealing system; (2) plywood with an impervious sealing system applied to the inside face; (3) carefully applied sprayed urethane finished with mastic, which also serves as a fire retardant.

A room is considered sufficiently sealed if, under uniform temperature and barometric conditions, 1 h after the room is pressurized to 250 Pa (gage), 25 to 50 Pa remains. A room with external gas generation is considered satisfactorily sealed if it loses pressure at double the above rate, and the test prescribed for a room with product-generated atmosphere is about one air change of the empty room in a 30 day period.

Extreme care in all details of construction is required to obtain a seal that passes these tests. Doors are well sealed and have sills that can be bolted down; electrical conduits and special seals around pipe and hanger penetrations must allow some movement while keeping the hermetic seal intact. Structural penetrations through the seal must be avoided, and the structure must be stable. Controlled atmosphere rooms in multifloor buildings, where the structure deflects appreciably under load, are extremely difficult to seal.

Gas seals are normally applied at the cold side of the insulation, so that they may be easily maintained and points of leakage can be detected. However, this placement causes some moisture entrapment, and the insulation materials must be carefully selected so that this moisture causes minimal damage. In some installations, cold air with a dew point lower than the inside surface temperature is circulated through the space between the gas seal and the insulation to provide drying of this area. Chapters 21 and 23 have additional information on conditions required for storage of various commodities in controlled atmosphere storage rooms.

Automated Warehouses

Automated warehouses usually contain tall, fixed rack arrangements with stacker cranes under fully automatic, semiautomatic, or manual control. The control systems can be tied into a computer system to retain a complete inventory of product and location.

The following are some of the advantages of automation:

- First-in, first-out inventory can be maintained.
- Enclosure structure is high, requiring a minimum of floor space and favorable cost per cubic metre.
- Product damage and pilfering are minimized.
- Direct material handling costs are minimized.

The following are some of the disadvantages:

- First cost of the racking system and building are very high compared to conventional designs.
- Access may be slower, depending on product flow and locations.
- Cooling equipment may be difficult to access for maintenance, unless installed in a penthouse.
- Air distribution must be carefully evaluated.

Refrigerated Rooms

Refrigerated rooms may be appropriate for long-term storage at temperatures other than the temperature of the main facility, for bin storage, for controlled atmosphere storage, or for products that deteriorate with active air movement. Mechanically cooled walls, floors, and ceilings may be economical options for controlling the temperature. Embedded pipes or air spaces through which refrigerated air is recirculated can provide the cooling; with this method, leakage is absorbed into the walls and prevented from entering the refrigerated space.

The following must be considered in the initial design of the room:

- Initial cool-down of the product, which can impose short-term peak loads
- Service loads when storing and removing product
- Odor contamination from products that deteriorate over long periods
- Product heat of respiration

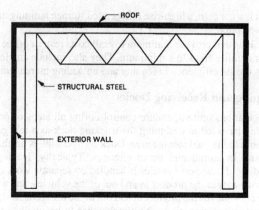

Fig. 2 Total Exterior Vapor Retarder System

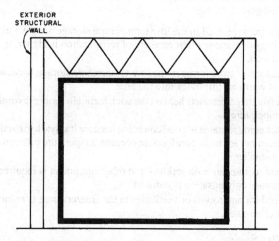

Fig. 3 Entirely Interior Vapor Retarder System

Supplementary refrigeration or air-conditioning units in the refrigerated room that operate only as required can usually alleviate such problems.

CONSTRUCTION METHODS

Cold storages, more than most construction, require correct design, quality materials, good workmanship, and close supervision. Design should ensure that proper installation can be accomplished under various adverse job site conditions. Materials must be compatible with each other. Installation must be made by careful workers directed by an experienced, well-trained superintendent. Close cooperation between the general, roofing, insulation, and other contractors increases the likelihood of a successful installation.

Enclosure construction methods can be classified as (1) insulated structural panel, (2) mechanically applied insulation, or (3) adhesive or spray-applied foam systems. These construction techniques seal the insulation within an airtight, moisturetight envelope that must not be violated by major structural components.

Three methods are used to achieve an uninterrupted vapor retarder/insulation envelope. The first and simplest is total encapsulation of the structural system by an **exterior** vapor retarder/insulation system under the floor, on the outside of the walls, and over the roof deck (Figure 2). This method offers the least number of penetrations through the vapor retarder, as well as the lowest cost. The second method is an entirely **interior** system in which the vapor retarder envelope is placed within the room, and insulation is added to the walls, floors, and suspended ceiling (Figure 3). This technique is used where walls and ceilings must be washed, where an existing structure is converted to refrigerated space, or for smaller rooms that are located within large coolers or unrefrigerated facilities or are part

Refrigerated Facility Design

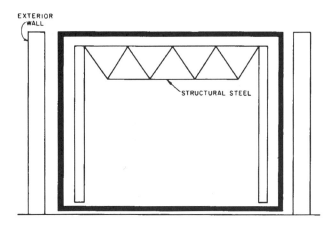

Fig. 4 Interior-Exterior Vapor Retarder System

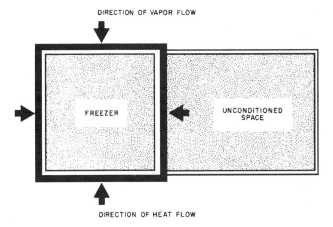

Fig. 5 Separate Exterior Vapor Retarder Systems for Each Area of Significantly Different Temperature

of a food-processing facility. The third method is **interior-exterior** construction (Figure 4), which involves an exterior curtain wall of masonry or similar material tied to an interior structural system. Adequate space allows the vapor retarder/insulation system to turn up over a roof deck and be incorporated into a roofing system, which serves as the vapor retarder. This construction method is a viable alternative, although it allows more interruptions in the vapor retarder than the exterior system.

The total exterior vapor retarder system (Figure 2) is best because it has the fewest penetrations and the lowest cost. Areas of widely varying temperature should be divided into separate envelopes to retard heat and moisture flow between them (Figure 5).

Space Adjacent to Envelope

Condensation at the envelope is usually caused by high humidity and inadequate ventilation. Poor ventilation occurs most often within a dead air space such as a ceiling plenum, hollow masonry unit, through-metal structure, or beam cavity. All closed air spaces should be eliminated, except those large enough to be ventilated adequately. Ceiling plenums, for instance, are best ventilated by mechanical vents that move air above the envelope.

If possible, the insulation envelope and vapor retarder should not be penetrated. All steel beams, columns, and large pipes that project through the insulation should be vapor-sealed and insulated with a 1200 mm wrap of insulation. Conduit, small pipes, and rods should be insulated at four times the regular wall insulation thickness. In both cases, the thickness of insulation on the projection should be half that on the regular wall or ceiling. Voids within metal projections should be filled. Where practicable, the wrap insulation should be located, and the metal projection sealed, on the warm side. Vapor sealing on the inside of conduits prevents moisture flow through the conduits.

Air/Vapor Treatment at Junctions

Air and vapor leakage at wall/roof junctions is perhaps the predominant construction problem in cold storage facilities.

When a cold room of interior-exterior design (Figure 4) is lowered to operating temperature, the structural elements (roof deck and insulation) contract and can pull the roof away from the wall. Negative pressure in the space of the wall/roof junction can cause warm, moist air to leak into the room and form frost and ice. Therefore, proper design and construction of the air/vapor seal is critical.

An air/vapor flashing sheet system is best for preventing leakage. This is a transition from the roof system vapor retarder to the exterior wall system vapor retarder. A good corner flashing sheet must be flexible, tough, airtight, and vaportight. Proper use of flexible insulation at overlaps, mastic adhesive, and a good mastic sealer ensure leak-free performance. To remain airtight and vaportight during the life of the facility, a properly constructed vapor retarder should

- Be flexible enough to withstand building movements that may occur at operating temperatures
- Allow for thermal contraction of the insulation as the room is pulled down to operating temperature
- Be constructed with a minimum of penetrations that might cause leaks (Wall ties and structural steel that extend through the corner flashing sheet may eventually leak no matter how well sealed during construction; keep these to a minimum, and make them accessible for maintenance.)
- Have corner flashing sheet properly lapped and sealed with adhesive and mechanically fastened to the wall vapor retarder
- Have corner flashing sealed to roof without openings
- Have floor to exterior vapor retarders that are totally sealed

The interior-exterior design (Figure 4) is likely to be unsuccessful at the wall/roof junction because of extreme difficulty in maintaining an airtight or vaportight environment.

The practices outlined for the wall/roof junction apply for other insulation junctions. The insulation manufacturer and the designer must coordinate the details of the corner flashing design.

Poor design and shoddy installation cause moist air leakage into the facility, resulting in frost and ice formation, energy loss, poor appearance, loss of useful storage space, and, eventually, expensive repairs.

Floor Construction

Refrigerated facilities held above freezing need no special underfloor treatment. A below-the-floor vapor retarder is needed in facilities held below freezing, however. In these facilities, the subsoil eventually freezes; any moisture in this soil also freezes and causes floor frost heaving. In moderate climates, underfloor tubes vented to ambient air are sufficient to prevent heaving. Artificial heating, either by air circulated through underfloor ducts or by glycol circulated through plastic pipe, is the preferred method to prevent frost heaving. Electric heating cables installed under the floor can also be used to prevent frost formation. The choice of heating method depends on energy cost, reliability, and maintenance requirements. Air duct systems should be screened to keep rodents out and sloped for drainage to remove condensation.

Future facility expansion must be considered when underfloor heating systems are being designed. Therefore, a system including artificial heating methods that do not require building exterior access is preferred.

Surface Preparation

When an adhesive is used, the surface against which the insulating material is to be applied should be smooth and dust-free. Where room temperatures are to be below freezing, masonry walls should be leveled and sealed with cement back plaster. Smooth poured concrete surfaces may not require back plastering.

No special surface preparation is needed for a mechanical fastening system, assuming that the surface is reasonably smooth and in good repair.

The surface must be warm and dry for a sprayed foam system. Any cracks or construction joints must be prepared to prevent projection through the sprayed insulation envelope. All loose grout and dust must be removed to ensure a good bond between the sprayed foam insulation and the surface. Very smooth surfaces may require special bonding agents.

No special surface preparation is needed for insulated panels used as a building lining, assuming the surfaces are sound and reasonably smooth. Grade beams and floors should be true and level where panels serve as the primary walls.

Finishes

Insulated structural panels with metal exteriors and metal or reinforced plastic interior faces are prevalent for both coolers and freezers. They keep moisture from the insulation, leaving only the joints between panels as potential areas of moisture penetration. They are also available with surface finishes that meet government requirements.

For sanitary washdowns, a scrubbable finish is sometimes required. Such finishes generally have low permeance; when one is applied on the inside surface of the insulation, a lower permeance treatment is required on the outside of the insulation.

All insulated walls and ceilings should have an interior finish. The finish should be impervious to moisture vapor and should not serve as a vapor retarder, except for panel construction. The permeance of the in-place interior finish should be significantly greater than the permeance of the in-place vapor retarder.

To select an interior finish to meet the in-use requirements of the installation, the following factors should be considered: (1) fire resistance, (2) washdown requirements, (3) mechanical damage, (4) moisture and gas permeance, and (5) government requirements. All interior walls of insulated spaces should be protected by bumpers and curbs wherever there is a possibility of damage to the finish.

Suspended Ceilings

Suspended (hung) insulated ceilings perform well when they are surfaced with a vapor retarder that is attached with flashing to the wall insulation vapor retarder on the top or warm side. The space above the ceiling must be ventilated at a minimum of six air changes per hour to minimize the possibility of condensation. Roof-mounted exhaust fans and uniformly spaced vents around the perimeter of the plenum may be used. The mechanical ventilators should be thermostat and/or humidistat controlled to turn off when the outside temperature is below 10°C and the outside humidity is above 60% rh. At these conditions, condensation rarely occurs, and the ventilation only reduces the insulating effect of the dead air space.

Suspended ceilings should be designed for light foot traffic for inspection and maintenance of piping and electrical wiring. Fastening systems for ceiling panels include spline, U channel, and camlock.

Floor Drains

Floor drains should be avoided if possible, particularly in freezers. If they are necessary, they should have short, squat dimensions and be placed high enough to allow the drain and piping to be installed above the insulation envelope.

Electrical Wiring

Electrical wiring should be brought into a refrigerated room through as few locations as possible (preferably one), piercing the wall vapor retarder and insulation only once. Plastic-coated cable is recommended for this service where codes permit. If codes require conduit, the last fitting on the warm side of the run should be explosionproof and sealed to prevent water vapor from entering the cold conduit. The light fixtures in the room should not be vapor-sealed but should allow free passage of moisture. Care should be taken to maintain the vapor seal between the outside of the electrical service and the cold room vapor retarder.

Tracking

Cold room meat tracking, wherever possible, should be erected and supported within the insulated structure, entirely independent of the building itself. This places the entire weight of the tracking on the cold room floor, eliminates flexure of the roof structure or overhead members, and simplifies maintenance.

Cold Storage Doors

Doors should be strong yet light enough for easy opening and closing. Hardware should be of good quality, so that it can be set to compress the gasket uniformly against the casing. All doors to rooms operating below freezing should be equipped with heaters.

In-fitting doors are not recommended for rooms operating below freezing unless they are provided with heaters, and they should not be used at temperatures below −18°C with or without heaters.

See the subsection on Doors in the section on Applying Insulation for more information.

Hardware

All metal hardware, whether within the construction or exposed to conditions that will rust or corrode the base metal, should be heavily galvanized, plated, or otherwise protected.

Refrigerated Docks

The purpose of the refrigerated facility (e.g., distribution, in-transit storage, or seasonal storage) will dictate the loading dock requirements. Shipping docks and corridors should provide liberal space for (1) movement of goods to and from storage, (2) storage of pallets and idle equipment, (3) sorting, and (4) inspecting. The dock should be at least 9 m wide. Commercial use facilities usually require more truck dock space than specialized storage facilities due to the variety of products handled.

Floor heights of refrigerated vehicles vary widely but are often greater than those of unrefrigerated vehicles. Rail dock heights and building clearances should be verified by the railroad serving the plant. A dock height of 1370 mm above the rail is typical for refrigerated rail cars. Three to five railroad car spots per 30 000 m^3 of storage should be planned.

Truck dock heights must comply with the requirements of fleet owners and clients, as well as the requirements of local delivery trucks. Trucks generally require a 1370 mm height above the pavement, although local delivery trucks may be much lower. Some reefer trucks are up to 1470 mm above grade. Adjustable ramps at some of the truck spots will partly compensate for height variations. If dimensions permit, seven to ten truck spots per 30 000 m^3 should be provided in a public refrigerated facility.

Refrigerated docks maintained at temperatures of 2 to 7°C require about 190 W of refrigeration per square metre of floor area. Cushion-closure seals around the truck doorways reduce infiltration of outside air. An inflatable or telescoping enclosure can be extended to seal the space between a railcar and the dock. Insulated doors for docks must be mounted on the inside walls. The relatively high costs of doors, cushion closures, and refrigeration influence dock size and number of doors.

Refrigerated Facility Design

Schneider System

The Schneider system, and modifications thereof, is a cold storage construction and insulation method primarily used in the western United States, with most of the installations in the Pacific Northwest. It is an interior-exterior vapor retarder system, as illustrated in Figure 4. The structure features the use of concrete tilt-up walls and either glue-laminated wood beams or bowstring trusses for the roof. Fiberglass batts coupled with highly efficient vapor retarders and a support framework are used to insulate the walls and roof. The floor slab construction, insulation, and underfloor heat are conventional for refrigerated facilities.

The key to success for the Schneider system is an excellent vapor retarder system that is professionally designed and applied, with special emphasis on the wall/roof junction. Fiberglass has a high permeability rating and loses its insulating value when wet. It is therefore absolutely essential that the vapor retarder system perform at high efficiency. Typical wall vapor retarder materials include aluminum B foil and heavy-gage polyethylene, generously overlapped and adhered to the wall with a full coating of mastic. The roofing materials act as a vapor retarder for the roof. The vapor retarder at the wall/roof junction is usually a special aluminum foil assembly installed to perform efficiently in all weather conditions.

The fiberglass insulation applied to the wall is usually 250 to 300 mm thick for freezers and 150 to 200 mm for coolers. It is retained by offset wood or fabricated fiberglass-aluminum sheathed studs on 600 mm centers. Horizontal girts are used at intervals for bracing. The inside finish is provided by 25 mm thick perforated higher density fiberglass panels that can breathe to allow any moisture that passes through the vapor retarder to be deposited as frost on the evaporator coils.

The fiberglass insulation applied to the roof structure is usually 300 to 350 mm thick for freezers and 200 to 250 mm for coolers and is applied between 50 mm by 300 mm or 50 mm by 350 mm joists that span the glue-laminated wood beams, purlins, or trusses. The exterior finish is the same as described for walls. Battens attached to the underside of the joists hold the finish panels and insulation in place.

Advantages of the Schneider system over insulated panels, assuming equal effectiveness of the vapor retarder/insulation envelopes over time, include lower first cost for structures over 3700 m^2, lower operating cost, and fewer interior columns. Disadvantages include a less clean appearance, unsuitability where washdown is required, impracticality where a number of smaller rooms are required, and a smaller number of capable practitioners (i.e., architects, engineers, contractors) available.

REFRIGERATION SYSTEMS

Types of Refrigeration Systems

Refrigeration systems can be broadly classified as unitary or applied. In this context, **unitary** refers to units that are designed by manufacturers, assembled in factories, and installed in a refrigerated space as prepackaged units. Heat rejection and compression equipment is either within the same housing as the low-temperature air-cooling coils or separated from the cooling section. Such units normally use hydrochlorofluorocarbon (HCFC) refrigerants.

Applied units denote field-engineered and -erected systems and form the vast majority of large refrigerated (below freezing) facility systems. Installations generally have a central machinery room or series of machinery rooms convenient to electrical distribution services, cooling tower, outside service entrance, and so forth. Essentially made to order, applied systems are generally designed and built from standard components obtained from one or more suppliers. Key elements are compressors, fan-coil units, receivers, pump circulation systems, controls, refrigerant condensers (evaporative, shell-and-tube, liquid-cooled, or air-cooled), and other pressure vessels.

The selection of the refrigeration system for a refrigerated facility should be made in the early stages of planning. If the facility is a single-purpose, low-temperature storage building, most types of systems can be used. However, if commodities to be stored require different temperatures and humidities, a system must be selected that can meet the demands using isolated rooms at different conditions.

Using factory-built packaged unitary equipment may have merit for the smallest structures and also for a multiroom facility that requires a variety of storage conditions. Conversely, the central compressor room has been the accepted standard for larger installations, especially where energy conservation is important.

Multiple centrally located, single-zone condensing units have been used successfully in Japan and other markets where high-rise refrigerated structures are used or where local codes drive system selection.

Direct refrigeration, either a flooded or pumped recirculation system serving fan-coil units, is a dependable choice for a central compressor room. Refrigeration compressors, programmable logic controllers, and microprocessor controls complement the central engine room refrigeration equipment.

Load Determination

The refrigeration loads of refrigerated facilities of the same capacity vary widely. Many factors, including building design, indoor and outdoor temperatures, and especially the type and flow of goods expected and the daily freezing capacity, contribute to the load. Therefore, no simple design rules apply. Experience from comparable buildings and operations is valuable, but an analysis of any projected operation should be made. Compressor and room cooling equipment should be designed for maximum daily requirements, which will be well above any monthly average.

The load factors to be considered include

- Heat transmission through insulated enclosures
- Heat and vapor infiltration load from warm air passing into refrigerated space and improper air balance
- Heat from pumps or fans circulating refrigerant or air, power equipment, personnel working in refrigerated space, product-moving equipment, and lights
- Heat removed from goods in lowering their temperatures from receiving to storage temperatures
- Heat removed in freezing goods received unfrozen
- Heat produced by goods in storage
- Other loads, such as office air conditioning, car precooling, or special operations inside the building
- Refrigerated shipping docks
- Heat released from automatic defrost units by fan motors and defrosting, which increases overall refrigerant system requirements
- Blast freezing or process freezing

High humidity, warm temperatures, or manual product handling may dramatically affect design, particularly that of the refrigeration system.

A summation of the average proportional effect of the load factors is shown in Table 1 as a percentage of total load for a facility in the southern United States. Both the size and the effect of the load factors are influenced by the facility design, usage, and location.

Heat leakage or transmission load can be calculated using the known overall heat transfer coefficient of various portions of the insulating envelope, the area of each portion, and the temperature difference between the lowest cold room design temperature and the highest average air temperature for three to five consecutive days at the building location. For freezer storage floors on ground, the average yearly ground temperature should be used.

Heat infiltration load varies greatly with the following variables: size of room, number of openings to warm areas, protection on

Table 1 Refrigeration Design Load Factors for Typical 10 000-m² Single-Floor Freezer

Refrigeration Load Factors	Long-Term Storage Cooling Capacity		Short-Term Storage Cooling Capacity		Distribution Operation Cooling Capacity	
	kW	Percent	kW	Percent	kW	Percent
Transmission losses	343	49	343	43	343	36
Infiltration	35	5	70	9	140	15
Internal operation loads	175	25	196	24	217	22
Cooling of goods received	24	3	53	6	105	11
Other factors	123	18	143	18	158	16
Total design capacity	700	100	805	100	963	100

Note: Based on a facility located in the southern United States using a refrigerated loading dock, automatic doors, and forklift material handling.

openings, traffic through openings, and cold and warm air temperatures and humidities. Calculation should be based on experience, remembering that most of the load usually occurs during daytime operations. Chapter 12 presents a complete analysis of refrigeration load calculation.

Heat from goods received for storage can be approximated from the quantity expected daily and the source. Generally, 5 to 10 K of temperature reduction can be expected, but for some newly processed items and for fruits and vegetables direct from harvesting, 35 K or more temperature reduction may be required. For general public cold storage, the load may range from 0.5 to 1 W of cooling capacity per cubic metre to allow for items received direct from harvest in a producing area.

The freezing load varies from zero for the pure distribution facility, where the product is received already frozen, to the major portion of the total load for a warehouse near a producing area. The freezing load depends on the commodity, the temperature at which it is received, and the method of freezing. More refrigeration is required for blast freezing than for still freezing without forced-air circulation.

Heat is produced by many commodities in cooler storage, principally fruits and vegetables. Heat of respiration is a sizable factor, even at 0°C, and is a continuing load throughout the storage period. Refrigeration loads should be calculated for maximum expected occupancy of such commodities.

Manual handling of product may add 30 to 50% more load to a facility in tropical areas due to constant interruption of the cold barriers at doors and on loading docks.

Unit Cooler Selection

Fan-Coil Units. These units may have direct-expansion, flooded, or recirculating liquid evaporators with either primary or finned-coil surfaces or a brine spray coolant. Storage temperature, packaging method, type of product, and so forth, must be considered when selecting a unit. The coil surface area, temperature difference between the refrigerant coil and the return air, and volumetric airflow depend on the application.

Fans are normally of the axial propeller type, but may be centrifugal if a high static discharge loss is expected. In refrigerated facilities, the fan-coil units are usually draw-through. That is, the room air is drawn through the coil and discharged through the fan. Blow-through units are used in special applications, such as fruit storages where refrigerant and air temperatures must be close. Heat from the motor is absorbed immediately by the coil on a blow-through unit and does not enter the room. Motor heat must be added to the room load with draw-through units. Figure 6 illustrates fan-coil units commonly used in refrigerated facility construction.

It is important when selecting fan-coil units to consider the **throw**, or distance air must travel to cool the farthest area. Failure to properly consider throw can result in areas of stagnant air and hot spots in the refrigerated space. Consult manufacturers' recommendation in all cases. Do not rely on guesses or "rules of thumb" in selecting units with proper airflow. Units vary widely in terms of fan type, design of the diffuser leaving the fan-coil, and coil air pressure drop.

Defrosting. Although all fan-coils occasionally require defrosting if operated below the room dew-point conditions, units located above entrances to a refrigerated space tend to draw in warm, moist air from adjacent spaces and frost the coil quickly. If this occurs, more frequent defrosting will be required to maintain the efficiency of the cooling coil.

A properly engineered and installed system can be automatically defrosted successfully with hot gas, water, electric heat, or continuously sprayed brine. The sprayed brine system has the advantage of producing the full refrigeration capacity at all times; however, it does require a supply and return pipe system with a means of boiling off the absorbed condensed moisture and can be subject to contamination with odors, biological pollution, or airborne dust.

Valve Selection. Refer to Chapter 45 and manufacturers' literature for specific information on control valve type and selection (sizing).

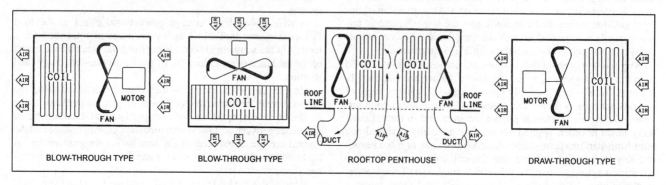

Fig. 6 Fan-Coil Units for Refrigerated Facilities

Refrigerated Facility Design

Valving Arrangements. Critical to the performance of all fan-coil units are proper refrigerant feed valve, block valve, and defrosting valve arrangements. When coils defrost, condensate that has formed as ice or frost on the coils melts. This new condensate is collected in a pan beneath the coil and drained into collection drains outside of the freezer space. Because the space is cold, condensate pans are connected to the hot-gas defrost system or otherwise heated to prevent ice formation. Likewise, all condensate drain lines must be wrapped in heat-tracing tape and trapped outside of the refrigerated space to ensure that condensate can drain unrestricted.

A variety of valve piping schemes are used. See Chapter 3 for typical piping arrangements.

Valve Location. Vital to convenient maintenance of control valves and service block valves is the location of these valves. The owner/designer has some options in most plants. If penthouse units are used, all valves are generally located outside the penthouse and are accessible from the roof. Fan-coil units mounted in the refrigerated space are generally hung from the ceiling and must be accessed via ladder or personnel lift cage on a forklift or other service vehicle. It is recommended that valve stations be located outside the freezer storage area if possible in order to facilitate access.

System Considerations. For refrigerated temperatures below −32°C, two-stage compression is generally used. Compound compressors having capacity control on each stage may be used. For variable loads, separate high- and low-stage (or booster) compressors, each with capacity control or of different capacities, may provide better operation. If blast freezers are included, pipe connections should be arranged so that sufficient booster capacity for the blast freezers can be provided by the low-stage suction pressure compressor, while the other booster is at higher suction pressure for the freezer room load. Interstage pressure and temperatures are usually selected to provide refrigeration for loading dock cooling and for rooms above 0°C.

In a two-stage system, the liquid refrigerant should be precooled at the high-stage suction pressure (interstage) to reduce the low-stage load. An automatic purger to remove air and other noncondensable gases is essential. Ammonia plants should have oil traps at all points where oil may accumulate in the system (i.e., on the compressor discharge, condensers, intercoolers, and evaporators), so that oil can be removed periodically without loss of ammonia. Chapters 1 through 7 have further information on refrigeration systems.

The use of commercial, air-cooled packaged or factory-assembled units is common, especially in smaller plants. These units have lower initial cost, smaller space requirements, and no need for a special machinery room or operating engineer. However, they use more energy, have higher operating and maintenance costs, and have a shorter life expectancy for components (usually compressors) than central refrigeration systems.

Multiple Installations. To distribute air without ductwork, installations of multiple fan-coil units have been used. The quantity of air circulated per unit refrigeration capacity may be higher than that required for sprayed unit systems, so proper application is important. For single-story buildings, sprayed units installed in penthouses with ducted air distribution have been used to make full use of floor space in the storage area (Figure 7). Either prefabricated or field-erected refrigeration systems or cooling units connected to a central plant can be incorporated in penthouse design.

Unitary cooling units are located in a penthouse, with distributing ductwork projected through the penthouse floor and under the insulated ceiling below. Return air passes up through the penthouse floor grille. This system avoids the interference of fan-coil units hung below the ceiling in the refrigerated chamber and facilitates maintenance access.

Defrost water drain piping passes through the penthouse insulated walls and onto the main storage roof. Refrigerant mains and electrical conduit can be run over the roof on suitable supports to the central compressor room or to packaged refrigeration units on the adjacent roof. Temperature control thermostats and electrical equipment can be housed in the penthouse.

A personnel access door to the penthouse is required for convenient equipment service. The inside insulated penthouse walls and ceiling must be vaportight to keep condensation from deteriorating the insulation and to maintain the integrity of the building vapor retarder. Some primary advantages of penthouses are

- Cooling units, catwalks, and piping do not interfere with product storage space and are not subject to physical damage from stacking truck operations.
- Service to all cooling equipment and controls can be handled by one individual from an essential floor or roof deck location.
- Maintenance and service costs are minimized.
- Main piping, control devices, and block valves are located outside the refrigerated space.
- If control and block valves are located outside the penthouse, any refrigerant leaks will occur outside the refrigerated space.

Freezers

Freezers within refrigerated facilities are generally used to freeze products or to chill products from some higher temperature to storage temperature. Failure to properly cool the incoming product transfers the product cool-down load to the facility, greatly increasing facility operating costs. Of perhaps greater concern is that dormant storage in a cold area may not cool the product fast enough to prevent bacterial growth, which causes product deterioration. In addition, other stored, already frozen products may be damaged by providing some or all of the refrigerating effect to the incoming products.

For this reason, many refrigerated facilities have a **blast freezer** that producers can contract to use. Blast freezing ensures that the products are properly frozen in minimum time before they are put into storage and that their quality is maintained. Modern control systems allow sampling of inner core product temperatures and printout of records that customers may require. The cost of blast freezer service can be properly apportioned to its users, allowing higher efficiency and lower cost for other cold storage customers.

Although there are many types of freezers, including belt, tray, contact plate, spiral, and other packaged types, the most common arrangement used in refrigerated facilities is designed to accept pallets of products from a forklift. The freezing area is large and free from obstructions and has large doors. See Chapter 15, Industrial Food Freezing Systems.

Figure 8 illustrates a typical blast freezer used in a refrigerated facility. Temperatures are normally about −35°C but may be higher or lower depending upon the product being frozen. Once the room is filled to design capacity, it is sealed and the system is started. The refrigeration process time can be controlled by a time clock, by manual termination, or by measuring internal product temperature and stopping the process once the control temperature is reached. The last method gives optimum performance. Once the product is

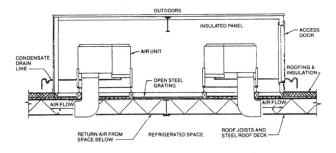

Fig. 7 Penthouse Cooling Units

frozen, the pallets are transferred to general refrigerated storage areas.

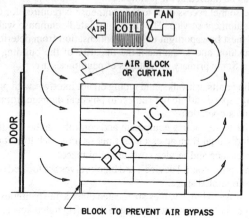

Fig. 8 Typical Blast Freezer

Because the blast freezer normally operates intermittently, freezer owners should endeavor to operate it during the times when energy cost is lowest. Unfortunately, food products must be frozen as quickly as possible, and products are usually delivered during times of peak electrical rates. Alternative power sources, such as natural gas engine or diesel drives, should be considered. Although these normally have first cost and maintenance cost premiums, they are not subject to time-varying energy rates and may offer savings.

Defrost techniques for blast freezers are similar to normal defrost methods for refrigerated facility fan-coil units. Coils can often be defrosted after the product cooling cycle is completed or while the freezer is being emptied for the next load.

Pumped refrigerant recycling systems and flooded surge drum coils have both been used with success. Direct-expansion coils may be used, but the designer should be careful with expansion valve systems to address coil circuitry, refrigerant liquid overfeed, oil return, defrost, shutdown liquid inventory management, and so forth. Conventional oil removal devices should be supplied on flooded coil and pumped systems because the blast freezer is normally the lowest temperature system in the facility and may accumulate oil over time. Materials of construction can be similar to the rest of facility. Floor heating may be convenient if the products are damp or wet during loading.

Most blast freezers are accessed from a refrigerated space, so that (1) products can be moved directly from the freezer to storage racks. Also, the blast freezers can be used for storage when not operating.

Controls

The term **controls** refers to any mechanism or device used to start, stop, adjust, protect, or monitor the operation of a moving or functional piece of equipment. Controls for any system can be as simple as electromechanical devices such as pressure switches and timer relays or as complex as a complete digital control system with analog sensors and a high-speed communications network connected to a supervisory computer station. Because controls are required in every industry, there is a wide variety from which to choose. In recent years, the industrial refrigeration industry has migrated away from the use of electromechanical devices and toward the use of specialized microprocessors, programmable logic controllers, and computers for unit and system control.

Electromechanical devices for control will continue to be used for some time and may never be replaced entirely for certain control functions (e.g., use of relays for electrical high current isolation and float switches for refrigerant high level shutdowns). See Chapter 45 for more information.

Microprocessor controls and electronic sensors generally offer the following advantages over electromechanical control:

- More accurate readings and therefore more accurate control
- Easier operation
- Greater flexibility due to adjustable set points and operating parameters
- More information concerning operating conditions, alarms, failures, and trouble-shooting
- Capability for interfacing with remote operator stations

There are four main areas of control in all refrigerated facility systems with a central compressor room:

1. **Compressor package control.** Minimum requirements: orderly start-up, orderly shutdown, capacity control to maintain suction pressure, alarm monitoring, and safety shutdown.
2. **Condenser control.** Minimum requirements: fan and water pump start and stop to maintain a reasonable constant or floating refrigerant discharge pressure.
3. **Evaporator control.** Minimum requirements: control of the air unit fans and refrigerant liquid feed to maintain room air temperatures and staging of the air unit refrigerant valve stations to provide automatic coil defrosts.
4. **Refrigerant flow management.** Minimum requirements: maintenance of desired refrigerant levels in vessels, control of valves and pumps to transfer refrigerant as needed between vessels and air units in the system, and proper shutdowns in the event of refrigerant overfeed or underfeed.

Other areas of control, such as refrigerant leak detection and alarm, sequencing of multiple compressors for energy efficiency, and underfloor warming system control, may be desired.

Because of the wide variety of control components and systems available and the fast-changing capabilities of those components, it is impossible to define or recommend an absolute component list. However, it is possible to provide guidelines for the design and layout of the overall control system, regardless of the components used or the functions to be controlled. This design or general layout can be termed a **control system architecture**.

All control systems consist of four main building blocks:

1. **Controller(s)**—Microprocessor with control software.
2. **Input/Outputs (I/Os)**—Means of connecting devices or measurements to the controller.
3. **Operator interface(s)**—Means of conveying information contained in the controller to a human being.
4. **Interconnecting media**—Means of transferring information between controllers, I/Os, and operator interfaces.

The control system architecture defines the quantity, location, and function of these basic components. The architecture determines the reliability, expandability, operator interface opportunities, component costs, and installation costs of a control system. Therefore, the architecture should be designed before any controls component manufacturers or vendors are selected.

The following are the basic steps in designing a refrigerated facility control system:

Step 1. Define the **control tasks**. This step should provide a complete and detailed I/O listing, including quantity and type. With this list and a little experience and knowledge of available hardware, the type, quantity, and processing power of the necessary controllers can be determined.

Step 2. Determine the physical locations of the controlled devices and the measurements to be taken. If remote I/Os or multiple controllers are located close to the devices and sensors, field wiring installation costs can be reduced. In order to avoid extra costs or impracticalities, the environments of the various locations must be compared with the environmental specifications of the hardware

Refrigerated Facility Design

to be placed in them. Maintenance requirements can also affect the selection of physical location of the I/Os and controller.

Step 3. Determine the control task integration requirements. Control tasks that require and share the same information (such as a discharge pressure reading for starting both a condenser fan and a condenser water pump) must be accomplished either with the same controller or with multiple controllers that share information via interconnecting media. Tasks that do not share information can be performed by separate controllers. The use of multiple controllers minimizes the chance of catastrophic control failures. With multiple controllers that share information, the interconnecting media must be robust. In particular, the speed of data transfer between controllers must be suitable to maintain the control accuracy required. If the task is critical, the chance of failure of the interconnecting media must be minimal.

Step 4. Determine the operator interface requirements. This includes noting which controllers must have a local operator interface, which controllers must have a remote operator interface, and how many remote operator interface stations are required, and defining the hardware and software requirements of the local and remote operator interfaces.

Step 5. Select the interconnecting media between controllers and their remote I/Os, between controllers and controllers, and between controllers and operator interfaces. The interconnecting medium to the remote I/Os is typically defined by the controller manufacturer; it must be robust and high-speed because controllers' decisions depend on real-time data. The interconnecting medium between controllers themselves is also typically defined by the controller manufacturer; the speed requirements depend upon the tasks being performed with the shared information. For the interconnecting media between the controllers and the operator interfaces, speed is typically not as critical because the control continues even if the connection fails. For the operator interface connection, the speed of accessing a controller's data is not as critical as having access to all the available data from the controller.

Step 6. Evaluate the architecture for technical merit. The first five steps should produce a list of controllers, their locations, their operator interfaces, and their control tasks. Once the list is complete, the selected controllers should be evaluated for both processor memory available for programming and processor I/O capacity available for current and future requirements. The selected interconnecting media should be evaluated for distance and speed limitations. If any weaknesses are found, a different model, type, or even manufacturer of the component should be selected.

Step 7. Evaluate the architecture for software availability. The best microprocessor is of little good if no software exists to make it operate. It must be ascertained that software exists or can easily be written to provide (1) information transfer between controllers and operator interfaces; (2) the programming functions needed to perform the control tasks; and (3) desired operator interface capabilities such as graphics, historical data, reports, and alarm management. Untested or proprietary software should be avoided.

Step 8. Evaluate the architecture for failure conditions. Determine how the system will operate with a failure of each controller. If the failure of a particular controller would be catastrophic, more controllers can be used to further distribute the control tasks, or electromechanical components can be added to allow manual completion of the tasks. For complex tasks that are impossible to control manually, it is essential that spare or backup control hardware be in stock and that operators be trained in the trouble-shooting and reinstallation of control hardware and software.

Step 9. Evaluate the proposed architecture for cost, including field wiring, components, start-up, training, downtime, and maintenance costs. All these costs must be considered together for a fair and proper evaluation. If budgets are exceeded, then Steps 1 through 8 must be repeated, removing any nonessential control tasks and reducing the quantity of controllers, I/Os, and operator interfaces.

Once the control system architecture is designed, specifics of software operation should be determined. This includes items such as set points necessary for a control task, control algorithms and calculations used to determine output responses, graphic screen layouts, report layouts, alarm message wording, and so forth. More detail is necessary, and more time could be spent determining the details of software operation than is needed and spent defining the system architecture. However, if the system architecture is solid, the software can always be modified as needed. With improper architecture, functional additions or corrections can be costly, time consuming, and sometimes impossible.

INSULATION TECHNIQUES

The two main functions of an insulation envelope are to reduce the refrigeration requirements for the refrigerated space and to prevent condensation.

Vapor Retarder System

The primary concern in the design of a low-temperature facility is the vapor retarder system, which should be as close to 100% effective as is practical. *The success or failure of an insulation envelope is due entirely to the effectiveness of the vapor retarder system in preventing water vapor transmission into and through the insulation.*

The driving force behind water vapor transmission is the difference in vapor pressure across the vapor retarder. Once water vapor passes a vapor retarder, a series of detrimental events begins. The water migrating into the insulation diminishes the thermal resistance of the insulation and eventually destroys the envelope. Ice formation inside the envelope system usually grows and physically forces the building elements apart to the point of failure.

Another practical function of the vapor retarder is to stop air infiltration. The driving force behind infiltration can be atmospheric pressure or ventilation.

After condensing or freezing, some of the water vapor in the insulation revaporizes or sublimes. This vapor is eventually drawn to the refrigeration coil and disposed of via the condensate drain, but the amount removed is usually not sufficient to dry out the insulation unless the vapor retarder discontinuity is located and corrected.

The vapor retarder must be located on the warm side of the insulation. Each building element inside the prime retarder must be more permeable than the last to permit moisture to move through it, or it becomes a site of condensation or ice. This precept is abandoned for the sake of sanitation at the inside faces of coolers. However, the inside faces of freezers are usually permitted to breathe by leaving the joints uncaulked in the case of panel construction or by using less permeable surfaces for other forms of construction. Factory-assembled insulation panels endure this double vapor retarder problem better than other types of construction.

In walls with insufficient insulation, the temperature at the inside wall surface may during certain periods reach the dew point of the migrating water vapor, causing condensation and freezing. This can also happen to a wall that originally had adequate insulation but, through condensation or ice formation in the insulation, lost part of its insulating value. In either case, the ice deposited on the wall will gradually push the insulation and protective covering away from the wall until the insulation structure collapses.

It is extremely important to properly install vapor retarders and seal joints in the vapor retarder material to ensure continuity from one surface to another (i.e., wall to roof, wall to floor, or wall to ceiling). The failure of vapor retarder systems for refrigerated facilities is almost always due to poor installation. The contractor must be competent and experienced in the installation of vapor retarder systems to be able to execute a vaportight system.

Condensation on the inside of the cooler is unacceptable because (1) the wet surface provides the culture base for bacteria growth,

Table 2 Recommended Insulation R-Values

Type of Facility	Temperature Range, °C	Thermal Resistance R, m²·K/W		
		Floors	Walls/Suspended Ceilings	Roofs
Cooler[a]	4 to 10	Perimeter insulation only	4.4	5.3 to 6.2
Chill cooler[a]	−4 to 2	3.5	4.2 to 5.6	6.2 to 7.0
Holding freezer	−23 to −29	4.8 to 5.6	6.2 to 7.0	7.9 to 8.8
Blast freezer[b]	−40 to −46	5.3 to 7.0	7.9 to 8.8	8.8 to 10.6

Note: Because of the wide range in the cost of energy and the cost of insulation materials based on thermal performance, a recommended R-value is given as a guide in each of the respective areas of construction. For more exact values, consult a designer and/or insulation supplier.

[a] If a cooler has the possibility of being converted to a freezer in the future, the owner should consider insulating the facility with the higher R-values from the freezer section.

[b] R-values shown are for a blast freezer built within an unconditioned space. If the blast freezer is built within a cooler or freezer, consult a designer and/or insulation supplier.

and (2) any dripping onto the product gives cause for condemnation of the product in part or in whole.

Stagnant or dead air spots behind beams or inside metal roof decks can allow localized condensation. This moisture can be from within the cooler or freezer (i.e., not necessarily from a vapor retarder leak).

The reality is that there is no 100% effective vapor retarder system. A system is successful when the rate of moisture infiltration is equal to the rate of moisture removal by refrigeration, with no detectable condensation.

Types of Insulation

Rigid Insulation. Insulation materials, such as polystyrene, polyisocyanurate, polyurethane, and phenolic material, have proven satisfactory when installed with the proper vapor retarder and finished with materials that provide protection from fire and form a sanitary surface. Selection of the proper insulation material should be based primarily on the economics of the installed insulation, including the finish, sanitation, and fire protection.

Panel Insulation. The use of prefabricated insulated panels for insulated wall and roof construction is widely accepted. These panels can be assembled around the building structural frame or against masonry or precast walls. Panels can be insulated at the factory with either polystyrene or urethane. Other insulation materials do not lend themselves to panelized construction.

The basic advantage, besides economics, of using insulated panel construction is that repair and maintenance are simplified because the outer skin also serves as the vapor retarder and is accessible. This is of great benefit if the structure is to be enlarged in the future. Proper vapor retarder tie-ins then become practical.

Foam-in-Place Insulation. This application method has gained acceptance as a result of developments in polyurethane insulation and equipment for installation. Portable blending machines are available with a spray or frothing nozzle for feeding insulation into the wall, floor, or ceiling cavities to fill without joints the space provided for monolithic insulation construction. This material has no usable vapor resistance, and its application in floor construction should be limited.

Precast Concrete Insulation Panels. This specialized form of construction has been successful when proper vapor retarder and other specialized elements are incorporated. As always, vapor retarder continuity is the key to a successful installation.

Insulation Thickness

The R-value of insulation required varies with the temperature held in the refrigerated space and the conditions surrounding the room. Table 2 shows generally recommended R-values for different types of facilities. The range in R-values is due to variations in cost of energy, insulation materials, and climatic conditions. For more exact values, consult a designer and/or insulation supplier. Insulation with R-values lower than those shown should not be used.

APPLYING INSULATION

The method and materials used to insulate roofs, ceilings, walls, floors, and doors need careful consideration.

Roofs

The suspended ceiling method of construction is preferred for attaining a complete thermal and vapor envelope. Insulating materials may be placed on the roof or floor above the refrigerated space rather than adhered to the structural ceiling. If this type of construction is not feasible, and the insulation must be installed under a concrete or other ceiling, then the vapor retarder, insulation, and finish materials should be mechanically supported from the structure above rather than relying on adhesive application only. Suspending a wood or metal deck from the roof structure and applying insulation and a vapor retarder to the top of the deck is another method of hanging ceiling insulation. Skill of application and attention to positive air and vapor seals are essential to continued effectiveness.

Suspended insulated ceilings, whether built-up or prefabricated, should be adequately ventilated to maintain near-ambient conditions in the plenum space; this minimizes both condensation and deterioration of vapor retarder materials. A minimum of six air changes per hour is adequate in most cases. Permanent sealing around insulating hanger rods, columns, conduit, and other penetrations is needed.

The structural designer usually includes roofing expansion joints when installing insulation on top of metal decking or concrete structural slabs for a building larger than 30 m by 30 m. Because the refrigerated space is not normally subject to temperature variations, structural framing is usually designed without expansion or contraction joints if it is entirely enclosed within the insulation envelope. Board insulation laid on metal decking should be installed in two or more layers with the seams staggered. An examination of the coefficients of linear expansion for typical roof construction materials illustrates the need for careful attention to this phase of the building design.

Although asphalt built-up roofs have been used, loosely laid membrane roofing has become popular and requires little maintenance.

Walls

Wall construction must be designed so that as few structural members as possible penetrate the insulation envelope. Insulated panels applied to the outside of the structural frame prevent conduction through the framing. Where masonry or concrete wall construction is used, structural framing must be independent of the exterior wall. The exterior wall cannot be used as a bearing wall unless suspended insulated ceiling is used.

Where interior insulated partitions are required, a double column arrangement at the partition prevents structural members from penetrating the wall insulation. For satisfactory operation and long life of the insulation structure, envelope construction should be used wherever possible.

Refrigerated Facility Design

Governing codes for fire prevention and sanitation must be followed in selecting a finish or panel. For conventional insulation materials other than prefabricated panels, a vapor retarder system should be selected.

Abrasion-resistant membranes, such as 0.25 mm thick black polyethylene film with a minimum of joints, are suitable vapor retarders. Rigid insulation can then be installed dry and finished with plaster or sheet finishes, as the specific facility requires. In refrigerated facilities, contraction of the interior finish is of more concern than expansion because temperatures are usually held far below installation ambient temperatures.

Floors

Freezer buildings have been constructed without floor insulation, and some operate without difficulty. However, the possibility of failure is so great that this practice is seldom recommended.

Underfloor ice formation, which causes heaving of floors and columns, can be prevented by heating the soil or fill under the insulation. Heating can be by air ducts, electric heating elements, or pipes through which a liquid is recirculated.

The air duct system works well for smaller storages. For a larger storage, it should be supplemented with fans and a source of heat if the pipe is more than 30 m long. The end openings should be screened to keep out rodents, insects, and any material that might close off the air passages. The ducts must be sloped for drainage to remove condensed moisture. Perforated pipes should not be used.

The electrical system is simple to install and maintain if the heating elements are run in conduit or pipe so they can be replaced; however, operating costs may be very high. Adequate insulation should be used because it directly influences energy consumption.

The pipe grid system, shown in Figure 9, is usually best because it can be designed and installed to warm where needed and can later be regulated to suit varying conditions. Extensions of this system can be placed in vestibules and corridors to reduce ice and wetness on floors. The underfloor pipe grid also facilitates future expansion. A heat exchanger in the refrigeration system, steam, or gas engine exhaust can provide a source of heat for this system. The temperature of the recirculated fluid is controlled at 10 to 21°C, depending on design requirements. Almost universally, the pipes are made of plastic.

The pipe grid system is usually placed in the base concrete slab directly under the insulation. If the pipe is metal, a vapor retarder should be placed below the pipe to prevent corrosion. The fluid should be an antifreeze solution such as propylene glycol with the proper inhibitor.

The amount of warming for any system can be calculated and is about the same for medium-sized and large refrigerated spaces regardless of ambient conditions. The calculated heat input requirement is the floor insulation leakage based on the temperature difference between the 4.4°C underfloor earth and room temperature (e.g., 27.4 K temperature difference for a −23°C storage room). The flow of heat from the earth, about 4.1 W per square metre of floor, serves as a safety factor.

Doors

The selection and application of cold storage doors are a fundamental part of cold storage facility design and have a strong bearing on the overall economy of facility operation. The trend is to have fewer and better doors. Manufacturers offer many types of doors supplied with the proper thickness of insulation for the intended use. Four basic types of doors are swinging, horizontal sliding, vertical sliding, and double acting. Door manufacturers' catalogs give detailed illustrations of each. Doors used only for personnel cause few problems. In general, a standard swinging personnel door, 0.9 m wide by 2 m high and designed for the temperature and humidity involved, is adequate.

The proper door for heavy traffic areas should provide maximum traffic capacity with minimum loss of refrigeration and require minimum maintenance.

The following are factors to consider when selecting cold storage doors:

- Automatic doors are a primary requirement with forklift and automatic conveyor material handling systems.
- Careless forklift operators are a hazard to door operation and effectiveness. Guards can be installed but are effective only when the door is open. Photoelectric and ultrasonic beams across the

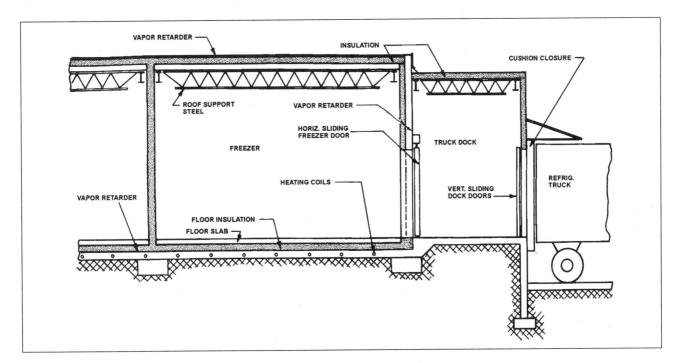

Fig. 9 Typical One-Story Construction with Underfloor Warming Pipes

doorway or proximity loop control on both sides of the doorway can provide additional protection by monitoring objects in the door openings or approaches. These systems can also control door opening and closing.

- The selection of automatic door systems to suit traffic requirements and the building structure may require experienced technical guidance.
- To ensure continuous door performance, the work area near the doors must be supervised, and the doors must have planned maintenance.
- Cooled or refrigerated shipping platforms increase door efficiency and reduce door maintenance because the humidity and temperature difference across the doorway is lower. Icing of the door is lessened, and fogging in trafficways is reduced.

Bi-parting and Other Doors. Air curtains, plastic or rubber strip curtains, and bi-parting doors, give varied effectiveness. Strip curtains are not accepted by USDA standards if open product moves through the doorway. Often the curtain seems to the forklift driver to be a substitute for the door, so the door is left open with a concurrent loss of refrigerated air. Quick-operating powered doors of fabric or rigid plastic are beneficial for draft control.

Swinging and Sliding Doors. A door with hinges on the right edge (when observed from the side on which the operating hardware is mounted) is called a right-hand swing. A door that slides to the right to open (when observed from the side of the wall on which it is mounted) is called a right-slide door.

Vertical Sliding Doors. These doors, which are hand or motor operated with counterbalanced spring or weights, are used on truck receiving and shipping docks.

Refrigerated Room Doors. Doors for pallet material handling are usually automatic horizontal sliding doors, either single-slide or biparting.

Metal or Plastic Cladding. Light metal cladding or a reinforced plastic skin protects most doors. Areas of abuse must be further protected by heavy metal, either partial or full-height.

Heat. To prevent ice formation and resultant faulty door operation, doors are available with automatic electrical heat—not only in the sides, head, and sill of the door or door frame, but also in switches and cover hoods of power-operated units. Such heating elements are necessary on all four edges of double-acting doors in low-temperature rooms. Safe devices that meet electrical codes must be used.

Bumpers and Guard Posts. Power-operated doors require protection from abuse. Bumpers embedded in the floor on both sides of the wall and on each side of the passageway help preserve the life of the door. Correctly placed guard posts protect sliding doors from traffic damage.

Buck and Anchorage. Effective door operation is impossible without good buck and anchorage provisions. The recommendations of the door manufacturers should be coordinated with wall construction.

Door Location. Doors should be located to accommodate safe and economical material handling. Irregular aisles and blind spots in trafficways near doors should be avoided.

Door Size. A hinged insulated door opening should provide at least 300 mm clearance on both sides of a pallet. Thus, 1.8 m should be the minimum door width for a 1.2 m wide pallet load. Double-acting doors should be 2.4 m wide. Specific conditions at a particular doorway can require variations from this recommendation. A standard height of 3 m accommodates all high-stacking forklifts.

Sill. A concrete sill minimizes the rise at the door sill. A thermal break should be provided in the floor slab at or near the plane of the front of the wall.

Power Doors. Horizontal sliding doors are standard when electric operation is provided. The two-leaf biparting unit keeps opening and closing time to a minimum, and the door is out of the way

and protected from damage when open. Also, because leading edges of both leaves have safety edges, personnel, doors, trucks, and product are protected. A pull cord is used for opening, and a time-delay relay, proximity-loop control, or photoelectric cell controls closing. Potential for major door damage may be reduced by proper location of pull-cord switches. Doors must be protected from moisture and frost with heat or baffles. Automatic doors should have a preventive maintenance program to check gaskets, door alignment, electrical switches, safety edges, and heating circuits. Safety releases on locking devices are necessary to prevent entrapment of personnel.

Fire-Rated Doors. Available in both swinging and sliding types, fire-rated doors are also insulated. Refrigerated buildings have increased in size, and their contents have increased in value, so insurance companies and fire authorities are requiring fire walls and doors.

Large Door Openings. Door openings that can accommodate forklifts with high masts, two-pallet-high loads, and tractor-drawn trailers are large enough to cause appreciable loss of refrigeration. Infiltration of moisture is objectionable because it forms as condensate or frost on stacked merchandise and within the building structure. Door heights up to 3 to 3.7 m are frequently required, especially where drive-through racks are used. Refrigeration loss and infiltration of moisture can be particularly serious when doors are located in opposite walls of a refrigerated space and crossflow of air is possible. It is important to reduce infiltration with enclosed refrigerated loading docks and, in some instances, with one-way traffic vestibules.

OTHER CONSIDERATIONS

Temperature Pulldown

Because of the low temperatures in freezer facilities, contraction of structural members in these spaces will be substantially greater than in any surrounding ambient or cooler facilities. Therefore, contraction joints must be properly designed to prevent structural damage during facility pulldown.

The first stage of temperature reduction should be from ambient down to 2°C at whatever rate of reduction the refrigeration system can achieve.

The room should then be held at that temperature until it is dry. Finishes are especially subject to damage when temperatures are lowered too rapidly. Portland cement plaster should be fully cured before the room is refrigerated.

If there is a possibility that the room is airtight (most likely for small rooms, 6 m by 6 m maximum), swinging doors should be partially open during pulldown to relieve the internal vacuum caused by the cooling of the air, or vents should be provided. Permanent air relief vents are needed for continual operation of defrosts in small rooms with only swinging doors.

The concrete slab will contract during pulldown, causing slab/wall joints, contraction joints, and other construction joints to open. At the end of the holding period (i.e., at 2°C), any necessary caulking should be done.

An average time for achieving dryness is 72 h. However, there are indicators that may be used, such as watching the rate of frost formation on the coils or measuring the rate of moisture removal by capturing the condensation during defrost.

After the refrigerated room is dry, the temperature can then be reduced again at whatever rate the refrigeration equipment can achieve until the operating temperature is reached. Rates of 6 K per day have been used in the past, but if care has been taken to remove all the construction moisture in the previous steps, faster rates are possible without damage.

Material Handling Equipment

Both private and public refrigerated facilities can house high-volume, year-round operations with fast-moving order pick areas

Refrigerated Facility Design

backed by in-transit bulk storage. Distribution facilities may carry 300 to 3000 items or as many as 30 000 lots. Palletized loads stored either in bulk or on racks are transported by forklifts or high-rise storage/retrieval machines in a −18 to −29°C environment. Standard, battery-driven forklifts that can lift up to 7.6 m can service one-deep, two-deep reach-in, drive-in, drive-through, or gravity flow storage racks. Special forklifts can lift up to 18 m.

Automated storage/retrieval machines make better use of storage volume, require fewer personnel, and reduce the refrigeration load because the facility requires less roof and floor area. This equipment operates in a height range of 7 to 30 m to service one-deep, two-deep reach-in, two- to twelve-deep rollpin, or gravity flow pallet storage racks. On-line computers and bar code identification allow a system to automatically control the retrieval, transfer, and delivery of products. In addition, these systems can record product location and inventory and load several delivery trucks simultaneously from one order pick conveyor and sorting device.

A refrigerated plant may have two or more material handling systems if the storage area contains fast- and slow-moving reserve storage, plus slow-moving order pick. Fast-moving items may be delivered and order picked by a conventional forklift pallet operation with up to 9 m stacking heights. In the fast-moving order pick section, the storage room internal height is raised to accommodate storage/retrieval machines; reserve pallet storage; order pick slots; multilevel palletizing; and the infeed, discharge, and order pick conveyors. Mezzanines may be considered in order to provide maximum access to the order pick slots. Intermediate level fire protection sprinklers may be required in the high rack or mezzanine areas above 4.3 m high.

Fire Protection

Ordinary wet sprinkler systems can be applied to refrigerated spaces above freezing. In rooms below freezing, entering water will freeze if a sprinkler head malfunctions or is mechanically damaged. If this occurs, the affected piping must be removed. In lower temperature spaces, a dry air or nitrogen-charged system should be selected.

The design of a dry sprinkler system operating in areas below 0°C requires special knowledge and should not be undertaken without expert guidance. Freezer storage with rack storages 9 m high or higher may require special design, and the initial design should be shown to the insuring company.

Local regulations may require ceiling isolation smoke curtains and smoke vents near the roof in large refrigerated chambers. These features allow smoke to escape and help the fire-fighting crew locate the fire. If the building does not have a sprinkler system, central reporting or warning systems are available for hazardous areas.

Inspection and Maintenance

Buildings can change in dimension due to settling, temperature change, and other factors; therefore, cold storage facilities should be inspected regularly to spot problems early, so that preventive maintenance can be performed in time to avert serious damage.

Inspection and maintenance procedures fall into two areas: **basic system** (floor, wall, and roof/ceiling systems); and **apertures** (doors, frames, and other access to cold storage rooms).

Basic System

- Stack pallets at a sufficient distance (460 mm) from walls or ceiling to permit air circulation.
- Examine walls and ceiling at random every month for frost buildup. If buildup persists, locate the break in the vapor retarder.
- For insulated ceilings below a plenum, inspect the plenum areas for possible roof leaks or condensation.
- If condensation or leaks are detected, make repairs immediately.

Apertures

- Remind personnel to close doors quickly to reduce frosting in rooms.
- Check the rollers and door travel periodically to ensure that the seal at the door edge is effective. If leaks are detected, adjust the door to restore a moisture- and airtight condition.
- Check doors and door edges to detect damage from forklifts or other traffic. Repair any damage immediately to prevent door icing or motor overload due to excessive friction.
- Lubricate doors according to the maintenance schedule from the door manufacturer to ensure free movement and complete closure.
- Periodically check seals around openings for ducts, piping, and wiring in the walls and ceiling.

BIBLIOGRAPHY

Aldrich, D.F. and R.H. Bond. 1985. Thermal performance of rigid cellular foam insulation at subfreezing temperatures. In: *Thermal performance of the exterior envelopes of buildings III*, pp. 500-09. ASHRAE.

Baird, C.D., J.J. Gaffney, and M.T. Talbot. 1988. Design criteria for efficient and cost effective forced air cooling systems for fruits and vegetables. *ASHRAE Transactions* 94(1):1434-54.

Ballou, D.F. 1981. A case history of a frost heaved freezer floor. *ASHRAE Transactions* 87(2):1099-1105.

Beatty, K.O., E.B. Birch, and E.M. Schoenborn. 1951. Heat transfer from humid air to metal under frosting conditions. *Journal of the ASRE* (December):1203-7.

Cole, R.A. 1989. Refrigeration load in a freezer due to hot gas defrost and their associated costs. *ASHRAE Transactions* 95(2).

Coleman, R.V. 1983. Doors for high rise refrigerated storage. *ASHRAE Transactions* 89(1B):762-65.

Corradi, G. 1973. Air cooling units for the refrigerating industry and new equipment. *Revue generale du froid* 1(January):45-51.

Courville, G.E., J.P. Sanders, and P.W. Childs. 1985. Dynamic thermal performance of insulated metal deck roof systems. In: *Thermal performance of the exterior envelopes of buildings III*, pp. 53-63. ASHRAE.

Crawford, R.R., J.P. Mavec, and R.A. Cole. 1992. Literature survey on recommended procedures for the selection, placement, and type of evaporators for refrigerated warehouses. *ASHRAE Transactions* 98(1).

D'Artagnan, S. 1985. The rate of temperature pulldown. *ASHRAE Journal* 27(9):36.

Downing, C.G. and W.A. Meefert. 1993. Effectiveness of cold storage infiltration protection devices. *ASHRAE Transactions* 99(2).

FDA. 1997. *Food code*. U.S. Food and Drug Administration. Available from National Technical Information Service, Springfield, VA. No. PB97-141204LEU.

Hampson, G.R. 1981. Energy conservation opportunities in cold storage warehouses. *ASHRAE Transactions* 87(2):845-49.

Hendrix, W.A., D.R. Henderson, and H.Z. Jackson. 1989. Infiltration heat gains through cold storage room doorways. *ASHRAE Transactions* 95(2).

Holske, C.F. 1953. Commercial and industrial defrosting: General principles. *Refrigerating Engineering* 61(3):261-62.

Kerschbaumer, H.G. 1971. Analysis of the influence of frost formation on evaporators and of the defrost cycles on performance and power consumption of refrigerating systems. Proceedings of the 13th International Congress of Refrigeration, 1-12.

Kurilev, E.S. and M.Z. Pechatnikov. 1966. Patterns of airflow distribution in cold storage rooms. Bulletin of the International Institute of Refrigeration, Annex 1966-1, 573-79.

Lehman, D.C. and J.E. Ferguson. 1982. A modified jacketed cold storage design. *ASHRAE Transactions* 88(2):228-334.

Lotz, H. 1967. Heat and mass transfer and pressure drop in frosting finned coils; Progress in refrigeration science and technology. Proceedings of the 12th International Congress of Refrigeration, Madrid 2:49-505.

Niederer, D.H. 1976. Frosting and defrosting effects on coil heat transfer. *ASHRAE Transactions* 82(1):467-73.

Paulson, B.A. 1988. Air distribution in freezer areas. International Institute of Ammonia Refrigeration 10th Annual Meeting, No. 10, 261-71.

Powers, G.L. 1981. Ambient gravity air flow stops freezer floor heaving. *ASHRAE Transactions* 87(2):1107-16.

Sainsbury, G.F. 1985. Reducing shrinkage through improved design and operation in refrigerated facilities. *ASHRAE Transactions* 91(1B):726-34.

Sastry, S.K. 1985. Factors affecting shrinkage of foods in refrigerated storage. *ASHRAE Transactions* 91(1B):683-89.

Shaffer, J.A. 1983. Foundations and superstructure systems for stacker crane high-rise freezers and coolers. *ASHRAE Transactions* 89(1).

Sherman, M.H. and D.T. Grimsrud. 1980. Infiltration-pressurization correlation: Simplified physical modeling. *ASHRAE Transactions* 86(2):778-807.

Soling, S.P. 1983. High rise refrigerated storage. *ASHRAE Transactions* 89(1B):737-61.

Stoecker, W.F. 1960. Frost formations on refrigeration coils. *ASHRAE Transactions* 66:91-103.

Stoecker, W.F., J.J. Lux, and R.J. Kooy. 1983. Energy considerations in hot-gas defrosting of industrial refrigeration coils. *ASHRAE Transactions* 89(2A):549-68.

Treschel, H.R., P.R. Achenbach, and J.R. Ebbets. 1985. Effect of an exterior air infiltration barrier on moisture condensation and accumulation within insulated frame wall cavities. *ASHRAE Transactions* 91(2A):545-59.

Tye, R.P., J.P. Silvers, D.C. Brownell, and S.E. Smith. 1985. New materials and concepts to reduce energy losses through structural thermal bridges. In: *Thermal performance of the exterior envelopes of buildings III*, pp. 739-50. ASHRAE.

Voelker, J.T. 1983. Insulating considerations for stacker crane high rise freezers and coolers. *ASHRAE Transactions* 89(1B):766-68.

Wang, I.H. and Touber. 1987. Prediction of airflow pattern in cold stores based on temperature measurements. Proceedings of Commission D, International Congress of Refrigeration, Vienna, 52-60.

CHAPTER 14

METHODS OF PRECOOLING FRUITS, VEGETABLES, AND CUT FLOWERS

Product Requirements 14.1	Package Icing 14.5
Calculation Methods 14.1	Vacuum Cooling 14.5
Cooling Methods 14.3	Selecting a Cooling Method 14.8
Hydrocooling 14.3	Cooling Cut Flowers 14.9
Forced-Air Cooling 14.4	Symbols 14.9

COOLING is generally considered the removal of field heat from freshly harvested products to inhibit spoilage and to maintain preharvest freshness and flavor. The term precooling implies the removal of heat before the product is shipped to a distant market, processed, or stored. Some products are slowly cooled in the room in which they are stored. Precooling is generally done in a separate facility within a few hours or even minutes. Therefore, room cooling is not considered precooling.

PRODUCT REQUIREMENTS

Product physiology, in relation to harvest maturity and ambient temperature at harvest time, largely determines precooling requirements and methods. Some products are highly perishable and must be cooled as soon as possible after harvest. Vegetables in this category include: asparagus, snap beans, broccoli, cauliflower, sweet corn, cantaloupes, summer squash, vine-ripened tomatoes, leafy vegetables, globe artichokes, brussels sprouts, cabbage, celery, carrots, snow peas, and radishes. Vegetables such as white potatoes, sweet potatoes, winter squash, pumpkins, and mature green tomatoes may need to be cured or ripened at some temperature higher than desirable for holding more perishable produce. Cooling of these products is not as important; however, some cooling is necessary if ambient temperature is high during harvest. Vegetables not listed may or may not be cooled because of lack of economic importance or susceptibility to chilling injury—for example, kiwi fruits.

Commercially important fruits that need to be precooled immediately after harvest include: apricots, avocados, all of the berries except cranberries, tart cherries, peaches and nectarines, plums and prunes, and tropical and subtropical fruits such as guavas, mangos, papayas, and pineapples. Tropical and subtropical fruits of this group are susceptible to chilling injury and thus need to be cooled according to individual temperature requirements. Sweet cherries, grapes, pears, and citrus fruit have a longer postharvest life than the former fruits, yet prompt cooling is essential to maintain high quality during the holding period. Bananas require special ripening treatment and therefore are not precooled. Chapter 8 lists recommended storage temperatures for many products. Quality is maintained better if products are cooled to these temperatures quickly after harvest, although some products may be damaged if cooled too quickly or if cooled below the recommended temperature.

CALCULATION METHODS

Heat Load

The refrigeration capacity needed for precooling is much greater than that required for holding a product at a constant temperature or for slow cooling of a product. Therefore, the heat load on a precooling system should be determined as accurately as possible. While it is imperative to have an adequate amount of refrigeration for effective precooling, it is uneconomical to have more refrigerating capacity available than is normally needed.

On jobs where refrigeration is needed only during a regular 8- to 12-h workday, ice builder equipment can be installed to reduce the size of the high-pressure side (compressor and condenser) of the refrigeration equipment. Also, ice building equipment can reduce electrical cost where off-peak power rates are available.

The total heat load comes from the product, surroundings, air infiltration, containers, and heat-producing devices such as motors, lights, fans, and pumps. Product heat accounts for the major portion of total heat load on a precooling system. Product heat load depends on product temperature and cooling rate, amount of product cooled in a given time, and specific heat of the product. Heat from respiration is part of the product heat load, but it is generally small. Chapter 12 discusses how to calculate the refrigeration load in more detail.

Mass-Average Temperature. Product temperature must be determined accurately in order to calculate heat load accurately. During rapid heat transfer, a temperature gradient develops within the product, with faster cooling causing larger gradients. This gradient is a function of product properties, surface heat transfer parameters, and cooling rate. Initially, for example, hydrocooling rapidly reduces the temperature of the exterior of a product while the center temperature may not change at all. Most of the product mass is in the outer portion. Thus, calculations based on center temperature would show little heat removal while, in fact, substantial heat has been extracted. For this reason, the product mass-average temperature must be used for product heat load calculations (Smith and Bennett 1965). A mass-average temperature denotes the single value from the transient temperature distribution that would become the uniform product temperature when held for a period under adiabatic conditions.

Figure 1 can be used to determine the mass-average temperature t_{ma} of peaches during hydrocooling. Subsequently, the product cooling load can be calculated as

$$q = mc_p(t_i - t_{ma}) \qquad (1)$$

Figure 1 can be applied to products other than peaches, when temperature ratio with respect to cooling time and product size is known. Figure 2 illustrates the comparative relationship of fractional temperature difference, or temperature ratio Y, to cooling time for 62 mm diameter peaches, 70 mm diameter apples and citrus fruit, and 50 mm diameter sweet corn when hydrocooled under ideal conditions of negligible surface resistance to heat transfer.

Cooling Rate

Cooling rate data are commonly presented in terms of **cooling coefficient** C and **half-cooling time** Z. The **temperature ratio** Y is the unaccomplished temperature change at any time in relation to

---

The preparation of this chapter is assigned to TC 10.9, Refrigeration Application for Food and Beverages.

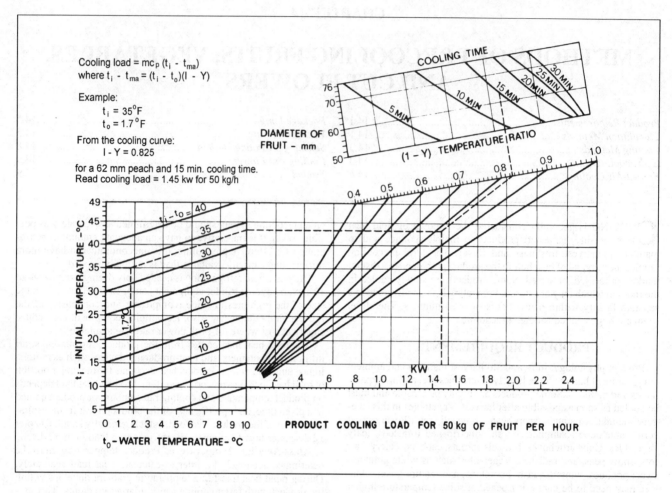

Fig. 1 Nomograph to Determine Product Heat Load of Hydrocooled Peaches

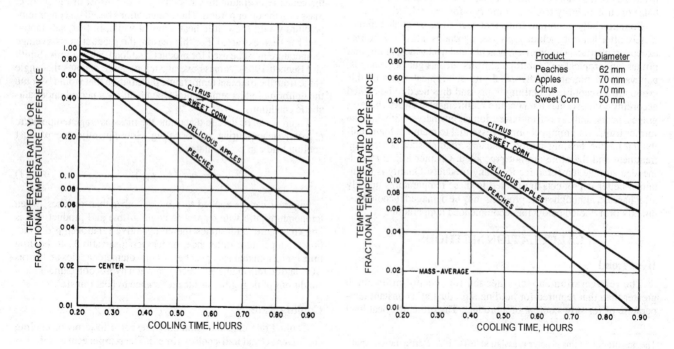

Fig. 2 Time-Temperature Response for Hydrocooled Produce

the total temperature change possible for the cooling condition. It is calculated by the equation

$$Y = \frac{t - t_o}{t_i - t_o} \qquad (2)$$

The inverted temperature ratio denotes the percent accomplished temperature change. The value of the cooling coefficient is the same in either case.

The cooling coefficient denotes the change in product temperature per unit change of cooling time for each degree temperature difference between the product and its surroundings. In most cases, a logarithmic transformation of the temperature ratio provides a fairly accurate straight line fit to the data points. If Newtonian heat transfer occurs (a negligible temperature gradient within the product during cooling) and the straight line asymptote intercepts the temperature ratio at unity for zero time, the cooling coefficient may be calculated by the simplified formula

$$C = \ln Y / \theta \qquad (3)$$

When produce is rapidly cooled, the conditions for Newton's law are seldom satisfied because a considerable temperature gradient often develops within the product, depending on its properties and rate of surface heat transfer. Also, the zero asymptote intercepts the temperature ratio at some value either greater or less than unity. In this case, the constant cooling coefficient is calculated by the equation

$$C = \frac{\ln Y_1 - \ln Y_2}{\theta_1 - \theta_2} \qquad (4)$$

where the subscripts 1 and 2 denote points along the straight line asymptote to the cooling curve.

The half cooling time is the time required to reduce the temperature difference between the product and its surroundings by one-half. Mathematically, it is expressed as

$$Z = \ln(0.5)/C \qquad (5)$$

where C is a negative value.

The previous solutions treat the product as an individual unit in a specified environment and isolated from the influence of surrounding material, which is rarely the case. Often, products are cooled in bulk or in packages of several layers, where the surrounding material influences the cooling response, particularly in air-cooling systems. Thus the individual approach usually gives poor estimates for cooling of bulk loads.

Baird and Gaffney (1976) developed a numerical technique for calculating cooling rates in bulk loads of fruits or vegetables. This procedure, applicable to spherical products, uses finite difference solutions of the differential equations that describe heat transfer both within individual fruits and at different levels in bulk loads. The models can be used to calculate temperatures at various points within individual products and at different depths in bulk loads at regular time intervals during cooling. Required inputs to the models include product size and shape, thermal properties of the product, the convective coefficient, and flow rate of the cooling medium.

COOLING METHODS

The principal methods of precooling are hydrocooling, forced-air cooling, forced-air evaporative cooling, package icing, and vacuum cooling. Most cooling is done at the packinghouse or in central cooling facilities. Some products can be cooled by any of these methods without suffering any adverse effects. For these products, the cooling method chosen is often determined more by such factors as economy, convenience, relation of the cooling equipment to the total packing operation, and personal preference.

HYDROCOOLING

Because of its simplicity, economy, and effectiveness, hydrocooling is a popular precooling method. When a film of cold water flows briskly and uniformly over the surface of a warm substance, the surface temperature of the substance becomes essentially equal to that of the water. Rate of internal cooling is limited by the size and shape (volume in relation to surface area) and thermal properties of the substance being cooled. For example, Stewart and Lipton (1960) showed a substantial difference in half-cooling time for sizes 36 and 45 cantaloupes. A weighted average of temperatures taken at different depths showed that 20 min was required to half-cool size 36 melons and only 10 min for size 45.

Freestone peaches, clingstone peaches for canning, tart cherries, and cantaloupes are hydrocooled. Few apples and citrus fruits are hydrocooled. Hydrocooling is not popular for citrus fruit because it has a long marketing season and good postharvest holding ability. Citrus fruits are also susceptible to increased peel injury and to decay and loss of quality and vitality after hydrocooling. Apples are usually cooled in the storage rooms. Apples and pears, however, are often dumped from field bins into a water dump at the packing house. Cold water, then, provides some cooling.

Many vegetables are successfully hydrocooled. The more important of these are sweet corn, celery, radishes, and carrots.

Types of Hydrocoolers

Hydrocooling is accomplished by flooding, spraying, or immersing the product in an agitated bath of chilled water.

Flood-type and **bulk-type hydrocoolers** are used to cool freestone peaches. The flood-type hydrocooler cools the packaged product by flooding as it is conveyed through a cooling tunnel. Adaptations consist of conveying the product through the cooling tunnel in loose bulk or in bulk bins. The bulk-type cooler uses combined immersion and flood cooling. Loose fruit, dumped into cold water, remains immersed for half of its travel through the cooling tunnel. An inclined conveyor gradually lifts the fruit out of the water and moves it through an overhead shower. The bulk-type cooler permits greater packaging flexibility than the flood-type.

Most vegetables are moved by fork truck in unit loads of as many as 40 packages. In one system, chilled water sprays on stacks of packed vegetables placed in a refrigerated room. Water is collected in floor drains leading to a sump and recirculated over the product. One nozzle spraying approximately 5.7 L/s of water is located over each stack, which may contain from 30 to 40 crates (Grizzell and Bennett 1966).

Other unit load systems convey the stacks through cooling tunnels. Either spray nozzles or a flood pan may be used to deliver up to 25 L/s of chilled water to each stack. Most commercial hydrocoolers provide overhead showering at a rate of 7 to 10 L/s per square metre of top face area. This type of hydrocooler requires less space than the batch type, but is not as flexible. With the batch system, chilled water can be sprayed on the product indefinitely, depending on the season and the incoming product temperature. Also, the crates can be left in place after cooling until they are shipped.

Spray or waterfall hydrocoolers with conveyors are used to cool vegetables prior to packaging. The water is cooled by flooded refrigerated plates or pipe coils located over or adjacent to flood pans above the mesh belt conveyor. The flood pans deliver 1°C water evenly over the produce as it is conveyed below. The water passes through the mesh conveyor, is filtered, and returned by pump and piping to the chiller. In some cases, particularly where produce has a high initial temperature, it may be worthwhile to augment the basic hydrocooler with ambient water or air cooling sections prior to the refrigerated hydrocooler to reduce refrigeration and energy costs.

Hydraircooling uses a mixture of refrigerated air and water in a fine mist spray that is circulated around and through the stack by forced convection (Henry et al. 1976). It has the advantage of reduced water requirements, the potential for improved sanitation, and the capability of adapting to fiberboard containers of the type that cannot be used in conventional hydrocooling systems. Cooling rates equal to, and in some cases better than, those obtained in conventional unit load hydrocoolers are possible.

Hydrocooler Design and Operation

The rate of heat transfer q is directly proportional to the surface heat transfer coefficient h, the total surface area A, and the difference in temperature between the surface and its surroundings Δt. To determine refrigeration requirements, the product mass-average temperature, which was described previously, should be used.

If the fluid velocity is sufficient, as in gravity flow or forced convection over fruit and vegetable products, the resistance to heat transfer at the surface is negligible. In this case, the heat is removed as fast as it comes to the surface, and the temperature difference across the surface boundary layer is small (0.5 K or less). In ideal hydrocooling, the optimum average film coefficient of heat transfer is roughly 680 W/(m²·K); the average temperature difference across the boundary layer is about 0.4 K. On this basis, the rate of heat transfer per unit surface area is $h = 680 \times 0.4 = 270$ W per square metre of product area. The value of h varies substantially, depending on many conditions. However, because of limitations in water temperature and product thermal properties and similarity among products, this is about as fast as any fruit or vegetable can be cooled in water. The ideal (maximum h) time-temperature response curves for hydrocooling select fruit and vegetable products are shown in Figure 2, based on both center and mass-average temperatures.

Flooded ammonia systems are often used to cool the water for hydrocoolers in large packing houses. Some use a secondary coolant. In packing houses with central plants cooling coils are contained in a tank through which water is rapidly circulated. Refrigerant temperature inside the cooling coils is approximately $-2°C$, except in systems with ice building coils, which may operate at a lower temperature.

Solar load, radiation from hot surfaces, convection from ambient air, or conduction from surroundings can affect the load. Protecting the hydrocooler from these sources of heat gain will enhance efficiency. Refrigeration capacity in excess of needs and cooling to a temperature below that required also reduces efficiency.

When hydrocooler water is recirculated, as is usually the case in mechanically refrigerated units, decay-producing organisms accumulate. Mild disinfectants such as chlorine will reduce buildup of bacteria and fungus spores, but they will not kill infections already in the products or sterilize either the water or the product surfaces. Brown rot and rhizopus decay spores on the surface and under the skin of peaches can be destroyed by soaking the fruit in water at 52°C for 2 to 3 min. (Smith and Redit 1968), or by treating the fruit with a chemical fungicide, which is usually put into the hydrocooling water. The principal problem with chemicals is maintenance of uniform concentrations, particularly in ice-refrigerated equipment because of the constant dilution from melting ice. In addition, hydrocooler shower pans and/or trash screens need regular (daily) cleaning to provide maximum efficiency.

FORCED-AIR COOLING

Theoretically, air cooling at a rate comparable to hydrocooling can be obtained by providing certain conditions of product exposure and air temperature. In air cooling, the optimum value of the surface heat transfer coefficient is considerably smaller than in cooling with water. However, Pflug et al. (1965) showed that apples moving through a cooling tunnel on a conveyer belt cool faster with air at $-6.7°C$ approaching the fruit at a velocity of 3 m/s than they would in a water spray at 1.7°C. For this condition of air cooling, they calculated an average film coefficient of heat transfer of 41 W/(m²·K). They note that the advantage of air is its lower temperature and that if the water were reduced to 1°C, the time for water-cooling would be less.

In tests to evaluate film coefficients of heat transfer for anomalous shapes, Smith et al. (1970) obtained an experimental value of 37.8 W/(m²·K) for a single Red Delicious apple in a cooling tunnel with air approaching at 8 m/s. At this rate of airflow, the logarithmic mean surface temperature of a single apple cooled for 0.5 h in air at $-6.7°C$ is approximately 1.7°C. The average temperature difference across the surface boundary layer is, therefore, 8.4 K and the rate of heat transfer per square metre of surface area is:

$$q/A = 37.8 \times 8.4 = 318 \text{ W/m}^2$$

For these conditions, the cooling rate compares favorably with that obtained in ideal hydrocooling. However, these coefficients are based on single specimens isolated from surrounding fruit. Had the fruit been in a packed bed at equivalent flow rates, the values would have been less because less surface area would have been exposed to the cooling fluid. Also, the rate of evaporation from the product surface significantly affects the cooling rate.

Because of physical characteristics, mostly geometry, various fruit and vegetable products respond differently to similar treatments of airflow and air temperature. For example, peaches cool faster than potatoes when they are cooled in a packed bed under similar conditions of airflow and air temperature, for example.

Surface coefficients of heat transfer are sensitive to the physical conditions involved among objects and their surroundings. Experimental surface coefficients ranging from 50 to 68 W/(m²·K) were obtained by Soule et al. (1966) for bulk lots of Hamlin oranges and Orlando tangelos with air approaching the fruit at velocities ranging from 1.1 to 1.8 m/s. Bulk bins containing 450 kg of 72 mm diameter Hamlin oranges were cooled from 27°C to a final mass-average temperature of 8°C in 1 h with air approaching the fruit at 1.7 m/s (Bennett et al. 1966). Surface heat transfer coefficients for these tests averaged slightly above 62 W/(m²·K). On the basis of a log mean air temperature of 6.7°C, the calculated half-cooling time was 970 s.

By correlating data from experiments on cooling 70 mm diameter oranges in bulk lots with results of a mathematical model, Baird and Gaffney (1976) found surface heat transfer coefficients of 8.5 and 51 W/(m²·K) for approach velocities of 0.055 and 2.1 m/s, respectively. A Nusselt-Reynolds heat transfer correlation representing data from six experiments on air cooling of 70 mm diameter oranges and seven experiments on 107 mm diameter grapefruit, with approach air velocities ranging from 0.025 to 2.1 m/s, gave the relationship $Nu = 1.17 Re^{0.529}$, with a correlation coefficient of 0.996.

Ishibashi et al. (1969) constructed a stage-type forced-air cooler that exposed bulk fruit to air at a progressively declining temperature (10, 0, and $-10°C$) as the fruit was conveyed through the cooling tunnel. Air approached the fruit at 3.6 m/s. With this system, 65 mm diameter citrus fruit cooled from 25 to 5°C in 1 h. Their half-cooling time of 0.32 h compares favorably with a half-cooling time of 0.30 h for similarly cooled Delicious apples at an approach air velocity of 2 m/s (Bennett et al. 1969). Perry et al. (1968) obtained a half-cooling time of 0.5 h for potatoes in a bulk bin with air approaching at 1.3 m/s, as compared to 0.4 h for similarly treated peaches and 0.38 h for apples. Optimum approach velocity for this type of cooling is in the range of 1.5 to 2 m/s, depending on conditions and circumstances.

Methods of Precooling Fruits, Vegetables, and Cut Flowers

Commercial Methods

Produce can be satisfactorily cooled (1) with air circulated in refrigerated rooms adapted for that purpose, (2) in rail cars or highway vans using special portable cooling equipment that cools the load before it is transported, (3) with air forced through the voids of bulk products moving through a cooling tunnel on continuous conveyors, (4) on continuous conveyors in wind tunnels, or (5) by the forced-air method of passing air through the containers by pressure differential. Each of these methods is used commercially, and each is suitable for certain commodities when properly applied.

In circumstances where air cannot be forced directly through the voids of products in bulk, a container type and a load pattern that permits air to circulate through the container and reach a substantial part of the product surface is beneficial. Examples of this are (1) small products such as grapes and strawberries that offer appreciable resistance to airflow through the voids in bulk lots, (2) delicate products that cannot be handled in bulk, and (3) products that are packed in the shipping containers before they are precooled.

Forced-air or pressure cooling involves definite stacking patterns and the baffling of stacks so that cooling air is forced through, rather than around, individual containers. Successful use of the method requires a container with vent holes placed in the direction in which the air will move and a minimum of packaging materials that would interfere with free movement of air through the containers. Under these conditions, a relatively small pressure differential between the two sides of the containers results in good air movement and excellent heat transfer. Differential pressures in use are about 60 to 750 Pa, with airflows ranging from 1 to 3 L/s per kilogram of product.

Because the cooling air comes in direct contact with the product being cooled, cooling is much faster than with conventional room cooling. This gives the advantage of rapid product movement through the cooling plant, and the size of the plant is one-third to one-fourth that of an equivalent cold room type of plant.

Mitchell et al. (1972) observed that forced-air cooling usually cools in one-fourth to one-tenth the time needed for conventional room cooling, but it still takes two to three times longer than hydrocooling or vacuum cooling.

A proprietary direct-contact heat exchanger cools air and maintains high humidities using chilled water as a secondary coolant and a continuously wound polypropylene monofilament packing. It contains about 24 km of filament per cubic metre of packing section. Air is forced up through the unit while chilled water flows downward. The dew-point temperature of the air leaving the unit equals the entering water temperature. Chilled water can be supplied from coils submerged in a tank. Buildup of ice on the coils provides an extra cooling effect during peak loads. This design also allows an operator to add commercial ice during long periods of mechanical equipment outage.

In one portable, forced-air method, refrigeration components are mounted on flat bed trailers and the warm, packaged produce is cooled in refrigerated transport trailers. Usually the refrigeration equipment is mounted on two trailers—one holds the forced-air evaporators and the other holds compressors, air-cooling condensers, a high-pressure receiver, and electrical gear. The loaded produce trailers are moved to the evaporator trailer and the product is cooled. After cooling, the trailer is transported to its destination.

Effects of Containers and Stacking Patterns

The accessibility of the product to the cooling medium, essential to rapid cooling, may involve both access to the product in the container and to the individual container in a stack. This effect is evident in the cooling rate data of various commodities in various types of containers reported by Mitchell et al. (1972). Parsons et al. (1972) developed a corrugated paperboard container venting pattern for palletized unit loads that produced cooling rates equal to those from conventional register stacked patterns. Fisher (1960) demonstrated that spacing apple containers on pallets reduced cooling time by 50% as compared with pallet loads stacked solidly.

Palletization is essential for shipment of many products, and pallet stability is improved if cartons are packed closely together. Thus, cartons and packages should be designed to allow ample airflow though the stacked products. Amos (1993) and Parsons (1972) showed the importance of vent sizes and location to obtain good cooling in palletized loads without reducing the strength of the container. Some operations wrap the palletized products in polyethylene to increase stability. In this case, the product may need to be cooled before it is palletized.

Computer Solution

Baird et al. (1988) developed an engineering economic model for designing forced-air cooling systems. Figure 3 shows the type of information that can be obtained from the model. By selecting a set of input conditions (which varies with each application) and varying the approach air velocity, the entering air temperature, or some other variable, the optimum (minimum cost) value can be determined. The curves in Figure 3 show that the selection of air velocity for containers is critical, whereas the selection of entering air temperature is not as critical until the desired final product temperature of 4°C is approached. The results shown are for four cartons deep with a 4% vent area in the direction of airflow, and they would be quite different if the carton vent area was changed. Other design parameters that can be optimized using this program are the depth of product in direction of airflow and the size of evaporators and condensers.

PACKAGE ICING

Finely crushed ice placed in shipping containers can effectively cool products that are not harmed by contact with ice. Spinach, collards, kale, brussels sprouts, broccoli, radishes, carrots, and onions are commonly packaged with ice (Hardenburg et al. 1986). Cooling a product from 35 to 2°C requires melting ice equal to 38% of the product's mass. Additional ice must melt to remove heat leaking into the packages and to remove heat from the container. In addition to removing field heat, package ice can keep the product cool during transit.

Top icing, or placing ice on top of packed containers, is used occasionally to supplement another cooling method. Because corrugated containers have largely replaced wooden crates, the use of top ice has decreased in favor of forced-air and hydrocooling. Wax-impregnated corrugated containers, however, have allowed the use of icing and hydrocooling of products after packaging.

Pumping **slush ice** or liquid ice into the shipping container through a hose and special nozzle that connect to the package is another method used for cooling some products. Some systems can ice an entire pallet at one time.

VACUUM COOLING

Principle

Vacuum cooling of fresh produce by the rapid evaporation of water from the product works best with vegetables having a high ratio of surface area to volume. In vacuum refrigeration, water, as the primary refrigerant, vaporizes in a flash chamber under low pressure. The pressure in the chamber is lowered to the saturation point corresponding to the lowest required temperature of the water.

Vacuum cooling is a batch process. The product to be cooled is loaded into the flash chamber, the system is put into operation, and the product is cooled by reducing the pressure to the corresponding saturation temperature desired. The system is then shut down, the product removed, and the process repeated. Since the product is normally at ambient temperature before it is cooled, vacuum cooling can be thought of as a series of intermittent operations of a vacuum

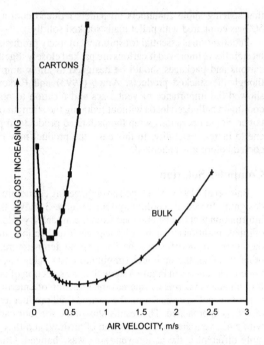
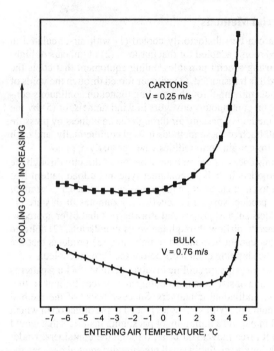

Fig. 3 Engineering-Economic Model Output for a Forced-Air Cooler

refrigeration system in which the water in the flash chamber is allowed to come to ambient temperature before each start. The functional relationships for determining refrigerating capacity are the same in each case.

Cooling is achieved by boiling water, mostly off the surface of the product to be cooled. The heat of vaporization required to boil the water is furnished by the product, which is cooled accordingly. As the pressure is further reduced, cooling continues to the desired temperature level. The saturation pressure for water at 100°C is 101.3 kPa. At 0°C, the saturation pressure is 0.610 kPa. Commercial vacuum coolers normally operate in this range.

Although the cooling rate of lettuce could be increased without danger of freezing, by reducing the pressure to 0.517 kPa, corresponding to a saturation temperature of -2°C, most operators do not reduce the pressure below the freezing temperature of water because of the extra work involved and the freezing potential.

Pressure, Volume, and Temperature

In a vacuum cooling operation, the thermodynamic process is assumed to take place in two phases. In the first phase, the product is assumed to be loaded into the flash chamber at ambient temperature, and the temperature in the flash chamber remains constant until saturation pressure is reached. At the onset of boiling, the small remaining amount of air in the chamber is replaced by the water vapor, the first phase ends, and the second phase begins simultaneously. The second phase continues at saturation until the product has cooled to the desired temperature.

If the ideal gas law is applied for an approximate solution in a commercial vacuum cooler, the pressure-volume relationships are:

Phase 1 $pv = 8.697$ kN·m/kg

Phase 2 $pv^{1.056} = 16.985$ kN·m/kg

where p is the absolute pressure and v is the specific volume.

The pressure-temperature relationship is determined by the value of ambient and product temperature. Based on 30°C for this value, the temperature in the flash chamber theoretically remains constant at 30°C as the pressure is reduced from atmospheric to saturation,

after which it declines progressively along the saturation line. These relationships are illustrated in Figure 4. The product temperature would respond similarly but would vary depending on where temperature is measured in the product, the physical characteristics of the product, and the amount of product surface water available. While it is possible for some vaporization to occur within the intercellular spaces beneath the product surface, most of the water is vaporized off the surface. The heat required to vaporize this water is also taken off the product surface where it flows by conduction under the thermal gradient produced. Thus, the rate of cooling depends on the relation of surface area to volume of the product and the rate at which the vacuum is drawn in the flash chamber.

Because water is the sole refrigerant, the amount of heat removed from the product depends on the mass of water vaporized m_v and its latent heat of vaporization L. Assuming an ideal condition, with no heat gain from surroundings, total heat Q removed from the product is:

$$Q = m_v L \qquad (6)$$

The amount of moisture removed from the product during vacuum cooling, then, is directly related to the specific heat of the product and the amount of temperature reduction accomplished. A product with a specific heat capacity of 4 kJ/(kg·K) would theoretically lose 1% moisture for each 6°C reduction in temperature. In a study of vacuum cooling of 16 different vegetables, Barger (1963) showed that cooling of all products was proportional to the amount of moisture evaporated from the product. Temperature reductions averaged 5 to 5.5 K for each 1% of mass loss, regardless of the product cooled. This mass loss may lower the amount of money the grower receives and reduce the turgor and crispness of the product. Some vegetables are sprayed with water before cooling to reduce this loss.

Commercial Systems

The four types of vacuum refrigeration systems that use water as the refrigerant are: (1) steam ejector, (2) centrifugal, (3) rotary, and (4) reciprocating. A schematic of the vacuum-producing mechanism of each is illustrated in Figure 5.

Methods of Precooling Fruits, Vegetables, and Cut Flowers

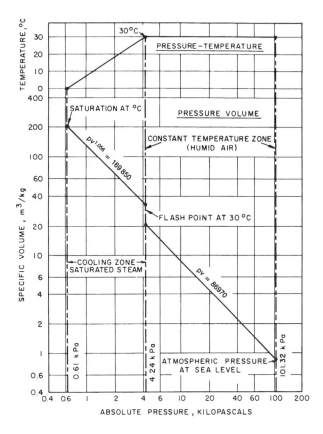

Fig. 4 Pressure, Volume, and Temperature in a Vacuum Cooler Cooling Product from 30 to 0°C

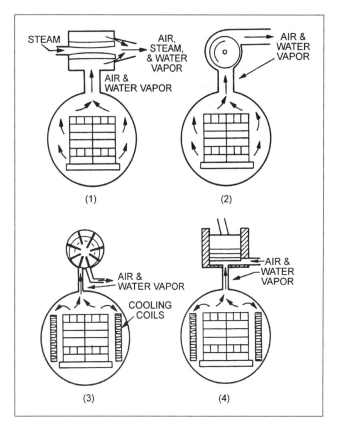

Fig. 5 Schematic Cross Sections of Vacuum-Producing Mechanisms

Of these, the steam ejector type is best suited for displacing the extremely high volumes of water vapor encountered at the low pressures needed in vacuum cooling. It also has the advantage of having few moving parts, thus requiring no compressor to condense the water vapor. High-pressure steam is expanded through a series of jets or ejectors arranged in series and condensed in barometric condensers mounted below the ejectors. Cooling water for condensing is accomplished by means of an induced-draft cooling tower. In spite of these advantages, few steam ejector vacuum coolers are used today, due to the inconvenience of using steam and the lack of portability. Instead, vacuum coolers mounted on semitrailers are used to follow the seasonal crops.

The centrifugal compressor is also a high-volume pump and can be adapted to water vapor refrigeration. However, its use in vacuum cooling is limited because of inherent mechanical difficulties at the high rotative speeds required to produce the low pressures needed.

Both rotary and reciprocal vacuum pumps are capable of producing the low pressures needed for vacuum cooling, and they also have the advantage of portability. Being positive-displacement pumps, however, they have low volumetric capacity; therefore, vacuum coolers using rotary or reciprocating pumps have separate refrigeration systems to condense much of the water vapor that evaporates off the product, thus substantially reducing the volume of water vapor passing through the pump. Ideally, when it can be assumed that all of the water vapor is condensed, the required refrigeration capacity is equal to the amount of heat removed from the product during cooling.

The condenser must contain adequate surface to condense the large amount of vapor removed from the produce in a few minutes. Refrigeration is furnished from cold brine or a direct-expansion system. A very large peak load occurs from rapid condensing of so much vapor. Best results are obtained if the refrigeration plant is equipped with a large brine or icemaking tank having enough stored refrigeration to smooth out the load. A standard three-tube plant, with capacity to handle three cars per hour, has a peak refrigeration load of at least 900 kW.

To increase cooling effectiveness and reduce product moisture loss, the product is sometimes wetted before cooling begins. However, lettuce is rarely prewetted. A proprietary modification of vacuum cooling continuously circulates chilled water over the product throughout the vacuum cooling process. Among the chief advantages are increased cooling rates and residual refrigeration that is stored in the chilled water following each vacuum process. It also prevents water loss from products that show objectionable wilting after conventional vacuum cooling.

Applications

Because vacuum cooling is generally more expensive, particularly in capital cost, than other cooling methods, its use is primarily restricted to products for which vacuum cooling is much faster or more convenient. Lettuce is ideally adapted to vacuum cooling. The numerous individual leaves provide a large surface area and the tissues release moisture readily. It is possible to freeze lettuce in a vacuum chamber if pressure and condenser temperatures are not carefully controlled. However, even lettuce does not cool entirely uniformly. The fleshy core, or butt, releases moisture more slowly than the leaves. Temperatures as high as 6°C have been recorded in core tissue when leaf temperatures were down to 0.5°C (Barger 1961).

Other leafy vegetables such as spinach, endive, escarole, and parsley are also suitable for vacuum cooling. Vegetables that are less suitable but adaptable by wetting are asparagus, snap beans, broccoli, brussels sprouts, cabbage, cauliflower, celery, green peas, sweet corn, leeks, and mushrooms. Of these vegetables, only cauliflower, celery, cabbage, and mushrooms are commercially vacuum

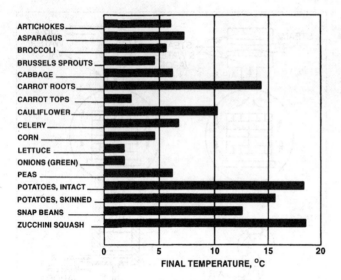

Vacuum-Cooling Conditions

Initial Product Temp.	= 20 to 22°C
Minimum Pressure	= 530 to 610 Pa
Condenser Temperature	= −1.7 to 0°C
Time in Vacuum Tank	= 0.42 to 0.5 h

Fig. 6 Comparative Cooling of Vegetables Under Similar Vacuum Conditions

cooled in California. Fruits are generally not suitable, except some berries. Cucumbers, cantaloupes, tomatoes, dry onions, and potatoes cool very little because of their low surface-to-mass ratio and relatively impervious surface. The final temperatures of various vegetables when vacuum cooled under similar conditions are illustrated in Figure 6.

The rate of cooling and the final temperature attained by vacuum cooling are largely affected by the ratio of the surface area of the commodity to its mass and the ease with which the product gives up water from its tissues. Consequently, the adaptability of fruits and vegetables varies tremendously for this method of precooling. For products that have a low surface-to-mass ratio, high temperature gradients occur. To prevent the surface from freezing before the desired mass-average temperature is reached, a procedure referred to as "bouncing" is practiced. This is accomplished by switching the vacuum pump off and on to keep the saturation temperature above freezing.

Mechanical vacuum coolers have been designed in several sizes. Most installations use cylindrical or rectangular retorts, sized to hold either a half- or a quarter-car of produce. A few are large enough to hold an entire refrigerator carload. For portability, some vacuum coolers and associated refrigeration equipment have been placed on flat bed trailers.

SELECTING A COOLING METHOD

Packing house size and operating procedures, response of product to the cooling method, and market demands largely dictate the cooling method used. Other factors considered are whether the product is packaged in the field or in a packinghouse, the product mix being cooled, length of cooling season, and comparative costs of dry versus water-resistant cartons. In some cases, there is little question about the type of cooling to be used. For example, vacuum cooling is most effective on lettuce and other similar vegetables. Peach packers in the southeastern United States and some vegetable

Table 1 Cooling Methods Suggested for Horticultural Commodities

	Size of Operation	
Commodity	Large	Small
Tree fruits		
Citrus	R	R
Deciduous [a]	FA, R, HC	FA
Subtropical	FA, R	FA
Tropical	FA, R	FA
Berries	FA	FA
Grapes [b]	FA	FA
Leafy vegetables		
Cabbage	VC, FA	FA
Iceberg lettuce	VC	FA
Kale, collards	VC, R, WV	FA
Leaf lettuces, spinach, endive, escarole, Chinese cabbage, bok choy, romaine	VC, FA, WV, HC	FA
Root vegetables		
with tops [c]	HC, PI, FA	HC, FA
topped	HC, PI	HC, PI, FA
Irish potatoes, sweet potatoes [d]	R w/evap coolers, HC	R
Stem and flower vegetables		
Artichokes	HC, PI	FA, PI
Asparagus	HC	HC
Broccoli, brussels sprouts	HC, FA, PI	FA, PI
Cauliflower	FA, VC	FA
Celery, rhubarb	HC, WV, VC	HC, FA
Green onions, leeks	PI, HC	PI
Mushrooms	FA, VC	FA
Pod vegetables		
Beans	HC, FA	FA
Peas	FA, PI, VC	FA, PI
Bulb vegetables		
Dry onions [e]	R	R, FA
Garlic	R	
Fruit-type vegetables [f]		
Cucumbers, eggplant	R, FA, FA-EC	FA, FA-EC
Melons		
cantaloupes, muskmelons, honeydew, casaba	HC, FA, PI	FA, FA-EC
crenshaw	FA, R	FA, FA-EC
watermelons	FA, HC	FA, R
Peppers	R, FA, FA-EC, VC	FA, FA-EC
Summer squashes, okra	R, FA, FA-EC	FA, FA-EC
Sweet corn	HV, VC, PI	HC, FA, PI
Tomatillos	R, FA, FA-EC	FA, FA-EC
Tomatoes	R, FA, FA-EC	
Winter squashes	R	R
Fresh herbs		
not packaged [g]	HC, FA	FA, R
packaged	FA	FA, R
Cactus		
leaves (nopalitos)	R	FA
fruit (tunas or prickly pears)	R	FA
Ornamentals		
Cut flowers [h]	FA, R	FA
Potted plants	R	R

R = Room cooling   WV = Water spray vacuum cooling
HC = Hydrocooling   PI = Package icing
FA = Forced-air cooling   FA-EC = Forced-air evaporative cooling
VC = Vacuum cooling

[a] Apricots cannot be hydrocooled.
[b] Grapes require rapid cooling facilities adaptable to sulfur dioxide fumigation.
[c] Carrots can be vacuum cooled.
[d] With evaporative coolers, facilities for potatoes should be adapted to curing.
[e] Facilities should be adapted to curing onions.
[f] Fruit-type vegetables are chilling sensitive but at varying temperature.
[g] Fresh herbs can be easily damaged by water beating in hydrocooler.
[h] When cut flowers are packaged, only use forced-air cooling.

Reprinted with permission from A.A. Kader, *Post Harvest Technology of Horticultural Crops*, The Regents of the University of California Division of Agriculture and Natural Resources, 1992.

and citrus packers are satisfied with hydrocooling. Air (room) cooling is used for apples, pears, peaches, plums, nectarines, sweet cherries, strawberries, and apricots. In other cases, choice of cooling method is not so clearly defined. Celery and sweet corn are usually hydrocooled, but they may be vacuum cooled as effectively. Cantaloupes may be satisfactorily cooled by several methods.

When more than one method can be used, cost becomes a major consideration. While rapid, forced-air cooling is more costly than hydrocooling, if the product does not require rapid cooling, a forced-air system can operate almost as economically as hydrocooling. In a study to evaluate costs of hypothetical precooling systems for citrus fruit, Gaffney and Bowman (1970) found that the cost for forced-air cooling in bulk lots was 20% more than that for hydrocooling in bulk and that forced-air cooling in cartons costs 45% more than hydrocooling in bulk.

Table 1 is a summary of precooling and cooling methods suggested for various commodities.

COOLING CUT FLOWERS

Because of their high rates of respiration and low tolerance to heat, deterioration in cut flowers is rapid at field temperatures. Refrigerated highway vans do not have the capacity to remove the field heat in sufficient time to prevent some deterioration from occurring (Farnham et al. 1979). Forced-air cooling is commonly used by the flower industry. As with most fruits and vegetables, the cooling rate of cut flowers varies substantially among the various types. Rij et al. (1979) found that the half-cooling time for packed boxes of gypsophila was about 3 min compared to about 20 min for chrysanthemums at airflows ranging from 38 to 123 L/s per box. Within this range, cooling time was proportional to the reciprocal of airflow but varied less with airflow than with flower type.

SYMBOLS

A = product surface area, m^2
c_p = specific heat of product, kJ/(kg·K)
C = cooling coefficient, reciprocal of hours
G = geometry index
G' = mass rate of airflow, kg/(s·m^2)
h = surface heat transfer coefficient, W/(m^2·K)
k = thermal conductivity, W/(m·K)
l = characteristic length, m
L = heat of vaporization, kJ/kg
m = mass of product, kg
m_v = mass of water vaporized, kg
M_1 = first root of transcendental function
p = pressure, Pa
q = cooling load or rate of heat transfer, W
Q = total heat, kJ
t = temperature of any point in product, °C
t_i = initial uniform product temperature, °C
t_o = surrounding temperature, °C
t_{ma} = mass-average temperature, °C
v = specific volume of water vapor, m^3/kg
Y = temperature ratio $(t - t_o)/(t_i - t_o)$
Z = half-cooling time, h
α = thermal diffusivity, m^2/s
θ = cooling time, h
μ = dynamic viscosity, Pa·s

REFERENCES

Amos N.D., D.J. Cleland, and N.H. Banks. 1993 Effect of pallet stacking arrangement on fruit cooling rates within forced-air pre-coolers. *Refrigeration Science and Technology* 3: 232-241.

Baird, C.D. and J.J. Gaffney. 1976. A numerical procedure for calculating heat transfer in bulk loads of fruits or vegetables. ASHRAE *Transactions* 82(2):525.

Baird, C.D., J.J. Gaffney, and M.T. Talbot. 1988. Design criteria for efficient and cost effective forced air cooling systems for fruits and vegetables. ASHRAE *Transactions* 94(1):1434.

Barger, W.R. 1961. Factors affecting temperature reduction and weight loss of vacuum-cooled lettuce. USDA, *Marketing Research Report*.

Barger, W.R. 1963. Vacuum precooling—A comparison of cooling of different vegetables. USDA, *Marketing Research Report* No. 600.

Bennett, A.H., J. Soule, and G.E. Yost. 1966. Temperature response of citrus to forced-air precooling. ASHRAE *Journal* 8(4):48.

Bennett, A.H., J. Soule, and G.E. Yost. 1969. Forced-air precooling for Red Delicious apples. USDA, Agricultural Research Service ARS 52-41.

Farnham, D.S., et al. 1979. Comparison of conditioning, precooling, transit method, and use of a floral preservative on cut flower quality. Proceedings, *Journal of American Society of Horticultural Science* 104(4):483.

Fisher, D.V. 1960. Cooling rates of apples packed in different bushel containers and stacked at different spacing in cold storage. ASHRAE *Journal* (July):53.

Gaffney, J.J. and E.K. Bowman. 1970. An economic evaluation of different concepts for precooling citrus fruits. ASHRAE Symposium on Precooling Fruits and Vegetables, San Francisco (January).

Grizell, W.G. and A.H. Bennett. 1966. *Hydrocooling stacked crates of celery and sweet corn.* USDA, Agricultural Research Service, ARS 52-12.

Hardenburg, R.E., A.E. Watada, and C.Y. Wang. 1986. The commercial storage of fruits, vegetables, and florist and nursery stocks. USDA *Agricultural Handbook* No. 66.

Henry, F.E. A.H. Bennett, and R.H. Segall. 1976. Hydraircooling—A new concept for precooling pallet loads of vegetables. ASHRAE *Transactions* 82(2):541.

Ishibashi, S., R. Kojima, and T. Kaneko. 1969. Studies on the forced-air cooler. JSAM, Japan 31(2), September.

Kader, A.A., R.F. Kasmire, and J.F. Thompson. 1992. *Cooling horticultural commodities.* University of California Division of Agricultural and Natural Resources *Publication* 3311.

Mitchell, F.G., R. Guillou, and R.A. Parsons. 1972. *Commercial cooling of fruits and vegetables.* Manual 43, Division of Agricultural Sciences, University of California, Berkeley.

Parsons, R.A., F.G. Mitchell, and G. Mayer. 1972. Forced-air cooling of palletized fresh fruit. *Transactions of the ASAE* 15(4):729.

Pflug, I.J., J.L. Blaisdell, and I.J. Kopelman. 1965. Developing temperature-time curves for objects that can be approximated by a sphere, infinite plate, or infinite cylinder. ASHRAE *Transactions* 71(1):238.

Rij, R.E., J.F. Thompson, and D.S. Farnham. 1979. *Handling, precooling, and temperature management of cut flower crops for truck transportation.* USDA-SEA Western Series No. 5 (June).

Smith, R.E. and A.H. Bennett. 1965. Mass-average temperature of fruits and vegetables during transient cooling. *Transactions of the ASAE* 8(2):249.

Smith, R.E., A.H. Bennett, and A.A. Vacinek. 1970. Convection film coefficients related to geometry for anomalous shapes. *Transactions of the ASAE* 13(2).

Smith, R.E., G.L. Nelson, and R.L. Henrickson. 1967. Analyses on transient heat transfer from anomalous shapes. *Transactions of the ASAE* 10(2):236.

Smith, R.E., G.L. Nelson, and R.L. Henrickson. 1968. Applications of geometry analysis of anomalous shapes to problems in transient heat transfer. *Transactions of the ASAE* 11(2):296.

Smith, W.L. and W.H. Redit. 1968. Postharvest decay of peaches as affected by hot-water treatments, cooling methods, and sanitation. USDA, *Marketing Research Report* No. 807.

Soule, J., G.E. Yost, and A.H. Bennett. 1966. Certain heat characteristics of oranges, grapefruit and tangelos during forced-air precooling. *Transactions of the ASAE* 9(3):355.

Stewart, J.K. and W.J. Lipton. 1960. Factors influencing heat loss in cantaloupes during hydrocooling. USDA Marketing Research *Report* 421.

BIBLIOGRAPHY

Ansari, F.A. and A. Afaq. 1986. Precooling of cylindrical food products. *International Journal of Refrigeration* 9(3):161-63.

Arifin, B.B. and K.V. Chau. 1988. Cooling of strawberries in cartons with new vent hole designs. *ASHRAE Transactions* 94(1):1415-26.

Bennett, A.H. 1962. Thermal characteristics of peaches as related to hydrocooling. USDA *Technical Bulletin* No. 1292.

Bennett, A.H., W.G. Chace, Jr., and R.H. Cubbedge. 1969. Heat transfer properties and characteristics of Appalachian area Red Delicious apples. *ASHRAE Transactions* 75(2):133.

Bennett, A.H., R.E. Smith, and J.C. Fortson. 1965. Hydrocooling peaches—A practical guide for determining cooling requirements and cooling times. USDA *Agriculture Information Bulletin* No. 298 (June).

Bennett, A.H., W.G. Chace, Jr., and R.H. Cubbedge. 1970. Thermal properties and heat transfer characteristics of marsh grapefruit. USDA, *Technical Bulletin* No. 1413.

Burton, K.S., C.E. Frost, and P.T. Atkey. 1987. Effect of vacuum cooling on mushroom browning. *International Journal of Food Science & Technology* 22(6):599-606.

Chau, K.V. 1994. Time-temperature-humidity relations for the storage of fresh commodities. *ASHRAE Transactions* 100(2):348-353.

Gariepy, Y., G.S.V. Raghavan, and R. Theriault. 1987. Cooling characteristics of cabbage. *Canadian Agricultural Engineering* 29(1):45-50.

Hackert, J.M., R.V. Morey, and D.R. Thompson. 1987. Precooling of fresh market broccoli. *Transactions of the ASAE* 30(5): 1489-93.

Hayakawa, K. 1978. Computerized simulation for heat transfer and moisture loss from an idealized fresh produce. *Transactions of the ASAE* 21(5):1015-24.

Hayakawa, K. and J. Succar. 1982. Heat transfer and moisture loss of spherical fresh produce. *Journal of Food Science* 47(2):596-605.

Isenberg, F.M.R., R.F. Kasmire, and J.E. Parson. Vacuum cooling vegetables. *Information Bulletin* 186, Cornell University Cooperative Extensive Service, Ithaca, NY.

Rohrbach, R.P., R. Ferrell, E.O. Beasley, J.R. Fowler. 1984. Precooling blueberries and muscadine grapes with liquid carbon dioxide. *Transactions of the ASAE* 27(6):1950-55.

Stewart, J.K. and H.M. Couey. 1963. Hydrocooling vegetables—A practical guide to predicting final temperatures and cooling times. USDA *Marketing Research Report* No. 637.

Thompson, J.F., Y.L. Chen, and T.R. Rumsey. 1987. Energy use in vacuum coolers for fresh market vegetables. *Applied Engineering in Agriculture* 3(2):196-99, American Society of Agricultural Engineers, St. Joseph, MI.

CHAPTER 15

INDUSTRIAL FOOD FREEZING SYSTEMS

Freezing Methods 15.1
Blast Freezers 15.1
Contact Freezers 15.4
Cryogenic Freezers 15.5
Cryomechanical Freezers 15.5
Other Freezer Selection Criteria 15.5
Refrigeration Systems 15.6

FREEZING is a widely used method of food preservation that slows the physical changes and chemical and microbiological activity that cause deterioration in foods. Reducing temperature slows molecular and microbial activity in food, thus extending its useful storage life. Although every product has an individual ideal storage temperature, most frozen food products are stored at −18 to −35°C. Chapter 10 has frozen storage temperatures for specific products.

Freezing reduces the temperature of a product from ambient to storage level and changes most of the water in the product to ice. Figure 1 shows the three phases of freezing: (1) cooling, which removes sensible heat, reducing the temperature of the product to the freezing point; (2) removal of the product's latent heat of fusion, changing the water to ice crystals; and (3) continued cooling below the freezing point, which removes more sensible heat, reducing the temperature of the product to the desired or optimum frozen storage temperature. Values for specific heats, freezing points, and latent heats of fusion for various products are given in Chapter 10.

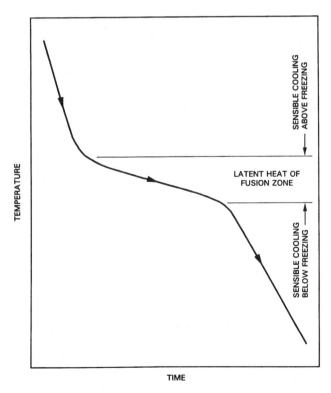

Fig. 1 Typical Freezing Curve

The preparation of this chapter is assigned to TC 10.9, Refrigeration Application for Foods and Beverages.

The longest part of the freezing process is the removal of the latent heat of fusion as water turns to ice. Many food products are sensitive to freezing rate, which affects yield (dehydration), quality, nutritional value, and sensory properties. The freezing method and system selected can thus have substantial economic impact.

The following factors should be considered in the selection of freezing methods and systems for specific products: special handling requirements, capacity, freezing times, quality consideration, yield, appearance, first cost, operating costs, automation, space availability, and upstream/downstream processes.

This chapter covers general freezing methods and systems. Additional information on freezing of specific products is covered in Chapters 13, 16 through 19, and 24 through 28. Related information can be obtained in Chapters 8 and 9, which cover thermal properties of foods as well as their cooling and freezing times. Information on refrigeration system practices is given in Chapters 1, 2, and 3.

FREEZING METHODS

Freezing systems can be grouped by their basic method of extracting heat from food products:

Blast freezing (convection). Cold air is circulated over the product at high velocity. The air removes heat from the product and releases it to an air/refrigerant heat exchanger before being recirculated.

Contact freezing (conduction). Food, packaged or unpackaged, is placed on or between cold metal surfaces. Heat is extracted by direct conduction through the surfaces, which are cooled by a circulating refrigerated medium.

Cryogenic freezing (convection and/or conduction). Food is exposed to an environment below −60°C, which is achieved by spraying liquid nitrogen or liquid carbon dioxide into the freezing chamber.

Cryomechanical freezing (convection and/or conduction). Food is first exposed to cryogenic freezing and then finish frozen through mechanical refrigeration.

Special freezing methods, such as immersion of poultry in chilled brine, are covered under the specific product chapters.

BLAST FREEZERS

Blast freezers use air as the heat transfer medium and depend on contact between the product and the air. Sophistication in airflow control and conveying techniques varies from crude blast freezing chambers to carefully controlled impingement-style freezers.

The earliest blast freezers consisted of cold storage rooms with extra fans and a surplus of refrigeration. Improvements in airflow control and mechanization of conveying techniques have made heat transfer more efficient and product flow less labor-intensive.

While **batch freezing** is still widely used and appropriate for certain production lines, more sophisticated freezers allow integrate the freezing process into the production line. This integration, known as **process-line freezing**, has become essential for large-volume, high-quality, cost-effective operations. A wide range of blast freezer systems are available, including:

Batch
- Cold storage room
- Stationary blast cell
- Push-through trolley

Continuous/Process-line
- Straight belt (two-stage, multipass)
- Fluidized bed
- Fluidized belt
- Spiral belt
- Carton (carrier)

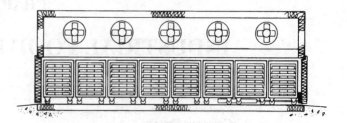

Fig. 3 Push-Through Trolley Freezer

Cold Storage Room

Although a cold storage room is not considered a freezing system, it is sometimes used for this purpose. Because a storage room is not designed to be a freezer, it should only be used for freezing in exceptional cases. Freezing is so slow that the quality of most products suffers. The quality of the already frozen products stored in the room is jeopardized because the temperature of the frozen products may rise considerably due to the excess refrigeration load. Also, flavors from the warm products may be transferred.

Stationary Blast Cell Freezing Tunnel

The stationary blast cell (Figure 2) is the simplest freezer that can be expected to produce satisfactory results for most products. It is an insulated enclosure equipped with refrigeration coils and axial or centrifugal fans that circulate the air over the products in a controlled way. Products are placed on trays, which are then placed into racks in such a way that an air space is left between adjacent layers of trays. The racks are moved in and out of the tunnel manually. It is important that the racks be placed so that air bypass is minimized. The stationary blast cell is a universal freezer—almost all products can be frozen in a blast cell. Vegetables and other products (e.g., spinach, bakery items, meat patties, fish fillets, and prepared foods) may be frozen either in cartons or unpacked and spread in a layer on trays. In some instances, this type of freezer is also used to reduce to −18°C or below the temperature of palletized, cased products that have previously been frozen through the latent heat of fusion zone by other means. The flexibility of a blast cell makes it suitable for small quantities of varied products; however, labor requirement is relatively high and product flow is cumbersome.

Push-Through Trolley Freezer

The push-through trolley freezer (Figure 3), in which the racks are fitted with wheels, incorporates a moderate degree of mechanization. The racks, or trolleys, are usually moved on rails by a pushing mechanism, which can be hydraulic. This type of freezer is similar to the stationary blast cell, except that labor costs and product handling time are decreased. This system is widely used to **crust-freeze** (quick chill) wrapped packages of raw poultry.

Straight Belt Freezer

The first mechanized blast freezers consisted of a wire mesh belt conveyor in a blast room, which satisfied the need for continuous product flow. A disadvantage to these early systems was the poorly controlled airflow and resulting inefficient heat transfer. Current versions use controlled vertical airflow, which forces cold air up through the product layer, thereby creating good contact with the product particles. Straight belt freezers are generally used with fruits, vegetables, French fried potatoes, cooked meat toppings (e.g., diced chicken), and cooked shrimp.

The principal design is the **two-stage belt freezer** (Figure 4), which consists of two mesh conveyor belts in series. The first belt initially precools or crust-freezes an outer layer or crust to condition the product before transferring it to the second belt for freezing and sensible cooling to −18°C or below. To ensure uniform cold air contact and effective freezing, products should be distributed uniformly over the entire belt. Two-stage freezers are generally operated at −10 to −4°C refrigerant temperature in the precool section and at −32 to −40°C in the freezing section. Capacities range from 0.9 to 18 Mg of product per hour, with freezing times from 3 to 50 min.

When products to be frozen are hot (e.g., French fries from the fryer at 80 to 95°C), another cooling section is added ahead of the normal precool section. This section supplies either refrigerated air at approximately 10°C or filtered ambient air to cool the product and congeal the fat. Refrigerated air is preferred because filtered ambient air has greater temperature variations and may contaminate the product.

Multipass Straight Belt Freezer

For larger products with longer freezing times (up to 60 min) and higher capacity requirements (0.5 to 5.4 Mg/h), a single straight belt freezer would require a very large floor space. Required floor space can be reduced by stacking belts above each other to form either (1) a single-feed/single-discharge multipass system (usually three passes) or (2) multiple single-pass systems (multiple infeeds and discharges) stacked one on top of the other. The multipass (triple-pass) arrangement (Figure 5) provides another benefit in that the product, after being surface frozen on the first (top) belt, may be stacked more deeply on the lower belts. Thus, the total belt area required is reduced, as is the overall size of the freezer. This system has a potential for product damage and product jams at the belt transfers.

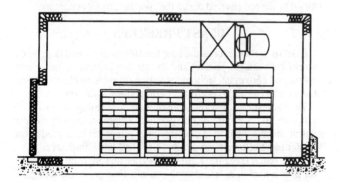

Fig. 2 Stationary Blast Cell

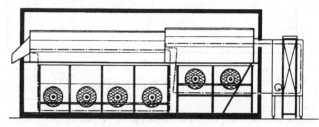

Fig. 4 Two-Stage Belt Freezer

Industrial Food Freezing Systems

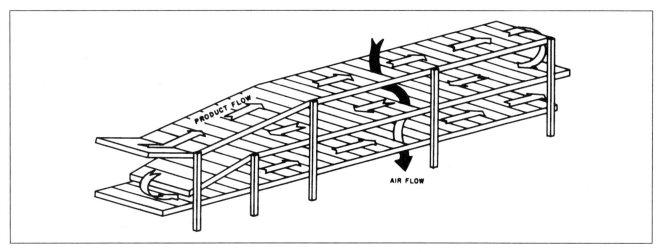

Fig. 5 Multipass, Straight Belt Freezer

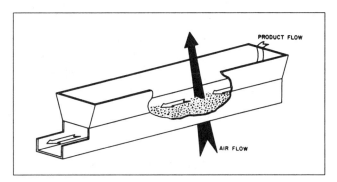

Fig. 6 Fluidized Bed Freezer

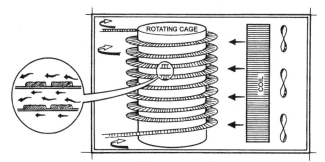

Fig. 7 Horizontal Airflow Spiral Freezer

Fluidized Bed Freezers

This freezer uses air both as the medium of heat transfer and for transport; the product flows through the freezer on a cushion of cold air (Figure 6). This proprietary design is well suited for small, uniform-sized particulate products such as peas and diced carrots.

The high degree of fluidization improves the heat transfer rate and allows good use of floor space. The technique is limited to well-dewatered products of uniform size that can be readily fluidized and transported through the freezing zone. Because the principle depends on rapid crust-freezing of the product, the operating refrigerant temperature must be −40°C or lower. Fluidized bed freezers are normally manufactured as packaged, factory-assembled units with capacity ranges of 0.9 to 4.5 Mg/h. The particulate products generally have a freezing time of 3 to 11 min.

Fluidized Belt Freezers

A hybrid of the two-stage belt freezer and the fluidized bed freezer, the fluidized belt freezer has a fluidizing section in the first belt stage. An increased air resistance is designed under the first belt to provide fluidizing conditions for wet incoming product, but the belt is there to assist in the transport of heavier, less uniform products that do not fluidize fully. Once crust-frozen, the product can be loaded deeper for greater efficiency on the second belt. Two-stage fluidized belt freezers operate at −34 to −37°C refrigerant temperature and in capacity ranges from 1 to 14 Mg/h.

Spiral Belt Freezer

This freezer is advantageous for products with a long freezing time (generally 10 min to 3 h), and for products that require gentle handling during freezing. An endless conveyor belt that can be bent laterally is wrapped cylindrically, one tier below the last; this is a configuration that requires minimal floor space for a relatively long belt. The original spiral belt principle uses a spiraling rail system to carry the belt, while more recent designs use a proprietary self-stacking belt requiring less overhead clearance. The number of tiers in the spiral can be varied to accommodate different capacities. In addition, two or more spiral towers can be used in series for products with long freezing times. Spiral freezers are available in a range of belt widths and are manufactured as packaged, modular, and field-erected models to accommodate various upstream processes and capacity requirements.

Air flow varies from little more than an open, unbaffled spiral conveyor to flow through extensive baffling and high-pressure fans. Horizontal airflow is applied to spiral freezers (Figure 7) by axial fans mounted along one side. The fans blow air horizontally across the spiral conveyor with minimal baffling limited to two portions of the spiral circumference. The rotation of the cage and belt produces a rotisserie effect, with product moving past the high-velocity cold air near the discharge. This aids in providing uniform freezing.

Several proprietary designs are available to control airflow. One design (Figure 8) has a mezzanine floor that separates the freezer into two pressure zones. Baffles around the outside and inside of the belt form an air duct so that air flows up or down around the product as the conveyor moves the product. The controlled airflow reduces freezing time for some products.

Another design (Figure 9) splits the airflow so that the coldest air contacts the product both as it enters and as it leaves the freezer. The coldest air introduced on the incoming, warm product may increase surface heat transfer and freeze the surface more rapidly, which may reduce product dehydration.

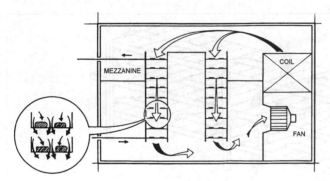

Fig. 8 Vertical Airflow Spiral Freezer

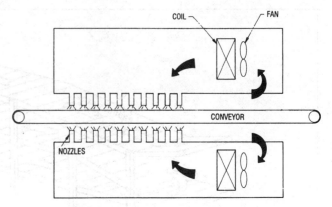

Fig. 10 Impingement-Style Freezer

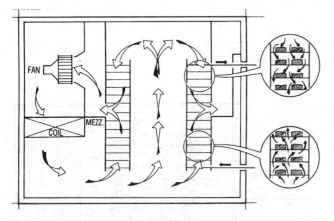

Fig. 9 Split Airflow Spiral Freezer

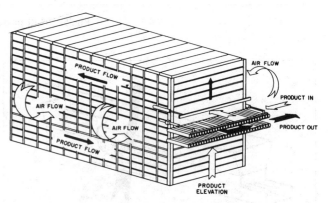

Fig. 11 Carton (Carrier) Freezer

Typical products frozen in spiral belt freezers include raw and cooked meat patties, fish fillets, chicken portions, pizza, and a variety of packaged products. Spiral freezers are available in a wide range of capacities, from 0.5 to 4.5 Mg/h. They dominate today's frozen food industry and account for the majority of unpackaged nonparticulate frozen food production.

Impingement-Style Freezers

In this design (Figure 10), cold air flows perpendicular to the product's largest surfaces at a relatively high velocity. Thousands of air nozzles with corresponding return ducts are mounted above and below the conveyors. The airflow constantly interrupts the boundary layer that surrounds the product, enhancing the surface heat transfer rate. The technique may therefore reduce the freezing time of products with large surface-to-mass ratios (thin hamburger patties, for example). Impingement-style freezers are designed with single-pass or multipass straight belts. Freezing times are 1 to 4 min. Application is limited to thin food products (less than 25 mm thick).

Carton Freezers

The carton (or carrier) freezer (Figure 11) is a very high capacity freezer (4.5 to 18 Mg/h) for medium to large cartons of such products as beef, poultry, and ice cream.

In the top section of the freezer, a row of loaded product carriers is pushed toward the rear of the freezer, while on the lower section they are returned to the front. Elevating mechanisms are located at both ends. A carrier is similar to a bookcase with shelves. When it is indexed in the loading/discharge end of the freezer, the already-frozen product is pushed off each shelf one row at a time onto a discharge conveyor. When the carrier is indexed up, this shelf aligns with the loading station, where new products are continuously pushed onto the carrier before it is moved once again to the rear of the freezer. Refrigerated air is circulated over the cartons by forced convection.

CONTACT FREEZERS

A contact freezer's primary means of heat transfer is by conduction; the product or package is placed in direct contact with a refrigerated surface. Contact freezers can be categorized as follows:

Batch
- Manual horizontal plate
- Manual vertical plate

Process-line
- Automatic plate
- Contact belt (solid stainless steel)
- Specialized design

The most common type of contact freezer is the **contact plate freezer**, in which the product is pressed between metal plates. Refrigerant is circulated inside channels in the plates, which ensures efficient heat transfer. This is reflected in short freezing times, provided that the product is a good conductor of heat, as in the case of fish fillets, chopped spinach, or meat offal. However, packages or cavities should be well filled, and if metal trays are used, they should not be distorted.

Manual and Automatic Plate Freezers

Contact plate freezers (Figure 12) are available in horizontal or vertical arrangements with manual loading/unloading. Horizontal plate freezers are also available in an automatic loading/unloading version, which generally accommodates higher capacities and continuous operation. The advantage of good heat transfer in contact plate freezers is gradually reduced with increasing product thickness. For this reason, thickness is often limited to 50 to 80 mm. Contact plate freezers operate efficiently because they require no fans, and they are very compact. An advantage with packaged products is that pressure from the plates minimizes any bulging that may occur.

Industrial Food Freezing Systems

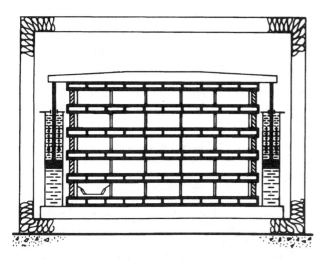

Fig. 12 Plate Freezer

Thus, packages are even and square within close tolerances. Automatic plate freezers accommodate up to 200 packages per minute, with freezing times from 10 to 150 min. When greater capacities are required, the freezers are placed in series with associated conveyor systems to handle the loading and unloading of packages.

Specialized Contact Freezers

A combination of air and contact freezing is used for wet fish fillets and similar soft, wet products with relatively large, flat surfaces. The continuous, solid stainless steel belt is typically 1.2 to 2 m wide and may be 30 m long. Product is loaded onto the belt at one end of the freezer and then travels in a fixed position through the freezing zone to the discharge end. Freezing is usually accomplished both by conduction through the belt to a cooling medium below the belt and by convection through controlled airflow above the belt. This freezer design produces attractive product, but a drawback is the physical size of the freezer. Capacities for typical products are generally limited to 0.9 to 2.3 kg/h, with a freezing time of less than 30 min.

Another specialized contact freezer conveys food products on a continuous plastic film over a low-temperature ($-40°C$) refrigerated plate. Upon contacting the film, the bottom surface of the products is frozen approximately 1 mm deep in about 1 min. This equipment is used to eliminate product deformation or wire mesh belt markings on products that are flat, wet and sticky, soft, or in need of hand shaping before freezing. Another benefit of contact prefreezing is that it reduces dehydration losses in the subsequent freezing step. Examples of products suitable for contact prefreezing are marinated, boneless chicken breasts and thin fish fillets.

CRYOGENIC FREEZERS

Cryogenic (or gas) freezing is often an alternative for (1) small-scale production, (2) new products, (3) overload situations, or (4) seasonal products. Cryogenic freezers use liquid nitrogen or liquid carbon dioxide (CO_2) as the refrigeration medium, and the freezers may be batch cabinets, straight belt freezers, spiral conveyors, or liquid immersion freezers.

Liquid Nitrogen Freezers

The most common type of liquid nitrogen freezer is a straight-through, single-belt, process-line tunnel. Liquid nitrogen at $-195°C$ is introduced at the outfeed end of the freezer directly onto the product; as the liquid nitrogen vaporizes, those cold vapors are circulated toward the infeed end, where they are used for precooling and initial freezing of the product. The "warmed" vapors (typically $-45°C$) are then discharged to the atmosphere. The low temperature of the liquid and vaporous nitrogen provides rapid freezing, which can improve quality and reduce dehydration for some products. However, the freezing cost is high, and the surface of products having a high water content may crack if precautions are not taken.

Consumption of liquid nitrogen is in the range of 0.9 to 2.0 kg of nitrogen per kilogram of product, depending on the water content and temperature of the product. While this translates into relatively high operating costs, the small initial investment makes liquid nitrogen freezers cost-effective for some applications.

Carbon Dioxide Freezers

A similar cryogenic freezing method places boiling (subliming) CO_2 in direct contact with foods frozen in a straight belt or spiral freezer. Carbon dioxide boils at approximately $-79°C$, and the system operates like a liquid nitrogen freezing system, consuming cryogenic liquid as it freezes product. Applications for CO_2 freezing include producing individual quick frozen (IQF) diced poultry, pizza toppings, and seafood.

CRYOMECHANICAL FREEZERS

Although the technique is not new, cryomechanical freezing (combination of cryogenic and blast freezing) applications are increasing. High-value, sticky products, such as IQF shrimp, and wet, delicate products, such as IQF strawberries and IQF cane berries, are common applications for these systems.

A typical cryomechanical freezer has an initial immersion step in which the product flows through a bath of liquid nitrogen to set the product surface. This rapid surface freezing reduces dehydration and improves the handling characteristics of the product, thus minimizing sticking and clumping. The cryogenically crust-frozen product is then transferred directly into a mechanical freezer, where the remainder of the heat is removed and the product temperature is reduced to $-18°C$ or lower. The cryogenic step is sometimes retrofitted to existing mechanical freezers to increase their capacity. The use of the mechanical freezing step makes operating costs lower than for cryogenic-only freezing.

OTHER FREEZER SELECTION CRITERIA

Reliability

Due to the harsh operating conditions, the freezing system is probably the most vulnerable equipment in a process line. A process line usually incorporates only one freezer, which makes reliability a major concern.

In order to achieve normal equipment life expectancy, freezing systems must be designed and constructed with adequate safety factors for electrical/mechanical components and with materials that can withstand harsh environments and rugged usage.

Hygiene

Cleanability and sanitary design are as important as reliability. Freezing systems should (1) have a minimum number of locations where the product can hang up, (2) be constructed of noncorrosive, safe materials, and (3) be equipped with manual and/or automatic sanitation systems for wash-down and cleanup. If the equipment is not or cannot be properly cleaned and sanitized, product contamination can result. These features are particularly important for chilled, partially cooked, and fully cooked products that may not be fully reheated or properly prepared prior to consumption.

Quality

The quality of processed food products is affected by physical changes and by rates of microbiological activity and chemical reactions, each of which is influenced by the rate of temperature change. The freezing process physically changes the food product; the rate of physical change or the freezing time determines the size of ice crystals produced. See Chapter 11 for more information on food microbiology.

At a slow freezing rate, initially formed ice crystals can grow to a relatively large size; fast freezing forces more crystals to be seeded with a smaller average size. However, different sized ice crystals are formed because the product's surface freezes faster than its inner parts. Large ice crystals may puncture the cell walls of the product, usually increasing loss of juices during thawing. For some products, this can greatly affect the texture and flavor of the remaining product tissue.

The influence of freezing time is more apparent in some products than in others. For strawberries, a shorter freezing time can significantly reduce drip loss. Drip loss is 20% for strawberries frozen in 12 h but only 8% for strawberries frozen in 15 min. Cryogenic systems perform the same freezing function in 8 min or less, reducing drip loss to less than 5%.

A well-applied mechanical freezer or a cryogenic freezer can crust-freeze products rapidly, minimizing loss of natural juices, aromatics, and flavor essences and maintaining higher, more marketable product quality. Also, lower storage temperature and fewer, less severe temperature fluctuations tend to help preserve quality.

Economics

Ironically, freezing equipment is considered both the most expensive and the least expensive link in the modern processing chain. Although the freezer frequently represents the single largest investment in a line, its operating costs are usually only 3 to 5% of the total. Packaging costs may vary widely but generally are several times greater than total freezing cost.

One essential factor to consider when choosing freezing equipment is the loss in product mass that occurs during freezing. The cost of this loss may be about the same as the operating cost of the freezer for inexpensive products such as peas; the loss is even more significant for expensive products such as seafood.

Loss of product mass during freezing may be caused by mechanical losses, downgrading, and dehydration. **Mechanical losses** include juice dripping or products dropping to the floor or sticking to conveyor belts, all of which are specific to each processing plant. A modern freezer should produce minimal losses in this category. **Downgrading losses** refer to product damage, breakage, and other occurrences that render the product unsalable at the top-quality price. For most products, a modern freezing system should incur minimal losses from damage and breakage.

Dehydration losses occur in any freezing system. The evaporation of water vapor from unpackaged products during freezing becomes evident as frost builds up on evaporator surfaces. Frost is also caused by excessive infiltration of warm, moist air into the freezer. Still air inside a diffusiontight carton often creates larger dehydration losses than individual quick freezing of unpackaged products. Heat transfer is poor because no circulation of air occurs within the package. The resulting evaporation of moisture can be significant; however, the frost stays inside the carton.

A poorly designed freezing system operates with dehydration losses of easily 3 to 4%, while well-designed mechanical or cryogenic freezing systems can be built to operate with losses near 0.5%. Liquid nitrogen tunnels normally operate with a dehydration loss of about 0.4 to 1.25%, which occurs when the nitrogen gas is circulated over the product at the infeed end of the freezer. Infeed circulation is sometimes needed to temper the product and to use the heat capacity in the nitrogen most efficiently. Nitrogen immersion freezers have lower dehydration losses but use more liquid nitrogen. A CO_2 freezer using jet impingement operates with a dehydration loss of about 0.5 to 1.25%.

REFRIGERATION SYSTEMS

Most mechanical food freezers use ammonia as the refrigerant and are equipped with either liquid overfeed or gravity flooded evaporators. The choice of evaporator system depends on freezer size and configuration, space limitations, freezer location, existing plant systems (where applicable), relative cost, and end user preference. For systems with three or more evaporators, a liquid overfeed system is usually less costly to install and operate.

The evaporators may be defrosted with water, hot gas, or a combination of both. The defrost systems can be manual, manual start/automatic run, or fully automatic. Coil defrost can take place at a shift change or be sequential, so that the freezer remains in continuous operation for long periods. Selection of a defrost system depends on plant and product requirements, water supply and disposal situation, sanitation regulations, and end user preference.

With a liquid overfeed system, close consideration should be given to refrigeration line sizing and potential static pressure penalties if the liquid overfeed recirculator is remote from the freezer. See Chapter 1 for design considerations.

A design evaporator temperature for the freezer should be selected to achieve the lowest overall capital and operating cost possible for the freezer and the other high- and low-side refrigeration components while still remaining consistent with product requirements and other plant operating conditions.

Operation

Modern conveyor freezers are equipped with programmable logic controls (PLCs) and/or computer control systems that can monitor and control key elements of freezer operation to maximize productivity, product quality, and safety. Items to be monitored and controlled include belt speeds, air and refrigerant temperatures, air and refrigerant pressures, evaporator defrost cycles, belt washers and dryers, amperage for electric motors, safety and alarm functions, and other variables specific to the products being frozen.

The presence of electronic controls alone does not guarantee freezer performance; human operators are still needed. The number and specialty of operators required depends on the size of the plant and the quantity of freezers. A small plant may have a combination operator covering belt production and refrigeration. In larger plants, a freezer operator may oversee production while a specialist attends to the refrigeration cycle. Long-term success requires well-trained, knowledgeable operators who make the proper adjustments as changes occur in the process.

Maintenance

Freezing systems operate in a harsh environment in which some of the components are hidden from view by the enclosure and product. Many freezers operate 5000 to 7000 h per year. The best freezers are ruggedly constructed and well designed for ease of maintenance. Nevertheless, a well-run maintenance program is essential to productivity and safety.

Freezer manufacturers supply operation and maintenance manuals with key instructions, information on components, parts lists, and suggestions regarding maintenance and safety inspections and tasks on a planned-frequency basis. Manufacturers also provide training programs for maintenance technicians.

It is important for plants to have a sufficient number of properly trained maintenance technicians to maintain all systems. Duties include prescribed inspections, routine maintenance tasks, troubleshooting, and required maintenance during nonproduction periods. Depending on plant size, the technicians may be individual mechanics, electricians, and refrigeration specialists or combinations of the three.

CHAPTER 16

MEAT PRODUCTS

Sanitation 16.1
Carcass Chilling and Holding 16.2
Processed Meats 16.12
Frozen Meat Products 16.15
Shipping Docks 16.16
Energy Conservation 16.17

AROUND the world about 4 to 5 million (0.4 million in the United States) four legged animals such as hogs, cattle, calves, buffalo, water buffalo, lamb, mutton, goats, and venison are slaughtered each day to supply the demand for red meats and their products. The majority of these animals are slaughtered in commercial slaughterhouses (abattoirs) under supervision, while a small portion (0.08% in the U.S.) are still killed on the farm. The slaughter process from live animals to packaged meat products is described in Figure 1.

SANITATION

Sound sanitary practices should be applied at all stages of food processing, not only to protect the public but to meet aesthetic requirements. In this respect, meat processing plants are no different from other food plants. The same principles apply regarding sanitation of buildings and equipment; provision of sanitary water supplies and wash facilities; disposal of waste materials; insect and pest control; and proper use of sanitizers, germicides, and fungicides. All U.S. meat plants operate under regulations set forth in inspection service orders. For detailed sanitation guidelines to be followed in all plants producing meat under federal inspection, refer to *Agriculture Handbook* No. 570, available from the U.S. Department of Agriculture, the Food Safety and Inspection Service (FSIS), and Marriott (1994).

Proper safeguards and good manufacturing practices should minimize bacterial contamination and growth. This involves using clean raw materials, clean water and air, sanitary handling throughout, good temperature control (particularly in coolers and freezers), and scrupulous between-shift cleaning of all surfaces in contact with the product.

Precooked products present additional problems because favorable conditions for bacterial growth exist after the product has cooled to below 55°C. In addition, potential pathogens may experience enhanced growth because their competitor organisms were destroyed during cooking. Any delay in processing at this stage allows surviving microorganisms to multiply, especially when the cooked and cooled meat is handled and packed into containers prior to processing and freezing. Creamed products afford especially favorable conditions for bacterial growth. Filled packages should be removed immediately on filling and quickly chilled. Fast chilling not only reduces the time for growth, but can also reduce the number of bacteria.

It is even more important during processing to avoid any opportunity for the growth of pathogenic bacteria that may have entered the product. While these organisms do not grow as quickly at temperatures below 5°C, they can survive freezing and prolonged frozen storage.

Storage at a temperature of about −4°C will permit the growth of psychrophilic spoilage bacteria, but at −10°C these, as well as all other bacteria, are dormant. Even though some cells of all bacteria types die during storage, activity of the survivors is quickly renewed with rising temperature. The processor should recommend safe preparation practices to the consumer. The best procedure is to provide instructions for cooking the food without preliminary thawing. In the freezer, sanitation is confined to keeping physical cleanliness and order, preventing access of foreign odors, and maintaining the desired temperatures.

Role of HACCP

Many of the procedures for the control of microorganisms are managed by the Hazard Analysis and Critical Control Point (HACCP) system of food safety, which is described in Chapter 11. It is a logical process of preventive measures that can control food safety problems. HACCP plans are required by the USDA in all large plants as of January 1998, and smaller plants will be required to have plans within two years. One aspect of the plan recommends

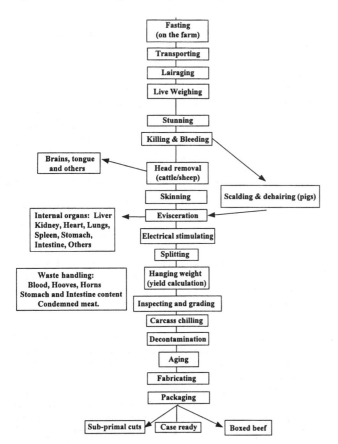

Fig. 1 Steps of Meat Processing

The preparation of this chapter is assigned to TC 10.9, Refrigeration Application for Foods and Beverages.

that red meat carcasses and variety meats be chilled to 5°C within 24 h, and that this temperature be maintained during storage, shipping, and product display.

CARCASS CHILLING AND HOLDING

A hot carcass cooler removes live animal heat as rapidly as possible. Side effects such as cold shortening, which can reduce tenderness, must be considered. Electrical stimulation can minimize cold shortening. Rapid temperature reduction is important in reducing the growth rate of microorganisms that may exist on carcass surfaces. Conditions of temperature, humidity, and air motion must be considered to attain desired meat temperatures within the time limit and to prevent excessive shrinkage, bone taint, sour rounds, surface slime, mold, or discoloration. The carcass must be delivered with a bright, fresh appearance.

Spray Chilling Beef

Spraying cold water intermittently on beef carcasses for 3 to 8 h during chilling is currently the normal procedure in commercial beef slaughter plants (Johnson et al. 1988). Basically, this practice reduces evaporative losses and speeds chilling. Regulations do not allow the chilled carcass to exceed the prewashed hot carcass mass. The carcass is chilled to a large extent by evaporative cooling. As the carcass surface tissue dries, moisture migrates toward the surface, where it evaporates. Eventually, an equilibrium is reached when the temperature differential narrows and reduces the evaporative loss.

When carcasses were shrouded, a once frequently used method for reducing mass loss (shrink), typical evaporative losses ranged from 0.75 to 2.0% for an overnight chill (Kastner 1981). Allen et al. (1987) found that spray-chilled beef sides lost 0.3% compared with 1.5% for non-spray-chilled sides. Those authors stated that although variation in carcass shrink of spray-chilled sides was influenced by carcass spacing, other factors, especially those affecting the dynamics of surface tissue moisture, may be involved. Carcass washing, length of spray cycle, and carcass fatness also influence the variation in shrink. With sufficient care, however, carcass cooler shrink can be nearly eliminated.

Loin eye muscle color and shear force are not affected by spray chilling, but fat color can be lighter in spray-chilled compared to nonspray chilled sides. Over a 4-day period, color changes and drip losses in retail packs for rib steaks and round roasts were not related to spray chilling (Jones and Robertson 1989). Those authors also concluded that spray-chilling could provide a moderate reduction in carcass shrinkage during cooling without having a detrimental influence on muscle quality.

Vacuum-packaged inside rounds from spray-chilled sides had significantly more purge, i.e., air removed, (0.4 kg or 0.26%) than those from conventionally chilled sides. Spacing treatments where foreshanks were aligned in opposite directions and where they were aligned in the same direction but with 150 mm between sides both result in less shrink during a 24-h spray-chill period than the treatment where foreshanks were aligned in the same direction but with all sides tightly crowded together (Allen et al. 1987). Some studies with both beef (Hamby et al. 1987) and pork (Greer and Dilts 1988) indicated that bacterial populations of conventionally and spray-chilled carcasses were not affected by chilling method (Dickson 1991). However, Acuff (1991) and others showed that use of a sanitizer (chlorine, 200 ppm, or organic acid, 1 to 3%) significantly reduces carcass bacterial counts.

Chilling Time

Although certain basic principles are identical, beef and hog carcass chilling differs substantially. The massive beef carcass is only partially chilled (although shippable) at the end of the standard overnight period; the average hog carcass may be fully chilled (but not ready for cutting) in 8 to 12 h, while the balance of the period accomplishes only temperature equalization.

The beef carcass surface retains a large amount of wash water, which provides much evaporative cooling in addition to that derived from actual shrinkage; but evaporative cooling of the hog carcass, which retains little wash water, occurs only through actual shrinkage. A beef carcass, without skin and destined largely for sale as fresh cuts, must be chilled in air temperatures sufficiently high to avoid freezing and damage to appearance. Although it must subsequently be well tempered for cutting and scheduled for in-plant processing, a hog carcass, including the skin, can tolerate a certain amount of surface freezing. Beef carcasses can be chilled with an overnight shrinkage of 0.5%, whereas equally good practice on hog carcasses will result in 1.25 to 2% shrinkage.

The bulk (16 to 20 h) of beef chilling is done overnight in high humidity chilling rooms with a large refrigeration and air circulation capacity. The balance of the chilling and temperature equalization occurs during a subsequent holding or storage period that averages one day, but can extend to 2 or 3 days, usually in a separate holding room with a low refrigeration and air circulation capacity.

Some packers load for shipment the day after slaughter, since some refrigerated transport vehicles have ample capacity to remove the balance of the internal heat in round or chuck beef during the first two days in transit. This practice is most important in rapid delivery of fresh meat to the marketplace. Carcass beef that is not shipped the day after slaughter should be kept in a beef-holding cooler at temperatures of 1 to 2°C with minimum air circulation to avoid excessive color change and mass loss.

Refrigeration Systems for Coolers

Refrigeration systems commonly used in carcass chilling and holding rooms are operated with ammonia as the primary refrigerant and are of three general types: dry coils, chilled brine spray, and sprayed coil.

Dry Coil Refrigeration. Dry coil systems comprise most chilling and holding room installations. Dry coil systems usually include unit coolers equipped with coils, defrosting equipment, and fans for air-vapor circulation. Because the coils are operated without continuous brine spray, eliminators are not required. Coils are usually finned, with fins limited to 6 to 8 mm spacing or with variable fin spacing to avoid icing difficulties. The units may be mounted on the floor, overhead on the rail beams, or overhead on converted brine spray decks.

Dry coil systems operated at surface temperatures below 0°C build up a coating of frost or ice, which ultimately reduces the airflow and cooling capacity. Coils must therefore be defrosted periodically, normally every 4 to 24 h for coils with 6 to 8 mm fin spacing, to maintain capacity. The rate of buildup, and hence the defrosting frequency, decreases with large coil capacity and high evaporating pressure.

Defrosting may be done either manually or automatically by the following:

- **Hot-gas defrost** is accomplished, with the fans off, by introducing hot gas directly from the system compressors into the evaporator coils. The evaporator suction is throttled to maintain a coil pressure of about 400 to 500 kPa (gage) (at approximately 5 to 10°C). The coils then act as condensers and supply the heat for melting the ice coating. Other evaporators in the system must supply the load for the compressors during this period. Hot-gas defrost is rapid, normally requiring 10 to 30 min for completion. See Chapter 3 for further information about hot-gas defrost piping and control.

- **Coil spray defrost** is accomplished (with the fans turned off) by spraying the coil surfaces with water, which supplies the heat required to melt the ice coating. Suction and feed lines are closed off, with pressure relief from the coil to the suction line to minimize the refrigeration effect. Enough water at 10 to 25°C must be

Meat Products

used to avoid freezing on the coils, and care must be taken to ensure that drain lines do not freeze. The sprayed water tends to produce some fog in the refrigerated space. Coil spray defrost may be more rapid than hot-gas defrost.

- **Room air defrost** (for rooms 2°C or higher) is accomplished with the fans running while suction and feed lines are closed off (with pressure relief from coil to suction line), to permit buildup of coil pressure and melting of the ice coating by transfer of heat out of the air flowing across the coils. Refrigeration therefore continues during the defrosting period, but at a drastically reduced rate. Room air defrost is slow; the time required may vary from 30 min to several hours if the coils are undersized for dry coil operation.
- **Electric defrost** is accomplished with electric heaters with fans either on or off. During defrost, refrigerant flow is interrupted.

Unit coolers may be defrosted by any one or combinations of the first three methods. All methods involve a reduction in chilling capacity, which varies with time loss and heat input. Hot-gas and coil spray defrost interrupt the chilling only for short periods, but they introduce some heat into the space. Room air defrost severely reduces the chilling rate for long periods, but the heat required to vaporize the ice is obtained entirely from the room air.

Evaporator controls customarily employed in carcass chilling and holding rooms include refrigerant feed controls, evaporator pressure controls, and air circulation control.

Refrigerant feed controls are designed to maintain, under varying loads, as high a liquid level in the coil as can be carried without excessive liquid spillover into the suction line. This is accomplished by using an expansion valve that throttles the liquid from supply pressure [typically 1 MPa (gage)] to evaporating pressure [usually 140 kPa (gage) or higher]. The throttling of the liquid flashes some of it to gas, which chills the remaining liquid to saturation temperature at the lower pressure. If it does not bypass the coil, the flashed gas tends to reduce flooding of the interior coil surface, thus lowering coil efficiency.

The valve used may be a hand-controlled expansion valve supervised by operator judgment alone, a thermal expansion valve governed by the degree of superheat of the suction gas, or a float valve (or solenoid valve operated by a float switch) governed by the level of feed liquid in a surge drum placed in the coil suction line. This surge drum suction trap permits the ammonia flashed to gas in the throttling process to flow directly to the suction line, bypassing the coil. The trap may be small and placed just high enough so that its level governs that in the coils by gravity transfer. Or, as in the ammonia recirculation system, it may be placed below coil level so that the liquid is pumped mechanically through the coils in much greater quantity than is required for evaporation. In the latter case, the trap is sized large enough to carry its normal operating level plus all the liquid flowing through the coils, thus effectively preventing liquid spillover to the compressors. Nevertheless, it is necessary in all cases to provide further protection at the compressors' liquid return.

Present practice strongly favors liquid ammonia recirculation systems, mainly because of the greater coil heat transfer rates with the resultant greater refrigerating capacity over other systems (see Chapter 3). Some have coils mounted above the rail beams with 1.2 to 1.8 m of ceiling head space. Air is forced through the coils, sometimes using two-speed fans.

Manual and thermal expansion valves do not provide good coil flooding under varying loads and do not bypass the flashed feed gas around the coils. As a result, evaporators so controlled are usually rated 15 to 25% less in capacity than those controlled by float valve or ammonia recirculation.

Evaporator pressure controls regulate coil temperature, and thereby the rate of refrigeration, by varying evaporating pressure within the coil. This is accomplished by using a throttling valve in the evaporator suction line downstream from the surge drum suction trap. All such valves impose a definite loss on the refrigeration system, and the amount of the loss varies directly with the pressure drop through the valve. This increases the work of compression for a given refrigeration effect.

The valve used to control evaporating pressure may be a manual suction valve set solely by operator judgment, a back pressure valve actuated by coil pressure or temperature, or a back pressure valve actuated by a temperature-sensing element somewhere in the room. Manual suction valves require excessive attention when loads fluctuate. The coil-controlled back pressure valve seeks to hold a constant coil temperature but does not control room temperature unless the load is constant. Only the room-controlled compensated back pressure valve responds to room temperature.

Air circulation control is frequently used when an evaporator must handle separate load conditions differing greatly in magnitude, such as the load in chilling rooms that are also used as holding rooms or for the negligible load on weekends. The use of two-speed fan motors (operated at reduced speed during the periods of light load) or turning the fans off and on can control air circulation to a degree.

Chilled Brine Spray Systems. These systems are generally being abandoned in favor of other systems due to such a system's large required building space, inherent low capacity, brine carryover tendencies, and difficulty of control.

Sprayed Coil Systems. These consist of unit coolers equipped with coils, brine spray banks, eliminators to prevent brine carryover, and fans for air-vapor circulation. The units are usually mounted (without ductwork) either on the floor or overhead on converted brine spray decks. Refrigeration is supplied by the primary refrigerant in the coils. Chilled or nonchilled recirculated brine is continually sprayed over the coils, thus eliminating ice formation and the need for periodic defrosting.

The brine predominantly used is sodium chloride, with caustic soda or another additive for controlling pH. Because sodium chloride brine is corrosive, bare-pipe coils (without fins) generally see service. The brine is also highly corrosive to the rail system and other cooler equipment.

Propylene glycol with added inhibitor complexes is another coil spray solution used in place of sodium chloride. As with sodium chloride brine, propylene glycol is constantly diluted by moisture condensed out of the spaces being refrigerated and must be concentrated by evaporating water from it. The reconcentration process requires special equipment designed to minimize glycol losses. Sludge that accumulates in the concentrator may become an operating problem; to avoid it, additives must be selected and pH closely controlled. Finned coils are usually used with propylene glycol.

Because of its noncorrosive nature in comparison to sodium chloride, propylene glycol greatly reduces the cost of unit cooler construction as well as maintenance of space equipment.

Other Systems. Considerable attention is being directed to system designs that will reduce the amount of evaporative cooling at the time of entrance into the cooler and eliminate ceiling rail and beam condensation and drip. Good results have been achieved by using low-temperature blast chill tunnels before entrance into the chill room. The volume of ceiling condensate is reduced because the rate of evaporative cooling is reduced in proportion to the degree of surface cooling. Room condensation has been reduced by the addition of heat above carcasses (out of the main air stream), fans, minimized hot water usage during cleanup, better dry cleanup, timing of cleanup, and using wood rail supports.

Grade and yield sorting, with its simultaneous filling of several rooms, has shortened the chilling time available if refrigeration is kept off during the filling cycle. Its effect has to be offset by more chill rooms and more installed refrigeration capacity. If full refrigeration is kept at the start of filling, the peak load is reduced to the rooms being filled. Hot carcass cutting has been started with only a short chilling time. Cryogenic chilling has also been tested for hot carcass chilling.

Beef Cooler Layout and Capacity

Carcass halves or sides are supported by hooks suspended from one-wheel trolleys running on overhead rails. The trolleys are generally pushed from the dressing floor to the chilling room by powered conveyor chains equipped with fingers that engage the trolleys, which are then distributed manually over the chilling and holding room rail system. Chilling and holding room rails are commonly placed on 0.9 to 1.2 m centers in the holding rooms, with pullout or sorting rails between them. The rails must be placed a minimum of 0.6 m from the nearest obstruction, such as a wall or building column, and the tops of the rails must be at least 3.4 m above the floor. The supporting beams should be placed a minimum of 1.8 m below the ceiling for optimum air distribution. Applicable to new construction in plants engaged in interstate commerce, regulations for some of these dimensions are issued by the Meat Inspection Division of the FSIS.

To assure effective air circulation, carcass sides should be placed on the rails in both chilling and holding coolers so that they do not touch each other. Required spacing varies with the size of the carcass and averages 750 mm per two sides of beef. In practice, however, sides are often more crowded.

A chilling room should be of such size that the last carcass loaded into it does not materially retard the chilling of the first carcass. While size is not as critical as in the case of the hog carcass chill room (because of the slower chill), to better control shrinkage and condensation, it is desirable to limit chill cooler size to hold not more than 4 h of the daily kill. Holding coolers may be as large as desired because of their ability to maintain more uniform temperature and humidity.

While overall plant chilling and holding room capacities vary widely, chilling coolers generally require a capacity equal to the daily kill; holding coolers require 1 to 2 times the daily kill.

Beef Carcasses. Dressed beef carcasses, each split into two sides, range in mass from approximately 140 to 450 kg, averaging about 300 kg per head. Specific heats of carcass components range from 2.1 kJ/(kg·K) for fat to 3.3 kJ/(kg·K) or more for lean muscle, averaging about 3.1 kJ/(kg·K) for the carcass as a whole.

The body temperature of an animal at slaughter is about 39°C. After slaughter, physiological changes occur that generate heat and tend to increase carcass temperature, while heat loss from the surface tends to lower it.

The largest part of the carcass is the round, and at any given stage of the chilling cycle its center has the highest temperature of all carcass parts. This *deep round* temperature (about 41°C when the carcass enters the chilling cooler) is therefore universally used as a measure of chilling progress. If it is to be significant, the temperature must be taken accurately. Incorrect techniques will show temperatures as much as 5 K lower than actual deep round temperature. An accurate technique that yields consistent results is shown in Figure 2. The technique applies a fast-reacting, easily read stem dial thermometer, calibrated before and after tests, inserted upward to the full depth through the hole in the aitchbone.

At the time of slaughter, the water content of beef muscle is approximately 75% of the total mass. Thereafter, a gradual drying of the surface takes place, resulting in mass loss or shrinkage. Shrinkage and its measurement are greatly affected by the final operations of the dressing process: weighing and washing. Weighing must be done prior to washing if the masses are to reflect actual product shrinkage.

A beef carcass retains large amounts of wash water on its surface, which it carries into the cooler. The loss of this water, occurring in the form of vapor, does not constitute actual product loss. However, it must be considered when estimating the system capacity since the

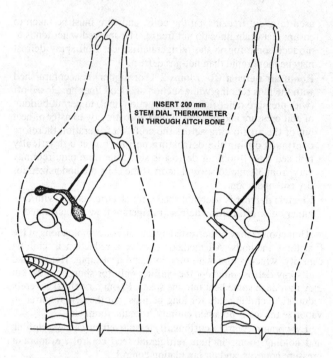

Fig. 2 Deep Round Temperature Measurement in Beef Carcass

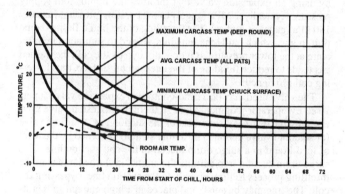

Fig. 3 Beef Carcass Chill Curves

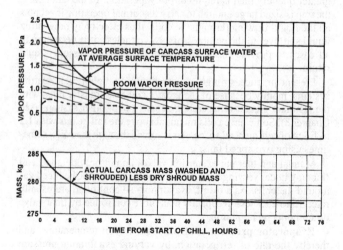

Fig. 4 Beef Carcass Shrinkage Rate Curves

Meat Products

vapor must be condensed on the coils, thus constituting an important part of the refrigeration load.

The amount of wash water retained by the carcass depends on its condition and on washing techniques. A carcass typically retains 3.6 kg, part of which is lost by drip and part by evaporation. Water pressures used in washing vary from 0.3 to 2.1 MPa (gage), and temperatures from 15 to 46°C.

To minimize spoilage, a carcass should be reduced to a uniform temperature of about 1.5°C as rapidly as possible. In practice, deep round temperatures of 15°C (measured as in Figure 2), with surface temperatures of 1.5 to 7°C, are common at the end of the first day's chill period.

To prevent formation of surface slime most carcasses are cut, vacuum packaged, and boxed within 24 to 72 h. Otherwise, a carcass surface needs to be a certain dryness during storage. Exposed beef muscle chilled to an actual temperature of 2°C will not slime readily if dried at the surface to a water content of 90% of dry mass (47.4% of total mass). Such a surface is in vapor pressure equilibrium with a surrounding atmosphere at the same temperature (2°C) and 96% rh. In practice, a room atmosphere at 0 to 1°C and approximately 90% rh will maintain a well-chilled carcass in good condition without slime (Thatcher and Clark 1968).

Chilling-Drying Process. Curves of carcass temperature during a chilling-holding cycle are shown in Figure 3. Note that some heat loss occurs before a carcass enters the chilling cooler. The evaporative cooling of surface water dominates in the initial stages of hot carcass chilling. As chilling progresses, the rate of losses by evaporative surface cooling diminishes and the sensible transfer of heat from the carcass surface increases. Note that the time-temperature rates of change are subject to variations between summer and winter ambient conditions, which influence system capacity.

The rate of transfer is increased both by more rapid circulation of air and lower air temperature, but these are limited by the necessity of avoiding surface freezing.

Estimated differences in vapor pressure between surface water (at average surface temperature) and atmospheric vapor during a typical chilling-holding cycle, and the corresponding shrinkage curve for an average carcass, are shown in Figure 4. Note the tremendous vapor pressure differences during the early part of the chill cycle when the carcass is warm. The evaporative loss could be reduced by beginning the chill with room temperature high, then lowering it slowly to minimize the pressure difference between carcass surface water and room vapor at all times. However, this slows the chill and prolongs the period of rapid evaporation. The quick chill practice is favored; but the cold shortening effect and bacterial growth must be considered in carcass quality and keeping time.

Evaporation from the warm carcass in cool air is nearly independent of room relative humidity because the warm carcass surface generates a much higher vapor pressure than the cooler vapor surrounding the carcass. If the space surrounding a warm carcass is saturated, evaporation forms fog, which can be observed at the beginning of any chill.

Evaporation from the well-chilled carcass with surface temperature at or near room temperature is different. The spread between surface and room vapor pressures approaches zero when room air is near saturation. Evaporation proceeds slowly, without forming fog. Evaporation does not cease when a room is saturated; it ceases only if the carcass is chilled through to room temperature, and no heat transfer is taking place.

The ultimate disposition of water condensed on the coils depends on the temperature of the coil surface and the method of coil operation. In continuous defrost (sprayed coil) operation, condensed and trapped water dilute the solution sprayed over the coil. In nonfrosting dry-coil operation, condensed water falls to the evaporator pan and drains to the sewer. Water frozen on the coil is lost to the sewer if removed by hot-gas or coil spray defrost. Periodic room air defrost, however, vaporizes part of the ice and returns it to room

Table 1 Mass Changes in Beef Carcass

Chilling Cooler	Mass, kg
Initial dry mass	279
Wash water pickup	3.6
Initial wet mass	283
Spray chill water use	7.3
Drip (not evaporated)	4.5
Mass at maximum (8 h postmortem)	287
Mass loss 8-24 h postmortem	8.1
Net mass loss (ideal)	0
Holding Cooler	
Mass loss/day	1.4
Final mass (48 h postmortem)	278

atmosphere, while losing the remainder to drain. This method of defrost is not normally used in beef chill or holding coolers because temperatures are not suitable, and thus the defrost period is excessive, resulting in abnormal room temperature variations. Most chill and holding evaporators are automatically defrosted with water or hot gas on selected time cycles. The massmass changes that take place in an average beef carcass are given in Table 1.

Chilling of the beef carcass is not completed in the chill cooler but continues at a reduced rate in the holding cooler. A carcass well chilled when it enters the holding cooler shows minimum holding shrinkage; a poorly chilled one shows high holding shrinkage.

If shrinkage values are to have any significance, they must be carefully derived. Actual product loss must be determined by first weighing the dry carcass prior to washing and then weighing it out of the cooler. In-motion masses are not sufficiently precise; carcasses must be weighed at rest. Scales must be accurate, and, if possible, the same scale should be used for both weighings. If shrinkage is to have any comparison value, it must be measured on carcasses chilled to the same temperature, since chilling occurs largely by evaporative mass loss.

Design Conditions and Refrigeration Load. Equipment selection should be based on conditions at peak load, when product loss is greatest. Room losses, equipment heat, and carcass heat add up to a total load that varies greatly—not only in magnitude but in proportion of sensible to total heat (sensible heat ratio)— throughout the chill. As the chill progresses, the vapor load decreases and the sensible load becomes more predominant.

Under peak chilling load, excess moisture condenses into fog—enough to warm the air-vapor-fog mixture to the sensible heat ratio of the heat removal process. The heat removal process of the coil therefore underestimates the actual rate of water removal by the amount of vapor condensed to fog (Table 2).

Fog does not generally form under later chilling room loads and all holding room loads, although it may form locally and then vaporize. Sensible heat ratios of air vapor heat gain and air vapor heat removal are then equal (Table 3).

Beef chilling rooms generally have evaporator capacity sufficient to hold room temperature under load approximately as shown in Figure 3. This results in an increase in room temperature to 2 to 5°C, with gradual reduction to 0 to 1°C. However, many installations provide greater capacity, particularly dry coil systems, which thereby avoid excessive coil frosting. In batch-loaded coolers, room temperature may be as low as −4°C under peak load, provided it is raised to −1°C as the chill progresses, without surface freezing of the beef. The shrinkage improvement affected by these lower temperatures, however, tends to be less than expected (in beef chilling) because of the relatively small part played by sensible transfer of heat.

Standard practice in the holding room calls for providing enough evaporator capacity to keep the room temperature at 0 to 1°C at all times. Holding room coils sized at peak load, low air

Table 2 Load Calculations for Beef Chilling

Cooler size, m:	18.6 m by 22.2 m by 5.55 m
Cooler capacity:	476 carcasses
Average carcass mass:	284 kg
Beef mass per second for first 4h:	3.75 kg/s
Assumed chill rate:	28 K in 20 h
Assumed air circulation:	79 m³/s
Loading time:	3.3-h maximum
Assumed air to coil:	0.6°C, 100% rh
Assumed fan motive power:	26.8 kW
Specific heat of beef:	3.14 kJ/(kg·K)
Density of air:	1.28 kg/m³

Heat Gain—Room Load	Loads, kW		
	Sensible	Latent	Total
1. Transmission, infiltration, personnel, fan motor, lights, and equipment heat	47.5	1.2	48.7
2. Product heat (average, first 4 h:			
a. 3.75 × 3.14 × 27.8	326.9		
b. 0.11 kg/s × 2489 kJ/kg[a]	−276.3	276.3	326.9
3. Total heat gain (room load), kW (Items 1 + 2a + 2b)	98.1	277.5	375.6
Heat Removal—Coil Load			
4. Air circulation, dry air, kg/s 79.0 × 1.28 = 101.1 kg/s	—	—	—
5. Heat removed per kg of dry air, kilojoules (Item 3)/(Item 4) = 3.71 kJ/kg	—	—	—
6. Air-vapor enthalpy, kJ/kg dry air:			
a. Air to coil, 0.6°C, 100% rh	18.4	9.9	28.3
b. Kilojoules removed, temperature drop 2.06°C	−2.2	−1.6	−3.7
c. Air from coil, −2.12°C, 100% rh	16.2	8.3	24.6
7. Cool air-vapor heat removal, kilowatts (Item 4)(Item 6b)	218.3	157.4	375.6
8. Room vapor condensed to fog (Item 7) − (Item 3)	120.1	−120.1	
9. Water (ice) removed by coil 0.111 kg/s × 335 kJ/kg[b]		37.4	37.4
10. Total heat removal (coil load), kW (Items 3 + 8 + 9)	218.2	194.8	413.0

[a] Heat of vaporization   [b] Heat of fusion

Table 3 Load Calculations for Beef Holding

Cooler size:	30.5 m by 41.5 m by 5.55 m
Cooler capacity, one day's kill:	1120 carcasses
Average carcass mass:	276.7 kg
Beef mass chilled per second for first 4h:	4.3 kg/s
Assumed chill rate:	4.17°C in 24 h
Assumed air circulation:	43 m³/s
Assumed air to coil:	1.1°C, 96% rh
Assumed fan motive power:	17.9 kW
Specific heat of beef:	3.14 kJ/(kg·K)
Air density:	1.28 kg/m³

Heat Gain—Room Load	Loads, kW		
	Sensible	Latent	Total
1. Transmission, infiltration, personnel, fan motor, lights, and equipment heat	71.8	2.9	74.9
2. Product heat (average, first 4 h:			
a. 4.3 × 3140.1 × 4.17	56.3		
b. 0.0152 kg/s × 2489 kJ/kg[a]	−37.8	37.8	56.3
3. Total heat gain (room load), kW (Items 1 + 2a + 2b)	90.3	40.7	131.0
Heat Removal—Coil Load			
4. Air circulation, dry air, kg/s 43 × 1.28 = 55.0 kg/s	—	—	—
5. Heat removed per kg of dry air, kJ (Item 3)/(Item 4) = 2.38 kJ/kg	—	—	—
6. Air-vapor enthalpy, kJ/kg dry air:			
a. Air to coil, 1.1°C, 96% rh	19.0	9.8	28.8
b. kJ removed, temperature drop 1.61°C	−1.6	−0.8	−2.4
c. Air from coil, −0.51°C, 100% rh	17.4	9.0	26.4
7. Cool air-vapor heat removal, kW	85.8	41.4	131.2
8. Room vapor condensed to fog	—	—	—
9. Water (ice) removed by coil 0.0152 × 3,350 kJ/kg[b]		5.1	5.1
10. Total heat removal (coil load), W (Items 3 + 8 + 9)	90.3	45.8	136.1

[a] Heat of vaporization   [b] Heat of fusion

vapor circulation rate, and a coil temperature 5 K below room temperature tend to maintain the approximately 90% rh that avoids excessive shrinkage and prevents surface sliming.

From the average temperature curve shown in Figure 4 and the shrinkage curve in Figure 5, certain generalizations useful in calculating carcass chilling load may be made. In the chilling cooler, the average carcass temperature is reduced approximately 28 K, from about 39°C to about 8°C, in 20 h. Simultaneously, about 6.5 kg of water is vaporized for each 284 kg carcass entering the chill; only 2.0 kg of this is actual shrinkage. The losses of sensible heat and water occur at about the same rate. In the sample load calculations, this is calculated at an average of 10% for the first 4 h of chill for sensible heat and 13% for the evaporation of moisture, which roughly agrees with the curves of Figures 4 and 5. This is the maximum rate of chill and is used for sizing the refrigeration equipment and piping.

In the holding cooler, the average carcass temperature is reduced from 8 to 4°C in 24 h. Simultaneously, about 0.8 kg of water is vaporized per carcass. Where spray chilling is employed, this shrinkage approaches zero. Here also, the losses of sensible heat and water occur at about the same rate. The sample load calculations are figured at a 5% average for the first 4 h for sensible heat and 6% for latent heat.

Under peak chilling and holding room conditions, water trapped and condensed out by the coils imposes a further latent load on the evaporators. This occurs in the form of heat extracted to freeze condensed water into ice or of heat removed to chill the returning warmed and strengthened spray solution. In the absence of a more complex evaluation, this load may be considered equal to the latent heat of fusion (334 kJ/kg) of the water removed.

Based on the data just mentioned, cooler loads may be calculated as illustrated in Tables 2 and 3. Transmission, infiltration, personnel, and equipment loads are estimated by standard methods such as those discussed in Chapter 12.

The complete calculation is made to illustrate the heat removal process associated with the chilling-drying of the carcass. In particular, it illustrates that the sensible heat ratio of the heat transfer in the coil cannot be used to measure the amount of water removed from the space when fog is involved.

Evaporator Selection. Evaporator selection is a procedure of approximation only, because of the inaccuracies of load determination on the one hand and of predicting sustained field performance of coils on the other. Furthermore, there is rarely complete freedom of specification; for example, the air vapor circulation rate for a given coil may be limited to avoid spray solution carryover or excessive fan power.

Sprayed and dry coil systems perform equally well with respect to shrinkage, provided compressor capacity is adequate and the evaporators are correct for the system selected. Evaporator

Meat Products

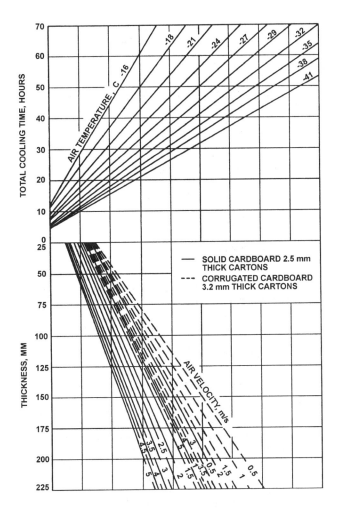

Fig. 5 Freezing Times of Boneless Meat

Table 4 Sample Evaporator Installations for Beef Chilling[a]

Cooler size, m:	18.6 m × 22.2 m × 5.55 m
Cooler capacity:	476 carcasses
Deep-round chill:	to 10°C in 20h
Design load:	413 kW
Coil operation:	liquid recirculation
Loading time:	3.3 h
Average carcass mass:	283.5 kg
Assumed air to coil:	0.6°C, 100% rh
Sensible heat ratio:	53%

	Dry Coil
Coil Description:	
Type of coil	Finned
Fin spacing, center to center	6 mm
Coil depth, number of pipe rows	8
Coil face area	1.9 m²
Coil surface area, total	207.9 m²
Fan Description, Airflow:	
Type of fan	Centrifugal
Flow through coil	4390 L/s
Flow per square metre coil face area	2320 L/s
Fan motive power, kW	1.5 kW
Unit Rating[b] (Total Heat):	
TD for capacity rating[c]	5.6°C
Chilling capacity	23.7 kW
Temperature drop, air through coil	2.1 K
Equipment for 520 Carcasses:	
Number units required	18
Total motive power, fans and pumps	26.85 kW
Coil surface per carcass	8.175 m²
Airflow per carcass	165 L/s

[a]While data describe actual successful installations, other successful installations may be different.
[b]Ratings shown are estimated from performance of actual systems. Dry coil ratings are at average frost conditions, with airflow reduced by frost obstruction. While this example describes actual installations, it is not to be interpreted as an accepted standard. Other installations, employing both more and less equipment, are also successful.
[c]TD is temperature difference between refrigerant and air.

requirements vary widely with the type of system. Comparative evaporator data on a typical, successful flooded coil installation in the chilling cooler is presented in Table 4.

The coil overall heat transfer coefficient and airflows shown describe sustained field performance under actual chilling conditions and loads; they should not be confused with clean coil test ratings. The heat transfer coefficient varies greatly with the character of the coil and its operation and is influenced by such variables as (1) the ratio of extended-to-prime surface, which may range from 7-to-1 to 21-to-1 in standard dry coils; (2) coil depth, which typically ranges from 8 to 12 rows in sprayed coils and from 4 to 10 rows in dry coils; (3) fin spacing, which may be 6 to 8 mm in typical dry coils; (4) condition of the surface, either continuously defrosted or generally coated with frost; and (5) airflow, which may vary from 1.3 to 3.8 m³/s over the coil face area.

Greater temperature differences (TD) than those shown are sometimes used, but a higher TD is valid only at high room temperatures. The lower TD (5 K) shown for dry coils is desirable to limit frosting. Many dry coil evaporators have higher ratios of extended-to-prime surface and higher airflows per unit face area than shown.

The difficulties involved in obtaining accurate shrinkage figures on carcasses chilled to a specified degree cause wide differences of opinion as to the coil capacity required for good chilling. While data describe actual successful installations, other successful installations may differ.

Boxed Beef

The majority of the output of beef slaughterhouses is in the form of prefabricated sections of the carcass, vacuum-packed in plastic bags and shipped in corrugated boxes. Standard cuts can be sold at cost savings to the market. The shipping density is much greater, with easier material handling, and the bones and fat are removed where their value as a byproduct is greatest. Customers purchase only the sections they need, and the trim loss at final processing to primal cuts is minimized.

Vacuum-packaging with added carbon dioxide, nitrogen, or a combination of gases has the following advantages:

- Creates anaerobic conditions, preventing the growth of mold (which is aerobic and requires the presence of oxygen for growth)
- Provides more sanitary conditions for carcass breaking
- Retains moisture, retards shrinkage
- Excludes bacteria entry, extends shelf life
- Retards bloom until opened

After normal chilling, a carcass is broken into primal cuts, vacuum packed, and boxed for shipment. Temperatures are usually held at −2°C to prevent the development of pathogenic organisms. Aging of the beef continues after vacuum-packaging and during shipment, because the exclusion of oxygen or the addition of gases does not slow enzymatic action in the muscle.

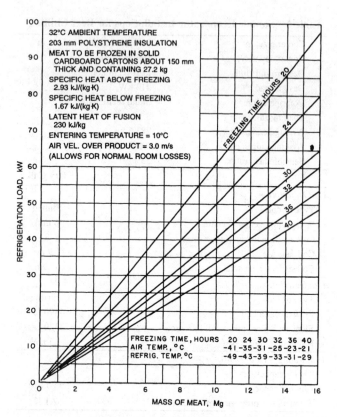

Fig. 6 Blast Freezer Loads

Freezing Times of Boneless Meat. The cooling of boneless meat from 10 to −12°C requires the removal of about 310 kJ/kg of lean meat (74% water), most of which is latent heat liberated when the liquid water in the meat changes to ice. Most of the time needed to freeze meat is spent in cooling from −1 to −4°C.

For boneless meat in cartons, the rate of freezing depends on the temperature and velocity of the surrounding air and on the thickness and thermal properties of the carton and the meat itself.

Figure 5 shows the effects of the first three factors on cooling times for lean meat in two carton types. For example, the chart shows a total cooling time of about 31 h for solid cardboard cartons 125 mm thick at an air temperature of −32°C and a velocity of 2 m/s. The corresponding air temperatures and cooling times may be found for any specified thickness of carton and air velocity. Conversely, the chart can be used to find combinations of air velocity and temperature needed to freeze cartons of a particular thickness in a specified time (see Figure 6).

Accuracy of the estimated freezing times is about 3% for air velocities greater than 2 m/s. Calculations are based on Plank's Equation as modified by Earle. A latent heat of 249 kJ/kg and an average freezing point of −2°C are assumed for lean meat.

Increasing the fat content of meat reduces the water content and hence the latent heat load. Thermal conductivity of the meat is reduced at the same time, but the overall effect is for freezing times to drop as the percentage of fat rises. Actual cooling times for mixtures of lean and fatty tissue should therefore be somewhat less than the times obtained from the chart. For meat with 15% fat, the reduction is about 17%.

Hog Chilling and Tempering

The internal temperature of hog carcasses entering the chill coolers from the killing floor varies from 38 to 41°C. The specific heat shown in Chapter 8 is 2.60 kJ/(kg·K), but in practice 2.9 to 3.1 kJ/(kg·K) is used because changed feeding techniques have created leaner hogs. The dressed mass varies from 40 to 200 kg approximately, the average being near 80 kg. Present practice requires dressed hogs to be chilled and tempered to an internal ham temperature of 3 to 4°C overnight. This limits the chilling and tempering time to 12 to 18 h.

Cooler and refrigeration equipment must be designed to chill the hogs thoroughly with no frozen parts at the time the carcasses are moved to the cutting floor. Carcass crowding, reducing exposure to circulated chilled air, and excessively high peak temperatures are all detrimental to proper chilling.

The following hog cooler design details will provide:

- Sufficiently quick chilling to retard bacterial development and prevent deterioration
- A cooler shrinkage from 0.1 to 0.2%
- Firm carcasses that are dry and bright without frozen surface or internal frost, suitable for efficient cutting

Hog Cooler Design. The capacity of hog coolers is set by the dressing rate of hogs and the planned hours of operation. However, on a one-shift basis, it is economically sound to provide cooler hanging capacity for 10-h dressing in order to properly handle the chilling of large sows that require more than 24-h exposure in the chill room; handle increased dressing volumes when market conditions warrant overtime operation; and have some flexibility in unloading and loading the cooler during normal operations. On a two-shift basis, extra cooler capacity for overtime operation is not necessary.

The rail height should be 2.75 m to provide both good air circulation and adequate clearance between the floor and the largest dressed hog. (In the United States, this is a requirement of the FSIS and most state regulations.) Rails should be spaced at a minimum of 760 mm on centers to provide sufficient clearance for the hanging hogs and to prevent contact between carcasses.

The spacing of hogs on the rail varies according to the size of the hog. Hogs should be spaced so that there is at least 40 to 50 mm between the flank of one carcass and the back of the carcass immediately in front of it. The rail spacing of 330 mm on centers is normal for 80 kg dressed hogs.

Many meat-packing refrigeration engineers maintain that several hog chill coolers with a capacity of 2-h loading for 300 to 600 kill/h or 4-h loading for lower killing capacities is more economical than one large chill cooler.

The hog cooler should be designed on the following basis:

1. Total amount of hanging rail should be equal to:
   a. (One-shift operation) (10 h × Rate of kill × 0.33 m/hog).
   b. (Two-shift operation) (16 h × Rate of kill × 0.33 m/hog).
   c. For combination carcass loading, hogs and cattle, calves, or sheep, the figures should be modified accordingly.
2. The rail height should be 2.75 m. It may be 3.35 m for combination beef and hog coolers.
3. The rail spacing should be a minimum of 760 mm.
4. The inside building height will vary depending on the type of refrigeration equipment installed. A clear height of 1.8 m above the rail support is adequate for space to install units, piping, and controls; it provides sufficient plenum over the rails to ensure even air distribution over the hog carcasses.

Preliminary, intensive batch, inline chilling is practiced in some plants, resulting in smaller shrinkages with the variations in chilling systems.

Selecting Refrigeration Equipment. Both floor units and units installed above the rail supports are used. Floor units, with a top discharge outlet equipped with a short section of duct to discharge air in a space over the rail supports, are used by a few pork processors. A few brine spray units equipped with ammonia coils, and water or hot-gas defrost units are being used.

Meat Products

Two types of units are available for installation above the rail supports. One uses a blower fan to force air below and through a horizontally placed coil with the air discharging horizontally from the front or top of the unit. The other consists of a vertical coil with axial fans or blowers to force the air through the unit. Both units are designed for various types of liquid feed control.

The horizontal units are equipped for both hot-gas and water defrosting. All have finned surfaces designed for hog chill cooler operation. Evaporator controls should be provided as described in the previous section on beef carcass coolers.

Careful selection of units and use of automatic controls, including liquid recirculation, provides an air circulation, temperature, and humidity balance that chills hogs with minimum shrinkage in the quickest time.

The temperature control should be set to provide an opening room temperature of −3 to −2°C. As the cooler is loaded, the suction temperature decreases to provide the additional refrigeration effect required to handle increased refrigeration load and maintain room temperature at −1 to 0°C. Ample compressor and unit capacity is required to achieve these results.

Dry coils selected to maintain a 5.5 to 6.5 K temperature difference (refrigerant to air) at peak operation provides adequate coil surface and a TD of 0.5 to 3 K prior to opening and about 10 h after closing the cooler. This low-temperature difference results in economical high-humidity conditions during the entire chilling cycle. Cutting practices typically use a high initial chilling TD and a lower TD at the end of the chilling cycle.

Sample Calculation. The hog chilling time-temperature curves (Figure 7) are composite curves developed from several operation tests. The relation of the room temperature and ammonia suction gas temperature curves show that the refrigeration load decreased about 9 h after closing the cooler. After about 9 h, the room temperature is increased and the hog is tempered to an internal ham temperature of 3 to 4°C.

Table 5 was prepared using empirical calculations to coordinate product and unit refrigeration loads. A shortcut method for determining hog chill cooler refrigeration loads is presented in Table 6. The latent heat of the product has been neglected, since the latent heat of evaporation is equal to the reduction of sensible heat load of the product. Total sensible heat was used in all calculations.

Example 1: Select cooling units for 600 hogs/h at 82 kg average dressed mass using 2-h loading time cooler.

Four coolers minimum requirement (five desirable)
Each cooler:
Capacity 1200 hogs = 39.7 kW (Table 6)
Product peak load = 6 × 49.3 = 295.8 kW (Table 5)
Total = 335.5 kW

Select 18 units of 18.7 kW at 5.7 K temperature difference per cooler
Approximately 93.4 m³/s
Air changes per second = 93.4/1926 = 0.048 or 2.9 per minute

The refrigeration of the hog cutting room, where the carcass is cut up into its primal parts, is an important factor in maintaining the quality of the product. A maximum dry-bulb temperature of 10°C should be maintained. This level is low enough to prevent excessive rise in temperature of the product during its relatively short stay in the cutting room and also complies with USDA-FSIS requirements.

Chilled carcasses entering the room may have surface temperatures as low as −1°C. Unless the dew point of the air in the cutting room is maintained at a temperature below −1°C, moisture will condense on the surface of the product, providing an excellent medium for bacterial growth.

Floor, walls, and all machinery on the cutting floor must be thoroughly cleaned at the end of each day's operation. Cleaning releases a large amount of vapor in the room that, unless quickly removed, will condense on the walls, ceiling, floor, and machinery surfaces. When the outside dew-point temperature is less than the room temperature, vapor removal can be accomplished by installing fans to continuously exhaust the room during cleanup.

These fans should operate only during the cleanup period and as long as required to remove the vapor produced by cleaning. Exhaust facilities equivalent to five air changes per hour should be satisfactory. When the room temperature is lower than the outside dew-point temperature, this method cannot be used. The water vapor must then be condensed out on the evaporator surfaces.

Many people work in the hog cutting room. Attention should be given to the heat given off by personnel, normally 290 W per

Table 5 Product Refrigeration Load, kW

	Cooler Loading Time, h			
Hours	1	2	4	8
1	25.3[a]	25.3	25.3	25.3
2	23.9	49.3[a]	49.3	49.3
3	22.8	46.8	72.1	72.1
4	21.8	44.6	94.6[a]	93.9
5	20.9	42.7	89.5	114.7
6	20.1	41.0	85.6	134.8
7	19.6	39.7	82.3	154.4
8	19.1	38.7	79.7	173.6[a]
9	18.9	38.1	77.8	167.1
10	—	18.9	57.7	139.7
11	—	—	38.1	120.4
12	—	—	18.9	98.7
13	—	—	—	77.8
14	—	—	—	57.7
15	—	—	—	38.1
16	—	—	—	18.9

[a]Values are for peak load:
100 hogs/h at 82 kg dressed average.
Chilled from 38.9 to 3.3°C in 16 h.
Based on operating test, not laboratory standards.

Table 6 Average Chill Cooler Loads Exclusive of Product

Cooler Capacity	Room Dimensions, m	Floor Area, m²	Room Volume, m³	Refrigeration kW[a]
1200 Hogs	12.2 × 30.5 × 5.2	370.3	1926	39.7
2400 Hogs	24.4 × 30.5 × 5.2	740.7	3852	79.5
3600 Hogs	36.6 × 30.5 × 5.2	1111.0	5777	119.2
4800 Hogs	30.5 × 48.8 × 5.2	1481.3	7703	159.0
6000 Hogs	30.5 × 61.0 × 5.2	1851.7	9629	198.7

[a]Based on room volume of 48 m³/kW (or use detailed calculations for building heat gains, infiltration, people, lights, and average unit cooler motors from Chapter 28 in the 1997 *ASHRAE Handbook—Fundamentals*).

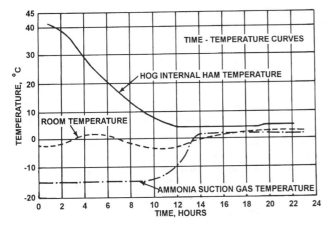

Fig. 7 Composite Hog Chilling Time-Temperature Curves

person. Heat given off by electric motors must also be included in the refrigeration load.

Special consideration should also be given to the latent heat load caused by knife boxes, wash water, workers, and infiltration. If the sensible heat load is not sufficient, this high latent heat load must be offset by reheat at the refrigeration units in order to maintain the desired low relative humidity.

The quantity of air circulated is influenced by the amount of sensible heat to be picked up and the relative humidity to be maintained, usually between 7 and 12 complete air changes each hour. The air distribution pattern requires careful attention to prevent drafts on the workers.

Forced air units are satisfactory for refrigerating cutting floors. Ceiling height must be sufficient to accommodate the units. A wide selection of forced air units may be applied to these rooms. They can be floor or ceiling mounted, with either dry-coil or wetted-surface units arranged for flooded, recirculated, or direct-expansion refrigerant systems.

Suction pressure regulators should be provided with both the flooded and the direct-expansion units. Automatic dry- and wet-bulb controls are essential for best operation.

Pork Trimmings

Pork trimmings come from the chilled hog carcass, principally from the primal cuts: belly, plate, back fat, shoulder, and ham.

Trimmings per hog average 1.8 to 3.6 kg. Only trimmings used in sausage or canning operations are discussed here.

In the cutting or trimming room, trimmings are usually at a temperature between 3 to 7°C; an engineer must design for the higher temperature. The product requires only moderate chilling to be in proper condition for grinding, if it is to be used locally in sausage or canning operations. If it is to be stored or shipped elsewhere, hard chilling is required. Satisfactory final temperature for local processing is −2°C. This is the average temperature after tempering and should not be confused with surface temperature immediately after chilling. It is possible for the trimmings to have a much lower surface temperature than internal temperature immediately after chilling, especially if they have been quick-chilled.

Good operating practice requires rapid chilling of pork trimmings as soon as possible after removal from the primal cuts. This retards enzymatic action and microbial growth, which are responsible for poor flavor, rancidity, loss of color, and excessive shrinkage.

The choice of chilling method depends largely on local conditions and consists of a variation of air temperature, air velocity, and method of achieving contact between air and meat. Continuous belt equipment using low-temperature air fluids or one of the cryogenic fluids to obtain lower shrinkage is available.

Truck Chilling. Economic conditions may require existing chilling or freezer rooms to be used. Some may require an overnight chill, others less than an hour. The following methods make use of existing facilities:

- Trimmings are often chilled with CO_2 snow prior to grinding to prevent excessive temperature rise during grinding and blending. Additionally, finished products, such as sausage or hamburger in chubs, may be crusted or frozen in a glycol/brine liquid contact chiller.
- Trimmings are put on truck pans to a depth of 50 to 100 mm and held in a suitable cooler kept at −1°C. This method requires a short chill time and results in a near uniform temperature (−1°C) of the trimmings.
- Trimmings are spread 100 to 125 mm deep on truck pans in a −20°C freezer and held for 5 to 6 h, or until the meat is well stiffened with frost. Use of temperatures lower than −20°C (with or without fans) will expedite chilling if time is limited. After trimmings are hardchilled, they are removed from the metal pans and tightly packed into suitable containers. They are held in a −3 to −2°C room until shipped or used. Average shrinkage using this system is 0.5% up to the time they are put in the containers.
- Trimmings from the cutting or trimming room are put in a meat truck and held in a cooler with a temperature from −2 to −1°C. This method usually requires an overnight chill and is not likely to reach a temperature of 0°C in the center of the load.
- If trimmings are not to be used within one week, they should be frozen immediately and held at −23°C or lower.

Fresh Pork Holding

Fresh pork cuts are usually packed on the cutting floor. If they are not shipped the same day that they are cut and packed, they should be held in a cooler with a temperature of −7 to −2°C.

Forced-air cooling units are frequently used for holding room service because they provide better air circulation and more uniform temperatures throughout the room, minimize ceiling condensation caused by air entering doorways from adjacent warmer areas (because of traffic), and eliminate the necessity of coil scraping or drip troughs if hot-gas defrost is used.

Cooling units may be the dry type with hot-gas defrost, or wetted surface with brine spray. Units should have air diffusers to prevent direct air blast on the products. Unless the room shape is unusually odd, discharge ductwork should not be necessary.

Since the product is boxed and wrapped and the holding period is short, humidity control is not too important. Various methods of automatic control may be used. CO_2 has been applied within boxes of pork cuts. Care must be taken to maintain the ratio of pounds of CO_2 to pounds of meat for the retention period. The enclosures must be relieved and ventilated in the interest of life safety.

Calf and Lamb Chilling

Dry coils are typically used for calf and lamb chilling. These can be either the between-the-rail type, the suspended type above the rail, or floor units. The same type of refrigerating units used for pork may be used for lamb, with some modifications. For example, in the chilling of lambs and calves, it is desirable to use reduced air changes over the carcass. This is accomplished by two-speed motors, using the higher speed for the initial chill and reducing the rate of air circulation when the carcass temperatures are reduced, approximately 4 to 6 h after the cooler is loaded.

Lambs usually have a mass of 18 to 36 kg, with an approximate average dressed carcass mass of 23 kg. Sheep have a mass up to an average of approximately 57 kg and readily take refrigeration. Adequate coil surface should be installed to maintain a room temperature below −1°C and a relative humidity of 90 to 95% in the loading period. The evaporating capacity should be based on an average 5 K temperature differential between refrigerant and room air temperature, with an opening room temperature of 0°C.

Compensating back pressure regulating valves, which vary the evaporator pressure as the room temperature changes, should be used. As the room and carcass temperatures drop, the temperature differential is reduced, thus holding a high relative humidity (40 to 45%). At the end of a 4- to 6-h chill period, the air over the carcass may be reduced to help keep the bloom and color of the product.

Carcasses should not touch each other. They enter the cooler at 37 to 39°C, with the carcass temperature taken at the center of the heavy section of the rear leg. The specific heat of a carcass is 2.9 kJ/(kg·K). Air circulation for the first 4 to 6 h should be approximately 50 to 60 changes per hour, reduced to 10 to 12 changes per hour. The carcass should reach 1 to 2°C internally in about 12 to 14 h and should be held at that point with 85 to 90% rh room air until shipped or otherwise processed. This gives the least possible shrinkage and prevents excess surface moisture.

In calf chilling coolers, approximately the same procedure is acceptable, with carcasses hung on 300 to 380 mm centers. The dressed mass varies at different locations, with an approximate 39 to

Meat Products

41 kg average in dairy country and a 90 to 160 kg (sometimes heavier) average in beef-producing localities. The same time and temperature relationship and air velocities previously given for chilling lambs are used for chilling calves, except when calves are chilled with the hide on. Also, the time may be extended for air circulation. Air circulation need not be curtailed in hide-on chilling in this application because rapid cooling gives better color to these carcasses after they are skinned.

The refrigerating capacity for lamb and calf chill coolers is calculated the same as for other coolers, but additional capacity should be added to permit reduced air circulation and maintain close temperature differential between room air and refrigerant.

Chilling and Freezing Variety Meats

The temperature of variety meats must be lowered rapidly to −2 to −1°C to reduce spoilage. Large boxes are particularly difficult to cool. For example, a 125 mm deep box containing 32 kg of hot variety meat can still have a core temperature as high as 16°C 24 h after it enters a −30°C freezer. Variety meats in boxes or packages more than 75 mm thick may be chilled very effectively during freezing by adding dry ice to the center of the box.

For design calculations, variety meat has an initial temperature of about 38°C. Specific heats of variety meats vary with the percentage of fat and moisture in each. For design purposes, a specific heat of 3.1 kJ/(kg·K) should be used. used method consists of quick chilling at lower temperatures and higher air velocities, using the same type of truck equipment as in the overnight chilling method. This method is also used for chilling trimmings. Care must be exercised in the design of the quick chilling cabinet or room to provide for the refrigeration load imposed by the hot product. One industry survey shows that approximately 50% of large establishments are using quick-chill in their variety meat operations.

The quick-chilling cabinet or room should be designed to operate at an air temperature of approximately −30 to −40°C with air velocities over the product of 2.5 to 5 m/s. During initial loading, the air temperature may rise to −20°C. In the design of quick-chilling units, refrigeration coils are used with axial flow fans for air circulation.

The recommended defrosting method is with water and/or hot gas, except where units with continuous defrost are used. The product is chilled to the point where the outside is frosted or frozen and a temperature of −2 to −1°C is obtained when the product later reaches an even temperature throughout in the packing or tempering room. The time required to chill the product by this method depends on the depth of product in the pans, the size of individual pieces, air temperature, and velocity. Normally, 0.5 to 4 h is a satisfactory chill period to attain the required −2 to −1°C temperature. In addition to the obvious savings in time and space, an important advantage of this method is the low total shrinkage, averaging only 0.5 to 1%. These values were obtained in the same survey as those in Table 7.

Another method of handling variety meats involves packaging the product before chilling as near as possible to the killing floor. The packed containers are placed on platforms and frozen in a freezer. Separators should permit air circulation between packages.

This method is used in preparing products for frozen shipment or freezer storage. The internal temperature of the product should reach −4°C for prompt transfer to a storage freezer. For immediate shipment, the internal temperature of the product must be reduced to −20°C; this may be done by longer retention in the quick freezer. Here package material and size, particularly package thickness, largely determine the rate of freezing. For example, a 125 mm thick box will take from 16 h upward to freeze, depending on the type of product, package material, size, and loading method.

The dry-bulb air temperature in these freezers is maintained at −40 to −30°C, with air velocities over the product at 2.5 to 5 m/s. The time required to reach the desired internal temperature is a combination of refrigeration capacity, size of largest package, insulating properties of package material, and so forth. A generous safety factor should be used in sizing evaporator coils for this type of service. These freezers are best incorporated within refrigerated rooms. Defrosting is by the water or hot-gas method, except where units with continuous defrost are used. Shrinkage varies in the range of only 0.5 to 1%.

The initial freezing equipment cost and design load can be reduced if carbon dioxide is included in the packaging phase as part of the operational plan. Another efficient cooling method uses plate freezers to form blocks of product that can be loaded on pallets with minimal packaging.

Table 7 Storage Life of Meat Products

	Months			
	Temperature, °C			
Product	−12	−18	−23	−29
Beef	4 to 12	6 to 18	12 to 24	12+
Lamb	3 to 8	6 to 16	12 to 18	12+
Veal	3 to 4	4 to 14	8	12
Pork	2 to 6	4 to 12	8 to 15	10
Chopped beef	3 to 4	4 to 6	8	10
Pork sausage	1 to 2	2 to 6	3	4
Smoked ham and bacon	1 to 3	2 to 4	3	4
Uncured ham and bacon	2	4	6	6
Beef liver	2 to 3	2 to 4		
Cooked foods	2 to 3	2 to 4		

Packaging and Storage

Packages for variety meats do not yet have any standard sizes or dimensions. Present requirements are a package that will stand shipping, with sizes to suit individual establishments. The importance of the package is brought more into focus in the case of the hot pack freezing method. Standardization of sizes and package materials promotes faster chilling and more economical handling.

Storage of variety meats depends on its end use. For short storage (under one week) and local use, −2 to −1°C is considered a good internal product temperature. If stored for shipping, the internal temperature of the product should be kept at −20°C or below. Recommended length of storage is controversial; type of package, freezer temperature and relative humidity, amount of moisture removed in original chill, and the variations of the products themselves all affect storage life.

Packers' recommendations regarding storage time vary from 2 to 6 months and longer, since variety meats pick up rancidity on the surface and soft muscle tissue dehydrates while freezing. More rapid freezing and vapor-proof packaging are important in increasing storage life.

Refrigeration Load Computations

The average evaporator refrigerating load for a typical chilling process above freezing may be computed by:

$$q_r = m_m c_m (t_1 - t_2) + m_t c_t (t_1 - t_2) + q_w + q_i + q_m \qquad (1)$$

where

q_r = refrigeration load, kW
m_m = mass of meat, kg/s
m_t = mass of trucks, kg/s
c_m = specific heat of meat, kJ/(kg·K)
c_t = specific heat of truck, containers, or platforms (0.50 for steel), kJ/(kg·K)
t_1 = average initial temperature, °C
t_2 = average final temperature, °C
q_w = heat gain through room surfaces, kW
q_i = heat gain from infiltration, kW
q_m = heat gain from equipment and lighting, kW

The following example illustrates the method of computing the refrigeration load for a quick-chill operation.

Example 2. Find the refrigeration load for chilling six trucks of offal from a maximum temperature of 38 to 1 °C in 2 h. Each truck has a mass of 180 kg empty and holds 330 kg of offal. The specific heat is 0.50 kJ/(kg·K) for the truck and 3.14 kJ/(kg·K) for the offal. The room temperature is to be held at −18 °C, with an outdoor temperature of 4 °C and a rh of 70%. The walls, ceiling, and floor gain 0.426 W/(m²·K) in 24 h and have an area of 88 m². The volume of the room is 53 m³ and 12 air changes in 24 h are assumed.

Solution: Values for substitution in Equation (1) are as follows:

$m_m = 6 \times 330/2 = 990$ kg/h $= 0.275$ kg/s
$c_m = 3.14$ kJ/kg·K
$m_t = 6 \times 180/2 = 540$ kg/h $= 0.15$ kg/s
$c_t = 0.50$ kJ/(kg·K)
$q_w = 88 \times 0.426 [4 - (-18)] = 0.82$ kW

From a pychrometric chart or table the heat removed from the infiltrating air is 41.6 kJ/m³. Then:

$q_i = 41.6 \times 53 \times 12/(24 \times 3600) = 0.31$ kW
$q_m = 7.5 + 0.2 + 7.7$ kW where
7.5 = assumed fan and motor kilowatts
0.2 = lights in kilowatts

Substituting in Equation (1):

$$q_r = 0.275 \times 3.14(38-1) + 0.15 \times 0.50(38-1)$$
$$+ 0.82 + 0.31 + 7.7 = 43.6 \text{ kW}$$

Good practice is to add 10 to 25% to the computed refrigeration load.

PROCESSED MEATS

Prompt chilling, handling, and storage under controlled temperatures help in the production of mild and rapidly cured and smoked meats. The product is usually transferred directly from the smokehouse to a refrigerated room, but sometimes a drop in temperature of 5 to 15 K can occur if the transfer time is appreciable.

Since the day's production is not usually removed from the smokehouses at one time, the refrigeration load is spread over nearly 24 h. Table 8 outlines temperatures, relative humidities, and time required in refrigerated rooms used in handling smoked meats.

Prechilling of smoked meat reduces drips of moisture and fat, thus increasing yield. Meats can be chilled at higher temperatures, with air velocities of up to 2.5 m/s (Table 8). At lower temperatures, air velocities of 5 m/s and higher are used. Chilling in the hanging room or in the wrapping and packaging rooms results in slow chilling and high temperatures when packing. Slow chilling is not desirable for a product that is to be stored or shipped a considerable distance.

Meats handled through smoke and into refrigerated rooms are hung or racked on cages that are moved on an overhead track or mounted on wheels. Sometimes the product is transferred from suspended cages to wheel-mounted cages between smoking and subsequent handling.

Smoked hams and picnic meats must be chilled as rapidly as possible through the incubation temperature range of 40 to 10 °C. A product requiring cooking before eating is brought to a minimum internal temperature of 60 °C to destroy possible live trichinae, while a product not requiring cooking before eating is brought to a minimum internal temperature of 68 °C.

A maximum storage room temperature of 5 °C dry bulb should be used when delivery from the plant to retail outlets is made within a short time. A room dry-bulb temperature of −2 to 0 °C is desirable when delivery is to points considerably removed from the plant and transfer is made through controlled low humidity rooms, docks, cars, or trucks, keeping the dew point below that of the product.

Bacon usually reaches a maximum temperature of 52 °C in the smokehouse. Since most smoked bacon in a sliced form is packaged, it may be transferred directly to the chill room if it has

Table 8 Room Temperatures and Relative Humidities for Smoking Meats

	Room Conditions			
	°C Dry-Bulb	% Relative Humidity	Final Product Temp., °C	Time, h
Prechill method				
Hams, picnics, etc.				
High temperature	3 to 4	80	15	8 to 10
Low temperature	−3 to −2	80	15	2 to 3
Derind bacon				
Normal	−3 to −2	80	−2	8 to 10
Blast	−18 to −12	80	−3	2 to 3
Hanging or tempering				
Ham, picnics, etc.	7 to 10	70	10 to 13	
Derind bacon	−3 to −2	70	−3 to −2	
Wrapping or packaging				
Hams, picnics, etc.	7 to 10	70		
Storage	−2 to 7	70		

been skinned before smoking. If the bacon is to be skinned after smoking, it is usually allowed to hang in the smokehouse vestibule for 2 to 4 h, until it drops to a temperature of 32 °C before skinning.

Bacon is usually molded and sliced at temperatures just below −2 °C. Chill rooms are usually designed to reduce the internal temperature of the bacon to −3 °C in 24 h or less, requiring a room dry-bulb temperature of −8 to −7 °C. A tempering room (which also serves as storage for stock reserve), held at the exact temperature at which bacon is sliced, is often used.

Bacon molding can be done either after tempering, in which case it is moved directly to the slicing machines, or after the initial hardening, and then be transferred to the tempering room. In the latter case, care should be taken that none of the slabs is below −4 °C so that the product will not crack during molding. Bacon cured by the pickle injection process generally shows less pickle pockets if it is molded after hardening, placed no more than eight slabs high on pallets, and held in the tempering room.

In any of the rooms mentioned, air distribution must be uniform. To minimize shrinkage, the air supply from floor-mounted unit coolers should be delivered through slotted ducts or by means of closed ducts supplying properly spaced diffusers directed so that no high velocity airstreams impinge on the product itself. The exception would be in blast chill rooms where high air velocities are needed, but in reality the product is subjected to the condition for only a short time.

Refrigeration may be supplied by floor or ceiling-mounted dry or wet coil units. If the latter are selected, water, hot-gas, or electric defrost must also be used.

Many processors use three methods of chilling smoked meats: rapid blast chilling, brine chilling by direct contact spraying, and cryogenic chilling. Direct contact spraying is especially emphasized, because it minimizes shrinkage, increases shelf life, and provides more uniform chilling. This method is usually carried out in special enclosures designed to combat the detrimental effects of salt brine. Color and salt taste may need close monitoring in contact spraying.

The product should enter the slicing room chilled to a uniform internal temperature not to exceed 10 °C for beef rounds and 5 to 7 °C for other fresh carcass parts, depending on the individual packer's temperature standard. Internal temperatures below −3 °C tend to cause shattering of products such as bacon during slicing and slow fabrication. For that type product, temperatures above 0 °C cause improper shingling from the slicing machine.

The slicing and packaging room temperature and air movement are usually the result of a compromise between the physical comfort

Meat Products

demands of the operating personnel and the requirements of the product. The design room dry-bulb temperature should be below 10°C, according to USDA-FSIS regulations.

An objectionable amount of condensation on the product may occur. To guard against this, the coil temperature should be maintained below the temperature of the bacon entering the room, thus keeping the dew point of the room air below the product temperature. The product should be exposed to the room air for the shortest possible time.

Bacon Slicing and Packaging Room

Exhaust ventilation should remove smoke and fumes from the sealing and packaging equipment and comply with OSHA occupancy regulations. Again, heat exchangers should be considered for reducing the resulting increased refrigeration load.

Refrigeration for this room may be supplied by forced-air units—floor or ceiling mounted, dry or wet coil—or finned-tube ceiling coils. Dry coils should have defrost facilities if coil temperatures are to be kept at more than several degrees below freezing. Air discharge and return should be evenly distributed, using ductwork if necessary. To avoid drafts on personnel, air velocities in the occupied zone should be in the range of 0.13 to 0.18 m/s. The temperature differential between primary air and room air should not exceed 5 K to assure personnel comfort. To provide optimum comfort and dew-point control, reheat coils are necessary.

Where ceiling heights are adequate, multiple ceiling units can be used to minimize the amount of ductwork. Automatic temperature and humidity controls are desirable in cooling units.

To provide draft-free conditions, drip troughs with suitable drainage should be added to finned-type ceiling coils. However, it is difficult, if not impossible, to maintain a room air dew point low enough to approach the product temperature. Some installations operating with relative humidities of 60 to 70% do not have product condensation problems.

One control method consists of individual coil banks connected to common liquid and suction headers. Each bank is equipped with a thermal expansion valve. The suction header has an automatically operated back-pressure valve. The thermostatically controlled dual back-pressure regulator and liquid-header solenoid are both controlled by a single thermostat. This arrangement provides a simple automatic defrosting cycle.

Another system uses fin coils with glycol sprays. Humidity is controlled by varying the concentration strength of the glycol and the refrigerant temperature.

Sausage Dry Rooms

Refrigeration or air conditioning is an integral part of year-round sausage dry rooms. The purpose of these systems is to produce and control the air conditions for proper removal of moisture from the sausage.

A variety of dry sausage is manufactured, for the most part uncooked. Its keeping qualities depend on curing ingredients, spices, and removal of moisture from the product by drying.

This sausage is generally of two distinct types—smoked and not smoked.

FSIS regulates the minimum temperature and amount of time that the product must be held after stuffing and prior to release, depending on the method of production. The dry room temperature shall not be lower than 7°C and the length of time the product shall be held in the dry room prior to release is dependent on the diameter of the sausage after stuffing and the method of preparation used.

After stuffing, the sausages are held at a temperature of 16 to 24°C and 75 to 95% rh in the sausage greenroom to develop the cure. The sausages are suspended from sticks at the time of stuffing and may be held on the trucks or railed cages or be transferred to racks in the greenroom. Sausages in 90 to 100 mm diameter casings are generally spaced about 150 mm on centers on the sausage sticks. The length of time the sausages are held in the greenroom depends on the method of preparation, the type and dimensions of the sausage, the operator, and the judgment of the sausage maker regarding proper flavor, pH, and other characteristics.

Those sausage varieties that are not smoked are then transferred to the sausage dry room; those that are to be smoked are transferred from the greenroom to the smokehouse and then to the dry room.

In the dry room, approximately 30% of the moisture is removed from the sausage, to a point at which the sausage will keep for a long time, virtually without refrigeration. The drying period required depends on the amount of moisture to be removed to suit trade demand, type of sausage, and type of casing. The moisture transmission characteristics of synthetic casings vary widely and greatly influence the rate of drying. The diameter of the sausage is probably the most important factor influencing the drying rate.

Small-diameter sausages, such as pepperoni, have more surface in proportion to the mass of material than do large-diameter sausages. Furthermore, the moisture from the sausage interior has to travel a much shorter distance to reach the surface, where it can evaporate. Thus, the drying time for small-diameter sausages is much shorter than that for large-diameter sausages.

Typical conditions in the dry room are approximately 7 to 13°C and 60 to 75% rh. Some sausage makers favor the lower range of temperatures for unsmoked varieties of dry sausage and the higher range for smoked varieties.

In processing dry sausage, moisture can only be removed from the product at the rate at which the moisture comes to the surface of the casing. Any attempt to speed the drying rate results in overdrying the surface of the sausage, a condition known as case hardening. This condition is identified by a dark ring inside the casing, close to the surface of the sausage, which precludes any further attempt to remove moisture from the interior of the sausage. On the other hand, if sausage is dried at too slow a rate, excessive mold occurs on the surface of the casing, sometimes leading to an unsatisfactory appearance. An exception is the Hungarian salami, which requires a high humidity so that a prolific mold growth can occur and flourish.

As with any other cool or refrigerated space, sausage dry rooms should be properly insulated to prevent temperatures in adjoining spaces from influencing the temperature in the dry room. Ample insulation is especially important in the case of dry rooms located adjacent to rooms of much lower temperature or rooms on the top floor, where the ceiling may be exposed to relatively high temperatures in summer and low temperatures in winter.

Insulation should be adequate to prevent the inner surface of the walls, floor, and ceiling of the dry room from differing more than a degree or two from the average temperature in the room. Otherwise, there is the possibility of condensation due to the high relative humidity in these rooms, which leads to mold growth on the surfaces themselves and, in some cases, on the sausages as well.

The sticks of sausage are generally supported on permanent racks built into the dry room. In the past, these were frequently made of wood; however, sanitary requirements have virtually outlawed the use of wood for this purpose in new construction. The uprights and rails for the racks are now made of either galvanized pipe, hot dip galvanized steel, or stainless steel. The rails for supporting the sausage sticks should be spaced vertically at a distance that leaves ample room for air circulation below the bottom row and between the top row and the ceiling. Generally, a spacing of not less than 300 to 600 mm between rails depends on the length of sausage stick used by the individual manufacturer.

The horizontal spacing between sausages should be such that they do not touch at any point, to prevent mold formation or improper development of color. Generally, with the large 100 mm diameter sausages, a spacing of 150 mm on centers is adequate.

Dry-Room Equipment. In general, two types of refrigeration equipment are currently used for attaining the required conditions in a dry room. The most common is a refrigeration-reheat system, in which the room air is circulated either through a brine spray or over a refrigerated coil and sufficiently cooled to reduce the dew point to the temperature required in the room. The other type involves the use of a hygroscopic liquid sprayed over a refrigerating coil in the dehumidifier, thus condensing the moisture from the air without the severe overcooling usually required by the refrigeration-reheat type of system. The chief advantage of this arrangement is that refrigeration and heating loads are greatly reduced.

The use of any type of liquid, brine or hygroscopic, requires periodic tests and adjusting the pH to minimize corrosion of the equipment. Although most systems depend on a type of liquid spray to prevent frost buildup on the refrigerating coils, some successful rooms use dry coils with hot-gas or water defrost.

The air for conditioning the dry room is normally drawn through the refrigerating and dehumidifying systems by a suitable blower fan (or fans) and discharged into the distribution ductwork.

Rooms used exclusively for small-diameter products with a rapid drying rate may actually have air leaving the room to return to the conditioning unit at a lower dry-bulb temperature and greater density than the point at which it is introduced. A dry-room designer needs to know what the room will be used for to determine the natural circulation of the room air. Supply and return ducts can then be arranged to take advantage of and accelerate this natural circulation to provide thorough mixing of incoming dry air with the air in the room.

Regardless of the location of the supply and return ducts, care should be taken to prevent strong drafts or high velocity airstreams from impinging on the product; this will result in local overdrying and unsatisfactory products.

A study of air circulation within the product racks will show that as air passes over the sausages and moisture evaporates from them, this air becomes cooler and heavier, and thus has a tendency to drop toward the bottom of the room, creating a vertical downward air movement within the sausage racks. This natural tendency must be considered in designing the duct installation if uniform conditions are to be achieved.

An example of the calculation involved in designing a sausage dry room follows. These calculations apply to a room used for an assortment of sausages, with an average drying time of approximately 30 days. They would not be directly applicable to a room used primarily for very large salami (which has a much longer drying period) or a room used primarily for small-diameter sausage.

In the latter case, use of the air-circulating rate shown in this example will allow the air to absorb so much moisture in passing through the room that it will be difficult to obtain uniform conditions throughout the space. Furthermore, the amount of refrigeration required to lower the air temperature enough to produce the required low inlet air dew point would be excessive. An air circulation rate of 12 air changes per hour should therefore be considered average for use in average rooms. The actual circulating rate should be adjusted to obtain the best compromise of refrigeration load and air uniformity for the particular type of product handled.

Example 3. Air conditioning for sausage drying room
Room Dimensions:
  12.2 m × 10.2 m × 3.5 m
  Floor space: 124.4 m²
  Volume: 435.5 m³
  Outdoor wall area: 91 m²
  Partition wall area: 71.5 m²
Hanging Capacity:
  Number of racks: 12
  Length of racks: 8.25 m
  Number of rails high: 5
  Spacing of sticks: 0.15 m
  Number of pieces of sausage per stick: 7
  Average mass per sausage: 1.8 kg
  Total mass: 1.8 × 12 × 5 × 7(8.25/0.15) = 41 580 kg
  Assume 42 000 kg green hanging capacity
  Loading per day: 700 kg

Assumed Outdoor Conditions (Summer):
  35°C db; 24°C wb, h = 72 kJ/kg, W = 14.35 g/kg
Dry-Room Conditions Desired:
  13°C db; 10°C wb, h = 29 kJ/kg, W = 6.4 g/kg
Sensible Heat Calculations:
  Walls [U = 0.57 W/(m²·K)]:
    0.57 × 91(35 − 13)/1000 = 1.14 kW
  Partition [U = 0.38 W/(m²·K)]:
    0.38 × 71.5(35 − 13)/1000 = 0.60 kW
  Floor and ceiling [U = 0.57 W/(m²·K)]:
    0.57 × 124.4 × 2(13 − 13)/1000 = none
  Infiltration (Assume ρc_p = 1.20 kJ/(m³·K) and 0.5 air changes/h)
    1.20(435.5 × 0.5/3600)(35 − 13) = 1.60 kW
  Lights = 0.60 kW
  Motors = 3.75 kW
  Daily product load
    3.35[700/(24 × 3600)](35 − 13) = 0.60 kW
    Total sensible heat gain = 8.29 kW

Latent Heat Calculations:
  Product
    700/(24 × 3600) × 0.30 × 2470 kJ/kg = 6.00 kW
    6.00/2470 = 0.00243 kg/s
  Infiltration
    1.2 × 435.5(0.5/3600)(72 − 29) = 3.12 kW
    (3.12 − 1.60)/2470 = 0.00062 kg/s

Total moisture = 0.00243 + 0.00062 = 0.00305 kg/s

Assume 12 air changes per hour with an empty room volume of 435.5 m³ or 1.2 kg/m³ × 435.5 × 12/3600 = 1.74 kg/s. Then each kilogram of air must absorb 0.00305/1.74 = 0.00175 kg of moisture. Air at the desired room condition contains 0.0064 kg/kg, so the entering air must contain 0.0064 − 0.00175 = 0.00465 kg/kg, corresponding to about 5.0°C db and 4.0°C wb (h = 16.6 kJ/kg).

Temperature rise due to sensible heat gain [air specific heat = 1.0 kJ/(kg·K)]:

  8.29/(1.75 × 1.0) = 5.0 K

Temperature drop due to evaporative cooling from latent heat of product only:

  6.00/(1.74 × 1.0) = 3.4 K

Net temperature rise in the room = 5.0 − 3.4 = 1.6 K, or the air entering the room must be 13 − 1.6 = 11.4°C db and 7.2°C wb (0.00465 kg/kg).

Refrigerating load = 1.74(29 − 16.6) = 21.5 kW
Reheat load = 1.74(23.5 − 16.6) = 12.0 kW
Room load = 1.74(29 − 23.5) = 9.6 kW

Lard Chilling

In federally inspected plants, the USDA FSIS designates the types of pork fats which, when rendered, are classified as lard. Other pork fats, when rendered, are designated as rendered pork fats. The rendering process requires considerable heat, and the subsequent temperature of the lard at which refrigeration is to be applied may be as high as 50°C. The following data for refrigeration requirements may be used for either product type.

The fundamental requirement of the FSIS is good sanitation through all phases of handling. The use of copper or copper-bearing alloys that come in contact with lard should be avoided, because minute traces of copper lower the stability of the product.

Lard has the following properties:

Density at	−20°C	= 990 kg/m³
	20°C	= 930 kg/m³
	70°C	= 880 kg/m³
Heart of solidification		= 112 kJ/kg
Melting begins at −37 to −40°C.		
Melting ends at 43 to 46°C.		
Point of half fusion is around 4°C.		
Specific heat in solid state:	−80°C	= 1.18 kJ/(kg·K)
	−40°C	= 1.42 kJ/(kg·K)
Specific heat in liquid state	40°C	= 2.09 kJ/(kg·K)
	100°C	= 2.18 kJ/(kg·K)

Meat Products

In the production of lard, refrigeration is applied so that the final product will have enough texture and a firm consistency. The finest possible crystal structure is desired.

Calculations for chilling 500 kg of lard per hour are:

Initial temperature:	50°C
Final temperature:	25°C
Heat of solidification:	112 kJ/kg
Specific heat:	2.09 kJ/(kg·K)

$$S_f = 100\frac{t_e - t_f}{t_e - t_b} = 100\frac{46 - 25}{46 - (-40)} = 24.4\%$$

where

S_f = percent solidification at final temperature
t_e = temperature at which melting ends
t_f = final temperature
t_b = temperature at which melting begins

Latent heat of solidification:

$$112(24.4/100) = 27.3 \text{ kJ/kg}$$

Sensible heat removed:

$$2.09(50 - 25) = 52.2 \text{ kJ/kg}$$

Total heat removed:

$$(27.3 + 52.2)500/3600 = 11.0 \text{ kW of refrigeration}$$

Assuming a 15% loss because of radiation, for example, in the process, the required refrigeration for application to chill 500 kg lard per hour would be $1.15 \times 11.0 = 12.7$ kW.

Filtered lard at 50°C can be chilled and plasticized in compact internal swept surface chilling units, which use either ammonia or halogenated hydrocarbons. A refrigerating capacity of about 23 W per kilogram of lard handled per hour for the product only should be provided. Additional refrigeration for the requirements of heat equivalent to the work done by the internal swept surface chilling equipment will also be needed.

When operating this type of equipment, it is essential to keep the refrigerant free of oil and other impurities so that the heat transfer surface will not have a film of oil on it to act as insulation and cut down the unit's capacity. Some installations have oil traps connected to the liquid refrigerant leg on the floor below to provide an oil accumulation drainage space.

The safety requirements for this type of chilling equipment are described in ASHRAE *Standard* 15. Note that such units are pressure vessels and, as such, require properly installed and maintained safety valves.

The recommended storage temperature for packaged refined lard is −1 to 1°C. The storage temperature required for prime steam lard in metal containers is 5°C or below for up to a 6-month storage period. Lard stored for a year or more should be kept at −20°C.

Blast and Storage Freezers

The standard method of sharp-freezing a product destined for storage freezers comprises freezing the product directly from the cutting floor in a blast freezer until its internal temperature reaches the holding room temperature. The product is then transferred to holding or storage freezers.

The product to be sharp-frozen may be bagged, wrapped, or boxed in cartons. Individual loads are usually placed on pallets, dead skids, or in wire basket containers. In general, the larger the ratio of surface exposed to blast air to the volume of either the individual piece or the product's container, the greater the rate of freezing. Product loads should be placed in a blast freezer to ensure that each load is well exposed to the blast air and to minimize possible short circuiting of the airflow. Each layer on a load should be separated by 50 mm spacers to give the individual pieces as much exposure to the blast air as possible.

The most popular types of blast equipment are the self-contained air handling or cooling units that consist of a fan, evaporator, and other elements in one package. These units are usually used in multiples and placed in the blast freezer to provide the optimum blast air coverage. The unit fans should be capable of high air velocity and volumetric flow; two air changes per minute is the accepted minimum.

The coils of the evaporator may have either a wet or a dry surface. See Chapter 42 for information on defrosting.

Blast chill design temperatures vary throughout the industry. Most designs are within a −30 to −40°C range. For low temperatures, booster compressors that discharge through a desuperheater into the general plant suction system are used.

Blast freezers require sufficient insulation and good vapor barriers. If possible, a blast freezer should be located so that temperature differentials between it and adjacent areas are minimized in order to decrease insulation costs and refrigeration losses.

Blast freezer entrance doors should be power operated. Suitable vestibules should also be provided as air locks to decrease infiltration of outside air.

Besides normal losses, heat calculations for a blast freezer should include the loads imposed by such material handling equipment as electric trucks, skids, and spacers, together with the packaging materials for the product. Some portion of any heat added under the floor to prevent frost heaving must also be added to the room load.

Storage freezers are usually maintained at −18 to −26°C. If the plant operates with several high and low suction pressures, the evaporators can be tied to a suitable plant suction system. The evaporators can also be tied to a booster compressor system; if the booster system is operated intermittently, provisions must be made to switch to a suitable plant suction system when the booster system is down. Storage freezer coils can be defrosted by hot gas, electricity, or water. Emphasis should be placed on not defrosting too fast with hot gas (because of pipe expansion) and on providing well-insulated, sloped, heat traced drains and drain pans to prevent freezeups.

Direct Contact Meat Chilling

Continuous processes for smoked and cooked wieners use direct sodium chloride brine tanks or deluge tunnels to chill the meat as soon as it comes out of the cooker. Everyday, the brine is prepared fresh in 2 to 13% solutions depending on the chilling temperature and the salt content of the meat.

Cooling is usually done on sanitary stainless steel surface coolers, which are either refrigerated coils or plates in cabinets. Using this type of unit allows coolant temperatures near the freezing point without damaging the cooler; damage may occur when brine is confined in a tubular cooler. The brine is circulated in quantities necessary to fully wet the surface cooler and fill the distribution troughs of the deluge.

Another type of continuous process uses a brine tank into which the wieners drop out of a conveyor that has carried them through the smoking and cooking process. The pumped brine moves the product to the end of the tank, where it is removed by hand and inserted into peeling and packaging lines.

FROZEN MEAT PRODUCTS

The handling and selling of consumer portions of frozen meats has many potential advantages compared with merchandising fresh meat. The preparation and packaging can be done at the packinghouse, allowing economies of mass production, byproduct savings, lower transportation costs, and flexibility in meeting market demands. At the retail level, frozen meat products reduce space and investment requirements and labor costs.

Freezing Quality of Meat

After an animal is slaughtered, physiological and biochemical reactions continue in the muscle until the complex system supplying energy for work has run down and the muscle goes into rigor. These changes continue for up to 32 h postmortem in major beef muscles. Hot boning with electrical stimulation renders meat tender on a continuous basis without conventional chilling. Freezing meat or cutting carcasses for freezing before the completion of these changes cause cold-shortening and thaw-shortening, which render meat tough. The best time to freeze meat is either after rigor has passed or later, when natural tenderization is more or less complete. Natural tenderization is completed during 7 days of aging in most major beef muscles. Where flavor is concerned, freezing as soon as tenderization is complete is desirable.

For frozen pork, the age of the meat before freezing is even more critical than it is for beef. Pork loins aged 7 days before freezing deteriorate more rapidly in frozen storage than loins aged 1 to 3 days. In tests, a difference could be detected between 1- and 3-day-old loins, favoring those only 1 day old. With frozen pork loin roasts from carcasses chilled for 1 to 7 days, the flavor of lean and fat in the roasts was progressively poorer with longer holding time after slaughter.

Effect of Freezing on Quality

Freezing affects the quality—including color, tenderness, and amount of drip—of meat.

Color. The color of frozen meat depends on the rate of freezing. Tests in which prepackaged, steak-size cuts of beef were frozen by immersion in liquid or exposure to an air blast at between -30 and $-40°C$ revealed that airblast freezing at $-30°C$ produced a color most similar to that of the unfrozen product. An initial meat temperature of $0°C$ was necessary for best results (Lentz 1971).

Flavor and Tenderness. Flavor does not appear to be affected by freezing per se, but tenderness may be affected depending on the condition of the meat and the rate and end-temperature of freezing. Faster freezing to lower temperatures was found to increase tenderness; however, consensus on this effect has not been reached.

Drip. The rate of freezing generally affects the amount of drip, and meat nutrients, such as vitamins, are lost from cut surfaces after thawing. Faster freezing tends to reduce the amount of drip, although many other factors, such as the pH of meat, also have an effect on drip.

Changes in Fat. Fat of pork changes significantly in 112 days at $-21°C$, whereas beef shows no change within 260 days at this temperature. At -30 and $-35°C$, no measurable change occurs in either meat in one year.

The relationship of fat rancidity and oxidation flavor has not been clearly established for frozen meat, and the usefulness of antioxidants in reducing flavor changes during frozen storage is doubtful.

Storage and Handling

Pork remains acceptable for a shorter storage period than beef, lamb, and veal due to the differences in fatty acid chain length and saturation in the different species. Storage life is also related to storage temperature. Because animals within a species vary greatly in nutritional and physiological backgrounds, their tissues differ in susceptibility to change when stored. As a result of the differences between meat animals, packaging methods, and acceptability criteria, a wide range of storage periods is reported for each type of meat (Table 7).

Lentz (1971) found that appreciable changes in the color and flavor of frozen beef occur at storage temperatures down to $-40°C$ in 1 to 90 days (depending on temperature) for samples held in the dark. Changes were much more rapid (1 to 7 days) for samples exposed to light. Color changes were less pronounced after thawing than when frozen.

Reports on the effect of different storage temperatures on fat oxidation and palatability of frozen meats indicate that a temperature of $-20°C$ or lower is desirable. Cuts of pork back fat held at $-6, -12, -18$ and $-23°C$ show increases in peroxide value; free fatty acid is most pronounced at the two higher temperatures. For a storage period of 48 weeks, $-18°C$ or lower is essential to avoid fat changes. Pork rib roasts of $-18°C$ showed little or no flavor change up to 8 months, whereas at $-12°C$, fat was in the early stages of rancidity in 4 months. Ground beef and ground pork patties stored at $-12, -18,$ and $-23°C$ indicate that meats must be stored at $-18°C$ or lower to retain good quality for 5 to 8 months. For longer storage, $-30°C$ is desirable.

The desirable flavor in pork loin roasts stored at -21 to $-22°C$, with maximum fluctuations of 3 to 4 K, decreased slightly, apparently without significant difference between treatments. Fluctuations from -18 to $-12°C$ did not harm quality.

Storage temperature is perhaps more critical with meat in frozen meals because of the differing stability of the various individual dishes included. Frozen meals show marked deterioration of most of the foods after 3 months at -11 to $-9°C$.

Storage and Handling Practices. Surveys of practices in the industry indicate why some product reaches the consumer in poor condition. One unpublished survey indicated that 10% of frozen foods may be at $-14°C$ or higher in warehouses, $-9°C$ or higher in assembly rooms, $-6°C$ or higher during delivery, and $-8°C$ or higher in display cases. All these temperatures should be maintained at $-18°C$ for complete protection of the product.

Packaging

At the time of freezing, a package or packaging material serves to hold the product and prevent it from losing moisture. Other functions of the wrapper or box become important as soon as the storage period begins. Ideal packaging material in direct contact with meat should have: low moisture vapor transmission rate; low gas transmission rate; high wet strength; resistance to grease; flexibility over a temperature range including subfreezing; freedom from odor, flavor, and any toxic substance; easy handling and application characteristics adaptable to hand or machine use; and reasonable price. Individually or collectively, these properties are desired for good appearance of the package, protection against handling, preventing dehydration, which is unsightly and damages the product, and keeping oxygen out of the package.

Desiccation through use of unsuitable packaging material is one of the major problems with frozen foods. Another problem is that of distorted or damaged containers due either to lack of expansion space for the product in freezing or to selection of low-strength box material.

Whenever free space is present in a container of frozen food, ice sublimes and condenses on the film or package. Temperature fluctuation increases the severity of frost deposition.

SHIPPING DOCKS

A refrigerated shipping dock can eliminate the need for assembling orders on the non-refrigerated dock or other area, or using a more valuable storage space for this purpose. This is especially true in the case of freezer operations. Some businesses do not really need a refrigerated order assembly area. One example is a packing plant that ships out whole carcasses or sides in bulk quantities and does not need a large area in which to assemble orders. Many such plants are constructed without any dock at all, simply having the load-out doors lead directly into the carcass-holding cooler, which is satisfactory, provided adequate refrigerating capacity is installed in the vicinity of the shipping doors to prevent undue temperature rise in the coolers during shipping.

A refrigerated shipping dock can perform a second function of reducing the refrigeration load, which is most important in the case of freezers; although it serves almost as valuable a function with

Meat Products

coolers. Even with cooler operations, the installation of a refrigerated dock greatly reduces the load on the cooler's refrigerating units and assures a more stable temperature within the cooler. At the same time, it is possible to only provide refrigeration to maintain dock temperatures on the order of 5 to 7°C, so that the refrigerating units can be designed to operate with a wet coil. In this way, frost buildup on the units is avoided and the capacity of the units themselves substantially increased, making it unnecessary to install as many or as large units in this area.

In the case of freezers, the units should be designed and selected to maintain a dock temperature slightly above freezing, usually about 1.5°C. With this dock temperature, orders may be assembled and held prior to shipment without the risk of defrosting the frozen product, and the workers can assemble orders in a much more comfortable space than the freezer. The design temperature should be low enough that the dew point of the dock atmosphere is below the product temperature. Condensation on the surface of the products is one step in developing off-condition product.

With a dock temperature of 1.5°C, the temperature difference between the freezer itself and the outdoor summer condition is split roughly in half. Since the airflow through the loading doors or other openings is proportional to the square root of the temperature difference, this results in an approximate 30% reduction in airflow through the doors—both the doors into the dock itself and the doors from the dock into the freezer. At the same time, by cooling the outdoor air to approximately 1.5°C, in most cases about 50% of the total heat in the outdoor air is removed by the refrigerating units on the dock.

Since using a refrigerated dock reduces the airflow through the door into the freezer by approximately 30% and 50% of the heat in the air that does pass through this door is removed, the net effect is to reduce the infiltration load on the units within the freezer itself by about 65%. This is not a net gain; since an equal number of these units operate at a much higher temperature, the power required to remove the heat on the dock is substantially lower than it would be if the heat were allowed to enter the freezer.

The infiltration load from the shipping door, whether it opens directly into a cooler or freezer or into a refrigerated dock, is extremely high. Even with well-maintained foam or inflatable door seals, a great deal of warm air leaks through the doors whenever they are open. Such air infiltration may be calculated approximately by:

$$V = CHW\theta(H)^{0.5}(t_1 - t_2)^{0.5} \qquad (2)$$

where

- V = air volume, m³/s at the higher temperature condition
- C = 0.017 = empirical constant selected to account for the contraction of the airstream as it passes through the door and for the obstruction created by a truck parked at the door with only nominal sealing
- H = height of the door, m
- W = width of the door, m
- θ = time the door is open, decimal part of an hour
- t_1 = outdoor air temperature or the air at the higher temperature, °C
- t_2 = temperature of air in the dock or cooler, °C

Time θ is estimated, based on the time the door is assumed to be obstructed or partially obstructed. If the doors are equipped with good seals and these are well-maintained so that the average truck will be tightly sealed to the building, this time would be assumed as only the time necessary to spot the truck at the door and complete the air seal.

The unit cooler providing refrigeration for the dock area should be ceiling suspended with a horizontal air discharge. Each unit should be aimed toward the outer wall and above each of the truck loading doors, if possible, so that cold air strikes the wall and is deflected downward across the door. This downward airflow just inside the door tends to oppose the natural airflow of the entering warm air, thus helping reduce the total amount of infiltration.

In general, a between-the-rails unit cooler has proved most successful for this purpose, since it distributes the air over a fairly wide area and at low outlet velocity. Such an airflow pattern does not create severe drafts in the working area and is more acceptable to employees working in the refrigerated space. The above comments and equation for determining air infiltration also apply to shipping doors, which open directly into storage or shipping coolers.

ENERGY CONSERVATION

Water, a utility previously considered free, frequently has the most rapid rate increases. Coupled with high sewer rates, it is the largest single-cost item in some plants. If fuel charges are added to the hot water portion of water usage, water is definitely the most costly utility. Costs can be reduced by better dry cleanup, use of heat exchangers, use of filters and/or settling basins to collect solids and greases, use of towers and/or evaporative condensers, elimination of water as a means of product transport, and an active conservation program.

Air is needed for combustion in steam generators, sewage aeration, air coolers or evaporative condensers, and blowing product through lines. Used properly in conjunction with heat exchangers, air can reduce other utility costs—fuel, sewage, water, and electricity. Nearly all plants need close monitoring of valves either leaking through or left open in product conveying. Low-pressure blowers are frequently used in place of high-pressure air, reducing initial investment and operating costs of driving equipment.

Steam generation is a source of large savings through efficient boiler operation (fuel and water sides). Reduced use of hot water and sterilizer boxes, and proper use of equipment in plants with electric and steam drives, should be promoted. Sizable reductions can be realized by scavenging heat from process-side steam and hot water and by systematically checking steam traps. In some plants, excess hot water and low energy heat can be recovered using heat exchangers and better heat balances.

Electrical energy needs can be reduced by:

- Properly sizing, spacing, and selecting light fixtures and an energy program of keeping lights off (lights comprise 25 to 33% of an electric bill)
- Monitoring and controlling the demand portion of electric usage
- Checking and sizing motors to their actual loads for operation within the more efficient ranges of their curves
- Adjusting the power factor to reduce initial costs in transformers, switchgear, and wiring
- Lubricating properly to cut power demands

Although refrigeration is not a direct utility, it involves all or some of the factors just mentioned. Energy use in refrigeration systems can be reduced by:

- Operating with lower condenser and higher compressor suction pressures
- Properly removing oil from the system
- Purging noncondensable gases from the system
- Adequately insulating floors, ceilings, walls, and hot and cold lines
- Using energy exchangers on exhaust and air makeup
- Keeping doors closed to cut humidity or prevent an infusion of warmer air
- Installing high-efficiency motors
- Maintaining compressors at peak efficiency
- Keeping condensers free of scale and dirt
- Using proper water treatment in the condensing system
- Operating with a microprocessor based management system

Utility savings are also possible when usage is considered with product line flows and storage space. A strong energy conservation program not only saves total energy but frequently results in greater product yields and product quality improvements, and therefore increased profits. Prerigor or hot processing of pork and beef products is being used to greatly reduce the energy required for postmortem chilling. Removal of waste fat and bone prior to chilling reduces the amount of chilling space by 30 to 35% per beef carcass.

REFERENCES

Acuff, G.R. 1991. Acid decontamination of beef carcasses for increased shelf life and microbiological safety. *Proceedings of the Meat Industry Resources Conference*, Chicago.

Allen, D.M., M.C. Hunt, A.L. Filho, R.J. Danler, and S.J. Goll. 1987. Effects of spray chilling and carcass spacing on beef carcass cooler shrink and grade factors. *Journal of Animal Science* 64:165.

ASHRAE. 1994. Safety code for mechanical refrigeration. ANSI/ASHRAE Standard 15-1994.

Dickson, J.S. 1991. Control of *Salmonella typhimurium, Listeria monocytogenes*, and *Escherichia coli* O157:H7 on beef in a model spray chilling system. *Journal of Food Science* 56:191.

Earle, R.L. Physical aspects of the freezing of cartoned meat. *Bulletin* 2, Meat Industry Research Institute of New Zealand.

Greer, G.G. and B.D. Dilts. 1988. Bacteriology and retail case life of spray-chilled pork. *Canadian Institute of Food Science Technology Journal* 21:295.

Hamby, P.L., J.W. Savell, G.R. Acuff, C. Vanderzant, and H.R. Cross. 1987. Spray-chilling and carcass decontamination systems using lactic and acetic acid. *Meat Science* 21:1.

Heitter, E.F. 1975. Chlor-chill. *Proceedings of the Meat Industry Resources Conference AMIF*, Arlington, VA. 31-32.

Johnson, R.D., M.C. Hunt, D.M. Allen, C.L. Kastner, R.J. Danler, and C.C. Schrock. 1988. Moisture uptake during washing and spray chilling of Holstein and beef-type carcasses. *Journal of Animal Science* 66:2180.

Jones, S.M. and W.M. Robertson. 1989. The effects of spray-chilling carcasses on the shrinkage and quality of beef. *Meat Science* 24:177-88.

Kastner, C.L. 1981. Livestock and meat: Carcasses, primal and subprimals. In CRC *Handbook of Transportation and Marketing in Agriculture*, edited by E.E. Finney, Jr. CRC Press, Inc., Boca Raton, FL. 239-58.

Lentz, C.P. 1971. Effect of light and temperature on color and flavor of prepackaged frozen beef. *Canadian Institute of Food Technology Journal* 4:166.

Marriott, N.G. 1994. *Principles of food sanitation*, 3rd edition. Chapman and Hall, New York.

CHAPTER 17

POULTRY PRODUCTS

Poultry Processing 17.1	Poultry Plant Sanitation 17.7
Chilling 17.1	Tenderness Control 17.8
Decontamination of Broiler Carcasses 17.3	Distribution and Retail Holding Refrigeration 17.8
Further Processing of Poultry 17.3	Preserving Quality in Storage and Marketing 17.8
Freezing 17.4	Thawing and Use 17.9
Packaging 17.7	

POULTRY and broilers in particular are the most widely grown farm animal on earth. Two major challenges currently face the poultry industry: (1) keeping food safe from human pathogens carried by poultry in small numbers that could multiply sometimes to dangerous levels during processing, handling, and meal preparation; and (2) developing environmentally sound, economical waste management facilities. Innovative engineering and refrigeration are a part of the solutions for these issues.

POULTRY PROCESSING

Poultry processing is composed of three major segments:

- **Dressing**, where the birds are placed on moving line, killed, and defeathered.
- **Eviscerating**, where the viscera are removed, the carcass is chilled, and the birds are inspected and graded.
- **Further processing**, where the largest portion of the carcasses are cut up, deboned, and processed into various products. The products are packaged and stored chilled or frozen.

A schematic poultry processing flowsheet is described in Figure 1. Today's highly automated poultry processing plant processes 1 to 3 million birds per week. Mountney (1976) shows a schematic of a plant that was standard in the United States in the 1970s and now is in operation around the globe. It processes 120,000 birds per week (1500 birds per hour, 2 shifts, 5 days). Brant et al. (1982), USDA/UC Davis (1975) (turkeys), and USDA/ Univ. of Georgia (1970) (broilers) provide guidelines for design and operation of poultry processing plants.

CHILLING

Poultry products in the United States may be chilled to −3.5°C or frozen to lower than −3.5°C. Means of refrigeration include ice, mechanically cooled water or air, dry ice (carbon dioxide sprays), and liquid nitrogen sprays. Continuous chilling and freezing systems, with various means for conveying the product, are common. According to USDA regulations (1990), poultry carcasses with a mass of less than 1.8 kg should be chilled to 4.5°C or below in less than 4 h, carcasses of 1.8 to 3.6 kg in less than 6 h, and carcasses of more than 3.6 kg in less than 8 h.

Slow air chilling was considered adequate for semiscalded, uneviscerated poultry in the past. But with the transformation to eviscerated, ready-to-cook poultry, sometimes subscalded, air chilling was replaced by chilling in tanks of slush ice. Immersion chilling is more rapid than air chilling, it prevents dehydration, and it affects a net absorption of water of from 4% up to 12%. Objections to this mass gain from external water, a concern that water chillers can be recontamination points, and the high cost of disposing of the

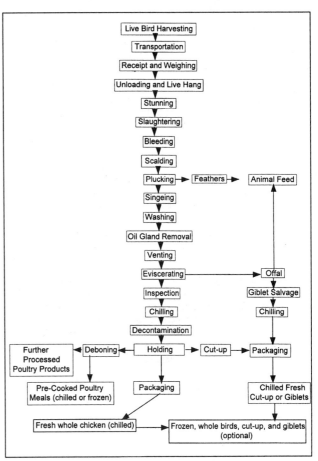

Fig. 1 Processing Sequence of Fresh Poultry

waste water in an environmentally sound manner have encouraged some operators to consider a return to air chillers.

Continuous immersion slush ice chillers, which are fed automatically from the end of the evisceration conveyor line, have replaced slush ice tank chilling, a batch process. In general, tanks are only used to hold chilled carcasses in an iced condition prior to cutting up, or to age prior to freezing.

The following types of continuous chillers are used:

- **Continuous drag chillers.** Suspended carcasses are pulled through troughs containing agitated cool water and ice slush.
- **Slush ice chillers.** Carcasses are pushed by a continuous series of power-driven rakes.
- **Concurrent tumble systems.** Free-floating carcasses pass through horizontally rotating drums suspended in tanks of successively cool water and ice slush. The movement of the carcasses is regulated by the flow rate of recirculated water in each tank.

The preparation of this chapter is assigned to TC 10.9, Refrigeration Application for Foods and Beverages.

- **Counterflow tumble chillers.** Carcasses are carried through tanks of cool water and ice slush by horizontally rotating drums with helical flights on the inner surface of the drums
- **Rocker vat systems.** Carcasses are conveyed by the recirculating water flow and agitation is accomplished by an oscillating, longitudinally oriented paddle. Carcasses are removed automatically from the tanks by continuous elevators.

These chillers are capable of reducing the internal temperature of broilers from 32 to 4.5°C in 20 to 40 min, at processing speeds of 5000 to 10 000 birds/h. USDA regulations require a minimum overflow of 1.9 L of water per broiler from continuous immersion chillers and 3.8 L per turkey. Chilling water must not be higher than 18°C in the warmest part of the chilling system.

Adjuncts and replacements for continuous immersion chilling should be used, if available, because immersion chilling is believed to be a major cause of bacterial contamination. Water spray chilling, air blast chilling, carbon dioxide snow, or liquid nitrogen spray are alternatives, but with the following limitations:

- Liquid water has a much higher heat transfer coefficient than any gas at the same temperature of cooling medium, so water immersion chilling is more rapid and efficient than gas chilling.
- Water spray chilling, without recirculation, requires much greater amounts of water than immersion chilling.
- Product appearance should be equivalent for water immersion or spray chilling, but inferior for air blast, carbon dioxide, or nitrogen chilling, due to surface dehydration.
- Gas chilling without packaging could cause a 1 to 2% loss of moisture, while water immersion chilling permits from 4 to 15% moisture uptake, and water spray chilling up to 4% moisture uptake. (Carpenter et al. (1979) reported on the effect of salt brine chilling on dripless ice-packed broiler carcasses.)

The temperature of the coolant and the degree of contact between the coolant and the product are most important in transferring heat from the poultry carcass surface to the cooling water. The heat transfer coefficient between the carcass and the water can be as high as 2000 W/(m²·K). Mechanical agitation, injection of air, or both can improve the heat transfer rate (VeerKamp 1995). VeerKamp and Hofmans (1974) expressed heat removed from poultry carcasses by the following empirical relationship.

$$\frac{\Delta Q}{\Delta Q_i} = (-0.009 \log h + 0.73) \log \theta$$
$$- (0.194 \log h - 0.187) \log m + 0.564 \log h - 2.219 \quad (1)$$

where
h = apparent heat transfer coefficient, W/(m²·K)
m = mass of the carcass, kg
θ = cooling time, s
ΔQ_i = maximum heat removal, J

Figure 2 shows time-temperature curves in a commercial counterflow chiller and compares calculated and measured values.

With adequately washed carcasses and adequate chiller overflow in a direction counterflow to the carcasses, the bacterial count on carcasses should be reduced by continuous water immersion chilling. However, incidence of a particular low level contaminant, such as *Salmonella*, may increase during continuous water immersion chilling. Such transfer of microbes can be controlled by chlorinating the chill water.

Spray chilling without recirculation has reduced bacterial surface counts 85 to 90% (Peric et al. 1971). Microbe transfer by spray chilling is unlikely. Chilling with air, carbon dioxide, or nitrogen presents no obvious microbiological hazards, although good sanitary practices are essential. If the surface of the carcass freezes as a part of the chilling process, bacterial load may be reduced as much as 90%.

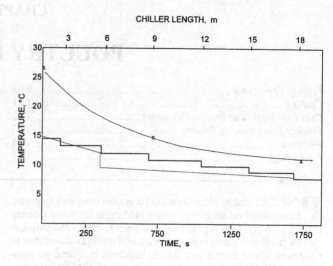

x product temperature
— — water temperature in relation to length of chiller
——— theoretical water temperatures
(assuming six perfectly mixed tanks in cascade)

Fig. 2 Broiler and Coolant Temperatures in a Countercurrent Immersion Chiller

Air or gas chilling is commonly used in Europe. In air-blast chilling and evaporative chilling, heat is conducted partly by the air-to-carcass contact and partly by evaporation of moisture from the carcass surface. The amount of water removed by evaporation depends on the carcass temperature, but even at −10°C it is about 1%. The apparent heat transfer coefficient ranges from 50 to 200 W/(m²·K). Major disadvantages of air chilling are slow cooling, dripping from one bird to another in multitiered chillers, and loss of mass during chilling.

Cryogenic gases are generally used in long insulated tunnels through which the product is conveyed on an endless belt. Some freezing of the outer layer (crust freezing) usually occurs, and the temperature is allowed to equilibrate to the final, intended chill temperature. Some plants use a combination of continuous water immersion chilling to reach 2 to 5°C and a cryogenic gas tunnel to reach −2°C. The water-chilled poultry, either whole or cut up, is generally packaged before gas chilling to prevent dehydration.

Water absorption by the carcass in water immersion or spray chilling, and possible loss of water during gas chilling, are important economic and quality factors. USDA regulations set maximum levels of water absorption and retention varying with mass and type of product and specify testing procedures and bases for rejection or retention of product. The set values range from 4.3% mass increase for 12 kg and over consumer-packaged whole turkeys to 12% for ice-packed poultry at the end of the drip line. Of the water absorbed, very little appears in the muscle; most is held in the connective tissue in the skin and between skin and flesh. For example, increases in percentage of moisture due to water immersion chilling were reported from 0.5 to 2% for muscle and 10 to 12% for skin.

Kotula et al. (1960) compared moisture uptake and drainage loss for air-agitated slush ice tank chilling and for four mechanical continuous immersion slush ice chillers. Immediately after chilling, the total moisture uptake including 2 to 3% by prechill washing ranged from 7% for 2 to 4 h of tank chilling to 17% for a counterflow tumble continuous immersion water-ice chiller. After draining, shipping, and a final 30-min drain, all uptakes ranged from 5 to 7%. Initial uptakes were much greater in a counterflow tumble system for carcasses with thigh skin area cut compared to uncut, i.e., 18 versus 12%. After shipping and draining, the residual uptakes were 7 and 5%.

Poultry Products

Although soluble solids are lost from the carcass to the chill water in immersion chilling, no evidence exists that, under accepted chilling practices, any appreciable loss of flavor or other desirable qualities occurs.

A major disadvantage of water chilling is extensive dripping in storage. Frozen carcasses, in particular, drip significantly during thawing. Also, water chilled poultry looses more moisture during cooking and has poorer eating quality than evaporatively cooled poultry. The European Union requires that water uptake in broilers not to exceed 6% and dripping loss not to exceed 5.2%. In the United States, USDA allows up to 8% uptake of water.

Ice requirements per bird for continuous immersion chilling depend on entering carcass temperatures and mass, entering water temperature, and exit water and carcass temperature. For a counterflow system, 15°C entering water and 18°C exit water, 0.25 kg of ice per kilogram of carcass is a reasonable estimate. This may be compared to a requirement of 0.5 to 1 kg of ice per kilogram of poultry for static ice slush chilling in tanks. For continuous counterflow water immersion chillers, if the plant water temperature is considerably above 18°C, it may be economical to use a heat exchanger between incoming plant water and exiting (overflow) chill water.

Ice production for chilling is usually a complete in-plant operation, with large piping and pumps to convey the small crystalline ice or ice slush to the point of use. To reduce the use of ice, some immersion chillers are double-walled and depend on circulating refrigerant to chill the water in the chiller. The chiller has an ammonia or refrigerant lubricant between the outer and inner jacket with the inner jacket serving as the heat transfer medium. Agitation or a defrost cycle must be provided during periods of slack production to prevent the chiller from freezing up.

Chilling and holding to about −2°C, the point of incipient freezing, gives the product a much longer shelf life compared with a product held at ice-pack temperatures (Stadelman 1970).

DECONTAMINATION OF BROILER CARCASSES

The contamination of poultry meat by foodborne pathogens during processing can be potentially dangerous if microbes are allowed to multiply and reach critical numbers and/or produce poisonous toxins (Zeidler 1996, 1997). The Hazard Analysis of Critical Control Points (HACCP) system (described in Chapter 11) was specifically developed for each food to eliminate or keep pathogen levels very low so food-related illnesses cannot break out. Appropriate refrigeration and strict temperature control throughout the food channel is vital to suppress microbial growth in high moisture perishable foods and meats in particular.

Decontamination steps are now being added at the end of the processing line just before packaging and/or after packaging. Numerous methods have developed and some are implemented (Bolder 1997 and Mulder 1995), among them **lactic acid** (1%), **hydrogen peroxide** (0.5%), and **trisodium phosphate** (TSP) sprays. **Ozone** (O_3) is a strong oxidizer and can be used to decontaminate chiller and scalding water; however, it is very corrosive.

Gamma irradiation of poultry is approved in many countries, including the United States; and products are available for sale in few outlets. The public's fear of this technique limit sales. However, the threat of food poisoning is reducing objections to irradiated foods because irradiation is a very effective method that can kill 95.5% of non-spore forming pathogens (Stone 1995). A dose of 25 kGy is the most suitable for poultry.

Steam under vacuum effectively kills 99% of the surface bacteria on beef and pork carcasses and has started to be used commercially. In this continuous system, the carcass is carried on a rail to a chamber. A vacuum is pulled and steam at 143°C is applied for 25 ms. Upon breaking of the vacuum, the carcass surface is cooled to prevent the surface from cooking. USDA engineers developed steam equipment for poultry in 1996.

FURTHER PROCESSING OF POULTRY

Most chickens and turkeys, for both chilled and frozen distribution, are cut up in the processing plant. More than 90% of the broilers in the United States are sold as cut-up products produced at the processing plant. The cutting procedure is almost fully automatic as the carcass is reduced to the various parts. Assembly of the parts into a final product is still largely manual as employees place the appropriate parts into a foam plastic tray.

A popular cut for broilers is a nine-piece cut, usually two drumsticks, two thighs, two wings, two breast halves, and back. However, there are at least eight different cutting patterns, involving greater subdivision of breast and leg portion. Backs and necks are often mechanically deboned, giving a comminuted slurry that is frozen in rectangular flat cartons containing about 27 kg. Turkey breasts, legs, and drumsticks are available as separate film-packaged parts, and turkey thigh meat is marketed as a ground product resembling hamburger. Partial cooking and breading and battering of broiler parts is done in poultry processing plants.

Unit Operations

The following types of equipment used for further processing of poultry products are also used in red meat facilities.

Size Reduction and Mixing Machines. Several types of size-reduction and mixing equipment are available.

- **Grinding.** Meat is conveyed by an auger and forced through a grinding plate.
- **Flaking** is done by cutting blades locked at a specific angle on a rotating drum. Flaking does not extensively break muscle cells as in grinding and moisture loss and dripping are limited. Product texture resembles muscle texture.
- **Chopping** is generally conducted in a silent cutter equipment. The meat is placed in a rotating bowl with ice, which is used to keep the temperature low. Vertical rotating blades keep chopping the moving meat. The length of the chopping time determines the particle size. The end product is used in hot dogs and sausages.
- **Mixing and tumbling and injecting machines** produce a uniform product out of the various meats and non-meat ingredients such as salt, sugar, dairy or egg proteins, spices, and flavorings. Together with salt, mixing also helps extract myosin, which acts likes a glue in holding the product together.
- **Injection machines** are used to insert an accurate and repeatable volume of liquid that contain salt and flavorings to large chunks of muscle meats such as turkey breasts or whole turkey carcasses. The procedure disperses these ingredients better and faster than soaking in brine and marinade. It also protects the meat from drying during cooking, especially at home.
- **Tumblers** shaped like concrete mixers tumble injected large meat chunks mostly under vacuum. The tumbling helps distribute injected brine and spices throughout the meat.

Shaping Forms and Dimension. These machines establish the form, size, and desired mass of the size-reduced poultry meats.

- **Stuffing machines** are used to make hot dogs and sausages by stuffing meat emulsion into the casing. Modern stuffing machines operate under vacuum to eliminate bubbles and other textural defects. In recent years, dough products or muscle meats are stuffed with other meats, fruit or vegetable pieces, etc. These products are produced by equipment that was originally designed to stuff doughnuts with jelly.
- **Forming machines** are used to make hamburgers and nuggets. They are basically presses that force the meat through a plate with holes of various sizes and shapes.
- **Metal molds.** Many products such as turkey rolls and luncheon meats are made from many meat chunks which are placed into metal molds and cooked to produce a restructured log. The meat is chilled in the molds before being released.

- **Coating.** Batter and breading give the product a uniform shape as well as higher palatability and mass. The coating is conducted by carrying the product on belts through ingredients that coat the products. The products are fried immediately after.

  Cooking Techniques. Many meat products are produced as ready-to-eat meals that need warming only or are eaten cold. These products are fully cooked in the plant by various methods. Other products are produced as ready-to-cook and skip the cooking step.

- **Smoking/cooking** is the most popular method of meat cooking. Here, smoke from slow-burning wood outside the cooking chamber flows over the hanging product. To eliminate some smoke carcinogenic compounds and to accelerate the process, liquid smoke is used to treat the product before cooking (Lazar 1997). Smoking is done best on a dry, uncooked surface, which better absorbs the smoke ingredients. Smokehouses are generally the bottleneck of the process and their high capital cost and large size limits the number of units in the plant. Every product is cooked to a specific internal temperature, commonly ranged between 63 and 80°C, followed by immediate chilling by water showers from sprinklers located in the cooking chamber.
- **Continuous hot air ovens** are used to cook hamburgers and chicken breast products. These ovens accelerate cooking and reduce labor compared to batch-type equipment. Wireless, solid-state temperature monitoring devices that travel with the product optimize the cooking process and provide cooking records.
- **Cooking in water bath** is a fast and low-cost way to cook meats due to better heat transfer than in air cooking. The product is protected from the water by waterproof plastic packaging. Most operations are batch-type.
- **Frying** provides higher palatability at the cost of increasing fat content. Frying provides crispiness as the hot oil above 100°C replaces the water in the skin, batter, and breading. Frying is a fast method of cooking as the heat transfer of hot oil is higher than that of water
- **Microwave heating** is used in poultry plants to thaw incoming frozen products rapidly. Cooking by microwave is limited despite the advantage of short cooking times due to the inability of the microwave oven to crisp and brown meats.
- **Rotisserie** is a fast growing cooking technique because it does not add fat and provides superior taste. Its main disadvantages are moisture loss, slow cooking, and flavor deterioration a short time after cooking.

FREEZING

Effect on Product Quality

Rapid freezing is essential to obtain satisfactorily light appearance in certain types of frozen poultry (van den Berg and Lentz 1958). Birds scalded at 60°C and above completely lose their outer layer of skin during normal machine picking and become particularly susceptible to the development of a dark frozen appearance if not frozen rapidly. Also, immature fryer-roaster turkeys, which have a thin, practically fat-free skin, are naturally dark and require a rapid rate of freezing to assure a light, pleasing frozen appearance.

Air blast tunnels operating at air temperatures ranging from −30 to −40°C and at air velocities of 2.5 m/s or more provide rapid freezing. In an evaluation of various factors contributing to freezing rate and frozen appearance, Klose and Pool (1956) compared freezing on open shelves versus freezing in boxes and found that lower air blast temperature and higher air blast velocity were important, in that order. At an air blast temperature of −30°C or below, increasing air blast velocity beyond 3 m/s had little beneficial effect. Also at 6.5 m/s, decreasing air-blast temperature from −30 to −35°C produced almost no additional improvements in frozen appearance. Birds placed in the blast freezer immediately after evisceration and packaging and while still warm did not develop an appreciably darker frozen appearance than those chilled in ice slush before packaging and blast freezing. However, this procedure has the serious disadvantage that the duration of holding above freezing temperatures may be reduced to an amount inadequate for optimum tenderization. This concern is eliminated using the MTPS described in the Tenderness Control section. The factor of finish, or amount of fat in the skin layer, can exert a much greater beneficial effect on frozen appearance than any possible changes in processing or freezing practices.

USDA regulations now define frozen poultry as cooled to −3.5°C or lower. This rule prevents the practice of cooling meat to above −20°C, thawing it in destination, and selling it as fresh. Poultry that is frozen to less than −20°C is now called deep frozen.

The freezing rate of diced cooked chicken meat does affect the quality of the frozen meat. Hamre and Stadelman (1967a) reported that cryogenic freezing procedures were desirable as the resulting color was lighter, but too rapid a freezing rate resulted in a shattering of the meat cubes. The freeze-drying rates for rapidly frozen material were slower than for products frozen by slower methods. Hamre and Stadelman (1967b) indicated that tenderness of freeze-dried chicken after rehydration was affected by freezing rate prior to drying. Liquid nitrogen spray or carbon dioxide snow freezing were selected as preferred methods for overall quality of diced cooked chicken meat to be freeze-dried.

Freezing Methods

Poultry may be frozen between refrigerated double-walled plates, in a blast of refrigerated air, by immersion in a refrigerated liquid, or by a shower of liquefied gas such as nitrogen. IQF (individual quick frozen) freezing with CO_2 is also used, particularly in preparing for freeze drying. Rectangularly packaged, cut-up poultry can be frozen in multiplate freezers within several hours. Chapter 15 describes commercial freezing equipment.

Air Blast Tunnel Freezers. Most whole, ready-to-cook birds are frozen in air blast tunnel freezers, with air temperatures ranging from −25 to −40°C and air velocities of 1.5 to 5 m/s and up. It is desirable to have air temperatures at −35°C or below during operation and air velocities over the product surfaces of at least 3 m/s. To obtain high air velocity over the product, the blast tunnel should be completely loaded across its cross section, with proper spacing of the units of the product to assure airflow around all sides and no large openings to permit bypassing of the airstream. In some cases, the whole bird may be packed into cartons or boxes, and the cartons stacked on pallets in the blast tunnel, with spacing between layers and between cartons in the same layer to permit adequate airflow and freezing rate.

However, as mentioned, freezing rates required to give a sufficiently light frozen appearance are so high for birds scalded at elevated temperatures and immature turkeys that modified freezing methods are necessary. Boxes may be left open during freezing, or they may be constructed with holes or cutouts on the ends, or, best of all, the individual, bagged birds may be frozen on open racks or shelves to provide maximum airflow and freezing rate. If freezing rates in excess of those possible in air blast tunnels are required, immersion freezing may be considered.

Freezing by Direct Contact. Low-temperature brine or other liquids such as glycols are frequently used to freeze poultry by direct contact. Esselen et al. (1954) found that the freezing time per 0.5 kg of packaged ready-to-cook birds immersed in −29°C brine was in the range of 20 to 30 min. In −29°C calcium chloride brine, times for the interior to reach −9°C for warm eviscerated, packaged broilers, 5.5 kg turkeys, and 11.4 kg turkeys were approximately 1.5, 5, and 7 h, respectively. Lentz and van den Berg (1957) have evaluated factors affecting freezing time and frozen appearance of poultry. Immersion freezing at −29°C produced a lightness of appearance that could be matched by air blast freezing only at a temperature of −75°C. Combinations of immersion freezing and air blast freezing were tested. A minimum immersion time at −29°C of 20 min for chickens and 40 min for turkeys was necessary to obtain a white appearance, typical of immersion freezing, in birds subsequently frozen on open shelves in a −29°C, 1.5 to 2.5 m/s air blast. The light-colored chalk-like appearance of immersion frozen poultry was maintained indefinitely in −29°C storage, but darkened after 6 weeks at −18°C or 2 weeks at −6.7°C.

Comparison of various liquids (methanol, salt brines, and glycols) that might be used in immersion freezing revealed that freezing times depended mainly on viscosity of the liquids and were maximum in glycerol and minimum in calcium chloride brine. Increased agitation decreased freezing times in the surface layers by as much as 50%. Spraying the liquid was as effective in reducing freezing times as the highest rate of agitation used. Advantages of immersion freezing are lighter, more pleasing appearance; high rate of heat transfer that makes a line operation feasible; lower initial and upkeep costs than air blast; and possible use in combination with subsequent air freezing. Possible disadvantages include lack of uniformity of frozen appearance, development of leaks due to pinpoint holes or cracks in packaging film, the question of acceptance of the chalk-white birds, and lack of versatility for freezing all types of material.

Typical freezing rate curves are shown in Figures 3 to 7, which illustrate the effect of bird size and depth below skin surface, temperature and velocity of the cooling medium, and type of medium (air versus liquid). Pflug (1957) reviews the relative efficiencies and requirements for air and liquid immersion freezing and the critical importance of the conductance of the coolant-product interfacial film, citing a film heat transfer coefficient of 17 W/(m^2·K) for air at −30°C and 2.5 m/s, compared to 170 W/(m^2·K) for sodium chloride brine at −20°C and a velocity of 0.025 m/s. Tables 1 and 2 list physical properties of poultry related to refrigeration.

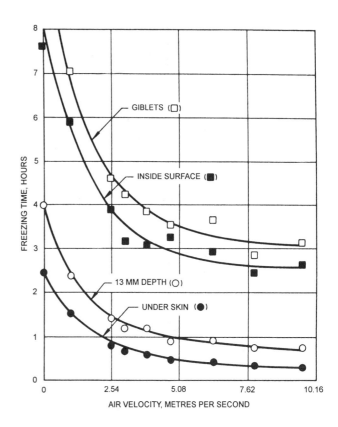

Fig. 4 Relation Between Freezing Time and Air Velocity
(van den Berg and Lentz 1958)

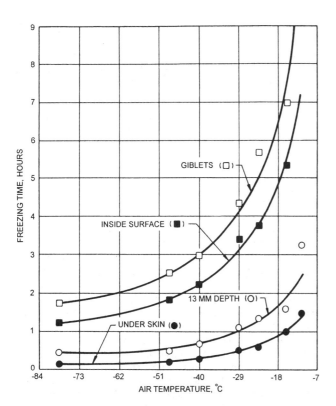

Fig. 3 Relation Between Poultry Freezing Time and Air Temperature
(van den Berg and Lentz 1958)

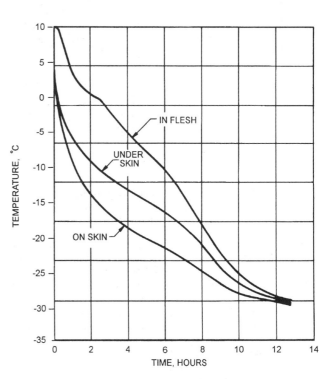

Fig. 5 Temperature During Freezing of Packaged, Ready-to-Cook Turkeys
(Klose and Pool 1956)

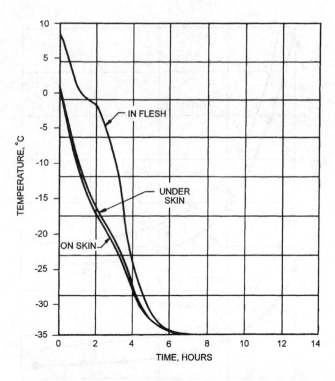

Fig. 6 Temperature During Freezing of Packaged, Ready-to-Cook Turkeys

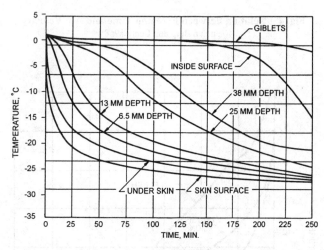

Fig. 7 Temperatures at Various Depths in Breast of 6.8 kg Turkeys During Immersion Freezing at −30°C
(Lentz and van den Berg 1957)

Liquid Nitrogen. A liquid nitrogen shower system has been used to freeze the outer shell of products to −75°C or below, after which the entire product may be equilibrated in a holding room to a temperature around −25°C. Temperature in the freezer box where the liquid nitrogen is released in a shower may be as low as −160°C.

The freezing rate of precooked chicken affects the quality of product. Breaded precooked drumsticks frozen with liquid nitrogen are susceptible to cracking and separation of the meat from the bone. Precooked chicken that is lightly breaded (7 to 10% breading) and frozen by cryogenic procedures is susceptible to developing small white freezer burn areas next to the surface. These areas may or may not show up immediately after freezing, but will show up almost immediately after reconstituting with hot oil. In some instances, the white areas will cover the entire surface of the piece

Table 1 Thermal Properties of Ready-to-Cook Poultry

Property	Value	Reference
Specific heat, above freezing	2.94 kJ/(kg·K)	Pflug (1957)
Specific heat, below freezing	1.55 kJ/(kg·K)	Pflug (1957)
Latent heat of fusion	247 kJ/kg	Pflug (1957)
Freezing point	−2.8°C	Pflug (1957)
Average density		
Poultry muscle	1070 kg/m^3	
Poultry skin	1030 kg/m^3	
Thermal conductivity, kJ/(m·K)		
Broiler breast muscle =	0.42 at 27°C	Walters and May (1963)
Broiler breast muscle ⊥	0.50 at 20°C	Sweat et al. (1973)
Broiler breast muscle ⊥	1.380 at −20°C	Sweat et al. (1973)
Broiler breast muscle ⊥	1.51 at −40°C	Sweat et al. (1973)
Broiler dark muscle ⊥	1.56 at −40°C	Sweat et al. (1973)
Turkey breast muscle ⊥	1.26 at −20°C	Sweat et al. (1973)
Turkey breast muscle =	1.61 at −20°C	Sweat et al. (1973)
Turkey leg muscle ⊥	1.44 at −20°C	Lentz (1961)

⊥ indicates heat flow perpendicular to the muscle fibers.
= indicates heat flow parallel to the muscle fibers.

Table 2 Thermal Conductivity of Broiler Carcasses With and Without Cardboard Box Packaging

	Thermal Conductivity, W/(m·K)	
Temperature, °C	Unpackaged Carcass	Packaged Carcass
−30	1.89	0.93
−20	1.78	0.88
−10	1.64	0.81
−5	1.49	0.73
0	0.47	0.26
5	0.47	0.26
10	0.48	0.26

From Veerkamp (1995)

of chicken. Altering the fat content of thighs does not reduce the severity of this problem.

Predicting Freezing or Thawing Times. The following equation can be used to predict freezing and thawing time with an accuracy of about 10% (Calvelo 1981; and Cleland et al. 1982, 1984).

$$\theta_f = \rho \frac{\Delta H}{\Delta t}\left(\frac{d}{6h} + \frac{d^2}{24k}\right) \quad (2)$$

where
- θ_f = freezing time, s
- ρ = product density, kg/m^3
- d = equivalent diameter of product, m
- ΔH = enthalpy difference, kJ/kg
- Δt = temperature difference between air and mean freezing temperature, K
- h = heat transfer coefficient, W/(m^2·K)
- k = thermal conductivity at mean freezing temperature, W/(m^2·K)

Individual Quick Frozen (IQF) Products. This method of freezing creates a crust on the bottom of the product, which moves on thin, disposable plastic sheets. IQF works well for marinated bones, chicken breast, and chicken tenders because they are moist and softer than other parts and tend to stick to the freezer belts. The plastic sheet keeps the product from sticking.

Freezer Conveyors. Automated units may be designed to handle packages, cartons of birds, or unwrapped pieces of chicken or turkey. Product may be transported through the freezing chamber on belts or trays. One such system is adaptable to all sizes of whole birds, packages, or cartons. The system automates the freezing operation from the point birds are placed in cartons until they are frozen and ready for a carton top to be put on. A typical design handles nearly 1 bird per second with about 70 Mg total capacity.

Refrigeration coils and fans are located at the side of the machine to give a high-velocity two-pass airflow that applies the coldest air to the warmest product. Frost or ice buildup is minimized because the shelves never come outside the freezer.

A tray system is available that automatically loads trays from a moving belt, conveys them through the air blast freezing chamber, empties the trays into a holding bin in the freezing chamber, and conveys the product onto a belt, which carries it from the freezer to a packaging station. This system is particularly useful for cut-up chicken parts. Belt systems, using refrigerated air blast or cryogenic gases, are used for small parts, packages, and particle size poultry such as precooked diced chicken.

PACKAGING

Most packaged poultry is now tray packed, either for frozen or chilled distribution. All-plastic packages and automated packaging lines using plastic film have been engineered. Changes in packaging methods and materials are so rapid that the best sources of information on this subject are the companies that fabricate films and packages and distribute the materials. They are listed in the most recent Encyclopedia Issue of *Modern Packaging*.

Packages for frozen, whole, and ready-to-cook poultry consist principally of plastic film bags that are tough and reasonably impermeable to moisture vapor and air. The commonly used polyvinylidene chloride, polyethylene, and polyester films are sufficient barriers to water vapor and air to give adequate protection for normal commercial times and temperatures. Turkeys, ducks, and geese are packaged mostly in the whole, ready-to-cook form, while frozen chickens appear whole and in packaged, cut-up form.

Large fiberboard cartons or containers for holding and shipping from 2 to 12 individually packaged birds should be rectangular in shape to facilitate palletizing and should be strong enough to support 5 m high stacked loads common in refrigerated warehouses. If freezing is to be accomplished for material such as fryer turkeys that need to be frozen rapidly, holes or cutaway sections in the sides and ends are needed to permit a rapid airflow across the poultry surfaces in the air blast freezer.

POULTRY PLANT SANITATION

Poultry meat is highly perishable as it composed of nutrients that are ideal for microbial growth. During processing, excessive amounts of meat and drippings soil equipment and floors. If not spotlessly cleaned and sanitized, it becomes a source of bacterial growth that can recontaminate incoming new meats. Therefore, specific cleaning teams clean the plant at the end of the working day using steam, soap and sanitizing agents. In many instances work is stopped and certain equipment is cleaned every few hours.

In January 1997, the rules for meat inspection changed dramatically (USDA 1996). The processing plants are required to (1) inspect their own processes by writing their own Sanitation Standard Operation Procedures (SSOP) and implementing them into their operations, (2) monitor the processes, and (3) take corrective action when necessary. Precise records should be kept in a format ready for instant review by purchasers.

Proper sanitation should be addressed when the structure, processing equipment, and refrigeration systems are designed. The plant structure should be designed to prevent pests such as mice, rats, cockroaches, and birds from entering the facility and finding places to hide that cannot be reached. This includes drainage, sewage, windows, vents, etc. Equipment should be designed for easy cleaning and easy assembly and disassembly. It should not have any areas on which product particles can accumulate. Refrigeration systems should be designed to restrict airflow from raw to cooked meat areas and to eliminate possible condensation and dripping into the product or into drip pans that cannot be reached for easy cleaning.

Clearly written procedures, constant training of employees, and adequate numbers of employees are essential for successful implementation of the program. Also, constant management commitment is vital.

HACCP Systems in Poultry Processing

HACCP is a logical process of preventative measures that can control food safety problems. HACCP is a process control system designed to identify and prevent and microbial and other hazards in food production. It is designed to prevent problems before they occur and to correct deviations as soon as they are detected. This method of control emphasizes a preventative approach rather than a reactive approach, which can reduce the dependence on final product testing. The fundamentals of HACCP is described in Chapter 11.

HACCP systems are being designed and implemented in the poultry meat processing industry to improve the safety of fresh meats and their products. HACCP programs are required by the USDA in all large plants (establishments with 500 or more employees) as of January 1998. Medium and small plants will be required to comply over the next two years.

Poultry is associated with numerous microbial pathogens that occur naturally in wild birds, rats, mice, and cockroaches. Poultry is contaminated by feed containing feces of these pests. They are also transferred to the meat during processing from unclean equipment, processing water, air, and from human hands, hair or clothing. Strict temperature control throughout the system will strongly suppress microbial growth, keeping pathogen levels too low to generate food poisoning outbreaks. **In most outbreaks, temperature control breakdown or temperature abuse is involved** (Zeidler 1996).

The major pathogens associated with poultry are various types of *Salmonella* and *Campylobacter jejuni*, which recently became the leading pathogen in poultry meat. HACCP programs cover production farms, processing plant, and shipping trucks. In the processing plant, critical control points (CCP) are placed in the receiving and killing, scalding and defeathering, evisceration, inspection, and chilling areas. Water baths as in chilling and scalding could easily spread pathogens and the circulating water must be treated. The aerosol, places where condensation may accumulate, back-up of sewage, and used processing water are also CCPs. Reducing human touch, bird-to-bird contact, and dripping from bird to bird during air chilling, as well as increased automation help reduce contamination. Appropriate temperature control throughout the system is vital as food-borne disease outbreaks always involve temperature abuse. Therefore all measured temperatures are CCPs.

An example HACCP program approach to receiving of chilled poultry meat is illustrated in Figure 8.

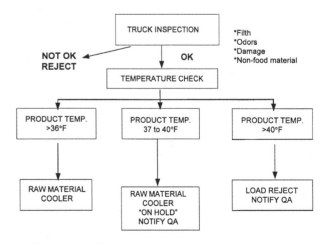

Fig. 8 HACCP Approach to Receiving Chilled Poultry
(Guelph Food Technology Center 1997)

TENDERNESS CONTROL

Tenderness in cooked poultry meat is a prerequisite to acceptability. Relative tenderness decreases as birds mature, and this toughness has always been considered in the recommendations for cooking birds of various ages. However, another type of toughness depends primarily on the length of time that the carcass is held in an unfrozen state before cooking. Birds cooked before they have time to pass through rigor are very tough. Normal tenderization after slaughter is arrested by freezing. For birds held at 5°C, complete tenderization occurs for all muscles within 24 h and for many muscles in a much shorter time.

Other factors that interfere with normal tenderization are immersion in 60°C water and cutting into the muscle. Formerly, birds were held unfrozen for sufficient time in the normal channels of processing and use to permit adequate tenderization. Shorter chilling periods, more rapid freezing, and cooking without a preliminary thawing period have shortened the period during which tenderization can occur to such an extent that toughness has become a potential consumer complaint.

Hanson et al. (1942) observed a rapid increase in tenderness within the first 3 h of holding and a gradual increase thereafter. Shannon et al. (1957), working with hand-picked stewing hens, demonstrated an increased toughness because of increased scalding temperature or increased scalding time in the ranges of 120 to 195°F and 5 to 160 s. However, the differences in toughness that occurred within the limits of temperature and time, necessary or practical in commercial plants, were quite small.

In addition to the beneficial effect of increasing holding time above freezing, tenderness is increased by lower scalding temperatures, by reducing the scalding time, and particularly by reducing the extent of beating received by the birds during picking operations. Additionally, turkey fryers should be held at least 12 h above freezing to develop optimum tenderness. Holding fryers at −18°C for 6 months and longer has no tenderizing effect, but holding in a thawed state (2°C) after frozen storage has as much tenderizing effect as an equal period of chilling before freezing. Turkeys frozen 1 h after slaughter are adequately tenderized by holding for 3 days at −2°C, a temperature at which the carcass is firm and no important quality loss occurs for the period involved. Behnke et al. (1973) confirmed this effect for Leghorn hens.

Overall processing efficiency is improved by cutting up the carcass directly from the end of the eviscerating line, packaging the parts, and then chilling the still-warm packaged product in a low-temperature air blast or cryogenic gas tunnel. Webb and Brunson (1972) reported that cutting the breast muscle and removing a wing at the shoulder joint before chilling significantly decreased tenderness of treated muscles, though cut carcasses were aged in ice slush before cooking. Klose et al. (1972) found that under commercial plant conditions, making an eight-piece hot-cut before chilling and aging significantly reduced tenderness of breast and thigh muscles, compared to cutting after chilling. Smith et al. (1966) indicated that too-rapid chilling of poultry might have a toughening effect, similar to cold shortening observed in red meats.

Electrical Stimulation. Electrical stimulation is used routinely to tenderize beef and pork, but not yet for poultry. Cladfelter and Webb (1987) describe successful electrical stimulation of broiler chickens shortly after slaughter in order to tenderize the meat. To meet the goals described by Ingling (1978), Webb et al. (1989) modified the system using electrical stunning in a minimum time process system (MTPS). An evaluation of breast muscle tenderness from the MTPS indicates that tenderness levels achieved in 40 min post mortem are equivalent to those achieved in about 6 h of normal processing. The MTPS not only shortens meat aging time, but it allows straight-through processing by eliminating the need for chillers. MTPS broilers can be processed, cut up, cooked, packaged, and frozen as a continuous line operation.

DISTRIBUTION AND RETAIL HOLDING REFRIGERATION

Chilled poultry handled under proper conditions is an excellent product. However, there are limitations in its marketability because of the relatively short shelf life caused by bacterial deterioration. Bacterial growth on poultry flesh, as on other meats, has a high temperature coefficient. Studies based on total bacterial counts have shown that birds held at 2°C for 14 days are equivalent to those held at 10°C for 5 days or 24°C for 1 day. Spencer and Stadelman (1955) found that birds at −0.6°C had 8 days additional shelf life over those at 3.3°C.

The generation time of psychrophilic organisms isolated from chickens was 10 to 35 h at 0°C, depending on the species studied (Ingraham 1958). Raising the temperature to 2°C reduced generation time to 8 to 14 h, again depending on the species.

Frequent cleaning of processing equipment, as well as thorough washing of the eviscerated carcasses, is essential. Goresline et al. (1951) reported a substantial decrease in bacterial contamination and an increase in shelf life by the use of 20 ppm of chlorine in processing and chilling water. Water is routinely chlorinated in the United States, but chlorine is not allowed to touch poultry meat in some European countries.

Because shelf life is limited considerably by bacterial growth (slime formation) on the skin layer, it is reasonable to assume that drastic changes in the skin surface, such as the removal of the epidermal layer by high temperature scalding, might appreciably affect shelf life. Ziegler and Stadelman (1955) reported approximately 1 day more chilled shelf life for 53°C scalded birds than for 60°C scalded ones.

Chickens, principally broilers, are sold as whole, ready-to-cook or cut-up, ready-to-cook. Poultry is shipped in wax-coated corrugated containers. A number of precooked poultry meat products are being sold in wholesale and retail markets as refrigerated, nonfrozen products. Such items are usually vacuum packaged or packaged in either a carbon dioxide or nitrogen gas atmosphere. The desired temperature for such products is also −2 to −1°C.

PRESERVING QUALITY IN STORAGE AND MARKETING

Important qualities of frozen poultry include appearance, flavor, and tenderness. Optimum quality requires care in every phase of the marketing sequence, from the frozen storage warehouse, through transportation facilities, wholesaler, retailer, and finally to the frozen food case or refrigerator in the home.

Tissue Darkening. Darkening of the bones is a condition that occurs in immature chickens and has become more prevalent as broilers are marketed at younger and younger ages. During chilled storage or during freezing and defrosting, some of the pigment normally contained inside the bones of particularly young chickens leaches out and discolors adjacent tissues. This discoloration does not affect the palatability of the product. Brant and Stewart (1950) found that development of dark bones was greatly reduced by a combination of freezing and storage at −35°C and immediate cooking after rapid thawing. Aside from this combination, freezing rate, temperature and length of storage, and temperature fluctuations during storage were not found to have a significant effect.

Further research suggested that freezing and thawing not only liberated hemoglobin from the bone marrow cells but modified the bone structure to permit penetration by the released pigment. Roasting pieces of chicken 0.5 h prior to freezing reduced discoloration of the bone. Ellis and Woodroof (1959) found that heating legs and thighs to 82°C before freezing effectively controlled meat darkening. Methods of preheating, in order of preference, include microwave oven, steam, radiant heat oven, and deep fat frying.

Dehydration. During storage, poultry may become dehydrated, causing a condition known as **freezer burn**. Dehydration can be controlled by humidification, lowering of storage temperatures, or by packaging the product adequately. Aside from adversely affecting the appearance of the product, dehydration, unless severe, does not impair quality. When freezer burn is extensive, quality is decreased because of toughening and development of oxidative rancidity of the affected area. If a storage temperature of −20°C or lower is maintained, freezer burn is usually the factor limiting the length of time that poultry can be held in storage without adequate packaging.

Willis et al. (1948) found that the appearance of poultry suffered greatly when stored at −7°C. The most serious defects were microbiological changes, desiccation, and development of a stale, rancid or storage odor. Serious changes in flavor and juiciness occurred in poultry that had been frozen 3 to 9 months at −12°C.

Storage Temperature Variations. Klose et al. (1955) evaluated moisture losses, chemical changes, and palatability after 6-, 12-, and 18-month storage times. Under average commercial conditions of frozen storage (moisture impermeable package and temperature range of −23 to −12°C), the only factor for which a periodic temperature fluctuation is inferior to the mean temperature is the accumulation of frost in the package. Frost formation, which influences appearance but not eating quality, increased with storage temperature, and for the −23 to −12°C fluctuation was considerably greater than for the highest (−12°C) constant temperature. Results after 12-month storage indicated a definite superiority to −23°C storage over −12 and −18°C but no detectable taste superiority of −35 to −23°C.

Rancidity. Poultry fat becomes rancid during very long storage periods or at extremely high storage temperatures. Rancidity in frozen, eviscerated whole poultry stored for 12 months is not a serious problem if the bird is packaged in essentially impermeable film and held at −20°C or below. Danger of rancidification is greatly increased when poultry is cut up before freezing and storage, because of the increased surface exposed to atmospheric oxygen.

Length of Storage. Klose et al. (1959) studied quality losses in frozen, packaged, and cut-up frying chickens over temperatures of −35 to −7°C and storage periods from 1 month to 2 years. All commercial-type samples examined were acceptable after a storage period at −18°C of at least 6 months, and some were stable for more than a year. In a comparison of a superior (moisture-vapor-proof) commercial package with a fair commercial package, increased adequacy of packaging resulted in as much extension in storage life as a decrease in storage temperature of about −7°C. The results indicate that no statement on storage life can have general value unless the packaging condition is accurately specified.

Frozen storage tests by Klose et al. (1960) on commercial packs of ready-to-cook ducklings and ready-to-cook geese established that these products have frozen storage lives similar to other commercial forms of poultry. Ducks and geese should be stored at −20°C or below to maintain their high quality for 8 to 12 months.

Incorporation of polyphosphates into poultry meat by adding it to the chilling water has been shown to increase shelf life in frozen or refrigerated storage and to control loss of moisture in refrigerated storage and during thawing and cooking.

Storage of Precooked Poultry. Studies on frozen fried chicken indicated that precooking produces a product much less stable than a raw product. Rancidity development is the limiting factor and it is detected in the meat slightly sooner than in the skin and fatty coating of the fried product. The marked beneficial effect of oxygen (air)-free packaging was demonstrated in tests in which detectable off-flavors were observed at −18°C in air-packed samples after 2 months, while nitrogen-packed samples developed no off-flavors for periods exceeding 12 months.

Cooling the precooked parts in ice water prior to breading was found to reduce the TBA (thiobarbituric acid) values of precooked parts (Webb and Goodwin 1970). In this study, no difference in rancidity was noted for chicken stored 6, 8, or 10 months. By removing the skin from precooked broilers, the TBA values were lower, but yield and tenderness were reduced. No difference was detected in the TBA values of the thighs frozen in liquid refrigerant with or without skin. Chicken parts that were blast frozen without skin were less rancid than those frozen with skin. Precooked frozen chicken parts browned for 120 s at 200°C were less rancid than those parts browned at 150°C (Love and Goodwin 1974).

In contrast to a loosely packed product such as frozen fried chicken, Hanson and Fletcher (1958) reported that a solid-pack product such as chicken and turkey pot pies, in which the cooked poultry is surrounded by sauce or gravy, with consequent exclusion of air, had a storage life at −18°C of at least 1 year. As is the case with raw poultry, turkey products have less fat stability than chicken products, but the stability can be increased by substituting more stable fats in the sauces or by using antioxidants. A quality defect found in precooked frozen products containing a sauce or gravy is a liquid separation and curdled appearance of the sauce or gravy when thawed for use. This separation is extremely sensitive to storage temperature. Sauces can be stored at least five times as long at −18°C as at −12°C before separation takes place. Hanson et al. (1951) established that the flour in the sauce was the cause of the separation, and found among a large number of alternative thickening agents that waxy rice flour produced superior stability. Sauces and gravies prepared with waxy rice flour are completely stable for about a year at −20°C.

Since precooked frozen foods are not apt to be sterilized in the reheating process in the home, the processor has an added responsibility to keep bacterial counts in the product well below hazardous levels. Extra precautions should be taken in general plant sanitation, in rapid chilling and freezing of the cooked products, and in seeing that the products do not reach a temperature that will permit bacterial growth at any time during storage or distribution.

THAWING AND USE

Under ordinary conditions, poultry should be kept frozen until shortly before its consumption. The general procedure is to defrost in air or in water. No significant difference has been found in palatability between thawing in oven, refrigerator, room, or water.

For turkeys that have been scalded at high temperatures and fast frozen to give a light appearance, the temperature in retail storage and display must be kept as low as possible (−20°C is reasonable) to prevent darkening. Thawing in the package minimizes darkening.

The safest procedure for thawing turkeys is to hold the turkey in the refrigerator (2 to 5°C) for 2 to 4 days depending on the size of the bird. Other methods are immersion in cool water in the bag for 4 to 6 h or holding in a paper bag or styrofoam chest for 12 to 36 h at room temperature. When using these nonrefrigerated thawing techniques, care must be taken to keep the bird's surface cool to inhibit microbiological growth.

Some retail stores allow frozen poultry to start thawing in the chilled (1 to 3°C) section of the meat display case where poultry is for sale on the particular day. This is an advantage to the consumer who wants to cook the poultry that night. However, this is a safe practice only if careful, constant control is maintained over the chilled inventory so that the product is not held beyond its overall shelf life in store and home. Freezing and thawing in itself does not reduce the refrigerated shelf life of the product. Elliott and Straka (1964) found that frozen-thawed chicken had a shelf life at 2°C about equal to unfrozen counterparts at 2°C, as measured by total counts of psychrophilic bacteria and by odor tests.

Ready-to-cook turkeys in a frozen, prestuffed raw form have been marketed. Extreme care should be exercised in producing and consuming this type of product to assure that the original bacterial count in the birds and stuffing is at a minimum and that, in roasting, the internal temperature reaches a value high enough to provide a safe product.

REFERENCES

Behnke, J.R., O. Fennema, and R.W. Haller. 1973. Quality changes in prerigor poultry at –3°C *Journal of Food Science* (38):275.

Bolder, N.M.. 1997. Decontamination of meat and poultry carcasses. *Trends in Food Sci & Technol.* 8:221-227.

Brant, A.W. and G.F. Stewart. 1950. Bone darkening in frozen poultry. *Food Technology* (4):168.

Brant, A.W., J.W. Goble, J.A. Hamann, C.J. Wabeck, and R.E. Walters. 1982. Guidelines for establishing and operating broiler processing plants. *USDA Agr. Handbook No.* 581.

Calvelo, B., 1981. Recent studies on meat freezing. In *Development in meat sciences* - 2 (R. Laurie, ed.), pp 125-158. Appl. Sci. Pub., London.

Carpenter, M.D., D.M. Janky, A.S. Arafa, J.L. Oblinger, and J.A. Koburger. 1979. The effect of salt brine chilling on drip loss of icepacked broiler carcasses. *Poultry Science* (58):369.

Clatfelter, K.A. and J.E. Webb. 1987. Method of eliminating aging step in poultry processing. U.S. Patent No. 4,675,947, June 30.

Cleland, A.C. and R.L. Earle. 1984. Assessment of freezing time prediction formula. *J. Food Sci.* (49) 1034-1042.

Cleland, A.C., R.L. Earle, and D.J. Cleland. 1982. The effect of freezing rate on the accuracy of numerical freezing calculations. *Int. J. of Refrigeration* 5: 294-301.

Elliott, R.P. and R.P. Straka. 1964. Rate of microbial deterioration of chicken meat at 2°C after freezing and thawing. *Poultry Science* (43):81.

Ellis, C. and J.G. Woodroof. 1959. Prevention of darkening in frozen broilers. *Food Technology* (13):533.

Esselen, W.B., A.F. Lexine, I.J. Pflug, and L.L. Davis. 1954. Brine immersion cooling and freezing of packaged ready-to-cook poultry. *Refrigerating Engineering* 62(7):61.

Goresline, H.E., M.A. Howe, E.R. Baush, and M.F. Gunderson. 1951. Inplant chlorination does a 3-way job. *U.S. Egg and Poultry Magazine* (4):12.

Hamre, M.L. and W.J. Stadelman. 1967a. Effect of various freezing methods on frozen diced chicken. *Quick Frozen Foods* 29(4):78.

Hamre, M.L. and W.J. Stadelman. 1967b. The effect of the freezing method on tenderness of frozen and freeze dried chicken meat. *Quick Frozen Foods* 30(8):50.

Hanson, H.L. and L.R. Fletcher. 1958. Time-temperature tolerance of frozen foods. Part XII, Turkey dinners and turkey pies. *Food Technology* (12):40.

Hanson, H.L., A. Campbell, and H. Lineweaver. 1951. Preparation of stable frozen sauces and gravies. *Food Technology* (5):432.

Hanson, H.L., G.F. Stewart, and B. Lowe. 1942. Palatability and histological changes occurring in New York dressed broilers held at 1.7°C (35°F). *Food Research* (7):148.

Ingraham, J.L. 1958. Growth of psychrophilic bacteria. *Journal of Bacteriology* (6):75.

Klose, A.A., A.A. Campbell, and H.L. Hanson. 1960. Stability of frozen ready-to-cook ducks and geese. *Poultry Science* (39):1136.

Klose, A.A., M.F. Pool, M.B. Wiele, H.L. Hanson, and H. Lineweaver. 1959. Time-temperature tolerance of frozen foods. Ready-to-cook cut-up chicken. *Food Technology* (13):477.

Klose, A.A., M.F. Pool, and H. Lineweaver. 1955. Effect of fluctuating temperatures on frozen turkeys. *Food Technology* (9):372.

Kotula, A.W., J.E. Thomson, and J.A. Kinner. 1960. Water absorption by eviscerated broilers during washing and chilling. USDA, Agricultural Marketing Service, *Marketing Research Report* No. 438 (October).

Lazar, V. 1997. Natural vs. liquid smoke. *Meat Processing* 36(9): 28-31.

Lentz, C.P. and L. van den Berg. 1957. Liquid immersion freezing of poultry. *Food Technology* (11):247.

Love, B.E. and T.L. Goodwin. 1974. Effects of cooking methods and browning temperatures on yields of poultry parts. *Poultry Science* (53):1391.

Mountney, G.J. 1976. Plant layout. In *Poultry products technology*, 2nd ed. pp 116-131. AVI Pub. Westport, Comm.

Mulder, R.W. A. W. 1995. Decontamination of broiler carcasses. *Misset World Poultry* 11(3): 39-43.

Peric, M., E. Rossmanith, and L. Leistner. 1971. Verbesserung der microbiologischen Qualität von Schlachthähnchen durch die Sprühkühlung. *Die Fleischwirtschaft* (April):574.

Pflug, I.J. 1957. Immersion freezing found to improve poultry appearance. *Frosted Food Field* (June):17.

Shannon, W.G., W.W. Marion, and W.J. Stadelman. 1957. Effect of temperature and time of scalding on the tenderness of breast meat of chicken. *Food Technology* (11):284.

Smith, M.C., Jr., M.D. Judge, and W.J. Stadelman. 1966. A cold shortening effect in avian muscle *Journal of Food Science* (31):450.

Spencer, J.V. and W.J. Stadelman. 1955. Effect of certain holding conditions on shelf life of fresh poultry meat. *Food Technology* (9):358.

Stadelman, W.J. 1970. 28 to 32°F temperature is ideal for preservation, storage and transportation of poultry. *ASHRAE Journal* 12(3):61.

Stone, D.R., 1995. Can irradiation zap consumer resistance? *Poultry Marketing and Technology* 3(2): 20.

Sweat, V.E., C.G. Haugh, and W.J. Stadelman. 1973. Thermal conductivity of chicken meat at temperatures between –75 and 20°C. *Journal of Food Science* (38):158.

USDA and U.C. Davis. 1975. Guidelines for turkey processing plant layout. *USDA Marketing Research Report* #1036. Washington, D.C.

USDA and Univ. of Georgia, 1970. Guidelines for poultry processing plant layout. *USDA Marketing Research Report* #878. Washington D.C.

USDA/FSIS. 1990. Poultry products inspection regulations. Chapter 3, Sub-Chapter C, Part 381. Washington, D.C.

USDA/FSIS. 1996. Sanitation standard operation procedures (SSOP) reference guide.

van den Berg, L. and C.P. Lentz. 1958. Factors affecting freezing rate and appearance of eviscerated poultry frozen in air. *Food Technology* (12):183.

Veerkamp, C.H. 1995. Chilling, freezing and thawing. In *Processing of poultry* (G.C. Mead, ed.). pp 103-125, Chapman & Hall, London.

Veerkamp, C.H. and G. J. P. Hofmans. 1974 Factors influencing cooling of poultry carcasses. *J. Food Sci.* (39): 980-984.

Walters, R.E. and K.N. May. 1963. Thermal conductivity and density of chicken breast, muscle and skin. *Food Technology* (17):808.

Webb, J.E. and C.C. Brunson. 1972. Effects of eviscerating line trimming on tenderness of broiler breast meat. *Poultry Science* (51):200.

Webb, J.E., R.L. Dake, and R.E. Wolfe. 1989. Method of eliminating aging step in poultry processing. U.S. Patent No. 4,860,403. August 29.

Webb, J.E. and T.L. Goodwin. 1970. Precooked chicken. Effect of cooking methods and batter formula on yields and storage conditions on 2-thiobarbituric acid values. *British Poultry Science* (11):171.

Wells, F.E., J.V. Spencer, and W.J. Stadelman. 1958. Effect of packaging materials and techniques on shelf life of fresh poultry meat. *Food Technology* (12):425.

Willis, R., B. Lowe, and G.F. Stewart. 1948. Poultry storage at subfreezing temperatures—Comparisons at –10 and +10°F. *Refrigerating Engineering* (56):237.

Zeidler, G. 1997. New light on foodborne and waterborne diseases. *Misset World Poultry* 13(9): 490-53.

Zeidler, G. 1996. How can food-borne microorganisms make you ill. *Misset World Poultry* 12(X)

Ziegler, F. and W.J. Stadelman. 1955. The effect of different scald water temperatures on the shelf life of fresh, non-frozen fryers. *Poultry Science* (34):237.

BIBLIOGRAPHY

Barbut, S. 1992. Poultry processing and product technology, in *Encyclopedia of food science and technology* (Hui, Y.H., ed.). pp 2145-56. John Wiley & Sons, New York.

Bowers, P. 1997. In-plant irradiation emerges. *Poultry Marketing and Technology* 5(4): 18.

Bowers, P. 1997. Hot off the bone. *Poultry Marketing and Technol.* 5(4) 14.

Hoggins, J., 1986. Chilling broiler chicken: An overview. In *Proceedings* of Recent Advances and Development in the Refrigeration of Meat by Chilling. pp. 133-147. Inter. Inst. of Refrigeration, Paris.

Evans, T. 1997. Watt poultry statistical yearbook. *Poultry Inter.* 36 (9).

Herwill, J. 1986. What to do before meat hits your boning line. *Broiler Industry* 19 (1): 124-128.

Mogens, J., 1986. Chilling broiler chicken: An overview. In *Proceedings* of Recent Advances and Development in the Refrigeration of Meat by Chilling, pp. 133-141. Inter. Inst. of Refrigeration, Paris.

Stadelman, W.J., V.M. Olson, G.A. Shemwell, and S. Pasch. 1988. Scalding. In *Egg and Poultry Meat Processing*, pp 127-128. Ellis Horwood Int. Pub. In *Sci & Technol.* Chichester, England

Todd, E.C.D. 1980. Poultry associated foodborne diseases—Its occurrence, cost, source and prevention. *J. Food Protection* 43:129-139.

Zeidler, G. 1997. Changes in consumer behaviors and in economic and demographic trends in the US as reflected in successful new poultry product introductions. In *Poultry Meat Quality* (J. Kijowski and J. Piskell, eds.) pp 43-14-32.

CHAPTER 18

FISHERY PRODUCTS

FRESH FISHERY PRODUCTS 18.1	FROZEN FISHERY PRODUCTS 18.4
Care Aboard Vessels 18.1	Packaging 18.4
Shore Plant Procedure and Marketing 18.3	Freezing Methods 18.5
Packaging Fresh Fish 18.3	Storage of Frozen Fish 18.7
Fresh Fish Storage 18.3	Transportation and Marketing 18.9

THE major types of fish and shellfish harvested from North American waters and used for food include the following:

1. Groundfish (haddock, cod, whiting, flounder, and ocean perch), lobster, clams, scallops, snow crab, shrimp, capelin, herring, and sardines from New England and Atlantic Canada.
2. Oysters, clams, scallops, striped bass, and blue crab from the Middle and South Atlantic.
3. Shrimp, oysters, red snapper, clams, and mullet from along the Gulf Coast.
4. Lake herring, chubs, carp, buffalofish, catfish, yellow perch, and yellow pike from the Mississippi Valley and the Great Lakes region.
5. Alaska pollock, Pacific pollock, tuna, halibut, salmon, Pacific cod, various species of flatfish, king and Dungeness crab, scallops, shrimp, and oysters from the Pacific Coast and Alaska.
6. Catfish, salmon, trout, oysters, and mussels from aquaculture operations in various parts of North America.

Fish harvested from tropical waters are reported to have a substantially longer shelf life than fish harvested from cold waters. This may be due to the bacterial flora naturally associated with the fish. The bacteria associated with fish from tropical waters are mainly gram-negative mesophiles. The bacteria that cause spoilage of tropical or other fish during refrigerated storage are usually gram-negative psychrophiles. The time required for this bacterial population shift (from mesophiles to psychrophiles) after refrigeration may account for the increased shelf life.

The major industrial fish used for fish meal and oil is menhaden from the Atlantic and Gulf coasts. In addition, the parts of fish not used for human consumption are often used to manufacture fish meal and oil.

Fish meal and oil are the principal components of the feed used in the aquaculture of trout and salmon. Meal also is a component of the diets of poultry and pigs. Fish oil is used in margarine, in paints, and in the tanning industry. It is also being refined for pharmaceutical purposes.

This chapter deals with the preservation and processing of fresh and frozen fishery products; the care of fresh fish aboard vessels and ashore; the technology of freezing fish; and present commercial trends in the freezing, frozen storage, and distribution of seafood.

See Chapter 26 for additional information regarding fishery products for precooked and prepared foods.

HACCP System. Many of the procedures for the control of microorganisms are managed by the Hazard Analysis and Critical Control Point (HACCP) system of food safety. Each food manufacturing site should have a HACCP team to develop and implement its HACCP plan. See Chapter 11 for additional information on sanitation.

FRESH FISHERY PRODUCTS

CARE ABOARD VESSELS

After fish are brought aboard a vessel, they must be promptly and properly cared for to assure maximum quality. Trawl-caught fish on the New England and Canadian Atlantic coasts, such as haddock and cod, are usually eviscerated, washed, and then iced down in the pens of the vessel's hold. The offshore Canadian fleet and the fleets of Iceland, the United Kingdom, and other European countries have been icing the fish in boxes for optimum quality. Because of their small size, other groundfish (e.g., ocean perch, whiting, and flounder) are not eviscerated and are not always washed. Instead, they are iced down directly in the hold of the vessel.

Crustaceans, such as lobsters and many species of crabs, are usually kept alive on the vessel without refrigeration. Warm-water shrimp are beheaded, washed, and stored in ice in the hold of the vessel; on some vessels, however, the catch is frozen either in refrigerated brine or in plate freezers. Cold water shrimp are stored whole in ice or in chilled sea water, or they may be cooked in brine, chilled, and stored in containers surrounded with ice.

Freshwater fish in the Great Lakes and Mississippi River areas are caught in trap nets, haul seines, or gill nets. They are sorted according to species into 23 or 45 kg boxes, which are kept on the deck of the vessel. In most cases, fishermen carry ice aboard their vessels, and the fish are landed the day they are caught.

Freshwater fish in the lakes of Canada are iced down in the summertime and stored at collecting stations on the lakes, where they are picked up by a collecting boat with a refrigerated hold. Winter-caught Canadian freshwater fish and Arctic saltwater fish are usually weather frozen on the ice immediately after catching and are marketed as frozen fish.

Line-caught fish of the Pacific Northwest, such as halibut caught largely by bottom long-line gear and salmon caught by trolling gear, are eviscerated, washed, and iced in the pens of the vessel. Pacific salmon caught by seines and gill nets for cannery use are usually stored whole for several days, either aboard vessels or ashore in tanks of seawater refrigerated to −1°C. A small but significant volume of halibut is held similarly in refrigerated seawater aboard vessels. Tuna caught offshore by seiners or clipper vessels are usually brine-frozen at sea. However, tuna caught inshore by the smaller trollers or seiners are often iced in the round or refrigerated with a brine spray.

Fish raised by aquaculture farms are usually harvested and sold as required by the fresh fish market. They are usually shipped in containers in which they are surrounded by ice.

Icing of Fish

Fish lose quality because of bacterial or enzymatic activity or both. Reduction of storage temperature retards these activities significantly, thus delaying spoilage and autolytic deterioration.

Low temperatures are particularly effective in delaying growth of psychrophilic bacteria, which are primarily responsible for the spoilage of nonfatty fish. The shelf life of species such as haddock

---

The preparation of this chapter is assigned to TC 10.9, Refrigeration Application for Foods and Beverages.

and cod is doubled for each 4 to 5.5 K decrease in storage temperature within the range of 16 to −1°C.

To be effective, ice must be clean when used aboard a vessel. Bacteriological tests on ice in the hold of a fishing vessel showed bacterial counts as high as 5 billion per gram of ice. These results indicate that (1) chlorinated or potable water should be used in making the ice at the ice plant, (2) ice should be stored under sanitary conditions, and (3) unused ice should be discarded from a vessel at the end of each trip.

Both flake ice and crushed block ice are used aboard fishing vessels, although flake ice is more common because it is cheaper to produce and easier to handle mechanically.

The amount of ice used aboard vessels varies with the particular fishery and vessel; however, it is essential to provide sufficient ice around the fish to obtain a proper cooling rate (see Figure 1). A common ratio of ice to fish used in bulk icing on New England trawlers is one part ice to three parts fish. Experiments on English trawlers in boxing fish at sea with one part ice to two parts fish demonstrated improved quality in the landed fish, and, as ice has become more plentiful and less costly relative to the value of fish, the ratio of ice to fish continues to increase. Mechanical refrigeration is employed in some vessels to retard the melting of ice en route to the fishing grounds; however, the hold temperature must be controlled after fish are taken to allow the ice to melt for effective cooling of the fish.

Saltwater Icing

Iced fish storage temperatures must be maintained close to the freezing point of fish. To obtain lower ice temperatures, the freezing point may be depressed by adding salt to the water from which the ice is made. Adequate amounts of ice made from a 3% solution of sodium chloride brine will maintain a storage environment of about −1°C. Tests conducted on the storage of haddock in saltwater ice aboard a fishing vessel showed that, under parallel conditions, fish iced with saltwater ice cooled faster and to a lower temperature than fish iced with plain ice. However, the saltwater ice melted faster than the plain ice because of its lower latent heat and its greater temperature differential. Therefore, once the saltwater ice melted, the fish stored in this ice rose to a higher temperature than those stored in plain ice. Since it is not always possible to replenish ice on fish at sea, sufficient quantities of saltwater ice must be used initially to make up for its faster melting rate.

In making ice from water containing a preservative, rapid freezing, use of a stabilizing dispersant, or both is essential to prevent migration of the additive to the center of the ice block. This problem is not encountered in flake ice because flake ice machines freeze water rapidly into thin layers of ice, thus fixing additives within the ice flakes. Chapter 33 describes the manufacture of flake ice in more detail.

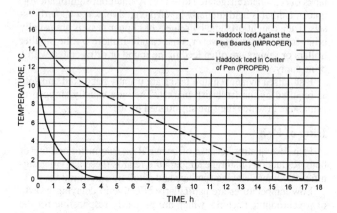

Fig. 1 Cooling Rate of Properly and Improperly Iced Haddock

Use of Preservatives

In the United States and Canada, the use of antibiotics in ice or in dips for treatment of whole or gutted fish, shucked scallops, and unpeeled shrimp is prohibited by regulation.

Storage of Fish in Refrigerated Sea Water

Refrigerated seawater (RSW) is used commercially for preserving fish. On the Pacific Coast, substantial quantities of net-caught salmon are stored in RSW aboard barges and cannery tenders for delivery to the canneries. On the East and Gulf coasts, RSW installations on fishing vessels are used for chilling and holding menhaden and industrial species needed for production of meal, oil, and pet food. On the east and west coasts of Canada, RSW installations are used for chilling and holding herring and capelin, which are processed on shore for their roe. Other more limited applications of RSW include holding Pacific halibut and Gulf shrimp aboard a vessel; chilling and holding Maine sardines in shore tanks for canning, and short-term holding of Pacific groundfish in shore tanks for later filleting.

With groundfish and shrimp, RSW works well for short-term storage (2 to 4 days), but it is not suitable for longer periods because of the excessive salt uptake, accelerated rancidity, poorer texture, and increased bacterial spoilage that may result. These problems can be partially overcome by introducing carbon dioxide (CO_2) gas into the RSW; holding in RSW saturated with CO_2 can increase the storage life of some species of fish by about 1 week. Additional benefits of RSW are (1) the reduction of handling that results from the bulk storage of the fish and (2) the reduction of pressure on the fish as a result of buoyancy, faster cooling, and lower storage temperature.

In many RSW systems, the refrigeration effect is provided by ammonia flowing through external chillers or pipe coils located within the tanks. Best results have been achieved with external chillers.

Boxing at Sea

There are many advantages to using containers or boxes instead of bulk storage aboard fishing vessels. Known as boxing at sea, the use of containers reduces pressure on the fish while they are stowed in a vessel's hold. Because significant reductions in handling during and subsequent to unloading are possible, mechanical damage and product temperature rise may be virtually eliminated and handling costs may be reduced. Fish can be sorted into boxes by size and species as soon as they are caught. Boxed fish lend themselves more readily to mechanized handling, such as machine filleting, because they are generally firmer and of more uniform shape; fillet yields are generally better than they are with bulk-stored fish.

Boxing at sea is not generally practiced in the United States, except by some inshore vessels. The principal problems associated with converting a fishing vessel from bulk storage to boxed storage are the increased labor required by the crew for handling the boxes, the reduced hold capacity, and a relatively large investment for boxes. Many fisheries have difficulties working out the logistics for assuring the prompt return of properly cleaned boxes to the vessel. Most of these problems have been solved in European fleets, the Canadian offshore fleet, and South American hake fishing fleets. The use of nonreturnable containers for boxing at sea simplifies logistics and reduces initial capital outlay; it has proved justifiable in some U.S. fisheries.

Reusable containers for boxing at sea are usually made of plastic. Careful icing is necessary to minimize the surface area of fish in contact with the box. Plastic boxes provide more heat transfer resistance than aluminum boxes in vessels with uninsulated fish holds and for in-plant storage prior to processing.

All fish boxes must be equipped with drains, preferably directed outside the boxes on the bottom of a stack.

Fishery Products

Table 1 Organoleptic Quality Criteria for Fish

Factor	Good Quality	Poor Quality
Eyes	Bright, transparent, often protruding	Cloudy, often pink, sunken
Odor	Sweet, fishy, similar to seaweed	Stale, sour, presence of sulfides, amines
Color	Bright, characteristic of species, sometimes pearlescent at correct light angles	Faded, dull
Texture	Firm, may be in rigor, elastic to finger pressure	Soft, flabby, little resilience, presence of fluid
Belly	Walls intact, vent pink, normal shape	Often ruptured, bloated, vent brown, protruding
Organs (including gills)	Intact, bright, easily recognizable	Soft to liquid, gray homogeneous mass
Muscle tissue	White or characteristic of species and type	White flesh pink to gray, spreading of blood color around backbone

SHORE PLANT PROCEDURE AND MARKETING

Proper use of ice and adherence to good sanitary practices ensure maintenance of iced fish freshness during unloading from the vessel, at the shore plant, during processing, and throughout the distribution chain. Fish landed in good quality will spoil rapidly if these practices are not carried out.

Fish unloaded from the vessel are usually graded by the buyer for species, size, and minimum quality specification. A price is based in part on the quality in relation to market requirements. Fish also may be inspected by local and federal regulatory agencies for wholesomeness and sanitary condition. Organoleptic criteria are most important for evaluating quality; however, there is a growing acceptance, particularly in Canada and some European countries, of objective chemical and physical tests as indexes of quality loss or spoilage. Organoleptic quality criteria vary somewhat among species, but the information in Table 1 can be used as a general guide in judging the quality of whole fish.

In New England and the Canadian Atlantic provinces, groundfish unloaded from the vessel may be placed in boxes and trucked to the shore plant or conveyed directly from the hold or deck to the shore plant. Single- or double-wall insulated boxes are normally used for transporting fish. Wooden boxes are rarely used because they are a source of microbiological contamination. Ice should be applied generously to each box of fish, even if the period prior to processing is only a few hours. Fish in the plant awaiting processing for longer than a few hours should be iced heavily and stored in insulated containers or in single-wall boxes in a chill room refrigerated to 2°C. If refrigerated facilities are not available, the boxes of fish should be kept in a cool section of the plant that is clean and sanitary and has adequate drainage.

Large boxes of resin-coated plywood or reinforced fiberglass that hold up to 45 kg of fish and ice are used by some plants in preference to icing fish overnight on the floor. These **tote boxes** are moved and stacked by forklift, can be used for trucking fish to other plants, and make better use of plant floor space. Generally, fish awaiting processing should not be kept longer than overnight.

Fresh fish are marketed in different forms: fillets, whole fish, dressed-head on, dressed-headed (head removed), and, in some instances, steaks. The method of preparing fish for marketing depends largely on the species of fish and on consumer preference. For example, groundfish such as cod and haddock are usually marketed as fillets or as dressed-headed fish. Freshwater fish such as catfish and bullheads are usually dressed and skinned; lake trout are not skinned, but are merely dressed; and lake herring are marketed in dressed, round, or filleted form.

PACKAGING FRESH FISH

Most fresh fish is packaged in institutional containers of 2 to 16 kg capacity at the point of processing. Polyethylene trays, steel cans, aluminum trays, plastic-coated solid boxes, wax-impregnated corrugated fiberboard boxes, foamed polystyrene boxes, and polyethylene bags are used.

Fresh fish is often packaged while it still contains process heat from wash water. In these cases, it is advantageous to use a packaging material that is a good heat conductor. The fresh fish industry makes little use of controlled prechilling equipment in packaging. As a result, product temperatures may never reach the optimum level subsequent to packaging. Traditionally, institutional fresh fish travels packed in wet ice; in this case, it may cool to the proper level in transit even if process heat is initially present. However, there is a trend toward the use of leaktight shipping containers for fresh fish because modern transportation equipment is not designed to handle wet shipments. Also, some customers want to avoid the cost of transporting ice yet demand a product that is uniformly chilled to 0 to 2°C when it reaches their door. Shippers who use leaktight shipping containers have to upgrade their product temperature control systems to ensure that the fish reaches ice temperature prior to packaging. Rapid prechilling systems that result in crust freezing can be applied to some fresh seafood products, but this practice must be used with discretion since partial freezing harms quality.

Some general requirements for institutional containers that hold products such as fillets, steaks, and shucked shellfish are (1) sufficient rigidity to prevent pressure exerted on the product, even when containers are stacked or heavily covered with ice; and (2) measures to prevent ice-melt water from contaminating the product. Some containers have drains permitting the drip associated with the fish itself to run off. Others are sealed and may be gastight, which increases shelf life. One problem associated with sealed containers is the emission of a strong odor when the package is first opened. Although this odor may be foul, it soon dissipates and has no adverse effect on quality. Dressed or whole fish may be placed in direct contact with ice in a gastight container.

Leaktight shipping containers are used with nonrefrigerated transportation systems, such as air freight, and consequently require insulation. Foamed polystyrene is particularly suited to this application. For typical air freight shipments, the most economical thickness of insulation is between 25 and 50 mm. To maintain product temperature in transit, shippers use either dry ice, packaged wet ice, packaged gel refrigerant, or wet ice with absorbent padding in the bottom of the container. Foamed polystyrene containers may be of molded construction or of the composite type, in which foam inserts and a plastic liner are used with a corrugated fiberboard box.

At the retail level, fresh fish may be handled in two ways. Stores with service counters display fish in unpackaged form. However, markets without service counters sometimes package fish prior to displaying for sale. Both types of outlets receive the product in institutional containers. If the fish is prepackaged at the market, high labor and packaging costs may be incurred, and the temperature of the product is likely to rise. Often, relatively warm fish is placed in a foam tray, wrapped, and displayed in a meat case, the temperature of which may be 4°C or more. This drastically reduces the shelf life of the fish. Centralized prepackaging at the point of initial processing appears to have many important advantages over the present system. A number of retail chains have their suppliers prepackage the product under controlled temperature and sanitary conditions.

FRESH FISH STORAGE

The maximum storage life of fish varies with the species. In general, the storage life of East and West Coast fish, properly iced and stored in refrigerated rooms at 2°C, is 10 to 15 days. This depends on the condition of the fish when it is unloaded from the boat. Generally, freshwater fish properly iced in boxes and stored in refrigerated rooms may be held for only 7 days. Both of the above time limitations refer to the period between when the fish is landed and processed to when it is consumed.

Cold storage facilities for fresh fish should be maintained at about 2°C with over 90% relative humidity. Air velocity should be limited to control ice loss. Temperatures less than 0°C retard ice melting and can result in excessive fish temperatures. This is particularly important when storing round fish such as herring, which generate heat from autolytic processes.

Floors should have adequate drainage with ample slopes toward drains. All inside surfaces of a cold storage room should be easy to clean and able to withstand the corrosive effects of frequent washings with antimicrobial compounds.

Radiopasteurization of Fresh Seafood

Ionizing radiation can double or triple the normal shelf life of refrigerated, unfrozen fish and shellfish stored at 1°C (see Table 2). No off-odors, adverse nutritional effects, or other changes are imparted to the product as a result of the radiation treatment. However, irradiation of fish is still not common and is not permitted in some jurisdictions.

Modified Atmosphere Packaging

A product environment with modified levels of nitrogen, CO_2, and oxygen can curtail the growth of bacteria and extend shelf life of fresh fish. For example, whole haddock stored in a 25% CO_2 atmosphere from the time it is caught keeps about twice as long as it would in air. However, a modified atmosphere does not inhibit all microbes, and spoilage bacteria, because of their great number, usually restrict the growth of the few pathogenic bacteria present. Traditionally, the obvious signs of spoilage serve as the safeguard against eating fish that may have dangerous levels of pathogenic bacteria.

Because modified atmosphere packaging can be a safety hazard, it is being introduced slowly in several countries under close monitoring by regulatory agencies. This type of packaging requires complete knowledge of regulations and a good control system that maintains proper temperature and sanitation levels.

Table 2 Optimal Radiation Dose Levels and Shelf Life at 1°C for Some Species of Fish and Shellfish

Species	Optimal Radiation Dose, kGy Air Packed	Shelf Life, Weeks
Oysters—shucked, raw	2.0	3 to 4
Shrimp	1.5	4
Smoked chub	1.0	6
Yellow perch	3.0	4
Petrale sole	2.0	2 to 3 (4 to 5 when vac pac)
Pacific halibut	2.0	2 (4 when vac pac)
King crabmeat	2.0	4 to 6
Dungeness crabmeat	2.0	3 to 6
English sole	2 to 3	4 to 5
Soft-shell clam meat	4.5	4
Haddock	1.5 to 2.5	3 to 4
Pollock	1.5	4
Cod	1.5	4 to 5
Ocean perch	2.5	4
Mackerel	2.5	4 to 5
Lobster meat	1.5	4

FROZEN FISHERY PRODUCTS

The production of frozen fishery products varies with geographical location and includes primarily the production of groundfish fillets, scallops, breaded precooked fish sticks, breaded raw fish portions, fish roe, and bait and animal food in the north-eastern states and in Atlantic Canada; round or dressed halibut and salmon, halibut and salmon steaks, groundfish fillets, surimi, herring roe, and bait and animal food in the northwestern states and in British Columbia; halibut, groundfish fillets, crab, salmon, and surimi in Alaska; shrimp, oysters, crabs, and other shellfish and crustacea in the Gulf of Mexico and South Atlantic states; and round or dressed fish in the areas bordering on the Great Lakes.

The fish obtained from these areas differ considerably in both physical and chemical composition. For example, cod or haddock are readily adaptable to freezing and have a comparatively long storage life, but other fatty species, such as mackerel, tend to become rancid during frozen storage and, therefore, have a relatively short storage life. The differences in composition and marketing requirements of many species of fish necessitate consideration of the specific product with regard to quality maintenance and methods of packaging, freezing, cold storage, and handling.

Temperature is the most important factor limiting the storage life of frozen fish. At temperatures below freezing, bacterial activity as a cause of spoilage is limited. However, even fish frozen within a few hours of catching and stored at −29°C will deteriorate very slowly until it becomes unattractive and unpleasant to eat.

Fish proteins are permanently altered during freezing and cold storage. This denaturation occurs quickly at temperatures not far below freezing, even at −18°C fish deteriorates rapidly. Badly stored fish is easily recognized; the thawed product is opaque, white, and dull, and juice is easily squeezed from it. While the properly stored product is firm and elastic, poorly stored fish is spongy, and in very bad cases, the flesh breaks up. Instead of the succulent curdiness of cooked fresh fish, cooked denatured samples have a wet and sloppy consistency at first and, on further chewing, become dry and fibrous.

Among other factors that determine how quickly quality deteriorates in cold storage are the initial quality and composition of the fish, the protection of the fish from dehydration, the freezing method, and the environment during storage and transport. These factors are reflected in four principal phases of frozen fish production and handling—packaging, freezing, cold storage, and transportation.

Today, many species are brought from warm and tropical waters where parasites and toxins could infect them. In addition, food dishes that use raw seafood, such as sushi and sashimi, have gained wide popularity, making them a potential health risk. Parasites are not life threatening but can cause pain and inconvenience. They are easily destroyed by cooking or by deep freezing (−40°C). Marine toxins could be deadly and are not affected by temperature. Susceptible species should not be eaten during periods when toxins could be developed.

PACKAGING

Materials for packaging frozen fish are similar to those for other frozen foods. A package should (1) be attractive and appeal to the consumer, (2) protect the product, (3) allow rapid, efficient freezing and ease of handling, and (4) be cost-effective.

Package Considerations in Freezing

Refrigeration equipment and packaging materials are frequently purchased without considering the effect of the package size on freezing rate and efficiency. For example, a thin consumer package results in a faster rate of product freezing, lower total freezing costs, higher handling costs, and higher packaging material costs; a

Fishery Products

thicker institutional-type package results in a slower rate of product freezing, higher freezing costs, lower handling costs, and lower packaging material costs.

Tests indicate that the time required to freeze packaged fish fillets in a plate freezer is directly proportional to the square of the package thickness. Thus, if it takes 3 h to freeze packaged fish fillets 50 mm thick, it takes about 4.7 h to freeze packaged fish fillets 65 mm thick. The insulating effect of the packaging material, the fit of the product in the package, and the total surface area of the package must be considered. A packing material with low moisture-vapor permeability has an insulating effect, which increases freezing time and cost.

The rate of heat transfer through packaging material is inversely proportional to its thickness; therefore, packaging material should be (1) thin enough to produce rapid freezing and an adequate moisture-vapor barrier in frozen storage and (2) thick enough to withstand heavy abuse. Aluminum foil cartons and packages offer an advantage in this regard.

Proper fit of package to product is essential; otherwise, the insulating effect of the air space formed reduces the freezing rate of the product and increases freezing cost. The surface area of the package is also important because of its relation to the size of the freezer shelves or plates. Maximum use of freezer space can be obtained by designing the package so that it fits the freezer properly. Often, however, these factors cannot be changed and still meet customer requirements for a specific package.

Package Considerations for Frozen Storage

Fish products lose considerable moisture and become tough and fibrous during frozen storage unless a package with low moisture-vapor permeability is specified. The package in contact with the product must also be resistant to oils or moisture exuded from the product, or rancidity of the oils and softening of the package material will occur. The package must fit the product tightly to minimize air spaces and thereby reduce moisture migration from the product to the inside surfaces of the package.

Unless temperatures are very low or special packaging is used, the oils in fish will oxidize in frozen storage, producing an off flavor. One effective type of packaging is to replace the air surrounding the frozen fish with pure nitrogen and to seal the fish in a leak-proof bag made of a material that is a barrier to the passage of oxygen.

Types of Packages

Packaging materials consist of either paperboard cartons coated with various waterproofing materials or cartons laminated with moisture-vapor-resistant films and heat-sealable overwrapping materials with a low moisture-vapor permeability. Paperboard cartons are usually made of a bleached kraft stock, coated with a suitable fortified wax, polyethylene, or other plastic material.

Overwrapping materials should be highly resistant to moisture transmission, inexpensive, heat sealable, adaptable to machinery application, and attractive in appearance. Various types of hot melt coated waxed paper, cellophane, polyethylene, and aluminum foil are available in different forms and laminate combinations to best suit each product.

Consumer Packages. These usually hold less than 500 g and are generally printed, bleached paperboard coated with wax or polyethylene and closed with adhesive. Fish sticks and portions, shrimp, scallops, crabmeat, and precooked dinners and entrees are packaged in this way. In the case of dinners and entrees, rigid plastic, pressboard, or aluminum trays are used inside the printed paperboard package. Rigid plastic or pressboard packages are becoming more common because they are better for microwave cooking. The packaging of these products is normally mechanized.

Materials such as polyethylene combined with cellophane, polyvinylidene chloride, or polyester and combinations of other plastic materials are used with high-speed automatic packaging machines to package shrimp, dressed fish, fish fillets, fish portions, and fish steaks prior to freezing. In some instances, tearing of the wrapping material by fins protruding from the fish has been a problem. Otherwise, this method of packaging is satisfactory and affords the product considerable protection against dehydration and rancidity at a comparatively low cost. This packaging method has also created new markets for merchandising frozen fish products. Boil-in-bag pouches made of polyester-polyethylene and combinations of foil, polyethylene, and paper are used for packaging shrimp, fish fillets, and entrees. These packages are also suitable for microwave cooking.

Institutional Packages. The 2 kg and larger cartons used in the institutional trade are commonly constructed of bleached paperboard that has been waxed or polyethylene coated. Folding cartons with self-locking covers, full-telescoping covers, or glued closures are used. Often the cartons are packaged inside a corrugated master carton or are shrink-wrapped in polyethylene film.

Products such as fish fillets and steaks are individually wrapped in cellophane or another moisture-vapor-resistant film and then packed in the carton. Fish, such as headed and dressed whiting and scallop meats, are packed into the carton and covered with a sheet of cellophane. The cover is then put in place and the package is frozen upside down in the freezer. Raw, unbreaded products, such as shrimp, scallops, fillets, and steaks, are sometimes individually quick frozen (IQF) prior to packaging. When IQF, they can be glazed to enhance moisture retention. This method is preferred over freezing after packaging because it leads to a product that is more convenient to handle and sometimes obviates the need to thaw the fish prior to cooking.

For institutional frozen fish, the current trend is toward printed paperboard folding cartons coated with moisture-vapor-resistant materials instead of waxed paper or cellophane overwrap, though "shatter pack" bulk is also common. Some frozen fish products and seafood entrees destined for institutional markets are packaged in aluminum trays or in rigid plastic trays so they may be heated within the package.

FREEZING METHODS

Product characteristics, such as size and shape, freezing method, and rate of freezing, affect the quality, appearance, and cost of production.

Quick freezing of fish offers the following advantages:

- Chills the product rapidly, preventing bacterial spoilage
- Facilitates rapid handling of large quantities of product
- Makes use of conveyors and automatic devices practical, thus materially reducing handling costs
- Promotes maximum use of the space occupied by the freezer
- Produces a packaged product of uniform appearance, with a minimum of voids or bulges

For further information, see Chapters 8, 9, and 15.

Blast Freezing

Blast freezers for fishery products are generally small rooms or tunnels in which cold air is circulated by one or more fans over an evaporator and around the product to be frozen, which is on racks or shelves. A refrigerant such as ammonia, a halocarbon, or brine flowing through a pipe coil evaporator furnishes the necessary refrigeration effect.

Static pressure in these rooms is considerable, and air velocities average between 2.5 and 7.5 m/s, with 6 m/s being common. Air velocities between 2.5 and 5 m/s give the most economical freezing.

Lower air velocities slow down product freezing, and higher velocities increase unit freezing costs considerably.

Some factories have blast freezers in which conveyors move fish continuously through a blast room or tunnel. These freezers are built in a number of configurations, including (1) a single pass through the tunnel, (2) multiple passes, (3) spiral belts, and (4) moving trays or carpets. The configuration and type of conveyor belt or freezing surface depend on the type and quantity of the product to be frozen, the space available to install the equipment, and the capital and operating costs of the freezer.

Batch loaded blast freezers are used for freezing the following: shrimp, fish fillets, steaks, scallops, and breaded precooked products in institutional packages; round, dressed, and panned fish; and shrimp, clams, and oysters packed in metal cans.

Conveyor-type blast freezers are widely used to freeze products prior to packaging. These products include all types of breaded, precooked seafoods; IQF fillets, loins, tails, steaks, scallops, and shrimp; and raw, breaded fish portions. In the case of portions, which are sliced or sawed from blocks, the function of the blast freezer is to harden the batter and breading prior to packaging and to again lower the temperature of the frozen fish for storage if it has been tempered for slicing.

Dehydration of product, or freezer burn, may occur in freezing unpackaged whole or dressed fish in blast freezers unless the velocity of air is kept to about 2.5 m/s and the period of exposure to the air is controlled. Consumer packages of fish fillets or fish-fillet blocks requiring close dimensional tolerances undergo bulging and distortion during freezing unless restrained. In blast rooms or tunnels, the product can be frozen on specially designed trucks, enabling distribution of pressure on the surfaces of the package and remedying this condition. It is difficult to control the expansion of the product on conveyor installations.

Freezing times for various sizes of packaged fishery products are shown in Figure 2.

Plate Freezing

In the multiplate freezer, freezing is accomplished by refrigerant flowing through connected passageways in horizontal movable plates stacked vertically within an insulated cabinet or room. The plate freezer is used extensively in the freezing of fishery products packaged in consumer cartons and in 2 and 5 kg institutional cartons. Fish to be plate frozen should be properly packaged to minimize air spaces. Spacers should be used between the plates during freezing to prevent crushing or bulging of the package. For most products, the thickness of the spacers should be about 0.8 to 1.5 mm less than that of the package.

Where very close package tolerances are required, as in the manufacture of fish fillet blocks, a metal frame or tray is used to hold the packages of fish during freezing. The frame or tray is generally the same width as the package and the length of one or two blocks. It must be rigid enough to prevent bulging and to hold the fish block's dimensions. This is sometimes accomplished with rigid spacers that limit the mass and cost of the tray.

Fish blocks are available in two common sizes: 7.5 kg (480 mm by 250 mm by 60 mm) and 8.4 kg (480 mm by 290 mm by 60 mm). Other blocks are sized for special applications. The fish can be packed in the block with the long dimensions of the fillets along the length of the frame (long-pack) or along the width of the frame (cross-pack). The orientation depends on the eventual cutting pattern and type of cutting used to convert the block into a finished product.

A tray is not necessary for other packaged seafoods such as shrimp, fillets, fish sticks, or scallops, where close package tolerances are not as essential. Therefore, an automatic continuous plate freezer with properly sized spacers is satisfactory for these products.

The plate freezer provides rapid and efficient freezing of packaged fish products. The freezing time and energy required for freezing packaged fish sticks is greater than that for fish fillets because heat transfer is slowed by the air space within the package. Energy required to freeze a unit mass of product increases with thickness. The freezing times of consumer and institutional size packages of fish fillets and fish sticks are shown in Figure 3.

Immersion Freezing

Immersion in low-temperature brine was one of the first methods used for quick-freezing fishery products. A number of direct

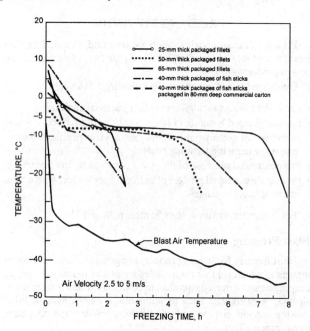

Fig. 2 Freezing Time of Fish Fillets and Fish Sticks in Tunnel-Type Blast Freezer (Air Velocity 2.5 to 5 m/s)

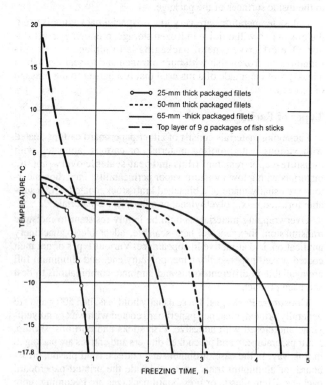

Fig. 3 Freezing Time of Fish Fillets and Fish Sticks in Plate Freezer

immersion freezing machines were developed for whole or panned fish. These machines were generally unsuitable for packaged fish products, which make up the bulk of frozen fish production, and have been replaced by methods employing air cooling, contact with refrigerated plates or shelves, and combinations of these methods.

Immersion freezing is used primarily for the freezing of tuna at sea and, to a lesser extent, for freezing shrimp, salmon, and Dungeness crab. Extensive research has been conducted on brine freezing of groundfish aboard vessels, but this method is not in commercial use.

An important consideration in immersion freezing of fish is selection of a suitable freezing medium. The medium should be nontoxic, acceptable to public health regulatory agencies, easy to renew, and inexpensive; it should also have a low freezing temperature and viscosity. It is difficult to obtain a freezing medium that meets all these requirements. Sodium chloride brine and a mixture of glucose and salt in water are acceptable media. The glucose reduces salt penetration into the fish and provides a protective glaze.

Liquid nitrogen spray and CO_2 are coming into wider use for IQF seafood products such as shrimp. Although the cost per unit mass is high, fish frozen by these methods is of good quality, there is virtually no mass loss from dehydration, and there are space and equipment savings. The fish should not be directly immersed in the liquid nitrogen since this will cause the flesh to shatter and rupture.

Immersion Freezing of Tuna. Most tuna harvested by the United States fleet is brine-frozen aboard the fishing vessel. Freezing at sea enables the vessel to make extended voyages and return to port with a full payload of high-quality fish.

Tuna are frozen in brine wells, which are lined with galvanized pipe coils on the inside. Direct expansion of ammonia into the evaporator coils provides the necessary refrigeration effect. The wells are designed so that tuna can be precooled and washed with refrigerated seawater and then frozen in an added sodium chloride brine. After the fish are frozen, the brine is pumped overboard, and the tuna is kept in −12°C dry storage. Prior to unloading, the fish are thawed in −1°C brine. In some cases, the fish are thawed in tanks at the cannery. If the fish are thawed ashore, thawing on the vessel is not required beyond the stage needed to separate those fused together in the vessel's wells.

Sometimes tuna are held in the wells for a long period prior to freezing or are frozen at a very slow rate because of high well temperatures caused by overloading, insufficient refrigeration capacity, or inadequate brine circulation. These practices have a detrimental effect on product quality, especially for smaller fish, which are more subject to salt penetration and quality changes. Tuna that is not promptly and properly frozen may undergo excessive changes, absorb excessive quantities of salt, and possibly be bacteriologically spoiled when landed. Some freezing times for tuna of various sizes are shown in Figure 4.

Specialized Contact Freezers. Fish frozen by this method are placed on a slowly moving, solid stainless steel belt. This belt conveys the fish fillets through a tunnel, where they are frozen not only by an air blast but also by direct contact between the conveyor belt and a thin layer of glycol pumped through the plates that support the belt. A refrigerant, such as ammonia or a halocarbon, also flows through separate channels in the plates. This provides the refrigeration effect with minimal temperature difference between the evaporating refrigerant and the product.

Freezing Fish at Sea

Freezing fish at sea has found increasing commercial application in leading fishery nations such as Japan, Russia, the United Kingdom, Norway, Spain, Portugal, Poland, Iceland, and the United States. Including freezer trawlers, factory ships, and refrigerated transports in fisheries, hundreds of large freezer vessels are operating throughout the world. United States factory freezer trawlers, factory surimi trawlers, and floating factory ships supplied by

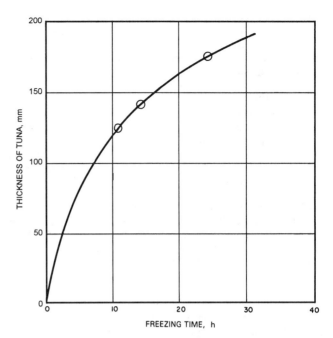

Fig. 4 Freezing Time for Tuna Immersed in Brine

catcher vessels operate off Alaska. These vessels process mainly Alaskan pollock, cod, and flounder, although they do process other species.

Freezing groundfish at sea is uncommon in the northeastern United States largely because fresh fish commands a better price than frozen fish. For the same reason, East Coast United States producers avoid putting their product into frozen packs if they can sell it fresh. Hence, much of the frozen fish used in the United States, with the exception of Alaskan fish, is imported from other countries.

Where used, the factory vessel is equipped to catch, process, and freeze the fish at sea and to use the waste material in the manufacture of fish meal and oil. A large European factory vessel measures 85 m in length, displaces 3400 t, and is equipped to stay at sea for about 80 days without being refueled. About 65 to 100 people are required to operate the vessel and to process and handle the fish. On most vessels of this type, contact-plate freezers are used. The freezers can freeze about 27 t of fish per day, and the total capacity of the frozen fish hold may be as high as 680 t.

Because the factory trawler stays at sea for long periods, it can fully use its space for storing fish. However, due to limited available labor, frozen packs are generally of the less labor-intensive types.

The freezer trawler was designed to resolve the disadvantages associated with factory freezer vessels. It is smaller and equipped to freeze fish in bulk for later thawing and processing ashore. Freezer trawlers use vertical plate freezers to freeze dressed fish in blocks of about 50 kg.

Some countries use freezer trawlers to supply raw material to shore-based processing plants producing frozen fish products. This allows the trawlers to fill their holds in distant waters and transport the fish to home base, where it becomes frozen raw material that is held in storage until required for processing. In some cases, trawlers have been designed as dual fisheries, fishing and freezing groundfish blocks during part of the year and catching, processing, and freezing Northern shrimp for the rest of the year.

STORAGE OF FROZEN FISH

Fishery products may undergo undesirable changes in flavor, odor, appearance, and texture during frozen storage. These changes are attributable to dehydration (moisture loss) of the fish, oxidation of the oils or pigments, and enzyme activity in the flesh.

The rate at which these changes occur depends on (1) the composition of the species of fish, (2) the level and constancy of storage room temperature and humidity, and (3) the protection afforded the product through the use of suitable packaging materials and glazing compounds.

Composition

The composition of a particular species of fish affects its frozen storage life considerably. Fish having a high oil content, such as some species of salmon, tuna, mackerel, and herring, have a comparatively short frozen storage life because of the development of rancidity as a result of the oxidation of the oils and pigments in the flesh. Certain fish, such as sablefish, are quite resistant to oxidative deterioration in frozen storage, despite their high oil content. The development of rancidity is less pronounced in fish with a low oil content. Therefore, lean fish such as haddock and cod, if handled properly, can be kept in frozen storage for many months without serious loss of quality. The relative susceptibility of various species of fish to oxidative changes during frozen storage is shown in Table 3.

Temperature

The quality loss of frozen fish in storage depends primarily on temperature and duration of storage. Fish stored at −29°C has a shelf life of more than a year. In Canada, the Department of Fisheries recommends a storage temperature of −26°C or lower. Storage above −23°C, even for a short period, results in rapid loss of quality. Time-temperature tolerance studies show that frozen seafoods have memory; that is, each time they are subjected to high temperatures or poor handling practices, the loss in quality is recorded. When the product is finally thawed, the total effect of each exposure to mistreatment is reflected in the quality of the product at the consumer level. Continuous storage at temperatures lower than −26°C reduces oxidation, dehydration, and enzymatic changes, resulting in longer product shelf life. From the time they are frozen until they reach the consumer, frozen seafoods should be kept at temperatures as close to −26°C as possible. The shelf life of frozen fish products stored at different temperatures is given in Table 4. Note the increase in shelf life at the lowest temperatures.

For many years, it was thought too costly to operate refrigerated warehouses at temperatures lower than −23°C. However, improvements in the design and operation of refrigeration equipment have made such temperatures economically possible. The production of surimi by West Coast-based factory ships has resulted in the construction of ultracold rooms for its storage. Japanese standards call for this product to be kept at −30°C.

Humidity

A high relative humidity in the cold storage room tends to reduce the evaporation of moisture from the product. The relative humidity of air in the refrigerated room is directly affected by the temperature difference between room cooling coils and room temperature. A large temperature difference results in decreased relative humidity and an accelerated rate of moisture withdrawal from the frozen product. A small temperature difference between the air and evaporator cooling coils results in high relative humidity and reduced moisture loss from the product.

The relative humidity in commercial cold storages is 10 to 20% higher than that of an empty cold storage because of constant evaporation of moisture from the product. In a cold storage operating at −20°C, with a 70% rh and pipe coil temperature of −25°C, the moisture-vapor pressure of the air within the package (in direct contact with the frozen fish) would be 109 Pa. The air in the cold storage would have a vapor pressure of 91 Pa, and the moisture-vapor pressure at the coils would be 64 Pa. These differences in moisture-vapor pressure will result in considerable moisture loss from the product unless it is adequately protected by suitable packaging materials or glazing compounds. The evaporator coils in the freezer should be sized properly so that the desired high relative humidities can be obtained. However, because of material costs and space limitations, a temperature difference of 5 K between evaporator coils and room air is the most practical.

Packaging and Glazing

Adequate packaging of fishery products is important in preventing product dehydration and consequent quality loss. The packaging, which, in most instances, occurs prior to freezing, has been described. Individual fish, whether frozen in the round or dressed, cannot usually be suitably packaged; therefore, they must be protected by a glazing compound.

A glaze acts as a protective coating against the two main causes of deterioration during storage (i.e., dehydration and oxidation). It protects against dehydration by preventing moisture from leaving the product and against oxidation by mechanically preventing air

Table 3 Relative Susceptibility of Representative Species of Fish to Oxidative Changes in Frozen Storage

Severe	Moderate	Minor	Very Slight
Pink salmon	Chum salmon	Cod	Yellow pike
Rockfish	Coho salmon	Haddock	Yellow perch
Lake chub	King salmon	Flounder	Crab
Whiting	Halibut	Sole	Lobster
	Ocean perch	Sablefish	
	Herring	Oysters	
	Mackerel		
	Tuna		
	Lake herring		
	Sheepshead		
	Lake trout		

Table 4 Effect of Storage Temperature on Shelf Life of Frozen Fishery Products

Product	Temperature, °C	Shelf Life, Months
Packaged haddock fillets	−12	4 to 5
	−18	11 to 12
	−29	Longer than 12
Packaged cod fillets	−12	5
	−18	6
	−23	10 to 11
Packaged pollock fillets[a]	−7	1
	−12	2
	−18	8
	−23	11
	−29	24
Packaged ocean perch fillets	−9	1.5 to 2
	−12	3.5 to 4
	−18	6 to 8
	−23	9 to 10
Packaged striped bass fillets	−9	4
	−18	9
Glazed whole halibut	−12	3
	−18	6
	−23	9
	−29	12
Whole blue fin tuna	−12	4
	−18 to −20	8
	−29	12
Glazed whole herring	−18	6
	−25	9
Packaged mackerel fillets	−9	2
	−18	3
	−23	3 to 5

[a]Prepared from 1-day-old iced fish.

Fishery Products

contact with the product. It may also minimize these changes chemically with an antioxidant.

Maximum storage life of fishery products can be obtained by employing the following procedures:

- Select only high-quality fish for freezing.
- Use moisture-vapor-resistant packaging materials and fit package tightly around product, or use a modified atmosphere and oxygen-barrier package.
- Freeze fish immediately after processing or packaging.
- Glaze frozen fish prior to packaging.
- Glaze round, unpackaged fish prior to cold storage.
- Put fish in frozen storage immediately after freezing and glazing, if required.
- Store frozen fish at −26°C or lower.
- Renew glaze on round, unpackaged fish as required during frozen storage.

The recommended protection and expected storage life for various species of fish at −18°C are shown in Table 5.

Space Requirements

Packaged products such as fillets and steaks are usually packed in cardboard master cartons for storage and shipment. These master cartons are stacked on pallets and transferred to various areas of the cold storage room by forklift. The master cartons are strong enough to support one or two pallet loads placed on the shelf of each rack in the cold storage. In cold storages without racks, cartons should be stacked to a height that does not cause crushing of the bottom cartons. Cartons for products in packages that contain a lot of air, such as IQF fillets, must be stronger than those for solid packages of fish in order to resist crushing during storage.

Whole or dressed fish frozen in blocks in metal pans, such as mackerel, chub, or whiting, are removed from the pans after freezing, glazed, and then packaged in wooden boxes lined with wax-impregnated paper or in cardboard cartons.

Round fish stored in wooden boxes can be easily reglazed at periodic intervals during frozen storage. The space requirements for the storage of fishery products are shown in Table 6.

Thawing Frozen Fish. Frozen fishery products can be thawed by circulating air or water. In thawing, the fish should not be allowed to rise above refrigerated temperatures; otherwise, rapid deterioration may occur. Thawing is a slower and more difficult process than freezing when done to ensure that quality is maintained. Each application should be carefully designed.

TRANSPORTATION AND MARKETING

Conditions of temperature and humidity recommended for frozen storage should also be applied during transportation and marketing to minimize product quality loss. Shipment in nonrefrigerated or improperly refrigerated carriers, exposure to high ambient temperatures during transfer from one environment to another, improper loading of common carriers or display cases, equipment failure, and other poor practices lead to increased product temperature and, consequently, to quality loss.

Frozen fish is transported under mechanical refrigeration in trucks, railroad cars, or ships. Most of these vehicles are capable of maintaining temperatures of −18°C or lower. Additional information on equipment used in the transportation and marketing of frozen fish and other foods is given in Chapters 10, 12, 13, 29-31, and 47 of this volume and in Chapter 28 of the 1994 *ASHRAE Handbook—Refrigeration*.

To minimize quality loss during transportation and marketing, the following procedures should be adhered to:

1. Transport frozen fish in refrigerated carriers (mechanical or dry ice systems) with ample capacity to maintain a temperature of −18°C over long distances.
2. Precool refrigerated carriers to at least −12°C before loading.
3. Remove frozen products from the warehouse only when the carrier is ready to be loaded. Load directly into the refrigerated carrier; do not allow the product to sit on the dock.
4. Check the frozen fish temperature with a thermometer before loading.
5. Do not stack frozen fish directly against floors or walls of the carrier. Provide floor and wall racks or strips to permit air circulation around the entire load.
6. Continuously record the temperature of the refrigerated carrier during transit. Use an alarm to warn of equipment failure.
7. Measure the temperature of the product when it is removed from the common carrier at its destination.
8. If products are shipped in an insulated container, apply sufficient dry ice to maintain temperatures of −18°C or lower for the duration of the trip.
9. Maintain food delivery or breakup rooms at −18 to −12°C. Do not hold products in breakup rooms any longer than necessary.
10. When received at the retail store, place the product in a −18°C storage room immediately.
11. Hold display cases in retail stores at −18°C or lower.
12. Do not overload display cases, especially above the frost line.
13. Record the temperature of the display cases. Provide an alarm to warn of an excessive rise in temperature.

Table 5 Storage Conditions and Storage Life of Frozen Fish

Fish	Recommended Protection[a]	Storage Life (−18°C), Months
Chub, pink salmon	Ice glazing and packaging	4-6
Mackerel, sea herring, pollock, chub, smelts	Ice glazing and packaging	5-9
Pacific sardines, tuna	Packaging	4-6
Buffalofish, flounder, halibut, ocean perch, rockfish, sablefish, red, sockeye, silver or coho salmon, whiting, shrimp	Packaging	7-12
Haddock, blue pike, cod, hake, lingcod	Packaging	Over 12

[a]All packaging should be with moisture-resistant films.

Table 6 Space Requirements for Frozen Fishery Products

Commodity	Product Package	Container for Storage	Space Required, kg/m³
Fish sticks, breaded shrimp, breaded scallops	225 or 275 g	Corrugated master containers	400 to 480
Fish fillets, fish steaks, small dressed fish	0.5, 2.5, or 5 kg	Corrugated master containers	800 to 960
Shrimp	1.0 and 2.5 kg	Corrugated master containers	550
Panned, frozen fish (mackerel, herring, chub)	None	Wooden or fiberboard boxes	550
Round halibut	None	Wooden box	480 to 550
		Stacked loose	600
Round groundfish (cod, etc.)	None	Stacked loose	500
Round salmon	None	Stacked loose	525 to 550

14. Because of the accelerated deterioration of frozen fish products in the distribution and retail chain, hold products in these areas for as short a period as possible.

BIBLIOGRAPHY

Barnett, H.J, R.W. Nelson, P.J. Hunter, S. Bauer, and H. Groninger. 1971. Studies on the use of carbon dioxide dissolved in refrigerated brine for the preservation of whole fish. *Fishery Bulletin* 69(2).

Bibek, R. 1996. *Fundamental food microbiology.* CRC Press, Boca Raton, FL.

Charm, S.E. and P. Moody. 1966. Bound water in haddock muscle *ASHRAE Journal* 8(4):39.

Dassow, J.A. and D.T. Miyauchi. 1965. Radiation preservation of fish and shellfish of the Northeast Pacific and Gulf of Mexico; Ronsivalli, L.J., M.A. Steinberg, and H.L. Seagran. Radiation preservation of foods. National Academy of Science *Publication* No. 1273. Washington, D.C.

Feiger, E.A. and C.W. du Bois. 1952. Conditions affecting the quality of frozen shrimp. *Refrigerating Engineering* (September):225.

Holston, J. and S.R. Pottinger. 1954. Some factors affecting the sodium chloride content of haddock during brine freezing and water thawing. *Food Technology* 8(9):409.

Kader, A.A., ed. 1992. *Postharvest technology of horticultural crops,* 2nd ed. University of California Division of Agriculture and Natural Resources.

NACMCF. 1992. Hazard Analysis and Critical Control Point System. *International Journal of Food Microbiology* 16:1-23.

Nelson, R.W. 1963. Storage life of individually frozen Pacific oyster meats glazed with plain water or with solutions of ascorbic acid or corn syrup solids. *Commercial Fisheries Review* 25(4):1.

Peters, J.A. 1964. Time-temperature tolerance of frozen seafood. *ASHRAE Journal* 6(8):72.

Peters, J.A., E.H. Cohen, and F.J. King. 1963. Effect of chilled storage on the frozen storage life of whiting. *Food Technology* 17(6):109.

Peters, J.A. and J.W. Slavin: Comparative keeping quality, cooling rates, and storage temperatures of haddock held in fresh water ice and salt water ice. *Commercial Fisheries Review* 20(1):6.

Ronsivalli, L.J. and J.W. Slavin. 1965. Pasteurization of fishery products with gamma rays from a cobalt 60 source. *Commercial Fisheries Review* 27(10):1.

Stansby, M.E., ed. 1976. *Industrial fishery technology,* 2nd ed. Robert E. Krieger Publishing Co., Huntington, NY.

Tressler, D.K, W.B. van Arsdel, and M.J. Copley, eds. 1968. *The freezing preservation of foods,* 4th ed. AVI Publishing, Westport, CT.

Wagner, R.L, A.F. Bezanson, J.A. Peters. Fresh fish shipments in the BCF insulated leakproof container. *Commercial Fisheries Review* 31(8 and 9):41.

CHAPTER 19

DAIRY PRODUCTS

Milk Production and Processing 19.1
Butter Manufacture 19.6
Cheese Manufacture 19.9
Frozen Dairy Desserts 19.13
Ultrahigh-Temperature (UHT) Sterilization and Aseptic Packaging 19.19
Evaporated, Sweetened Condensed, and Dry Milk 19.21

RAW milk is either processed for beverage milks, creams, and related milk products for marketing, or is used for the manufacture of dairy products. Milk is defined in the United States Code of Federal Regulations, Title 21, Section 131.110. This definition includes goat's milk as defined in the Grade A Pasteurized Milk Ordinance, Part I, Section 1, A.1 of the 1991 revision. Milk products are defined in the Code of Federal Regulations, Title 21, Parts 131 through 135. Public Law 519 defines butter. Note that there are many nonstandard dairy-based products that may be processed and manufactured by the equipment described in this chapter. Dairy plant operations include receiving of raw milk; purchase of equipment, supplies, and services; processing of milk and milk products; manufacture of frozen dairy desserts, butter, cheeses, and cultured products; packaging; maintenance of equipment and other facilities; quality control; sales and distribution; engineering; and research.

Farm cooling tanks and most dairy processing equipment manufactured in the United States meet the requirements of the *3-A Sanitary Standards* (IAMFES). These standards set forth the minimum design criteria acceptable for composition and surface finishes of materials in contact with the product, construction features such as minimum inside radii, accessibility for inspection and manual cleaning, criteria for mechanical cleaning in place (CIP), insulation of nonrefrigerated holding and transport tanks, and other factors which may adversely affect the quality and safety of the product or the ease of cleaning and sanitizing the equipment. Also available are *3-A Accepted Practices*, which deal with construction, installation, operation, and testing of certain systems rather than individual items of equipment.

The *3-A Sanitary Standards* and *Accepted Practices* are developed by the 3-A Standards Committees, which are composed of conferees representing state and local sanitarians, the U.S. Public Health Service, dairy processors, and equipment manufacturers. Compliance with the *3-A Sanitary Standards* is voluntary, but a manufacturer who complies and has authorization from the 3-A Symbol Council may affix to his equipment a plate bearing the 3-A Symbol, which indicates to regulatory inspectors and purchasers that the equipment meets the pertinent sanitary standards.

MILK PRODUCTION AND PROCESSING

Handling Milk at the Dairy

Most dairy farms have bulk milk tanks to receive, cool, and hold the milk. Tank capacity ranges from 0.8 to 19 m³, with a few larger tanks. As the cows are mechanically milked, the milk flows through sanitary pipelines to an insulated stainless steel bulk tank. An electric agitator stirs the milk, and mechanical refrigeration begins to cool it even during milking.

When operated with a condensing unit of the minimum capacity given on the nameplate, a tank must have sufficient refrigerated surface at the first milking to cool 50% of its capacity in an everyday pickup tank, or 25% of its capacity in a tank for every-other-day pickup, from 32.2 to 10°C within the first hour, and from 10 to 4.4°C within the next hour. During the second and subsequent milking, there must be sufficient refrigerating capacity to prevent the temperature of the blended milk from rising above 7.2°C. The nameplate must state the maximum rate at which the milk may be added and still meet the cooling requirements of the *3-A Sanitary Standards*.

Automatic controls maintain the desired temperature within a preset range in conjunction with the agitation. Some dairies continuously record the milk temperature in the tank, which is required in some states. Since the milk is picked up from the farm tank daily or every other day, the milk from the additional milkings generally flows into the reservoir cooled from the previous one. Some large dairy farms may use a plate or tubular heat exchanger to rapidly cool the milk. Cooled milk may be stored in an insulated silo tank (a vertical cylinder 3 m or more in height).

Milk in the farm tank is pumped into a stainless steel tank on a truck for delivery to the dairy plant or receiving station. The tanks are well insulated to alleviate the need for refrigeration of the milk during transportation. Temperature rise when testing the tank full of water should not be more than 1.1 K in 18 h, when the average temperature difference between the water and the atmosphere surrounding the tank is 16.7 K.

The most common grades of raw milk are Grade A and Manufacturing Grade. The former is that used for market milk and related products such as cream. Surplus Grade A milk is used for ice cream and/or manufactured products. To produce Grade A milk, the dairy farmer must meet state and federal standards. In addition to the state requirements, a few municipal governments also have raw milk regulations.

For raw milk produced under the provisions of the Grade A Pasteurized Milk Ordinance (PMO) recommended by the U.S. Public Health Service, the dairy farmer must have healthy cows, adequate facilities (barn, milkhouse, and equipment), maintain satisfactory sanitation of these facilities, and have milk with a bacteria count of less than 100 000 per mL for individual producers. Commingled raw milk can not have more than 300 000 counts per mL. The milk should not contain pesticides, antibiotics, sanitizers, and so forth. However, current methods detect even minute traces, and total purity is difficult. Current regulators require no positive results on drug residue. Milk should be free of objectionable flavors and odors.

Receiving and Storage of Milk

A milk plant receives, standardizes, processes, packages, and merchandises milk products that are safe and nutritious for human consumption. Most dairy plants either receive raw milk in bulk from a producer organization or arrange for pickup directly from dairy farms. The milk level in a farm tank is measured with a dipstick or a direct-reading gage, and the volume is converted to mass. Fat test

---

The preparation of this chapter is assigned to TC 10.9, Refrigeration Application for Foods and Beverages.

and mass are common measures used to base payment to the farmer. A few organizations and the state of California include the percent of nonfat solids and protein content.

Some plants determine the amount of milk received by weighing the tanker, metering the milk as it is pumped from the tanker to a storage tank, and using load cells on the storage tank or other methods associated with the amount in the storage tank.

Milk is generally received more rapidly than it is processed, so ample storage capacity is needed. A holdover supply of raw milk at the plant may be needed for start-up before arrival of the first tankers in the morning. Storage may be required for nonprocessing days and emergencies. Storage tanks vary in size from 4 to 230 m^3. The tanks have a stainless steel lining and are well insulated.

The *3-A Sanitary Standards* for silo-type storage tanks specify that the insulating material should be of a nature and an amount sufficient to prevent freezing, or an average 18-h temperature change of no more than 1.6 K in the tank filled with water when the average temperature differential between the water and the surrounding air is 16.7 K. Inside tanks should have a minimum insulation R-value of 1.76 whereas partially or wholly outside tanks have minimum R-value of 2.64. R-value units are m$^2 \cdot$K/W. For horizontal storage tanks, the allowable temperature change under the same conditions is 1 K.

Agitation is essential to maintain uniform milk fat distribution. Milk held in such large tanks as the silo type is continuously agitated with a slow-speed propeller driven by a gearhead electric motor or with filtered compressed air. The tank may or may not have refrigeration, depending on the temperature of the milk flowing into it and the maximum holding time. Some plants pass the milk through a plate cooler to maintain 4.4°C or less on all milk directed into the storage tanks.

If cooling is provided for milk in a storage tank, it may be by refrigeration of the surface around the lining. This cooling surface may be an annular space from a plate welded to the outside of the lining for direct refrigerant cooling or circulation of chilled water or propylene glycol solution. Another system provides a distributing pipe at the top for the chilled liquid to flow down the lining and drain from the bottom. Direct refrigerant cooling must be carefully applied to prevent milk from freezing on the lining. This limits the evaporator temperature to approximately −4 to −2°C.

Separation and Clarification

Before pasteurizing, milk and cream are standardized and blended to control the milk fat content with legal and practical limits. Nonfat solids may also need to be adjusted for some products; some states require added nonfat solids, especially for low-fat milk such as 2% (fat) milk. Table 1 shows the approximate legal milk fat and nonfat solids requirements for milks and creams in the United States.

One means of obtaining the desired fat standard is by separating a portion of the milk. The required amount of cream or skim milk is returned to the milk. Milk with an excessive fat content may be processed through a standardizer-clarifier that removes fat to a predetermined percentage and clarifies it at the same time. This machine can be adjusted to remove 0.1 to 2.0% of the fat in milk. To increase the nonfat solids, condensed skim milk or low-heat nonfat dry milk may be added.

Milk separators are enclosed and fed with a pump. Separators designed to separate cold milk, usually not below 4.4°C, have increased capacity and efficiency as the milk temperature is increased. Capacity of a separator is doubled as milk temperature is raised from 4.4 to 32.2°C. The efficiency of fat removal with a cold milk separator decreases as temperature decreases below 4.4°C. The maximum efficiency for fat removal is attained at approximately 7 to 10°C or above. Milk is usually separated at 20 to 33°C, but not above 38°C in warm milk separators. If raw, warmed milk or cream is to be held for more than 20 min before pasteurizing, it should be immediately recooled to 4.4°C or below after separation. The pump supplying milk to the separator should be adjusted to pump the milk at the desired rate without causing a partial churning action.

At an early stage between receiving and before pasteurizing, the milk or the resulting skim milk and cream should be filtered or clarified. An optimum time to effectively filter is during the transfer from the pickup tanker into the plant equipment. A clarifier removes extraneous matter and leucocytes, thus improving the appearance of homogenized milks.

Pasteurization and Homogenization

There are two systems of pasteurization—batch and continuous. The minimum feasible continuous operation is about 250 g/s. Therefore, batch pasteurization is used for relatively small quantities of liquid milk products. The product is heated in a stainless steel-lined vat to not less than 62.8°C and held at that temperature or above for not less than 30 min. The Grade A PMO requires that means be provided and used in batch or vat pasteurizers to keep the atmosphere above the product at a temperature not less than 2.8 K higher than the minimum required temperature of pasteurization during the holding period. Whole milk, low-fat milk, half-and-half, and coffee cream are cooled, usually in the vat, to 54°C and then homogenized. Cooling is continued in a heat exchanger (e.g., a plate or tubular unit) to 4.4°C or lower and then packaged.

Plate coolers may have two sections, one using plant water and the second using chilled water or propylene glycol. The temperature of the product leaving the cooler depends on the flow rates and temperature of the cooling medium. Pasteurizing vats are heated with hot water or with steam vapor in contact with the outer surface of the lining. One heating method consists of spraying the heated water around the top of the lining. It flows to the bottom where it drains into a sump, is reheated by steam injection, and again is pumped through the spray distributor. Steam-regulating valves control the temperature of the hot water.

Most pasteurizing vats are constructed and installed so that the plant's cold water is used for initial cooling of the product after pasteurization. For final vat cooling, refrigerated water or propylene glycol is recirculated through the jacket of the vat to attain a product temperature of 4.4°C or less. Cooling time to 4.4°C should be less than 1 h.

Table 1 U.S. Requirements for Milk Fat and Nonfat Solids in Milks and Creams

Product	Legal Minimum Milk Fat, %			Legal Minimum Nonfat Solids, %		
	Federal	Range	Most Often	Federal	Range	Most Often
Whole milk	3.25	3.0 to 3.8	3.25	8.25	8.0 to 8.7	8.25
Low-fat milk	0.5	0.5 to 1.9	2.0	8.25	8.25 to 10.0	8.25
Skim milk	0.5[a]	0.1 to 0.5	0.5[a]	—	8.25 to 9.0	8.25
Flavored milk	—	2.8 to 3.8	3.25	—	7.5 to 10.0	8.25
Half-and-half	10.5	10.0 to 18.0[a]	10.5	—	—	—
Light (coffee) cream	18.0	16.0 to 30.0[a]	18.0	—	—	—
Light whipping cream	30.0	30.0 to 36.0[a]	30.0	—	—	—
Heavy cream	36.0	36.0 to 40.0	36.0	—	—	—
Sour cream	18.0	16.0 to 20.0	18.0	—	—	—

[a]Maximum

Dairy Products

High temperature, short time (HTST) pasteurization is a continuous process in which the milk is heated to a temperature of not less than 71.7°C and held at this temperature for at least 15 s. The complete pasteurizing system usually consists of a series of heat exchange plates contained in a press, a milk balance tank, one or more milk pumps, a holder tube, flow diversion valve, automatic controls, and sources of hot and chilled water or propylene glycol for heating and cooling the milk. Homogenizers are used in many HTST systems as timing pumps used to process Grade A products. The heat exchanger plates are arranged so that milk to be heated or cooled flows between two plates, and the heat exchange medium flows in the opposite direction between alternate pairs of plates.

Ports in the plates are arranged to direct the flow where desired, and gaskets are arranged so that any leakage will be to the outside of the press. Terminal plates are inserted to divide the press into three sections (heating, regenerating, and cooling) and arranged with ports for inlet and outlet of milk, hot water, or steam for heating, and chilled water or propylene glycol for cooling. To provide a sufficient heat-exchange surface for the temperature change desired in a section, the milk flow is arranged for several passes through each section. The capacity of the pasteurizer can be increased by arranging several streams for each pass made by the milk. The capacity range of a complete HTST pasteurizer is 13 g/s to about 13 kg/s. A few shell-and-tube and triple-tube HTST units are in use, but the plate type is by far the most prevalent.

Figure 1 shows one example of a flow diagram for an HTST plate pasteurizing system. Raw product is first introduced into a constant level (or balance) tank from a storage tank or receiving line by either gravity or a pump. A uniform level is maintained in this tank by means of a float-operated valve or similar device. A booster pump is often used to direct the flow through the regeneration section. The product may be clarified and/or homogenized or simply directly pumped to the heating section by means of a timing pump. From the heating section, the product continues through a holding tube to the flow diversion valve. If the product is at or above the preset temperature, it passes back through the opposite sides of the plates in the regeneration section and then through the final cooling section. The flow diversion valve is set at 72°C or above; if the product is below this minimum temperature, it is diverted back into the balance tank for repasteurization. The exchange of heat in the regeneration section causes the cold raw milk to be heated by the heated pasteurized milk going downstream from the heater section. According to the PMO, the pasteurized milk pressure must be maintained at least 14 kPa above the raw. The flow rate of both products is the same, and the temperature change is about the same.

Most HTST heat exchangers have 80 to 90% regeneration. The cost of additional equipment to achieve more than 90% regeneration should be compared with savings in the increased regeneration to determine feasibility. The percentage of regeneration may be calculated as follows:

$$\frac{59°C \text{ (regeneration)} - 4°C \text{ (raw product)}}{72°C \text{ (pasteurization)} - 4°C \text{ (raw product)}} = \frac{55}{68} = 81\%$$

The temperature of a product going into the cooling section can be calculated if the percent regeneration is known and the raw product and pasteurizing temperatures are determined. If they are 80%, 7°C, and 72°C, respectively,

$$(72 - 7) \times 0.80 = 52 \text{ K}$$
$$72 - 52 = 20°C$$

The product should be cooled to at least 4.4°C, preferably even lower, to compensate for the increase while in the sanitary pipelines or package (including filling, sealing, casing, and transfer into cold storage). The average temperature increase of milk between the time of discharge from the cooling section of the HTST unit and arrival at the cold storage in various containers is as follows: glass bottles, 4.4 K; preformed paperboard cartons, 3.3 K; formed paperboard, 2.8 K; and semirigid plastic, 2.2 K.

Many plate-type pasteurizing systems are equipped with a cooling section containing propylene glycol solution to cool the milk or milk product to temperatures lower than are practical by circulating only chilled water. This requires an additional section in the plate heat exchanger, a glycol chiller, a pump for circulating the glycol solution, and a product temperature actuated control to regulate the flow of glycol solution and prevent freezing of the product. Also, some plants use propylene glycol exclusively to cool the milk products, thus avoiding the use of chilled water and the requirement for two separate cooling sections. Milk is usually cooled with propylene glycol to approximately 1°C, then packaged. The lower temperature allows the milk to absorb heat from the containers and still maintain a low enough temperature for excellent shelf life. Milk should not be cooled to temperatures between 0.8°C and freezing because of the tendency toward increased foaming in this range. The propylene glycol is usually chilled to approximately −2 to −1°C for circulation through the milk cooling section.

The product flow rate through the pasteurizer may be more or less than the filling rate of the packaging equipment. Pasteurized product storage tanks are generally used to hold the product until it is packaged.

The number of plates in the pasteurizing unit is determined by the volume of product needed per unit of time, the desired percentage of regeneration, and the temperature differentials between the product and the heating and cooling media. The heating section will usually have ample surface so that the temperature of the hot water entering the section will be no more than 1 to 3 K higher than the pasteurizing, or outlet, temperature of the product. On larger units, steam may be used for the heater section instead of hot water. The cooling section is usually sized so that the temperature of the pasteurized product leaving the section is about 2 to 3 K higher than the entering temperature of the chilled water or propylene glycol.

The holder tube size and length is selected so that not less than 15 s will elapse for the product to flow from one end of the tube to the other. An automatic, power-actuated, flow diversion valve, controlled by a temperature recorder-controller, is located at the outlet end of the holder tube and diverts the flow back to the raw product constant level tank as long as the product is below the minimum set pasteurizing temperature. The product timing pump is a variable speed, positive displacement, rotary type that can be sealed by the

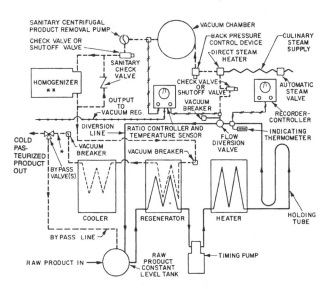

Fig. 1 Flow Diagram of Plate HTST Pasteurizer with Vacuum Chamber

local government milk plant inspector at a maximum speed and volume. This assures a product hold of not less than 15 s in the holder tube.

As a means of reducing undesirable flavors and odors in milk (usually caused by specific types of dairy cattle feed), some plants use a vacuum process in addition to the usual pasteurization. In this process, milk from the flow diversion valve passes through a direct steam injector or steam infusion chamber where the milk is heated with culinary steam to a temperature of 82 to 93°C. The milk is then immediately sprayed into a vacuum chamber, where it is cooled by evaporation to the pasteurizing temperature and promptly pumped to the regeneration section of the pasteurizing unit. The vacuum in the evaporating chamber is automatically controlled so that the same amount of moisture will be removed as was added by steam condensate. Noncondensable gases are removed by the vacuum pump, and vapor from the vacuum chamber is condensed in a heat exchanger which is cooled by the plant water.

The vacuum chamber can be installed with any type of HTST pasteurizer. In some plants, after preheating in the HTST system, the product is further heated by direct steam infusion or injection. It then is deaerated in the vacuum chamber. The product is pumped from the chamber by a timing pump through final heating, holder, flow diversion valve, and regenerative and cooling sections. Homogenization may occur either immediately after preheating for pasteurization or after the product passes through the flow diversion valve. If the product is heated by direct steam injection and deaerated, the preferred practice is to homogenize after deaeration.

Where volatile weed and feed taints in the milk are mild, some processors use only a vacuum treatment for reduction of the off-flavor. The main objection to vacuum treatment alone is that, to be effective, the vacuum must be low enough to cause some evaporation, and the moisture so removed constitutes a loss of product. With this type of treatment, the vacuum chamber may be installed immediately after preheating, where it effectively deaerates the milk prior to heating or immediately after the flow diversion valve where it is more effective in removing the volatile taints.

Nearly all milk processed in the United States is homogenized to improve stability of the milk fat emulsion, thus preventing creaming (concentration of the buoyant milk fat in the upper portion of the containerized milk) during normal shelf life. The homogenizer is a high-pressure, reciprocating pump with 3 to 7 pistons, fitted with a special homogenizing valve. There are several types of homogenizing valves in use, all of which cause the melted fat globules in the stream of milk to be subjected to enough shear to be divided into several smaller globules. Homogenizing valves may either be single or two in series.

For effective homogenization of whole milk, the fat globules should be 2 μm or less in diameter. The usual temperature range is from 54 to 82°C, and the higher the temperature within this range, the lower the pressure required for satisfactory homogenization. The homogenizing pressure for a single-stage homogenizing valve ranges from about 8 to 17 MPa for milk; for a two-stage valve, from 8 to 14 MPa on the first stage plus 2 to 5 MPa on the second, depending on the design of the valve and the product temperature and composition. To conserve energy, use the lowest homogenizing pressure consistent with satisfactory homogenization; the higher the pressure, the greater the power requirements.

Packaging of Milk Products

Cold product from the pasteurizer cooling section flows to the packaging machine and/or a surge tank 4 to 38 m^3 or larger. These tanks are stainless steel, well insulated, and have agitation and usually refrigeration.

Milk and related products are packaged for distribution in paperboard, plastic, or glass containers in various sizes. Fillers vary in design. Gravity flow is used, but positive piston displacement is used on paper machines. Filling speeds range from roughly 16 to 250 units/min, but vary with container size. Some fillers handle only one size, while others may be adjusted to automatically fill and seal several size containers. Paperboard cartons are usually formed ahead of filling, but may be preformed prior to delivery to the plant. Semirigid plastic containers may be blow-molded in-plant ahead of the filler or preformed. Plastic pouches (called bags) arrive at the plant ready for filling and sealing. The filling of dispenser cans and bags is a semimanual operation.

The paperboard carton for milk consists of a 0.41-mm thick kraft paperboard from virgin paper with a 0.025-mm polyethylene film laminated onto the inside and a 0.019-mm film onto the outside. Gas or electric heaters supply heat for sealing while pressure is applied.

Blow-molded plastic milk containers are fabricated from high density polyethylene resin. The resin temperature for blow-forming varies from 170 to 218°C. The molded 4 L has a mass of approximately 60 to 70 g, and the 2 L, about 45 g. Most equipment uses direct expansion water chillers or another cooling medium to cool the mold head and clutch. The refrigeration demand is sufficiently large to require the cooling load to be included in planning a plastic blow-molded operation. The blow-molded equipment manufacturer should be contacted for the refrigeration requirements of a specific machine. Some plants combine the blow-molding refrigeration with the central refrigeration system to achieve better overall efficiency and utilization.

Packages containing the product may be placed into cases mechanically. Stackers place the cases 5 or 6 high, and conveyors transfer the stacks into the cold storage area.

Most milk processing activities from receiving to storage can be automated. This necessitates the installation and operation of control panels and associated computers. Automation uses a meter-based system that controls the separation, fat and/or nonfat solids content, and ingredient addition for a variety of common products. If the initial fat tests fed into the computer are correct, the accuracy of the fat content of the standardized product is ±0.01%. Added ingredients (e.g., flavoring and sweetening products) must be in liquid form for a computerized operation.

Equipment Cleaning

Several systems for automatic CIP are used in milk processing plants. These may involve the holding and reuse of the detergent solution or the preparation of a fresh solution (single-use) each day. The means of programming the automatic control of each cleaning and sanitizing step also varies. Tanks, vats, and other large equipment can be cleaned by using spray balls and similar devices that assure complete coverage of soiled surfaces. Tubing, HTST units, and equipment with relatively low volume may be cleaned by the full-flood system. The solutions should have a velocity of not less than 1.5 m/s and must be in contact with all soiled surfaces. Surfaces used for heating milk products, such as in batch or HTST pasteurization, are more difficult to clean than the other equipment surfaces. Other surfaces difficult to clean are those in contact with high-fat products, products containing added solids and/or sweeteners, and highly viscous products. The usual cleaning steps for this equipment are a warm water rinse, hot acid solution wash, rinse, hot alkali solution wash, and rinse. The variables of time, temperature, concentration, and velocity may need to be adjusted for effective cleaning. Just before use, the product surfaces should be sanitized with chemical solution, hot water, or steam.

Milk Storage and Distribution

Cases containing packaged products are conveyed into a cold storage room or directly to delivery trucks for wholesale or retail distribution. The temperature of the storage area should be 0.6 to 4.4°C, and for improved keeping quality, the product temperature in the container on arrival in storage should be 4.4°C or less.

Dairy Products

The refrigeration load for cold storage areas includes transmission through insulation; product and packaging materials temperature reduction; internally generated loads (lights, motors, personnel); infiltration air load; and refrigeration equipment-related load. See Chapter 12 for refrigeration load calculation.

The moisture load in these storage areas is generally high, which can lead to high humidity or wet conditions if the evaporators are not selected properly. These applications usually require higher temperature differences between refrigerant and return air to achieve lower humidity. In addition, supply air temperatures should be controlled to prevent product freezing. The use of reheat coils to provide humidity control is not recommended, as bacteriological growth on these surfaces could be rapid. Evaporators for these applications should be equipped with automatic coil defrost to remove the rapidly forming frost periodically as required. Defrost cycles add to the refrigeration load and should be considered in the design.

A proprietary system used in some plants sprays coils continuously with glycol to prevent frost from forming on the coil. These fan-coil units eliminate defrosting, can control humidity to an acceptable level with less danger of product freezing, and reduce bacteriological contamination. The glycol absorbs the water, which is continuously reconcentrated in a separate apparatus. A separate load calculation and analysis is required for these systems.

The floor space required for cold storage depends on product volume, height of stacked cases, kind of package (glass requires more space than paperboard), whether mechanized or manual handling, and the number of processing days per week. A 5-day processing week requires a capacity for holding product supply for 2 days. A very general estimate is that 490 kg of milk product in paperboard cartons can be stored per square metre of area. Approximately one-third more area should be allowed for aisles. Some automated, racked storages are used for milk products. They can be more economical than manually operated storages.

Milk product may be transferred by conveyor from storage room to dock for loading onto delivery trucks. In-floor drag-chain conveyors are commonly used, especially for retail trucks. Refrigeration losses are reduced if the load-out doorway has the protrusion of cushioning material to contact the doorway frame of the truck as it is backed to the dock.

Distribution trucks need refrigeration to protect quality and extend the keeping quality of milk products. Refrigeration capacity must be sufficient to maintain Grade A products at 7.2°C or less. Many plants use insulated truck bodies with integral refrigerating systems powered by the truck engine or one that can be plugged into a remote electric power source when it is parked. In some facilities, cold plates in the truck body are connected to a coolant source in the parking space. These refrigerated trucks can also be loaded when convenient and held over at the connecting station until the next morning.

Half-and-Half and Cream

Half-and-half is standardized to 10.5 to 12% milk fat and, in most states, to about the same percent nonfat milk solids. Coffee cream should be standardized to 18 to 20% milk fat. Both are pasteurized, homogenized, cooled, and packaged similarly to milk. Milk fat content of whipping cream is adjusted to 30 to 35%. Care must be taken during processing to preserve the whipping properties; this includes the omission of homogenization.

Buttermilk, Sour Cream, and Yogurt

Retail buttermilk is not from the butter churn but is rather a cultured product. To reduce the microorganisms to a low level and improve the body of the resulting buttermilk, skim milk is pasteurized at 82°C or higher for 0.5 to 1 h and cooled to 21 to 22°C. One percent of a lactic acid culture (starter) specifically for buttermilk is added and the mixture incubated until firmly coagulated by the correct lactic acid production (pH 4.5). The product is cooled to 4.4°C or less with gentle agitation to inhibit serum separation subsequent to packaging and distribution. Salt and/or milk fat (0.5 to 1.0%) in the form of cream or small fat granules may be added. Package equipment and containers are the same as for milk. Pasteurizing, setting, incubating, and cooling are usually accomplished in the same vat. Rapid cooling is necessary, so chilled water is used. If a 2 m^3 vat is used, as much as 90 to 110 kW of refrigeration may be needed. Some plants have been able to cool buttermilk with a plate heat exchanger without causing a serum separation problem (wheying off).

Cultured half-and-half and cultured sour cream are manufactured similarly to the method for cultured buttermilk. Rennet may be added at a rate of 1.3 mL (diluted in water) per 100 L cream. Care must be exercised to use an active lactic culture and to prevent postpasteurization contamination by bacteriophage, bacteria, yeast, or molds. An alternate method consists of packaging the inoculated cream, incubating, and then cooling by placing packages in a refrigerated room.

Skim milk may be used, or milk fat standardized to within the range of 1 to 5%, and a 0.1 to 0.2% stabilizer may be added to yogurt. Either vat pasteurization at 66 to 93°C for 0.5 to 1 h or HTST at 85 to 140°C for 15 to 30 s can be used. For yogurt to have optimum body, the milk homogenization is at 54 to 66°C and 3.5 to 14 MPa. After cooling to between 38 to 43°C, the product is inoculated with a yogurt culture. Incubation for 1.5 to 2.0 h is necessary; the product is then cooled to about 32°C, packaged, incubated 2 to 3 h (acidity 0.80 to 0.85%), and chilled to 4.4°C or below in the package. Varying yogurt cultures and yogurt manufacturing procedures should be selected on the basis of consumer preferences. Numerous flavorings are used (fruit is quite common), and sugar is usually added. The flavoring material may be added at the same time as the culture, after incubation, or ahead of packaging. In some dairy plants, a fruit (or sauce) is placed into the package before filling with yogurt.

Refrigeration

The choice of refrigerant for product plants is primarily ammonia (R-717). Some small plants may use halocarbons, and some large plants may use halocarbons for special, small applications.

Product plants use single-stage compression, and new applications are equipped with rotary screw compressors with microprocessors and automatic control. Older plants may be equipped with reciprocating compressors, but added capacity is generally with rotary screw compressors.

Most refrigerant condensing is accomplished with evaporative condensers. Freeze protection is required in cold climates, and materials of construction are an important consideration in subtropical climates. Water treatment is usually required.

The selection of a system requires an analysis of each application to determine the overall best economics as well as quality factors. Some plants use 0.5 to 1°C chilled water for cooling milk products in the various processes as well as blow molding if part of the central plant. Where chilled water is used, it is often in combination with falling film water chillers and ice builder chillers. Ice builder chilling should be evaluated versus falling film chilling for each application, considering both initial capital and operating costs. By building ice on evaporator surfaces during periods when chilled water is not required, the ice builder system permits the use of a refrigeration system with considerably less capacity than is required for the peak cooling load. When chilled water is required, the melting ice adds cooling capacity to that supplied by the refrigeration system. Additional information on ice storage is found in Chapter 40 of the 1995 *ASHRAE Handbook—Applications*.

Other plants use propylene glycol at −2 to −1°C for process cooling requirements. With this system, the propylene glycol is cooled in a welded plate heat exchanger or a shell-and-tube heat exchanger. The ammonia feed system is either gravity flooded or liquid overfeed. Advantages to this system are a lower ammonia charge (especially with a plate heat exchanger) and a lower fluid temperature to achieve lower milk product temperatures. This system may have a higher operating cost as there is no stored refrigeration.

In addition, there are combination systems whereby chilled water is used for most of the process requirements and a separate, smaller propylene glycol system is used in final cooling sections to provide lower milk product temperatures.

The evaporators or cooling units for milk storage areas are either direct ammonia evaporators (flooded or liquid overfeed) or propylene glycol. In choosing new systems, an evaluation is recommended that involves capital requirements, operating costs, ammonia charges, and plant safety.

Some plants also refrigerate chilled water and/or propylene glycol for plant air conditioning with the central ammonia refrigeration system. The choice between chilled water and propylene glycol depends on the plant winter climate conditions.

Most new or expanded plants rely on automated operation and computer controls for operating and monitoring the refrigeration systems. There also is a trend to use welded plate heat exchangers for water and propylene glycol cooling in milk product plants and to reduce or eliminate direct ammonia refrigeration in the plant process areas. This approach may add somewhat to the capital and operating costs, but it substantially reduces the ammonia charge in the system and confines the ammonia to the refrigeration machine room area.

BUTTER MANUFACTURE

Much of the butter production is in combination butter-powder plants. These plants get the excess milk production after current market needs for milk products, frozen dairy desserts, and, to some extent, cheeses are met. Consequently, seasonal variation in the volume of butter manufactured is large; spring is the period of highest volume, fall the lowest.

Separation and Pasteurization

After separation of the milk, the cream with 30 to 40% fat is either pumped to the pasteurizer or cooled to 7°C and held for later pasteurization. Cream from cold milk separation does not need to be recooled except for extended storage. Cream is received, weighed, sampled, and, in some plants, graded according to flavor and acidity. It is pumped to a refrigerated storage vat and cooled to 7.2°C if held for a short period or overnight. Cream with developed acidity is warmed to 27 to 32°C, and neutralized to 0.12 to 0.15% titratable acidity just prior to pasteurization. If the acidity is above 0.40%, it is neutralized with a soda-type compound in aqueous solution to about 0.30% and then to the final acidity with aqueous lime solution. Sodium neutralizers include $NaHCO_3$, Na_2CO_3, and $NaOH$. Limes are $Ca(OH)_2$, MgO, and CaO.

Batch pasteurization is usually at 68 to 79°C for 0.5 h, depending on intended storage temperature and time. HTST continuous pasteurization is at 85 to 121°C for at least 15 s. HTST systems may be plate or tubular. After pasteurization, the cream is immediately cooled. The temperature depends on the time that the cream will be held before churning, whether or not it is ripened, the season (higher in winter due to fat composition), and the churning method. The range is 4 to 13°C. The ripening process consists of adding a flavor-producing lactic starter to tempered cream and holding until acidity has developed to 0.25 to 0.30%. The cream is cooled to prevent further acid development and warmed to the churning temperature just before churning. Ripening cream is not a common practice in the United States, but is customary in some European countries such as Denmark. First, tap water is used to reduce the temperature to between 25 to 35°C. Refrigerated water or brine is then used to reduce the temperature to the desired level. The cream may be cooled by passing the cooling medium through a revolving coil in the vat or through the vat jacket, or by using a plate or tubular cooler.

If the temperature of 500 kg of cream is to be reduced by refrigerated water from 40 to 4°C, and the specific heat is 3.559 kJ/(kg·K), the heat to be removed would be

$$500(40-4)3.559 = 64\,062 \text{ kJ}$$

This heat can be removed by 64 062/335 = 191 kg of ice at 0°C plus 10% for mechanical loss.

The temperature of the refrigerated water commonly used for cooling cream is 0.6 to 1.1°C. The ice builder system is efficient for this purpose. Brine is not currently used. About 1000 L of cream can be cooled from 37.7 to 4.4°C in a vat using refrigerated water in an hour.

After a vat of cream has been cooled to the desired temperature, the temperature increases during the following 3 h. It may increase several degrees depending on the rapidity with which the cream was cooled, the temperature to which it was cooled, the richness of the cream, and the properties of the fat. The rise in temperature is due to liberation of heat when fat is changed from a liquid to a crystal form.

Rishoi (1951) presented data in Figure 2 that show the thermal behavior of cream heated to 75°C followed by rapid cooling to 30°C and to 10.4°C, as compared with cream heated to 50°C and cooled rapidly to 31.4°C and to 12°C. The curves indicate that when cream is cooled to a temperature at which the fat remains liquid, the

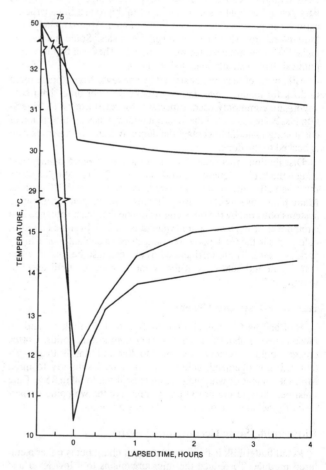

Fig. 2 Thermal Behavior of Cream Heated to 75°C Followed by Rapid Cooling to 30°C and to 10.4°C; Comparison with Cream Heated to 50°C, then Rapid Cooling to 31.4°C and to 12°C

cooling rate is normal, but when the cream is cooled to a temperature at which some fractions of the fat have crystallized, a spontaneous temperature rise takes place after cooling.

Rishoi also determined the amount of heat liberated by the part of the milk fat that crystallizes in the temperature range of 29 to 0.6°C. The results are shown in Figure 3 and Table 2.

Table 2 shows that, at a temperature below 10°C, about one-half of the liberated heat evolved in less than 15 s. The heat liberated during fat crystallization constitutes a considerable portion of the refrigeration load required to cool fat-rich cream. Rishoi states:

"If we assume an operation of cooling cream containing 40% fat from about 65 to 4°C, heat of crystallization evolved represents about 14% of the total heat to be removed. In plastic cream containing 80% fat it represents about 30% and in pure milk fat oil about 40%."

Churning

To maintain the yellow color of butter from cream that came from cows on green pasture in spring and early summer, yellow coloring is added to the cream in the amount needed to produce the color that is obtained naturally during other periods of the year. After cooling, pasteurized cream should be held a minimum of 2 h and preferably overnight. It is tempered to the desired batch churning temperature, which varies with the season and feed of the cows but ranges from 7°C in early summer to 13°C in winter, to maintain a churning time 0.5 to 0.75 h. Lower churning time results in soft butter that is more difficult (or impossible) to work into a uniform composition.

Today most butter is churned by continuous churns, but batch units remain in use, especially in smaller butter factories. Batch churns are usually made of stainless steel, although a few aluminum ones are still in use. They are cylinder, cube, cone, or double cone in shape. The inside surface of metal churns is sandblasted during fabrication to reduce or prevent butter from sticking to the surface. Metal churns may have accessories to draw a partial vacuum or introduce an inert gas (e.g., nitrogen) under pressure. Working the butter under a partial vacuum reduces the air in the butter. Churns have two or more speeds, with the faster rate for churning. The speed should provide maximum agitation of the cream, usually within 0.25 to 0.5 rev/s.

When churning, temperature is adjusted and the churn is filled 40 to 48% of capacity. The churn is revolved until the granules break out and attain a diameter of 5 mm or slightly larger. The buttermilk is drained and should have no more than 1% milk fat. The butter may or may not be washed. The purpose of washing is to remove buttermilk and temper the butter granules if they are too soft for adequate working. Wash water temperature is adjusted to 0 to 6 K below churning temperature. The preferred procedure is to spray wash water over granules until it appears clear from the churn drain vent. The vent is then closed, and water is added to the churn until the volume of butter and water is approximately equal to the former amount of the cream. The churn is revolved slowly 12 to 15 times and drained or held for an additional 5 to 15 min for tempering so granules will work into a mass of butter without becoming greasy.

The butter is worked at a slow speed until free moisture is no longer extruded. The free water is drained, and the butter is analyzed for moisture content. The amount of water needed to obtain the desired content (usually 16.0 to 18.0%) is calculated and added. Salt may be added to the butter. The percent is standardized between 1.0 and 2.5% according to customer demand.

Salt may be added in dry form either to a trench formed in the butter or spread over the top of the butter. It also may be added in moistened form using the water required for standardizing the composition to not less than 80.0% fat. Working continues until the granules are completely compacted and the salt and moisture droplets are uniformly incorporated. The moisture droplets should become invisible by normal vision with adequate working. Most churns have ribs or vanes which cause tumbling and folding of the butter as the churn revolves. The butter passes between the narrow slit of shelves attached to the shell and the roll. A leaky butter is inadequately worked, possibly leading to economic losses due to

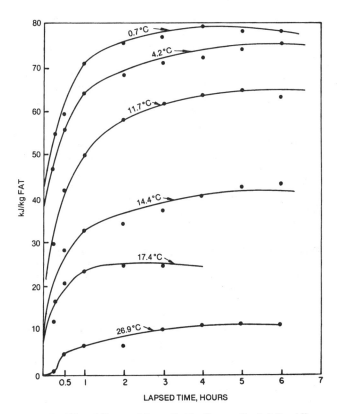

Fig. 3 Heat Liberated from Fat in Cream Cooled Rapidly from Approximately 30°C to Various Temperatures
(Rishoi 1951)

Table 2 Heat Liberated from Fat in Cream Cooled Rapidly from about 30°C to Various Temperatures

Calculated temperature for zero time, °C:	0.8	4.2	11.7	14.4	17.4	26.9	29.8[a]
First observed temperature:	2.3	6.5	12.5	14.8	17.7	26.9	29.8
Final equilibrium temperature:	4.1	7.8	14.4	15.9	18.5	28.1	29.8
Lapsed time, min	**Heat Liberated, kJ/kg**						
0.25	42.6	38.8	18.0	6.7	5.3	0	0
15	54.9	46.5	29.8	16.7	12.1	0.8	0
30	59.3	55.6	41.9	28.4	20.9	4.9	0
60	70.7	64.0	49.8	32.6	23.7	6.3	0
120	75.4	68.2	58.2	34.2	24.7	6.7	0
180	76.5	70.2	61.9	37.2	24.7	10.0	0
240	79.1	71.9	63.3	40.7	24.7	11.4	
300	78.2	74.0	64.9	42.8		11.6	
360	78.2	75.6	63.3	43.7		11.6	

Percent heat liberated at zero time compared with that at equilibrium: 54.5, 51.3, 27.7, 20.7, 21.7.
Percent total heat liberated compared with that liberated at about 0°C: 100.0, 95.7, 82.0, 55.0, 31.0, 12.5, 0.
Iodine values of three samples of butter produced while these tests were in progress were: 28.00, 28.55, and 28.24.
[a] Cooled in an ice water bath.

reduction of mass and shorter keeping quality. The average composition of U.S. butter on the market has these ranges:

Fat	80.0 to 81.2%	Moisture	16.0 to 18.0%
Salt	1.0 to 2.5%	Curd, etc.	0.5 to 1.5%

Cultured skim milk is added to unsalted butter as part of the moisture and thoroughly mixed in during working. Cultured skim milk increases acid flavor and the diacetyl content associated with butter flavor.

Butter may be removed manually from small churns, but it is usually emptied by a mechanized procedure. One method is to dump the butter from the churn directly into a stainless steel boat on casters or a tray that has been pushed under the churn with the door removed. Butter in boats may be augered to the hopper for printing (forming the butter into retail sizes) or pumping into cartons 27 to 30 kg in size. The bulk cartons are held cold before printing or shipment. Butter may be stored in the boats or trays and tempered until printing. A hydraulic lift may be used for hoisting the trays and dumping the butter into the hopper. Cone-shaped churns with a special pump can be emptied by pumping butter from churn to hopper.

Continuous Churning

The basic steps in two of the continuous buttermaking processes developed in the United States are (1) fat emulsion in the cream is destabilized and the serum separated from the milk fat; (2) the butter mix is prepared by thoroughly mixing together the correct amount of milk fat, water, salt, and cultured skim milk; (3) this mixture is worked and chilled at the same time; and (4) butter is extruded at 3 to 10°C with a smooth body and texture.

Several European continuous churns consist of a single machine that directly converts the cream to butter granules, drains off the buttermilk, and washes and works the butter, incorporating the salt in continuous flow. Each brand of continuous churn may vary in equipment design and specific operation details for obtaining the optimum composition and quality control of the finished product. Figure 4 shows a flow diagram of a continuous churn.

In one system, milk is heated to 43.3°C and separated to cream with 35 to 50% fat and skim milk. The cream is pasteurized at 95°C for 16 s, cooled to a churning temperature of 8 to 14°C, and held for 6 h. The cream enters the balance tank and is pumped to the churning cylinder where it is converted to granules and serum in less than 2 s by vigorous agitation. Buttermilk is drained off and the granules are sprayed with tempered wash water while being agitated.

Next, salt, in the form of 50% brine prepared from microcrystalline sodium chloride, is fed into the product cylinder by a proportioning pump. If needed, yellow coloring may be added to the brine. The high-speed agitators work the salt and moisture into the butter in the texturizer section and then extrude it to the hopper for packaging into bulk cartons or retail packages. The cylinders on some designs have a cooling system to maintain the desired temperature of the butter from churning to extrusion. The butterfat content is adjusted by fat test of the cream, churning temperature of the cream, and flow rate of product.

Continuous churns are designed for CIP. The system may be automated or a cream tank may be used to prepare the detergent solution prior to circulation through the churn after the initial rinsing.

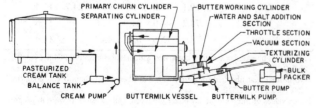

Fig. 4 Flow Diagram of Continuous Butter Manufacture

Packaging Butter

"Printing" refers to the process of forming (or cutting) butter into retail sizes. Each print is then wrapped with parchment or parchment-coated foil. The wrapped prints may be inserted in paperboard cartons or overwrapped in cellophane, glassine, and so forth, and heat-sealed. For institutional uses, butter may be extruded into slabs. These are cut into patties, embossed, and each slab of patties wrapped in parchment paper. Most common numbers of patties are 105 to 158 per kg.

Butter keeps better if stored in bulk. If the butter is intended to be stored for several months, the temperature should not be above −18°C, and preferably below −30°C. For short periods, 0 to 4°C is satisfactory for bulk or printed butter. Butter should be well protected to prevent absorption of off-odors during storage and loss of mass due to evaporation, and to minimize surface oxidation of the fat.

The specific heat of butter and other dairy products at temperatures varying from 0 to 60°C is given in Table 3. The temperature of the butter when removed from the churn ranges from 13 to 16°C. Assuming a temperature of 15°C of the packed butter, the heat that must be removed from 500 kg to reduce the temperature to 0°C is

$$500(15-0)2.18/1000 = 16.4 \text{ MJ}$$

It is assumed that the average specific heat at the given range of temperatures is 2.18 kJ/(kg·K). Heat to be removed from the butter containers and packaging material should be added.

Deterioration of Butter in Storage

The development of an undesirable flavor in butter during storage may be caused by (1) growth of microorganisms (proteolytic organisms causing putrid and bitter off-flavors); (2) absorption of odors from the atmosphere; (3) fat oxidation; (4) catalytic action by metallic salts; (5) activity of enzymes, principally from microorganisms; and (6) low pH (high acid) of salted butter.

Normally, microorganisms do not grow below 0°C; if salt-tolerant bacteria are present, their growth will be slow below 0°C. Growth of microorganisms does not take place at −18°C or below, but some may survive in the butter held at this temperature. It is important to store butter in a room where the atmosphere is free of odors. Butter readily absorbs odors from the atmosphere or from odoriferous materials with which it comes into contact.

Oxidation causes a stale, tallowy flavor. Chemical changes take place slowly in butter held in cold storage, but are hastened by the presence of metals or metallic oxides.

With almost 100% replacement of tinned copper equipment with stainless steel equipment, a tallowy flavor is not as common as in the past. Factors that favor oxidation are light, high acid, high pH, and metal.

Table 3 Specific Heats of Milk and Milk Derivatives, kJ/(kg·K)

	0°C	15°C	40°C	60°C
Whey	4.095	4.086	4.078	4.070
Skim milk	3.936	3.948	3.986	4.932
Whole milk	3.852	3.927	3.984	3.844
15% Cream	3.140	3.864	3.764	3.768
20% Cream	3.027	3.936	3.684	3.710
30% Cream	2.818	4.116	3.567	3.601
45% Cream	2.537	4.254	3.295	3.320
60% Cream	2.345	4.409	3.019	3.086
Butter	(2.114)[a]	(2.207)[a]	2.328	2.438
Milk fat	(1.863)[a]	(1.955)[a]	2.093	2.219

[a] For butter and milk fat, values in parentheses were obtained by extrapolation, assuming that the specific heat is about the same in the solid and liquid states.

Dairy Products

Enzymes present in raw cream are inactivated by current pasteurization temperatures and holding times. The only enzymes that may cause deterioration of butter are those produced by microorganisms that gain entrance to the pasteurized cream and butter or survive pasteurization. The chemical changes caused by enzymes present in butter are retarded by a lowering of the storage temperature.

A fishy flavor may develop in salted butter during cold storage. The development of the defect is favored by high acidity (low pH) of the cream at the time of churning and by metallic salts. With the use of stainless steel equipment and the proper control of the butter's pH, this defect now occurs very rarely. For salted butter to be stored for several months, even at $-23°C$, it is advisable to use good-quality cream; avoid exposing the milk or cream to strong light, copper, or iron; and adjust any acidity developed in the cream so that the butter serum has a pH of 6.8 to 7.0.

Total Refrigeration Load

Some dairy plants that manufacture butter also process and manufacture other products such as ice cream, fluid milk, and cottage cheese. A common refrigeration system is used. The method of determining the refrigeration load is illustrated by the following example:

Example 1. Determine the product refrigeration load for a plant manufacturing butter from 6000 kg of 30% cream per day in three churnings.

Solution: The refrigeration requirement is obtained in the following steps. Assume that refrigeration is accomplished with chilled water from an ice builder.

The cream would be cooled in steps A and B. The butter would then have to be cooled through steps C, D, and E. Refrigerated water is normally used as a cooling medium in steps A, B, and C. The ice builder system is used to produce 2°C water, and the load should be expressed in kilograms of ice that would have to be melted to handle steps A, B, and C. This load would be added to the refrigerated water load from the various other products such as milk, cottage cheese, and so forth, in sizing the ice builder.

A. If the cream is separated in the plant rather than on the farm, it would have to be cooled from 32°C separating temperature to 4°C for holding until it is processed.

$$\frac{6000(32-4)3.56}{335} = 1790 \text{ kg ice/day}$$

B. After pasteurization, the temperature of the cream is reduced to approximately 38°C with city water, then down to 4°C with refrigeration.

$$\frac{6000(38-4)3.56}{335} = 2170 \text{ kg ice/day}$$

C. After churning, the 15°C butter wash water (city water) is usually cooled to 7°C, then used to wash the butter granules. A mass of water equal to the mass of cream churned may be used.

$$\frac{6000(15-7)4.187}{335} = 600 \text{ kg ice/day}$$

Total ice load	4560 kg ice/day
Plus 10% mechanical loss	460 kg ice/day
Total ice required	5020 kg ice/day

D. Approximately 2250 kg of butter would be obtained. (6000 kg cream × 30% fat = 1800 kg of fat. If butter contains approximately 80% fat, 1800 kg divided by 80% equals approximately 2250 kg of butter.) The butter temperature going into the refrigerated storage room is usually about 17°C and must be cooled down to 4°C during the following 16 h. (For long-term storage, the butter is held at -23 to $-17°C$.) The average specific heat for butter over this range is 2.30 kJ/(kg·K).

2250(17 − 4)2.30	=	67.3 MJ
135 kg (metal container) × (24 − 4)0.50	=	1.4 MJ
Total/24 h		68.7 MJ

E. After 24 h or longer, the butter is removed from the cooler to be cut and wrapped in 450 g or smaller units. During this process, the butter temperature will rise to approximately 13°C, which constitutes another product load in the cooler when it goes back for storage.

2250(13 − 4)2.30	=	46.6 MJ
90 kg (paper container) × (24 − 4)1.38	=	2.5 MJ
Total/24 h		49.1 MJ

Total of Steps D and E, Product Load in Cooler:

$$\frac{1000(68.7 + 49.1)}{16 \text{ h} \times 3600} = 2.05 \text{ kW}$$

Whipped Butter

To whip butter by the batch method, the butter is tempered to 17 to 21°C, depending on such factors as the season, type of whipper, and so forth. The butter is cut into slabs for placing into the whipping bowl. The whipping mechanism is activated, and air is incorporated until the desired overrun is obtained, usually between 50 and 100%. The whipped butter is packaged mechanically or manually into semirigid plastic containers.

With one continuous system, butter directly from cooler storage is cut into pieces and augered until soft. However, it can be tempered and the augering step omitted. The butter is then pumped into a cylindrical continuous whipper that uses the same principles as those for incorporating air in ice cream. Air or nitrogen is incorporated until the desired overrun is obtained. Another continuous method (used less commercially) is to melt butter or standardize butter oil to the composition of butter with moisture and salt. The fluid product is pumped through a chiller-whipper. Metered air or nitrogen provides overrun control. Whipped butter is pumped in a soft state to the hopper of the filler and packaged in rigid or semirigid containers, such as plastic. It is chilled and held in storage at 0 to 4.4°C.

CHEESE MANUFACTURE

Approximately 800 cheeses have been named, but there are only 18 distinctly different types. A few of the more popular types in the United States are cheddar, cottage, Roquefort or blue, cream, ricotta, mozzarella, Swiss, Edam, and Provolone. Such details of manufacture as setting (starter organisms, enzyme, milk or milk product, temperature, and time), cutting, heating (cooking), stirring, draining, pressing, salting, and curing (including temperature and humidity control) are varied to produce a characteristic variety and its optimum quality.

The production of cheddar cheese in the United States far exceeds the other cured varieties; cottage cheese production is much greater than that of the other uncured types. Another trend in the cheese industry is large factories. These plants may have sufficient curing facilities for the total production. If not, the cheese is shipped to central curing plants.

The physical shape of cured cheese varies considerably. Barrel cheese is common; it is cured in a metal barrel or similar impervious container in units of approximately 225 kg. Cheese may also be cured in rectangular metal containers holding 900 kg.

The microbiological flora of cured cheese are important in the development of flavor and body. Heating the milk for cheese is general practice. The milk may be pasteurized at the minimum HTST conditions or be given a subpasteurization treatment that results in a positive phosphatase test. This is possible when the milk quality is good (low level of spoilage microorganisms and pathogens). Such treatments of milk give the cheese some of the characteristics of raw-milk cheese in curing, such as production of higher flavors, in a shorter time. Pasteurization to produce phosphatase-negative milk is practiced in the making of soft, unripened varieties of cheese and some of the more perishable of the ripened types such as Camembert, Limburger, and Munster.

The standards and definitions of the Federal Food and Drug Administration and of most state regulatory agencies require that cheese that is not made from phosphatase-negative milk must be cured for not less than 60 days at not less than 1.7°C. Raw-milk cheese contains not only lactic acid producing organisms such as *Streptococcus lactis*, which are added to the milk during the cheese making process, but also the heterogeneous mixture of microorganisms present in the raw milk, many of which may produce gas and off-flavors in the cheese. With milk pasteurization, some control of the bacterial flora of the cheese is possible.

Freshly manufactured cheese of the cured types is rubbery in texture and has little flavor; perhaps the more characteristic flavor is slightly acid. The presence of definite flavor(s) in freshly made cheese indicates poor quality, probably resulting from off-flavored milk. On curing under proper conditions, however, the body of the cheese breaks down, and the nut-like, full-bodied flavor characteristic of aged cheese develops. These changes are accompanied by certain chemical and physical changes during the curing process. The calcium paracaseinate of cheese gradually changes into proteoses, amino acids, and ammonia. These changes are a part of the ripening process and may be controlled by time and temperature of storage. As cheese cures, varying degrees of lipolytic activity also occur. In the case of blue or Roquefort cheese, this partial fat breakdown contributes substantially to the characteristic flavor.

During curing, the microbiological development produces changes according to the species and strains present. It is possible to predict from the microorganism data some of the usual defects in cheddar. In some cheeses (e.g., Swiss), gas production accompanies the desirable flavor development.

Cheese quality is evaluated on the basis of a score card. Flavor and odor, body and texture, and color and finish are principal factors. They are influenced by milk quality, the skill of manufacturing (including starter preparation), and the control effectiveness of maintaining optimum curing conditions.

Cheddar Cheese

Manufacture. Raw or pasteurized whole milk is tempered to 30 to 31°C and pumped to a cheese vat which is typically 10.7 m by 1.2 m by 1.2 m and holds approximately 18 Mg of milk. It is set by adding 0.75 to 1.25% active cheese starter and annatto yellow color, which depends on the market demand. After 15 to 30 min, 218 mL of single-strength rennet per 1000 kg milk is diluted in water 1:40 and slowly added with agitation of milk in the vat. After a quiescent period of 25 to 30 min, the curd should have developed proper firmness. The curd is cut into 6- to 10-mm cubes. After 15 to 30 min of gentle agitation, the cooking is initiated by heating water in the vat jacket by means of steam for 30 to 40 min. The curd and whey should increase 1°C per 5 min. Then a temperature of 38 to 39°C is maintained for approximately 45 min.

The whey is drained and the curd trenched along both sides of the vat, allowing a narrow area free of curd the length of the midsection of the vat. Slabs about 250-mm long are cut and inverted at 15-min periods during the cheddaring process. When acidity of the small whey drainage is at a pH of 5.3 to 5.2, the slabs are milled (cut into small pieces) and returned to the vat for salting and stirring, or the curd goes to a machine for the automatic addition of salt and its uniform incorporation into the curd. Weighed curd goes into hoops, which are placed into a press, and 140 kPa is applied. After 0.5 to 1 h, the hoops are taken out of the press, the bandage adjusted to remove wrinkles, and then the cheese is pressed overnight at 170 to 210 kPa or higher. Cheese may or may not be subjected to a vacuum treatment to improve body by reducing or eliminating air pockets. After the surface is dried, the cheese is coated by dipping into melted paraffin or wrapped with one of several plastic films, or oil with a plastic film, and sealed. Yield is about 10 kg per 100 kg of milk.

As cheese factories have grown larger, a change toward faster and more mechanized methods of making cheddar cheese has evolved. The stirred curd method (whereby the cheddaring step is omitted) is being used by more cheese makers. Deep circular or oblong cheese vats with special, reversible agitators and means for cutting the curd are becoming popular. The curd is pumped from these vats to draining and matting tables with sloped bottoms and low sides, then milled, slated, and hooped. In one method, the curd (except for Odenburg cheddar) is carried and drained by a draining-matting conveyor with a porous plastic belt to a second belt for cheddaring. The second belt carries the cheddared curd to the mill. The milled curd is then carried to a finishing table where it is salted, stirred, and moved out for hooping.

Another system, imported from Australia, is used in a number of cheddar cheese factories. This system requires a short method of setting. After the curd is cut and cooked, it is transferred to a series of perforated stainless steel troughs traveling on a conveyor where draining and partial fusion take place. The slabs are then transferred into buckets of a forming conveyor, transferred again to transfer buckets, and finally to compression buckets where cheddaring takes place. Cheddared slabs are discharged to a slatted conveyor which carries them to the mill and then to a final machine where the milled curd is salted, weighed, and hooped.

Curing. Curing temperature and time vary widely among cheddar plants. A temperature of 10°C cures the cheddar more rapidly than lower temperatures. The higher the temperature above 10°C to about 27°C, the more rapid the curing and the more likely that off-flavors will develop. At 10°C, 3 to 4 months are required for a mild to medium cheddar flavor. Six months or more are necessary for an aged (sharp) cheddar cheese. Relative humidity should be roughly 70%. Cheddar intended for processed cheese is cured in many plants at 21°C because of the economy of time. Some experts suggest that the cheddar, after its coating or wrapping, should be held in cold storage at approximately 4.4°C for about 30 days, then transferred to the 10°C curing room. During cold storage, the curd particles knit together, forming a close-bodied cheese. The small amount of residual lactose is slowly converted to lactic acid, along with other changes in optimum curing.

The maximum legal moisture content of cheddar is 39% and the fat must be not less than 50% of total solids. The amount of moisture directly affects the curing rate to some extent within the normal range of 34 to 39%. Cheese having a loose or crumbly body and a high acidity is less likely to cure properly. For best curing, the cheese should have a sodium chloride content of 1.5 to 2.0%. A lower percentage encourages off-flavors to develop, and higher amounts retard flavor development.

Moisture Losses. The loss in mass of cheese during curing is largely attributed to moisture loss. Paraffined cheddar cheese going into cure averages approximately 37% moisture. After a 12-month cure at 4.4°C, paraffined cheese averages approximately 33% moisture. This loss is a real loss to the cheese manufacturer unless the cheese is sold on the basis of total solids. Control of humidity can have an important role in moisture loss. Figure 5 shows the loss from paraffined longhorns in boxes held at 3.3°C and 70% rh over 12 months. The conditions were well controlled but the average loss was 7%. The high loss shown on the graph was influenced by the larger surface area in 5.5-kg longhorns, as compared to 31.8-kg cheddars. Curing the cheese within a good-quality sealed wrapper having a low moisture transmission (but some oxygen and carbon dioxide) largely eliminates loss of moisture.

Dairy Products

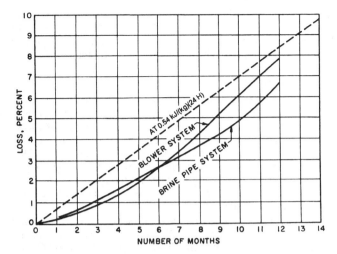

Fig. 5 Shrinkage of Cheese in Storage

Provolone and Mozzarella (Pasta Filata Types)

Provolone is an Italian plastic-curd-type cheese which represents a large group of pasta-filata-type cheeses. While these cheeses vary widely in size and composition, they are all manufactured by a similar method. After the curd has been matted, like cheddar, it is cut into slabs which are worked and stretched in hot water at 65 to 82°C. The curd is kneaded and stretched in the hot water until it reaches a temperature of about 57°C. The maker then takes the amount necessary for one cheese and folds, rolls, and kneads it by hand to give the cheese its characteristic shape and smooth, closed surface. Molding machines have been developed for large-scale operations to eliminate this hand labor. The warm curd of some varieties of pasta filata is placed in molds and submerged in or sprayed with 2°C cold water to harden into the desired shape. The hardened cheese is then salted in batch or continuous brine tanks for final cooling and salting depending on the size and variety. Some pasta filata cheese, such as mozzarella for pizza, is packaged for shipment with wrappers to protect it for the period it is held before use. This cheese may be sealed under vacuum in plastic bags for prolonged holding.

Provolone is salted by submersion in 24% sodium chloride solution at 7°C for 1 to 3 days, depending on the size. It is hung on a light rope to dry and then transferred to the smokehouse and exposed to hickory or other hardwood smoke for 1 to 3 days. The cheese is hung in a curing room for 3 weeks at 13°C and then for 2 to 10 months at 4.4°C. The size varies as well as the shape, but the most common in the United States is 6.4 kg and pear-shaped. The moisture content ranges from 37 to 45% and the salt from 2 to 4%. Milk fat usually comprises 46 to 47% of the total solids. The yield is roughly 9.5 kg per 100 kg of milk.

Swiss Cheese

One of the distinguishing characteristics of Swiss cheese is the eye formation during curing. These eyes result from the development of CO_2. Raw or heat-treated milk is tempered to 35°C and pumped to a large kettle or vat. One starter unit, consisting of 27 mL of *Propioni bacterium shermanii*, 165 mL of *Streptococcus thermophilus*, and 165 mL of *Lactobacillus bulgaricus*, is added per 500 kg of milk. After mixing, 77 mL of rennet per 500 kg is diluted 1:40 with water and slowly added with agitation of the milk. Curd is cut when firm (after 25 to 30 min) into very small granules. After 5 min, curd and whey are agitated for 40 min, and then the steam is released into the jacket without water. Curd is heated slowly to 50 to 54°C in 30 to 45 min. Without additional steam, the cooking is continued until curd is firm and has no tendency to stick when a group of particles is squeezed together (0.5 to 1 h and whey pH 6.3). Curd is dipped into

Table 4 Swiss Cheese Manufacturing Conditions

Processing Step	Temperature, °C	Relative Humidity, %	Time
Setting	35	—	0.4 to 0.5 h
Cooking	50 to 54.4	—	1.0 to 1.5 h
Pressing	26.7 to 29.4	—	12 to 15 h
Salting (brine)	10 to 11.1	—	2 to 3 days
Cool room hold	10 to 15.6	90	10 to 14 days
Warm room hold	21.1 to 23.9	80 to 85	3 to 6 weeks
Cool room hold	4.4 to 7.2	80 to 85	4 to 10 mos

hoops (73 kg) and pressed lightly for 6 h, redressing and turning the hoops every 2 h. Pressing is continued overnight. The next step consists of soaking the cheese in brine until it has about 1.5% salt. Table 4 shows temperature and time at which curing occurs.

Roquefort and Blue Cheese

Roquefort and blue cheese require a mold (*Penicillium roqueforti*) to develop the typical flavor. Roquefort is made from ewes' milk in France. Blue cheese in the United States is made from cow's milk. The equipment used for the manufacture and curing of blue cheese is the same as that used for cheddar with a few exceptions. The hoops are 190 mm in diameter and 150 mm high. They have no top or bottom covers and are thoroughly perforated with small holes. A manually or pneumatically operated device with 50 needles, which are 150 mm long and 3 mm in diameter, is used to punch holes in the blue curd wheels. An apparatus is also needed to feed moisture into the curing room to maintain at least 95% rh without causing a drip onto the cheese.

The milk may be raw or pasteurized and separated. The cream is bleached and may be homogenized at low pressure. Skim milk is added to the cream, and the milk is set with 2 to 3% active lactic starter. After 30 min, 90 to 120 mL of rennet per 500 kg is diluted with water (1:35) and thoroughly mixed into the milk. When the curd is firm (after 30 min), it is cut into 16-mm cubes. Agitation is begun 5 min later. After whey acidity is 0.14% (1 h), the temperature is raised to 33°C and held for 20 min. The whey is drained and trenched. Approximately 2 kg of coarse salt and 62 g of *P. roqueforti* powder are mixed into each 100 kg of curd.

The curd is transferred to stainless steel perforated cylinders (hoops). These hoops are inverted each 15 min for 2 h on a drain cloth, and curd matting is continued overnight. The hoops are removed and surfaces of the wheels covered with salt. The cheese is placed in a controlled room at 15°C and 85% rh and resalted daily for 4 more days (total 5 days). Small holes are punched through the wheels of cheese from top to bottom of the flat surfaces. The cheese is placed in racks on its curved edge in the curing room and held at 10 to 13°C and not less than 95% humidity. At the end of the month, the cheese surfaces are cleaned; the cheese is wrapped in foil and placed in a 2 to 4°C cold room for 2 to 4 months (Table 5). The surfaces are again scraped clean, and the wheels are wrapped in new foil for distribution.

Table 5 Typical Blue Cheese Manufacturing Conditions

Processing Step	Temperature, °C	Relative Humidity, %	Time
Setting	29 to 30	—	1 h
Acid development (after cutting)	29 to 30 / 33	— / —	1 h / 120 s
Curd matting	21 to 24	80 to 90	18 to 24 h
Dry salting	16	85	5 days
Curing	10 to 13	>95	30 days
Additional curing	2 to 4	80	60 to 120 days

Originally, Roquefort and blue cheese were cured in caves with high humidity and constant cool temperature. The refrigeration of insulated blue cheese curing rooms to the optimum temperature is not difficult. However, maintaining a uniform relative humidity of not less than 95% without excessive expense seems to be an engineering challenge, at least in some plants.

Cottage Cheese

Cottage cheese is made from skim milk. It is a soft, unripened curd and generally has a cream dressing added to it. There are small and large curd types, and they may or may not have added fruits or vegetables. The cheese plant equipment may consist of receiving apparatus, storage tanks, clarifier-separator, pasteurizer, cheese vats with mechanical agitation, curd pumps, drain drum, blender, filler, conveyors, and such accessory items as refrigerated trucks, laboratory testing facilities, and whey disposal equipment. The largest vats have a 20-Mg capacity. The basic steps are separation, pasteurization, setting, cutting, cooking, draining and washing, creaming, packaging, and distribution.

Skim milk is pasteurized at the minimum temperature and time of 71.7°C for 15 s to avoid adversely affecting the curd properties. If substantially higher heat treatment is practiced, the cottage cheese manufacturing procedure must be altered to obtain good body and texture quality and reduce curd loss in the whey. Skim milk is cooled to the setting temperature, which is 30 to 32°C for the short set (5 to 6 h) and 21 to 22°C for the overnight set (12 to 15 h). A medium set is used in a few plants. For the short set, 5 to 8% of a good cultured skim milk (starter) and 2.2 to 3.3 mL of rennet diluted in water are added per 1000 kg of skim milk. For the long set, 0.25 to 1% starter and 1 to 2 mL of diluted rennet per 1000 kg are thoroughly mixed into the skim milk. The use of rennet is optional. The setting temperature is maintained until the curd is ready to cut. The acidity of whey at cutting time depends on the total solids content of the skim milk (0.55% for 8.7% and 0.62% for 10.5%). The pH is typically 4.80, but it may be necessary to adjust for specific make procedures.

The curd is cut into 12-mm cubes for large curd and 6 mm for small curd cottage cheese. After the cut curd sets for 10 to 15 min, heat is applied to water in the vat jacket to maintain a temperature rise in the curd and whey of 1 K each 5 min. In very large vats, jacket heating is not practical, and superheated culinary steam in small jet streams is used directly in the vat; 20 to 30 min after cutting, very gentle agitation is applied. Heating rate may be increased to 1.5 to 2 K per 5 min as the curd firms enough to resist shattering. Cooking is completed when the cubes contain no whey pockets and have the desired firmness. The final temperature of curd and whey is usually 49 to 54°C, but some cheesemakers heat to 63°C when making the small type curd.

After cooking is completed, the hot water in both the jacket and the whey is drained. Wash water temperature is adjusted to about 21°C for the first washing and added gently to the vat to reduce curd temperature to 27 to 29°C. After gentle stirring and a brief hold, the water-whey mixture is drained. The temperature of the second wash is adjusted to reduce the temperature of curd to 10 to 13°C and to 4.4°C with a third wash. Water for the last wash may have 3 to 5 mg/kg of added chlorine. The curd is trenched for adequate drainage. The dressing is made from low-fat cream, salt, and usually 0.1 to 0.4% stabilizer based on cream mass. Salt averages 1%, and milk fat must be 4% or more in creamed cottage cheese or 2% in low-fat cottage cheese. The dressing is cooled to 4.4°C and blended into the curd.

A cheese vat can be reused sooner if the cheese pumps quickly convey the curd and whey at the completion of cooking to a special tank for drainage of the whey, washing, and blending of the dressing with the curd. Creamed cottage cheese is transferred mechanically to an automatic packaging machine. One type of filler employs an oscillating cylinder which holds a specific volume. Another type has a piston in a cylinder which discharges a definite volume. Cottage cheese is perishable and must be stored at 4.4°C or lower for prolonging the keeping quality to 2 or 3 weeks. A good yield is 15.5 kg of curd per 100 kg of skim milk with 9% total solids.

Other Cheeses

Table 6 presents data on a few additional common varieties of cheese in the United States. With the exception of the soft ripened cheeses such as Camembert and Liederkranz, freezing of cheese results in undesirable texture changes. This can be serious, as in the case of cream cheese, where a mealy, pebbly texture results. Other types, such as brick and Limburger, undergo a slight roughening of texture which is undesirable but which still might be acceptable to certain consumers. As a general rule, cheese should not be subjected to temperatures below −1.7°C.

When cured cheese is held above the melting point of milk fat, it becomes greasy because of the oiling off. The oiling-off point of all types of cheese except process cheese begins at 20 to 21°C. Consequently, storage should be substantially below the melting point (Table 7). Uncured cheese (i.e., cottage, cream) is highly perishable and thus should not be stored above 7°C and preferably at 1.7°C.

Processing protects cheese from oiling off. By heating the bulk cheese to temperatures of 60 to 82°C, and through the incorporation of emulsifying salts, a more stable emulsion is formed than in natural or nonprocessed cheese. Process cheese will not oil off even at melting temperatures. Because of the temperatures used in processing, process cheese is essentially a pasteurized product. Microorganisms causing changes in the body and flavor of the cheese during cure are largely destroyed; hence there will be practically no further flavor development. Consequently, the maximum permissible temperature for the storage of process cheese is considerably higher than any of the other types. Table 7 shows the maximum temperatures of storage for cheese of various types.

Table 6 Curing Temperature, Humidity, and Time of Some Cheese Varieties

Variety	Curing Temperature, °C	Relative Humidity, %	Curing Time
Brick	15 to 18	90	60 days
Romano	10 to 15	85	5 to 12 mos
Mozzarella	21	85	24 to 72 h
Edam	10 to 15	85	3 to 4 mos
Parmesan	13 to 15	85 to 90	14 mos
Limburger	10 to 15	90	2 to 3 mos

Table 7 Temperature Range of Storage, Common Types of Cheese

Cheese	Ideal Temperature, °C	Maximum Temperature, °C
Brick	−1 to 1	10
Camembert	−1 to 1	10
Cheddar	−1 to 1	15
Cottage	−1 to 1	7
Cream	0 to 1	7
Limburger	−1 to 1	10
Neufchatel	0 to 1	7
Process American	4 to 7	24
Process brick	4 to 7	24
Process Limburger	4 to 7	24
Process Swiss	4 to 7	24
Roquefort	−1 to 1	10
Swiss	−1 to 1	15
Cheese foods	4 to 7	13

Dairy Products

Refrigeration of Cheese Rooms

Cheeses that are to be dried prior to wrapping or waxing enter the cooler at approximately room temperature. Sufficient refrigerating capacity must be provided to reduce the cheeses to drying room temperature. The product load may be taken as 1 kW for each 1500 to 1900 kg per day. Product load in a cheese-drying room is usually small compared to total room load. Extreme accuracy in calculating product load is not warranted.

When determining the peak refrigerating load in a cheese drying room, the factor to be remembered is that peak cheese production may coincide with periods of high ambient temperature. In addition, these rooms normally open directly into the cheese-making room, where both temperature and humidity are quite high. Also, traffic in and out of the drying room may be heavy; therefore, ample allowance for door losses should be made. Two to three air changes per hour are quite possible during the flush season. See Chapter 12 for information on load calculations.

To maintain the desired humidity, refrigerating units for the cheese-drying room should be sized to handle the peak summer load with not more than a 10 K difference between the return air temperature and evaporator temperature. Units operated from a central refrigerating system should be equipped with suction pressure regulators.

Temperature control may be obtained through a room thermostat controlling a solenoid valve in the liquid supply to the unit or units. Fans should be allowed to run continuously. A modulating suction pressure regulator is not a satisfactory temperature control for a cheese-drying room because it causes undesirable variations in humidity.

Air circulation should only be sufficient to assure uniform temperature and humidity throughout the room. Strong drafts or air currents should be avoided because they cause uneven drying and cracking of the cheeses. The most satisfactory refrigerating units are the ceiling-suspended between-the-rails type or the penthouse type. One unit for each 37 to 46 m^2 of floor area usually assures uniform conditions. One unit should be placed near the door to the room to cool the warm moist air before it has a chance to spread over the ceiling. Otherwise, condensation dripping from the ceiling or mold growth will result.

Humidity control during winter may present problems in cold climates. Since most of the refrigeration load (during the peak season) is due to insulation losses and warm air entering through the door, refrigerating units may not operate enough during cold weather to take up the moisture given up by the cheese, resulting in excess humidity and improper drying of the cheese. Within certain limits, this can be overcome by reducing the speed of the unit fans, lowering the suction pressure, and increasing the number of defrosts. If there are several units in the room, the refrigeration may be turned off on some. Fans should be left running to assure uniform conditions throughout the room. If these adjustments are not sufficient, or if automatic control of humidity is desired, it is necessary to use reheat coils in the airstream leaving the units. These may be electric heaters, steam or hot water coils, or hot gas from the refrigerating system. A heating capacity of 15 to 20% of the refrigerating capacity of the units is usually sufficient.

A humidistat may be used to operate the heaters when the humidity rises above the desired level. The heater should be wired in series, with a second room thermostat set to shut it off if the room temperature becomes excessive. Because of variation in size and shape of drying rooms, it is impossible to generalize about air velocities and capacities. Airflow should be regulated so that the cheese feels moist for the first 24 h and then becomes progressively drier and firmer.

Calculation of the product refrigeration load for a cheese curing room involves a simple computation of heat to be removed from the cheese at the incoming temperature in order to bring it to curing temperature, using 2.72 kJ/(kg·K) as the specific heat of cheese. For most varieties, the heat given off during curing is negligible.

While the fermentation of lactose to lactic acid which occurs in cheese is an exothermic reaction, this process is substantially completed in the first week after cheese is made; hence further heat given off during curing is of no significance. Assuming that average conditions for American cheese curing are approximately 7°C and 70% rh, if −1 to 1.7°C refrigerant is used in the cooling system, a humidity of about 70% will be maintained.

FROZEN DAIRY DESSERTS

Ice cream is the most common frozen dairy dessert. Legal guidelines for the composition of frozen dairy desserts generally follow federal standards. The amount of air incorporated during freezing is controlled for the prepackaged products by the standard specifying the minimum density and/or a minimum density of food solids.

The basic dairy components of frozen dairy dessert are milk, cream, and condensed or nonfat dry milk. Some plants also use butter, butter oil, buttermilk (liquid or dry), and dry or concentrated whey of the sweet type. The acid-type whey (e.g., from cottage cheese) can be used for sherbets.

Ice Cream

Milk fat content (called butterfat by some standards) is one of the principal factors in the legal standards for ice cream. The fat in other ingredients such as eggs, nuts, cocoa, or chocolate do not satisfy the legal minimum. Federal standards set the minimum milk fat content at 8% for bulky flavored ice cream mixes (e.g., chocolate) and 10% or above for the other flavors (e.g., vanilla). Manufacturers, however, usually make two or more grades of ice cream, one being competitively priced with the minimum legal fat content, and the others richer in fat, higher in total solids, and lower in overrun for a special trade. Such ice cream may be made with a fat content of 16 or 18%, although most ice cream is made with a fat content ranging from 10 to 12%.

Serum solids content designates the nonfat solids from milk. The chief components of milk serum are lactose, milk proteins (casein, albumin, and globulin), and milk salts (sodium, potassium, calcium, and magnesium as chlorides, citrates, and phosphates). The following average composition for serum solids is useful for general calculations: lactose, 54.5%; milk proteins, 37.0%; and milk salts, 8.5%.

The serum solids in ice cream produce a smoother texture, better body, and better melting characteristics. Because serum solids are relatively inexpensive compared with fat, they are used liberally. The total solids content usually is kept below 40%.

The lower limit in the serum solids content, 6 to 7%, is found in a homemade type of ice cream, where the only dairy ingredients are milk and cream. Ice creams with an unusually high fat content are also kept near this serum solids content so that the total solids content will not be excessive. Most ice cream, however, is made with condensed or nonfat dry milk added to bring the serum solids content within the range of 10 to 11.5%. The upper extreme of 12 to 14% serum solids can be used only where rapid product sales turnover or other special means avoid sandiness.

The sugar content of ice cream is of special interest because of its effect on the freezing point of the mix and its hardening behavior. The extreme range of sugar content encountered in ice cream is from 12 to 18%, with 16% being most representative of the industry. The chief sugar used is sucrose (cane or beet sugar), either in granulated or liquid form. Many manufacturers use dextrose and corn syrup solids to replace part of the sucrose. Some manufacturers prefer sucrose in liquid form, or in a mixture with syrup, because of lower cost and easier handling in tank car lots. In some instances, 50% of the sucrose content has been replaced by other sweetening agents. A more common practice is to replace one-fourth to one-third of the sucrose with dextrose or corn syrup solids, or a combination of the two.

Practically all ice cream is made with a stabilizer to help maintain a smooth texture, especially under the conditions that prevail in retail cabinets. Manufacturers who do not use stabilizers offset this omission by a combination of factors such as a high fat and solids content, the use of superheated condensed milk to aid in smoothing the texture and imparting body, and a sales program designed to provide rapid turnover. The stabilizing substances most commonly added are carboxymethylcellulose (CMC) and sodium alginate, a product made from giant kelp gathered off the coast of California. Gelatin is used for some ice cream mixes that are to be batch pasteurized. Other stabilizers are locust or carob bean gum, gum arabic or acacia, gum tragacanth, gum Karaya, psyllium seed gum, and pectin. The amount of stabilizer commonly used in ice cream ranges from 0.20 to 0.35% of the mass of the mix.

Many plants now combine an emulsifier with the stabilizer to produce a smoother and richer product. The emulsifier reduces the surface tension between the water and fat phase to produce a drier-appearing product.

Egg solids in the form of fresh whole eggs, frozen eggs, or powdered whole eggs or egg yolks are used by some manufacturers. While flavor and color may motivate this choice, the most common reason for selecting them is to aid the whipping qualities of the mix. The amount required is about 0.25% egg solids, with 0.50% being about the maximum content for this purpose. To obtain the desired result, the egg yolk should be in the mix at the time it is being homogenized.

In frozen custards or parfait ice cream, the presence of eggs in liberal amounts and the resulting yellow color are identifying characteristics. Federal standards specify a minimum 1.4% egg yolk content for these products.

Ice Milk

Ice milk commonly contains 3 to 4% fat and 13 to 15% serum solids; formulations with respect to sugar and stabilizers are similar to those for ice cream. The sugar content in ice milk is somewhat higher in order to build up the total solids content. The stabilizer content is also higher in proportion to the higher water content of ice milk. The overrun is approximately 70%.

Soft Ice Milk or Ice Cream

Machines that serve freshly frozen ice cream are common in roadside stands, retail ice cream stores, and restaurants. These establishments must meet sanitation requirements and have facilities for proper cleaning of the equipment, but very few blend and process the ice cream mix used. The mix is usually supplied either from a plant specializing in producing ice cream mix only or from an established ice cream plant. The mix should be cooled to about 1.7°C at the time of delivery, and the ice cream outlet should have ample refrigerated space to store the mix until it is frozen. To be served in a soft condition, this ice cream mix is usually frozen stiffer than would be customary for a regular plant operation with a 30 to 50% overrun. Some mixes are prepared only for soft serve. They are 1 to 2% greater in serum solids and have 0.5% stabilizer-emulsifier to aid in producing a smooth texture. The overrun is limited to 30 to 40% during freezing to the soft-serve condition.

Frozen Yogurt

Hard- or soft-serve frozen yogurt is similar to low-fat ice cream in composition and processing. The significant exception is the presence of a live culture in the yogurt.

Sherbets

Sherbets are fruit (and mint) flavored frozen desserts characterized by their high sugar content and tart flavor. They must contain between 1 and 2% milk fat and not more than 5% by mass of total milk solids. While the milk solids can be supplied by milk, the general practice is to supply them by using ice cream mix. Typically, a solution of sugars and stabilizer in water is prepared as a base for sherbets of various flavors. To 70 kg of sherbet base, 20 kg of flavoring and 10 kg of ice cream mix are added. The sugar content of sherbets ranges from 25 to 35%, with 28 to 30% being most common. One example of a sherbet formula is: milk solids, 5%, of which 1.5 is milk fat; sugar, 13%; corn syrup solids, 22%; sherbet stabilizer, 0.3%; and flavoring, acid, and water, 59.7%.

In sherbets, and even more so in ices, a high overrun is not desirable because the resulting product will appear foamy or spongy under serving conditions. The overrun should be kept within 25 to 40%. This fact and the problem of preventing bleeding (the leakage of syrup from the frozen product) emphasize the importance of the choice of stabilizers. If gelatin is selected as the stabilizer in sherbets, the freezing conditions must be managed to avoid an excessive overrun. The gums added to ice cream are commonly used as the stabilizer in sherbets and ices.

Ices

Ices contain no milk solids, but closely resemble sherbets in other respects. To offset the lack of solids from milk, the sugar content of ices is usually slightly higher than in sherbets; it is usually 30 to 32%. A combination of sugars should be used in order to prevent crusty sugar crystallization, just as in the case of sherbets. The usual procedure is to make an ice-base solution of the sugars and stabilizer, from which different flavored ices may be prepared by adding the flavoring material in the same general manner as mentioned for sherbets.

Ices contain few ingredients with lubricating qualities and often cause extensive wear on scraper blades in the freezer. Frequent resharpening of the blades is necessary. Where a number of freezers are available, and the main production is ice cream, it is desirable to confine the freezing of ices and sherbets to a specific freezer or freezers, which should then receive special attention to resharpening.

Making Ice Cream Mix

The chosen composition for a typical ice cream would be

| Fat | 12.5% | Sugar | 15.0% |
| Serum solids | 10.5% | Stabilizer | 0.3% |

Mixing and Pasteurizing. Generally, the liquid dairy ingredients are placed in a vat equipped with suitable means of agitation, especially to keep the sugar in suspension until it is dissolved. The dry ingredients are then added, with suitable precautions to prevent lump formation of such products as stabilizers, nonfat dry milk solids, powdered eggs, and cocoa. Gelatin should be added while the temperature is still low to allow time for the gelatin to imbibe water before its dissolving is promoted by heat. Dry ingredients that tend to form lumps may be successfully added by first mixing them with some of the dry sugar so that moisture may penetrate freely. Where vat agitation is not fully adequate, sugar may be withheld until the liquid portion of the mix is partly heated so that promptness of solution avoids settling out.

The mix is pasteurized to destroy any pathogenic organisms, to lower the bacterial count so as to enhance the keeping quality of the mix and comply with bacterial count standards, to dissolve the dry ingredients, and to provide a temperature suitable for efficient homogenization. A pasteurizing treatment of 68.3°C maintained for 0.5 h is the minimum allowed. The mix should be homogenized at the pasteurizing temperature. Vat batches should be homogenized in 1 h and preferably less.

Many ice cream plants use continuous pasteurization methods that use plate-type heat exchange equipment for heating and cooling the mix. If some solid ingredients are selected, such as skim milk

Dairy Products

powder and granulated sugar, a batch is made up in a mixing tank at a temperature of 38 to 60°C. This preheated mix is then pumped through a heating section of the plate unit where it is heated to a temperature of 79.4°C or higher, and held for 25 s while passing through a holding tube. The mix is then homogenized and pumped to the precooling plate section using city, well, or cooling tower water as the cooling medium. Final cooling of the mix may be done in an additional plate section, using chilled water as the cooling medium, or the mix may be cooled through a separate mix cooler. A propylene glycol medium is sometimes used for cooling to temperatures just above the freezing point.

Large plants generally use all liquid ingredients, especially if the production is automated and computerized. The ingredients are blended at 4.4 to 15°C. The mix passes through the product-to-product regeneration section of a plate-type heat exchanger with about 70% regeneration during preheating. The mix is HTST heated to not less than 79.4°C, homogenized, and held for 25 s. Greater heat treatment is common, and 104.4°C for 40 s is not unusual. The final heating may be accomplished with plate equipment, a swept-surface heat exchanger, or a direct steam injector or infusor.

Steam injection and infusion equipment may be followed by vacuum chamber treatment, whereby the mix is flash cooled to 82 to 88°C by a partial vacuum. It is further cooled through a regenerative plate section and additionally cooled indirectly to 4.4°C or less with chilled water. The chief advantage of the vacuum treatment is the flavor improvement of the mix if prepared from raw materials of questionable quality.

Homogenizing the Mix. Homogenization disperses the fat in a very finely divided condition so it will not churn out during freezing. Most of the fat in milk and cream is in globules 3 to 7 μm in diameter. Some of the globules are as large as 12 μm or larger in diameter, especially if there has been some churning incidental to handling. In a properly homogenized mix, globules are seldom over 2 μm in diameter.

Cooling and Holding Mix. Methods of final cooling of ice cream mix after pasteurization depend on the equipment used and the final mix temperature desired. The mix should be as cold as possible, to about −1°C minimum for greater capacity and less refrigeration load on the ice cream freezers. Smaller plants generally use the vat holding system of pasteurization with either a Baudelot-type surface cooler or a plate-type heat exchanger, both with precooling and final cooling sections. The precooling may be done with city, well, or cooling tower water, and mix leaving the precooling section is about 6 K warmer than the entering water temperature. The Baudelot-type cooler may be arranged for final cooling with chilled water, propylene glycol, or direct expansion refrigerant. A final mix temperature of −1 to 0.5°C can be obtained over the surface cooler using propylene glycol or refrigerant. Final mix temperature when using chilled water will be about 4.4°C.

For larger ice cream plants, where low mix temperature is desired and where plate-type pasteurizing equipment is installed, it may be desirable to use separate equipment for the final cooling. Where the mix is preheated to about 60°C, it will be precooled to about 6 K warmer than the entering precooling water temperature; final cooling can be done in a remote cooler. An ammonia jacketed, scraped surface chiller is often selected. Where cold liquid mix is used through a continuous pasteurizing, high-heat vacuum system with regeneration at about 70%, the temperature of the mix to the final cooling unit will be 29°C, assuming 4.4°C original mix temperature and 88°C temperature of mix returning through the regenerating section.

Where plants have ample ice cream mix holding tank capacity (enabling the mix to be held overnight), part of the final mix cooling may be accomplished by means of a refrigerated surface built into the tanks. Using refrigerated mix holding tanks, the average rate of cooling may be estimated at 0.6 K/h. Mixes made up with gelatin as a stabilizer should be aged 24 h to allow time for the full set of the gelatin to develop. Mixes made with sodium alginate or other vegetable-type stabilizers develop maximum viscosity on being cooled, and can be used in the freezer immediately.

Freezing Process

The ice cream freezer freezes the mix to the desired consistency and whips in the desired amount of air in a finely divided condition. The aim is to conduct the freezing and later hardening to obtain the smoothest possible texture.

Freezing an ice cream mix means, of course, freezing a mixed solution. The solutes that determine the freezing point are the lactose and soluble salts contained in the serum solids and the sugars added as sweetening agents. The other constituents of the mix affect the freezing point only indirectly, by displacing water and affecting the in-water concentration of the solutes mentioned. Leighton (1927) developed a reliable method for computing the freezing points of ice cream mixes from their known composition. He added the lactose and sucrose content of the mix, expressed their concentration in terms of parts of sugar per 100 parts of water, and determined the freezing-point depression due to the sugars by reference to published data for sucrose. This computation is justified because lactose and sucrose have the same molecular mass.

$$\% \text{ lactose in mix } = 0.545(\% \text{ serum solids})$$

$$\frac{(\% \text{ lactose} + \% \text{ sucrose})100}{\% \text{ water in mix}} = \frac{\text{parts lactose} + \text{sucrose}}{\text{per 100 parts water}}$$

To the freezing point depression caused by these sugars, he added the depression that will be caused by the soluble milk salts. The depression caused by the salts is computed as follows:

Freezing-point depression caused by salt solids in °C

$$= \frac{2.37(\% \text{ serum solids})}{\% \text{ water in mix}}$$

Table 8 presents the freezing points of various ice creams and a typical sherbet and an ice, as computed by Leighton's method. The freezing point represents the temperature at which freezing commences. As in the case of all solutions, the unfrozen portion becomes more concentrated as the freezing progresses, and the freezing temperature therefore decreases as the freezing progresses. In a simple solution, containing only one solute, this trend will progress until the unfrozen portion represents a saturated solution of the solute, and thereafter the temperature will remain constant until the freezing has been completed. This temperature is known as the cryohydric point of the solute. In a mixed solution such as ice cream, which contains several sugars and a number of salts, no such point can be recognized.

Table 8 Freezing Points of Typical Ice Creams, Sherbet, and Ice

	Composition of the Mix, %					Freezing Point, °C
	Fat	Serum Solids	Sugar	Stabilizer	Water	
Ice cream	8.5	11.5	15	0.4	64.6	−2.45
	10.5	11.0	15	0.35	63.15	−2.46
	12.5	10.5	15	0.30	61.7	−2.47
	14.0	9.5	15	0.28	61.22	−2.40
	16.0	8.5	15	0.25	60.25	−2.33
	10.5	8.4	{ S 12 D 4 }	0.40	64.7	−2.56
Sherbet	1.2	1.0	{ S 22 D 8 }	0.50	67.3	−3.35
Ice	0	0	{ S 23 D 9 }	0.50	67.5	−3.51

S = Sucrose   D = Dextrose

Table 9 Freezing Behavior of Typical Ice Cream[a]

Water Frozen to Ice, %	Freezing Point of Unfrozen Portion, °C	Water Frozen to Ice, %	Freezing Point of Unfrozen Portion, °C
0	−2.47	40	−4.22
5	−2.58	45	−4.65
10	−2.75	50	−5.21
15	−2.90	55	−5.88
20	−3.11	60	−6.78
25	−3.31	70	−9.45
30	−3.50	80	−14.92
35	−3.87	90	−30.16

[a]Composition, %: fat, 12.5; serum solids, 10.5; sugar, 15; stabilizer, 0.30; water, 61.7.

On the contrary, the sugars remain in solution in a supersaturated state in the unfrozen portion of the product. This is due to the fact that by the time the saturation point has been reached, the temperature is so low and the viscosity so high that essentially a glass state exists. In a mixed solution, however, the temperature required for complete freezing must be somewhat below the cryohydric point of that solute with the lowest cryohydric point. In ice cream, that solute is calcium chloride, contained as a component of serum solids. The cryohydric point of calcium chloride is −51°C. Therefore, the ice cream ranges from 0 to 100% frozen within the approximate range of −2.5 to −55°C.

Therefore, the temperature to which the ice cream has been frozen becomes a measure of the degree to which it has been frozen, as illustrated by Table 9. In the table, the freezing points of the unfrozen portions of the third ice cream listed in Table 8 have been computed when 0 to 90% of the original water has been frozen out as ice.

Refrigeration Requirements. Exact calculation of refrigeration requirements is complicated by the number of factors involved. The specific heat of the mix varies with its composition. According to Zhadan (1940), the specific heat of food products may be computed by assuming the following specific heats in kJ/(kg·K) for the chief components: carbohydrates, 1.42; proteins, 1.55; fats, 1.67; and water, 4.18. Salts are normally not included. Where they are present in significant amounts, as in ice cream (8.5% of the serum solids), a specific heat of 0.84 is accurate. The value given by Zhadan for fats is apparently for solid fats. For milk fat in a liquid condition, Hammer and Johnson (1913) found the specific heat to be 2.18. In addition, their data clearly show that the latent heat of fusion of fats becomes involved. From their data, the latent heat of fusion of milk fat is about 81.4 kJ/kg.

The change from liquid to solid fat occurs over a wide temperature range, approximately 27 to 5°C; in changing from solid to liquid fat, the range is approximately 10 to 40°C. This wide discrepancy between solidifying and melting behavior is apparently because milk fat is a mixture of glycerides, and mutual solubility of the glycerides is involved. In any case, the latent heat of fusion of the fat is involved in cooling the mix from the pasteurizing and homogenizing temperature down to the aging temperature of 3.3 to 4.4°C. Instead of making detailed calculations, a specific heat of 3.35 kJ/(kg·K) is assumed for ice cream mix, which is high for mixes ranging from 36 to 40% total solids.

In calculating the refrigeration required for freezing and hardening, a single value of a specific heat for frozen ice cream cannot be chosen. As shown in Table 9, any change in temperature in freezing and hardening involves some latent heat of fusion of the water, as well as the sensible heat of the unfrozen mix and the ice. Near the initial freezing point, much more latent heat of fusion is involved per degree temperature change than in well-hardened ice cream, e.g., at −23 to −24°C. For this reason, instead of using an overall value of specific heat, freezing load may be computed as follows:

1. First determine the temperature to which the freezing is to occur; then determine (by calculations such as those used to develop Table 9) how much water will be converted to ice. The heat to be removed is the product of the heat of fusion of ice and the mass of water frozen.

2. Compute the sensible heat that must be removed in the desired temperature change, by treating the product as a mix; that is, use the specific heat for ice cream mix. The temperature change times the mass of product times 3.35 = sensible heat to be removed.

In such a calculation, the water present is treated as though it all remained in a liquid form until the desired temperature had been reached, although ice was forming progressively. Because ice has a specific heat of 2.060 kJ/(kg·K) instead of 4.187 kJ/(kg·K) as for water, this calculation will err in the direction of generous refrigeration. To offset this, the freezer agitation develops friction heat. Approximately 80% of the energy input in the motor of the freezer is converted to heat in the product. Where the product is frozen to a stiff consistency, power requirements increase, and should be added to the load calculation.

A litre of ice cream mix has a mass of from 1.08 kg for mixes with a high fat content, to 1.11 kg for mixes with a low fat content and a high content of serum solids and sugar. The mass of a unit volume of ice cream varies with the mix and overrun according to the following relationship:

$$\text{Percentage overrun} = \frac{100(\text{density of mix} - \text{density of ice cream})}{\text{density of ice cream}}$$

Freezing Ice Cream. Both batch and continuous ice cream freezers are in general use. Both are arranged with a freezer cylinder having either an annular space or coils around the cylinder, where cooling is accomplished by direct refrigerant cooling, either in a flooded arrangement with an accumulator or controlled by a thermostatic expansion valve. The freezer cylinder has a dasher, which revolves within the cylinder. Sharp metal blades on the dasher scrape the inner surface of the cylinder to remove the frozen film of ice cream as it forms. Some batch freezers use plastic dashers and blades.

Batch freezers range in size from 2 to 40 L of ice cream per batch, the smaller sizes being used for retail or soft ice cream operations, and the 40-L size used in small commercial ice cream plants or in large plants for running small special-order quantities. Batch freezers larger than the 40-L size have not been used extensively since the development of the continuous freezer.

In operation, a measured quantity of mix is placed in the freezer cylinder and the required flavor, fruit, or nuts are added as freezing of the mix progresses. Freezing is continued until the desired consistency is obtained in the judgment of the operator or by the indication of a meter showing an increase in the current drawn by the motor as the partly frozen mix becomes stiffer. At the desired point of freezing, the refrigeration is cut off from the freezer cylinder, usually by closing the refrigerant suction valve. The dasher continues operating until enough air has been taken into the mix by the whipping action to produce the desired overrun. The overrun is checked by taking and weighing a sample from the freezer. When the desired overrun is obtained, the entire batch is discharged from the freezer cylinder into cans or cartons, and the machine is then ready for a new batch of mix.

The output of a batch-type freezer varies with the sharpness of the blades, the refrigeration supplied, and the overrun desired. The average maximum output for commercial batch freezers is 8 batches per hour. This schedule allows 3 to 4 min to freeze, 2 to 3 min to whip, and about 1 min to empty the ice cream and refill with mix. For this time schedule, it is assumed the ice cream is drawn from the freezer at not over 100% overrun, at a temperature of about −4°C and at a refrigerant temperature around the freezer cylinder of about −26°C.

Dairy Products

Continuous ice cream freezers range in size from 40 to 2800 mL/s at 100% overrun, and they are used almost exclusively in commercial ice cream plants. Where large capacities are required, multiple units are installed with the ice cream discharge from several machines connected together to supply the requirements of automatic or semiautomatic packaging or filling machines. In operation, the ice cream mix is continuously pumped to the freezer cylinder by a positive displacement rotary pump. Air pressure within the cylinder is maintained from 140 kPa to more than 690 kPa (gage), supplied by either a separate air compressor or drawn in with the mix through the mix pump. The mix entering the rear of the freezer cylinder becomes partly frozen and takes on the overrun because of air pressure and the agitation of the dasher and freezer blades as it moves to the front of the cylinder and is discharged.

The output capacity of most continuous freezers can be varied from 50 to 100% rated capacity by regulating the variable speed control supplied for the mix pump. Continuous freezers can be used for nearly every flavor of ice cream, ice milk, sherbets, and ices. Where flavors requiring nuts, whole fruits, or candy pellets are run, the base or unflavored mix is run through the continuous freezer and then passed through a fruit feeder which automatically feeds and mixes the flavor particles into the ice cream. Ice cream can be discharged from continuous freezers at temperatures of −4°C, as required for ice cream bar (novelty) operations, up to a very stiff consistency at −6.7°C, as required for automatic filling of small packages.

Special low-temperature ice cream freezers are available to produce very stiff ice cream for extruded shapes, stickless bars, and sandwiches. Ice cream temperatures as low as −9°C can be drawn with some mixes. When ample refrigerating effect is supplied, a variation of ice cream discharge temperature can be obtained by regulating the evaporator temperature around the freezer cylinder by a suction pressure regulating valve. For filling cans and cartons, the average discharge temperature from the continuous freezer is about −5.6°C, when operating with ammonia in a flooded system at about −32°C.

To calculate accurately the refrigeration requirement for freezing the ice cream mix in the freezer, the mass of the mix per gal and the amount of water should be known. This can be checked by weighing, knowing the percentage of water, or by calculating the mass from the mix formula, as in Example 2.

Example 2. Find the density of mix for the following composition (by percent): milk fat, 12.0; serum solids, 10.5; sugar, 16.0; stabilizer, 0.25; egg solids, 0.25; and water, 61.0.

Solution: The density of the mix is

$$\frac{100}{\left(\dfrac{\%\text{ milk fat}}{930} + \dfrac{\%\text{ solids, not fat}}{1580} + \dfrac{\%\text{ water}}{1000}\right)}$$

$$= \frac{100}{(12/930) + (27/1580) + (61/1000)} = 1099 \text{ kg/m}^3$$

The overrun in ice cream varies from 60 to 100%, which affects the required refrigeration. For a continuous freezer, the required refrigeration may be calculated as in Example 3.

Example 3. Assume a typical ice cream mix as listed in Example 2 with 100% overrun. The mix contains 61% water and goes to the freezer at a temperature of 4.4°C. Freezing would start in this mix at about −3°C, and 48% of the water would be frozen at −5.5°C.

The mass of mix required to produce 1000 L of ice cream is

$$\frac{100}{100 + \%\text{ overrun}} \times 1000 \text{ L} \times \text{Mix density}$$

For the ice cream being considered, the mass of mix required for 1 m³ (1000 L) would be

$$\frac{100}{100 + 100} \times 1 \times 1099 = 550 \text{ kg}$$

Calculations of capacity required to freeze 1000 L/h (or 0.153 kg/s) of ice cream are as follows:

$$\begin{aligned}
\text{Sensible heat of mix: } 0.153[4.4 - (-3)]3.35 &= 3.79 \text{ kW} \\
\text{Latent heat: } 0.153 \times 0.61 \times 0.48 \times 335 &= 15.01 \text{ kW} \\
\text{Sensible heat of slush: } 0.153[(-3) - (-5.5)]2.72 &= 1.04 \text{ kW} \\
\text{Heat from motors: Assume 10 kW} &= 10 \text{ kW} \\
\text{Total} &= 29.8 \text{ kW} \\
\text{5\% losses from freezer and piping (estimated)} &= 1.5 \text{ kW} \\
\text{Total refrigeration} &= 31.3 \text{ kW}
\end{aligned}$$

In continuous freezer operations, the heat gain from motors and losses from freezer and piping would remain about the same at all levels of overrun, but the necessary refrigerating effect would vary with the mass of mix required to produce a litre of ice cream, as shown in Table 10.

Table 10 Continuous Freezing Loads for Typical Ice Cream Mix

Overrun, %	Ammonia Refrigeration at 21 kPa (gage) Suction Pressure, kJ/L
60	135
70	130
80	125
90	121
100	117
110	113
120	110

Hardening Ice Cream. After leaving the freezer, the ice cream is in a semisolid state and must be further refrigerated to become solid enough for storage and distribution. The ideal serving temperature for ice cream is about −13°C; it is considered hard at −18°C. To retain a smooth texture in hardened ice cream, the remaining water content must be frozen rapidly, so that the ice crystals formed will be small. For this reason, most hardening rooms are maintained at −29°C and some as low as −35°C. Most modern hardening rooms have forced-air circulation, usually from fan-coil evaporators. With the ice cream containers arranged so that air will circulate around them, the hardening time is about one-half that of rooms having overhead coils or coil shelves and gravity circulation. With forced-air circulation in the hardening room and average plant conditions, ice cream in 10- or 20-L containers (or smaller packages in wire basket containers), all spaced to allow air circulation, will harden in about 10 h. Hardening rooms are usually sized to allow space for a minimum of three times the daily peak production and for a stock of all flavors, with the sizing based on 400 L/m² of floor area in a 2.7-m high room when stacked loose, which includes aisles.

Some larger plants use ice cream hardening carton (carrier) freezers, which discharge into a low-temperature storage room. Because of the various size packages to be hardened, most tunnels are the air-blast type, operating at temperatures of −34 to −40°C and, in some cases, as low as −46°C. Containers under 2 L are usually hardened in these blast tunnels in about 4 h.

Contact-plate hardening machines are also used. They must continuously and automatically load and unload to introduce the packages from the filler without delay. Compared to carton (carrier) freezers, contact-plate hardeners save space and power and eliminate package bulging. They are limited to packages of uniform thickness having parallel flat sides. These freezers are described in Chapter 15.

Temperature in the storage room is held at about −30°C. Space in storage rooms can be estimated at 1000 L/m² when palletized and stacked solid 1.8 m high, including space for aisles. Many storage rooms today use pallet storage and racking systems. Such rooms may be 10 m tall or more, some using stacker-crane automation. Freezer storages are described in Chapter 13.

Refrigeration required to harden ice cream varies with the temperature from the freezer and the overrun. The following example calculates the refrigeration required to harden a typical ice cream mix.

Example 4. Assume ice cream with 100% overrun enters the hardening room at a temperature of −4°C. At this temperature, approximately 30% of the water would be frozen in the ice cream freezer with the remainder to be frozen in the hardening room. The mass of 1 L of ice cream at 100% overrun, from a mix with a density of 1099 kg/m³, is 0.55 kg. The mix is assumed to contain 61% water, and the hardening room is at −29°C. Calculate the refrigeration required to harden the ice cream in kJ/L.

Solution:

$$
\begin{aligned}
\text{Latent heat of hardening: } 0.55 \times 0.61 \times 0.70 \times 335 &= 78.7 \text{ kJ/L} \\
\text{Sensible heat: } 0.55 \times 2.09[-4-(-29)] &= 28.7 \text{ kJ/L} \\
\text{Total} &= 107.4 \text{ kJ/L} \\
\text{Loss due to heat of container and exposure to outside air, assumed 10\%} &= 10.7 \text{ kJ/L} \\
\text{Total to harden} &= 118.1 \text{ kJ/L}
\end{aligned}
$$

Percent overrun, when calculated on the basis of the quantity of ice cream delivered by the freezer or the quantity placed in the hardening room, would affect the refrigeration required, as shown in Table 11.

Table 11 Hardening Loads for Typical Ice Cream Mix

Overrun, %	Hardening Load, kJ/L
60	148
70	139
80	130
90	125
100	118
110	113
120	108

Example 5. Calculate the refrigeration load in an ice cream hardening room, assuming 4000 L of ice cream at 100% overrun are to be hardened in 10 h in a forced-air circulation room at a temperature of −29°C. The hardening room, for three times this daily output, should have 30 m² of floor area measuring approximately 5 m by 6 m by 3 m high. The total insulated surface of 126 m² requires 100 to 150 mm of urethane or equivalent. For this example, the heat conductance through the insulated surface is selected at 0.227 W/(m²·K). The average ambient temperature is assumed to be 32°C.

Solution:

$$
\begin{aligned}
\text{Heat leakage: } 126 \times 0.227[32-(-29)] &= 1.74 \text{ kW} \\
\text{Heat from fan motor (assumed)} &= 1.5 \text{ kW} \\
\text{Heat from lights (600 W assumed)} &= 0.6 \text{ kW} \\
\text{Air infiltration and persons in room} \\
\text{(approximately 20\% leakage)} &= 0.35 \text{ kW} \\
\text{Hardening 4000 L ice cream in} \\
\text{10 h (0.111 L/s) at 118.1 kJ/L} &= 13.12 \text{ kW} \\
\text{Total} &= 17.31 \text{ kW}
\end{aligned}
$$

Additional refrigeration load calculation information is located in Chapter 12.

Other products, such as sherbets, ices, ice milk, and novelties, usually represent a small percentage of the total output of the plant, but should be included in the total requirement of the hardening room.

Ice Cream Bars and Other Novelties

Ice cream plants may manufacture and merchandise a limited number of the many novelties. The most common are chocolate-coated ice cream bars, popsicles, fudgesicles, drumsticks, ice cream sundae cups, ice cream sandwiches, and so forth. Small plants freeze most of these products, especially those with sticks, in metal trays each containing 24 molds, which are submerged in a special brine tank with a built-in evaporator surface and brine agitation. The product mix is prepared in a tank and cooled to 1.5 to 4.5°C. A controlled quantity of mix is poured into the tray molds or measured in with a dispenser. The tray molds are placed in the brine tank for complete freezing. Brine temperature is −34 to −38°C. The freezing rate should be rapid to result in small ice crystals, but it varies with the product and generally is 15 to 20 min. The frozen product is loosened from the molds by momentary melting of the outer layers of the product in a water bath. It is immediately removed from the molds; each is separately wrapped or put in a novelty bag and promptly placed in frozen storage for distribution.

Example 6. Calculate the refrigeration required to freeze 1200 popsicles per hour based on a mix containing 85% water. Each popsicle has a mass of 0.085 kg and a density of 1060 kg/m³.

Solution:

$$
\begin{aligned}
\text{Estimate mass flow of mix: } 1200 \times 0.085/3600 &= 0.0283 \text{ kg/s} \\
\text{Cool mix from 5°C to freezing at } -3°C: \\
0.0283[5-(-3)]3.64 &= 0.83 \text{ kW} \\
\text{Freezing load: } 0.0283 \times 0.85 \times 335 &= 8.07 \text{ kW} \\
\text{Subcool to } -35°C: 0.0283[-3-(-35°C)]2.09 &= 2.25 \text{ kW} \\
\text{Cooling trays (50/h or 0.138/s at 15°C):} \\
0.0138 \times 3.6 \text{ kg/tray} \times [15-(-35)]0.50 &= 1.25 \text{ kW} \\
\text{Heat from agitator (750 W)} &= 0.75 \text{ kW} \\
\text{Leakage through 1 m by 4 m by 1 m tank} &= 0.22 \text{ kW} \\
\text{Loss, top of tank and piping (assumed)} &= 2.19 \text{ kW} \\
\text{Total refrigeration load} &= 15.6 \text{ kW}
\end{aligned}
$$

In making ice cream, ice milk, and similar kinds of bars, the mix is processed through the freezer and is extruded in a viscous form at about −6°C. Using similar calculations, the estimated refrigeration load to freeze 100 dozen would be 7.7 kW for 85-g ice cream bars with 100% overrun.

The equipment to make and package novelties in large plants is available in several designs and capacities. Some are limited to the manufacture of one or a few kinds of similar novelties. Other machines have more versatility; for example, they can be used to make novelties with or without sticks, coated or uncoated, and of numerous sizes, shapes, and flavor combinations. Some of these machines include packaging in a bag or wrap, plus placement and sealing in a carton in units of 6, 8, 12, 14, 18, 24, or 48. In other plants, a separate packaging unit may be required. Some units harden the product by air at a temperature within the range of −37 to −43°C. Brine, usually calcium chloride, with a specific gravity of 1.275 or more and a temperature of −33 to −39°C may be the hardening medium. The capacity of novelty makers varies with the shape and size of the specific product. Common production of a novelty machine is generally within the range of 3500 to 35 000 or more per hour. Novelty equipment in plants may be semiautomatic or automatic in performance of the necessary functions.

An example of a simple novelty machine is one that has two parallel conveyor chains on which the mold strips are fastened. The molds are conveyed through filling, stick inserting, freezing, and defrosting stages. The extractor conveyor removes the frozen

Dairy Products

product from the mold cups and carries it to packaging or through the dipping operation; it is then discharged at packaging. In the meantime, the molds go through a wash and rinse and back to be filled. The novelty is either bagged or wrapped by machine, grouped and placed into cartons, and conveyed to cold storage.

Refrigeration Compressor Equipment Selection and Operation

Counter-type freezers are usually designed for use with halocarbon refrigerants and may be arranged in a self-contained cabinet with the condensing unit mounted under the freezer. Nearly all commercial ice cream plants, particularly the larger plants, use ammonia. Some smaller plants operate continuous ice cream freezers and refrigerate hardening rooms to acceptable temperatures with single-stage refrigerant compressors. In most cases, these plants operate reciprocating compressors at conditions above the maximum compression ratio recommended by the manufacturer.

For economy of operation, within reasonable limits of compression ratio, ice cream plants normally use multistage compression. For freezing ice cream, producing frozen novelties, and refrigerating an ice cream hardening room to −30°C, one or more booster compressors may be used at the same suction pressure, discharging into second-stage compressors, which also handle the mix cooling and ingredient cold storage room loads. If a carton freezer is used at a temperature of −40°C or below for ice cream hardening, two low suction pressure systems should be used, the lower one for the carton freezer and the higher one for the ice cream freezers and storage room. Both low suction pressure systems discharge into the second-stage compressor system. For plants with carton freezers arranged for large volumes, an analysis of operating costs may indicate savings in using three-stage compression with the low-temperature booster compressors used for the tunnel, discharging into the second-stage booster compressor system used for freezers and storage, and then the second-stage booster compressor system discharging into the third or high stage compressor system.

High-temperature loads in an ice cream plant usually consist of refrigeration for cooling and holding cream, cooling ice cream mix after pasteurization, cooling for mix holding tanks, and refrigeration for the ingredient cold storage room. If direct refrigerant cooling is used for the high-temperature loads, then compressor selections for the high stage can be made at about −7°C saturated suction temperature and combined with the compressor capacity required to handle the booster discharge load. Approximately the same high suction temperature can be estimated if ice cream mix and mix holding tanks are cooled by chilled water from a falling film water chilling system. If an ice builder supplies chilled water for cooling pasteurized ice cream mix, it may be desirable to provide a separate compressor system to handle this refrigerating load.

Refrigeration is a significant and important cost in an ice cream plant because of the relatively large refrigeration capacity required at low suction pressure (temperature). It is imperative to use efficient two-stage or three-stage compression systems at the highest suction pressures and lowest discharge pressures practical to achieve the desired product temperatures.

The effectiveness of the heat transfer surfaces is reduced by oil films, excessive ice and frost, scale, noncondensable gases, abnormal temperature differentials, clogged sprays, improper liquid circulation, poor airflow, and foreign materials in the system. Adequate operations and maintenance procedures for all components and systems should be in force to assure maximum performance and safety.

Process operation performance is also critical to the effectiveness of the refrigeration system. Items that adversely affect ice cream freezing rates include dull scraper blades, mix viscosity, high mix inlet temperatures, low ice cream discharge temperatures, overrun below specifications, and incorrect mix composition.

Rooms and storage areas should be well maintained to preserve the integrity of the insulation. This includes doors and passageways, which may be a major source of air infiltration load.

New and updated ice cream plants should be equipped with microprocessor compressor controls and an overall computerized control system for operations and monitoring. When properly used, these controls help to provide safe, efficient operation of the refrigeration system.

ULTRAHIGH-TEMPERATURE (UHT) STERILIZATION AND ASEPTIC PACKAGING

Ultrahigh-temperature (UHT) sterilization of liquid dairy products destroys viability of microorganisms with a minimum adverse effect on sensory and nutritional properties. **Aseptic packaging (AP)** containerizes the sterilized product without recontamination. Sterilization, in the true sense, is the destruction or elimination of all viable microorganisms. In industry, however, the term sterilized may refer to a product that does not deteriorate microbiologically, but in which viable organisms may have survived the sterilization process. In other words, heat treatment renders the product safe for consumption and imparts an acceptable shelf life microbiologically.

Sterilization Methods and Equipment

Retort sterilization of milk products has been a commercial practice for many years. It consists of sterilizing the product after hermetically sealing it in a metal or glass container. The heat treatment is sufficiently severe to cause a definite cooked off-flavor in milk and to decrease the heat-labile nutritional constituents of milk products. UHT-AP has the advantage of causing less cooked flavor, color change, and loss of vitamins while producing the same sterilization effect as the retort method.

UHT-AP has been applied to common fluid milk products (whole milk, 2% milk, skim milk, and half-and-half), various creams, flavored milks, evaporated milk, and such frozen dessert mixes as ice cream, ice milk, milk shakes, soft-serve, and sherbets. UHT-sterilized dairy foods include eggnog, salad dressings, sauces, infant preparations, puddings, custards, and nondairy coffee whiteners and toppings.

UHT sterilization is accomplished by rapid heating of the product to the sterilizing temperature, holding the temperature for a definite number of seconds, and then rapid cooling. The methods have been classified as direct steam or indirect heating. The following are some of the advantages of the direct methods: (1) heating is faster, (2) processing intervals between equipment cleanings are longer, and (3) the flow rate is easier to change. Among the indirect methods, advantages are the following: (1) regeneration potential is greater, (2) potable steam is not necessary, and (3) viscous products and those with small pieces of solids can be processed with the scraped-surface unit.

The direct steam method is subdivided into injection or infusion. In direct injection, steam is forced into the product, preferably in small streamlets, with sufficient turbulence to minimize localized overheating of the milk surfaces that the steam initially contacts. In the infusion system, the product is sprayed into a steam chamber. Advantages of infusion over injection are (1) slightly less steam pressure required (there are exceptions), (2) less localized overheating of a portion of the product, and (3) more flexibility for change of the product flow rate.

The three important indirect systems are tubular, plate, and cylinder with mechanical agitation. In the tubular type, the tube diameter must be small and the velocity of flow high to maximize heat transfer into the product.

The essential components for direct steam injection are storage or balance tank, timing pump, preheater (tubular or plate), steam injector or infuser unit, holding unit, flow-diversion valve, vacuum chamber, aseptic pump, aseptic homogenizer, plate or tubular cooler, and control instruments. The minimum items of equipment for steam infusion are the same, except that the infuser is used to heat the product from the preheat to the sterilization temperature.

The necessary equipment for the indirect systems is similar: storage or balance tank, timing pump, preheater (tubular or plate type, and preferably regenerative), homogenizer, final plate or tubular heater, holding tube or plate, flow-diversion valve, cooler (1 to 3 stages), and control instruments. The mechanically agitated heat exchanger replaces the tubes or plates in the final heating stage. Otherwise, the same items of equipment are used for this system of sterilization.

In addition to the basic equipment, many combinations of essential and supplemental items of UHT equipment are available. For example, one deviation in the indirect system is to use the pump portion of the homogenizer as a timing pump when it is installed after the balance tank. The first stage of homogenization may occur after preheating, and the second stage may occur after precooling. A vacuum chamber may be placed in the line after preheating, for precooling after sterilization, or installed in both locations. A condenser in the vacuum chamber allows the advantages of deaeration without moisture losses that otherwise would occur in the indirect system. In Europe, some indirect systems have a hold of several minutes, after preheating, to reduce the rate of solids accumulation on the final heating surfaces of the tubes or plates. A bactofuge may be included in the line after preheating to reduce a high microbiological content, especially of the bacterial spores.

Self-acting controls and other instrumentation are available to assure automatic operation in nearly every respect. Particularly important is automatic control of the temperatures for preheating, sterilizing, and precooling in the vacuum chamber, and to some extent, of the final temperature before packaging. This may include temperature-sensing elements to control heating and cooling and pressure-sensing elements for operating pneumatic valves. The cleaning cycle may be automated, beginning with a predetermined solids accumulation on specific heating surfaces. Timers regulate the various cleaning and rinsing steps.

In some systems, one or more aseptic surge tanks are installed between the UHT sterilizer and the AP equipment. Surge tanks permit continuation of the sterilizer should the operation of the AP equipment be interrupted, or the continuation of the AP equipment should sterilization be interrupted. It also makes the use of two or more AP units easier than direct flow from the UHT sterilizer to the AP machines.

When aseptic surge tanks are used, they must be constructed to withstand the steam pressure required for equipment sterilization and be provided with a sterile air venting system. Aseptic surge tanks may be unloaded by applying sterile air to force product out to the AP equipment. The pressure for air unloading can be controlled at a constant value, making uniform filling possible even when one of several AP machines is removed from service.

Aseptic surge tanks make it possible to hold bulk product, even for several days, until it is convenient to package it.

Basic Steps. After the formula is prepared and the product standardized, the processing steps are (1) preheat to 65 to 75°C by a plate or tubular heat exchanger; (2) heat to a sterilization temperature of 140 to 150°C; (3) hold for 1 to 20 s at sterilizing temperature; and (4) cool to 4.4 to 38°C, depending on product keeping quality needs. Cooling may be by one to three stages; generally two are used. The direct steam method requires at least two cooling stages. The first is flash cooling in a vacuum chamber to 65 to 75°C to remove moisture equal to the steam injected during sterilization. The second stage reduces the temperature to within 10 to 38°C. A third stage is required in most plants if the temperature is lowered additionally to 2 to 10°C.

The products with fat are homogenized to increase stability of the fat emulsion. The direct method requires homogenization after sterilization and precooling. Homogenization may follow preheating or precooling, but usually follows preheating in the indirect method. Efficient homogenization is very important in delaying the formation of a cream layer during storage.

Sterilized plain milks (such as whole, 2%, and skim milk) are most vulnerable to having a cooked off-flavor. Consequently, the aim is to have low sterilization temperature and time consistent with satisfactory keeping quality. The total cumulative heat treatment is directly related to the intensity of the cooked off-flavor. The total processing time from preheating to cooling varies widely among systems. Most operations in the United States range from 30 to 200 s; in European UHT processes, it may be much longer.

Several factors influence the minimum sterilization temperature and time needed to control adverse effects on flavor and physical, chemical, and nutritional changes. The type of product, initial number of spores and their heat resistance, total solids of the product, and pH are the most important factors. Obviously, the relationship is direct for the number and the heat resistance of the spores. Total solids also have a direct relationship, but for an acid pH, it is inverse.

Several terms are used in the designation of the influence of UHT on the microbiological population. **Decimal reduction** refers to a reduction of 90% (e.g., 100 to 10, or one log cycle). An example of a three-decimal reduction is 10 000 to 10. **Decimal reduction time**, or D **value**, is the time required to obtain a 90% decrease. **Sterilizing effect**, or **bactericidal effect**, is the number of decimal reductions obtained and expressed as a logarithmic reduction ($\log_{10}$ initial count minus $\log_{10}$ final count). A sterilizing effect of six means one organism remaining from a million per mL (10^6), and seven would be one remaining in 10 mL (a final count of 10^{-1}).

The Z **value** is the temperature increase required to reduce the D value by one log cycle (90% reduction of microorganisms with the time held constant). The F **value** (thermal death time) is the time required to reduce the number of microorganisms by a stated amount or to a specific number. For example, assuming D value of 36 s for *Bacillus substilis* spores at 120°C and a need to reduce the spores from 10^6 per mL to < 1 per mL, the thermal treatment time would be 6 × 36 s = 216 s (F value).

Aseptic Packaging

Aseptic fillers are available for coated metal cans, glass bottles, plastic-paperboard-foil cartons, thermoformed plastic containers, blow-molded plastic containers, and plastic pouches. The aseptic can equipment includes a can conveyor and sterilizing compartment, filling chamber, lid sterilizing compartment, sealing unit, and instrument controls. The procedure sterilizes the cans with steam at 290°C as they are conveyed, fills the cans by continuous flow, simultaneously sterilizes the lids, places the lids on the cans, and seals the lids onto the cans. Pressure control apparatus is not used for entry or exit of cans.

A similar system is used for glass bottles or jars. The jars are conveyed into a turret chamber; air is removed by vacuum; the jars are then sterilized for 2 s with wet steam at 400 kPa (gage) and moved into the filler. The temperature of the glass equalizes to 50°C and the filling takes place. Next, the transfer is to the capper for placement of sterile caps, which are screwed onto the jars. The filling and capping space is maintained at 260°C.

Several aseptic blowmold forming and filling systems have been developed. Each system is different, but the basic steps using molten plastic are (1) extruded into a parison, (2) extended to the bottom of the mold, (3) mold closed, (4) preblown with compressed air that inflates the plastic film into a bottle shape, (5) parison cut and the neck pinched, (6) final air application, (7) bottle filled and foam removed, (8) top sealed, and (9) mold opened and filled bottle ejected.

The basic steps in the manufacture of aseptic, thermoformed plastic containers are as follows: (1) a sheet of plastic (e.g., polystyrene) is drawn from a roll through the heating compartment and then

Dairy Products

multistamped into units, which constitute the containers; (2) these units are conveyed to the filler, which is located in a sterile atmosphere, and are filled; (3) a sheet of sterilized foil is heat sealed to the container tops; and (4) each container is separated by scoring and cutting.

One of the two aseptic systems for the plastic-paperboard-foil cartons draws the material from a roll through a concentrated hydrogen peroxide bath to destroy the microorganisms. The peroxide is removed by drawing the sheet between twin rolls, by exposure to ultraviolet light and hot air, or by superheated, sterilized air forced through small slits at high velocity. The packaging material is drawn downward in a vertical, sterile compartment for forming, filling by continuous flow, sealing, separation, and ejection.

In the other plastic-paperboard-foil aseptic system, the prepared, flat blanks are formed and the bottoms are heat sealed. In the next step, the inside surfaces are fogged with hydrogen peroxide. Sterilized hot air dissipates the peroxide. The cartons are conveyed into the aseptic filling and then into top-sealing compartments. The air forced into these two areas is rendered devoid of microorganisms by high efficiency filters.

Operational Problems. Aseptic operational problems are reduced by careful installation of satisfactory equipment. The equipment should comply with *3-A Sanitary Standards*. Milk and milk products that are processed to be commercially sterile and aseptically packaged must also meet the Grade A Pasteurized Milk Ordinance and be processed in accordance with 21 CFR Part 113, "Thermally processed low-acid foods packaged in hermetically sealed containers." Generally, the simplest system, with a minimum of equipment for product contact surfaces and processing time, is desirable. It is specifically important to have as few pumps and nonwelded unions as possible, particularly those with gaskets. The gaskets and O-rings in unions, pumps, and valves are much more difficult to clean and sterilize than are the smooth surfaces of chambers and tubing. Automatic controls, rather than manual attention, is generally more satisfactory.

Complete cleaning and sterilizing of the processing and packaging equipment are essential. Milk solids accumulate rapidly on heated surfaces; therefore, cleaning is necessary after 0.5 h of processing for the tubular or plate UHT heat exchangers, although cleaning after 3 to 4 h is more common. The cleaning practice for the sterilizer, filler, and accessory equipment usually involves the CIP method for the rinse and alkali cleaning cycle, rinse, acid cleaning cycle, and rinse. Some plants only periodically acid clean the storage tanks and packaging equipment, e.g., once or twice a week. Steam sterilization just before processing is customary. At 160 to 170 kPa (absolute) of wet steam, 1.5 to 2.0 h (or a shorter time at higher steam pressure) may be required. Water sterilized by steam injection or the indirect method can be used for rinsing and for the cleaning solution.

Survival of spores during UHT processing, or subsequent recontamination of the product before the container is sealed, is a constant threat. Inadequate sealing of the container also may be troublesome with certain types of containers. Another source of poststerilization contamination is airborne microorganisms. These may contact the product through inadequate sterilization of air that enters the storage vat for the processed product or through air leaks into the product upstream of the sterilized product pumps or homogenizer, if a reduced pressure is created. During packaging, air may contaminate the inside of the container or the product itself during filling and sealing.

Quality Control

Poor quality of raw materials must be avoided. The higher the spore count of the product before sterilization, the larger the spore survival number at a constant sterilization temperature and time. Poor quality can also contribute to other product defects (off-flavor, short keeping quality) because of sensory, physical, or chemical changes. Heat stability of the raw product must be considered.

A good quality sterilized product has a pleasing, characteristic flavor and color that are similar to the pasteurized samples. The cooked flavor should be slight, or negligible, with no unpleasant aftertaste. The product should be free of microorganisms and adulterants such as insecticides, herbicides, and peroxide or other container residues. It should have good physical, sensory, and keeping quality.

Deterioration in storage may be evaluated by holding samples at 21, 32, 37 or 45°C for 1 or 2 weeks. The number of samples for storage testing should be selected statistically and should include samplings of the first and last of each product packaged during the processing day. In order to identify the source of microbiological spoilage, continuous aseptic sampling into standard size containers after sterilization and/or just ahead of packaging may be practiced. Sampling rate should be set to change containers each hour.

The rate of change in storage of sterilized milk products is directly related to the temperature. Commercial practice varies with storage ranging from 1.7°C to room temperature, which may go as high as 35°C or more. In plain milks, the cooked flavor may decrease the first few days, and then remain at its optimum for 2 to 3 weeks at 21°C before gradually declining. When the milk is held at 21°C, a slight cream layer becomes noticeable in approximately 2 weeks and slowly continues until much of the fat has risen to the top. Thereafter, the cream layer becomes increasingly difficult to reincorporate or reemulsify.

Viscosity increases slightly the first few weeks at 21°C and then remains fairly stable for 4 to 5 months. Thereafter, gelation gradually occurs. However, milks vary in stability to gelation depending on such factors as feeds, stage of lactation, preheat treatment, and homogenization pressure. The addition of sodium tetraphosphate to some milks causes gelatin to develop more slowly.

Occasionally, some sterilized milk products develop a sediment on the bottom of the container because of crystallization of complex salt compounds or sugars. Browning can also occur during storage. Usually, the off-flavors develop more rapidly and render the product unsalable before the off-color becomes objectionable.

Heat-Labile Nutrients

The results reported by researchers on the effects of UHT sterilization on the heat-sensitive constituents of milk products lack consistency. The variability may be attributed to the analytical methods and to the difference in total heat treatment among various UHT systems, especially in Europe. In a review, Van Eekelen and Heijne (1965) summarized the effect of UHT sterilization on milk as follows: slight or none for vitamins A, B_2, and D, carotene, pantothenic acid, nicotinic acid, biotin, and calcium; and no decrease in biological value of the proteins. The decreases were: 3 to 10%, thiamine; 0 to 30%, B_6; 10 to 20%, B_{12}; 25 to 40%, C; 10%, folic acid; 2.4 to 66.7%, lysine; 34%, linoleic acid; and 13%, linolenic acid. Protein digestibility was decreased slightly. A substantial loss of vitamins C, B_6, and B_{12} occurred during a 90-day storage. Brookes (1968) reported that Puschel found that babies fed sterilized milk averaged a gain of 27 g per day, compared to 20 g for the control group.

EVAPORATED, SWEETENED CONDENSED, AND DRY MILK

Evaporated Milk

Raw milk intended for processing into evaporated milk should have a heat stability quality with little and preferably no developed acidity. As the milk is received it should be filtered and held cold in a storage tank. The milk fat is standardized to nonfat solids at the ratio of 1:2.2785. The milk is preheated to 93 to 96°C for 10 to 20 min or 115 to 127°C for 60 to 360 s to reduce product denaturation during the sterilizaton process. Moisture is removed by batch or

Table 12 Inversion Times for Cases of Evaporated Milk in Storage

Storage Temperature, °C	Time
32	1 month initially and each 15 days
27	1 to 2 months
21	2 to 3 months
15	3 to 6 months

Table 13 Typical Steam Requirements for Evaporating Water from Milk

No. of Evaporating Effects	Steam Required, MJ/kg water
Single	2.9 to 2.3
Double	1.4 to 1.1
Triple	0.90 to 0.80
Quadruple	0.68 to 0.56

(usually) continuous evaporation until the total solids have been concentrated to 2.25 times the original content.

The condensed product is pumped from the evaporator and, with or without additional heating, is homogenized at 14 to 21 MPa and 49 to 60°C. It is cooled to 7°C and held in storage tanks for restandardization to not less than 7.9% milk fat and 25.9% total solids. The product is pumped to the packaging unit for filling the cans made from tin-coated sheet steel. The filled cans are conveyed continuously through a retort, whereby the product is rapidly heated with hot water and steam to 118°C and held for 15 min to complete the sterilization. Rapid cooling with water to 27 to 32°C follows. The evaporated milk is agitated while in the retort by the can movement. Application of labels and placement of cans in shipping cartons are done automatically.

Storage at room temperature is common, but deterioration of flavor, body, and color is decreased by lowering the storage temperature to 10 to 15°C. Relative humidity should be less than 50% to reduce can and label deterioration. The recommended inversion of cases during storage to reduce fat separation is shown in Table 12.

Sweetened Condensed Milk

Sweetened condensed milk is manufactured similarly to evaporated milk in several aspects. One important difference, however, is that added sugar replaces heat sterilization to extend storage life. Filtered cold milk is held in tanks and standardized to 1:2.2942 (fat to nonfat solids). The milk is preheated to 63 to 71°C, homogenized at 17 MPa, and heating is continued to 82 to 93°C for 5 to 15 min or to 116 to 149°C for 30 s to 5 min. The milk is condensed in a vacuum pan to slightly higher than a 2:1 ratio. Liquid sugar (pasteurized) is added at the rate of 18 to 20 kg/100 kg of condensed milk.

As the mixture is pumped from the vacuum pan, it is cooled through a heat exchanger to 30°C and held in a vat with an agitator. Nuclei for proper lactose crystallization are provided by adding finely powdered lactose (200-mesh). The product is cooled slowly, taking an hour to reach 24°C with agitation. Then cooling is continued more rapidly to 15°C. Improper crystallization forms large crystals, which cause sandiness (gritty texture). The sweetened condensed milk is pumped to a packaging unit for filling into retail cans and sealing. Labeling of cans and placement in cases is mechanized, similar to the process used for evaporated milk. The product is usually stored at room temperature, but the keeping quality is improved if the storage temperature is below 21°C.

Condensing Equipment. Both batch and continuous equipment are used to reduce the moisture content of fluid milk products. The continuous types have single, double, triple, or more evaporating effects. The improvement in efficiency with multiple effects is shown in Table 13 by the reduction in steam required to evaporate 1 kg of water.

A simple evaporator is the horizontal tube. In this design, the tubes are in the lower section of a vertical chamber. During operation, water vapor is removed from the top and the product, from the bottom of the unit. For the vertical short-tube evaporator, the chamber design may be similar to the horizontal tube. The long-tube vertical unit may be designed to operate with a rising or falling film in the tubes. The latter is common. For the falling film, product Reynolds number should be greater than 2000 for good heat transfer. Falling-film units may have a high k-factor at low temperature differentials, resulting in low steam requirements per mass of water evaporated per area of heating surface. This type (falling film) has a rapid start-up and shutdown. Thermocompressing and mechanical compressing evaporators have the advantage of operating efficiently at lower temperatures, thus reducing the adverse effect on heat-sensitive constituents. Vapors removed from the product are compressed and used as a source of heat for additional evaporation.

Plate-type evaporators are also in common use. They are similar to plate heat exchangers used for pasteurization in that they have a frame and a number of plates gasketed to carry the product in a passage between two plates and the heating medium in adjacent passages. They differ in that, in addition to ports for product, they have large ports to carry vapor to a vapor separator. Vapors flow from the separator chamber to a condenser similar to those used for other types of evaporators. Plate evaporators require less head space for installation than other types, may be enlarged or decreased in capacity by a change in the number of plates, and offer a very efficient heat exchange surface.

Equipment Operation. Positive pumps of the reciprocating type are often used to obtain 20 kPa (absolute) in the chamber. Steam jet ejectors may be used for 17 kPa (absolute), for one stage; two stages permit 6.5 kPa (absolute); and three stages, 0.4 kPa (absolute). Condensers between stages remove the heat and may reduce the amount of vapor for the following stage. Either a centrifugal or reciprocating pump may be used to remove water from the condenser. A barometric leg may also be placed at the bottom of a 10.3 m or longer condenser to remove the water by gravity.

Dry Milk and Nonfat Dry Milk

There are two important methods of drying milk—spray and drum. Each has modifications, such as the foam spray and the vacuum drum drying methods. Spray drying exceeds by far the other methods for drying milk, and the largest volume of dried dairy product is skim milk.

In the manufacture of spray-dried nonfat dry milk (NDM), cold milk is preheated to 32°C and separated, and the skim milk for low-heat NDM is pasteurized at 71.7°C for 15 s or slightly higher and/or longer. It is condensed with caution to restrict total heat denaturation of the serum protein to less than 10%. This requires using a low-temperature evaporator or operating the first effect of a regular double effect evaporator at a reduced temperature. After increasing the total solids to 40 to 45%, the condensed skim milk is continuously pumped from the evaporator through a heat exchanger to increase the temperature to 63°C. The concentrated skim milk is filtered and enters a positive pump operating at 21 to 28 MPa, which forces the product through a nozzle with a very small orifice, producing a mist-like spray in the drying chamber. Hot air of 143 to 204°C or higher dries the milk spray rapidly. Nonfat dry milk with 2.5 to 4.0% moisture is conveyed from the drier by pneumatic or mechanical means, then cooled, sifted, and packaged. Packages for industrial users are 22.7- or 45.5-kg bags.

High-heat nonfat dry milk is used principally for bread and other bakery products. The manufacturing procedure is the same as for low-heat NDM except that (1) the pasteurization temperature is well above the minimum (e.g., 79.4°C for 20 s or higher); (2) after pasteurization, the skim milk is heated to 85 to 91°C for 15 to 20 min,

Dairy Products

condensed; and (3) the concentrate is heated to 71 to 74°C ahead of filtering and then is spray dried, similar to the process for low-heat NDM. Storage of low- or high-heat NDM is usually at room temperature.

Dry Whole Milk. The raw whole milk in storage tanks is standardized at a ratio of fat to nonfat solids of 1:2.769. The milk is preheated to 71°C, filtered or clarified, and homogenized at 71°C and 21 MPa on the first stage and 5 MPa on the second stage. The heating continues to 93°C with a 180-s hold. The milk is drawn into the evaporator and the total solids are condensed to 45%. The product is continuously pumped from the evaporator, reheated in a heat exchanger to 71°C, and spray dried to 1.5 to 2.5% moisture. Dry whole milk (DWM) is cooled (not below dew point) and sifted through a 12-mesh screen. For industrial use within 2 or 3 months, the dry whole milk is packaged in 22.7-kg bags and held at room temperature or, preferably, well below 21°C.

In order to retard oxidation, the dry whole milk may be containerized in large metal drums or in customer-size cans unsealed and subjected to 6.5 kPa (absolute). Less than 2% oxygen in the head space of the package after a week of storage is a common aim. The oxygen desorption from the entrapped content in lactose is slow, and two vacuum treatments may be necessary with a 7- to 10-day interval between them. Warm DWM directly from the drier has a faster oxygen desorption rate than after it has cooled. Nitrogen is used to restore atmospheric pressure after each vacuum treatment. After the hold period for the first vacuum treatment, the DWM in the drums is dumped into a hopper, mechanically packaged into retail size metal cans, and given the second vacuum treatment.

Foam spray drying permits the total solids to be increased to 50 to 60% in the evaporator prior to drying. Gas, compressed air, or nitrogen is distributed, by means of a small mixing device, into the condensed product between the high-pressure pump and the spray nozzle. A regulator and needle valve are used to adjust the gas flow into the product. Gas usage is approximately 3.7 L/L of concentrated product. Otherwise, the procedure is the same as for regular drying. Foam spray-dried NDM has poor sinkability but good reconstitutability in water. The density is roughly half that of regular spray-dried NDM. The additional equipment for foam spray drying is limited to a compressor, storage drum, pressure regulator, and a few accessory items. The cost is relatively small, especially if compressed air is used.

Spray driers are made in various shapes and sizes with one or many spray nozzles. The horizontal spray driers may be box shaped or a teardrop design. The vertical spray driers are usually cone or silo shaped.

Heat Transfer Calculations. The typical atomization in U.S. spray-drying plants is produced by a high-pressure pump that forces the liquid through a small orifice in a nozzle designed to give a spreading effect as it emerges from the nozzle. Single-nozzle driers have an orifice opening diameter of 2.7 to 4.5 mm. The diameter for multinozzle driers is 0.64 to 1.32 mm. In Europe, the spinning disk is the most common means of atomizing in milk drying plants. Droplet sizes of 50 to 250 μm in diameter are usual. Droplet size has an inverse relationship to the rate of drying at a uniform hot air temperature. Larger droplets require a higher air drying temperature and/or longer exposure than the smaller ones.

Other essential steps in spray drying are (1) moving, filtering, and heating the air; (2) incorporating the hot air with the product droplets; and (3) removing the moisture vapors and separating the moist air from the product particles. After passing through a rough or intermediate filter, the air is heated indirectly by steam coils or directly with a gas flame to 120 to 260°C. During the short drying exposure time, the air temperature drops to 70 to 93°C.

Thermal efficiency is the percentage of the total heat used to evaporate the water during the drying process. The efficiency is improved by recovery of heat from the exhaust air, decreased radiation loss, and high drying air temperature versus a low outlet air temperature. Roughly 5.0 to 7.2 MJ of steam are needed to evaporate 1 kg of moisture in the drier.

$$\text{Thermal efficiency} = \frac{(1 - R/100)(t_1 - t_2)}{t_1 - t_0}$$

where

R = radiation loss, percentage of temperature decrease in drier
t_1 = inlet air temperature, °C
t_2 = outlet air temperature, °C
t_0 = ambient air temperature, °C

Most of the dried particles are separated from the drying air by gravity and fall to the bottom of the drier or the collectors. The fine particles are removed by directing the air-powder mixture through bag filters or a series of cyclone collectors. Air movement in the cyclone is designed to provide a centrifugal force for separation of the product particles. In general, several small-diameter cyclones with a fixed pressure drop will be more efficient for removal of the fines than two large units.

The drier has sensing elements to continuously record the hot air (inlet) temperature and the moist air (outlet) temperature. Adjustments of either of these temperatures during drying is done with a steam valve or gas inlet valve.

Drum Drying

Relatively little skim milk or whole milk has been drum dried in the last few years. Drum-dried products, when reconstituted, have a cooked or scorched flavor compared with the spray-dried products. The heat treatment during drying denatures the protein and results in a high insolubility index. In preparation for drying, the skim milk is separated or the whole milk is standardized to 1:2.769 (e.g., 3.2 fat and 8.86 SNF). The product is filtered or clarified, homogenized after preheating, and pasteurized. If the resulting dry product is intended for bakery purposes, the milk is heated to approximately 85°C for 10 min. The fluid product may or may not be concentrated by moisture evaporation to not more than 2 to 1. The product is then dried on the drum(s)—skim milk to not more than 4.0% and whole milk to not more than 2.5% moisture. A blade pressed against the drum scrapes off the sheet of dried product. An auger conveys the dry material to the hammer mill, where it is pulverized and sifted through an 8-mesh screen. Drum dried milks are usually packaged at the sifter into 23- to 45-kg kraft bags with a plastic liner.

A double-drum drier is more common than a single drum for drying milk. Cast iron is used more often in drum construction than stainless steel or alloy steel and chrome plate steel. The knife metal must be softer than the drum. In the double drum unit, the drums are spaced from 0.5 to 1.1 mm apart. End plates on the drums create a reservoir into which the product, at 85°C, is sprayed the length of the drums. The steam-heated drums boil the product continuously as a thin film adheres to the revolving drums. After about 0.875 of one revolution, the film of product is dry and is scraped off. Drums normally revolve between 0.2 to 0.32 r/s. The steam pressure inside the drums is approximately 500 to 600 kPa, as indicated by the pressure gage at the inlet of the condensate trap.

The steam pressure is adjusted for drying the product to the desired moisture content. Superheated steam will cause scorching of the product. Condensate inside the drums must be continuously removed, while the exterior vapors from the product are exhausted from the building with a hood and fan system. Capacity, dried product quality, and moisture content depend on many factors. Some important ones are: steam pressure in drums, rotation speed of drums, total solids of product, smoothness of drum surface and sharpness of the knives, properly adjusted gap between the two drums, liquid level in drum reservoir, and product temperature as it enters the reservoir.

REFERENCES

Brookes, H. 1968. New developments in longlife milk and dairy products. *Dairy Industries* (May).

Burdick, R. 1991. Salt brine cooling systems in the cheese industry. IIAR, 1991 Annual Meeting Technical Papers.

Hammer, B.W. and A. R. Johnson. 1913. The specific heat of milk and milk derivatives. *Research Bulletin* No. 14, Iowa Agricultural Experiment Station.

IAMFES. 3-A *Sanitary Standards*. International Association of Milk, Food, and Environmental Sanitarians, Ames, IA.

Leighton, A. 1927. On the calculation of the freezing point of ice cream mixes and of the quantities of ice separated during the freezing process. *Journal of Dairy Science* 10:300.

Rishoi, A.H. 1951. *Physical characteristics of free and globular milk fat*. American Dairy Science Association, Annual Meetings (June).

Van Eeckelen, M. and J.J.I.G. Heijne. 1965. Nutritive value of sterilized milk. In *Milk sterilization*. Food and Agricultural Organization of the United Nations, Rome.

Zhadan, V.Z. 1940. *Specific heat of foodstuffs in relation to temperature*. Kholod'naia Prom. 18 (4):32. Cited from Stitt and Kennedy (Russian).

BIBLIOGRAPHY

Arbuckle, W.S. 1972. *Ice cream*, 2nd ed. AVI Publishing, Westport, CT.

Farrall, A.W. 1963. *Engineering for dairy and food products*. John Wiley and Sons, New York.

Griffin, R.C. and S. Sacharow. 1970. *Food packaging*. AVI Publishing, Westport, CT.

Hall, C.W. and T.I. Hedrick. 1971. *Drying of milk and milk products*. AVI Publishing, Westport, CT.

Henderson, F.L. 1971. *The fluid milk industry*. AVI Publishing, Westport, CT.

Judkins, H.F. and H.A. Keener. 1960. *Milk production and processing*. John Wiley and Sons, New York.

Kosikowski, F.V. 1966. *Cheese and fermented milk foods*. Published by author, Ithaca, NY.

Reed, G.H. 1970. *Refrigeration*. Hart Publishing, New York.

Sanders, G.P. Cheese varieties and descriptions. *Agriculture Handbook* No. 54. USDA, U.S. Government Printing Office, Washington, D.C.

Webb, B.W. and E. A. Whittier. 1970. *Byproducts from milk*. AVI Publishing.

Wilcox, G. 1971. *Milk, cream and butter technology*. Noyes Data Corporation, Park Ridge, NJ.

Wilster, G.H. 1964. *Practical cheesemaking*, 10th ed. Oregon State University Bookstore, Corvallis, OR.

CHAPTER 20

EGGS AND EGG PRODUCTS

SHELL EGGS 20.1	EGG PRODUCTS 20.8
Egg Structure and Composition 20.1	Egg Breaking Process 20.9
Egg Quality and Safety 20.2	Refrigerated Liquid Egg Products 20.10
Shell Egg Processing 20.5	Frozen Egg Products 20.11
Effect of Refrigeration on Egg Quality and Safety 20.5	Dehydrated Egg Products 20.11
Packaging 20.7	Egg Product Quality 20.12
Transportation 20.8	Sanitary Standards and Plant Sanitation 20.12

ABOUT 72% of the table eggs produced in the United States are sold as shell eggs. The remainder are further processed into liquid, frozen, or dehydrated egg products that are used in food service or as an ingredient in food products. Small amounts of further processed eggs are converted to retail egg products, mainly mayonnaise and salad dressings. Shell egg processing includes cleaning, washing, candling for interior and exterior quality, sizing, and packaging. Further processed eggs require shell removal, filtering, blending, pasteurization, and possibly freezing or dehydration.

Following processing, shell eggs intended for use within several weeks are stored at 7 to 13°C and relative humidities of 75 to 80%. These conditions reduce the evaporation of water from the egg, which reduces the egg's mass and hastens the breakdown of the albumen—an indicator of quality and grade. Shell eggs are also refrigerated during transportation, during short- and long-term storage, in retail outlets, and at the institutional and consumer levels.

Research has shown that microbial growth can be curtailed by holding eggs at less than 5°C. The USDA has proposed that processed eggs be kept in an ambient temperature below 7°C until they reach the consumer. This requirement is intended to prevent the growth of *Salmonella* (see October 27, 1992, United States Federal Register). The proposal is expected to include the retail trade, manufacturers, institutional users, and restaurants. If so, major changes in the way eggs are handled will have to be implemented. Storage and display areas not now refrigerated will have to be corrected. All egg storage areas will have to be able to maintain ambient temperatures at 7.2°C.

SHELL EGGS

EGG STRUCTURE AND COMPOSITION

Physical Structure

The parts of an egg are shown in Figure 1, and physical properties of eggs are given in Table 1.

The **shell** is about 11% of the egg mass and is deposited on the exterior of the outer shell membrane. It consists of a mammillary layer and a spongy layer. The shell contains large numbers of pores (approximately 17 000) that allow water, gases, and small particles, such as microorganisms, to move through the shell. A thin, clear film (cuticle) on the exterior of the shell covers the pores. This material is thought to retard the passage of microbes through the shell and serves to prevent moisture loss from the egg's interior. The shape and structure of the shell provide enormous resistance to pressure stress, but very little resistance to breakage caused by impact.

Tough **fibrous shell membranes** surround the albumen. As the egg cools, ages, and looses moisture, an air cell develops on the large end of the egg between these two membranes. The size of the air cell is an indirect measure of the egg's age and is used to evaluate interior quality.

The **white** (albumen) constitutes about 58% of the egg mass. The white consists of a thin, inner chalaziferous layer of firm protein containing fibers that twist into chalazae on the polar ends of the yolk. These structures anchor the yolk in the center of the egg (Figure 1). The albumen consists of inner thin, thick, and outer thin layers.

The **yolk** constitutes approximately 31% of the egg mass. It consists of a yolk (vitelline) membrane and concentric rings of six yellow layers and narrow white layers (Figure 1). In the intact egg, these layers are not visible. Most of the egg's lipids and cholesterol are bounded into a lipoprotein complex that is found more in the white layers. The yolk contains the germinal disc, which consists of about 20 000 cells if the egg is fertile. However, eggs produced for human consumption are not fertile because the hens are raised without roosters.

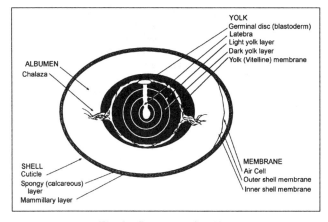

Fig. 1 Structure of an Egg

Table 1 Physical Properties of Chicken Eggs
(Burley and Vadehra 1989)

Property	Whole Egg	Albumen	Yolk
Solids, %	26.4	11.5	52.5
pH (fresh eggs)		7.6	6.0
Density, kg/m³	1080	1035	1035
Surface tension, Pa			4.4
Freezing point, °C		−0.42	−0.57
Specific heat, kJ/(kg·K)	3.23		
Viscosity, mPa·s			
Thick white		164	
Thin white		4	
Electrical conductivity, mS/m		82.5	0.7
Water activity, % relative humidity		97.8	98.1

The preparation of this chapter is assigned to TC 10.9, Refrigeration Application for Foods and Beverages. The chapter was rewritten in 1994 by G. Zeidler, R.R. Ernst, J.F. Thompson, D.D. Bell, and A.J. King.

Table 2 Composition of Whole Egg

Egg Component	Protein, %	Lipid, %	Carbohydrate, %	Ash, %	Water, %
Albumen	9.7-10.6	0.03	0.4-0.9	0.5-0.6	88.0
Yolk	15.7-16.6	31.8-35.5	0.2-1.0	1.1	51.1
Whole egg	12.8-13.4	10.5-11.8	0.3-1.0	0.8-1.0	75.5

Note: Shell is not included in above percentages.

	Percent of Egg	Calcium Carbonate	Magnesium Carbonate	Calcium Phosphate	Organic Matter
Shell	11	94.0	1.0	1.0	4.0

Source: Stadelman and Cotterill (1990).

Chemical Composition

The mass of the chicken egg varies from 35 to 80 g or more. The main factors affecting mass and size are the bird's age, breed, and strain. Nutritional adequacy of the ration and ambient temperature of the laying house also influence egg size. Size affects the composition of the egg in that the proportion of the parts changes as egg mass increases. For example, small eggs laid by young pullets just coming into production will have relatively more yolk and less albumen than eggs laid by older hens. Table 2 presents the general composition of a typical egg weighing 60 g.

The shell is low in water content and high in inorganic solids. The solids are mainly calcium carbonate as calcite crystals plus small amounts of phosphorus and magnesium and some trace minerals. Most of the shell's inorganic matter is protein. It is found in the matrix fibers closely associated with the calcite crystals and in the cuticle layer covering the shell surface. Protein fibers are also present in the pore canals and in the two shell membranes. The membranes contain keratin, a protein that makes the membranes tough even though they are very thin.

Egg albumen, or egg white, is a gel-like substance consisting of ovomucin fibers and globular-type proteins in an aqueous solution. Ovalbumin is the most abundant protein in egg white. When heated to about 60°C, coagulation occurs and the albumen becomes firm. Several fractions of ovoglobulins have been identified by electrophoretic and chromatographic analyses. These proteins impart excellent foaming and beating qualities to egg white when making cakes, meringues, candies, etc. Ovomucin is partly responsible for the viscous characteristic of raw albumen and also has a stabilizing effect on egg-white foams, an important property in cakes and candy.

Egg white contains a small amount of carbohydrates. About half is present as free glucose and half as glycoproteins containing mannose and galactose units. In dried egg products, glucose interacts with other egg components to produce off-colors and off-flavors during storage; therefore, glucose is enzymatically digested before drying.

The yolk comprises one third of the edible portion of the egg. Its major components are water (48 to 52%), lipids (33%), and proteins (17%). The yolk contains all of the fatty material of the egg. The lipids are very closely associated with the proteins. These very complex lipoproteins give yolk special functional properties, such as emulsifying power in mayonnaise and foaming and coagulating powers in sponge cakes and doughnuts.

Nutritive Value

Eggs are a year-round staple in the diet of nearly every culture. The composition and nutritive value of eggs differ among the various avian species. However, only the chicken egg is considered here, as it is the most widely used for human foods.

Eggs contain high-quality protein, which supplies essential amino acids that cannot be produced by the body or that cannot be synthesized at a rate sufficient to meet the body's demands. Eggs are also an important source of minerals and vitamins in the human diet. Although the white and yolk are low in calcium, they contain substantial quantities of phosphorus, iron, and trace minerals. Except for vitamin C, one or two eggs daily can supply a significant portion of the recommended daily allowance for most vitamins, particularly the vitamins A and B_{12}. Eggs are second only to fish liver oils as a natural food source of vitamin D.

The fatty acids in the egg yolk are divided into saturated and unsaturated in a ratio of 1:1.8, with the latter further subdivided into mono- and polyunsaturated fatty acids in a ratio of 1:0.3. Eggs are a source of oleic acid, a monosaturated fatty acid; they also contain polyunsaturated linoleic acid, an essential fatty acid. The fatty acid composition of eggs and the balance of saturated to unsaturated fatty acids can be changed by modifying the hen's diet. Several commercial egg products with modified lipids have been marketed.

EGG QUALITY AND SAFETY

Quality Grades and Mass Classes

In the United States, the Egg Products Inspection Act of 1970 requires that all eggs moving in interstate commerce be graded for size and quality. The USDA standards for quality of individual shell eggs are shown in Table 3. The quality of shell eggs begins to decline immediately after the egg is laid. Water loss from the egg causes a thinning of the albumen and an increase in the size of the air cell. Carbon dioxide migration from the egg results in an increase in albumen pH and a decrease in vitelline membrane strength.

Classes for shell eggs are shown in Table 4. The average mass of shell eggs from commercial flocks varies with age, strain, diet, and environment. Practically all eggs produced on commercial poultry farms are processed mechanically. They are washed, candled, sized, oiled, then packed. Although eggs are sold by units of 6, 12, 18, or 30 per package, the packaged eggs must maintain a minimum mass that relates to the egg size.

Table 3 United States Standards for Quality of Shell Eggs

Quality Factor	AA Quality	A Quality	B Quality
Shell	Clean	Clean	Clean to slightly stained[a]
	Unbroken	Unbroken	Unbroken
	Practically normal	Practically normal	Abnormal
Air Cell	3 mm or less in depth	5 mm or less in depth	Over 5 mm in depth
	Unlimited movement and free or bubbly	Unlimited movement and free or bubbly	Unlimited movement and free or bubbly
White	Clear	Clear	Weak and watery
	Firm	Reasonably firm	Small blood and meat spots present[b]
Yolk	Outline slightly defined	Outline fairly well defined	Outline plainly visible
	Practically free from defects	Practically free from defects	Enlarged and flattened
			Clearly visible germ development but no blood
			Other serious defects

For eggs with dirty or broken shells, the standards of quality provide two additional qualities. These are:

Dirty	Check
Unbroken. Adhering dirt or foreign material, prominent stains, moderate stained areas in excess of B quality.	Broken or cracked shell but membranes intact, not leaking.[c]

[a] Moderately stained areas permitted (1/32 of surface if localized, or 1/16 if scattered).
[b] If they are small (aggregating not more than 3 mm in diameter).
[c] Leaker has broken or cracked shell and membranes, and contents are leaking or free to leak.

Source: *Federal Register*, CFR7, Part 56, May 1, 1991. USDA *Agriculture Handbook* No. 75, p. 18.

Eggs and Egg Products

Table 4 United States Egg Classes for Consumer Grades

Size or Mass Class	Minimum Net Mass per Dozen, g	Minimum Net Mass per 30-Dozen Case, kg	Minimum Mass for Individual Eggs, g
Jumbo	850	25.4	68.5
Extra Large	765	22.9	61.4
Large	680	20.4	54.3
Medium	595	17.9	47.3
Small	510	15.4	40.2
Peewee	425	12.7	

Quality Factors

Besides legal requirements, egg quality encompasses all the characteristics that affect an egg's acceptability to a particular user. The specific meaning of egg quality may vary. To a producer, it might mean the number of cracked or loss eggs that cannot be sold, or the percentage of undergrades on the grade-out slip. The processor associates quality with prominence of yolk shadow under the candling light and the resistance of the shell to damage on the automated grading and packing lines. The consumer looks critically at shell texture and cleanliness and the appearance of the broken-out egg and considers these factors in their relationship to a microbially safe product.

Shell Quality. Strength, texture, porosity, shape, cleanliness, soundness, and color are factors determining shell quality. Of these, shell soundness is the most important. It is estimated that about 10% of all eggs produced are cracked or broken between oviposition and retail sale. Eggs that have shell damage can only be salvaged for their liquid content, but eggs that have both shell and shell membrane ruptured are regarded as a loss and cannot be used for human consumption. Shell strength is highly dependent on shell thickness and crystalline structure, which is affected by genetics, nutrition, length of continuous lay, disease, and environmental factors.

Eggs with smooth shells are preferred over those with a sandy texture or prominent nodules that detract from the egg's appearance. Eggs with rough or thin shells or other defects are often weaker than those with smooth shells. Although shell texture and thickness deteriorate as the laying cycle progresses, the exact causes of these changes are not fully understood. Some research suggests that debris in the oviduct collects on the shell membrane surface, resulting in rough texture formation (nodules).

The number and structure of pores are factors in microbial penetration and loss of carbon dioxide and water. Eggs without a cuticle or with a damaged cuticle are not as resistant to water loss, water penetration, and microbial growth as those with this outer proteinaceous covering. External oiling of the egg shell provides additional protection.

Eggs have an oval shape with shape indexes (breadth/length × 100) ranging from 70 to 74. Eggs that deviate excessively from this norm are considered less attractive and break more readily in packaging and in transit.

Shells with visible soil or deep stains are not allowed in a high-quality pack of eggs. Furthermore, soil usually contains a heavy load of microorganisms that may penetrate the shell, get into the egg contents, and cause spoilage.

Shell color is a breed characteristic. Brown shells owe their color to a reddish-brown pigment, ooporphyrin, which is derived from hemoglobin. The highest content of the pigment is near the surface of the shell. White shells contain a small amount of ooporphyrin, too, but it is degraded soon after laying by exposure to light. Brown shelled eggs tend to vary in color.

Albumen Quality. Egg white viscosity differs in various areas of the egg. A dense layer of albumen is centered in the middle and is most visible when the egg is broken out onto a flat surface. Raw albumen has a yellowish-green cast. In high-quality eggs, the white should stand up high around the yolk with a minimum spreading of the outer thin layer of the albumen. The quality of thick albumen in the freshly laid egg is affected by genetics, duration of continuous production, and environmental factors. Albumen quality generally declines as age progresses, especially in the last part of the laying cycle. Breakdown of thick white is a continuing process in eggs held for food marketing or consumption. The rate of quality loss depends on holding conditions.

Intensity of the color is associated with the amount of riboflavin in the ration. The albumen of top-quality eggs should be free of any blood or meat spots. Incidence of non-meat spots such as blood spots and related problems has been reduced to such a low level by genetic selection that it is no longer a serious concern.

The chalazae may be very prominent in some eggs and can create a negative reaction from consumers who are unfamiliar with these structures (Figure 1). The twisted, rope-like cords are merely extensions of the chalaziferous layer surrounding the yolk and are a normal part of the egg. The chalazae stabilize the yolk in the center of the egg.

Yolk Quality. Shape and color are the principal characteristics of yolk quality. In a freshly laid egg, the yolk is nearly spherical, and when the egg is broken out onto a flat surface, the yolk stands high with little change in shape. Shell and albumen tend to decline in quality as the hen ages. However, yolk quality, as measured by shape, remains relatively constant throughout the laying cycle.

Yolk shape depends on the strength of the vitelline membrane and the chalaziferous albumen layer surrounding the yolk. After oviposition, these structures gradually undergo physical and chemical changes that decrease their ability to keep the yolk's spherical shape. These changes alter the integrity of the vitelline membrane so that water passes from the white into the yolk, increasing the yolk's size and weakening the membrane.

Color as a quality factor of yolk depends on the desires of the user. Most consumers of table eggs favor a light-to-medium yellow color, but some prefer a deeper yellowish-orange hue. Processors of liquid, frozen, and dried egg products generally desire a darker yolk color than users of table eggs because these products are used in making mayonnaise, doughnuts, noodles, pasta, and other foods that depend on eggs for their yellowish color. If laying hens are confined, yolk color is easily regulated by adjusting the number of carotenoid pigments supplied in the hen's diet. Birds having access to growing grasses and other plants usually produce deep-colored yolks of varying hues.

Yolk defects that detract from their quality include blood spots, embryonic development, and mottling. Blood on the yolk can be from (1) hemorrhages occurring in the follicle at the time of ovulation, or (2) embryonic development that has reached the blood-forming stage. The second source is a possibility only in breeding flocks where males are present.

Yolk surface mottling or discoloration can be present in the fresh egg or may develop during storage and marketing. Very light mottling, resulting from an uneven distribution of moisture under the surface of the vitelline membrane, can often be detected on close examination, but this slight defect usually passes unnoticed and is of little concern. Certain coccidiostats (nicarbazin) and wormers (piperazine citrate and dibutylin dilaurate) have been reported to cause mottled yolks and should not be used above recommended levels in layer rations. More serious are the olive-brown mottled yolks produced by rations containing cottonseed products with excessive amounts of free gossypol. This fat-soluble compound reacts with iron in the yolk to give the discoloration. Cottonseed meal may also have cyclopropanoid compounds that increase vitelline membrane permeability. When iron from the yolk passes through the membrane and reacts with the conalbumen of the white, a pink pigment is formed in the albumen. Cyclopropanoid compounds also cause yolks to have a higher proportion of saturated fats than normal, giving the yolks a pasty, custard-like consistency when they are cooled.

Flavors and Odors. When birds are confined and fed a standard ration, eggs have a uniform and mild flavor. Off-flavors can be caused by rations with poor-quality fish meal containing rancid oil or by birds having access to garlic, certain wild seeds, or other materials foreign to normal poultry rations. Off-flavors or odors from rations are frequently found in the yolk, as many compounds imparting off-flavors are fat-soluble. Once eggs acquire off-flavor during storage, their quality is unacceptable to consumers. Eggs have a great capacity to absorb odors from the surrounding atmosphere (Carter 1968). Storage should be free from odor sources such as apples, oranges, decaying vegetable matter, gasoline, and organic solvents (Stadelman and Cottermill 1990). If this cannot be avoided, odors can be controlled with charcoal absorbers or periodical ventilation.

Control and Preservation of Quality

Egg quality is evaluated by shell appearance, air-cell size, and the apparent thickness of the yolk and white. A low storage temperature and shell oiling slow down the escape of carbon dioxide and moisture and prevent shrinkage and thinning of the white. Clear white mineral oil sprayed on the shell after washing partially protects the egg; but its use in commercial operations is diminishing. Rapid cooling will also reduce moisture loss.

Egg quality loss is slowed by maintaining egg temperatures near the freezing point. Albumen freezes at 0.4°C, and the yolk freezes at −0.6°C. Stadelman et al. (1954) and Tarver (1964) found that eggs stored for 15 or 16 days at 7 to 10°C had significantly better quality than eggs stored at 13 to 16°C.

Stadelman and Cotterill (1990) recommend that storage and humidity should be controlled and maintained between 75 and 80%. As a rule, eggs lose about 1% of their mass per week in storage. When large amounts of eggs are palletized, humidity in the center of the pallet may be higher than that of the surrounding air. Therefore, air flow through the eggs is needed to remove excess humidity above 95% to prevent mold growth and decay.

Albumen quality loss is associated with carbon dioxide loss from the egg. Quality losses can be reduced by increasing carbon dioxide levels around the eggs. Controlled atmosphere storage and modified atmosphere packaging have been studied, but they are not used commercially because eggs typically do not need long-term storage. Oiling also helps retard carbon dioxide and moisture loss.

Egg Spoilage and Safety

Microbiological Spoilage. Shell eggs deteriorate in three distinct ways: (1) decomposition by bacteria and molds, (2) changes from chemical reactions, and (3) changes because of absorption of flavors and odors from the environment. Dirty or improperly cleaned eggs are the most common source of bacterial spoilage. Dirty eggs are contaminated with bacteria. Improper washing caused by using water colder than the eggs or water with high iron content increases the possibility of contamination, although it removes evidence of dirt. Most improperly cleaned eggs spoil during long-term storage. Therefore, extremely high sanitary standards are required when washing eggs that will go into long-term storage.

Eggs contaminated with certain microorganisms spoil quickly, resulting in black-, red-, or green-rot, crusted yolks, mold, etc. However, eggs occasionally become heavily contaminated without any outward manifestations of spoilage. Clean, fresh eggs are seldom contaminated internally. However, if the shell becomes wet or sweaty, particularly under conditions of fluctuating temperatures, bacteria and/or molds may cause spoilage.

Preventing Microbial Spoilage. Egg quality can be severely jeopardized by the invasion of microorganisms that cause off-odors and off-flavors. With frequent gathering, proper cleaning, and refrigeration, sound-shell eggs that move quickly through market channels have few spoilage problems.

Sound-shell eggs have a number of defenses against microbial attack, some of which are mechanical and some chemical. Although most of the shell pores are too large to impede bacterial movement, the cuticle layer, and possibly materials within the pores, offer some protection, especially if the shell surface remains dry. Bacteria successful in penetrating the shell are next confronted by a second physical barrier, the shell membrane.

Microorganisms reaching the albumen find it unfavorable for growth. Movement is retarded by the viscosity of the egg white. Also, most bacteria prefer a pH near neutral; but the pH of egg white, initially at 7.6 when newly laid, increases to 9.0 or more after several days, providing a deterring alkaline condition.

Conalbumen, which is believed to be the main microbial defense system of albumen, complexes with iron, zinc, and copper, thus making these elements unavailable to the bacteria and restricting their growth. The chelating potential increases with the rise in albumen pH.

Eggs are capable of warding off a limited quantity of organisms, but should be handled in a manner that minimizes contamination. Egg washing must be done with care. Proper overflow, maintenance of a minimum water temperature of 32°C (as required by USDA regulations), and use of a sufficient quantity of approved detergent-sanitizer are important for effective cleaning. The wash water should be at least 11 K warmer than the internal temperature of the eggs to be washed. Likewise, the rinse water should be a few degrees higher than the wash water. Under these conditions, the contents of the eggs expand to create a positive pressure, which tends to repel penetrations of the shell by microorganisms.

Regular changes of the wash water, as well as thorough daily cleaning of the washing machine, are very important. When the wash water temperature exceeds the egg temperature by more than 30 K, an inordinate number of cracks in the shells, called thermal cracks, occur. Excessive shell damage also occurs if the washer and its brushes are not properly adjusted. Most egg processors use wash waters at temperatures of 43 to 52°C.

Some changes that take place during storage are caused by chemical reaction and temperature effects. As the egg ages, the pH of the white increases, the thick white thins, and the yolk membrane thins. Ultimately, the white becomes quite watery, although total protein content changes very little. Some coincidental loss in flavor usually occurs, although it develops more slowly.

HACCP Plan for Shell Eggs

Many of the procedures for the control of microorganisms are managed by the Hazard Analysis for Critical Control Points (HACCP), which is currently being implemented in United States egg farms, egg packaging sites, egg processing facilities, and the distribution system. Information on the fundamentals of the HACCP system can be found in Chapter 11.

HACCP systems in the egg industry is focused mainly on the prevention of *Salmonella* food poisoning. In the past, *S. typhimurium* was the leading strain in food poisoning related to eggs. However, since 1985 *S. enteritidis* has taken the leading role in egg related salmonellosis illnesses (about 25%).

Salmonella is found naturally in the intestines of mice, rats, snakes and wild birds and not in domesticated chickens. Chicken feed, which attracts the rodents and birds, is the main source of chicken intestine contamination. Unfortunately, *S. enteritidis* can invade the hen ovaries and contaminate the developing yolks, thus being transferred into the egg interior. There it is unreachable for sanitizing agents and the technology for the pasteurization of eggs in the shell is not yet commercially implemented. Fortunately, only a very small portion of eggs are internally contaminated. As the number of internal bacteria is very small, immediate cooling to 7°C and preferably to 5°C will suppress bacterial growth to below the hazard level until the egg is consumed, normally 10 to 30 days after being laid.

Eggs and Egg Products

SHELL EGG PROCESSING

Off-Line and In-Line Processing

Poultry farms either send eggs to a processing plant or package them themselves. On commercial farms, the hens reside in cages with sloped floors. Eggs immediately roll onto a gathering tray or conveyor where they are either gathered by hand, packed on flats, and stored for transport to an processing line (off-line); or the eggs are conveyed directly from the poultry house to a packing machine (in-line) operation. Machines have the ability to package both in-line and off-line eggs, thereby increasing the flexibility of the operation (Figure 2). Off-line operations have coolers both for incoming eggs and for outgoing finished product (Figure 3). An in-line operation has only one cooler for the outgoing finished product (Figure 4).

Figure 5 illustrates material flow during egg packaging in an off-line facility. Egg packaging machines wash the eggs by brushing with warm detergent solution followed by rinsing with warm water and sanitizing with an approved sanitizing agent. Sodium hydrochloride is most commonly used.

The eggs are then dried by air and moved by conveyer, which rotates the egg as they enter the candling booth. There a strong source of light under the conveyor illuminates the eggs' internal and shell defects. Two operators (candlers) remove the defective eggs. The eggs are then weighed and sized automatically and the different sizes are packaged into cartons (12 eggs) or flats (20 or 30 eggs).

One trend in packaging equipment is to increase output to 450 to 500 cases per hour. This development could be possible if automatic candling is developed, as the human eye can only efficiently respond to about 250 cases per hour.

Kuney et al. (1992) demonstrated the high cost of good eggs overpulled in error by candlers. Machine speed was the major factor related to overpulling. Packaging is another area that could be automated because feeding packaging materials, packaging cartons or flats into cases, and palletizing are still largely manual operations.

EFFECT OF REFRIGERATION ON EGG QUALITY AND SAFETY

Refrigeration is the most effective and practical means for preserving quality of shell eggs. It is widely used in farm holding rooms, processing plants, and in marketing channels. Refrigeration conditions for shell eggs to prevent quality loss during short- and long-term storage are as follows:

Temperature, °C	Relative Humidity, %	Storage Period
10 to 16	75 to 80	2 to 3 weeks
7	75 to 80	2 to 4 weeks
−1.5 to −0.5	85 to 92	5 to 6 months

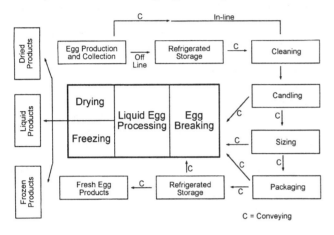

Fig. 2 Unit Operations in Off-Line and In-Line Egg Packaging

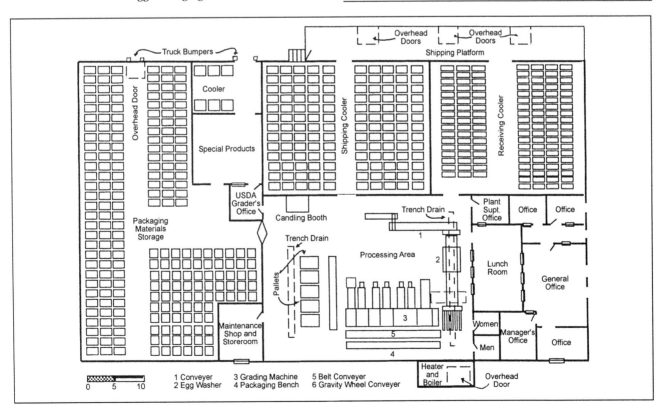

Fig. 3 Off-Line Egg Processing Operation
(Goble 1980)

A relative humidity of 75 to 80% in egg storage rooms must be maintained to prevent moisture loss with a subsequent loss of egg mass. Too high a relative humidity causes mold growth, which can penetrate the pores of the shell and contaminate the egg contents. Mold will grow on eggs when the relative humidity is above 90%.

For long-term storage, eggs should be kept just above their freezing point, −2.8°C. However, long-term storage is seldom used because most eggs are consumed within a short period. Low temperatures can cause sweating, i.e., condensation of moisture on the shell, which causes microbial spoilage.

Refrigeration Requirement Issues

Temperature has a profound effect on *Salmonella enteritidis* on and within eggs. Research has shown that the growth rate of *S. enteritidis* in eggs is directly proportional to the temperature at which the eggs were stored. Holding eggs at 4 to 8°C reduced the heat resistance of *S. enteritidis* and suggested that not only does refrigeration reduce the level of microbial multiplication in shell eggs, but it lowers the temperature at which the organism is killed during cooking.

At present, most shell eggs in the United States are refrigerated to 10 to 13°C after processing. Commonly they are transported in refrigerated trucks and displayed in refrigerated retail displays. Proposed legal changes may require that eggs be maintained in an ambient temperature of 7°C after being packaged and during transport. If implemented, this temperature requirement will cause major changes in the way eggs are now handled.

Condensation on Eggs

Moisture often condenses on the shell surface when cold eggs are moved from the cool storage into hot and humid outside conditions or if the temperature varies widely inside the cooler. Sweating results in a wet egg, and the ability of any microbes present on the shell to penetrate the shell is increased. In addition, wet eggs are more likely to become stained when handled.

Plastic wrapped around the pallets to stabilize the load for shipping can also prevent moisture loss and increase the humidity in the load, which can cause mold problems when eggs are held too long in this condition.

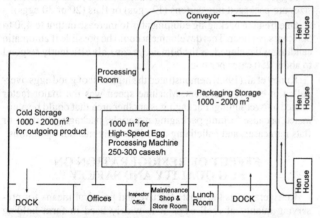

Fig. 4 Typical In-Line Processing Operation
(Zeidler and Riley 1993)

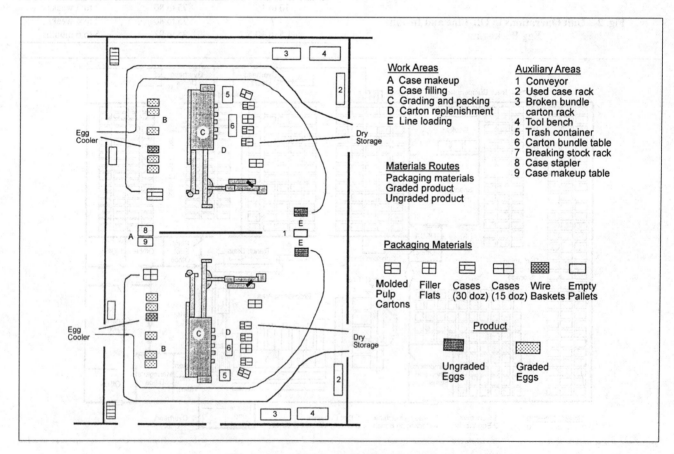

Fig. 5 Material Flow in Off-Line Operation
(Hamann et al. 1978)

Eggs and Egg Products

Table 5 Ambient Conditions When Moisture Condenses on Cold Eggs

Outside Temperature, °C	Outside Relative Humidity, %	
	Egg Temperature	
	7°C	13°C
12	72	—
15	60	—
18	50	73
21	40	60
24	34	50
27	28	42
30	24	35
33	20	30

Condensation or sweating can be predicted from a psychrometric chart. Table 5 lists typical conditions in which sweating may occur.

Initial Egg Temperatures

Cooling requirements for shell eggs obviously vary with the mass of eggs to be cooled and their initial temperature. Anderson et al. (1992) showed that incoming egg temperature depends on the type of processing operation and the time of year. In off-line plants, eggs typically arrive at the plant with internal temperatures ranging from 16.5 to 20°C. Before processing, the eggs are placed in a cooler, which is held between 10 and 15°C. In in-line plants, eggs are conveyed directly from poultry houses to the packing area. Anderson measured incoming egg temperatures ranging from 31 to 36°C. Czarick and Savage (1992) reported that incoming egg temperature in an in-line system reached equilibrium with the layer house temperature. House temperatures are often maintained at 24 to 27°C, however, 32°C sometimes occurs.

Egg Temperatures After Processing

Experience has shown that quality defects are more readily detected when the eggs are allowed to age. Thus, in off-line processing, eggs from production units are usually stored overnight at 13 to 16°C before processing. With present cooling methods, eggs would need to be kept in cold storage about 48 h to cool completely.

The cooling of eggs prior to processing is limited by the temperature rise the shells can tolerate without cracking, which is about 34 K. Most processors wash eggs in warm water ranging from 46 to 52°C (Anderson et al. 1992). This wash temperature would cause shell cracking if eggs are initially cooled to the minimum temperature prescribed by USDA (5°C). Therefore, the lowest egg temperature acceptable before processing would be about 15°C. In contrast, eggs in in-line operations are commonly processed while still warm from the house and are packaged in a warm state.

Hillerman (1955) reported that wash water kept at 46°C would increase internal egg temperature by 0.2 K per second. Anderson (1993) showed that the internal temperature of eggs continued to rise due to the high temperatures during washing resulting in a 4.5 to 6.5 K internal temperature increase above their starting temperature. As a result, egg temperature after washing and packing in in-line systems can typically reach 24 to 29°C and in rare cases may reach nearly 38°C.

Cooling Rates

Henderson (1957) showed that air rates of 0.5 to 3 m/s flowing past an individual egg caused it to cool within one hour to 90% of the difference between the initial egg temperature and the temperature of the cooling air. Eggs packed in filler flats required 4 to 5 h to achieve 90% of total possible temperature drop. Bell and Curley (1966) reported that 13°C air forced around fiber flats in vented corrugated fiberboard boxes cooled eggs from 32 to 16°C in 2 to 5 h. Unvented cartons with the same pack required more than 30 h to cool.

The Canadian Agriculture Department placed eggs with an internal temperature of 27 to 38°C either on fiber flats and stacked six high or in egg flats placed in 6-flat (15 dozen) fiber cases. The eggs were then placed in a 10°C cooler. Eggs in the outermost cells of the cased flats cooled to 10°C in 9 h and all eggs in the fiber flats cooled to 10°C in 24 h. However, eggs at the center of the cases had not reached 10°C after 36 h. Czarick and Savage (1992) found that it took more than 5 days for a pallet of eggs in cases to cool from 29 or 32°C to 7°C in a 7°C cold room.

Egg moisture loss is not increased by rapid cooling. Funk (1935) found that mass loss was the same for eggs in wire baskets cooled in 1 h with circulating air or 15 h with still air.

Cooling for Storage

With current practices of handling, packed eggs require more than one week of storage before they reach the temperature of the storage room. This slow cooling results in egg temperatures in the optimal growth range for *S. enteritidis* from 24 to 72 h after processing. The packaging materials effectively insulate the eggs from the surrounding environment, especially in the center of the pallet. In addition, pallets are often stacked touching each other and may be wrapped in plastic, which further insulates the inner cases and reduces airflow. Also, most eggs are moved from storage within hours of processing, so they are barely cooled. But delaying shipment to allow the eggs to cool results in less-than-fresh products being delivered to the consumer and interior quality suffers.

Adequate air flow through a box requires that the box be vented. In a study done for fruits and vegetables, Baird et al. (1988) showed that cooling cost increases rapidly when carton face vent area decreased below that of 4% of the total area. Other packing materials, such as liners, wraps, flats, or cartons, must not prevent the air that enters the box from contacting the eggs. Also, cases must be stacked to allow air to circulate freely around the pallets.

Because of the inefficiency of cooling eggs in containers, it would seem best to cool them before packing. Eggs could be cooled between washing and packing just prior to being placed into cases, perhaps in a space-saving spiral belt cooler such as that used for food freezing. This would allow the use of current packaging. However, existing equipment is not designed to incorporate this procedure.

Accelerated Cooling Methods

Forced-Air Cooling. Henderson (1957) showed that forced ventilation of palletized eggs produced cooling times close to that of cooling individual eggs. Knutson et al. (1997) arranged a 30-case pallet of eggs so that a 0.47 m^3/s fan drew 4.5°C air through openings in the cases. The eggs were cooled to less than 7°C within 1 to 3 h. This cooling method can be used in an existing refrigerated storage room with little additional investment.

Cryogenic. Curtis et al. (1995) showed that eggs exposed to a −51°C carbon dioxide environment for 3 min continued to cool after packaging and 15 min later were at 7°C. The process maintained egg quality and did not increase the incidence of shell cracking.

PACKAGING

Shell eggs are packaged for the individual consumer or the institutional user. Consumer packs are usually a one dozen carton or variations of it. The institutional user usually receives shell eggs in 30 dozen cases on twelve 30-egg filler-flats.

Consumer cartons are generally made of paper pulp or foam plastic. Some cartons have openings in the top designed for viewing the eggs, which also facilitates cooling. Cartons are generally delivered to the retailer in corrugated containers that hold 15 to 30 dozen eggs, in wire or plastic display baskets that hold 15 dozen eggs, or on rolling display carts. The wire baskets and the rolling racks allow more rapid cooling, but are also more expensive and take up more space in storage and in transport.

TRANSPORTATION

Shell eggs are transported from the off-line egg production site to the egg processing plants, and from the processing plant to local or regional retail and food service outlets. Less frequently eggs are transported from one state to another or overseas. Truck transport is most common and refrigerated trucks will be mandatory in the United States, with possible exemption for small quantities delivered to short-distance customers.

Cases and baskets are generally stored and transported on wooden pallets in 30-case lots (five cases high with six cases per layer). The common carrier for local and long-distance hauling is the refrigerated tractor/trailer combination. Trailers carry 24 to 26 thirty-case pallets of eggs, often of one size category. A typical load of 720 to 780 cases has a mass of about 20 Mg. Some additional cases may be added when small or medium eggs are being transported. Eggs are not generally stacked above six cases high to allow the cold air to travel to the rear of the trailer and to minimize crushing of the lower level cases.

The interregional shipment of eggs is quite common with production and consumption areas often 2500 km apart. Such shipments usually require two to three days using team driving.

Local transportation of eggs may be with similar equipment, especially when delivered to retailer warehouses. Smaller trucks with capacities of 250 to 400 cases are often used when multiple deliveries are required. Local deliveries are commonly made directly to retail or institutional outlets. Individual store deliveries require a variety of egg sizes to be placed on single pallets. This assembly operation in the processing plant is very labor-intensive. Local delivery may involve multiple short stops and considerable opening and closing of the storage compartment with resultant loss of cooling. Many patented truck designs are available to protect cargo from temperature extremes during local delivery, yet none has been adopted by the egg industry.

A 1993 USDA survey found that over 80% of the trucks used to deliver eggs were unsuitable to maintain 7°C. Douglas et al. (1991) in a survey of three egg transport companies in Tampa and Dallas found the average temperature of trailers during non-stop warehouse deliveries was 8°C. The front of the trailer averaged below 7°C 20 to 25% of the time while the back of the trailer was below 7°C 65% of the time. The loads were below 7°C 37% of the time while the reefer discharge was below 5°C.

Trailers used for store-door deliveries had temperatures averaging approximately 7°C at the start of the route; however, some areas only reached a low of 9.2°C. As the deliveries continued and the volume of eggs decreased, temperatures increased and temperature recovery never occurred.

EGG PRODUCTS

Egg products are classified into four groups according to the American Egg Board Guidelines (1981):

1. Refrigerated egg products
2. Frozen egg products
3. Dried egg products
4. Specialty egg products (including hard-cooked eggs, omelets, scrambled eggs, egg substitutes)

Most of these products are not seen at the retail level, but are used as further processed ingredients by the food processing industry for such products as mayonnaise, salad dressing, pasta, quiches, bakery products, and eggnog. Other egg products such as deviled eggs, Scottish eggs, frozen omelets, egg patties, and scrambled eggs are prepared for fast food and institutional food establishments, hotels, and restaurants. In recent years, several products such as egg substitutes (which are made from egg whites) and scrambled eggs have appeared. Yet to be developed are large volume items such as aseptically filled, ultrapasteurized, chilled liquid egg and low-cholesterol chilled liquid eggs.

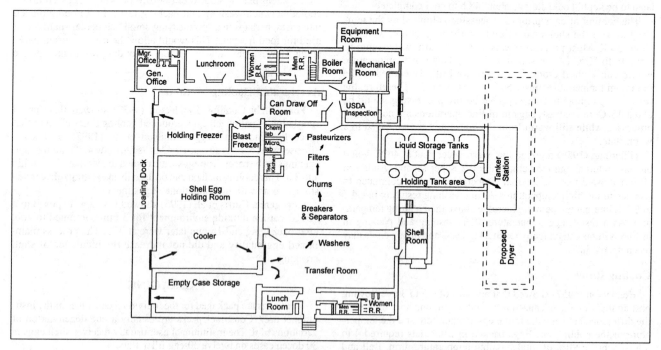

Fig. 6 Floor Plan and Material Flow in Large Egg Breaking Plant
(Courtesy of Seymour Food)

Eggs and Egg Products

Table 6 Minimum Cooling and Temperature Requirements for Liquid Egg Products
(Unpasteurized product temperature within 2 h from time of breaking)

Product	Liquid (Other than salt product) Held 8 h or less	Liquid (Other than salt product) Held in excess of 8 h	Liquid Salt Product	Temperatures within 2 h after pasteurization	Temperatures within 3 h after stabilization
Whites (not to be stabilized)	12.8°C or lower	7°C or lower	—	7°C or lower	—
Whites (to be stabilized)	21°C or lower	12.8°C or lower	—	12.8°C or lower	a
All other products (except product with 10% or more salt added)	7°C or lower	4.4°C or lower	If held 8 h or less, 7°C or lower. If held in excess of 8 h, 4.4°C or lower.		
Liquid egg product (10% or more salt added)	—	If to be held 30 h or less, 18.3°C or lower. To be held in excess of 30 h, 7°C or lower.	—	18.3°C or lower[b]	—

Source: Regulations governing the inspection of eggs and egg products (7C Federal Register Part 59) May 1, 1991.
[a]Stabilized liquid whites shall be dried as soon as possible after the removal of glucose. The storage of stabilized liquid whites shall be limited to that necessary to provide a continuous operation.
[b]The cooling process shall be continued to assure that any salt product held in excess of 24 h is cooled and maintained at 7°C or lower.

EGG BREAKING PROCESS

The egg breaking process transforms shell eggs into liquid products: whole egg, egg white, and yolk. Liquid egg products are chilled, frozen, or dried. These items can be used as is or are processed as an ingredient in food products. Only a few products, such as hard-cooked eggs, do not use the breaking operation system. Dried egg powder, which is the oldest processed egg product, lost ground as a proportion of total egg products, whereas chilled egg products are booming due to superior flavor, aroma, pronounced egg characteristics, and convenience. Most liquid egg products (about 44% of all egg products) must be consumed in a relatively short time because of their short storage life. Frozen or dried egg products may be stored considerably longer.

Surplus eggs, small eggs, and cracked eggs are the major supply source for egg breaking operations. Those eggs must be cleaned in the same manner as shell eggs. Washing and loading of eggs to be broken must be conducted in a separate room from the breaking operation (Figure 6). Eggs with broken shell membrane (leakers) or blood spots are not allowed to be broken for human consumption. Most egg breaking operations are close to egg production areas and in many cases are merely a separate area of a shell egg processing and packaging facility. An egg breaking operation usually receives its eggs from several processing plants in the area that do not have breaking equipment. Storage and transport of eggs, and especially of cracked eggs, reduces the quality of the end product.

Two types of egg breaking equipment are available:

1. **Basket centrifuge.** Shell eggs are dumped into a centrifuge and a whole egg liquid is collected. Several states and some local health authorities ban this equipment for breaking eggs for human consumption due to the high risk of contamination. Similar centrifuges are used to extract liquid egg residue from the discarded egg shells. This inedible product is used mostly for pet food.
2. **Egg breaker and separator.** These machines can process up to 100 cases per hour (36 000 eggs), which is still a slow process compared to up to 500 cases per hour (180 000 eggs) handled by modern table egg packaging equipment.

Holding Temperatures

The prepasteurization holding temperatures required by USDA for out-of-shell liquid egg products are listed in Table 6.

Pasteurization

In the United States, the USDA requires all egg products made by the breaking process to be pasteurized and free of salmonella and requires all plants to be inspected. The minimum required temperatures and holding times for pasteurization of each type of egg product are listed in Table 7.

Table 7 Pasteurization Requirements of Various Egg Products

Liquid Egg Products	Minimum Temperature, °C	Minimum Holding Time, minutes
Albumen (without use of chemicals)	56.7	3.5
	55.6	6.2
Whole egg	60.0	3.5
Whole egg blends (less than 2% added non-egg ingredients)	61.1	3.5
	60.0	6.2
Fortified whole eggs and blends (24 to 38% egg solids, 2 to 12% non-egg ingredients)	62.2	3.5
	61.1	6.2
Salt whole egg (2% salt added)	63.3	3.5
	62.2	6.2
Sugar whole egg (2 to 12% sugar added)	61.1	3.5
	60.0	6.2
Plain yolk	61.1	3.5
	60.0	6.2
Sugar yolk (2% or more sugar added)	63.3	3.5
	62.2	6.2
Salt yolk (2 to 12% salt added)	63.3	3.5
	62.2	6.2

Source: Regulations governing the inspection of eggs and egg products (7C Federal Register Part 59) May 1, 1991.

Plate heat exchangers, commonly used for pasteurization of milk and dairy products, are also commonly used for liquid egg products. Prior to entering the heat exchanger, the liquid egg is moved through a clarifier that removes solid particles such as vitelline (the yolk membrane) and shell pieces.

Egg white solids may be made *Salmonellae* negative by heat treatments. Spray-dried albumen is heated in closed containers so that the temperature throughout the product is not less than 54.4°C for not less than 7 days, until it is free of *Salmonellae*. For pan-dried albumen, the requirement is 51.7°C for 5 days until it is free of *Salmonellae*. For the dried whites to be labeled pasteurized, the USDA requires that each lot be sampled, cultured, and found to contain no viable *Salmonellae*.

Temperature, time, and pH affect the pasteurization of liquid eggs. Various countries specify different pasteurization time, temperature, and pH, but all specifications provide the same pasteurization effects (Table 8). Higher pH requires lower pasteurization temperature and pH 9.0 is most commonly used for egg whites (Figure 7). Various egg products demonstrate different destruction curves (Figure 8), and therefore, different pasteurization conditions were set for these products (Table 8).

Egg whites are more sensitive to higher temperatures than whole eggs or yolk, and therefore will coagulate. Thus, lactic acid is added to adjust the pH to 7.0 to allow the egg whites to withstand 61 to 62°C. Egg whites can be pasteurized at 52°C for 1.5 min if, after the

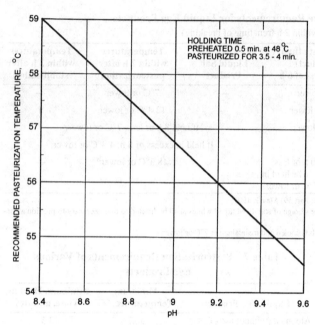

Fig. 7 Effect of pH on Pasteurization Temperature of Egg White

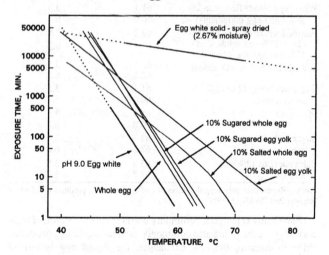

Fig. 8 Thermal Destruction Curves of Several Egg Products
(Stadelman and Cotterill 1990)

Table 8 Minimum Pasteurization Requirements in Various Countries

Country	Temperature, °C	Time (minutes)
Great Britain	64.4	2.5
Poland	66.1-67.8	3
China (PRC)	63.3	2.5
Australia	62.5	2.5
Denmark	65-69	1.5-3
USA	60	3.5

Source: Stadelman et al. (1988)

heat treatment, 0.075 to 0.1% hydrogen peroxide is added for 2 min, followed by its elimination with the enzyme catalase. Liquid yolk, on the other hand requires higher temperatures for pasteurization than liquid whole eggs (62°C for 3.5 min).

Yields

The ratios of white, yolk, and shell vary with the size of the egg. During the laying cycle, the hens lay small, medium, and large size eggs, which have different proportions of yolk and white. Therefore, the distribution of egg sizes that the breaking plant receives during the year varies with season, breed, egg prices, and surplus sizes. As a result, the processing yields of white, yolk, and shell vary accordingly (Table 9).

Table 9 Liquid and Solid Yields From Shell Eggs

Constituent	Liquid (% by weight)		Solids (%)
	Mean	Range	
Shell	10.5	7.8 to 13.6	99.0
Whites	58.5	53.1 to 68.9	11.5
Yolk	31.0	24.8 to 35.5	52.5
Edible whole egg	89.5	86.4 to 92.2	24.5

Source: Shenstone (1968)

REFRIGERATED LIQUID EGG PRODUCTS

Liquid egg products are extremely perishable and should be cooled immediately after pasteurization to below 5°C and kept cool at 1 to 5°C during storage. Refrigerated liquid egg products are convenient to use, do not need defrosting, and can be delivered in bulk tank trucks, totes, or pails, which reduces packaging costs. However, shelf-life at +1 to −1°C is about 2 to 3 weeks; therefore, this product is used mostly as an ingredient in further food processing and manufacturing.

Extending the shelf life of liquid egg products is difficult because egg proteins are much more heat-sensitive than dairy proteins. As a result, ultrapasteurized liquid eggs must be kept under refrigeration whereas ultrapasteurized milk can be kept at room temperature. Ball et al. (1987) used ultrapasteurization and aseptic packaging to extend the shelf life of refrigerated whole eggs to 24 weeks.

Chilled Egg Products

Chilled or **Frozen Liquid** whole egg, yolk, and whites are the major high volume products.

Stabilized Egg Products. Additives are added to yolk products before freezing to prevent coagulation during thawing. Ten percent salt is added to yolks used in mayonnaise and salad dressings, and 10% sugar is added to yolks used in baking, ice cream, and confectionery manufacturing. Whole egg products are also fortified with salt or sugar according to finished product specifications. However, egg whites are not fortified as they do not have gelation problems during defrosting.

UHT Products. High temperature processing (UHT) was initially aimed at producing sterile milk with superior palatability and shelf life by replacing conventional sterilization at 120°C for about 12 to 20 min. with 135°C for 2 to 5 s. UHT treatment of liquid eggs is more complicated, as egg proteins are more sensitive to heat treatment; therefore, UHT liquid eggs must be kept under refrigeration.

In one study, researchers applied aseptic processing and packaging technology to extend the shelf life of liquid egg products to several months under refrigerated (4.5°C) conditions. According to the USDA, the process conditions for extended shelf life liquid whole egg is about 64°C for 3.5 min. Ultrapasteurized, aseptically filled, chilled, whole liquid egg product is now limited to institutional food establishments in the United States; although retail products are available in some European countries.

Egg Substitutes. Substitutes are made from egg whites, which do not contain cholesterol or fat. The yolk is replaced with vegetable oil, food coloring, gums, and nonfat dry milk. Recent formulations have reduced the fat content to almost zero. These products are packaged in cardboard containers and sold frozen or chilled in numerous formula variations. Aseptic packaging extends the shelf life of the refrigerated product.

Eggs and Egg Products

Low-Cholesterol Eggs. Many techniques have been developed to remove cholesterol from eggs, yet no commercial product is currently available.

FROZEN EGG PRODUCTS

Egg products are usually frozen in cartons, plastic bags, 30-lb plastic cans, or 200 L drums (for bulk shipment). Table 3 in Chapter 8 lists thermal properties involved in freezing egg products. Freezing is usually accomplished by air blasts at temperatures ranging from −20 to −40°C. Pasteurized products designated for freezing must be frozen solid or cooled to a temperature of at least −12°C within 60 h after pasteurization. Newer freezing techniques for products containing cooked white (deviled eggs or egg rolls) include individual quick freezing at very low temperatures (−20 to −150°C).

Defrosting. Frozen eggs may be defrosted below 7°C in approved metal tanks in 40 to 48 h. If defrosted at higher temperatures (up to 10°C), the time cannot exceed 24 h. Running water can be used for defrosting. When the frozen mass is crushed by crushers, all sanitary precautions must be followed.

DEHYDRATED EGG PRODUCTS

Spray drying is the most commonly used method for egg dehydration. However, other methods are used for specific products such as scrambled eggs, which are made by freeze drying, and egg white products, which are usually made by pan drying to produce a flake-like product. In spray drying (Figure 9), the liquid is atomized by nozzles which operate at 3.5 to 4 MPa. The centrifugal atomizer, in which a spinning disc or rod rotates at 3500 to 50 000 rpm, creates a hollow cone pattern for the liquid, which enters the drying chamber. The atomized droplets meet a 120 to 230°C hot air cyclone, which is created and driven by a fan blowing in the opposite direction. Because the surface area of the atomized liquid is so large, the moisture evaporates very rapidly. The dry product is separated from the air, cooled, and, in many cases, sifted before being packaged into fiber drums lined with vapor retarder liners. Military specifications usually call for gas-packaging in metal cans. Moisture level in dehydrated products is usually around 5%, whereas in pan dryer products moisture level is around 2%.

Spray dryers are classified as vertical or horizontal types. However, large variations exist in methods of atomizing, drying air movement, and powder separation.

Whole egg, egg white, and yolk products naturally contain reduced sugar. In order to extend the shelf life of these products and to prevent color change through browning (Mylard reaction), the glucose in the egg is removed by baker's yeast, which consumes the glucose within 2 to 3 h at 30°C. Many commercial firms are replacing the baker's yeast method with a glucose oxidase-catalase enzyme process because it is more controllable. The enzyme treated liquid is then pasteurized in continuous heat exchangers at 61°C for 4 min and dried. Whole egg and yolk powder have excellent emulsifying, binding, and heat coagulating properties, whereas egg white possesses whipping capabilities.

Different dry egg products are used in different baking products such as sponge cakes, layer cakes, pound cakes, doughnuts, and cookies. Numerous dry products exist because it is possible to dry eggs together with other ingredients such as milk, other dairy products, sucrose, corn syrup, and other carbohydrates.

Common Dried Products. Figure 9 shows the processing for several dried egg products. Some common dried egg products include

- Pan-dried egg whites, spray-dried egg white solids, whole egg solids, yolk solids
- Stabilized (desugared) whole egg, stabilized yolk

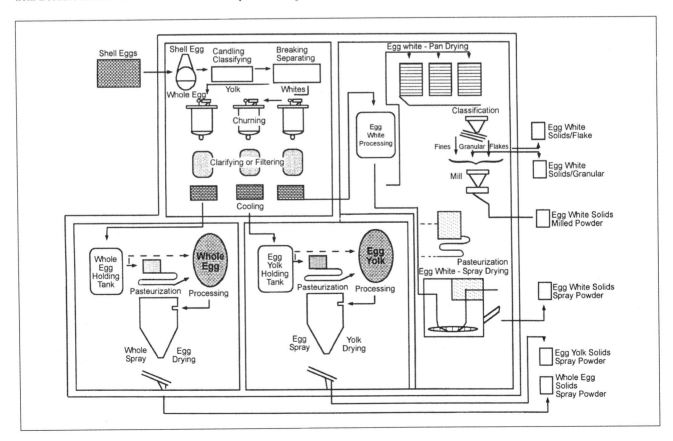

Fig. 9 Steps in Egg Product Drying

- Free-flowing (sodium silicoaluminate) whole egg solids, free flowing yolk solids
- Dry blends: whole egg or yolk with carbohydrates such as sucrose, corn syrup
- Dry blends with dairy products such as scrambled egg mix

EGG PRODUCT QUALITY

Criteria usually used in evaluating egg product quality are odor, yolk color, bacteria count, solids and fat content (for yolk and whole egg), yolk content (for whites), and performance. All users want a wholesome product with a normal odor that performs satisfactorily in the ways it will be used. For noodles, a high solids content and color are important. Bakers are particular about performance—whites do not perform well in angel food cake if excessive yolk is present. They test the foaming performance of whites based on the height and volume of angel food cake and meringue. Performance is also critical for candy (using whites). Salad dressing and mayonnaise are used to evaluate the performance of the yolk as an emulsifier, and emulsion stability is tested.

SANITARY STANDARDS AND PLANT SANITATION

In the United States, *Egg 3-A Sanitary Standards and Accepted Practices* are formulated by the cooperative efforts of the U.S. Public Health Service; the U.S. Department of Agriculture; the Poultry and Egg Institute of America; the Dairy Industry Committee; International Association of Milk, Food, and Environmental Sanitarians; and the Dairy Food Industries Supply Association. The Standards are published by the *Journal of Food Protection* (formerly the *Journal of Milk and Food Technology*).

Egg processing facilities and equipment require daily cleaning and sanitation. Plastic egg flats should be sanitized after each use to avoid microbial contamination of eggs. Chlorine or quaternary-based sanitizers are often used for egg washing and for cleaning equipment, egg flats, floor, walls, etc. Water for egg washing should have low iron content (below 2 ppm) to prevent bacterial growth.

Filters in forced-air egg drying equipment should be cleaned a minimum of once per week. Egg processing rooms should be well ventilated. Inlet air filters should be cleaned weekly. Egg cooling rooms should be kept clean and free from dust or molds.

HACCP Program for Egg Products. Liquid egg products are more susceptible to spoilage as well as being a carrier of pathogens. Therefore, careful temperature control, strict sanitation, good manufacturing practices (GMP), and a HACCP program are essential for the production of high quality, safe egg products.

REFERENCES

Anderson, K.E., F.T. Jones, and P.A. Curtis. 1992. Heat loss from commercially packed eggs in post-processing coolers. Extension Report in Poultry Science, North Carolina Cooperative Extension Service.

Ball, H.R., Jr., M. Hamid-Samirin, P.M. Foegeding, and K.R. Swartzel. 1987. Functional and microbial stability of ultrapasteurized aseptically packaged refrigerated whole egg. *J. Food Sci.* 52:1212-18.

Baird, C.D., J.J. Gafney, and M.T. Talbot. 1988. Design criteria for efficient and cost effective forced air cooling systems for fruits and vegetables. *ASHRAE Transactions* 94(1):1434-54.

Bell, D.D. and R.G. Curley. 1966. Egg cooling rates affected by containers. *California Agriculture* 20(6):2-3.

Burley, R.W. and D.V. Vadehra. 1989. *The avian egg: Chemistry and biology*. John Wiley and Sons, New York.

Carter, T.C. (ed). 1968. *Egg Quality: A study of the hen's egg*. Oliver & Boyd, Edinburgh.

Czarick, M. and S. Savage. 1992. Egg cooling characteristics in commercial egg coolers. *J. Appl. Poultry Res.* 1:258-70.

Douglas, C.R., B.L. Damron, and R.D. Jacobs. 1992. Characterization of temperature patterns in transport vehicles. Southeastern International Poultry Exposition.

Funk, E.M. 1935. The cooling of eggs. Missouri Ag. Exp. Sta. Bull. 350.

Henderson, S.M. 1957. On-the-farm egg processing: Cooling. *Agricultural Engineering* 38(8):598-601, 605.

Knutson, J., J.F. Thompson, and G. Zeidler. 1997. Forced air cooling of shell eggs. *California Poultry Letter*, August.

Kuney, D.R., S. Bokhari, G. Zeidler, R. Ernst, and D. Bell. 1992. Factors affecting candling errors. Proceedings of the 1992 Egg Processing, Packaging and Marketing Seminar, San Bernardino and Modesto, California.

Shenstone, F.S. 1968. The gross composition, chemistry, and physico-chemical basis of organization of the yolk and white. In: *Egg quality: A study of the hen's egg*. T.C. Carter (ed). Oliver and Boyd, Edinburgh.

Stadelman, W.J. 1992. Eggs and egg products. In: *Encyclopedia of Food Science and Technology*, Vol. 2. Y.H. Hui (ed). J. Wiley & Sons, Inc., New York.

Stadelman, W.J., E.L. Baum, J.G. Darroch, and H.G. Walkup. 1954. A comparison of quality in eggs marketing with and without refrigeration. *Food Technology* 8:89-102.

Stadelman, W.J. and O.J. Cottermill (eds). 1990. *Egg science and technology*, 3rd ed. Food Products Press, Binghamton, New York.

Stadelman, W.J., V.M. Olson, G.A. Shemwell, and S. Pasch. 1988. *Egg and poultry meat processing*. Ellis Harwood, Ltds., Chickester, England.

Tarver, F.R. 1964. The influences of rapid cooling and storage conditions on shell egg quality. *Food Technology* 18(10):1604-06.

BIBLIOGRAPHY

Cottrill, O.J. 1981 (Revised in 1990). A scientist speaks about egg products. An American Egg Board publication, Park Ridge, IL.

Dawson, L.E. and J.A. Davidson. 1951. Farm practices and egg quality: Part III. Egg-holding conditions as they affect decline in quality. Quarterly Bulletin, Michigan Agricultural Experiment Station 34(1):105-44.

Goble, J.W. 1980. Designing egg shell grading and packaging plants. USDA Marketing Research *Report* No. 1105.

Hamann, J.A., R.G. Walters, E.D. Rodda, G. Serpa, and E.W. Spangler. 1978. Shell egg processing plant design. USDA/ARS Market Research *Report* No. 912.

Henderson, S.M. 1958. On-the-farm egg processing: Moisture loss. *Agricultural Engineering* 39(1):28-30, 34.

Rhorer, A.R. 1991. What every producer should know about refrigeration. *Egg Industry* (May/June):16-25.

USDA. 1990. Egg grading manual. *Agriculture Handbook* No. 75. U.S. Department of Agriculture, Agricultural Marketing Service.

USDA. 1991. Criteria for shelf-life of refrigerated liquid egg products. U.S. Department of Agriculture, Agricultural Marketing Service.

Van Rest, D.J. 1967. Operations research on egg management. *Transactions of the American Society of Agricultural Engineers*, St. Joseph, MI (Dec.):752-55.

Wells, R.G. and C.G. Belyavin. 1984. *Egg quality: Current problems and recent advances*. Butterworth's, London.

Zeidler, G. and D. Riley. 1993. The role of humidity in egg refrigeration. *Proceedings* 1993 Egg Processing, Packaging, and Marketing Seminar.

CHAPTER 21

DECIDUOUS TREE AND VINE FRUIT

Fruit Storage and Handling Considerations 21.1	Peaches and Nectarines 21.8
Apples 21.1	Apricots 21.9
Pears 21.4	Berries 21.9
Grapes 21.5	Strawberries 21.9
Plums 21.7	Figs 21.10
Sweet Cherries 21.8	Supplements to Refrigeration 21.10

THE most obvious losses from marketing fruit crops are caused by mechanical injury, decay, and aging. Losses in moisture, vitamins, and sugars are less obvious, but they adversely affect quality and nutrition. Rough handling and holding at undesirably high or low temperatures increases loss. Loss can be substantially reduced by greater care in handling and by following recommended storage practices.

FRUIT STORAGE AND HANDLING CONSIDERATIONS

Quality and Maturity

Maximum storage life can be obtained only by storing high-quality commodities soon after harvest. Different lots of fruit may vary greatly in their storage behavior due to variety, climate, soil and cultural conditions, maturity, and handling practices. When fruit is transported from a distance, is grown under unfavorable conditions, or is in a deteriorated state, proper storage allowance should be made.

Fresh fruit for storage should be as free as possible from skin breaks, bruises, and decay. These defects reduce the value of the product and may cause rapid deterioration not only of the damaged fruit, but also of the stored fruit nearby. This process occurs because damaged fruit often exhibits increased ethylene production, which can cause rapid ripening of many types of climacteric fruit. For the same reason, it is unwise to store fruit or vegetables having different storage characteristics together; some may emit ethylene, causing a more sensitive crop to ripen prematurely. Natural cooling in well-ventilated storage slows down or halts these processes.

The amount of incipient decay infection, which influences storage potential of grapes and apples, can be predicted in the early storage period. Only lots with good storage potential should be held for late-season marketing.

The maturity of the fruit at harvest time determines the refrigerated storage life and quality of the product. For any given produce, there is a maturity best suited for refrigerated storage. Undermature produce will not ripen or develop good quality during or following refrigerated storage. For many crops, excessively overmature produce deteriorates quickly during storage, although there are some exceptions for late-harvested fruit (in particular, late-harvested kiwifruit). Determination of maturity can be a complex problem. A number of measurements are used depending on the crop; these include penetromer firmness, color, degree-days since flowering or fruit set, soluble solids, or other physical, chemical or biological tests. In critical cases, a combination of tests may be used.

Handling and Harvesting

Rising handling costs have encouraged the use of bulk handling and large storage bins for many kinds of fruit. Moving, loading, and stacking bins by forklift trucks must be done carefully to maintain proper ventilation and refrigeration of the product. Bins should not be so deep that excessive weight damages the produce near the bottom.

Mechanical harvesters for fruit frequently cause some bruising. This damage can materially reduce the quality of the produce.

Transportation

As in storage, losses from deterioration during distribution are affected by temperature, moisture, diseases, and mechanical damage. Gradual aging and deterioration are continuous after harvest. Time in transit may represent a large portion of postharvest life for some commodities, such as cherries and strawberries. Thus, the environment during this period largely determines produce salability when it reaches the consumer.

To prevent undue warming and condensation of moisture, which promote decay and deterioration, fruit-handling systems must be well designed to minimize rewarming and moisture condensation on the product. For example, fruit should not be removed from cool storage and left unattended for significant periods of time before loading and transport in refrigerated vehicles. When the product is removed from cool storage, deterioration in flavor and condition may be accelerated after long periods of storage; therefore, the product should be consumed as quickly as possible or retained at low temperature.

Details on storage and handling of common fruit are given in the following sections. For more information on storage requirements and physical properties of specific commodities, see Chapter 10. Table 1 shows recommended controlled atmosphere (CA) and modified atmosphere (MA) conditions (Kader et al. 1992).

APPLES

Apples are not only the most important fruit stored on a tonnage basis, but their average storage period is considerably greater than that of any other fruit. The length of storage may be short for early varieties and those going into processing, but cold storage is critical to proper handling and marketing.

Recommended storage temperature depends on the cultivar. For most apple varieties, cool storage at 0 to 1°C is recommended. Specific recommendations for each commercial cultivar are usually available from marketing organizations [or see Kader et al. (1992)].

Storage life of apple varieties depends on harvest maturity, elapsed time and temperature between harvest and storage, cooling rate in storage, and sometimes cultural factors. The best storage potential is usually found in apples that are mature but have not yet attained their peak of respiration when harvested. However, the grower is inclined to sacrifice storage quality for the better color often gained in red varieties by holding them longer on the tree. Even if harvesting begins at the proper time, the fruit picked last may be at an advanced stage of maturity. Such late-harvested apples do not have good storage characteristics; neither do those harvested on the immature side, but this is seldom a problem with apples intended for storage before marketing. Harvest at proper maturity,

---

The preparation of this chapter is assigned to TC 10.9, Refrigeration Application for Foods and Beverages.

Table 1 Recommended Controlled Atmosphere or Modified Atmosphere Conditions During Transport and/or Storage of Deciduous Tree Fruit

Commodity	Temperature,[a] °C	Controlled Atmosphere[b] % Oxygen	Controlled Atmosphere[b] % Carbon Dioxide	Potential for Benefit[c]	Remarks[d]
Apple	0 to 5	1 to 3	1 to 5	A	About 50% of production is stored under CA
Apricot	0 to 5	2 to 3	2 to 3	C	No commercial use
Cherry, sweet	0 to 5	3 to 10	10 to 15	B	Some commercial use
Fig	0 to 5	5 to 10	15 to 20	B	Limited commercial use
Grape	0 to 5	2 to 5	1 to 3	C	Incompatible with SO_2 fumigation
Kiwifruit	0 to 5	1 to 2	3 to 5	A	Some commercial use; C_2H_4 must be maintained below 20 ppb
Nectarine	0 to 5	1 to 2	3 to 5	B	Limited commercial use
Peach	0 to 5	1 to 2	3 to 5	B	Limited commercial use
Pear, Asian	0 to 5	2 to 4	0 to 1	B	Limited commercial use
Pear, European	0 to 5	1 to 3	0 to 3	A	Some commercial use
Persimmon	0 to 5	3 to 5	5 to 8	B	Limited commercial use
Plum and prune	0 to 5	1 to 2	0 to 5	B	Limited commercial use
Raspberry and other cane berries	0 to 5	5 to 10	15 to 20	A	Increasing use during transport
Strawberry	0 to 5	5 to 10	15 to 20	A	Increasing use during transport
Nuts and dried fruit	0 to 25	0 to 1	0 to 100	A	Effective insect control method

Source: Kader et al. (1992), Table 5.1. Reprinted by permission.
[a] Usual and/or recommended range. A relative humidity of 90% to 95% is recommended.
[b] Best CA combination may vary among cultivars and according to storage temperature and duration.
[c] A = excellent, B = good, C = fair.
[d] Comments about use refer to domestic market only; many of these commodities are shipped under MA for export marketing.

Table 2 Storage Periods for Certain Apple Cultivars and Their Susceptibility to Storage Disorders

Cultivar	Months of Storage Normal	Months of Storage Maximum[a]	Storage Scald Susceptibility	Other Disorders Likely to Occur in Storage[b]
Baldwin	4 to 5	6 to 7	Moderate	Bitter pit, brown core
Cortland	3 to 4	6 to 7	Very high	Senescent breakdown
Delicious	5 to 6	8 to 11	Moderate	Bitter pit, senescent breakdown, soft scald
Empire	4 to 5	8 to 9	Slight	Chilling injury, brown core, senescent breakdown
Golden Delicious	5 to 6	7 to 11	Slight	Shriveling, bitter pit, senescent breakdown
Gravenstein	2	3	Moderate	Bitter pit, Jonathan spot
Granny Smith	5 to 6	7 to 9	High	Bitter pit, brown core, senescent breakdown
Idared	5 to 6	8 to 9	Slight	Breakdown, Jonathan spot
Jonathan	3 to 4	6 to 8	Moderate	Breakdown, Jonathan spot, soft scald
McIntosh	4 to 5	7 to 8	Moderate	McIntosh breakdown, brown core
Northern Spy	4 to 5	7 to 8	Slight	Senescent breakdown, Spy spot, bitter pit
Rome Beauty	5 to 6	7 to 8	Very high	Jonathan spot, soft scald
Spartan	5 to 6	7 to 10	Slight	Spartan breakdown, brown core
Stayman Winesap	4 to 5	7 to 8	Very high	Senescent breakdown
Winesap	5 to 6	8 to 9	High	Senescent breakdown
Yellow Newtown	5 to 6	8 to 9	High	Internal browning, senescent breakdown, bitter pit
York Imperial	4 to 5	6 to 7	Very high	Cork spot

Source: Hardenburg et al. (1986).
[a] For maximum storage, cultivars must be harvested at optimum maturity, stored under ideal temperature and humidity, and in most cases in the recommended controlled atmosphere. Some fruit may be stored 1 to 2 months longer than shown.
[b] Water core not listed for Delicious, Jonathan, Winesap, Stayman, and others, as it is present at harvest and does not develop in storage.

careful handling, and prompt storage after harvest are conducive to long storage life.

Chilling injury is the term commonly applied to disorders that occur at low storage temperatures where freezing is not a factor. The exact mutual relationship of the many types of chilling injury is unknown. The principal disorders classed as chilling injuries in apples are (1) soft scald; (2) soggy breakdown; (3) brown core; and (4) internal browning. Varieties susceptible to one or more of these disorders are Rome Beauty, Braeburn, Jonathan, Golden Delicious, Empire, Grimes Golden, McIntosh, Rhode Island Greening, and Yellow Newtown. In addition to variable susceptibility by variety, there are also yearly variations related to climate, fruit size, and cultural factors.

Table 2 lists the range in storage life at −1°C of several apple varieties. The following practices affect the condition of apples held for both conventional and controlled atmosphere storage:

Maturity. Because there is no reliable maturity index, growers must use personal experience of the variety, area, or orchard to decide when the crop is mature. Availability of labor, size of operation and crop, weather, storage facilities, and intended length of storage also affect the time of harvest.

Handling to Storage. For optimum storage, apples should be cooled within one or two days of harvest because they can deteriorate as much during one day at field temperatures as during one week at proper storage temperature. If other factors prevent final packaging, fruit can be cooled and stored in the field bins used for

Deciduous Tree and Vine Fruit

harvest. In this case, no grading will have been done to remove substandard product. Subsequent grading may result in an increased level of bruising, especially if the fruit is still cold when handled.

Normally, apples are placed in storage and cooled by the room refrigeration equipment to about 0°C in 1 to 3 days. Hydrocooling is sometimes used, but it requires careful disease control. It also interferes with scald inhibitors, which must be applied to the warm fruit.

Controlled Atmosphere Storage

Controlled atmosphere (CA) storage offers important gains in extending the market life of certain apple varieties. Chilling injury is eliminated in some varieties by elevating the storage temperature to about 4°C and altering the composition of the atmosphere.

Only apples of good quality and high storage potential should be placed in CA storage. Harvest maturity and handling practices are crucial; only fruit harvested at proper maturity should be considered. In any one district, this limits the apple harvest for CA storage to only a few days. Immature apples or those retained on the tree to gain better color, as is often the case with Delicious and McIntosh, are equally undesirable.

The rooms for gas storage must be gastight, and there must be provision to remove excess carbon dioxide (CO_2) from the air. Carbon dioxide may be removed by circulating the air through (1) washers or scrubbers filled with sodium hydroxide, ethanolamine, or plain water; or (2) a cabinet or room containing bags of hydrated lime.

Modified atmosphere (MA) storage may be obtained by filling the room with fruit, sealing it, and allowing the respiration of the fruit to provide the desired proportions. These proportions are maintained by operation of the scrubber and by ventilation. A simple pressure equalization system is vital for the safe operation of CA storage. However, in CA storage, the atmosphere is established and maintained by using controlled supplies of gas from cylinders.

Table 3 lists approximate temperature and atmospheric requirements for CA storage of several varieties. Since these may vary from region to region, more precise requirements should be obtained from authorities within a region. For example, Jonathan, Rome Beauty, and Stayman Winesap are reported best at 2% CO_2 in New York, whereas a somewhat wider range is acceptable in Washington.

Disadvantages of gas storage include the difficulty of making the storage room gastight, the danger of suffocation to persons entering the room, and the impossibility of entering the room to examine the fruit without losing the desired atmosphere. Some of these disadvantages have been overcome by a method that passes the air entering the storage room through a generator, which reduces the oxygen (O_2) level and raises the CO_2 content to desired levels. The process is continuous, so airtight rooms are not essential. Also, the desired atmosphere may be restored quickly after the rooms are opened for inspection.

Storage Diseases and Deterioration

Storage problems in apples may be caused either by invading microorganisms or by the fruit's own physiological processes. Physiological disorders, although sometimes resembling rots, are related to biochemical processes within the fruit. Susceptibility to such disorders is often a variety characteristic, but it may be influenced by cultural and climatic factors and storage temperature.

Alternaria Rot. Dark brown to black, firm, fairly dry to dry storage decay centering at wounds, in skin cracks, in core area, or in scald patches—one of the blackest of storage decays. *Control*: Cultural practices that produce apples of good finish and prevent skin diseases and injuries that open the way for infection.

Ammonia Gas Discoloration. Circular spots centering at lenticels; dull green on unblushed side and brown to black on blushed side. Injury may disappear from slightly affected fruit. *Control*: Ventilate as soon as possible. Examine fruit for injury at various points in the room because some sections may escape.

Bitter Pit. Many small, sunken bruise-like spots, usually on the calyx half of the fruit. Masses of brown, spongy tissue occur adjacent to surface pits or may be found deeper in the flesh. In storage, spongy tissue near surface loses moisture and tends to become hollow. New areas may appear and develop in storage. *Control*: Apply boron and calcium, as recommended, in the orchard. Follow cultural practices that promote regular bearing and stabilize moisture. Store fruit of proper maturity and cool promptly to 0°C. Maintain humidity high enough to prevent moisture loss.

Blue Mold Rot (*Penicillium*). Spots of various sizes with decayed tissue that is soft and watery and can be readily scooped out of the surrounding healthy flesh. Rot usually as deep as wide. Advanced stages have white tufts of mold that turn bluish-green as spores are produced under moist conditions. Affected tissue has moldy or musty flavor or odor. Most prevalent type of storage decay of apples. *Control*: Handle carefully to prevent skin breaks. Cool promptly to 0°C. Use fungicides in wash treatments. Keep picking boxes, packing house, and storage room sanitary. Whitewash walls and ceiling.

Brown Core. No external symptoms. First appears as slight browning or discoloration of core tissue between the seed cavities. Later, part or all of the flesh between the seed cavities and the core line may become brown. Serious in McIntosh and other susceptible varieties stored for long periods at −1°C. *Control*: Pick at proper stage of maturity. Use CA storage at 3°C. A disorder with similar symptoms has been reported as a result of exposure to excessive concentrations of CO_2.

Freezing Injury. Water-soaked, rubbery condition of large areas or of entire apple. Vasculars (water-conducting strands) brown. Bruised areas in frozen apples large, with wrinkled gray to light brown surface. Moisture lost rapidly from affected areas. In refrigerator cars, most prevalent on floor and at doorways; in storage rooms, most injury in bottom layer boxes, near coils, or against walls next to freezer storage. *Control*: Heat car during subfreezing weather. Prevent cold pockets in storage rooms by adequate air circulation. Minimize handling of fruit while frozen. Thaw at 5 to 10°C. Move thawed fruit into trade channels promptly; do not allow it to become overripe.

Internal Breakdown. Mealy breakdown of internal tissue in overripe fruit. Flesh soft. Surface often duller and darker than normal. Hastened by too high storage temperature, freezing, bruising, or presence of water core, which it often follows. *Control*: Pick before overmature. Cool promptly at temperatures as near 0°C as

Table 3 Requirements for Controlled Atmosphere Storage of Apples

Cultivar	Carbon Dioxide, %	Oxygen, %	Temperature, °C
Cortland	5	2 to 3	2.2
	2 to 3	2 to 3	0
Delicious	1 to 2	1.5 to 2[a]	−0.5 to 0
Golden Delicious	1 to 3	1.5 to 2[a]	−0.5 to 0
Granny Smith	1 to 3	2 to 3	−0.5 to 0
Idared	2 to 3	2 to 3	−0.5 to 0
McIntosh	2 to 3 one month, then 5	2.5 to 3	2.2
	2 to 3 one month, then 5	2	3.3
Rome Beauty	1 to 3	2 to 3	−0.5 to 0
Stayman Winesap	2 to 5	2 to 3	−0.5 to 0
Yellow Newtown			
California	8	3	4.4
Oregon	5 to 6	3	2.2

Source: Hardenburg et al. (1986).
[a] 1.5% oxygen not recommended for Delicious or Golden Delicious in New York because less than 2% oxygen is injurious.

possible for varieties that tolerate that temperature. Watch ripening rate, particularly of fruit with water core.

Internal Browning. No abnormal skin appearance. Sometimes appears only around core; the apple's outer fleshy portion remains normal in appearance. Occasionally only outer flesh is involved; but when internal browning develops in the outer fleshy portion, it is usually accompanied by browning around the core. Disease develops uniformly throughout tissue. Occurs in firm, sound apples. *Control*: Use CA at 3°C for Empire and other susceptible varieties.

Jonathan Spot. Slate-brown to black, entirely superficial or very slightly sunken, skin-deep spots in color-bearing cells of skin. In some varieties, spots center at lenticels. *Control*: Refrigerate promptly as this disease is greatly aggravated by delayed storage. Use CA storage.

Lenticel Rots. Bullseye rot (*Neofabrabraea*): most common of group; of importance only in apples from Northwest; spots fairly firm, pale centers, decay mealy, may penetrate nearly as deep as wide. Fisheye rot (*Corticum*): tough leathery spot, often follows scab; decayed tissue stringy. Side rots (*Phialophora*): spots shallow with tender skin, decayed tissue wet, slippery. *Control*: Harvest at prime maturity; store and cool promptly; use forecasting technique for bullseye rot to determine potential keeping quality.

Scab (*Venturia*). Occasionally, active scab spots on fruit at time of storage will enlarge. Fruit may be infected in orchard but show no disease at the time of storage. Disease may subsequently develop in storage as small brown or jet black spots in peel, often without breaking cuticle of fruit. *Control*: Follow recommended orchard spray schedule.

Scald. Diffuse browning and killing of skin of fruit stored for several months. Ordinarily most prevalent on immature fruit or on green portions of fruit. *Control*: Pick apples when well matured. Treat with effective scald-inhibiting chemicals. Scald develops less on fruit in controlled atmospheres.

Soft Scald. Sharply defined or slightly sunken ribbon-like areas in the skin. Affected tissue shallow and rubbery. Most severe on Jonathan, Golden Delicious, and Wealthy. *Control*: Store promptly. Use recommended controlled atmospheres, temperatures, and lengths in storage for each variety.

Soggy Breakdown. Light brown, moist, rubbery, definitely delimited areas in cortex of apple. Not visible on surface. Worst in Grimes Golden, Wealthy, and Golden Delicious. *Control*: Same as for soft scald.

Water Core. Hard, glassy, water-soaked regions in flesh of apple at core or under skin. Decreases in extent during storage but predisposes fruit to internal breakdown. *Control*: Pick as soon as mature. Watch fruit in storage and move before it becomes overripe.

PEARS

Bartlett is the most important pear variety, exceeding the total of all others by a wide margin. Other Pacific Coast varieties are Hardy, Comice, Anjou, Bosc, and Winter Nelis. The eastern states have limited varieties due to the severe problem of fire blight and primarily grow the Kieffer variety. Although most Bartlett and Hardy and many Winter Nelis pears are canned, cold storage prior to ripening for canning is the usual procedure. A 10 day to 2 week cold storage period for Bartlett pears is commonly used by canners because it improves uniform ripening. Substantial quantities may also be stored for periods approaching maximum storage life of the variety to better use processing facilities.

Maturity at harvest has a very important bearing on subsequent storage life, as it does for apples. However, unlike apples, pears do not ripen on the tree, nor do most varieties ripen at cold storage temperatures. If harvested too early, they are subject to excessive water loss in storage. If permitted to become overmature on the tree, their storage life is shortened, and they may be highly susceptible to scald and core breakdown. Flesh firmness as measured by a pressure tester is perhaps the best measure of potential storage life of pears from any single orchard. For the Bartlett variety, a firmness of 85 to 75 N, measured with a Magnus-Taylor pressure tester or similar device, using an 8 mm plunger head, indicates best storage quality. If average firmness is as low as 67 N, storage for any prolonged period is hazardous. Pressure test information for each lot of pears going into storage may be very helpful to both the fruit owner and the cold storage operator in determining the storage program.

Careful harvesting and handling are essential to good storage quality. Bruises and skin breaks are likely sites for infection by microorganisms. Varieties such as Winter Nelis and Bosc are highly susceptible to punctures caused by stems broken in the harvesting operation. Comice is also easily damaged because of its very tender skin. Many pears are now being harvested into pallet bins holding about 450 kg of fruit. Care in dumping fruit from a picking container is important in keeping mechanical damage to a minimum.

For best storage quality, rapid cooling after harvest is essential. Pears ripen rapidly at elevated temperatures but do not soften or change color in the early ripening stages. Therefore, a considerable part of the storage life may be used up without a visible change in the fruit. If cold storage rooms do not have adequate refrigeration and ventilation capacity for the rapid cooling of fruit, precooling in special rooms (or hydrocooling) prior to placing in the storage room should be considered. When warm fruit is placed in a room with cold fruit, the loading arrangements should be such that the temperature of the cold fruit is not elevated.

Pears are very sensitive to temperature and should be stored at −1°C and 90 to 95% rh. Recommendations as low as −1.7°C have been made, but the risk of freezing injury is great unless the temperature in all parts of the room can be controlled precisely. Pears are not subject to chilling injury as are some apple varieties, so elevated storage temperatures are not required. The stacking arrangement recommended for apples in the cold storage room also applies for pears.

Since pears lose water more readily than most apple varieties, good humidity conditions in the storage room must be maintained. For long storage, 90 to 95% rh is recommended. Perforated film box liners are excellent for moisture loss control.

The approximate storage life of pears at −1°C is shown in Table 4. These values assume an additional time for transportation and marketing. If Bartlett pears for canning are harvested at the best stage of maturity and quickly cooled to −1°C, their safe storage life may be as long as 4 months, since marketing involves only ripening for processing. However, quality deteriorates during storage, particularly as the maximum storage life is approached.

After removal from storage, best dessert quality is attained if pears are ripened in a controlled temperature range of about 16 to 21°C. This applies to fruit for the cannery and for fresh use. For cannery fruit, ripening at 20 to 22°C is more practical than lower temperatures because the shorter time involved reduces overhead costs with no measurable difference in quality.

Controlled Atmosphere Storage

The practice of CA storage of pears is promising. The storage life of Bartlett pears can be extended to 5 to 6 months at −1°C for fruit of desirable maturity (75 to 89 N firmness) in an atmosphere

Table 4 Controlled Atmosphere Storage Life of Pear Varieties at −1°C

Variety	Storage Life, Months
Bartlett, Hardy, and Kieffer	2 to 3
Bosc, Comice, and Seckel	3 to 4
Anjou	6 to 7
Packham	5 to 6
Winter Nelis	7 to 8

Deciduous Tree and Vine Fruit

containing 2 to 2.5% oxygen and 0.8 to 1% CO_2. Bartlett pears of advanced maturity are intolerant to elevated CO_2 and develop core and flesh browning within a few weeks. They are tolerant to low O_2, but have less storage potential than pears of desirable maturity. Chronological age and pressure test (under 71 N firmness) are evidence of advanced maturity.

Commercial storage of pears in a controlled atmosphere has not been considered as necessary as it is in the apple industry. Since no low-temperature disorders have been recognized, there has not been a need for further extension of the storage life.

Many pears from western states, when packed for storage before marketing, have perforated polyethylene liners in the container. While such liners give excellent protection against water loss, there is no agreement as to their value in modifying the atmosphere within the container.

Storage Diseases and Deterioration

The principal storage disorders in pears are (1) core breakdown; (2) scald and failure to ripen; and (3) fungus rots.

Core Breakdown. Often accompanies scald. Soft, brown breakdown in core area having acrid, disagreeable odor of acetaldehyde. *Control*: Do not allow pears to become overmature on tree. Cool promptly. Store at $-1°C$. Ripen fruit between 18 and 24°C.

Core breakdown is associated with overmaturity at harvest. This problem has become more important because of growth regulator sprays used to keep pears from dropping during the harvest season. Pressure test information on late-harvested fruit is helpful in locating lots susceptible to core breakdown. Records of the time lapse between harvest and storage are important, since pressure test information may not be a true measure of relative storage quality where storage is delayed. Pear color, particularly in California Bartletts, is a very poor measure of potential storage life because of great variability among pears from different districts.

Scald. Often accompanies core breakdown. Brown to black softening of large areas of skin and tissues immediately beneath skin. Affected areas slough off readily. Acetaldehyde odor and flavor prominent. *Control*: Pick before overmature. Cool promptly. Store only for proper period. Oiled paper wraps do not control scald.

Pear scald is associated with pears that have been stored too long and have lost their capacity to ripen. It is not related to apple scald and cannot be controlled by any supplemental treatments. The problem develops progressively earlier as the temperature is raised above $-1°C$. Yellowing of the fruit is the principal storage symptom; Bartlett and Bosc are the two most susceptible varieties. Anjou and Comice may not develop scald but do lose their capacity to ripen. Periodic inspection is desirable to be sure that green pear varieties are removed from storage before yellowing progresses to the danger point. Yellow pears may show no scald in storage but may develop scald on removal to a ripening temperature. If pear scald does show in storage, the pears have been kept too long and may be worthless.

Anjou Scald. Anjou pears are often affected with a surface browning more superficial than common scald and distinct from it, resembling apple scald. Anjou scald is controlled by oiled paper wraps and effective scald-inhibiting chemicals.

Gray Mold Rot (*Botrytis*). Extensive, firm, dull brown, water-soaked decay with bleached border. Dirty white to gray extensive mycelium forming nests of decayed fruit. *Control*: Wrap fruit in copper-impregnated paper. Use fungicide in spray or wax on packing line. Cool promptly to $-1°C$.

Gray mold rot caused by *Botrytis cinerea* grows at cold storage temperatures and can be a serious threat to long-stored winter varieties such as Anjou and Winter Nelis. Without control measures, the disease may spread from one fruit to another by contact.

Alternaria Rot. Surface is dark brown to black. Decayed tissue is gray to black, dry in center, gelatinous at edge, easily removable as core from surrounding flesh. Found late in storage season, usually at punctures. *Control*: Prevent skin breaks. Remove from storage at first appearance of trouble.

Brown Core. Anjou and Bartlett pears stored in sealed, polyethylene-lined boxes with inadequate permeability may show various degrees of pithy brown core and desiccated air pockets. Prolonged storage in concentrations of 4% or more CO_2 often produces brown core, particularly when pears are harvested at advanced maturity or are cooled slowly after packing. *Control*: Harvest at proper maturity. Cool promptly. Store at $-1°C$. Use perforated film liners to maintain CO_2 level at 1 to 3%.

Freezing Injury. Bartlett and Anjou pears exposed for 4 to 6 weeks just below their freezing point develop glassy, water-soaked external appearance with tan pithy area around core. Pears frozen sharply may break down completely or show abruptly sunken large pits where slightly bruised while frozen. *Control*: Keep transit and storage temperature above $-1°C$.

GRAPES

Grapes are widely grown in the United States, but over 90% are grown in California. This state produces grapes of the *Vitis vinifera* species almost exclusively. This species can withstand the rigors of handling, transport, and storage required of table grapes for wide distribution over a long marketing period. Almost all of this fruit is precooled and much of it stored for varying periods before consumption. On the other hand, for fresh use, the fruit of the species *Vitis labrusca* (Eastern type) is largely limited to local market distribution.

Grapes grow relatively slowly and should be mature before harvest because all of their ripening occurs on the vine. **Mature** here means that stage of physiological development when the fruit appears pleasing to the eye and can be eaten with satisfaction. However, grapes should not be overripe, as this predisposes them to two serious postharvest disorders: (1) weakening of the stem attachment in some varieties, such as Thompson seedless, which causes the berries to separate from the pedicel attachment; and (2) progressively greater susceptibility to invading decay organisms. Danger of fruit decay is increased with exposure to rain or excessively damp weather before harvest (conditions favorable for the inception of field infections by *Botrytis cinerea Pers*).

Cooling and Storage

Grapes are vulnerable to the drying effect of the air because of their relatively large surface-to-volume ratio, especially that of the stems. Stem condition is an important quality factor and an excellent indicator of the past treatment of the fruit. Stems should be maintained in a fresh green condition not only for appearance but because they become brittle when dry and are apt to break. The stem of a grape cluster, unlike that of other fruit, is the handle by which the fruit is carried; if breakage (shatter) occurs, the fruit is lost for all practical purposes even though the shattered berries may still be in excellent condition. Therefore, careful attention should be paid to those operations that minimize moisture loss.

The rate of water loss is especially high before and during precooling because grapes are normally harvested under hot, dry conditions. Field heat should be removed promptly after the fruit is picked to minimize the exposure of grapes to low vapor pressure conditions. Volume and temperature of the precooling air, velocity of the air past or through the containers (lugs), and accessibility of the fruit to this air are significant factors in the rate of heat removal. These factors are drastically influenced by the location and amount of venting of the containers, alignment of the containers (air channels), and packing materials such as curtains, cluster wraps, and pads.

Two general systems of air handling used for precooling table grapes are (1) the velocity or conventional system, and (2) the pressure system.

In the **velocity system**, air is forced through channels parallel to the long axis of the containers. Heat transfer is effected by conduction through the packaging materials and by penetration of the cold air to the fruit from turbulence in and around the vents. Satisfactory precooling rates can be attained if (1) the containers are aligned so that there are no obstructed channels; (2) the velocity of the air through these channels is at least 0.5 m/s; and (3) air of not more than 1.5°C can be supplied at the rate of at least 0.17 L/s per kilogram of fruit. Very humid air helps to reduce the rate of water loss from the fruit. Large cooling surfaces can be maintained, or atomized water can be added to the airstream. However, the rate of removal of field heat is the significant factor; humidifying techniques that retard precooling are likely to be a liability.

In the **pressure system**, a pressure gradient is set up so that there is a positive flow of cold air through the fruit from one vented side of the container to the other. The containers are arranged so that the air must pass *through* the containers before returning to the refrigeration surface. Precooling time may be as little as one-fifth that of the velocity system if the pressure differential across the packages is equivalent to at least 60 Pa and cold air is supplied at the rate of at least 1 L/s per kilogram of fruit.

The recommended storage temperature for *Vitis vinifera* (European or California type) grapes is −1°C. The relative humidity should be 90 to 95%. Although temperatures as low as −1.7°C have not been injurious to well-matured fruit of some varieties, other varieties having low sugar content have been reported damaged by exposure to −0.5°C. Grape storage plants in California should provide uniform air circulation in the rooms. Some have precooling rooms where the grapes are cooled to about 4°C in 6 to 24 h before storage. In some plants, all of the cooling is done in the storage rooms, but only a few have sufficient air movement to cool the fruit as quickly as desired. After the fruit has been precooled, the air velocity should be reduced to a rate that maintains uniform temperatures throughout the room (no more than 0.05 to 0.1 m/s in the channels between the lugs). Ventilation is required only to exhaust sulfur dioxide and air following fumigation.

The greatest change that takes place in grapes in storage is loss of water. The first noticeable effect is drying and browning of stems and pedicels. This effect becomes evident with a loss of only 1 to 2% of the mass of the fruit. When the loss reaches 3 to 5%, the fruit loses its turgidity and softens.

Maintaining a relative humidity of 90 to 95% in grape storage is often a problem, especially at the beginning of the storage season when the rooms are being filled with dry lugs. Each lug will absorb 0.15 to 0.3 kg of water over a month, and, unless moisture is supplied to the room, this water must come from the fruit. Spray humidification is an effective method of supplying water to minimize shrinkage. With proper balance of water and air pressure and the correct type of nozzle, a fine spray can be obtained that will vaporize readily even at −0.5°C.

Fumigation

Vinifera grapes must be fumigated with sulfur dioxide (SO_2) after they are packed to prevent or retard the spread of decay. The treatment surface-sterilizes the fruit, particularly wounds made during handling.

Fumigation with SO_2 in storage prevents new infections of the fruit but does not control infections that have already occurred in the vineyard. Frequently, these have not developed far enough to be detected at harvest and consequently are the primary cause of decay in storage. A method of measuring field infection has been developed and used to forecast decay during storage. The forecast indicates the lots that are sound and can be safely stored and also those that are likely to decay and should be marketed early (Harvey 1955, 1984).

It has become common practice to accumulate packed fruit in the precooler during the daily packing and to fumigate the fruit in the evening. In this way, precooling is not delayed, and fumigation can be done after most of the working crew has left. This initial treatment often becomes the responsibility of the refrigeration personnel.

Amount of Sulfur Dioxide. Other commodities should not be treated along with the grapes or even held where the fumigant can reach them, as most of them are very easily injured by the gas. Because grapes also can be injured, they should be exposed to the minimum quantity of SO_2 required, which depends on the following:

- Decay potential and condition of the fruit
- Amount of fruit to be treated
- Type of containers and packing materials
- Air velocity and uniformity of air distribution
- Size of the room
- Losses from leakage and sorption on walls

Under favorable conditions, a basic SO_2 concentration of 0.5% by volume for 20 min is adequate. To keep the concentration at this level, consider the absorptive capacity of the containers and fruit as well as their volume. The dosage can then be calculated from the following equation:

$$W_s = \frac{AB}{E} + CD$$

where

W_s = quantity of SO_2 required, kg
A = concentration of SO_2 to be used, %
B = free volume of room (total volume minus 0.014 m³ for each container), m³
C = number of carloads of 12.7 kg lugs (1000 lugs/car)
D = quantity of SO_2 absorbed by each carload, kg
E = volume occupied by SO_2 gas at 0°C = 0.34 m³/kg

For factor D, 0.5 kg per car is adequate when the fruit is sound, air velocities are maintained past both sides of every container at 0.25 m/s or more (0.4 to 0.5 m/s if the fruit has curtains over it or the clusters are wrapped), and the room is relatively gastight with no opportunity for the fumigant to be lost on refrigeration surfaces. Conversely, a higher value of 1 kg/car would be used when these factors are less favorable.

Grapes must be fumigated weekly in storage to prevent *Botrytis cinerea* from spreading from infected fruit to adjacent sound fruit. The amount of SO_2 needed depends on the same factors as for the initial treatment. However, a basic concentration of 0.1% for 30 min is adequate. Also, an absorptive factor in the range of 0.15 to 0.3 kg of SO_2 per carload should be used.

Gas Distribution Procedure. The gas must be distributed quickly and evenly to all parts of the room. This can be done by spacing special nozzles 1800 mm apart along the ceiling in the room. If the outlet is placed in front of a fan, there should be one for each fan, or the air from the single fan should be distributed evenly across the room through a plenum.

The same requirements of proper container alignment, adequate fan capacity, and uniform air distribution apply here as for the initial treatment. The lugs should be oriented parallel to the airflow, and 20 to 40 mm channels should be provided on both sides and kept completely unobstructed through the stacked fruit. The fruit should be stacked as near the ceiling as possible so that curtains provided over the fruit will prevent air from passing over the fruit and thus bypassing the channels. The working distance between pallets should be kept to an absolute minimum to avoid wide channels, and no holes should be left in the wall of lugs when pallets of fruit are withdrawn.

The **hot-gas method** of delivery may be used if the room requires 4.5 kg or less of gas. The steel cylinder containing the liquid SO_2 is first connected to the gas inlet, and the valve is then opened. The cylinder should then be placed in a pot of boiling water

Deciduous Tree and Vine Fruit

to vaporize the fumigant as rapidly as possible. Only about 7.5 g/s can be delivered this way.

For larger quantities, the **cold-gas method** is usually more practical. A riser extends to the bottom of the cylinder through which the liquid SO_2 rises and flows through the delivery line. Every precaution must be taken to provide enough air volumetric flow and velocity to vaporize and mix the gas thoroughly with the air before it reaches the fruit. Up to 45 kg of the material can be released in 2 to 3 min. After 30 min, the room should be purged of the gas-laden air until personnel can remain in the space without excessive discomfort.

In plants that are devoted entirely to the storage of grapes, the gas is sometimes released into the air ducts of the plant, thus using the air-cooling system for even distribution and good circulation in the rooms. With a brine spray system of refrigerating the air, a bypass around the spray chamber prevents the gas from contacting the wet metal surfaces, since it readily forms a corrosive acid in combination with water. For the same reason, SO_2 should be cleared from the air of the rooms before the damper is turned and the air is circulated through the spray. It is advisable to check the acidity of the brine frequently to guard against corrosion.

Precautions. Sulfur dioxide has certain properties that demand care in its use as a fumigant in cold storage plants. The concentrations recommended for the fumigation of grapes in storage can cause respiratory spasms and death if the victim cannot escape the fumes. When working in even weak concentrations of SO_2, goggles to protect against injury to the eyes and a gas mask fitted with canister for acid gases (not the usual canister for ammonia gas) should be worn. Concentrations as low as 30 to 40 mg/kg can be detected by smell. It requires several times these concentrations to cause discomfort.

Because a small segment of the population may experience severe allergic reactions to sulfites, the U.S. Environmental Protection Agency has proposed a 10 mg/kg tolerance for sulfite residues in table grapes (EPA 1989, Harvey et al. 1988). Fruit with residues exceeding the tolerance cannot be marketed.

Another precaution about SO_2 that cannot be overemphasized is its injurious effects on other produce. For this reason, care must be taken that only grapes are stored in the room that is to be fumigated and that there are no leaks through walls or halls to adjacent rooms storing other produce.

Periodic inspection of the fruit is recommended to check whether the SO_2 gas is reaching the center of the stacks or whether some grapes are being overtreated. If the pedicels and stems retain a yellow or green color and broken berries show no mold and appear to be dried or seared, the gas has reached the fruit in question and is having the desired effect. Serious bleaching on unbroken grapes means too high a concentration or too long an exposure; there should be better distribution of the gas, lower concentration, or shorter fumigation periods.

Diseases

Blue Mold Rot (*Penicillium*). Watery, mushy condition. Early production of typical bluish-green spores on berries and stems. Moldy odor and flavor. *Control*: Prevent deterioration in fruit by careful handling and prompt refrigeration, preferably to 0°C. Fumigate with SO_2 in storage.

Cladosporium Rot. Black, firm, shallow decay which produces an olive-green surface mold. Common on stored grapes harvested early in the season. Infections occur on small growth cracks at the blossom end and sides of the grape. *Control*: Precool and store grapes promptly at 0°C. After harvest, fumigate with SO_2 to reduce spread.

Gray Mold Rot (*Botrytis*). Early stage: slip skin with no mold growth. Later, nest of fairly firm decay covered with abundant fine gray mold and grayish-brown, velvety spore masses. *Control*: Cull out decay when packing. Fumigate grapes with SO_2. For storage, cool grapes rapidly to $-1°C$. Use forecasting technique to determine safe storage periods. Use short storage period for grapes harvested in rainy periods or after slight freezes.

Table 5 Storage Life of California Table Grapes at 0°C

Variety	Storage Life, Months
Emperor, Ohanez, Ribier	3 to 5
Malaga, Red Malaga, Cornichon	2 to 3
Thompson seedless, Tokay	1 to 2.5
Muscat, Cardinal	1 to 1.5

Table 6 Storage Life of *Labrusca* Grapes at 0°C

Variety	Storage Life, Weeks
Catawba	5 to 8
Concord, Delaware	4 to 7
Niagara, Moore	3 to 6
Worden	3 to 5

Rhizopus Rot. Soft, mushy, leaky decay causing staining of lugs. Coarse extensive mycelium and black sporangia develop under moist conditions. *Control*: Prevent skin breaks. Cool promptly to below 10°C.

Sulfur Dioxide Injury. Bleached sunken areas on berry at skin breaks or cap-stem attachment. Decolorized portions have disagreeable astringent flavor. Does not appear in full severity until cool grapes are warmed. *Control*: Apply proper concentration and distribution of gas for recommended period.

Storage Life

The normal storage life of the principal varieties of California table grapes at $-1°C$ is shown in Table 5. Under exceptional conditions, sound fruit will keep longer than indicated; for example, Emperor grapes have been held in good condition for 7 months, and Thompson seedless for 4 months.

The storage life of grapes is affected most by the attention given to selecting and preparing the fruit. Grapes should be picked at the best maturity for storage, especially Thompson seedless and Ohanez. Stems and pedicels should be well developed, and the fruit should be firm and mature. Soft and weak fruit should not be stored. The display lug is a satisfactory package for storage since it can be cooled and fumigated easily.

Cooling to 4 to 7°C is advised for grapes that are to be in transit a day or two before reaching storage. Special care should be taken during transit so that decay does not start. It is not good practice to delay fumigation until the grapes reach a distant storage plant, for in the picking and packing of grapes, many berries are injured sufficiently to permit mold to begin unless the fruit is fumigated promptly.

For *labrusca* (Eastern type) grapes, a storage temperature of 0°C and humidity of 85% are recommended. Care in packing and handling the fruit, a minimum of delay before storage, and prompt cooling are important for best results with these varieties, as they are with the *vinifera* grapes. The Eastern varieties are not fumigated with SO_2 due to their susceptibility to injury from it. The storage life of the important commercial varieties at 0°C is shown in Table 6.

PLUMS

Plums are not suitable for long storage. Among the major shipping varieties, well-matured Santa Rosa, El Dorado, Nubiana, Queen Ann, Laroda, Late Santa Rosa, and Casselman are sometimes stored for short periods. The Italian Prune can be held for no more than 2 weeks before marketing begins.

Plums intended for storage should be harvested at a high soluble solids level for the variety, although doing so may delay the harvest beyond the normal picking date. Harvested fruit should be carefully graded to remove disease, defects, and injuries before packing in the shipping container.

The fruit should be thoroughly cooled before storage. Cooling may be done in the 400 to 450 kg bulk bins that are used for harvest. The shipping containers are normally vented to aid cooling after packing. While most fruit is air cooled in conventional room coolers, some shippers use forced air to cool fruit quickly in bulk bins or shipping containers.

Plums can usually be stored for 1 month at −1 to 0°C with 90 to 95% rh. Results of storage life tests have been variable, with some lots in certain seasons remaining in good condition even after 4 to 5 months in storage. Other lots in some seasons have not been held satisfactorily beyond 2 months. Fruit with the highest soluble solids has consistently shown the longest storage life, even when harvested several weeks after the completion of commercial harvest. Some plum varieties benefit from CA storage.

Storage Diseases and Deterioration

Plum deterioration appears as changes in appearance and flavor. A poststorage holding period should be used in judging the condition of stored fruit. Fruit that appears bright and flavorful in storage can show severe deterioration when removed to room temperature for 2 to 3 days.

Some flesh softening and a gradual loss of varietal flavor and tartness occur even at low storage temperatures. The first visual sign of deterioration is the development of translucence, first around the pit, then extending outward through the fruit. Translucence is followed by the development of progressively more severe flesh browning following the same pattern. The first noticeable loss in flavor is generally associated with the first symptoms of translucence in the tissue. It is necessary to cut through the fruit to judge condition, since fruit held under good storage conditions may appear sound from the outside while being seriously deteriorated internally. See the section on Sweet Cherries for diseases. See the section on Diseases under Peaches and Nectarines for information on cold storage and sulfur dioxide injuries.

SWEET CHERRIES

Harvesting Techniques

Sweet cherries for storage must be harvested with stems attached. Mechanical harvesting takes most of the fruit without stems; consequently, the cherries must be processed or otherwise used immediately.

Cooling

Rapid cooling to −1°C is essential if the fruit is to be stored. Hydrocooling has been used successfully, and the wetting is tolerable as long as the fruit remains cold. Fungicidal postharvest sprays or dips are helpful in reducing decay during storage.

Forced-air or pressure cooling can be used to quickly cool the fruit without the problem of wetting. Moisture loss and stem drying can be minimized by rapid movement from the field to the cooler, rapid cooling, and maintaining low temperatures and high humidity during cooling and storage.

Storage

When sweet cherries are stored, they are normally held in shipping containers, often with polyethylene liners. These liners permit an increase in CO_2 gas surrounding the fruit, which tends to reduce decay rates and increase storage time. Cherries should be stored at −1°C and may be held 2 weeks after harvest and still retain enough quality for shipment to market.

Controlled atmospheres with 20 to 25% CO_2 or 0.5 to 2% O_2 help maintain firmness and bright, full color during storage (Hardenburg et al. 1986). Polyethylene liners can extend the market life. The liner must be perforated when removed from storage. Modified atmosphere bags are now used commercially.

Diseases

Alternaria and Cladosporium Rot. Light brown, dry, firm decay lining skin breaks that can be removed easily from surrounding healthy tissue. Mycelium on the area are fine and white above and dark green below. *Control*: Sort out cherries with cracks and other skin breaks at packing. Use fungicide in spray or sizer on packing line.

Blue Mold Rot (*Penicillium*). Circular, flat spots covering conical, soft, mushy decay that can be scooped out cleanly from surrounding healthy flesh. White fungus tufts turning to bluish green develop on surface. Musty odor and flavor. *Control*: Prevent skin breaks. Use fungicide in spray or sizer on the packing line. Market promptly. Refrigerate promptly to 0°C.

Brown Rot (*Monilinia*). See the section on Diseases under Peaches and Nectarines. *Control*: Follow recommended orchard spray practices. Use fungicide in spray or sizer on packing line. Refrigerate promptly to 0°C. Package cherries in polyethylene bags to reduce desiccation of stem and fruit, preserve color, and reduce decay development.

Gray Mold Rot (*Botrytis*). Light brown, fairly firm, watery decay covered with extensive delicate, dirty-white mycelium. On completely decayed cherries, grayish-brown velvety spores may be found. *Control*: Handle carefully. Use fungicide in spray or sizer on packing line. Refrigerate promptly to 0°C.

Rhizopus Rot. Extensive soft, leaking decay with little change from normal color. Coarse mycelium and black spore heads are prominent under moist conditions. More prevalent in upper-layer packages in refrigerator car. *Control*: Rhizopus develops very slowly at temperatures below 10°C, so storage at recommended temperature keeps decay in check.

PEACHES AND NECTARINES

This discussion relates primarily to peaches but also applies to nectarines in many respects.

Storage Varieties

Peaches do not adapt well to prolonged storage. However, if they are sound and well matured, most freestone varieties can be stored for up to 2 weeks (some freestone and most clingstone for up to 4 weeks) without any noticeable deterioration in flavor, texture, or appearance. Storage life appears to depend on the harvest season. Early varieties, particularly the freestone peaches grown in Florida and the early clingstones grown in the Southeast, have an extremely short storage life and should be used as soon as possible after harvest. However, some late season varieties can be safely stored for up to 6 weeks. In the West, the Rio Oso Gem is consistently stored for 4 to 6 weeks before being marketed.

Harvest Techniques

Peaches for fresh consumption must be in a condition to survive a postharvest holding period of several days to several weeks. The fruit must be sound and bruise-free and must be handled delicately during the harvesting and packing operations. With the widespread use of bulk bins or pallet boxes, hand-picked fruit requires extra careful handling. Hydrodumpers are generally employed for dumping the pallet bins. With proper care, pallet boxes cause less bruising than small field boxes.

Cooling

Cooling peaches to 4°C soon after harvesting is essential to postharvest retention of quality and control of decay. Peaches begin to soften and decay in a few hours without proper temperature management. All peaches shipped out of the Southeast are hydrocooled. Originally, the fruit was cooled in flood-type hydrocoolers as a final operation after it was packed in containers. In the West, most fresh

Deciduous Tree and Vine Fruit

peaches are air cooled in pressure coolers to remove the field heat rapidly for the postharvest holding period. With forced-air or pressure cooling, peaches or nectarines in two-layer plastic tray packs with 6% side-vented corrugated containers will cool 80% in about 6 h with an airflow of 0.2 L/s per kilogram of fruit.

Storage

Peaches are normally stored in corrugated or tray pack shipping containers.

An environment of −0.5°C and 90 to 95% rh with very low air movement is best for peaches. Under these conditions, peaches can be held for 2 to 6 weeks, depending on variety.

The same storage conditions may be used for nectarines; however, they are somewhat more susceptible to shrivel than are peaches. Air velocity in the storage room should be as low as possible but still maintain proper storage temperatures. Frequent checks should be made of the fruit at the edge of alleyways, for example, to detect the first signs of shrivel.

Good experimental CA results have been obtained with peaches and nectarines held in 1% O_2 with 5% CO_2 at 0°C. Extended storage of 6 to 9 weeks is possible. The fruit ripens or softens with good flavor and is juicy on removal. Low-temperature breakdown, which is usually encountered with lengthy storage, is controlled by CA. While CA reduces decay, it does not completely control it; thus, a fungicide is needed for extended storage.

Diseases

Brown Rot (*Monolinia*). Extensive firm, brown, unsunken areas turning dark brown to black in the center and generally covered with yellowish-gray spore masses. Skin clings tightly to center of old lesions. *Control*: Follow recommended field and postharvest control measures involving use of heat treatments and fungicides. Refrigerate promptly to as near 0°C as feasible.

Cold Storage Injury. Fruit loses flavor, becomes dry and mealy. Breakdown starting around pit is grayish brown, water-soaked, or mealy. *Control*: Refrigerate promptly to 0°C. Breakdown appears earlier at 3°C. Store for only 2 to 4 weeks, depending on variety.

Pustular Spot (*Coryneum*). Common on peaches from the West, occasionally on Eastern fruit. At first small purplish-red spots, later up to 13 mm in diameter, brown, sunken with white center. *Control*: Treat with orchard sprays. Cool harvested fruit to below 7°C.

Rhizopus Rot. Extensive, fairly firm, watery decay with uniformly brown surface color. Skin slips readily from center of lesions. Coarse mycelium, black spherical sporangia develop. *Control*: Store cannery peaches at 0°C before ripening. Prevent skin breaks. Follow recommended field and postharvest control measures. Refrigerate promptly to as near 0°C as feasible.

Sulfur Dioxide Injury. Bleached and pitted areas on fruit surface. After removal from refrigeration, injured areas of peaches are brown, dry, and collapsed. Skin may slough off. *Control*: Avoid SO_2 contact of peaches (and other stone fruit) in storage or in transit with grapes.

Sour Rot. An unfamiliar postharvest disease in peaches noticed in some packing sheds in the southeastern states. First signs of the infection may be peaches that are easily skinned by the brushes and belts on the packing line. Affected peaches then develop softened and sunken brown lesions that eventually become covered with a white or creamy exudation. The infected areas generally emit a vinegar-like, sour odor. *Control*: Chlorination of dump tank water, chlorination of hydrocooling water, and careful culling of all overripe, bruised, and damaged fruit. In short, good shed sanitation and quality control are the keys to eliminating sour rot.

APRICOTS

Apricots are not stored for a prolonged time but may be held for 2 or 3 weeks if they are picked firm enough that they will not bruise. Unfortunately, this maturity does not yield good dessert-quality fruit. Care must be used in sizing and packing the fruit going into storage, as small surface bruises can become infected with disease-producing organisms. Chapter 10 has further details.

Apricots for short-term storage are harvested in much the same way as freestone peaches, precooled, and placed in storage promptly. Storage temperature should be 0°C with 90 to 95% rh.

Diseases and Deterioration

See the section on Peaches and Nectarines.

BERRIES

Blackberries, raspberries, and related berries cannot be stored for more than 2 or 3 days even at −0.5°C with a relative humidity of 90 to 95%. An atmosphere with 20 to 40% CO_2 will increase storage life by 3 or 4 days by inhibiting fungal rots.

As they come from the field, cranberries are stored in field boxes at 2 to 4°C and 90 to 95% rh. They are usually not stored longer than 2 months. Storage at −1 to 0°C causes chilling and physiological breakdown. Modified atmospheres have not extended the storage life of fresh cranberries beyond that attainable in conventional storage.

Diseases

Cladosporium Rot. Surfaces of berries covered with olive to olive-green mold. In raspberries, the mold is most abundant on inside or cup of berry. *Control*: Avoid bruising; pack and ship promptly. Refrigerate to 0°C.

Gray Mold Rot. Causes soft, watery rot. Fruit may be covered with dense, dusty gray growth of fungus, which spreads rapidly in package, forming nests. *Control*: Avoid bruising. Refrigerate to 0°C in transit.

Anthracnose (*Gloeosporium sp.*). Berries may be completely rotted and show masses of spores glistening in salmon-colored droplets on fruit. *Control*: Refrigerate to 0°C.

Alternaria Rot. Affected berries remain firm and show gray-white woolly fungal growth from injured cap-stem areas. Nesting occurs in tight clusters scattered throughout containers. *Control*: Refrigerate to 0°C.

Chilling Injury. Berries held for 4 or more weeks become tough and rubbery; surfaces are dull in appearance, red in color throughout. *Control*: Hold fruit at 2.8°C.

Fungus Rots (Several Fungi). Limited portions or entire berries are brown, soft, or collapsed. Some berries turn into water bags. *Control*: Spray in field. Handle carefully. Reduce temperature to 3.3°C after harvest.

STRAWBERRIES

Diseases

Gray Mold Rot (*Botrytis*). Brown, fairly firm, fairly dry decay. Dirty-gray mold and grayish-brown velvety spore masses present. Nesting common. *Control*: Apply recommended fungicides in field. Handle carefully to prevent skin breaks. Cull out all diseased berries. Cool promptly to 4°C or below.

Leather Rot (*Phytophthora*). Large, slightly discolored tough areas with indefinite purplish margins. Vascular system browned, flavor bitter. *Control*: Mulch plants to keep berries from contact with infested soil. Cool promptly to 4°C or below.

Rhizoctonia Rot. Hard dark brown decay on one side of berry, usually small quantities of soil adhering. Develops only a little after harvest. *Control*: Mulch plants to keep berries from contact with infested soil. Cull thoroughly.

Rhizopus Rot. Mushy, leaky collapse of berries associated with coarse black mycelium and sporangia. Extensive red staining of containers from leaking juice. *Control*: Reduce temperature promptly to 4°C or below. Handle carefully to prevent skin breaks.

FIGS

Diseases

Alternaria Spot. White fungal growth on surfaces that soon darkens. As fungus spots enlarge, tissue beneath becomes slightly sunken. *Control*: Cool promptly after harvest, hold at 7°C in transit.

Black Mold Rot (*Aspergillus*). Disease first appears as a dirty-white to pink color of the skin and pulp. White mold growth develops within fig. Cavities formed in fruit become lined with black spore masses. *Control*: Store fresh figs at 0°C and 85 to 90% rh.

SUPPLEMENTS TO REFRIGERATION

Antiseptic Washes

Many types of fruit are washed before packing to remove dirt and improve appearance. In some cases, hydrocoolers are used to remove field heat. If the water is recirculated, it may become heavily contaminated with decay-producing bacteria and fungi. Chlorine can be added to the water at 50 to 100 mg/kg to control the buildup of these organisms. Other fungicides may also be used, but they must be legally registered for the specific application.

Protective Packaging

Proper packaging protects against bruising, moisture loss, and spread of disease. Packaging materials may also contain chemicals to control spoilage. Packages must have good stacking strength for palletizing and must also perform under high-humidity conditions.

Selective Marketing

The potential storage life of grapes and apples can be predicted within a few weeks after they are stored. Thus, those with a short storage life may be marketed while still in good condition, and longer lived products can be stored for late-season marketing. Samples taken from each lot placed in storage are kept for a few weeks at temperatures and relative humidities that favor rapid development of decay. Grapes that will not keep long can be detected in about 2 weeks and apples in about 60 days. Since both kinds of fruit may be stored for several months, knowing their potential storage life can significantly reduce spoilage losses.

Heat Treatment

Heat treatments to reduce decay also kill insects on and microorganisms near the surface of the fruit without leaving a residue. For example, brown rot and rhizopus rot of peaches are reduced by exposing the fruit for 1.5 min in 55°C water or for 3 min in 49°C water.

Fungicides

Fungicides may be applied during cleaning, brushing, or waxing of some fruit. Only fungicides registered for the particular fruit and application may be used.

Irradiation

Gamma radiation has effectively controlled decay in some products. High dosages can cause discoloration, softening, or flavor loss. Commercial application of gamma radiation is limited due to the cost and size of equipment needed for the treatment and to uncertainty about the acceptability of irradiated foods to the consumer (Hardenburg et al. 1986).

Ultraviolet lamps are sometimes used to control bacteria and mold in refrigerated storage. While ultraviolet light kills bacteria and fungi that are sufficiently exposed to the direct rays, it does not reduce decay of packaged fruit in storage. Even ultraviolet light directed on fruit as it passed over a grader did not control decay.

REFERENCES

EPA. 1989. Interim policy for sulfiting agents on grapes; Pesticide tolerance for sulfur dioxide. *Federal Register* 54(3):382-85.

Hardenburg, R.E., A.E. Watada, and C.Y. Wang. 1986. The commercial storage of fruit, vegetables, and florist and nursery stocks. USDA *Agriculture Handbook* No. 66. U.S. Department of Agriculture.

Harvey, J.M. 1955. A method of forecasting decay in California storage grapes. *Phytopathology* 45:229-32.

Harvey, J.M. 1984. *Instructions for forecasting decay in table grapes for storage*. ARS-7. U.S. Department of Agriculture.

Harvey, J.M., C.M. Harris, T.A. Hanke, and P.L. Hartsell. 1988. Sulfur dioxide fumigation of table grapes: Relative sorption of SO_2 by fruit and packages, SO_2 residues, decay, and bleaching. *American Journal of Enology and Viticulture* 39:132-36.

Kader, A.A., R.F. Kasmire, and J.F. Thompson. 1992. *Cooling horticultural commodities: Selecting a cooling method*. Pub. 3311. University of California Division of Agriculture and Natural Resources.

BIBLIOGRAPHY

Chau, K.V., C.D. Baird, P.C. Talasila, and S.A. Sargent. 1992. Development of time-temperature-humidity relations for fresh fruits and vegetables. Final report for ASHRAE Research Project 678-RP.

Nelson, K.E. 1979. Harvesting and handling California table grapes for market. University of California *Publication* 4095.

Ryall, A.L. and W.T. Pentzer. 1982. *Handling, transportation, and storage of fruits and vegetables*, 2nd ed. Vol. 2, *Fruits and tree nuts*. AVI Publishing Co., Westport, CT.

CHAPTER 22

CITRUS FRUIT, BANANAS, AND SUBTROPICAL FRUIT

CITRUS FRUIT 22.1	Diseases and Deterioration 22.5
Maturity and Quality 22.1	Exposure to Excessive Temperatures 22.5
Harvesting and Packing 22.1	Wholesale Processing Facilities 22.5
Transportation 22.3	SUBTROPICAL FRUIT 22.8
Storage 22.3	Avocados 22.8
Storage Disorders and Control 22.4	Mangoes 22.8
BANANAS 22.5	Pineapples 22.8
Harvesting and Transportation 22.5	

THIS chapter covers the harvesting, handling, processing, storage requirements, and possible disorders of fresh market citrus fruit grown in Florida, California, Texas, and Arizona; of bananas; and of subtropical fruit grown in California, Florida, Hawaii, and Puerto Rico.

CITRUS FRUIT

MATURITY AND QUALITY

The degree of citrus fruit ripeness at the time of harvest is the most important factor determining eating quality. Oranges and grapefruit do not improve in palatability after harvest. They contain practically no starch and do not undergo marked composition changes after they are picked from the tree (as do apples, pears, and bananas), and their sweetness comes from the natural sugars they contain when picked.

The ripening of citrus fruit is a slow, gradual process closely related to increases in diameter and mass. Citrus fruit must be of high quality when harvested to assure quality during storage and shelf life.

Quality is often associated with the fruit rind's appearance, firmness, thickness, texture, freedom from blemishes, and color. However, quality determination should be based on the texture of the flesh, juiciness, soluble solids (principally sugars), total acid, aromatic constituents, and vitamin and mineral content. Age is also important. Immature fruit is usually coarse and very acid or tart and has an internal texture that is ricey or coarse. Overripe fruit held on the tree too long may become insipid, develop off-flavors, and possess short transit, storage, and shelf life. The importance of having good quality fruit at harvest cannot be overemphasized. The main objective thereafter is to maintain quality and freshness.

HARVESTING AND PACKING

Picking

Citrus fruit is harvested in the United States throughout the year, depending on the growing area and kind of fruit. Figure 1 shows the approximate commercial shipping seasons for Florida, California-Arizona, and Texas citrus. Trained crews from independent packing houses or large associations conduct the picking operations. These organizations schedule picking to meet market demands. The fruit that is not handled through cooperatives is normally sold on the tree to the shippers or processors and is picked at the latter's discretion.

The pickers carefully removed the fruit from the trees, either by hand or with special clippers, and then place the fruit in picking bags that are emptied into field boxes. An increasing amount of fruit is

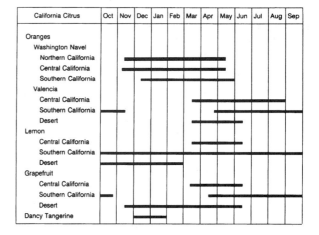

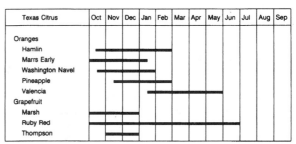

Fig. 1 Approximate Commercial Shipping Season for U.S. Citrus

---

The preparation of this chapter is assigned to TC 10.9, Refrigeration Application for Foods and Beverages.

handled in bulk, so the pickers put the fruit into pallet boxes or wheeled carts. In some cases, especially when fruit is picked for processing, it is loaded loose into open truck trailers. In Florida, over 90% of the oranges and slightly more than 50% of the grapefruit and specialty fruit are processed, while in California and Arizona less than 35% of all citrus fruit is processed.

At the beginning of the season, the fruit is often spot-picked; only the riper, larger, or outside fruit is harvested. Later, the trees are picked clean. In California, lemons usually are picked for size with the aid of sizing rings.

Various labor-saving devices have been tested, including mechanical platforms and positioners, tree shakers with catch frames, and air blasts for fruit removal. Mechanical harvesting, however, is limited to a very small percentage of the total crop. Because of damage incurred, fruit intended for processing only is mechanically harvested. Preharvest sprays have also been developed to improve the color and to loosen the fruit in order to facilitate harvest.

Handling

After the fruit is received at the packinghouse, it is removed from the boxes or bulk containers carefully to prevent damage to the fruit. It is then presized to remove fruit that is oversized or undersized. Before washing, the fruit may be floated through a soak tank, which usually contains a detergent for cleaning and an antiseptic for decay control.

The washer is generally equipped with transverse brushes that revolve up to 120 rpm. If not applied at the soak tank, soap or antiseptic may be dribbled or foamed on the first series of brushes. A fresh water spray then rinses the fruit.

The fruit then passes under fans that circulate warm air through the moving pieces. When dried, the fruit is polished and waxed. Since water waxes are normally used, a second drying operation follows the waxer. The fruit then passes over roller conveyor grading tables. After grading, it is conveyed to sizing equipment that separates the pieces into the standard packing sizes; the pieces are then dropped at stations for hand packing or conveyed to automatic or semiautomatic box-filling or bagging machines. Electronic sizing, based on machine vision, is used extensively.

The packinghouse handling of California lemons for fresh market is interrupted by an extended storage period. After washing, the fruit is conveyed to a sorting table for color separation by electronic means or by human eye. Usually, the separation process recognizes four colors, which are designated as dark green, light green, silver, and yellow. The dark green is a full green; the light green is a partially colored green (a green with color well broken); the silver is fully colored with a green tip (stylar end); and the yellow is fully colored and mature with no green showing. The normal storage life for dark green fruit is 4 to 6 months; for yellow fruit, it is 3 to 4 weeks. These periods are approximate, as the storage (or keeping) quality of fruit varies considerably with season and grove. A light concentration of water-wax emulsion is usually applied to lemons before they are put into storage. The section on Storage has more information on storing the different varieties of citrus fruit.

After storage, lemons are waxed, then sized and packed. Poststorage washing to remove mold soilage is desirable but requires a washer incorporating very soft roller brushes.

Shipping and storage containers vary considerably for the different types of citrus fruit. The 28 L fiberboard cartons have become the standard in California and Florida. In addition, over 15% of Florida fresh fruit is consumer-packed in mesh and polyethylene bags that are normally shipped in 18 kg master cartons. After the packages are filled and closed, they are conveyed either to precooling rooms to await shipment or directly to standard refrigerator cars or trucks. The containers are stacked so that air distribution is uniform throughout the load. Either slipsheets or wooden pallets are commonly used for palletized handling.

Accelerated Coloring or Sweating

All varieties of citrus fruit must be mature before they are picked. Maturity standards are based on internal attributes of soluble solids, acid percentage, and juice percentage. Color is not always a criterion of maturity. The natural change of color in oranges from dark green to deep orange is a gradual process while the fruit remains on the tree. The fruit remains dark green from its formation until it is nearly full size and approaching maturity; then the color may change very rapidly. The color change is influenced greatly by temperature variations. A few cold nights followed by warm days may turn very green oranges to deep orange. The color changes in lemons and grapefruit similarly turn the fruit yellow. Unfavorable weather conditions may delay coloring even after maturity.

Up to a certain point, the natural color changes in Valencia oranges follow the trend described, but complete or nearly complete orange color generally develops some time before the fruit is mature. Some regreening of Valencias may occur after the fruit has reached its prime. Navel oranges in California, as well as the Florida varieties of Hamlin, Parson Brown, and Pineapple harvested in late fall and early winter, may be mature and of good eating quality even if the rind is green in color. Grapefruit, lemons, tangerines, tangelos, and other specialty fruit may also be sufficiently mature for eating before they are fully colored. Because the consumer is accustomed to fruit of characteristic color, poorly colored fruit is put through a coloring or **degreening** process in special rooms, bulk bins, or trailer degreening equipment.

These units are equipped to maintain temperature and humidity at desired levels. Approximately 5 mg/kg ethylene is maintained in the air. The concentration of ethylene and the duration of the degreening periods depends on the variety of fruit and the amount of chlorophyll to be removed. During degreening, fresh air is introduced into the room, and a relative humidity of 88 to 92% is maintained. In Florida, temperatures of 28 to 29°C are recommended, while in California temperatures of 18 to 21°C are used. Lemons are usually degreened at 16°C without added ethylene. In California, the process is called sweating instead of coloring or degreening.

Oranges, grapefruit, and specialty citrus fruit requiring ethylene treatment are frequently degreened upon delivery to the packinghouse, but they may receive a fungicide drench prior to degreening. Lemons are washed and graded or color-separated before being degreened.

Color-Added Treatment

A high percentage of Florida's early and midseason varieties of oranges and tangelos receive color-added treatment with a certified food dye that causes the rind of pale fruit to take on a brighter and more uniform orange color. This is usually in addition to degreening with ethylene gas. In color-added treatment, the fruit is subjected for 2 to 3 min to the dye solution, which is maintained at 49°C. Treatment can be in an immersion tank filled with vegetable dye solution, or the dye can be flooded on the fruit as it passes on a roller conveyor. The immersion tank is between the washer and the wax applicator. Oranges with the desired color at harvest time, as well as tangerines and grapefruit, bypass the dye tank, or the flow of dye may be cut off as the fruit passes over the equipment. Standards for maturity are slightly higher in Florida for oranges given color-added treatment. California oranges are not artificially colored.

Cooling

After the fruit is packed, it is cooled. The efficiency of the cooling rooms depends on the following conditions:

1. Cooling air rate per railcar load (at least 1400 L/s)
2. Relative humidity of supply air (95% or above)
3. Temperature of supply air entering room (no more than 1 K below the selected cooling temperature)

Citrus Fruit, Bananas, and Subtropical Fruit

The fruit may also be cooled in a refrigerated truck trailer or container after it has been loaded.

In California, air is used to cool oranges but not lemons or grapefruit. In Florida, specialty fruit such as Temple oranges, tangerines, and tangelos may be cooled. Chapter 14 discusses cooling practices and equipment used for various commodities in more detail.

TRANSPORTATION

Fruit packed in piggybacks, trucks, ship vans, or rail cars should be stowed in appropriate modifications of the spaced bonded block to assure good air circulation, uniform temperature, and stable load. No dunnage is required. Such stowing provides continuous air channels through the interior of the load and improves the likelihood of sound arrival. Trailers and containers that circulate air from the bottom provide uniform temperatures throughout the load with a regular bonded-block stow.

In Florida, the present quarantine treatment for the Caribbean fruit fly, *Anastrepha suspensa*, is to subject an export load of citrus to specified temperatures for up to 24 days (Ismail et al. 1986). This treatment may be implemented in containers or in a ship's hold.

A uniform sample of 1500 fruit is withdrawn from a shipment before ship or container loading and is held at 27°C or higher. These fruit are then examined after a 10 day incubation period. When an infestation of *A. suspensa* is found, the entire load must undergo the long treatment process. Table 1 details temperature and time schedules.

Table 1 Quarantine Treatment of Citrus Fruit for the Caribbean Fruit Fly

	Temperature,[a] °C	Days
Short Treatment[b]		
	0.6	10
	1.1	12
	1.7	14
	2.2	17
Long Treatment[b]		
	0.6	14
	0.8	16
	1.1	17
	1.4	19
	1.7	20
	1.9	22
	2.2	24

[a]These temperatures are required center pulp temperatures.
[b]To avoid chilling injury, a conditioning period of 7 days at 15°C is recommended before initiation of the cold treatment process.

STORAGE

The performance of any citrus storage facility depends on three conditions: (1) the provision of sufficient capacity for peak loads; (2) an evaporator and secondary refrigerating surface sufficient to permit operation at high back pressures, which prevents low humidity and lowers operating costs; and (3) efficient air distribution, which assures velocities high enough to effect rapid initial cooling and volumetric flows great enough to permit operation during storage with only a small temperature rise between delivery and return air. Chapters 10, 12, and 13 have further information on storage design.

Oranges

Valencia oranges grown Florida and Texas can be stored successfully for 8 to 12 weeks at 0 to 1°C with a relative humidity of 85 to 90%. The same requirements apply to Pope's Summer orange, a late-maturing Valencia-type orange. A temperature range of 4 to 7°C for 4 to 6 weeks is suggested for California oranges. Arizona Valencias harvested in March store best at 9°C, but fruit harvested in June store best at 3°C.

Table 2 Heat of Respiration of Citrus Fruit

Temp., °C	Heat of Respiration, W/Mg of Fruit					
	Oranges			Grapefruit		Lemons
	Florida	Calif. Navels	Calif. Valencias	Florida	Calif. Marsh	Calif. Eureka
0	9	12	5	7	7	9
4	19	19	13	15	11	15
10	36	40	35	20	27	34
16	62	67	38	38	35	47
21	90	81	52	47	52	67
27	105	107	62	57	65	77

Source: Haller et al. (1945)

Oranges lose moisture rapidly, so high humidity should be maintained in the storage rooms. For storage longer than the usual transit and distribution periods, 85 to 90% relative humidity is recommended.

Florida and Texas oranges are particularly susceptible to stem end rots. Citrus fruit from all producing areas are subject to blue and green mold rot. These decays develop in the packinghouse, in transit, in storage, and in the market, but they can be greatly reduced if fruit is properly treated. Proper temperature is effective in reducing decay. However, once storage fruit is removed to room temperature, decay develops rapidly.

Prolonged holding at relatively low temperatures may cause the development of physiological rind disorders not ordinarily encountered at room temperature. This possibility often complicates the storage of oranges. Aging, pitting, and watery breakdown are the most prevalent rind disorders caused by low storage temperatures. California and Arizona oranges are generally more susceptible to low-temperature rind disorders than Florida oranges.

Successful long storage of oranges requires harvest at the proper maturity, careful handling of fruit, good packinghouse methods, fungicidal treatments, and prompt storage after harvest.

The rate of respiration of citrus fruit is usually much lower than that of most stone fruit and green vegetables and somewhat lower than that of apples. Navel oranges have the highest respiration rate, followed by Valencia oranges, grapefruit, and lemons. The heat from respiration is a relatively small part of the heat load. Table 2 shows heat generated through respiration.

Grapefruit

Florida and Texas grapefruit is frequently placed in storage for 4 to 6 weeks without serious loss from decay and rind breakdown. The recommended temperature is 10°C. A temperature range of 14.5 to 16°C is recommended for the storage of California and Arizona grapefruit.

A relative humidity of 85 to 90% is usually recommended for the storage rooms containing grapefruit. Loss of mass and water occurs rapidly and can be avoided by maintaining the correct humidity and taking the additional precaution of a wax coating.

Long storage of grapefruit may cause decay and rind breakdown during storage or following removal from storage. Proper prestorage treatments with fungicides greatly reduces these problems. Also, periodic inspections of stored fruit for the least symptom of rind pitting or excessive decay should be made so that storage can be terminated if necessary.

Export may require 10 days to 4 weeks of storage in a refrigerated hold and present problems similar to those encountered in refrigerated storage. Marsh Seedless and Ruby Red grapefruit picked before January retain appearance best when stored at 15°C. With riper fruit, 10 to 13°C is better for export shipments. Very ripe fruit harvested in April and May, however, develops excessive decay following storage at 10 to 15°C.

Lemons

Most of the lemon crop is picked during the period of least consumption and stored until consumer demand justifies shipment. Lemons are generally stored near the producing areas rather than the consuming areas.

All lemons, except the relatively small percentage that are ripe when harvested, must be conditioned or cured and degreened before shipping. When lemons are stored prior to shipment, the curing and degreening processes occur during storage. These lemons are usually stored at 11 to 13°C and 86 to 88% rh. Local conditions may suggest slight modifications of these values.

Lemons picked green but intended for immediate marketing, (e.g., most lemons grown in the desert portions of Arizona and California) are degreened and cured for 6 to 10 days at 22 to 26°C and 88 to 90% rh. The thin-skinned Pryor strain of Lisbon lemons degreens in about 6 days, whereas the thick-skinned old-line Lisbon requires as long as 10 days.

Lemon storage rooms must have accurately controlled temperature and relative humidity; the air should be clean and uniformly circulated to all parts of the room. Ventilation should be sufficient to remove harmful metabolic products. Air-conditioning equipment is necessary to provide satisfactory storage conditions, as natural atmospheric conditions are not suitable for the necessary length of time.

A uniform storage temperature of 10 to 15°C is important. Fluctuating or low temperatures cause lemons to develop an undesirable color or bronzing of the rind. Temperatures 10°C and lower cause a staining or darkening of the membranes dividing the pulp segments and may affect the flavor. Temperatures above 12.8°C shorten storage life and promote the growth of decay-producing organisms.

A relative humidity of 86 to 88% is generally considered satisfactory for lemon storage, although a slightly lower humidity may be desirable in some locations. Higher humidities prevent proper curing of the lemons, encourage mold growth on walls and containers, and hasten decay of the fruit; much lower humidities cause excessive shrinkage.

Proper stacking of the fruit containers in storage rooms is important to secure uniform air circulation and temperature control. The stacks should be at least 50 mm apart, and the rows should be 100 mm apart; trucking aisles at least 2 m wide should be provided at intervals.

Specialty Citrus Fruit

In Florida, small amounts of various specialty citrus fruit are grown commercially. These types of fruit, which are usually channeled to fresh market, include tangerines, tangerine hybrids (Murcott Honey oranges, Temple oranges, tangelos), King oranges, and other mandarin-type fruit.

Careful handling during picking and packing is especially necessary for these types of fruit. Because of their perishable nature and limited shelf life, specialty citrus fruit should not be stored longer than required for orderly marketing (2 to 4 weeks). A temperature of about 4°C at 90 to 95% rh is recommended. Adequate precooling and continuous refrigeration during transit are required.

Tahiti or Persian limes are also grown in southern Florida. This is the only citrus fruit marketed while it is green in color. The fully ripe (yellow) fruit lacks consumer appeal and is undesirable for fresh market. Limes should be picked while still green, but after the fruit has lost the dimpled appearance around the blossom end. Good-quality fruit may be stored satisfactorily for 6 to 8 weeks at 9 to 10°C. However, mature fruit gradually turns yellow at this temperature. Prevention of desiccation is very important, as is a relative humidity above 85%. Pitting occurs at temperatures below 7°C, and temperatures above recommended permit the development of stem end rot.

Controlled Atmosphere Storage

Tests of modified or controlled atmosphere (CA) storage for oranges, grapefruit, lemons, and limes have obtained minor benefits but have not shown that CA storage extends storage or market life. For this reason, CA storage is not generally recommended for citrus fruit. Atmospheres used for storage of apples and other deciduous fruit are unsatisfactory for citrus fruit and lead to rind injuries, off-flavors, and decay.

STORAGE DISORDERS AND CONTROL

Postharvest Diseases

Citrus fruit often carries incipient fungus infections when harvested. Decay organisms may also enter minor injuries caused during harvesting and handling. The major postharvest diseases (with symptoms) encountered in storage include the following:

Alternaria rot. Usually a black, dry, deeply penetrating decay at stylar end of navel oranges. A slimy, leaden-brown storage decay of core starting at stem end in other citrus fruit. *Control*: Provide optimum growing conditions. Harvest oranges before they are overripe. Do not store tree-ripe lemons. Restrict storage period for other lots known to be weak. Green buttons indicate strong fruit.

Anthracnose (*Colletotrichum*). Leathery, dark brown, sunken spots or irregular areas. Internal affected tissues dark gray, fading through pink to normal color. Most serious with degreened early season tangerines, tangerine hybrids, and long-stored oranges, and long-stored grapefruit. *Control*: Use recommended postharvest fungicide. Avoid long storage and move promptly.

Blue (and green) mold rot (*Penicillium*). Soft, watery, decolorized lesions that, under moist conditions, become quickly covered with blue or olive-green powdery spores. *Control*: Prevent skin breaks. Use recommended fungicides in washes. Cool fruit to as near to 0°C as practicable.

Brown rot (*Phytophthora*). Extensive firm, brown decay having a penetrating rancid odor. Chiefly on fruit from California and Arizona. *Control*: Orchard spraying and good sanitation. Submerge fruit at packing for 2 min in 45.5°C water.

Sour rot (*Geotrichum*). Soft, watery rot with sour smell following peel injuries. Similar to early stages of mold rot, except that no powdery spores are formed. Most serious on lemons and mandarin-type fruit. *Control*: Avoid peel injuries at harvest; refrigerate at lowest practical temperature. Approved fungicides are of little or no value.

Stem end rot (*Diplodia*; *Phomopsis*). Pliable, fairly firm, extensive, brown decay starting at stem. Sour, pungent odor. Prevalent in Florida and found occasionally in Arizona and California fruit. *Control*: Treat harvested fruit promptly in recommended fungicides and cool promptly below 10°C.

The U.S. Environmental Protection Agency (EPA) approves several chemical fungicides for post-harvest use on citrus fruit. These include thiabendazole (TBZ), orthophenylphenol (OPP or SOPP), and imazalil. These materials are applied after washing and before waxing or are incorporated in the wax coating. Under certain conditions, it is beneficial to use a combination of these materials since all are not equally effective against the same organism. Strains of the blue and green molds (*Penicillium*), which are resistant to certain fungicides, have developed in citrus storage houses, so care must be taken in selecting the fungicide and the time of application.

Physiological Disturbances

Various physiological conditions can also cause defects. The use of fruit at prime maturity and proper handling after harvest can eliminate these defects. Proper temperature and humidity levels are required during handling, storage, and transit. The following are the physiological disorders and symptoms:

Stem end rind breakdown. Small to large sunken, drying, discolored, firm areas in skin around stem button or on the upper part of fruit. *Control*: Pick before overmature. Avoid overheating in packinghouse treatments. Wax fruit. Store for limited period only in fairly high relative humidity (85 to 90%). Follow storage temperatures recommended for variety and growing area.

Freezing injury. Field freezing is found scattered through boxes. Transit and storage freezing are worse in exposed fruit in bottom-layer boxes or those nearest cooling coils. Affected fruit may show water-soaked areas in rind. The internal tissue is disorganized, water-soaked, and milky and has rind flavor. Frozen fruit loses moisture, causing drying, separation of juice vesicles, and buckling of segment walls. The freezing point of citrus fruit is about −2°C.

Internal decline. In lemons, core tissues near the stylar end break down and dry, becoming pink. *Control*: Maintain optimum moisture conditions in grove.

Pitting. This physiological disease is manifested by depressed areas of 3 to 20 mm diameter in the peel of citrus fruit. Affected tissues collapse and may appear bleached or brown. Pits occur anywhere on the fruit and may coalesce to form large irregular areas. The cause is not fully understood. In general, pitting is a low-temperature disorder. However, lack of immediate cold storage for Florida grapefruit has accentuated pitting. *Control*: Follow storage temperatures recommended for cultivar and growing area.

BANANAS

HARVESTING AND TRANSPORTATION

Bananas do not ripen satisfactorily on the plant; even if they did, deterioration of ripe fruit is too rapid to allow shipping from tropical growing areas to distant markets. Bananas are harvested when the fruit is mature but unripe, with dark green peels and hard, starchy, inedible pulps. Each banana plant produces a single stem of bananas that contains from 50 to 150 individual pieces of fruit (or fingers). The stem is cut from the plant as a unit with fingers attached and transported to nearby boxing stations.

Bananas are removed from the stem, washed, and cut into consumer-sized cluster units of four or more fingers. The clusters are packed in protective fiberboard cartons that contain 18 kg of fruit. The cartons move by rail from the tropical boxing stations to port and then are loaded into the holds of refrigerated ships. On the ship, the fruit is cooled to the optimum carrying temperature, usually about 14°C, depending on variety.

Bananas are unloaded still green and unripe at seaboard and transported under refrigeration at a holding temperature of 14.5°C to interior wholesale distribution centers by both truck and railcar. The objective is to maintain the product in an optimal environment and move it to its destination as quickly as possible to minimize postharvest deterioration.

DISEASES AND DETERIORATION

Bananas are subject to various diseases and physiological disorders. Proper temperature and moisture during storage and careful handling slow aging and development of decay.

Anthracnose (ripe rot) (*Gloeosporium*). Shallow black spots on stems of ripening fruit. Under moist conditions, pink spore masses cover center of spots. Dark discoloration of skin may extend from stem end over entire fruit. *Control*: Protect fruit from mechanical injury; damage is reduced if fruit is transported in corrugated boxes. Schedule ripening so that fruit can be marketed and consumed before appearance of defect.

Black rot (*Ceratocystis*). Transmitted from wounds via fibrovascular system of plant. Progresses into crowns and stem ends of fingers. Produces brownish-black areas in peel at fruit ends. As fruit ripens, skin becomes grayish-black in color and water-soaked. Pulp is rarely affected. *Control*: In the tropics, dip or spray freshly cut tips and bunches with fungicides before boxing. Avoid mechanical injury and maintain sanitation program from tropics to ripening room.

Chilling injury. Dull gray skin color with increased tendency to darken on slight bruising. Latex in green fruit does not bleed freely and will be clear rather than cloudy. Subsurface peel tissue streaked with brown. Turning or ripe bananas are more susceptible to injury than green fruit. *Control*: Avoid temperatures below 13°C. Moving air accelerates chilling.

Fungus rots (several fungi). Extensive soft rot of scarred, split, or broken fruit. Affected skin and flesh moist and brown to black. Under high humidity, the surface is often covered with mold. *Control*: Handle fruit carefully to avoid bruising and mechanical injury. Cool stored fruit rapidly to 13.5°C.

EXPOSURE TO EXCESSIVE TEMPERATURES

Fruit pulp temperatures only a few degrees below optimum holding temperatures, although considerably above the actual freezing point of bananas, can cause chilling injury, as previously described. The severity of chilling injury varies directly with the duration of exposure and indirectly with temperature. It is primarily a peel injury in which certain surface cells of the banana peel are killed. The contents of the dead cells eventually darken due to oxidation and give the fruit a dull appearance. Both green and ripe bananas are susceptible to chilling injury; severely chilled green bananas never ripen properly. Fruit pulp temperatures only a few degrees above the optimum holding temperatures can cause the fruit to ripen prematurely in transit.

Once the bananas arrive at wholesale distribution centers, they are unloaded and placed in specially equipped processing rooms for controlled ripening. As soon as the bananas have ripened to an edible state, they are rushed to retail because ripening cannot be stopped. Even under ideal refrigeration, ripe bananas eventually progress to the point where they are too ripe to sell.

WHOLESALE PROCESSING FACILITIES

Wholesale banana processing facilities are distinguished from general wholesale produce storage facilities by special banana ripening rooms. The ripening room controls initiation and completion of fruit ripening, a natural physiological phenomenon. Figure 2 shows a typical banana room. The ability to properly ripen bananas is so critically linked to the design of the ripening rooms that major banana importers maintain technical staffs that specialize in banana room design. These technical staffs provide free, nonobligatory consultation to wholesalers, architects, engineers, contractors, and others involved in banana ripening facility design, construction, and operation.

A typical banana processing facility consists of a bank of five or more individual ripening rooms. For design purposes, one complete turnover per week is assumed; therefore, the combined capacity of

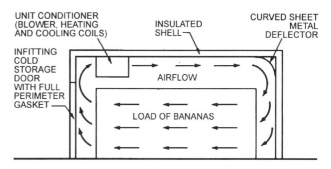

Fig. 2 Banana Room (Side View)

all rooms should approximately equal total weekly volume, allowing for seasonal variations.

Each load is scheduled for optimum ripeness on a particular day. Fruit shipped from this load a day ahead of schedule will be underripe; fruit shipped a day late will be overripe. Therefore, shipping for several days out of one room is not practical. There should be at least as many rooms as there are retail shipping days per week.

Because bananas cannot be processed on a continuous flow basis, individual room capacities are multiples of carlots, usually one-half or one carlot. As the capacity of transportation equipment has increased in recent years, the design capacity of processing rooms has also increased. It is generally cheaper to build one or two large rooms than several small rooms with equivalent total capacity. However, minimizing construction cost is not the pertinent consideration; having all bananas in each particular processing room reach optimum ripeness for shipment to retail at the same time is more important.

Airtightness

Exposing the fruit to ethylene gas, introduced into the room from cylinders, initiates banana ripening. The dose is 1 m^3 of ethylene gas per 1000 m^3 of room air space. Ethylene is explosive in air at concentrations between 2.75 and 28.6%. Many ethylene systems gas the fruit automatically over a 24 h period.

To be effective, the gas must be confined to the ripening room for 24 h, so banana rooms must be airtight. Floor drains must be individually trapped to prevent gas leakage. Special care should be taken to seal all penetrations in room walls where refrigerant piping, plumbing lines, and the like enter rooms. Doors should have single-seal gaskets all around and sweep gaskets at the floor line.

Refrigeration

A direct-expansion halocarbon system is recommended for use in banana rooms. Because of ammonia's harmful effect on bananas, direct-expansion ammonia systems should not be used. Malfunctioning refrigeration equipment during processing could cause heavy product losses, so each ripening room should have a completely separate system despite high initial installation costs.

For maintenance-free operation in the high-humidity environment of processing rooms, evaporator coils should have a fin spacing of 6.4 mm. Coils should be amply sized and capacity rated at a design temperature difference of 8 K with a refrigerant temperature of 4.5°C. Air temperatures used during processing range from 7 to 18°C. Because of the danger of banana chilling, refrigerant temperatures below 4.5°C are not recommended. With programmers, suction pressure control devices or hot-gas bypass systems must be installed.

Refrigeration Load Calculations

These calculations are based on the same methods used for other fresh fruit. A typical half-carlot-capacity banana room holds approximately 432 boxes of fruit. Approximate outside dimensions for a three-tier forklift-type room, shown in Figure 3, are 9.2 m long by 1.8 m wide by 6.7 m high. Pallets are 1.2 m by 1.0 m and are stacked 6 pallets deep in each of 3 tiers, totaling 18 pallets per room. Boxes for use in banana rooms are approximately 250 mm high by 400 mm wide by 560 mm long and are stacked 4 boxes per each of 6 layers, totaling 24 boxes per pallet. With 18 pallets per room, 432 boxes of bananas can be stored (18 pallets by 24 boxes). Each box has a net mass of 19 kg and a gross mass of 21.3 kg.

Transmission load is calculated in the normal manner; the air change load is negligible. The electrical load is based on the continuous operation of multi-kilowatt fan motor(s). The peak heat of respiration is 0.3 W/kg multiplied by the total net mass of bananas in the ripening room. For product cooling, the specific heat of bananas is 3.55 kJ/(kg·K) multiplied by the total net mass of the bananas, plus the total tare mass of the cartons multiplied by 1.7 kJ/(kg·K), the specific heat of fiberboard. The total calculated load is thus

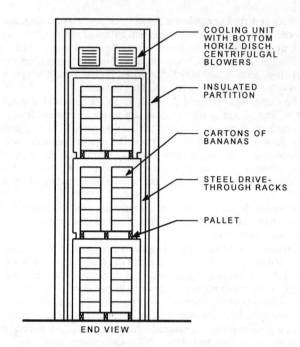

Fig. 3 Three-Tier Forklift Banana Room (End View)

approximately 17 W per box. A pulldown rate of 0.5 K/h is assumed. Total system design capacity is calculated by assuming simultaneous peak respiration and pulldown load.

Heating

Heat is not required during most ripening cycles. However, occasional loads may come in at temperatures below desired levels for treatment with ethylene gas, making heating necessary. Many banana room refrigeration units come with electrical heating elements as an integral part of the unit. If electrical heating strips are used, they should be enclosed in a corrosion-resistant sheath and have a surface temperature of not more than 425°C in dead still air; this temperature limitation is necessary because of the proximity of the heating strips to refrigerant coils and the inherent danger should leakage occur. Portable plug-in electric heaters are also used. Heating system capacity should be sufficient to raise load temperature at a rate of 0.5 K/h. Open-flame gas heaters should never be used in banana rooms for two reasons: (1) ethylene gas used during ripening is explosive at certain concentrations; and (2) the necessary room airtightness could easily result in the open-flame heaters' consuming the available oxygen within the space, thereby extinguishing the flame and permitting raw gas to enter the room.

Air Circulation

Table 3 shows the fruit pulp temperature schedules for 4 to 8 day ripening cycles. A temperature variation of only a few degrees considerably alters the rate of fruit ripening. For even ripening, fruit temperatures must be uniform throughout the room, so the volumetric airflow circulated throughout the entire load must be comparatively large. Centrifugal fans are necessary. They are installed for bottom horizontal discharge, so that the top boxes are not chilled immediately in front of the unit. Fan air output should be rated at 150 Pa external static pressure. Because of heat of respiration, heat must be continually withdrawn from the product even when it is held at a constant temperature. A temperature variation in the load is therefore inevitable, with warmer fruit downstream relative to the circulated air. Unit conditioners at the front of the room over the door discharge toward the rear of the room. This arrangement leaves riper fruit near the door to be shipped first.

Citrus Fruit, Bananas, and Subtropical Fruit

Table 3 Fruit Temperatures for Banana Ripening

Ripening Schedule	Temperature, °C							
	1st Day	2nd Day	3rd Day	4th Day	5th Day	6th Day	7th Day	8th Day
Four days	18	18	17	16				
Five days	17	17	17	17	16			
Six days	17	17	16	16	16	14		
Seven days	16	16	16	16	16	14	14	
Eight days	14	14	14	14	14	14	14	14

For improved air distribution, a sheet metal (or other suitable material) air deflector curved to a 90° arc is mounted full width on the back room wall. This deflector reduces turbulence and directs the air downward for return through the load.

Airflow Requirements

Air volumetric flow requirements are calculated on the basis of conditions required at the end of the pulldown period. Assume a maximum allowable fruit temperature variation of 1 K, an air temperature drop through the cooling unit of 1 K, and product temperature reduction proceeding at a rate of 0.1 K/h. During the initial pulldown, the air quantity so calculated will give about a 3 K drop through the cooling unit. The general equations are

$$q_t = q_r + q_p \quad (1)$$

$$q_t = mc_p \Delta t \quad (2)$$

where

q_t = total heat removed, W
q_r = heat of respiration, W
q_p = pulldown load, W
m = mass flow rate of air, kg/s
c_p = specific heat of air = 1000 J/(m·K)
Δt = temperature change of air, K

The values of q_r and q_p can be determined using the heat of respiration and specific heat values given in the section on Refrigeration Load Calculations. q_p is calculated on the basis that at the end of the pulldown period, temperature reduction is proceeding at the rate of 0.1 K/h.

q_r = 0.3 W/kg × 19 kg/box = 5.7 W/box

q_p = 0.1 K/h[3350 J/(kg·K) × 19 kg/box
    + 1700 J/(kg·K) × 2.3 kg/box]/3600 s/h = 1.9 W/box

q_t = 5.7 + 1.9 = 7.6 W/box

At equilibrium, the air temperature Δt equals the fruit temperature Δt, and

$m = q_t/c_p\Delta t$ = (7.6 W/box)/[1000 J/(kg·K) × 1 K]
  = 0.0076 kg$_{air}$/s per box

Volumetric flow rate = [0.0076 kg$_{air}$/(s·box)]/1.2 kg$_{air}$/m^3
  = 0.0063 m^3/s per box = 6.3 L/s per box

For a room with 432 boxes, the airflow should be 432 × 6.3 = 2720 L/s at 150 Pa external static pressure.

Humidity

A high relative humidity around the fruit is important during banana ripening. Bananas ripened under low-humidity conditions are more susceptible to handling damage. When bananas were ripened on the stem, naked fruit was directly exposed to the moving airstream, and automatic room humidifiers were used to prevent excessive fruit dehydration. However, banana room humidifiers are not required with tropical boxing. The fiberboard carton shields the fruit from the moving airstream. In addition, ample sizing of evaporator coils keeps the temperature difference across the coil to 6 to 8 K, thereby limiting dehumidification. Both natural transpiration of the fruit and airtight room design also contribute to high room humidity.

Controls

Ripening room air temperatures are varied frequently during banana processing. Temperatures should be controlled by remote bulb-type thermostats, with bulbs for heating and cooling mounted in the return airstream within the ripening room to prevent short-cycling of equipment. Thermostats should be mounted on the exterior of the ripening room and have a range of 7 to 21°C, calibrated in 0.5 K increments, with no more than a 1 K differential. The thermostats are best mounted on a control panel having a selector switch providing heating and cooling with continuous fan operation.

Automatic temperature controllers or programmers are installed in new facilities. Bananas produce heat continuously, but the rate of heat production varies considerably during the ripening cycle. Figure 4 shows a generalized heat-of-respiration curve. Although more complex, the removal of heat of respiration from the load can be viewed, for the purpose of analysis, as a simple conduction process. Applying the general conduction equation,

$$q = \frac{1000kA\Delta t}{L} \quad (3)$$

$$\frac{1000k}{L} = \frac{1}{R}$$

where

q = heat of respiration, W
k = thermal conductivity of packaging material, W/(m·K)
A = surface area of packaging material, m^2
Δt = banana temperature minus air temperature, K
L = thickness of packaging material, mm
R = thermal resistance of packaging material, m^2·K/W

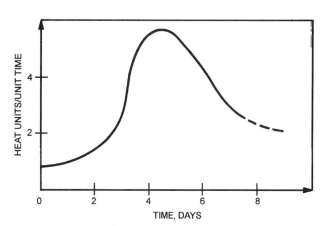

Fig. 4 Heat of Respiration During Banana Ripening

If kA/L is a constant, Δt must vary as q. Assuming the banana temperature is also constant, the room air temperature must be set lower as q increases during ripening. However, the exact value of q at any particular point in the ripening cycle is unknown.

With the conventional, manually adjustable air-sensing thermostat-control system, fruit temperatures are taken manually with a pulp thermometer, and thermostat settings are continually adjusted to follow ripening schedules. This is essentially a trial-and-error procedure.

By contrast, the temperature programmer has a remote bulb, which is placed in a box of bananas in the load. Since the bulb senses fruit temperature directly, heat of respiration is compensated for during ripening. Fruit temperature is automatically adjusted to follow preset cycles.

SUBTROPICAL FRUIT

AVOCADOS

Avocado cultivars grown in California are not grown commercially in Florida, and vice versa. Florida cultivars tend to be larger fruited than those from California. In California, the Fuerte variety accounts for 75% of the annual crop and is available from October through March. Hass (black skin) is available from April through September. In order of importance, Florida cultivars are Booth 8, Lula, Waldin, and Booth 7. Waldin appears on the market in August, followed by Booth 8 in September, and Booth 7 and Lula from October to February.

The best storage temperature for cold-tolerant Florida avocado cultivars such as Booth 8 and Lula is 4.5°C. All Florida avocado cultivars produced during the summer, such as Waldin, are cold-intolerant and store best at about 12.5°C. A few cultivars, such as Fuerte, store best at 7°C. Cold-tolerant cultivars can be held in storage a month or longer, but storage of cold-intolerant cultivars is usually limited to 2 weeks because of their susceptibility to softening and chilling injury. The best ripening temperature for avocados is 15°C, but temperatures from 13 to 24°C are usually satisfactory. Temperatures above 26°C frequently cause off-flavor, skin discoloration, uneven ripening, and increased decay.

Storage Disorders

Anthracnose (*Colletotrichum*) are scattered black spots covering firm decayed tissue that can be removed easily from surrounding flesh. Pink spore masses form on the spots under moist conditions. Control requires preventing blemishes and other breaks in the skin.

Chilling injury is typified by small to large sunken pits in the skin that turn brown or black and are often accompanied by general browning of the skin and light smoky streaks in the flesh, which develop independently.

MANGOES

The important early cultivars Tommy Atkins and Irwin mature during June and July, followed by such midseason cultivars as the Kent and Palmer, which mature during July and August. The most important cultivar produced in Florida is the large-fruited Keitt, which is late-season and matures during August and September.

Optimum storage temperature for mangoes is 12 to 13°C for 2 to 3 weeks, although 10°C is adequate for some cultivars for shorter periods. Mangoes are subject to chilling injury at temperatures below 10°C. The best ripening temperatures for mangoes are from 21 to 24°C, but temperatures of 15 to 18°C are also satisfactory under certain conditions. At 15 to 18°C, the fruit develops a bright and most attractive skin color, but the flavor is usually tart and requires an additional 2 to 3 days at 21 to 24°C to attain sweetness. Mangoes ripened at 27°C and higher frequently have a strong flavor and mottled skin.

Storage Disorders

Anthracnose (*Colletotrichum*) are large scattered black spots in the skin of ripening fruit. Under moist conditions, pink spore masses develop in spots. The disease is controlled by regular spraying on the tree and by use of hot water treatment (55°C for 450 s) after harvest.

Chilling injury causes pitting of the skin, which sometimes develops a gray cast. Fruit with chilling injury usually does not ripen uniformly. Control requires storing at proper temperatures.

PINEAPPLES

Fresh pineapples are available throughout the year, but the supply is much larger from March through June. Only three pineapple cultivars are commercially important in the United States: the Smooth Cayenne from Hawaii and the Red Spanish and Smooth Cayenne from Puerto Rico.

Pineapples harvested at the half-ripe stage can be held for 2 weeks at 7 to 13°C and still have a shelf life of about 1 week. Continuous maintenance of storage temperature is as important as the specific storage temperature. Ripe fruit should be held at 7 to 8°C. Harvesting at the mature green stage is not recommended because some fruit would be too immature to ripen. Mature green fruit is especially susceptible to chilling injury at temperatures below 10°C.

Storage Disorders

Black rot (*Ceratocystis*) causes extensive damage to tissues, which become soft and leaky, ranging from normal to jet black in color. To control, treat freshly cut stem parts with benzoic acid-talc dust, prevent bruising, and cool to 10°C.

Brown rot (*Penicillium; Fusarium*) causes brown, firm decay starting at eyes or cracks; it is common on overripe fruit. To control, provide good growing conditions and market before the fruit is overripe.

Chilling injury causes the fruit to take on a dull hue, the flesh to develop water soaking, and the core to darken. To control, hold at recommended temperatures.

REFERENCES

Haller, M.H., D.H. Rose, J.M. Lutz, and P.L. Harding. 1945. Respiration of citrus fruits after harvest. *Journal of Agricultural Resources* 71:327.

Ismail, M.A., T.T. Hatton, D.J. Dezman, and W.R. Miller. 1986. In transit cold treatment of Florida grapefruit shipped to Japan in refrigerated van containers: Problems and recommendations. *Proceedings of the Florida State Horticultural Society* 99:117-121.

BIBLIOGRAPHY

Chace, W.G., Jr., J.J. Smoot, and R.H. Cubbedge. 1979. Storage and transportation of Florida citrus fruits. *Florida Citrus Industry* 51:16.

Chau, K.V., C.D. Baird, P.C. Talasila, and S.A. Sargent. 1992. Development of time-temperature-humidity relations for fresh fruits and vegetables. Final report for ASHRAE Research Project 678-RP.

Hardenburg, R.E., A.E. Watada, and C.Y. Wang. 1986. The commercial storage of fruits, vegetables, and florist and nursery stocks. USDA *Handbook* No. 66. U.S. Department of Agriculture, Washington, D.C.

McCornack, A.A., W.F. Wardowski, and G.E. Brown. 1976. Postharvest decay control recommendations for Florida citrus fruit. *Florida Cooperative Extension Service Circular* 359-A.

Smoot, J.J., L.G. Houck, and H.B. Johnson. 1971. Market diseases of citrus and other subtropical fruits. USDA *Handbook*, 398.

Wardowski, W.F., S. Nagy, and W. Grierson. 1986. *Fresh citrus fruits*. AVI Publishing, Westport, CT.

CHAPTER 23

VEGETABLES

Product Selection and Quality Maintenance 23.1
In-Transit Preservation 23.2
Preservation in Destination Facilities 23.3
Refrigerated Storage Considerations 23.4
Storage of Various Vegetables 23.4

LOSSES (shrinkage) in marketing fresh vegetables (harvesting, handling, packing, storing, and retailing) are caused, in part, by overly high temperatures during handling, storage, and transport, which increase ripening, decay, and the loss of edible quality and nutrient values. In some cases, freezing or chilling injury from overly low temperatures may be involved. Other serious losses are caused by mechanical injury from careless or rough handling and by shrinkage or wilting because of moisture loss. Shrinkage can be reduced substantially by following recommended handling, cooling, transport, and storage practices. Improved packaging, refrigerated transport, and awareness of the role of refrigeration in maintaining quality throughout marketing have made it possible to transport vegetables in field-fresh condition to distant markets.

This chapter covers postharvest handling, cooling, packaging, in-transit preservation, and storage at destination locations for fresh vegetables. It also gives storage requirements for specific vegetables, including potential product deterioration due to improper handling and storage conditions. Vegetable precooling is covered in Chapter 14, and vegetable processing and freezing is covered in Chapter 26.

PRODUCT SELECTION AND QUALITY MAINTENANCE

The principal hazards to quality retention during marketing include

1. Metabolic changes (composition, texture, color) associated with respiration, ripening, and senescence (aging)
2. Moisture loss with resultant wilting and shriveling
3. Bruising and other mechanical injuries
4. Parasitic diseases
5. Physiological disorders
6. Freezing and chilling injury
7. Flavor and nutritional changes
8. Growth (sprouting, rooting)

Fresh vegetables are living tissues and have a continuing need for oxygen for respiration. During respiration, stored food such as sugar is converted to heat energy, and the product loses quality and food value. In maintaining commodity temperatures during storage or transportation, some of the refrigeration load can be attributed to respiration. For example, a 10-Mg load of asparagus cooled to 4°C can produce enough heat of respiration during a cross-country trip to melt 4 Mg of ice.

Vegetables that respire the fastest often have greater handling problems because they are the most perishable. Variations are due to the type of plant part involved. For example, root crops such as carrots and radish have lower respiration rates than fruit vegetables (cucumber, pepper) and sprouts (asparagus). Refrigeration is the best method of slowing respiration and other life processes. Chapters 8 and 10 give more information on the respiration rates of many vegetables.

Vegetables are usually covered with natural populations of microorganisms which will cause decay given the right conditions. Deterioration because of decay is probably the greatest source of spoilage during marketing. When mechanical injuries break the skin of the produce, decay organisms will enter. If it is then exposed to warm temperatures, especially under humid conditions, infection usually increases. Adequate refrigeration is the best method of controlling decay because low temperatures control the growth of most microorganisms.

Many color changes associated with ripening and aging can be delayed by refrigeration. For example, broccoli may show some yellowing in one day on a nonrefrigerated counter, while it remains green at least 3 to 4 days in a refrigerated display.

Refrigeration can retard deterioration caused by chemical and biological reactions. Sweet corn may lose 50% of its initial sugar content in a single day at 21°C, while only about 5% will be lost in one day at 0°C (Appleman and Arthur 1919). Also, freshly harvested asparagus will lose 50% of its vitamin C content in one day at 20°C, whereas it takes 4 days at 10°C or 12 days at 0°C to lose this amount (Lipton 1968).

Loss of moisture with consequent wilting and shriveling is one of the obvious ways to lose freshness. **Transpiration** is the loss of water vapor from living tissues. Moisture losses of 3 to 6% are enough to cause a marked loss of quality for many kinds of vegetables. A few commodities may lose 10% or more in moisture and still be marketable, although some trimming may be necessary, such as for stored cabbage. For more on transpiration, see Chapter 8.

Postharvest Handling

After harvest, most highly perishable vegetables should be removed from the field as rapidly as possible and placed under refrigeration, or they should be graded and packaged for marketing. Since aging and deterioration of vegetables continues after harvest, the marketable life depends greatly on the temperature and care in physical handling.

The effects of rough handling of vegetables are cumulative. Several small bruises on a tomato can produce an off-flavor. Bruising also stimulates the ripening rate of products such as tomatoes and thereby shortens potential storage and shelf life. Mechanical damage increases moisture loss; skinned potatoes may lose 3 to 4 times as much mass as nonskinned ones.

Care should be exercised in stacking bulk bins in storage so that proper ventilation and refrigeration of the product is maintained. Bins should not be of such depth that excessive mass damages the product near the bottom.

Quality maintenance is further aided by the following procedures:

1. Harvest at optimum maturity or quality
2. Handle carefully to avoid mechanical injury
3. Handle rapidly to minimize deterioration
4. Provide protective containers and packaging

The preparation of this chapter is assigned to TC 10.9, Refrigeration Application for Foods and Beverages.

5. Use preservative chemical, heat, or modified-atmosphere treatments
6. Enforce good plant sanitation procedures while handling
7. Precool to remove field heat
8. Provide high relative humidity to minimize moisture loss
9. Provide proper refrigeration throughout marketing

Cooling

Rapid cooling of a commodity after harvest, before or after packaging or before it is stored or moved in transit, reduces deterioration of the more perishable vegetables. The faster field heat is removed after harvest, the longer produce can be maintained in good marketable condition. Cooling slows natural deterioration, including aging and ripening; slows growth of decay organisms (and thereby the development of rot); and reduces wilting, since water losses occur much more slowly at low temperatures than at high temperatures. After cooling, produce should be refrigerated continuously at recommended temperatures. If warming is allowed, much of the benefit of prompt precooling may be lost.

Types of cooling include hydrocooling, vacuum cooling, air cooling, and cooling with contact ice and top ice. These methods are discussed in detail in Chapter 14. The choice of cooling method depends on factors such as refrigeration sources and costs, volume of product shipped, and compatibility with the product.

Protective Packaging and Waxing

Vegetables for transit and destination storage should be packed in containers with adequate cushioning materials, stacking strength, and durability to protect against crushing and high-humidity conditions. Bulging crates should be stacked on their sides or stripped between layers to keep mass off the commodity. Many vegetables are now stored or shipped in fiberboard or corrugated containers, but the weakening of fiberboard materials by moisture absorption at high storage humidities is frequently a serious problem.

Manufacturers have improved the strength and reduced the degree of moisture absorption by fiberboard materials. Special treatment of fiberboard permits the use of certain cartons during hydrocooling and with package and top ice. Cartons may be strengthened for stacking by using dividers, wooden corner posts, and full telescoping covers.

Produce is often consumer-packaged at production locations using many types of trays, wraps, and film bags, which may present special transit and handling problems when master containers for these packages lack stacking strength.

Desiccation often can be minimized by using moisture retentive plastic packaging materials. Polyethylene film box liners, pallet covers, and tarpaulins may be helpful in reducing moisture loss from vegetables. Plastic films, if sealed or tightly tied, may restrict the transfer of carbon dioxide, oxygen, and water vapor, leading to harmful concentrations of these respiratory gases; and films restrict heat transfer, which retards the rate of cooling (Hardenburg 1971).

Waxes are applied to rutabagas, cucumbers, mature green tomatoes, and cantaloupes, and to a lesser extent to peppers, turnips, sweet potatoes, and certain other crops. With products such as cucumbers and root crops, waxing reduces moisture loss and thus retards shriveling. With some products, an improved glossy appearance is the main advantage. Thin wax coatings may give little if any protection against moisture loss; coatings which are too heavy may increase decay and breakdown. Waxing alone does not control decay, but waxing combined with fungicides may be beneficial. Waxing is not recommended for potatoes either before or after storage (Hardenburg et al. 1959).

IN-TRANSIT PRESERVATION

Good equipment is available to transport perishable commodities to market under refrigeration by rail, trucks, piggyback trailers, and containers. A high relative humidity (rh) of about 95% is desirable for most vegetables to prevent moisture loss and wilting. Many vegetables benefit from 95 to 100% rh. Humidity in both iced and mechanically refrigerated cars and trailers is usually high. Top ice and package ice, used most often for leafy vegetables, provide refrigeration and increased moisture.

Cooling Vehicle and Product

Vehicles used to ship vegetables that require low temperatures during transit should have their interiors cooled before loading. With mechanical refrigeration, the units should be operated with the doors closed until the interior temperature of the vehicle is reduced to the approximate transit temperature.

Generally, vegetables that require a low temperature during shipping should be cooled before they are loaded into transport vehicles. Cooling produce in tightly loaded refrigerator cars or trailers is a slow process, and the portion of the load exposed to the cold air discharge may be frozen when the interior of the load is still warm. It is also uneconomical for vehicles to provide refrigeration capacity to remove field heat.

Shipping, Packaging, Loading, and Handling

Containers must protect the commodity, permit heat exchange as necessary, and serve as an appropriate merchandising unit with sufficient strength to withstand normal handling. Freight container tariffs describe approved containers and loading procedures.

Containers should be loaded to take advantage of their maximum strength and to permit adequate stripping, or use of spacers to hold the load in alignment. Proper vertical alignment of containers is essential to obtain their maximum stacking strength. Although maximum stacking is frequently incompatible with providing channels for air circulation, these channels must be maintained even at the sacrifice of some capacity and resistance of the load to shifting. If there are any ventilation openings in containers, they should have the greatest possible exposure to channels or flues.

When different types of containers are used in the same load, stacks should be separated so that one type will not damage another. If separation of stacks is impossible, containers made of lighter material, such as fiberboard, should always be loaded on top of heavier wood containers.

Providing Refrigeration and Air Circulation

Safe transit temperatures for various vegetables and suggested temperatures for mechanically refrigerated cars and trailers are given in Table 1. For safety, the suggested thermostat settings for cool season vegetables are usually 1 to 2 K above the freezing point. The various means for obtaining specific transit temperatures in the United States are provided under the regulations of the Perishable Protective Tariffs issued by the National Freight Committee and the Railway Express Agency.

With the many kinds of refrigeration, heating, and ventilating services now available, the shipper has only to specify the desired transport temperature. Generally, the shipper or the receiver is responsible for selecting the protective service for his commodity in transit. The various protective services are described in detail in Ashby et al. (1987) and Redit (1969).

Protection from Cold

During the winter, vegetables must be protected from freezing. Mechanically refrigerated freight cars and trucks, equipped to handle the full range of both fresh and frozen commodities, are also designed to provide heat for cold weather protection. The heat is supplied by electric heating elements or by reverse-cycle operation of the refrigerating unit (hot gas from the compressor is circulated in the cooling coils). The change from cooling to heating is done automatically by thermostatic controls.

Vegetables

In-transit freezing is a major problem in moving late crop potatoes from storage to market. Cars and trailers should be warmed before loading, protected during loading with canvas door shields or loading tunnels, and loaded properly.

Checking and Cleaning Equipment

If the thermostat is out of adjustment by a few degrees, products may be damaged by freezing or by insufficient refrigeration. Thermostats should be calibrated at regular intervals to assure that the proper amount of refrigeration is furnished. Trailers and railcars should be cleaned carefully before loading. Debris from previous shipments may contaminate loads and should be swept from floors and floor racks.

Modified Atmospheres in Transit

A variety of systems provide controlled or modified atmospheres in trucks, piggyback trailers, railcars, and seavans. Though the label on each process may differ, all systems alter the levels of oxygen, carbon dioxide, and nitrogen surrounding the produce.

The oxygen concentration in air is usually lowered to 1 to 5% because this level tends to depress the respiration rate.

No single modified atmosphere can be expected to benefit more than a few commodities; crop requirements and tolerances are quite specific (Harvey 1965). Also, certain vegetables may tolerate an atmosphere at one temperature but not at another, or they may tolerate a modified atmosphere for only a limited time.

The load compartment must be fairly tight to maintain desired atmospheres. Modified atmosphere equipment is predominantly installed in vehicles used to transport chilled perishables in long haul movements (over five days transit time). Atmosphere systems used in transport either vaporize liquid nitrogen or generate nitrogen by passing heated compressed air through hollow fibers.

Lettuce is the main vegetable shipped under a modified atmosphere. The physiological disorder known as **russet spotting** is reduced when lettuce is shipped in low oxygen atmospheres with good refrigeration (Stewart and Ceponis 1968). Lettuce is damaged by accumulated carbon dioxide (Stewart et al. 1970), so some fresh hydrated lime is usually placed in the cargo space to absorb carbon dioxide.

Temperature is critical in determining the benefits of modified atmospheres. If temperatures are higher than recommended, decay and other deterioration may be increased rather than reduced. Modified atmospheres should never be used as a substitute for good temperature management.

Use of controlled atmospheres is discussed later in this chapter. For further information on refrigerated transport by truck, railway car, ship, and air, see Chapters 29 through 31 in this volume and Chapter 28 of the 1994 *ASHRAE Handbook—Refrigeration*.

PRESERVATION IN DESTINATION FACILITIES

Wholesale warehouse facilities usually do not have a whole range of controlled temperature rooms to provide optimum storage conditions for each kind of produce, and this is not necessary for short holding. About one-half the produce handled can be stored at 0°C. Enough refrigeration capacity should be available to maintain a year-round temperature of 0°C with 90 to 95% rh. Higher temperatures and lower humidities accelerate quality loss and increase waste for more perishable vegetables. Enough refrigerated coil surface should be provided to allow a differential of only a few degrees between the coil and return air temperatures while still providing adequate refrigeration. A difference of as little as 1 K is desirable to maintain optimum humidity conditions.

Desirable air temperature and humidity cannot be maintained if excessive air exchange occurs between the warehouse cold storage

Table 1 Optimal Transit Temperatures for Various Vegetables

Vegetable	Desirable Transit Temperature, °C	Suggested Thermostat Setting[a], °C	Highest Freezing Point[b], °C
Artichokes	0	0.6	−1.2
Asparagus	0-2	1.7	−0.6
Beans, lima	3-5	3	−0.6
Beans, snap	4-7	7	−0.7
Beets, topped	0	1	−0.9
Broccoli	0	1	−0.6
Brussels sprouts	0	1	−0.8
Cabbage	0	1	−0.9
Cantaloupes	2-5	3	−1.2
Carrots, topped	0	0.6	−1.4
Cauliflower	0	1	−0.8
Celery	0	1	−0.5
Corn, sweet	0	1	−0.6
Cucumbers	10-13	10	−0.5
Eggplant	8-12	10	−0.8
Endive and escarole	0	1	−0.1
Greens, leafy	0	1	—
Honeydew melon	7-10	7	−0.9
Lettuce	0	1	−0.2
Onions, dry	0-4	1.7	−0.8
Onions, green	0	1	−0.9
Peas, green	0	1	−0.6
Peppers, sweet	7-10	8	−0.7
Potatoes:			
Early crop	10-16	10	−0.6
Late crop	4-10	4.4	−0.6
For chipping:			
Early crop	18-21	18-21	−0.6
Late crop	10-16	10-16	−0.6
Radishes	0	1	−0.7
Spinach	0	1	−0.3
Squash, summer	5-10	5	−0.6
Squash, winter	10-13	10	−0.8
Sweet potatoes	13-16	13	−1.3
Tomatoes:			
Mature green	13-21	13	−0.6
Pink	10	10	−0.8
Watermelons	10-16	10	−0.4

Sources: USDA *Agricultural Handbook* No. 195 (Redit 1969), USDA *Marketing Research Report* No. 196 (Whiteman 1957), and USDA *Agricultural Handbook* No. 66 (Hardenburg et al. 1986).

[a] For U.S. shipments of vegetables in mechanically refrigerated cars under Rule 710 and in trailers under Rule 800 of the Perishable Protective Tariff.

[b] Highest temperature at which freezing occurs.

Table 2 Recommended Temperatures for Maintaining Quality of Fresh Vegetables in Wholesale Warehouses

Store at 0°C and 90 to 95% rh		Store at 10°C and 80 to 85% rh
Artichokes	Horseradish	Cucumbers
Asparagus	Kohlrabi	Eggplants
Beans, lima	Lettuce	Garlic, dry
Beets	Mushrooms	Melons
Broccoli	Onions, green	Okra
Brussels sprouts	Parsnips	Onions, dry
Cabbage	Peas, green	Peppers
Carrots	Radishes	Potatoes
Cauliflower	Rhubarb	Pumpkins
Celery	Rutabagas	Squash, hard shell
Corn, sweet	Spinach	Squash, summer
Endive	Turnips	Sweet potatoes
Escarole	Beans, green	Tomatoes, ripe

Source: Adapted from Bogardus and Lutz (1961).

room and warmer areas. Operators should consider the use of air curtains, flap doors, or other infiltration reduction devices whenever doors to cold rooms must be opened often or for prolonged periods.

Some vegetables should not be stored at 0°C because of the danger of chilling injury. Other less perishable vegetables, such as dry onions, can be held satisfactorily for short periods at warmer temperatures. For these vegetables, controlled storage conditions during wholesaling are still desirable. Table 2 shows which vegetables should be held at 0°C and which should be held at 10°C during wholesaling.

Mature green tomatoes (for ripening) usually need separate, controlled temperature rooms where 13 to 21°C temperatures with 85 to 90% rh can be maintained to delay or speed ripening as desired.

REFRIGERATED STORAGE CONSIDERATIONS

The refrigeration requirement of any storage facility must be based on peak refrigeration load. This peak usually occurs when outside temperatures are high and warm produce is being moved into the plant for cooling and storage. The peak refrigeration load depends on the amount of commodity received each day, the temperature of the commodity at the time it is placed under refrigeration, the specific heat of the commodity, and the final temperature attained. Information on general aspects of cold storage design and operation can be found in Chapters 10, 12, and 13.

Sprout Inhibitors

Sprout inhibitors are used when cold storage facilities are lacking or if low temperatures might injure the vegetable or affect its processing quality. In-storage sprouting of onions, potatoes, and carrots can be inhibited by spraying the plants a few weeks before harvest with a solution of maleic hydrazide. Potatoes are also sprayed or dipped in a solution that inhibits sprouting. Vaporized nonanol alcohol is circulated through ducts to suppress potato sprouting in Great Britain (Burton 1958).

Gamma irradiation suppresses sprouting of onions, sweet potatoes, and white potatoes at dosages of 0.05 to 0.15 kGy. Dosages above 0.15 kGy cause breakdown and increased decay in white potatoes (Kader et al. 1984).

Controlled and Modified Atmosphere Storage

Refrigeration is most effective in retarding respiration and lengthening storage life. For some products, reducing the oxygen level in the storage air and/or increasing the carbon dioxide level as a supplement to refrigeration can provide extended storage life. Careful control of the concentration of oxygen and carbon dioxide is essential. If all of the oxygen is used, produce will suffocate and may develop an alcoholic off-flavor in a few days. Carbon dioxide given off in respiration or from dry ice may accumulate to injurious levels.

Modified or controlled atmospheres (CA) have received little application for vegetable storage, in contrast to their wide use for apples. Many atmospheres tested on vegetables were injurious or produced only minor benefits (Dewey et al. 1969). If commercial use of CA for vegetables increases, it is likely to be with the use of external generators to create desired atmospheres or with the addition of nitrogen gas or dry ice, rather than by using product respiration in a gastight room.

Danish cabbage has kept better during a 4 to 5 month period at 0°C in gas mixtures with 2.5 to 5% oxygen and carbon dioxide (with the balance nitrogen) than it has in air (Isenberg and Sayles 1969). Tests have indicated that mature green tomatoes may keep predominantly green for 5 to 6 weeks at 13°C in an atmosphere of 3% oxygen with 97% nitrogen. After removal to air at 18°C, these tomatoes ripened normally with acceptable flavor (Parsons and Anderson 1970).

Asparagus and mushrooms in refrigerated storage have kept better for short periods in atmospheres with 5 to 10% carbon dioxide than they have in air. The carbon dioxide inhibits soft rot of asparagus (Lipton 1965), and it retards cap opening and inhibits mold in mushrooms. Some experimental data showed improved quality retention during CA short storage have also been obtained for head lettuce, broccoli, brussels sprouts, green onions, and radishes.

Hypobaric storage, or storage at reduced atmospheric pressure, is another supplement to refrigeration that involves principles similar to those in controlled-atmosphere storage. At atmospheric pressures 0.1 or 0.2 of normal, several kinds of produce have an extended storage life. The benefits are attributed both to the low oxygen level maintained and to the continuous removal of ethylene, carbon dioxide, and possibly other metabolically active gases. Their rates of production under hypobaric ventilation are also lower.

Positive hypobaric ventilation is absolutely required to achieve the desired low ethylene concentration within and around produce to retard ripening. The continuous flow of water-saturated air at a low pressure flushes away emanated gases and prevents loss of mass.

Despite benefits, the costs of hypobaric storage structures are not commercially viable.

Injury

Chilling injury may be defined as an injury caused by exposure to low but nonfreezing temperatures, often between 0 to 10°C. At these temperatures, vegetables are weakened because they are unable to carry on normal metabolic processes. Often, vegetables that are chilled look sound when removed from low temperatures. However, symptoms of chilling (pitting or other skin blemishes, internal discoloration, or failure to ripen) become evident in a few days at warmer temperatures (Morris and Platenius 1938). Vegetables that have been chilled may be particularly susceptible to decay. **Alternaria rot** is often severe on tomatoes, squash, peppers, and cantaloupes that have been chilled. Tomatoes that have been severely chilled usually ripen slowly and rot rapidly. Figure 1 shows the increasing extent of rot in mature green tomatoes held at chilling temperatures.

Both time and temperature are involved in chilling injury. Damage may occur in a short time if temperatures are considerably below the safe storage temperatures, but a product may withstand a few degrees in the danger zone for a longer time. Also, the effects of chilling are cumulative. Low temperatures in the field before harvest, in transit, and in storage all contribute to the total effects of chilling. A list of vegetables susceptible to chilling injury together with the symptoms and the lowest safe temperatures are shown in Table 3.

Table 1 shows the **freezing points** of various vegetables. The freezing point is the highest temperature at which ice crystal formation in the tissues has been recorded experimentally. Most vegetables have a freezing point between -2.2 to -0.6°C (Whiteman 1957). Different vegetables vary widely in their susceptibility to freezing injury.

Tissues injured by freezing generally appear water-soaked. Even though some vegetables are somewhat tolerant to freezing, it is desirable to avoid freezing temperatures because they shorten storage life (Parsons and Day 1970).

To minimize damage, fresh commodities should not be handled while frozen. Fast thawing damages tissues, but very slow thawing (0 to 1°C), permits ice to remain in the tissues too long. Thawing at 4°C is suggested (Lutz 1936).

STORAGE OF VARIOUS VEGETABLES

In the following sections, the temperatures and relative humidity recommendations (shown in parentheses) are the optimum for maximum storage in the fresh condition. For short storage, higher temperatures may be satisfactory for some commodities. Temperature requirements represent commodity temperature levels that should

Vegetables

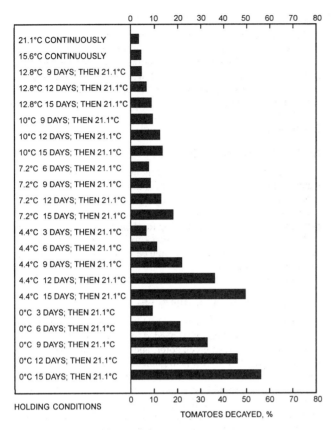

Fig. 1 Effect of Different Temperatures and Holding Periods on Rot, Principally *Alternaria*, in Tomatoes Ripened at 21°C
(McColloch and Worthington 1954)

Table 3 Vegetables Susceptible to Chilling Injury at Moderately Low but Nonfreezing Temperatures

Commodity	Approximate Lowest Safe Temperature, °C	Character of Injury Between 0°C and Safe Temperature[a]
Beans (snap)	7	Pitting and russeting
Cucumbers	10	Pitting; watersoaked spots, decay
Eggplants	7	Surface scald; *Alternaria* rot
Melons:		
Cantaloupes	a	Pitting; surface decay
Honeydew	7 to 10	Pitting; failure to ripen
Casaba	7 to 10	Pitting; surface decay
Crenshaw and Persian	7 to 10	Pitting; surface decay
Watermelons	4	Pitting; objectionable flavor
Okra	7	Discoloration; watersoaked areas; pitting; decay
Peppers, sweet	7	Sheet pitting, *Alternaria* rot on pods and calyxes
Potatoes	3 to 4	Mahogany browning (Chippewa and Sebago); sweetening[b]
Pumpkins and hard-shell squash	10	Decay, especially *Alternaria* rot
Sweet potatoes	13	Decay; pitting; internal discoloration
Tomatoes:		
Ripe	10[a]	Watersoaking and softening; decay
Mature green	13	Poor color when ripe; *Alternaria* rot

Sources: From USDA *Agricultural Handbook* No. 66 (Hardenburg et al. 1986).
[a]See text.
[b]Often these symptoms appear only after removal to warm temperatures, as in marketing.

be maintained. Much of this information is taken from USDA *Agricultural Handbook* No. 66 (Hardenburg et al. 1986).

The quality of each lot of produce should be determined at the time of storage and regular inspections should be made during storage. Such vigilance will permit early detection of disease so that the affected commodity can be moved out of storage before serious loss occurs.

Artichokes, Globe (0°C and 95 to 100% rh)

Globe artichokes are seldom stored, but for temporary holding a temperature of about 0°C is recommended. A high rh of at least 95% will help prevent wilting. This product should keep for 2 weeks in storage if buds are uninjured and wilting is prevented. Perforated polyethylene liners with 250 6-mm holes/m² help retard moisture loss. Hydrocooling or room cooling to 4°C on the day of harvest reduces deterioration.

Gray Mold Rot (Botrytis): The most common decay of harvested artichokes. Reddish brown to dark brown firm rot (see Table 4, Note 2). *Control*: Practice sanitation in the field. Refrigerate promptly.

Asparagus (0 to 2°C and 95 to 100% rh)

Asparagus deteriorates very rapidly at temperatures above 2°C and especially at room temperature. It loses sweetness, tenderness, and flavor, and decay develops later. If the storage period is 10 days or less, 0°C is recommended; asparagus is subject to chilling injury if held longer at this temperature. Asparagus is not ordinarily stored except temporarily, but at 2°C with a high relative humidity, it can be kept in salable condition for 3 weeks. However, after a long haul to market, even under refrigeration, it cannot be expected to keep longer than 1 to 2 weeks.

Asparagus should be cooled immediately after cutting. Hydrocooling is the usual method. During transit or storage, the butts of asparagus should be placed on some moist absorbent material to prevent loss of moisture and to maintain freshness of the spears. Sometimes asparagus bunches are set in shallow pans of water in storage.

Bacterial Soft Rot: Mushy, soft, watersoaked areas on tips and cut ends of asparagus (see Table 4, Note 1). *Control*: Avoid excessive bruising of tips; cool to 4°C.

Fusarium Rot: Watersoaked areas changing through yellow to brown, chiefly on asparagus tips; white to pink delicate mold. *Control*: Cool and ship at temperatures of 4°C or lower; handle promptly. Keep tips dry on way to market.

Phytophthora Rot: Large, watersoaked, or brownish lesions at the side of cut asparagus stalks. Lesions later are extensively shriveled. *Control*: Cool to 4°C. Maintain low transit temperatures. Market promptly.

Beans, Green or Snap (4 to 7°C and 95% rh)

Green or snap beans are probably best stored at 4 to 7°C, where they may keep for about 1 week. Even these recommended temperatures cause some chilling injury, but are best for short storage. When stored at 4°C or below for 3 to 6 days, surface pitting and russet discoloration may appear in a day or two following removal for marketing. The russeting will be especially noticeable in the centers of the containers where condensed moisture remains. *Contact icing is not recommended*. To prevent wilting, the rh should be maintained at 90 to 95%. Beans for processing can be stored up to 10 days at 4°C. Containers of beans should be stacked to allow abundant air circulation; otherwise, the temperature may rise from the

Table 4 Notes on Diseases of General Occurrence

Note 1. Bacterial Soft Rot

Occurs on various vegetables as dark green, greasy or watersoaked soft spots and areas in leaves and stems. Soft, mushy, yellowish spots or soupy areas on stems, roots, and tubers of vegetables. Frequently accompanied by repulsive odor from secondary invaders. *Control*: Use sanitation practices during picking and packing to reduce contamination of harvested product. Use bactericides in postharvest wash treatments. Where possible avoid bruising and injury. Shade harvested produce in the field and reduce temperature promptly to 4°C or lower for commodities that can withstand low temperatures.

Note 2. Gray Mold Rot (Botrytis)

Decayed tissues are fairly firm to semiwatery. Watersoaked grayish tan to brownish in color. Gray mold and grayish brown, with conspicuous velvety spore masses. *Control*: Use sanitation practices during harvesting and packing. Avoid wounds as much as possible. Use storage and transit temperatures as low as otherwise practicable because decay progresses even at 0°C.

Note 3. Rhizopus Rot

Decayed tissues are watersoaked, leaky, and softer than those with gray mold rot or watery soft rot. Coarse mycelium and black spore heads develop under moist conditions. Nesting is common. *Control*: Insofar as possible, avoid injury and bruising. Reduce temperature promptly and maintain below 10°C for commodities that can withstand low temperatures.

Note 4. Watery Soft Rot (Sclerotinia)

Decayed tissues are watersoaked, with slightly pinkish or brownish tan borders; very soft and watery in later stages, accompanied by development of fine white to dingy cottony mold and black to brown mustard seed-like bodies called *sclerotia*. Nesting is common. *Control*: Use sanitary practices in harvesting and packing. Cull out specimens with discolored or dead portions. Maintain temperature as low as practicable because rot progresses even at 0°C. Do not store commodities known to have watery soft rot at harvest time.

---

heat of respiration. Beans stored too long or at too high a temperature are subject to such decays as watery soft rot, slimy soft rot, and rhizopus rot.

Anthracnose (Colletotrichum): Circular or oval sunken spots; reddish brown around border with tan centers that frequently bear pink spore mounds. *Control*: Use resistant varieties. Plant disease-free seed. Refrigerate harvested beans promptly to 7°C.

Bacterial Blight (Pseudomonas): Small, greasy-appearing, watersoaked spots in pod. Older spots show red at the center, with watersoaked area surrounding and penetrating to the seed. *Control*: Keep the field sanitary. Use disease-free seed. Reduce the transit temperature promptly to 7°C.

Bacterial Soft Rot: See Table 4, Note 1.

Cottony Leak (Pythium): Pods with large, watersoaked spots, accompanied by abundant white cottony mold. *Control*: Sort out diseased pods in packing. Reduce transit temperature promptly to 7°C.

Freezing Injury: Slight freezing results in watersoaked mottling on surface of exposed pods. Severely frozen beans become completely watersoaked, limp, and dry out rapidly. Snap beans freeze at about −0.5°C.

Russeting: Chestnut brown or rusty, diffuse surface discoloration on both sides of pods. *Control*: Permit no surface moisture on warm beans. Cool promptly; avoid temperatures lower than 7°C.

Soil Rot (Rhizoctonia): Large, reddish brown, sunken, decayed spots on pods. Cream-colored or brown mycelium and irregular chocolate-colored sclerotia may develop. Nesting is common. *Control*: Maintain transit temperatures of 7 to 10°C.

Watery Soft Rot: Presence of large black sclerotia in white mold helps to separate this from cottony leak (see Table 4, Note 4).

Beans, Lima (3 to 5°C and 95% rh)

For best quality, lima beans should not be stored. Unshelled lima beans can be stored for about 1 week at 3 to 5°C. They should be used promptly after removal because the pods discolor rapidly at room temperature. Even with only 1 week of refrigerated storage, the pods may develop rusty brown or brown specks, spots, and larger discolored areas that reduce salability of the beans. The pod discoloration will increase sharply during an additional day at 21°C following storage.

Beets (0°C and 98 to 100% rh)

Topped beets are subject to wilting because of the rapid loss of water when the storage atmosphere is too dry. When stored at 0°C with at least 95% rh, they should keep for 4 to 6 months. Before beets are stored, they should be topped and sorted to remove all beets that are diseased and showing mechanical injury. Bunch beets under the same conditions will keep 1 to 2 weeks. Contact icing is recommended. The containers should be well ventilated and stacked to allow air circulation.

Bacterial Soft Rot: See Table 4, Note 1.

Broccoli (0°C and 95 to 100% rh)

Italian or sprouting broccoli is highly perishable and is usually stored for only a brief period as needed for orderly marketing. Good salable condition, fresh green color, and the vitamin C content are best maintained at 0°C. If it is in good condition and is stored with adequate air circulation and spacing between containers to avoid heating, broccoli should keep satisfactorily 10 to 14 days at 0°C. Longer storage is undesirable because leaves discolor, buds may drop off, and tissues soften. The respiration rate of freshly harvested broccoli is high comparable to that of asparagus, beans, and sweet corn. This high rate of respiration should be considered when storing broccoli, especially if it is held without package ice.

Brussels Sprouts (0°C and 95 to 100% rh)

Brussels sprouts can be stored in good condition for a maximum of 3 to 5 weeks at the recommended temperature of 0°C. Deterioration, yellowing of the sprouts, and discoloration of the stem are rapid at temperatures of 10°C and above. Rate of deterioration is twice as fast at 4°C as at 0°C. Loss of moisture through transpiration is rather high even if the relative humidity is kept at the recommended level. Film packaging is useful in preventing moisture loss. As with broccoli, sufficient air circulation and spacing between packages is desirable to allow good cooling and to prevent yellowing and decay.

Cabbage (0°C and 98 to 100% rh)

A large percentage of the late crop of cabbage is stored and sold during the winter and early spring, or until the new crop from southern states appears on the market. If it is stored under proper conditions, late cabbage should keep for 3 to 4 months. The longest keeping varieties belong to the Danish class. Early crop cabbage, especially southern grown, has a limited storage life of 3 to 6 weeks.

Cabbage is successfully held in common storage in northern states, where a fairly uniform inside temperature of 0 to 2°C is maintained. An increasing quantity of cabbage is now held in mechanically refrigerated storage, but in some seasons its value does not justify the expense. Use of controlled atmospheres to supplement refrigeration aids quality retention. An atmosphere with 2.5 to 5% oxygen and 2.5 to 5% carbon dioxide can extend the storage life of late cabbage. Cabbage should not be stored with fruits emitting ethylene. Concentrations of 10 to 100 mg/kg of ethylene cause leaf abscission and loss of green color within 5 weeks at 0°C.

Pallet bins are used as both field and storage containers, so the cabbage requires no handling from the time of harvest until preparation for shipment. Before the heads are stored, all loose leaves

Vegetables

should be trimmed away; only 3 to 6 tight wrapper leaves should be left on the head. Loose leaves interfere with ventilation between heads which is essential for successful storage. When removed from storage, the heads should be trimmed again to remove loose and damaged leaves.

Chinese cabbage can be stored for 2 to 3 months at 0°C with 95 to 100% rh.

Alternaria Leaf Spot: Small to large spots bearing brown to black mold. This spotting opens the way for other decays. *Control*: Avoid injuries. Maintain 0 to 1°C temperature in transit and storage. Practice sanitation in storage rooms.

Bacterial Soft Rot: This slimy decay frequently starts in the pith of a cut stem or in leaf spots caused by other organisms (see Table 4, Note 1).

Black Leaf Speck: Small, sharply sunken, brown or black specks occurring anywhere on outer leaves or on leaves throughout the head. Occurs under refrigeration in transit and storage, and in association with sharp temperature drops. *Control*: No effective control measures are known.

Freezing Injury: Heads frozen slightly may thaw without apparent injury. Freezing injury first appears as brown streaks in the stem, then as light brown watersoaking of the heart leaves and stem. *Control*: Prevent any extended exposure to temperatures below −0.5°C.

Watery Soft Rot: See Table 4, Note 4.

Carrots (0°C and 98 to 100% rh)

Carrots are best stored at 0°C with a very high rh. Like beets, they are subject to rapid wilting if the humidity is low. For long storage, carrots should be topped and free from cuts and bruises. If they are in good condition when stored and promptly cooled after harvest, mature carrots should keep 5 to 9 months. Carrots lose moisture readily and wilting results. Humidity should be kept high, but condensation or dripping on the carrots should be avoided, since this causes decay.

Most carrots for the fresh market are not fully mature. Immature carrots are prepackaged in polyethylene bags either at the shipping point or in terminal markets. They are usually moved into marketing channels soon after harvest, but they can be stored for a short period to avoid a market glut. If the carrots are cooled quickly and all traces of leaf growth are removed, they can be held 4 to 6 weeks at 0°C.

In Texas, immature carrots are often stored in clean 23-kg burlap sacks. The sacks of carrots should be stacked in such a manner that at least one surface of each sack is in contact with top ice at all times. Top ice provides some of the necessary refrigeration and prevents dehydration. Bunched carrots may be stored 10 to 14 days at 0°C. Contact icing is recommended.

Bitterness in carrots, which may develop in storage, is due to abnormal metabolism caused by the ethylene emitted by apples, pears, and some other fruits and vegetables. It can also be caused by other sources such as internal combustion engines. Bitterness can be prevented by storing carrots away from products that give off ethylene.

Bacterial Soft Rot: See Table 4, Note 1.

Black Rot (Stemphylium): Fairly firm black decay at the crown, on the side, or at the tips of harvested roots. *Control*: Avoid bruising. Store at 0°C.

Crater Rot (Rhizoctonia): Circular brown craters with white to cream-colored mold in the center. Develops under high humidity in cold storage. *Control*: Field sanitation measures. Avoid surface moisture on storage roots.

Freezing Injury: Roots are flabby and on cutting show radial cracks in the flesh of the central part and tangential cracks in the outer part. *Control*: Prevent exposure to temperatures below −1°C.

Gray Mold Rot: See Table 4, Note 2.

Rhizopus Rot: See Table 4, Note 3.

Watery Soft Rot: See Table 4, Note 4.

Cauliflower (0°C and 95% rh)

Cauliflower may be stored for 3 to 4 weeks at 0°C with about 95% rh. Successful storage depends on retarding the aging of the head or curd, preventing decay marked by spotting or watersoaking of the white curd, and preventing yellowing and dropping of the leaves. When it is necessary to hold cauliflower temporarily out of cold storage, packing in crushed ice will aid in keeping it fresh. Freezing causes a grayish brown discoloration, softening of the curd, and a watersoaked condition. Affected tissues are rapidly invaded by soft rot bacteria.

Much of the cauliflower now marketed has closely trimmed leaves and is prepackaged in perforated cellophane overwraps that are then packed in fiberboard containers. Vacuum cooling is a fairly efficient method of cooling prepackaged cauliflower. In general, use of controlled atmospheres with cauliflower has not been promising. Atmospheres containing 5% carbon dioxide or higher are injurious to cauliflower, although the damage may not be apparent until after cooking.

Bacterial Soft Rot: See Table 4, Note 1.

Brown Rot (Alternaria): Brown or black spotting of the curd. *Control*: Use seed treatment and field spraying. Keep the curds dry. Maintain low transit temperatures. Store at 0°C.

Celery (0°C and 98 to 100% rh)

Celery is a relatively perishable commodity, and for storage of 1 to 2 months, it is essential that a commodity temperature of 0°C be maintained. The rh should be high enough to prevent wilting (98 to 100%). Considerable heat is given off because of respiration, and for this reason the stacks of crates should be separated and dunnage used to allow cold air circulation under and over the crates and between the bottom crates and the floor. Forced-air circulation should be provided; otherwise there may be a 1 to 2 K temperature differential between the top and the bottom of the room.

Celery can be cooled by forced-air cooling, hydrocooling, or vacuum cooling. Hydrocooling is the most common cooling method; temperatures should be brought to as near 0°C as possible. In practice, temperatures are reduced to 4 to 7°C. Vacuum cooling is widely used for celery packed in corrugated cartons for long-distance shipment.

Bacterial Soft Rot: See Table 4, Note 1.

Black Heart: Brown or black discoloration of tips or all of the heart leaves. Affected celery should not be stored because of rapid development of bacterial soft rot. *Control*: Good cultural practices with special attention to available calcium. Harvest promptly after celery is mature.

Early Blight (Cercospora): Circular pale yellow spots on leaflets. In advanced stages, spots coalesce and become brown to ashen gray. No spots develop in storage, but affected lots lose moisture and their fresh appearance. *Control*: Control early blight in the field by spraying or dusting.

Freezing Injury: Characteristic loosening of the epidermis is best demonstrated by twisting the leaf stem. Severe freezing causes celery to become limp and to dry out rapidly. Freezing may cause watery soft rot and bacterial soft rot. The freezing point is about −0.5°C.

Late Blight (Septoria): Small (3 mm or less), yellowish, indefinite spots in the leaflet and elongated spots on the leaf stalk bearing black fruiting bodies of pinpoint size on the surface and surrounding green tissue. Development of blight in storage is probably negligible, but it opens the way for storage decays. *Control*: Control late blight in the field with sanitary measures and fungicides. Store infected lots for short periods only.

Mosaic: Leaflets are mottled; the stalk shows brownish sunken streaks. Badly affected stalks shrivel. *Control*: Eradicate weeds that carry the virus. Grade out all discolored stalks at packing time.

Watery Soft Rot (Pink Rot): This is the principal decay of celery. It is often severe on field-frozen stock and on celery harvested after prolonged cool, moist weather. In early stages, it often has a pink color. *Control*: Grade out all discolored stalks at packing time. Storage at 0°C will retard but not prevent the disease.

Corn, Sweet (0°C and 95 to 98% rh)

Sweet corn is highly perishable and is seldom stored except to temporarily protect an excess supply. Corn, as it usually arrives on the market, should not be expected to keep for more than 4 to 8 days even in 0°C storage.

The sugar content, which so largely determines quality in corn and decreases rapidly at ordinary temperatures, decreases less rapidly if the corn is kept at about 0°C. The loss of sugar is about 4 times as rapid at 10°C as it is at 0°C.

Sweet corn should be cooled promptly after harvest. Usually, corn is hydrocooled, but vacuum cooling is also satisfactory if the corn is prewetted and top-iced after cooling. Where precooling facilities are not available, corn can be cooled with package and top ice. Sweet corn should not be handled in bulk unless it is copiously iced because of its tendency to heat throughout the pile. Sweet corn ears should be trimmed to remove most shank material before shipment to minimize moisture loss and prevent kernel denting. A loss of 2% moisture from sweet corn can result in objectionable kernel denting.

Cucumbers (10 to 13°C and 95% rh)

Cucumbers can be held only for short periods of 10 to 14 days at 10 to 13°C with a relative humidity of 95%. Cucumbers held at 7°C or below for longer periods develop surface pitting or dark-colored watery areas. These blemishes indicate chilling injury. Such areas soon become infected and decay rapidly upon removal of the cucumbers to warmer temperatures. Slight chilling may develop in 2 days at 0°C and severe chilling injury within 6 days at 0°C. The susceptibility of cucumbers to chilling injury does not preclude their exposure to temperatures below 10°C for short intervals as long as they are used immediately after removal from cold storage. Chilling symptoms develop rapidly only at higher temperatures. Thus, 2 days at 0°C or 4 days at 4°C are both harmless. Waxing is of some value in reducing loss of mass and in giving a brighter appearance. Shrink wrapping with polyethylene film can also prevent the loss of turgidity.

At temperatures of 10°C and above, cucumbers ripen rather rapidly as the green color changes to yellow. Ripening is accelerated if they are stored in the same room with ethylene-producing crops for more than a few hours.

Anthracnose (Colletotrichum): Circular, sunken, watersoaked spots that soon produce pink spore masses in the center. Later, the spots turn black. *Control*: Use disease-free seed and fungicidal applications in the field.

Bacterial Soft Rot: See Table 4, Note 1.

Bacterial Spot (Pseudomonas): Small, circular, watersoaked spots, later chalky or moist with gummy exudate. *Control*: If possible, avoid shipping infected cucumbers. Pack them dry and maintain temperatures as near optimum as practicable.

Black Rot (Mycosphaerella): Irregular, brownish, watersoaked spots of varying size, later nearly black. Black fruiting bodies are sometimes present. *Control*: Exclude infected cucumbers from the pack, if possible. Reduce carrying temperatures to about 10°C.

Chilling Injury: Numerous sunken and slightly watersoaked areas in the skin of cucumbers after removal from storage, found on cucumbers stored for longer than a week at temperatures below 7°C. *Control*: Store cucumbers at temperatures between 10 and 13°C, for no longer than 2 weeks.

Cottony Leak (Pythium): Large, greenish, watersoaked lesions. Luxuriant, white, cottony mold over wet decay. *Control*: Exclude infected cucumbers from the pack, if possible. Reduce carrying temperatures to about 10°C.

Freezing Injury: Large areas in cucumbers that are soft, flabby, watersoaked, and wrinkled, especially toward the stem end. *Control*: Prevent exposure to temperatures below −0.5°C.

Watery Soft Rot: See Table 4, Note 4.

Eggplants (8 to 12°C and 90 to 95% rh)

Eggplants are not adapted to long storage. They cannot be kept satisfactorily even at the optimum temperatures of 8 to 12°C for over a week and still retain good condition during retailing. Eggplants are subject to chilling injury at temperatures below 7°C. Surface scald or bronzing and pitting after sand scarring are symptoms of chilling injury. Eggplants that have been chilled are subject to decay by *Alternaria* when they are removed from storage. Exposure to ethylene for 2 or more days hastens deterioration.

Cottony Leak (Pythium): Decayed areas are large, bleached, discolored (tan), wrinkled, moist, and soft; later they exhibit abundant cottony mold. *Control*: Reduce carrying temperatures to about 8°C.

Fruit Rot (Phomopsis): Numerous somewhat circular brown spots that later coalesce over much of the fruit with pycnidia dotting the older lesions. This is a very common decay of eggplant. *Control*: Use fungicide sprays in the field. Reduce carrying temperature to about 8°C. Move the fruit promptly if decay is evident.

Endive and Escarole (0°C and 95 to 100% rh)

Endive and escarole are leafy vegetables not adapted to long storage. Even at 0°C, which is considered the best storage temperature, they will not keep satisfactorily for more than 2 or 3 weeks. They should keep somewhat longer if they are stored with cracked ice in or around the packages. Some desirable blanching usually occurs in endive held in storage.

Garlic, Dry (0°C and 65 to 70% rh)

If it is in good condition and is well cured when stored, garlic should keep at 0°C for 6 to 7 months. Garlic cloves sprout most rapidly at 4 to 18°C; therefore, prolonged storage at this temperature should be avoided. In California, it is frequently put in common storage, where it can be held 3 to 4 months or sometimes longer if the building can be kept cool, dry, and well ventilated.

Blue Mold Rot (Penicillium): Soft, spongy, or powdery dry decay of cloves. Affected cloves finally break down completely into gray or tan powdery masses. *Control*: Prevent bruising; keep garlic dry.

Waxy Breakdown: Yellow or amber waxy translucent breakdown of the outside cloves. *Control*: No control measures have been developed.

Greens, Leafy (0°C and 95 to 100% rh)

Leafy greens such as collards, chard, beet and turnip greens are very perishable and should be held as close to 0°C as possible. At this temperature, they can be held 10 to 14 days. They are commonly shipped with package and top-ice to maintain freshness and are handled like spinach. Kale packed with polyethylene crate liners should keep at least 3 weeks at 0°C or 1 week at 4°C. Vitamin content and quality are retained better when wilting is prevented.

Lettuce (0°C and 95 to 100% rh)

Lettuce is highly perishable. To minimize deterioration, it requires a temperature as close to its freezing point as possible without actually freezing it. Lettuce will keep about twice as long at 0°C as at 3°C. If it is in good condition when stored, lettuce should keep 2 to 3 weeks at 0°C with a high rh. Most lettuce is packed in cartons and vacuum-cooled to about 1 or 2°C soon after harvest. It should

Vegetables

then be immediately loaded into refrigerated cars or trailers for shipment or placed in cold storage rooms for holding prior to shipping.

An increasing quantity of lettuce is shipped in modified atmospheres to aid quality retention. Modified atmospheres are a supplement to proper transit refrigeration but are not a substitute for refrigeration. Lettuce is not tolerant of carbon dioxide and is injured by concentrations of 2 to 3% or higher. Romaine and leaf lettuce tolerate slightly higher carbon dioxide levels than head lettuce. Romaine is injured by 10% carbon dioxide, but not by 5% at 0°C.

Since excess wrapper leaves are usually trimmed off before sale or use, it is suggested that lettuce be trimmed to 2 wrapper leaves before packaging (rather than the usual 5 or 6) to save space and mass. The extra wrapper leaves are not needed to maintain quality.

Bacterial Soft Rot: The most common cause of spoilage in transit and storage. Often, it starts on bruised leaves. This decay normally is the controlling factor in determining the storage life of lettuce and is much less serious at 0°C than at higher temperatures (see Table 4, Note 1).

Brown Stain: Lesions that are typically tan, brown, or even black and about 6 mm wide and 12 mm long with distinct margins that are darker than the slightly sunken centers. The margins give a halo effect. The lesions develop on head leaves just under the cap leaves. The heart and wrapper leaves are not affected. Brown stain is caused by carbon dioxide accumulation from normal product respiration in railcars or trailers. *Control*: Ventilate to keep carbon dioxide below 2% in transport vehicles by keeping one water drain open. Enclose bags of hydrated lime (in vehicles shipped under a modified atmosphere) to absorb carbon dioxide.

Gray Mold: See Table 4, Note 2.

Pink Rib: Characterized by diffuse pink discoloration near the bases of the midribs of the outer head leaves. In heads with severe symptoms, all but the youngest head leaves may be pink and discoloration may reach into large veins. The cause has not been identified but shipment in low oxygen atmospheres at undesirably high temperatures (10°C) can accentuate the disease. It is most common in hard to overmature lettuce. *Control*: Store and ship lettuce at recommended low temperatures.

Russett Spotting: This occasionally causes serious losses. Small tan or rust-colored pitlike spots appear mostly on the midrib but possibly develop on other parts of leaves. Exposure to ethylene and to storage or transport temperatures above 3°C are the main causes of this disorder. Hard lettuce is more susceptible to it than firm lettuce. *Control*: Avoid storing or shipping lettuce with apples, pears, or other products that give off ethylene. Precool lettuce adequately to 1 to 3°C and refrigerate it continuously. Shipment in a low oxygen atmosphere (1 to 8%) allows effective control.

Rusty Brown Discoloration: A serious market disorder of western head lettuce; a diffuse discoloration which tends to follow the veins but also spreads to adjacent tissue. The disorder starts on the outer head leaves but in severe cases may affect all leaves. *Control*: There is no known control method.

Tipburn: Dead brown areas along the edges and tips of inner leaves. This is considered to be of field origin, but occasionally the severity of the disease increases after harvest. *Control*: Keep the affected stock well cooled and market it promptly after unloading to avoid secondary bacterial rots.

Watery Soft Rot: See Table 4, Note 4.

Melons

Persians should keep at 7 to 10°C for up to 2 weeks; **honeydews** for 2 to 3 weeks; and **casabas** for 4 to 6 weeks. These melons will definitely be injured in 8 days at temperatures as low as 0°C. Honeydews are usually given an 18- to 24-h ethylene treatment (5000 mg/kg) to obtain uniform ripening. Pulp temperature should be 21°C or above during treatment. Honeydews must be mature when harvested; immature melons fail to ripen even if treated with ethylene.

Cantaloupes harvested at the hard-ripe stage (less than full slip) can be held about 15 days at 2 to 3°C. Lower temperatures may cause chilling injury. Full-slip hard-ripe cantaloupes can be held for a maximum of 10 to 14 days at 0 to 2°C. They are more resistant to chilling injury. Cantaloupes are precooled by hydrocooling or forced-air cooling before loading or by top-icing after loading in railcars or trucks.

Watermelons are best stored at 10 to 15°C and should keep for 2 to 3 weeks. Watermelons decay less at 0°C than at 4°C, but they tend to become pitted and have an objectionable flavor after 1 week at 0°C. At low temperatures, they are subject to various symptoms of chilling injury—loss of flavor and fading of red color. Watermelons should be consumed within 2 to 3 weeks after harvest, primarily because of the gradual loss of crispness.

Alternaria Rot: Irregular, circular, brownish spots, sometimes with concentric rings, later covered with black mold. Often found on melons that have been chilled. *Control*: Avoid chilling temperatures. If cold melons are to be held at room temperature, they should be so stacked that condensed moisture will evaporate readily. Market melons promptly.

Anthracnose (Colletotrichum): Numerous greenish, elevated spots with yellow centers, later sunken and covered with moist pink spore masses. *Control*: Apply recommended field control measures.

Chilling Injury: Honeydew and honeyball melons stored for 2 weeks or longer at temperatures of 0 to 1°C sometimes show large, irregular, water soaked, sticky areas in the rind. *Control*: Store melons at 7 to 10°C.

Cladosporium Rot: Small black shallow spots later covered with velvety green mold. On cantaloupes, this rot is evident on extensive shallow areas at the stem ends or at points of contact between melons and it can be rubbed off easily. *Control*: Control measures are the same as for *Alternaria rot*.

Fusarium Rot: Brown areas on white melons; white or pink mold over indefinite spots on green melons. Affected tissue is spongy and soft with white or pink mold. *Control*: Avoid mechanical injuries; reduce carrying temperatures to 7°C.

Phytophthora Rot: Brown slightly sunken areas; later water-soaked and covered with a wet, appressed, whitish mold. *Control*: Cull out the affected fruits during packing. Reduce carrying temperatures to 7°C.

Rhizopus Rot: The affected melon is soft but not soupy and leaky as it is in similar decay on other vegetables. Coarse fungus strands may be demonstrated in decayed tissue (see Table 4, Note 3).

Stem End Rot (Diplodia): Fairly firm brown decay usually starting at the stem end and affecting a large part of the watermelon. Black fruiting bodies develop later. *Control*: At the time of loading in cars, recut the stems and treat them with Bordeaux paste or another recommended fungicide.

Mushrooms (0°C and 95% rh)

Mushrooms are usually sold in a retail market within 24 to 48 h after harvesting. They keep in good salable condition at 0°C for 5 days, at 4°C for 2 days, and at 10°C or above for about 1 day. A rh of 95% is recommended during storage. While being transported or displayed for retail sale, mushrooms should be refrigerated. Deterioration is marked by brown discoloration of the surfaces, elongation of the stalks, and opening of the veils. Black stems and open veils are correlated with dehydration.

Controlled-atmosphere storage can prolong the shelf life of mushrooms held at 10°C, if the oxygen concentration in the atmosphere is 9% or the carbon dioxide concentration is 25 to 50%.

Moisture-retentive film overwraps of caps are usually beneficial in reducing moisture loss.

Okra (7 to 10°C and 90 to 95% rh)

Okra deteriorates rapidly and is normally stored only briefly before marketing or processing. It has a very high respiration rate at warm temperatures. Okra in good condition can be kept satisfactorily for 7 to 10 days at 7 to 10°C. A rh of 90 to 95% is desirable to prevent wilting. At temperatures below 7°C, okra is subject to chilling injury which is shown by surface discoloration, pitting, and decay. Holding okra for 3 days at 0°C may cause pitting. Contact or top-ice causes water spotting in 2 or 3 days at all temperatures.

Fresh okra bruises easily; blackening of the damaged areas occurs within a few hours. A bleaching type of injury may also develop when okra is held in hampers for more than 24 h without refrigeration.

Onions (0°C and 65 to 70% rh)

A comparatively low rh is essential in the successful storage of dry onions. However, humidities as high as 85% and forced-air circulation have given satisfactory results. At higher humidities, at which most other vegetables keep best, onions develop root growth and decay; at too high a temperature, sprouting occurs. Storage at 0°C with 65 to 70% rh is recommended to keep them dormant.

Onions should be adequately cured in the field, in open sheds, or by artificial means before, or in, storage. The most common method of curing in northern areas is by forced ventilation in storage. Onions are considered cured when the necks are tight and the outer scales are dried until they rustle. If they are not cured, onions are likely to decay in storage.

Onions are stored in 23-kg bags, in crates, in pallet boxes that hold about 450 kg of loose onions, or in bulk bins. Bags of onions are frequently stored on pallets. Bagged onions should be stacked to allow proper air circulation.

In the northern onion-growing states, onions of the globe type are generally held in common storage because average winter temperatures are sufficiently low. They should not be held after early March unless they have been treated with maleic hydrazide in the field to reduce sprout growth.

Refrigerated storage is often used to hold onions for marketing in late spring. Onions to be held in cold storage should be placed there immediately after curing. A temperature of 0°C will keep onions dormant and reasonably free from decay, provided the onions are sound and well cured when stored. Sprout growth indicates too high a storage temperature, poorly cured bulbs, or immature bulbs. Root growth indicates the relative humidity is too high.

Globe onions can be held for 6 to 8 months at 0°C. Mild or Bermuda types can usually be held at 0°C for only 1 to 2 months. Onions of the Spanish type are often stored; if well matured, they can be held at 0°C at least until January or February. In California, onions of the sweet Spanish type are held at 0°C until April or May.

Onions are damaged by freezing, which appears as watersoaking of the scales when cut after thawing. If allowed to thaw slowly and without handling, onions that have been slightly frozen may recover with little perceptible injury. When onions are removed from storage in warm weather, they may sweat due to condensation of moisture. This may favor decay. Warming onions gradually (for example to 10°C over 2 to 3 days) with good air movement should avoid the difficulty. Onions should not be stored with other products that tend to absorb odors.

Onion sets require practically the same temperature and humidity conditions as onions, but because they are smaller in size they tend to pack more solidly. They are handled in approximately 11-kg bags and should be stacked to allow the maximum air circulation.

Green onions (scallions) and green shallots are usually marketed promptly after harvest. They can be stored 3 to 4 weeks at 0°C with 95% rh. Crushed ice spread over the onions aids in supplying moisture. Packaging in polyethylene film also aids in preventing moisture loss. Storage life of green onions at 0°C can be extended to 8 weeks by (1) packaging them in perforated polyethylene bags or in waxed cartons and (2) holding them in a controlled atmosphere of 1% oxygen with 5% carbon dioxide.

Ammonia Gas Discoloration: Exposure of onions to 1% ammonia in air for 24 h causes the surface of yellow onions to turn brown, red onions to turn deep metallic black, and white onions to turn greenish yellow. *Control*: Ventilate storage rooms as soon as possible after exposure.

Bacterial Soft Rot: This decay often affects one or more scales in the interior of the bulb. Decayed tissue is more mushy than gray mold rot (see Table 4, Note 1).

Black Mold (Aspergillus): Black powdery spore masses on the outermost scale or between outer scales. *Control*: Store onions at just above 0°C and at 65% rh.

Freezing Injury: A watersoaked, grayish yellow appearance of all the outer fleshy scales results from a slight freezing injury. All scales are affected and become flabby with severe injury. Opaque areas appear in affected scales. *Control*: Prevent exposure to −1°C temperatures and lower. Thaw frozen onions at 4°C.

Fusarium Bulb Rot: Semiwatery to dry decay progressing up the scales from the base. Decay is usually covered with dense, low-lying white to pinkish mold. *Control*: Do not store badly affected lots. Pull out infected bulbs in slightly affected lots. Store onions at 0°C.

Gray Mold Rot: This is the most common type of onion decay; it usually starts at the neck, affecting all scales equally. Decay often is pinkish (see Table 4, Note 2). *Control*: Cure onions thoroughly. Protect them from rain. Store them at just above 0°C.

Smudge: Black blotches or aggregations of minute black or dark green dots on the outer drying scales of white onions. Under moist conditions sunken yellow spots develop on fleshy scales. *Control*: Protect onions from rain after harvest. Store them just above 0°C.

Translucent Scale: The outer 2 or 3 scales are gray and watersoaked, as in freezing. The entire scale may not be discolored; no opaque area is noticeable. Sometimes translucent scale is found in the field. *Control*: No control is known. Store onions at 0°C after curing.

Parsley (0°C and 95 to 100% rh)

Parsley should keep 1 to 2.5 months at 0°C and for a somewhat shorter period at 2 to 3°C. High humidity is essential to prevent desiccation. Package icing is often beneficial.

Bacterial Soft Rot: See Table 4, Note 1.
Watery Soft Rot: See Table 4, Note 4.

Parsnips (0°C and 98 to 100% rh)

Topped parsnips have similar storage requirements to topped carrots and should keep for 2 to 6 months at 0°C. Parsnips held at 0 to 1°C for 2 weeks after harvest attain a sweetness and high quality equal to that of roots subjected to frosts for 2 months in the field. Ventilated polyethylene box or basket liners can aid in preventing moisture loss. Parsnips are not injured by slight freezing while in storage, but they should be protected from hard freezing. They should be handled with great care while in a frozen condition. The main storage problems with parsnips are decay, surface browning, and their tendency to shrivel. Refrigeration and high rh will retard deterioration.

Bacterial Soft Rot: See Table 4, Note 1.

Canker (Itersonilia sp.): An organism enters through fine rootlets and through injuries. The surface of the infected parsnip is first brown to reddish; later, it turns black where a depressed canker is formed. *Control*: Follow the recommended field spray program. Practice crop rotation.

Gray Mold Rot: See Table 4, Note 2.
Watery Soft Rot: See Table 4, Note 4.

Vegetables

Peas, Green (0°C and 95 to 98% rh)

Green peas lose part of their sugar rapidly if they are not refrigerated promptly after harvest. They should keep in salable condition for 1 to 2 weeks at 0°C. Top icing is beneficial in maintaining freshness. Peas keep better unshelled than shelled.

Bacterial Soft Rot: See Table 4, Note 1.
Gray Mold Rot: See Table 4, Note 2.
Watery Soft Rot: See Table 4, Note 4.

Peas, Southern (4 to 5°C and 95% rh)

Freshly harvested southern peas at the mature-green stage should have a storage life of 6 to 8 days at 4 to 5°C with high rh. Without refrigeration, they remain edible for only about 2 days, the pods yellowing in 3 days and showing extensive decay in 4 to 6 days.

Peppers, Dry Chili or Hot

Chili peppers, after drying to a moisture content of 10 to 15%, are stored in nonrefrigerated warehouses for 6 to 9 months. The moisture content is usually low enough to prevent fungus growth. A relative humidity of 60 to 70% is desirable. Polyethylene-lined bags are recommended to prevent changes in moisture content. Manufacturers of chili pepper products hold part of their raw material supply in cold storage at 0 to 10°C, but they prefer to grind the peppers as soon as possible and to store them in the manufactured form in airtight containers.

Peppers, Sweet (7 to 13°C and 90 to 95% rh)

Sweet peppers can be stored for a maximum of 2 to 3 weeks at 7 to 13°C. They are subjected to chilling injury if they are stored at temperatures below 7°C. The symptoms of this injury are surface pitting and discoloration near the calyx, which develops a few hours after removal from storage. At temperatures of 0 to 2°C peppers usually develop pitting in a few days. When stored at temperatures above 13°C, ripening (red color) and decay develop rapidly. Rapid cooling of harvested sweet peppers is essential in reducing marketing losses. It can be done by forced-air cooling, hydrocooling, or vacuum cooling. Forced air cooling is the preferred method. Peppers are often waxed commercially, which reduces in-transit chafing and moisture loss.

Bacterial Soft Rot: See Table 4, Note 1.
Freezing Injury: The outer wall is soft, flabby, watersoaked, and dark green in color. The core and seeds turn brown with severe freezing. Sweet peppers freeze at about −0.5°C.
Gray Mold Rot: See Table 4, Note 2.
Rhizopus Rot: See Table 4, Note 3.

Potatoes (Temperature, see following; 90 to 95% rh)

The proper potato storage environment will promote the most rapid healing of bruises and cuts, reduce rot penetration to a minimum, allow the least loss of mass and other storage losses to occur, and reduce to a minimum the deleterious quality changes that might occur during storage.

Early-crop potatoes are usually only stored during congested periods. They are more perishable and do not keep as well or as long as late-crop tubers. Refrigerated storage at 40°C, following a curing period of 4 or 5 days at 21°C, is recommended; or they can be stored for about 2 months at 10°C without curing. If early-crop potatoes are to be used for chipping or French frying, storage at 21°C is recommended. Holding these potatoes in cold storage (even at moderate temperatures of 10 to 13°C) for only a few days causes excessive accumulation of reducing sugars, which results in production of dark-colored chips.

Late-crop potatoes produced in the northern half of the United States are usually stored. The greater part of the crop is held in nonrefrigerated commercial and farm storages, but some potatoes are held in refrigerated storages. Potatoes in nonrefrigerated storages are usually held in bulk bins 2.4 to 6.1 m deep. Shallower bins are used in milder climates. Some potatoes are stored in pallet boxes. In refrigerated warehouses, potatoes can be stored in sacks, pallet boxes, or bulk.

Late-crop potatoes should be cured immediately after harvest by being held at 10 to 16°C and high relative humidity for about 10 to 14 days to permit suberization and wound periderm formation (healing of cuts and bruises). If properly cured, they should keep in sound dormant condition at 3 to 4°C with 95% rh for 5 to 8 months. A lower temperature is not desirable, except in the case of seed stock for late planting. For this purpose, 3°C is best. At 3°C or below, Irish potatoes tend to become sweet. For ordinary table use, potatoes stored at 4°C are satisfactory, but they will probably be unsatisfactory for chipping or French frying unless desugared or conditioned at about 18 to 21°C for 1 to 3 weeks prior to use. However, conditioning may be costly and good results are often uncertain.

Potatoes will remain dormant at 10°C for 2 to 4 months; since tubers from this temperature are more desirable for both table use and processing than those from 4°C, late-crop potatoes intended for use within 4 months should be stored at 10°C and those for later use at 4°C. All potatoes should be stored in the dark to prevent greening.

A storage temperature of 10 to 13°C is recommended for most cultivars of potatoes intended for chip manufacture. At these temperatures, they usually remain in satisfactory condition if their reducing-sugar content is low enough when they are initially stored. Storage at 16 to 18°C is less desirable because shrinkage, internal and external sprout growth, and decay are greater at these temperatures than at 10 to 13°C. Russet Burbank potatoes for table stock or for chipping are stored at 7°C with 95% rh.

Potatoes usually do not sprout until 2 to 3 months after harvest, even at 10 to 16°C. However, after 2 to 3 months of storage, sprouting occurs in potatoes stored at temperatures above 4°C and particularly at temperatures around 16°C. Although limited sprouting does not affect potatoes for food purposes, badly sprouted stock shrivels and is difficult to handle and market.

Certain growth-regulating chemicals have been approved by the U.S. Food and Drug Administration to control or reduce sprouting on potatoes. Potatoes treated with chemical sprout inhibitors should not be stored in the same warehouse with seed potatoes. If 90% or slightly higher rh is maintained, potatoes will have the best quality and the least amount of shrinkage. Cunningham et al. (1971) recommend 95% or higher rh for late-crop potatoes. Condensation on the ceiling and resultant moisture drip is sometimes a problem when very high humidity is maintained.

Ventilation or air circulation in potato storage is needed to provide and maintain optimum temperature and relative humidity throughout storage and the tubers it contains. In northern states, where average outdoor temperatures during storage are low, little circulation or ventilation is needed. Shell or perimeter circulation is extensively used in these areas for seed and table stock potatoes. Forced circulation through the potatoes is required for the higher temperature storage of processing potatoes and for table and seed stock in the warmer parts of the late-crop area. Rapid air circulation may lower the relative humidity of the air immediately surrounding the potatoes; it is conducive to drying and loss of mass, which may be desirable if there are disease problems but undesirable with sound potatoes because of increased shrinkage. For late-crop Idaho potatoes, a uniform airflow, which does not have to be continuous, of 47 L/s is recommended. With this ventilation, Russet Burbank potatoes stored at 7°C with 95% rh should keep in good condition for 10 months or longer.

Potatoes should not be kept in the same room with fruits, nuts, eggs, or dairy products because of the objectionable flavor they may impart. Also, potatoes may absorb odors from cheese or from volatile chemicals.

Bacterial Ring Rot: Yellow, soft, cheesy decay of the thin layer of tissue in the vascular ring. The outer 6 mm of tuber and the inner part may appear normal. *Control*: Use disease-free seed; store promptly at 4°C.

Bacterial Soft Rot (see Table 4, Note 1): This disease probably causes more loss in the early and intermediate crops than all other potato diseases combined.

Freezing Injury: If frozen solidly, tubers become soft and cream-colored and exude moisture. Slightly frozen tubers show darkening of the vascular ring and dull gray to black areas in the flesh. Potatoes freeze at about −1°C.

Fusarium Rot: Brown to black, spongy, and fairly dry; white or pink mold inside cavities of stored potatoes. *Control*: Avoid cutting and bruising during harvesting. After proper curing, maintain well-ventilated storage at 4°C.

Late Blight (Phytophthora): Reddish brown to black granular discoloration of the outer 3 to 6 mm of tuber. The affected tissue is firm to rock hard. *Control*: Apply recommended fungicides in the field. Kill vines prior to harvesting tubers or keep tubers away from blighted vines at harvest. Keep them dry; store at 4°C; market promptly.

Leak (Pythium): A large, gray to black, moist, decayed area starting at bruises or the stem end of the tuber. The internal tissue is granular and cream-colored at first, turning through reddish brown to inky black. *Control*: Prevent bruising. Refrigerate tubers to 4°C and keep them dry.

Mahogany Browning: Reddish brown patches or blotches in the flesh of tubers. Chippewa and Katahdin varieties are most susceptible. This differs from flesh discoloration caused by freezing in being reddish brown instead of gray. *Control*: Store at 4°C or above because lower temperatures cause the discoloration.

Net Necrosis: Dark brown vascular ring and vascular netting of the flesh, most prominent at the stem end, but extends well toward the bud end; increases during storage. *Control*: Reduce storage temperature promptly to 4°C; the infected tubers show symptoms earlier at higher temperatures.

Scald and Surface Discoloration: On early potatoes, this appears as sunken injured areas; later, it turns black and sticky and is followed by bacterial rots. *Control*: Move potatoes promptly to market; cool to 4°C.

Southern Bacterial Wilt: Moist, sticky exudation from the vascular ring when the tuber is cut. Sometimes there is advanced mushy decay in the center of the tuber. *Control*: Avoid shipping infected tubers; market promptly.

Stem End Browning: Dark brown to black vascular tissue, occurs in streaks that extend 10 to 25 mm into the flesh from the stem end; develops during storage. *Control*: After curing, reduce the storage temperature promptly to 4°C. Higher temperatures allow rapid development in susceptible lots.

Tuber Rot (Alternaria): Black to purplish, slightly sunken, shallow, irregularly shaped lesions, 6 to 25 mm in diameter, developing during storage. *Control*: Apply recommended fungicides in the field. Keep the tubers away from blighted vines as much as possible at harvest time. If the tubers are damp, inspect them frequently during storage and use forced-air ventilation to dry up excess moisture.

Pumpkins and Squash

Hard shell winter squash, such as the Hubbards, can be successfully stored for 6 months or longer at 10 to 13°C with a relative humidity of 60 to 75%. Dry storage is needed for quality retention. All specimens should be well-matured, carefully handled, and free from injury or decay when stored. Hubbard and other dark-green skinned squashes should not be stored near apples, as the ethylene from apples may cause the skin to turn orange-yellow. Most varieties of **pumpkins** do not keep in storage for as long as the usual storage varieties of squash. Such varieties as Connecticut Field and Cushaw do not keep well and cannot be kept in good condition for more than 2 to 3 months. **Acorn squash** can be stored satisfactorily for 5 to 8 weeks at 10°C. **Butternut squash** should keep at least 2 to 3 months at 10°C with 50% rh.

Summer squash, such as yellow crookneck and giant straight-neck, are harvested at an immature stage for best quality. The skin is tender and these varieties are easily wounded and perishable. The storage temperature range for summer squash is 5 to 10°C with 95% rh. A temperature of 5°C is best for zucchini squash stored up to 2 weeks, since they are sensitive to chilling injury.

Black Rot (Mycosphaerella): Hard, dry, black decay, dotted with minute black pimplelike fruiting bodies that occurs at stem ends or sides of the fruit. *Control*: Avoid skin breaks on the fruit; handle promptly.

Dry Rots (Alternaria; Cladosporium; Fusarium): Small, deep, dry, decayed areas. The decayed portion is easily lifted out of the surrounding healthy tissue. The surface mold is low-growing and either greenish black or pinkish white in color. *Control*: Prevent skin breaks. Do not store hard shell squashes below 10°C.

Rhizopus Rot: See Table 4, Note 3.

Radishes (0°C and 95 to 100% rh)

After harvest, topped spring radishes should be precooled quickly, often by hydrocooling, to 5°C or below. If they are then packaged in polyethylene bags, radishes can usually be held 3 to 4 weeks at 0°C and for a somewhat shorter time at 4°C.

Bunched radishes with tops are more perishable. They can be stored at 0°C and a rh of 95 to 100% for 1 to 2 weeks.

Rhubarb (0°C and 95% rh)

Fresh rhubarb stalks wilt and decay rapidly. Rhubarb in good condition can be stored 2 to 4 weeks at 0°C with a 95% rh or above. Moisture loss during holding or storage can be minimized by using nonsealed polyethylene crate liners or by film wrapping consumer size bunches. Removing and discarding leaf blades at harvest is desirable, because it not only reduces the possibility of decay and loss of mass but also reduces shipping mass and package size by one-third. Rhubarb is usually marketed with about 10 mm of the leaf blade attached to the petiole. Splitting of the petiole is more serious if the entire leaf is removed.

Fresh rhubarb cut into 25-mm pieces and packaged in 0.5-kg perforated polyethylene bags can be held 2 to 3 weeks at 0°C with high relative humidity. Splitting of cut ends and curling of these pieces in film bags may be a problem if marketing is at warm temperatures.

Gray Mold Rot (Botrytis): Grayish, smoke-colored growths and grayish brown spore masses on stalks. *Control*: Refrigerate to 0°C.

Phytophthora Rots (Phytophthora): Watery, greenish brown, sunken lesions starting at the base of the leafstalk and causing brown decay. *Control*: Decay is retarded with transit and storage temperatures below 4°C.

Rutabagas (0°C and 98 to 100% rh)

Rutabagas lose moisture and shrivel readily if they are not stored under high humidity conditions. A hot paraffin wax coating, often given to rutabagas, is effective in preventing wilting and loss of mass; it also slightly improves appearance. Too heavy a wax coating may produce severe injury from internal breakdown caused by suboxidation. Rutabagas in good condition, when stored, should keep 4 to 6 months at 0°C.

Freezing Injury: Rare, because the commodity can stand slight freezing without injury. Severe freezing causes watersoaking and light browning of the flesh, a mustard odor, and fermentation. *Control*: Prevent repeated slight freezing or severe freezing.

Gray Mold Rot: See Table 4, Note 2.

Vegetables

Spinach (0°C and 95 to 98% rh)

Spinach is very perishable and can be stored for only short periods of 10 to 14 days at 0°C with 95 to 98% rh. It will deteriorate rapidly at higher temperatures. Spinach is commercially vacuum cooled and forced-air cooled. If it is thoroughly cooled, it can be held for 10 to 14 days at 0°C without the addition of any package ice prior to storage. When precooling facilities are not available, crushed ice should be placed in each package to provide rapid cooling and to take care of the heat of respiration. Top-ice is also beneficial.

Bacterial Soft Rot: See Table 4, Note 1.

Downy Mildew (Peronospora): A field disease, commonly found at the marketing stage as pale yellow irregular areas in the leaves. Downy gray mold is present on the lower surface. *Control*: Control it in the field; market promptly.

White Rust (Albugo): Slight yellowing of areas in the leaf above white blisterlike pustules filled with white masses of spores. *Control*: Control it in the field.

Sweet Potatoes (13 to 16°C, 85 to 90% rh)

Most sweet potatoes are stored in nonrefrigerated commercial or farm storages. Preliminary curing at 29°C and 90 to 95% rh for 4 to 7 days is essential in the healing of injuries received in harvesting and handling and in preventing the entrance of decay organisms. After curing, the temperature should be reduced to 13 to 16°C, usually by ventilating the storage, and the relative humidity should be retained at 85 to 90%. Most varieties will keep satisfactorily for 4 to 7 months under these conditions. Loss of mass of 2 to 6% can be expected during curing and about 2% a month during subsequent storage.

Usually, sweet potatoes will not keep satisfactorily if they have been subjected to excessively wet soil conditions just before harvest or chilled before or after harvest by exposure to temperatures of 10°C or below. Short periods at temperatures as low as 10°C need not cause alarm; but after a few days at lower temperatures, sweet potatoes may develop discoloration of the flesh, internal breakdown, increased susceptibility to decay, and off-flavors when cooked.

Temperatures above 16°C stimulate development of sprouting (especially at high humidities), pithiness, and internal cork (a virus disease). Refrigeration is frequently used in large sweet potato storages to extend the marketing season into warm weather when ventilation will not maintain low enough temperatures.

Sweet potatoes are usually stored in slatted crates or bushel baskets. Palletization of crates and use of pallet boxes facilitates handling. Sweet potatoes are usually washed and graded and are sometimes waxed before being shipped to market. They may be treated with a fungicide to reduce decay during marketing.

Black Rot (Ceratocystis): Greenish black decay, frequently fairly shallow, and sometimes circular in outline at the surface. *Control*: Follow recommended field and postharvest control measures. Heat treatment of seed roots at 41 to 43°C for 24 h will prevent development of black rot.

Chilling Injury: Brown tinged with black discolored areas scattered or associated with vasculars. The interior becomes pithy. Chilling injury is often produced by exposure to lower temperatures for only a few days. Uncured roots are more sensitive than cured ones. *Control*: Store sweet potatoes at 13 to 16°C.

Freezing Injury: Soft, leaky condition of the flesh. The outer layer of the potato is dark brown. *Control*: Do not subject potatoes to low temperatures; sweet potatoes may freeze at −1°C.

Rhizopus Rot: See Table 4, Note 3. *Control*: Cure potatoes for 4 to 7 days at 29°C before storage. Follow recommended field and postharvest control measures.

Tomatoes (Mature Green, 13 to 21°C; Ripe, 10°C; 90 to 95% rh)

Mature green tomatoes cannot be successfully stored at temperatures that greatly delay ripening, even at a temperature of 13°C, which is considered to be a nonchilling temperature. Tomatoes held for 2 weeks or longer at 13°C may develop an abnormal amount of decay and fail to reach as intense a red color as tomatoes ripened promptly at 18 to 21°C. Temperatures of 20 to 22°C, and a relative humidity of 90 to 95% are probably used most extensively in commercial ripening of mature green tomatoes. At temperatures above 21°C, decay is generally increased. A temperature range of 14 to 16°C is probably the most desirable for slowing ripening without increasing decay problems. At this temperature, the more mature fruit will ripen enough to be packaged for retailing in 7 to 14 days. Tomatoes should be kept out of cold, wet rooms because, in addition to potential chilling injury, extended refrigeration damages the ability of fruit to develop desirable fresh tomato flavor.

Ethylene gas is sometimes used to hasten and give more even ripening to mature green tomatoes. In ripening rooms, a concentration of one part ethylene per 5000 parts of air daily for 2 to 4 days will usually shorten the ripening period by about 2 days at 20 to 22°C. Some tomatoes are gassed with ethylene in loaded railcars prior to shipping. Adding ethylene has little or no effect on tomatoes just before or after they have started to turn pink. Tomatoes themselves give off considerable ethylene as they ripen. Interest is increasing in the commercial use of low oxygen atmospheres of 3 to 5% during storage or transport to retard ripening and decay.

Storage temperatures below 10°C are especially harmful to mature green tomatoes; these chilling temperatures make the fruit susceptible to *Alternaria* decay during subsequent ripening. Increased decay during ripening occurs following 6 days' exposure to 0°C, or 9 days at 4°C (see Figure 1).

Firm ripe tomatoes may be held at 7 to 10°C with a relative humidity of 85 to 90% overnight or over a holiday or weekend. Tomatoes showing 50 to 75% of the surface colored (the usual ripeness when packed for retailing) cannot be successfully stored for more than 1 week and be expected to have a normal shelf life during retailing. Such fruits should also be held at 7 to 10°C and 85 to 90% rh. A storage temperature of 10 to 13°C is recommended for pink-red to firm red tomatoes raised in greenhouses.

When it is necessary to hold firm ripe tomatoes for the longest possible time, consistent with immediate consumption on removal from storage, such as on board a ship or for an overseas military base, they can be held at 0 to 2°C for up to 3 weeks, with some loss in quality. Mature green, turning, or pink tomatoes should be ripened before storage at this low temperature.

Alternaria Rot: Decayed area is brown to black, with or without a definite margin. Lesions are firm; rot extends into the flesh of the fruit. Dense, velvety, olive green or black spore masses frequently grow over affected surfaces. *Control*: Avoid mechanical injuries at packing time. Avoid temperatures below 10°C in green fruit.

Bacterial Soft Rot: See Table 4, Note 1.

Cladosporium Rot: Thin, brownish blemishes or black shiny spots of shallow decay, later covered by green, velvety mold. *Control*: Take care in harvesting and packing. Ship high quality tomatoes free of field chilling injury under protective services that will provide temperatures of 13 to 20°C.

Late Blight Rot (Phytophthora): Greenish brown to brown, roughened areas with a rusty tan margin. *Control*: Apply recommended field control measures. Cull tomatoes carefully before packing.

Phoma Rot: Slightly sunken, moderately penetrating, black areas at the edge of the stem scar and elsewhere on the fruit. Black pimplelike fruiting bodies develop later. Decayed tissues are firm and brown to black in color. Phoma rot is found in eastern-grown tomatoes. *Control*: Apply field control measures. Exercise care in harvesting and packing. Avoid temperatures below 13°C.

Rhizopus Rot: See Table 4, Note 3.

Soil Rot (Rhizoctonia): Small circular brown spots, frequently with concentric ring markings; later, large, brown, and fairly firm lesions. In advanced stages, under warm conditions, cream-colored or brown mycelium and irregular sclerotia may develop. *Control*: Before packing, sort out tomatoes with early lesions if the disease is prevalent.

Turnips (0°C and 95% rh)

Turnips in good condition can be expected to keep 4 to 5 months at 0°C with 90 to 95% rh. At higher temperatures (5°C and above), decay will develop much more rapidly than at 0°C. Injured or bruised turnips should not be stored. Store turnips in slatted crates or bins and allow good circulation around containers.

Dipping turnips in hot melted paraffin wax gives them a glossy appearance and is of some value in reducing moisture loss during handling. However, waxing is primarily to aid in marketing and is not recommended prior to long-term storage.

Turnip greens are usually stored for only short periods (10 to 14 days). They should keep about as well as spinach at 0°C with crushed ice in the packages.

REFERENCES

Agricultural Statistics. 1988. U.S. Department of Agriculture, Washington, D.C.

Appleman, C.O., and J.M. Arthur. 1919. Carbohydrate metabolism in green sweet corn. *Journal of Agricultural Research* 17:137.

Ashby, H.B., R.T. Hinsch, L.A. Risse, W.G. Kindya, W.L. Craig, Jr., and M.T. Turczyn. 1987. Protecting perishable foods during transport by truck. USDA *Agricultural Handbook* No. 669.

Bogardus, R.K., and J.M. Lutz. 1961. Maintaining the fresh quality in produce in wholesale warehouses. *Agricultural Marketing* 6(12):8.

Burton, W.G. 1958. Suppression of potato sprouting in buildings. *Agriculture* 65:299.

Cunningham, H.H., M.V. Zaehringer, and W.C. Sparks. 1971. Storage temperature for maintenance of internal quality in Idaho Russet Burbank potatoes. *American Potato Journal* 48:320.

Dewey, D.H., R.C. Herner, and D.R. Dilley. 1969. Controlled atmospheres for the storage and transport of horticultural crops. *Horticultural Report* No. 9, Michigan State University (July).

Hardenberg, R.E. 1971. Effect of in-package environment on keeping quality of fruits and vegetables. *HortScience* 6(3):198.

Hardenberg, R.E., H. Findlen, and H.W. Hruschka. 1959. Waxing potatoes—Its effect on weight loss, shrivelling, decay, and appearance. *American Potato Journal* 36:434.

Hardenburg, R.E., A.E. Watada, and C.Y. Wang. 1986. The commercial storage of fruits, vegetables, and florist and nursery stocks. USDA *Agriculture Handbook* No. 66.

Harvey, J.M. 1965. Nitrogen—Its strategic role in produce freshness. *Produce Marketing* 8(7):17.

Isenberg, F.M. and R.M. Sayles. 1969. Modified atmosphere storage of Danish cabbage. *Journal of the American Society for Horticultural Science* 94(4):447.

Kader, A.A. et al. 1984. Irradiation of plant products. Comments from CAST 1984-1. Council of Agricultural Science and Technology, Ames, IA. ISSN 0194-4096.

Lipton, W.J. 1965. Post-harvest responses of asparagus spears to high carbon dioxide and low oxygen atmospheres. *Proceedings of the American Society for Horticultural Science* 86:347.

Lipton, W.J. 1968. Effect of temperature on asparagus quality. *Proceedings*, Conference on Transportation of Perishables, Davis, CA, 147.

Lutz, J.M. 1936. The influences of rate of thawing on freezing injury of apples, potatoes and onions. *Proceedings of the American Society for Horticultural Science* 33:227.

McColloch, L.P., and J.T. Worthington. Ways to prevent chilling mature green tomatoes. *PrePack-Age* 7(6):22.

Morris, L.L. and H. Platenius. 1938. Low temperature injury to certain vegetables after harvest. *Proceedings of the American Society for Horticultural Science* 36:609.

Parsons, C.S. and R.E. Anderson. 1970. Progress on controlled-atmosphere storage of tomatoes, peaches and nectarines. *United Fresh Fruit and Vegetables Association Yearbook*, 175.

Parsons, C.S. and R.H. Day. 1970. Freezing injury of root crops—Beets, carrots, parsnips, radishes, and turnips. USDA *Marketing Research Report* No. 866.

Redit, W.H. 1969. Protection of rail shipments of fruits and vegetables. USDA *Handbook* No. 195.

Stewart, J.K. and M.J. Ceponis. 1968. Effects of transit temperatures and modified atmospheres on market quality of lettuce shipped in nitrogen-refrigerated and mechanically refrigerated trailers. USDA *Marketing Research Report* No. 832 (December).

Stewart, J.K., M.J. Ceponis, and L. Beraha. 1970. Modified atmosphere effects on the market quality of lettuce shipped by rail. USDA *Marketing Research Report* No. 863.

Watada, A.E. and L.L. Morris. 1966. Effect of chilling and nonchilling temperatures on snap bean fruits. *Proceedings of the American Society for Horticultural Science* 89:368.

Whiteman, T.M. 1957. Freezing points of fruits, vegetables and florist stocks. USDA *Marketing Research Report* No. 196 (December).

CHAPTER 24

FRUIT JUICE CONCENTRATES AND CHILLED JUICE PRODUCTS

ORANGE JUICE 24.1	Pure Fruit Juice Powders 24.5
Orange Concentrate 24.1	OTHER CITRUS JUICES 24.6
Cold Storage 24.3	NONCITRUS JUICES 24.6
Concentration Methods 24.3	Pineapple Juice 24.6
Quality Control 24.4	Apple Juice 24.6
Chilled Juice 24.4	Grape Juice 24.7
Refrigeration 24.5	Strawberry and Other Berry Juices 24.7

CITRUS products, especially orange juice, comprise the largest percentage of the total volume of juices sold in the United States. Much of the technology used in processing noncitrus juices was developed from citrus processing.

ORANGE JUICE

ORANGE CONCENTRATE

Processed orange juice is sold in four principal forms:

1. Frozen concentrate (3-plus-1 concentration, in which three volumes of water are added to one volume of concentrate for reconstitution) in a variety of package sizes. These are the familiar retail products.
2. Concentrate in bulk at 65° Brix. This is an intermediate product that is bought and sold daily as futures on the Commodity Exchange. Most of this product will ultimately be sold in one of the other forms.
3. Chilled orange juice, which is ready to drink when poured from the carton. It is either reconstituted concentrate or nonconcentrated juice. By law, these two products must be labeled "from concentrate" or "not from concentrate."
4. Institutional or restaurant concentrates in special packaging at 4-plus-1 or higher concentrations.

After processing, frozen citrus concentrates in retail (3-plus-1) packages must be stored at −18°C. Highly concentrated bulk juice (65° Brix) may be satisfactorily stored at about −9°C. Chilled single-strength juices are stored at about −1 to 0°C.

Figure 1 shows a schematic flow diagram of citrus processing. The **Brix scale** is a hydrometer scale that indicates the percentage by mass of sugar in a solution at a specified temperature.

Selecting, Handling, and Processing Fresh Fruit

Selection. Fruit is selected for proper quality and maturity. Some fruit that is blemished but sound in quality, referred to as packinghouse eliminations, is used. A major portion of the crop is taken directly from the grove to the processing plant. To be mature, the fruit must have the proper Brix-acid ratio, and the juice content and Brix must be above specified values. Fruit should be handled without delay because no real maturing occurs after harvesting; instead, the temperature and condition of the fruit determine the rate of deterioration. Citrus fruit is sufficiently rugged to withstand mechanical handling on conveyors, elevators, and belts, provided that the fruit is processed within a day or two after picking. Samples are taken mechanically as fruit enters the bins, and records of chemical analyses are maintained. Usually fruit from two or more bins is used simultaneously to improve uniformity. The fruit passes over inspection tables both before and after temporary storage in bins, and damaged or deteriorated fruit is removed.

Washing. Prior to juice extraction, the fruit is wetted by sprays. The wetting agent is dispensed onto the fruit as it travels over rotating brushes. Water sprays near the end of the washer unit rinse the fruit. A sanitizing solution may be used to sanitize conveyors and elevators.

Juice Extraction. Individual high-speed mechanical juice extractors handle from 300 to 700 pieces of fruit per minute. Some machines halve the fruit and ream or squeeze the juice from the half. Other machines insert a tube through the middle of the fruit and squeeze the juice through the fine holes into the tube, at the same time sieving away the seeds and large pieces of membrane. After the juice has been extracted, it passes to finishers that remove the remaining seeds, pieces of peel, and excess cell or fruit membrane. In the past, this was a comparatively simple process involving one or two stages, but it has become complicated in recent years and varies extensively from plant to plant. Usually, one or two stages of screw-type finishers separate most of the pulp from the juice.

Pulp washing, a process during which soluble solids in separated segment and cell walls are recovered by countercurrent extraction with water, is permitted in Florida, provided that the resultant extract is not used in frozen orange concentrates. It may be used in other formulated products permitted by the Federal Standard of Identity for frozen concentrated orange juice.

The juice or pulp wash liquor from the finishers may require treatment in high-speed desludging centrifuges that remove suspended matter before transferring the juice to the evaporator. These centrifuges have peripheral discharges that open and close at intervals to discharge a thick suspension of pulp cells. This operation decreases the viscosity of the juice in the evaporator, improves the efficiency of evaporation, and improves the appearance of the final product. Special means are used to classify orange pulp for inclusion in products with a high pulp content.

Heat Treatment. When frozen concentrated orange juice was first developed, heat treatment was avoided in order to maintain optimum flavor. Such concentrate, if prepared from good sound fruit, remains stable for a considerable time at −18°C and for nearly a year at −15°C. However, with large-scale production, it is not possible to assure storage below −15°C. Concentrates originally of good quality tend to gel or clarify rapidly during storage. Heat treatment inactivates enzymes responsible for the development of these defects during improper storage. While earlier methods employed plate-type, steam-heated pasteurizers, heat treatment is now almost universally included as an integral part of heat conservation in the evaporation process.

The preparation of this chapter is assigned to TC 10.9, Refrigeration Application for Foods and Beverages.

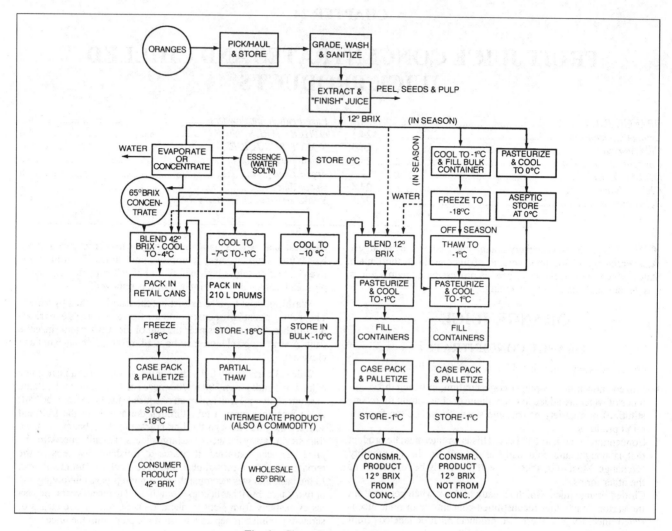

Fig. 1 Citrus Processing Schematic

Flavor Fortification. Originally, frozen concentrated orange juice was overconcentrated, and fresh juice (**cutback juice**) was added to reduce the concentration to the desired level and provide fresh flavor in the final product. This process is used extensively in frozen concentrates, but flavor levels cannot be standardized by this method alone.

Essential oil from orange peel does not by itself supply completely balanced fresh flavor, but it is now used extensively to control flavor intensity. Some **peel oil** is found in the cutback juice, but little remains in the juice from the evaporator. The addition of peel oil to the finished concentrate, at approximately 0.014% by volume in the reconstituted juice, has become standard practice.

Several variations have been introduced to supplement the use of peel oil and cutback fresh juice for flavor fortification. In most cases, the condensate from the first stage of concentration, which uses evaporator distillers, is used to produce an **essence** that is restored to the final concentrate in order to enhance flavor. Although cutback fresh juice and cold-pressed peel oil are commonly added, most operations depend heavily on essence recovery and its incorporation into the final product.

Blending, Packaging, and Freezing. The final step in processing is the blending of concentrate with flavor-enhancing components such as cold-pressed orange oil and liquid essence. The 65° Brix concentrate is then reduced to about 42° Brix, the requirement for 3-plus-1 product, which reconstitutes to about 12° Brix when used by the consumer. The availability of cold-pressed oil and essence when no fresh fruit is available makes blending and packaging possible year-round.

Most retail product is packaged in fiber-foil cans, although aluminum and tinned steel cans may be used. Sizes range from 180 to 1900 mL. Gable-top milk cartons and specialized large disposable plastic reservoirs for dispensers are also used.

Prior to filling, the product is cooled to about −4 to 2°C. This is usually done in the swept-surface cold-wall tank in which the concentrate is blended, but it may also be done in a bladed heat exchanger or, preferably, in a plate-type unit. Alternatively, the 65° Brix concentrate from the evaporator may be precooled (unblended) and packed in 210 L open-top drums or delivered to large bulk storage tanks.

The filled cans pass through blast freezers, where their temperature is reduced to −18°C or below in 45 to 90 min. The cans are transported on link belts, and the air handlers are arranged so that air at about −32°C is forced down through the loaded belt.

Thermal Properties. In calculating cooling and freezing requirements, concentrates may be considered to approximate sucrose solutions of the same concentration. In the range of 60 to 65° Brix, the specific heat is about 2.85 kJ/(kg·K); for 42° Brix, the value is about 3.10 kJ/(kg·K). Thermal conductivity in the liquid state is in the range of 0.29 to 0.31 W/(m·K). Regarding the heat of fusion for freezing tunnel design, the value should be 163 kJ/kg in the range of −1 to −18°C; however, experience has shown that a value of about

Fruit Juice Concentrates and Chilled Juice Products

230 kJ/kg should be used. This allows for extraneous heat gains, coil defrosting, and so forth.

COLD STORAGE

Cold storage facilities for citrus processing can be divided into three categories according to temperature requirements of −18, −10, and −1°C.

Finished goods for retail and institutional markets are stored at −18°C in insulated, refrigerated buildings. Bulk 65° Brix product packed in drums is also stored at −18°C. The refrigerated buildings range from a few hundred square metres to a hectare or more. Other than the usual insulation requirements, two factors are critical to the design: (1) the vapor retarder outside the insulation must be as close to hermetic as possible, and (2) irrespective of the insulation in the floor, a heat supply must be installed beneath the insulation in order to maintain the temperature below the floor at about 0°C. Otherwise, the floor will ultimately heave from ice formation below the floor.

The −10°C buildings are used for bulk storage of 65° Brix concentrate. In a typical installation, a −10°C building would house several large stainless steel tanks. Tank sizes range from about 10 to 760 m³ each. At the stated temperature, the product is barely pumpable, requiring sanitary positive-displacement pumps. Because the temperature is virtually impossible to change after the product is in the tank, the product must be cooled to storage temperature before it is introduced to the tanks. Cooling usually occurs in a plate heat exchanger.

Finally, −1°C storage rooms are used largely for chilled single-strength juice in retail packages or for the not-from-concentrate bulk tank systems. This product is discussed in the sections on Chilled Juice and Refrigeration.

CONCENTRATION METHODS

The three major methods for producing concentrates are (1) high-temperature, single pass, multiple-effect evaporators; (2) freeze concentration with mechanical separation; and (3) low-temperature, recirculatory, high-vacuum evaporators.

Thermally Accelerated Short-Time Evaporator (TASTE)

At present, the TASTE is the standard evaporator of the citrus industry. This unit has also been successfully applied to grape, apple, and other juices. TASTEs produce at least 90% of all juice concentrate in the western hemisphere. The first cost of this unit is among the lowest of all alternatives, and it has excellent thermal efficiency. The flavor quality and storage stability of juice processed in this unit compare favorably with those of juice from alternate methods of concentration.

The TASTE includes all standard methods of heat conservation. It is multiple-effect, uses concurrent vapor for staged preheating, and flashes condensate to discharge condensed vapors at the lowest possible temperature. Residence time is minimal because all stages are single pass. Total residence time is 2 to 8 min, varying directly with evaporator capacity.

Figure 2 shows a schematic diagram of a typical TASTE. A triple-effect unit illustrates the basic design. Typical units are offered in four to seven effects (five to nine stages), and capacity ranges from about 1.3 to 11.3 kg/s water removal. The vapor-to-steam ratio is approximately the number of effects minus 0.5. In Figure 2, note that the juice enters the spray nozzles atop each stage at a temperature well above the saturation temperature at which that body operates. The resultant flashing of vapor, combined with the spray effect of the nozzle, distributes the juice on the vertical tube walls. Thus, the juice film and the water vapor flow concurrently down the tube; the juice flows down the wall as the vapor flows in the central area of the tube. The centrifugal pumps that transfer the juice to the succeeding stage operate continuously in a cavitating, or almost cavitating, condition.

As juice is transferred from stage 3 to stage 4, it must be reheated to a higher temperature (60°C) to effect sufficient flashing. Reheating is typical for any effect that has more than one stage. These stage divisions are necessary when water removal renders the remaining

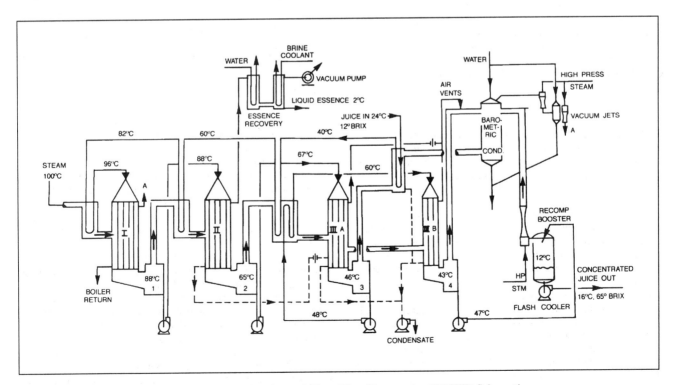

Fig. 2 Thermally Accelerated Short-Time Evaporator (TASTE) Schematic

juice insufficient to wet the total required surface for that effect. Then flash cooling quickly reduces the concentrate from 47°C to the pump-out temperature of 16°C.

On a TASTE, the vapor flow is fixed and virtually unalterable. Although the juice flow shown in Figure 2 is concurrent with the vapor, in many cases it is mixed flow, in which the juice may first be introduced to an intermediate effect before being delivered to the first effect. In all cases, however, the essence-bearing material must be recovered from the vapor stream that is produced where the juice first enters the evaporator.

Figure 2 also shows an essence recovery method. Several different systems for essence recovery are used, some of which are proprietary. The refrigeration load for essence recovery is relatively small, usually 50 to 140 kW, according to evaporator capacity.

Freeze Concentration

In the freeze concentration system, juice is introduced and pumped rapidly through a swept-surface heat exchanger, in which ice nuclei are formed at about −2°C. This slurry is delivered to a recrystallizer, in which small crystals are melted to form larger crystals. The slurry of larger crystals, together with the resultant concentrate, is delivered to the wash column, where the ice rises. As it does, chilled water (melted ice) is introduced at the top to wash the ice crystals; these continue to rise and melt as they reach the top of the column. The concentrate (now ice-free) is drawn off the bottom of the wash column.

At first, centrifuges were used to separate ice crystals from concentrate, but due to poor separation and attendant problems with crystal washing, the losses of orange soluble solids were unacceptable. The proprietary wash-column process drastically reduces these losses. The output concentration of commercial freeze concentration equipment is presently limited to under 50° Brix.

When first-quality oranges are used to prepare the juice, freeze concentration makes a concentrate that is indistinguishable from fresh orange juice. However, first costs for equipment and installation are relatively high, and operating costs at best are the same as for modern evaporators. The cost is high despite the 334 kJ/kg required to freeze water, as opposed to about 2300 kJ/kg for evaporation. A contributing factor is the relative cost of electricity (compressors) versus steam from fossil fuel. Another factor is that, by using multiple-effect evaporators, the 2300 kJ/kg can be divided by the number of effects in the evaporator, thus reducing the energy requirement to the range of 350 to 580 kJ/kg, depending on the number of effects.

QUALITY CONTROL

The most important factor in quality control is the use of good sound fruit, but the quality must also be checked during processing and again in the final product. Standards for the various concentrates describe the quality and serve as a basis for grading. Concentrates are checked for Brix value, Brix-acid ratio, peel oil content, and other factors. Additionally, tests may be run to ensure that other requirements of a particular brand are met. At periodic intervals, bacteriological samples are taken during various stages in the plant, plated on orange serum agar, and examined after incubation. Although the necessary sanitation measures are known, the bacteriological samples serve as a check and indicate whether the procedures have been effective.

When the weather is warm, cleanups must be more stringent and more frequent than during cold weather. TASTEs can run for only 6 to 12 h before substantial loss of evaporation or the release of hesperidin crystals forces a cleanup cycle. These cycles are normally spaced to coincide with other required cleaning in the juice extraction system and often take no more than 30 min of production time. The thermal conditions in a properly operated TASTE are so adverse to microorganisms and enzymes that these do not influence the cleaning cycle. Normally, low-temperature evaporators are cleaned at least every 24 h; however, in some instances when arrangements are favorable, they may run as long as 7 days between cleanups. Total plate counts in final product are generally maintained well below 10^6 organisms per millilitre of reconstituted juice. Counts of a few thousand per millilitre are common and indicate a high degree of plant cleanliness. Sanitation is based on asepsis rather than antisepsis. The natural acidity and high sugar content of citrus juice concentrate normally inhibit rapid growth of organisms. Generally, counts tend to decrease in storage.

CHILLED JUICE

Chilled juice is usually packed either in fiber cartons or in plastic bottles or jugs. While the ideal storage temperature is about −1°C, it is frequently handled in the retail trade at 4 to 7°C and is typically stored in household refrigerators. Normal shelf life is 3 to 4 weeks.

Chilled juice is marketed in two basic forms: "from-concentrate" and "not-from-concentrate." Due to higher overall costs, the not-from-concentrate product is higher priced. Figure 1 shows the processing steps for each of the two products. The sole difference in the two products is the source of the juice.

From-Concentrate Juice. Bulk concentrate, taken either from partially thawed drums or from bulk storage, is mixed in a blending tank with water, essence, and cold-pressed oil so that it is reconstituted to about 12° Brix. This juice is then processed in a three-stage pasteurizer. First, the juice is preheated in a regenerative section that recovers heat from juice leaving the pasteurizing section. It then flows to the pasteurizer, where it is steam-heated to 82 to 88°C. Then it flows back through the regenerative section, where it is partially cooled by the incoming juice. Finally, it passes through the cooling section, where it is cooled to about −1°C. A refrigerated glycol/water solution usually serves as the chilled brine for this purpose. These pasteurizer units are usually plate-type units, but tubular equipment is also used. From this point onward, every effort is made to maintain the juice at −1°C as it is packaged and placed in −1°C storage.

Not-from-Concentrate Juice. When fruit is mature and in season, fresh juice may be taken directly from extraction to pasteurizing and packaging. No blending is required, and the remaining process duplicates that for the from-concentrate product. However, unless special means are used in the extraction process, the peel-oil content of the juice may be excessive. A low-oil extraction method or heating and flashing the juice in a special deoiler prior to pasteurizing can remove excess oil.

Because not-from-concentrate chilled juice is in demand year-round and mature fruit is only available a maximum of 8 months per year, two storage methods are commonly used to ensure year-round supply: (1) store the juice in large bulk tanks in an aseptic environment at a temperature just above the freezing point or (2) freeze the fresh juice and store it in a solid form at about −18°C.

The aseptic system usually involves a number of large stainless steel storage tanks in a cold room. The entire system must be carefully sanitized before use and must be so maintained during use. The juice is pasteurized and cooled before introduction to the tanks.

Freezing fresh juice predates the aseptic liquid method. The freezing method always includes cooling the juice to about −1°C prior to placing it in a container for freezing. The choice of container is related to (1) the rate at which the juice will be frozen and (2) the method used to thaw the product and remove it from the container for later processing. In some instances, the advantage for freezing may be counterbalanced by disadvantages for thawing and removal.

For example, juice may be cooled, placed in an open-top drum, and immediately transferred to −23°C storage, where it will slowly freeze. The product quality will be satisfactory, but thawing and dumping the product from the large drum is difficult. In another method, the juice is frozen and stored in a specially constructed

Fruit Juice Concentrates and Chilled Juice Products

container. Later, it is fully thawed and pumped from the container. In still another method, the juice is encased in a plastic bag and then quick-frozen in a blast freezer room. Thawing and removal methods are similar to that for drums. Some juice is also stored in open ice blocks from which it is easier to recover, but the juice is more susceptible to contamination or losses.

A third distinct method of providing fresh juice in the off-season is to store fresh oranges under refrigeration for extraction of juice as required. This method requires large cold rooms held at about −1 to 0°C. In many cold storages, moisture must be added to keep the air at saturation. Added moisture, however, increases the refrigeration load because all condensate from the cooling coils must be recycled to the conditioned space. Because of the cost and other limitations, fresh fruit is seldom stored for juice.

REFRIGERATION

Refrigeration Equipment

The choice of refrigerant for juice processing is almost universally R-717 (ammonia); a few small systems use R-22. R-22 is sometimes used in freeze concentration systems, although ammonia has also been used for this purpose.

High-stage compressors have been of the reciprocating type, but to meet greater demands, there is a trend toward rotary screw compressors.

Most condensers are evaporatively cooled. Exposure to subtropical climatic conditions makes materials of construction an important consideration. In Florida, water conditioning is critical because the available makeup water has a high mineral content.

The most common refrigerant evaporators in cold rooms and blast freezing tunnels are finned or plain coils. In larger installations, a single, low-pressure receiver serves a multitude of coils. Liquid refrigerant is pumped through the coils at two to three times the evaporation rate, and the liquid/gas mixture is returned to the low-pressure receiver, from which the compressors take their suction. At refrigerant temperatures below 0°C, coils must be defrosted on a regular basis, generally by hot gas from the compressor discharge manifold. Some smaller ammonia installations employ air units, each with its own surge drum and controls for flooded operation. Small installations using R-22 are usually direct-expansion, and some use electric or water defrost.

To solve coil defrosting problems, some cold rooms and freezing tunnels operate with continuous defrost, in which a strong solution of propylene glycol is sprayed continuously over the coils. The weaker glycol solution is removed, and the acquired water is driven off in an external concentrator. Unless extreme care is used in designing the eliminators, the glycol solution may be entrained and deposited on the containers. The resultant appearance of the containers is unacceptable.

Another common evaporator is a shell-and-tube unit that chills a secondary coolant, typically propylene glycol. Ethylene glycol should not be used for food processing because it is toxic. Plate-type juice processors require large volumes of chilled polypropylene glycol. Although the refrigerant can be supplied directly to a gasketed plate heat exchanger, most heat exchangers do not satisfy the pressure requirements of the refrigerant.

Freezing tunnels using an alcohol-water solution chilled in shell-and-tube equipment have been constructed. The chilled solution was sprayed over the containers for quick freezing. These tunnels are now largely obsolete and have been replaced with blast units.

Swept-surface heat exchangers for concentrate precooling preparatory to can filling were also used. These units operated as flooded refrigerant systems, but most have been replaced with plate units, which also require glycol coolant. Today, application of swept-surface units is generally limited to freeze concentration systems.

A typical installation has a number of small loads, including

- Essence recovery (condensing of water vapor laden with essence). This load is typically handled with shell-and-tube equipment either by direct expansion or with a secondary coolant (glycol).
- Winterizing of cold-pressed oil in order to form precipitates of undesired materials in a quiescent storage vessel. The process goes forward at about −18 to −40°C and requires months of storage. Winterizing may be done in 210 L drums in a cold room or, preferably, in a jacketed tank using direct refrigerant expansion or very cold glycol.
- Storage of essence. Tanks of essence may be stored in a 0°C cold room or in an outdoor insulated storage tank in which the heat gain is removed by continuous recirculation of the essence through an external plate heat exchanger cooled by glycol brine.

Refrigeration Loads

Refrigeration loads vary not only with plant throughput but also according to the relative rates at which different products are processed. For a given amount of consumable product, the volumes of chilled juice are higher than those for concentrates by a factor of 4:1 to 6:1. Thus, the refrigeration load depends largely on the proportions of concentrates and chilled juices being processed.

To illustrate the magnitude of the refrigeration loads, the following fixed set of processing conditions is assumed. All percentages are percentage of total throughput of **soluble solids**.

- The plant has an operating evaporator capacity of 12.6 kg/s of water removal.
- 10% is packed as not-from-concentrate chilled juice. Of this, half is packaged directly for immediate sale; the remainder is frozen for storage and use during the off-season.
- 90% is processed into concentrate, which is made up of 85.5% fed to the evaporator and 4.5% reserved for use as cutback to make 42° Brix retail packs.
- Of the concentrate from the evaporator, 30% is packed in bulk for later use as from-concentrate chilled juice; an additional 24% is similarly packed for later use as retail frozen concentrate.
- The remaining 31.5% from the evaporator is blended with the 4.5% cutback (to total 36%). This is packed into retail containers at 42° Brix and immediately frozen.
- The chilled juice line processes 30% of the concentrate (i.e., same as quantity from the evaporator stored for this purpose).

Given these conditions, the plant refrigeration loads would be approximately as follows:

	Refrigeration, kW	
	Low-Stage	High-Stage
Process	880	2250
Cold storage	530	985
Miscellaneous loads	—	350
Total	1410	3585

Compressor Manifolding

In most cases, low-stage compressors have their own isolated high-stage machines, with an intermediate pressure of 205 to 240 kPa (gage). The remaining high-stage compressors are usually paralleled to handle all other high-stage loads at a somewhat lower suction pressure [105 to 140 kPa (gage)] to optimize power requirements.

Frequently, these systems operate as central systems, with all condensers manifolded using a common receiver.

PURE FRUIT JUICE POWDERS

One manufacturer has successfully produced a vacuum dried orange juice powder. In this **puff-drying** process, concentrate of about 58° Brix is introduced into a vacuum chamber, where it is dried on a moving stainless steel belt. The dried powder is flavored with a locked-in orange oil prepared by dispersing orange oil in a mixture of molten sugars, extruding, and cooling rapidly. The oil is thus kept out of contact with the powdered orange juice until water is added to reconstitute the juice.

Another process, known as **foam-mat drying**, has been under investigation for citrus juices. In this process, a small amount of foam stabilizer is added (0.5% of dry solids content), and the chilled concentrate of about 50° Brix is beaten into a foam. This foam is laid out in a sheet about 3 mm thick on perforated trays. An air blast clears the foam from the holes, and the stacked trays are conveyed up a column while hot air passes up through them and reduces the moisture to about 1.25% in 12 min. The product is then chilled until it hardens and is scraped from the trays. In a variation of the process, a thin layer of the foam is placed on a polished stainless steel belt, dried in a stream of hot air, and finally stripped from the belt in dried form with a doctor blade.

Both the puff-drying and foam-mat drying processes were developed through U.S. Department of Agriculture (USDA) research.

OTHER CITRUS JUICES

Grapefruit Juice

In the production of frozen concentrated grapefruit juice, essentially the same equipment is used as in the production of frozen concentrated orange juice. Some adjustments at the extractors are necessary to accommodate grapefruit. Because bittering is considered a defect, debittering systems may also be used to improve the flavor.

Both sweetened and unsweetened frozen concentrate grapefruit juices are prepared (the sweetened product in greater quantities). The final unsweetened product may vary from 38 to 42° Brix. The sweetened product must contain at least 416 g of soluble grapefruit solids per litre exclusive of added sweetening ingredients. The final Brix may vary from 38 to 48°. In Grade A unsweetened concentrate, the Brix-acid ratio may vary from 9:1 to 14:1; in the sweetened product, the ratio may vary from 10:1 to 13:1. All the types mentioned are 3-plus-1 concentrate. Either seedless or seeded grapefruit can be used, but the seeded varieties, such as Duncan, are generally preferred.

Blended Grapefruit and Orange Juice

The same procedures are used in producing a grapefruit and orange juice blend as are used in preparing the separate products. USDA grade standards recommend no less than 50% orange juice in the mixture and as much as 75% orange juice when it is light in color. Military specifications require 60 to 75% orange juice. USDA grades require 40 to 44° Brix in unsweetened concentrates. In sweetened concentrates, the Brix must be at least 38° before sweetening and 40 to 48° after sweetening. For Grade A, the Brix-acid ratios in the packed concentrate may vary from 10:1 to 16:1 if unsweetened, and from 11:1 to 13:1 if sweetened.

Tangerine Juice

Tangerines require different methods of handling during picking, hauling, and storage at the plant. While the grapefruit and orange are generally round, quite firm, and able to withstand considerable rough handling, the tangerine is somewhat flat and irregular in shape and has a loose, tender skin that is easily broken. If the skin is broken and the fruit bruised, bacteria and yeasts readily attack the fruit, and undesirable enzyme actions occur. Thus, tangerines cannot be handled in orange bins but must be handled in boxes or loose in trucks to a depth of no more than 600 mm.

The processes and equipment used in manufacturing concentrated tangerine juice are practically the same as those used with oranges. Because the fruit is smaller, the yield of juice from a given number of extractors is smaller, and about twice the extracting equipment is required to furnish enough juice to keep the evaporators operating at full capacity. Generally, the values for Brix-acid ratio, peel oil content, and concentration have followed those prescribed for the orange product. Recently, more of the pack has been sweetened, and there has been a trend toward packing at a higher concentration. A Brix of 44° is common for a 3-plus-1 concentrate.

NONCITRUS JUICES

PINEAPPLE JUICE

Pineapple juice is prepared from small fruit and the parts of larger pineapples that are unsuitable for packing as fruit pieces. The main sources are the cores, the layer of flesh between the shell and cylinder that is cut for the preparation of pineapple slices, and the juice that drains from crushed pineapple. The juice material amounts to about one-third of the mass of the fresh fruit. It is inspected in order to remove pieces of shell and spoiled flesh. The juice is extracted by passing through disintegrators and screw presses. It is then centrifuged to remove heavy foreign material and excessive insoluble solids. Juice processing up to this point is the same for both single-strength and concentrate.

Pineapple concentrate is produced from single-strength juice in equipment similar to that used to produce orange and other fruit juice concentrates. The first step in the concentrating operation is to strip out the volatile flavoring materials. These are separated as about a 100-fold concentrate and added back to the final concentrate. The concentration takes place in multiple-effect evaporators.

Pineapple concentrate is produced either as a 3-to-1 product with a Brix of about 46.5° or as a 4-1/2-to-1 product with a Brix of about 61°. The 3-to-1 concentrate is produced in both a sterile form and a frozen form. However, even the sterile product is stored and sold under refrigeration in order to preserve quality. The 4-1/2-to-1 concentrate is also produced in both a sterile form and a frozen form. It may be held for short periods without refrigeration, but it should be stored at 4°C or lower. The frozen 61° Brix concentrate is packaged in polyethylene bags and held in 25 L fiber containers. This product is stored under refrigeration.

Bulk pineapple concentrate is used principally as an ingredient for mixing with citrus concentrate in the production of frozen juice blends. Pineapple concentrate is also used as an ingredient in many types of canned fruit drinks.

The composition of pineapple juice varies greatly. Brix varies between 12 and 18°, with an average of about 13.5 to 14°. Brix-acid ratio ranges from 12:1 to over 20:1 and usually averages between 16:1 and 17:1. Because pineapple concentrate is produced at a standard Brix, variation in composition shows up only in the acidity and the Brix-acid ratio.

APPLE JUICE

Evaporative procedures include some method of essence (ester) recovery for incorporating the volatile components of apple flavor into the final concentrate. These procedures are designed to take advantage of the fact that in the distillation of apple juice, most of the volatile flavors are found in the first 10% of the distillate. This portion of the distillate is passed through a fractionating column to obtain the volatile flavors in concentrated form, usually about 100-fold compared to the fresh juice. The remaining 90% of the original juice, now stripped of its volatile flavors, is then concentrated under vacuum to somewhat more than the concentration desired for the final product.

For example, 100 L of apple juice prepared in a conventional manner, may be treated to yield 1 L of 100-fold essence and 24 L of concentrated stripped juice. The combination of these two fractions

Fruit Juice Concentrates and Chilled Juice Products

yields full-flavored fourfold concentrate. If a higher concentrate is desired, concentration of the stripped juice is carried out to a greater extent. In such instances, however, the juice must be depectinized to avoid excessive viscosity and gelation of the highly concentrated juice.

One report shows that fourfold apple juice (depectinized) concentrate was essentially unchanged after one year of storage at $-18°C$. Similar information on concentrate prepared without depectinization is not available.

GRAPE JUICE

Concord Grapes

Most of the concentrated grape juice marketed in North America is prepared from Concord grapes (*Vitis labrusca*) grown in New York, Michigan, Washington, Pennsylvania, Ohio, Arkansas, and Ontario. The grapes are harvested when the soluble solids reach a concentration of 15 to 16%. This varies with maturity and is influenced by cultural and climatic factors.

After washing, Concord grapes are conveyed to the stemmer, which consists of a perforated, slowly revolving (20 rpm) horizontal drum inside of which several beaters, revolving at a much faster speed (200 rpm), knock the berries off the cluster and partially crush them before they are discharged through the drum perforations. The cluster stems are expelled from the open end of the drum. The crushed fruit is then pumped through a tubular heat exchanger, where it is heated to 60 to 63°C for good extraction of the pigments and juice. The hot pulp then goes to hydraulic presses, where the juice is removed in the same manner as apple juice. The expressed juice may be clarified in a centrifuge or filter press. With a filter press, 1 to 2% diatomaceous earth is used to maintain a high filtering rate and remove a substantial amount of suspended matter. Under normal operating conditions, 790 to 805 L of juice are obtained from a megagram of grapes. In some plants, screw presses are used to remove juice from all or part of the crushed grapes, but this increases the amount of suspended matter to be removed later.

The clarified juice is pasteurized in tubular or plate heat exchangers to a temperature of 82 to 88°C and cooled immediately to $-1°C$ before storage in tanks in refrigerated rooms maintained at $-2°C$. The juice is usually cooled in two or more steps. Some heat exchange systems begin with a regeneration cycle, in which the hot juice leaving the pasteurizer preheats entering juice. The cooling water discharged from the heat exchangers may be piped to the washers in order to heat the water applied to the incoming grapes.

The method of handling the cooled juice depends on the intended use of the concentrate. If it is to be used later in jelly manufacture, the juice is stored at $-2°C$ for 1 to 6 months to permit settling of the **argols**; these consist of potassium acid tartrate, tannins, and some colored materials that would give a gritty texture to the jelly or detract from its clarity. The clear juice is siphoned off the precipitate in the storage tanks and may be refiltered. If the concentrate is to be sold as a blend formed by mixing in sugar and ascorbic acid before canning and freezing, the cold storage tank merely serves as a surge tank because the juice is pumped out to the concentrator within a few hours. A polishing filter is used before the concentrator to minimize fouling of the evaporator tubes. Concentrates for both jelly manufacture and blended juice may be stored in tanks at $-3°C$ prior to processing. Whenever single-strength juice is bulk-stored at $-3°C$ prior to concentration, spoilage by fermentation is a danger. To minimize yeast growth during storage, all pipelines and equipment from the pasteurizer to the cold room should be designed for ready and frequent cleaning. The interior surfaces of storage tanks must be relatively smooth and free from crevices, and the tank should be thoroughly cleaned before use.

Concentration involves two steps. In the first, the volatile flavoring materials are stripped from the juice, and the stripped juice is then concentrated to the desired density. The volatile components are removed by heating the single-strength juice to 104 to 110°C for a few seconds in a heat exchanger, flashing a percentage of the liquid into a vapor in a jacketed tube bundle, and discharging the liquid and vapor tangentially through an orifice into a separator. The separator should be large enough to reduce the vapor velocity to 3 m/s or less for minimal entrainment. Twenty to 30% by mass of the original juice flashes off as vapor, which is led into the base of a fractionating column filled with ceramic saddles or rings. A reflux condenser on the vapor line from the column and a reboiler section at the base of the column provide the necessary reflux ratio. The vent gases from the reflux condenser are then chilled in a heat exchanger, and the condensate containing the essence is collected at a rate equivalent to 1/150 of the volume of entering flavoring material.

Methyl anthranilate, which has a boiling point of 267°C and is only slightly soluble in water, is an important flavor component of Concord grape juice. The high boiling point and low solubility cause losses of methyl anthranilate when the efficiency of the stripping column is low. These losses may be reduced by increasing the vaporizing temperature.

In a typical formulation, grape juice is concentrated to a little over 34° Brix, and essence or fresh cutback juice is added to reduce it to 34° Brix. Sucrose is added to achieve 47° Brix, and citric acid is added until the total acidity is 1.8% calculated at tartaric acid. The concentrate may be cooled to -7 to $-1°C$ in a heat exchanger or cold wall tank, filled into cans, sealed, cased, and allowed to freeze in storage below $-18°C$. When diluted with an equal quantity of water, this concentrate yields the equivalent of sweetened single-strength juice; however, for a more palatable beverage, three parts of water are added. The product is labeled as concentrated, sweetened grape juice.

Muscadines

Muscadines (*Vitis rotundifolia*), typical of the southeastern United States, are processed into juice as follows: grapes are harvested, transported in trucks to the receiving station, dumped into hoppers, and crushed with potassium persulfate ($K_2S_2O_8$) added to give 50 mg/kg free sulfite. These grapes differ from *V. labrusca* in that they grow not in clusters but as individual berries, so there is no need to destem them.

After crushing, the grapes are conveyed into a pneumatic press (without any heating), and pressure up to 500 kPa is exerted. Once the juice is extracted, it is quickly pasteurized as it passes through a plate heat exchanger that heats it to about 85°C. The juice is partially cooled, and a mixture of pectinases and cellulases is added to clarify it. After 1 to 2 h, the juice is passed through an ultrafiltration (UF) tubular unit with 2.0 μm pores for filtering. The juice is then repasteurized and cooled to 7°C in plate heat exchangers and sent to large refrigerated storage tanks for removal of tartrate and sediment at $-2°C$. Some of the juice is concentrated to about 65° Brix in a triple-effect evaporator with an essence recovery system. This concentrate is field-frozen and used later for bottle juice. When ready to bottle, juice is pumped through the UF unit, bottled and capped, and passed through a pasteurizer/cooler in a conveyor belt.

STRAWBERRY AND OTHER BERRY JUICES

Frozen strawberry juice concentrate, a sevenfold concentrate with separately packed concentrated (100-fold) essence, is used for manufacturing, especially jellies. Availability of sevenfold frozen concentrate also allows the marketing of high-quality strawberry juice solids. Concentrates of red raspberry, black raspberry, and blackberry juices are also available in limited quantities.

The preparation of strawberry and other berry juice concentrates involves essence recovery in which 12 to 20% of the juice is separated by a stripping process using a steam injection heater. Vapors containing volatile flavors are concentrated in a fractionating column to the desired degree. The juice remaining after the essence recovery step is concentrated under vacuum three- to sevenfold by volume. For strawberry juice, a maximum temperature of 38°C for 2.5 h should not be exceeded, whereas temperatures up to 54°C may be used in preparing boysenberry concentrate in a batch-type operation.

Preparation of juice for concentration involves chopping or coarse milling of cold, sound berries and mixing with pectic enzymes and filter aid. After several hours (4 to 5 h at room temperature), juice is expressed with a bag press or rack and clothes press. The cloudy juice is clarified in a filter press. Recovered essences are concentrated and packaged separately so that the jelly manufacturer can incorporate the essence in the jelly just before filling. This procedure greatly reduces the amount of essence lost by volatilization. The essence can also be incorporated in the concentrate to make a full-flavored product for shipping as a single unit.

Concentrated juice without essence can be packed in enamel-lined containers that need only be liquid-tight. Concentrated essence should be kept in carefully sealed cans to avoid loss of the highly volatile flavor. Both juice concentrate and essence are kept frozen for proper quality retention.

CHAPTER 25

BEVERAGES

BREWERIES 25.1	Heat Treatment of Red Musts 25.9
Malting 25.1	Juice Cooling 25.9
Process Aspects 25.1	Heat Treatment of Juices 25.9
Processing 25.3	Fermentation Temperature Control 25.9
Pasteurization 25.6	Potassium Bitartrate Crystallization 25.9
Carbon Dioxide 25.6	Storage Temperature Control 25.10
Heat Balance 25.7	Chill-Proofing Brandies 25.10
Common Refrigeration Systems 25.7	CARBONATED BEVERAGES 25.10
Vinegar Production 25.8	Beverage and Water Coolers 25.10
WINE MAKING 25.8	Size of Plant 25.11
Must Cooling 25.8	Liquid Carbon Dioxide Storage 25.11

THIS chapter discusses the processes and use of refrigeration in breweries, wineries, and carbonated beverage plants.

BREWERIES

MALTING

Malt is the primary raw ingredient used in the brewing of beer. While adjuncts such as corn grits and rice contribute considerably to the composition of the extract, they do not possess the necessary enzymatic components required for the preparation of the wort. They lack nutrients (amino acids) required for yeast growth and contribute little to the flavor of beer. Malting is the initial stage in preparing raw grain to make it suitable for mashing. Traditionally, this operation was carried out in the brewery, but in the past century, this phase has become so highly specialized that it is now almost entirely the function of a separate industry.

Various grains such as wheat, oats, rye, and barley can be malted; however, barley is the predominate grain used in the preparation of mash because it has a favorable ratio of protein to starch. It has the proper enzyme systems required for conversion, and the barley hull provides an important filter bed during lautering. Also, barley is readily available in most of the world.

There are three steps to malting barley. The first step is the **steeping** phase. The raw grain is soaked or steeped in 4 to 18°C water for two to three days. The moisture content of the barley kernel increases from 12% to approximately 45%. The water is changed frequently and the grain is aerated. After two or three days, the kernels start to germinate and the white tips of rootlets appear at the end of the kernels. At this time the water is drained and the barley is transferred to where it is germinated.

Germination is the second step of malting. The growth of the kernel continues during this 4 to 5 day process. The green malt is constantly turned over to assure uniform growth of the kernels. Slowly revolving drums can be used to turn over the growing malt. In a compartment system, slowly moving, mechanically driven plow like agitators are used for mixing. Cool (10 to 18°C) saturated moist air is used to maintain temperature and green malt moisture levels. At the desired stage in its growth, the green malt is transferred to a kiln.

Kilning, the final step, stops the growth of the barley kernel by reducing its moisture level. Warm (49 to 66°C) dry air is used to remove the moisture from the green malt. The kilning is usually done in two stages. First, the moisture content of the malt is reduced to approximately 8 to 14%; then, the heat is increased until the moisture is further reduced to about 4%. Using this heating procedure reduces excessive destruction of enzymes. The desired color and aroma are obtained by controlling the final degree of heat.

After kilning, the malt is cleaned to separate dried rootlets from the grain, which is then stored for future use. The finished malt differs from the original grain in several significant ways. The hard endosperm was modified and is now chalky and friable. The enzymatic activity has been greatly increased, especially alpha amylase which is not present in unmalted barley. The moisture content is reduced, making it more suitable for storing and subsequent crushing. It now has a distinctive flavor and aroma, and the starch and diastase are readily extractable in the brewhouse.

PROCESS ASPECTS

Two distinct types of chemical reactions (mashing and fermentation) are used in brewing beer. The first, **mashing**, is carried out in the brewhouse. The starches in the malted grain are hydrolyzed into sugars and complex proteins are broken down into simpler proteins, polypeptides, and amino acids. These reactions are brought about by first crushing the malt and suspending it in warm (38 to 50°C) water by means of agitation in the mash tun. When adjuncts are used, a portion of the malt is cooked separately with the adjunct, usually corn grits or rice. After boiling, this mixture is combined with the main mash, which has been so proportioned that a combining temperature generally in the range of 63 to 72°C results. Within this temperature range, the alpha and beta amylases degrade the starch to mono-, di-, tri-, and higher saccharides. By suitably choosing a time and temperature regimen, the brewer controls the amount of fermentable sugars produced. The enzyme diastase (essentially a mixture of alpha and beta amylase), which induces this chemical reaction, is not consumed but acts merely as a catalyst. Some of the maltose is subsequently changed by another enzyme, maltase, into a fermentable monosaccharide, glucose.

Mashing is complete when the starches are converted to iodine-negative sugars and dextrins. At this point, the temperature of the mash is raised to a range from 75 to 77°C, which is the "mashing-off" temperature. This stops the amylolytic action and fixes the ratio of fermentable to nonfermentable sugars. The **wort** is separated from the mash solids using a lauter tub, a mash filter, or other proprietary equipment (MBAA 1981). Hot water (76 to 77°C) is then "sparged" through the grain bed to recover additional extract. Wort and sparge water are added to the brew kettle and boiled with hops, which may be in the form of pellets, extract, or whole cones. After boiling, the brew is quickly cooled and transferred to the starting cellar where yeast is added to induce fermentation. Figure 1 shows a double-gravity system with the grains stored at the top of the

---

The preparation of this chapter is assigned to TC 10.9, Refrigeration Application for Foods and Beverages.

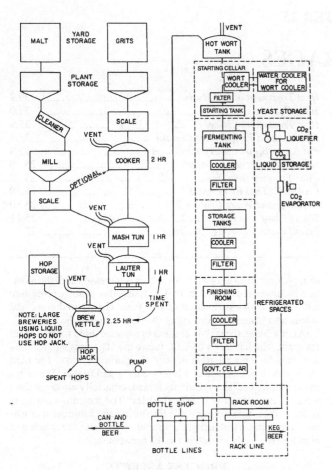

Fig. 1 Brewery Flow Diagram

Table 1 Total Solids in a Cubic Metre of Wort at 10°C

% Solids[a]	Total Density, kg/m³	Mass of Solids, kg	Specific Heat, kJ/(kg·K)
0	999.6	0.0	4.19
1	1003	10.0	4.16
2	1007	20.1	4.13
3	1011	30.3	4.10
4	1015	40.6	4.07
5	1019	51.0	4.04
6	1023	61.4	4.01
7	1027	71.9	3.98
8	1031	82.5	3.95
9	1036	93.2	3.92
10	1040	104.0	3.89
11	1044	114.8	3.86
12	1048	125.8	3.84
13	1052	136.8	3.81
14	1057	147.9	3.78
15	1061	159.1	3.75
16	1065	170.4	3.72
17	1070	181.8	3.69
18	1074	193.3	3.66
19	1078	204.9	3.63
20	1083	216.6	3.60

[a]Saccharometer readings.

brewhouse. As the processing continues, gravity creates a downward flow. The hot wort from the bottom of the brewhouse is then pumped to the top of the stockhouse, where it is cooled and again proceeds by gravity through fermentation and lagering.

After cooling the wort, yeast and sterile air are injected into it. The yeast is pumped in as a slurry at a rate of 4 to 12 g of slurry per litre of wort. Normally, oil-free compressed air is filtered and treated with ultraviolet light and then added to the wort, which is nearly saturated with approximately 11 mg/kg of oxygen. However, the wort may also be oxygenated with pure oxygen.

The **fermentation** process takes place in two phases. During the first phase, called the respiratory or aerobic phase, the yeast consumes the oxygen present. It uses a metabolic pathway, preparing it for the anaerobic fermentation to follow. The process typically lasts 6 to 8 h.

The depletion of the oxygen present signals the yeast to start anaerobically metabolizing the sugars present in the extract, releasing heat and producing CO_2 and ethanol as metabolic by-products.

During the early fermentation phase, the yeast multiplies rapidly then more slowly as it consumes the available sugars. Normal multiplication for the yeast is 5 to 6 times. A representative value for the heat released during fermentation is 650 kJ/kg of extract (sugar) fermented.

Wort is measured by the saccharometer (measures sugar content), which is a hydrometer calibrated to read the percentage of maltose solids in solution with water. The standard instrument is the Plato saccharometer, and the reading is referred to as percentage of solids by saccharometer, or degrees Plato (°P). Table 1 illustrates the various data deducible from reading the saccharometer.

The same instrument is used to check the progress of the fermentation. While it still gives an accurate measure of density of the fermenting liquid, it is no longer a direct indicator of dissolved solids because the solution now contains alcohol, which is less dense than water. This saccharometer reading is called the apparent extract, which is always less than the real extract (apparent attenuation is calculated from the hydrometer reading of apparent extract and the original extract). In engineering computations, 81% of the change in apparent extract is considered a close approximation of the change in real extract. Thus, 81% of the difference between the solids shown in Table 1 for saccharometer readings before and after fermentation would represent the mass of maltose fermented. This mass (in kilograms per cubic metre of wort at 10°C) times 650 kJ/kg gives the heat of fermentation. The difference between the original solids and the mass of fermented solids gives the residual solids per metre. It is assumed that there is no change in the volume because of fermentation. The specific heat of beer is assumed to be the same as that of the original wort, but the mass per unit volume is decreased according to the apparent attenuation.

Bottom fermentation-type yeast (*Saccharomyces uvarium*, formerly *carlsbergensis*) is used in fermenting lager beer. Top fermentation-type yeast (*Saccharomyces cerevisiae*) is used in making ale. They are so called because, after fermentation, one settles to the bottom and the other rises to the top. A more significant difference between the two types is that in the top fermentation type, the fermenting liquid is allowed to attain a higher temperature before a continued rise is checked. The following are the characteristics of brewing ale that cause it to differ from brewing lager:

- A more highly kilned darker malt is used.
- Malt forms a greater proportion of the total grist (less adjunct).
- Infusion mashing is used and a wort of higher original specific gravity is generally produced.
- More hops are added during the kettle boil.
- A different yeast and temperature of fermentation are used.

Therefore, ale may have a somewhat higher alcohol content and a fuller, more bitter flavor than lager beer. With the bottom fermentation yeasts, fermentation is generally carried out between 7 and 18°C and most commonly between 10 and 16°C. Ale fermentations are generally carried out at somewhat higher temperatures, often peaking in the range of 21 to 24°C. In either type, the temperature during fermentation would continue to rise above that desired if not checked by cooling coils or attemperators through which a cooling

Beverages

medium such as propylene glycol, ice water, brine, or ammonia is circulated. In the past, these attemperators were manually controlled, but more recent installations are automatic.

PROCESSING

Wort Cooling

To prepare the boiling wort from the kettle for fermentation, it must first be cooled to a temperature of 7 to 13°C. To avoid contamination with foreign organisms that would adversely affect the subsequent fermentation, this cooling must be done as quickly as possible, especially through the temperatures around 38°C. Besides the primary function of wort cooling, other beneficial effects accrue that are essential to good fermentation, including precipitation, coagulation of proteins, and oxidation because of natural or induced aeration depending on the type of cooler used.

In the past, the Baudelot cooler was almost universally used because it is inherently easy to clean and affords the necessary aeration of the wort. However, the traditional open-type Baudelot cooler was replaced by one consisting of a series of swinging leaves encased within a removable enclosure into which sterilized air was introduced for aeration. This modified form, in turn, has virtually been replaced by the totally enclosed heat exchanger. Air for aeration is admitted under pressure into the wort steam, usually at the discharge end of the cooler. The air is first filtered and then irradiated to kill bacteria, or it can be sterilized by heating in a double-pipe heat exchanger with steam. By injecting 40 L of air per cubic metre of wort, which is the amount necessary to saturate the wort, a normal fermentation should result. The quantity can be accurately increased or diminished as the subsequent fermentation indicates.

The coolant section of wort coolers is usually divided into two or three sections. For the first section, a potable source of water is used. The heated effluent goes to hot water tanks where, after additional heating, it is used for succeeding masking and sparging brews. Final cooling is done in the last section, either by direct expansion of the refrigerant or by means of an intermediate coolant such as chilled water or propylene glycol. Between these two, a third section may be used from which the warm water can be recovered and stored in a wash-water tank for later use in various washing and cleaning operations around the plant.

Closed coolers save on space and money for expensive cooler room air-conditioning equipment. They also permit a faster cooling rate and provide accurate control of the degree of aeration. The problem of cleaning plate coolers is solved by having a spare set of plates ready for weekly replacement. The cleaning between successive brews is accomplished by circulating cleaning or sterilizing solutions, or both, through the cooler.

In selecting a wort cooler, the following should be considered:

- The cooling rate should allow the contents of the kettle can to be cooled in less than 2 h.
- The heat transfer surfaces to be apportioned between the first section, using an available water supply, and the second section, using refrigeration, should be such that the most economical use is made of each of these resources. Cost of water, its temperature, and its availability should be balanced against the cost of refrigeration. Usual design practice is to cool the wort in the first section to within 6 K of the available water.
- Usable heat should be recovered (the effluent from the first section is a good source of preheated water). After additional heating, it can be used for succeeding brews and as wash waste in other parts of the plant. At all times, the amount of heat recovered should be consistent with the overall plant heat balance.
- Meticulous sanitation and maintenance costs are important factors.

The size of the wort coolers is determined by the rate of cooling desired, the rate of water flow, and the temperature differences used. A brew, which may vary in size from 5 to 100 m^3 and over, is ordinarily cooled in less than 1 to 4 h. Open-type coolers are made in stands up to about 6 m long. Where more length is needed, two or more stands are operated in parallel.

Open coolers are best operated with a wort flow of 10.7 to 12 L/s per metre of stand. As the flow increases beyond this rate, an increasingly larger part of the wort splashes from the top tube of the cooler and drops directly into the collecting pan below without contacting the cooler surfaces. An increased amount of wort, which flows over the surfaces, must be subcooled to offset what has been bypassed.

In the plate-type cooler, this bypassing does not occur, and wort velocities can be increased to a point where the friction pressure through the cooler approaches the maximum design pressure of the press and gasketing. The number of passes and streams per pass afford the designer much latitude in selecting the most favorable parameters for optimum performance and economical design. This design is based on (1) the specific heat of wort, (2) its initial temperature and the range through which it is to be cooled, (3) the temperature of the available water supply, and (4) the ratio of the quantity of cooling water to wort that is to be used. Design and operating features of a typical plate cooler are as follows:

Specifications

Quantity of wort to be cooled	2.14 kg/s
Temperature of hot wort	98°C
Temperature of cooled wort	4°C
Temperature of available water (maximum)	21°C
Water used, not to exceed	4.28 kg/s
Temperature of water leaving cooler	60°C
Temperature of wort leaving first section of cooler	27°C
Temperature of incoming recirculated chilled water	1°C

Plate cooler (first section)

Number of plates	40
Heat transfer surface per plate	0.4 m^2
Heat transfer surface in first section	16 m^2
Number of passes	5
Number of streams per pass	4
Water flow rate	4.28 kg/s
Wort flow rate	2.14 kg/s

Plate cooler (second section)

Number of plates	24
Heat transfer surface per plate	0.4 m^2
Heat transfer surface in second section	9.6 m^2
Number of passes	3
Number of streams per pass	4
Chilled water flow rate	6.5 kg/s

A shell-and-tube or plate-type cooler with two stages of cooling can cool the wort efficiently. In the first (hot) stage, potable water is used counterflow to the wort, and the usual discharge temperature is about 76°C. This hot water is then used in the following brews at various blended temperatures. Excess is used in the brewery's general operations.

The second stage of wort cooling is accomplished at about 2°C by a closed system of refrigerated water through a closed cooler, which cools the wort to 10°C or lower, depending on the brewer. Lower temperature water (1°C) may be used in open units where no danger of freezing exists.

Wort cooling may be accomplished in one stage, depending on the potable water temperature available and plant refrigeration capacity. This might be a somewhat simpler and less expensive arrangement, but sewerage cost may be an important factor where water saving is mandatory. These heat exchangers have many benefits, including easier cleaning, uniform temperature control, uniform control of the sterile air injected for aeration, reduction of overall steam requirements for brewing, and minimal water wasted. The water consumption has been reduced from 17 to 20 m³ per cubic metre of beer to 6 to 10 m³ per cubic metre of beer.

Fermenting Cellar

After cooling, the wort is pitched with yeast and collected in a fermenting tank where respiration and fermentation occur according to the chemical reaction previously discussed. The daily rate of fermentation varies depending on the operating procedure adopted in each plant. On the first day, a representative rate might be 8 kg of converted maltose per cubic metre of wort. The rise in temperature caused by fermentation and by the growth and changing physiology of the yeast increases this rate to 27 kg/m³ on the second day. By now, the maximum desired temperature has been attained, and a further rise is checked by an attemperator, so that on the third day another 27 kg/m³ is converted. This rate continues through the fourth day. Two examples of the fermentation rate follow; one is for normal gravity brewing, and the other is for high (heavy) gravity brewing.

Example 1 Normal Gravity Brewing

Fermentation Day	°Plato	Real Extract	Mass of Extract, kg/m³	Extract Fermented, kg/m³
0	11	11.0	114.5	—
1	10	10.2	105.8	8.70
2	8	8.6	88.7	17.12
3	5	6.1	62.3	26.40
4	3	4.5	45.7	16.66
5	2.5	4.1	41.6	4.10
				72.98

Example 2 High Gravity Brewing

Fermentation Day	°Plato	Real Extract	Mass of Extract, kg/m³	Extract Fermented, kg/m³
0	16	16.0	170.0	—
1	15	15.2	161.0	9.01
2	12	12.8	134.2	26.71
3	7	8.7	89.3	44.49
4	4	6.3	64.4	25.36
5	3.5	5.9	60.2	4.17
				109.74

By now, the amount of unconverted maltose remaining in the beer is greatly diminished. Because alcohol, carbon dioxide, and other products of fermentation inhibit further yeast propagation, the action nearly stops on the fifth day when only about 12 kg/m³ are converted. At this stage the yeast begins to flocculate (clump together) and either settles to the bottom of the fermenter (bottom yeast) or rises to the top (top yeast). Because of the reduced fermentation rate, the temperature of the beer begins to fall, either as the result of increased attemperation applied to the tank itself, heat loss from the tank to the surrounding area, or a combination of both. Many fermentation programs call for the beer to be cooled to a temperature ranging from 1 to 8°C at this time. This period of more rapid cooling aids in settling the yeast. At the completion of this cooling period, the fermentation rate is essentially zero, and the beer is ready to be transferred off the settled yeast. Complete fermentation generally occurs in about 7 days. The introduction of new types of beers (i.e., reduced calorie, reduced alcohol) and the more general use of high gravity brewing have led to the use of a variety of fermentation programs both between brewers making the same product and within the same brewery for different products.

While complete fermentation can be accomplished in less than 7 days, most brewers take 7 to 14 days or more for the fermentation and subsequent cooling. The length of the process depends on original gravity, whether a secondary fermentation is used, and available cooling capacity. Most brewers cool the beer to between 7 and 3°C after ending fermentation or after the final days of quick cooldown in the fermenting tank. In addition, the long rest allows time for the yeast to settle. Some brewers agitate the beer in cylindrical fermenters, which enables them to ferment the beer faster and then to separate out the yeast by centrifuge. Most brewers cool the beer to the desired −2 to 7°C temperature before it goes into storage for resting and settling between fermentation and final aging.

Fermenting Cellar Refrigeration

The agitation necessary for heat exchange between the attemperator and the beer is provided partly by convection resulting from temperature gradients in the beer. Agitation is principally by the ebullition caused by the carbon dioxide (CO_2) bubbles rising to the surface of the liquid. In estimating the heat transfer surface required, a heat transfer rate range from 85 to 170 W/(m²·K) is reasonable. The heat loss from tank walls and the surface of the liquid may be disregarded when calculating the attemperator coil surface requirements. However, if the room temperature is allowed to drop appreciably below 10°C, the heat dissipated through the metal tank walls becomes important. Depending on the degree of heat dissipation, fermentation may be retarded or even inhibited. In such instances, wooden fermenting tanks or insulated tank walls and bottoms are indicated so that the control of heat removal remains in the attemperator.

Refrigeration requirements are based on the maximum volume of wort being fermented, as illustrated by Example 3.

Example 3. Figure 2 illustrates the volume of wort production based on a 60 m³/day production rate. The days are represented by the abscissa, and the kilograms of solids converted per cubic metre of beer, by the ordinate. The individual brews in fermentation on any particular day are additive. For example, on the fifth day, Brew No. 1 is finishing with a conversion rate of 12 kg/m³ for that day; Brew No. 5, which is just beginning the fermentation cycle, is fermenting at the rate of 8 kg/m³; and Brews No. 2, 3, and 4 are each at the maximum rate of 28 kg/m³ per day. The total solids fermented on this day is 104 kg/m³ for the 300 m³ in fermentation, totaling 6240 kg of solids converted per day (0.0722 kg/s). Since the heat of fermentation is 650 kJ/kg, the refrigeration load is

$$0.0722 \times 650 = 46.9 \text{ kW}$$

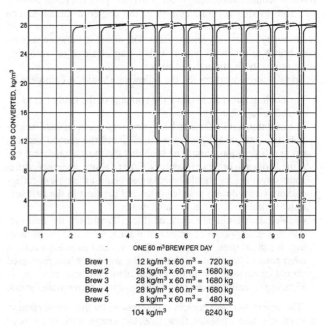

Brew 1	12 kg/m³ × 60 m³ =	720 kg
Brew 2	28 kg/m³ × 60 m³ =	1680 kg
Brew 3	28 kg/m³ × 60 m³ =	1680 kg
Brew 4	28 kg/m³ × 60 m³ =	1680 kg
Brew 5	8 kg/m³ × 60 m³ =	480 kg
	104 kg/m³	6240 kg

Fig. 2 Solids Conversion Rate

Beverages

Calculations for sizing attemperators must consider (1) the internal dimensions of the fermenting tank and its capacity; (2) the temperature difference between the coolant and the fermenting beer; (3) the maximum daily sugar conversion rate; and (4) the heat evolved, which is at the rate of 650 kJ/kg of fermentable sugar converted.

Assuming a square fermenting tank 4 m per side to hold a brew of 60 m^3 and allowing 0.25 m between the tank wall and the attemperator for easy cleaning, a 3.5 m square attemperator can be used, giving 14 m of tubing.

From Figure 2, the maximum daily conversion rate is shown to be 28 kg/m^3. Calculating for 60 m^3 per day at 650 kJ/kg of sugar converted

$$28 \times 60 \times 650/(24 \times 3600) = 12.6 \text{ kW}$$

Assuming a 10°C fermenting beer temperature and a −7°C brine temperature, a temperature difference of 17 K, and a heat transfer rate of 85 W/($m^2 \cdot$K) for the attemperator, the surface area required would be:

$$12.6 \times 1000/(85 \times 17) = 8.72 \text{ m}^2$$

Considering 100-mm OD tubing with an external area of 0.314 m^2/m, the length required would be:

$$8.72/0.314 = 27.8 \text{ m}$$

Two attemperators (each 3.5 m square) would give 28 m of tubing, which is adequate for the conditions outlined.

The amount of CO_2 generated per litre of beer fermented depends on the original gravity of the beer and the degree of fermentation. A little more than half of the generated CO_2 is collected and liquefied for later carbonating use in the brewery; for counterpressure in transfer operations; and for the bottle, can, and keg lines.

Adequate fresh air must be provided in an open-tank fermenting cellar to safely dilute the CO_2 emanating from the fermenters. The amount permissible should be such that there is no health hazard to the employees working in these spaces. Concentrations below 0.5% are considered safe. Increasing amounts reduce a worker's efficiency, and 4% concentrations make protracted work periods untenable. Since it is heavier than air, CO_2 tends to settle to the floor, from where it may be withdrawn by scupper connections at the floor level. Fresh air may be introduced through the ceiling, the openings being located so as not to remove or disturb the layer of protective CO_2 gas on top of the fermenting beer.

Old attemperators usually consisted of one or more rings of 75 or 100 mm copper or stainless tubing, concentric with the walls of the tank and supported at about two-thirds of the height of the liquid. Later designs consist of a partial exterior jacket at about the same height as the old coils. The jacket is a multipass of three or more baffled passes, or a dimple plate design, providing good flow and heat transfer. A glycol solution or liquid ammonia may be circulated through the cooling jackets. These tank changes, dictated by automation and economics, allow easier in-place cleaning of tanks and provide more cooling effect in fermenting.

Stock Cellar

The stock cellar is a refrigerated room containing tanks into which the cooled beer from the fermenter is transferred for the purpose of aging and settling. The period of beer retention in the stock cellar is the brewer's discretion and may be one week to several months. The beer is stored in the presence of some yeast remaining from the original fermentation. Under these conditions, slow, subtle, but important changes take place and contribute to the flavor characteristic of the beer, including the coagulation of protein, which might produce haze in the finished product.

While the CO_2 pollution of air in the stock cellar is far less than in the fermenting room, adequate provision must be made to supply fresh air in sufficient amounts to keep the CO_2 concentration below 0.5%. Air-conditioning equipment, using chemical dehumidification and refrigeration, is generally used to maintain dry conditions

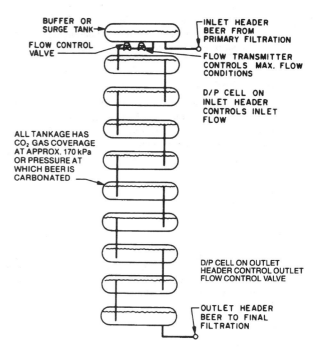

Fig. 3 Continuous Aging Gravity Flow

such as 0°C and 50% rh in storage areas. This decreases mold growth and rusting of steel girders and other steel structures. To maintain lower CO_2 concentration in tightly closed cellars and to reduce operational cost, heat exchange sinks and thermal wheels are used to cool incoming fresh air and to exhaust cold stale air.

Air compressor systems commonly use air driers with refrigerated aftercoolers, 0°C glycol coolers, desiccant drying, or a combination. This is necessary if lines pass through areas below 0°C.

A continuous aging process used in multistory buildings, all gravity flow, is shown in Figure 3. The process is better for larger operations that principally produce one brand of beer.

Kraeusen Cellar

Instead of carbonating the beer during the finishing step, some brewers prefer to carbonate by the Kraeusen method. In this procedure, the fully fermented beer is moved from the fermenting tank to a tank capable of holding about 140 kPa (gage). A small percentage of actively fermenting beer is added. The tank is allowed to vent freely for 24 to 48 h, then it is closed and the CO_2 pressure is allowed to build. Since the amount of CO_2 retained in the beer is a function of temperature and pressure, the brewer can achieve the desired carbonation level by controlling either or both pressure and temperature. At the conclusion of the Kraeusen fermentation, generally a week or more, the beer may be moved to another storage tank. However, the brewer can accomplish the same effect by leaving the beer in the Kraeusen tank and cooling the beer either by space cooling, tank coils, or both.

Heat is generated by this secondary fermentation, but the temperature of the liquid does not rise as high as it did in the fermenter because the fermentable sugars are only available from the small percentage of actively fermenting beer, added as Kraeusen. Furthermore, the bulk of the liquid may have a lower starting temperature than in primary fermentation. Typically, a temperature of 4 to 10°C may be reached at the peak, after which the liquid cools to the ambient temperature of the room. This cooling can be accelerated in the tanks by means of attemperators through which a cooling liquid, such as propylene glycol, is circulated. Since heat is generated during the Kraeusen fermentation, refrigeration load calculations must include removal of this heat by transfer to the air in the cellar, by means of tank coils, or by a combination of both. Furthermore, if the tank is to be used as a storage tank, the calculation must include the necessary heat removal to reduce the beer temperature to the desired level.

Finishing Operations

After flavor maturation and clarification in the storage tanks, the beer is ready for finishing. Finishing includes the processes of carbonation, stabilization, standardization, and clarification.

Carbonation. Any of the following processes are used to raise the CO_2 concentration from 1.2 to 1.7 volumes/volume to about 2.7 volumes/volume:

- Kraeusen
- In-line
- In-tank with stones
- Saturator
- Aging train

Stabilization. The formation of colloidal haze, caused by soluble proteins and tannins forming insoluble protein-tannin complexes, is reduced by any of the following materials:

- Enzymes (papain)
- Tannic acid
- Tannin absorbents
- Protein absorbents, silica gel, bentonite

Standardization. Chilled, deaerated, and carbonated water are added to adjust original density from high-density level (14 to 16°Brix) down to normal package levels (10 to 12°Brix or lower for low-calorie beer).

Clarification. In the finishing cellar, the beer is polished by filtration and is then carbonated by any of several methods. The filtering is usually done through a series of cellulose pulp filters and/or diatomaceous earth filters. The number of filters used depends on the brilliance desired in the finished product. After this final processing, beer is transferred to the government cellar and held until it is needed for filling kegs in the racking room or bottles or cans in the packaging plant. In some breweries, initial clarification is accomplished using centrifuges. This reduces the load on the filtration system, allowing higher flow rates and longer filter runs.

Outdoor Storage Tanks

Some breweries are using vertical outdoor fermenting and holding tanks (similar to those popular with dairies). These tanks have working capacities of 250 to 1200 m^3. The geometry of these tanks includes a conical bottom and height-to-diameter ratios from 1:1 to 5:1. The tanks are jacketed and use propylene glycol or the direct expansion of ammonia for cooling. Insulation is usually100 to 150 mm thick polyurethane foam with an aluminum cladding. They may be built as fermenters, as aging tanks, or in many cases, fermenting and aging are completed in the same tank with no beer transfer.

Hop Storage

Hops should be stored at a temperature of 0 to 1°C with 55 to 65% rh and very little air motion to prevent excessive drying. Sweating of the bales should not be permitted because this would carry off the light aromatic esters and deteriorate the fine hop character. Nothing else should be stored in the hops cellar as foreign odors may be absorbed by the hops, which results in off-flavors in the beer.

Yeast Culture Room

In the yeast culture room, yeast is propagated to be used in reseeding and replacing yeast that has lost its viability. Normal fermentation of aerated wort also propagates yeast. The amount of yeast will roughly triple during fermentation depending on the degree of aeration. A portion of this yeast is repitched (reused) in later fermentation, and the balance is discarded as waste yeast which is sometimes sold for other purposes. Clean yeast, usually the middle layer of the yeast deposit that remains in a fermenting tank after removal of the beer, is selected for repitching.

Repitched yeast is carefully handled to avoid contamination with bacteria and is stored in the yeast room as a liquid slurry (yeast balm) in suitable vats. If open vats are used, 80% rh is required to prevent the yeast from hardening on the vat walls. The CO_2 blanket on top of the vats should not be disturbed by excessive air motion. There is considerable variation in yeast handling and recycling practices.

PASTEURIZATION

Plate pasteurizers heat the beer to a temperature sufficient for proper pasteurization (15 s at 71°C or 10 s at 74°C) and then cool the pasteurized product with incoming cold beer. Plate pasteurizers and microfiltration are used to produce a beer that is similar to draft beer but does not require refrigeration to prevent spoilage. It is distributed in bottles and cans that can be of slightly lighter construction since they do not have to withstand the high pressure created in tunnel pasteurizers.

CARBON DIOXIDE

Collection

Carbon dioxide gas, produced as a by-product of fermentation, can be collected from closed fermenters, compressed, and stored in pressure tanks for later use. It may be used for final carbonation, counterpressure in storage and finishing tanks, transfer, and the bottling and canning of the product. In the past, the CO_2 was stored in the gaseous state at about 1.8 MPa. However, in most medium and large breweries, the gas is collected and, after thorough washing and purification, it is liquefied and stored. Carbon dioxide stored in the liquid state would occupy about 2% of the volume of an equal mass of gas at the same pressure at room temperature.

As an example, from each cubic metre of wort fermented, about 50 kg of CO_2 are generated over a period of five days, though not at a constant daily rate. Therefore, brews must be carefully scheduled to provide the necessary CO_2 gas, thereby minimizing storage requirements. As a general rule, only about 50 to 60% of the total gas generated is collected. The gas generated at the beginning and end of the fermentation cycle is discarded because of excessive air content and other impurities.

From the fermenting tank, the gas is piped through a foam trap to a gas pressure booster. Surplus gas is discharged to the outside from a water-column safety relief tank, which also protects the fermenting tank from excessive gas pressure. To compensate for friction pressure loss in the long lines to the compressor and to increase its capacity, the booster raises the pressure from as low as 0.25 to 30 to 40 kPa (gage).

Compressors, which in the past were of the two-stage type with water injection, are being replaced by nonlubricated compressors that use carbon or nonstick fluorocarbon rings. Today, lubricated screw and reciprocating compressors are used for food and beverage-grade CO_2 production in commercial CO_2 plants. These may be single two-stage compressors or two compressors comprising individual high and low stages. A complete collection system consists of suction and foam trap; rotary boosters, where required; scrubber; deodorizer; compressor (or compressors); intercoolers and aftercoolers; dehumidifying tower; condenser (with refrigeration from plant cascade system or separate compressor); liquid storage tanks; and vaporizers, all interconnected and automatically controlled.

Beverages

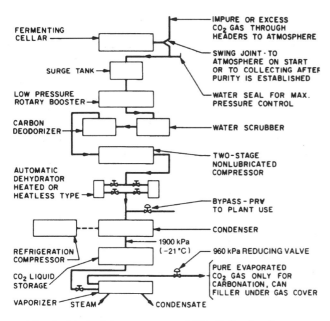

Fig. 4 Typical Arrangement of CO_2 Collecting System

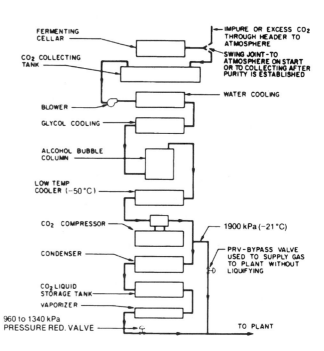

Fig. 5 Special CO_2 Collecting and Liquefaction System

Liquefaction

The condensing pressures of carbon dioxide at several temperatures are:

−30°C	1.43 MPa
−25°C	1.68 MPa
−20°C	1.97 MPa

The latent heat at saturation temperature is about 280 kJ/kg. The refrigerant for liquefying the compressed gas should be about −30°C to effectively condense the CO_2. Most of the moisture must be removed from the compressed gas; this may be done by passing the gas through a horizontal-flow finned coil (located in a 2°C cellar) which condenses out about 80% of the moisture, i.e., the condensate is drained from the system. Also, this is done effectively with refrigerant-cooled precoolers, intercoolers, and aftercoolers. Further removal is effected by passing the gas through desiccant driers. The emerging gas has a slightly higher temperature, but has a dew point around −57°C or lower.

Under these conditions, the gas is liquefied when it comes in contact with the liquefying surfaces, which stay ice-free because of the low moisture content (−40°C dew point) of the gas, thus assuring continuous service. The driers are installed in duplicate with automatic timing for regeneration of the desiccant material. Desiccant driers usually rely on liquid CO_2 as the refrigeration gas. An earlier method used dual sets of double-pipe driers which froze out moisture and retained it in the heat exchanger.

Liquefiers are vertical shell-and-tube, inclined double-pipe, or shell-and-tube types. The refrigerant side is operated fully flooded, with the refrigerant being supplied from the main plant and a booster compressor discharging into the plant suction main. Carbonating systems have undergone changes with all-closed fermenters, refrigerated condensing systems, and large liquid CO_2 holding tanks. See Figures 4 and 5 for collecting and liquefaction system flow diagrams.

CO_2 Storage and Reevaporation

The condensed CO_2 drains into a storage tank, which is usually designed for a working pressure of 2.1 MPa and varying storage capacities of 4.5 to 54 Mg each. The vessel is enclosed in an insulated box and is equipped with equalizing connections, safety valves, liquid-level indicators, and electric heating units. Gas purity tests are regularly conducted from samples withdrawn from above the liquid level.

As liquid is withdrawn from the tank, it is introduced to a steam-heated liquid evaporator, which is automatically controlled to give the desired superheat to the reevaporated gas. This type of storage tank is gradually replacing other types because of its ability to control the temperature of the reevaporated gas closely. Then the evaporated gas is stored temporarily in buffer tanks until it is needed in the brewhouse and packaging plant.

HEAT BALANCE

Most of the steam required for processing, heating water, and general plant heating can be obtained as a by-product. Because the manufacture of beer is a batch process with various peaks occurring at different times, the study of the best heat balance possible is difficult. In a given plant, it depends on many variables, and a comprehensive study of all factors is necessary.

In brewery plant locations where electric energy costs are high, the installation of cogeneration facilities can be favorable. However, in plants that produce in excess of 350 000 m³ annually, the steam turbine as a prime mover often comes into prominence. A bleeder-type steam turbine operating at 2.8 MPa can be used to drive a refrigeration compressor, electrical generator, or both; steam bled from it can be used for process and other needs requiring low-pressure steam. In smaller plants, less favorable heat balances must be accepted in line with more economical plant investment programs. Each brewery requires individual study to procure the most economical program.

COMMON REFRIGERATION SYSTEMS

Absorption Machine for Heat Balance (especially for air conditioning and water cooling for wort cooling). The unit requires a careful heat balance study to determine if it is economical.

Halocarbon Refrigerant Cascade System. Oil-free ammonia as brine can be pumped at a 4:1 ratio. The system requires pumps, but more often is expanded on a 1:1 ratio if connected with a compression system. The ratio of motive power to refrigeration at 170 and 1270 kPa (gage) is 0.32 kW/kW.

Direct Centrifugal. This is often an oil-free ammonia system with a probable 1:1 ratio, or it can be recirculated. The unit requires pumps or a pressure transfer system.

Oil-Sealed Screw Compressors. Ammonia is circulated at a 1:1 ratio in this system. The units require pumps or a pressure transfer system. The ratio of motive power to refrigeration is 0.29 kW/kW.

Oil-Free Compressors, Screw Type. The ratio of motive power to refrigeration is 0.32 kW/kW.

Large Balanced Opposed Horizontal Double-Acting Reciprocating Compressor. While the system is not oil-free, good oil separation equipment minimizes this problem. It can use recirculation or direct expansion. The ratio of motive power to refrigeration is 0.26 kW/kW.

Further automation has been accomplished by programming the flow of materials in the brewhouse, as well as the entire brewing operation. The newest brewing operations are fully automated.

Where necessary, a cooling tower may be used to reduce thermal pollution or to conserve water in the pasteurizing phase. Ecology plays an important part in the brewery; stacks are monitored for particulates, effluent is checked, and heat from kettle vents and others is recovered. Water usage is more closely regulated, and water-saving equipment, including evaporative condensers, is used in the refrigeration systems.

VINEGAR PRODUCTION

Vinegar is produced from any liquid capable of first being converted to alcohol (such as wine, cider, and malt) and syrups, glucose, molasses, and the like.

First Stage: Conversion of sugar to alcohol by yeast (anaerobic)

$$C_6H_{12}O_6 \rightarrow 2CH_3CH_2OH + 2CO_2$$

Second Stage: Conversion of alcohol to acetic acid by bacterial action (aerobic)

$$CH_3CH_2HO + O_2 \rightarrow CH_3COOH + H_2O$$

Bacteria are active only at the surface of the liquid where air is available. Two methods are used to increase the air-to-vinegar surface:

The **packed generator** or Frings generator is a vertical cylinder with a perforated plate and is filled with oak shavings or other inert support material intended to increase column surface area. The weak alcohol and the vinegar culture are introduced, and the solution is continuously circulated through a sparger arm, with air being introduced through drilled holes in the top of the tank. A heat exchanger is used to remove the heat generated and to maintain the solution at 30°C. This is a batch process requiring 72 h.

In the **submerged fermentation process**, the air is distributed to the bacteria by continuously disbursing air bubbles through the mash in a tank that is practically filled with cooling coils to maintain the 30°C temperature. This also is a batch process requiring 39 h.

Concentration is best accomplished by removing some of the water in the form of ice, which achieves a 12 to 40% increase in the concentration of the acid. In freezing out the water, a rotator is often used. About −18 to −12°C is required on the evaporating surface to produce the best crystals; the ice is separated in a centrifuge. The vinegar is then stored 30 days before filtering. Effective concentration can also be achieved by distillation (as is done for distilled white vinegar).

WINE MAKING

The use of refrigeration to control the rates of various physical, chemical, enzymatic, and microbiological reactions in commercial wine making is well established. Periods at elevated temperatures, followed by rapid cooling, can be used to denature oxidative enzymes and proteins in grape juices, to retain desirable volatile constituents of grapes, to enhance the extraction of color pigments from the skins of red grapes, to modify the aroma of the juices from certain white grape cultivars, and to inactivate the fungal populations of mold-infected grapes. Reduced temperatures can slow the growth rate of natural yeast and of the enzymatic oxidation of certain phenolic compounds, assist in the natural settling of grape solids in juices, and favor the formation of certain by-products during fermentation. Also, reduced temperatures can be used to enhance the nucleation and crystallization of potassium bitartrate from wines, to slow the rate of aging reactions during storage, and to promote the precipitation of wood extractives of limited solubility from aged brandies.

The extent to which refrigeration is used in these applications depends on such factors as the climatic region in which the grapes have been grown, the grape cultivars used, the physical condition of the fruit at harvest, the styles and types of wines being produced, and the discretion of the winemaker.

Presently, the wine industry in the United States is heavily committed to the production of table wines (ethanol content less than 14% by volume). Considerably less emphasis is being placed on the production of dessert wines and brandies than in the past. Additionally, the recent growth in wine cooler popularity has significantly altered winery operations where they are produced. A variety of enological practices and winery equipment can be found between the batch emphasis of small wineries (crushing tens of tonnes per season) and the continuous emphasis of large wineries (crushing hundreds or thousands of tonnes per season).

The applications of refrigeration will be classified and considered in the following order:

1. Must cooling
2. Heat treatment of red musts
3. Juice cooling
4. Heat treatment of juices
5. Control of fermentation temperature
6. Potassium bitartrate crystallization
7. Control of storage temperatures
8. Chill-proofing of brandies

MUST COOLING

Must cooling is the cooling of crushed grapes prior to the separation of the juice from the skins and seeds. White wine grape musts will often be cooled prior to being introduced to a juice-draining system or a skin-contacting tank; this is done to reduce the rate of oxidation of certain juice components, as well as to prevent the onset of spontaneous fermentation by wild and potentially undesirable organisms. Must cooling can be used when grapes are delivered to the winery at excessively high temperatures or when they have been heated to aid in pressing or extracting red color pigments.

In general, tube-in-tube or spiral heat exchangers are used for this application. Tubes of at least 100 mm internal diameter with detachable end sections of large-radius return bends are necessary to reduce the possibility of blockage by any stems that might be left in the must after the crushing-destemming operation. The cooling medium can be chilled water, a glycol solution, or a directly expanding refrigerant. Overall heat transfer coefficients for must cooling range between 400 and 700 W/(m²·K), depending on the proportions of juice and skins, with the must side providing the controlling resistance. In small wineries, jacketed draining tanks and fermenters are often used to cool musts in a relatively inefficient batch procedure in which the overall coefficients are on the order of 10 to 30 W/(m²·K) because the must is stationary and, therefore, rate controlling.

to three weeks. The crystallization at these temperatures can be increased dramatically by the introduction of nuclei, either potassium bitartrate powder or other neutral particles, and subsequent agitation. In modern wineries, it is particularly important to supply nuclei for crystallization, since unlike older wooden cooperage, stainless steel tanks do not offer convenient sites for rapid growth. The holding times can be reduced to 1 to 4 h by these methods. Several continuous and semicontinuous processes have been developed, most incorporating an interchange of the cold exit stream with the warmer incoming wine (Riese and Boulton 1980). Dessert wines can be stabilized in the same manner, except that the congealing temperatures are usually in the range of −11 to −14°C. It is usual for the stabilization to take place some time after the harvest period and for the suction temperatures of the refrigeration compressors to be adjusted in favor of low coolant temperatures rather than refrigeration capacity.

STORAGE TEMPERATURE CONTROL

The control of storage temperature is perhaps the most important aspect of the postfermentation handling of wines, particularly generic white wines. The transfer of wine from a fermenter to a storage vessel generally results in at least a partial saturation with oxygen. The rates at which oxidative browning reactions (and the associated development of acetaldehyde) advance depend on the wine, its pH and free sulfur dioxide level, and its storage temperature. Berg and Akiyoshi (1956) indicate that in the oxidation of white wines, for temperatures below ambient, the rate was reduced to one-fifth its value for each 10 K reduction in temperature. Similar studies of the hydrolysis of carboxylic esters (Ramey and Ough 1980) produced during low-temperature fermentations indicate that the rate was more than halved for each 10 K reduction in temperature. These latter data suggest that on average, the esters have half-lives of 380, 600, and 940 days when wine is stored at 15, 10, and 5°C, respectively. As a result, wines should generally be stored at temperatures between 5 and 10°C if oxidation and ester hydrolysis are to be reduced to acceptable levels. The importance of cold bulk wine storage will likely increase as vintners strive to reduce the amount of sulfur dioxide used to control undesirable yeasts.

The cooling requirements during storage are easily calculated using the vessels' dimensions and construction materials as well as the thickness of the insulation used.

CHILL-PROOFING BRANDIES

In the production of brandies, refrigeration is used in the chill-proofing step, just prior to bottling. When the proof is in the range of 100 to 120 proof (50 to 60% v/v ethanol), aged brandies will contain polysaccharide fractions extracted from the wood of the barrels. When the proof is reduced to 80 proof for bottling, some of these components with limited solubility will become unstable and precipitate, often as a dispersed haze. These components are removed by rapidly chilling the diluted brandy to a temperature in the −18 to 0°C range and filtering while cold. This is generally done by using a plate heat exchanger and a pad filter. The outgoing filtered brandy is then used to precool the incoming stream, thus reducing the cooling load. Calculations can be made by using the properties of equivalent ethanol-water mixtures, with particular attention to the viscosity effects on the heat transfer coefficient.

CARBONATED BEVERAGES

Refrigeration equipment is used in many carbonated beverage plants. The refrigeration load varies with plant and production conditions; small plants may use 500 kW of refrigeration and large plants may require over 5000 kW.

Dependency on refrigeration equipment has diminished in carbonated beverage plants using modern deaerating, carbonating, and high-speed beverage container-filling equipment. In facilities that use refrigeration, the product water is often deaerated before cooling to aid carbonation. In addition, cooling the product at this stage of production (1) facilitates carbonation to obtain maximum stability of the carbonated beverage during filling (reduces foaming), (2) permits reducing the pressure at which the beverage is filled into the container (minimizing glass bottle breakage at filler), and (3) reduces overall filling equipment size and investment.

Immediately before filling operations, beverage product preparation requires the use of equipment for proportioning, mixing, and carbonating so that the finished beverage has the proper release of carbon dioxide gas when it is served. The equipment for these functions is frequently found as an integrated apparatus (device), often called a mixer-carbonator or a proportionator.

Table 2 lists the volume of carbon dioxide dissolved per volume of water at various temperatures. Table 2 indicates that at 15.6°C and atmospheric pressure, a given volume of product water will absorb an equal volume of carbon dioxide gas. If the carbon dioxide gas is supplied to the product water under a pressure of approximately 205 kPa (absolute), it will absorb two volumes. For each additional 100 kPa, one additional volume of gas is absorbed by the water. Reducing the temperature of the product water to 0°C increases the absorption rate to 1.7 volumes. Therefore, at 0°C product temperature, each increase of 100 kPa in CO_2 pressure results in the absorption of an additional 1.7 volumes instead of one volume as when the product water temperature is 15.6°C. Carbonated levels for different products vary from less than 2.0 volumes to around 5.0 volumes.

BEVERAGE AND WATER COOLERS

The main sanitation requirements for beverage and/or water coolers are hygiene and ease of cleaning, particularly if the beverage is cooled rather than the water. The key point is that water freezes easily, and the design of cooler equipment needs to avoid this. The early Baudelot tank solved the problem by not forming ice; however, since sanitation of such systems is a problem, their use is not recommended.

If a cooler is needed, most plants choose plate heat exchangers and careful control of temperature. Plate heat exchangers reduce ice formation through high turbulence, which reduces thermal gradients. Furthermore, they are hygienic and easy to clean. Such heat

Table 2 Volume of CO_2 Gas Absorbed in One Volume of Water

Temperature °C	Pressure in Bottle, kPa (absolute)										
	101.3	170.3	239.2	308.2	377.1	446.1	515.0	584.0	652.9	721.9	790.8
0	1.71	2.9	4.0	5.2	6.3	7.4	8.6	9.7	10.9	12.2	13.4
4.4	1.45	2.4	3.4	4.3	5.3	6.3	7.3	8.3	9.2	10.3	11.3
10.0	1.19	2.0	2.8	3.6	4.4	5.2	6.0	6.8	7.6	8.5	9.5
15.6	1.00	1.7	2.3	3.0	3.7	4.3	5.0	5.7	6.3	7.1	7.8
21.1	0.85	1.4	2.0	2.5	3.1	3.7	4.2	4.8	5.4	6.1	6.6
26.7	0.73	1.2	1.7	2.2	2.7	3.2	3.6	4.1	4.6	5.2	5.7
32.2	0.63	1.0	1.5	1.9	2.3	2.7	3.2	3.6	4.0	4.5	4.9
37.8	0.56	0.9	1.3	1.7	2.0	2.4	2.8	3.2	3.5	3.9	4.3

HEAT TREATMENT OF RED MUSTS

Most red grapes have white or greenish-white flesh (pulp) and juice. The coloring matter or pigments (anthocyanins) reside in the skins. Color can be rapidly extracted from these varieties by heat treating the musts so that the pigment-containing cells are disrupted prior to actual fermentation. This is done in several countries throughout the world when grape skins are low in color or when color extraction during fermentation is poor. Heating the must is necessary to produce the desirable flavor profile of some varieties, most notably Concord, when manufacturing juices, jellies, and wines. The series of operations is referred to as thermovinification. The must is heated to temperatures in the range 55 to 75°C, generally by draining off some of the juice, condensing steam, and returning the hot juice to the skins for a given contact time, often 30 min. The complete must can be cooled prior to separation and pressing, or the colored juice can be drawn off and cooled prior to the fermentation.

Heat treatment of red grapes can also be used to inactivate the more active oxidative enzymes found in red grapes infected by the mold *Botrytis cinerea*. It can further aid the action of pectic enzyme preparations added to facilitate the pressing of some cultivars. In all cases, the temperature-time pattern used is a compromise between desirable and undesirable reactions. The two most undesirable reactions are caramelization and accelerated oxidation of the juice. Condensing steam and tube-in-tube exchangers are generally used for these applications, with design coefficients similar to those given previously for must cooling.

JUICE COOLING

Juices separated from the skins of white grapes are usually cooled to between 5 and 20°C to aid in the natural settling of suspended grape solids, to retain volatile components in the juice, and to prepare it for a cool fermentation. Tube-in-tube, shell-and-tube, and spiral exchangers and small jacketed tanks are used either with direct expansion refrigerant, propylene glycol solution, or chilled water as the cooling medium. Overall coefficients for juice cooling range between 550 and 850 $W/(m^2 \cdot K)$ for the exchanger and 25 to 50 $W/(m^2 \cdot K)$ for small jacketed tanks.

The small jacketed tank values can be improved significantly by juice agitation. Transport and thermal properties of 24% (by mass) sucrose solutions can be used for grape juices. There is a general tendency for medium and large wineries to use continuous-flow juice cooling arrangements of jacketed tanks.

HEAT TREATMENT OF JUICES

Juices from sound grapes can be exposed to a high-temperature, short-time (HTST) treatment to denature grape proteins, reduce the number of unwanted microorganisms, and, some winemakers believe, enhance the varietal aroma of certain juices. The denaturation of proteins reduces the need for their removal (by absorptive clays such as bentonite) from the finished wine. However, turbid juices and wines can result from this treatment, presumably because of modification of the pectin and polysaccharide fraction. Clarified juices from mold-infected grapes can be treated in a similar way to denature oxidative enzymes and to inactivate the molds. Most wineries in the United States rely on "pure culture" fermentation to achieve consistently desirable results, hence the value of HTST treatment in the control of microbial populations.

A typical program would include rapidly raising the juice temperature to 90°C, holding it for 2 s, and rapidly cooling it to 15°C. Plate heat exchangers are used for this treatment because of their thin film paths and high overall coefficients. Grape pulp and seeds can cause problems in this equipment if they are not completely removed beforehand.

FERMENTATION TEMPERATURE CONTROL

The anaerobic conversion of grape sugars to ethanol and carbon dioxide by yeast cells is exothermic, although the yeast is capturing a significant quantity of the overall energy change in the form of high-energy phosphate bonds. Experimental values of the heat of reaction range between 83.7 and 100.5 kJ/mole with the value of 99.6 kJ/mole being generally accepted for fermentation calculations (Bouffard 1985).

One litre of juice at 262 kg/m^3 sucrose (24°Brix) will produce approximately 60 L of carbon dioxide during fermentation. Allowing for the enthalpy lost by this gas, with its saturation levels of water and ethanol vapors, the corrected heat release value is 95.9 kJ/mole at 15°C and 88.3 kJ/mole at 25°C. The adiabatic temperature rise of the 262 kg/m^3 juice would then be 48.3°C, based on the 15°C value, and 45.6°C, based on the 25°C value. Whether or not a fermentation will approach these adiabatic conditions depends on the difference between the rate of heat generation by fermentation and the rate of heat removal by the cooling system. For constant temperature fermentations, which are the most common type of temperature control practiced, these rates must be equal. Red wine fermentations are generally controlled at temperatures between 24 and 32°C, while white wines are fermented at 10, 15, 20°C, depending on the cultivar and the type of wine being produced. The more rapid fermentations of red wines are used in the cooling load calculations of individual fermenters; a more involved composite calculation, allowing for both red and white fermentations staggered in time, is necessary for the overall daily fermentation loads.

At 20°C, red wines have average fermentation rates in the range of 40 to 50 kg/m^3 per day, which correspond to heat release rates of approximately 217 to 270 W/m^3 per hour. The peak fermentation rate is generally 1.5 times the average, leading to values of 325 to 405 W/m^3 per hour. This value, multiplied by the volume of must fermenting, provides the maximum rate of heat generation. The heat transfer area of the jacket or external exchanger can then be calculated from the average coolant temperature and the overall heat transfer coefficient.

The largest volume of a fermenter of given proportions, whose fermentation can be controlled by jacket cooling alone, is a function of the maximum fermentation rate and the coolant temperature. The limitation occurs because the volume (and hence the heat generation rate) increases with the diameter cubed, while the jacket area (and hence the cooling rate) only increases with the diameter squared. Similarly, the temperature rise in small fermenters, cooled only by ambient air, depends on the fermenter's volume and shape and the ambient air temperature (Boulton 1979a).

The development of a kinetic model for wine fermentations (Boulton 1979b) has made it possible to predict the daily or hourly cooling requirements of a winery. The many different fermentation temperatures, volumes, and starting times can now be incorporated into algorithms that predict future demands and schedule off-peak electricity usage, allowing for optimal control of refrigeration compressors.

POTASSIUM BITARTRATE CRYSTALLIZATION

Freshly pressed grape juices are usually saturated solutions of potassium bitartrate. The solubility of this salt decreases as alcohol accumulates so that newly fermented grape wines are generally supersaturated. The extent of supersaturation and even the solubility of this salt depends on the type of wine. Young red wines can hold almost twice the potassium content at the same tartaric acid level as young white wines, with other effects due to pH and ethanol concentration. Since salt solubility also decreases with temperature, wines will usually be cold stabilized so that the crystallization occurs in the tank rather than in the bottle if the finished wine is chilled. In the past, this was done by holding the wine close to its congealing temperature (usually −4 to −6°C for table wines) for two

Beverages

exchange devices are normally fed by a brine (not direct refrigeration) and are protected against brine leakage, for example, by ensuring that the brine pressure is lower than the beverage product/water pressure.

In many existing beverage plants, coolers with patented direct-expansion refrigeration equipment are used to achieve system security, hygiene, ease of cleaning, etc., using a Baudelot-type system. However, this equipment is only used for water cooling (not product), making it easier to clean. This is achieved by the equipment manufacturer's long experience with a proprietary system and its attention to detailed equipment design.

When coolers are necessary, it is recommended that this component of the refrigeration system be located adjacent to, or integrated with, the proportioner-mixer-carbonator. Usually these devices are physically positioned next to the beverage container filler. Normally, the refrigeration plant itself should be housed separately from the product processing and filling areas, preferably located together with the other plant utilities (boilers, hot water heater, air compressors, etc.).

It is most important to keep the beverage free from contamination by foreign substances or organisms picked up from the atmosphere or from metals dissolved in transit. Consequently, coolers are designed for ease of cleaning and freedom from water stagnation. The coolers and product water piping are fabricated of corrosion-resistant nontoxic metal (preferably stainless steel); however, certain plastics are usable. For example, acrylonitrile-butadiene-styrene (ABS) is used in the beverage industry for raw water piping.

Refrigeration Plant

Halogenated hydrocarbons or ammonia refrigerants are commonly used for those plants requiring beverage product and/or water coolers. Refrigeration compressors vary from the two-cylinder vertical units to the larger, multicylinder V-style compressors.

The refrigeration plant should be centralized in larger production facilities. With the multiplicity of product sizes, production speeds, and other factors affecting the refrigeration load, an automatically controlled central plant conserves energy, reduces electrical energy costs, and improves opportunities for a preventive maintenance program.

Makeup water and electrical energy costs have encouraged careful selection and use of compressors, air-cooled condensers, evaporative condensers, and cooling towers. Some plants have economized by using the spent water from empty can and bottle rinsers as makeup water for the evaporative condensers and cooling towers.

As indicated earlier, the temperature to which the product must be cooled depends on the type of filling machinery used, as well as the deaerating-mixing-proportioning-carbonating equipment used. These cooling needs may be divided into three general categories: (1) those that use water at supply temperature or less, (2) those that operate with water at 7 to 13°C, and (3) those that require water at 5°C or lower. The exact temperature to which the product should be cooled depends on the specific requirements of the beverage product and the needs of the particular plant. These requirements are primarily based on product preparation, production equipment availability, and capital costs versus operating costs.

Refrigeration Load

The refrigeration load is determined by the amount of water being cooled per unit of time. This is derived from the maximum fluid output of the beverage filler. Most cooling units are of the instantaneous type; they must furnish the desired output of cold water continuously, without relying on storage reserve.

Knowing the water temperature from the supply source, the temperature to which the water is to be cooled, and the water demand, the refrigeration load can be determined by

$$q_R = Q c_p (t_s - t_c) \rho$$

where

q_R = cooling load, kW
Q = water flow rate, m³/s
t_s = supply water temperature, °C
t_c = cold water temperature, °C
c_p = specific heat of water = 4.19 kJ/(kg·K)
ρ = density of water = 1000 kg/m³

In the computation of the refrigeration load, one of the most troublesome values to determine is the highest temperature the incoming supply water can be expected to reach. This temperature usually occurs during the hottest summer period. Moreover, allowance should be made for additional warming of the supply water from its flowing through the piping and water treatment equipment in the beverage plant.

SIZE OF PLANT

The output of each plant depends on the beverage-filling capacity of the plant production equipment. Small, individual filling units turn out approximately 600 cases of 24 beverage containers (approximately 0.25 L capacity) per hour or 240 containers per minute (240 cpm); intermediate units turn out up to 1200 cases per hour (480 cpm); and high-speed, fully automatic machines begin at approximately 1200 cases per hour and go through several increases in size up to the largest units, which approach 5000 cases per hour (2000 cpm).

Operation of these filling machines, which also determines the demand on refrigeration machinery, usually exceeds 8 h per day, especially during summer months when market demands are highest.

An arbitrary classification of beverage plants may be (1) small plants that produce under 1.25 million cases per year, (2) intermediate plants that produce about 2.5 million cases per year, and (3) large plants that produce 15 million or more cases per year. The latter require installation of multiple-filling lines.

The usual distribution area of the finished beverages is within the metropolitan area of the city in which the plant is located. Some plants have built such a reputation for their goods, that they ship to warehouses several hundred kilometres away. Local distribution is made from there. A few nationally known products are shipped long distances from producing plants to specialized markets.

In the warehouse, cans and nonreturnable bottles filled with precooled beverage are commonly warmed to a temperature exceeding the dew-point temperature to prevent condensation and resulting package damage. Bottled goods should be protected against excessive temperature and direct sunlight while in storage and transportation. At the point of consumption, the carbonated beverage is often cooled to temperatures close to 0°C.

LIQUID CARBON DIOXIDE STORAGE

Liquefied carbon dioxide, used to carbonate water and beverages, is truck delivered in bulk to the beverage plant. The liquid is piped from the trucks to large outdoor storage-converter tanks equipped with mechanical refrigeration and electrical heating. The typical tank unit is maintained at internal temperatures not exceeding −18°C, so that the equilibrium pressure of the carbon dioxide does not exceed 2 MPa and the storage tanks need not be built for excessively high pressures. Full-controlled equipment heats or refrigerates, and safety relief valves discharge sufficient carbon dioxide to relieve excess pressure.

REFERENCES

Berg, H.W. and M. Akiyoshi. 1956. Some factors involved in the browning of white wines. *American Journal of Enology* 7:1.

Bouffard, A. 1985. Determination de la chaleur degagée dans la fermentation alcoölique. Comptes rendus hebdomadaires des séances, Academie des Sciences, Paris 121:357. *Progres agricole et viticole* 24:345.

Boulton, R. 1979a. The heat transfer characteristics of wine fermentors. *American Journal of Enology and Viticulture* 30:152.

Boulton, R. 1979b. A kinetic model for the control of wine fermentations. *Biotechnology and Bioengineering Symposium* No. 9:167.

MBAA. 1981. *The practical brewer*, 2nd ed. Master Brewer's Association of the Americas, Madison, WI.

Ramey, D.R. and C.S. Ough. 1980. Volatile ester hydrolysis or formation during storage of model solutions and wine. *Journal of Agriculture and Food Chemistry* 28:928.

Riese, H. and R. Boulton. 1980. Speeding up cold stabilization. *Wines and Vines* 61(11):68.

CHAPTER 26

PROCESSED, PRECOOKED, AND PREPARED FOODS

Main Dishes, Meals	26.1
Vegetables	26.3
Fruits	26.4
Potato Products	26.5
Other Prepared Foods	26.6
Long-Term Storage	26.6

ALTHOUGH there are many categories of processed, precooked and prepared foods, this chapter covers primarily main dishes, meals, fruits, vegetables, and potato production from the perspective of the refrigeration and air-conditioning concerns of facilities producing these categories of food.

To ensure the safety of food products, the Hazard Analysis and Critical Control Point (HACCP) system is now accepted by food manufacturers and regulators. It is a preventive system that builds safety control features into the food product's design and the process by which it is produced. The HACCP system is used to manage physical and chemical hazards as well as biological hazards.

Each food manufacturing site should have a HACCP team to develop and implement its HACCP plan. HVAC&R practioners are a part of that team. Areas of concern include HVAC&R equipment and the associated air distribution system, sanitation equipment and procedures, and processing (automation, heating, cooling, temperature control). Chapter 11 includes additional information on HACCP.

MAIN DISHES, MEALS

Main dishes constitute the largest category of processed, precooked and prepared foods. They are primarily frozen products, but many are also refrigerated. Most of these dishes can be prepared in a conventional or microwave oven. Many contain sauces and/or gravies. Consequently, sauce and/or gravy kitchens are an integral part of these production facilities. A principal characteristic of main dishes is the requirement for a substantial number of ingredients, several unit operations, an assembly-type packaging line, and subsequent cooling or freezing in individual cartons or cases. Examples of these products include:

- Soups and chowders
- Main dishes of meat, poultry, fish, and pasta items
- Complete dinners, each with a main dish usually with sauce and/or gravies, a vegetable, and dessert
- Lunches and breakfasts
- Ethnic main dishes and dinners, particularly Italian, Mexican, and Oriental styles
- Low caloric or diet versions of many of the above
- Snack foods such as pizza, fish sticks, and breaded items

General Plant Characteristics

The plant facilities for preparing, processing, packaging, cooling, freezing, casing, and storing these products vary widely. Since the variety of products is diverse, and it is beyond the scope of this chapter to cover all the operations involved in all the products, product formulations, food chemistry, and process details are not discussed. This chapter also does not cover the basic process details for preparing meat, poultry, and fishery products. For information on these subjects, refer to Chapters 16, 17, and 18, respectively.

A prepared foods plant has the following production areas and/or rooms: receiving areas; storage areas for packaging materials and supplies; storage areas for ambient, refrigerated, and frozen ingredients; rooms and/or areas for thawing and defrosting; refrigerated rooms for in-process storage; areas for mixing, cutting, chopping, and assembly; sauce and/or gravy kitchens; areas for cooking and cooling rice, other starches, and pasta; unit operations for preparing main dishes such as meat patties, ethnic foods, and poultry items; dough manufacturing for pies and pizzas; assembly, filling, and packaging areas; cooling and freezing facilities; casing and palletizing operations; finished goods storage rooms for refrigerated and frozen products; and shipping areas for outbound finished goods by truck or rail.

In addition, these operations have support and utilities areas for equipment and utensil sanitation; personnel facilities; and areas for utilities such as refrigeration, steam, water, wastewater disposal, electric power, natural gas, air, and vacuum.

Plants and equipment for the production of prepared foods should be constructed and operated to provide for minimum bacteriological contamination and ease of cleanup and sanitation. Sound sanitary practices should be adhered to at all stages of production, not only to protect the end user but also to provide wholesome, healthy products. All meat and poultry plants in the United States, and hence most prepared food plants, operate under U.S. Department of Agriculture (USDA) regulations. For detailed information and guidelines, refer to *Agricultural Handbook* No. 570 from the USDA, Food Safety Inspection Service.

Preparation, Processing, Unit Operations

The initial steps in the production of prepared foods involve the preparation, processing, and unit manufacture of items for assembly and filling on the packaging line. These generally include scheduling ingredients; thawing or defrosting frozen ingredients, where applicable; manufacturing sauces and gravies; cooking and cooling rice, other starches, and/or pasta; unit operations for manufacturing meat patties and ethnic foods, such as burritos; mixing vegetables and/or vegetables and rice, other starches, or pasta; manufacturing dough; and cooling, storage, and transport operations prior to packaging.

The above-mentioned processes require refrigeration for controlled tempering in refrigerated rooms; plate heat exchangers or swept surface heat exchangers using chilled water or propylene glycol for preparation of sauces and gravies in processors; chilled water for pasta, other starches, or rice cooling; refrigerated preparation rooms for meat and poultry products; in-line coolers or freezers for meat patties, burritos, etc.; ice for dough manufacture; and cooler rooms for in-process storage and inventory control.

The refrigeration loads for each of these categories should be calculated individually using the methods described in Chapter 12. In addition, the loads should be tabulated by time-of-day and classified by evaporator temperature, chilled water, ice, and propylene glycol

---

The preparation of this chapter is assigned to TC 10.9, Refrigeration Application for Foods and Beverages.

requirements. An assessment should be made whether an ice builder is appropriate for chilled water requirements to reduce refrigeration compressor capacity.

The refrigerated rooms should be amply sized, be capable of maintaining temperatures from 0 to 10°C as required for the specific application, have power-operated doors equipped with infiltration reduction devices, and have evaporators that are easily sanitized. Evaporators for use in rooms where personnel are working should be equipped with gentle airflow to minimize drafts. Rooms with temperature maintained at 3°C or below should have evaporators equipped for automatic coil defrost. Temperature controls for such rooms should be tamperproof.

Proper safeguards and good manufacturing practices during preparation and processing minimize bacteriological contamination and growth. This involves use of clean raw materials, clean water and air, sanitary handling of product throughout, proper temperature control, and thorough sanitizing of all product contact services during clean-up periods. Sauces, gravies, and cooked products must be cooled quickly to prevent conditions favorable for microbial growth. Refer to Chapter 11 for further information on control of microorganisms, cleaning, and sanitation.

Assembly, Filling, and Packaging

These activities cover transporting the components to packaging lines; preparing and depositing doughs for pies, ethnic dishes and pizzas; filling or placing components into containers; placing containers into packages; closing and checking the packages; and transporting the packages to cooling and freezing equipment.

Items that are pumpable, such as gravies and sauces, are usually pumped to a hopper or tank adjacent to the packaging lines. Items such as free-flowing individually quick frozen (IQF) vegetables may be transported to the lines in bulk boxes via lift truck directly from cold storage facilities. For example, mixes of vegetables, rice, other starches, and/or pastas that have been prepared at the plant may be wheeled to the packaging lines in portable tanks or vats. Tempered, prefrozen meat or poultry rolls may be placed on special carts and wheeled to the packaging lines.

A typical packaging line for a meals product consists of a timed conveyor system with equipment for dispensing containers; a filler or fillers for components that can be filled volumetrically; volumetric timed dispensers for sauces, gravies, and desserts; net weigh filler systems for components and mixes that cannot be placed volumetrically; slicing and dispensing apparatus for placing components such as tempered meat or poultry rolls and similarly configured items; line space for personnel to place components that must be placed manually; liquid dispensers for placing spices and flavorings; and a sealing mechanism for placing a sealed plastic sheet over the containers. This system may be two or three compartments wide for high-volume production.

The timed conveyor system is followed by a single filer to align the containers in single file prior to indexing into a cartoner. The containers are automatically inserted into cartons, after which they are coded and sealed closed. Subsequent to the cartoner, the filled cartons are automatically checked for underweights and tramp metal, and then conveyed to cooling or freezing equipment. Other types of packaging lines may include some or all of the above apparatus; they are then usually specific for the prepared food items being filled and packaged.

Previous comments regarding product and equipment safeguards, good manufacturing practices, and actions to minimize microbial growth apply to packaging activities as well.

It is good practice to air condition filling and packaging areas, particularly when these areas are subject to ambient temperatures and humidities that can affect product quality and significantly increase potential bacteriological exposure. In addition, workers are more productive in air conditioned areas. Packaging area air conditioning is usually supplied through air-handling units that also provide ventilation, heat, and positive pressure to the area. Some applications use separate refrigeration systems; others use chilled water or propylene glycol available from systems installed for product chilling. Generally, no other significant refrigeration is required for the filling and packaging areas.

Cooling, Freezing, Casing

The cartons from packaging are cooled or frozen in different ways depending on package sizes and shapes, speed of production, cooling or freezing time, inlet temperatures, plant configuration, available refrigeration systems, and labor costs relative to production requirements. Refer to Chapter 15 for additional information.

In small plants, stationary air-blast or push-through trolley freezers are used when flexibility is required for a variety of products. In these cases, fully mechanized in-line freezers are not economically justified. In larger plants where production rates are high, mechanized freezers such as automatic plate freezers, carrier freezers, and spiral belt freezers are used extensively. These freezers significantly reduce labor costs and provide for in-line freezing.

The prepared foods business has many line extension additions and changes. Each product change results in component differences that may change the freezing load and/or freezing time due to inlet temperatures, latent heat of freezing, and/or package size (particularly depth). Each product should be checked to assure that production rates for the line extension or different products match the current freezing capacity.

Over time, packaging lines tend to become more efficient and capable of higher production output per unit of time due to experience as well as advances in packaging line technology. These advances often occur with existing cooling or freezing capacity, which is relatively constant. This may result in freezer exit temperatures above −20°C if reserve freezing capacity is not available or cannot be physically added due to space limitations. For new or expanded plants, allow space and/or reserve freezer capacity. Increases in packaging line speed and efficiency of 25 to 50% are reasonable to expect. Each case should be individually evaluated.

Casing the cartoned product follows freezing with in-line production of meals and main dishes. Manual casing is used in small plants and/or with low production rates. High-speed production lines use semiautomatic and automatic casing methods to attain high productivity and to lower labor costs. Inspection at this point is necessary to ensure that freezing has been satisfactory and that the cartons are properly sealed and not disfigured. It is also important to ensure that cartons do not become defrosted due to conveyor hangups or line stoppages.

Product palletizing follows product casing. Manual palletizing is used for small plants and/or low production rates, and automatic palletizers are used for higher production rates. Most palletizing is done adjacent to or near cold storage facilities and the pallets are transported to the cold storage rooms via lift truck. Some manual palletizing occurs inside cold storage rooms, particularly for slow production lines, to prevent product warm-up, but it is more costly as labor rates are higher for workers in cold storage rooms.

Some plants are equipped with air-blast freezing cells to augment in-line freezing capacity. These are used primarily for cased products when existing in-line freezers are overloaded and to reduce temperatures of products that have been frozen through the latent zone but have not been sufficiently pulled down for placement in the cold storage rooms. These products are usually loosely stacked on pallets with sufficient air space around the sides of the cases to achieve rapid pull-down to −20°C or lower.

Finished Goods Storage and Shipping

Larger prepared foods operations usually have sufficient cold storage space in which to store the necessary refrigerated and frozen ingredients and at least 72 h of finished goods production. This

Processed, Precooked, and Prepared Foods

volume of space provides for proper inventory control, adequate scheduling of ingredients for production, and sufficient control of finished goods to ensure that the product is −20°C or lower prior to shipment and that it meets the criteria established for product quality and bacteriological counts.

The refrigeration loads for these production warehouses are calculated as suggested in Chapters 12 and 13. Special attention, however, should be given to product pull-down loads and infiltration. Liberal allowances should be made for product pull-down as freezing problems do occur, and some plants are under negative pressure at times during the year because more air is exhausted than supplied through ventilation. This can result in infiltration by direct inflow, which is a serious refrigeration load (see Chapter 12) that should be corrected. This problem is not only costly in energy use, but it also makes it difficult to maintain proper storage temperatures.

Refrigerated trucks and/or rail cars are generally used for shipping. Shipping areas are usually refrigerated to 2 to 7°C, and the truck loading doors are equipped with cushion-closure seals to reduce infiltration of outside air. Refer to Chapter 13 for additional information on refrigerated docks. Where refrigerated docks are not provided, special care should be taken to ensure that the frozen product is rapidly handled to prevent undue warming.

Refrigeration Loads

Refrigeration loads cover a wide range of evaporator temperatures and different types of equipment. Most plants cover these with two or three basic saturated suction temperatures. Where two suction temperatures are provided, they are usually at −37 to −43°C for freezing and cold storage and −12 to −7°C for cooling loads. Where provided, a third suction temperature is usually from −29 to −23°C for frozen product storage rooms and other medium-low temperature loads. The third suction temperature is advantageous with relatively large frozen product storage loads to reduce energy costs.

Refrigeration loads should be tabulated by time of day, season of year, evaporator temperature, and equipment type or function. This record should be made periodically for existing operations; it is essential for new and/or expanded plants. These tabulations reveal the diversity of the loadings and provide guidance for existing operations and for equipment sizing for added capacity or new plants.

In addition, loads should be tabulated for off-shift production, weekends, and holidays to provide proper equipment sizing and for economic operation for these relatively small loads. Refer to Chapter 12 for information on load calculation procedures.

Refrigeration Systems

Most refrigeration systems for prepared foods use ammonia as the refrigerant. Two-stage compression systems are dominant, because compression ratios are high when freezing is involved and energy savings warrant the added expense and complexity. Evaporative condensers are extensively used for condensing the refrigerant. Evaporators are designed for full-flooded or liquid overfeed operation, depending on the equipment or application. Direct-expansion evaporators are not used extensively.

New plant designs limit the exposure of plant employees to large quantities of ammonia. This can be accomplished by locating evaporators, such as propylene glycol chillers, water chillers, and ice builders, in or near machine rooms and away from production employees. Freezers can be located in isolated clusters near machine rooms so that the low-pressure receivers are also in or near the machine room. These practices limit exposure and provide a means for close supervision by trained, competent operators.

Some plants use glycol chillers to circulate propylene glycol to evaporators located in production areas. This results in an energy penalty because of the secondary heat transfer, but it is deemed affordable because of less potential exposure to employees and product in the event of an ammonia spill.

Other plants have a major portion of the ammonia main lines installed on the roof to reduce internal exposure to personnel and product and provide accessibility. These mains require close monitoring and inspection, as an ammonia spill can still result in an injury to internal personnel both inside and outside the plant as well as damage to products.

Machine room equipment should be sized not only for the full refrigeration loads imposed, but it should also be able to handle the relatively small off-shift and weekend loads without using large compressors and components at low capacity levels.

Many plants take advantage of energy savings measures, such as floating head pressure controls with oversized evaporative condensers coupled with two-speed fans, single-stage refrigeration for small areas and loads; variable-speed pumps for glycol chiller systems; ice builders to compensate for peak loads; door infiltration protection devices; added insulation; and computer-based control systems for monitoring and controlling the system. These measures should be considered for existing plants that are not so equipped and should be included in the design of plant expansions and new plants.

All refrigeration systems for prepared food plants should comply with applicable codes and standards (see Chapter 51).

Ventilation

Adequate ventilation and positive air pressure relationships are important in processed and prepared food plants to minimize cross-contamination between raw and cooked products and to prevent condensation. Each plant has to be analyzed separately to provide proper differential pressures and to overcome air and steam exhaust quantities. Many plants are under negative pressure for at least part of the year; this condition should be corrected.

VEGETABLES

Frozen vegetables are prepared foods in that they are essentially precooked and require minimal preparation. Chapters 14 and 23 have further information on handling, cooling, and storing fresh vegetables. Most vegetables to be frozen are received directly from harvest. Some are cooled and stored to smooth out production and others are processed directly. They are cleaned, washed, and graded; cut, trimmed, or chopped (if necessary); and then, blanched, cooled, and inspected prior to freezing. At this point some vegetables are filled or packed into cartons prior to freezing, whereas others are frozen and then filled into packages, bags, cases, or bulk bins.

Products cartoned prior to freezing generally can only be manually packed and/or have to be check weighed prior to closing the cartons. Examples include broccoli spears, asparagus spears, leaf spinach, French cut green beans, okra, cauliflower florets, and some volumetrically machine-filled vegetables that are cartoned prior to freezing such as peas, cut corn, and cut green beans. Products that are cartoned prior to freezing are usually frozen in manual plate freezers, automatic plate freezers, stationary airblast tunnels, and push-through trolley freezers.

Products that are frozen prior to packaging are included in the free-flowing or individually quick frozen (IQF) categories. These include true IQF products such as peas, cut corn, cut green beans, diced carrots, and lima beans, as well as products that are more difficult to IQF such as broccoli florets, cauliflower florets, sliced carrots, sliced squash, and chopped onions.

The easily IQF products are usually frozen in straight belt freezers, fluidized bed freezers, or fluidized belt freezers. The more difficult products to IQF are usually frozen in fluidized bed freezers and fluidized belt freezers. Cryomechanical freezers are sometimes used where high value, sticky products are frozen. Hydrocooling of these products to temperature of 7 to 15°C prior to freezing reduces the freezing load and overall energy requirements.

The products from IQF freezers are packed either directly into cartons or polyethylene bags, into cases for bulk shipment, or into tote bins for repacking at a later date or shipment to other customers for repack or use in prepared foods. The products packed into tote bins for repacking into cartons or polyethylene bags are used for single products, products of various vegetable mixes, and products with butter or cheese sauces. Repacking operations rather than direct in-season packing are preferred for most companies, as they allow the packer to produce directly for orders, which saves buying finished goods packaging material until required.

Products in tote bins for repacking are placed in dumpers in a cold storage room adjacent to the packaging lines. The products are metered onto a conveyor in proportion to the end mix required and pneumatically or mechanically conveyed to the filler hopper for volumetrically filled mixes or single products. Products and product mixes that require weighing are generally conveyed to net weigh filler systems. After packaging, the products are usually cased semiautomatically or automatically and returned to the cold storage immediately. The products are rapidly handled to minimize warm-up and clustering.

Corn-on-the-cob is prepared in the same manner as other vegetables, but because of its bulk and to retain quality, it is usually cooled with refrigerated water. After cooling, the product is either packaged in polyethylene bags and frozen in stationary blast cells, push-through trolleys, or manual plate freezers; or frozen bare in this same equipment or in straight belt freezers or multipass straight belt freezers, and packed into cases for institutional use or tote bins for repacking throughout the year.

Raw, whole onions for French fried onion rings are cleaned, sliced, and prepared prior to being coated in breading machines; after this, they are fried in an oil fryer, cooled, and frozen. Almost all production is IQF for the restaurant and food service markets. The product is frozen in various types of belt freezers. Refrigerated precoolers or precooler sections coupled with the IQF freezers are common. Product handling must be gentle, and the product should not be cooled below −15°C prior to discharge from the freezer as it can become brittle and fractured, resulting in some product downgrading.

Changes in the Prepared Vegetable Industry

Several significant changes have occurred in the prepared vegetable industry that affect refrigeration and production in the United States. These include a shift in production of certain vegetables to Mexico and Central and South America and an increased use of vegetables in other prepared foods and in the food service markets. The vegetables being processed in Mexico and Central America and exported to the United States include asparagus, broccoli, cauliflower, Brussels sprouts, okra, and fruits, such as strawberries.

The freezing equipment used for these products is primarily operated manually as automation cannot be financially justified except for some types of belt freezers. Most of the production in the southern hemisphere is delivered in tote bins and packaged in the consuming country.

The principal products involved from the southern hemisphere are those with a short season of production in the United States (e.g., peas, green beans, and corn). If an entire year's estimated requirement is based on one short harvest season, it must be stored in freezer warehouses at considerable expense. On the other hand, if a portion of the estimated requirements is produced approximately six months later, the storage costs can be reduced significantly, and the requirements can be estimated more accurately.

Frozen vegetables for retail prepared foods and food service applications is most significant and ever-increasing. All the production is in bulk, either in cases or in tote bins. Some products are used frozen as a main vegetable or as part of a mix. In the case of mixes, some are used frozen and others are defrosted to mix with additional items prior to becoming a portion of a meal. Much could be done to improve product use, improve the quality of finished products, and reduce costs through cooperative relationships between vegetable processors and prepared food processors.

Refrigeration Loads and Systems

The principal refrigeration loads for vegetable operations are raw product cooling and storage, product cooling after blanching, freezing, process equipment located in freezer storage facilities, and freezer storage warehouses. Not all vegetable operations have all of these loads; each plant has unique conditions. Raw product cooling and storage are covered in Chapters 14 and 23.

Cooling after blanching is done with fresh water, with refrigerated water, and with evaporation cooling. If fresh water is used, it is usually well water at 13 to 16°C that is reused once or twice prior to blanching for product washing, cleaning, or waste product transfer. If refrigerated water is used, it is often used in combination with well or municipal water. Refrigerated water at 2 to 4°C reduces freezing loads and, for some products, enhances quality (e.g., cut green beans).

Freezing loads and corresponding freezing capacity vary widely between different vegetables depending on the initial product temperature and the latent heat of fusion. Special attention must be paid to the variety and size of the particular vegetables to be frozen. A belt freezer designed to freeze 4 500 kg/h of lima beans may only freeze 3 400 kg/h of 25 mm cut green beans. In addition, the freezing time will be approximately twice as long for cut green beans due to the differences in shape and bulk. Specific and latent heats of fusion for vegetables are listed in Chapter 8.

Freezer warehouse loads are calculated as suggested in Chapters 12 and 13. Freezer storages located at vegetable processing plants have three additional potential loads to consider: (1) the extra capacity reserve needed for product pull-down during peak processing; (2) the negative pressure almost all vegetable plants are under, which can substantially increase infiltration by direct flow-through; and (3) the process machinery load (particularly pneumatic conveyors) associated with repack operations.

Almost all vegetable freezing operations use ammonia as the refrigerant. Two-stage compression systems are in general use, even in those with short peak seasons. The product cooling load is done at the intermediate suction pressure. Freezing and freezer warehouse loads use the first-stage suction. Design saturated suction temperatures vary from −35 to −40°C in the first stage to −12 to −7°C in the second stage. Design saturated condensing temperatures vary from 30 to 35°C.

A unique feature of vegetable plant refrigeration is a lack of spare equipment and redundancy for those that operate for short periods at peak capacity—usually 1 500 to 2 500 h/year. In these applications spare capacity cannot be justified financially. Extra care and maintenance are usually provided prior to the peak season to ensure that full capacity is available and to minimize downtime.

Flooded evaporators are often used for product cooling and cold storage facilities. Liquid overfeed systems are used for freezing apparatus and for cold storage facilities and product cooling. Direct-expansion evaporators are rarely used.

Condensing is usually done with evaporative condensers. Some older plants use shell-and-tube condensers with once-through water usage. The warmed water is reused for product washing and cleaning prior to blanching. This method provides some risk if ammonia leaks into the water stream and should not be used in new installations.

FRUITS

Frozen fruits are processed foods that are thawed prior to serving except for specialties such as fruit pies, which are cooked. This section covers fruits that can be successfully frozen. Chapters 14, 21

Processed, Precooked, and Prepared Foods

and 22 have further information on handling, cooling, and storing fruits prior to processing.

Most fruits to be frozen are received directly from harvest. Some are cooled and stored to smooth out production, and others are processed directly. Fruits are typically cleaned, washed, and graded; cut, trimmed, or sliced (if required); and then inspected prior to freezing. At this point some fruits are packaged into containers with sugar or syrup prior to freezing, whereas others are frozen and then filled into polyethylene bags, cases, and bulk bins.

Usually no special step is taken to kill pathogens in processed fruit. Hence, it is imperative that strict sanitary practices and standards be imposed to minimize the presence of pathogenic organisms. The acidity of most fruits is a bacteriological deterrent, but it is not a guarantee of bacteriological safety.

Products cartoned prior to freezing include products such as sliced strawberries mixed with sugar in a 4 to 1 ratio and whole strawberries, other berries, mixed fruits, and melon balls in a sugar syrup. The cartons and containers are liquid tight to prevent spillage. These products are usually frozen in manual or automatic plate freezers, stationary air-blast tunnels, and push-through trolley freezers. These types of products are losing market share to other forms such as fresh fruit.

Products that are frozen prior to packaging are usually included in the free flowing or individually quick frozen (IQF) categories. These include whole fruits such as strawberries, cherries and grapes; and fruit pieces (slices, dices, halves, balls), or fruits such as apples, peaches, melons, pineapples, and citrus.

The products for IQF are usually frozen in straight belt freezers, fluidized belt freezers, fluidized bed freezers, and cryomechanical freezers. Many fruits and fruit pieces are sticky, fragile, and have a relatively high latent heat value. There is a trend toward the use of cryomechanical freezers for those applications in both new installations and in retrofits, as they often provide a superior IQF product at a reasonable cost. Chapter 15 has further information regarding freezers. The products from IQF freezers are packed and handled in a similar manner as described under the Vegetables section.

Fruit pies are a specialty pack. The IQF or fresh fruits are deposited in a dough shell and a top dough sheet is added on an assembly, filling, and packaging line as described in the previous section, Main Dishes, Meals. Fruit pies are usually frozen in automatic plate freezers or spiral belt freezers.

Refrigeration Loads and Systems

The principal refrigeration load calculations for fruits are similar to those as described under the Vegetables section except there is no cooling after blanching. The Vegetables section commentary with respect to freezing, freezer warehouse loads and refrigeration systems applies to fruits as well. Fruits do, however, respond better to a lower storage temperature ($-24°C$ versus $-20°C$) due to their higher sugar content.

POTATO PRODUCTS

The primary frozen potato products include various types of French fried potatoes for fast food restaurant, regular restaurant, and institutional uses. They include regular, shoestring, crinkle cut, and curly fries. Potato sales for retail consumption are far less than those for institutional use. Other potato products include potato puffs, tots, and wedges which are a coproduct or made from waste-stream potatoes and rejected fries. Specialty potato products include hash browns, twice-baked potatoes, and refrigerated French fries. The refrigerated product has a relatively short shelf life and requires close control of handling, shipping, and storage.

French Fries

French fried potatoes and coproducts are processed almost year-round. In the northern United States, product is received in bulk directly from the field for 1-1/2 to 2 months in the fall and thereafter from storage. The storage of fresh potatoes for processing is covered in Chapter 23.

French fried potato lines are usually of large capacity, often with a raw potato capacity of 23 000 kg/h with a finished capacity of 11 000 kg/h plus 900 kg/h of coproduct.

The bulk potatoes are metered into the processing lines, after which they are conveyed through a destoner to provide initial washing and to remove stones and other debris. They are then graded, and small potatoes are diverted to the coproduct line. Some processors also grade out large potatoes to be sold fresh as baking potatoes. The potatoes for French fries are then steam peeled and trimmed to remove blemishes. (French fries made with skins are not peeled.) Whole potatoes are cut into the desired fry shapes, and the slivers are automatically graded out and diverted to the coproduct line.

The fry shapes are then processed in one of two ways. The first method is to blanch, cool, blanch, mini-dry, and oil fry the potatoes, producing a product of approximately 28 to 30% solids. This product has a high moisture content and is more difficult to freeze because of its higher latent heat of fusion. The second method is to blanch, process the product through a two- or three-stage drier, and then oil fry the potatoes, which produces a product of approximately 34 to 36% solids, a so-called "dry fry."

After frying, the French fries are frozen on a straight belt freezer system that has three separate conveyors. In the first section, filtered ambient air or refrigerated air cools the product from 82 to 93°C to approximately 21°C to drain and/or congeal the excess fryer oil on the surface of the fries. The refrigerated air is provided either by a direct flooded ammonia evaporator, by an ammonia thermosyphon system, or by water coils where plant water is heated for reuse in processing. The second and third sections, which are two mesh conveyors in series, function as a two-stage belt freezer. The first section is used for precooling and/or crust freezing to condition the fries before total freezing and sensible cooling in the third section. French fries are usually only frozen to -15 to $-12°C$, as they become brittle below those temperatures and may be damaged in subsequent handling.

After freezing, the fries are size graded with the longest lengths slated for institutional markets, the shorter lengths for retail, and the slivers and pieces relegated for use as cattle feed. The grading occurs in an area usually maintained at 21°C. The product from the graders is collected in tote bins for subsequent packaging, with some of the totes going directly to the packaging operation and the balance to freezer storage.

The filling of fries is accomplished with net weigh fillers. The containers used include retail cartons, retail poly bags, and institutional polyethylene bags. The institutional volume is dominant. The packaging area is usually air conditioned to minimize product warm-up and clustering.

Potato Coproducts

Puffs, tots, and wedges are the coproducts manufactured from small whole potatoes that were graded out prior to peeling and slivers that were graded out after cutting the whole potatoes into shapes from the French fry line. The graded out product is about 7% of the French fried potato production. If more coproducts are desired, small whole potatoes are used.

The graded out product and/or whole potatoes that have been steam peeled or diced are inspected to remove blemishes and are then blanched. The product is then conveyed to a retrograde cooler, where the product temperature is reduced to approximately 2°C, and the product is partially dehydrated. From the cooler, the potatoes are conveyed to other equipment where they are chopped into small pieces, mixed with flavorings and condiments, and formed into the desired shapes. From the formers, the product is conveyed to the oil fryers and then to the freezer.

These products are usually frozen on spiral belt freezers, as the freezing time is relatively long due to the bulky shape and the high product inlet temperature of approximately 80°C. These products are usually not cooled in a refrigerated cooler prior to entering the spiral freezer.

The products can be packaged directly from the freezer or stored in bins for later packaging. They are distributed in both retail and institutional markets and primarily packaged in polyethylene bags.

Hash Brown Potatoes

Hash browns are manufactured from steam-peeled whole potatoes that are too small for French fries. They are then blanched and cooled conventionally or by retrograde cooling in bins, after which they are conveyed to a slicer and sliced into strips.

They are placed on a conveyor belt and formed into shapes that are scored and pulled apart to make discrete groupings of patties. They are conveyed to a straight line belt freezer and frozen to −20°C or below. The product is generally packed into polyethylene bags for either retail or institutional distribution.

Other Potato Products

This category includes twice-baked potatoes, potato skins, and boiled potatoes with or without skins for frozen meals or main dishes.

Refrigeration Loads and Systems

The refrigeration loads result primarily from cooling and freezing products and the associated freezer storage for in-process and finished goods storage. As noted previously, the production rates and total capacity of these plants are high, to take advantage of the economics of scale.

Belt freezer refrigeration loads are the major component, and a careful analysis should be made to ensure that adequate performance is achieved to meet the capacity requirements of the various product forms to be frozen. Refrigeration loads and capacity levels for the same freezing apparatus change significantly due to differences in latent heats of fusion, inlet and outlet temperatures, specific heats, and the size and shape of individual product pieces. Additional information regarding freezing times and refrigeration loads for specific foods may be found in Chapters 8 and 9.

Freezer warehouse loads are calculated as suggested in Chapters 12 and 13. Freezer storage located at potato processing plants has additional refrigeration requirements due to product pull-down loads. French fries are often discharged from belt freezers at −15 to −12°C to reduce or eliminate product breakage due to brittleness at lower temperatures. This discharge temperature coupled with subsequent packaging can result in product inlet temperature to the freezer warehouse of −9°C. The product temperature should be lowered promptly in the freezer storage to −20°C or below prior to shipping. In addition, the product quantities for a typical plant can be several hundred thousand pounds per day. Also, some freezer warehouses may be attached to plants under negative pressure, which can substantially increase infiltration. This should be corrected, as it is very difficult and costly to offset with refrigeration and infiltration reduction devices.

Potato freezing plants primarily use ammonia as a refrigerant. Two-stage compression systems are in general use, but some single-stage plants are in operation, particularly in the western United States. In two-stage systems, the product cooling is done at the intermediate pressure. Freezing and freezer warehouse loads use the first stage. Design saturated suction temperatures vary from −36 to −40°C in the first stage and −12 to −7°C in the second stage. Design saturated condensing temperatures vary from 30 to 35°C.

In single-stage systems, separate compressors are used for the −12 to −7°C product cooling and liquid refrigerant precooling. Separate compressors are used for freezing and freezer storage with design saturated suction temperatures of −33 to −36°C. The higher design suction temperature is achieved with more evaporator surface. Design saturated condensing temperatures are usually 30°C in the low wet-bulb design temperature areas where these plants are located. As the wet-bulb temperatures are even lower during the early production months in the fall, winter, and spring, the compression ratios are tolerable. Some firms find inadequate or borderline financial justification in electric power savings for the extra capital associated with two-stage systems.

Single-stage systems are simple and have a low first cost. Most of these plants are in rural areas, and it is easier and usually more satisfactory to train operators and mechanics for single-stage systems.

Condensing is usually done with evaporative condensers. New plants are often designed for 30°C condensing temperature with floating condensing pressures for lower wet-bulb temperatures and partial loads.

Potato processing plants have heavy refrigeration loads and operate for 6000 to 7000 h per year. Some spare machine room equipment may be needed, but it is not generally provided. Major maintenance is performed during periods of lower production as well as during downtime periods of four to six weeks per year.

The French fried potato freezers function under heavy loads and severe duty. Features include modular, rugged construction for ease of installation, evaporators of aluminum or hot-dipped galvanized tubing with variable fin spacing, axial or centrifugal fans with updraft air flow, noncorrosive materials for product contact parts, regular or sequential water defrost, belt washing apparatus, catwalks for access, and insulated panel housings with interiors constructed to withstand periodic wash-down. A few of these freezers are designed to operate continuously to provide full capacity at all times. Most are designed to be defrosted every 7 to 7.5 h, between shifts, to maintain capacity. In the latter case, the freezer should provide full capacity at the end of the shift. Spiral freezers for coproducts and straight belt freezers for hash browns are similar to those described in Chapter 15.

OTHER PREPARED FOODS

Several other types of prepared foods, including appetizers, sandwiches, breads, rolls, cakes, cookies, fruit pies, toppings, ice creams, sherbets, yogurts, and frozen novelties are covered in other chapters. Bakery products are covered in Chapter 27, and ice cream products are covered in Chapter 19.

LONG-TERM STORAGE

Most prepared foods are not produced with long-term frozen storage as an objective. Inventories are closely supervised, and production and sales are closely linked to minimize inventory. Profitability is reduced by having finished goods in storage and in the distribution chain.

One exception is some vegetables and fruits that can be processed and frozen only during the harvest season. Even here, steps are usually taken to maximize in-process storage of bulk products and to package them as required by sale orders and projections.

Some components and ingredients for prepared foods, however, must withstand long-term frozen storage if they are only produced annually or infrequently. These products require close monitoring to ensure that the quality still meets standards when used.

Regardless of the length of storage, it is important that ingredients, components, and finished goods are stored at −20°C or below with minimal temperature fluctuations.

CHAPTER 27

BAKERY PRODUCTS

Ingredient Storage 27.1	Bread Cooling 27.4
Mixing 27.2	Slicing and Wrapping 27.5
Fermentation 27.3	Bread Freezing 27.5
Bread Makeup 27.3	Freezing Other Bakery Products 27.6
Final Proof 27.3	Retarding Doughs and Batters 27.6
Baking 27.4	Choice of Refrigerants 27.6

THIS chapter addresses refrigeration and air conditioning as applied to bakery products, including items distributed (1) at ambient temperature, (2) refrigerated but unfrozen, and (3) frozen. Refrigeration plays an important part in modern bakery production.

Some of the major uses of refrigeration in the baking industry are

- Ingredient cooling and dough mixing
- Freezing bread for later holding, thawing, and sale
- Refrigerating unfrozen dough products
- Freezing dough for the food service industry and supermarkets
- Freezing fried and baked products for sale to consumers

Refrigeration equipment needed to accomplish these refrigeration assignments includes

- Normal air conditioning
- Dough mixers with jackets through which chilled water, low-temperature antifreeze or direct-expansion refrigerant passes
- Cooling tunnels with refrigerated air flowing counterflow to the product
- Medium-temperature cool rooms for storage of refrigerated dough products
- Freezing tunnels for bread
- Spiral freezer chambers in which the product remains for about 20 min to 1 h and is subjected to air at about −34°C for freezing

Total plant air conditioning is used more and more in new plant construction except in areas immediately surrounding ovens, in final proofers, and in areas where cooking vessels prepare fruit fillings and hot icings. Total plant cooling was first used in plants producing Danish pastry, croissants, puff pastry, and pies and has expanded to new-construction facilities for frozen dough operations and for general production. Flour dust in the air should be filtered out because it fouls air passages in the air-conditioning equipment and seriously reduces heat transfer rates.

INGREDIENT STORAGE

Raw materials are generally purchased in bulk quantities except in small operations. Deliveries are made by truck or railcar and stored in bins or tanks with required temperature protection while in transit and storage.

Flour. Flour is stored in bins at ambient temperature. Some bakeries locate these bins outside their buildings; however, inside storage is recommended where outside temperatures vary greatly. This improves control of product temperature and decreases the risk of moisture condensation inside the storage bins. Pneumatic conveyance and subsequent sifting of the flour before use generally increase flour temperature a few degrees. Smaller quantities of different types of flour, such as clear, rye, and whole wheat, are usually received in bags and stored on pallets.

Sugars and Syrups. Sugar is handled in both dry and liquid bulk forms by many large production bakeries. Although most prefer liquid, many cake and sweet goods plants produce their own powdered sugar for icings by passing granulated sugar through a pulverizer. Refrigeration dehumidifiers are sometimes used to minimize caking or sticking of the powdered sugar for proper pneumatic handling. Liquid sucrose (cane or beet sugar), generally having a solids content of 66 to 67%, is stored at ambient temperature; however, it can be cooled to as low as 7°C without crystallizing out of solution. Corn syrups and various blends of sucrose and corn syrups should be stored at 32 to 38°C to improve fluidity and pumpability. Unlike sucrose, corn syrups become more viscous when cooled. High-fructose corn syrups are best handled at 27 to 32°C. Lower storage temperatures cause the sugars to crystallize, and higher temperatures accelerate caramelizing. Dextrose (corn sugar) solutions containing 65 to 67% solids must be stored in heated tanks at 54°C to prevent crystallization. Many bakeries use high-fructose corn syrups. Because these syrups are stored at a lower temperature than conventional syrup, less thermal input is required during storage, and the refrigeration load during mixing is significantly reduced. Smaller volume and specialized sugars are received in poly-lined bags and stored at ambient temperature.

Shortenings. Shortenings are stored in heated tanks or a "hot room," where the temperature is maintained at 6 K above the American Oil Chemists' Society (AOCS) capillary closed-tube melting point of the fat. Lard, for example, should be stored at 49°C to be totally liquid. Other shortenings need slightly higher temperatures. Fluid shortenings and oils are stored at room temperature, but fluid shortenings need constant slow-speed agitation to prevent hard fats from separating to the bottom of the tanks.

Yeast. Fresh yeast comes in 0.5 kg blocks packaged in cartons of various sizes, in crumbled form in 25 kg bags, and in liquid cream form handled in bulk tanks. Refrigerated storage temperatures ranging from 7°C to the freezing point of the product are required. For maximum storage life, 1 to 2°C is considered best. Active dry and instant dry forms of yeast are available that do not need refrigeration.

Egg Products. Liquid egg products (whole eggs, egg whites, egg yolks and fortified eggs) are commonly used in small retail and large cake and sweet goods bakeries. They generally come frozen in 15 kg containers that must be thawed under refrigeration or cold-water baths. Where large quantities are needed, liquid bulk refrigerated handling can be an economic advantage. Storage temperatures for liquid egg products should be less than 4.5°C, with 2 to 3°C being the ideal storage temperature range. Dried egg solids, which need no refrigeration, are also used. A shelf-stable whole egg that requires no refrigeration has been introduced. This stability was achieved by removing two-thirds of the water from the eggs and replacing it with sugar, thereby lowering the water activity to the point that most organisms cannot grow.

Other. Dried milk products, cocoa spices, and other raw ingredients in baking are usually put into dry storage, ideally at 21°C. Ideal storage is rarely achieved under normal bakery conditions. Refrigerated storage is sometimes used where longer shelf lives are

---

The preparation of this chapter is assigned to TC 10.9, Refrigeration Application for Foods and Beverages.

desired or high storage temperatures are the norm. This decreases flavor loss and change, microbial growth, and insect infestation.

MIXING

Bread, buns, and sweet rolls are the most important yeast-leavened baked products in terms of production volume. After scaling the ingredients, mixing is the next active step in production. Proper development of the flour's gluten proteins is what gives doughs their gas-retaining properties, which affect the volume and texture of the baked products. Temperature control during mixing is essential. Refrigeration is required because of the generation of heat and the necessity of controlling dough temperature at the end of mixing.

Yeast metabolism is materially affected by the temperatures to which the yeast is exposed. During dough mixing, the following heat factors are encountered: (1) **heat of friction**, by which the electrical energy input of the mixer motor is converted to heat; (2) **specific heat** of each ingredient; and (3) **heat of hydration**, generated when a dry material absorbs water. If ice is used for temperature control, **heat of fusion** is involved. Finally, the temperature of the dough ingredients must be considered. Yeast acts very slowly below 7.2°C. It is extremely active in the presence of water and fermentable sugars at 27 to 38°C, but yeast cells are killed at about 60°C and at a lower but sustained pace below its freezing point of −3.3°C. Precise temperature control is essential at all stages of storage and production, especially during mixing, because of its impact on downline processing.

Mixers

The three most common styles of mixers are the horizontal, vertical or planetary, and spiral. **Horizontal mixers** are primarily used by wholesale bakeries and are designed with the agitator bars in a horizontal position. They range in capacity from 90 to 1350 kg. Because of the large dough sizes, these mixers are generally jacketed with some form of cooling. **Vertical mixers** are more common in retail bakeries and are categorized by their largest bowl capacity. Bowls range in size from 11 to 19 L for tabletop models to as high as 320 L in some large wholesale plants. The bowls have no refrigeration jacket and can be removed from the mixers. The hook or agitator revolves as it travels around the inside of the bowl. **Spiral mixers** are somewhat newer and are gaining in popularity with retail and specialty bread bakers. Here the bowl revolves, bringing the ingredients to the off-center spiral agitator. The bowls for the smaller models (23 to 180 kg) are not removable, but those for the larger models (up to 450 kg) can be removed. Like vertical mixers, spiral mixers are not jacketed.

Dough Systems

The three principal types of batch dough mixes are the straight dough, no-time dough, sponge dough, and liquid ferment. These methods are called **dough systems** or **dough process** in the baking industry. The type of dough system determines the stages in the process, the equipment needed, and the general processing parameters (times and temperatures). **Straight dough** and **no-time dough** systems are the two most common in retail bakeries. All of the ingredients are mixed at one time. Straight doughs require fermentation, whereas no-time doughs do not. These systems are sometimes found in wholesale production.

The more common systems used in wholesale bakeries are the sponge dough and liquid ferment or liquid sponge. The **sponge dough** process requires more equipment and longer fermentation times than the straight and no-time dough systems. Here, only a part of the total amount of flour and water required are mixed with all of the yeast and yeast food. The resulting mixture, or sponge, is then fermented for a period of time before it is given the final mix or remix with the remaining ingredients. This is done just before makeup, improving both tolerance of schedule disruptions and dough machinability.

The principal heat generated during mixing comes from hydration as the flour absorbs water and from the friction of the mixer. To absorb this excess heat and maintain the dough at 25.6 to 27.8°C, the ingredient water is usually supplied to the mix at 2 to 4°C, and horizontal mixers are generally jacketed to circulate a cooling medium around the bowl.

The **liquid sponge** process has gained popularity with wholesale bakeries because of the excellent temperature control and acceptable product quality. Liquid sponge ingredients can be incorporated and fermented in special equipment either in batches or on an uninterrupted basis. To render the mixture pumpable, more water is incorporated than the actual amount used in a sponge to be fermented in a trough. On completion of the predetermined fermentation time, the liquid sponge (at required pH and titratable acid levels) is chilled through heat exchange equipment from about 27 to 31°C to about 7 to 13°C. The cold liquid sponge is then maintained at the required temperature in a storage or feed tank until it is weighed or metered and pumped to the mixer, where it is combined with the remaining ingredients prior to being remixed into a dough. Regular sponges come back for remixing at about 29°C. Often, it is necessary to use the refrigerated surface on the mixer jacket in conjunction with ice water or ice to achieve the required dough temperature after mixing.

Positive dough temperature control is achieved using a cold liquid sponge at 7 to 13°C, which limits the need for the refrigeration jacket and eliminates using ice in the doughs. In cold weather, remaining dough water temperatures of 38°C or higher are often required.

Specially designed equipment is needed for making liquid sponge doughs by the continuous mix process. A brew, or liquid sponge, is formed using 0 to 100% of the flour, 10 to 15% of the sugar, 25 to 50% of the salt, and all of the yeast and yeast food in about 85% of the water. The liquid sponge is fed into an incorporator, where the remaining ingredients are added. The resultant thicker batter finally passes through a developer-mixer from which the finished dough, in loaf size, is extruded into the baking pans.

Properly formulated dough can also be deposited on floured belts, rounded up, and then conveyed through conventional makeup and molding equipment. Hot dog and hamburger rolls are also produced in quantity by pumping bun dough from the continuous mix developer directly to the hopper of the makeup equipment. A coating of flour on outside dough surfaces affects gas development and retention, which in turn leads to a grain/texture more closely resembling that of sponge dough products. External symmetry and crust characteristics are similarly changed.

Dough Cooling

Some dough mixers are cooled by direct-expansion refrigerant, but the most common means of cooling is with chilled water or an antifreeze such as propylene glycol. The temperature of the evaporating refrigerant or antifreeze supplied may in many cases be as low as −1°C to maintain the dough at the desired temperature. When the dough mixers are cooled by evaporating refrigerant, the condensing unit is usually located close to the mixers. Table 1 lists the sizes of condensing units commonly selected for mixers. The ingredient water is cooled in separate liquid chillers, and when the mixer bowl is cooled by antifreeze, the liquid chiller may be remotely located.

When large batches of dough are handled in the mixers, the required cooling is sometimes greater than the available heat transfer

Table 1 Size of Condensing Units for Various Mixers

Dough Capacity, kg	Condensing Unit Size, kW
300	3
450	7
600	10
750	13
900	15

Bakery Products

surface can produce at −1°C, and the refrigerant temperature must be lowered. A refrigerant temperature below −1°C can, however, cause a thin film of frozen dough to form on the jacketed surface of the mixer. This frozen film effectively insulates the surface and impairs heat transfer from the dough to the refrigerant.

Formulas, batch sizes, mixing times, and almost every other part of the mixing process vary considerably from bakery to bakery and must be determined for each application. These variations usually fall within the following limits:

Final batch dough mass	Up to 1500 kg
Ratio of sponge to final dough mass	50 to 75%
Ratio of flour to final dough mass	50 to 65%
Ratio of water to flour	50 to 65%
Sponge mixing time	240 to 360 s
Final dough mixing time	480 to 720 s
Number of mixes per hour	2 to 5
Continuous mix production rates	Up to 1 kg/s

The total cooling load is the sum of the heat that must be removed from each ingredient to bring the homogeneous mass to the desired final temperature plus the generated heat of hydration and friction. In large batch operations, the sponge and final dough are mixed in different mixers, and refrigeration requirements must be determined separately for each process. In small operations, a single mixer is used for both sponge and final dough. The cooling load for final dough mix is greater than that for sponge mix and is used to establish refrigeration requirements.

FERMENTATION

After completion of the sponge mix, the sponge is placed in large troughs that are rolled into an enclosed conditioned space for a fermentation period of 3 to 5 h, depending on the dough formula. The sponge comes out of the mixer at a temperature of 22 to 24.5°C. During the fermentation period, the sponge temperature rises 3 to 6 K as a result of the heat produced by the yeast and fermentation.

To equalize the temperature substantially throughout the dough mass, the room temperature is maintained at the approximate mean sponge temperature of 27°C. Even temperature throughout the batch produces even fermentation action and a uniform product.

Water makes up a large part of the sponge, and uncontrolled evaporation causes significant variations in the quality and mass of the bread. The rate of evaporation from the surface of the sponge varies with the relative humidity of the ambient air and the airflow rate over the surface. The rate of movement of moisture from the inside of the sponge to the surface does not react similarly to external conditions, and the net result can be the drying of the surface and formation of a crust. Crusted dough is inactive and does not develop. When folded into the dough mass later, undeveloped portions produce hard, dark streaks in the finished bread.

To control the evaporation rate from the surface of the sponge, the air is maintained at 75% rh, and the conditioned air is moved into, through, and out of the process room without producing crust-forming drafts.

In calculating the room cooling load, the product is not considered because the air temperature is maintained at approximately the mean of the various dough temperatures in the room. Transmission heat loss through walls, ceiling, and floor is the principal load source. Infiltration is estimated as 1.5 times the room volume per hour. The lighting requirement for a fermentation room is usually about 2 W per square metre of floor area. The conditioning units are often placed within the conditioned space, and the full motor heat load must be considered.

The only source of latent heat is the approximate 0.5% loss of mass in the sponge. Under full operating load, this could account for a 1 K increase in dew-point temperature. For conditions of 27°C dry bulb and 75% rh, the dew-point temperature would be 21.9°C, and the supply air would be introduced into the conditioned space at 22.2°C dry bulb and 21°C dew point.

In large rooms, sufficient air volume can be introduced to pick up the sensible heat load with a 4.5 K rise in air temperature. In smaller rooms, a latent heat load may need to be added by spraying water directly in the room through compressed air atomizing nozzles. Water sprays have been most successful when a relatively large number of nozzles is spaced around the periphery of the room.

Because a high removal ratio of sensible to latent heat is desirable, the condensing unit is usually specified for operation as close to 15°C evaporator temperature as is practicable with the temperature and quantity of condensing water available.

BREAD MAKEUP

After the dough is mixed using conventional mixers and fermented, it is placed into the hopper of the divider. The two types of dividers used in wholesale bakeries are the ram and piston and the rotary (extrusion). With the **ram and piston divider**, the dough is forced into cylinders; the pistons adjust the cylinder opening, which controls the unit mass. The **rotary divider** extrudes the dough through an opening using a metering pump; a rotating knife then cuts off the dough. The unit mass is adjusted by the speed of the metering pump and/or the speed of the knife. Because dough density changes with time and the dividers work on the principle of volume, the baker must routinely check scaling mass and adjust the dividers.

Next, the irregularly shaped units are rounded into dough balls for easier handling in the subsequent processing steps. This **rounding** is done on drum, cone, or bar/belt rounders. The dough pieces are well floured, and a smooth skin is established on the outer surface to reduce sticking.

A short resting or relaxation period follows rounding. The equipment used is referred to as an intermediate proofer. Flour-dusted trays or belts hold the dough pieces and carry them to the next pieces of equipment via a transfer shoot. Residency time is 3 to 8 min.

The dough is next formed into a loaf by the **sheeter** and the **molder**. The relaxed dough ball is sheeted (e.g., to about 3 mm thick for white pan bread) by passing through sets of rollers. This reduces the size of gas cells and multiplies them, producing a finer grain in the baked bread. Next the dough goes to the molder, where it receives its final shape. There are three styles of molders: (1) straight grain, (2) cross grain, and (3) tender-curl. The molder produces the final length of the dough piece before it is placed into the greased baking pan.

FINAL PROOF

After it is formed into loaves and placed in pans, the dough is placed in the **proofer** for 50 to 75 min. The proofer is an insulated enclosure having a controlled atmosphere in which the dough receives its final fermentation or proof before it is baked. To stimulate the fermentative ability of the yeast, the temperature is maintained at 35 to 46°C, depending on the exact formula, the prior intensity of dough handling, and the desired character of the baked loaf.

For proper crust development during the bake, the exposed surface of the dough must be kept pliable by maintaining the relative humidity of the air in the range of 75 to 95%. Some bakeries find it necessary to compromise on a lower humidity range because of the effect on the flow of dough.

With dew points inside the proofer at about 33 to 43°C and ambient conditions as low as 18°C at times, the proofer must be adequately insulated to keep the inside surface warm enough to prevent condensation from forming because mold growth can be a serious problem on warm, moist surfaces. A thermal conductance of $C = 0.68$ W/(m^2·K) is adequate for most conditions.

In small and midsized bakeries, one of the more commonly used proofers for this process consists of a series of aisles with doors at both ends. Frequent opening of the doors to move racks of panned dough in and out makes control of the conditioned air circulation an unusually important engineering consideration. The air is recirculated at 90 to 120 changes per hour and is introduced into the room to temper the infiltration air and cause a rolling turbulence throughout the enclosure. This brings the newly arrived racks and pans up to room temperature as soon as possible.

The problem of air circulation is somewhat simpler when large automatically loaded and unloaded tray and conveyer or spiral proofers are used. These proofers have only minimal openings for the entrance and exit of the pans, and the thermal load is reduced by elimination of the racks moving in and out.

BAKING

The fully developed loaves are baked in the oven at temperatures around 200 to 230°C for 18 to 30 min, depending on the variety and size of product. Several styles of ovens are available.

Deck ovens are found in some smaller retail establishments, specialty cookie and hearth bread bakeries, pizzerias, and restaurants. The baking surfaces are stationary, requiring manual loading and unloading of product. Uneven heat distribution, little air movement, and low clearance between decks make the deck oven unsuitable for some products. Newer designs have made this oven the choice for retail baking of the denser, heavy, European-type hearth breads.

Rack ovens are very popular in in-store supermarket bakeries, freestanding retail bakeries, and some wholesale bakeries. These ovens have a small chamber into which one or two racks of panned product can be rolled at one time. Using forced-air convection, baking can be accomplished at lower temperatures for shorter baking times. These ovens are highly efficient in heat transfer and easy to operate. They generate their own steam without the use of boilers and are well suited for crusty breads.

There are two popular styles of tray ovens, which are loaded and unloaded at the same end. The **reel-type tray oven**, which looks like a Ferris wheel inside a heated baking chamber, is one of the most popular ovens in retail bakeries and is used by some large wholesalers. This design, with its manual loading and unloading, allows different products to be baked simultaneously. Because of the revolving action, hot spots and uneven baking are less likely than in deck ovens.

The **single-lap traveling tray oven** is the other type of tray oven. It is used principally in large production bakeries and is usually equipped with pan loaders and unloaders for efficient product handling and continuous baking. There is also a double-lap design, which travels back and forth twice, but it is not popular in today's bakeries.

Wholesale bakeries are being equipped with either tunnel (traveling hearth) ovens or conveyor-style ovens. **Tunnel ovens**, as the name might suggest, have the product loaded in at one end and unloaded at the other. This permits ideal temperature control for the entire baking time. **Conveyor-style ovens** have a continuous spiral climbing to the top of a heated chamber and then back down again, with each pan carried on its own section of the conveyor and every product passing through the same spot within the baking chamber. Another style of tunnel oven is the **impingement oven**. Its directional flow of heated air to the top and bottom of the product provides excellent baking speed and efficiency for pizza and other low-profile products.

Ovens are usually gas fired, but oil firing may be more economical in some areas. Where gas interruption is a problem, combination oil and gas burners are an alternative. While both indirect- and direct-fired gas systems are used, the indirect firing is generally more popular because of its greater flexibility. Ribbon burners running across the width of the oven are operated as an atmospheric system at pressures of 1.5 to 2.0 kPa.

The burners are located directly beneath the path of the hearth or trays, and the flames may be adjusted along the length of the burner to equalize heat distribution across the oven. The oven is divided into zones, with the burners zone-controlled so that the heat can be varied for different periods of the bake.

Controlled air circulation inside the oven ensures uniform heat distribution around the product, which produces desired crust color and thickness. When baking hearth breads, ovens are equipped with steam ejection. Low-pressure (14 to 35 kPa), high-volume wet steam is used for the first few minutes of baking to control loaf expansion and crust crispness and shine. This is accomplished with a series of perforated steam tubes located in the first zone of baking in place of the air-recirculating tubes. Heating calculations are based on 1050 kJ per kilogram of bread baked. Steam requirements run approximately 620 kJ per kilogram of bread baked.

BREAD COOLING

Baked loaves come out of the oven with an internal temperature of 90 to 99°C because of the evaporative cooling effect of the moisture that is driven off during baking. The crust temperature is closer to the oven baking temperature, 230°C.

The loaves are then removed from the pans and allowed to cool to an internal temperature of 32 to 35°C. The cooling of a hygroscopic material takes place in two phases, which are not distinct periods. When the bread first comes out of the oven, the vapor pressure of the moisture in the loaf is high compared to the vapor pressure of the moisture in the surrounding air. Moisture is rapidly evaporated with a resultant cooling effect. This **evaporative cooling**, combined with **heat transmission** from bread to air due to a relatively high temperature differential, causes rapid cooling in the early cooling stage, as indicated by the steepness of the temperature versus time curve (Figure 1). As the vapor pressure approaches equilibrium, heat transfer is mainly by transmission, and the cooling curve flattens rapidly.

While small bakeries still cool bread on racks standing on the open floor for 1.5 to 3 h, depending on air conditions, many large operations cool the bread while it is moving continuously on belt-, spiral-, or tray-type conveyors. Cooling is mostly atmospheric even on these conveyors. However, to ensure a uniform final product, cooling is often handled in air-conditioned enclosures with a conventional counterflow movement of air in relation to the product.

An internal temperature of 32 to 35°C stabilizes moisture in the bread enough to reduce excessive condensation inside the wrapper, thus discouraging mold development. Approximately 50 to 75 min of cooling time is required to bring the internal loaf temperature to 32°C, as shown in Figure 1.

In counterflow cooling, the optimum temperature for air introduced into the cooling tunnel is about 24°C. Air at 80 to 85% rh controls moisture loss from the bread during the latter stage of cooling and improves the keeping quality of the bread.

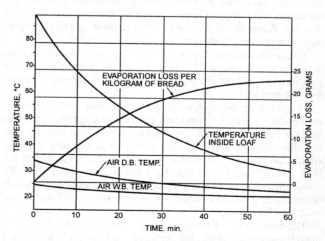

Fig. 1 Moisture Loss and Air Temperature Rise in Counterflow Bread-Cooling Tunnel

Bakery Products

Attempting to shorten the cooling time by forced-air cooling with air temperatures below 24°C is not satisfactory. The rate of heat and moisture loss from the surface becomes much greater than that from the interior to the surface. This causes the crust to shrivel and crack.

Product heat load is calculated assuming sensible heat transfer based on a specific heat of 2.93 kJ/(kg·K) for bread and a temperature reduction from 82°C to 32°C. For calculating purposes, evaporation of moisture from the bread may be considered responsible for reducing the bread temperature to 82°C.

Under normal conditions, the conditioning of the air can best be accomplished by evaporative cooling in an air washer. For periods when the outdoor wet-bulb temperature exceeds 22°C, which is the maximum allowable based on 24°C dry bulb and 85% saturation, refrigeration is required.

The amount of air required is rather high because the maximum temperature rise of the air passing over the bread is usually about 11 K; consequently, the refrigeration load is comparatively high. In localities where wet-bulb temperatures above 22°C are rare or of short duration, the expense of refrigeration is not considered justified, and the bread is sliced at a higher than normal temperature.

SLICING AND WRAPPING

Bread from the cooler goes through the slicer. The slicer's high-speed cutting blades (similar to band saw blades) cut cleanly through properly cooled bread. If the moisture evaporation rate from the surface and the replacement rate of the surface moisture from the interior of the loaf are not kept in balance, the bread develops a soggy undercrust that fouls the blades, causing the loaf to crush during slicing. A brittle crust may also develop, which leads to excessive crumbing during the slicing. From the slicer, the bread moves automatically into the bagger.

BREAD FREEZING

The bagged bread normally moves into the shipping area for delivery to various markets by local route trucks or by long-distance haulers. Part of the production may go into a quick-freezing room and then into cold storage.

Two important problems face bakers in the freezing of bread and other bakery products. The first is connected with the short work week. Most bread and roll production bakeries are inoperative on Saturday; thus, production near the end of the week is much larger than for the earlier part of the week. The problem increases for bakeries on a 5 day week, with Tuesday usually being the second day off. Freezing a portion of each day's production for distribution on the days off would enable a more even daily production schedule.

A second problem is the increased demand for variety breads and other products. The daily production run of each variety is comparatively small, so the constant setup change is expensive and time-consuming. Running a week's supply of each variety at one time and freezing it to fill daily requirements can reduce operating cost.

Both problems concern the staleness of bread, which is a perishable commodity. After baking, starch from the loaf progressively crystallizes and loses moisture until a critical point of moisture loss is reached. A tight wrap helps keep the moisture content high over a reasonable time. Starch crystallization, when complete, produces the crumbly texture of stale bread. This is a spontaneous action that increases in rate both with decreased moisture and with decreased temperature. Starch crystallization is accelerated as the product passes through a critical temperature zone of 10°C to the freezing point of the product. The rate then decreases until the temperature reaches −18°C, where moisture loss seems to be somewhat arrested.

Bread freezes at −9 to −7°C (Figure 2). It is necessary to cool the bread through the freezing or latent heat removal phase as fast as possible to preserve the cell structure. Because the moisture loss rate increases with reduced temperature, the bread should be cooled rapidly through the entire range from the initial temperature down

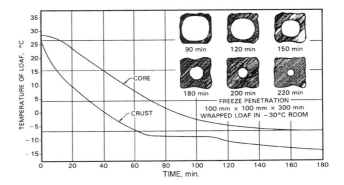

Fig. 2 Core and Crust Temperatures in Freezing Bread

to, and through, the freezing points. Successful freezing has been reported in room temperatures of −18, −23, −29, and −34°C.

In U.S. Department of Agriculture (USDA) laboratory tests, core temperatures of loaves of wrapped bread placed in 3.5 m/s cold air blasts were brought down from 21°C to −10°C in the times given:

Freezer Air Temperature, °C	Time, h	Bread Core Temperature at End of 2 h, °C
−40	2	−9.4
−34.4	2.25	−8.9
−28.9	3	−7.2
−23.3	3.75	−6.1
−17.8	5	−5.5

Changes in air velocity from 1.0 to 6.5 m/s caused relatively little change in the cooling of wrapped bread.

Cooling from −18 to −29°C is 10 to 30 min faster for unwrapped bread than for wrapped bread. However, the wrapper's value in retaining the moisture content of the product during freezing and thawing makes freezing in the wrapped condition advisable.

Some commercial installations freeze wrapped bread in corrugated shipping cartons. The additional insulation provided by the carton increases the freezing time considerably and causes a wide variation in freezing time between variously located loaves. One test found that a corner loaf reached −9°C in 5.5 h while the center loaf required 9 h.

Most freezers are batch-loaded rooms in which the bread is placed on the wire shelves of steel racks. One of the principal disadvantages of this arrangement is that the racks must be moved in and out of the room manually. Continuous-type freezers with wire belt- or tray-type conveyors permit a steady flow and do not expose personnel to the freezer temperature. See Chapter 15 for freezer descriptions.

For air freezing temperatures of −29°C or below, two-stage compression systems should be considered for overall economy of operation.

In addition to the primary air blowers designed to handle about 10 L/s per kilogram of bread frozen per hour, a series of fans is employed to ensure good air turbulence in all parts of the room. Heat load calculations are based on the specific and latent heat values in Table 2.

After the quick-freeze, the bread is moved into a −23°C holding room where the temperature throughout the loaf equalizes. The bread is often placed in shipping cartons after freezing and stacked tightly on pallets.

Defrosting. Frozen bread must be thawed or defrosted for final use. Slow, uncontrolled defrosting requires only that the frozen item be left to stand, usually in normal atmospheric conditions. For quality control, the defrosting rate is just as critical as the freezing rate. Passing the product rapidly through the critical temperature range of 10°C to the freezing point of the product yields maximum crumb softness. Too high a relative humidity causes excessive condensation on the wrapper with some resultant susceptibility to handling

Table 2 Important Heat Data for Baking Applications

Specific heat	
Baked bread (above freezing)	2.93 kJ/(kg·K)
Baked bread (below freezing)	1.42 kJ/(kg·K)
Butter	2.39 kJ/(kg·K)
Dough	2.51 kJ/(kg·K)
Flour	1.76 kJ/(kg·K)
Ingredient mixture	1.68 kJ/(kg·K)
Lard	1.88 kJ/(kg·K)
Milk (liquid whole)	3.98 kJ/(kg·K)
Liquid sponge (50% flour)	2.93 kJ/(kg·K)
Heat of friction per kilowatt of motive power of mixer motor	1 kW
Heat of hydration of dough or sponge	15.1 kJ/kg
Latent heat of baked bread	109.3 kJ/kg
Specific heat of steel	0.50 kJ/(kg·K)

damage. The product will defrost in about 1.75 h when placed in air at 49°C and 50% rh or less. Good air movement over the entire product surface at 1.0 m/s or higher helps minimize the condensation and make the defrosting rate more uniform.

FREEZING OTHER BAKERY PRODUCTS

Retail bakeries freeze many products to meet fluctuating demand. Cakes, pies, sweet yeast dough products, soft rolls, and doughnuts are all successfully frozen. A summary of tests and commercial practice shows that these products are less sensitive to the freezing rate than are bread and rolls. Freezing at −18 to −12°C apparently produces just as satisfactory results as freezing at −23 to −29°C.

Storage at −23 to −29°C or below keeps packaged dinner rolls and yeast-raised and cake doughnuts satisfactorily fresh for 8 weeks. Cinnamon rolls keep for only about 3 weeks, apparently because of the raisins, which absorb moisture from the crumb of the roll.

Pound, yellow layer, and chocolate layer cakes can be frozen and held at −12°C for 3 weeks without significantly affecting their quality. Sponge and angel food cakes tend to be much softer as the freezing temperature is reduced to −18°C. Layer cakes with icing freeze well, but condensation on exposed icing during thawing ruins the gloss; therefore, these cakes are wrapped before freezing.

Unlike cakes, which can be satisfactorily frozen after baking, pies frozen after baking have an unsatisfactory crust color, and the bottom crust of fruit pies becomes soggy when the pie is thawed. Freezing of unbaked fruit pies is highly successful. Freezing time has little, if any, effect on product quality, but storage temperature does have an effect. Frozen pies stored at temperatures above −18°C develop badly soaked bottom crusts after 2 weeks, and the fillings tend to boil out during baking, possibly due to moisture migration from the starch-based syrups. Freezing of baked or fried products is generally a high-production operation carried out in freezing tunnels or spiral freezers.

One of the fastest growing applications of refrigeration is to freeze dough for in-store bakeries. A few frozen dough products are sold in supermarkets directly to consumers, but the major portion of frozen dough is destined either for the food service industry or for supermarket bakeries.

Danish and sweet dough products are frozen baked or unbaked, depending on how quickly they will be required for sale after they are removed from the freezer. Custard and chiffon pie fillings have not had uniformly good results, but some retail bakeries have achieved satisfactory results by carefully selecting starch ingredients. Meringue toppings made with proper stabilizers stand up very well, and whipped cream seems to improve with freezing. Cheesecake, pizza, and cookies also freeze well.

Although some products are of better quality if frozen at −23°C while others are improved by freezing at −29°C, variety shops must compromise on a single freezer temperature so that all products can be placed in one freezer. The freezer is usually maintained between −23 and −29°C. Freezing time is not a factor because the products are kept in storage in the freezer. Freezers range from large reach-in refrigerators in retail shops to walk-in boxes in wholesale shops.

RETARDING DOUGHS AND BATTERS

Freezing is usually employed if the products are to be held for 3 days to 3 weeks. For shorter holding periods, such as might be required in order to have freshly baked products all day from one batch mix, a temperature just cold enough to retard fermentation action in the dough is applied. Retarder temperatures of 0 to 4.5°C slow the yeast action sufficiently to allow holding for 3 h to 3 days.

Doughs to be retarded are sometimes made up into final shaped units ready for proofing and baking. Cold slabs of dough can also be stored, with the units for baking being made up after thawing. This method is especially satisfactory for Danish pastry dough and other doughs with rolled-in shortening, such as pie crust. Chilling to retarded temperatures improves the flakiness of these products. Refrigeration load calculations should be made following the recommendations included in Chapter 12.

In the retarding refrigerator, a relative humidity of about 85% is required to prevent the product from drying out. Condensation on the product is undesirable. Complete batch loading is usually used, so refrigeration calculations must be based on the introduction of the batch over a short period of time. The refrigeration equipment must have the capacity to absorb the product and carrier heat load in 0.75 to 3 h, depending on cabinet size and handling technique. The products most commonly handled in this manner are Danish pastry, dough for sweet rolls and coffee cake, cookies, layer cake mixes, pie crust mixes, and bun doughs.

Temperatures required for retarding are very similar to those required for storing ingredients, and the refrigerators are usually designed to handle both ingredient storage and dough retarding.

CHOICE OF REFRIGERANTS

The most popular refrigerants serving the baking industry in the past have been R-12, R-22, R-502, and ammonia (R-717). R-12 is a chlorofluorocarbon (CFC) and can no longer be used. The most successful replacement for R-12 has been R-134a. In general, any R-12 system can be refitted with R-134a along with the substitution of the proper lubricant. R-22 is a hydrochlorofluorocarbon (HCFC), so it is destined for phaseout in several decades. R-502 is an azeotrope containing the CFC R-115, so no new R-502 systems are being installed. Hydrofluorocarbon (HFC) substitutes for R-502 and R-22 are available, but they cost much more. Therefore, the preferred configuration of the system may be different than for the traditional refrigerants.

For water chillers, R-134a is becoming popular. For chilling antifreeze to lower temperatures, the HFC replacements for R-22 and R-502 are possibilities. For large freezing facilities (e.g., spiral freezers), ammonia dominates due to its excellent low-temperature performance. Some producers also choose expendable refrigerants, such as carbon dioxide (CO_2) or nitrogen (N_2), which many believe provides a superior frozen product because of the low temperature of the freezing medium. Hybrid systems can be used in which CO_2 and N_2 rapidly freeze a crust on the product surface and then vapor compression, with its lower operating cost, completes the freezing process.

BIBLIOGRAPHY

American Institute of Baking. 1995. *Baking science and technology bread lecture book*. American Institute of Baking, Manhattan, KS.

Kulp, K., K. Lorenz, and J. Brümmer. 1995. *Frozen and refrigerated doughs and batters*. American Association of Cereal Chemists, St. Paul, MN.

Matz, S.A. 1987. *Formulas and processes for bakers*. Pan-Tech International, McAllen, TX.

Matz, S.A. 1988. *Equipment for bakers*. Pan-Tech International.

Pyler, E.J. 1988. *Baking science and technology*, 3rd ed. Sosland Publishing Company, Merriam, KS.

CHAPTER 28

CANDIES, NUTS, DRIED FRUITS, AND DRIED VEGETABLES

CANDY MANUFACTURE	28.1
Milk and Dark Chocolate	28.1
Hand Dipping and Enrobing	28.2
Bar Candy	28.2
Hard Candy	28.3
Hot Rooms	28.3
Cold Rooms	28.3
Cooling Tunnels	28.3
Coating Kettles or Pans	28.4
Packing Rooms	28.4
Refrigeration Plant	28.4
STORAGE	28.5
Candy	28.5
Nuts	28.7
Dried Fruits and Vegetables	28.7
Controlled Atmosphere	28.8

CANDY MANUFACTURE

AIR conditioning and refrigeration are essential for successful candy manufacturing. Proper atmospheric control increases production, lowers production costs, and improves and standardizes product quality.

One or more of several standardized spaces or operations is encountered in every plant. These spaces include hot rooms; cold rooms; cooling tunnels; coating kettles; packing, enrobing, or dipping rooms; and storage.

Sensible heat must be absorbed by air-conditioning and refrigeration equipment, which includes the air distribution system, plates, tables, and cold slabs in tunnels or similar coolers. In calculating the loads, such sensible heat sources as people, power, lights, sun effect, transmission losses, infiltration, steam and electric heating apparatus, and the heat of the entering product must be considered. See Chapter 12 for more information. Table 1 summarizes the optimum design conditions for refrigeration and air conditioning.

Two of the basic ingredients in candy are sucrose and corn syrup. These change easily from a crystalline form to a fluid, depending on temperature, moisture content, or both. The surrounding temperature and humidity must be controlled to prevent moisture gain or loss, which affects the product's texture and storage life. Temperature should be relatively low, generally below 21°C. The relative humidity should be 60% or less, depending on the type of sugar used. For chocolate coatings, temperatures of 18°C or less are desirable, with relative humidity 50% or less.

MILK AND DARK CHOCOLATE

Cocoa butter is either the only fat or the principal fat in chocolate, constituting 25 to 40% or more of various types. Cocoa butter is a complex mixture of triglycerides of high relative molecular mass fatty acids, mostly stearic, oleic, and palmitic. Because cocoa butter is present in such large amounts in chocolate, anything affecting cocoa butter affects the chocolate product as well.

Because cocoa butter is a mixture of triglycerides, it does not act as a pure compound. Its physical properties, melting point, solidification point, latent heat, and specific heat affect the mixture. Cocoa butter softens over a wide temperature range, starting at about 27°C and melting at about 34°C. It has no definite solidification point; this varies from just below its melting point to 27°C or lower, depending on the quantity of cocoa butter and the time it is held at various temperatures. The presence of milk fat in milk chocolate lowers both the melting point and the solidification point of the cocoa butter. High-quality milk chocolate remains fluid for easy

Table 1 Optimum Design Air Conditions[a]

Department or Process	Dry-Bulb Temperature, °C	Relative Humidity, %
Chocolate pan supply air	13 to 17	55 to 45
Enrober room	27 to 29	30 to 25
Chocolate cooling tunnel supply air	4 to 7	85 to 70[b]
Hand dipper	17	45
Molded goods cooling	4 to 7	85 to 70
Chocolate packing room	18	50
Chocolate finished stock storage	18	50
Centers tempering room	24 to 27	35 to 30
Marshmallow setting room	24 to 26	45 to 40
Grained marshmallow (deposited in starch) drying	43	40
Gum (deposited in starch) drying	52 to 66	25 to 15
Sanded gum drying	38	25 to 40
Gum finished stock storage	10 to 18	65
Sugar pan supply air (engrossing)	29 to 41	30 to 20
Polishing pan supply air	21 to 27	50 to 40
Pan rooms	24 to 27	35 to 30
Nonpareil pan supply air	38 to 49	20
Hard candy cooling tunnel supply air	16 to 21	55 to 40
Hard candy packing	21 to 24	40 to 35
Hard candy storage	10 to 21	40
Caramel rooms	21 to 27	40
Raw Material Storage		
Nuts (insect)	7	60 to 65
Nuts (rancidity)	1 to 3	85 to 80
Eggs	−1	85 to 90
Chocolate (flats)	18	50
Butter	−7	
Dates, figs, etc.	4 to 7	75 to 65
Corn syrup[c]	32 to 38	
Liquid sugar	24 to 27	40 to 30
Comfort air conditions	24 to 27	60 to 50

Note: Conditions given in this table are intended as a guide and represent values found to be satisfactory for many installations. However, specific cases may vary widely from these values because of such factors as type of product, formulas, cooking process, method of handling, and time. Acceleration or deceleration of any of the foregoing will change the temperature, humidity, or both to some degree.

[a]Temperature and humidity ranges are given in respective order (i.e., first temperature corresponds to first humidity).
[b]Optimum conditions.
[c]Depends on removal system. With higher temperatures, coloration and fluidity are greater.

The preparation of this chapter is assigned to TC 10.9, Refrigeration Application for Foods and Beverages.

handling at temperatures as low as 30 to 31°C. Sweet chocolate remains fluid at temperatures as low as 32 to 33°C.

Chocolate can be subcooled below its melting point without crystallization. In fact, it does not crystallize en masse but rather in successive stages, as solid solutions of a very unstable crystalline state are formed under certain conditions. The latent heat of crystallization (or fusion) is a direct function of the manner in which the chocolate has been cooled and solidified. Once crystallization has started, it continues until completion, taking from several hours to several days, depending on exposure to cooling, particularly to low temperatures (subcooling).

The latent heat of solidification of the grades of chocolate commonly used in candy manufacture varies from approximately 84 to 93 kJ/kg. Average values for the specific heat of chocolate may be taken as 2.3 kJ/(kg·K) before solidification and 1.3 kJ/(kg·K) after solidification. In calculating the cooling load, a margin of safety should be added to these figures.

Cocoa butter's cooling and solidification properties exist in five polymorphic forms: one stable form and four metastable or labile ones. Cocoa butter usually solidifies first in one of its metastable forms, depending on the rate and temperature at which it solidifies. In solidified cocoa butter, the lower melting labile forms change rapidly to the higher melting forms. The higher melting labile forms change slowly, and seldom completely, to the stable form.

Commercial chocolate blocks are cast in metal molds after the tempering process. During this process, it is desirable to cool the chocolate in the molds as quickly as possible, thus requiring the shortest possible cooling tunnel. However, cooling the blocks too quickly (particularly large commercial blocks, which are standard 5 kg cakes) may cause checking or cracking, which, while not serious to quality, adversely affects appearance. Depositing the chocolate into metal molds at 29 to 32°C is common.

Dark chocolate should be cooled very slowly at 32 to 33°C; milk chocolate, at 30 to 31°C. Air entering the cooling tunnel, where the goods are unmolded, may be 4°C. The air may be 17°C where the goods enter the tunnel. After the chocolate is deposited in the mold, it can be moved into a cooling tunnel for a continuous cooling process, or the molds can be stacked up and placed in a cooling room with forced-air circulation. In either case, temperatures of 5 to 10°C are satisfactory. The discharge room from the cooling tunnel or the room to which the molds are transferred for packing should be maintained at a dew point low enough to prevent condensation on the cooled chocolate. In load calculations for the cooling or cold room, it is necessary to account for transmission and infiltration losses, any load derived from further cooling of the molds, and the sensible and latent heat cooling loads of the chocolate itself.

The tunnel is designed to introduce 4°C air countercurrent to the flow of chocolate; the coldest air enters the tunnel where the cooled chocolate leaves the tunnel. Because the tunnel air warms up on its way out, the warmest air leaves the tunnel at the point where the warmest molten chocolate enters. The leaving chocolate is markedly cooler than the entering chocolate, and the subcooling is greatly reduced. This in turn reduces the large temperature difference between the chocolate and the cooling air along the entire tunnel length.

For any particular application, only testing will determine the length of time the chocolate should remain in the tunnel and the subsequent temperature requirements. Good cooling is generally a function of tunnel length, belt speed, and the actual time the product contacts the cooling medium.

HAND DIPPING AND ENROBING

The candy centers of chocolate-coated candies are either formed by hand or cast in starch or rubber molds. They are then dipped by hand or enrobed mechanically. The chocolate supply for hand dipping is normally kept in a pan maintained at the lowest temperature that still secures sufficient fluidity for the process. Because this temperature is higher than the dipping room temperature, a heat source, such as electrically heated **dipping pans** with thermostatic controls, is required. The dipped candy is placed either on trays or on belts while the chocolate coating sets.

Setting is controlled by conditioning the air in the dipping room. A dry-bulb temperature of 2 to 4°C best promotes rapid setting and provides a high gloss on the finished goods. However, the temperature in the dipping room is raised for human comfort. The suggested conditions for hand-dipping rooms are 18°C dry bulb and a relative humidity not exceeding 50 to 55%. The principal aim is to achieve uniform air distribution without objectionable drafts. The loads for this room include transmission, lights, and people, as well as the heat load from the chocolate and the heat used to warm dipping pans.

In high-speed production of bar candy, the chocolate coating is applied in an **enrober machine**, which consists essentially of a heated and thermostatically controlled reservoir for the fluid chocolate. This chocolate is pumped to an upper flow pan that allows it to flow in a curtain down to the main reservoir. An open chain-type belt carries the centers through the flowing chocolate curtain, where they pick up the covering. At the same time, grooved rolls pick up some chocolate and apply it to the bottom of the centers. The centers should be cooled to 24 to 27°C to assist solidification and retention of the proper amount of coating.

The coated pieces are transferred from the enrober to the **bottomer slab** and then pass into the enrober cooling tunnel. The function of the bottomer slab is to set the bottom coating as rapidly as possible in order to form a firm base for the pieces as they pass through the enrober tunnel. The bottomer slab is often a flat-plate heat exchanger fed with chilled water or propylene glycol or directly supplied with refrigerant. The belt carrying the candy passes directly over this plate, and heat transfer must take place from the candy through the belt to the surfaces of the bottomer slab. The bottomer slab is sometimes located before the enrober to create a good bottom prior to full coverage.

The **enrober cooling tunnel** sets the balance of the chocolate coating as rapidly as is consistent with high quality and good appearance of the finished pieces. The discharge end of the enrober tunnel is normally in the packing room, where the finished candy is wrapped and packed.

Although not absolutely necessary, air conditioning the enrobing room is desirable. Because the coating is exposed to the room atmosphere, the atmosphere should be clean to prevent contamination of the coating with foreign material. It is advisable to maintain conditions of 24 to 27°C dry bulb and 50 to 55% rh; that is, low enough to prevent the centers from warming up and to assist in the setting of the chocolate after it is applied.

BAR CANDY

The production of bar candy calls for high-speed semiautomatic operations to minimize production costs. From the kitchen, the center material is either delivered to spreaders, which form layers on tables, or cast in starch molds. Depending on the composition of the center, the hot material may be delivered at temperatures as high as 70 to 80°C. Successive layers of different color or flavor may be deposited to build up the entire center material. These layers usually consist of nougat, caramel, marshmallow whip, or similar ingredients, to which peanuts, almonds, or other nuts may be added. Because each of the ingredients requires a different cooking process, each separate ingredient is deposited in a separate operation. Thus, a 3 mm layer of caramel may be deposited first, then a layer of peanuts, followed by a 20 mm layer of nougat. Except for nuts, it is necessary to allow time for each successive layer to set before the next is applied. If the candy is spread in slabs, the slab must be

Candies, Nuts, Dried Fruits, and Dried Vegetables

cooled and then cut with rotary knives into pieces the size of the finished center.

HARD CANDY

Manufacturing hard candy with high-speed machinery requires air conditioning to maintain temperature and humidity. Candy made of cane sugar has somewhat different requirements from that made partly with corn syrup. For example, a dry-bulb temperature of 24 to 27°C with 40% rh is satisfactory for corn syrup (as the corn syrup percentage increases, the relative humidity must decrease), whereas the same temperature with 50% rh is satisfactory for cane sugar.

Where relative humidity is to be maintained at 40% or less, standard dehydrating systems employing such chemicals as lithium chloride, silica gel, or activated alumina should be used. A combination of refrigeration and dehydration is also used.

The amount of air required is a direct function of the sensible heat of the room. Approximate rules indicate that the quantity should be between 7.5 and 12.5 L/s per square metre of floor area, with a minimum of 15% outdoor air or 15 L/s per person. The sensible heat in hard candy, which is at a high temperature to keep it pliable during forming, must also be taken into account.

If concentrations of the finished product in containers or tubs are located in the general conditioned area, the quantity of air must be increased to prevent the product from sticking to the container.

Unitary air conditioners employing dry coils are satisfactory if they have a sufficient number of rows and adequate surface. A central station apparatus employing cooling and dehumidifying coils of similar design may also be used. Good filtration is essential for air purity as well as for preventing dirt accumulation on cooling coils. Reheat is required for some temperature and humidity conditions. Air distribution should be designed to provide uniform conditions and to minimize drafts.

HOT ROOMS

Such products as jellies and gums can best be dried in air-conditioned hot rooms. These products are normally cast into starch molds. The molds are contained in a tray approximately 890 mm by 380 mm by 40 mm with an extra 13 to 20 mm blocking at the bottom for air circulation. These trays are racked up on trucks, with the number of trays per truck (usually 25 to 30) determined by the method of loading. The trucks are loaded into the hot room where the actual drying is accomplished.

Normal drying conditions average between 50 and 65°C dry bulb with 15 to 20% rh. While humidity is important, close humidity control is not necessary.

Some operators prefer manual humidity control, which requires frequent inspection of actual conditions; others prefer automatic humidity control by instruments calibrated to maintain desired dry-bulb temperature and relative humidity in the hot room regardless of the moisture from the candy. With full automation, the supply air system should provide dry air to the unit and also cool the air usually needed in hot weather to purge the hot room after the drying cycle.

For proper air distribution in the hot room, the maximum amount of air must be in contact with the product. Providing space between trays is one means of accomplishing this. The trucks within the hot room must be placed to ensure continuous airflow from truck to truck with the shortest airflow path. Space must also be maintained at the entering and leaving air sides to ensure flow from the top to the bottom tray for each truck. A large air quantity is required to secure uniformity over the entire product zone.

One device for achieving uniformity is the **ejector nozzle system**. This system consists of a supply header fitted with conical ejector nozzles designed to have a tip velocity of 10 to 25 m/s with a static pressure behind the nozzle ranging as high as 3 kPa. Nozzles arranged in this way induce a flow of air about three times that actually supplied by the nozzles. This ratio gives the most economical balance between air quantity supplied and fan power. The ejector system causes the primary and induced airstreams to mix over the product and the space between the top tray and the ceiling. Sufficient ceiling height must be allowed for this mixing, which rapidly decreases the differential between the air supply temperature and the actual room temperature. The high airflow thus created decreases the temperature drop across the product. Because the temperature drop is proportional to the heat pickup, a greater airflow has a lower temperature drop. Thus, the spread between the air temperatures entering and leaving the product zone is reduced, promoting uniform drying.

When drying is completed, the product must be cooled rapidly to facilitate unloading. This quick cooling is provided by a second outdoor air intake, which bypasses the heating coil or air-conditioning unit. When this bypass intake is activated, drop dampers are opened in the bottom of the ejector header, so the air also bypasses the ejector nozzles and removes the heat from the room. A ceiling exhaust fan is also recommended for removal of the rising heated air.

The equipment for this operation consists of a fan and heating coil located outside the hot room. No electric motors should be in the room because of the hazard of sparking. This unit has outdoor air intakes, ejector headers, return air dampers, and dampers for the outdoor air intake. A recording controller to maintain an accurate record of each batch is recommended. The controller simply regulates the flow of steam to the heating coil to maintain the desired room temperature. Control switches should be provided to position the outdoor and return air dampers because a rise in humidity requires more outdoor air and a drop in humidity requires more return air. In some cases, this function can be achieved automatically with a humidity control. An end position can be included on the control switches for the cooling-down period to start the exhaust fan when the outdoor air damper is opened wide.

COLD ROOMS

Many confectionary products (e.g., marshmallows, certain types of bar centers, and cast cream centers) require chilling and drying but cannot withstand high temperatures. Drying conditions of about 24°C and 45% rh are required in the cold room, and the drying period varies from 24 to 48 h. An ejector-type system similar to that for hot rooms is used, the difference being that cooling coils are provided in the unit. The sensible and latent heat components of the load must be carefully determined so that the actual air quantity, together with the air supply and refrigerant temperatures, can be calculated.

Controlling relative humidity is an inherent part of the system design. The control system is similar to the one used for the hot room, except that its recording regulator must control the flow of steam to the heating coil in winter and regulate the flow of refrigerant to the cooling coil in summer. Flushing dampers and a cooling-down cycle are necessary. One precaution in connection with starch or sugar dust picked up in the return airstream must be observed. During the cooling cycle, condensate forms on the cooling coils, so that any starch or sugar dust deposits on the coils as a paste. This reduces capacity and necessitates frequent maintenance and cleaning of the equipment. Thus, air filters should always be used for the outdoor and return air entering the cooling coil. In addition, a coil wash system should be provided to aid in cleaning the coils.

COOLING TUNNELS

Various candy plant cooling requirements can best be handled in a cooling tunnel, including the cooling of (1) coated centers after they leave the enrober, (2) cast chocolate bars, and (3) hard candy. These operations are usually set up for a continuous flow of high-rate production. The product is normally conveyed on belts through either the enrober or the casting machine and then through a cooling chamber.

A cooling tunnel consists of an insulated box placed around the conveyor so that the product may travel through it in a continuous flow. Refrigerated air is supplied to this enclosure to cool the product. To achieve maximum heat transfer between the air and product, air should flow counter to the material flow. In general, air supply temperatures of 2 to 7°C with air velocities up to 2.5 m/s have been found satisfactory.

The actual size of the tunnel is determined by the size of the conveyor belt and the air quantity. The air quantity depends on the heat load and the desired rate of cooling. The rise in air temperature through the tunnel should be limited to 8 to 11 K maximum. Each tunnel generally has one refrigerated air handler, which normally consists of a fan and coil with the necessary duct connections to and from the unit. An outdoor air intake is advisable in appropriate climates because cooling can at times be accomplished with outdoor air (without operating the refrigeration plant). The outside air intake should be equipped with suitable air filters to minimize contamination. The tunnel should be made as tight as possible, and the entrance and exit openings for the candy should be as small as possible to limit air loss from the tunnel or air infiltration into the tunnel; in some cases, it is advisable to use a flexible canvas curtain to control airflow. As some loss is unavoidable, it is practicable in some applications to take a small amount of outdoor air or air from adjoining spaces to provide a slight excess pressure in the tunnel.

For chocolate-enrobing work, the condition of the air is the paramount factor in securing the best possible luster and most even coating. The best results are obtained with rather slow cooling, but this requires either a low production rate or excessively long tunnels; the final design is a compromise. The coating must be in the proper condition when it is poured over the centers because improper temperature at this point causes blushing or loss of luster. Proper temperature, however, is a function of the enrober machine and its operation; no amount of correction in the tunnel can compensate.

A variation of the standard single-pass counterflow tunnel has been used for enrobing. The tunnel is divided horizontally by an uninsulated sheet metal partition. The belt carrying the candy rides directly on this partition, and the return belt is brought back through the space below the partition. Cold air is supplied to the lower chamber near the enrober, progresses to the opposite end of the tunnel, is transferred to the top chamber, progresses back to a point near the enrober, and is then returned to the cooling equipment. This tunnel has two important advantages: (1) it chills the return belt so that the belt can act as a bottomer slab to quickly set the base of the coated piece, and (2) the uninsulated partition, with the coldest air below it, assists in this bottoming operation. Thus, the air supply has already absorbed some of its heat load by the time it is actually introduced to the product-cooling zone. The method approaches the advantage of slow cooling but keeps the tunnel at a minimum length.

For some applications, a spiral belt cooling tunnel can be used to save floor space. Products that are sensitive to injury or abrasion due to lateral movement of the belt underneath are generally unsuitable for a spiral conveyor. Typical cooling tunnel configurations are described and illustrated in Chapter 15.

COATING KETTLES OR PANS

The application of air conditioning to revolving coating kettles or pans is advantageous.

Originally these pans were merely supplied with warm air, ranging from 27 to 52°C. This air was then exhausted from the kettle to the room, creating a severe nuisance from sugar dust blowing out of the kettle. Another difficulty was that a portion of the energy required to rotate the kettle was converted to heat in the centers being coated, causing enough expansion to produce cracking or checking. To mitigate this problem, the following practice evolved: a portion of the coating is applied, the product is withdrawn for a seasoning period of up to 24 h, and the material is then returned to the kettles for additional coating.

Some installations overcome most of the sugar dust problem by providing a conditioned air supply to the kettles and positive exhaust from the kettles. The wet- and dry-bulb temperature of the air supplied to the kettles is controlled so that the rate of coating evaporation and the drying are uniform, at a high production rate and reduced labor cost. Evaporation of the moisture in the coating material tends to take place at the wet-bulb temperature of the air supplied to the kettle. A large portion of the heat of crystallization entering the product is absorbed, and the centers are not overheated. This eliminates the need for a seasoning period and permits continuous operation.

With air conditioning applied to coating kettles, the number of rejects caused by splitting, cracking, uneven coating, or doubles can be reduced considerably. Sugar dust recovery is accomplished with cyclone-type dust collecting devices.

PACKING ROOMS

Manufacturers spend considerable time and effort in the design and application of packaging materials due to their effect on product keeping quality. Important packaging considerations are moisture-proof containers, vapor retarders for abnormally high humidity conditions, and protection against freezing or extreme heat. Controlling the air in the packing room is essential for proper packing.

For example, air surrounding products packed in a room at 30°C dry bulb and 60% rh, which is not unusual in a normal summer, would have a 21°C dew point. If this sealed package were subjected to temperatures below 21°C, the air in the package would become supersaturated and moisture would condense on the surfaces of the container and the product. If the product were then subjected to a higher temperature, the moisture would reevaporate. In the process, chocolates would lose their luster or show sugar bloom, and marshmallows would develop either a sticky or a grained surface, depending on the formula used.

In practice, packing room conditions of 18°C dry bulb and 50% rh have been found most practicable. Results improve if the relative humidity is reduced several points. In the case of hard candy, which is intensely hygroscopic, the relative humidity should be reduced to 35 to 40% at 21 to 24°C.

REFRIGERATION PLANT

Large candy manufacturers often use a central refrigeration plant for cooling water and/or a secondary coolant for circulation throughout the plant to meet the various load requirements. Propylene glycol is the current secondary coolant of choice for food plants. See Chapter 4 for design and application information regarding secondary coolants.

The central refrigeration plant for cooling water and/or propylene glycol may use ammonia or one of the environmentally suitable halocarbon refrigerants. Heat transfer equipment both for new plants and for retrofits for cooling water or propylene glycol may be welded-plate heat exchangers or extended-surface shell-and-tube heat exchangers to increase plant efficiency and to reduce the refrigerant charge. They are usually piped for gravity-flooded or liquid overfeed operation. Compressors vary with the refrigerant used, plant size, initial and operating costs, and plans for future expansion. All installations should adhere to applicable refrigeration codes and regulations.

Secondary coolant temperatures usually vary between −2 and 0°C for central refrigeration plants. These temperatures seldom require an artificial defrost system for the evaporator coils. Some cooling tunnel applications may require lower coolant temperatures and associated defrost cycles.

A secondary coolant distribution system makes it feasible to connect all service points to one source of refrigeration and, with

Candies, Nuts, Dried Fruits, and Dried Vegetables

control systems, to maintain dry-bulb and dew-point temperatures precisely. In addition, this system can be extended to comfort cooling in offices and other nonproduction areas.

Smaller manufacturing plants may be better served by multiple condensing units and direct-expansion evaporators using an environmentally suitable refrigerant such as R-134a. These condensing unit/evaporator combinations may be grouped at appropriate refrigeration or air-conditioning load points as dictated by the plant layout. This arrangement provides flexibility to expand or retrofit a plant a portion at a time at a reasonable investment and operating cost.

STORAGE

CANDY

Most candies are held for 1 week to over a year between manufacture and consumption. Storage may be in the factory, in warehouses during shipping, or in retail outlets. It is important that the candy maintain quality during that time.

Low-temperature storage does not produce undesirable results if the following conditions are met:

- The candies are made of proper ingredients for refrigerated storage.
- The packages have a moisture barrier.
- The storage room is held at equilibrium humidity with desirable moisture conditions for preserving the candy.
- The candy is brought to room temperature before the packages are opened.

The storage period depends on (1) the marketing season of the candy; (2) the stability of the candy; and (3) the storage temperature and humidity (see Table 2).

The shelf life of a candy is determined by the stability of its individual ingredients. Common candy ingredients are sugar (including sucrose, dextrose, corn syrups, corn solids, and invert syrups), dark and milk chocolates, nuts (including coconut, peanuts, pecans, almonds, walnuts, and others), fruits (including cherries, dates, raisins, figs, apricots, and strawberries), dried milk and milk products, butter, dried eggs, cream of tartar, gelatin, soybean flour, wheat flour, starch, and artificial colors.

Refrigerated storage of candy ingredients is especially advantageous for seasonal products, such as peanuts, pecans, almonds, cherries, coconut, and chocolate. Ingredients with delicate flavors and colors, such as butter, dried eggs, and dried milk, retain quality more evenly year-round if kept properly refrigerated. Otherwise, ingredients containing fats or proteins may lose considerable flavor or develop off-flavors before being used.

Candies are semiperishable: the finest candies or candy ingredients may be ruined by a few weeks of improper storage. This includes many candy bars and packaged candies and some choice bulk candies. Unless refrigeration is provided from time of manufacture through the retail outlet, the types of candies offered for sale must be greatly reduced in the summer.

Benefits from refrigerated storage of candies, especially during the summer, are the following:

- Insects are rendered inactive at temperatures below 9°C.
- The tendency to become stale or rancid is reduced.
- Candies remain firm as an assurance against sticking to the wrapper or being smashed.
- Loss of color, aroma, and flavor is reduced.
- Candies can be manufactured year-round and accumulated for periods of heavy sales.

Color

Many colors used in hard candies, hard creams, and bonbons gradually fade during storage at room temperature, especially in the light. However, the most marked effect of storage temperature on color occurs with chocolates. In candies high in protein and nuts, there is a gradual darkening of color, especially at higher storage temperatures.

Temperatures of 30 to 35°C cause graying of chocolates in only a few hours and darkening of nuts within a month. **Graying or fat blooming** is caused by crystals of fat on the surface of the chocolate coating. While this condition is usually associated with old candies, new candies can become gray after one day under adverse storage conditions. Chilling of chocolates following exposure to high temperatures produces graying very quickly, but chilling without previous heat exposure does not.

Sugar blooming of chocolate looks similar to graying or fat blooming and is caused by crystallized sugar deposited on the surface from condensation of moisture following removal of the candy from refrigerated storage without proper tempering. Experiments have shown that chocolate tempered by gradually raising the temperature to normal without opening the package did not incur sugar blooming even after storage at −20°C or lower. Sugar bloom may also occur due to storage in overhumid air and migration of moisture from the centers to the surface.

Flavor

Keeping candy fresh is one of the chief reasons for refrigerated storage. Most flavors added to candies, including peppermint, lemon, orange, cherry, and grape, are distinctive and stable during storage. Less pronounced flavors such as those of butter, milk, eggs, nuts, and fruits are more sensitive to high temperatures.

Low temperatures retard the development of staleness and rancidity in fats, preserve flavors in fruit ingredients, and prevent staleness and other off-flavors in candies containing such semiperishable ingredients as milk, eggs, gelatin, nuts, and coconut. Candies containing fruit become strong in flavor when they are stored at room temperature or higher for more than a few weeks; those containing nuts become rancid. There are no specific critical temperatures at which undesirable changes occur, but the lower the temperature, the more slowly they take place.

Table 2 Expected Storage Life for Candy

Candy	Moisture Content, %	Relative Humidity, %	Storage Life, Months Storage Temperature, °C			
			20	9	0	−18
Sweet chocolate	0.36	40	3	6	9	12
Milk chocolate	0.52	40	2	2	4	8
Lemon drops	0.76	40	2	4	9	12
Chocolate-covered peanuts	0.91	40 to 45	2	4	6	8
Peanut brittle	1.58	40	1	1.5	3	6
Coated nut roll	5.16	45 to 50	1.5	3	6	9
Uncoated peanut roll	5.89	45 to 50	1	2	3	6
Nougat bar	6.14	50	1.5	3	6	9
Hard creams	6.56	50	3	6	12	12
Sugar bonbons	7.53	50	3	6	12	12
Coconut squares	7.70	50	2	3	6	9
Peanut butter taffy kisses	8.20	40	2	3	5	10
Chocolate-covered creams	8.09	50	1	3	6	9
Chocolate-covered soft creams	8.22	50	1.5	3	5	9
Plain caramels	9.04	50	3	6	9	12
Fudge	10.21	65	2.5	5	12	12
Gumdrops	15.11	65	3	6	12	12
Marshmallows	16.00	65	2	3	6	9

Texture

Candy becomes increasingly soft at high temperatures and increasingly hard and brittle at low temperatures, reaching an optimum for eating at about 21°C. Changes in texture are reversible from below −20°C to 27°C, enabling refrigerated candies to be returned satisfactorily to any desired temperature for eating. This is extremely important because the texture of most candies subjected to very low temperatures (or even shipped in contact with dry ice at about −80°C) is not permanently changed.

Most candies are manufactured at controlled temperatures. Their texture (except in hard candies and hard creams) is maintained best at temperatures below 20°C.

Insects

Candies containing fruit, chocolate, nuts, or coconut are favorite hosts for insects. Because fumigation and insect repellants are seldom permissible, refrigeration is used to inactivate insects in candy and candy ingredients.

Common insects become active at about 10°C, and activity increases as the temperature is raised to 38°C. Although common cold storage temperatures do not kill many insects, temperatures below 10°C do inactivate them. Both adults and eggs may exist for months at above-freezing temperatures without feeding or propagating. Candies with insect eggs on either the product or wrappers may be refrigerated for long periods with no apparent damage, but when they are warmed up, a serious insect infestation may develop.

Both adults and eggs are killed at storage temperatures of about −20°C. Storing candies at −20°C for a few weeks usually destroys all forms of insect life. Lower temperatures and long storage periods are lethal to insects.

Storage Temperature

Regarding their effect on candies, storage temperature is difficult to separate from humidity, but the latter is more important. There are no specific critical refrigerated temperatures at which certain types of candy must be held. In general, the lower the temperature, the longer the storage life, but the greater the problem of moisture condensation on removal.

Air Conditioning (20 to 21°C). Because the storage of candies begins in the tempering room of the manufacturing plant, 20°C and 50% rh is desirable to prevent pieces from being packaged when they are soft and sticky. Under these conditions, all candies remain firm and there is little or no graying of chocolates; their original luster can be held.

Hard candies and candies containing only sugar ingredients keep in good condition for more than 6 months at 20 to 21°C. Other types become stale, lose flavor and luster, and darken in color.

Under prolonged storage, candy containing nuts or chocolate becomes musty or rancid, and even the colors and flavors of some hard candies may fade. Only **summer candies** should be held for more than a few weeks at 20 to 21°C or higher. Unless precautions are taken, the temperature of truck or rail shipments may rise to the melting point of semisoft candies or to the graying point of chocolates. Candies temporarily stored in the sunshine or in warm places in buildings may suffer severe loss of shape, luster, and color. Some companies use refrigerated trucks and railroad cars for hauling candies. Portable refrigerated containers provide a viable means of ensuring that candies maintain their quality from manufacturer to retail outlet.

Cool Storage (9 to 10°C). Candies stored at this temperature remain firm and retain good texture and color; only those containing nuts, butter, cream, or other fats become stale or rancid within 4 months. Candies that remain practically fresh for 4 months are fudge, caramels, sugar bonbons, gumdrops, marshmallows, lemon drops, hard creams, and semisweet chocolates; they are wrapped in aluminum foil to give added protection. Candies that become stale at this temperature are peanut butter taffy kisses, peanut brittle, uncovered peanut rolls, chocolate-covered peanut rolls, and nougat bars.

Cold Storage (0 to 1°C). Most candies can be successfully held in cold storage for at least one year, and many for much longer. Only those containing nuts, coconut, chocolate, or other fatty materials become stale or rancid.

Freezer Storage (−20°C). The need and economic justification for freezing candy is the same as for any other food—better preservation for a longer time. This method of preservation is suitable for candies (1) in which high quality standards must be maintained; (2) in which a longer shelf life is desired than is accomplished from other methods of storage; (3) that are normally manufactured 6 to 9 months in advance of consumption; and (4) that are especially suitable for retailing as frozen items. Because of their high sugar content and low moisture content, little ice formation occurs in candies at −20°C.

One of the chief reasons for freezing candies is to hold them in an unchanged condition for as long as 9 months, then thaw and sell them as fresh candies. Experience shows that this is not only possible but also practical if the manufacturer (1) freezes only those candies that would lose quality when held at a higher temperature; (2) eliminates the few kinds that crack during freezing; (3) packages the candies in moistureproof containers similar to those used for other frozen foods; and (4) thaws the candies in the unopened packages to avoid condensation of moisture on the surface.

Moistureproof Packaging. Experiments show that candies for freezing require more protection than those for common storage because the storage period is usually longer and condensation on removal is more likely. A single layer of moistureproof material (e.g., aluminum foil, plastic film, or glassine) affords adequate protection. Candies not fully protected from desiccation become hard and grainy and lose flavor.

Adequate protection is provided when the moisture barrier is in contact with the candy in the form of a sealed, individual wrapper. Inner liners for the boxes protect candy, provided they are sealed (which is difficult and seldom accomplished). Moisture barriers are usually applied as overwraps for the boxes, chiefly because overwraps are easiest to apply and seal by machines. Overwraps provide less protection than wraps for individual pieces of candy because of the larger amount of air enclosed in the box. Also, boxes with extended edges offer less protection.

Thawing. Frozen candies should be thawed in the unopened packages. While freezing itself affects only a few types of candies, the manner of removal from storage affects all candies, especially those unprotected by special coatings or individual wraps.

Improvement. Candies that are improved in freshness, mellowness, or textural smoothness by freezing include those with high moisture content and without protective coatings or individual wrapping. Usually these are candies ordinarily subject to surface drying. Marshmallows, jellies, caramels, fudges, divinities, coconut macaroons, fruit loaves, coconut bonbons, panned Easter eggs, malted milk balls, and chocolate puffs are in this group.

Stability. The stability of candies after freezing is good. Candies may be held frozen for 6 months or more, carefully thawed, and then sold as fresh. Some candies are prepared especially for freezing. These are made of low melting point fat, have more flavor and softer texture than most candies, and should be eaten while they are cold.

Humidity Requirements

Sugar ingredients are stable over a wide range of storage temperatures, but they are sensitive to high or low humidity. The initial moisture content of candies largely determines the optimum relative humidity of the storage atmosphere. Candies with a moisture content of 12 to 16%—marshmallows, gumdrops, coconut sticks, jelly beans, and fudge—should be stored at 60 to 65% rh to avoid becoming (1) sticky, runny or moldy or (2) hard and crusty. Candies with

Candies, Nuts, Dried Fruits, and Dried Vegetables

a moisture content of 5 to 9%—most fine candies, nougat bars, nut bars, hard and soft creams, bonbons, and caramels—should be stored at 50 to 55% rh to retain their original mass, finish, and texture. Candies with a moisture content below 2%—milk chocolate bars, chocolate covered nuts, and all kinds of hard candies—should be stored at 45% rh or lower.

The hygroscopicity of the ingredients also determines the relative humidity at which candies must be held in order to retain their original firmness and finish. Candies with a high proportion of invert syrups, such as taffy kisses, must be kept very dry, even though their moisture content is not extremely low. Other candies containing high proportions of invert syrup, honey, or corn syrup must be held in a drier atmosphere than the moisture content indicates.

Candies stored at low temperatures have a wider range of critical relative humidities than those stored at high temperatures. For example, nougat bars stored at 65% rh (10% too high) become sticky within a few days at room temperature, but at 5°C or lower, stickiness might not develop for many weeks. Similarly, marshmallows stored in a room with 55% rh (10% too low) become dry and crusty within a few days at room temperature, but at 5°C or lower, they show little change for several weeks. Refrigeration retards the ill effects of storage under improper humidity conditions. Humectants, such as sorbitol, glycerine, and high conversion corn syrup, are advantageous for maintaining original moisture content of certain candies.

NUTS

Commonly refrigerated nuts include peanuts, walnuts, pecans, almonds, filberts, chestnuts, and imported cashews and Brazil nuts. The advantages of refrigerated storage of nuts are the following:

- Marketable life is increased as much as ten times.
- Natural texture, color, and flavor are retained almost perfectly from one season to the next.
- Staleness, rancidity, and molding are retarded for more than 2 years, depending on the temperature.
- Insect activity is arrested at temperatures below 9°C.

With optimum temperature, humidity, atmospheric conditions, and packaging, good-quality nuts may be successfully stored for up to 5 years.

Temperature

Other conditions being equal, the lower the temperature, the longer the storage life of nuts. Storage life may be doubled or tripled with each 10 K drop in temperature. The freezing points of nuts, depending on the moisture content, are about −5°C for chestnuts; −10°C for walnuts, pecans, and filberts; and −10.6°C for peanuts. Normal moisture content for stored nuts is as follows: chestnuts, 30%; peanuts, 6%; walnuts, 4.5%; pecans, 4%; and filberts, 3.5%.

Shelled nuts to be stored from one harvest season to the next without appreciable loss in quality must be held at 2°C or lower; those to be stored for 6 to 9 months must be kept at 9°C or lower; and all nuts stored for 4 to 6 months should be held below 20°C. Storage life at a given temperature doubles if the nuts are unshelled.

Relative Humidity

Although the storage temperature of nuts may range from 20 to −30°C or lower, the relative humidity must remain between 65 and 75%. This is to maintain the optimum moisture content for desired texture, color, flavor, and stability. If the moisture content rises as much as 2% above normal, the nuts (except chestnuts) darken, become stale, and may become moldy. If the moisture drops more than 2% below normal, the nuts become objectionably hard and brittle.

When the relative humidity is suitable, nuts that are too high or too low in moisture may be safely stored with the assurance that rapid air circulation will bring the moisture content to a safe level. In this sense, the storage room acts as a conditioning room.

Atmosphere

All nuts (again, except chestnuts) contain 45% or more oil and readily absorb odors and flavors from the atmosphere and surrounding products. Certain gases, particularly ammonia, react with tannin in the seed coats of nuts, causing them to turn black. Therefore, the atmosphere in the nut storage room must be free of all odors. This includes the containers, walls of the room, pallets, and other stored products.

Products that can be safely stored with nuts include dried fruits, candies, rice, and goods packaged in cans, bottles, or barrels. Commodities that should not be stored with nuts are onions, meats, cheese, chocolate, fresh fruits, and other products having an odor or a high moisture content.

Packaging

The storage life of nuts may be greatly influenced by the choice of package. Nuts become bruised when they are shelled, and oil *crawls* over the surface and onto the package in a very thin film. Unless this crawling is retarded by a package that acts as a barrier, contains an antioxidant, or removes air by vacuum, the nuts become stale and rancid. Furthermore, some packaging materials (e.g., polyethylene) should be avoided because they impart an undesirable odor to the nuts.

DRIED FRUITS AND VEGETABLES

Dried fruits and vegetables differ from each other in the following ways:

1. Fruits contain sugars that render them more hygroscopic, harder to dry, and greater absorbers of moisture during storage.
2. Moisture in dried fruits may range from 32% with sorbate treatment to as low as 2%, while that in vegetables ranges from 7% to a low of 0.3%.
3. Dried fruits are acid, more highly colored, and more stable during storage than vegetables, which are nonacid.
4. Fruits are generally dried raw with active enzyme and respiration systems, while vegetables are blanched or precooked, with no active enzymes (thus, dried fruits are more responsive to storage temperature and humidity conditions than are vegetables).
5. The high sugar and acid content of dried fruits provides an adverse physical environment for bacteria, thus making their growth almost impossible even though the moisture level is higher than that in dried vegetables.

Dried fruits and vegetables maintain quality longer when stored at low temperatures. The storage life of dehydrated vegetables has been extended by packing them in a nitrogen or carbon dioxide atmosphere in the presence of a desiccant to achieve further reduction in moisture content. Staleness (charred flavor or off-flavor) and other deteriorative changes in dehydrated vegetables are inhibited by packing in nitrogen. In air-packed samples, the rate of staling is reduced at low temperature. Cut fruits and dried vegetables are widely treated with sulfite to retain color and extend storage life. A light coating of laundry starch applied to diced carrots prior to dehydration has achieved excellent results in retention of color and other quality factors during storage in cellophane at 29°C without sulfite.

Refrigerated storage at 5 to 10°C or lower retards and controls insect infestation. Substantial killing occurs with exposure at 0°C for 6 months or longer, and a temperature of −20°C kills insects within a few hours. An alternative method is fumigation, which is generally used in commercial practice.

Nonenzymatic browning (browning in products that have been scaled or blanched adequately to inactivate enzymes) is reduced in dehydrated vegetables at low temperature. Cold storage offers protection for several years.

Increasing the temperature of dried apricots accelerates oxygen consumption, carbon dioxide production, disappearance of sulfur dioxide, and darkening.

Molds and yeasts do not grow in dried fruits that have an adequate sulfur dioxide content or less than 25% moisture. At 0°C, relative humidity is less important than at higher temperatures.

Other methods of dehydration include the following:

Dehydrofreezing. In this process, the raw, prepared product is dried to about 50% in mass, followed by freezing. This method has yielded excellent fruit and vegetable products when storage was at −20°C or lower. Concentration of juices by low temperature and high vacuum followed by preservation of the concentrate by freezing (Chapter 24) is another application of the process.

Freeze-Drying. Products to be freeze-dried are usually frozen slowly to form larger ice crystals to increase process efficiency while retaining quality. The frozen product is placed in freeze-dry chambers, where the moisture is removed by sublimation through the application of vacuum and low heat such that product porosity is preserved for subsequent reconstitution. This method is successfully used with products of high value, high protein, low fat, and low sugar content. While refrigerated storage is not necessary to prevent freeze-dried food from spoiling, it is necessary in order to preserve maximum flavor and natural color.

Dried Fruit Storage

Refrigeration is beneficial in augmenting drying as a means of preserving fruits. The optimum conditions for holding most dried fruits are about 55% rh and just above the individual fruit's freezing temperature. Because the sugar content of these products is high, the freezing point varies from about −6°C to −3°C. Refrigerated storage is beneficial in retaining natural flavor, ascorbic acid, carotene, and sulfur dioxide and in controlling browning, insects, rancidity, and molding. Other than for insect control, low humidity is more important than low temperature for storing dried fruit. Packaged, sulfured cut fruits keep adequately at higher humidities.

Although most dried fruits are adversely affected by softening and injured by freezing, dates are held best by freezing. Before storage, dried fruits should be brought to the desired moisture content.

When practical, dried fruits should be packed in moistureproof containers made of metal or foil, which not only ensure a constant moisture content in storage, but also prevent injury from moisture condensation on removal from storage. The permeability of the packaging medium is extremely important due to the adverse effects of storage humidity. The following are dried fruit storage recommendations:

Raisins. At 0 to 4°C and 50 to 60% rh, sugaring is prevented for one year, provided the moisture content of the dried fruit is not unusually high. Raisins contain 15 to 18% moisture; for extremely long storage, the lowest possible moisture content should be maintained.

Figs. These may be held for a year at 0 to 4°C and 50 to 60% rh. A temperature of 13°C or lower prevents darkening for more than 5 months, and low humidity controls sugaring.

Prunes. These may be held for a year at 0 to 4°C and 50 to 60% rh. For storage of 4 to 5 months, a relative humidity of 75 to 80% is not detrimental.

Apples. At 0 to 4°C and 55 to 65% rh, dried apples retain excellent color and texture for more than a year. A relative humidity of 70 to 80% is not objectionable at 0°C, but at 4°C enough moisture may be gained to cause the fruit to mold within 8 months. Browning develops gradually at 4°C and above.

Pears. Same as for apples.

Peaches. Sun-dried freestone peaches are harder to store than most dried fruits. Therefore, the temperature should be held close to 0°C with a relative humidity of 55 to 65%. At 4°C and moderate humidity, the moisture pickup causes rapid molding and browning. Clingstone peaches (dehydrated after steam scalding) should be stored at 0 to 4°C and 55 to 75% rh. Sun-dried peaches tolerate a slightly higher humidity than dehydrated peaches.

Apricots. Dried apricots are easy to keep in refrigerated storage at 0 to 4°C and 55 to 65% rh. They remain in excellent condition for more than a year. At 4°C and moderate humidity, there is enough gain in moisture content to cause molding.

Dates (sucrose or hard type). For storage of 6 months or less, dates may be held at 0°C and 70 to 75% rh, but for longer storage they should be stored at −4 to −3°C. Usually it is more convenient to store them at 0°C, at which temperature they can be stored for over a year.

Soft or invert sugar-type dates may be held for 6 months at −2 to 0°C, but if storage is for 9 to 12 months, the temperature should be −18 to −12°C. Uncured dates should be stored at −18 to −12°C.

Dried Vegetable Storage

Few specific recommendations for dehydrated vegetable storage temperatures are available. Decrease in storage temperature retards deterioration, but cold storage is considered necessary only for long storage periods. Among the advantages are (1) control of insects at 7°C or lower; (2) preservation of natural colors; and (3) retention of initial flavors and vitamins.

Because most dried vegetables have very low moisture content, are well packaged, and are usually surrounded by an atmosphere of nitrogen or rarefied air, refrigeration is less essential than it is for fresh vegetables.

CONTROLLED ATMOSPHERE

Low oxygen in the storage atmosphere suppresses the growth of insects and molds, retards rancidity and staleness, and reduces oxidative changes in flavors, odors, and colors. In small packages, oxygen can be reduced by vacuum; in storage and large shipping containers, oxygen can best be flushed out with nitrogen. Excellent results have been obtained by substituting up to 98% of the atmosphere with nitrogen. Nitrogen is preferred to carbon dioxide, ethylene, or other gases for storage of low-moisture products; it greatly extends the shelf life even with refrigeration.

CHAPTER 29

TRUCKS, TRAILERS, AND CONTAINERS

Types of Equipment	29.1
Body Design and Construction	29.1
Auxiliary Equipment	29.3
Types of Refrigeration Systems	29.4
Mechanical Refrigeration Components	29.5
Calculation of Cooling Loads	29.6

THE transport of perishable commodities may be as simple as a direct delivery from farm to market by a small truck. However, they are more likely to travel to distant markets intermodally, that is, by some combination of highway, ocean, and railroad. This chapter describes the vehicles and refrigeration equipment that preserve perishable commodities as they travel. Additional details may be found in the section on Bibliography.

TYPES OF EQUIPMENT

Refrigerated transport equipment can be broadly classified by type of operation—highway and intermodal equipment, or straight trucks. Intermodal equipment often includes special provisions for marine service.

Highway and Intermodal Vehicles

Refrigerated semitrailers for overland use are up to 16.8 m long, 2590 mm wide, and 4270 mm overall height. Where permitted, double or triple trailers may be pulled by one tractor, or a trailer may be pulled by a refrigerated truck. USDA *Agricultural Handbook* 669 (Ashby et al. 1995) gives recommended loading methods and temperatures for many perishable products.

Highway trailers, uncoupled from their tractors, are carried "piggyback" on railroad flatcars (trailer on flatcar, or TOFC). Trailers with tractors—and trucks—are driven onto ships (roll-on roll-off, or RORO). Containers are carried on trailer and truck chassis, on railroad flatcars (container on flatcar, or COFC), and above and below deck on container ships.

Containers differ from trailer bodies primarily in the hardware for attachment to the chassis, and in structural hardware required for lifting and stacking. In a typical trailer-ship transfer operation, a container is lifted from a trailer chassis and placed either on deck or in the hold. Special fittings permit these containers to be stacked six-high in ship holds (and at shoreside terminals). On-deck carriage of containers may involve stacking two refrigerated containers, with a dry cargo van on top. The corner posts must be designed to carry the floor load and added vertical load applied by lashing gear. The American National Standards Institute (ANSI) and the International Organization for Standardization (ISO) have developed standards for refrigerated containers for intermodal use (ANSI *Standard* MH5.1.1.5, ISO *Standard* 668).

For long-haul land transportation, by either highway or rail, each vehicle requires an independent means for refrigeration. Containers interchanged between marine and land use can be refrigerated with independent systems. Alternatively, they can be connected to a central refrigerated air system on specially equipped ships and to clip-on refrigeration units on land. Two or more containers are sometimes connected end-to-end to form a single unit for long-haul operation and are separated for local delivery.

Some railroad yards transfer trailer bodies from the trailer chassis to gondola or flatcars in a manner similar to those for the trailer-to-ship operation. Various container transfer machines are used to move, stack, or load containers. Finger-lift, pincer-lift, or cable-lift systems each apply characteristic stresses that must be considered in structural design. Many truck and rail carriers apply standard piggyback techniques (i.e., the trailer is driven or lifted onto a special railroad flatcar and attached to the car for travel).

Trucks

Refrigerated trucks are used primarily for short-haul wholesale delivery within or between population centers. Body styles have been developed to suit the character and distribution needs of perishable products. Trucks are used in operations that may require more door openings than trailers. Penney and Phillips (1967) measured the air exchange and cooling load caused by door openings of typical refrigerated trucks.

Various devices such as power lift gates and conveyors, which facilitate loading and unloading, are built into the body. Trucks for retail delivery may be furnished for either walk-in or reach-in service. Wheeled racks, which can be rolled into place in the truck, are used to expedite loading.

Multitemperature Vehicles

Trucks and trailers may be partitioned into several compartments that can be held at different temperatures. This enables a variety of products to be carried in a single vehicle, making food distribution more efficient. For example, multitemperature vehicles can carry ice cream at −23°C, frozen vegetables at −18°C, meat at −1°C, and fresh vegetables at 1°C. This is usually accomplished via a thermostatically controlled fan-coil unit in each compartment. These fan-coil units are supplied with refrigerant and electricity from a single engine-driven condensing unit; otherwise, they function independently.

BODY DESIGN AND CONSTRUCTION

An insulated body, when matched with a suitable refrigeration system, should provide economical temperature control for the commodity being transported or stored. The necessary features are determined by the type of service intended. Trucks may have doors on the two sides rather than at the rear, but most trailers and containers have full opening rear doors large enough to permit forklift trucks to be driven into the vehicle. Many trailers are also equipped with a curbside door. Meat rails are provided if the vehicle is to be used for hanging meat. Temporary or permanent discharge ducts from the refrigerating unit can be used to improve air movement through the cargo.

Insulation

The goal of body builders is to construct a light body, having sufficient strength, in which the insulation will stay dry and maintain

---

The preparation of this chapter is assigned to TC 10.6, Transport Refrigeration.

its original insulating value. Insulation should have moderate cost, low density, low thermal conductivity, low moisture permeability and retention, ease of application, uniformity, resistance to breakdown at temperature extremes, and fire resistance. The insulation should resist cracking, crumbling, shifting, and packing from the shock, vibration, and flexing of the body structure.

Bodies for low-temperature use (about −18°C or lower) have been insulated with 50 to 100 mm of urethane foam having a k-factor of about 0.022 W/(m·K). Legal requirements and other factors limit the outside width, height, and length of vehicles. Therefore, the designer must establish an insulation thickness to obtain optimum cargo space and operating performance. For example, increasing insulation thickness from 75 to 100 mm in a 12.2 m long trailer decreases cargo space by 2.8 m^3, or 5%. Decreased insulation thickness, however, requires an increase in refrigerating or heating capacity. In determining the optimum insulation thickness, the transport body and its refrigeration unit must be considered as one system.

The heat transferred from outside to inside the vehicle, excluding heat that enters when the door is open for loading or unloading cargo, includes heat transferred (1) through insulation, (2) through structural members, and (3) by air and moisture leakage. Air leakage can affect overall heat transfer significantly, so some buyers specify a maximum rate of air leakage with a given air pressure imposed in the cargo space.

Floor Insulation

Floor loads are frequently supported on rigid insulating material to eliminate floor beams. Most truck, trailer, and container floors must support forklift trucks. Floors should be watertight, and if separate floor racks are not furnished, the floor surface should allow adequate air movement under the load. Several types of formed or extruded floor surfaces are available, but not all of these permit enough air circulation under the load.

Galvanized or treated steel and aluminum are used widely for floors in ice cream trucks, while aluminum is most often used for trailer floors. Expanded polystyrene and polyurethane foam are used for floor insulation. A metal skirt reaching at least 150 mm up the walls should be bonded to the floor so that water running down the wall or collecting on the floor will not enter the insulation. Drains, if used, must be self-closing. Plywood, aluminum, other metals, and certain plastics are used for interior wall surfaces. Glass-reinforced plastic materials are being used for both interior and exterior surfaces.

Reducing Heat Transmission

Often the buyer specifies the maximum allowable heat transfer rate, usually at 38°C and 50% rh outside, and −18°C inside. Bodies for low temperature (−18 to −29°C) typically have 50 to 100 mm of polyurethane insulation. For temperatures above freezing, 25 to 65 mm of polyurethane is applied. At the option of the customer, the insulation thickness is usually nonuniform on the different surfaces, with the front wall generally thicker for structural reasons. For multistop delivery operations, trailers and trucks are sometimes provided with roll-up insulated doors.

Moisture Penetration

All exterior surfaces of an insulated body must be made as airtight and water vaportight as possible. Water vapor will pass through any opening or nonvaporproof barrier in the outer shell and will condense in the insulation itself or in cavities in it. Some ice cream truck bodies have increased more than 225 kg over a period of use, and large semitrailers in frozen food service have gained up to 680 kg. Since the k-factor of wet insulation may be much higher than its original value, the required low temperature may not be obtained and a much greater load is placed on the refrigerating equipment. Use of closed-cell insulation in place of fibrous types minimizes moisture buildup problems.

Air Leakage

Eby and Collister (1955) pointed out the serious heat gain penalty imposed by air entering the insulation space through cracks in the front of the vehicle during vehicle travel. They showed that this driving force is 300 Pa at 80 km/h, and that a 2-m length of unsealed seam can permit 11.3 L/s of air to enter. At ambient conditions of 38°C and 50% rh, and a trailer temperature of −18°C, the heat gain caused by this amount of infiltration air could be 1 350 W, assuming that the leakage air left the vehicle saturated at −18°C.

Refrigerated vehicles leak air even when they are stationary, probably because of the stack effect of the temperature difference between inside and outside. This driving force for air infiltration for a body 2.4 m high and a temperature difference of 55 K, is about 7 Pa (Phillips et al. 1960). Openings with an aggregate area of 700 mm^2 each at the top and bottom of the exterior skin would permit an air leakage of about 1.0 L/s, if they function as thin plate orifices. For a nominal 11-m trailer, observed leakage rates of more than 5.5 L/s and an accumulation of nearly 0.13 g/s of ice at ambient laboratory conditions of 38°C at 50% rh with a −18°C trailer temperature, illustrate the value of improving the vaportightness of the exterior shell.

The need for better sealing is even more significant at road speeds. For example, if a 10-cm^2 opening were located where it was subjected to 300 Pa ram air pressure at 80 km/h, approximately 14 L/s of air could be driven into the insulation space of the vehicle. At ambient temperature and humidity conditions of 38°C and 50%, the heat gain from this leakage would be about 1.73 kW, if the air was cooled to the trailer temperature of −18°C.

Road tests of nominal 11-m commercial trailers by Phillips et al. (1960) have shown air leakage rates as high as 12.5 L/s at 80 km/h. Figure 1 shows the heat gain resulting from air leaking into a vehicle with an interior temperature of −18°C, at various ambient temperatures and relative humidities.

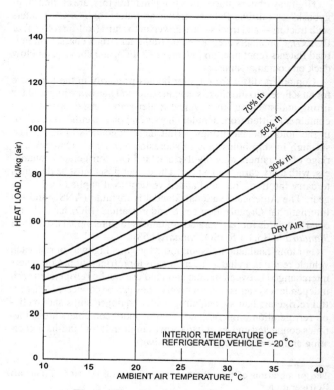

Fig. 1 Heat Load from Air Leakage

Trucks, Trailers, and Containers

When moist air moves through the insulated space, the water vapor left behind causes trouble in several ways:

- Increased heat gain caused by the latent heat of vaporization and fusion
- Loss of payload because of gain in mass
- Corrosion
- Increased heat gain through insulation
- Coating of coils with frost, resulting in loss of refrigerating effect
- More frequent defrosting
- Physical damage to insulation
- Rotting of wood members
- Odor

Additional sealing of the metal skin of a truck or trailer is required to make it leakproof. Three methods are popular: (1) use of foamed-in-place plastic insulation; (2) lining the inside of the exterior skin with a nonpermeable vapor barrier, such as aluminum foil coated with a plastic binder, which can be sealed at the joints; or (3) coating the interior surface of the exterior skin with some type of vapor-sealing compound such as neoprene. The vapor seal must not be destroyed where wiring, piping, or frames penetrate the barrier.

Typical trailer specifications call for air leakage of less than 1 L/s at a negative pressure of 125 Pa in the trailer. The interior skin need not, and probably should not, be vaportight; it must be water resistant so that cleaning does not readily soak the insulating material, however. All doors must fit properly and be gasketed to reduce air and heat leakage. Hardware for these doors must be substantial enough to withstand severe conditions.

The interior walls of trailers, which may be used in vacuum cooling of fresh produce, must be vented, preferably into the cargo space, to prevent pressure damage to the insulated spaces during the vacuum cooling operation.

Insulated bodies with one-piece molded plastic exterior shells are now being produced. Types and sizes range from small retail meat trucks to maximum legal length trailers and containers. High impact, thermal resistance, and good vapor-sealing characteristics are claimed for this type of construction.

Air Circulation Systems

Air circulation is one of the most important factors in protecting refrigerated loads of perishable foods. Inadequate air distribution is probably the principal cause of improper cargo refrigeration during transport. Satisfactory product temperatures cannot be maintained unless the load is surrounded by proper air and/or surface temperatures.

Products that are to be cooled in transit, or loads that produce heat (e.g., fresh fruits, vegetables, and flowers) must be arranged so that refrigerated air can move not only around the load but through it as well. Non-heat-producing loads, placed in the vehicle at the proper temperature, need only be blanketed with air in areas where heat could enter the product.

The two major methods of circulating air in refrigerated vehicles are shown in Figure 2. The overhead (or top air) delivery system is prevalent in over-the-road vehicles. In this method, the conditioned air flows into the space between the cargo and vehicle ceiling. The bottom-air delivery system has been employed extensively in seagoing containers for several decades, but it has been used only to a limited extent in highway trailers. In trucks that use overhead eutectic plates, either forced or gravity circulation is used.

Most systems using fans can move an adequate air supply over the top of the load. When light density loads approach the ceiling, or when structure interferes with the throw of air, supply ducts, either permanent or temporary, can facilitate the delivery of air to the rear of the vehicle. In addition to those that are obvious in Figure 2, air channels must be provided on the sides and rear of the vehicle via batten strips, or similar means. Attention must be paid to the movement of air along the floor under the load. As mentioned

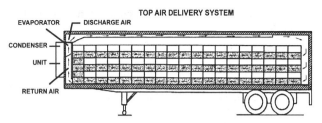

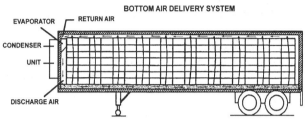

Fig. 2 Section of Vehicle Showing Air Circulation

in the section on Floor Insulation, racks or formed metal floors are used to enable air to flow the length of the vehicle under the load.

AUXILIARY EQUIPMENT

Heating Systems (Heaters)

Most trailer and truck refrigeration units are also capable of heating the cargo space. They use the evaporator to desuperheat hot gas from the compressor, and the evaporator fan circulates the heated air. On occasion, insulated vehicles that do not have refrigeration units require heat. This is accomplished by diesel engine-powered heating units or by direct combustion heaters.

Engine-powered heating units mount on the front of a truck or trailer (like a refrigeration unit). The engine drives a fan that circulates air through an engine-coolant heated finned-tube coil to the cargo space. Some recently designed units use microprocessors to control their operation.

Direct combustion heaters burn alcohol, kerosene, butane, propane, or charcoal. Because of the hazards involved, these heaters must be appropriately designed and carefully used.

Controlled and Modified Atmospheres

Systems have been developed to help maintain fresh produce quality during transport by controlling atmospheric chemistry inside the refrigerated vehicle. Commodity respiration, or the combining of natural sugars with oxygen, can be slowed by reducing the ambient oxygen level. Increasing the carbon dioxide level further slows respiration and helps prevent senescence (aging). Specific combinations of both gases will control, and in some cases eliminate, certain pathogens and insects. Also, texture and color can be maintained by limiting ethylene, a natural ripening hormone.

Each commodity has unique physiological characteristics which dictate specific oxygen and carbon dioxide levels. Monitoring and control of these levels is critical, since excursions outside acceptable ranges can have detrimental effects, causing spoilage and potential shipment loss.

Controlled atmosphere is generally defined as continual monitoring and adjustment of gas levels. Systems include nitrogen generators used to displace oxygen and ethylene, combined with sophisticated sensors and controls. Oxygen can be increased by venting to outside air. Carbon dioxide is supplied from bottles, or as a by-product of cargo respiration. **Modified** atmosphere is typically a one-time surrounding of the commodity with an appropriate gas or mixture, combined with unique packaging.

The advantages of controlled and modified atmospheres are proportional to the length of exposure time. Accordingly, the benefits in shipboard transportation far outweigh those in overland shipments.

Thermometers

Refrigerated vehicles usually have an indicating or recording thermometer to monitor cargo space temperature. Many systems are equipped with microprocessors that indicate, control, and record cargo space temperature.

TYPES OF REFRIGERATION SYSTEMS

During the development of transport refrigeration, various methods of cooling perishable cargoes were tried. These include ventilation, product storage effect, ice, ice and salt, CO_2 (solid and liquid) and N_2 (liquid), eutectic plates, and engine-driven mechanical refrigeration units; the last one dominates now. However, the others are still used, and they are synopsized in this section.

Ventilation

Control of temperature in an insulated vehicle with outside air is the least dependable of any system. Obviously, it is limited to areas or times of the year in which outdoor conditions are appropriate. Ventilation is usually accomplished by leaving small doors open at the front and rear of the vehicle. Pressure caused by forward motion of the vehicle and the adjustment of doors by the operator control the airflow. Ventilation can be used with mechanical refrigeration to remove field heat or reduce the concentration of ripening gases generated by a commodity.

Storage Effect

Some refrigerated commodities are transported without providing refrigerating equipment on the vehicle by using the heat-absorbing capacity of the product itself. Precooled milk is transported in tank trucks and trailers, some as large as 20 m^3 capacity. Chilled orange juice and hot liquids are also transported in this way.

For example, 20 m^3 of milk absorbs 79.8 MJ/K temperature rise. If the tank truck has a heat leakage of 1900 W, the milk could be transported 2600 km an average speed of 65 km/h with a temperature rise of less than 3.5 K without any refrigeration during the trip. Orange juice shipped in bulk from Florida to New York, a distance of nearly 2300 km, has warmed less than 5 K with no en route refrigeration. Product quality and other factors must be considered, however, before relying on this method for shipments of perishable products.

Water Ice

This means for transport cooling has largely been replaced. A layer of ice is sometimes placed on top of loads to supplement mechanical refrigeration in the removal of field heat. It is important to recognize that airflow (see the section on Air Circulation Systems) through top-iced cargo can be severely restricted if the evaporator leaving air temperature is low enough to keep the ice frozen.

Dry Ice (Carbon Dioxide)

In the past, dry ice was used to maintain frozen commodities in delivery trucks. Currently it is used for shipping small quantities of specialty items or items requiring very low temperature. Dry ice sublimes at $-78°C$ and absorbs about 560 kJ/kg in the process.

Liquid Nitrogen or Liquid Carbon Dioxide Spray

Liquid nitrogen has a boiling point of $-196°C$ and, with superheating to $-18°C$, has a total refrigerating effect of about 373 kJ/kg. Liquid carbon dioxide, with a boiling point of $-78°C$ and superheating to $-18°C$, has a refrigerating effect of about 280 kJ/kg.

For road operation, the refrigerant is carried in storage vessels located either in or out of the refrigerated space. To eliminate the need for high-pressure nitrogen tanks, the storage vessels are insulated and equipped with relief valves set for about 170 kPa (gage) and vented into the cargo space. Liquid carbon dioxide must be stored at pressures in excess of 416 kPa (gage) and at temperatures below 31°C. For cooling cargo after loading, some terminals use liquid refrigerant sprays fed from dock-mounted storage tanks. In this operation, the distributing tube can be introduced through a partially open door. Caution must be taken to prevent refrigerant from being released when personnel are in the cargo space; there is risk of freezing and/or suffocation. Liquid carbon dioxide systems must be designed to prevent nozzle blockage by snow formation.

Guilfoy (1973) describes a liquid nitrogen clip-on unit for land refrigeration of intermodal containers. During shipboard operation, the containers are refrigerated either by immersion in a temperature-regulated hold or by connecting ducts to a central system.

Eutectic Plates for Holdover

Station Charging. Some retail service trucks use a holdover plate arrangement to maintain low temperatures. A holdover plate consists of a coil for the primary refrigerant mounted inside a thin tank filled with a eutectic solution with a freezing temperature sufficiently low to meet the required conditions. Holdover, or eutectic, plates vary in size from 450 to 920 mm wide, 1000 to 3100 mm long, and 25 to 75 mm thick. The refrigerating capacity and mass for a given size plate vary with the thickness. For example, a 760 mm by 1680 mm plate of 67 mm thickness has a refrigerating capacity of about 17.9 MJ (at -22 to $-23°C$ eutectic temperature) and a mass of about 136 kg, while the same size plate of 38 mm thickness has a capacity of about 9500 kJ and a mass of about 102 kg. Standard eutectic plates operating at temperatures of -50, -34, -26, -24, -23, -22, -21, -8, -5, and $-3°C$ are available.

Several of these plates are mounted on the walls, ceiling, or both, or are used as shelves or compartment dividers in the vehicle to provide the required refrigerating capacity while the vehicle is away from the garage or station where the plates are refrozen. When the truck is garaged, flexible connections from the stationary refrigerating plant are attached to connecting devices provided on the truck. A charging island, where several trucks can be connected at one time, can be used for fleet operation. Any refrigerant can be used—the most common being Refrigerant 717 or, more recently, Refrigerant 22. Expansion valves may be mounted on the truck or on the garage wall, and various types of automatic-closure quick connecting devices are available for attaching the flexible lines. In plants using a chilled brine circuit, brine may be circulated through the plates to freeze the eutectic solution.

To provide proper cooling capacity, the plates must be mounted so that air can move freely on both sides, usually 40 to 50 mm from walls with top edges 150 to 200 mm from the ceiling. On ceilings, plates should be sloped at least 85 mm/m across the shortest dimension and should be at least 50 mm below the ceiling at the higher edge. Hangers for plates must be securely mounted, preferably by fasteners attached to the vehicle structure. (As with other similar fasteners, these must not compromise the vapor barrier or extend to the exterior.) A drip trough may be desirable under each plate, although many vehicles are defrosted by scraping the frost without warming the plates. If necessary, to protect plates from damage, guard rails should be placed at least 25 mm away. Eutectic plate systems usually circulate air by gravity, but some have fans with ducts and dampers for temperature control.

Vehicle-Mounted Condensing Unit. Some trucks are equipped with condensing units powered by electric motors. These units operate on commercial ac power to refreeze the eutectic plates when the vehicle is stopped. The comparative lengths of the time for road operation and for refreezing are factors in determining the required size of the condensing unit and the amount of eutectic solution (Guilfoy and Mongelli 1971). Variations of truck-mounted

Trucks, Trailers, and Containers

condensing unit systems include: separate engine drive, compressor driven by the truck engine, and nose or chassis mounting.

Plastic Plates or Tubes. Several European manufacturers have developed holdover systems that contain the eutectic solution in either rectangular or round cross-sectional tubes. Plastic parts replace relatively heavy steel ones.

Mechanical Refrigeration

Independent Engine or Electric Motor-Driven. Many styles of independent engine and/or electric motor-driven mechanical refrigerating units are available and constitute the most popular application for both trucks and trailers. The one-piece, plug-type, self-contained unit mounts in an opening in the front of the vehicle. The condensing section (on the outside) and the evaporator section (on the inside) are separated by insulation. A structure attaches to the vehicle wall and supports the various parts of the unit. For trailers, usually the plug units are thin enough to mount between the tractor and trailer. Others are short and deep and extend over the truck roof. For intermodal containers, the most widely used refrigerating unit is electric motor-driven, usually hermetic, with or without a companion engine-driven generator. The refrigeration unit essentially forms most of the front wall of the container.

The refrigerating capacity of self-contained, independent engine-driven plug-type trailer units in an ambient temperature of 38°C range as follows: 6 to 14 kW at 2°C trailer temperature; 3.5 to 10 kW at −18°C trailer temperature; and 1.8 to 7.3 kW at −29°C trailer temperature. These units have a mass of 360 to 770 kg. Similar self-contained units, which have a mass of 230 to 460 kg and have a lower cooling capacity, are available for use on trucks.

Some units have two belt-driven compressors; others have one compressor that is either belt- or direct-driven by the engine. Engines may operate on fuel from the truck tank, from a separate gasoline or diesel fuel supply on trailers, or on liquefied petroleum gas. In addition to the engine, many refrigeration units are equipped with an electric motor for standby operation. Most units can be used for heating, and they defrost automatically; both are done by the hot-gas method [see the section on Heating Systems (Heaters)]. Most units are thermostatically controlled, starting and stopping and/or reducing speed as refrigeration need requires. Condenser and evaporator fans are driven by the unit engine. Tight-fitting dampers close off evaporator airflow to permit defrosting without unduly warming the vehicle interior. Alternatively, means to stop the evaporator fan during defrost may be provided. One-piece construction facilitates unit replacement in the event of breakdown or removal for maintenance.

Two-piece units are also used for both trucks and trailers. One style uses an engine-driven generator mounted under the trailer to provide electric power to operate one or two independent refrigerating systems contained in a plug-type unit mounted in the front wall. This unit consists of a self-contained, hermetic compressor system, and the housing under the trailer contains an engine-driven alternator. The compressor can be operated from commercial ac power during standby. Another style uses an evaporator section, usually with electric motor-driven fans, mounted in the nose of the vehicle, and an engine-driven condensing unit installed under the body. In this case, permanent piping is run between the condensing unit and evaporator, and a generator on the engine provides electric power to operate the evaporator fan. Some units are designed to cycle on thermostatic control, others operate continuously, but change engine speed in response to the need for refrigeration. Some units are designed for automatic defrosting; others are defrosted by manual operation of hot-gas valves or electric heaters.

Power Takeoff from Vehicle Engine or Transmission. For most cases, this type of refrigerating system is limited to trucks, although an axle-driven unit has been devised for trailers. Several types of power takeoffs are available. One manufacturer has an alternator, belt-driven from the engine crankshaft, which produces a regulated alternating current voltage that is rectified to drive direct current motors for the compressor and fans in the body. Temperature control is accomplished by cycling the compressor. Refrigeration is produced only when the truck engine is in operation; full refrigerating capacity is claimed for all engine speeds above 500 rpm.

One manufacturer uses an all-electric, self-contained hermetic unit to provide refrigeration for medium-temperature truck bodies. A specially regulated alternator produces an ac power supply of variable voltage and frequency at essentially constant current, with engine speeds from idle to 4000 rpm. The alternator is mounted on the truck engine and belt-driven from the engine crankshaft pulley. Standby operation is provided by electrical connection to commercial ac power. A two-speed centrifugally operated gear shift is available to provide more suitable generator speed if the truck engine speed range is too wide.

Power takeoff systems of the mechanical type sometimes have the condenser mounted in front of the truck radiator, with flexible tubing to connect to the unit in the cargo space. The most common system uses an engine-mounted compressor, which is belt-driven from the crankshaft pulley. One manufacturer has a flexible shaft drive for the unit compressor and condenser fan, with a belt-driven electric clutch at the forward end of the flexible shaft. In this system, of course, the compressor speed is a function of engine speed. A variation of the power transmission mechanism for use with a belt-driven flexible shaft includes a standby electric motor to drive the compressor. The flexible shaft drives an electric clutch attached to a universal joint. An overriding clutch idles the electric motor when the compressor is being driven by the flexible shaft. When the electric motor operates, the electric clutch is disengaged. The flexible shaft is driven by a belt from the engine crankshaft, and the electric clutch is used for temperature control.

Hydraulic power transmission is a special case of power takeoff from the truck transmission or engine. The hydraulic pump is either mounted in the truck engine compartment and belt-driven from the engine crankshaft, or is direct-driven from a power takeoff in the truck transmission. Hydraulic drive systems may be designed to provide a near constant output speed from a variable input speed.

MECHANICAL REFRIGERATION COMPONENTS

The major components of a transport unit are its power source, compressor, condenser and fan, evaporator and fan, and controls. Some of the fundamental aspects of system and component design for a transport refrigeration system follow.

- All components must withstand shock and vibration. Avoid flexing short, rigid refrigerant lines. In particular, copper can cold work and fail. Heavy objects in refrigerant lines, such as dryers and control valves, should be individually anchored.
- Account for temperature extremes. Transport refrigeration units are usually expected to start up, and cool or heat in ambients that range from −40 to 60°C (and even higher transients). This affects most components, especially engines, compressors, heat exchangers, and safety devices.
- Consider water and dirt. Protect all components from corrosion, especially those for intermodal containers which are exposed to sea water. Shield bearings and shaft seals from water, dirt, or abrasive material penetration. Protect motors and other electrical devices from condensate, rain, and splashes.

Power Source

Refrigeration systems for short trucks that require moderate cooling typically use the truck engine for power. A belt on the engine crankshaft drives the compressor, and the truck alternator powers the fans. An optional drive motor can be plugged into commercial ac power. If the cooling demand is higher (as for frozen food, a large truck body, or frequent door openings), the system is usually powered by a small engine. (Most are diesels, with a 410 to 575 mL displacement.) Eutectic systems with 3.7 to 5.6 kW motors may also meet the needs of this application.

Large diesel engines, up to about 1.3 L, provide power for units for large trucks (up to 8.5 m) that have high service load or carry frozen products. Most truck units driven by their own engine are available with an optional electric motor drive. Trailer refrigeration units are typically driven by diesel engines with 1.6 to 2.3 L displacement. Optional electric standby may be available in some models.

Compressor

Small compressors, driven by the truck engine, are available from the automobile air-conditioning industry. Designs include in-line or V-piston, rotary vane, swash-plate piston, and scroll. Large truck compressors, driven by an engine in the unit, are generally reciprocating or scroll. Trailer units use reciprocating or screw-type compressors. Compressor selection criteria include: thermodynamic performance, physical size, mass, vibration, sound level, and reliability.

Condensers

Truck engine-powered unit condensers may be mounted in front of the truck engine radiator, above the truck cab with a dc motor-driven fan, or below the truck body. The condenser and fans for large truck and trailer units are normally packaged with the compressor, engine, and other components. Condensers are designed for forced-convection heat transfer and generally have aluminum fins and copper tubes.

Evaporator Systems

Most transport refrigeration systems use plate-fin and tube evaporation coils; fans or blowers provide forced-convection heat transfer. During the defrost cycle, airflow to the cargo is stopped either by stopping the fan(s) or by closing dampers. The heat required for defrost is provided by hot refrigerant gas or electric heaters. Eutectic plates were discussed previously in the section on Eutectic Plates for Holdover.

Control Requirements and Devices

Frozen and fresh produce must be maintained close to its optimum temperature for best retention of food value and appearance. The quality of some frozen foods is adequately maintained at $-18°C$, while others, such as ice cream specialty products, may require a temperature of $-29°C$. Compartmentalized trailers, or trucks with two or three temperatures, are often used for wholesale delivery of frozen, fresh, or dry commodities in the same vehicle (see the section on Multitemperature Vehicles). Certain pharmaceuticals, films, or electronic devices may require humidity control combined with temperature control.

Chapter 10 presents a table of preferred temperatures for long-term storage of perishable products. USDA *Agricultural Handbooks* 669 (Ashby et al. 1995) and 668 (McGregor 1987) discuss many aspects of maintenance of quality of fruits, vegetables, plants, and flowers during transport. The Association of Food and Drug Officials has prepared a model code that outlines storage and transportation temperature limits for frozen foods.

The rigid requirements for maintaining product temperature dictate the following:

- A suitably designed and maintained truck or trailer with provisions for perimeter airflow
- A satisfactory loading pattern
- An adequate supply of conditioned air held within a suitable temperature differential
- A precise thermostat with a proper sensor location

When the control sensor is located in the return airstream of a refrigerated load, all points upstream of the sensor (i.e., all of the load space) are subject to temperatures lower than the sensor. In this case, the possibility exists for overcooling part of the load. Conversely, if the system is heating a load, all of the space is subject to temperatures higher than at the sensor. If the control sensor is positioned in the supply airstream, the temperature relationship between sensor and return air is reversed.

In general, little difficulty is experienced with the sensor located in the return air sensor for frozen loads or precooled fresh produce. However, the thermostat should be set at a higher set-point temperature to avoid overchilling products requiring cooling in transit—normally 2 to 3 K above the product chill or freeze point. For a product with a high specific heat cooling load, the thermostat setting should be increased to 4 to 5 K above the chill or freeze point.

The use of various capacity reduction or modulation devices tends to reduce the air temperature differences in the cargo space. These devices can also help to minimize dehydration of fresh cargoes by bringing evaporator surface temperature closer to the desired product temperature.

Control Devices. Electronic (microprocessor) thermostats operate over a wide control range with a high degree of accuracy. Most temperature controls are multistep and/or modulating for both heating and cooling modes. Auxiliary features include temperature indication, multipoint data logging, and out-of-range warming. Recent developments include communication by satellite to locate a vehicle or determine its load temperature.

Other operational or safety controls include lubricating oil safety devices for engines and compressors, devices for automatic engine starting and stopping, and automatic defrost initiation and termination. Microprocessor equipped units ordinarily sense a number of pressures and temperatures, and use these data to control and record system operating parameters.

Many mechanical refrigeration units are designed to operate from −30 to 20°C in the cargo space and to pull down from a higher temperature at the beginning of every trip. Various means are used to limit the power required during these transient pulldown periods. These include compressor crankcase pressure-limiting controls, compressor unloading, and thermostatic expansion valves with outlet pressure control.

CALCULATION OF COOLING LOADS

Phillips and Penney (1967) describe a method of rating refrigerated trucks. In an earlier study, Phillips et al. (1960) describe a rating method for refrigerated trailers. The Truck-Trailer Manufacturers Association developed a method for rating heat transmission in refrigerated vehicles (TTMA 1989). The International Standards Organization has prepared thermal test standards (ISO *Standard* 1496-2).

Although suitable standards are not in general use for rating the performance of all types of refrigerating systems used for trucks and trailers, the Air-Conditioning and Refrigeration Institute has developed ARI *Standard* 1110, which deals with speed-governed and variable-speed transport refrigeration units using forced circulation air coolers. Until standards are available for all types of systems, comparisons of refrigerating units should be based on net refrigerating capacity at specified conditions of air temperature at the condenser and evaporator inlets. For example, compare using 38°C ambient temperature and −18°C trailer interior temperature.

The Refrigerated Transport Foundation developed a classification system based on the combined thermal performance of both trailers and refrigeration units (CGTFL 1988). The trailer thermal rating is based on a practice recommended by the Truck Trailers Manufacturers Association (TTMA 1989). The refrigeration unit cooling capacity is determined by ARI *Standard* 1110. By this

Trucks, Trailers, and Containers

classification system, the combined trailer and unit can be classified in one or more of four temperature ranges:

DF for deep frozen, –29°C
F for frozen, −18°C
C35 for chilled temperature, 2°C
C65 for chilled temperature, 18°C

To qualify for the C35 and C65 classification, sufficient heat must be available to achieve the classification at an ambient of −18°C. The classification is voluntary for all parties—the trailer manufacturer, the unit manufacturer, and the carrier or owner. Trailers that are classified display a permanent plate or decal that lists (1) temperature range(s), (2) trailer U-factor, (3) data on area provided for airflow around the periphery of the loading space, (4) type of bulkhead, (5) presence or absence of air distribution ducts on the ceiling, (6) amount of excess unit cooling capacity beyond the steady-state requirements at the coldest range chosen while at 38°C ambient, and (7) airflow available from the unit.

Insulated trailers range from 11 to 16 m in length, and steady-state heat gain values (U-factor) range from 40 to 130 W/K. The refrigeration system must have an additional cooling capacity beyond that needed for steady-state heat transfer to satisfy deterioration factors, solar radiation, air infiltration, door openings, and product loads. Extra capacity also reduces the pulldown time prior to loading, which improves equipment use.

Phillips and Penney (1967) found that solar radiation can increase the cooling load of stationary vehicles more than 20% for several hours during a sunny day. Penney and Phillips (1967) also noted that the cooling load caused by door usage can be five times the body heat transmission when a truck is used in multistop operation.

Manufacturers of refrigeration units for trucks and trailers have devised computer programs or charts to match the unit to the body for the specific operational details planned by the customer. The inputs for the calculations include body U-factors, losses from door openings, product data factors, ambient conditions, and refrigeration unit capabilities.

Sample Calculation for Trailer Cooling Load

Assume a trailer is loaded with 17 Mg of peaches at an average pulp temperature of 12°C. The load will be delivered 72 h later at an average product temperature of 1°C. Average ambient temperature is 30°C.

The trailer U-factor is 80 W/K as determined either from data on the trailer or estimated from information on insulation properties, size, and condition of the trailer.

The specific heat of peaches (above freezing) is 3.82 kJ/(kg·K). To estimate the heat of respiration, the average temperature of peaches during transit is assumed to be 5°C. The heat of respiration of peaches at 5°C is an average of 23 W/Mg. A more precise calculation of heat of respiration would recognize that it varies with product temperature, which decreases during the pulldown period.

Total heat load = Product heat load + Heat of respiration + Heat transmitted
Product heat load = (Specific heat) (Mass) (Temperature change)
= (3.82) (17) (12 − 1) = 714 MJ
Heat of respiration = (Heat of respiration factor) (Mass) (Seconds)
= (23 W/Mg) (17 Mg) (72 × 3600)
= 101 MJ
Heat transmitted = (U-factor) (Seconds) (Ambient temperature − Thermostat set temperature)
= 80 × 72 × 3600 (30 − 2)/10^6

Total heat load for 72 h = 714 + 101 + 581 = 1396 MJ
Average load/h = 1396 × 1000/(72 × 3600)
= 5.39 kW

If ice were used instead of mechanical refrigeration, the amount of ice melted would be about 1396 × 1000/335 = 4170 kg. Liquid nitrogen absorbs about 407 kJ/kg during evaporation and warming to 0°C. Thus, if liquid nitrogen were used in this example, the amount of liquid nitrogen required would be 1396 × 1000/407 = 3430 kg.

The preceding calculation neglects initial cooling requirements for the trailer structure, air in the trailer, and cargo packing materials.

Perishables should be fully cooled prior to loading. However, some cooling is usually required in transit for several reasons. Product temperature can rise during loading. During periods of high production, cooling facilities can be overloaded and insufficient time is allowed to fully cool the center of a product.

Cooling during transit can reduce the overall time from harvest to consumption. However, airflow through the load is essential to cool the product in transit successfully. The section on Control Requirements and Devices also discusses product cooling.

REFERENCES

ANSI. 1990. Freight containers—Road/rail dry van containers. ANSI *Standard* MH5.1.1.5-90 (R1997). American National Standards Institute, New York.

ARI. 1992. Mechanical transport refrigeration units. *Standard* 1110-92. Air-Conditioning and Refrigeration Institute, Arlington, VA.

Ashby, H.B., R.T. Hinsch, L.A. Risse, W.G. Kindya, W.L. Craig, Jr., and M.T. Turczyn. 1995. Protecting perishable foods during transport by truck. *Agricultural Handbook* 669. U.S. Department of Agriculture, Washington, D.C.

CGTFL. 1988. Refrigerated transportation foundation method classification of controlled temperature vehicles. *Recommended Practice* No. 1-89. California Grape and Tree Fruit League, Fresno, CA.

Eby, C.W. and R.L. Collister. 1955. Insulation in refrigerated transportation body design. *Refrigerating Engineering* (July):51.

Guilfoy, R.F., Jr. 1973. Refrigeration systems for transporting frozen foods. *ASHRAE Journal* (May):58.

Guilfoy, R.L., Jr. and R.C. Mongelli. 1971. A method for measuring cost and performance of refrigeration systems in local delivery vehicles. ARS *Publication* 52-64. USDA, Agricultural Research Service.

ISO. 1995. Series I freight containers—Classifications, dimensions and ratings, 4th ed. *Standard* 668:1995. International Organization for Standardization, Geneva.

ISO. 1996. Series I freight containers—Specifications and testing—Part 2: Thermal containers, 3rd ed. *Standard* 1496-2:1996.

McGregor, B. 1987. Tropical products transport handbook. *Agriculture Handbook* Number 668. USDA.

Penney, R.W. and C.W. Phillips. 1967. Refrigeration requirements for truck bodies—Effects of door usage. Agricultural Research Service Technical Bulletin No. 1375. USDA.

Phillips, C.W., et al. 1960. A rating method for refrigerated trailer bodies hauling perishable foods. Marketing Research Report No. 433. Agricultural Marketing Service, USDA.

Phillips, C.W. and R.W. Penney. 1967. Development of a method for testing and rating refrigerated truck bodies. USDA Agricultural Research Service, Technical Bulletin No. 1376.

TTMA. 1989. Method for rating heat transmissions of refrigerated vehicles. Recommended Practice No. 38-89. Truck-Trailer Manufacturers' Association, Washington, D.C.

BIBLIOGRAPHY

AFDO. Code of recommended practices for the handling of frozen foods. Association of Food and Drug Officials, York, PA.

ASHRAE. 1971. Refrigeration systems for perishable food delivery vehicles. *Symposium Bulletin* WA-71-4.

ASHRAE. 1972. Long-haul transportation of respiring perishable commodities in refrigerated containers. *Symposium Bulletin* NO-72-7.

ATA. Summary of size and weight limits. American Trucking Association, Alexandria, VA.

ATP. 1970. Agreement on the international carriage of perishable foodstuffs and on the special equipment to be used for such carriage. United Nations, New York.

Chau, K.V., R.A. Romero, C.D. Baird, and J.J. Gaffney. 1987. Transportation coefficients of fruits and vegetables in refrigerated storage. Final Report for Project 370-RP. Department of Agricultural Engineering, University of Florida, Gainesville, FL.

Hardenburg, R.E., A.E. Watada, and C.Y. Yang. 1986. The commercial storage of fruits and vegetable and florist and nursery stocks. *Handbook 66* (revised). U.S. Department of Agriculture (USDA), Washington, D.C.

Heap, R.D. 1989. Design and performance of insulated and refrigerated ISO intermodal containers. *International Journal of Refrigeration* 12:139-45.

IIR. 1974. Transport of perishable produce in refrigerated vehicles and containers.

IIR. 1984. Refrigeration and the quality of fresh vegetables.

IIR. 1985. Long distance refrigerated transport—Land and sea.

IIR. 1985. Technology advances in refrigerated storage and transport.

IIR. 1986. Recommendations for the processing and handling of frozen foods.

IIR. 1990. Progress in the science and technology of refrigeration in food engineering.

IIR. 1991. Proceedings of the 18th International Congress of Refrigeration.

IIR. 1993. Compression cycles for environmentally acceptable refrigeration, air conditioning and heat pump system.

IIR. 1994. Cold store guide.

IIR. 1994. New applications of refrigeration to fruit and vegetables processing.

IIR. 1995. Guide to refrigerated transport.

IIR. 1995. Proceedings of the 19th International Congress of Refrigeration. Volume I.

IIR. 1996. New developments in refrigeration for food safety and quality.

IIR. 1996. Refrigeration, climate control and energy conservation.

IIR. 1996. Research, design and construction of refrigeration and air conditioning equipment in Eastern European countries.

Mercantila Publishers. 1989. Guide to food transport—Fruit and vegetables. July.

Mercantila Publishers. 1990. Guide to food transport—Fish, meat and dairy products. September.

Postharvest Technology of Horticultural Crops, 2nd ed. *Publication* 3311. University of California, Division of Agriculture and Natural Resources.

Ryall, A.L. and W.J. Lipton. 1972. Handling transportation, and storage of fruits and vegetables, Vol. 1. AVI Publishing, Westport, CT.

Ryall, A.L. and W.T. Pentzer. 1982. Handling transportation, and storage of fruits and vegetables, Vol. 2—Fruits and Tree Nuts, 2nd ed. AVI Publishing, Westport, CT.

UN. 1970. Agreement on the international carriage of perishable foodstuff and the special equipment to be used for such carriage (ATP). Economic Commission for Europe, Inland Transport Committee, United Nations.

CHAPTER 30

MARINE REFRIGERATION

CARGO REFRIGERATION 30.1	Storage Areas 30.8
Built-In Refrigerators 30.1	Ship Refrigerator Design 30.9
Refrigeration System 30.3	FISHING VESSELS 30.10
Refrigeration Distribution 30.4	Refrigeration System Design 30.10
Refrigeration Load 30.6	Refrigeration with Ice 30.13
Cargo Containers 30.7	Refrigeration with Seawater 30.14
SHIPS' REFRIGERATED STORES 30.7	Tuna Seiners 30.14
Commodities 30.8	Freezer Trawlers, Factory Vessels, and Mother Ships 30.15

CARGO REFRIGERATION

MARINE transport is an interim operation between preshipment storage of indeterminate duration and early distribution at destination ports. Frequently, the marine transport period is equal to or even exceeds the full high-quality life of the perishables being transported. Good design, therefore, requires that shoreside criteria be applied to the floating cold storage plant.

The increased use of various sizes of cargo containers has effected savings, mostly by providing faster unloading and reloading of the vessels. Most refrigerated cargo can be containerized. However, containers will never completely supplant the ship's built-in cargo refrigerator for such services as those provided by passenger ships, logistic supply ships, all-refrigerated fruit carriers, and special service vessels. These services will continue to require the insulation and refrigeration of the principal structural compartments of a ship.

The physical and mechanical aspects of both containerized services, as well as the break-bulk (or individual package) type of operation are discussed in this section.

BUILT-IN REFRIGERATORS

The location and arrangement of insulated compartments and compartment subdivisions within the hull should reflect the trade in which the vessel will serve. The volume of trade, the scheduling of ports of call, and the efficient and speedy handling of produce with minimum exposures are all factors that will influence these arrangements. Perishables should be the last cargo loaded and the first to be discharged.

Arrangement and Utility

Limitations of dimension and arrangement are due to the ship's structure, compliance with floodability compartmentation of the hull, and the fire-resistant regulations to which many ships are subject.

The refrigerators should not be designed exclusively for high-temperature cargo services unless it is certain that the vessel will always remain in that limited trade. Otherwise, all compartments should be readily convertible to any temperature service from below −29 to 15°C, thus allowing flexibility in segregating cargo according to temperature requirements and ports of call.

In a vessel in which only a portion is refrigerated and designed for late loading and early discharge, the compartments are usually located under the topmost deck to the hull. The naval architect considers the effect on trim of the vessel in normal loading conditions. Generally, the added mass of the insulation and the perishable cargo is not as influential as the mass of dry cargoes, which are usually of greater density. Often these compartments are empty or lightly loaded during some of the voyage. Trim conditions generally require the refrigerators to be placed forward of the midlength of the vessel or grouped about the middle section, where the moment arms for trim are of lesser consequence. Almost without exception, refrigerators are arranged symmetrically about the ship's longitudinal centerline.

The rooms or areas where the refrigerating control valves are installed should be arranged and located so that the apparatus or controls are accessible to the operating personnel through trunks or passageways at all times. When the boundary bulkheads parallel hatches, they should clear the openings by 1 m as a safety measure, providing adequate room for handling hatch beams and covers.

The greater the number of subdivisions in the refrigerated compartments, the greater the loss to the ship's revenue spaces, due to the volumes occupied by insulated partitions, cooling apparatus, insulated piping, and access areas. On this basis, the all-refrigerated ship, with only the main structural boundaries insulated, make the most efficient use of a ship's costly enclosures. On the other hand, the conventional all-refrigerated ship, with insulated hatch plugs at the weather deck, has several undesirable features, such as the difficulty of providing uniform compartment temperatures and of refrigerating hatch areas far from the refrigeration apparatus.

With large-volume rooms, the extended period in which the hatch is open to the outside atmosphere for loading and discharge is not conducive to a good environment for frozen or fresh cargoes. Seldom are frozen goods sufficiently cooled to tolerate an extended loading interval. Fresh products often are not precooled before loading, and the accumulation and delayed extraction of respiratory heat in the compartment is detrimental. During long discharge intervals in ports, atmospheric moisture causes sweating or frosting of products, some of which may suffer from this exposure. With partial delivery at a port of call, such exposure may unfavorably affect the remaining cargo. Also, harbor workers may refuse to work in spaces with the refrigerating equipment in operation. Air curtains help substantially to keep heat from entering the refrigerated area.

A more suitable arrangement of the all-refrigerated ship can be made by fitting the insulated compartments in the 'tween decks with access doors in hatch-side bulkheads, and by means of various combinations of insulated plug hatches at the lower hold and 'tween deck levels. Hatch areas may be used for general cargo or treated as separately refrigerated compartments. Many ships are fitted with side port doors at the 'tween deck level, through which cargo is handled by conveyors which are independent of the ship's overall cargo gear.

The central refrigeration machinery plant should be located in, or immediately adjacent to, the main propulsion machinery room, where the ship's responsible watch officers are in constant attendance. A central location for the refrigeration machinery usually

---

The preparation of this chapter is assigned to TC 10.6, Transport Refrigeration.

results in economy of space and close connections for power and pumping facilities.

Insulation and Construction

Insulation. Moisture-vapor and water-resistant insulation is of particular importance on board ship because of frequent and extreme temperature cycles due to intermittent refrigeration. On termination of refrigeration at discharging ports, insulation will be at lower temperatures than the open room, and often the room surfaces stand dripping wet with atmospheric moisture, which enters through the open door or hatch. Both the warm and the cold side should be moisture-sealed equally, and cold-side breather ports are not recommended. Other common sources of water in ships' refrigerators are melting of ice (packed with vegetables) and defrosting of cool surfaces.

Severe service conditions, which subject the insulation to injury or change by mechanical damage or vibration, and intermittent refrigeration, place exacting requirements on insulation for ships' refrigerators. The ideal shipboard composite insulation should have the following characteristics:

- High insulating value
- Imperviousness to moisture from any source
- Low mass
- Flexibility and resilience to accommodate ships' stresses and loading
- Good structural strength
- Resistance to infiltrating air
- Resistance to disintegration or deterioration
- Fire resistance or fireproof self-extinguishing requirement
- Odorless
- Not conducive to harboring rodents or vermin
- Reasonable installation cost
- Workability in construction

In the United States, the properties of the insulation and the details of construction must meet approval by the U.S. Coast Guard and the U.S. Public Health Service. For information on insulation materials and moisture barriers, see Chapters 22 and 24 of the 1997 *ASHRAE Handbook—Fundamentals*.

Construction. The three principal parts to the refrigerator boundary include the envelope (or basic structure), the insulating material, and the room lining.

The envelope is usually partly composed of the ship's hull, the watertight decks, or watertight main bulkheads whose members resist the entry of vapor from the warm side. The inboard boundaries outlining the refrigerators should have an equal ability to resist moisture. A continuous steel internal bulkhead with lap seams and welded stiffeners provides a boundary of adequate strength and tightness. Details of design may accommodate dimensioned insulators or facilitate means of fastening these materials. Doorway main bucks of steel channel provide good structure but are usually a source of sweating on low-temperature rooms because of heat gain through the metal. Wooden door bucks minimize sweating but are a retreat from efforts to eliminate concealed wooden structure.

Partitional bulkheads may be of similar detail, but airtight sealing is less important. Some installations are framed with angle-bar grids, between which the insulator is installed. In passenger vessels (over 12 passengers), Coast Guard regulations governing fire-resistant construction restrict wood assembly. Under no circumstances should wood be a part of the floor assembly since it deteriorates rapidly under the prevailing conditions.

The assembled boundary of a ship's refrigerator must withstand heavy floor loads and several wall thrusts of cargo when the vessel rolls or pitches in a heavy sea; it must be able to flex with the hull structure being stressed in any angle. The assembly must resist vibration caused by the propelling machinery, the sea, and the careless handling of cargo. The vapor seal of all surfaces must remain intact.

Only in extreme cases should voids in the insulation assembly be concealed. Filling such volumes with insulating material is cheaper and more effective than constructing internal framing. The exceptions to this rule are in the deep volumes formed by bilge brackets, deck brackets, and open box girders. Solid filling results in more insulation thickness than is needed for a heat barrier in the overhead and the ship's side where beams and frames are deep.

The third part of the boundary assembly is the room lining. This surface must be sturdy enough to withstand the impact of frequent cargo loading and handling. On passenger vessels, U.S. Coast Guard regulations require that the lining be fire-resistant. Tongue-and-groove lumber is considered obsolete for any ship, and other linings are employed. On freight vessels where wood is permissible, exterior grade plywood is sometimes applied. A few installations have been made either of laminated plastic sheets or wood fiber hardboard; both are satisfactory when properly supported. Steel linings are costly, impractical, and difficult to maintain and repair. The favored lining is the cement and fire-resistant fiber hardboard panel with aluminum sandwich lamination. When the insulator is secured with adhesives containing volatile solvents, the aluminum laminations should be of perforated metal or mesh.

The U.S. Public Health Service requires all linings to be rat-proofed. Vulnerable linings should have an underlay of 6-mm, 16-gage galvanized wire mesh for rat-proofing.

Applying Insulation

In the application of panels with adhesives over block insulators, the butted joints should be separated sufficiently for the adhesive to extrude, provide a moisture-sealed joint, and accommodate movement of the panels by flexing of the ship's structure. The joints may be covered by cargo battens, which are also secured by adhesive and with brass screws. The walls of all refrigerators should be fitted with vertical cargo battens on 380- to 460-mm centers to hold the cargo clear of the insulated wall. This spacing permits circulation of air and prevents the contacting package from assuming the part of a room insulator.

Container vessels of cellular-type construction move the containers along guiding columns; the boundaries of the hold, the hatch coamings excepted, are not subject to mechanical damage. Here the insulation may be of the simplest, low-cost, rat-proofed form without fitted hard panels or cargo battens.

Urethane foam floor insulation requires a restricting surface to confine and distribute the expanding material. In one method, plywood is secured over foam spacer blocks, and the material is injected into the space through properly spaced holes. The membrane and the wearing surface covering are laid over the plywood. But wood in the floor is not good practice, nor is the wood a suitable base for a heavy-duty wearing surface material. An alternate method calls for laying expanded board or blocks in an approved adhesive. The use of the more resilient corkboard laid in adhesives may be good practice in some applications.

The greatest weakness in ship refrigerator construction is the floor covering. Research and practice have not brought forth a totally satisfactory covering. Unlike a warehouse, a tightly packed cargo refrigerator must have floor gratings both to assure the circulation of air and to protect the bottom tier of products from heat leakage. The grating supports carry the cargo mass to concentrated load areas of the floor. The room may be filled with warm general cargo in alternate service, and a thermoplastic covering may be punctured by the supports.

The floor covering must be flexible enough to withstand the flexing of a ship or extreme temperature fluctuations, as well as maintain a moistureproof cover over the insulation. The most satisfactory material is a mastic composed of emulsified asphalt, sand, and cement. This material is applied cold; on setting, it has good load-bearing qualities, is impervious to water, has a small degree of ductility, and may be used in less thickness than concrete. It should

Marine Refrigeration

always be reinforced, and expansion joints should be included to accommodate shrinkage and adjustment to movements of the ship.

All rigid or semirigid floor coverings should have rubber-base composition expansion joints capable of bonding to the edges of the floor slabs or wall and not subject to shrinking from age. The expansion joints should trace the periphery of the room, the line of all underdeck girder systems, bulkhead offsets to pillars, and similar lines of anticipated ship stress.

Water from ice-packed vegetables and defrost water draining onto the floor covering requires it to be impervious to moisture in the slab as well as the joints. If the floor insulation is wetted, it will deteriorate, lose its efficiency as an insulator, and give off odors. If wet floor insulation freezes, it will lift the deck covering and destroy it. If water penetrates to the steel, the ships' structure will corrode unnoticed.

All floor coverings will crack or become damaged during the ship's life. As a secondary security against water, a membrane should be laid between the insulator and the floor covering. The membrane need not be flashed to the wall if suitable sealing expansion joints are fitted at the juncture of the wall panel and the floor covering.

When an attempt is made to seal insulation with waterproof paper, it should be applied in double thickness, and the laps and perforations should be cement-sealed. Wall and ceiling panels attached with suitable adhesives do not require waterproof paper inner linings.

Floors and Doors

The floor is the weakest point in the ship's refrigerator. The weakest element in the floor is the floor drain, or scupper, because of the difficulty in bonding the floor covering with the metal drain fitting. Water will often find its way between the covering and the insulation. The conventional floor drain is fitted with a perforated plate flush with the floor covering and hidden by the floor gratings. If the perforations become clogged by debris, water will accumulate at the scupper. A drain fitting near a wall or corner will, on most occasions, create a weak section in the floor covering and will develop cracks running to the wall or across the corner. Figure 1 shows a satisfactory scupper fitting. It is flush with the top of the gratings, has a liftout cover for easy cleaning, and is bonded to the floor covering with expansion joint material. Drains between decks should be omitted wherever practicable.

Refrigerator doors are generally a manufactured product. They should have generously designed steel hardware, and the door and frame should be metal sheathed and have a flat sill and double gasket. Very large or double doors should have additional dogs to assure proper sealing when closed.

Sliding doors should be installed wherever possible to reduce interferences and conserve adjacent revenue space. When used, brackets should be installed to support portable horizontal spars inside the doors to prevent cargo from falling against the doors in a seaway. Molded glass fiber swinging doors insulated with urethane foams poured in place are available. They are strong, have low density, are easily handled, and can be fitted with low-density hardware.

In finishing, wooden surfaces should be varnished rather than shellacked, since the latter material has little protective penetration. Manufactured nonmetallic surfaced materials may be painted or varnished, but if they are nonhygroscopic, their original surface will usually present a good appearance for longer than a painted coating.

Low-temperature apparatus or piping should be inspected carefully during installation. All joints and surfaces should be generously sealed to keep out atmospheric moisture. Special attention should be given to pipe covering ends, valves, and bulkhead penetrations. On subzero services, special composition adhesives should be used. The smallest omission or breach of a seal will allow progressive destruction of the covering.

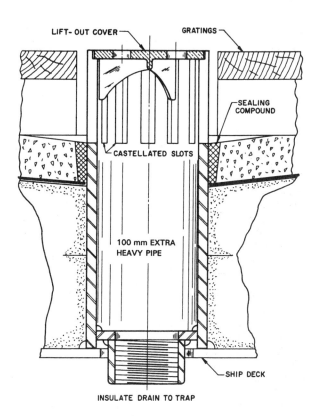

Fig. 1 Floor Drain Fitting

On shipboard, where piping systems are relatively short, the function of the insulation is more that of preventing sweating or frosting of cold surfaces than preventing heat gain.

REFRIGERATION SYSTEM

Refrigerants

Refrigerants for new shipboard equipment must meet the environmental regulations that apply to land-based systems. The choice of chemicals is similar, but special attention should be given to the availability of both the refrigerant and the compressor lubricant at all ports of call. Shipboard safety is also a consideration; it is addressed in ASHRAE *Standard* 26, Mechanical Refrigeration and Air-Conditioning Installations Aboard Ship. Currently, new centrifugal systems may use R-22, R-123, or R-134a. New systems with other types of compressors may use R-22, R-134a, or one of several hydrofluorocarbon blends designed to replace R-22 or R-502. R-717 (ammonia) is widely used on fishing and seafood processing vessels; however, the scope of this section is confined to the other refrigerant types mentioned.

Compressors

In addition to centrifugal compressors, which are usually associated with large comfort cooling systems, reciprocating, helical screw, and other types of rotating compressors may be used in marine systems. Chapter 34 of the 1996 *ASHRAE Handbook—Systems and Equipment* describes compressors in detail and discusses their application and control. Much of this is identical for stationary and shipboard systems. The following aspects differ.

Parallel operation of reciprocating halocarbon compressors is usually avoided because vessel motion, list, and trim cause an oil-level control problem. A similar restriction on parallel operation usually applies to other types of compressors. If multiple compressors are required, each usually has a separate refrigerant circuit.

Table 1 Operating and Reserve Capacities of Condensing Units

No. of Units, 100% Load	Additional or Reserve Unit, %	Total No. of Units
1	100	2
2	50	3
3	33⅓	4
4	25	5
5 or more	20	6 or more

Intermittent operation of compressors should be minimized because shipboard electrical systems tend to be sensitive to operating transients. One or more appropriate means of compressor capacity control (cylinder unloaders, inlet guide vanes, etc.) should be used to help match capacity and load. Oversized compressors should be avoided. Both of these measures help minimize temperature variations.

The shafts of equipment with significant rotating inertia are usually oriented fore-and-aft to minimize the gyroscopic bearing loads that occur when a ship rolls. Also, compressor lubrication systems must be designed to function properly under roll, pitch, list, and trim conditions.

Ships may include compartments with a minimum design temperature of −29°C. This can require compound compression or a cascade system; details are beyond the scope of this section.

Reserve compressor capacity and spare part lists are specified by the American Bureau of Shipping (ABS). This agency recommends two condensing units, one of which, running continuously, should be capable of maintaining full cargo lading in tropical waters. The rules further state that when refrigerated spaces are of 425 m^3 total capacity or less, consideration should be given to the installation of a single condensing unit with adequate spare parts. Experienced system designers usually ignore the ABS rule provision and require a reserve condensing unit. Table 1 suggests reserve capacity for various installations.

Condensers and Coolers

Halocarbon condensers are of the conventional shell-and-tube design and, since these refrigerants are noncorrosive to all commercial metals commonly used in refrigeration, a wide variety of materials is available. High heat transfer characteristics and resistance to corrosion/erosion by circulating seawater should be considered in selection and design.

The refrigerant is usually between the tubes and shell of the condenser, and seawater circulates through the tubes in a multiple-pass arrangement. Steel is usually used for shells and 90-10 cupronickel for tubes and tube sheets. Bronze water boxes are preferred. If used, cast iron water boxes should have zinc anodes and be coated with a protective material such as epoxy. The tendency to substitute high velocities of circulating water for cooling surface in heat exchange should be resisted to provide long life. Water velocity of 1.8 m/s in the tubes is considered high, and water box velocities above 0.6 m/s are conducive to turbulence and accelerated erosion of tubes. Cooling surfaces are designed for maximum conditions.

Shell-and-tube condensers need double drains if installed level; vertical outlet lines long enough to ensure a liquid seal at the expected angle of list or trim are essential. A single outlet may be used if the condenser is inclined at an angle greater than the expected angle of list or trim. Horizontal receivers must be drained in a similar manner. Vertical receivers with a single outlet are an option. Receivers may be constructed of welded steel and should be arranged to assure submergence of the liquid outlet under all sea conditions. The receiver should have sufficient capacity to hold a complete charge of the system, plus 20% reserve volume. Means should be provided for operators to observe the liquid level.

Brine cooler specifications are similar to those of the condensers except that the materials must resist the corrosive effects of brine. Steel tubes and tube sheets are generally used. In conventional design, the refrigerant circulates through the tubes to reduce the refrigerant charge to a minimum and provides maximum wetting of cooling surfaces. Large coolers may have two or more independent tube groups in parallel, each with its own expansion valve. Multiple parallel expansion results in improved flexibility, increased reliability, and better control at partial loads.

REFRIGERATION DISTRIBUTION

Refrigeration may be distributed by direct-expansion (DX) systems or by brine as the secondary refrigerant. Because the refrigerated compartments are located far from the machinery, direct-expansion systems have many shortcomings. The extended return lines require a large total pressure drop to produce high enough velocities at low loads to return the oil to the machinery. A full load on the same line will reduce the capacity of the compressor cylinders because of the resultant rarefied vapor.

When the evaporators or the return lines in DX systems are lower than the compressor suction, oil traps are possible; these may, under certain conditions, cause alternate starving and slugging of the compressors. Small dual risers (the bottom one of which forms a trap) provide a suitable oil lift when the vapor velocity in vertical return lines falls below 7.6 m/s when one compressor is operating. The long low-side lines are generally concealed by insulation or are inaccessible for repair at sea or when the ship is loaded. The cargo may be seriously damaged if vibrations, the flexing of the ship, or damages cause the lines to develop leaks. In addition, the extended systems require large charges of refrigerant with many more scattered points where moisture or air could enter the system.

Brine has many advantages over DX systems. The full charge of the primary system of such a plant is small in quantity, and, in case of accident to the piping or the components, escaping refrigerant is confined to the refrigerating machinery room. The short runs of piping result in small total pressure drops and permit an efficient layout with fewer compressor units. This facilitates the return of oil to the compressors.

Brine cushions abrupt thermal changes and eliminates the unwanted sensitivity of control that is characteristic of DX latent heat systems. Even though the primary system should be automatically controlled, excellent results in cargo space conditions are obtainable with manual control of return brine flow. Brine systems provide great flexibility in plant operations by the capabilities of compounding or sharing refrigeration loads. Total centralized control is feasible.

Space Cooling

The refrigerators in ships are tightly stowed, without aisles or clearances, and the method of stowage has more influence on cooling than the design of the refrigeration equipment. Marine refrigerators are of countless variations in size, dimension, proportion, and configuration. As a result, conventional tests have little significance as guides to operations.

Because cargoes are so tightly stowed, wall coil refrigeration or gravity air circulation is obsolete. Circulating air removes heat from the cargo, but does accelerate dehydration of vulnerable cargo. A long air path requires a higher velocity airstream than a short path for the same air temperature rise.

One method of refrigeration, used often in banana and other fruit carriers, uses external coil bunkers and fan systems to circulate chilled air through ducts and ported false bulkheads. Typically, brine-cooled serpentine coils of pipe have been used. This system supplies air to one side of the compartment and fans exhaust it from the opposite side. Some installations permit reversal of the direction of the airflow. Refrigeration effectively reaches the entire volume if it is properly stowed and the ports are correctly adjusted. The usual distribution results in low air velocities and relatively low moisture

Marine Refrigeration

pickup. European builders often employ vertical airstreams between floor and ceiling plenums, but this system has not been adopted by American operators.

Cooling Units—Diffusers

These units are composed of three sections: a bottom air inlet, the middle encasing the cooling coils, and the top housing the motor-driven fans, which are usually in multiple arrangement on a common shaft. Large units may be arranged horizontally in similar sequence. Diffusers are packaged units that are easily and cheaply installed. These units are fitted with distributing ducts that direct the air along the ceiling and down the bulkheads between the cargo battens.

The cooling coils are, in some cases, small tube DX evaporators with closely spaced fins. Each vertically arranged tube pass is fed liquid from the distributing head of a thermoexpansion valve. The outlet consists of a connecting manifold at the base of the coils. When air flows horizontally, the evaporator tubes are horizontal and the connecting manifold vertical. Other types employ prime surface tubes or coils cooled by DX or brine.

The manufacturers of the units will usually specify cooling performance as refrigeration capacity divided by the air-to-refrigerant temperature difference. The ratio provides a ready comparison of competitive designs or proposals.

Cooling units or diffusers are installed behind guards against which the cargo may be piled and behind which access is provided for maintenance. The cargo space lost to these units and their associated ducts increases proportionately as room volume becomes smaller. Fans should have output adjustment in steps to half-rated speed, and the motors should be watertight.

Defrosting Systems

Most marine refrigerator cooling surfaces operate below 0°C, so they rapidly accumulate frost. Prime surface coils require less frequent defrosting than do extended surfaces of equivalent capacity rating. The problem becomes most severe during initial cooling. Maintaining a small temperature difference between the evaporator and room air is the principal means of providing desirable high humidities for fresh perishable cargo.

All cooling surfaces should have means to melt accumulated frost and remove the resulting water. Direct-expansion systems, if not too far from the compressors, may be defrosted by hot gas or, alternatively, with warm water spray. Because hot gas defrosting is complex, most plants defrost with heated fresh or seawater. When using heated water, its temperature should not exceed 32°C to avoid high refrigerant pressures in the evaporator.

Brine-cooled surfaces may be defrosted by passing heated brine through the coils. A steam-heated, thermostatically controlled brine heater may be installed in a central location for the purpose.

In some long voyage chill services, such as for bananas, citrus fruits, and apples, replacement air is necessary to remove excessive quantities of carbon dioxide and other gases. Continuous airflow maintains uniform conditions within the room and is preferred over intermittent airflow. Outside air should first pass through the cooling unit before distribution.

Citrus fruits should have replacement air at the rate of 2 to 3% per minute based on the gross volume of the refrigerator; bananas require as much as 5% replacement air. When replacement ventilation is provided, accessible, efficient, and well-insulated dampers should be fitted to cut off this air supply when frozen food is carried.

Plant Layout and Piping

Plant layout aboard ship should be as simple as possible without sacrificing reliability. In addition, the machinery plant should be closely associated with the main power plant to provide short piping and power connections and facilitate close supervision.

The personnel changes of engineers and refrigeration crew members demand that strangers to the installation be able, on short notice, to trace well-labeled systems and place the plant in operation or maintain it without undue hazards to the machinery or cargo. Even at the expense of valuable revenue space, an uncrowded machinery room should be provided that will give ample space for operations, maintenance, and the repair of apparatus or the ships's structure, and proper clearances should be allowed for the insulation of low-temperature parts.

All machinery should have sturdy foundations, and all parts should be secured against vibrations set up either by themselves or the main propulsion plant. High-speed machinery should be mounted on fore-and-aft centerlines, and all gravity feeds, drains, and tanks should be designed or installed with full consideration of the effects of trim, roll, or pitch of the vessel.

Each refrigeration system must work independently of other systems in the primary refrigerant side. Table 1 lists the recommended number of units for a plant. The entire plant should be cross-connected to back up the load from any system normally assigned to a specific use.

Multiple compartmentalized refrigerator arrangements operating with DX can become a complex central plant, particularly with hot gas defrosting. Remotely located evaporators result in decentralized control with apparatus scattered over the ship.

Halocarbon return piping may be designed on the basis of a 1 K total temperature drop between the evaporator and the compressor. Oil trapping should be avoided. Minimum vapor velocities in horizontal lines of 3.5 to 4 m/s are recommended for oil return flow. The liquid refrigerant should be subcooled immediately after it leaves the receiver, because flashing is easier to prevent than recondensation. This requires placing the liquid-suction heat interchanger close to the condenser and receiver.

Liquid lines should be arranged to prevent line flashing either from high temperature or from excessive pressure drop in the static head of risers. In situations that require a high lift, a refrigerant-cooled subcooler may be needed in place of a liquid-suction heat interchanger. Liquid lines should be fitted with efficient driers and strainers.

Two-Temperature Brine Plant

Brine refrigerant plants require a condensing unit for each evaporator, but the brine system may be cross connected. Most marine plants operate from subzero to ambient temperatures with any or all compartments being interchangeable. A two-temperature brine plant allows high-temperature brine to refrigerate fresh products with minimum desiccation effect, and low-temperature brine to refrigerate frozen cargoes (see Figure 2). A two-temperature brine plant suggests two circulating piping systems served by two pumps and a standby pump available for either system. The desired brine can be selected by setting valves at the refrigerators.

Brine returns from each cooling unit should run to the central plant return-valve twin manifold where room temperatures may be remotely controlled by brine flow. By controlling return flow, the systems are kept under pressure, and air binding will be at a minimum. The returning brine is next diverted to its respective collector tank. Two types of systems are in use: the open system, in which the return brine streams are in view of the operator; and the closed return system, in which a static head is maintained in the system by a vented overhead gravity tank. Brine systems use a suitable heat transfer fluid; consult Chapter 20 of the 1997 *ASHRAE Handbook—Fundamentals*.

A simple two-temperature brine system can be made by including either a closed high-temperature brine loop or a cooling unit with a recirculating pump. The selective brine temperature is maintained by a thermostatically operated pilot valve which feeds low-temperature brine to the loop or cooler as required. The central plant thus becomes a one-temperature installation.

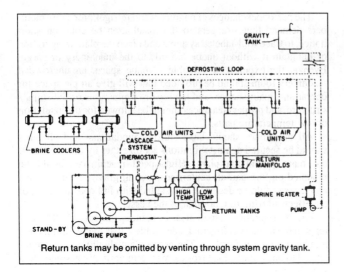

Fig. 2 Two-Temperature Brine System

With the use of a secondary refrigerant such as brine, an additional temperature differential results between the primary evaporating refrigerant and the compartment air temperature. Cascading of a refrigeration plant is desirable for a room temperature of −29°C. In this operation, the high-temperature brine circulates through the condenser in lieu of seawater and reduces pressure and compression ratios over those of single-stage compression.

For defrosting with hot brine, an independent small pipe loop having its own pump and heater is suggested. In this operation, the cold brine valves are closed at the refrigerator, and the cooling coils are placed in series with the pump and heater. Only the brine content of the coil is displaced; the entire pipeline is not heated.

Thermometers

The thermometer is the principal indicator of how a refrigerator plant is functioning. Room temperature depends on the proper placement of the sensor within that space. In rooms operating with wide variation below or above the freezing temperature of the product, the sensor can be placed almost anywhere, but in rooms operating only slightly above the commodity's freezing temperature, care must be taken to place the sensor where it will read the temperature accurately and avoid freezing. The sensor should not be placed in direct contact with warm-side surfaces or refrigerating equipment. Rooms with cold air units should have the sensors located in the supply airstream, that being the point of critical temperatures for chill cargo. When 0 to 0.5°C supply air is distributed, the heat of the load is removed as rapidly as possible with no danger of freezing the product.

Recording thermometers are essential to proper operation and management of cargo refrigerators. Indicating thermometers should have the dial installed outside and near the access doors in each room. All sensors used for comparative readings should be installed adjacent to each other. Mercury thermometers will, because of their design, be near the refrigerator.

Indicating thermometers inserted in walls through the insulation are not satisfactorily accurate or representative. Indicating thermometers should have the dial installed externally and near the access doors in each room. All sensors used for comparative readings should be installed adjacent to each other. Electronic thermometers should be installed with the display located at the log desk.

Recording thermometers are essential to proper operation and management of cargo refrigerators. Mercury actuated instruments will, because of system limitations, have the instrument near the refrigerator. Electronic recording instruments are best located in the machinery room. In some cases, entire cargoes are electronically monitored and controlled from a central location.

REFRIGERATION LOAD

Specifications

The refrigeration specifications should set forth the extreme operating conditions of loading, ambient and sea temperatures, and rates of pulldown. In the all-purpose installation, each compartment should be designed for the refrigeration of warm, fresh products from the field or orchard; for the overall condition, a percentage division of chill and freezer cargo with simultaneous and total loading should be stated.

Typical conditions include the following:

1. Arrangement and net volume capacities of the refrigerated compartments
2. Thicknesses and kinds of insulation
3. Ambient temperatures

Weather surfaces	38°C
Adjacent machinery spaces	38°C
Other adjacent spaces	29°C
Sea temperatures	29°C

4. Overall stowage factor — 2.2 m³/Mg
5. Percentage total loading as chill — 75%
   Percentage total loading as freezer — 25%
6. Receiving temperature, chill cargo — 27°C
   Receiving temperature, frozen cargo — −4°C
7. Carrying temperature, chill cargo — 1°C
   Carrying temperature, frozen cargo — −18°C
8. Initial period of cargo heat removal (equivalent) — 72 h
9. Replacement air at 29°C db, 24°C wb — 3%

The owner should describe the kind of refrigeration system to be installed and specify the number of compressors and other auxiliary parts or apparatus together with sources of emergency pumping and water facilities. All equipment and installations must be specified as complying with the rules and regulations of the Classification Societies (American Bureau of Shipping, Lloyd's Register, and others), the U.S. Coast Guard, the U.S. Public Health Service, ASHRAE *Standard* 15, and ASHRAE *Standard* 26.

The specification writer should specify machinery performance only, unless the writer wishes to assume full responsibility for functioning of the plant and all its parts. The specifier should, however, obtain from the vendors full descriptions, details, capacities, and specifications of the equipment proposed to permit comparative analysis.

Completion tests should be required to determine workmanship and functional performance. Performance guarantees should cover operations under loaded service conditions.

Calculations

The following method of refrigeration load calculation for a general service plant carrying heterogeneous chill cargo may appear simplified, but it is justified by the great range of conditions common to marine installations. Refrigeration loads for freezer cargo may be calculated in a similar manner using the same stowage factor, a specific heat of 1.67 kJ/kg·K, and an equivalent pulldown period of 72 h. The loads for respiration heat, replacement air, or latent heat of fusion will not be present.

Specialized service in known ambient conditions may be calculated more precisely, but an arbitrary 10% margin should still be added to the results to compensate for aging and unforeseen heat gains.

With general calculations, the following operating conditions should be assumed:

1. *Weather ambient conditions*: Up to 38°C.
2. *Ambient sea conditions*: Up to 29°C.

Marine Refrigeration

3. *Conductivity of insulation*: According to standards given for the material, urethane foam with an installed *k*-value of about 0.022 W/(m·K) is suggested.
4. *Resistivity of outer boundaries, inner linings, and surface films*: These factors should be ignored because boundaries and linings are usually dense and have high conductivity values.
5. *Infiltration and open door leakage*: For cargo refrigeration installations, such losses at sea are nil. Port exposures during loading and discharge reestablish pulldown conditions.
6. *Ventilation or replacement air*: This factor is often omitted when carrying heterogeneous cargo for short to medium length voyages. For specialized service, it may be as much as 300% of the gross room volume per hour.
7. *Electrical energy conversion*: The energy load from fans and brine pumps will be on demand load rather than connected load. An arbitrary assumption of 85% efficiency may be used
8. *Product load*: This factor ranges widely for heterogeneous cargo. Its volume will vary from 270 to 800 kg/m^3 and its specific heat will vary from 0.92 to 3.98 kJ/(kg·K). The gross mass of the package and the specific heat of the product should be used. An average density is 460 kg/m^3.
9. *Receiving temperature of cargo*: Chill cargo will range from carrying temperature to ambient. Frozen cargo will range from −29 to −2°C.
10. *Carrying temperature of cargo*: Ranges from 0 to 13°C for chill cargo, and −20 to −29°C for frozen foods.
11. *Respiratory heat of chill cargo*: Meat products, eggs, and dairy products have no respiratory heat. Chapter 8 lists heat of respiration of many horticultural products at various storage temperatures.

CARGO CONTAINERS

The early programs for marine refrigerated containerization borrowed from developments in land transport, particularly the highway vehicular concept. Currently called the conventional container, the shortcomings of its precursor have been inherited along with the merits. Highway transport has developed on the premise of an interim transport of relatively short duration between source and delivery by a single carrier. For fresh horticultural products, initial transport is assumed to occur when the products are hardiest and most likely to endure poor environments. These influences are reflected in the conventional marine container. However, the use of interchangeable (or intermodal) transport of containerized products demanding excellent conditions is increasing. For the usual interval of an overseas voyage, which is often added to storage and land haul, good cold storage is essential.

Standards

As surface transport equipment often passes through many countries, the International Organization for Standardization (ISO) has published container standards for overall dimensions, fastenings (lashings), handling, and lifting (ISO 1996). The American Bureau of Shipping has issued *Rules for Certification of Cargo Containers*. Lloyd's Register of Shipping has published rules covering refrigerated cargo containers.

Container Design

Most containers are 2.6 m high, 2.4 m wide, and either 6.1 or 12.2 m long. Some 2.7- or 2.9-m high containers are also in use. Container insulation is urethane foam because of its low density, its closed-cell characteristics, and its high bonding properties. For containers capable of carrying frozen foods, the urethane foam is 50 to 100 mm thick. Considering aging, skin densities, possible imperfections of installation, and the inevitable penetrations and heat bridges, an average *k*-value of 0.22 W/(m·K) is recommended in the calculation of transmitted heat loads. Floors are of extruded aluminum; linings are aluminum or stainless steel. All must have cargo battens or corrugations on the walls to prevent surface contact by the cargo and to assure ventilation.

The methods of assembly, the calculation of refrigeration loads, and the testing procedures for marine containers are the same as for highway vehicles.

Refrigeration Equipment

The front wall of refrigerated containers is formed by a flush-mounted refrigeration unit. It may be powered by a dismountable diesel generator for land-based operation. The unit is thermostatically controlled and includes automatic defrosting. In new containers, the cold air flows between the T-bars that form the deck and up through the cargo. Microprocessor-based control of return air temperature and high airflow can maintain the product temperature within ±0.5 K throughout the container for frozen loads. It is usual to control supply air temperature to within ±0.25 K for perishable loads.

Because of the considerable rejected heat of the machines and cargo, many containers are carried on the weather decks. Many container refrigeration units are equipped with both a water-cooled and an air-cooled condenser. When it is operated in the hold, the water-cooled unit is connected to a cooling water source, and the fan on the air-cooled condenser is turned off.

Other Heat Removal Equipment

Liquid nitrogen is economically feasible for some services. In such arrangements, the supply tanks and control apparatus may be carried by the transporting chassis, and the attachments to the spray header and the sensing bulb are separable.

Modified Atmospheres in Transport

Supplementing refrigeration, modified atmospheres slow the metabolism of fresh horticultural products during transport. Low-oxygen atmospheres are avoided in conventional refrigerated holds because of the hazard to personnel. This restriction does not apply to containers as they are not entered while in transit. Additional information appears in Chapter 29.

Criteria for Marine Carriage

Products intended for export must be of the highest quality and in prime condition. They must be packed in packages that will not be crushed during transport and will allow heat removal; these are responsibilities of the shipper and should be subject to requirements of the marine carrier. All products carried in cargo containers should be cooled to carrying temperature before original departure, and the container should be expertly loaded to allow air movement through the stack, yet remain secure during transport. Uniformity of spacing and product temperature are requisites of good refrigeration for frozen foods as well as fresh horticultural products.

Even with orderly initial stacking within the container, disarrangement may occur en route because of the ship's motion. Vibrations caused by propulsion machinery can cause improperly secured packages to creep. Shifting of packages can lock air passages and effectively retard refrigeration. The shipper's method of stowage, like the quality of the products, is concealed from the marine carrier who assumes the responsibility for good delivery at destination.

SHIPS' REFRIGERATED STORES

Most vessels carry enough provisions for long voyages without replenishing en route. The refrigerating equipment must operate under extreme ambient conditions. Storage of frozen foods, packaging, humidity, air circulation, and space requirements are important factors.

Table 2 Classifications for Ships' Refrigeration Services

Service	Temp., °C	Passenger Vessels	Freight Vessels
Freezer rooms			
Meats/poultry	−29	X	X
Frozen foods	−29	X	X
Ice cream	−29	X	X
Fish	−29	X	X
Ice	−2	X	
Bread	−18	X	
Chill rooms			
Fresh fruit/vegetables	1	X	X
Dairy products/eggs	0	X	X
Thaw rooms	4 to 7	X	X
Wine rooms	9	X	
Bon voyage packages	4	X	
Service boxes			
Main galley			
Cooks' boxes	4	X	X
Butchers' boxes	4	X	
Bakers' boxes	4	X	
Salad pantry refrigerator	4	X	
Coffee pantry refrigerator	4	X	
Ice cream cabinet	−12	X	
Mess rooms or pantries	4	X	X
Deck pantries	4	X	
Wine stewards' box	4	X	
Bars/fountains	Various	X	
Miscellaneous			
Ice cube freezers	See text	X	
Ice cream freezers	See text	X	
Biologicals	4	X	
Drinking water systems	See text	X	X
Ventilated stores			
Hardy root vegetables	See text	X	X
Flour/cereals	See text	X	X

Perishable foods can be fresh, dehydrated, canned, smoked, salted, and frozen. For most, refrigeration is necessary; for some it may be omitted if the storage period is not too long. Space aboard ship is costly and limited; many rooms at different temperatures cannot be provided. Suitable product storage can be obtained by providing conditions outlined in Table 2 and described in the following sections.

COMMODITIES

Meats and Poultry

Substantial savings in space and preparation labor and better quality can be obtained with precut, boned, frozen meat, and poultry packed in moisture-vaporproof cartons and wrappers. For this reason, increased −20°C storage space should be anticipated. Fresh meats are less suitable because of their relatively short storage life. Also, the space required for fresh meats is two to four times more than that needed for prepackaged meats.

Fish, Ice Cream, and Bread

Good quality fish, properly prepared and packaged, will remain odorless and palatable for a long time.

Commercially prepared ice cream is nearly always available and used to a great extent for both passenger and freight vessels. Ice cream for immediate use should be kept at a slightly higher temperature in an ice cream cabinet in the galley or pantry. Ice cream-making equipment may be desired, in which case provision must be made for hardening the ice cream, ices, and sherbets.

Excellent results can be obtained by purchasing freshly baked bread, sealing it in moistureproof wrappers, and storing it at −18°C. This supply may be supplemented by bread and other bakery items made on the ship. Frozen bread may be thawed in its wrapper in a few hours.

Fruits and Vegetables

Packaged frozen fruits, fruit juices or concentrates, and vegetables may be stored in any freezer room. All packaged frozen products may be held in a common −18°C storage space. However, improved accessibility, especially on large passenger vessels, may justify separate refrigerated spaces for some products.

In some cases, fresh-grown product is desired. These items may be stored in a common chill room, but some compromises with their optimum storage conditions must be expected.

Dairy Products, Ice, and Drinking Water

All dairy products may be stored in a single room, following customary shoreside practice. Strong cheeses with odors that might be adsorbed by other foods should be stored in a tightly enclosed chest or cabinet placed in the dairy refrigerator. Eggs may be processed by oil dipping or heat stabilization to make them less sensitive to unfavorable humidity conditions or odors. Large passenger vessels should be fitted with a separate egg storage room. Butter for reserve supply should come aboard frozen and be kept in a freezer. Frozen homogenized milk has been perfected to a degree that it can be carried for reasonably long periods. Aseptically canned whole milk may be stored without benefit of refrigeration, but this product has some limitations because of the detectable cooked flavor.

Flake ice machines and automatic ice cube makers are common on passenger and freight vessels. Chilled drinking water is piped to many parts of a ship. The water is cooled in closed-system scuttlebutts, and the necessary circulating lines serve living and machinery spaces where drinking fountains and carafe-filling taps are installed. Remote stations that would require unusually long insulated piping runs are better served by independent refrigerated drinking fountains.

STORAGE AREAS

Many borderline perishables, such as potatoes and onions, are satisfactorily stored with ventilation only. Hardy root vegetables are carried on freight vessels not destined for winter zones in slatted bins on a protected weather deck. Flour and cereals must be stored in cool, well-ventilated spaces to minimize conditions conducive to the propagation of weevils and other insects.

Storage Space Requirements

Space requirements for refrigerated ships' stores can be approximated by formulas. However, catering officials and supervising stewards have specific ideas regarding the total volume and subdivisions, and these sometimes vary greatly. A freight ship in ordinary scheduled service seldom exceeds 45 days between replenishment of stores and passenger vessels, considerably less. In addition, deliveries en route are possible.

The space provided should allow suitable working floor areas for good storekeeping. When possible, stable piling 1.8 m high is good practice; and, if the clear height of the room is less than 2.1 m, allowances must be made for air circulation. A storage factor of 2.8 m^3/Mg should suffice and allow floor working area. In the absence of a directing caterer or steward for consultation, Equation (1) may be used to estimate the total refrigerated storage space for merchant vessels. Ice storage is not included because of the various methods used in supplying it.

Marine Refrigeration

$$V = \frac{NDPF}{1000} \quad (1)$$

where

V = total volume of refrigerated storage (not including ice), m^3
N = number of crew and passengers
D = number of days between re-storing
P = mass of refrigerated perishables per person per day, kg
  = 4.5 kg for freighters
  = 6.1 kg for passenger vessels
F = stowage factor (approximately 2.8 m^3/Mg)

For example, for a freighter on a 45-day voyage with a crew of 53 and 12 passengers,

$$V = \left(\frac{65 \times 45 \times 4.5}{1000}\right)28 = 36.9 \text{ m}^3$$

With 1.8-m high stowage, the net floor area would be 20.5 m^2. With 2.4-m high ceilings, the gross volume would be 49.2 m^3.

Gross volume represents actual space available for storage of foods up to ceiling height and does not include the space occupied by cooling units, coils, gratings, or other equipment.

Stores' Arrangement and Location

Next to the arrangement of ships to meet their major purpose, the planning of ships' housekeeping facilities is most important. Efficient operation by culinary workers requires not only well-arranged working spaces, but also convenient supply stores. Storerooms are usually located in spaces least suitable for living quarters or revenue-earning volumes and in areas adjacent to the main galley and pantry. The arrangement should provide easy access, which generally places the reserve storage refrigerators on the deck below the galley.

Aboard freight vessels, the refrigerators serve for daily issue as well as reserve storage. Aboard passenger vessels, the reserve storerooms are less frequently entered, and greater use is made of the service or work boxes.

Passenger vessels carry a corps of steward's storekeepers, who should have an issue counter and office located within sight of the exits serving this area. The storerooms should extend to the ship's sides or have passageways reaching to sideport doors through which the stores may be loaded directly into the ship. However, the arrangement of passenger ship stores will likely be compromised because of the interferences of structure, machinery or access hatches, and ventilation trunks.

In addition to the requirements for reserve storage of perishable foods, refrigerators (often referred to as working boxes) must be provided for the galley and pantry crew. On cargo vessels, a large domestic-type refrigerator will suffice. When more space is needed, a commercial walk-in box can be used. On passenger vessels, larger boxes are built-in like reserve refrigerators. The capacity of the passenger ship refrigerators will be governed by the number of passengers carried, the variety of the menu, and the arrangement of the galley and pantries.

Ice cream stored in the reserve boxes is too hard for serving, hence a dry or closed type of serving cabinet that will maintain temperatures from −15 to −12°C must be installed in the pantry. Passenger vessels may require ice cream-making machines as well as bar and soda fountain equipment, the latter being fitted with commercial, independent refrigerating units.

SHIP REFRIGERATOR DESIGN

Marine refrigeration equipment for off-shore vessels should be designed, selected, and applied so that it will function properly under extreme conditions with a minimum dependence on expert servicing.

Refrigerated Room Construction

Free water that might enter the insulation through faulty floor or wall surfaces is the most harmful element to ships' refrigerators. Room linings and floor coverings should be made of the materials and have the surface character that will give life-long resistance to the absorption of water by the insulation and the adherence of moisture on the room's interior surfaces. The construction of reserve and built-in refrigerators should follow details similar to those of the conventionally designed cargo refrigerators described in the section on Cargo Refrigeration.

Adequate floor drains of the type that may be cleaned without lifting floor gratings should be provided and located so that, with the probable stowage plan, the scuppers will be accessible for cleaning without moving shelving or excessive weights of stores.

All details must be in compliance with the regulations of the U.S. Public Health Service, which also emphasizes rat-proofing. American ships are also subject to strict fire-resistance regulations.

All doors and frames should be of sturdy construction to resist frequent slamming and should have metal sheathing or reinforced glass fiber doors. They should be large enough to facilitate the loading of stores. The locking device should permit release of its fastenings from the inside by a person accidentally locked in.

All rooms should have galvanized or stainless steel racks or shelves to meet storekeeping needs, and they should be easily removable from their supports to facilitate rapid and thorough cleaning. The meat room and thaw room should have a single fore-and-aft meat rail for miscellaneous uses and thawing, respectively. Floor gratings or duckboards, fitted to each room, should be of a size and mass to facilitate removal and cleaning of gratings and the room.

The refrigerators should be fitted with waterproof lighting fixtures well guarded from damage by storing operations. The wiring should be bronze basket-weave cables, surface mounted, and well secured. Lighting switches should be mounted inside each room at the door with indicating lights in the outer passageway. Each room should have an efficient audible alarm for the use of any person inadvertently locked inside.

Remote reading thermometers, from which room temperatures can be read in the outside passageway, are essential to good operations. The sensor should be located in a representative location in the room—generally the geometric center at the ceiling. A large passenger installation justifies a duplicate electronic thermometer, with the instrument located in the refrigeration machinery room.

Service boxes in the galley and pantries should be constructed with a minimum amount of wood. The linings and shelving should be made of materials and have a surface character that facilitates thorough cleaning. Service refrigerators should not have raised door sills, and the floor should present a flush surface that is easily drained and cleaned. The cooling surfaces should be totally accessible for cleaning. Small units should be mounted without floor clearance on elevated bases, or be provided with at least an 200-mm clearance to facilitate scrubbing underneath.

Methods of Refrigeration

Freighters that carry no refrigerated cargo or that normally carry perishables in one direction, will only, in most cases, use direct-expansion systems for stores. Freight ships normally carrying refrigerated cargoes round trip may be arranged to use brine refrigeration for the combined service. Passenger vessels may have brine circulating systems, individual condensing units on each fixture, or a suitable multiplex arrangement.

The location of the condensing units for ships' stores, particularly for galley service, bars, and soda fountains, often presents a plumbing problem with the circulating water. This may make air-cooled units advisable for such applications. When air-cooled units are used, an ample supply of the coolest available air must be assured, and the hot air leaving the condenser must be expelled if the units are located in small rooms. Ventilating fans are mandatory for air-cooled condenser installations. Condensing units should not be installed in hot machinery rooms. Standby units for the more important refrigerating system components are advisable to ensure against extended interruptions of service at sea.

Direct-Expansion Systems. These systems may be classed as central plant, multiplexed, and unit installations. The central plant installation uses two condensing units, one as a standby. If both low- and high-temperature refrigerators are to be served by the single operating unit, it must be selected and based on the lowest refrigerant temperature to be used in the system; this entails some sacrifice of compressor capacity. For successful operation, the smallest cooling unit must balance with the condensing unit at a saturated suction temperature above the low-pressure safety cut-out switch setting. If this condition is not met, hot gas bypass and desuperheating valves will be needed.

Multiplexed Systems. In these systems, the low-temperature fixtures are grouped on one condensing unit, the medium-temperature services on another, and water cooling and high-temperature loads on another. If the refrigerant temperatures required by the various fixtures on one group differ greatly, the higher temperature evaporators should be supplied with evaporator pressure regulators for automatic control. Assuming a series of four refrigerators for ships' foods—one for dairy products, one for fruits and vegetables, one for meats, and another for frozen foods—each would have its own condensing unit with associated cooling units. This provides maximum protection of perishables in case of mechanical failure. A better arrangement provides one or more standby condensing units. With proper piping and valving, the standby unit can be connected to substitute for any unit that has failed.

Unit Installations. These systems apply a single condensing unit to each fixture—an arrangement preferred by some engineers.

Cooling Units. Packaged forced-air cooling units with finned tubes speed up and reduce the cost of the installation. Automatic defrosting may be used, where average room temperatures are about 1°C and above. Air should be directed above the food; it should then be allowed to diffuse down and return to the cooling unit for recirculation. One or more of the units may be installed in a room after considering air distribution. If finned-tube, forced air units are used for temperatures below 1°C, some positive means of defrosting must be provided. Either hot gas or electric defrost is successful if properly applied and used.

When wall coils are used, paneled baffles should be installed over them both to prevent the produce from contacting the low-temperature surfaces and to promote air circulation. To cool products in larger spaces, ceiling-mounted fans should be installed for positive air distribution. The fan is generally located in the center of the room. In designing direct-expansion wall coils, the circuits must not be so long that excessive superheat of the expanded refrigerant results. On the other hand, a shortage of cooling surface results in excessive differences between refrigerant and room air temperature, with consequent dehydration of stored products.

Wall coils with an odd number of horizontal passes simplify series connection. Flow should be from top to bottom to minimize oil hang-up. Intermediate risers must be sized to ensure proper velocity for transport of the oil and unevaporated refrigerant leaving that coil section. Each riser may need to be preceded by a close-coupled trap to minimize oil hang-up in the bottom pass. The final riser should be sized as described in Chapter 2. Double risers may be needed to cope with reduced capacity operation.

Suction lines must be connected to overhead mains so oil cannot drain into idle or low-side equipment. Suction lines should run level and above the compressor suction inlets. If a riser must be used, a suction line accumulator may be needed. If a double suction riser is used, a suction line accumulator is essential to avoid oil slugging.

Controls. The most reliable system is the simplest one with a minimum of automatic apparatus. For marine applications, liquid controls should be of the automatic expansion type. Thermostatic valves are suitable for most applications. Rolling, pitching, trim, and list of vessels make float valve operation difficult. Expansion valves, being spring operated, are not affected by these conditions. Capillary tubes are commonly used in unit refrigerators and will function on vessels. If a float-type control is unavoidable, its vertical plane of operation should be fore-and-aft rather than athwartship. Float controls are unsuitable in vessels with propulsion machinery aft or which normally trim aft.

Temperature controls using thermostats, low-pressure controls, evaporator-regulating valves, and solenoid valves perform satisfactorily in ships' service systems. Such equipment, however, must be unaffected by vibration or ships' motion. Mercury bulb controls should not be used.

Direct-expansion controls should be located adjacent to their respective units, be installed clear of damage hazards caused by handling stores, have strong guards, and be fitted with locked covers to prevent unauthorized persons from tampering with them. Controls should be installed outside the refrigerators; on closely coupled systems using a central plant, they may be installed in the machinery room without the benefit of guards.

The calculation of refrigerant loads of a ship's service refrigerator system should follow the same procedure indicated for cargo systems in the section on Cargo Refrigeration. A larger margin for miscellaneous heat gain, such as frequently opened doors, should be allowed. An arbitrary figure of 10 to 25% of the total load may be used.

FISHING VESSELS

All vessels harvesting fish products, from the small open gill net boats used in inland lakes to the large factory processing ships on the high seas, use some form of refrigeration, whether it is ice picked up each day sufficient for the day's catch, or a version of the most advanced blast freezing system.

REFRIGERATION SYSTEM DESIGN

When designing a refrigeration system, the following issues should be considered:

- The vessel owner and the design engineer must be aware of the monetary value of a fully loaded fish hold. The money saved by selecting and using substandard equipment may be a needless and expensive gamble.
- The vessel may be several hundred kilometres from a qualified service technician and have very limited resources on board for emergency repair. In the event of a system failure, effective initial design may maintain temperatures longer, thus preserving the product.
- Marine refrigeration systems are subjected to severe conditions, including high engine room temperatures, low ambient temperatures, electrolysis (corrosion), impacts, and vibrations. In some cases, these conditions are compounded by little or no maintenance, or even worse, abusive maintenance.
- The system should be well laid out and designed in such a way as to allow new operators to adapt to the system quickly.
- All safety and operating controls should be employed. In the event of a component failure, a backup system should be available, or, ideally, built into the system.
- On completion, the vessel should be provided with all wiring and refrigerant flow diagrams, an operator's manual, and a supply of spare parts.

Marine Refrigeration

In the initial planning, the designer must know the following:

- For what fisheries the vessel is being equipped and in what area of the world the vessel will operate.
- In what future fisheries the vessel may be required to work. (At this point, such considerations will probably add little or no cost to the system.) Necessary alterations may be as few as increasing the spacing in the freezing racks.

Hold Preparation

On any vessel presently being refrigerated or being fitted for future refrigeration installations, 150 mm of insulating spray-on urethane is required. Special attention must be given to insulating areas of high heat such as engine rooms, bulkheads, and the underside of the main decks. High heat sources such as the hatch coaming, shaft log, and fuel tanks with fuel returns from the engines, must also be insulated. The insulation must be protected to prevent moisture from destroying its insulating quality. In the northern Pacific, laid up fiberglass is commonly used because of its strength, low density, and versatility. Pen board guides, mounting brackets, and plate racks are at times fiberglassed into the liner, and thus become a very secure part of the vessel. Fiberglass has the advantage of being easily cleaned and sanitized.

Case Study

In this case study, the owner of a 15-m north Pacific trawler requested plate-type freezing surfaces secured to the front and rear bulkheads and under the deck to control heat gain through insulation, and a freezing rack made up of plates on which fish would be laid during freezing. All equipment uses a single compressor.

Heat Load Calculations. Depending on the sizes of the vessels, the available spaces, and budgets, most vessels employ one or more systems. U.S. law requires the product to be core frozen to $-29°C$ and stored at or below that temperature.

1. Precalculated load from all heat gain sources (not including product load), including a 10% safety factor, is 2.3 kW.
2. Saturated suction temperatures: $-40°C$
3. Required hold temperature: $-29°C$
4. Temperature difference (Δt): $-40 - (-29) = 11$ K
5. The U-factor, or the amount of heat absorbed by the plate surface, is 11 W/(m²·K) (below 0°C in still air). The U-factor can be increased by increasing airflow over the plates. The heat transfer per square metre of plate surface then equals:

$$U\Delta t = 11 \times 11 = 120 \text{ W/m}^2$$

6. Plate heat transfer surface required: $2300/120 = 19$ m²
7. On a visit to the vessel, it was found that 560 mm by 1200 mm plates would be best suited to this vessel. The number of plates required (assuming both sides of the plate provide cooling in this application) are as follows:

$0.56 \text{ m} \times 1.2 \text{ m} \times 2 \text{ sides} = 1.34 \text{ m}^2$ cooling surface per plate
$19/1.34 = 15$ plates required

8. The hatch coaming heat load has been considered in the calculation. The hatch is heavily insulated, has a small entry port, and an insulated floor, again with a small entry port into the hold. The hatch should be kept at above 0°C for bulk fresh food storage. It is equipped with two additional plates and an EPR valve (see Figure 3).
9. A typical method of mounting under the deck plate surfaces is shown in Figure 4.

Freezer Cell Loading. A typical freezing cell (Figure 5) is usually located across the front bulkhead or in another suitable space. It is constructed with 100- to 130-mm spaces between the plates and

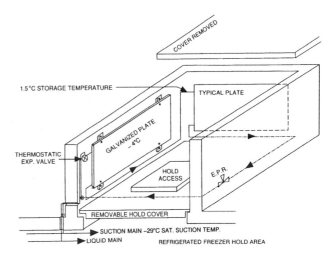

Fig. 3 Cross Section of Product Store

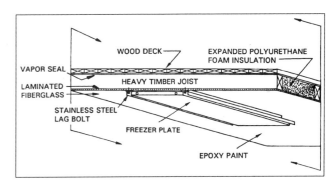

Fig. 4 Typical Under-Deck Freezer Plate Installation

with 2 or 3 spaces at 130 to 180 mm for salmon. Tuna requires spacing of 150 mm and greater. The freezing cell is usually constructed of noncorrosive materials, such as aluminum, with stainless bolts.

1. The example vessel expects to handle an average catch of 140 kg of salmon daily, and to load its freezing cell twice with 700 kg per load. On average, 29 to 34 kg of fish per square metre should be used for proper freezing. The surface area required is

$$\frac{700 \text{ kg}}{29 \text{ kg/m}^2} \times 1.1 \text{ safety factor} = 26.4 \text{ m}^2$$

Note: Only one side of a plate can be used.

2. A 1.2 m by 1.2 m freezing surface allows easy access to 3 sides and requires a minimum of deck area. It also permits the use of storage pens on both sides. The number of freezing surfaces required are as follows:

$1.2 \times 1.2 = 1.4 \text{ m}^2/\text{surface}$
Number of freezing surfaces $= 26.4/1.4 = 18.9$, or
19 freezing surfaces

3. The covering plate not used for loading can be used as a heat transfer surface and must be considered in these calculations. Total heat transfer surface available is

$19 \times 1.2 \times 1.2 \times 2 \text{ sides} = 54.7 \text{ m}^2$
Heat absorption capacity $= UA\Delta t$
$= 11 \times 54.7 \times 11 = 6.6 \text{ kW}$

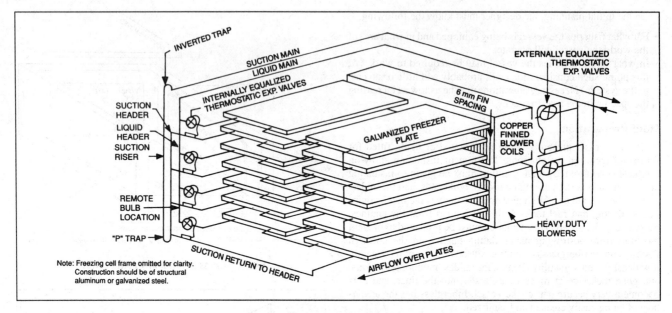

Fig. 5 Typical Marine Freezing Cell

4. Total product load for 700 kg of salmon frozen to −29°C in 12 h is

$$q = \frac{m[C_1(t_1 - t_f) + h_{if} + C_2(t_f - t_2)]}{\theta}$$

where

q = freezing load, W
m = mass of product, kg
C_1 = specific heat of product above freezing
= 3.6 kJ/(kg·K) for salmon
C_2 = specific heat of product below freezing
= 1.63 kJ/(kg·K) for salmon
t_1 = initial temperature of product = 10°C
t_2 = final temperature of product = −29°C
h_{if} = latent heat of product
= 285.2 kJ/kg of salmon
θ = time period, h

$$q = 700\{3.6(10 - 0) + 285.2 + 1.63[0 - (-29)]\}/(12 \times 3600)$$
$$= 6.0 \text{ kW}$$

Assuming a 10% safety factor

$$q = 6.0 \times 1.1 = 6.6 \text{ kW}$$

Note that the plate absorption capacity and product bond are closely correlated in still air.

Plate capacity may be increased in two ways. One way is to increase the airflow across the product, which will increase the U-factor of the plate to as high as 23 W/(m²·K). The other way is to install a compressor that runs the system at a greater TD. This method has a higher initial cost, however. In addition, problems associated with lower suction temperatures and pressures may arise. In most cases, a fan forcing 100 to 200 L/s over the plate will increase capacity by about 20%. The fan should be turned off when freezing is no longer required.

Total Evaporator Loading. The total evaporator load, then, is the heat absorption capacity of the freezing plates and the heat gain from other sources.

$$q_e = 2.3 + 6.6 = 8.9 \text{ kW}$$

As an alternative to plates, forced-air cooling units may be used. They should be constructed with not more than 6.4 mm fin spacing and with no dissimilar metals to help prevent corrosion. If an all-aluminum coil is used, it should be constructed using marine grade material. Copper-aluminum construction should not be used.

The most effective blast freezing system is a plate freezing cell, sized and constructed as previously described, with hot gas defrost and equipped with a blast coil to force air over the plates. The blast coil should have about 6.4 mm fin spacing and deliver an air blast of 5000 L/s per square metre of coil face area. The air blast discharges across one-half of the plate freezing cell and returns across the other half (Figure 5).

Compressor and Drive Selection. The compressor may have two duties: (1) to hold the frozen product, and (2) to freeze the fresh product entering the hold. When not required to freeze fresh product, the evaporator surface assists in pulling the hold down to a selected temperature.

The compressor chosen for this application has a capacity rating of 11.6 kW at −40°C saturated suction temperature (SST) and +32°C saturated discharge temperature (SDT) at 1750 rpm. At 1450 rpm, the rating is 9.7 kW, which is closer to the load of this application and is preferred for longer compressor life. Manufacturers' ratings should be consulted to ensure that the compressor will operate within all specifications (i.e., above minimum rpm to ensure proper lubrication and head and oil cooling, if required).

Mounting Frame. The mounting frame should prevent flexing and vibration. Special consideration should be given to diesel-driven units to prevent harmonic vibration due to their higher compression ratios. The combination of high compression ratios and the great mass of most diesel engines precludes flimsy frame construction. The material used in the frame construction of the example was 9.5 mm by 200 mm channel, well cross-braced. This base construction should be considered a minimum for all marine units.

To save space, most marine units are direct driven. The coupling must be aligned to the manufacturer's tolerances, or better, to prevent compressor damage. A good dial gage and knowledge of its use is required for this procedure.

Since a vessel has many pieces of equipment operating at various speeds and under varying load conditions, harmonic vibrations may be created. These high-frequency vibrations cannot be detected in some cases. To prevent damage, all piping and tubing connected to the compressor should have vibration eliminators. Gages mounted

Marine Refrigeration

on the compressor should be oil-filled and equipped with antipulsation dampers. When possible, all controls, gages, switch gear, and so forth should be mounted on a remote panel away from the compressor or unit-mounting skid. A remote panel gives greater access for servicing the compressor drive system while protecting delicate components. The high noise levels in the engine room require the engineer to use ear protection devices, which makes it necessary to include an alarm system to indicate when a system shuts down. For operator convenience, the alarm, if not audible in the wheelhouse, should be equipped with a flashing indicator light. This light, along with operating pressure gages and remote reading thermometers, should be mounted in a control console within easy view of the operator.

Condenser Selection. See the section on Refrigeration System under Cargo Refrigeration for further information. On a small vessel, a keel cooler may be fixed to the underside of the vessel, and the refrigerant is piped directly to it. However, this system is not in wide use because, if the vessel hits a submerged object and the keel cooler is ruptured, refrigerant is lost and the system fills with seawater. Secondly, as this system is primarily used for economic reasons, they are at times undersized and do not provide enough heat transfer surface. Keel coolers are inefficient when the vessel is tied up because no water flows past the hull; they are too small to create adequate natural circulation around their surface. This situation causes a higher operating pressure and temperature and, often, a complete system shutdown.

The most widely used system uses shell-and-tube receiver condensers or tube-in-tube condensers. Regardless of which condenser is selected, it must have removable heads that are accessible inside the vessel to permit inspection and cleaning. The water system should have at least two pickup points, so if one becomes plugged, the other may be of use. These intakes should be located so that the pumps will not cavitate during rough sea conditions. Easily cleaned seawater filters should be included and the operating pressure should be controlled with a 3-way water-regulating valve.

The receiver-condenser should have sufficient capacity to hold the refrigerant charge. It should be installed in the most stable position available, with a liquid outlet at both ends. If the receiver condenser does not have sufficient holding capacity, both outlets should be piped to a vertical receiver of sufficient size to hold the additional charge. Vertical receivers with a bottom outlet always ensure that a liquid refrigerant column is available to the expansion valves to prevent flash gas regardless of the vessel's attitude. A vertical receiver should be designed into a system with a tube-in-tube condenser with no pumpdown capacity.

Another condensing system widely used in medium-sized vessels employs an oversized keel cooler to permit full operation at all times. An antifreeze (propylene glycol) and water solution is circulated through the keel cooler and to a land-based or marine condenser. Operating pressures may be controlled with a 3-way water regulator, thus bypassing excess solution around the condenser back into the keel cooler. Natural water movement allows the refrigeration system to operate at full capacity at all times, even while stationary and even over kelp beds, which may plug the water intakes of salt water cooling systems.

Oil Separators. Oil separators should be used with all low-temperature systems, especially with plate or pipe-style evaporators, because they may oil log if the suction gas velocity is restricted. Crankcase pressure regulators are required to prevent compressor motors from overloading during start-up.

Suction Accumulators. Properly designed suction accumulators prevent compressor wash-outs by liquid refrigerant. The accumulator should be sized to hold at least the low-temperature evaporator operating charge. Accumulators equipped with liquid boiling coils not only help to boil off liquid refrigerant in these accumulators, but also provide some measure of liquid subcooling. Boiling coils are simply several loops of the liquid line inside the accumulator. These accumulators, used in conjunction with properly sized liquid suction heat exchangers, increase system capacity through liquid subcooling.

Liquid Pumps. Liquid pumps, available in a full range of voltages suitable for marine refrigeration, are used on some vessels. This addition allows the system to operate at lower head pressures, while supplying a full column of liquid refrigerant to the expansion valves, thus preventing flash gas.

Mechanical Subcooling. Mechanical subcooling, which uses an additional refrigeration system to cool the liquid of the main system, provides up to 30% increase in capacity. Where an existing system is undersized and cannot be increased for reasons such as space limitations or power limitations of the main system, subcooling becomes an important addition.

Liquid Line Driers. Liquid line driers are usually of the cartridge style and are installed with bypass valves to allow quick and easy changing of the drier blocks. The drier should be placed in a visible location, preferably upstream from the liquid moisture indicator and in an area with minimum vibration.

Other Considerations. Refrigerant flow to the freezing plates must be accurately controlled. There are a few methods to accomplish this. The most common practice is to use a 3.5-kW, internally equalized expansion valve, feeding between 2.8 and 4.6 m^2 of plate surface area. The valve chosen should operate below –40°C.

A liquid header of 22 mm or larger, to which all expansion valves are attached, ensures a full column of liquid to all valves. Isolation valves should be used wherever possible to isolate various sections of the vessel's refrigerating system. Isolating areas such as the port, starboard, under-deck plates, and other sections allows partial use of the system in the event of component failures. Use of these valves should be included in the overall piping plan. All refrigerant lines should be well protected from damage and properly secured to prevent damage through vibration and various vessel movements. If clamps of dissimilar metal to the piping are to be used, the pipe must be protected from direct contact with the clamp, using a good quality rubberized gasket material. This material should be wrapped around the pipe before the clamp is secured to prevent electrolysis that would corrode the copper.

All flow control devices must be of the best possible quality, ensuring that replacement parts are available for future service. All controls, gages, and thermometers should be waterproof and capable of withstanding strenuous use. All electrical wiring must be completed by a qualified marine electrician, while all refrigeration equipment must be installed by a qualified refrigeration technician.

REFRIGERATION WITH ICE

Ice is commonly used for the preservation of groundfish, shrimp, halibut, and most other commercial species. Bin or pen boards are installed to divide the hold as desired (Figure 6). Ice is usually stored in alternate bins so that it is handy for packing around the fish as the fish is loaded into the adjacent bin. The crushed ice varies in size up to 120 mm lumps. As the fish are stowed with crushed ice, each pen is generally divided horizontally by the insertion of boards so that the bottom fish will not be crushed. The compartmentalized sections should not be more than 760 mm high if undesirable crushing and bruising are to be avoided.

The approximate amount of ice required is 1 Mg for each 2 Mg of fish in summer, and 1 Mg to each 3 Mg of fish in winter, based on a voyage of about 8 days. Less ice is needed if the ship has supplemental refrigeration.

The method of stowing the fish in ice is very important to the keeping quality. The depth of ice on the floor of the pen should be a minimum of 50 mm at the end of the voyage. This is obtained by having the initial bedding of ice 25 mm thick for each day of the voyage. In stowing, one or two layers of fish are laid on the bedding ice so that the ice is just completely covered. In no case should the layer of fish exceed 300 mm in thickness. The top covering layer of ice is

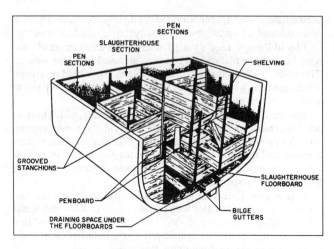

Fig. 6 Typical Layout of Pens in Hold

about 230 mm thick, heaped up higher in the center than along the sides. This method of stowing permits the pile to adjust itself to melting and settling and results in good drainage of water and fish slime.

Most small fishing vessels are constructed of wood; the holds are not insulated. The large vessels are of steel construction with insulated holds lined with wood or aluminum. Mechanical refrigeration is used on some vessels to keep the ice from melting fast and to maintain lower temperatures. The most common mechanical system uses DX cooling coils under the deckhead of the hold and sometimes around the entire shell. Another system uses both a double-wall construction with cold air circulated around the entire hold between the insulated hull and deck and an inner metal lining.

REFRIGERATION WITH SEAWATER

In the United States, recirculated refrigerated seawater is commonly used instead of ice for holding industrial fish (i.e., menhaden and salmon) in satisfactory condition. When used for food fish, it is most suitable for larger fish where salt penetration is not a problem. The seawater is continuously pumped either around the fish and over baffled cooling coils placed along one side of the insulated tank or through external brine coolers and then sprayed over the fish. Welded reinforced aluminum tanks and aluminum cooling coils have been used successfully. The smaller trollers require about 1 kW refrigeration capacity for each 500 kg of fish capacity, while the larger fishing vessels require about 1 kW refrigeration capacity for each 1.3 to 1.5 Mg fish capacity.

Refrigerated seawater is used to preserve the highly perishable menhaden. In one commercial method of preservation, seawater is chilled at the rate of 90 mL/s per kilowatt of refrigeration from 10 to 7°C with 3.1°C suction at the compressor, based on cooling 180 Mg of menhaden to 1.7°C in 24 h at the rate of 3300 fish per megagram. The calculated load of 151 kW of refrigeration includes 10% for insulation losses, so a plant of 200 kW capacity provides considerable margin for contingencies. The chilled water is sprayed over the partially filled fish hold and returned from the bottom of the hold to the chiller.

Distant water vessels are usually equipped for freezing and handling the catch at sea because they stay out for periods of 3 to 9 months, making storage with ice unfeasible. Although a large number of different vessels and freezing systems are used, the general types can be classified as either those freezing large whole fish, such as tuna and halibut, and those freezing processed and semiprocessed fishery products, such as fish blocks or groundfish in bulk lots.

The method of freezing is determined by the physical and biochemical characteristics of the fish and the desired end product. For the most part, large fish such as tuna, which are eventually canned and somewhat resistant to salt intake, are conveniently frozen in brine wells where space savings and ease of handling offer convincing benefits. Cod, haddock, hake, and similar demersal or midwater species, which are more delicate than tuna, are usually frozen rapidly either in vertical or horizontal plate freezers, as described in the previous section, or in air blast freezers.

TUNA SEINERS

Three main factors distinguish the tuna seine fishery from almost all others. These are the big hauls that may be made at any one time, the great size of the larger fish with its corresponding slow heat removal, and the high temperatures of landed tropical tunas.

A system suitable for freezing tuna must be designed to take account of these factors and also meet other general requirements. Physically, it must be reliable and permit large quantities of tuna to be moved in and out in a short time. It must be compact, accessible for service, and compatible with the other space, mass, and trim requirements of the vessel. The cost of installation and maintenance must be low enough to permit an adequate return on the investment. Finally, the system must be able to chill, hold, freeze, store, and thaw tuna under conditions that assure high quality when delivered to the cannery. Some of these factors are now considered in more detail.

Purse-seining is the predominant catch method used in the eastern Pacific tuna fishery. The vessel capacity may vary from 400 to 2000 Mg of fish. For tuna in brine wells, the maximum storage density is approximately 800 kg/m^3. Maximum storage density for tuna in dry refrigerated holds is approximately 580 kg/m^3. Enough refrigeration capacity is needed to freeze fish from 21 to −18°C in 2 to 3 days and to store the frozen fish at −18°C in dry well.

Vessels can range several thousand kilometres and may be required to hold fish for 12 weeks or more, although the average trip time is nearer 6 to 8 weeks. The average catch rate varies considerably. Tuna landed for canning vary in size from 2 to 60 kg. The cross-sectional dimensions of these tuna at the point of greatest girth are approximately 180 mm for 12-kg fish and 330 mm for 60-kg fish. Tropical tunas are fished in waters up to 30°C. The temperatures of the landed fish are higher, reaching as much as 32°C. The required rate of unloading fish from the vessel is 360 Mg/day.

Brine Wells

A tuna seiner is over 75 m long, has a cruising range of 7400 km, and a hold capacity of over 1800 Mg of frozen tuna. To accommodate catches of up to 1800 Mg, a large seiner may have 20 brine wells (2400 m^3) lined with galvanized pipe coils using ammonia direct expansion refrigeration. A total of 1400 kW of refrigeration is required for fish storage, ship service, cold storage room, and air conditioning. Each brine well is insulated on all sides with 150-mm foamed-in-place polyurethane. Basic refrigeration for the brine well evaporated coils is supplied by five reciprocating ammonia compressors with capacity reduction controls, horizontal oil separators, and automatic oil return.

The hold is divided into steel wells or tanks arranged on both sides of the shaft alley. The well insulation is about 125 to 150 mm thick. The wells are constructed of 125-mm wooden planking except on sides adjacent to the skin of the ship where 50-mm planking is used. The shaft alley accommodates pipelines, control valves, brine pumps, and other mechanical devices all closely arranged in a very limited space. The tanks are lined with cooling coils, and each is equipped with a brine circulating pump, seawater inlet and outlet, and connections to the brine transfer lines. Additional fuel oil is carried in some fish wells, since the permanent fuel tanks cannot carry enough fuel oil for the long trip.

The wells on large seiners are prepared for receiving fish by being filled with fresh seawater and being chilled to −1.7°C with the refrigeration system. Since seawater freezes at about −2.2°C, it is not practical to cool below −1.7°C in the preliminary chilling operation. Before the first well is completely filled with fish, the second well is filled with fresh seawater, cooled down, and made ready for loading. With the refrigeration plant operating and the brine recirculating, the fish are chilled down to −1°C internal temperature in 24 to 72 h. This preliminary chilling time varies with the operator, some chilling the fish as rapidly as possible and others preferring a three-day precooling period, feeling that the longer time for chilling seals the pores of the fish better and prevents excessive salt penetration, which may increase freezing time when dense brine is used in the later freezing process.

The next operation strengthens the brine so that fish may be frozen at a lower temperature. This is accomplished by dumping salt directly into the well. It requires about 50 kg of salt for each megagram of fish capacity of the well. This makes dense brine of the seawater, the dense brine having a 1126 kg/m^3, a specific heat of 3.49 kJ/(kg·K), and a freezing point of −13.2°C. A cubic metre of the dense brine and fish mixture has a mass of 1130 kg, of which 800 kg is fish and 330 kg is brine.

Within 2 or 3 days of operation with the dense brine and additional refrigeration, the internal temperature of the fish reaches −7°C or lower as the brine temperature is maintained at about −8°C. The fish become rigid because about 75% or more of their water content is frozen. The dense brine solution is then pumped to another well for reuse, although in some cases it may be pumped overboard if badly contaminated with fish slime and blood. After brine freezing, the well is drained of dense brine and the temperature of the fish may be further reduced by the refrigeration coils only.

The fish are maintained in a dry frozen condition at −8°C or lower until port is almost reached. About one day from port, the circulation of seawater may be started through some of the wells so that the tuna will be sufficiently loosened from each other to start unloading. The fish must be thoroughly thawed before entering the cannery processing line. Steel buckets are lowered into the wells, and the fish are manually thrown into the buckets. Power winches lift the buckets to the wharf. The fish are unloaded into flumes, which carry them to the weighing tank and then into the cannery.

Refrigeration Equipment

For this type of refrigeration duty, each tuna vessel generally has three or more refrigeration compressors, driven by electric motors, and three different suction lines—one for −2°C brine, one for −8°C brine, and one for holding. Each fish well and tank is connected to each of the three suction lines, and the ammonia compressors are cross-connected so that any or all of them may operate on any of the three different types of loads. Ammonia is generally used because of the desirability of parallel operation of compressors and because oil problems prevent halocarbon compressors from being operated interconnected aboard ship.

For a large tuna seiner (see Figure 7), duplicate 250-mm bait pumps, each capable of delivering 145 L/s of seawater against 65 kPa, and one 75-mm brine transfer pump, capable of handling 19 L/s of seawater against 150 kPa, are provided. For emergency operation, the brine transfer pump may be cross-connected on the suction and discharge sides with the ship's general service pump.

Each brine tank and each of the three bait tanks, which are also arranged for freezing fish, are provided with a 50-mm brine circulating pump with shutoff valves on inlet and outlet. Each pump circulates 12.6 L/s of seawater against 65 kPa. Pumps are mounted in the shaft alley. These pumps draw from the bottoms of the tanks and discharge into the hatches above the tanks through 65-mm galvanized piping. This circulation of the brine improves the heat transfer of the cooling coils and makes possible the rapid, uniform cooling and subsequent freezing of the fish.

FREEZER TRAWLERS, FACTORY VESSELS, AND MOTHER SHIPS

On a worldwide basis, the largest quantity of fish frozen at sea comprises such species as cod, hake, pollock, redfish, and other demersal or midwater fish. The fish are caught by bottom or midwater trawls and may be either frozen in bulk on the vessel or processed into fish fillets or fish fillet-type blocks and then frozen. The catch rate varies with the particular fishing operation, but may be as high as 55 to 65 Mg per day with averages well below 20 Mg per day in most northern fisheries and 32 to 36 Mg per day on an overall average. The vessels vary in size from 50 m for the smaller bulk freezer trawlers, to over 60 m for factory freezer vessels, and over 90 m for the larger mother ships.

In bulk freezing aboard vessels in the United Kingdom, the fish are landed on the vessel, eviscerated, and then frozen in bulk in vertical plate freezers using R-22. A typical British bulk freezer trawler will have a length of 65 m with a capacity of about 1600 Mg gross mass. A vessel with this classification has a 10 to 12 station top loading, side unloading, vertical plate freezer, producing blocks of whole frozen whitefish, 1060 mm long by 530 mm wide by 100 mm thick, each block weighing approximately 45 kg. The storage rate is 1.7 to 1.9 m^3/Mg. Total freezing capacity is about 32 Mg of fish per day.

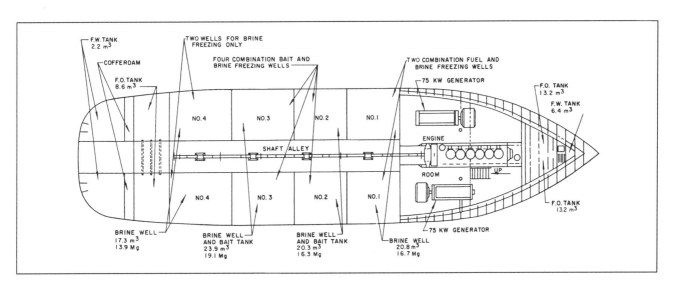

Fig. 7 Plan of Hold for Tuna Seiner

In a typical operation, fish are landed on the vessel, sorted and gutted by hand, washed in a rotating cylindrical washer, and conveyed to storage bins alongside the two rows of freezers that run fore and aft on either side of the factory space. The fish are packed between the pairs of freezer plates and are reduced to a temperature of about −21°C in about 3.5 h. Hot trichlorethylene is then circulated through the evaporated plates both to help release the blocks from the freezer and to ensure that the wet fish do not stick to the plates when reloading begins. The discharged frozen blocks of fish are chuted through the insulated hatch in the factory deck to the cold storage room below, which operates at −29°C and has a capacity of about 450 Mg. Trawlers of this type make voyages of 30 to 60 days, depending on the catching rate, and fish principally off Greenland, Newfoundland, and Labrador.

Factory trawlers are equipped to catch, fillet, package, and freeze the product on the vessel in a manner very similar to shoreside processing operations. A large vessel with a length of about 75 m has refrigeration capacity sufficient to freeze 27 Mg of fish fillets a day and to store 540 Mg at −7°C. After being landed, the fish are stored in ice or chilled seawater and then are headed, filleted, and skinned by machine. The fillets are packed in trays and frozen in blocks, usually in horizontal plate freezers, and then transferred to the cold storage room. This pattern of operations is similar in most factory freezing vessels. Russian, Polish, and German companies have emphasized use of blast freezers, and some of the larger vessels can freeze at a rate of 100 Mg a day. Some vessels may produce a combination of factory-finished products, bulk frozen fish, and iced fish that is the last day's catch.

The mother ship is a processing vessel with no catching capacity. It is the central part of a total fishing complex consisting of a fleet of fishing trawlers and fuel supply vessels that operate at considerable distances from home port. The fishing trawlers, which fish for several days, transfer their catch to the mother ship, where the fish are processed, frozen, and stored in much the same manner as on the factory vessel.

REFERENCES

ABS. 1987. Rules for certification of cargo containers. *Document* 13-87. American Bureau of Shipping, New York.

ASHRAE. 1994. Safety code for mechanical refrigeration. ANSI/ASHRAE *Standard* 15-1994.

ASHRAE. 1996. Mechanical refrigeration and air-conditioning installations aboard ship. ANSI/ASHRAE *Standard* 26-1996.

ISO. 1996. Series 1 freight containers—Specification and testing—Part 2: Thermal containers. *Standard* 1496/2:1996. International Organization for Standardization, Geneva.

CHAPTER 31

AIR TRANSPORT

Perishable Air Cargo 31.1
Perishable Commodity Requirements 31.2
Design Considerations 31.2
Shipping Containers 31.3
Transit Refrigeration 31.3
Ground Handling 31.4

AIR freight service is provided by all-cargo carriers and passenger airlines. The latter companies also have all-cargo aircraft. Wide-body aircraft have a passenger and cargo mix on the main deck, increasing cargo capacity (Figures 1 and 2). All lines maintain regularly scheduled flights so shippers may adequately plan delivery time. Special charter flights are also available from regular terminals and from airports located close to the producing areas. Payload range comparisons of wide-body jets are shown in Figure 2.

6-Pallet Arrangement

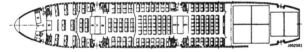

8-Pallet Arrangement

10-Pallet Arrangement

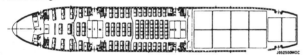

Lower Deck Arrangement

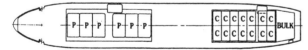

Fig. 1 Flexible Passenger/Cargo Mix

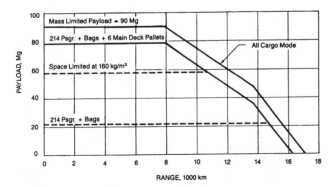

Fig. 2 Payload-Range Comparison for Wide-Body Jet

The preparation of this chapter is assigned to TC 10.6, Transport Refrigeration.

Prospective shippers should contact the airlines serving their locality to obtain specific details for handling perishable shipments.

PERISHABLE AIR CARGO

Some aircraft have cargo compartment temperature control with setting options ranging from just above freezing to normal room temperature. Most compartments have a single temperature control. The control is achieved by balancing the skin heat loss with the supply of expended passenger cabin air and, when necessary, the introduction of hot jet engine bleed air through eductors. Skin heat exchangers are used to assist in maintaining the lower temperatures at high (cold) altitudes. This mode of refrigeration is not available at low altitudes or on the ground where skin temperatures can exceed the compartment temperature significantly. Refrigeration techniques for aircraft rely primarily on precooling, insulated containers, dry ice charged containers, quick handling, and short-time exposure to adverse conditions. Airports seeking to expand cargo operations are adding refrigerated warehouses internationally. The availability of refrigerated warehouses is generally the result of specific market demands and competition.

Fruits and vegetables, flowers and nursery stock, poultry and baby chicks, hatching eggs, meats, seafoods, dairy products, live animals, whole blood, body organs, and drugs (biologicals) are transported by air. Items so shipped are generally of such a perishable nature that slower modes of transportation result in excessive deterioration in transit, making air movement the only possible means of delivery. Certain early season and specialty fruits and vegetables can be flown to distant markets economically because of the high market prices when there is a short supply. Some items, such as cut flowers and papayas, arrive at distant markets in better condition than they would otherwise, so the extra transportation cost is justified. Flowers are shipped on a regular basis from Hawaii to the mainland United States and from California and Florida to the large midwestern and eastern cities. Air movement of strawberries has increased tremendously, including direct shipments to global destinations. Papayas are shipped from Hawaii almost exclusively by air.

When carefully handled, ice cream is shipped successfully to overseas markets from the United States; however, some unsuccessful shipments have occurred because customs inspectors have opened containers for inspection and have taken too much time. The lowering of trade barriers has reduced the risk.

Fruits and Vegetables

All fresh fruits, vegetables, and cut flowers are alive and remain living throughout their entire salable period. Being alive, they respond to their environment and have definite limitations regarding the conditions they can tolerate. They remain alive by the process of respiration, which breaks down stored foods into energy, carbon dioxide, and water, with the uptake of atmospheric oxygen. Respiration, together with accompanying chemical changes, results in quality changes and the eventual death of the commodity. These internal changes associated with life cannot be stopped but should be retarded if a high level of quality is to be retained for a prolonged period.

Seafood

Seafood and fish also benefit from the speed of air freight. The abundance of fresh fish at restaurants and markets throughout the United States is the result of air shipment.

Animals

Design of aircraft cargo compartments for animals is based on SAE *Standard* AIR 1600 and the U.S. Code of Federal Regulations, Title 9. Temperature and ventilation regulations as well as recommendations for birds and animals of all sizes are included in these documents. Air transportation limits exposure to the extremes that would otherwise require special handling and additional cost for animal safety in accordance with the regulations.

PERISHABLE COMMODITY REQUIREMENTS

The justification for the air movement of perishable commodities is based on (1) the time element; and (2) the delivery of a higher quality product than is possible by other modes of transportation. Better delivered quality increases returns to the shipper. This not only offsets the added transportation costs but increases consumer demand and acceptance as well. The market quality of perishable items is definitely controlled by a time and temperature relationship. Temperature cannot be ignored even for the few hours now required for transcontinental air movement. Proper temperature and humidity must be maintained at all times.

Pentzer et al. (1958) lists desirable transit environments for most perishable horticultural commodities. Figure 3 shows the result of a test of air shipments of strawberries from California to Chicago in a refrigerated but uninsulated container. The shipments were exposed to high ambient temperatures during ground handling at origin, resulting in fruit temperatures ranging from 10 to 15°C instead of the desired 0 to 1°C. These berries were compared with those shipped by rail in 4.5 days with temperatures averaging 3°C for the transit period. Appearance and decay on delivery were about the same for both lots. Thus, the advantage of the short 22-h air movement was offset by a loss in quality arising from the unfavorable temperature.

Top quality of many of the most perishable commodities can be significantly reduced by only a few hours' exposure to unfavorably high temperatures. Many drugs (biologicals) and other items, such as whole blood, can be rendered completely ineffective or toxic if not kept at the specified low temperature.

Some flowers, fruits, and vegetables respond favorably to reduced oxygen levels, increased amounts of carbon dioxide, or both, which could be maintained by gastight packaging or containers.

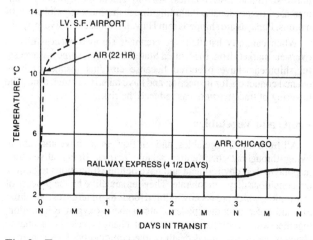

Fig. 3 Temperature of Strawberries Shipped by Air and Rail

The maintenance of temperatures near freezing is not desirable for all products because some are subject to chilling injury when they are exposed to temperatures well above this point. Chilling injury is most pronounced in tropical products, such as bananas, tomatoes, cucumbers, avocados, and orchids. Temperatures above 13°C are usually safe for cold-sensitive commodities. Other items, such as cut flowers, require temperatures between 0 and 13°C.

Certain fruits and vegetables require humidity control. Humidity should be kept between 85 and 95% to prevent wilting and general loss of water. The relative humidity in the cabin of an airplane flying at about 12 000 m is generally less than 10%. However, the respiration of fruits and vegetables, placed in closed containers with recirculated cooling air, should produce the required humidity level with no additional water added.

Certain vegetables, such as peas, broccoli, lettuce, and sweet corn, have high respiration rates and the heat produced may amount to the equivalent of 110 kg or more of ice meltage per tonne of vegetables per day at 15°C. In designing the refrigeration systems for aircraft containers, the additional evaporator capacity required to handle the heat of respiration should be considered.

DESIGN CONSIDERATIONS

A refrigeration or air-conditioning system for a cargo airplane or airborne cargo containers has conflicting design temperature requirements, depending on the type of cargo to be carried, which makes it difficult to use one optimum refrigeration system for all kinds of cargo. For example, frozen foods should have a temperature of −18°C or lower, fresh meat and produce −1 to 7°C, and live animals generally require temperatures in the same comfort range as for humans. Today, many of the commercial jet cargo planes operate with the main cabin divided between cargo compartments and passenger compartments, and they are supplied by a single air system controlled to the comfort of the human occupants. In this case, perishable cargo must be packed in containers, insulated, and iced or precooled.

The design ambient temperatures that an airplane will experience in flight are given in Chapter 9 of the 1995 *ASHRAE Handbook—Applications*. A cargo jet cruising close to Mach 0.9, has an increase in skin temperature over ambient of about 28 K. With an all-cargo load, the basic air-conditioning systems are capable of maintaining main cargo compartment temperatures on a design hot day from 4°C at 9000 m to −1°C at 12 000 m.

The air-conditioning systems are equipped with controls to prevent freezing of moisture condensed from the air at low altitudes. With the extremely dry air prevailing at cruise altitudes, an override of these anti-icing controls would permit an even lower cabin temperature, although it is doubtful that storage temperatures of frozen goods could be met. Thus, some insulation would still be required in frozen food containers. Further, the airplanes are often required to hold at relatively low altitudes of 6000 m or less (because of heavy traffic at the busier airports) for periods of 30 min or more.

Permanent attachment of a mechanical refrigeration system to a cargo container may not be desirable for several reasons: increased load, reduction in usable volume, and difficulty in rejecting the condensing unit heat load overboard. These objections are particularly applicable to containers carried in the main cargo hold. On the other hand, permanently attached units would permit refrigeration of just part of the cargo load while the remainder could be held at temperatures in the normal human or animal comfort zone. Temperature control of products requiring widely differing transit and storage temperatures would be more feasible with onboard refrigeration units.

SHIPPING CONTAINERS

Fruits and vegetables are generally shipped in the same containers used for surface transportation: wooden boxes, veneer crates of various types, or fiberboard cartons. Most flower containers are constructed of either plain or corrugated fiberboard materials. Wooden cleats are used for bracing material, generally as dividers or corner braces inside the cargo box. Where lading may be exposed to very cold surfaces, external cleats may be used as spacers to prevent direct contact. Certain flowers such as gardenias and orchids may be packaged in individual cellophane-wrapped boxes or trays and placed in a master container. Any tightly sealed film wrap must be perforated by at least one small hole to permit release of air from the container during ascent to high altitudes.

Containers, built on pallets and shaped to make maximum use of the interior airplane volume, are in use. Airline containers presently in use are described in the IATA Register. Containers for aircraft, except for the belly cargo holds, are not shaped to make maximum use of the interior volume of the airplane. One reason for this is that the individual packages that will fill the containers are generally rectangular in shape anyway. Another reason is to permit easier intermodal transport, e.g., from motor truck to the airplane and vice versa. Because of the size of aircraft loading doors and irregular aircraft cross-sections compared to surface vehicles and vessels, containerization may require this compromise.

Containerization is a system of moving goods in sealed, reusable freight containers too large for manual handling and which do not have wheels permanently attached. The advantages include far less cargo damage and pilferage, lower packaging costs, minimized handling, and lower shipping rates. Presently, these containers may be loaded at the air freight terminal or loaded at the shipper's facility and transported by flatbed truck-trailer or railroad, or both, to the air terminal.

The critical condition for design of insulation and refrigeration systems (detachable plug-in type or permanently installed) for cargo containers is the time that the container is on the dock in the hot sun waiting for shipment. For this condition, an ambient temperature of 38°C db is assumed. The average outside skin temperature of an unpainted metal container is about 45°C.

Under the conditions just mentioned, the 2.4 m by 2.4 m by 3.0 m container with 13 mm of high efficiency insulation (recirculating the air and considering no latent load) requires about 5 kW of refrigeration to maintain 1.7°C inside and about 7 kW to maintain −18°C inside. For quick pulldown to these temperatures of the container and fresh perishable contents (assuming prefreezing of frozen products), the capacities should increase by about 50%.

Fresh fish, shrimp, and oysters may be packed in boxes, barrels, or special containers. Proper precaution must be taken to prevent drippage from melted ice into the cargo space. Live lobsters are packed in insulated containers with salt water seaweed. Frozen foods are always packed in insulated containers. Whole blood is shipped in specially developed containers. Insulated bags are also used.

The configurations and dimensions of two insulated containers are shown in Figure 4. Insulated with closed-cell, rigid plastic foam, the containers are a fabricated sandwich structure and are sized to fit conventional pallets and materials handling systems. The heat transfer rate for the entire standard container is 15 W/K and 17 W/K for the commercial size. More recent aircraft such as the MD-11 depicted in Figure 1, use pallets 3.2 m wide by 1.6 m, 2.2 m, and 2.4 m which may be loaded to 1.6 m high and retained by straps. Load capacities are 3000 kg, 4500 kg, and 5000 kg, respectively. Containers are LD-3 half width and LD-6 full width, the latter having twice the capacity and width. Both are 1.5 m deep and 1.6 m high. The LD-3 width is 2 m, the volume is 4.5 m³, and the capacity is 1500 kg. A plug-in portable mechanical refrigerating unit may be positioned in the doorway for standby operation. Tight construction permits controlled atmosphere application. A smaller shipping unit,

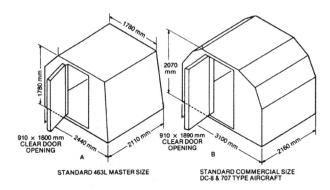

Fig. 4 Insulated Containers Designed to Fit Configuration of Cargo Aircraft

insulated with a foamed plastic, has inside dimensions of 1.1 m by 0.5 m by 0.6 m, i.e., a total area of 3.0 m³ and a capacity of 0.33 m². The heat transfer rate of the entire container is 1.5 W/K.

TRANSIT REFRIGERATION

Many commodities must receive refrigeration in transit. In most cases, this is accomplished by a refrigerant in the package. Water ice, dry ice, and certain proprietary refrigerants are used. Because no method of transit refrigeration can economically cool a warm commodity to its desired transit temperature, all perishable items must be cooled before shipment. Flowers are generally wrapped in several layers of paper or light insulating material and kept cool by water ice. The ice (solid, chopped, or flaked) is placed in a plastic bag or wrapped in many layers of paper and tied to one of the cleats of the container. In some instances, the block of ice may be chilled to −20°C in a freezer before putting it in the package, thereby obtaining a slightly greater refrigeration capacity. Newspapers are sometimes wadded up, thoroughly wetted, and then frozen to −20°C or lower. In all cases, the paper helps absorb the ice meltage water and reduces the chances for leakage into the cargo space.

Some voids should be left in the containers to permit air circulation and uniform cooling. Boxes should be sealed to prevent air exchange. Placing the ice or water for freezing in sealed plastic bags eliminates drippage, but melting ice in open containers increases humidity, which is particularly desirable for cut flowers. A packaged refrigerant with no escape of free liquid must be used with commodities that would be damaged by water. Water ice acts as refrigerant in special blood containers.

Dry ice is used extensively with frozen products and fresh strawberries, the amount depending on the type of container and the length of the journey. Sometimes it is placed in with water ice, not only for its own refrigerating value, but also to slow down meltage of the water ice and extend its value to the end of the transit period. Dry ice alone is seldom used for flowers because its very low temperature may cause freezing damage to adjacent blooms if not properly spaced or insulated. Unless proper ventilation of the cargo compartment is provided, the use of large amounts of dry ice may cause a buildup of carbon dioxide gas in concentrations dangerous to humans and animals.

When the heat transfer rate of insulated shipping containers is known, the amount of refrigeration required can be estimated with reasonable accuracy from

$$Q = 3.6 \, HD\Delta t$$

where

Q = total heat transfer, kJ
H = heat transfer rate of entire container, W/K

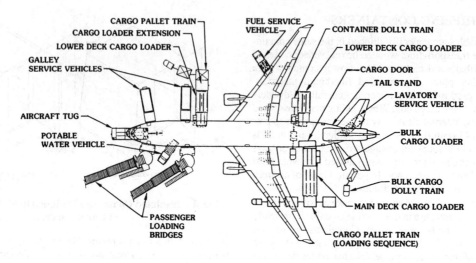

Fig. 5 Ground Service Equipment Arrangement

Δt = difference between ambient temperature and that at which product is to be carried, K

D = duration of transit, h

For example, assume the small container previously described holds 15 standard strawberry trays, each holding 6 kg of berries (90 kg total), with the fruit cooled to, and carried at, 2°C at an ambient temperature of 24°C for a transit time of 24 h:

$$H = 1.5 \text{ W/K}$$

$$\Delta t = 24 - 2 = 22 \text{ K}$$

$$D = 24 \text{ h}$$

$$Q = 3.6 \times 1.5 \times 22 \times 24 = 2850 \text{ kJ}$$

The heat of respiration generated by the berries at 2°C is about 54 W/Mg (see Table 9 in Chapter 8). For 90 kg of berries 24 h in transit:

$$3.6 \times 54 \times (90/1000) \times 24 = 420 \text{ kJ}$$

The ice required to absorb this heat is

$$(2850 + 420)/335 = 9.8 \text{ kg}$$

The amount of dry ice required would be about 5.5 kg.

These simple calculations can be made only when the thermal conductance or heat-transfer rate of the container is known. It would therefore be of considerable value to all concerned—shipper, carrier, and receiver—to have this factor determined and clearly displayed for all insulated shipping containers. Such ratings have been made on truck-trailer bodies, as discussed in Chapter 29.

When package refrigeration is not available, rapid warm-up can be retarded by insulated containers or blanket insulation over stacks or pallet loads. This method has been satisfactory with some of the less perishable fruits such as peaches. Temperatures can be maintained for several in-flight hours with the proper use of these blankets. Care must be exercised in loading to ensure that insulating material is wrapped completely around the cargo and that containers are not in direct contact with hot or cold surfaces. Some cargo compartments on passenger aircraft may be cooled by the air-conditioning system, but the temperature will not be in the optimum range for most perishable commodities.

GROUND HANDLING

All the advantages of speed can be lost if the shipper, carrier, and receiver do not follow the good handling practices that keep deterioration to a minimum.

Ground handling can amount to over 70% of the total elapsed time from shipper to receiver. To reduce this ground time, load palletization and special pallet carriers and loaders, in conjunction with improved load-handling systems aboard the aircraft, are used. Air freight terminals are now designed and built to use these new handling techniques. Combination cargo/passenger jets present unique loading techniques. A typical ground service equipment arrangement is shown in Figure 5.

Fast pickup and delivery are also essential. Because of the generally high ground temperatures at shipping point terminals and intermediate points, most perishable agricultural commodities should be cooled as soon after harvesting as possible and delivered to the air terminal in properly refrigerated vehicles, particularly if they are shipped in uninsulated containers. At the terminal, these shipments must be held at proper temperatures if prompt loading from the pickup vehicle is not possible. Holding rooms, refrigerated mechanically or by ice, should be provided. During seasonal loading peaks, refrigerated trucks or trailers may be used as temporary holding rooms. During hot weather, cargo space must be cooled before loading. Portable air-conditioning equipment, such as that used for passenger aircraft, is used.

The airlines have developed rules for handling various perishable commodities. These include the temperatures desired in transit, the amount of seasonal protection needed, loading methods for various types of containers, and other factors involved in proper handling.

REFERENCES

International Airport Association. Updated annually. IATA *Register of containers and pallets*. Montreal.

Pentzer, W.T., Jr., et al. 1958. Air transportation of fruits, vegetables and cut flowers: Temperature and humidity requirements and perishable nature. AMS *Report* No. 280. USDA.

SAE. 1985. Animal environment in cargo compartments. ANSI/SAE *Standard* AIR 1600-85. SAE International, Warrendale, PA.

Taylor, W.P. 1990. In Proceedings of the 25th Intersociety Energy Conversion Engineering Conference 4:285-87. IEEE Catalog No. 90CH2942-1.

Tyree, L., Jr. 1973. Refrigerated containerized transport for "jumbo" jets. *Progress in Refrigeration Science and Technology* 4:515-25. AVI Publishing Company, Westport, CT.

USDA. 1993. *Animal Welfare*. In Chapter 1 of Code of Federal Regulations, Title 9, Animals and Animal Products.

CHAPTER 32

INSULATION SYSTEMS FOR REFRIGERANT PIPING

Insulation Properties at Below-Ambient Temperatures	32.1
The Insulation System	32.2
Installation Guidelines	32.9
Maintenance of Insulation Systems	32.10

THIS chapter is a guide to specifying insulation systems for piping, fittings, and vessels typically required at below-ambient temperatures ranging from +10°C to −70°C.

The success of an insulation system for cold piping is contingent on such factors as the following:

- Correct refrigeration system design
- Correct specification of insulating system
- Correct specification of insulation thickness
- Correct installation of the insulation and related materials such as vapor retarders
- Quality of the installation
- Adequate maintenance of the insulating system

A variety of insulation materials are used for steam and hot water lines. These lines are insulated for the following reasons:

- Heat conservation and/or freeze prevention
- Process control
- Personnel protection
- Fire protection

Refrigeration lines are insulated for much different reasons than high-temperature lines. Some of these reasons are to:

- Minimize heat gain to the refrigerant suction gas and liquid line. This improves efficiency which in turn reduces energy consumption and operating costs.
- Control surface condensation
- Reduce noise
- Protect personnel and prevent unsafe ice formations

The design features for typical refrigeration insulation applications recommended in this chapter may be followed unless building codes dictate otherwise. For insulation and insulation accessories, specific manufacturer instructions override the recommendations in this chapter. A qualified engineer may be consulted to specify both the insulation material and the insulation thickness (see Tables 2 through 11) based on specific design conditions. All fabricated pipe, valve and fitting coverings should have dimensions and tolerances in accordance with ASTM *Standards* C 585 and C 450. The installation of all materials used for thermal insulation should be carried out in accordance with the Midwest Insulation Contractors Association's (MICA) National Commercial and Industrial Insulation Standards.

INSULATION PROPERTIES AT BELOW-AMBIENT TEMPERATURES

Below-ambient conditions affect several properties that may affect insulation performance. These properties include water absorption, water vapor permeability, thermal expansion and contraction, and wicking of water.

Water absorption is the ability of a material to absorb and hold liquid water. Water absorption is important where systems are exposed to water. This water may come from external sources such as rain, surface condensation or wash down water. The property of water absorption is especially important for outdoor systems.

Water vapor permeability is a measure of the property of a material that allows water vapor to pass through it. The lower the permeability, the higher is the resistance of the material to water vapor intrusion. Water vapor permeability can be critical in design because water vapor can penetrate materials that are unaffected by water in the liquid form. Water vapor intrusion is a particular concern for insulation subjected to a thermal gradient. Pressure differences between ambient conditions and the colder operating conditions of the piping drive water vapor into the insulation. There it may be retained as water vapor, condense to liquid water, or condense and freeze to form ice. Thermal properties of insulation materials are negatively affected as the moisture or vapor content of the insulation material increases.

Thermal expansion and contraction is a concern both for insulation that is continuously below ambient or that cycles between below ambient and elevated temperatures. Thermal contraction of insulation materials may be substantially different from those of the metal pipe. A large difference in the contraction between the insulation and the piping may open joints in the insulation, which not only creates a thermal short circuit at that point, but may also affect the integrity of the entire system. Insulation materials that have large contraction coefficients and do not have a high enough tensile or compressive strength to compensate may shrink and subsequently crack. At the high temperature end of a cycle the reverse is a concern. High thermal expansion coefficients may cause warping or buckling that for some insulation materials is permanent. In this instance, the possible stress on an external vapor retarder or weather barrier should be considered.

Wicking is the tendency of an insulation material to absorb liquid due to capillary action. Wicking is measured by partially submerging a material and measuring both the mass of liquid that is absorbed and the volume that the liquid has filled within the insulation material.

Open cell or fibrous materials tend to absorb more water and have a higher water vapor permeability and wicking than do closed-cell materials. The amount of closed cell structure compared to open cell structure may give some indication of the performance of the material; however some closed cell materials may still have high permeance, so all specific physical properties should be compared.

HVAC and chilled water systems are often insulated to conserve energy and prevent surface condensation. When an open-cell material is used and the vapor retarder system fails, water vapor may move into the insulation, condense, and eventually saturate the insulation material. This may lead to partial or complete failure of the insulation system. The problem becomes more severe at lower operating temperatures and when operating continuously at cold temperatures. The driving forces are greater in these cases and water vapor will condense and freeze on or in the insulation. As more water vapor is absorbed, the thermal conductivity of the insulation material increase, which leads to lower surface temperatures. These lower surface temperatures lead to more condensation, which may cause the insulation material to pop off due to ice formation. With

---

This preparation of this chapter is assigned to TC 10.3, Refrigerant Piping, Controls, and Accessories.

refrigeration equipment operating at −60°C or lower, the problem may be severe.

THE INSULATION SYSTEM

The elements of a below-ambient temperature insulation system include:

- Pipe preparation
- Insulation material
- Insulation joint sealant/adhesive
- Vapor retarders
- Weather barrier/jacketing

Pipe Preparation

Before any insulation is applied, all equipment and pipe surfaces **must** be dry and clean from contaminants and rust. Corrosion of any metal under any thermal insulation can occur for a variety of reasons. The outer surface of the pipe should be properly prepared prior to the installation of the insulation. With any insulation, the pipe can be primed to minimize the potential for corrosion. Careful consideration at the time of the insulation system design stage is essential.

Insulated carbon steel surfaces that operate continuously below −5°C do not present major corrosion problems. However, equipment or piping operating either steadily or cyclically at or above these temperatures may have significant corrosion problems. These problems are aggravated by inadequate insulation thickness, improper insulation material, improper insulation system design, and improper installation of insulation.

Common flaws are:

- Incorrect insulation materials, joint sealants/adhesives or vapor retarders used on below-ambient temperature systems, such as open-cell, wicking or high permeance materials
- Improper specification of insulation materials by generic type rather than specifying specific material properties required for the intended service
- Improper or unclear application methods

Carbon Steel. Carbon steel corrodes not because it is insulated, but because it is contacted by aerated water and/or a water-borne corrosive chemical. For corrosion to occur, water must be present. Under the right conditions, corrosion can occur under all types of insulation. Examples of improperly insulated systems that create conditions that may promote corrosion include:

- Annular space or crevice for the retention of water
- Insulation material that may wick or absorb water
- Insulation material that may contribute contaminants that may increase the corrosion rate

The corrosion rate of carbon steel depends on the temperature of the steel surface and the contaminants in the water. The two primary sources of water are infiltration of liquid water from external surfaces and condensation of water vapor on cold surfaces.

Infiltration occurs when water from external sources enters an insulated system through breaks in the vapor retarder or breaks in the insulation itself. The breaks may be the result of inadequate design, incorrect installation, abuse, or poor maintenance practices. Infiltration of external water can be reduced or prevented.

Condensation results when the metal temperature or the insulation surface temperature is lower than the dew point. Insulation systems cannot always be made completely vapor tight, so condensation must be recognized in the system design.

The main contaminants found in insulation are chlorides and sulfates. These contaminants may have been introduced during the manufacture of the insulation or from external sources. These contaminants may hydrolyze in water to produce free acids that are highly corrosive.

Table 1 lists a few of many protective coating systems that can be used for carbon steel. For other systems or for more details, contact the coating manufacturer.

Copper. External stress corrosion cracking (ESCC) is a type of localized corrosion of various metals, notably copper. For ESCC to occur in refrigeration piping, the copper must undergo the combined effects of sustained stress and a specific corrosive species. During ESCC, copper degrades so that localized chemical reactions occur often at the grain boundaries in the copper. The localized corrosion attack creates a small crack that advances into the metal because of the tensile stress. The common form of ESCC in copper results from grain boundary attack (i.e. intergranular ESCC). Once the advancing crack extends through the metal, the pressurized refrigerant leaks from the line.

ESCC occurs in the presence of the following four conditions:

1. The presence of oxygen (air)
2. The presence of tensile stress, either residual or applied. In copper, stress can be put in the metal at the time of manufacture (residual) or during installation (applied) of a refrigeration system.
3. The presence of a chemical corrosive
4. The presence of water (or moisture) to allow the copper corrosion process to occur

The following precautions reduce the risk of ESCC in refrigeration systems:

- Properly seal all seams and joints of the insulation to prevent condensation between the insulation and the copper tubing.
- Avoid introducing applied stress to copper during installation. Applied stress can be caused by any manipulation, direct or

Table 1 Protective Coating Systems for Piping

Substrate	Temperature Range	Surface Prep.[d]	Surface Profile	Prime Coat[a]	Intermediate Coat[a]	Finish Coat[a]
Carbon Steel System No.1	−45 to 60°C	NACE No. 2	50 to 75 μm	125 μm high-build (HB) epoxy	N/A	125 μm HB epoxy
Carbon Steel System No.2	−45 to 60°C	NACE No. 2	50 to 100 μm	180 to 250 μm metallized aluminum	13 to 20 μm of MIL-P-24441/1[b] epoxy polyamide (EPA) followed by 75 μm of MIL-P-24441/1[c] EPA	75 μm of MIL-P-24441/2[c] EPA
Carbon Steel System No.3	93°C max.	NACE No. 2	50 to 75 μm	50 to 75 μm moisture-cured urethane aluminum primer	50 to 75 μm moisture-cured micaceous aluminum	Two 75 μm coats of acrylic urethane
Carbon Steel System No.3	−45 to 150°C	NACE No. 2	50 to 75 μm	150 μm epoxy/phenolic or high-temperature rated amine-cured coal tar epoxy	N/A	150 μm epoxy/phenolic or high-temperature rated amine-cured coal tar epoxy

[a] Coating thicknesses are typical dry film values.
[b] General Specification for Epoxy-Polyamide Paint. MIL-P-24441, Part 1.
[c] General Specification for Epoxy-Polyamide Paint. MIL-P-24441, Part 2.
[d] NACE No. 2/SSPC-SP 10, Near-White Metal Blast Cleaning. Standard of the National Association of Corrosion Engineers International, Houston, TX and Steel Structures Painting Council, Pittsburgh, PA.

Insulation Systems for Refrigerant Piping

- indirect, that stresses the copper tubing; for example, applying stress to align a copper tube with a fitting or physically damaging the copper prior to installation.
- Never use chlorinated solvents such as 1,1,1-trichloroethane to clean refrigeration equipment. Such solvents have been linked to rapid corrosion.
- Use no acidic materials such as citric acid or acetic acid (vinegar) on copper. Such acids are found in many cleaners.
- Make all soldered connections gas-tight because a leak could cause the section of insulated copper tubing to fail. A gas-tight connection prevents self-evaporating lubricating oil, and even refrigerants, from reacting with moisture to produce corrosive acidic materials such as acetic acid.
- Choose the appropriate thickness of insulation for the environment and the operating condition to avoid condensation on the copper tubing.
- Never mechanically constrict or adhere the insulation to the copper. An example of mechanical constriction is the use of wire ties to compress the insulation. This operation may cause water to pool between the insulation and the copper tubing.
- Prevent extraneous chemicals or chemical-bearing materials such as corrosive cleaners containing ammonia and/or amine salts, wood smoke, nitrites, and ground or trench water, from contacting the insulation or copper.
- Prevent water from entering between the insulation and the copper. Where the layout of the system is such that condensation may form and run along uninsulated copper by gravitational force, completely adhere and seal the beginning run of insulation to the copper or install vapor stops.
- Use copper that complies with ASTM *Standard* B 280. Buy copper from a reputable manufacture.
- When pressure testing copper tubing, take care not to exceed the specific yield point of the copper.
- When testing copper for leaks, use only a commercial refrigerant leak detector solution specifically designed for that purpose. Assume that all commercially available soap and detergent products contain ammonia or amine-based materials, all of which contribute to the formation of stress cracks.
- Replace any insulation that has become wetted or saturated with refrigerant lubricating oils. Such oils can react with moisture to form corrosive materials.

Stainless Steel. Certain grades of stainless steel piping are susceptible to ESCC. ESCC occurs in austenitic steel piping and equipment when chlorides in the environment or insulation material are transported in the presence of water to the hot surface of stainless steel and are then concentrated by evaporation of the water. This situation occurs most commonly beneath thermal insulation, but the presence of insulation is not a requirement. Thermal insulation simply provides a medium to hold and transport the water with its chlorides to the metal surface.

Most ESCC failures occur when the metal temperature is in the range of 50 to 150°C. Failures are less frequent when the metal temperature are outside of this range. Below 50°C the reaction rate is slow and the evaporative concentration mechanism is not significant. Equipment that cycles through the dew point is particularly susceptible. Water that is present at the low temperature evaporates at the higher temperature. During the high temperature cycle the chloride salts dissolved in the water concentrate on the surface.

As with copper, in order for ESCC to develop, sufficient tensile stress must be present in the stainless steel. Most mill products, such as sheet, plate, pipe and tubing, contain enough residual processing tensile stresses to develop cracks without additional applied stress. When stainless steel is used coatings may be applied to prevent ESCC. A metallurgist should be consulted to avoid catastrophic piping system failures.

Insulation Materials

All insulation must be stored in a cool, dry location and be protected from the weather before and during application. Vapor retarders and weather barriers must be installed over dry insulation. The insulation itself should be a low thermal conductivity material with low water vapor permeability and it should be nonwicking.

The following insulation materials are commonly used in refrigeration applications. Table 2 summarizes some of their physical properties and Tables 3 through 12 list recommended thicknesses for pipe insulation.

- **Cellular glass** has excellent compressive strength, but it is rigid. Density varies between 100 and 140 kg/m^3. Density does not greatly affect the thermal performance of cellular glass. When installed on applications that are subject to excessive vibration, the inner surface of the material may need to be coated. The coefficient of thermal expansion for this material is relatively close to that of carbon steel. When installed on refrigeration systems, provisions for expansion and contraction of the insulation are usually only recommended for applications that cycle from below ambient to high temperatures. For outdoors or direct buried applications, a jacketing or mastic coating is recommended.
- **Flexible elastomerics** are soft and flexible. This material is suitable for use on non-rigid tubing. It has a low density and low vapor permeability.
- **Closed-cell phenolic foam insulation** has a very low thermal conductivity. It is able to provide the same thermal performance as other insulations at a reduced thickness.
- **Polyisocyanurate insulation** has low thermal conductivity, low density, and excellent compressive strength.
- **Polystyrene insulation** has a low density and good compressive strength. This product does not meet a smoke developed index of 50 or less.

Table 2 Properties of Insulation Materials

	Cellular Glass	Flexible Elastomeric	Closed-Cell Phenolic	Polyisocyanurate	Polystyrene
Standard that specifies material and temperature requirements	ASTM C 552	ASTM C 534	ASTM C 1126	ASTM C 591	ASTM C 578
Suitable temp. range, °C	−270 to 430	−30 to 80[d]	−180 to 120	−183 to 150	−55 to 75
Flame spread rating[a]	5	≤25	≤25	≤25	≤25
Smoke developed rating[a]	0	≤50	≤50	≤50	115
Water vapor permeability[b], ng/(s·m·Pa)	≤0.007	≤0.15	≤3.0	≤6.5	≤2.2
Thermal conductivity[c], W/(m·K)					
At −20°C mean temperature	0.039	0.038	—	0.027	—
At +25°C mean temperature	0.045	0.040	0.019	0.027	0.035
At +50°C mean temperature	0.048	0.043	0.022	0.030	0.037

[a] Tested in accordance with ASTM E 84
[b] Tested in accordance with ASTM E 96, Procedure A.
[c] New insulation tested in accordance with ASTM C 177 or C 518.
[d] Around −30°C flexible elastomeric insulation becomes stiff, but this does not affect its thermal performance or water vapor permeability.

Table 3 Cellular Glass Insulation Thickness for Indoor Design Conditions
(32°C ambient temperature, 80% relative humidity, 0.9 emittance, 0 km/h wind velocity.)

Nominal Pipe Size, mm	Pipe Operating Temperature, °C							
	+5	−7	−20	−30	−40	−50	−60	−70
15	25	25	40	40	50	50	50	65
20	25	40	40	50	50	50	65	65
25	25	40	40	50	50	50	65	65
40	25	40	40	50	65	65	75	75
50	25	40	40	50	65	65	75	75
65	25	40	50	65	65	75	75	75
75	25	40	50	65	65	75	75	75
100	25	40	50	65	65	75	75	90
125	40	40	50	65	65	75	75	90
150	40	50	50	65	75	75	90	90
200	40	50	50	65	75	75	90	90
250	40	50	50	65	75	90	90	100
300	40	50	50	65	75	90	90	100
350	40	50	65	75	75	90	100	100
400	40	50	65	75	90	90	100	115
450	40	50	65	75	90	90	100	115
500	40	50	65	75	90	90	100	115
600	40	50	65	75	90	100	100	115
700	40	50	65	75	90	100	100	115
750	40	50	65	75	90	100	100	115
900	40	50	65	75	90	100	115	115

Notes:
1. Insulation thickness is chosen to either prevent or minimize condensation on the outside pipe surface or to limit heat gain to 25 W/m^2, whichever thickness is greater.
2. All thicknesses are in millimetres.
3. These values do not include a safety factor or aging factor. Actual operating conditions may vary. Consult a design engineer for an appropriate recommendation for your specific system.
4. Data calculated using NAIMA 3E Plus program.

Table 4 Cellular Glass Insulation Thickness for Outdoor Design Conditions
(38°C ambient temperature, 90% relative humidity, 0.4 emittance, 12 km/h wind velocity.)

Nominal Pipe Size, mm	Pipe Operating Temperature, °C							
	+5	−7	−20	−30	−40	−50	−60	−70
15	40	50	65	75	90	90	100	100
20	50	65	75	90	90	90	90	100
25	50	65	65	75	90	100	100	115
40	65	75	75	90	100	115	115	125
50	50	65	75	90	100	115	115	125
65	65	75	90	100	115	125	125	140
75	65	75	90	100	115	125	125	140
100	65	75	90	100	115	125	140	150
125	65	90	100	115	125	140	150	165
150	65	90	100	115	125	140	150	165
200	75	90	115	125	140	150	165	180
250	75	100	115	140	150	180	180	190
300	75	100	115	140	150	180	190	205
350	90	100	125	140	165	180	190	205
400	90	115	125	150	165	180	190	215
450	90	115	125	150	165	190	205	215
500	90	115	125	150	180	190	205	215
600	90	115	125	150	180	190	205	230
700	90	115	140	165	180	205	215	230
750	90	115	140	165	180	205	215	230
900	90	115	140	165	190	205	230	240

Notes:
1. Insulation thickness is chosen to either prevent or minimize condensation on the outside pipe surface or to limit heat gain to 25 W/m^2, whichever thickness is greater.
2. All thicknesses are in millimetres.
3. These values do not include a safety factor or aging factor. Actual operating conditions may vary. Consult a design engineer for an appropriate recommendation for your specific system.
4. Data calculated using NAIMA 3E Plus program.

Insulation Systems for Refrigerant Piping

Table 5 Flexible Elastomeric Insulation Thickness for Indoor Design Conditions
(32°C ambient temperature, 80% relative humidity, 0.9 emittance, 0 km/h wind velocity.)

Nominal Pipe Size, mm	Pipe Operating Temperature, °C							
	+5	−7	−20	−30	−40	−50	−60	−70
15	25	25	40	40	50	50	50	50
20	25	25	40	50	50	50	65	65
25	25	25	40	50	50	50	65	65
40	25	25	40	50	50	65	65	75
50	25	25	50	50	50	65	75	75
65	25	40	50	50	65	65	75	75
75	25	40	50	50	65	65	75	75
100	25	40	50	65	65	75	75	75
125	40	40	50	65	65	75	90	90
150	40	50	50	65	75	75	90	90
200	40	50	50	65	75	75	90	90
250	40	50	50	65	75	90	90	90
300	40	50	50	65	75	90	100	100
350	40	50	65	65	75	90	100	100
400	40	50	65	65	90	90	100	100
450	40	50	65	65	90	90	100	115
500	40	50	65	75	90	90	100	115
600	40	50	65	75	90	100	100	115
700	40	50	65	75	90	100	100	115
750	40	50	65	75	90	100	100	115
900	40	50	65	75	90	100	115	115

Notes:
1. Insulation thickness is chosen to either prevent or minimize condensation on the outside pipe surface or to limit heat gain to 25 W/m^2, whichever thickness is greater.
2. All thicknesses are in millimetres.
3. These values do not include a safety factor or aging factor. Actual operating conditions may vary. Consult a design engineer for an appropriate recommendation for your specific system.
4. Data calculated using NAIMA 3E Plus program.

Table 6 Flexible Elastomeric Insulation Thickness for Outdoor Design Conditions
(38°C ambient temperature, 90% relative humidity, 0.4 emittance, 12 km/h wind velocity.)

Nominal Pipe Size, mm	Pipe Operating Temperature, °C							
	+5	−7	−20	−30	−40	−50	−60	−70
15	40	50	65	65	65	75	75	75
20	50	65	65	65	75	75	90	90
25	50	65	65	75	75	90	90	100
40	50	65	75	75	75	90	100	100
50	50	75	75	75	90	100	100	115
65	65	75	75	75	90	100	100	115
75	65	75	90	90	100	115	115	125
100	65	75	90	100	115	115	125	125
125	65	90	100	100	115	125	125	140
150	65	90	100	115	115	125	140	150
200	75	90	115	115	125	140	150	165
250	75	100	115	125	140	150	165	180
300	75	100	115	140	140	150	165	180
350	90	100	125	140	150	165	165	180
400	90	115	125	140	150	165	180	190
450	90	115	125	140	150	165	180	190
500	90	115	125	140	150	165	180	190
600	90	115	125	140	165	180	190	205
700	90	115	140	150	165	180	190	205
750	90	115	140	150	165	180	190	205
900	90	115	140	150	180	180	190	205

Notes:
1. Insulation thickness is chosen to either prevent or minimize condensation on the outside pipe surface or to limit heat gain to 25 W/m^2, whichever thickness is greater.
2. All thicknesses are in millimetres.
3. These values do not include a safety factor or aging factor. Actual operating conditions may vary. Consult a design engineer for an appropriate recommendation for your specific system.
4. Data calculated using NAIMA 3E Plus program.

Table 7 Closed-Cell Phenolic Foam Insulation Thickness for Indoor Design Conditions
(32°C ambient temperature, 80% relative humidity, 0.9 emittance, 0 km/h wind velocity.)

Nominal Pipe Size, mm	Pipe Operating Temperature, °C							
	+5	−7	−20	−30	−40	−50	−60	−70
15	25	25	25	25	40	40	40	40
20	25	25	25	40	40	40	40	40
25	25	25	25	40	40	40	40	40
40	25	25	25	40	40	40	40	40
50	25	25	25	40	40	40	40	40
65	25	25	25	40	40	40	40	40
75	25	25	25	40	40	50	50	50
100	25	25	40	40	40	50	50	50
125	25	25	40	40	40	50	50	65
150	25	25	40	40	50	50	50	65
200	25	25	40	40	50	50	50	65
250	25	25	40	40	50	50	50	65
300	25	25	40	40	50	50	50	65
350	25	25	40	40	50	50	50	65
400	25	25	40	40	50	50	65	65
450	25	25	40	40	50	50	65	65
500	25	25	40	40	50	50	65	65
600	25	25	40	40	50	50	65	65
700	25	25	40	40	50	50	65	65
750	25	25	40	40	50	50	65	65
900	25	25	40	50	50	50	65	65

Notes:
1. Insulation thickness is chosen to either prevent or minimize condensation on the outside pipe surface or to limit heat gain to 25 W/m^2, whichever thickness is greater.
2. All thicknesses are in millimetres.
3. These values do not include a safety factor or aging factor. Actual operating conditions may vary. Consult a design engineer for an appropriate recommendation for your specific system.
4. Data calculated using NAIMA 3E Plus program.

Table 8 Closed-Cell Phenolic Foam Insulation Thickness for Outdoor Design Conditions
(38°C ambient temperature, 90% relative humidity, 0.4 emittance, 12 km/h wind velocity.)

Nominal Pipe Size, mm	Pipe Operating Temperature, °C							
	+5	−7	−20	−30	−40	−50	−60	−70
15	25	25	40	40	40	50	50	50
20	25	40	40	40	50	50	50	65
25	25	40	40	40	50	50	50	65
40	25	40	40	40	50	50	50	65
50	25	40	40	40	50	50	65	65
65	25	40	40	40	50	50	65	65
75	25	40	50	50	65	65	75	75
100	40	40	50	65	65	75	75	75
125	40	50	50	65	65	75	75	90
150	40	50	50	65	75	75	90	90
200	40	50	65	65	75	75	90	100
250	40	50	65	65	75	90	90	100
300	40	50	65	75	75	90	100	100
350	40	50	65	75	90	90	100	115
400	40	50	65	75	90	100	100	115
450	40	65	65	75	90	100	100	115
500	50	65	65	75	90	100	100	115
600	50	65	75	75	90	100	115	125
700	50	65	75	75	90	100	115	125
750	50	65	75	90	90	100	115	125
900	50	65	75	90	90	100	115	125

Notes:
1. Insulation thickness is chosen to either prevent or minimize condensation on the outside pipe surface or to limit heat gain to 25 W/m^2, whichever thickness is greater.
2. All thicknesses are in millimetres.
3. These values do not include a safety factor or aging factor. Actual operating conditions may vary. Consult a design engineer for an appropriate recommendation for your specific system.
4. Data calculated using NAIMA 3E Plus program.

Insulation Systems for Refrigerant Piping

Table 9 Polyisocyanurate Foam Insulation Thickness for Indoor Design Conditions

(32°C ambient temperature, 80% relative humidity, 0.9 emittance, 0 km/h wind velocity.)

Nominal Pipe Size, mm	Pipe Operating Temperature, °C							
	+5	−7	−20	−30	−40	−50	−60	−70
15	25	25	40	40	40	40	50	50
20	25	25	40	40	40	50	50	50
25	25	25	40	40	40	50	50	50
40	25	25	40	40	40	50	50	50
50	25	25	40	40	40	50	50	65
65	25	25	40	40	40	50	50	65
75	25	25	40	40	50	65	65	65
100	25	25	40	40	50	65	65	75
125	25	40	40	50	50	65	65	75
150	25	40	40	50	50	65	65	75
200	25	40	40	50	50	65	65	75
250	25	40	40	50	50	75	75	90
300	25	40	40	50	65	75	75	90
350	25	40	40	50	65	75	75	90
400	25	40	50	50	65	75	75	90
450	25	40	50	50	65	75	90	90
500	25	40	50	50	65	75	90	90
600	25	40	50	50	65	75	90	90
700	25	40	50	50	65	75	90	100
750	25	40	50	50	65	75	90	100
900	25	40	50	50	65	75	90	100

Notes:
1. Insulation thickness is chosen to either prevent or minimize condensation on the outside pipe surface or to limit heat gain to 25 W/m^2, whichever thickness is greater.
2. All thicknesses are in millimetres.
3. These values do not include a safety factor or aging factor. Actual operating conditions may vary. Consult a design engineer for an appropriate recommendation for your specific system.
4. Data calculated using NAIMA 3E Plus program.

Table 10 Polyisocyanurate Foam Insulation Thickness for Outdoor Design Conditions

(38°C ambient temperature, 90% relative humidity, 0.4 emittance, 12 km/h wind velocity.)

Nominal Pipe Size, mm	Pipe Operating Temperature, °C							
	+5	−7	−20	−30	−40	−50	−60	−70
15	25	40	40	50	50	65	65	65
20	25	40	50	50	65	65	65	75
25	25	40	50	50	65	65	75	90
40	40	40	50	50	65	65	75	90
50	40	40	50	65	75	75	90	100
65	40	40	50	65	75	75	90	100
75	40	50	65	75	75	90	100	115
100	40	50	65	75	90	90	100	115
125	40	50	65	75	90	100	115	125
150	50	65	75	75	90	100	115	125
200	50	65	75	90	100	115	125	140
250	50	65	75	90	100	115	125	150
300	50	65	75	90	115	125	140	150
350	50	65	90	100	115	125	140	150
400	50	75	90	100	115	125	150	165
450	50	75	90	100	115	140	150	165
500	50	75	90	100	115	140	150	165
600	50	75	90	100	125	140	150	180
700	50	75	90	100	125	140	150	180
750	65	75	90	100	125	140	165	180
900	65	75	90	100	125	140	165	180

Notes:
1. Insulation thickness is chosen to either prevent or minimize condensation on the outside pipe surface or to limit heat gain to 25 W/m^2, whichever thickness is greater.
2. All thicknesses are in millimetres.
3. These values do not include a safety factor or aging factor. Actual operating conditions may vary. Consult a design engineer for an appropriate recommendation for your specific system.
4. Data calculated using NAIMA 3E Plus program.

Table 11 Polystyrene Foam Insulation Thickness for Indoor Design Conditions
(32°C ambient temperature, 80% relative humidity, 0.9 emittance, 0 km/h wind velocity.)

Nominal Pipe Size, mm	Pipe Operating Temperature, °C							
	+5	−7	−20	−30	−40	−50	−60	−70
15	25	40	40	50	50	50	65	65
20	40	40	40	50	50	65	65	65
25	40	40	40	50	50	65	65	65
40	40	40	50	50	50	65	65	65
50	40	40	50	50	65	65	65	75
65	40	40	50	50	65	65	65	75
75	40	50	50	65	65	75	75	90
100	40	50	50	65	75	75	75	90
125	40	50	65	65	75	75	90	90
150	40	50	65	65	75	90	90	90
200	40	50	65	65	75	90	90	100
250	40	50	65	75	75	90	100	100
300	40	50	65	75	90	90	100	100
350	40	50	65	75	90	100	100	100
400	40	50	65	75	90	100	100	115
450	40	50	65	75	90	100	100	115
500	40	65	75	75	90	100	100	115
600	40	65	75	90	90	100	100	115
700	40	65	75	90	90	100	115	115
750	40	65	75	90	90	100	115	115
900	50	65	75	90	100	100	115	125

Notes:
1. Insulation thickness is chosen to either prevent or minimize condensation on the outside pipe surface or to limit heat gain to 25 W/m^2, whichever thickness is greater.
2. All thicknesses are in millimetres.
3. These values do not include a safety factor or aging factor. Actual operating conditions may vary. Consult a design engineer for an appropriate recommendation for your specific system.
4. Data calculated using NAIMA 3E Plus program.

Table 12 Polystyrene Foam Insulation Thickness for Outdoor Design Conditions
(38°C ambient temperature, 90% relative humidity, 0.4 emittance, 12 km/h wind velocity.)

Nominal Pipe Size, mm	Pipe Operating Temperature, °C							
	+5	−7	−20	−30	−40	−50	−60	−70
15	40	50	65	65	65	75	75	75
20	40	50	65	65	75	75	90	90
25	40	50	65	75	75	90	90	100
40	50	50	65	75	75	90	100	100
50	50	65	75	75	90	100	100	115
65	50	65	75	75	90	100	100	115
75	65	75	90	90	100	115	115	125
100	65	75	90	100	115	115	125	125
125	65	75	90	100	115	125	125	140
150	65	90	90	115	115	125	140	150
200	65	75	115	115	125	140	150	165
250	75	90	115	125	140	150	165	180
300	75	90	115	125	140	150	165	180
350	75	100	115	140	150	165	165	180
400	75	100	125	140	150	165	180	190
450	90	100	125	140	150	165	180	190
500	90	100	125	140	150	165	180	190
600	90	100	125	140	165	180	190	205
700	90	100	125	150	165	180	190	205
750	90	100	125	150	165	180	190	205
900	90	115	125	150	165	180	190	205

Notes:
1. Insulation thickness is chosen to either prevent or minimize condensation on the outside pipe surface or to limit heat gain to 25 W/m^2, whichever thickness is greater.
2. All thicknesses are in millimetres.
3. These values do not include a safety factor or aging factor. Actual operating conditions may vary. Consult a design engineer for an appropriate recommendation for your specific system.
4. Data calculated using NAIMA 3E Plus program.

Insulation Systems for Refrigerant Piping

Insulation Joint Sealant/Adhesive

All insulation materials that operate in below ambient conditions should be protected by a joint sealant. The sealant should resist liquid water and water vapor, and it should bond to the specific insulation surface. The joint sealant is applied as a full bedding coat to all sealant joints. A properly designed and constructed insulation/sealant/insulation joint retards liquid water and water vapor migration through the insulation system.

Vapor Retarders

Closed-cell insulation materials, as such, do not absorb water. However, this is not exactly the case as most insulations will absorb a certain amount of water. The insulation materials should either have a low water vapor permeability of less than 0.15 ng/(s·m·Pa) or the material should be protected by a continuous and effective vapor retarder.

The service life of the insulation and pipe depends primarily on the in-place water vapor permeance of the vapor retarder. Therefore, the vapor retarder must be free of discontinuities and penetrations. It must be installed so that expansion and contraction can occur without compromising the integrity of the vapor retarder. The manufacturer should have specific design and installation instructions for their products.

Vapor retarders may be of the following types:

- **Metallic foil or all service jacket (ASJ) retarders** are applied to the surface of the insulation by the manufacturer or field applied. This type of jacket has a low water vapor permeance under ideal conditions [0.03 ng/(s·m·Pa)]. These jackets have longitudinal joints and butt joints, so achieving low permeability depends on complete sealing of all joints and seams. The jackets may be sealed with a contact adhesive applied to both overlapping surfaces. Manufacturers' instructions must be strictly followed during the installation. Butt joints are sealed in a similar fashion using metallic faced ASJ material and contact adhesive. ASJ jacketing when used outdoors with metal jacketing may be damaged by the metal jacketing, so extra care should be taken when installing it. Self-seal lap joints and butt joints may be acceptable, but they must be perfectly sealed.
- **Coatings**, **mastics**, and **heavy, paint-type products** are available for covering insulation by trowel, brush, or spraying. The permeability of the materials is a function of the thickness applied. Some products are recommended for indoor use only while others are available for indoor or outdoor use. These products may impart odors and manufacturers instructions should be meticulously followed. Care should also be taken to insure that the mastics used are chemically compatible with the insulation system.

    Mastics should be applied in two coats (with an open weave fiber reinforcing mesh) to obtain a total dry film thickness as recommended by the manufacturer. The mastic should be applied as a continuous monolithic retarder and extend at least 50 mm over any membrane where applicable. This is typically done only at valves and fittings. Mastics must be tied to the rest of the insulation or bare pipe at the termination of the insulation preferably with a 50 mm overlap to maintain continuity of the retarder.
- A **laminated membrane retarder** with a rubber bitumen adhesive on a polyethylene or PVC film has been used successfully. This type of retarder has a very low permeance of 0.9 ng/(s·m²·Pa). Some solvent based adhesives can attack this vapor retarder. All joints should have a 50 mm overlap to insure adequate sealing. Other types of finishes may be appropriate depending upon environmental or other factors.

Weather Barrier/Jacketing

Jacketing on insulated pipes and vessels protects the vapor retarder and insulation. Various plastic and metallic products are available for this purpose. Some specifications suggest that the jacketing should function to preserve and protect the "egg-shell like" vapor retarder over the insulation. This being the case, bands must be used to secure the jacket. Pop rivets, sheet metal screws, staples, or any other item that punctures should not be used because they will compromise the vapor retarder. The use of such materials may indicate that the installer does not understand the vapor retarder concept and corrective education steps should be taken.

Protective jacketing is designed to be installed over the vapor retarder and insulation to prevent weather and abrasion damage. The protective jacketing must be installed independently and in addition to any factory or field applied vapor retarder. Ambient temperature cycling will cause the jacketing to expand and contract. The manufacturer's instructions should show how to install the jacketing to permit this expansion and contraction.

Metal jacketing may be smooth, textured, embossed, or corrugated aluminum or stainless steel with a continuous moisture retarder. Metallic jackets are recommended for exposed, roof-mounted piping. As an alternative, PVC jacketing or other finishes may be appropriate, in some situations. The PVC should be at least 0.75 mm thick, smooth, ultra-violet inhibited, and in precurled rolls.

Protective jacketing is required whenever piping is exposed to washing, physical abuse, or traffic. White PVC is popular inside buildings where degradation from sunlight is not a factor. Colors can be obtained at little, if any, additional cost. All longitudinal and circumferential laps should be seal welded using a solvent welding adhesive. The laps should be located at the 10:00 o'clock or 2:00 o'clock positions. A sliding lap (PVC) expansion/contraction joint should be located near each end point and at immediate joints no more than 6 m apart. Where very heavy abuse and/or hot, scalding washdowns are encountered, a CPVC material is required. These materials can withstand temperatures as high as 110°C, where standard PVC will warp and disfigure at 60°C.

Roof piping should be jacketed with a minimum 0.41 mm aluminum (embossed or smooth finish depending on aesthetic choice). On pitched lines, this jacketing should be installed with a minimum 50 mm overlap arranged to shed any water in the direction of the pitch. Only stainless steel bands should be used to install this jacketing (13 mm wide by 0.50 mm thick 304 stainless) and spaced every 300 mm. Jacketing on valves and fittings should match that of the adjacent piping.

INSTALLATION GUIDELINES

Preliminary Preparation. Corrosion of any metal under any thermal insulation can occur for many reasons. With any insulation, the pipe can be primed to minimize the potential for corrosion. Prior to installing the insulation:

- Complete all welding and other hot work
- Complete hydrostatic and other performance testing
- Remove oil, grease, loose scale, rust, and foreign matter from surfaces to be insulated. Surface must also be dry and free from frost.
- Complete site touch-up of all shop coating including preparation and painting at field welds

Insulating Fittings and Joints. Insulation for fittings, flanges, and valves should be the same thickness as the insulation of the pipe and must be fully vapor sealed.

- If the valve design allows, valves should be insulated to the packing glands.
- Stiffener rings where provided on vacuum equipment and/or piping, should be insulated with the same thickness and type of insulation as specified for that piece of equipment or line. The rings should be fully independently insulated.
- Where multiple layers of insulation are used, all joints should be staggered or beveled where appropriate.

- Insulation should be applied with all joints fitted to eliminate voids. Large voids should not be filled with vapor sealant or fibrous insulation, but eliminated by refitting or replacing the insulation.
- All joints, with the exception of contraction joints and the inner layer of a double layer system, should be sealed with either the proper adhesive or a joint sealer during installation.
- Each line should be insulated as a single unit. Adjacent lines must not be enclosed within a common insulation cover.

Planning Work. Open cell or high permeance insulations may require special protection during installation. All insulation applied in one day should have at least one coat of the vapor retarder mastic applied the same day. If impractical to apply the first coat of vapor retarder mastic, the insulation must be temporarily protected with a moisture retarder, such as an appropriate polyethylene film, and sealed to the pipe or equipment surface. All exposed insulation terminations should be protected before work ends for the day.

Vapor Stops. Vapor stops should be installed using either sealant or the appropriate adhesive at all directly attached pipe supports, guides, and anchors, and at all locations requiring potential maintenance, such as valves, flanges, and instrumentation connections to piping or equipment. If valves or flanges must be left uninsulated until after the plant starts up, temporary vapor stops should be installed using either sealant or the appropriate adhesive.

Securing Insulation. When applicable, the innermost layer of insulation should be applied in two half sections and secured with 19 mm wide pressure sensitive filament tape banding spaced a maximum of 230 mm apart and applied with a 50% overlap. Single and outer layers more than 450 mm in diameter and inner layers with radiused and beveled segments should be secured by 9.5 mm wide stainless steel bands spaced on 230 mm maximum centers. The bands shall be firmly tensioned and sealed.

Applying Vapor Retarder. Irregular surfaces and fittings should be vapor sealed by applying a thin coat of vapor retarder mastic or finish with a minimum wet film thickness as recommended by the manufacturer. While the mastic or finish is still tacky, an open weave glass fiber reinforcing mesh should be laid smoothly into the mastic or finish and should be thoroughly embedded in the coating. Care should be taken not to rupture the weave. The fabric should be overlapped a minimum of 50 mm at joints to provide strength equal to that maintained elsewhere.

Before the first coat is completely dry, a second coat should be applied over the glass fiber reinforcing mesh with a smooth unbroken surface. The total thickness of the mastic or finish should be in accordance with the coating manufacturer's recommendation.

Pipe Supports and Hangers. When possible, the pipe support should be located outside of the insulation. Supporting the pipe outside of the protective jacketing eliminates the need to insulate over the pipe clamp, hanger rods, or other attached support components. This method minimizes the potential for vapor intrusion, and thermal bridges because a continuous envelope surrounds the pipe.

ASME *Standard* B31 establishes basic stress allowances for piping material. The loading on the insulation material is a function of its compressive strength. Table 13 suggests spacing for pipe supports. Related information in this table is also in Chapter 40 of the 1996 *ASHRAE Handbook—Systems and Equipment*.

The insulation material may or may not have the compressive strength to support loading at these distances. Therefore, the force due to the load of the piping and contents on the bearing area of the insulation should be calculated. In refrigerant piping, typically bands or clevis hangers are used with rolled metal shields or cradles. Although the shields are typically rolled to wrap the outer diameter of the insulation in an arc of 180°, the bearing area is calculated over a 120° arc of the outer circumference of the insulation multiplied by the shield length. If the insulated pipe is subjected to point loading such as where it rests on a beam or a roller, the bearing area arc is reduced to 60° and multiplied by the shield length. In this case, rolled plate may be more suitable than sheet metal. Provisions should be made to secure the shield on both sides of the hanger (metal band), and the shield should be centered in the support. Table 14 lists widths and thicknesses for pipe shields.

Expansion Joints. Some installations require an expansion or contraction joint. These joints are normally required in the innermost layer of insulation. Expansion/contraction joints may be constructed in the following manner:

- Make a 25 mm break in the insulation
- Tightly pack the break with fibrous insulation material
- Secure insulation on either side of the joint with stainless steel bands that have been hand tightened
- Cover the joint with the an appropriate vapor retarder and seal properly

MAINTENANCE OF INSULATION SYSTEMS

Periodic inspections are needed to determine the presence of moisture, which will lower the insulation thermal efficiency, often destroying the insulation system. The frequency of inspection should be determined by the critical nature of the process, the external environment, and the age of the insulation. During a routine inspection:

Table 13 Suggested Pipe Support Spacing for Straight Horizontal Runs

Nominal Pipe Size, mm	Std. Steel Pipe[a,b]	Copper Tube
	Support Spacing, m	
15	1.8	1.5
20	1.8	1.5
25	1.8	1.8
40	3.0	2.4
50	3.0	2.4
65	3.3	2.7
75	3.6	3.0
100	4.2	3.6
150	4.9	4.2
200	4.9	4.9
250	4.9	5.5
300	4.9	
350	4.9	
400	4.9	
450	4.9	
500	4.9	
600	4.9	

Source: Adapted from MSS *Standard* SP-69 and ASME B31.1
[a] Spacing does not apply where span calculations are made or where concentrated loads are placed between supports such as flanges, valves, specialties, etc.
[b] Suggested maximum spacing between pipe supports for horizontal straight runs of standard and heavier pipe.

Table 14 Shield Dimensions for Insulated Pipe and Tubing

Nominal Pipe Size, mm	Shield Length, mm	Shield Thickness, gage (mm)
15 to 90	300	18 (1.22)
100	300	16 (1.52)
125 to 150	450	16 (1.52)
200	600	14 (1.90)

Source: Adapted from MSS *Standard* SP-69
Note: Protection shield gages listed are for use with band type hangers only. For point loading, increase shield thickness and length.

Insulation Systems for Refrigerant Piping

- Look for signs of moisture or ice on the lower part of horizontal pipe, at the bottom elbow of a vertical pipe, and around pipe hangers/saddles as moisture may migrate to low areas.
- Look for jacketing penetrations, openings, or separations.
- Check the jacketing to determine if the banding is loose.
- Look for bead caulking failure especially around flange and valve covers.
- Look for jacketing integrity and open seams around all intersecting points such as pipe transitions, branches and tees.
- Look for cloth visible through the mastic or finish if the pipe is protected by a reinforced mastic weather barrier.

During an extensive inspection of an insulation system:

- Use thermographic equipment to isolate areas of concern.
- Design a method to repair, close and seal any cut into the insulation or vapor retarder so that a positive seal may be maintained throughout the entire system.
- Examine the pipe surface for corrosion if the insulation is physically wet.

The extent of moisture present in the insulation system and (or) the corrosion of the pipe will determine the need to replace the insulation. All wet parts of the insulation must be replaced.

REFERENCES

ASME. 1992. Power Piping. *Standard* B31.1-92.

ASTM. 1993. Test method for steady-state heat flux measurements and thermal transmission properties by means of the guarded hot-plate apparatus. *Standard* C 177. American Society for Testing and Materials, West Conshohocken, PA.

ASTM. 1994. Practice for prefabrication and field fabrication of thermal insulating fitting covers for NPS piping, vessel lagging, and dished head segments. *Standard* C 450.

ASTM. 1990. Test method for steady-state heat flux measurements and thermal transmission properties by means of the heat flow meter apparatus. *Standard* C 518.

ASTM. 1997. Specification for preformed flexible elastomeric cellular thermal insulation in sheet and tubular form. *Standard* C 534.

ASTM. 1991. Specification for cellular glass thermal insulation. *Standard* C 552.

ASTM. 1995. Specification for rigid, cellular polystyrene thermal insulation. *Standard* C 578.

ASTM. 1990. Practice for inner and outer diameter of rigid thermal insulation for nominal sizes of pipe and tubing (NPS System). *Standard* C 585.

ASTM. 1994. Specification for unfaced preformed rigid cellular polyisocyanurate thermal insulation. *Standard* C 591.

ASTM. 1996. Specification for faced and unfaced rigid cellular phenolic thermal insulation. *Standard* C 1126.

ASTM. 1995. Test method for surface burning characteristics of building materials. *Standard* E 84.

ASTM. 1995. Test methods for water vapor transmission of materials. *Standard* E 96.

Hedlin, c.p. 1977. Moisture gains by foam plastic roof insulations under controlled temperature gradients. *Journal of Cellular Plastics* Sept./Oct., 313-326.

Lenox, R.S. and P.A. Hough. 1995. Minimizing corrosion of copper tubing used in refrigeration systems. *ASHRAE Journal* 37:11.

Kumaran, M.K. 1989. Vapor transport characteristics of mineral fiber insulation from heat flow meter measurements. ASTM STP 1039, p. 19-27. American Society for Testing and Materials, West Conshohocken, PA.

Kumaran, M.K., M. Bomberg, N.V. Schwartz. 1989. Water vapor transmission and moisture accumulation in polyurethane and polyisocyanurate foams. ASTM STD 1039, p. 63-72. American Society for Testing and Materials, West Conshohocken, PA.

Malloy, J.F. 1969. *Thermal insulation.* Van Nostrand Reinhold Co., New York.

MIL-P-24441. General specification for paint, epoxy-polyamide. Naval Publications and Forms Center, Philadelphia, PA.

MICA. 1993. National commercial and industrial insulation standards. Midwest Insulation Contractors Association, Omaha, NE.

MSS. 1991. Pipe hangers and supports—selection and application. *Standard* SP-69-91.

National Association of Corrosion Engineers. Near white metal blast cleaning. Pub. No. 2/SSPC-SP10. NACE International, Houston, TX.

National Association of Corrosion Engineers. 1997. *Corrosion under insulation.* NACE International, Houston, TX.

SofTech[2]. 1996. NAIMA 3E Plus. Grand Junction, CO.

CHAPTER 33

ICE MANUFACTURE

Ice Makers	33.1
Thermal Storage	33.4
Ice Storage	33.4
Delivery Systems	33.5
Commercial Ice	33.7
Ice-Source Heat Pumps	33.7

MOST commercial ice production is done with ice makers that produce three basic types of fragmentary ice which vary according to the type and size required for a particular application. The basic types of fragmentary ice are flake, tubular, and plate. Chapter 54 of the 1978 *ASHRAE Handbook and Product Directory—Applications* includes information on block ice manufacturing. Among the many applications for manufactured ice are

- Processing: Fish, meat, poultry, dairy, bakery products, and hydrocooling
- Storage and transportation: Fish, meat, poultry, and dairy products
- Manufacturing: Chemicals and pharmaceuticals
- Others: Retail consumer ice; concrete mixing and curing; and off-peak thermal storage

ICE MAKERS

Flake Ice

Flake ice is produced by applying water to the inside or outside of a refrigerated drum or to the outside of a refrigerated disk. The drum is either vertical or horizontal and may be either stationary or fixed. The disk is vertical and rotates about a horizontal axis.

Ice removal devices fracture the thin layer of ice produced on the freezing surface of the ice maker, breaking it free from the freezing surface and allowing it to fall into an ice bin, which is generally located below the ice maker.

The thickness of the ice produced by flake ice machines can be varied by adjusting the speed of the rotating part of the machine, varying evaporator temperature, or regulating the water flow on the freezing surface. Flake ice is produced continuously, unlike tubular and plate ice, which are produced in an intermittent cycle or harvest operation. The resulting thickness ranges from 1 to 4.5 mm. A continuous operation (without a harvest cycle) requires less refrigeration capacity to produce a kilogram of ice than any other type of ice manufacture with similar makeup water and evaporating temperatures. The exact amount of refrigeration required varies by the type and design of the flake ice machine. Typical flake ice machines are shown in Figures 1 and 2.

All water used by flake ice machines is converted into ice; therefore, there is no waste or spillage. Flake ice makers are usually operated at a lower evaporating temperature than tube or plate ice makers, and the ice is colder when it is removed from the icemaking surface. The surface of flake ice is not wetted by thawing during removal from the freezing surface, as is common with other types of ice. Since it is produced at a colder temperature, flake ice is most adaptable to automated storage, particularly when low-temperature ice is desired.

The rapid freezing of water on the freezing surface entrains air in the flake ice, giving it an opaque appearance. For this reason, flake ice is not commonly used for applications where clear ice is important. Where rapid cooling is important, such as in chemical processing and concrete cooling, flake ice is ideal because the flakes present the maximum amount of cooling surface for a given amount of ice.

When used as ingredient ice in sausage making or other food grinding and mixing, flake ice provides rapid cooling while minimizing mechanical damage to other ingredients and wear on mixing/cutting blades.

Some flake ice machines can produce salty ice from seawater. These are particularly useful in shipboard applications. Other flake ice machines require adding trace amounts of salt to the makeup water to enhance the release of ice from the refrigerated surface. In rare cases, the presence of salt in the finished product may be objectionable.

Tubular Ice

Tubular ice is produced by freezing a falling film of water either on the outside of a tube with evaporating refrigerant on the inside, or on the inside of tubes surrounded by evaporating refrigerant on the outside.

Outside Tube. When ice is produced on the outside of a tube, the freezing cycle is normally from 8 to 15 min, with the final ice thickness from 5 to over 13 mm following the curvature of the tube. The refrigerant temperature inside the tube continually drops from an initial suction temperature of about −4°C to the terminal suction temperature in the range of −12 to −26°C. At the end of the freezing cycle, the circulating water is shut off, and hot discharge gas is introduced to harvest the ice. To maintain proper harvest temperatures, typical discharge gas pressure is 1.1 MPa. This drives the liquid refrigerant in the tube up into an accumulator and melts the inside of the tube of ice, which slides down through a sizer and mechanical breaker, and finally down into storage. The defrost cycle is normally about 30 s. The unit returns to the freezing cycle by returning the liquid refrigerant to the tube from the accumulator.

This type of ice maker operates with refrigerants R-717, R-404A, R-507, and R-22. R-12 may be found in some older units. Higher capacity units of 9 Mg per 24 h and larger usually use R-717. The capacity of the unit increases as the terminal suction pressure decreases. A typical unit with 21°C makeup water and R-717 as the refrigerant produces 17.5 Mg of ice per 24 h with a terminal suction pressure of 265 kPa and requires 126 kW of refrigeration. This equates to 7.2 kW of refrigeration per megagram of ice. The same unit produces 37.7 Mg of ice per 24 h with a terminal suction pressure of 145 kPa and requires 280 kW of refrigeration. This equates to 7.5 kW of refrigeration per megagram of ice. Figure 3 shows the physical arrangement for an ice maker that makes ice on the outside of the tubes.

Inside Tube. When ice is produced inside a tube, it can be harvested as a cylinder or as crushed ice. The freezing cycle ranges from 13 to 26 min. The tube is usually 20 to 50 mm in diameter, producing a cylinder that can be cut to desired lengths. The refrigerant temperature outside the tube is continually dropping, with an initial

---

The preparation of this chapter is assigned to TC 10.2, Automatic Icemaking Plants and Skating Rinks.

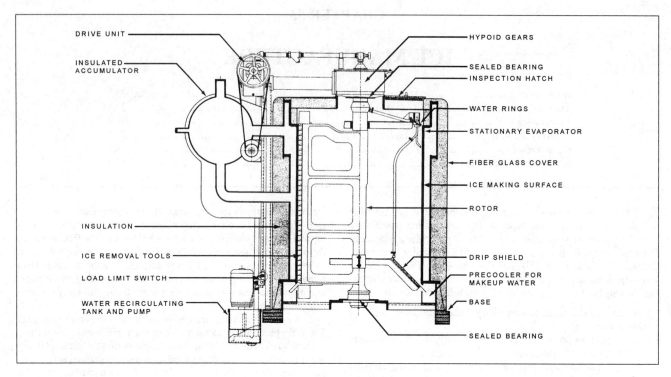

Fig. 1 Flake Ice Maker

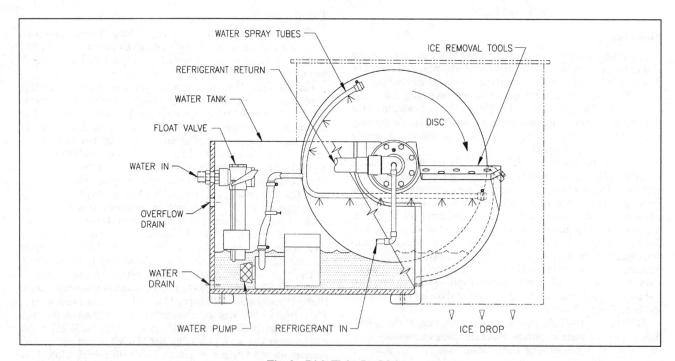

Fig. 2 Disk Flake Ice Maker

temperature of −4°C and a terminal suction temperature ranging from −7 to −20°C. At the end of the freezing cycle, the circulating water is shut off and the ice is harvested by introducing hot discharge gas into the refrigerant in the freezing section. To maintain gas temperature, typical discharge gas pressure is 1.2 MPa. This releases the ice from the tube; the ice descends to a motor-driven cutter plate that can be adjusted to cut the ice cylinders to the length desired (up to 40 mm). At the end of the defrost cycle, the discharge gas valve is closed and water circulation resumes.

These units can use refrigerants R-717 and R-22; R-12 may be found in older units. Again, the capacity increases as the terminal suction pressure decreases. A typical unit with 21°C makeup water and R-717 as the refrigerant produces 39 Mg of ice per 24 h with a terminal suction pressure of 275 kPa and requires 262 kW of refrigeration. This equates to 6.7 kW of refrigeration per megagram of ice. The same unit produces 60 Mg of ice per 24 h with a terminal suction pressure of 210 kPa and requires 475 kW of refrigeration. This equates to 7.9 kW of refrigeration per megagram of ice.

Ice Manufacture

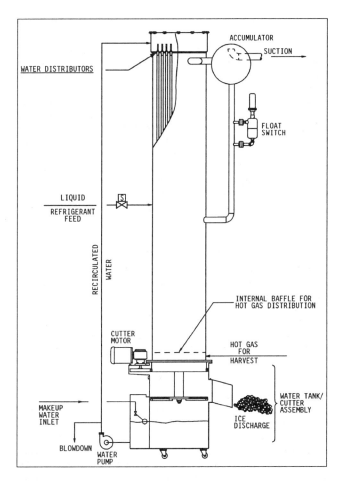

Fig. 3 Tubular Ice Maker

Tubular ice makers are advantageous because they produce ice at higher suction pressures than other types of ice makers. They can make relatively thick and clear ice, with curvatures that help prevent bridging in storage. Tubular ice makers have a greater height requirement for installation than do plate or flake ice makers, but a smaller footprint. Provision must be made in the refrigeration system high side to accommodate the volume of refrigerant required for the proper amount of harvest discharge gas. Ice temperatures are generally higher than the temperature of flake ice makers.

Supply Water. Supply water temperature has a great effect on the capacity of either type of tubular ice maker. If the supply water temperature is reduced from 21 to 4°C, the ice production of the unit increases approximately 18%. In larger systems, the economics of precooling the water in a separate water cooling system with higher suction pressures should be considered.

Plate Ice

Plate ice makers are commonly defined as those that build ice on a flat vertical surface. Water is applied above freezing plates and flows by gravity over the freezing plates during the freeze cycle. Liquid refrigerant at a temperature between −21 and −7°C is contained in circuiting inside the plate. The length of the freezing cycle governs the thickness of ice produced. Ice thicknesses in the range of 6 to 20 mm are quite common, with freeze cycles varying from 12 to 45 min. Figure 4 shows a flow diagram of a plate ice maker using water for harvest. All plate ice makers use a sump and recirculating pump concept, whereby an excess of water is applied to the freezing surface. Water not converted to ice on the plate is collected in the sump and recirculated as precooled water for ice making.

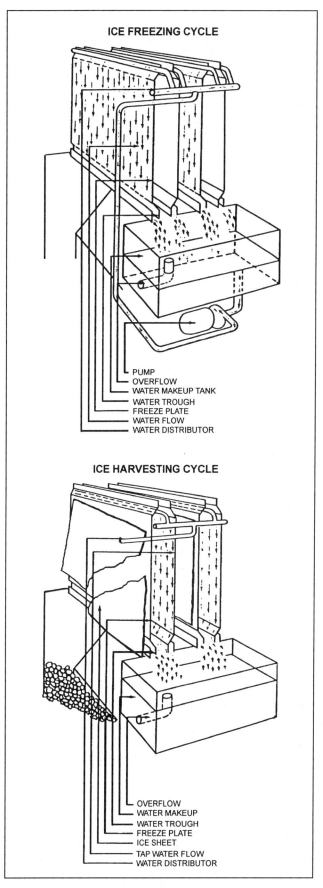

Fig. 4 Plate Ice Maker

Ice is harvested from plate ice makers by one of two methods. One method involves the application of hot gas to the refrigerant circuit to warm the plates to 4 to 10°C, causing the ice surface touching the plate to reach its melting point and thereby release the ice from the plate. The ice falls by gravity to the storage bin below or to a cutter bar or crusher that further reduces the ice to a more uniform size. Plate ice makers using the hot gas method of harvesting are capable of producing ice on one or two sides of the plate, depending on the design.

In the second method of harvesting ice, warm water flows on the back side of the plate. This heats the refrigerant inside the plate above the ice melting point, and the ice is released. Ice makers using the water warming harvest principle manufacture ice on one side of a plate only. Harvest water is chilled by passing over the plates. It is then collected in the sump and recirculated to become precooled water for the next batch of ice.

For plate ice makers, the freezing time, harvest time, water, pump, and refrigeration are controlled by adjustable electromechanical or electronic devices. Using the wide variety of thicknesses and freezing times available, plate ice makers can produce clear ice. Thus, the plate ice maker is commonly used in applications requiring clear ice.

Because of the harvest cycle involved, plate ice makers require more refrigeration per unit mass of ice produced than flake ice makers. This disadvantage is offset by the capability of plate ice makers to operate at higher evaporating temperatures; thus, connected motor power per kilowatt refrigeration is usually less than that of flake ice makers. During the harvest cycle, the suction pressure rises considerably depending on the design of the ice maker. When a common refrigeration system is used for multiple refrigerated requirements, a stable suction pressure can be maintained for all the refrigeration loads by using a dedicated compressor for the ice machine, or by using a dual pressure suction regulator at each ice machine to minimize the load placed on the suction main during harvest. This may occur in large processing plants, refrigerated warehouses, and so forth. Large plate ice makers can be arranged such that only sections or groups of plates are harvested at one time. Properly adjusting the timing of harvesting each section can reduce the fluctuation in suction pressure.

Plate ice makers using the water harvest principle rely on the temperature of the water for harvesting. A minimum of 18°C is usually recommended to minimize both harvest cycle time and harvest water consumption. For installations in cold water areas, or where wintertime inlet water temperatures are low, it is advisable to provide auxiliary means of warming the inlet water to 18°C.

Ice Builders

Ice builders comprise various types of apparatus that produce ice on the refrigerated surfaces of coils or plates submerged in insulated tanks of water. This equipment is commonly known as an ice bank water chiller. The ice built on the freezing coils is not used as a manufactured ice product but rather as a means of cooling water circulating through the tank as the ice melts from the coils. The ice builder is most often used for thermal storage applications with high peak and intermittent cooling loads that require chilled water. See Chapter 40 of the 1995 *ASHRAE Handbook—Applications* for more information.

Scale Formation

The performance of all ice makers is affected by the characteristics of the inlet water used. Impurities and excessive hardness can cause a scale to be deposited on the freezing surface of the ice maker. The deposit reduces the heat transfer capability of the freezing surface, thereby reducing ice-making capacity. Deposited scale may further reduce ice-making capacity by causing the ice to stick on the freezing surface during the harvest process. The rated capacity of all ice makers is based on the substantial release of all the ice from the freezing surface during the removal period. Because the process of freezing water into ice tends to freeze a greater proportion of pure water on the ice maker freezing surface, impurities tend to remain in the excess or recirculated water. A blowdown, or bleedoff, whereby a portion of the recirculated water is bled off and discharged, can be installed. The bleedoff system can control the concentration of chemicals and impurities in the recirculated water. The necessity of a bleedoff system and the effectiveness of this concept for controlling scale deposits depend on local water conditions. Some refrigeration system loss is experienced because the recirculated water that is bled off to drain is precooled. Water that is bled off may be passed through a heat exchanger to precool incoming makeup water. Water conditions, water treatment, and related water problems in ice making are covered in Chapter 44 of the 1995 *ASHRAE Handbook—Applications*.

THERMAL STORAGE

Interest in energy conservation has renewed interest in the ice storage concept of providing thermal storage of cooling capacity for air-conditioning or process applications. The ice is produced and stored using lower off-peak and weekend power rates. During the day, stored ice provides refrigeration for the chilled water system. The design and features of thermal storage equipment are covered in Chapter 40 of the 1995 *ASHRAE Handbook—Applications*.

ICE STORAGE

Fragmentary ice makers can produce ice either on a continuous basis or in a constant number of harvest cycles per hour. The use of the ice is generally not at a constant rate but on a batch basis. Batches vary greatly based on user requirements. The ice must be stored and recovered from storage on demand. Labor savings, economics, quantity of ice to be stored, amount of automation desired, and user delivery requirements must all be considered in ice storage and storage bin design.

Ice makers can produce ice 24 h a day. By making ice during off-shifts and weekends, as well as during work shifts, considerable savings in total ice-making and refrigeration system requirements can be achieved. In addition, by using electrical power during off-peak hours, peak loads on the power system are reduced during the day. Many power companies offer reduced rates during off-peak hours.

Ice storages vary in type from short- to prolonged-term. Degree of automation for filling and discharge ranges from manual shoveling to a completely automatic rake system.

Short-term storage generally requires provision for one day's ice production. The ice maker is mounted over a bin, and ice falls by gravity into the bin. The bin is an insulated, airtight enclosure with one or more insulated doors for access. Ice is removed from the bin by shoveling or scooping. In such storage, the subcooling effect of the ice generally offsets the heat loss through the insulated bin walls without excessive melting. In most situations, it is not necessary to provide refrigeration units in an ice storage bin where ice production is being used on a daily basis and ambient temperatures are reasonable.

Prolonged ice storage requires a refrigerated, airtight, insulated storage bin. Some designs provide for false walls and floor, which produce an envelope effect that allows cold air to circulate completely around the mass of ice in storage. If wet ice is placed in a bin refrigerated to a temperature below 0°C, it will freeze together and may be difficult to remove.

Time and pressure affect the storage quality of fragmentary ice. Even though a bin is refrigerated to a temperature well below 0°C, pressure can cause local melting near the bottom. Thus, there are limits to the size and configuration of a gravity-filled storage bin. The ice falling from an ice maker forms a cone directly underneath the drop

Ice Manufacture

in the bin. With slight variation because of the type of ice, the angle of repose is approximately 30°. Fusion of ice under pressure limits the practical ice storage depth to 3 to 3.5 m. To use the volume of the bin more efficiently, a leveling screw mounted in the overhead can be used to carry the ice away from the top of the ice cone.

There is also a practical limit to the size of a storage bin from which ice can be manually removed through refrigerator doors. The simplest device used to remove ice from a bin is a screw conveyor with a trough at floor level, which is equipped with gratings and removable sectional covers. The removable covers protect the screw from ice blockage when the conveyor is not running. The gratings are for the protection of personnel.

Ice Rake and Live Bottom Bins

The ice rake system is used for larger and fully automated storages. These storages generally have a 10 to 300 Mg capacity for a single rake system. Depending on plant demands, combinations of rake systems can be developed into an integrated production, storage, and delivery system. Such a mechanism could have capabilities of up to 1000 Mg of storage with multiple screw or pneumatic conveying delivery systems. A range of delivery rates, up to 60 Mg of ice per hour, can be achieved. The advantages of such rake systems are the elimination of labor for storing and transporting ice, longer and more effective distribution systems, faster delivery and termination of ice flow, less waste of ice, and the elimination of physical contamination.

Storages that incorporate rake systems are of two basic types. One type encloses the ice storage and rake system in an arrangement of steel framework and panels with the complete unit installed in a refrigerated room. The second type involves the construction of an insulated enclosure around the ice bin and rake system. This type can be installed inside a building or outside, depending on the weather protection provided. For either type, the ice makers are mounted outside of the refrigerated space. Figure 5 shows the arrangement of components for a typical rake system.

Once fragmentary ice comes to rest in the storage bin, it will not flow freely, and a mechanical force is necessary to start the ice moving. The deeper the ice is stored, the greater the pressure on the ice near the bottom. The ice near the bottom tends to fuse together faster. Rake systems work from the top of the ice; they continuously level and fill the storage bin, as well as automatically remove the ice on demand. The systems operate in nonrefrigerated or refrigerated bins; since most users of large storages also want ice to be dry for ease of handling, large automated storages are usually refrigerated.

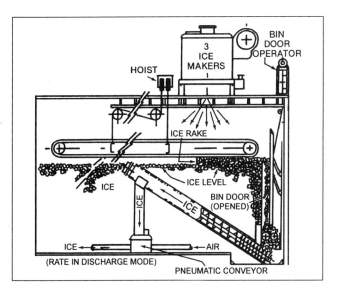

Fig. 5 Ice Rake System

The ice rake itself consists of a structural steel mechanism with drive. It operates similar to the tracks of a crawler tractor. By means of a hoist and timer, the rake is raised or lowered to automatically maintain its position suspended over, and in close contact with, the ice level. Wide scraper-type conveyors, mounted across the tracks along the full width of the bin, spread the ice out and drag it toward the back of the bin during the filling mode. To dispense the ice at delivery, the scraper conveyors reverse direction and drag the ice to the opposite end of the bin, dropping it into a screw conveyor mechanism. From this point, the ice is transferred to the external delivery system of screw conveyors.

Some rake systems have features that allow ice deliveries from the bin to be remotely controlled and volumetrically metered. Ice deliveries can be recorded on digital counters at the storage bin, remote stations, and control centers. Accuracy is in the range of ±2%. Another method of metering ice from a rake system storage involves the screw conveyor delivering the ice to a weigh belt. As the ice passes along the moving belt, it is electronically weighed and the mass is recorded. Selection of the belt material carrying the ice is critical in preventing ice from sticking to the belt. The weigh belt is often installed in a refrigerated area adjacent to the ice storage.

Another type of ice storage with delivery system capabilities is the **live bottom** type with a multiplicity of screws arranged in various configurations on the bottom of the storage bin. Because of ice fusion, these bins are limited to short-term storage. The success of this type of bin depends on the type and quality of fragmentary ice being stored and the ability of the design to overcome particle fusion. Particle fusion may result in ice bridges forming over the top of the screws; then the screws will bore holes in the ice rather than empty the bin.

Primarily for the consumer bagged-ice industry, a bin and automatic storage system is used in which the entire floor moves, carrying the ice load into slowly rotating beaters. As the ice breaks loose, it drops to a screw conveyor, which feeds an ice bagger. This type of bin is located in a refrigerated room, and the ice makers are located away from the bin. The ice makers must be shut off so that no ice can flow into the storage during the discharge and bagging process.

The **ice silo** is used for long- or short-term storage with capacities in the range of 20 to 100 Mg. The silo tank comprises a cylindrical part and a tapered, conical part leading the ice to the outlet at the bottom of the tank. From this point, the ice is transported by a screw delivery system. A rotating flexible chain arrangement is provided in the silo to assist in ice removal and to partially overcome the fusion problem. The ice maker is mounted over the top of the silo, and the ice falls into the storage. No leveling of ice is required in the bin because the diameter of the silo is sized to be compatible with the ice maker. The larger the ice storage, the higher the silo. As a result, proper ice discharge becomes more critical in the design when considering the fusion of ice and the fact that the ice must finally pass through a relatively small opening at the bottom of a tapered zone.

DELIVERY SYSTEMS

The location of ice manufacture is rarely the location of ice usage. Usually it is necessary to move ice from the ice machine or storage bin to some other area where it will be used; thus, a conveying system is required. Most conveyor applications use screws, belts, or pneumatic systems. Great care must be taken in selecting the size and type of conveyor to be used because no matter what type of fragmentary ice is being handled, problems such as fines, freeze-up, and ice jams can be encountered with an improperly designed system. Fines, or snow, are small particles of ice that chip off the larger pieces during harvesting, crushing, or conveying operations.

Screw and Belt Conveyors

Screw conveyors are the most popular of all the conveyances used for transporting ice. Screw conveyors are manufactured in sizes of 100 mm diameter and up, as well as in various screw pitches. Most ice-conveying operations use 150, 230, and 300 mm diameter screws.

The sizing and drive power requirements of screw conveyors are determined by the ice delivery rate, the inclination of the conveyor, and the conveyor screw pitch. With fragmentary ice, the selection of an undersized conveyor will result in excessive conveyor speed or require that the conveyor run too full of ice. These conditions can produce excessive fines.

When screw conveyors transport ice through high ambient inside areas or outside in the weather (e.g., in icing fishing vessels), it is advisable to insulate the screw conveyor trough and provide the conveyor with insulated covers. Rain is as problematic as sunshine for contributing to ice meltage and delivery difficulties. For this reason, most screw conveyors operating in the weather are provided with sectional and removable covers.

Belt conveyors are often used when excess moisture has to be removed from the ice or to minimize the fines. The mesh-type belts allow snow and excess water to fall through. Stainless steel, galvanized steel, or high-density polyethylene are commonly used for belting.

Pneumatic Ice Conveying

Pneumatic ice conveying systems have proved desirable, economical, and practical for transporting fragmentary ice distances of 30 m or more and when multiple delivery stations must be served. A pneumatic system is advantageous when delivery stations are in different directions or at different elevations, when delivery through a pressure hose is needed, or when flexibility is required for future changes or addition of delivery stations.

The basic principle of conveying ice by a pneumatic system involves a rotary blower, which delivers air to a rotary air-lock valve or conveying valve. Ice is fed into the conveying valve, and compressed air conveys the mixture of air and ice at high velocity through thin-walled tubing (aluminum, stainless steel, or plastic). Figure 6 shows the diagrammatic arrangement of pneumatic system components.

Delivery rates between 2.5 and 10 kg/s and conveying distances up to 180 m are common. Delivery distances exceeding 180 m can be achieved at reduced delivery rates, with the maximum practical distance being approximately 300 m. Conveying pressures range from 30 to 70 kPa (gage), depending on the delivery rate and the maximum distance the ice is to be conveyed. The air velocity required to keep the ice in suspension in the conveying line will vary among the different types of fragmentary ice. A pneumatic system cannot satisfactorily convey all sizes of fragmentary ice. The manufacturer of the ice-making equipment should be consulted for recommended line velocities. Sometimes, storage bins for ice plants using a pneumatic delivery system are refrigerated to ensure cold, free-flowing ice with minimum moisture. Because the ice remains in the tubing a very short time, tubing insulation is seldom needed or used. However, in warm climates, shading the conveying line reduces the solar load. The tubing typically used has a diameter of 100 to 200 mm and requires minimum support, making installation easy and economical.

Multiple delivery points are served by automatic Y-type diverter valves or multiple-way slide valves, either air or electrically operated. Pneumatically blown ice can be delivered under pressure out of the end of a hose. An alternate method of delivery is a cyclone receiver, which takes the ice at high velocity, dissipates the air, and drops the ice by gravity. Combinations of hose stations and cyclone delivery stations in the same system are common.

When a pneumatic conveying system is used in areas of high ambient and wet-bulb temperatures on an application requiring higher conveying pressures, a heat exchanger is often used to cool and dehumidify the pneumatic air before it enters the conveying valve. The heat exchanger is provided with a cooling coil, either refrigerant or chilled-water cooled, a demister or other means of separating moisture from the air, and a condensate trap to expel the

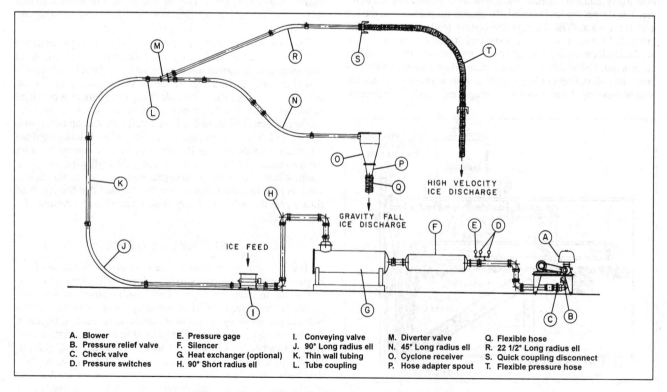

A. Blower	E. Pressure gage	I. Conveying valve	M. Diverter valve	Q. Flexible hose
B. Pressure relief valve	F. Silencer	J. 90° Long radius ell	N. 45° Long radius ell	R. 22 1/2° Long radius ell
C. Check valve	G. Heat exchanger (optional)	K. Thin wall tubing	O. Cyclone receiver	S. Quick coupling disconnect
D. Pressure switches	H. 90° Short radius ell	L. Tube coupling	P. Hose adapter spout	T. Flexible pressure hose

Fig. 6 Typical Flake Ice Pneumatic Conveying System

entrapped moisture. Geographical location, system pressure, and the quality and use of the ice at the delivery point must be considered when determining whether to use a heat exchanger.

Slurry Pumping

A mixture of particle ice and water can be pumped as a slurry. This method has some advantages for transporting ice. Generally, the slurry mix is approximately 50% water and 50% ice. For specialized application, mixtures of up to 80% ice and 20% water can be successfully pumped. Delivery distances of 250 m have been achieved with delivery rates of 15 kg/s of slurry mix. This practice has been extensively used in the produce industry and has potential in concrete cooling, chemical processing, and other ice or chilled-water related applications. Ice slurry mixes are of particular interest where there is a need for low-temperature chilled water at or near 0°C. In converting ice to water, the absorption of latent heat at the usage point enables more cooling to be done with a slurry mix than with straight chilled-water cooling. Pumping volumes and line sizes are minimized, and ice meltage during mixing and pumping does not normally exceed 1 to 3%. The system is capable of automation for continuous operation.

The basic system for slurry pumping includes a mixing tank, in which the ice and water are mixed. The ice is carried by any of the conventional conveying methods from the ice storage bin to the slurry mix tank. Agitators in the tank operate continuously to maintain a mixture with uniform consistency. Pumps discharge the slurry mix through pipelines to the point of use. The pumps are of the centrifugal type, modified for pumping slurry. When the icing cycle at the usage points is intermittent, a recirculating system returns unused slurry to the mixing tank. Thus the slurry is kept moving at all times, and the possibility of ice blockage in the lines is minimized. The temperature of the slurry solution in the tank is maintained at 0°C.

The fresh produce industry offers a unique application for slurry mixes. Body icing of the fresh produce, which is generally of nonuniform size and configuration, is achieved by applying the slurry mix to the dry packed product. Drain holes are provided in the shipping container to remove water. Because of its suspension in water, ice is carried to all parts of the container. The ice solidifies as the water drains from the container. The product is then completely surrounded with ice, the voids are filled, and potential hot spots are eliminated. The drained water can be collected and returned to the mixing tank.

COMMERCIAL ICE

Commercial ice is primarily used for human consumption. It is also called **packaged** or **consumer** ice and is used for cooling beverages and for other applications in restaurants, hotels, and similar institutions. This use requires packaging at the ice plant for storage and eventual distribution. In bagged form, commercial ice is also available for sale in grocery stores and in automatic, coin-operated vending machines. When the ice is to be used in beverages, ice produced by plate or tube ice makers is preferred because of the clear appearance and the fact that it can be made in greater thicknesses. Rake systems are often used to store packaged ice and convey it into the packaging system.

Packaging. A packaging system normally comprises an ice bagger and a bag closer. These components are available from ice packaging equipment manufacturers in various types and sizes. In the bagging process, the ice is fed from the ice storage bin into the bagging machine by a screw or belt conveyor. The bagging machine meters ice into a bag placed below its discharge chute. The amount of ice measured into the bag can be determined by mass, volume, or sight approximation, depending on the equipment used.

When packaging by mass, the ice bag is placed on a weighing table on the bagging machine. Ice is then dispensed into the ice bag until the desired mass is in the bag. At this point, a switch mounted on the scale stops the flow of ice.

The volumetric bagging machine deposits the ice in a rotating chamber, which is adjustable in volume. After a predetermined volume of ice enters the chamber, the ice is discharged into the bag below. Because the shape and size of the ice is not constant, the volumetric chamber is usually set to produce a 3 to 5% overage by volume. Therefore, the proper minimum mass of ice in the bag is assured.

The bag closer consists of a mechanical unit that ties and seals the top of the bag with a wire ring, wire twist tie, or plastic clip. Smaller bagging operations do not use a bag closer, and the bags are manually closed with plastic ties, wire rings, staples, and so forth. In more elaborate systems, the bag is formed from roll stock, filled with ice, automatically removed from the bagging machine, and automatically closed before it is dropped onto a conveyor, which carries the bagged ice to the refrigerated storage room. The degree of automation for the bagging and closing operation is determined by the number of bags of ice to be produced per day, the size of the bags, and the cost-benefit relationship between automated equipment and reduced labor costs.

Ice bags are made of plastic, most often polyethylene, or, very rarely, heavy moisture-resistant paper. Plastic bags are used in most modern plants.

Storage. Packaged ice must be stored in a refrigerated warehouse or room prior to distribution. The ice storage area is sized to meet the daily production of the plant and the distribution requirements. Generally, a bag ice storage facility can store 3 to 7 days' production. Although the ice will not melt at a storage temperature below 0°C, it is important that the storage be maintained at a temperature between -12 and -4°C. The lower temperature subcools the ice and avoids meltage during distribution. Depending on the type and quality of the ice, the bagged ice can contain some water. The percentage of water can range from 0 to 5%. For this reason, provision is made in the storage room refrigeration system for the product load of refreezing the water.

ICE-SOURCE HEAT PUMPS

Ice-making systems can be configured to provide heating alone, or heating and cooling, for a building or process. The conversion of water to ice occurs at a relatively high evaporator temperature and coefficient of performance compared to air-source heat pumps operating at low ambient temperatures. Systems can provide necessary heating, with the resulting ice disposed of by melting with low-grade heat, such as solar. In addition, ice can be used for useful cooling through daily, weekly, or seasonal storage.

The concept was originally considered mainly for residential heating and cooling, but installations are proving feasible for larger structures, such as office buildings. Energy consumption savings resulting from the coefficient of performance of a conventional heat pump system are achieved. Using off-peak night and weekend rates can reduce power costs, and the ice produced is used for building cooling requirements. As a result, a system can be developed that consumes less energy at a lower utility rate.

Ice-source heat pumps follow two basic approaches. The first involves using the ice builder principle, with coils in a large tank, as the evaporator component of the heat pump. The second approach uses a fragmentary type of ice maker as the evaporator, with ice being stored in a tank as a mixture of ice and water. Many variations, combinations, and adaptations may be developed from these basic systems. The requirement for thermal storage of a large quantity of ice dictates new planning in architectural building design. Heating and cooling system designs for the building are also influenced.

BIBLIOGRAPHY

Dorgan, C.E. 1985. Ice-maker heat pumps operation and design. *ASHRAE Transactions* 91(1B): 856-62.

Dorgan, C.E., G.C. Nelson, and W.F. Sharp. 1982. Ice-maker heat pump performance—Reedsburg Center. *ASHRAE Transactions* 88(1): 1271-78.

CHAPTER 34

ICE RINKS

Applications 34.1	General Rink Floor Design 34.6
Refrigeration Requirements 34.1	Building, Maintaining, and Planing Ice Surfaces 34.8
Ice Rink Conditions 34.4	Rink Fog and Ceiling Dripping 34.8
Equipment Selection 34.4	Imitation Ice-Skating Surfaces 34.9

ANY level sheet of ice made by refrigeration (the term **artificial ice** is sometimes used) is referred to in this chapter as an ice rink regardless of use and whether it is located indoors or outdoors.

The freezing of an ice sheet is usually accomplished by the circulation of a heat transfer fluid through a network of pipes or tubes located below the surface of the ice. The heat transfer fluid is predominantly a secondary coolant such as glycol, methanol, or calcium chloride (see Chapter 20 of the 1997 ASHRAE Handbook—Fundamentals).

R-22 and R-717 are most frequently used for chilling secondary coolants for ice rinks. R-12 and R-502 have also been used; however, due to the phaseout of the CFC refrigerants, they should no longer be considered for use. R-22 will also be phased out in the future, so for new rink equipment selection, R-22 and CFC replacements should be evaluated according to status and availability.

In some rinks, R-22, and R-717 to a lesser degree, have been applied as a direct coolant for freezing. The direct refrigerant rinks operate at higher compressor suction pressures and temperatures, thus achieving an increased COP, compared to secondary coolants. However, due to emissions regulations, the projected R-22 phaseout, building codes, and fire regulations, R-22 and R-717 should not be used to freeze ice directly in rinks.

APPLICATIONS

Most ice surfaces are used for a variety of sports, although some are constructed for specific purposes and are of specific dimensions. Usual rink sizes include:

Hockey. The accepted North American hockey rink size is 26 m by 61 m. Radius corners of 8.5 m are recommended by professional and amateur rules. The Olympic and international hockey rink size is 30 m by 60 m, with 6 m radius corners. Many rinks are considered adequate with dimensions of 26 m by 56.4 m, 24.4 m by 54.9 m, and 21.3 m by 51.8 m. In substandard size rinks, a corner radius of not less than 6 m should be provided to permit the use of mechanical resurfacing equipment.

Curling. Regulation surface for this sport is 4.3 m by 45 m; however, the width of the ice sheet is often increased to allow space for installation and dividers between the sheets, particularly at the circles. Most are laid out on ice sheets measuring 4.5 m by 46 m.

Figure Skating. School or compulsory figures are generally done on a patch approximately 5 m by 12 m. Freestyle and dance routines generally require an area of 18 m by 36 m or more.

Speed Skating. Indoor speed skating has traditionally been on hockey-size rinks. The Olympic-size outdoor speed skating track is a 400 m oval, 10 m wide with 112 m straightaways and curves with an inner radius of 25 m. Most speed skating ovals are outdoors; however, some recently constructed speed skating rinks are full size and indoors.

Recreational Skating. Recreational skating can be done on any size or shape rink, as long as it can be efficiently resurfaced.

The preparation of this chapter is assigned to TC 10.2, Automatic Icemaking Plants and Skating Rinks.

Generally, 2.8 m^2 is allowed for each person actually skating; 2.3 m^2 per skater is acceptable, except where a large number of preteens are skating. A 26 m by 61 m hockey rink with 8.5 m radius corners has an area of 1517 m^2 and will accommodate a mixed group of about 650 skaters.

Public Arenas, Auditoriums, and Coliseums

Public arenas, auditoriums, field houses, etc., are designed primarily for spectator events. They are used for ice sports, ice shows, and recreational skating, as well as for non-ice events, such as basketball, boxing, tennis, conventions, exhibits, circuses, rodeos, and stock shows. The refrigeration system can be designed so that, with adequate personnel, the ice surface can be produced within 12 to 16 h. However, general practice is to leave the ice sheet in place and to hold other events on an insulated floor placed on the ice. This approach saves significant time, labor, and energy.

REFRIGERATION REQUIREMENTS

The heat load factors considered in the following section include type of service, length of season, usage, type of enclosure, radiant load from roof and lights, and geographic location of the rink with associated wet- and dry-bulb temperatures. In the case of outdoor rinks, the sun effect and weather conditions must also be considered.

A fairly accurate estimate of refrigeration requirements can be made based on data from a number of rink installations with the pipes covered by not more than 25 mm of sand or concrete and not more than 40 mm of ice—a total of 65 mm sand or concrete and ice.

The refrigeration load may be estimated either by: (1) calculating the refrigeration necessary to freeze the ice to required conditions in a specified time, or (2) calculating the refrigeration necessary to maintain the ice surface and temperature during the most severe usage and operating conditions that coincide with the maximum ambient environmental conditions.

In the time-to-freeze method, the quantity of ice required (rink surface area multiplied by thickness) is calculated first. Then the refrigeration is determined to: (1) reduce the water from application temperature to 0°C, (2) freeze the water to ice, (3) reduce the ice to the required temperature, and (4) handle the heat loads and system losses during the freezing period. The total requirement is divided by system efficiency and freezing period to determine the required refrigeration.

Example 1. Calculate the refrigeration required to build 25 mm thick ice on a 1500 m^2 rink in 24 hours.

Assume the following material properties and conditions:

Material	Specific Heat, kJ/kg·K	Temperature, °C Initial	Temperature, °C Final	Density or Mass
150 mm concrete slab	0.67	2	−6	2400 kg/m^3
Supply water	4.18	11	0	1000 kg/m^3
Ice	2.04	0	−4	—
Ethylene glycol, 35%	3.5	5	−9	14 000 kg

Latent heat of freezing water = 334 kJ/kg
Building and pumping heat load = 170 kW of refrigeration
System losses = 15%
Mass of water = 1500 m² × 0.025 m × 1000 kg/m³ = 37 500 kg
Mass of concrete = 1500 m² × 0.150 m × 2400 kg/m³ = 540 000 kg
Then:

$$q_R = (\text{Sys. losses})(q_F + q_C + q_{SR} + q_{HL})$$

where

q_R = refrigeration requirement
q_F = water chilling and freezing
q_C = concrete chilling load
q_{SR} = refrigeration to cool secondary coolant
q_{HL} = building and pumping heat load

$$q_F = \frac{37\,500\,\text{kg}\{4.18(11-0) + 334\,\text{kJ/kg} + 2.04[0-(-4)]\}}{24\,\text{h} \times 3600\,\text{s/h}}$$

$$= 168.5\,\text{kW}$$

$$q_C = \frac{540\,000 \times 0.67[2-(-6)]}{24 \times 3600} = 33.5\,\text{kW}$$

$$q_{SR} = \frac{14\,000 \times 3.5[5-(-9)]}{24 \times 3600} = 7.9\,\text{kW}$$

$$q_R = 1.15(168.5 + 33.5 + 7.9 + 170) = 437\,\text{kW}$$

When no time restrictions apply, the estimated refrigeration is the amount needed to offset the usage loads plus the coincidental heat loads during the most severe operating conditions. Table 1 lists approximate refrigeration requirements for various rinks with controlled and uncontrolled atmospheric conditions. Table 1 should only be used to check the calculated refrigeration requirements. Table 2 shows the distribution of various load components for basic construction and the estimated potential load reductions that may be obtained when energy-conserving design and operating techniques are used.

Table 1 Range of Ice Rink Refrigeration

4 to 5 Winter Months, Above 37° Latitude	
	m²/kW (refrigeration)
Outdoors, unshaded	2 to 8
Outdoors, covered	3 to 5
Indoors, uncontrolled atmosphere	5 to 8
Indoors, controlled atmosphere	4 to 9
Curling rinks, indoors	5 to 10
Year-Round (Indoors) (Controlled Atmosphere)	
	m²/kW (refrigeration)
Sports arena	3 to 4
Sports arena, accelerated ice making	1 to 3
Ice recreation center	3 to 5
Figure skating clubs and studios	3 to 5
Curling rinks	4 to 6
Ice shows	2 to 3

Heat Loads

Energy and operating costs for ice rinks are very significant, and these costs should be analyzed during design. A good estimate of required refrigeration can be calculated by summing the heat load components at design operating conditions. The heat loads for ice rinks consist of conductive, convective, and radiant components. Connelly (1976) collected the performance data summarized in Tables 2 and 3. The amount of control over each load sour reduction of the load that is possible through effective design and operation.

Conductive Loads. If a rink is uninsulated, **heat gain from the ground** below the rink and at the edges averages 2 to 4% of the total heat load. Permafrost may accumulate and frost heaving, which is detrimental to both the rink and the piping, may result. Heaving also makes it more difficult to maintain a usable ice surface.

Table 2 Ice Rink Heat Loads, Indoor Rinks

Load Sources	Approx. Max. Percentage of Total Load[a]	Max. Reduction through Design and Operation, %
Conductive loads:		
Ice resurfacing	12	60
System pump work	15	80
Ground heat	4	80
Header heat gain	2	40
Skaters	4	0
Convective loads:		
Rink air temperature	13	50
Rink humidity	15	40
Radiant loads:		
Ceiling radiation	28	90
Lighting radiation	7	40
Total	100	

[a]Load distribution for basic rink without insulation below rink floor.

Table 3 Ice Rink Heat Loads, Outdoor Rinks

Load Sources	Approx. Max. Percentage of Total Load[a]	Max. Reduction through Design and Operation, %
Conductive loads:		
Ice resurfacing	9	50
System pump work	12	80
Ground heat	2	40
Header heat gain	1	30
Skaters	1	0
Convective loads:		
Air velocity	0 to 15	10
Air temperature	0 to 15	0
Humidity	0 to 15	0
Radiant loads:		
Solar load	10 to 30	60
Total	100	

[a]Load distribution for basic rink without insulation below rink floor.

The heat gain from the ground and perimeter is highest when the system is first placed in operation; however, it decreases as the temperature of the mass beneath the rink decreases and permafrost accumulates. Ground heat gain is reduced substantially with insulation. Chapter 24 of the 1997 *ASHRAE Handbook—Fundamentals* gives details on computing heat gain with insulation.

Heat gain to the piping is normally about 2 to 4% of the total refrigeration load depending on length of piping, surface area, and ambient temperatures. The ice and frost that naturally accumulate on headers reduce the heat gain. Insulation can be applied to reduce the heat gain to the piping and keep ice from accumulating. A header designed for balanced circuit flow to the freezing grid, may, with precautions and the use of steel headers and piping, be imbedded within the rink floor. The imbedded headers contribute to the ice freezing and eliminate the trench to rink floor piping penetrations.

A circuit loop should be placed around the rink perimeter to prevent soft ice from developing at the edges (see the section on Rink Piping and Pipe Supports).

Heat gain from coolant circulating pumps can represent up to 12% of the refrigeration load. The pumps normally operate 24 h per day. The pumping heat load is the pump power plus adjustment for the pump and motor operating efficiency. Energy consumption from pump operation can be reduced by using pump cycling, two-speed motors, multiple pumps, or variable-speed motors with the appropriate controls. Proprietary variable motor speed controls are also available. The coolant flow should be sufficient at all times for acceptable chiller operation and to maintain a balanced flow through the piping grid.

The refrigeration system auxiliaries, such as condenser pumps, condensers, cooling tower or evaporative condensers, and condenser fans consume substantial electrical energy. Appropriate design and control of the system and good equipment selection should keep these auxiliary electric loads reasonable.

Ice resurfacing represents a significant operating heat load. Water is flooded onto the ice surface, normally at temperatures between 55 and 80°C, to restore the ice surface condition. The heat load resulting from the flood water application may be calculated as follows:

$$Q_f = 1000 V_f [4.2(t_f - 0) + 334 + 2.0(0 - t_i)]$$

where

Q_f = heat load per flood, kJ
V_f = flood water volume (typically 0.4 to 0.7 m³ for a 30 m by 60 m rink), m³
t_f = flood water temperature, °C
t_i = ice temperature, °C

The resurfacing water temperature affects the load and time required to freeze the flood water. Maintaining good water quality through proper treatment may permit the use of lower flood water temperature and less volume.

Convective Loads. The convective load from the air to the ice may represent as much as 28% or more of the total heat load to the ice (Tables 2 and 3). The convective heat load is affected by air temperature, relative humidity, and air velocity near the ice surface. Precautions should be taken to minimize the influence of air movement across the ice surface in the design of the rink heating and dehumidification air distribution system. The convection heat load may be estimated using the procedure from Appendix 5 in the publication, "Energy Conservation in Ice Skating Rinks" (DOE 1980). The estimated convective heat transfer coefficient can be calculated using the formula:

$$h = 3.41 + 3.55 V$$

where

h = convective heat transfer coefficient, W/(m²·K)
V = air velocity over the ice, m/s

The effective heat load (including the latent heat effect of convective mass transfer) is given by the following equation:

$$Q_{cv} = h(t_a - t_i) + [K(X_a - X_i)(2852 \text{ kJ/kg})(18 \text{kg/mole})]$$

where

Q_{cv} = convective heat load, W/m²
K = mass heat transfer coefficient
t_a = air temperature, °C
t_i = ice temperature, °C
X_a = mole fraction of water vapor in air, kg mol/kg mol
X_i = mole fraction of water in saturated ice, kg mol/kg mol

When the mole fraction of air is calculated using a relative humidity of 80% and a dry bulb of 3.3°C, X_a is approximately 6.6 × 10⁻³, and X_i for saturated ice at 100% and a temperature of −6.1°C is 3.6 × 10⁻³. On the basis of the Chelton Colburn analogy, $K \approx 0.23$ g/(s·m²) (DOE/TIC 1980).

In locations with high ambient wet-bulb temperatures, dehumidification of the building interior should be considered. This process lowers the load on the icemaking plant and reduces condensation and fog formation. Traditional air conditioners are inappropriate because the large ice slab tends to maintain a lower than normal dry-bulb temperature.

Radiant Loads. Indoor ice rinks create a unique condition where a large, relatively cold plane (the ice sheet) is maintained beneath an equally warm plane (the ceiling). The ceiling is warmed by conductive heat flow from the outside and by normal stratification of arena air. Up to 35% of the heat load on the ice sheet comes from radiant sources. On outdoor rinks radiant sources are the sun or a warm cloud cover. Vertical hanging cloth suspended from east-west horizontal overhead wires has been used to reduce the winter sun load.

In indoor and covered rinks, lighting is the major source of radiant heat to the ice sheet. The actual quantity depends on the type of lighting and how the lighting is applied. The direct radiant heat component of the lighting can be as much as 60% of the kilowatt rating of the luminaires. A radiant heating system can be another source of radiant heat gain to the ice. If radiant heat is used to maintain the comfort level in the promenade or spectator area, the radiant heaters should be located and directed to avoid direct radiation to the ice surface. The infrared components of the lighting can be estimated from manufacturers' data.

The infrared heat gain component from the ceiling and building structure, which is warmer than the ice surface, can be calculated by applying the Stefan-Boltzmann equation as follows:

$$q_r = A f_{ci} \sigma (T_c^4 - T_i^4)$$

$$f_{ci} = \left[\frac{1}{F_{ci}} + \left(\frac{1}{\varepsilon_c} - 1 \right) + \frac{A_c}{A_i} \left(\frac{1}{\varepsilon_i} - 1 \right) \right]^{-1}$$

where

q_r = radiant heat load, W/m²
A_c = ceiling area, m²
A_i = ice area, m²
ε = emissivity
f_{ci} = gray body configuration factor, ceiling to ice surface
F_{ci} = angle factor, ceiling to ice interface (from Figure 1)
T = temperature, K
σ = Stefan-Boltzmann constant = 5.67 × 10⁻⁸ W/(m²·K⁴)

Example 2. An ice rink has the following conditions:

Ice dimension: 26 m × 60 m = 1560 m²
Ice temperature: −4°C (269 K), $\varepsilon_i = 0.95$
Ceiling radiating area: 28 m × 60 m = 1680 m²
Ceiling mid-height: 7.6 m

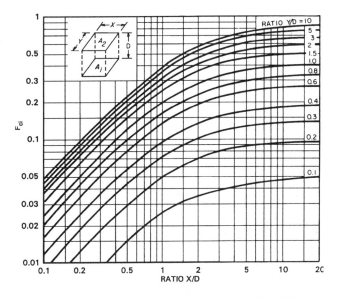

Fig. 1 Angle Factor for Radiation between Parallel Rectangles F_{ci}

Ceiling temperature: 16°C (289 K), $\varepsilon_c = 0.90$

$$x/d = 28/7.6 = 3.6$$
$$y/d = 60/7.6 = 7.9$$

From Figure 1, $F_{ci} = 0.68$

$$f_{ci} = \left[\frac{1}{0.68} + \left(\frac{1}{0.90} - 1\right) + \frac{1680}{1560}\left(\frac{1}{0.95} - 1\right)\right]^{-1} = 0.610$$

Then:

$$q_r = 1680 \times 0.610 \times 5.67 \times 10^{-8}(289^4 - 269^4)/1000$$
$$= 101 \text{ kW}$$

The ceiling radiant heat load can be reduced by lowering the temperature of the ceiling, keeping warm air away from the ceiling, increasing the roof insulation, and, more significantly, by lowering the emissivity of the ceiling material to shield the ice from the building structure.

Ceiling and roof materials and exposed structural members have an emissivity that may be as high as 0.9. Special aluminum paint can lower the emissivity to between 0.5 and 0.2. Polished metal such as polished aluminum or aluminum foil have an emissivity of 0.05.

Also because a low-emissivity ceiling is cooled very little by radiant loss, most of the time its temperature remains above the dew point of the rink air. Thus, condensation and dripping is substantially reduced or eliminated.

Low emissivity fabric or tiled ceilings are frequently incorporated into new and existing rinks to reduce radiation loads, decrease condensation problems, and reduce the overall lighting required.

Radiant heat gain to the ice, especially in outdoor rinks, can be further controlled by painting the ice about 25 mm below the surface with whitewash or slaked lime. Commercial paints with a low solar absorptivity, which are generally water based, are also available.

ICE RINK CONDITIONS

Properly designed indoor rinks, as well as properly designed renovated rinks, can be operated year-round without shut down. However, some indoor rinks operate from 6 to 11 months and shut down for various reasons including maintenance, rink construction, inability to control indoor conditions, or unprofitable operation during part of the year. Outdoor, uncovered rinks generally operate from early November to mid-March above 40° North latitude. However, if sufficient refrigeration capacity is provided, the ice can be maintained for a longer period.

Indoor rinks are operating successfully even in warm tropical climates. Relative humidity, temperature, and ceiling radiant losses must be controlled in these climates to prevent fog, ceiling dripping, and high operating cost.

Steel frame, brick, concrete, and various forms of plastic have been used to enclose ice skating rinks. Rinks have also been built under air-supported structures for seasonal use and are usually over a multipurpose surface.

Arena heating is frequently provided for skater and/or spectator comfort and can be provided in conjunction with a dehumidification system. Heat recovery from the refrigeration system may be used for limited heating, supplementing the heating system, or dehumidification reheat. Ice rink temperatures are usually maintained between 5 and 15°C; however, for skater or spectator comfort, higher temperatures are sometimes preferred. The relative humidity maintained in the arena depends on factors such as building construction, indoor temperature, and outdoor wet bulb.

The system should be designed to reduce fogging and ice surface condensation. Relative humidity at or below 80% with rink temperatures between 5 and 15°C is usually sufficient to eliminate fogging; however, condensation can occur on the ceiling or roof structure due to radiation from the building structure to the ice. Low relative humidity is needed to reduce this condition when a high emissivity ceiling is exposed to the ice surface.

Ventilation should be the minimum required for the building occupancy so that the humidity introduced with outdoor air is kept as low as is feasible; but enough outdoor air must enter to maintain acceptable indoor air quality (see ASHRAE *Standard* 62). Gas engine resurfacing machines should be equipped with catalytic exhaust convertors to reduce carbon monoxide emissions. The makeup air or ventilation air in humid climates should be dehumidified prior to being supplied to the arena.

Carbon monoxide and nitrogen dioxide are pollutant emissions from gasoline- or propane-fueled ice resurfacers. The concentration of these chemicals can reach dangerously high levels if they are not controlled or eliminated. In some areas, regulations require sensors to detect and alarm at unsafe chemical concentrations. Check the health regulations for local requirements.

Each rink user group has its own preference for the type of ice used. Hockey players and curlers prefer hard ice; figure skaters prefer softer (i.e., warmer) ice so they can clearly see the tracings of their skates; and recreational skaters prefer even softer ice, which minimizes the buildup of shavings and scrapings.

Since ice surface temperature can not be measured easily, the ice condition is customarily controlled either from a predetermined coolant average temperature or from the ice temperature measured by a thermocouple embedded beneath the ice surface. With approximately 7°C air temperature and one 25-mm ice thickness, ice at −6.5 to −5.5°C is satisfactory for hockey, −4 to −3°C for figure skating, and −3 to −2°C for recreational skating. A 0.5 K higher ice temperature may be feasible when water with a low mineral content is used for resurfacing. To achieve these ice temperatures, the coolant temperature is maintained about 3 to 6 K lower than the ice temperature. The temperature of the coolant must be lowered to maintain the same ice conditions when there are higher wet-bulb temperatures or abnormally high loads, such as when television lighting is used.

EQUIPMENT SELECTION

Compressors

Two or more refrigeration compressors should be used in an ice rink system. When two compressors are used, one compressor should be specified with ample capacity to maintain the ice sheet under normal load and operating conditions. When greater capacity is required during the initial ice freezing or under high heat loads, the second compressor picks up the load. In multiple compressor installations, a multistage thermostat microprocessor control and/or a motorized sequence control may be used to control the operation of the compressors. The multiple compressors serve as backups; they maintain the ice in the event of compressor failure or a service requirement.

Compressors and evaporators should operate at a suction pressure corresponding to a 6 K mean temperature difference between the coolant and primary refrigerant in systems operating with secondary coolants, or between the ice and the refrigerant in direct refrigerant rinks.

Condensers and Heat Recovery

Wells, lakes, or rivers can be good sources of condenser cooling water, if they are available. Capacity is easy to regulate and the low coolant temperature maintains low condensing pressures, which saves energy. But, condensers require high quality water, which may need treatment to prevent scale formation, fouling, or corrosion in the condenser tubes.

Cooling towers used with water-cooled condensers, evaporative condensers, or air-cooled condensers are alternatives. When

Ice Rinks

selecting a cooling tower or evaporative condenser, not only the maximum expected wet-bulb temperature during the skating season should be considered, but also suitable controls to cover the wide range in capacities and protection against freezeup needed in cold weather. A water treatment specialist should also be consulted.

Air-cooled condensers are used in northern climates, particularly where the rink is used only in the winter. They can be economically sized and require no water, so that the possibility of freezeup is eliminated. This type of condenser, however, is not economical for year-round operation, and for seasonal operation it must have wide-range capacity control. Heat rejected by the condensers can be recovered and used with water or air-cooled condensing systems.

When a cooling tower system is selected, heat from condensers can be used for such energy-saving applications as arena heating, subfloor heating, domestic water heating, and snow melting. This is generally done either by circulating the condenser cooling water through heat exchangers or with fan coil units. The circulated cooling water can also be used in conjunction with a heat pump as a heat sink/source for heating or cooling various areas within the rink building and for water heating.

Ice Temperature Control

Ice temperature may be controlled by various methods. Thermostats that sense the return coolant temperature or the differential temperature between the supply and return coolant can be used to control the refrigeration system. They may also be used in controlling operation of the coolant pump. To be effective, a differential sensor should sense a small temperature difference. The return coolant temperature can be sensed by multistage sensors that sense a larger temperature difference. Another strategy varies coolant flow by controlling the pump with a temperature sensor buried in the ice. Direct refrigerant systems can be controlled by regulating compressor operation with a sensor in the ice. This method has been used with a direct refrigerant impulse pumping system. Compressor capacity and pump operation may be controlled from the low-pressure receiver when refrigerant pumps are used to circulate the refrigerant.

Rink Piping and Pipe Supports

High flow rate secondary systems use standard mild steel pipe 20, 25, or 32 mm in diameter; thin-walled polyethylene plastic pipe 25 mm in diameter; or UHMW (ultrahigh molecular mass) polyethylene plastic pipe 25 mm in diameter. These are placed at 90- or 100-mm centers on the rink floor. A proprietary low flow rate secondary coolant system uses 6-mm tubing made of flexible plastic with tube spacing averaging 20 mm or one dual tube every 40 mm. Direct refrigerant rinks generally use 16 to 22 mm steel tubing, which is placed on 75-mm centers for outdoor rinks and 100-mm centers for indoor rinks.

The pipe grid must be maintained as close to level as possible, regardless of the rink piping system used. When a pipe rink surface is of the open type with sand fill around and over the pipes, the conduit usually rests on pressure-treated sleepers set level with the subbase; however, the sleepers can be omitted in a rink that is to be operated year-round. The piping is then spaced with clips, plastic stripping, or punched metal spacers.

In permanent concrete floors, the pipe or steel tubes are supported on notched iron supports or welded chair supports. The latter must be used in the case of plastic pipe.

Headers and Expansion Tanks

Secondary coolant rinks using large-diameter pipe generally run the piping lengthwise, with the supply and return headers across one end. Small-diameter tubing rinks generally run crosswise, with the supply and return headers along one side. Direct refrigerant rinks generally run lengthwise, with the supply header at one end and the return header at the opposite end in a balanced system. The header must be sized to assure an even distribution of coolant through every pipe. The systems are generally designed with low coolant velocities, which do not need balancing valves. If at all possible, the return header should be placed at the same elevation as the rink piping, with a minimum of two air vents to eliminate the trapping of air.

The three-pipe reversed return header and distribution arrangement (Figure 2) is commonly used. However, a properly sized two-pipe header system (Figure 3) is frequently applied and gives nearly uniform circuit flow with no discernible differences in the ice surface. To allow for thermal contraction and expansion, headers and main piping should be free to move without producing excessive stress.

Polyethylene distribution headers should only be used with proper allowances for expansion and contraction. The coefficient of thermal expansion for steel is relatively low and very close to that of concrete, while the polyethylene pipe expansion coefficient is much higher. Pipe clamp connections must remain accessible for inspection and tightening. Clamps are not considered permanent joints.

A closed secondary coolant system requires an expansion tank to safely accommodate the expansion and contraction of the coolant resulting from fluid temperature changes. The expansion tank must be installed so that it cannot be isolated from the system.

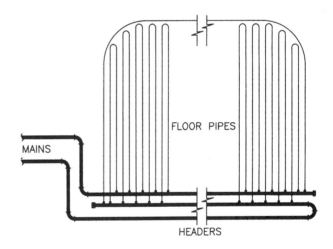

Fig. 2 Reverse Return System of Distribution

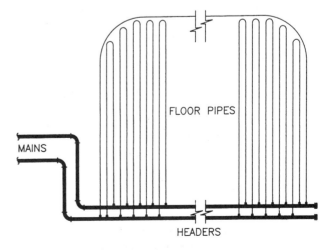

Fig. 3 Two Pipe Header and Distribution

Coolant Equipment

The coolant circulating pumps must be sized for the particular type of rink and system involved. Large-diameter pipe rinks require 180 to 270 mL/s per kilowatt of refrigeration to maintain the required 1 to 2 K temperature differential between incoming and outgoing coolant. These operate at approximately 170 kPa (gage). Low flow rate dual tubing or mat rinks use about 40 mL/s per kilowatt. Differentials of 2 to 3 K are normal, but 6 to 7 K differentials can be experienced in high load conditions with no reduction in ice quality. Uniform temperatures are achieved in mat rinks by temperature averaging between closely spaced, adjoining counterflow tubes operating at approximately 280 to 350 kPa pressure.

For auditoriums and sports arenas, the rink surface should have provision for deicing in less than 4 h. In this operation, the floor is heated to about 10°C so that the ice can be peeled off by first breaking the bond between the floor and ice and then breaking the ice and removing it with power tractors.

A standard heat exchanger can be used, with piping so arranged that all the coolant can be pumped through the heater, with the coolant flowing in the tubes and the steam or water in the shell. Approximately 1100 W per square metre of rink surface is needed to heat the coolant in the system enough to warm the floor and break the ice bond.

In sports arena rinks designed for frequent deicing, ice may be required every other day, and 16 mm ice or more must be frozen in 12 h to allow the ice to temper before being skated on the following day. In such cases, a cold coolant accumulator keeps the size of the refrigeration equipment and connected power within reasonable limits. This accumulator is a coolant storage tank bypassed from the thawing cycle. When refrigeration on the ice surface is not required, a large volume of coolant in the accumulator may be cooled to approximately −32°C and be ready to be pumped into the rink piping when needed. The cold coolant accumulator should store a sufficient volume to cool the entire cooling system coolant volume from 18 to −18°C. This cold coolant tank usually holds more than three times the volume of the cooling system's charge.

With the increase in construction and operating costs, the use of accumulators has been declining. In place of accumulators, ice-making equipment is sized to handle the demand loads, and arenas are programmed to eliminate the need for quickly making and removing the ice. The use of phase change materials may reduce the accumulator size by 6 to 8 times, eliminating high demand charges and allowing total off-peak operation at lower electric rates.

Energy Consumption

Energy consumption for an ice rink facility is somewhat unique. Maintenance of internal conditions is affected by the cold ice sheet. The lighting, ventilation, heating, and dehumidification systems depend on the use and occupancy of the facility. The energy consumed by the refrigeration equipment is affected by construction, operation, water quality, and the various use factors. Methods to reduce heat load and energy consumption should be considered in both the design and operation of an ice rink. These include:

- Install low emissivity ceilings to reduce refrigeration and lighting loads and to permit compressors to operate at a higher saturated suction temperature
- Reclaim the refrigerant superheat to preheat shower water, heat the ice resurfacing water, melt ice shavings, heat the subfloor, etc.
- Select a pumping system and controls to reduce coolant flow during part load conditions
- Install an energy management system
- Insulate the subfloor and header piping
- Control the temperature and humidity in the arena to reduce sensible and latent heat gain to the ice
- Install high efficiency luminaires
- Use demineralized water or water with a very low mineral content for the ice and resurfacing

GENERAL RINK FLOOR DESIGN

Generally, five types of rink surface floors are used (Figure 4):

- Open or sand fill type, for plastic or metal piping or tubing
- Permanent, general-purpose type, with piping or tubing embedded in concrete on grade
- All-purpose type, with piping or tubing embedded in concrete with floor slab insulated on grade
- All-purpose floors, supported on piers or walls
- All-purpose floor with reheat; use this type when the water table and moisture are severe problems or when the rink is to operate for more than six months

The open sand fill floor is the least expensive type of rink floor. The cooling pipes rest on wood sleepers over a bed of crushed stone or other fill. The clean washed sand is filled in around the cooling pipes. Curling rink floors, as well as hockey and skating rinks, where first cost is a factor and the building is not intended for other uses, are usually constructed in this manner. Clay or cinders should never be used in the bed or for fill around the pipes. Tubing rinks do not need supports or sleepers; the tubes are laid on accurately leveled sand.

Rinks using 25 mm plastic pipe or the mat type are usually covered with sand to a depth of 13 to 25 mm to provide additional strength to the ice surface and to reduce cracking. Many portable outdoor rinks have used this arrangement for laying the plastic pipes or tubing mats on top of existing sodded areas, black top, or concrete. More permanent installations of outdoor semiportable rinks have used this same arrangement where recreational area is at a premium. Such an installation consists of steel pipes supported on notched steel sleepers, which in turn are supported on concrete piers down to solid ground.

To obtain a better return on investment, most indoor rinks that operate with an ice surface for only a portion of the year have a permanent general-purpose concrete floor with subfloor insulation and heat pipes so that the floor may be used for other purposes when the skating season is over. The floor should withstand the average street load and is usually designed with 25 or 30 mm steel or plastic pipe embedded in a steel-reinforced concrete slab 100 to 150 mm thick, depending on the anticipated loading and coolant pipe diameter.

In sports arenas, where the ice is removed and the floor made ready for other sports and entertainment, the ice floor must be constructed to withstand the frequent change from hot to cold. The refrigerating machinery must be of sufficient capacity to freeze a sheet of ice 16 mm thick in 12 h. This type of floor is always insulated.

Subfloor insulation must be installed when quick changeovers are desired, when there is a high moisture content in the subsoil, when the floor is elevated, or when the rink is in continuous use for more than 9 months. This subfloor insulation reduces the refrigeration load on ice-making equipment and slows down, but does not eliminate, the cooling of the subsoil on surfaces installed on grade.

Drainage

The suitability of an ice rink's subsoil has a great influence on the rink's success. Complete ice surfaces have had to be rebuilt because of poor drainage and the ultimate heaving of the ice surface. Thus, skating rinks should not be built on swampy or low-lying land unless adequate drainage is provided.

Moist subsoil will freeze in the ground to a depth of 1200 mm or more. The frozen water will heave the ice surface when freezing takes place at a depth of 150 mm or more. Heaving creates an uneven skating surface; moves and raises walls, piers, and header trenches; cracks walls and piping; and necessitates the eventual drainage and rebuilding of the rink floor.

Not only should there be a complete drainage system around the footings of the rink to prevent seepage, but there should also be one

Ice Rinks

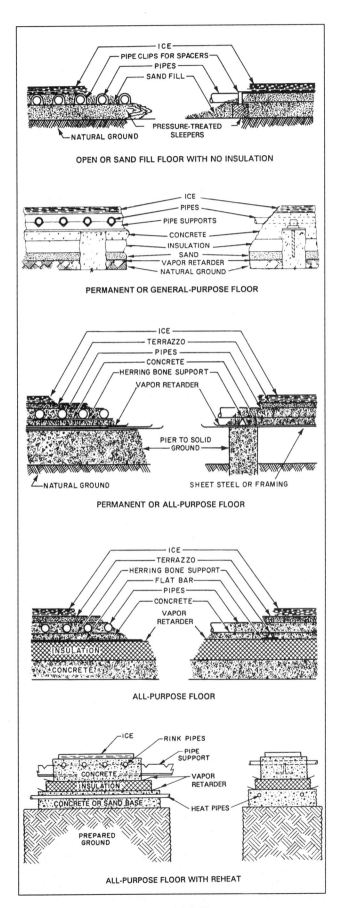

Fig. 4 Ice Rink Floors

under the rink surface itself. This is particularly important when a sand fill floor is used; a good system will assure that the ice melted after the skating season will completely drain away and the sand will dry out as quickly as possible.

Subfloor Heating for Freeze Protection

Subfloor heating, by electrical heating cables or a pipe or tubing recirculating system using a warm antifreeze solution, is found in most new rinks to prevent below floor permafrost development and the resultant heaving. Pipes or tubing are on 300 to 600-mm centers located under 50 to 100 mm of insulation. They are generally installed in sand rinks, which are used year-round, although they may be poured into a concrete base slab with insulation between the base slab and the rink slab (see Figure 4).

Alternatively, the heating pipes may be laid directly in the subfoundation below the rink pipe or insulation. However, an installation not equipped with insulation requires a greater depth between the heating pipes and the ice-making pipes to prevent an increased load.

Neither water nor warm air should be used for subfloor heating. Water, if inadvertently allowed to freeze, cannot be readily melted out. In time, warm air ducts become filled with frost and ice because of high rink humidity and air duct leakage.

Usually, the same fluid used for the coolant in the ice-making system is used for subfloor heating; it can be heated to the necessary 4 to 10°C in a heat exchanger warmed by compressor waste heat. Subfloor insulation should be of a rigid moistureproof board, such as high-density polystyrene foam, and be completely enveloped in a polyethylene vapor retardant.

Preparation for Rink Floor

When building on natural ground, regardless of whether a sand fill or a permanent general-purpose floor is intended, proper preparation of the bed is important unless the rink is built on elevated sand and gravel subsoil. If the rink is to be built on clay, part clay, or rock subsoil, water should be prevented from collecting in low areas. Either the clay or rock should be excavated or the rink level should be built up with crushed stone and gravel to a height of about 1200 mm, after which it should be well rolled. Water should not be used for settling the fill.

In the case of sand fill rinks, quickly draining the melted ice at the end of the skating season ensures rapid drying of the sand and rink piping and results in a longer life for the steel piping. Cinders should never be used as fill in open sand fill rinks because of the possibility of sulfur in the cinders which, when damp, accelerates corrosion of steel piping.

Care must be taken to assure a level surface over the entire rink with no more than ±3 mm in any 1 m^2 area and ±6 mm overall.

Permanent General-Purpose Rink Floor

When constructing a permanent general-purpose floor, the same subsoil precaution must be taken as for a sand fill rink. The concrete floor should withstand, at a minimum, the average road pavement load.

When local conditions make it advisable, the rink floor should be insulated. Insulation may be laid on a level concrete or sand base.

The concrete mixture should have a 28-day strength of 140 to 240 kPa and be put in place in a quality manner (a concrete engineer is recommended to specify concrete, its placement, and curing). Suitable cross-reinforcing and pipe supports are necessary.

Concrete floors with mat-type tubing are poured in two courses. A first course is poured and leveled; the mats are then rolled out and positioned. A 150 mm by 150 mm wire mesh is laid on top of the mats; then a second course, with grouting between it and the first, is poured on top of the first course, mats, and wire. Water pressure should be kept in the tubing to spot any leaks or cuts that may develop. Once started, the pouring of each course of the concrete floor should be continuous with interruptions not to exceed 15 min.

General-purpose rink floors should not be defrosted too frequently. When a rink constructed with a general-purpose floor is to be used during the ice season for purposes that require an ice-free floor, it is preferable to place an insulated portable-section wood floor over the ice for each occasion.

All-Purpose Floors

If a rink floor as used in sports arenas is to withstand both the expansion and contraction of frequent frosting and defrosting and thermal shock because of the circulation of very low-temperature coolant, then extra precautions must be taken in its construction, such as provisions for the free movement of the freezing slab with respect to the subfloor.

Header Trench

A well-constructed header trench of sufficient size to house the headers and connections and the subfloor heating system, if applicable, is essential unless the steel distribution headers are cast into the concrete slab as part of the rink. Provisions for movement of pipes due to thermal expansion and contraction should be incorporated into the design. This trench should be equipped with removable covers and be well-drained to facilitate drying out. The headers and piping in the trench are not usually insulated, which allows for periodic inspection and painting of the piping. However, unless a large trench is provided, consideration should be given to insulating the headers on rinks that operate year-round because of the massive buildup of frost. Provision must be made for purging air from the rink piping and header system.

Snow Pit

A snow-melting pit should be provided at a suitable point, usually at one end of the rink. It should be of sufficient size to handle the scraped off snow and the ice accumulated during planing, or it may be made large enough to accommodate the complete ice removal.

Discharge water from shell-and-tube condensers, a waste heat recovery system, or some other heat source should be provided to melt the snow and ice. An average load for a snow melting pit from a mechanical ice resurfacer is between 40 and 50 kW for a 1500 m^2 rink. A large drain with overflow, as well as a large removable screen to filter out trash, should be provided.

BUILDING, MAINTAINING, AND PLANING ICE SURFACES

Regardless of the type of rink floor used, when the plant is first placed in operation, the equipment should be operated long enough for a sharp frost to appear on the surface. Then the entire surface should be uniformly covered with a fine spray. This process should be repeated until a 13 mm thickness of ice is built, or until the surface is level. After applying a layer of water base white paint, another 10 mm thick layer of ice is built before painting the red and blue lines. Red and blue lines are available in plasticized paper; however, they need to be covered with a minimum of 13 mm of ice to protect against damage. It is essential that sand floors be thoroughly wet before freezing because dry sand has poor conductivity. The surface should not be frozen any colder than required after this buildup so as to allow the ice to temper before it is used for skating and also to deter cracking.

To maintain an ice surface, it is customary to scrape off the snow after each skating session or hockey period. In all but the smallest rinks, this is done by a motorized resurfacer. On small rinks, the scraping is done manually with a wide hardened-steel scraper blade. The most satisfactory method of resurfacing the ice between sessions is to wheel a sprinkler tank filled with hot water over the ice. The sprinkler has an adjustable valve to control the quantity of water, which is sprayed into a terry cloth bag that wipes the fine snow off the ice surface and fills the crevices cut by the skaters. In this manner, the least amount of water is added, reducing the ice buildup and refrigeration load.

By far the most common method is the use of automatic resurfacing machines. Mounted on four-wheel drive chassis, the machines plane the ice, pick up the snow, and lay down a new ice surface using hot or cold water. Hot water generally gives harder ice, since air bubbles are removed, but high energy costs have led many rinks to alternate hot and cold water resurfacings. Rink corners should be at least a 6.1-m, preferably 8.5-m, radius for effective use of this equipment. Smaller equipment is available for studio and small rinks.

Because of inattentive ice making, improper sprinkling equipment, or deep cutting of the ice during public skating, the ice may become uneven and excessively thick. There may be a fairly slight variation in the ice thickness across the rink, but more serious is the resulting variation in the condition of the ice. In any case, the low spots on the ice must be built up, increasing the thickness and refrigeration requirements.

For example, under assumed conditions, where $-8°C$ coolant would be cold enough to hold a 38 mm thickness of ice, calculations show that $-21°C$ coolant would be required if the ice were permitted to build up to 150 mm, with a corresponding decrease in effective refrigeration capacity and an increase in operating costs. In other words, every additional 25 mm of ice thickness required from the refrigeration system increases 8 to 15%, depending on system heat load (DOE 1980).

Since ice of 13 to 25 mm thickness is satisfactory for skating and is the most economical thickness to freeze and hold, the ice should be periodically planed to maintain this desired thickness.

Water Quality

The quality of the water affects energy consumption and ice quality. Water contaminants, such as minerals, organic matter, and dissolved air, can affect both the freezing temperature and the ice thickness necessary to provide satisfactory ice conditions. Proprietary treatment systems for arena flood water are available. When these treatments are properly applied, they reduce or eliminate the effects of contaminants and improve ice conditions.

RINK FOG AND CEILING DRIPPING

During mild weather, particularly in early fall and late spring in the northern United States and Canada, condensation often drips from the roofs and roof supports of rinks (especially curling rinks), due to construction, internal conditions, or insufficient internal heat loads. The condensate dropping on the ice ruins the curling surface and the fog obstructs the view. These conditions cannot be solved by ventilation because the introduction of outdoor air only aggravates the problem when the weather is mild and humid. Insulating the roof also aggravates the drip during mild outside weather conditions. Low-emissivity ceilings stay warmer and thus reduce condensation and drip.

Under these conditions, to prevent condensation in the roof space and to clear the fog, a six-sheet curling rink, with 1160 m^2 of ice, would require the removal of 15 kg of moisture per hour, necessitating about 21 kW of refrigeration.

Units using the coolant from the rink piping as a cooling medium avoid frosting by recirculating with a small bypass pump to keep coil inlet brine above 0°C. Reheat coils are often included and frequently use back waste heat from condenser cooling water to counter the cooling effect of the dehumidifier coil. Self-contained, air-cooled, compressor-type packaged dehumidifying units, as well as desiccant drier types with gas or electric regeneration, are available.

Various dehumidification and defogging systems should be evaluated in an owning and operating cost analysis. Unlike the normal behavior of moist air, air with high moisture content does not rise in an ice rink. Instead, the air remains near the ice surface because it is colder than the surrounding air. The air circulation system should remove the cold, moist air away from the ice with minimum draft on the ice surface. Increased air velocity near the ice surface increases the heat load to the ice and can cause surface wetness. Air circulation is also important in removing carbon monoxide from resurfacing equipment; the carbon monoxide tends to remain below the top of the dasher boards and near the ice surface.

IMITATION ICE-SKATING SURFACES

A number of different imitation ice-skating surfaces have been marketed; these use semiporous plastic panels dressed with a synthetic lubricant. The coefficient of friction of ice is approximately 0.03 at $-3°C$ and is even less because of the film of water produced by pressure under the skate. In considering the use of imitation surfaces, the actual friction coefficients of these surfaces—both when freshly lubricated and after a period of usage—should be investigated.

BIBLIOGRAPHY

Albern, W.F. and J.J. Seals. 1983. Heat recovery in an ice rink? They did it at Cornell University. *ASHRAE Journal* 25(9):38-39.

ASHRAE. 1968. Ice skating rinks. *Symposium* at ASHRAE meeting in Columbus, OH.

Banks. N.J. 1990. Desiccant dehumidifiers in ice arenas. *ASHRAE Transactions* 96(1):1269-72.

Blades, R.W. 1992. Modernizing and retrofitting ice skating rinks. *ASHRAE Journal* 34(4):34-42.

Brauer, M., J.D. Spengler, K. Lee, and Y. Yanagisana. 1992. Air pollutant exposures inside hockey rinks: Exposure assessment and reduction strategies. *Proceedings* Second International Symposium on Safety in Ice Hockey, Pittsburgh, PA.

Canadian Electrical Associates. 1992. Potential electricity savings in ice arenas and curling rinks through improved refrigeration plant. CEA No. 9129-858 Manbek Resource Consul Book.

Connelly, J.J. 1976. ASHRAE *Seminar* on Ice Rinks (February), Dallas, TX.

DOE. 1980. Energy conservation in ice skating rinks. Prepared by B.K. Dietrich and T.J. McAvoy. U.S. Department of Energy.

Matus, S.E. et al. 1988. Carbon monoxide poisoning at an indoor ice skating facility. Proceedings ASHRAE IAQ 88 Conference, pp. 275-283.

Minnesota Department of Health. 1990. Indoor air quality unit: Regulating air quality in ice arenas.

Rein, R.G. and C.M. Burrows. 1981. Basic concepts of frost heaving. *ASHRAE Transactions* 87(2):1087-97.

CHAPTER 35

CONCRETE DAMS AND SUBSURFACE SOILS

CONCRETE DAMS 35.1	SOIL STABILIZATION 35.4
Methods of Temperature Control 35.1	Thermal Design 35.4
System Selection Parameters 35.3	Passive Cooling 35.4
CONTROL OF SUBSURFACE WATER FLOW 35.3	Active Systems 35.6

REFRIGERATION is one of the more important tools of the heavy construction industry, particularly in the temperature control of large concrete dams. It is also used to stabilize both water-bearing and permanently frozen soil. This chapter briefly describes some of the cooling practices that have been used for these purposes.

CONCRETE DAMS

Without the application of mechanical refrigeration during construction of massive concrete dams, much smaller construction blocks or monoliths would have to be used, which would slow construction. By removing unwanted heat, refrigeration can speed construction, improve the quality of the concrete, and lower the overall cost.

METHODS OF TEMPERATURE CONTROL

Temperature control of massive concrete structures can be achieved by (1) selecting the type of cement, (2) replacing part of the cement with pozzolanic materials, (3) using embedded cooling coils, or (4) precooling the materials. The measures used depend on the size and type of structure and on the time permitted for its construction.

Cement Selection and Pozzolanic Admixtures

The temperature rise that occurs after concrete is placed is due principally to the cementing materials' heat of hydration. This temperature rise varies directly with cement content per unit volume and, more significantly, with the type of cement. Ordinary Portland cement (Type I) releases about 420 kJ/kg, half of which is typically generated in the first day after the concrete is placed. Depending on specifications, Type II cement may generate slightly less heat. Type IV is a low heat cement that generates less heat at a slower rate.

Pozzolanic admixtures may be used in lieu of part of the cement. Such pozzolans include fly ash, calcined clays and shales, diatomaceous earths, and volcanic tuffs and pumicites. The heat-generating characteristics of these pozzolans vary, but are generally about one-half that of cement.

When determining the system refrigeration load, heat release data for the cement being used should be obtained from the manufacturer.

Cooling with Embedded Coils

In the early to mid 1900s, the heat of curing on large concrete structures was removed by embedded cooling coils for glycol or water recirculation. Further, they lowered the temperature of the structure to its final state during construction. This is desirable where volumetric shrinkage of a large mass is necessary during construction, for example, to allow the contraction joint grouting of intermediate abutting structures to be completed.

In the past, thin wall tubing was placed as a grid-like coil on top of each 1.5 to 2.3 m lift of concrete in the monoliths. Chilled water was then pumped through the tubing, using a closed loop system to remove the heat. A typical system used 25-mm OD tubing with a flow of about 0.25 L/s through each embedded coil. While the number of coils in operation at any time varies with the size of the structure, 150 coils is not uncommon in larger dams. Initially, the temperature rise in each coil can be as much as 4 to 6 K, but it later becomes 1.5 to 2 K. An average temperature rise of 3 K is normal. When sizing the refrigeration equipment, the heat gain of all the circuits is added to the heat gain through the headers and connecting piping.

For a typical system with 150 coils based on a design temperature rise of 3 K in the embedded coils and a total heat loss of 3 K elsewhere, the size of the refrigeration plant would be about 1000 kW. Figure 1 shows a flow diagram of a typical embedded coil system.

Cooling with Chilled Water and Ice

The actual temperature of the mix at the time of placement has a greater effect on the overall temperature changes and subsequent contraction of the concrete than any change caused solely by varying the heat-generating characteristics of the cementing materials.

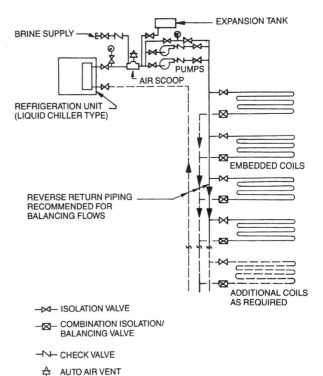

Fig. 1 Flow Diagram of Typical Embedded Coil System

The preparation of this chapter is assigned to TC 10.1, Custom Engineered Refrigeration Systems.

Further, placing the concrete at a lower temperature normally results in a smaller overall temperature change than that obtained with embedded coil cooling. Because of these inherent advantages, precooling measures have been applied to most concrete dams.

Glen Canyon Dam illustrates the installation required. The concrete was placed at a maximum placing temperature of 10°C during summer months when the aggregate temperature was about 31°C, cement temperature was as high as 65°C, and the river water temperature was about 29°C. Maximum air temperatures averaged over 38°C during the summer months. The selected system included cooling aggregates with 1.7°C water jets on the way to the storage bins, adding refrigerated mix water at 1.7°C, and adding flaked ice for part of the cold water mix. Subsequent cooling of the concrete to temperatures varying from 4°C at the base of the dam to 13°C at the top was also required. The total connected brake power of the ammonia compressors in the plant was 4600 kW, with a refrigeration capacity equivalent to making 5400 Mg of refrigeration.

The maximum amount of chilled water that may be added to the concrete mix is determined by subtracting the amount of surface water from the total mix water, which is free water. Frequently, if a chemical admixture is specified, some water must be added to dissolve the admixture—usually about 20% of the total free water. This limits the amount of ice that can be added to the remaining 80% of free water available. After the amount of ice that is needed for cooling is determined, the size of the ice-making equipment can be fixed. When determining the capacity of the equipment, allowances should also be made for cleaning, service time, and storage for the ice during nonproductive times.

When calculating the actual heat removal, the ice is considered to be 0°C when introduced into the mixer. The chilled water is assumed to be 4°C entering the mixer, even though the chiller may supply the water at a lower temperature.

Cooling by Inundation

The temperatures specified today cannot be achieved solely by adding ice to the mix. In fact, it is not possible on heavy construction of this type (in view of low cement content and low water-cement ratio specified) to put a sufficient amount of ice in the mix to obtain the specified temperatures. As a result, inundation (deluging or overflowing) of aggregates in refrigerated water was developed and was one of the first uses of refrigeration in dams.

When aggregates are cooled by inundation with water, generally the three largest sizes are placed in large cylindrical tanks. Normally two tanks are used for each of the three aggregate sizes to provide backup capacity and a constant flow of materials. The cooling tanks, loaders, unloaders, chutes, screens, and conveyor systems from the tanks into the concrete plant should be enclosed and cooled from 7 to 4°C by refrigeration units, with blowers placed at appropriate points in the housing around the tanks and conveyors.

Peugh and Tyler (House 1949) determined the inundation (or soaking) time required by calculations and actual tests. Pilot tests corroborated their computations of aggregates' cooling times as indicated in Table 1.

This study indicated that an immersion period of about 40 min will bring the aggregate down to an average temperature of 4.4°C. Theoretically, smaller sizes can be brought to the desired temperature in less time. Considerations, such as the rate at which cooling water can be pumped, make it unlikely that a cooling period less than 30 min should be considered. Any excess cooling will provide the needed safety factor. However, the limiting factor on the overall cycle is the cooling time for the largest aggregate, which is nearly 45 min, plus about 15 min for loading and unloading. Backup capacity should be considered for maintaining a constant flow of materials.

Air Blast Cooling

Following the inundation method's development, air-blast cooling was developed. This method of cooling aggregates does not require particular changes to handling the materials or additional tanks for inundation; the aggregate is cooled by blowing cold air through the aggregate in the batching bins above the concrete mixers. Also, the air cycle used in cooling can be used in heating aggregates during cold weather. The aggregate is cooled during the final stage of handling; this does not increase the moisture content.

The compartmented bins where air-blast cooling is usually accomplished are generally sized so that if any supply breakdowns occur, the mixing plant will not have to shut down before a particular pour can be completed. On this basis, the average concrete octagonal bin above the mixing plant on a large job holds at least 460 m³. The size is usually more than adequate to allow time for air cooling. However, certain minimum requirements must be considered. If possible, based on a one-hour loading and cooling schedule, at least 2 h of storage volume should be provided for each size aggregate that air will cool. The minimum volume should be 1.5 h plus cycling time, based on the tables shown for cooling 150-mm aggregate.

The bin compartment analysis shown in Table 2 may be used as a starting point in determining the air refrigeration loads and static pressures. This type of analysis should give approximately equal storage periods for each aggregate size used. In practice, after air cycle cooling is calculated, more air volume is needed in the smaller aggregate compartments, and less in the sand and aggregate, to obtain maximum cooling. This is because of the higher air resistance in the smaller aggregate sections and the fact that air does not cool the sand compartment as effectively.

Table 1 Temperature of Various Size Aggregates Cooled by Inundation

Time, min.	Aggregate Size, mm				
	150	75	40	20	10
1	29.4°C	20.6°C	9.4°C	3.9°C	3.3°C
2	25.0	15.0	5.0	3.3 (5.6)	
5	18.9	7.8 (7.2)	3.3 (5.6)		
10	13.3	4.4 (5.6)			
15	10.0	3.3			
20	7.8				
25	6.7				
30	5.6				
40	4.4				
50	3.9				

Source: Peugh, V.L. and I. Tyler. 1934. "Mathematical theory of cooling concrete aggregates." In House (1949). Numbers in parentheses are from tests by R. McShea.
Note: The temperatures listed are at the center of a cobble with an assumed thermal diffusivity of 1.8 mm²/s. The aggregate initial temperature is 32°C and the cooling water temperature is 1.7°C.

Table 2 Bin Compartment Analysis for Determining Refrigeration Loads and Static Pressures

Material Size, mm	kg/m³	% of Total	No. of Bin Compartments	Bin %
Stones				
150 to 75	480	21.33	4	25.00
75 to 40	420	18.67	3	18.75
40 to 20	390	17.33	2	12.50
20 to 6	360	16.00	3	18.75
Sand	600	26.67	4	25.00
Total	2250	100	16	100

Note: In practice, these relative sizes will vary from time to time; the amounts shown are for an assumed design mix of the principal classes of concrete. On any given job, several design mixes requiring different amounts for each size are needed.

Concrete Dams and Subsurface Soils

Computing Air-Blast Cooling Loads. To calculate the required cooling, several assumptions must be made:

Assumption 1. Normally, the lowest temperature of air leaving the cooling coils is 3 to 4°C. While lower air temperatures may be achieved, these should not be trusted because a temperature lower than 2°C usually causes rapid frosting of the coils.

Assumption 2. The heat transfer between the aggregate and air will be only 80 to 90% effective. To allow for air temperature rise in the ducts, heat leakage, pressure drop, etc., an empirical factor of 85% may be used. Thus, with 26°C aggregate and 4°C cooling air, the effective temperature differential would be 0.85(26 − 4) = 18.7 K.

Assumption 3. The rise in temperature of air passing through the aggregate compartment normally will not exceed 80% of the difference between the entering aggregate and the entering air temperatures. Thus with 26°C aggregate and 4°C air, the temperature rise of the air is 0.80(26 − 4) = 17.6 K. The maximum temperature of the return air is 4 + 17.6 = 21.6°C.

Assumption 4. An allowance should be included for heat leakage into the air ducts on the aggregate bin sides—normally about 2% of a total air-blast cooling load.

Another important consideration is the static pressure against air flowing through a body of aggregates. The resistance to air flowing through a bin varies as the square of the air volume or velocity; this is summarized in Table 3. The resistance pressure listed is for a unit height of aggregate. To use the values shown in Table 3, the cross-sectional area of the aggregate compartment and the height of the aggregate column must be known. The manufacturers of concrete mixing bins and mixing equipment can supply this information.

Table 3 Resistance Pressure

Aggregate Size, mm	Velocity, m/s	Resistance Pressure, Pa/m of Height
150 to 75	1.5	240
75 to 40	1.0	260
40 to 20	0.5	222
20 to 6	0.3	246

Other Cooling Methods

As specifications require lower and lower placing temperatures, direct methods of cooling sand and cement have been initiated to obtain or lower the heat removal required. No method has been proven to cool cement. Sand cooling methods that have been tried and found to be unsuccessful are

1. Water inundation. The increase in free moisture content of the sand when batched makes correctly proportioning the mix difficult.
2. Moving the sand through screw conveyors with hollow flights. Chilled water was pumped through the flights, but because components were cooled below the dew point resulting in condensation, serious handling problems occurred.
3. Vacuum systems, which evaporate the surface moisture to reduce aggregate temperature. The unreliability of the equipment and the batch nature of the process precluded success.

An alternate method of cooling sand, air cooling, is now being tried on some projects, but it is too early for final results of this method.

For small pours, liquid nitrogen is sometimes used to reduce the temperature of the mixture to the final pour temperature. Nitrogen is used because the initial capital costs are considerably lower than a mechanical system, but because of the high cost of manufacturing nitrogen, the operation cost is much greater than a mechanical system. The cost comparison must be done on a case by case basis.

SYSTEM SELECTION PARAMETERS

For larger installations, plant selection depends on a number of factors, including the following:

1. Normal pouring rate, m^3/h (contractor or contract specified).
2. Maximum pouring rate, m^3/h (contractor or contract specified).
3. Total allowable mixing water, kg/m^3 (usually contract specified).
4. Required concrete placement temperature (usually contract specified).
5. Concrete temperature when coming from the mixer (to be determined considering materials handling to placement site, time in transit, and storage at placement site).
6. Average ambient temperature and aggregate temperature during period of maximum placement. The average ambient temperature of the aggregates is assumed to be the mean ambient temperature—including night and day—during the period of storage, which is determined by the amount of storage capacity provided by the contractor and the rate of concrete placement. If the minimum live storage is, for example, 100 Gg of aggregates and the pouring rate is 2000 m^3/day, the consumption would be approximately 4.5 Gg of aggregates per day. If the job is working a 5-day week, this would provide 22 days of storage. Unless weather conditions are unusual, the temperature of the rock delivered to the reclaiming tunnel is assumed to equal the average ambient temperature for the 22 days preceding the delivery.
7. Specific heat of materials. The specific heat of the aggregates and sand may vary with project location. Typical values include: sand = 0.444 kJ/(kg·K); aggregates = 0.50 kJ/(kg·K); water = 4.18 kJ/(kg·K); cement = 0.50 kJ/(kg·K).
8. Heat release rate for cement (material specifications).

Where the aggregate cooling range from initial to final temperature of the mix is relatively small (8 to 11 K), or where the required pour temperature is relatively high (18°C or more), chilled water plus ice in the mix or chilled water plus air blast on larger aggregates may handle the entire cooling load. When the overall temperature reduction is greater than this, or when lower pour temperatures are specified, such as 10°C or less, a combination of all three types of cooling will probably be required because only a limited amount of heat removal can be obtained by one of these methods alone.

Cooling by air blast alone is limited by the entering air temperature, which must be maintained high enough to prevent coil frosting. Cooling by inundation, although it requires large inundation tanks, offers the most positive and sure method of cooling. Ice can also be added to the mix to remove the remainder of the heat. The result is a very satisfactory blending of the aggregates and exact control of the amount of water in the mix.

CONTROL OF SUBSURFACE WATER FLOW

Refrigeration has been used successfully since 1880 to freeze moisture in unstable and water-bearing soil and to stop underground flows of water in pervious material or gravelly stream beds. Other common methods of stopping the flow of water include sheet piling, cement grout, chemicals, and well points. In many cases, freezing has been the last resort after other methods are unsuccessful. In a number of cases, a combination of well points, grouting, and freezing has solved the problem.

Applications include

- Deep excavations for building foundations (to prevent water from seeping into the excavation before the foundation is poured)
- Deep ditches for laying pipe (to keep the banks from sliding in)
- Large dam excavations (to stop water seepage until the dam footings are poured)
- Mining (to temporarily stabilize areas affected by water seepage)

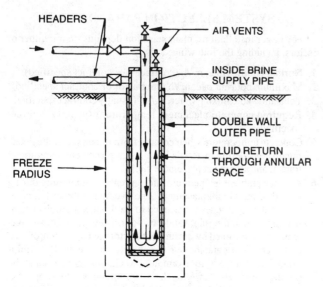

Fig. 2 Typical Freezing Point

Using refrigeration, the common practice is to lay a series of concentric pipes in a line or arch pattern in the path of the subsurface water flow. A wall of earth is frozen by pumping cold brine down the inside pipe and letting it flow back through the annular space between the inside and outside pipe. The growth of frozen soil on the outside of the concentric pipe proceeds until it connects with the frozen cylinder formed on the adjacent pipe.

Spacing between pipes can vary, depending on the time available to complete the wall of ice, but spacings of 600 to 1200 mm on center have been used successfully. Freezing pipes have been successfully used to control subsurface water in both dam and mining projects. Figure 2 shows a typical freezing pipe and system diagram.

The brine or refrigerant and method of containment should be carefully selected in system design. If a leak develops in the system, groundwater contamination could result or the soil saturated with the refrigerant may have to be excavated and cleaned, depending on the type of refrigerant used. Double-wall piping, or an environmentally safe refrigerant that will vaporize when exposed to the air, should be considered.

Ammonia is generally used as the basic refrigerating medium. Then brine, chilled by the ammonia, is circulated through the freezing pipes. The brine is commonly calcium chloride ($CaCl_2$), but magnesium chloride ($MgCl_2$) is recommended because it has less of a tendency to precipitate and clog the piping at low temperatures. To monitor the progress of freezing, thermocouples are normally located throughout the area to be frozen. Brine temperatures may range from −10 to −30°C depending on the state of the freezing and the amount of time available to complete the task.

Liquid nitrogen has been used for small projects.

SOIL STABILIZATION

In northern latitudes where areas of permanently frozen earth, or permafrost, prevail, methods of soil stabilization for building and equipment foundations are required. These methods range from providing a non-frost-susceptible gravel pad to rigid below-grade insulation plus an active or passive refrigeration system to freeze the soil and keep it frozen. Since many of these systems are in remote areas, simplicity and reliability become major factors in system design.

THERMAL DESIGN

Piling Design

Frost heave has long been a problem for the designers of piles and buildings in the arctic or subarctic. Uplift forces as high as 40 Mg/m of perimeter must be taken into account when designing nonthermal piling (Long and Yarmak 1982). Designers have used sleeves, greases, waxes, and plastics to reduce the adfreeze bond in the active layer and lower frost heave forces. Almost all the methods for reducing heave forces have proven to be only temporary (Long and Yarmak 1982). Increased pile embedment remains the only sure method of preventing frost heaving of conventional piling. The use of thermopiles can effectively eliminate frost heaving forces (Long and Yarmak 1982). The thermopile freezes the active layer radially from the pile. As a result, active layer temperatures adjacent to the pile remain at the same temperature as the rest of the pile.

Slab-on-Grade Buildings, Outdoor Slabs, and Equipment Pads

When a slab-on-grade building, outdoor slab, or equipment pad is constructed on a permafrost area, the resulting soil thawing or thaw bulb must be considered. Thermopiles, thermoprobes, or an active refrigerated foundation system may be required to stabilize the structure, slab, or pad.

Design Considerations

Soil properties have a pronounced effect on the capacity requirements for passive or active refrigeration systems. The soil within the radius of influence is an integral part of the refrigeration system. Highly conductive soils will increase the radius of influence of the system and allow more heat to be pumped out of the subsoils. Conversely, poorly conductive soils will cause the radius of influence to be small; the thermal lag through the soils will be high, and the heat transfer rate, low. Soil moisture contents and soil classifications are valuable for estimating thermal conductivities. Backup capabilities for active and passive systems should be considered to avoid foundation failure.

PASSIVE COOLING

The three processes used by passive systems for heat removal are air convection, liquid convection, and two-phase liquid/vapor convection. All passive refrigeration systems rely on the temperature differential between the soil and the winter air to operate. When the temperature of the soil in contact with the refrigeration system is lower than the air temperature, the system is dormant.

Air Convection Systems

Air convection has been used to provide subgrade cooling below on-grade and pile-supported structures. For a passive air convection system to work, two criteria must be met in the design—the air outlet must be higher than the inlet to promote convection, and the air distribution system must be designed so that the friction imposed by the distribution system is low enough to allow convection to begin. Air convection systems are usually designed to take advantage of the prevailing winds at a specific site. The wind can push air through a distribution system at greater velocities than convection would allow; however, the wind in the arctic frequently carries large quantities of snow. Distribution ducts can be blocked by snow and ice causing the system to fail (Long and Yarmak 1982). Ducting fans are built into many air convection systems for active refrigeration backup.

Liquid Convection Systems

Liquid convection has been used to provide subgrade cooling below on-grade structures. Radiator and heat absorber portions of the system are normally connected by either a double or single pipe with a flow splitter to decrease frictional losses in the system and to provide maximum cooling of the working fluid. Examples of some working fluids are trichloroethylene, kerosene, and methanol and water. Frictional losses within the liquid system are high and limit heat transfer rates. Large-diameter pipes may be used to overcome frictional losses within liquid systems so that high heat transfer rates can be achieved. Some liquid convection systems allow an option for mechanical circulation of the working fluid as an active refrigeration backup. Liquid systems must be sealed to avoid leakage into the subsoils. Introducing the working fluid into the permafrost subsoils could depress the soil freezing point and increase the probability of foundation failure.

Two-Phase Systems (Heat Pipes)

Two-phase liquid/vapor convection systems are the most widely used passive refrigeration systems for permafrost foundations and earth stabilization. A typical two-phase unit is constructed of pipe enclosed at both ends and charged with a passive refrigerant gas. The radiator (aboveground condenser) portion of the unit can have a bare or finned surface, depending on heat transfer requirements (Figure 3). The evaporator portion of the unit can have any configuration as long as a slope remains between the evaporator and the radiator (Figure 4).

Refrigeration of the subgrade occurs when the radiator has a lower temperature than the soil in contact with the bottom of the evaporator, where the liquid portion of the refrigerant is pooled (Yarmak and Long 1982). Condensation occurs in the radiator, initiating evaporation of the refrigerant in the evaporator. The condensate wets the walls of the unit and flows down to the evaporator. Reevaporation of refrigerant condensate with subsequent cooling occurs where the soil in contact with the evaporator is warmer than the soil adjacent to the liquid pool of refrigerant at the bottom of the unit. Then the entire evaporator unit is reduced in temperature, cooling the surrounding soil.

A two-phase system will start with a temperature differential of as little as 0.005 K between the radiator and the evaporator. Liquid and air convection systems may require temperature differentials of 2 to 8 K before they start.

Propane, butane, halocarbons, anhydrous ammonia, and carbon dioxide have been used as the refrigerant gas in two-phase systems. Choice of refrigerant gas depends primarily on the allowable internal pressure capabilities of the vessel containing the gas, the quality of available gases, the molecular stability of the gas, and the preference of either the customer or manufacturer of the system. Relatively low-pressure systems using propane, halocarbons, or anhydrous ammonia have been known to gas lock, that is, gases other than the refrigerant gas accumulate in the radiator portion of the unit and prevent the refrigerant gas from condensing. Purging or venting of noncondensable gases may be required on a system following start-up.

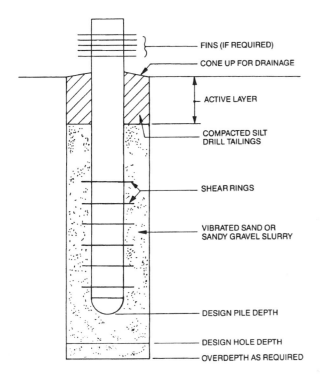

Fig. 3 Thermo Ring Pile Placement

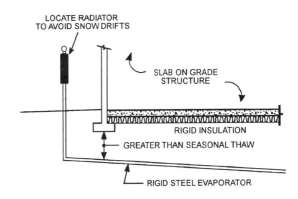

Fig. 4 Typical Thermo-Probe Installation

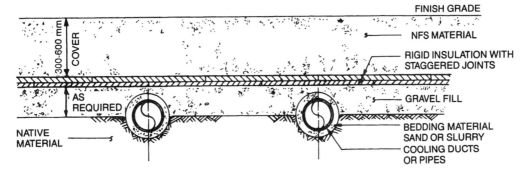

Note: Cooling duct center-to-center distance and pipe size calculated from fluid temperature, soil temperature, time to solidify soil, and annual weather pattern.

Fig. 5 Active Ground Stabilization System

ACTIVE SYSTEMS

Active ground-freezing refrigeration systems are used to keep building or equipment foundations stable when a passive system will not maintain the required stability. The system is normally a network of underground ductwork or piping through which a cooling fluid flows. Heat is removed using a heat exchanger or cooling coil, with refrigeration provided by medium- or low-temperature refrigeration units. System design can include provisions to bypass the refrigeration units during cold weather. Outside air should not be used directly in systems that use air as a cooling fluid due to the moisture introduced into the system in the form of snow and humidity. A defrost system for the cooling coil in a circulating air system should be considered. Figure 5 shows a typical system installation.

REFERENCES

House, R.F. 1949. The inundation method of cooling concrete aggregates at Bull Shoals Dam. USA Corps of Engineers, Little Rock, AR.

Long, E.L. and E. Yarmak, Jr. 1982. Permafrost foundations maintained by passive refrigeration. Petroleum Division, Ocean Engineering Division—ASME (March).

Yarmak, E., Jr. and E.L. Long. 1982. Some considerations regarding the design of two-phase liquid/vapor convection type passive refrigeration systems. Petroleum Division, Ocean Engineering Division—ASME (March).

BIBLIOGRAPHY

Kinley, F.B. 1955. Refrigeration for cooling concrete mix. *Air Conditioning, Heating, and Ventilating* (March):94.

Long, E.L. 1963. The long thermopile, permafrost. Proceedings of International Conference, National Academy of Sciences, Washington, D.C., 487-91.

Townsend, C.L. Control of cracking in mass concrete structures. *Engineering Monograph* No. 34. U.S. Department of Interior, Bureau of Reclamation.

CHAPTER 36

REFRIGERATION IN THE CHEMICAL INDUSTRY

Flow Sheets and Specifications 36.1	*Equipment Characteristics* 36.3
Refrigeration—Service	*Start-Up and Shutdown* 36.4
or Utility 36.1	*Refrigerants* 36.5
Load Characteristics 36.2	*Refrigeration Systems* 36.5
Safety Requirements 36.2	*Refrigeration Equipment* 36.6

CHEMICAL industry refrigeration systems range in capacity from a few kilowatts of refrigeration to thousands of kilowatts. Temperature levels range from those associated with chilled water through the cryogenic range. The degree of sophistication and interrelation with the chemical process varies from that associated with comfort air conditioning of laboratories or offices to that where the production of reliable refrigeration is vital to product quality or to the safety of the operation.

However, two significant characteristics identify most chemical industry refrigeration systems. Almost exclusively, they are engineered one-of-a-kind systems, and equipment used for normal commercial application may be unacceptable for chemical plant service.

This chapter gives guidance to refrigeration engineers in working with chemical plant designers so they can design an optimum refrigeration system. Refrigeration engineers must be familiar with the chemical process for which the refrigeration facilities are being designed. An understanding of the overall process is also desirable. Computer programs are also available that can calculate cooling loads based on the gas chromatographic analysis of a process fluid. These programs accurately define not only the thermodynamic performance of the fluid to be chilled, but also the required heat transfer characteristics of the chiller.

Occasionally, because the process is secret, refrigeration engineers may have limited access to process information. In such cases, chemical plant design engineers must be aware of the restrictions this may place on providing a satisfactory refrigeration system.

FLOW SHEETS AND SPECIFICATIONS

The starting point in attaining a sound knowledge of the chemical process is the flow sheet. Flow sheets serve as a road map to the unit being designed. They include such information as heat and material balances around the major system components and pressures, temperatures, and composition of the various streams within the system. Flow sheets also include refrigeration loads, the temperature level at which refrigeration is to be provided, and the manner in which refrigeration is to be provided to the process (such as via a primary refrigerant or via a secondary coolant). They indicate the nature of the chemicals and processes to be anticipated in the vicinity in which the refrigeration system is to be installed. This information should indicate the need for special safety considerations in the design of the refrigeration system or for construction materials that resist corrosion by process materials or process fumes.

Different portions of a process flow sheet may be developed by different process engineers; consequently, the temperature levels at which refrigeration is specified may vary by only a few degrees. A study of such a situation might reveal that a single temperature is satisfactory for several or even all of the users of refrigeration, which could reduce project cost by eliminating multilevel refrigeration facilities.

Most process flow sheets indicate the design maximum refrigeration load required. The refrigeration engineer should also know the minimum design load. Process loads in the chemical industry tend to fluctuate through a wide range, creating potential operational problems.

Flow sheets also indicate the significance of the refrigeration system to the overall process and the desirability of providing redundant systems, interlocking systems, and so forth. In some cases, refrigeration is mandatory to ensure safe control of a process chemical reaction or to achieve satisfactory product quality control. In other cases, loss or malfunction of the refrigeration system has much less significance.

Other sources of information are also valuable. A properly prepared set of specifications and process data expands on flow sheet information. These generally cover the proposed process design in much more detail than the flow sheets and may also detail the mechanical systems. Information regarding the design principles, including continuity of operation, safety hazards, degree of automation, and special start-up requirements, is generally found within the specifications. Equipment capacities, design pressures and temperatures, and materials of construction may be included. Specifications for piping, insulation, instrumentation, electrical, pressure vessels and heat exchangers, painting, and so forth are normally issued as part of the design package available to a refrigeration engineer.

It is imperative that refrigeration engineers establish effective communication with chemical process engineers. The refrigeration engineer must know what information to request and what information to give to the chemical process engineer for design optimization. The following sections outline some of the significant characteristics of chemical industry refrigeration systems. A full understanding of these peculiarities is of value in achieving effective communication with chemical plant designers.

REFRIGERATION—SERVICE OR UTILITY

Refrigeration engineers unfamiliar with the chemical industry must understand that unless the chemical process is cryogenic in nature, chemical plant designers probably consider refrigeration merely as a service or utility of the same nature as steam, cooling water, compressed air, and the like. Chemical engineers expect the reliability of the refrigeration to be of the same quality as other services. When a steam valve is opened, chemical engineers expect steam to be available instantly, in whatever quantities demanded. When steam is no longer required, the engineer expects to be able to shut off the steam supply at any time without adversely affecting any other steam user or the source of steam. The same response from the refrigeration system will be expected. This high degree of reliability is usually so strongly implied that no specific mention of it may be made in specifications.

Because refrigeration is frequently considered a service, process designers spend insufficient time analyzing temperature levels, potential load combinations, energy recovery potentials, and the like. The potential for minimizing the size of the refrigeration system, the total plant investment, or both, by providing refrigeration at

---

The preparation of this chapter is assigned to TC 10.1, Custom Engineered Refrigeration Systems. This chapter last received a major revision in 1971.

a minimum number of temperature levels, is frequently not investigated by process engineers. Likewise, the potential for power recovery is frequently overlooked.

Part of the reason for this attitude is that refrigeration facilities represent only a minor part of the total plant investment. The entire utilities installation for the chemical industry usually falls in the range of 5 to 15% of the total plant investment, with the refrigeration system only a small portion of the utilities investment. Process requirements may be overruling, but process engineers must recognize legitimate process necessities and avoid unnecessary and costly restrictions on the refrigeration system design.

LOAD CHARACTERISTICS

Flow sheet values generally indicate direct process refrigeration requirements and do not include heat gains from the equipment or piping. Flow sheet peaks and average loads generally do not allow for unusual start-up conditions or off-normal process operation that may impose unusual refrigeration loads. This information must be gained by a thorough understanding of the process and by discussing the potential effect on refrigeration system design for off-normal process conditions with the process engineer.

Once the true peak loads are established, the duration and frequency of the peaks must be considered. For some simple processes this information is fairly straightforward; however, if the plant is designed for both batch operation and multiproduct manufacture, this can be an enormous task. Computer simulation of such a combination of processes has led to optimization of not only the refrigeration equipment but also the process equipment. Computer simulation assures that the refrigeration machine, secondary coolant storage tanks, and circulating pumps are properly sized to handle both average and peak loads with a minimum investment. Few applications require computer simulation, but a thorough understanding of the relationship of peak loads to average loads and their influence on refrigeration system component sizing is vital to good design.

Sometimes unusually light load conditions must be met. If the process is cyclic and an on-off operation is undesirable, the refrigeration system may run for significant periods under a very light load or even a no-load condition. Light loads often require special design of the system controls, such as multistep unloading and hot gas bypass with reciprocating compressors, a combination of suction throttling and hot gas bypass with centrifugal compressors, and slide valve unloading and hot gas bypass with screw compressors. A secondary coolant system might require a bypass arrangement.

The investment in refrigeration equipment can be kept to a minimum when only a few levels of refrigeration are required. Checking the specified temperatures to be sure they are based on some process requirement may show that fewer temperature levels than shown on the flow sheet are necessary. If multiple levels of refrigeration are required, a compound system should be evaluated. The evaluation must consider the limited ability of a compound system to provide the precise temperatures required of some processes.

Production Philosophy

The flow sheet and specifications generally indicate whether a chemical process is in continuous or batch operation, but further research may be required to understand the required continuity of service for the refrigeration facilities. The chemical industry frequently requires a high degree of continuity; plant production rates are often based on 8000 h or more per year. In general, the refrigeration equipment is worked extremely hard, with no off-peak period because of seasonal changes, and any unscheduled interruption of refrigeration service may create large production losses. In most cases, scheduled maintenance shutdowns are not only infrequent but also highly vulnerable to cancellation or delays because of the press of production requirements.

As a result, reliability is key in the design of chemical industry systems. Equipment that is satisfactory in commercial or light industrial service is frequently unfit where high service rates and minimum availability for maintenance are the rule. In some cases, duplicate systems are justified. More often, multiple part-capacity units are installed so that a refrigeration system breakdown will not create a total process production loss. Major equipment and hardware items that require minimal maintenance or that permit maintenance with the refrigeration system in operation should be selected (for instance, dual lube oil filters or a bypass line that permits temporary reversal of condenser water flow for cleaning). Particular attention should be paid to equipment layout, so that adequate access, tube pullout space, and laydown space are available to minimize refrigeration system maintenance time. In some cases, overhead steel supports for rigging heavy equipment, or permanent monorails, are justifiable.

Flexibility Requirements

The chemical industry constantly develops new processes; consequently, the usual chemical plant undergoes constant modification. On occasion, total processes are rendered obsolete and scrapped before design production rates are ever reached. Thus flexibility should be designed into the refrigeration system so it may be adaptable to some process modification. Designing for optimum flexibility is difficult; however, a study of the potentials or probabilities of process modifications and of the expected life of the process facility will help in making design decisions.

SAFETY REQUIREMENTS

Most chemical processes require special design to ensure safe operation. Many raw materials, intermediates, or finished products are themselves corrosive or toxic or are potential fire or explosion hazards. Frequently, the chemical reactions involved in the process generate extremes of pressure or temperature that must be properly contained for safe operation. Refrigeration engineers must be aware of these potential hazards as well as any abnormal hazards that may develop during start-up, unscheduled shutdown, or other upset within the chemical process. When designing modifications or expansions for an existing facility, the possibility that certain construction or maintenance techniques may be safety hazards must be considered.

Corrosion

The shell, tubes, tube sheets, gaskets, packing, O rings, seal materials, and components of instrument or control hardware must be properly specified. The potential hazards of leakage between the normal process side and the refrigeration side of heat exchange equipment must be investigated, because an undesirable chemical reaction may occur between the process material and the refrigerant.

An additional corrosive hazard may result from leaks, spills, or upsets within the process area. Safe chemical plant design must anticipate the unusual as well as the usual hazards. For example, if refrigerant piping will run adjacent to a flanged piping system containing a highly corrosive material, special materials for the piping and insulation systems may be justified.

Toxicity

If the refrigeration system indirectly contacts a toxic material via heat exchange equipment, flanges and such elements as gaskets, packing, and seals in direct contact with the toxic material must be designed carefully. The possibility of a leak in equipment that might allow refrigerants or secondary coolants to mix with process chemicals and cause a toxic or otherwise dangerous reaction must also be considered. In some cases, the potential for toxic leaks may be so high that a special ventilation system may be required.

Refrigeration in the Chemical Industry

Even though containment and ventilation can handle toxic materials under normal process conditions, toxic material may need to be vented in abnormal situations to avoid the hazards of fire or explosion. In such cases, the toxic material is frequently vented or diluted through a tall stack so that ground or operating level concentrations do not reach toxic limits. Refrigeration engineers should evaluate the desirability of locating the refrigeration equipment itself or its controls outside the operating areas. An alternate solution is to ventilate either the refrigeration system or its controls with a system that has a remote air intake.

The hazards of toxicity are not confined to the chemical process itself. Some common refrigeration chemicals are toxic in varying degrees. Because of the chemical industry's interest in safety, refrigeration engineers are required to treat some of the common refrigerants and secondary coolants with much more caution than required for the usual commercial or industrial system.

Fire and Explosion

Plant specifications normally define the area classifications, which determine the need for special enclosures for electrical equipment. For areas designated as being in an explosion-proof environment, standard components are normally grouped in a single large explosion-proof cabinet. Intrinsically safe control systems, which eliminate the need for explosion-proof enclosures at all end devices, are also used. Intrinsically safe shutdown systems eliminate arcing at the end device by using low-voltage signals controlled by a microprocessor. Control equipment can also be mounted in standard or weatherproof cabinets equipped with an inert gas purge system.

Items other than the electrical system may require special consideration in areas of high fire or explosion hazard. The use of flammable materials should be carefully reviewed. Stainless steel instrument tubing should be chosen rather than unprotected plastic tubing, for example. Both insulation materials and insulation finish systems should minimize flame spread in the event of a fire. In some extremely hazardous areas, nonflammable refrigerants or secondary coolants may be required, rather than flammable materials that would provide otherwise superior performance.

In areas of high explosion potential, modification of the usual refrigeration system design may be required. Design pressures for refrigeration vessels or piping may be determined by process considerations as well as refrigeration system requirements. Special pressure relief systems, such as rupture disk in series with relief valves, dual full-sized relief valves, or a parallel relief valve and rupture disc with transfer valves between them, are often required.

Refrigeration System Malfunction

A malfunction can itself be a significant safety hazard, whether the upset is caused by an internal failure of the system or by fire, explosion, or some other catastrophe. Most process areas designated as being in a hazardous environment are protected by automatic gas and flame detection systems that shut down the refrigeration system when explosive mixtures or fire are detected in an area. Loss of refrigeration may permit a process reaction to run out of control and cause loss of product, fire, explosion, or the release of toxic material in an area remote from the original source of trouble. Thus, one or several degrees of redundancy may be needed to minimize the consequence of refrigeration system malfunction.

Storage of cold coolant, ice, or cold eutectic, or an alternate emergency supply of cooling water may be necessary to meet emergency peaks. Alternate sources of electric power to the refrigeration system may be desirable. Dual drive capability by either electric motor or steam turbine might even be justified. Uninterruptible power systems may also be used for control systems.

Frequently, special facilities are used to protect against the extreme consequences of refrigeration system malfunction. One example is the quenching of the process reaction via an inhibiting or neutralizing chemical introduced to the process in the event of an unusual pressure or temperature rise. Another simpler and more common example requires closing one or more process valves in the event of a loss of refrigeration.

Maintenance

In an operating chemical plant, because of hazards frequently encountered, the normal maintenance procedures may not be permitted. Because welding, burning, or the use of an open flame is often prohibited throughout large areas of a chemical plant, maintenance flanges or screwed connections are used to permit replacement of piping and equipment. Sometimes extra access space or handling facilities (such as monorails) are provided to permit efficient removal of machinery to an area where welding, burning, and the like, is permitted.

EQUIPMENT CHARACTERISTICS

Automation

In chemical plant operations, instrumentation represents a significant percentage of the plant investment. For this reason, most chemical plant designers insist on standardization of instrumentation throughout the plant. These requirements may not include familiar refrigeration components and may create problems in the refrigeration design. Therefore, it is vital that the refrigeration engineer and the chemical or instrumentation engineer must agree on the instrument requirements early in the design phase.

A second concept frequently adopted is the control and monitoring of all plant operation from a single central control room. It is not unusual to operate a multimillion dollar processing facility with two central control room operators and perhaps a single roving operator. This concept influences the refrigeration system in several ways. Refrigeration system controllers and alarm and shutdown lights are mounted on the central control room (CCR) panel, as are recorders or indicators that display refrigeration system temperatures, pressures, flows, and so forth.

Even with central control room operation, a local panel is required for the display or recording of additional information that can aid in troubleshooting an emergency shutdown. Frequently, start-up control is available only at this local panel, to assure that start-up is not attempted unless an operator is present to witness the operation of major items of refrigeration equipment. The CCR-mounted hardware would certainly conform to the standards of the process instrumentation to minimize operator confusion either in reading informative devices or in operating control devices. Most plants now use centralized computer or microprocessor controls. In some cases, the refrigeration unit is controlled from the main computer or microprocessor.

When applying a CCR concept, it is important to determine exactly how much information and control are to be provided at the control room and how much are to be provided locally. Transmission of unnecessary information to the CCR can be costly, but, if sufficient information is not available, serious process upsets are inevitable. Most process operators are not trained to understand the intricacies of refrigeration machinery. The CCR system must permit monitoring of the refrigeration system performance and control of that performance to suit process needs.

Operators require alarms to indicate abnormal conditions for which they can make corrections, either at the CCR or in the field, and to indicate a system breakdown or shutdown. They also require a locally mounted manual shutdown station in the event of an emergency. Devices such as sequencing alarms and lube oil or bearing temperature recorders, which are either troubleshooting aids or can be checked or logged by the roving operator, are best installed locally.

The concept of designing for a minimum of operator attention with personnel that are not refrigeration specialists is yet another reason that a high degree of reliability is required for a chemical industry refrigeration system.

Outdoor Construction

Another chemical industry characteristic is outdoor construction. The chemical industry installs sophisticated process and auxiliary facilities outdoors all over the world. Whether the problems are those imposed by low temperatures, heavy snows and freezing rain, dust storms, baking heat, or hurricane force winds with salt-laden rains, they are generally unfamiliar to the uninitiated refrigeration engineer. For example, explosion-proof electrical construction is not necessarily weather resistant. Lube oil heaters and a prestart-up circulation of heated lube oil may be required for compressors or other rotating machinery. In areas with high winds, special attention must be paid to the detailed installation instructions for insulation jacketing applied to pipe or vessels.

Winter operation of cooling towers may require multiple-cell construction, two-speed or reverse rotation cooling-tower fans, or even facilities for steam heating the cooling-tower basin. Instrument air for transmission of signals or power to pneumatic operators should be dried to a dew point (under pressure conditions) lower than anticipated ambient conditions to avoid condensate or ice from forming in the instrument air lines or in the instruments themselves. In fact, it has become almost standard that the instrument air provided is both oil-free and dried to a low dew point. The effect of ambient temperature, as well as radiation from the sun, should be considered in determining system design pressures, especially when equipment may be idle. Ambient temperatures up to 50°C and vessel skin temperatures of 75°C can be experienced in hot climates. To determine whether or not purchased equipment meets the requirements for outdoor installation, a detailed check of vendors' drawings, specifications, descriptive literature, and vendor-procured components is required.

Energy Recovery

Both the installed cost and the operating costs for the refrigeration system of some chemical processes can be reduced by the intelligent use of energy recovery techniques. If the process requires large quantities of low-pressure steam, the use of back pressure turbines to drive the refrigeration compressors could significantly reduce refrigeration system energy costs. On the other hand, if the process generates an excess of low-pressure steam, an absorption system may provide an overall saving. For processes with an excess of low-pressure steam only during the summer months when heating requirements are at a minimum, a condensing turbine may be economical if it is sized to operate at the low steam pressure when the excess is available and at a higher pressure during the heating season.

Other energy imbalances occurring within a chemical process can be advantageous. A waste heat boiler installed in a high-temperature gas stream may provide a source for low-cost steam. If the gas stream is at a moderate pressure as well as a high temperature, the possibility of a gas-driven power recovery turbine should be considered.

Another means by which operating costs frequently can be reduced is the reuse of once-through cooling water. Frequently, turbine-driven centrifugal refrigeration machines can use cooling water from the refrigerant condenser to condense the turbine exhaust steam, either in a shell-and-tube or a low-level jet condenser. Another possibility is the reuse of refrigeration system cooling water in process heat exchangers. Again, a thorough understanding of the process is a prerequisite for understanding the energy recovery potential, and a flow sheet should be helpful.

Performance Testing

Frequently, a requirement for performance testing of the refrigeration facilities is included within the contract for a chemical industry processing unit. Agreement should be reached as early as possible between the owner and the contractor regarding the exact procedure to be used for testing. If the test is to be run at some condition other than design conditions, both parties must agree on the methods of converting the test results to design conditions. Approximation techniques, such as those outlined in Air-Conditioning and Refrigeration Institute (ARI) *Standard* 550, are usually unacceptable in the chemical industry. The refrigeration engineer must be assured that adequate facilities for an equitable test are designed into the refrigeration system. This may require additional flow-metering devices or more accurate temperature-measuring devices than are required for normal plant operation.

Insulation Requirements

The service conditions imposed by the chemical industry on both piping and equipment insulation are frequently more exacting than those experienced in the usual commercial or industrial installation. Not only must the initial integrity be as near perfect as possible, but it must also resist the high degree of both physical and chemical abuse that it is likely to incur during its lifetime. To achieve both a minimum permeability and a maximum resistance to abuse, multi-component finish systems may be required. In some cases, a vapor barrier mastic coating system (which may include reinforcing cloth) is covered with aluminum, stainless steel, or an epoxy-coated carbon steel jacket to protect against physical and chemical abuse. Piping and equipment insulated under ideal shop working conditions must be designed to withstand loading and unloading and erection into position on the job site without damaging the vapor barrier.

Because the fire hazard is generally high and the potential loss of personnel and investment resulting from a fire is prohibitive, a strict limit is usually placed on insulation finish systems having a high flame-spread rating, particularly in indoor construction.

The frequent use of stainless steel in the chemical industry for piping and equipment creates another problem with regard to insulation system design. Many stainless steels fail when they are exposed to chlorides. Stress corrosion cracking can occur in a matter of hours. Consequently, chloride-bearing insulation materials must not contact stainless steels even in minute quantities and should not be used anywhere in the system unless a valid vapor retarder is interposed. Chapter 23 of the 1997 *ASHRAE Handbook—Fundamentals* covers the general subject of thermal insulation and water vapor barriers.

Design Standards and Codes

Relatively few suppliers of refrigeration equipment regularly manufacture to meet codes or standards that apply to the chemical industry. Another variation from commercial or industrial design practice is the use of company standards. For the usual commercial or industrial plant, the client will at best provide performance specifications and a statement of what is to be done, leaving the preparation of detailed specifications for equipment, piping, ducting, insulation, and painting to the designer. However, many such items are covered by company standards, which, though established primarily for use in the process cycles, can yield corporate benefits if they are also used in design of the refrigeration systems. A request for all applicable company standards at the start of the design and an effort to use them will avoid costly rework of the design following a review by the client.

START-UP AND SHUTDOWN

Processes are most hazardous during start-up and shutdown. Though present in batch or discontinuous processing units, the

Refrigeration in the Chemical Industry

problem is usually more severe in continuously operating units for the following reasons:

1. Instrumentation and control must be designed for the normal condition, and the cost of features intended for use only during start-up or shutdown often cannot be justified. Frequently, conditions at start-up or shutdown fall outside the range of the operating instruments and control, so that manual control is necessary.
2. The same argument holds for much of the process equipment, so that extraordinary measures, such as severely throttled flow, minimum-flow bypasses, and recycling may be needed.
3. The operators go through these conditions only infrequently and may have forgotten the techniques of operation at the time they are most needed.
4. The process conditions at start-up and shutdown are usually not recorded on flow sheets or in descriptions because they occur so infrequently and usually vary continually as the units are brought on and off stream. As a consequence, designers have a tendency to overlook them and to concentrate on the conditions in the operating range.

Refrigeration engineers must inquire whether start-up or shutdown is likely to impose any special conditions on the refrigeration system. Start-up and shutdown is a special burden in this respect, since operators are particularly busy with the processing equipment and cycle during these times and generally cannot monitor or adjust the operation of what they regard as a service system. Therefore, process engineers must be made thoroughly aware of the precise limitations that start-up or shutdown of the refrigeration system may impose on process operation.

REFRIGERANTS

Such factors as flammability, toxicity, and compatibility with proposed construction materials may influence the final selection of a refrigerant more than in other applications. Special attention should be paid to the consequences of leakage between the process materials and the refrigerant or the secondary coolant. Chapters 18 and 20 of the 1997 *ASHRAE Handbook—Fundamentals* discuss refrigerants and secondary coolants in detail.

Because of their nontoxic, nonflammable properties, halogenated hydrocarbons have been used predominantly, but recent environmental concerns have reduced their application throughout the chemical industry. Hydrocarbons such as methane, propane, ethane, propylene, ethylene, or ammonia are used in many cases where the process stream involves them as constituents. In the petrochemical industry, the use of these materials within the process is common.

Of the secondary coolants, calcium and sodium chloride brines have been used most often, although glycols and such halocarbons as methylene chloride, trichloroethylene, R-11, and R-12 also have been frequent choices. Again, environmental concerns predicate against the use of R-11 and R-12 in new facilities.

Many of the same factors that influence refrigerant selection must be considered in choosing a secondary coolant. Corrosivity, toxicity, and stability are of special significance in determining suitability for chemical plant service.

REFRIGERATION SYSTEMS

An indirect system, in which brine or chilled water is circulated to air washers, cooling coils, and process heat exchangers from a central refrigeration plant, is much more prevalent in the chemical industry than in the food industry or in residential or light commercial comfort air conditioning. This is particularly true where large capacities or low temperature levels are involved. An indirect system permits centralization of the refrigeration equipment and associated auxiliaries, which may offer significant advantages in operation and maintenance, particularly if remote location of the refrigeration equipment permits design, operation, and maintenance in a nonhazardous location. It also may permit the installation of a minimum number of large units rather than many small units located in remote areas. For low-temperature systems of significant capacity, an indirect brine cooling system installed in the process area close to the process users is common.

Where the number of process heat exchangers requiring cooling and the length of piping can be kept to a minimum, a direct system, which uses the refrigerant in the process heat exchange equipment, often is the optimum design, particularly for small or medium loads. Since an indirect heat exchanger is not required in this case, a higher operating suction pressure and consequent lower operating and investment costs may be possible. Direct systems are also used when a refrigerant is involved in the manufacturing process stream, as in the production of ammonia or many petrochemicals. Here, the length of refrigerant lines, with possible high refrigerant losses because of leakage, is a less significant factor in system selection. Direct systems are usually of the dry expansion or flooded evaporator design; flooded coil systems of the gravity feed or pumped liquid overfeed design are relatively uncommon.

Direct systems for larger capacity, low-temperature, multiple user service have advantages and disadvantages that must be considered for proper system selection. Some of the disadvantages are the following:

1. Maintaining an extensive refrigeration piping system free of leaks is difficult. Leakage from piping for secondary coolants is frequently less objectionable than the refrigerant gases. Checking for refrigerant leaks or repairing them in certain high explosion hazard process areas can be a problem because halide and electronic leak detectors may not be permitted and burning or welding may not be possible without a plant shutdown. If air or moisture leaks into a system operating at vacuum conditions, extensive icing and corrosion problems can result.
2. Higher piping costs are often involved when all items are considered, including large and expensive vapor and liquid-control valves at individual heat exchangers, particularly for the halocarbon refrigerants. Generous refrigerant knockout separators are necessary at each stage. Although both refrigerant and secondary coolant lines require insulation to prevent capacity losses, sweating, and icing, refrigerant lines, especially on the vapor return to the compressor, can greatly exceed the size of those required for brine recirculation.
3. No system reserve capacity is available as is the case with a secondary coolant, particularly if the latter is designed as a storage system. Process upsets can directly and suddenly increase the load on the refrigeration unit, causing cycling of the equipment and possible damage. For some processes, meeting short, sharp load peaks is of paramount importance to avoid off-standard production or unsafe operating conditions.
4. Constant temperature control is often more difficult or more costly to maintain with direct refrigeration than with a secondary coolant.
5. Initial testing of an extensive direct refrigeration system may be a significant problem. Testing must be done pneumatically rather than hydrostatically to prevent problems associated with water left in the refrigerant system. Pneumatic testing is a hazardous operation and is forbidden in some chemical plants. The alternative of extensive posttesting dehydration is usually both expensive and time consuming.
6. The initial cost for refrigerants is usually much higher in an extensive direct refrigeration system than in a secondary coolant system operating at temperatures most frequently encountered in the chemical industry. In the case of system leaks, the costs of makeup coolant are generally less than the costs of makeup primary refrigerant.

Some of the advantages offered by direct refrigeration systems include the following:

1. Careful control of corrosion inhibitors may be necessary to keep secondary coolants stable so that they do not cause extensive equipment damage.
2. Less equipment and maintenance may be required; secondary coolant circulation and control or coolant mixing and makeup facilities are not needed.
3. Power costs are generally lower because of higher suction pressures and, in some designs, because pumps are not required.
4. Damage because of equipment freezing is not likely. Such damage can occur in a secondary coolant system if the coolant condition or the refrigeration plant is not properly operated.

Thus, the broad scope of refrigeration applications within the chemical industry permits the use of virtually any refrigeration system under the proper process conditions.

REFRIGERATION EQUIPMENT

For the most part, the refrigeration equipment used in the chemical industry is identical to, or closely parallels, the equipment used in other industries. The chemical industry is unique, however, in the wide variety of applications, the large temperature ranges covered by these applications, the diversity of equipment usage, and the variation of mechanical specifications required. Where possible, the chemical industry uses standard equipment, but this is frequently impossible because of the particularly rigorous demands of chemical plant service. Therefore, this section only briefly describes the application and modification of refrigeration equipment for chemical plant service.

Compressors

Refrigeration engineers may find difficulty in applying conventional refrigeration compressors to chemical plant service. Most process engineers are familiar with heavy duty, forged steel, high pressure, single or double throw reciprocating gas compressors; they are uncomfortable with the high speed, cast iron, or steel compressors that are standard to the refrigeration industry. Another difference between commercial and chemical plant usage is the greater use of open-drive equipment in the chemical plant.

The large capacities and low temperatures frequently encountered in chemical plant duty have led to wide use of either centrifugal compressors or high capacity rotary or screw compressors. These large machines vary from standard commercial equipment principally in the amount and complexity of controls or other auxiliaries provided. Load control devices such as multistep unloaders or hot-gas bypass systems are often required to permit a compressor turndown to 10% of full load or, in some cases, to permit no-load operation without either compressor surge problems or on-off operation. Most systems with large multistage centrifugal compressors use economizers to minimize power and suction volume requirements. Compressor lube oil systems are often provided with auxiliary oil pumps, dual oil filters, dual oil coolers, and the like, to permit routine maintenance without shutdown and to minimize shutdown frequency. Compressor control and alarm systems are frequently tied into central control room panel boards to permit monitoring and/or control of compressors.

Compressors for hydrocarbon gas refrigerants find their greatest use within the chemical industry, particularly in the field of petrochemicals. The relatively low cost and ready availability of pure hydrocarbons and hydrocarbon mixtures frequently dictate their use. Many offer the additional advantage of positive pressure operation throughout the entire refrigeration cycle.

Since refrigeration systems within the chemical industry are often required to operate for a year or more without shutdown, standby compression equipment is frequently installed. Even the larger refrigeration loads sometimes require 100% standby protection. Special controls may be required to provide rapid and automatic start-up of the standby equipment. The main drive is commonly an electric motor and the standby drive may be either a steam turbine or an internal combustion engine. Provisions must be made via nonelectric drivers or emergency generating equipment to keep all necessary auxiliaries and controls operative during the electrical outage. Oversized crankcase heaters may be required, as well as electric or steam tracing of various lubricant system components.

High in-service requirements, plant standardization, explosion hazards, and corrosive atmospheres all require special controls. Often, the copper instrument tubing normally used on commercial equipment must be replaced with steel or stainless steel tubing more suitable to the proposed plant atmosphere. Lubricant piping must be stainless steel pipe with nonferrous valves, coolers, and filters. This requirement is primarily to minimize expensive delays in initial plant start-up that may result when rust or scale within lubricant systems causes damage to bearings or seals in high-speed centrifugal or screw compressors.

Absorption Equipment

Chapter 1 of the 1997 *ASHRAE Handbook—Fundamentals* and Chapter 41 of this volume discuss absorption equipment in detail. Absorption equipment has seen little use in chemical plants, even though plant waste heat is available to operate it.

By special design and reselection of materials, hot streams of many fluids can be used as the energy source instead of using hot water or steam. Direct fired units are available. Hot condensable vapors can also be used as the energy source.

Commercial absorption machine equipment must be modified to permit outdoor operation. Manufacturers' recommendations should be followed regarding changes necessary to prevent freezing on the water side and solution crystallization on the absorption side of the equipment, particularly during shutdown.

Condensers

Water-Cooled Condensers. Units for chemical plant service require relatively minor design changes from those provided for commercial installations. Since cooling water is frequently of low quality, special materials of construction may be required throughout the tube side. An example is the necessity to switch from copper to a cupronickel when cooling water comes from a tidal source that is high in chlorides. If the cooling water is high in mud or silt content, it is sometimes justifiable to install piping and valving that will permit backflushing the condenser without requiring a refrigeration machine shutdown. Chemical plant requirements normally dictate shell-and-tube condensers of the replaceable tube type. Process engineers may insist on conservative tube-side velocities (2.4 m/s or less as a maximum for copper tubes) and replaceable tube bundles. The long hours of operation without opportunity for cleaning and the types of cooling water used frequently require that a higher waterside fouling factor be assumed than on commercial installations.

Air-Cooled Condensers. With increasing restrictions on the use of water for condensing, air-cooled condensing systems have been used in several instances, even in larger centrifugal-type plants. These have usually been installed in warmer locations where the increase in condensing pressure (temperature) over that from the use of cooling tower water or once-through water systems is minimal.

Air-cooled condensers for chemical plant service are normally fabricated to one of the API standards for forced-convection coolers. Care must be taken when specifying these coolers so that the manufacturer understands the type of duty associated with a condensing refrigerant. The service required of an air-cooled condenser in a chemical plant atmosphere dictates either the use of more expensive alloys in the tube construction or conventional materials of greater wall thickness, to give acceptable service life. Air coolers

Refrigeration in the Chemical Industry

may be more difficult to locate because recirculation of hot discharge air or fouling by hot process exhaust gases must be avoided.

Evaporative Condensers. Evaporative condensers, particularly for smaller refrigerating loads, are used extensively in the chemical industry, and they should become more prevalent as more emphasis is placed on the reduction of thermal contamination of rivers, lakes, and streams. In a few larger installations, the combination of an air cooler and an evaporative condenser operating in series satisfies the condensing requirements.

In most cases, the commercial evaporative condenser is totally unsuitable for chemical plant service, but satisfactory results can be obtained if this equipment is carefully specified. The major items of concern are the atmospheric conditions to which such equipment may be exposed and the long in-service requirements of the chemical plant. The chemical plant atmosphere, which may abound in vapors or dusts that are corrosive in themselves, can be an even more serious problem when these vapors and dusts are passed over surfaces that are constantly being wetted. Another problem is that dusts from nearby raw material storages or grinding operations may infiltrate the water recirculating system and plug the spray nozzles. The problems of water treatment and winter freeze protection are usually much more severe in chemical plant service because of the lower quality water that is frequently available and the demand for both year-round operation and a high turndown ratio. Light load operation in freezing weather calls for extreme care in design to avoid freezeup.

Two other areas of commercial evaporative condenser design that must frequently be strengthened for chemical plant duty are the electrical equipment, which must be satisfactory for the plant environment, and the fans, dampers, and recirculating pumps, which must be suitable for long-life low-maintenance service.

Evaporators

The general familiarity of chemical plant design personnel with heat exchanger design and application may sometimes lead them to suggest that refrigeration evaporators for the chemical plant should be designed similarly to evaporators in nonrefrigeration service. While the general laws of heat transfer apply in either case, there are special requirements for evaporators in refrigeration service which are not always present in other types of heat exchanger design. Refrigeration engineers must coordinate the process engineers' experience with the special requirements of a refrigeration evaporator. To do so, the standards of the Tubular Exchanger Manufacturers Association (TEMA) should be consulted to ensure that the end product is familiar to the plant engineer while still performing efficiently as a refrigeration chiller.

Paramount in these special requirements are the proper treatment of oil circulation in the refrigeration evaporator and proper evaluation of liquid submergence as it may affect low-temperature evaporator performance. When the evaporator in chemical plant service is being used with reciprocating and rotary screw compression equipment, continuous oil return from the evaporator must normally be provided. If continuous oil return is not possible, an adequate oil reservoir for the compression equipment, with periodic transfer of oil from the low side of the system, may be needed. On evaporators used with centrifugal compression equipment, continuous oil return from the evaporators is not necessary. In general, centrifugal compressors pump very little oil, so oil contamination of the low side of the system is not as serious as with positive displacement equipment. However, even with centrifugal equipment, the low side evaporators eventually become contaminated with oil, which must be removed. Most centrifugal systems operate for several years before oil accumulation in the evaporator adversely affects evaporator performance. Newer tube surfaces with porous coatings may be more sensitive to the presence of oil in the refrigerant than would be conventional finned surfaces. For newer surfaces, a continuous oil return system may be essential for centrifugal systems.

Flooded shell refrigeration evaporators operating at extremely low temperatures and low suction pressures may build up an excessive liquid pressure, which can create higher evaporating pressures and temperatures at the bottom of the evaporator than at the top. Spray-type evaporators with pump recirculation of refrigerant eliminate this static pressure penalty.

Special materials for evaporator tubes and shells of particularly heavy wall thickness are frequently dictated to cool process streams of a highly corrosive nature. Corrosion allowances in evaporator design, which are seldom a factor in the commercial refrigeration field, are often required in chemical service. Ranges of permissible velocities are frequently specified to prevent sludge deposits or erosion at tube ends.

Process side construction suitable for high pressures seldom encountered in usual refrigeration applications is frequently necessary. Choice of process side scale factors must also be made carefully without overstatement.

Differences between process inlet and outlet temperatures of 50 K or more are not uncommon. For this reason, special consideration must be given to thermal stresses within the refrigerant evaporator. U-tube or floating tube sheet construction is frequently specified in chemical plant service, but minor process side modifications may permit the use of less expensive standard fixed tube sheet design. The refrigerant side of the evaporator may be required to withstand pressures resulting from maximum process temperature or the evaporator must be able to bypass the process stream under certain high-temperature conditions (e.g., in a refrigeration system failure).

Relief devices and safety precautions common to the refrigeration field normally meet chemical plant needs but should be reviewed against individual plant standards and local statutory requirements. Forged steel relief valves are becoming more common as they meet the applicable refinery piping codes. In hazardous service, relief valves are sized for emergency discharge in the event of fire. The effect of chemical vapors on the downstream (outlet) internal parts of relief valves may call for special materials or trapped outlet piping with isolating liquid seals.

Process requirements frequently call for sudden or unexpected load changes on the refrigeration evaporator. Possible thermal shocks, with attendant stresses, must be evaluated, and the evaporator must be designed to meet any such conditions.

Evaporators in chemical plant duty normally require inspection and cleaning on an annual basis. For this reason, they should be located for accessibility and ease of tube replacement. Possible contamination of the process stream or the refrigerant side, because of leakage, should be evaluated. Special means of leak detection from one side of the evaporator to the other may be justified on occasion.

Low-temperature refrigeration in the chemical industry often creates extremely high viscosities on the process side of the equipment. Special evaporator designs may be needed to minimize pressure drops on the process side and to maintain optimum heat-transfer performance. Small tube diameters may not be compatible with the process stream because certain processes may call for extra large tubes. For extremely low temperature and high viscosity duties, evaporators are sometimes provided with rotating internal wall scrapers to assure flow of high viscosity fluids through the evaporators. Similarly, jacketed process vessels are used to cool highly viscous materials, while rotary scrapers keep the vessel walls clean.

For proper process flow, evaporators usually are remote from the other refrigeration equipment to minimize piping and pumping costs. Because remotely located evaporators place special emphasis on proper refrigerant piping practices, secondary coolant systems may be used. Chemical plants frequently use flooded refrigeration systems, which pump refrigerant from the central compressor station to remote evaporators. The use of these systems often reduces the design difficulties in assuring adequate oil return, and special provisions must be made at the central refrigeration station to protect the compressor against liquid carryover in the suction gas. The system must have an adequate accumulator to assure dry gas to the compressor.

Standard air-side evaporators may require modification, mainly to solve special corrosion problems in handling air or process gases that attack standard coil materials. Occasionally, process requirements demand coil designs that do not match standard commercial air-side pressure drops, air-side design temperature range, or both. Coils of special depth and special finning may be required and coil casings and fan casings of alloy steel are common.

Instrumentation and Controls

Since the heart of the chemical plant is its instrument control system, it follows that the instrumentation and control is much more advanced in the chemical industry than in commercial or the usual industrial refrigeration applications. As previously discussed, chemical industry refrigeration instrumentation hardware is much more sophisticated in design, particularly in regard to providing increased safety, reliability, and compatibility with process instrumentation devices. This sophistication extends to the application and design of individual hardware items. The chemical industry seldom settles for integral control devices such as self-contained pressure regulators or capillary-actuated thermal control valves. The usual chemical industry control loop consists of a sensing device, a transmitter, a recorder/controller, a positioner, and an operator, all pneumatically or electrically interconnected. Many plants use central computer and microprocessor controls. Interfacing between the refrigeration system and control system may be necessary.

Cooling Towers and Spray Ponds

In a refrigeration system that uses water-cooled rather than air-cooled or evaporative condensers, heat may be rejected to once-through cooling water, spray ponds, or cooling towers. The chemical industry uses mechanical draft towers almost exclusively. These are generally of the induced draft design and are about evenly divided between crossflow and counterflow operation. Although a familiarity with these items is necessary, chemical plant engineers are usually responsible for their design.

Miscellaneous Equipment

Pumps. Refrigeration system pumps are usually of a high quality centrifugal design, the primary exception being small positive displacement pumps for compressor lube oil systems. In the past, heavy duty design was the rule rather than the exception, and secondary coolant and chilled water units were usually of a horizontal split-case design, patterned after boiler-house or water plant construction. Chemical process designers have advocated standard chemical plant pump designs, which usually have a vertically split case and an end suction. If the selection is made carefully, this design is successful in many applications, and the resultant savings in pump costs, space requirements, and spare parts stocking requirements make it economically attractive.

For pumped materials difficult to contain, such as most refrigerants and many secondary coolants, mechanical shaft seals of various designs are frequently used. As a result of their highly successful use in pumping difficult process fluids, canned or sealless pumps are used in such applications as liquid overfeed systems using halocarbons. Because the pumping of difficult fluids is a common problem, chemical process designers can be of invaluable assistance.

Piping. As a consequence of several factors, including low fluid temperatures, large pipe sizes, congested pipe alley space, and the industry's reluctance to use expansion joints for high duty service and in corrosive atmospheres, piping flexibility problems are much more complex. Expansion joints are frequently prohibited, which increases space requirements dramatically. Secondly, piping and valve standards that apply to both process and service facilities are frequently established by the process designer. The engineer who is accustomed to using carbon steel piping systems with tongue and groove flanging and valves may find that plant standards call for a welded nickel steel system with raised face flanges, spirally wound stainless steel gaskets, and cast steel valving.

Most piping construction problems resulting from the difference between expectations of the process engineer and the experience of the refrigeration engineer can be resolved by constructing the system to meet ASME *Standard* B31.3. The ASME B16 series of standards that defines the flanges and fittings of the process industry should also be followed. Chapters 1 through 4 also discuss piping sizing for various refrigeration systems.

Tanks. Chemical plants use storage tanks for both refrigerants and secondary coolants more frequently than most commercial or industrial plants. In chilled water or brine circulation systems, storage tanks often serve a dual purpose: (1) to store secondary coolants during operation to provide a reserve capacity and thus smooth out short-term peak requirements and (2) to store secondary coolants during a maintenance shutdown of process evaporators. In some cases, brine mix and storage facilities are provided, so that any brine lost due to leakage or unusual maintenance demands can be quickly replaced, thus minimizing unscheduled process outages. In many cases, refrigerant pumpout compressors and storage receivers can minimize loss of the refrigerant and unscheduled outage time because of refrigeration system failures on the refrigerant side.

The chemical industry designs all pressure vessels in accordance with the ASME *Boiler and Pressure Vessel Code*, in particular Section VIII, Division 1, for unfired pressure vessels, regardless of local government regulations requiring such design. In most plants, standards are established regarding such items as pressure relief devices, manhole design, insulation supports, and tank supports. A thorough knowledge of the plant standards to be applied should be gained before specifications and design details are established for refrigeration system tankage.

REFERENCES

ARI. 1992. Centrifugal and rotary screw water-chilling packages. ANSI/ARI *Standard* 550-92. Air-Conditioning and Refrigeration Institute, Arlington, VA.

ASME. 1995. Rules for construction of pressure vessels. ANSI/ASME *Boiler and pressure vessel code*, Section VIII-95, Division 1. American Society of Mechanical Engineers, New York.

ASME. 1996. Process piping. ANSI/ASME *Standard* B31.3-96.

CHAPTER 37

ENVIRONMENTAL TEST FACILITIES

Cooling Systems 37.1
Heating Systems 37.2
System Control and Instrumentation 37.4
Design Calculations 37.5
Chamber Construction 37.6

ENVIRONMENTAL test facilities are used to simulate an environment or combination of environments under laboratory controlled conditions that duplicate or exaggerate the effects found in actual service. They assist the engineer and scientist in exploring the effects of equipment and in developing equipment for resistance to the many environmental forces.

The acceptance of and demand for environmental simulation facilities result from the following factors: (1) parallel and reproducible tests can be made; (2) equipment being tested can usually be observed and analyzed during testing; and (3) supporting equipment requirements are reduced to a minimum. Field testing and product development costs are reduced, lead time required for completion of product development is shortened, and most desirable reliability features can be incorporated in the original manufacture of the product.

Environmental test facilities are used not only to determine the performance of mechanical and electrical equipment, but for certain tests on personnel as well. Personnel testing includes (1) checking protective equipment and clothing; (2) altitude and space procedures indoctrination; and (3) studying physiological and psychological effects on the human body and mind.

Environmental testing is usually divided into two general classifications—climatic and dynamic. The **climatic tests** of primary interest include the following:

Temperature chambers. These are used for (1) temperature soaks at high and low extremes; (2) temperature shock testing in which the part is subjected to rapid high and low temperature cycling; and (3) programmed cycling in which the parts are subjected to repetitive expansion and contraction stresses and breathing.

Humidity chambers. These may involve simply exposing the equipment to a constant humidity level, or **cycling**, wherein the temperature and relative humidity are varied. Cycle testing induces breathing and condensation within the parts tested. Subcooling may also be used to produce icing conditions.

Salt spray chambers. These are used for study of the corrosive action of salt vapor on components, usually at constant high-humidity test conditions.

Fungus chambers. These are used to stimulate the growth of fungus on equipment, primarily electrical, being studied to assure protection against tropical climates.

Miscellaneous climatic chambers. These include chambers for the simulation of desert sand and dust with high-velocity air movement, sunshine, snow, and rain, as well as chambers for proof testing of explosion-resistant wiring devices.

Altitude and space chambers. Altitude chambers have been used for some time in testing aircraft equipment. In recent years, space chambers have been developed for missiles, rockets, and space vehicle development work.

Combined environment testing. This type of test facility combines two or more of the above environments in one system, with all the complexity implied. Current thinking is that this type of testing is the only satisfactory means of proving that a piece of equipment will stand up in actual service.

Dynamic or nonclimatic tests include vibration, physical shock, acceleration, mechanical stress, nuclear radiation, cosmic radiation, micrometeorite bombardment, and many other types of stress.

The air-conditioning and refrigerating engineer may be directly concerned with design of test chambers for the simulation of most climatic environments. Some suggested methods of designing of test equipment for the production and control of various environments are given; the dynamic tests, however, are beyond the scope of this text. Because of the wide variety of tests that must be produced, detailed descriptions are not possible within this chapter.

Many techniques used in the design of environmental test equipment are the same as those employed for low-temperature refrigeration (Chapter 39), biomedical applications (Chapter 40), and cryogenic equipment (Chapter 38).

Many existing federal specifications outline basic environmental test specifications and the design method for some of the chambers to be used in this work.

COOLING SYSTEMS

Temperature reduction in test chambers can be accomplished by both mechanical refrigeration systems and the use of expendable refrigerants. Both methods can be used directly or in conjunction with secondary heat transfer fluids.

Primary Refrigerants

Any refrigerant suitable for mechanical refrigeration can be used as the primary refrigerant. See Chapter 39 for information on low-temperature refrigerants and refrigeration systems. Selection requires an evaluation of equipment size, cooling load, type of evaporator, temperatures to be produced, method of condensing, hazards, lubrication, and special operating requirements. The use of chlorofluorocarbons is being eliminated. Although single-stage and cascade systems are the most common designs presently used, a two- or three-stage compound system may be selected for certain load applications.

Test chambers are frequently required to operate at high as well as low temperatures. When using primary refrigerants in evaporators, which may be subjected to temperatures above 150°C, thermal isolation and cooling of the evaporator are necessary to prevent lubricant decomposition or other possible deterioration of the refrigerant circuit. Refrigerant filter cartridges should be changed frequently in systems where the evaporator is subjected to high temperatures. A secondary refrigerant heat exchange fluid, which is pumped out of the cooling coil at some safe, predetermined temperature, is another method for protection against overheating of a primary refrigerant coil. However, it is more common to use expendable refrigerants for high-temperature equipment, even with the disadvantage of higher operating costs at low temperature.

The preparation of this chapter is assigned to TC 9.2, Industrial Air Conditioning.

Expendable Refrigerants

Expendable refrigerants, such as dry ice, liquid carbon dioxide, liquid nitrogen, liquid helium, and others discussed in Chapter 38, are suitable for producing low temperatures in environmental chambers. Sublimation of dry ice within the chamber, chilling brine for circulation through heat exchangers, or direct expansion of the liquids within the test space are common expendable refrigerant methods. Liquid nitrogen, liquid helium, and other cryogenic liquids work particularly well below the temperature range of mechanical refrigeration systems.

The advantages of expendable refrigerants in environmental test equipment are reduced initial cost, basic simplicity, reduced mass and size of the chamber, and the ability to produce very rapid pull-down rates to low temperatures. Disadvantages include higher operating costs, the need for a reliable source of expendable refrigerants, the potential personnel hazard resulting from the absence of oxygen when the air is displaced in the test space, and the possible detrimental effect of submerging the tested product in the gas or liquid of an expendable refrigerant which is expanded directly into the test space. Indirect brine systems frequently approach the first cost of mechanical refrigeration systems. Direct expansion of expendable refrigerants in altitude chambers is impracticable.

Chambers using dry ice as the expendable refrigerant have, in most cases, been replaced with direct injection of liquid carbon dioxide or liquid nitrogen, which eliminates the need for a dry ice compartment. Also, the chamber is smaller and the temperature more easily controlled. Only on-off cycling is possible for control of direct injection liquid CO_2 systems because of the triple point of CO_2. Solenoid actuated two-way valves and pneumatically or electrically actuated ball valves have been used with much success. If reduced capacity operation is desirable, the control valve should be time-pulsed or more than one valve should be used with small and large expansion orifices. Solenoid valves with built-in orifices designed for this specific application are available.

Liquid CO_2 is used at two pressures; low-pressure bulk liquid is stored in refrigerated receivers at approximately −20°C and 2 MPa. Its latent heat is approximately 280 kJ/kg. High-pressure liquid is stored at room temperature at pressures ranging from 5 to 7 MPa, depending on ambient temperatures. Its latent heat varies from approximately 115 to 175 kJ/kg, depending on initial temperatures. High-pressure liquid is used only for very small or infrequent cooling loads because the latent heat is low, the liquid fills only 60 to 70% of a high-pressure cylinder, and the cost is considerably higher than for low-pressure liquid. For increased efficiency when high-pressure liquid CO_2 is selected, an economizer that precools incoming liquid by heat exchange to the exhaust cold vapors can be used to reduce the CO_2 requirement by 15 to 30%.

Liquid nitrogen can be applied either directly or indirectly to produce temperatures cheaply and easily down to its approximate −196°C boiling point (at 101.3 kPa). Control is accomplished either by on-off cycling of special solenoid actuated valves or by specially designed flow-metering valves. The liquid is stored in Dewar flasks or vacuum-insulated tanks. Transfer is accomplished either by self-pressurizing the storage vessel or by using dry air or nitrogen, properly pressure regulated, to pressurize the flask and force the liquid from a discharge tube.

Piping, valves, heat exchangers, and so forth should be fabricated of nonferrous metals, stainless steel, or high nickel steel. Few plastics are suitable seals in this temperature range. Although its latent heat of evaporation at 101.3 kPa pressure is only 198 kJ/kg, another 130 kJ/kg is available from the superheating of gas to −73°C. Therefore, it has a greater capacity of heat absorption than does low-pressure liquid CO_2 when operating at a −73°C or higher control point. Suitable precautions should be taken against the low-temperature hazards to operating personnel. Oxygen monitoring systems should be located in the test cells to warn personnel if oxygen depletion occurs. Carbon dioxide and nitrogen gases should be vented to the outside atmosphere after they have been used in the test chamber.

Other cryogenic liquids such as liquid helium and liquid hydrogen provide test temperatures close to absolute zero. These are too expensive for routine work at temperatures which can be obtained by more economical means. Special storage and control valves are required. More detailed information on cryogenic liquids, their handling and piping, may be found in Chapter 38.

Secondary Coolants

Low-temperature heat transfer fluids are used where (1) control flexibility is better accomplished by their use; (2) large central cooling systems have been chosen; (3) the high temperatures experienced will rule out primary refrigerants because of complex mechanical design problems; (4) expendable refrigerants are used and direct cooling of air or a product is not possible; (5) the wall of a temperature chamber or the shroud of a space simulator is to be conditioned and a close temperature gradient is required; or (6) thermal shock by alternately immersing the test device in liquids of different temperatures is required.

Halocarbons, liquid hydrocarbons, alcohol, some primary refrigerants, silicone fluids, and aqueous glycol solutions are used as secondary coolants. Thermal considerations include viscosity, specific heat, density, thermal conductivity, freezing and boiling points, and coefficient of expansion. Other important considerations are flammability, toxicity, corrosiveness, vapor pressure, and water solubility.

To minimize evaporation losses, the more volatile fluids are used in closed systems with suitable expansion tanks or accumulators and are frequently pressurized by an inert gas such as nitrogen. However, some secondary refrigerants dissolve in nitrogen. In that case, other methods must be used to pressurize the system such as an expansion tank with a diaphragm separator. Also, certain fluids must be kept refrigerated to prevent boiling off under standby conditions.

Secondary coolants are cooled in insulated sublimation tanks by spraying the liquid over dry ice or directly injecting liquid CO_2 or liquid N_2 into the secondary coolant. Such systems require a pump for recirculation of the secondary coolant. Common operating difficulties in sublimation systems include both cavitation in pumps because of the release of dissolved CO_2 or liquid N_2 gas and foaming in the sublimation tank with resultant carryover of the secondary heat transfer liquid with the vented gases. Positive displacement pumps and positive suction pressures are recommended.

In a system with a wide temperature range, the pump and shaft seal must be carefully selected. Magnetic drive and canned pumps have no external seals, which eliminates the shaft seal problem.

Secondary coolants can also be cooled by a mechanical refrigeration system or by a heat exchanger cooled by liquid nitrogen. These systems avoid carryover, cavitation, and moisture problems. The evaporator or heat exchanger must be designed to avoid freeze-up. In Chapter 20 of the 1997 *ASHRAE Handbook—Fundamentals*, more detail is included and additional coolants and their properties are discussed.

HEATING SYSTEMS

Electric heat is most commonly used in environmental chambers. Prime or extended surface, tubular, or strip heaters are suitable for circulating airstream systems. This method is a conduction or convection type of heating used to duplicate conditions in storage, in transportation, and when a protective housing is provided around equipment. With proper precautions, open nichrome, strip, or coil wire heaters can be used; they offer the advantages of a rapid response because of their low thermal mass. All types of electric heaters require proper insulation and protection against moisture. Generally speaking, temperature test chambers (dry-bulb control) may use open wire resistance heaters,

Environmental Test Facilities

although condensation at low temperatures may produce excessive moisture and corrosion on the heaters.

Some specifications prohibit direct radiation from chamber heaters to the test object. In this case, heater baffling may be required.

Salt-spray chambers are heated indirectly, usually by circulating heated water or air around the outside of the chamber shell outside the salt-fog atmosphere. Heaters for sand and dust chambers must use sheathed heaters that have good resistance against the erosion of the high-velocity sand and dust. Explosion-proof testing chambers contain an explosive fuel and air mixture. Heaters must have temperature-limiting devices on the sheath with terminals extended to the outside to prevent ignition of the mixture.

Altitude and space chambers present special heat problems. Arcing between terminals under vacuum conditions must be guarded against. General practice brings the terminals to the outside of the vacuum space through a vacuum-tight fitting. The reduced convective heat transfer of a heater under vacuum must be considered in the design to prevent burnout. Thermal and mechanical bonding of sheathed heaters to the exterior wall of the vacuum structure has been successfully used to avoid some of these problems. Internal forced convection heating is usually suitable up to about 15 km of altitude, approximately 10 kPa.

Indirect heating is also suitable for environmental chambers. Hot water, steam, brines, and oils can be used in coils in various configurations. These heating systems are particularly applicable where close tolerance must be provided within the system because complete modulation is possible. Modulation control is also possible in electrical resistance heating systems by using power proportioning equipment. Hot water recirculating systems and steam systems can be used with relatively standard design approaches. Complete drainage of the coils and piping within the chamber must be assured if the chamber will be operated below freezing. Proper water treatment should be provided.

Many brines and heat-exchange fluids that are not limited to the temperature ranges of water or steam are available. No one fluid for the commonly required wide ranges of environmental test equipment, such as −75 to 150°C, is available at present. Most chemical brines present some problems with viscosity, toxicity, flammability, chemical reaction with the piping, and other limitations. Any heat-exchange fluid must be carefully evaluated for limitations.

Another type of heating system is the radiant type, which is required to simulate such situations as sun exposure where both the heat-flux density and the wavelength are important; more commonly, radiant sources are used in equipment exposed to radiation from some high-temperature source. Radiant sources in environmental chambers vary from exposed nichrome wire heaters to specialized equipment such as arc lamps and mercury xenon lamps used to simulate sunshine.

Since the wavelength of the source is a function of its temperature, the conditions to be simulated must be understood before a suitable radiant heat source can be designed. Where the problem is simply to produce high power density on the product to be tested, bare nichrome wires, sheathed heaters, carbon and silicon carbide rod heating elements, and similar devices are available. Tubular quartz heat lamps are often suitable for very rapid heat-up conditions. Sunshine at sea level is usually simulated by the use of mercury vapor lamps or combinations of tungsten filament heat lamps to produce the proper wavelength and power density. Radiant heating systems may be required to produce heat densities in excess of 1300 kW/m^2. Careful design of the chamber structure, proper support of the heating elements, and correct location of the heating elements are required to prevent mechanical and thermal stress problems resulting from large temperature differentials and rapid heat-up and cool-down cycles.

Under altitude simulation conditions, removal of heat from a test object is possible by convection up to about 30 km (1.1 kPa). To accomplish this, the air mover should be of the largest practical size so that the residual air in the chamber is passed at high velocity through the cooling coils and directed on and around the test object. An envelope of cooled, rarified air, within which temperature gradients will be reasonably small, is thus produced around the test object.

Large gradients may occur throughout the rest of the chamber but will not affect the test results. Airflow should be reduced at low temperatures to prevent fan motor overload and minimize the addition to the cooling load. Cold wall construction reduces the heat load to be transferred to the cooling coils within the vacuum space, provides surface for radiant transfer, and overcomes some of the losses through the structure. Use a pumped secondary refrigerant to prevent high wall temperature gradients.

Evaporators should be maintained at the lowest practical temperature to provide the maximum temperature difference for both radiant and convective heat transfer. In simulated altitudes over 30 km, a radiation heat transfer system should be designed into the chamber to produce the desired temperature-control conditions on the test part. Treating the walls of the chamber to produce high emissivity is necessary in this case; close control of the entire wall surface is generally advisable. The limitation of radiant heat transfer in the low-temperature range must be considered because the amount of heat transferred from one surface to another is a function of the ratio of the square of the absolute temperatures. With temperatures in the −75°C range, the heat transfer rate is small. The detailed calculation of this effect is described in Chapter 3 of the 1997 *ASHRAE Handbook—Fundamentals*, where the Stefan-Boltzmann equation is discussed.

Air Movement

All types of fans and air movers are used in test chambers. Propeller, axial flow, and centrifugal fans are generally selected. Positive displacement or multistage turbine-type blowers provide the necessary total pressure for ram air simulation under altitude conditions. Drives for internally mounted fans must be located externally when extreme conditions are encountered in the airstream. To eliminate corrosion and minimize heat transmission, stainless steel shafts are commonly used. In almost all instances, a vapor seal is required to minimize or eliminate transmission to or from the ambient air. Internal bearings must withstand the full range of environmental conditions with a reasonable life expectancy. Altitude chamber fan shafts must be equipped with vacuum-tight shaft seals.

Air distribution within the work space of the test equipment is important in producing both uniform gradient conditions and system response. Frequently, close temperature gradients specified for air circulation rate and distribution will complicate the control problem. Air densities may vary almost three-to-one in temperature test chambers and more than fifty-to-one in altitude test chambers. Therefore, motor sizing and speed control of fans require careful attention. The airflow for a specific gradient is directly proportional to the net heat gain or loss in the work space and inversely proportional to the permissible temperature gradient and density of the air.

Usually the air volume must be increased at elevated temperatures because of the lower density. The circuit pressure drop will be approximately proportional to the air density. Therefore, the system balance points should be determined from the fan characteristic curves for various operating range conditions in the chamber.

Humidification

Common means of raising the relative humidity within a test chamber include directly introducing low-pressure steam, vaporizing water by electric immersion heaters in an open evaporation pan, and directly atomizing water sprays alone or in combination with electric heaters to aid vaporization of the spray.

Steam and vapor generators are best for increasing humidity as well as temperatures, since considerable sensible heat is introduced.

Control of steam may be modulating or on-off. With the latter, control set point anticipation is suggested to minimize overrun because of too rapid a rise in moisture content. When self-contained steam generators are used, they should include a low water cut-off device, suitable pressure relief valves, and other safety devices. Makeup water should be provided from a distilled or demineralized source.

Vapor generators operating at atmospheric pressure are commonly selected when it is desirable to produce test conditions of 95 ± 5% rh at temperatures between 40 and 70°C. Protection against immersion heater burnout is mandatory for good design.

For production of high humidities at temperatures near and below ambient, some form of atomized water spray is recommended because of the adiabatic cooling effect on evaporation. Although water spray systems do not have as rapid a response as steam, this is no disadvantage at lower temperatures where the humidity ratio is low. By controlling the temperature of the sprayed water, both humidification and dehumidification can be achieved with a single system. Heating and cooling means may be located either inside the conditioning spray plenum or externally in the recirculating water circuit. Close control of humidities is possible with such a design. Distilled or demineralized water is suggested, since the hardness of tap water may be corrosive or cause contamination. Recirculating spray systems should be periodically flushed out because of possible airstream contamination of equipment or test parts. For the system to operate properly, careful consideration of the construction material must be exercised.

Dehumidification

Low humidity conditions with controlled dew points above freezing are usually produced by mechanical refrigeration cooling and electric reheating because most environmental chambers require mechanical refrigeration for extended dry-bulb temperature range control. For the control of humidity conditions requiring dew points below freezing and continuous operation, either alternately defrosted dual evaporators or automatically regenerating desiccant dehumidifiers are used. If an enclosure is sufficiently vaportight and there is no internal latent load, an evaporator sufficiently large to handle the frost buildup because of initial latent load may be an acceptable solution.

Ram air coolers that simulate the cooling air supply for electronic airborne equipment at controlled temperature, humidity, and altitude conditions, present special dehumidification problems. If 100% outdoor air is required, it is best handled by a wet coil which removes a large part of the moisture just above freezing, followed by a set of parallel alternately defrosted evaporators. For ram air simulators in a closed-loop arrangement, simpler systems are possible if the cooling of air does not involve dehumidification.

When using certain liquid or solid desiccants in very high or low dry-bulb ranges, mechanical refrigeration cooling may be required for precooling or aftercooling. In such systems, a portion of the chamber air is continuously withdrawn by an auxiliary blower and passed through the drying agent and any necessary precooling or aftercooling coils.

Some tests require the conditioned air dew point to remain sufficiently low to prevent condensation within the test article during dry-bulb control temperature cycling. This requirement can be a problem during rapid heat pull-up and may require a desiccant dehumidifier.

SYSTEM CONTROL AND INSTRUMENTATION

Control Tolerances

Environmental test facilities generally require minimal control tolerances over a wider range than that required for ordinary HVAC systems. Control of temperatures as close as ±0.3 to 0.5 K are commonly specified. Relative humidity control tolerance is at most ±5% and more often ±3% or even less. To meet these requirements, the control and instrumentation system must have an even better accuracy since operating tolerances often exceed control tolerances.

When, in addition, the test chamber is designed to provide a wide range of test conditions, the HVAC equipment must be oversized for most of those conditions. The control system is then required to modulate equipment output to a small percentage of capacity, with all of the difficulties inherent in controlling oversized equipment.

For these reasons, the control devices used with test chambers must be of the highest quality. In general this will be "industrial quality," as used in process control applications. Most systems operate in a modulating mode, using proportional plus integral (PI) or, occasionally, proportional plus integral plus derivative (PID). Proportional only control is not acceptable due to its inherent deviation from set point. For discussions of control theory, operating modes and typical systems, see Chapter 42 of the 1995 *ASHRAE Handbook—Applications* and Haines (1987).

Temperature Controls

Systems for temperature control are of the types discussed in Chapter 42 of the 1995 *ASHRAE Handbook—Applications* and Haines (1987). Controllers should operate in PI mode, although PID may be needed when rapid cycling over a wide range of set points is required.

Temperature sensors may be thermocouples, thermistors, or resistance temperature detectors (RTDs). Thermistors and RTDs should be specified to have an absolute accuracy of ±0.3 K and a sensitivity of at least 0.06 K. Most good RTD and thermistor sensors can meet these requirements. Thermistors must be specified with factory-certified performance and a guarantee of not more than 0.5 K drift per year so that they can be readily replaced. Thermistors do drift and must, therefore, be recalibrated or replaced at frequent intervals, depending on the tolerance specifications. Wound wire RTDs may be specified with no drift over time. The best-quality wound wire RTDs are platinum; these may be obtained with certified accuracies, traceable to industry standards. Platinum RTDs using solid-state deposition techniques are also available. These are accurate and have a fast response but are subject to drift over time. Sensitivity and accuracy may not be suitable for close tolerances when bulb and capillary-type sensors are used with pneumatic controllers. RTD sensors used with pneumatic controllers can greatly improve accuracy.

Sensors must be suitable for the temperature ranges required. Consult the sensor manufacturer for this information. Where wide ranges are required, two or more instruments may provide better accuracy over the entire range.

Recorders may be digital or of the strip or circular chart type. They may use the same sensors as are used by the controllers. If separate sensors are used they must be calibrated together. For accurate calibration, laboratory-type mercury thermometers should be used, depending on the range.

Humidity Controls

Accurate humidity sensing is very difficult. Materials that vary dimensionally with changes in relative humidity are not suitable for close control because of hysteresis and drift. Electrolytes and wet-bulb/dry-bulb sensors require essentially continuous maintenance to ensure accuracy. For any system, accuracy at extremely high or below-freezing temperatures may be questionable. For reasonably close control (±5% rh), solid-state deposition-type sensors of the capacitance or resistance type are satisfactory and have a fast response. They must be calibrated regularly—the best have a drift of about 1% per year. For very close control, for calibration, and for continuous accurate sensing, the chilled-mirror dew-point sensor should be used. This device is available in several packaging arrangements and, if needed, can be obtained with a certification of

Environmental Test Facilities

accuracy traceable to an agency standard. Its response is somewhat slow for use with rapidly changing conditions. In some cases, it has been used to back up and check the solid-state instruments which are actually used for control.

Controllers must operate in the PI or PID mode, and recorders may use sensors in common with the controllers.

Pressure Controls

An altitude measurement and control system may consist of a simple manometer for indicating pressure and a hand-throttling or bleed valve to control the level; or the system may be completely automatic, operated by a mechanical or electronic sensing device that actuates modulating bleed controls to automatically regulate the vacuum. The U-tube manometer shows relative pressure between the site barometric pressure and the internal vacuum of the chamber. Absolute vacuum measuring manometers are recommended for altitude chambers to avoid the need to correct for local barometric pressure variations.

Opposed-bellows absolute pressure controllers and electrical instruments using strain gage pressure transducers with bridge circuit electronic controllers are available. These instruments are suitable for altitudes to 60 km. They provide on-off control or proportioning output signals and can control various functions of the vacuum system. Electronic vacuum-sensing devices measure the change in interference across an air gap between two electronic devices because of the reduction in gas molecules in the space. Coupled with amplifiers, this effect initiates action in control circuits at a set point.

One instrument system has an alpha emitter and measures the amount of energy reaching the grid to determine the vacuum level. Another system uses ionization of a filament. A thermocouple measures the change in the heat exchange rate as a function of vacuum. Most electronic vacuum instruments are available with microamp or millivolt output that may be coupled to a standard potentiometer to provide both a record and an output signal for controlling the vacuum system.

The most common method of automatically controlling a vacuum system is to bleed air into the chamber through a modulating valve. This floods the pump to the desired capacity so that good control can be maintained with the proper size bleed valve. Frequently, two or more valves are required to control the pressure in the chamber over a wide range.

Altitude chambers can be controlled by cycling a solenoid valve in the pump suction line or simply turning the pump on and off. The latter measure is not recommended because of the excessive strain on the motor and the drive mechanism. These two methods are not practical for diffusion pumping systems.

Valves and Dampers

Haines (1987) discusses the factors to be addressed in selecting control valves and dampers. In test chambers where a wide range of conditions must be simulated, the system gains are aggravated by the need to size the HVAC equipment for the most severe conditions, which means that it is oversized for every other condition, sometimes extremely so. Oversizing the valve or damper inevitably leads to poor control. It follows that valves and dampers should be undersized rather than oversized if close control is to be obtained. Undersizing penalizes the air or water system hydraulics but may be necessary to obtain the desired results. Use of multiple small dampers should be considered.

Computers as Controllers

Typically, direct digital control (DDC) with computers is used. The computer simply takes over the controller function. It can be programmed to any kind of operating and timed sequence desired. It can also provide visual and printed output of system status, either in real time or as history, including graphic and statistical presentations. The engineer should select the functions needed to satisfy the design parameters.

Accessories

To provide satisfactory operation, a control system must include other accessories besides sensing elements and controllers. A pneumatic system should include a pressure-reducing valve, a filter-drier, and a surge tank; all are standard with such systems. For systems operating below freezing, nitrogen or dried compressed air should be considered as the pneumatic medium rather than compressed air if the system is to run within the chamber.

The selection of safety and other interlock relays and switches should be carefully considered not only to provide safe operation but to control the system and reduce the number of switch settings to be made by the operator. Vacuum switches are frequently used to cut out the heaters or switch them to series wiring whenever an altitude chamber is under vacuum, thus preventing burnout because of low convection transfer. Often there is a temperature limit controller to prevent damage that may occur because of excessive heating or cooling of the test module.

In addition to the conventional control systems for temperature, humidity, and vacuum, the designer may encounter requirements for measuring or controlling many other variables. These variables include: hydrogen ion concentration (pH); fluid velocity and pressure, dust, and smoke density; explosive fuel and air mixtures; ozone, oxygen, and carbon dioxide levels; and others.

DESIGN CALCULATIONS

Cooling and Heating Loads

Because the rate of temperature change is usually important in the operation of test chambers, both the steady-state and transient load must be calculated to size the refrigeration and heating system. The steady-state cooling load includes heat transferred through the insulation, framing, windows, sleeves and penetrations, fan shafts and other conducting materials, and the internal live heat loads of lighting, product, fans, and personnel. The sum of these loads that occur simultaneously at the lowest air temperature in the chamber is the minimum design capacity. The equipment needs to be sized to handle this design capacity, plus a factor of safety, at the refrigerant evaporating temperature required and as determined by the evaporator size.

Transmission losses must also be determined to calculate the total steady-state heating load. The live loads, such as blowers and lights, will decrease the heating load, but usually they are not deducted from the transmission losses. The heaters are then sized with the load at the high temperature and derated for low voltage and operating temperature.

For transient conditions, the thermal mass of all items, including chamber air, liner and bracing, insulation and framing, evaporators, heat exchangers, heaters, blower, brackets, shelves or other gear, and the test specimen must be determined. The product of the mean specific heat and the mass gives the average thermal mass. The product of the time rate of temperature change and the thermal mass gives the instantaneous load due to temperature transients. The total load may include transmission and live loads as well.

Under transient conditions, a gradient will exist between the room ambient and the heat source or sink. For small chambers, conservative design assumes that the temperature of all items directly contacting the chamber air changes at the same rate as the air temperature. In cases with very rapid temperature changes and high mass loads, the surface areas, air velocity, and point of temperature control must be considered in designing the necessary refrigerating or heating system. A larger system cooling requirement may be calculated by this method rather than by one based on the total load changing with the air temperature—especially if a linear rate of change is required and the final temperature is low on the capacity curve of the refrigeration equipment. This larger capacity is also required for heating, except electric heating, where the capacity is essentially uniform at all temperatures.

Relative Humidity

Equipment selection for relative humidity control is similar to normal air-conditioning applications, only more complex in that widely varied control conditions must be produced. A typical control range is 2 to 85°C dry-bulb temperature and from 20 to 98% rh above 2°C dew point. Very low dew points may be required for special test programs. Sensible cooling loads may be calculated in a manner similar to that for low-temperature loads, but must also account for such additional factors as the sensible heat of steam when used for humidification.

Since dehumidification is normally accomplished by mechanically cooling and reheating the air, the most common approach is to take a portion of the total recirculating air, cool it to the required dew point, then reheat it as required to keep it at the desired dry-bulb temperature level. If the total air volume were cooled and then reheated, rather large dehumidification and reheat capacities would result. When long-term tests are required at controlled dew points near freezing, heat transfer cooling surfaces must be amply sized to operate without freezing.

Where large internal heat gains are present and high humidities are desired, spray-wetted surfaces producing adiabatic saturation may be used. By atomizing the spray and controlling the temperature of the recirculating spray water, systems may be used for simultaneous cooling and humidification.

The sizing of heaters for increasing dry-bulb temperature is the same as for normal chamber heating. Atmospheric pressure vapor generators and direct steam injection are alternate methods of humidification. The amount of water that must be added to achieve the desired increase in absolute humidity for the known volume of air within the chamber determines the rate of moisture addition. Both of these methods add sensible heat, as well as latent heat, which must be added to the cooling load to obtain the total load.

CHAMBER CONSTRUCTION

Inner liners of test chambers for high- and low-temperature and humidity are most commonly made of type 302 or 304, heliarc welded stainless steel. The outside liner is made of aluminum or mild steel, welded or sealed to provide a vapor-tight structure for the insulation. For walk-in rooms, prefabricated, urethane insulated, foamed-in-place modular panels are popular. Special care must be taken in the sealing of the panel joints.

A means must be provided to prevent an excessive pressure differential between the inside and outside of the chamber during rapid changes of temperature. Even opening and closing a door, when test space conditions are different from ambient, may change the air temperature and consequent pressure enough to buckle the liner of a well-sealed chamber if it is not properly vented.

The structure must be designed to provide for the considerable expansion and contraction that take place during wide-range temperature cycling. Doors are a particularly difficult problem. Any fan baffles, coil mountings, or other accessories must be carefully planned so that moving parts are not interfered with during temperature cycling runs. Shock loads induced by liquid nitrogen and other refrigerants can cause violent stress concentrations. Humidity chambers must be designed so that any condensate formed on the coils, walls, or floor will drain rapidly from the chamber. This feature is important in programmed humidity chambers in which operation ranges from high- to low-humidity conditions.

Altitude Chambers

Generally, a chamber that simulates altitude to 75 km (2.1 Pa) is considered an altitude chamber; above that altitude, the chamber is considered a space simulator. Interior liners of chambers that simulate altitude are usually constructed of aluminum, mild steel, or stainless steel, with exterior structural steel reinforcing designed to withstand a vacuum or 100 kPa (gage).

Depending on the desired test space configuration, chamber liners may be rectangular or cylindrical. Design of the vessel to withstand the exterior pressure should be in accordance with the latest edition of the American Society of Mechanical Engineers' (ASME) *Boiler and Pressure Vessel Code*.

Altitude chambers that control temperature or humidity or both, in combination with altitude simulation, are built with the pressure liner on the inside or outside of the insulation space. The inside liner, whether or not it is the pressure liner, should be made of a corrosion resistant material such as stainless steel. The outside liner may be mild steel. If the outside liner is the pressure liner, it must be of sufficient thickness or reinforced to withstand a vacuum or 100 kPa pressure differential. It must also be treated for corrosion resistance due to equalization with the test space, which will introduce various combinations of temperature and (high) humidity into the liner surfaces and insulation.

If the inner liner is the pressure liner, it must be stressed for the temperature range involved as well as the pressure. The inner liner should be the pressure liner if high altitude use is a major consideration, because the insulation space is not exposed to the vacuum. However, because of the liner and reinforcing mass, the cooling and heating load during temperature transition is greater than with an outside pressure liner. With an outside pressure liner, the insulation space must be equalized with the test space to prevent inner liner collapse. Achieving high vacuum on the insulation and insulation space is very difficult unless it is dry; therefore, this fact must be considered during temperature or temperature-humidity runs, or when diving the chamber to atmospheric pressure after an altitude test.

Seam welds of the vacuum vessel should be on the vacuum side and should be of one continuous pass with skip welding, if required, on the outside. This method of welding prevents virtual leaks and allows for leak checking.

Insulation

Chapters 22, 23, and 24 of the 1997 *ASHRAE Handbook—Fundamentals* discuss insulation fundamentals and applications in detail.

Doors

Both overlap and plug-type doors are used, the latter most frequently on both high-humidity chambers (because of internal condensation) and high-temperature chambers, to prevent or minimize warping. They must be rugged and well fitted, particularly for altitude chambers. Vapor sealing with multiple gaskets is important, particularly on smaller chambers, since defrosting means are seldom provided. Suitable gasket materials would be natural rubber, synthetic rubber, and silicone. Plastics or light-gage stainless steel are recommended for thermal breaker strips; pressed wood hardboard materials generally have insufficient resistance to physical damage, high temperatures, or high humidities. Low wattage heater cables, installed under the breaker strip, prevent larger doors from freezing closed. Safety measures are essential for walk-in freezers, including doors easily opened from inside.

For altitude chambers, the door assembly must register against a gasket that will easily vacuum seal. Vacuum gaskets must be properly retained to prevent their being drawn into the chamber.

Environmental Test Facilities

Windows

Special hermetically sealed multilight window assemblies are manufactured for special conditions. They are suitable for wide-range temperature and humidity tests and, if provided with an inner pane of sufficient strength, for altitude tests as well. Tempered plate is required for the inner panes so that temperature shocks will not cause failure. Because of the pressures that build up, sealed assemblies for higher temperatures require careful design. For temperatures up to 500°C, windows of tempered glass are required and hermetic sealing is difficult, but research indicates a possible range of −180 to 500°C. The number of panes and design for a given task are usually specified by the window manufacturer. A typical −75 to 150°C window contains six lights of glass enclosing dry gas spaces.

Accessories

Interior lighting is mostly incandescent, in vaportight and sometimes explosionproof fixtures. Ample illumination will be given usually by 60 to 100 W/m^2 of floor space. Lamps are not recommended for use inside chambers above 300°C.

Power leads may handle normal alternating current or special aircraft voltages and frequencies.

Thermocouples in almost any number may be called for. Bare iron-constantan should not be installed in a high-humidity chamber. For the operator's convenience, thermocouples are frequently connected to plugs and outlets.

Turning shafts, if small and manually operated, may be coupled to adjustable devices on the item being tested. These shafts should be stainless steel tubes. O rings on the outside of the chamber may be used for pressure sealing.

Power shafts, provided for driving rotating equipment, must have vacuumtight shaft seals for vacuum chambers.

Sleeves and plugs are required for general purpose testing. Holes of various sizes may be specified. On altitude chambers, threaded or flanged caps must be provided.

Special lines for pneumatically or hydraulically operated equipment may require pressure pipes with special connections at the ends.

Protective panels are used to prevent damage to the chamber if an internal explosion or sudden release of pressure occurs.

Window wipers are installed on windows to remove condensation if vision is required when temperature adjustments are made within the chamber.

Reach-in ports, with or without gauntlets, may be provided for minor work within a test space without opening doors or disturbing internal conditions.

BIBLIOGRAPHY

ASME. 1995. *Boiler and pressure vessel code.* American Society of Mechanical Engineers, New York.

Haines, R.W. 1987. *Control systems for heating, ventilating and air conditioning*, 4th ed. Van Nostrand Reinhold, New York.

Holladay, W.L. 1950. Low temperature test chamber design. *Refrigerating Engineering* (July):656.

Missimer, D.J. 1956. Cascade refrigeration systems for ultra low temperatures. *Refrigerating Engineering* (February).

Missimer, D.J. 1972. Mechanical system can reach −140°C. *Research/Development* (July):40.

Missimer, D.J. 1973. Ultra low-temp systems—A practical summary. *Refrigeration Service and Contracting* (December):18.

Missimer, D.J. and W.L. Holladay. 1967. Cascade refrigerating systems—State-of-the-art. *ASHRAE Journal* (April):70.

U.S. Government Printing Office. 1990. Environmental test methods and engineering guidelines. MIL-Std-810E. Washington, D.C. (February).

CHAPTER 38

CRYOGENICS

General Applications 38.1	Equipment 38.16
Low-Temperature Properties 38.1	Low-Temperature Insulations 38.19
Refrigeration and Liquefaction 38.3	Storage and Transfer Systems 38.22
Cryocoolers 38.8	Instrumentation 38.23
Separation and Purification of Gases 38.12	Hazards with Cryogenic Systems 38.24

CRYOGENICS is a term normally associated with low temperatures. However, the location on the temperature scale at which refrigeration generally ends and cryogenics begins has never been well defined. Most scientists and engineers working in this field restrict cryogenics to a temperature below 125 K because the normal boiling points of most of the permanent gases (e.g., helium, hydrogen, neon, nitrogen, argon, oxygen, and air) occur below this temperature. In contrast, most of the common refrigerants have boiling points above this temperature.

Cryogenic engineering therefore is involved with the design and development of low-temperature systems and components. In such activities the designer must be familiar with the properties of the fluids used to achieve these low temperatures as well as the physical properties of the components used to produce, maintain, and apply such temperatures.

GENERAL APPLICATIONS

The application of cryogenic engineering has become extensive. In the United States, for example, nearly 30% of the oxygen produced by cryogenic separation is used by the steel industry to reduce the cost of high-grade steel, while another 20% is used in the chemical process industry to produce a variety of oxygenated compounds. Liquid hydrogen production has risen from laboratory quantities to a level of over 2.1 kg/s. Similarly, liquid helium demand has required the construction of large plants to separate helium from natural gas by cryogenic means. The demands for energy likewise have accelerated the construction of large base-load liquefied natural gas (LNG) plants. Some of the current applications include the development of high-field magnets and sophisticated electronic devices that use the phenomenon of **superconductivity** of materials at low temperatures. Space simulation requires **cryopumping**, the freezing of residual gases in a chamber on a cold surface, to provide the ultrahigh vacuum representative of conditions in space. This concept has also been used in commercial high-vacuum pumps.

The food industry uses large amounts of liquid nitrogen to freeze the more expensive foods like shrimp and to maintain frozen food during transport. Liquid nitrogen cooled containers are used to preserve whole blood, bone marrow, and animal semen for extended periods. Cryogenic surgery is performed to cure such involuntary disorders as Parkinson's disease. Medical diagnosis uses magnetic resonance imaging (MRI), which requires cryogenically cooled superconducting magnets. Finally, the chemical processing industry relies on cryogenic temperatures to recover the more valuable heavy components or upgrade the heat content of the fuel gas from natural gas, to recover useful components like argon and neon from air, to purify various process and waste streams, and to produce ethylene from a mixture of olefin compounds.

LOW-TEMPERATURE PROPERTIES

Test data are necessary because properties at low temperatures are often significantly different from those at ambient temperature. For example, the phenomenon of superconductivity, the vanishing of specific heats, and the onset of ductile brittle transitions in carbon steel cannot be inferred from property measurements obtained at ambient temperatures.

Fluid Properties

Some thermodynamic data for cryogenic fluids are given in Chapter 19 of the 1997 *ASHRAE Handbook—Fundamentals*. Computer compiled tabulations include those of MIPROPS prepared by NIST and GASPAK, HEPAK, and PROMIX developed by Cryodata (Arp 1998). Some unique properties of several cryogens are discussed in the following paragraphs.

Helium exists in two isotopic forms, the most common being helium 4. The more rare form, helium 3, exhibits a much lower vapor pressure, which has been exploited in the development of the helium dilution refrigerator to attain temperatures as low as 0.02 to 0.05 K. Whenever helium is referenced without isotopic designation, it can be assumed to be helium 4.

As a liquid, helium exhibits two unique phases—liquid helium I and liquid helium II (Figure 1). Helium I is labeled as the normal fluid while helium II is designated as the superfluid because, under

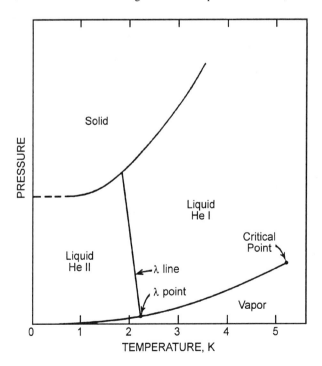

Fig. 1 Phase Diagram for Helium 4

This preparation of this chapter is assigned to TC 10.4, Ultra-Low Temperature Systems and Cryogenics.

certain conditions, the fluid exhibits no viscosity. The phase transition between these two liquids is identified as the lambda (λ) line. Intersection of helium II with the vapor pressure curve is known as the λ point. Immediately to the right of the λ line, the specific heat of helium I increases to a large value as the temperature approaches this line. Once below the λ line the specific heat of helium II rapidly decreases to zero. The thermal conductivity of helium I, on the other hand, decreases with decreasing temperature. However, once the transition to helium II has been made, the thermal conductivity of the liquid can increase in value by many orders of magnitude over that for helium I. Also, helium 4 has no triple point and requires a pressure of 2.5 MPa or more to exist as a solid below a temperature of 3 K.

A property of **hydrogen** that sets it apart from other substances is that it can exist in two molecular forms: orthohydrogen and parahydrogen. These forms differ by having parallel (orthohydrogen) or opposed (parahydrogen) nuclear spins associated with the two atoms forming the hydrogen molecule. At ambient temperatures the equilibrium mixture of 75% orthohydrogen and 25% parahydrogen is designated as normal hydrogen. With decreasing temperatures the thermodynamics shifts to 99.79% parahydrogen at 20.4 K, the normal boiling point of hydrogen. The conversion from normal hydrogen to parahydrogen is exothermic and evolves sufficient energy to vaporize ~1% of the stored liquid per hour assuming negligible heat leak into the storage container. The fractional rate of conversion is given by

$$\frac{dx}{d\theta} = -kx^2 \qquad (1)$$

where x is the orthohydrogen fraction at time θ in hours and k is the reaction rate constant, 0.0114/h. The fraction of liquid remaining in a storage dewar at time θ is then

$$\ln\frac{m}{m_o} = \frac{1.57}{1.33 + 0.0114\theta} - 1.18 \qquad (2)$$

Here m_o is the mass of normal hydrogen at $\theta = 0$ and m is the mass of remaining liquid at time θ. If the original composition of the liquid is not normal hydrogen at $\theta = 0$, a new constant of integration based on the initial orthohydrogen concentration can be evaluated from Equation (1). Figure 2 provides a summary of the calculations.

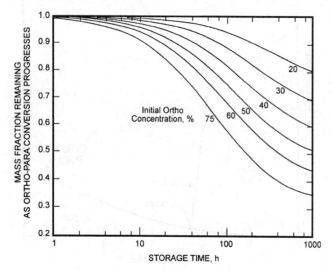

Fig. 2 Fraction of Liquid Hydrogen Evaporated due to Ortho-Parahydrogen Conversion as Function of Storage Time

To minimize such losses in the commercial production of liquid hydrogen, a catalyst is used to hasten the conversion from normal hydrogen to the thermodynamic equilibrium concentration during the liquefaction process. Hydrous iron oxide, Cr_2O_3 on an Al_2O_3 gel carrier or NiO on an Al_2O_3 gel are used as catalysts. The latter combination is about 90 times as rapid as the others and is therefore the preferred choice.

In contrast to other cryogenic fluids, **liquid oxygen** (LOX) is slightly magnetic. Its paramagnetic susceptibility is 1.003 at its normal boiling point. This characteristic has prompted the use of a magnetic field in a liquid oxygen dewar to separate the liquid and gaseous phases under zero gravity conditions.

Both gaseous and liquid oxygen are chemically reactive, particularly with hydrocarbon materials. Since oxygen presents a serious safety problem, systems using liquid oxygen must be maintained scrupulously clean of any foreign matter. Liquid oxygen cleanliness in the space industry has come to be associated with a set of elaborate cleaning and inspection specifications representing a near ultimate in large scale equipment cleanliness.

Liquid nitrogen is of considerable importance as a cryogen because it is a safe refrigerant. Because it is rather inactive chemically and is neither explosive nor toxic, liquid nitrogen is commonly used in hydrogen and helium liquefaction cycles as a precoolant.

Liquefied natural gas (LNG) is the liquid form of natural gas consisting primarily of methane, a mixture of heavier hydrocarbons, and other impurities such as nitrogen and hydrogen sulfide. Liquefying natural gas reduces its specific volume by a factor of approximately 600 to 1, which makes handling and storage economically possible despite the added cost of liquefaction and the need for insulated transport and storage equipment.

Thermal Properties

Specific heat, thermal conductivity, and thermal expansivity are the thermal properties of major interest at low temperatures. The specific heat can be predicted fairly accurately by mathematical models through the use of statistical mechanics and quantum theory. For solids, the Debye model gives a satisfactory representation of the specific heat with changes in temperature. However, difficulties are encountered when the Debye theory is applied to alloys and compounds.

Several computer programs provide thermal data for many of the metals used in low-temperature equipment. METALPAK (Arp 1998), for example, is a reference program for the thermal properties of thirteen metals used in low-temperature systems.

The specific heat of cryogenic liquids generally decreases in a manner similar to that observed for crystalline solids as the temperature is lowered. At low pressures the specific heat decreases with a decrease in temperature. However, at high pressures near the critical, humps in the specific heat curve are observed for all of the cryogenic fluids.

Adequate predictions of the thermal conductivity of pure metals at low temperatures can be made with the use of the Wiedeman-Franz law, which states that the ratio of the thermal conductivity to the product of the electrical conductivity and the absolute temperature is a constant. This ratio for high-conductivity metals extrapolates close to the Sommerfeld value of 2.445×10^{-8} W·Ω/K^2 at absolute zero; but this ratio is considerably below this value at higher temperatures. However, high-purity aluminum and copper exhibit peaks in thermal conductivity between 20 and 50 K, but these peaks are rapidly suppressed with increased impurity levels or cold work of the metal. Aluminum alloys, for example, show a steady decrease in thermal conductivity with a decrease in temperature. Other structural alloys such as Monel, Inconel and stainless steel, exhibit similar thermal conductivity properties and thus are helpful in reducing heat leak into a cryogenic system.

All cryogenic fluids, except for helium and hydrogen, exhibit increasing thermal conductivities with a decrease in the temperature.

Cryogenics

The thermal conductivity of the two exceptions decrease as the temperature is decreased. The kinetic theory of gases predicts the decrease in thermal conductivity of all gases as the temperature is lowered.

The expansion coefficient of a solid can be estimated with an approximate thermodynamic equation of state for solids that equates the volumetric thermal expansion coefficient with the quantity $\gamma c_v \rho/B$, where γ is the Grüneisen dimensionless ratio, c_v the specific heat of the solid, ρ the density of the material, and B its bulk modulus. For face centered cubic metals the average value of the Grüneisen constant is approximately 2.3. However, this constant has a tendency to increase with atomic number.

Electrical and Magnetic Properties

The ratio of the electrical resistivity of most pure metallic elements at ambient temperature to that at moderately low temperatures is approximately proportional to the ratio of the absolute temperatures. However, at very low temperatures the electrical resistivity (with the exception of superconductors) approaches a residual value almost independent of temperature. Alloys, on the other hand, have resisitivities much higher than those of their constituent elements and exhibit resistance temperature coefficients that are quite low. Electrical resistivity as a consequence is largely independent of temperature and may often be of the same magnitude as the ambient temperature value.

The insulating quality of solid electrical conductors usually improves with a lowering of the temperature. However in the temperature region from 1 to 5 K, the electrical resistivity of many semiconductors increases quite rapidly with a small decrease in temperature. This has formed the basis for the development of numerous sensitive semiconductor resistance thermometers for very low temperature measurements.

The phenomenon of superconductivity is described as the disappearance of electrical resistance in certain materials that are maintained below a characteristic temperature, electrical current and magnetic field, and the appearance of perfect **diamagnetism**. This phenomenon is the most distinguishing characteristic of superconductors.

More than twenty metallic elements have been shown to be superconductors at low temperatures. In fact, the materials whose superconducting properties have been identified extends into the thousands. Bednorz and Müller's (1986) discovery of high-temperature superconductors and the intensive research to extend the upper temperature limit above 135 K has further increased this list of superconductors.

The properties that are affected when a material becomes superconducting include specific heat, thermal conductivity, electrical resistance, magnetic permeability, and thermoelectric effect. As a consequence, the use of superconducting materials in the construction of equipment subjected to operating temperatures below their critical temperatures needs to be evaluated carefully.

Mechanical Properties

Mechanical properties of interest for the design of a facility subjected to low temperatures include ultimate strength, yield strength, fatigue strength, impact strength, hardness, ductility, and elastic moduli. Chapter 39 and Wigley (1971) include information on some of these properties.

The mechanical properties of metals at low temperature are classified most conveniently by their lattice symmetry. The face centered cubic (fcc) metals and their alloys are most often used in the construction of cryogenic equipment. Al, Cu, Ni, their alloys, and the austenitic stainless steels of the 18-8 type are fcc and do not exhibit an impact ductile to brittle transition at low temperatures. As a general rule, the mechanical properties of these metals improve as the temperature is reduced. The yield strength at 20 K is considerably larger than at ambient temperature; Young's modulus is 5 to 20% larger at the lower temperatures, and fatigue properties, with the exception of 2024 T4 aluminum, are improved at the lower temperature. Because annealing can affect both the ultimate and yield strengths, care must be exercised under these conditions.

The body centered cubic (bcc) metals and alloys are normally undesirable for low-temperature construction. This class includes Fe, the martenistic steels (low carbon and the 400 series of stainless steels), Mo, and Nb. If not brittle at room temperature, these materials exhibit a ductile to brittle transition at low temperatures. Hard working of some steels, in particular, can induce the austenite to martensite transition.

The hexagonal close packed (hcp) metals exhibit mechanical properties intermediate between those of the fcc and bcc metals. For example, Zn undergoes a ductile to brittle transition whereas Zr and pure Ti do not. The latter and its alloys, having a hcp structure, remain reasonably ductile at low temperatures and have been used for many applications where mass reduction and reduced heat leakage through the material have been important. However, small impurities of O, N, H, and C can have detrimental effect on the low-temperature ductility properties of Ti and its alloys.

Plastics increase in strength as the temperature decreases, but this increase in strength is also accompanied by a rapid decrease in elongation in a tensile test and decrease in impact resistance. Nonstick, fluorocarbon resins and glass-reinforced plastics retain appreciable impact resistance as the temperature is lowered. The glass-reinforced plastics also have high strength to mass and strength to thermal conductivity ratios. All elastomers, on the other hand, become brittle at low temperatures. Nevertheless, many of these materials including rubber, polyester film, and nylon can be used for static seal gaskets provided they are highly compressed at room temperature, prior to cooling.

The strength of glass under constant loading also increases with decrease in temperature. Because failure occurs at a lower stress when the glass surface contains surface defects, the strength can be improved by tempering the surface.

REFRIGERATION AND LIQUEFACTION

A refrigeration or liquefaction process at cryogenic temperatures usually involves ambient compression of a suitable fluid with heat rejected to a coolant. During the compression and cooling process, the enthalpy and entropy of the fluid decrease. At the cryogenic temperature where heat is absorbed, the enthalpy and entropy of the fluid increase. The reduction in temperature of the process fluid is usually accomplished by heat exchange with a returning colder fluid and then followed with an expansion of the process fluid. This expansion may take place using either a throttling device approximating an isenthalpic expansion with only a reduction in temperature or a work-producing device approximating an isentropic expansion in which both the temperature and the enthalpy are decreased.

Normal commercial refrigeration generally uses a vapor compression process. Temperatures down to about 200 K can be obtained by cascading vapor compression processes in which refrigeration is obtained by liquid evaporation in each stage. Below this temperature, isenthalpic or isentropic expansion are generally used either singly or in combination. With few exceptions, refrigerators using these methods also absorb heat by vaporization of the liquid. If no suitable liquid exists to absorb the heat by evaporation over a temperature range, a cold gas must be available to absorb the heat. This is generally accomplished by using a work-producing expansion engine.

In a continuous refrigeration process, no refrigerant accumulates in any part of the system. This is in contrast with a liquefaction system, where liquid accumulates and is continuously withdrawn.

Thus, in a liquefaction system the total mass of the returning cold gas that is warmed prior to compression is less than the mass of gas that is to be cooled, creating an unbalance of flow in the heat exchangers. In a refrigerator, the mass flow rates of the warm and cold gas streams in the heat exchangers are usually equal, unless a portion of the warm gas flow is diverted through a work-producing expander. This condition of equal flow is usually called a **balanced flow condition** in the heat exchangers.

Isenthalpic Expansion

The thermodynamic process identified either as the simple **Linde cycle** or the **Joule-Thomson cycle** (J-T cycle) is shown schematically in Figure 3A. In this cycle, the gaseous refrigerant is compressed essentially isothermally at ambient temperature by rejecting heat to a coolant. The compressed refrigerant is then cooled in a heat exchanger with the cold gas stream returning to the compressor intake. Joule-Thomson cooling accompanying the expansion of the gas exiting the heat exchanger further reduces the temperature of the refrigerant until, under steady-state conditions, a small fraction of the refrigerant is liquefied. For a refrigerator, the liquid fraction is vaporized by absorbing the heat Q that is to be removed, combined with the unliquefied fraction, and, after warming in the heat exchanger, returned to the compressor. Figure 3B shows the ideal process on a temperature-entropy diagram.

Applying the first law to this refrigeration cycle, while assuming no heat leaks to the system as well as negligible kinetic and potential energy changes in the refrigeration fluid, the refrigeration effect per unit mass of refrigeration is simply the difference in enthalpies of streams 1 and 2 on Figure 3A. Thus, the coefficient of performance (COP) of the ideal J-T cycle is given by

$$\text{COP} = \frac{Q}{W} = \frac{h_1 - h_2}{T_1(s_1 - s_2) - (h_1 - h_2)} \qquad (3)$$

where Q is the refrigeration effect, W the work of compression, h_1 and s_1 are the specific enthalpy and entropy at point 1, and h_2 and s_2 the specific enthalpy and entropy at point 2 of Figure 3A.

For a liquefier, the liquefied portion is continuously withdrawn from the liquid reservoir and only the unliquefied portion of the fluid is warmed in the heat exchanger and returned to the compressor. The fraction y that is liquefied is determined by applying the first law to the heat exchanger, J-T valve, and liquid reservoir. This results in

$$y = \frac{h_1 - h_2}{h_1 - h_f} \qquad (4)$$

where h_f is the specific enthalpy of the saturated liquid being withdrawn. The maximum liquefaction occurs when the difference between h_1 and h_2 is maximized.

To account for any heat leak Q_L into the system, Equation (4) needs to be modified to

$$y = \frac{h_1 - h_2 - Q_L}{h_1 - h_f} \qquad (5)$$

resulting in a decrease in the fraction liquefied. The work of compression is identical to that determined for the J-T refrigerator. The figure of merit (FOM) is defined as $(W/m_f)_i/(W/m_f)$, where $(W/m_f)_i$ is the work of compression per unit mass liquefied for the ideal liquefier and (W/m_f) is the work of compression per unit mass liquefied for the J-T liquefier. The FOM reduces to

$$\text{FOM} = \left[\frac{T_1(s_1 - s_f) - (h_1 - h_f)}{T_1(s_1 - s_2) - (h_1 - h_2)}\right]\left[\frac{h_1 - h_2}{h_1 - h_f}\right] \qquad (6)$$

Liquefaction by this cycle requires that the inversion temperature of the refrigerant be above ambient temperature to provide cooling as the process is initiated. Auxiliary refrigeration is required if the J-T cycle is to be used to liquefy fluids whose inversion temperature is below ambient. This condition occurs when liquefaction of helium, hydrogen, and neon is attempted. Liquid nitrogen is the optimum auxiliary refrigerant for hydrogen and neon liquefaction systems, while liquid hydrogen accompanied by liquid nitrogen are the normal auxiliary refrigerants for helium liquefaction systems. Upper operating pressures for the J-T cycle are often as high as 20 MPa.

To reduce the work of compression in the previous cycle, a two-stage or dual-pressure cycle may be used in which the pressure is reduced by two successive isenthalpic expansions as shown in Figure 4. Since the work of compression is approximately proportional to the logarithm of the pressure ratio, and the Joule Thomson cooling is roughly proportional to the pressure difference, compressor work reduces more than refrigeration performance. Hence, the

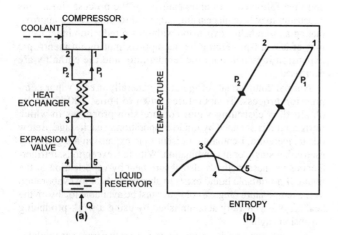

Fig. 3 (a) Schematic and (b) Temperature-Entropy Diagram for Simple Joule-Thomson Cycle Refrigerator

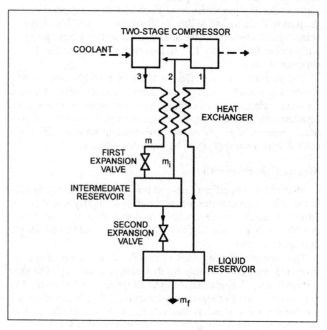

Fig. 4 Dual Pressure Joule-Thomson Cycle Used as Liquefier

Cryogenics

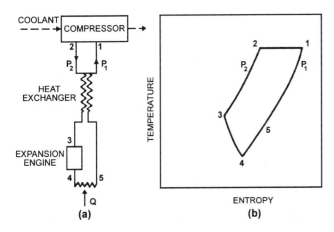

Fig. 5 (a) Schematic for Cold Gas Expansion Refrigerator
(b) Temperature-Entropy Diagram for Cycle

dual-pressure process produces the same quantity of refrigeration with less energy input than the simple J-T process.

The theoretical liquid yield for the dual-pressure J-T cycle is obtained by making an energy balance around the cold heat exchanger, lower J-T valve, and liquid reservoir as follows

$$(m - m_f - m_i)h_1 + m_i h_2 + m_f h_f - m h_3 = 0 \quad (7)$$

where m, m_f, and m_i are the mass flow rates of the streams designated on Figure 4. Solving for the liquid yield y for the ideal dual-pressure cycle results in the following relationship:

$$y = \frac{h_1 - h_3}{h_1 - h_f} - \frac{m_i}{m}\left[\frac{h_1 - h_2}{h_1 - h_f}\right] \quad (8)$$

The intermediate pressure in this cycle must be optimized. For a cycle with an upper pressure of 20 MPa, the intermediate optimum pressure generally occurs between 3 to 6 MPa.

Isentropic Expansion

Because the temperature always decreases in a work-producing expansion, cooling does not depend on being below the inversion temperature prior to expansion. In large industrial refrigerators, the work produced during the expansion is conserved. In small refrigerators, the energy from the expansion is usually dissipated in a gas or hydraulic pump or other suitable device.

A schematic of a refrigerator using the work-producing expansion principle and the corresponding temperature-entropy diagram are shown in Figure 5. Gas compressed isothermally at ambient temperature is cooled in a heat exchanger by gas being warmed on its return to the compressor intake. Further cooling takes place in the expansion engine. In practice, this expansion is never truly isentropic, as shown by the path 3–4 on the temperature-entropy diagram. The refrigerator shown produces a cold gas that absorbs heat during the path 4–5 and provides a method of refrigeration that can be used to obtain temperatures between those of the boiling points of the lower boiling cryogenic fluids.

The coefficient of performance of an ideal gas refrigerator with varying refrigerator temperature can be obtained from the relation

$$COP = \frac{T_5 - T_4}{T_1 \ln(T_5/T_4) - (T_5 - T_4)} \quad (9)$$

where the absolute temperatures refer to the designated points shown in Figure 5 and T_1 is the ambient temperature.

Combined Isenthalpic and Isentropic Expansion

To take advantage of the increased reversibility gained by the use of a work-producing expansion engine while at the same time minimizing the problems associated with the formation of liquid in such a device, Claude developed a process that combined both expansion processes in the same cycle; a schematic of this cycle along with the corresponding temperature-entropy diagram is shown in Figure 6.

An energy balance for this system, which incorporates the heat exchangers, expansion engine, J-T expansion valve, and liquid reservoir permits evaluation of the refrigeration effect or, in the case of a liquefier, allows evaluation of the liquid yield. For the refrigerator, the refrigeration effect is

$$Q = m(h_1 - h_2) + m_e(h_3 - h_e) \quad (10)$$

where m_e and h_e are the mass flow rate through the expander and the specific enthalpy of the stream leaving the expander, respectively.

For a liquefier, the energy balance yields

$$m h_2 - (m - m_f) h_1 - m_e(h_3 - h_e) - m_f h_f = 0 \quad (11)$$

Collecting terms and using previously adopted symbols gives the liquid fraction obtained as

$$y = \frac{h_1 - h_2}{h_1 - h_f} - \frac{m_e}{m}\left[\frac{h_3 - h_e}{h_1 - h_f}\right] \quad (12)$$

For a nonideal expander the h_e term would be replaced with $h_{e'}$.

One modification of the **Claude cycle** that has been used extensively in high-pressure liquefaction plants for air is the **Heylandt cycle**. In this cycle the first warm heat exchanger in Figure 5 is eliminated, thereby allowing the inlet of the expander to operate at ambient temperatures, which minimizes many of the lubrication problems that are often encountered at lower temperatures.

Other modifications of the basic Claude cycle are the use of a "wet" expander operating in the two-phase region to replace the throttling valve and the addition of a saturated vapor compressor following the liquid reservoir. The two-phase expander is used with systems involving helium as the working fluid because the thermal capacity of the compressed gas is, in many cases, larger than the latent heat of the liquid phase. Experience has shown that operation of the expander with helium in the two-phase region has

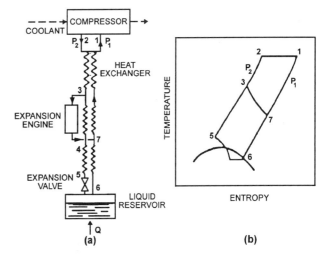

Fig. 6 (a) Schematic for Claude Cycle Refrigerator
(b) Temperature-Entropy Diagram for Cycle

little effect on expander efficiency as experienced with expanders handling either nitrogen or air. The addition of the saturated vapor compressor actually improves the thermodynamic performance of the system.

Another modification of the basic Claude cycle is the dual-pressure Claude cycle, similar in principle to the Linde dual-pressure system shown in Figure 4. In this dual-pressure cycle, only the gas that flows through the expansion valve is compressed to the high pressure; this reduces the work requirement per unit mass of gas liquefied. Barron (1985) compared the Claude dual-pressure cycle with the Linde dual-pressure cycle and showed that in the liquefaction of air the liquid yield can be doubled while the work per unit mass liquefied can be halved when the Claude dual-pressure cycle is selected.

Still another extension of the Claude cycle is the **Collins helium liquefier**. Depending on the helium inlet pressure, from two to five expansion engines are used in this system. The addition of a liquid nitrogen precooling bath results in a two- to threefold increase in liquefaction performance.

Mixed Refrigerant Cycle

With the advent of large natural gas liquefaction plants, the mixed refrigerant cycle (MRC) is used exclusively in LNG production. This cycle, in principle, resembles the cascade cycle occasionally used in LNG production as shown in Figure 7. After purification, the natural gas stream is cooled by the successive vaporization of propane, ethylene, and methane. These gases have each been liquefied in a conventional refrigeration loop. Each refrigerant may be vaporized at two or three different pressure levels to increase the natural gas cooling efficiency, but at a cost of considerably increased process complexity.

Cooling curves for natural gas liquefaction by the cascade process are shown in Figures 8a and 8b. The cascade cycle efficiency can be improved considerably by increasing the number of refrigerants used. For the same flow rate the actual work required for the nine-level cascade cycle is approximately 80% of that required by the three-level cascade cycle. This increase in efficiency is achieved by minimizing the temperature difference throughout the cooling curve.

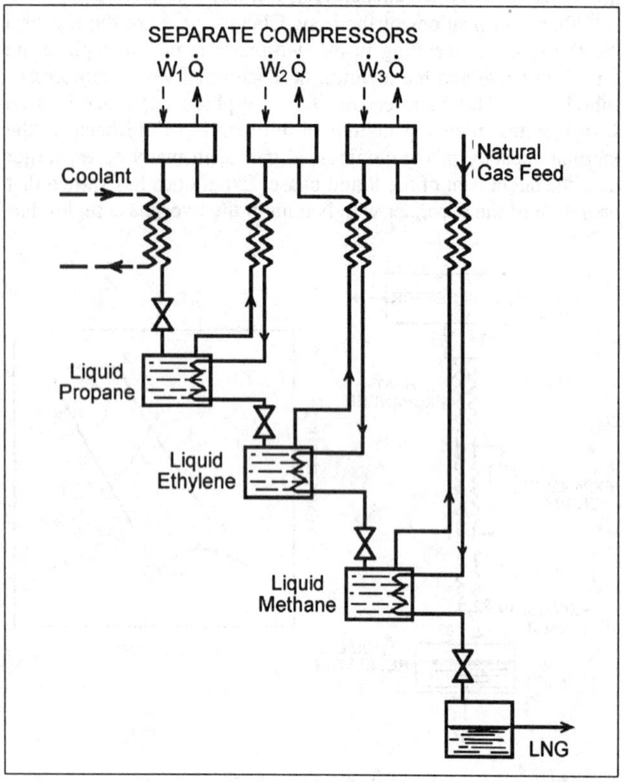

The cascade system can be designed to approximate any cooling curve; that is, the quantity of refrigeration provided at the various temperature levels can be chosen so that the temperature differences between the warm and cold streams in the evaporators and heat exchangers approach a practical minimum (smaller temperature differences equate to lower irreversibilities and therefore lower power consumption).

The simplified version of the mixed refrigerant cycle shown in Figure 9 is a variation of the cascade cycle described previously. This version involves the circulation of a single mixed refrigerant system, which is repeatedly condensed, vaporized, separated, and expanded. As a result, such processes require more sophisticated design approaches and more complete knowledge of the thermodynamic properties of gaseous mixtures than expander or cascade cycles. Nevertheless, simplifying compression and heat exchange in such cycles offers potential for reduced capital expenditure over conventional cascade cycles. Note the similarity of the temperature versus enthalpy diagram shown in Figure 10 for the mixed

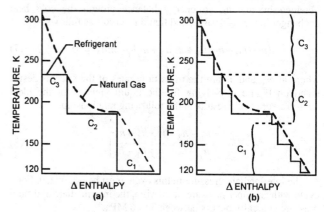

Fig. 8 (a) Three-Level and (b) Nine-Level Cascade Cycle Cooling Curves for Natural Gas

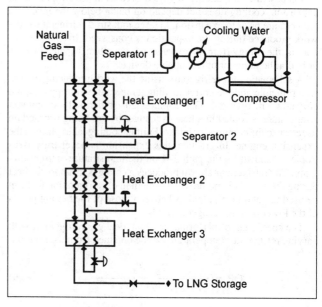

Fig. 9 Mixed Refrigerant Cycle Used for Liquefaction of Natural Gas

Fig. 7 Classical Cascade Compressed Vapor Refrigerator

Cryogenics

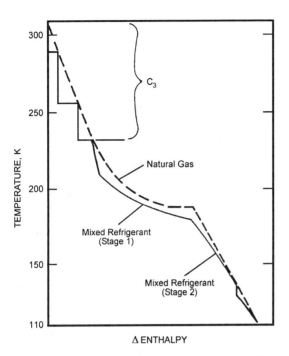

Fig. 10 Propane Precooled Mixed Refrigerant Cycle Cooling Curve for Liquefaction of Natural Gas

refrigerant cycle with the corresponding diagram for the nine-level cascade cycle in Figure 8b.

Proprietary variations of the mixed refrigerant have developed the technology. In one commercial process, for example, the gas mixture is obtained by condensing part of the natural gas feed. This has the advantage of requiring no fluid input other than the natural gas itself. On the other hand, this procedure can cause a slow start up because of the refrigerant that must be collected and the time required to adjust its composition. Another version uses a multi-component refrigerant that is circulated in a completely separate flow loop. The mixture is prepared from pure gaseous components to obtain the desired composition. Thereafter only occasional makeup gases are added, usually from cylinders. This process is simpler and allows for rapid start up. On the other hand, the refrigerant mixture must be stored when the process is shut down. Accordingly, suction and surge drums must be provided with a resulting increase in capital expenditure. Gaumer (1993) presents a good review of the mixed refrigerant cycle.

Comparison of Refrigeration and Liquefaction Systems

The measure of the thermodynamic quality associated with either a piece of equipment or an entire process is its reversibility. The second law, or more precisely the entropy increase, is an effective guide to this degree of irreversibility. However, to obtain a clearer picture of what these entropy increases mean, it is convenient to relate such an analysis to the additional work that is required to overcome these irreversibilties. The fundamental equation for such an analysis is

$$W = W_{rev} + T_o \sum m \Delta s \quad (13)$$

where the total work is the sum of the reversible work W_{rev} plus a summation of the losses in availability for various steps in the analysis. Here T_o is the reference temperature (normally ambient), m the flow rate through each individual process step, and Δs the change in entropy across these same process steps.

Caution must be exercised before accepting comparisons made in the literature because it is difficult to put all processes on a comparable basis. Many assumptions must be made in the course

Table 1 Comparison of Several Liquefaction Systems Using Air as the Working Fluid

Air Liquefaction System (Inlet conditions of 294.4 K and 101.3 kPa)	Liquid Yield $y = m_f/m$	Work per Unit Mass Liquefied, kJ/kg	Figure of Merit
Ideal reversible system	1.000	715	1.000
Simple Linde system, $p_2 = 20$ MPa, $\eta_c = 100\%$, $\varepsilon = 1.0$	0.086	5240	0.137
Simple Linde system, $p_2 = 20$ MPa, $\eta_c = 70\%$, $\varepsilon = 0.95$	0.061	10620	0.068
Simple Linde system, observed	—	10320	0.070
Precooled simple Linde system, $p_2 = 20$ MPa, $T_3 = 228$ K, $\eta_c = 100\%$, $\varepsilon = 1.00$	0.179	2240	0.320
Precooled simple Linde system, $p_2 = 20$ MPa, $T_3 = 228$ K, $\eta_c = 70\%$, $\varepsilon = 0.95$	0.158	3700	0.194
Precooled simple Linde system, observed	—	5580	0.129
Linde dual-pressure system, $p_3 = 20$ MPa, $p_2 = 6$ MPa, $i = 0.8$, $\eta_c = 100\%$, $\varepsilon = 1.00$	0.060	2745	0.261
Linde dual-pressure system, $p_3 = 20$ MPa, $p_2 = 6$ MPa, $i = 0.8$, $\eta_c = 70\%$, $\varepsilon = 0.95$	0.032	8000	0.090
Linde dual-pressure system, observed	—	6340	0.113
Linde dual-pressure system, precooled to 228 K, observed	—	3580	0.201
Claude system, $p_2 = 4$ MPa, $x = m_e/m = 0.7$, $\eta_c = \eta_e = 100\%$, $\varepsilon = 1.00$	0.260	890	0.808
Claude system, $p_2 = 4$ MPa, $x = m_e/m = 0.7$, $\eta_c = 70\%$, $\eta_{e,ad} = 80\%$, $\eta_{e,m} = 90\%$, $\varepsilon = 0.95$	0.189	2020	0.356
Claude system, observed	—	3580	0.201
Cascade system, observed	—	3255	0.221

η_c = compressor overall efficiency
η_e = expander overall efficiency
$\eta_{e,ad}$ = expander adiabatic efficiency
$\eta_{e,m}$ = expander mechanical efficiency
ε = heat exchanger effectiveness
$i = m_i/m$ = mass in intermediate stream divided by mass through compressor
$x = m_e/m$ = mass through expander divided by mass through compressor

of the calculations and these can have considerable effect on the conclusions. The chief factors upon which assumptions generally have to be made include heat leak, temperature differences in the heat exchangers, efficiencies of compressors and expanders, number of stages of compression, fraction of expander work recovered, state of flow, etc. For this reason, differences in power requirements of 10 to 20% can readily be due to differences in assumed variables and can negate the advantage of one cycle over another. A comparison that demonstrates this point is made in Table 1 of some of the more common liquefaction systems described earlier using air as the working fluid with a compressor inlet gas temperature and pressure of 294.4 K and 0.1 MPa, respectively.

An analysis that avoids these pitfalls is one that compares the minimum power requirements to produce a unit of refrigeration. Table 2 provides such data for some of the more common cryogens in addition to the minimum power requirements for liquefaction. The warm temperature in all cases is fixed at 300 K while the cold temperature is assumed to be the normal boiling temperature T_{bp} of the fluid. The specific power requirements increase rapidly as the boiling point of the fluids decrease.

Table 2 Reversible Power Requirements

		Ideal Power Input for		
Fluid	T_{bp}, K	1 W Refrig. Capacity, W/W	1 L Liquid, W/L	1 W Refrig. Capacity from Liquid[a], W/W
Helium	4.2	70.4	236	326
Hydrogen	20.4	13.7	278	31.7
Neon	27.1	10.1	447	15.5
Nitrogen	77.4	2.88	173	3.87
Fluorine	85.0	2.53	238	3.26
Argon	87.3	2.44	185	2.95
Oxygen	90.2	2.33	195	2.89
Methane	111.5	1.69	129	2.15

[a] Obtained by dividing ideal liquefaction power requirement by heat of vaporization of fluid.

Table 2 also shows that more power is required to produce a given amount of cooling with the liquid from a liquefier than is needed for a continuously operating refrigerator. This greater power is needed because the refrigeration effect of the effluent vapor is not available for cooling the feed gas stream when the liquid is evaporated at another location (other than the source of liquefaction itself).

An ideal helium liquefier requires a power input of 236 W to produce liquid at the rate of 1 L/h (850 MJ/m³). Because the heat of vaporization of helium is low, 1 W will evaporate 1.38 L/h. Thus, 326 W would be required to ideally operate a liquefier whose liquid product was used to absorb 1 W of refrigeration at 4.2 K. An ideal refrigerator, on the other hand, would only require 70.4 W of input power to produce the same quantity of refrigeration also at 4.2 K. This difference in power requirement does not mean that a refrigerator will always be chosen over a liquefier for all cooling applications; in some circumstances a liquefier may be the better or the only choice. For example, a liquefier is required when a constant temperature bath is used in a low-temperature region.

Another comparison of low-temperature refrigeration examines the ratio of W/Q for a Carnot refrigerator with the ratio obtained for the actual refrigerator. This ratio indicates the extent to which an actual refrigerator approaches ideal performance. (The same ratio can be formed for liquefiers using the values from Table 2 and the actual power consumption per unit flow rate.) This comparison with ideal performance is plotted in Figure 11 as a function of refrigeration capacity for actual refrigerators and liquefiers. The capacity of the liquefiers included in this comparison was obtained by determining the percent of Carnot performance that these units achieved as liquefiers and then calculating the refrigeration output of a refrigerator operating at the same efficiency with the same power input.

The data for these low-temperature refrigerators cover a wide range of capacity and temperatures. The more efficient facilities are the larger ones that can more advantageously use complex thermodynamic cycles. The same performance potential may exist in the smaller units, but costs are prohibitive and the savings in electrical power have generally not justified the greater capital expenditure.

CRYOCOOLERS

Small, low-temperature refrigerators that provide no more than a few watts of cooling are generally called **cryocoolers**. As noted in the previous section, these units are less efficient. However, other problems of unreliability, size, mass, vibration, and cost have also been major concerns for cryocooler developers. The seriousness of any one of these problems depends on the application of the cryocooler. The largest application of cryocoolers has been in the cooling of infrared sensors in satellites by the military. Stirling cryocoolers with a refrigeration capacity of about 0.25 W at 80 K

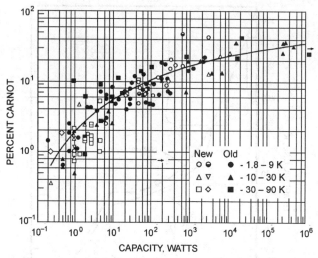

Fig. 11 Efficiency as Percent of Carnot Efficiency
(For low-temperature refrigerators and liquefiers as function of refrigeration capacity)

have primarily been used for this application. However, the mean-time-to-failure of about 4000 h has been inadequate to meet the required needs of space operations. This inadequate reliability of the Stirling cryocooler has provided the impetus for major innovations for such units and the rapid growth of research and development of pulse tube refrigerators.

The largest application of cryocoolers in the commercial area has been in cryopumping for the manufacture of semiconductors. Gifford-McMahon cryocoolers providing a few watts of refrigeration at a temperature of about 15 K are the most popular devices for this area. This choice may change, however, because of the vibration characteristics of this cryocooler. Semiconductor manufacturers are forced to achieve the narrower line widths in their computer chips to provide more compact packaging of semiconductor circuits.

Even though cryocoolers can be classified by the thermodynamic cycle that is followed, generally there are only two types; **recuperative** or **regenerative units** (Walker 1983). A cryocooler is a recuperative type if only recuperative heat exchangers are used in the unit, and it is a regenerative type if at least one regenerative heat exchanger (defined as a regenerator) is used in the unit.

Because the recuperative heat exchangers provide two separate flow channels for the refrigerant, the refrigerant flow is always continuous and in one direction, analogous to a dc electrical system. This heat exchanger requires either valving with reciprocating compressors and expanders or rotary or turbine type compressors and expanders. In regenerative cycles the refrigerant flow oscillates, analagous to an ac electrical system. This oscillatory effect allows the regenerator matrix to store energy for the first half of the cycle in the matrix and release the energy during the next half of the cycle from the matrix. To be effective, the solid matrix material placed within the regenerator must have high heat capacity and good thermal conductivity. Some of the recent advances that have occurred in both types of cryocoolers are reviewed in the following sections.

Recuperative Systems

The J-T and the **Brayton cycles**, compared schematically in Figure 12, are two recuperative systems.

Joule-Thomson Cryocoolers. The J-T effect of achieving cooling by throttling a nonideal gas is one of the oldest but least efficient methods for attaining cryogenic temperatures. However, significant improvements in J-T cryocoolers have been achieved by applying novel methods of fabrication, incorporating more complex cycles, and using special gas mixtures as refrigerants. These developments

Cryogenics

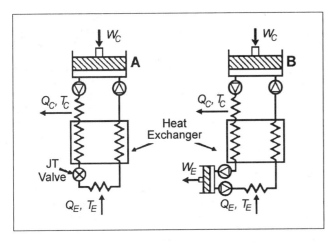

Fig. 12 Schematic of Joule-Thomson and Brayton Cycles
(Two recuperative types of cycles used in cryocoolers)

have made the J-T cryocooler competitive in many applications even when compared with cryocoolers that generally exhibit more efficient cooling cycles. Their relative simplicity combined with their small size, low mass, and absence from mechanical noise or vibration have provided additional advantages for these small refrigerators. In the past, J-T cryocoolers have been fabricated by winding a finned, capillary tube on a mandril, attaching an expansion nozzle at the end of the capillary tube, and inserting the entire unit in a tightly fitting tube closed at one end with inlet and exit ports at the other end. Little (1989) introduced a method of fabricating J-T cryocoolers using a photolithographic manufacturing technique in which gas channels for the heat exchangers, expansion capillary, and liquid reservoir are etched on thin planar, glass substrates that are fused together to form a sealed unit. These miniature cryocoolers have been fabricated in a wide range of sizes and capacities. A cryocooler operating at 80 K with a refrigeration capacity of 250 mW uses a heat exchanger whose channels have a width of 200 μm and a depth of 30 μm. The channels etched on the glass substrate must be controlled to a tolerance of ±2 μm, and the bond between the different substrates must withstand pressures on the order of 15 to 20 MPa. This cryocooler, used in spot cooling of electronic systems, has an overall dimension not including the compressor of only 75 mm × 14 mm × 2 mm.

Fabrication of miniature J-T cryocoolers by this approach has made it much simpler to use more complex refrigeration cycles and multistage configurations. The dual-pressure J-T cycle in which the refrigerant pressure is reduced by two isenthalpic expansions provides either a lower spot cooling temperature or a higher coefficient of performance for the same power input. Attainment of 20 K requires two separate refrigeration systems in which a nitrogen J-T refrigeration stage provides the precooling to 77 K of a hydrogen J-T refrigeration stage. For 4 K, an additional helium stage of refrigeration cooled by the hydrogen stage is required. Use of multistage units in these miniature cryocoolers requires an order of magnitude better dimensional control of the etching process in order to match the desired flows and capacities specified for the heat exchangers, expansion capillaries, and liquid reservoirs.

To attain temperatures of 77 K, pure nitrogen has been used as the refrigerant in J-T cryocoolers. At 300 K, nitrogen must be compressed to a very high pressure (10 to 20 MPa) to achieve any significant enthalpy change. The high pressure required, leads to a low compression efficiency with high stresses on compressor components, while the small enthalpy change results in a low cycle efficiency. Alfiev et al. (1973) using a gaseous mixture of 30 mol% nitrogen, 30 mol% methane, 20 mol% ethane, and 20 mol% propane achieved a temperature of 78 K using a 50:1 pressure ratio. The system efficiency with this gas mixture was 10 to 12 times better than when pure nitrogen gas was used as the refrigerant. Temperatures below 70 K were obtained by adding neon, hydrogen, or helium to the mixture.

Little (1984) established that the addition of the fire retardant, CF_3Br, to the nitrogen-hydrocarbon mixture was sufficient to render the mixture nonflammable and was retained in the resulting liquid solution down to 77 K or lower without precipitation because of the excellent solvent properties of the mixture. As a result, a series of nitrogen-hydrocarbon gas mixtures that are reasonably safe to use and provide refrigeration efficiencies approaching 50% of Carnot efficiency (excluding compressor losses) are now available.

An example highlights this advance. Consider the temperature-entropy diagram shown in Figure 13 for a hydrocarbon mixture of 27% methane, 50% ethane, 13% propane, and 10% butane on a volumetric basis. A throttling process for this gas mixture, initially at 300 K and 4.5 MPa can ideally (constant enthalpy) achieve an exit temperature of 200 K at a final exit pressure of 0.1 MPa. Pure nitrogen gas undergoing a similar throttling process from the same inlet conditions to the same final exit pressure only achieves an exit gas temperature of 291 K. Thus, there can be as much as an eleven-fold increase in the temperature drop of the refrigerant after the throttling process by using the gas mixture instead of the pure nitrogen gas over these pressure and temperature conditions. That is, refrigeration performance comparable to that using pure nitrogen gas at 12 to 15 MPa inlet pressures can be achieved with specific gas mixtures at pressure as low as 3 to 5 MPa.

The effect of various gas mixture concentrations on the efficiency of any cycle can be analyzed by evaluating the coefficient of performance of the cycle. The refrigeration effect Q of a J-T refrigerator using such gas mixtures is given by

$$Q = n(h_{lp} - h_{hp})_{min} = n\Delta h_{min} \qquad (14)$$

where n is the molar flow rate of the gas mixture, h_{lp} is the molar enthalpy of the low-pressure stream, and h_{hp} is the molar enthalpy of the high-pressure stream at the location in the recuperative heat exchanger that provides a minimum difference in molar enthalpies

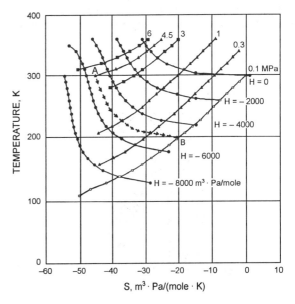

Fig. 13 Isenthalpic Expansion of Multicomponent Gaseous Mixture from A to B
(27% methane, 50% ethane, 13% propane, and 10% butane)

Δh_{min} between the two streams. The ideal work of compression is evaluated from

$$W_{ideal} = n[(h_2 - h_1) - T_o(s_2 - s_1)] = n\Delta g_o \quad (15)$$

where s_1 and s_2 are the molar entropies at the inlet and outlet from the compressor, respectively, at a constant compression temperature of T_o. The Δg_o is the change in the molar Gibbs free energy also at T_o. The ideal COP of the refrigerant cycle is then

$$COP = Q/W_{ideal} = \Delta h_{min}/\Delta g_o \quad (16)$$

This indicates that a maximum efficiency in the J-T cycle is achieved when the value of $\Delta h_{min}/\Delta g_o$ is maximized for the refrigerant mixture in the temperature range of interest.

Brayton Cryocoolers. The use of an expansion engine to carry out the expansion of the refrigerant in the Brayton cycle leads to higher cycle efficiencies than are attainable with J-T cryocoolers. The Brayton cycle is commonly used in large liquefaction systems accompanied by a final J-T expansion. The units have a high reliability due to the use of turboexpanders operating with gas bearings. For small cryocoolers the challenge has been in fabricating the miniature turboexpanders while maintaining a high expansion efficiency and minimizing heat leakage. Swift and Sixsmith (1993) addressed this challenge by developing a single-stage Brayton cryocooler with a small turboexpander (rotor diameter of 3.2 mm) providing 5 W of refrigeration at 65 K with neon as the working fluid. The compressor also uses gas bearings with an inlet pressure of 0.11 MPa and a pressure ratio of 1.6. The unit operates between 65 and 280 K with a Carnot efficiency of 7.7%. However, the present cost of such cryocoolers limits their service to space applications, which require high reliability, high thermodynamic efficiency, and low vibration.

Regenerative Systems

Stirling Cryocoolers. The Stirling refrigerator, which boasts the highest efficiency of all cryocoolers, is the oldest and most common of the regenerative systems. The elements of the Stirling refrigerator normally include two variable volumes at different temperatures, coupled together through a regenerative heat exchanger, a heat exchanger rejecting the heat of compression and a refrigerator absorbing the refrigeration effect. These elements can be arranged in a wide variety of configurations and operate as either single- or double-acting systems. The single-acting units are either two-piston or piston-displacer systems as shown in Figure 14.

The ideal Stirling cycle consists of four processes. Step one involves an isothermal compression of the refrigerant in the compression stage at ambient temperature by rejecting heat Q_c to the surroundings. This is followed by a constant volume regenerative cooling where heat is transferred from the working fluid to the regenerator matrix. The reduction in temperature at constant volume causes a reduction in the pressure. The third step consists of an isothermal expansion in the expansion space at the refrigeration temperature T_E. Heat Q_E in this step is absorbed from the surroundings of the expansion space. The process is completed by a constant volume regenerative heating in which heat is transferred from the regenerator matrix to the working fluid. The increase in temperature at constant volume increases pressure back to the initial conditions. Successful operation of the cycle requires that the volume variations in the expansion space lead those in the compression space.

Thousands of small, single-stage Stirling cryocoolers have been manufactured. Capacities range from about 10 mW to 1 W at 80 K. The largest units (not including the compressor) are generally no more than 150 mm with a mass of less than 3 kg. Power inputs range from 40 to 50 W per watt of refrigeration, which is equivalent to an efficiency of 6 to 7% of Carnot.

Many of the recent developments in Stirling cryocoolers have been directed towards improved reliability. In most applications, for example, linear motor drives have replaced rotary drives to eliminate many of the moving parts as well as reduce the side forces existing between the piston and cylinder. Lifetimes of about 4000 h are the norm with the linear compressors; however, lifetimes greater than 15,000 h have been achieved with the use of improved materials for the rubbing contact. Longer lifetimes have been achieved with piston devices by using flexure, gas, or magnetic bearings to center the piston and displacer in the cylinder housing. Davey (1990) reviews the development of these cryocoolers.

As noted earlier, space applications require lifetimes of 10 to 15 years, low mass, and low energy consumption. These considerations have directed research on the orifice pulse tube refrigerator shown schematically in Figure 15. This unit is a variation of the Stirling cryocooler in which the moving displacer is replaced by a pulse tube, orifice, and reservoir volume. Radebaugh (1990) gives a detailed review of pulse tube refrigerators.

Orifice Pulse Tube Refrigerators. The orifice pulse tube refrigerator operates on a cycle similar to the Stirling cycle except that the proper phasing between mass flow and pressure is established by the passive orifice rather than by the moving displacer. In this cycle, a low-frequency compressor raises the pressure of the helium refrigerant gas to a level between 0.5 and 2.5 MPa during the first half of

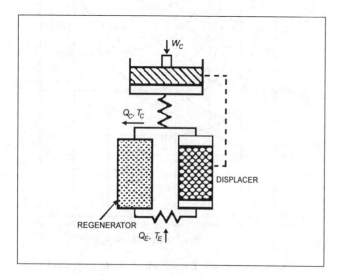

Fig. 14 Schematic of Stirling Cryocooler

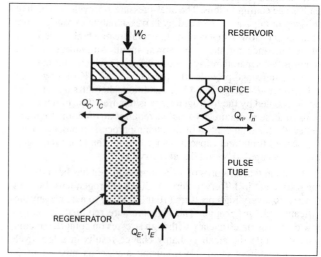

Fig. 15 Schematic for Orifice Pulse Tube Cryocooler

a sinusoidal compression cycle. (The oscillating pressure for the OPTR can be provided either by a compressor similar to that used in the Stirling refrigerator or by a Gifford-McMahon compressor that has been modified with appropriate valving to achieve the required oscillating pressure.) The high-pressure gas, after being cooled in the regenerator, adiabatically compresses the gas in the pulse tube.

Approximately one third of the compressed gas originally in the pulse tube flows through the orifice to the reservoir volume with the heat of compression being removed in the hot exchanger. During the latter half of the sinusoidal cycle, the gas in the pulse tube expands adiabatically, which causes a cooling effect. The cold, expanded gas is forced past the cold heat exchanger as a buffer volume of gas that allows a temperature gradient to exist between the hot and cold ends of the pulse tube. Some mixing or turbulence occurs in the buffer volume because the time averaged enthalpy flow which represents the gross refrigeration capacity is only 55 to 85% of the ideal enthalpy flow.

By assuming simple harmonic pressure, mass flow, and temperature oscillations in the entire pulse tube refrigerator as well as adiabatic operation in the pulse tube itself, researchers at NIST developed an analytical model that reasonably predicts refrigeration performance. Thermoacoustic theories include a linear approximation with higher harmonics and realistic heat transfer and viscous effects between the gas and pulse tube wall. The losses accounted for in these models result in a time-averaged enthalpy flow that agrees closely with experimental values.

The orifice concept for the pulse tube refrigerators achieves a refrigeration temperature of 60 K with only one stage. Further refinement permits achieving temperatures below 40 K However, the improved Stirling refrigerator remains the choice for most space applications because its Carnot efficiency is presently higher than those obtained from the OPTR.

Zhu et al. (1990) improved efficiencies for pulse tube refrigerators with higher operating frequencies by adding a second orifice, as shown in Figure 16. The addition of this orifice, identified as the double-inlet concept, permits the gas flow needed to compress and expand the gas at the warm end of the pulse tube to bypass the regenerator and pulse tube. The reduced mass flow through the regenerator reduces the regenerator losses, particularly at high frequencies where these losses become large. Addition of the second orifice can reduce the refrigerator temperature by at least 15 to 20 K in a well-designed pulse tube operating at frequencies of 40 to 60 Hz. This was substantiated in 1992 with a temperature of 28 K, the lowest temperature achieved to date with a single-stage, double-inlet arrangement.

A comparison of the percent of Carnot efficiency obtained for the improved pulse tube refrigerators with the Stirling refrigerators is shown in Figure 17. The shaded area represents the efficiency range obtained for most of the recent Stirling refrigerators while the circles represent the individual efficiencies obtained from recent pulse tube refrigerators. The highest power and highest efficiency unit is the pulse tube refrigerator described by Radebaugh (1995). This unit provided 31.1 W of refrigeration at 80 K with a rejection temperature of 316 K equivalent to a relative Carnot efficiency of 13%. The average operating pressure was 2.5 MPa while the operating frequency was maintained at 4.5 Hz. The other two circles with lower efficiencies in the 65 to 80 K range represent the efficiencies obtained for the same unit but with different input powers and different cold end temperatures. The efficiency shown in the 30 to 35 K range is for a small unit developed by Burt et al. (1995). Even though the data for pulse tube refrigerators are limited, the efficiencies of the most recent pulse tube refrigerators is becoming quite competitive with the best Stirling refrigerators of comparable size.

Two or more pulse tube refrigerator stages are normally used to maintain high efficiency when temperatures below about 80 K are desired. Purposes of the staging are to provide net cooling at an intermediate temperature and to intercept regenerator and pulse tube losses at a higher temperature. Three methods exist for the staging arrangement. The first uses a **parallel arrangement** of a separate regenerator and pulse tube for each stage with the warm end of each pulse tube located at ambient temperature. In the second method, shown in Figure 18, the warm end of the lower stage pulse tube is thermally anchored to the cold end of the next higher stage in a **series configuration**. The third method uses a third orifice to permit a fraction of the gas removed from an optimized location in the regenerator to enter the pulse tube at an intermediate temperature. This staging configuration, identified as the **multi-inlet arrangement**, maintains the simple geometrical arrangement of a single pulse tube, although it would normally require a change in diameter at the tube junction with the pulse tube to maintain a constant gas velocity in the pulse tube. The lowest temperature attained with a three-stage parallel arrangement of pulse tube refrigerators is 3.6 K.

Gifford-McMahon Refrigerator. As noted earlier, J-T cryocoolers using pure gas refrigerants require very high operating pressures; therefore most commercial closed-cycle cryocoolers use one or more expanders to achieve part of the cooling effect. One of the most widely used regenerative cryocoolers is the Gifford-McMahon refrigerator schematically shown in Figure 19. These units can achieve temperatures of 65 to 80 K with one stage of expansion and 15 to 20 K with two stages of expansion. Precooling of the expansion stage is accomplished with regenerators using carefully selected matrix materials. Because regenerators essentially store energy, the matrix materials must possess a high heat capacity as

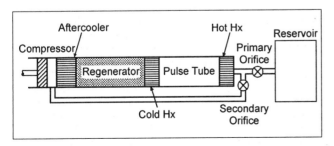

Fig. 16 Schematic of Double-Inlet Pulse Tube Refrigerator Using Secondary Orifice

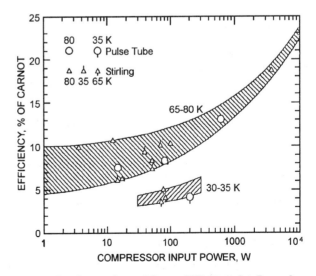

Fig. 17 Comparison of Carnot Efficiency for Several Recent Pulse Tube Cryocoolers with Similarly Powered Stirling Cryocoolers

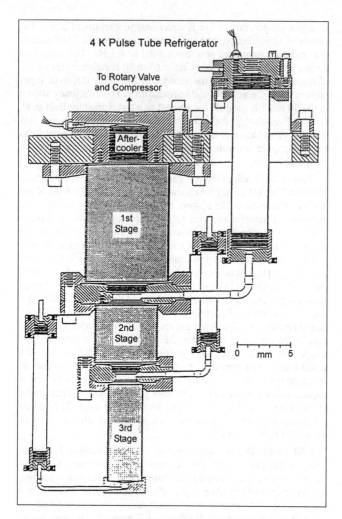

Fig. 18 Three-Stage Series Orifice Pulse Tube Cryocooler for Liquefying Helium

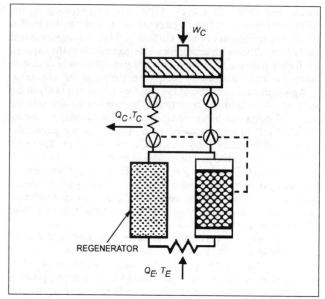

Fig. 19 Schematic for Single-Stage Gifford-McMahon Refrigerator

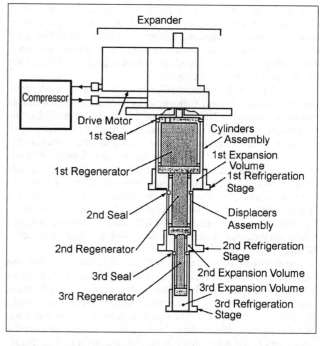

Fig. 20 Cross Section of Three-Stage Gifford-McMahon Refrigerator

well as a good thermal conductivity. Lead shot has been the regenerator material selected for most regenerative cryocooler operation between 10 to 65 K. However, its heat capacity becomes ineffective below this temperature range.

Without a suitable matrix material, addition of a third stage with its accompanying regenerator has made it impossible for any regenerative cryocooler of achieving a temperature below 10 to 12 K. Attaining a temperature of 4.2 K to reliquefy helium boil-off has required a two-stage Gifford-McMahon refrigerator equipped with a J-T loop using a compact countercurrent heat exchanger. However, the availability of new matrix materials (rare earth compounds) has made it possible to use a three-stage Gifford-McMahon refrigerator to provide sufficient cooling to reliquefy helium boil-off from the superconducting magnets serving MRI units. Nagao et al. (1994) described such a device, shown in Figure 20. It uses $Er_{1.5}Ho_{1.5}Ru$ as the matrix material in the third-stage regenerator and provides a refrigeration capacity of more than 150 mW at 4.2 K. This unit, which is smaller than the conventional 4 K Gifford-McMahon refrigerator, also has greater reliability as well as lower operating costs.

SEPARATION AND PURIFICATION OF GASES

The major application of low-temperature processes in industry involves the separation and purification of gases. Much of the commercial oxygen and nitrogen, and all of the neon, argon, krypton, and xenon are separated from air. Pressure swing adsorption processes account for the oxygen and nitrogen production that is not obtained by cryogenic separation of air. Commercial helium is separated from helium-bearing natural gas by a low-temperature process. Cryogenics has also been used commercially to separate hydrogen from various sources of impure hydrogen. Even the valuable low-boiling components of natural gas (i.e., methane, ethane, ethylene, propane, propylene, and others) are recovered and purified by various low-temperature schemes. The separation of these gases is dictated by the **Gibbs phase rule**. The degree to which they separate is based on the physical behavior of the liquid and vapor phases. This behavior is governed, as at ambient temperatures, by the laws of Raoult and Dalton.

Cryogenics

The energy required to reversibly separate gas mixtures is the same as the work needed to isothermally compress each component in the mixture from its own partial pressure in the mixture to the final pressure of the mixture. This reversible isothermal work per unit mass is given by the relation

$$(W/m)_i = T_1(s_1 - s_2) - (h_1 - h_2) \quad (17)$$

where s_1 and h_1 refer to the entropy and enthalpy before the separation and s_2 and h_2 refer to the entropy and enthalpy after the separation. For a binary system of components A and B, and assuming an ideal gas for both components, Equation (17) simplifies to

$$(W/n_T)_i = -RT\left[n_A \ln \frac{p_T}{p_A} + n_B \ln \frac{p_T}{p_B}\right] \quad (18)$$

in which n_A and n_B are the moles of components A and B in the mixture, p_A and p_B are the partial pressures of these two components in the mixture, and p_T is the total pressure of the mixture.

The figure of merit for a separation system is defined in a manner similar to that for a liquefaction system, namely

$$\text{FOM} = \frac{(W/m)_i}{(W/m)_{act}} \quad (19)$$

The number of stages or plates to effect a low-temperature separation is determined by the same procedures as developed for normal separations. A computer is programmed to make interactive mass and energy balances around each plate in a separation column to determine the number of plates required to effect a desired separation. To make these computations meaningful requires accurate thermodynamic data for mixtures and an appreciation for the efficiency of separation that can be expected on each plate. Experience is an important factor in selecting appropriate efficiency factors because they vary from 65 to 100%.

Air Separation

Figure 21 provides a simplified schematic of the Linde single-column originally used for air separation and presently used commercially to produce low-impurity **nitrogen** gas for inerting purposes. The separation scheme shown uses the simple J-T liquefaction cycle considered earlier but with a rectification column substituted for the liquid reservoir. (Any other liquefaction cycle could have been used in place of the J-T cycle since it is immaterial as to how the liquefied air is furnished to the column). As shown here, purified compressed air is precooled in a three-channel heat exchanger if gaseous oxygen is the desired product. (If liquid oxygen is recovered from the bottom of the column, a two-channel heat exchanger is used for the compressed air and waste nitrogen streams.) The precooled air then flows through a coil in the boiler of the rectifying column, where it is further cooled to saturation while simultaneously serving as the heat source to vaporize the liquid in the boiler. After leaving the boiler, the compressed fluid expands essentially to atmospheric pressure through a throttling valve and enters the top of the column as reflux for the separation process. Rectification in the column occurs in a manner similar to that observed in ambient temperature columns. If oxygen gas is to be the product, the air must be compressed to pressures of 3 to 6 MPa; if it is to be liquid oxygen, pressures of 10 to 20 MPa are necessary.

Although the oxygen product purity is high from a simple single-column separation scheme, the nitrogen effluent stream always contains about 6 to 7 mol% oxygen. This means that approximately one third of the oxygen liquefied as feed to the column appears in the nitrogen effluent. This loss in oxygen product is not only undesirable but wasteful in terms of compression requirements. This problem was solved by the introduction of the Linde double-column gas-separation system in which two columns are placed one on top of the other as shown in Figure 22.

In this system liquid air is introduced at an intermediate point in the lower column. A condenser-evaporator at the top of the lower column provides the reflux needed for both columns. Since the condenser must condense nitrogen vapor in the lower column by evaporating liquid oxygen in the upper column, it is necessary to operate the lower column at a higher pressure between 0.5 and 0.6 MPa, while the upper column is operated just above 0.1 MPa. This

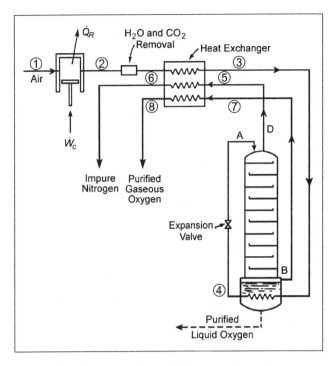

Fig. 21 Linde Single-Column Gas Separator

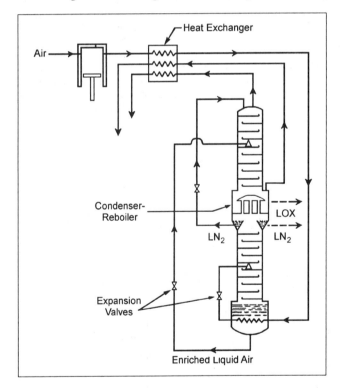

Fig. 22 Traditional Linde Double-Column Gas Separator

requires throttling the overhead nitrogen and the ~45 mol% oxygen products from the lower column as they are transferred to the upper column. The reflux and rectification process in the upper column produce high-purity oxygen at the bottom and high-purity nitrogen at the top of the column, provided that argon and the rare gases have previously been removed.

Figure 23 illustrates the scheme for removing and concentrating the **argon**. The upper column is tapped at a level where the argon concentration is highest in the column. This gas is fed to an auxiliary column where a large fraction of the argon is separated from the oxygen and nitrogen mixture, which is returned to the appropriate level in the upper column. The crude argon product normally contains about 45 mol% argon, 50 mol% oxygen, and 5 mol% nitrogen. The oxygen is readily removed by chemical reduction or adsorption. For high-purity argon, the nitrogen must be removed with the aid of another separation column.

Since helium and neon have boiling points considerably below that of nitrogen, these components from the air feed stream collect on the nitrogen side of the condenser-reboiler unit. These gases are recovered by periodically removing a small portion of the gas in the dome of the condenser and sending the gas to a small nitrogen refrigerated condenser rectifier. The resulting crude helium and neon are further purified to provide high-purity neon.

Atmospheric air contains only very small concentrations of **krypton** and **xenon**. As a consequence, very large amounts of air must be processed to obtain appreciable amounts of these rare gases. Because the krypton and xenon tend to collect in the oxygen product, the liquid oxygen from the reboiler of the upper column is first sent to an auxiliary condenser-reboiler to increase the concentration of these two components. The enriched product is further concentrated in another separation column before being vaporized and passed through a catalytic furnace to remove any remaining hydrocarbons with oxygen. The resulting water vapor and carbon dioxide are removed by a caustic trap and the krypton and xenon absorbed in a silica gel trap. The krypton and xenon are finally separated either with another separation column or by a series of adsorptions and desorptions on activated charcoal.

Figure 24 shows a schematic of the double-column gas-separation system presently used to produce **gaseous oxygen**. Such a column has both theoretical and practical advantages over the Linde double-column discussed above. A second law analysis for the two columns shows that the modern double-column has fewer irreversibilities than are present in the Linde double-column. This results in lower power requirements. From a practical standpoint, only two pressure levels are needed in the present column instead of the three required in the Linde double-column. A further advantage of the double-column in Figure 24 is that it does not require a reboiler in the bottom of the lower column thereby simplifying the heat transfer process for providing the needed vapor flow in this column. The cooled gaseous air from an expansion turbine provides additional feed to the upper column. High-purity oxygen vapor is available from the vapor space above the liquid in the reboiler of the upper column if impurities in the air stream are removed at the appropriate locations in the column as noted earlier with the traditional Linde double-column gas separation system. Further details on modifications made to modern cryogenic air separation plants are given by Grenier et al. (1986).

Helium Recovery

The major source of helium in the United States comes from helium-rich natural gas. Because the major constituents of natural gas have boiling points considerably higher than that of helium, the separation can be accomplished with condenser evaporators rather than with the more expensive separation columns.

A typical scheme pioneered by the U.S. Bureau of Mines for separating helium from natural gas is shown in Figure 25. In this scheme, the natural gas is treated to remove impurities and compressed to approximately 4.2 MPa. The purified and compressed natural gas stream is then partially condensed by the returning cold

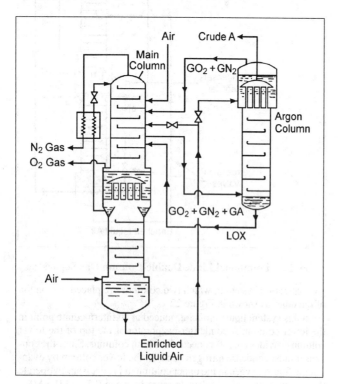

Fig. 23 Argon Recovery Subsystem

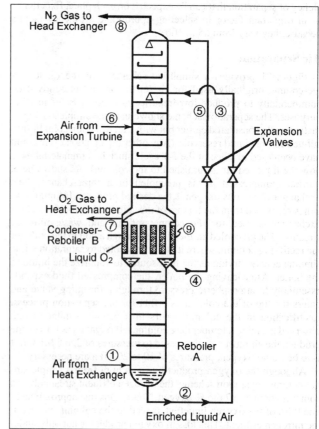

Fig. 24 Contemporary Double-Column Gas Separator

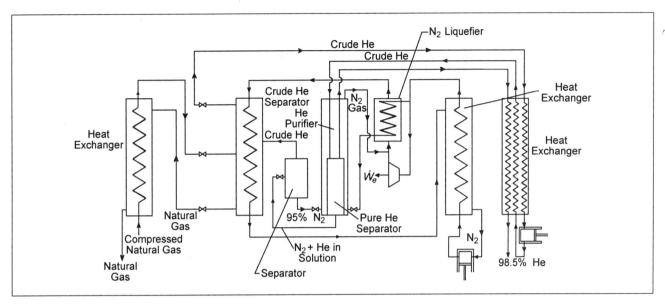

Fig. 25 Schematic of U.S. Bureau of Mines Helium Separation Plant

low-pressure natural gas stream, throttled to a pressure of 1.7 MPa, and further cooled with cold nitrogen vapor in a heat exchanger separator where 98% of the gas is liquefied. The cold nitrogen vapor, supplied by an auxiliary refrigeration system, not only provides necessary cooling but also results in some rectification of the gas phase in the heat exchanger, thereby increasing the helium concentration. The remaining vapor phase, consisting of about 60 mol% helium and 40 mol% nitrogen with a very small amount of methane is warmed to ambient temperature for further purification. The liquid phase, having been depleted of its helium content, furnishes the refrigeration required to cool and partially condense the incoming high-pressure gas. The process is completed by recompressing the stripped natural gas and returning it to the natural gas pipeline with a higher heating value.

Purification of the crude helium is accomplished by compressing the gas to 18.6 MPa and cooling it first in a heat exchanger and then in a separator that is immersed in a bath of liquid nitrogen. Nearly all of the nitrogen in the crude helium gas mixture is condensed in the separator and removed as a liquid. The latter contains some dissolved helium which is released and recovered when the pressure is reduced to 1.7 MPa. Helium gas from the separator has a purity of about 98.5 mol%. Final purification to 99.995% is accomplished by sending the cold helium through charcoal adsorption purifiers to remove the nitrogen impurity.

Natural Gas Processing

The need to obtain greater recoveries of the light hydrocarbons in natural gas has resulted in the expanded use of low-temperature processing of these streams. The cryogenic processing of natural gas brings about a phase change and involves the physical separation of the newly formed phase from the main stream. The lower the temperature for a given pressure, the greater is the selectivity of the phase separation for a particular component.

Most natural gas processing at low temperatures uses the turboexpander cycle to recover light hydrocarbons. Feed gas is normally available from 1 to 10 MPa. The gas is first dehydrated to dew points of 200 K and lower. After dehydration, the feed is cooled with cold residue gas. Liquid produced at this point is separated before entering the expander and sent to the condensate stabilizer. The gas from the separator flows to the expander. The expander exhaust stream can contain as much as 20 mass % liquid. This two-phase mixture is sent to the top section of the stabilizer which separates the two phases. The liquid is used as reflux in this unit while the cold gas exchanges heat with fresh feed and is recompressed by the expander driven compressor. Many variations to this cycle are possible and have been used in actual plants.

Purification Procedures

The nature and concentration of impurities to be removed depends entirely on the type of process involved. For example, in the production of large amounts of oxygen, various impurities must be removed to avoid plugging of the cold process lines or to avoid buildup of hazardous contaminants. The impurities in air that would contribute most to plugging are water and carbon dioxide. Helium, hydrogen, and neon, on the other hand, accumulate on the condensing side of the oxygen reboiler and will reduce the rate of heat transfer unless removed by intermittent purging. The buildup of acetylene, however, can be dangerous even if the feed concentration of the air is no greater than 0.04 mg/kg.

Refrigeration purification is a relatively simple method for removing water, carbon dioxide, and certain other contaminants from a process stream by condensation or freezing. (As noted later, either regenerators or reversing heat exchangers may be used for this purpose because a flow reversal is periodically necessary to reevaporate and remove the solid deposits.) The effectiveness of this method depends on the vapor pressure of the impurities relative to that of the major components of the process stream at the refrigeration temperature. Thus, assuming ideal gas behavior, the maximum impurity content in a gas stream after refrigeration would be inversely proportional to its vapor pressure. However, at higher pressures the impurity content can be significantly greater than that predicted for the ideal situation. Data of this behavior are available as **enhancement factors**, defined as the ratio of the actual molar concentration to the ideal molar concentration of a specific impurity in a given gas.

Purification by a solid adsorbent is one of the most common low-temperature methods for removing impurities. Materials such as silica gel, carbon, and synthetic zeolites (molecular sieves) are widely used as adsorbents because of their extremely large effective surface areas. Most of the gels and carbon have pores of varying sizes in a given sample, but the synthetic zeolites are manufactured with closely controlled pore size openings ranging from 0.4 to about 1.3 nm. This pore size makes them even more selective than other adsorbents because it permits separation of gases on the basis of molecular size.

The equilibrium adsorption capacity of the gels and carbon is a function of temperature, the partial pressure of the gas to be adsorbed, and the properties of the gas. An approximation generally exists between the amount adsorbed per unit of adsorbent and volatility of the gas being adsorbed. Thus, carbon dioxide would be adsorbed to a greater extent than nitrogen under comparable conditions. In general, the greater the difference in the volatility of the gases, the greater will be the selectivity for the more volatile component.

The design of low-temperature adsorbers requires knowledge of the equilibrium between the solid and the gas and the rate of adsorption. Equilibrium data for the common systems generally are available from the suppliers of such material. The rate of adsorption is usually very rapid and the adsorption is essentially complete in a relatively narrow zone of the adsorber. If the concentration of the adsorbed gas is more than a trace, then the heat of adsorption may also be a factor of importance in the design. (The heat of adsorption is usually of the same order or larger than the normal heat associated with a phase change.) Under such situations, it is generally advisable to design the purification process in two steps; that is, first removing a significant portion of the impurity either by condensation or chemical reaction and then completing the purification with a low-temperature adsorption system.

In normal plant operation, at least two adsorption units are employed—one is in service while the other is being desorbed of its impurities. In some cases a third adsorbent unit offers some advantage, one adsorbs, one desorbs, and one is cooled to replace the adsorbing unit as it becomes saturated. Adsorption units are generally cooled by using some of the purified gas to avoid adsorption of additional impurities during the cooling period.

Low-temperature adsorption systems are used for many applications. For example, such systems are used to remove last traces of carbon dioxide and hydrocarbons in air separation plants. Adsorbents are also used in hydrogen liquefaction to remove oxygen, nitrogen, methane, and other trace impurities. They are also used in the purification of helium suitable for liquefaction (Grade A) and for ultrapure helium (Grade AAA, 99.999% purity). Adsorption at 35 K will, in fact, yield a helium with less than 2 µg/kg of neon, which is the only detectable impurity in the helium after this treatment.

Even though most chemical purification methods are not carried out at low temperatures, they are useful in several cryogenic gas separation systems. Ordinarily, water vapor is removed by refrigeration and adsorption methods. However, for small-scale purification, the gas can be passed over a desiccant, which removes the water vapor as water of crystallization. In the krypton-xenon purification system, carbon dioxide is removed by passage of the gas through a caustic, such as sodium hydroxide, to form sodium carbonate.

When oxygen is an impurity it can be removed by reacting with hydrogen in the presence of a catalyst to form water. The latter then is removed by refrigeration or adsorption. Palladium and metallic nickel have proved to be effective catalysts for the hydrogen-oxygen reaction.

EQUIPMENT

The production and use of low temperatures require the use of highly specialized equipment including compressors, expanders, heat exchangers, pumps, transfer lines, and storage tanks. As a general rule, design principles applicable at ambient temperature are also valid for low-temperature design. However underlying each aspect of such a design must be a thorough understanding of temperature on the properties of the fluids being handled and the materials of construction being selected.

Compression Systems

Compression power accounts for more than 80% of the total energy required to produce industrial gases and liquefy natural gas.

The three major types of compressors used today are reciprocating, centrifugal, and screw. No particular type of compressor is generally preferred for all applications. The final selection ultimately depends on the specific application, the effect of plant site, available fuel source and its reliability, existing facilities, and power structure.

The key feature of reciprocating compressors is their adaptability to a wide range of volumes and pressures with high efficiency. Some of the largest units for cryogenic gas production range up to 11 MW. They use the balanced-opposed machine concept in multistage designs with synchronous motor drive. When designed for multistage, multiservice operation, these units incorporate manual or automatic, fixed or variable, volume clearance packets and externally actuated unloading devices where required. Balanced opposed units not only minimize vibrations, resulting in smaller foundations, but also allow compact installation of coolers and piping, further increasing the savings.

Air compressors for constant speed service normally use piston-type suction valve loaders for low-pressure lubricated machines. Nonlubricated units require diaphragm operated unloaders. Medium-pressure compressors for argon and hydrogen often use this type of unloader as well. The trend towards nonlubricating machines has led to piston designs using glass-filled PTFE rider rings and piston rings, with cooled packing for the piston rods.

Larger units operate as high as 277 rpm with piston speeds for air service up to 4.3 m/s. Larger compressors with provision for multiple services reduce the number of motors or drivers and minimize the accessory equipment, resulting in lower maintenance cost.

Nonlubricated compressors used in oxygen compression have carbon- or bronze-filled PTFE piston rings and piston rod packing. The suction and discharge valves are specially constructed for oxygen service. The distance pieces that separate the cylinders from the crankcase are purged with an inert gas such as nitrogen, to preclude the possibility of high concentrations of oxygen in the area in the event of excessive rod packing leakage. Compressors for oxygen service are characteristically operated at lower piston speeds of the order of 3.3 m/s. Maintenance of these machines requires rigid control of cleaning procedures and inspection of parts to ensure the absence of oil in the working cylinder and valve assemblies.

Variable-speed engine drives can generally operate over a 10 to 100% range in the design speed with little loss in operating efficiency because compressor fluid friction losses decrease with lower revolutions per minute.

Technological advances achieved in centrifugal compressor design have resulted in improved high-speed compression equipment with capacities exceeding 280 m^3/s in a single unit. Discharge pressure of such units is usually between 0.4 and 0.7 MPa. Large centrifugal compressors are generally provided with adjustable inlet guide vanes to facilitate capacity reductions of up to 30% while maintaining economical power requirements. As a consequence of their high efficiency, better reliability, and design upgrading, centrifugal compressors have become accepted for low-pressure cryogenic processes such as air separation and base load LNG plants.

Separately driven centrifugal compressors are adaptable to low-pressure cryogenic systems because they can be coupled directly to steam turbine drives, are less critical from the standpoint of foundation design criteria, and lend themselves to gas turbine or combined cycle applications. Isentropic efficiencies of 80 to 85% are usually obtained.

Most screw compressors are oil lubricated. They are either semihermetic (the motor is located in the same housing as the compressor) or have an open-drive (the motor is located outside of the compressor housing and thus requires a shaft seal). The only moving parts in screw compressors are two intermeshing helical rotors. Since rotary screw compression is a continuous positive-displacement process, no surges are created in the system.

Screw compressors require very little maintenance because the rotors turn at conservative speeds and they are well lubricated with

Cryogenics

a cooling lubricant. Fortunately, most of the lubricant can easily be separated from the gas in screw compressors. Typically, only small levels of impurities of between 1 and 2 mg/kg remain in the gas after separation. Charcoal filters can be used to reduce the impurities below this level.

A major advantage of screw compressors is that they permit high pressure ratios to be attained in a single mode. To handle these same large volumes with a reciprocating compressor requires a double-stage unit. Because of this and other advantages, screw compressors are now preferred over reciprocating compressors for helium refrigeration and liquefaction applications. They are competitive with centrifugal compressors in other applications as well.

Expansion Devices

The primary function of a cryogenic expansion device is to reduce the temperature of the gas to provide useful refrigeration for the process. In expansion engines the temperature is reduced by converting part of the energy of the high-pressure gas stream into mechanical work. This work in large cryogenic facilities is recovered and used to reduce the overall compression requirements of the process. A gas can also be cooled by expanding the gas through an expansion valve (provided that its initial temperature is below the inversion temperature of the gas). The cooling here converts part of the energy of the high-pressure gas stream into kinetic energy. No mechanical work is obtained from such an expansion.

Expanders are of either the reciprocating or the centrifugal type. Centrifugal expanders have gradually displaced the reciprocating type in large plants. However, the reciprocating expander is still popular for those processes where the inlet temperature is very low, such as for hydrogen or helium gas. Units up to 2700 kW are in service for nitrogen expansion in liquid hydrogen plants, while nonlubricated expanders with exhausts well below 33 K are used in liquid hydrogen plants developed for the space program.

For reciprocating expanders, efficiencies of 80% are normally quoted and values of 85% are quoted for high-capacity centrifugal types. The latter are generally identified as turboexpanders. Generally, reciprocating expanders are selected when the inlet pressure and pressure ratio are high and when the volume of gas handled is low. The inlet pressure to expansion engines used in air separation plants varies between 4 to 20 MPa, while capacities range from 0.1 to 3 m^3/s.

The design features of reciprocating expanders used in low-temperature processes include rigid, guided cam-actuated valve gears, renewable hardened valve seats, helical steel or air springs, and special valve packing that eliminates leakage. Cylinders are normally steel forgings effectively insulated from the rest of the structure. Removable nonmetallic cylinder liners and floating piston design offer wear resistance and good alignment in operation. Piston rider rings serve as guides for the piston. Nonmetallic rings are used for nonlubricated service. Both horizontal and vertical design, and one- and two-cylinder versions, have been used successfully.

Nonlubricated reciprocating expansion engines are generally used whenever possible oil contamination is unacceptable or where extremely low operating temperatures preclude the use of cylinder lubricants. This type of expansion engine is found in hydrogen and helium liquefaction plants and in helium refrigerators.

Reciprocating expanders in normal operation should not accept liquid in any form during the expansion cycle. However, the reciprocating device can tolerate some liquid for short periods if none of the constituents freeze in the expander cylinder and cause serious mechanical problems. Inlet pressure and temperature must be changed to eliminate any possibility of entering the liquid phase and especially the triple point range on expansion during normal operation.

Turboexpanders are classified as either axial or radial. Most turboexpanders built today are of the radial type because of their generally lower cost and reduced stresses for a given tip speed. This design allows them to run at higher speeds with higher efficiencies and lower operating costs. On the other hand, axial flow expanders are more suitable for multistage expanders because these units provide an easier flow path from one stage to the next. Where low flow rates and high enthalpy reductions are required, an axial-flow two-stage expander is generally used with nozzle valves controlling the flow. For example, in the processing of ethylene, gas leaving the demethanizer is normally saturated, and expansion conditions cause a liquid product to exit from the expander. Since up to 15 to 20% liquid at the isentropic end point can be handled in axial-flow impulse-turbine expanders, recovery of ethylene is feasible by the procedure. Depending on the initial temperature and pressure entering the expander and the final exit pressure, good flow expanders are capable of reducing the enthalpy of an expanded fluid by between 175 to 350 kJ/kg, and this may be multistaged. The change in enthalpy drop can be regulated by turbine speed.

Today's highly reliable and efficient turboexpanders have made large capacity air separation plants and base load LNG facilities a reality. Notable advances in turboexpander design center on improved bearings, lubrication, and wheel and rotor design to permit nearly ideal rotor assembly speeds with good reliability. Pressurized labyrinth sealing systems use dry seal gas under pressure mixed with cold gas from the process to provide seal output temperatures above the frost point. Seal systems for oxygen compressors are more complex than for air or nitrogen and prevent lubricant carryover to the processed gas. By the combination of variable area nozzle grouping or partial admission of multiple nozzle grouping, efficiencies up to 85% have been obtained with radial turboexpanders.

Turboalternators were developed to improve the efficiency of small cryogenic refrigeration systems. This is accomplished by converting the kinetic energy in the expanding fluid to electrical energy, which in turn is transferred outside of the system where it can be converted to heat and dissipated to an ambient heat sink.

The **expansion valve** (often called the J-T valve) is an important component in any liquefaction system, although not as critical a component as the others mentioned in this section. This valve resembles a normal valve that has been modified to handle the flow of cryogenic fluids. These modifications include exposing the high-pressure stream to the lower part of the valve seat to reduce sealing problems and a valve stem that has been lengthened and surrounded by a thin walled tube to reduce heat transfer.

Heat Exchangers

One of the more critical components of any low-temperature liquefaction and refrigeration system is the heat exchanger. This point is demonstrated by considering the effect of heat exchanger effectiveness on the liquid yield of nitrogen in a simple J-T liquefaction process operating between a lower and upper pressure of 0.1 to 20 MPa. The liquid yield under these conditions is zero if the effectiveness of the heat exchanger is less than 85%. (**Heat exchanger effectiveness** is defined as the ratio of the actual heat transfer to the maximum possible heat transfer in the heat exchanger).

With the exception of helium II, the behavior of most cryogens may be predicted by using the principles of mechanics and thermodynamics that apply to many fluids at room temperature. This behavior has permitted the formulation of convective heat transfer correlations for low-temperature designs of heat exchangers similar to those used at ambient conditions and ones that use Nusselt, Reynolds, Prandtl, and Grashof numbers.

However, the need to operate more efficiently at low temperatures has made the use of simple exchangers impractical in many cryogenic applications. Some of the important advances in cryogenic technology are the development of complex but very efficient heat exchangers. Some of the criteria that have guided the development of these units for low-temperature service are (1) small temperature differences at the cold end of the exchanger to enhance efficiency, (2) large heat exchange surface area to heat exchanger volume ratios to minimize heat leak, (3) high heat transfer rates to reduce surface area, (4) low mass to minimize start up time, (5) multichannel capability to minimize the number of exchangers, (6) high pressure capability to provide design flexibility, (7) low or reasonable pressure drops in the exchanger to minimize compression requirements, and (8) minimum maintenance to minimize shutdowns.

Minimizing the temperature difference at the cold end of the exchanger has some problems, particularly if the specific heat of the cold fluid increases with increasing temperature as demonstrated by hydrogen. In such cases a temperature pinch or a minimum temperature difference between the two streams in the heat exchanger can occur between the warm and cold ends of the heat exchanger. This problem is generally alleviated by adjusting the mass flow of the key stream into the heat exchanger. In other words, the capacity rate is adjusted by controlling the mass flow to offset the change in specific heats. Problems of this nature can be avoided by balancing enthalpy in incremental steps from one end of the exchanger to the other.

The selection of an exchanger for low-temperature operation is normally determined by process design requirements, mechanical design limitations, and economic considerations. The principal industrial exchangers finding use in cryogenic applications are coiled-tube, plate-fin, reversing, and regenerator units.

Construction. A large number of aluminum tubes are wound around a central core mandrel of a **coiled-tube exchanger**. Each exchanger contains many layers of tubes, both along the principal and radial axes. Pressure drops in the coiled tubes are equalized for each specific stream by using tubes of equal length and carefully varying the spacing of these tubes in the different layers. A shell over the outer tube layer together with the outside surface of the core mandrel form the annular space in which the tubes are nested. Coiled-tube heat exchangers offer unique advantages, especially for those low-temperature conditions where simultaneous heat transfer between more than two streams is desired, a large number of heat transfer units is required, and high operating pressures in various streams are encountered. The geometry of these exchangers can be varied to obtain optimum flow conditions for all streams and still meet heat transfer and pressure drop requirements.

Optimization of the coiled-tube heat exchanger involves such variables as tube and shell flow velocities, tube diameter, tube pitch, and layer spacing. Other considerations include single-phase and two-phase flow, condensation on either the tube or shell side, and boiling or evaporation on either the tube or shell side. Additional complications occur when multicomponent streams are present, as in natural gas liquefaction, because mass transfer accompanies the heat transfer in the two-phase region.

The largest coiled-tube exchangers contained in one shell have been constructed for LNG base load service. These exchangers handle liquefaction rates in excess of 28 m^3/s with a heat transfer surface of 25,000 m^2, an overall length of 60 m, a maximum diameter of 4.5 m, and a mass of over 180 Mg.

Plate-and-fin heat exchangers are fabricated by stacking alternate layers of corrugated, high-uniformity, die-formed aluminum sheets (fins) between flat aluminum separator plates to form individual flow passages. Each layer is closed at the edge with aluminum bars of appropriate shape and size. Figure 26 illustrates the elements of one layer and the relative position of the components before being joined by a brazing operation to form an integral structure with a series of fluid flow passages. These flow passages are combined at the inlet and exit of the exchanger with common headers. Several sections can be connected to form one large exchanger. The main advantage of this type of exchanger is that it is compact (about nine times as much surface area per unit volume as conventional shell and tube exchangers), yet permits wide design flexibility, involves minimum mass, and allows design pressures to 7 MPa from 4.2 to 340 K.

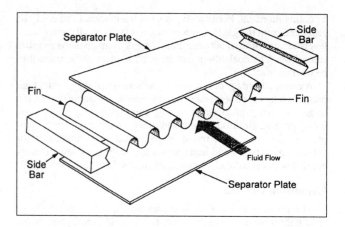

Fig. 26 Enlarged View of One Layer of Plate-and-Fin Heat Exchanger Before Assembly

The fins for these heat exchangers which are typically 10 mm in height can be manufactured in a variety of configurations that can significantly alter the heat transfer and pressure drop characteristics of the exchanger. Various flow patterns can be developed to provide multipass or multistream arrangements by incorporating suitable internal seals, distributors, and external headers. The type of headers used depends on the operating pressures, the number of separate streams involved, and, in the case of counterflow exchangers, whether reversing duty is required.

Plate-and-fin exchangers can be supplied as single units or as manifolded assemblies that consist of multiple units connected in parallel or in series. Sizes of single units are presently limited by manufacturing capabilities and assembly tolerances. Nevertheless, the compact design of brazed aluminum plate-and-fin exchangers makes it possible to furnish more than 35 000 m^2 of heat transfer surface in one manifolded assembly. These exchangers find application in helium liquefaction, helium extraction from natural gas, hydrogen purification and liquefaction, air separation, and low-temperature hydrocarbon processing. Design details for plate fin exchangers are available in most heat exchanger texts.

Removal of Impurities. Continuous operation of low-temperature processes requires that impurities in feed streams be removed almost completely prior to cooling the streams to very low temperatures. The removal of impurities is necessary because their accumulation in certain parts of the system creates operational difficulties or constitutes potential hazards. Under certain conditions the necessary purification steps can be accomplished by using reversing heat exchangers.

A typical arrangement of a reversing heat exchanger for an air separation plant is shown in Figure 27. Channels A and B constitute the two main reversing streams. Operation of such an exchanger is characterized by the cyclical changeover of one of these streams from one channel to the other. The reversal normally is accomplished by pneumatically operated valves on the warm end and by check valves on the cold end of the exchanger. The warm end valves are actuated by a timing device which is set to a period such that the pressure drop in the feed channel is prevented from increasing beyond a certain value because of the accumulation of impurities. Feed enters the warm end of the exchanger and as it is progressively cooled, impurities are deposited on the cold surface of the exchanger. When the flows are reversed, the return stream reevaporates the deposited impurities and removes them from the system.

Proper functioning of the reversing exchanger will depend on the relationship between the pressures and temperatures of the two streams. Because pressures are normally fixed by other considerations, the purification function of the exchanger is usually controlled by the proper selection of temperature differences

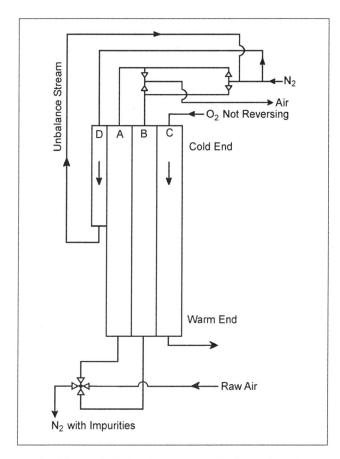

Fig. 27 Typical Flow Arrangement for Reversing Heat Exchanger in Air Separation Plant

throughout the exchanger. These differences must be such that at every point in the exchanger where reevaporation takes place, the vapor pressure of the impurity must be greater than the partial pressure of the impurity in the scavenging stream. Thus, a set of critical values for the temperature differences exists depending on the pressures and temperatures of the two streams. Since ideal equilibrium concentrations can never be attained in an exchanger of finite length, allowances must be made for an exit concentration in the scavenging stream sufficiently below the equilibrium one. Generally a value close to 85% of equilibrium is selected.

The use of **regenerators** was proposed by Frankl in the 1920s for the simultaneous cooling and purification of gases in low-temperature processes. In contrast to the reversing heat exchangers in which the flows of the two fluids are continuous and countercurrent during any period, the regenerator operates periodically by storing heat in a high heat capacity packing in one half of the cycle and then releasing this stored heat to the fluid in the other half of the cycle. Such an exchanger, shown in Figure 28, consists of two identical columns packed with typical matrix materials such as metal screens or lead shot through which a cyclical flow of gases is maintained. In the process of cooldown, the warm feed stream deposits impurities on the cold surface of the packing. When the streams are switched, the impurities are reevaporated as the cold stream is warmed while simultaneously cooling the packing. Thus, the purifying action of the regenerator is based on the same principles as for the reversing exchanger, and the same limiting critical temperature differences must be observed if complete reevaporation of the impurities is to take place.

Regenerators frequently are selected for those applications in which the heat transfer effectiveness, defined as Q_{actual}/Q_{ideal}, must be greater than 0.98. A high regenerator effectiveness requires a

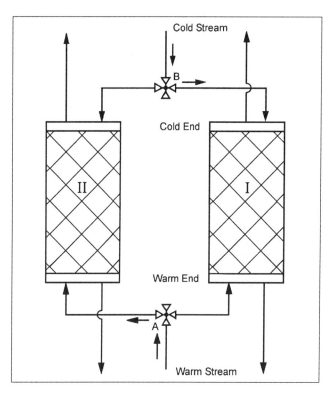

Fig. 28 Flow Arrangement in Regenerator Operation

matrix material with a high heat capacity per unit volume and also a large surface area per unit volume. Until the early 1990s, recuperative heat exchangers rather than regenerators were used in cryocoolers because the heat capacity of typical matrix materials rapidly decreases to a negligible value below 10 K. Because an increase in specific heat of a material can only occur when there is a physical transition that occurs in the material, studies have been directed to heavy rare earth compounds that exhibit a magnetic phase transition at these low temperatures. Some of the experimental results are shown in Figure 29. Hashimoto et al. (1992) determined that specific heats of the $ErNi_{1-x}Co_x$ system are more than twice the values obtained for Er_3Ni at 7 K. Kuriyama et al. (1994) used layered rare earth matrix materials with higher heat capacities than Er_3Ni by itself in the cold end of the second stage of a Gifford-McMahon refrigerator and increased the refrigeration power of the refrigerator by as much as 40% at 4.2 K.

The low cost of the heat transfer surface along with the low-pressure drop are the principal advantages of regenerators. However, the contamination of fluid streams by mixing caused by periodic flow reversals and the difficulty of designing a regenerator to handle three or more fluids has restricted its use and favored the adoption of the plate-and-fin exchangers for air separation plants.

LOW-TEMPERATURE INSULATIONS

As noted, the effectiveness of a liquefier or refrigerator depends to a large extent on heat leaking into a system. Because heat removal becomes more costly as temperature is reduced (the Carnot limitation), most cryogenic systems include some form of insulation to minimize the effect. Cryogenic insulations generally are divided into five general categories: high-vacuum, multilayer insulation, powder, foam, and special insulations. The type of insulation chosen for a given cryogenic use depends on the specific application. An optimum insulation system should not only possess maximum insulation effectiveness, but also minimum mass, ease of fabrication and handling, adequate service life, and reasonable cost. Generally, the selection is aided by a knowledge

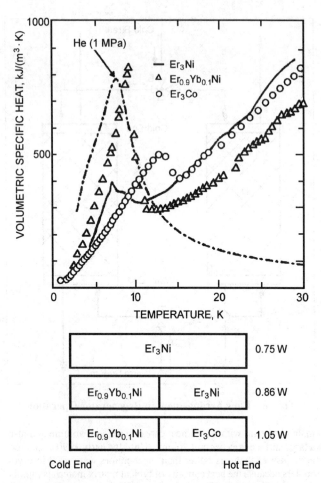

Fig. 29 Specific Heat of Several Rare Earth Matrix Materials
(Courtesy of T. Kuriyama, Toshiba, Kawasaki, Japan)

of the properties of the particular insulation, such as thermal conductivity, emissivity, moisture content, evacuability, porosity, and flammability.

Heat flows through an insulation by solid conduction, gas conduction, and radiation. Because these heat transfer mechanisms operate simultaneously and interact with each other, an apparent thermal conductivity k_a is used to characterize the insulation. The value of k_a is measured experimentally during steady-state heat transfer and evaluated from the basic one-dimensional Fourier equation. Typical k_a values for a variety of insulations used in cryogenic service are presented in Table 3.

High-Vacuum Insulation

The mechanism of heat transfer prevailing across an evacuated space (0.13 mPa or less) is by radiation and conduction through the residual gas. Radiation is generally the more predominant mechanism and can be approximated by

$$\frac{q_r}{A_1} = \sigma(T_2^4 - T_1^4)\left[\frac{1}{\varepsilon_1} + \frac{A_1}{A_2}\left(\frac{1}{\varepsilon_2} - 1\right)\right]^{-1} \quad (20)$$

where q_r/A_1 is the radiant heat flux, σ the Stefan-Boltzmann constant, and ε the emissivity of the surface. The subscripts 1 and 2 refer to the cold and warm surfaces, respectively. The bracketed term on the right is the emissivity for an evacuated space with diffuse radiation between spheres or cylinders (with length much less than diameter).

Table 3 Apparent Thermal Conductivity of Selected Insulations

Type of Insulation	Apparent Thermal Conductivity k_a (Between 77 K and 300 K), W/(m·K)	Bulk Density ρ, kg/m³
Pure gas (at 100 kPa, 180 K)		
n-H$_2$	0.1216	0.138
N$_2$	0.0163	1.91
Pure vacuum (0.13 mPa or less)	0.0170	Nil
Foam insulation		
Polystyrene foam	0.026	32 to 48
Polyurethane foam	0.033	80 to 128
Glass foam	0.035	144
Evacuated powder		
Perlite (0.13 Pa)	0.001 to 0.002	64 to 144
Silica (0.13 Pa)	0.0017 to 0.0021	64 to 96
Multilayer insulation		
Aluminum foil and fiberglass		
(12–28 layers/cm, 0.13 mPa)	3.5×10^{-5} to 7×10^{-5}	64 to 112
(30–60 layers/cm, 0.13 mPa)	1.75×10^{-5}	120
Aluminum foil and nylon net		
(32 layers/cm, 0.13 mPa)	3.5×10^{-5}	89

When the mean free path of gas molecules becomes large relative to the distance between the walls of the evacuated space as the gas pressure in the evacuated space rises, free molecular conduction is encountered. The gaseous heat conduction under free molecular conditions for most cryogenic applications is given by

$$\frac{q_{gc}}{A_1} = \frac{\gamma+1}{\gamma-1}\left(\frac{R}{8\pi MT}\right)^{1/2}\alpha p(T_2 - T_1) \quad (21)$$

where α, the overall accommodation coefficient, is defined by

$$\alpha = \frac{\alpha_1 \alpha_2}{\alpha_2 + \alpha_1(1-\alpha_2)(A_1/A_2)} \quad (22)$$

and γ is the ratio of the heat capacities, R the molar gas constant, M the relative molecular mass of the gas, and T the temperature of the gas at the point where the pressure p is measured. The subscripted A_1 and A_2, T_1 and T_2, and α_1 and α_2 are the areas, temperatures, and accommodation coefficients of the cold and warm surfaces, respectively. The accommodation coefficient depends on the specific gas surface combination and the surface temperature. Table 4 presents accommodation coefficients of three gases at several temperatures.

Heat transfer across an evacuated space by radiation can be reduced significantly by inserting one or more low-emissivity floating shields within the evacuated space. Such shields provide a reduction in the emissivity factor. The only limitation on the number of floating shields used is one of complexity and cost.

Table 4 Accommodation Coefficients for Several Gases

Temp., K	Helium	Hydrogen	Air
300	0.29	0.29	0.8 to 0.9
77	0.42	0.53	1
20	0.59	0.97	1

Multilayer Insulations

Multilayer insulation provides the most effective thermal protection available for cryogenic storage and transfer systems. It consists of alternating layers of highly reflecting material, such as aluminum foil or aluminized polyester film, and a low-conductivity spacer material or insulator, such as fiberglass mat or paper, glass fabric, or nylon net, all under high vacuum. (The desired vacuum of 0.13 mPa or less is maintained by using a getter such as activated charcoal to adsorb gases that desorb from the surfaces in the vacuum space.) When properly applied at the optimum density, this type of insulation can have an apparent thermal conductivity as low as 17 to 70 μW/(m·K) between 20 and 300 K. The very low thermal conductivity of multilayer insulations can be attributed to the fact that all modes of heat transfer are reduced to a minimum.

The apparent thermal conductivity of a highly evacuated (pressures on the order of 0.13 mPa or less) multilayer insulation can be determined from

$$k_a = \frac{1}{N/\Delta x}\left\{h_s + \frac{\sigma e T_2^3}{2-\varepsilon}\left[1 + \left(\frac{T_1}{T_2}\right)^2\right]\left[1 + \frac{T_1}{T_2}\right]\right\} \quad (23)$$

where $N/\Delta x$ is the number of complete layers (reflecting shield plus spacer) of insulation per unit thickness, h_s the solid conductance for the spacer material, σ the Stefan-Boltzmann constant, ε the effective emissivity of the reflecting shield, and T_2 and T_1 the absolute temperatures of the warm and cold surfaces of the insulation, respectively. Equation (23) indicates that apparent thermal conductivity is inversely proportional to the number of complete layers used in the evacuated space. However, as the multilayer insulation is compressed, the increase in solid conductivity outweighs the decrease in radiative heat, thereby establishing an optimum layer density.

The effective thermal conductivity values generally obtained with actual cryogenic storage and transfer systems often are at least a factor of two greater than the thermal conductivity values shown in Figure 30, which were obtained under carefully controlled conditions. This degradation in insulation thermal performance is caused by the combined presence of edge exposure to isothermal boundaries, gaps, joints, or penetrations in the insulation blanket required for structural supports, fill and vent lines, and the high lateral thermal conductivity of these insulation systems.

Powder and Fibrous Insulations

The difficulties encountered with applying multilayer insulation to complex structural storage and transfer systems can be minimized by using evacuated powder or fibrous insulation. This substitution in insulation materials, however, decreases the overall thermal effectiveness of the insulation system by ten fold. Nevertheless, in applications where this is not a serious factor and investment cost is a major factor, even unevacuated powder insulation may be the proper choice of insulating material. Such is the case for large LNG storage facilities.

A **powder insulation** system consists of a finely divided particulate material such as perlite, expanded SiO_2, calcium silicate, diatomaceous earth, or carbon black inserted between the surfaces to be insulated. When used at 0.1 MPa gas pressure (generally with an inert gas), the powder reduces both convection and radiation and, if the particle size is sufficiently small, can also reduce the mean free path of the gas molecules. The apparent thermal conductivity of gas-filled powders is given by the expression

$$k_a = \left[\frac{V_r}{k_s} + \frac{1}{k_g/(1-V_r) + 4\sigma T^3 d/V_r}\right] \quad (24)$$

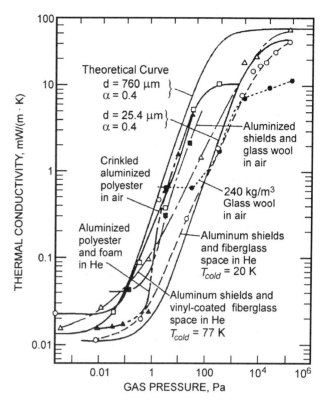

Fig. 30 Effect of Residual Gas Pressure on Apparent Thermal Conductivity of Multilayer Insulation

where V_r is the ratio of the solid powder volume to the total volume, k_s the thermal conductivity of the powder, k_g the thermal conductivity of the residual gas, σ the Stefan-Boltzmann constant, T the mean temperature of the insulation, and d the mean diameter of the powder.

The insulating value of powders is increased considerably by removing the interstitial gas. Thus, when powders are used at pressures of 0.13 mPa or less, gas conduction is negligible and heat transport is mainly by radiation and solid conduction. Figure 31 shows the apparent thermal conductivity of several powders as a function of interstitial gas pressure.

The radiation contribution for evacuated powders near room temperature is larger than the solid conduction contribution to the total heat transfer rate. On the other hand, the radiant contribution is smaller than the solid conduction contribution for temperatures between 77 and 20 or 4 K. Thus, evacuated powders can be superior to vacuum alone (for insulation thicknesses greater than 0.1 m) for heat transfer between ambient and liquid nitrogen temperatures. Conversely, since solid conduction becomes predominant at lower temperatures, vacuum alone is usually better for reducing heat transfer between two cryogenic temperatures.

Evacuated fibrous insulations have performance levels comparable to evacuated powders, but usually are more expensive and require higher vacuum levels to attain their best performance. Both their costs and sensitivity to vacuum level are generally related to fiber size and density; better performance and higher cost are usually associated with small fiber diameter and high density.

Foam Insulations

The apparent thermal conductivity of foam depends on the bulk density of the foamed material, the gas used as the foaming agent, and the temperature levels to which the insulation is exposed. Heat transport across a foam is determined by convection and radiation in the cells of the foam and by conduction in the solid structure.

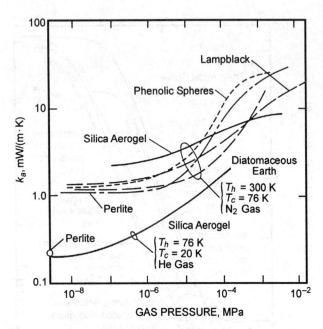

Fig. 31 Apparent Thermal Conductivity of Several Powder Insulations as Function of Residual Gas Pressure

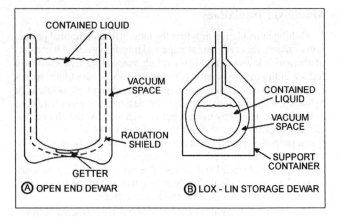

Fig. 32 Laboratory Storage Dewars for Liquid Oxygen and Nitrogen

Evacuation of a foam effectively reduces its thermal conductivity, indicating a partially open cellular structure, but the resulting values are still considerably higher than either multilayer, evacuated powder or fibrous insulations. The opposite effect, diffusion of atmospheric gases into the cells, can cause an increase in the apparent thermal conductivity. In particular, the diffusion of hydrogen and helium into the cells can increase the apparent thermal conductivity of the foam by a factor of three or four. Of all the foams, polyurethane and polystyrene have received the widest use at low temperatures.

The major disadvantage of foams is not their relatively high thermal conductivity compared to other insulations, but rather their poor thermal behavior. When applied to cryogenic systems, they tend to crack on repeated cycling and lose their insulation value.

Special Insulations

No single insulation has all the desirable thermal and strength characteristics required in many cryogenic applications. Consequently, numerous composite insulations have been developed. One such insulation consists of a polyurethane foam, reinforcement materials to provide adequate compressive strength, adhesives for sealing and securing the foam to the container, enclosures to prevent damage to the foam from external sources, and vapor barriers to maintain a separation between the foam and atmospheric gases. Another external insulation system for space applications uses honeycomb structures. Phenolic-resin reinforced fiberglass cloth honeycomb is most commonly used. Filling the cells with a low-density polyurethane foam further improves the thermal effectiveness of the insulation.

STORAGE AND TRANSFER SYSTEMS

Storage vessels range from low-performance containers where the liquid in the container boils away in a few hours to high-performance containers and dewars where less than 0.1% of the fluid contents is evaporated per day.

Storage Systems

The essential elements of a storage vessel consist of an inner vessel, which encloses the cryogenic fluid to be stored, and an outer vessel, which contains the appropriate insulation and serves as a vapor barrier to prevent water and other condensables from reaching the cold inner vessel. The value of the cryogenic liquid stored dictates whether or not the insulation space is evacuated. In small laboratory dewars designed for storage of liquid nitrogen and oxygen, as shown in Figure 32, the insulation is obtained by coating the two surfaces facing the insulation space with low-emissivity materials and then evacuating the space to a pressure of 0.13 mPa or lower. Laboratory storage of liquid hydrogen and liquid helium, on the other hand, requires multilayer insulation. In larger vessels, insulations such as powders, evacuated powders, or multilayer insulations are used, depending on the type of fluid and storage applications. Table 5 provides a brief tabulation of storage systems based on the preceding criteria.

The transport of cyrogens for more than a few hundred metres generally requires specially built transport systems for truck, railroad, or airline delivery. Volumes from 0.02 to more than 100 m³ have been transported successfully by these carriers. Large barges and ships built specifically for shipment of cryogens, particularly LNG, have increased the volume transported manyfold.

Several requirements must be met in the design of the inner vessel. The material of construction must be compatible with the stored

Table 5 Insulation Selection for Various Cryogenic Storage Vessels

Fluid	Application	Volume, m³	Insulation
Liquid natural gas	Liquefier storage	Up to 1600	Powder
Liquid natural gas	Sea transport	Up to 40 000	Powder and foam with N_2 purge
Liquid oxygen or nitrogen	Laboratory use	Up to 0.1	Vacuum
Liquid oxygen or nitrogen	Transport	Up to 0.2	Multilayer
Liquid oxygen or nitrogen	Liquefier storage	Up to 50	Evacuated powder
Liquid oxygen or nitrogen	Liquefier storage	Above 50	Powder
Liquid hydrogen or helium	Laboratory use	Up to 0.15	Multilayer
Liquid hydrogen	Truck transport	Up to 100	Multilayer
Liquid hydrogen	Liquefier storage	Up to 3800	Evacuated powder
Liquid helium	Truck transport	Up to 35	Multilayer
Liquid helium	Liquefier storage	Up to 115	Multilayer

Note: This tabulation should only serve as a guideline as economics will determine the final selection.

Cryogenics

cryogen. Nine percent nickel steels are acceptable for high-boiling cryogens ($T > 75$ K), while many aluminum alloys and austenitic steels are usually structurally acceptable throughout the entire temperature range. Economic and cooldown considerations dictate that the inner shell be as thin as possible. Accordingly, the inner container is designed to withstand only the internal pressure and bending forces, while stiffening rings are used to support the mass of the fluid. The minimum thickness of the inner shell for a cylindrical vessel under such a design arrangement is given in Section VIII of the ASME *Boiler and Pressure Vessel Code*.

The outer shell of the storage vessel, on the other hand, is subjected to atmospheric pressure on the outside and evacuated conditions on the inside. Such a pressure difference requires an outer shell of sufficient material thickness with appropriately placed stiffening rings to withstand collapsing or buckling. Here again, specific design charts can be found in the ASME code.

Heat leaks into cryogen storage vessel by radiation and conduction through the insulation and by conduction through the inner shell supports, piping, instrumentation leads, and access ports. Conduction losses are reduced by making long heat-leak paths, by making the cross sectional areas for heat flow small, and by using materials with low thermal conductivity. Radiation losses, a major factor in the heat leak through insulations, are reduced by using radiation shields (such as multilayer insulation), boil-off vapor-cooled shields, and opacifiers in powder insulation.

Most storage vessels for cryogens are designed for a 90% liquid volume and a 10% vapor or ullage volume. The latter permits reasonable vaporization of the liquid contents due to heat leakage without incurring too rapid a buildup of pressure in the vessel. This, in turn, permits closing the container for short periods either to avoid partial loss of the contents or to permit the safe transport of flammable or hazardous cryogens.

Transfer Systems

A cryogen is transferred from the storage vessel by one of three methods: self pressurization of the container, external gas pressurization, and mechanical pumping. Self pressurization involves removing some of the fluid from the container, vaporizing the extracted fluid, and then reintroducing the vapor into the ullage space to displace the contents of the container. External gas pressurization uses an external gas to displace the container contents. In the mechanical pumping method, the contents of the storage vessel are removed by a cryogenic pump located in the liquid drain line.

Several types of pumps have been used with cryogenic fluids. In general, positive displacement pumps are best suited for low-flow rates at high pressures. Centrifugal or axial flow pumps are generally best for the high-flow applications. Centrifugal or axial flow pumps have been built and used for liquid hydrogen with flow rates of up to 3.8 m^3/s and pressures of more than 6.9 MPa. Cryogen subcooling, thermal contraction, lubrication, and compatibility of materials must be considered carefully for successful operation.

Cryogenic fluid transfer lines are generally classified as one of three types: uninsulated, foam-insulated lines, and vacuum-insulated lines. The latter may entail vacuum insulation alone, evacuated powder insulation, or multilayer insulation. A vapor retarder must be applied to the outer surface of foam-insulated transfer lines to minimize the degradation of the insulation that occurs when water vapor and other condensables diffuse through the insulation to the cold surface of the lines.

Cooldown of a transfer line always involves two-phase flow. Severe pressure and flow oscillations occur as the cold liquid comes in contact with the successive warm sections of the line. Such instability continues until the entire transfer line is cooled down and filled with liquid cryogen.

INSTRUMENTATION

Cryogenic instrumentation is used primarily to determine the condition or state of cryogenic fluids, such as pressure and temperature. Such information is typically required for process optimization and control. In addition, the question of quantity and quality transferred or delivered has become commercially important. Accordingly, the instrumentation system must also be able to accurately indicate liquid level, density, and flow rate.

Pressure Measurement

Pressure in cryogenic systems has been measured by simply attaching gage lines from the points where the pressure is to be measured to some convenient location at ambient temperature where a suitable pressure-measuring device is available. This method works well for many applications, but it can have problems of independent frequency response and thermal oscillations. In addition, heat leak, uncertainties in hydrostatic pressure in gage lines, and fatigue failure of gage lines can be significant in some applications. Such problems can be eliminated by installing pressure transducers at the point of measurement. Many pressure transducers used in the cryogenic environment are similar to those used at ambient conditions, such as strain gage and capacitance transducers. However, because pressure sensing devices often behave quite differently under cryogenic conditions, a systematic testing program can identify the device most suitable for the specific application.

Thermometry

Most low-temperature temperature measurements are made with metallic resistance thermometers, nonmetallic resistance thermometers, or thermocouples. Vapor pressure thermometers find limited application, but they can provide convenient temperature check points. The selection of a thermometer for a specific application must consider such factors as absolute accuracy, reproducibility, sensitivity, heat capacity, self heating, heat conduction, stability, simplicity and convenience of operation, ruggedness, and cost.

The resistivity of a metallic element or compound varies with a change in temperature. While many metals are suitable for resistance thermometry, platinum occupies a predominant position, mainly because it is chemically inert, easy to fabricate, sensitive down to 20 K, and very stable.

Many semiconductors, such as germanium, silicon, and carbon, also have useful thermometric properties at low temperatures. Carbon, though not strictly a semiconductor, is included in this group because of its similarity in behavior to semiconductors.

Thermocouples can be very small, so that the disturbance to the object being sensed is very slight and the response time very fast. However, such devices generate rather small voltages, and these become smaller as the temperature is reduced. Copper and constantan is the most commonly used thermocouple pair for low-temperature thermometry. Such a thermocouple has an accuracy of ±0.1 K and a sensitivity of 40 µV/K at room temperature, 17 µV/K at liquid oxygen temperatures, and 5 µV/K at liquid hydrogen temperatures. Even though other thermocouple combinations such as gold (2.1% atomic cobalt) and copper, have larger thermoelectric powers at low temperatures, copper-constantan is still favored because it suffers less from the inhomogenieties common with other thermocouples.

Liquid-Level Measurements

Liquid-level is one of several measurements needed to establish the contents of a cryogenic container. Other measurements may include volume as a function of depth, density as a function of physical storage conditions, and, sometimes, useful contents from total contents. Of these measurements, liquid level is presently the most advanced; it can be as accurate and precise as thermometry, and often with greater simplicity.

The operation of cryogenic liquid-level sensors generally depends on a large property change that occurs at the liquid-vapor interface (e.g., a significant change in density). This change may not occur if the fluid is stored near its critical point or if it has stratified after prolonged storage. A convenient way to classify such liquid-level sensors is according to whether the output is discrete (point sensors) or continuous. Point level sensors are tuned to detect sharp property differences and give an on-off signal.

The **capacitance type** of **liquid-level sensor** recognizes the differences in dielectric constant between the liquid and the vapor, which is closely related to the fact that the liquid is more dense than the vapor. The **thermal** or **hot wire sensor** detects the large difference in heat transfer between the liquid and vapor phases. The **optical sensor**, on the other hand, detects the change in refractive index between the liquid and the vapor, which is also related to the dielectric constant and density. Several **acoustic** and **ultrasonic devices** operate on the principle that the damping of a vibrating member is greater in liquid than in a vapor.

Continuous liquid-level sensors take many forms. Some liquid-level sensors determine mass, while others are just continuous analogs of some point sensor and merely follow the liquid-vapor interface. Mass sensors include direct weighing, nuclear radiation attenuation, and radio frequency techniques to detect mass. Differential pressure sensors, capacitance, and acoustic devices are used to follow the vapor-liquid interface.

Density Measurements

Measurements of liquid density are closely related to quantity and liquid-level measurements because both are often required simultaneously to establish the mass contents of a tank. The same physical principle may often be used for either measurement, because liquid-level detectors sense the steep density gradient at the liquid vapor interface. Density may be detected by direct weighing, differential pressure, capacitance, optical, acoustic, and nuclear radiation attenuation. In general, the various liquid-level principles apply to density measurement techniques as well.

Two exceptions are noteworthy. In the case of homogeneous pure fluids, density can usually be determined more accurately by an indirect measurement. That is, density can be calculated from accurate thermophysical properties data by the measurement of pressure and temperature.

Nonhomogeneous fluids are quite different. LNG, for example, is often a mixture of five or more components whose composition and, hence, density vary. Accordingly, temperature and pressure measurements alone will not suffice to determine density. In that case, a dynamic, direct measurement, using liquid-level measurement, is required.

Flow Measurement

Three types of flow meters are useful for liquid cryogens: pressure drop or head type, turbine type, and momentum type.

The **pressure drop** or **head meter** embodies the oldest method of measuring flowing fluids. Its distinctive feature is a restriction that is used to reduce the static pressure of the flowing fluid. This static pressure difference between the pressures upstream and downstream sides of the restriction is measured. These meters are simple, and they do not need calibration if proper design, application theory, and practices are followed. Orifice accuracy is generally within ±3% of full scale and repeatability is ±1%. However, transient fluctuation in flow can cause erratic readout particularly during cooldown and warmup. Also static liquid pressures within the meter should be well above the fluid saturation pressure to avoid an erroneous flow measurement due to possible cavitation in the flowing stream. In spite of these shortcomings, head meters are widely used and quite reliable.

The turbine-type, **volumetric flow meter** is probably the most popular of the various flow-measuring instruments because of its simple mechanical design and demonstrated repeatability. This meter consists of a freely spinning rotor having a number of blades. The rotor is supported in guides or bearings mounted in a housing that forms a section of the pipeline. The primary requirement is that the angular velocity of the rotor be directly proportional to the volumetric flow rate or, more correctly, to some average velocity of the fluid in the pipe.

Mass-reaction or **momentum-flow meters** are primarily of three types. In one type an impeller imparts a constant angular momentum to the fluid stream, which is measured as a variable torque on a turbine. In a second type, an impeller is driven at a constant torque and the variable angular velocity of the impeller is measured. In the third type a loop of fluid is driven at either a constant angular speed or a constant oscillatory motion and the mass reaction is measured. Several of these momentum mass flow meters provide liquid hydrogen mass-flow measurement accuracies on the order of ±0.5%. However, degradation of mass flow measurement accuracy occurs whenever two-phase flow is encountered.

HAZARDS WITH CRYOGENIC SYSTEMS

Hazards can best be classified as those associated with the response of the human body and the surroundings to cryogenic fluids and their vapors, and to those associated with reactions between certain of the cryogenic fluids and their surroundings.

Physiological Hazards

Exposure of the human body to cryogenic fluids or to surfaces cooled by cryogenic fluids can cause severe **cold burns** because damage to the skin or tissue is similar to that caused by an ordinary burn. The severity of the burn depends on the contact area and the contact time—prolonged contact results in deeper burns. Severe burns are seldom sustained if withdrawal is rapid. Cold gases may not be damaging if the turbulence in the gas is low, particularly since the body can normally withstand a heat loss of nearly 95 W/m^2 for an area of limited exposure. If a burn is inflicted, the only first aid treatment is to liberally flood the affected area with lukewarm water. The affected area should not be massaged because additional tissue damage can occur.

Protective clothing, including safety goggles, gloves, and boots, is imperative for personnel that handle liquid cryogens. Such operations, should only be attempted when sufficient personnel are available to monitor the activity.

Since nitrogen is a colorless, odorless, inert gas, personnel must be aware of associated respiratory and asphyxiation hazards. Whenever the oxygen content of the atmosphere is diluted due to spills or leaks of nitrogen, there is the danger of **nitrogen asphyxiation**. In general, the oxygen content of air for breathing purposes should not be below 16% or greater than 25%.

An **oxygen-enriched atmosphere**, on the other hand, produces exhilarating effects when breathed. However, lung damage can occur if the oxygen concentration in the air exceeds 60%, and prolonged exposure to an atmosphere of pure oxygen may initiate bronchitis, pneumonia, or lung collapse. An additional threat of oxygen-enriched air can come from the increased flammability and explosion hazards.

Construction and Operations Hazards

Most failures of cryogenic equipment can be traced to an improper selection of construction materials or a disregard for the change of some material property at low temperatures. For example, low temperatures make some construction materials brittle or less ductile. This behavior is further complicated because some materials become brittle at low temperatures but still can absorb considerable impact strength. (See Chapter 39 for additional

Cryogenics

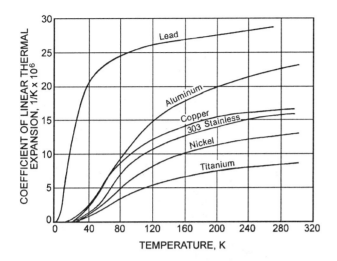

Fig. 33 Coefficient of Linear Expansion for Several Metals as Function of Temperature

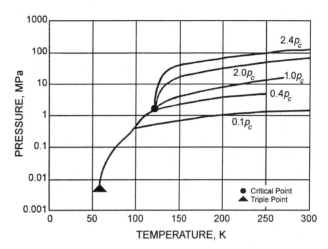

Fig. 34 Pressure Developed During Warming of Liquid Nitrogen in Closed Container

(Courtesy of Air Products and Chemicals, Inc., Allentown, PA)

details). Brittle fracture can cause almost instantaneous failure. Such a failure can cause shrapnel damage if the system is under pressure, and release of a fluid such as oxygen can cause a fire or an explosion.

Low-temperature equipment can also fail because thermal stresses cause thermal contraction of the materials. Figure 33 provides information for evaluating the thermal contraction exhibited by several metals used widely in low-temperature construction. Solder in joints must be able to withstand stresses caused by differential contraction where two dissimilar metals are joined. Contraction in long pipes is also a serious problem; for example, a stainless steel pipeline 30 m long will contract approximately 57 mm when filled with liquid oxygen or nitrogen. Provisions must be made for this change in length during both cooling and warming of the pipeline by using bellows, expansion joints, or flexible hose. Pipe anchors, supports, and so on, must also be carefully designed to permit contraction and expansion to take place. The primary hazard of failure due to thermal contraction is spillage of the cryogen and the possibility of fire or explosion.

Overpressure. All cryogenic systems should be protected against overpressure due to the phase change from liquid to gas. Systems containing liquid cryogens can reach bursting pressures, if not relieved, simply by trapping the liquid in an enclosure. For example, Figure 34 shows the pressure buildup in a closed vessel of liquid nitrogen. The rate of the pressure rise in storage containers for cryogens depends on the rate of heat transfer into the liquid. In uninsulated systems the liquid is vaporized rapidly and pressure in the closed system can rise very rapidly. The more liquid that is originally in the tank before it is sealed off, the greater will be the resulting final pressure.

Heat leakage and vaporization of cryogenic fluid trapped in valves, fittings, and sections of piping can cause excessive pressure buildup and possible rupture of the equipment. For instance, liquid hydrogen expands about 850 times its volume when warmed to ambient temperature. Relief valves and burst disks are normally used to relieve piping systems at a pressure slightly above the operating design pressure of the equipment. Such relief needs to be provided between valves, on tanks, and at all locations of possible (though perhaps unintentional) pressure rise in a piping system.

Overpressure in cryogenic systems can also occur in a more subtle way. Vent lines without appropriate rain traps can collect rainwater and freeze closed. So can exhaust tubes on relief valves and burst disks. Small necked, open mouth dewars can collect moisture from the air and freeze closed. Entrapment of cold liquids or gases can occur by freezing water or other condensables in some portion of the cold system. If this occurs in an unanticipated location, the relief valve or burst disk may be isolated and afford no protection.

Another source of system overpressure that is frequently overlooked results from cooldown surges. If a liquid cryogen is admitted to a warm line with the intention of transferring the liquid from one point to another, severe pressure surges will occur. These pressure surges can be ten times the operating or transfer pressure and can even cause backflow into the storage container. Protection against such overpressure must be included in the overall design and operating procedures for the transfer systems.

Release of Cryogens. Although several cryogenic liquids produce vapors of lower molecular mass than air, the lower temperatures result in a more dense vapor. Released vapors travel along the ground and collect in low places. Exposure to these vapors is hazardous. Exposure to oxygen vapors has caused clothing or any equipment with oil lubricated parts to become oxygen enriched and cause fires.

In making an accident or safety analysis, the possibility of encountering even more serious secondary effects from any cryogenic accident should be considered. For example, any one of the failures discussed previously (brittle fracture, contraction, overpressure, etc.) may release sizable quantities of cryogenic liquids or cold gases causing a severe fire or explosion hazards, asphyxiation possibilities, further brittle fracture problems, or shrapnel damage to other flammable or explosive materials. In this way the accident can rapidly and progressively become much more serious.

Flammability and Detonability Hazards

Almost any flammable mixture will, under favorable conditions of confinement, support an explosive flame propagation or even a detonation. A fuel-oxidant mixture of a composition favorable for high-speed combustion first loses its capacity to detonate when it is diluted with an oxidant, fuel, or inert substance. Further dilution causes it to lose its capacity to burn explosively. Eventually, the lower or upper flammability limits will be reached and the mixture will not maintain its combustion temperature and will automatically extinguish itself. These principles apply to the combustible cryogens, hydrogen and methane. The flammability and detonabiltiy limits for these two cryogens with either air or oxygen are presented in Table 6. Because the flammability limits are rather broad, great care must be exercised to exclude oxygen from these cryogens. This is particularly true with hydrogen since even trace

Table 6 Flammability and Detonability Limits of Hydrogen and Methane Gas

Mixture	Flammability Limits, mol%	Detonability Limits of Mixture, mol%
H_2-Air	4–75	20–65
H_2-O_2	4–95	15–90
CH_4-Air	5–15	6–14
CH_4-O_2	5–61	10–50

amounts of oxygen will condense, solidify, and build up with time in the bottom of the liquid hydrogen storage container. Eventually the upper flammability limits will be reached. Then some ignition source, such as a mechanical or electrostatic spark, may initiate a fire or possibly an explosion.

Liquid hydrogen and LNG spills from a vapor blanket that includes zones of combustible mixtures that could ignite the entire spilled fuel. Both these fluids burn clean; hydrogen produces a nearly invisible flame. Compared to a flammability limit of 2 to 9% (by volume) for jet fuel in air, hydrogen has a flammability limit of 4 to 75% and LNG from 5 to 15%. The ignition of explosive mixtures of hydrogen with oxygen or air occurs with low energy input, about one tenth that of a gasoline-air mixture. All ignition sources should be eliminated, and all equipment and connections should be grounded. Lightning protection in the form of lightning rods, aerial cables, and ground rods suitably connected, should be provided at all preparation, storage, and use areas for these flammable cryogenic fluids.

The methane-oxygen-nitrogen flammability limits diagram of Figure 35 can be used to analyze a fuel-oxidant-diluent. The diagram is typical of any system where methane represents any fuel, oxygen represents any oxidant, and nitrogen represents any diluent. Gas mixtures that lie outside the flammability envelope will not burn or detonate because insufficient heat is released upon combustion to attain a temperature at which combustion is supported. The flame temperatures are at a minimum around the edges of the envelope but increase toward the center of the envelope. The line drawn from a 21% O_2/79% N_2 binary to pure CH_4 represents the compositions that can result from mixing methane with air. Flammability occurs from about 5 to 15% methane in air. If an LNG tank is purged with air, the gas mixture resulting would at first be too rich in methane for combustion. However, as purging continues, flammable mixtures would develop, resulting in a hazardous condition.

Such predictability of behavior, however, depends on the homogenous composition of the fuel-oxidant-diluent mixture achievable in gases. Where local concentration gradients exist, the combustion behavior is less certain. In such cases, the flame velocity depends on such factors as particle size, droplet size, and heat capacities of components in the system. Flame arresters, which are inserted in a line or vessel, prevent the propagation of a flame front by absorbing energy from the combustion process and lowering the flame temperature below a level that supports combustion. The minimum ignition energy, however, varies considerably, being as much as an order of magnitude lower for a hydrogen air mixture than for a methane air mixture.

Flammable cryogens are best disposed of in a burnoff system in which the liquid or gas is piped to a remote area and burned with air in multiple burner arrangements. Such a system should include pilot ignition methods, warning systems in case of flameout, and means for purging the vent line.

Because of its chemical activity, oxygen also presents a safety problem in handling. Liquid oxygen is chemically reactive with hydrocarbon materials. Ordinary hydrocarbon lubricants are even dangerous to use in oxygen compressors and vacuum pumps exhausting gaseous oxygen. In fact, valves, fittings, and lines used

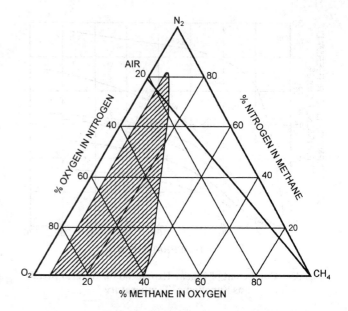

Fig. 35 Flammable Limits for O_2-N_2-CH_4 System
(Courtesy of Air Products and Chemicals, Inc. Allentown, PA)

with oil pumped gases should never be used with oxygen. Serious explosions have resulted from the combination of oxygen and hydrocarbon lubricants. To ensure against such unwanted chemical reactions, systems using liquid oxygen must be kept scrupulously clean of any foreign matter.

Liquid oxygen equipment must also be constructed of materials incapable of initiating or sustaining a reaction. Only a few polymeric materials can be used in such equipment because most will react violently with oxygen under mechanical impact. Also, reactive metals such as titanium and aluminum should be used cautiously, because they are potentially hazardous. Once the reaction is started, an aluminum pipe containing oxygen burns rapidly and intensely. With proper design and care, however, liquid oxygen systems can be operated safely.

Even though nitrogen is an inert gas and will not support combustion, in some subtle ways a flammable or explosive hazard may develop. Cold traps or open-mouth dewars containing liquid nitrogen can condense air and cause oxygen enrichment of the liquid nitrogen. The composition of air as it condenses into the liquid nitrogen container is about 50% oxygen and 50% nitrogen. As the liquid nitrogen evaporates, the liquid oxygen content steadily increases so that the last portion of liquid to evaporate will have a relatively high oxygen concentration. The nitrogen container must then be handled as if it contained liquid oxygen; fire, compatibility and explosive hazards all apply to this oxygen enriched liquid nitrogen.

Because air condenses at a temperature above the normal boiling point of liquid nitrogen, uninsulated pipelines transferring liquid nitrogen will condense air. This oxygen-enriched condensate can drip on combustible materials, causing an extreme fire hazard or explosive situation. The oxygen-rich condensate can saturate clothing, rags, wood, asphalt pavement, etc. and cause the same problems associated with the handling and spillage of liquid oxygen.

Hazard Evaluation Summary

The best-designed facility is no better than the detailed attention that has been paid to every aspect of safety. Such attention cannot be considered once and forgotten. Rather, it is an ongoing activity that requires constant review of every conceivable hazard that might be encountered.

REFERENCES

Alfiev, V.N., V.M. Brodyansky, V.M. Yagodin, V.A. Nikolsky, and A.V. Invantso. 1973. UK Patent 1,336,892.

Anderson, J.E., D.A. Fester, and A.M. Czysz. 1990. Evaluation of long term cryogenic storage system requirements. *Advances in Cryogenic Engineering* 35:1725. Plenum Press, New York.

Arp, V. 1997. Electrical and thermal conductivities of elemental metals below 300 K. *Proceedings of the 13th Symposium on Thermophysical Properties*, Boulder, CO.

Arp, V. 1998. A summary of fluid properties including near-critical behavior. *Proceedings of the International Conference on Cryogenics and Refrigeration*, Hangzhou, China.

Augustynowicz, S.D., J.A. Demko, and V.I. Datskov. 1994. Analysis of multilayer insulation between 80 K and 300 K. *Advances in Cryogenic Engineering* 39:1675. Plenum Press, New York.

Bednorz, J.G. and K.A. Müller. 1986. *Z. Phys.* B64:189.

Burt, W.W. and C.K. Chan. 1995. Demonstration of a high performance 35 K pulse-tube cryocooler. *Cryocoolers* 8:313. Plenum Press, New York.

Davey, G. 1990. Review of the Oxford cryocooler. *Advances in Cryogenic Engineering* 35:1423. Plenum Press, New York.

Gaumer, L.S. 1986. LNG processes. *Advances in Cryogenic Engineering* 31:1095. Plenum Press, New York.

Gistan, G.M., J.C. Villard, and F. Turcat. 1990. Application range of centrifugal compressors. *Advances in Cryogenic Engineering* 35:1031. Plenum Press, New York.

Grenier M. and P. Petit. 1986. Cryogenic air separation: The last twenty years. *Advances in Cryogenic Engineering* 31:1063. Plenum Press, New York.

Hashimoto, T., K. Gang, H. Makuuchi, R. Li, A. Oniski, and Y. Kanazawa. 1994. Improvement of refrigeration characteristics of the G-M refrigerator using new magnetic regenerator material $ErNi_{1-x}CO_x$ system. *Advances in Cryogenic Engineering* 37:859. Plenum Press, New York.

Hashimoto, T., M. Ogawa, A. Hayaski, M. Makino, R. Li, and K. Aoki. 1992. Recent progress on rare earth magnetic regenerator materials. *Advances in Cryogenic Engineering* 37:859. Plenum Press, New York.

Haynes, W.M. and D.G. Friend. 1994. Reference data for thermophysical properties of cryogenic fluids. *Advances in Cryogenic Engineering* 39:1865. Plenum Press, New York.

Johnson, P.C. 1986. Updating LNG plants. *Advances in Cryogenic Engineering* 31:1101. Plenum Press, New York.

Kittel, P. 1992. Ideal orifice pulse tube refrigerator performance. *Cryogenics* 32:843.

Kun, L.C. 1988. Expansion turbines and refrigeration for gas separation and liquefaction. *Advances in Cryogenic Engineering* 33:963. Plenum Press, New York.

Kuriyama, T., M. Takahashi, H. Nakagoma, T. Hashimoto, H. Nitta, and M. Yabuki. 1994. Development of 1 watt class 4K GM refrigerator with magnetic regenerator materials. *Advances in Cryogenic Engineering* 39:1335. Plenum Press, New York.

Lee, J.M., P. Kittel, K.D. Timmerhaus, and R. Radebaugh. 1993. Flow patterns intrinsic to the pulse tube refrigerator. *Proc. 7th International Cryocooler Conference*, 125. Air Force *Rept.* PL CP 93 1001.

Lemmon, E.W., R.T. Jacobsen, S.G. Penoncello, and S.W. Beyerlein. 1994. Computer programs for the calculation of thermodynamic properties of cryogens and other fluids. *Advances in Cryogenic Engineering* 39:1891. Plenum Press, New York.

Little, W.A. 1990. Advances in Joule Thomson cooling. *Advances in Cryogenic Engineering* 35:1305. Plenum Press, New York

Little, W.A. 1984. Microminiature refrigeration. *Rev. Sci Instrum.* 55:661.

Longsworth, R.C. 1988. 4 K Gifford-McMahon/Joule-Thomson cycle refrigerators. *Advances in Cryogenic Engineering* 33:689. Plenum Press, New York.

Nagao, M., T. Anaguchi, H. Yoshimura, S. Nakamura, T. Yamada, T. Matsumoto, S. Nakagawa, K. Moutsu, and T. Watanabe. 1994. 4K three stage Gifford-McMahon cycle refrigerator for MRI magnet. *Advances in Cryogenic Engineering* 39:1327. Plenum Press, New York.

Radebaugh, R. 1995. Recent developments in cryocoolers. *Proc. XIX International Congress of Refrigeration* IIIb:973. IIR, Paris.

Radebaugh, R. 1990. A review of pulse tube refrigeration. *Advances in Cryogenic Engineering* 35:1191. Plenum Press, New York.

Rao, M.G. and R.G. Scurlock. 1986. Cryogenic instrumentation with cold electronics: A review. *Advances in Cryogenic Engineering* 31:1211. Plenum Press, New York.

Sarwinski, R.E. 1988. Cryogenic requirements for medical instrumentation. *Advances in Cryogenic Engineering* 33:87. Plenum Press, New York.

Scurlock, R.G. 1992. A brief history of cryogenics. *Advances in Cryogenic Engineering* 37:1. Plenum Press, New York.

Shuh, Q.S., R.W. Fast, and H.L. Hart. 1988. Theory and technique for reducing the effect of cracks in multilayer insulation from room temperature to 77 K. *Advances in Cryogenic Engineering* 33:291. Plenum Press, New York.

Sixsmith, H., R. Hasenbein, J.A. Valenzuela, J.C. Theilacker, and J. Fuerst. 1990. A miniature wet turboexpander. *Advances in Cryogenic Engineering* 35:989. Plenum Press, New York.

Spradley, I.E., T.C. Nast, and D.J. Frank. 1990. Experimental studies of MLI systems at very low boundary temperatures. *Advances in Cryogenic Engineering* 35:477. Plenum Press, New York.

Swift, W.L., and H. Sixsrnith. 1993. Performance of a long life reverse Brayton cryocooler. *Proc. 7th International Cryocooler Conference*, 84. Air Force *Rept.* PL CP 93 1001.

Vinen, W.F. 1990. Fifty years of superfluid helium. *Advances in Cryogenic Engineering* 35:1. Plenum Press, New York

Wigley, D.A. 1971. *Mechanical properties of materials at low temperatures*. Plenum Press, New York.

Zhu, S., P. Wu, and Z. Chen 1990. Double inlet pulse tube refrigerators: An important improvement. *Cryogenics* 30:514.

BIBLIOGRAPHY

Barron, R.F. 1985. *Cryogenic systems*, 2nd ed. Oxford University Press, London and New York.

Edeskuty, F.J., and W.F. Stewart. 1986. *Safety in the handling of cryogenic fluids*. Plenum Press, New York.

Edeskuty, F.J. and K.D. Williamson. 1983. *Liquid cryogens*, Vols 1 and 2. CRC Press, Boca Raton, Florida.

Frey, H. and R.A. Haefer. 1981. *Tieftemperatur Technologie*. VDI Verlag, Berlin, Germany.

Jacobsen, R.T. and S.G. Penoncello. 1997. *Thermodynamic properties of cryogenic fluids*. Plenum Press, New York.

Pavese, F. and G. Molinas. 1992. *Modern gas-based temperature and pressure measurements*. Plenum Press, New York.

Timmerhaus, K.D. and T.M. Flynn. 1989. *Cryogenic process engineering*. Plenum Press, New York.

Van Sciver, S.W. 1991. *Helium cryogenics*. Plenum Press, New York.

Walker, G. 1983. *Cryocoolers*. Plenum Press, New York.

CHAPTER 39

LOW-TEMPERATURE REFRIGERATION

Autocascade Systems 39.1	Low-Temperature Materials 39.6
Custom-Designed and Field-Erected Systems 39.2	Insulation 39.9
Single-Refrigerant Systems 39.2	Heat Transfer 39.9
Cascade Systems 39.3	Secondary Coolants 39.11

LOW-TEMPERATURE refrigeration is defined here as refrigeration in the temperature range of −50 to −100°C. What is considered low temperature for an application depends on the temperature range for that specific application. Low temperatures for air conditioning are around 0°C; for industrial refrigeration, −35 to −50°C; and for cryogenics, approaching 0 K. Applications such as freeze-drying, as well as the pharmaceutical, chemical, and petroleum industries, use refrigeration in the temperature range designated low in this chapter.

The −50 to −100°C temperature range is treated separately because design and construction considerations for systems that operate in this range differ from those encountered in the two fields bracketing it, namely industrial refrigeration and cryogenics. Designers and builders of cryogenic facilities are rarely active in the low-temperature refrigeration field. One major type of low-temperature system is the **packaged type**, which often serves such applications as environment chambers. The other major category is **custom-designed and field-erected** systems. Industrial refrigeration practitioners are the group most likely to be responsible for these systems, but they may deal with low-temperature systems only occasionally; the experience of a single organization does not accumulate rapidly. The objective of this chapter is to bring together available experience for those whose work does not require daily contact with low-temperature systems.

The refrigeration cycles presented in this chapter may be used in both standard packaged and custom-designed systems. Cascade systems are emphasized, both autocascade (typical of the packaged units) and two-refrigerant cascade (encountered in custom-engineered low-temperature systems).

AUTOCASCADE SYSTEMS

An autocascade refrigeration system is a complete, self-contained refrigeration system in which multiple stages of cascade cooling effect occur simultaneously by means of vapor-liquid separation and adiabatic expansion of various refrigerants. Physical and thermodynamic features, along with a series of counterflow heat exchangers and an appropriate mixture of refrigerants, make it possible for the system to reach low temperature.

Autocascade refrigeration systems offer many benefits, such as a low compression ratio and relatively high volumetric efficiency. However, system chemistry and heat exchangers are complex, refrigerant compositions are sensitive, and compressor displacement is large.

Operational Characteristics

Components of an autocascade refrigeration system typically include a vapor compressor, an external air- or water-cooled condenser, a mixture of refrigerants with descending boiling points, and a series of insulated heat exchangers. Figure 1 is a schematic diagram of a simple system illustrating a single stage of autocascade effect.

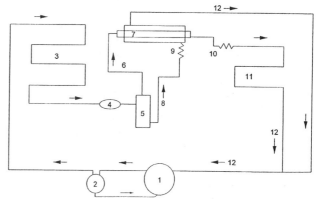

1. Compressor
2. Oil separator
3. Condenser
4. Filter drier
5. Phase separator
6. Vapor tube to second condenser
7. Heat exchanger (evaporator of first refrigerant, cooling second refrigerant)
8. First refrigerant liquid line to metering device
9. First refrigerant metering device feeding heat exchanger, cooling second refrigerant
10. Second refrigerant metering device
11. Evaporator
12. Suction lines to compressor

Fig. 1 Simple Autocascade Refrigeration System

In this system, two refrigerants (A and B), which have significantly different boiling points, are compressed and circulated by one vapor compressor. Assume that refrigerant A is R-23 (normal boiling point, −82°C), while refrigerant B is R-404a (normal boiling point, −46.7°C). Assume that ambient temperature is 25°C and that the condenser is infinitely efficient.

With properly sized components, this system should be able to achieve −60°C in the absorber while the compression ratio is maintained at 5.1 to 1. As the refrigerant mixture is pumped through the main condenser and cooled to 25°C at the exit, compressor discharge pressure is maintained at 1524 kPa (gage). At this condition, virtually all R-404a is condensed at 35°C and then further chilled to subcooled liquid. Although R-23 molecules are present in both liquid and vapor phases, the R-23 is primarily vapor due to the large difference in the boiling points of the two refrigerants. A phase separator located at the outlet of the condenser therefore collects the liquid by gravitational effect, and the R-23-rich vapor is removed from the outlet of the phase separator to the heat exchanger.

At the bottom of the phase separator, expansion device 1 adiabatically expands the collected R-404a-rich liquid such that the outlet of the device produces a low temperature of −19°C at 220 kPa (gage) (Weng 1995). This cold stream is immediately sent back to the heat exchanger in a counterflow pattern such that the R-23-rich vapor is condensed to liquid at −17°C and 1524 kPa (gage). The R-23-rich liquid is then adiabatically expanded by expansion device 2 to −60°C. As it absorbs an appropriate amount of heat in the absorber, the R-23 mixes with the expanded R-404a and evaporates in the heat exchanger, providing a cold source for condensing R-23 on the high-pressure side of the heat exchanger. Leaving the heat exchanger at superheated conditions, the vapor mixture then returns to the suction of the compressor for the next cycle.

The preparation of this chapter is assigned to TC 10.4, Ultra-Low Temperature Systems and Cryogenics.

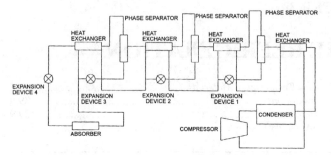

Fig. 2 Four-Stage Autocascade System

As can be seen from this simple example, the autocascade effect derives from a **short cycle** of the refrigerant circuit within the system that performs only internal work to condense the lower boiling point refrigerant.

The concept of the single-stage cycle can be extended to multiple stages. Figure 2 shows the flow diagram of a four-stage system. The condensation and subsequent expansion of one refrigerant provides the cooling necessary to condense the next refrigerant in the heat exchanger downstream. This process continues until the last refrigerant with the lowest boiling point is expanded to achieve extremely low temperature.

Design Considerations

Compressor Capacity. As can be seen from Figures 1 and 2, a significant amount of compressor work is used for internal evaporating and condensing of refrigerants. The final gain of the system is therefore relatively small. Compressor capacity must be sufficient to produce an appropriate amount of final refrigerating effect.

Heat Exchanger Sizing. Because there is a significant amount of refrigerant vapor in each stage of the heat exchanger, the overall heat transfer coefficients on both the evaporating side and the condensing side are rather small compared to those of pure components at phase-changing conditions. Therefore, generous heat transfer area should be provided for energy exchange between refrigerants on the high- and low-pressure sides.

Expansion Devices. Each expansion device is sized to provide sufficient refrigerating effect for the adjacent downstream heat exchanger.

Compressor Lubrication. General guidelines for lubrication of refrigeration systems should be adopted.

CUSTOM-DESIGNED AND FIELD-ERECTED SYSTEMS

If the refrigeration assignment is to maintain a space at a low temperature to store a modest quantity of product in a chest or cabinet, the packaged low-temperature system is probably the best choice. Prefabricated walk-in size environmental chambers are also practical solutions when they can accommodate space needs. When the required refrigeration capacity exceeds that of packaged systems, or when a fluid must be chilled, a custom-engineered system should be considered.

The refrigeration requirement may be to chill a certain flow rate of a given fluid from one temperature to another. Part of the design process is to choose the type of system, which may be a multistage plant using a single refrigerant or a two-circuit cascade system using a high-pressure refrigerant for the low-temperature circuit. The compressor(s) and condenser(s) must be selected, and the evaporator and interstage heat exchanger (in the case of the cascade system) must be either selected or custom designed.

The design process includes selection of (1) metal for piping and vessels and (2) insulating material and method of application. The product to be refrigerated may actually pass through the evaporator, but in many cases a secondary coolant transfers heat from the final product to the evaporator. Brines and antifreezes that perform satisfactorily at higher temperatures may not be suitable at low temperatures. Compressors are subjected to unusual demands when operating at low temperatures, and because they must be lubricated, oil selection and handling must be addressed.

SINGLE-REFRIGERANT SYSTEMS

Single-refrigerant systems are contrasted with the cascade system, which consists of two separate but thermally connected refrigerant circuits, each with a different refrigerant (Stoecker and Jones 1988).

In the industrial refrigeration sector, the traditional refrigerants have been R-22 and ammonia (R-717). Because R-22 will ultimately be phased out, various hydrofluorocarbon (HFC) refrigerants and blends are proposed as replacements. Two that might be considered are R-507 and R-404a.

Two-Stage Systems

In systems where the evaporator operates below about −20°C, two-stage or compound systems are widely used. These systems are explained in Chapter 3 of this volume and in Chapter 1 of the 1997 *ASHRAE Handbook—Fundamentals*. Advantages of two-stage compound systems that become particularly prominent when the evaporator operates at low temperature include

- Improved energy efficiency due to the removal of flash gas at the intermediate pressure and desuperheating of the discharge gas from the low-stage compressor before it enters the high-stage compressor.
- Improved energy efficiency because the two-stage compressors are more efficient operating against discharge-to-suction pressure ratios that are lower than for a single-stage compressor.
- Avoidance of high discharge temperatures typical of single-stage compression. This is important in reciprocating compressors but of less concern with oil-injected screw compressors.
- Possibility of a lower flow rate of liquid refrigerant to the evaporator because the liquid is at the saturation temperature of the intermediate pressure rather than the condensing pressure, as is true of single-stage operation.

Refrigerant and Compressor Selection

The compound, two-stage (or even three-stage) system is an obvious possibility for low-temperature applications. However, at very low temperatures, certain limitations of the refrigerant itself appear. These limitations include the freezing point, the pressure ratios required of the compressors, and the volumetric flow at the suction of the low-stage compressor per unit refrigeration capacity. Table 1 shows some key values for four candidate refrigerants, thereby illustrating some of the concerns that arise when considering refrigerants that are widely applied in industrial refrigeration systems. Refrigerants not included in Table 1 are hydrocarbons (HCs), which are candidates particularly in the petroleum and petrochemical industry, where the entire plant is geared toward working with flammable gases.

Table 1 Low-Temperature Characteristics of Several Refrigerants at Three Evaporating Temperatures

Refrigerant	Freezing Point	Pressure Ratio with Two-Stage System			Volumetric Flow of Refrigerant, L/s per kW		
		Evaporating Temp.			Evaporating Temp.		
		−50°C	−70°C	−90°C	−50°C	−70°C	−90°C
R-22	−160°C	4.6	8.14	17.9	1.57	4.81	19.8
R-507	<−100°C	4.4	7.8	16.1	1.29	4.08	17.5
R-717	−77.7°C	5.75	11.1	25.4	2.05	7.24	—
R-404a	<−100°C	4.36	7.58	15.2	1.34	4.13	16.75

Low-Temperature Refrigeration

The first factor shown in Table 1 is the **freezing point.** It is not a limitation for the halocarbon refrigerants, but ammonia freezes at −78°C, so its use must be restricted to temperatures safely above that temperature.

The next data shown are **pressure ratios** against which the compressors must operate in two-stage systems. A condensing temperature of 35°C is assumed, with the intermediate pressure being the geometric mean of the condensing and evaporating pressures. Many low-temperature systems may be small enough that a reciprocating compressor would be favorable, but the limiting pressure ratio with reciprocating compressors is usually about 8, a value that is chosen to limit the discharge temperature. An evaporating temperature of −70°C is about the lowest permissible for systems using reciprocating compressors. For evaporating temperatures lower than −70°C, a three-stage system should be considered. An alternative to the reciprocating compressor is the screw compressor, which operates with lower discharge temperatures because it is oil flooded. The screw compressor can therefore operate against larger pressure ratios than the reciprocating compressor; in larger systems, it is the favored type of compressor.

The third characteristic shown in Table 1 is the required **volumetric pumping capacity** of the compressor, measured at the compressor suction. This value is an indicator of the physical size of the compressor; the values become huge at the −90°C evaporating temperature.

The following are some conclusions from Table 1:

- A single-refrigerant, two-stage system can adequately serve a plant in the higher temperature portion of the range considered here, but it becomes impractical in the lower temperature portion.
- Ammonia, which has so many favorable properties for industrial refrigeration, has little appeal for low-temperature refrigeration because of its relatively high freezing point and pressure ratios.

Special Multistage Systems

Special high-efficiency operations to recover volatile compounds such as hydrocarbons use the **reverse Brayton cycle.** This consists of one or two conventional compressor refrigeration cycles with the lowest stage ranging from −60 to −100°C. This final stage is achieved by using a turbo compressor/expander and enables the collection of liquefied hydrocarbons (Emhö 1997, Jain and Enneking 1995, Enneking and Priebe 1993).

CASCADE SYSTEMS

The cascade system, illustrated in Figure 3, confronts some of the problems of single-refrigerant systems. The system consists of two separate circuits, each using a refrigerant appropriate for its temperature range. The two circuits are thermally connected by the cascade condenser, which is the condenser of the low-temperature circuit and the evaporator of the high-temperature circuit.

Refrigerants typically chosen for the high-temperature circuit include R-22, ammonia, R-507, R-404a, and so forth. For the low-temperature circuit, a high-pressure refrigerant possessing a high vapor density (even at low temperatures) is chosen. For many years, R-503, an azeotropic mixture of R-13 and R-23, has been a popular choice, but R-503 is no longer available because R-13 is an ozone-depleting chlorofluorocarbon (CFC). R-23 could be and has been used alone, but R-508b, an azeotrope of HFC-23 and R-116, has superior properties. R-508b is discussed further in the section on Refrigerants for Low-Temperature Circuit.

The cascade system possesses some of the thermal advantages of two-stage, single-refrigerant systems in that it approximates the flash gas removal process and also permits each compressor to take a share of the total pressure ratio between the low-temperature evaporator and the condenser. The cascade system has the thermal disadvantage of needing to provide an additional temperature lift in the cascade condenser because the condensing temperature of the low-temperature refrigerant is higher than the evaporating temperature of the high-temperature refrigerant. There is an optimum operating temperature of the cascade condenser for minimum total power requirement, just as there is an optimum intermediate pressure in two-stage, single-refrigerant systems.

Shown in Figure 3 is a fade-out vessel, which limits the pressure in the low-temperature circuit when the system shuts down. At room temperature, the pressure of R-23 or R-508b in the system would exceed 4000 kPa if liquid were present. The entire low-temperature system must be able to accommodate this pressure. The process taking place in the fade-out vessel is at constant volume, as shown on the pressure-enthalpy diagram of Figure 4. When the system operates at low temperature, the refrigerant in the system is a mixture of liquid and vapor, indicated by point A. When the system shuts down, the refrigerant begins to warm and follows the constant-volume line, with the pressure increasing according to the saturation curve. When the saturated vapor line is reached at point B, further increases in temperature result in only a slight increase in pressure because the refrigerant is superheated vapor.

The larger cascade systems are field engineered, but packaged systems are also available. Figure 5 shows a two-stage system with the necessary auxiliary equipment. Figure 6 shows a three-stage cascade system.

Refrigerants for Low-Temperature Circuit

CFC-503 is no longer available for general use. HFC-23 can be used, but R-508b, an azeotropic mixture of two non-ozone-depleting refrigerants, HFC-23 and R-116, is superior. It is nonflammable and

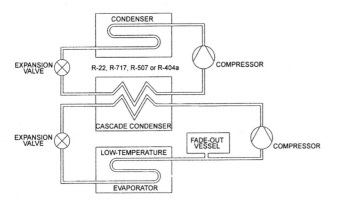

Fig. 3 Simple Cascade System

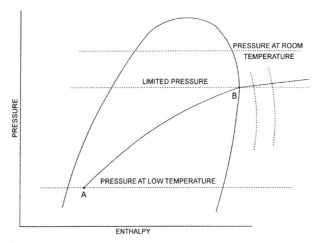

Fig. 4 Simple Cascade Pressure-Enthalpy Diagram

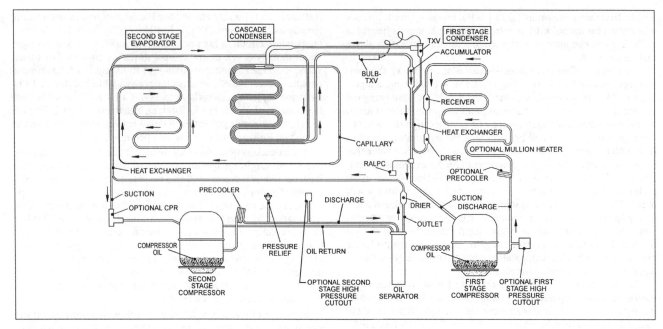

Fig. 5 Two-Stage Cascade System

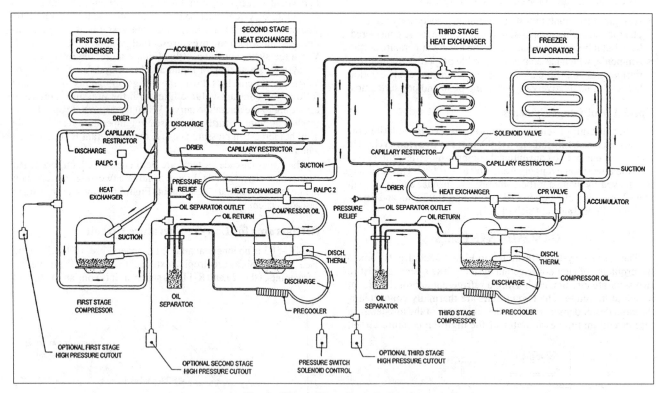

Fig. 6 Three-Stage Cascade System

has zero ozone depletion potential (ODP). Table 2 lists some properties of R-508b.

R-508b offers excellent operating characteristics compared to R-503 and R-13. Capacity and efficiency values are nearly equivalent to R-503 and superior to R-13. The compressor discharge temperature is lower than for R-23; lower discharge temperatures may equate to longer compressor life and better lubricant stability. The estimated operating values of a cascade system running with R-508b are shown in Table 3. Performance parameters of R-503, R-13, and R-23 are shown for comparison.

Table 4 shows calculated data for R-23 and R-508b for two operating ranges. The volumetric efficiency is 100%. Actual compressor performance varies with pressure ratios and would yield lower capacity and efficiencies and higher discharge temperatures and flow requirements than shown.

Compressor Lubrication

The criteria for selecting a lubricant for use with R-508b in an existing low-temperature system should include (1) refrigerant/

Low-Temperature Refrigeration

Table 2 Properties of R-508b

Boiling point (101.325 kPa)	−88°C
Critical temperature	13.7°C
Critical pressure	3935 kPa
Latent heat of vaporization at boiling point	168.2 kJ/kg
Ozone depletion potential (R-12 = 1)	0
Flammability	Nonflammable
Exposure limit (8- and 12-hour) [a]	1000 ppm

[a] The exposure limit is a calculated limit determined from the DuPont airborne exposure limit (AEL) of the individual components. The AEL is the maximum amount to which nearly all workers can be repeatedly exposed during a working lifetime without adverse effects.

Table 3 Theoretical Performance of a Cascade System Using R-13, R-503, R-23, or R-508b

	R-503	R-13	R-23	R-508b
Capacity (R-503 = 100)	100	71	74	98
Efficiency (R-503 = 100)	100	105	95	103
Discharge pressure, kPa	999	717	848	1013
Suction pressure, kPa	110	83	90	110
Discharge temperature, °C	107	92	136	85

Note: Operating conditions are −84.4°C evaporator, −35°C condenser; 5.8 K subcooling; −17.8°C suction temperature; 70% isentropic compression efficiency, 4% volumetric clearance.

Table 4 Theoretical Compressor Performance Data for Two Different Evaporating Temperatures

Refrigerant	Evaporating Temperature, °C	Pressure Ratio	Discharge Temperature, °C	Volumetric Flow, L/s per kW
R-23	−80	7.49	58	1.01
R-508b	−80	6.51	32	0.857
R-23	−100	26.88	72	3.85
R-508b	−100	21.88	39	2.95

Basis: −35°C condensing temperature; compressor efficiency of 70%; volumetric efficiency of 100%; 10 K subcooling, and 50 K suction superheat.

lubricant miscibility, (2) chemical stability, (3) materials compatibility, and (4) refrigeration system design. Original equipment manufacturers and compressor suppliers should be consulted.

It has been long-standing practice in the low-temperature industry to use additives to enhance system performance, and this practice may be applied to R-508b. The miscibility of R-508b with certain polyol esters (POEs) is slightly better than the limited miscibility of R-13 and R-503 with mineral oil and alkylbenzene, which helps oil circulation at the low evaporator temperatures. Even with the increased miscibility, additives may enhance performance. Certain POE oils designed for use in very low temperature systems have been used successfully with R-508b in equipment retrofits. Compressor manufacturers and suppliers should be consulted before a final decision on the lubricants and any additives is made.

Compressors

Larger cascade systems typically use standard, positive-displacement compressors in the dual refrigeration system. The evaporator of the higher temperature refrigerant system serves as the condenser for the lower temperature refrigerant system. This permits rather normal application of the compressors on both systems in relation to the pressures, compression ratios, and oil and discharge temperatures within the compressors. However, several very important items must be considered for both the high and low sides of the cascade system in order to avoid operational problems. Commercially available compressors must be analyzed for both sides of the cascade system to determine the best combination for a suitable intermediate high-side evaporator/low-side condenser and minimum (or economical) system power usage.

High-Temperature Circuit. The higher temperature system is generally a single- or two-stage system using a commercial refrigerant (R-134a, R-22, R-404a, or R-717); evaporating temperature is approximately −23 to −45°C, and condensing is at normal ambient conditions. Commercially available reciprocating and screw compressors are suitable for this duty. If the compressor evaporating temperature is below −45°C, it becomes necessary to provide a suction-line heat exchanger to superheat the compressor suction gas to at least −43°C to avoid the low-temperature metal brittleness associated with lower temperatures at the compressor suction valve and body. Suction piping materials and the evaporator/condenser (evaporator side) must also be suitable for these low temperatures per American Society of Mechanical Engineers (ASME) code requirements.

The lubricant for this higher temperature system must be compatible with the refrigerant employed and suitable for the type of system, considering oil carryover and return from the evaporator and the low-temperature conditions within the evaporator.

Low-Temperature Circuit. The compressor for this duty may also be a standard refrigeration compressor, provided that the compressor suction gas is superheated to at least −43°C to avoid low-temperature metal brittleness. Operation is typically well within standard pressure, oil temperature, and discharge temperature limits. The refrigerant is usually R-23 or R-508b. Because the temperatures in the low side are below −45°C, all piping, valves, and vessels must be of materials in compliance with the ASME codes pertaining to these temperatures.

It may be difficult to obtain compressor rating data for low-temperature applications with these refrigerants because little actual test data are available, and the manufacturer may be reluctant to be specific. Therefore, the low side should not be designed too close to the required specification. A good practice is to calculate the actual volumetric flow rate to be handled by the compressor (at the expected superheat) to be certain that it can perform as required.

The capacity loss from the high superheat is more than recouped (for a net capacity gain) from the liquid subcooling obtained by the suction-line heat exchanger. In rare cases, it may be necessary to inject a small quantity of hot gas into the suction to ensure maximum suction temperature.

The lubricant selected for the low side of the cascade must be compatible with the specific refrigerant employed and also suitable for the low temperatures expected in the evaporator. It is important that adequate coalescing oil separation (5 ppm) is provided to minimize oil carryover from the compressor to the evaporator.

In direct-expansion evaporators, any oil is forced through the tubes, but the velocity in the return lines must be high enough to keep this small amount of oil moving back to the compressor. If the system has capacity control, then multiple suction risers or alternate design procedures may be required to prevent oil logging in the evaporator and ensure oil return. At these ultralow temperatures, it is imperative to select a lubricant that remains fluid and does not plate out on the evaporator surfaces, where it can foul heat transfer.

Choice of Metal for Piping and Vessels

The usual construction metal for use with thermal fluids is carbon steel. However, at temperatures below −29°C, carbon steel should not be used due to its loss of ductility. The designer should consider the use of 304 or 316 stainless steel because of their good low-temperature ductility. Another alternative is to use carbon steel that has been manufactured specifically to retain good ductility at low temperatures (Dow Corning USA 1993). For example,

Carbon steel	Down to −29°C
SA - 333 - GR1	−29 to −46°C
SA - 333 - GR7	−46 to −73°C
SA - 333 - GR3	−59 to −101°C
SA - 333 - GR6	To −46°C

LOW-TEMPERATURE MATERIALS

Choosing material for a specific low-temperature use is often a compromise involving several factors:

- Cost
- Stress level at which the product will operate
- Manufacturing alternatives
- Operating temperature
- Ability to weld and stress-relieve welded joints
- Possibility of excessive moisture and corrosion
- Thermal expansion and modulus of elasticity characteristics for bolting and connection of dissimilar materials
- Thermal conductivity and resistance to thermal shock

Effect of Low Temperature on Materials. When the piping and vessels are to contain refrigerant at low temperature, special materials must usually be chosen because of the effect of low temperature on material properties. Chemical interactions between the refrigerant and containment material must also be considered. Mechanical and physical properties, fabricability, and availability are some of the important factors to consider in selecting a material for low-temperature applications. Few generalizations can be made, except that a decrease in temperature increases hardness, strength, and modulus of elasticity. The effect of low temperatures on ductility and toughness can vary considerably between materials. With a decrease in temperature, some metals show an increase in ductility, while others show an increase to some limiting low temperature followed by a decrease at lower temperatures. Still other metals show a decrease in toughness and ductility as the temperature is decreased below room temperature.

The effect of temperature reduction on polymers depends on the type of polymer. Thermoplastic polymers, which soften when heated above their glass transition temperature T_g, become progressively stiffer and finally brittle at low temperatures. Thermosetting plastics, which are highly cross-linked and do not soften when heated, are brittle at ambient temperatures and remain so at lower temperatures. Elastomers (rubbers) are lightly cross-linked and stiffen like thermoplastics as the temperature is lowered, becoming fully brittle at very low temperatures.

Although polymers become brittle and may crack at low temperatures, their unique combination of properties—excellent thermal and electrical insulation capability, low density, low heat capacity, and nonmagnetic character—make them attractive for a variety of lower temperature applications. At extremely low temperatures, all plastic materials are very brittle and have low thermal conductivity and low strength relative to metals and composites, so selection and use must be carefully evaluated.

Fiber composite materials have gained widespread use at low temperatures, despite their incorporation of components that are often by themselves brittle. A factor that must be considered in the use of composites is the possibility of **anisotropic** behavior, in which they exhibit properties with different values when measured along axes in different directions. Composites with aligned fibers are highly anisotropic.

Metals

The relation of tensile strength to temperature for metals commonly used for structural applications at low temperatures is shown in Figure 7 (Askeland 1994). The slopes of the curves indicate that the increase in strength with decrease in temperature varies among the different metals. However, tensile strength is not the best criterion for determining the suitability of a material for low-temperature service because most failures result from a loss of ductility.

Lower temperatures can have a dramatic effect on the **ductility** of metal; the effect depends to a large extent on crystal structure. Metals and alloys that are face-centered cubic (FCC) and ductile at ambient temperatures remain ductile at low temperatures. Metals in

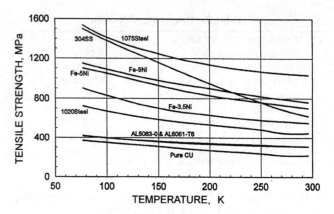

Fig. 7 Tensile Strength Versus Temperature of Several Metals

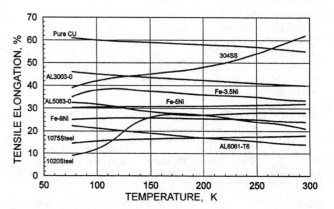

Fig. 8 Tensile Elongation Versus Temperature of Several Metals

this category include aluminum, copper, copper-nickel alloys, nickel, and austenitic stainless steels. Metals and alloys that are body-centered cubic (BCC), such as pure iron, carbon steel, and many alloy steels, become brittle at low temperatures. Many BCC metals and alloys exhibit a ductile-to-brittle transition at lower temperatures (see 1020 steel in Figure 8). This loss of ductility comes from a decrease in the number of operating slip systems, which accommodate dislocation motion. Hexagonal close-packed (HCP) metals and alloys occupy an intermediate place between FCC and BCC materials and may remain ductile or become brittle at low temperatures. Zinc becomes brittle, whereas pure titanium (Ti) and many Ti alloys remain ductile.

Ductility values obtained from the **static tensile test** may give some clue to ductility loss, but the **notched-bar impact test** gives a better indication of how the material performs under dynamic loading and how it reacts to complex multidirectional stress. Figure 8 shows ductility, as measured by percent elongation in the tensile test, in relation to temperature for several metals. As temperature drops, the curves for copper and aluminum show an increase in ductility, while AISI 304 stainless steel and Ti-6%Al-4%V show a decrease.

Aluminum alloys are used extensively for low-temperature structural applications because of cost, weldability, and toughness. Although their strength is considered modest to intermediate, they remain ductile at lower temperatures. Typical mechanical properties at −196°C are listed in Table 5. Property values between ambient (21°C) and −196°C are intermediate between those at these temperatures.

Aluminum 1100 (relatively pure at 99% Al) has a low yield strength but is highly ductile and has a high thermal conductivity. It

Low-Temperature Refrigeration

Table 5 Several Mechanical Properties of Aluminum Alloys at −196°C

Aluminum Alloy	Modulus of Elasticity, GPa	Yield Strength, MPa	Ultimate Tensile Strength, GPa	Percent Elongation at Failure, %	Plane Strain Fracture Toughness, MPa·m$^{1/2}$
1100-0	78	50	190	—	—
2219-T851	85	440	568	14	45
5083-0	80	158	434	32	61
6061-T651	77	337	402	23	42

finds use in nonstructural applications such as thermal radiation shields. For structural purposes, alloys 5083, 5086, 5454, and 5456 are often employed. Alloys such as 5083 have a comparatively high strength in the annealed (0) condition and can be readily welded with little loss of strength in the heat-affected zone; post-welding heat treatments are not necessary. These alloys find use in the storage and transportation areas. Alloy 3003 is widely used for plate-fin heat exchangers because it is easily brazed with an Al-7%Si filler metal. Aluminum-magnesium alloys (6000 series) are used as extrusions and forgings for such components as pipes, tubes, fittings, and valve bodies.

Copper alloys are rarely used for structural applications because of joining difficulties. Copper and its alloys behave similarly to aluminum alloys as temperature decreases. Strength is typically inversely proportional to impact resistance; high-strength alloys have low impact resistance. Silver soldering and vacuum brazing are the most successful methods for joining copper. Brass is useful for small components and is easily machined.

Nickel and nickel alloys do not exhibit a ductile-to-brittle transition as the temperature decreases and can be welded successfully, but their high cost limits use. High-strength alloys can be used at very low temperatures.

Iron-based alloys that are body-centered cubic usually exhibit a ductile-to-brittle transition as the temperature is lowered. The BCC phase of iron is ferromagnetic and easily identified because it is attracted to a magnet. Extreme brittleness is often observed at lower temperatures. Thus, BCC metals and alloys are not normally used for structural applications at lower temperatures. A notable exception is iron alloys having a high nickel content.

Nickel and manganese are added to iron to stabilize the austenitic phase (FCC), promoting low-temperature ductility. Depending on the amount of Ni or Mn added, a great deal of low-temperature toughness can be developed. Two notable high-nickel alloys for use below ambient temperature include 9% nickel steel and austenitic 36% Ni iron alloy. The 9% alloy retains good ductility down to 100 K (−173°C). Below 100 K, ductility decreases slightly, but a clear ductile-to-brittle transition does not occur. Iron containing 36% Ni possesses the unusual feature of nearly zero thermal contraction during cooling from room temperature to near absolute zero. It is therefore an attractive metal for subambient use where the thermal stress associated with differential thermal contraction is to be avoided. Unfortunately, this alloy is quite expensive and therefore sees limited use.

Lesser amounts of Ni can be added to Fe to lower cost and depress the ductile-to-brittle transition temperature. Iron with 5% Ni can be used down to 150 K (−123°C), and Fe with 3.5% Ni remains ductile to 170 K (−102°C). The high-nickel steels are usually heat treated before use by water quenching from 800°C followed by tempering at 580°C. The 580°C heat treatment tempers martensite formed during quenching and produces 10 to 15% stable austenite, which is responsible for the improved toughness of the product.

The austenitic stainless steels (300 series) are widely used for low-temperature applications. Many retain high ductility down to 4 K (−269°C) and below. Their attractiveness is based on good strength, stiffness, toughness, and corrosion resistance, but cost is high compared to Fe-C alloys. A stress relief heat treatment is generally not required after welding, and impact strengths vary only slightly with decreasing temperature. A popular, readily available steel with moderate strength for low-temperature service is AISI type 304, with the low-carbon grade preferred. Where higher strengths are needed and welding can be avoided, strain-hardened or high-nitrogen grades are available. Castable austenitic steels are also available; a well-known example (14-17%Cr, 18-22%Ni, 1.75-2.75%Mo, 0.5%Si max, and 0.05%C max) retains excellent ductility and strength to extremely low temperatures.

Titanium alloys are characterized by high strength, low density, and poor thermal conductivity. Two alloys often used at low temperatures are Ti-5%Al-2.5%Sn and Ti-6%Al-4%V. The Ti-6-4 alloy has the higher yield strength, but it loses ductility below about 80 K (−193°C). The low-temperature properties are dramatically affected by oxygen, carbon, and nitrogen content. Higher levels of these interstitial elements increase strength but decrease ductility. Extra low interstitial (ELI) grades containing about half the normal levels are usually specified for low-temperature applications. Both Ti-6-4 and Ti-5-2.5 are easily welded but expensive and difficult to form. They find application where a high strength-to-mass or strength-to-thermal conductivity ratio is attractive. Titanium alloys are not recommended for applications where an oxidation hazard exists.

Thermoplastic Polymers

Reducing the temperature of thermoplastic polymers restricts molecular motion (bond rotations and molecules sliding past one another), so that the material becomes less deformable. A rapid change in behavior normally takes place over a narrow temperature range beginning at the material's glass transition temperature T_g.

Figure 9 shows the general mechanical response of linear amorphous thermoplastics to temperature. At or above the melting temperature T_m, bonding between polymer chains is weak, the material flows easily, and the modulus of elasticity is nearly zero. Just below T_m, the polymer becomes rubbery; with applied stress, the material deforms by elastic and plastic strain. The combination of these deformations is related to the applied stress by the shear modulus. At still lower temperatures, the polymer becomes stiffer, exhibiting "leathery" behavior and a higher stress at failure. Many

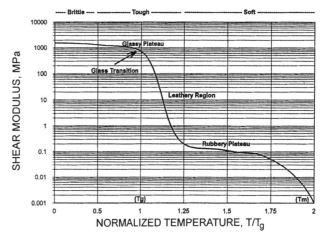

Fig. 9 Shear Modulus Versus Normalized Temperature (T/T_g) for Thermoplastic Polymers

commercially available polymers (e.g., polyethylene) are used in this condition. T_g is located at the transition between the leathery and glassy regions. T_g is usually 0.5 to 0.75 times the absolute melting temperature T_m. Table 6 lists T_g and T_m values for common polymers. In the glassy state, at temperatures below T_g, the polymer is hard, brittle, and glass-like. Although polymers in the glassy region have poor ductility and formability, they are strong, stiff, and creep-resistant.

For thermoplastic polymers, the temperature at which stress is applied and the rate of stress application are interdependent based on time-temperature superposition. This relationship allows different types of tests, such as creep or stress relaxation, to be related through a single curve that describes the viscoelastic response of the material to time and temperature. Applying stress more rapidly has an effect equivalent to applying stress at lower temperatures.

Figure 10 shows tensile strength versus temperature for plastic and polymer composites.

Thermoplastic polymers such as polyethylene and polyvinyl chloride (PVC) may be used for plastic films and wire insulation but are not generally suitable for structural applications because of their brittle nature at temperatures below T_g. The only known polymer that exhibits appreciable ductility at temperatures substantially below T_g is polytetrafluoroethylene (PTFE). Because of their large thermal contraction coefficients, thermoplastic polymers should not be restrained during cool-down. Large masses should be cooled slowly to ensure uniform thermal contraction; the coefficient of thermal contraction decreases with temperature. Contractions of 1 to 2% in cooling from ambient to $-196°C$ are common. For instance, nylon, PTFE, and polyethylene contract 1.3, 1.9, and 2.3%, respectively. These values are large compared to those for metals, which contract 0.2 to 0.3% over the same temperature range.

Thermosetting Plastics

Thermosetting plastics such as epoxy are relatively unaffected by changes in temperature. The materials as a class are brittle and generally used in compression and not tension. Care must be taken in changing their temperature to avoid thermally induced stress, which could lead to cracking. Adding particulate fillers such as silica (SiO_2) to thermosetting resins can increase elastic modulus and decrease strength. The main reasons for adding fillers are to reduce the coefficient of thermal expansion and to improve thermal conductivity. Filled thermosetting resins such as epoxy and polyester can be made to have coefficients of thermal expansion that closely match those of metal. Such materials may be used as insulation and spacers but are not generally used for load-bearing structural applications.

Fiber Composites

Nonmetallic filamentary reinforced composites have gained wide acceptance for low-temperature structural applications because they have good strength, low density, and low thermal conductivity. Nonmetallic insulating composites are usually formed by laminating together layers of fibrous materials in a liquid thermosetting resin such as polyester or epoxy. The fibers are often but not necessarily continuous; they can take the form of bundles, mats, yarns, or woven fabrics. The most frequently used fiber materials include glass, aramid, and carbon. The reinforcing fibers add considerable mechanical strength to otherwise brittle matrix material and can lower the thermal expansion coefficient to a value comparable to that of metals. High figures of merit (ratio of thermal conductivity to elastic modulus or strength) can reduce refrigeration costs substantially from those obtained with fully metallic configurations. Nonmetallic composites can be used for tanks, tubes, struts, straps, and overlays in low-temperature refrigeration systems. These materials perform well in high loading environments and under cyclic stress; they do not degrade chemically at low temperatures.

Different combinations of fiber materials, matrices, loading fractions, and orientations yield a range of properties. Material properties are often anisotropic, with maximum properties in the fiber direction. Failure of composites is caused by cracking in the matrix layer perpendicular to the direction of stress. The cracking may propagate along the fibers but does not generally lead to debonding. Maximum elongations at failure for glass-reinforced composites are usually 2 to 5%; the material is generally elastic all the way to failure.

A major advantage of using glass fibers with a thermosetting binder matrix is the ability to match thermal contraction of the composite to that of most metals. Aramid fibers produce laminates with lower density but higher cost. With carbon fibers, it is possible to produce components that show virtually zero contraction on cooling.

Typical tensile mechanical property data for glass-reinforced laminates are given in Table 7. Under compressive loading, strength and modulus values are generally 60 to 70% of those for tensile loading because of shrinkage of the matrix away from fibers and microbuckling of fibers.

Table 6 Approximate Melting and Glass Transition Temperatures for Common Polymers

Polymer	Melting Temperature T_m		Glass Transition Temperature T_g	
	K	°C	K	°C
Addition polymers				
Low-density polyethylene	388	115	153	−120
High-density polyethylene	413	140	153	−120
Polyvinyl chloride	—	—	363	90
Polypropylene	443	170	258	−15
Polystyrene	—	—	378	105
Polytetrafluoroethylene	603	330	—	—
Polymethyl methacrylate (acrylic)	—	—	368	95
Condensation polymers				
6-6 Nylon	538	265	323	50
Polycarbonate	—	—	418	145
Polyester	528	255	348	75
Elastomers				
Silicone	—	—	148	−125
Polybutadiene	393	120	183	−90
Polychloroprene	353	80	223	−50
Polyisoprene	303	30	203	−70

Source: Askeland (1994). Derived from Table 15-2, p. 482.

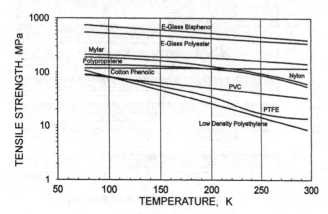

Fig. 10 Tensile Strength Versus Temperature of Plastics and Polymer Matrix Laminates

Table 7 Tensile Properties of Unidirectional Fiber-Reinforced Composites

Composite	Test Temperature, °C	Tensile Strength, MPa	Tensile Modulus, GPa
E Glass (50%)			
Longitudinal	22	1050	41
	−196	1340	45
Transverse	22	9	11
	−196	8	12
Aramid fibers (63%)			
Longitudinal	25	1130	71
	−196	1150	99
Transverse	25	4.2	2.5
	−196	3.6	3.6

Source: Hands (1996). Table 11.3.

Adhesives

Adhesives for bonding composite materials to themselves or to other materials include epoxy resins, polyurethanes, polyimides, and polyheterocyclic resins. Epoxy resins, modified epoxy resins (with nylon or polyamide), and polyurethanes apparently give the best overall low-temperature performance. The joint must be properly designed to account for the different thermal contractions of the components. It is best to have the adhesives operate under compressive loads. Before bonding, surfaces to be joined should be free of contamination, have uniform fine scale roughness, and preferably be chemically cleaned and etched. An even bond gap thickness of 0.1 to 0.2 mm is usually best.

INSULATION

Refrigerated pipe insulation, by necessity, has become an **engineered** element of the refrigeration system. The complexity and cost of this element now rival that of the piping system. This is especially true of systems operating at the depressed temperatures considered in this chapter.

Some factory-assembled, close-coupled systems that operate intermittently can function with a relatively simple installation of flexible sponge/foam rubber pipe insulation. Larger systems that operate continuously require much more investment in design and installation. Higher technology materials and techniques, which are sometimes waived (at risk of invested capital) for systems operating at warmer temperatures, are quite critical for low-temperature operation. Also, the nature of the application does not usually permit shutdown for repair.

Pipe insulation systems are distinctly different from cold room construction. Cold room construction vapor leaks can be neutral if they reach equilibrium with the dehumidification effect of the refrigeration unit. Moisture entering the pipe insulation can only accumulate and form ice, destroying the insulation system. At these low temperatures, it is proper to have redundant vapor retarders (e.g., reinforced mastic plus membrane plus sealed jacket). Insulation should be multilayer to allow expansion and contraction, with inner plies allowed to slide and the outer ply joint sealed. Sealants are placed in the warmest location because they may not function properly at the lower temperature of inner plies. Insulation should be thick enough to prevent condensation (above dew point) at the outside surface.

The main components of a low-temperature refrigerated pipe insulation system are shown in Table 8.

HEAT TRANSFER

The heat transfer coefficients of boiling and condensing refrigerant and the convection heat transfer coefficients of secondary coolants are the most critical heat transfer issues in low-temperature refrigeration. In a cascade system, for example, the heat transfer coefficients in the high-temperature circuit are typical of other refrigeration applications at those temperatures. In the low-temperature circuit, however, the lower temperatures appreciably alter the properties of the refrigerant and therefore the boiling and condensing coefficients.

The expected changes in properties with a decrease in temperature are as follows. **As the temperature drops,**

- Density of liquid increases
- Specific volume of vapor increases
- Enthalpy of evaporation increases
- Specific heat of liquid decreases
- Specific heat of vapor decreases
- Viscosity of liquid increases
- Viscosity of vapor decreases
- Thermal conductivity of liquid increases
- Thermal conductivity of vapor decreases

Table 8 Components of a Low-Temperature Refrigerated Pipe Insulation System

Insulation System Component	Primary Roles	Secondary Roles	Typical Materials
Insulation	Efficiently insulate pipe	Limit water movement toward pipe	Polyurethane-modified polyisocyanurate foams
	Provide external hanger support	Reduce rate of moisture/vapor transfer toward pipe	Extruded polystyrene foams
		Protect vapor retarder from external damage	Cellular glass
Elastomeric joint sealant	Limit liquid water movement through insulation cracks		Synthetic rubbers
	Reduce rate of moisture/vapor transfer toward pipe		Resins
Vapor retarder	Severely limit moisture transfer toward pipe		Mastic/fabric/mastic
	Eliminate liquid water movement toward pipe		Laminated membranes
Protective jacket	Protect vapor retarder from external damage	Reduce moisture/vapor transfer toward pipe	Aluminum
	Limit water movement toward pipe		Stainless steel
			PVC
Protective jacket joint sealant	Prevent liquid water movement through gaps in protective jacket	Limit rate of moisture/vapor transfer toward pipe	
Vapor stops	Isolate damage caused by moisture penetration		Mastic/fabric/mastic

In general, increases in liquid density, enthalpy of evaporation, specific heats of liquid and vapor, and thermal conductivity of liquid and vapor cause an increase in the boiling and condensing heat transfer coefficients. Increases in specific volume of vapor and viscosities of liquid and vapor cause a decrease in these heat transfer coefficients.

Data from laboratory tests or even field observations are scarce for low-temperature heat transfer coefficients. However, heat transfer principles indicate that in most cases the lowering of the temperature level at which the heat transfer occurs reduces the coefficient. The low-temperature circuit in a custom-engineered cascade system encounters lower temperature boiling and condensation than are typical of industrial refrigeration. In some installations, refrigerant boiling is within the tubes, while in others, it is outside the tubes. Similarly, the designer must decide whether condensation at the cascade condenser takes place inside or outside the tubes.

Some relative values based on correlations in Chapter 4 of the 1997 *ASHRAE Handbook—Fundamentals* may help the designer determine which situations call for conservative sizing of heat exchangers. The values in the following subsections are based on changes in properties of R-22 because data for this refrigerant are available down to very low temperatures. Other halocarbon refrigerants used in the low-temperature circuit of the cascade system are likely to behave similarly. Predictions are complicated by the fact that in a process inside tubes, the coefficient changes constantly as the refrigerant passes through the circuit. For both boiling and condensing, temperature has a more moderate effect when the process occurs outside the tubes than when it occurs inside the tubes.

A critical factor in the correlations for boiling or condensing inside the tubes is the mass velocity G in $g/(s \cdot m^2)$. The relative values given in the following subsections are based on keeping G in the tubes constant. The result is that G drops significantly because the specific volume of vapor experiences the greatest relative change of all the properties. As the vapor becomes less dense, the linear velocity can be increased and still maintain a tolerable pressure drop of the refrigerant through the tubes. So G would not drop to the extent used in the comparison below, and the reductions shown for tube-side boiling and condensing would not be as severe as shown.

Condensation Outside Tubes. Based on Nusselt's film condensation theory, the condensing coefficient at $-20°C$, a temperature that could be encountered in a cascade condenser, would actually be 17% higher than the condensing coefficient in a typical condenser at $30°C$. The increase is due to higher latent heat, liquid density, and thermal conductivity. The penalizing influence of the increase in specific volume of vapor is not present because this term does not appear in the Nusselt equation.

Condensation Inside Tubes. Using the correlation of Ackers and Rosson (Table 3, Chapter 4 of the 1997 *ASHRAE Handbook—Fundamentals*) with a constant velocity and thus decreasing the value of G by 1/5, the condensation coefficient at $-20°C$ is one-fourth that at $30°C$.

Table 9 Overview of Some Secondary Coolants

Coolant	Flash Point, °C	Freezing Point, °C	Boiling Point, °C	Temperature at Which Viscosity > 10 mm²/s
Polydimethyl-siloxane	46.7	−111.1	175	−60
d-Limonene	46.1	−96.7	154.4	−80
Diethylbenzene[a]	58	<−84	181	−80
Diethylbenzene[a]	57	−75	181	−70
Hydrofluoroether	not flamm.	−130	60	−30
Ethanol	12	−117	78	−60
Methanol	11	−98	64	−90

[a]Two proprietary versions containing different additives.

Table 10 Refrigerant Properties of Some Low-Temperature Secondary Coolants

Temperature, °C	Viscosity, mPa·s	Density, kg/m³	Heat Capacity, kJ/(kg·K)	Thermal Conductivity, W/(m·K)
Polydimethylsiloxane[a]				
−100	78.6	978	1.52	0.1341
−90	33.7	968	1.54	0.1324
−80	20.1	958	1.56	0.1306
−70	13.3	948	1.58	0.1288
−60	9.4	937	1.60	0.1269
−50	6.4	927	1.62	0.1251
d-Limonene[b]				
−80	1.8	929.7	—	0.139
−70	1.7	921.0	—	0.137
−60	1.6	912.3	—	0.135
−50	1.5	903.6	1.39	0.133
Diethylbenzene[c, d]				
−90		Below Freezing Point		
−80	10	933.6	1.571	0.1497
−70	7.11	926.8	1.593	0.1475
−60	5.12	920.0	1.615	0.1454
−50	3.78	913.0	1.635	0.1435
Hydrofluoroether[e]				
−100	21.226	1814	0.933	0.093
−90	10.801	1788	0.953	0.091
−80	6.412	1762	0.973	0.089
−70	4.235	1737	0.993	0.087
−60	3.017	1711	1.013	0.085
−50	2.268	1686	1.033	0.083
Ethanol[f]				
−100	47	717.0	1.884	0.199
−90	28.3	726.0	1.918	0.198
−80	18.1	735.1	1.943	0.197
−70	12.4	744.1	1.964	0.195
−60	8.7	753.1	1.985	0.194
−50	6.4	762.2	2.011	0.192
Methanol[f]				
−100	16	720	2.178	0.224
−90	8.8	729	2.203	0.223
−80	5.7	738	2.228	0.222
−70	40.2	747	2.253	0.221
−60	2.98	756	2.278	0.220
−50	22.6	765	2.303	0.219
Acetone				
−94		Freezing point		
−90	—	—	1.996	0.150
−80	1.487	—	2.002	0.148
−70	1.22	—	2.010	0.146
−60	0.984	—	2.020	0.145
−50	0.807	—	2.031	0.143
20	—	791	—	—

Sources:
[a]Dow Corning USA (1993)
[b]Florida Chemical Co. (1994)
[c]Dow Corning USA (1993)
[d]Therminol LT (1992)
[e]3M Company (1996)
[f]Raznjevic (1997)

Low-Temperature Refrigeration

Boiling Inside Tubes. Using the correlation of Pierre [Equation (1) in Table 2, Chapter 4 of the 1997 *ASHRAE Handbook—Fundamentals*] and maintaining a constant velocity, when the temperature drops to −70°C the boiling coefficient drops to 46% of the value at −20°C.

Boiling Outside Tubes. In a flooded evaporator with refrigerant boiling outside the tubes, the heat transfer coefficient also drops as the temperature drops. Once again, the high specific volume of vapor is a major factor, restricting the ability of liquid to be in contact with the tube, which is essential for good boiling. Figure 4 in Chapter 4 of the 1997 *ASHRAE Handbook—Fundamentals*, based on data of Stephan, shows that the heat flux has a dominant influence on the coefficient. For the range of temperatures presented for R-22, the boiling coefficient drops by 12% as the boiling temperature drops from −15°C to −41°C.

SECONDARY COOLANTS

Secondary coolant selection, system design considerations, and applications are discussed in Chapter 4 of this volume; properties of brines, inhibited glycols, halocarbons, and nonaqueous fluids are given in Chapter 20 of the 1997 *ASHRAE Handbook—Fundamentals*. The focus here is on secondary coolants for low-temperature applications in the range of −50 to −100°C.

An ideal secondary coolant should

- Have favorable thermophysical properties (high specific heat, low viscosity, high density, and high thermal conductivity)
- Be nonflammable, nontoxic, environmentally acceptable, stable, noncorrosive, and compatible with most engineering materials
- Possess a low vapor pressure

There are just a few fluids that meet such criteria, especially in the entire −50 to −100°C range. Some of these fluids are hydrofluoroether (HFE), diethylbenzene, d-limonene, polydimethylsiloxane, trichloroethylene, and methylene chloride. Table 9 provides an overview of these coolants. Table 10 gives refrigerant properties for the coolants at various low temperatures.

Polydimethylsiloxane, known as silicon oil, is environmentally friendly, nontoxic, and combustible and can operate in the whole range. Due to its high viscosity (greater than 10 mPa·s), its flow pattern is laminar at lower temperatures, which limits heat transfer.

d-Limonene is optically active terpene ($C_{10}H_{16}$) extracted from orange and lemon oils. This fluid can be corrosive and is not recommended for contact with some important materials (polyethylene, polypropylene, natural rubber, neoprene, nitrile, silicone, and PVC). Some problems with stability, such as increase of viscosity with time, are also reported. Contact with oxidizing agents should be avoided. The values listed are based on data provided by the manufacturer in a limited temperature range. d-Limonene is a combustible liquid with a flash point of 46.1°C.

The synthetic aromatic heat transfer fluid group includes **diethylbenzene.** Different proprietary versions of this coolant contain different additives. In these fluids, the viscosity is not as strong a function of temperature. Freezing takes place by crystallization, similar to water.

Hydrofluoroether (1-methoxy-nonafluorobutane, $C_4F_9CH_3$), is a new fluid, so there is limited experience with its use. It is nonflammable, nontoxic, and appropriate for the whole temperature range. No ozone depletion is associated with its use, but its global warming potential is 500 and its atmospheric lifetime is 4.1 years.

The alcohols—**methanol and ethanol**—possess suitable low-temperature physical properties, but they are flammable and methanol is toxic, so their application is limited to industrial situations where these characteristics can be accommodated.

Another possibility for a secondary coolant is **acetone** (C_3H_6O).

REFERENCES

Askeland, D.R. 1994. *The science and engineering of materials*, 3rd ed. PWS Publishing Company, Boston.

Dow Corning USA. 1993. Syltherm heat transfer fluids. The Dow Corning Corporation, Midland, MI.

Emhö, L.J. 1997. HC-recovery with low temperature refrigeration. Presented at ASHRAE Annual Meeting, Boston, MA, June 30.

Enneking, J.C. and Priebe, S. 1993. Environmental application of Brayton cycle heat pump at Savannah River Project. Meeting Customer Needs with Heat Pumps, Conference/Equipment Show.

Florida Chemical Co. 1994. d-Limonene product and material safety data sheets. Winter Haven, FL.

Hands, B.A. 1986. *Cryogenic engineering*. Academic Press, New York.

Jain, N.K. and Enneking, J.C. 1995. Optimization and operating experience of an inert gas solvent recovery system. Air and Waste Management Association Annual Meeting and Exhibition, San Antonio, June 18-23.

Raznjevic, K. 1997. *Heat transfer*. McGraw-Hill, New York.

Stoecker, W.F. and J.W. Jones. 1982. *Refrigeration and air conditioning*, 2nd ed. McGraw-Hill, New York.

Therminol LT. 1992. *Technical Bulletin* No. 9175. Monsanto, St. Louis.

3M Company. 1996. Performance Chemicals and Fluids Laboratory, St. Paul, MN.

Wark, K. 1982. *Thermodynamics*, 4th ed. McGraw-Hill, New York.

Weng, C. 1990. Experimental study of evaporative heat transfer for a non-azeotropic refrigerant blend at low temperature. Master's degree thesis, Ohio University.

Weng, C. 1995. Non-CFC autocascade refrigeration system. U.S. Patent 5,408,848 (April).

CHAPTER 40

BIOMEDICAL APPLICATIONS OF CRYOGENIC REFRIGERATION

Preservation Applications	40.1
Research Applications	40.6
Clinical Applications	40.7
Refrigeration Hardware for Cryobiological Applications	40.8

THE controlled exposure of biological materials to subzero (i.e., potentially freezing) states has multiple practical applications, which have been rapidly multiplying in recent times. Primary among these applications are the long-term preservation of cells and tissues, the selective surgical destruction of tissue by freezing, the preparation of aqueous specimens for electron microscopy imaging, and the study of biochemical mechanisms used by a multitude of living species to withstand the rigors of extreme environmental cold. Some of the applications are restricted to the research laboratory, but clinical and commercial environments are increasingly frequent venues for activities in low-temperature biology. The success of much of this work depends on the design and availability of an apparatus that can control temperatures and thermal histories. This apparatus can be adapted and programmed to meet the specific needs of particular applications.

This chapter briefly describes many of the principles driving the present growth and development of low-temperature biological applications. An understanding of these principles is required to optimally design practical apparatus for executing low-temperature biological processes. Although this field is growing in both breadth and sophistication, this chapter is restricted to processes that involve temperatures below which ice formation is normally encountered, i.e., 0°C, and to an overview of the state-of-the-art.

PRESERVATION APPLICATIONS

Principles of Biological Preservation

Successful cryopreservation of living cells and tissues is coupled to control the thermal history during exposure to below freezing temperatures. The objective of cryopreservation is to reduce the specimen's temperature to such an extent that the rates of chemical reactions that control processes of degeneration become very small, creating a state of effective suspended animation. An Arrhenius analysis (Benson 1982) shows that temperatures must be maintained well below freezing to reduce reaction kinetics enough to store specimens injury free for an acceptable time (usually measured in years). Consequently, one of two types of processes is typically encountered: either the specimen freezes or it undergoes a transition to a glassy state (vitrification). Although both of these phenomena may lead to irreversible injury, most of the destructive consequences of cryopreservation can be avoided.

A change in chemical composition occurs with freezing as the water segregates in the solid ice phase, leaving a residual solution that is rich in electrolytes. This process occurs progressively as the solidification process proceeds through a temperature range that defines a "mushy zone" between the ice nucleation and eutectic states (Körber 1988). If this process follows a series of equilibrium states, the liquidus line on the solid/liquid phase diagram for a system of the chemical composition of the specimen defines the relationship between the system temperature and the solute concentration. The fraction of total water that is solidified increases as the temperature is reduced, according to the function defined by applying the lever rule to the phase diagram liquidus line for the initial composition of the specimen (Prince 1966). This relationship has been worked out for a simple binary model system of water and sodium chloride and has been used to calculate the thermal history of a specimen of defined geometry during cryopreservation (Diller et al. 1985). As explained later, the osmotic stress on the cells with a concurrent efflux of intracellular water results from chemical changes. The critical range of states over which this process occurs corresponds closely to the temperature extremes defined by the mushy zone. At higher temperatures there is no phase change so osmotic stress does not exist. At lower temperatures the permeability of the cell plasma membrane is reduced significantly (as described via an Arrhenius function), and the membrane transport impedance is so high that no significant efflux can occur. Thus, the specimen's chemical history and osmotic response are coupled to its thermal history as defined by the phase diagram properties.

The property of a cell that dictates the response to freezing is the permeability of the plasma membrane to water and permeable solutes. The permeability determines the mass exchange between a cell and its environment when an osmotic stress develops during cryopreservation. The magnitude of the permeability decreases exponentially with the absolute temperature. Thus, the resistance to the movement of chemical species in and out of the cell becomes much larger as the temperature is reduced during a freezing process. Since the osmotic driving force also increases as the temperature decreases, in general, the balance between the osmotic force and resistance determines the extent of mass transfer that will occur during freezing. At high subfreezing temperatures (defined generally by the mushy zone), the osmotic force dominates and extensive transport occurs. At low subfreezing temperatures, the resistance dominates and the chemical species are immobilized either inside or outside the cells. The amount of mass exchanged across the membrane is a direct function of the amount of time spent in states for which the osmotic force dominates the resistance. Thus, at slow cooling rates, the cells of a sample experience extensive dehydration, and at rapid cooling rates, very little net transport occurs. The absolute magnitude of the cooling rate that defines the slow and rapid regimes for a specific cell depends on the plasma membrane permeability. A cell with a large permeability requires a rapid cooling rate to prevent extreme transport. The converse holds for cells having a small membrane permeability; they require prolonged high-temperature exposure to effect significant accumulated transport.

When very little transport occurs before low temperatures are reached during cooling, water becomes trapped within the cell in a subcooled state. Chemical equilibration is achieved with extracellular ice by the intracellular nucleation of ice. This phenomenon is referred to as intracellular freezing. In this process, a substantial degree of liquid subcooling occurs prior to nucleation, so the resulting ice structure is dominated by numerous very small crystals. Further, at the low temperatures, the extent of subsequent

---

The preparation of this chapter is assigned to TC 10.4, Ultra-Low Temperature Systems and Cryogenics.

recrystallization is minimal and the intracellular solid state surface energy is high.

At slow cooling rates and at high subfreezing temperatures, both extensive dehydration of cells and an extended period of exposure to concentrated electrolyte solutions occur. Clear evidence shows that some combination of the dehydration and the exposure to concentrated solutes leads to irreversible injury (Mazur 1970, Meryman et al. 1977). Mazur (1977) has also demonstrated that freezing at cooling rates that are rapid enough to cause intracellular ice formation causes a second mechanism of irreversible cell injury. These processes are illustrated in Figure 1, which shows that each extreme of the cooling process during freezing produces a potential for damaging cells. Figure 1 also implies that an intermediate cooling rate should minimize the aggregate effects of these injury processes and define the conditions at which optimum recovery from cryopreservation can be achieved.

Experimental data has been obtained for the survival of a large number of cell types for freezing and thawing as a function of the cooling rate. Nearly without exception the survival function follows an inverted V profile when plotted against cooling rate (Figure 2). This plot has been described as the survival signature of a cell. It illustrates the tradeoff between competing heat and mass transfer processes that govern the cryopreservation process. Solution concentration/osmotic effects lead to slow cooling rate injury. In this state there is adequate time for transport of water out of the cell before sufficient heat transport occurs to lower the temperature enough to drive the membrane permeability to nearly zero. Conversely, at rapid cooling rates the cell temperature is lowered so quickly that there is insufficient time for dehydration, and injury is caused by the formation of intracellular ice. The magnitude of the optimum intermediate cooling rate is a function of the magnitude of the membrane transport permeability. Higher permeabilities result in higher optimum cooling rates. Thus, the optimum thermal history for any cell type must be tailored for its unique constitutive properties.

For most cell types, the band width of cooling rates for optimum cryopreservation survival is small, and the highest achievable survival is unacceptably low. Fortunately, for practical clinical applications, the spectrum of working cooling rates can be broadened and the maximum survival increased by adding a **cryoprotective agent (CPA)** to the sample prior to freezing. Although a wide range of chemicals exhibit cryoprotective properties, the most commonly used include glycerol, dimethyl sulfoxide (DMSO), and polyethylene glycol. Numerous theories have been postulated to explain the action of CPAs. In simplest terms, they modify the processes of solute concentration and/or intracellular freezing (e.g., Lovelock 1954, Mazur 1970). Introducing CPAs to cell systems results in a major modification of the phase diagram for the system (Fahy 1980). In particular, the rate of electrolyte concentration with decreasing temperature may be reduced by nearly ten times, and the eutectic state depressed by as much as 60 to 80 K. These consequences greatly extend the regime of the mushy zone during solidification (Cocks et al. 1975, Jochem and Körber 1987).

Although phase diagrams provide much information for understanding the possible states that may occur during the cryopreservation of living tissues, their interpretation is limited due to two major factors. First, the chemical complexity of living systems is far greater than the simple binary, ternary, or quaternary mixtures that are used to model their behavior. Second, and more importantly, the thermal data used to generate phase diagrams is usually obtained for near equilibrium conditions. In contrast, most cryopreservation is executed under conditions far from the equilibrium state. For some situations, the goal is to maintain a state of disequilibrium; this domain includes vitrification methods that are applied to reach a solid glassy state that avoids ice crystal formation, latent heat effects, and solute concentration effects. In many cases, the degree of thermodynamic equilibrium reached for the low-temperature storage state may differ significantly between the intracellular and extracellular volumes (Mazur 1990). The equilibration can be controlled by manipulating the thermal boundary conditions of the cryopreservation protocol and by alternating the system's chemical composition prior to initiating cooling. Many of the same chemicals used for cryoprotection may be added at higher concentrations to decrease the probability of ice crystal formation at subzero temperatures and elicit vitrification (Fahy 1988).

In addition to the thermal history of the interior of a specimen, the thermodynamic relations determining the release of the latent heat of fusion as a function of temperature in the mushy zone (Hayes et al. 1988) must be considered. A specimen of finite dimension has a distribution of thermal histories within it during the freezing process (Meryman 1966). The pattern assumed for modeling the evolution of latent heat during freezing has a large effect on the cooling rates predicted as a function of local position in a specimen. Consequently, the anticipated spatial distribution of cell survival as a consequence of the preservation process may depend strongly on

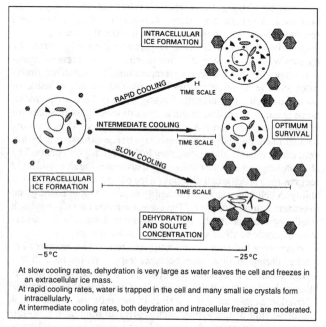

Fig. 1 Schematic of Response of Single Cell During Freezing as Function of Cooling Rate

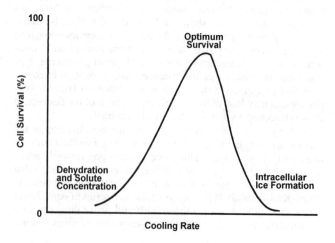

Fig. 2 Generic Survival Signature Indicating Independent Injury Mechanisms Associated with Extremes of Slow and Rapid Cooling Rates During Cell Freezing

Biomedical Applications of Cryogenic Refrigeration

the model chosen for the thermodynamic coupling between the system's thermal and osmotic properties. Hartman et al. (1991) applied this principle to evaluate how to choose the optimum location for a thermal sensor to record the most representative thermal history during the freezing of a specimen of finite dimensions. Hartman et al.'s analysis indicated that the geometric center of a sample is a poor selection for positioning the sensor. A position approximately one third of the distance from the center to the periphery more accurately represents the integrated thermal history experienced by the mass during freezing.

Preservation of Biological Materials by Freezing

Biological materials are primarily cryopreserved by freezing them to deep below freezing temperatures. Among clinical and commercial tissue banks, freezing is the predominant method for preservation. Following the discovery of the cryoprotective properties of glycerol (Polge et al. 1949) and other CPAs, procedures for cryopreservation were developed for storing a variety of cells and tissues. Table 1 summarizes the types of tissue frozen by standard cryopreservation techniques as of 1993.

A typical protocol for cryopreservation consists of the following steps:

1. Place specimen in an appropriate container.
2. Add CPA by sequential increments at reduced temperatures.
3. Cool to subzero temperatures.
4. Possibly induce extracellular ice nucleation followed by a controlled period of thermal and osmotic equilibration.
5. Cool through high subfreezing temperatures with the greatest degree of thermal control invoked for the entire process then quench to storage temperature, usually in liquid nitrogen.
6. Store for extended periods.
7. Warm relatively rapidly by immersion in a heated water bath.
8. Serially dilute to remove CPA using nonpenetrating solutes to control the intracellular/extracellular osmotic balance.
9. Harvest the specimen for its intended application.

Table 1 Spectrum of Various Types of Living Cells and Tissues Commonly Stored by Freezing (as of 1993)

Tissue	Comments	References
Blood vessels	DMSO used for CPA; cooling rate < 1 K/min.	Gottlob et al. (1982)
Bone marrow stem cells	DMSO is usual CPA; widely used in cancer therapy.	McGann et al. (1981)
Cornea	DMSO is usual CPA.	Armitage (1991)
Erythrocytes	Usual CPA is glycerol; high concentrations for slow cooling and low concentrations for rapid cooling; wide spread clinical use.	Turner 1970, Valeri (1976), AATB (1985), Huggins (1985)

Embryos
  Mouse
  Rat
  Goat
  Sheep
  Rabbit
  Bovine
  Drosophila
  Human
	Many species of mammalian embryos have been cryopreserved successfully. The most common CPAs for these applications are glycerol and DMSO. 1,2-Propanediol is used for humans. Variations in the required thermal protocol and processing steps exist among the different species. Ice crystal nucleation is usually controlled by seeding.	CIBA (1977), Zeilmaker (1981)
		Whittingham et al. (1972), Wilmut (1972)
		Whittingham (1975)
		Bilton and Moore (1976)
		Willadsen et al. (1976)
		Bank and Maurer (1974)
		Wilmut and Rowson (1973)
		Mazur et al. (1992)
		Troundson and Mohr (1983)

Heart valves	DMSO is usual CPA; cooling rate of 1 to 2 K/min.	Angell et al. (1987)
Hepatocytes	DMSO is usual CPA; cooling rate of 2.5 K/min.	Fuller and Woods (1987)
Islets	DMSO used for CPA; cooling rate < 1 K/min.	Rajotte et al. (1983), Taylor and Benton (1987)
Lymphocytes	DMSO is usual CPA; primary application is in clinical testing.	Knight (1980), Scheiwe et al. (1981)
Microorganisms	DMSO and glycerol used for CPAs.	James (1987)

Oocytes
  Hamster
  Mouse
  Primate
  Rabbit
  Rat
  Human
	Many species of mammalian oocytes have been cryopreserved successfully. The most common CPAS are glycerol and DMSO. Variations in the required thermal protocol and processing steps exist among the different species. Clinical applications in humans have been difficult to achieve.	Bernard (1991)
		Critser et al. (1986)
		Whittingham (1977)
		DeMayo et al. (1985)
		Diedrick et al. (1986)
		Kasai et al. (1979)
		Van Uem et al. (1987)

Parathyroid	DMSO used for CPA; cooling rate of 1 K/min.	Wells et al. (1977)
Periosteum	DMSO used for CPA; cooling rate of 1 K/min.	Kreder et al. (1993)
Plants	Selected plants are cold hardy; some germplasm is cryopreserved.	Grout (19870, Withers (1987)
Platelets	Best success is with DMSO as CPA; high sensitivity to freezing and osmotic injury.	Schiffer et al. (1985) Sputtek and Körber (1991)
Skin	Both glycerol and DMSO used as CPA.	Aggarwal et al. (1985)

Sperm
  Animal
  Human
	First mammalian cells frozen successfully. Broad applications for animals and humans using glycerol as CPA.	Polge (1980)
		Sherman (1973)

Details vary among individual tissue types; the references in Table 1 give sources of specific parameter values for individual tissues. The refrigeration requirements may vary considerably among different tissues, but basic principles and processes of the cryopreservation processes are generally consistent. This chapter's bibliography and the references in Table 1 identify appropriate introductory references.

Most of the initial applications of cryobiology were in clinical, research, and nonprofit (e.g., the Red Cross) venues. However, the commercial sector has recently adopted cryopreservation methods (McNally and McCaa 1988). As the arsenal of practical cryopreservation methods has grown, the profit potential of freezing tissues for prolonged storage is being recognized and exploited. Thus, an added set of incentives and motives is driving the development of techniques which make use of challenging refrigeration schemes.

Preservation of Biological Materials by Freeze Drying

Freeze drying extends long-term storage at ambient temperatures without the threat of product deterioration. This process removes water from the specimen while it is frozen by sublimation. As a result, no thawing process occurs during rewarming. Thus, none of the decay processes associated with the presence of water in the liquid state are active. Freeze drying has been applied widely in the food processing industry where the product need not be rehydrated in the living state. Other applications, such as taxidermy, also avoid this stringent requirement. The list of biological materials that are frequently processed by freeze drying is extensive and encompasses various microorganisms, protein solutions, pharmaceuticals, and bone. Rowe (1970) reviewed the early state-of-the-art of the physical and engineering aspects of freeze drying; more recently, Franks (1990) has reviewed the physical and chemical principles that govern the freeze drying process from the perspective of achieving an optimal process design.

Franks (1990) characterized the processing steps for freeze drying as summarized in Figure 3. In the figure, note the alternate process pathways from the native to the stored state. The path can be controlled by equipment design and operator intervention. The material shown initially is in the native state from which cooling is initiated. As subfreezing temperatures are reached, either the material remains subcooled in the liquid state or ice crystals form either by spontaneous nucleation or by active seeding a substrate on which a solid phase may form. The material will vitrify if sufficiently subcooled. Ice crystals will grow in a nucleated material, with the rate of temperature change determining the structure and size distribution. Simultaneously, the solute becomes concentrated until the eutectic state is reached. At this state an additional solid phase may form, or the liquid solution may become supersaturated as the temperature is reduced further.

The material is dried, either in the crystalline or vitreous state, by drawing a vacuum on the system at low temperature. Finally, the dried material may be stored at ambient temperatures, although the material is often stored at high subfreezing temperatures to minimize the probability of product deterioration by the activity of residual water. Production methods have been developed mainly by empirical experience and art. However, Franks (1990) argues for the need of a stronger scientific base to more effectively increase the process productivity and quality.

During freeze drying, many phase transitions either never occur or are precipitated at states far from equilibrium, and the slow kinetics of subsequent diffusion processes at low temperatures limit the system from moving toward equilibrium. Even if water crystallizes, the residual solutes are likely to never crystallize fully, if at all. As a supersaturated solution is cooled, the viscosity becomes so large that crystallization processes become undetectable. The so called glass transition temperature is defined by the intersection of the liquidus curve on the phase diagram and an isoviscosity curve for which the mechanical properties of the material are glass-like. These states are illustrated on a state diagram, as shown in Figure 4. By definition, the state diagram does not represent a locus of the system's equilibrium states, but it provides a map of the temperature and composition combinations of defined kinetic properties (Franks 1985).

The intersection of the liquidus curve, when extended past the eutectic temperature t_e into the supersaturated region, and the glass transition curve defines a specific glass transition temperature t_g. This property depends on the combination of the solute concentration and composition. At states above the glass transition threshold, the material behaves as a viscoelastic medium, which is unacceptable for long-term storage. An important aspect of the state diagram is that the slope of the glass transition curve is very steep at high

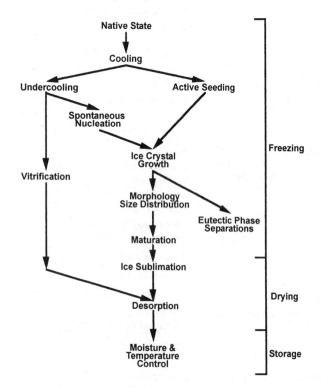

Fig. 3 Key Steps in Freeze Drying Process

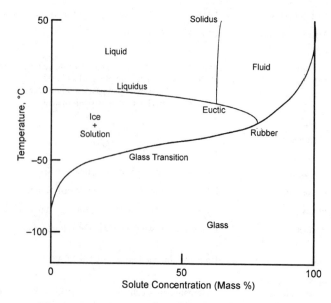

Fig. 4 Generic State Diagram for Aqueous Solution

concentrations of solute. As a consequence, the glass transition temperature t_g is well above 0°C for a pure solute, thus providing for stable storage conditions. However, since small amounts of residual water in the system can significantly lower the glass transition temperature, it is important to check and control the moisture content of a freeze dried product. To this end, Levine and Slade (1988) provided extensive data on the glass transition temperature and unfreezable water fraction of many molecular solutions of interest in the design of freeze drying processes. It is most important that the water is removed from the material at a state temperature lower than the glass transition value at the local solute concentration value. If salts do not precipitate into a solid phase, then unfrozen water remains, which can affect the freeze-drying kinetics (Murase et al. 1991).

Preservation of Biological Materials by Vitrification

The vitrified state has been noted as playing an integral, albeit partial and secondary, role in the preservation of tissues by freezing and by freeze drying. Vitrification may also be used as a storage technique in its own right. Since solidification is avoided, problems associated with the freezing concentration of solutes are avoided. In addition, there are no complications due to latent heat removal from the specimen at a moving phase front. However, water cannot achieve a vitreous state simply by cooling a solution of physiological composition. Therefore, vitrification is achieved only with the prior addition of high concentrations of solutes (i.e., CPAs) to alter the kinetics of the crystallization process and the locus of the liquidus curve and the glass transition curve on the state diagram. Although the CPA concentration that must be achieved prior to cooling is higher for vitrification than for freezing, the vitrification process produces no subsequent solution concentration phenomenon as does freezing. Therefore, if the higher initial CPA concentrations can be tolerated without injury at above 0°C where the addition occurs, vitrification may present a distinct benefit as an approach to long-term cryopreservation.

Fahy (1988) presents an extensive summary of various constitutive properties of candidate CPAs for vitrification, as well as empirical data for the crystallization properties of solutes in aqueous solutions. Other important sources of data for the design of vitrification processes are publications of Boutron, who has measured the thermal and glass forming properties of solutions of particular relevance to cryopreservation. For example, Boutron (1993) deals with the glass forming tendency and stability of the amorphous (glassy) state of 2,3—butanediol in physiological solutions of varying chemical complexity.

In addition to modifying a system to be cryopreserved by adding a CPA, Fahy et al. (1984) has explored modifying the state behavior of tissues by cooling under high pressures. Pressures of up to 100 MPa were used during cooling to reduce the melting temperature of water to about −9°C and the homogeneous nucleation temperature to −54°C, which is equivalent to the reduction in phase change state achieved by introducing a 3 molar concentration of a common CPA. However, limiting factors associated with thermodynamic properties and design of apparatus must be solved before this technique can be considered for practical applications.

The growth of submicroscopic (light) ice crystals, primarily during warming, has been hypothesized to be injurious to vitrified cells and tissues. Several approaches have been pursued to control this process. Rapid warming through the region of sensitive temperatures where crystal nucleation and growth are most probable is used to reduce the time of exposure to these processes (e.g., Marsland 1987). Problems with this technique have included insuring a homogeneous temperature throughout the tissue and matching the hardware to the impedance properties of the specimen, especially for large organs composed of heterogeneous tissues. Alternatively, Rubinsky et al. (1992) have used biological antifreezes from polar fishes (which adsorb to specific faces of ice crystals to inhibit crystal growth) as a CPA constituent to reduce the susceptibility of mammalian tissues to injury. Accordingly, antifreeze glycopeptides have been added to the vitrifying solution to increase the post thaw viability of vitrified porcine oocytes and embryos.

Vitrification of tissues has been most successful with physically small specimens. Freezing cryopreservation has also been most successful with specimens having this same characteristic, e.g., suspensions of isolated cells and small multicellular tissues. One of the major anticipated advantages for vitrification techniques is in processing whole organs for cryopreservation. To date this potential has not been realized, due in part to difficulty in solving engineering problems associated with the processing. The specimen must cool rapidly throughout to prevent significant numbers of ice crystals from forming in any portion of the tissue volume, which could then later propagate into other areas. Unfortunately, boundary conditions and heat transfer characteristics of relatively large organs do not allow such rapid cooling. The threshold cooling rates can be altered as a function of the tissue's chemical composition by adding a CPA; that is, the most promising approach to resolving this limitation is likely to be chemical rather than thermal. Nonetheless, more effective control of the thermal boundary conditions could be beneficial.

The cooling process also produces a second problem that is in direct conflict with satisfying the threshold cooling rate requirement. As progressively larger temperature gradients are created within the specimen in order to boost the cooling rate, corresponding internal thermal stresses are generated. In the glassy state the elastic strength of the vitrified tissue can easily be exceeded, causing mechanical fracture of the tissue to be preserved (Fahy 1990). This phenomenon is obviously irreversible and totally unacceptable. Thus, cooling must be designed to reduce the temperature fast enough to avoid ice nucleation but slow enough to avoid mechanical fracture. Fortunately, some possible solutions to this quandary have been tested (i.e., annealing stages at appropriate thermal states) and hold promise for vitrification of large organs.

Preservation of Biological Materials by Undercooling

An option for cryopreservation in the undercooled state has found a limited range of applications. This technique avoids heterogeneous nucleation of ice crystals in subcooled water and maintains the storage temperature above the value at which homogeneous nucleation occurs (Franks 1988).

The undercooling method of storage is based on the fact that aqueous solutions can be cooled to temperatures substantially below the equilibrium phase change state without the nucleation of ice crystals. The temperature of spontaneous homogeneous nucleation for pure water is approximately −40°C. Thus, if externally induced heterogeneous nucleation can be blocked, a substantial window of subzero temperatures can be used for storage of biological materials. This approach avoids the injurious effects of ice formation and the freeze concentration of solutes as well as the need to add and remove chemical CPAs from the specimen; although the temperature range available is not low enough to insure long-term storage without product deterioration. Because the physical basis of the undercooling process is much different from the alternate methods described previously, the strategy for developing effective storage is also substantially different.

The key to undercooled storage is the ability to control (prohibit) the nucleation of ice in the specimen. Although the homogeneous nucleation temperature is about 40 K below the equilibrium freezing state, in practice it is difficult to reach even −20°C due to heterogeneous nucleation by particulate matter in the specimen. Further, the presence of just a single ice crystal nucleus is adequate to feed the growth of ice throughout a large volume of aqueous medium. However, because heterogeneous nucleation occurs in the extracellular subvolume of a cell suspension, Franks (1983) suspended the biological material in a medium of innocuous oil formed into microdroplets, thereby dispersing the bulk aqueous suspending solution. In effect, the material, such as cells, was suspended in a very thin film of aqueous solution, which dramatically depresses the ice nucleating tendency of the extracellular matrix. By this method, living cells may be undercooled to nearly the homogeneous nucleation temperature (Franks et al. 1983). Subsequently, many different types of cells have been undercooled in water-in-oil dispersions to $-20°C$ or lower without injury (Mathias et al. 1985).

A similar approach has been developed for the storage of biochemicals. For example, an aqueous protein solution can be dispersed in an oil carrier formulated to form a gel, thereby trapping the biological material in very small isolated droplets in the inert matrix. Each of the microdroplets is unable to communicate with any neighboring droplets, thus preventing local ice nuclei from providing a substrate for ice growth in the material. Challenges of this process involve creating microdroplet dispersions for effective storage that recover when returned to ambient temperatures. The temperature must be precisely controlled to avoid both homogeneous nucleation by becoming too cold and accelerated product deterioration by becoming too warm. Typical storage temperatures are around $-20°C$.

RESEARCH APPLICATIONS

Electron Microscopy Specimen Preparation

Freezing has been adopted widely as a method of preparing specimens for examination by electron microscopy. The advantages of freezing are that it need not involve chemical modification of the specimen in the active liquid state and that the physical substructure of components may be preserved. Conversely, the cooling process may cause ice crystals to form, which would alter or mask the structure to be imaged and which could concentrate the solute locally and cause internal osmotic flows that would produce image artifacts. Thus, control of the thermal history during cooling is critical in obtaining a high quality preparation for viewing on the microscope. Cooling rates of 10^5 to 10^6 K/s or higher are desirable to minimize osmotic dehydration of cells and to avoid ice crystal nucleation and growth. Cooling removes heat from the surface of the specimen, and in most cases, the highest cooling rates occur at the boundary of the specimen. Thus the quality of preparation may vary significantly as a function of position, so the specimen should be mounted so that the dimension normal to the primary direction of heat transfer is as small as possible.

Bald (1987) analyzed factors that govern the cooling process during the cryopreparation of specimens. In each case, the objective is to cool the specimen as rapidly as possible. Three different approaches have been developed for cryofixation; these are classified as slamming, plunging, and spraying. Cooling by slamming is effected by mechanically driving the specimen and its mounting holder onto the surface of a cryogenically refrigerated solid block, which has a large thermal inertia in comparison with the specimen. The impact velocity of the specimen against the cold block is high to achieve as rapid a change in the thermal boundary conditions as possible. The drive mechanism is spring loaded to maintain continuous contact with the block following impact so that thermal resistance to the specimen is minimized.

The plunging technique uses a liquid rather than a solid refrigeration sink. As in slamming, the specimen is driven into a relatively large volume of cryogenic liquid. In common practice, the liquid is prepared in a subcooled or supercritical state so that heat transport from the specimen is not limited by a boiling boundary layer at the interface (Bald 1984). It is also important to eliminate a stratified layer of chilled vapor above the liquid through which the specimen would pass during plunging. Such a vapor layer would cool the specimen somewhat before contact with the liquid cryogen in the vapor medium, but because it has a relatively low convective coefficient the effective cooling rate is substantially reduced.

For the spraying method of cryofixation the specimen is held in a stationary mount, and a jet of liquid cryogen is directed onto the specimen. Heat is removed by a combination of evaporation and convection of the cryogen.

Analysis by Bald (1987) indicated that slamming is potentially the most effective method of rapid cooling for cryofixation. The velocity of the specimen during plunging must be 20 m/s or greater to reach thermal performance levels characteristic of slamming. In general, it is easier to design apparatus to achieve the velocities required for satisfactory performance by spraying than plunging. Further, high plunge velocities are more likely to damage the specimen than are equivalent spray velocities. The most effective cryogen for both plunging and spraying is subcritical ethane.

After the temperature is reduced, further preparation for viewing on the electron microscope may involve mechanical fracture of the specimen, chemical substitution of one constituent such as water (Hunt 1984), or removal of a chemical constituent such as by vacuum sublimation of water (Linner and Livesey 1988, Livesey and Linner 1988, Echlin 1992). Sectioning and fracturing techniques are used to expose internal structure and constituents of a specimen. This approach to preparation is particularly appropriate at cryogenic temperatures since biological materials become quite brittle and very little plastic deformation occurs that would alter the morphology. The exposed internal surfaces may be either imaged directly or modified by mechanical or chemical means. Echlin (1992) presented a comprehensive summary of the cryoprocessing of materials for electron microscopy.

Cryomicroscopy

Initial investigations that made use of cryomicroscopy were conducted in the early 1800s and have been pursued ever since. From its earliest adoption, cryomicroscopy made it possible to obtain useful information about the behavior of living tissues at subfreezing temperatures, but the rigor and breadth of application has been limited primarily by the difficulty in controlling the refrigeration applied to the specimen.

Diller and Cravalho (1970) designed a cryomicroscope in which independently controlled refrigerating and heating sources controlled the specimen temperature and its time rate of change during both cooling and heating. Heating was produced by applying a variable voltage across a transparent, electrically resistive thick film coating deposited on the underside of a glass plate on which the biological specimen was mounted. The local temperature was monitored via a microthermocouple positioned in direct contact with the specimen, and this signal was applied as the input to the electronic control system. By miniaturizing the thermal masses of all components of the system, much higher rates of temperature change were achieved with this system than previously possible (cooling rates approaching 10^5 K/min). This system was cooled by circulating a chilled refrigerant fluid through a closed chamber directly beneath the plate on which the specimen was mounted.

Subsequently, the design was modified by McGrath et al. (1976) to eliminate the flow of refrigerant fluid from passing through the optical path of the microscope. Rather, heat was conducted away from the specimen via a thin radial plate that was chilled on its periphery by a refrigerant. This design is mechanically more satisfactory and offers a thinner working cross-section through the optical path, but the lateral temperature gradients are much higher. These two designs are known as convection and conduction cryomicroscopes, respectively (Diller 1988). The former has been adapted to allow for the simultaneous alteration of both the chemical and thermal environments of the specimen (Walcerz and Diller 1990), and the latter has been commercially marketed with a computer control system (McGrath 1987).

The operation of these cryomicroscopes is based on modulating the temperature over time at the site where the specimen is mounted on the microscope to create the desired thermal history for an experimental trial. The dimensions of the specimen are limited by the field of view of the microscope optics, since the specimen is stationary during a trial. An alternative approach has been adapted to study the control of a different set of variables. In this system a steady state temperature gradient is established across the viewing area of the microscope, and the specimen is moved in time through the gradient to produce the desired temperature history (Rubinsky 1984, Körber 1988). Advantages of this system are that specimens of macroscopic size may be frozen, as has been adapted to controlled thermal preparation of specimens for electron microscopy (Bischoff et al. 1990); and the cooling rate applied to a specimen can be investigated as defined by the product of the spatial temperature gradient and the velocity of advance of the phase interface (Beckmann et al. 1990). A similar gradient stage was built by Koroush and Diller (1983) for analysis of solidification processes. This system included feedback control of the temperatures at the ends of the gradient to view a stationary specimen. Gradient cryomicroscopes have seen little application for cryobiology owing to the limited range of cooling rates that can be achieved.

Cryomicrotome

The refrigerated microtome maintains tissue specimens at a subfreezing temperature in a mechanically rigid state so that very thin sections may be cut in preparation for viewing by electron microscopy. The degree of rigidity required is a function of the thickness of the specimen to be cut; thinner sections require greater rigidity, which is achieved by lower temperatures. Stumpf and Roth (1965) have determined that temperatures above −30°C are adequate to obtain sections 1 µm thick while temperatures below −70°C facilitate the cutting of sections thinner than 1 µm. Thus, the apparatus must produce both a wide range of temperatures and accurate thermal control during the processing. The apparatus must also be designed to exclude environmental moisture that could contaminate the specimen, and to isolate the refrigeration apparatus from the sectioning chamber in order to minimize mechanical vibrations that could compromise the dimensional integrity of the delicate cutting process.

CLINICAL APPLICATIONS

Hypothermia

Although accidental hypothermia is the most widely encountered clinical condition of lowered body core temperature, induced hypothermia has been developed as a method of reducing the metabolic rate of selected organs, such as the heart and brain, during surgical procedures. This procedure is of particular benefit in neonatal patients for whom the blood vessels and surgical field are too small to effectively apply standard cardiac bypass procedures for maintaining peripheral circulation during surgery. If the temperature can be reduced to a suitably low level (12 to 20°C), then it is possible to stop the heart and to pursue surgical procedures (in the absence of blood perfusion) without incurring irreversible injury. The period for which the body can be subjected to the absence of perfused oxygenated blood is a function of the hypothermic temperature and may last as long as an hour. These procedures require (1) the temperature of the organ to be within tolerances that limit tissue damage and (2) the ability to quickly lower and raise the temperature to provide the maximum fraction of the low temperature period for the surgical procedure. For example, Eberhart addressed the challenge of achieving a suitably rapid rate of cooling for the brain by perfusion through the vascular network with a chilled solution (Olsen et al. 1985).

The most effective approach to cooling an internal organ is to circulate the blood through a heat exchanger outside the body. The blood is then perfused through the vascular system of the organ, which acts as a physiological heat exchanger. Weinbaum and Jiji (1989) demonstrated the efficacy of thermal equilibration between various components of the vascular tree and the local embedding tissue. Earlier procedures relied primarily on surface cooling to chill internal organs, which is significantly less effective than perfusion in most applications. The results of Olsen et al. (1985) indicate that the brain can be very rapidly cooled to a hypothermic state by an infusion of cold arterial blood. However, when blood circulation was stopped for cardiac surgical procedures, a gradual, but significant rewarming of the brain occurred due to parasitic heat flow from surrounding structures that had not been cooled. Thus, a combination of cold perfusion through the vascular system and surface cooling seems to provide the best control of the body core temperature during hypothermic surgical procedures.

Cryosurgery

In contrast with the previous applications in which the objective is to maximize the survival of tissues exposed to freezing and thawing, cryosurgery has the goal of selective total destruction of a targeted area of tissue within the body. Cryosurgery is applied to destroy and/or excise tissue that is either dead or diseased. It is usually one of several treatment alternatives and has risen and fallen in favor as a method of treating various types of lesions. In general, it has been most effective in treating lesions for which there is direct or easy external access to allow mechanical placement of a cryoprobe or the spray of a cryogenic fluid. The most commonly accepted uses of cryosurgery include the treatment of skin, mucosal and gynecological lesions, liver cancer, and in cardiac surgery for treatment of tachyarrhythmias (Gage 1992). Other uses that have demonstrated efficacy but not such broad adoption are the treatment of hemorrhoids; oral, prostate and anorectal cancer; bone tumors; vertigo; retinal detachment; and visceral tumors (Gage 1992).

Primary advantages of cryosurgery are that (1) it provides a bloodless approach to surgery, (2) in some applications it reduces the rate of death, and (3) the extent of destruction inside the affected area can be imaged with noninvasive methods (Gilbert et al. 1985). This latter process makes use of a continuous ultrasonic scan of the freezing zone to monitor the interface between the solid and liquid phases as it grows into the targeted tissue. Experimental evidence indicates that a close correlation exists between the extent of phase interface propagation and the boundary of the zone of tissue destruction (Rubinsky et al. 1990), and these results may be explained in large part by a model for the mechanism of destruction of the freezing process (Rubinsky and Pegg 1988). The model asserts that during freezing of tissue, ice forms preferentially in the vascular network. The ice also propagates through the vessels as the solidification front advances. The cells near the vascular network dehydrate due to osmotic stress, and this water then freezes in the vascular lumina. As a result, vessels may expand by as much as a factor of two (for electron micrographs see Rubinsky et al. 1990) causing irreversible injury. Thus, the primary action of freezing in destroying tissue during cryosurgery may be by rendering the vascular system nonfunctional rather than by causing direct cryoinjury. Without an active microcirculatory blood flow, the thawed tissue will die rapidly.

Tissue is best destroyed by using a true cryogenic fluid, which is most commonly liquid nitrogen at −196°C. The size of the probe and the flow rate of cryogen through it determine the volume of tissue that may be frozen. For example, a 9-mm diameter probe will produce in tissue an ice ball with a diameter as large as 25 mm (Dilley et al. 1993). Frequently, tumors exceed the capacity of a single probe, but at present, commercial multiprobe cryosurgery systems do not exist. As a result, multiple systems are used, which are both hardware intensive and compromise control over the freezing process (Onik and Rubinsky 1988). Thus, opportunities exist to improve cryosurgical apparatus.

Recent innovations have included operating the refrigerant system under vacuum thereby creating liquid phase heat transfer with the active heat transfer surface of the probe, which has a considerably lower thermal resistance than a boiling interface (Baust et al. 1992). This approach to enhancing thermal performance is similar to that used to cool specimens for electron microscopy rapidly (Bald 1987).

Other problems in the design of cryosurgical equipment remain to be solved. For example, parasitic heat leakage along the probe stem to the cold tip extends the active surface capable of causing tissue damage away from the area designed for destruction. This leakage is particularly compromising to the surgical procedure for treating malignant diseases in locations other than on the body surface (Onik and Rubinsky 1988). The simple and convenient interchangeability of probe tips, having various geometries and thermal capacities, would enhance the flexibility of cryosurgical apparatus. Further, the increasing incidence of sexually transmitted diseases dictates the need for a cryosurgical probe that may either be effectively sterilized (Evans 1992) or be disposable (Baust 1993).

REFRIGERATION HARDWARE FOR CRYOBIOLOGICAL APPLICATIONS

In general, two classes of refrigeration sources have been adapted successfully to biological applications: vapor compression cycle cooling and boiling of liquid cryogens. Also two types of thermal performance standards may be required of these refrigeration sources. As indicated in the previous sections, the thermal history during cooling is very often a critical factor in determining the success of a cryobiological procedure. The refrigerating apparatus must achieve a critical cooling rate within a specimen and regulate the cooling rate within specified tolerances over a designated range of temperatures. If the refrigeration apparatus is designed for general applications, these criteria will be demanded for a large variety of procedures.

A second important performance standard is the minimum specimen temperature that can be maintained in the system. Many biological applications depend on continuously holding the specimen at a temperature below a value at which significant process kinetics may occur. Of most importance are (1) control of the nucleation of ice or other solid phases in vitrified materials, and (2) limitation of recrystallization of small ice crystals that form during cooling. Many cryopreservation procedures require that the specimen be

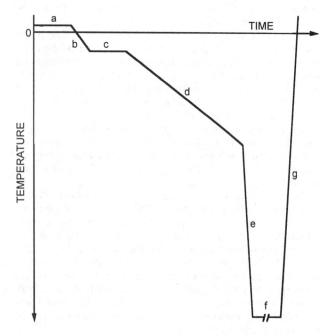

Fig. 5 Generic Thermal History for Example Cryopreservation Procedure

warmed from the stored state as rapidly as possible to avoid the above phenomena for which the kinetics are most favorable at higher subfreezing temperatures. For long-term storage of biological materials, temperatures below −120°C are generally considered to be safe from the effects of devitrification and crystal growth. This state pushes the limits of refrigeration that can be produced by mechanical means, however.

An example of a generic cooling, storage, and warming protocol for cryopreservation is shown in Figure 5. The protocol is divided into seven steps. The first (a) consists of adding a cryoprotective agent at a temperature slightly above freezing. This operation is usually executed with the specimen held in a constant temperature circulating bath. The mixing and osmotic equilibration process may occur in several serial steps and last for half an hour or longer. The specimen is then immersed into a second constant temperature bath held at a high subfreezing temperature (such as −10°C). The cooling rate during this process (b) is uncontrolled, being governed by the inherent heat transfer characteristics of the container and the refrigerant fluid. This constant temperature holding period (c) enables nucleation of ice in the specimen at a predetermined thermodynamic state and provides time for release of the latent heat of fusion and for osmotic equilibration between the intracellular and extracellular volumes. Subsequently, the specimen is placed into a controlled rate refrigerator to execute process (d) during which the temperature is reduced at a rate that maintains a balance between an acceptable osmotic state of the cells and avoids intracellular ice formation. As discussed earlier, the absolute magnitude of this cooling rate depends on the properties of the subject cell, and it may vary over several orders of magnitude for different specimen types. When the specimen reaches a temperature where kinetic rate processes approach zero (e.g., −80°C), the specimen may be plunged (e) into a liquid nitrogen bath for long-term storage (f). Finally, the specimen is warmed and thawed by removing it from the refrigerator and immersing it directly in a water bath (g).

In practice, many variations exist on the cryopreservation scheme shown in Figure 5. One of the most frequent simplifications is to eliminate one or more of the steps (b through d). Whether or not such a simplification is acceptable depends on the sensitivity of the specimen to variations in the thermal history. This sensitivity is a function of the properties of the cells, the physical geometry of the specimen and its packaging for cryopreservation, and chemical modifications performed during step (a).

As the scientific basis for understanding and designing optimal protocols for processes in cryobiology has been strengthened, the specificity and sophistication of the associated refrigeration apparatus has likewise progressed. Therefore, considerable opportunity for improvements in cryobiology hardware remains. The decade of the 1980s witnessed the founding of many new commercial ventures with the objective of exploiting this potential. A common theme was an effective link to the scientific and/or medical community to insure that equipment was designed to address the needs of the customers.

REFERENCES

Aggarwal, S.J., C.R. Baxter, and K.R. Diller. 1985. Cryopreservation of skin: An assessment of current clinical applicability. *J. Burn Care Rehabil.* 6:469-76.

AABB. 1985. *Technical manual of the American Association of Blood Banks.* American Association of Blood Banks, Arlington, VA.

Angell, W.W., J.D. Angell, J.H. Oury, J.J. Lamberti, and T.M. Greld. Long-term follow-up of viable frozen aortic homografts. A viable homograft valve bank. *J. Thorac. Cardiovasc. Surg.* 93:815-22.

Armitage, W.J. 1991. Preservation of viable tissues for transplantation. In *Clinical applications of cryobiology.* B.J. Fuller and B.W.W. Grout, editors. CRC Press, Boca Raton, FL, 170-89.

Bald, W.B. 1984. The relative efficiency of cryogenic fluids used in the rapid quench cooling of biological samples. *J. Micros.* 134:261-70.

Bald, W.B. 1987. *Quantitative cryofixation.* Adam Hilger, Bristol, England.

Bank, H. and R.R. Maurer. 1974. Survival of frozen rabbit embryos. *Exp. Cell Res.* 89:188-96.

Baust, J.G. 1993. Cautions in cryosurgery. *Cryo-Letters* 14:1-2.

Baust, J.G., Z. Chang, and T.C. Hua. 1992. Emerging technology in cryosurgery. *Cryobiology* 29:777.

Beckmann, J., Ch. Körber, G. Rau, A. Hubel, and E.G. Cravalho. 1990. Redefining cooling rate in terms of ice front velocity and thermal gradient: First evidence of relevance to freezing injury of lymphocytes. *Cryobiology* 27:279-87.

Benson, S.W. 1982. *The foundation of chemical kinetics.* Robert E. Kreiger, Malabar, FL.

Bernard, A.G. 1991. Freeze preservation of mammalian reproductive cells. In *Clinical applications of cryobiology.* B.J. Fuller and B.W.W. Grout, editors. CRC Press, Boca Raton, FL, 149-68.

Bilton, F.J. and N.M. Moore. 1976. In vitro culture, storage and transfer of goat embryos. *Aust. J. Biol. Sci.* 29:125-29.

Bischoff, J., C.J. Hunt., B. Rubinsky, A. Burgess, and D.E. Pegg. 1990. Effects of cooling rate and glycerol concentration on the structure of the frozen kidney: Assessment by cryo-scanning electron microscopy. *Cryobiology* 27:301-10.

Boutron, P. 1993. Glass-forming tendency and stability of the amorphous state in solutions of a 2,3—butanediol containing mainly the levo and dextro isomers in water, buffer, and Euro-Collins. *Cryobiology* 30:86-97.

CIBA Foundation. 1977. *The freezing of mammalian embryos.* North Holland/Elsevier, Amsterdam, Holland.

Cocks, F.H., W.H. Hildebrandt, and M.L. Shepard. 1975. Comparison of the low temperature crystallization of glasses in the ternary systems H_2O-NaCl-dimethyl sulfoxide and H_2O-NaCl-glycerol. *J. Appl. Phys.* 46:359-72.

Critser, J.K., B.W. Arneson, D.V. Aaker, and G.D. Ball. 1986. Cryopreservation of hamster oocytes: Effects of vitrification or freezing on human sperm penetration of zona-free hamster oocytes. *Fertil. Steril.* 46:277-84.

DeMayo, F.J., R.G. Rawlins, and W.R. Dukelow. 1985. Xenogenous and in vitro fertilisation of frozen/thawed primate oocytes and blastomere separation of embryos. *Fertil. Steril.* 43:295-300.

Diedrick, K., S. al-Hasani, H. Van der Ven, and D. Krebs. 1986. Successful in vitro fertilisation of frozen thawed rabbit and human oocytes. *Journal of In Vitro Fertilization and Embryo Transplantation* 3:65.

Diller, K.R. Cryomicroscopy. In *Low temperature biotechnology: Emerging applications and engineering contributions.* J.J. McGrath and K.R. Diller, editors. ASME, New York, 347-62.

Diller, K.R. and E.G. Cravalho. 1970. A cryomicroscope for the study of freezing and thawing processes in biological cells. *Cryobiology* 7:191-99.

Diller, K.R., L.J. Hayes, and M.E. Crawford. 1985. Variation in thermal history during freezing with the pattern of latent heat evolution. *AIChE Symposium Series* 81:234-39.

Dilley, A.V., D.Y. Dy, A. Warlters, S. Copeland, A.E. Gillies, R.W. Morris, D.B. Gibb, T.A. Cook, and D.L. Morris. 1993. Laboratory and animal model evaluation of the Cryotech LCS 2000 in hepatic cryotherapy. *Cryobiology* 30:74-85.

Echlin, P. 1992. *Low-temperature microscopy and analysis.* Plenum Press, New York.

Evans, D.T.P. 1992. In search of an optimum method for the sterilization of a cryoprobe in a sexually transmittable diseases clinic. *Genitorin. Med.* 68:275-76.

Fahy, G.M. 1980. Analysis of "solution effects" injury: Equations for calculating phase diagram information for the ternary systems NaCl-dimethylsulfoxide-water and NaCl-glycerol-water. *Biophys. J.* 32:837-50.

Fahy, G.M. 1988. Vitrification. In *Low temperature biotechnology: Emerging applications and engineering contributions.* J.J. McGrath and K.R. Diller, editors. ASME, New York, 113-46.

Fahy, G.M., D.R. MacFarlane, C.A. Angell, and H.T. Meryman. 1984. Vitrification as an approach to cryopreservation. *Cryobiology* 21:407-26.

Fahy, G.M., J. Saur, and R.J. Williams. 1990. Physical problems with the vitrification of large biological systems. *Cryobiology* 27:465-71.

Franks, F. 1985. *Biophysics and biochemistry at low temperatures.* Cambridge University Press, Cambridge.

Franks, F. 1988. Storage in the undercooled state. In *Low temperature biotechnology: Emerging applications and engineering contributions.* J.J. McGrath and K.R. Diller, editors. ASME, New York, 107-12.

Franks, F. 1990. Freeze drying: From empiricism to predictability. *Cryo-Letters* 11:93-110.

Franks, F., S.F. Mathias, P. Galfre, S.D. Webster, and D. Brown. 1983. Ice nucleation and freezing in undercooled cells. *Cryobiology* 20:298-309.

Fuller, B.J. and R.J. Woods. 1987. Influence of cryopreservation on uptake of 99m Tc Hida by isolated rat hepatocytes. *Cryo-Letters* 8:232-37.

Gage, A. 1992. Progress in cryosurgery. *Cryobiology* 29:300-4.

Gilbert, J.C., G.M. Onik, W.K. Hoddick, and B. Rubinsky. 1985. Real time ultrasonic monitoring of hepatic cryosurgery. *Cryobiology* 22:319-30.

Gottlob, R., L. Stockinger, and G.F. Gestring. 1982. Conservation of veins with preservation of viable endothelium. *J. Cardiovasc. Surg.* 23:109-16.

Grout, B.W.W. 1987. Higher plants at freezing temperatures. In *The effects of low temperatures on biological systems.* B.W.W. Grout and G.J. Morris, editors. Edward Arnold, London, 293-314.

Hartman, U., B. Nunner, Ch. Körber, and G. Rau. 1991. Where should the cooling rate be determined in an extended freezing sample? *Cryobiology* 28:115-30.

Hayes, L.J., K.R. Diller, H.J. Chang, and H.S. Lee. 1988. Prediction of local cooling rates and cell survival during the freezing of cylindrical specimens. *Cryobiology* 25:67-82.

Huggins, C.E. 1985. Preparation and usefulness of frozen blood. *Ann. Rev. Med.* 36:499-503.

Hunt, C.J. 1984. Studies on cellular structure and ice location in frozen organs and tissues: The use of freeze-substitution and related techniques. *Cryobiology* 21:385-402.

James, E. 1987. The preservation of organisms responsible for parasitic diseases. In *The effects of low temperatures on biological systems.* B.W.W. Grout and G.J. Morris, editors. Edward Arnold, London, 410-31.

Jochem, M. and Ch. Körber 1987. Extended phase diagrams for the ternary solutions H_2O-NaCl-glycerol and H_2O-NaCl-hydroxyethylstarch (HES) determined by DSC. *Cryobiology* 24:513-36.

Kasai, M., A. Iritani, and B.C. Chang. 1979. Fertilisation in vitro of rat ovarian oocytes after freezing and thawing. *Biol. Reprod.* 21:839-44.

Körber, Ch. 1988. Phenomena at the advancing ice-liquid interface: Solutes, particles and biological cells. *Quart. Rev. Biophysics* 21:229-98.

Knight, S.C. 1980. Preservation of leukocytes. In *Low temperature preservation in medicine and biology.* M.J. Ashwood-Smith and J. Farrant, editors. University Park Press, Baltimore, 121-28.

Kreder, H.J., F.W. Keeley, and R. Salter. 1993. Cryopreservation of periosteum for transplantation. *Cryobiology* 30:107-12.

Levine, H. and L. Slade. 1988. Principles of "Cryostabilization" technology from structure/property relationships of carbohydrate/water systems. *Cryo-Letters* 9:21-63.

Lipton, J.M. 1985. Thermoregulation in pathological states. In *Heat transfer in biology and medicine: Volume 1 analysis and applications.* A. Shitzer and R.C. Eberhart, editors. Plenum Press, New York, 79-105.

Linner, J.G. and S.A. Livesey. 1988. Low temperature molecular distillation drying of cryofixed biological samples. In *Low temperature biotechnology: Emerging applications and engineering contributions.* J.J. McGrath and K.R. Diller, editors. ASME, New York, 117-58.

Livesey, S.A. and J.G. Linner. 1988. Cryofixation methods for electron microscopy. In *Low temperature biotechnology: Emerging applications and engineering contributions.* J.J. McGrath and K.R. Diller, editors. ASME, New York, 159-74.

Lovelock, J.E. 1954. The protective action by neutral solutes against haemolysis by freezing and thawing. *Biochem. J.* 56:265-70.

Marsland, T.P. 1987. The design of an electomagnetic rewarming system for cryopreserved tissue. In *The biophysics of organ cryopreservation.* D.E. Pegg and A.M. Karow, Jr., editors. Plenum Press, New York, 367-85.

Mathias, S.F., F. Franks, R.H.M. Hatley. 1985. Preservation of viable cells in the undercooled state. *Cryobiology* 22:537-46.

Mazur, P. 1963. Kinetics of water loss from cells at subzero temperature and the likelihood of intracellular freezing. *J. Gen Physiol.* 47:347-69.

Mazur, P. 1970. Cryobiology: The freezing of biological systems. *Science* 168:939-49.

Mazur, P. 1977. The role of intracellular freezing in the death of cells cooled at supraoptimal rates. *Cryobiology* 14:251-72.

Mazur, P. 1990. Equilibrium, quasi-equilibrium and nonequilibrium freezing of mammalian embryos. *Cell Biophysics* 17:53-92.

Mazur, P., K.W. Cole, J.W. Hall, P.D. Schreuders, and A.P. Mahowald. 1992. Cryobiological preservation of *Drosophila* embryos. *Science* 258:1932-35.

McGann, L.E., A.R. Turner, M.J. Allalunis, and J.M. Turc. 1981. Cryopreservation of human peripheral blood stem cells: Optimal cooling and warming conditions. *Cryobiology* 18:469-72.

McGrath, J.J., E.G. Cravalho, and C.E. Huggins. 1975. An experimental comparison of intracellular ice formation and freeze-thaw survival of hela S-3 cells. *Cryobiology* 12:540-50.

McGrath, J.J. 1987. Temperature-controlled cryogenic light microscopy—An introduction to cryomicroscopy. In *The effects of low temperatures on biological systems*. B.W.W. Grout and G.J. Morris, editors. Edward Arnold, London, 234-67.

McNally, R.T. and C. McCaa. 1988. Cryopreserved tissues for transplant. In *Low temperature biotechnology: Emerging applications and engineering contributions*. J.J. McGrath and K.R. Diller, editors. ASME, New York, 91-106.

Meryman, H.T. 1966. The interpretation of freezing rates in biological materials. *Cryobiology* 2:165-70.

Meryman, H.T., R.J. Williams, and M. St J. Douglas. 1977. Freezing injury from "Solution" effects and its prevention by natural or artificial cryoprotection. *Cryobiology* 14:287-302.

Murase, N., P. Echlin, and F. Franks. 1991. The structural states of freeze-concentrated and freeze-dried phosphates studied by scanning electron microscopy and differential scanning calorimetry. *Cryobiology* 28:364-75.

O'Brien, M.F., G. Stafford, M. Gardner, P. Pohlener, D. McGriffin, N. Johnston, A. Brosna, and P. Duffy. 1987. The viable cryopreserved Allograft aortic valve. *J. Cardiac. Surg.* 2:153-67.

Olson, R.W., L.J. Hayes, E.H. Wissler, H. Nikaidoh, and R.C. Eberhart. 1985. Influence of hypothermia and circulatory arrest on cerebral temperature distributions. *Trans. ASME, J. Biomech. Engr.* 107:354-60.

Onik, G. and B. Rubinsky. 1988. Cryosurgery: New developments in understanding and technique. In *Low temperature biotechnology: Emerging applications and engineering contributions*. J.J. McGrath and K.R. Diller, editors. ASME, New York, 57-80.

Polge, C. 1980. Freezing of spermatozoa. In *Low temperature preservation in medicine and biology*. M.J. Ashwood-Smith and J. Farrant, editors. University Park Press, Baltimore, 45-64.

Polge, C., A.U. Smith, and A.S. Parkes. 1949. Revival of spermatazoa after vitrification and dehydration at low temperatures. *Nature* (London) 49:666.

Prince, A. 1966. *Alloy phase equilibria*. Elsevier Publishing Company, Amsterdam, Holland.

Rajotte, R.V., G.L. Warnock, L.C. Bruch, and A.W. Procyshyn. 1983. Transplantation of cryopreserved and fresh rat islets and canine pancreatic fragments: Comparison of cryopreservation protocols. *Cryobiology* 20:169-84.

Rall, W.F. and G.M. Fahy. 1985. Ice free cryopreservation of mouse embryos at −196°C by vitrification. *Nature* 313:573-75.

Rowe, T.W.G. 1970. Freeze-drying of biological materials: Some physical and engineering aspects. In *Current trends in cryobiology*. A.U. Smith, editor. Plenum Press, New York, 61-138.

Rubinsky, B., A. Arav, and A.L. DeVries. 1992. The cryoprotective effect of antifreeze glycopeptides from antarctic fishes. *Cryobiology* 29:69-79.

Rubinsky, B. and M. Ikeda. 1985. A cryomicroscope using directional solidification for the controlled freezing of biological material. *Cryobiology* 22:55-68.

Rubinsky, B., C.Y. Lee, L.C. Bastacky, and G. Onik. 1990. The process of freezing and the mechanism of damage during hepatic cryosurgery. *Cryobiology* 27:85-97.

Rubinsky, B. and D.E. Pegg. 1988. A mathematical model for the freezing process in biological tissue. *Proc. Royal Soc. London B* 234:343-58.

Scheiwe, M.W., Z. Pusztal-Markos, U. Essers, R. Seelis, G. Rau, Ch. Körber, K.H. Stürner, H. Jung, and B. Liedtke. 1981. Cryopreservation of human lymphocytes and stem cells (CFU-c) in large units for cancer therapy—A report based on the data of more than 400 frozen units. *Cryobiology* 18:344-56.

Schiffer, C.A., J. Aisner, and J.P. Dutcher. 1985. Platelet cryopreservation using dimethyl sulfoxide. *Ann. N.Y. Acad Sci.* 459:353-61.

Sherman, J.K. 1973. Synopsis of the use of frozen human semen since 1964: State of the art of human semen banking. *Fertil. Steril.* 24:397-412.

Sputtek, A. and Ch. Körber. 1991. Cryopreservation of red blood cells, platelets, lymphocytes, and stem cells. In *Clinical applications of cryobiology*. B.J. Fuller and B.W.W. Grout, editors. CRC Press, Boca Raton, FL, 95-147.

Stumpf, W.F. and L.J. Roth. 1965. Frozen sectioning below −60°C with a refrigerated microtome. *Cryobiology* 1:227-32.

Taylor, M.J. and M.J. Benton. 1987. Interaction of cooling rate, warming rate and extent of permeation of cryoprotectant in determining survival of isolated rat islets of langerhans during cryopreservation. *Diabetes* 36:59-65.

Troundson, A. and L. Mohr. 1983. Human pregnancy following cryopreservation, thawing and transfer of an 8-cell embryo. *Nature* 305:707-9.

Turner, A.R. 1970. *Frozen blood—A review of the literature 1949-1968*. Gordon and Breach, London.

Valeri, C.R. 1976. *Blood banking and the use of frozen blood products*. CRC Press, Boca Raton, FL.

Van Uem, J.F.H.M., D.R. Siebzehnruebele, B. Schuh, R. Koch, S. Trotnow, and N. Lang. 1987. Birth after cryopreservation of unfertilized oocytes. *Lancet* 1:752-53.

Walcerz, D.B. and K.R. Diller. 1991. Quantitative light microscopy of combined perfusion and freezing processes. *J. Microscopy* 161:297-311.

Weinbaum, S. and L.M. Jiji. 1989. The matching of thermal fields surrounding countercurrent microvessels and the closure approximation in the Weinbaum—Jiji Equation. *Trans. ASME, J. Biomech. Engr.* 111:234-37.

Wells, S.A., J.C. Gunnells, R.A. Gutman, J.D. Shelburne, S.G. Schneider, and L.M. Sherwood. 1977. The successful transplantation of frozen parathyroid tissue in man. *Surgery* 81:86-91.

Whittingham, D.G. 1975. Survival of rat embryos after freezing and thawing. *J. Reprod. Fertil.* 43:575-78.

Whittingham, D.G. 1977. Fertilisation in vitro and development to term of unfertilised mouse oocytes previously stored at −196°C. *J. Reprod. Fertil.* 49:89-94.

Whittingham, D.G., P. Mazur, and S.P. Leibo. 1972. Survival of mouse embryos frozen to −196°C and −269°C. *Science* 178:411-14.

Willadsen, S.M., C. Polge, L.E.A. Rowson, and R.M. Moor. 1976. Deep freezing of sheep embryos. *J. Reprod. Fertil.* 46:151-54.

Wilmut, I. 1972. The effect of cooling rate, cryoprotective agent and stage of development on survival of mouse embryos during freezing and thawing. *Life Sci* 11:1071-79.

Wilmut, I. and L.E.A. Rowson. 1973. Experiments on the low-temperature preservation of cow embryos. *Vet. Rec.* 93:686-90.

Withers, L.A. 1987. The low temperature preservation of plant cell, tissue and organ cultures and seed for genetic conservation and improved agricultural practice. In *The effects of low temperatures on biological systems*. B.W.W. Grout and G.J. Morris, editors. Edward Arnold, London, 389-409.

Zeilmaker, G. (editor) 1981. *Frozen storage of laboratory animals*. Gustav Fischer, Stuttgart.

BIBLIOGRAPHY

Primary literature: The main English language sources for general literature on cryobiology are found in two archival journals: *Cryobiology* and *Cryo-Letters*. In addition, the *Bulletin of the International Institute of Refrigeration* provides a timely listing of the world literature in low-temperature biology. Other references are distributed among a large number of journals that are either more general or are oriented toward specific physiological or applications areas.

Monographs: A number of monographs have been written on the principles and applications of low-temperature biology. In general, these have been edited works in which a number of contributing authors provide a series of expositions in focused areas of expertise. Over the last forty years, they have appeared rather consistently. A list of selected monographs follows:

Bald, W.B. 1987. *Quantitative cryofixation*. Adam Hilger, Bristol.

Diller, K.R. 1992. Modeling of bioheat transfer processes at high and low temperatures. In *Advances in heat transfer: Bioengineering heat transfer* 22. Y.I. Cho, editor. Academic Press, Boston, 157-357.

Franks, F. 1985. *Biophysics and biochemistry at low temperatures*. Cambridge University Press, Cambridge.

Fuller, B.J. and B.W.W. Grout, eds. 1991. *Clinical applications of cryobiology*. CRC Press, Boca Rotan, FL.

Grout, B.W.W. and G.J. Morris, eds. 1987. *The effects of low temperatures on biological systems*. Edward Arnold Publishers, Ltd, London.

McGrath, J.J. and K.R. Diller, eds. 1988. *Low temperature biotechnology: Emerging applications and engineering contributions*. ASME, New York.

Meryman, H.T., ed. 1966. *Cryobiology*. Academic Press, New York.

Pegg, D.E. and A.M. Karrow (editors). 1987. *The biophysics of organ cryopreservation*. Plenum Press, New York.

Smith, A.U. 1961. *Biological effects of freezing and supercooling*. Williams and Wilkins, Baltimore.

Smith, A.U., ed. 1970. *Current trends in cryobiology*. Plenum Press, New York.

CHAPTER 41

ABSORPTION COOLING, HEATING, AND REFRIGERATION EQUIPMENT

Terminology 41.1	*Ammonia-Water-Hydrogen Cycle* 41.9
Basic Cycles 41.2	*Special Applications* 41.11
Water-Lithium Bromide Absorption Technology 41.2	*Evolving Concepts* 41.11
Ammonia-Water Absorption Technology 41.9	*Alternative Absorption Working Fluids* 41.12

THIS chapter describes both the standard absorption cooling and heating units available and new concepts that are evolving. Absorption units offer several advantages. They reduce the use of CFC refrigerants and eliminate concerns about lubricants in refrigerants. In addition, particularly with direct-fired systems, they level the demand for natural gas year-round, which promotes efficient maintenance of pipeline systems and lowers costs.

Refrigeration with absorption is possible at evaporator temperatures ranging from 10°C to as low as –60°C with a variety of cycles and fluids. For heat pumps and heat transformers, evaporator fluid inlet temperatures can easily reach 100°C.

Absorption equipment, which is heat driven, can both cool and heat for human comfort and process control. Heat sources for driving an absorption unit include the following:

- Direct gas and oil firing
- Indirect steam heating from boilers
- Steam turbine exhaust
- Gas turbine exhaust
- Waste process steam
- Solar-heated or diesel or gas engine-heated hot water
- Cogeneration heat recovery steam or hot water
- Hot process fluids
- Heat recovery from process streams and flue gases

With larger absorption units, water is usually cooled by passing it through the tubes of the evaporator. This cooled water then flows through air coils to air condition spaces or through other exchangers to cool process fluids. The concept can be applied to directly cool any process fluid passed through the tubes of the absorption unit evaporator, as long as the materials of construction are compatible with the fluid and the temperatures required are attainable by the refrigerant. Smaller absorption units can be used to directly cool air or gases that are passed over finned evaporator coils.

TERMINOLOGY

The following terms refer to the basic absorption cycle and equipment:

Absorption. The process in which refrigerant vapor is absorbed into a concentrated (strong) solution. The heat of condensation of the water and the heat of mixing are released into the fluid by the absorption process. The fluid must be cooled (usually with cooling tower water) to allow sufficient refrigerant to be continuously absorbed into solution while maintaining a low-pressure condition.

Coefficient of performance (COP). For a chiller, the ratio of cooling capacity to required heat energy input. For a heat pump or heat transformer, the ratio of heat energy output capacity to the required heat energy input.

Concentrated (strong) solution. Solution with a relatively high mass fraction of sorbent, and, accordingly, a relatively low mass fraction of refrigerant.

Condensation. Liquefaction of the refrigerant vapor caused when it gives up any superheat (if present) and its latent heat to an available heat sink.

Crystallization. The freezing or conversion of the pumpable liquid solution into a solid or slushy mass when it is excessively strong in sorbent at a given temperature. Not all solutions experience this problem; water-lithium bromide does but ammonia-water does not, except at very low temperatures.

Desorption (generation). The process in which heat is added to a dilute (weak) solution to drive refrigerant vapor from the solution, thus concentrating the solution and permitting the refrigerant to be recycled for reuse at the evaporator.

Dilute (weak) solution. Solution with a relatively low mass fraction of sorbent, and, accordingly, a relatively high mass fraction of refrigerant.

Effect versus Stage. The terms single-stage and single-effect are used interchangeably as are multiple-effect and multistage and two-stage and double-effect.

Evaporation. Vaporization of liquid refrigerant by heat supplied from a heat source, thus producing a cooling effect on the heat source.

Heat pump. A Type 1 heat pump is an absorption machine that uses heat from a high temperature source to elevate the quantity and temperature of heat from available sources and moves this energy to another location at an intermediate temperature. A Type 2 heat pump, which is usually called a heat transformer, is an absorption machine that substantially elevates the temperature of energy from available sources at reduced quantities, using the inherent latent heats and chemical heat of mixing for the fluids of the cycle.

Performance additive. A surfactant (usually one of the octyl alcohols) that is added in minute quantities to the lithium bromide solution. It reduces surface tension and triggers a violent convection at the interface between refrigerant vapor and solution (the Marangoni Effect). The effect considerably enhances the rate of absorption of water vapor by the solution.

P-T-x equilibrium diagram. The plot of equilibrium properties for a given solution relating pressure (P), temperature (T), and concentration (x).

Rectification. Thermally induced mass transfer in which vapor of a volatile sorbent is stripped from the refrigerant vapors driven from the dilute solution in the desorption (generation) process. Not all sorbents are sufficiently volatile to require rectification. For example, in a water-lithium bromide working fluid, water is the refrigerant, and lithium bromide is a nonvolatile sorbent that does not need rectification. However, in an ammonia-water working fluid, ammonia is the refrigerant, and water, the sorbent, is sufficiently volatile to require rectification.

The preparation of this chapter is assigned to TC 8.3, Absorption and Heat Operated Machines.

Refrigerant. A volatile substance that (1) leaves the solution at the desorber or generator, (2) performs the refrigeration process at an evaporator, and (3) is reabsorbed at the absorber to complete the thermodynamic cycle.

Solution. A liquid of at least two components, one soluble in the other. The combination is called the working fluid or *fluid pair* of an absorption system. Chapter 1 of the 1997 *ASHRAE Handbook—Fundamentals* discusses the characteristics of practical fluid pairs, and Chapter 17 lists thermodynamic property data for the two commonly used solutions, water-lithium bromide and ammonia-water.

Sorbent. The portion of the solution that transports the absorbed refrigerant through the processes that make up the thermochemical compressor.

Thermal recuperation. Heat exchange usually between strong and weak solutions within the cycle that improves the overall thermal efficiency.

BASIC CYCLES

Single-Effect Cycle for Cooling

Figure 1 shows the absorption cycle in two portions: a refrigeration portion and a thermochemical compressor portion, thus establishing a parallel with vapor compression refrigeration. A volatile liquid refrigerant, such as water or ammonia, which is suitable for the operating condition, evaporates in the evaporator vessel at a low pressure, thus producing cooling at a low temperature.

Instead of the refrigerant vapor being drawn into a compressor at low pressure and being physically compressed as in a mechanical refrigeration cycle, the refrigerant vapor is absorbed by a separate adjacent absorber into a solution that has had its vapor pressure reduced while it was being cooled. The heat of condensation and heat of mixing caused by the high chemical affinity of the absorbent for the refrigerant is removed and rejected to a heat sink by a second fluid such as water or ambient air to prevent the evaporator and absorber pressure from rising. Usually this heat is rejected by a water-cooled, an air-cooled, or an evaporative condenser.

The diluted sorbent solution is pumped to the higher temperature and pressure generator or desorber side of the system, where the solution can be reconcentrated to complete the cycle.

A recuperative energy exchange usually takes place between the hot concentrated solution leaving the generator and the cooler dilute solution being pumped from the absorber back to the generator.

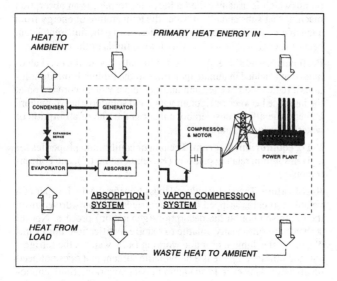

Fig. 1 Similarities between Absorption and Vapor Compression Systems

Heat Transformer

The single-effect absorption heat pump cycle is sometimes called a Type 1 heat pump, and a heat transformer is referred to as a Type 2 heat pump. The single-effect cycle involves sinks and sources at three temperature levels: the highest temperature is for the heat input to the generator, the intermediate temperature is for the heat rejection from both the condenser and the absorber (usually near ambient temperatures), and the lowest temperature is for the heat addition to the evaporator. For a cooling device, the useful result is the cooling of the load by the evaporator, and the rejection is to the ambient atmosphere.

For a Type 1 heat pump, the useful result is the rejection heat, which is transferred to the load being heated, and the atmospheric reference is the heat source for the evaporator. In both cases, the driving energy input is the heat to the generator. A COP (heating) of 1.3 to 1.8 is routinely achieved for respective temperature rises of 40 to 20 K with corresponding waste heat source temperatures of 40 to 90°C and driving heat input source temperatures of 143 to 151°C using water-lithium bromide machines.

In the Type 2 heat pump (heat transformer), the heat sink of a Type 1 system becomes a heat source. The object is to take heat at an intermediate temperature that has no value (because it is not hot enough) and generate a smaller amount of heat at a higher temperature that can be used. The pressure level of the evaporator/absorber pair is arranged to be higher than that of the condenser/generator pair. The highest temperature usable output is from the absorber, and the lowest temperature is the heat rejected to the ambient heat sink of the condenser. Heat at the intermediate temperature level is added to both the evaporator and the generator. In a simple heat transformer, the amount of heat produced at the higher temperature absorber is about half the heat added at the intermediate level evaporator and regenerator. Temperature rises of 25 to 50 K are easily achieved for processes with corresponding intermediate source temperatures of 80 to 95°C using water-lithium bromide machines.

Multiple-Effect Systems for Cooling

Since the thermochemical compressor is governed by the limits of the Carnot cycle (defined by the absolute temperatures of the heat sources and the heat sinks), its efficiency may be improved by increasing the temperature of the generator. Constraints of system assemblies and the characteristics of the chosen working fluids determine the ability to make thermodynamic improvements.

In single-effect systems, the change in concentration and the difference between equilibrium temperatures of the boiling dilute and the outgoing strong solutions for the generator are small. The solution flow rates relative to the flow of refrigerant are large. Significant thermodynamic cycle losses are tied to any inefficiency of the weak solution/strong solution heat exchanger. Temperature approaches should be as small as practical.

The term **double effect** is generally restricted to systems in which a second generator/condenser pair is added to the system; its input is from the heat-driving energy source, and the heat rejection from its condenser is the heat input to the original generator. A given amount of external heat into the first-effect generator provides about 50 to 80% more cooling in a double-effect system than in a single-effect system. The attractiveness of additional condenser/desorber stages is countered by limits of pressure, temperature, corrosion, increases in first costs, and the tendency to generate noncondensables.

WATER-LITHIUM BROMIDE ABSORPTION TECHNOLOGY

Absorption equipment using water as the refrigerant and lithium bromide as the absorbent is classified by the method of heat input to the primary generator (the firing method) and whether the absorption cycle is single- or multiple-effect. Single- and double-effect

Absorption Cooling, Heating, and Refrigeration Equipment

absorption chillers are described in the section on Absorption Machinery.

Machines using steam or hot liquids as a heat source are **indirect fired**, while those using direct combustion of fossil fuels as a heat source are **direct fired**. Machines using clean, hot waste gases as a heat source are also classified as indirect fired but are often referred to as **heat-recovery chillers**.

Components

Solution recuperative heat exchangers, also referred to as economizers, are typically shell-and-tube or plate heat exchangers. They transfer heat between hot and cold absorbent solution streams, thus recycling energy. The material of construction is mild steel or stainless steel.

Condensate subcooling heat exchangers, a variation of solution heat exchangers, are used on steam-fired double-effect machines and on some single-effect, steam-fired machines. These heat recovery exchangers use the hot condensed steam to add heat to the solution entering the generator.

Indirect-fired generators are usually of the shell-and-tube type, with the absorbent solution either flooded or sprayed outside the tubes, and the heat source (steam or hot fluid), inside the tubes. The absorbent solution boils outside the tubes, and the resulting intermediate or strong concentration absorbent solution flows from the generator through an outlet pipe. The refrigerant vapor evolved passes through a vapor/liquid separator consisting of baffles and eliminators and then flows to the condenser section. Ferrous materials are used for absorbent containment; copper, copper-nickel alloys, stainless steel, or titanium are used for the tube bundle.

Direct-fired generators consist of a fire-tube section, a flue-tube section, and a vapor/liquid separation section. The fire tube is typically a double-walled vessel with an inner cavity large enough to accommodate a radiant or open-flame fuel oil or natural gas burner. Dilute solution flows in the annulus between the inner and outer vessel walls and is heated by contact with the inner vessel wall. The flue-tube section is typically a tube- or plate-type heat exchanger connected directly to the fire tube.

Heated solution from the fire-tube section flows on one side of the heat exchanger, and flue gases flow on the other side. Hot flue gases further heat the absorbent solution and cause it to boil. The flue gases leave the generator, while the partially concentrated absorbent solution and refrigerant vapor mixture pass to a vapor and liquid separator chamber. This chamber separates the absorbent solution from the refrigerant vapor. Materials of construction are mild steel for the absorbent containment parts and mostly stainless steel for the flue gas heat exchanger.

Secondary or second-stage generators are used only in double- or multi-stage machines. They are both a generator on the low pressure side and a condenser on the high pressure side. They are usually of the shell-and-tube type and operate similarly to indirect-fired generators of single-effect machines. The heat source, which is inside the tubes, is high-temperature refrigerant vapor from the primary generator shell. Materials of construction are mild steel for absorbent containment and usually copper-nickel alloys or stainless steel for the tubes. Droplet eliminators are typically stainless steel.

Evaporators are heat exchangers, usually of the shell-and-tube type, over which liquid refrigerant is dripped or sprayed and evaporated. The liquid to be cooled passes through the inside of the tubes. Evaporator tube bundles are usually copper or a copper-nickel alloy. Refrigerant containment parts are mild steel. Eliminators and drain pans are typically stainless steel.

Absorbers are tube bundles over which strong absorbent solution is sprayed or dripped in the presence of refrigerant vapor. The refrigerant vapor is absorbed into the absorbent solution, thus releasing heat of dilution and heat of condensation. This heat is removed by cooling water that flows through the tubes. Dilute absorbent solution leaves the bottom of the absorber tube bundle. Materials of construction are mild steel for the absorbent containment parts and copper or copper-nickel alloys for the tube bundle.

Condensers are tube bundles located in the refrigerant vapor space near the generator of a single-effect machine or the second-stage generator of a double-effect machine. The water-cooled tube bundle condenses refrigerant from the generator on the surface of the tubes. Materials of construction are mild steel, stainless steel, or other corrosion resistant materials for the refrigerant containment parts and copper for the tube bundle. For special waters, the condenser tubes can be copper-nickel, which derates the performance of the unit.

High-stage condensers are found only in double-effect machines. This type of condenser is typically the inside of the tubes of the second-stage generator. Refrigerant vapor from the first-stage generator condenses inside the tubes, and the resulting heat is used to concentrate absorbent solution in the shell of the second-stage generator when heated by the outside surface of the tubes.

Pumps move absorbent solution and liquid refrigerant in the absorption machine. Pumps can be configured as individual (one motor, one impeller, one fluid stream) or combined (one motor, multiple impellers, multiple fluid streams). The motors and pumps are hermetic or semihermetic. Motors are cooled and bearings lubricated either by the fluid being pumped or by a filtered supply of liquid refrigerant. Impellers are typically brass, cast iron, or stainless steel; volutes are steel or impregnated cast iron, and bearings are babbitt-impregnated carbon journal bearings.

Refrigerant pumps (when used) recirculate liquid refrigerant from the refrigerant sump at the bottom of the evaporator to the evaporator tube bundle in order to effectively wet the outside surface and enhance heat transfer.

Dilute solution pumps take dilute solution from the absorber sump and pump it to the generator.

Absorber spray pumps recirculate absorbent solution over the absorber tube bundle to assure adequate wetting of the absorber surfaces. These pumps are not found in all equipment designs. Some designs use a jet eductor for inducing concentrated solution flow to the absorber sprays. Another design uses drip-type distributors fed by gravity and the pressure difference between the generator and absorber.

Purge systems are required on lithium bromide absorption equipment to remove noncondensables (air) that leak into the machine or hydrogen (a product of corrosion) that is produced during equipment operation. Noncondensable gases, present even in small amounts, can cause reduction in chilling capacity and even lead to solution crystallization. Purge systems for larger sizes above 359 kW of refrigeration typically consist of these components:

- Vapor pickup tube(s) usually located at the bottom of large absorber tube bundles
- Noncondensable separation and storage tank(s), located in the absorber tube bundle or external to the absorber/evaporator vessel
- A vacuum pump or valving system using solution pump pressure to periodically remove noncondensables collected in the storage tank

Some variations include jet pumps (eductors), powered by pumped absorbent solution and placed downstream of the vapor pickup tubes to increase the volume of sampled vapor, and water-cooled absorbent chambers to remove water vapor from the purged gas stream.

Because of their size, smaller units have fewer leaks, which can be more easily detected during manufacture. As a result, small units may use variations of solution drip and entrapped vapor bubble pumps plus purge gas accumulator chambers.

Palladium cells, found in large direct-fired and small indirect-fired machines, continuously remove the small amount of hydrogen gas that is produced by corrosion. These devices operate on the principle that thin membranes of heated palladium are permeable to hydrogen gas only.

Corrosion inhibitors, typically lithium chromate, lithium nitrate, or lithium molybdate, protect machine internal parts from the corrosive effects of the absorbent solution in the presence of air. Each of these chemicals is used as a part of a corrosion control system. Acceptable levels of contaminants and the correct solution pH range must be present for these inhibitors to work properly. Solution pH is controlled by adding lithium hydroxide or hydrobromic acid.

Performance additives are used in most lithium bromide equipment to achieve design performance. The heat- and mass-transfer coefficients for the simultaneous absorption of water vapor and cooling of lithium bromide solution have relatively low values that must be enhanced. A typical additive is one of the octyl alcohols.

Refrigerant flow control between condensers and evaporators is typically achieved with orifices (suitable for high- or low-stage condensers) or liquid traps (suitable for low-stage condensers only).

Solution flow control between generators and absorbers is typically achieved with flow control valves (primary generator of double-effect machines), variable-speed solution pumps, or liquid traps.

Absorption Machines

Single-Effect Chillers. Figure 2 is a schematic diagram of a commercially available, single-effect indirect-fired liquid chiller, Table 1 lists typical characteristics of this chiller. showing one of several configurations of the major components. During operation, heat is supplied to tubes of the **generator** in the form of a hot fluid or steam, causing dilute absorbent solution on the outside of the tubes to boil. This evolved refrigerant vapor (water vapor) flows through eliminators to the **condenser**, where it is condensed on the outside of tubes that are cooled by a flow of water from a heat sink (usually a cooling tower). Both the boiling and condensing processes take place in a vessel that has a common vapor space at a pressure of about 6 kPa.

The condensed refrigerant passes through an orifice or liquid trap in the bottom of the condenser and enters the evaporator. In the **evaporator**, the liquid refrigerant boils as it contacts the outside surface of tubes that contain a flow of water from the heat load. In

Table 1 Characteristics of Typical Single-Effect, Indirect-Fired, Water-Lithium Bromide Absorption Chiller

Performance characteristics	
Steam input pressure	60 to 80 kPa (gage)
Steam consumption (per kilowatt of refrigeration)	1.48 to 1.51 kW
Hot fluid input temp.	115 to 132°C, with as low as 88°C for some smaller machines for waste heat applications
Heat input rate (per kilowatt of refrigeration)	1.51 to 1.54 kW, with as low as 1.43 kW for some smaller machines
Cooling water temp. in	30°C
Cooling water flow (per kilowatt of refrigeration)	65 mL/s, with up to 115 mL/s for some smaller machines
Chilled water temp. off	6.7°C
Chilled water flow (per kilowatt of refrigeration)	43 mL/s, with 47 mL/s for some smaller international machines
Electric power (per kilowatt of refrigeration)	3 to 11 W with a minimum of 1 W for some smaller machines
Physical characteristics	
Nominal capacities	180 to 5800 kW, with 18 to 35 kW for some smaller machines
Length	3.3 to 10 m, with as low as 0.9 m for some smaller machines
Width	1.5 to 3.0 m, with 0.9 m minimum for some smaller machines
Height	2.1 to 4.3 m, with 1.8 m for some smaller machines
Operating mass	5 to 50 Mg, with 320 kg for some smaller machines

this process, the water in the tubes is cooled as it releases the heat required to boil the refrigerant. Refrigerant that does not boil is collected at the bottom of the evaporator, flows to a **refrigerant pump**, is pumped to a distribution system located above the evaporator tube bundle, and is sprayed over the evaporator tubes again.

The dilute absorbent solution that enters the generator increases in concentration (percentage of sorbent in the water) as it is boiled and releases water vapor. The resulting strong absorbent solution leaves the generator and flows through one side of a **solution heat exchanger** where it cools as it heats a stream of dilute absorbent solution passing through the other side of the solution heat exchanger on its way to the generator. This increases the efficiency of the machine by reducing the amount of heat from the primary heat source that must be added to the dilute solution before it begins to boil in the generator.

The cooled, strong absorbent solution then flows (in some designs via a jet eductor or solution spray pumps) to a solution distribution system located above the **absorber tubes** and drips or is sprayed over the outside surface of the absorber tubes. The absorber and evaporator share a common vapor space at a pressure of about 0.7 kPa. This allows refrigerant vapor, which is evolved in the evaporator, to be readily absorbed into the absorbent solution flowing over the absorber tubes. This absorption process releases heat of condensation and heat of dilution, which are removed by cooling water flowing through the absorber tubes. The resulting dilute absorbent solution flows off the absorber tubes and then to the absorber sump and **solution pump**. The pump and piping convey the dilute absorbent solution to the heat exchanger, where it accepts heat from the strong absorbent solution returning from the generator. From there, the dilute solution flows into the generator, thus completing the cycle.

These machines are typically fired with low-pressure steam or medium-temperature liquids. Several manufacturers in the United States have machines with capacities ranging from 180 to 5840 kW of refrigeration. Machines of capacities 18 to 35 kW are also available from international sources.

Typical COPs for large single-effect machines at ARI (Air Conditioning and Refrigeration Institute) rating conditions are 0.7 to 0.8.

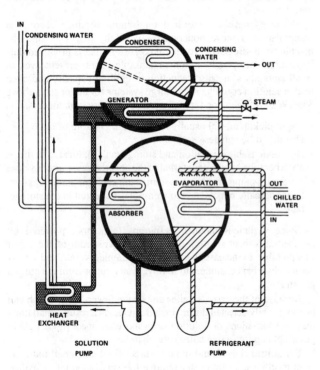

Fig. 2 Two-Shell Lithium Bromide Cycle Water Chiller

Absorption Cooling, Heating, and Refrigeration Equipment

Single-Effect Heat Transformer. Figure 3 shows a schematic of a single-effect heat transformer (or Type 2 heat pump). All major components are similar to the single-effect, indirect-fired liquid chiller. However, the absorber/evaporator is located above the desorber (generator)/condenser due to the higher pressure level of the absorber and evaporator compared to the desorber/condenser pair, which is the opposite of a chiller.

High pressure refrigerant liquid enters the top of the evaporator and heat released from a waste hot water stream converts it to a vapor. The vapor travels to the absorber section where it is absorbed by the incoming rich solution. The heat released during this process is used to raise the temperature of a secondary fluid stream to a useful level.

The diluted solution leaves the bottom of the absorber shell and flows through a solution heat exchanger. There it released heat in counterflow to the rich solution. Following the solution heat exchanger, the dilute solution flows through a throttling device where its pressure is reduced before it enters the generator unit. In the generator, heat from a waste hot water system generates low pressure refrigerant vapor. The rich solution leaves the bottom of the generator shell where a solution pump pumps it to the absorber.

The low pressure refrigerant vapor flows from the generator to the condenser coil, where it releases heat to a secondary cooling fluid and condenses. The condensate flows by gravity to a liquid storage sump and is pumped into the evaporator. Liquid refrigerant, which is not evaporated, collects at the bottom of the evaporator and flows back into the storage sump below the condenser. Measures must be taken to control the refrigerant pump discharge flow and to prevent vapor from blowing back from the higher pressure evaporator into the condenser during start-up or during any other operational event that causes low condensate flow. Typically, a column of liquid refrigerant is used to seal the unit to prevent blowback and a float operated valve controls the refrigerant flow to the evaporator.

Excess refrigerant flow is maintained to adequately distribute the liquid with only fractional evaporation.

Double-Effect Chillers. Figure 4 is a schematic of a commercially available, double-effect indirect-fired liquid chiller. Table 2 lists typical characteristics of this chiller. All major components are similar to the single-effect chiller except for an added generator (first-stage or primary generator), condenser, heat exchanger, and optional condensate subcooling heat exchanger.

Operation of the double-effect absorption machine is similar to that for the single-effect machine. The primary generator receives heat from the external heat source, which boils dilute absorbent solution. The pressure in the vapor space of the primary generator is about 100 kPa. This vapor flows to the inside of tubes in the second-effect generator. At this pressure the refrigerant vapor has a condensing temperature high enough to boil and concentrate absorbent solution on the outside of these tubes, thus creating additional refrigerant vapor with no additional primary heat input.

The extra solution heat exchanger (high-temperature heat exchanger) is placed in the intermediate and dilute solution streams flowing to and from the primary generator to preheat the dilute solution. Because of the relatively large pressure difference between the vapor spaces of the primary and secondary generators, a mechanical solution flow control device is required at the outlet of the high-temperature heat exchanger to maintain a liquid seal between the two generators. A valve at the heat exchanger outlet that is controlled by the liquid level leaving the primary generator can maintain this seal.

One or more condensate heat exchangers may be used to remove additional heat from the primary heat source steam by subcooling the steam condensate. This heat is added to the dilute or intermediate solution flowing to one of the generators. The result is a reduction in the quantity of steam required to produce a given refrigeration effect; however, the required heat input remains the same. The COP is not improved by condensate exchange.

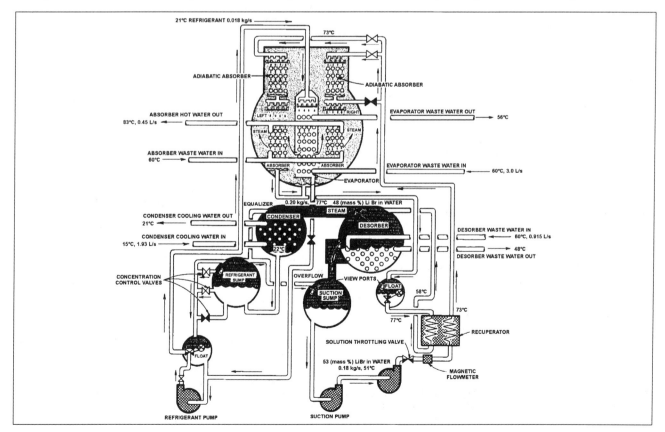

Fig. 3 Single-Effect Heat Transformer

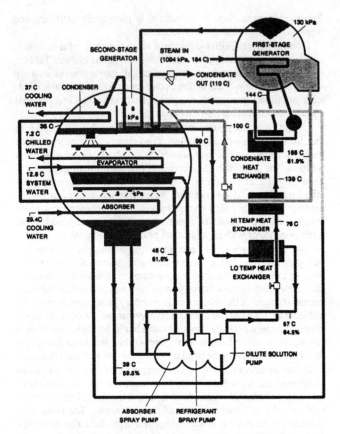

Fig. 4 Double-Effect Indirect-Fired Chiller

As with the single-effect machine, the strong absorbent solution flowing to the absorber can be mixed with dilute solution and pumped over the absorber tubes or can flow directly from the low-temperature heat exchanger to the absorber. Also, as with the single-effect machines, the four major components can be contained in one or two vessels.

The solution flow shown in Figure 4 is a series flow and is one of several solution flow cycles that is used in double-effect absorption equipment. The following solution flow cycles may be used:

Series flow. All solution leaving the absorber runs through a pump and then flows sequentially through the low-temperature heat exchanger, high-temperature heat exchanger, first-stage generator, high-temperature heat exchanger, second-stage generator, low-temperature heat exchanger, and absorber.

Parallel flow. Solution leaving the absorber is pumped through appropriate portions of the combined low- and high-temperature solution heat exchanger and is then split between the first- and second-stage generators. Both solution flow streams then return to appropriate portions of the combined solution heat exchanger, are mixed together, and flow to the absorber.

Reverse parallel flow. All solution leaving the absorber is pumped through the low-temperature heat exchanger and then to the second-stage generator. Upon leaving this generator, the solution flow is split, with a portion going to the low-temperature heat exchanger and on to the absorber. The remainder goes sequentially through a pump, the high-temperature heat exchanger, the first-stage generator, and the high-temperature heat exchanger. This stream then rejoins the solution from the second-stage generator; both streams flow through the low-temperature heat exchanger and to the absorber, as shown in Figure 5.

These machines are typically fired with medium-pressure steam of 550 to 990 kPa or hot liquids of 150 to 200°C. Typical operating

Table 2 Characteristics of Typical Double-Effect, Indirect-Fired, Water-Lithium Bromide Absorption Chiller

Performance Characteristics	
Steam input pressure	790 kPa (gage)
Steam consumption (with condensate saturated conditions) (per kilowatt of refrigeration)	780 to 810 W
Hot fluid input temperature	188°C
Heat input rate (per kilowatt of refrigeration)	0.83 kW
Cooling water temperature in	30°C
Cooling water flow (per kilowatt of refrigeration)	65 to 80 mL/s
Chilled water temperature off	7°C
Chilled water flow (per kilowatt of refrigeration)	43 mL/s
Electric power (per kilowatt of refrigeration)	3 to 11 W
Physical Characteristics	
Nominal Capacities	350 to 6000 kW
Length	3.1 to 9.4 m
Width	1.8 to 3.7 m
Height	2.4 to 4.3 m
Operating mass	7 to 60 Mg

COPs are 1.1 to 1.2. These machines are available commercially from several manufacturers and have capacities ranging from 350 to 6000 kW of refrigeration.

Figure 5 is a schematic of a commercially available, double-effect direct-fired chiller with a reverse parallel flow cycle. Table 3 lists typical characteristics of this chiller. All major components are similar to the double-effect indirect-fired chiller except for substitution of the direct-fired primary generator for the indirect-fired primary generator and elimination of the steam condensate subcooling heat exchanger. Operation of these machines is identical to that of the double-effect indirect-fired machines. The typical direct-fired, double-effect machines can be ordered with a heating cycle. Also available on some units is a simultaneous cycle. This cycle provides about 80°C water, via a heat exchanger, and chilled water, simultaneously. The combined load is limited by the maximum burner input.

These machines are typically fired with natural gas or fuel oil (most have dual fuel capabilities). Typical operating COPs are 0.92 to 1.0 on a fuel input basis. These machines are available commercially from several manufacturers and have capacities ranging from 350 to 5300 kW. Machine capacities of 70 to 350 kW are also available from international sources.

Operation

Modern water-lithium bromide chillers are trouble free and easy to operate. As with any equipment, careful attention should be paid to operational and maintenance procedures recommended by the manufacturer of the equipment. The following characteristics are common to all types of lithium bromide absorption equipment.

Operational Limits. Chilled water temperature leaving the evaporator should normally be between 4 and 15°C. The upper limit is set by the pump lubricant and is somewhat flexible. The lower limit exists because the refrigerant (water) freezes at 0°C.

Cooling water temperature entering the absorber tubes is generally limited to between 12 and 43°C, although some machines limit the entering cooling water temperatures to between 21 and 35°C. The upper limit exists because of hydraulic and differential pressure limitations between the generator-absorber, the condenser-evaporator, or both, and to reduce absorbent concentrations and corrosion effects. The lower temperature limit exists because, at excessively low cooling water temperature, the condensing pressure drops too low and excessive vapor velocities carry over solution to the refrigerant in the condenser. Sudden lowering of cooling water temperature at high loads will also promote crystallization; therefore, some manufacturers will dilute the solution with refrigerant liquid to help

Absorption Cooling, Heating, and Refrigeration Equipment

Table 3 Characteristics of Typical Double-Effect, Direct-Fired, Water-Lithium Bromide Absorption Chiller

Performance Characteristics	
Fuel consumption (high heating value of fuel) (per kilowatt of refrig.)	1 to 1.1 kW
COP (high heating value)	0.92 to 1.0
Cooling water temperature in	30°C
Cooling water flow (per kilowatt of refrig.)	79 to 81 mL/s
Chilled water temperature off	7°C
Chilled water flow (per kilowatt of refrig.)	43 mL/s
Electric power (per kilowatt of refrig.)	3 to 11 W
Physical Characteristics	
Nominal capacities	350 to 5300 kW
Length	3.0 to 10.4 m, with minimum of 1.5 m for some machines
Width	1.5 to 6.5 m, with minimum of 1.2 m for some machines
Height	2.1 to 3.7 m
Operating mass	5 to 80 Mg, with a minimum of 1.5 Mg for some machines

prevent crystallization. The supply of refrigerant is limited, however, so this dilution is done in small steps.

Operational Controls. Modern absorption machines are equipped with electronic control systems. The primary function of the control system is to safely operate the absorption machine and modulate its capacity in order to satisfy the load requirements placed upon it.

The temperature of the chilled water leaving the evaporator is set at a desired value. Deviations from this set point indicate that the machine capacity and the load applied to it are not matched. Machine capacity is then adjusted as required by modulation of the heat input control device. Modulation of heat input results in changes to the concentration of absorbent solution supplied to the absorber if the pumped solution flow remains constant.

Some equipment uses solution flow control to the generator(s) in combination with capacity control. The solution flow may be reduced with modulating valves or solution pump speed controls as the load decreases (which reduces the required sensible heating of solution in the generator to produce a given refrigeration effect), thereby improving part-load efficiency.

Operation of lithium bromide machines with low entering cooling water temperatures or a rapid decrease in cooling water temperature during operation can cause liquid carryover from the generator to the condenser and possible crystallization of absorbent solution in the low-temperature heat exchanger.

For these reasons, most machines have a control that limits heat input to the machine based on entering cooling water temperature. Since colder cooling water enhances machine efficiency, the ability of machines to use colder water, when available, is important.

Use of electronic controls with advanced control algorithms have improved part-load and variable cooling water temperature operation significantly compared to older pneumatic or electric controls. Electronic controls have also made chiller setup and operation simpler and more reliable.

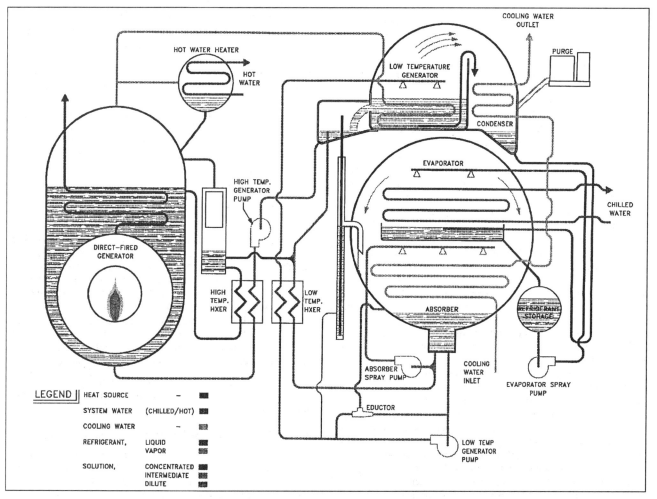

Fig. 5 Double-Effect, Direct-Fired Chiller

These steps are involved in a typical start-run-stop sequence of an absorption chiller with chilled and cooling water flows pre-established (this sequence may vary from one product to another):

1. Cooling required signal is initiated by building control device or in response to rising chilled water temperature.
2. All chiller unit and system safeties are checked.
3. Solution and refrigerant pumps are started.
4. Heat input valve is opened or burner is started.
5. Chiller begins to meet the load and controls chilled water temperature to desired set point by modulation of heat input control device.
6. During operation, all limits and safeties are continually checked. Appropriate action is taken, as required, to maintain safe chiller operation.
7. Load on chiller decreases below minimum load capabilities of chiller.
8. Heat input device is closed.
9. Solution and refrigerant pumps continue to operate for several minutes to dilute the absorbent solution.
10. Solution and refrigerant pumps are stopped.

Limit and Safety Controls. In addition to capacity controls, these chillers require several protective devices. Some controls keep the units operating within safe limits and others stop the unit before damage occurs due to a malfunction. Each limit and safety cutout function usually uses a single sensor when electronic controls are used. The following limits and safety features are normally found on absorption chillers:

Low-temperature chilled water control/cutout. Allows the user to set the desired temperature for chilled water leaving the evaporator. Control then modulates the heat input valve to maintain this set point. This control incorporates chiller start and stop by water temperature. A safety shutdown of the chiller is invoked if a low-temperature limit is reached.

Low-temperature refrigerant limit/cutout. A sensor in the evaporator monitors refrigerant temperature. As the refrigerant low-limit temperature is approached, the control limits further loading, then prevents further loading, then unloads, and finally invokes a chiller shutdown.

Chilled water, chiller cooling water, and pump motor coolant flow. Flow switches trip and invoke chiller shutdown if flow stops in any of these circuits.

Pump motor over-temperature. A temperature switch in the pump motor windings trips if safe operating temperature is exceeded and shuts down the chiller.

Pump motor overload. Current to the pump motor is monitored, and the chiller shuts down if the current limit is exceeded.

Absorbent concentration limit. Key solution and refrigerant temperatures are sensed during chiller operation and used to determine the temperature safety margin between solution temperature and solution crystallization temperature. As this safety margin is reduced, the control first limits further chiller loading, then prevents further chiller loading, then unloads the chiller, and finally invokes a chiller shutdown.

In additional to this type of control, most chiller designs incorporate a built-in overflow system between the evaporator liquid storage pan and the absorber sump. As the absorbent solution concentration increases in the generator/absorber flow loop, the refrigerant liquid level in the evaporator storage pan increases. The initial charge quantities of solution and refrigerant are set such that liquid refrigerant will begin to overflow the evaporator pan when maximum safe absorbent solution concentration has been reached in the generator/absorber flow loop. The liquid refrigerant overflow goes to the absorber sump and prevents further concentration of the absorbent solution.

Burner fault. Operation of the burner on direct-fired chillers is typically monitored by its own control system. A burner fault indication is passed on to the chiller control and generally invokes a chiller shutdown.

High-temperature limit. Direct-fired chillers typically have a temperature sensor in the liquid absorbent solution near the burner fire tube. As this temperature approaches its high limit, the control first limits further loading, then prevents further loading, then unloads, and finally invokes a chiller shutdown.

High-pressure limit. Double-effect machines typically have a pressure sensor in the vapor space above the first-stage generator. As this pressure approaches its high limit, the control first limits further loading, then prevents further loading, then unloads, and finally invokes a chiller shutdown.

Machine Setup and Maintenance

Large capacity lithium bromide absorption water chillers are generally put into operation by factory-trained technicians. Proper procedures must be followed in order to insure that the machines will function as designed and continue to function in a trouble-free manner for their intended design life (20 plus years). Steps required to set up and start a lithium bromide absorption machine include:

1. Level the unit so that internal pans and distributors can function properly.
2. Isolate the unit from foundations with pads if it is located near noise-sensitive areas.
3. Confirm that factory leaktightness has not been compromised.
4. Charge the unit with refrigerant water (distilled or deionized water is required) and lithium bromide solution.
5. Add corrosion inhibitor to the absorbent solution if required.
6. Calibrate all control sensors and check all controls for proper function.
7. Start the unit and bring it slowly to design operating condition while adding performance additive (usually one of the octyl alcohols).
8. If necessary to obtain design conditions, adjust absorbent and/or refrigerant charge levels. This procedure is known as trimming the chiller, and, if done correctly, will allow the chiller to operate safely and efficiently over its entire operating range.
9. Fine tune control settings.
10. Check purge operation.

Recommended periodic operational checks and maintenance procedures typically include:

- Purge operation and air leaks. Confirm that the purge system operates correctly and that the unit does not have chronic air leaks. Continued leakage of air into an absorption chiller will deplete the corrosion inhibitor, cause corrosion of internal parts, contaminate the absorbent solution, reduce chiller capacity and efficiency, and may cause crystallization of the absorbent solution.
- Sample absorbent and refrigerant periodically and check for contamination, pH, corrosion-inhibitor level, and performance additive level. Use these checks to adjust the levels of additives in the solution and as an indicator of internal machine malfunctions.

Mechanical systems such as the purge, solution pumps, controls, and burners all have periodic maintenance requirements recommended by the manufacturer.

Chiller Performance at Other Than Design Rating

The performance of lithium bromide absorption machines is affected by the operating conditions and the heat transfer surface chosen by the manufacturer. Manufacturers of this equipment can provide detailed performance information for their equipment at specific alternate operating conditions.

AMMONIA-WATER ABSORPTION TECHNOLOGY

Figure 6 shows a typical schematic of an ammonia-water machine, which is available as a direct-fired air-cooled liquid chiller in capacities of 10 to 18 kW. Table 4 lists physical characteristics of this chiller. Ammonia-water equipment varies from water-lithium bromide equipment to accommodate three major differences:

1. Water (the absorbent) is also volatile, so the regeneration of weak absorbent to strong absorbent is a fractional distillation process.
2. Ammonia (the refrigerant) causes the cycle to operate at condenser pressures of about 2100 kPa (absolute) and at evaporator pressures of approximately 480 kPa (absolute). As a result, vessel sizes are held to a diameter of 150 mm or less to avoid construction code requirements on small systems, and positive-displacement solution pumps are used.
3. Air cooling requires condensation and absorption to occur inside the tubes so that the outside can be finned for greater air contact.

Components

Generator. The vertical vessel is finned on the outside to extract heat from the combustion products. Internally, a system of analyzer plates creates intimate counterflow contact between the vapor generated, which rises, and the absorbent, which descends. Atmospheric gas burners depend on the draft of the condenser air fan to sustain adequate combustion airflow to fire the generator. The exiting flue products mix with the air that has passed over the condenser and absorber.

Heat Exchangers. Heat exchange between strong and weak absorbents takes place partially within the generator-analyzer. A tube bearing strong absorbent spirals through the analyzer plates and in the solution-cooled absorber, where strong absorbent metered from the generator through the solution capillary passes over a helical coil bearing weak absorbent. In the solution-cooled absorber, the strong absorbent absorbs some of the vapor from the evaporator, thus retaining its heat of absorption within the cycle to improve its COP. The strong absorbent and unabsorbed vapor continue from the solution-cooled absorber into the air-cooled absorber, where absorption is completed and the heat of absorption is rejected to the air.

Table 4 Physical Characteristics of Typical Ammonia-Water Absorption Chiller

Cooling capacities	10 to 18 kW
Length	1020 to 1230 mm
Width	740 to 850 mm
Height	960 to 1170 mm
Mass	250 to 350 kg

The **rectifier** is a heat exchanger that consists of a spiral coil through which weak absorbent from the solution pump passes on its way to the absorber and generator. Some type of packing is included to assist counterflow contact between condensate from the coil (which is refluxed to the generator) and the vapor (which continues on to the air-cooled condenser). The function of the rectifier is to concentrate the ammonia in the vapor from the generator by cooling and stripping out some of the water vapor.

Absorber and Condenser. These finned tubes are arranged so that most of the incoming air flows over the condenser tubes and most of the exit air flows over the absorber tubes.

Evaporator. The liquid to be chilled drips over a coil bearing evaporating ammonia, which absorbs the refrigeration load. On the chilled-water side, which is at atmospheric pressure, a pump circulates the chilled liquid to the load source. Refrigerant to the evaporator is metered from the condenser through restrictors. A tube-in-tube heat exchanger provides the maximum refrigeration effect per unit mass of refrigerant. The tube-in-tube design is particularly effective in this cycle because water present in the ammonia produces a liquid residue that evaporates at increasing temperatures as the amount of residue decreases.

Solution Pumps. The reciprocating motion of a flexible sealing diaphragm moves solution through suction and discharge valves. Hydraulic fluid pulses delivered to the opposite side of the diaphragm by a hermetic vane or piston pump at atmospheric suction pressure impart this motion.

Capacity Control. A thermostat usually cycles the machine on and off. A chilled-water switch shuts the burners off if the water temperature drops close to freezing. Units may also be underfired by 20% to derate to a lower load.

Protective Devices. Typical protective devices include (1) flame ignition and monitor control, (2) a sail switch that verifies airflow before allowing the gas to flow to the burners, (3) a pressure relief valve, and (4) a generator high-temperature switch.

Equipment Performance and Selection

Ammonia absorption equipment is built and rated to meet ANSI *Standard* Z21.40.1, Gas-fired Absorption Summer Air Conditioning Appliances, Sixth Edition, for outdoor installation. The rating conditions are ambient air at 35°C dry bulb and 24°C wet bulb and chilled water delivered at the manufacturer's specified flow at 7.2°C. A COP of about 0.5 is realized, based on the higher heating value of the gas.

Although most units are piped to a single furnace, duct, or fan coil and operated as air conditioners, multiple units supplying a multicoil system for process cooling and air conditioning are common. Also, chillers are packaged with an outdoor boiler and can supply chilled or hot water as the cooling or heating load requires.

AMMONIA-WATER-HYDROGEN CYCLE

Domestic absorption refrigerators use a modified absorption cycle with ammonia, water, and hydrogen as working fluids. Wang and Herold (1992) reviewed the literature on this cycle. These units are popular for recreational vehicles because they can be dual-fired

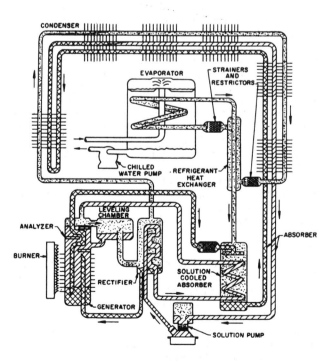

Fig. 6 Ammonia-Water Direct-Fired Air-Cooled Chiller

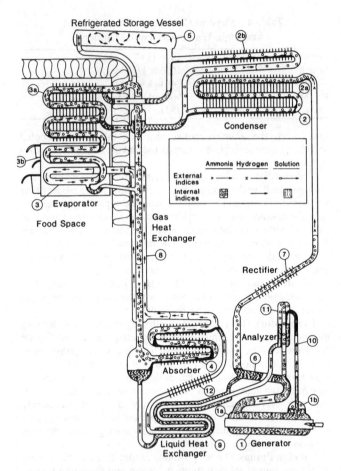

Fig. 7 Domestic Absorption Refrigeration Cycle

by gas or electric heaters. They are also popular for hotel rooms because they are silent. The refrigeration unit is hermetically sealed. All spaces within the system are open to each other and, hence, are at the same total pressure, except for minor variations caused by fluid columns used to circulate the fluids.

The key elements of the system shown in Figure 7 include a generator (1), a condenser (2), an evaporator (3), an absorber (4), a rectifier (7), a gas heat exchanger (8), a liquid heat exchanger (9), and a bubble pump (10). The following three distinct fluid circuits exist in the system: (I) an ammonia circuit, which includes the generator, condenser, evaporator, and absorber; (II) a hydrogen circuit, which includes the evaporator, absorber, and gas heat exchanger; and (III) a solution circuit, which includes the generator, absorber, and liquid heat exchanger.

Starting with the generator, a gas burner or other heat source applies heat to expel ammonia from the solution. The ammonia vapor generated then flows through an analyzer (6) and a rectifier (7) to the condenser (2). The small amount of residual water vapor in the ammonia is separated by atmospheric cooling in the rectifier and drains to the generator (1) through the analyzer (6).

The ammonia vapor passes into section (2a) of the condenser (2), where it is liquified by air cooling. Fins on the condenser increase the cooling surface. The liquified ammonia then flows into an intermediate point of the evaporator (3). A liquid trap between the condenser section (2a) and the evaporator prevents hydrogen from entering the condenser. Ammonia vapor that does not condense in the condenser section (2a) passes to the other section (2b) of the condenser and is liquified. It then flows through another trap into the top of the evaporator.

The evaporator has two sections. The upper section (3a) has fins and cools the freezer compartment directly. The lower section (3b) cools the refrigerated food section.

Hydrogen gas, carrying a small partial pressure of ammonia, enters the lower evaporator section (3) and, after passing through a precooler, flows upward and counterflow to the downward flowing liquid ammonia, increasing the partial pressure of the ammonia in the vapor as the liquid ammonia evaporates. While the total pressures in the evaporator and the condenser are the same, typically 2000 kPa, substantially pure ammonia is in the space where condensation takes place, and the vapor pressure of the ammonia essentially equals the total pressure. In contrast, the ammonia partial pressures entering and leaving the evaporator are typically 100 and 300 kPa, respectively.

The gas mixture of hydrogen and ammonia leaves the top of the evaporator and passes down through the center of the gas heat exchanger (8) to the absorber (4). Here, ammonia is absorbed by liquid ammonia-water solution, and hydrogen, which is almost insoluble, passes up from the top of the absorber, through the external chamber of the gas heat exchanger (8), and into the evaporator. Some ammonia vapor passes with the hydrogen from the absorber to the evaporator. Because of the difference in molecular mass of ammonia and hydrogen, the gas circulation is maintained between the evaporator and absorber by natural convection.

Countercurrent flow in the evaporator permits placing the box cooling section of the evaporator at the top of the food space, which is the most effective location. Also, the gas leaving the lower temperature evaporator section (3b) can pick up more ammonia at the higher temperature in the box cooling evaporator section (3a), thus increasing capacity and efficiency. In addition, the liquid ammonia flowing to the lower temperature evaporator section is precooled in the upper evaporator section. The dual liquid connection between the condenser and the evaporator permits extending the condenser below the top of the evaporator to provide more surface, while maintaining gravity flow of liquid ammonia to the evaporator. The two-temperature evaporator partially segregates the freezing function from the box cooling function, thus giving better humidity control.

In the absorber, the strong absorbent flows counter to and is diluted by direct contact with the gas. From the absorber, the weak absorbent flows through the liquid heat exchanger (9) to the analyzer (6) and then to the weak absorbent chamber (1a) of the generator (1). Heat applied to this chamber causes vapor to pass up through the analyzer (6) and to the condenser. The solution passes through an aperture in the generator partition into the strong absorbent chamber (1b). Heat applied to this chamber causes vapor and liquid to pass up through the small-diameter bubble pump (10) to the separation vessel (11). While liberated ammonia vapor passes through the analyzer (6) to the condenser, the strong absorbent flows through the liquid heat exchanger (9) to the absorber. The finned air-cooled loop (12) between the liquid heat exchanger and the absorber precools the solution further. The heat of absorption is rejected to the surrounding air.

The refrigerant storage vessel (5), which is connected between the condenser outlet and the evaporator circuit, is a reservoir for the refrigerant to compensate for changes in load and the heat rejection air supply temperature.

Controls

Burner Ignition and Monitoring Control. These controls are either electronic or thermomechanical. Electronic controls ignite, monitor, and shut off the main burner as required by the thermostat. For thermomechanical control, a thermocouple monitors the main flame. The low-temperature thermostat then changes the input to the main burner in a two-step mode. A pilot is not required because the main burner acts as the pilot on low fire.

Low-Temperature Thermostat. This thermostat monitors the temperature in the cabinet and controls the gas input.

Absorption Cooling, Heating, and Refrigeration Equipment

Safety Device. Each unit has a fuse plug to relieve pressure in the event of fire. Gas-fired installations require a flue exhausting to outside air. Nominal operating conditions are as follows:

Ambient temperature	35°C
Freezer temperature	−12°C
Input	1.0 kW/m³ of cabinet interior

SPECIAL APPLICATIONS

A single-stage indirect-fired lithium bromide absorption chiller as a bottoming cycle for an exhausting steam turbine that drives a centrifugal water chiller uses the steam twice. As a result, thermal efficiencies and costs are comparable to those of the double-effect absorption chiller. The chilled water flows in series through the absorption unit, then out of the centrifugal chiller. The cooling water can flow through the centrifugal unit condenser and then the absorption unit, or the cooling water can be piped in parallel. Approximately two-thirds of the cooling duty is carried by the absorption unit and one-third by the centrifugal unit.

The exhausting steam turbine drive for the centrifugal compressor is cheaper than the usual condensing turbine required for high efficiency, and the absorption generator becomes the main steam condenser. The steam turbine uses superheated steam at 700 kPa (gage) and exhausts the steam at between 60 and 80 kPa (gage) to the absorption chiller generator.

Single-stage absorption units have been used to condense hot hydrocarbon process vapors from the top of a distillation column while cooling solvents for chemical plants. They have also been installed to use heat extracted directly from hot gasoline in the tubes of the generator while directly cooling lean oil for oil refinery absorbers. These applications took considerable engineering development and are difficult to justify today.

Single-stage absorption units have used solar-heated and diesel or gas engine-heated hot water at 82°C to cool buildings for comfort.

An absorption unit could be added to condense waste steam to subcool refrigerant from an existing conventional single-stage refrigeration plant and boost the available cooling capacity of the existing compressors. Also, with heat recovery features of absorption units provided by some manufacturers, hot exhaust gases supplemented by gas or oil firing could provide comfort and process cooling and heating.

EVOLVING CONCEPTS

The absorption chiller and heat pump industry has changed in recent years. The following systems and cycles are being studied for commercial use.

Triple-Effect Cycles

Triple-effect absorption cooling can be classified as single-loop and dual-loop cycles. Single-loop triple-effect cycles are basically double-effect cycles with an additional generator and condenser. The resulting system with three generators and three condensers operates similarly to the double-effect system. Primary heat (from a natural gas or fuel oil burner) concentrates absorbent solution in a first-stage generator at about 230°C. A fluid pair other than water-lithium bromide has to be used for the high-temperature cycle. The refrigerant vapor produced is then used to concentrate additional absorbent solution in a second-stage generator at about 150°C. Finally, the refrigerant vapor produced in the second-stage generator concentrates additional absorbent solution in a third-stage generator at about 93°C. The usual internal heat-recovery devices (solution heat exchangers) can be used to improve cycle efficiency. As with the double-effect cycles, several variations of solution flow paths through the generators are possible.

Theoretically, the COP obtainable with these triple-effect cycles is about 1.7 (not taking into account burner efficiency). Difficulties with these cycles include the following:

- High solution temperatures pose problems to solution stability, performance additive stability, and material corrosion
- High pressure in the first-stage generator vapor space requires costly pressure vessel design and high-pressure solution pump(s)

A double-loop triple-effect cycle consists of two cascaded single-effect cycles. One cycle operates at normal single-effect operating temperatures and the other at higher temperatures. The smaller high-temperature topping cycle is direct-fired with natural gas or fuel oil and has a generator temperature of about 230°C. A fluid pair other than water-lithium bromide must be used for the high-temperature cycle. Heat is rejected from the high-temperature cycle at 93°C and is used as the energy input for the conventional single-effect bottoming cycle. Both the high-and low-temperature cycles remove heat from the cooling load at about 7°C.

Theoretically, the overall COP obtainable with this triple-effect cycle is about 1.8 (not taking into account burner efficiency).

As with the single-loop triple-effect cycle, high temperatures create problems with solution and additive stability and material corrosion. Also, the use of a second loop requires additional heat exchange vessels and additional pumps. However, both loops operate below atmospheric pressure and, therefore, do not require costly pressure vessel designs.

GAX (Generator-Absorber Heat Exchange) Cycle

The air-cooled absorption air-conditioning equipment presently available operates at gas-fired cooling COPs of just under 0.5 at ARI rating conditions. The absorber heat exchange cycle of past air conditioners had a COP of about 0.67 at the rating conditions. In recent years, several projects have been initiated around the world to develop generator-absorber heat exchange (GAX) cycle systems. The best known programs have been directed toward cycle COPs of about 1.0.

The GAX cycle is a heat-recuperating cycle in which absorber heat is used to heat the lower temperature section of the generator as well as the rich ammonia solution being pumped to the generator. This cycle, like others capable of higher COPs, is more complex and difficult to develop than the ammonia single-stage and absorber heat exchange cycles, but its potential gas-fired COPs of 0.9 in cooling mode and 1.8 in heating mode make it capable of significant energy savings on an annual basis. In addition to providing a more effective use of heat energy than the most efficient furnaces, the GAX heat pump is able to supply all the heat a house requires to outdoor temperatures below −18°C without the use of supplemental heat.

Solid-Vapor Sorption Systems

Solid-vapor heat pump technology is being developed for zeolite, silica-gel, activated-carbon, and coordinated complex adsorbents. The cycles are periodic in that the refrigerant is transferred periodically between two primary vessels. Several concepts providing quasi-continuous refrigeration have been developed. One advantage of solid-vapor systems is that no solution pump is needed. The main challenge in designing a competitive solid-vapor heat pump is to package the adsorbent in such a way that good heat and mass transfer are obtained in a small volume. A related constraint is that good thermal performance of periodic systems requires that the thermal mass of the vessels be small to minimize cyclic heat-transfer losses. Research in the area of solid-vapor technology can be found in the proceedings of two recent conferences (IIR 1991 and 1992).

Liquid Desiccant-Absorption Systems

In efforts to reduce a building's energy consumption, designers have successfully integrated liquid desiccant equipment with standard absorption chillers. These applications have been building specific and are sometimes referred to as application hybrids. In a more general approach, the absorption chiller is modified so that rejected heat from its absorber can be used to help regenerate the liquid desiccant. Only liquid desiccants are appropriate for this integration because they can be regenerated at lower temperatures than solid desiccants.

The desiccant dehumidifier dries ventilation air sufficiently that, when it is mixed with return air, the building's latent load is satisfied. The desiccant drier is cooled by cooling tower water so that a significant amount of the cooling load is transferred directly to the cooling water. Consequently, the absorption chiller size is significantly reduced, potentially to as little as 60% of the size of the chiller in a conventional installation.

Because the air handler is restricted to sensible load, the evaporator in the absorption machine will run at higher temperatures than normal. Consequently, a machine operating at normal concentrations in its absorber will reject heat at higher temperatures. To permit convenient regeneration of the liquid desiccant, only moderate increases in solution concentration are required. These are subtle but significant modifications to a standard absorption chiller.

Such combined systems seem to work best when about one-third of the supply air comes from outside the conditioned space. These systems do not require 100% outside air for ventilation, so they should be applicable to conventional buildings as newly mandated ventilation standards are accommodated. Because they always operate in a form of economizer cycle, they are particularly effective during shoulder seasons (spring and fall). As lower cost liquid desiccant systems become available, reduced first costs may join the advantages of decreased energy use, better ventilation, and improved humidity control.

ALTERNATIVE ABSORPTION WORKING FLUIDS

The fluids chosen for liquid-vapor absorption cycles are usually water-lithium bromide or ammonia-water. These fluids currently exhibit the best combination of properties for many typical applications. However, these choices limit the machine design in several ways. Limiting characteristics of water-lithium bromide include (1) evaporator temperature must be above 0°C to avoid freezing the water, (2) coolant temperatures must be relatively low to avoid crystallization of the lithium bromide (making air cooling difficult), and (3) corrosion is a major factor at temperatures above 177°C. Limiting characteristics of ammonia-water include (1) the toxicity of ammonia and (2) its high vapor pressure.

Many fluid alternatives have been considered by researchers. Several systems appear to be approaching commercial use. An aqueous mixture of nitrate salts has good high-temperature properties for industrial absorption heat pumps and topping cycles. These salts have been used previously as corrosion inhibitors in water-lithium bromide systems.

A large effort is being expended to find corrosion inhibitors for lithium bromide systems that will allow high-temperature operation. In particular, a lithium bromide-based triple-effect cycle will require such inhibitors or will require metals having greater corrosion resistance.

BIBLIOGRAPHY

Alefeld, G. 1985. Multi-stage apparatus having working-fluid and absorption cycles, and method of operation thereof. U.S. Patent No. 4,531,374.

Bogart, M. 1981. *Ammonia absorption refrigeration in industrial processes.* Gulf Publishing Co., Houston.

Eisa, M.A.R., S.K. Choudhari, D.V. Paranjape, and F.A. Holland. 1986. Classified references for absorption heat pump systems from 1975 to May 1985. *Heat Recovery Systems* 6:47-61. Pergamon Press Ltd., Great Britain.

Hanna, W.T. and W.H. Wilkinson. 1982. Absorption heat pumps and working pair developments in the U.S. since 1974, New working pairs for absorption processes, pp. 78-80. *Proceedings of Berlin Workshop* by the Swedish Council for Building Research, Stockholm, Sweden.

Huntley, W.R. 1984. Performance test results of a lithium bromide-water absorption heat pump that uses low temperature waste heat. Oak Ridge National Laboratory *Report* No. ORNL/TM9702, Oak Ridge, TN.

IIR. 1991. *Proceedings of the XVIIIth International Congress of Refrigeration* (August 10-17, 1991, Montreal, Canada), Volume III. International Institute of Refrigeration, Paris, France.

IIR. 1992. *Proceedings of Solid Sorption Refrigeration Meetings of Commission B1* (November 18-20, 1992, Paris, France). International Institute of Refrigeration, Paris, France.

Niebergall, W. 1981. *Handbuch der Kältetechnik,* Volume 7: Sorptionsmaschinen. R. Plank, ed. Springer Verlag, Berlin.

Phillips, B.A. 1990. Development of a high-efficiency, gas-fired, absorption heat pump for residential and small-commercial applications: Phase I Final *Report*: Analysis of advanced cycles and selection of the preferred cycle. ORNL/Sub/86-24610/1, September.

Scharfe, J., F. Ziegler, and R. Radermacher. 1986. Analysis of advantages and limitations of absorber-generator heat exchange. *International Journal of Refrigeration* 9:326-33.

Vliet, G.C., M.B. Lawson, and R.A. Lithgow. 1982. Water-lithium bromide double-effect cooling cycle analysis. *ASHRAE Transactions* 88(1):811-23.

Wang, L. and K.E. Herold. 1992. Diffusion-absorption heat pump. Annual *Report* to Gas Research Institute, GRI-92/0262.

Wilkinson, W. H. 1991. A simplified high efficiency DUBLSORB system. *ASHRAE Transactions* 97(1).

CHAPTER 42

FORCED-CIRCULATION AIR COOLERS

Types of Forced-Circulation Air Coolers 42.1
Components 42.2
Air Movement and Distribution 42.2
Rating Units 42.3

FORCED-CIRCULATION unit coolers and product coolers are designed to operate continuously within refrigerated enclosures. An evaporator coil and motor-driven fan make up the basic components of these coolers. Any unit, such as a blower coil, unit cooler, product cooler, cold diffuser unit, or air-conditioning air handler is considered a forced-air cooler when operated under refrigeration conditions. Many design and construction choices are available, including: (1) various coil types and fin spacing; (2) electric, gas, air, or water defrosting; (3) discharge air velocity; (4) centrifugal or propeller fans, either belt- or direct-driven; (5) ducted or nonducted, and/or (6) freestanding or ceiling suspended.

The fan in these units directs air over a refrigerated coil. For nearly all applications, the coil lowers the temperature of this air below its dew point, which causes condensate or frost to form on the coil surface. However, the normal refrigeration load is a sensible heat load, therefore the coil surface is considered dry. Rapid and frequent defrosting on a timed cycle can maintain this dry-surface condition, or the coil and airflow can be designed to minimize the frost accumulation and its effect on the refrigeration capacity.

TYPES OF FORCED-CIRCULATION AIR COOLERS

Sloped Front Unit Coolers

These units, which are also called reach-in units, range from 125 mm to 25 mm high. Their distinctive sloped fronts are designed for horizontal top mounting as a single unit, or for installation as a group of parallel connected units. Direct-drive fans are sloped to fit within the restricted return air stream, which rises past the access doors and across the ceiling of the enclosure. Airflows are usually less than 70 L/s per fan. Commonly, these units are installed in back-bar and under-the-counter fixtures, as well as in vertical, self-serve, glass door reach-in enclosures.

Low Air Velocity Unit Coolers

These units feature a long, narrow profile. They have dual coils and two or more fans. These units are used in meat-cutting rooms and in carcass and floral walk-in holding rooms. These units are designed to maintain high humidity in the enclosure. For this reason the coils have amply spaced fins to allow air to pass through when frost accumulates between defrost cycles. Discharge air velocities at the coil face range from 0.4 to 1.0 m/s.

Medium Air Velocity Unit Coolers

These units have a half-round appearance, although the long, narrow, dual-coil units with higher volume fans also fit in this category. They are used in vegetable preparation rooms, wrapped fresh meat walk-in rooms, and dairy coolers. These units normally extract more moisture from the room than the low velocity units. Discharge air velocities at the coil face range from 1 to 2 m/s.

The preparation of this chapter is assigned to TC 8.4, Air-to-Refrigerant Heat-Transfer Equipment.

Standard Air Velocity Unit Coolers

Low silhouette units are 300 to 380 mm high. Medium or mid-height units are 450 to 900 mm high. Those over 900 m high are classified as high silhouette unit coolers. The air velocity at the coil face can be as high as 3 m/s. Outlet air velocities range from 5 to 10 m/s when the unit is equipped with a cone-shaped air discharge to extend the throw of the air jet.

High Air Velocity Product Coolers

These units are used for blast tunnel freezing and for cooling of products that are not adversely affected by moderate dehydration during rapid cooling. They generally draw air through the coil at over 3 m/s and discharge at velocities over 10 m/s.

Sprayed Coil Product Coolers

Spray coils feature a saturated coil surface that can cool the processed air closer to the coil surface temperature than can a regular (non-sprayed) coil. In addition, the spray continuously defrosts the low-temperature coil. Unlike unit coolers, spray coolers are usually floor mounted and discharge air vertically. The unit sections include a drain pan/sump, coil with spray section, moisture eliminators, and fan with drive. The eliminators prevent airborne spray droplets from discharging into the refrigerated area. Typically, belt-driven centrifugal fans draw air through the coil at 3 m/s or less.

Water can be used as the spray medium for coil surfaces with temperatures above freezing. For coil surfaces with temperatures below freezing, a suitable material must be added to the water to lower the freezing point to −11°C or lower than the coil surface temperature. Some suitable recirculating solutions include:

- **Sodium chloride** solution is limited to a room temperature of −12°C or higher. Its minimum freezing point is −21°C.
- **Calcium chloride** solution can be used for enclosure temperatures down to about −23°C, but its use may be prohibited in enclosures containing food products.
- **Aqueous glycol** solutions are commonly used in water and/or sprayed coil coolers operating below freezing. Food-grade propylene glycol solutions are commonly used because of their low toxicity, but they generally become too viscous to pump below −25°C. Ethylene glycol solutions may be pumped at temperatures down to −40°C. Because of its toxicity, sprayed ethylene glycol in other than sealed tunnels or freezers (no human access allowed during process) is usually prohibited by most jurisdictions. When a glycol mix is sprayed in food storage rooms, any carryover of the spray must be maintained within the limits prescribed by all applicable regulations.

All brines are hygroscopic; that is, they absorb condensate and become progressively weaker. This dilution can be corrected by continually adding salt to the solution to maintain a sufficient below-freezing temperature. Salt is extremely corrosive, so it must be contained in the sprayed coil unit with suitable corrosive-resistant materials or coatings, that must be inspected and maintained.

Sprayed coil units are usually installed in refrigerated enclosures requiring high humidity, e.g., chill coolers. Paradoxically, the same sprayed coil units can be used in special applications requiring low relative humidity. For such dehydration applications, both a high brine concentrate (near its eutectic point) and a large difference between the process air and the refrigerant temperature are maintained. Process air is reheated downstream from the sprayed coil to correct the dry-bulb temperature.

COMPONENTS

Draw-Through and Blow-Through Airflow

Unit fans may draw air through the cooling coil and discharge it through the fan outlet into the enclosure; or the fans may blow air through the cooling coil and discharge it from the coil face into the enclosure. Blow-through units have a slightly higher thermal efficiency because heat from the fan is removed from the forced airstream by the coil. Draw-through fan energy adds to the heat load of the refrigerated enclosure, but neither load from the small (less than 1 kW) or small three-phase integral fan motors is significant. Selection depends more on a manufacturer's design features for the unit size required, the air throw required for the particular enclosure, and accessibility of the coil for cleaning.

The blow-through unit has a lower discharge air velocity because the entire coil face area is usually the discharge opening (grilles and diffusers not considered). An air throw of 10 m or less is common for the average air velocity from a blow-through unit. Greater throw, in excess of 30 m, is normal for draw-through, centrifugal fan units. The propeller fan in the high silhouette, draw-through unit cooler is popular for intermediate ranges of air throw.

Fan Assemblies

Direct-drive propeller fans (motor plus blade) are popular because they are simple, economical, and can be installed in multiple assemblies in a unit cooler housing. Additionally, they require less motor power for a given airflow capacity.

The centrifugal fan assembly includes belts, bearings, sheaves, and coupler drives along with their inherent maintenance problems. Yet, this design is necessary for applications having high air distribution static pressure losses. These applications include enclosures with ductwork runs, tunnel conveyors, densely stacked products, and produce ripening rooms.

Casing

Casing materials are selected for compatibility with the enclosure environment. Aluminum, either coated or uncoated, or steel, either galvanized or suitably coated, are typical casing materials. Stainless steel is also used in food storage or preparation enclosures where sanitation must be maintained. On larger cooler units, internal framing is fabricated of sufficiently substantial material, such as galvanized steel, and casings are usually made with similar material. Some plastic casings are used in small unit coolers, while some large, ceiling-suspended units may feature all aluminum construction to reduce weight.

Coil Construction

Coil construction varies from uncoated (all) aluminum tube and fin to hot-dipped galvanized (all) steel tube and fin, depending on the type of refrigerant used and the environmental exposure of the coil. The most popular unit coolers have coils with copper tubes and aluminum fins. ammonia refrigerant evaporators are never constructed with copper because ammonia corrodes copper. Also, sprayed coils are not constructed with aluminum fins they are protected with a baked on phenolic dip coating or similar protection applied after fabrication.

Fin spacings vary from 3 to 4 mm between fins for coils with surfaces above 0°C when latent loads are insignificant. Otherwise, 4 to 8 mm between fins is the accepted spacing for coil surfaces below 0°C, with a 6-mm fin spacing when latent loads exceed 15% of the total load. Fin spacings of 25 and 12 mm are used when defrosting is set for once a day, such as in low-temperature supermarket display cases. Staged fins in a rows of coils, such as a 24-12-6 mm fin spacing combination greatly reduces fin blockage due to frost accumulation (Ogawa et al. 1993).

Even distribution of the refrigerant flow to each circuit of the coil is vital for attaining maximum performance. distributor assemblies are used for direct expansion halocarbon refrigerants and occasionally for large, medium-temperature ammonia units. Application information provided by the manufacturer should be closely followed, particularly as to orifice sizing and assembly orientation. For liquid pumped recirculating systems, orifice disks are used in lieu of a distributor assembly. These disks are sized and installed by the coil manufacturer. They fit in the inlet (supply) header, at the connection to each coil circuit.

Headers and their piping connections are part of the coil assembly. Usually header lengths equal the coil height dimension; therefore each header is sized to the coil capacity for the application, based on flow velocities and not on the temperature equivalent of the saturated suction temperature drop. Velocities of approximately 7.5 m/s are used to compute the size of the return gas header and its connection size. In the field this size is often mistaken to be the recommended return line size. But the size of lines installed in the field should be based on the suction drop (see Chapter 1).

Controls

In the simplest form, electromechanical controls cycle the refrigeration components to maintain the desired enclosure temperature and defrost cycle. Pressure responsive modulating control valves such as evaporator pressure regulators and head pressure controls are also used. A temperature control could be a thermostat mounted in the enclosure that cycles either the compressor on and off or a liquid line solenoid valve that allows liquid refrigerant to flow to the evaporator. A suction pressure switch at the compressor can substitute for the wall-mounted thermostat.

Electronic controls have made electromechanical controls obsolete, except on very small installations. Microprocessor controllers mounted at the compressor receive and process signals from one or more temperature diode(s) and/or pressure transducer(s). These signals are converted to coordinate control of the compressor and the suction, discharge, and liquid line flow control valves. Defrost cycling, automatic callout for service, and remote site operation checks are standard options on this type of controller. For large warehouses and supermarkets, an electronically based energy management system (EMS) can incorporate several compressors into one controller.

AIR MOVEMENT AND DISTRIBUTION

The direction of the air and air throw should be such that air moves where there is a heat gain. This principle implies that the air sweep the enclosure walls and ceiling as well as to the product. Unit coolers should be placed (1) so they do not discharge air at any doors or openings; (2) away from doors that do not incorporate an entrance vestibule or pass to another refrigerated enclosure in order to keep from inducing additional infiltration into the enclosure; and (3) away from the airstream of another unit to avoid defrosting difficulties.

The velocity and relative humidity of air passing over an exposed product affect the amount of surface drying and mass loss. Air velocities of 2.5 m/s over the product are typical for most freezer applications. Higher velocities require additional fan power and, in many cases, only slightly decrease the cooling time. For example, air velocities in excess of 2.5 m/s for freezing plastic-wrapped bread reduce freezing time very little. However, increasing the air velocity from 2.5 to 5.0 m/s over unwrapped pizza reduces the freezing time and product exposure by almost half. This variation shows that

Forced-Circulation Air Coolers

product testing is necessary to design the special enclosures intended for blast freezing and/or automated food processing. Sample tests should yield the following information: ideal air temperature, air velocity, product mass loss, and dwell time. With this information, the proper unit or product coolers, as well as the supporting refrigeration equipment and controls, can be selected.

RATING UNITS

Currently, no industry standard exists for rating unit and product coolers. Part of the difficulty in developing a workable standard is that many variables are encountered. Cooler coil performance and capacities should be based on a fixed set of conditions, and they greatly depend on (1) air velocity, (2) refrigerant velocity, (3) circuit configuration, (4) refrigerant blend glide, (5) temperature difference, (6) frost condition, and (7) superheating adjustment. The most significant items are refrigerant flow rate, as related to refrigerant feed through the coil, and frost condition defrosting in low-temperature applications. The following sections discuss performance differences relative to some of the available unit cooler variations.

Refrigerant Velocity

Depending on the commercial refrigerant feed method used, both the capacity ratings of the cooler and the refrigerant flow rates vary. The following feed methods are used.

Dry Expansion. In this system, a thermostatically controlled, direct-expansion valve allows just enough liquid refrigerant into the cooling coil to ensure that it vaporizes at the outlet. In addition, 5 to 15% of the coil surface is used to superheat the vapor. The direct expansion (DX) coil flow rates are usually the lowest of the various feed methods.

Recirculated Refrigerant. This system is similar to a dry expansion feed except it includes a recirculated refrigerant drum, i.e. a low pressure receiver, and a liquid refrigerant pump connected to the coil. It also has a hand expansion valve, which is a metering device used to control the flow of the entering liquid refrigerant. The coil is intentionally overfed by the pump, such that complete coil flooding eliminates superheating of the refrigerant in the evaporator. The amount of liquid refrigerant pumped through the coil may be two to six times greater than that passed through a dry DX coil. As a result, this coil's capacity is higher than that for a dry expansion feed (see Chapter 1 for further information).

Flooded. This system has a liquid reservoir (surge drum or accumulator) located next to each unit or set of units. The surge drum is filled with a subcooled refrigerant and connected to the cooler coil. To ensure gravity flow of the refrigerant and a completely wet internal coil surface, the liquid level in the surge drum must be equal to the top of the coil. The capacity of gravity-recirculated feed is usually the highest attainable, in part because large coil tubes (≥ 25 mm o.d.) are required so that virtually no pressure drop exists.

Brine. In this chapter this term encompasses any liquid or solution that absorbs heat in the coil without a change in state. Ethylene glycol in water and propylene glycol in water are well accepted and thus most often used. Food grade propylene glycol should be used in refrigerated food processing applications. Calcium chloride in water or sodium chloride in water, for extra low temperature applications, and R-30, can be used only under tightly controlled and monitored conditions. For corrosion protection, most of these solutions must be neutralized or inhibited (preferably by the chemical manufacturer) prior to use.

The capacity rating for a brine coil depends on such variables as flow, viscosity, specific heat, and density. This rating is obtained by special request from the coil manufacturer. Generally, coils handling secondary refrigerants have about 11% less capacity at low temperatures and 14% less capacity at medium temperatures than comparable direct-expansion halocarbon refrigerants. The glycol temperature must run about 5 K lower than the comparable saturated suction temperature of a comparable DX coil to obtain the same capacity.

Frost Condition

Frost accumulation on the coil and defrosting are perhaps the most indeterminate variables that affect the capacity rating of forced-air coolers. Ogawa et al. (1993) showed that a light frost accumulation slightly improves the heat transfer of the coil. A continuous accumulation has a varying result, depending on the airflow. Performance suffers when airflow through the coil is reduced because of an increase in the air-side static pressure (e.g., the propeller fan arrangement in a unit cooler). But if the airflow through a frosting coil is maintained (e.g., a variable speed fan arrangement), the frost will reduce the capacity by about 2 to 10% (Kondepudi and O'Neal 1990, Rile and Crawford 1991). The thermal resistance of the frost (ice) varies with time and temperature, and ice pack growth is a product of having a surface temperature below the air dew point. Ultimately, defrosting is the only way to return to rated performance. This is usually initiated when unit performance drops to 75 to 80% of rated.

Controlled lab tests also showed that frost growth on a finned surface is not uniform with coil depth. Fin spacing and, for DX type coils, location of the superheat region in the coil had the most effect on uniformity. Oskarsson et al. (1990) discussed the effect of the length of time of frosting on uniformity. Industry generally considers that ice formation is uniform through a coil with a wide fin spacing, i.e., < 5 mm. This spacing is used to determine an inter-fin free-air area to estimate the air static pressure drop through a coil operating under frosting conditions.

Defrosting

The defrost cycle may be initiated and terminated in many ways. The microprocessor control, which has largely replaced the mechanical time clock, has reduced energy use and helped to maintain product quality (by reducing the temperature rise during the defrost cycle). Accurate defrosting also provides better protection of the refrigeration equipment. That is, improper and/or incomplete defrosting can damage the evaporator coil, to the extent that refrigerant leaks develop when ice is allowed to build up and crush the coil tube(s). The following defrosting methods are in use.

Enclosure Air Temperature Above 2°C. Enclosure air that is 2°C or slightly warmer can be used to defrost the coil. Fan(s) are left on and defrosting occurs during the compressor off cycles. However, some of the moisture on the coil surface evaporates into the air, which is undesirable for a low-humidity application. The following methods of control are commonly used.

1. If the refrigeration cycle is interrupted by a defrost timer, the continually circulating air melts the coil frost and ice. The timer can operate either the compressor or a liquid-line solenoid valve.
2. An oversized unit cooler controlled by a wall thermostat defrosts during its normal *off* and *on* cycling. The thermostat can control a refrigeration solenoid in a multiple-coil system or the compressor in a unitary installation. *Note*: An oversized unit is able to handle a 24 h cooling load in 16 h.
3. Pressure control can be used for slightly oversized unitary equipment. A low-pressure switch connected to the compressor suction line is set at a cut-out point such that the design suction pressure corresponds to the saturated temperature required to handle the maximum enclosure load. The suction pressure at the compressor drops and causes the compressor motor to stop as the enclosure load fluctuates or as the oversized compressor overcomes the maximum loading.

The thermostatic expansion valve on the unit cooler controls the evaporator temperature by regulating liquid refrigerant flow, which varies with the load. The cut-in point, which starts the compressor motor, should be set at the suction pressure that corresponds to the equivalent saturated temperature of the desired enclosure temperature. The pressure differential between the cut-in and cut-out points corresponds to the temperature difference between the enclosure air and the coil temperatures. The pressure setting should allow for the pressure drop in the suction line.

Enclosure Air Temperature Below 2°C. Whenever the enclosure air is below 2°C, supplementary heat must be introduced into the enclosure to defrost the coil surface and drain pan. Unfortunately, some of this defrost heat remains in the enclosure until the unit starts operation after the defrost cycle. The following supplemental heat sources are used for defrosting.

1. **Gas defrosting** can be the fastest and most efficient method if an adequate supply of hot gas is available. Besides performing the defrost function, the hot refrigerant gas clears the coil and drain pan tube assembly of accumulated compressor oil. This aids in returning the oil back to the compressor. Gas defrosting is used for small, commercial single and multiplex units, and for large, industrial central plants for virtually all low-temperature applications. Hot-gas defrosting also increases the capacity of a large, continuously operating compressor system because it removes some of the load from the condenser as it alternately defrosts the multiple evaporators. This method of defrost puts the least amount of heat into the enclosure, especially when latent gas (sometimes called cool gas) from the receiver is used.
2. **Electric defrost** effectiveness depends on the location of the electric heating elements. The elements can be either attached to the finned coil surface or inserted inside special fin holes or dummy tubes in the coil element. Electric defrost can be efficient and rapid. It is simple to operate and maintain, but it does dissipate the most heat into the enclosure.
3. **Heated air** may be circulated around unit coolers that are constructed to isolate the frosted coil from the cold enclosure air. Once the coil is isolated, the air around it is heated by hot gas or electric heating elements and is circulated to hasten the defrost. This heated air also must heat a drain pan, which is needed in all enclosures at temperatures of 1°C or less. Some units have specially constructed housings and ducting to run warm air from adjoining areas.
4. **Water defrost** is the quickest method of defrosting a unit. It is efficient and effective for rapid cleaning of the complete coil surface. Water defrost can be performed manually or on an automatic cycle. This method becomes less desirable as the enclosure temperature decreases much below freezing, but it has been successfully used in cases as low as −40°C. Water defrost is used more for large units used for cooling ndustrial products than for small ceiling-suspended units.
5. **Hot brine** can be used to defrost brine-cooled coils by heating the brine for the defrost cycle. This system heats from within the coil and is as rapid as hot-gas defrost. The heat source can be steam, electric resistance elements, or condenser water.

Defrost Control. For the most part, the fan is turned off when defrosting with supplemental heat. Defrosting for either too little or too much time can degrade overall performance; thus, a defrost cycle is best ended by monitoring temperature. A thermostat may be mounted in the cooler coil to sense a rise in the temperature of the finned or tubed surface. A temperature of at least 7°C indicates the removal of frost and automatically returns the unit to cooling.

Fan operation is delayed, usually by the same thermostat, until the coil surface temperature approaches its normal operating level. This practice prevents unnecessary heating of the enclosure after defrost. It also prevents drops of defrost water from being blown off the coil surface, which avoids icing of the fan blade, guard, and orifice ring. In some applications, fan delay prevents a rapid buildup of air pressure, which could structurally damage the enclosure.

Initiation of defrost can be automated by time clocks, running time monitors, air pressure differential controls, or by monitoring the air temperature difference through the coil (which increases as frost accumulation reduces the airflow). Adequate supplementary heat for the drain pan and drain lines should be considered. Often two methods are run simultaneously (i.e., hot-gas and electric) to simplify drain pan defrosting and shorten the defrost cycle. Drain lines should be properly pitched, insulated, and trapped outside the freezer, preferably in a warm area.

Basic Cooling Capacity

Most rating tables state gross capacity and assume that the fan assembly or defrost heat is included as part of the enclosure load calculation. Some manufacturers' cooler coil ratings may appear as sensible capacity, while some may be listed as total capacity, which includes both sensible and latent capacities. Some ratings include reduction factors to account for frost accumulation in low-temperature applications or for some unusual condition. Others include capacity multiplier factors for various refrigerants (i.e. R-12 = 1.00 and R-134a = 0.97).

The rating, defined as the basic cooling capacity, is based on the temperature difference between the inlet air and the refrigerant in the coil [watt per degree TD (temperature difference)]. The coil inlet air temperature is considered to be the same as the enclosure air temperature, and the refrigerant temperature is assumed to be the temperature equivalent to the saturated pressure at the coil outlet. For heavy use such as for a blast freezer or an assembly line, manufacturer's ratings should be applied on the average of the coil inlet-to-outlet temperatures. This is regarded as the average enclosure temperature.

The TD necessary to obtain the unit cooler capacity varies with the application. The TD can be related to the humidity desired in the enclosure—a smaller TD maintains a higher humidity. The following TDs for above −4°C saturated suction are suggested:

- For a very high relative humidity (about 90%), a temperature difference of 4 to 5 K is common.
- For a high relative humidity (approximately 80%), a temperature difference of 6 to 7 K is recommended.
- For a medium relative humidity (approximately 75%), a temperature difference of 7 to 9 K is recommended.

Temperature differences beyond these limits usually result in low enclosure humidities, which dry the product. However, for packaged products and workrooms, a TD of 14 to 16 K is not unusual. Paper storage or similar products also require a low humidity level. Here, a TD of 11 to 16 K may be necessary.

For low-temperature applications below −4°C saturated suction, the temperature difference is generally kept below 8 K because of system economics and frequency of defrosting rather than for humidity control.

BIBLIOGRAPHY

ARI. 1989. Unit coolers for refrigeration. *Standard* 420. Air Conditioning and Refrigeration Institute, Arlington, VA.

ASHRAE. 1990. Method of testing forced and natural convection air coolers for refrigeration. *Standard* 25.

Kondepudi, S.N. and D.L. O'Neal. 1990. The effect of different fin configurations on the performance of finned-tube heat exchangers under frosting conditions. *ASHRAE Transactions* 96(2):439-444.

Ogawa, K., N. Tanaka, M. Takashita. 1993. Performance improvement of plate fin-and-tube heat exchangers under frosting conditions. *ASHRAE Transactions* 99(1):762-771.

Oskarsson, S.P., K.I. Krakow, and S. Lin. 1990. Evaporator models for operation with dry, wet, and frosted finned surfaces—Part II: Evaporator models and verification. *ASHRAE Transactions* 96(1):381-392.

Rile, R.W. and R.R. Crawford. 1991. The effect of frost accumulation on the performance of domestic refrigerator freezer finned-tube evaporator coils. *ASHRAE Transactions* 97(2):428-437.

CHAPTER 43

LIQUID CHILLING SYSTEMS

GENERAL CHARACTERISTICS 43.1	CENTRIFUGAL LIQUID CHILLERS 43.8
Principles of Operation 43.1	Equipment Description 43.8
Common Liquid Chilling Systems 43.1	Performance and Operating Characteristics 43.9
Equipment Selection 43.3	Method of Selection 43.10
Control 43.4	Control Considerations 43.10
Standards 43.5	Auxiliaries and Special Applications 43.11
Methods of Testing 43.5	Operation and Maintenance 43.12
General Maintenance 43.5	SCREW LIQUID CHILLERS 43.13
RECIPROCATING LIQUID CHILLERS 43.6	Equipment Description 43.13
Equipment Description 43.6	Performance and Operating Characteristics 43.13
Performance Characteristics and Operating Problems 43.6	Method of Selection 43.14
Method of Selection 43.7	Control Considerations 43.14
Control Considerations 43.7	Auxiliaries and Special Applications 43.15
Special Applications 43.8	Maintenance 43.15

A LIQUID chilling system cools water, brine, or other secondary coolant for air conditioning or refrigeration. The system may be either factory assembled and wired or shipped in sections for erection in the field. The most frequent application is water chilling for air conditioning, although both brine cooling for low-temperature refrigeration and chilling of fluids in industrial processes are also common.

The basic components of a vapor-compression, liquid chilling system include a compressor, a liquid cooler (evaporator), a condenser, a compressor drive, a liquid refrigerant expansion or flow-control device, and a control center; the system may also include a receiver, an economizer, an expansion turbine, and/or a subcooler. In addition, certain auxiliary components may be used, such a lubricant cooler, lubricant separator, lubricant-return device, purge unit, lubricant pump, a refrigerant transfer unit, refrigerant vents, and/or additional control valves.

GENERAL CHARACTERISTICS

PRINCIPLES OF OPERATION

Liquid (usually water) enters the cooler, where it is chilled by liquid refrigerant evaporating at a lower temperature. The refrigerant vaporizes and is drawn into the compressor, which increases the pressure and temperature of the gas so that it may be condensed at the higher temperature in the condenser. The condenser cooling medium is warmed in the process. The condensed liquid refrigerant then flows back to the evaporator through an expansion device. A fraction of the liquid refrigerant changes to vapor (flashes) as the pressure drops between the condenser and the evaporator. Flashing cools the liquid to the saturated temperature at the evaporator pressure. It produces no refrigeration effect in the cooler. The following modifications (sometimes combined for maximum effect) reduce flash gas and increase the net refrigeration effect per unit of power consumption.

Subcooling. Condensed refrigerant may be subcooled to a temperature below its saturated condensing temperature in either the subcooler section of a water-cooled condenser or a separate heat exchanger. Subcooling reduces the amount of flashing and increases the refrigeration effect in the chiller.

Economizing. This process can occur either in a direct-expansion (DX), an expansion turbine, or a flash-type system. In a **DX system**, the main liquid refrigerant is usually cooled in the shell of a shell-and-tube heat exchanger, at condensing pressure, from the saturated condensing temperature to within several degrees of the intermediate saturated temperature. Before cooling, a small portion of the liquid flashes and evaporates in the tube side of the heat exchanger to cool the main liquid flow. Although subcooled, the liquid will still be at the condensing pressure.

An **expansion turbine** extracts rotating energy as a portion of the refrigerant vaporizes. As in the DX system, the remaining liquid is supplied to the cooler at the intermediate pressure.

In a **flash-type system**, the entire liquid flow is expanded to the intermediate pressure in a vessel that supplies liquid to the cooler at the saturated intermediate pressure; however, the liquid is at the intermediate pressure.

In any case, the flash gas enters the compressor at either an intermediate stage of a multistage centrifugal compressor, at the intermediate stage of an integral two-stage reciprocating compressor, at an intermediate pressure port of a screw compressor, or at the inlet of a high-pressure stage on a multistage reciprocating or screw compressor.

Liquid Injection. Condensed liquid is throttled to the intermediate pressure and injected into the second-stage suction of the compressor to prevent excessively high discharge temperatures and, in the case of centrifugal machines, to reduce noise. In the case of screw compressors, condensed liquid is injected into a port fixed at slightly below discharge pressure to provide lubricant cooling.

COMMON LIQUID CHILLING SYSTEMS

Basic System

The refrigeration cycle of a basic system is shown in Figure 1. Chilled water enters the cooler at 12°C, for example, and leaves at 7°C. Condenser water leaves a cooling tower at 30°C, enters the condenser, and returns to the cooling tower near 35°C. Condensers may also be cooled by air or through evaporation of water. This system, with a single compressor and one refrigerant circuit with a water-cooled condenser, is used extensively to chill water for air conditioning because it is relatively simple and compact.

These liquid chillers are available with hermetic, semihermetic, and open-type compressors with various means of starting.

The preparation of this chapter is assigned to TC 8.1, Positive Displacement Compressors, and TC 8.2, Centrifugal Machines.

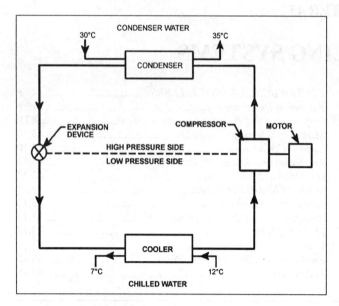

Fig. 1 Equipment Diagram for Basic Liquid Chiller

Multiple Chiller System

A multiple chiller system has two or more chillers connected by parallel or series piping to a common distribution system. Multiple chillers offer operational flexibility, standby capacity, and less disruptive maintenance. The chillers can be sized to handle a base load and increments of a variable load to allow each chiller to operate at its most efficient point.

Multiple chiller systems offer some standby capacity if repair work must be done on one chiller. Starting in-rush current is reduced, as well as power costs at partial-load conditions. Maintenance can be scheduled for one chilling machine during part-load times, and sufficient cooling can still be provided by the remaining unit(s). These advantages require an increase in installed cost and space, however.

Water should flow constantly through the chillers for stable control. Load variation is temperature-related and is easily detected by temperature controls. In contrast, when water flow varies the load becomes flow related. Because a temperature control system cannot sense a variation in flow, it is unable to maintain stable control. However, some applications do have variable water flow through the cooling coils. In this case, a decoupled system is typically used to separate the distribution pumping from the production pumping. It allows variable flow through the cooling coils but maintains constant water flow through the chillers, allowing good control of the multiple chillers.

A typical decoupled system is shown in Figure 2. The multiple pumps are connected by a bypass pipe that connects the return and supply headers. Each chiller-pump combination operates independently from the remaining chillers. Capacity control is simplified and as if each chiller operated alone. Instead of using temperature as an indicator of demand, relative flow is the indicator. If greater flow is demanded than that supplied by the chiller-pumps, return water is forced through the bypass into the supply header. This flow indicates a need for additional chiller capacity and another chiller-pump starts. Bypass flow in the opposite direction indicates overcapacity and the chiller-pumps are turned off.

Two basic multiple chiller systems are used: **parallel** and **series chilled water flow**. In the **parallel arrangement**, liquid to be chilled is divided among the liquid chillers; the multiple chilled streams are combined again in a common line after chilling. As the cooling load decreases, one unit may be shut down. Unless water flow is stopped through the inoperative chiller, the remaining unit(s) provide colder-than-design chilled liquid. The combined

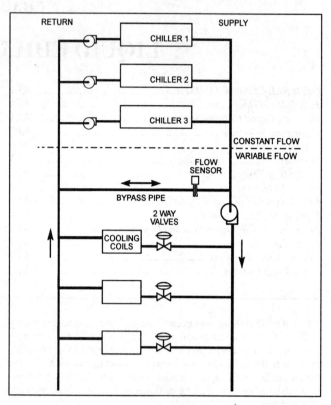

Fig. 2 Decoupled System

streams (including one from the idle chiller) then supply the chilled water at the design temperature in the common line.

When the design chilled water temperature is above about 7°C, all units should be controlled by the combined exit water temperature or by the return water temperature (RWT), since overchilling will not cause dangerously low water temperature in the operating machine(s). Chilled water temperature can be used to cycle one unit off when it drops below a capacity that can be matched by the remaining units.

When the design chilled water temperature is below about 7°C, each machine should be controlled by its own chilled water temperature, both to prevent dangerously low evaporator temperatures and to avoid frequent shutdowns by the low-temperature cutout. In this case, the temperature differential setting of the RWT must be adjusted carefully to prevent short cycling caused by the step increase in chilled water temperature when one chiller is cycled off. These control arrangements are shown in Figures 3 and 4.

In the **series arrangement**, the chilled liquid pressure drop may be higher if shells with fewer liquid-side passes or baffles are not available. No overchilling by either unit is required, and compressor power consumption is lower than it is for the parallel arrangement at partial loads. Because the evaporator temperature never drops below the design value (because no overchilling is necessary), the chances of evaporator freezeup are minimized. However, the chiller should still be protected by a low-temperature safety control.

Water cooled condensers in series are best piped in a counterflow arrangement so that the lead machine is provided with warmer condenser and chilled water and the lag machine is provided with colder entering condenser and chilled water. Refrigerant compression for each unit is nearly the same. If about 55% of design cooling capacity is assigned to the lead machine and about 45% to the lag machine, identical units can be used. In this way, either machine can provide the same standby capacity if the other is down, and lead and lag machines may be interchanged to equalize the number of operating hours on each.

Liquid Chilling Systems

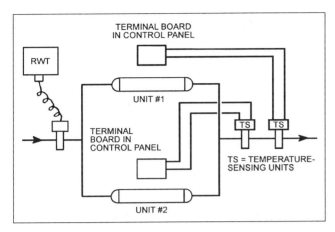

Fig. 3 Parallel Operation High Design Water Leaving Coolers (Approximately 7°C and Above)

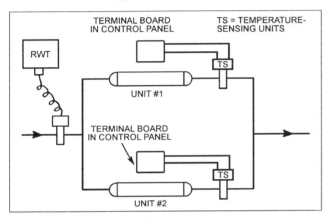

Fig. 4 Parallel Operation Low Design Water Leaving Coolers (Below Approximately 7°C)

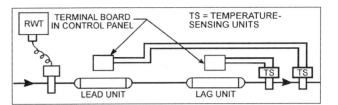

Fig. 5 Series Operation

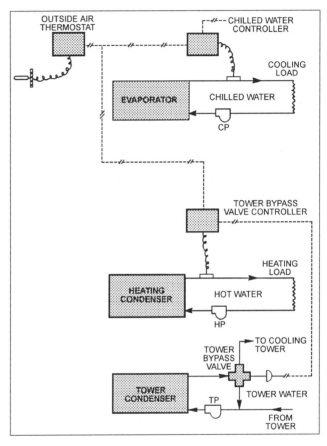

Fig. 6 Heat-Recovery Control System

A control system for two machines in series is shown in Figure 5. (On reciprocating chillers, RWT sensing is usually used instead of leaving water sensing because it allows closer temperature control.) Both units are modulated to a certain capacity; then, one unit shuts down, leaving less than 100% load on the operating machine.

One machine should be shut down as soon as possible, with the remaining unit carrying the full load. This not only reduces the number of operating hours on a unit, but also leads to less total power consumption because the COP tends to decrease below the full load value when unit load drops much below 50%.

Heat-Recovery Systems

Any building or plant requiring the simultaneous operation of heat-producing and cooling equipment has the potential for a heat-recovery installation. Heat-recovery equipment should be considered for all new or retrofit installations. In some cases, the installed cost may be less because of the elimination or reduction of both heating equipment and the space required for it.

Heat-recovery systems extract heat from chilled liquid and reject some of that heat, plus the energy of compression, to a warm-water circuit for reheat or heating. Air-conditioned spaces thus furnish heating for other spaces in the same building. During the full-cooling season, all heat must be rejected outdoors, usually by a cooling tower. During spring or fall, some heat is required inside, while a portion of the heat extracted from the air-conditioned spaces must be rejected outside simultaneously.

Heat recovery offers the user low heating costs and reduces space requirements for mechanical equipment. A control system must be designed carefully, however, to take the greatest advantage of the recovered heat and to maintain proper temperature and humidity in all parts of the building. Chapter 8 of the 1996 *ASHRAE Handbook—Systems and Equipment* covers balanced heat-recovery systems.

Since cooling tower water is not satisfactory for heating coils, a separate, closed warm-water circuit with another condenser bundle or auxiliary condenser, in addition to the main water chiller condenser, must be provided. In some cases, it is economically feasible to use a standard condenser and a closed-circuit water cooler.

A suggested control scheme is shown in Figure 6. The heating water temperature is controlled by a cooling tower bypass valve, which modulates the flow of condenser cooling water to the tower. An outside air thermostat resets the hot water control point upward as the outdoor temperature drops and resets the chilled water temperature control point upward on colder days. In this way, extra power is not consumed unnecessarily by the compressor in attempting to maintain summer design coil temperatures during dry, cold outdoor conditions.

EQUIPMENT SELECTION

The largest factor that determines total liquid chiller owning cost is the cooling load size; therefore, the total required chiller capacity should be calculated accurately. The practice of adding 10 to 20% to load estimates is unnecessary because of the availability of accurate load estimating methods, and it proportionately increases costs related to equipment purchase, installation, and the poor efficiency resulting from wasted power. Oversized equipment can also cause operational difficulties such as frequent on-off cycling or surging of centrifugal machines at low loads. The penalty for a small underestimation of cooling load, however, is not serious. On the few design load days of the year, an increase in chilled liquid temperature is often acceptable. However, for some industrial or commercial loads, a safety factor can be added to the load estimate.

The life-cycle cost as discussed in Chapter 33 of the 1995 *ASHRAE Handbook—Applications* should be used to minimize the overall purchase and operating costs. Total owning cost is composed of the following:

- **Equipment Price.** Each machine type and/or manufacturer's model should include all the necessary auxiliaries such as starters and vibration mounts. If these are not included, their price should be added to the base price. Associated equipment, such as condenser water pump, tower, and piping, should be included.
- **Installation Cost.** Factory-packaged machines are both less expensive to install and usually considerably more compact, resulting in space savings. The cost of field assembly of field-erected chillers must also be evaluated.
- **Energy Cost.** Using an estimated load schedule and part-load power consumption curves furnished by the manufacturer, a year's energy cost should be calculated.
- **Water Cost.** With water cooled towers, the cost of acquisition, water treatment, tower blowdown, and overflow water should be included.
- **Maintenance Cost.** Each bidder may be asked to quote on a maintenance contract on a competitive basis.
- **Insurance and Taxes.**

For package chillers that include heat recovery, system cost and performance should be compared in addition to equipment costs. For example, the heat-recovery chiller installed cost should be compared with the installed cost of a chiller plus a separate heating system. The following factors should also be considered: (1) energy costs, (2) maintenance requirements, (3) life expectancy of equipment, (4) standby arrangement, (5) relationship of heating to cooling loads, (6) effect of package selection on sizing, and (7) type of peripheral equipment.

Condensers and coolers are often available with either **liquid heads**, which require the water pipes to be disconnected for tube access and maintenance, or **marine-type water boxes**, which permit tube access with water piping intact. The liquid head is considerably lower in price. The cost of disconnecting piping must be greater than the additional cost of marine-type water boxes to justify their use. Typically, an elbow and union or flange connection is installed only to facilitate the removal of heads.

These methods are helpful in choosing between equipment types and makes. Chapter 33 of the 1995 *ASHRAE Handbook—Applications* has further cost information.

The following types of liquid chillers are generally used for air conditioning:

Up to 90 kW	— Reciprocating or scroll
90 to 280 kW	— Screw, reciprocating, or scroll
280 to 1600 kW	— Screw, reciprocating, or centrifugal
700 to 3500 kW	— Screw or centrifugal
3500 kW	— Centrifugal

For air-cooled condenser duty, brine chilling, or other high pressure applications from 280 to about 700 kW, reciprocating and screw liquid chillers are more frequently installed than centrifugals. Centrifugal liquid chillers (particularly multistage machines), however, may be applied quite satisfactorily at high pressure conditions.

Factory packages are available to about 8400 kW and field-assembled machines to about 35 MW.

CONTROL

Liquid Chiller Controls

The **chilled liquid temperature sensor** sends an air pressure (pneumatic control) or electrical signal (electronic control) to the control circuit, which then modulates compressor capacity in response to leaving or return chilled liquid temperature change from its set point.

Compressor capacity adjustment is accomplished differently on the following liquid chillers:

Reciprocating chillers use combinations of cylinder unloading and on-off compressor cycling of single or multiple compressors.

Centrifugal liquid chillers, driven by electric motors, commonly use adjustable prerotation vanes, which are sometimes combined with movable diffuser walls. Turbine and engine drives and inverter-driven, variable-speed electric motors allow the use of speed control in addition to prerotation vane modulation, reducing power consumption at partial loads.

Screw compressor liquid chillers include a slide valve that adjusts the length of the compression path. Inverter-driven, variable-speed electric motors and turbine and engine drives can also modulate screw compressor speed to control capacity.

In air-conditioning applications, most centrifugal and screw compressor chillers modulate from 100% to approximately 10% load. Although relatively inefficient, hot-gas bypass can be used to reduce capacity to nearly 0% with the unit in operation.

Reciprocating chillers are available with simple on-off cycling control in small capacities and with multiple steps of unloading down to 12.5% in the largest multiple compressor units. Most intermediate sizes provide unloading to 50, 33, or 25% capacity. Hot-gas bypass can reduce capacity to nearly 0%.

The **water temperature controller** is a thermostatic device that unloads or cycles the compressor(s) when the cooling load drops below minimum unit capacity. An *antirecycle timer* is sometimes used to limit starting frequency.

On centrifugal or screw compressor chillers, a **current limiter** or **demand limiter** limits compressor capacity during periods of possible high power consumption (such as pulldown) to prevent current draw from exceeding the design value; such a limiter can be set to limit demand, as described in the section on Centrifugal Liquid Chillers.

Controls That Influence the Liquid Chiller

Condenser cooling water may need to be controlled to regulate condenser pressure. Normally, the temperature of the water leaving a cooling tower can be controlled by fans, dampers, or a water bypass around the tower. Bypass around the tower allows the water velocity through the condenser tubes to be maintained, which prevents low-velocity fouling.

A flow-regulating valve is another common means of control. The orifice of this valve modulates in response to condenser pressure. For example, a reduction in pressure decreases the water flow, which, in turn, raises the condenser pressure to the desired minimum level.

For air-cooled or evaporative condensers, compressor discharge pressure can be controlled by cycling fans, shutting off circuits, or flooding coils with liquid refrigerant to reduce the heat transfer.

A reciprocating chiller usually has a thermal expansion valve, which requires a restricted range of pressure to avoid starving the evaporator (at low pressure).

An expansion valve(s) usually controls a screw compressor chiller. Cooling tower water temperature can be allowed to fall with

Liquid Chilling Systems

decreasing load from the design condition to the chiller manufacturer's recommended minimum limit.

Screw compressor chillers above 500 kW may use flooded-type evaporators and evaporator liquid refrigerant controls similar to those used on centrifugal chillers.

A thermal expansion valve may control a centrifugal chiller at low capacities, while higher capacity machines employ a high-pressure float, orifice(s), or even a low-side float valve to control refrigerant liquid flow to the cooler. These latter types of controls allow relatively low condenser pressures, particularly at partial loads. Also, a centrifugal machine may surge if pressure is not reduced when cooling load decreases. In addition, low pressure reduces compressor power consumption and operating noise. For these reasons, in a centrifugal installation, cooling tower water temperature should be allowed to fall naturally with decreasing load and wet-bulb temperature, except that the liquid chiller manufacturer's recommended minimum limit must be observed.

Safety Controls

Some or all of the cutouts listed below may be provided in a liquid chilling package to stop the compressor(s) automatically. Cutouts may be manual or automatic reset.

- **High Condenser Pressure.** This pressure switch opens if the compressor discharge pressure exceeds the value prescribed in ASHRAE *Standard* 15.
- **Low Refrigerant Pressure (or Temperature).** This device opens when evaporator pressure (or temperature) reaches a minimum safe limit.
- **High Lubricant Temperature.** This device protects the compressor if loss of lubricant cooling occurs or if a bearing failure causes excessive heat generation.
- **High Motor Temperature.** If loss of motor cooling or overloading because of a failure of a control occurs, this device shuts down the machine. It may consist of direct-operating bimetallic thermostats, thermistors, or other sensors embedded in the stator windings; it may be located in the discharge gas stream of the compressor.
- **Motor Overload.** Some small, reciprocating compressor hermetic motors may use a directly operated overload in the power wiring to the motor. Some larger motors use pilot-operated overloads. Centrifugal and screw compressor motors generally use starter overloads or current-limiting devices to protect against overcurrent.
- **Low Lubricant Sump Temperature.** This switch is used either to protect against a lubricant heater failure or to prevent starting after a prolonged shutdown before the lubricant heaters have had time to drive off refrigerant dissolved in the lubricant.
- **Low Lubricant Pressure.** To protect against clogged lubricant filters, blocked lubricant passageways, loss of lubricant, or a lubricant pump failure, a switch shuts down the compressor when lubricant pressure drops below a minimum safe value or if sufficient lubricant pressure is not developed shortly after the compressor starts.
- **Chilled Liquid Flow Interlock.** This device may not be furnished with the liquid chilling package, but it is needed in the external piping to protect against a cooler freezeup in case the liquid stops flowing. An electrical interlock is typically installed.
- **Condenser Water Flow Interlock.** This device, which is similar to the chilled liquid flow interlock, is sometimes used in the external piping.
- **Low Chilled Liquid Temperature.** Sometimes called **freeze protection**, this cutout operates at a minimum safe value of leaving chilled liquid temperature to prevent cooler freezeup in the case of an operating control malfunction.
- **Relief Valves.** In accordance with ASHRAE *Standard* 15, relief valves, rupture disks, or both, set to relieve at the shell design working pressure, must be provided on most pressure vessels or on piping connected to the vessels. Fusible plugs may also be used in some locations. Pressure relief devices should be vented outdoors or to the low-pressure side, in accordance with regulations or the standard.

STANDARDS

ARI *Standards* 550, Centrifugal and Rotary Screw Water-Chilling Packages, and 590, Positive Displacement Compressor, provide guidelines for the rating of centrifugal and reciprocating liquid chilling machines, respectively.

The design and construction of refrigerant pressure vessels are governed by the ASME *Boiler and Pressure Vessel Code*, Section VIII, except when design working pressure is 103 kPa or less (as is usually the case for R-123 liquid chilling machines). The water-side design and construction of a condenser or cooler is not within the scope of the ASME code unless the design pressure is greater than 2.1 MPa or the design temperature is greater than 99°C.

ASHRAE *Standard* 15, Safety Code for Mechanical Refrigeration, applies to all liquid chillers and new refrigerants on the market. New standards for equipment rooms are included. Methods for the measurement of unit sound levels are described in ARI *Standard* 575, Method of Measuring Machinery Sound Within an Equipment Space.

METHODS OF TESTING

All tests of reciprocating liquid chillers for rating or verification of rating should be conducted in accordance with ASHRAE *Standard* 30, Methods of Testing Liquid Chilling Packages. Centrifugal or screw liquid chiller ratings should be derived and verified by test in accordance with ARI *Standard* 550.

GENERAL MAINTENANCE

The following general maintenance specifications apply to reciprocating, centrifugal, and screw chillers. The equipment should be neither overmaintained nor neglected. A preventive maintenance schedule should be established; the items covered can vary with the nature of the application. The list is intended as a guide; in all cases, the manufacturer's specific recommendation should be followed.

Continual Monitoring

- Condenser water treatment—treatment is determined specifically for the condenser water used.
- Operating conditions—daily log sheets should be kept to indicate trends and provide an advanced notice of deteriorating chillers.
- Brine quality for concentration and corrosion inhibitor levels.

Periodic Checks

- Leak check
- Purge operation
- System dryness
- Lubricant level
- Lubricant filter pressure drop
- Refrigerant quantity or level
- System pressures and temperatures
- Water flows
- Expansion valves operation

Regularly Scheduled Maintenance

- Condenser and lubricant cooler cleaning
- Evaporator cleaning on open systems
- Calibrating pressure, temperature, and flow controls
- Tightening wires and power connections
- Inspection of starter contacts and action
- Safety interlocks

- Dielectric checking of hermetic and open motors
- Tightness of hot gas valve
- Lubricant filter and drier change
- Analysis of lubricant and refrigerant
- Seal inspection
- Partial or complete valve or bearing inspection, as per manufacturer's recommendations
- Vibration levels

Extended Maintenance Checks

- Compressor guide vanes and linkage operation and wear
- Eddy current inspection of heat exchanger tubes
- Compressor teardown and inspection of rotating components
- Other components as recommended by manufacturer

RECIPROCATING LIQUID CHILLERS

EQUIPMENT DESCRIPTION

Components and Their Function

The reciprocating compressor described in Chapter 34 of the 1996 *ASHRAE Handbook—Systems and Equipment* is a positive-displacement machine that maintains fairly constant volume flow rate over a wide range of pressure ratios. The following types of compressors are commonly used in liquid chilling machines:

- Welded hermetic, to about 90 kW chiller capacity
- Semihermetic, to about 700 kW chiller capacity
- Direct-drive open, to about 1600 kW chiller capacity

Open motor-driven liquid chillers are usually more expensive than hermetically sealed units, but they can be more efficient. Hermetic motors are generally suction gas cooled; the rotor is mounted on the compressor crankshaft.

Condensers may be evaporative, air- or water-cooled. Water-cooled versions may be either tube-in-tube, shell-and-coil, shell-and-tube, or plate type heat exchangers. Most shell-and-tube condensers can be repaired, while others must be replaced if a refrigerant-side leak occurs.

Air-cooled condensers are much more common than evaporative condensers. Less maintenance is needed for air-cooled heat exchangers than for the evaporative type. Remote condensers can be applied with condenserless packages. (Information on condensers can be found in Chapter 35 of the 1996 *ASHRAE Handbook—Systems and Equipment*.)

Coolers are usually direct-expansion, in which refrigerant evaporates while it is flowing inside tubes and chilled liquid is cooled as it is guided several times over the outside of the tubes by shell-side baffles. Flooded coolers are sometimes used on industrial chillers. Flooded coolers maintain a level of refrigerant liquid on the shell side of the cooler, while the liquid to be cooled flows through tubes inside the cooler. Tube-in-tube coolers are sometimes used with small machines; they offer low cost when repairability and installation space are not important criteria. (A detailed description of coolers can be found in Chapter 37 of the 1996 *ASHRAE Handbook—Systems and Equipment*.)

The **thermal expansion** valve, capillary, or other expansion device modulates refrigerant flow from the condenser to the cooler to maintain enough suction superheat to prevent any unevaporated refrigerant liquid from reaching the compressor. Excessively high values of superheat are avoided so that unit capacity is not reduced. (For additional information, see Chapter 45.)

Lubricant cooling is not usually required for air conditioning. However, if lubricant cooling is necessary, a refrigerant-cooled coil in the crankcase or a water-cooled cooler may be used. Lubricant coolers are often used in conjunction with low suction temperatures or high-pressure ratio applications when extra lubricant cooling is needed.

Capacities and Types Available

Available capacities range from about 7 to 1600 kW. Multiple reciprocating compressor units have become popular for the following reasons:

- The number of capacity increments is greater, resulting in closer liquid temperature control, lower power consumption, less current in-rush during starting, and extra standby capacity.
- Multiple refrigerant circuits are used, resulting in the potential for limited servicing or maintenance of some components while maintaining cooling.

Selection of Refrigerant

R-12 and R-22 have been the primary refrigerants used in chiller applications. CFC-12 has been replaced with HFC-134a, which has similar properties. However, R-134a requires synthetic lubricants because it is not miscible with mineral oils. R-134a is suitable for both open and hermetic compressors.

R-22 provides greater capacity than R-134a for a given compressor displacement. R-22 is used for most open and hermetic compressors, but as an HCFC, it is scheduled for phaseout in the future. R-717 (ammonia) has similar capacity characteristics to R-22, but, because of odor and toxicity, R-717 is subject to restrictions for use in public or populated areas. However, R-717 chillers are becoming more popular because of bans on CFC and HCFC refrigerants. R-717 units are open-drive compressors and are piped with steel because copper cannot be used in ammonia systems.

PERFORMANCE CHARACTERISTICS AND OPERATING PROBLEMS

A distinguishing characteristic of the reciprocating compressor is its pressure rise versus capacity. Pressure rise has only a slight influence on the volume flow rate of the compressor, and, therefore, a reciprocating liquid chiller retains nearly full cooling capacity, even on above design wet-bulb days. It is well suited for air-cooled condenser operation and low-temperature refrigeration. A typical performance characteristic is shown in Figure 7 and compared with

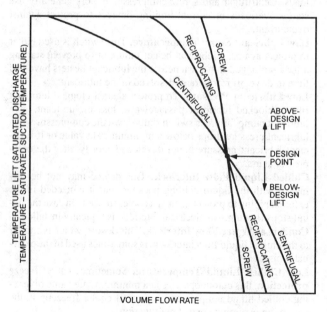

Fig. 7 Comparison of Single-Stage Centrifugal, Reciprocating, and Screw Compressor Performance

Liquid Chilling Systems

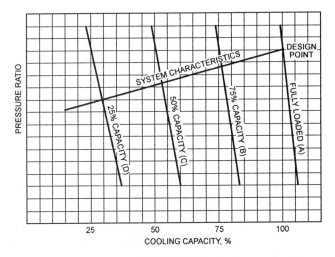

Fig. 8 Reciprocating Liquid Chiller Performance with Three Equal Steps of Unloading

the centrifugal and screw compressors. Methods of capacity control are furnished by the following:

- Unloading of compressor cylinders (one at a time or in pairs)
- On-off cycling of compressors
- Hot-gas bypass
- Compressor speed control
- Combination of the previous methods

Figure 8 illustrates the relationship between system demand and performance of a compressor with three steps of unloading. As cooling load drops to the left of the fully loaded compressor line (A), compressor capacity is reduced to that represented by line (B), which produces the required refrigerant flow. Since cooling load varies continuously while machine capacity is available in fixed increments, some compressor on-off cycling or successive loading and unloading of cylinders is required to maintain fairly constant liquid temperature. In practice, a good control system minimizes the load-unload or on-off cycling frequency while maintaining satisfactory temperature control.

METHOD OF SELECTION

Ratings

Two types of ratings are published. The first, for a packaged liquid chiller, lists values of capacity and power consumption for many combinations of leaving condenser water and chilled water temperatures (ambient dry-bulb temperatures for air-cooled models). The second type of rating shows capacity and power consumption for different condensing temperatures and chilled water temperatures. This type of rating permits selection with a remote condenser that can be evaporative, water, or air cooled. Sometimes the required rate of heat rejection is also listed to aid in selection of a separate condenser.

Power Consumption

With all liquid chilling systems, power consumption increases as condensing temperature rises. Therefore, the smallest package, with the lowest ratio of input to cooling capacity, can be used when condenser water temperature is low, the remote air-cooled condenser is relatively large, or when leaving chilled water temperature is high. The cost of the total system, however, may not be low when liquid chiller cost is minimized. Increases in cooling tower or fan coil cost will reduce or offset the benefits of reduced compression ratio. Life-cycle costs (initial cost plus operating expenses) should be evaluated.

Fouling

A fouling allowance of 0.044 $m^2 \cdot K/kW$ is included in manufacturers' ratings in accordance with ARI *Standard* 590. However, fouling factors greater than 0.044 should be considered in the selection if water conditions are other than ideal.

CONTROL CONSIDERATIONS

A reciprocating chiller is distinguished from centrifugal and screw compressor-operated chillers by its use of increments of capacity reduction rather than continuous modulation. Therefore, special arrangements must be used to establish precise chilled liquid temperature control while maintaining stable operation free from excessive on-off cycling of compressors or unnecessary loading and unloading of cylinders.

To help provide good temperature control, return chilled liquid temperature sensing is normally used by units with steps of capacity control. The resulting flywheel effect in the chilled liquid circuit damps out excessive cycling. Leaving chilled liquid temperature sensing has the advantage of preventing excessively low leaving chilled liquid temperatures if chilled liquid flow falls significantly below the design value. It may not provide stable operation, however, if rapid changes in load are encountered.

An example of a basic control circuit for a single compressor-packaged reciprocating chiller with three steps of unloading is shown in Figure 9. The unit is started by moving the on-off switch to the *on* position. The programmed timer will start operating. Assuming that the flow switch, field interlocks, and chiller safety

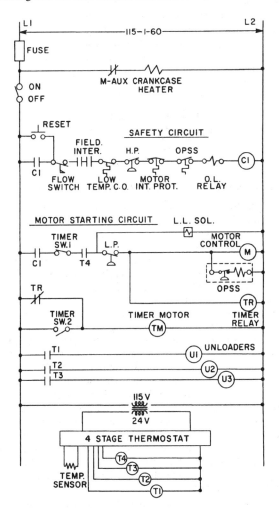

Fig. 9 Reciprocating Liquid Chiller Control System

devices are closed, pressing the momentarily closed reset button will energize control relay C1, locking in the safety circuit and the motor starting circuit. When the timer completes its program, timer switch 1 closes and timer switch 2 opens. Timer relay TR energizes, stopping the timer motor. When timer switch 1 closes, the motor starting circuit is completed and the motor contactor holding coil is energized, starting the compressor.

The four-stage thermostat controls the capacity of the compressor in response to the demand of the system. Cylinders are loaded and unloaded by de-energizing and energizing the unloader solenoids. If the load is reduced so that the return water temperature drops to a predetermined setting, the unit shuts down until the demand for cooling increases.

Opening a device in the safety circuit will de-energize control relay C1 and shut down the compressor. The liquid line solenoid is also de-energized. Manual reset is required to restart. The crankcase heater is energized whenever the compressor is shut down.

If the automatic reset, low-pressure cutout opens, the compressor shuts down, but the liquid line solenoid remains energized. The timer relay (TR) is de-energized, causing the timer to start and complete its program before the compressor can be restarted. This prevents rapid cycling of the compressor under low-pressure conditions. A time delay low-pressure switch can also be used for this purpose with the proper circuitry.

SPECIAL APPLICATIONS

For multiple chiller applications and a 5 K chilled liquid temperature range, the use of a parallel chilled liquid arrangement is common because of the high cooler pressure drop resulting from the series arrangement. For a large (10 K) range, however, the series arrangement eliminates the need for overcooling during operation of one unit only. Special coolers with low water pressure drop may also be used to reduce total chilled water pressure drop in the series arrangement.

CENTRIFUGAL LIQUID CHILLERS

EQUIPMENT DESCRIPTION

Components and Their Function

The centrifugal compressor is described in Chapter 34 of the 1996 *ASHRAE Handbook—Systems and Equipment*. Since it is not a constant displacement machine, it offers a wide range of capacities continuously modulated over a limited range of pressure ratios. By altering built-in design items (including number of stages, compressor speed, impeller diameters, and choice of refrigerant), it can be used in liquid chillers having a wide range of design chilled liquid temperatures and design cooling fluid temperatures. Its ability to vary capacity continuously to match a wide range of load conditions with nearly proportional changes in power consumption makes it desirable for both close temperature control and energy conservation. Its ability to operate at greatly reduced capacity allows it to run most of the time with infrequent starting.

The hour of the day for starting an electric-drive centrifugal liquid chiller can often be chosen by the building manager to minimize peak power demands. It has a minimum of bearing and other contacting surfaces that can wear; this wear is minimized by providing forced lubrication to those surfaces prior to startup and during shutdown. Bearing wear usually depends more on the number of startups than the actual hours of operation. Thus, by reducing the number of startups, the life of the system is extended, and maintenance costs are reduced.

Both open and hermetic compressors are manufactured. Open compressors may be driven by steam turbines, gas turbines or engines, or electric motors, with or without speed-changing gears. (Engine and turbine drives are covered in detail in Chapter 7 and electric motor drives in Chapter 39 of the 1996 *ASHRAE Handbook—Systems and Equipment*.)

Packaged electric-drive chillers may be of the open- or hermetic-type and use two-pole, 50- or 60-Hz polyphase electric motors, with or without speed-increasing gears. Hermetic units use only polyphase induction motors. Speed-increasing gears and their bearings, in most open- and hermetic-type packaged chillers, operate in a refrigerant atmosphere, and the lubrication of their contacting surfaces is incorporated in the compressor lubrication system.

Magnetic and SCR (silicon controlled rectifier) motor controllers are used with packaged chillers. When purchased separately, the controller must meet the specifications of the chiller manufacturer to ensure adequate equipment safety. When timed step starting methods are used, the time between steps should be long enough for the motor to overcome the relatively high inertia of the compressor and attain sufficient speed to minimize the electric current drawn immediately after transition.

Flooded coolers are commonly used, although direct-expansion coolers are employed by some manufacturers in the lower capacity ranges. The typical flooded cooler uses copper tubes or copper alloy that are mechanically expanded into the tube sheets, and, in some cases, into intermediate tube supports, as well.

Because liquid refrigerant that flows into the compressor increases power consumption and may cause internal damage, mist eliminators or baffles are often used in flooded coolers to minimize refrigerant liquid entrainment in the suction gas. (Additional information on coolers for liquid chillers can be found in Chapter 37 of the 1996 *ASHRAE Handbook—Systems and Equipment*.)

The condenser is generally water cooled, with refrigerant condensing on the outside of copper tubes. Large condensers may have refrigerant drain baffles, which direct the condensate from within the tube bundle directly to the liquid drains, reducing the thickness of the liquid film on the lower tubes.

Air-cooled condensers can be used with units that use higher pressure refrigerants, but with considerable increase in unit energy consumption at design conditions. Operating costs should be compared with systems using cooling towers and condenser water circulating pumps.

System modifications, including subcooling and economizing (described under Principles of Operation) are often used to conserve energy. Some units combine the condenser, cooler, and refrigerant flow control in one vessel; a subcooler may also be incorporated. (Additional information concerning thermodynamic cycles can be found in Chapter 1 of the 1997 *ASHRAE Handbook—Fundamentals*. For information on condensers and subcoolers, see Chapter 35 of the 1996 *ASHRAE Handbook—Systems and Equipment*.)

Capacities and Types Available

Centrifugal packages are currently available from about 300 to 8500 kW at nominal conditions of 7°C leaving chilled water temperature and 35°C leaving condenser water temperature. This upper limit is continually increasing. Field-assembled machines extend to about 35 MW. Single-stage and two-stage internally geared machines and two- and three-stage direct-drive machines are commonly used in packaged units. Electric motor-driven machines constitute the majority of units sold.

Units with hermetic motors, cooled by refrigerant gas or liquid, are offered from about 300 to 7000 kW. Open-drive units are not offered by all hermetic manufacturers in the same size increments but are generally available from 300 kW to 35 MW.

Selection of Refrigerant

In choosing refrigerant, the manufacturer considers coefficient of performance, operating pressures, heat-transfer properties, flow

Liquid Chilling Systems

rate, stability, toxicity, flammability, cost, and availability for their machine designs. Halogenated hydrocarbons are normally used as refrigerants because they offer reasonable characteristics with respect to the application. ASHRAE *Standards* 15 and 34 cover the use, handling, and numbering of refrigerants.

The centrifugal compressor is particularly suitable for handling relatively large flows of suction vapor. As the volumetric flow of suction vapor increases with higher capacities and lower suction temperatures, the higher pressure refrigerants, for example, R-134a and R-22, are used. The physical size and weight of the refrigerant piping and, often, other components of the refrigeration system are reduced by the use of higher pressure refrigerants. In order of decreasing volumetric flow and increasing pressures are refrigerants R-113, R-123, R-11, R-114, R-134a, R-12 or R-500, and R-22. The CFC refrigerants R-11, R-12, R-113, R-114, and R-500 have been phased out.

Pressure vessels for use with R-123 usually have a design working pressure of 103 kPa on the refrigerant side. The vessel shells are usually stronger than necessary for this requirement to ensure sufficient rigidity and prevent collapse under vacuum.

The thermal stability of the refrigerant and its compatibility with materials it contacts are also important. Selection of elastomers and electrical insulating materials require special attention because many of these materials are affected by the refrigerants. (Additional information concerning refrigerants can be found in Chapters 18 and 19 of the 1997 *ASHRAE Handbook—Fundamentals*.)

PERFORMANCE AND OPERATING CHARACTERISTICS

Figure 10 illustrates a compressor's performance at constant speed with various inlet guide vane settings. Figure 11 illustrates that compressor's performance at various speeds with open inlet guide vanes. Capacity is modulated at constant speed by automatic adjustment of prerotation vanes that whirl the refrigerant gas at the impeller eye. This effect matches system demand by shifting the compressor performance curve downward and to the left (as shown in Figure 10). Compressor efficiency, when unloaded in this manner, is superior to suction throttling. Some manufacturers also automatically reduce diffuser width or throttle the impeller outlet with decreasing load.

Speed control for a centrifugal compressor offers even lower power consumption. Down to about 50% capacity or at off design conditions, the speed may be reduced gradually without surging. Control is transferred to the prerotation vanes for operation at lower loads. While capacity is related directly to a change in speed, the pressure produced is proportional to the square of the change in speed. Therefore, the pressure produced by reducing the speed may be less than that required by the load. Combined use of gas bypass, prerotation vanes, etc., would then be necessary.

A **gas bypass** allows the compressor to operate down to zero load. This feature is a particular advantage for such intermittent industrial applications as the cooling of quenching tanks. Bypass vapor obtained by either method maintains the power consumption at the same level attained just prior to starting bypass, regardless of load reductions. At light loads, some bypass vapor, if introduced into the cooler below the tube bundle, may increase the evaporating temperature by agitating the liquid refrigerant and thereby more thoroughly wetting the tube surfaces.

Figure 12 shows how **temperature lift** varies with load. A typical reduction in entering condenser water temperature of 5 K helps to reduce temperature lift at low load. Other factors producing lower lift at reduced loads are

- Reduction in condenser cooling water range (the difference between entering and leaving temperatures, resulting from decreasing heat rejection)
- Decrease in temperature difference between condensing refrigerant and leaving condenser water
- Similar decrease between evaporating refrigerant and leaving chilled liquid temperature

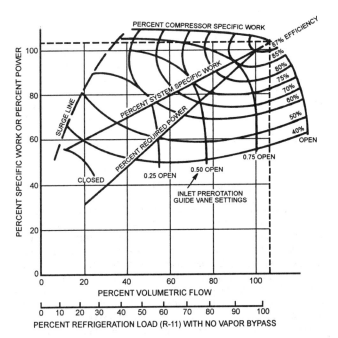

Fig. 10 Typical Centrifugal Compressor and System at Constant Speed

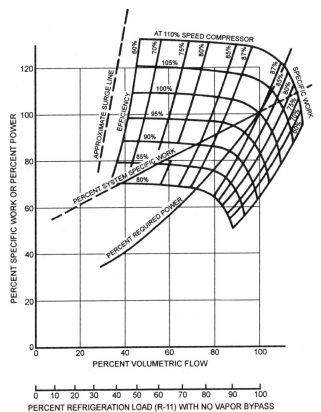

Fig. 11 Typical Centrifugal Compressor Performance at Various Speeds

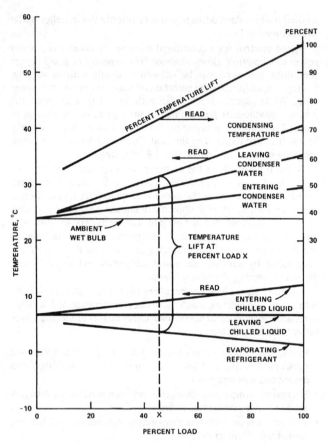

Fig. 12 Temperature Relations in a Typical Centrifugal Liquid Chiller

In many cases, the actual reduction in temperature lift is even greater because the wet-bulb temperature usually drops with the cooling load, producing a greater decrease in entering condenser water temperature.

As stated earlier, speed control is usually used from 100% down to about 50% load or at off-design conditions; below 50%, inlet vane control is used. Power consumption is reduced when the coldest possible condenser water is used, consistent with the chiller manufacturer's recommended minimum condenser water temperature. In cooling tower applications, minimum water temperatures should be controlled by a cooling tower bypass and/or by cooling tower fan control, not by reducing the water flow through the condenser. Maintaining a high flow rate at lower temperatures minimizes fouling and power requirements.

Surging occurs when the system specific work becomes greater than the compressor developed specific work or above the surge line indicated in Figures 10 and 11. Excessively high temperature lift and corresponding specific work commonly originate from

- Excessive condenser or evaporator water-side fouling beyond the specified allowance
- Inadequate cooling tower performance and higher-than-design condenser water temperature
- Noncondensables in the condenser, which increase condenser pressure.

METHOD OF SELECTION

Ratings

A refrigeration machine with specified details is chosen from selection tables for given capacities and operating conditions, or through computer-generated selection or performance programs. Rating tables differ from selection tables in that they list the capacities and operating data for each refrigeration machine under various operating conditions, often with specific details for the listed conditions. The details specified for centrifugal systems include the number of passes and the water-side pressure drop in each of the heat exchangers, the required power input, electrical characteristics, and part-load performance.

The maximum number of condenser and cooler water passes should be used, without producing excessive water pressure drop. The greater the number of water-side passes, the less the power consumption. Sometimes a slight reduction in condenser water flow (and slightly higher leaving water temperature) allows a better selection (lower power consumption or smaller model) than will the choice of fewer water passes when a rigid pressure-drop limit exists.

Fouling

In accordance with ARI *Standard* 590, a fouling allowance of 0.044 $m^2 \cdot K/kW$ is included in manufacturers' ratings. (Chapter 35 of the 1996 *ASHRAE Handbook—Systems and Equipment* has further information about fouling factors.) To reduce fouling, a minimum water velocity of about 1 m/s is recommended in coolers or condensers. Maximum water velocities exceeding 3.3 m/s are not recommended because of potential erosion problems.

Proper water treatment and regular tube cleaning are recommended for all liquid chillers to reduce power consumption and operating problems. See Chapter 44 of the 1995 *ASHRAE Handbook—Applications* for water treatment information.

Continuous or daily monitoring of the quality of the condenser water is desirable. Checking the quality of the chilled liquid is also desirable. The intervals between checks become greater as the possibilities for fouling contamination become less—for example, an annual check should be sufficient for closed-loop water-circulating systems for air conditioning. Corrective treatment is required, and periodic, usually annual, cleaning of the condenser tubes usually keeps fouling within the specified allowance. In applications where more frequent cleaning is desirable, an on-line cleaning system may be economical.

Noise and Vibration

The chiller manufacturer's recommendations for mounting should be followed to prevent transmission or amplification of vibration to adjacent equipment or structures. Auxiliary pumps, if not connected with flexible fittings, can induce vibration of the centrifugal unit, especially if the rotational speed of the pump is nearly the same as either the compressor prime mover or the compressor. Flexible tubing becomes less flexible when it is filled with liquid under pressure and some vibration can still be transmitted. General information on noise, measurement, and control may be found in Chapter 7 of the 1997 *ASHRAE Handbook—Fundamentals*, Chapter 43 of the 1995 *ASHRAE Handbook—Applications*, and ARI *Standard* 575.

CONTROL CONSIDERATIONS

In centrifugal systems, the **chilled liquid temperature sensor** is usually placed in thermal contact with the leaving chilled water. In electrical control systems, the electrical signal is transmitted to an electronic control module, which controls the operation of an electric motor(s) positioning the capacity controlling inlet guide vanes. A current limiter is usually included on machines with electric motors. An electrical signal from a current transformer in the compressor motor controller is sent to the electronic control module. The module receives indications of both the leaving chilled water temperature and the compressor motor current. The portion of the electronic control module responsive to motor current is called the current limiter. It overrides the demands of the temperature sensor.

Liquid Chilling Systems

The **inlet guide vanes**, independent of the demands for cooling, do not open more than the position that results in the present setting of the current limiter. Pneumatic capacity controls operate in a similar manner. The chilled liquid temperature sensor provides a pneumatic signal. The controlling module receives both that signal and the motor current electrical signal and controls the operation of a pneumatic motor(s) positioning the inlet guide vanes. Both controlling systems have sensitivity adjustments.

The **current limiter** on most machines can limit current draw during periods of high electrical demand charges. This control can be set from about 40 to 100% of full load current. Whenever power consumption is limited, cooling capacity is correspondingly reduced. If cooling load is only 50% of the full value, the current (or demand) limiter can be set at 50% without loss of cooling. By setting the limiter at 50% of full current draw, any subsequent high demand charges are prevented during pulldown after startup. Even during periods of high cooling load, it may be desirable to limit electrical demand if a small increase in chiller liquid temperature is acceptable. If the temperature continues to decrease after the capacity control has reached its minimum position, a low-temperature control stops the compressor and restarts it when a rise in temperature indicates the need for cooling. Manual controls may also be provided to bypass the temperature control. Provision is included to ensure that the capacity control is at its minimum position when the compressor starts to provide an unloaded starting condition.

Additional operating controls are needed for appropriate operation of lubricant pumps, lubricant heaters, purge units, and refrigerant transfer units. An **antirecycle timer** should also be included to prevent frequent motor starts. Multiple unit applications require additional controls for capacity modulation and proper sequencing of units. (See the Multiple Chiller System section.)

Safety controls protect the unit under abnormal conditions. Safety cutouts that may be required are for high condenser pressure, low evaporator refrigerant temperature or pressure, low lubricant pressure, high lubricant temperature, high motor temperature, and high discharge temperature. Auxiliary safety circuits are usually provided on packaged chillers. At installation, the circuits are field wired to field-installed safety devices, including auxiliary contacts on the pump motor controllers and flow switches in the chilled water and condenser water circuits. Safety controls are usually provided in a lockout circuit, which will trip out the compressor motor controller and prevent automatic restart. The controls reset automatically, but the circuit cannot be completed until a manual reset switch is operated and the safety controls return to their safe positions.

AUXILIARIES AND SPECIAL APPLICATIONS

Auxiliaries

Purge units are required for centrifugal liquid chilling machines using R-123, R-11, R-113, or R-114 because evaporator pressure is below atmospheric pressure. If a purge unit were not used, air and moisture would accumulate in the refrigerant side. Noncondensables collect in the condenser during operation, reducing the heat-transfer coefficient and increasing condenser pressure as a result of both their insulating effect and the partial pressure of the noncondensables. Compressor power consumption increases, capacity is reduced, and surging may occur.

Moisture may build up as free moisture once the refrigerant becomes saturated. Acids produced by a reaction between the free moisture and the refrigerant will then cause internal corrosion. A purge unit prevents the accumulation of noncondensables and ensures internal cleanliness of the chiller. However, a purge unit does not reduce the need to check for leaks and the need to repair them, which is required maintenance for any liquid chiller. Purge units may be manual or automatic, compressor-operated, or compressorless. To reduce the potential for air leaks when chillers are off, the chillers may be heated externally to pressurize them to atmospheric pressure.

ASHRAE *Standard* 15 requires purge units and rupture disks to be vented outdoors. Because of environmental concerns and the increasing cost of refrigerants, high efficiency (air to refrigerant) purges are available that reduce refrigerant losses during normal purging operations.

Lubricant coolers may be water cooled, using condenser water when the quality is satisfactory, or chilled water when a small loss in net cooling capacity is acceptable. These coolers may also be refrigerant or air cooled, eliminating the need for water piping to the cooler.

A **refrigerant transfer unit** may be provided for centrifugal liquid chillers. The unit consists of a small reciprocating compressor with electric motor drive, a condenser (air- or water-cooled), a lubricant reservoir and separator, valves, and interconnecting piping. Refrigerant transfers in three steps:

1. **Gravity Drain.** When the receiver is at the same level as or below the cooler, some liquid refrigerant may be transferred to the receiver by opening valves in the interconnecting piping.

2. **Pressure Transfer.** By resetting valves and operating the compressor, refrigerant gas is pulled from the receiver to pressurize the cooler, forcing refrigerant liquid from the cooler to the storage receiver. If the chilled liquid and condenser water pumps can be operated to establish a temperature difference, the migration of refrigerant from the warmer vessel to the colder vessel can also be used to assist in the transfer of refrigerant.

3. **Pump-Out.** After the liquid refrigerant has been transferred, valve positions are changed and the compressor is operated to pump refrigerant gas from the cooler to the transfer unit condenser, which sends condensed liquid to the storage receiver. If any chilled liquid (water, brine, etc.) remains in the cooler tubes, pump-out must be stopped before cooler pressure drops below the saturation condition corresponding to the freezing point of the chilled liquid.

If the saturation temperature corresponding to cooler pressure is below the chilled liquid freezing point when recharging, refrigerant gas from the storage receiver must be introduced until the cooler pressure is above this condition. The compressor can then be operated to pressurize the receiver and move refrigerant liquid into the cooler without danger of freezing.

Water-cooled transfer unit condensers provide fast refrigerant transfer. Air-cooled condensers eliminate the need for water, but they are slower and more expensive.

Special Applications

Heat-Recovery. Instead of rejecting all heat extracted from the chilled liquid to a cooling tower, a separate, closed condenser cooling water circuit is heated by the condensing refrigerant for such purposes as comfort heating, preheating, or reheating. Some factory packages include an extra condenser water circuit, either in the form of a double-bundle condenser or an auxiliary condenser.

A centrifugal heat-recovery package is controlled as follows:

- **Chilled liquid temperature** is controlled by a sensor in the leaving chilled liquid line signaling the capacity control device.

- **Hot water temperature** is controlled by a sensor in the hot water line that modulates a cooling tower bypass valve. As the heating requirement increases, hot water temperature drops, opening the tower bypass slightly. Less heat is rejected to the tower, condensing temperature increases, and hot water temperature is restored as more heat is rejected to the hot water circuit.

The hot water temperature selected has a bearing on the installed cost of the centrifugal package, as well as on the power consumption while heating. Lower hot water temperatures of 35 to 40°C result in a less expensive machine that uses less power. Higher temperatures require a greater compressor motor output, perhaps higher pressure condenser shells, sometimes extra compression stages, or a cascade arrangement. Installed cost of the centrifugal heat-recovery machine is increased as a result.

Another concern in the design of a central chilled water plant with heat-recovery centrifugal compressors is the relative size of the cooling and heating loads. The heating and cooling loads should be equalized on each machine so that the compressor may operate at optimum efficiency during both the full cooling and full heating seasons. When the heating requirement is considerably smaller than the cooling requirement, multiple packages will lower operating costs and allow standard air-conditioning centrifugal packages of lower cost to be used for the remainder of the cooling requirement. In multiple packages, only one unit is designed for heat recovery and carries the full heating load.

Free Cooling. Cooling without operating the compressor of a centrifugal liquid chiller is called free cooling. When a supply of condenser water is available at a temperature below the needed chilled water temperature, the chiller can operate as a thermal siphon. Low-temperature condenser water condenses refrigerant, which is either drained by gravity or pumped into the evaporator. Higher-temperature chilled water causes the refrigerant to evaporate, and vapor flows back to the condenser because of the pressure difference between the evaporator and the condenser. Free cooling is limited to about 10 to 30% of the chiller design capacity. Free cooling capacity depends on chiller design and the temperature difference between the desired chilled water temperature and the condenser water temperature. Free cooling is also available using either direct or indirect methods as described in Chapter 36 of the 1996 *ASHRAE Handbook—Systems and Equipment*.

Air-Cooled System. Two types of air-cooled centrifugal systems are prevalent. One consists of a water-cooled centrifugal package with a closed-loop condenser water circuit. The condenser water is cooled in a water/air heat exchanger. This arrangement results in higher condensing temperature and increased power consumption. In addition, winter operation requires the use of glycol in the condenser water circuit, which reduces the heat-transfer coefficient of the unit.

The other type of unit is directly air-cooled, which eliminates the intermediate heat exchanger and condenser water pumps, resulting in lower power requirements. However, the condenser and refrigerant piping must be kept leak free.

Because a centrifugal machine will surge if it is subjected to a pressure appreciably higher than design, the air-cooled condenser must be designed to reject the required heat. In common practice, the selection of a reciprocating air-cooled machine is based on an outside dry-bulb temperature that will be exceeded 5% of the time. A centrifugal may be unable to operate during such times because of surging, unless the chilled water temperature is raised proportionately. Thus, the compressor impeller(s) and/or speed should be selected for the maximum dry-bulb temperature to ensure that the desired chilled water temperature will be maintained at all times. In addition, the condenser coil must be kept clean.

An air-cooled centrifugal chiller should allow the condensing temperature to fall naturally to about 21°C during colder weather. The resulting decrease in compressor power consumption is greater than that for reciprocating systems controlled by thermal expansion valves.

During winter shutdown, precautions must be taken to prevent freezing of the cooler liquid caused by a free cooling effect from the air-cooled condenser. A thermostatically controlled heater in the cooler, in conjunction with a low refrigerant pressure switch to start the chilled liquid pumps, will protect the system.

Other Coolants. Centrifugal liquid chilling units are most frequently used for water chilling applications. But centrifugals are also used with such coolants as calcium chloride, methylene chloride, ethylene glycol, and propylene glycol. (For a complete description of secondary coolants, see Chapter 20 of the 1997 *ASHRAE Handbook—Fundamentals*.) Coolant properties must be considered in calculating heat-transfer performance and pressure drop. Because of the greater temperature rise, higher compressor speeds and possibly more stages may be required for cooling these coolants. Compound and/or cascade systems are required for low-temperature applications.

Vapor Condensing. Many process applications condense vapors such as ammonia, chlorine, or hydrogen fluoride. Centrifugal liquid chilling units are used for these applications.

OPERATION AND MAINTENANCE

Proper operation and maintenance are essential for reliability, longevity, and safety. See Chapter 35 of the 1995 *ASHRAE Handbook—Applications* for general information on principles, procedures, and programs for effective maintenance. The manufacturer's operation and maintenance instructions should also be consulted for specific procedures. In the United States, Environmental Protection Agency regulations require (1) certification of service technicians, (2) a statement of minimum pressures necessary during evacuation of the system, and (3) definition of when a refrigerant charge must be removed prior to opening a system for service. All service technicians or operators maintaining systems must be familiar with these regulations.

Normal operation conditions should be established and recorded at initial startup. Changes from these conditions can be used to signal the need for maintenance. One of the most important items is to maintain a leak free unit.

Leaks on units operating at subatmospheric pressures allow air and moisture to enter the unit, which increases the condenser pressure. While the purge unit can remove noncondensables sufficiently to prevent an increase in condenser pressure, continuous entry of air and attendant moisture into the system promotes refrigerant and lubricant breakdown and corrosion. Leaks from units that operate above atmospheric pressure may release environmentally harmful refrigerants. Regulations require that annual leakage not exceed a percentage of the refrigerant charge. It is good practice, however, to find and repair all leaks.

Periodic analysis of the lubricant and refrigerant charge can also be used to identify system contamination problems. High condenser pressure or frequent purge unit operation indicate leaks that should be corrected as soon as possible. With positive operating pressures, leaks result in loss of refrigerant and such operating problems as low evaporator pressure. A leak check should also be included in preparation for a long-term shutdown. (See Chapter 6 for a detailed discussion of the harmful effects of air and moisture.)

Normal maintenance should include periodic lubricant and refrigerant filter changes as recommended by the manufacturer. All safety controls should be checked periodically to ensure that the unit is protected properly.

Cleaning of inside tube surfaces may be required at various intervals, depending on the water condition. Condenser tubes may only need annual cleaning if proper water treatment is maintained. Cooler tubes need less frequent cleaning if the chilled water circuit is a closed loop.

If the refrigerant charge must be removed and the unit opened for service, the unit should be leak-checked, dehydrated, and evacuated properly before recharging. Chapter 46 has information on dehydrating, charging, and testing.

Liquid Chilling Systems

SCREW LIQUID CHILLERS

EQUIPMENT DESCRIPTION

Components and Their Function

Single- and twin-screw compressors are both positive-displacement machines with nearly constant flow performance. Compressors for liquid chillers can be both lubricant-injected and lubricant-injection-free. (Chapter 35 of the 1996 *ASHRAE Handbook—Systems and Equipment* describes screw compressors in greater detail.)

The cooler may be flooded or direct-expansion. No particular design has a cost advantage over the other. The flooded cooler is more sensitive to freezing, requires more refrigerant, and requires closer evaporator pressure control, but its performance is easier to predict and it can be cleaned. The direct-expansion cooler requires closer mass flow control, is less likely to freeze, and returns lubricant to the lubricant system rapidly. The decision to use one or the other is based on the relative importance of these factors on a given application.

Screw coolers have the following characteristics: (1) high maximum working pressure, (2) continuous lubricant scavenging, (3) no mist eliminators (flooded coolers), and (4) distributors designed for high turndown ratios (direct-expansion coolers). A suction gas, high-pressure liquid heat exchanger is sometimes incorporated into the system to provide subcooling for increased thermal expansion valve flow and reduced power consumption. (For further information on coolers, see Chapter 37 of the 1996 *ASHRAE Handbook—Systems and Equipment*.)

Flooded coolers were once used in units with a capacity larger than about 1400 kW. Direct-expansion coolers are also used in larger units up to 2800 kW with a servo-operated expansion valve having an electronic controller that measures evaporating pressure, leaving secondary coolant temperature, and suction gas superheat.

The condenser may be included as part of the liquid chilling package when water-cooled, or it may be remote. Air-cooled liquid chilling packages are also available. When remote air-cooled or evaporative condensers are applied to liquid chilling packages, a liquid receiver generally replaces the water-cooled condenser on the package structure. Water-cooled condensers are the cleanable shell-and-tube type (see Chapter 35 of the 1996 *ASHRAE Handbook—Systems and Equipment*).

Lubricant cooler loads vary widely, depending on the refrigerant and application, but they are substantial because lubricant injected into the compressor absorbs a portion of the heat of compression. Lubricant is cooled by one of the following methods:

- Water-cooled using condenser water, evaporative condenser sump water, chilled water, or a separate water- or glycol-to-air cooling loop
- Air-cooled using a lubricant-to-air heat exchanger
- Refrigerant-cooled (where lubricant cooling load is low)
- Liquid injection into the compressor
- Condensed refrigerant liquid thermal recirculation (thermosyphon), where appropriate, compressor head pressure is available.

The latter two methods are the most economical both in first cost and overall operating cost because cooler maintenance and special water treatment are eliminated.

Efficient lubricant separators are required. The types and efficiencies of these separators vary according to refrigerant and application. Field built systems require better separation than complete factory-built systems. Ammonia applications are most stringent because no appreciable lubricant returns with the suction gas from the flooded coolers normally used in ammonia applications. However, separators are available for ammonia packages, which do not require the periodic addition of lubricant that is customary on other ammonia systems. The types of separators in use are centrifugal, demister, gravity, coalescer, and combinations of these.

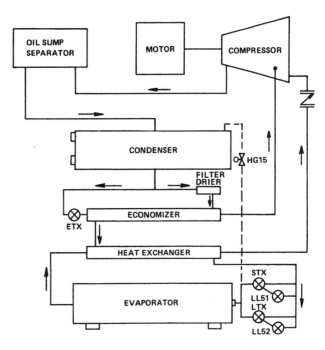

Fig. 13 Refrigeration System Schematic

Hermetic compressor units may use a centrifugal separator as an integral part of the hermetic motor while cooling the motor with discharge gas and lubricant simultaneously. A schematic of a typical refrigeration system is shown in Figure 13.

Capacities and Types Available

Screw compressor liquid chillers are available as factory-packaged units from about 100 to 4400 kW. Both open and hermetic styles are manufactured. Packages without water-cooled condensers, with receivers, are made for use with air-cooled or evaporative condensers. Most factory-assembled liquid chilling packages use R-22 and some use R-134a.

Additionally, compressor units, comprised of a compressor, hermetic or open motor, and lubricant separator and system, are available from 70 to 7000 kW. These are used with remote evaporators and condensers for low, medium, and high evaporating temperature applications. Condensing units, similar to compressor units in range and capacity but with water-cooled condensers, are also built. Similar open motor-drive units are available for ammonia, as are booster units.

Selection of Refrigerant

The refrigerants most commonly used with screw compressors on liquid chiller applications are R-22, R-134a, and R-717. The active use of R-12 and R-500 has been discontinued for new equipment.

PERFORMANCE AND OPERATING CHARACTERISTICS

The screw compressor operating characteristic shown in Figure 7 is compared with reciprocating and centrifugal performance. Additionally, since the screw compressor is a positive-displacement compressor, it does not surge. Because it has no clearance volume in the compression chamber, it pumps high volumetric flows at high pressure. Because of this, screw compressor chillers suffer the least capacity reduction at high condensing temperatures.

The screw compressor provides stable operation over the whole working range because it is a positive-displacement machine. The working range is wide because the discharge temperature is kept low and is not a limiting factor because of lubricant injection into the compression chamber. Consequently, the compressor is able to operate single-stage at high pressure ratios.

An economizer can be installed to improve the capacity and lower the power consumption at full-load operation. An example of such an economizer arrangement is shown in Figure 13, where the main refrigerant liquid flow is subcooled in a heat exchanger connected to the intermediate pressure port in the compressor. The evaporating pressure in this heat exchanger is higher than the suction pressure of the compressor.

Lubricant separators must be sized for the size of compressor, type of system (factory-assembled or field-connected), refrigerant, and type of cooler. Direct-expansion coolers have less stringent separation requirements than do flooded coolers. In a direct-expansion system, the refrigerant evaporates in the tubes, which means that the velocity is kept so high that the lubricant rapidly returns to the compressor. In a flooded evaporator, the refrigerant is outside the tubes, and some type of external lubricant-return device must be used to minimize the concentration of lubricant in the cooler. Suction or discharge check valves are used to minimize backflow and lubricant loss during shutdown.

Because the lubricant system is on the high-pressure side of the unit, precautions must be taken to prevent lubricant dilution. Dilution can also be caused by excessive floodback through the suction or intermediate ports; and unless properly monitored, it may go unnoticed until serious operating or mechanical problems are experienced.

METHOD OF SELECTION

Ratings

Screw liquid-chiller ratings are generally presented similarly to those for centrifugal-chiller ratings. Tabular values include capacity and power consumption at various chilled water and condenser water temperatures. In addition, ratings are given for packages without the condenser that list capacity and power versus chilled water temperature and condensing temperature. Ratings for compressors alone are also common, showing capacity and power consumption versus suction temperature and condensing temperature for a given refrigerant.

Power Consumption

Typical part-load power consumption is shown in Figure 14. Power consumption of screw chillers benefits from reduction of

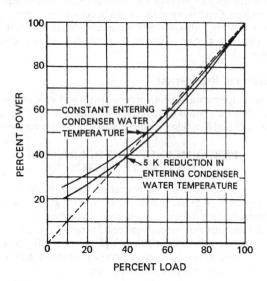

Fig. 14 Typical Screw Compressor Chiller Part-Load Power Consumption

condensing water temperature as the load decreases, as well as operating at the lowest practical pressure at full load. However, because direct-expansion systems require a pressure differential, the power consumption saving is not as great at part load as shown.

Fouling

A fouling allowance of 0.044 m²·K/kW is incorporated in screw compressor chiller ratings. Excessive fouling (above the design value) increases power consumption and reduces capacity. Fouling of water-cooled lubricant coolers results in higher than desirable lubricant temperatures.

CONTROL CONSIDERATIONS

Screw chillers provide continuous capacity modulation, from 100% capacity down to 10% or less. The leaving chilled liquid temperature is sensed for capacity control. Safety controls commonly required are (1) lubricant failure switch, (2) high discharge pressure cutout, (3) low suction pressure switch, (4) cooler flow switch, (5) high lubricant and discharge temperature cutout, (6) hermetic motor inherent protection, (7) lubricant pump and compressor motor overloads, and (8) low lubricant temperature (flood back/dilution protection). The compressor is unloaded automatically (slide valve driven to minimum position) before starting. Once it starts operating, the slide valve is controlled hydraulically by a temperature-load controller that energizes the load and unload solenoid valves.

The current limit relay protects against motor overload from higher than normal condensing temperatures or low-voltage and also allows a demand limit to be set, if desired. An antirecycle timer is used to prevent overly frequent recycling. Lubricant sump

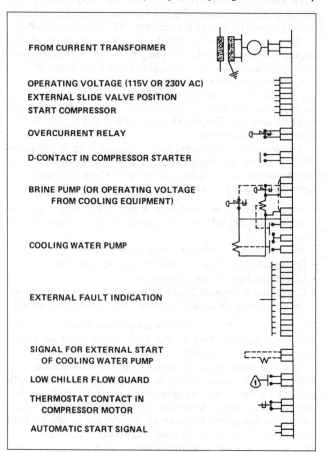

Fig. 15 Typical External Connections for Screw Compressor Chiller

Liquid Chilling Systems

heaters are energized during the off cycle. A hot gas capacity control is optionally available and prevents automatic recycling at no-load conditions such as is often required in process liquid chilling. A suction to discharge starting bypass sometimes aids starting and allows the use of standard starting torque motors.

Some units are equipped with electronic regulators specially developed for the screw compressor characteristics. These regulators include PI control (Proportional-Integrating) of the leaving brine temperature and such functions as automatic/manual control, capacity indication, time circuits to prevent frequent recycling and to bypass the lubricant pressure cutout during startup, switch for unloaded starting, etc. (Typical external connections are shown in Figure 15.)

AUXILIARIES AND SPECIAL APPLICATIONS

Auxiliaries

A **refrigerant transfer unit** is similar to the unit described in the section on Auxiliaries under Centrifugal Liquid Chillers. It is designed for R-22 operating pressure. Its flexibility is increased by including a reversible liquid pump on the unit. It is available as a portable unit or mounted on a storage receiver.

A **lubricant-charging pump** is useful for adding lubricant to the pressurized lubricant sump. Two types are used: a manual pump and an electric motor-driven positive-displacement pump.

Acoustical enclosures are available for installations that require low noise levels.

Special Applications

Because of the screw compressor's positive-displacement characteristic and lubricant-injected cooling, its use for high differential applications is limited only by power considerations and maximum design working pressures. Therefore, it is being used for many special applications because of reasonable compressor cost and no surge characteristic. Some of the fastest growing areas are:

- Heat-recovery installations
- Air-cooled
  Split packages with field-installed interconnecting piping
  Factory-built rooftop packages
- Low-temperature brine chillers for process cooling
- Ice rink chillers
- Power transmission line lubricant cooling

High temperature compressor and condensing units are being used increasingly for air conditioning because of the higher efficiency of direct air-to-refrigerant heat exchange resulting in higher evaporating temperatures. Many of these installations have air-cooled condensers.

MAINTENANCE

Manufacturer's maintenance instructions should be followed, especially because some items differ substantially from reciprocating or centrifugal units. Water-cooled condensers must be cleaned of scale periodically (see the General Maintenance section). If the condenser water is also used for the lubricant cooler, this should be considered in the treatment program. Lubricant coolers operate at higher temperatures and lower flows than condensers, so it is possible that the lubricant cooler may have to be serviced more often than the condenser.

Because large lubricant flows are a part of the screw compressor system, the lubricant filter pressure drop should be monitored carefully and the elements changed periodically. This is particularly important in the first month or so after startup of any factory-built package and is essential on field-erected systems. Since the lubricant and refrigeration systems merge at the compressor, much of the loose dirt and fine contaminants in the system eventually find their way to the lubricant sump, where they are removed by the lubricant filter. Similarly, the filter-drier cartridges should be monitored for pressure drop and moisture during initial start and regularly thereafter. Generally, if a system reaches an acceptable dryness level, it stays that way unless it is opened.

It is good practice to check the lubricant for acidity periodically, using commercially available acid test kits. The lubricant does not need to be changed unless it is contaminated by water, acid, or metallic particles. Also, a refrigerant sample should be analyzed yearly to determine its condition.

Certain procedures that should be followed on a yearly basis or during a regularly scheduled shutdown. These include checking and calibrating all operation and safety controls, tightening all electrical connections, inspecting power contacts in starters, dielectric checking of hermetic and open motors, and checking the alignment of open motors.

Leak testing of the unit should be performed regularly. A water-cooled package used for summer cooling should be leak tested annually. A flooded unit with proportionately more refrigerant in it, used for year-round cooling, should be tested every four to six months. A process air-cooled chiller designed for year-round operation 24 hours per day should be checked every one to three months.

Based on 6000 operating hours per year and depending on the above considerations, a typical inspection or changeout timetable is as follows:

Shaft seals	1.5 to 4 yr	Inspect
Hydraulic cylinder seals	1.5 to 4 yr	Replace
Thrust bearings	4 to 6 yr	Check preload via shaft end play every 6 months and replace as required
Shaft bearings	7 to 10 yr	Inspect

CHAPTER 44

COMPONENT BALANCING IN REFRIGERATION SYSTEMS

Refrigeration System 44.1
Components 44.1
Selecting Design Balance Points 44.2
Energy and Mass Balances 44.2
System Performance 44.4

THIS chapter describes the methods and components used in balancing a primary refrigeration system. A refrigerant is a fluid used for heat transfer in a refrigeration system. The fluid absorbs heat at a low temperature and pressure and transfers heat at a higher temperature and pressure. The heat transfer process can involve either a complete or partial change of state in the case of a primary refrigerant. Energy transfer is a function of the heat transfer coefficients; the temperature differences; and the amount, type, and configuration of the heat transfer surface and, hence, the heat flux on either side of the heat transfer device.

REFRIGERATION SYSTEM

A typical basic direct-expansion refrigeration system includes an evaporator, which vaporizes the entering refrigerant as it absorbs heat, increasing the refrigerant's heat content or enthalpy. A compressor pulls the vapor from the evaporator through suction piping and compresses the refrigerant gas to a higher pressure and temperature. The refrigerant gas then flows through the discharge piping to a condenser, where it is condensed by rejecting its heat to a coolant (e.g., other refrigerants, air, water, or air/water spray). The condensed liquid is supplied to a device that reduces pressure, cools the liquid by flashing vapor, and meters the flow. The cooled liquid is supplied back to the evaporator. For more information on the basic refrigeration cycle, see Chapter 1 of the 1997 ASHRAE Handbook—Fundamentals.

Compression of the gas theoretically follows a line of constant entropy. In practice, adiabatic compression cannot occur due to friction and other inefficiencies of the compressor. Therefore, the actual compression line deviates slightly from the theoretical. Power to the shaft of the compressor is added to the refrigerant, and compression increases the pressure, temperature, and enthalpy of the refrigerant.

In applications involving a large compression ratio (such as low-temperature freezing, multitemperature applications, etc.), multiple compressors in series are used to completely compress the refrigerant gas. In multistage systems, interstage desuperheating of the lower stage compressor's discharge gas protects the high-stage compressor. Liquid refrigerant can also be subcooled at this interstage condition and delivered to the evaporator for improved efficiencies.

An intermediate-temperature condenser can serve as a cascading device. A low-temperature, high-pressure refrigerant condenses on one side of the cascade condenser surface by giving up heat to a low-pressure refrigerant that is boiling on the other side of the surface. The vapor of this boiling refrigerant transfers the energy to the next compressor (or compressors); heat of compression is added and, at a higher pressure, the last refrigerant is condensed on the final condenser surface.

Heat is rejected to air, water, or water spray. The saturation temperatures of evaporation and condensation throughout the system fix the terminal pressures that the single or multiple compressors must operate against.

Generally, the smallest differential between the saturated evaporator and the saturated condensing temperatures results in the lowest energy requirement for compression. Liquid refrigerant cooling or subcooling should be used where possible to improve efficiencies and minimize energy consumption. Liquid refrigerant cooling or subcooling should be used to minimize energy consumption.

Where intermediate pressures have not been specifically set for the system operation, the compressors automatically balance at their respective suction and discharge pressures as a function of their relative displacements and compression efficiencies depending on the load and temperature requirements. This chapter covers the technique used to determine the balance points for a typical brine chiller, but the theory can be expanded to apply to single- and two-stage systems with a different types of evaporators, compressors, and condensers.

COMPONENTS

Evaporators may have flooded, direct-expansion, or liquid overfeed cooling coils with or without fins. Evaporators are used to cool air, gases, liquids, and solids; condense volatile substances; and freeze products.

Ice builder evaporators accumulate ice to store cooling energy for later use. Embossed plate evaporators are available (1) to cool a falling film of liquid; (2) to cool, condense, and/or freeze out volatile substances from a fluid stream; or (3) to cool or freeze a product by direct contact. Brazed- and welded-plate fluid chillers can be used to improve efficiencies and reduce refrigerant charge.

Ice, wax, or food products are frozen and scraped from some freezer surfaces. Electronic circuit boards, mechanical products, or food products (where permitted) are being flash-cooled by direct immersion in boiling refrigerants. These are some of the diverse applications demanding innovative configurations and materials that perform the function of an evaporator.

Compressors take the form of positive displacement, reciprocating piston, rotary vane, scroll, single and double dry and lubricant flooded screw devices, and single-stage or multistage centrifugals. They can be operated in series or in parallel with each other, in which case special controls may be required.

The drivers for the compressors can be direct hermetic, semihermetic, or open with mechanical seals on the compressor. In hermetic and semihermetic drives, the motor inefficiencies are added to the refrigerant as heat. Open compressors are driven with electric motors, fuel-powered reciprocating engines, or steam or gas turbines. Intermediate gears, belts, and clutch drives may be included in the drive.

Cascade condensers are used with high-pressure, low-temperature refrigerants (such as R-23) on the bottom cycle, and high-temperature refrigerants (such as R-22, azeotropes, and refrigerant blends or zeotropes) on the upper cycle. Cascade condensers are manufactured in many forms, including shell-and-tube, embossed plate, submerged, direct-expansion double coils, and brazed or welded plate heat exchangers. The high-pressure refrigerant from the compressor(s) on the lower cycle condenses at a given intermediate temperature. A separate, lower-pressure refrigerant evaporates on the other

---

The preparation of this chapter is assigned to TC 10.1, Custom Engineered Refrigeration Systems.

side of the surface at a somewhat lower temperature. The vapor formed from the second refrigerant is compressed by the higher cycle compressor(s) until it can be condensed at an elevated temperature.

Desuperheating of suction gas at intermediate pressures where multistage compressors balance is essential to reduce discharge temperatures of the upper-stage compressor. Desuperheating also helps reduce oil carryover and reduces energy requirements. Subcooling improves the net refrigeration effect of the refrigerant supplied to the next lower temperature evaporator and reduces the system energy requirements. The total heat is then rejected to a condenser.

Subcoolers can be of shell-and-tube, shell-and-coil, welded-plate, or tube-in-tube construction. Friction losses reduce the liquid pressure that feeds refrigerant to an evaporator. Subcoolers are used to improve the efficiency of the system and to prevent refrigerant liquid from flashing due to pressure loss caused by friction and the vertical rise in lines. Refrigerant blends (zeotropes) can take advantage of temperature glide on the evaporator side with a direct-expansion-in-tube serpentine or coil configuration. In this case, the temperature glide from the bubble point to the dew point promotes increased efficiency and lower surface requirements for the subcooler. A flooded shell for the evaporating refrigerant requires use of only the higher dew-point temperature.

Lubricant coolers remove friction heat and some of the superheat of compression. Heat is usually removed by water, air, or a direct-expansion refrigerant.

Condensers that reject heat from the refrigeration system are available in many standard forms. These include water or brine cooled shell-and-tube, shell-and-coil, plate-and-frame, or tube-in-tube condensers; water cascading over or sprayed over plate or coil serpentine models; and air-cooled, fin-coil condensers. Special heat pump condensers are available in other forms such as tube-in-earth and submerged tube bundle, or as serpentine and cylindrical coil condensers that heat baths of boiling or single-phase fluids.

SELECTING DESIGN BALANCE POINTS

The refrigeration load at each designated evaporator pressure, the refrigerant properties, the liquid refrigerant temperature feeding each evaporator, and evaporator design determine the required flow rate of refrigerant in a system. The additional flow rates of refrigerant that provide refrigerant liquid cooling, desuperheating, and compressor lubricant cooling, where used, depend on the established liquid refrigerant temperatures and intermediate pressures.

For a given refrigerant and flow rate, the suction line pressure drop, suction gas temperature, pressure ratio and displacement, and volumetric efficiency determine the required size and speed of rotation for a positive displacement compressor. At low flow rates, particularly at very low temperatures and in long suction lines, heat gain through insulation can significantly raise the suction temperature. Also, at low flow rates a large, warm compressor casing and suction plenum can further heat the refrigerant before it is compressed. These heat gains increase the required displacement of a compressor. The compressor manufacturer must recommend the superheating factors to apply. The final suction gas temperature due to suction line heating is calculated by an iterative process.

Another concern is that more energy is required to compress the refrigerant to a given condenser pressure as the suction gas gains more superheat. This can be seen by examining a pressure-enthalpy diagram for a given refrigerant such as R-22, which is shown in Figure 4 in Chapter 19 of the 1997 *ASHRAE Handbook—Fundamentals*. As suction superheat increases along the horizontal axis, the slopes of the constant entropy lines of compression decrease. This means that a greater enthalpy change must occur to produce a given pressure rise. For a given flow, then, the power required for compression is increased. With centrifugal compressors, pumping capacity is related to wheel diameter and speed, as well as to volumetric flow and acoustic velocity of the refrigerant at the suction entrance. If the thermodynamic pressure requirement becomes too great for a given speed and volumetric flow, the centrifugal compressor will experience periodic backflow and surging.

Figure 1 shows an example of a system of curves needed to represent the maximum refrigeration capacities for a brine chilling plant. The example shows only one type of positive displacement compressor using a water-cooled condenser in a single-stage system operating at a steady-state condition. The figure is a graphical method of expressing the first law of thermodynamics with an energy balance applied to a refrigeration system.

One set of nearly parallel curves (A) represents the capacity of the cooler at various brine temperatures versus saturated suction temperature (a pressure condition) at the compressor, allowing for suction line pressure drops. The (B) set of curves represents the capacities of the compressor as the saturated suction temperature is varied and the saturated condenser temperature (a pressure condition) is varied. The (C) set of curves represents the heat transferred to the condenser by the compressor. It is calculated by adding the heat input at the evaporator to the energy imparted to the refrigerant by the compressor. The (D) set of curves represents the condenser performance at various saturated condenser temperatures as the inlet temperature of a fixed quantity of cooling water is varied.

The (E) set of curves represents the combined compressor and condenser performance as a "condensing unit" at various saturated suction temperatures for various cooling water temperatures. These curves were cross-plotted from the (C) and (D) curves back to the set of brine cooler curves as indicated by the dotted construction lines for the 27 and 33°C cooling water temperatures. Another set of construction lines (not shown) would be used for the 30°C cooling water. The number of construction lines used can be increased as necessary to adequately define curvature (usually no more than three per condensing-unit performance line).

The intersections of curves (A) and (E) represent the maximum capacities for the entire system at those conditions. For example, these curves show that the system develops 528 kW when cooling the brine to 7°C at 2.4°C (saturated) suction and using 27°C cooling water. At 33°C cooling water, the capacity drops to 473 kW if the required brine temperature is 6°C and the required saturated suction temperature is 1.7°C. The corresponding saturated condensing temperature for 6°C brine with an accompanying suction temperature of 2.4° and using 27°C water is graphically projected on the brine cooler line with a capacity of 528 kW to meet a newly constructed 2.4°C saturated suction temperature line (parallel to the 1°C and 3°C lines). At this junction, draw a horizontal line to intersect the vertical saturated condensing temperature scale at 34.2°C. The condenser heat rejection is apparent from the (C) curves at a given balance point.

The equation at the bottom of Figure 1 may be used to determine the shaft power required at the compressor for any given balance point. A sixth set of curves could be drawn to indicate the power requirement as a function of capacity versus saturated suction and saturated condensing temperatures.

The same procedure can be repeated to calculate the performance of cascade systems. The rejected heat at the cascade condenser would be treated as the chiller load in making a cross-plot of the upper cycle, high-temperature refrigeration system.

For cooling air at the evaporator(s) and for condenser heat rejection to ambient air or evaporative condensers, the same procedures would be used. The performance of coils and expansion devices such as thermostatic expansion valves may also be graphed, once the basic concept of heat and mechanical energy input equivalent combinations is recognized. Chapter 1 of the 1997 *ASHRAE Handbook—Fundamentals* has further information.

This graphical plotting method finds the natural balance points of compressors operating at their maximum capacities. For multiple stage loads at several specific operating temperatures, the usual way of controlling compressor capacities is with a suction pressure

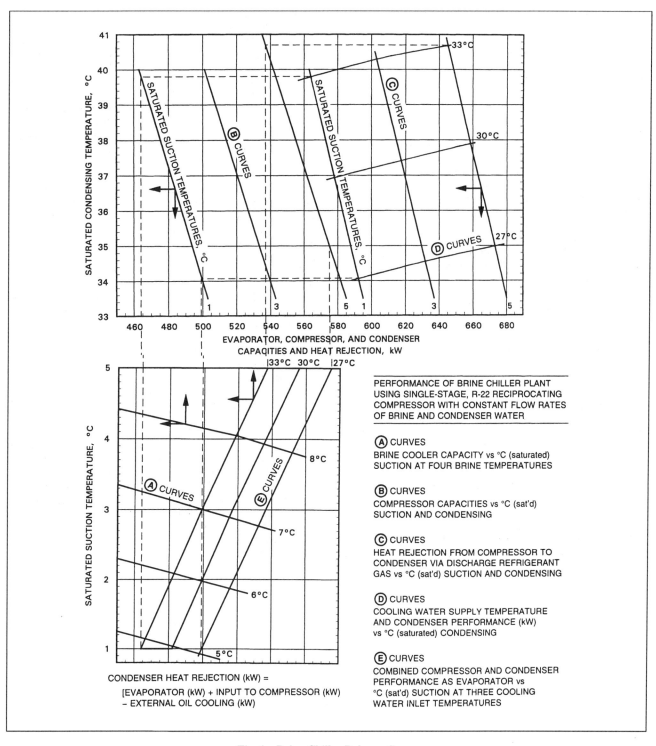

Fig. 1 Brine Chiller Balance Curve

control and compressor capacity control device. This control then accommodates any mismatch in the pumping capabilities of multistage compressors instead of allowing each compressor to find its natural balance point.

Computer programs could be developed to determine balance points of complex systems. However, because applications, components, and piping arrangements are so diverse, many designers use available capacity performance data from vendors and plot the balance points for chosen components. Individual computer programs may be available for specific components, which will speed the process.

ENERGY AND MASS BALANCES

A systematic, point-to-point flow analysis of the system (including the piping) is essential in accounting for pressure drops and heat gains, particularly in long suction lines. Air-cooled condensers, in particular, can have large pressure drops, which must be included in the analysis to estimate a realistic balance. Making a flow diagram of the system with designated pressures and temperatures, loads, enthalpies, flow rates, and energy requirements helps in identifying all important factors and components.

An overall energy and mass balance for the system is also essential to avoid mistakes. The overall system represented by the complete flow diagram should be enclosed by a dotted line envelope. Any energy inputs to or outputs from the system that directly affect the heat content of the refrigerant itself should cross the dotted envelope line and must enter the energy balance equations. Accurate estimates of the ambient heat gains through insulation and heat losses from discharge lines where they are significant improves the comprehensiveness of the energy balance and accuracy of equipment selections.

Cascade condenser loads and subcooler or desuperheating loads carried by a refrigerant are internal to the system and thus do not enter into the overall energy balance. The total energy entering the system equals the total energy leaving the system. If calculations do not show an energy balance within reasonable tolerances for the accuracy of data used, then an omission occurred or a mistake was made and should be corrected.

The dotted envelope technique can be applied to any section of the system, but all energy transmissions must be included in the equations including the enthalpies and mass flow rates of streams that cross the dotted line.

SYSTEM PERFORMANCE

Rarely are sufficient sensors and instrumentation devices available, nor are conditions proper at a given job site to permit the calculation of a comprehensive, accurate energy balance for an operating system. Water-cooled condensers and oil coolers for heat rejection and the use of electric motor drives, where motor efficiency and power factor curves are available, offer the best hope for estimating the actual performance of the individual components in a system. The evaporator heat loads can be derived from the measured heat rejection and derived mechanical or measured electrical energy inputs. A comprehensive flow diagram assists in a field survey.

Various coolant flow detection devices are available for direct measurement inside a pipe and for measurement from outside the pipe with variable degrees of accuracy. Sometimes flow rates may be estimated by simply weighing or measuring an accumulation of coolant over a brief time interval.

Temperature and pressure measurement devices should be calibrated and be of sufficient accuracy. Digital calibrated scanning devices for comprehensive simultaneous readings are best. Electrical power meters are not always available, so voltage and current at each leg of a motor power connection must be measured. Voltage drops for long power leads must be calculated when the voltage measurement points are far removed from the motor. The motor load versus efficiency and power factor curves must be used to determine motor output to the system.

Gears and belt or chain drives have friction and windage power losses that must be included in any meaningful analysis.

Stack gas flows and enthalpies for engine or gas turbine exhausts as well as air inputs and speeds must be included. In this case, heavy reliance must be placed on the performance curves issued by the vendor to estimate the energy input to the system.

Calculating steam turbine performance requires measurements of turbine speed, steam pressures and temperatures, and condensate mass flow coupled with confidence that the vendor's performance curves truly represent the current mechanical condition. Plant personnel normally experience difficulty in obtaining operating data at specified performance values.

Heat rejection from air-cooled and evaporative condensers or coolers is extremely difficult to measure accurately because of changing ambient temperatures and the extent and scope of airflow measurements required. Often, one of the most important issues is the wide variation or cycling of process flows, process temperatures, and product refrigeration loads. Hot gas false loading and compressor continuous capacity modulations complicate any attempt to make a meaningful analysis.

The prediction and measurement of the performance of systems using refrigerant blends (zeotropes) are especially challenging because of the temperature variations between bubble points and dew points.

Nevertheless, ideal conditions of nearly steady-state loads and flows with a minimum of cycling sometimes occur frequently enough to permit a reasonable analysis. Computer controlled systems can provide the necessary data for a more accurate system analysis. Several sets of nearly simultaneous data at all points over a short time enhance the accuracy of any calculation concerning the performance of a given system. In all cases, the proper purging of condensers and elimination of excessive lubricant contamination of the refrigerant at the evaporators are essential to determine system capabilities accurately.

CHAPTER 45

REFRIGERANT-CONTROL DEVICES

CONTROL SWITCHES	45.1
Pressure Control Switches	45.1
Pressure Transducers	45.2
Temperature Control Switches	45.2
Fluid Flow-Sensing Switches	45.2
Differential Control Switches	45.2
Float Switches	45.3
CONTROL VALVES	45.4
Thermostatic Expansion Valves	45.4
Electric Expansion Valves	45.9
Constant Pressure Expansion Valves	45.9
Evaporator Pressure and Temperature Regulators	45.10
Suction Pressure Regulators	45.12
Condenser Pressure Regulators	45.13
High-Side Float Valves	45.13
Low-Side Float Valves	45.14
Solenoid Valves	45.14
Condensing Water Regulators	45.18
Check Valves	45.19
Relief Devices	45.19
DISCHARGE-LINE LUBRICANT SEPARATORS	45.21
CAPILLARY TUBES	45.21
SHORT TUBE RESTRICTORS	45.27

THE control of refrigerant flow is essential in any refrigeration system. The section on Control Switches details control switches, including (1) pressure control switches, (2) pressure transducers, (3) temperature control switches, (4) fluid flow-sensing switches, (5) differential control switches, and (6) float switches.

The section on Control Valves addresses the operation, selection, and application of control valves, including (1) thermostatic expansion valves, (2) electric expansion valves, (3) constant pressure expansion valves, (4) evaporator pressure and temperature regulators, (5) suction pressure regulators, (6) condenser pressure regulators, (7) high-side float valves, (8) low-side float valves, (9) solenoid valves, (10) condensing water regulators, (11) check valves, and (12) relief devices.

The third section discusses discharge-line lubricant separators; the fourth, capillary tubes; and the fifth, short tube restrictors. Most examples, references, and capacity data in this chapter refer to the more common refrigerants. For further information on automatic control, see Chapter 42 of the 1995 *ASHRAE Handbook—Applications*.

CONTROL SWITCHES

The **control switch** operates one or more sets of electrical contacts, which are used, for example, to open or close water or refrigerant valves, engage and disengage compressor clutch coils and relays, active timers, and thermostats. Control switches respond to a variety of physical changes such as pressure, temperature, liquid level, flow velocity, and proximity.

Pressure- and temperature-responsive controls have one or more power elements, which may use bellows, diaphragms, snap disks, or bourdon tubes to produce the force needed to operate the mechanism. Level-responsive controls may use floats, mercury balance tubes, or electronic probes to operate (directly or indirectly) one or more sets of electrical contacts.

Refrigeration controls may be categorized into three basic groups: (1) operating, (2) primary, and (3) limit. Operating controls such as thermostats, turn systems on and off. Primary controls, such as floats, provide safe continuous operation. Limit controls, such as the high-pressure cut-out switch, protect a refrigeration system from unsafe operation.

The preparation of this chapter is assigned to TC 8.8, Refrigerant System Controls and Accessories.

PRESSURE CONTROL SWITCHES

Commercial Applications

The refrigerant pressure is applied directly to the element, which moves against a spring that can be adjusted to control any operation at the desired pressure. If the control is to operate in the subatmospheric (or vacuum) range, the bellows or diaphragm force is sometimes reversed to act in the same direction as the adjusting spring. To counteract this, an additional spring may be needed to overcome the reversed bellows force; consequently, the force of both the bellows and the vacuum spring oppose the adjusting spring. However, the controls are sometimes built without any spring and use the spring force of the bellows or diaphragm in place of the adjusting spring.

The force available for doing work in this control switch depends on the pressure in the system and on the area of the bellows or diaphragm. With proper area, a sufficient force can be produced to operate heavy switches, water valves, or refrigerant valves. In heavy-duty controls, the minimum differential is large because of the large size of the bellows and the opposing spring system.

Automotive Applications

Pressure control switches in most automotive refrigeration systems are used primarily to control electrical engagement of the compressor clutch or condenser fan. These switches may be installed in various locations (e.g., mounted on a line, the accumulator/dehydrator or the receiver/dehydrator, or in the compressor body). The following is a list of types of pressure control switches with their corresponding functions:

Type	Function
High-pressure cut-out (HPCO)	Disengages compressor clutch when excessive pressure occurs
High-side low-pressure (HSLP)	Disengages compressor clutch under low ambient and/or low-pressure conditions
High-side fan cycling (HSFC)	Cycles condenser fan on and off to provide proper condenser pressure
Low-side low-pressure (LSLP)	Disengages compressor clutch when low charge or system blockage occurs
Low-side clutch cycling (LSCC)	Cycles compressor clutch on and off to provide proper evaporator pressure

The automotive pressure control switch incorporates one or more steel snap disks, which provide positive pressure at the electrical contacts. Use of the snap action disk construction also assures consistent differential pressure between on and off settings. (See Figure 1.) One very important benefit of this construction over the

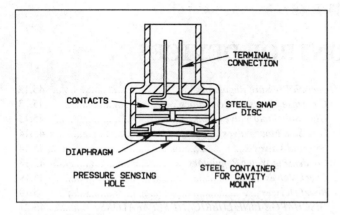

Fig. 1 Pressure Control Switch (Compressor Mounted)

earlier direct contact or creep switch is the substantial reduction in electrical contact bounce or flutter, which can be very damaging to compressor clutch assemblies, fan relays, and electronic control modules (ECM).

PRESSURE TRANSDUCERS

Pressure transducers are most often used as a substitute for one or more pressure switches. The pressure transducer is typically an analog device that produces a continuous low voltage output signal proportional to the pressure applied. It requires a three pin connection: input voltage (constant reference), output voltage (variable), and ground.

The primary advantages of the pressure transducer in an automotive air-conditioning application include the following:

- **Enhanced engine load management.** Based on a correlation between refrigerant pressure and compressor torque, the vehicle engine controller utilizes the transducer signal to vary the idle air flow as required when the compressor engages and disengages. This results in lower engine idle speeds, more economic fuel consumption and elimination of the "power drain" that is experienced when the compressor clutch is engaged.
- **Component/cost reduction.** A single transducer output can be used in place of multiple switches for sensing such parameters as low and high pressure cut off pressures and designated switch points for condenser fan control. This integration increases the accuracy of the overall system while reducing overall system control costs.
- **Performance.** A properly designed transducer provides a highly reliable operation with low hystersis (<1% V supply) and minimal drift over life (<2.5% V supply).
- **Noise/vibration/harshness (NVH) reduction.** With multi-speed fans the pressure transducer can be used to minimize NVH caused from on/off switching of the cooling fan.

Three key points to consider in deciding if transducer integration would be practical for a given application are as follows: (1) the transducer is most effective when it performs multiple functions and continuously monitors air-conditioning refrigerant pressure; (2) transducers, unlike switches, require supporting components, amplifiers, relays, and control modules; and (3) separate transducers are needed for low and high side applications due to different transducer sensitivity and pressure range requirements.

TEMPERATURE CONTROL SWITCHES

An **indirect temperature control switch** is a pressure control switch in which the pressure-responsive element is replaced by a temperature-responsive element. The exact temperature-pressure or temperature-volume relationship of the fluid used in the element

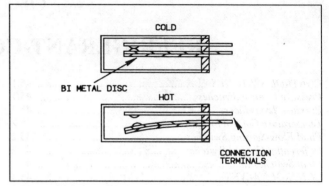

Fig. 2 Direct Temperature Control Switch

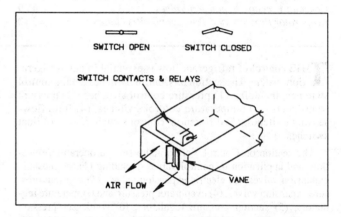

Fig. 3 Air Flow Switch

allows the temperature of the bulb to control the switch accurately. Operation of the control switch results from these changes in pressure or volume.

A **direct temperature control switch** generally contains bimetallic disks that activate electrical contacts when the temperature increases or decreases. As the temperature of a bimetallic disk or blade in the switch increases or decreases, the attached electrical contacts engage or disengage. This switch is often used for thermal limit control because a simple on/off control is not precise enough for most primary refrigerant control. The same principle is used in thermostat construction in which a semicoiled strip of bimetallic material is used to activate electrical contacts. (See Figure 2.)

FLUID FLOW-SENSING SWITCHES

Both airflow and liquid flow-sensing switches are available. Airflow switches (Figure 3) are used where it is important to detect air movement, for example, in refrigeration air ducts. This switch may be used to shut down fan blower motors or the total system if the airflow stops. The spring-loaded vanes of this switch are installed in a regulated, high-velocity air stream. The pressure of the air perpendicular causes the vane to deflect and make electrical contact.

Liquid flow switches work in a similar manner (Figure 4). Water or refrigerant flow applied perpendicular to the vane causes it to deflect and make electrical contact. Liquid flow switches are often used to indicate sufficient water or refrigerant flow through chillers or condensers.

DIFFERENTIAL CONTROL SWITCHES

Differential control switches maintain a given difference in pressure or temperature between two pipe lines or spaces. An example is the lubricant safety switch used with reciprocating compressors with forced-feed pressure lubrication. These controls have two

Refrigerant-Control Devices

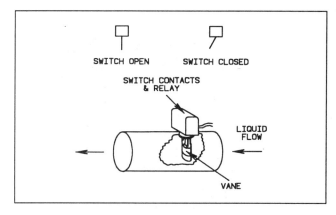

Fig. 4 Liquid Flow Switch

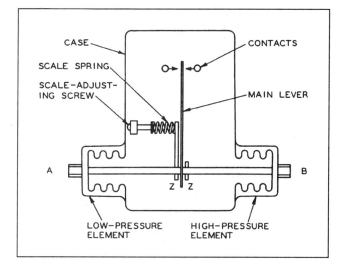

Fig. 5 Differential Pressure Control Switch

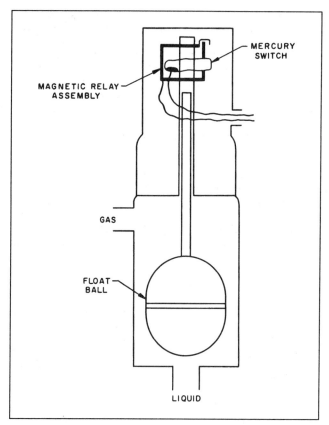

Fig. 6 Magnetic Float Switch

elements (either pressure- or temperature-sensitive) because they sense conditions in two locations. Figure 5 is a schematic of a differential pressure control switch that uses bellows as power elements. As shown, the two elements are rigidly connected by a rod so that motion of one causes motion of the other. On the connecting rod, a power takeoff operates either single-pole double-throw contacts (as shown) or valves. A compression spring permits the setting of the differential pressure at which the device operates. The sum of the forces developed by the low-pressure bellows and the spring equals the force developed by the high-pressure bellows at the control point.

Instrument differential is the difference in pressure between the low- and the high-pressure elements for which the instrument is adjusted. In the case of a temperature element, this difference is expressed in degrees. **Operating differential** is the difference in pressure or temperature required to open or close the switch contacts. It is actually the change in instrument differential from cut-in to cut-out for any setting. Operating differential can be varied by a second spring that acts in the same direction as the first and takes effect only at the cut-in or cut-out point without affecting the other spring. A second method is the adjustment of the distance between collars Z-Z on the connecting rod. The greater the distance between them, the greater the operating differential.

If a constant instrument differential is required on a temperature-sensitive differential control switch throughout a large temperature range, one element usually has a different fill than the other, if both are of the vapor-charged type. The alternative is to use liquid filled elements.

A second type of differential-temperature control uses two sensing bulbs and capillaries connected to one bellows with a liquid fill. This is known as a **constant-volume fill**, because the operating point corresponds to a constant volume of the two bulbs, capillaries, and bellows. If the two bulbs have equal volume, a rise in the temperature of one bulb requires an equivalent fall in the temperature of the other to maintain the operating point.

FLOAT SWITCHES

A **float switch** has a float that operates one or more sets of electrical contacts as the level of a liquid changes. It is connected by equalizing lines to the vessel in which the liquid level is to be maintained or indicated.

Operation and Selection

Some float switches (see Figure 6) operate from the movement of a magnetic armature located in the field of a permanent magnet. Others use solid-state circuits in which a variable signal is generated by liquid contact with a probe that replaces the float. The latter methods are adapted to remote-controlled applications and are preferred for ultralow temperature applications. Switches that have mercury-tube contacts are usually not recommended for installation in an ambient temperature lower than −32°C, because mercury freezes at approximately −39°C.

Application

The float switch can maintain or indicate the level of a liquid, operate an alarm, control the operation of a pump, or perform other functions. A float switch, solenoid liquid valve, and hand expansion valve combination can control the refrigerant level on the high- or low-pressure side of the refrigeration system in the same way that high- or low-side float valves are used. The hand expansion valve, located in the refrigerant liquid line immediately downstream of the solenoid valve, is initially adjusted to provide a refrigerant flow rate at maximum load to keep the solenoid liquid valve in the open position 80 to 90% of the time; it need not be adjusted thereafter. From the outlet side of the hand expansion valve, the refrigerant passes through a line and enters either the evaporator or the surge drum, depending on design.

When the float switch is used for low-pressure level control, precaution must be taken to provide a quiet liquid level that falls in response to an increase in evaporator load and rises with a decrease in evaporator load. The same recommendations for insulation of the body and liquid leg of the low-pressure float valve apply to the float switch when it is used for refrigerant-level control on the low-pressure side of the refrigeration system. To avoid floodback in this application, controls should be wired to prevent the opening of the solenoid liquid valve when the solenoid suction valve closes or the compressor stops.

CONTROL VALVES

Control valves are used to start, stop, direct, and modulate the flow of refrigerant to satisfy system requirements in accordance with load requirements. To ensure satisfactory performance, valves should be protected from foreign material, excessive moisture, and corrosion by properly sized strainers, filters, and/or filter-driers. Either a diaphragm or a bellows are used for the various types of refrigerant flow-control valves.

THERMOSTATIC EXPANSION VALVES

The thermostatic expansion valve controls the flow of liquid refrigerant entering the evaporator in response to the superheat of the gas leaving it. It functions to keep the evaporator active without permitting liquid to return through the suction line to the compressor. This is done by controlling the mass flow of refrigerant entering the evaporator so it equals the rate at which it can be completely vaporized in the evaporator by the absorption of heat. Because this valve is operated by superheat and responds to changes in superheat, a portion of the evaporator must be used to superheat the refrigerant gas.

Unlike the constant pressure valve, the thermostatic expansion valve is not limited to constant load applications. It is used for controlling refrigerant flow to all types of direct-expansion evaporators in air-conditioning, and commercial, low-temperature, and ultralow-temperature refrigeration applications.

Operation

A schematic cross section of the typical thermostatic expansion valve, with the principal components identified, is shown in Figure 7. The following forces govern thermostatic expansion valve operation:

P_1 = the pressure of the thermostatic element (a function of the bulb's charge and the bulb temperature), which is applied to the top of the diaphragm and acts to open the valve

P_2 = the evaporator pressure, which is applied under the diaphragm through the equalizer passage and acts in a closing direction

P_3 = the pressure equivalent of the superheat spring force, which is applied underneath the diaphragm and is also a closing force

At any constant operating condition, these forces are balanced and $P_1 = P_2 + P_3$.

An additional force, which is small and not considered fundamental, arises from the unbalanced pressure across the valve port. To a degree, it can affect thermostatic expansion valve operation.

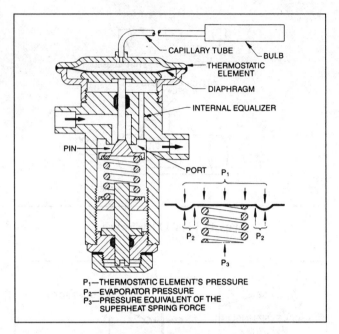

Fig. 7 Typical Thermostatic Expansion Valve

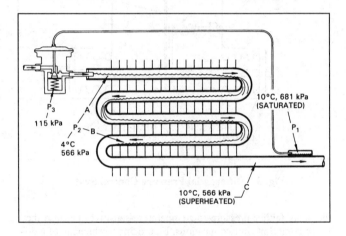

Fig. 8 Thermostatic Expansion Valve Controlling Flow of Liquid R-22 Entering Evaporator, Assuming R-22 Charge in Bulb

For the configuration shown in Figure 7, the force resulting from port imbalance is the product of the pressure drop across the port and the difference in area of the port and the pin; it is an opening force. In other designs, depending on the direction of flow through the valve, the port imbalance may result in a closing force.

The principal effect of port imbalance is on the stability of the valve control. As with any modulating control, if the ratio of the diaphragm area to the port is kept large, the unbalanced port effect is minor. However, depending on this ratio or system operating conditions, valves are made with balanced port construction.

Figure 8 shows an evaporator operating with R-22 at a saturation temperature of 4°C [566 kPa (gage)]. Liquid refrigerant enters the expansion valve, is reduced in pressure and temperature at the valve port, and enters the evaporator at Point A as a mixture of saturated liquid and vapor. As flow continues through the evaporator, more of the refrigerant is evaporated. Assuming there is no pressure drop, the refrigerant temperature remains at 4°C until the liquid is entirely evaporated at Point B. From this point, additional heat absorption increases the temperature and superheats the refrigerant gas, while the pressure remains constant at 566 kPa (gage), until, at Point C

Refrigerant-Control Devices

(the outlet of the evaporator), the refrigerant gas temperature is 10°C. At this point, the superheat is 6 K (10 to 4°C).

An increase in the heat load on the evaporator increases the temperature of the refrigerant gas leaving the evaporator. The bulb of the valve senses this increase and the thermostatic charge pressure P_1 increases and causes the valve to open wider. The increased flow results in a higher evaporator pressure P_2 and a balanced control point is again established. Conversely, a decrease in the heat load on the evaporator decreases the temperature of the refrigerant gas leaving the evaporator and causes the thermostatic expansion valve to start closing.

The new control point, following an increase in valve opening, is at a slightly higher operating superheat because of the spring rate of the diaphragm and superheat spring. Conversely, a decrease in load results in an operating superheat slightly lower than the original control point.

These superheat changes in response to load changes are illustrated by the gradient curve of Figure 9. Superheat at no load A, **static superheat**, ensures sufficient spring force to keep the valve closed during equipment shutdown. An increase in valve capacity or load is approximately proportional to superheat until the valve is open fully. The **opening superheat**, represented by the distance AB, may be defined as the superheat increase required to open the valve to match the load. **Operating superheat** is the sum of static superheat and opening superheat.

Capacity

The **factory superheat setting** (static superheat setting) of thermostatic expansion valves is made when the valve starts to open. Valve manufacturers establish capacity ratings on the basis of opening superheat from 2 to 4 K, depending on valve design, valve size, and application. Full-open capacities usually exceed rated capacities by 10 to 40% to allow a reserve, represented by the distance BC in Figure 9, for manufacturing tolerances and application contingencies.

A valve should not be selected on the basis of its reserve capacity, which is available only at higher operating superheat. The added superheat may have an adverse effect on performance. Because valve gradients used for rating purposes are selected to produce

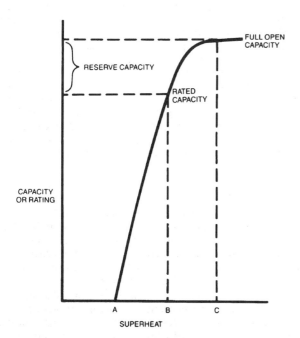

Fig. 9 Typical Gradient Curve for Thermostatic Expansion Valves

optimum modulation for a given valve design, manufacturers' recommendations should be followed.

Thermostatic expansion valve capacities are normally published for various evaporator temperatures and valve pressure drops. (See ASHRAE *Standard* 17 and ARI *Standard* 750 for testing and rating methods.) Nominal capacities apply at 4°C evaporator temperature. Capacities are reduced at lower evaporator temperatures. These reductions in capacity are the result of the change in the refrigerant pressure-temperature relationship at lower temperatures. For example, if R-22 is used, the change in saturated pressure between 4 and 7°C is 65.6 kPa, whereas between −30 and −27°C the change is 21.8 kPa. Although the valve responds to pressure changes, capacities are based on superheat gradients. Thus, the valve opening and, consequently, the valve capacity, is less for a given superheat change at lower evaporator temperatures.

Pressure drop across the valve port is always the net pressure drop available at the valve, rather than the difference between compressor discharge and compressor suction pressures. Allowances must be made for the following:

1. Pressure drop through condenser, receiver, liquid lines, fittings, and liquid line accessories, such as filters, driers, solenoid valves, etc.
2. Static pressure in a vertical liquid line. If the thermostatic expansion valve is at a higher level than the receiver, there will be a pressure loss in the liquid line because of the static pressure of liquid.
3. Distributor pressure drop.
4. Evaporator pressure drop.
5. Pressure drop through suction line and accessories, such as evaporator pressure regulators, solenoid valves, accumulators etc.

Variations in valve capacity related to changes in system conditions are approximately proportional to the following relationship:

$$q \approx C\sqrt{\rho \Delta p}\,(h_g - h_f) \tag{1}$$

where

q = heat flow
C = constant related to thermostatic expansion valve design
ρ = entering liquid density
Δp = valve pressure difference
h_g = enthalpy of vapor exiting evaporator
h_f = enthalpy of liquid entering the thermostatic expansion valve

Thermostatic expansion valve ratings are based on vapor-free liquid entering the valve. If flash gas is present in the entering liquid, the valve capacity is reduced substantially because the gas must be handled along with the liquid. Flashing of the liquid refrigerant may be caused by pressure drop in the liquid line, the filter-drier, the vertical lift, or a combination of these. If the refrigerant subcooling at the receiver outlet is not adequate to prevent the formation of flash gas, additional subcooling means must be used to remove it. Liquid-to-suction heat exchange provides a moderate degree of subcooling, but for extreme requirements, a separate liquid-cooling coil may be necessary.

Thermostatic Charges

Each type of thermostatic charge has certain advantages and limitations. The principal types of thermostatic charges and their characteristics are described below.

Gas Charge. Conventional gas charges are limited liquid charges that use the same refrigerant in the thermostatic element that is used in the refrigeration system. The amount of charge is such that, at a predetermined temperature, all of the liquid has vaporized and any temperature increase above this point results in virtually no increase in element pressure. Figure 10 shows the pressure-temperature relationship of the R-22 gas charge in the

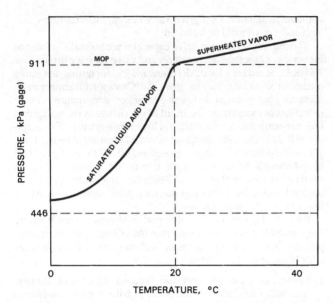

Fig. 10 Pressure-Temperature Relationship of R-22 Gas Charge in Thermostatic Element

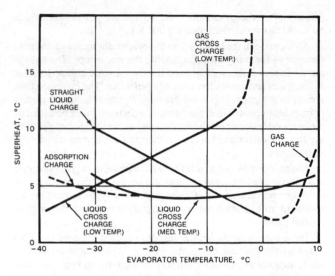

Fig. 11 Typical Superheat Characteristics of Common Thermostatic Charges

thermostatic element. Because of the characteristic pressure-limiting feature of its thermostatic element, the gas-charged valve can provide compressor motor overload protection on some systems by limiting the maximum operating suction pressure (MOP). It also helps prevent **floodback** (the return of refrigerant liquid to the compressor through the suction line) on starting. Increasing the superheat setting lowers the maximum operating suction pressure; decreasing the superheat setting raises it because the superheat spring, together with the evaporator pressure, act to balance the element pressure through the diaphragm.

Gas-charged valves must be carefully applied to avoid loss of control from the bulb. If either the diaphragm chamber or the capillary tube becomes colder than the bulb, the small amount of charge in the bulb condenses at the coldest point. This results in the valve throttling or closing as detailed in the section on Application.

Liquid Charge. Straight liquid charges use the same refrigerant in the thermostatic element that is used in the refrigeration system. The volumes of the bulb, capillary tubing, and diaphragm chamber are so proportioned that the bulb contains some liquid under all temperature conditions. Therefore, the bulb always controls the valve operation, even with a colder diaphragm chamber or capillary tube.

The characteristics of the straight liquid charge (see Figure 11) result in an increase in operating superheat as the evaporator temperature decreases. This usually limits the use of the straight liquid charge to moderately high evaporator temperatures. The valve setting required for a reasonable operating superheat at a low evaporator temperature may cause floodback during cooling from normal ambient temperatures.

Liquid Cross Charge. Liquid cross charges, unlike the conventional liquid charges, use a liquid in the thermostatic element that is different than the refrigerant in the system. Cross charges have flatter pressure-temperature curves than the system refrigerants with which they are used. Consequently, their superheat characteristics differ considerably from those of the straight liquid or gas charges.

Cross charges in the commercial temperature range generally have superheat characteristics that are nearly constant or that deviate only moderately through the evaporator temperature range. This charge, also illustrated in Figure 11, is generally used in the evaporator temperature range of 4 to −20°C or slightly below.

For evaporator temperatures substantially below −20°C, a more extreme cross charge may be used. At high evaporator temperatures, the valve controls at a high superheat. As the evaporator temperature falls to the normal operating range, the operating superheat also falls to normal. This characteristic prevents floodback on starting, reduces the load on the compressor motor after start-up, and permits a rapid pulldown of suction pressure. To avoid floodback, valves with this type of charge must be set for the optimum operating superheat at the lowest evaporator temperature expected.

Gas Cross Charge. Gas cross charges combine the features of the gas charge and the liquid cross charge. They use a limited amount of liquid, thereby providing a maximum operating pressure. The liquid used in the charge is different than the refrigerant in the system and is chosen to provide superheat characteristics similar to those of the liquid cross charges (low temperature). Consequently, they provide both the superheat characteristics of a cross charge and the maximum operating pressure of a gas charge (Figure 11). While a commercial (medium-temperature) gas cross charge is possible, its uses are limited.

Adsorption Charge. Typical adsorption charges depend on the property of an adsorbent, such as silica gel or activated carbon, that is used in an element bulb to adsorb and desorb a gas such as carbon dioxide, with accompanying changes in temperature. The amount of adsorption or desorption changes the pressure in the thermostatic element. Because adsorption charges respond primarily to the temperature of the adsorbent material, they are essentially unaffected by cross-ambient conditions. The comparatively slow response time of the adsorbent results in a charge characterized by its stability. Superheat characteristics can be varied by using different charge materials, adsorbents, and/or charge pressures. The pressure-limiting feature of the gas or gas cross charges is not available with the adsorption-type element.

Type of Equalization

Internal Equalizer. When the refrigerant pressure drop through a single-circuit evaporator is equivalent to not more than a 1 K change in evaporator temperature, a thermostatic expansion valve that has an internal equalizer may be used. Internal equalization describes valve outlet pressure transmitted through an internal passage to the underside of the diaphragm (see Figure 7).

The pressure drop in many evaporators is greater than the 1 K equivalent. When a refrigerant distributor is used, the pressure drop across the distributor will cause the pressure at the outlet of the expansion valve to be considerably higher than the pressure at the evaporator outlet. As a result, an internally equalized valve will control at an abnormally high superheat. Under these conditions, the

Refrigerant-Control Devices

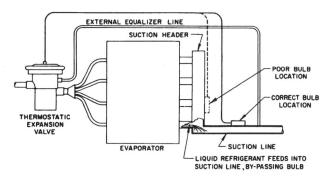

Fig. 12 Bulb Location for Thermostatic Expansion Valve

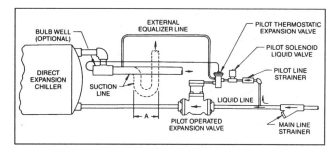

Fig. 13 Pilot-Operated Thermostatic Expansion Valve Controlling Liquid Refrigerant Flow to Direct-Expansion Chiller

evaporator does not perform efficiently because it is starved for refrigerant. Furthermore, the distributor pressure drop is not constant but varies with evaporator load and cannot be compensated for by adjusting the superheat setting of the valve.

External Equalizer. Because evaporator and/or refrigerant distributor pressure drop produces poor system performance with an internal equalizer valve, a valve that has an external equalizer is used. Instead of the internal communicating passage shown in Figure 7, an external connection to the underside of the diaphragm is provided. The external equalizer line is connected either to the suction line, as shown in Figure 12, or into the evaporator at a point downstream from the major pressure drop.

Alternate Types of Construction

Pilot-operated thermostatic expansion valves are used on large systems in which the required capacity per valve is beyond the range of direct-operated valves. The pilot-operated valve consists of a piston-type pilot-operated regulator, which is used as the main expansion valve, and a low-capacity thermostatic expansion valve, which serves as an external pilot valve. The small pilot thermostatic expansion valve supplies pressure to the piston chamber or, depending on the regulator design, bleeds pressure from the piston chamber in response to a change in the operating superheat. Pilot operation permits the use of a characterized port in the main expansion valve to provide good modulation over a wide loading range. Therefore, the pilot-operated valve performs well on refrigerating systems that have some form of compressor capacity reduction, such as cylinder unloading. Figure 13 illustrates such a control valve applied to a large capacity, direct-expansion chiller.

The auxiliary pilot controls should be sized to handle only the pilot circuit flow. For example, in Figure 13 a small solenoid valve in the pilot circuit, installed ahead of the thermostatic expansion valve, converts the pilot-operated valve into a stop valve when the solenoid valve is closed.

Equalization Features. When the compressor stops, a thermostatic expansion valve usually moves to the *closed* position. This movement sustains the difference in refrigerant pressures in the evaporator and the condenser. Low-starting torque motors require that these pressures be equalized to reduce the torque needed to restart the compressor. One way to provide pressure equalization is to add, parallel to the main valve port, a small fixed auxiliary passageway, such as a slot or drilled hole in the valve seat or valve pin. This opening permits a limited fluid flow through the control, even when the valve is closed and allows the system pressures to equalize on the *off* cycle. The size of such a fixed auxiliary passageway must be limited so its flow capacity is not greater than the smallest flow that must be controlled in normal system operation.

Another more complex control is available for systems requiring shorter equalizing times than can be achieved with the fixed auxiliary passageway. This control incorporates an auxiliary valve port, which bypasses the primary port and is opened by the element diaphragm as it moves toward and beyond the position at which the primary valve port is closed. The flow capacity of such an auxiliary valve can be considerably larger than that of the fixed auxiliary passageway, so that pressures can equalize more rapidly.

Flooded System. Thermostatic expansion valves are seldom applied to flooded evaporators because superheat is necessary for proper valve control; only a few degrees of suction vapor superheat in a flooded evaporator incurs a substantial loss in capacity. If the bulb is installed downstream from a liquid-to-suction heat exchanger, a thermostatic expansion valve can be made to operate at this point on a higher superheat. Valve control is apt to be poor because of the variable rate of heat exchange as flow rates change (see the section on Application).

Expansion valves with modified thermostatic elements are available in which electrical heat is supplied to the bulb. The bulb is inserted in direct contact with refrigerant liquid in a low-side accumulator. The contact of cold refrigerant liquid with the bulb overrides the artificial heat source and throttles the expansion valve. As the liquid falls away from the bulb, the valve feed increases again. Although similar in construction to a thermostatic expansion valve, it is essentially a modulating liquid level control valve.

Desuperheating Valves. Thermostatic expansion valves with special thermostatic charges are used to reduce gas temperatures (superheat) on various air conditioning and refrigeration systems. Suction gas in a single-stage system can be desuperheated by injecting liquid directly into the suction line. This cooling may be required with or without discharge gas bypass used for compressor capacity control. The line upstream of the valve bulb must be long enough so the injected refrigerant can mix adequately with the gas being desuperheated. On compound compression systems, liquid may be injected directly into the interstage line upstream of the valve bulb to provide intercooling.

Application

Hunting is alternate overfeeding and starving of the refrigerant feed to the evaporator. It produces sustained cyclic changes in the pressure and temperature of the refrigerant gas leaving the evaporator. Extreme hunting reduces the capacity of the refrigeration system because the mean evaporator pressure and temperature are lowered and the compressor capacity is reduced. If overfeeding of the expansion valve causes intermittent flooding of liquid into the suction line, the compressor may be damaged.

Although hunting is commonly attributed to the thermostatic expansion valve, it is seldom solely responsible. One reason for hunting is that all evaporators have a time lag. When the bulb signals for a change in refrigerant flow, the refrigerant must traverse the entire evaporator before a new signal reaches the bulb. This lag or time lapse may cause continuous overshooting of the valve both opening and closing. In addition, the thermostatic element, because of its mass, has a time lag that may be in phase with the evaporator lag and amplify the original overshooting.

It is possible to alter the response rate of the thermostatic element by either using thermal ballast or changing the mass or heat capacity of the bulb, thereby damping or even eliminating the hunting. A change in valve gradient may produce the same result.

Extremely high refrigerant velocity in the evaporator can also cause hunting. Liquid refrigerant under these conditions moves in waves, called **slugs**, that fill a portion of the evaporator tube and erupt into the suction line. These unevaporated slugs chill the bulb and temporarily reduce the feed of the valve, resulting in intermittent starving of the evaporator.

On multiple-circuit evaporators, a lightly loaded or overfed circuit will also flood into the suction line, chill the bulb, and throttle the valve. Again, the effect is intermittent; when the valve feed is reduced, the flooding ceases, and the valve reopens.

Hunting can be minimized or avoided by the following actions:

1. Select the proper valve size from the valve capacity ratings rather than nominal valve capacity; oversized valves aggravate hunting.
2. Change the valve adjustment. A lower superheat setting usually (but not always) increases hunting (Huelle 1972, Wedeking & Stoecker 1966, Stoecker 1966).
3. Select the correct thermostatic element charge. Cross-charged elements have inherent antihunt characteristics.
4. Design the evaporator section for even refrigerant and airflow. Uniform heat transfer from the evaporator is only possible if refrigerant is distributed by a properly selected and applied refrigerant distributor and air distribution is controlled by a properly designed housing. (Air-cooling and dehumidifying coils, including refrigerant distributors, are detailed in Chapter 21 of the 1996 *ASHRAE Handbook—Systems and Equipment*.)
5. Size and arrange suction piping correctly.
6. Locate and apply the bulb correctly.
7. Select the best location for the external equalizer line connection.

Bulb Location. Most installation requirements are met by strapping the bulb to the suction line to obtain good thermal contact between them. Normally, the bulb is attached to a horizontal line upstream of the external equalizer connection (if used) at a 3 or 9 o'clock position as close to the evaporator as possible. While the bulb is not normally placed near or after suction line traps, some designers test and prove locations that differ from these recommendations. A good moisture-resistant insulation over the bulb and suction line eliminates any adverse effect of varying ambient temperatures at the bulb location.

Occasionally, the bulb of the thermostatic expansion valve is installed downstream from a liquid-suction heat exchanger to compensate for a capacity shortage due to an undersized evaporator. While this procedure seems to be a simple method of obtaining maximum evaporator capacity, installing the bulb downstream of the heat exchanger is undesirable from a control standpoint. As the valve modulates, the liquid flow rate through the heat exchanger changes, causing the rate of heat transfer to the suction vapor to change. An exaggerated valve response follows, resulting in hunting. It may be possible to find a bulb location downstream from the heat exchanger that reduces the hunt considerably. However, the danger of floodback to the compressor normally overshadows the need to attempt this method.

Certain installations require increased **bulb sensitivity** as a protection against floodback. The bulb, if located properly in a well in the suction line, has a rapid response feature because of its intimate contact with the refrigerant stream. The bulb sensitivity can be increased by the use of a bulb smaller than is normally supplied. However, the use of the smaller bulb is limited to gas-charged valves. Good piping practice also effects expansion valve performance.

Figure 14 illustrates the proper **piping arrangement** when the suction line runs above the evaporator. A lubricant trap that is as short as possible is located downstream from the bulb. The vertical

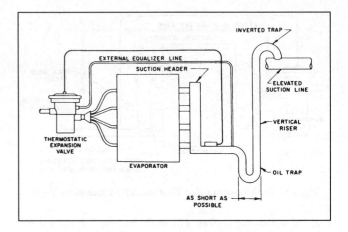

Fig. 14 Bulb Location When Suction Main is Above Evaporator

riser(s) must be sized to produce a refrigerant velocity that ensures continuous return of lubricant to the compressor. The terminal end of the riser(s) enters the horizontal run at the top of the suction line; this avoids both interference from the overfeeding of any other expansion valve or any drainback during the off cycle.

If circulated with lubricant-miscible refrigerant, a heavy concentration of lubricant elevates boiling temperature of the refrigerant. The response of the thermostatic charge of the expansion valve is related to the saturation pressure and the temperature of pure refrigerant. In an operating system, the false pressure-temperature signals of lubricant-rich refrigerants cause both floodback or operating superheats considerably lower than indicated and quite often cause erratic valve operations. To keep the lubricant concentration at an acceptable level, either the lubricant pumping rate must be reduced or an effective lubricant separator must be used.

The **external equalizer** line is ordinarily connected at the evaporator outlet, as shown in Figure 12. It may also be connected at the evaporator inlet or at any other point in the evaporator downstream of the major pressure drop. On evaporators with long refrigerant circuits that have inherent lag, hunting may be minimized by changing the connection point of the external equalizer line.

The ambient temperature at the valve does not ensure a corresponding temperature of the diaphragm chamber because a certain amount of heat is conducted to the cooler valve body. It is not unusual for the diaphragm-chamber temperature to fall below the bulb temperature. When this fall occurs with a gas-charged valve, the entire thermostatic charge can condense in the diaphragm chamber, and the refrigerant feed can then be controlled from that point. Extreme starving of the evaporator, to the point of complete cessation of feed, is characteristic of this condition. For this reason, gas-charged valves are normally used only when sufficient pressure drop exists between the outlet of the valve and the bulb location. On multiple-circuit evaporators, the refrigerant distributor serves this purpose. The pressure drop through the distributor elevates the valve body temperature above that of the evaporator and assists in maintaining control at the valve bulb. Straight liquid and liquid cross-charged valves operate properly when the temperature of the space in which they are located is temperate. However, consideration must be given to the location of the valve when the temperature of the space is extreme because of its adverse effect on the pressure in the bulb, capillary tube, and diaphragm-chamber assembly. Adsorption-charged valves are unaffected by the surrounding ambient temperature.

Direct-expansion chillers located in a heated environment are among the few applications in which gas-charged valves operate successfully without refrigerant distributors. This is possible because operation occurs at considerably lower superheats (2 to 3 K) and narrower ranges of temperatures than other types of direct-expansion evaporators. When this type of chiller is exposed to a cold ambient temperature, action must be taken to prevent the charge from migrating to the diaphragm housing of the valve element. The adsorption charge can be used for these applications because it is unaffected by the colder ambients. If a pressure-limiting feature is necessary, special hydraulic elements can use the gas cross charge in such a way as to prevent charge migration.

ELECTRIC EXPANSION VALVES

The electronically controlled expansion valve is a liquid refrigerant flow-control device that is similar to the thermostatic expansion valve. These valves may be classified into the following types:

- Heat motor operated
- Magnetically modulated
- Pulse width modulated (*on-off* type)
- Step motor driven

Heat motor valves may be either of two types. In one type, one or more bimetallic elements are heated electrically, causing them to deflect. The bimetal elements are linked mechanically to a valve pin or poppet. In a second type of heat motor expansion valve, a volatile material charge is contained within an electrically heated chamber so that the charge temperature (and pressure) is controlled by electrical power input to the heater. The charge pressure is made to act on a diaphragm or bellows, which is balanced against either ambient air pressure or refrigeration system suction pressure. The diaphragm is linked to a pin or poppet.

In a **magnetically modulated valve**, a direct current electromagnet modulates smoothly, while an armature, or plunger, compresses a spring progressively as a function of coil current. The modulating electromagnet plunger may be connected to a valve pin or poppet directly or may be used as the pilot element in a servo loop to operate a much larger valve. When the modulating plunger operates a pin or poppet directly, the valve may be of a pressure-balanced port design so that pressure differential has little or no influence on valve opening.

The **pulse width modulated valve** is an *on-off* solenoid valve with special features that allow it to function as an expansion valve through a life of millions of cycles. Even though the valve is either fully opened or closed, it operates as an infinitely variable metering device by pulsing the valve open regularly. The duration of each opening, or pulse, is regulated by the electronics. For example, a valve may be pulsed every six seconds. If 50% flow is needed, the valve would be held open three seconds and closed for three seconds.

A **step motor** is an electronically commutated, multiphase motor capable of running continuously in forward or reverse, or it can be discretely positioned in increments of a small fraction of a revolution. Step motors are used in instrument drives, plotters, and other applications where accurate positioning is required. Step motors require electronics, such as an integrated circuit, to switch windings in proper sequence. When used to drive expansion valves, a lead screw changes the rotary motion of the rotor to a linear motion suitable for moving a valve pin or poppet. The lead screw may be driven directly from the rotor, or a reduction gearbox may be placed between the motor and lead screw. The motor may be hermetically sealed within the refrigerant environment, or the motor and gearbox can be sealed to operate in air.

Electric expansion valves may be controlled by either digital or analog electronic circuits. Electronic control gives the additional flexibility to consider control schemes that are impossible with conventional valves.

CONSTANT PRESSURE EXPANSION VALVES

The constant pressure expansion valve is operated by evaporator or valve-outlet pressure; it regulates the mass flow of liquid refrigerant entering the evaporator and maintains this pressure at a constant value. Although this valve was first used as a liquid refrigerant expansion valve, other applications are also addressed.

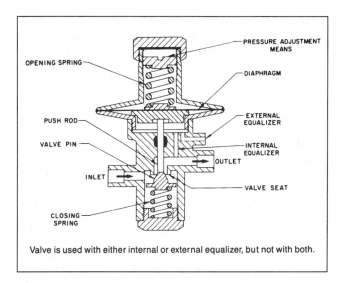

Fig. 15 Constant Pressure Expansion Valve

Operation

Figure 15 shows a cross section of a constant pressure expansion valve. The valve has both an adjustable spring, which exerts its force on top of the diaphragm in an opening direction, and a spring beneath the diaphragm, which exerts its force in a closing direction. Evaporator pressure is admitted beneath the diaphragm, through either the internal or external equalizer passage, and the combined forces of the evaporator pressure and the closing spring act to counterbalance the opening spring pressure.

With the valve set and refrigerant flowing at a given pressure, a small increase in the evaporator pressure forces the diaphragm up, which restricts the refrigerant flow and limits the evaporator pressure. When the evaporator pressure drops below the valve setting due to a decrease in load, the top spring pressure moves the valve pin in an open direction. As a result, the refrigerant flow increases and raises the evaporator pressure to the balanced valve setting. This valve controls evaporation of the liquid refrigerant in the evaporator at a constant pressure and temperature.

Constant pressure expansion valves automatically adjust the flow of liquid refrigerant to the evaporator to balance compressor pumping capacity. When this balance is established, evaporator pressure remains constant during the remainder of the running phase of the refrigeration cycle. This low-side pressure balancing point is selected by turning the valve adjuster to the desired pressure setting. To avoid floodback to the compressor during low heat load periods, the valve setting should be selected to use evaporator surface effectively when there is a minimum heat load on the system. When there is a maximum heat load on the system, the increase in evaporator pressure causes the valve to close partially. This prevents the pressure from rising any further and overloading the compressor.

For reasons of operation and adjustment, constant pressure expansion valves are most effective on applications with a constant heat load. When the compressor stops at the end of the running phase of the cycle, standard constant pressure expansion valves close to prevent off-cycle refrigerant flow. The rapid equalization features described in the section on Thermostatic Expansion Valves can also be built into constant pressure expansion valves to provide off-cycle pressure equalization for use with low-starting torque compressor motors.

Selection

The constant pressure expansion valve for liquid refrigerant expansion service should be selected for the required capacity and system refrigerant at the lowest expected pressure drop across the valve and should have an adjustable pressure range to provide the required evaporator (valve outlet) pressure. The system designer should decide whether the valve should be a bleed-type or a standard expansion valve.

Application

Because the constant pressure expansion valve, when applied as a liquid refrigerant expansion valve, is suitable only for constant load applications, its use is limited. Ordinarily, only one such liquid expansion valve is used on each system. When applied to a variable load, this valve starves the evaporator at the high load and overfeeds it at the low load, with possible compressor damage resulting in the latter case. Because of the difference in the operating principles of the constant pressure expansion valve and the thermostatic expansion valve, both cannot be used in the same system as liquid-refrigerant expansion valves.

The constant pressure expansion valve is used as a pilot valve in some large suction pressure regulating valves. It may also be used to bypass compressor discharge gas. Such applications may arise either when compressor capacity must be reduced or where a low-limit control of either the evaporator, the suction pressure, or both is required. After passing through the valve, the hot gas may be introduced directly into the suction line, in which case some additional desuperheating means may be required, or it may be introduced at the evaporator inlet, where it mixes with the cold refrigerant and is desuperheated before it reaches the compressor. Constant pressure expansion valves are used in drink dispensers, food dispensers, water coolers, ice cream freezers, and self-contained room air-conditioners.

EVAPORATOR PRESSURE AND TEMPERATURE REGULATORS

The **evaporator pressure regulator** (back-pressure regulator) regulates the evaporator pressure (pressure entering the regulator) at a constant value. It is used in the evaporator outlet or suction line wherever low-limit control of the evaporator pressure or temperature is required.

Operation

As shown in Figure 16, the inlet pressure acts on the bottom of the seat disk and opposes the adjusting spring. The outlet pressure acts on the bottom of the bellows and the top of the seating disk. Because the effective areas of the bellows and the port are equal, the two forces cancel each other, and the valve responds to inlet pressure only. When the evaporator pressure rises above the force exerted by the spring, the valve begins to open. When the evaporator pressure drops below the spring force, the valve closes. In operation, the valve assumes a throttling position to balance the load.

This change in the pressure, which acts on and operates the diaphragm or bellows, is called differential. The **total valve differential** is the difference between the pressure at which the valve operates at rated stroke and the pressure at which the valve starts to open. **Pressure drop** is the difference between the pressure at the valve inlet and the pressure at the valve outlet. The difference between pressure differential and pressure drop must be understood when determining the proper valve size for a given load condition.

Pilot-operated evaporator pressure regulators are either self-contained or high-pressure driven. The self-contained regulator (Figure 17) starts to operate when the evaporator pressure rises above the pressure setting of the diaphragm spring; the diaphragm starts to open, which increases the pressure above the piston. This increase moves the piston down, thereby forcing the main valve to open. As a result, the evaporator pressure drops back to the pressure setting of the pilot. When the evaporator pressure drops, the pilot valve starts to close, allowing the pressure above the piston to decrease. Then, the main spring forces the main valve to start closing, preventing the evaporator pressure from falling below the pressure setting of the pilot. In operation, the pilot valve and the main valve assume either intermediate or throttling positions, depending on the load.

Many pilot-operated regulators are of a normally open design and require high-pressure liquid or gas to provide a closing force. The advantage of this regulator is that it requires no minimum operating suction pressure drop to operate. When the valve inlet pressure

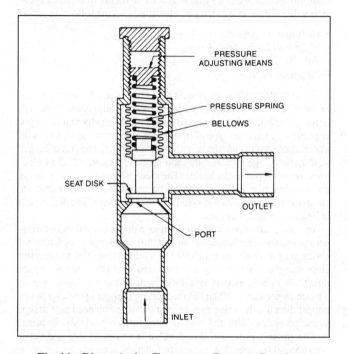

Fig. 16 Direct-Acting Evaporator Pressure Regulator

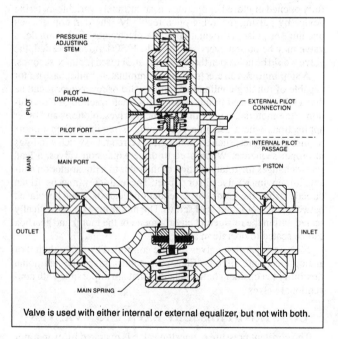

Fig. 17 Pilot-Operated Evaporator Pressure Regulator (Self-Contained)

Refrigerant-Control Devices

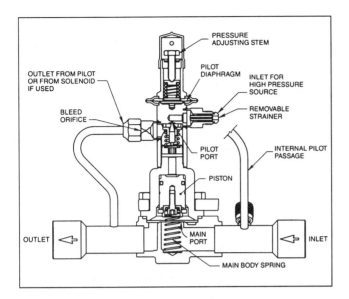

Fig. 18 Pilot-Operated Evaporator Pressure Regulator
(High Pressure Driven)

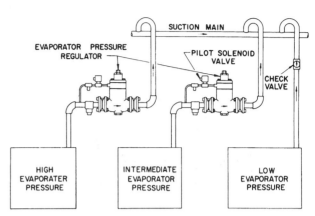

Fig. 19 Evaporator Pressure Regulators in Multiple System

increases above set point, the pilot valve diaphragm lifts against the adjustment spring load (Figure 18). When the diaphragm moves up in response to the increasing inlet pressure, the pilot valve pin closes the pilot valve port. Gas or liquid from the high-pressure side of the system is throttled by the pilot, allowing the main piston spring to move the valve to a more open position. Because the top of the piston chamber bleeds to the downstream side of the valve through a bleed orifice, a continuous flow of high-pressure liquid or gas through the pilot valve drives the piston down to a closed or partially closed position. A solenoid valve may be used to drive the piston down, either by porting pressure directly to the top of the piston or by closing the bleed orifice, thus closing the valve for defrost operation.

Selection

Unless a system has special requirements, the evaporator pressure regulator should be selected for the required capacity and refrigerant at the required evaporator (regulator inlet) pressure and the lowest pressure drop across the regulator that are consistent with good regulator performance and economical compressor operation.

Application

Evaporator pressure regulators are used on finned-coil evaporators where frosting should be prevented or evaporator pressure must be maintained constant and higher than suction pressure to prevent dehumidification. These regulators are used on water and brine chiller applications to prevent chillers from freezing under low-load conditions.

On **multiple evaporator installations**, as shown in Figure 19, evaporator pressure regulators (direct-acting or pilot) may be installed to control evaporator pressure and temperature in each unit. The regulators maintain the desired evaporator temperature in the warmer units, while the compressor continues to operate to satisfy the coldest unit. With such an installation, the compressor may be controlled by either a low-pressure switch or by thermostats installed in the individual units.

The evaporator pressure regulator, with internal pilot passage, receives its source of pressure for pilot operation at the regulator inlet connection. In contrast, the regulator with the external pilot connection not only permits a choice of pressure source for pilot operation, but also permits the use of a remote pressure pilot or pilot valves for other purposes.

If the regulator inlet pressure is unsteady and adversely affects regulator performance, a source of pressure steadier than that available at the regulator inlet may be obtained by connecting the external pilot line to a surge drum or an enlarged section of suction line upstream from the regulator.

A remote pressure pilot installed in the external pilot line facilitates adjustment of the pressure setting when the regulator must be installed in an inaccessible location. A pressure pilot on the self-contained design with a pneumatic connection permits resetting and controlling of the evaporator pressure and temperature by a pneumatic control system.

In Figure 19, a pilot solenoid valve installed in the external pilot line allows the regulator to function as a suction stop valve as well as an evaporator pressure regulator. This feature is particularly useful on a flooded evaporator to prevent pumping refrigerant out of the evaporator, which occurs when the thermostat is satisfied and the compressor continues to operate to cool other evaporators.

Two pressure pilots may be combined to make a dual-pressure regulator. The lower pressure pilot is piped through a pilot solenoid valve and, when energized, controls the lower pressure. The higher pressure, which occurs when the pilot solenoid valve is deenergized, is generally used for a warmer evaporator requirement or for pressure above 0°C saturation for defrosting.

In some instances, it is desirable to control, at a constant temperature, the air or liquid entering or leaving the evaporator. By modulating the evaporator pressure and temperature according to load demand, the evaporator temperature is decreased during heavy loads and increased during lighter loads. Modulation can be accomplished with an evaporator pressure regulator modified in one of the following ways:

1. By using a temperature-actuated pilot in the external pilot line in place of the pressure pilot and by placing the bulb of the temperature pilot where it will respond to the temperature of the air or liquid being cooled. The temperature pilot modulates the position of the regulator according to the load requirement and maintains the air or liquid at a constant temperature.

2. By using a pneumatic thermostat and a pressure pilot with a pneumatic connection so that the evaporator pressure and temperature are raised by an increase in air pressure. This increase is supplied to the pressure pilot at reduced loads and is lowered by a decrease in air pressure supplied to the pressure pilot at increased loads, as required to maintain the temperature of the pneumatic thermostat bulb. This bulb is placed to respond to the temperature of the air or liquid being cooled. Thus, the air or liquid being cooled is maintained at a constant temperature.

3. By using a pressure pilot that is adaptable to a reversible electric motor drive, which resets both the pressure pilot and a potentiometer-type thermostat to control the reversible electric motor. The evaporator pressure and temperature are raised at reduced

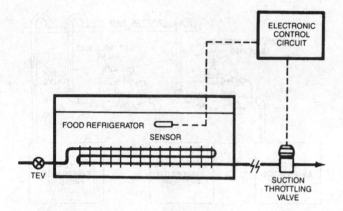

Fig. 20 Electronically Controlled, Electrically Operated Suction Throttling Regulator

loads and lowered at increased loads, as required to maintain the temperature of the thermostat bulb, which is placed to respond to the temperature of the air or liquid being cooled. The air or liquid being cooled is maintained at a constant temperature.

In addition to the normal applications, these regulators, when equipped with either high-pressure pilots and suitable valve-seat material or a higher range pressure spring, such as in the direct-acting regulator, are used in several applications with air-cooled condensers. The regulators are used for maintaining low-limit control during cold weather operation of the compressor discharge, the liquid line pressure, or both.

Electronically controlled, electrically operated, suction throttling valves have been developed to control temperature in a food merchandising refrigerator or other refrigerated space (Figure 20). This valve is a form of evaporator pressure regulator, although it responds only to temperature in the space, rather than pressure in the evaporator or suction line. The system consists of a temperature sensor, an electronic control circuit, and a suction throttling valve. A temperature setting is made by turning a calibrated potentiometer or rotary switch normally located on the control circuit or through communication software. The valve responds to the difference between set point temperature and the prevailing temperature. A temperature above the set point drives the valve further open, while a temperature below set point modulates the valve in the closing direction. During defrost, the control circuit usually closes the valve.

By modulating suction gas flow in response to the measured temperature, the refrigerated space may be held close to the set point regardless of variations in heat load or compressor suction pressure. An electronic regulator is particularly useful on refrigerators containing fresh meat or other products where close control of temperature is important. In some instances, an energy savings may be realized, because a temperature regulating control reduces refrigeration during low heat load conditions more effectively than an evaporator pressure regulator.

SUCTION PRESSURE REGULATORS

The suction pressure regulator (holdback valve or crankcase pressure regulator) limits the compressor suction pressure (regulator outlet pressure) to a maximum value. This type of regulator should be used in the suction line at the compressor on any refrigeration installation in which the liquid expansion valve cannot limit the suction pressure and compressor motor overload would otherwise exist because of the following:

- Excessive starting load
- Excessive suction pressure following the defrost cycle
- Prolonged operation at excessive suction pressure
- Low-voltage and high-suction pressure conditions

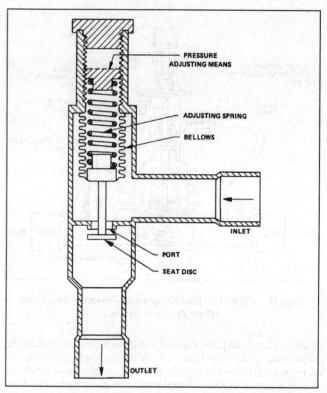

Fig. 21 Direct-Acting Suction Pressure Regulator

Operation

Direct-acting suction pressure regulators sense only their outlet or downstream pressure (compressor crankcase or suction pressure). As illustrated in Figure 21, the inlet pressure acts on the underside of the bellows and the top of the seating disk. Since the effective areas of the bellows and port are equal, these forces cancel each other and do not affect valve operation. The valve outlet pressure acts on the bottom of the disk and exerts a closing force, which is opposed by the adjustable spring force. When the outlet pressure drops below the equivalent force exerted by the spring, the valve moves in an opening direction. If the outlet pressure rises, the valve moves in a closing direction and throttles the refrigerant flow to maintain the set point of the valve.

Externally and internally pilot-operated suction pressure regulators are available for larger system applications. Although their design is more complex, due to pilot operation, their method of operation is similar to that described above.

Selection

The suction pressure regulator should be selected for the required capacity, system refrigerant, required regulator inlet pressure, and lowest practical pressure drop across the regulator to minimize any loss in system capacity.

Application

In addition to the normal applications, these regulators—when equipped with high-pressure pilots, suitable valve-seat material, or a higher range pressure spring such as in the direct-acting regulator—are used in several methods of application in refrigeration systems with air-cooled condensers. They are used for maintaining low-limit control of the compressor discharge, liquid line pressure, or both, during cold weather operation. In addition, such modified regulators are used to bypass compressor discharge gas

Refrigerant-Control Devices

in refrigeration systems, as described in the section on Application under Constant Pressure Expansion Valves.

CONDENSER PRESSURE REGULATORS

Various condenser pressure-regulating valves are used to maintain sufficient condensing pressure to allow air-cooled condensers to operate properly during the winter. Both single- and two-valve arrangements have been used for this purpose. See Chapter 2 of this volume and Chapter 35 of the 1996 *ASHRAE Handbook—Systems and Equipment* for more information.

The two-valve arrangement often uses a valve that is constructed and operates similarly to the evaporator pressure-regulating valve shown in Figures 16 and 17. This control is installed either in the liquid line between the condenser and receiver or in the discharge line. It throttles when the condenser or discharge pressure falls as a result of a low ambient condition.

The second valve in the two-valve arrangement bypasses discharge gas around the condenser to the receiver to mix with cold liquid and maintain adequate high-side pressure. Several bypass valves are available, some of which are similar to the suction pressure-regulating valve shown in Figure 21. This valve responds to outlet pressure (receiver pressure). When receiver pressure decreases as a result of a decrease in ambient temperature, the valve opens and bypasses discharge bypass gas to the receiver. Figure 22 shows another device that responds to changes in pressure between its inlet and outlet. As the differential pressure increases, the valve opens. Thus, when the other valve in this two-valve arrangement throttles and restricts liquid flow, a differential is created, and this bypass device opens.

It is sometimes an advantage to substitute a single three-way condenser pressure-regulating valve for the two-valve arrangement described previously. The three-way valve (Figure 23) simultaneously holds back liquid in the condenser and passes compressor discharge into the receiver to maintain pressure in the liquid line. The lower side of a metal diaphragm is exposed to system high-side pressure, while the upper side is exposed to a noncondensable gas charge (usually dry nitrogen). A pushrod connects the diaphragm to the valve poppet, which seats on either the upper or lower port and throttles either the discharge gas or the liquid from the condenser, respectively.

During system start-up in extremely cold weather, the poppet may be tight against the lower seat, stopping all liquid flow from the condenser and bypassing discharge gas into the receiver until adequate system pressure is developed. During stable operation in cold weather, the poppet modulates at an intermediate position, with liquid flow from the condensing coil mixing with compressor discharge gas within the valve and flowing to the receiver. During warm weather, the poppet seats tightly against the upper port, allowing free flow of liquid from the condenser but preventing flow of discharge gas.

Three-way condenser pressure-regulating valves are not usually adjustable by the user. The pressure setting is established by the pressure of the gas charge placed in the dome above the diaphragm during manufacture.

HIGH-SIDE FLOAT VALVES

Operation

A high-side float valve controls the mass flow of refrigerant liquid entering the evaporator so it equals the rate at which the refrigerant gas is pumped from the evaporator by the compressor. Figure 24 shows a cross section of a typical valve. The refrigerant liquid flows from the condenser into the high-side float valve body, where it raises the float and moves the valve pin in an opening direction, permitting the liquid to pass through the valve port, expand, and flow into the evaporator. Most of the system refrigerant charge is contained in the evaporator at all times. The high-side float system is a flooded system.

Selection

For acceptable performance, the high-side float valve is selected for the refrigerant and a rated capacity neither excessively large nor too small. The orifice is sized for the maximum required capacity

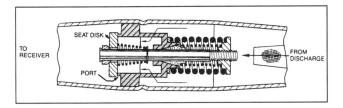

Fig. 22 Condenser Bypass Valve

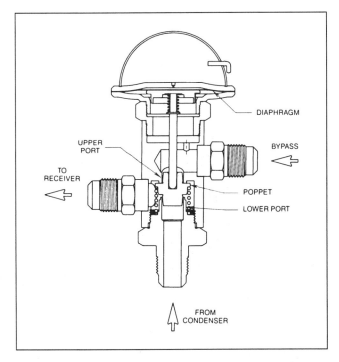

Fig. 23 Three-Way Condenser Pressure-Regulating Valve

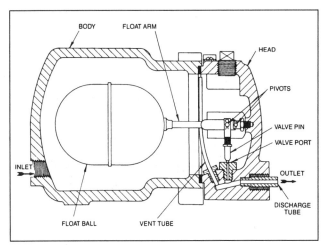

Fig. 24 High-Side Float Valve

with the minimum pressure drop across the valve. The valve operated by the float may be a pin-and-port construction (Figure 24), a butterfly valve, a balanced double-ported valve, or a sliding gate or spool valve. The internal bypass vent tube allows installation of the high-side float valve near the evaporator and above the condenser without danger of the float valve becoming gas bound. Some large-capacity valves use a high-side float valve for pilot operation of a diaphragm or piston-type spring-loaded expansion valve. This arrangement can provide improved modulation over a wide range of load and pressure-drop conditions.

Application

A refrigeration system in which a high-side float valve is used consists ordinarily of a single evaporator, compressor, and condenser. The operating receiver or a liquid sump at the condenser outlet can be quite small. A full-sized receiver is required for pumping out the flooded evaporator. Under certain conditions, the high-side float valve may be used to feed more than one evaporator; consequently, additional control valves are required. The amount of refrigerant charge is critical with a high-side float valve. An excessive charge causes floodback, while an insufficient amount reduces system capacity.

LOW-SIDE FLOAT VALVES

Operation

The low-side float valve performs the same function as the high-side float valve, but it is connected to the low-pressure side of the system. When the evaporator liquid level drops, the float opens the valve. Refrigerant liquid then flows from the liquid line through the valve port and directly into the evaporator or surge drum. In another valve design, the refrigerant liquid flows through the valve port, passes through a remote feed line, and enters the evaporator through a separate connection. (A typical direct-feed valve construction is shown in Figure 25.) The low-side float system is a flooded system.

Selection

Low-side float valves are selected in the same manner as the high-side float valves discussed previously.

Application

In the low-side float valve system, the refrigerant charge is not critical. The low-side float valve can be used with multiple evaporators such that some evaporators may be controlled by other low-side float valves and some by thermostatic expansion valves.

Depending on its design, the float valve is mounted either directly in the evaporator or surge drum or in an external chamber connected to the evaporator or surge chamber by equalizing lines, i.e., a gas line at the top and a liquid line at the bottom. In the externally mounted type, the float valve is separated from the float chamber by a gland that maintains a quiet level of liquid in the float chamber for steady actuation of the valve.

In evaporators with high boiling rates or restricted liquid and gas passages, the boiling action of the liquid raises the refrigerant level during operation. When the compressor stops or the solenoid suction valve closes, the boiling action of the refrigerant liquid ceases, and the refrigerant level in the evaporator drops. Under these conditions, the high-pressure liquid line supplying the low-side float valve should be shut off by a solenoid liquid valve to prevent overfilling of the evaporator. Otherwise, excess refrigerant will enter the evaporator on the *off* cycle, which can cause floodback when the compressor starts or the solenoid suction valve opens.

When a low-side float valve is used, precautions must be taken that the float is in a quiet liquid level that falls properly in response to an increase in evaporator load and rises with a decrease in evaporator load. In low-temperature systems particularly, it is important that the equalizer lines between the evaporator and either the float chamber or the surge drum be generously sized to eliminate any reverse response of the refrigerant liquid level in the vicinity of the float. Where the low-side float valve is located in a nonrefrigerated room, the equalizing liquid and gas lines and the float chamber must be insulated to provide a quiet liquid level for the float.

SOLENOID VALVES

A solenoid valve is closed by gravity, pressure, or spring action and opened by a plunger actuated by the magnetic action of an electrically energized coil, or vice versa. Figures 26 and 27 show cross sections of solenoid valves with their principal components identified.

Because solenoid valves are actuated electrically, they may be conveniently operated in remote locations by any suitable electric switch. These valves are always fully open or fully closed, in contrast to motorized valves, which can modulate flow. Solenoid valves may be used to control the flow of many different fluids if the pressures and temperatures involved, the viscosity of the fluid, and the suitability of the materials used in the valve construction are carefully considered.

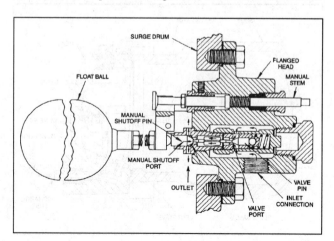

Fig. 25 Low-Side Float Valve

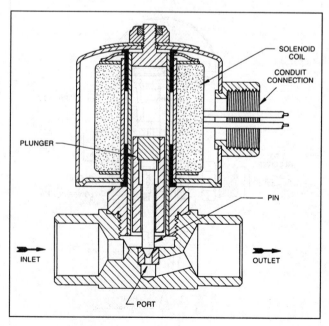

Fig. 26 Normally Closed Direct-Acting Solenoid Valve with Hammer-Blow Feature

Refrigerant-Control Devices

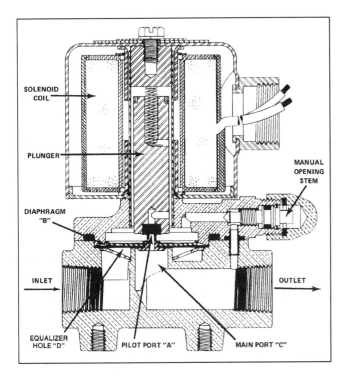

Fig. 27 Normally Closed Pilot-Operated Solenoid Valve with Direct-Lift Feature

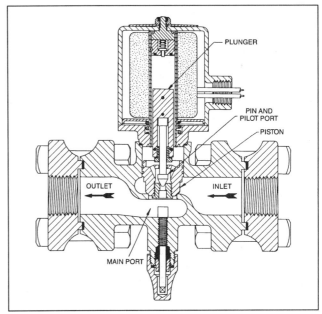

Fig. 28 Normally Closed Pilot-Operated Solenoid Valve with Hammer-Blow and Mechanically Linked Piston-Pin Plunger Features

Solenoid valves can be divided into the following general types:

1. **Normally closed solenoid valves**, in which the closure member moves away from the port to open the valve when the coil is energized, e.g., a two-way solenoid valve.
2. **Normally open solenoid valves**, in which the closure member moves to the port to close the valve when the coil is energized, e.g., a two-way solenoid valve.
3. **Multiaction solenoid valves**, which combine (in one body) the action of one or more normally open and one or more normally closed solenoid valves, e.g., a three-way solenoid valve or a four-way solenoid valve.

Operation

While all types of solenoid valves are used, the normally closed type is used far more extensively. In the normally closed direct-acting solenoid valve shown in Figure 26, the solenoid coil, acting on the plunger, pulls the valve pin away from and off the valve port, thereby opening it directly. Because this valve depends on the power of the solenoid coil for operation, its port size for a given operating pressure differential is limited by the limitations of solenoid coil size.

Figure 27 shows a medium-sized, normally closed pilot-operated solenoid valve. In this valve, the solenoid coil, acting on the plunger, does not open the main port directly but opens the pilot Port A. Pressure trapped on top of Diaphragm B is released through the pilot port, thus creating a pressure imbalance across the diaphragm, forcing it upward and opening the main Port C. When the solenoid coil is de-energized, the plunger drops and closes pilot Port A. Then the pressures above and below the diaphragm equalize again through the equalizer Hole D, and the diaphragm drops and closes the main port. In some pilot-operated solenoid valve designs, a piston is used for the main closing member instead of a diaphragm. In medium-sized valves, the pilot port is usually located in the main closing member.

Such pilot-operated valves depend on a certain minimum pressure drop across the valve (approximately 3.5 kPa or more) to hold the piston or diaphragm in the *open* position. If a valve is oversized such that inadequate pressure drop is developed, the valve may chatter or fail to open fully. Valves should be sized by the capacity tables provided by the manufacturer, rather than by pipe or tube size or port diameter. Where it is desirable to keep the valve open without this pressure drop penalty, such as on refrigeration suction lines, the piston or diaphragm may be linked mechanically to the solenoid valve pin and plunger, as shown in Figure 28. The opening and closing actions are the same as before. However, the increased pulling force of the plunger as it approaches its stop position in the coil is used to hold the piston in the open position, without requiring valve pressure drop.

To obtain the maximum operating pressure differential for a given solenoid pulling power, many valves leave the plunger free to gain momentum before it knocks the valve pin out of the valve port or pilot port by impact (see Figures 26 and 28). Solenoid valves with this impact feature must have the full rated voltage, within the customary tolerance of +10 to −15%, applied instantaneously to their coils so that the valves open under rated conditions. Otherwise, they will fail. If, after the valve is in the open position, the line voltage drops below the hold-in voltage value but not to zero, the plunger drops and the valve closes. After the line voltage builds up again, the valve will not reopen, whether it is used on alternating or direct current, because the effect of the impact is lost. However, in the case of an alternating current valve, the coil will overheat and may burn out under this condition because the inherent high inrush current continues to flow through the coil when the plunger is not pulled all the way into the coil to close the air gap.

Direct-lift solenoid valves (see Figure 27), in which the valve pin is an integral part of the plunger, can be designed to open fully at rated voltage, +10 to −15%, whether the voltage is applied gradually or instantaneously. Normally open solenoid valves are usually held in the *open* position by gravity or a spring force. When energized, the power of the solenoid coil, acting on the plunger, pulls the valve pin on the valve port or pilot port to close it and, therefore, reverses the valve action.

Multiaction solenoid valves are available to accommodate many different flow configurations, such as (1) a common inlet three-way valve, which directs flow from a common inlet connection to one of two outlet connections; or (2) the four-way valve described in the section on Refrigerant-Reversing Valves. Design compromises usually result in the ports of these direct-acting valves being smaller (or the maximum operating pressure differential lower) than the equivalent normally closed two-way solenoid valves. As a result, multiaction pilot-operated solenoid valves are used in most cases. Their use warrants careful application analysis because the pressure differences required to shift and hold the valves may not exist under all operating conditions.

Selection

When solenoid valves are selected, the following factors should be considered:

1. Basic flow configuration, such as two-way and three-way normally closed.
2. Type of fluid to be handled.
3. Temperature and pressure conditions of the entering fluid.
4. Allowable fluid flow pressure drop across the valve needed to establish the port size for the required capacity.
5. Capacity in appropriate terms; do not size for less pressure drop than is needed to open a piloted valve. Use capacity tables for sizing to ensure that pressure drop at least equals the minimum pressure drop required for operation as specified by the manufacturer. If a piloted valve is too far oversized, it may chatter or fail to open fully. If no minimum pressure drop is specified, 3.5 kPa is normally safe. If the valve is advertised to have a **zero pressure drop** opening feature, minimum pressure drop need not be considered.
6. Maximum operating pressure differential under which the valve will be required: (*a*) to open, for the normally closed valve; and (*b*) to close, for the normally open valve. Three- and four-way valves require additional information on the operating conditions for which they are intended.
7. Maximum rated pressure should not be confused with the pressure difference under which the valve is required to open.
8. Type and size of line connections.
9. Electrical characteristics for the solenoid coil. Voltage and frequency must be specified for alternating current, but only voltage is specified for direct current.
10. Ambient temperature in which the valve will be located.
11. Cycling rate of valve.
12. Hazard of location, which may make explosion-proof coil housings necessary.

Application

Spring-loaded solenoid valves (see Figure 27) usually can be installed in vertical lines or any other position. The solenoid valves in Figure 26 should be installed upright in horizontal lines.

Solenoid valves must be used with the correct current characteristics. Momentary overvoltage is not harmful, but sustained overvoltage of more than 10% may cause coils to burn out. Undervoltage is harmful to alternating current-operated valves if it reduces operating power enough to prevent the valve from opening when the coil is energized. This condition may cause burnout of an alternating current coil, as discussed previously.

When the solenoid valve is energized by a control transformer of limited capacity, the transformer must be able to provide proper voltage during the inrush load. As the inrush alternating current may be several times the holding current, it is useless to check the voltage at the coil leads when only holding current is being supplied. For such applications, the inrush current in amperes, multiplied by the rated coil voltage, gives the necessary volt-ampere capacity that must be provided by the transformer for each solenoid valve simultaneously actuated. The inrush and holding currents of a direct current solenoid valve are equal.

Fuses protecting electrical lines for solenoid valves should be sized according to holding current and preferably should be slow blowing type. To protect the coil insulation and the controlling switch, a capacitor or other device may be wired across the coil leads of a high-voltage, direct current solenoid valve to absorb the counter-voltage surge generated by the coil when the circuit is broken.

Whenever a solenoid valve is reassembled after installation, the magnetic coil sleeves (if required) must be replaced in their correct positions to operate the valve properly. Leaving the coil sleeves out of an alternating current valve may cause the coil to burn out.

To avoid valve failure because of low voltage, the solenoid valve coil should not be energized by the same contacts or at the same instant that a heavy motor load is connected to the electrical supply line. The solenoid coil can be energized immediately before or after the heavy motor load is connected to the line.

Solenoid valves are used for the following applications:

1. **Refrigerant Liquid.** To prevent flow of refrigerant liquid to the evaporator, a solenoid valve is installed in the liquid line just ahead of the expansion valve to (1) prevent flow of refrigerant liquid to the evaporator when the compressor is idle, (2) provide individual temperature control in each room of a multiple system, or (3) control the number of evaporator sections used as the load varies on a central air-conditioning installation.
2. **Refrigerant Suction Gas.** In commercial multiple systems, especially those having evaporators containing a large amount of refrigerant, solenoid valves are often provided in the suction line from each unit or room, as well as in the liquid line, to isolate each evaporator completely. Otherwise, refrigerant gas can migrate from one evaporator to another through the suction line during the *off* cycle. This migration causes uneven performance and possible floodback when the compressor starts again.

   In applications 1 and 2, the solenoid valves are operated by thermostats. The compressor may be operated by a low-pressure switch or directly by a thermostat. The compressor may or may not be operated on a pumpdown cycle, either a continuous one or a pumpdown and lockout cycle.
3. **Refrigerant Discharge Gas.** In many hot-gas defrost applications, a solenoid valve, installed in a line connected to the discharge line between the compressor and the condenser, feeds the evaporator with hot gas, which provides heat for the defrosting operation. The solenoid valve remains closed, except during the defrosting operation. A solenoid valve installed in a bypass around one or more compressor cylinders provides compressor-capacity control. This valve may be used to bypass the entire compressor output and reduce the compressor starting load where required.
4. **Water and Other Liquids.** Solenoid valves are used to control the flow of water and many other liquids. Water is one of the more harmful liquids because it deposits solids on the internal solenoid valve surfaces, causing corrosion; therefore, solenoid valves for water service should be easy to dismantle and clean.
5. **Air.** Many air systems rely on solenoid valves to operate controls or actuators. Since rapid cycling is often required, solenoid valves for air service should be selected for endurance.
6. **Steam.** The application of solenoid valves in industrial steam systems is quite varied. Because of the continuous high steam temperatures involved, special high-temperature solenoid coils are usually required. The ambient temperatures in which the valves are located are also a concern.

Pilot Solenoid Valve Application

Two-way, normally closed, direct-acting pilot solenoid valves are used with (1) a large, piston-type, spring-loaded expansion valve to provide refrigerant liquid shutoff service, as shown in Figure 13; (2) either an evaporator-pressure regulator to provide

Refrigerant-Control Devices

shutoff service or a pressure pilot selector; (3) a large piston-type, spring-loaded regulator to provide refrigerant gas shutoff service; and (4) four-way refrigerant switching valves on heat-pump systems to accomplish cooling, heating, and defrosting.

Three-way, direct-acting, pilot solenoid valves are used to operate (1) a cylinder unloading mechanism for compressor capacity reduction; and (2) three- and four-way reversing valves on heat-pump systems to accomplish cooling, heating, and defrosting.

Refrigerant-Reversing Valves

These are three- or four-way two-position valves that are usually operated by pilot solenoid valves and designed for reversing or changing the direction of refrigerant flow through certain parts of a refrigeration system. They are used on refrigeration and year-round air-conditioning (heat-pump) systems to control cooling, heating, and defrosting operations.

Operation. Reversing valves may be operated by either a two-, three-, or four-way pilot solenoid valve, which may be an integral part of the reversing valve or a separate valve connected to the reversing valve by tubes.

Pilot solenoid valves can be divided into poppet valve and slide valve groups. The reversing valve is usually a slide-type valve, with the slide classified by the spool design or shell design. Figures 29 and 30 show a slide-type pilot valve with a shell design slide component. The slide component F moves across a flat seat containing ports K, L, and M, which connect to tubes of the refrigeration system. With a spool slide component, a spool with flow passages moves across ports inside the cylindrical reversing valve body. The following principles of operation apply to most valves, although some designs slightly modify the operation.

The **four-way reversing valve** may be connected so the refrigeration system is fail-safe on either the heating or cooling cycle if the solenoid valve coil fails. The valve is connected by interchanging the two lines on the reversing valve, which connect to the inside and outside heat-exchange coils. Figure 29 shows refrigeration flow through a four-way slide reversing valve in the cooling (or defrosting) cycle.

During the cooling cycle shown in Figure 29, the pilot valve is energized and pilot Port A opens to bleed high-pressure refrigerant into Chamber C. Simultaneously, pilot Port B is connected to low pressure and bleeds refrigerant from Chamber D. The pressure difference across Piston G develops enough force on Piston G to cause Pistons G and H and the connecting structural member to move Slide F from encompassing Ports K and L to encompassing Ports M and L. When G reaches the end of its stroke, the valve has reversed, and the inside coil is connected to low-pressure cool refrigerant, while the outside coil is connected to high-pressure hot refrigerant.

To reverse to the heating cycle (Figure 30), the pilot valve is de-energized. This opens pilot Port B, which bleeds high-pressure refrigerant into Chamber D and connects pilot Port A to low pressure, which bleeds refrigerant from Chamber C. The valve slide moves in the same manner as before, and the inside coil is connected to high-pressure hot refrigerant, while the outside coil is connected to low-pressure cool refrigerant.

A similar reversing action of Pistons G and H can be produced by a three-way pilot valve. The connection J to the high-pressure tube of the main valve is omitted, and the pilot valve operates to connect either Chamber C or D to the low-pressure source. The high pressure in Chamber E acts on the Pistons G and H and moves the Slide F toward the chamber with the low pressure. In this case, a bleed hole in each piston allows refrigerant to fill Chambers C and/or D when these chambers expand in volume. Also, the piston must seal the main valve end-cap ports to prevent leakage through the pilot valve to the suction Tube S of the main valve.

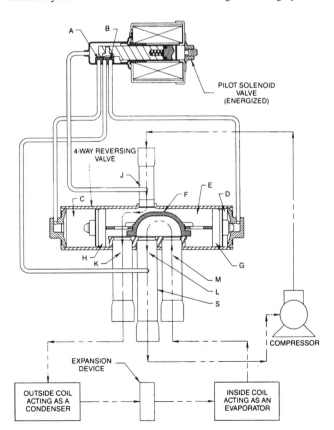

Fig. 29 Four-Way Slide-Type Refrigerant-Reversing Valve Used in Cooling (or Defrosting) Cycle of Refrigeration System

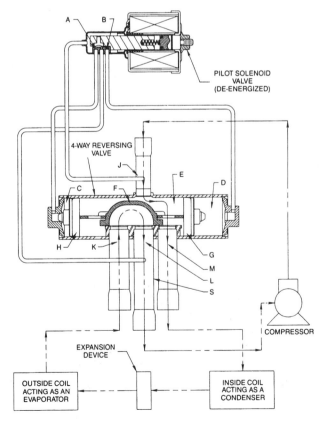

Fig. 30 Four-Way Slide-Type Refrigerant-Reversing Valve Used in Heating Cycle of Refrigeration System

Reversing valves operate well with most fluorinated refrigerants, but the type of refrigerant used affects the size of the valve selected for a given capacity. The valve should have minimum pressure drops through the valve passages in both the discharge and suction gas paths so that the compressor capacity is not reduced.

Application. In addition to their year-round air-conditioning (heat-pump) application on residential or commercial systems, four-way reversing valves are used in refrigeration systems in motor trucks, trailers, and railway refrigerator cars for the transportation of perishable cargo. The transport systems can be arranged to operate automatically and to provide cooling, heating, and defrosting.

While the four-way valves are sometimes adapted for other applications, three-way reversing valves are designed specifically for commercial refrigeration systems to defrost the evaporator or for heat reclaim using an auxiliary condenser.

Regardless of the type of installation to which the reversing valve is added (e.g., window unit, residential unit, commercial unit, transportation unit, or customer-built unit), the system may be made to operate automatically. If a dual-action thermostat is used, cooling automatically occurs when the temperature rises above a preset value, and heating automatically occurs when the temperature drops below a preset value.

CONDENSING WATER REGULATORS

Two-Way Regulators

The condensing water regulator modulates the quantity of water passing through a water-cooled refrigerant condenser in response to the condensing pressure. This regulator is used on a vapor-cycle refrigeration system to maintain a condensing pressure that loads but does not overload the compressor motor. The regulator automatically modulates to correct for both variations in temperature or pressure of the water supply and variations in the quantity of refrigerant gas that the compressor is sending to the condenser.

Operation

The condensing water regulator consists of a valve and an actuator that are linked together, as shown in Figure 31. The actuator consists of a metallic bellows and adjustable spring combination connected to the system condensing pressure. For equipment with a large water flow, a small condensing water regulator is used as pilot valve for a diaphragm-type main valve.

After a compressor starts, the condensing pressure begins to rise. When the opening pressure setting of the regulator spring is reached, the bellows moves to open the valve disk or the slide gradually. The regulator continues to open as the condensing pressure rises until the water flow balances the required heat rejection. At this point the condensing pressure is stabilized. When the compressor stops, the continuing water flow through the regulator causes the condensing pressure to drop gradually and the regulator becomes fully closed when the opening pressure setting of the regulator is reached.

Selection

The condensing water regulator should be selected from the manufacturer's data on the basis of maximum required flow, minimum available pressure drop, and water temperature. Also, while one standard bellows operator can sometimes handle several refrigerants, special springs or bellows may be required for very high- or low-pressure refrigerants. For example, R-717 (ammonia), requires the use of stainless steel rather than brass bellows.

The water flow required depends on condenser performance, the temperature of available water, the quantity of heat that must be rejected to the water, and the allowable exit water temperature. For a given opening of the valve seat, which corresponds to a given pressure rise above the regulator opening point, the flow rate handled by a given size water regulator is a function of the available water-pressure drop across the valve seat. Available water-pressure drop is determined for the required flow by deducting condenser water-pressure drop, pipeline pressure drop, and static pressure losses from the pressure of the water at its supply point.

Application

To avoid hunting, a regulator should not be oversized. When a two-way condensing water regulator (see Figure 31) is used on a recirculating cooling tower, the throttling action of the valve during cold weather reduces circulation through the pump and tower, which is undesirable for that equipment.

Three-Way Regulators

A three-way condensing water regulator should be used on a cooling tower requiring individual condensing pressure control (see Figure 32). These regulators are similar in construction to two-way regulators, but they have an additional port, which opens to bypass water around the condenser as the port controlling water flow to the condenser closes. Thus, the tower decking or sprays and the circulating pump receive a constant supply of water, although the water supply to individual condensers is modulated for control.

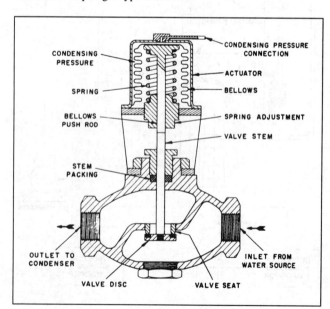

Fig. 31 Two-Way Condensing Water Regulator

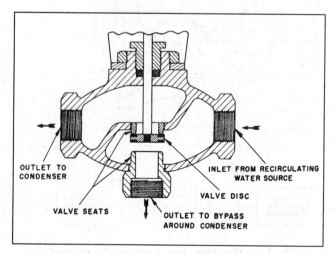

Fig. 32 Three-Way Condensing Water Regulator

Refrigerant-Control Devices

Three-way condensing water regulators must be supplemented by other means if cooling tower systems are to be operated in freezing weather. An indoor sump is usually required, and a temperature-actuated three-way water control valve is used to divert periodically all of the condenser leaving water directly to the sump whenever the water becomes too cold.

Unlike the recommended and accepted practice pertaining to the application of refrigerant-control valves, a strainer is not ordinarily used with a water regulator because a strainer usually requires more cleaning and servicing than a regulator without a strainer.

CHECK VALVES

Refrigerant check valves are normally used in refrigerant lines in which pressure reversals can cause undesirable reverse flows. A check valve is usually opened by a portion of the pressure drop. Closing usually occurs either when a reversal of pressure takes place or when the pressure drop across the check valve is less than the minimum opening pressure drop in the normal flow direction.

The conventional, large check valve uses piston construction in a globe valve body, while in-line designs are common for 50 mm or smaller valves. Either design may be include a closing spring; a heavier spring gives more reliable and tighter closing but require greater pressure to open. Although conventional check valves may be designed to open at less than 7 kPa, they may not be reliable below −32°C because the light closing springs can not overcome the viscous lubricant.

Other designs are used for various special functions not covered here, such as the **excess flow check valve**, which closes only when flow exceeds the maximum desired rate; the **electrically lifted check valve**, which requires no pressure drop to remain open; the **remote pressure-operated check valve**, which is normally open but closes when supplied with a higher pressure source of refrigerant; and the **pressure differential valve**, which maintains a uniform pressure differential between components for special functions in a refrigeration system.

Seat Materials

Although precision metal seats may be manufactured nearly bubble-tight, they are not economical for refrigerant check valves. Seats made of synthetic elastomers provide excellent sealing at medium and high temperatures, but may leak at low temperatures due to their lack of resilience. Because high temperatures deteriorate most elastomers suitable for refrigerants, plastic materials have become more widely used, despite being susceptible to damage by large pieces of foreign matter.

Applications

In compressor discharge lines, check valves are used to prevent flow from the condenser to the compressor during the *off* cycle or to prevent flow from an operating compressor to an idle compressor. While a 15 to 40 kPa pressure drop is tolerable, the check-valve must resist pulsations cause by the compressor and the temperature of discharge gas. Also, the valve must be bubble-tight to prevent liquid refrigerant from accumulating at the compressor discharge valves or in the crankcase.

In liquid lines, a check valve prevents reverse flow through the unused expansion device on a heat-pump or prevents backup into the low-pressure liquid line of a recirculating system during a defrosting. While a 15 to 40 kPa pressure drop is usually acceptable, the check-valve seat must be bubble-tight.

In the suction line of a low-temperature evaporator, a check valve may be used to prevent the transfer of refrigerant vapor to a lower temperature evaporator on the same suction main. In this case, the pressure drop must be less than 14 kPa, the valve seating must be reasonably tight, and the check valve must be reliable at low temperatures.

Normally, open pressure-operated check valves are used to close suction lines, gas, or liquid legs in gravity recirculating systems during defrost.

In hot-gas defrost lines, check valves may be used in the branch hot-gas lines connecting the individual evaporators to prevent crossfeed of refrigerant during the cooling cycle when the defrost operation is not taking place. In addition, check valves are used in the hot-gas line between the hot-gas heating coil in the drain pan and the evaporator, to prevent pan coil sweating during the refrigeration cycle. Tolerable pressure drop is typically 14 to 41 kPa, seating must be nearly bubble-tight, and seat materials must withstand high temperatures.

To prevent chatter or pulsation, check valves should be sized for the particular pressure drop that ensures that they are in the wide-open position at the desired flow rate.

RELIEF DEVICES

A refrigerant relief device has either a safety or functional use. A safety relief device is designed to relieve positively at its set pressure for one crucial occasion without prior leakage. The relief may be to the atmosphere or to the low-pressure side.

A functional relief device is a control valve that may be called on to open, modulate, and close with repeatedly accurate performance. Relief is usually from a portion of the system at higher pressure to a portion at lower pressure. Design refinements of the functional relief valve usually make it unsuitable or uneconomical as a safety relief device.

Safety Relief Valves

These valves are most commonly a pop-type design, which open abruptly when the inlet pressure exceeds the outlet pressure by the valve setting pressure (see Figure 33). Seat configuration is such that once lift begins, the resulting increased active seat area causes the valve seat to pop wide open against the force of the setting spring. Because the flow rate is measured at a pressure of 10% above the setting, the valve must open within this 10% increase in pressure.

This relief valve operates on a fixed pressure differential from inlet to outlet. Because the valve is affected by back pressure, a rupture disk must not be installed at the valve outlet.

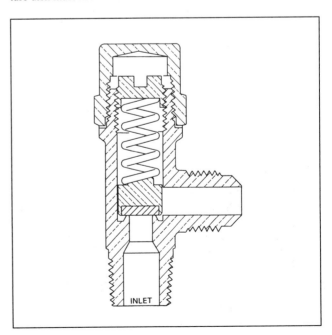

Fig. 33   Pop-Type Safety Relief Valve

Relief valve seats are made of metal, plastic, lead alloy, or synthetic elastomers. Elastomers are commonly used because they have greater resilience and, consequently, reseat more tightly than other materials. Some valves that have lead-alloy seats have an emergency manual reseating stem that permits reforming the seating surface by tapping the stem lightly with a hammer. The advantages of the pop-type relief valve are simplicity of design, low initial cost, and high-discharge capacity.

Other Safety Relief Devices

The **fusible plug** and the **rupture disk** (Figure 34) provide similar safety relief. The former contains a fusible member that melts at a predetermined temperature corresponding to the safe saturation pressure of the refrigerant, but is limited in application to pressure vessels with internal gross volumes of 0.085 m^3 or less and internal diameters of 150 mm or less. The rupture member contains a preformed disk designed to rupture at a predetermined pressure. These devices may be used as a stand alone device or installed at the inlet to a safety relief valve.

When these devices are installed in series with a safety relief valve, the chamber created by the two valves must have a pressure gage or other suitable indicator. A rupture disk will not burst at its design pressure if back pressure builds up in the chamber.

The rated relieving capacity of a relief valve alone must be multiplied by 0.9 when it is installed in series with a rupture disk (unless the relief valve has been rated in combination with the rupture disk).

Discharge Capacity. The minimum required discharge capacity of the pressure relief device or fusible plug for each pressure vessel is determined by the following formula, specified by ASHRAE Standard 15:

$$C = fDL \qquad (2)$$

where

C = minimum required air discharge capacity of relief device, kg/s
D = outside diameter of vessel, m
L = length of vessel, m
f = factor dependent on refrigerant, as shown in Table 1

Capacities of pressure relief valves are determined by test in accordance with the provisions of the ASME *Boiler and Pressure Vessel Code*. Relief valves approved by the National Board of Boiler and Pressure Vessel Inspectors are stamped with the code symbol, which consists of the letters UV in a clover leaf design with the letters NB stamped directly below this symbol. In addition, the pressure setting and capacity are stamped on the valve.

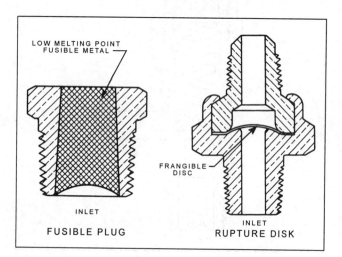

Fig. 34 Safety Relief Devices

Table 1 Values for f for Discharge Capacity of the Pressure Relief Devices

Refrigerant	Value of f
When used on the low side of a limited-charge cascade system:	
R-170, R-744, R-1150	0.082
R-13, R-13B1, R-503	0.163
R-14	0.203
Other applications:	
R-717	0.041
R-11, R-40, R-113, R-123, R-142b, R-152a, R-290, R-600, R-600a, R-611, R-764	0.082
R-12, R-22, R-114, R-134a, R-C 318, R-500, R-1270	0.163
R-115, R-502	0.203

Notes:
1. The values of f listed do not apply if fuels are used within 6 m of the pressure vessel. In this case, the methods in API RP 520 shall be used to size the pressure relief device.
2. When one pressure relief device or fusible plug is used to protect more than one pressure vessel, the required capacity must be the sum of the capacities required for each pressure vessel.
3. For refrigerants not listed, consult ASHRAE *Standard* 15.

When relief valves are used on pressure vessels of 0.28 m^3 internal gross volume or more, a relief system consisting of a three-way valve and two relief valves in parallel is required.

The rated discharge capacity of a rupture member or fusible plug that discharges to the atmosphere under critical flow conditions is determined by calculation, using the formula provided in ASHRAE *Standard* 15.

Pressure Setting. The maximum pressure setting for a relief device is limited by the design working pressure of the vessel to be protected. Pressure vessels normally have a safety factor of 5. Therefore, the minimum bursting pressure is five times the rated design working pressure. The relief device must have enough discharge capacity to prevent the pressure in the vessel from rising more than 10% above its design pressure. Since the capacity of a relief device is measured at 10% above its stamped setting, the setting cannot exceed the design pressure of the vessel.

To prevent loss of refrigerant through pressure relief devices during normal operating conditions, the relief setting must be substantially higher than the system operating pressure. For rupture members, the setting should be 50% above a static system pressure and 100% above a maximum pulsating pressure. Failure to provide this margin of safety causes fatigue of the frangible member and rupture well below the stamped setting.

For relief valves, the setting should be 25% above maximum system pressure. This safety factor provides sufficient spring force on the valve seat to maintain a tight seal and still allow for setting tolerances and other factors that cause settings to vary. Although relief valves are set at the factory to be close to the stamped setting, the variation may be as much as 10% after the valves have been stored or placed in service.

Discharge Piping. The size of the discharge pipe from the pressure relief device or fusible plug must not be less than the size of the pressure relief device or fusible plug outlet. The maximum length of the discharge piping is provided in a table or may be calculated from the formula provided in ASHRAE *Standard* 15.

Selection and Installation. The following factors should be considered when selecting and installing a relief device:

- Select a relief device with sufficient capacity for code requirements and one suitable for the type of refrigerant used.
- Use the proper size and length of discharge tube or pipe.
- Do not discharge the relief device prior to installation or when pressure testing the system.
- For systems containing large quantities of refrigerant, use a three-way valve and two relief valves.

Refrigerant-Control Devices

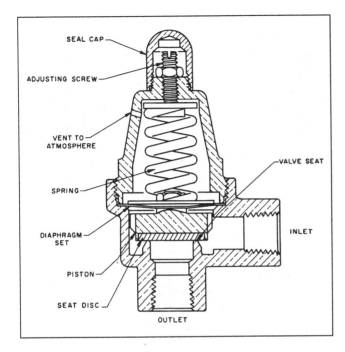

Fig. 35 Diaphragm-Type Relief Valve

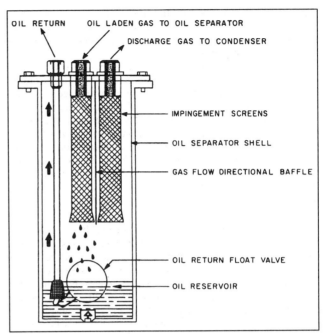

Fig. 36 Discharge-Line Lubricant Separator

- Install a pressure vessel that permits the relief valve to be set at least 25% above the maximum system pressure.

Functional Relief Valves

Functional relief valves are usually diaphragm types in which the system pressure acts on a diaphragm that lifts the valve disk from the seat (Figure 35). The other side of the diaphragm is exposed to both the adjusting spring and atmospheric pressure. The ratio of effective diaphragm area to seat area is high, so the outlet pressure has little effect on the operating point of the valve.

Because the lift of the diaphragm is not great, the diaphragm valve is frequently built as the pilot or servo of a larger piston-operated main valve to provide both sensitivity and high flow capacity. Construction and performance are similar to the previously described pilot-operated evaporator pressure regulator, except that diaphragm valves are constructed for higher pressures. Thus, the valves are suitable for use as defrost relief from evaporator to suction pressure, as large-capacity relief from a pressure vessel to the low side, or as a liquid refrigerant pump relief from pump discharge to the accumulator to prevent excessive pump pressures when some evaporators are valved closed.

DISCHARGE-LINE LUBRICANT SEPARATORS

The **discharge-line lubricant separator** removes lubricant from the discharge gas of helical rotary (screw) and reciprocating compressors. Lubricant is separated by (1) reducing gas velocity, (2) changing direction of flow, (3) impingement on baffles, (4) mesh pads or screens, and (5) centrifugal force. The separator reduces the amount of lubricant reaching the low-pressure side, helps maintain the lubricant charge in the compressor sump, and muffles the sound of the gas flow.

Figure 36 shows a small separator incorporating inlet and outlet screens and a high-side float valve. A space below the float valve allows for dirt or carbon sludge. When lubricant accumulates it raises the float ball. Lubricant then passes through a needle valve and returns to the low-pressure crankcase. When the level falls, the needle valve closes, preventing the release of hot gas into the crankcase. Insulation and electric heaters may be added to prevent the refrigerant from condensing when the separator is exposed to low temperatures. A wide variety of horizontal and vertical flow separators is manufactured with one or more of such elements as centrifuges, baffles, wire mesh pads, or cylindrical filters.

Selection

Separators are usually given capacity ratings for several refrigerants at several suction and condensing temperatures. Another rating method gives the capacity in terms of compressor displacement volume. Some separators also show a marked reduction in separation efficiency at some stated minimum capacity. Because the compressor capacity increases when the suction pressure is raised or the condensing pressure is lowered, the system capacity at its lowest compression ratio should be the criterion for selecting the separator.

Application

A discharge-line lubricant separator is best for ammonia or hydrocarbon refrigerants to reduce fouling in the evaporator. With lubricant-soluble halocarbon refrigerants, only certain flooded systems, low-temperature systems, or systems with long suction lines or other lubricant return problems need lubricant separators. (See Chapter 2 for more information about lubricant separators.)

CAPILLARY TUBES

Every refrigerating unit requires a pressure-reducing device to meter the flow of refrigerant from the high pressure side to the low pressure side of the refrigerating system according to load demand. The capillary tube is especially popular for smaller unitary hermetic equipment such as household refrigerators and freezers, dehumidifiers, and room air conditioners. Capillary tube use also extends to larger units such as unitary air conditioners in sizes up to 35 kW capacity.

The capillary operates on the principle that liquid passes through it much more readily than vapor. It is a length of drawn copper tubing with a small inner diameter. When used for controlling refrigerant flow, it connects the outlet of the condenser to the inlet of the evaporator. Despite its small inner diameter, the term capillary tube is a misnomer because the inner bore is much too large to allow capillary action. In some applications, the capillary tube is soldered to the suction line and the combination is called a **capillary tube-suction line heat exchanger**.

A high pressure-side liquid receiver is not normally used with a capillary tube; consequently, a less refrigerant charge is needed. In a few applications, such as household refrigerators, freezers, room air conditioners, and heat pumps, a small, low-pressure-side accumulator may be used. Because the capillary tube allows pressure to equalize when the refrigerator is off, a compressor motor with a low starting torque may be used. A capillary tube does not operate as efficiently over as wide a range of conditions as does a thermostatic expansion valve; however, a capillary tube is less expensive and generally performs nearly as well.

Theory

A capillary tube passes liquid much more readily than vapor due to the increased friction with the vapor; as a result, it is a practical metering device. When a capillary tube is sized to permit the desired flow of refrigerant, the liquid seals its inlet. If the system becomes unbalanced, some vapor (uncondensed refrigerant) enters the capillary tube. This vapor reduces the mass flow of refrigerant considerably, which increases condenser pressure and causes subcooling at the condenser exit and capillary tube inlet. The result is an increase of the mass flow of refrigerant through the capillary tube. If properly sized for the application, the capillary tube compensates automatically for load and system variations and gives acceptable performance over a wide range of operating conditions.

A common flow condition is to have subcooled liquid at the entrance to the capillary tube. Bolstad and Jordan (1948) described the flow behavior from temperature and pressure measurements along the tube as follows:

"With subcooled liquid entering the capillary tube, the pressure distribution along the tube is similar to that shown in the graph (see Figure 37). At the entrance to the tube, section 0-1, a slight pressure drop occurs, usually unreadable on the gauges. From point 1 to point 2, the pressure drop is linear. In the portion of the tube 0-1-2, the refrigerant is entirely in the liquid state, and at point 2, the first bubble of vapor forms. From point 2 to the end of the tube, the pressure drop is not linear, and the pressure drop per unit length increases as the end of the tube is approached. For this portion of the tube, both the saturated liquid and saturated vapor phases are present, with the percent and volume of vapor increasing in the direction of flow. In most of the runs, a significant pressure drop occurred from the end of the tube into the evaporator space.

With a saturation temperature scale corresponding to the pressure scale superimposed along the vertical axis, the observed temperatures may be plotted in a more efficient way than if a uniform temperature scale were used. The temperature is constant for the first portion of the tube 0-1-2. At point 2, the pressure has dropped to the saturation pressure corresponding to this temperature. Further pressure drop beyond point 2 is accompanied by a corresponding drop in temperature, the temperature being the saturation temperature corresponding to the pressure. As a consequence, the pressure and temperature lines coincide from point 2 to the end of the tube."

Mikol (1963) and Li et al. (1990) showed that generation of the first vapor bubble does not occur at the point where the liquid pressure reaches the saturation pressure (Point 2 on Figure 37), but rather the refrigerant remains in the liquid phase for some limited length past Point 2, reaching a pressure below the saturation pressure. This delayed evaporation, often referred to as a metastable or superheated liquid condition, must be accounted for in analytical modeling of the capillary tube, or the mass flow rate of refrigerant will be underestimated (Kuehl and Goldschmidt 1991, Wolf et al. 1995).

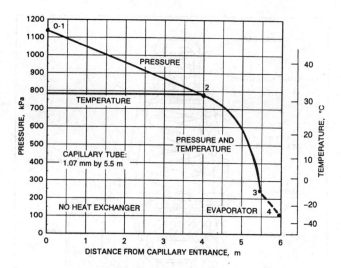

Fig. 37 Pressure and Temperature Distribution Along Typical Capillary Tube
(Bolstad and Jordan 1948)

The rate of refrigerant flow through a capillary tube always increases with an increase in inlet pressure. Flow rate also increases with a decrease in external outlet pressure down to a certain critical value, below which the flow does not change (choked flow). Figure 37 illustrates a case in which the outlet pressure inside the capillary tube has reached the critical value, which is higher than the external pressure. Such a condition is typical for normal operation. The point at which the first gas bubble appears, is called the **bubble point**. The preceding portion of capillary tube is called the **liquid length**, and that following is called the **two-phase length**.

System Design Factors

A capillary tube must be compatible with other components. In general, once the compressor and heat exchangers have been selected to meet the required design conditions, the capillary tube size and system charge are determined. However, detailed design considerations may be different for different applications (domestic refrigerator, window air conditioner, residential heat pump).

Capillary tube size and system charge together are used to determine subcooling and superheat for a given design. Performance at off-design conditions should also be checked. Capillary tube systems are generally much more sensitive to the amount of refrigerant charge than expansion valve systems.

The high-pressure side must be designed for use with a capillary tube. To prevent rupture in case the capillary tube becomes blocked, the high-side volume should be sufficient to contain the entire refrigerant charge. A sufficient refrigerant storage volume may also be needed to protect against excessive discharge pressures during high-load conditions.

Unloading during the off period is another concern in designing the high-side. When the unit stops, refrigerant continues to pass through the capillary tube from the high side to the low side until pressures are equal. If liquid is trapped in the high side, it will evaporate there during the off cycle, pass to the low side as a warm gas, condense, and add latent heat to the evaporator. Therefore, good drainage of the liquid to the capillary tube during this unloading interval should be provided. Liquid trapping may also increase the time for the pressure to equalize after the compressor stops operating. If this interval is too long, the compressor may not be sufficiently unloaded to start easy, especially in domestic refrigerators.

The maximum quantity of refrigerant is in the evaporator during the off cycle and the minimum during the running cycle. The suction piping should be arranged to reduce the adverse effects of the

Refrigerant-Control Devices

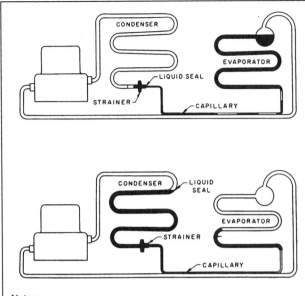

Notes:
1. Capillary selected for capacity balance conditions. Liquid seal at capillary inlet but no excess liquid in condenser. Compressor discharge and suction pressures normal. Evaporator properly charged.
2. Too much capillary resistance—liquid refrigerant backs up in condenser and causes evaporator to be undercharged. Compressor discharge pressure may be abnormally high. Suction pressure below normal. Bottom of condenser subcooled.

Fig. 38 Effect of Capillary Tube Selection on Refrigerant Distribution

variable-charge distribution. A suitable liquid accumulator is sometimes needed.

In some systems (e.g., refrigerators), a capillary tube is soldered to the suction line. The capillary tube heats the suction line to keep water vapor from the ambient air from condensing on the suction line. This capillary tube/suction line heat exchanger may also increase efficiency (Domanski et al. 1994).

When a capillary tube-suction line heat exchanger is used, the excess capillary tube length may be coiled and placed at either end of the heat exchanger. Although more heat is exchanged when the excess is coiled at the evaporator, the system stability is improved if some of the capillary tube coil is located at the condenser. Coils and bends must be formed carefully to avoid local restrictions. The effect of coils and bends on the restriction should be considered when specifying the capillary tube.

Capacity Balance Characteristic

The selection of a capillary tube depends on the application and anticipated range of operating conditions. One approach to the problem involves the concept of **capacity balance**. A refrigeration system operate s at capacity balance when the resistance of the capillary tube is sufficient to maintain a liquid seal at its entrance without excess liquid accumulating in the high side (see Figure 38). Only one such capacity balance point exists for any given compressor discharge pressure. A curve through the capacity balance points for a range of compressor discharge pressures is called the capacity balance characteristic of the system. Such a curve is shown in Figure 39. Ambient temperatures are drawn on the chart for a typical air-cooled system. A given set of compressor discharge and suction pressures is associated with fixed condenser and evaporator pressure drops; these pressures establish the capillary tube inlet and outlet pressures.

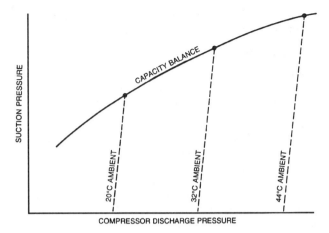

Note: Operation below this curve results in a mixture of liquid and vapor entering the capillary. Operation above the capacity balance points causes liquid to back up in the condenser and elevate its pressure.

Fig. 39 Capacity Balance Characteristic of Capillary System

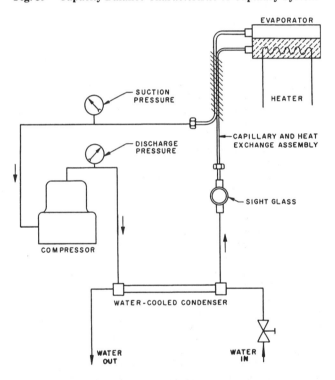

Fig. 40 Test Setup for Determining Capacity Balance Characteristic of Compressor, Capillary, and Heat Exchanger

The capacity balance characteristic curve for any combination of compressor and capillary tube may be determined experimentally by the arrangement shown in Figure 40. This test arrangement makes it possible to vary independently the suction and discharge pressures until capacity balance is obtained. The desired suction pressure may be obtained by regulating the heat input to the low side, usually by electric heaters. The desired discharge pressure may be obtained by a suitably controlled water-cooled condenser. A liquid indicator is located at the entrance to the capillary tube. The usual test procedure is to hold the high-side pressure constant and, with gas bubbling through the sight glass, slowly increase the suction pressure until a liquid seal forms at the capillary tube entrance. Repeating this procedure at various discharge pressures determines the capacity balance characteristic curve similar to that shown in

Figure 39. This equipment may also be used as a calorimeter to determine the capacity of the refrigerating system.

Optimum Selection and Refrigerant Charge

Whether the initial capillary tube selection and charge are optimum for the unit is always questioned, even in such simple applications as a room air-conditioning unit. The refrigerant charge in the unit can be varied using a small refrigerant bottle (valved off and sitting on a scale) connected to the circuit. The interconnecting line must be flexible and arranged so that it is filled with vapor instead of liquid. The charge is brought in or removed from the unit by heating or cooling the bottle.

The only test for varying the capillary tube restriction is to remove the element, install a new capillary tube, and determine the optimum charge, as outlined previously. A method occasionally used is to pinch the capillary tube to determine whether or not increased resistance is needed.

The refrigeration unit should be operated through its expected range to determine power and cooling capacity for any given selection and charge combination.

Application

Processing and Inspection. To prevent mechanical clogging by foreign particles, a strainer should be installed ahead of the capillary tube. Also, all parts of the system must be evacuated adequately to eliminate water vapor and non-condensable gases, which may cause clogging by corrosion. The lubricant should be free from wax separation at the minimum operating temperature.

The interior surface of the capillary tube should be smooth and uniform in diameter. Although plug-drawn copper is more common, wire-drawn or sunk tubes are also available. Life tests should be conducted at low evaporator temperatures and high condensing temperatures to check the possibility of corrosion and plugging. Material specifications for seamless copper tube are given in ASTM *Standard* B 75. Similar information on hand-drawn copper tubes can be found in ASTM *Standard* B 360.

A procedure should be established to ensure uniform flow capacities, within reasonable tolerances, for all capillary tubes used in product manufacture. This procedure may be conducted as follows. The final capillary tube, determined from tests, is removed from the unit and given an airflow capacity rating, using the wet-test meter method described in ASHRAE *Standard* 28. Master capillary tubes are then produced, by using the wet-test meter airflow equipment, to provide the maximum and minimum flow capacities for the particular unit. The maximum flow capillary tube has a flow capacity equal to that of the test capillary tube, plus a specified tolerance. The minimum flow capillary tube has a flow capacity equal to that of the test capillary tube, less a specified tolerance. One sample of the maximum and minimum capillary tubes is sent to the manufacturer of capillary tubes to be used as tolerance guides for elements supplied for a particular unit. Samples are also sent to the inspection group for quality control.

Considerations. In the selection of a capillary tube for a specific application, practical considerations influence the length. For example, the minimum length is determined by such geometric considerations as the physical distance between the high side and low side and the length of capillary tube required for optimum heat exchange. It may also be dictated by exit velocity, noise, and the possibility of plugging with foreign materials. The maximum length may be determined primarily by cost. It is fortunate, therefore, that the flow characteristics of a capillary tube can be adjusted independently by varying either its bore or its length. Thus, it is feasible to select the most convenient length independently and then (within certain limits) select a bore to give the desired flow. An alternate procedure is to select a standard bore and then adjust the length, as required.

ASTM *Standard* B 360 lists standard diameters and wall thicknesses for capillary tubes. Many nonstandard tubes are also used, resulting in nonuniform interior surfaces and variations in flow.

Capillary Tube Selection

Wolf et al. (1995) developed refrigerant-specific rating charts to predict refrigerant flow rates through adiabatic capillary tubes. The methodology involves determining two quantities from a series of curves. They are similar to rating charts for R-12 and R-22 developed by Hopkins (1950) and Whitesel (1957). The two quantities necessary for the prediction of the refrigerant flow rate through the adiabatic capillary tube are a flow rate through a reference capillary tube and a flow factor f, which is a geometric correction factor. These two quantities are multiplied together to calculate the flow rate.

Figures 41 and 42 are rating charts for pure R-134a through adiabatic capillary tubes. Figure 41 plots the capillary tube flow rate as a function of the inlet condition and inlet pressure for a reference capillary tube geometry of 0.034 in. I.D. and 130 in. long. Figure 42 is a geometric correction factor. Using the desired capillary tube geometry, a flow factor f may be determined. This flow factor is then multiplied with the flow rate from Figure 41 to determine the predicted capillary tube flow rate. *Note*: For quality inlet conditions for R-134a, an additional correction factor of 0.95 is necessary to obtain the proper results.

Figures 43, 44, and 45 present the rating charts for pure R-410A through adiabatic capillary tubes. The method of selection from these charts is identical to that previously presented for R-134a with the exception that an additional correction factor for quality inlet conditions is not used. A separate flow factor chart (Figure 45), however, is provided to determine the value of the geometric correction factor for quality inlet conditions. The same methodology using Figures 46, 47, and 48 can be used to determine the mass flow rate of R-22 through adiabatic capillary tubes. Wolf et al. (1995) developed additional rating charts for R-152a.

Wolf et al. (1995) also presented limited performance results for refrigerants R-134a, R-22, and R-410A with 1.5% lubricant. While these results did indicate a 1% to 2% increase in the refrigerant mass flow through the capillary tubes, this magnitude was considered insignificant.

Generalized Prediction Equations. Based on tests with R-134a, R-22, and R-410A, Wolf et al. (1995) developed a general method for predicting refrigerant mass flow rate through a capillary tube. In this method the Buckingham pi theorem was applied to the physical factors and fluid properties that affect capillary tube flow rate. The physical factors include capillary tube diameter and length, capillary tube inlet pressure, and the refrigerant inlet condition, while the fluid properties included specific volume, viscosity, surface tension, specific heat, and enthalpy of formation. The result of this analysis was a group of eight dimensionless pi-terms shown in Table 2.

All fluid properties for respective pi terms, both liquid and vapor, are evaluated at the saturation state using the capillary tube inlet temperature. Separate π_5 terms are presented for subcooled and two-phase refrigerant conditions at the capillary tube entrance. *Note*: The bubble formation term π_3 is statistically insignificant and, therefore, does not appear in prediction Equations (3) and (4).

A regression analysis of the refrigerant flow rate data for the subcooled inlet results produced the following equation for subcooled inlet conditions.

$$\pi_8 = 1.8925 \pi_1^{-0.484} \pi_2^{-0.824} \pi_4^{1.369} \pi_5^{0.0187} \pi_6^{0.773} \pi_7^{0.265} \quad (3)$$

where $\pi_5 = d_c^2 c_p \Delta t_{sc}/v_f^2 \mu_f^2$ and $1 \text{ K} < \Delta t_{sc} < 17 \text{ K}$

The following equation includes the π_5 term that only includes the quality x.

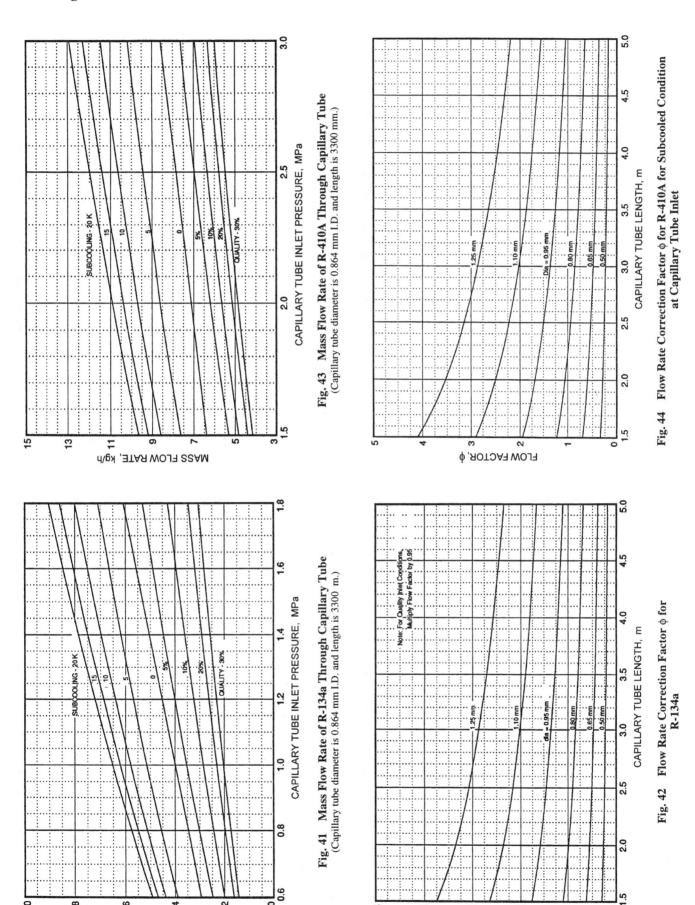

Fig. 41 Mass Flow Rate of R-134a Through Capillary Tube
(Capillary tube diameter is 0.864 mm I.D. and length is 3300 m.)

Fig. 42 Flow Rate Correction Factor φ for R-134a

Fig. 43 Mass Flow Rate of R-410A Through Capillary Tube
(Capillary tube diameter is 0.864 mm I.D. and length is 3300 mm.)

Fig. 44 Flow Rate Correction Factor φ for R-410A for Subcooled Condition at Capillary Tube Inlet

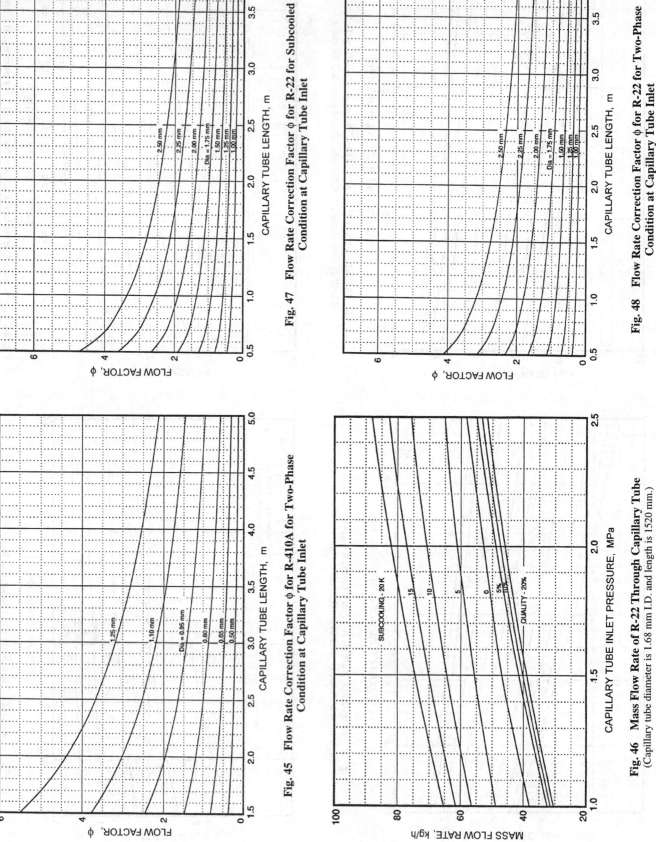

Fig. 47 Flow Rate Correction Factor φ for R-22 for Subcooled Condition at Capillary Tube Inlet

Fig. 48 Flow Rate Correction Factor φ for R-22 for Two-Phase Condition at Capillary Tube Inlet

Fig. 45 Flow Rate Correction Factor φ for R-410A for Two-Phase Condition at Capillary Tube Inlet

Fig. 46 Mass Flow Rate of R-22 Through Capillary Tube
(Capillary tube diameter is 1.68 mm I.D. and length is 1520 mm.)

Refrigerant-Control Devices

Table 2 Capillary Tube Dimensionless Parameters

Pi-term	Definition	Description
π_1	L_c/d_c	Geometry effect
π_2	$d_c^2 h_{fg}/v_f^2 \mu_f^2$	Vaporization effect
π_3	$d_c \sigma / v_f \mu_f^2$	Bubble formation
π_4	$d_c^2 p_{in}/v_f^2 \mu_f^2$	Inlet pressure
π_5 (subcooled)	$d_c^2 c_p \Delta t_{sc}/v_f^2 \mu_f^2$	Inlet condition
π_5 (quality)	x	Inlet condition
π_6	v_g/v_f	Density effect
π_7	$(\mu_f - \mu_g)/\mu_g$	Viscous effect
π_8	$\dot{m}/d_c \mu_f$	Flow rate

where
L_c = capillary tube length
d_c = capillary tube diameter
h_{fg} = enthalpy of formation
v_f = liquid specific volume
v_g = vapor specific volume
μ_f = liquid viscosity
μ_g = vapor viscosity
σ = surface tension
p_{in} = capillary tube inlet pressure
c_p = liquid specific heat
Δt_{sc} = degree of subcooling
x = quality (decimal)
$\dot{m}$ = mass flow rate

$$\pi_8 = 187.27 \pi_1^{-0.635} \pi_2^{-0.189} \pi_4^{0.645} \pi_5^{-0.163} \pi_6^{0.213} \pi_7^{-0.483} \quad (4)$$

where π_5 = quality x from $0.03 < x < 0.25$

Wolf et al. (1995) compared these equations to R-134a results by Dirik et al. (1994) and R-22 results by Kuehl and Goldschmidt (1990). Wolf et al. also compared experimental results with R-152a and predictions by Equations (3) and (4) and found agreement within 5%. In addition, less than 1% of the R-134a, R-22, and R-410A experimental results used by Wolf et al. to develop correlation Equations (3) and (4) were outside ±5% of the flow rates predicted by the equations.

Sample Calculations

Example 1: Determine the mass flow rate of R-134a through a capillary tube of 1.10 mm ID, 3.0 m long, operating without heat exchange at 1.4 MPa inlet pressure and 15 K subcooling.

Solution: From Figure 41, at 1.4 MPa and 15 K subcooling, the flow rate of HFC-134a for a capillary tube 0.864 mm ID and 3.302 m long is 7.5 kg/h. The flow factor ϕ from Figure 42 for a capillary tube of 1.10 mm ID and 3.0 m long is 2.0. The predicted R-134a flow rate is then 7.5 × 2.0 = 15.0 kg/h.

Example 2: Determine the mass flow rate of R-134a through a capillary tube of 0.65 mm ID, 2.5 m long, operating without heat exchange at 1.2 MPa inlet pressure and 5% vapor content at the capillary tube inlet.

Solution: From Figure 41, at 1.2 MPa and 5% quality, the flow rate of R-134a for a capillary tube 0.864 mm ID and 3.302 m long is 4.0 kg/h. The flow factor ϕ from Figure 42 for a capillary tube of 0.65 mm ID and 2.5 m long is 0.5. The predicted R-134a flow rate is then 0.95 × 4.0 × 0.5 = 1.9 kg/h, where 0.95 is the additional correction factor for R-134a quality inlet conditions.

Example 3: Determine the mass flow rate of R-410A through a capillary tube of 0.95 mm ID, 3.5 m long, operating without heat exchange, 2.5 MPa inlet pressure, and 20 K subcooling at the capillary tube inlet.

Solution: From Figure 43, at 2.5 MPa and 20 K subcooling, the flow rate of R-410A for a capillary tube of 0.864 mm ID and 3.302 m long is 12.0 kg/h. The flow factor ϕ from Figure 44 for a capillary tube of 0.95 mm ID and 3.5 m long is 1.25. The predicted R-410A flow rate is then 12.0 × 1.25 = 15.0 kg/h.

Example 4: Determine the mass flow rate of R-22 through a capillary tube of 1.75 mm ID, 2.0 m long, operating without heat exchange, 1.5 MPa inlet pressure, and 10% vapor content at the capillary tube inlet.

Solution: From Figure 46, at 1.5 MPa and 10% quality, the flow rate of R-22 for a capillary tube of 1.676 mm ID and 1.524 m long is 40 kg/h. The flow factor ϕ from Figure 48 for a capillary tube of 1.75 mm ID and 2.0 m long is 1.0. The predicted R-22 flow rate is then 40 × 1.0 = 40 kg/h.

SHORT TUBE RESTRICTORS

Application

Short tube restrictors are widely used in residential air conditioners and heat pumps. They offer the advantages of low cost, high reliability, ease of inspection and replacement, and potential elimination of check valves in the design of a heat pump. Due to their pressure-equalizing characteristics, short tubes allow the use of a low-starting-torque compressor motor.

Short tube restrictors, as used in residential systems, are typically 10 to 13 mm in length, with a length-to-diameter ratio of 3 L/D < 20. Short tubes are also called plug orifices or orifices, although the latter is reserved for restrictors having an L/D ratio less than 3. Capillary tubes have an L/D ratio much greater than 20.

An **orifice tube**, a type of short tube restrictor, is commonly used in automotive air conditioners. The L/D ratio of an orifice tube falls between that of a capillary tube and a short tube restrictor. Most automotive applications use orifice tubes with L/D ratios between 21 and 35 and inside diameters from 1 and 2 mm. An orifice tube allows the evaporator to operate in a flooded condition, which improves performance. To prevent liquid from flooding the compressor, an accumulator/dehydrator is installed to separate liquid from vapor and to meter a small amount of lubricant-rich refrigerant to the compressor. The accumulator/dehydrator does cause a pressure drop penalty on the suction side, however.

There are two basic designs for short tube restrictors: stationary and movable. Movable short tube restrictors consist of a piston that moves within its housing (see Figure 49). A movable short tube restricts refrigerant flow in one direction. In the opposite direction, the refrigerant pushes the restrictor off its seat, opening a larger area for the flow. The stationary design is used in units that cool only, while movable short tubes are used in heat pumps that require different flow restrictions for the cooling and heating. Two movable

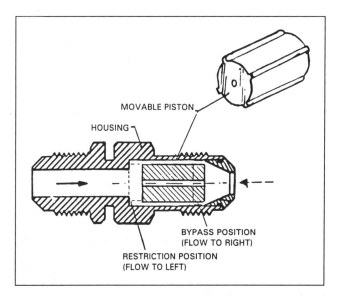

Fig. 49 Schematic of a Movable Short Tube Restrictor

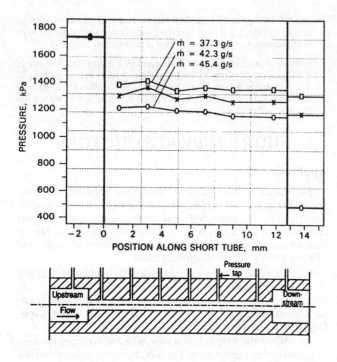

Fig. 50 Pressure Profile at Various Downstream Pressures with Constant Upstream Conditions:
$L = 12.7$ mm, $D = 1.35$ mm, subcooling 13.9 K
(Adapted from Aaron and Domanski 1990)

short tubes, installed in series and faced in opposite directions, eliminate the need for check valves, which are needed for capillary tubes and thermostatic expansion valves.

The refrigerant mass flow rate through a short tube depends strongly on upstream subcooling and upstream pressure. For a given inlet pressure, inlet subcooling, and downstream pressure below the saturation pressure corresponding to the inlet temperature, the flow has a very weak dependence on the downstream pressure, indicating a nearly choked flow. This flow dependence is shown in Figure 50, and represents test data obtained on a 12.7 mm long, laboratory-made short tube at three different downstream pressures and the same upstream pressure.

A significant drop in downstream pressure from approximately 1170 to 480 kPa produces a smaller increase in the mass flow rate than does a modest change of downstream pressure from 1310 to 1170 kPa. The pressure drops only slightly along the length of the short tube. The large pressure drop at the entrance is due to the rapid fluid acceleration and the inlet losses. The large pressure drop in the exit plane, typical for heat pump operating conditions and represented in Figure 50 by the bottom pressure line, indicates that choked flow has nearly occurred.

Among geometric parameters, the short tube diameter has the strongest influence on the mass flow rate. Chamfering the inlet of the short tube may increase the mass flow rate by as much as 25%, depending on the length-to-diameter ratio and chamfer depth. Chamfering the exit causes no appreciable change in the mass flow rate.

Although refrigerant flow inside a short tube is different than flow inside a capillary tube, choked flow is common for both, making both types of tubes suitable as metering devices. Systems equipped with short tubes, as with capillary tubes, must be precisely charged with the proper amount of refrigerant. Inherently, a short tube does not operate as efficiently over a wide range of operating conditions as does a thermostatic expansion valve. However, its performance is generally good in a properly charged system.

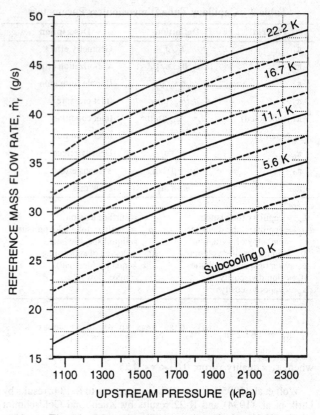

Fig. 51 Mass Flow Rate versus Condenser Pressure for Reference Short Tube. $L = 12.7$ mm, $D = 1.35$ mm, sharp-edged.

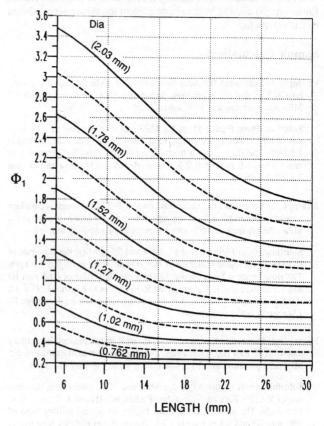

Fig. 52 Correction Factor for Short Tube Geometry

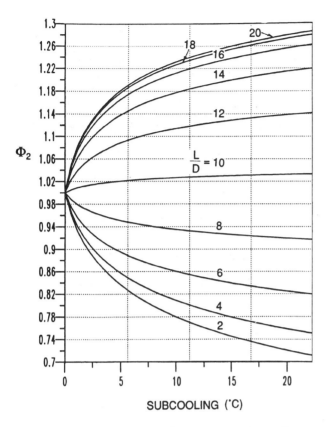

Fig. 53 Correction Factor for *L/D* Versus Subcooling

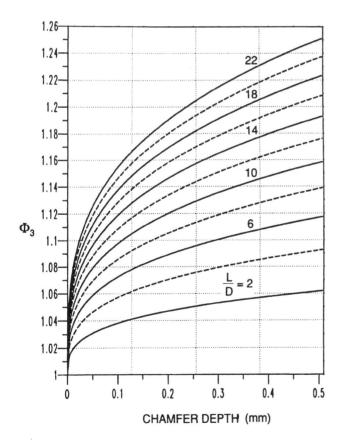

Fig. 54 Correction Factor for Inlet Chamfering

Selection

Figures 51 through 54, from Aaron and Domanski (1990), can be used for a preliminary evaluation of the mass flow rate of R-22 at a given inlet pressure and subcooling in air-conditioning and heat pump applications (i.e., applications in which the downstream pressure is below the saturation pressure of refrigerant at the inlet). The method requires reading mass flow rate for the reference short tube from Figure 51 and modifying the reading with multipliers that account for the short tube geometry according to the equation

$$\dot{m} = \dot{m}_r \phi_1 \phi_2 \phi_3 \qquad (5)$$

where

$\dot{m}$ = mass flow rate for the short tube
$\dot{m}_r$ = mass flow rate for the reference short tube from Figure 51
ϕ_1 = correction factor for tube geometry from Figure 52
ϕ_2 = correction factor for *L/D* versus subcooling from Figure 53
ϕ_3 = correction factor for chamfered inlet from Figure 54

Aaron and Domanski (1990) also provide a more accurate correlation than the graphical method. Neglecting downstream pressure on the graphs may introduce an error in the prediction as compared to the correlation results; however, this discrepancy should not exceed 3% due to the choked-flow condition at the tube exit. *Note:* The lines for 0 and 22.2 K subcooling in Figure 51 were obtained by extrapolation beyond the test data and may carry a large error.

Example 5: Determine the mass flow rate of R-22 through a short tube restrictor 9.5 mm long, of 1.52 mm inside diameter, and chamfered 0.25 mm deep at an angle of 45°. The inlet pressure is 1600 kPa, and subcooling is 5.6 K.

Solution: From Figure 51, for 1600 kPa and 5.6 K subcooling, the flow rate for the reference short tube, $\dot{m}_r$, is 30 g/s. The value for ϕ_1 from Figure 52 is 1.62. The value for ϕ_2 from Figure 53 for 5.6 K subcooling and L/D = 9.5/1.52 = 6.25 is 0.895. The value of ϕ_3 from Figure 54 for L/D = 6.25 and a chamfer depth of 0.025 mm is 1.10. Thus, the predicted mass flow rate through the restrictor is 30 × 1.62 × 0.895 × 1.10 = 47.8 g/s.

REFERENCES

Aaron, D.A. and P.A. Domanski. 1990. Experimentation, analysis, and correlation of Refrigerant-22 flow through short tube restrictors. *ASHRAE Transactions* 96(1):729-42.

API. 1993. Sizing, selection, and installation of pressure-relieving devices in refineries Part I—Sizing and selection, Sixth edition. *Recommended Practice* RP 520 PT I 1993. American Petroleum Institute, Washington, DC.

API. 1994. Sizing, selection, and installation of pressure-relieving devices in refineries Part II—Installation, Fourth edition. *Recommended Practice* RP 520 PT II 1994. American Petroleum Institute, Washington, DC.

ARI. 1987. Thermostatic refrigerant expansion valves. *Standard* 750. Air-Conditioning and Refrigeration Institute, Arlington, VA.

ASHRAE. 1986. Method of Testing for capacity rating of thermostatic refrigerant expansion valves. *Standard* 17-1986, reaffirmed 1990.

ASHRAE. 1996. Method of testing flow capacity of refrigerant capillary tubes. *Standard* 28-1996.

ASHRAE. 1994. Safety code for mechanical refrigeration. *Standard* 15-1994.

ASTM. 1995. Standard specification for seamless copper tube. *Standard* B 75 Rev A-95. American Society for Testing and Materials, West Conshohocken, PA.

ASTM. 1995. Standard specification for hand-drawn copper capillary tube for restrictor applications. *Standard* B 360-1995. American Society for Testing and Materials.

Bolstad, M.M. and R.C. Jordan. 1948. Theory and use of the capillary tube expansion device. *Refrigerating Engineering* (December):519.

Dirik, E., C. Inan, and M.Y. Tanes. 1994. Numerical and experimental studies on adiabatic and non-adiabatic capillary tubes with HFC-134a. *Proceeding* Intl. Refrigeration Conf., Purdue Univ., W. Lafayette, IN.

Domanski, P.A., D.A. Didion, and J.P. Doyle. 1994. Evaluation of suction line-liquid line heat exchange in the refrigeration cycle. *Intl. J. of Refrigeration* 17(7):487-493.

Hopkins, N.E. 1950. Rating the restrictor tube. *Refrigerating Engineering* (November):1087.

Kuehl, S.J. and V.W. Goldschmidt. 1991. Modeling of steady flows of R-22 through capillary tubes. *ASHRAE Transactions* 97(1):139-48.

Kuehl, S.J. and V.W. Goldschmidt. 1990. Steady flow of R-22 through capillary tubes: Test data. *ASHRAE Transactions* 96(1):719-728.

Li, R.Y., S. Lin, Z.Y. Chen, and Z.H. Chen. 1990. Metastable flow of R-12 through capillary tubes. *Intl. J. of Refrigeration* 13(3):181-186.

Mikol, E.P. 1963. Adiabatic single and two-phase flow in small bore tubes. *ASHRAE Journal* 5(11): 75-86.

Whitesel, H.A. 1957a. Capillary two-phase flow. *Refrigerating Engineering* (April):42.

Whitesel, H.A. 1957b. Capillary two-phase flow—Part II. *Refrigerating Engineering* (September):35.

Wolf, D.A., R.R. Bittle, and M.B. Pate. 1995. Adiabatic capillary tube performance with alternative refrigerants. *Final Report* ASHRAE 762-RP.

BIBLIOGRAPHY

Bittle, R.R., D.A. Wolf, and M.B. Pate. 1998. A generalized performance prediction method for adiabatic capillary tubes. *HVAC&R Research J.* 4(1):27-43.

Bittle, R.R. and M.B. Pate. 1996. A theoretical model for predicting adiabatic capillary tube performance with alternative refrigerants. *ASHRAE Transactions* 102(2): 52-64.

Bolstad, M.M. and R.C. Jordan. 1949. Theory and use of the capillary tube expansion device. Part II—Nonadiabatic flow. *Refrigerating Engineering* (June):577.

Koizumi, H. and K. Yokoyama. 1980. Characteristics of refrigerant flow in a capillary tube. *ASHRAE Transactions* 86(2):19-27.

Kuehl, S.J. and V.W. Goldschmidt. 1990. Transient response of fixed area expansion devices. *ASHRAE Transactions* 96(1):743-50.

Lathrop, H.F. 1948. Application and characteristics of capillary tubes. *Refrigerating Engineering* (August):129.

Marcy, G.P. 1949. Pressure drop with change of phase in a capillary tube. *Refrigerating Engineering* (January):53.

Pate, M.B. and D.R. Tree. 1987. An analysis of choked flow conditions in a capillary tube-suction line heat exchanger. *ASHRAE Transactions* 93(1):368-80.

Schulz, U. 1985. State of the art: The capillary tube for, and in, vapor compression systems. *ASHRAE Transactions* 91(1):92-105.

Wijaya, H. 1991. An experimental evaluation of adiabatic capillary tube performance for HFC-134a and CFC-12, pp. 474-83. *Proceedings* of the International CFC and Halon Alternatives Conference. The Alliance of Responsible CFC Policy, Arlington, VA.

CHAPTER 46

FACTORY DEHYDRATING, CHARGING, AND TESTING

Dehydration (Moisture Removal) 46.1
Moisture Measurement 46.3
Charging 46.4
Testing for Leaks 46.4
Performance Testing 46.5

PROPER dehydration, charging, and testing of packaged refrigeration systems and components (compressors, evaporators, and condensing coils) help ensure proper performance and extend the life of refrigeration systems. This chapter covers the methods used to perform these functions. The chapter does not address criteria such as allowable moisture content, refrigerant quantity, and performance, which are specific to each machine.

DEHYDRATION (MOISTURE REMOVAL)

Factory dehydration may be feasible only for certain sizes of equipment. On large equipment, which is open to the atmosphere when connected in the field, factory treatment is usually limited to purge and backfill, with an inert holding charge of nitrogen. In most instances, this equipment is stored for short periods only, so this method suffices until total system evacuation and charging can be done at the time of installation.

Excess moisture in refrigeration systems may lead to freeze-up of the capillary tube or expansion valve. Contaminants can cause valve breakage, motor burnout, and bearing and seal failure. Chapter 6 has more information on moisture and other contaminants in refrigerant systems.

These effects, with the exception of freeze-up, are not normally detected by a standard factory test. Therefore, it is important to use a dehydration technique that yields a safe moisture level without adding foreign elements or solvents. In conjunction with dehydration, an accurate method of moisture measurement must be established. Many factors, such as the size of the unit, its application, and the type of refrigerant, determine the acceptable moisture content. Table 1 shows moisture limits recommended by various manufacturers for particular refrigeration system components.

Sources of Moisture

Moisture in refrigerant systems can be (1) retained on the surfaces of metals; (2) produced by combustion of a gas flame; (3) contained in liquid fluxes, oil, and refrigerant; (4) absorbed in the hermetic motor insulating materials; (5) derived from the factory ambient at the point of unit assembly; and (6) provided by free water. The moisture contained in the refrigerant has no effect on the dehydration of the component or unit at the factory. However, because the refrigerant is added after the dehydration process, it must be considered in determining the overall moisture content of the completed unit. The moisture in the oil may or may not be removed in the dehydration process, depending on when the oil is added to the component or system.

Bulk mineral oils, as received, have 20 to 30 mg/kg of moisture. Synthetic polyol ester lubricants have 50 to 85 mg/kg; they are highly hygroscopic, so they must be handled carefully to prevent moisture contamination. Refrigerants have an accepted commercial tolerance of 10 to 15 mg/kg on bulk shipments. Controls at the factory are needed to ensure the maintenance of these moisture levels in the oils and refrigerant.

Newer insulating materials in hermetic motors retain much less moisture compared to the old rag paper and cotton-insulated motors. However, tests by several manufacturers have shown that the stator, with its insulation, is still the major source of moisture in compressors.

Dehydration by Heat, Vacuum, or Dry Air

Heat may be applied by placing components in an oven or by using infrared heaters. Oven temperatures of 80 to 170°C are usually maintained. The oven temperature should be selected carefully to prevent damage to the synthetics used and to avoid breakdown of any residual run-in oil that may be present in compressors. The air in the oven must be maintained at low humidity. When dehydrating by heat alone, the time and escape area are critical; therefore, the size of parts that can be economically dehydrated by this method is restricted.

The **vacuum** method reduces the boiling point of water below the ambient temperature. The moisture then changes to vapor, which is pumped out by the vacuum pump. Table 3 in Chapter 6 of the 1997 *ASHRAE Handbook—Fundamentals* shows the relationship of temperature and pressure for water at saturation.

Vacuum is classified according to the following absolute pressure ranges:

Low Vacuum	101.325 to 3.4 kPa
Medium Vacuum	3.4 kPa to 130 mPa
High Vacuum	130 to 0.13 mPa
Very High Vacuum	130 to 0.13 μPa
Ultra High Vacuum	0.13 μPa and below

The degree of vacuum achieved and the time required to obtain the specified moisture level are a function of (1) the type and size of vacuum pump used, (2) the internal volume of the component or system, (3) the size and composition of water-holding materials in the system, (4) the initial amount of moisture in the volume, (5) piping and fitting sizes, (6) the shape of the gas passages, and (7) the external temperatures maintained. The pumping rate of the vacuum pump is critical only if the unit is not evacuated through a conductance-limiting orifice such as a purge valve. Excessive moisture content, such as a pocket of puddled water, takes a long time to remove due to the volume expansion to vapor. At absolute pressures below 610 Pa, the water freezes, and removal via sublimation without heat is minimal.

Vacuum measurements should be taken directly at the equipment (or as close to it as possible) rather than at the vacuum pump. Small tubing diameters or long tubing runs between the pump and the equipment should be avoided because line/orifice pressure drops reduce the actual evacuation level at the equipment.

If **dry air or nitrogen** is drawn or blown through the equipment for dehydration, it removes moisture by becoming totally or partially saturated. In systems with several passages or blind passages, flow may not be sufficient to dehydrate. The flow rate should obtain

---

The preparation of this chapter is assigned to TC 8.1, Positive Displacement Compressors.

Table 1 Typical Factory Dehydration and Moisture-Measuring Methods for Refrigeration Systems

Component	Dehydration Method	Moisture Audit	Moisture Limit
Coils and tubing	121°C oven, −57°C dry air sweep	Dew point recorder	10 mg
Evaporator coils			
Small	−57°C dew point dry air sweep, 240 s	P_2O_5	25 mg
Large	−57°C dew point dry air sweep, 240 s	P_2O_5	65 mg
Evaporators/condensers	149°C oven, 1 h, dry air sweep	Cold trap	200 mg
Evaporators/condensers	Dry air sweep	Nesbitt tube	90 mg/m² surf. area
Condensing unit (1 to 25 kW)	Purchase dry	P_2O_5	25 to 85 mg
Condensing unit	Dry air sweep	Nesbitt tube	90 mg/m² surf. area
Air-conditioning unit	Evacuate to 240 μm Hg	P_2O_5	35 mg/kg
Air-conditioning unit	3 h winding heat, 0.5 h vacuum	Refrigerant moisture check	25 mg/kg
Refrigerator	121°C oven, dc winding heat, vacuum	Cold trap	200 mg
Freezer	−57°C dew point dry air ambient, −40°C dew point air sweep	P_2O_5	10 mg/kg
Compressors			
	DC Winding Heat		
	0.5 h dc winding heat 177°C, 0.25 h vacuum/repeat	Cold trap	200 mg
	DC winding heat 88°C, 0.5 h vacuum	Cold trap	1200 mg
7 to 210 kW semihermetic	DC winding heat, 30 min, evacuation, N_2 charge	Cold trap	1000 to 3500 mg
	Oven Heat		
	121°C oven, 4 h vacuum	Cold trap	180 mg
	121°C oven, 5.5 h at −51°C dew point air	Cold trap	200 mg
2 to 40 kW hermetic	149°C oven 4 h, −57°C dew point air 3.5 min	Cold trap	150 to 400 mg
175 to 350 kW	Oven at 132°C, 4 h evacuate to 1000 μm Hg	Cold trap	750 mg
5 to 20 kW hermetic	171°C oven, −73°C dew point dry air, 1.5 h	Cold trap	100 to 500 mg
7 to 140 kW semihermetic	121°C oven, −73°C dew point dry air, 3.5 h	Cold trap	100 to 1100 mg
20 to 525 kW open	79°C oven, evacuate to 1 mm Hg	Cold trap	400 to 2700 mg
Scroll 7 to 35 kW hermetic	149°C oven 4 h, 50 s evacuation and 10 s −57°C dew point air charge/repeat 7 times	Cold trap	300 to 475 mg
	Hot Dry Air, N_2		
10 to 20 kW	Dry air at 135°C, 3 h	Cold trap	250 mg
25 to 55 kW	Dry air at 135°C, 0.5 h vacuum	Cold trap	750 mg
70 to 140 kW	Dry N_2 sweep at 135°C, 3.5 h evacuate to 200 μm Hg	Cold trap	750 mg
	Dry N_2 Flush		
Reciprocating, semihermetic	N_2 run, dry N_2 flush, N_2 charge	—	—
Screw, hermetic/semihermetic	R-22 run, dry N_2 flush, N_2 charge	—	—
Screw, open	N_2 run, dry N_2 flush, N_2 charge	—	—
	Evacuation Only		
Screw, open, 175 to 5300 kW	Evacuate <1500 μm Hg, N_2 charge		
Refrigerants	As purchased	Electronic analyzer	Typically 10 mg/kg
Lubricants			
Mineral oil	As purchased	Karl Fischer method	25 to 35 mg/kg
Mineral oil	As purchased and evacuation	Hygrometer	10 mg/kg
Synthetic polyol ester	As purchased	Karl Fischer method	50 to 85 mg/kg

optimum moisture removal, and its success depends on the overall system design and temperature.

Combination Methods

Each of the following methods can be effective if controlled carefully, but a combination of two or even three of the methods is preferred because of the shorter drying time and more uniform dryness of the treated system.

Heat and Vacuum Method. The heat of this combination method drives deeply sorbed moisture to the surfaces of materials and removes it from walls; the vacuum lowers the boiling point, making the pumping rate more effective. The heat source can be an oven, infrared lamps, or an ac or dc current circulating through the internal motor windings of semihermetic and hermetic compressors. Combinations of vacuum, heat, and then vacuum again can also be used.

Heat and Dry Air Method. The heat drives the moisture from the materials. The dry air picks up this moisture and removes it from the system or component. The dry air used should have a dew point between −40 and −73°C. Heat sources are the same as those mentioned previously. Heat can be combined with a vacuum to accelerate the process. The heat and dry air method is effective with open, hermetic, and semihermetic compressors. The heating temperature should be selected carefully to prevent damage to the compressor parts or breakdown of any residual oil that may be present.

The advantages and limitations of the various methods depend greatly on the system or component design and the results expected. Goddard (1945) considers double evacuation with an air sweep between vacuum applications the most effective method, while Larsen and Elliot (1953) believe the dry air method, if controlled carefully, is just as effective as the vacuum method and much less expensive, although it incorporates a 1.5 h evacuation after the hot

Factory Dehydrating, Charging, and Testing

air purge. Tests by one manufacturer show that a 138°C oven bake for 1.5 h, followed by a 20 min evacuation, effectively dehydrates compressors that use newer insulating materials.

MOISTURE MEASUREMENT

Measuring the correct moisture level in a dehydrated system or part is important but not always easy. Table 1 shows measuring methods used by various manufacturers, and others are described in the literature. Few standards are available, however, and the moisture limits accepted by different manufacturers vary.

Cold Trap Method. This common method of determining residual moisture monitors the production dehydration system to ensure that it produces equipment that meets the required moisture specifications. An equipment sample is selected after completion of the dehydration process, placed in an oven, and heated at 65 to 135°C (depending on the limitations of the sample) for 4 to 6 h. During this time, a vacuum is drawn through a cold trap bottle immersed in an acetone and dry-ice solution (or an equivalent), which is generally held at about −73°C. Vacuum levels are between 1.3 and 13 Pa, with the lower levels preferred. Important factors are leaktightness of the vacuum system and cleanliness and dryness of the cold trap bottle.

Vacuum Leakback. Measuring the rate of vacuum leakback is another means of checking components or systems to ensure that no water vapor is present. This method is used primarily in conjunction with a unit or system evacuation that removes the noncondensables prior to final charging. This test allows a check of each unit, but too rapid a pressure buildup may signify a leak, as well as incomplete dehydration. The time factor may be critical in this method and must be examined carefully. Blair and Calhoun (1946) show that a small surface area in connection with a relatively large volume of water may only build up vapor pressure slowly. This method also does not give the actual condition of the charged system.

Dew Point. When dry air is used, a reasonably satisfactory check for dryness is a dew point reading of the air as it leaves the part being dried. If the airflow is relatively slow, there should be a marked difference in dew point between the air entering and the air leaving the part, followed by a decrease in dew point of the leaving air until it eventually equals the dew point of the entering air. As is the case with all systems and methods described in this chapter, the values considered acceptable depend on the size, usage, and moisture limits desired. Different manufacturers use different limits.

Gravimetric Method. In this method, described by ASHRAE *Standard* 35, a controlled amount of refrigerant is passed through a train of flasks containing phosphorous pentoxide (P_2O_5), and the mass increase of the chemical (caused by the addition of moisture) is measured. Although this method is satisfactory when the refrigerant is pure, any oil contamination produces inaccurate results. This method must be used only in a laboratory or under carefully controlled conditions. Also, it consumes considerable time and cannot be used when production quantities are high. Furthermore, the method is not effective in systems containing only small charges of refrigerant because it requires 200 to 300 g of refrigerant for accurate results. If it is used on systems where withdrawal of any amount of refrigerant changes the performance, recharging is required.

Aluminum Oxide Hygrometer. This sensor consists of an aluminum strip that is anodized by a special process to provide a porous oxide layer. A very thin coating of gold is evaporated over this structure. The aluminum base and the gold layer form two electrodes that essentially form an aluminum oxide capacitor.

In the sensor, the water vapor passes through the gold layer and comes to equilibrium on the pore walls of the aluminum oxide in direct relation to the vapor pressure of water in the ambient surrounding the sensor. The number of water molecules absorbed in the oxide structure determines the sensor's electrical impedance, which modulates an electrical current output that is directly proportional to the water vapor pressure. This device is suitable for both gases and liquids over a temperature range of 70 to −110°C and a pressure range of about 1 Pa to 34.5 MPa. The **Henry's Law constant** for each fluid must be determined. This constant is the saturation parts per million by mass of water for the fluid divided by the saturated vapor pressure of water at a constant temperature. For many fluids, this constant must be corrected for the operating temperature at the sensor.

Christensen Moisture Detector. The Christensen moisture detector is used for a quick check of uncharged components or units on the production line. In this method, dry air is blown first through the dehydrated part and then over a measured amount of calcium sulfate ($CaSO_4$). The temperature of the $CaSO_4$ rises in proportion to the quantity of water it absorbs, and desired limits can be set and monitored. One manufacturer reports that coils were checked in 10 s with this method. Moisture limits for this detector are 2 to 60 mg. Corrections must be made for variations in the desiccant grain size, the quantity of air passed through the desiccant, and the difference in instrument and component temperatures.

Karl Fischer Method. In systems containing refrigerant and oil, moisture may be determined by (1) measurement of the dielectric strength or (2) the Karl Fischer method (Reed 1954). In this method, a sample is condensed and cooled in a mixture of chloroform, methyl alcohol, and Karl Fischer reagent. The refrigerant is then allowed to evaporate as the solution warms to room temperature. When the refrigerant has evaporated, the remaining solution is titrated immediately to a **dead stop** electrometric end point, and the amount of moisture is determined. This method requires a sample of 50 to 60 g and takes about 1 h to perform. It is generally considered inaccurate below 15 mg/kg; however, as the method does not require that oil be boiled off the refrigerant, it can be used for checking complete systems. Reed points out that additives in the oil, if any, must be checked to ensure that they do not interfere with the reactions of the method. The Karl Fischer method may also be used for determining moisture in oil alone (Reed 1954, ASTM *Standard* D 117, Morton and Fuchs 1960).

Electrolytic Water Analyzer. Taylor (1956) describes an electrolytic water analyzer designed specifically to analyze moisture levels in a continuous process, as well as in discrete samples. The device passes the refrigerant sample, in vapor form, through a sensitive element consisting of a phosphoric acid film surrounding two platinum electrodes; the acid film absorbs the moisture. When a dc voltage is applied across the electrodes, the water absorbed in the film is electrolyzed into hydrogen and oxygen, and the resulting dc current, in accordance with **Faraday's first law of electrolysis**, flows in proportion to the mass of the products electrolyzed. Liquids and vapor may be analyzed because the device has an internal vaporizer. This device handles the popular halocarbon refrigerants, but the samples must be free of oils and other contaminants. In tests on desiccants, this method is quick and accurate with R-22.

Sight-Glass Indicator. In fully charged halocarbon systems, a sight-glass indicator can be used in the refrigerant lines. This device consists of a colored chemical button, visible through the sight glass, that indicates excessive moisture by a change in color. This method requires that the system be run for a reasonable length of time to allow moisture to circulate over the button. This method compares moisture only qualitatively to a fixed standard. Sight-glass indicators have been used on factory-dehydrated split systems to ensure that they are dry after field installation and charging and are in common use in conjunction with filter driers to monitor moisture in operating systems.

Special Considerations. Although all the methods described in this section can effectively measure moisture, their use in the factory requires certain precautions. Operators must be trained in the use of the equipment or, if the analysis is made in the laboratory, the proper method of securing samples must be understood. Sample flasks must be dry and free of contaminants; lines must be clean, dry, and properly purged. The procedures for weighing the sample, the time during the cycle, and the location of the sample part should be clearly defined and followed carefully. Checks and calibrations of the equipment must be made on a regular basis if consistent readings are to be obtained.

CHARGING

The accuracy required when charging refrigerant or oil into a unit depends on the size and application of the unit. Charging equipment must also be adapted to the particular conditions of the plant; equipment may be manual or automatic. Standard charging is used where extreme accuracy is not necessary or the production rate is not high. Fully automatic charging boards check the vacuum in the units, evacuate the charging line, and meter the desired amount of oil and refrigerant into the system. These devices are accurate and suitable for high production.

Refrigerant and oil must be handled carefully during charging; the place and time of oil and refrigerant charging have a great bearing on the life of a system. If a complete unit is charged prior to performance testing, the presence of liquid refrigerant in the crankcase can cause damage because of slugging. If the oil is added after the refrigerant is already in the crankcase, excessive foaming and oil vapor lock may cause bearing damage. Refrigerant lines must be dry and clean, and all charging lines must be kept free of moisture and noncondensable gases. Also, new containers must be connected with proper purging devices. Carelessness in observing these precautions may lead to excess moisture and noncondensables in the refrigeration system.

Oil drums must lie on their sides in storage to prevent dirt and water, which might get into the oil as soon as a drum is opened, from accumulating on the top of the drum. Oil drums should be opened at the last moment before charging, even on applications requiring that the oil be degassed. Regular checks for moisture or contamination must be made at the charging station to ensure that the oil and refrigerant delivered to the unit meet specifications. Compressors charged with oil for storage or shipment must be charged with dry nitrogen. Compressors without oil may be charged with dry air.

TESTING FOR LEAKS

Extended warranties and critical refrigerant charges add to the importance of proper leak detection prior to charging.

Basically, the **allowable leakage rate** depends entirely on the system or component characteristics. Any leak on the low-pressure side of a system operating below atmospheric pressure is dangerous, no matter how large the refrigerant charges. A system that has 100 to 200 g of refrigerant and a 5 year warranty must have virtually no leak (30 g in 10 years or more), whereas in a system that has 4.5 to 9 kg of refrigerant, the loss of 30 g of refrigerant in 1 year is not critical. Before any leak testing is done, the component or system should be strength tested at a pressure considerably higher than the leak test pressure. This test ensures safety when the unit is being tested under pressure in an exposed condition. Applicable design test pressures for high- and low-side components have been established by Underwriters Laboratories (UL), the American Society of Mechanical Engineers (ASME), the American National Standards Institute (ANSI), and ASHRAE. Units or components using composition gaskets as joint seals should have the final leak test after dehydration. A final torquing of this type of joint after dehydration can be beneficial in reducing leaks.

Leak Detection Methods

Water Submersion Test. The most popular method of leak and strength testing is the water submersion test. The unit or component is pressurized to the specified positive pressure and submerged in a well-lighted tank filled with clean water. A long time may be needed to obtain the leak test sensitivity desired.

Pressure Testing. The unit is sealed off under pressure or vacuum, and a decrease or rise in pressure is noted over time. The disadvantages of this test are the time involved, the lack of sensitivity, and the inability to determine the exact location of the leak.

Soap Bubble Leak Test. A pressurized system's suspected leak areas are brushed with a soapy solution. Bubbles form in the immediate leak zone.

Halide Leak Testing. The halide torch is used on systems charged with a halogenated refrigerant. This test can detect a leakage rate of approximately 30 g per year. The gas, drawn across a faintly bluish flame, turns the flame a greenish blue that varies in intensity with the size of the leak. Each joint or area can easily be probed in order to locate the leak. The sensitivity of halide torches is reduced by refrigerant contamination; therefore, testing should be done in well-ventilated areas or chambers. Large leaks, even in well-ventilated areas, may cause contamination levels so high that small leaks are not detected. **Caution:** Some of the newer halocarbon refrigerant blends contain flammable components; the use of a halide torch may be dangerous, especially if the leak is large (see ASHRAE *Standard* 34).

Electronic Leak Testing. The electronic leak detector consists of a probe that draws air over a platinum diode, the positive ion emission of which is greatly increased in the presence of a halogen gas. This increased emission is translated into a visible or audible signal. Electronic leak testing shares with halide torches the disadvantages that every suspect area must be explored and that contamination makes the instrument less sensitive; however, it does have some advantages. The main advantage is increased sensitivity. With a well-maintained detector, it is possible to identify leakage at a rate of 10^{-3} mm^3/s (standard), which is roughly equivalent to the loss of 30 g of refrigerant in 100 years. Another advantage is that the detector can be calibrated in many ways, so that a leak can be measured quantitatively. The instrument also can be desensitized to the point that leaks below a predetermined rate are not found. Some models have an automatic compensating feature to accomplish this.

The problem of contamination is more critical with improved sensitivity, so the unit under test is placed in a chamber slightly pressurized with outside air, which keeps contaminants out of the production area and carries contaminating gas from leaky units. An audible signal allows the probe operator to concentrate on probing, without having to watch a flame or dial. Equipment maintenance presents a problem because the sensitivity of the probe must be checked at short intervals. Any exposure to a large amount of refrigerant causes loss of probe sensitivity. A rough check (e.g., air underwater testing) is frequently used to find large leaks prior to use of the electronic device.

Fluorescent Leak Detection. This system involves infusing a small quantity of a fluorescent additive into the oil/refrigerant charge of an operating system. Leakage is observed as a yellow-green glow under an ultraviolet (UV) lamp. This method is suitable for halocarbon systems.

Mass Spectrometer. The most sensitive leak detection method is probably the mass spectrometer. The unit to be tested is evacuated and then surrounded by a helium and air mixture. The vacuum is sampled through a mass spectrometer; any trace of helium indicates one or more leaks. The mass spectrometer is extremely sensitive, able to detect leaks of 10^{-7} mm^3/s. Effective test levels in the manufacturing environment, though, are closer to 10^{-2} mm^3/s. The helium for testing is normally kept inside a chamber that is completely closed except at the bottom. The unit to be tested is simply raised into the lighter-than-air helium atmosphere.

This method, in addition to being extremely sensitive, has the advantage of measuring all leaks on all joints simultaneously. Therefore, a quick test is possible. However, the cost of the equipment and of the helium is high, the instrument must be maintained carefully, and a method of locating individual leaks must be developed.

Factory Dehydrating, Charging, and Testing

The required concentration of helium depends on the maximum leak permissible, the configuration of the system under test, the time the system can be left in the helium atmosphere, and the vacuum level in the system; the lower the vacuum level, the higher the helium readings. The longer a unit is exposed to the helium atmosphere, the lower the concentration necessary to maintain the required sensitivity. If, due to the shape of the test unit, a leak is distant from the point of sampling, a good vacuum must be drawn, and sufficient time must be allowed for traces of helium to appear on the mass spectrometer.

In general, a helium concentration of more than 10% is costly. The inherently high diffusion rate causes it to disperse to the atmosphere, no matter how effectively the chamber is designed. As in the case of other methods described in this chapter, the best testing procedure in using the spectrometer is to locate calibrated leaks at extreme points of the test unit and to adjust exposure time and helium concentration in the most economical manner. One manufacturer of refrigeration equipment found leaks of 1.4 g of refrigerant per year by using a 10% concentration of helium and exposing the tested system for 10 min.

The sensitivity of the mass spectrometer method can be limited by the characteristics of the tested system. Because only the total leakage rate is found, it is impossible to tell whether a leakage rate of, for example, 30 g per year, is caused by one fairly large leak or a number of small leaks. If the desired sensitivity rejects units outside of the sensitivity range of tests listed earlier in this chapter, it is necessary to use a **helium probe** for the location of leaks. In this method, the component or system to be probed is fully evacuated to clear it of helium; then, while the system is connected to the mass spectrometer, a fine jet of helium is sprayed over each joint or suspect area. With large systems, a waiting period is necessary because some time is required for the helium to pass from the leak point to the mass spectrometer. To save time, isolated areas (e.g., return bends on one end of a coil) may be hooded and sprayed with helium to determine whether the leak is in the region.

Special Considerations

Two general categories of leak detection may be selected: one group furnishes a leak check before refrigerant is introduced into the system, and the other group requires refrigerant. The methods that do not use refrigerant have the advantage that heat applied to repair a joint has no harmful effects. On units containing refrigerant, the refrigerant must be removed and the unit vented before any welding, brazing, or soldering is attempted. This practice avoids refrigerant breakdown and pressure buildup, which would prevent the successful completion of a sound joint.

All leak-testing equipment must be calibrated frequently to ensure maximum sensitivity. The electronic leak detector and the mass spectrometer are usually calibrated with equipment furnished by the manufacturer. Mass spectrometers are usually checked by a flask containing helium. A glass orifice in the flask allows the helium to escape at a known rate; the operator calibrates the spectrometer by comparing the measured escape rate with the standard.

The effectiveness of the detection system can best be checked with calibrated leaks made of glass, which can be bought commercially. These leaks can be built into a test unit and sent through the normal leak detection cycles to evaluate the effectiveness of the detection method. Care must be taken that the test leakhole does not become closed; the leakage rate of the test leak must be determined before and after each system audit.

From a manufacturing standpoint, the use of any leak detection method should be secondary to leak prevention. Improper brazing and welding techniques, unclean parts, untested sealing compounds or improper fluxes and brazing materials, and poor workmanship result in leaks that occur in transit or later. Careful control and analysis of each joint or leak point make it possible to concentrate tests on areas where leaks are most likely to occur. If operators must scan hundreds of joints on each unit, the probability of finding all leaks is rather small, whereas concentration on a few suspect areas reduces field failures considerably.

PERFORMANCE TESTING

Because there are many types and designs of refrigeration systems, this section only presents specific information on reciprocating compressor testing and covers some important aspects of performance testing of other components and complete systems.

Compressor Testing

The two prime considerations in compressor testing are power and capacity. Secondary considerations are leakback rate, low-voltage starting, noise, and vibration.

Testing Without Refrigerant. A number of tests measure compressor power and capacity before the unit is exposed to refrigerant. In cases where excessive power is caused by friction of running gear, **low-voltage tests** spot defective units early in assembly. In these tests, voltage is increased from a low or zero value to the value that causes the compressor to **break away**, and this value is compared with an established standard. When valve plates are accessible, performance can be tested by using an air pump for **leakback tests**. Air at fixed pressure is put through the unit to determine the flow rate at which valves open properly. The air pressure exerted against the closing side of the valve indicates its efficiency. This method is effective only when the valves are reasonably tight and is difficult to use on valves that must be run in before seating properly.

Extreme care should be taken when a compressor is used to pump air because the combination of oil, air, and high temperatures caused by compression can result in a diesel effect or an explosion.

In a common test using the compressor as an air pump, the discharge airflow is measured through a flowmeter, orifice, or other flow-measuring device. When the volumetric efficiency of the compressor with refrigerant is known, the flow rate that can be expected with air at a given pressure may be calculated. Because this test adiabatically compresses the air, the head pressure must be low to prevent overheating of discharge lines and oil oxidation if the test lasts longer than a few minutes. [The temperature of adiabatic compression is 140°C at 250 kPa (gage), but 280°C at 850 kPa (gage).] When the compressor is run long enough to stabilize temperatures, both power and flow can be compared with established limits. Both temperature readings at discharge and rpm measurements aid in analyzing defective units. If a considerable amount of air is discharged or trapped, the air used in the test must be dry enough to prevent condensation from causing rust or corrosion on the discharge side.

Another method of determining compressor performance requires the compressor to pump from a free air inlet into a fixed volume. The time required to reach a given pressure is compared against a maximum standard acceptable value. The pressure used in this test is approximately 860 kPa (gage), so that a reasonable time spread can be obtained. The time needed for measuring the capacity of the compressor must be sufficient for accurate readings but short enough to prevent overheating. Power readings can be recorded at any time in the cycle. By shutting off the compressor, the leakback rate can be measured as an additional check. In addition to the pump-up and leakback tests noted above, a vacuum test should also be performed.

The **vacuum test** should be performed by closing off the suction side with the discharge open to the atmosphere. The normal vacuum obtained under these conditions is 6.5 to 10 kPa (absolute). Abrupt closing of the suction side also allows the oil to serve as a check on the priming capabilities of the pump because of the suppression of the oil and attempt to deaerate. This test also checks for porosity and leaking gaskets. To establish reasonable pump-up times, leakback rates, and suctions, a large number of production units must be tested to determine the range of production variation.

In any capacity test using air, only clean, dry air should be used in order to prevent compressor contamination,.

The acceptance test limits described are best established by taking compressors of known capacity and power and observing their performance during the test. Precautions should be taken to prevent oil used repeatedly for the lubrication of many compressors from becoming acidic or contaminated.

Testing with Refrigerant. The most common test method is the **runaround cycle**. Successful variations and modifications of this cycle are described in ASHRAE *Standard* 23. This method requires an expansion device and a condenser large enough to handle the heat of compression. The gas compressed by the compressor is flash-cooled until its enthalpy is the same as that at suction conditions. It is then expanded back to the suction state. This method eliminates the need for an evaporator and uses a condenser about one-fifth the size normally used with the compressor. On compressors of small capacity, a piece of tubing that connects discharge to suction and has a hand expansion valve can be used effectively. The measure of performance is usually the relationship of suction and discharge pressures to power. When a water-cooled condenser is used, the head pressure is usually known, and the water temperature rise and flow are used as capacity indicators.

As a further refinement, flow-measuring devices can be installed in the refrigerant lines. This system is charge-sensitive if predetermined head and suction pressures and temperatures are to be obtained. This is satisfactory when all units have the same capacity and one test point is acceptable because the charge desired can be determined with little experimentation. When a variety of sizes is to be tested, however, or more than one test point is desired, a liquid sump or receiver after the condenser can be used for full-liquid expansion.

The refrigerant must be free of contamination, inert gases, and moisture; the tubing and all other components should be clean and sealed when they are not in use. In the case of hermetic and semi-hermetic systems, a motor burnout on the test stand makes it imperative not to use the stand until it has been thoroughly flushed and is absolutely acid-free. In all tests, oil migration must be observed carefully, and the oil must be returned to the crankcase.

The length of a compressor performance test depends on various factors. Stabilization of conditions is a prerequisite if accuracy is to be obtained. If oil pump or oil charging problems are inherent, the compressor should be run long enough to ensure that all defective units are detected. Most manufacturers use test periods of approximately 30 min to 1 h. A check of the test system is usually made by running a sample of the production units on a calorimeter under controlled conditions.

Testing of Complete Systems

In a factory, testing of any system may be done at a controlled ambient temperature or at an existing shop ambient temperature. In both cases, tests must be run carefully, and any necessary corrections must be made. Because measuring air temperature and flow is difficult, production-line tests are usually more reliable when secondary conditions are used as capacity indicators. Measurements of water temperature and flow, power, cycle time, refrigerant pressures, and refrigerant temperatures are reliable capacity indicators.

When testing self-contained air conditioners, for example, a fixed load may be applied to the evaporator using any air source and either a controlled ambient or shop ambient temperature. As long as the load is relatively constant, its absolute value is not important. For water-cooled units, in which water flow can be absolutely controlled, capacity is best measured by the heat rejected from the condenser. Suction and discharge pressures can be measured for the analysis.

Suction and discharge pressures can be used as a direct measure of capacity in units having air-cooled condensers. As long as the load is relatively constant, the absolute value is not important. Air distribution, velocity, or temperature over the coil of the test unit must be kept constant during the test, and the performance of the test unit must then be correlated with the performance of a standard unit. Power measurements supplement the suction and discharge pressure readings. As a rule, suction and discharge temperatures are useful in determining unsatisfactory operation of the unit and are particularly important when the evaporator or condenser loads are not reasonably stable. In such cases, simultaneous readings of suction temperature and pressure throughout an entire cycle permit the experienced observer to judge the performance of the unit.

The length of the test run depends on the test used, but in general, stabilization should be achieved in approximately 30 min. For units in which air flows over an evaporator, the conditions should load the test unit with a dry coil. This reduces the time necessary to balance the test system. The test requires preheating of the air temperature to a level that maintains the test unit evaporator saturated temperature above the dew point of the supplied air. For units with air-cooled condensers, the air leaving the condenser can be recirculated back to the evaporator for use as the load. This second arrangement is simple and inexpensive, but it may cause wide variation in performance if the system is not controlled carefully.

The primary function of the factory performance test is to ensure that a unit is constructed and assembled properly. Therefore, all equipment must be compared to a standard unit. This standard unit should be typical of the unit used to pass the Air Conditioning and Refrigeration Institute (ARI) and Association of Home Appliance Manufacturers (AHAM) certification programs for compressors and other units. ARI and AHAM provide rating standards with applicable maximum and minimum tolerances. Several ASHRAE and International Organization for Standardization (ISO) standards specify applicable rating tests.

Normal causes of malfunction in a complete refrigeration system are overcharging, undercharging, presence of noncondensable gases in the system, blocked capillaries or tubes, and excessive power. To determine the validity and sensitivity of any test procedure, it is best to use a unit with known characteristics and then establish limits for deviations from the test standard. If the established limits for charging are ±30 g of refrigerant, for example, the test unit is charged first with the correct amount of refrigerant and then with 30 g more and 30 g less. If this procedure does not establish clearly defined limits, it cannot be considered satisfactory and new values must be established. This same procedure should be followed regarding all variables that influence performance and cause deviations from established limits. All equipment must be maintained carefully and calibrated if tests are to have any significance. Gages must be checked at regular intervals and protected from vibration. Capillary test lines must be kept clean and free of contamination. Power leads must be kept in good repair to eliminate high-resistance connection, and electrical meters must be calibrated and protected to yield consistent data.

In plants where component testing and manufacturing control have been so well managed that the average unit performs satisfactorily, units are tested only long enough to find major flaws. Sample lot testing is sufficient to ensure product reliability. This approach is sound and economical because complete testing taxes power and plant capacity and is not necessary.

When the evaporator load is static (e.g., for refrigerators or freezers), time, temperature, and power measurements are used to measure performance. Performance is determined by the time elapsed between start and first compressor shutoff or by the average on-and-off period during a predetermined number of cycles in a controlled or known ambient temperature. Also, concurrent suction and discharge temperatures in connection with power readings are used to establish conformity to standards. On units where the necessary

Factory Dehydrating, Charging, and Testing

connections are available, pressure readings may be taken. Such readings are usually possible only on units where refrigerant loss is not critical because some loss is caused by gages.

Units with complicated control circuits usually undergo an operational test to ensure that controls function within design specifications and operate in the proper sequence.

Testing of Components

Component testing must be based on a thorough understanding of the use and purpose of the component. Pressure switches may be calibrated and adjusted with air in a bench test and need not be checked again if there is no danger of blocked passages or pulldown tripout during the operation of the switch. However, if the switch is brazed into the final assembly, precautions are needed to prevent blocking of the switch capillary.

Capillaries for refrigeration systems are checked by air testing. When the capillary limits are known, it is relatively easy to establish a flow rate and pressure drop test for eliminating crimped or improperly sized tubing. When several capillaries are used in a distributor, a series of water manometers check for unbalanced flow and can find damaged or incorrectly sized tubes.

In plants with good manufacturing control, only sample testing of evaporators and condensers is necessary. Close control of coils during manufacture leads to the detection of improper expansion, poor bonding, split fins, or uneven spacing. Proper inspection eliminates the need for costly test equipment. In testing the sample, either a complete evaporator or condenser or a section of the heat transfer surface is tested. Because liquid-to-liquid is the most easily and accurately measurable method of heat transfer, a tube or coil can be tested by passing water through it while it is immersed in a bath of water. The temperature of the bath is kept constant, and the capacity is calculated by measuring the coil flow rate and the temperature differential between water entering and leaving the coil.

REFERENCES

ASHRAE. 1992. Method of testing desiccants for refrigerant drying. ANSI/ASHRAE *Standard* 35-1992.

ASHRAE. 1992. Number designation and safety classification of refrigerants. ANSI/ASHRAE *Standard* 34-1992.

ASHRAE. 1993. Methods of testing for rating positive displacement refrigerant compressors and condensing units. ANSI/ASHRAE *Standard* 23-93.

ASTM. 1996. Guide for sampling, test methods, specifications, and guide for electrical insulating oils of petroleum origin. *Standard* D 117-96. American Society for Testing and Materials, West Conshohocken, PA.

Blair, H.A. and J. Calhoun. 1946. Evacuation and dehydration of field installations. *Refrigerating Engineering* (August):125.

Goddard, M.B. 1945. Moisture in Freon refrigerating systems. *Refrigerating Engineering* (September):215.

Larsen, L.W. and J. Elliot. 1953. Factory methods for dehydrating refrigeration compressors. *Refrigerating Engineering* (December):1325.

Morton, J.D. and L.K. Fuchs. 1960. Determination of moisture in fluorocarbons. *ASHRAE Transactions* 66:434.

Reed, F.T. 1954. Moisture determination in refrigerant oil solutions by the Karl Fischer method. *Refrigerating Engineering* (July):65.

Taylor, E.S. 1956. New instrument for moisture analysis of "Freon" fluorinated hydrocarbons. *Refrigerating Engineering* (July):41.

CHAPTER 47

RETAIL FOOD STORE REFRIGERATION AND EQUIPMENT

Display Refrigerators 47.1	*Condensing Methods* 47.10
Meat Processing 47.7	*Methods of Defrost* 47.13
Storage Refrigerators 47.7	*Refrigerant Lines* 47.14
Refrigerators and Systems 47.8	*Effects of New Refrigerants* 47.14
Compressor Configurations 47.9	*Interaction Between Refrigeration and Air Conditioning* 47.14

THE modern retail food store is a high-volume sales outlet with maximum inventory turnover. Almost half of retail food sales are of perishable or semiperishable foods requiring refrigeration. These foods include fresh meats, dairy products, perishable produce, frozen foods, ice cream and frozen desserts, and various special items such as bakery and deli products and prepared meals. These foods are displayed in highly specialized and flexible storage, handling, and display apparatus.

These food products must be kept at safe temperatures during transportation, storage, and processing, as well as during display. The back room of a food store is both a processing plant and a warehouse distribution point. It includes specialized refrigerated rooms, which must be coordinated during construction planning because of the interaction between the store's environment and the refrigeration equipment (Figure 1). Chapter 2 of the 1995 *ASHRAE Handbook—Applications* also covers the importance of coordination.

Refrigeration equipment used in retail food stores may be broadly grouped into display refrigerators, storage refrigerators, processing refrigerators, and mechanical refrigeration machines. Chapter 48 presents food service and general commercial refrigeration equipment.

DISPLAY REFRIGERATORS

Each category of perishable food has its own physical characteristics, handling logistics, and display requirements that dictate specialized display shapes and flexibility required for merchandising. Also, the same food product requires different display treatment in different locations, depending on such things as neighborhood preferences, neighborhood income level, size of store, volume of sales, and local availability of food items by type.

Open display refrigerators for medium and low temperatures are widely used in food markets. However, glass door multideck models have also rapidly gained in popularity. **Deck** is an industry term for a shelf, pan, or rack that supports the displayed product. Many operators combine single-deck and multideck models in most departments where perishables are displayed and sold. **Closed-service refrigerators** are used to display unwrapped fresh meat, delicatessen food, and, frequently, fish on crushed ice instead of or supplemented by mechanical refrigeration. A store employee assists the customer by obtaining product out of the service-type case. More complex layouts of display refrigerators have been developed as new or remodeled stores strive to be distinctive and more attractive. Refrigerators are allocated in relation to expected sales volume in each department. Thus, floor space is allocated to provide balanced stocking of merchandise and smooth flow of traffic in relation to expected peak volume periods.

Small stores accommodate a wide variety of merchandise in a limited floor space. Thus, display refrigerators installed in small and

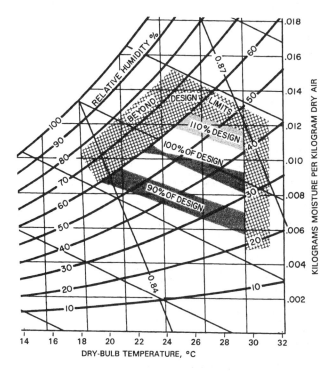

Fig. 1 Open Display Refrigerator Load Factors Superimposed on Psychrometric Chart
Reprinted by permission of Tyler Refrigeration.

medium-sized stores tend to be the glass door reach-in type, which can display 50 to 100% more quantity and variety of merchandise than single-deck refrigerators in the same amount of floor space. This concentration of large refrigeration loads in a small space makes year-round space temperature and humidity control essential.

Product Temperatures

Display refrigerators are designed to merchandise food to maximum advantage while providing short-term (24 to 72 h) protection. Table 1 lists air temperatures for the maintenance of the product. Codes established with maximum product temperature requirements for the case display period may require different case air temperatures from those listed in Table 1.

Display refrigerators are not designed to cool the product; they are designed to maintain product temperature. The merchandise, when put into the case, should be at or near the proper temperature. Food placed directly into the refrigerator or into another adequately refrigerated storage space on delivery to the store should come from properly refrigerated trucks. Little or no delay in transferring perishables from storage or trucks to the display refrigerator or storage space should be permitted.

The preparation of this chapter is assigned to TC 10.7, Commercial Food and Beverage Cooling, Display and Storage.

Table 1 Air Temperatures in Display Refrigerators

Type of Fixture	Air Discharge Temperatures, °C[a]	
	Minimum	Maximum
Dairy		
Multideck	1.1	3.3
Produce, packaged		
Single-deck	1.7	3.3
Multideck	1.7	3.3
Meat, unwrapped (closed display)		
Display area	2.2[b]	3.3[b]
Deli smoked meat		
Multideck	0	2.2
Meat, wrapped (open display)		
Single-deck	−4.5	−3.3
Multideck	−4.5	−3.3
Frozen food		
Single-deck	c	−25[c]
Multideck, open	c	−23[c]
Glass door reach-in	c	−20[c]
Ice cream		
Single-deck	c	−31[c]
Glass door reach-in	c	−24[c]

[a]These air temperatures are measured with the thermometer in the outlet of the refrigerated airstream and not in contact with the product displayed.
[b]Unwrapped fresh meat should only be displayed in a closed, service-type display case. The meat should be cooled to 2.2°C internal temperature prior to placing on display. The case air temperature should be adjusted to keep the internal meat temperature at 2.2°C or lower for minimum dehydration and optimum display life. Display case air temperature varies with manufacturer.
[c]Minimum temperatures for frozen foods and ice cream are not critical (except for energy conservation); maximum temperature is important for proper preservation of product quality. The differences in display temperatures among the three different styles of frozen food and ice cream display cases are due to the orientation of the refrigeration air curtain and the size and style of the opening. The single-deck case has a horizontal air curtain and opening of approximately 760 to 1070 mm. The multideck, open case has a vertical air curtain and an opening of about 1070 to 1270 mm. The glass door reach-in case has a vertical air curtain protected by a multiple-pane insulated glass door.

Display refrigerators should be loaded properly. The product on display should never be piled so high that it is out of the refrigerated zone or be stacked so that circulation of refrigerated air is blocked. The load line recommendations of the manufacturer must be followed to obtain good refrigeration performance. Proper refrigerator design and loading minimize energy use, maximize the efficiency of the refrigeration equipment, and minimize product loss.

Store Ambient Effect

Display fixtures are affected significantly by the temperature, humidity, and movement of surrounding air. These display cases are designed primarily for supermarkets, virtually all of which are air conditioned.

Table 2 summarizes a study of ambient conditions in retail food stores. Individual store ambient readings showed that only 5% of all readings (including those when the air conditioning was not working or not turned on) exceeded 24°C dry bulb or 10.2 g of moisture per kilogram of dry air. Based on these data, the industry has chosen 24°C dry bulb and 18°C wet bulb (55% rh, 14.2°C dew point) as summer design conditions. This is the ambient condition at which refrigeration load for food store display cases is normally rated.

When store ambient relative humidity is different from that at which the cases were rated, the energy requirements for case operation will vary. Howell (1993a, 1993b) concludes that compared to operation at 55% store rh, display case energy savings at 35% rh range from 5% for glass door reach-in cases to 29% for multideck

Table 2 Average Store Conditions in the United States

Season	Dry Bulb, °C	Wet Bulb, °C	Grams Moisture per Kilogram Dry Air	rh, %
Winter	20.6	12.2	5.4	36
Spring	21.1	14.4	7.9	50
Summer	21.7	16.1	9.1	56
Fall	21.1	14.4	7.9	50

Store Conditions Survey conducted by Commercial Refrigerator Manufacturers' Association from December 1965 to March 1967. Approximately 2000 store readings in all parts of the country, in all types of stores, during all months of the year reflected the above ambient store conditions.

Table 3 Relative Refrigeration Requirements with Varying Store Ambient Conditions

Case Model	21°C Dry-Bulb Relative Humidity, %					26°C Dry-Bulb Relative Humidity, %		
	30	40	55	60	70	50	55	65
Multideck dairy	0.90	0.95	1.00	1.08[a]	1.18[b]	0.99	1.08[a]	1.18[b]
Multideck low-temperature	0.90	0.95	1.00	1.08[a]	1.18[b]	0.99	1.08[a]	1.18[b]
Single-deck low-temperature	0.90	0.95	1.00	1.08[a]	1.15	0.99	1.05	1.15
Single-deck red meat	0.90	0.95	1.00	1.08[a]	1.15	0.99	1.05	1.15
Multideck red meat	0.90	0.95	1.00	1.08[a]	1.18[b]	0.99	1.08[a]	1.18[b]
Low-temperature reach-in	0.90	0.95	1.00	1.05[a]	1.10	0.99	1.05[a]	1.10

Note: Package warm-up may be more than indicated. Standard flood lamps are clear PAR 38 and R-40 types.
[a]More frequent defrosts required.
[b]More frequent defrosts required plus internal condensation (not recommended).

deli cases. Table 3 lists correction factors for the effect of store relative humidity on display case refrigeration requirements when the dry-bulb temperature is 21°C and 26°C.

Manufacturers sometimes publish ratings for open refrigerators at lower ambient conditions than the standard because the milder conditions may significantly reduce the heat load on the refrigerators. In addition, lower ambient conditions may permit both reductions in anticondensate heaters and fewer defrosts, allowing substantial energy savings on a complete store basis.

The application engineer needs to verify that the year-round store ambient conditions are within the performance ratings of the various cases selected for the store. Because relative humidity varies throughout the year, the dew point for each period should be analyzed. The sum of these case energy requirements provides the total annual energy consumption. In a store designed for a maximum relative humidity of 55%, the air-conditioning system will dehumidify only when the relative humidity exceeds 55%.

In climates where the outdoor air temperature is low in winter, infiltration of outdoor air can cause store humidity to drop below 55% rh. Separate calculations need to be done for periods during which mechanical dehumidification is used and periods when it is not required. As an example, in Boston, Massachusetts, mechanical dehumidification is required for only about 3-1/2 months of the year, while in Jacksonville, Florida, it is required for almost 7-1/2 months of the year. Also, in Boston, there are 8-1/2 months when the store relative humidity is below 40%, while Jacksonville has these conditions for only 4-1/2 months. The engineer must weigh the savings at lower relative humidity against the cost of the mechanical equipment required to maintain relative store humidity levels at, for example, below 40% instead of 55%.

Additional savings can be achieved by controlling anticondensate heaters and reducing defrost frequency at ambient relative humidities below 55%. Energy savings credit for reduced use of display case anticondensate heaters can only be taken if the display cases are equipped with humidity-sensing controls that reduce the

Retail Food Store Refrigeration and Equipment

amount of power supplied to the heaters as the store dew point decreases. Also, defrost savings can be considered when defrost frequency or duration is reduced. Controls can reduce the frequency of defrost as the store relative humidity decreases. Individual manufacturers give specific anticondensate and defrost values for their equipment at stated store conditions. Less defrosting is needed as store temperature and humidity decrease from the design conditions.

Attention should be given to the opposite condition, in which store environments higher than the industry standard dramatically raise the refrigeration requirements and consequently the energy demand. Higher store ambient temperatures may also significantly affect the performance of the refrigeration equipment—particularly open, self-service models.

Adverse Heat Sources

Lighting. Refrigeration performance and food temperatures are adversely affected by display lighting and other forms of radiant heat. High-intensity lighting raises product temperatures several degrees, as shown in Figure 2, and can discolor meats. Recently, T8 fluorescent lamps with electronic ballasts have been used to produce less heat load in the case. Case shelf ballasts are now often located out of the refrigerated space to further reduce case load.

The heat-rejecting lamp is questionable, and conventional incandescent floodlights and supplemental incandescent lighting should never be used to display retail meats.

Ceiling Temperatures. The radiant energy coming into the meat case from the adjacent ceiling can cause product temperature control problems that cannot be cured by lowering case air temperature. Proper ceiling and light fixture design minimizes this problem.

Packaging. The product surface temperature of a loosely wrapped package of meat with an air space between the film and surface may be 1 to 2 K above the ambient temperature. This is called the greenhouse effect.

Load Lines and Loading. The following observations indicate the effect of loading:

- Voids in the display raise the surface temperature of a package in front of a void 1 to 3 K.
- For all makes and styles of display refrigerators, keeping the top of the product 50 mm below the maximum load line lowers the surface temperatures 1 to 3 K.

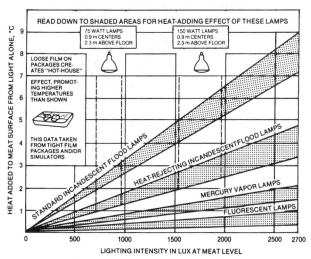

Note: Package warm-up may be more than indicated. Standard flood lamps are clear PAR 38 and R-40 types. Blue, pink, or other hues in these types reduce the amount of light, but not the heat load.

Fig. 2 Heating Effect of Lighting on Packaged Meat in Open Display Refrigerators

Refrigerator Construction

Commercial refrigerators for market installations are usually of the endless construction type, which allows a continuous display as refrigerators are joined. Clear plastic panels are often used to separate case interiors when adjacent cases are connected to different refrigeration circuits. Separate end sections are provided for the first and last units in a continuous display. Methods of joining self-service refrigerators vary, but they are usually bolted or cam locked together.

All refrigerators are constructed with surface zones of transition between the refrigerated area and the room atmosphere. Thermal breaks of various designs separate the zones to minimize the amount of refrigerator surface that is below the dew point. Surfaces that may be below the dew point include (1) in front of discharge air nozzles, (2) sometimes on the nose of the shelving, and (3) sometimes at the front rails or center flue of the refrigerator. In glass door reach-in freezers or medium-temperature refrigerators, the frame jambs and the glass can be below the dew point. In these locations, resistance heat is used effectively to raise the exterior surface temperature above the dew point to prevent accumulation of condensation.

With the current emphasis on energy efficiency, designers have developed means other than resistance heat to raise the surface temperatures above the dew point. However, when no other technique is known, resistance heating becomes a necessity. Control by cycling and/or proportional controllers to vary heat with store ambient changes can reduce energy consumption.

Store designers can do a great deal to promote energy efficiency. Not only does controlling the atmosphere within a store reduce the refrigeration requirements, it also reduces the need to heat the surfaces of refrigerators. This heat not only consumes energy; it also places added demand on the refrigeration load.

Evaporators and air distribution systems for display refrigerators are highly specialized and are usually fitted precisely into the particular display refrigerator. As a result, they are inherent in the fixture and are not standard independent evaporators. The design of the air circuit system, the evaporator, and the means of defrosting are the result of extensive testing to produce the particular display results desired.

Cleaning and Sanitizing Equipment

Because the evaporator coil is the most difficult part to clean, the operator should consider the judicious use of high-pressure, low liquid volume sanitizing equipment. This type of equipment enables personnel to spray cleaning and sanitizing solutions into the duct, grill, coil, and waste outlet areas with minimum disassembly and maximum effectiveness. However, this equipment must be used carefully because the high-pressure stream can easily displace sealing and caulking materials. High-pressure streams should not be directed toward electrical devices. Hot liquid can also break the glass on models with glass fronts and on closed-service fixtures.

Dairy Display

Dairy products include items with significant sales volume, such as fresh milk, butter, eggs, and margarine. They also include a myriad of small items such as fresh (and sometimes processed) cheeses, special above-freezing pastries, and other perishables. The available display equipment includes the following:

1. Full-height, fully adjustable shelved display units without doors in back for use against a wall (Figure 3); or with doors in back for rear service or for service from the rear through a dairy cooler. The effect of rear service openings on the surrounding refrigeration must be considered.

2. Closed-door displays built in the wall of a walk-in cooler with adjustable shelving behind doors. Shelves are located and stocked in the cooler (Figure 4).

3. A variety of other special display units, including single-deck and island-type display units, some of which are self-contained and reasonably portable for seasonal, perishable specialties.
4. A refrigerator, similar to Item 1, but able to receive either conventional shelves and a base shelf and front or premade displays on pallets or carts. This version comes with either front-load capability only or rear-load capability only (Figure 5). These are called front roll-in or rear roll-in display cases.

Produce Display

Wrapped and unwrapped produce is often intermixed in the same display refrigerator. Ideally, unwrapped produce should have low-velocity refrigerated air forced up through the loose product. Water is usually also sprayed on the leafy vegetables, either by manually operated spray hoses or by automatic misting systems, to retain their crispness and freshness. Produce is often displayed on a bed of ice for freshness appeal. However, packaging prevents this air from circulating through wrapped produce and requires higher velocity air. The equipment available for displaying both packaged and unpackaged produce is usually a compromise between these two desired features and is suitable for both types of product. The equipment available includes the following:

1. Wide or narrow single-deck display units with or without mirrored superstructures.
2. Two- or three-deck display units, similar to the one in Figure 6, usually for multiple-case lineups near the above display refrigerators.
3. Due to the nature of produce merchandising, a variety of nonrefrigerated display units of the same family design are usually designed for connection in continuous lineup with the refrigerators.
4. A refrigerator, like Item 2, but able to receive either conventional shelves and a base shelf and front or premade displays on pallets and carts. This version comes with either front-load capability only or rear-load capability only (Figure 5).

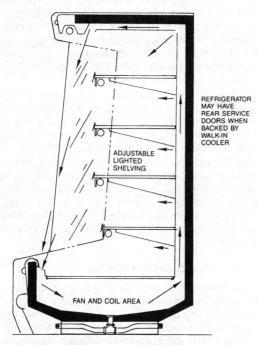

Fig. 3 Multideck Dairy Display Refrigerator

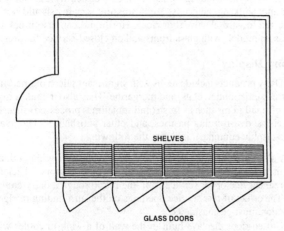

Fig. 4 Typical Walk-In Cooler Installation
Reprinted by permission of Bangor Refrigeration Corporation.

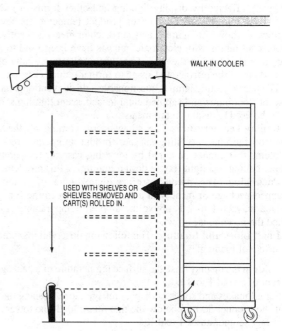

Fig. 5 Vertical Rear-Load Dairy (or Produce) Refrigerator with Roll-In Capability

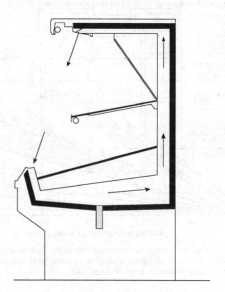

Fig. 6 Multideck Produce Refrigerator
Reprinted by permission of Hussmann Corporation.

Retail Food Store Refrigeration and Equipment

This produce equipment is generally available with a variety of merchandising and other accessories, including bag compartments, sprayers for wetting the produce, night covers, scale racks, sliding mirrors, and other display shelving and apparatus.

Frozen Food and Ice Cream Display

To display frozen foods most effectively (depending on varied need), many types of display refrigerators have been designed and are available. These include the following:

1. Single-deck well-type refrigerators for one-side shopping. Many types of merchandising superstructures for related nonrefrigerated foods are available. Configurations of these refrigerators are designed for matching lineup with fresh meat refrigerators, and there are similar refrigerators for matching lineup of ice cream refrigerators with their frozen food counterparts. These refrigerators are offered with or without glass fronts (Figure 7).
2. Single-deck island for shop-around (Figure 8). These are available in widths ranging from the single-deck refrigerators in Item 1 to refrigerators of double width, with various sizes in between. Some across-the-end increments are available with or without various merchandising superstructures for selling related non-refrigerated food items to complete the shop-around configuration.
3. Freezer shelving in two to six levels with many refrigeration system configurations (Figure 9). Multideck self-service frozen food and ice cream fixtures are generally more complex in design and construction than single-deck models. Because they have wide, vertical display compartments, they are more affected by ambient conditions in the store. Generally, open multideck models have two or three air curtains to maintain product temperature and shelf life requirements.
4. Glass door, front reach-in refrigerators, usually of a continuous lineup design. This style allows for maximum inventory volume and variety in minimum floor space. These advantages must be balanced against the barrier produced by the doors and the greater labor requirement in restocking. The front-to-back interior dimension of these cabinets is usually about 600 mm. Greater attention must be given to the back product to provide the desired rotation. While these refrigerators generally consume less energy in their operation than open multideck low-temperature refrigerators, specific comparisons by model should be made to determine original and operating costs (Figure 10).
5. Spot merchandising refrigerators, usually self-contained and sometimes arranged for quick change from nonfreezing to freezing temperature to allow for promotional items of either type (i.e., fresh asparagus or ice cream).

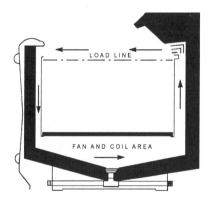

Fig. 7 Single-Deck Well-Type Frozen Food Refrigerator

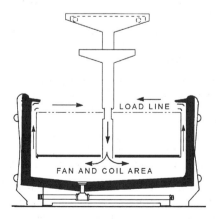

Fig. 8 Single-Deck Island Frozen Food Refrigerator

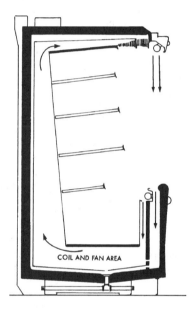

Fig. 9 Multideck Frozen Food Refrigerator

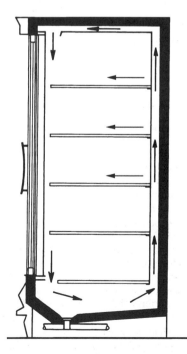

Fig. 10 Glass Door, Frozen Food Reach-In Refrigerator

6. Versions of most of the above items for ice cream, usually with modified defrost heaters and other changes necessary for the approximately 5.5 K colder required temperature. As display temperature decreases to below −18°C (product temperature), the problem of frost and ice accumulation in flues and in the product zone increases dramatically. Proper product rotation and restocking frequently minimize frost accumulation.

Meat Display

Prepackaged meat is becoming more popular. Most meat is sold in packaged form. Some of this product is cut and packaged on the store premises. Control of temperature, time, and sanitation from the truck to the checkout counter is important. Surface temperature on the meat in excess of 4.4°C shortens the salable life of meat products significantly and increases the rate of discoloration.

The design of open fresh meat display refrigerators, either well-type single-deck or vertical multideck, is limited by the freezing point of meat. Ideally, the refrigerators are set to operate as cold as possible without freezing the meat. The temperatures are maintained with a minimum of fluctuations (with the exception of defrost) to ensure the coldest possible stable internal and surface meat temperatures.

Sanitation is also important. If all else is kept equal, good sanitation can as much as double the salable life of meat in a display refrigerator. In this chapter, sanitation includes the control of the time during which meat is exposed to temperatures above 4.4°C. If meat has been handled in a sanitary manner until it is placed in the display case, elevated temperatures can be more tolerable. When the meat surfaces are contaminated by dirty knives, meat saws, table tops, and so forth, even optimum display temperatures will not prevent premature discoloration and subsequent deterioration of the meat. See the section on Meat Processing for information about the refrigeration requirements of the meat wrapping area.

Along with molds and natural chemical changes, bacteria discolor meat. With good control of sanitation and refrigeration, experiments in stores have produced meat shelf life of one week and more. Bacterial population is greatest on the exposed surface of displayed meat because the surface is warmer than the interior. Although the cold airflow refrigerates each package, the surface temperature (and thus bacterial growth) is cumulatively increased by the following:

- Infrared rays from lights
- Infrared rays from the ceiling surface
- High stacking of meat products
- Voids in display
- Store drafts that disturb refrigerator air

Where these factors are handled improperly, the surface temperature of the meat is often above that allowed by the food handling codes. It takes great care in every building and equipment detail, as well as in refrigerator loading, to maintain meat surface temperature below 4.4°C. However, the required diligence is rewarded by excellent shelf life, improved product integrity, higher sales volume, and less scrap or spoilage.

The surface temperature rises during defrost. Tests have compared matched samples of meat—one goes through normal defrost, and the other is removed from the refrigerator during its defrosting cycles. Although defrosting characteristics of refrigerators vary, such tests have shown that the effects on shelf life of properly handled defrosts are negligible. Tests for a given installation can easily be run to prove the effects of defrosting on shelf life for that specific set of conditions.

Self-Service Meat Refrigerators. Self-service meat products are displayed in packaged form. The meat department planner can select from a wide variety of available meat display possibilities:

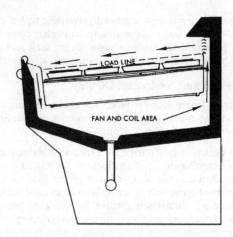

Fig. 11 Single-Deck Meat Display Refrigerator

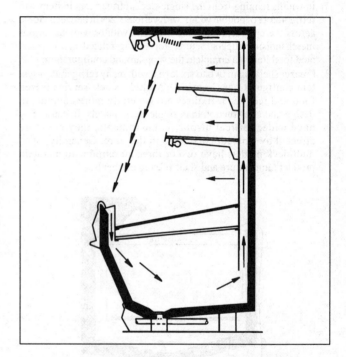

Fig. 12 Multideck Meat Refrigerator

1. Single-deck refrigerators with or without rear or front access storage doors (Figure 11)
2. Multideck refrigerators with or without rear access (Figure 12)
3. Either of the above with or without glass fronts
4. Processed meat versions of the above, often designed for somewhat higher temperatures but including special merchandising shelves or accessories

All of the above refrigerators are available with a variety of lighting, superstructures, shelving, and other accessories tailored to special merchandising needs. Storage compartments are rarely used in self-service meat refrigerators.

Closed-Service Meat or Deli Refrigerators. Service meat products are generally displayed in bulk, unwrapped. Generally, closed refrigerators are definable in one of the following categories:

1. Fresh red meat, with or without storage compartment (Figure 13)
2. Deli and smoked or processed meats, with or without storage
3. Fresh fish and poultry, usually without storage but designed to display the fresh fish and/or poultry on a bed of cracked ice

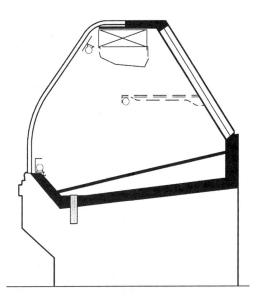

**Fig. 13 Closed-Service Display Refrigerator
(Gravity Coil Model with Curved Front Glass)**
Reprinted by permission of Hussmann Corporation.

Closed-service meat display refrigerators are offered in a variety of configurations. They are available with gravity or forced-convection coils, and their fronts may be nearly vertical or angled up to 20° from vertical in flat or curved glass panels, either fixed or hinged. Gravity coils are usually preferred for more critical products, but forced-air coil models using various forms of humidification systems are also common.

These service refrigerators typically have sliding rear access doors, which are sometimes removed during busy periods. This practice is not recommended by manufacturers, however, because the internal product display zone temperature and humidity levels are affected.

Lighting. The elimination of excessive heat due to lighting, high ceiling temperatures, and poor display practices should be required in addition to strict sanitation practices.

There is a conflict between the need for highlighting the product in a closed refrigerator and the need for maintaining a high moisture level in the atmosphere within the refrigerator. Even the lighting within the store is involved. Bright store lighting necessitates even brighter refrigerator lighting to highlight the product in the refrigerator. Lights in the refrigerator produce heat. Reheating refrigerated air increases the total refrigeration load on the fixture.

Light also discolors smoked, cured, and table-ready meats. No more than 750 lx of display lighting at the product level should be used on smoked, cured, and table-ready meats unless they will be sold within 24 h. Discoloration is directly proportional to both light intensity and length of time displayed.

The best solution for this problem is to subdue ambient lighting around the refrigerators so that minimal lighting within the refrigerator is not dominated. As mentioned in the section on Adverse Heat Load, fluorescent lamps with electronic ballasts or ballasts located out of the refrigerated space have been used to reduce heat load in the case. Meat and deli refrigerators usually maintain a long shelf life when the air temperature in the refrigerator is between 1.5 and 2.5°C. At the same time, the velocity of the refrigerated air over the product should be minimized. High relative humidity and a stable temperature should be maintained in the product zone for maximum shelf life. Use of an evaporator pressure control is an often recommended method for refrigeration control. A minimum TD thermostat is used as a backup for times when the load is small. This control method minimizes product dehydration.

MEAT PROCESSING

In a self-service meat market, the operations of cutting, wrapping, sealing, weighing, and labeling involve precise production control and scheduling to meet varying sales demands. The faster the processing, the less critical the temperature and corresponding refrigeration demand.

The wrapping room should not be too dry, but condensation on the meat, which provides a medium for bacterial growth, should be avoided by maintaining a dew point temperature within a few degrees of the sensible temperature. Fan-coil units should be selected with a maximum of 6 K temperature difference (TD) between the entering air and the evaporator temperature. Low-velocity fan-coil units are generally used to reduce the drying effect on exposed meat. Gravity coils are also available and have the advantage of lower room air velocities.

The meat wrapping area is generally cooled to about 7 to 13°C, which is desirable for the personnel, but not low enough for meat storage. Thus, meat should be held in that room only for the cutting and packaging operation; then, as soon as possible, it should be moved to a packaged product storage cooler held at −2 to 0°C. The meat wrapping room may be a refrigerated room adjacent to the meat storage cooler or one compartment of a two-compartment cooler. In such a cooler, one compartment is refrigerated at about −2 to 0°C and used as a meat storage cooler, while the second compartment is refrigerated at 7 to 13°C and used as a cutting and packaging room. Best results are attained when the meat is cut and wrapped to minimize exposure to temperatures above −2 to 0°C.

STORAGE REFRIGERATORS

Wrapped Meat Storage

At some point between the wrapping room and the display refrigerator, refrigerated storage for the wrapped cuts of meat must be provided. Without this space, a balance cannot be maintained between the cutting/packaging rate and the selling rate for each particular cut of meat. Display refrigerators with refrigerated bottom storage compartments, equipped with racks for holding trays of meats, offer one solution to this problem. However, the amount of stored meat is not visible, and the inventory cannot be controlled at a glance.

A second option is a pass-through, reach-in cabinet. This cabinet has both front and rear insulated glass doors and is located between the wrapping room and the display refrigerators. After wrapping, the meats are passed into the cabinet for temporary storage at −2 to 0°C and then are withdrawn from the other side for restocking to the display refrigerator. Because these pass-through cabinets have glass doors, the inventory of wrapped meats is visible and therefore controllable.

The third and most common option involves a section of the back room walk-in meat storage cooler or a completely separate packaged meat storage cooler. The cooler is usually equipped with rolling racks into which the trays of meat can be slid. This method of storage also offers visible inventory control. This method provides convenient access to both the wrapping room and the display cases.

The overriding philosophy in successful meat wrapping and merchandising can be summarized thus: keep it clean, keep it cold, and keep it moving.

Walk-In Storage Coolers

Each category of displayed food product that requires refrigeration for preservation is usually backed up by storage in the back room. This storage usually consists of refrigerated rooms with sectional walls and ceilings equipped with the necessary storage racks for a particular food product. Walk-in coolers are required for the storage of meat, some fresh produce, dairy products, frozen food, and ice cream. Medium and large stores have separate produce and dairy coolers usually in the 2 to 4°C range. Meat coolers are used in all food stores, with storage conditions between −2 and 0°C.

Unwrapped meat, fish, and poultry should each be stored in separate coolers to prevent odor transfer. Walk-in coolers, which serve the dual purpose of storage and display, are equipped with either sliding or hinged glass doors on the front. These door sections are often prefabricated and set into an opening in the front of the cooler. In computing the refrigeration load, allowance must be made for the extra service load.

Moisture conditions must be confined to a relatively narrow range because an excessive humidity level encourages bacteria and mold growth, which leads to sliming. A level that is too low leads to excessive dehydration.

Air circulation must be maintained at all times to prevent stagnation, but it should not be so rapid as to cause drying of an unwrapped product. Forced-air blasts must not be permitted to strike products; therefore, low-velocity coils are recommended.

For optimum humidity control, unit coolers should be selected at about a 6 K TD between entering air temperature and evaporator temperature. Note that the published ratings of commercial unit coolers do not reflect the effect of frost accumulation on the evaporator. The unit cooler manufacturer can determine the correct frost derating factor for its published capacity ratings. From experience, a minimum correction multiplier of 0.80 is typical.

A low-temperature storage capacity equivalent to the total volume of the low-temperature display equipment in the store is satisfactory. Storage capacity requirements can be reduced by frequent deliveries.

Generally, forced-air coils are selected for low-temperature coolers where humidity is not critical for packaged products. For low-temperature coolers, gas or electric defrost is required. Off-cycle defrosts are used in produce and dairy coolers. Straight time or time-initiated, time- or temperature-terminated gas or electric defrosts are generally used for meat coolers.

REFRIGERATORS AND SYSTEMS

Food stores sell all types of perishable foods that require a variety of refrigeration systems to best preserve and most effectively display each product. Moreover, the refrigerating system must be highly reliable because it must operate 24 h per day for 10 or more years, to protect the large investment in highly perishable foods. The various refrigerators have different evaporator pressure/temperature requirements ranging from the highest for the meat processing room to the lowest for the ice cream level. Produce and meat wrapping rooms may approach the suction pressures used in air-conditioning applications. Open ice cream display cases may have suction pressures corresponding to temperatures as low as $-40°C$. All other cases and coolers fall between these extremes.

Temperature controls also vary greatly, from a produce preparation room (which may operate with a wet coil) requiring no defrost to the ice cream case requiring induced heat to defrost the coil periodically. Electronic sensor control is the most accurate and can also provide a temperature alarm to prevent food loss. Defrost methods include (1) off-time defrost, (2) gas defrost, (3) electric defrost, and (4) defrost using ambient air induced into the refrigerator.

Various Refrigeration System Design Solutions

The selection of refrigeration equipment used to operate display refrigerators and storage rooms for food stores involves the following considerations: (1) cost/space limitations, (2) reliability, (3) maintainability and complexity, and (4) operating efficiency. Solutions span the very simple (one compressor and associated controls on one refrigerator) to the complex (central refrigeration plant operating all refrigerators in a store).

Condensing unit systems, which use single compressors with multiple cases or coolers, provide a popular, suitable approach. This method places produce, dairy, meat, or frozen food refrigerators in groups, each on a single compressor that may have its own condenser or may have an air- or water-cooled remote condenser.

Parallel Systems. Another common refrigeration technique couples two compressors in parallel. Oil distribution, temperature control at more than one suction pressure, and defrosting must be considered. Parallel units usually operate connected to one or more large condensers. The condensers are usually remote air-cooled or remote evaporative cooled, but they can also be built as part of the compressor rack assembly.

Carrying the sequence one step further leads to a large assembly tying three or more compressors in parallel. The needs defined for two units in parallel apply to the larger styles as well. However, efficiency may be sacrificed when greatly differing suction pressures are grouped in one suction pressure designed system. Split suction manifolds separate load groups on a rack. To match changing evaporator loads, rack capacity is varied by cycling compressors, varying the speed of one or more compressors, and/or unloading compressor cylinders by closing valves or moving ports on screw compressors.

Running all the refrigeration equipment at the low suction pressure required for ice cream on low-temperature systems and for meat on medium-temperature systems is inefficient. In large parallel systems, the ice cream and meat refrigeration are frequently isolated. Single compressor satellites tied into the parallel compressors are often used for small ice cream or meat loads, while split-suction parallel equipment is used for larger loads. Different suction pressures are obtained, but all compressors discharge into a common header.

Load Versus Ratings

Food store refrigerator manufacturers publish case load ratings to match the proper condensing unit with the fixture load. For single compressor applications only, the ratings can be stated, for selection convenience, as the capacity the condensing unit must deliver at an arbitrary suction pressure (evaporator temperature). In general, manufacturers of display refrigerators use ASHRAE *Standards* 72 and 117, which specify standard methods of testing open and closed refrigerators for food stores. These standards establish refrigeration load requirements at rated ambient conditions of 24°C and 55% rh in the sales area. Similar display refrigerators vary between manufacturers, so manufacturers' recommendations must be followed to achieve proper results in both efficiency and product integrity.

Multiplexed Systems

When dissimilar display case refrigerator loads are connected to the same condensing system, the manufacturer should be consulted to determine the true maximum suction pressure at the fixture refrigerant line outlet for each system and the true load that such a system adds to the total. The condensing unit must deliver the total of all the loads at a suction pressure at the machine no higher than the lowest system pressure requirement less the suction line pressure drop. The other systems must have suction line restriction (by evaporator pressure regulators or some other sure means) to prevent higher temperature evaporators from adding unnecessarily to the load. The pressure regulator then ensures that the condensing unit will pull down to the pressure necessary for the lowest evaporator temperature required.

Electronically or mechanically actuated pilot-operated evaporator pressure regulator (EPR) valves control evaporator pressures in multiplexed systems. These valves cause little or no measurable pressure drop when they are in the full-open position. Another design uses liquid or discharge pressure to open and close the EPR valves, so the pressure drop is negligible. Another design has no pilot valve but typically requires a minimum of 6.9 kPa pressure drop for operation.

The suction gas temperature leaving display fixtures is often superheated. Particularly on low-temperature fixtures, the suction line gas temperature increase from heat gained from the store ambient can be substantial. This increase, which adversely affects condensing unit capacity and compressor discharge gas temperature, must be considered for system design; both capacity effects and the cooling needs of compressors are involved.

Retail Food Store Refrigeration and Equipment

One solution to excessively high superheat is to run the suction line and liquid line tightly together from the fixture refrigerant line outlet, with the pair insulated together for a distance of 10 to 20 m from the fixture outlet. This technique cannot be used with gas defrost or refrigerants requiring low suction superheat at the compressor suction (for example, low-temperature single-stage R-22 systems). Most manufacturers produce suction line to liquid line heat exchangers installed in the display fixture. This technique converts a loss into a gain by allowing the suction gas to pick up heat from the liquid instead of the store ambient. Under all conditions, the suction line should be insulated from the point where it leaves the display case to the suction service valve on the compressor. The insulation and its installation must be vapor resistant.

To ensure proper thermostatic expansion valve operation, the engineer should verify that the liquid entering the fixture is subcooled. Some case and/or system designs require liquid line insulation. This is very important when ambient outdoor air or mechanical subcooling is being used to improve system efficiency.

COMPRESSOR CONFIGURATIONS

Matching Refrigerator Requirements with Compressor Capacity

The designer matches requirements of refrigerator lineups to the capacity of a single compressor unit or divides them into manageable circuits for parallel compressor systems. The parallel systems adapt easily to gas defrost. Gas directly from the compressor discharge, or, in some instances, from the top of the receiver at a lower temperature, flows through a manifold. Electric defrost, reverse air defrost, and off-cycle defrost can also be used on both parallel and single-unit systems. Liquid and/or suction line solenoid valves control the circuits for defrosting. Often, a suction stop is combined with the temperature-controlling EPR. The entire system with its individual circuits is controlled by a multicircuit time clock. Temperature control on the branch circuits is also achieved by refrigerator thermostats actuating liquid-line solenoid valves. Shutoff valves isolate each circuit for service convenience. Refrigerator requirements are now often given as a refrigeration load per unit length, with a lower value sometimes allowed for parallel systems. The rationale for this lower value is that peak loads are less with programmed defrost. Refrigerator temperature recovery after a defrost period is less of a strain than it would be to a single-compressor system.

The published refrigerator load requirements allow for extra capacity for temperature pulldown after defrost, per ASHRAE *Standard* 72. The industry considers a standard store condition to be 24°C and 55% rh, which should be maintained with air conditioning. Much of this air-conditioning load is carried by the open refrigerators, and credit for the heat removed by them should be considered in sizing the air-conditioning system.

Typical Parallel Compressor Energy Efficiency

A typical supermarket includes one or more medium-temperature parallel systems for meat, deli, dairy, and produce refrigerators and the medium-temperature walk-in coolers. The system may have a separate compressor for the meat or deli refrigerators, or all units may have a single compressor. Energy efficiency ratios typically run from 2.3 to 2.6 W/W for the main load. Low-temperature refrigerators and coolers are grouped on one or more parallel systems with ice cream refrigerators on a satellite or on a single compressor. EERs range from 1.2 to 1.5 W/W for the frozen food units and as low as 1.0 to 1.2 W/W for the ice cream units. Cutting and preparation rooms are most economically placed on a single unit because the refrigeration EER is at nearly 2.9 W/W. Air-conditioning compressors are also separate because their EERs can range up to 3.2 W/W (Figure 14).

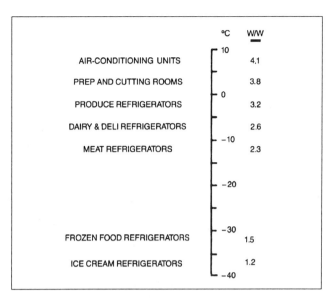

Fig. 14 Typical Single-Stage Compressor Efficiency

Capacity Control of Parallel Systems

This system must be designed to maintain proper refrigerator temperatures under peak summer load. During the remainder of the year, store conditions can be easily maintained at a more ideal condition, and refrigeration load will be lower. In the past, refrigeration systems were operated at 32°C condensing conditions or above to maintain enough high-side pressure to feed the refrigerated display fixture expansion valves properly. When outdoor ambient conditions allow, today's technology permits the condensing temperature to follow the ambient down to about 21°C or less. When proper liquid-line piping practices are observed, the expansion valves feed the evaporators properly under these low condensing temperatures. Therefore, at partial load, the system has excessive capacity to perform adequately.

Multiple compressors may be controlled or staged based on a drop in system suction pressure. If the compressors are equal in size, a mechanical device can turn off one compressor at a time until only one is running. The suction pressure will be perhaps 34 kPa or more below optimum. Newer, solid-state control devices can cycle units; the run time for each compressor motor can be equalized while the system remains at one economical pressure range. Satellite compressors can be controlled accurately with one control that also monitors other components, such as oil pressure and alarm functions. Rack capacity can also be varied by compressor cylinder unloading.

Staging Unequally Sized Compressors

Unequally sized compressors can be staged to obtain more steps of capacity than the same number of equally sized compressors. Figure 15 shows seven stages of capacity from a 5, 7, and 10 kW compressor parallel arrangement. Unloaders on multibank compressors provide greater flexibility, as does variable speed (usually on the lead compressor) for capacity control.

Screw Compressor Central Plant Units

Some supermarkets have begun to use the screw compressor to replace the multireciprocating compressor rack. Precise temperature and power control is available if microprocessor controls are used to control the screw compressor(s). Microprocessors offer the option of remote control and system operation for all types of compressors.

Use of the Economizer Port on Screw Compressors

Small walk-in coolers and preparation rooms are higher temperature applications than the medium-temperature cases. The side center port of the screw compressor is being used for these and other

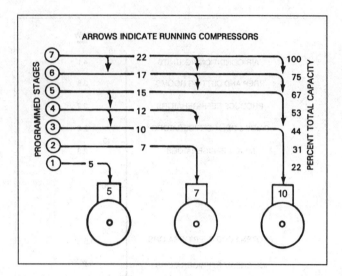

Fig. 15 Stages with Mixed Compressors
Reprinted by permission of Hussmann Corporation.

types of loads where a higher suction pressure is needed. This reduces the number of compressors needed for the variety of load requirements. Planning the load management and the sizing of the screw compressors is very important to a successful refrigeration installation.

Factory-Assembled Equipment

Factory assembly of the necessary compressor systems with either a direct air-cooled condenser or any style of remote condenser is common practice. Both single and parallel systems can be housed, prepiped, and prewired at the factory. The complete unit is then delivered to the job site for placement on the roof or beside the store.

Prefabricated Equipment Rooms

Many supermarkets have elected to purchase the compressor equipment installed in a housing to reduce real estate costs for the building. The time requirements for installation of piping and wiring may also be reduced with prefabrication. Most of the rooms are modular and prewired and include some refrigeration piping. Their fabrication in a factory setting should offer good quality control of the assembly. They are usually put into operation quickly upon arrival at the site.

Subcooling Liquid Refrigerant

Allowing refrigerant to subcool in cool weather as it returns from the remote condenser saves energy. Subcooling can be done by several means. The most common is to flood the condenser and allow the liquid refrigerant to cool to close to the ambient temperature.

Mechanical subcooling may also be economical in many areas. A subcooling compressor can be combined on a parallel system. Another method is to have a separate parallel system with branches to subcoolers on the other parallel and single systems in the store. The mechanical subcooling would be set to operate when the ambient air cools the refrigerant to above the desired subcooled temperature. The advantage to this method is that the mechanical subcooling compressor can run at twice the efficiency of the main system, thus saving energy through year-round liquid temperature control.

Heat Recovery

Heat recovery may be an important feature of virtually every compressor system, parallel or single. A heat recovery coil is simply a second condenser coil placed in the air handler. If the store needs heat, this coil is energized and run in series with the regular condenser (Figure 17). The heat recovery coil can be sized for a 17 to 28 K TD depending on the capacity in cool weather. Lower pressures in parallel systems permit little heat recovery. However, when heat is required in the store, simple controls create a higher pressure for heat recovery. Compared with the cost of auxiliary gas or electric heat, the higher energy consumption of the compressor system may be compensated for by the value of the heat gained.

On a large, single unit, water can be heated by a desuperheater; on two-stage or compound R-22 parallel systems, water is commonly heated by the interstage desuperheater.

CONDENSING METHODS

Many commercial refrigeration installations use air-cooled condensers. Alternatively, evaporative condensers or water-cooled condensers with cooling towers may be specified. To obtain the lowest operating costs, equipment should operate at the lowest pressure possible. The minimum allowable pressures are determined by other design and component considerations incorporated by the designer.

Techniques that permit a system to operate satisfactorily with lower condensing temperatures include (1) insulating liquid lines and/or receiver tank, (2) subcooling the liquid refrigerant by design, and (3) connecting the receiver as a surge tank with appropriate valving. Condensing pressure must still be controlled, at least to the lower limit required by the expansion valve, gas defrosting, and heat reclaim. The typical thermostatic expansion valve is capable of feeding the evaporator properly, assuming that a solid liquid column of refrigerant is always supplied to the expansion valve and that there is sufficient pressure drop across the valve. Balanced-port thermostatic expansion valves have enhanced the opportunity for floating pressures down with varying ambient temperatures below the design point.

Air-Cooled Machine Room

The arrangement of standard air-cooled condensing units in a separate air-cooled machine room is still used in some supermarkets. Dampers, which may be powered or gravity operated, supply air to the room; fans or blowers controlled by room temperature at a thermostat exhaust the air.

A complete air-cooled condensing unit located indoors requires ample, well-distributed ventilation. Ventilation requirements vary depending on maximum summer conditions and evaporator temperature, but 500 to 600 L/s per condensing unit kilowatt has given proper results. Exhaust fans should be spaced for an even distribution of air (see Figure 16).

Rooftop air intake units should be sized for 3.8 m/s velocity or less to keep airborne moisture from entering the room. When condensing units are stacked (as shown in Figure 16), the ambient air design should provide upper units with adequate ventilation. Rooftop intakes are preferred because they are not as sensitive to wind as side wall intakes, especially in winter in cold climates. Butterfly dampers installed in upblast exhaust fans, which are controlled by a thermostat in the compressor room, exhaust warm air from the space.

The air baffle helps prevent intake air from short-circuiting to the exhaust fans (Figure 16). Because air recirculation is needed around the condensers for proper winter control, the intake air should not be baffled to flow only through the condensers.

Ventilation fans for air-cooled machine rooms normally do not have a capacity equal to the total of all the individual condenser fans. Therefore, if the air is baffled to flow only through the condenser during maximum ambient temperatures, the condensers will not receive full free air volume when all or nearly all condensing units are in operation. Also, during winter operation, tight baffling of the air-cooled condenser prevents recirculation of condenser air,

Retail Food Store Refrigeration and Equipment

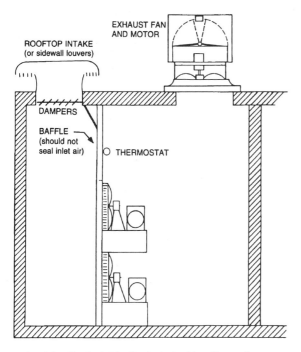

Fig. 16 Typical Air-Cooled Machine Room Layout

which is essential to maintaining sufficiently high room temperature for proper refrigeration system performance.

Machine rooms that are part of the building need to be airtight so that air from the store is not drawn by the exhaust fans into the machine room. Additional load is placed on the store air-conditioning system if the compressor machine room, with its large circulation of outside air, is not isolated from the rest of the store.

Condenser Sizing

To minimize energy consumption, condensers should be more generously sized than they are for typical air-conditioning applications. Typical condenser selection is based on the temperature difference between the air entering the condenser and the saturated condensing temperature. Generally accepted TDs for refrigeration condenser sizing are 8 K for medium-temperature systems and 6 K for low-temperature systems.

Remote Air-Cooled Condenser

The remote condenser may be placed outdoors, or it may be placed indoors to heat portions of the building in winter. Regardless of the arrangement of the remote condenser or the indoor-outdoor mechanical package, the following design points are relevant. The air-cooled condenser may be either a single-circuit or a multiple-circuit condenser. The manufacturer's heat rejection factors should be followed to ensure that the desired TD is accommodated.

Pressure must be controlled on most outdoor condensers. Fan cycle controls work well down to 10°C on condensers with single compressors or with parallel groups of compressors. Below 10°C, condenser flooding (with the refrigerant) can be used alone or with fan controls. Flooding requires a larger refrigerant charge and larger liquid receiver. In contrast, split condensers with solenoid valves in the hot-gas lines and bleedoff in the liquid return lines can reduce the condenser surface during cold weather. Natural subcooling can be integrated into the design to save energy.

Fans are controlled by pressure controls, liquid-line thermostats, or a combination of both. Ambient control of condenser fans is commonly used; however, it may not give the degree of condensing temperature control required in systems designed for high-efficiency gain. Ambient control of condenser fans is not recommended except in mild climates down to 10°C. Sometimes, pressure switches, in conjunction with gravity louvers, cycle the condenser fan. This system requires no refrigerant flooding charge.

The sizing of the receiver tank on the high-pressure side, especially for remote condensers, must be considered. Remote condenser installations, particularly when associated with heat recovery, have substantially higher internal high-side volume than other types of systems. Much of the high side is capable of holding liquid refrigerant, particularly if runs are long and lines are large.

Roof-mounted condensers should have at least 0.9 m of space between the roof deck and the bottom of the condenser slab to minimize the radiant heat load from the roof deck to the condenser surface. Also, free airflow to the condenser should not be restricted. Remote condensers should be placed at least 0.9 m from any wall, parapet, or other airflow restriction. Two side-by-side condensers should be placed at least 1.8 m from each other. In Chapter 15 of the 1997 *ASHRAE Handbook—Fundamentals*, the problems of locating equipment for proper airflow are discussed.

Evaporative Condenser Arrangements

Evaporative condensers are also available as single- or multiple-circuit condensers. Manufacturer conversion factors for operating at a given condensing temperature and wet-bulb temperature must be applied to determine the required size of the evaporative condenser.

In cold climates, the condenser must be installed to guard against freezing during winter. Evaporative condensers demand a regular program of maintenance and water treatment to ensure uninterrupted operation. The receiver tank should be capable of storing the extra liquid refrigerant during warm months. Line sizing must be considered to keep a reasonable tank size.

Closed water condenser/evaporative cooler systems are used frequently. In this arrangement, an evaporative condenser cools water instead of refrigerant. This water flows in a closed, chemically stabilized circuit with a regular water-cooled condenser (a two-stage heat transfer system). Heat from the condensing refrigerant transfers to the closed water loop in the regular water-cooled condenser. The warmed water then passes to the evaporative cooler.

The water-cooled condenser and evaporative cooler must be selected considering the temperature difference (1) between the refrigerant and the circulating water and (2) between the circulating water and the available wet-bulb temperature. The double temperature difference results in higher pressures than when the refrigerant is condensed in the evaporative condenser. On the other hand, this arrangement causes no corrosion inside the refrigerant condenser itself because the water flows in a closed circuit and is chemically stabilized.

The extremely high temperature of the entering discharge gas is the prime cause of evaporative condenser deterioration. The severity of deterioration can be substantially reduced by using the closed water condensing arrangement. The extent of this reduction in deterioration relates directly to the reduction in the difference between the high discharge gas temperatures experienced even with generously sized evaporative condensers and the design entering water temperature for the closed water circuit.

Water flow in the closed water circuit can be balanced between multiple condensers on the same evaporative cooler circuit with water-regulating valves. Usually, low pressures are prevented by temperature control of the closed water circuit. Three-way valves provide satisfactory water distribution control between condensers.

Cooling Tower Arrangements

Few supermarkets use water-cooled condensing units because the trend is toward air cooling. Nearly all water-cooled condensing units are installed with a water-saving cooling tower because of the high cost of water and sewage disposal.

The engineering of water cooling towers for perishable foods is different than for air conditioning because (1) the hours of operation required are much greater than for space conditioning; (2) refrigeration is required year-round; and (3) in some applications, cooling towers must survive severe winters. A thermostat must control the tower fan for year-round control of the condensing pressure. The control is usually set to turn off the fan when the water temperature drops to a temperature that produces the lowest desired condensing pressure. Water-regulating valves are sometimes used in a conventional manner. Dual-speed fan control is also used.

Some engineers use balancing valves for water flow control between condensers and rely on water temperature control to avoid low pressure. Proper bleedoff is required to ensure the satisfactory performance and full life of the cooling towers, condensers, water pumps, and piping. Water treatment specialists should be consulted because each locality has different water and atmospheric conditions. A regular program of water treatment is mandatory.

Condensing Units

Air-Cooled Condensing Unit. Single-unit compressors with air-cooled condenser systems can be mounted in racks up to three high to save space. These units may have condensers sized so that the TD is in the 6 to 14 K range. Optionally available next larger size condensers are often used to achieve lower TDs and higher energy efficiency ratios (EERs) in some supermarkets, convenience stores, and other applications. Single compressors with heated crankcases and heated insulated receivers and other suitable outdoor controls are assembled into weatherproof racks for outdoor installations. Sizes range from 0.4 to 22 kW.

Water-Cooled Condensing Unit. Water-cooled units range in size from 0.4 to 22 kW and are best for hot, dry climates. The city water cooled condensing unit is usually no longer economical due to the high cost of water and sewer fees. A cooling towers, which cools the water for all compressors, has been used instead. Closed water systems are also used in areas that can use an evaporative water cooler.

Remote Condenser-Compressor Unit. Remote units operate efficiently with minimum condenser maintenance. Evaporative condensers are also used in some areas. Sizes range from 0.4 to 22 kW.

Recommended sizing for remote air-cooled condensers is 6 K TD for low-temperature application and 8 K TD for medium-temperature application. Remote water-cooled condensers are often used in areas with abundant water.

Single-Compressor Control. Single-compressor units make up half of the supermarket compressor equipment presently used. A solid-state pressure control for single units can help control excess capacity when the ambient temperature drops. The control senses the pressure and adjusts the cutout point to eliminate short cycling, which ruins many compressors in low-load conditions. This control also saves energy by maintaining a higher suction pressure than would otherwise be possible and by reducing overall running time.

Parallel Compressors. Multiple compressors may be operated in parallel with a large receiver and multistation manifolds for liquid, suction, and discharge gas. These systems usually have a remote condenser and frequently recover heat from a secondary condenser coil in the air handler. A separate compressor for ice cream refrigerators on a low-temperature system or for meat refrigerators on a medium-temperature system can be physically mounted on the rack and piped so that only the heat removed from the lower temperature refrigerators is supplied at the less efficient rate. Parallel operation is also applied in two-stage or compound systems for low-temperature applications. Two-stage compression includes interstage gas cooling before the second stage of compression to avoid excessive discharge temperatures. Parallel compressors may be equal or unequal in size and may have unloaders or variable-speed drives.

Remote Air-Cooled Condenser and Heat Recovery Coil. Remote air-cooled condensers are popular for use with parallel compressors. The condenser coil TD is in the range of 6 K for low-temperature and 8 K for medium-temperature applications. For energy conservation, generous sizing of the condenser with a lower TD is recommended. Figure 17 diagrams a basic parallel system with an air-cooled condenser and heat recovery coil.

Evaporative Condenser. Evaporative condensers have a coil, in which refrigerant is condensed or closed-loop circulating water is cooled, and a means of supplying air and water over its external surfaces. Heat is transferred from the condensing refrigerant inside the coil or the water in the loop to the external wetted surface and then into the moving airstream principally by evaporation. In areas where the wet-bulb temperature is about 17 K below the dry bulb, the condensing temperature can be 6 to 17 K above the wet-bulb temperature. This lower condensing temperature saves energy; one

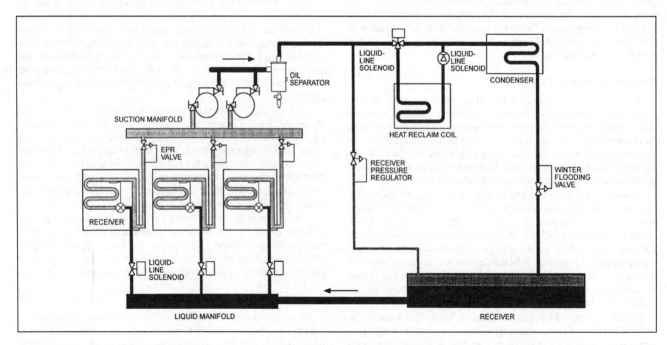

Fig. 17 Basic Parallel System with Remote Air-Cooled Condenser and Heat Recovery

Retail Food Store Refrigeration and Equipment

evaporative condenser can be installed for the entire store. Chapter 35 of the 1996 *ASHRAE Handbook—Systems and Equipment* gives more details.

Condensing Unit Noise

Air-cooled condensing units located outdoors, either as single units with weather covers or grouped in prefabricated machine rooms, produce sounds that must be evaluated. The largest source of noise is usually propeller-type condenser fans. Other sources are compressor and fan motors, high-velocity refrigerant gas, general vibration, and amplification of sound where vibration is transmitted to mounting structures (this is most critical in roof-mounted units).

A fan-speed or fan-cycle control helps control fan air noise by ensuring that only the amount of air necessary to maintain proper pressure is generated. Care should be taken not to restrict discharge air. When possible, it should be discharged vertically upward.

Resilient mountings for fan motors and small compressors and isolation pads for larger motors and compressors are helpful in reducing noise transmission. Proper discharge line sizing and mufflers are the best solution for high-velocity gas noise. Lining enclosures with sound-absorbing material is of minimal value. Isolation pads can help on roof-mounted units, but even more important is choosing the right location with respect to the supporting structure, so that structural member vibration does not amplify the noise.

If sound levels are still excessive after the previously mentioned controls have been implemented, location becomes the greatest single factor. Distance from a sensitive area is most important in choosing a location; each time the distance from the source is doubled, the noise level is halved. Direction is also important. Condenser air intakes should face parking lots, open fields, or streets zoned for commercial use. In sensitive areas, ground level installation close to building walls should be avoided because walls reflect sound.

When it is impossible to meet requirements by adjusting location and direction, barriers can be used. Although a masonry wall is an effective barrier, it may be objectionable because of cost and weight. If a barrier is used, it must be sealed at the bottom because any opening will allow sound to escape. Barriers also must not restrict condenser entering air. Keep the open area at the top and sides at least equal to the condenser face area.

When noise is a consideration, (1) purchase equipment designed to operate as quietly as possible (e.g., 850 rpm condenser fan motors instead of higher speed motors); (2) choose the location carefully; and (3) use barriers when the first two steps do not meet requirements.

See Chapter 43 of the 1995 *ASHRAE Handbook—Applications* for information on outdoor sound criteria, equipment sound levels, sound control for outdoor equipment, and vibration isolation.

METHODS OF DEFROST

Defrosting is accomplished by latent heat reverse-cycle gas defrosting, selective ingestion of store air, electric heaters, or cycling the compressor. In the defrost operation, particularly in low-temperature equipment, frost in the air flues and around the fan blades must be melted and completely drained.

Defrosting is usually controlled by a variety of clocks. They are often part of a compressor controller system. Electronic systems often have communication capabilities outside the store.

Gas Defrost

Gas defrost uses heat from the compressor's discharge gas to defrost the evaporators. To remove the coil frost, the discharge pressure gas is directed to the case evaporator. One method tempers the impact of the hot gas on the cold suction line by beginning the defrost cycle with saturated gas from the top of the liquid in the receiver. Occasionally, supplemental electric case heaters are added to ensure rapid and reliable defrosting. A timer generally terminates the defrost cycle, although temperature termination is sometimes used.

Air Defrost

Air defrost moves ambient air from the store into the refrigerator. A variety of systems are used; some use supplemental electric heat to ensure reliability. The heat content of the store ambient air during the winter is critical for good results from this method.

Electric Defrost

Electric defrost methods usually apply heat externally to the evaporator and require a longer defrost period than the gas defrost method, usually up to 1.5 times longer. The heating element may be in direct contact with the evaporator, depending on conduction for defrost; or it may be located between the evaporator fans and the evaporator, depending on convection or a combination of conduction and convection for defrost. In both instances, the manufacturer generally installs a temperature-limiting device on or near the evaporator to prevent excessive temperature rise if any controlling device fails to operate.

The electric defrost method simplifies the installation of low-temperature fixtures. The controls used to automate the cycle usually include one or more of these devices: (1) defrost timer, (2) solenoid valve, (3) electrical contactor, and (4) evaporator fan delay switch. Some applications of open low-temperature cases may operate the fans during the defrost cycle.

Condensing Unit Off-Time

This method simply shuts off the unit and allows it to remain off until the evaporator reaches a temperature that permits defrosting and gives ample time for condensate drainage. Because this method obtains its heat from the air circulating in the display fixture, it is slow and limited to open fixtures operating at temperatures of 1°C or above. Methods of defrost control include (1) suction pressure control (no time clock required); (2) time clock initiation and termination; (3) time clock initiation and suction pressure termination; (4) time clock initiation and temperature termination; and (5) demand defrost or proportional defrost.

Suction Pressure Control. This control is adjusted for a cut-in pressure high enough to allow defrosting during the off cycle. This method is usually used in fixtures maintaining temperatures from 3 to 6°C. Air is circulated over the coil, melting the frost on the coil. This provides some cooling effect to the product during defrost. If an excessive heat or humidity load occurs, the evaporator frost becomes ice. When the evaporator pressure is lowered to the cutout point of the control, the control initiates a defrost cycle to clear the evaporator.

However, condensing units and/or suction lines may, at times, be subjected to ambient temperatures below the evaporator's temperature. This prevents the buildup of suction pressure to the cut-in point, and the condensing unit will remain off for prolonged periods. In such instances, fixture temperatures may become excessively high, and displayed product temperatures will increase.

A similar situation can exist if the suction line from a fixture is installed in a trench or conduit with many other cold lines. The other cold lines may prevent the suction pressure from building to the cut-in point of the control.

Methods (2), (3), and (4) of controlling off-cycle defrosting use defrost time clocks to break the electrical circuit to the condensing unit initiating a defrost cycle. The difference lies in the manner in which the defrost period is terminated.

Time Initiation and Termination. A timer initiates and terminates the defrost cycle after the selected time interval. The length of the defrost cycle must be determined and the clock set accordingly.

Time Initiation and Suction Pressure Termination. This method is similar to the first method, except that suction pressure terminates the defrost cycle. The length of the defrost cycle is automatically adjusted to the condition of the evaporator, as far as frost and ice are concerned. However, to overcome the problem of the suction pressure not rising because of the defrost cut-in pressure previously described, the timer has a fail-safe time interval to terminate the defrost cycle after a preset time, regardless of suction pressure.

Time Initiation and Temperature Termination. This method is also similar to the time initiation and termination method, except that temperature terminates the defrost cycle. The length of the defrost cycle varies depending on the amount of frost on the evaporator or in the air stream leaving the evaporator. A temperature sensor is located on a tube of the evaporator or in the airstream leaving the evaporator. The timer also has a fail-safe setting in its circuit to terminate the defrost cycle after a preset time, regardless of the temperature.

Demand Defrost or Proportional Defrost. This system initiates defrost based on demand (need) or in proportion to humidity or dew point. Techniques vary from measuring change in the temperature spread between the air entering and leaving the coil, to changing the defrost frequency based on store relative humidity. Other systems use a device that senses the frost level on the coil.

REFRIGERANT LINES

Sizing liquid and suction refrigerant lines is critical in the average refrigeration installation due to the typically long horizontal runs and frequent use of vertical risers. Correct liquid line sizes are essential to ensure a full feed of liquid to the expansion valve. Oversizing of liquid lines must be avoided to prevent system pumpdown or defrost cycles from operating improperly in single-compressor systems.

Proper suction line sizing is required to ensure adequate oil return to the compressor without excessive pressure drop. Oil separates in the evaporator and moves toward the compressor at a lower velocity than the refrigerant. Unless the suction line is properly installed, the oil can accumulate at low places, causing problems such as compressor damage from liquid slugging or insufficient lubrication. Excessive pressure drop and reduced system capacity can also result from undersized or improperly installed suction lines. To prevent these problems, horizontal suction lines must pitch down as the gas flows toward the compressor, the bottom of all suction risers must be trapped, and the refrigerant velocity in suction risers must be maintained according to piping practices described in Chapters 2 and 3. To overcome a large pressure drop, the suction lines may be oversized on long runs, but they still must pitch down toward the compressor.

Manufacturers' recommendations and appropriate line sizing charts should be followed to avoid adding heat to either the suction or the liquid lines. In large stores, both suction and liquid lines can be insulated profitably, particularly if subcooling is present.

EFFECTS OF NEW REFRIGERANTS

Selection of a suitable refrigerant for food stores has become complex due to international concern about the ozone-depleting effect of some refrigerants. International treaties no longer allow developed nations to manufacture equipment that uses these refrigerants. In addition, as these refrigerants are phased out, recharging old equipment with these refrigerants will become too expensive to remain feasible. Hydrochlorofluorocarbon R-22 is still popular while its price remains low and availability is assured for a reasonable time. Current alternatives in use in the United States include R-404A, R-134a and R-507. Other refrigerants are listed in ASHRAE *Standard* 34.

Compressor performance and material compatibility are two major concerns in selecting new refrigerants. Close consultation with equipment manufacturers must be maintained in order to stay current on this most important choice because it is still a changing issue.

INTERACTION BETWEEN REFRIGERATION AND AIR CONDITIONING

Open display equipment often extracts enough heat from the store ambient air to reduce the air temperature in the customer aisles to as much as 9 K below the desired level. The air-conditioning return duct system or fans can be used to move chilled air from the floor in front of the cases back to the store air handler. Lack of attention to this element can substantially reduce sales in these areas.

Heat Reclamation

Heat reclaim condensers and related controls operate as alternates to or in series with the normal refrigeration condensers. They can be used in winter to return most of the refrigeration and compressor heat to the store. They may also be used in mild spring and fall weather when some heating is needed to overcome the cooling effect of the refrigeration system itself. Another use is for cooling coil reheat when needed for humidity control in spring, summer, and fall. Excess humidity in the store must be avoided because it can increase the display case refrigeration load as much as 20% at the same dry-bulb temperature. Also, heat reclamation can be used to heat water for store use. The section on Supermarkets in Chapter 2 of the 1995 *ASHRAE Handbook—Applications* has more detailed information on the interrelation of the store environment and the refrigeration equipment.

Environmental Equipment and Control

Major components of common store environmental equipment include (1) central air handler with fresh makeup air mixing box, (2) air-cooling coils, (3) heat recovery coils, (4) supplemental heat equipment, (5) connecting ducts, and (6) termination units such as air diffusers and return grilles.

Environmental control is the heart of energy management. Control panels designed for the unique heating, cooling, and humidity control requirements for food stores provide several stages of heating (up to eight) and cooling (up to three), plus a dehumidification stage. When high humidity exists in the store, cooling is activated to remove moisture. The controller receives input from temperature and dew point sensors that are located in the sales area.

Some controllers include night setback for cool climates and night setup for warm climates. This feature may save energy by turning the air handler off and allowing the store temperature to fluctuate several degrees above or below the set point temperature. However, store warm-up practices impose an energy use penalty to the display case refrigeration systems and affect display case performance, particularly open models.

Following are rules for good air distribution in food stores:

Air Circulation. Operate 100% of the time the store is open at a volumetric flow of 3 to 5 L/s per square metre of sales area. Air supply and return grilles should be located so they do not disturb the air in open display cases and negatively affect case performance.

Fresh Air. Introduce whenever the air handler is operating. Supply should meet the required indoor air quality or equal the total for all exhaust fans, whichever is greater.

Discharge Air. Discharge most or all of the air in areas where heat loss or gain occurs. This load is normally at the front of the store and around glass areas and doors.

Return Air. Locate return air registers as low as possible. With low registers, return air temperature may be 10 to 13°C. Low returns reduce heating and cooling requirements and temperature stratification. A popular practice, where store construction allows, is to return air under case ventilated bases and through trenches.

REFERENCES

ASHRAE. 1983. Method of testing open refrigerators for food stores. *Standard* 72-1983.

ASHRAE. 1992. Methods of testing closed refrigeration. ANSI/ASHRAE *Standard* 117-1992.

ASHRAE. 1992. Number designation and safety classification of refrigerants. ANSI/ASHRAE *Standard* 34-1992.

Howell, R.H. 1993a. Effects of store relative humidity on refrigerated display case performance. *ASHRAE Transactions* 99(1).

Howell, R.H. 1993b. Calculation of humidity effects on energy requirements of refrigerated display cases. *ASHRAE Transactions* 99(1).

Tyler Refrigeration. 1972. *Trade talk* (February).

CHAPTER 48

FOOD SERVICE AND GENERAL COMMERCIAL REFRIGERATION EQUIPMENT

Reach-In and Specialty Cabinets 48.1
Roll-In Cabinets 48.3
Food Freezers 48.3
Walk-In Coolers/Freezers 48.3

FOOD service requires refrigerators that meet a variety of needs. This chapter covers refrigerators available for restaurants, fast-food restaurants, cafeterias, commissaries, hospitals, schools, convenience stores, and other specialized applications.

Many of the refrigeration products used in food service applications are self-contained, and the corresponding refrigeration systems are conventional. Some systems, however, do use ice for fish, salad pans, or specialized preservation and/or display. Chapters 47, 49, and 50 have further information on some of these products.

Generally, electrical and sanitary requirements of refrigerators are covered by criteria, standards, and inspections of Underwriters Laboratories (UL), NSF International, and the U.S. Public Health Service.

Frame construction is usually of metal. Occasionally, wood is used in minor amounts, but it is rarely visible in the final assembly. Nonmetallic materials are used increasingly for interior liners, decoration, finish, trim, thermal break strips, and gaskets.

Stainless steel is the most common sheet metal used for these refrigerators. Other sheet metals used are aluminum and carbon steel, either hot- or cold-rolled. Stainless steel usually has a grained or polished finish when exposed in final assembly. Aluminum is generally unfinished, although it may occasionally be lacquered or coated with some other organic finish. Carbon steel may be black; plated; aluminized; or coated with either high-solids powder paint, enamel, or another special finish.

Insulation is usually one of the following: (1) foam polyurethane poured in place, (2) slab polyurethane, (3) slab polystyrene, or (4) glass fiber batts. Foam polyurethane is advantageous because it conforms to various cavity shapes and provides added structural integrity through adherence bonding to cabinet liners.

REACH-IN AND SPECIALTY CABINETS

The **reach-in refrigerator** or freezer is an upright, box-shaped cabinet with straight vertical front(s) and hinged or sliding doors (Figure 1). It is usually about 750 to 900 mm deep and 1800 m high and ranges in width from about 900 to 3000 mm. Capacities range from about 500 to 2500 L. Undercounter models 900 mm high with the same dimensions are also available. These capacities and dimensions are standard from most manufacturers.

The typical reach-in cabinet (Figure 1) is available in many styles and combinations, depending on its intended application. Other shapes, sizes, and capacities are available on a custom basis from some manufacturers (Figure 2).

There are many varied adaptations of refrigerated spaces for storing perishable food items. Reach-ins, by definition, are medium- or low-temperature refrigerators small enough to be moved into a building. This definition also includes refrigerators and freezers built for special purposes, such as mobile cabinets or refrigerators on wheels and display refrigerators for such products as beverages, pies, cakes, and bakery goods. The latter cabinets usually have glass doors. Candy refrigerators are also specialized in size, shape, and temperature.

Refrigerated vending machines satisfy the general definition of reach-ins, but they also receive coins and dispense products individually. Generally, the full product load of a vending machine is not accessible to the customer. Beverage-dispensing units dispense a measured portion into a cup rather than in a bottle or can.

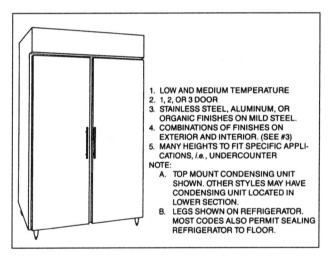

Fig. 1 Reach-In Food Storage Cabinet Features

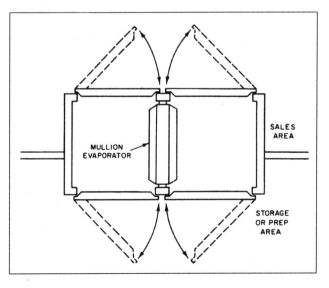

Fig. 2 Pass-Through Styles Facilitate Some Handling Situations

---

The preparation of this chapter is assigned to TC 10.7, Commercial Food and Beverage Cooling, Display and Storage.

Types of Construction

Reach-in refrigerators are available in two basic types of construction. The older style is a wood frame substructure clad with a metal interior and exterior. The newer style is a welded assembly of exterior panels with insulation and liner inserts.

Materials used on exteriors and interiors are stainless steel, painted steel, aluminum-coated steel, aluminum, and vinyl-clad steel with wood grain or other patterns. The requirements are for a material that (1) matches or blends with that used on nearby equipment; (2) is easy to keep clean; (3) is not discolored or etched by commonly used cleaning materials; (4) is strong enough to resist denting, scratching, and abrasion; and (5) provides the necessary frame strength. The material chosen by an individual purchaser depends a great deal on layout and budget.

Temperature Ranges

Reach-in refrigerators are available for medium- or low-temperature ranges. The medium-temperature range has a maximum of 7°C and a minimum of 0°C internal product temperature, with the most desirable average temperature close to 3°C. The low-temperature range need not be below −23.5°C and should not exceed −12°C internal product temperature. The desirable average is −18°C for frozen foods and −20.5°C internal product temperature for ice cream. Both temperature ranges are available in cabinets of many sizes, and some cabinets combine both ranges.

Refrigeration Systems

Remote refrigeration systems are often used if cabinets are installed in a hot or otherwise unfavorable location where the noise or heat of the condensing units would be objectionable. Other special circumstances may also make remote refrigeration desirable.

Self-contained systems, in which the condensing unit and controls are built into the refrigerator structure, are of two general types and are usually air-cooled. The first type has the condensing unit beneath the cabinet; in some designs it takes up the entire lower part of the refrigerator, while in others it occupies only a corner at one lower end. The second type has the condensing unit on top.

There is no advantage to locating a self-contained condensing unit beneath the refrigerator; although the air near the floor is generally cooler, and thus beneficial to the condensing unit, it is usually dirtier. Putting the condensing unit on top of the cabinet allows full use of cabinet space, and although the air passing over the condenser may be warmer, it is cleaner and more abundant. In addition, top mounting of both condensing unit and coil offers physical and constructional advantages. Having the condensing unit and coil in the same location gives a refrigeration unit that can be removed, serviced, and replaced in the field as a whole. Servicing can then be done at an off-site repair facility.

Styles

Reach-in refrigerators have doors on the front. Refrigerators that have doors on both front and rear are called **pass-through or reach-through refrigerators**. Doors are either full height (one per section) or half height (two per section). Doors may have windows or be solid, hinged, or sliding.

Interiors

Shelves are standard interior accessories and are usually furnished three or four per full-height section. Generally, various types of shelf standards are used to provide vertical shelf adjustment.

Modifications and Adaptations

Food Service. These applications often require extra shelves or tray slides, pan slides, or other interior accessories to increase food-holding capacity or make operation more efficient. Because certain stored foods create a corrosive atmosphere in the enclosure, the evaporator coil may have special coatings or fin materials to prevent oxidation. With increasing use of foods prepared off-premises, specialization of on-premise storage cabinets is growing. This is developing an increasing pressure for designs that consider new food shapes, as well as in-and-out handling and storage.

Beverage Service. If reach-ins are required, standard cabinets are used except when glass doors and special interior racks are needed for chilled product display.

Meal Factories. These applications, which include airline or central feeding commissaries, require rugged, heavy-duty equipment, often fitted for bulk in-and-out handling.

Retail Bakeries. Special requirements of bakeries are the dough retarder refrigerator and the bakery freezer, which permit the baker to spread the work load over the entire week and to offer a greater variety of products. The recommended temperature for a dough retarder is 2 to 4.5°C. The relative humidity should be in excess of 80% to prevent crusting or other undesirable effects. In the freezer, the temperature should be held at −18°C. All cabinets or wheeled racks should be equipped with racks to hold the 460 mm by 660 mm bun pans, which are standard throughout the baking industry.

Retail Stores. Stores use reach-ins for many different nonfood items. Drugstores often have refrigerators with special drawers for storage of biological compounds. (See the section on Nonfood Installations.)

Retail Florists. Florists use reach-in refrigerators for displaying and storing flowers. Although a few floral refrigerator designs are considered conventional in the trade, the majority are custom built. The display refrigerator located in the sales area at the front of the shop may include a picture window display front and have one or more display-type access doors, either swinging or sliding. A variety of open refrigerators may also be used.

For the general assortment of flowers in a refrigerator, most retail florists have found best results at temperatures from 4.5 to 7°C. The refrigeration coil and condensing unit should be selected to maintain a high relative humidity. Some florists favor a gravity cooling coil because the circulating air velocity is low. Others, however, choose forced-air cooling coils, which develop a positive but gentle airflow through the refrigerator. The forced-air coil has an advantage when the in-and-out service is especially heavy because it provides quick temperature recovery during these peak conditions.

Nonfood Installations. A variety of applications use a wide range of reach-ins, some standard except for accessory or temperature modifications and some completely special. Examples are (1) biological and pharmaceutical cabinets; (2) blood bank refrigerators; (3) low- and ultralow-temperature cabinets for bone, tissue, and red-cell storage; and (4) specially shaped refrigerators to hold column chromatography and other test apparatus.

Blood bank refrigerators for whole blood storage are usually standard models, ranging in size from under 500 L to 1300 L, with the following modifications:

- Temperature is controlled at 3 to 5°C.
- Special shelves and/or racks are sometimes used.
- A temperature recorder with a 24 h or 7 day chart is furnished.
- An audible and/or visual alarm system is supplied to warn of unsafe blood temperature variation.
- Sometimes an additional alarm system is provided to warn of power failure.

Biological, laboratory, and mortuary refrigerators involve the same technology as refrigerators for food preservation. Most biological serums and vaccines require refrigeration for proper preservation and to retain highest potency. In hospitals and laboratories, refrigerator temperatures should be 1 to 3.5°C. The refrigerator should provide low humidity and should not freeze. Storage in

Food Service and General Commercial Refrigeration Equipment

mortuary refrigerators is usually short-term, normally 12 to 24 h at 1 to 3.5°C. Refrigeration is provided by a standard air- or water-cooled condensing unit with a forced-air cooling coil.

Products in biological and laboratory refrigerators are kept in specially designed stainless steel drawers sized for convenient storage, labeled for quick and safe identification, and perforated for proper air circulation.

Mortuary refrigerators are built in various sizes and arrangements, the most common being two- and four-cadaver self-contained models. The two-cadaver cabinet has two individual storage compartments, one above the other. The condensing unit compartment is above and indented into the upper front of the cabinet; ventilation grills are on the front and top of this section. The four-cadaver cabinet is equivalent to two two-cadaver cabinets set together; the storage compartments are two cabinets wide by two cabinets high with the compressor compartment above. Six- and eight-cadaver cabinets are built along the same lines. The two-cadaver refrigerator is approximately 965 mm wide by 2390 mm deep by 1956 mm high and is shipped completely assembled.

Each compartment contains a mortuary rack consisting of a carriage supporting a stainless steel tray. The carriage is telescoping, equipped with roller bearings so that it slides out through the door opening and is self-supporting even when extended. The tray is removable. Some specifications call for a thermometer to be mounted on the exterior front of the cabinet to show the inside temperature.

ROLL-IN CABINETS

These cabinets are very similar in style and appearance to reach-in cabinets. Roll-ins (Figure 3) are usually part of a food-handling or other special-purpose system (Figure 4). Pans, trays, or other specially sized and/or shaped receptacles are used to serve a specific system need, such as the following:

- Food-handling for schools, hospitals, cafeterias, and other institutional facilities
- Meal manufacturing
- Bakery processing
- Pharmaceutical products
- Body parts preservation (i.e., blood)

Fig. 3 Open and Enclosed Roll-In Racks

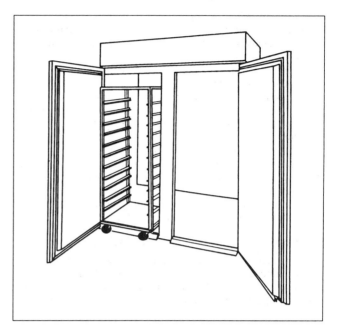

Fig. 4 Roll-In Cabinet —Usually Part of a Food-Handling or Other Special-Purpose System

The roll-in differs from the reach-in in the following ways:

1. The inside floor is at about the same level as the surrounding room floor, so wheeled racks of product can be rolled directly from the surrounding room into the cabinet interior.
2. Cabinet doors are full height, with drag gaskets at the bottom.
3. Cabinet interiors have no shelves or other similar accessories.

The racks that roll in and out of these cabinets are generally fitted with slides to handle 460 mm by 660 mm pans, although some newer systems call for either 300 mm by 500 mm or 300 mm by 460 mm steam table pans. Racks designed for special applications are available, but usually custom designed.

Manufacturers and contractors offer various methods of insulating the floor area. This is important if the roll-in is to hold frozen food.

FOOD FREEZERS

Some hospitals, schools, commissaries, and other mass-feeding operations use on-premises freezing to level work loads and operate kitchens efficiently on normal schedules. Industrial freezing equipment is usually too large for these applications, so operators use either regular frozen food storage cabinets for limited amounts of freezing or special reach-ins that are designed and refrigerated to operate as batch-type blast freezers.

Chapter 15 covers the industrial freezing of food products.

WALK-IN COOLERS/FREEZERS

This type of commercial refrigerator is a factory-made, prefabricated, sectional version of the built-in, large-capacity cooling room. It closely matches the reach-in type in meeting a wide variety of applications.

Its function is to store foods and other perishable products in larger quantities and for longer periods than the reach-in refrigerator. Good refrigeration practice dictates that dissimilar unpackaged foods be stored in separate rooms because they require different temperature and humidity and because odors from some foods are absorbed by others. Large food operations may have three rooms: one for fruits and vegetables, one for meats and poultry, and one for dairy products. A fourth room, at −18°C, may be added for frozen foods. Smaller food operations that use appropriate food packaging may require only two rooms; one for medium-temperature refrigeration and one for frozen storage.

Foam plastic materials have improved thermal insulation in both self-contained and remotely refrigerated sectional coolers. Polyurethane foam-in-place insulation between two skins of metal makes a light, water-resistant panel. Additionally, the foam is a very efficient insulator, allowing slimmer panels for equal insulation value compared to most other insulations.

The sectional cooler offers flexibility over the built-in type. It is easily erected and easily moved, and it can be readily altered to meet changing requirements, uses, or layouts by adding standard sections. Also, the sectional walk-in cooler can be erected outside a building, providing more refrigerated storage with no building costs except for footings and an inexpensive roof supported by the cooler.

Self-Contained Sectional Walk-In Coolers

The versatility of sectional walk-in coolers was greatly increased by the introduction of self-contained models. There are various methods of application. Figure 5 shows one arrangement; the top-mount refrigeration unit is labeled (a), the straddle mount unit is (b), and the side-mount unit is (c). These self-contained units use complete refrigeration systems, usually air-cooled, in a single compact package. The units are installed in the sectional cooler/freezer wall or ceiling panels.

Walk-In Floors

Sectional walk-in coolers termed floorless by the supplier are furnished with floor splines to fasten to the existing floor to form a base for the wall sections. Models with an insulated floor are also available.

A medium-temperature (above-freezing) cooler can be erected on an uninsulated concrete floor on the ground. Generally, floor losses are considered small.

Level entry is becoming more important as the use of hand trucks and electric trucks increases. The advantage and convenience of level entry afforded by a floorless cooler can also be obtained by recessing a sectional insulated floor.

Design Characteristics

The factory-made walk-in cooler consists of standard top, bottom, wall, door, and corner sections, which are shipped to the user and erected on the site. The frames are filled with insulation and are covered on the inside and outside with metal. The edges of these frames are usually of tongue-and-groove construction and either fitted with a gasket material or provided with suitable caulking material to ensure a tight vapor seal when assembled. These sections are assembled on the site with either lag screws or hooks operated from inside the cooler.

Exterior and interior surfaces may be painted and are made of one or more of the following:

- Galvanized steel
- Aluminum
- Aluminum-coated steel
- Stainless steel
- Vinyl-clad steel

Coolers may be used to hold sides or quarters of beef, lamb carcasses, crates of vegetables, and other bulky items. Food operations now rarely use such items. If they do, the items are broken down, trimmed, or otherwise processed before entering refrigerated storage. The modern cooler is not a storage room for large items, but a temporary place for quantities of small, partially or totally processed products.

The food cooler, therefore, is likely to be equipped with sturdy, adjustable shelving about 460 mm deep and arranged in tiers, three or four high, around the inside walls. Alternatively, the cooler is often provided with rolling racks that are actually shelving on wheels. These racks are rolled directly into and out of the cooler.

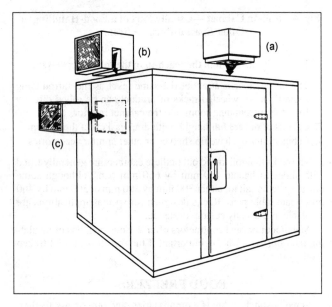

Fig. 5 Refrigeration Equipment Added to Make a Walk-In Cooler Self-Contained

CHAPTER 49

HOUSEHOLD REFRIGERATORS AND FREEZERS

Primary Functions 49.1
Performance Characteristics 49.1
Safety Requirements 49.2
Durability and Service 49.2
Cabinets 49.2
Refrigerating Systems 49.5
Evaluation 49.9

THIS chapter covers the design and construction of full-sized household refrigerators and freezers, the most common of which are illustrated in Figure 1. Small portable and secondary refrigerators are not specifically addressed here. Some of these small refrigerators use absorption systems, special forms of compressors, and, in some cases, **thermoelectric** refrigeration. Applications for water-ammonia **absorption** systems have developed for recreational vehicles, picnic coolers, and hotel room refrigerators, where noise is an issue.

The section on Refrigerating Systems only covers the **vapor-compression** cycle, which is almost universally used for full-sized household refrigerators and freezers. In these applications, where several hundred watts are pumped through temperature differentials from freezer to room temperature in excess of 55 K, other **electrically powered** systems compare unfavorably to vapor-compression systems in terms of manufacturing and operating costs. Typical operating efficiencies of the three most practical refrigeration systems are as follows for a $-18°C$ freezer and $32°C$ ambient:

Thermoelectric   Approximately 0.09 W/W
Absorption   Approximately 0.44 W/W
Vapor-compression   Approximately 1.61 W/W

An absorption system may operate from **gas** at a lower cost per unit of energy, but the initial cost, size, and mass have made it unattractive to use gas systems for major appliances where electric power is available. Because of its simplicity, thermoelectric refrigeration could replace other systems if (1) an economical thermoelectric material were developed and (2) design issues such as the need for a direct current (dc) power supply and an effective means for transferring heat from the module were addressed.

PRIMARY FUNCTIONS

Providing food storage space at reduced temperature is the primary function of a refrigerator or freezer, with ice making an essential secondary function. For the preservation of fresh food, a general storage temperature between 0 and 4°C is desirable. Higher or lower temperatures or a humid atmosphere are more suitable for storing certain foods. A discussion of special-purpose storage compartments designed to provide these conditions may be found in the section on Cabinets. Food freezers and combination refrigerator-freezers used for long-term storage are designed to hold temperatures near $-18°C$ and always below $-13°C$ during steady-state operation. In single-door refrigerators, the frozen food space is usually warmer than this and is not intended for long-term storage. Optimum conditions for food preservation are addressed in more detail in Chapters 8 through 28.

PERFORMANCE CHARACTERISTICS

A refrigerator or freezer must maintain desired temperatures and have reserve capacity to cool to these temperatures when started on a hot summer day. Most models cool down within hours in a 43°C ambient at rated voltage.

Overall system efficiency has become important both because rising energy costs have driven operating costs upward and because federal energy standards dictate consumption limits. The challenge for the designer to control noise and vibration has been made more complex with the need for fans for forced-air circulation and compressors with higher efficiencies and capacities. The need for increased storage volumes and better insulation efficiency has resulted in almost universal use of foam insulation, which is less acoustically absorbent than glass fiber. Vibrations from running or stopping the compressor must be isolated to prevent mechanical transmission to the cabinet or to the floor and walls, where it may cause additional vibration and noise.

Energy Efficiency Standards

Under the National Appliance Energy Conservation Act (NAECA) in the United States, the U.S. Department of Energy (DOE) is required to set efficiency standards for residential

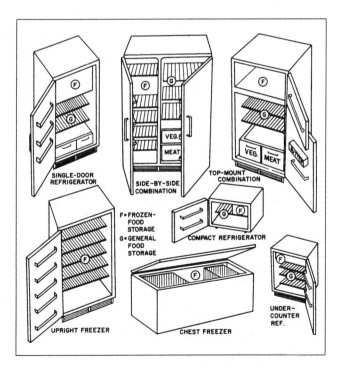

Fig. 1 Configurations of Contemporary Household Refrigerators and Freezers

---

The preparation of this chapter is assigned to TC 7.1, Residential Refrigerators and Food Freezers.

appliances. For refrigerators and freezers, these standards are set in terms of the maximum annual electric energy consumption, which is measured according to a prescribed test procedure. Maximum energy consumption varies with cabinet volume and by product class. The current energy test procedure and the maximum allowable energy consumption are documented in the *Code of Federal Regulations* (10 CFR Part 430, Subparts B and C, respectively). Energy standard levels took effect in 1993 and are subject to periodic review and revision.

SAFETY REQUIREMENTS

American manufacturers comply with Underwriters Laboratories (UL) *Standard* 250, Household Refrigerators and Freezers, which protects the user from electrical shock, fire dangers, and other hazards under normal and some abnormal conditions. Specific areas addressing product safety focus on motors, hazardous moving parts, grounding and bonding, stability (cabinet tipping), door-opening force, door-hinge strength, shelf strength, component restraint (shelves and pans), glass strength, cabinet and unit leakage current, leakage current from surfaces wetted by normal cleaning, high-voltage breakdown, ground continuity, testing and inspection of polymeric parts, and uninsulated live electrical parts accessible with an articulated probe.

DURABILITY AND SERVICE

Studies show that the average refrigerator or freezer will belong to its original owner for 10 to 15 years and will continue to give satisfactory service for a considerably longer period. There is a wide variation in life span, however; some refrigerators run for over 30 years. Their reliability, particularly that of the older and simpler refrigerators, has led consumers to expect a 15 to 20 year life from the hermetic unit; in many respects, the appliance must be designed to protect itself over this period. Motor overload protectors are normally incorporated, and an attempt is made to design fail-safe circuits so that the hermetic motor of the compressor will not be damaged by failure of a minor external component, unusual voltage extremes, or voltage interruptions.

Customer-operated devices must withstand frequent use. For example, a refrigerator door may be opened and shut over 300,000 times in its lifetime. To protect the customer against the cost of premature failure, most manufacturers will replace faulty parts within one year and repair or replace a faulty hermetic system within five years or longer at no charge for materials. Beyond these warranties, the terms vary among manufacturers; warranties are sometimes combined with an extended service contract.

In the design of refrigerators and freezers, provisions must be made for economical and effective servicing if damage or malfunction occurs in the field.

CABINETS

Good cabinet design achieves the optimum balance of the following objectives:

- Maximum food storage volume for the floor area occupied by the cabinet
- Maximum utility, performance, convenience, and reliability
- Minimum heat gain
- Minimum cost to the consumer

Use of Space

The fundamental factors in cabinet design are usable food storage capacity and external dimensions. Food storage volume has increased considerably without a corresponding increase in the external dimensions of the cabinet due to the use of thinner but more effective insulation and to reduction of the space occupied by the compressor and condensing unit. The method of computing storage volume and shelf area is described in Association of Home Appliance Manufacturers (AHAM) *Standard* HRF-1.

Frozen Food Storage

The increased use of frozen foods necessitates refrigerators with much larger frozen food storage compartments. In **single-door models**, the frozen food storage volume is provided by a freezing compartment across the top of the general food storage compartment. An insulated baffle beneath the evaporator allows it to operate at the low temperatures required for short-term frozen food storage, while maintaining temperatures above freezing in the general food storage compartment. Sufficient air passes around the baffle to cool the general storage compartment.

In larger refrigerators, the frozen food space often represents a large part of the total volume, and it usually has a separate exterior door or drawer and a lower temperature capability; in this case, the model is classified as a **combination refrigerator-freezer**. The frozen food compartment in these combinations is most often positioned across the top, but on the largest models, side-by-side arrangements are not uncommon. A few bottom-mounted models have reappeared on the market.

Two-door cabinets are sometimes built with two separate inner liners housed in a single outer shell. Others use only a single liner divided into two sections by an insulated panel that separates the frozen food storage from the fresh food storage. This panel may be integral with the liner or installed as a separate piece. Located beneath the evaporator, it allows the unit to operate at the low temperatures required for short-term frozen food storage, while maintaining temperatures above freezing in the general food storage compartment. The insulation must be sufficiently thick to prevent excessive amounts of condensation from forming on the fresh food side of the panel. Reduced heat transfer through the panel increases the demand for cool dry air in the fresh food compartment. This dry air improves performance in high-humidity environments when doors are opened for normal usage.

Food freezers are offered in two forms: **upright** (vertical) and **chest**. Locks are often provided on the lids or doors as protection against accidental door opening or access by children. A power supply indicator light or a thermometer with an external dial may be provided to warn of high storage temperatures.

Special-Purpose Compartments

Special-purpose compartments provide a more suitable environment for storage of specific foods. For example, a warmer compartment for maintaining butter at spreading temperature is often found in the refrigerator door. Some refrigerators have a meat storage compartment that can maintain storage temperatures just above freezing and may include an independent temperature adjustment feature. Some models have a special compartment for fish, which is maintained at approximately −1°C. High-humidity compartments for storage of leafy vegetables and fresh fruit are found in practically all refrigerators. These drawers, located within the food compartment, are generally tight-fitting to protect vulnerable foods from the desiccating effects of dry air circulating in the general storage compartment. The desired conditions are maintained in the special storage compartments and drawers by (1) enclosing them to prevent air exchange with the general storage area and (2) surrounding them with cold air to maintain the desired temperature.

Ice and Water Service

Through a variety of manual and automatic means, refrigerators provide ice. For **manual** operation, ice trays are placed in the freezing compartment in a stream of air that is substantially below 0°C or placed in contact with a directly refrigerated evaporator surface.

Automatic Ice Makers. Automatic ice-making equipment in household refrigerators is increasingly common. Almost all

Household Refrigerators and Freezers

automatic defrost refrigerators either include factory-installed automatic ice makers or can accept field-installable ice makers.

The ice maker mechanism is located in the freezer section of the refrigerator and requires attachment to a water line. The ice-freezing rate is primarily a function of the system design. Most ice makers are in no-frost refrigerators, and the water is frozen by refrigerated air passing over the ice mold. Because the ice maker must share the available refrigeration capacity with the freezer and food compartments, the ice production rate is usually limited by design to 2 to 3 kg per 24 h. An ice production rate of about 2 kg per 24 h, coupled with an ice storage container capacity of 3 to 5 kg, is adequate for most users.

In the design of an ice maker, the various methods of accomplishing the basic functions must be evaluated to determine whether they meet the design objectives. The basic functions are as follows:

1. **Initiating** the ejection of the ice as soon as the water is frozen is necessary to obtain a satisfactory production rate. Ejection before complete freezing causes wet cubes to freeze together in the storage container and may cause the ice mold to overfill. One method is to initiate ejection in response to the temperature of a selected location in the mold that indicates complete freezing. Another successful method is to initiate ejection based on the time required to freeze the water under normal freezer temperatures. In either method, the temperature or time required may vary in different applications, depending on the cooling air temperature and the rate and direction of the airflow.

2. **Ejecting** the ice from the mold must be a reliable operation. In several designs, ejection is accomplished by freeing the ice from the mold with an electric heater and pushing it from the tray into an ice storage container. In other designs, water is frozen in a plastic tray by passing refrigerated air over the top so that the water freezes from the top down. The natural expansion that takes place during this freezing process causes the ice to partially freeze free from the tray. Through twisting and rotation of the tray, the ice can be completely freed and ejected into a container.

3. **Driving** the ice maker is done in most designs by a gear motor, which operates the ice ejection mechanism and may also be used to time the freezing cycle and the water-filling cycle and to operate the stopping means.

4. **Filling** the ice mold with a constant volume of water, regardless of the variation in line water pressure, is necessary to ensure uniform-sized ice cubes and prevent overfilling. This is done by timing a solenoid flow-control valve or by using a solenoid-operated, fixed-volume slug valve.

5. **Stopping** is necessary after the ice storage container is filled until some ice is used. This is accomplished by using a feeler-type ice level control or a weight control.

Ice service has become more convenient in some models that dispense ice through the door. In one design, the ice is ejected into a storage container, which is accessible from the outside of the freezer door as a tilt-out compartment. In another design, ice is delivered through a trap door in the freezer door by an auger mechanism operating in the ice storage container. The auger motor is energized when a push-button switch is contacted by the action of placing a glass under the trap door. On some designs, an additional selector switch is available to actuate an ice crusher as the cubes pass through the door. Chilled water and/or juice dispensing are provided on yet other designs.

Thermal Considerations

The total heat load imposed on the refrigerating system comes from both external and internal heat sources. The relative values of the basic or predictable components of the heat load (those that are independent of usage) are shown in Figure 2. A large portion of the peak heat load may result from door openings, food loading, and ice making, which are variable and unpredictable quantities dependent on customer use. As the beginning point for the thermal design of the cabinet, the significant portions of the heat load are normally calculated and then confirmed by test.

The major predictable heat load is the heat passing through the cabinet walls. Table 1 shows the insulating values of fibrous and foam insulations commonly used to insulate the cabinet; for further information, see Chapter 22 of the 1997 *ASHRAE Handbook—Fundamentals*.

External sweating can be avoided by keeping exterior surfaces warmer than the dew point of the environment. Condensation is most likely to occur around the hardware, on door mullions, along the edge of door openings, and on any cold refrigerant tubing that may be exposed outside the cabinet. In a 32°C room, no external surface temperature on the cabinet should be more than 3 K below the room temperature. If it is necessary to raise the exterior surface temperature to avoid sweating, this can be done either by routing a loop of the condenser tubing under the front flange of the cabinet outer shell or by locating low-wattage wires or ribbon heaters behind the critical surfaces. Most refrigerators that incorporate electric heaters have power-saving electrical switches that allow the user to deenergize these electrical heaters when the environmental conditions do not require their use.

Temporary condensation on internal surfaces may occur with frequent door openings, so the interior of the general storage compartment must be designed to avoid objectionable accumulation or drippage.

Figure 2 shows the design features of the throat section where the door meets the face of the cabinet. On products with metal liners, metal-to-metal contact between inner and outer panels is prevented by thermal breaker strips. Because the air gap between the breaker strip and the door panel provides a low-resistance heat path to the door gasket, the clearance should be kept as small as possible and the breaker strip as wide as practical. When the inner liner is made of plastic rather than steel, there is no need for separate plastic breaker strips because these are an integral part of the liner.

Cabinet heat leakage can be reduced by using door gaskets with more air cavities to reduce conduction or by using internal secondary gaskets. Care must be taken so that the maximum door opening force of 67 N (as specified in United States *Federal Register* Vol. 38, No. 242, Tuesday, December 18, 1973) is not exceeded.

Structural supports, used to support and position the food compartment liner from the outer shell of the cabinet, are usually constructed of a combination of steel and plastics to provide adequate strength with maximum thermal insulation.

Table 1 Effect of Thermal Insulation on Cabinet Wall Thickness

		Wall Thickness, mm			
	Thermal Conductivity, W/(m·K)	For Threshold of External Sweating in 32°C at 75% rh		Common Practice	
Insulation		−18°C	3°C	−18°C	3°C
Mineral or glass fiber, air filled	0.032 to 0.040	50 to 70	33 to 44	75 to 90	60 to 70
Foam-in-place urethane foam, heavy gas filled	0.019	32	22	48	40
Foamed slab urethane foam, heavy gas filled	0.023	38	25	50	38

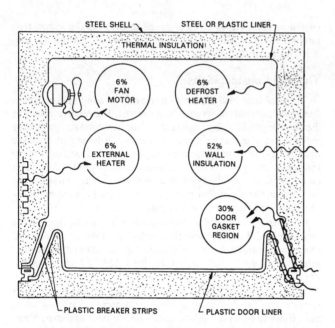

Fig. 2 Cabinet Cross Section Showing Typical Contributions to Total Basic Heat Load

Internal heat loads that must be overcome by the system's refrigerating capacity are generated by ice makers, lights, timers, fan motors used for air circulation, and heaters used to prevent undesirable internal cabinet sweating or frost buildup.

Structure and Materials

The external shell of the cabinet is usually a single fabricated steel structure that supports the inner food compartment liner, the door, and the refrigeration system. The space between the inner and outer walls of the cabinet is usually filled with foam slabs or foam-in-place insulation. In general, the door and breaker strip construction is similar to that shown in Figure 2, although breaker strips and food liners formed of a single plastic sheet are also common. The doors cover the whole front of the cabinet, and plastic sheets become the inner surface for the doors, so no separate door breaker strips are required. The door liners are usually formed to provide an array of small door shelves and racks. Cracks and crevices are avoided, and edges are rounded and smooth to facilitate cleaning. Interior lighting is from incandescent lamps controlled by mechanically operated switches actuated by the opening of the refrigerator door(s) or chest freezer lid. Table 2 summarizes the most common materials and manufacturing methods used in the construction of refrigerator and freezer cabinets.

The cabinet design must provide for the special requirements of the refrigerating system. For example, it may be desirable to refrigerate the freezer section by attaching evaporator tubing directly to the food compartment liner. Also, it may be desirable, particularly with food freezers, to attach the condenser tubing directly to the shell of the cabinet to prevent external sweating. Both designs influence the cabinet heat leakage and the amount of insulation required.

The method of installing the refrigerating system into the cabinet is also important. Frequently, the system is installed in two or more component pieces and then assembled and processed in the cabinet. Unitary installation of a completed system directly into the cabinet allows the system to be tested and charged beforehand. The cabinet design must be compatible with the method of installation chosen. In addition, systems using forced air frequently require ductwork in the cabinet or insulation spaces.

The overall structure of the cabinet must be strong enough to withstand shipping, in which case it will be strong enough to withstand daily usage. However, additional support is typically provided in packaging material. Plastic food liners must withstand the thermal stresses they are exposed to during shipping and usage, and they must be unaffected by common contaminants encountered in a kitchen environment. Shelves must be designed not to deflect excessively under the heaviest anticipated load. Refrigerator doors and associated hardware must withstand a minimum of 300 000 door openings.

Foam-in-place insulation has had an important influence on cabinet design and assembly procedures. Not only is the wall thickness reduced due to the foam's superior thermal conductivity, but the rigidity and bonding action of the foam usually eliminate the need for structural supports. The foam is normally expanded directly into the insulation space, adhering to the food compartment liner and the outer shell. Unfortunately, this precludes disassembly of the cabinet for service or repairs. Alternative constructions that overcome this restriction use prefoamed slabs or expand the foam against an inner mold, which is later withdrawn and replaced with the food compartment liner. In either case, the liner is not bonded to the foam, and it can be easily removed if required. However, the foam no longer provides the structural tie between the liner and the outer shell.

Outer shells of refrigerator and freezer cabinets are now typically of prepainted steel, thus reducing the volatile emissions that accompany the finishing process and providing a consistently durable finish to enhance product appearance and avoid corrosion.

Use of Plastics. As much as 7 to 9 kg of plastic is incorporated in a typical refrigerator or freezer, and the use of plastic is increasing due to the following characteristics:

- Wide range of physical properties
- Good bearing qualities
- Electrical insulating ability
- Moisture and chemical resistance
- Low thermal conductivity
- Ease of cleaning
- Pleasing appearance with or without an applied finish
- Potential of multifunctional design in a single part
- Transparency, opacity, and colorability
- Ease of forming and molding
- Lower cost

Table 2 Cabinet Materials and Manufacturing Methods

Component	Material	Common Thickness, mm	Manufacturing Method	Finish
Outer cabinet assembly			Welded assembly	
Wrapper and top	Low-carbon cold-rolled steel	0.61 to 0.91	Roll form and bend	Organic finish (polyester, alkyd, acrylic, etc.)
Back	Low-carbon cold-rolled steel	0.56 to 0.84	Draw and stamp	
Bottom	Low-carbon cold-rolled steel	0.41 to 0.64	Draw and stamp	
Inner cabinet liner	Enameling iron	0.61 to 0.91	Bend and weld	Vitreous enamel
	Low-carbon cold-rolled steel	0.61 to 0.91	Bend and weld	Organic finish
	Aluminum	0.61 to 0.91	Bend and weld	Anodized or organic finish
	Plastic	1.3 to 5.1	Injection molded or vacuum formed	None
Inner door liner	Plastic	1.9 to 2.4	Vacuum formed or injection molded	None
Breaker strips	Plastic	1.9 to 2.4	Extruded or injection molded	None

Household Refrigerators and Freezers

A few examples illustrate the versatility of plastics. High-impact polystyrene and acrylonitrile butadiene styrene (ABS) plastics are used for inner door liners and food compartment liners. In these applications, no applied finish is necessary. These and similar thermoplastics such as polypropylene and polyethylene are also selected for evaporator doors, baffles, breaker strips, drawers, pans, and many small items. The phenolics are used for decorative door panels, terminal boards, and terminal covers and as a binder for the glass fiber insulation. The good bearing qualities of nylon and acetal are used to advantage in such applications as hinges, latches, and rollers for sliding shelves. Gaskets, both for the refrigerator and for the evaporator doors, are generally made of vinyl or rubber.

Many items (such as ice cubes and butter) readily absorb odors and tastes from materials to which they are exposed. Accordingly, manufacturers take particular care to avoid using any plastics or other materials that will impart an odor or taste in the interior of the cabinet.

Moisture Sealing

For the cabinet to retain its original insulating qualities, the insulation must be kept dry. Moisture may get into the insulation through leakage of water from the food compartment liner, through the defrost water disposal system, or, most commonly, through vapor leaks in the outer shell.

The outer shell is generally crimped, seam welded, or spot welded and carefully sealed against vapor transmission with mastics and hot-melt asphaltic or wax compounds at all joints and seams. In addition, door gaskets, breaker strips, and other parts should provide maximum barriers to vapor flow from the room air to the insulation. When refrigerant evaporator tubing is attached directly to the food compartment liner, as is generally done in chest freezers, moisture will not migrate from the insulation space, and special efforts must be made to vapor-seal this space.

Although urethane foam insulation tends to inhibit moisture migration, it does have a tendency to trap water when migrating vapor reaches a temperature below its dew point. The foam then becomes permanently wet, and its insulation value is decreased. For this reason, a vaportight exterior cabinet is equally important with foam insulation.

Door Latching

Latching of doors is accomplished by mechanical or magnetic latches that act to compress relatively soft compression gaskets made of extruded rubber or vinyl compounds. Gaskets with magnetic materials embedded in the gasket are generally used. Chest freezers are sometimes designed so that the mass of the lid acts to compress the gasket, although most of the mass is counterbalanced by springs in the hinge mechanism.

In 1956, the Refrigerator Safety Act, Public Law 84-930, was enacted prohibiting shipment in interstate commerce of any household refrigerator that was not equipped with a device or system permitting the door to be opened from the inside. Although freezers are not included under PL 84-930, UL *Standard* 250 has adopted the requirements of PL 84-930 to prevent entrapment in freezers.

In addition, because freezers are sometimes located in areas that have public access, they are often equipped with key locks to prevent pilferage. Underwriters Laboratories requires that these key locks be the non-self-engaging type, and the key must be self-ejected from the slot when not in place to prevent entrapment caused by accidentally locking the door.

Cabinet Testing

Specific tests necessary to establish the adequacy of the cabinet as a separate entity include (1) structural tests, such as repeated twisting of the cabinet and door; (2) door slamming test; (3) tests for vapor-sealing of the cabinet insulation space; (4) odor and taste transfer tests; (5) physical and chemical tests of plastic materials; and (6) heat leakage tests. Cabinet testing is also discussed later in the section on Evaluation.

REFRIGERATING SYSTEMS

The vapor-compression refrigerating systems used with modern refrigerators vary considerably in capacity and complexity, depending on the refrigerating application. They are hermetically sealed and normally require no replenishment of refrigerant or oil during the useful life of the appliance. The components of the system must provide optimum overall performance and reliability at minimum cost. In addition, all safety requirements of UL *Standard* 250 must be met. The fully halogenated refrigerant R-12 was used in household refrigerators for many years. However, due to its strong ozone depletion property, appliance manufacturers have replaced R-12 with environmentally acceptable R-134a or isobutane.

The design of refrigerating systems for refrigerators and freezers has improved due to new refrigerants and oils, wider use of aluminum, smaller and more efficient motors and compressors, universal use of capillary tubes, and simplified electrical components. These refinements have kept the vapor-compression system in the best competitive position for household application.

Refrigerating Circuit

Figure 3 shows the refrigerant circuit for a vapor-compression refrigerating system. The refrigeration cycle is as follows:

1. Electrical energy supplied to the motor drives a positive displacement compressor, which draws cold, low-pressure refrigerant vapor from the evaporator and compresses it.
2. The resulting high-pressure, high-temperature discharge gas then passes through the condenser, where it is condensed to a liquid while the heat is rejected to the ambient air.
3. The liquid refrigerant passes through a metering (pressure reducing) capillary tube to the evaporator, which is at low pressure.
4. The low-pressure, low-temperature liquid in the evaporator absorbs heat from its surroundings, evaporating to a gas, which is again withdrawn by the compressor.

Note that energy enters the system through the evaporator (heat load) and through the compressor (electrical input). Thermal energy is rejected to the ambient by the condenser and the compressor shell. A portion of the capillary tube is usually soldered to the suction line for heat exchange. Capacity and efficiency are increased by cooling the refrigerant in the capillary tube with the suction gas.

A strainer-drier is usually placed ahead of the capillary tube to remove foreign material and moisture. Refrigerant charges of 250 g or less are common. A thermostat (or cold control) cycles the compressor to provide the desired temperatures within the refrigerator. During the off cycle, the capillary tube permits the pressures to equalize throughout the system.

Materials used in refrigeration circuits are selected for (1) their mechanical properties, (2) their compatibility with the refrigerant

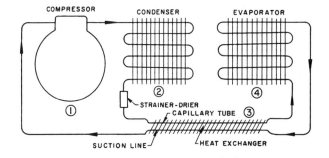

Fig. 3 Refrigeration Circuit

and oil on the inside, and (3) their resistance to oxidation and galvanic corrosion on the outside. Evaporators are usually made of aluminum tubing, either with integral extruded fins or with extended surfaces mechanically attached to the tubing. Condensers are usually made of steel tubing with an extended surface of steel sheet or wire. Steel tubing is used on the high-pressure side of the system, which is normally dry, and copper is used for suction tubing, where condensation can occur. Because of its ductility, corrosion resistance, and ease of brazing, copper is used for capillary tubes and often for small connecting tubing. Wherever aluminum tubing comes in contact with copper or iron, it must be protected against moisture to avoid electrolytic corrosion.

Defrosting

Manual Defrost. A few manufacturers still use the manual defrost method in which the cooling effect is generated by natural convection of air over a refrigerated surface (evaporator) located at the top of the food compartment. The refrigerated surface forms some of the walls of a frozen food space, which usually extends across the width of the food compartment. Defrosting is typically accomplished by manually turning off the temperature control switch.

Cycle Defrosting (Partial Automatic Defrost). Combination refrigerator-freezers sometimes use two separate evaporators for the fresh food and freezer compartments. The fresh food compartment evaporator defrosts during each off cycle of the compressor, with the energy for defrosting provided mainly by the heat leakage into the fresh food compartment. The cold control senses the temperature of the fresh food compartment evaporator and cycles the compressor on when the evaporator surface is above 0°C. The freezer evaporator requires infrequent manual defrosting.

No-Frost Systems (Automatic Defrost). Most combination refrigerator-freezers and some upright food freezers are refrigerated by air that is fan-blown over a single evaporator concealed from view. Because the evaporator is colder than the freezer compartment, it collects practically all of the frost, and there is little or no permanent frost accumulation on the frozen food or on exposed portions of the freezer compartment. The evaporator is defrosted automatically by electric heat or hot refrigerant gas, and the defrosting period is short in order to limit food temperature rise. The resulting water is disposed of automatically by draining to the exterior, where it is evaporated in a pan located in the warm condenser compartment. Defrosting is usually initiated by a timer at intervals of up to 24 h. If the timer operates only when the compressor runs, the accumulated time tends to reflect the probable frost load.

Adaptive Defrost. Developments in electronics have allowed the introduction of microprocessor-based control systems to some household refrigerators. An adaptive defrost function is usually included in the software. Various parameters are monitored so that the period between defrosts varies according to actual conditions of use. Adaptive defrost tends to reduce energy consumption and improve food preservation.

Forced Heat for Defrosting. All no-frost systems add heat to the evaporator to accelerate melting during the short defrosting cycle. The most common method uses a 300 to 1000 W electric heater. The typical defrost cycle is initiated by a timer, which stops the compressor and energizes the heater.

When the evaporator has melted all the frost, a defrost termination thermostat opens the heater circuit. In most cases, the compressor is not restarted until the evaporator has drained for a few minutes and the system pressures have stabilized; this reduces the applied load for restarting the compressor. Commonly used defrost heaters include metal-sheathed heating elements in thermal contact with evaporator fins and radiative heating elements positioned to radiate heat to the evaporator.

Evaporator

The **manual defrost** evaporator is usually a box with three or four sides refrigerated. Refrigerant may be carried in tubing brazed to the walls of the box, or the walls may be constructed from double sheets of metal that are brazed or metallurgically bonded together with integral passages for the refrigerant. In this construction, the walls are usually aluminum, and special attention is required to avoid (1) contamination of the surface with other metals that would promote galvanic corrosion and (2) configurations that may be easily punctured during use.

The **cycle defrost** evaporator for the fresh food compartment is designed for natural defrost operation and is characterized by its low thermal capacity. It may be either a vertical plate, usually made from bonded sheet metal with integral refrigerant passages, or a serpentine coil with or without fins. In either case, the evaporator should be located near the top of the compartment and be arranged for good water drainage during the defrost cycle. In some designs, this cooling surface has been located in an air duct remote from the fresh food space, with air circulated continuously by a small fan.

The **no-frost forced-convection** evaporator is usually a forced-air fin-and-tube arrangement designed to minimize the effect of frost accumulation, which tends to be relatively rapid in a single evaporator system. The coil is usually arranged for airflow parallel to the long dimension of the fins.

The fins may be more widely spaced at the air inlet to provide for preferential frost collection and to minimize the air restriction effects of the frost. All surfaces must be heated adequately during the defrost cycle to ensure complete defrosting, and provision must be made for draining and evaporating the defrost water outside the food storage spaces.

Freezers. Evaporators for chest freezers usually consist of tubing that is in good thermal contact with the exterior of the food compartment liner. The tubing is generally concentrated near the top of the liner, with wider spacing near the bottom to take advantage of natural convection of the air inside. Most upright food freezers usually have refrigerated shelves and a refrigerated surface at the top of the food compartment. These are commonly connected in series with an accumulator at the exit end. No-frost freezers usually incorporate a fin-and-tube evaporator and an air-circulating fan as used in the no-frost combination refrigerator-freezers.

Condenser

The condenser is the main heat-rejecting component in the refrigerating system. It may be cooled by natural draft on freestanding refrigerators and freezers or fan-cooled on larger models and on models designed for built-in applications.

The **natural-draft condenser** is located on the back wall of the cabinet and is cooled by natural air convection under the cabinet and up the back. The most common form of natural-draft condenser consists of a flat serpentine of steel tubing with steel cross wires welded on 6 mm centers on one or both sides perpendicular to the tubing. Tube-on-sheet construction may also be used.

The **hot wall condenser**, another natural-draft arrangement, is used principally with food freezers. It consists of condenser tubing attached to the inside surface of the cabinet shell. The shell thus acts as an extended surface for heat dissipation. With this construction, external sweating is seldom a problem.

The **forced-draft condenser** may be of fin-and-tube, folded banks of tube-and-wire, or tube-and-sheet construction. Various forms of condenser construction are used to minimize clogging caused by household dust and lint. The compact, fan-cooled condensers are usually designed for low airflow rates because of noise limitations. Air ducting is often arranged to use the front of the machine compartment for the entrance and exit of air. This makes the cooling air system largely independent of the location of the refrigerator and permits built-in applications.

Household Refrigerators and Freezers

A portion of the condenser may be located under the defrost water evaporating pan to promote water evaporation. The condenser may also incorporate a section for **compressor cooling**; from here the partially condensed refrigerant is routed to an oil-cooling loop in the compressor, where the liquid refrigerant, still at high pressure, absorbs heat and is reevaporated. The vapor is then routed through the balance of the condenser, to be condensed in the normal manner. In some designs, as noted previously, a portion of the condenser tubing is routed internally in contact with the outer case in place of anticondensation heaters.

Condenser performance may be evaluated directly on calorimeter test equipment similar to that used for compressors. However, the final design of the condenser must be determined by performance tests on the refrigerator under a variety of operating conditions.

Generally, the most important design requirements for a condenser include (1) sufficient heat dissipation at peak-load conditions, (2) storage volume that prevents excessive pressures during pulldown or in the event of a restricted or plugged capillary tube, (3) good refrigerant drainage to minimize the off-cycle losses and the time required for equalization of system pressures, (4) an external surface that is easily cleaned or designed to avoid dust and lint accumulation, and (5) an adequate safety factor against bursting.

Capillary Tube

The most commonly used refrigerant metering device is the capillary tube, a small-bore tube connecting the outlet of the condenser to the inlet of the evaporator. The regulating effect of this simple control device is based on the principle that a given mass of liquid passes through a capillary more readily than the same mass of gas at the same pressure. Thus, if uncondensed refrigerant vapor enters the capillary, the mass flow will be reduced, giving the refrigerant more cooling time in the condenser. On the other hand, if liquid refrigerant tends to back up in the condenser, the condensing temperature and pressure rise, resulting in an increased mass flow of refrigerant. Under normal operating conditions, a capillary tube gives good performance and efficiency. Under extreme conditions, the capillary either passes considerable uncondensed gas or backs liquid refrigerant well up into the condenser. Figure 4 shows the typical effect of capillary refrigerant flow rate on system performance.

A capillary tube has the advantage of extreme simplicity and no moving parts. It also lends itself well to being soldered to the suction line for heat exchange purposes. This positioning prevents sweating of the otherwise cold suction line and increases the refrigerating capacity and efficiency. Another advantage is that the pressure equalizes throughout the system during the off cycle and reduces the starting torque required of the compressor motor. The capillary is the narrowest passage in the refrigerant system and the place where low temperature first occurs. For that reason, a combination strainer-drier is usually located directly ahead of the capillary to prevent it from being plugged by ice or any foreign material circulating through the system (see Figure 3).

Selection. The optimum metering action can be obtained by variations in either the diameter or the length of the tube. Such factors as the physical location of the system components and the heat exchanger length (900 mm or more is desirable) may help determine the optimum length and bore of the capillary tube for any given application.

Capillary tube selection is covered in detail in Chapter 45. Once a preliminary selection is made, an experimental unit can be equipped with three or more different capillaries that can be activated independently. System performance can then be evaluated by using in turn capillaries with slightly different flow characteristics.

Final selection of the capillary requires an optimization of performance under both no-load and pulldown conditions, with maximum and minimum ambient and load conditions. The optimum refrigerant charge can also be determined during this process.

Compressor

While a more detailed description of compressors can be found in Chapter 34 of the 1996 *ASHRAE Handbook—Systems and Equipment*, a brief discussion of the small compressors used in household refrigerators and freezers is included here.

These products use positive displacement compressors in which the entire motor-compressor is hermetically sealed in a welded steel shell. Capacities range from about 90 W to about 600 W when measured at the usual rating conditions of −23°C evaporator, 54°C condenser, 32°C ambient, with the suction gas superheated to 32°C and the liquid subcooled to 32°C.

Design emphasis is placed on ease of manufacturing, reliability, low cost, quiet operation, and efficiency. Figure 5 illustrates the two reciprocating piston compressor mechanisms and two types of rotary compressors that are used in virtually all conventional refrigerators and freezers; no one type is much less costly than the others. While rotary compressors are somewhat more compact than reciprocating compressors, a greater number of close tolerances is involved in their manufacture.

These compressors are directly driven by two-pole, 3450 rpm squirrel cage induction motors, although some four-pole, 1750 rpm motors are also used. Field windings are insulated with special wire enamels and plastic slot and wedge insulation; all are chosen for their compatibility with the refrigerant and oil. During continuous runs at rated voltage, motor winding temperatures may be as high as 120°C when tested in a 43°C ambient temperature. In addition to maximum operating efficiency at normal running conditions, the motor must provide sufficient torque at the anticipated extremes of line voltage for starting and temporary peak loads due to start-up and pulldown of a warm refrigerator and for the load conditions associated with defrosting.

Starting torque is provided by a split-phase winding circuit, which in the larger motors may include a starting capacitor. When the motor comes up to speed, an external electromagnetic relay or positive temperature coefficient (PTC) device disconnects the start winding. A run capacitor may be employed for greater motor efficiency. Motor overload protection is provided by an automatic resetting switch, which is sensitive to a combination of motor current and compressor case temperature or to internal winding temperature.

The compressor is cooled by rejecting heat to the surroundings. This is easily accomplished with a fan-cooled system. However, an oil-cooling loop carrying partially condensed refrigerant may be necessary when the compressor is used with a natural-draft condenser and in some forced-draft systems above 300 W.

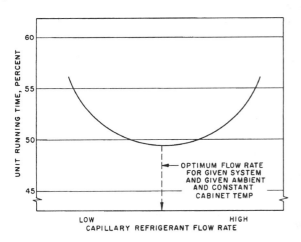

Fig. 4 Typical Effect of Capillary Tube Selection on Unit Running Time

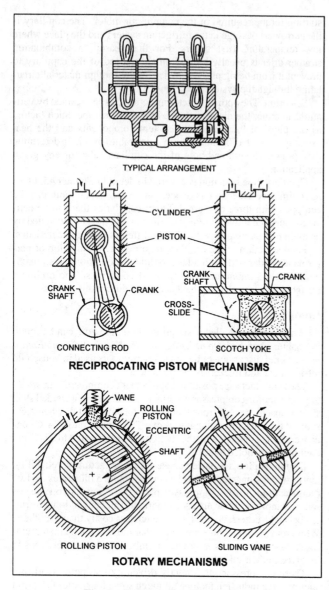

Fig. 5 Refrigerator Compressors

Temperature Control System

The temperature control thermostat is generally an electromechanical switch actuated by a temperature-sensitive power element that has a condensable gas charge, which operates a bellow or diaphragm. At operating temperature, this charge is in a two-phase state, and the temperature at the gas-liquid interface determines the pressure on the bellows. To maintain temperature control at the bulb end of the power element, the bulb must be the coldest point at all times.

The thermostat must have an electrical switch rating for the inductive load of the compressor and other electrical components that is carried through the switch. The thermostat is usually equipped with a shaft and knob for adjustment of the operating temperature.

In the simple gravity-cooled system, the sensing bulb of the thermostat is normally clamped to the evaporator. The location of the bulb and the degree of thermal contact are selected to produce both a suitable cycling frequency for the compressor and the desired refrigerator temperature. Small refrigerators sold in Europe are sometimes equipped with a manually operated push button to prevent the control from coming on until defrost temperatures are reached; afterward, normal cycling is resumed.

In a combination refrigerator-freezer with a split air system, the location of the thermostat sensing bulb depends on whether an automatic damper control is used to regulate the airflow to the fresh food compartment. When such an auxiliary control is used, the sensing bulb is usually located to sense the temperature of the air leaving the evaporator. In manual damper controlled systems, the sensing bulb is usually placed in the cold airstream to the fresh food compartment. The sensing bulb location is frequently related to the damper effect on the airstream. Depending on the design of this relationship, the damper may become the freezer temperature adjustment or it may serve the fresh food compartment, with the thermostat being the adjustment for the other compartment. The temperature-sensing bulb should be located to provide a large enough temperature differential to drive the switch mechanism, while avoiding (1) excessive cycle length; (2) short cycling time, which can cause compressor starting problems; and (3) annoyance to the user from frequent noise level changes.

In some refrigerators, microprocessor-based control systems have replaced the electromechanical thermostat switch; in some cases, both compartment controls use thermistor sensing devices that relay electronic signals to the microprocessor. Electronic control systems provide a higher degree of independence in temperature adjustments for the two main compartments.

System Design and Balance

A principal design consideration is the selection of components that will operate together to give the optimum system performance and efficiency when the total cost is considered. Normally, a range of combinations of values for these components meets the performance requirements, and the lowest cost is only obtained through a careful analysis or a series of tests—usually both. For instance, for a given cabinet configuration, food storage volume, and temperature, the following can be traded off against one another: (1) insulation thickness and overall shell dimensions, (2) insulation material, (3) system capacity, and (4) individual component performance (e.g., fan, compressor, and evaporator). Each of these variables affects the total cost, and most of them can be varied only in discrete steps.

The experimental procedure involves a series of tests. Calorimeter tests may be made on the compressor and condenser, separately or together, and on the compressor and condenser operating with the capillary tube and heat exchanger. Final selection of the components requires performance testing of the system installed in the cabinet. These tests also determine the refrigerant charge, airflows for the forced-draft condenser and evaporator, temperature control means and calibration, necessary motor protection, and so forth. The section on Evaluation covers the final evaluation tests to be made on the complete refrigerator. The interaction between components is further addressed in Chapter 44. This experimental procedure assumes a knowledge (equations or graphs) of the performance characteristics of the various components, including the heat leakage of the cabinet and the heat load imposed by the customer. The analysis may be performed manually point by point. If enough component information exists, it can be entered into a computer simulation program capable of responding to various design conditions or statistical situations. Although the available information may not always be adequate for an accurate analysis, this procedure is often useful, although it must be followed by confirming tests.

Processing and Assembly Procedures

All parts and assemblies that are to contain refrigerant are processed to avoid unwanted substances or remove them from the final sealed system and to charge the system with refrigerant and oil. Each component should be thoroughly cleaned and then stored in a clean, dry condition prior to assembly. The presence of free water in stored parts produces harmful compounds such as rust and

Household Refrigerators and Freezers

aluminum hydroxide, which are not removed by the normal final assembly process. Procedures for dehydration, charging, and testing may be found in Chapter 46.

Assembly procedures are somewhat different, depending on whether the sealed refrigerant system is completed as a unit before being assembled to the cabinet, or components of the system are first brought together on the cabinet assembly line. With the unitary installation procedure, the system may be tested for its ability to refrigerate and then be stored or delivered to the cabinet assembly line.

EVALUATION

Once the unit is assembled, laboratory tests, supplemented by field testing, are necessary to determine actual performance. The following aspects are considered in this section:

- Test facilities required
- Established test procedures published by standard, technical, and industry organizations
- Special performance testing
- Materials testing
- Life testing of components
- Field testing

Environmental Test Rooms

Controlled temperature and humidity test rooms are essential for performance testing of refrigerators and freezers. AHAM *Standard* HRF-1 describes the environmental conditions to be maintained. The rooms should be capable of providing ambient temperatures ranging from 21 to 45°C accurate to within 0.5 K of the desired value. The temperature gradient and the air circulation within the room should also be maintained closely. To provide more flexibility in testing, it may be desirable to have an additional test room that can cover the range between −18 and 21°C. At least one test room should be capable of maintaining a desired relative humidity within a tolerance of ±2% up to 85% rh.

All instruments should be calibrated at regular intervals. The instruments for adequate performance testing of a refrigerator or freezer are described in AHAM *Standard* HRF-1. Instrumentation should have accuracy and response capabilities of sufficient quality to measure the dynamics of the systems tested.

Computerized data acquisition systems that record power, current, voltage, temperature, and pressure are used in testing refrigerators and freezers. Refrigerator test laboratories have developed automated means of control and data acquisition (with computerized data reduction output) and automated test programming.

Standard Performance Test Procedures

AHAM *Standard* HRF-1 describes tests for determining the performance of refrigerators and freezers. It specifies the standard ambient conditions, power supply, and means for selecting samples and measuring temperatures. Test procedures include the following:

No-Load Pulldown Test. This tests the ability of the refrigerator or freezer in a 43°C ambient temperature to pull down from a stabilized warm condition to design temperatures within an acceptable period.

Simulated Load Test (*Refrigerators*) or **Storage Load Test** (*Freezers*). This test determines the electrical energy (kWh) consumption rate per 24 h period, the percent operating time of the compressor motor, and temperatures at various locations within the cabinet at 21, 32, and 43°C ambient for a range of temperature control settings. The cabinet doors remain closed during the test. The freezer compartment is loaded with filled frozen packages. Each test point may take 8 h or more to ensure steady-state condition and accuracy of data. The data taken are usually plotted as shown in Figure 6 for a combination refrigerator-freezer with only a fresh food

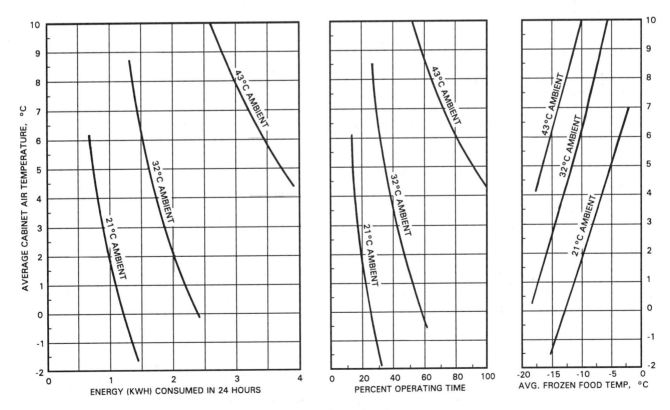

Fig. 6 Sample Plot of Simulated Load Test Results

temperature control. If there is a separate control for freezer temperature, these graphs can carry additional curves for high and low freezer control settings.

Freezers are tested similarly, but in a 32°C ambient. Under actual operating conditions in the home, with frequent door openings and ice making, the performance may not be as favorable as that shown by this test. However, the test indicates general performance, which can serve as a basis for comparison.

Ice-Making Test. This test, performed in a 32°C ambient, determines the rate of making ice with the ice trays or other ice-making equipment furnished with the refrigerator.

External Surface Condensation Test. This test determines the extent of moisture condensation on the external surfaces of the cabinet in a 32°C, high-humidity ambient when the refrigerator or freezer is operated at normal cabinet temperatures. Although AHAM *Standard* HRF-1 calls for this test to be made at a relative humidity of 75 ± 2%, it is customary to determine the sweating characteristics through a wide range of relative humidity up to 85%. This test also determines the need for, and the effectiveness of, anticondensation heaters in the cabinet shell and door mullions.

Internal Moisture Accumulation Test. This dual-purpose test is also run under high-temperature, high-humidity conditions. First, it determines the effectiveness of the moisture sealing of the cabinet in preventing moisture from getting into the insulation space and degrading the performance and life of the refrigerator. Secondly, it determines the rate of frost buildup on refrigerated surfaces, the expected frequency of defrosting, and the effectiveness of any automatic defrosting features, including defrost water disposal.

This test is performed in ambient conditions of 32°C and 75% rh with the cabinet temperature control set for normal temperatures. The test extends over a 21 day period with a rigid schedule of door openings over the first 16 h of each day. The test calls for 96 door openings per day for a general refrigerated compartment and 24 per day for a freezer compartment and for food freezers.

Current Leakage Test. This test determines the electrical current leakage through the entire electrical insulating system under severe operating conditions. To eliminate the possibility of a shock hazard, a current leakage of 0.75 mA at rated line voltage, in accordance with UL *Standard* 250, is considered the industry maximum.

Handling and Storage Test. As with most other major appliances, it is during shipping and storage that a refrigerator is exposed to the most severe impact forces, to vibration, and to extremes of temperature. When packaged, it should withstand without damage a drop of several centimetres onto a concrete floor, the impact experienced in a freight car coupling at 4.5 m/s, and jiggling equivalent to a trip of several thousand kilometres by rail or truck.

The widespread use of plastic parts makes it important to select materials that also withstand the high and low temperature extremes that may be experienced. This test determines the ability of the cabinet, when packaged for shipment, to withstand handling and storage conditions in extreme temperatures. It involves raising the crated cabinet 150 mm off the floor and suddenly releasing it on one corner. This is done for each of the four corners. This procedure is carried out at stabilized temperature conditions—first in a 60°C ambient temperature, and then in a −18°C ambient. At the conclusion of the test, the cabinet is uncrated and operated, and all accessible parts are examined for damage.

Special Performance Testing

To ensure customer acceptance, several additional performance tests are customarily performed.

Usage Test. This is similar to the internal moisture accumulation test, except that additional performance data are taken during the test period, including (1) electrical energy consumption per 24 h period, (2) percent running time of the compressor motor, and (3) cabinet temperatures. These data give an indication of the reserve capacity of the refrigerating system and the temperature recovery characteristics of the cabinet.

Low Ambient Temperature Operation. It is customary to conduct a simulated load test and an ice-making test at ambient temperatures of 13°C and below. This test determines performance under unusually low temperature conditions.

Food Preservation Tests. This test determines the food-keeping characteristics of the general refrigerated compartment and is useful for evaluating the utility of special compartments such as vegetable crispers, meat keepers, high-humidity compartments, and butter keepers. This test is made by loading the various compartments with food, as recommended by the manufacturer, and periodically observing the condition of the food.

Noise Tests. The complexity and increased size of refrigerators has made it difficult to keep the sound level within acceptable limits. Thus, sound testing is important to ensure customer acceptance.

A meaningful evaluation of the sound characteristics may require a specially constructed room with a background sound level of 30 dB or less. The wall treatment may be reverberant, semireverberant, or anechoic, with the reverberant construction usually favored in making an instrument analysis. A listening panel is most commonly used for the final evaluation, and most manufacturers strive to correlate instrument readings with the judgment of the panel.

High- and Low-Voltage Tests. The ability of the compressor to start and pull down the system after an ambient soak is tested with applied voltages that are at least 10% above and below the rated voltage. The starting torque is reduced at low voltage; the motor tends to overheat at high voltage.

Special Functions Tests. Refrigerators and freezers with special features and functions may require additional testing. In the absence of formal procedures for this purpose, test procedures are usually improvised as required.

Determining Energy Consumption. This is a special 32°C ambient, closed-door test using the test procedure for electric refrigerators, refrigerator-freezers, and freezers published by the DOE in the United States *Federal Register* (Vol. 54, No. 221, Friday, November 17, 1989). This test procedure specifies a statistical sampling plan, which must be followed to establish the estimated annual cost of energy for labeling under the U.S. Federal Trade Commission's Energyguide Program, as well as conformance with DOE energy standards. The DOE test procedure presently references AHAM *Standard* HRF-1 for methods of testing.

Materials Testing

The materials used in a refrigerator or freezer should meet certain test specifications. All materials in contact with foods must meet U.S. Food and Drug Administration requirements. Metals, paints, and surface finishes may be tested according to procedures specified by the American Society for Testing and Materials and others. Plastics may be tested according to procedures formulated by the Society of the Plastics Industry appliance committee. In addition, the following tests on materials, as applied in the final product, are assuming importance in the refrigeration industry [*Federal Specification* AA-R-00211H(GL)].

Odor and Taste Contamination. This test determines the intensity of odors and tastes imparted by the cabinet air to uncovered, unsalted butter stored in the cabinet at operating temperatures.

Stain Resistance. The degree of staining is determined by coating the cabinet exterior surfaces and the surface of plastic interior parts with a typical staining food (e.g., prepared cream salad mustard).

Environmental Cracking Resistance Test. This tests the cracking resistance of the plastic inner door liners and breaker strips at operating temperatures when coated with a 50/50 mixture of oleic acid and cottonseed oil. The cabinet door shelves are loaded with weights, and the doors are slammed on a prescribed schedule extending over an 8 day test period. The parts are then examined for cracks and crazing.

Breaker Strip Impact Test. This test determines the impact resistance of the breaker strips at operating temperatures when coated with a 50/50 mixture of oleic acid and cottonseed oil. The breaker strip is impacted by a 0.9 kg dart dropped from a prescribed height. The strip is then examined for cracks and crazing.

Component Life Testing

Various components of a refrigerator and freezer cabinet are subject to continual use by the consumer throughout the life of the product; they must be adequately tested to ensure their durability for at least a 10 year life. Some of these items are (1) hinges, (2) latch mechanism, (3) door gasket, (4) light and fan switches, and (5) door shelves. These components may be checked by an automatic mechanism, which opens and closes the door in a prescribed manner. A total of 300,000 cycles is generally accepted as the standard for design purposes. Door shelves should be loaded as they would be for normal home usage. Several other important characteristics may be checked during the same test: (1) retention of door seal, (2) rigidity of door assembly, (3) rigidity of cabinet shell, and (4) durability of inner door panels.

Life tests on the electrical and mechanical components of the refrigerating system may be made as required.

Field Testing

Additional information may be obtained from a program of field testing in which test models are placed in selected homes for observation. Because high temperature and high humidity are the most severe conditions encountered, the Gulf Coast of the United States is a popular field test area. Laboratory testing has limitations in the complete evaluation of a refrigerator design, and field testing can provide the final assurance of customer satisfaction.

Field testing is only as good as the degree of policing and the completeness and accuracy of reporting. However, if the testing is done properly, the data collected are important, not only in product evaluation, but also in providing criteria for more realistic and timely laboratory test procedures and acceptance standards.

REFERENCES

AHAM. 1988. Household refrigerators, combination refrigerator freezers, and household freezers. ANSI/AHAM *Standard* HRF-1-1988. Association of Home Appliance Manufacturers, Chicago.

Code of Federal Regulations. 10 CFR, Part 430, Subparts B and C.

Federal Register. 1973. Vol. 38, No 242. Dated Tuesday, December 18, 1973.

Federal Register. 1989. Vol. 54, No. 221. Dated Friday, November 17, 1989, pp. 47916-47945.

Federal Specification. 1977. Refrigerators, mechanical, household (electrical, self-contained). AA-R-00211H(GL).

Refrigeration Safety Act. 1956. *Public Law* 84-930.

UL. 1994. Household refrigerators and freezers, 10th ed. *Standard* 250-93. Underwriters Laboratories, Northbrook, IL.

CHAPTER 50

AUTOMATIC ICE MAKERS

Terminology 50.1
Ice Maker Construction 50.2
Application 50.5

THIS chapter addresses commercial-size automatic ice makers —their construction, operation, and application. Specifically omitted are ice makers used in domestic refrigerators (covered in Chapter 49), large plant units requiring trained personnel (covered in Chapter 33), and ice machines used for thermal storage (see Chapter 40 in the 1995 *ASHRAE Handbook—Applications*). This chapter is concerned with machines that make small, fairly uniform pieces of ice, such as ice flakes or ice cubes, up to approximately 60 g in mass. These units usually include a bin or ice storage facility and are arranged and controlled to maintain a certain stated amount of product in storage. Unit capacities range from production rates of 7 kg/24 h to several hundred kilograms a day.

Fundamentally, ice means spot cooling. Water resulting from meltage is often used in applications where products, especially vegetables, tend to lose mass (through dehydration) and become less attractive in appearance. Today, the largest markets for ice makers are bars, restaurants, cafeterias, soft drink parlors, motels, hotels, hospitals, fish and vegetable markets, concession stands, fast-food services, and prepackaged ice for home use.

Although ice can be made from seawater (with use usually confined to fresh fish preservation) or almost any other water, ice in the automatic ice maker industry is customarily made from potable water and is kept relatively uncontaminated. Further classifications are **clear ice** and **cloudy ice**; there are different degrees of cloudiness, depending on the ice maker design.

Machines using a batch process make clear ice particles or chunks; machines using a continuous process either produce cloudy flakes or they compact the flakes into more or less opaque pieces of various geometric shapes. An appreciation of the characteristics and source material (water) of product ice is important not only to the design engineer but also to anyone wishing to apply, specify, or purchase ice-making equipment for specific applications.

Fundamentally, ice is crystalline in structure and follows the physical laws of crystalline material. However, water is seldom pure H_2O and, depending on its source, is generally made up of a complex mixture of layers of tiny water crystals interspersed with layers of an amorphous matrix of other chemicals. This composition accounts not only for the clarity (or lack of it), but also for a number of other qualities of ice, such as hardness or, particularly, the shearing quality.

Factors influencing the actual character of the ice formed are not only the water of which it is made but also the freezing rate and degree of washing of the interface between the already frozen ice and the water to be frozen.

The process of removing heat from a body of water to turn it into ice is normally a concentrating process that tends to freeze pure H_2O and leave the remaining liquid water with a higher percentage of extraneous chemicals than it had at the beginning of the process. This concentrating occurs at the interface between the newly formed ice and the water surrounding it. If the freezing rate is too fast, the rejected chemicals are frozen into a matrix surrounding the pure H_2O crystals, and the remaining water is consumed or converted into ice at the original concentration. If, however, freezing is slow enough or the ice interface is washed by a flow of water, these rejected chemicals can be removed and the resulting ice is significantly purer than the water from which it was made.

Continuous ice makers make ice that contains most of the chemicals in the original water, although some rejection of extraneous chemicals may result if sufficient bleeding of the makeup water mixture is continuously maintained. Units operating on a batch cycle, however, tend to produce ice that is purer than its source water.

Since water easily dissolves many substances, many impurities are always present in natural water, the most common of which are compounds of sodium, calcium, magnesium, and iron. In addition, the water may carry such suspended impurities as fine clay, sand, and fragments of vegetation, as well as entrapped air and microscopic organisms, including bacteria.

Calcium and magnesium salts make water hard and tend to come out of solution and deposit in an ice maker, eventually making the ice maker inoperative if periodic cleaning and lime removal are not performed conscientiously. Typhoid, cholera, and dysentery are primarily spread by infected water supplies, so only water that is bacteriologically and chemically safe must be used for product ice that is to be used in beverages or is to come in direct contact with food.

Practically all domestic water systems have different mixtures of the same common chemicals. Whenever the total chemical content exceeds 400 mg/kg, or if especially objectionable gases or compounds are present, auxiliary water treatment is indicated.

TERMINOLOGY

The following terms may be used in the ice-making industry:

Kilograms of corrected products, or **334 kJ ice.** Describes an imaginary product that absorbs 334 kJ/kg while melting at 0°C.

To convert the product of any particular machine to 334 kJ ice, a sample of the actual product as it leaves the evaporator is tested in a calorimeter. The reading in kJ/kg of product is then divided by 334, which gives a multiplier to be used with that machine to establish an equitable comparison of kilograms of ice per 24 h among all types of machines.

Cube ice. Normally refers to a fairly uniform product that is hard, solid, usually clear, and generally weighs less than 60 g per piece—as distinguished from flake, crushed, or fragmented ice.

Flake ice or **flaked ice.** Made in a thin sheet approximately 1.5 to 4.5 mm thick. The sheet may be flat or curved, but the thin ice is generally broken into random-sized flakes when harvested. The term also refers to machines that produce ice in this manner and, additionally, compress or extrude a product in larger chunks that either resemble pebbles or are in irregular shapes but fairly uniform sizes.

Crushed ice. Made in hard masses that are later crushed into a smaller size. This ice is characterized by the amount of fine or slush ice mixed in with the more uniform larger chunks.

Blowdown or **bleedoff.** The rejection of a certain amount of recirculated or ingredient water in order to control the amount of chemicals that are present because of the concentrating effect of water frozen into ice.

The preparation of this chapter is assigned to TC 10.2, Automatic Icemaking Plants and Skating Rinks.

Harvesting or **harvest cycle** (sometimes called the **defrost cycle**). The removal or separation of the manufactured ice from the evaporator.

Auger. Originally referred to the scraper or helical wedging device that rotated inside a cylindrical evaporator, removed the ice from the interior wall, and pushed the separated ice up and out of the evaporator section. Later, the term was applied to the ice remover (usually helical), even when the ice was made on the outside of a small, vertical, cylindrical evaporator.

Density. As used in this chapter, refers to the mass per unit volume (kg/m^3) of a sample of the product from a given machine and not to the density of an individual particle of product. The definition is useful in determining the amount of ice that can be stored in a bin having a known usable storage volume.

ICE MAKER CONSTRUCTION

Types of Product and Evaporators

Flake ice is uniformly thin, randomly shaped in its perimeter, and sometimes clear but more often cloudy or actually white. Evaporators currently used to produce flake ice are of four main forms: cylinders, flat plates, flexible belt, and disks.

The most common evaporator is a cylinder, usually brass, with refrigerated tubing wrapped on the outside and bonded to it. The evaporator may also be two concentric cylinders with water on either the outside of the outside cylinder or the inside of the inside cylinder, and with the annular space between cylinders occupied by the refrigerant. The simplest form, however, is a cylinder with the refrigerant on the inside and the water to be frozen on the outside. These evaporators require augers, helixes, or equivalent components to remove the ice.

Both cylinder types can be made with an extended auger section in which the ice flakes removed are compressed and the resultant product extruded and ejected. This compressed product differs considerably in both density and appearance from the uncompressed flake ice and can be binned and dispensed far more readily.

The cylinder evaporator is subject to many variations. Both clear and opaque ice can be produced, depending on the means of distributing the water. The cylinder can also be stationary or rotated, as well as placed in a vertical or horizontal position. The harvesting means is also open to many variations, depending on whether ice is formed on the inside or the outside of the cylinder.

The cylinder evaporator has been produced to meet capacities from 50 kg to 20 Mg or more per day, using the whole range of refrigerants from the halogenated fluorocarbons and methyl chloride to ammonia. Larger flexible-cylinder machines have also been refrigerated with circulated cold brine. In general, if the type of ice produced is satisfactory for the intended application, this evaporator form can be the least expensive to produce and the most efficient to operate thermodynamically. These machines are almost exclusively operated on a continuous cycle rather than in a batch process, although batch-process machines are available when clear ice is sought. Cracked ice can be produced on a flat evaporator, using a batch process and either an ice crusher or some other ice-breaking feature.

The **belt-ice** method is used mostly in industrial-size machines. Belt ice is produced from a metal belt turning over two pulleys—the belt sliding upward across a slightly-curved inclined refrigerated surface. Flowing water strikes the belt near the upper end of its length and flows downward forming ice as it flows. The ice thus formed continues past the point of water impact and leaves the belt in a ribbon or sheet as the belt passes over the upper pulley.

Cube ice is produced from a great variety of evaporators. Most of these machines fall into five main types: flat plates (using either the whole surface or selected spots), multiple cells or molds, tube machines, rod machines, and channel machines.

Flat-plate evaporators may be vertical or in an inclined position; they may have water flowing on either the top or the underside of the evaporator. The main characteristics are that the ice is formed in a rectangular slab during the freezing cycle, and it is usually quite clear. During the harvest cycle, the evaporator plate is heated, and the ice drops or slides off. This slab is then cut into square or rectangular pieces, most often with a grid of either electrically heated wires or small tubes carrying a heated fluid. The slab thickness can usually be varied at will, but the cube sizes are determined by the previously selected grid pattern.

There are several other versions of the flat-plate evaporator in which grids of metal or plastic are pressed against the flat evaporator and a third movable element closes the side opposite the evaporator plate, thereby forming individual cells. Recirculated water is supplied to each cell. In these versions, the individual cubes are ejected during the harvest cycle, and no cutting is required.

Cell- or **mold-type evaporators** make the individual cubes without any additional cutting. The cups or cells may be any shape from round to square to polysided. They may be inverted, and then water is sprayed up into the cavities to produce clear ice. The cells are usually perpendicular to the vertical plate. Water then flows by gravity from the top of the evaporator and washes into the cells enroute to a recirculating water pan. During the harvest cycle, the water is shut off, and the cavities are heated in various ways to free the ice cubes. The cubes fall onto a grill that directs the ice to a chute or bin below.

Tube machines make ice inside the tubes, with the refrigerant outside the tubes. The ice is made in a long cylindrical shape, usually with a hole in the center, and is broken into short lengths as it emerges from the evaporator. The ice is kept clear by making sure that an excess of water is continuously passing through the tubes.

In one version, tube sheets within a vertical shell hold a series of tubes in place. The shell and tubes act as a flooded evaporator during the freezing process and as a condenser during the harvesting cycle. Water is pumped to the top of the shell and flows by gravity down the inside walls of the freezing tubes.

Another version is a single tube within a tube. The ice is formed on the inside of the inner tube, and the refrigerant is in the annular space between the two tubes. Water is pumped through the inner tube until the resistance to flow becomes greatly increased, at which time hot gas is introduced into the annular space. When ice is free, it is forced out by water pump pressure and cut or broken into lengths of two or three diameters.

A third cube machine of the tube type forms separate ice bodies in square, vertical, stainless steel tubes. These tubes are banded at uniformly spaced intervals with copper heat conductors, which, in turn, are refrigerated. Water flows down the inside of the tubes and turns to ice in the region of the bands, thereby producing clear ice cubes with an hourglass-shaped hole through the center. Hot-gas harvesting is used, and the cubes fall by gravity into a storage bin below.

Rod units usually employ a series of short refrigerated rods (or, in some cases, short tubes) sealed at the bottom, which drop into a tank of flowing or agitated water. Clear, thimble-like pieces of ice form, and at a predetermined time, the water tank is removed (or the fingers are raised). When the hot gas thaws the ice loose, the ice falls free and is guided into an opening that leads to an ice bin. Both the length of the ice piece and its outer diameter can be varied.

Channel-type evaporators are made by forming thin, stainless steel sheets into a series of channels. These sheets are then mounted upright so that the channels are vertical. On one side, water flows down the sides and bottom. On the opposite side and at right angles to the channels is a refrigerant-carrying serpentine tube. This tube is bonded to the stainless steel at equally spaced intervals, thus forming a series of cold spots over which the water flows. When hot gas replaces the cold refrigerant, individual pieces of ice are freed and fall into a bin below.

Almost as many types of evaporators and ice machines have been built as there are makes of automobiles. The surviving

Automatic Ice Makers

machines have the fewest number of moving parts, are economical to manufacture, and have controls as close to fail-safe as possible. As long as the ice is compatible with its ultimate use, the acceptance of any specific design is based as much on reliability as on efficiency or even first cost.

Complete Ice Maker Packages

Automatic ice-making equipment can be purchased in a wide variety of models, with some companies offering as many as three or four different ice shapes. Some units are made of separate modules capable of being assembled into complete units to satisfy a wide range of applications. These modules often include an automatic ice-making section for either cubes or flakes; the section shuts off when it is set on a container or bin of some kind. Crusher sections are offered that can be set to produce crushed ice from ice cubes and direct uncrushed ice into one bin section and crushed into another. Various bin sizes with different access ports are often available. Additionally, sanitary dispensing units are offered as an alternate to a simple ice bin. Other add-on accessories include water stations for filling drinking glasses with ice and water and for soda fountain dispensing units.

In addition to the modular approach, units are specifically designed for many special applications. These special units may be designed to conserve floor space or fit under counters. Hospital units must protect the product from contamination, so the units must be able to withstand thorough cleaning and sanitizing by cleaning personnel.

Coin-operated machines that make, bag, and vend ice (all automatically) are not economical because the cost and complication of the machines are too great for the yearly volume of ice sold. However, vending machines that simply vend bags of ice are gaining acceptance. In this case, the ice is made and bagged elsewhere. This operation is especially economical in connection with merchandising requiring refrigerated holding rooms for storing backup supplies.

Performance and Operating Characteristics

ARI (1990) shows typical water use and energy input for ice cube machines tested in accordance with ASHRAE *Standard* 29 as follows:

	Air Cooled	Water Cooled
Potable water used, L/kg ice	1.10 to 5.27	1.10 to 6.32
Condenser water used, L/kg ice	na	9.0 to 25.1
Energy input, kJ/kg ice	430 to 1790	370 to 1130

System Design

All automatic ice-making systems can be broken down into the following subsystems:

1. Refrigeration circuit
2. Water (for ice making) circuit
3. Ice removal and/or harvest system
4. Unit electrical controls
5. Ice storage and dispensing systems, where included
6. General safety and sanitation codes as they apply to each of the above systems
7. Special application and specifications that may bear on individual system designs

Refrigeration Circuit. Any refrigeration system is a balance of the compressor capacity, the condenser capacity, and the evaporator capacity. In the previous sections, the types of evaporators in current production were reviewed. The thinner the ice produced, the more thermally efficient the production method; as a corollary, the more prime refrigerant surface per mm^3 of ice, the more efficient the heat transfer. Also, a continuous process without a harvesting and a freezing cycle is thermally more efficient than a batch process.

The determining factors in any evaporator design are (1) production of the kind of ice required for the specific end use, (2) manufacturing economics, and (3) the ease of cleaning and extent of service required to keep the evaporator running for at least eight to ten years.

The best heat-transfer metals are the most desirable for an evaporator. The surface on which ice is made, however, must be corrosion resistant, nontoxic, nonporous, and readily cleanable; most important of all, the refrigerant passages must be clean and clear of all oxides and corrosive fluxes. The cross-sectional areas must be uniform so gas velocities do not fall below those necessary to prevent oil separation or accumulation.

The high side (compressor, condenser, and receiver) is generally made up by the ice maker manufacturer from separately selected components suitable for the specific design, rather than from stock high-side units furnished by compressor manufacturers.

Both air- and water-cooled condensers have their particular application advantages. The air-cooled units are the simplest; however, proper ventilation of the installation area may not be practical, or the rejected heat may be too much of a load for the air-conditioning system if the unit is installed in an air-conditioned space. Thus, water-cooled units may be the simplest way of conducting all the rejected heat from the installation area. An alternative is a remotely located air-cooled condenser, which rejects heat away from the air-conditioned space and yet retains the economy of the air-cooled system.

Commercial high sides are usually low-temperature units for flaked-ice machines and medium-temperature units for cube-ice machines (batch process systems where hot gas is used for defrost). The compressor motors are generally of high-torque design.

The liquid control means may be (1) an automatic expansion valve (especially good if the evaporator load is fairly steady), (2) a fixed superheat thermal expansion valve (the easiest to apply), or (3) a capillary restrictor tube (the least expensive to provide, but the most difficult to apply).

If a thermal expansion valve is used, a liquid receiver generally is required, but if an automatic expansion valve or a capillary tube is used for liquid control, a suction line accumulator is generally used.

Unless the manufacturer has sophisticated process equipment and a reliable quality-control system is in operation, all systems should have strainer-driers in the high side. Ice-making systems are essentially low-temperature systems and are therefore very sensitive to excess moisture in the refrigerant circuit.

For most cubers and other batch-process units, a hot-gas solenoid valve bypasses hot compressor discharge gas around the condenser and moves it directly into the evaporator during the harvest cycle. This is normally an electrically operated closed valve, which must be large enough to (1) provide as short a harvest cycle as practical without creating such a thermal shock that the ice tends to break up and (2) prevent excessive compressor overload.

High- and low-pressure controls may or may not be used in the refrigerant circuit. Most water-cooled condensing units use automatic-reset high-pressure controls. Depending on the extent of system safety, the low-pressure cutout is most often a manual reset control; this arrangement ensures that the unit will be investigated to determine cause when an excessively low pressure develops.

Ingredient Water Circuit. The potable water used to make ice is normally a part of the city water supply. In the majority of flaked-ice units a float valve maintains a constant water volume in the system. This mechanically operated valve is usually mounted in a separate float chamber.

A flexible line runs from this float chamber to the evaporator section. A constant and fixed water level is maintained in the evaporator because of the vertical relationship between it and the water level in the float chamber.

Where clear ice is made, a water pump usually circulates water, taken from a water tank or sump, through the evaporator, from which the water returns to the sump. A float valve frequently maintains a constant water level in the sump. In this case, the float valve may be either mounted directly in the sump or have a separate float chamber.

The float valve and its enclosure must meet sanitary code requirements so that no back siphonage can occur to contaminate the primary water supply. This is usually accomplished by ensuring that the valve-outlet orifice is at least 25 mm above the float chamber or any possible flood level within the machine proper. Also, if such a flood develops (for instance, because of a failure of the float valve to shut off), the exposed electrical equipment must not be shorted out by splash or free-running water.

Unit protection from water failure with air-cooled condensers is normally limited to a low-pressure cutout in the refrigerant circuit. Not all units use water pumps. In some machines, a stirring device keeps water moving. Compressed air has also been used to agitate water adjacent to the ice.

The trend in ice makers is to have a minimum quantity of water in the water system at any one time. If the unit is of the continuous-process type and is making clear or fairly clear ice, a constant bleed-off must be used to control the mineral concentration. With this type of operation, no relationship exists between quantity of bleedoff and total system water quantity. In batch processes, where clear ice is made, either a constant bleedoff or batch dumping a certain amount of water at the end of each cycle is used. As stated earlier, keeping the water quantity to a minimum maintains a good flushing action at the end of each cycle. This dumping can be obtained by a siphon effect, initiated by stopping the water pump and allowing all the system water to flow back into the water sump, thus increasing the water level to the point where the siphon starts. Alternatively, it may be simply a matter of energizing a solenoid valve in the water sump drain line, allowing any given amount of water to escape during the harvest cycle.

In all cases, drainage must be provided at the lowest point in the water system so the unit can be drained completely after any routine cleaning operation, before shipping, and when preparing for winter storage. All materials that come into contact with the ingredient water or the ice must be nontoxic, noncorrosive, smooth, impervious, nonodor imparting, cleanable, and durable. Minimum requirements for sanitary design are specified in NSF *Standard* 12; which is almost mandatory.

Ice Removal and/or Harvesting Systems. Usually an auger removes the ice from flaked-ice continuous-process machines. The design of most of these machines is based on breaking the ice away from the evaporator at the interface between the ice and the metal freezing surface. However, this bond is not consistent in shear strength; this, as well as the irregular pattern of the ice leaving the freezing surface, causes a highly irregular torque requirement on the auger drive shaft. As a result, side thrusts on the auger vary considerably. As designed, cutting or breaking the bond is balanced equally between cutters approximately 180° apart. When ice breaks loose irregularly, this balancing effect is lost. Side thrusts become of major proportion; thus, if close clearances are to be maintained between cutter and evaporator, the evaporator shell and auger must be quite sturdy.

The strength of the primary bond or fracture plane between evaporator and ice is influenced by many variables. Strength can vary with the temperature of the ice at the bonding plane, the water composition, the conditions of the evaporator surface, and the angle and shape of the cutting blade. These conditions can change from day to day or season to season. Thus, the ice-harvesting design becomes crucial, and this factor alone has been the downfall of many promising evaporator designs.

Most ice-cube machines are batch-process units that use hot gas to melt the bond between the ice and the evaporator. If the ice is made in cups, the adhesion or surface tension can create a vacuum and prevent the ice from falling. Either vacuum-breaking holes or some means of exerting pressure on the ice may be required to release the ice. When ice forms on ice rods, it must reach a critical mass to overcome the adhesive forces at the interface between evaporator and ice so it will drop off. Ice is abrasive, and this fact must be considered on all ice-handling components.

Unit Electrical Controls. The control systems for continuous-process machines are relatively simple, requiring only that the unit be shut down manually or automatically when an ice bin has been filled. To protect the machine, controls must shut the unit down whenever the water supply fails, an auger overloads its drive motor, excessively high discharge pressures are experienced, or abnormally low suction pressures are encountered.

A batch ice maker, which goes through a freezing and harvesting cycle, requires a more complicated control system than one that freezes continuously. A batch cycle must terminate the freezing at the proper time to start the thawing cycle and to resume the freezing operation when harvesting is complete. Timers, pressure-operated switches, ambient-compensated thermostats, water-overflow actuators, two-element thermostats, ice thickness feelers, and combinations of these have been used. Selection depends on what is most suitable for the particular evaporator design. Additional control is needed to shut off the unit when the storage bin is full of ice; a temperature-sensing element, which can be contacted by the ice, is often used. Mechanical feelers and photoelectric devices have also been applied successfully.

All controls must be effective over a range of ambient air temperatures from 4 to 43°C and supply water temperatures from 4 to 38°C.

Numerous safety devices help prevent damage to the apparatus or injury to personnel. These include (1) high-pressure cutout, (2) low-pressure cutout, (3) motor overload protector, (4) overfreeze protection, (5) fusible plugs in receiver shells, (6) safety switches interlocking with access panels, and (7) other devices that a particular design may require.

Ice Storage and Ice-Dispensing Systems. Most automatic ice makers provide an ice storage bin designed to hold approximately a 10- to 12-hour ice production. A certain demand does exist for units of a medium-to-large capacity of both flake ice and ice cubes with no integral bins. These units are ideal for applications where high demands occur less frequently than every 12 hours. For example, country clubs have high weekend demands. Also, supermarkets often use ice beds to display vegetables, meats, or fish; these beds are remade only once or twice a week. Often, a standard bin is made by the ice maker manufacturer and holds the usual 12-hour production of ice. Bins made by independent ice bin manufacturers are better for larger quantities of ice for special applications.

Ice bins are usually built to NSF construction standards, with from 40 to 75 mm of insulation on the sides and top and from 50 to 100 mm of insulation on the bottom. These bins are seldom refrigerated because the ice may refreeze together, therefore drainage must be provided. Drains should never be less than 12 mm and preferably should be 20 mm in internal diameter. A strainer should be included at the drain inlet. Access to the ice can be achieved at the top, bottom, or side of the storage bin.

Ice is quite impervious to outside attack. However, as air near the ice cools, moisture and any contaminants present in the air are condensed onto the ice surface. Besides unsanitary areas, the greatest source of contamination lies in scoops, shovels, or other instruments introduced into the ice and then withdrawn and left in other unsanitary areas.

Many sanitary dispensers are available, but the first cost has held sales down, except in those applications where sanitation is crucial. Such an ice dispenser (1) must protect the ice from outside contamination while it is in storage, (2) must be readily cleanable so that frequent cleaning can be done by regular housekeeping help, and (3) should cost less than the ice machine. Many current designs meet these objectives, and sanitary ice dispensers are an increasing part of the standard product line for major manufacturers. A need for storing and dispensing uncontaminated ice continues to be of major interest to hospitals, motels, restaurants, and other operations where ice may come in direct contact with food or beverages.

Safety and Sanitary Standards. The following standards apply to ice-making equipment.

- NSF *Standard* 12, Automatic Ice Making Equipment.
- Underwriters Laboratories (UL) *Standard* 563, Ice Makers, covers electrical safety.
- Canadian Standards Association (CAN/CSA) *Standard* C22.2 No. 120 covers electrical safety of ice maker products for the Canadian market.
- Air-Conditioning and Refrigeration Institute (ARI) *Standard* 810, Automatic Commercial Ice Makers, details the conditions and methods of rating. These ratings are based on ASHRAE *Standard* 29, Method of Testing Automatic Ice Makers.
- Other standards, such as the ASME *Boiler and Pressure Vessel Code*, may apply, as well as individual state and city plumbing and sanitary codes.

Special Concerns. Noise and vibration, while of little concern in a busy kitchen, should be considered for equipment installed in a dining room or quiet hospital area. A noise level equal to that of a window air conditioner is acceptable for most applications, although some specifications may quote specific NC sound power levels.

Condensation or sweating on the bin or ice-dispensing part of the equipment may be a concern. Generally, equipment is satisfactory for most applications if no moisture drips from the unit when it operates for 4 h in an ambient of 32°C dry bulb, 25.5°C wet bulb.

Installation, operation, and owner maintenance determines the eventual life of the unit, no matter how well it is designed and built. Therefore, the designer should make the unit as simple and easy to clean as possible. In addition, operating and maintenance instructions that are clear and positive will help the owner extend the life of the machine.

Special Applications and Specifications. Several special applications require additional nonstandard design features. Shipboard application usually requires that units continue to produce ice when they are subjected to a 15° pitch and a 30° roll; for naval vessels, however, 15° permanent list and high shock are additional requirements. Also, a no-radio interference requirement is often added.

Explosion-proof units are sometimes required, but this is a special feature that normally is not a production item. Usually, odd electrical current characteristics can be handled by transformers, since direct current is now rarely encountered. For the international market, many manufacturers provide specific models that meet specific local voltage and frequency requirements. Occasionally, measures are required to prevent fungus and high humidity.

APPLICATION

Many ice maker designs, capacities, and ice shapes are available. The primary factors in selecting the best equipment for any specific application include the following considerations:

1. What type (or types) of ice best satisfies the application?
2. What storage or dispensing methods will be required?
3. Is the cost of ice of primary importance?
4. Is the cost of electricity and/or water of primary importance?
5. What is the quality of the available water supply?
6. Are there noise-level restrictions that must be met?
7. Are reliability and long life of primary importance?
8. Are service and maintenance facilities readily available?
9. What are the space limitations (if any) for the installation?
10. Are there any general environmental conditions or restrictions that must be met?

Choice of Ice Types

Differentiating among the types of ice, two general physical characteristics are apparent: (1) ice can be clear, white, or have varying degrees of opacity; and (2) the individual ice pieces can come in a myriad of sizes and shapes. Generally, the shape description is cubes, flakes, crushed, or aggregate. This last designation is not generally used, but it is meant to cover the more or less uniform ice that results from thin ice flakes that have been compressed and extruded and then broken again as the extrusion is forced against a stop of some kind. (Trade designations are **pebble ice**, **granular ice**, **nugget ice**, or **cracked ice**.) This ice differs from crushed ice in that it has relatively few fines and is neither entirely clear nor entirely white, but has various degrees of cloudiness. Also, ice that is stored for a few hours may vary greatly from ice just coming out of the ice maker discharge chute. The stored ice, unless cubed, is said to cure. In addition to having been drained of most of its excess water, it has changed its structure so that both its density and its thermal capacity (kJ/kg) have changed. This change can be important when figuring bin capacities.

Sometimes, maximum ice size is important. For instance, it must be small enough to readily enter the opening of a patient's carafe for hospital use. It must not have so many fines that it clogs straws or glass drinking tubes.

Shape can also be important; for instance, ice flakes are suited for packing flowers because they do not bruise the petals. This shape is also ideal for fish because it tends to conform to the surface on which it lies. On the other hand, an egg shape has long-lasting qualities because it presents a minimum of outside surface to the liquid to be cooled. Also, various forms of ice react differently in bins. Some do not pack well, and some do not lend themselves to deep bins. Ice cubes generally flow more easily.

Probably the most important consideration for the selection of one ice form over another is user preference. In many applications, one type of ice is clearly superior to others, but, frequently, any form will do the job. When ice is to be delivered to specification, however, all ice-making capacities must be quoted from ASHRAE *Standard* 29, which converts all product mass to 334 kJ/kg ice per 24 h at stated ambient and water conditions. Also, ice bin capacities generally are understood to mean kilograms of cured ice and not 334 kJ/kg ice, unless the latter is specifically mentioned. Normally, these are not significant points, but they become important if strict specifications are involved or actual cooling ability is guaranteed.

In the beverage industry, the ice must not impart any flavor to the beverage. If the local water contains gases or solids that impart a distinctive odor or taste, a flake-ice machine may not be satisfactory, whereas a cube ice maker would eliminate most of the objectionable flavor. On the other hand, proper water pretreatment may make flake ice acceptable.

Ice Storage

Most automatic ice makers not only produce ice, they also store it and provide some means of dispensing it. Storage is expected to be an insulated container with 50 to 75 mm of insulation designed for periodic cleaning or sanitizing. (Ideally, it will meet NSF International standards). Additionally, it must be designed so that ice can be discharged from the bin with a minimal hazard of withdrawal affecting the sanitary conditions of the remaining ice. (This ideal prevents the use of scoops or shovels that touch the ice remaining in storage.)

This design also precludes any openings into the ice-containing area in which packages or bottles can be left to cool and possibly contaminate the ice in storage. Whether or not the equipment is maintained in a sanitary condition is entirely up to the owner. The equipment supplier, however, must ensure that the equipment is capable of being maintained in a sanitary condition when straightforward cleaning instructions supplied by the manufacturer are followed.

Ice-storage capacities are commonly stated in terms of how much ice, or product, can be accommodated when the bin is full. Since ice has a natural angle of repose, the machine generally turns off before reaching the full bin capacity. This factor should be considered when selecting equipment. In any application, ice bin and machine production capacities must balance to fit the ice-demand cycle. Equipment must also be selected for maximum demand, and machine capacities at lowest production rates must be compatible with the installation environment.

Ice Costs

Ice is a relatively inexpensive commodity. If ice is used in small quantities, the cost of making it is secondary to the space the ice maker occupies, the convenience of keeping the ice as close as possible to the point of use, or the original first cost. This relative importance must be kept in mind when choosing the ice form or size and when considering reliability. Thus, two small machines are safer than one large one, and the machine that makes ice for the least amount of money is the logical choice.

Equipment Life

Unit life depends on installation, environment, and regular cleaning and maintenance. Unless these factors are considered at the time of installation, ice-making equipment must be classified as relatively short-lived. Where ice is added to food or beverages, the regular cleaning and inspection required for sanitary purposes

Table 1 Ice Maker Application Data[a]

Classification	Application	Preferred Form of Ice	Use Cycle[b]	Consumption[c]
Food and Drink Service				
Fast-food drive-in	Water, soft drinks	Crushed/flake	7 days; 150% 2 consec. days	0.1 to 0.2 kg/customer
Night clubs	Food-drinks	Cube	7 days; 200% 2 consec. days	0.9 to 1.8 kg/seat
Bar cocktail lounges	Drinks			
Carry home	Retail stores			
	Recreation areas	Cube/crushed	7 days; 200% 3 consec. days	Varies by season, available in 2.3 and 4.6 kg size bags
	Marinas, fishing piers			
Caterers	Banquets	All	Varies	0.46 kg per meal
	Truck (mobile)	Flaked/crushed	7 days; 140% 5 consec. days	90 to 180 kg/truck
	In-plant feeding	All	7 days; 140% 5 consec. days	0.23 to 0.46 kg/meal
Auditoriums, stadiums		Crushed/flake	Varies	0.23 kg/customer
Churches	Banquets	Cube	7 days; 200% 2 consec. days	0.23 kg/meal
Theaters	Snack bar	Crushed/flake	7 days; 200% 2 consec. days	0.23 kg/snack bar patron
Hotel-motel, resorts	Banquet, meeting room	Cube	7 days; 200% 2 consec. days	0.46 kg/meal
	Dining room	Cube/flake	7 days; 200% 2 consec. days	0.46 kg/meal
	Coffee shop	Cube/flake	Daily, 100%	0.46 kg/meal
	Guest room service	Cube	Daily, 100%	2.3 kg/room
	Kitchen	Flake/crushed	Daily, 100%	0.23 kg/meal
	Buffets-display	Flake/crushed	Varies Daily, 100%	49 kg/m^2 of display area
Restaurants, cafeterias	Dining room	All	7 days; 150% 2 consec. days	0.23 kg/meal
	Serving line display	Flake/crushed	7 days; 100%	49 kg/m^2 of display
	Kitchen	Flake/crushed	7 days; 100%	0.23 kg/meal
Military bases	Mess halls	Flake/crushed	7 days; 100%	0.23 kg/meal
	Post exchange clubs	Cube/crushed	Daily, 100%	0.23 kg/customer
Prisons	Dining halls	Flake/crushed	Daily, 100%	0.33 lb/meal
Airlines, airline caterers	In-flight feeding	Cube	7 days; 150% 2 consec. days	1.0 lb/meal
Clubs, country and private	Banquet, meeting room	Cube	7 days; 200% 2 consec. days	0.23 kg/meal
	Dining room	Cube	7 days; 200% 2 consec. days	0.23 kg/meal
	Bar, cocktail lounge	Cube	7 days; 200% 2 consec. days	0.23 to 0.46 kg/customer
	Kitchen	Crushed/flake	1 day; 100%	0.23 kg/meal
	Golf course	All	1 day; 100%	0.1 to 0.5 kg per litre of water cooler capacity
Food Preservation				
Seafood markets	Merchandising (display)	Crushed/flake	7 days; 140% 5 consec. days	24.5 kg/m^2 of display
Florists	Shipping	Flakes	Varies	0.9 kg/box of flowers
Groceries	Display	Crushed/flake	Daily	25 kg/m^2 of display
Medical				
Hospitals	Dietary	Crushed/flake	Daily, 100%	0.23 kg/meal
	Nursing service	Crushed/flake	Daily, 100%	2.3 kg/bed
Nursing homes	Dietary	Crushed/flake	Daily, 100%	0.15 kg/meal
	Nursing service	Crushed/flake	Daily, 100%	1.4 kg/bed
Athletic fields	Ice packs and drinking water	Crushed/flake	Daily, 100%	1.4 to 2.3 kg/person per day
Construction	Drinking water coolers	Crushed/cubes	Daily, 100%	1.4 to 2.3 kg/person per shift

[a] Additional application information other than food, beverages, or display counters may be found under general refrigeration requirements by industry or product.
[b] These values assist in balancing bin capacity and ice-making capability to obtain the most economical combinations. The length of the normal use cycle is shown first. The next figure is an approximation of how much ice might be consumed during a peak period in the use cycle; it is expressed as a percentage of the average daily production. The last figure indicates how long the high consumption might last.
[c] Ice consumption is generally cyclic. Proper storage capacity helps ensure an adequate supply of ice during high-consumption periods. The values shown are average consumption for the indicated applications.

should promote longer life. The necessity for regular cleaning, not only of the ice-storage facilities but also of the ice-making parts of the ice maker, must be impressed on the user. Neglected equipment in areas of poor makeup water may have a life as short as three years, whereas if water treatment is used and the equipment is regularly cleaned and maintained, it could extend to a ten-year life.

Installation

Proper installation is the foremost requirement of economical, maintenance-free operation. Manufacturer's recommendations should be followed closely.

Often, the necessary drain and its pitch are overlooked when a location is chosen. The drain must take care of blowdown, ice meltage, and in the case of water-cooled condensers, waste water. Bin drains should be insulated to prevent moisture from condensing on their cold exteriors; this condensation is often mistaken for a leaking bin.

Bad water conditions can create innumerable problems for ice-making equipment. When melted, the ice may leave a scum on water in a glass, or it may have an objectionable taste. Bad water also may have deleterious effects on the equipment. When the local water conditions are known to cause problems, water treatment must be considered at the outset.

Water treatment firms in the vicinity of the installation are the best source of water treatment recommendations. In general, demineralization offers the best overall water conditioning preparation for ice-making equipment. Softening by ion exchange (sodium cycle zeolite) eliminates most scaling problems, but when the dissolved solids content of makeup water exceeds 400 mg/kg, cloudy mushy ice is apt to result. Treating makeup water with polyphosphates only reduces the scaling tendencies, but this lengthens equipment life.

Ambient conditions have an important influence on performance and maintenance requirements. By avoiding locations where temperatures are extreme—either high or low—better performance may be ensured. Also, ease of accessibility for cleaning and maintenance helps prolong life. Most ice makers create some noise, so it is important to locate them where noise is not objectionable.

Table 1 lists ice maker application data and is a digest of the experiences of many individuals and firms recommending and installing ice-making equipment in diverse areas of the country under various local conditions. Therefore, the figures are averages, and on any particular application they must be modified up or down according to local conditions.

Environmental Considerations

Many areas experience severe shortages of electric power and supply water, so it is important for ice-making systems to operate with maximum efficiency. Some measures of operating efficiency are as follows:

1. Kilograms of ice per cubic metre of supply water
2. Kilograms of ice per megajoule (or kWh) of power consumed by the ice-making system
3. Kilograms of ice per cubic metre of condensing water
4. Kilograms of ice per square metre of floor space

BIBLIOGRAPHY

ARI. 1995. Automatic commercial ice makers. *Standard* 810-1995. Air-Conditioning and Refrigeration Institute, Arlington, VA.

ARI. Annual. Directory of certified automatic commercial ice cube machines and ice storage bins.

ASHRAE. 1988. Method of testing automatic ice makers. ASHRAE *Standard* 29.

CSA. 1991. Refrigeration equipment. CAN/CSA *Standard* C22.2 No. 120-M91.

NSF. 1992. Automatic ice making equipment. NSF *Standard* 12-1992. NSF International, Ann Arbor, MI.

Proctor, W.T. 1977. Ice vending—A growing market. *American Automatic Merchandisers* (May):36.

UL. 1995. UL Standard for safety, Ice makers, seventh ed. UL *Standard* 563-1995. Underwriters Laboratories Inc., Northbrook, IL.

CHAPTER 51

CODES AND STANDARDS

THE Codes and Standards listed in Table 1 represent practices, methods, or standards published by the organizations indicated. They are valuable guides for the practicing engineer in determining test methods, ratings, performance requirements, and limits applying to the equipment used in heating, refrigerating, ventilating, and air conditioning. *Copies of the publications can usually be obtained from the organizations listed in the Publisher column.* Addresses of the organizations are given at the end of the chapter.

Table 1 Codes and Standards Published by Various Societies and Associations

Subject	Title	Publisher	Reference
Air Conditioners	Installation Techniques for Perimeter Heating and Cooling, 11th ed.	ACCA	ACCA Manual 4
	Methods of Testing for Rating Ducted Air Terminal Units	ASHRAE	ANSI/ASHRAE 130-1996
	Ducted Air-Conditioners and Air-to-Air Heat Pumps—Testing and Rating for Performance	ISO	ISO 13253:1995
	Non-Ducted Air Conditioners and Heat Pumps—Testing and Rating for Performance	ISO	ISO 5151:1994
	Heating and Cooling Equipment (1995)	UL	UL 1995
		CSA	CAN/CSA-C22.2 No. 236-95
Central	Performance Standard for Split-System Central Air-Conditioners and Heat Pumps	CSA	CAN/CSA-C273.3-M91
	Performance Standard for Single Package Central Air-Conditioners and Heat Pumps	CSA	CAN/CSA-C656-M92
	Performance Standard for Rating Large Air Conditioners and Heat Pumps	CSA	CAN/CSA-C746-93
	Air Conditioners, Central Cooling (1982)	UL	ANSI/UL 465-1984
Gas-Fired	Gas-Fired Absorption Summer Air Conditioning Appliances (with 1982 addenda)	AGA	ANSI Z21.40.1-1994
	Requirements for Gas-Fired, Engine-Driven Air Conditioning Appliances	AGA	4-89
	Requirements for Gas-Fired Desiccant Type Dehumidifiers and Air Conditioners	AGA	9-90
Packaged Terminal	Packaged Terminal Air Conditioners	ARI	ARI 310-93
	Packaged Terminal Heat Pumps	ARI	ANSI/ARI 380-90
	Standards for Packaged Terminal Air-Conditioners and Heat Pumps	CSA	C744-93
Room	Room Air Conditioners	AHAM	ANSI/AHAM RA C-1-1992
	Method of Testing for Rating Room Air Conditioners and Packaged Terminal Air Conditioners	ASHRAE	ANSI/ASHRAE 16-1983 (RA 88)
	Method of Testing for Rating Room Air Conditioner and Packaged Terminal Air Conditioner Heating Capacity	ASHRAE	ANSI/ASHRAE 58-1986 (RA 90)
	Methods of Testing for Rating Room Fan-Coil Air Conditioners	ASHRAE	ANSI/ASHRAE 79-1984 (RA 91)
	Performance Standard for Room Air Conditioners	CSA	CAN/CSA-C368.1-M90
	Room Air Conditioners	CSA	C22.2 No. 117-1970 (R 1992)
	Room Air Conditioners (1993)	UL	ANSI/UL 484
Unitary	Commercial Load Calculation, 4th ed. (1988)	ACCA	ACCA Manual N
	Application of Sound Rated Outdoor Unitary Equipment	ARI	ARI 275-84
	Commercial and Industrial Unitary Air-Conditioning Equipment	ARI	ANSI/ARI 360-93
	Sound Rating of Outdoor Unitary Equipment	ARI	ARI 270-84
	Unitary Air-Conditioning and Air-Source Heat Pump Equipment	ARI	ANSI/ARI 210/240-89
	Method of Rating Computer and Data Processing Room Unitary Air Conditioners	ASHRAE	ANSI/ASHRAE 127-1988
	Method of Rating Unitary Spot Air Conditioners	ASHRAE	ANSI/ASHRAE 128-1989
	Methods of Testing for Rating Heat Operated Unitary Air-Conditioning Equipment for Cooling	ASHRAE	ANSI/ASHRAE 40-1980 (RA 92)
	Methods of Testing for Rating Unitary Air-Conditioning and Heat Pump Equipment	ASHRAE	ANSI/ASHRAE 37-1988
	Methods of Testing for Seasonal Efficiency of Unitary Air Conditioners and Heat Pumps	ASHRAE	ANSI/ASHRAE 116-1995
Air Conditioning	Commercial Low Pressure, Low Velocity Duct System Design (1990)	ACCA	Manual Q
	Residential Duct Systems (1995)	ACCA	ACCA Manual D-R
	Residential Load Calculation, 7th ed. (1986)	ACCA	ACCA Manual J
	Environmental System Technology (1984)	NEBB	NEBB
	Installation of Air Conditioning and Ventilating Systems	NFPA	ANSI/NFPA 90A-1993
	Standard of Purity for Use in Mobile Air-Conditioning Systems	SAE	SAE J 1991-1989
	HVAC Systems—Applications, 1st ed. (1986)	SMACNA	SMACNA
	HVAC Systems—Duct Design (1990)	SMACNA	SMACNA
	Installation Standards for Residential Heating and Air Conditioning Systems (1988)	SMACNA	SMACNA
	Heating and Cooling Equipment (1995)	UL	UL 1995
		CSA	CAN/CSA-22.2 No. 36-95
Aircraft	Air Conditioning of Aircraft Cargo	SAE	SAE AIR 806A-1978 (R 1992)
	Air Conditioning of Subsonic Aircraft at High Altitudes	SAE	SAE AIR 795

51.1

Table 1 Codes and Standards Published by Various Societies and Associations (*Continued*)

Subject	Title	Publisher	Reference
Aircraft (*continued*)	Air Conditioning Systems for Subsonic Airplanes	SAE	ANSI/SAE ARP 85E-1991
	Aircraft Fuel Weight Penalty Due to Air Conditioning	SAE	SAE AIR 1168/8-1989
	Aircraft Ground Air Conditioning Service Connection	SAE	SAE AS 4262
	Air Cycle Air Conditioning Systems for Military Air Vehicles	SAE	ANSI/SAE ARP 4073-1993
	Control of Excess Humidity in Avionics Cooling	SAE	SAE ARP 987-1970 (R 1992)
	Engine Bleed Air Systems for Aircraft	SAE	ANSI/SAE ARP 1796-1987
	Guide for Qualification Testing of Aircraft Air Valves	SAE	ANSI/SAE ARP 986B-1980
	Nomenclature, Aircraft Air-Conditioning Equipment	SAE	SAE ARP 147C-1978 (R 1992)
	Testing of Commercial Airplane Environmental Control Systems	SAE	SAE ARP 217B-1973 (R 1992)
Automotive	Automotive Air-Conditioning Hose	SAE	ANSI/SAE J 51-1989
	Design Guidelines for Air Conditioning Systems for Off-Road Operator Enclosures	SAE	SAE J 169-1985
	Extraction and Recycle Equipment for Mobile Automotive Air-Conditioning Systems	SAE	SAE J 1990-1992
	Guide to the Application and Use of Passenger Car Air-Conditioning Compressor Face Seals	SAE	SAE J 1954-1990
	Information Relating to Duty Cycles and Average Power Requirements of Truck and Bus Engine Accessories	SAE	SAE J 1343-1988
	Rating Air Conditioner Evaporator Air Delivery and Cooling Capacities	SAE	ANSI/SAE J 1487-1985
	Service Hose for Automotive Air Conditioning	SAE	SAE J 2196-1992
	Test Method for Measuring Power Consumption of Air Conditioning and Brake Compressors for Trucks and Buses	SAE	ANSI/SAE J 1340-1990
Ships	Mechanical Refrigeration and Air-Conditioning Installations Aboard Ship	ASHRAE	ASHRAE 26-1996
Air Curtains	Air Curtains for Entranceways in Food and Food Service Establishments	NSF	ANSI/NSF-37-1992
	Air Distribution Basics for Residential and Small Commercial Buildings (1989)	ACCA	ACCA Manual T
	Residential Equipment Selection, 2nd ed. (1995)	ACCA	ACCA Manual S
	Test Methods for Air Curtain Units	AMCA	AMCA 220-91
	Air Terminals	ARI	ANSI/ARI 880-94
	Method of Testing for Rating the Performance of Air Outlets and Inlets	ASHRAE	ANSI/ASHRAE 70-1991
	Standard Methods for Laboratory Air Flow Measurement	ASHRAE	ANSI/ASHRAE 41.2-1987 (RA 92)
	Rating the Performance of Residential Mechanical Ventilating Equipment	CSA	CAN/CSA-C260-M90
	Direct Gas-Fired Door Heaters	AGA	ANSI Z83.17-1990; Z83.17a-1991; Z83.17b-1992
Air Diffusion	Test Code for Grilles, Registers and Diffusers	ADC	ADC 1062:GRD-84
	Method of Testing for Rating the Performance of Air Outlets and Inlets	ASHRAE	ANSI/ASHRAE 70-1991
	Method of Testing for Room Air Diffusion	ASHRAE	ANSI/ASHRAE 113-1990
Air Filters	Method for Measuring Performance of Portable Household Electrical Cord Connected Room Air Cleaners	AHAM	ANSI/AHAM AC-1-1988
	Commercial and Industrial Air Filter Equipment	ARI	ARI 850-93
	Residential Air Filter Equipment	ARI	ARI 680-93
	Gravimetric and Dust-Spot Procedures for Testing Air-Cleaning Devices Used in General Ventilation for Removing Particulate Matter	ASHRAE	ANSI/ASHRAE 52.1-1992
	Method for Sodium Flame Test for Air Filters	BSI	BS 3928
	Particulate Air Filters for General Ventilation—Requirements, Testing Marking	BSI	BS EN 779:1993
	Electrostatic Air Cleaners (1995)	UL	UL 867
	High-Efficiency, Particulate, Air Filter Units (1990)	UL	ANSI/UL 586-1990
	Test Performance of Air Filter Units (1994)	UL	UL 900
Air-Handling Units	Commercial Low Pressure, Low Velocity Duct System Design (1990)	ACCA	ACCA Manual Q
	Residential Duct Systems (1995)	ACCA	ACCA Manual D-R
	Central Station Air-Handling Units	ARI	ANSI/ARI 430-89
	Direct Gas-Fired Make-Up Air Heaters	AGA	ANSI Z83.4-1991; Z83.4a-1992
Air Leakage	Air Leakage Performance for Detached Single-Family Residential Buildings	ASHRAE	ANSI/ASHRAE 119-1988 (RA 94)
	A Method of Determining Air Change Rates in Detached Dwellings	ASHRAE	ANSI/ASHRAE 136-1993
	Standard Practices for Air Leakage Site Detection in Building Envelopes	ASTM	ASTM E 1186-87 (R 1992)
	Test Method for Determining Air Change in a Single Zone by Means of a Tracer Gas Dilution	ASTM	ASTM E 741-95
	Test Method for Determining Air Leakage Rate by Fan Pressurization	ASTM	ASTM E 779-87 (R 1992)
	Test Method for Determining the Rate of Air Leakage Through Exterior Windows, Curtain Walls, and Doors Under Specified Pressure and Temperature Differences Across the Specimen	ASTM	ASTM E 1424-91
	Test Method for Determining the Rate of Air Leakage Through Exterior Windows, Curtain Walls, and Doors Under Specified Pressure Differences Across the Specimen	ASTM	ASTM E 283-91
	Test Method for Field Measurement of Air Leakage Through Installed Exterior Window and Doors	ASTM	ASTM E 783-93
Boilers	A Guide to Clean and Efficient Operation of Coal Stoker-Fired Boilers	ABMA	ABMA
	Boiler Water Limits and Steam Purity Recommendations for Watertube Boilers	ABMA	ABMA

Codes and Standards

Table 1 Codes and Standards Published by Various Societies and Associations (*Continued*)

Subject	Title	Publisher	Reference
Boilers *(continued)*	Boiler Water Requirements and Associated Steam Purity—Commercial Boilers	ABMA	ABMA
	Fluidized Bed Combustion Guidelines	ABMA	ABMA
	Guidelines for Industrial Boiler Performance Improvement	ABMA	ABMA
	Matrix of Recommended Quality Control Requirements	ABMA	ABMA
	Operation and Maintenance Safety Manual	ABMA	ABMA
	Recommended Design Guidelines for Stoker Firing of Bituminous Coals	ABMA	ABMA
	(Selected) Summary of Codes and Standards of the Boiler Industry	ABMA	ABMA
	Thermal Shock Damage to Hot Water Boilers as a Result of Energy Conservation Measures	ABMA	ABMA
	Commercial Applications, Systems and Equipment (1993)	ACCA	ACCA Manual CS
	Methods of Testing for Annual Fuel Utilization Efficiency of Residential Central Furnaces and Boilers	ASHRAE	ANSI/ASHRAE 103-1993
	Boiler and Pressure Vessel Code (11 sections) (1995)	ASME	ASME
	Boiler, Pressure Vessel, and Pressure Piping Code	CSA	CSA B51-95
	Testing and Rating Standard for Heating Boilers (1989)	HYDI	IBR
	Prevention of Furnace Explosions/Implosions in Multiple Burner Boilers	NFPA	ANSI /NFPA 8502-1995
	Heating, Water Supply, and Power Boilers—Electric (1995)	UL	UL 834
Gas or Oil	Gas-Fired Low-Pressure Steam and Hot Water Boilers	AGA	ANSI Z21.13-1991; Z21.13a-1993; Z21.13b-1994
	Gas Utilization Equipment in Large Boilers	AGA	ANSI Z83.3-1971; Z83.3a-1972; Z83.3b-1976 (R 1989)
	Requirements for High Pressure Steam Boilers	AGA	AGA 3-89
	Control and Safety Devices for Automatically Fired Boilers	ASME	ANSI/ASME CSD.1-1992
	Industrial and Commercial Gas-Fired Package Boilers	CGA	CAN1-3.1-77 (R 1985)
	Oil-Fired Steam and Hot-Water Boilers for Residential Use	CSA	B140.7.1-1976 (R 1991)
	Oil-Fired Steam and Hot-Water Boilers for Commercial and Industrial Use	CSA	B140.7.2-1967 (R 1991)
	Prevention of Furnace Explosions/Implosions in Multiple Burner Boilers	NFPA	ANSI/NFPA 8502-1995
	Single Burner Boiler Operations	NFPA	ANSI/NFPA 8501-1992
	Commercial-Industrial Gas Heating Equipment (1994)	UL	UL 795
	Oil-Fired Boiler Assemblies (1990)	UL	ANSI/UL 726-1990
	Standards and Typical Specifications for Deaerators, 5th ed. (1992)	HEI	HEI
	Method and Procedure for the Determination of Dissolved Oxygen, 2nd ed. (1963)	HEI	HEI
Building Codes	ASTM Standards Used in Building Codes	ASTM	ASTM
	BOCA National Building Code, 12th ed. (1993)	BOCA	BOCA
	BOCA National Property Maintenance Code, 4th ed. (1993)	BOCA	BOCA
	National Building Code of Canada (1995)	NRCC	NRCC
	One- and Two-Family Dwelling Code (1995)	CABO	CABO
	Model Energy Code (1995)	CABO	BOCA/ICBO/SBCCI
	Uniform Building Code (1994)	ICBO	ICBO
	Uniform Building Code Standards (1994)	ICBO	ICBO
	Directory of Building Codes and Regulations, State and City Volumes (1996 ed.)	NCSBCS	NCSBCS
	Standard Building Code (1994)	SBCCI	SBCCI
Mechanical	Safety Code for Elevators and Escalators (plus two yearly supplements)	ASME	ANSI/ASME A 17.1-1993
	Natural Gas Installation Code	CGA	CAN/CGA-B149.1-M95
	Propane Installation Code	CGA	CAN/CGA-B149.2-M91
	Safety Code for Elevators	CSA	CAN/CSA-B44-94
	BOCA National Mechanical Code, 8th ed. (1993)	BOCA	BOCA
	Uniform Mechanical Code (1994) (with Uniform Mechanical Code Standards)	ICBO	ICBO
		IAPMO	IAPMO
	Standard Gas Code (1994)	SBCCI	SBCCI
	Standard Mechanical Code (1994)	SBCCI	SBCCI
Burners	Guidelines for Burner Adjustments of Commercial Oil-Fired Boilers	ABMA	ABMA
	Domestic Gas Conversion Burners	AGA	ANSI Z21.17-1991; Z21.17a-1993; Z21.17b-1994
	Installation of Domestic Gas Conversion Burners	AGA	ANSI Z21.8-1994
	General Requirements for Oil Burning Equipment	CSA	CAN/CSA-B140.0-M87 (R 1991)
	Installation Code for Oil Burning Equipment	CSA	CAN/CSA-B139-M91
	Oil Burners; Atomizing-Type	CSA	CAN/CSA-B140.2.1-M90
	Pressure Atomizing Oil Burner Nozzles	CSA	B140.2.2-1971 (R 1991)
	Replacement Burners and Replacement Combustion Heads for Residential Oil Burners	CSA	B140.2.3-M1981 (R 1991)
	Vapourizing-Type Oil Burners	CSA	B140.1-1966 (R 1991)
	Commercial/Industrial Gas and/or Oil-Burning Assemblies with Emission Reduction Equipment (1993)	UL	UL 2096
	Commercial-Industrial Gas Heating Equipment (1994)	UL	UL 795
	Oil Burners (1994)	UL	ANSI/UL 296-1995
Chillers	Methods of Testing Liquid Chilling Packages	ASHRAE	ASHRAE 30-1995
	Absorption Water-Chilling and Water Heating Packages	ARI	ARI 560-92
	Centrifugal and Rotary Screw Water-Chilling Packages	ARI	ANSI/ARI 550-92
	Positive Displacement Compressor Water-Chilling Packages	ARI	ANSI/ARI 590-92

Table 1 Codes and Standards Published by Various Societies and Associations (*Continued*)

Subject	Title	Publisher	Reference
Chillers (continued)	Performance Standard for Rating Packaged Water Chillers	CSA	C743-93
Chimneys	Design and Construction of Masonry Chimneys and Fireplaces	CSA	CAN/CSA-A405-M87
	Chimneys, Fireplaces, Vents, and Solid Fuel-Burning Appliances	NFPA	ANSI/NFPA 211-1992
	Chimneys, Factory-Built, Medium Heat Appliance (1994)	UL	UL 959
	Chimneys, Factory-Built, Residential Type and Building Heating Appliance (1994)	UL	ANSI/UL 103-1995
Clean Rooms	Procedural Standards for Certified Testing of Cleanrooms (1988)	NEBB	NEBB
	Standard Practice for Continuous Sizing and Counting of Airborne Particles in Dust-Controlled Areas and Clean Rooms Using Instruments Capable of Detecting Single Sub-Micrometre and Larger Particles	ASTM	ASTM F 50-92 (R 1996)
Coils	Forced-Circulation Air-Cooling and Air-Heating Coils	ARI	ANSI/ARI 410-91
	Methods of Testing Forced Circulation Air Cooling and Air Heating Coils	ASHRAE	ASHRAE 33-1978
Comfort Conditions	Thermal Environmental Conditions for Human Occupancy	ASHRAE	ANSI/ASHRAE 55-1992 with Addendum 55a-1994
	Ergonomics—Determination of Metabolic Heat Production	ISO	ISO 8996:1990
	Ergonomics of the Thermal Environment—Estimation of the Thermal Insulation and Evaporative Resistance of a Clothing Ensemble	ISO	ISO 9920:1995
	Hot Environments—Estimation of the Heat Stress on Working Man, Based on the WBGT Index (Wet Bulb Globe Temperature)	ISO	ISO 7243:1989
	Moderate Thermal Environments—Determination of the PMV and PPD Indices and Specification of the Conditions for Thermal Comfort	ISO	ISO 7730:1994
Compressors	Compressors and Exhausters (reaffirmed 1986)	ASME	ANSI/ASME PTC 10-1965 (R 1992)
	Displacement Compressors, Vacuum Pumps and Blowers	ASME	ANSI/ASME PTC 9-1974 (R 1992)
	Safety Standard for Air Compressor Systems	ASME	ANSI/ASME B19.1-1990
	Safety Standard for Compressors for Process Industries	ASME	ANSI/ASME B19.3-1991
	Compressed Air and Gas Handbook, 5th ed. (1988)	CAGI	CAGI
Refrigerant	Ammonia Compressor Units	ARI	ANSI/ARI 510-93
	Method for Presentation of Compressor Performance Data	ARI	ARI 540-91
	Positive Displacement Refrigerant Compressors, Compressor Units and Condensing Units	ARI	ANSI/ARI 520-90
	Methods of Testing for Rating Positive Displacement Refrigerant Compressors and Condensing Units	ASHRAE	ANSI/ASHRAE 23-1993
	Safety Code for Mechanical Refrigeration	ASHRAE	ANSI/ASHRAE 15-1994
	Testing of Refrigerant Compressors	ISO	ISO 917:1989
	Refrigerant Compressors—Presentation of Performance Data	ISO	ISO 9309:1989
	Hermetic Refrigerant Motor-Compressors (1996)	UL	UL 984
		CSA	CAN/CSA-C22.2 No.140.2-M91
Computers	Method of Rating Computer and Data Processing Room Unitary Air Conditioners	ASHRAE	ANSI/ASHRAE 127-1988
	Protection of Electronic Computer/Data Processing Equipment	NFPA	ANSI/NFPA 75-1995
Condensers	Commercial Applications, Systems and Equipment (1993) (for equipment selection only)	ACCA	ACCA Manual CS
	Remote Mechanical-Draft Air-Cooled Refrigerant Condensers	ARI	ARI 460-87
	Remote Mechanical Draft Evaporative Refrigerant Condensers	ARI	ANSI/ARI 490-89
	Water-Cooled Refrigerant Condensers, Remote Type	ARI	ARI 450-87
	Methods of Testing for Rating Remote Mechanical-Draft Air-Cooled Refrigerant Condensers	ASHRAE	ASHRAE 20-1970
	Methods of Testing Remote Mechanical-Draft Evaporative Refrigerant Condensers	ASHRAE	ANSI/ASHRAE 64-1995
	Methods of Testing for Rating Water-Cooled Refrigerant Condensers	ASHRAE	ANSI/ASHRAE 22-1992
	Safety Code for Mechanical Refrigeration	ASHRAE	ANSI/ASHRAE 15-1994
	Steam Condensing Apparatus	ASME	ANSI/ASME PTC 12.2-1983 (R 1988)
	Standards for Steam Surface Condensers, 9th ed. (1995)	HEI	HEI
	Standards for Direct Contact Barometric and Low Level Condensers, 6th ed. (1995)	HEI	HEI
Condensing Units	Commercial Applications, Systems and Equipment (1993)	ACCA	ACCA Manual CS
	Residential Equipment Selection, 2nd ed. (1995)	ACCA	ACCA Manual S
	Commercial and Industrial Unitary Air-Conditioning Condensing Units	ARI	ANSI/ARI 365-94
	Methods of Testing for Rating Positive Displacement Refrigerant Compressors and Condensing Units	ASHRAE	ANSI/ASHRAE 23-1993
	Heating and Cooling Equipment (1995)	UL	UL 1995
		CSA	CAN/CSA-C22.2 No. 236-M95
	Refrigeration and Air-Conditioning Condensing and Compressor Units (1987)	UL	ANSI/UL303-1988
Contactors	Definite Purpose Contactors for Limited Duty	ARI	ANSI/ARI 790-86
	Definite Purpose Magnetic Contactors	ARI	ANSI/ARI 780-86
Containers	Requirements for Closed Van Cargo Containers	ASME	ASME MH5.1.1M
	Series I Freight Containers—Classifications, Dimensions, and Ratings (1995)	ISO	ISO 688

Codes and Standards

Table 1 Codes and Standards Published by Various Societies and Associations (*Continued*)

Subject	Title	Publisher	Reference
Containers (*continued*)	Series I Freight Containers—Specifications and Testing—Part 2: Thermal Containers (1996)	ISO	ISO 1496-2
	Animal Environment In Cargo Compartments	SAE	SAE AIR 1600-85
Controls	Control Systems (1989)	AABC	National Standards, CH 25
	Quick-Disconnect Devices for Use with Gas Fuel	AGA	ANSI Z21.41-1989;Z21.41a-1990 Z21.41b-1992
	Energy Management Control Systems Instrumentation	ASHRAE	ANSI/ASHRAE 114-1986
	BACnet™—A Data Communication Protocol for Building Automation and Control Networks	ASHRAE	ANSI/ASHRAE 135-1995
	Performance Requirements for Electric Heating Line-Voltage Wall Thermostats	CSA	C273.4-M1978 (R 1992)
	Temperature-Indicating and Regulating Equipment	CSA	C22.2 No. 24-93
	Control Centers for Changing Message Type Electric Signals (1991)	UL	ANSI/UL 1433-1995
	Limit Controls (1994)	UL	ANSI/UL 353-1995
	Primary Safety Controls for Gas- and Oil-Fired Appliances (1994)	UL	ANSI/UL 372-1994
	Solid State Controls for Appliances (1994)	UL	ANSI/UL 244A-1995
	Temperature-Indicating and -Regulating Equipment (1994)	UL	UL 873
	Tests for Safety-Related Controls Employing Solid-State Devices (1995)	UL	UL 991
Commercial and Industrial	Industrial Control and Systems General Requirements	NEMA	ANSI/NEMA ICS 1-1993
	Industrial Control and Systems, Controllers, Contactors, and Overload Relays Rated Not More than 2000 Volts AC or 750 Volts DC	NEMA	ANSI/NEMA ICS 2-1993
	Instructions for the Handling, Installation, Operation and Maintenance of Motor Control Centers	NEMA	NEMA ICS 2.3-1983 (R 1990)
	Preventive Maintenance of Industrial Control and Systems Equipment	NEMA	NEMA ICS 1.3-1986
	Industrial Control Equipment (1993)	UL	UL 508
Residential	Automatic Gas Ignition Systems and Components	AGA	ANSI Z21.20-1993; Z21.20a-1994
	Gas Appliance Pressure Regulators	AGA	ANSI Z21.18-1993; Z21.18a-1994
	Gas Appliance Thermostats	AGA	ANSI Z21.23-1993; Z21.23a-1994
	Manually Operated Gas Valves for Appliances, Appliance Connector Valves and Hose End Valves	AGA	ANSI Z21.15-1992
	Manually-Operated Piezo Electric Spark Gas Ignition Systems and Components	AGA	ANSI Z21.77-1989; Z21.77a-1993
	Hot-Water Immersion Controls	NEMA	NEMA DC-12-1985 (R 1991)
	Line-Voltage Integrally Mounted Thermostats for Electric Heaters	NEMA	NEMA DC 13-1991
	Residential Controls—Quick Connect Terminals	NEMA	NEMA DC 2-1982 (R 1988)
	Residential Controls—Electrical Wall-Mounted Room Thermostats	NEMA	NEMA DC 3-1989
	Residential Controls—Surface Type Controls for Electric Storage Water Heaters	NEMA	NEMA DC 5-1989
	Residential Controls—Temperature Limit Controls for Electric Baseboard Heaters	NEMA	NEMA DC 10-1983 (R 1989)
	Residential Controls—Class 2 Transformers	NEMA	NEMA DC 20-1992
	Safety Guidelines for the Application, Installation, and Maintenance of Solid State Controls	NEMA	NEMA ICS 1.1-1984 (R 1988)
	Electrical Quick-Connect Terminals (1995)	UL	UL 310
Coolers	Refrigeration Equipment	CSA	CAN/CSA-C22.2 No. 120-M91
	Unit Coolers for Refrigeration	ARI	ANSI/ARI 420-94
Air	Methods of Testing Forced Convection and Natural Convection Air Coolers for Refrigeration	ASHRAE	ANSI/ASHRAE 25-1990
	Commercial Bulk Milk Dispensing Equipment	NSF	ANSI/NSF 20-1992
Drinking Water	Self-Contained, Mechanically-Refrigerated Drinking-Water Coolers	ARI	ANSI/ARI 1010-94
	Methods of Testing for Rating Drinking-Water Coolers with Self-Contained Mechanical Refrigeration Systems	ASHRAE	ANSI/ASHRAE 18-1987 (RA 97)
	Drinking-Water Coolers (1993)	UL	ANSI/UL 399-1992
Food and Beverage	Methods of Testing and Rating Bottled and Canned Beverage Vendors and Coolers	ASHRAE	ANSI/ASHRAE 32-1986 (RA 90)
	Refrigerated Vending Machines (1995)	UL	UL 541
	Manual Food and Beverage Dispensing Equipment	NSF	ANSI/NSF 18-1990
Liquid	Refrigerant-Cooled Liquid Coolers, Remote Type	ARI	ANSI/ARI 480-87
	Methods of Testing for Rating Liquid Coolers	ASHRAE	ANSI/ASHRAE 24-1989
	Liquid Cooling Systems	SAE	ANSI/SAE AIR 1811-1985 (R 1992)
Cooling Towers	Special Systems: Cooling Tower Performance Tests (1989)	AABC	National Standards, CH 22
	Commercial Applications, Systems and Equipment (1993)	ACCA	ACCA Manual CS
	Atmospheric Water Cooling Equipment	ASME	ANSI/ASME PTC 23-1986 (R 1992)
	Water-Cooling Towers	NFPA	ANSI/NFPA 214-1992
	Acceptance Test Code for Spray Cooling Systems (1985)	CTI	CTI ATC-133
	Acceptance Test Code for Water Cooling Towers: Mechanical Draft, Natural Draft Fan Assisted Types, Evaluation of Results, and Thermal Testing of Wet/Dry Cooling Towers (1990)	CTI	CTI ATC-105
	Certification Standard for Commercial Water Cooling Towers (1991)	CTI	CTI STD-201
	Code for Measurement of Sound from Water Cooling Towers (1981)	CTI	CTI ATC-128
	Fiberglass-Reinforced Plastic Panels for Application on Industrial Water-Cooling Towers (1986)	CTI	CTI STD-131
	Nomenclature for Industrial Water-Cooling Towers (1983)	CTI	CTI NCL-109
	Recommended Practice for Airflow Testing of Cooling Towers (1994)	CTI	CTI PFM-143

Table 1 Codes and Standards Published by Various Societies and Associations (*Continued*)

Subject	Title	Publisher	Reference
Crop Drying	Density, Specific Gravity, and Mass-Moisture Relationships of Grain for Storage	ASAE	ANSI/ASAE D241.4-1993
	Moisture Measurement—Forages	ASAE	ASAE S358.2-1993
	Moisture Measurement—Unground Grain and Seeds	ASAE	ASAE S352.2-1992
	Moisture Relationships of Grains	ASAE	ASAE D245.4-1994
	Resistance to Airflow of Grains, Seeds, Other Agricultural Products, and Perforated Metal Sheets	ASAE	ASAE D272.2-1994
Dehumidifiers	Commercial Applications, Systems and Equipment (1993)	ACCA	ACCA Manual CS
	Dehumidifiers	AHAM	ANSI/AHAM DH-1-1992
	Dehumidifiers	CSA	C22.2 No. 92-1971 (R 1992)
	Dehumidifiers (1993)	UL	ANSI/UL 474-1992
Desiccants	Method of Testing Desiccants for Refrigerant Drying	ASHRAE	ANSI/ASHRAE 35-1992
Dryers	Method of Testing Liquid Line Refrigerant Driers	ASHRAE	ANSI/ASHRAE 63.1-1995
	Liquid-Line Driers	ARI	ANSI/ARI 710-86
Ducts and Fittings	Fibrous Glass Duct Liner Standards (1994)	NAIMA	NAIMA AH 124
	Hose, Air Duct, Flexible Nonmetallic, Aircraft	SAE	SAE AS 1501C-1994
	Ducted Electric Heat Guide for Air Handling Systems (1971)	SMACNA	SMACNA
	Factory-Made Air Ducts and Air Connectors (1994)	UL	UL 181
	Marine Rigid and Flexible Air Ducting (1986)	UL	ANSI/UL 1136-1986
Construction	Preferred Metric Sizes for Flat Metal Products	ASME	ANSI/ASME B32.3M-1984 (R 1994)
	Sheet Metal Welding Code	AWS	ANSI/AWS D90.1-90
	Pipes, Ducts and Fittings for Residential Type Air Conditioning Systems	CSA	B228.1-1968
	Fibrous Glass Duct Construction Standards, 2nd ed. (1993)	NAIMA	NAIMA AH 116
	Fibrous Glass Duct Construction with 1-1/2" Duct Boards (1994)	NAIMA	NAIMA AH 120
	Fibrous Glass Residential Duct Construction Standards (1993)	NAIMA	NAIMA AH 119
	Fibrous Glass Duct Construction Standards, 6th ed. (1992)	SMACNA	SMACNA
	HVAC Duct Construction Standards—Metal and Flexible, 2nd ed. (1995)	SMACNA	SMACNA
	Rectangular Industrial Duct Construction (1980)	SMACNA	SMACNA
	Round Industrial Duct Construction (1977)	SMACNA	SMACNA
	Thermoplastic Duct (PVC) Construction Manual, 2nd ed. (1994)	SMACNA	SMACNA
Installation	Flexible Duct Performance and Installation Standards	ADC	ADC-91
	Installation of Air Conditioning and Ventilating Systems	NFPA	ANSI/NFPA 90A-1993
	Installation of Warm Air Heating and Air-Conditioning Systems	NFPA	ANSI/NFPA 90B-1993
Materials Specifications	Specification for General Requirements for Flat-Rolled Stainless and Heat-Resisting Steel Plate, Sheet, and Strip	ASTM	ASTM A 480/A 480M-96
	Specification for General Requirements for Steel, Sheet, Carbon, and High-Strength, Low-Alloy, Hot-Rolled and Cold-Rolled	ASTM	ASTM A 568/A 568M-95
	Specification for General Requirements for Steel Sheet, Metallic-Coated by the Hot-Dip Process	ASTM	ASTM A 924/A 924M-95A
	Specification for Steel, Carbon (0.15 Maximum Percent), Hot-Rolled Sheet and Strip Commercial Quality	ASTM	ASTM A 569/A 569M-91A (R 1993)
	Specification for Steel, Sheet, Carbon, Cold-Rolled, Commercial Quality	ASTM	ASTM A 366/A 366M-91 (R 1993)
	Specification for Steel Sheet, Zinc-Coated (Galvanized) or Zinc-Iron Alloy-Coated (Galvannealed) by the Hot-Dip Process	ASTM	ASTM A 653/A 653M-95
System Design	Commercial Low Pressure, Low Velocity Duct System Design (1990)	ACCA	ACCA Manual Q
	Residential Duct Systems (1995)	ACCA	ACCA Manual D-R
	Closure Systems for Use with Rigid Air Ducts and Air Connectors (1994)	UL	UL 181A
Testing	Special Systems: Duct Testing (1989)	AABC	National Standards, CH 23
	Flexible Air Duct Test Code	ADC	ADC FD-72 (R 1979)
	Test Method for Measuring Acoustical and Airflow Performance of Duct Liner Materials and Prefabricated Silencers	ASTM	ASTM E 477-90
	HVAC Air Duct Leakage Test Manual (1985)	SMACNA	SMACNA
	HVAC Duct Systems Inspection Guide (1989)	SMACNA	SMACNA
Electrical	Voltage Ratings for Electrical Power Systems and Equipment	ANSI	ANSI C84.1-1989
	Test Method for Bond Strength of Electrical Insulating Varnishes by the Helical Coil Test	ASTM	ASTM D 2519-96
	Canadian Electrical Code, Part I (17th ed.)	CSA	C22.1-1994
	Canadian Electrical Code, Part II—General Requirements	CSA	CAN/CSA-C22.2 No. 0-M91
	Application Guide for Ground Fault Circuit Interrupters	NEMA	NEMA 280-1990
	Application Guide for Ground Fault Protective Devices for Equipment	NEMA	ANSI/NEMA PB 2.2-1988
	Enclosures for Electrical Equipment (1000 Volts Maximum)	NEMA	ANSI/NEMA 250-1991
	Industrial Control and Systems Enclosures	NEMA	ANSI/NEMA ICS 6-1993
	General Requirements for Wiring Devices	NEMA	NEMA WD 1-1983 (R 1989)
	Low Voltage Cartridge Fuses	NEMA	ANSI/NEMA FU 1-1986
	Molded Case Circuit Breakers and Molded Case Switches	NEMA	NEMA AB 1-1993
	Industrial Control and Systems Terminal Blocks	NEMA	NEMA ICS 4-1993
	National Electrical Code	NFPA	ANSI/NFPA 70-1996
	Compatibility of Electrical Connectors and Wiring	SAE	ANSI/SAE AIR 1329A-1988
	Class T Fuses (1988)	UL	ANSI/UL 198H-1987

Codes and Standards

Table 1 Codes and Standards Published by Various Societies and Associations (*Continued*)

Subject	Title	Publisher	Reference
Electrical (continued)	Enclosures for Electrical Equipment (1995)	UL	ANSI/UL 50-1995
	Fuseholders (1993)	UL	ANSI/UL 512-1992
	High-Interrupting-Capacity Fuses, Current-Limiting Types (1986)	UL	ANSI/UL 198C-1986
	Molded-Case Circuit Breakers and Circuit-Breaker Enclosures (1991)	UL	ANSI/UL 489-1994
	Terminal Blocks (1993)	UL	UL 1059
	Thermal Cutoffs for Use in Electrical Appliances and Components (1994)	UL	UL 1020
Energy	Air Conditioning and Refrigerating Equipment Nameplate Voltages	ARI	ANSI/ARI 110-90
	Energy Conservation in Existing Buildings (Supersedes *Standards* 100.2, 100.3, 100.4, 100.5, and 100.6)	ASHRAE	ASHRAE/IESNA 100-1995 with Addendum 100a
	Energy-Efficient Design of New Low-Rise Residential Buildings	ASHRAE	ASHRAE/IESNA 90.2-1993
	Energy Efficient Design of New Buildings Except Low-Rise Residential Buildings	ASHRAE	ASHRAE/IESNA 90.1-1997
	Standard Methods of Measuring and Expressing Building Energy Performance	ASHRAE	ANSI/ASHRAE 105-1984 (RA 90)
	Model Energy Code (1995)	CABO	BOCA/ICBO/SBCCI
	Uniform Solar Energy Code (1994)	IAPMO	IAPMO
	Model Energy Code, Thermal Envelope Compliance Guide	NAIMA	NAIMA BI407
	Energy Management Guide for Selection and Use of Polyphase Motors	NEMA	NEMA MG 10-1994
	Energy Management Guide for Selection and Use of Single-Phase Motors	NEMA	NEMA MG 11-1977 (R 1992)
	Building Systems Analysis and Retrofit Manual (1995)	SMACNA	SMACNA
	Energy Conservation Guidelines (1984)	SMACNA	SMACNA
	Energy Recovery Equipment and Systems, Air-to-Air (1991)	SMACNA	SMACNA
	HVAC Commissioning Manual (1994)	SMACNA	SMACNA
	Retrofit of Building Energy Systems and Processes (1982)	SMACNA	SMACNA
	Energy Management Equipment (1994)	UL	ANSI/UL 916-1993
Exhaust Systems	Return and Exhaust Air Systems (1989)	AABC	National Standards, CH 20
	Commercial Low Pressure, Low Velocity Duct System Design (1990)	ACCA	ACCA Manual Q
	Fundamentals Governing the Design and Operation of Local Exhaust Systems	ANSI	ANSI/AIHA Z9.2-1979 (R 1991)
	Laboratory Ventilation	ANSI	ANSI/AIHA Z9.5-1992
	Open-Surface Tanks—Ventilation and Operation	ANSI	ANSI/AIHA Z9.1-1991
	Safety Code for Design, Construction, and Ventilation of Spray Finishing Operations	ANSI	ANSI/AIHA Z9.3-1985
	Abrasive Blasting Operations—Ventilation and Safe Practices	ANSI	ANSI/AIHA Z9.4-1985
	Method of Testing Performance of Laboratory Fume Hoods	ASHRAE	ANSI/ASHRAE 110-1995
	Compressors and Exhausters	ASME	ANSI/ASME PTC 10-1965 (R 1992)
	Mechanical Flue-Gas Exhausters	CSA	CAN 3-B255-M81
	Exhaust Systems for Air Conveying of Materials	NFPA	ANSI/NFPA 91-1995
	Draft Equipment (1993)	UL	UL 378
Expansion Valves	Thermostatic Refrigerant Expansion Valves	ARI	ANSI/ARI 750-94
	Method of Testing for Capacity Rating of Thermostatic Refrigerant Expansion Valves	ASHRAE	ANSI/ASHRAE 17-1986 (RA 90)
Fan-Coil Units	Room Fan-Coil and Unit Ventilators	ARI	ANSI/ARI 440-89
	Methods of Testing for Rating Room Fan-Coil Air Conditioners	ASHRAE	ANSI/ASHRAE 79-1984 (RA 91)
	Fan-Coil Units and Room Fan-Heater Units (1986)	UL	ANSI/UL 883-1986
Fans	Commercial Low Pressure, Low Velocity Duct System Design (1990)	ACCA	ACCA Manual Q
	Residential Duct Systems (1995)	ACCA	ACCA Manual D-R
	Designation of Rotation and Discharge of Centrifugal Fans	AMCA	AMCA 99-2406-83
	Drive Arrangements for Centrifugal Fans	AMCA	AMCA 99-2404-78
	Drive Arrangements for Tubular Centrifugal Fans	AMCA	AMCA 99-2410-82
	Recommended Safety Practices for Users and Installers of Industrial and Commercial Fans	AMCA	AMCA 410-90
	Inlet Box Positions for Centrifugal Fans	AMCA	AMCA 99-2405-83
	Motor Positions for Belt or Chain Drive Centrifugal Fans	AMCA	AMCA 99-2407-66
	Industrial Process/Power Generation Fans: Site Performance Test Standard	AMCA	AMCA 803-94
	Standards Handbook	AMCA	AMCA 99-86
	Fans and Blowers	ARI	ARI 670-90
	Methods for the Measurement of Noise Emitted by Small Air-Moving Devices	ASA	ANSI S12.11-1987 (R 1993)
	Laboratory Methods of Testing Fans for Rating	ASHRAE AMCA	ANSI/ASHRAE 51-1985 ANSI/AMCA 210-85
	Laboratory Method of Testing In-Duct Sound Power Measurement Procedure for Fans	ASHRAE AMCA	ANSI/ASHRAE 68-1997 ANSI/AMCA 330-86
	Methods of Testing Fan Vibration—Blade Vibrations and Critical Speeds	ASHRAE	ANSI/ASHRAE 87.1-1992
	Fans	ASME	ANSI/ASME PTC 11-1984 (R 1990)
	Fans and Ventilators	CSA	C22.2 No. 113-M1984 (R 1993)
	Rating the Performance of Residential Mechanical Ventilating Equipment	CSA	CAN/CSA-C260-M90
	Acoustics—Method for the Measurement of Airborne Noise Emitted by Small Air-Moving Devices	ISO	ISO 10302:1996

Table 1 Codes and Standards Published by Various Societies and Associations (*Continued*)

Subject	Title	Publisher	Reference
Fans *(continued)*	Electric Fans (1994)	UL	UL 507
Ceiling	AC Electric Fans and Regulators	ANSI	ANSI-IEC Pub. 385
Fenestration	Specification for Sealed Insulated Glass Units	ASTM	ASTM E 774-92
	Standard Practice for Calculation of Photometric Transmittance and Reflectance of Materials to Solar Radiation	ASTM	ASTM 971-88 (R 1996)
	Standard Practice for Determining the Minimum Thickness and Type of Glass Required to Resist a Specific Load	ASTM	ASTM E 1300-94
	Standard Tables for Terrestrial Direct Normal Solar Spectral Irradiance for Air Mass 1.5	ASTM	ASTM E 891-87 (R 1992)
	Standard Tables for Terrestrial Solar Spectral Irradiance at Air Mass 1.5 for a 37° Tilted Surface	ASTM	ASTM E 892-87 (R 1992)
	Test Method for Seal Durability of Sealed Insulated Glass Units	ASTM	ASTM E 773-88
	Test Method for Solar Absorptance, Reflectance, and Transmittance of Materials Using Integrating Spheres	ASTM	ASTM E 903-96
	Test Method for Solar Photometric Transmittance of Sheet Materials Using Sunlight	ASTM	ASTM E 972-96
	Test Method for Solar Transmittance (Terrestrial) of Sheet Materials Using Sunlight	ASTM	ASTM E 1084-86 (R 1996)
	Energy Performance Evaluation of Swinging Doors	CSA	CSA-A453-95
	Energy Performance Evaluation of Windows and Sliding Glass Doors	CSA	CAN/CSA-A440.2-93
	Windows	CSA	CAN/CSA-A440-M90
Filters	Flow-Capacity Rating and Application of Suction-Line Filters and Filter Driers	ARI	ANSI/ARI 730-86
	Specification for Octave-Band and Fractional-Octave-Band Analog and Digital Filters	ASA	ANSI S1.11-1986 (R 1993)
	Method of Testing Flow Capacity of Suction Line Filters and Filter Driers	ASHRAE	ANSI/ASHRAE 78-1985 (RA 97)
	Method of Testing Liquid Line Filter-Drier Filtration Capability	ASHRAE	ANSI/ASHRAE 63.2-1996
	Exhaust Hoods for Commercial Cooking Equipment (1990)	UL	ANSI/UL 710-1992
	Grease Filters for Exhaust Ducts (1979)	UL	UL 1046
Fireplaces	Factory-Built Fireplaces (1988)	UL	ANSI/UL 127-1992
	Fireplace Stoves (1995)	UL	ANSI/UL 737-1995
Fire Protection	Standard Method for Fire Tests of Building Construction and Materials	ASTM	ASTM E 119-95A
	Method of Test of Surface Burning Characteristics of Building Materials	ASTM	ASTM E 84-95A
		NFPA	NFPA 255-1996
	BOCA National Fire Prevention Code, 9th ed. (1993)	BOCA	BOCA
	Uniform Fire Code (1994)	IFCI	IFCI
	Uniform Fire Code Standards (1991)	IFCI	IFCI
	Fire-Resistance Tests—Elements of Building Construction Part 1: General Requirements	ISO	ISO 834:1975
	Fire-Resistance Tests—Methods of Test of Fire Doors and Shutters	ISO	ISO 3008:1976
	Reaction to Fire Tests—Ignitability of Building Products Using a Radiant Heat Source	ISO	ISO 5657:1986
	Fire-Resistance Tests—Service Installations In Buildings—Fire-Resisting Ducts	ISO	ISO6944:1985
	Interconnection Circuitry of Noncoded Remote-Station Protective Signalling Systems	NEMA	NEMA SB 3-1969 (R 1989)
	Fire Doors and Fire Windows	NFPA	ANSI/NFPA 80-1995
	Fire Hazard Properties of Flammable Liquids, Gases, and Volatile Solids	NFPA	ANSI/NFPA 325-1994
	Fire Prevention Code	NFPA	ANSI/NFPA 1-1992
	Fire Protection for Laboratories Using Chemicals	NFPA	ANSI/NFPA 45-1996
	Fire Protection Handbook, 18th ed. (1996)	NFPA	NFPA
	Flammable and Combustible Liquids Code	NFPA	ANSI/NFPA 30-1993
	Health Care Facilities	NFPA	ANSI/NFPA 99-1996
	Installation of Sprinkler Systems	NFPA	NFPA 13-1996
	Life Safety Code	NFPA	ANSI/NFPA 101-1994
	National Fire Codes (issued annually)	NFPA	NFPA
	Methods of Fire Tests of Door Assemblies	NFPA	ANSI/NFPA 252-1995
	Standard Fire Prevention Code (1994 ed. with 1995 revisions)	SBCCI	SBCCI
	Fire, Smoke and Radiation Damper Installation Guide for HVAC Systems (1992)	SMACNA	SMACNA
	Fire Dampers (1995)	UL	UL 555
	Fire Tests of Building Construction and Materials (1992)	UL	UL 263
	Fire Tests of Door Assemblies (1993)	UL	UL 10B
	Fire Tests of Through-Penetration Firestops (1994)	UL	UL 1479
	Heat Responsive Links for Fire-Protection Service (1993)	UL	ANSI/UL 33-1995
Smoke Management	Commissioning Smoke Management Systems	ASHRAE	ASHRAE Guideline 5-1994
	Recommended Practice for Smoke Control Systems	NFPA	ANSI/NFPA 92A-1993
	Guide for Smoke Management Systems in Malls, Atria, and Large Areas	NFPA	ANSI/NFPA 92B-1995
	Leakage Rated Dampers for Use in Smoke Control Systems	UL	ANSI/UL 555S-1993
Freezers	Capacity Measurement and Energy Consumption Test Methods for Refrigerators, Combination Refrigerator-Freezers, and Freezers	CSA	CAN/CSA-C300-M91
	Refrigeration Equipment	CSA	CAN/CSA-C22.2 No. 120-M91
Commercial	Dispensing Freezers	NSF	ANSI/NSF 6-1989

Codes and Standards

Table 1 Codes and Standards Published by Various Societies and Associations (*Continued*)

Subject	Title	Publisher	Reference
Commercial (*continued*)	Food Service Refrigerators and Storage Freezers	NSF	NSF 7-1990
	Ice Cream Makers (1993)	UL	ANSI/UL 621-1992
	Commercial Refrigerators and Freezers (1992)	UL	ANSI/UL 471-1991
	Ice Makers (1995)	UL	UL 563
Household	Household Refrigerators, Combination Refrigerator-Freezers and Household Freezers	AHAM	ANSI/AHAM HRF-1-1988
	Household Refrigerators and Freezers (1993)	UL	UL 250
		CSA	C22.2 No. 63-93
Fuels	Standard Classification of Coals by Rank	ASTM	ANSI/ASTM D975-96
	Standard Specification for Diesel Fuel Oils	ASTM	ANSI/ASTM D 975-96
	Standard Specification for Fuel Oils	ASTM	ANSI/ASTM D 396-95
	Standard Specification for Gas Turbine Fuel Oils	ASTM	ANSI/ASTM D 2880-96
Furnaces	Commercial Applications, Systems and Equipment (1993)	ACCA	ACCA Manual CS
	Residential Equipment Selection, 2nd ed. (1995)	ACCA	ACCA Manual S
	Methods of Testing for Annual Fuel Utilization Efficiency of Residential Central Furnaces and Boilers	ASHRAE	ANSI/ASHRAE 103-1993
	BOCA National Mechanical Code, 8th ed. (1993)	BOCA	BOCA
	Prevention of Furnace Explosions/Implosions in Multiple Burner Boilers	NFPA	ANSI /NFPA 8502-1995
	Standard Mechanical Code (1994)	SBCCI	SBCCI
	Heating and Cooling Equipment (1995)	UL	UL 1995
		CSA	CAN/CSA-C22.2 No. 236-95
	Residential Gas Detectors (1991)	UL	UL 1484
	Single and Multiple Station Carbon Monoxide Detectors (1992)	UL	UL 2034
Gas	Direct Vent Central Furnaces	AGA	ANSI Z21.64-1990
	Gas-Fired Central Furnaces	AGA	ANSI Z21.47-1993
		CGA	CAN/CGA-2.3-M93
	Gas-Fired Duct Furnaces	AGA	ANSI Z83.9-1990; Z83.9a-1992
	Gas-Fired Gravity and Fan Type Direct Vent Wall Furnaces	AGA	ANSI Z21.44-1995
	Gas-Fired Gravity and Fan Type Floor Furnaces	AGA	Z21.48-1992
	Gas-Fired Gravity and Fan Type Vented Wall Furnaces	AGA	Z21.49-1992
	Gas-Fired Duct Furnaces	CGA	CAN/CGA-2.8-M86
	Gas-Fired Gravity and Fan Type Direct Vent Wall Furnaces	CGA	CAN1-2.19-M81
	Gas-Fired Gravity and Fan Type Vented Wall Furnaces	CGA	CAN/CGA-2.5-M86
	Industrial and Commercial Gas-Fired Package Furnaces	CGA	CGA 3.2-1976
	Standard Gas Code (1994)	SBCCI	SBCCI
	Commercial-Industrial Gas Heating Equipment (1973)	UL	UL 795
Oil	Standard Specification for Fuel Oils	ASTM	ANSI/ASTM D 396-95
	Test Method for Smoke Density in Flue Gases from Burning Distillate Fuels	ASTM	ANSI/ASTM D 2156-94
	Oil Burning Stoves and Water Heaters	CSA	B140.3-1962 (R 1991)
	Oil-Fired Warm Air Furnaces	CSA	B140.4-1974 (R 1991)
	Installation of Oil-Burning Equipment	NFPA	ANSI/NFPA 31-1992
	Oil-Fired Central Furnaces (1986)	UL	UL 727-1986
	Oil-Fired Floor Furnaces (1987)	UL	ANSI/UL 729-1987
	Oil-Fired Wall Furnaces (1987)	UL	ANSI/UL 730-1986
Solid Fuel	Standard Classification of Coals by Rank	ASTM	ANSI/ASTM D975-96
	Installation Code for Solid-Fuel-Burning Appliances and Equipment	CSA	CAN/CSA-B365-M91
	Solid-Fuel-Fired Central Heating Appliances	CSA	CAN/CSA-B366.1-M91
	Solid-Fuel and Combination-Fuel Central and Supplementary Furnaces (1991)	UL	ANSI/UL 391-1991
Heaters	Gas-Fired Infrared Heaters	AGA	ANSI Z83.6-1990; Z83.6a-1992; Z83.6b-1993
	Requirements for Gas-Fired Infrared Patio	AGA	5-90
	Requirements for Residential Radiant Tube Heaters	AGA	7-89
	Air Heaters	ASME	ANSI/ASME PTC 4.3-1968 (R 1991)
	Direct Gas-Fired Non-Recirculating Make-up Air Heaters	CGA	CAN1-3.7-77
	Gas-Fired Infra-Red Heaters	CGA	CAN1-2.16-M81
	Electric Air Heaters	CSA	C22.2 No.46-M1988
	Electric Duct Heaters	CSA	C22.2 No. 155-M1986 (R 1992)
	Portable Kerosine-Fired Heaters	CSA	CAN 3-B140.9.3 M86
	Standards for Closed Feedwater Heaters, 5th ed. (1992)	HEI	HEI
	Electric Air Heaters (1980)	UL	ANSI/UL 1025-1991
	Electric Dry Bath Heaters (1994)	UL	ANSI/UL 875
	Electric Heating Appliances (1987)	UL	ANSI/UL 499-1987
	Electric Oil Heaters (1991)	UL	ANSI/UL 574-1990
	Oil-Burning Stoves (1993)	UL	UL 896
	Oil-Fired Air Heaters and Direct-Fired Heaters (1993)	UL	UL 733
Combination	Requirements for Gas-Fired Combination Space Heating/Water Heating Appliances	AGA	11-90
Engine	Electric Engine Preheaters and Battery Warmers for Diesel Engines	SAE	SAE J 1310-1993
	Fuel Warmer—Diesel Engines	SAE	ANSI/SAE J 1422-1989

Table 1 Codes and Standards Published by Various Societies and Associations (*Continued*)

Subject	Title	Publisher	Reference
Engine (*continued*)	Selection and Application Guidelines for Diesel, Gasoline, and Propane Fired Liquid Cooled Engine Pre-Heater	SAE	SAE J 1350-1988
Nonresidential	Direct Gas-Fired Industrial Air Heaters	AGA	ANSI Z83.18-1990; Z83.18a-1991; Z83.18b-1992
	Gas-Fired Construction Heaters	AGA	ANSI Z83.7-1990; Z83.7a-1991; Z83.7b-1993
	Gas-Fired Unvented Commercial and Industrial Heaters	AGA	ANSI Z83.16-1982; Z83.16a-1984; Z83.16b-1989
	Requirements for Direct Gas-Fired Circulating Heaters for Agricultural Buildings	AGA	5-88
	Requirements for High Pressure LP Infrared Poultry and Livestock Heating Systems	AGA	4-87
	Portable Industrial Oil-Fired Heaters	CSA	B140.8-1967 (R 1991)
	Fuel-Fired Heaters—Air Heating—For Construction and Industrial Machinery	SAE	ANSI/SAE J 1024-1989
	Commercial-Industrial Gas Heating Equipment (1994)	UL	UL 795
	Electric Heaters for Use in Hazardous (Classified) Locations (1991)	UL	ANSI/UL 823-1990
Room	Gas-Fired Room Heaters, Vol. I, Vented Room Heaters	AGA	ANSI Z21.11.1-1991; Z21.11.1a-1993; Z21.11.1b-1995
	Gas-Fired Room Heaters, Vol. II, Unvented Room Heaters	AGA	ANSI Z21.11.2-1992 Z21.11.2a-1993; Z21.11.2b-1995
	Gas-Fired Unvented Catalytic Room Heaters for Use with Liquefied Petroleum (LP) Gases	AGA	ANSI Z21.76-1994
	Requirements for Gas-Fired Vented Catalytic Type Room Heaters	AGA	1-81
	Requirements for Unvented Room Heaters Equipped with Oxygen Depletion Safety Shutoff Systems	AGA	2-79
	Fixed and Location-Dedicated Electric Room Heaters (1992)	UL	UL 2021
	Movable and Wall- or Ceiling-Hung Electric Room Heaters (1994)	UL	UL 1278
	Room Heaters, Solid Fuel-Type (1994)	UL	UL 1482
	Unvented Kerosene-Fired Room Heaters and Portable Heaters (1993)	UL	UL 647
Pool	Gas-Fired Pool Heaters	AGA	ANSI Z21.56-1994
	Oil-Fired Service Water Heaters and Swimming Pool Heaters	CSA	B140.12-1976 (R 1991)
Transport	Heater, Aircraft Internal Combustion Heat Exchanger Type	SAE	SAE AS 8040-1988
	Heater, Airplane, Engine Exhaust Gas to Air Heat	SAE	SAE ARP 86A-1952 (R 1992)
	Installation, Heaters, Airplane, Internal Combustion Heater Exchange Type	SAE	SAE ARP 266-1952 (R 1992)
	Motor Vehicle Heater Test Procedure	SAE	SAE J 638-1993
Unit	Gas Unit Heaters	AGA	ANSI Z83.8-1990; Z83.8a-1990; Z83.8b-1992
	Gas Unit Heaters	CGA	CAN/CGA-2.6-M86
	Oil-Fired Unit Heaters (1995)	UL	UL 731
Heat Exchangers	Remote Mechanical-Draft Evaporative Refrigerant Condensers	ARI	ANSI/ARI 490-89
	Method of Testing Air-to-Air Heat Exchangers	ASHRAE	ANSI/ASHRAE 84-1991
	Standard Methods of Test for Rating the Performance of Heat-Recovery Ventilators	CSA	CAN/CSA-C439-88
	Standards for Power Plant Heat Exchangers, 2nd ed. (1990)	HEI	HEI
	Standards of Tubular Exchanger Manufacturers Association, 7th ed. (1988)	TEMA	TEMA
Heating	Commercial Applications, Systems and Equipment (1993)	ACCA	ACCA Manual CS
	Installation Techniques for Perimeter Heating and Cooling, 11th ed.	ACCA	ACCA Manual 4
	Residential Equipment Selection, 2nd ed. (1995)	ACCA	ACCA Manual S
	Determining the Required Capacity of Residential Space Heating and Cooling Appliances	CSA	CAN/CSA-F280-M90
	Automatic Flue-Pipe Dampers for Use with Oil-Fired Appliances	CSA	B140.14-M1979 (R 1991)
	Heater Elements	CSA	C22.2 No.72-M1984 (R 1992)
	Advanced Installation Guide for Hydronic Heating Systems (1991)	HYDI	IBR 250
	Heat Loss Calculation Guide	HYDI	IBR H-21 (1984), IBR H-22 (1989)
	Installation Guide for Residential Hydronic Heating Systems, 6th ed. (1988)	HYDI	IBR 200
	Radiant Floor Heating (1993)	HYDI	IBR 400
	Environmental System Technology (1984)	NEBB	NEBB
	Pulverized Fuel Systems	NFPA	ANSI/NFPA 8503-1992
	Aircraft Electrical Heating Systems	SAE	ANSI/SAE AIR 860-1965 (R 1992)
	Performance Test for Air-Conditioned, Heated, and Ventilated Off-Road Self-Propelled Work Machines	SAE	ANSI/SAE J 1503-1986
	Heating Value of Fuels	SAE	SAE J 1498-1990
	HVAC Systems—Applications, 1st ed. (1986)	SMACNA	SMACNA
	Installation Standards for Residential Heating and Air Conditioning Systems (1988)	SMACNA	SMACNA
	Electric Baseboard Heating Equipment (1994)	UL	ANSI/UL 1042-1995
	Electric Central Air Heating Equipment	UL	ANSI/UL 1096-1986
	Heating and Cooling Equipment (1995)	UL	UL 1995
		CSA	CAN/CSA-C22.2 No. 236-95
Heat Meters	Method of Testing Thermal Energy Meters for Liquid Streams in HVAC Systems	ASHRAE	ANSI/ASHRAE 125-1992
Heat Pumps	Commercial Applications, Systems and Equipment (1993)	ACCA	ACCA Manual CS

Table 1 Codes and Standards Published by Various Societies and Associations (*Continued*)

Subject	Title	Publisher	Reference
Heat Pumps (*continued*)	Heat Pump Systems, Principles and Applications (Commercial and Residence), 2nd ed. (1984)	ACCA	ACCA Manual H
	Residential Equipment Selection, 2nd ed. (1995)	ACCA	ACCA Manual S
	Commercial and Industrial Unitary Heat Pump Equipment	ARI	ANSI/ARI 340-86
	Ground Source Closed-Loop Heat Pumps	ARI	ARI 330-93
	Ground Water-Source Heat Pumps	ARI	ANSI/ARI 325-93
	Water-Source Heat Pumps	ARI	ANSI/ARI 320-93
	Methods of Testing for Rating Unitary Air-Conditioning and Heat Pump Equipment	ASHRAE	ANSI/ASHRAE 37-1988
	Methods of Testing for Seasonal of Unitary Air-Conditioners and Heat Pumps	ASHRAE	ANSI/ASHRAE 116-1995
	Installation Requirements for Air-to-Air Heat Pumps	CSA	C273.5-1980 (R 1991)
	Performance Standard for Split-System Central Air-Conditioners and Heat Pumps	CSA	CAN/CSA-C273.3-M91
	Heating and Cooling Equipment (1995)	UL	UL 1995
		CSA	CAN/CSA C22.2 No. 236-95
	Heat Pumps (1985)	UL	ANSI/UL 559-1985
Gas-Fired	Requirements for Gas-Fired, Absorption and Adsorption Heat Pumps	AGA	10-90
Heat Recovery	Gas Turbine Heat Recovery Steam Generators	ASME	ANSI/ASME PTC 4.4-1981 (R 1992)
	Energy Recovery Equipment and Systems (1991)	SMACNA	SMACNA
	Requirements for Heat Reclaimer Devices for Use with Gas-Fired Appliances	AGA	ANSI Z21.40.1-1994
Humidifiers	Method for Measuring Performance of Appliance Humidifiers	AHAM	ANSI/AHAM HU-1-1987
	Central System Humidifiers for Residential Applications	ARI	ANSI/ARI 610-89
	Commercial and Industrial Humidifiers	ARI	ARI 640-90
	Self-Contained Humidifiers for Residential Applications	ARI	ANSI/ARI 620-89
	Humidifiers (1993)	UL	UL 998
		CSA	C22.2 No. 104-93
Ice Makers	Automatic Commercial Ice Makers	ARI	ARI 810-95
	Ice Storage Bins	ARI	ANSI/ARI 820-95
	Methods of Testing Automatic Ice Makers	ASHRAE	ANSI/ASHRAE 29-1988
	Refrigeration Equipment	CSA	CAN/CSA-C22.2 No. 120-M91
	Standards for Safety, Ice Makers	CSA	CSA 133-1964
	Performance of Automatic Ice-Makers and Ice Storage Bins	CSA	CAN/CSA-C742-94
	Automatic Ice-Making Equipment	NSF	ANSI/NSF 12-1992
	Ice Makers (1995)	UL	UL 563
Incinerators	Large Incinerators	ASME	ANSI/ASME PTC 33-1978 (R 1991)
	Incinerators and Waste and Linen Handling Systems and Equipment	NFPA	ANSI/NFPA 82-1992
	Residential Incinerators (1993)	UL	UL 791
Indoor Air Quality	Standard Practice for Continuous Sizing and Counting of Airborne Particles in Dust-Controlled Areas and Clean Rooms Using Instruments Capable of Detecting Single Sub-Micrometre and Larger Particles	ASTM	ASTM F 50-92 (R 1996)
	Standard Practices for Referencing Suprathreshold Odor Intensity	ASTM	ASTM E 544-75 (R 1993)
	Ambient Air—Determination of Mass Concentration of Nitrogen Dioxide—Modified Griess-Saltzman Method	ISO	ISO 6768:1985
	Format for the Exchange of Air Quality Data—Part 1: General Data Format	ISO	ISO 7168-1:1985
	Format for the Exchange of Air Quality Data—Part 2: Condensed Data Format	ISO	ISO 7168-2:1985
Induction Units	Room Air-Induction Units	ARI	ANSI/ARI 445-87
	Frame Assignments for Alternating Current Integral-Horsepower Induction Motors	NEMA	NEMA MG 13-1984 (R 1990)
Industrial Duct	Rectangular Industrial Duct Construction (1980)	SMACNA	SMACNA
	Round Industrial Duct Construction (1977)	SMACNA	SMACNA
Insulation	Classification for Rating Sound Insulation	ASTM	ASTM E 413-87 (R 1994)
	Classification of Potential Health and Safety Concerns Associated with Thermal Insulation Materials and Accessories	ASTM	ASTM C 930-92
	Guide for Sampling, Test Methods, Specifications, and Guide for Electrical Insulating Oils of Petroleum Origin	ASTM	ASTM D117-96
	Practice for Prefabrication and Field Fabrication of Thermal Insulating Fitting Covers for NPS Piping, Vessel Lagging, and Dished Head Segments	ASTM	ASTM C450-1994
	Specification for Adhesives for Duct Thermal Insulation	ASTM	ASTM C 916-85 (R 1990)
	Specification for Thermal and Acoustical Insulation (Glass Fiber, Duct Lining Material)	ASTM	ASTM C 1071-91
	Specification for Preformed Flexible Elastomeric Cellular Thermal Insulation in Sheet and Tubular Form	ASTM	ASTM C534-1997
	Specification for Cellular Glass Thermal Insulation	ASTM	ASTM C552-1991
	Specification for Rigid, Cellular Polystyrene Thermal Insulation	ASTM	ASTM C578-1995
	Specification for Unfaced Preformed Rigid Cellular Polyisocyanurate Thermal Insulation	ASTM	ASTM C591-1994
	Specification for Faced and Unfaced Rigid Cellular Phenolic Thermal Insulation	ASTM	ASTM C1126-1996
	Standard Practice for Determination of Heat Gain or Loss and the Surface Temperature of Insulated Pipe and Equipment Systems by the Use of a Computer Program	ASTM	ASTM C 680-89 (R 1995)

Table 1 Codes and Standards Published by Various Societies and Associations (Continued)

Subject	Title	Publisher	Reference
Insulation (continued)	Standard Practice for Inner and Outer Diameters of Rigid Thermal Insulation for Nominal Sizes of Pipe and Tubing (NPS System))	ASTM	ASTM C 585-90
	Standard Practice for Thermographic Inspection of Insulation Installations in Envelope Cavities of Frame Buildings	ASTM	ASTM C 1060-90
	Standard Terminology Relating to Thermal Insulating Materials	ASTM	ASTM C 168-90
	Test Method for Steady-State Heat Flux Measurements and Thermal Transmission Properties by Means of the Guarded Hot Plate Apparatus	ASTM	ASTM C 177-85 (R 1993)
	Test Method for Steady-State Heat Flux Measurements and Thermal Transmission Properties by Means of the Heat Flow Meter Apparatus	ASTM	ASTM C 518-91
	Test Method for Steady-State Heat Transfer Properties of Horizontal Pipe Insulations	ASTM	ASTM C 335-95
	Test Method for Steady-State and Thermal Performance of Building Assemblies by Means of a Guarded Hot Box	ASTM	ASTM C 236-89 (R 1993)
	Thermal Insulation, Mineral Fibre, for Buildings	CSA	A101-M1983
	Thermal Insulation—Definition of Terms	ISO	ISO 9229:1991
	National Commercial and Industrial Insulation Standards	MICA	MICA 1993
Louvers	Test Methods for Louvers, Dampers and Shutters	AMCA	AMCA 500-89
Lubricants	Method of Testing the Floc Point of Refrigeration Grade Oils	ASHRAE	ANSI/ASHRAE 86-1994
	Practice for Calculating Viscosity Index from Kinematic Viscosity at 40 and 100°C	ASTM	ASTM D 2270-93
	Practice for Conversion of Kinematic Viscosity to Saybolt Universal Viscosity or to Saybolt Furol Viscosity	ASTM	ASTM D 2161-93
	Test Method for Estimation of Molecular Weight (Relative Molecular Mass) of Petroleum Oils from Viscosity Measurements	ASTM	ANSI/ASTM D 2502-92
	Test Method for Separation of Representative Aromatics and Nonaromatics Fractions of High-Boiling Oils by Elution Chromatography	ASTM	ANSI/ASTM D 2549-91 (R 1995)
	Classification of Industrial Fluid Lubricants by Viscosity System	ASTM	ANSI/ASTM D 2422-86 (R 1993)
	Test Method for Carbon-Type Composition of Insulating Oils of Petroleum Origin	ASTM	ASTM D 2140-91
	Test Method for Dielectric Breakdown Voltage of Insulating Liquids Using Disk Electrodes	ASTM	ASTM D 877-87 (R 1995)
	Test Method for Dielectric Breakdown Voltage of Insulating Oils of Petroleum Origin Using VDE Electrodes	ASTM	ASTM D 1816-84A (R 1990)
	Test Method for Mean Molecular Weight of Mineral Insulating Oils by the Cryoscopic Method	ASTM	ASTM D 2224-78 (R 1983)
	Test Method for Relative Molecular Mass (Molecular Weight) of Hydrocarbons by Thermoelectric Measurement of Vapor Pressure	ASTM	ASTM D 2503-92
	Test Methods for Pour Point of Petroleum Oils	ASTM	ASTM D 97-93
	Petroleum Industry—Corrosiveness to Copper—Copper Strip Test	ISO	ISO 2160:1985
	Semiconductor Graphite	NEMA	NEMA CB 4-1989
Measurements	A Standard Calorimeter Test Method for Flow Measurement of a Volatile Refrigerant	ASHRAE	ANSI/ASHRAE 41.9-1988
	Engineering Analysis of Experimental Data	ASHRAE	ASHRAE Guideline 2-1986(RA 96)
	Methods of Measuring Solar-Optical Properties of Materials	ASHRAE	ANSI/ASHRAE 74-1988
	Standard Method for Measurement of Proportion of Lubricant in Liquid Refrigerant	ASHRAE	ASHRAE 41.4-1996
	Standard Method for Measurement of Moist Air Properties	ASHRAE	ANSI/ASHRAE 41.6-1994
	Standard Methods of Measuring and Expressing Building Energy Performance	ASHRAE	ANSI/ASHRAE 105-1984 (RA 90)
	Measuring Air Change Effectiveness	ASHRAE	ASHRAE 129-1997
	Measurement of Industrial Sound	ASME	ANSI/ASME PTC 36-1985
	Measurement of Rotary Speed	ASME	ANSI/ASME PTC 19.13-1961 (R 1986)
	Measurement Uncertainty	ASME	ANSI/ASME PTC 19.1-1985 (R 1990)
	Method for Establishing Installation Effects on Flowmeter	ASME	ANSI/ASME MFC-10M-1994
	Procedure for Bench Calibration of Tank Level Gaging Tapes and Sounding Rules	ASME	ANSI/ASME MC88.2-1974 (R 1987)
	Specification and Temperature-Electromotive Force (EMF) Tables for Standardized Thermocouples	ASTM	ASTM E 230-96
	Standard Practice for Continuous Sizing and Counting of Airborne Particles in Dust-Controlled Areas and Clean Rooms Using Instruments Capable of Detecting Single Sub-Micrometre and Larger Particles	ASTM	ASTM F 50-92 (R 1996)
	American National Standard for Use of International System of Units (SI)—The Modern Metric System	IEEE/ASTM	ANSI/IEEE/ASTM SI 10-97
	Test Methods for Water Vapor Transmission of Materials	ASTM	ASTM E 96-95
	Ergonomics—Determination of Metabolic Heat Production	ISO	ISO 8996:1990
	Ergonomics of the Thermal Environment—Estimation of the Thermal Insulation and Evaporative Resistance of a Clothing Ensemble	ISO	ISO 9920:1995
	Thermal Environments—Instruments and Methods for Measuring Physical Quantities	ISO	ISO 7726:1985
Fluid Flow	Standard Methods of Measurement of Flow of Liquids in Pipes Using Orifice Flowmeters	ASHRAE	ANSI/ASHRAE 41.8-1989
	Application of Fluid Meters	ASME	ASME 19.5-72

Codes and Standards

Table 1 Codes and Standards Published by Various Societies and Associations (*Continued*)

Subject	Title	Publisher	Reference
Fluid Flow (*continued*)	Fluid Flow in Closed Conduits—Connections for Pressure Signal Transmissions Between Primary and Secondary Devices	ASME	ANSI/ASME MFC-8M-1988
	Glossary of Terms Used in the Measurement of Fluid Flow in Pipes	ASME	ANSI/ASME MFC-1M-1991
	Measurement of Fluid Flow by Means of Coriolis Mass Flowmeters	ASME	ANSI/ASME MFC-11M-1989 (R 1994)
	Measurement of Fluid Flow in Pipes Using Orifice, Nozzle, and Venturi	ASME	ASME MFC-3M-1989
	Measurement of Fluid Flow in Pipes Using Vortex Flow Meters	ASME	ASME/ANSI MFC-6M-1987
	Measurement of Fluid Flow Using Small Bore Precision Orifice Meters	ASME	ANSI/ASME MFC-14M-1995
	Measurement of Liquid Flow in Closed Conduits by Weighting Method	ASME	ANSI/ASME MFC-9M-1988
	Measurement of Liquid Flow in Closed Conduits Using Transit-Time Ultrasonic Flowmeters	ASME	ANSI/ASME MFC-5M-1985 (R 1994)
	Measurement Uncertainty for Fluid Flow in Closed Conduits	ASME	ANSI/ASME MFC-2M-1983 (R 1988)
	Measurement of Fluid Flow in Closed Conduits—Velocity Area Method Using Pitot Static Tubes	ISO	ISO 3966:1977
Gas Flow	Standard Methods for Laboratory Air Flow Measurement	ASHRAE	ANSI/ASHRAE 41.2-1987 (RA 92)
	Standard Method for Measurement of Flow of Gas	ASHRAE	ANSI/ASHRAE 41.7-1984 (RA 91)
	Measurement of Gas Flow by Means of Critical Flow Venturi Nozzles	ASME	ASME/ANSI MFC-7M-1987 (R 1992)
	Measurement of Gas Flow by Turbine Meters	ASME	ANSI/ASME MFC-4M-1986 (R 1990)
Pressure	Standard Method for Pressure Measurement	ASHRAE	ANSI/ASHRAE 41.3-1989
	Gauges—Pressure Indicating Dial Type—Elastic Element	ASME	ANSI/ASME B40.1-91
	Guide for Dynamic Calibration of Pressure Transducers	ASME	ANSI MC88-1-1972 (R 1987)
	Pressure Measurement	ASME	ANSI/ASME PTC 19.2-1987
Temperature	Standard Method for Temperature Measurement	ASHRAE	ANSI/ASHRAE 41.1-1986 (RA 91)
	Temperature Measurement	ASME	ANSI/ASME PTC 19.3-1974 (R 1986)
	Total Temperature Measuring Instruments (Turbine Powered Subsonic Aircraft)	SAE	SAE AS 793-1966 (R 1991)
Thermal Properties	Standard Practice for Determining Thermal Resistance of Building Envelope Components from In-Situ Data	ASTM	ASTM C 1155-95
	Standard Practice for In-Situ Measurement of Heat Flux and Temperature on Building Envelope Components	ASTM	ASTM C 1046-95
	Test Method for Steady-State Heat Flux Measurements and Thermal Transmission Properties by Means of the Guarded Hot Plate Apparatus	ASTM	ASTM C 177-85 (R 1993)
	Test Method for Steady-State Heat Flux Measurements and Thermal Transmission Properties by Means of the Heat Flow Meter Apparatus	ASTM	ASTM C 518-91
	Test Method for Thermal Performance of Building Assemblies by Means of a Calibrated Hot Box	ASTM	ASTM C 976-90 (R 1996)
Mobile Homes and Recreational Vehicles	Residential Load Calculation, 7th ed. (1986)	ACCA	ACCA Manual J
	Recreational Vehicle Cooking Gas Appliances	AGA	ANSI Z21.57-1993
	Mobile Homes	CSA	CAN/CSA-Z240 MH Series-92
	Mobile Home Parks	CSA	Z240.7.1-1972
	Oil-Fired Warm Air Heating Appliances for Mobile Housing and Recreational Vehicles	CSA	B140.10-1974 (R 1991)
	Park Model Trailers	CSA	CAN/CSA-Z41 Series-92
	Recreational Vehicle Parks	CSA	Z240.7.2-1972
	Recreational Vehicles	CSA	CAN/CSA-Z240 RV Series-M86 (R 1992)
	Gas Supply Connectors for Manufactured Homes	IAPMO	IAPMO TSC 9-1992
	Manufactured Home Installations	NCSBCS	ANSI A225.1-1994
	Recreational Vehicles	NFPA	NFPA 501C-1993 (ANSI A119.2)
	Plumbing System Components for Manufactured Homes and Recreational Vehicles	NSF	ANSI/NSF 24-1988
	Gas Burning Heating Appliances for Mobile Homes and Recreational Vehicles (1995)	UL	UL 307B
	Gas-Fired Cooking Appliances for Recreational Vehicles (1993)	UL	UL 1075
	Liquid Fuel-Burning Heating Appliances for Manufactured Homes and Recreational Vehicles (1995)	UL	UL 307A
	Low Voltage Lighting Fixtures for Use in Recreational Vehicles (1994)	UL	UL 234
	Roof Jacks for Manufactured Homes and Recreational Vehicles (1994)	UL	ANSI/UL 311-1995
	Roof Trusses for Manufactured Homes (1995)	UL	UL 1298
	Shear Resistance Tests for Ceiling Boards for Manufactured Homes (1992)	UL	UL 1296
Motors and Generators	Steam Generating Units	ASME	ANSI/ASME PTC 4.1-1964 (R 1991)
	Testing of Nuclear Air-Treatment Systems	ASME	ANSI/ASME N510-1989
	Nuclear Power Plant Air Cleaning Units and Components	ASME	ANSI/ASME N509-1989
	Test Methods for Film-Insulated Magnet Wire	ASTM	ANSI/ASTM D 1676-95
	Energy Efficiency Test Methods for Three-Phase Induction Motors (Efficiency Quoting Method and Permissible Efficiency Tolerance)	CSA	C390-93
	Motors and Generators	CSA	C22.2 No. 100-95
	Energy Management Guide for Selection and Use of Polyphase Motors	NEMA	NEMA MG 10-1994
	Energy Management Guide for Selection and Use of Single-Phase Motors	NEMA	NEMA MG 11-1977 (R 1992)

Table 1 Codes and Standards Published by Various Societies and Associations (*Continued*)

Subject	Title	Publisher	Reference
Motors and Generators (*continued*)	Magnet Wire	NEMA	NEMA MW 1000-87
	Motion/Position Control Motors, Controls, and Feedback Devices	NEMA	NEMA MG 7-1993
	Motors and Generators	NEMA	NEMA MG 1-1993
	Electric Motors (1994)	UL	UL 1004
	Electric Motors and Generators for Use in Division 1 Hazardous (Classified) Locations (1994)	UL	ANSI/UL 674-1993
	Impedance-Protected Motors (1994)	UL	UL 519
	Thermal Protectors for Motors (1991)	UL	ANSI/UL 547-1991
Operation and Maintenance	Preparation of Operating and Maintenance Documentation for Building Systems	ASHRAE	ASHRAE Guideline 4-1993
Pipe, Tubing, and Fittings	Process Piping	ANSI	ANSI/ASME B31.3-96
	Building Services Piping	ASME	ANSI/ASME B31.9-1988
	Pipe Threads, General Purpose (Inch)	ASME	ANSI/ASME B1.20.1-1983 (R 1992)
	Power Piping	ASME	ANSI/ASME B31.1-1995
	Refrigeration Piping	ASME	ASME/ANSI B31.5-1992
	Scheme for the Identification of Piping Systems	ASME	ANSI/ASME A13.1-1996
	Standard Practice for Obtaining Hydrostatic or Pressure Design Basis for "Fiberglass" (Glass-Fiber-Reinforced Thermosetting-Resin) Pipe and Fittings	ASTM	ANSI/ASTM D 2992-96
	Standards of the Expansion Joint Manufacturers Association, Inc., 6th ed. (1993)	EJMA	EJMA
	Guideline for Quality Piping Installation	MCAA	MCAA
	Pipe Hangers and Supports—Materials, Design and Manufacture	MSS	MSS SP-58-93
	Pipe Hangers and Supports—Selection and Application	MSS	MSS SP-69-91
	Welding Procedure Specifications	NCPWB	NCPWB
	Electrical Nonmetallic Tubing (ENT)	NEMA	NEMA TC 13-1993
	Filament-Wound Reinforced Thermosetting Resin Conduit and Fittings	NEMA	NEMA TC 14-1984 (R 1986)
	National Fuel Gas Code	NFPA	ANSI/NFPA 54-1992
		AGA	ANSI Z223.1-1992; Z223.1a-1994
	Refrigeration Tube Fittings	SAE	ANSI/SAE J 513-1994
	Seismic Restraint Manual Guidelines for Mechanical Systems (1991)	SMACNA	SMACNA
	Tube Fittings for Flammable and Combustible Fluids, Refrigeration Service, and Marine Use (1993)	UL	UL 109
Plastic	Specification for Acrylonitrile-Butadiene-Styrene (ABS) Plastic Pipe, Schedules 40 and 80	ASTM	ASTM D 1527-89
	Specification for Chlorinated Polyvinyl Chloride (CPVC) Plastic Pipe, Schedules 40 and 80	ASTM	ASTM F 441/F 441M-96A
	Specification for Plastic Insert Fittings for Polybutylene (PB) Tubing	ASTM	ASTM F 845-95
	Specification for Polybutylene (PB) Plastic Hot- and Cold- Water Distribution Systems	ASTM	ASTM D 3309-96
	Specification for Polybutylene (PB) Plastic Pipe (SIDR-PR) Based on Controlled Inside Diameter	ASTM	ASTM D 2662-96
	Specification for Polybutylene (PB) Plastic Pipe (SDR-PR) Based on Outside Diameter	ASTM	ASTM D 3000-95A
	Specification for Polybutylene (PB) Plastic Tubing	ASTM	ASTM D 2666-96
	Specification for Polyethylene (PE) Plastic Pipe, Schedule 40	ASTM	ASTM D 2104-90
	Specification for Polyvinyl Chloride (PVC) Plastic Pipe, Schedules 40, 80, and 120	ASTM	ASTM D 1785-96A
	Test Method for Obtaining Hydrostatic Design Basis for Thermoplastic Pipe Materials	ASTM	ASTM D 2837-92
	Corrugated Polyolefin Coilable Plastic Utilities Duct	NEMA	NEMA TC 5-1990
	Corrugated Polyvinyl-Chloride (PVC) Coilable Plastic Utilities Duct	NEMA	NEMA TC 12-1991
	Electrical Plastic Tubing (EPT) and Conduit Schedule EPC-40 and EPC-80	NEMA	NEMA TC 2-1990
	Extra-Strength PVC Plastic Utilities Duct for Underground Installation	NEMA	NEMA TC 8-1990
	Fittings for ABS and PVC Plastic Utilities Duct for Underground Installation	NEMA	NEMA TC 9-1990
	PVC and ABS Plastic Utilities Duct for Underground Installation	NEMA	NEMA TC 6-1990
	Smooth-Wall Coilable Polyethylene Electrical Plastic Duct	NEMA	ANSI/NEMA TC 7-1990
	Plastics Piping Components and Related Materials	NSF	ANSI/NSF 14-1990
	Rubber Gasketed Fittings for Fire-Protection Service (1993)	UL	UL 213
Metal	Welded and Seamless Wrought Steel Pipe	ASME	ASME/ANSI B36.10M-1995
	Specification for Pipe, Steel, Black and Hot-Dipped, Zinc-Coated, Welded and Seamless	ASTM	ASTM A 53-96
	Specification for Seamless Carbon Steel Pipe for High-Temperature Service	ASTM	ASTMA 106-95
	Specification for Seamless Copper Pipe, Standard Sizes	ASTM	ASTM B 42-89
	Specification for Seamless Copper Tube	ASTM	ASTM B 75 Rev A-95
	Specification for Hand-Drawn Copper Capillary Tube for Restrictor Applications	ASTM	ASTM B 360-95
	Specification for Seamless Copper Tube for Air Conditioning and Refrigeration Field Service	ASTM	ASTM B 280-95
	Specification for Seamless Copper Water Tube	ASTM	ASTM B 88-96
	Specification for Welded Copper and Copper Alloy Tube for Air Conditioning and Refrigeration Service	ASTM	ASTM B 640-90
	Thickness Design of Ductile-Iron Pipe	AWWA	ANSI/AWWA C150/A21.50-91

Codes and Standards

Table 1 Codes and Standards Published by Various Societies and Associations (*Continued*)

Subject	Title	Publisher	Reference
Metal (*continued*)	Fittings, Cast Metal Boxes, and Conduit Bodies for Conduit and Cable Assemblies	NEMA	NEMA FB 1-1993
	Polyvinyl-Chloride (PVC) Externally Coated Galvanized Rigid Steel Conduit and Intermediate Metal Conduit	NEMA	NEMA RN 1-1989
Plumbing	BOCA National Plumbing Code, 9th ed. (1993)	BOCA	BOCA
	Uniform Plumbing Code (1994) (with IAPMO Installation Standards)	IAPMO	IAPMO
	Safety Requirements for Plumbing	IAPMO/ MCAA/ NAPHCC	ANSI A40-1993 (1996 Pending)
	International Plumbing Code, 1st ed. (1995)	ICC	BOCA/ICBO/SBCCI
	International Private Sewage Disposal Code	ICC	BOCA/ICBO/SBCCI
	National Standard Plumbing Code (NSPC)	NAPHCC	NSPC 1996
	Standard Plumbing Code (1994 ed. with 1995 revisions)	SBCCI	SBCCI
Pumps	Centrifugal Pumps	ASME	ASME PTC 8.2-1990
	Displacement Compressors, Vacuum Pumps and Blowers	ASME	ANSI/ASME PTC 9-1970 (R 1992)
	Liquid Pumps	CSA	CAN/CSA C.22.2 No. 108-M89
	Performance Standard for Liquid Ring Vacuum Pumps, 1st ed. (1987)	HEI	HEI
	Centrifugal Pumps	HI	ANSI/HI 1.1-1.5 (1994)
	Vertical Pumps	HI	ANSI/HI 2.1-2.5 (1994)
	Rotary Pumps	HI	ANSI/HI 3.1-3.5 (1994)
	Sealless Rotary Pumps	HI	ANSI/HI 4.1-4.6 (1994)
	Sealless Centrifugal Pumps	HI	ANSI/HI 5.1-5.6 (1994)
	Reciprocating Power Pumps	HI	ANSI/HI 6.1-6.5 (1994)
	Controlled Volume Pumps	HI	ANSI/HI 7.1-7.5 (1994)
	Direct Acting (Steam) Pumps	HI	ANSI/HI 8.1-8.5 (1994)
	Pumps—General Guidelines	HI	ANSI/HI 9.1-9.6 (1994)
	Engineering Data Book, 2nd ed. (1990)	HI	HI
	Circulation System Components and Related Materials for Swimming Pools, Spas/Hot Tubs	NSF	ANSI/NSF 50-1992
	Swimming Pool Pumps, Filters and Chlorinators (1993)	UL	UL 1081
	Motor-Operated Water Pumps (1991)	UL	ANSI/UL 778-1991
	Pumps for Oil-Burning Appliances (1993)	UL	ANSI/UL 343-1992
Radiators	Testing and Rating Standard for Baseboard Radiation, 6th ed. (1990)	HYDI	IBR
	Testing and Rating Standard for Finned-Tube (Commercial) Radiation (1990)	HYDI	IBR
Receivers	Refrigerant Liquid Receivers	ARI	ARI 495-93
Refrigerant-Containing Components	Refrigerant-Containing Components for Use in Electrical Equipment	CSA	C22.2 No. 140.3-M1987 (R 1993)
	Refrigerant-Containing Components and Accessories, Non-Electrical (1993)	UL	ANSI/UL 207-1993
Refrigerants	Refrigerant Recovery Recycling Equipment	ARI	ARI 740-95
	Specifications for Fluorocarbon and Other Refrigerants	ARI	ARI 700-93
	Format for Information on Refrigerants	ASHRAE	ASHRAE Guideline 6-1996
	Method of Testing Flow Capacity of Refrigerant Capillary Tubes	ASHRAE	ANSI/ASHRAE 28-1996
	Methods of Testing Discharge Line Refrigerant-Oil Separators	ASHRAE	ANSI/ASHRAE 69-1990
	Number Designation and Safety Classification of Refrigerants	ASHRAE	ANSI/ASHRAE 34-1992 with addenda
	Reducing Emission of Halogenated Refrigerants in Refrigeration and Air-Conditioning Equipment and Systems	ASHRAE	ASHRAE Guideline 3-1996
	Refrigeration Oil Description	ASHRAE	ANSI/ASHRAE 99-1981 (RA 87)
	Sealed Glass Tube Method to Test the Chemical Stability of Material for Use Within Refrigerant Systems	ASHRAE	ANSI/ASHRAE 97-1983 (RA 89)
	Test Method for Acid Number of Petroleum Products by Potentiometric Titration	ASTM	ASTM D 664-95
	Test Method for Concentration Limits of Flammability of Chemicals	ASTM	ASTM E 681-94
	Refrigerants-Number Designation	ISO	ISO 817:1974
	Recommended Service Procedure for the Containment of HFC-134a	SAE	SAE J 2211-1991
	HFC-134a Recycling Equipment for Mobile Air-Conditioning Systems	SAE	SAE J 2210-1991
	CFC-12 (R-12) Extraction Equipment for Mobile Automotive Air-Conditioning Systems	SAE	SAE J 2209-1992
	HFC-134a (R-134a) Service Hose Fittings for Automotive Air-Conditioning Service Equipment	SAE	SAE J 2197-1992
	Standard of Purity for Recycled HFC-134a for Use in Mobile Air Conditioning Systems	SAE	SAE J 2099-1991
	Recommended Service Procedure for the Containment of R-12	SAE	SAE J 1989-1989
	Procedure for Retrofitting CFC-12 (R12) Mobile Air Conditioning Systems to HFC 134-a (R134a)	SAE	ANSI/SAE J 1661-1993
	Field Conversion/Retrofit of Products to Change to an Alternate Refrigerant—Construction and Operation (1993)	UL	ANSI/UL 2170-1995
	Field Conversion/Retrofit of Products to Change to an Alternate Refrigerant—Insulating Material and Refrigerant Compatibility (1993)	UL	ANSI/UL 2171-1995
	Field Conversion/Retrofit of Products to Change to an Alternate Refrigerant—Procedures and Methods (1993)	UL	ANSI/UL 2172-1995
	Refrigerant Recovery/Recycling Equipment (1989)	UL	ANSI/UL 1963-1991

Table 1 Codes and Standards Published by Various Societies and Associations (*Continued*)

Subject	Title	Publisher	Reference
Refrigerants (continued)	Refrigerants (1994)	UL	UL 2182
Refrigeration	Safety Code for Mechanical Refrigeration	ASHRAE	ANSI/ASHRAE 15-1994
	Mechanical Refrigeration Code	CSA	B52-95
	Refrigeration Equipment	CSA	CAN/CSA-C22.2 No. 120-M91
	Equipment, Design and Installation of Ammonia Mechanical Refrigerating Systems	IIAR	ANSI/IIAR 2-1992
	Refrigerated Medical Equipment (1993)	UL	UL 416
Refrigeration Systems	Refrigerating Systems—Test Methods Part 1: Testing of Systems for Cooling Liquids and Gases Using a Positive Displacement Compressor	ISO	ISO 916-1
	Refrigerating Systems—Test Methods Part 2: Refrigerating Units Using a Positive Displacement Compressor, Condensing Units and Evaporator-Compressor	ISO	ISO 916-2
	Refrigerating Systems—Test Methods Part 3: Systems and Units for Cooling Liquids and Gases Using a Turbo-Compressor	ISO	ISO 916-3
Steam Jet	Ejectors	ASME	ASME PTC 24-1976 (R 1982)
	Standards for Steam Jet Vacuum Systems, 4th ed. (1988)	HEI	HEI
Transport	Mechanical Transport Refrigeration Units	ARI	ARI 1110-92
	Mechanical Refrigeration and Air-Conditioning Installations Aboard Ship	ASHRAE	ASHRAE 26-1996
	General Requirements for Application of Vapor Cycle Refrigeration Systems for Aircraft	SAE	SAE ARP 731A-1973 (R 1992)
	Safety and Containment of Refrigerant for Mechanical Vapor Compression Systems Used for Mobile Air-Conditioning Systems	SAE	SAE J 639-1994
Refrigerators	Method of Testing Open Refrigerators for Food Stores	ASHRAE	ANSI/ASHRAE 72-1983
	Methods of Testing Closed Refrigerators	ASHRAE	ANSI/ASHRAE 117-1992
Commercial	Energy Performance Standard for Commercial Refrigerated Display Cabinets and Merchandise	CSA	CAN/CSA-C657-95
	Food Carts	NSF	ANSI/NSF 59-1986
	Food Equipment	NSF	ANSI/NSF 2-1992
	Food Service Refrigerators and Storage Freezers	NSF	ANSI/NSF 7-1990
	Soda Fountain and Luncheonette Equipment	NSF	NSF 1-1984
	Commercial Refrigerators and Freezers (1992)	UL	ANSI/UL 471-1991
	Refrigerating Units (1994)	UL	UL 427
	Refrigeration Unit Coolers (1993)	UL	ANSI/UL 412-1992
Household	Refrigerators Using Gas Fuel	AGA	ANSI Z21.19-1990; Z21.19a-1992
	Household Refrigerators, Combination Refrigerator-Freezers and Household Freezers	AHAM	ANSI/AHAM HRF-1-1988
	Capacity Measurement and Energy Consumption Test Methods for Refrigerators, Combination Refrigerator-Freezers, and Freezers	CSA	CAN/CSA C300-M91
	Household Refrigerators and Freezers (1993)	UL	UL 250
		CSA	CAN/CSA C22.2 No. 63-93
Retrofitting			
Building	Retrofit of Building Energy Systems and Processes (1982)	SMACNA	SMACNA
	Building Systems Analysis and Retrofit Manual (1995)	SMACNA	SMACNA
Refrigerant	Procedure for Retrofitting CFC-12 (R12) Mobile Air Conditioning Systems to HFC-134a (R134a)	SAE	ANSI/SAE J 1661-1993
	Field Conversion/Retrofit of Products to Change to an Alternate Refrigerant—Construction and Operation (1993)	UL	ANSI/UL 2170-1995
	Field Conversion/Retrofit of Products to Change to an Alternate Refrigerant—Insulating Material and Refrigerant Compatibility (1993)	UL	ANSI/UL 2171-1995
	Field Conversion/Retrofit of Products to Change to an Alternate Refrigerant—Procedures and Methods (1993)	UL	ANSI/UL 2172-1995
Roof Ventilators	Commercial Low Pressure, Low Velocity Duct System Design (1990)	ACCA	ACCA Manual Q
	Power Ventilators (1984)	UL	ANSI/UL 705-1984
Safety Devices	Safety Devices for Protection Against Excessive Pressure—Part 2: Bursting Disc Safety Devices	ISO	ISO 4126-2:1981
	Safety Devices for Protection Against Excessive Pressure—Part 3: Safety Valves and Bursting Disc Safety Devices in Combination	ISO	ISO 4126-3: 1981
Solar Equipment	Method of Measuring Solar-Optical Properties of Materials	ASHRAE	ANSI/ASHRAE 74-1988
	Methods of Testing to Determine the Thermal Performance of Flat-Plate Solar Collectors Containing a Boiling Liquid	ASHRAE	ANSI/ASHRAE 109-1986 (RA 96)
	Methods of Testing to Determine the Thermal Performance of Solar Collectors	ASHRAE	ANSI/ASHRAE 93-1986 (RA 91)
	Methods of Testing to Determine the Thermal Performance of Solar Domestic Water Heating Systems	ASHRAE	ASHRAE 95-1981 (RA 87)
	Methods of Testing to Determine the Thermal Performance of Unglazed Flat-Plate Liquid-Type Solar Collectors	ASHRAE	ANSI/ASHRAE 96-1980 (RA 89)
	Reference Solar Spectral Irradiance at the Ground at Different Receiving Conditions—Part 1: Direct Normal and Hemispherical Solar Irradiance for Air Mass 1.5	ISO	ISO 9845-1:1992
	Solar Energy—Calibration of a Pyranometer Using a Pyrheliometer	ISO	ISO 9846:1993

Codes and Standards   51.17

Table 1 Codes and Standards Published by Various Societies and Associations (*Continued*)

Subject	Title	Publisher	Reference
Solar Equipment (*continued*)	Solar Heating—Domestic Water Heating Systems—Part 2: Outdoor Test Methods for System Performance Characterization and Yearly Performance Prediction of Solar-Only Systems	ISO	ISO 9459-2:1995
	Solar Heating—Swimming Pool Heating Systems—Dimensions, Design and Installation Guidelines	ISO	ISO 12596:1995
	Solar Water Heaters—Elastomeric Materials for Absorbers, Connecting Pipes and Fittings—Method of Assessment	ISO	ISO 9808:1990
	Test Methods for Solar Collectors—Part 2: Qualification Test Procedures	ISO	ISO 9806-2:1995
	Test Methods for Solar Collectors—Part 3: Thermal Performance of Unglazed Liquid Heating Collectors (Sensible Heat Transfer Only) Including Pressure Drop	ISO	ISO 9806-3:1995
Solenoid Valves	Solenoid Valves for Use with Volatile Refrigerants	ARI	ANSI/ARI 760-94
Sound Measurement	Method for the Calibration of Microphones (reaffirmed 1986)	ASA	ANSI S1.10-1966 (R 1986)
	Specification for Sound Level Meters	ASA	ANSI S1.4-1983; ANSI S1.4A-1985 (R 1990)
	Test Method for Laboratory Measurement of Airborne Sound Transmission Loss of Building Partitions	ASTM	ASTM E 90-90
	Test Method for Measuring Acoustical and Airflow Performance of Duct Liner Materials and Prefabricated Silencers	ASTM	ASTM E 477-90
	Sound and Vibration Design and Analysis (1994)	NEBB	NEBB
Fans	Methods for Calculating Fan Sound Ratings from Laboratory Test Data	AMCA	AMCA 301-90
	Reverberant Room Method for Sound Testing of Fans	AMCA	AMCA 300-94
	Methods for the Measurement of Noise Emitted by Small Air-Moving Devices	ASA	ANSI S12.11-1987 (R 1993)
	Laboratory Method of Testing In-Duct Sound Power Measurement Procedure for Fans	ASHRAE	ANSI/ASHRAE 68-1997
		AMCA	ANSI/AMCA 330-86
	Acoustics—Method for the Measurement of Airborne Noise Emitted by Small Air-Moving Devices	ISO	ISO 10302:1996
Other Equipment	Application of Sound Rated Outdoor Unitary Equipment	ARI	ARI 275-84
	Method of Measuring Machinery Sound Within Equipment Space	ARI	ARI 575-94
	Method of Measuring Sound and Vibration of Refrigerant Compressors	ARI	ANSI/ARI 530-89
	Rating the Sound Levels and Sound Transmission Loss of Packaged Terminal Equipment	ARI	ANSI/ARI 300-88
	Sound Rating of Large Outdoor Refrigerating and Air-Conditioning Equipment	ARI	ARI 370-86
	Sound Rating of Non-Ducted Indoor Air-Conditioning Equipment	ARI	ARI 350-86
	Sound Rating of Outdoor Unitary Equipment	ARI	ARI 270-84
	Statistical Methods for Determining and Verifying Stated Noise Emission Values of Machinery and Equipment	ASA	ANSI S12.3-85 (R 1996)
	Sound Level Prediction for Installed Rotating Electrical Machines	NEMA	NEMA MG 3-1974 (R 1990)
Techniques	Methods for Measurement of Sound Emitted by Machinery and Equipment at Workstations and Other Specified Positions	ANSI	ANSI S12.43-1997
	Methods for Calculation of Sound Emitted by Machinery and Equipment at Workstations and Other Specified Positions from Sound Power Level	ANSI	ANSI S12.44-1997
	Criteria for Evaluating Room Noise	ASA	ANSI S12.2-1995
	Engineering Method for the Determination of Sound Power Levels of Noise Sources Using Sound Intensity	ASA	ANSI S12.12-1992
	Engineering Methods for the Determination of Sound Power Levels of Noise Sources for Essentially Free-Field Conditions over a Reflecting Plane	ASA	ANSI S12.34-1988 (R 1993)
	Guidelines for the Use of Sound Power Standards and for the Preparation of Noise Test Codes	ASA	ANSI S12.30-1990
	Measurement of Sound Pressure Levels in Air	ASA	ANSI S1.13-1995
	Methods for Determination of Insertion Loss of Outdoor Noise Barriers	ASA	ANSI S12.8-1987
	Methods for the Determination of Sound Power Levels of Noise Sources in a Special Reverberation Test Room	ASA	ANSI S12.33-1990
	Precision Methods for the Determination of Sound Power Levels of Broad-Band Noise Sources in Reverberation Rooms	ASA	ANSI S12.31-1990 (R 1996)
	Precision Methods for the Determination of Sound Power Levels of Discrete-Frequency and Narrow-Band Noise Sources in Reverberation Rooms	ASA	ANSI S12.32-1990
	Precision Methods for the Determination of Sound Power Levels of Noise Sources in Anechoic and Hemi-Anechoic Rooms	ASA	ANSI S12.35-1990 (R 1996)
	Preferred Frequencies, Frequency Levels, and Band Numbers for Acoustical Measurements	ASA	ANSI S1.6-1984 (R 1990)
	Procedure for the Computation of Loudness of Noise	ASA	ANSI S3.4-1980 (R 1992)
	Procedures for Outdoor Measurement of Sound Pressure Level	ASA	ANSI S12.18-1994
	Reference Quantities for Acoustical Levels	ASA	ANSI S1.8-1989
	Survey Methods for the Determination of Sound Power Levels of Noise Sources	ASA	ANSI S12.36-1990
	Measurement of Industrial Sound	ASME	ASME/ANSI PTC 36-1985
	Test Method for Evaluating Masking Sound in Open Offices Using A-Weighted and One-Third Octave Band Sound Pressure Levels	ASTM	ASTM E 1573-93
	Test Method for Measurement of Sound in Residential Spaces	ASTM	ASTM E 1574-95
	Acoustics–Measurement of Sound Insulation in Buildings–Part 1: Requirements for Laboratory Test Facilities with Suppressed Flanking Transmission	ISO	ISO 140-1: 1990
	Acoustics–Measurement of Sound Insulation in Buildings and of Building Elements–Part 4: Field Measurements of Airborne Sound Insulation Between Rooms	ISO	ISO 140-4: 1978

Table 1 Codes and Standards Published by Various Societies and Associations (*Continued*)

Subject	Title	Publisher	Reference
Techniques (*continued*)	Acoustics—Measurement of Sound Insulation in Buildings and of Building Elements—Part 5: Field Measurements of Airborne Sound Insulation of Facade Elements and Facade	ISO	ISO 140-5: 1990
	Acoustics—Measurement of Sound Insulation in Buildings and of Building Ele-ments—Part 6: Laboratory Measurements of Impact Sound Insulation of Floors	ISO	ISO 140-6: 1978
	Acoustics—Measurement of Sound Insulation in Buildings and of Building Elements—Part 7: Field Measurements of Impact Sound Insulation of Floors	ISO	ISO 140-7: 1990
	Acoustics—Measurement of Sound Insulation in Buildings and of Building Elements—Part 8: Laboratory Measurements of the Reduction of Transmitted Impact Noise by Floor Coverings on a Solid Standard Floor	ISO	ISO 140-8: 1978
	Acoustics—Determination of Sound Power Levels of Noise Sources Using Sound Intensity—Part 1: Measurement at Discrete Points	ISO	ISO 9614-1:1993
	Acoustics—Determination of Sound Power Levels of Noise Sources Using Sound Intensity—Part 2: Measurement by Scanning	ISO	ISO 9614-2:1996
	Acoustics—Method for Calculating Loudness Level	ISO	ISO 532:1975
	Procedural Standards for the Measurement and Assessment of Sound and Vibration (1994)	NEBB	NEBB
Terminology	Acoustical Terminology	ASA	ANSI S1.1-1994
	Standard Terminology Relating to Environmental Acoustics	ASTM	ASTM C 634-89
Space Heaters	Method of Testing for Rating Combination Space-Heating and Water-Heating Appliances	ASHRAE	ANSI/ASHRAE 124-1991
	Electric Air Heaters	CSA	C22.2 No. 46-M1988
	Electric Air Heaters (1980)	UL	ANSI/UL 1025-1991
	Fixed and Location-Dedicated Electric Room Heaters (1992)	UL	UL 2021
	Movable and Wall- or Ceiling-Hung Electric Room Heaters (1994)	UL	UL 1278
	Gas-Fired Room Heaters, Vol. I, Vented Room Heaters	AGA	ANSI Z21.11.1-1991; Z21.11.1a-1993; Z21.11.1b-1995
	Gas-Fired Room Heaters, Vol. II, Unvented Room Heaters	AGA	ANSI Z21.11.2-1992 Z21.11.2a-1993; Z21.11.2b-1995
Symbols	Graphic Electrical/Electronic Symbols for Air-Conditioning and Refrigeration Equipment	ARI	ARI 130-88
	Graphic Symbols for Heating, Ventilating, and Air Conditioning	ASME	ANSI/ASME Y32.2.4-1949 (R 1993)
	Graphic Symbols for Pipe Fittings, Valves and Piping	ASME	ANSI/ASME Y32.2.3-1949 (R 1994)
	Graphic Symbols for Plumbing Fixtures for Diagrams used in Architecture and Building Construction	ASME	ANSI/ASME Y32.4-1977 (R 1994)
	Symbols for Mechanical and Acoustical Elements as used in Schematic Diagrams	ASME	ANSI/ASME Y32.18-1972 (R 1993)
	Graphic Symbols for Electrical and Electronic Diagrams	IEEE	ANSI/IEEE 315-1975 (R 1994)
	Abbreviations for Use on Drawings and in Text	ASME	ANSI/ASME Y1.1-1989
	Glossary of Terms Concerning Letter Symbols	ASME	ANSI/ASME Y10.1-72
	Letter Symbols and Abbreviations for Quantities Used in Acoustics	ASME	ASME Y10.11-84
	Letter Symbols for Illuminating Engineering	ASME	ASME Y10.18-67 (RA 1987)
	Safety Color Code	NEMA	ANSI/NEMA Z535.1-1991
Terminals, Wiring	Residential Controls—Quick Connect Terminals	NEMA	NEMA DC 2-1982 (R 1988)
	Electrical Quick-Connect Terminals (1995)	UL	UL 310
	Equipment Wiring Terminals for Use with Aluminum and/or Copper Conductors (1994)	UL	ANSI/UL 486E-1994
	Splicing Wire Connectors (1991)	UL	ANSI/UL 486C-1990
	Wire Connectors and Soldering Lugs for Use with Copper Conductors (1991)	UL	ANSI/UL 486A-1990
	Wire Connectors for Use with Aluminum Conductors (1991)	UL	ANSI/UL 486B-1990
Testing and Balancing	Industrial Process/Power Generating Fans: Site Performance Test Standard	AMCA	AMCA 803-94
	The HVAC Commissioning Process	ASHRAE	ASHRAE Guideline 1-1989
	Practices for Measurement, Testing, Adjusting, and Balancing of Building Heating, Ventilating, Air-Conditioning, and Refrigeration Systems	ASHRAE	ANSI/ASHRAE 111-1988
	Centrifugal Pump Test	HI	ANSI/HI 1.6-1994
	Vertical Pump Tests	HI	ANSI/HI 2.6-1994
	Rotary Pump Tests	HI	ANSI/HI 3.6-1994
	Reciprocating Pump Tests	HI	ANSI/HI 6.6-1994
	Pumps—General Guidelines (Including "Measurement of Airborne Sound")	HI	HI 9.1-9.6-1994
	Procedural Standards for Certified Testing of Cleanrooms (1988)	NEBB	NEBB
	Procedural Standards for Testing, Adjusting, Balancing of Environmental Systems, 5th ed. (1991)	NEBB	NEBB
	Building Systems Analysis and Retrofit Manual (1995)	SMACNA	SMACNA
	HVAC Systems—Testing, Adjusting and Balancing (1993)	SMACNA	SMACNA
Thermal Storage	Method of Testing Active Sensible Thermal Energy Storage Devices Based on Thermal Performance	ASHRAE	ANSI/ASHRAE 94.3-1986 (RA 96)
	Method of Testing Active Latent Heat Storage Devices Based on Thermal Performance	ASHRAE	ANSI/ASHRAE 94.1-1985 (RA 91)

Codes and Standards 51.19

Table 1 Codes and Standards Published by Various Societies and Associations (*Continued*)

Subject	Title	Publisher	Reference
Thermal Storage (*continued*)	Methods of Testing Thermal Storage Devices with Electrical Input and Thermal Output Based on Thermal Performance	ASHRAE	ANSI/ASHRAE 94.2-1981 (RA 96)
	Practices for Measurement, Testing, Adjusting, and Balancing of Building Heating, Ventilation, Air-Conditioning, and Refrigeration Systems	ASHRAE	ANSI/ASHRAE 111-1988
Turbines	Steam Turbines	ASME	ANSI/ASME PTC 6-1996
	Wind Turbines	ASME	ANSI/ASME PTC 42-1988
	Standard Specification for Gas Turbine Fuel Oils	ASTM	ANSI/ASTM D 2880-96
	Land Based Steam Turbine Generator Sets	NEMA	NEMA SM 24-1991
	Steam Turbines for Mechanical Drive Service	NEMA	ANSI/NEMA SM 23-1991
Valves	Methods of Testing Nonelectric, Nonpneumatic Thermostatic Radiator Valves	ASHRAE	ANSI/ASHRAE 102-1983 (RA 89)
	Face-to-Face and End-to-End Dimensions of Valves	ASME	ANSI/ASME B16.10-1992
	Pressure Relief Devices	ASME	ANSI/ASME PTC 25-1994
	Valves—Flanged Threaded, and Welding End	ASME	ANSI/ASME B16.34-1988
	Control Valve Capacity Test Procedure	ISA	ANSI/ISA S75.02-88
	Flow Equations for Sizing Control Valves	ISA	ANSI/ISA S75.01-85 (R 1995)
	Industrial Valves—Part-Turn Valve Actuator Attachments—Part 1	ISO	ISO 5211-1:1977
	Industrial Valves—Part-Turn Valve Actuator Attachments—Part 2	ISO	ISO 5211-2:1979
	Industrial Valves—Part-Turn Valve Actuator Attachments—Part 3	ISO	ISO 5211-3:1982
	Metal Valves for Use in Flanged Pipe Systems—Face-to-Face Dimensions	ISO	ISO 5752:1982
	Safety Devices for Protection Against Excessive Pressure Part 1: Safety Valves	ISO	ISO 4126-1:1991
	High Pressure Oxygen System Filler Valve	SAE	SAE AS 1225
	Electrically Operated Valves (1994)	UL	UL 429
	Pressure Regulating Valves for LP-Gas (1994)	UL	UL 144
	Safety Relief Valves for Anhydrous Ammonia and LP-Gas (1993)	UL	UL 132R
	Valves for Anhydrous Ammonia and LP-Gas (Other than Safety Relief) (1993)	UL	UL 125
	Valves for Flammable Fluids (1993)	UL	ANSI/UL 842-1992
Gas	Automatic Gas Valves for Gas Appliances	AGA	ANSI Z21.21-1993
	Combination Gas Controls for Gas Appliances	AGA	ANSI Z21.78-1992; Z21.78a-1993; Z21.78b-1994
	Manually Operated Gas Valves for Appliances, Appliance Connection Valves, and Hose End Valves	AGA	ANSI Z21.15-1992
	Relief Valves and Automatic Gas Shutoff Devices for Hot Water Supply Systems	AGA	ANSI Z21.22-1986; Z21.22a-1990
	Requirements for Automatic Non-Shutoff Modulating Gas Valves	AGA	1-92
	Requirements for Gas Operated Valves for High Pressure Natural Gas	AGA	3-93
	Requirements for Manually Operated Gas Valves for Use in House Piping Systems	AGA	3-88
	Requirements for Manually Operated Valves for High Pressure Natural Gas	AGA	2-93
	Large Metallic Valves for Gas Distribution (Manually Operated, NPS-2 1/2 to 12, 125 psig Maximum)	ASME	ANSI/ASME B16.38-1985 (R 1994)
	Manually Operated Metallic Gas Valves for Use in Gas Piping Systems up to 125 psig	ASME	ANSI/ASME B16.33-1990
	Manually Operated Thermoplastic Gas Shutoffs and Valves in Gas Distribution Systems	ASME	ANSI/ASME B16.40-1985 (R 1994)
Refrigerant	Refrigerant Access Valves and Hose Connectors	ARI	ANSI/ARI 720-94
	Refrigerant Pressure Regulating Valves	ARI	ARI 770-84
	Solenoid Valves for Use with Volatile Refrigerants	ARI	ANSI/ARI 760-94
	Thermostatic Refrigerant Expansion Valves	ARI	ANSI/ARI 750-94
Vapor Retarders	Standard Practice for Selection of Vapor Retarders for Thermal Insulation	ASTM	ASTM C 755-85 (R 1990)
	Standard Practice for Determining the Properties of Jacketing Material for Thermal Insulation	ASTM	ASTM C 921-89
	Standard Specification for Flexible, Low Permeance Vapor Retarders for Thermal Insulation	ASTM	ASTM C 1136-92
	Test Method for Water Vapor Transmission Rate of Flexible Barrier Materials Using an Infrared Detection Technique	ASTM	ASTM F 372-94
Vending Machines	Methods of Testing Pre-Mix and Post-Mix Soft Drink Vending and Dispensing Equipment	ASHRAE	ANSI/ASHRAE 91-1976 (RA 89)
	Vending Machines	CSA	CAN/CSA-C22.2 No.128-95
	Vending Machines for Food and Beverages	NSF	ANSI/NSF 25-1990
	Vending Machines (1995)	UL	UL 751
	Refrigerated Vending Machines (1995)	UL	UL 541
Vent Dampers	Automatic Vent Damper Devices for Use with Gas-Fired Appliances	AGA	ANSI Z21.66-1994
	Vent or Chimney Connector Dampers for Oil-Fired Appliances (1994)	UL	UL 17-1995
Ventilation	Commercial Low Pressure, Low Velocity Duct System Design (1990)	ACCA	ACCA Manual Q
	Guide for Testing Ventilation Systems	ACGIH	ACGIH
	Industrial Ventilation: A Manual of Recommended Practice, 22nd ed. (1995)	ACGIH	ACGIH
	Measuring Air Change Effectiveness	ASHRAE	ASHRAE 129-1997
	A Method of Determining Air Change Rates in Detached Dwellings	ASHRAE	ANSI/ASHRAE 136-1993
	Method of Testing for Room Air Diffusion	ASHRAE	ANSI/ASHRAE 113-1990
	Ventilation for Acceptable Indoor Air Quality	ASHRAE	ANSI/ASHRAE 62-1989

Table 1 Codes and Standards Published by Various Societies and Associations (*Continued*)

Subject	Title	Publisher	Reference
Ventilation (*continued*)	Design of Ventilation Systems for Poultry and Livestock Shelters	ASAE	ASAE D270.5-1991
	Residential Mechanical Ventilation Systems	CSA	CAN/CSA F326-M91
	Installation of Air Conditioning and Ventilating Systems	NFPA	ANSI/NFPA 90A-1993
	Parking Structures	NFPA	ANSI/NFPA 88A-1995
	Repair Garages	NFPA	ANSI/NFPA 88B-1991
	Ventilation Control and Fire Protection of Commercial Cooking Operations	NFPA	ANSI/NFPA 96-1994
	Food Equipment	NSF	ANSI/NSF 2-1992
	Class II (Laminar Flow) Biohazard Cabinetry	NSF	NSF 49-1992
	Test Procedure for Battery Flame Retardant Venting Systems	SAE	SAE J 1495-1992
	Heater, Airplane, Engine Exhaust Gas to Air Heat Exchanger Type	SAE	SAE ARP 86A-1952 (R 1992)
	Aerothermodynamic Systems Engineering and Design	SAE	SAE AIR 1168/3-1990
Venting	Draft Hoods	AGA	ANSI Z21.12-1990; Z21.12a-1993; Z21.12b-1994
	National Fuel Gas Code	AGA	ANSI Z223.1-1992; Z223.1a-1994
		NFPA	ANSI/NFPA 54-1992
	Requirements for Electrically Operated Automatic Combustion and Ventilation Air Control Devices for Use with Gas-Fired Appliances	AGA	1-88
	Requirements for Mechanical Venting Systems	AGA	6-90
	Chimneys, Fireplaces, Vents and Solid Fuel-Burning Appliances	NFPA	ANSI/NFPA 211-1992
	Explosion Prevention Systems	NFPA	ANSI/NFPA 69-1992
	Smoke and Heat Venting	NFPA	ANSI/NFPA 204M-1991
	Guide for Steel Stack Design and Construction (1996)	SMACNA	SMACNA
	Draft Equipment (1993)	UL	UL 378
	Gas Vents (1994)	UL	UL 441
	Low-Temperature Venting Systems, Type L (1994)	UL	ANSI/UL 641-1995
Vibration	Mechanical Vibration of Rotating and Reciprocating Machinery—Requirements for Instruments for Measuring Vibration Severity	ASA	ANSI S2.40-1984 (R 1992)
	Methods for Analysis and Presentation of Shock and Vibration Data	ASA	ANSI S2.10-1971 (R 1990)
	Selection of Calibrations and Tests for Electrical Transducers Used for Measuring Shock and Vibration	ASA	ANSI S2.11-1969 (R 1986)
	Techniques of Machinery Vibration Measurement	ASA	ANSI S2.17-1980 (R 1986)
	Vibrations of Buildings—Guidelines for the Measurement of Vibrations and Evaluation of Their Effects on Buildings	ASA	ANSI S2.47-1990
	Evaluation of Human Exposure to Whole-Body Vibration—Part 2: Continuous and Shock-Induced Vibrations in Buildings (1 to 80 Hz)	ISO	ISO 2631-2:1989
	Guidelines for the Evaluation of the Response of Occupants of Fixed Structures, Especially Buildings and Off-Shore Structures, to Low-Frequency Horizontal Motion (0.063 to 1 Hz)	ISO	ISO 6897:1984
	Procedural Standards for the Measurement and Assessment of Sound and Vibration (1994)	NEBB	NEBB
	Sound and Vibration Design and Analysis (1994)	NEBB	NEBB
Water Heaters	Gas Water Heaters, Vol. I, Storage Water Heaters with Input Ratings of 75,000 Btu per Hour or Less	AGA	ANSI Z21.10.1-1993; Z21.10.1a-1993; Z21.10.1b-1994
	Gas Water Heaters, Vol. III, Storage, with Input Ratings Above 75,000 Btu per Hour, Circulating and Instantaneous Water Heaters	AGA	ANSI Z21.10.3-1993; Z21-10.3a-1993; Z21.10.3b-1994
	Requirements for Non-Metallic Dip Tubes for Use in Gas-Fired Water Heaters	AGA	1-89
	Requirements for Indirect Water Heaters for Use with External Heat Source	AGA	1-91
	Desuperheater/Water Heaters	ARI	ARI 470-87
	Method of Testing for Rating Commercial Gas, Electric, and Oil Water Heaters	ASHRAE	ANSI/ASHRAE 118.1-1993
	Method of Testing for Rating Residential Water Heaters	ASHRAE	ANSI/ASHRAE 118.2-1993
	Methods of Testing for Efficiency of Space-Conditioning/Water-Heating Appliances that Include a Desuperheater Water Heater	ASHRAE	ANSI/ASHRAE 137-1995
	Methods of Testing to Determine the Thermal Performance of Solar Domestic Water Heating Systems	ASHRAE	ANSI/ASHRAE 95-1981 (RA 87)
	Method of Testing for Rating Combination Space-Heating and Water-Heating Appliances	ASHRAE	ANSI/ASHRAE 124-1991
	Construction and Test of Electric Storage-Tank Water Heaters	CSA	CAN/CSA-C22.2 No. 110-M90
	Performance of Electric Storage Tank Water Heaters	CSA	CAN/CSA-C191 Series-M90
	Oil Burning Stoves and Water Heaters	CSA	B140.3-1962 (R 1991)
	Oil-Fired Service Water Heaters and Swimming Pool Heaters	CSA	B140.12-1976 (R 1991)
	Water Heaters, Hot Water Supply Boilers, and Heat Recovery Equipment	NSF	NSF 5-1992
	Commercial-Industrial Gas Heating Equipment (1994)	UL	UL 795
	Electric Booster and Commercial Storage Tank Water Heaters (1995)	UL	ANSI/UL 1453-1987
	Household Electric Storage Tank Water Heaters (1995)	UL	UL 174
	Oil-Fired Storage Tank Water Heaters (1995)	UL	UL 732
Wood-Burning Appliances	Installation Code for Solid Fuel Burning Appliances and Equipment	CSA	CAN/CSA-B365-M91
	Solid-Fuel-Fired Central Heating Appliances	CSA	CAN/CSA-B366.1-M91
	Chimneys, Fireplaces, Vents, and Solid-Fuel-Burning Appliances	NFPA	ANSI/NFPA 211-1992
	Commercial Cooking, Rethermalization and Powered Hot Food Holding and Transport Equipment	NSF	ANSI/NSF 4-1992

Codes and Standards   51.21

ABBREVIATIONS AND ADDRESSES

AABC	Associated Air Balance Council, 1518 K Street NW, Washington, D.C. 20005
ABMA	American Boiler Manufacturers Association, 950 North Glebe Road, Suite 160, Arlington, VA 22203-1824
ACCA	Air Conditioning Contractors of America, 1712 New Hampshire Avenue, NW, Washington, D.C. 20009
ACGIH	American Conference of Governmental Industrial Hygienists, 1330 Kemper Meadow Drive, Cincinnati, OH 45240
ADC	Air Diffusion Council, 11 South Lasalle, Suite 1400, Chicago, IL 60603
AGA	American Gas Association, 1515 Wilson Boulevard, Arlington, VA 22209
	Also available through International Approval Services U.S., Inc., 8501 East Pleasant Valley Road, Cleveland, OH 44131
AHAM	Association of Home Appliance Manufacturers, 20 North Wacker Drive, Suite 1600, Chicago, IL 60606
AIHA	American Industrial Hygiene Association, 2700 Prosperity Avenue, Suite 250, Fairfax, VA 22031
AMCA	Air Movement and Control Association, Inc., 30 West University Drive, Arlington Heights, IL 60004-1893
ANSI	American National Standards Institute, 11 West 42nd Street, 13th floor, New York, NY 10036-8002
ARI	Air-Conditioning and Refrigeration Institute, 4301 North Fairfax Drive, Suite 425, Arlington, VA 22203
ASA	Acoustical Society of America, Standards Secretariat, 120 Wall Street, 32nd floor, New York, NY 10005-3993
	For ordering publications: Standards and Publications Fulfillment Center, P.O. Box 1020, Sewickley, PA 15143-9998
ASAE	American Society of Agricultural Engineers, 2950 Niles Road, St. Joseph, MI 49085-9659
ASHRAE	American Society of Heating, Refrigerating and Air-Conditioning Engineers, Inc., 1791 Tullie Circle, NE, Atlanta, GA 30329
ASME International	The American Society of Mechanical Engineers, 345 East 47 Street, New York, NY 10017-2392
	For ordering publications: ASME Marketing Department, P.O. Box 2350, Fairfield, NJ 07007-2350
ASTM	American Society for Testing and Materials, 100 Barr Harbor Drive, West Conshohocken, PA 19428-2959
AWS	American Welding Society, Inc., 550 N.W. LeJeune Road, Miami, FL 33126
AWWA	American Water Works Association, 6666 W. Quincy Avenue, Denver, CO 80235
BOCA	Building Officials and Code Administrators International, Inc., 4051 West Flossmoor Road, Country Club Hills, IL 60478-5795
BSI	British Standards Institution, 389 Chiswick High Road, London W4 4AL, England
CABO	Council of American Building Officials, 5203 Leesburg Pike, Suite 708, Falls Church, VA 22041
CAGI	Compressed Air and Gas Institute, 1300 Sumner Avenue, Cleveland, OH 44115-2851
CGA	Canadian Gas Association, 55 Scarsdale Road, Toronto, ON M3B 2R3, Canada
CSA	Canadian Standards Association, 178 Rexdale Boulevard, Etobicoke (Toronto), ON M9W 1R3, Canada
CTI	Cooling Tower Institute, P.O. Box 73383, Houston, TX 77273
EJMA	Expansion Joint Manufacturers Association, Inc., 25 North Broadway, Tarrytown, NY 10591-3201
HEI	Heat Exchange Institute, 1300 Sumner Avenue, Cleveland, OH 44115-2851
HI	Hydraulic Institute, 9 Sylvan Way, Parsippany, NJ 07054-3802
HYDI	Hydronics Institute, 35 Russo Place, P.O. Box 218, Berkeley Heights, NJ 07922
IAPMO	International Association of Plumbing and Mechanical Officials, 20001 Walnut Drive South, Walnut, CA 91789-2825
ICBO	International Conference of Building Officials, 5360 Workman Mill Road, Whittier, CA 90601
ICC	International Code Council, 5360 Workman Mill Road, Whittier, CA 90601
IEEE	Institute of Electrical and Electronics Engineers, 445 Hose Lane, P.O. Box 1331 Piscataway, NJ 08855-1331
IESNA	Illuminating Engineering Society of North America, 120 Wall Street, 17th floor, New York, NY 10005-4001
IFCI	International Fire Code Institute, 5360 Workman Mill Road, Whittier, CA 90601-2298
IIAR	International Institute of Ammonia Refrigeration, 1200 19th Street, NW, Suite 300, Washington, DC 20036-2412
ISA	ISA—The International Society for Measurement and Control, P.O. Box 12777, Research Triangle Park, NC 27709
ISO	International Organization for Standardization, 1, rue de Varembé, Case postale 56, CH-1211 Genève 20, Switzerland
	Publications available in the U.S. from ANSI, 11 West 42nd Street, 13th floor, New York, NY 10036-8002
MCAA	Mechanical Contractors Association of America, 1385 Piccard Drive, Rockville, MD 20850-4329
MICA	Midwest Insulation Contractors Association, 2017 South 139th Circle, Omaha, NE 68144
MSS	Manufacturers Standardization Society of the Valve and Fittings Industry, Inc., 127 Park Street, N.E., Vienna, VA 22180
NAIMA	North American Insulation Manufacturers Association, 44 Canal Center Plaza, Suite 310, Alexandria, VA 22314
NAPHCC	National Association of Plumbing-Heating-Cooling Contractors, P.O. Box 6808, Falls Church, VA 22040
NCPWB	National Certified Pipe Welding Bureau, 1385 Piccard Drive, Rockville, MD 20850
NCSBCS	National Conference of States on Building Codes and Standards, 505 Huntmar Park Drive, Suite 210, Herndon, VA 22070
NEBB	National Environmental Balancing Bureau, 8575 Grovemont Circle, Gaithersburg, MD 20877-4121
NEMA	National Electrical Manufacturers Association, 1300 North 17th Street, Suite 1847, Rosslyn, VA 22209
NFPA	National Fire Protection Association, 1 Batterymarch Park, P.O. Box 9101, Quincy, MA 02269-9101
NRCC	National Research Council of Canada, Client Services, M-20, 1200 Montreal Road, Ottawa, ON K1A 0R6, Canada
NSF	NSF International, P.O. Box 130140, Ann Arbor, MI 48113-0140
SAE	Society of Automotive Engineers, 400 Commonwealth Drive, Warrendale, PA 15096-0001
SBCCI	Southern Building Code Congress International, Inc., 900 Montclair Road, Birmingham, AL 35213-1206
SMACNA	Sheet Metal and Air Conditioning Contractors' National Association, 4201 Lafayette Center Drive, Chantilly, VA 22021-1209
TEMA	Tubular Exchanger Manufacturers Association, Inc., 25 North Broadway, Tarrytown, NY 10591-3201
UL	Underwriters Laboratories Inc., 333 Pfingsten Road, Northbrook, IL 60062-2096

ASHRAE HANDBOOK

ADDITIONS AND CORRECTIONS

This section includes additional information and notes technical errors found in the SI edition of the *ASHRAE Handbooks*. Changes marked with a * were also published in the Additions and Corrections section of the 1996 *HVAC Systems and Equipment* volume. Occasional typographical errors and nonstandard symbol labels will be corrected in future volumes.

The authors and editor encourage you to notify them if you find other technical errors. Please send corrections to: Handbook Editor, ASHRAE, 1791 Tullie Circle NE, Atlanta, GA 30329, or e-mail bparsons@ashrae.org.

1995 HVAC Applications

*p. 7.6, 2nd column.** In the 3rd line of the paragraph on "Intensive Care Unit," the temperature range should read 24°C to 27°C.

p. 14.2, Equation (1). The engine heat release is in kW, not W.

p. 20.9, Table 3. The second type of new construction should read, Single glass lapped (laps sealed). The number of air changes is 1.0.

p. 27.1, 2nd column. The sentence and equation below the small table should read

Moisture pickup is 22 − 10 = 12 g/kg (dry air). The rate of evaporation in the dryer is

41(228 − 32)/100 = 80.36 kg/h = 22.3 g/s

*p. 42.5, 2nd column, 5th para.** Replace paragraph with

A *flow coefficient* (defined in Chapter 34) is sometimes used to compare valve capacities.

p. 42.32, 1st column. The first sentence in the section, Size of Controlled Area, should read in part:

No individually controlled area should exceed 500 m^2 ....

*p. 44.8, 2nd column.** Delete the last paragraph in the WATER TREATMENT General Considerations section and replace with the following paragraph.

Some users have reported success with field installations of various nonchemical equipment in side-by-side comparative tests with conventional water treatment methods. Investigations of nonchemical treatment have been mixed, with some positive (Brecevic and Kralj 1989; Busch et al. 1985, 1986; Craine 1984; Donaldson and Grimes 1988; Lin 1989: Lin and Benguigui 1985; Quinn 1989; Saam 1980) and some negative (Alleman 1985; Gruber and Carda 1981; Hasson and Bramson 1985; Meckler 1974a; Puckorius 1981; Rosa 1988; Sohnel 1988). (See the Additions and Corrections section in the 1996 ASHRAE Handbook for the list of references.)

*p. 45.11, Table 7.** The units for Maximum Daily hot water demand for Type A and Type B Food service establishments should be litres per max. meals per day.

p. 45.19, 1st column, 8th line up. The area of the oval is

$$\pi R^2 + LW$$

p. 45.19, Equation (12). Equation must be divided by 3600 s/h for units shown.

p. 45.20, 2nd column. The reference for Figures 23 and 24 is

Hunter, R.B. 1941. Water distributing systems for buildings. National Bureau of Standards *Report* BMS 79.

p. 48.10, 1st column, 2nd line. The equation for *B* should read:

$$B = 3460[1/(273 + t_o) − 1/(273 + t_s)]$$

1996 HVAC Systems and Equipment

p. 6.14. Figure 20 shows earth banked up to the top of the foundation wall. Most building codes require some clearance between the earth and any wood framing.

p. 12.4, Equation (13). A brace in the equation is misplaced. The equation should read

$$V_t = 2V_s\left[\left(\frac{v_2}{v_1} − 1\right) − 3\alpha\Delta t\right]$$

p. 28.2, 1st column. The operating efficiencies of the listed furnace categories should be exchanged. That is, Category I furnaces operate at a steady-state efficiency less than 83%, Category II at greater than 83%, Category III at less than 83%, and Category IV at greater than 83%

p. 30.6, Equation (12). The right side of the equation should be multiplied by ρ_m, which is the mean gas density calculated from Equation (2).

p. 30.12, 2nd column. In step 6 of Example 2, the first calculation for Δp should be divided by 2. The equation should then read

$$\Delta p = (3.65)(0.291)\frac{(21.4)^2}{2} = 243 \text{ Pa flow losses}$$

1997 Fundamentals

p. 1.8, Equation (39). The equation for *h* should read $h_3 = h_4$ instead of $h_1 = h_2$.

p. 6.2, Equation (5). The exponent in the definition of C_3 should be E−03, so its value is −0.009 677 843.

p. 6.15, 1st column, 13th line. Replace "pound" with "kilogram".

p. 7.2, 2nd column, 13th line up. Delete entire parenthetic sentence, (The SI units are used....).

p. 7.11, 1st column. Third line of the section on Enclosures and Barriers: delete "weight or".

p. 8.8, 1st column, 10th line up. Replace 14.7 psia with 101.325 kPa.

p. 8.26, 3rd column. Units for V_{CO2} and V_{O2} should be L/s.

p. 10.3, 2nd column, section on Cyclic Conditions. Paragraph 3, 8th line: 1 to 2 footcandles should read 10 to 20 lx. Paragraph 4, 4th line: 1 to 3 footcandles should read 10 to 30 lx.

p. 10.8, 2nd column. In the section on Heat Production, 7th line, add a minus sign to 40°C.

p. 11.13, 2nd column, 4th line up in section on Wheat and Barley. Replace "bushel" with "grain".

p. 14.3, Equation 1. Delete *J* from equation and nomenclature. Delete "k" from units of specific heat c_p; units should be J/(kg·K).

A.1

p. 15.11, 1st column, 4th line up. The AIHA standard number should read Z9.5.

p. 17.9, 1st column, 4th paragraph, 5th line. Replace "pound" with "kilogram".

p. 18.8, Table 8. Delete data for R-11 and R-123 in Section F and Section G because the superheat values for these refrigerants are unrealistically large. Also, correct the last column of the table to read the following values:

Refrigerant No.	Power Consumption, kW	Refrigerant No.	Power Consumption, kW
Section E		Section H	
717	0.309	717	0.147
134a	0.328	134a	0.150
123	0.297	123	0.142
11	0.290	11	0.141
Section F		Section I	
717	0.360	717	0.137
134a	0.392	134a	0.142
		123	0.133
Section G		11	0.131
717	0.333	290	0.144
134a	0.313		

p. 22.15, 2nd column. Delete third sentence in last paragraph that begins, "For example …"

p. 23.12, 1st column, last line. Gravel pad thickness should read 10 mm.

pp. 24.4-7, Table 4. Change caption in the $1/k$ resistance column to read, "Per Metre Thickness," not "Per Inch Thickness."

p. 24.19, 2nd column, 5th line up. Replace foot with metre.

pp. 24.21-22, Table 13. In last line of footnote, replace inches with mm.

p. 26.3, 1st column. The extreme value distribution is a **double** exponential distribution, so the equation for F at the bottom of the column should read

$$F = -\frac{\sqrt{6}}{\pi}\left[0.5772 + \ln\ln\left(\frac{n}{n-1}\right)\right]$$

pp. 26.6-52, Tables 1A, 2A, and 3A. In the MWS/MWD to DB columns, the columns headed PWD should read MWD.

pp. 26.6-52, Tables 1A, 2A, and 3A. The humidity ratio (HR) units are grams of water vapor per kilogram of dry air.

pp. 26.7-24. The temperature values in Tables 1B and 2A are shown to the nearest degree Celsius. If you need these tables with values to the nearest 0.1°C, contact Handbook Editor at ASHRAE (bparsons@ashrae.org) for a printed copy.

p. 26.9. DP/MDB and HR data for Hartford, Brainard Field, Connecticut, should be corrected to read as follows:

DP/MDB AND HR								
0.4%			1%			2%		
DP	HR	MDB	DP	HR	MDB	DP	HR	MDB
22.5	17.3	27.2	21.8	16.5	26.3	20.9	15.7	25.6

p. 26.26. The latitude and longitude for the Salta Airport weather station in Argentina are S and W, respectively.

p. 26.52. The latitude and longitude for the Treinta Y Tres weather station in Uruguay are S and W, respectively.

p. 27.4, Figure 1. To reflect the revised weather data in Chapter 26, the caption on the x-axis of the graph should read, 1% DESIGN HUMIDITY RATIO (W).

p. 27.12, Table 16. The values in the first and third kelvin-day column headings have been switched. The first value should read 1640, and the value in the last column should read 4130.

p. 28.30. Replace Table 25 with the following table, which includes SI units.

Table 25 Room Transfer Functions: v_0 and v_1 Coefficients[a]

Heat Gain Component	Room Envelope Construction[b]	v_0	v_1
		Dimensionless	
Solar heat gain through glass[c] with no interior shade; radiant heat from equipment and people	Light	0.224	$1 + w_1 - v_0$
	Medium	0.197	$1 + w_1 - v_0$
	Heavy	0.187	$1 + w_1 - v_0$
Conduction heat gain through exterior walls, roofs, partitions, doors, windows with blinds or drapes	Light	0.703	$1 + w_1 - v_0$
	Medium	0.681	$1 + w_1 - v_0$
	Heavy	0.676	$1 + w_1 - v_0$
Convective heat generated by equipment and people, and from ventilation and infiltration air	Light	1.000	0.0
	Medium	1.000	0.0
	Heavy	1.000	0.0

Heat Gain from Lights[d]				
Furnishings	Air Supply and Return	Type of Light Fixture	v_0	v_1
Heavyweight simple furnishings, no carpet	Low rate; supply and return below ceiling ($V \leq 25$)[e]	Recessed, not vented	0.450	$1 + w_1 - v_0$
Ordinary furnishings, no carpet	Medium to high rate, supply and return below or through ceiling ($V \geq 25$)[e]	Recessed, not vented	0.550	$1 + w_1 - v_0$
Ordinary furnishings, with or without carpet on floor	Medium to high rate, or induction unit or fan and coil, supply and return below, or through ceiling, return air plenum ($V \geq 25$)[e]	Vented	0.650	$1 + w_1 - v_0$
Any type of furniture, with or without carpet	Ducted returns through light fixtures	Vented or free-hanging in air-stream with ducted returns	0.750	$1 + w_1 - v_0$

[a] The transfer functions in this table were calculated by procedures outlined in Mitalas and Stephenson (1967) and are acceptable for cases where all heat gain energy eventually appears as cooling load. The computer program used was developed at the National Research Council of Canada, Division of Building Research.
[b] The construction designations denote the following:
Light construction: such as frame exterior wall, 50-mm concrete floor slab, approximately 150 kg of material per square metre of floor area.
Medium construction: such as 100-mm concrete exterior wall, 100-mm concrete floor slab, approximately 540 kg of building material per square metre of floor area.
Heavy construction: such as 150-mm concrete exterior wall, 150-mm concrete floor slab, approximately 630 kg of building material per square metre of floor area.
[c] The coefficients of the transfer function that relate room cooling load to solar heat gain through glass depend on where the solar energy is absorbed. If the window is shaded by an inside blind or curtain, most of the solar energy is absorbed by the shade, and is transferred to the room by convection and long-wave radiation in about the same proportion as the heat gain through walls and roofs; thus the same transfer coefficients apply.
[d] If room supply air is exhausted through the space above the ceiling and lights are recessed, such air removes some heat from the lights that would otherwise have entered the room. This removed light heat is still a load on the cooling plant if the air is recirculated, even though it is not a part of the room heat gain as such. The percent of heat gain appearing in the room depends on the type of lighting fixture, its mounting, and the exhaust airflow.
[e] V is room air supply rate in litres per second per square metre of floor area.

Additions and Corrections

p. 28.14, Table 9B, 1st column. In last two entries, replace 17 in. with 430 mm.

p. 28.18, Table 10, 2nd column. In definition of Q beneath Equation (20), replace cfm with m^3/s.

pp. 28.26–27, Tables 18 and 19. The heading in the first column of these tables should read Wall Group.

p. 28.28, 1st column, last para., 3rd line. Replace Btu/(h·ft^2) with W/m^2.

p. 28.33, Example 5, Solution, 2nd line. Replace 400 ft^2 with 40 m^2.

p. 28.38, 1st column, 2nd equation. Delete "Btu/h".

p. 28.41, 1st column, 2nd line up. Replace "feet" with "metres".

pp. 28.47-48. The headings for the tables should read:

Table 33B Wall Types, Mass Evenly Distributed, for Use with Table 32

Table 33C Wall Types, Mass Located Outside Insulation, for Use with Table 32

p. 28.49, Example 7, 8th line. Area should be 10 m^2, not 10 m^3.

p. 28.55, 2nd column, infiltration table. In the last column, the units of q should be W, not Btu/h.

p. 28.57, 1st column. In Equation (49), the letter l should be a λ.

p. 28.57, 2nd column. Under "Power," delete "horse" from definition of P. Under "Appliances," change energy input to power input in definition of q_{input}.

p. 29.28, 1st column, 15th line up. Replace 45°F with 7°C.

p. 29.28, 2nd column. The first sentence in the third paragraph up should refer to Chapter 26, not Chapter 24.

p. 30.12, 1st column, equation nomenclature. Change definition of c to read, c = fluid capacity, kg/s.

p. 32.9, Figure 9. The shaded area in the friction chart shown on page A.4 is the suggested range of air velocity and friction loss for design.

p. 33.2, Example 1. Solution equation should be

$$\Delta p = 1.5 \times 1000 \times 1^2/2 = 750 \text{ Pa}$$

p. 33.7, 2nd column, 18th line. Replace "gallons per minute" with "litres per second".

p. 33.9, 2nd column, 15th line. Replace 4 mm with 40 mm.

p. 33.17, 1st column. 1st para., last line: Change "specific gravity" to "density". 2nd para., 5th line: Change "Btu input" to "energy input". 3rd para., 2nd line: Replace "specific gravity of 0.60 with "density of 0.735 kg/m^3".

p. 34.4, 1st column, Letter Symbols. Delete J, mechanical equivalent of heat.

p. 35.2, last line of Table 2. Conversion of Fahrenheit to °R should read $x + 459.67$.

p. 39.1. The page numbers in the table of contents should read 39.#, not 38.#.

p. 28.30. Replace Table 25 with the following table.

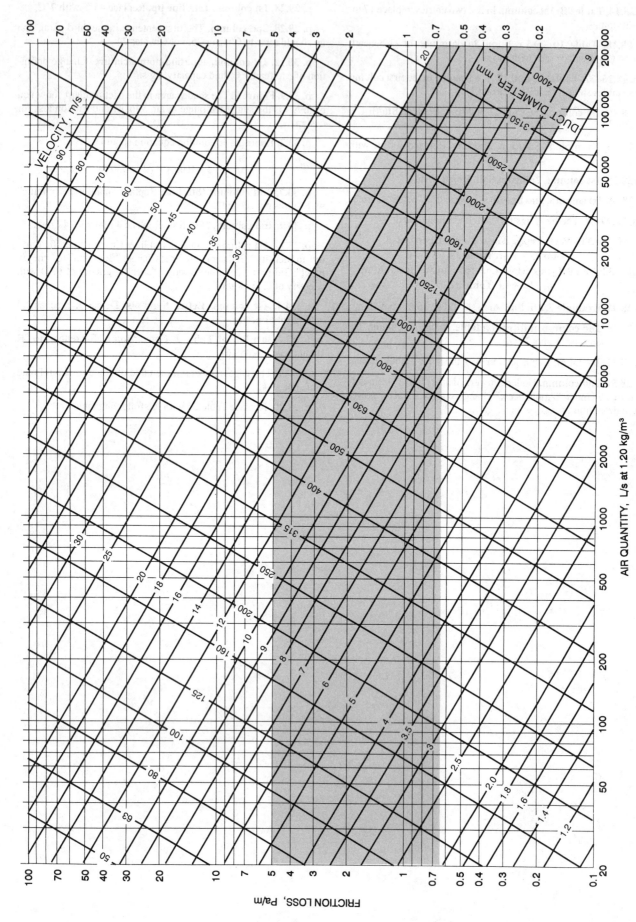

Fig. 9 Friction Chart for Round Duct ($\rho = 1.20$ kg/m³ and $\varepsilon = 0.09$ mm)

COMPOSITE INDEX
ASHRAE HANDBOOK SERIES

This index covers the current Handbook series published by ASHRAE. The four volumes in the series are identified as follows:

A = 1995 Applications
S = 1996 Systems and Equipment
F = 1997 Fundamentals
R = 1998 Refrigeration

Alphabetization of the index is letter by letter; for example, **Heaters** precedes **Heat exchangers**, and **Zones** precedes **Zone systems**.

The page reference for an index entry includes the book letter and the chapter number, which may be followed by a decimal point and the beginning page or page range within the chapter. For example, the page number R31.4 means the information may be found in the 1998 Refrigeration volume, Chapter 31, beginning on page 4.

Each Handbook is revised and updated on a four-year cycle. Because technology and the interests of ASHRAE members change, some topics are not included in the current Handbook series but may be found in the earlier Handbook editions cited in the index.

Abbreviations, F34
 computer programming, F34.1
  table, F34.2
 drawings, F34.1
  table, F34.2
 letter symbols, F34.1
  table, F34.4
 text, F34.1
  table, F34.2
Absorbents
 liquid, F21.3
Absorptance
 fenestration, F29.17
 table, F3.8
Absorption, F21.1
 dehumidification, S22.1, 8
 gaseous contaminant control, A41.9
 industrial gas cleaning
  gaseous contaminant control, S25.17
 refrigeration cycles, F1.14
  ammonia-water, F1.18; F19.1
  lithium bromide-water, F1.16; F19.1
  solar cooling, A30.21
Absorption refrigeration, R41
 ammonia-water-hydrogen cycle, R41.9
 ammonia-water technology, R41.9
 basic cycles, R41.2
  heat transformer, R41.2
  multiple-effect, R41.2
  single-effect, R41.2
 evolving technologies, R41.11
  generator-absorber heat exchange cycle, R41.11
  liquid desiccant absorption, R41.12
  solid-vapor sorption, R41.11
  triple-effect cycle, R41.11
 terminology, R41.1
 water-lithium bromide technology, R41.2
  components, R41.3
  control, R41.7
  machines, R41.4
   double-effect chillers, R41.5
   single-effect chillers, R41.4
   single-effect heat transformers, R41.5
  maintenance, R41.8
  operation, R41.6
ACH. *See* **Air changes per hour (ACH)**
Acoustics
 computer applications, A36.8-9
Activated charcoal
 gaseous contaminant control, A41.9
Adjusting (*see also* **Testing, adjusting, and balancing**)
 HVAC systems, A34

ADPI. *See* **Air diffusion performance index (ADPI)**
Adsorbents
 solid, F21.4
Adsorption, F21.1
 dehumidification, S22.1, 9
 gaseous contaminant control, A41.9-11
 industrial gas cleaning, S25.17, 23
  applications
   odor control, S25.25
   solvent recovery, S25.24
  equipment, S25.24
   fixed beds, S25.24
   fluidized beds, S25.24
   moving beds, S25.24
  gaseous contaminant control, S25.23
  impregnated adsorbents, S25.24
Aeration
 farm crops, A21.4, 10-12; F11.11
Affinity laws
 pumps, centrifugal, S38.7
Age of air, F25.6
Agitated-bed drying, A27.6
Air
 emittance (table), F24.2
 liquefaction, R38.5
 lubricant solubility, R7.20
 moist. *See* **Moist air**
 properties, F6.16
 separation, R38.13
 standard, F28.15
 thermal resistance (table), F24.2, 3
Air cargo. *See* **Air transport**
Air changes per hour (ACH), F25.3
Air cleaners (*see also* **Filters, air; Gas-cleaning equipment**), S24
 absorbers, A41.9
 adsorbers, A41.9-11
 aerosol, S24.1
 air washers, S19.7
 atmospheric dust, S24.1
 chemisorbers, A41.11
 collection mechanisms, S24.5
  diffusion, S24.5
  direct interception, S24.5
  electrostatic effects, S24.5
  inertial deposition, S24.5
  straining, S24.5
 combination, S24.6
 electronic, S24.6, 8; S28.2
 fibrous media, S24.5
 industrial exhaust systems, A26.10
 industrial ventilation, S24.1
 maintenance, S24.9
 ratings, S24.2

 renewable media, S24.6, 7
 safety requirements, S24.11
 selection, S24.9
 small forced-air systems, S9.1
 standards, S24.2, 5
 synthetic dust, S24.2, 3
 test methods, S24.2
  arrestance, S24.2, 3
  DOP penetration, S24.3
  dust-holding capacity, S24.2, 3
  dust-spot efficiency, S24.2, 3
  environmental, S24.5
  fractional efficiency, S24.2
  leakage (scan), S24.4
  particle size efficiency, S24.2, 4
  penetration, S24.2, 3
 types, S24.5
Air cleaning (*see also* **Gas-cleaning equipment**)
 absorption, A41.9
 adsorption, A41.9-11
 evaporative cooling, A47.3
 gaseous contaminant removal, A41.8-12
 incineration, A41.11-12
Air conditioners
 central systems, S1.3
  air distribution, S1.6
   ceiling plenums, S1.6
   controls, S1.6
   ductwork, S1.6
   insulation, S1.6
   room terminals, S1.6
   sound control, S1.6
   vibration control, S1.6
  applications, S1.3
   environmental control, S1.3
   large areas, S1.3
   precision control, S1.3
   primary sources, S1.3
   uniform loads, S1.3
  components, S1.4
   air-conditioning units, S1.4
   automatic dampers, S1.4
   cooling coils, S1.5
   filters, S1.5
   humidifiers, S1.6
   mixing plenums, S1.5
   outdoor air intakes, S1.5
   preheat coils, S1.5
   reheat coils, S1.5
   relief air fans, S1.4
   relief openings, S1.4
   return air dampers, S1.5
   return air fans, S1.4
   supply air fans, S1.6
  performance, S1.3

I.1

Air conditioners *(cont.)*
  central systems *(cont.)*
    primary systems, S1.2, 7
      absorption chillers, S1.8
      air-cooled condensers, S1.8
      centrifugal compressors, S1.8
      cooling towers, S1.8
      evaporative condensers, S1.8
      fuels, S1.8
      heating equipment, S1.7
      instrumentation, S1.9
      piping, S1.9
      pumps, S1.8
      reciprocating compressors, S1.8
      refrigeration equipment, S1.8
      rotary compressors, S1.8
    secondary systems, S1.2, 3
    smoke management, S1.3
    space requirements, S1.9
      equipment access, S1.10
      fan room, S1.9
      interior shafts, S1.10
      mechanical, electrical, plumbing, S1.9
  system selection and design, S1
    constraints, S1.1
      architectural, S1.2
      cooling load, S1.1
      heating and ventilation, S1.1
      zoning requirements, S1.1
    goals, S1.1
    narrowing the choice, S1.2
    selection report, S1.2
  packaged terminal, S5.2; S45.1
    classifications, S45.1
    design, S45.1, 2
    sizes, S45.1
    testing, S45.3
  residential, A1.4-6
  room, S43.1
    codes, S43.4
    design, S43.1
      compressors, S43.2
      condensers, S43.2
      control, S43.2
      evaporators, S43.2
      fans, S43.2
      restrictors, S43.2
    filters, S43.4
    installation/service, S43.5
    noise, S43.4
    performance, S43.2
      efficiency, S43.2
      energy conservation, S43.3
      sensible heat ratio, S43.3
    sizes/classifications, S43.1
    special features, S43.3
    standards, S43.4
  small forced-air systems, S9.1
  unitary, S44; S44.6
    accessories, S44.7
    air-handling systems, S44.7
    application, S44.1
    certification, S44.5
    codes, S44.5
    desuperheaters, S44.4
    efficiency, S44.5
    electrical design, S44.7
    energy conservation, S44.5
    installation, S44.1
    maintenance, S44.3
    mechanical design, S44.7
    refrigerant circuit, S44.6
    residential, A1.4-5
    space conditioning/water heating, S44.4

    split system, S44.1, 6
    standards, S44.5
    through-the-wall, S5.2
    types, S44.3, 4
    window-mounted, S5.2
Air conditioning
  air-and-water systems, S3
    changeover temperature, S3.4
    description, S3.1
    exterior zone loads, S3.1
      external loads, S3.1
      internal loads, S3.1
    fan-coil systems, S3.3
    four-pipe systems, S3.8
      description, S3.8
      evaluation, S3.8
      room control, S3.8
      zoning, S3.8
    induction systems, S3.2
      advantages, S3.2
      disadvantages, S3.2
    performance under varying load, S3.3
    primary air, S3.1, 3
    refrigeration load, S3.4
    secondary water, S3.1, 8
    ships, A10.4
    three-pipe systems, S3.7; 1976 Systems, Chapter 4, pp. 8-9, 13
      (See explanation on first page of Index)
    two-pipe systems, S3.5
      changeover temperature, S3.6
      description, S3.5
      design, S3.5
      electric heat, S3.7
      evaluation, S3.7
      nonchangeover design, S3.6
      room control, S3.7
      zoning, S3.6
  aircraft, A9
  airports, A3.12
  all-air systems, S2
    advantages, S2.1
    control, S2.9
      constant volume, S2.9
      dual-duct, S2.10
      variable-air-volume, S2.9
    cooling, S2.2
    costs, S2.5
    dehumidification, S2.3
    design, S2.3
      distribution systems, S2.3
      zoning, S2.4
    disadvantages, S2.1
    dual-duct, S2.7
      constant volume, S2.7
      variable-air-volume, S2.7
    energy conservation, S2.5
    equipment, S2.9
    heating, S2.2
    humidification, S2.2
    multizone, S2.8
    single-duct, S2.5
      constant volume, S2.5
      variable-air-volume, S2.6
  all-water systems, S4
    bare pipe, S4.1
    baseboard radiation, S4.1
    convectors, S4.1
    fan-coil units, S4.1
      advantages, S4.4
      applications, S4.4
      capacity control, S4.2
      disadvantages, S4.4
      location, S4.1

      maintenance, S4.3
      piping, S4.2
      selection, S4.2
      types, S4.1
      ventilation, S4.3
      water distribution, S4.3
      wiring, S4.2
    panel systems, S4.1
    radiators, S4.1
    unit ventilators, S4.4
  ammonia systems, R3.1
  animal buildings, A20.3
  arenas, A4.4-5
  atriums, A4.9
  auditoriums, A4.3-4
  automobiles, A8.1-7
  bakeries, R27
  bars, A3.5-6
  bowling centers, A3.10
  buses, A8.7-9
  bus terminals, A3.12
  candy manufacturing plants (table), R28.1
  central systems
    ships, A10.3-4
  clean spaces, A15
  colleges, A6.3-4
  commercial buildings, A3
  communications centers, A3.11
  concert halls, A4.4
  convention centers, A4.5-6
  cooling load calculations, F28
  data processing areas, A16
  dehumidification for, S22.6
  domiciliary facilities, A5
  dormitories, A5.3
  dual duct systems
    ships, A10.4-5
  educational facilities, A6
  engine test facilities, A14
  exhibition centers, A4.5-6
  fairs, A4.8
  fixed guideway vehicles, A8.11
  gymnasiums, A4.5
  health care facilities, A7
    disease prevention/treatment, A7.1-2
    outpatient, A7.10-11
  hospitals, A7.2-10
  hotels, A5.3-4
  houses of worship, A4.3
  industrial environment, A11; A24
  kitchens, A3.5, 6; A28
  laboratories, A13
  mines, A25
  motels, A5.3-4
  natatoriums, A4.6-8
  nightclubs, A3.6
  nuclear facilities, A23
  nursing homes, A7.11-12
  office buildings, A3.6-8
  outpatient health care facilities, A7.10-11
  panel systems, S6
  paper products facilities, A22.2-4
  photographic processing/storage areas, A19
  places of assembly, A4
  plant growth facilities, A20.15
  printing plants, A17
  public buildings, A3
  railroad cars, A8.9-11
  reheat systems
    ships, A10.4
  residential, A1
  restaurants, A3.5
  retail stores, A2
  retrofitting
    contaminant control, R6.12

A = 1995 Applications   S = 1996 Systems and Equipment   F = 1997 Fundamentals   R = 1998 Refrigeration

Composite Index

I.3

Air conditioning *(cont.)*
  schools, A6.2-3
  ship docks, A3.12
  ships, A10
  stadiums, A4.4-5
  temporary exhibit buildings, A4.8
  textile processing, A18.4-6
  theaters, A4.3, 4
  thermal storage, A40.4-6
  transportation centers, A3.11-13
  unitary systems, S5
    characteristics, S5.1
      advantages, S5.1
      disadvantages, S5.1
    economizers, S5.7
      airside, S5.7
      waterside, S5.7
    heat pumps
      air-to-air, S5.3
      through-the-wall, S5.2
      window-mounted, S5.2
    indoor equipment, S5.4
      description, S5.4
      design considerations, S5.4
    outdoor equipment, S5.3
      advantages, S5.4
      design considerations, S5.3
      disadvantages, S5.4
    self-contained, S5.5
      acoustics, S5.6, 7
      advantages, S5.6, 8
      commercial, S5.6
      controls, S5.8
      disadvantages, S5.6, 8
    through-the-wall, S5.2
      advantages, S5.3
      controls, S5.3
      design considerations, S5.2
      disadvantages, S5.3
    window-mounted, S5.2
      advantages, S5.2
      design considerations, S5.2
      disadvantages, S5.2
  unit ventilators, S31.1
  universities, A6.3-4
  warehouses, A3.13
  warm humid climates, F23.9
  wood products facilities, A22.1
Air contaminants. *See* **Contaminants, gaseous; Contaminants, particulate**
Air coolers. *See* **Coolers**
Air cooling
  evaporative. *See* **Evaporative cooling**
Aircraft air conditioning, A9
  air distribution, A9.3-4
  air quality, A9.2-3
  air source, A9.4
  control, A9.8
  design conditions, A9.1-3
    air conditioning performance, A9.1-2
    load determination, A9.3
    operational recommendations, A9.2
    pressurization performance, A9.2
    temperature/humidity/pressure, A9.1
    ventilation, A9.2
  pressurization
    control, A9.7
    performance, A9.2
  refrigeration systems, A9.4-6
    air cycle, A9.4-6
    direct expansion, A9.6
    vapor cycle, A9.6
  temperature control, A9.6-7
  typical system, A9.7-8

Air curtains, S17.7
  industrial environment, A24.14-16
Air diffusers, S17
  air curtains, S17.7
  air distribution, S17.6
  ceiling-mounted, F31.15
  industrial environment, A24.12-13
  K factors, A34.2
  ships
    merchant, A10.3
    naval, A10.7
  supply air outlets, S17.1
    selection, S17.1
      smudging, S17.2
      sound levels, S17.2
      surface effects, S17.2
      temperature differential, S17.2
      variable-air-volume, S17.2
    types, S17.3
      ceiling diffusers, S17.4
      grilles, S17.3
      registers, S17.3
      slots, S17.3
Air diffusion, F31
  air jets, F31.8
    angle of divergence, F31.8
    Archimedes number, F31.11
    centerline velocity, F31.9
      example, F31.10
    classification, F31.8
    Coanda effect, F31.12
    entrainment ratios, F31.11
    isothermal free axial jets, F31.8
    isothermal radial flow jets, F31.11
    multiple jets, F31.12
    nonisothermal free jets, F31.11
    surface jets, F31.12
    throw, F31.7, 10
  ceiling mounted diffusers, F31.15
    interior spaces, F31.15
    perimeter spaces, F31.15
  duct approaches, F31.12
  evaluation, F31.6
    air diffusion performance index (ADPI), F31.6
    air velocity, F31.6
    draft, F31.6
    effective draft temperature, F31.6
    temperature gradient, F31.6
  exhaust inlets, F31.13
  methods, F31.1
    displacement ventilation, F31.5
    localized ventilation, F31.5
    mixing systems, F31.1
    unidirectional ventilation, F31.5
  outlets, F31.2
    Group A, F31.2, 14
    Group B, F31.3, 14
    Group C, F31.3, 14
    Group D, F31.4, 14
    Group E, F31.4, 14
    location, F31.13
    performance, F31.4
    selection, F31.13
  return air design, F31.15
  return inlets, F31.13
  system design, F31.12
    design procedure, F31.13
    example, F31.13
  terminology, F31.1
Air diffusion performance index (ADPI), F31.6
Air distribution
  aircraft air conditioning, A9.3-4
  animal environments, A20.2, 4-5

bowling centers, A3.10
bypass boxes, S17.7
ceiling induction boxes
  air-to-air, S17.7
  fan-assisted, S17.7
central system, S1.6
communications centers, A3.11
data processing areas, A16.3-5
dual-duct boxes, S17.7
dump boxes, S17.7
fixed guideway vehicles, A8.11
industrial environment, A24.9, 12-13
kitchens, A28.9-10
laboratories, A13.9
modeling, F30.10
natatoriums, A4.7
places of assembly, A4.2-3
plant growth facilities, A20.8-18
railroad cars, A8.10
reheat boxes, S17.6
retail stores, A2
ships
  merchant, A10.5
  naval, A10.7
small forced-air systems, S9
  commercial systems, S9.13
  components, S9.1
    air cleaners, S9.1
    controls, S9.2
    cooling units, S9.1
    ducts, S9.1
    heating units, S9.1
    humidifiers, S9.1
    return grilles, S9.2
    supply outlets, S9.2
  design, S9.3
    airflow requirements, S9.4
    box plenum/flexible duct, S9.12
    distribution design, S9.4
    embedded loop duct systems, S9.12
    equipment selection, S9.4
    heating/cooling loads, S9.4
    outlets/ducts location, S9.3
    zone control, S9.7
    zone damper systems, S9.12
  duct design, S9.5
    branch supply air sizing, S9.6
    equal friction sizing, S9.5
    equivalent lengths, S9.5
    return air sizing, S9.7
    supply trunk sizing, S9.6
    velocity reduction sizing, S9.7
  grille/register selection, S9.13
static pressure control, S17.7
terminal boxes, S17.6
testing, adjusting, balancing, A34.3-6
textile processing, A18.4-6
Air exchange rate
  modeling, F25.21
    empirical, F25.21
    multizone, F25.21
    single-zone, F25.21
  multizone measurement, F25.6
  time constants, F25.3
  tracer gas measurement, F25.5
    constant concentration method, F25.6
    constant injection method, F25.6
    decay method, F25.5
Air filters. *See* **Filters, air**
Airflow
  around buildings, F15
    air intake contamination estimation, F15.9
      critical dilution, F15.10
      critical wind speed, F15.10

A = 1995 Applications   S = 1996 Systems and Equipment   F = 1997 Fundamentals   R = 1998 Refrigeration

Airflow (cont.)
 around buildings (cont.)
  air intake contamination estimation (cont.)
   minimum dilution, F15.9
   stretched-string distance, F15.9
   surface vents on flat roofs, F15.9
  air intake location, F15.15
  exhaust dilution, F15.8
   dilution factor, F15.8
  exhaust stack design, F15.10
   avoiding exhaust entrainment, F15.12
   critical dilution estimation, F15.13
   effective stack height, F15.9
   example, F15.14
   exhaust velocity, F15.11
   plume rise, F15.13
   standards, F15.11
  flow patterns, F15.1
  internal pressure, F15.8
  modeling/testing, F15.15
   similarity requirements, F15.15
  scaling length, F15.1
  wind data
   data sources, F15.5
   example, F15.5
   meteorological station, F15.3, 5
   remote site estimation, F15.6
  wind effects on system operation, F15.7
   fume hoods, F15.7
   pressure balance, F15.7
   system volume, F15.8
   ventilation, F15.7
  wind pressure, F15.3
   Bernoulli equation, F15.3
   coefficients, F15.4
 clean spaces, A15.3-5
  computer-aided modeling, A15.5
  nonunidirectional, A15.3
  unidirectional, A15.3-5, 8-9
 condensers, evaporative, S35.14
 control, A38.1-2
  dehumidification, S22.7
 displacement flow, F25.2
 energy recovery systems, S42.4
 entrainment flow, F25.2
 furnaces, S28.2
 kitchens
  balancing, A28.10-12
 measurement of, A34
 perfect mixing, F25.3
 smoke management, A48.5-6
 solar energy systems, A30.27-28
Airflow retarders, F22.16
 heating climates, F23.6
 mixed climates, F23.8
 warm humid climates, F23.9
Air jets, F31
 angle of divergence, F31.8
 Archimedes number, F31.11
 centerline velocity, F31.9
  example, F31.10
 classification, F31.8
 Coanda effect, F31.12
 entrainment ratios, F31.11
 throw, F31.7, 10
Air leakage (see also **Infiltration**), F25.1
 building components (table), F25.18
 building distribution, F25.16
 classes (table), F25.16
 commercial buildings, F25.19
 controlling, F25.17
  air-vapor retarder, F25.17
 leakage function, F25.11
 measurement, F25.15, 16
 residential buildings, F25.14

Air leakage area, F25.14, 15
Air pollutants. See **Contaminants, gaseous**; **Contaminants, particulate**
Airports, A3.12
Air quality (see also **Indoor air quality (IAQ)**)
 aircraft air conditioning, A9.2-3
 animal buildings, A20.2
 gaseous contaminant control, A41
 hospitals, A7.2-4
 kitchens, A28.9
 vehicular facilities, enclosed, A12
Airtightness
 measurement, F14.28
Airtightness rating, F25.15
Air transport, R31
 commodity requirements, R31.2
 design considerations, R31.2
 refrigeration, R31.3
 shipping containers, R31.3
Air washers, S19.5
 air cleaning, S19.7
 cooling, S19.7
 dehumidification, S19.7
 heat/mass simultaneous transfer, F5.13
 humidification, S19.6
 spray type, S19.5
 water treatment, A44.14
Alloys. See **Metals and alloys**
Altitude chambers, R37.6
Altshul-Tsal equation
 friction factor, F32.7
Ammonia
 air-conditioning systems, R3.1
 animal environments, A20.2
 refrigerant. See **Refrigerants**
 refrigeration systems. See **Refrigeration**
Ammonia-water
 absorption technology, R41.9
 properties, F19.1
 refrigeration cycle, F1.18
Ammonia-water-hydrogen
 absorption technology, R41.9
Anchor bolts
 seismic restraint, A50.4-5
Anemometers, A34.2, 3; F14.15
 cup, F14.15
 deflecting vane, F14.15
 laser Doppler, F14.16
 revolving vane, F14.15
 thermal, F14.15
Animal environments, A20.1-8; F10.1
 air contaminants, F10.4
 air distribution, A20.2, 4-5
 air ionization, F10.5
 air quality control, A20.2
 cooling
  air conditioning, A20.3
  earth tubes, A20.3
  evaporative, A20.3; A47.8-9
 disease control, A20.2
 emergency warning systems, A20.6
 evaporative cooling, A20.3; A47.8-9
 fans, A20.5-6
 gaseous contaminants, A20.2
 heating
  heat exchangers, A20.3
  supplemental, A20.4
 insulation, A20.4
 lighting, F10.3
 moisture control, A20.2
 physiological control, F10.2
  heat production, F10.2, 5
   genetics affecting, F10.5
  heat transfer, F10.3

 recommended practices, A20.6-8
  beef cattle, A20.6; F10.5
  chickens, F10.10
  dairy cattle, A20.6; F10.5
  laboratory animals, A20.8; F10.12
  poultry, A20.7-8
  sheep, F10.7
  swine, A20.6-7; F10.8
  turkeys, F10.10
 shelter degree, A20.2-3
 temperature control, A20.1-2; F10.3
  thermostats, A20.6
 ventilation management, A20.4-8
 ventilation systems, A20.4
  mechanical, A20.4
  natural, A20.4
Animals
 air transport, R31.2
 laboratory, A13.12-14
Antifreeze
 coolants, secondary, F20.4
 ethylene glycol, F20.4
 heat transfer, S12.16
 hydronic systems, S12.16
 propylene glycol, F20.4
Apartments
 service water heating, A45.11, 12, 15
Appliances
 cooling load, F28.10, 53
 heat gain (table), F28.11, 12
Aquifers
 thermal storage, A40.8
Archimedes number, F31.11
Arenas, A4.4-5
Argon
 recovery, R38.14
Asbestos, F9.7
Atriums, A4.9
Attics
 moisture control
  heating climates, F23.6
  mixed climates, F23.8
  warm humid climates, F23.10
 temperature, F27.8
 ventilated
  thermal resistance (table), F24.13
Auditoriums, A4.3-4
Automobile air conditioning, A8.1-7
 components, A8.3-6
  compressors, A8.3
  condensers, A8.3-4
  evaporator systems, A8.4
  expansion valves, A8.5
  filters/hoses, A8.4
  heater cores, A8.4
  lubricants, A8.3
  orifice tubes, A8.5
  receiver-dryer assembly, A8.4-5
  refrigerants, A8.3
   flow control, A8.5-6
  suction line accumulators, A8.5
 controls, A8.6-7
  electrical, A8.6
  manual, A8.6
  refrigerant flow, A8.5-6
  temperature, A8.6
  vacuum, A8.6
 environmental control, A8.1-2
  air conditioning, A8.1-2
  defrosting, A8.1
  heating, A8.1
  ventilation, A8.1
 general considerations, A8.2-3
Backflow-prevention devices, S41.12

A = 1995 Applications   S = 1996 Systems and Equipment   F = 1997 Fundamentals   R = 1998 Refrigeration

Composite Index

I.5

Bacteria
  food, growth in, R11.1
  humidifiers, S20.1
Bakery products, R27
  cooling, R27.4
  dough production, R27.2
  freezing, R27.5
  ingredient storage, R27.1
  thermal properties (table), R27.6
Balance point
  building, F30.17
  heat pumps, S44.8
Balancing (*see also* **Testing, adjusting, and balancing**)
  air distribution systems, A34.3-6
  central plant chilled water systems, A34.13
  cooling towers, A34.14-15
  dual-duct systems, A34.4
  hydronic systems, A34.6-14
  induction systems, A34.6
  kitchen ventilation systems, A28.10-12
  refrigeration systems, R44
  steam distribution systems, A34.14
  temperature control verification, A34.15-16
  variable-air-volume systems, A34.4-6
Barley
  drying, F11.13
Bars, A3.5-6
Baseboard units, S32.1
  application, S32.5
  finned-tube, S32.1
  nonstandard condition corrections, S32.3
  radiant, S32.1
  radiant-convector, S32.1
  rating, S32.3
Basements, F39.1, 3
  heat loss, F27.10
Beer. *See* **Beverages**
Beer's law, F3.11
Bernoulli equation, F2.2, 8; F15.3; F32.1
  kinetic energy factor, F2.2
  Bin method, F30.20
Beverage plants, R25.10
  liquid CO_2 storage, R25.11
  refrigeration systems, R25.11
Beverages, R25
  beer
    production, R25.1
    storage requirements, R10.5
  beverage coolers, R25.10
  carbonated, R25.10
  cooling time calculations, R9.1
  freezing time calculations, R9.7
  fruit juice, R24
  thermal properties, R8
  wine
    production, R25.8
    storage temperature, R25.10
Bioaerosols, F9.5; F12.1, 6
Blast freezers. *See* **Freezers**
BLC. *See* **Building loss coefficient (BLC)**
Blowers. *See* **Fans**
Boilers, S10.3; S27
  central plants, S11.2
  classifications, S27.1
    condensing/noncondensing, S27.2
    electric, S27.4
    fuel, S27.1
    material of construction, S27.1
    working pressure/temperature, S27.1
      high-pressure, S27.1
      low-pressure, S27.1
      medium-pressure, S27.1
      steam, S27.1
      water, S27.1

  codes, S27.5
  control, S27.6
  efficiency, S27.5
    combustion, S27.5
    overall, S27.5
    seasonal, S27.5
  gas-fired, S26.1
  material of construction, S27.1
    cast iron, S27.1
      dry-base, S27.1
      wet-base, S27.1
      wet-leg, S27.1
    copper, S27.2
    steel, S27.2
      fire-tube, S27.2
      scotch marine, S27.2
      water-tube, S27.2
  modeling, F30.13
  rating, S27.5
  residential, A1.3
  selection, S27.5
  service water heating, A45.21-22
  sizing, S27.5
  standards, S27.5
  steam, S27.1
  testing, S27.5
  water, S27.1
  water treatment, A44.15-16
Boiling, F4.1
  critical heat flux, F4.4
  evaporators, F4.1, 5
    flow mechanics, F4.5
    heat transfer, F4.6
  film boiling, F4.2, 4
  heat transfer equations (table), F4.3
  natural convection systems, F4.1
  nucleate, F4.1, 2
  points
    refrigerants (table), F18.3
    table, F36.1, 2
  pool boiling, F4.1
Bowling centers, A3.10
Brayton cycle, R38.8
Bread. *See* **Bakery products**
Breweries, R25
  carbon dioxide production, R25.6
  refrigeration
    fermenting cellar, R25.4
    Kraeusen cellar, R25.5
    refrigeration systems, R25.7
    stock cellar, R25.5
    wort cooler, R25.3
  storage tanks, R25.6
  vinegar production, R25.8
Bricks
  thermal storage, A40.7-8
Brines. *See* **Coolants, secondary**
Building commissioning, A39
  advantages, A39.1
  authority, A39.1-2
  basics, A39.1
  cost factors, A39.2
  phases, A39.2
  team, A39.1
Building energy monitoring, A37
  data analysis/reporting, A37.3
  design methodology, A37.3-10
    I—goals, A37.3-4
    II—data product/project output, A37.4
    III—experimental design approach, A37.4-5
    IV—data analysis/algorithms, A37.5-6
    V—field data monitoring points, A37.7-8
    VI—building characteristics, A37.8-9
    VII—data product accuracies, A37.9

    VIII—verification/quality assurance, A37.9-10
    IX—recording/data exchange formats, A37.10
  implementation/verification, A37.3
  planning, A37.3
  project types, A37.1-2
    building diagnostics, A37.2
    energy end use, A37.1
    specific technology assessment, A37.1-2
    whole-building energy use, A37.1
  retrofit performance monitoring, A37.10-13
    commercial, A37.11-13
    residential, A37.10-11
Building envelopes, F39
  ceilings, F39.5
  fenestration, F39.6
  foundations, F39.1
  roofs, F39.5
  thermal bridges, F39.8
    design recommendations, F39.10
    detrimental effects, F39.9
    mitigation of, F39.9
  thermal mass, F39.12
  wall/roof interface, F39.8
  wall/window interface, F39.7
  walls, F39.3
Building mass
  thermal storage, A40.8-9, 16-17
Building materials
  thermal properties (table), F24.4, 13
  vapor transmission properties (table), F24.16
Building operating dynamics/strategies, A38
  air-handling systems, A38.1-6
    airflow control, A38.1-2
    control response, A38.2-5
    set points, A38.1
    system energy effects, A38.5-6
  building dynamics, A38.14-15
    night setback recovery, A38.14
    precooling, A38.14-15
  chillers, A38.7-8
    multiple, A38.7-8
    sequencing, A38.8
  cooling towers, A38.9
  forecasting energy requirements, A38.18-20
    methods, A38.18-19
    simple algorithm, A38.19-20
  pumps, A38.9-10
  systems without storage, A38.6-14
    global optimization, A38.13-14
    load/ambient condition effects, A38.10-11
    near-optimal control, A38.7-9
    optimal control, A38.6-7
    versus others, A38.7
    simplified optimal control, A38.11-13
    variable-speed equipment, A38.11
  thermal storage systems, A38.15-18
    chiller-priority control, A38.16
    storage-priority control, A38.16-17
    versus optimal control, A38.17-18
Burners
  conversion, S26.1
  dual-fuel gas/oil, S26.7
  gas-fired, S26.1
    atmospheric, S26.2
    power-type, S26.2
  oil-fired, S26.4
  premix, S26.2
  pulse combustion, S26.2
  ring, S26.2
Bus air conditioning, A8.7-9
  interurban, A8.7-8
  controls, A8.8
  shock/vibration, A8.7-8

Bus air conditioning *(cont.)*
  urban, A8.8-9
    articulated coaches, A8.9
    compressors/drives, A8.9
    controls, A8.9
    maintenance, A8.9
    retrofit systems, A8.8-9
    system types, A8.8
Bus garages. *See* **Garages, bus**
Bus terminals, A3.12; A12.13-14
  operation areas, A12.14-15
  platforms, A12.14
Butter. *See* **Dairy products**
Bypass boxes
  air distribution, S17.7
Calcium chloride
  brines, F20.1
Candy, R28
  manufacture, R28.1
    coating, R28.4
    cooling, R28.3
    dipping, R28.2
    drying, R28.3
    refrigeration plant, R28.4
  storage, R28.5
    humidity, R28.6
    storage life (table), R28.5
    temperature, R28.6
  storage requirements (table), R10.3
Capillary tubes, R45.21
  capacity balance, R45.23
  selection, R45.24
Carbon dioxide
  animal environments, A20.2
  combustion, F17.1, 10
  greenhouse enrichment, A20.12
  liquefaction, R25.7
  measurement, F14.29
    amperometric electrochemical detectors, F14.29
    colorimetric detectors, F14.30
    nondispersive infrared detectors, F14.29
    photoacoustic detectors, F14.30
    potentiometric electrochemical detectors, F14.30
  storage, R25.11
  volume in water (table), R25.10
Carnot refrigeration cycle, F1.6
Cast iron
  boilers, S27.1
Cattle
  growth, F10.5
  heat and moisture production, F10.7
  lactation, F10.6
  recommended environment, A20.6
  reproduction, F10.6
Cavitation, S41.2; F2.5
  pumps, centrifugal, S38.8
Ceiling
  air movement, F39.6
  cathedral
    moisture control
      heating climates, F23.7
      mixed climates, F23.8
      warm humid climates, F23.10
  diffuser outlets, S17.4
  heat loss, F27.10
  induction boxes, S17.7
  moisture control, F39.6
  panels. *See* **Panel heating and cooling**
  plenums, S1.6
  R-values, F24.11
  sound correction, A43.24-25
  sound transmission, A43.23-24
  thermal performance, F39.5

Central plants, S11.2
  control, A42.28-30
Cetane number
  engine fuels, F17.6
Charging
  refrigeration systems, R46.4
Cheese. *See* **Dairy products**
Chemical industry refrigeration, R36
  equipment characteristics, R36.3
  load characteristics, R36.2
  refrigeration equipment, R36.6
  refrigeration systems, R36.5
  safety requirements, R36.2
Chickens *(see also* **Poultry***)*
  growth, F10.10
  heat and moisture production, F10.10
  reproduction, F10.10
Chilled water
  basic systems, S12.1, 11
    testing, adjusting, balancing, A34.7-8
  central plant
    testing, adjusting, balancing, A34.13
  cogeneration distribution, S7.29
  density (table), A40.11
  district heating and cooling, S11.23
  thermal storage, A40.9-11, 17
Chillers
  absorption, S1.8
    ammonia-water, R41.9
    heat-activated, S7.25
    water-lithium bromide, R41.4
  central plants, S11.2
  control of, A38.7-8; A42.28-30
  modeling, F30.15
  sound control, A43.9
  vapor compression
    modeling, F30.14
Chillers, liquid, R43
  centrifugal, R43.8
    air cooling, R43.12
    free cooling, R43.12
    heat recovery, R43.11
    purge units, R43.11
    refrigerant selection, R43.8
    refrigerant transfer units, R43.11
  control, R43.4
    regulating, R43.4
    safety, R43.5
  direct expansion
    piping, R2.21
  equipment selection, R43.3
  heat recovery, R43.3
  maintenance, R43.5
  multiple, R43.2
    parallel, R43.2
    series, R43.2
  reciprocating, R43.6
    capacity control, R43.7
    refrigerant selection, R43.6
  screw, R43.13
    capacity control, R43.14
    special applications, R43.15
Chilton-Colburn j-factor analogy, F5.8
Chimneys, S30
  accessories, S30.21
    draft fans, S30.23
    draft hoods, S30.21
    draft regulators, S30.22
    flue gas heat extractors, S30.23
    heat exchangers, S30.23
    vent dampers, S30.22
  air supply to fuel-burning equipment, S30.19
  altitude correction, S30.6
  capacity calculation examples, S30.10

  codes, S30.26
  design equations, S30.2
  draft, S30.1
    available, S30.1, 2, 6
    theoretical, S30.1, 2, 6
  fireplace, S30.18
  flow losses, S30.7
  flue gas, S30.1
  friction loss, S30.8
  functions, S30.1
  gas appliance venting, S30.15
    conversion to gas, S30.16
    masonry chimneys, S30.16
    prefabricated vents, S30.16
    vent connectors, S30.16
    without draft hoods, S30.16
  gas temperature, S30.5
  gas velocity, S30.7
  heat transfer, S30.5
  mass flow, S30.4
  materials, S30.20
  oil-fired appliance venting, S30.16
    condensation/corrosion, S30.17
    corrosion, S30.17
    equipment replacement, S30.18
    masonry chimneys, S30.18
    vent connectors, S30.18
  resistance coefficients, S30.8
    configuration effects, S30.9
    manifolding effects, S30.9
  standards, S30.21, 26
  terminations, S30.24
    caps, S30.24
    wind effects, S30.24
  wind effects, S30.2, 24
Chlorofluorocarbons (CFC). *See* **Refrigerants**
Chocolate, R28
Choking, F2.7
Churches. *See* **Houses of worship**
Clean spaces, A15
  air filters, A15.3
    HEPA filters, A15.3
    ULPA filters, A15.3
  airflow, A15.3-5
    computer-aided modeling, A15.5
    nonunidirectional, A15.3
    unidirectional, A15.3-5, 8-9
  applications, A15.2
  biomanufacturing, A15.6
  classes (table), A15.2
  construction/operation, A15.11-12
  cooling, A15.9
  energy conservation, A15.11
  exhaust, A15.9-10, 11
  fans, A15.11
  fire safety, A15.9-10
  humidity control, A15.10
  laminar (unidirectional) flow, A15.3-5, 8-9
  makeup air, A15.9, 11
  particulate control, A15.2-5
    particle sources, A15.2-3
  pharmaceutical, A15.6
    aseptic, A15.7
      qualification of, A15.7
      start-up of, A15.7
    nonaseptic, A15.8
  pressurization, A15.10
  semiconductor clean rooms, A15.8-9
    classes, A15.8
    unidirectional airflow, A15.8-9
  sound control, A15.11
  system sizing/redundancy, A15.10-11
  temperature control, A15.10
  terminology, A15.1-2

Composite Index

Clean spaces *(cont.)*
  testing, A15.5-6
  unidirectional (laminar) flow, A15.3-5, 8-9
  vibration control, A15.11
Climate. *See* Weather data
Climatic test chambers, R37
Clo, F8.8
Closed water systems, S12.1
Clothing
  insulation, F8.8
    table, F8.8, 9
  moisture permeability, F8.8
    table, F8.8
CNR. *See* Composite noise rating (CNR)
Coal, F17.7
  characteristics, F17.8
  classification (table), F17.7
  heating value, F17.8
Coanda effect
  air diffusion, F31.12
Codes, R51
  air conditioners, room, S43.4
  boilers, S27.5
  chimneys/fireplaces/gas vents, S30.26
  cogeneration, S7.46
  condensers
    evaporative, S35.17
    water-cooled, S35.7
  coolers, liquid, S37.4
  dehumidifiers, room, S43.6
  duct construction, S16.1
  engine drives, S7.46
  furnaces, S28.17
  heaters (in-space), S29.6
  makeup air units, S31.9
  motors, S39.1
  nuclear facilities, A23.8
  piping, S40.6
  service water heating, A45.22
  turbine drives, S7.46
  unitary systems, S44.5
Coefficient of performance (COP), F1.3
  air conditioners, room, S43.2
  compressors, S34.1
  economic
    cogeneration, S7.32
Cogeneration, S7
  codes, S7.46
  combined cycles, S7.33
  control and instruments, S7.37
  design/installation, S7.34
  distribution systems, S7.28
    geographical spread, S7.29
    media, S7.28
      chilled water, S7.29
      electrical, S7.28
      hot water, S7.29
      steam, S7.28
  economic feasibility, S7.42
    load duration curve, S7.42
      two-dimensional, S7.44
    simulation, S7.44
  electrical systems, S7.26
    generators, S7.26
    power quality, S7.27
    utilities, S7.26
  heat recovery, S7.18
    heat-activated chillers, S7.25
  maintenance, S7.38
  noise control, S7.38
  packaged systems, S7.31
  peak shaving, S7.32

  performance, S7.32
    economic coefficient of performance (ECOP), S7.32
    incremental heat rate, S7.32
  regulations, S7.27
    emissions, S7.28
    state laws, S7.28
    United States, S7.27
  standards, S7.46
  terminology, S7.45
  thermal energy storage, S7.32
  thermal output, S7.18
    heat/power ratio, S7.19
    heat quality, S7.19
  tracking/load profiling, S7.31
  unconventional systems, S7.33
    continuous duty standby, S7.33
    expansion engines/turbines, S7.34
    fuel cells, S7.33
  utilization systems, S7.30
    air, S7.30
    district heating/cooling, S7.31
    hybrid, S7.31
    hydronic, S7.30
    service hot water, S7.31
  vibration control, S7.38
    foundations, S7.38
Coils
  air-cooling, S1.5; S21
    airflow resistance, S21.6
    applications, S21.1, 3
    construction/arrangement, S21.1
    control, A42.16-17; S21.2
    dehumidifying performance, S21.9
    direct-expansion, S21.2
    fluid flow, S21.3
    glycol, aqueous, S21.1
    heat/mass simultaneous transfer, F5.15
    heat transfer, S21.6
    load determination, S21.14
    maintenance, S21.15
    modeling, F30.11
    performance/rating, S21.5, 7
    selection, S21.5
    ships, A10.2, 6
    water, S21.1
  air-heating, S1.5; S23
    applications, S23.2
    control, A42.15-16
    design, S23.1
    electric, S23.2
    flow arrangement, S23.2
    heat transfer, S23.3
      parametric effects, S23.3
    maintenance, S23.3
    pressure drop, S23.3
    selection, S23.3
    ships, A10.2-3, 6-7
    steam, S23.1
    water, S23.1
    water systems, medium/high temp., S14.8
  condensers
    air-cooled, S35.8
    evaporative, S35.14
  dehumidifying, S21
    heat/mass simultaneous transfer, F5.15
    load determination, S21.14
    maintenance, S21.15
    modeling, F30.11
    performance, S21.9
  desuperheating, S35.16
  finned-tube
    energy recovery loops, S42.12
  halocarbon refrigeration systems
    piping, R2.22

Colebrook equation
  friction factor, F32.7
Collectors, solar (*see also* Solar energy), A30.7-12, 16, 26, 27; S33.3
Colleges, A6.3-4
Combustion, F17
  air pollution, F17.15
  air required, F17.8
    table, F17.2, 9
  altitude compensation, F17.3
  analysis, F14.31
  calculations, F17.8
    air required, F17.8
    efficiency, F17.12
    example, F17.12
    flue gas dew point, F17.11
    flue gas produced, F17.10
    theoretical carbon dioxide, F17.10
    water vapor produced, F17.11
  coals, F17.7
    characteristics, F17.8
    classification (table), F17.7
    heating value, F17.8
  condensation, F17.16
  continuous, F17.2
  corrosion, F17.16
  diesel fuel, F17.6
  efficiency
    boilers, S27.5
  engine fuels, F17.6
    cetane number, F17.6
  excess air, F17.10
  fuel oils, F17.5
    characteristics, F17.5
    distillate oils, F17.5
    heating value, F17.6
    residual oils, F17.5
    sulfur content (table), F17.6
    viscosity, F17.5
  furnaces, S28.3
  gas burners, S26.3
  gaseous contaminant control, A41.11-12
  gaseous fuels, F17.4
    flammability limits, F17.1
      table, F17.3
    illuminants, F17.9
    liquefied petroleum gas, F17.4
    natural gas, F17.4
  gas turbine fuel, F17.6
  heating values, F17.3
    table, F17.3, 4, 6
  ignition temperature, F17.2
    table, F17.3
  industrial gas cleaning, S25.26
    applications, S25.27
    catalytic, S25.26
    thermal, S25.26
  liquefied petroleum gas, F17.4
  liquid fuels, F17.5
    engine fuels, F17.6
    fuel oils, F17.5
  natural gas, F17.4
  principles, F17.1
  products (table), F17.2
  pulse, F17.2
  reactions, F17.1
  solid fuels, F17.7
  soot, F17.16
  stoichiometric, F17.1
Comfort (*see also* Physiological principles, humans)
  comfort zones, F8.12, 18
    figure, F8.12

A = 1995 Applications   S = 1996 Systems and Equipment   F = 1997 Fundamentals   R = 1998 Refrigeration

Comfort *(cont.)*
  environmental indices, F8.19
    effective temperature, F8.19
    heat stress index, F8.19
    humid operative temperature, F8.19
    skin wettedness index, F8.20
    wet-bulb globe temperature, F8.20
    wet-globe temperature, F8.20
    wind chill index, F8.21
      table, F8.21
  environmental parameters, F8.10
    air velocity, F14.26
    asymmetric thermal radiation, F8.13
    draft, F8.13
    floor temperature, F8.14
    mean radiant temperature, F8.10; F14.26
    plane radiant temperature, F8.11; F14.26
    radiant temperature asymmetry, F8.12
    vertical air temperature difference, F8.14
  humidification, S20.1
  humidity, F14.26
  local discomfort, F8.13
  nonuniform conditions, F8.13
  predicted mean vote, F8.16; F14.26
  predicted percent dissatisfied, F8.17
  special environments, F8.21
    extreme cold, F8.24
    hot and humid environments, F8.23
    infrared heating, F8.21
      comfort equation, F8.23
  steady-state energy balance, F8.16
    two-node model, F8.17
  thermal sensation scale, F8.12
  zones, F8.12, 18; F9.14
    figure, F8.12
Commercial
  duct construction, S16.2
Commercial and public buildings, A3
  air leakage, F25.19
  airports, A3.12
  bars, A3.5-6
  bowling centers, A3.10
  bus terminals, A3.12
  characteristics (table), A32.4, 5
  communications centers, A3.11
  design criteria (table), A3.2-3
  dining and entertainment centers, A3.4-6
  energy consumption (table), A32.6
  garages, A3.12-13
  general criteria, A3.1-4
  indoor air quality, F25.4
  kitchens, A3.5, 6; A28
  libraries, A3.8-10
  load calculations, F28
  load characteristics, A3
  median electric demand/load factor (table), A32.6
  museums, A3.8-10
  nightclubs, A3.6
  office buildings, A3.6-8
  restaurants, A3.5
  service water heating, A45.10-12
  ship docks, A3.12
  transportation centers, A3.11-13
  ventilation rate procedure, F25.5, 23
  warehouses, A3.13
Commissioning
  automatic controls, F37.11
  buildings. *See* **Building commissioning**
  dehumidification equipment, S22.8
  desiccant systems, S22.8
  laboratories, A13.15-16
  makeup air units, S31.9
  pumps, centrifugal, S38.13
  thermal storage systems, A40.19

Communications centers, A3.11
  sound control, A3.11
Composite noise rating (CNR), A43.5-7
Compressors, S34
  air conditioners
    room, S43.2
  ammonia refrigeration systems, R3.2
  automobile air conditioning, A8.3
  bus air conditioning, A8.9
  centrifugal, S1.8; S7.39; S34.25
    angular momentum, S34.26
    capacity control, S34.31
    critical speed, S34.32
    design, S34.33
      accessories, S34.34
      bearings, S34.34
      casings, S34.33
      impellers, S34.33
    drivers, S34.33
    efficiency, S34.30
    isentropic analysis, S34.28
    lubrication, S34.34
    noise, S34.32
    paralleling, S34.33
    polytropic analysis, S34.28
    refrigeration cycle, S34.26
    surging, S34.31
    testing, S34.30
    vibration, S34.32
  chemical industry refrigeration, R36.6
  compressor drives, R3.2
  dynamic, S34.1
  halocarbon refrigeration systems
    piping, R2.19
  heat pumps, applied, S8.6
    centrifugal, S8.6
    floodback protection, S8.7
    reciprocating, S8.6
    rotary vane, S8.6
    screw, S8.6
    scroll, S8.6
  heat pumps, unitary, S44.11, 12
  liquid chiller, R43
  Mach number, S34.30
  modeling, F30.14, 15
  operation and maintenance, S34.34
  orbital (scroll), S34.21
    capacity control, S34.23
    efficiency, S34.24
    features, S34.23
    noise, S34.24
    vibration, S34.24
  positive-displacement, S34.1
    motors, S34.3
    performance, S34.1
      COP, S34.1
      efficiency, S34.2
      heat rejection, S34.2
      liquid subcooling, S34.2
    protective devices, S34.2
      liquid hazard, S34.3
      shock, S34.3
      testing, S34.3
      vibration, S34.3
  reciprocating, S1.8; S7.39; S34.4; R3.10
    application, S34.8
    crankcase, keeping liquid from, R2.28
    efficiency, S34.4
    features, S34.6
    lubrication, S34.8
    modeling, F30.14
    performance, S34.4
    special devices, S34.8
    types, S34.4
    valves, S34.7

refrigerant acoustic velocity, S34.30
retail food store refrigeration, R47.9
rotary, S1.8; S34.9
  rolling piston, S34.9
    efficiency, S34.9
    features, S34.9
    lubrication, S34.10
  rotary vane, S34.10; R3.12
    volume ratio, S34.11
  single-screw, S34.11
    capacity control, S34.14
    compression process, S34.11
    economizers, S34.13, 14
    efficiency, S34.15
    features, S34.11
    lubrication, S34.12
    noise, S34.16
    vibration, S34.16
    volume ratio, S34.14
  twin-screw, S34.16
    capacity control, S34.18
    compression process, S34.17
    economizers, S34.20
    efficiency, S34.21
    features, S34.17
    hermetic, S34.21
    lubrication, S34.20
    volume ratio, S34.18
  screw, S7.40; R3.13
    lubricant cooling, R3.13
  trochoidal (Wankel), S34.24
    performance, S34.25
Computer rooms. *See* **Data processing system areas**
Computers, A36
  abbreviations for programming, F34.1
    table, F34.2
  administrative uses, A36.10-11
    accounting, A36.11
    decision making, A36.10
    desktop publishing, A36.10
    employee records, A36.11
    integrity, A36.11
    job costing, A36.11
    management planning, A36.10
    project scheduling, A36.11
    security, A36.11
    specification writing, A36.10
    word processing, A36.10
  artificial intelligence, A36.14-15
    artificial neural networks, A36.14
    case-based reasoning, A36.15
    fuzzy logic, A36.14-15
    genetic algorithms, A36.15
    knowledge-based systems, A36.14
    natural language processing, A36.15
    optical character recognition, A36.15
    virtual reality, A36.15
  communications, A36.15-16
  computer-aided design (CAD), A36.12-13
    capabilities, A36.12-13
    hardware, A36.13
    software, A36.13
  controlling HVAC systems, A36.16
  data acquisition, A36.16
    hardware, A36.16
    software, A36.16
  design calculations, A36.4-10
    acoustic calculations, A36.8-9
    cooling loads, A36.4-5
    duct design, A36.7-8
    energy simulation, A36.5-7
    equipment selection/simulation, A36.9-10
    heating loads, A36.4-5

Composite Index

Computers *(cont.)*
  design calculations *(cont.)*
    piping design, A36.8
    system simulation, A36.5-7
  graphics, A36.13-14
  hardware options, A36.1-3
    laptops, A36.2
    mainframes, A36.1-2
    microcomputers, A36.2
    minicomputers, A36.2
    notebooks, A36.2
    personal digital assistants, A36.2
    personal organizers, A36.2-3
    programmable calculators, A36.2-3
  HVAC system control/monitoring, A36.16-17
  modeling, A36.13-14
  monitoring HVAC systems, A36.16
  productivity tools, A36.11-12
    advanced input/output, A36.12
    communications, A36.11
    database, A36.11-12
    graphics/imaging, A36.11
    special-purpose, A36.12
    spreadsheets, A36.11
  smoke control analysis, A48.12
  software options, A36.3-4
    acquisition, A36.3-4
      custom programs, A36.4
      public domain programs, A36.4
      purchased programs, A36.3-4
      royalty programs in time sharing, A36.4
  system architecture, A36.1
Concert halls, A4.4
Concrete
  cooling, R35.1
Condensate
  concealed, S20.3
  steam systems, S10.7; S11.20; F33.14
  visible, S20.2
  water treatment, A44.16
Condensation, F4.8
  combustion, F17.16
  cryogenic
    gaseous contaminant control, A41.12
  dropwise, F4.8
  fenestration, F29.12
  film, F4.8
  heat recovery equipment, S42.3, 11
  heat transfer coefficients (table), F4.9
  inside horizontal tubes, F4.10
  inside vertical tubes, F4.10
  noncondensable gases, F4.10
  oil-fired appliances, S30.17
  outside horizontal tubes, F4.10
  outside vertical tubes, F4.9
  prevention, F23.5
    dehumidification for, S22.6
  steam, F4.10
Condensers, S35
  air conditioners, room, S43.2
  air-cooled, S1.8; S35.7; R47.10
    application, S35.10
    coil construction, S35.8
    condensing unit, S35.9
    control, S35.11
    fans, S35.8
    heat transfer, S35.9
    installation, S35.12
    maintenance, S35.12
    noise, R47.13
    pressure drop, S35.9
    rating, S35.10
    remote from compressor, S35.9; R47.11
    total heat rejection, S35.10

  ammonia refrigeration systems, R3.2
    piping, R3.15
  automobile air conditioning, A8.3-4
  cascade, R44.1
  chemical industry refrigeration, R36.6
  evaporative, S1.8; S35.13; R3.17; R47.11
    blow-through, S35.14
    codes, S35.17
    coils, S35.14
      airflow, S35.14
      desuperheating, S35.16
      wetting method, S35.14
    control, S35.17
      dampers, S35.17
      fans, S35.17
      pony motors, S35.17
      two-speed motors, S35.17
      variable-speed motors, S35.17
    desuperheating coils, S35.16
    draw-through, S35.14
    eliminators, S35.14
    freeze prevention, S35.15
      remote sump, S35.15
    heat transfer, S35.13
    liquid subcooling, S35.16
    location, S35.14
    maintenance, S35.17
    multicircuiting, S35.15, 16
      trapped drop legs, S35.15
    noncondensable gases, S35.17
    purging, S35.17
    rating, S35.15
    standards, S35.17
    superheat of refrigerants, S35.16
    water
      consumption, S35.17
      treatment, S35.16
  halocarbon refrigeration systems
    piping, R2.18
    pressure control, R2.27
  modeling, F30.14
  retail food store refrigeration, R47.10
  water-cooled, S13; S35.1
    circuiting, S35.4
    codes, S35.7
    cooling tower systems, S13.1
      piping, S13.1
      pumps, S13.1
      water treatment, S13.2
    heat removal, S35.1
    heat transfer, S35.2
      fouling factor, S35.3
      overall coefficient, S35.2
      refrigerant-side film coefficient, S35.2
      surface efficiency, S35.3
      tube-wall resistance, S35.3
      water-side film coefficient, S35.2
    liquid subcooling, S35.4
    maintenance, S35.7
    noncondensable gases, S35.6
    once-through systems, S13.1
    pressure drop, S35.4
    standards, S35.7
    types, S35.5
      brazed plate, S35.6
      shell-and-coil, S35.5
      shell-and-tube, S35.5
      tube-in-tube, S35.6
Conductance, thermal, F3.3
Conduction
  steady-state, F3.1
Conduction drying, A27.3-4
Conductivity, thermal
  foods, R8.9
  industrial insulation (table), F24.18

  liquids (table), F36.2
  moist air
    figure, F6.16
  rocks (table), F24.15
  soils, S11.12; F24.14
    table, F24.15
  solids (table), F36.3
  vapors (table), F36.1
Consultants
  energy management, A32.2-3
Containers *(see also* **Trucks, trailers, and containers)**
  air transport, R31.3
  marine transport, R30.7
  surface transport, R29
Contaminants
  clean spaces, A15.2-5
  food, R11
  industrial environments, A11.5, 8-9
  refrigerant systems, R6
    dirt, R6.6
    generation by high temperature, R5.9
    lubricants, R6.7
    metallic, R6.6
    moisture, R6.1
    motor burnout, R6.7, 11
    noncondensable gases, R6.7
    sludge, tars, and wax, R6.6
    solvents, R6.7
Contaminants, gaseous
  ambient air standards (table), A41.4, 16; F9.2
  animal environments, A20.2; F10.4
  characteristics, A41.1
    table, A41.4
  chemical families (table), A41.2
  classification, F12.1
  combustion, F17.15
  control of, A41
    dehumidification for, S22.7
    economics, A41.16
    energy consumption, A41.15-16
    environmental effects on, A41.13
    system configurations, A41.12-13
    testing devices for, A41.14-15
  control techniques, A41.8-12
    absorption, A41.9
    adsorption, A41.9-11; S25.23
    catalysis, A41.11
    catalytic combustion, A41.12
    chemisorption, A41.11
    combustion, A41.11-12
    cryogenic condensation, A41.12
    general ventilation, A41.8-9
    hoods/local exhaust, A41.8
    incineration, S25.26
    masking, A41.12
    occupants, A41.12
      sorption (table), A41.6
    plants, A41.12
      adsorption (table), A41.12
    source elimination, A41.8
    spray dry scrubbing, S25.17
    wet-packed scrubbers, S25.18
  environmental tobacco smoke, F12.1
  fogs, F12.2
  gases, F12.2
    flammable, F12.4
  harmful effects, A41.2-5
    irritation, A41.3
    materials damage, A41.3-5
    odors, A41.3
    toxicity, A41.3
  industrial environments, A11.5, 8-9
  industrial gas cleaning, S25.17

Contaminants, gaseous *(cont.)*
  inorganic gases, F9.11
    exposure control, F9.12
    health effects, F9.12
    sources, F9.11
    standards, F9.12
      table, F9.12
  measurement, A41.13-14; F14.28
  mists, F12.2
  monitoring, A41.13-14
  nuclear facilities, A23.2-3, 4, 5
  odors, F13
  organic gases, F9.13
    exposure control, F9.13
    health effects, F9.13
    sources, F9.13
    standards
      table, F9.14
  permissible exposure limits (PEL), F9.1
  photographic processing, A19.1-2
  plant environments, F10.17
  polycyclic aromatic compounds (PAC), F9.8
  polycyclic aromatic hydrocarbons (PAH), F9.8
  radioactive, F12.6
  radon, F9.19
    action levels (table), F9.20
  recommended exposure limits (REL), F9.1
  smog, F12.2
  soil gases, F9.14
  sources, A41.5-7
    building materials/furnishings, A41.5
      pollutants from (table), A41.6; F25.4
    cleaning agents, A41.5
    equipment, A41.5-6
      pollutants from (table), A41.6
    occupants, A41.6
      pollutants from (table), A41.6
    outdoor air, A41.6-7
      compounds in (table), A41.7; F25.4
    tobacco smoke, A41.5; F9.8
      compounds in (table), A41.5
  textile processing, A18.6
  threshold limit values (TLV), F9.1
  vapors, F12.2
    flammable, F12.4
  ventilation pattern effects, A41.7-8
  volatile organic compounds (VOC), F9.9
    exposure control, F9.11
    health effects, F9.10
    sources, F9.10
Contaminants, particulate
  ambient air standards (table), F9.2
  animal environments, F10.4
  asbestos, F9.7
  classification, F12.1
  combustion, F17.15
  dusts, F9.7; F12.1
    combustible, F12.5
  environmental tobacco smoke, F12.1
  fumes, F12.1
  industrial gas cleaning, S25.2
    electrostatic precipitators, S25.7
    fabric filters, S25.11
    granular bed filters, S25.14
    HEPA filters, S25.2
    inertial collectors, S25.6
    mechanical collectors, S25.3
    principles of operation, S25.11, 15
      condensation, S25.15
      diffusion, S25.11, 15
      direct interception, S25.11
      electrostatic attraction, S25.11
      inertial impaction, S25.11, 15

  scrubbers, S25.14
  settling chambers, S25.3
  ULPA filters, S25.2
  measurement, F14.28
  particle size distribution, F12.2
  permissible exposure limits (PEL), F9.1
    table, F9.8
  pollen, F12.6
  polycyclic aromatic compounds (PAC), F9.8
  polycyclic aromatic hydrocarbons (PAH), F9.8
  radioactive, F12.6
  recommended exposure limits (REL), F9.1
  size, F12.2
  smokes, F12.1
  suspended particles, F12.2
  synthetic vitreous fibers, F9.8
  threshold limit values (TLV), F9.1
Continuity, F2.2
Control *(see also* **Controls, automatic***)*
  absorption units, R41.7, 10
  air-and-water systems, S3.7, 8
  air conditioners, room, S43.2
  aircraft
    air conditioning, A9.8
    pressurization, A9.7
    temperature, A9.6-7
  airflow, A38.1-2
    dehumidification, S22.7
  air handling systems, A42.19-23
    constant volume, A42.19-20
    dual-duct, A42.20-22
    makeup air, A42.22-23
    multizone, A42.22
    single-zone, A42.22
    variable-air-volume, A42.19
  all-air systems, S2.9
    constant volume, S2.9
    dual-duct, S2.10
    variable-air-volume, S2.9
  all-water systems, S4.2
  animal environments, A20.1-2, 6
  automobile air conditioning, A8.6-7
  boilers, S27.6
  bus air conditioning, A8.8, 9
  central plants, A42.28-30; S1.6
  central subsystems, A42.12-19
  chemical plants, R36.3
  chillers, A38.7-8; A42.28-30
  cogeneration systems, S7.37
  coils, air-cooling, S21.2
  condensers
    air-cooled, S35.11
    evaporative, S35.17
  cooling coils, A42.16-17
  cooling towers, A38.9; A42.30
  corrosion, A44
  data processing areas, A16.5
  duct heaters, A42.33
  educational facilities, A6.2
  energy recovery equipment, S42.5, 12, 15
  engine drives, S7.6
  environmental test facilities, R37.4
  fan-coil units, S4.2
  fans, A38.1-2; A42.13-15; S18.8
    constant volume control, A42.13
    duct static control, A42.13-14
    return fan control, A42.14
    sequencing, A42.14-15
    unstable operation, A42.15
  farm crop drying, A21.3
  fire, A48
  fixed guideway vehicle air conditioning, A8.11
  fuel-burning equipment, S26.12
  furnaces, S28.1, 7

  gaseous contaminants, A41
    dehumidification for, S22.7
  gas turbine drives, S7.10
  greenhouses, A20.12-13
  heaters (in-space), S29.2, 4
    electric
      thermostats, S29.4
    gas-fired
      thermostats, S29.2
      valves, S29.2
  heating coils, A42.15-16
    preheating, A42.15
    reheat/heating, A42.16
  heat pumps, A42.30
  heat pumps, applied systems, S8.8
  heat pumps, unitary, S44.12
  heat recovery, A42.30
  heat recovery systems, S8.19
  humidifiers, S20.8
    control location, S20.9
    electronic controllers, S20.8
    mechanical controls, S20.8
    variable-air-volume systems, S20.9
  humidity, A42.17-18; S22
    dehumidification, A42.17-18
    humidification, A42.18
  hydronic heating systems, A42.28; S11.6
  hydronic system capacity, S12.9
  infrared heaters, S15.4
  kitchen ventilation, A28.17
  laboratories, A13.11-12
  liquid chillers, R43.4, 7, 10, 14
  low temperature, R3.6
  makeup air units, S31.9
  moisture in insulation, F22; F23
    thermal in insulation, F22
  motors, S39.6
  outdoor air quantity, A42.12-13
  panel cooling systems, S6.19
  panel heating systems, S6.19
  paper manufacturing, A22.3
  plant growth environmental facilities, A20.15
  pressure, A42.19
  pumps, A38.9-10
  railroad air conditioning, A8.10-11
  refrigerant flow, R45
  residential heating/cooling equipment, A1.5
  retail store air conditioning, A2
  road tunnel ventilation, A12.7
  ship air conditioning
    merchant, A10.5-6
    naval, A10.7
  small forced-air systems, S9.2
  smoke, A48
  snow melting systems, A46.5, 8
  solar energy systems, A30.17-18, 28; S33.18
  sound, A43.1-33, 34-40, 41; F7.4
  space conditions, A42.18-19
  static pressure, S17.7
  steam systems, S10.13
  steam turbines, S7.12
  testing, adjusting, balancing, A34.15-16
  thermal in insulation, F23
  thermal storage systems, A38.15-18; A40.19; A42.30
    without storage, A38.6-14
  unitary air-conditioning systems
    self-contained, S5.8
    through-the-wall, S5.3
  unit heaters, S31.7
  unit ventilators, S31.3
  variable-air-volume, A38.1-2, 5-6, 7
  vibration, A43.1, 33-43
  water systems, medium/high temp., S14.8

Composite Index

Control *(cont.)*
  zone systems, A42.23-27
    perimeter units, A42.25-27
      fan-coil units, A42.26
      radiant panels, A42.27
      unit ventilators, A42.26-27
    terminal units
      constant volume, A42.25
      VAV, A42.23-24
Controlled atmosphere storage
  apples, R21.3
  pears, R21.4
  refrigerated facilities, R13.3
  vegetables, R23.4
Controls, automatic (*see also* **Control**), A42; F37
  advanced strategies, A38.1
  classification, A42.3; F37.3
    electric, A42.3; F37.3
      direct digital controller, A42.3; F37.3
    hybrid, A42.3; F37.4
    pneumatic, A42.3; F37.3
    self-powered, A42.3; F37.3
  closed loop (feedback), A42.1; F37.1
  commissioning, F37.11
  components, A42.4-12; F37.4
    auxiliary devices, A42.10-12; F37.10
    controlled devices, A42.4
      dampers, A42.6-7; F37.6
      positive positioners, A42.7; F37.7
      valves, A42.4-6; F37.4
    controllers, A42.9-10; F37.9
      direct digital, A42.10; F37.9, 12
      electric/electronic, A42.9; F37.9
      indicating, A42.9
      nonindicating, A42.9
      pneumatic receiver, A42.9; F37.9
      recording, A42.9
      thermostats, A42.10; F37.10
    sensors, A42.7-9, 32; F37.7
      flow rate, A42.9; F37.8
      humidity (hygrometers), A42.9; F37.8
      indoor air quality, F37.8
      lighting level, F37.8
      location, A42.32
      power transmission, F37.9
      pressure, A42.9; F37.8
      temperature, A42.8-9; F37.7
  computers for, A36.16-17
  control action types, A42.2-3; F37.2
    floating action, A42.2; F37.2
    proportional/integral (PI), A42.3; F37.3, 11
    proportional/integral/derivative (PID), A42.3; F37.3
    proportional-only (P), A42.3; F37.2
    timed two-position, A42.2; F37.2
    two-position, A42.2; F37.2
      heat anticipation, F37.2
  control response, A38.2-5
  dampers, A42.6-7; F37.6
    operators, A42.6-7; F37.6
    types, A42.6; F37.6
  design principles, A42.30-33
    building/mechanical subdivision, A42.31
    controlled area size, A42.32
    cost analysis, A42.32
    duct heater safety, A42.33
    energy conservation, A42.31
    explosive atmosphere concerns, A42.33
    limit controls, A42.33
    load matching, A42.32
    mechanical/electrical coordination, A42.31
    mobile effects, A42.32
    night setback, A42.32
    safety, A42.33
    sensor location, A42.32
    system selection, A42.31-32
  direct digital, A34.15-16; A42.3, 10, 35; F37.3, 9, 12
  feedback (closed loop), A42.1; F37.1
  feedforward (open loop), A42.1
  fundamentals, A42.1-4; F37
  open loop, F37.1
  open loop (feedforward), A42.1
  operation and maintenance, A42.33-34
    cost, A42.33
    documentation, A42.34
    flexibility, A42.33-34, 34
    maintainability, A42.34
    reliability, A42.34
  performance requirements, A42.3
  positive positioners, A42.7; F37.7
  proportional/integral (PI), A38.2-5; A42.3; F37.3
  proportional/integral/derivative (PID), A38.2-5; A42.3; F37.3
  proportional-only (P), A38.2-5; A42.3; F37.2
  refrigerant flow, R45
  start-up, A42.34-35
  supervisory, A38.1
  switches, R45.1
    differential control, R45.2
    float, R45.3
    fluid flow-sensing, R45.2
    pressure control, R45.1
    temperature control, R45.2
  terminology, A42.1-2; F37.1
  testing, A42.34-35
  transducers, pressure, R45.2
  tuning, A42.34-35; F37.11
    digital controllers, A42.35; F37.12
    PI controllers, A42.35; F37.11
  valves, A42.4-6; F37.4
    characteristics, A42.4-5; F37.4
    operators, A42.6; F37.5
    selection/sizing, A42.5-6; F37.5
Convection
  forced, F3.12; F4.5
  mass, F5.5
  natural, F3.11; F4.1
Convection drying, A27.4-6
  cabinet/compartment, A27.4-5
  rotary, A27.4
  spray, A27.5-6
  tunnel, A27.5
Convectors, S4.1; S32.1
  application, S32.5
  nonstandard condition corrections, S32.3
  rating, S32.2
Convention centers, A4.5-6
Conversion factors, F35
Coolants, secondary, F20
  acetone
    properties (table), R39.10
  brines, F20.1
    corrosion inhibition, F20.2
  calcium chloride solutions, F20.1
    density (figure), F20.1
    properties (table), F20.2
    specific heat (figure), F20.1
    thermal conductivity (figure), F20.3
    viscosity (figure), F20.2
  diethyl benzene
    properties (table), R39.10
  environmental test facilities, R37.2
  ethanol
    properties (table), R39.10
  ethylene glycol, F20.4
    density
      figure, F20.10
      table, F20.6
    properties (table), F20.4, 5
    specific heat
      figure, F20.10
      table, F20.6
    thermal conductivity
      figure, F20.10
      table, F20.7
    viscosity
      figure, F20.10
      table, F20.7
  halocarbons, F20.12
    properties (table), F20.12
  hydrofluoroether
    properties (table), R39.10
  inhibited glycols, F20.4
    corrosion inhibition, F20.11
    service considerations, F20.11
  limonene, F20.13
    properties (table), F20.12, 13; R39.10
  low-temperature refrigeration, R39.11
    properties (table), R39.10
  methanol
    properties (table), R39.10
  methylene chloride (R-30), F20.12
    properties (table), F20.12
  nonhalocarbon nonaqueous fluids, F20.12
  polydimethylsiloxane, F20.13
    properties (table), F20.12, 13; R39.10
  properties (table), R4.2
  propylene glycol, F20.4
    density
      figure, F20.10
      table, F20.8
    properties (table), F20.4, 5
    specific heat
      figure, F20.10
      table, F20.8
    thermal conductivity
      figure, F20.11
      table, F20.9
    viscosity
      figure, F20.11
      table, F20.9
  refrigeration systems, R4
  sodium chloride solutions, F20.1
    density (figure), F20.4
    properties (table), F20.3
    specific heat (figure), F20.3
    thermal conductivity (figure), F20.4
    viscosity (figure), F20.4
  trichloroethylene (R-1120), F20.12
    properties (table), F20.12
  water treatment, A44.15
Coolers (*see also* **Refrigerators**)
  beverage, R25.10
  cryocoolers, R38.8
  forced-circulation air, R42
    air distribution, R42.2
    components, R42.2
    controls, R42.2, 4
    cooling capacity, R42.4
    defrosting, R42.3
    refrigerant feed, R42.3
    types, R42.1
  retail food store, R47
  walk-in, R47.7; R48.3
  water, R25.10
Coolers, liquid, S37
  Baudelot, S37.2
  brazed (semiwelded) plate, S37.1

A = 1995 Applications   S = 1996 Systems and Equipment   F = 1997 Fundamentals   R = 1998 Refrigeration

Coolers, liquid (cont.)
  direct-expansion, S37.1
  freeze prevention, S37.5
  heat transfer, S37.2
    coefficients, S37.3
    fouling factors, S37.4
  maintenance, S37.5
  oil return, S37.5
  piping, R2.20
  pressure drop, S37.4
    fluid side, S37.4
    refrigerant side, S37.4
  refrigerant flow control, S37.5
  shell-and-coil, S37.2
  shell-and-tube, S37.2
  tube-in-tube, S37.1
  types, S37.1
  vessel design, S37.4
    chemical requirements, S37.4
      ammonia (R-717), S37.4
      brines, S37.5
      halocarbon refrigerants, S37.4
      water, S37.5
    electrical requirements, S37.5
    mechanical requirements, S37.4
      codes, S37.4
      standards, S37.4

Cooling
  absorption equipment, R41
  all-air systems, S2.2
  bakery products, R27.4
  concrete, R35.1
    air blast, R35.2
    chilled water, R35.1
    inundation, R35.2
  dehumidification, S22.5
  desiccant, S22.5
  design data, F26.2
  foods and beverages
    cooling time calculations, R9.1
  fruits and vegetables, R14
    forced-air cooling, R14.4
    hydrocooling, R14.3
    load calculation, R14.1
    package icing, R14.5
    rate of cooling, R14.1
    vacuum cooling, R14.5
  geothermal energy systems, A29.13
  greenhouses, A20.11-12
  panel systems, S6
  solar energy systems, A30.18-19, 28
  water systems, S12

Cooling load
  nonresidential, F28
    building design considerations, F28.4
    CLF
      hooded equipment (table), F28.53
      lighting (table), F28.52
      people (table), F28.51
      unhooded equipment (table), F28.51
    CLTD
      flat roofs (table), F28.42
      glass (table), F28.49
      walls (table), F28.43
    CLTD/SCL/CLF method, F28.3, 39
      appliances, F28.53
      example, F28.54
      glass load, F28.49
      glass load (table), F28.50
      humans, F28.51
      lighting, F28.52
      limitations, F28.55
      procedure (table), F28.40

    example, F28.33, 49, 52, 53
    heat balance, F28.2
      conduction transfer functions, F28.2
      space air energy balance, F28.3
    heat extraction rate and room temperature, F28.31
      example, F28.33
    heat gain, F28.5
      example, F28.6, 17
      fenestration, F28.7, 41
      infiltration, F28.11
      latent heat, F28.15
      partitions, walls, and ceilings, F28.28
      shading, F28.41
      sol-air temperature, F28.5, 6
      ventilation, F28.11
      walls and roofs, F28.17, 40
    heat sources, F28.7, 51
      appliances, F28.10, 53
      appliances (table), F28.11, 12
      humans, F28.7, 51
      humans (table), F28.8
      lighting, F28.7, 52
      motors, F28.9, 53
      office equipment, F28.11
      other sources, F28.16
    principles, F28.1
    room transfer function (RTF), F28.29, 31
      example, F28.31
    sol-air temperature, F28.5, 6
      table, F28.6
    space air transfer function, F28.32
    standard air, F28.15
      example, F28.15
    TETD/TA method, F28.3, 56
      example, F28.58, 59
      procedure (table), F28.57
    transfer function method (TFM), F28.3, 17
      example, F28.33
      procedure (table), F28.18
  residential calculations, F27
    block load, F27.1
      cooling load temperature difference (CLTD), F27.2
      multifamily residences (table), F27.2
      single-family residences (table), F27.2
    example, F27.5
    glass load factor (GLF), F27.2
      multifamily residences (table), F27.3
      single-family residences (table), F27.3
    latent factor, F27.5
    procedure, F27.5
    sensible heat factor, F27.5
    shade line factor (SLF), F27.2
    shading coefficient, F27.2
      table, F27.4

Cooling towers, S36
  approach to the wet bulb, S36.1
  capacity control, S36.8
    fan cycling, S36.8
    modulating dampers, S36.8
    two-speed fans, S36.8
  central systems, S1.8
  control of, A38.9; A42.30
  design conditions, S36.2
  drift, S36.10
    eliminators, S36.10
  fill, S36.2
  fogging, S36.10
  free cooling, S36.8
    direct, S36.9
    indirect, S36.9
  heat/mass simultaneous transfer, F5.14
  indirect evaporative coolers, S19.4

  Legionella pneumophila, S36.11
    inspections, S36.11
    maintenance, S36.12
    system cleanliness, S36.12
  maintenance, S36.10
  materials of construction, S36.6
    concrete, masonry, tile, S36.7
    graphite composites, S36.7
    metals, S36.6
    plastics, S36.7
    wood, S36.6
  modeling, F30.16
  number of transfer units (NTU), S36.17
  performance curves, S36.13
  piping, S36.8
  principle of operation, S36.1
  selection, S36.7
  siting, S36.7
  sound, S36.9
    attenuators, S36.10
  testing, S36.15
  testing, adjusting, balancing, A34.14-15
  theory, S36.15
    counterflow integration, S36.17
    crossflow integration, S36.17
  thermal capability, S36.2
    nominal capacity, S36.2
  thermal performance, S36.15
  tower coefficients, S36.18
  types, S36.2
    direct contact, S36.2, 3
      chimney towers (hyperbolic), S36.4
      counterflow heat transfer, S36.4
      crossflow heat transfer, S36.4
      forced draft, S36.4
      horizontal spray, S36.3
      induced draft, S36.4
      mechanical draft, S36.4
      nonmechanical draft, S36.3
      vertical spray, S36.3
      wet/dry, S36.5
    indirect contact, S36.2, 6
      closed-circuit fluid coolers, S36.6
      coil shed towers, S36.6
      mechanical draft, S36.6
  water treatment, S36.13
    dissolved solids, S36.13
    microorganisms, S36.13
    scale formation, S36.13
  winter operation, S36.9
    closed circulating water, S36.9
    open circulating water, S36.9
    sump water, S36.9
    freeze protection, S36.9
COP. *See* **Coefficient of performance (COP)**
Copper
  boilers, S27.2
  tube. *See* **Pipe**
Corn
  drying, A21.1-7; F11.12
  fungi (table), F11.2
Corrosion, A44
  accelerating factors, A44.2-3
    concentration differential, A44.2
    dissimilar metals, A44.2-3
    flow velocity, A44.3
    moisture, A44.2
    oxygen, A44.2
    pressure, A44.3
    solutes, A44.2
    stray current, A44.3
    stress, A44.3
    temperature, A44.3
  brines, F20.2

Composite Index

Corrosion *(cont.)*
  combustion, F17.16
  control, A44
    cathodic protection, A44.4-5, 5-6
      impressed current, A44.4-5
      sacrificial anodes, A44.4
    environmental change, A44.5
    geothermal energy systems, A29.8
    materials selection, A44.3-4
    protective coatings, A44.4, 6
    water treatment, A44.8-9
  energy recovery equipment, S42.3
  fire-side corrosion and deposits, A44.16
  inhibited glycols, F20.11
  mechanism, A44.1
  oil-fired appliances, S30.17
  secondary coolant systems, R4.5
  service water heating, A45.7
  solar energy storage tanks, S33.14
  terminology, A44.1
  underground structures, A44.5-6
    bacterial effects, A44.5
    control
      cathodic protection, A44.5-6
      protective coatings, A44.6
    insulation failure, A44.5
    soil effects, A44.5
  water-side corrosion/deposits, A44.6-16
    cooling systems, A44.7-8
    heating systems, A44.7-8
    water-caused problems, A44.8
    water characteristics, A44.6-7
      biological, A44.7
      chemical, A44.6-7
Costs *(see also* **Economics***)*
  all-air systems, S2.5
  automatic control, A42.33
  building commissioning, A39.2
  capital recovery factors (table), A33.7
  data summary (table), A33.2
  economic analysis techniques, A33.6-10
    cash flow, A33.10
    inflation, A33.7
    present value (worth), A33.6-7
      equal payment series, A33.6-7
      improved years to payback, A33.7
      single payment, A33.6
    simplified, A33.6
    uniform annualized costs, A33.8-10
  equipment service life (table), A33.4
  financing alternatives, A33.5-6
    cost sharing, A33.6
    leasing programs, A33.6
    low interest, A33.6
    shared savings, A33.6
  life-cycle, A33.6
    energy recovery equipment, S42.1
    operation and maintenance, A35.1
    piping insulation, S11.19
  life-cycle savings
    insulation, F22.10
  maintenance, A33.4-5
    estimating, A33.5
    refrigerant phaseout impact, A33.5
  operating, A33.3-4
    electrical energy, A33.3
    natural gas, A33.3-4
    other fuels, A33.4
  snow melting systems, A46.4, 5
  owning, A33.1-2
    analysis period, A33.1
    income taxes, A33.2
    initial cost, A33.1
    insurance, A33.1-2

    interest/discount period, A33.1
    property taxes, A33.2
Cotton
  drying, A21.8-9; F11.12
Crawl spaces, F39.1
  heat loss, F27.9
  moisture control, F23.11
  temperature, F27.9
Crops. *See* **Farm crops**
Cryogenics, R38
  biomedical applications, R40
    cryomicroscopy, R40.6
    cryopreservation, R40.1
    cryoprotective agents, R40.2
    cryosurgery, R40.7
    induced hypothermia, R40.7
    specimen preparation, R40.6, 7
  cryocoolers, R38.8
    recuperative
      Brayton, R38.10
      Joule-Thomson, R38.8
    regenerative
      Gifford-McMahon, R38.11
      orifice pulse tube, R38.10
      Stirling, R38.10
  cryogenic fluids, R38.1
    flammability, R38.25
    power requirements (table), R38.8
    storage vessels, R38.22
    transfer, R38.23
  equipment, R38.16
    compressors, R38.16
    expansion devices, R38.17
    heat exchangers, R38.17
    regenerators, R38.19
  freezers, industrial
    carbon dioxide, R15.5
    liquid nitrogen, R15.5
  hazards, R38.24
  instrumentation, R38.23
  insulation, R38.19
    selection (table), R38.22
    thermal conductivity (table), R38.20
  liquefaction of gases, R38.3
    system comparison (table), R38.7
  properties of materials, R38.1
  purification of gases, R38.15
  recovery of gases, R38.13, 14
  refrigeration cycles
    Brayton, R38.8
      figure, R38.9
    cascade, R38.6
    Claude, R38.5
      figure, R38.5
    Heylandt, R38.5
    isenthalpic expansion, R38.4
    isentropic expansion, R38.5
    Joule-Thomson, R38.4
      figures, R38.4, 9
    Linde, R38.4
    mixed refrigerant, R38.6
    Stirling, R38.10
      figure, R38.10
  separation of gases, R38.12
Dairy products, R19
  aseptic packaging, R19.20
  butter
    manufacture, R19.6
    refrigeration load, R19.9
  buttermilk, R19.5
  cheese
    cheese room refrigeration, R19.13
    manufacture, R19.9
  cream, R19.5

  display refrigerators, R47.3
  ice cream, R19.13
    freezing, R19.16
    freezing points (table), R19.15
    hardening, R19.17
    mix preparation, R19.14
    refrigeration equipment, R19.19
    refrigeration requirements, R19.16
  milk, dry, R19.22
  milk, evaporated, R19.21
  milk, fresh, R19.1
    processing, R19.2
    refrigeration system, R19.5
    storage, R19.4
  milk, sweetened condensed, R19.22
  storage requirements (table), R10.3
  thermal properties, R8
  UHT sterilization, R19.19
  yogurt, R19.5
Dampers
  automatic control with, A42.6-7; F37.6
  central systems, S1.4, 5
  cooling tower capacity control, S36.8
  fire and smoke control, A48.8
  furnaces, S28.2
  sound in, A43.11-12
  vehicular facilities, enclosed, A12.16-17
  vent, S30.22
Dams
  concrete cooling, R35.1
Darcy equation
  friction loss, F32.7
Darcy-Weisbach equation
  pressure drop, F2.8; F32.12; F33.1
Data processing system areas, A16
  air-conditioning systems, A16.2-3, 5-6
    central station supply, A16.3
      chilled water, A16.6
        packaged units, A16.3
      condensing methods, A16.6
      controls, A16.5
      humidification, A16.6
      packaged units, A16.3
      refrigeration, A16.5-6
  air distribution, A16.3-5
    ceiling plenum supply, A16.5
    overhead ducted supply, A16.5
    underfloor plenum supply, A16.4-5
    zoning, A16.4
  computer rooms, A16.1-2
    water-cooled equipment, A16.5
  cooling loads, A16.2
  design criteria, A16.1-2
    ancillary spaces, A16.2
    computer rooms, A16.1-2
  energy conservation, A16.7
  fire protection, A16.7
  heat recovery, A16.7
  humidification, S20.1
  instrumentation, A16.6-7
  return air, A16.5
  water-cooled computer equipment, A16.5
Definitions. *See* **Terminology**
Defrosting
  air coolers, forced circulation, R42.3
  air-source heat pump coils, S8.8
  ammonia liquid recirculation systems, R3.23
  heat pumps, unitary, S44.11
  household refrigerators and freezers, R49.6
  meat coolers, R16.2
  retail food store refrigerators, R47.13
Degree-day method, F30.16
  variable-base, F30.19

A = 1995 Applications   S = 1996 Systems and Equipment   F = 1997 Fundamentals   R = 1998 Refrigeration

Degree-days
U.S./Canada monthly and yearly, 1981
Fundamentals, Chapter 24, pp. 23-28
(See explanation on first page of Index)
Dehumidification, S22
absorption, S22.1, 8
adsorption, S22.1, 9
air filters, S22.7
airflow control, S22.7
air washers, S19.7
all-air systems, S2.3
applications, S22.1
atmospheric pressure, S22.6
air conditioning systems, S22.6
condensation prevention, S22.6
gaseous contaminant control, S22.7
industrial processes, S22.6
storage, S22.6
testing, S22.7
high pressures, S22.9
drying gases, S22.9
storage, S22.9
testing, S22.10
coils, S21
load determination, S21.14
maintenance, S21.15
performance, S21.9
commissioning, S22.8
cooling, S22.5
desiccant, S22.2; F21.1
liquid, S22.2; F21.3
solid, S22.3; F21.4
rotary, S22.3
design data, F26.2
ductwork, S22.7
equipment operation, S22.7
environmental test facilities, R37.4
evaporative cooling, A47.2-3
high pressures, S22.8
leakage, S22.7
methods, S22.1
mines, A25.3-4
Dehumidifiers
room, S43.5
capacity, S43.6
codes, S43.6
design, S43.5
efficiency, S43.6
energy conservation, S43.6
performance, S43.6
Dehydration
farm crops, A21; F11
industrial systems for, A27
refrigeration systems, R46.1
Density
chilled water, A40.11
fluids, F2.1
liquids (table), F36.2
solids (table), F36.3
vapors (table), F36.1
Desiccants, S22; F21
applications, F21.1
cooling, S22.5
cosorption of water vapor/air contaminants, F21.6
cycle, F21.1
dehumidification, S22.2
isotherms, F21.5
life, F21.5
liquid, S22.2
maintenance, S22.7
refrigerant system, R6.3, 5
equilibrium curves, R6.4
solid, S22.3
rotary, S22.3

types, F21.3
liquid absorbents, F21.3
solid adsorbents, F21.4
Desuperheaters
unitary systems, S44.4
Desuperheating
condensers, evaporative, S35.16
Dielectric drying, A27.4
Diffusers. *See* **Air diffusers**
Diffusion
coefficient, F5.2
eddy, F5.7
molecular, F5.1
Diffusivity, thermal
foods, R8.17
Dilution
exhaust, F15.8
critical, F15.10
minimum, F15.9
Dimensionless numbers
table, F3.2; F34.4
Dining and entertainment centers, A3.4-6
Discharge coefficients
valves, F2.11
District heating and cooling, S11
application, S11.1
central plant, S11.2
boiler, S11.2
chiller, S11.2
design considerations, S11.3
environmental equipment, S11.3
thermal storage, S11.3
components, S11.1
central plant, S11.1
distribution system, S11.1
user system, S11.1
consumer interconnects, S11.23
chilled water, S11.23
hot water, S11.23
steam, S11.23
distribution system, S11.8
aboveground, S11.8
conduit, S11.8
factory-prefabricated, S11.8
field-fabricated, S11.8, 11
insulation, S11.11
cellular glass, S11.11
economical thickness, S11.19
hydrocarbon powder, S11.11
hydrophobic powder, S11.11
life-cycle cost, S11.19
poured envelope, S11.11
thermal properties, S11.12
low-temperature, S11.9
plastic pipe, S11.8, 9
trench
deep-buried, S11.10
shallow, S11.9
tunnel, S11.9
economic benefits, S11.1
capital investment, S11.2
energy source, S11.2
expansion provisions, S11.19
anchors/guides/supports, S11.19
bends/loops, S11.19
cold springing, S11.20
joints, S11.19
heating design, S11.5
direct connection, S11.7
hot water, S11.5
indirect connection, S11.6
heat exchangers, S11.6
pressure, S11.6
steam, S11.5

temperature, S11.5
control, S11.6
heat transfer analysis, S11.13
pipes in air, S11.18
pipes in buried trenches/tunnels, S11.16
pipes in shallow trenches, S11.17
single buried pipe
in conduit with air space, S11.14
insulated, S11.14
uninsulated, S11.13
steady-state, S11.13
two pipes buried, S11.15
in conduit with air space, S11.15
hydraulic considerations, S11.20
condensate drainage/return, S11.20
drip leg, S11.20
drip station, S11.21
metering, S11.24
displacement meter, S11.24
propeller meter, S11.24
network calculations, S11.20
pressure losses, S11.20
valve vaults, S11.21
water hammer, S11.20
piping
layout, S11.7
materials, S11.7
sizing, S11.20
standards, S11.7
thermal design considerations, S11.12
soil conductivity, S11.12
Domestic hot water. *See* **Service water heating**
Domiciliary facilities, A5
air conditioning, A5
design criteria, A5.1, 3-4
dormitories, A5.3
energy efficient systems, A5.1-2
energy inefficient systems, A5.2
energy neutral systems, A5.2
hotels, A5.3-4
load characteristics, A5.1, 3
motels, A5.3-4
multiple-use complexes, A5.4
total energy systems, A5.2-3
Doors
U-factors, F24.12
table, F24.13; F29.10
Dormitories, A5.3
service water heating, A45.11, 12
Draft
air diffusion, F31.6
chimney, S30.1
available, S30.1, 2, 6
theoretical, S30.1, 2, 6
comfort affected by, F8.13
cooling towers, S36.3, 4, 6
fuel-burning equipment, S26.14
Drag, F2.5
Draperies, F29.43
double, F29.46
shading coefficients
example, F29.44, 45
table, F29.40
Driers (*see also* **Dryers**)
refrigerant systems, R6.5
Drip leg
steam systems, S11.20
Drip station
steam systems, S11.21
Drives
engine, S7.1
turbine, S7.8
variable-speed
geothermal energy systems, A29.9

Composite Index

Dryers (*see also* **Driers**)
    farm crops, A21
Drying
    farm crops, A21; F11
    gases
        dehumidification for, S22.9
    mechanism of, A27.1
Drying systems, industrial, A27
    agitated-bed, A27.6
    calculations, A27.2-3
    conduction, A27.3-4
    convection, A27.4-6
        cabinet/compartment, A27.4-5
        rotary, A27.4
        spray, A27.5-6
        tunnel, A27.5
    dielectric, A27.4
    drying time determination, A27.1-3
    flash, A27.6
    fluidized-bed, A27.6
    freeze drying, A27.6
    microwave, A27.4
    psychrometrics, A27.1-3
    radiant infrared, A27.3
    selection, A27.3
    superheated vapor, A27.6
    ultraviolet, A27.3
    vacuum, A27.6
Dual-duct systems
    air distribution boxes, S17.7
    all-air systems, S2.7
        constant volume, S2.7
        variable-air-volume, S2.7
    control, A42.20-22
    testing, adjusting, balancing, A34.4
DuBois equation
    body surface area, F8.3
Duct design, F32
    Bernoulli equation, F32.1
    computer applications, A36.7-8
    Darcy-Weisbach equation, F32.12
    design considerations, F32.14
        velocities for HVAC components, F32.17
        table, F32.18
    Duct Fitting Database, F32.11
    dynamic losses, F32.8
        Duct Fitting Database, F32.11
        local loss coefficients, F32.8
    examples
        heat transfer, F32.15
        industrial exhaust, F32.25
        procedures, F32.21
        thermal gravity effect (stack effect), F32.2, 3, 5
    fan-system interface, F32.12
        fan inlet and outlet conditions, F32.12
        fan system effect coefficients, F32.12
    fitting loss coefficients, F32.29
        Duct Fitting Database, F32.11
    friction losses, F32.7
        Altshul-Tsal equation, F32.7
        Colebrook equation, F32.7
        Darcy equation, F32.7
        flat oval ducts, F32.8
            round equivalent dimensions (table), F32.11
        friction chart, F32.8
            figure, F32.9
        noncircular ducts, F32.8
        rectangular ducts, F32.8
            round equivalent dimensions (table), F32.10
        roughness factors, F32.7
            table, F32.7

industrial exhaust systems, F32.22
    example, F32.25
insulation, F32.14
leakage, F32.15
    duct seal levels (table), F32.17
    leakage class, F32.16
    table, F32.16
methods, F32.18
    equal friction, F32.18
    static regain, F32.19
    T-method optimization, F32.19
    T-method simulation, F32.21
noise, F32.18
pressure, F32.2
    static pressure, F32.2
    total pressure, F32.2
    velocity pressure, F32.2
smoke management, F32.14
testing, adjusting, and balancing, F32.18
Ductility
    metals and alloys (figure), R39.6
Ducts
    air diffusers, F31.12
    central systems, S1.6
    classification, S16.1
        commercial, S16.1
        industrial, S16.1
        residential, S16.1
    cleaning, S16.2
    construction, S16
        codes, S16.1
        commercial, S16.2
            acoustical treatment, S16.4
            fibrous glass ducts, S16.3
            flat-oval ducts, S16.3
            flexible ducts, S16.4
            hangers, S16.4
            materials, S16.2
            plenums/apparatus casings, S16.4
            rectangular ducts, S16.2
            round ducts, S16.2
        grease-/moisture-laden vapors, S16.5
        industrial, S16.4
            construction details, S16.5
            hangers, S16.5
            materials, S16.4
            rectangular ducts, S16.5
            round ducts, S16.4
        master specifications, S16.6
        outside, S16.6
        residential, S16.2
        seismic qualification, S16.6
        sheet metal welding, S16.6
        standards, S16.1
            commercial, S16.2
            industrial, S16.4
            residential, S16.2
        thermal insulation, S16.6
        underground, S16.5
    dehumidification, S22.7
    equivalent diameter, F2.10
    equivalent lengths, S9.5
    flat oval ducts, F32.8
        round equivalent dimensions (table), F32.11
    fluid flow, F2
    friction chart, F32.8
    heaters for
        control, A42.33
    industrial exhaust systems, A26.7-10
    insulation, F23.18; F32.14
    laboratories
        leakage of, A13.9-10

    leakage, S16.2; F32.15
        duct seal levels (table), F32.17
        leakage class, F32.16
        table, F32.16
    noncircular ducts, F32.8
    plastic, S16.5
    rectangular ducts, F32.8
        round equivalent dimensions (table), F32.10
    road tunnels, A12.6
    roughness factors, F32.7
        table, F32.7
    small forced-air systems, S9.1, 5
    sound control, A43.11-23; F7.11
    velocity measurement in, F14.16
    vibration control, A43.40
Dump boxes
    air distribution, S17.7
Dynamic test chambers, R37
Earth
    stabilization, R35.3, 4
    ventilation tubes, A20.3
Earthquakes
    seismic restraint design, A50
Economic analysis techniques, A33.6-10
    cash flow, A33.10
    inflation, A33.7
    present value (worth), A33.6-7
        equal payment series, A33.6-7
        improved years to payback, A33.7
        single payment, A33.6
    simplified, A33.6
    uniform annualized costs, A33.8-10
Economics (*see also* **Costs**)
    cogeneration, S7.42
    district heating and cooling, S11.1
    energy management program, A32.3
    energy recovery equipment, S42.1
    evaporative cooling, A47.10-11
    gaseous contaminant control, A41.16
    insulation thickness, F22.22
    laboratories, A13.16
    owning and operating costs, A33
    pipe insulation thickness, S11.19
    steam turbines, S7.18
    thermal storage, A40.2-3
Economizers
    airside, S5.7
    compressors
        single-screw, S34.13, 14
        twin-screw, S34.20
    cycle, A38.5-6
    kitchen ventilation, A28.12
    nonresidential ventilation, F25.24
    waterside, S5.7
Eddy diffusivity, F5.7
Educational facilities, A6
    colleges, A6.3-4
    controls, A6.2
    design considerations, A6.1-2
    elementary schools, A6.2
    energy conservation, A6.4
    equipment selection, A6.4-5
    middle schools, A6.2-3
    preschools, A6.2
    secondary schools, A6.2-3
    sound control, A6.2
    system selection, A6.4-5
    universities, A6.3-4
    ventilation requirements, A6.4
EER. *See* **Energy efficiency ratio (EER)**
Effectiveness
    heat transfer, F3.4

Efficiency
  air conditioners, room, S43.2
  best efficiency point (BEP)
    pumps, centrifugal, S38.6
  boilers, S27.5
  combustion, F17.12
  compressors
    centrifugal, S34.30
    orbital (scroll), S34.24
    positive-displacement, S34.2
    reciprocating, S34.4
    rotary, S34.9
    single-screw, S34.15
    twin-screw, S34.21
  dehumidifiers, room, S43.6
  fins, F3.18
  furnaces, S28.5, 7, 9
  heating equipment, F30.18
  industrial gas cleaning, S25.3, 10
  infrared heaters, S15.4
  motors, S39.2
  pumps, centrifugal, S38.6
  unitary systems, S44.5
Eggs, R20
  egg products, R20.8
    dehydrated, R20.11
    frozen, R20.11
    processing, R20.9
    refrigerated, R20.10
  processing plant sanitation, R20.12
  shell eggs, R20.1
    processing, R20.5
    refrigeration, R20.5
    spoilage prevention, R20.4
    storage, R20.7
    transportation, R20.8
  storage requirements (table), R10.3
  thermal properties, R8
Electric
  boilers, S27.4
  coils
    air-heating, S23.2
  furnaces, residential, S28.4, 6, 8
  heaters (in-space), S29.4
    baseboard, S29.4
    control, S29.4
    panel, S29.4
    portable, S29.4
    radiant, S29.4
    wall/floor/toe space/ceiling, S29.4
  infrared heaters, S15.2
    metal sheath, S15.3
    quartz tube, S15.3
    reflector lamp, S15.3
    tubular quartz lamp, S15.3
  makeup air units, S31.9
  panel systems, S6.15
  service water heating, A45.2
  snow melting systems, A46.1-4, 8-13
  unit heaters, S31.4
  unit ventilators, S31.1
Electrical
  cogeneration, S7.26
    distribution, S7.28
  demand/load factor
    commercial buildings (table), A32.6
  hazards, F9.16
  insulation. *See* **Insulation, electrical**
Electricity
  measurement, F14.21
    ammeters, F14.21
    power-factor meters, F14.23
    voltmeters, F14.21
    wattmeters, F14.23

Electrostatic precipitators, S25.7
  components, S25.10
    electrical equipment, S25.10
    electrode cleaning, S25.11
    electrodes, S25.10
    safety interlocks, S25.11
  mathematical modeling, S25.10
    Anderson equation, S25.10
    Deutsch equation, S25.10
    efficiency, S25.10
  principles of operation, S25.8
    corona current, S25.8
    corona quenching, S25.9
    deposit resistivity, S25.9
    electric field, S25.8
    space charge, S25.8
Elevators
  smoke control, A48.11
Eliminators
  moisture
    condensers, evaporative, S35.14
Emissivity
  solids (table), F36.3
Emittance
  air spaces (table), F24.2
  table, F3.8
Energy
  accounting, A32.13
  audit field survey, A34.16-17
  audit levels, A32.7-10
  cost distribution chart, A32.10-11
  emergency use reduction, A32.13-14
  forecasting building needs, A38.18-20
  geothermal. *See* **Geothermal energy**
  monitoring. *See* **Building energy**
    monitoring
  simulation
    computer applications, A36.5-7
Energy balance
  comfort, F8.2, 16
  refrigeration systems, R44.3
Energy conservation
  air conditioners, room, S43.3
  all-air systems, S2.5
  automatic control for, A42.31
  buildings, A38
  clean spaces, A15.11
  data processing areas, A16.7
  dehumidifiers, room, S43.6
  educational facilities, A6.4
  farm crop drying, A21.4
  greenhouses, A20.14
  infrared heaters, S15.1
  inverse modeling, F30.26
  kitchen ventilation, A28.12-13, 18
  laboratories, A13.15
  measure (ECM), A32.2-13
  opportunity (ECO), A32.2-13
  pumps, centrifugal, S38.12
  textile processing, A18.6
  unitary systems, S44.5
Energy consumption
  buildings
    control effect on, A38.5-6
  commercial (table), A32.6
  emergency use reduction, A32.13-14
  gaseous contaminant control, A41.15-16
  humidifiers, S20.3
  residential (table), A32.8-9
  U.S., A31.4-6; F16.5
    per capita end-use, A31.5; F16.5
    projections, A31.5-6; F16.5
  world, A31.3-4; F16.3
    coal, A31.4; F16.4
    electricity, A31.4; F16.4

    natural gas, A31.3-4; F16.4
    per capita commercial, F16.4
    petroleum, A31.3; F16.3
Energy efficiency ratio (EER)
  air conditioners, room, S43.2
Energy estimating, F30
  balance point temperature, F30.17
  bin method, F30.20
  common elements, F30.1
  component modeling, F30.3
  control models, F30.21
  correlation methods, F30.20
  degree-day method, F30.16
  equation-based modeling, F30.21
  forecasting methods, A38.18-20
  forward modeling, F30.22
  furnaces
    electric, F30.19
    gas, F30.19
    oil, F30.19
  hybrid modeling, F30.28
  instantaneous sensible load calculation, F30.3
    heat balance method, F30.3
      example, F30.5
    thermal network method, F30.9
    weighting factor method, F30.8
  inverse modeling, F30.22, F30.25
    dynamic, F30.24
    energy conservation, F30.26
    energy management, F30.27
    selection (table), F30.28
    steady-state, F30.23
  models (table), F30.3, F30.27
  neural network modeling, F30.24
    commercial buildings, F30.24
    residential buildings, F30.25
  primary systems, F30.12
    boiler models, F30.13
    cooling tower models, F30.16
    modeling strategies, F30.12
      first principle models, F30.13
      HVAC1 Toolkit, F30.13
      part-load ratio, F30.13
      regression models, F30.12
    vapor compression chiller models, F30.14
      compressor models, F30.14, F30.15
      condenser models, F30.14
      evaporator models, F30.14
      other chiller models, F30.15
      reciprocating compressor models, F30.14
  seasonal efficiency, F30.18
  secondary systems, F30.10
    cooling/dehumidifying coils, F30.11
      effectiveness analysis, F30.11
    fans, pumps, distribution, F30.10
      HVAC2 Toolkit, F30.10
    heat and mass transfer, F30.11
  selecting a method, F30.2
  space load, F30.1
  system modeling, F30.16
    integration, F30.21
  variable-base degree-days, F30.19
Energy management, A32
  commercial buildings
    characteristics (table), A32.4, 5
    energy consumption (table), A32.6
    median electric demand/load factor (table), A32.6
  demand-side, A31.6; F16.6
  emergency energy use reduction, A32.13-14
  energy accounting, A32.13

A = 1995 Applications   S = 1996 Systems and Equipment   F = 1997 Fundamentals   R = 1998 Refrigeration

Composite Index

Energy management *(cont.)*
  energy audits, A32.7-12
    Level I, A32.10
    Level II, A32.10
    Level III, A32.10
  energy conservation measure (ECM), A32.2-13
  energy conservation opportunity (ECO), A32.2-13
  financing, A32.3
  implementing, A32.3-13, 14
    accomplishing measures, A32.13
    database compilation, A32.3-7
    maintaining measures, A32.13
    prioritizing, A32.7, 12-13
  inverse modeling, F30.27
  organizing, A32.1-3
    energy consultant, A32.2-3
    energy manager, A32.2
    motivation, A32.3
  residential buildings
    energy consumption (table), A32.8-9
Energy monitoring. *See* **Building energy monitoring**
Energy recovery (*see also* **Heat recovery**)
  chemical industry, R36.4
  industrial exhaust systems, A26.10
  laboratories, A13.15
  mines, A25.6
Energy resources, A31; F16
  characteristics, A31.1-2; F16.1
    electricity, A31.1-2; F16.1
    fossil fuels, A31.1-2; F16.1
    nonrenewable, A31.1; F16.1
    on-site energy forms, A31.1; F16.1
    renewable, A31.1; F16.1
  demand-side management, A31.6; F16.6
  integrated resource planning, A31.6; F16.6
  relationships to site energy, A31.7; F16.7
    intangible, A31.7
    local considerations, A31.7
    national/global considerations, A31.7
    quantifiable, A31.7
    site considerations, A31.7
  resource utilization factor (RUF), F16.7
  summary, A31.7
  U.S., A31.4-6; F16.5
    consumption projections, A31.4-6
      by resource, A31.5-6
      by sector, A31.6
      overall, A31.5
    outlook summary, A31.6; F16.6
    overall consumption projections, F16.5
    per capita consumption projections, F16.5
  use of, A32
  world, A31.2-4; F16.2
    consumption, A31.3-4; F16.3
    production, A31.2-3; F16.2
    reserves, A31.3; F16.3
Energy sources, F16.1
Engines, S7
  air-cooled engines, S7.36
  applications, S7.38
    centrifugal compressors, S7.39
    heat pumps, S7.40
    reciprocating compressors, S7.39
    screw compressors, S7.40
  codes, S7.46
  combustion air systems, S7.6, 36
  compressed air systems, S7.6
  control and instruments, S7.6
    alarm/shutdown, S7.6
    gas leakage prevention, S7.7
    governors, S7.7
    starting, S7.6

  design, S7.34
  ebullient cooling, S7.35
  exhaust gas, S7.35
  exhaust systems, S7.35
  fuels, S7.4; F17.6
    cetane number, F17.6
    heating value, S7.4
    selection, S7.4
  fuel systems, S7.4
    dual-fuel, S7.5
    fuel oil, S7.4
    multifuel, S7.5
    spark ignited gas, S7.5
  heat recovery, S7.20
  heat removal, S7.36
  jacket water, S7.5, 34
  lubrication, S7.5
  maintenance, S7.7
  noise control, S7.7
  performance characteristics, S7.3
  reciprocating, S7.1
    four-stroke, S7.2
    heat recovery, S7.20
      exhaust gas, S7.21
      jacket, S7.20
      lubricant, S7.21
      turbocharger, S7.21
    thermal output, S7.18
    two-stroke, S7.2
  sizing, S7.38
  standards, S7.46
  turbochargers, S7.6
  vibration control, S7.7
  water-cooled engines, S7.34
Engine test facilities
  air conditioning, A14
  engine heat release, A14.3
  ventilation, A14
Enthalpy
  calculation, F1.4
  definition, F1.1
  foods, R8.8
  recovery loop
    twin-tower, S42.15
  water vapor, F5.11
Entropy
  calculation, F1.4
  definition, F1.1
Environmental control
  animals. *See* **Animal environments**
  humans. *See* **Comfort**
  plants. *See* **Plant environments**
  retail food stores, R47.14
    store ambient effect, R47.2
Environmental health, F9
  biostatistics, F9.1
  cellular biology, F9.4
  epidemiology, F9.1
  ergonomics, F9.4
  genetics, F9.4
  industrial hygiene, F9.4
    short-term exposure, F9.5
    time-weighted average exposure, F9.5
  molecular biology, F9.3
  physical hazards, F9.14
    electrical hazards, F9.16
    electromagnetic radiation, F9.19
      ionizing radiation, F9.19
      nonionizing radiation, F9.20
      optical, F9.20
      radio waves, F9.21
      radon, F9.19
      radon action levels (table), F9.20
    noise, F9.18

    thermal comfort, F9.14
      comfort zones, F9.14
      diseases affected by, F9.15
      figure, F9.16
      hot and cold surfaces, F9.16
      hyperthermia, F9.15
      hypothermia, F9.15
    vibrations, F9.16
  standards, F9.1
    table, F9.2
  terminology, F9.1
  toxicology, F9.3
Environmental test facilities, R37
  chamber construction, R37.6
    altitude chambers, R37.6
  control, R37.4
  cooling systems, R37.1
  design calculations, R37.5
  heating systems, R37.2
  humidity, R37.3, 6
  refrigerants, R37.1
Ethylene glycol
  coolants, secondary, F20.4
  hydronic systems, S12.16
Eutectic plates
  refrigerated surface transport, R29.4
Evaporation
  in tubes
    equations (table), F4.7
    forced convection, F4.5
    natural convection, F4.1
Evaporative cooling, A47
  animal environments, A20.3; A47.8-9
  applications, commercial, A47.4-5
    two-stage, A47.5
  applications, general, A47.1-3
    air cleaning, A47.3
    cooling, A47.1-3
    dehumidification, A47.2-3
    heated recirculated water, A47.2
    humidification, A47.2
    preheating air, A47.2
    spray water, A47.2
  applications, industrial, A47.5-8
    area cooling, A47.7
    effective temperature, A47.5-7
    gas turbine engines, A47.8
    laundries, A47.8
    motors, A47.7
    paper products facilities, A47.8
    process cooling, A47.8
    spot cooling, A47.7
    wood products facilities, A47.8
  applications, other, A47.8-10
    animal environments, A47.8-9
    greenhouses, A47.9-10
    power generation facilities, A47.8
    produce storage, A47.9
      apples, A47.9
      citrus, A47.9
      potatoes, A47.9
  applications, residential, A47.4-5
  booster refrigeration, A47.4
  direct, A47.1; S19.1
  economic considerations, A47.10-11
    direct savings, A47.10
    indirect savings, A47.10-11
  equipment, S19
    air washers, S19.5
      air cleaning, S19.7
      dehumidification, S19.7
      high-velocity spray-type, S19.6
      humidification, S19.6, 7
      performance factor, S19.7
      spray-type, S19.5

A = 1995 Applications   S = 1996 Systems and Equipment   F = 1997 Fundamentals   R = 1998 Refrigeration

Evaporative cooling *(cont.)*
  equipment *(cont.)*
    direct, S19.1
      effectiveness of saturation, S19.1
      greenhouse, S19.3
      packaged rotary, S19.3
      random-media, S19.1
      remote pad, S19.3
      rigid-media, S19.2
      slinger packaged, S19.2
    hybrid, S19.4
    indirect, S19.3
      cooling tower/coil systems, S19.4
      effectiveness, S19.4
      heat pipes, S19.4
      heat wheels, S19.4
      packaged, S19.3
    indirect/direct combinations, S19.4
      makeup air pretreatment, S19.5
      precooling, S19.5
      three-stage, S19.5
      two-stage, S19.4
    maintenance, S19.7
  exhaust, A47.5
  gas turbines, S7.10
  greenhouses, A20.11-12; A47.9-10
    fan-and-pad systems, A20.11-12
    fog systems, A20.12
    unit systems, A20.12
  heat recovery equipment, S42.5
  hybrid, S19.4
  indirect, A47.2, 3-4, 4-5; S19.3; S42.5
  indirect/direct (staged) system, A47.4; S19.4
  industrial air conditioning, A11.8
  mines, A25.4, 5, 7
  precooling, A47.3-4; S19.5
    mixed air systems, A47.3-4
    outdoor air systems, A47.3
  psychrometrics, A47.1-2, 5, 11-12
  water treatment, S19.7
    Legionella pneumophila, S19.8
  water usage, S19.2
  weather considerations, A47.10

Evaporators
  air conditioners, room, S43.2
  ammonia refrigeration systems, R3.2
    piping, R3.19
  automobile air conditioning, A8.4
  chemical industry refrigeration, R36.7
  flooded, F4.5
  halocarbon refrigeration systems
    piping, R2.23
  liquid overfeed systems, R1.6
  modeling, F30.14

Exfiltration, F25.1

Exhaust
  animal buildings, A20.5-6
  cleaning. *See* **Gas-cleaning equipment**
  clean spaces, A15.9-10
  dilution, F15.8
  engine drives, S7.6, 35
    heat recovery, S7.21
  engine test facilities, A14.2, 4-5
  gas turbine drives, S7.9
  industrial air conditioning, A11.8, 9
  kitchens, A28.6-9, 18
  laboratories, A13.3-6, 8, 9-10
    stack height, A13.12
  photographic processing, A19.2-3
  stack design, F15.10
    minimizing reentrainment, F15.12
  vehicular facilities, enclosed, A12.17

Exhibition centers, A4.5-6

Expansion
  bends, S40.9
    L type, S40.9
    U type, S40.9
    Z type, S40.9
  joints, A43.40; S11.19; S40.10
    packed, S40.11
      flexible ball, S40.11
      slip, S40.11
    packless
      flexible hose, S40.12
      metal bellows, S40.11
      rubber, S40.12
  loops, S40.9
  tanks
    hydronic systems, S12.3, 7
      closed, S12.3
      diaphragm, S12.3
      hydraulic function, S12.7
      open, S12.3
      thermal function, S12.3
    secondary coolant systems, R4.3
    solar energy systems, A30.17
  valves. *See* **Valves**

Fairs, A4.8

Fan-coil units, S3.3; S4.1

Fans, S18
  air conditioners, room, S43.2
  animal environments, A20.5-6
  arrangements, S18.8
  axial flow, S18.1
  central systems
    relief air, S1.4
    return air, S1.4
    supply air, S1.6
  centrifugal, S18.1
  clean spaces, A15.11
  condensers
    air-cooled, S35.8
    evaporative, S35.17
  control of, A38.1; A41.14; A42.13-15; S18.8
  cooling tower capacity control, S36.8
  draft, S30.23
  ducts affecting, S18.6
  duct system interface, F32.12
  energy consumption, A38.2
  farm crop drying, A21.2
  furnaces, S28.2
  industrial exhaust systems, A26.10
  installation, S18.8
  isolation, S18.8
  kitchen exhaust, A28.8
  laws, S18.4
  modeling, F30.10
  noise, S18.8
  operating principles, S18.1
  parallel operation, S18.7
  plenum, S18.1
  plug, S18.1
  pressure relationships, S18.5
  rating, S18.4
  selection, S18.7
  ships
    merchant, A10.2
    naval, A10.6
  solar energy systems, A30.17
  sound control, A43.8-9, 9
  speed control, S35.17
  system effects, S18.7
  testing, S18.4
  types, S18.1
    axial flow, S18.1
    centrifugal, S18.1
    plenum, S18.1
    plug, S18.1
  vehicular facilities, enclosed, A12.15-16

Farm crops, A21; F11
  aeration, A21.4, 10-12; F11.11
  combination drying, A21.4
  deep bed drying, A21.4-7; F11.10
  dryeration, A21.4
  drying, A21.2-9; F11
    airflow resistance, F11.9
      constants (table), F11.10
    barley, F11.13
    corn, A21.1-7; F11.12
    cotton, A21.8-9; F11.12
    energy conservation, A21.4
    grain, F11.12
    hay, A21.8; F11.11
    peanuts, A21.9; F11.13
    physiological factors, F11
    rice, A21.9; F11.13
    seeds, F11.13
    soybeans, A21.7-8; F11.13
    theory, F11.8
    tobacco, F11.13
    wheat, F11.13
  drying equipment, A21
    batch dryers, A21.3
    column dryers, A21.3
    concurrent flow dryers, A21.3
    continuous flow dryers, A21.3-4
    controls, A21.3
    counterflow dryers, A21.3-4
    crossflow dryers, A21.3
    fans, A21.2
    heaters, A21.2-3
    rack-type dryers, A21.3
    recirculating/continuous flow bin dryers, A21.7
  full-bin drying, A21.4-6
  layer drying, A21.6-7
  microbial growth
    relative humidity affecting (table), A21.1
    temperature affecting (table), A21.1
  recirculation, A21.4
  shallow layer drying, A21.3-4; F11.8
  storing, A21.9-12; F11
    carbon dioxide, F11.3
    fungi, F11.6
      conditions promoting (table), F11.2
    grain aeration, A21.10-12; F11.11
    insects, F11.7
      conditions promoting (table), F11.3
    moisture content, F11.1
      equilibrium, F11.3
      measurement, F11.5
      safe levels (table), F11.2
    moisture migration, A21.9-10; F11.2
    oxygen, F11.3
    rodents, F11.7
    seeds, A21.12
    temperature, F11.3

Fenestration, F29; F39.1
  airflow windows, F29.24
  complex systems, F29.37
    absorptance calculation, F29.39
    solar-thermal separation, F29.39
    transmittance calculation, F29.39
  condensation resistance, F29.12
  cooling load, F28.7, 41
  daylighting from, F29.46
  doors, F29.10
  energy flow, F29.2
  energy performance, F29.46
  frames, F29.1, 4

Composite Index

Fenestration *(cont.)*
  glass, F29.47
    durability, F29.48
    example, F29.24
    occupant comfort, F29.48
    safety/strength, F29.48
    sound transmittance loss (table), F29.48
  glass block walls, F29.26
  glazing
    angular selectivity, F29.20
    glass, F29.47
    insulating glazing unit (IGU), F29.1
    plastic materials, F29.27
    properties, F29.19
    spectral selectivity, F29.20
    table, F29.48
  heat transfer coefficients, F29.4
    air space (table), F29.6
    glazing (table), F29.5
  infiltration, F29.12; F39.7
  optical properties, F29.17
    absorptance, F29.17
    angular dependence, F29.18
    reflectance, F29.17
    spectral dependence, F29.19
    transmittance, F29.17
    visible light transmittance (table), F29.25, 47
  shading, F29.2, 41
    computer calculations, F29.43
    draperies, F29.43
      double, F29.46
      example, F29.45
    example, F29.42, 43
    exterior, F29.41
    interior, F29.43
    louvers, F29.41
    partial shading, F29.42
    roller shades, F29.43
    roof overhangs, F29.42
    venetian blinds, F29.43
  shading coefficients (see also **Heat gain**), F29.23
    domed skylights (table), F29.26
    double glazing (table), F29.39
    glass (table), F29.25, 38, 39, 40
      with draperies (table), F29.40
      with roller shades (table), F29.38
      with venetian blinds (table), F29.38
    glass block walls (table), F29.27
    louvers (table), F29.38
    plastic (table), F29.27
  skylights (domed), F29.26
  solar angle, F29.16
  solar energy
    diffuse sky radiation, F29.16
    example, F29.16, 17
    extraterrestrial irradiance (table), F29.14
    flux, F29.14
    irradiance, F29.17
      table, F29.29, 30, 31, 32, 33, 34, 35
  solar heat gain, F39.7
    calculations, F29.22, 27, 28, 37
    example, F29.36
    frame, F29.23
    passive, F29.24
  solar heat gain coefficients (SHGC), F29.22
    table, F29.25
  solar heat gain factors (SHGF), F29.36, 37
    example, F29.37
    foreground surfaces (table), F29.36
    table, F29.29, 30, 31, 32, 33, 34, 35
  spacers, F29.1, 3
  Stefan-Boltzmann equation, F29.17
  thermal controls, F29.44
  U-factors, F29.2, 5
    center-of-glass, F29.3
    doors, F29.10
      table, F29.10
    edge-of-glass, F29.3
    example, F29.11
    fenestration products (table), F29.8
    frames, F29.4
  visual controls, F29.44
Fiberglass, A43.32
Fick's law
  molecular diffusion, F5.1
Field testing
  industrial ventilation, A24.6
Filters, air *(see also* **Air cleaners**), S24
  air conditioners, room, S43.4
  automobiles, A8.4
  central systems, S1.5
  clean spaces, A15.3
    HEPA, A15.3
    ULPA, A15.3
  dehumidification, S22.7
  electronic, S24.6, 8
    ozone, S24.9
    space charge, S24.9
  energy recovery equipment, S42.5
  furnaces, S28.2
  granular bed, S25.14
  HEPA, S24.2, 3, 5, 7, 9; S25.2
  hospitals, A7.2-3
  industrial air conditioning, A11.4, 8-9
  industrial gas cleaning, S25.11
    electrostatic augmentation, S25.12
    fabrics, S25.12
    pressure-volume relationships, S25.11
    types, S25.12
  installation, S24.9
  kitchens, A28.15-16
  laboratories, A13.8-9
    HEPA, A13.6-7, 8-9
  maintenance, S24.9
  makeup air units, S31.9
  natatoriums, A4.7
  nuclear facilities, A23.2-3, 5
    charcoal filters, A23.3, 5
    demisters, A23.3, 5
    dust filters, A23.3
    heaters, A23.3, 5
    HEPA filters, A23.3, 5
    sand filters, A23.3
  panel, S24.6
    dry, extended surface, S24.6
    viscous impingement, S24.6
  photographic processing, A19.2
  places of assembly, A4.2
  printing plants, A17.5
  pulse jet, S25.13
  ratings, S24.2
  renewable media, moving-curtain, S24.7
    dry media, S24.8
    performance, S24.8
    viscous impingement, S24.7
  residential, A1.5
  reverse flow, S25.13
  safety requirements, S24.11
  selection, S24.9
  shaker, S25.12
  ships
    merchant, A10.3
    naval, A10.7
  standards, S24.2, 5
  test methods
    arrestance, S24.2, 3
    DOP penetration, S24.3
    dust-holding capacity, S24.2, 3
    dust-spot efficiency, S24.2, 3
    environmental, S24.5
    fractional efficiency, S24.2
    leakage (scan), S24.4
    particle size efficiency, S24.2, 4
    penetration, S24.2, 3
    textile processing, A18.4-6
    types, S24.6
    ULPA, S24.4, 5, 7; S25.2
    unit heaters, S31.7
Filtration
  water treatment, A44.12-13
Finned tube
  coils
    energy recovery loops, S42.12
    heat transfer coils, F3.21
  units, S32.1
    application, S32.5
    nonstandard condition corrections, S32.3
    rating, S32.3
Fins
  efficiency, F3.18
  example, F3.19
  resistance, F3.21
Fireplaces
  chimneys, S3018
  duct design, F32.14
  heaters (in-space), S29.4
    factory-built, S29.5
    freestanding, S29.5
    simple, S29.4
Fire safety, A48
  clean space exhaust systems, A15.9-10
  data processing areas, A16.7
  kitchens, A28.13-16, 18
  laboratories, A13.10-11
  nuclear facilities, A23.2
Fish, R18
  fresh
    icing, R18.1
    packaging, R18.3
    refrigeration, R18.1
    storage, R18.3
    storage requirements (table), R10.3
  frozen
    freezing methods, R18.5
    packaging, R18.4
    storage, R18.7
    transport, R18.9
  thermal properties, R8
Fishing vessels. *See* **Ships**
Fittings, S40.2
  application data (table), S40.6
  Duct Fitting Database, F32.11
  effective length, F2.10
  elbow equivalents (table), F33.6
  halocarbon refrigeration systems, R2.9
    equivalent length (tables), R2.10
  loss coefficients, F2.10; F32.8
    table, F33.2
  pipe sizing, F33.1, 4
  tees, F33.6
  elbow equivalents (figure), F33.6
Fixed guideway vehicle air conditioning, A8.11
  (see also **Rapid transit systems**)
  air distribution, A8.11
  controls, A8.11
  fresh air, A8.11
  heating, A8.11
  refrigeration components, A8.11
  system types, A8.11
Fixture units, F33.7
Flammability limits
  fuels (table), F17.3

A = 1995 Applications   S = 1996 Systems and Equipment   F = 1997 Fundamentals   R = 1998 Refrigeration

Flash drying, A27.6
Floor coverings
  panel systems affected by, S6.6
  temperature
    comfort affected by, F8.14
Floors
  insulation, F23.3
  moisture control, F23.11
Floor slabs
  heat loss, F27.12
Flowers, cut
  air transport, R31.1, 3
  cooling, R14.9
  storage, R10.6
    conditions (table), R10.7
Flowmeters, F14.17
  flow nozzles, F14.18
  hoods, F14.21
  orifice plates, F14.18
  positive-displacement meters, F14.20
  rotameters, F14.20
  turbine meters, F14.20
  venturi meters, F14.18
Fluid dynamic modeling
  industrial ventilation, A24.6-7
Fluid flow, F2
  Bernoulli equation, F2.2, 8
    kinetic energy factor, F2.2
  boundary layer, F2.4
  cavitation, F2.5
  choking, F2.7
  compressible flow, F2.7, 11
    expansion factor, F2.7
  continuity, F2.2
  Darcy-Weisbach equation, F2.8
  drag, F2.5
    drag coefficient, F2.5
  example, F2.8
  fitting effective length, F2.10
  fitting loss coefficient, F2.10
  fluid properties, F2.1
    density, F2.1
    viscosity, F2.1
      absolute, F2.1
      dynamic, F2.1
      kinematic, F2.1
  incompressible flow, F2.11
  laminar flow, F2.3
  measurement, F2.12; F14.17
  noise, F2.14
  nonisothermal effects, F2.6
  pipe friction, F2.8
    equivalent diameter, F2.10
    friction factor, F2.8
    Moody chart (figure), F2.9
    roughness (table), F2.10
  pressure variation across flow, F2.2
  pressure variation along flow, F2.2
  Reynolds number, F2.3
  section change losses, F2.10
  sensors for, A42.9; F37.8
  separation, F2.4
    vena contracta, F2.4
  testing, adjusting, balancing, A34
  turbulent flow, F2.3
  two-phase, F4
    boiling, F4.1
    condensation, F4.8
    evaporation, F4.2, F4.5
  unsteady flow, F2.13
  valve losses, F2.10, 11
  wall friction, F2.3
    pipe factor, F2.3
Fluidized-bed drying, A27.6

Food
  cooling time calculations, R9.1
  freezing time calculations, R9.7
  industrial freezing methods, R15
  microbial growth, R11.1
    control, R11.3
  processing facilities, R26
    control of microorganisms, R11.3
    dairy, R19
    fruits, R26.4
    main dishes, R26.1
    meat, R16
    potato products, R26.5
    poultry, R17
    refrigeration systems, R26.3, 4, 6
    sanitation, R11.4
    vegetables, R26.3
  refrigeration
    dairy products, R19
    eggs and egg products, R20
    fishery products, R18.1
    fruits, fresh, R21; R22
    meat products, R16
    poultry products, R17
    vegetables, R23
  refrigerators
    commercial, R48
    retail food store, R47
  service (*see also* **Food service**)
    water heating, A45.11, 12, 16, 17
  storage
    retail walk-in coolers, R47.7
  storage requirements, R10
    canned foods, R10.5
    citrus fruit, R22.3
    dried foods, R10.5
    fruit, R21.1
    perishable foods (table), R10.2
  thermal properties, R8
    enthalpy, R8.8
      table, R8.10
    freezing point (table), R8.3
    heat of respiration, R8.17
      tables, R8.18, 20, 23
    heat transfer coefficient, surface, R8.24
      table, R8.25
    ice fraction, R8.2
    models (table), R8.2
    specific heat (table), R8.3
    thermal conductivity, R8.9
      tables, R8.12, 15
    thermal diffusivity, R8.17
      table, R8.18
    transpiration coefficient, R8.19
      tables, R8.24
    water content, R8.1
      table, R8.3
Food service
  refrigeration equipment, R48
Fouling factor
  condensers, water-cooled, S35.3
  coolers, liquid, S37.4
  heat exchangers, S42.3
Foundations
  insulation, F23.3
  moisture control, F39.2
  radon, F39.3
  thermal performance, F39.2
Fourier number
  table, F3.2
Four-pipe systems, S3.8; S4.3; S12.13
Frames
  fenestration, F29.1

Freeze drying, A27.6
  biological materials, R40.4
Freeze prevention
  condensers, evaporative, S35.15
    remote sump for, S35.15
  coolers, liquid, S37.5
  cooling tower sump water, S36.9
  heat recovery equipment, S42.3, 13
  hydronic systems, S12.15
  pipes, F23.16
  solar energy systems, A30.26; S33.1, 2, 19
Freezers
  blast, R13.9; R15.1; R16.15
  household, R49
    cabinet construction, R49.2
    defrosting, R49.6
    efficiency, R49.1
    refrigeration systems, R49.5
    testing, R49.9
  industrial, R15
    blast, R15.1
    contact, R15.4
    cryogenic, R15.5
    cryomechanical, R15.5
    selection, R15.5
  walk-in, R48.3
Freezing
  beverages
    freezing time calculations, R9.7
  biomedical applications, R40
  foods
    bakery products, R27.5
    egg products, R20.11
    fish, R18.5
    freezing time calculations, R9.7
    ice cream, R19.15
    industrial methods, R15
    meat products, R16.15
    poultry products, R17.4
    processed and prepared food, R26
  points
    foods (table), R8.3
    refrigerants (table), F18.3
    table, F36.2
  soil, R35.3, 4
Friction factor, F2.8
Friction losses, F2.8
  copper tube (figure), F33.5
  faucets/cocks (figure), F33.8
  fittings, F33.1, 4
  flat oval ducts, F32.8
  friction chart, F32.8
    figure, F32.9
  noncircular ducts, F32.8
  plastic pipe (figure), F33.5
  roughness factors, F32.7
    table, F32.7
  steel pipe (figure), F33.5
  valves, F33.1, 4
Fruit juice, R24
  chilled, R24.4
  concentrate
    concentration methods, R24.3
    processing, R24.1
    storage, R24.3
  powdered, R24.5
  refrigeration, R24.5
Fruits, dried
  thermal properties, R8
  storage, R28.7
Fruits, fresh, R21; R22
  air transport, R31.1
  apples, storage, A47.9; R21.1
  apricots, R21.9

Composite Index

Fruits, fresh *(cont.)*
  avocados, R22.8
  bananas, R22.5
    handling, R22.5
    ripening rooms, R22.5
  berries, R21.9
  cherries, sweet, R21.8
  citrus, A47.9; R22.1
    deterioration, R22.4
    diseases, R22.4
    handling, R22.1
    storage, R22.3
    transportation, R22.3
  cooling, R14
  desiccation, R10.1
  deterioration rate (table), R10.1
  display refrigerators, R47.4
  figs, R21.10
  frozen, R26.4
  grapes, R21.5
    fumigation, R21.6
  mangoes, R22.8
  nectarines, R21.8
  peaches, R21.8
  pears, R21.4
  pineapples, R22.8
  plums, R21.7
  respiration. *See* **Respiration**
  storage diseases, R21
  storage requirements (table), R10.2
  strawberries, R21.9
  thermal properties, R8
  transpiration. *See* **Transpiration**
Fuel-burning equipment, S26
  air supply, S30.19
  controls, S26.12
    combustion, S26.13
    drafts, S26.14
    flame quality, S26.13
    ignition, S26.13
    operations, S26.12
    programming, S26.13
    safety/interlocks, S26.13
  dual-fuel gas/oil, S26.7
    equipment selection, S26.7
  gas, S26.1
    industrial, S26.2
    residential, S26.1
  oil, S26.4
    industrial, S26.5
    residential, S26.4
  solid fuel, S26.9
Fuel oil. *See* **Oil, fuel**
Fuels, F17
  central systems, S1.8
  engine drives, S7.4
  gaseous, F17.4
  heating value, S7.4
  liquid, F17.5
  solid, F17.7
  turbine drives, S7.9
Fungi
  farm crop storage, F11.6
Furnaces, S28
  codes, S28.17
  commercial, S28.16
    design, S28.17
    ducted, S28.16
    selection, S28.17
    technical data, S28.17
    unducted space heaters, S28.16
  electric
    seasonal efficiency, F30.19
  floor (in-space), S29.2

  gas-fired, S26.1
    seasonal efficiency, F30.19
  installation practices, S28.17
  oil
    seasonal efficiency, F30.19
  regulating agencies, S28.17
  residential, A1.3; S28.1
    air cleaners, S28.2
    air filters, S28.2
    airflow variations, S28.2
      basement furnace, S28.3
      downflow furnace, S28.2
      gravity furnace, S28.3
      highboy furnace, S28.2
      horizontal furnace, S28.2
      lowboy furnace, S28.3
      multiposition furnace, S28.3
      upflow furnace, S28.2
    annual fuel utilization (AFUE), S28.5
    design, S28.5
      heating capacity, S28.5
    dynamic simulation model (HOUSE), S28.8
    electric, S28.4
      design, S28.6
      technical data, S28.8
    humidifiers, S28.2
    indoor/outdoor variations, S28.3
    liquefied petroleum gas (LPG), S28.4, 8
    natural gas, S28.1, 5, 6
      burners, S28.1
      capacity ratings, S28.5, 6
      casings, S28.1
      combustion variations, S28.3
      control, S28.1, 7
      efficiency ratings, S28.5, 7, 9
      fans/motors, S28.2
      heat exchangers, S28.1
      performance criteria, S28.5
    oil, S28.4
      technical data, S28.8
    performance, S28.8
      climate effect, S28.14
      examples, S28.11
      factors, S28.9
      furnace type effect, S28.14
      night setback effect, S28.14, 15
      size effect, S28.15
    selection, S28.5
    vent dampers, S28.2
    venting, S28.2
  standards, S28.18
  wall (in-space), S29.1
Garages, bus
  ventilation, A12.12-13
Garages, parking, A3.12-13
  ventilation, A12.11-12
Gas, natural. *See* **Natural gas**
Gas-cleaning equipment, industrial, S25
  auxiliary equipment, S25.27
    ducts, S25.27
    dust-slurry handlers, S25.28
      dust conveyors, S25.28
      dust disposal, S25.28
      hopper discharge, S25.28
      hoppers, S25.28
      slurry treatment, S25.28
    fans, S25.28
    temperature controls, S25.28
  controls, S25.2
  gaseous contaminant control, S25.17
    absorption, S25.17
    adsorption, S25.17, 23, 27
      equipment, S25.17
      fixed beds, S25.24

      fluidized beds, S25.24, 26
      impregnated adsorbents, S25.24
      moving beds, S25.24
      odor control, S25.25
      solvent recovery, S25.24
    incineration, S25.26, 27
      applications, S25.27
      catalytic, S25.26
      thermal, S25.26
    spray dry scrubbing, S25.17
      atomizers, S25.17
      cocurrent, S25.18
      countercurrent, S25.18
      equipment, S25.17
      mixed flow, S25.18
      principles of operation, S25.17
    wet-packed scrubbers, S25.18
      absorption efficiency, S25.20
      arrangements, S25.19
      countercurrent, S25.19
      crossflow, S25.19
      example, S25.23
      extended surface packings, S25.19
      gas film-controlled, S25.21
      general efficiency comparison, S25.23
      horizontal cocurrent, S25.19
      liquid effects, S25.23
      liquid film-controlled, S25.20
      packings, S25.18
      pressure drop, S25.20
      vertical cocurrent, S25.19
  gas flow distribution, S25.2
  measurement of particulate/gaseous components, S25.1
  monitors, S25.2
  operation and maintenance, S25.29
    corrosion, S25.29
    fires/explosions, S25.29
  particulate contaminant control, S25.2
    efficiency, S25.3, 10
    electrostatic precipitators, S25.7
      components, S25.10
      mathematical modeling, S25.10
      principles of operation, S25.8
    fabric filters, S25.11
      electrostatic augmentation, S25.12
      fabrics, S25.12
      pressure-volume relationships, S25.11
      pulse jet, S25.13
      reverse flow, S25.13
      shaker, S25.12
      types, S25.12
    fractional efficiency, S25.3, 6
    granular bed filters, S25.14
      electrostatic augmentation, S25.14
      principles of operation, S25.14
    HEPA filters, S25.2
    inertial collectors, S25.6
      baffles, S25.6
      cut size, S25.6
      cyclones, S25.6
      dry centrifugal collectors, S25.7
      louvers, S25.6
      multicyclones, S25.6
      superficial velocity, S25.6
    mechanical collectors, S25.3
    penetration, S25.3
    principles of operation, S25.11, 15
      condensation, S25.15
      diffusion, S25.11, 15
      direct interception, S25.11
      electrostatic attraction, S25.11
      inertial impaction, S25.11, 15

A = 1995 Applications   S = 1996 Systems and Equipment   F = 1997 Fundamentals   R = 1998 Refrigeration

Gas-cleaning equipment, industrial *(cont.)*
  particulate contaminant control *(cont.)*
    scrubbers, S25.14
      centrifugal-type collectors, S25.15
      electrostatic augmentation, S25.16
      impingement scrubbers, S25.15
      orifice-type collectors, S25.15
      principles of operation, S25.15
      spray towers, S25.15
      venturi scrubbers, S25.15
    settling chambers, S25.3
      fractional efficiency, S25.6
    ULPA filters, S25.2
  recirculated air, S25.1
  regulations, S25.1
  selection, S25.1
  stack sampling, S25.1
    program, S25.2
Gases
  liquefaction, R38.3
  purification, R38.12
  separation, R38.12
Gas-fired
  boilers, S26.1
  burners, S26.1
    altitude compensation, S26.3
    combustion/adjustments, S26.3
    industrial, S26.2
    residential, S26.1
    types, S26.2
      atmospheric, S26.2
      power-type, S26.2
    unitary heaters, S26.2
  equipment
    venting, S30.15
  furnaces, S26.1; S28.1, 5, 6
    codes, S28.17
    commercial, S28.16
    installation practices, S28.17
    residential, S28.1
    standards, S28.18
  heaters (in-space), S29.1
    control, S29.2
    efficiency (U.S. minimum), S29.2
    floor, S29.2
    room, S29.1
    sizing, S29.3
    venting, S29.3
    wall furnaces, S29.1
  infrared heaters, S15.1
    catalytic oxidation, S15.2
    indirect, S15.1
    porous matrix, S15.2
  makeup air units, S31.9
  service water heating, A45.1-2
  turbine drives, S7.8
  unit heaters, S31.4
  unit ventilators, S31.1
Generators
  absorption units, R41.3, 9, 11
  cogeneration, S7.26
Geothermal energy, A29
  characteristics, A29.3-4
    distance to usage, A29.3-4
    effluent disposal, A29.4
    fluid composition, A29.4
    load factor, A29.4
    load size, A29.4
    resource depth, A29.3
    resource life, A29.4
    resource temperature, A29.4
    temperature drop, A29.4
    well flow rate, A29.4

commercial/residential applications,
    A29.11-13
  service water heating, A29.13
  space cooling, A29.13
  space heating, A29.11-13
direct-use systems, A29.3-14
equipment, A29.7-11
  corrosion control, A29.8
  heat exchangers, A29.9-10
  materials, A29.7-8
  piping, A29.10-11
  pumps, A29.8-9
  valves, A29.10
  variable-speed drives, A29.9
fluids, A29.2
ground-source heat pump systems,
    A29.14-24; S8.4
  ground-coupled, A29.14-15, 16-20
    horizontal systems, A29.18-20
    vertical systems, A29.16-18
  groundwater, A29.15-16, 20-23
    direct systems, A29.21
    heat exchangers, A29.21
    indirect systems, A29.21-23
    well pumps, A29.21
  surface water, A29.16, 23-24
    closed-loop lake heat pumps,
      A29.23-24
    lake heat transfer, A29.23
    lake thermal patterns, A29.23
  terminology, A29.14
industrial applications, A29.14
residential applications, A29.11-13
resources, A29.1
temperatures, A29.1-2
uses, A29.2-3
water wells, A29.4-7
  flow testing, A29.5-6
  terminology, A29.4-5
  water quality testing, A29.6-7
Glass, F29
  block walls, F29.26
  CLTD (table), F28.49
  cooling load, F28.41
    table, F28.50
  daylight transmittance (table), F29.47
  durability, F29.48
  greenhouses, A20.10, 14
  load factors, F27.2
  occupant comfort, F29.48
  optical properties, F29.17
  safety/strength, F29.48
  selection, F29.47
  shading coefficients, F29.23
  sound transmittance, loss (table), F29.48
Glazing, F29
Glycols
  coolants, secondary, F20.4
  ethylene. *See* Ethylene glycol
  propylene. *See* Propylene glycol
Graetz number
  table, F3.2
Grain
  drying, F11.12
  insects (table), F11.3
  storage moisture (table), F11.2
Grashof number
  table, F3.2
Greenhouses *(see also* Plant environments),
    A20.8-14
  evaporative cooling, A47.9-10
  heating, F10.12, 14
  supplemental irradiance, F10.16

Grilles
  air outlets, S17.3
  small forced-air systems, S9.2
Ground-source energy. *See* Geothermal energy
Gymnasiums, A4.5
Halocarbons
  coolants, secondary, F20.12
  refrigerants. *See* Refrigerants
  refrigeration systems. *See* Refrigeration
Halons. *See* Refrigerants
Hay
  drying, A21.8; F11.11
Hazard Analysis and Critical Control Point
  (HACCP), R11.4
Hazen-Williams equation
  pressure drop, F33.1
Health
  environmental. *See* Environmental health
Health care facilities, A7
  disease prevention/treatment, A7.1-2
  hospitals. *See* Hospitals
  nursing homes. *See* Nursing homes
  outpatient. *See* Outpatient health care
    facilities
Heat
  animal production, F10.2, 5
  balance
    conduction transfer function, F28.2
    space air energy balance, F28.3
  latent
    respiratory loss, F8.4
    skin loss, F8.3, 10
  sensible
    respiratory, F8.4
    skin, F8.3
Heat conservation
  industrial environment, A24.17
Heaters
  automobiles, A8.1, 4
  codes, S29.6
  control, S29.2, 4
  direct-contact, S14.6
  electric, S29.4
  farm crop drying, A21.2-3
  fireplaces, S29.4
  gas-fired, S26.2; S29.1
  hydronic snow melting, A46.8
  infrared radiant, A49; S26.2
    electric, S15.2
    gas-fired, S15.1
    oil-fired, S15.3
  in-space, S29
  kerosene-fired, S29.3
  oil-fired, S29.3
  small forced-air systems, S9.1
  solid fuel, S29.4
  standards, S29.7
  stoves, S29.5
  testing, S29.7
  unducted space, S28.16
  unit, S31.3
    applications, S31.4
    control, S31.7
    filters, S31.7
    maintenance, S31.7
    piping, S31.7
    selection, S31.4
      heating medium, S31.4
      location, S31.6
      ratings, S31.6
      sound level, S31.6
    types, S31.6
      cabinet unit heaters, S31.6
      centrifugal fan units, S31.6

Composite Index

Heaters (*cont.*)
  unit (*cont.*)
    types (*cont.*)
      duct units, S31.6
      propeller fan units, S31.6
  water systems, medium/high temp., S14

Heat exchangers
  animal environments, A20.3-4
  chimneys, S30.23
  condensers. *See* **Condensers**
  coolers, liquid, S37.1
  counterflow, F3.4
  district heating and cooling, S11.6
  effectiveness, F3.4
  enhanced surfaces, F4.14
  external, S33.17
  fixed-plate, S42.9
  furnaces, S28.1
  geothermal energy systems, A29.9-10
    downhole, A29.10
    plate, A29.9-10
    shell-and-tube, A29.10
  halocarbon refrigeration systems, R2.23
  heat pipe, S42.14
  heat recovery, S42
  internal, S33.16
  mines, A25.7
  parallel flow, F3.4
  rotary, S42.11
  solar energy systems, A30.16-17; S33.16
  steam systems, S10.3
  thermosiphon, S42.16
  water systems, medium/high temp., S14.7

Heat flow
  apparent conductivity, F22.3
  insulation, F22.4
    calculations, F22.8
    performance, F22.4
    terminology, F22.1
  moisture affecting, F22.12
  series/parallel flow, F22.8
  surface conductance, F22.6
  surface temperature, F22.8
  thermal resistance, F22.7
  thermal transmittance, F22.6

Heat gain
  appliances, F28.10, 53
    table, F28.11, 12
  calculation, F28.5
  engine test facilities, A14.3
  fenestration, F28.7, 41
  humans, F28.7, 51
    table, F28.8
  infiltration, F28.11
  laboratories, A13.2-3
    equipment (table), A13.2
    lab animals (table), A13.13
  latent, F28.15
  lighting, F28.7, 52
  motors, F28.9, 53
  office equipment, F28.11
    table, F28.14
  panel systems, S6.7
  partitions, walls, and ceilings, F28.28
  sensible
    calculation, F30.3
  shading, F28.41
  solar
    fenestration, F29; F39.7
    frame, F29.23
    passive, F29.24
  solar heat gain coefficients (SHGC), F29.22
    glass (table), F29.25
  solar heat gain factors (SHGF), F29.36, 37
    example, F29.37
    foreground surfaces (table), F29.36
    table, F29.29, 30, 31, 32, 33, 34, 35
  ventilation, F28.11
  walls and roofs, F28.17, 40

Heating
  absorption equipment, R41
  all-air systems, S2.2
  all-water systems, S4
  animal environments, A20.3-4
  central system, S1.7
  design data, F26.2
  equipment, S1.7
    baseboard units, S32.1
    boilers, S27
    convectors, S32.1
    design, S32.3
      altitude corrections, S32.4
      enclosure/paint effect, S32.5
      mass effect, S32.4
      water velocity corrections, S32.3
    finned-tube units, S32.1
    furnaces, S28
    heat emission, S32.2
    in-space heaters, S29
    radiators, S32.1
    residential, A1.1-4
    space heaters, S14.8
  fan-coil units, S4.1
  geothermal energy systems, A29.11-13
  greenhouses, A20.9-11; F10.14
    framing type, A20.9-10
    heating systems, A20.10-11
    infrared heating, A20.11
    structural heat loss, A20.9
  hot-water (*see also* **Hydronic systems**)
  industrial environments, A11.6-7
  infrared radiant, A49; S15
  panel systems, S6
  plant growth facilities, A20.15
  residential, A1.1-4
  solar energy systems, A30.19-21, 28; S33
  spot, S15.1
  thermal storage, A40.7-8
  unit ventilators, S31.1
  water systems (*see also* **Hydronic systems**), S12
    low-temperature, S12.1, 10
    medium/high temp., S14
    nonresidential, S12.10
    solar energy, A30.13-15, 20-21
    water treatment for, A44.7-8

Heating load
  nonresidential, F28
    example, F28.39
    principles, F28.16
  residential calculations, F27
    attic temperature, F27.8
      example, F27.8
    crawl space heat loss, F27.9
    design conditions, F27.7
    example, F27.11
    infiltration heat loss, F27.13
    pickup load, F27.14
    procedure, F27.7
    transmission heat loss, F27.10
      example, F27.10
    unheated space temperature, F27.8
      example, F27.8

Heating values
  fuels, F17.3, 6, 8
    table, F17.3, 4, 6

Heat loss
  basement, F27.10
    design temperature, F27.11
    example, F27.11
    floors, F27.11
      table, F27.11
    walls, F27.10
      table, F27.11
    transient calculations, F27.12
  ceiling, F27.10
  crawl spaces, F27.9
    example, F27.9
  flat surfaces, F24.15
    example, F24.19
    table, F24.19
  floor slabs, F27.12
    heat loss coefficients (table), F27.12
    transient calculations, F27.12
  infiltration, F27.13
    air change method, F27.14
    crack length method, F27.13
    exposure factors, F27.14
    latent heat loss, F27.13
    sensible heat loss, F27.13
  panel systems, S6.7
  pipe, F24.15
    copper tube (table), F24.20
    example, F24.17, 19
    steel (table), F24.19
  roof, F27.10
  transmission, F27.10
    example, F27.10

Heat of fusion
  table, F36.2

Heat of vaporization
  table, F36.2

Heat pipes, S19.4
  heat exchangers, S42.14

Heat pumps
  absorption, R41.2, 5
  applied systems, S8.1
    components, S8.6
    compressors, S8.6
      centrifugal, S8.6
      floodback protection, S8.7
      reciprocating, S8.6
      rotary vane, S8.6
      screw, S8.6
      scroll, S8.6
      selection, S8.6
    controls, S8.8
    defrost, S8.8
    cycles, S8.1
      closed vapor compression, S8.1
      heat-driven Rankine, S8.2
      mechanical vapor recompression, S8.1
      open vapor recompression, S8.2
    heat sources/sinks, S8.2
      air, S8.2
      ground, S8.4
      solar energy, S8.4
      water, S8.4
    refrigeration components, S8.7
      check valves, S8.8
      expansion valves, S8.8
      refrigerant receivers, S8.8
      reversing valves, S8.8
    supplemental heating, S8.8
    types, S8.4
      air-to-air, S8.4
      air-to-water, S8.6
      ground-coupled, S8.6
      refrigerant-to-water, S8.6
      water-to-air, S8.4
      water-to-water, S8.6

A = 1995 Applications   S = 1996 Systems and Equipment   F = 1997 Fundamentals   R = 1998 Refrigeration

Heat pumps *(cont.)*
control, A42.30
reversing valves, R45.17
engine-driven, S7.40
ground-source
ground-coupled, A29.14-15, 16-20
groundwater, A29.15-16, 20-23
surface water, A29.16, 23-24
ice-source, R33.7
industrial process, S8.8
closed-cycle systems, S8.9
air-to-air, S8.9
air-to-water, S8.9
dehumidification, S8.9
heat pump water heaters, S8.9
process fluid-to-process fluid, S8.9
water-to-water, S8.9
open-cycle systems, S8.12
process emission-to-steam, S8.13
semi-open-cycle systems, S8.12
packaged terminal, S45.1
design, S45.2
operation, S45.3
testing, S45.3
residential air conditioning, A1.2-3, 5
unitary, S44
air-source, S44.1
air-to-air, S5.3
application, S44.1
balance point, S44.8
certification, S44.5
codes, S44.5
control, S44.12
desuperheaters, S44.4
efficiency, S44.5
energy conservation, S44.5
engine-driven, S44.9
components, S44.10
defrost cycle, S44.11
engine/compressor, S44.11
heat recovery, S44.11
maintenance, S44.9
selection, S44.9
standards, S44.10
installation, S44.1, 12
maintenance, S44.3
refrigerant circuit, S44.11
compressor selection, S44.12
defrost, S44.11
flow controls, S44.12
outdoor coil circuitry, S44.11
refrigerant charge, S44.12
water drainage, S44.11
selection, S44.8
space conditioning/water heating, S44.4
split system, S44.1
standards, S44.5
through-the-wall, S5.2
types, S44.3
window-mounted, S5.2
water heaters, A45.21
water-source, S45.3
certification, S45.4
design, S45.6
entering water temperatures, S45.4
testing, S45.4
types, S45.4
Heat recovery *(see also* **Energy recovery**), S8.14
applications, S8.21
balanced heat recovery, S8.19
cogeneration, S7.18
control, A42.30
data processing areas, A16.7

engine drives, S7.20
gas turbine drives, S7.24
heat-activated chillers, S7.25
heat balance
concept, S8.19
studies, S8.20
heat pumps, unitary, S44.11
heat redistribution, S8.19
industrial environment, A24.17
kitchen ventilation, A28.12
liquid chillers, R43.3, 11
multiple buildings, S8.21
retail food store refrigeration, R47.10, 14
service water heating, A45.3
steam systems, S10.3, 14
steam turbine drives, S7.24
supermarkets, A2.3-5
terminology, S8.1
waste heat, S8.14
water loop heat pump systems, S8.15
advantages, S8.19
controls, S8.19
design considerations, S8.17
limitations, S8.19
Heat recovery, air-to-air, S42
airflow arrangements, S42.4
applications, S42.6
coil energy recovery (runaround) loops, S42.12
characteristics, S42.13
construction materials, S42.13
cross-contamination, S42.13
effectiveness, S42.13
freeze prevention, S42.13
maintenance, S42.13
thermal transfer fluids, S42.14
comfort-to-comfort, S42.1
condensation, S42.3, 11
controls, S42.5, 12, 15
corrosion, S42.3
cross-leakage, S42.3
economic considerations, S42.1
evaporative air cooling, S42.5
examples, S42.7
face velocity, S42.4
filters, air, S42.5
fixed-plate exchangers, S42.9
condensation, S42.11
design, S42.10
differential pressure/cross-leakage, S42.10
hygroscopic, S42.11
performance, S42.10
fouling, S42.3
freeze prevention, S42.3, 13
heat pipe heat exchangers, S42.14
construction materials, S42.14
controls, S42.15
cross-contamination, S42.14
operating principles, S42.14
operating temperatures, S42.14
performance, S42.14
indirect evaporative cooling, S42.5
maintenance, S42.5, 12, 13, 16
performance rating, S42.2
precooling air reheater, S42.5
pressure drop, S42.4
process-to-comfort, S42.1
process-to-process, S42.1
rotary energy exchangers, S42.11
construction, S42.11
controls, S42.12
cross-contamination, S42.12
maintenance, S42.12
sensible vs. total recovery, S42.3

thermosiphon (two-phase) heat exchangers, S42.16
operating principles, S42.17
two-phase, S42.16
twin-tower enthalpy recovery loops, S42.15
cross-contamination, S42.16
design, S42.16
maintenance, S42.16
operating temperatures, S42.16
operation, S42.15
Heat storage. *See* **Thermal storage**
Heat stress, F8.23
index, F8.19
industrial environment, A24.2-3
mines, A25.1
Heat transfer, F3
air-heating coils, S23.3
air spaces, F22.7
animal environments, F10.3
antifreeze, S12.16
chimneys/fireplaces/gas vents, S30.5
coils
air-cooling, S21.6
dehumidifying, S21.9
condensers
air-cooled, S35.9
evaporative, S35.13
water-cooled, S35.2
conduction, F3.1
conductance, F3.3
resistance, F3.2
steady-state, F3.1
convection, F3.1
forced, F3.12
natural, F3.11
coolers, liquid, S37.2
dimensionless numbers (table), F3.2
effectiveness, F3.4
extended surfaces, F3.17
fins, F3.17
efficiency, F3.18
example, F3.19
finned-tube, F3.21
resistance, F3.21
fluids
solar energy systems, A30.16
insulation, F14.28
mass transfer analogy
convection, F5.6
eddy diffusion, F5.8
molecular diffusion, F5.3
mass transfer simultaneous with, F5.11
cooling coils, F5.15
modeling, F22.21; F30.11
effectiveness-NTU heat exchanger model, F30.11
number of transfer units (NTU), F3.4
overall, F3.2
coefficient, F3.2
panel surfaces, S6.2
pipes, S11.13
radiation, F3.1, F3.6
absorptances (table), F3.8
actual, F3.8
gray, F3.8
nonblack, F3.8
angle factor, F3.9
figure, F3.10
Beer's law, F3.11
blackbody, F3.6
emittances (table), F3.8
exchange between surfaces, F3.9
in gases, F3.11
Kirchoff's law, F3.9

Composite Index

I.25

Heat transfer *(cont.)*
  radiation *(cont.)*
    Lambert's law, F3.9
    radiosity, F3.9
    Stefan-Boltzmann law, F3.6
  snow melting systems, A46.1-4
  transient, F3.4
    cooling time estimation, F3.5
    cylinder, F3.5
    slab, F3.5
    sphere, F3.5
    temperature distribution, F3.5
  water, S12.2
Heat transfer coefficients, F29.4
  air space (table), F29.6
  condensation (table), F4.9
  convective, F8.8
  evaporative, F8.8
  foods, R8.24
  glazing (table), F29.5
  Lewis relation, F8.4
  low-temperature, R39.9
Heat transfer fluids
  snow melting systems, A46.5
Heat transmission, F24
  air spaces (table), F24.2, 3
  attics (table), F24.13
  basements, F27.10
  below-grade construction, F24.14
  building materials (table), F24.4
  ceilings, F27.10
  doors, F24.12
    table, F24.13
  floor slabs, F27.12
  insulating materials (table), F24.4
    industrial (table), F24.18
  pipe, F24.15
    buried, F24.23
    copper tube (table), F24.20
    steel (table), F24.19
  rocks (table), F24.15
  roofs, F27.10
  slab-on-grade construction, F24.14
  soils, F24.14
    table, F24.15
  windows, F24.12
Heat wheels, S19.4
Helium
  liquid, R38.1
  recovery, R38.14
Hoods
  draft, S30.21
  gaseous contaminant control, A41.8
  industrial exhaust systems, A26
    compound, A26.8-9
    hot process, A26.5-6
    jet-assisted, A26.6-7
    simple, A26.8
  kitchen exhaust, A28.2-6, 17-18
  laboratory fume, A13.3-6
    sound control, A43.27-28
Hospitals, A7.2-10
  air quality, A7.2-4
    air filters, A7.2-3
    air movement, A7.3-4
    exhaust outlets, A7.2
    outdoor intakes, A7.2
    pressure relationships and ventilation, A7.4
      general (table), A7.5
    smoke control, A7.4
  cooling, A7.10
  design criteria, A7.4-9
    administration, A7.10
    ancillary spaces, A7.7-8

    diagnostic and treatment, A7.8-9
    laboratories, A7.7-8
      bacteriology, A7.7
      infectious diseases/viruses, A7.7-8
      nuclear medicine, A7.8
    nursery, A7.6
    nursing, A7.6-7
    operating rooms, A7.4-6
    service areas, A7.9
    sterilizing and supply, A7.9
    surgery and critical care, A7.4
  energy, A7.10
  heating, A7.10
  hot water, A7.10
  infection sources/control, A7.2
  insulation, A7.10
  Legionella pneumophila, A7.2
  pressure relationships and ventilation, A7.4
    table, A7.5
  zoning, A7.9-10
Hotels, A5.3-4
Hot-gas bypass, R2.29
Hot water *(see also* **Heating, Hydronic systems,** and **Service water heating**)
  cogeneration distribution, S7.29
  makeup air units, S31.9
  unit heaters, S31.4
  unit ventilators, S31.1
Houses of worship, A4.3
Humans. *See* **Comfort; Physiological principles, humans**
Humidification
  air washers, S19.6
    effectiveness, S19.7
  all-air systems, S2.2
  data processing areas, A16.6
  environmental test facilities, R37.3
  evaporative cooling, A47.2
  rigid media, S19.7
Humidifiers, S20
  central systems, S1.6
  controls, S20.8
    control location, S20.9
    electronic, S20.8
    mechanical, S20.8
    variable-air-volume systems, S20.9
  enclosure characteristics, S20.2
    concealed condensation, S20.3
    vapor barriers, S20.2
    visible condensation, S20.2
  energy considerations, S20.3
    additional moisture loss, S20.4
    design conditions, S20.3
    internal moisture gains, S20.4
    load calculations, S20.3
    ventilation rate, S20.4
  environmental conditions, S20.1
    bacterial growth, S20.1
    comfort, S20.1
    data processing equipment, S20.1
    disease prevention/treatment, S20.1
    materials storage, S20.1
    process control, S20.1
    sound wave transmission, S20.2
    static electricity, S20.2
  furnaces, residential, S28.2
  industrial
    central air systems, S20.5
      atomizing, S20.8
      direct steam injection, S20.5
      electrically heated steam, S20.8
      heated pan, S20.5
      wetted media, S20.8

  residential, A1.5
    central air systems, S20.5
      atomizing, S20.5
      pan, S20.5
      wetted element, S20.5
    nonducted applications, S20.5
  small forced-air systems, S9.1
  supply water, S20.4
    scaling, S20.4
Humidity
  control, A42.17-18; S22; F21; F23.5
    clean spaces, A15.10
    greenhouses, A20.12
  drying, A27
  measurement, F14.9
  odors affected by, F13.5
  plant environments, F10.17
Hydrochlorofluorocarbons (HCFC). *See* **Refrigerants**
Hydrocooling
  fruits and vegetables, R14.3
Hydrofluorocarbons (HFC). *See* **Refrigerants**
Hydrogen
  liquefaction, R38.3
  liquid, R38.2
Hydrogen sulfide
  animal environments, A20.2
Hydronic systems, S12
  antifreeze, S12.16
    ethylene glycol, S12.16
    propylene glycol, S12.16
  capacity control, S12.9
  chilled water systems, S12.1, 11
  closed systems, S12.1
    basic components, S12.2
  control, A42.28
  control valve sizing, S12.10
  design considerations, S12.8
  distribution system, S12.5
    compound pumping, S12.7
    parallel pumping, S12.6
    primary-secondary pumping, S12.7
    pump curves, S12.5
    series pumping, S12.6
    standby pump, S12.7
  district heating and cooling, S11.5, 23
  dual-temperature, S12.1, 12
    four-pipe common load, S12.13
    four-pipe independent load, S12.13
    two-pipe, S12.12
  equipment layout, S12.15
  expansion tanks, S12.3, 7
    closed, S12.3
    diaphragm, S12.3
    hydraulic function, S12.7
    open, S12.3
    thermal function, S12.3
  freeze prevention, S12.15
    antifreeze, S12.16
  heating/cooling source, S12.3
    turndown ratio, S12.3
  hydraulic components, S12.5
  loads, S12.2
    dehumidification, S12.2
    heat transfer, S12.2
    pipe heat-carrying capacity, S12.3
    sensible, S12.2
  low-temperature heating systems, S12.10
  nonresidential heating systems, S12.10
  panel systems, S6.10
    ceiling, S6.12
    design considerations, S6.10
    floor, S6.14
    wall, S6.14

A = 1995 Applications   S = 1996 Systems and Equipment   F = 1997 Fundamentals   R = 1998 Refrigeration

Hydronic systems *(cont.)*
  pipe sizing, S12.15; F33.4
  piping circuits, S12.8
    diverting series, S12.8
    parallel, S12.9
    series, S12.8
  pressure drop determination, S12.15
  principles, S12.1
  radiators. *See* **Radiators**
  residential, A1.3, 5
  safety relief valves, S12.13
  snow melting, A46.1-4, 5-8
  temperature classification, S12.1
    chilled water, S12.1
    dual, S12.1
    high, S12.1
    low, S12.1
    medium, S12.1
  testing, adjusting, balancing, A34.6-14
  thermal components, S12.2
  water treatment, A44.15
Hygrometers, A42.9; F37.8
  calibration, F14.12
  capacitance, F14.11
  dew-point, F14.11
  electrolytic, F14.12
  gravimetric, F14.12
  impedance, F14.11
  mechanical, F14.11
  piezoelectric sorption, F14.12
  psychrometers, F14.9
  spectroscopic, F14.12
IAQ. *See* **Indoor air quality (IAQ)**
Ice
  delivery systems, R33.5
    conveyors, R33.6
    pneumatic conveying, R33.6
    slurry pumping, R33.7
  manufacture, R33
    flake ice, R33.1
    plate ice, R33.3
    tubular ice, R33.1
  packaged, R33.7
  storage, R33.4
  thermal storage, A40.11-15
Ice cream. *See* **Dairy products**
Ice makers
  commercial, R50
    construction, R50.2
    evaporators, R50.2
    ice storage, R50.5
    ice types, R50.5
    installation, R50.7
    safety and sanitary standards, R50.5
    system design, R50.3
    terminology, R50.1
  heat pumps, R33.7
  household refrigerator, R49.2
  large commercial, R33.1
  water treatment, A44.14-15
Ice rinks, R34
  conditions, R34.4
  dehumidification, R34.8
  energy conservation, R34.4, 6
  floor design, R34.6
  heat loads, R34.2
  refrigeration
    equipment, R34.4
    requirements, R34.1
  surface building and maintenance, R34.8
Ignition temperatures
  fuels, F17.2
    table, F17.3

Indoor air quality (IAQ) (*see also* **Air quality**), A41; F9.5
  bioaerosols, F9.5
    health effects, F9.5
    sampling, F9.6
    sources, F9.5
  environmental tobacco smoke (ETS), F9.8
  infiltration/ventilation, F25.4
  inorganic gases, F9.11
    exposure control, F9.12
    health effects, F9.12
    sources, F9.11
    standards, F9.12
      table, F9.12
  kitchens, A28.9
  organic gases, F9.13
    exposure control, F9.13
    health effects, F9.13
    sources, F9.13
    standards
      table, F9.14
  particulates, F9.7
    dust, F9.7
    exposure limits (table), F9.8
  polycyclic aromatic compounds (PAC), F9.8
  polycyclic aromatic hydrocarbons (PAH), F9.8
  radon, F9.19
    action levels (table), F9.20
  soil gases, F9.14
  standards, F9.1
    table, F9.2
  synthetic vitreous fibers, F9.8
  ventilation rate procedure, F25.5, 23
  volatile organic compounds (VOC), F9.9
    exposure control, F9.11
    health effects, F9.10
    sources, F9.10
    standards, F9.11
Induction
  air-and-water systems, S3.2
  ceiling boxes, S17.7
  supply air outlets, S17.1
  testing, adjusting, balancing, A34.6
Industrial
  duct construction, S16.4
  gas burners, S26.2
  gas cleaning. *See* **Gas-cleaning equipment, industrial**
  humidifiers, S20.5
  oil-fired burners, S26.5
  process heat pumps, S8.8
  process refrigeration, R36
  service water heating, A45.19-20
Industrial air conditioning, A11
  air filtration systems, A11.8-9
    contaminant control, A11.8-9
    exhaust air, A11.8
    exhaust systems, A11.9
  cooling systems, A11.7-8
    evaporative, A11.8
    refrigeration, A11.7-8
  design considerations, A11.5
    facilities checklist (table), A11.5
  employee requirements, A11.4-5
    contaminant levels, A11.5
    temperatures, A11.4-5
  equipment selection, A11.6-9
    air-handling, A11.6
    cooling, A11.7-8
    filtration, A11.8-9
    heating, A11.6-7
  heating systems, A11.6-7
    door heating, A11.7
    ducted heaters, A11.7

    floor, A11.6-7
    infrared, A11.7
    unit heaters, A11.7
  load calculations, A11.6
    fan heat, A11.6
    internal heat, A11.6
    makeup air, A11.6
    solar/transmission, A11.6
    stratification, A11.6
  maintenance, A11.9
  process and product requirements, A11.1-4
    air cleanliness, A11.4
    corrosion/rust/abrasion, A11.3
    moisture regain, A11.1
      conditioning/drying, A11.1
      hygroscopic materials, A11.1
    moisture regain (table), A11.4
      inorganic materials, A11.4
      natural textile fibers, A11.4
      organic materials, A11.4
      paper, A11.4
      rayons, A11.4
    product accuracy and uniformity, A11.3
    product formability, A11.4
    rates of biochemical reactions, A11.3
    rates of chemical reactions, A11.1
    rates of crystallization, A11.3
    static electricity, A11.3-4
    temperature/humidity requirements (table), A11.2-3
      abrasives, A11.2
      ceramics, A11.2
      distilling, A11.2
      electrical products, A11.2
      floor coverings, A11.2
      foundries, A11.2
      furs, A11.2
      gum, A11.2
      leather, A11.2
      lenses (optical), A11.2
      matches, A11.3
      paint application, A11.3
      photo studios, A11.3
      plastics, A11.3
      plywood, A11.3
      rubber-dipped goods, A11.3
      tea, A11.3
      tobacco, A11.3
  system selection, A11.6-9
    air-handling, A11.6
    cooling, A11.7-8
    filtration, A11.8-9
    heating, A11.6-7
Industrial environments, A24
  air curtains, A24.14-16
    design principles, A24.15-16
  air diffusers, A24.12-13
  air distribution, A24.9, 12-13
    selection methods, A24.9-11
  general comfort/dilution ventilation, A24.8-13
    air requirements, A24.8-9
  heat conservation, A24.17
  heat control, A24.1-5
    moderate environments, A24.2
      operative temperatures, A24.2
      PMV-PPD index, A24.2
    ventilation, A24.1-2
  heat exposure control, A24.5
    local exhaust ventilation, A24.5
    radiation shielding, A24.5
    source control, A24.5
  heat recovery, A24.17
  heat stress, A24.2-3
    wet-bulb globe temperature (WBGT), A24.2-3

Composite Index

Industrial environments *(cont.)*
  local discomfort, A24.3-5
    air speed, A24.3
    radiant asymmetry, A24.3
    spot cooling, A24.3-5
    temperature differences, A24.3
  natural ventilation, A24.13-14
  roof ventilators, A24.16-17
  spot cooling, A24.3-5, 11-12
  ventilation design principles, A24.5-8
    computer modeling, A24.7-8
    field testing methodology, A24.6
    fluid dynamic modeling, A24.6-7
    general ventilation, A24.5-6
    makeup air, A24.8
    natural ventilation, A24.13-14
Industrial exhaust systems *(see also* **Gas-cleaning equipment**), A26
  air cleaners, A26.10
  air-moving devices, A26.10
  duct design, F32.22
  energy recovery, A26.10
  exhaust stacks, A26.9-10
  fans, A26.10
  fluid mechanics, A26.1
  local exhaust, A24.5
  local exhaust system components, A26.1-9
    duct considerations, A26.7-8
      construction, A26.10
      losses, A26.9
      segment integration, A26.10
      size determination, A26.8
    hoods, A26.1-2
      capture velocities, A26.2-3
      compound hoods, A26.8-9
      design principles, A26.7
      entry loss, A26.8
      hot process hoods, A26.5-6
      jet-assisted hoods, A26.6-7
      simple hoods, A26.8
      special situations, A26.7
      volumetric flow rate, A26.3-4
  operation and maintenance, A26.11
  system testing, A26.11
Industrial hygiene, F9.4
Infiltration *(see also* **Air leakage**), F25
  air exchange rate, F25.3
    measurement, F25.5
  air leakage, F25.14, 16
    air-vapor retarder, F25.17
    building components (table), F25.18
    classes (table), F25.16
    controlling, F25.17
  climatic zones, F25.20
  commercial buildings, F25.19
  cooling load, F28.11
  driving mechanisms, F25.7
    combining, F25.10
    mechanical systems, F25.10
    stack pressure, F25.8
    wind pressure, F25.8
  examples, F25.22
  fenestration, F29.12; F39.7
  heat loss calculations
    air change method, F27.14
    crack length method, F27.13
    exposure factors, F27.14
    latent, F27.13
    sensible, F27.13
  indoor air quality, F25.4
  infiltration degree-days, F25.4
  leakage function, F25.11
  measurement, F14.28
  refrigerated facilities, R12.3
  residential buildings, F25.14
    examples, F25.22
  terminology, F25.1
  thermal loads, F25.3
  wind
    coefficients (table), F25.22
    shielding classes (table), F25.22
Infiltration degree-days, F25.4
Infrared
  greenhouse heating, A20.11
  heaters
    gas-fired, S26.2
  heating, A49; S15
    comfort, F8.21, 23
    industrial environments, A11.7
    low, medium, high intensity, A49.1, 8; S15.1
  snow melting systems, A46.12-13
Infrared drying, A27.3
Ionization
  animal environments, F10.5
Insects
  farm crop storage, F11.7
Instruments, F14
  central systems, S1.9
  cryogenic, R38.23
  data processing areas, A16.7
  electricity-measuring, F14.21
  flowmeters, F14.17
  hygrometers, F14.9
  infrared radiant heating, A49.6-7
  pressure-measuring, F14.12
  rotative speed-measuring, F14.23
  solar energy systems, A30.27
  sound-measuring, F7.7; F14.23
  testing, adjusting, balancing, A34
  thermometers, F14.3
  velocity-measuring, F14.14
  vibration-measuring, F14.24
Insulation, electrical
  refrigerant effects on, R5.4
Insulation, thermal, F22.2
  airflow retarders, F22.16
    heating climates, F23.6
    mixed climates, F23.8
    warm humid climates, F23.9
  animal environments, A20.4
  basic materials, F22.2
  central systems, S1.6
  clothing, F8.8
    table, F8.8, 9
  cryogenic, R38.19
    foam, R38.21
    high-vacuum, R38.20
    multilayer, R38.21
    powder, R38.21
  duct construction, S16.6
  ducts, F23.18; F32.14
  economic thickness, F22.9
    building envelopes, F22.10
    life-cycle savings, F22.10
    mechanical systems, F22.9
  environmental spaces, F23.19
  equivalent thickness, F22.22
    example, F22.23
  foundation and floors, F23.3
  heat flow, F22.4
  heat transfer, F14.28
  hospitals, A7.10
  industrial
    thermal conductivity (table), F24.18
  masonry construction, F23.2
  moisture control, F23
  performance, F23.4
  pipes, F23.15
    freeze prevention, F23.16
    table, F24.21
    underground, F23.16
  piping, S11.11, 19
  properties, F22.2
    health and safety, F22.3
    mechanical, F22.3
    sound control, F22.3
    thermal, F22.2
    table, F24.4
  refrigerant piping, R32
    installation, R32.9
    jacketing, R32.9
    joint sealant, R32.9
    maintenance, R32.10
    properties, R32.1
      table, R32.3
    thickness (tables), R32.4
    vapor retarders, R32.9
  refrigerated facilities, R12.1; R13.11
    application, R13.12
    recommended R-values (table), R13.12
  refrigerated rooms, F23.18
  roof deck construction, F23.3
  solar energy systems, S33.7, 14
  steel frame construction, F23.1
  tanks, vessels, equipment, F23.17
  thermal control, F23
  thermal storage systems, A40.18
    ice, A40.15
    water, A40.11
  transport vehicles, F23.19
  types, F22.2
  water vapor retarders, F22.17
  wood frame construction, F23.1
Intercoolers
  ammonia refrigeration systems, R3.3
Intermodal transport. *See* **Containers**
I-P unit conversion factors, F35
Iron pipe. *See* **Pipe**
Jets. *See* **Air jets**
Joule-Thomson cycle, R38.4
Juice, R24
Kelvin-day method. *See* **Degree-day method**
Kerosene-fired
  heaters (in-space), S29.3
Kirchoff's law, F3.9
Kitchens, A3.5, 6; A28
  air distribution, A28.9-10
  air filtration, A28.15-16
  air quality, A28.9
  balancing, A28.10-12
    multiple-hood systems, A28.11-12
  cooking equipment/processes, A28.1-2, 17
    appliance types, A28.1-2
    effluent generation, A28.1
    heat gain (table), F28.11, 12
    hot vs. cold, A28.1
    residential, A28.17
  energy conservation, A28.12-13, 18
    economizers, A28.12
    residential hoods, A28.18
  energy considerations, A28.12-13
  exhaust systems, A28.6-9, 18
    duct systems, A28.7-8
    effluent control, A28.7
    fan types, A28.8
    residential, A28.18
    terminations, A28.8-9
  fire safety, A28.13-16
    residential, A28.18
    spread prevention, A28.14-15
    suppression, A28.13-14

Kitchens (cont.)
  heat recovery, A28.12
  hoods, A28.2-6, 17-18
    recirculating, A28.6
    residential, A28.17, 18
    type I, A28.2
      exhaust flow rates, A28.3-5
      grease removal, A28.2-3
      makeup air options, A28.5-6
      sizing, A28.3
      static pressure, A28.6
      styles, A28.3
    type II, A28.6
  maintenance, A28.16-17, 18
    air systems, A28.17
    control systems, A28.17
    cooking equipment, A28.16
    exhaust systems, A28.16-17
    residential, A28.18
  makeup air systems, A28.9-10, 18
    indoor air quality, A28.9
    large buildings, A28.9
    non-code areas, A28.9
    non-HVAC system, A28.9
    residential, A28.18
    through HVAC system, A28.9
  operation, A28.16
  service water heating, A45.5
  system integration and balancing, A28.10-12
Krypton
  recovery, R38.14
Laboratories, A13
  air distribution, A13.9
  air intakes, A13.12
    dilution criteria, A13.12
  air supply systems, A13.8-9
    air distribution, A13.9
    filtration, A13.8-9
  animal labs, A13.12-14; A20.8
    animal heat generated (table), A13.13
    heat and moisture production, F10.12
  biological safety cabinets, A13.6-8
    Class I, A13.7
    Class II, A13.7
    Class III, A13.7-8
  clinical labs, A13.14-15
  commissioning, A13.15-16
  containment labs, A13.14
    biosafety level 1, A13.14
    biosafety level 2, A13.14
    biosafety level 3, A13.14
    biosafety level 4, A13.14
  controls, A13.11-12
    fume hoods, A13.11-12
    pressure, A13.11
    temperature, A13.11
  design parameters, A13.2-3
    architectural factors, A13.3
    heat gain, A13.2-3
  duct leakage rates, A13.9-10
  economics, A13.16
    life-cycle costs, A13.16
  energy conservation, A13.15
  energy recovery, A13.15
  exhaust devices, A13.8
  exhaust systems, A13.9-10
    construction/materials, A13.10
    ductwork leakage, A13.9-10
    types, A13.9
  filtration, A13.8-9
    HEPA filters, A13.6-7, 8-9
  fire safety, A13.10-11
  fume hoods, A13.3-6
    controls, A13.11-12
    performance, A13.4-6
    sash configurations, A13.4
    types, A13.4
  hazard assessment, A13.1-2
  heat gain, A13.2-3
    equipment (table), A13.2
    lab animals (table), A13.13
  hospitals, A7.7-8
  laminar flow clean benches, A13.8
  maintenance, A13.15
  nuclear facilities, A23.7-8
  operation, A13.15
  paper testing, A22.3
  radiochemistry labs, A13.15
  stack heights, A13.12
    dilution criteria, A13.12
  teaching labs, A13.14
  types, A13.1
    animal labs, A13.1
    biological labs, A13.1
    chemical labs, A13.1
    physical labs, A13.1
  ventilation, A13.8-12
    air supply systems, A13.8
    exhaust systems, A13.9-10
Lambert's law, F3.9
Laminar flow
  fluids, F2.3
Laundries
  evaporative cooling, A47.8
  service water heating, A45.18
Leakage function, F25.11
Leak detection
  refrigerants, F18.6
    ammonia, F18.9
    bubble method, F18.9; R46.4
    electronic detection, F18.6; R46.4
    fluorescent, R46.4
    halide torch, F18.9; R46.4
    mass spectrometer, R46.4
    pressure testing, R46.4
    sulfur dioxide, F18.9
    water submersion test, R46.4
Legionella pneumophila, A7.2; A44.12; A45.8
  air washers, S19.8
  cooling towers, S36.11
  evaporative coolers, S19.8
Lewis number, F5.11
Lewis relation, F5.11; F8.4
Libraries, A3.8-10
  sound and vibration control, A3.10
Light
  measurement, F14.25
Lighting
  animal environments, F10.3
  CLF (table), F28.52
  cooling load, F28.7, 52
  greenhouses, A20.13
  plant environments, F10.15, 16
  plant growth facilities, A20.15-18
Limonene
  coolants, secondary, F20.13
Linde cycle, R38.4
Lines, refrigerant. See **Piping**
Liquefied petroleum gas (LPG), F17.4
  furnaces, residential, S28.4, 8
Liquid
  chillers. See **Chillers, liquid**
  coolers. See **Coolers, liquid**
Liquid overfeed (recirculation) systems, R1
  ammonia refrigeration systems, R3.22
  circulating rate, R1.4
  evaporators, R1.6
  line sizing, R1.7
  liquid separators, R1.7
  overfeed rate, R1.4
  pump selection, R1.4
  receiver sizing, R1.7
  refrigerant distribution, R1.2
Lithium bromide-water (see also **Water-lithium bromide**)
  properties, F19.1
  refrigeration cycle, F1.16
  specific gravity (figure), F18.4
  specific heat (figure), F18.4
  viscosity (figure), F18.4
Load calculations
  air-and-water systems, S3.4
  coils, air-cooling/dehumidifying, S21.14
  computer applications, A36.4-5
  heat gain. See **Heat gain**
  humidification, S20.3
  hydronic systems, S12.2
  nonresidential, F28
    CLF
      hooded equipment (table), F28.53
      lighting (table), F28.52
      people (table), F28.51
      unhooded equipment (table), F28.51
    CLTD
      flat roofs (table), F28.42
      glass (table), F28.49
      walls (table), F28.43
    CLTD/SCL/CLF method, F28.3, 39
      appliances, F28.53
      example, F28.54
      glass load, F28.49
      glass load (table), F28.50
      humans, F28.51
      lighting, F28.52
      limitations, F28.55
      procedure (table), F28.40
    example, F28.33, 49, 52, 53
    heat balance, F28.2
      conduction transfer functions, F28.2
      space air energy balance, F28.3
      room transfer function (RTF), F28.29, 31
      example, F28.31
    TETD/TA method, F28.3, 56
      example, F28.58, 59
      procedure (table), F28.57
    transfer function method (TFM), F28.3, 17
      example, F28.33
      procedure (table), F28.18
  precooling, R14.1
  refrigerated facilities, R12
    equipment load, R12.5
    infiltration load, R12.3
    internal load, R12.2
    load factors (table), R13.8
    product load, R12.2
    transmission load, R12.1
  refrigerated trucks and trailers, R29.6
  residential, F27
    cooling, F27.1
      block load, F27.1
      cooling load temperature difference (CLTD), F27.2
      example, F27.5
      glass load factor (GLF), F27.2
      latent factor, F27.5
      procedure, F27.5
      sensible heat factor, F27.5
      shade line factor (SLF), F27.2
      shading coefficient, F27.2
    heating, F27.6
      attic temperature, F27.8
      crawl space heat loss, F27.9
      crawl space temperature, F27.9

Composite Index

Load calculations *(cont.)*
  residential *(cont.)*
    heating *(cont.)*
      design conditions, F27.7
      example, F27.8, 9, 10
      infiltration heat loss, F27.13
      pickup load, F27.14
      procedure, F27.7
      transmission heat loss, F27.10
      unheated space temperature, F27.8
Local exhaust
  gaseous contaminant control, A41.8
  industrial ventilation, A24.5; A26.1-9
Lorenz refrigeration cycle, F1.9
Loss coefficients
  control valves, F2.11
  Duct Fitting Database, F32.11
  fittings, F2.10
Louvers, F29.41
LPG. *See* **Liquefied petroleum gas (LPG)**
Lubricants (*see also* **Oil**), R7
  additives, R7.4
  air solubility, R7.20
  ammonia refrigeration systems
    lubricant management, R3.6
  automobile air conditioning, A8.3
  component characteristics, R7.3
  evaporator return, R7.14
  foaming, R7.20
  halocarbon refrigeration systems
    lubricant management, R2.9
    receivers, R2.27
    separators, R2.25
  lubricant cooling, R3.2
    screw compressors, R3.13
  moisture content, R46.1
  oxidation, R7.20
  properties, R7.4
    floc point, R7.15
    viscosity, R7.4, 5
  reactions with refrigerants, R5.7
    additives, R5.8
    polyalkylene glycols, R5.8
    polyol esters, R5.7
  refrigerant contamination, R6.7
  refrigerant solutions, R7.8
    density, R7.8
    miscibility, R7.9, 11, 12
    solubility, R7.8, 9, 13
    viscosity, R7.13
  requirements, R7.2
  separators, R45.21
  stability, R7.20
  synthetic lubricants, R7.3
  testing, R7.1
  water solubility, R7.19
  wax separation, R7.15
Lubrication
  compressors
    centrifugal, S34.34
    reciprocating, S34.8
    rotary, S34.10
    single-screw, S34.12
    twin-screw, S34.20
  engine drives, S7.5
    heat recovery, S7.21
  gas turbine drives, S7.9
  steam turbine drives, S7.12
Maintenance (*see also* **Operation and maintenance**)
  absorption units, R41.8
  air cleaners, S24.9
  air conditioners, room, S43.5
  air conditioners, unitary, S44.3

all-water systems, S4.3
bus air conditioning, A8.9
cogeneration systems, S7.38
coils
  air-cooling/dehumidifying, S21.15
  air-heating, S23.3
condensers
  air-cooled, S35.12
  evaporative, S35.17
  water-cooled, S35.7
cooking equipment, A28.16
coolers, liquid, S37.5
cooling towers, S36.10
costs, A33.4-5
documentation, A35.2-3
engine drives, S7.7
evaporative coolers, S19.7
fan-coil units, S4.3
filters, air, S24.9
gas turbine drives, S7.11
heat pumps, unitary, S44.3, 9
heat recovery equipment, S42.5, 12, 13, 16
industrial air conditioning, A11.9
infrared heaters, S15.5
kitchen ventilation systems, A28.16-17, 18
laboratory HVAC equipment, A13.15
large buildings, A35.4
liquid chillers, R43.5, 12, 15
makeup air units, S31.10
medium-sized buildings, A35.3-4
retail store air conditioners, A2
small buildings, A35.3
solar energy systems, A30.26-27
steam turbine drives, S7.17
terminology, A35.1
unit heaters, S31.7
Makeup air units, S31.8
  codes, S31.9
  commissioning, S31.9
  control, A42.22-23; S31.9
  door heating, S31.9
  filters, S31.9
  maintenance, S31.10
  selection, S31.8
    media, S31.9
      cooling, S31.9
      heating, S31.9
      location, S31.9
  spot cooling, S31.9
  standards, S31.9
Management
  life-cycle concept, A35.1-2
  operation and maintenance, A35
Manufactured homes
  heating/cooling systems, A1.6-7
Marine refrigeration. *See* **Refrigeration**
Masonry
  insulation, F23.2
Mass transfer, F5
  convection, F5.5
    coefficients for external flows, F5.5
    coefficients for internal flows, F5.5
    heat transfer analogy, F5.6
      example, F5.6
  eddy diffusion, F5.7
    Chilton-Colburn j-factor analogy, F5.8
    heat transfer analogy, F5.8
      example, F5.10
    Lewis relation, F5.11
    Reynolds analogy, F5.8
  heat transfer analogy
    convection, F5.6
    eddy diffusion, F5.8
    molecular diffusion, F5.3

heat transfer simultaneous with, F5.11
  air washers, F5.13
    example, F5.13
  cooling coils, F5.15
  cooling towers, F5.14
  dehumidifying coils, F5.15
  direct-contact equipment, F5.12
  enthalpy potential, F5.11
Lewis relation, F5.11; F8.4
molecular diffusion, F5.1
  coefficients, F5.2
  Fick's law, F5.1
  gas through a stagnant gas, F5.3
    example, F5.4
  heat transfer analogy, F5.3
  in liquids and solids, F5.4
Measurement (*see also* **Instruments**), F14
  air exchange rates, F25.5
  air infiltration, F14.28
  air leakage, F25.14, 16
  airtightness, F14.28
  carbon dioxide, F14.29
  combustion analysis, F14.31
  contaminants, F14.28
  data recording, F14.31
    chart recorders, F14.32
    data loggers, F14.32
    digital, F14.31
  electricity, F14.21
  fluid flow, F2.12; F14.17
  gaseous contaminants, A41.13-14
  heat transfer in insulation, F14.28
  humidity, F14.9
  light levels, F14.25
  moisture
    farm crops, F11.5
    refrigeration systems, R46.3
  moisture content, F14.27; R6.3
  moisture transfer, F14.27
  odors, F13.1, 5
  pressure, F14.12
  rotative speed, F14.23
  sound, F7.7; F14.23
  temperature, F14.3
  terminology, F14.1
  testing, adjusting, balancing, A34
  thermal comfort, F14.25
  uncertainty analysis, F14.2
  velocity, F14.14
  vibration, F7.14; F14.24
Meat, R16
  display refrigerators, R47.6
  frozen, R16.15
    freezing quality, R16.16
    storage and handling, R16.16
  processing facilities
    boxed beef, R16.7
    carcass coolers, R16.2
      beef, R16.2
      calves, R16.10
      hogs, R16.8
      lamb, R16.10
    energy conservation, R16.17
    pork trimmings, R16.10
    processed meats, R16.12
      bacon slicing rooms, R16.13
      freezing, R16.15
      lard chilling, R16.14
      sausage dry rooms, R16.13
    sanitation, R16.1
    shipping docks, R16.16
    variety meats, R16.11
  retail storage, R47.7
  storage requirements (table), R10.3
  thermal properties, R8

A = 1995 Applications   S = 1996 Systems and Equipment   F = 1997 Fundamentals   R = 1998 Refrigeration

Metabolic rate
  humans, F8.6
    activities (table), F8.6
Metals and alloys
  low-temperature properties, R39.6
Methylene chloride
  coolants, secondary, F20.12
Metric conversion. *See* **Conversion factors**
Microbiology
  foods, R11
Microwave drying, A27.4
Milk. *See* **Dairy products**
Mines, A25
  air cooling, A25.3-4
  dehumidification, A25.3-4
  equipment and applications, A25.4-6
    combination systems, A25.5
    components, A25.4-5
    cooling surface air, A25.4
    cooling surface water, A25.4
    energy recovery systems, A25.6
    evaporative cooling, A25.4, 5, 7
    water pressure reduction, A25.5-6
  evaporative cooling, A25.4, 5, 7
  heat exchangers, A25.7
    air vs. workplace cooling, A25.7
    cooling coils vs. fan position, A25.7
  heat sources, A25.1-3
    adiabatic compression, A25.1-2
    blasting, A25.3
    equipment, A25.2
    groundwater, A25.2
    wall rock, A25.2-3
  heat stress, A25.1
  mechanical refrigeration plants, A25.6-7
  spot cooling, A25.7
  wall rock heat flow, A25.2-3
  water sprays, A25.7
Modeling
  airflow, A15.5
    around buildings, F15.15
    similarity scaling, F15.15
  energy consumption, F30
  industrial ventilation, A24.6-8
  moisture in buildings, F22.21
Moist air, F6
  adiabatic mixing (example), F6.14
  adiabatic mixing of water (example), F6.14
  composition, F6.1
  cooling (example), F6.13
  heat absorption/moisture gain (example), F6.15
  heating (example), F6.12
  psychrometrics, F6
  thermal conductivity
    figure, F6.16
  thermodynamic properties, F6.2, 10
    example, F6.10, 12
    table, F6.3
  transport properties, F6.15
  viscosity
    figure, F6.16
Moisture
  animal buildings, A20.1-2
  animal production, F10.7, 8, 9, 10, 11, 12
  combustion, F17.11
  condensation
    concealed, S20.3
    visible, S20.2
  content/transfer, F14.27
  control, F23.5
    heating climates, F23.5
      example, F23.7

    mixed climates, F23.8
      example, F23.9
    warm humid climates, F23.9
      air conditioning, F23.9
      example, F23.10
      ventilation, F23.9
  corrosion affected by, A44.2
  diffusivity, F14.27
  eliminators, S35.14
  farm crops content, A21; F11.1
    measurement, F11.5
      direct methods, F11.5
      indirect methods, F11.6
  in attics
    heating climates, F23.6
    mixed climates, F23.8
    warm humid climates, F23.10
  in building materials, F22.13
  in buildings, F22.12; F23.4
    commercial and institutional, F23.14
    condensation prevention, F22.21; F23.5
    envelope component intersections, F23.13
    heat flow affected by, F22.12
    humidity control, F23.5
    modeling, F22.21
    moisture-related problems, F22.12
    steady-state design tools, F22.18
      dew-point method, F22.18
      example, F22.18, F22.20
      Kieper diagram, F22.20
  in cathedral ceilings
    heating climates, F23.7
    mixed climates, F23.8
    warm humid climates, F23.10
  in ceilings, F39.6
  in foundations, F23.11; F39.2
    crawl spaces, F23.11
    example, F23.12
    floor slabs, F23.11
  in insulation, F23
  in membrane roofs, F23.10
    inverted roofs, F23.11
    self-drying low-slope roofs, F23.11
  in refrigerant systems, R6
    desiccants, R6.3
    driers, R6.5
    drying methods, R6.2
    effects, R6.1
    factory dehydration, R46.1
    lubricant solubility, R7.19
    measurement, R6.3; R46.3
    sources, R6.1; R46.1
  in roofs, F39.6
  in walls, F39.4
  in windows, F39.7
  migration, F22.14; F23.4
    air leakage control, F23.4
    air movement, F22.15
    capillary suction, F22.14
    contact wetting angle, F22.15
    diffusion, F22.15
  paper content
    printing control of, A17.2
  permeability, F14.27
  permeance, F14.27
  refrigerators, cold pipes, freezers, F23.15
  regain
    products, A11.1, 4
  sorption isotherms, F14.27
  terminology, F22.1
  water vapor retarders, F22.17; F25.17
    heating climates, F23.6
    mixed climates, F23.8
    warm humid climates, F23.9

Montreal Protocol, F18.1
Moody chart (figure), F2.9
Motels, A5.3-4
  service water heating, A45.11
Motors, S39
  codes, S39.1
  compressors, S34.3
  control, S39.6
    control protection, S39.6
    motor protection, S39.6
    motor-starting methods, S39.7
  efficiency, S39.2
  electrical insulation
    refrigerant effects on, R5.4
  evaporative cooling, A47.7
  furnaces, residential, S28.2
  heat gain, F28.9, 53
    table, F28.10; R12.3
  hermetic, S39.4
    application, S39.4
  hermetic burnout
    cleanup, R6.11
    refrigerant contamination, R6.7, 11
  nonhermetic, S39.3
    application, S39.3
    types, S39.3
  pony, S35.17
  power supply (ac), S39.1
  pumps, centrifugal, S38.12
  standards, S39.1
  thermal protection, S39.5
  two-speed, S38.7, 11
Multiple-use complexes, A5.4
Multizone systems. *See* **Zone systems**
Museums, A3.8-10
  sound and vibration control, A3.10
Natatoriums (*see also* **Swimming pools**), A4.6-8
Natural gas, F17.4
  liquefaction, R38.6
    figure, R38.6
  liquefied, R38.2
  pipe sizing, F33.17
  processing, R38.15
  separation, R38.14
NC curves. *See* **Noise criterion (NC) curves**
Nightclubs, A3.6
Night setback
  automatic control, A42.32
  furnaces, residential, S28.14, 15
  recovery, A38.14
Nitrogen
  recovery, R38.13
Noise (*see also* **Sound**), F7.1, 3
  air conditioners, room, S43.4
  compressor
    centrifugal, S34.32
    orbital (scroll), S34.24
    single-screw, S34.16
  condensing units, R47.13
  fans, S18.8
  fluid flow, F2.14
  health effects, F9.18
  valves, S41.3
  water pipes, F33.3
Noise criterion (NC) curves, A43.5, 13-14; F7.5
Noncondensable gases
  condensers
    evaporative, S35.17
    water-cooled, S35.6
  refrigerant contamination, R6.7
Nonresidential. *See* **Commercial and public buildings**
NTU. *See* **Number of transfer units (NTU)**

Composite Index

Nuclear facilities, A23
  basic technology, A23.1-3
    criticality, A23.1
    filtration, A23.2-3, 5
      charcoal filters, A23.3, 5
      demisters, A23.3, 5
      dust filters, A23.3
      heaters, A23.3, 5
      HEPA filters, A23.3, 5
      sand filters, A23.3
    fire protection, A23.2
    radiation fields, A23.1
    regulation, A23.1
    safety design, A23.1-2
  codes and standards, A23.8
  Department of Energy facilities, A23.3-5
    confinement systems, A23.3-4
      air locks, A23.3
      differential pressures, A23.4
      zone pressure control, A23.3
      zoning, A23.3
    ventilation, A23.4-5
      control systems, A23.4
      radioactive effluents, A23.4-5
      requirements, A23.4
      systems, A23.4
  Nuclear Regulatory Commission facilities, A23.5-8
    boiling water reactor HVAC systems, A23.6
      primary containment, A23.6
      reactor building, A23.6
      turbine building, A23.6
    laboratories, A23.7-8
      fume hoods, A23.8
      glove boxes, A23.8
      radiobenches, A23.8
    medical/research reactors, A23.7
    nuclear power plants, A23.5
    other buildings/rooms, A23.6-7
    pressurized water reactor HVAC systems, A23.5-6
      air filtration, A23.5
      reactor containment building, A23.5-6

Number of transfer units (NTU)
  cooling towers, S36.17
  heat transfer, F3.4

Nursing homes, A7.11-12
  design concepts and criteria, A7.12
  pressure relationships and ventilation, A7.12
    table, A7.11
  service water heating, A45.11
  system applicability, A7.12

Nusselt number
  table, F3.2

Nuts
  storage, R28.7

Odor control
  industrial gas cleaning, S25.25, 27

Odors, F13
  analytical measurement, F13.5
  characteristics, A41.3
  control, A41.8-12
  factors affecting, F13.5
    humidity, F13.5
    sorption/release, F13.6
    temperature, F13.5
    ventilation, F13.6
  odor units, F13.4
  olf unit, F13.5
  sense of smell, F13.1
    anatomy/physiology, F13.1
    olfactory stimuli, F13.1
  sensory measurement, F13.1
    character, F13.2
    detectability, F13.1
    Dravnieks olfactometer, F13.2
    hedonics, F13.2
    odor acceptability, F13.4
    odor intensity, F13.2
    odor quality, F13.4
    olfactory acuity, F13.2
    threshold, F13.1, F13.2
      detection threshold, F13.1
      recognition threshold, F13.1
      table, F13.2
  sources, F13.5
  suprathreshold intensity, F13.3
    category scale, F13.3
    matching odorant intensity, F13.3
    perceived odor magnitude, F13.3
    psychophysical power law, F13.3
    ratio scaling, F13.3
    Stevens' law, F13.3
  threshold concentration, F13.1
  threshold limit value, F13.1

Odor units, F13.4

Office buildings, A3.6-8
  service water heating, A45.11, 12

Office equipment
  heat gain, F28.11
    table, F28.14

Oil (*see also* **Lubricants**)
  refrigerant systems, R7.2
  return
    coolers, liquid, S37.5

Oil, fuel, F17.5
  characteristics, F17.5
  distillate oils, F17.5
  handling, S26.8
  heating value, F17.6
  pipe sizing, F33.17
  preparation, S26.9
  residual oils, F17.5
  storage systems, S26.7
  sulfur content (table), F17.6
  viscosity, F17.5

Oil-fired
  burners, S26.4
    air atomizing, S26.5
    horizontal rotary cup, S26.6
    industrial, S26.5
    mechanical atomizing, S26.6
    pressure atomizing, S26.5
    residential, S26.4
    return-flow mechanical atomizing, S26.6
    return-flow pressure, S26.5
    steam atomizing, S26.6
  equipment
    venting, S30.16
  furnaces, residential, S28.4, 8
  heaters (in-space), S29.3
    powered atomizing, S29.3
    vaporizing pot oil, S29.3
  infrared heaters, S15.3
  service water heating, A45.1-2
  unit heaters, S31.4

Olf unit, F13.5

One-pipe diverting circuits, S12.8

Operating costs, A33.3-4

Operation and maintenance (*see also* **Maintenance**)
  automatic control systems, A42.33-34
  building types, A35.3
    large-sized, A35.4
    medium-sized, A35.3-4
    small-sized, A35.3
  compressors, S34.34
  desiccant systems, S22.7
  documentation
    maintenance manual, A35.3
    operation manual, A35.3
  gas-cleaning equipment, industrial, S25.29
  industrial exhaust systems, A26.11
  life-cycle management for, A35.1-2
  personnel knowledge/skills, A35.3
  responsibilities, A35.4
  system
    capability, A35.2
    dependability, A35.2
    effectiveness, A35.1-2
    maintainability, A35.2
    reliability, A35.2
  technology, A35.4
  terminology, A35.1

Outlets
  air diffusion, F31.2
    classification, F31.2
    location, F31.13
    performance, F31.4
    selection, F31.13
  small forced-air systems, S9.2
  supply air, S17.1
    ceiling diffusers, S17.4
    grilles, S17.3
    registers, S17.3
    selection, S17.1
    slot diffusers, S17.3
    sound level, S17.2
    VAV, S17.2

Outpatient health care facilities, A7.10-11
  diagnostic clinics, A7.10-11
  treatment clinics, A7.11

Outside air fraction, F25.3

Owning costs, A33.1-2

Oxygen
  liquid, R38.2
  recovery, R38.13

Ozone
  electronic air filters, S24.9

Paclet number
  table, F3.2

Panel heating and cooling, A49.1; S4.1; S6
  advantages, S6.1
  air-and-water systems, S3.3
  air-heated/cooled panels, S6.18
  applications, A49.8-9
  controls, S6.18
    cooling, S6.19
    electric heating slabs, S6.19
  design considerations, S6.5, 7
    floor covering effects, S6.6
    heat gain, S6.7
    heat loss, S6.7
    thermal resistance, S6.5
  distribution/layout, S6.13
    concrete slab-embedded, S6.14
    hydronic floor panels, S6.14
    hydronic wall panels, S6.14
    suspended floor piping, S6.14
  electrically heated systems, S6.15
    ceilings, S6.15
    floors, S6.17, 19
    walls, S6.17
  heat transfer, S6.2
    combined, S6.3
    convection, S6.2
    radiation, S6.2
  hydronic systems, S6.10
    ceiling, S6.12
    design considerations, S6.10

Panel heating and cooling *(cont.)*
  hydronic systems *(cont.)*
    floor, S6.14
    wall, S6.14
  new techniques, A49.9
Paper
  moisture content control, A17.2
Paper products facilities, A22.2-4
  evaporative cooling, A47.8
  finishing area, A22.3
  miscellaneous areas, A22.3
  motor control rooms, A22.3
  paper machine area, A22.2
  paper testing laboratories, A22.3
  process control rooms, A22.3
  system selection, A22.3
Parking garages. *See* **Garages, parking**
Peanuts
  drying, A21.9; F11.13
Permafrost
  stabilization, R35.4
Permeability, F14.27; F22.2
  building materials (table), F24.16
  clothing, F8.8
    table, F8.8
Permeance, F14.27; F22.2
  building materials (table), F24.16
Phase-change materials
  thermal storage, A40.11, 14, 18
Photographic materials, A19
  processing and printing, A19.1-3
    air filtration, A19.2
    exhaust requirements, A19.2
    preparatory operations, A19.1
    printing/finishing, A19.2
    processing operations, A19.1-2
    temperature control, A19.3
  storage
    processed materials, A19.3-4
      black/white prints, A19.4
      color film/prints, A19.4
      long-term storage, A19.3-4
      medium-term storage, A19.3
      nitrate base film, A19.4
    unprocessed materials, A19.1
Physiological principles, humans *(see also* **Comfort**)
  adaptation, F8.15
  age, F8.15
  body surface area (DuBois), F8.3
  clothing
    insulation, F8.8
      table, F8.8, 9
    moisture permeability, F8.8
      table, F8.8
  cooling load, F28.7, 51
  DuBois equation, F8.3
  energy balance, F8.2
  heat stress, F8.19, 23
  heat transfer coefficients, F8.7
    convective, F8.8
    evaporative, F8.8
    Lewis relation, F8.4
  latent heat loss, F8.3, 10
  mechanical efficiency, F8.6
  metabolic rate, F8.6
    activities (table), F8.6
  respiratory heat loss, F8.4
  seasonal rhythms, F8.15
  sensible heat loss, F8.3
  sex, F8.15
  skin heat loss, F8.5
    latent heat, F8.3, 10
    sensible heat, F8.3

  skin wettedness, F8.20
  thermoregulation, F8.1
Pipe, S40
  application data (table), S40.6
  buried, F24.23
  codes, S40.6
  cold springing, S40.10
  copper tube, S40.1
    friction loss (figure), F33.5
    heat loss (table), F24.20
    thermal expansion (table), S40.8
    tube data (table), S40.4
  expansion/flexibility, S11.19; S40.8
  expansion bends, S40.9
    L bends, S40.9
    U bends, S40.9
    Z bends, S40.9
  expansion joints, S40.10
    packed, S40.11
      flexible ball, S40.11
      slip, S40.11
    packless, S40.11
      flexible hose, S40.12
      metal bellows, S40.11
      rubber, S40.12
  expansion loops, S40.9
  fittings, S40.2
  fluid flow, F2
  freeze prevention, F23.16
  insulation, F23.15
    table, F24.21
  iron, S40.2
    cast, S40.2
    ductile, S40.2
  joining methods, S40.2
    brazing/soldering, S40.2
    flanges, S40.2
    flared/compression joints, S40.2
    reinforced outlet fittings, S40.5
    unions, S40.5
    welding, S40.5
  plastic, S11.8, 9; S40.12; F33.9
    allowable stresses, S40.12
    friction loss (figure), F33.5
    properties (table), S40.13
    selection, S40.12
  roughness (table), F2.10
  selection, S40.6
  special systems, S40.6
  standards, S40.2
  steel, S40.1
    friction loss (figure), F33.5
    heat loss (table), F24.19
    pipe data (table), S40.3
    thermal expansion (table), S40.8
  stress calculations, S40.7
  supporting elements, S11.19; S40.7
    anchors, S40.7
    guides, S40.7
    hangers, S40.7
    supports, S40.7
  thermal expansion (table), S40.8
  underground, F23.16
  wall thickness, S40.7
Pipe coils
  heat emission, 1988 Equipment, Chapter 28, Table 5, p. 3
  (See explanation on first page of Index)
Pipe sizing, F33
  fittings, F33.1, 4
    elbow equivalents (table), F33.6
    loss coefficients (table), F33.2
    tees, F33.6
      elbow equivalents (figure), F33.6

  fuel oil, F33.17
    suction line size (table), F33.18
  gas, F33.17
    flow rate (table), F33.17
  hydronic systems, S12.15; F33.4
    air separation, F33.4
    example, F33.4, 6
  pressure drop, F33.1
    copper tube (figure), F33.5
    Darcy-Weisbach equation, F33.1
    example, F33.2
    faucets/cocks vs. flow (figure), F33.8
    Hazen-Williams equation, F33.1
    plastic pipe (figure), F33.5
    steel pipe (figure), F33.5
  refrigerant
    ammonia systems, R3.9
      capacity tables, R3.8, 9
    halocarbon systems, R2.3
      tables, R2.4, 5, 6, 7, 8, 9
    liquid overfeed systems, R1.7
    retail food store refrigeration, R47.14
  service water, F33.6
    cold water sizing procedure, F33.9
    example, F33.9
  steam, F33.9
    condensate return, F33.14
      flow rate (table), F33.14, 15
      nonvented (closed), F33.15
      vented (open), F33.14
    example, F33.10, 13
    flow rate (figure), F33.11
    flow rate (table), F33.12, 13
    high-pressure, F33.13
    low-pressure, F33.12
    two-pipe systems, F33.14
    velocity (figure), F33.11
  valves, F33.1, 4
  water, F33.3
    fixture flow/pressure (table), F33.7
    fixture units, F33.7
      demand vs. (figure), F33.7
      demand weights (table), F33.7
    flow rate limitations, F33.3
      aging allowances, F33.3
      erosion, F33.3
      noise, F33.3
      water hammer, F33.4
    velocity (table), F33.3
Piping
  all-water systems, S4.2
  ammonia refrigeration systems, R3.7
    compressor piping, R3.10, 12, 13
    condenser and receiver piping, R3.15
    evaporator piping, R3.19
    materials, R3.7
  central systems, S1.9
  circuits
    hydronic systems, S12.8
  cooling towers, S36.8
  design
    computer applications, A36.8
  district heating and cooling, S11.8
    heat transfer, S11.13
    insulation thickness, S11.19
  fan-coil units, S4.2
  geothermal energy systems, A29.10-11
  halocarbon refrigeration systems, R2
    capacity tables, R2.4, 5, 6, 7, 8, 9
    compressor, R2.19
    condenser, R2.18
    defrost gas supply lines, R2.16
      capacity table, R2.16
    discharge lines, R2.14

Composite Index I.33

Piping *(cont.)*
   halocarbon refrigeration systems *(cont.)*
      evaporator, R2.23
      gas velocity, R2.1
      hot-gas bypass, R2.29
      insulation, R2.6
      liquid cooler, flooded, R2.20
      noise and vibration, R2.7
      pressure drop, R2.3
      suction risers (table), R2.12
   hydronic snow melting, A46.5-8
      metal, A46.5-8
      plastic, A46.5-8
   insulation
      refrigerant piping, R32
   panel systems, S6.13
   protective coating (table), R32.2
   railroad car air conditioning, A8.10
   refrigerant
      ammonia systems, R3
      halocarbon systems, R2
      insulation, R32
      jacketing, R32.9
      liquid overfeed systems, R1.7
      pipe preparation, R32.2
      supports and hangers, R32.10
      vapor retarders, R32.9
   service water heating, A45.3-6
   solar energy systems, A30.17; S33.8, 10
   sound control, A43.34-40
   sound transmission, A34.23-24
   standards, S11.7
   steam systems, S10.3, 5
   system identification, F34.12
   unit heaters, S31.7
   vibration control, A43.34-40
   vibration transmission, A34.23-24
   water systems, medium/high temp., S14.7
Pitot tube, A34.2; F14.16
Places of assembly, A4
   arenas, A4.4-5
      ancillary spaces, A4.5
      gymnasiums, A4.5
      load characteristics, A4.4-5
   atriums, A4.9
   auditoriums, A4.3-4
      concert halls, A4.4
      legitimate theaters, A4.4
      motion picture theaters, A4.3
      projection booths, A4.3-4
      stages, A4.4
   convention centers, A4.5-6
      load characteristics, A4.6
      system applicability, A4.6
   exhibition centers, A4.5-6
      load characteristics, A4.6
      system applicability, A4.6
   fairs, A4.8
      air cleanliness, A4.8
      design concepts, A4.8
      equipment, A4.8
      maintenance, A4.8
      occupancy, A4.8
      system applicability, A4.8
   general criteria, A4.1-2
      air conditions, A4.1-2
      filtration, A4.2
      lighting loads, A4.1
      sound control, A4.2
      ventilation, A4.1
      vibration control, A4.2
   gymnasiums, A4.5
   houses of worship, A4.3

   natatoriums, A4.6-8
      air distribution, A4.7
      design concepts, A4.6
      design criteria, A4.6-7
      filtration, A4.7
      load characteristics, A4.6
      noise levels, A4.7
      special considerations, A4.7-8
      system applicability, A4.6
   stadiums, A4.4-5
      ancillary spaces, A4.5
      enclosed stadiums, A4.5
      gymnasiums, A4.5
      load characteristics, A4.4-5
   system considerations, A4.2-3
      air conditioning, A4.2
      air distribution, A4.2-3
      ancillary facilities, A4.2
      mechanical equipment rooms, A4.3
      precooling, A4.2
      stratification, A4.2
   temporary exhibit buildings, A4.8
      air cleanliness, A4.8
      design concepts, A4.8
      equipment, A4.8
      maintenance, A4.8
      occupancy, A4.8
      system applicability, A4.8
Plant environments, A20.8-18; F10.12
   air composition, F10.17
   air contaminants, F10.17
   air movement, F10.18
   controlled environment rooms, A20.14-18
      air conditioning, A20.15
      controls, A20.15
      heating, A20.15
      lighting, A20.15-18
      location, A20.14
      luminaire height/spacing (table), A20.17
   energy balance, F10.14
   greenhouses, A20.8-14
      carbon dioxide enrichment, A20.12
      cooling, A20.11-12
      energy conservation, A20.14
      evaporative cooling, A20.11-12
      glazing, A20.14
      heating, A20.9-11; F10.12, 14
      heat loss reduction, A20.14
      humidity control, A20.12
      orientation, A20.9
      photoperiod control, A20.13
      radiant energy, A20.13
      sealants, A20.14
      shading, A20.11
      site selection, A20.8-9
      supplemental irradiance, F10.16
      supplemental lighting (table), A20.13
      ventilation, A20.11
   heating, F10.14
   humidity, F10.17
   lighting, F10.15
   other facilities, A20.18
   photoperiod control, F10.17
   plant growth facilities (chambers), A20.14-18
      air conditioning, A20.15
      controls, A20.15
      heating, A20.15
      lighting, A20.15-18
      location, A20.14
      luminaire height/spacing (table), A20.17
   supplemental irradiance, A20.13; F10.16
   temperature, F10.14
Plants
   gaseous contaminant control, A41.12

Plastic
   ducts, S16.5
   glazing, F29.27
   greenhouses, A20.10, 14
   pipe. *See* Pipe
   refrigerant swelling (table), F18.10
PMV. *See* Predicted mean vote (PMV)
Pollutants, gaseous. *See* Contaminants, gaseous
Polydimethylsiloxane
   coolants, secondary, F20.13
Ponds
   spray, S36.6
Positive positioners
   automatic control with, A42.7; F37.7
Potatoes
   products, frozen, R26.5
Poultry (*see also* Chickens, Turkeys)
   chilling, R17.1
   decontamination, R17.3
   freezing, R17.4
   processing, R17.1, 3
   processing plant sanitation, R17.7
   recommended environment, A20.7-8
   storage, R17.8
   storage requirements (table), R10.3
   thawing, R17.9
   thermal properties (table), R17.6
PPD. *See* Predicted percent dissatisfied (PPD)
Prandtl number
   table, F3.2
Precooling
   air reheater, S42.5
   buildings, A38.14-15
   flowers, cut, R14.9
   fruits and vegetables, R14
      load calculation, R14.1
      methods, R14.3
      rate, R14.1
   indirect evaporative cooling, A47.3-4
   places of assembly, A4.2
Predicted mean vote (PMV), F8.16; F14.26
Predicted percent dissatisfied (PPD), F8.17
Pressure
   absolute, F14.12
   aircraft, A9.2, 7
   clean spaces, A15.10
   control, A42.19
   corrosion affected by, A44.3
   differential, F14.13
   drying equipment, S22; S22.8
   dynamic, F14.13
   gage, F14.13
   measurement, A34.2; F14.12
      aneroid gages, F14.13
      Bourdon tubes, F14.13
      electromechanical transducers, F14.13
      manometers, F14.13
      McLeod gages, F14.13
      mechanical gages, F14.13
      standards, F14.13
   pipe sizing, F33
   sensors for, A42.9; F37.8
   smoke control, A48.5, 8
   stairwells
      smoke control, A48.8-9, 10-11
   static, F14.13
   steam systems, S10.5
   units, F14.13
   vacuum, F14.12
Pressure drop
   two-phase fluid flow, F4.12
      void fraction, F4.12

A = 1995 Applications   S = 1996 Systems and Equipment   F = 1997 Fundamentals   R = 1998 Refrigeration

Primary system, S1.2, 7
   air-and-water, S3.3
   energy estimating, F30.12
Printing plants
   air filtration for, A17.5
   binding, A17.5
   collotype printing, A17.4
   design criteria, A17.1-2
   flexography, A17.4
   letterpress, A17.3
   lithography, A17.3-4
   paper moisture content control, A17.2
   platemaking, A17.2-3
   relief printing, A17.3
   rotogravure, A17.4
   salvage systems, A17.4
   shipping, A17.5
Produce
   desiccation, R10.1
   deterioration rate (table), R10.1
   display refrigerators, R47.4
   storage requirements (table), R10.2
Properties
   boiling point (table), F36.1
   chilled water density (table), A40.11
   coolants, secondary, F20
   density
      liquids (table), F36.2
      solids (table), F36.3
      vapors (table), F36.1
   emissivity
      solids (table), F36.3
   freezing point (table), F36.2
   heat of
      fusion, F36.2
      vaporization, F36.2
   heat transmission, F24
   insulation, F22.2
      table, R32.3
   liquids (table), F36.2
   low-temperature
      fiber composites, R39.8
      metals and alloys, R39.6
         ductility (figure), R39.6
         tensile strength (figure), R39.6
      plastics and polymers, R39.7
   lubricants, R7.4
   pipe thermal expansion (table), S40.8
   plastic pipe (table), S40.13
   refrigerants
      physical properties, F18
         electrical properties (table), F18.5
         latent heat of vaporization, F18.5
         sound velocity, F18.4
         table, F18.3
         Trouton's rule, F18.5
      thermodynamic properties, F19
      transport properties, F19
   solids (table), F36.3
   specific heat
      liquids (table), F36.2
      solids (table), F36.3
      vapors (table), F36.1
   steam, S10.1
   thermal conductivity
      earth, S11.12
      liquids (table), F36.2
      solids (table), F36.3
      vapors (table), F36.1
   thermodynamic
      moist air, F6
      water at saturation, F6.2
   vapor pressure (table), F36.2
   vapors (table), F36.1

   viscosity
      liquids (table), F36.2
      vapors (table), F36.1
Propylene glycol
   coolants, secondary, F20.4
   hydronic systems, S12.16
Psychrometer, F6.9
Psychrometrics, F6
   air composition, F6.1
   charts, F6.10
      adiabatic mixing, F6.13
      adiabatic mixing of water, F6.14
      figure, F6.11
      heat absorption/moisture gain, F6.15
      moist air cooling, F6.13
      moist air heating, F6.12
      moist air properties, F6.12
   evaporative cooling systems, A47.1-2, 5, 11-12
   humidity parameters, F6.8
      absolute humidity, F6.8
      degree of saturation, F6.8
      dew-point temperature, F6.8, 9
      humidity ratio, F6.8
      relative humidity, F6.8
      saturation humidity ratio, F6.8
      specific humidity, F6.8
      wet-bulb temperature, F6.8, 9
   industrial drying, A27.1-3
   moist air
      thermal conductivity
         figure, F6.16
      thermodynamic properties, F6.2, 10
         example, F6.10, 12
         table, F6.3
      transport properties, F6.15
      viscosity
         figure, F6.16
   perfect gas equations, F6.8
   standard atmosphere U.S., F6.1
      altitudes (table), F6.1
   water at saturation
      thermodynamic properties, F6.2
Public buildings. *See* **Commercial and public buildings**
Pumps
   central systems, S1.8
   centrifugal, S12.5; S38
      affinity laws, S38.7
      arrangement, S38.9
         compound, S12.7
         distributed pumping, S38.11
         parallel pumping, S12.6; S38.9
         primary-secondary pumping, S12.7; S38.11
         series pumping, S12.6; S38.10
         standby pump, S12.7; S38.11
         two-speed motors, S38.11
         variable-speed pumping, S38.11
      best efficiency point (BEP), S38.6
      casing, S38.2
         diffuser, S38.2
         volute, S38.2
      cavitation, S38.8
      commissioning, S38.13
      components, S38.1
      efficiency, S38.6
      energy conservation, S38.12
      hydronic system curves, S38.4
      impeller trimming, S38.5, 7, 8
      installation, S38.13
      motive power, S38.12
      net positive suction, S38.8
      operating point, S38.4
      operation, S38.2, 13

      power, S38.6
      pump/hydronic system curves, S38.4
      pump performance curves, S38.3
         series curves, S38.4
      radial thrust, S38.8
      selection, S38.9
      troubleshooting, S38.13
      types, S38.2
         base-mounted horizontal (axial) split-case single-stage double-suction pump, S38.3
         base-mounted horizontal split-case multistage pump, S38.3
         circulator pump, S38.2
         close-coupled single-stage end-suction pump, S38.2
         frame-mounted end-suction pump on baseplate, S38.3
         vertical in-line pump, S38.3
         vertical turbine single- or multistage sump-mounted pump, S38.3
   control of, A38.9-10
   fluid flow indicators, A34.12-13
   geothermal energy systems, A29.8-9
      lineshaft, A29.8-9
      submersible, A29.9
   hydronic snow melting, A46.8
   liquid overfeed systems, R1.4
   modeling, F30.10
   pump performance curves, S12.5
   solar energy systems, A30.17
   water systems, medium/high temp., S14.6
Purge units
   centrifugal liquid chillers, R43.11
Radiant
   drying systems, A27.3
   energy transfer principles, S6.1
   panel cooling systems, S6
   panel heating systems, A42.27; S6; 29.4
   snow melting systems, A46.12-13
Radiant heating, A49; S15
   applications, A49.8-9
      considerations, A49.7-8
   asymmetry, A49.5
   beam heating design, A49.4; S15.5
      geometry, A49.4
   control, S15.4
   efficiency, S15.4
   energy conservation, S15.1
   design criteria, A49.2-4
      comfort chart, A49.2-4
   floor reradiation, A49.5
   heaters (in-space), S29.4
   maintenance, S15.5
   precautions, S15.4
   radiation patterns, A49.5-6
   reflectors, S15.4
   terminology, A49.1-2
      adjusted dry bulb temperature, A49.2
      angle factor, S15.5
      effective radiant flux, A49.2; S15.5
      fixture efficiency, S15.4
      flux distribution, S15.6
      mean radiant temperature, A49.1; S6.1; S15.4
      operative temperature, A49.1-2
      pattern efficiency, S15.4
      radiation-generating ratio, S15.4
      radiosity, S15.3
      skin/clothing absorptance/reflectance, S15.5
   test instrumentation, A49.6-7
      black globe thermometer, A49.6-7
      directional radiometer, A49.7
   total space heating design, A49.6

Composite Index

Radiation
  electromagnetic, F9.19
  optical waves, F9.20
  radio waves, F9.21
  radon, F9.19
  thermal, F3.1, 6
    angle factors, F3.9
      figure, F3.10
    blackbody, F3.6
    exchange between surfaces, F3.9
    gray, F3.8
    in gases, F3.11
    nonblack, F3.8
    transient, F3.4
      cylinder, F3.5
      slab, F3.5
      sphere, F3.5
Radiation shielding
  industrial environment, A24.5
Radiators, S4.1; S32
  application, S32.5
  nonstandard condition corrections, S32.3
  panel, S32.1
  rating, S32.2
  sectional, S32.1
  specialty, S32.1
  tubular, S32.1
Radiosity, F3.9
  radiant heating, S15.3
Radon, F9.19; F12.6; F39.3
Railroad cars
  air conditioning, A8.9-11
    air distribution, A8.10
    car construction, A8.9
    comfort conditions, A8.10
    controls, A8.10-11
    design limitations, A8.9-10
    equipment selection, A8.9
    future trends, A8.11
    piping design, A8.10
    system requirements, A8.10
    vehicle types, A8.9
  refrigerator cars, 1994 Refrigeration, Chapter 28
  (See explanation on first page of Index)
Rapid transit systems, A12.7-11
  (*see also* **Fixed guideway vehicle air conditioning**)
  design approach, A12.10-11
  station air conditioning, A12.9-10
  ventilation, A12.7-9
    emergency, A12.9
    mechanical, A12.7-9
    natural, A12.7
    smoke control, A12.9
Rating
  condensers
    air-cooled, S35.10
    evaporative, S35.15
RC curves. *See* **Room criterion (RC) curves**
Receivers
  ammonia refrigeration systems, R3.3, 15
  halocarbon refrigerant, R2.16
  liquid overfeed systems, R1.7
Recirculation systems (*see also* **Liquid overfeed (recirculation) systems**)
  water treatment for, A44.14, 15
Recycling
  refrigerants, R6.8, 9
Reflectance
  fenestration, F29.17
Reflectors
  radiant heating, S15.4

Refrigerant control devices, R45
  automobile air conditioning, A8.5-6
  capillary tubes, R45.21
  coolers, liquid, S37.5
  heat pumps, unitary, S44.12
  lubricant separators, R45.21
  pressure transducers, R45.2
  short tube restrictors, R45.27
  switches, R45.1
    differential control, R45.2
    float, R45.3
    fluid flow-sensing, R45.2
    pressure control, R45.1
    temperature control, R45.2
  valves, control
    check, S8.8; R45.19
    condenser pressure regulators, R45.13
    condensing water regulators, R45.18
    evaporator pressure regulators, R45.10
    expansion, constant pressure, R45.9
    expansion, electric, R45.9
    expansion, electronic, S8.8
    expansion, thermostatic, S8.8; R45.4
    float, high-side, R45.13
    float, low-side, R45.14
    pressure relief devices, R45.19
    reversing, S8.8; R45.17
    solenoid, R45.14
    suction pressure regulators, R45.12
Refrigerants, F18; F19
  absorption solutions, F19.1
  acoustic velocity, S34.30
  air conditioners, unitary, S44.6
  ammonia, F19.1
    chemical reactions, R5.6
    properties, R5.3
    refrigeration system practices, R3
  ammonia-water, F19.1
  automobile air conditioning, A8.3
  azeotropic, F1.6
  carbon dioxide, F19.1
  cascade refrigeration systems, R39.3
  compatibility with other materials, R5.4
    database on, R5.10
    elastomers, R5.6
    electrical insulation, R5.4
    plastics, R5.6
  construction materials affected by, F18.9
    elastomers, F18.9
      swelling (table), F18.10
    metals, F18.9
    plastics, F18.10
      swelling (table), F18.10
  contaminants in, R6
    generation by high temperature, R5.9
  cryogenic fluids, F19.1; R38.1
  environmental acceptability, R5.1
  environmental test facilities, R37.1
  halocarbons, F19.1
    azeotropic blends, F19.1
    ethane series, F19.1
    flow rate, R2.1
      figures, R2.2, 3
    hydrolysis, R5.6
    methane series, F19.1
    refrigeration system practices, R2
    thermal stability, R5.6
    zeotropic blends, F19.1
  heat pumps, unitary, S44.11
  hydrocarbons, F19.1
  inorganic refrigerants, F19.1
  leak detection, F18.6
    ammonia, F18.9
    bubble method, F18.9

    electronic detection, F18.6
    halide torch, F18.9
    sulfur dioxide, F18.9
  lithium bromide-water, F18.4; F19.1
    specific gravity (figure), F18.4
    specific heat (figure), F18.4
    viscosity (figure), F18.4
  low-temperature characteristics (table), R39.2
  lubricant solutions, R7.8
  moisture in, R6.1
  Montreal Protocol, F18.1
  number designation (table), F18.2
  performance, F18.5
    comparison (table), F18.7
  phaseout, F18.1
  phaseout costs, A33.5
  piping
    insulation, R32
    principles, R2.1
  properties, F18.1; R5.1
    atmospheric lifetime (table), R5.2
    electrical properties (table), F18.5
    flammability, R5.3
    global warming potential (tables), R5.2, 3
    latent heat of vaporization, F18.5
    ozone depletion potential (tables), R5.2, 3
    physical properties, F18.1
      table, F18.3
    sound velocity, F18.4
      table, F18.6
    Trouton's rule, F18.5
  reclamation, R6.8, 10
  recovery, R6.8
  recycling, R6.8, R6.9
  superheat, S35.16
  safety groups, F18.6
    table, F18.9
  thermodynamic properties, F19
    pressure-enthalpy diagrams, F19.1
    saturated liquid/vapor data, F19.1
  transport properties, F19
  water/steam, F19.1
  zeotropic, F1.6, 10
Refrigerant transfer units (RTU)
  centrifugal liquid chillers, R43.11
Refrigerated facilities, R13
  air circulation, R10.4
  air purification, R10.5
  automated, R13.4, 14
  construction methods, R13.4
    Schneider system, R13.7
  controlled atmosphere storage, R13.3
  controls, R10.4; R13.10
  design, R13.2
  freezers, R13.9
  insulation, R13.11
    minimum thickness (table), R12.1
  load calculations, R12
  refrigerated rooms, R13.4
  refrigeration systems, R13.7
  sanitation, R10.5
  secondary coolants. *See* **Coolants, secondary**
  temperature pulldown, R13.14
  vapor retarders, R13.5, 11
Refrigeration, F1.1
  absorption cycles (*see also* **Absorption refrigeration**), F1.14
    ammonia-water cycle, F1.18
    flow description, F1.14
    lithium bromide-water cycle, F1.16
      double-stage, F1.17
      single-stage, F1.16
    refrigerant-absorbent pairs, F1.15

A = 1995 Applications   S = 1996 Systems and Equipment   F = 1997 Fundamentals   R = 1998 Refrigeration

Refrigeration *(cont.)*
  air coolers, forced circulation, R42
  aircraft air conditioning, A9.4-6
    air cycle, A9.4-6
    direct expansion, A9.6
    vapor cycle, A9.6
  air transport, R31.3
  ammonia systems, R3
    compressors, R3.2
      cooling, R3.12, 13
        piping, R3.10, 12, 13
    condensers, R3.17
    controls, R3.6
    equipment, R3.2
    liquid recirculation (overfeed), R3.22
      hot-gas defrosting, R3.23
    lubricant management, R3.6
    multistage systems, R3.21
    piping, R3.7
    safety, R3.27
    system selection, R3.1
    valves, R3.10
  autocascade systems, R39.1
  beverage plants, R25.11
  biomedical applications, R40
  breweries, R25.3
  Carnot cycle, F1.6
  cascade systems, R39.3
    compressors, R39.5
    refrigerants, R39.3
  coefficient of performance (COP), F1.3
  chemical industry, R36
  compression cycles, F1.6
    actual systems, F1.12
      example, F1.12
    Carnot cycle, F1.6
      example, F1.7, F1.8
      single-stage theoretical, F1.8
    Lorenz cycle, F1.9
      example, F1.10
      single-stage theoretical, F1.10
    multistage, F1.10
      example, F1.11
    zeotropic mixture, F1.10
  concrete, R35.1
  coolers, liquid, S37
  cryogenic. *See* **Cryogenics**
  cycles, F1; R38.4
  food
    eggs and egg products, R20
    fish, R18.1
    fruits, fresh, R21; R22
    vegetables, R23
  food processing facilities, R26
    banana ripening rooms, R22.5
    control of microorganisms, R11.3
    meat plants, R16
  food service equipment, R48
  halocarbon systems
    accessories, R2.23
    fitting pressure losses (tables), R2.10
    heat exchangers, R2.23
    lubricant management, R2.9
    piping, R2.3
    refrigerant receivers, R2.16
    subcoolers, R2.24
    valves, R2.9
  heat reclaim
    service water heating, A45.3, 21
  ice rinks, R34.1
  insulation, F23.18; R32
  liquid overfeed systems, R1
  load calculations. *See* **Load calculations**
  low-temperature, R39
    autocascade systems, R39.1
    cascade systems, R39.3
    heat transfer, R39.9
    material selection, R39.6
    secondary coolants, R39.11
    single-refrigerant systems, R39.2
  marine, R30
    cargo, R30.1
    containers, R30.7
    fishing vessels, R30.10
    ships' stores, R30.7
  refrigerated facility design, R13
  retail food store systems, R47.8
  secondary coolant systems, R4
    coolant selection, R4.1
    corrosion prevention, R4.5
    design considerations, R4.2
  soils, subsurface, R35.3, 4
  storage facilities. *See* **Refrigerated facilities**
  systems
    charging, factory, R46.4
    chemical reactions, R5.6
    component balancing, R44
    contaminant control, R6
      retrofitting, during, R6.12
    copper plating, R5.9
    dehydration, factory, R46.1
    design balance points, R44.2
    energy and mass balance, R44.3
    evaluating system chemistry, R5.1
    moisture in, R46.1
    performance, R44.4
    testing, factory, R46.4
  thermal storage, A40.4-5, 18
  trucks and trailers, R29.4
  wineries, R25.8
Refrigerators
  commercial, R48
    reach-in, R48.1
    roll-in, R48.3
  cryocoolers, R38.8
  food service, R48
  household, R49
    absorption cycle, R41.9
    cabinet construction, R49.2
    defrosting, R49.6
    efficiency, R49.1
    ice makers, R49.2
    refrigeration systems, R49.5
    testing, R49.9
  mortuary, R48.3
  retail food store, R47
    display, R47.1
      temperatures (table), R47.2
    load variable with ambient (table), R47.2
    storage, R47.7
Registers
  air outlets, S17.3
  small forced-air systems, S9.2
Regulators *(see also* **Valves***)*
  condenser pressure, R45.13
  condensing water, R45.18
  draft, S30.22
  evaporator pressure, R45.10
  suction pressure, R45.12
Reheat boxes
  air distribution, S17.6
Residential
  air leakage, F25.14, 16
  climatic infiltration zones, F25.20
  duct construction, S16.2
  energy consumption (table), A32.8-9
  gas burners, S26.1
  heating/cooling systems, A1
  humidifiers, S20.5
  infiltration, F25.14
    examples, F25.22
  load calculations, F27
  oil burners, S26.4
  service water heating, A45.9-10
  ventilation, F25.20
Resistance, thermal *(see also* **R-values***)*, F3.21
  air spaces (table), F24.2, 3
  attics (table), F24.13
  building materials (table), F24.4
  insulating materials (table), F24.4
  panels, S6.5
Respiration
  fruits and vegetables, R8.17
  rate (table), R8.23
Restaurants, A3.5
Restrictors
  air conditioners, room, S43.2
Retail facilities
  air conditioning, A2
  convenience centers, A2.6
  department stores, A2.5-6
  design conditions, A2
  discount/outlet stores, A2.2
  load determination, A2
  multiple-use complexes, A2.7-8
  refrigeration, R47
  shopping centers, A2.6-7
  small stores, A2.1-2
  supermarkets, A2.3-5; A45.12
    refrigerators, R47
Retail food store refrigeration, R47
Retrofitting
  refrigerant systems
    contaminant control, R6.12
Reynolds number, F2.3
  table, F3.2
Rice
  drying, A21.9; F11.13
Road tunnels, A12.1-7
  allowable CO concentrations, A12.1
  carbon monoxide analyzers and recorders, A12.7
  carbon monoxide emission, A12.1
  control systems, A12.7
  pressure evaluation, A12.6
  ventilation, A12.1-6
    emergency, A12.6
    mechanical, A12.4-6
    natural, A12.3-4
    requirements, A12.1-3
    smoke control, A12.6
Rocks
  thermal conductivity (table), F24.15
Rodents
  farm crop storage, F11.7
Roofs
  air movement, F39.6
  CLTD (table), F28.42
  heat loss, F27.10
  insulation, F23.3
  moisture control, F23.10; F39.6
    inverted roofs, F23.11
    self-drying low-slope roofs, F23.11
  overhang shading, F29.42
  R-values, F24.11
  thermal performance, F39.5
  wall interface, F39.8
Roof ventilators
  industrial environment, A24.16-17
Room criterion (RC) curves, A43.3-4; F7.5

Composite Index

Rotative speed
  measurement, F14.23
    stroboscopes, F14.23
    tachometer-generators, F14.23
    tachometers, F14.23
Roughness
  pipe (table), F2.10
Roughness factors
  ducts, F32.7
R-values (*see also* **Resistance, thermal**)
  calculation, F24.2
    modified zone method, F24.11
    zone method, F24.10
      example, F24.10
  ceilings, F24.11
  masonry construction, F24.8
    example, F24.9
  roofs, F24.11
  steel frame construction, F24.9
    example, F24.10
  wood frame construction, F24.2
    example, F24.8
Safety
  air cleaners, S24.11
  air conditioners, room, S43.4
  automatic control, A42.33
  chemical plants, R36.2
  cryogenic equipment, R38.24
  filters, air, S24.11
  fuel-burning equipment, S26.13
  industrial gas cleaning
    fires/explosions, S25.29
  nuclear facilities, A23
  refrigerants, F18.6
    table, F18.9
  service water heating, A45.8
  solar energy systems, A30.26
  solid fuel heaters, S29.6
  water systems, medium/high temp., S14.9
Sanitation
  food production facilities, R11.4
    egg processing, R20.12
    HACCP, R11.4
    meat processing, R16.1
    poultry processing, R17.7
  refrigerated storage facilities, R10.5
Scale
  cooling towers, S36.13
  humidifiers, S20.4
  service water heating, A45.7
  water treatment control, A44.9-11
Schneider system, R13.7
Schools, A6.2-3
  service water heating
    elementary, A45.11
    high schools, A45.11, 15
Scrubbers
  industrial gas cleaning
    gaseous contaminant control
      spray dry scrubbing, S25.17
      wet-packed scrubbers, S25.18
    particulate contaminant control, S25.14
      centrifugal-type collectors, S25.15
      impingement scrubbers, S25.15
      orifice-type collectors, S25.15
      spray towers, S25.15
      venturi scrubbers, S25.15
Seasonal energy efficiency ratio. *See* **Energy efficiency ratio (EER)**
Secondary coolants. *See* **Coolants, secondary**
Secondary system, S1.2, 3
  air-and-water, S3.8
  energy estimating, F30.10

Seeds
  drying, F11.13
  storage, A21.12
SEER. *See* **Energy efficiency ratio (EER)**
Seismic restraint, A43.40-41; A50
  anchor bolts, A50.4-5
  design calculations, A50.1-2, 4
    dynamic analysis, A50.2
    examples, A50.6-9
    static analysis, A50.2, 4
  duct construction, S16.6
  installation problems, A50.9
  international seismic zone (table), A50.3
  snubbers, A50.6
  terminology, A50.1
  weld capacities, A50.5
Sensible heat ratio
  air conditioners, room, S43.3
Sensors
  automatic control with, A42.7-9, 32; F37.7
Separators
  lubricant, R45.21
Service life
  equipment (table), A33.4
Service water heating, A45
  codes, A45.22
  commercial needs/sizing, A45.10-12
    apartments, A45.11, 12, 15
    food service, A45.11, 12, 16, 17
    motels, A45.11
    office buildings, A45.11, 12
    supermarkets, A45.12
  corrosion, A45.7
  design considerations, A45.7
  distribution systems, A45.3-6
    commercial dishwashers, A45.5
    commercial kitchen water pressure, A45.5
    heat traced nonreturn system, A45.5
    manifolding, A45.6
    materials, A45.3-4
    pipe sizing, A45.4
    pressure differentials, A45.4
    return system, A45.4
    supply system, A45.4
    two-temperature service, A45.6
  geothermal energy systems, A29.13
  heat transfer equipment, A45.1-3
    blending injection, A45.3
    circulating tank, A45.3
    electric, A45.2
    gas-fired, A45.1-2
    heat recovery, A45.3
    indirect, A45.2-3, 21-22
    oil-fired, A45.1-2
    refrigeration heat reclaim, A45.3, 21
    semi-instantaneous, A45.3
    solar energy, A45.3
  hot water demand (table), A45.4, 11, 15, 21
  hot water requirements, A45.9-21
    commercial, A45.10-12
    institutional, A45.10-12
    residential, A45.9-10
  indirect water heating, A45.2-3, 21-22
    boilers for, A45.21-22
    control for, A45.22
  institutional needs/sizing, A45.10-12
    dormitories, A45.11, 12
    elementary schools, A45.11, 18
    high schools, A45.11, 15, 18
    military barracks, A45.11
    nursing homes, A45.11
  Legionella pneumophila, A45.8
  pipe sizing, F33.6

  residential needs/sizing, A45.9-10
    minimum water heater capacities (table), A45.9
  safety, A45.8
  scale, A45.7
  sizing, A45.9-21
    examples, A45.12-21
    instantaneous water heaters, A45.20-21
    refrigerant-based water heaters, A45.21
      heat pump, A45.21
      refrigeration heat reclaim, A45.21
    semi-instantaneous water heaters, A45.20-21
    storage equipment, A45.9-20
      concrete, ready-mix, A45.20
      examples, A45.12-21
      food service, A45.16-17
      industrial plants, A45.19-20
      laundries, coin-operated, A45.18
      laundries, commercial, A45.18
      schools, A45.18
      showers, A45.15-16
      swimming pools/health clubs, A45.18-19
      whirlpools/spas, A45.19
  solar energy systems, A30.15-18, 20-21, 28; A45.3
  standards, A45.22
  storage tank usable water, A45.8
  system planning, A45.1
    energy sources, A45.1
  temperature requirements, A45.8
  terminal usage devices, A45.6
  terminology, A45.6
  thermal storage, A40.7
  water heater placement, A45.8-9
  water quality, A45.7
Shading, F29.2, 41
  computer calculations, F29.43
  cooling load, F28.41
  draperies, F29.43
    double, F29.46
  exterior, F29.41
  interior, F29.43
  louvers, F29.41
  partial shading, F29.42
  roller shades, F29.43
  roof overhangs, F29.42
  venetian blinds, F29.43
Shading coefficients (*see also* **Heat gain**), F29.23
  domed skylights (table), F29.26
  double glazing (table), F29.39
  glass (table), F29.25, 38, 39, 40
    with draperies (table), F29.40
    with roller shades (table), F29.38
    with venetian blinds (table), F29.38
  glass block walls (table), F29.27
  louvers (table), F29.38
  windows (table), F27.4
Sheep
  growth, F10.7
  heat and moisture production, F10.8
  reproduction, F10.8
  wool production, F10.8
SHGC. *See* **Heat gain**
SHGF. *See* **Heat gain**
Ship docks, A3.12
Ships, A10
  air distribution
    merchant, A10.5
    naval, A10.7
  cargo refrigeration, R30.1
    built-in refrigerators, R30.1
    refrigeration load, R30.6
    refrigeration system, R30.3

A = 1995 Applications   S = 1996 Systems and Equipment   F = 1997 Fundamentals   R = 1998 Refrigeration

Ships *(cont.)*
  equipment selection
    merchant, A10.2-3
    naval, A10.6-7
  fishing vessels, R30.10
    fish freezing, R30.15
    fish refrigeration
      brine wells, R30.14
      icing, R18.1; R30.13
      refrigerated seawater, R18.2; R30.14
      system design, R30.10
  general criteria, A10.1
  merchant ships, A10.1-6
  naval surface ships, A10.6-7
  refrigerated stores, R30.7
Sick building syndrome, F13.5
Simulation. *See* **Modeling**
Single-duct systems
  all-air systems, S2.5
    constant volume, S2.5
    variable-air-volume, S2.6
SI unit conversion factors, F35
Skating rinks, R34
Skylights
  domed, F29.26
Slab-on-grade, F39.1, 2, 3
Slots
  air outlets, S17.3
Smoke management, A48
  acceptance testing, A48.12
  central systems, S1.3
  computer analysis, A48.12
  dampers, A48.8
  design parameters
    pressure differences, A48.8
    weather data, A48.8
  door-opening forces, A48.6
  duct design, F32.14
  elevators, A48.11
  flow areas, A48.6-8
    effective flow areas, A48.7-8
    open doors, A48.8
    symmetry, A48.8
  health care facilities, A7.4
  rapid transit system emergencies, A12.9
  road tunnel emergencies, A12.6
  smoke management, A48.4-6
    airflow, A48.5-6
    buoyancy, A48.6
    compartmentation, A48.4
    dilution near fire, A48.4-5
    pressurization, A48.5
    remote dilution, A48.4
  smoke movement, A48.1-4
    buoyancy, A48.2-3
    expansion, A48.3
    HVAC systems, A48.4
    stack effect, A48.1-2
    wind, A48.3-4
  stairwells, A48.8-11
    analysis, A48.9-10
    compartmentation, A48.9
    pressurization, A48.8-9, 10-11
      overpressurization relief, A48.10
      supply fan bypass, A48.10-11
  zones, A48.11-12
    acceptance testing, A48.12
    computer analysis, A48.12
Smudging
  supply air outlets, S17.2
Snow melting, A46
  control, A46.5, 8
    automatic, A46.5
    manual, A46.5

electric system design, A46.1-4, 8-13
  embedded wires, A46.11
  gutters and downspouts, A46.13
  heat density, A46.8
  infrared, A46.12-13
  installation, A46.11-12
  mineral insulated cable, A46.8-11
  switchgear/conduit, A46.8
free area ratio, A46.1
heating requirements, A46.1-4
  back/edge losses, A46.4
  free area ratio, A46.1
  heating equations, A46.1-3
hydronic system design, A46.1-4, 5-8
  controls, A46.8
  fluid heater/air control, A46.8
  heat transfer fluid, A46.5
  pipe system, A46.5-8
  pump selection, A46.8
  thermal stress, A46.8
idling load, A46.3
operating costs, A46.4, 5
pavement design, A46.4
performance classification, A46.3
  table, A46.3
snow detectors, A46.5
snowfall data, A46.3
  table, A46.2, 7
Snubbers
  seismic restraint, A50.6-9
Sodium chloride
  brines, F20.1
Soils
  corrosion affected by, A44.5
  stabilization, R35.3, 4
  thermal conductivity, S11.12; F24.14
    table, F24.15
Sol-air temperature, F28.5, 6
  table, F28.6
Solar angle, F29.16
Solar energy, A30; S33
  active systems, A30.20, 20, 22-24, 23-24
  collectors, A30.7-12, 16, 26, 27; S33.3
    array design, S33.10
      piping, S33.10
    concentrating, A30.9-10
    construction, S33.5
      absorber plates, S33.5
      glazing, S33.7
      housing, S33.6
      insulation, S33.7
    design/installation, A30.27
    evacuated heat pipe, S33.4
    evacuated tube, S33.3
    flat plate, A30.6, 7; S33.3
    glazing materials, A30.7-8
    integral storage, S33.3
    module design, S33.8
      piping, S33.8
      thermal expansion, S33.10
      velocity limits, S33.9
    mounting, A30.26
    operational results, S33.5
    performance, A30.10-12; S33.4
    plates, A30.8-9
    selection, S33.7
    testing, S33.4
    types, S33.3
      air-heating, S33.3
      liquid-heating, S33.3
      liquid-vapor, S33.3
  controls, A30.28; S33.18
    differential temperature controller, S33.18
    hot water dump, S33.19
    over-temperature protection, S33.18

cooling systems, A30.18-19, 28
  absorption refrigeration, A30.21
  nocturnal radiation/evaporation, A30.19
design, installation, operation checklist,
    A30.27-28
  airflow, A30.27-28
  collectors, A30.27
  controls, A30.28
  hydraulics, A30.27
  performance, A30.28
  thermal storage, A30.28
  uses, A30.28
diffuse sky radiation, F29.16
extraterrestrial irradiance (table), F29.14
fenestration, F29; F39.7
flux, F29.14
freeze protection, A30.26; S33.1, 2, 19
heating and cooling systems, A30.19-21, 28
  active, A30.20
  hybrid, A30.20
  passive, A30.19-20
heat exchangers, S33.16
  external, S33.17
  freeze protection, S33.19
  internal, S33.16
  performance, S33.17
  plate-and-frame, S33.17
  requirements, S33.16
  shell-and-tube, S33.17
  tube-and-tube, S33.17
heat gain. *See* **Heat gain**
heating systems, A1.4; A30.13-15, 20-21;
    A40.7; S33.1
  air systems, A30.15; S33.1, 11, 12
  direct circulation system, A30.14; S33.2
    drain-down, A30.14
  indirect systems, A30.14-15; S33.2
    drain-back systems, A30.15; S33.2
    freeze protection, S33.2
    nonfreezing fluid, S33.2
  integral collector, A30.15
  liquid systems, S33.1, 10, 12
    freeze protection, S33.2
  pool heating, A30.15
  residential, A1.4
  thermosiphon systems, A30.13-14
heat pump systems, applied, S8.4
hybrid systems, A30.20
installation guidelines, A30.25-27
  collector mounting, A30.26
  freeze protection, A30.26
  instrumentation, A30.27
  maintenance, A30.26-27
  over-temperature protection, A30.26
  performance monitoring, A30.27
  safety, A30.26
  start-up procedure, A30.26
irradiance, F29.17
  table, F29.29, 30, 31, 32, 33, 34, 35
passive systems, A30.19-20, 24-25
photovoltaic systems, S33.19
  applications, S33.21
  cells/modules, S33.20
  fundamentals, S33.19
  related equipment, S33.20
quality and quantity, A30.1-7
  flat plate collectors, A30.6
  incident angle, A30.3
  longwave atmospheric radiation, A30.6-7
  solar angles, A30.1-2
  solar constant, A30.1
  solar irradiation design values, A30.5-6
  solar position (tables), A30.2
  solar radiation at earth's surface, A30.4-5

A = 1995 Applications   S = 1996 Systems and Equipment   F = 1997 Fundamentals   R = 1998 Refrigeration

Composite Index I.39

Solar energy *(cont.)*
  quality and quantity *(cont.)*
    solar spectrum, A30.4
    solar time, A30.2-3
  service water heating, A30.15-18, 20-21, 28; A40.7; A45.3
    components, A30.15-18
      auxiliary heat sources, A30.17
      collectors, A30.16
      control systems, A30.17-18
      expansion tanks, A30.17
      fans, A30.17
      heat exchangers, A30.16-17
      heat transfer fluids, A30.16
      piping, A30.17
      pumps, A30.17
      thermal storage, A30.16
      valves/gages, A30.17
    load requirements, A30.18
    performance, A30.18
  sizing heating and cooling systems, A30.21-25
    active heating/cooling, A3022-24
    f-Chart method, A30.22-23
    passive systems, A30.24, 25
    simplified analysis, A30.21
    standard systems, A30.22
  thermal storage systems, A30.12-13, 16, 28; S33.11
    air systems, S33.12
    insulation, S33.14
    liquid systems, S33.12
      pressurized, S33.12
      unpressurized, S33.13
    packed rock beds, A30.12-13
    short circuiting, S33.15
    sizing, S33.15
      load matching, S33.15
      solar fraction, S33.15
      storage cost, S33.15
    stratification, S33.15
    tank construction, S33.14
      electrochemical corrosion, S33.14
      oxidation, S33.14
Solar heat gain coefficients (SHGC). *See* **Heat gain**
Solar heat gain factors (SHGF). *See* **Heat gain**
Solid fuel, F17.7
  fuel-burning equipment, S26.9
  heaters (in-space), S29.4
Sorbents, F21
Sound *(see also* **Noise**), F7.1
  air outlets, S17.2
  characteristics, F7.1
  cooling towers, S36.9
  loudness, F7.4, 6
    phons, F7.6
    sones, F7.6
  measurement, F7.7; F14.23
    instrumentation, F7.7
  sound power, F7.2, 8
    conversion to pressure, F7.9
    free-field method, F7.8
    progressive wave method, F7.9
    reverberant field method, F7.9
    sound intensity method, F7.9
  sources, F7.8
  terminology, F7.2
    bandwidths, F7.1, 9
    decibel, F7.2
    frequency spectrum, F7.1, 4
    loudness, F7.4
    phons, F7.6
    sones, F7.6
    sound intensity, F7.2

  sound power, F7.2, 8
  sound pressure, F7.2
  sound quality, F7.4
  tone, F7.4
  testing, F34.17-20
  transmission, F7.10
    airborne, F7.10
    flanking, F7.10
    humidity affecting, S20.2
    structure-borne, F7.10
  unit heaters, S31.6
  velocity
    refrigerants (table), F18.6
Sound control, A43
  absorption machines, A43.9
  A-weighted sound levels, F7.5
  barriers, F7.11
  ceiling system sound transmission, A43.23-24
    sound correction, A43.24-25
  central systems, S1.6
  chillers, A43.9
  clean spaces, A15.11
  cogeneration, S7.38
  communications centers, A3.11
  composite noise rating (CNR), A43.5-7
  cooling towers
    attenuators, S36.10
  design criteria, A43.1-3; F7.4, 7
    HVAC systems, A43.1-3, 29-32
    table, A43.5
  design software, A36.8-9
  ducts, A43.11-23; S16.4
    sound attenuation, A43.14-20; F7.11
      branch division, A43.19
      circular sheet metal (lined), A43.16-17
      circular sheet metal (unlined), A43.16
      end reflection loss, A43.19-20
      insulated flexible, A43.17
      plenum chambers, A43.14-15
      rect. sheet metal elbows, A43.17
      rect. sheet metal (lined), A43.15-16
      rect. sheet metal (unlined), A43.15
      silencers, A43.17-19
    sound generated in, A43.11-14
      air devices, A43.12-14
      dampers, A43.11-12
      duct velocities, A43.11
    wall sound radiation, A43.20-23
      breakout/breakin, A43.20-23
      rumble, A43.20
  educational facilities, A6.2
  enclosures, F7.11
  engine drives, S7.7
  engine test facilities, A14.5
  equipment, A43.8-11
    sound levels, A43.7-8
  fans, A43.8-9
    centrifugal, A43.8
    plug and plenum, A43.8
    point of operation, A43.8
    propeller, A43.8-9
    selection, A43.9
    sound, A43.9
      power prediction, A43.8
    vaneaxial, A43.8
  fiberglass products, A43.32
  fume hood duct system design, A43.27-28
  gas turbine drives, S7.11
  insertion loss, A43.15-29
  insulation, F22.3
  libraries, A3.10
  mechanical equipment rooms, A43.25-27
    chases/shafts, A43.26
    duct wall penetration, A43.25-26

    enclosed air cavity, A43.27
    floating floors, A43.26-27
    special walls, A43.26
  museums, A3.10
  noise criterion (NC) curves, A43.5, 13-14; F7.5
  outdoor equipment, A43.28-29
    sound barriers, A43.28-29
    sound propagation, A43.28
  piping systems, A43.34-40
    flow noise, A43.38
    connectors
      expansion joint/arched, A43.40
      flexible, A43.39-40
      hose, A43.40
    resilient hangers/supports, A43.38-39
  places of assembly, A4.2
  return air system sound transmission, A43.23
  rooftop air handlers, A43.10-11
  room criterion (RC) curves, A43.3-4; F7.5
  sound correction, A43.24-25
    ceiling sources, A43.24-25
    nonstandard rooms, A43.25
    point sources, A43.24
  sound criteria
    indoor, A43.3-5
      HVAC sound levels, A43.4-5
      room criterion (RC) curves, A43.3-4
    outdoor, A43.5-7
      composite noise rating (CNR), A43.5-7
      correction numbers (table), A43.6
  swimming pools (natatoriums), A4.7
  terminology, F7.10
    insertion loss, F7.10
    noise reduction, F7.10
    scattering, F7.11
    sound absorption coefficient, F7.11
    sound attenuation, F7.10
    spherical spreading, F7.11
    transmission loss, F7.10
  transmission loss, A43.14-28
  troubleshooting, A43.41
    noise problems, A43.41
    problem type, A43.41
    source, A43.41
  variable-air-volume systems, A43.9-10
    air modulation devices, A43.10
    design considerations, A43.9
    fan selection, A43.9
    terminal units, A43.10
Soybeans
  drying, A21.7-8; F11.13
Specific heat
  foods, R8.7
  liquids (table), F36.2
  solids (table), F36.3
Spot cooling
  evaporative, A47.7
  industrial environment, A24.3-5, 11-12
  makeup air units, S31.9
  mines, A25.7
Stack effect
  duct design, F32.2
  infiltration/ventilation, F25.8
  smoke movement, A48.1-2
Stacks
  design, F15.10
  exhaust velocity, F15.11
  height, F15.12
Stadiums, A4.4-5
Stairwells
  smoke control, A48.8-11
Standard atmosphere U.S., F6.1
  altitudes (table), F6.1

A = 1995 Applications   S = 1996 Systems and Equipment   F = 1997 Fundamentals   R = 1998 Refrigeration

Standards, R51
  air cleaners, S24.2, 5
  air conditioners, packaged terminal, S45.3
  air conditioners, room, S43.4
  air conditioners, unitary, S44.5
  air quality, U.S., A41.7
  boilers, S27.5
  chimneys/fireplaces/gas vents, S3021, 26
  cogeneration, S7.46
  condensers
    evaporative, S35.17
    water-cooled, S35.7
  coolers, liquid, S37.4
  dehumidifiers, room, S43.6
  duct construction, S16.1
    commercial, S16.2
    industrial, S16.4
    residential, S16.2
  engine drives, S7.46
  filters, air, S24.2, 5
  fittings, S40.2
  furnaces, S28.18
  heaters (in-space), S29.7
  heat pumps, packaged terminal, S45.3
  heat pumps, unitary, S44.5, 10
  heat pumps, water-source, S45.4
  ice makers, R50.5
  indoor air quality, F9.1, 11, 12
    table, F9.2, 12, 14
  liquid chillers, R43.5
  makeup air units, S31.9
  motors, S39.1
  nuclear facilities, A23.8
  pipe, S40.6
  piping, S11.7
  service water heating, A45.22
  turbine drives, S7.46
Stanton number
  table, F3.2
Static electricity
  humidity affecting, S20.2
Steam
  cogeneration distribution, S7.28
  coils
    air heating, S23.1
  condensation, F4.10
  distribution systems
    testing, adjusting, balancing, A34.14
  district heating and cooling, S11.5, 23
  heating systems, S10
    advantages, S10.1
    air, gas, water effects, S10.2
    boilers, S10.3; S27.1
      return piping, S10.3
      supply piping, S10.3
    condensate removal, S10.7
    convection systems, S10.12
      one-pipe, S10.12
      two-pipe, S10.12
    design pressure, S10.5
      high-pressure, S10.2, 5
      low-pressure, S10.2, 5
    fundamentals, S10.1
    heat exchangers, S10.3
    heat recovery, S10.3, 14
      direct recovery, S10.15
      flash steam, S10.14
    heat transfer, S10.2
    piping, S10.5
      return, S10.3, 6
      supply, S10.3, 5
      terminal equipment, S10.6
    steam distribution, S10.13
    steam source, S10.2
    steam traps, S10.7
      kinetic, S10.7, 9
      mechanical, S10.7, 9
      thermostatic, S10.7, 8
    temperature control, S10.13
    terminal equipment, S10.11
      forced-convection units, S10.11
      natural convection units, S10.11
      piping design, S10.6
    valves, pressure-reducing, S10.9
      installation, S10.9
      selection, S10.11
    waste heat boilers, S10.3
  makeup air units, S31.9
  one-pipe systems, 1993 Fundamentals,
    Chapter 33, pp. 18-19
    (See explanation on first page of Index)
  pipe sizing, F33.9
  properties, S10.1; F6.16
    refrigerant, F19.1
  turbine drives, S7.11
  unit heaters, S31.4
  unit ventilators, S31.1
Steam-jet refrigeration
  1983 Equipment, Chapter 13
  (See explanation on first page of Index)
Steel
  boilers, S27.2
  pipe. *See* **Pipe**
Stefan-Boltzmann equation, F3.6; F29.17
Stirling cycle, R38.10
Stokers
  capacity classification, S26.9
  types, S26.10
    chain (traveling) grate, S26.11
    spreader, S26.10
    underfeed, S26.11
    vibrating grate, S26.11
Storage
  apples, A47.9; R21.1
    controlled atmosphere, R21.3
  bakery ingredients, R27.1
  candy, R28.5
  carbon dioxide, R25.11
  citrus, A47.9; R22.3
  cold
    facility design, R13
    controlled atmosphere, R13.3
    cryogenic fluids, R38.22
  cut flowers, R10.6
    table, R10.7
  dehumidification, S22.6, 9
  eggs, R20.5
  farm crops, A21.9-12; F11
  fish, fresh, R18.3
  fish, frozen, R18.7
  food, canned, R10.5
  food, dried, R10.5
  food, perishable (table), R10.2
  fruit, dried, R28.7
  fruit, fresh, R21.1
  furs and fabrics, R10.5
  humidification, S20.1
  ice, R33.4; R50.5
  meat products, frozen, R16.16
  milk, fresh, R19.4
  nursery stock, R10.6
    table, R10.7
  nuts, R28.7
  photographic materials
    processed materials, A19.3-4
    unprocessed materials, A19.1
  potatoes, A47.9
  poultry products, R17.8
  refrigerated
    facility design, R13
    seeds, R10.6
    service water heating, A45.9-20
    thermal. *See* **Thermal storage**
    vegetables, R23.3
    vegetables, dried, R28.7
    wine, R25.10
    wood products, A22.1
Stores. *See* **Retail facilities**
Stoves, S29.5
  advanced design, S29.5
  conventional, S29.5
  fireplace inserts, S29.6
  pellet-burning, S29.6
Stratification
  places of assembly, A4.2
Subcoolers, R44.2
  condensers
    evaporative, S35.16
    water-cooled, S35.4
  two-stage, R2.24
Suction lines
  capacity tables, R2.8, 9
Suction risers, R3.25
  capacity table, R2.12
Sulfur
  fuel content (table), F17.6
Superheat
  refrigerants, S35.16
Superheated vapor drying, A27.6
Supermarkets. *See* **Retail facilities**
Surface transportation, A8
  automobiles, A8.1-7
  buses, A8.7-9
  fixed guideway vehicles, A8.11
  railroad cars, A8.9-11
Survival shelters
  1991 Applications, Chapter 11
  (See explanation on first page of Index)
Swimming pools (*see also* **Natatoriums**)
  water heating for, A45.18-19
  solar heating, A30.15
Swine
  growth, F10.8
  heat and moisture production, F10.9
  reproduction, F10.9
  recommended environment, A20.6-7
Symbols, F34
  graphical (table), F34.5
  letter, F34.1
    table, F34.4
  mathematical (table), F34.5
  subscripts (table), F34.5
Temperature
  adjusted dry bulb, A49.2
  aircraft, A9.6-7
  animal environments, F10.3
  average uncooled surface, S6.2
  clean spaces, A15.10
  corrosion affected by, A44.3
  dew-point, F6.8, 9
  effective, A47.5-7; F8.19
  geothermal fluids, A29.1-2
  horizontal/vertical differences, A24.3
  humid operative, F8.19
  mean radiant, A49.1; S6.1; S15.4; F8.10;
    F14.26
  measurement, F14.3
  odors affected by, F13.5
  operative, A49.1-2
  plane radiant, F8.11; F14.26
  plant environments, F10.14
  radiant asymmetry, A24.3; F8.12

Composite Index

I.41

Temperature *(cont.)*
  retail display refrigerators (table), R47.2
  sensors for, A42.8-9; F37.7
  vertical differences, F8.14
  wet-bulb, F6.8, 9
  wet-bulb globe, A24.2-3; F8.20
  wet-globe, F8.20
  wind chill index, F8.21
Temporary exhibit buildings, A4.8
Terminal boxes
  air distribution, S17.6
Terminology
  absorption refrigeration, R41.1
  air diffusion, F31.1
  automatic control, A42.1-2; F37.1
  clean spaces, A15.1-2
  cogeneration, S7.45
  corrosion control, A44.1
  ground-source heat pump systems, A29.14
  heat flow, F22.1
  heat recovery, S8.1
  ice makers, R50.1
  indoor environmental health, F9.1
  infiltration, F25.1
  measurement uncertainty, F14.1
  moisture transfer, F22.1
  operation and maintenance, A35.1
  radiant heating and cooling, A49.1-2
  seismic restraint design, A50.1
  service water heating, A45.6
  sound, F7.2
  sound control, F7.10
  testing, adjusting, balancing, A34.1
  thermal storage, A40.1
  thermodynamics, F1.1
  ventilation, F25.1
  vibration, F7.12
  water treatment, A44.1
  water wells, A29.4-5
Testing
  air cleaners, S24.2
  air conditioners, packaged terminal, S45.3
  air leakage
    fan pressurization, F25.15
  automatic control systems, A42.34-35
  clean spaces, A15.5-6
  compressors
    centrifugal, S34.30
    positive-displacement, S34.3
  condensers
    evaporative, S35.17
    water-cooled, S35.7
  cooling towers, S36.15
  dehumidification for, S22.7, 10
  desiccants, S22.7
  engine test facilities, A14
  environmental test facilities, R37
  fans, S18.4
  filters, air, S24.2
  freezers, household, R49.9
  gaseous contaminant control devices, A41.14-15
  geothermal fluids, A29.6-7
  heaters (in-space), S29.7
  heat pumps, packaged terminal, S45.3
  heat pumps, water-source, S45.4
  industrial exhaust systems, A26.11
  infrared heating, A49.6-7
  lubricants, R7.1
  refrigeration systems
    leak detection, R46.4
    performance testing, R46.5
  refrigerators, household, R49.9
  smoke control acceptance, A48.12

  solar collectors, S33.4
  sound, A34.17-20
    instrumentation, A34.18
    level criteria, A34.18
    procedure, A34.18-19
    transmission problems, A34.19-20, 23-24
  vibration, A34.20-24
    equipment, A34.21
      analyzing, A34.22-23
      measuring, A34.22
      procedure, A34.21-22
    instrumentation, A34.20
    isolation systems, A34.20-21
    piping transmission, A34.23
    procedure, A34.20
Testing, adjusting, and balancing *(see also* **Adjusting, Balancing),** A34
  air diffusers, A34.2
    K factors, A34.2
  air distribution systems, A34.3-6
    balancing procedure, A34.3
    equipment/system check, A34.3-4
    instrumentation, A34.3
    reporting results, A34.6
  airflow measurement, A34.2-3
    air diffuser K factor, A34.2
    duct flow, A34.2
    mixture plenums, A34.2
    pitot tube traverse, A34.2
    pressures, A34.2
    stratification, A34.2-3
  central plant chilled water systems, A34.13
  cooling towers, A34.14-15
    instrumentation, A34.14
    testing procedure, A34.14-15
  design considerations, A34.1-2
  dual-duct systems, A34.4
  duct design, F32.18
  energy audit field survey, A34.16-17
    building systems, A34.16
    data recording, A34.16
    energized subsystems, A34.16
    form for, A34.17
    instrumentation, A34.16
    process loads, A34.16
  hydronic systems, A34.6-14
    chilled water, A34.7-8
    flow vs. heat transfer, A34.7-8
    reduced flow rate heat transfer, A34.7
    water-side balancing, A34.8-14
      instrumentation, A34.8
      procedure (general), A34.10
      procedure (primary/secondary circuits), A34.10
      proportional method balancing, A34.9
      rated differential balancing, A34.9-10
      sizing balancing valves, A34.8
      temperature difference balancing, A34.8-9
      total heat transfer balancing, A34.10
  induction systems, A34.6
  instrumentation for, A34
  sound. *See* **Testing**
  steam distribution systems, A34.14
    balancing procedure, A34.14
    instrumentation, A34.14
  system components as flowmeters, A34.10-14
    differential pressure
      gage readout, A34.11
      head conversion of, A34.11
      mercury manometer readout, A34.11-12
    manufacturer data, A34.10-11
    other devices, A34.12
    pump as indicator, A34.12-13

  temperature control, A34.15-16
    verification procedure, A34.15-16
  terminology, A34.1
  variable-air-volume systems, A34.4-6
    air requirements, A34.5
    balancing procedure, A34.5-6
    diversity, A34.5
    pressure-dependent, A34.4
    pressure-independent, A34.4
    return air fans, A34.5
    static control, A34.4-5
    types, A34.5
  vibration. *See* **Testing**
Textile processing
  air-conditioning design, A18.4-6
    air distribution, A18.4-6
    collector systems, A18.4-5
    gaseous contaminants, A18.6
    health considerations, A18.6
  energy conservation, A18.6
  fabric making, A18.3-4
  fiber making, A18.1
  yarn making, A18.1-3
Theaters, A4.3, 4
Thermal bridges
  building envelopes, F39.8
Thermal comfort. *See* **Comfort**
Thermal conductance. *See* **Conductance, thermal**
Thermal conductivity. *See* **Conductivity, thermal**
Thermal diffusivity. *See* **Diffusivity, thermal**
Thermal gravity effect. *See* **Stack effect**
Thermal insulation. *See* **Insulation, thermal**
Thermal mass
  buildings, F39.12
Thermal resistance. *See* **Resistance, thermal**
Thermal storage, A40
  applications, A40.3-9
    control strategies, A40.3-4
    operating modes, A40.4
  benefits of, A40.2
  building mass effects, A40.8-9, 16-17
  closed systems, A40.18, 19
  cogeneration, S7.32
  commissioning, A40.19
  control of, A38.15-18; A40.19; A42.30
  district heating and cooling, S11.3
  economics, A40.2-3
  gas turbine air cooling, S7.10
  installation, A40.17-19
  insulation, A40.11, 15, 18
  off-peak air conditioning, A40.4-6
    air distribution, A40.5-6
    heat storage, A40.6
    refrigeration design, A40.4-5
  off-peak heating, A40.7-8
    service water heating, A40.7
    solar space/water heating, A40.7
    space heating, A40.7-8
  open systems, A40.19
  operation and maintenance, A40.17-19
  process cooling, A40.6-7
  refrigeration system, A40.4-5, 18
  retrofits, A40.6
  secondary coolant systems, R4.2
  solar energy systems, A30.12-13, 16, 28; A40.7; S33.11
  storage media, A40.1-2, 9-17
    aquifers, A40.8
    bricks, A40.7-8
    building mass, A40.8-9, 16-17
      charging/discharging, A40.17
      principles of, A40.16
      systems, A40.16-17

Thermal storage *(cont.)*
  storage media *(cont.)*
    electrically charged devices, A40.8, 15-16
      central, A40.15
      pressurized water, A40.8, 15-16
      room, A40.15
      underfloor, A40.16
    ice, A40.11-15
      encapsulated ice, A40.12-13, 17-18
      external melt ice-on-coil, A40.11-12, 17, 19
      harvesting system, A40.13-14, 18, 19
      internal melt ice-on-coil, A40.12, 17, 19
      system circuitry, A40.14-15
      tank insulation, A40.15
    phase-change materials, A40.11, 14, 18
    water, A40.8, 9-11, 17
      cool/warm water separation, A40.9-10
      density (table), A40.11
      performance, A40.10-11
      stratification diffuser design, A40.11
      tank insulation, A40.11
      temperature range/size, A40.9
  systems without
    control of, A38.6-14
  terminology, A40.1
  water systems, medium/high temp., S14.9
  water treatment, A40.19
Thermal transmission. *See* **Heat transmission**
Thermal transmittance. *See* **U-factors**
Thermodynamic properties
  moist air, F6
    example, F6.10, 12
    table, F6.3
  water at saturation, F6.2
    table, F6.5
Thermodynamics, F1
  absorption refrigeration cycles, F1.14
  calculations, F1.4
    enthalpy, F1.4
    entropy, F1.4
    phase equilibria, F1.5
  compression refrigeration cycles, F1.6
  equations of state, F1.3
  laws
    first, F1.2
    second, F1.2
  principles, F1.1
  properties
    calculation, F1.4
    definition, F1.1
    refrigerants, F19
    zeotropic mixture, F1.10
  refrigeration cycle analysis, F1.3
  terminology, F1.1
Thermoelectric cooling
  1981 Fundamentals, Chapter 1, pp. 27-33
  (See explanation on first page of Index)
Thermometers, F14.3
  error sources, F14.5
  infrared radiometers, F14.9
  infrared thermography, F14.9
  liquid-in-glass, F14.5
  resistance, F14.5
    resistance temp. devices (RTD), F14.5
    semiconductors, F14.7
    thermistors, F14.7
    thermocouples, F14.7
Thermosiphon
  heat exchangers, S42.16
  solar energy systems, A30.13-14
Thermostats, A42.10; F37.10
  heaters (in-space) control, S29.2, 4

Tobacco
  drying, F11.13
Tobacco smoke, A41.5; F9.8
Trailers. *See* **Trucks, trailers, and containers**
Transmittance
  fenestration, F29.17
    table, F29.25, 47
Transpiration
  fruits and vegetables, R8.19
    tables, R8.24
Transportation centers, A3.11-13
Transport properties
  refrigerants, F19
Traps
  ammonia refrigeration systems, R3.4, 17
  steam systems, S10.7
    kinetic, S10.7, 9
      disk traps, S10.9
      impulse traps, S10.9
      orifice traps, S10.9
      piston traps, S10.9
      thermodynamic traps, S10.9
    mechanical, S10.7, 9
      bucket traps, S10.9
      float and thermostatic (F&T) traps, S10.9
      inverted bucket traps, S10.9
    thermostatic, S10.7, 8
      bellows traps, S10.8
      bimetallic traps, S10.8
Trichloroethylene
  coolants, secondary, F20.12
Troubleshooting
  pumps, centrifugal, S38.13
  flow, S38.13
  noise, S38.13
Trouton's rule, F18.5
Trucks, trailers, and containers, R29
  air circulation, R29.3
  air leakage, R29.2
  heating systems, R29.3
  insulation, R29.1
  refrigeration systems, R29.4
    cooling load calculations, R29.6
    eutectic plates, R29.4
    mechanical refrigeration, R29.5
Tube. *See* **Pipe**
Tuning
  automatic control systems, A42.34-35; F37.11
Tunnels, vehicular. *See* **Road tunnels**
Turbines, S7
  codes, S7.46
  engine test facilities, A14.3
  gas, S7.8
    applications, S7.40
    combustion air systems, S7.9
      evaporative air coolers, S7.10
      online cooling, S7.10
      thermal energy storage, S7.10
    components, S7.8
    control and instruments, S7.10
    emissions, S7.11
    evaporative cooling, A47.8
    exhaust gas, S7.9
    fuels, S7.9
    gas turbine cycle, S7.8
    heat recovery, S7.24
    lubrication, S7.9
    maintenance, S7.11
    noise control, S7.11
    performance characteristics, S7.10
    starting, S7.9
    thermal output, S7.18
  standards, S7.46

  steam, S7.11
    applications, S7.41
    axial flow, S7.11
    control and instruments, S7.12
      automatic, S7.12
      governors, S7.12
      protective devices, S7.14
      starting, S7.12
    economics, S7.18
    heat recovery, S7.24
      extraction turbines, S7.24
      noncondensing turbines, S7.24
    lubrication, S7.12
    maintenance, S7.17
    performance characteristics, S7.14
    radial inflow, S7.12
    thermal output, S7.19
Turbochargers
  engine drives, S7.6
  heat recovery, S7.21
Turbulent flow
  fluids, F2.3
Turkeys *(see also* **Poultry***)*
  growth, F10.10
  heat and moisture production, F10.11
  reproduction, F10.11
Turndown ratio
  heating/cooling source, S12.3
Two-pipe systems, S3.5; S4.3; S12.12
U-factors
  center-of-glass, F29.3
  doors, F24.12; F29.10
    table, F24.13; F29.10
  edge-of-glass, F29.3
  fenestration, F29.5
    table, F29.8
  frame, F29.4
  U_o, F24.12
    example, F24.14
  windows, F24.12
Ultraviolet drying, A27.3
Uncertainty analysis
  measurement, F14.2
    precision uncertainty, F14.3
    systematic uncertainty, F14.3
Universities, A6.3-4
Vacuum cooling
  fruits and vegetables, R14.5
Vacuum drying, A27.6
Valves *(see also* **Regulators***),* S41
  ammonia refrigeration systems, R3.10
    float control, R3.20
  automatic control with, A42.4-6; F37.4
  automatic valves, S41.4
    actuators, S41.4
      electric, S41.5
      electrohydraulic, S41.5
      solenoids, S41.5
    applications, S41.9
    authority, S41.8
    butterfly valves, S41.7
    equal percentage, S41.7
    linear, S41.7
    quick opening, S41.7
    sizing, S41.8
    special purpose valves, S41.6
    thermostatic radiators, S41.6
    three-way valves, S41.6
    two-way valves, S41.6
  backflow-prevention devices, S41.12
    installation, S41.13
    selection, S41.12
  balancing, A34.8

Composite Index

Valves (*cont.*)
   balancing valves, S41.9
      automatic flow-limiting valves, S41.9
      selection, S41.9
   body styles, S41.3
   cavitation, S41.2
   check valves, S41.12; R45.19
   compressor
      reciprocating, S34.7
   control valves, S41.4; F2.10, 11
      discharge coefficients, F2.11
      sizing, S41.8
   expansion
      automatic, S43.2
      constant pressure, R45.9
      electric, R45.9
      thermostatic, S43.2; R45.4
   float control
      high-side, R45.13
      low-side, R45.14
   flow coefficient, S41.2, 8
   gaskets, S41.2
   geothermal energy systems, A29.10
   halocarbon refrigeration systems, R2.9
      equivalent lengths (table), R2.11
      float control, R2.21
   heaters (in-space) control, S29.2
   hydronic systems
      control, S12.10
      safety relief, S12.13
   manual valves, S41.3
      ball valves, S41.4
      butterfly valves, S41.4
      gate valves, S41.3
      globe valves, S41.3
      pinch valves, S41.4
      plug valves, S41.4
      selection, S41.3
   materials, S41.1
   multiple-purpose valves, S41.10
   noise, S41.3
   pressure drop, S41.2; F33.1, 4
   pressure-reducing valves, S41.12
      makeup water valves, S41.12
   pressure relief valves, S41.10; R45.19
   ratings, S41.1
   refrigerant control, R45.4
      check, S8.8
      expansion, S8.8
      reversing, S8.8
   reversing, R45.17
   seats, S41.2
   self-contained temperature control valves, S41.11
   solar energy systems, A30.17
   solenoid, R45.14
   steam systems, S10.9
   stems, S41.2
   stop-and-check valves, S41.12
   vaults, S11.21
   water hammer, S41.2
Vapor
   pressure (table), F36.2
Vapor retarders. *See* **Water vapor retarders**
Variable-air-volume
   all-air systems
      dual-duct, S2.7
      single-duct, S2.6
   control of, A38.1-2, 5-6, 7; A42.19, 23-24
   humidity control, S20.9
   outlets, S17.2
   sound control, A43.9-10
   testing, adjusting, balancing, A34.4-6
   ventilation, F25.24

Vegetables, R23
   air transport, R31.1
   cooling, R14
   desiccation, R10.1
   deterioration rate (table), R10.1
   display refrigerators, R47.4
   dried, storage, R28.7
   frozen, R26.3
   handling, R23.1
   packaging, R23.2
   potatoes, storage, A47.9
   respiration. *See* **Respiration**
   storage, R23.3
      chilling injury, R23.4
      controlled atmosphere, R23.4
      diseases, R23.4
      recommended temperatures (table), R23.3
      requirements (table), R10.2
   thermal properties, R8
   transpiration. *See* **Transpiration**
   transport, R23.2
      optimal temperatures (table), R23.3
Vena contracta, F2.4
Vehicular facilities, enclosed, A12
   bus garages, A12.12-13
   bus terminals, A12.13-15
   equipment, A12.15-17
   parking garages, A12.11-12
   rapid transit systems, A12.7-11
   road tunnels, A12.1-7
Velocity
   acoustical, of refrigerants, S34.30
   measurement, F14.14
      airborne tracers, F14.14
      anemometers, F14.15
      in ducts, F14.16
      pitot-static tubes, F14.16
Venetian blinds, F29.43
Ventilation, F25
   age of air, F25.6
   air change effectiveness, F25.6, 7
   aircraft, A9
   air exchange rate, F25.3
      measurement, F25.5
   airflow, F25.2
      displacement flow, F25.2
      entrainment flow, F25.2
      perfect mixing, F25.3
   all-water systems, S4.3
   animal environments
      management, A20.4-8
      mechanical systems, A20.4
      natural systems, A20.4
   automobiles, A8.1
   bus garages, A12.12-13
   bus terminals, A12.13-15
   CLF (table), F28.51
   climatic infiltration zones, F25.20
   commercial buildings, F25.23
   cooling load, F28.11
   displacement, F31.5
   driving mechanisms, F25.7
      combining, F25.10
      mechanical systems, F25.10
      stack pressure, F25.8
      wind pressure, F25.8
   economizers, F25.24
   educational facilities, A6.4
   effectiveness, F25.6
   enclosed vehicular facilities, A12
   engine test facilities, A14
   fan-coil units, S4.3
   forced, F25.1
   gaseous contaminant control, A41.7-8, 8-9

   greenhouses
      mechanical, A20.11
      natural, A20.11
   health care facilities, A7
   hospitals, A7.2-10
   indoor air quality, F25.4
      ventilation rate procedure, F25.5, 23
   industrial
      environments, A24
      exhaust systems, A26
   kitchens, A28
   laboratories, A13.8-12
   localized, F31.5
   leakage function, F25.11
   mechanical, F25.1
   mines, A25
   natural, F25.1, 12
      caused by stack effect, F25.13
      caused by wind, F25.12
      guidelines, F25.13
   nonresidential buildings, F25.23
      economizers, F25.24
      ventilation rate procedure, F25.5, 23
   nuclear facilities, A23.4-5
      control systems, A23.4
      radioactive effluents, A23.4-5
      requirements, A23.4
      systems, A23.4
   nursing homes, A7.11-12
   odor dilution, F13.6
   outpatient health care facilities, A7.10-11
   outside air fraction, F25.3
   parking garages, A12.11-12
   places of assembly, A4.1
   rapid transit systems, A12.7-11
   rate procedure, F25.5, 23
   residential, F25.20
   road tunnels, A12.1-7
   ships, A10
   terminology, F25.1
   thermal loads, F25.3
   unidirectional, F31.5
   unitary air-conditioning systems, S5
   variable-air-volume, F25.24
   warm humid climates, F23.9
   wind effect on, F15.7
Ventilators, S4.4
   unit, S31.1
      air conditioning, S31.1
      capacity, S31.1
      control, S31.3
      description, S31.1
      heating, S31.1
      capacity, S31.1
      location, S31.3
      selection, S31.3
Venting
   furnaces, S28.2
   gas appliances, S30.15
   heaters (in-space), S29.3
   oil-fired appliances, S30.16
Vibration
   compressors
      centrifugal, S34.32
      orbital (scroll), S34.24
      positive-displacement, S34.3
      single-screw, S34.16
   fans, S18.8
   health effects, F9.16
   measurement, F7.14; F14.24
      accelerometer, F7.14
   terminology, F7.12
   testing, A34.20-24

A = 1995 Applications   S = 1996 Systems and Equipment   F = 1997 Fundamentals   R = 1998 Refrigeration

Vibration control, A43.1, 33-43
  central systems, S1.6
  clean spaces, A15.11
  cogeneration, S7.38
  design criteria, A43.33-34
  ducts, A43.40
  engine drives, S7.7
  equipment vibration, A43.33
  isolator selection guide, A43.34-38
    table, A43.35-38
  libraries, A3.10
  museums, A3.10
  piping systems, A43.34-40
    connectors
      expansion joint/arched, A43.40
      flexible, A43.39-40
      hose, A43.40
    resilient hangers/supports, A43.38-39
  places of assembly, A4.2
  seismic restraint, A43.40-41; A50
  troubleshooting, A43.41-43
    problem type, A43.41
    source, A43.41
    vibration problems, A43.41-43
    vibration investigation, A43.41
Viscosity, F2.1
  absolute, F2.1
  dynamic, F2.1
  fuel oils, F17.5
  kinematic, F2.1
  liquids (table), F36.2
  lubricants, R7.4, 5
  moist air
    figure, F6.16
  vapors (table), F36.1
Volume ratio
  rotary vane compressors, S34.11
  single-screw compressors, S34.14
  twin-screw compressors, S34.18
Walls, F39.3
  air movement, F39.5
    air leakage, F39.5
    exfiltration, F39.5
    infiltration, F39.5
    intrusion, F39.5
  CLTD (table), F28.43
  masonry construction, F23.2; F39.4
  moisture control, F39.4
  roof interface, F39.8
    air leakage, F39.8
    moisture control, F39.8
  steel frame construction, F23.1
  window interface, F39.7
  wood frame construction, F23.1; F39.3
Warehouses, A3.13
  refrigerated. *See* **Refrigerated facilities**
Water
  air-and-water systems, S3
    secondary water systems, S3.8
  all-water systems, S4
    water distribution for, S4.3
      central equipment, S4.3
      four-pipe, S4.3
      two-pipe changeover, S4.3
  at saturation, F6.2
    table, F6.5
  characteristics, A44.6-7
  coils, S21.1
    air heating, S23.1
  coolers, R25.10

  heating
    boilers, S27.1
    geothermal energy systems, A29
    service, A45
    solar energy systems, A30.13-15, 20-21
    space conditioning combined with, S44.4
    water treatment for, A44.15
  humidifier supply, S20.4
  problems caused by, A44.8
  properties, S14.2; F6.16
    refrigerant, F19.1
  systems (*see also* **Hydronic systems**)
    condenser water, S13
    medium/high temp., S14
      air heating coils, S14.8
      basic system, S14.2
      cascade, S14.6
      characteristics, S14.1
      circulating pumps, S14.6
      control, S14.8
      design considerations, S14.2
      direct-contact heaters, S14.6
      direct-fired generators, S14.2
      expansion/pressurization, S14.4
      heat exchangers, S14.7
      piping, S14.7
      safety, S14.9
      space heaters, S14.8
      thermal storage, S14.9
      water treatment, S14.9
  pipe sizing, F33.3
  thermal storage, A40.8, 9-11, 17
  vapor transmission, F24
    building materials (table), F24.16
Water hammer, S41.2; F33.4
Water-lithium bromide (*see also* **Lithium bromide-water**)
  absorption technology, R41.2
Water treatment, A44
  biological control, A44.11-12
  condensers, evaporative, S35.16
  cooling towers, S36.13
  corrosion control, A44.8-9
  evaporative coolers, S19.7
  Langelier saturation index, A44.9-10
    nomograph, A44.10
  mechanical filtration, A44.12-13
  scale control, A44.9-11
  stability index, A44.9
  systems, A44.13-16
    air washers, A44.14
    boilers, A44.15-16
    brines, A44.15
    closed recirculating systems, A44.15
    ice machines, A44.14-15
    once-through systems, A44.14
    open recirculating systems, A44.14
    return condensate systems, A44.16
    sprayed coil units, A44.14
    water heating systems, A44.15
  terminology, A44.1
  thermal properties
    models (table), R8.2
  thermal storage, A40.19
  water characteristics, A44.6-7
    biological, A44.7
    chemical, A44.6-7
  water systems, medium/high temp., S14.9
Water vapor. *See* **Moisture**

Water vapor retarders, F22.17
  classification, F22.17
  heating climates, F23.6
  mixed climates, F23.8
  refrigerant piping insulation, R32.9
  refrigerated facilities, R13.5, 11
  warm humid climates, F23.9
Water wells, A29.4-7
Weather data
  annual extreme temperatures, F26.3
  cooling design conditions, F26.2
  dehumidifying design conditions, F26.2
  design conditions, F26
  heating design conditions, F26.2
  mean daily temperature range, F26.3
  other sources, F26.4
  representativeness, F26.3
  residential infiltration zones, F25.20
  snowfall, A46.3
    table, A46.2, 7
  sources, F26.2
  tables, F26.6, 22, 26
  uncertainty, F26.3
Welding
  sheet metal, S16.6
Wheat
  drying, F11.13
Wind
  chill index, F8.21
    table, F8.21
  coefficients (table), F25.22
  data sources, F15.5
  design data, F26.3
  effect on airflow around building, F15
  effect on chimneys, S30.2, 24
  effect on infiltration/ventilation, F25.8
  effect on system operation, F15.7
  pressure coefficients, F15.4
  pressure on buildings, F15.3
  shielding classes (table), F25.22
Windows
  air leakage, F39.7
  moisture control, F39.7
  shading, F29.2, 41
    coefficients, F29.23
  U-factors, F24.12; F29.2, 5
    table, F29.8
  wall interface, F39.7
Wine. *See* **Beverages**
Wineries, R25.8
  refrigeration, R25.8
  temperature control
    fermentation, R25.9
    storage, R25.10
Wood
  heaters (in-space), S29.5
  stoves, S29.5
Wood products facilities
  evaporative cooling, A47.8
  general operations, A22.1
  process area air conditioning, A22.1
  storage, A22.1
Xenon
  recovery, R38.14
Zones
  smoke control, A48.11-12
Zone systems
  all-air systems, S2.8
  control, A42.22, 23-27; S9.7
  residential, A1.3